Brooks/Cole Titles of Interest

Essentials of Pro/ENGINEER®
Yousef Haik and Mohammed Kilani
Florida State University
ISBN: 0-534-37864-1

Mastering Mechanical Desktop® Release 2: Surface, Parametric and Assembly Modeling
Ron K. C. Cheng
Hong Kong Polytechnic University
ISBN: 0-534-95760-9

Mastering AutoCAD®, Release 14
Ron K. C. Cheng
Hong Kong Polytechnic University
ISBN: 0-534-95761-7

Engineering Design Process
Yousef Haik
Florida State University
ISBN: 0-534-38014-X

Discrete-Time Control Problems Using MATLAB® and the Control System Toolbox
Joe H. Chow, Rensselaer Polytechnic Institute
Dean K. Frederick, Unified Technologies
Nicolas W. Chbat, General Electric CRD
ISBN: 0-534-38477-3

Contemporary Communication Systems Using MATLAB® and Simulink®
John G. Proakis, Northeastern University
Masoud Salehi, Northeastern University
Gerhard Bauch, DoCoMo Communication Laboratories
ISBN 0-534-40617-3

Pro/ENGINEER® Wildfire™

Louis Gary Lamit
De Anza College

With technical assistance provided by James Gee

Australia • Canada • Mexico • Singapore • Spain
United Kingdom • United States

Publisher/Executive Editor: *Bill Stenquist*
Editorial Assistant: *Julie Ruggiero*
Technology Project Manager: *Burke Taft*
Executive Marketing Manager: *Tom Ziolkowski*
Marketing Assistant: *Jennifer Gee*
Advertising Project Manager: *Vicki Wan*
Project Manager, Editorial Production: *Kelsey McGee*

Printed in the United States of America
3 4 5 6 7 07 06 05

Library of Congress Control Number: 2003106592

ISBN 0-534-40083-3

Print/Media Buyer: *Kristine Waller*
Permissions Editor: *Sommy Ko*
Production Service & Composition: *WestWords, Inc.*
Copy Editor: *Kamilla Storr*
Cover Design and Image: *Denise Davidson/Simple Design*
Cover Printing: *Phoenix Color Corp*
Printing and Binding: *Courier Corporation/Stoughton*

Brooks/Cole—Thomson Learning
10 Davis Drive
Belmont, CA 94002
USA

Asia
Thomson Learning
5 Shenton Way #01-01
UIC Building
Singapore 068808

Australia/New Zealand
Thomson Learning
102 Dodds Street
Southbank, Victoria 3006
Australia

Canada
Nelson
1120 Birchmount Road
Toronto, Ontario M1K 5G4
Canada

Europe/Middle East/Africa
Thomson Learning
High Holborn House
50/51 Bedford Row
London WC1R 4LR
United Kingdom

Latin America
Thomson Learning
Seneca, 53
Colonia Polanco
11560 Mexico D.F.
Mexico

Spain/Portugal
Paraninfo
Calle Magallanes, 25
28015 Madrid, Spain

About the Author

Louis Gary Lamit is currently an instructor at De Anza College in Cupertino, California, where he teaches computer-aided design.

Mr. Lamit has worked as a drafter, designer, numerical control (NC) programmer, technical illustrator, and engineer in the automotive, aircraft, and piping industries. A majority of his work experience is in the area of mechanical and piping design. He started as a drafter in Detroit (as a job shopper) in the automobile industry, doing tooling, dies, jigs and fixture layout, and detailing at Koltanbar Engineering, Tool Engineering, Time Engineering, and Premier Engineering for Chrysler, Ford, AMC, and Fisher Body. Mr. Lamit has worked at Remington Arms and Pratt & Whitney Aircraft as a designer, and at Boeing Aircraft and Kollmorgan Optics as an NC programmer and aircraft engineer. He also owns and operates his own consulting firm, and has been involved with advertising and patent illustrating.

Mr. Lamit received a BS degree from Western Michigan University in 1970 and did Masters' work at Wayne State University and Michigan State University. He has also done graduate work at the University of California at Berkeley and holds an NC programming certificate from Boeing Aircraft.

Since leaving industry, Mr. Lamit has taught at all levels (Melby Junior High School, Warren, Michigan; Carroll County Vocational Technical School, Carrollton, Georgia; Heald Engineering College, San Francisco, California; Cogswell Polytechnical College, San Francisco and Cupertino, California; Mission College, Santa Clara, California; Santa Rosa Junior College, Santa Rosa, California; Northern Kentucky University, Highland Heights, Kentucky; and De Anza College, Cupertino, California).

Textbooks, workbooks, tutorials, and articles by Louis Gary Lamit:

- ***Industrial Model Building*** (1981)
  Engineering Model Associates, Inc.

- ***Piping Drafting and Design*** (1981)
- ***Piping Drafting and Design Workbook*** (1981)
- ***Descriptive Geometry*** (1983)
- ***Descriptive Geometry Workbook*** (1983)
- ***Pipe Fitting and Piping Handbook*** (1984)
  Prentice-Hall

- ***Drafting for Electronics*** (3rd edition, 1998)
- ***Drafting for Electronics Workbook*** (2nd edition, 1992)
- ***CADD*** (1987)
  Charles Merrill (Macmillan-Prentice-Hall)

- ***Technical Drawing and Design*** (1994)
- ***Technical Drawing and Design Worksheets and Problem Sheets*** (1994)
- ***Principles of Engineering Drawing*** (1994)
- ***Fundamentals of Engineering Graphics and Design*** (1997)
- ***Engineering Graphics and Design with Graphical Analysis*** (1997)
- ***Engineering Graphics and Design Worksheets and Problem Sheets*** (1997)
- ***Basic Pro/ENGINEER® in 20 Lessons*** (Revision 18) (1998)
- ***Basic Pro/ENGINEER® (with references to PT/Modeler)*** (Revision 19 and PT/Modeler) (1999)
- ***Pro/ENGINEER 2000i®*** (Revision 2000i) (1999)
- ***Pro/E 2000i^{2}® (includes Pro/NC and Pro/SHEETMETAL)*** (Revision 2000i^{2}) (2000)
  Brooks/Cole

- ***IX Design*** (2001) CD text
  ImpactXoft

- ***Pro/ENGINEER® Wildfire™*** (Revision Wildfire) (2003)
  Brooks/Cole

Contents

Assembly Mode

Drawing Mode

Preface

Pro/ENGINEER is one of the most widely used CAD/CAM software programs in the world today. This book introduces you to the basics of the program and enables you to build on these basic commands to expand your knowledge beyond the scope of the book.

The book does not attempt to cover all available features, but rather to provide an introduction to the software, make you reasonably proficient in its use, and establish a firm basis for exploring and growing with the program as you use it in your career or classroom.

The book covers **Part Mode: Lessons 1-14, Assembly Mode: Lessons 15-16,** and **Drawing Mode: Lessons 17-20**. The basic premise of this book is that the more parts, assemblies, and drawings you create using Pro/ENGINEER Wildfire (Pro/E Wildfire), the better you learn the software. With this in mind, each lesson introduces a new set of commands, building on previous lessons. The parts created in Part Mode are used to create assemblies in Assembly Mode and to generate drawings in Drawing Mode This procedure allows you to work with actual completed parts, assemblies, and drawings in a short-lesson format, instead of building large complex projects where basic commands may be overshadowed and lost in a complicated process.

Every lesson introduces a new set of commands and concepts that are applied to a *part*, an *assembly*, or a *drawing*, depending on where in the book you are working.

Lessons involve creating a new part, an assembly, or a drawing, using a set of Pro/E commands that walk you through the process step by step. Each lesson starts with a list of objectives and ends with a lesson project. The lesson project consists of a part, assembly, or drawing that incorporates the lesson's new material and uses and expands on previously introduced material from other lessons.

COAch for Pro/ENGINEER®, has been referenced in the book's figures with the authorized use of illustrations. COAch is one of the best ways available for teaching and learning CAD/CAM software. ***A sampler CD can be requested from CADTRAIN®.*** This CD contains a sample of products offered in *CADTRAIN's COAch for Pro/ENGINEER®*.

For a small handling and shipping fee, a CD (with all Pro/E files used in this text) is available from the author for ***instructors*** who adopt this text.

A 100-page booklet on Pro/SHEETMETAL® (formerly Lesson 23) is available and includes a CD. To order, please go to the WEB site listed below and follow the instructions.

If you wish to contact the author concerning orders, questions, changes, additions, suggestions, comments, and so on, please send an email to one of the following:

Louis Gary Lamit & Lamit and Associates

Web Site: www.cad-resources.com

Email: lgl@cad-resources.com

Dedication

This book is dedicated to my daughter Corina and her husband Michael.

Om Mani Padme Hum

Acknowledgments

I want to thank the following people and organizations for the support and materials granted the author:

Ken Page, **Nick Maly** and **Larry Fire**	Parametric Technology Corporation
Dennis Stajic	CADTRAIN®
Bill Stenquist	Brooks/Cole
Thuy Dao Lamit	Lamit and Associates

In addition, I would like to thank the following for assistance in checking and editing the text: Gary Mahany, Tom Modrzejowski, Tracey Jones, Erika M. Shapiro and Dean Collins.

Resources

A variety of information, books, online products, and job opportunities (www.pejn.com) are available. We have listed some of the more useful and important resources. You can also search on the Web with **PTC, Pro/E,** and **Pro/ENGINEER** as keywords.

Parametric Technology Corporation www.ptc.com

Pro/ENGINEER Job Network www.pejn.com

A variety of services are available over the Internet: **Employers:** Announce your job openings on the Web site. Job seekers will fax, mail, or email their resumes directly to you. **Job Seekers:** Look over the job listings on the Web site, free of charge. Fax, mail, or email your resume directly.

CADTRAIN's COAch® for Pro/ENGINEER www.cadtrain.com

COAch® is a computer-based training (CBT) product designed to provide a comprehensive and affordable training program for Pro/ENGINEER users in their actual CAD environment. This self-paced onscreen training tool enables engineers, designers, drafters, and NC programmers to customize their training experience by following the learning sequence best suited to their individual needs.

FroTime® online tutorials for Pro/ENGINEER www.frotime.com

FroTime® is the leader in providing Online Web-Based Pro/ENGINEER training tutorials on a variety of subjects, including: Modeling, Detailing, Surfacing, Sheet Metal, Data Management, etc. In addition to training, FroTime® also provides several custom software applications to maximize efficiency for Pro/ENGINEER users.

Lamit and Associates (CAD-RESOURCES) tutorials for Pro/ENGINEER

www.cad-resources.com lgl@cad-resources.com

Booklets are now available on **Pro/SHEETMETAL** (100 pages, with CD). To order please go to the Web site and follow the ordering instructions.

Pro/ENGINEER Modes

Part Mode

Lesson 1 Introduction to Pro/ENGINEER Wildfire
Lesson 2 Extrusions
Lesson 3 Editing
Lesson 4 Holes and Rounds
Lesson 5 Datums, Layers, and Sections
Lesson 6 Revolved Protrusions and Revolved Cuts
Lesson 7 Chamfers and Threads
Lesson 8 Groups and Patterns
Lesson 9 Ribs, Relations, Failures, and Family Tables
Lesson 10 Drafts, Suppress, and Text Protrusions
Lesson 11 Shells, Reorder, and Insert Mode
Lesson 12 Sweeps
Lesson 13 Blends and Splines
Lesson 14 Helical Sweeps and 3D Model Notes

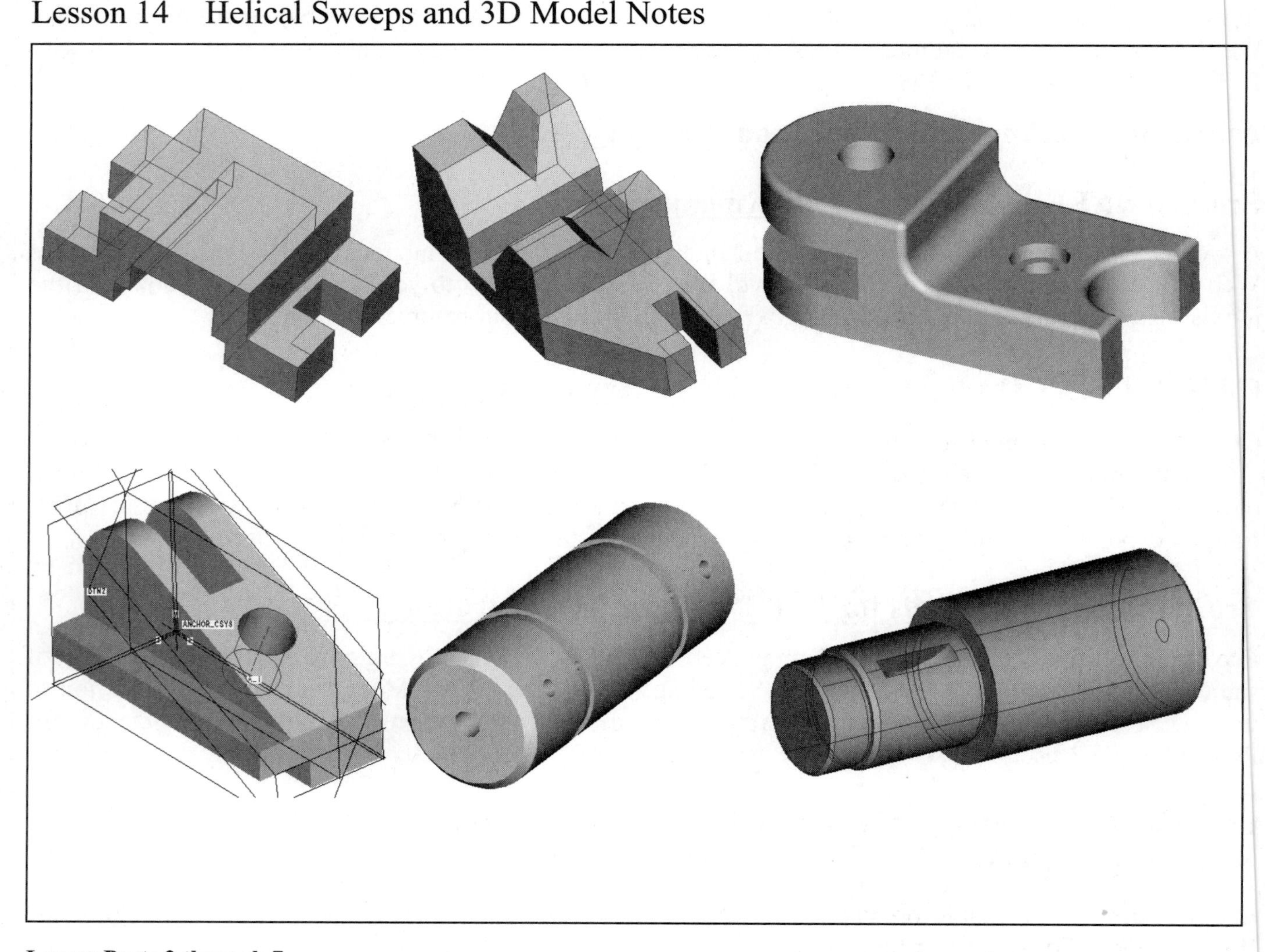

Lesson Parts 2 through 7

PARTS

The *design intent* of a feature, a part, or an assembly (and even a drawing) should be established before any work is done using Pro/ENGINEER. Skipping this step in the design process is a recipe for disaster.

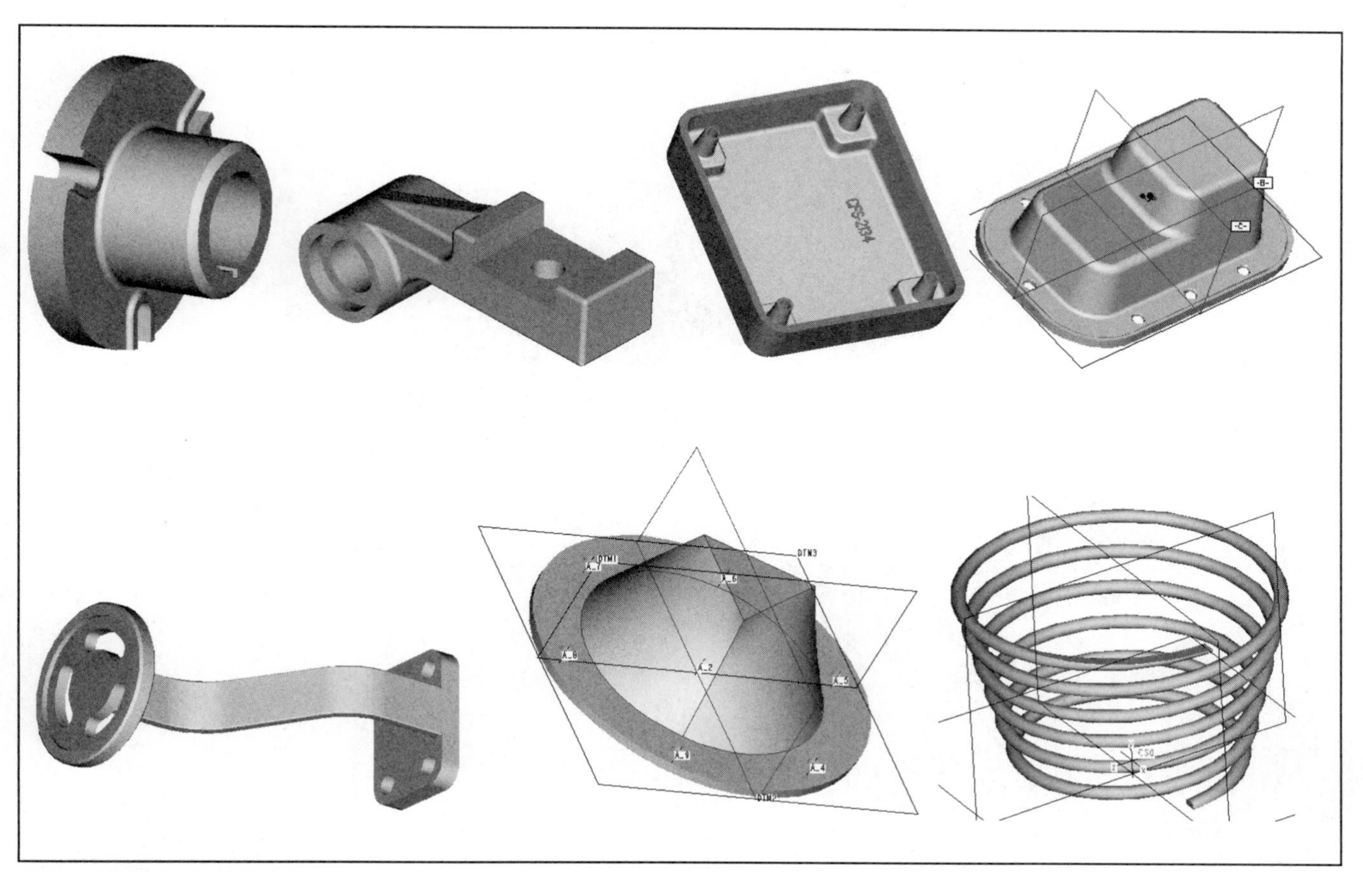

Lesson Parts 8 through 14

In industry, there are thousands of stories of how a designer created a *graphically correct* part (or assembly) that *"looked"* visually precise. Upon closer examination, the part (or assembly) had too many or too few datum planes, parent-child relationships that were glaring examples of the designer's incorrect use of Pro/E, and massive feature failures that resulted when minor ECOs were introduced after the original design was complete. Pro/E is only as good as the drafter, designer, or engineer using it.

Without proper process planning, organization, and well-defined design intent, the part model is useless. In most cases, such poor design habits result in the parts being remodeled, because it would take more time to reorder, modify, redefine, and reroute. In fact, most poor designs can't be fixed.

The **design intent** of a project must be understood before modeling geometry is started. Sketch and analyze your part before modeling.

The *dimensioning scheme* will establish the dimensions that are critical for the design: What dimensions on the part might be modified during an *ECO*? What dimensions are required for *manufacturing* the part economically and to the correct *tolerances*? Are there any dimensional *relationships* that must be established and maintained? Will the part be a member of a *family of similar parts?* How does the part relate to other *parts in the assembly?*

Assembly Mode

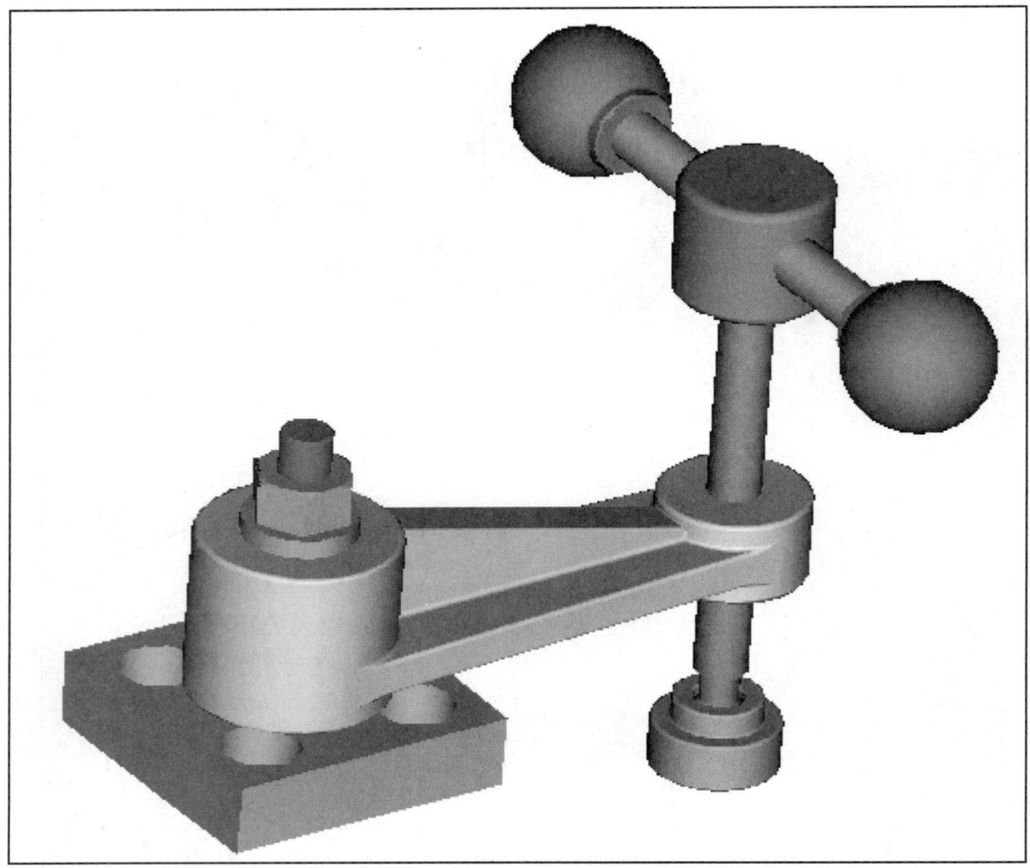

Swing Clamp Assembly

ASSEMBLIES

The **Assembly mode** allows you to place together component parts and subassemblies to form assemblies. Assemblies can then be modified, reoriented, analyzed, and documented. Assembly mode is used for the following functions:

- Placing components into assemblies (*bottom-up* assembly design)
- Altering the display settings for individual components
- Designing in Assembly mode (*top-down* assembly design)
- Part modification, including feature construction
- Analysis of assemblies
- Assemble component parts and subassemblies to form assemblies
- Delete or replace assembly components
- Modify assembly placement offsets, and create and modify assembly datum planes, coordinate systems, and sectional views
- Modify parts directly in Assembly mode
- Get assembly engineering information, perform viewing and layer operations
- Exploding views of assemblies

The process of creating an assembly is accomplished by adding models (parts/subassemblies) to a base component (parent part/subassembly) using a variety of constraints. A placement constraint specifies the relative position of a pair of surfaces on two components. The Mate, Align, and Insert commands and their variations are used to accomplish this task.

For approximately $60.00 US, the Swing Clamp can be purchased from CARRLANE at www.carrlane.com.

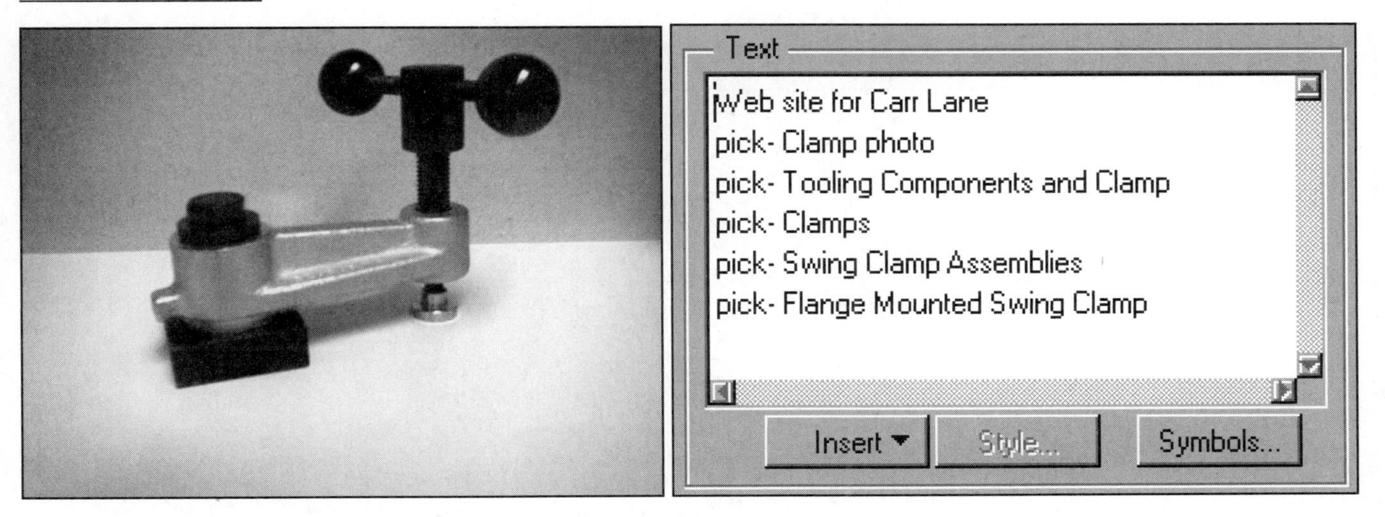

CARRLANE Swing Clamp http://www.carrlane.com/oncatfrm.html

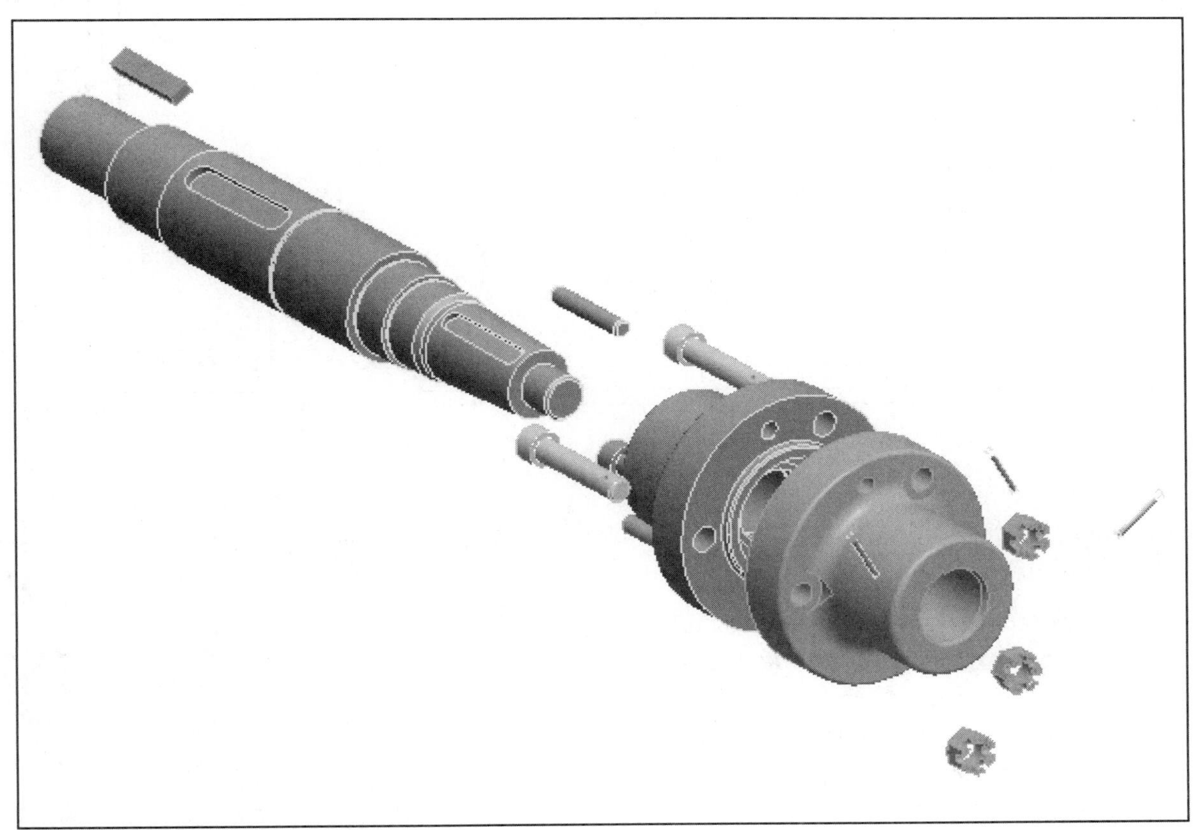

Exploded Coupling Assembly

Drawing Mode

Lesson 17 Formats, Title Blocks, and Views
Lesson 18 Detailing
Lesson 19 Sections and Auxiliary Views
Lesson 20 Assembly Drawings

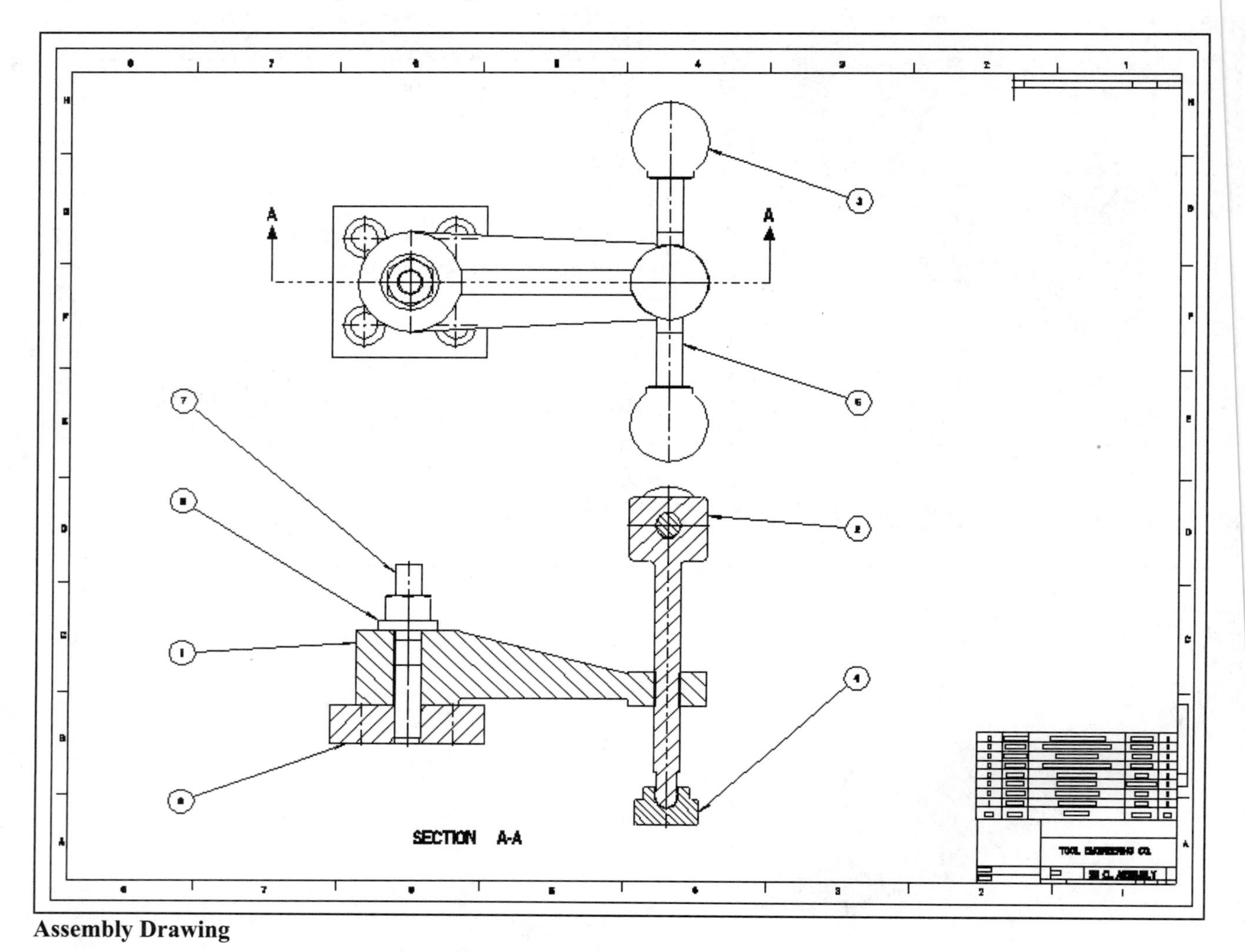

Assembly Drawing

DRAWINGS

The drawing functionality in Pro/E is used to create annotated drawings of parts and assemblies and has a variety of options, including:

- Use default views created on template drawings
- Add views of the part or assembly to the drawing
- Display existing design dimensions
- Create additional driven or reference dimensions
- Create and insert notes on the drawing
- Add views of additional parts or assemblies
- Add multiple sheets to the drawing
- Create a BOM and balloon the assembly
- Add draft entities to the drawing

Drawings can be created of parts and assemblies. Drawings can be multiview or pictorial and can include section, auxiliary, detailed, exploded, and broken views. With Pro/DETAIL, ANSI, ISO, DIN, or JIS, standard drawings can be created.

The **Drawing Mode** is designed to allow you to create drawings, add views, dimension, and document the part or assembly. Pro/E offers several methods for creating views on the drawing. All methods are based on the rules of orthographic projection. When you create a Drawing, Pro/E creates a new file to hold your drawing. Drawing files all have a .drw file extension. The new drawing is displayed in its own graphics window.

The Drawing Mode is very parameter-intensive. There are separate parameters, which control the display of drafting annotation, views, formats, etc. The user interface would become too cumbersome if all of these parameters were controlled by menu options, and because virtually all of these parameters are actually company standards (that you do not change constantly), the parameters are defined and modified in the Drawing Options.

All drawings have a Name, a Size (Height by Length), a Scale, a Projection Angle, and a Drawing Unit. The Name is the name of the file containing the drawing. The size defines, in Drawing Units, how much space is available on the drawing to place views and annotation. The size is also used to plot the drawing. The Scale establishes the default size at which views are placed on the drawing. This scale does not affect the size of annotation in the drawing. The Projection Angle controls how views are projected. Views are projected using one of two conventions: First-Angle or Third-Angle. The Projection Angle is controlled by a Drawing Options parameter called *projection_type*. The default value of this parameter is *third_angle.*

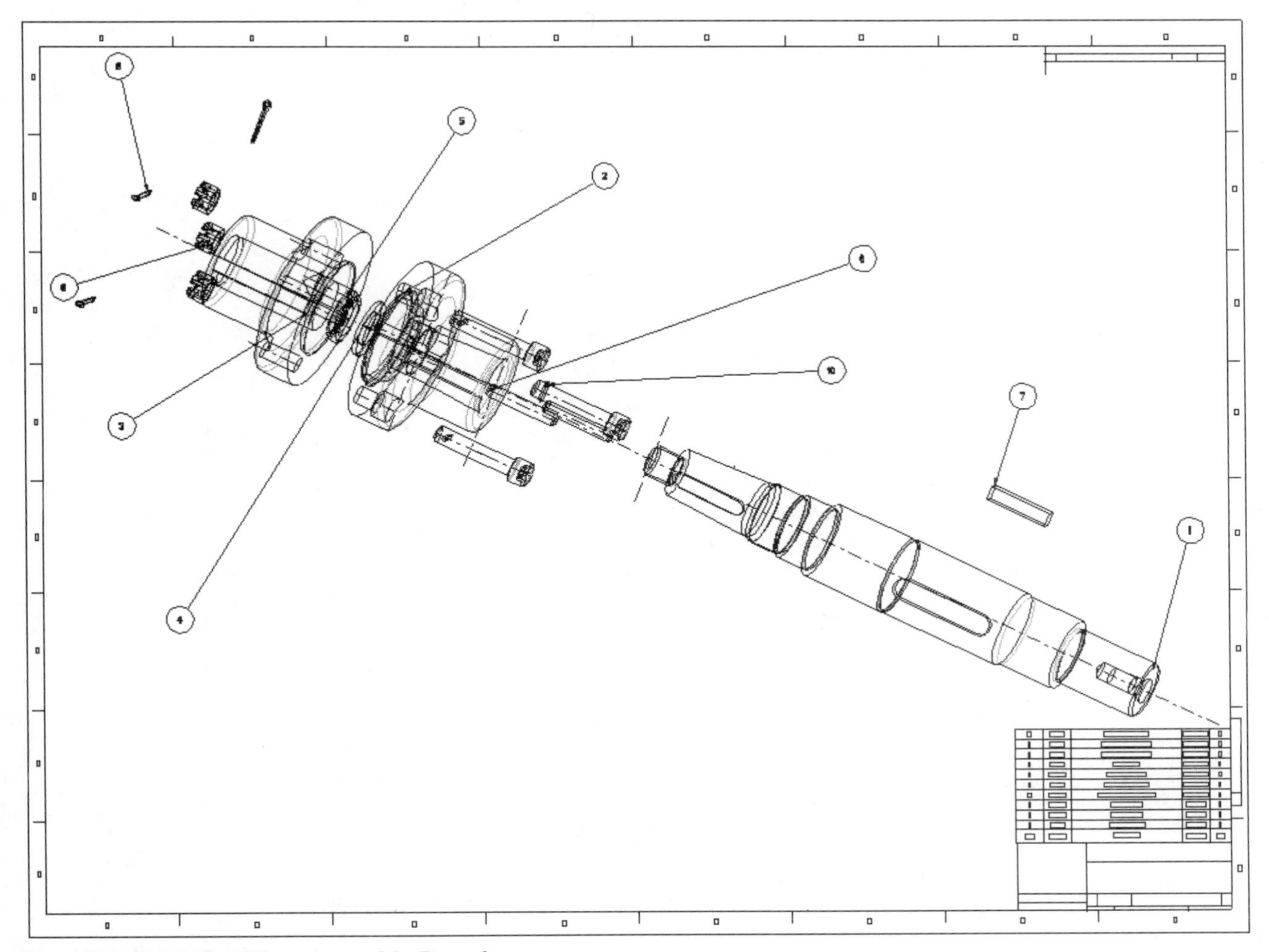

Coupling Exploded View Assembly Drawing

Pro/ENGINEER® Wildfire™

Introduction to Pro/ENGINEER® Wildfire™

This work text introduces the basic concepts of parametric design using Pro/ENGINEER (Pro/E®) to create and document individual parts, assemblies, and drawings. **Parametric** can be defined as *any set of physical properties whose values determine the characteristics or behavior of an object.* **Parametric design** enables you to generate a variety of information about your design: its mass properties, a drawing, or a base model. To get this information, you must first model your part design.

Parametric modeling philosophies used in Pro/E include the following:

Feature-Based Modeling Parametric design represents solid models as combinations of engineering features (Fig. 1).

Creation of Assemblies Just as features are combined into parts, parts may be combined into assemblies (Fig. 1).

Capturing Design Intent The ability to incorporate engineering knowledge successfully into the solid model is an essential aspect of parametric modeling.

Figure 1 Parts and Assembly Design

Parametric Design

Parametric design models are not drawn so much as they are *sculpted* from solid volumes of materials. To begin the design process, analyze your design. Before any work is started, take the time to ***tap*** into your own knowledge bank and others that are available. The acronym **TAP** can be used to remind yourself of this process: **T**hink, **A**nalyze, and **P**lan. These three steps are essential to any well-formulated engineering design process.

Break down your overall design into its basic components, building blocks, or primary features. Identify the most fundamental feature of the part to sketch as the first, or base, feature. A variety of **base features** can be modeled using *protrusion-extrude, revolve, sweep,* and *blend* tools.

Sketched features (*extrusions, sweeps, etc.)* and pick-and-place features called **referenced features** (*holes, rounds, chamfers, etc.*) are normally required to complete the design. With the **SKETCHER**, you use familiar 2D entities (points, lines, rectangles, circles, arcs, splines, and conics) (Fig. 2). There is no need to be concerned with the accuracy of the sketch. Lines can be at differing angles, arcs and circles can have unequal radii, and features can be sketched with no regard for the actual parts' dimensions. In fact, exaggerating the difference between entities that are similar but not exactly the same is actually a far better practice when using the SKETCHER.

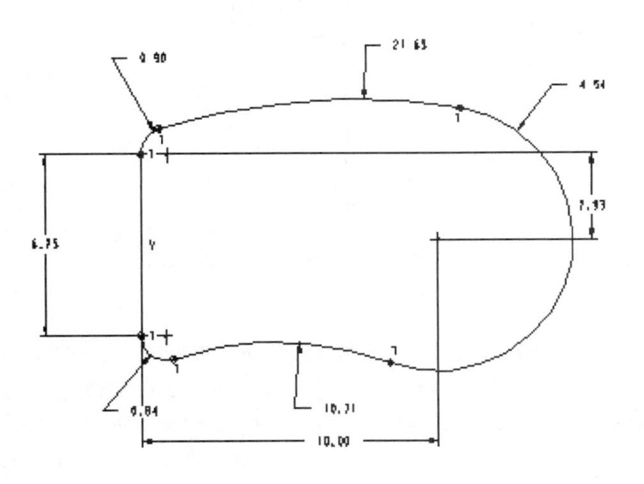

Figure 2 Sketching

Geometry assumptions and constraints close ends of connected lines, align parallel lines, and snap sketched lines to horizontal and vertical orthogonal orientations. Additional constraints are added by means of **parametric dimensions** to control the size and shape of the sketch.

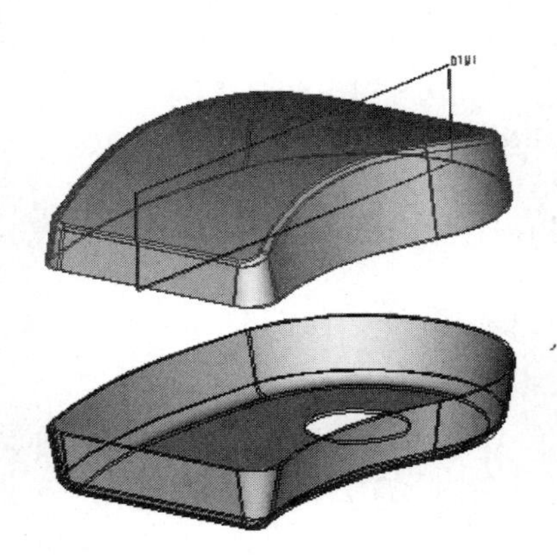

Figure 3 Feature Design

Features are the basic building blocks you use to create a part (Fig. 3). Features "understand" their fit and function as though "smarts" were built into the features themselves. For example, a hole or cut feature "knows" its shape and location and the fact that it has a negative volume. As you modify a feature, the entire part automatically updates after regeneration. The idea behind feature-based modeling is that the designer constructs a part so that it is composed of individual features that describe the way the geometry is supposed to behave if its dimensions change. This happens quite often in industry, as in the case of a design change. Feature-based modeling is diagramed in Figure 4.

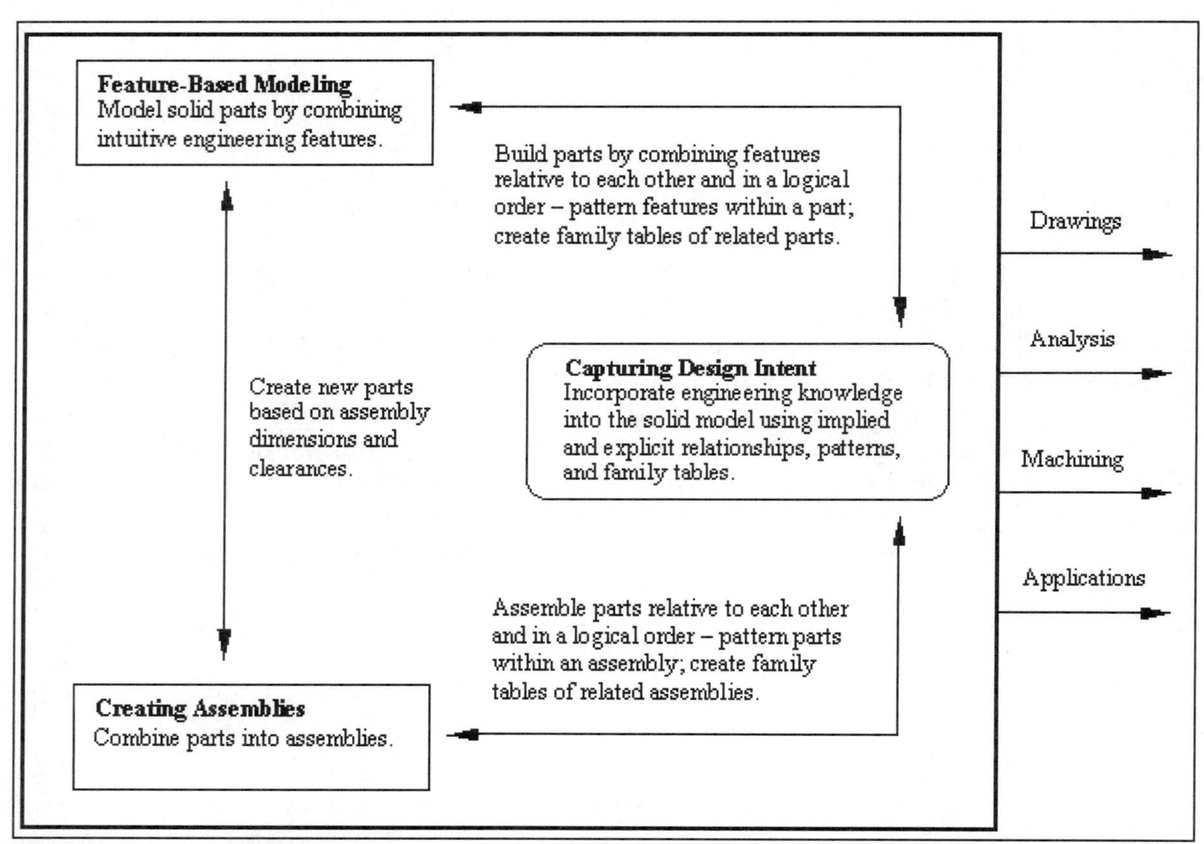

Figure 4 Parametric Design

Parametric modeling is the term used to describe the capturing of design operations as they take place, as well as future modifications and editing of the design. The order of the design operations is significant. Suppose a designer specifies that two surfaces be parallel, such that surface two is parallel to surface one. Therefore, if surface one moves, surface two moves along with it to maintain the specified design relationship. The second surface is a **child** of surface one in this example. Parametric modelers allow the designer to **reorder** the steps in the part's creation.

The various types of features are used as building blocks in the progressive creation of solid parts. Figure 5 shows base features, datum features, sketched features, and referenced features. The "chunks" of solid material from which parametric design models are constructed are called **features**.

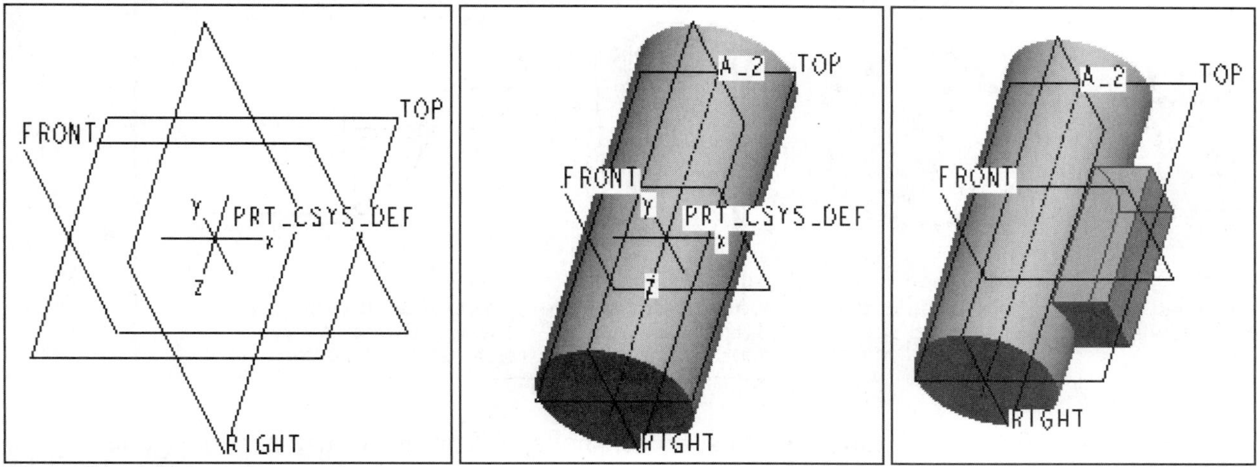

Figure 5(a-c) Features

Features generally fall into one of the following categories:

Base Feature The base feature is normally a set of datum planes referencing the default coordinate system. The base feature is important because all future model geometry will reference this feature directly or indirectly; it becomes the root feature. Changes to the base feature will affect the geometry of the entire model (Fig. 6).

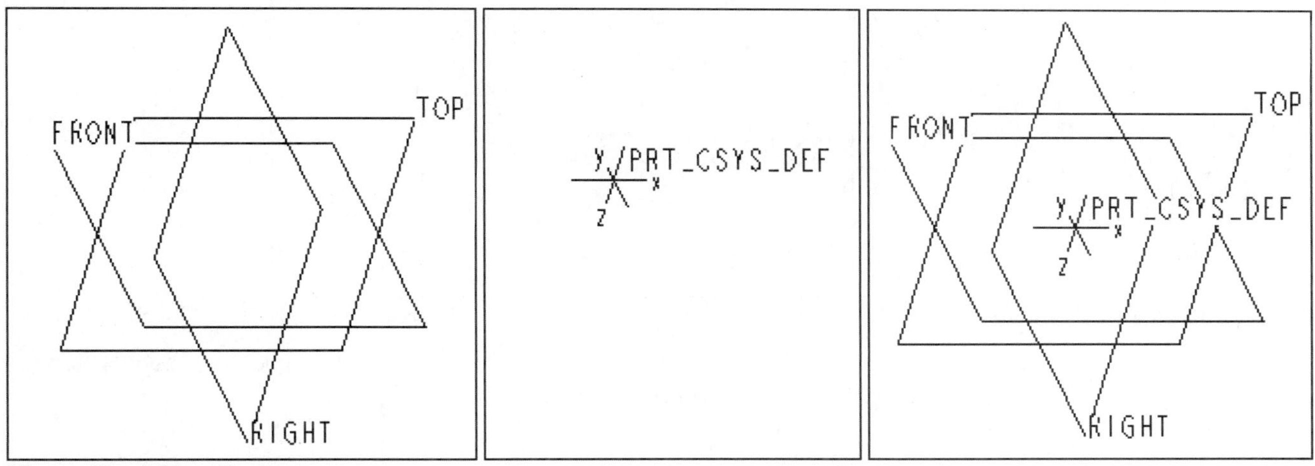

Figure 6(a-c) Base Features

Datum Features Datum features (lines, axes, curves, and points) are generally used to provide sketching planes and contour references for sketched and referenced features. Datum features do not have volume or mass and may be visually hidden without affecting solid geometry (Fig. 7).

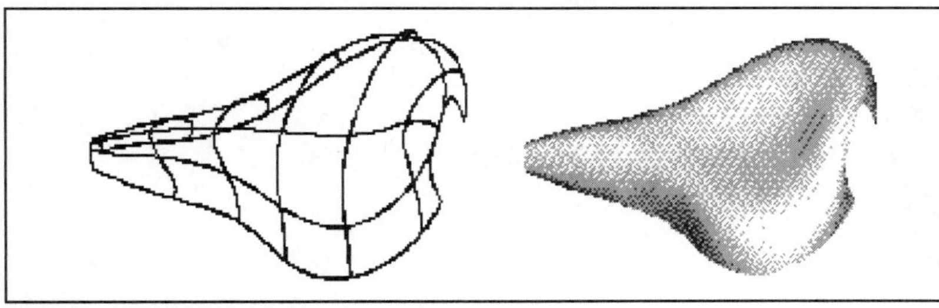

Figure 7 Datum Features

Sketched Features Sketched features are created by extruding, revolving, blending, or sweeping a sketched cross section. Material may be added or removed by protruding or cutting the feature from the existing model (Fig. 8).

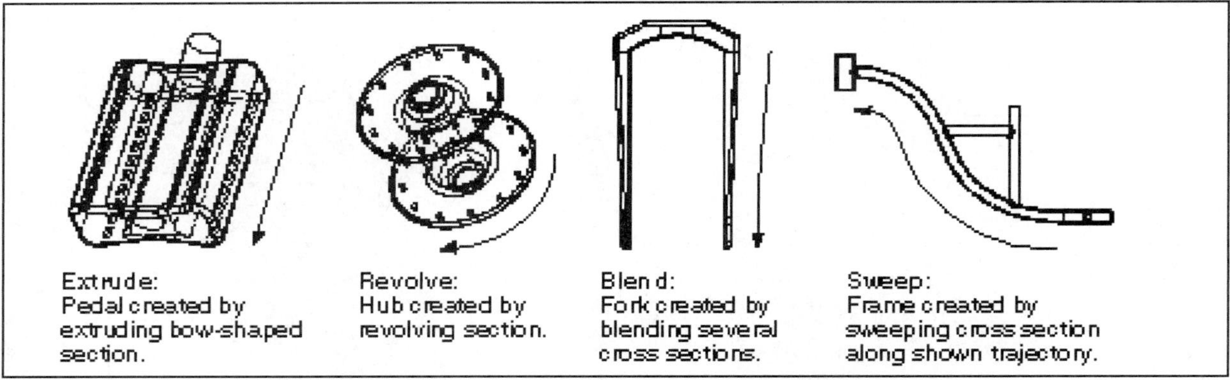

Figure 8 Sketched Features

Referenced Features Referenced features (rounds, holes, shells, and so on) utilize existing geometry for positioning and employ an inherent form; they do not need to be sketched (Fig. 9).

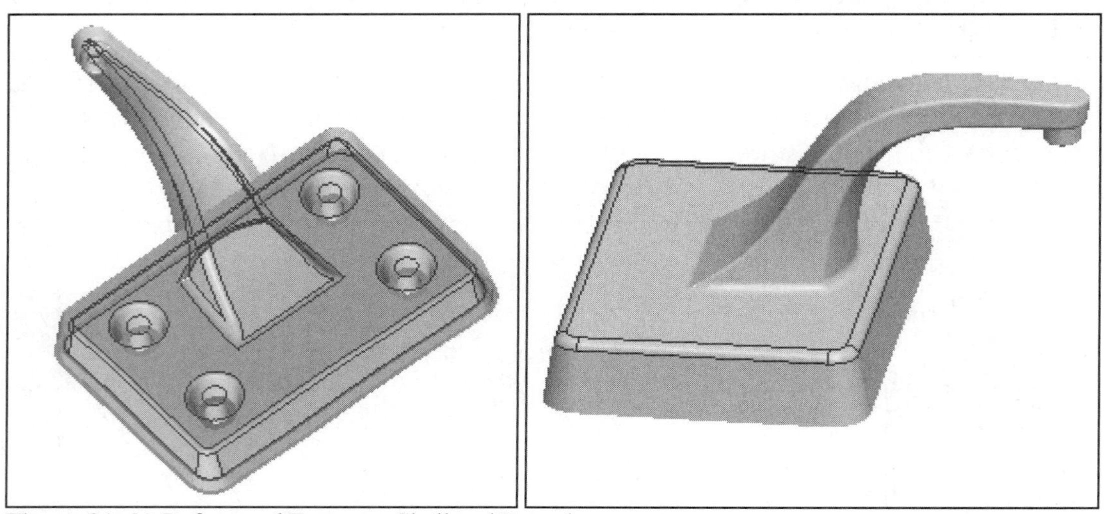

Figure 9(a-b) Referenced Features- Shell and Round

Wide varieties of features are available. These tools enable the designer to make far fewer changes by capturing the engineer's design intent early in the development stage (Fig.10).

Figure 10 Parametric Designed Part

Fundamentals

The design of parts and assemblies, and the creation of related drawings, forms the foundation of engineering graphics. When designing with Pro/ENGINEER, many of the previous steps in the design process have been eliminated, streamlined, altered, refined, or expanded. The model you create as a part forms the basis for all engineering and design functions. The part model contains the geometric data describing the part's features, but it also includes non-graphical information embedded in the design itself. The part, its associated assembly, and the graphical documentation (drawings) are parametric. The physical properties described in the part drive (determine the characteristics and behavior of) the assembly and drawing. Any data established in the assembly mode, in turn, determines that aspect of the part and, subsequently, the drawings of the part and the assembly. In other words, all the information contained in the part, the assembly, and the drawing is interrelated, interconnected, and parametric (Fig. 11).

Part Design

In many cases, the part will be the first component of this interconnected process. Therefore, in this text, the first set of Lessons 1-14 cover part design. The *part* function in Pro/E is used to design components.

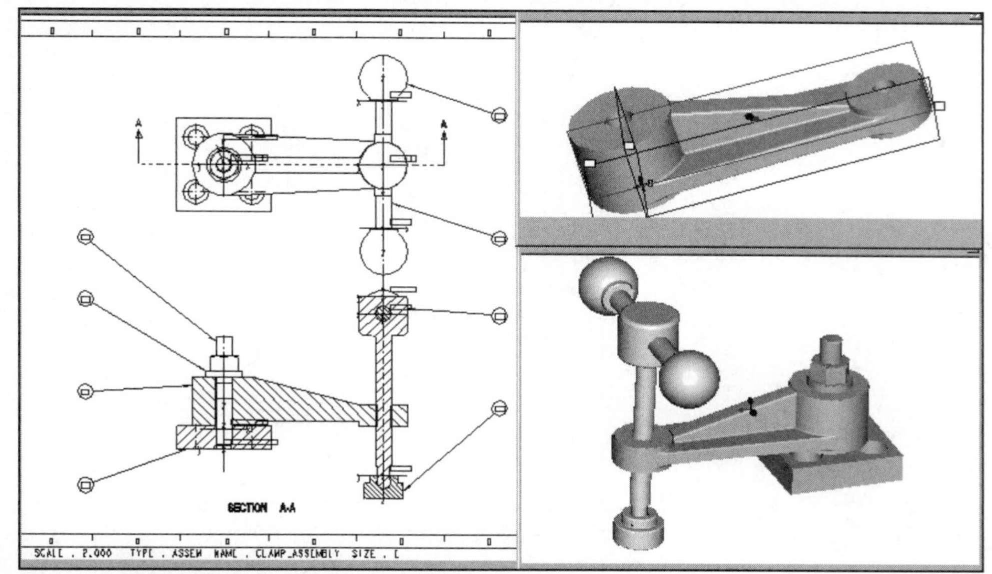

Figure 11 Assembly Drawing, Part, and Assembly

During part design (Fig. 12), you can accomplish the following:

- Define the base feature
- Define and redefine construction features to the base feature
- Modify the dimensional values of part features (Fig. 13)
- Embed design intent into the model using tolerance specifications and dimensioning schemes
- Create pictorial and shaded views of the component
- Create part families (family tables)
- Perform mass properties analysis and clearance checks
- List part, feature, layer, and other model information
- Measure and calculate model features
- Create detail drawings of the part

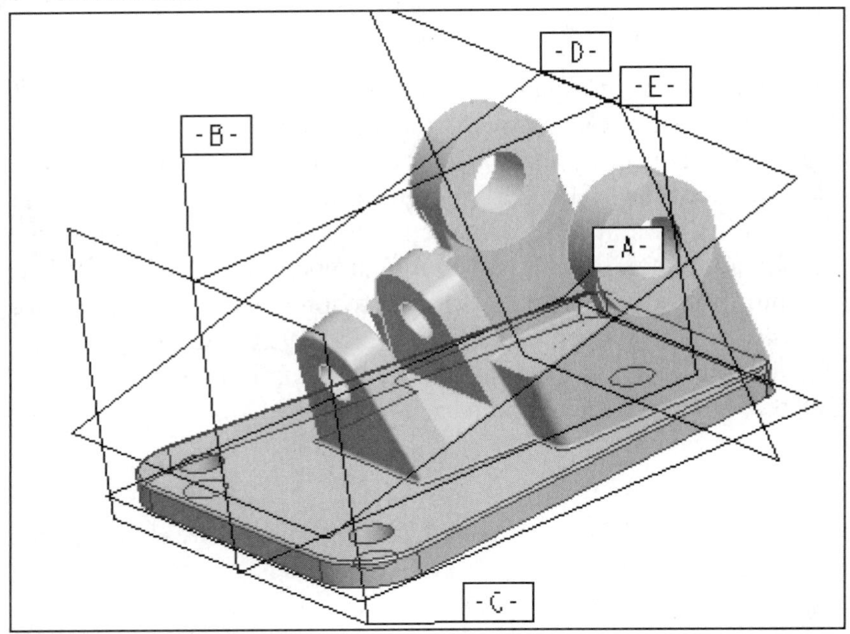

Figure 12 Part Design

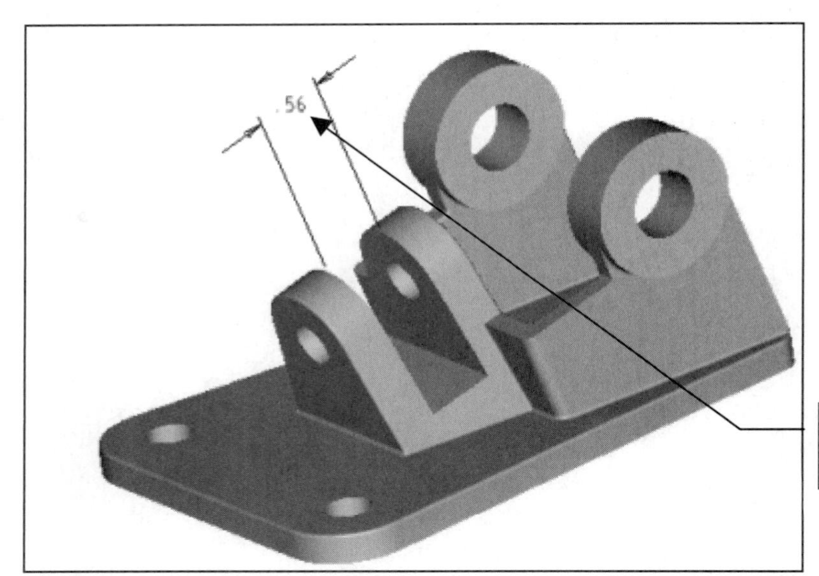

Modify **.56** by picking on the dimension and typing in a new value at the prompt

Figure 13 Modifying the Part Design

Establishing Features

The design of any part requires that the part be *confined*, *restricted*, *constrained*, and *referenced*. In parametric design, the easiest method to establish and control the geometry of your part design is to use three datum planes. Pro/E automatically creates the three **primary datum planes**. The default datum planes (**RIGHT**, **TOP**, and **FRONT**) constrain your design in all three directions.

Datum planes are infinite planes located in 3D model mode and associated with the object that was active at the time of their creation. To select a datum plane, you can pick on its name or anywhere on the perimeter edge. Datum planes are *parametric*--geometrically associated with the part. Parametric datum planes are associated with and dependent on the edges, surfaces, vertices, and axes of a part. For example, a datum plane placed parallel to a planar face and on the edge of a part moves whenever the edge moves and rotates about the edge if the face moves.

As you create parametric datum planes, relationships to the active part are determined by defining combinations of a placement option that link the datum plane to the part.

Datum planes are used to create a reference on a part that does not already exist. For example, you can sketch or place features on a datum plane when there is no appropriate planar surface. You can also dimension to a datum plane as though it were an edge.

In Figure 14, three **default datum planes** and a **default coordinate system** were created when a NEW part was started. Note that in the **Model Tree** window they have become the first four features of the part, which means that they will be the *parents* of the features that follow.

The three *default datum planes* and the *default coordinate system* appear in the Model Tree as the first four features of a new part **PRT0003.PRT**

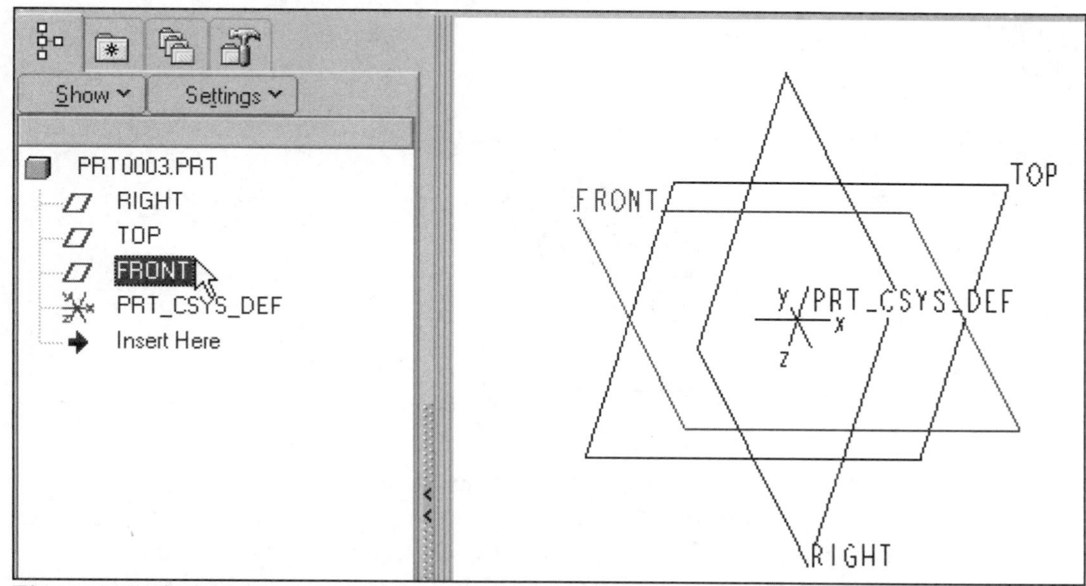

Figure 14 Default Datum Planes and Coordinate System

In order to see how datum planes work in the design of a part, try a simple exercise. Take a book or a box and put it on the floor of a room in your house or school. Choose the most important plane (flat side). You have now established **datum A**, the *primary datum plane* (FRONT in many, but not all, of the lessons and projects in this text). Choose the longest or second most important plane (flat side), and slide it up to and against a wall at the corner of the room. You have now established **datum B**, the *secondary datum plane* (TOP for many, but not all, of the lessons and projects in this text).

Finally, push the remaining side of the book or box against the other wall. You have now established **datum C**, the *tertiary datum plane* (RIGHT for many, but not all, of the lessons and projects in this text). The book/box is now constrained by these three planar surfaces. If the book/box were a real workpiece or stock material, you could secure the part to the floor with clamps (as though it were on a milling table) and machine it to the required design.

Although this exercise and description are simplified, and will not work for many parts, they do demonstrate the process of establishing your part in space using datum planes. In Pro/E, you can use any of the datum planes as sketching planes or, for that matter, any of the part planes for sketching geometry. Any number of other datum planes can be introduced into the part as required for feature creation, assembly operations, or manufacturing applications.

Datum Features

Datum features are planes, axes, and points you use to place geometric features on the active part. Datums other than defaults can be created at any time during the design process.

As we have discussed, there are three (primary) types of datum features (Fig. 14): **datum planes, datum axes**, and **datum points** (there are also ***datum curves*** and ***datum coordinate systems***). You can display all types of datum features, but they do not define the surfaces or edges of the part or add to its mass properties. In Figure 15, a variety of datum planes are used in the creation of the cell phone.

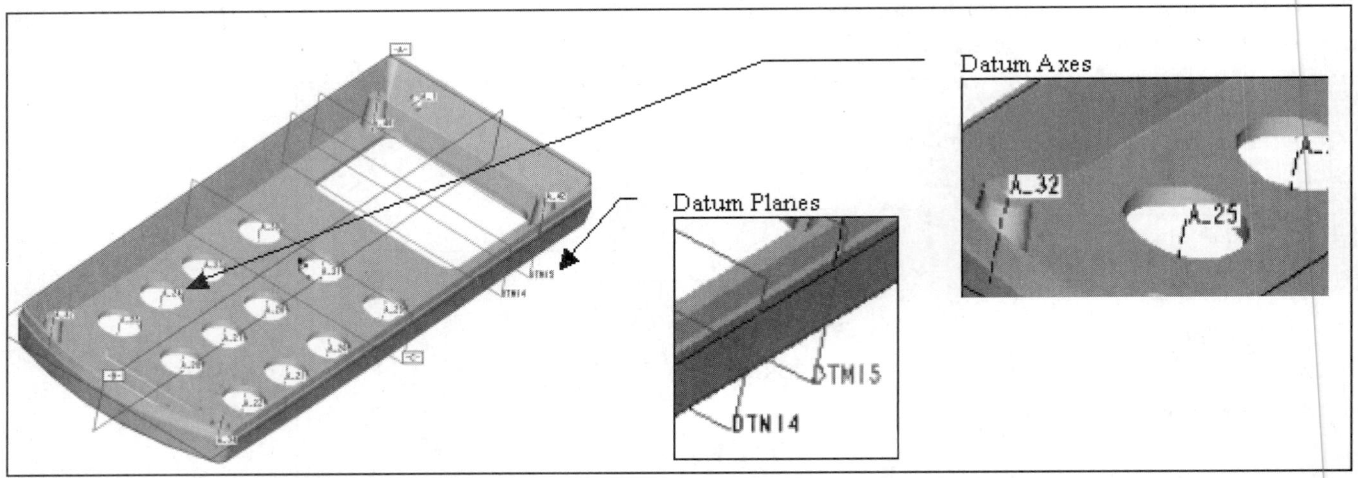

Figure 15 Datums in Part Design

Specifying constraints that locate it with respect to existing geometry creates a datum. For example, a datum plane might be made to pass through the axis of a hole and parallel to a planar surface. Chosen constraints must locate the datum plane relative to the model without ambiguity. You can also use and create datums in assembly mode.

Besides datum planes, datum axes and datum points can be created to assist in the design process. You can also automatically create datum axes through cylindrical features such as holes and solid round features by setting this as a default in your Pro/E configuration file. The part in Figure 16 shows **A_1** through the hole. **A_1** is the default axis of the circular cut.

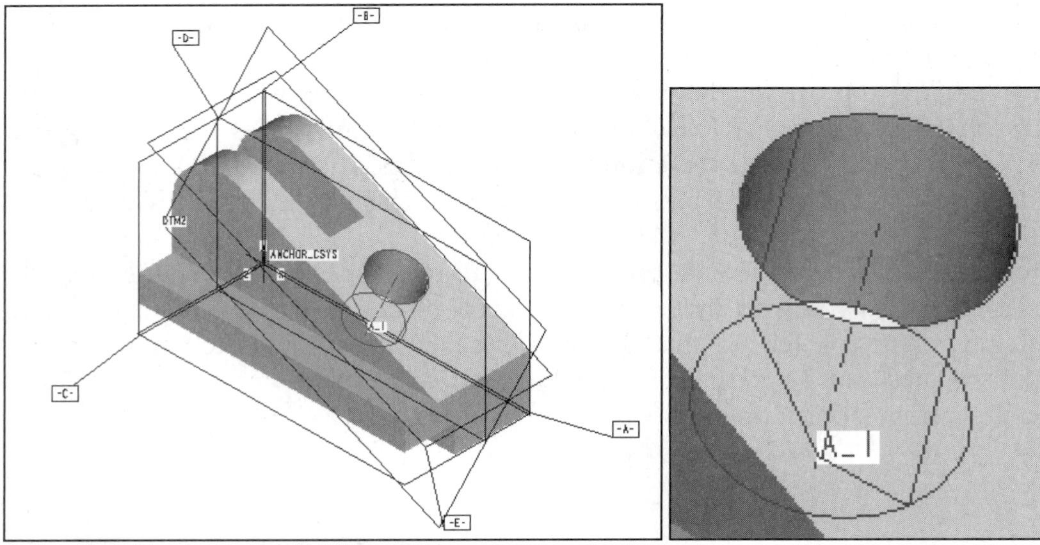

Figure 16(a-b) Feature Default Datum Axis

Parent-Child Relationships

Because solid modeling is a cumulative process, certain features must, of necessity precede others. Those that follow must rely on previously defined features for dimensional and geometric references. The relationships between features and those that reference them are termed ***parent-child relationships***. Because children reference parents, parent features can exist without children, but children cannot exist without their parents. This type of CAD modeler is called a history-based system. Using Pro/E's information command will list the models' information as shown in Figure 17.

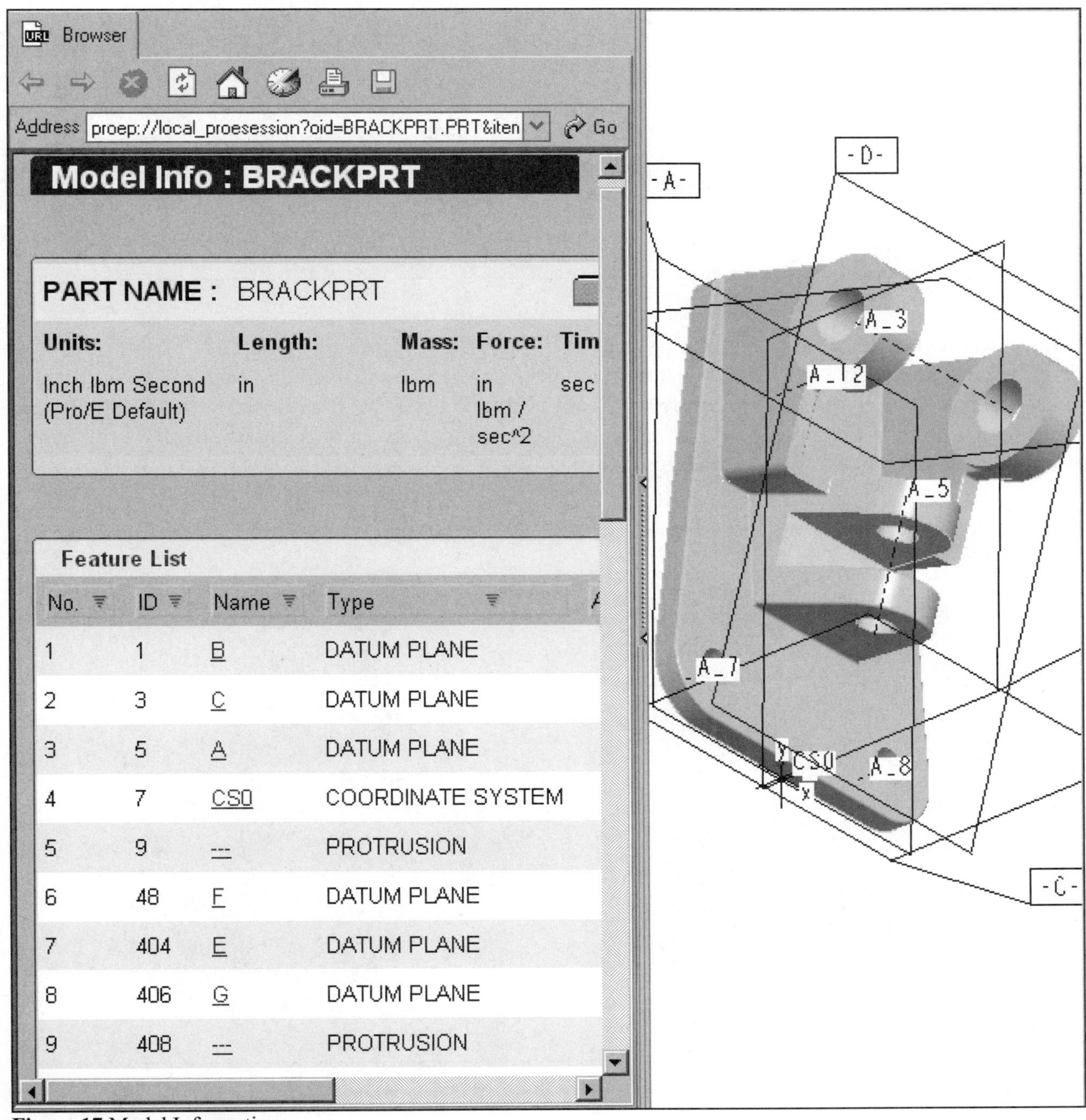

Figure 17 Model Information

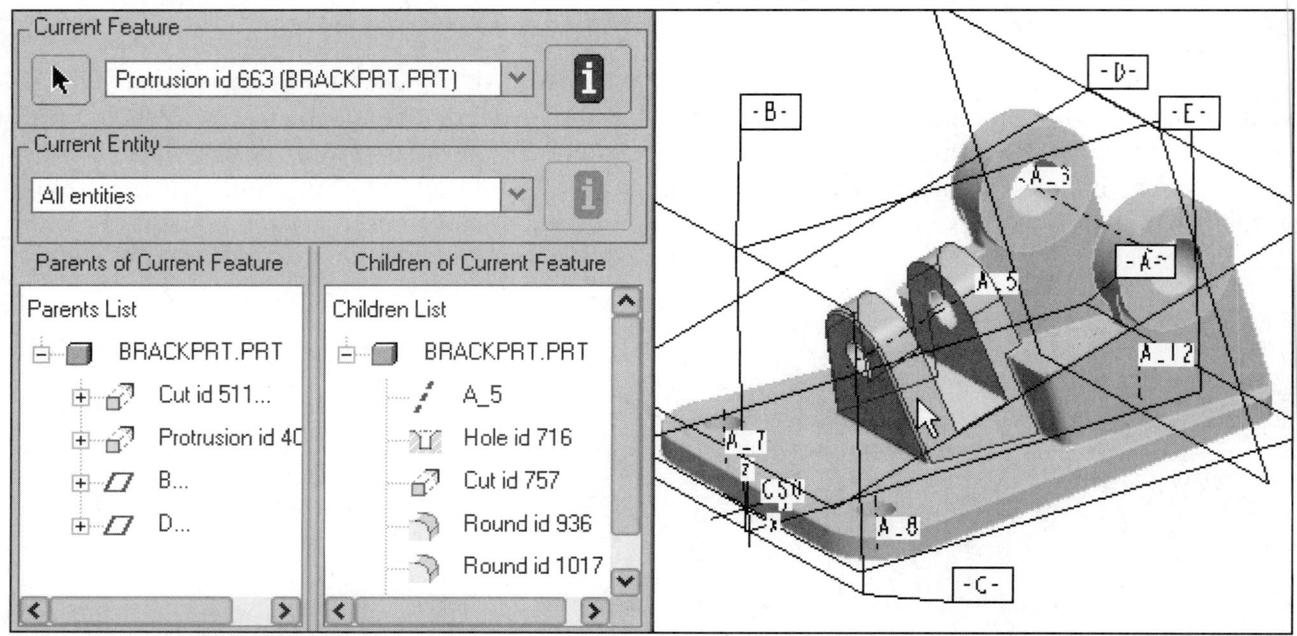

Figure 18 Parent-Child Information

The parent-child relationship is one of the most powerful aspects of parametric design. When a parent feature is modified, its children are automatically recreated to reflect the changes in the parent feature's geometry. It is essential to reference feature dimensions so that design modifications are correctly propagated through the model/part. Any modification to the part is automatically propagated throughout the model (Figure 19) and will affect all children of the modified feature.

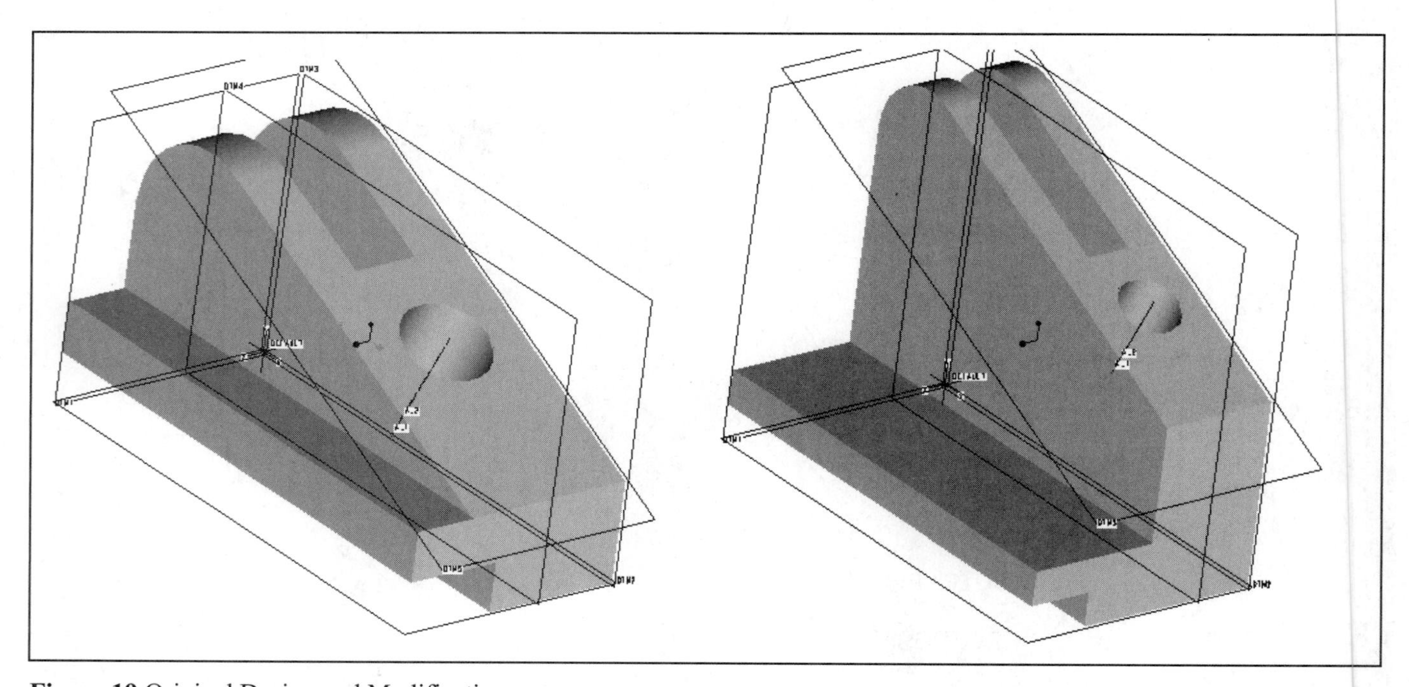
Figure 19 Original Design and Modification

Capturing Design Intent

A valuable characteristic of any design tool is its ability to *render* the design and at the same time capture its *intent* (Figure 20). Parametric methods depend on the sequence of operations used to construct the design. The software maintains a *history of changes* the designer makes to specific parameters. The point of capturing this history is to keep track of operations that depend on each other. Whenever Pro/E is told to change a specific dimension, it can update all operations that are referenced to that dimension.

For example, a circle representing a bolt hole circle may be constructed so that it is always concentric to a circular slot. If the slot moves, so does the bolt circle. Parameters are usually displayed in terms of dimensions or labels and serve as the mechanism by which geometry is changed. The designer can change parameters manually by changing a dimension or can reference them to a variable in an equation (**relation**) that is solved either by the modeling program itself or by external programs such as spreadsheets.

Features can also store non-graphical information. This information can be used in activities such as drafting, numerical control (NC), finite-element analysis (FEA), and kinematics analysis.

Capturing design intent is based on incorporating engineering knowledge into a model by establishing and preserving certain geometric relationships. The wall thickness of a pressure vessel, for example, should be proportional to its surface area and should remain so, even as its size changes.

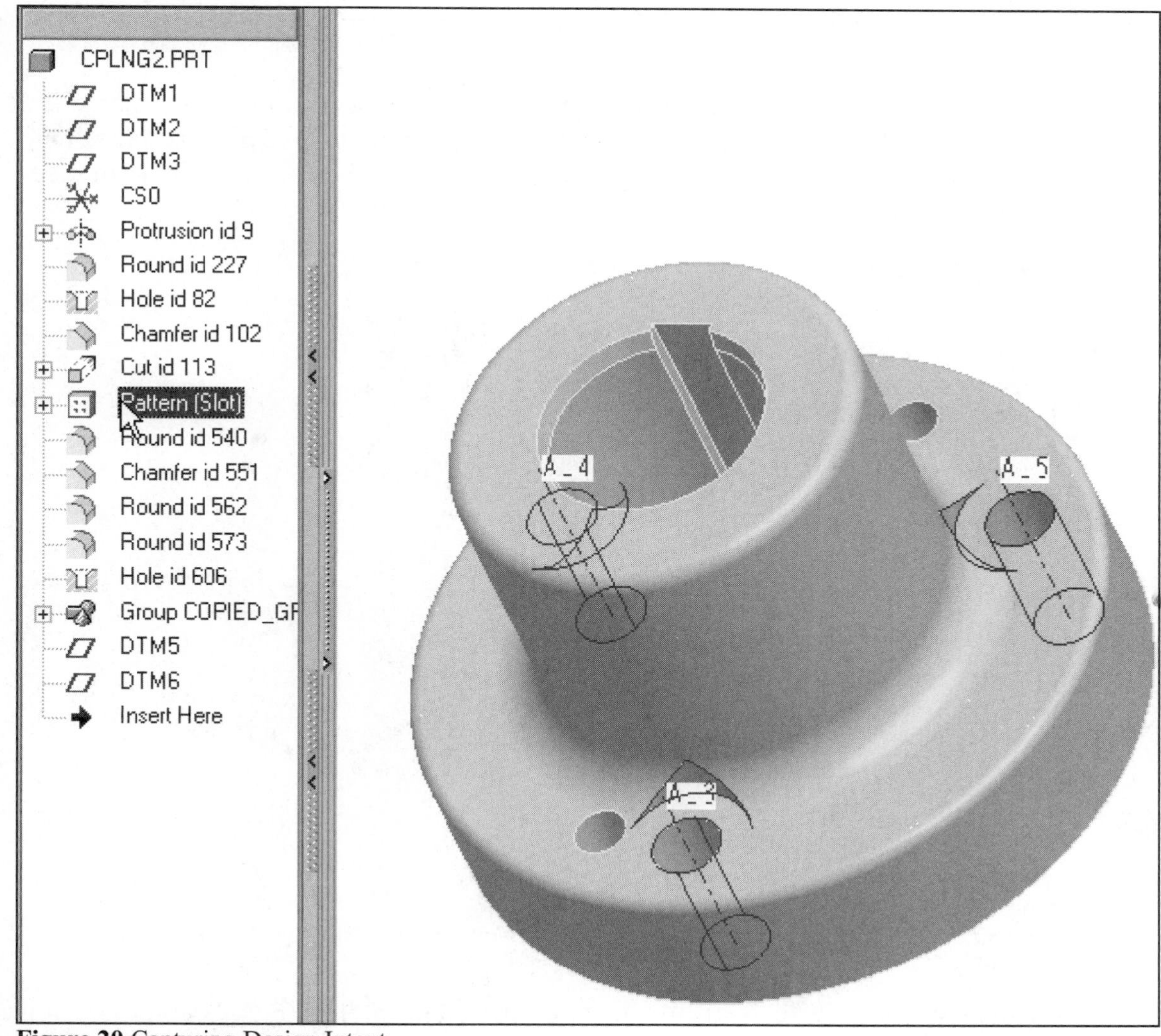

Figure 20 Capturing Design Intent

Parametric designs capture relationships in several ways:

Implicit Relationships Implicit relationships occur when new model geometry is sketched and dimensioned relative to existing features and parts. An implicit relationship is established, for instance, when the section sketch of a tire (Fig. 21) uses rim edges as a reference.

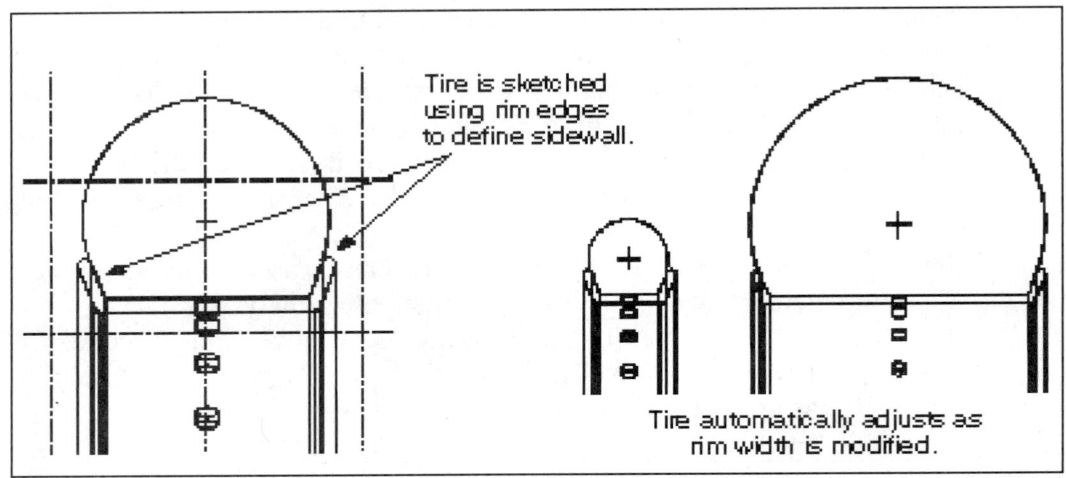

Figure 21 Tire and Rim

Patterns Design features often follow a geometrically predictable pattern. Features and parts are patterned in parametric design by referencing either construction dimensions or existing patterns. One example of patterning is a wheel hub with spokes (Fig. 22). First, the spoke holes are radially patterned. The spokes can then be strung by referencing this pattern.

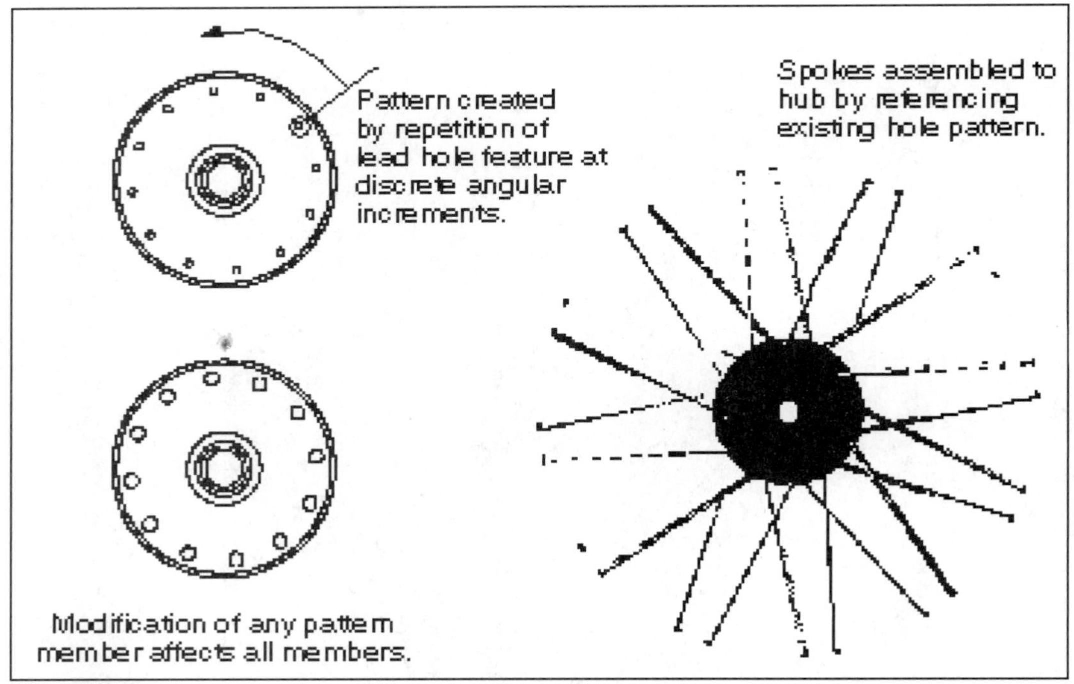

Figure 22 Patterns

Modification to a pattern member affects all members of that pattern. This helps capture design intent by preserving the duplicate geometry of pattern members.

Explicit Relations Whereas implicit relationships are implied by the feature creation method, the user mathematically enters an explicit relation. This equation is used to relate feature and part dimensions in the desired manner. An explicit relation (Fig. 23) might be used, for example, to control sizes on a model.

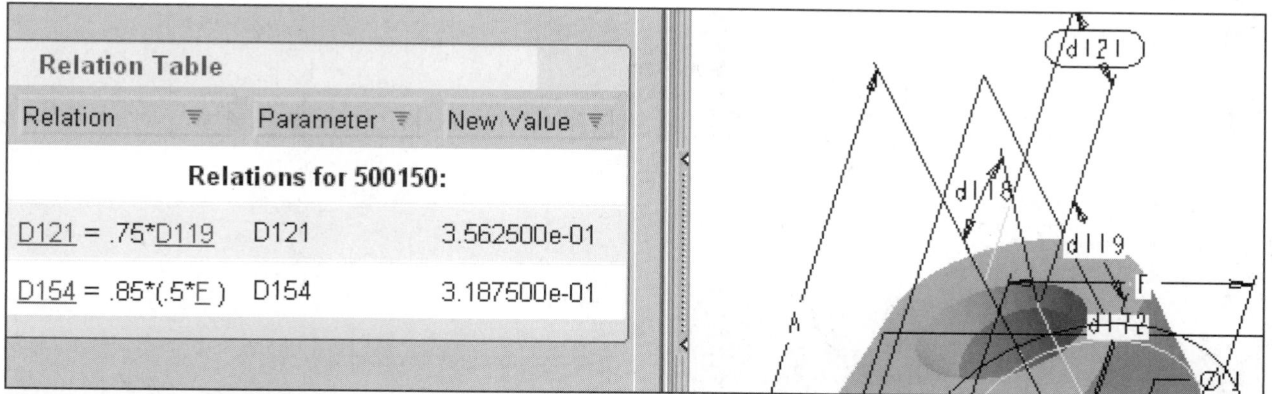

Figure 23 Adding Relations

Family Tables Family tables are used to create part families (Fig. 24) from generic models by tabulating dimensions or the presence of certain features or parts. A family table might be used, for example, to catalog a series of couplings with varying width and diameter as shown in Figure 25.

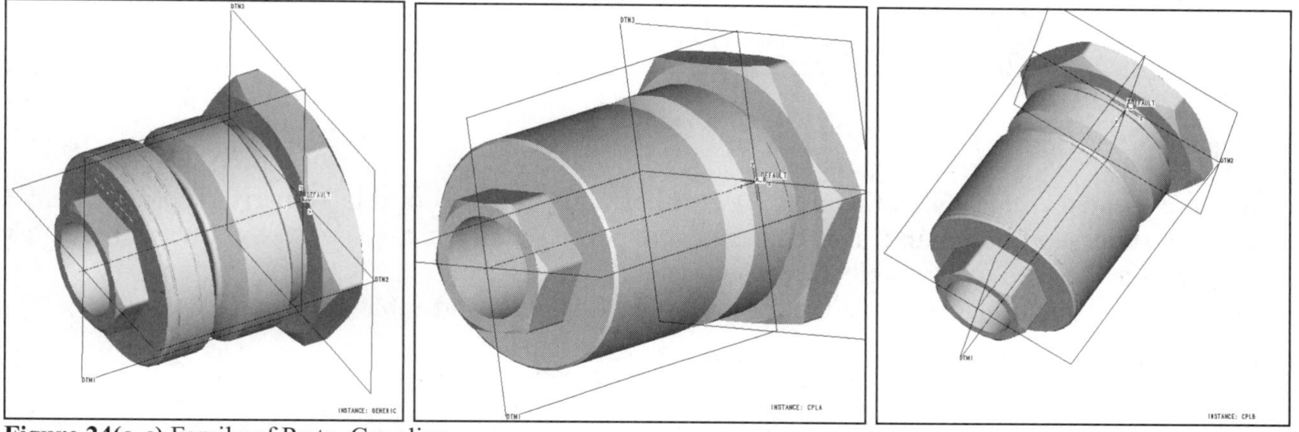

Figure 24(a-c) Family of Parts- Coupling

Family Table :COUPLING-FITTING

File Edit Insert Tools

Look In: COUPLING-FITTING

Type	Instance Name	d30	d46	F1068 [CUT]	F829 [SLOT]	F2620 COPIED_G...
	COUPLING-FITTI...	2.06	0.3130	Y	Y	Y
	CPLA	3.00	0.5000	N	N	N
	CPLB	3.25	0.6250	N	Y	N
	CPLC	3.50	0.7500	Y	N	Y

OK Open Cancel

Figure 25 Family Table for Coupling

The modeling task is to incorporate the features and parts of a complex design while properly capturing design intent to provide flexibility in modification. Parametric design modeling is a synthesis of physical and intellectual design (Fig. 26).

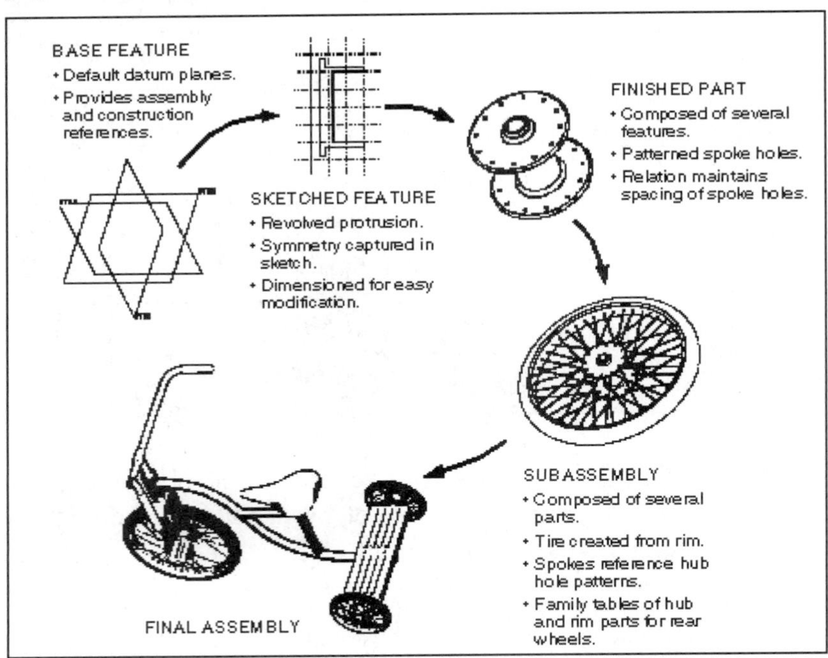

Figure 26 Relations

Assemblies

Just as parts are created from related features, **assemblies** are created from related parts. The progressive combination of subassemblies, parts, and features into an assembly creates parent-child relationships based on the references used to assemble each component (Fig. 27).

The *Assembly* functionality is used to assemble existing parts and subassemblies.

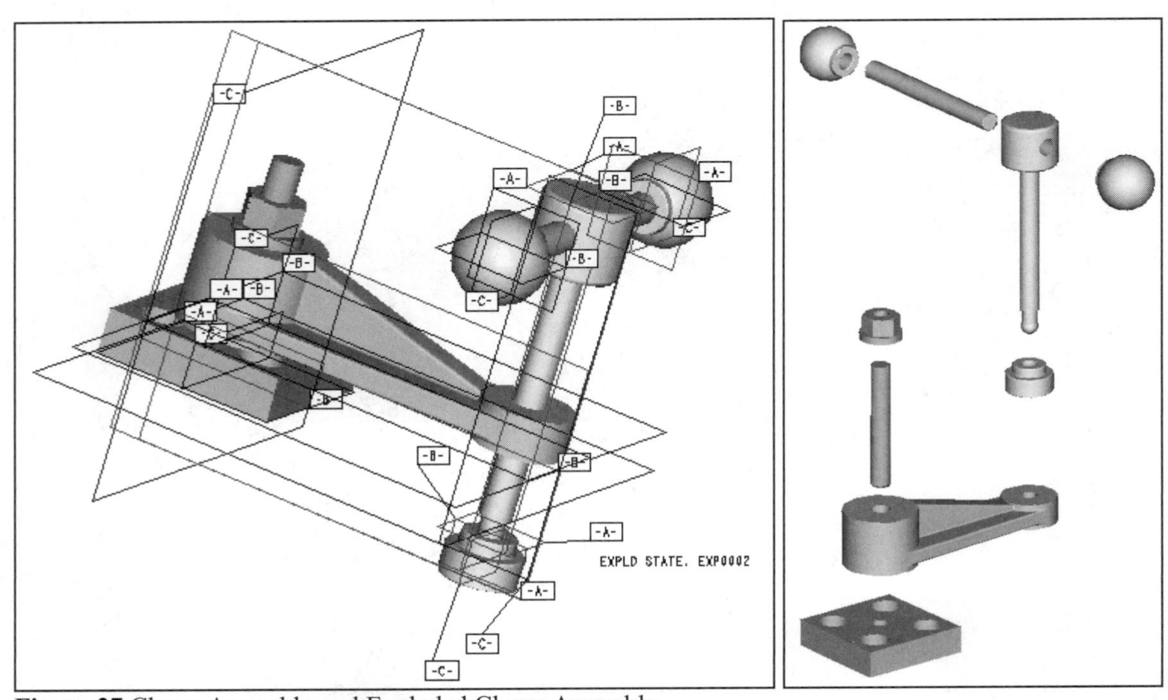

Figure 27 Clamp Assembly and Exploded Clamp Assembly

During assembly creation, you can:

- Simplify a view of a large assembly by creating a simplified representation
- Perform automatic or manual placement of component parts
- Create an exploded view of the component parts
- Perform analysis, such as mass properties and clearance checks
- Modify the dimensional values of component parts
- Define assembly relations between component parts
- Create assembly features
- Perform automatic interchange of component parts
- Create parts in Assembly mode
- Create documentation drawings of the assembly

Just as features can reference part geometry, parametric design also permits the creation of parts referencing assembly geometry. **Assembly mode** allows the designer both to fit parts together and to design parts based on how they should fit together.

In Figure 28, the assembly *Bill of Materials* report is generated.

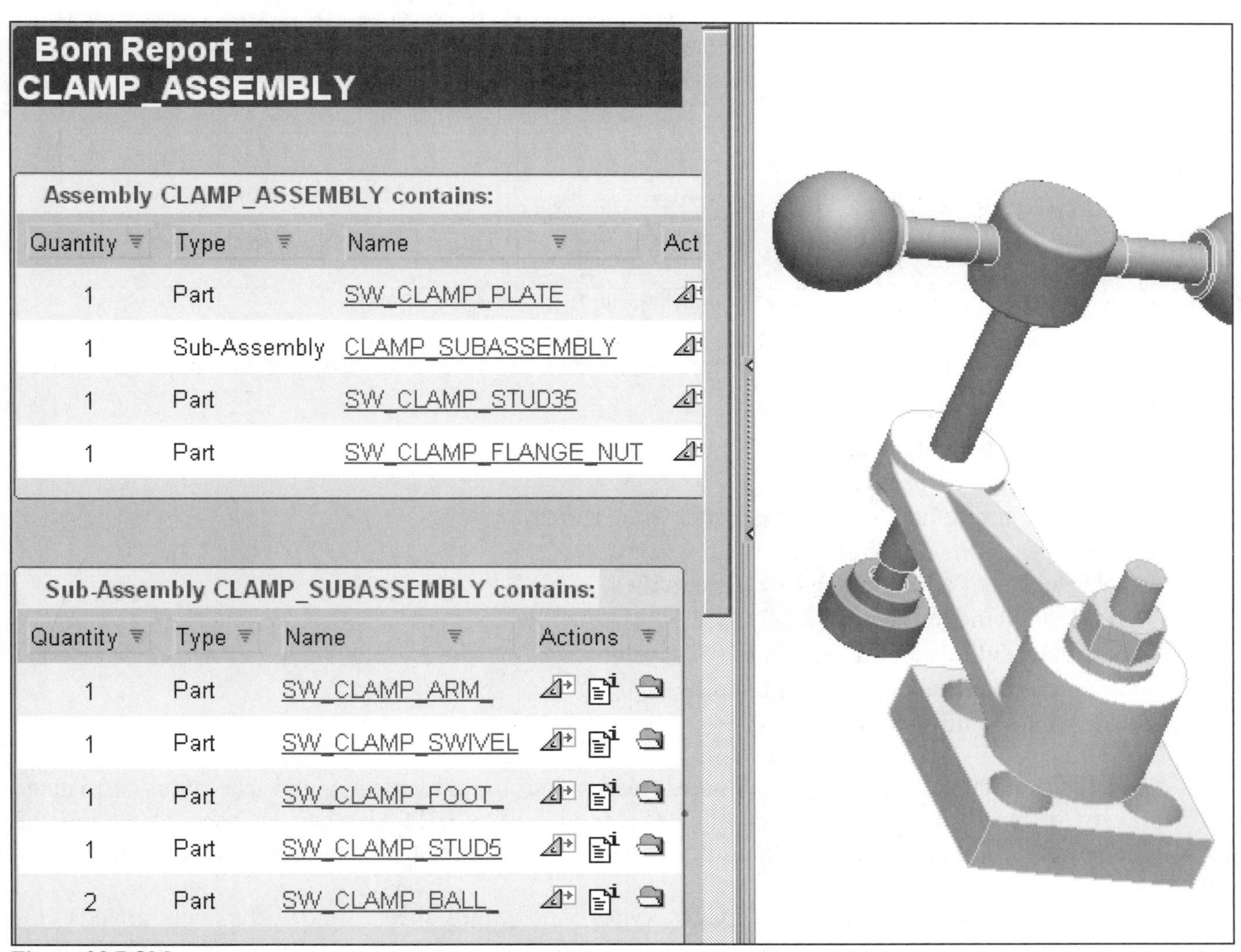

Bom Report : CLAMP_ASSEMBLY

Assembly CLAMP_ASSEMBLY contains:

Quantity	Type	Name	Act
1	Part	SW_CLAMP_PLATE	
1	Sub-Assembly	CLAMP_SUBASSEMBLY	
1	Part	SW_CLAMP_STUD35	
1	Part	SW_CLAMP_FLANGE_NUT	

Sub-Assembly CLAMP_SUBASSEMBLY contains:

Quantity	Type	Name	Actions
1	Part	SW_CLAMP_ARM_	
1	Part	SW_CLAMP_SWIVEL	
1	Part	SW_CLAMP_FOOT_	
1	Part	SW_CLAMP_STUD5	
2	Part	SW_CLAMP_BALL_	

Figure 28 BOM

Drawings

You can create drawings of all parametric design models (Fig. 29) or create or retrieve by importing files from other systems. All model views in the drawing are *associative:* if you change a dimensional value in one view, other drawing views update accordingly. Moreover, drawings are associated with their parent models. Any dimensional changes made to a drawing are automatically reflected in the model. Any changes made to the model (e.g., addition of features, deletion of features, dimensional changes, and so on) in Part, Sheet Metal, Assembly, or Manufacturing modes are also automatically reflected in their corresponding drawings.

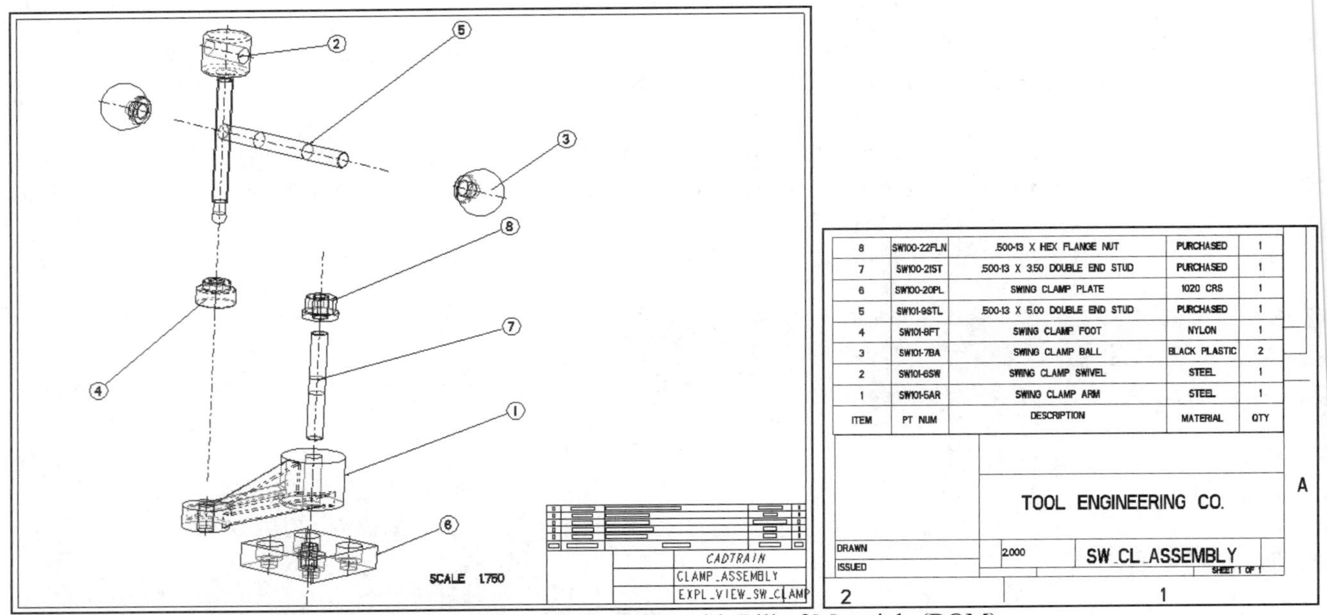

ITEM	PT NUM	DESCRIPTION	MATERIAL	QTY
8	SW100-22FLN	.500-13 X HEX FLANGE NUT	PURCHASED	1
7	SW100-21ST	.500-13 X 3.50 DOUBLE END STUD	PURCHASED	1
6	SW100-20PL	SWING CLAMP PLATE	1020 CRS	1
5	SW101-9STL	.500-13 X 5.00 DOUBLE END STUD	PURCHASED	1
4	SW101-8FT	SWING CLAMP FOOT	NYLON	1
3	SW101-7BA	SWING CLAMP BALL	BLACK PLASTIC	2
2	SW101-6SW	SWING CLAMP SWIVEL	STEEL	1
1	SW101-5AR	SWING CLAMP ARM	STEEL	1

Figure 29(a-b) Ballooned Exploded View Assembly Drawing with Bill of Materials (BOM)

The **Drawing** functionality is used to create annotated drawings of parts and assemblies. During drawing creation, you can:

- Add views of the part or assembly
- Show existing dimensions
- Incorporate additional driven or reference dimensions
- Create notes to the drawing
- Display views of additional parts or assemblies
- Add sheets to the drawing
- Create draft entities on the drawing
- Balloon components on an assembly drawing (Fig. 30)
- Create an associative BOM

You can annotate the drawing with notes, manipulate the dimensions, and use layers to manage the display of different items on the drawing. The module **Pro/DETAIL** can be used to extend the drawing capability or as a stand-alone module allowing you to create, view, and annotate models and drawings.

Pro/DETAIL supports additional view types and multi-sheets and offers commands for manipulating items in the drawing and for adding and modifying different kinds of textural and symbolic information. In addition, the abilities to customize engineering drawings with sketched geometry, create custom drawing formats, and make numerous cosmetic changes to the drawing are available.

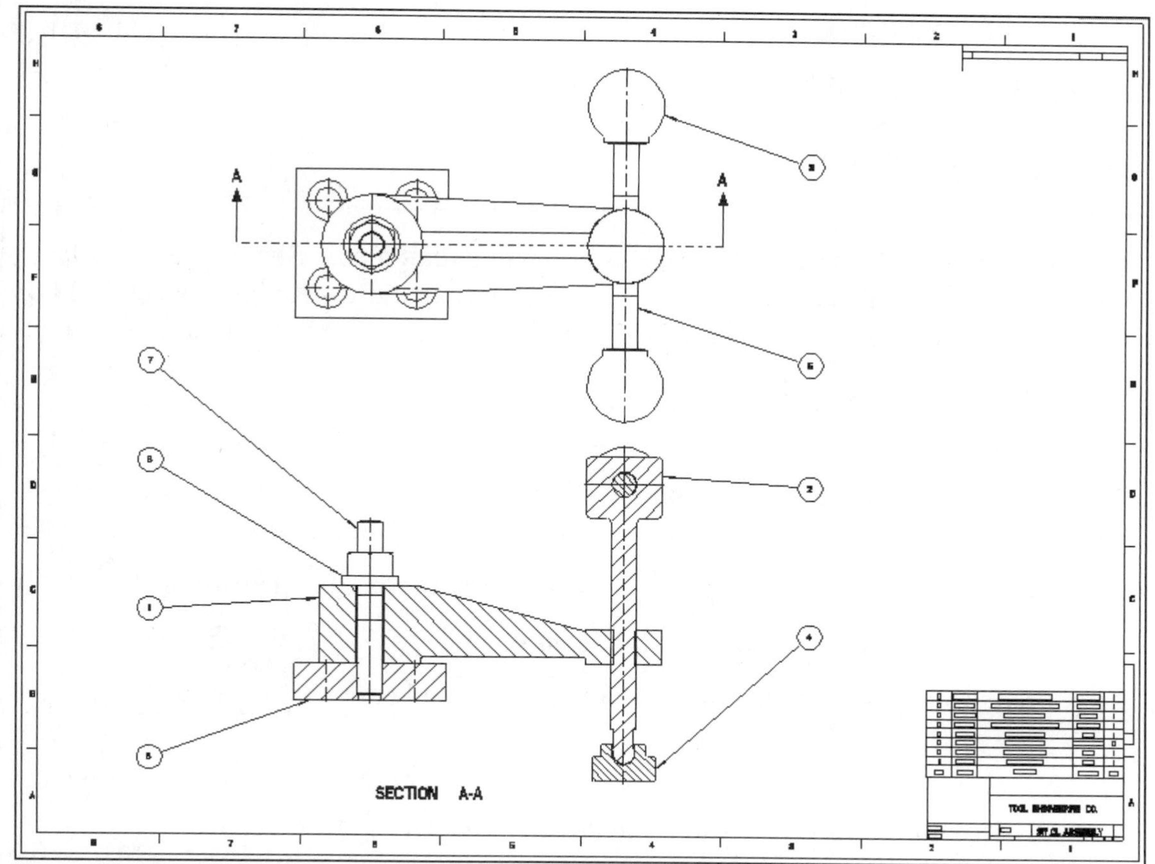

Figure 30 Assembly Drawing

Drawing mode in parametric design provides you with the basic ability to document solid models in drawings that share a two-way associativity (Fig. 31).

Changes that are made to the model in Part mode or Assembly mode will cause the drawing to update automatically and reflect the changes. Any changes made to the model in Drawing mode will be immediately visible on the model in Part and Assembly modes. The part shown in Figure 31 has been detailed. Basic Pro/E allows you to create drawing views of one or more models in a number of standard views with dimensions.

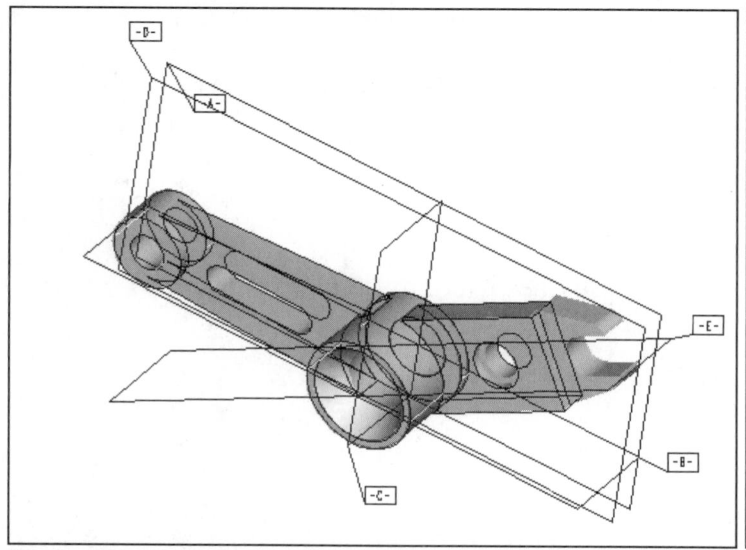

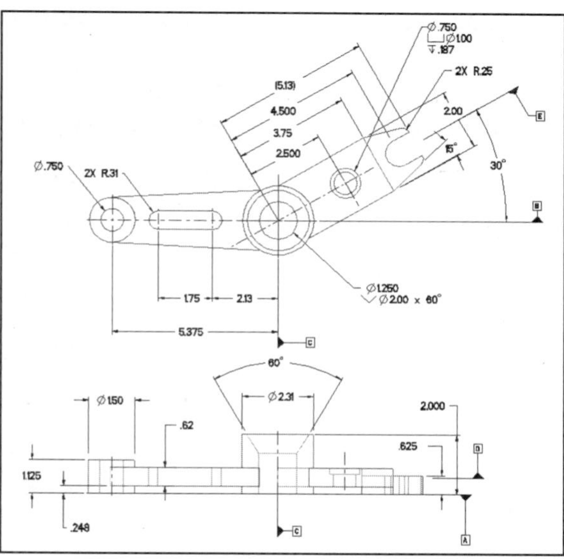

Figure 31(a-b) Angle Frame Model and Drawing

Using the Text

The following icons, symbols and conventions will be used throughout the text:

Command sequences are always in a box:

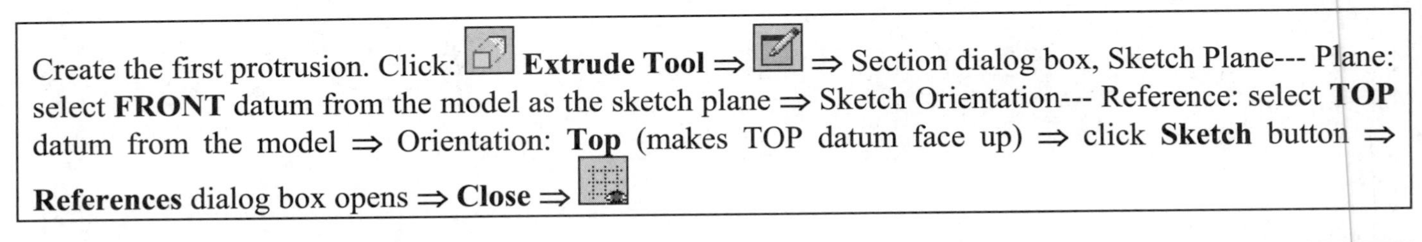

Create the first protrusion. Click: **Extrude Tool** ⇒ ⇒ Section dialog box, Sketch Plane--- Plane: select **FRONT** datum from the model as the sketch plane ⇒ Sketch Orientation--- Reference: select **TOP** datum from the model ⇒ Orientation: **Top** (makes TOP datum face up) ⇒ click **Sketch** button ⇒ **References** dialog box opens ⇒ **Close** ⇒

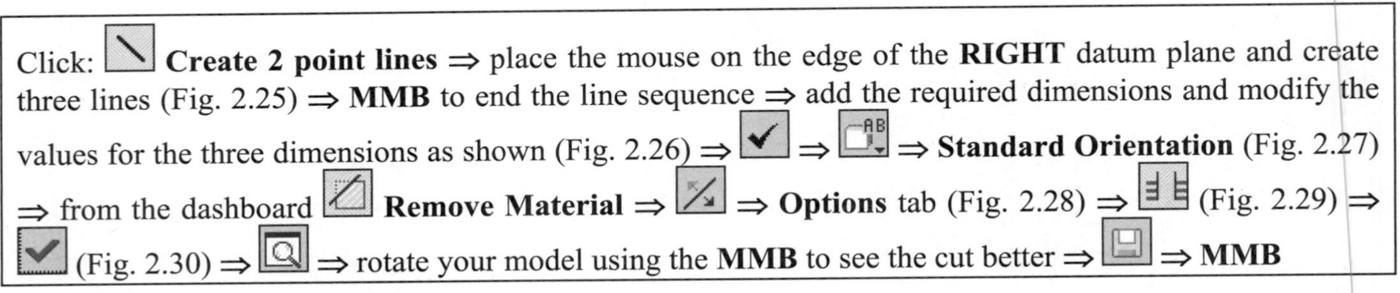

Click: **Create 2 point lines** ⇒ place the mouse on the edge of the **RIGHT** datum plane and create three lines (Fig. 2.25) ⇒ **MMB** to end the line sequence ⇒ add the required dimensions and modify the values for the three dimensions as shown (Fig. 2.26) ⇒ ⇒ ⇒ **Standard Orientation** (Fig. 2.27) ⇒ from the dashboard **Remove Material** ⇒ ⇒ **Options** tab (Fig. 2.28) ⇒ (Fig. 2.29) ⇒ (Fig. 2.30) ⇒ ⇒ rotate your model using the **MMB** to see the cut better ⇒ ⇒ **MMB**

Commands:

- ⇒ — Continue with command sequence or screen picks using LMB
- **Create lines** — icons (with description) indicate as command picks using LMB

Mouse or keyboard terms used in text:

- **LMB** — **L**eft **M**ouse **B**utton
  or "**Pick**" — term used to direct an action (i.e., "Pick the surface")
  or "**Click**" — term used to direct an action (i.e., "Click on the icon")
  or "**Select**" — term used to direct an action (i.e., "Select the feature")
- **MMB** — **M**iddle **M**ouse **B**utton (accept the current selection or value)
  or **Enter** — press **Enter** key to accept entry
  or — Click on icon to accept entry
- **RMB** — **R**ight **M**ouse **B**utton
  (toggles to next selection, or provides list of available commands)

Using Wildfire Pro/ENGINEER

The *Quick Reference Card* of the Pro/E Wildfire interface (Figs. 32-35). See www.ptc.com for PDF download.

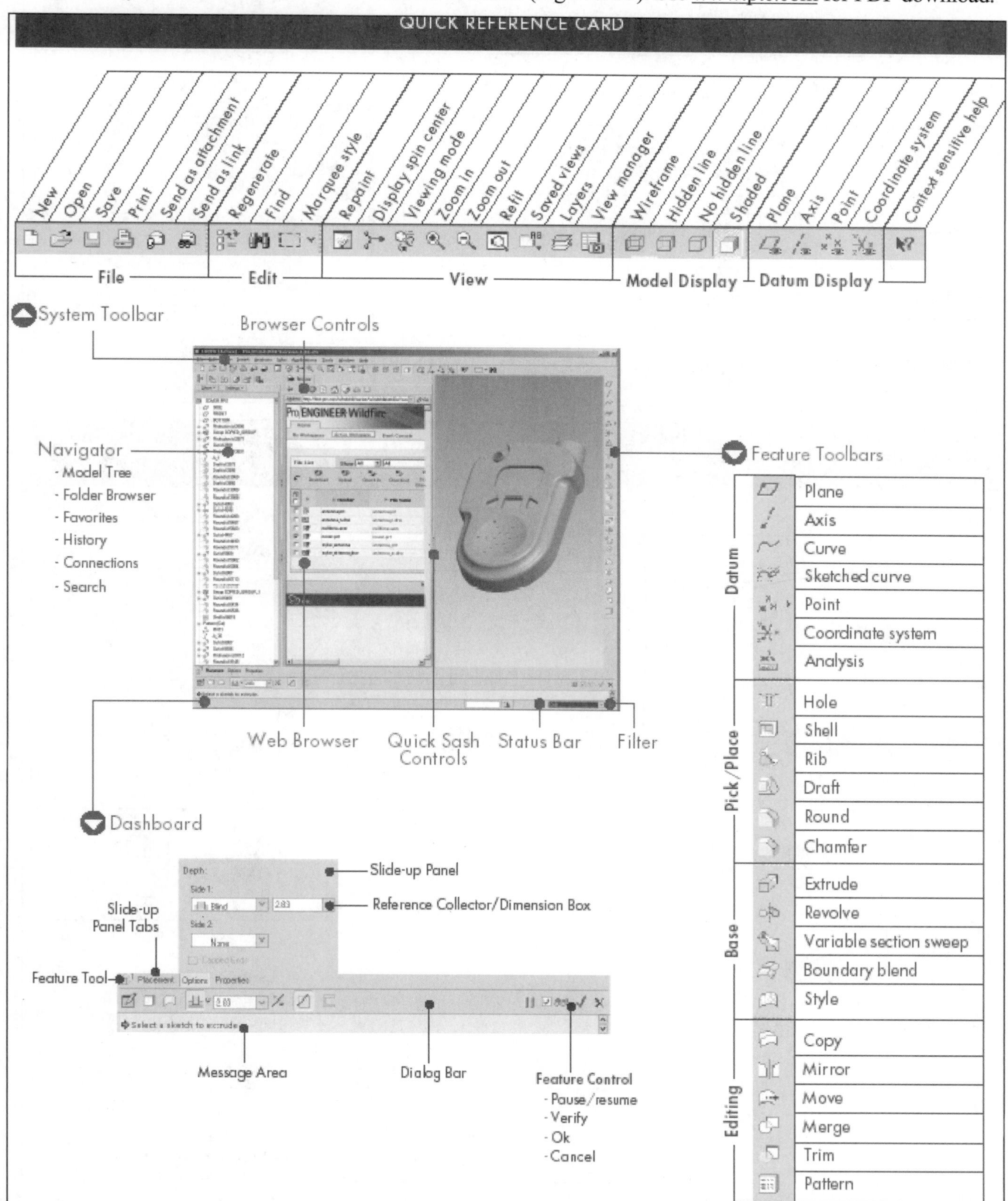

Figure 32 Quick Reference Card

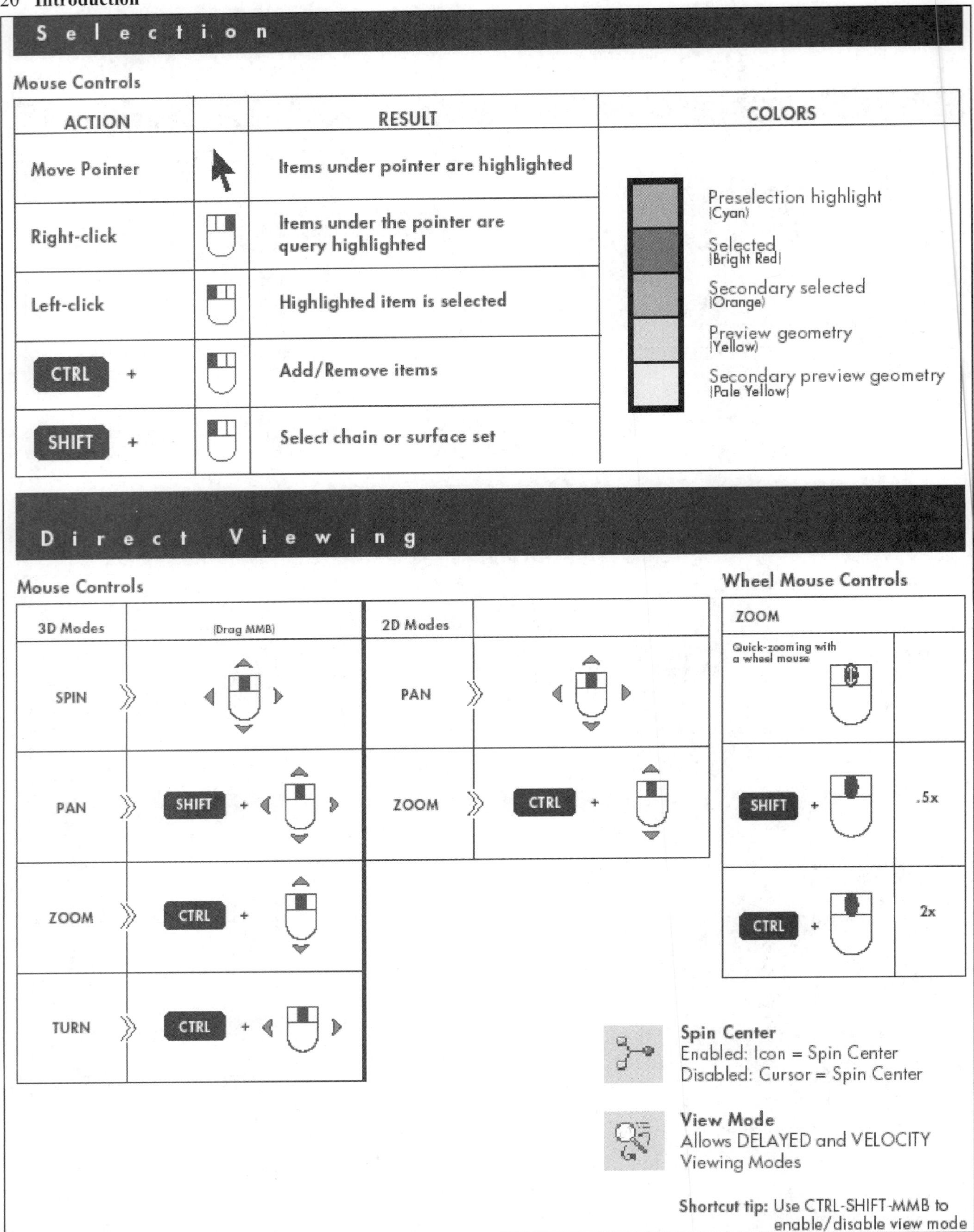

Figure 33 Selection and Direct Viewing

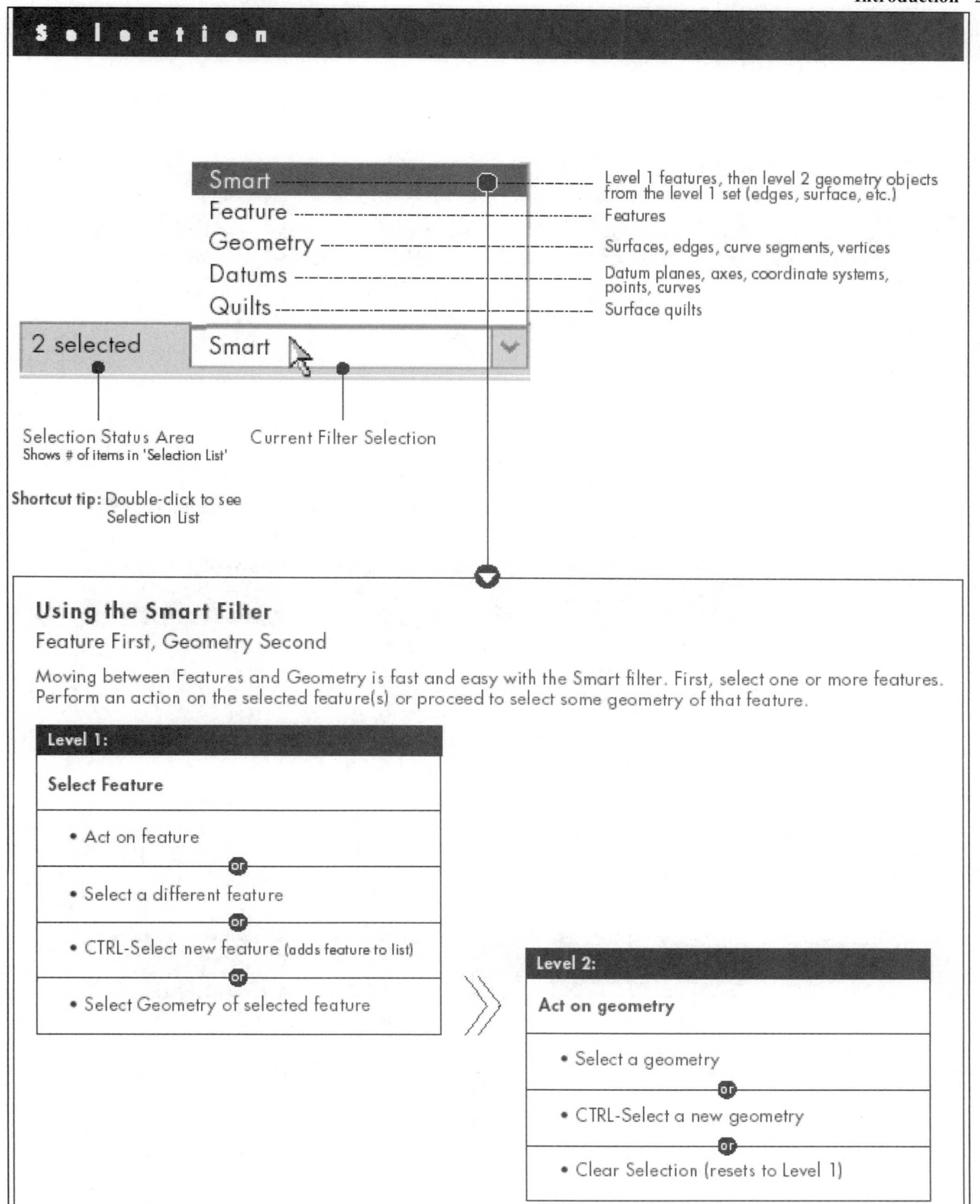

Figure 34 Selection and Filters

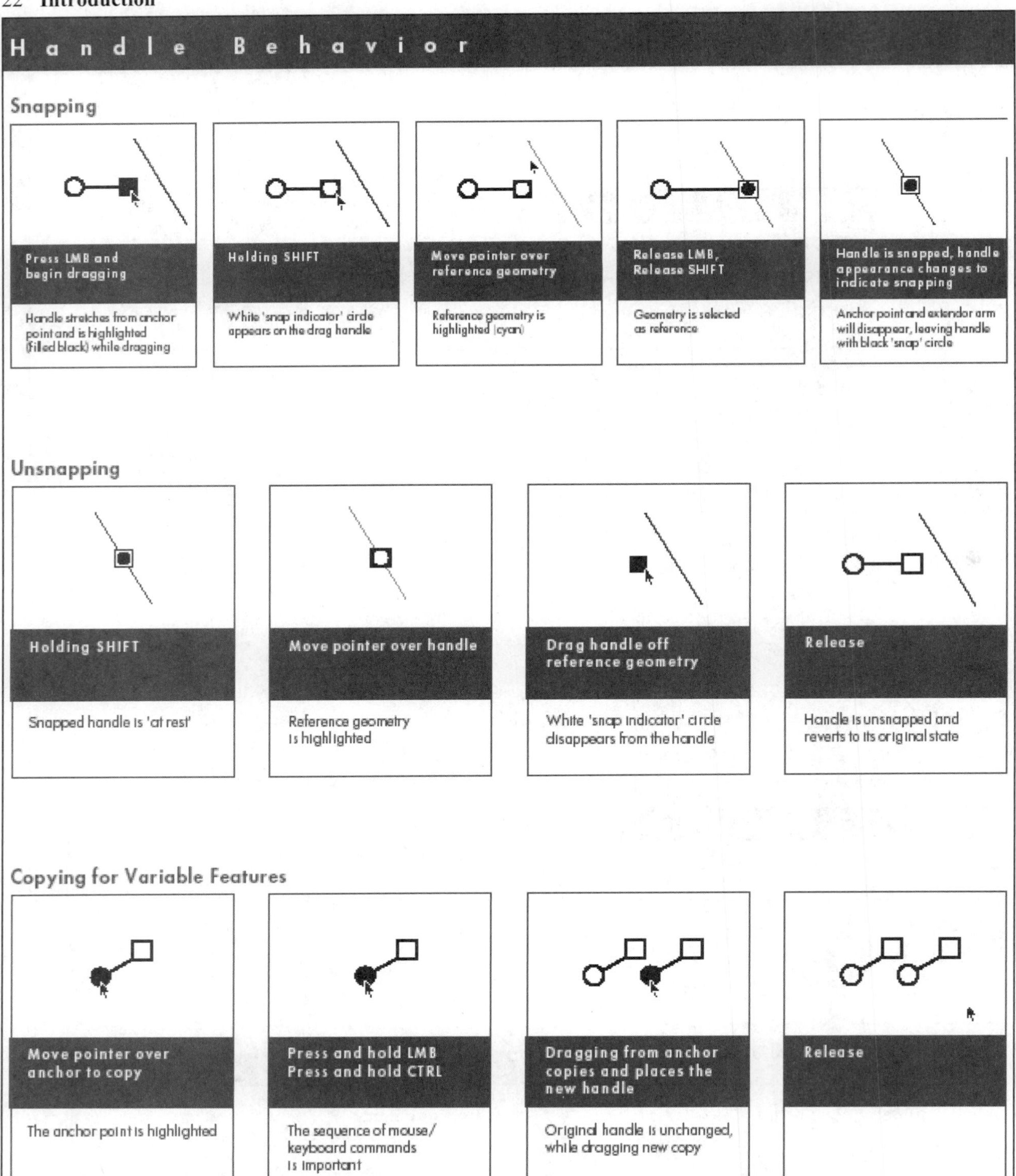

Figure 35 Handle Behavior

Lesson 1 Introduction to Pro/ENGINEER Wildfire

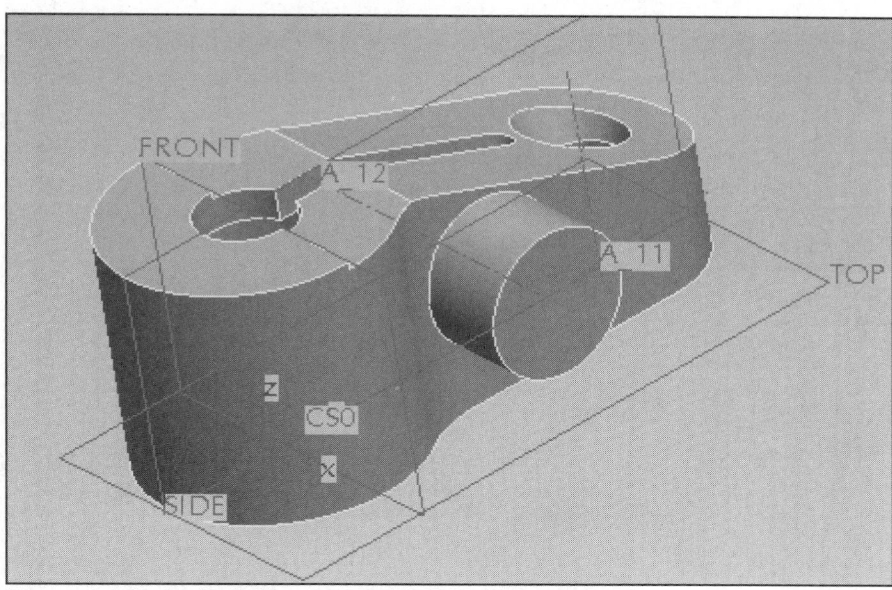

Figure 1.1 Swing/Pull Clamp Arm (SPX Fluid Power Part)

OBJECTIVES

- Understand the **User Interface (UI)**
- Download **Catalog Parts** using the **Browser**
- Master the **File Functions**
- Learn how to **Email** active Pro/ENGINEER objects
- Become familiar with the **Help** facility
- Be introduced to the **Display** and **View** capabilities
- Use **Mouse Buttons** to **Pan**, **Zoom**, and **Rotate** the object
- Change **Display Settings**
- Investigate an object with **Information Tools**
- Experience the **Model Tree** functionality

Pro/ENGINEER Wildfire

The first lesson will introduce you to Pro/ENGINEER's working environment. An existing part model (Fig. 1.1) will be downloaded from the Catalog and used to demonstrate the UI (user interface) and the general interaction required to master Pro/E.

You will be using a part available thru the **Catalog** using the **Browser**. If you are not connected to the Internet, your instructor will provide you with a simple start part. In addition, if you have Pro/Library installed, you may use any library part that you wish.

Pro/ENGINEER's Main Window

The Pro/ENGINEER user interface consists of a navigation area, an embedded Web browser, the menu bar, toolbars, information areas, and the graphics area (Fig. 1.2).

- **Navigation Area** Located on the left side of the Pro/E main window, and includes tabs for the Model Tree and Layer Tree, Folder Browser, Favorites, History and Connections.
- **Embedded Web Browser** The Web browser is located to the right of the navigation area. This standard browser provides you access to internal or external Web sites.

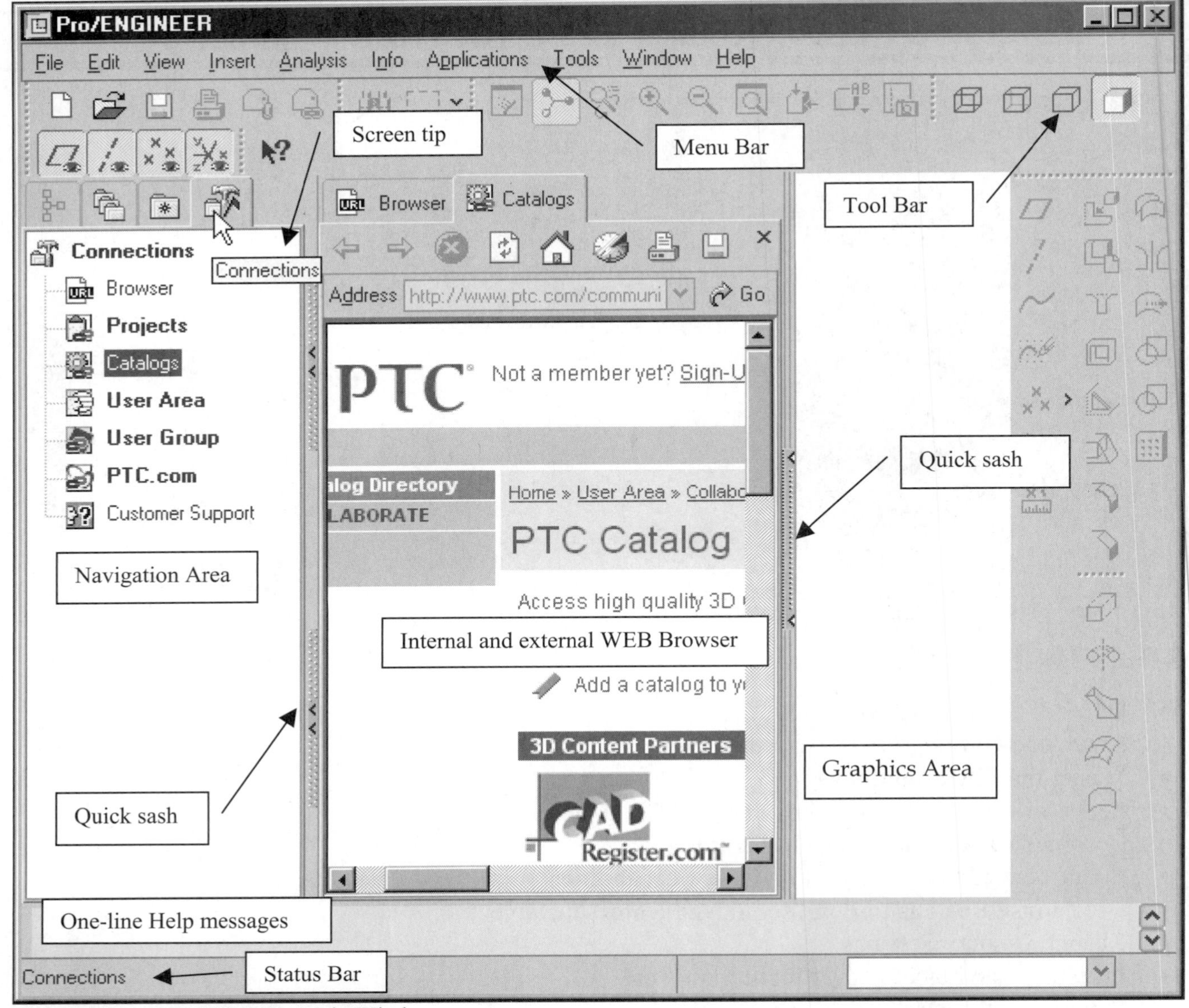

Figure 1.2 Pro/ENGINEER's Main Window

- **Menu Bar** The menu bar contains menus with options for creating, saving, and modifying models, and also contains menus with options for setting environment and configuration options.
- **Information Areas** Each Pro/E window has a message area near the top of and a status bar at the bottom of the window for displaying one-line Help messages.
  - **Message Area** Messages related to work performed are displayed here.
  - **Status Bar** At the bottom of the window is a one-line Help area that dynamically displays one-line context-sensitive Help messages. If you move your mouse pointer over a menu command or dialog box option, a one-line description appears in this area.
  - **Screen Tips** The status bar messages also appear in small yellow boxes near the menu option or dialog box item or toolbar button that the mouse pointer is passing over.
- **Toolbars** The toolbars contain icons to speed up access to commonly used menu commands. By default, the toolbars consist of a row of buttons located directly under the main menu bar. Toolbar buttons can be positioned on the top, left, and right of the window. Toolbar buttons can be added or removed from the Toolchest by customizing the layout.
- **Graphics Area** The graphics area is the main working space (main window) for modeling and is to the right of the embedded Browser. Normally the Browser is collapsed when modeling.

Lesson 1 STEPS

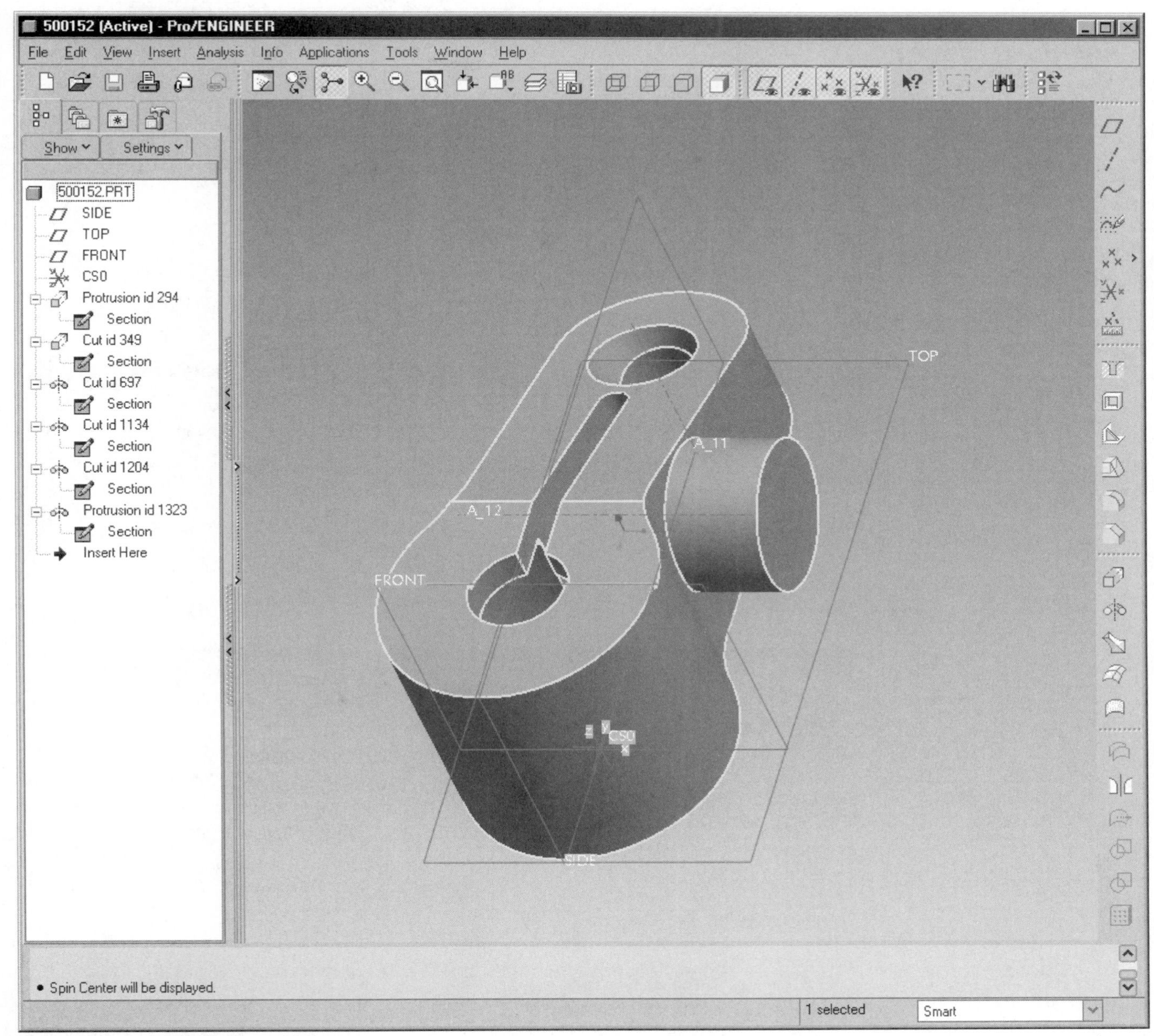

Figure 1.3 Swing/Pull Clamp Arm

Catalog Parts Swing/Pull Clamp

In order to see and use Pro/ENGINEER's UI we must have an active object (Fig. 1.3). Since you will not be modeling the part, you must download an existing model from the Catalog of parts from PTC. If you do not have access to the catalog, email the author at **lgl@cad-resources.com**.

Throughout the text, a box surrounds all commands and menu selections.

Open your Pro/ENGINEER program using a shortcut icon on your Desktop or with WINDOWS **Start** ⇒ **Programs** (Pro/E will open on your computer) ⇒ **Connections** tab ⇒ Catalogs (Fig. 1.4) ⇒ SPX HYTEC click Search the catalog

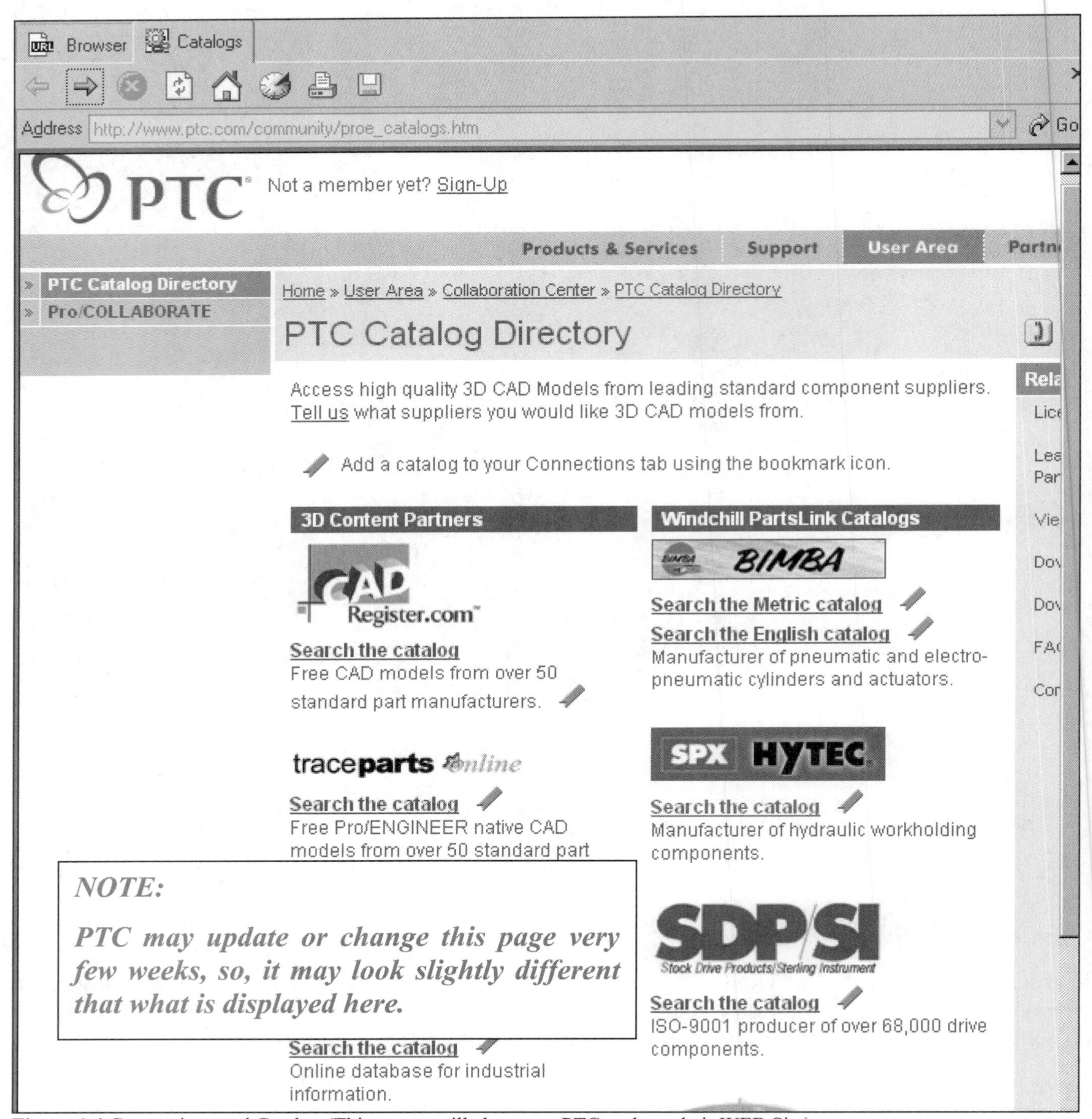

Figure 1.4 Connections and Catalog (This screen will change as PTC updates their WEB Site)

Click:  **Clamping** [Fig. 1.5(a)]

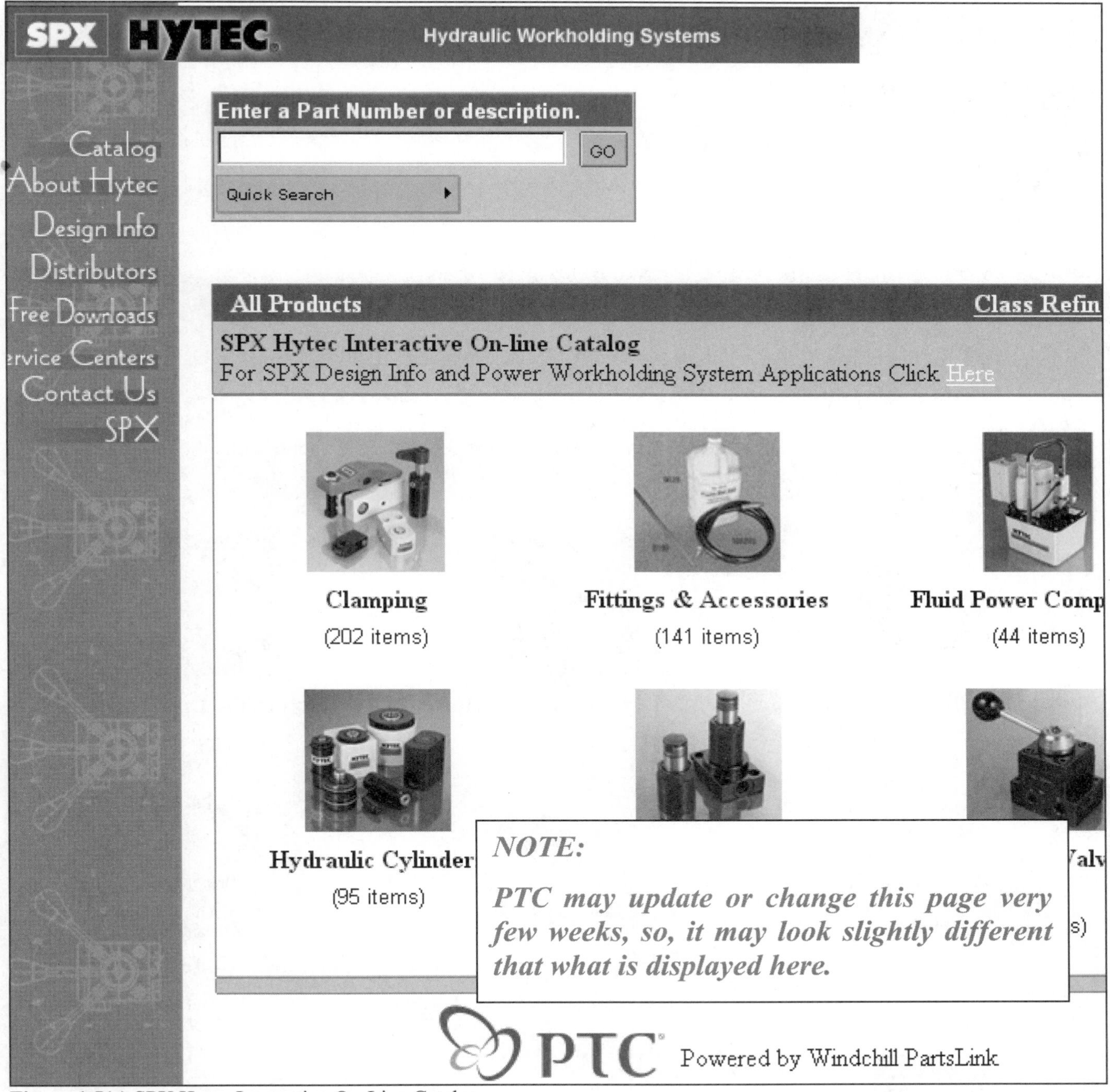

Figure 1.5(a) SPX Hytec Interactive On-Line Catalog

Click: 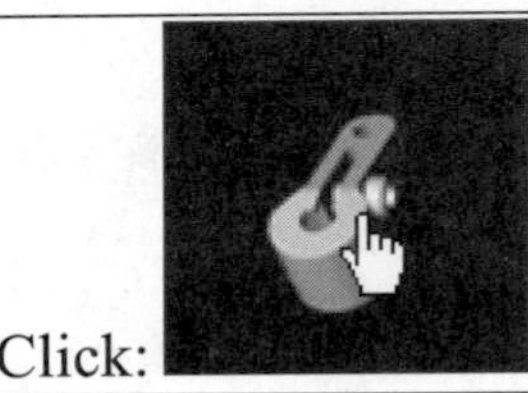 **Swing/Pull Clamp Arms** [Fig. 1.5(b)]

TEC Hydraulic Workholding Systems

Enter a Part Number or description.

GO

Quick Search

All Products / Clamping

Class Refine Search

Hytec's workholding devices include many types of hydraulic clamps that will handle most clamping applications. All of our hydraulic clamps are ideal for appications where it is neccassary for the clamping actuator to be moved away from the work piece. They perform the same function as clamping cylinders, but thier ability to swing or retract out of the way of cutters, plus the advantage of quick and easy part loading, makes them the perfect choice for the jobs with special workholding needs.

Swing/Pull Clamps

(162 items)

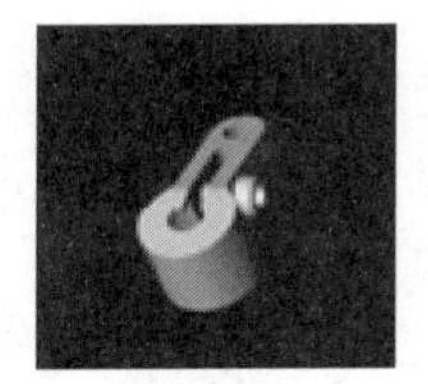

Swing/Pull Clamp Arms

(8 items)

Hydraulic Die Clamp

(1 items)

Hydraulic Edge Clamp

(2 items)

Hydraulic Retract Clamp

(6 items)

Hydraulic Uniforce Clamp

(15 items)

NOTE:

PTC may update or change this page very few weeks, so, it may look slightly different that what is displayed here.

Figure 1.5(b) SPX Hytec Catalog

Double-click on the entry for Product Number **500152** (use the slide bar to locate if needed) [Fig. 1.5(c)]:

4500 Lbs. i.0	Swing Clamp Arm	500152	SPX-HYTEC\CLAMPS-02	4	spx-spca-01.prt	500152

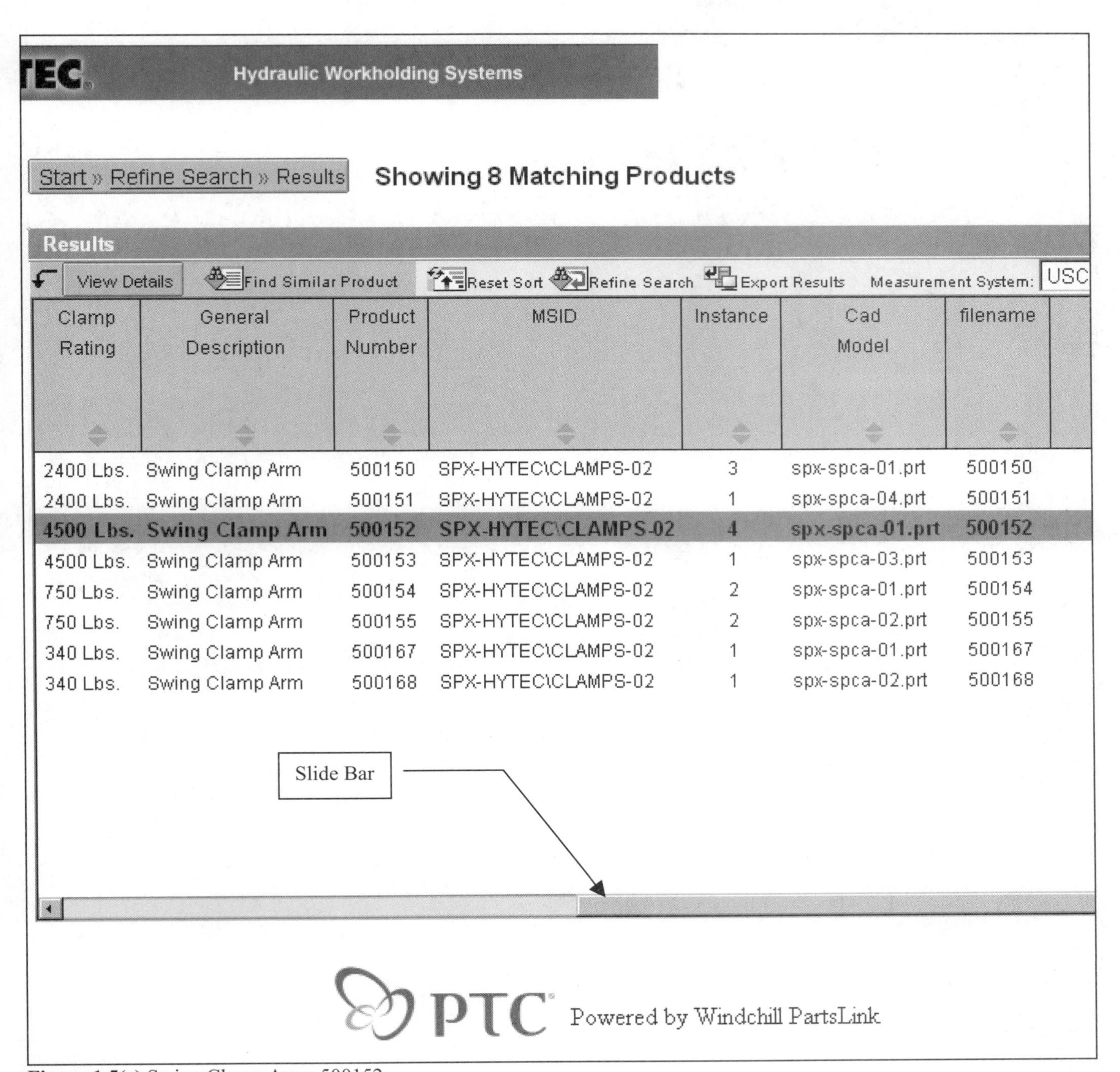

Figure 1.5(c) Swing Clamp Arms 500152

Click: 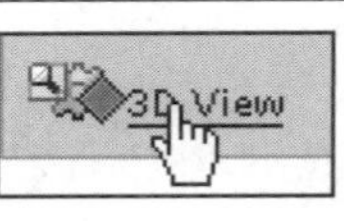 **3D View** [Fig. 1.5(d)]

Figure 1.5(d) Swing/Pull Clamp Arms Part Number 500152

ProductView Lite Edition

ProductView Lite is a data visualization tool that provides basic viewing capabilities for 3D models, including measurement and annotation. The Model Viewer lets you perform navigation tasks such as zoom, pan, rotate, fly-through, and exploding components of 3D models. You can also display models in wireframe, HLR (hidden lines removed), or shaded render modes. ProductView Lite provides the ability to view annotations that are associated with drawings or models. The annotations must be stored along with the file that you are viewing. Annotation sets appear along the left side of the ProductView Lite window as thumbnails. Click a thumbnail to view that annotation set.

You can use this tool to view annotations only; you cannot create annotations in ProductView Lite. In addition, 2D markups currently cannot be viewed with ProductView Lite; as a result, markups do not appear in the list of annotations.

The View Part window opens with the clamp displayed, click: Viewer Commands [Fig. 1.6(a-b)]

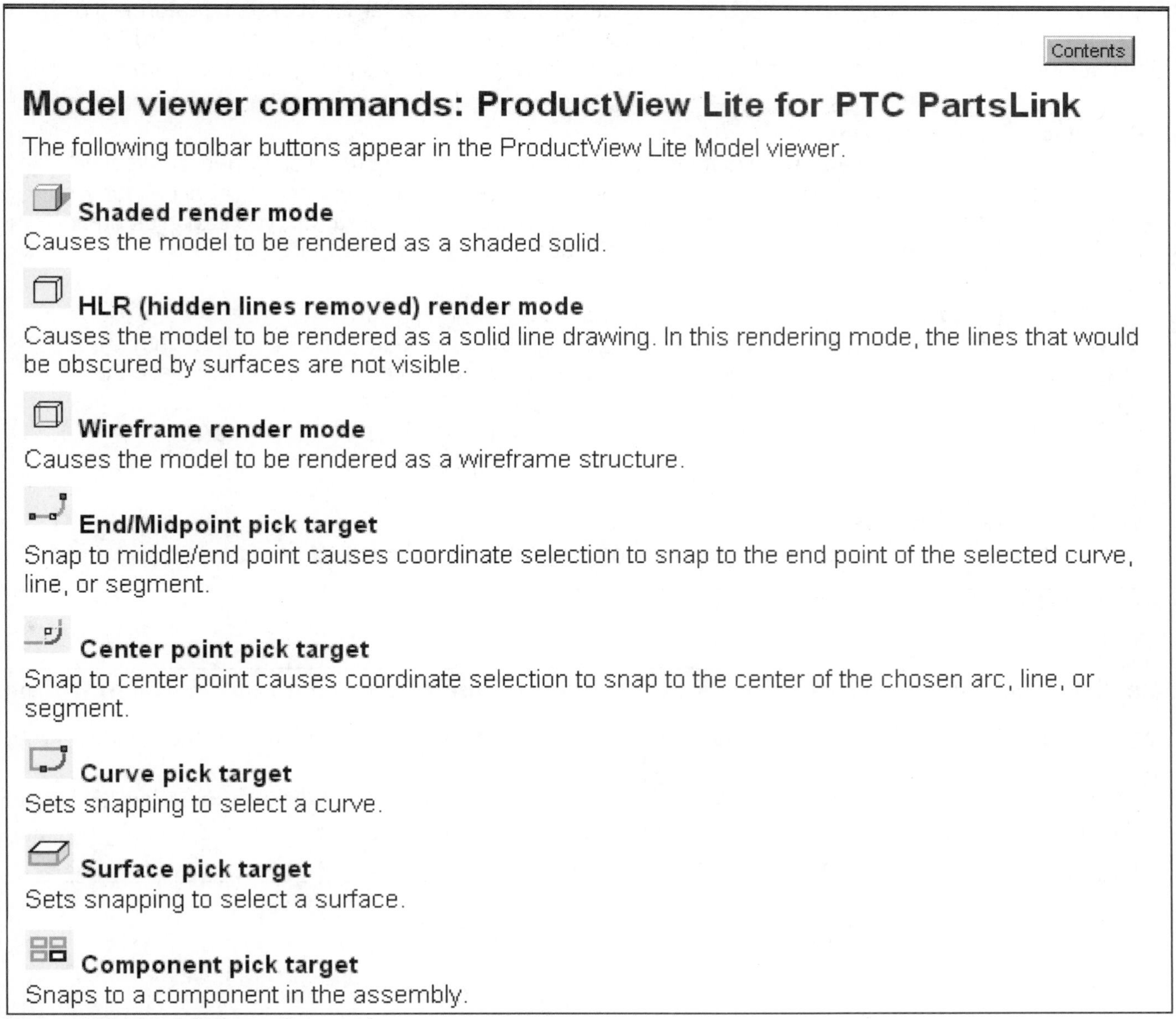

Contents

Model viewer commands: ProductView Lite for PTC PartsLink

The following toolbar buttons appear in the ProductView Lite Model viewer.

Shaded render mode
Causes the model to be rendered as a shaded solid.

HLR (hidden lines removed) render mode
Causes the model to be rendered as a solid line drawing. In this rendering mode, the lines that would be obscured by surfaces are not visible.

Wireframe render mode
Causes the model to be rendered as a wireframe structure.

End/Midpoint pick target
Snap to middle/end point causes coordinate selection to snap to the end point of the selected curve, line, or segment.

Center point pick target
Snap to center point causes coordinate selection to snap to the center of the chosen arc, line, or segment.

Curve pick target
Sets snapping to select a curve.

Surface pick target
Sets snapping to select a surface.

Component pick target
Snaps to a component in the assembly.

Figure 1.6(a) Model Viewer Commands

Zoom all
Adjusts the image size in the display to show the entire drawing within the window.

Zoom window
Magnifies the window to fill the screen. To use the Zoom Window tool, click and drag with the mouse to define a window. When you release the mouse button, the window is magnified to the maximum allowable size, while still showing all the areas within the zoom window borders.

Restore location
Allows you to move all parts back to their original positions. This command restores the location of selected components. If no components are selected, it restores all components.

Select
The Select command allows you to select a component. The selected component is highlighted in red in the graphical display. The selected component(s) become the target of certain features. To select a component, click the Select button. Press **Ctrl** to select multiple objects. Press **Shift** to indicate when selection is available for your current pick target (such as point, curve, or surface).

Dimension measurement
Select **Dimension measurement** to display measurements for a selected pick target. To display dimensions of a component, click on that component. The dimensions appear in a label callout. For example, you can click on a cylinder to display the height, radius, and area. Or, you can click on an arc to display the length and radius. For more information, see Taking measurements .

Distance measurement
Select **Distance measurement** to click and measure the distance between two pick targets. To measure distance between items, specify the pick target (snap) for item 1, click on it to select it, then specify the pick target for item 2, and click to select that item. In other words, you can change pick targets between clicks for distance measurement. For more information, see Taking measurements .

Clear measurements
Causes all currently displayed measurements and annotations to be cleared from the view.

Figure 1.6(b) Model Viewer Commands (continued)

From the Model Viewer Command window, click: **File** ⇒ **Close** ⇒ click on the model with **MMB** and "fly the shuttle icon" down to zoom out from the model until it is centered in the window [Fig. 1.6(c)] ⇒ click **RMB** to rotate the model (Fig. 1.7) ⇒ use both capabilities to position the model similar to that shown in Figure 1.7 ⇒ click on the model with **LMB** to turn on the enclosing box and coordinate system (Fig. 1.8) ⇒ reposition the model as desired

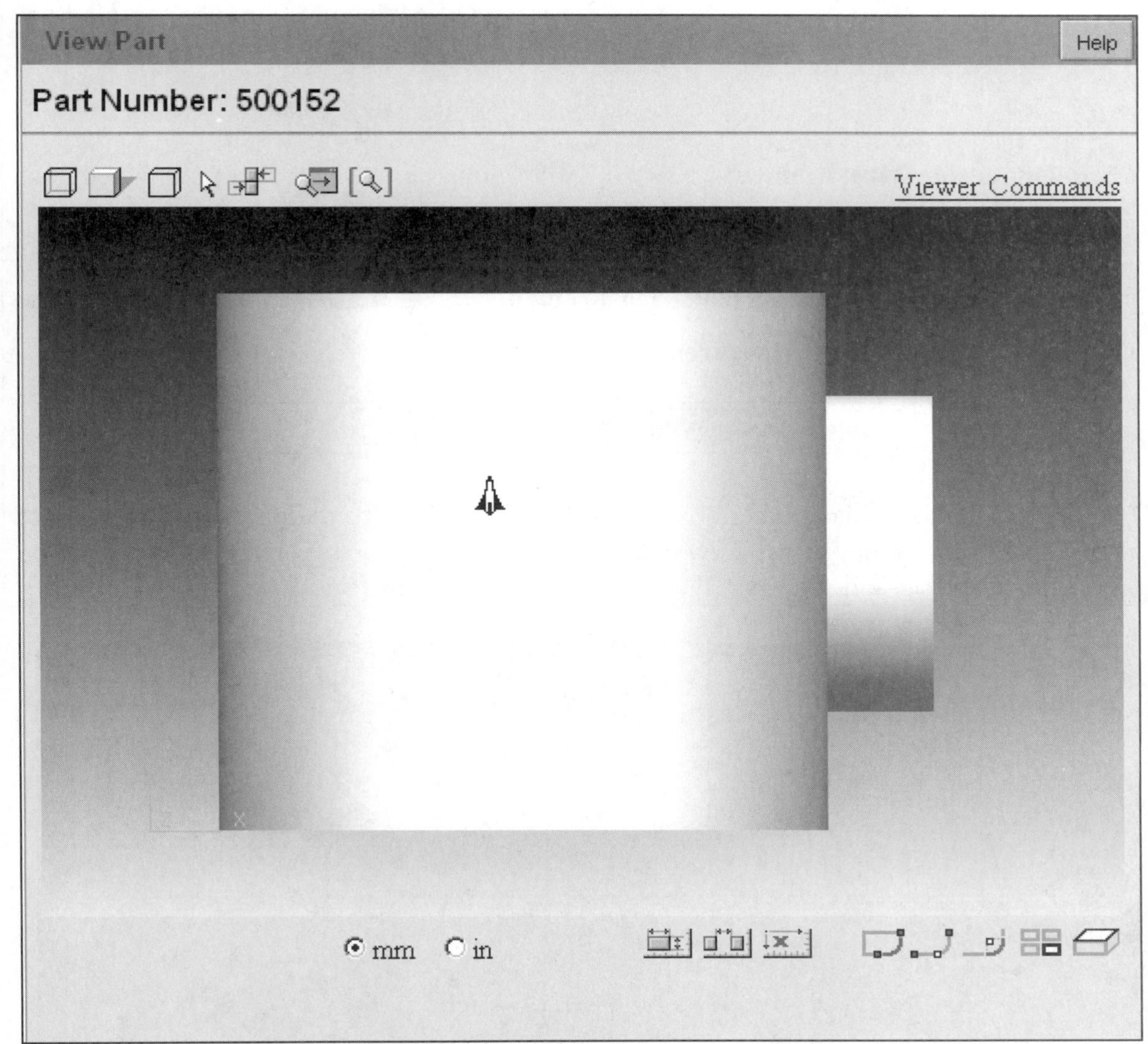

Figure 1.6(c) View Part Window with Clamp 500152 Displayed

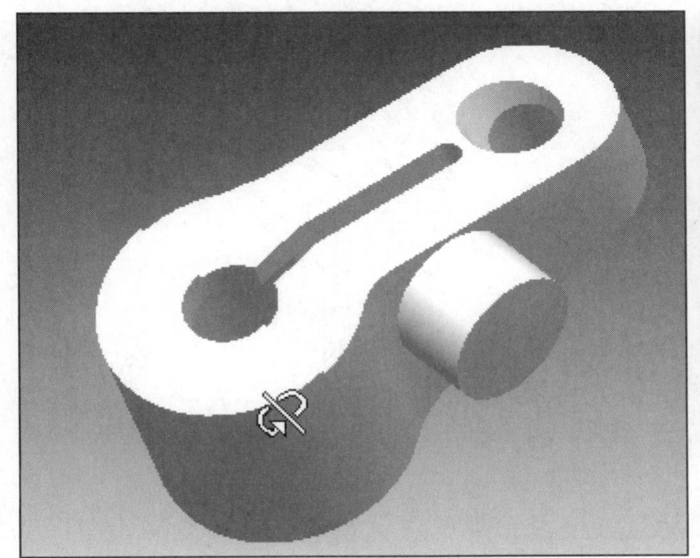

Figure 1.7 Fly and Rotate until Desired View is Obtained

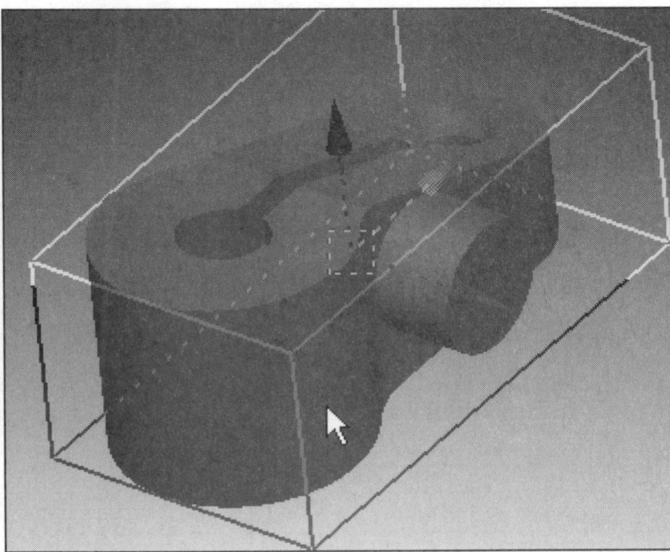

Figure 1.8 Pick on Model with LMB

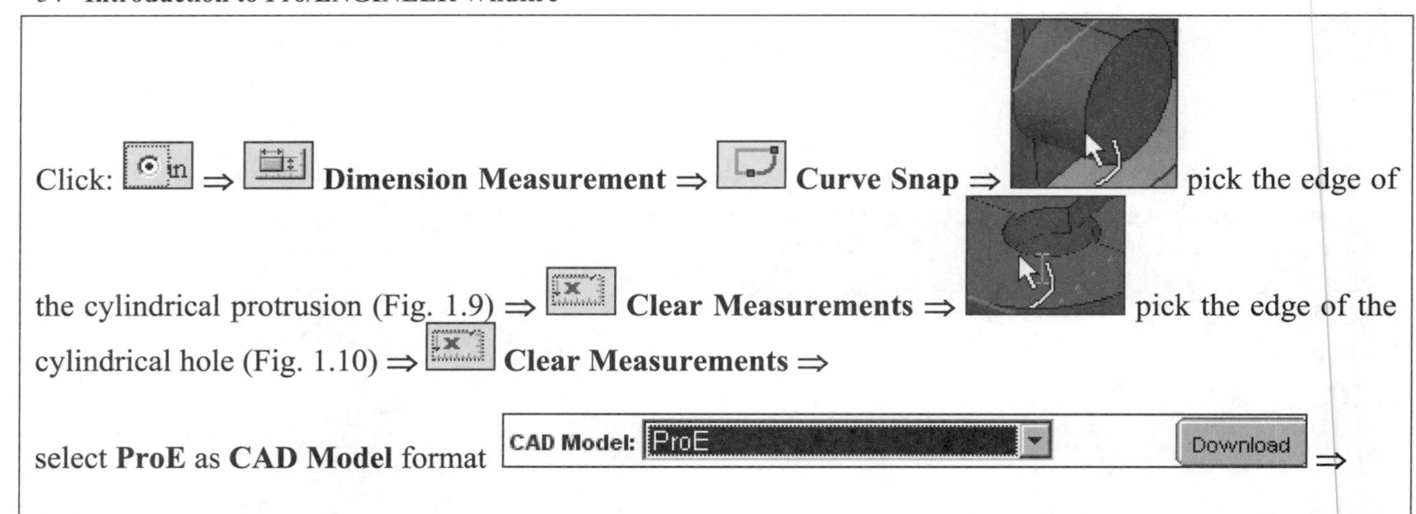

Click: ⇒ **Dimension Measurement** ⇒ **Curve Snap** ⇒ pick the edge of the cylindrical protrusion (Fig. 1.9) ⇒ **Clear Measurements** ⇒ pick the edge of the cylindrical hole (Fig. 1.10) ⇒ **Clear Measurements** ⇒

select **ProE** as **CAD Model** format ⇒

Download ⇒ **Open** from File Download dialog box ⇒ **I Agree** from **WinZip** dialog box (if you do not have WinZip you will have to download a free copy before this step at the prompt) (Fig. 1.11) ⇒ click on **500152.prt** [Fig. 1.12 (a)] and drag-and-drop into the Pro/E graphics window [Fig. 1.12 (b)]

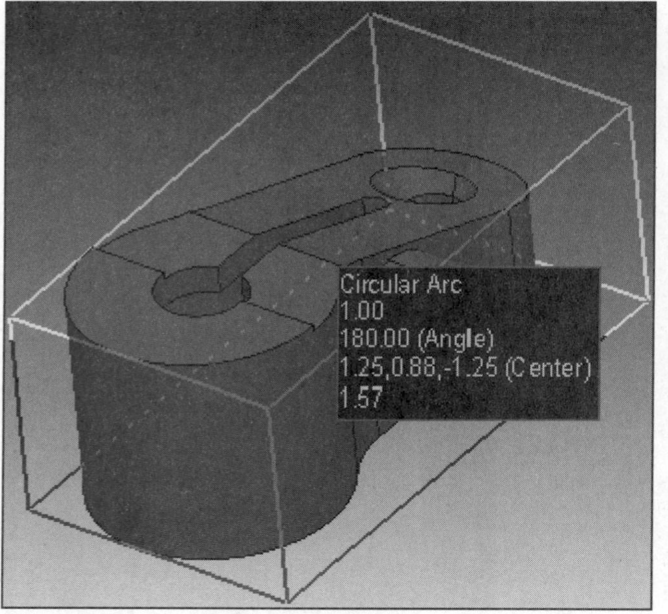

Figure 1.9 Cylindrical Protrusion Measurement

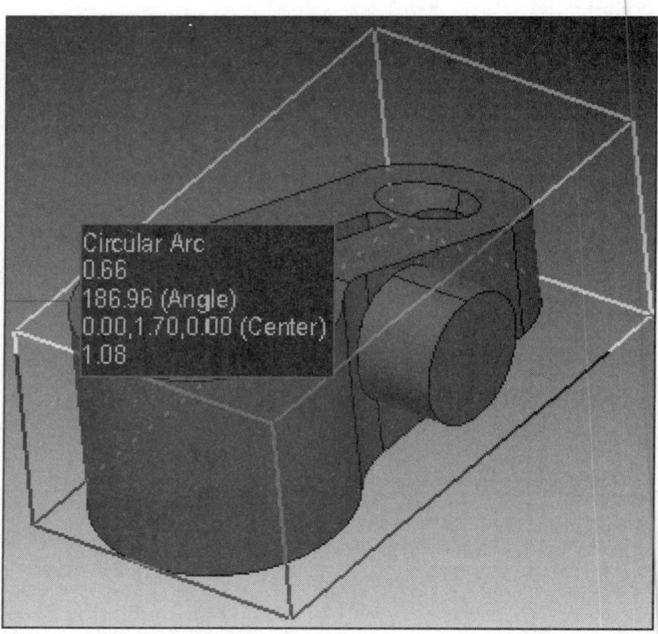

Figure 1.10 Hole Measurement

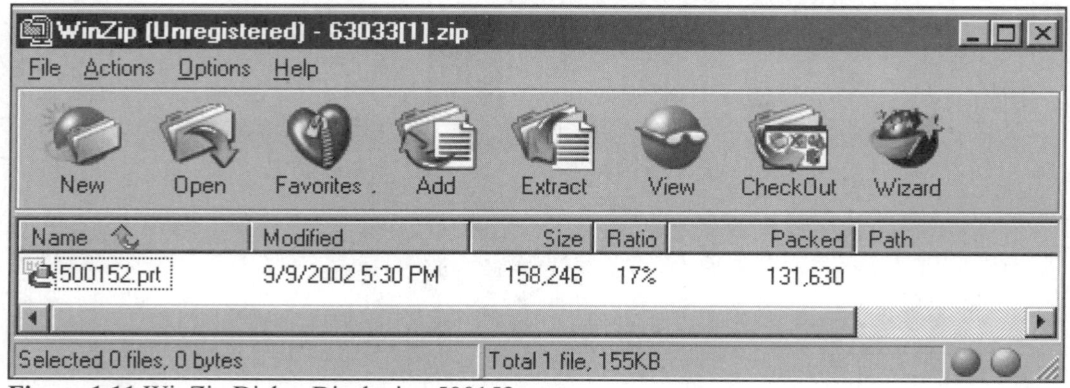

Figure 1.11 WinZip Dialog Displaying 500152.prt

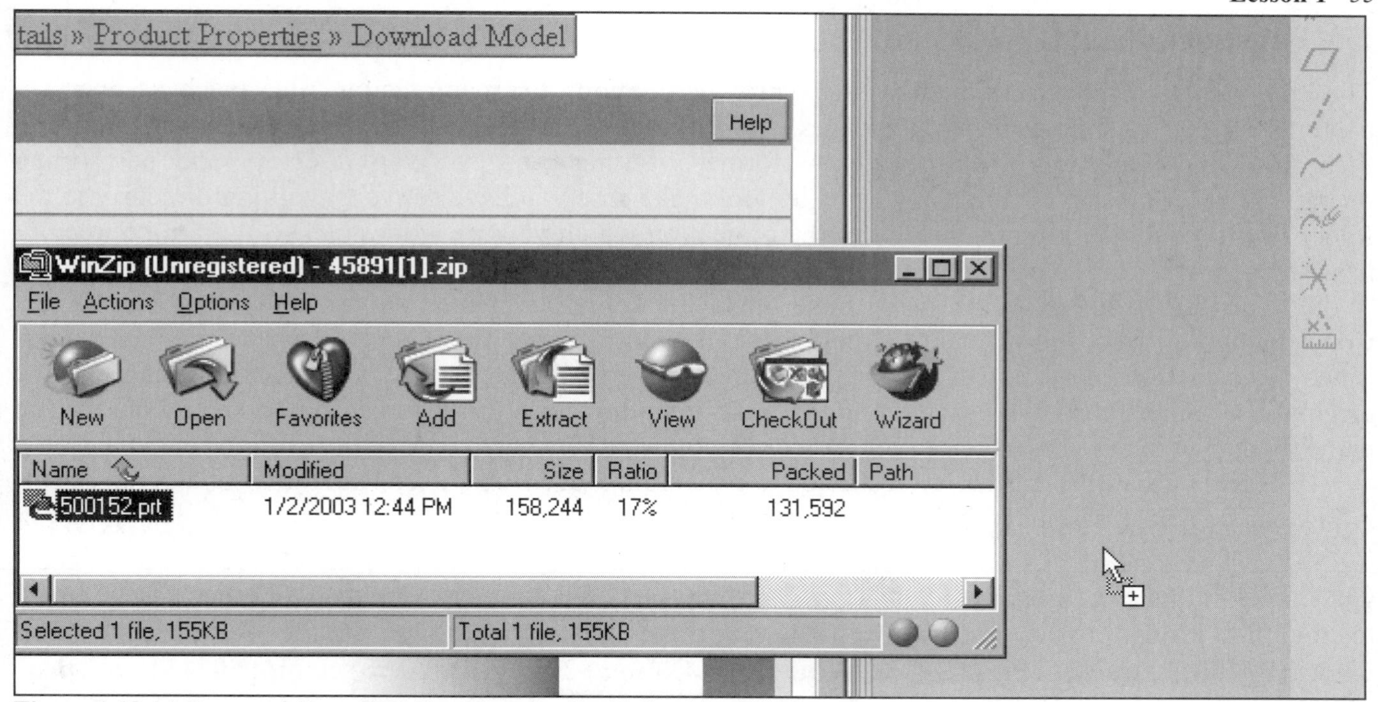

Figure 1.12 (a) Drag-and-Drop 500152.prt into the Pro/E Graphics Window

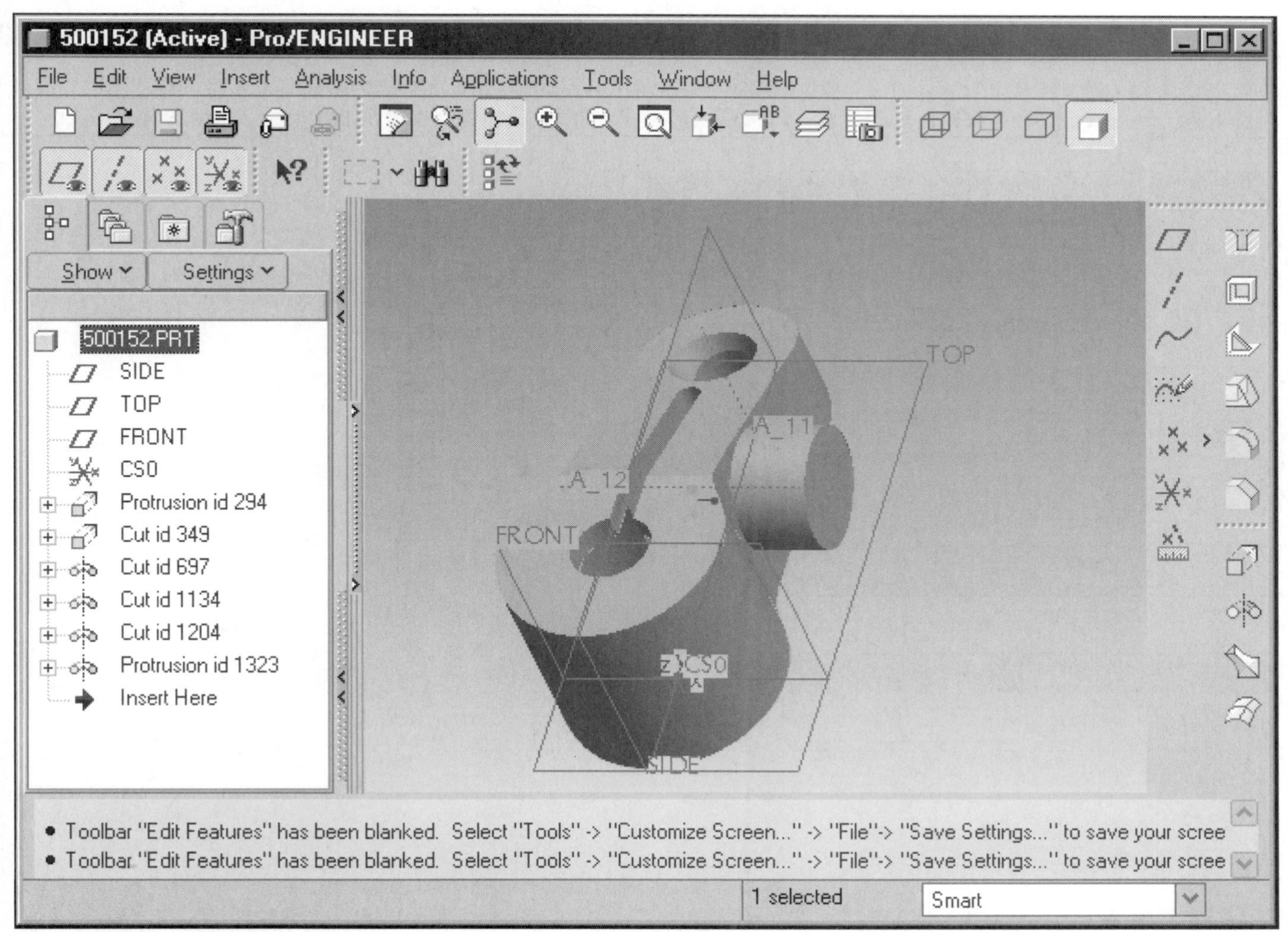

Figure 1.12 (b) Active Part 500152.prt Displayed in the Graphics Window and the Browser Automatically Closes

File Functions

The **File** menu on the Pro/E main window provides options for manipulating files (such as opening, creating, saving, renaming, backing up files, and printing). File functions include options for importing files from and exporting files to external formats, setting your working directory and performing operations on instances. Before using any File tool make sure you have set your working directory to the folder where you wish to save objects for the project on which you are working. The Working Directory is a directory that you set up to contain Pro/E files.

You can save Pro/E files using Save or Save a Copy commands on the File menu. The Save a Copy dialog box also allows you to export Pro/E files to different formats, and to save files as images. Since the name already exists in session, you cannot save or Rename a file using the same name as the original file name, even if you save the file in a different directory. Pro/E forces you to enter a unique file name by displaying the message: *An object with that name exists in session. Choose a different name.*

Names for all Pro/E files are restricted to a maximum of 31 characters and must contain no spaces. A File can be a part, assembly, or drawing. Each is considered an "object".

Click: **File ⇒ Set Working Directory** (Fig. 1.13) ⇒ select the directory you wish all objects to be saved to for this Pro/E project ⇒ **OK** ⇒ **File** ⇒ **Save a Copy** ⇒ type in the New Name field: **CLAMP_500152 ⇒ OK** (Fig. 1.14) ⇒ **Window ⇒ Close ⇒ File ⇒ Open ⇒ clamp_500152 ⇒ Open**

Depending on the computer you are using (yours or an organizations) the following may not work. If you have your mail server configured as the default then proceed, otherwise skip this step)

Send email with object in active window ⇒ email the object to yourself (or a friend with Pro/E software) ⇒ check on **Create a ZIP file ⇒ OK** (Fig. 1.15) ⇒ follow your email procedure as required

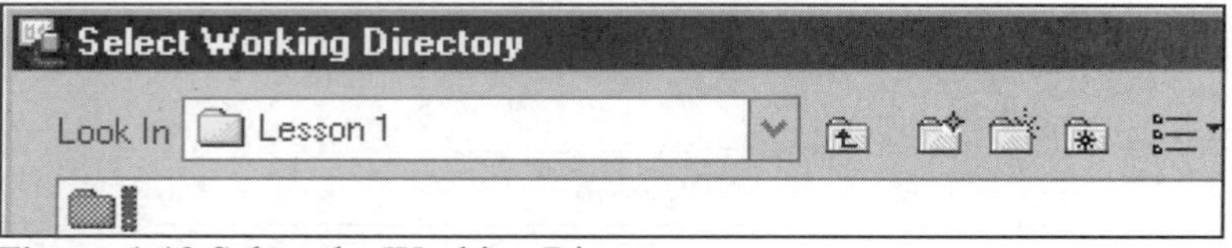

Figure 1.13 Select the Working Directory

Figure 1.14 Save a Copy

Figure 1.15 Create a Zip File and Send as Attachment

Help

Accessing the Help function is one of the best ways to learn CAD software. Use the Help tool as often as possible to understand the tool or command you are using at the time and to deepen your knowledge of the other capabilities provided by Pro/ENGINEER. Use the **Help** menu to gain access to online information, Pro/E release information, and customer service information. The following commands are available on the Pro/E **Help** menu in standard Part and Assembly modes.

- **Help Center** Displays the context-sensitive online help system. When you select this, your supported network browser opens to display a navigation tree and search tools to aid you in finding specific help topics. You can also access these topics by clicking for context-sensitive Help from windows, menu commands, and dialog boxes.
- **What's This?** Enables context-sensitive Help mode.
- **Release Notes** New in this release.
- **Technical Support Info** Displays product information, including the release level, license information, installation date, and customer support contact information.
- **About Pro/ENGINEER** Displays Pro/E copyright and release information.

Click: **Help ⇒ Help Center** (Fig. 1.16) ⇒ **Fundamentals ⇒ Pro/E Fundamentals** (Fig. 1.17)

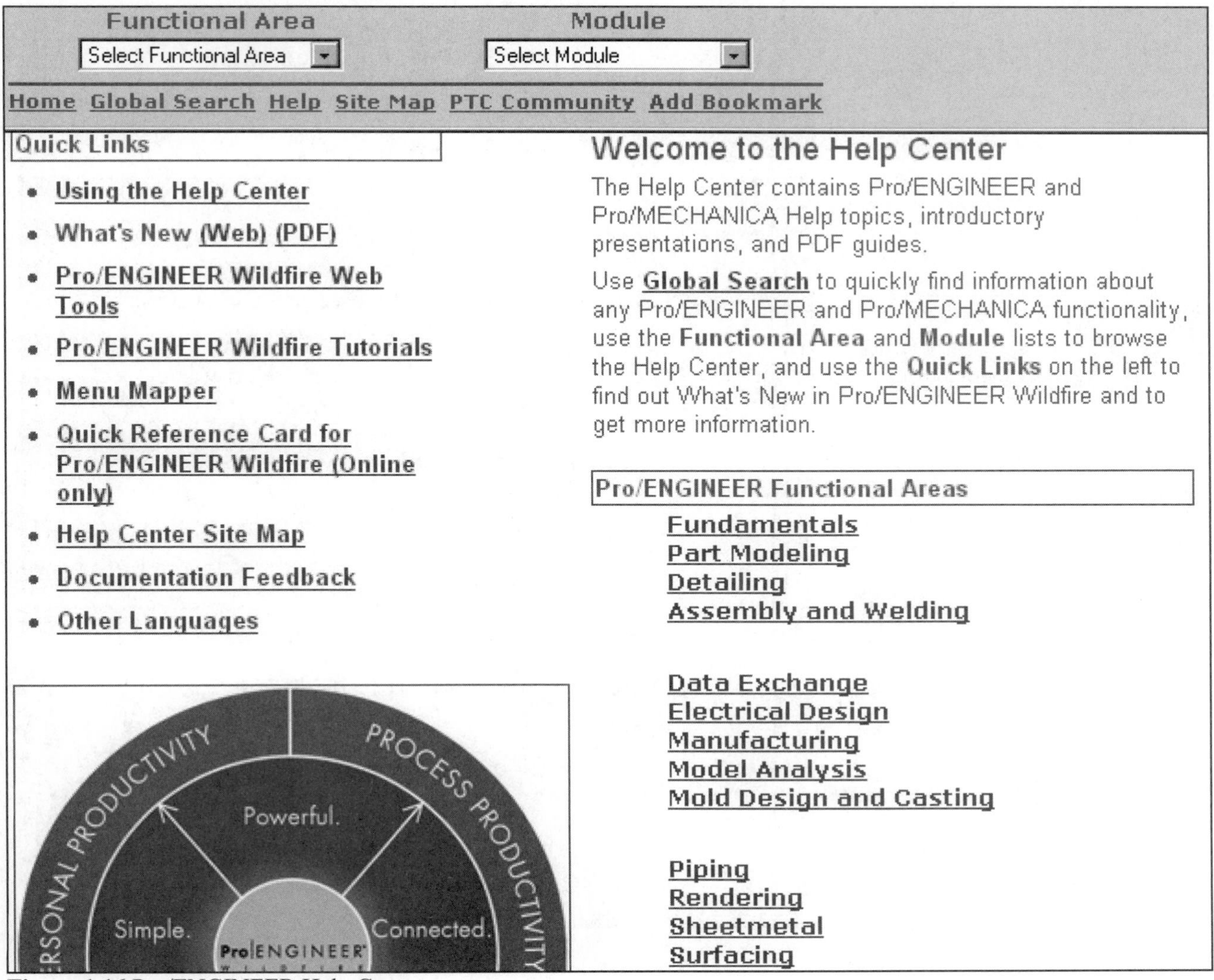

Figure 1.16 Pro/ENGINEER Help Center

Open the **Contents** choices as shown in Figure 1.17 (⊞ to expand or ⊟ to collapse). Explore the **View Menu**. Spend some time reading the documentation before continuing. ⇒

Click: Close ⇒ **Context sensitive help** from the Toolbar ⇒

Pick on the **Reorient View** button from the Toolbar ⇒

Read the documentation and then close the window.

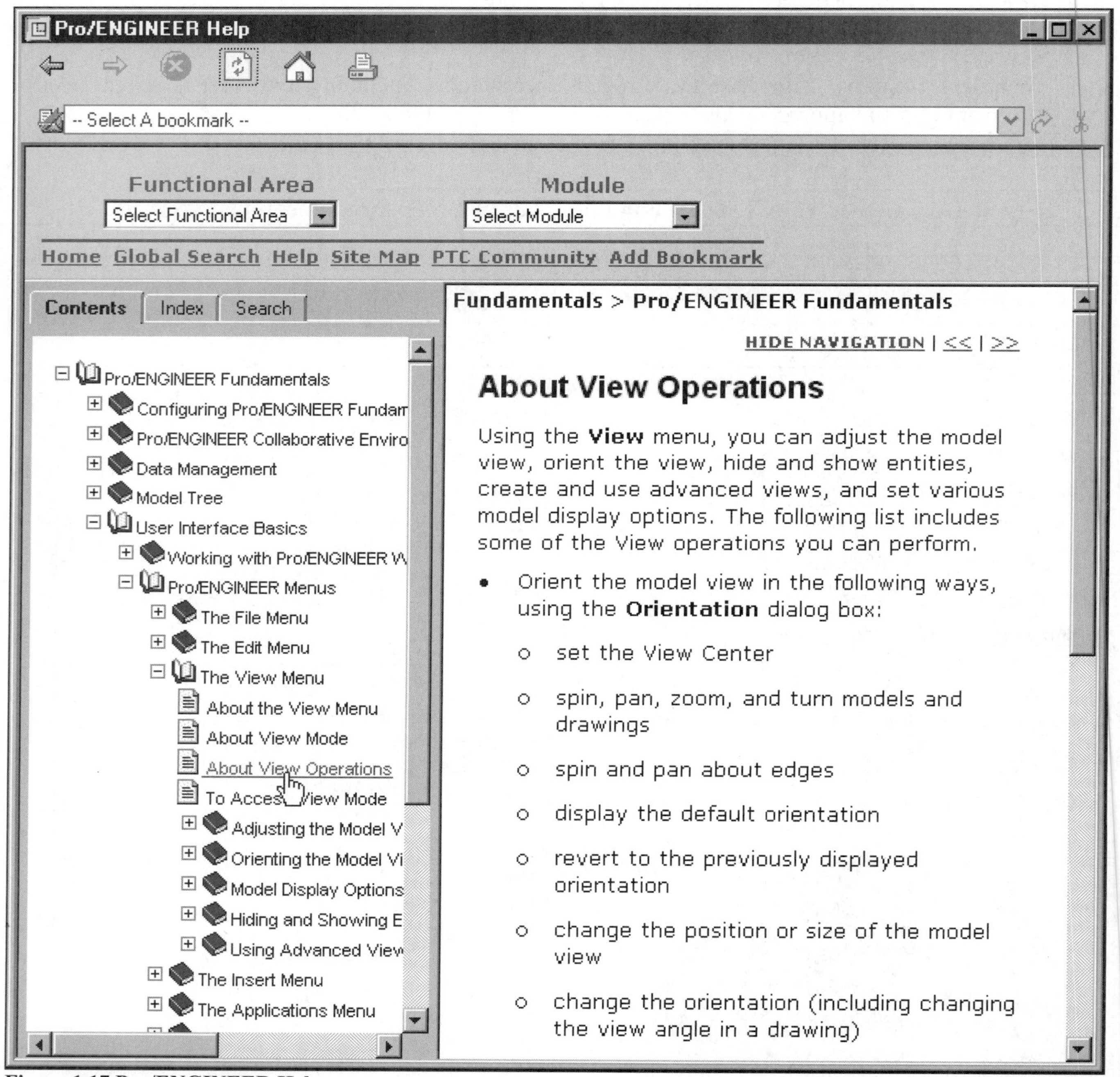

Figure 1.17 Pro/ENGINEER Help

View and Display Functions

Using the Pro/E **View** menu, you can adjust the model view, orient the view, hide and show entities, create and use advanced views, and set various model display options. The following list includes some of the View operations you can perform:

- Orient the model view in the following ways, using the **Orientation** dialog box: spin, pan, and zoom models and drawings, display the default orientation, revert to the previously displayed orientation, change the position or size of the model view, change the orientation (including changing the view angle in a drawing), and create new orientations.
- Temporarily shade a model by using cosmetic shading
- Show, dim, or remove hidden lines
- Highlight items in the graphics window when you select them in the Model Tree
- Explode or unexplode an assembly view
- Repaint the Pro/E graphics window
- Refit the model to the Pro/E window after zooming in or out on the model
- Update drawings of model geometry
- Hide and show entities, and hide or show items during spin or animation
- Use advanced views
- Add perspective to the model view

Click: **Wireframe** [Fig. 1.18(a)] ⇒ **Hidden Line** [Fig. 1.18(b)] ⇒ **No Hidden** [Fig. 1.18(c)] ⇒ **Shading**

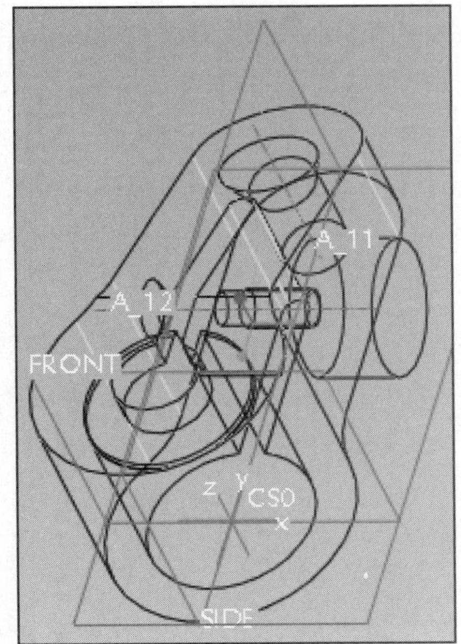

Figure 1.18(a) Wireframe

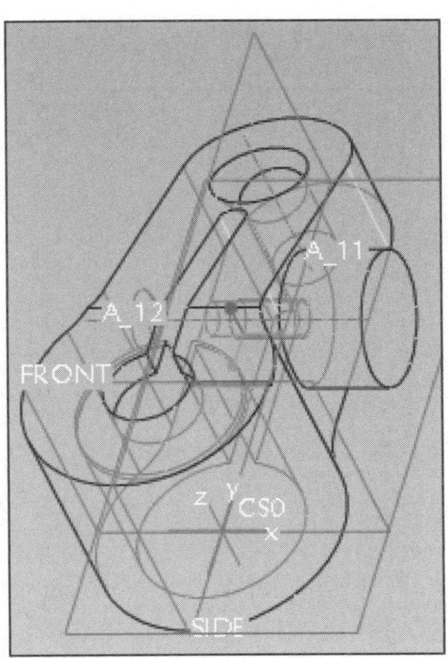

Figure 1.18(b) Hidden Line

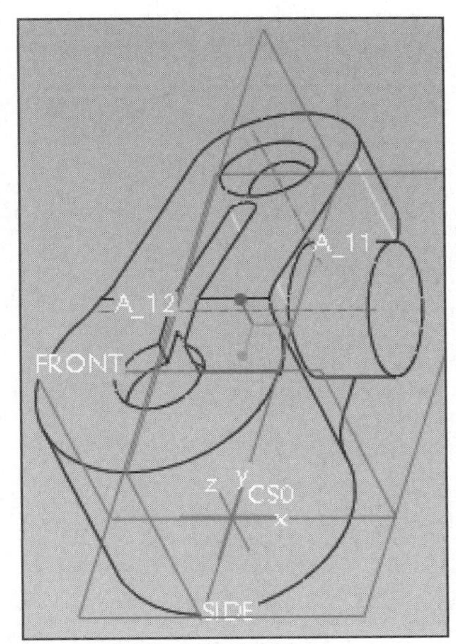

Figure 1.18(c) No Hidden

To see the standard views provided, click: ⇒ **Front** ⇒ ⇒ **Right** [Fig. 1.19(a-b)] ⇒ (try all the variations) ⇒ ⇒ **Standard Orientation**

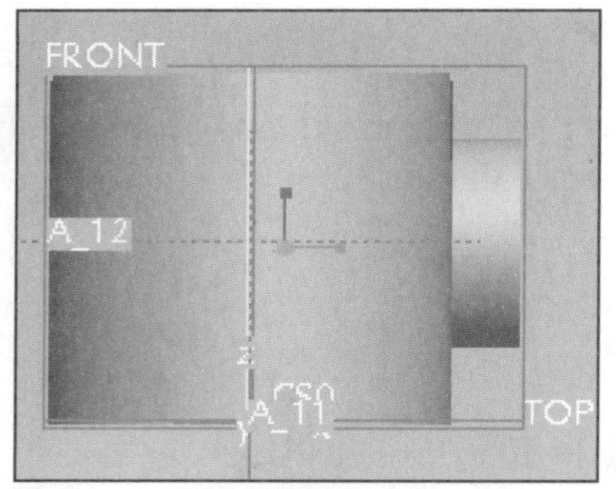

Figure 1.19(a) FRONT View

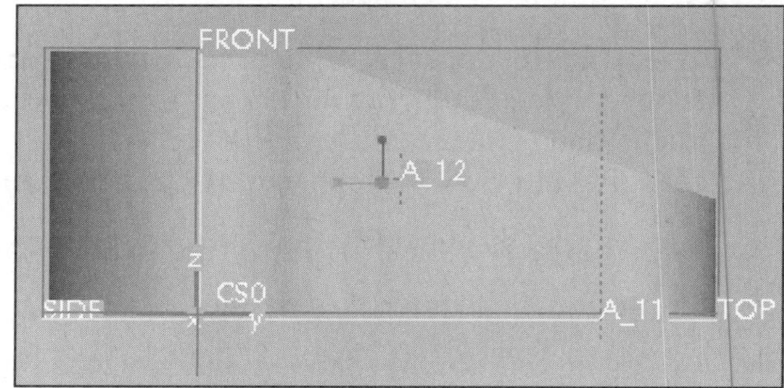

Figure 1.19(b) RIGHT View

View Tools

As with all CAD systems, Pro/E provides the typical view tools associated with CAD:

- **Zoom In** Use this tool to zoom in on a specific portion of the model. Pick two positions of a rectangular zoom box.
- **Zoom Out** Use this tool to reduce the view size of the model on the screen by 50%.
- **Refit object to fully display it on the screen** Use this tool to refit the model to the screen so that you can view the entire model. A refitted model fills 80% of the graphics window.

Click: **Zoom In** [Fig. 1.20(a)] ⇒ pick two positions about an area you wish to enlarge [Fig. 1.20(b)] ⇒ **Zoom Out** ⇒ **Refit** [Fig. 1.20(c)] ⇒ **Redraw the current view**

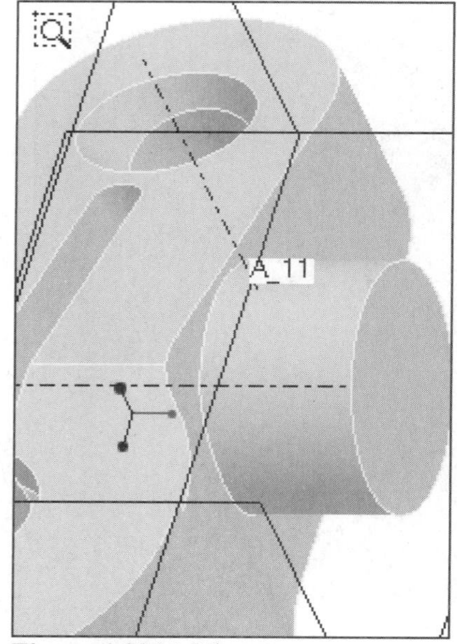

Figure 1.20(a) View

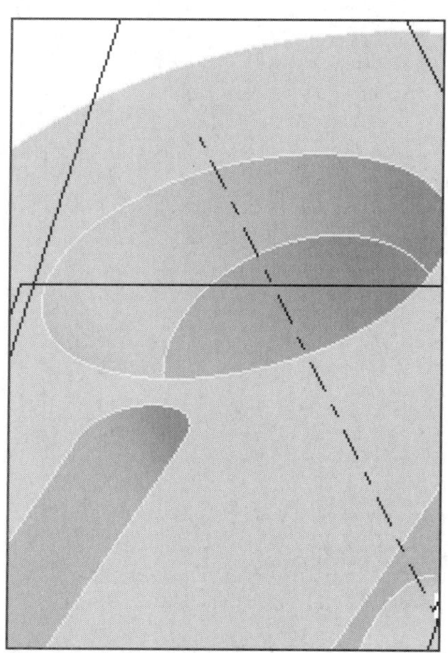

Figure 1.20(b) Zoom In

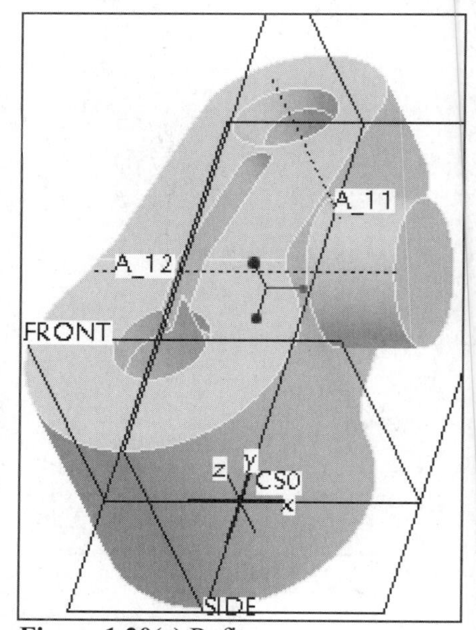

Figure 1.20(c) Refit

Using Mouse Buttons to Manipulate the Model

You can also dynamically reorient the model using the **MMB** by itself (**Rotate**) or in conjunction with the **Shift** key (**Pan**) or **Ctrl** key (**Zoom**).

Hold down **Ctrl** key and **MMB** in the graphics area near the model and move the cursor up (zoom out) [Fig. 1.21(a)] ⇒ hold down **Shift** key and **MMB** in the graphics area near the model and move the cursor about the screen (pan) [Fig. 1.21(b)] ⇒ hold down **MMB** in the graphics area near the model and move the cursor around (rotate) [Fig. 1.22(a-b)] ⇒ [icon] ⇒ **Standard Orientation**

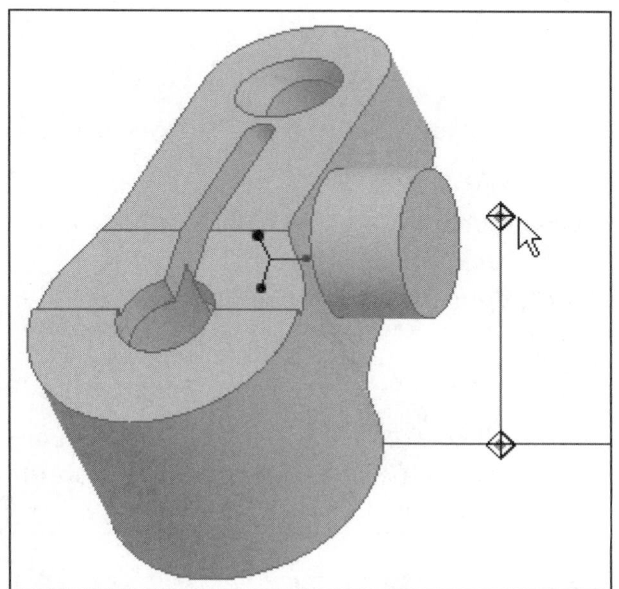

Figure 1.21(a) Zoom

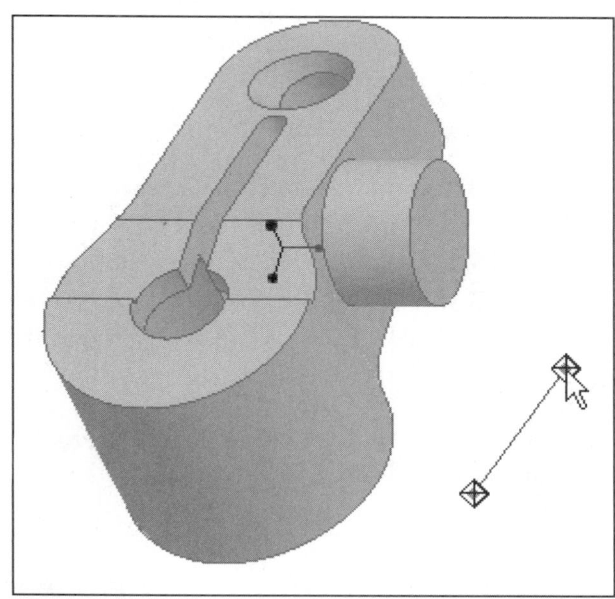

Figure 1.21(b) Pan

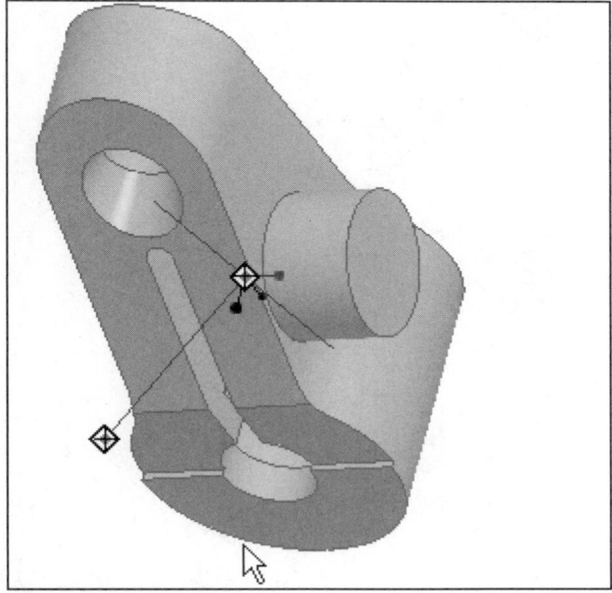

Figure 1.22(a) Rotate

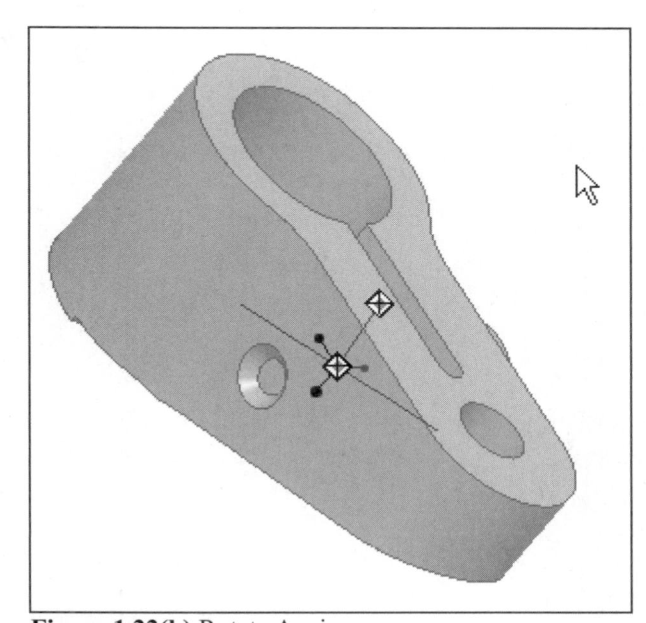

Figure 1.22(b) Rotate Again

You may have noticed that the illustrations of the text have changed since Figure 1.20. The default for the background and geometry colors has been changed so that the illustrations will capture and print clearer. In the next section, you will learn how to change the system display settings.

System Display Settings

You can make a number of changes to the default colors furnished by Pro/E, customizing them for your own use:

- Define, save, and open color schemes
- Customize colors used in the user interface
- Change your entire color scheme to a predefined color scheme (such as black on white)
- Change the top or bottom background colors
- Redefine basic colors used in models
- Assign colors to be used by an entity
- Store a color scheme so you can reuse it
- Open a previously used color scheme

The **Scheme** menu includes the following color schemes (text uses the **Black on White** selection):

- **Black on White** Black entities shown on a white background
- **White on Black** White entities shown on a black background
- **White on Green** White entities shown on a dark-green background
- **Initial** Reset the color scheme to the one defined by the configuration file settings
- **Default** Reset the color scheme to the system default

Click: **View** ⇒ **Display Settings** ⇒ **System Colors** [Fig. 1.23(a-c)] ⇒ Blended Background ⇒ **Scheme** ⇒ **Black on White** ⇒ experiment with different color schemes and system colors ⇒ **Scheme** ⇒ **Default** ⇒ **OK**

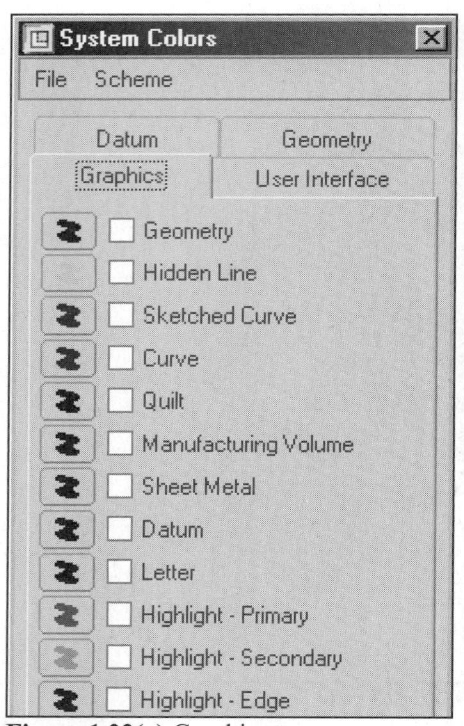

Figure 1.23(a) Graphics

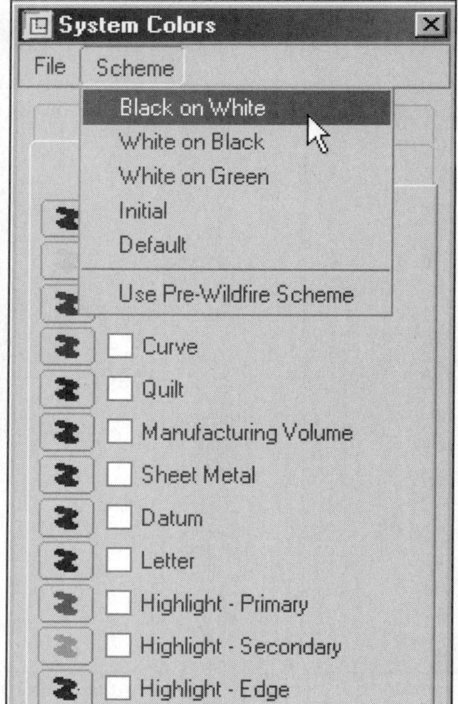

Figure 1.23(b) Scheme

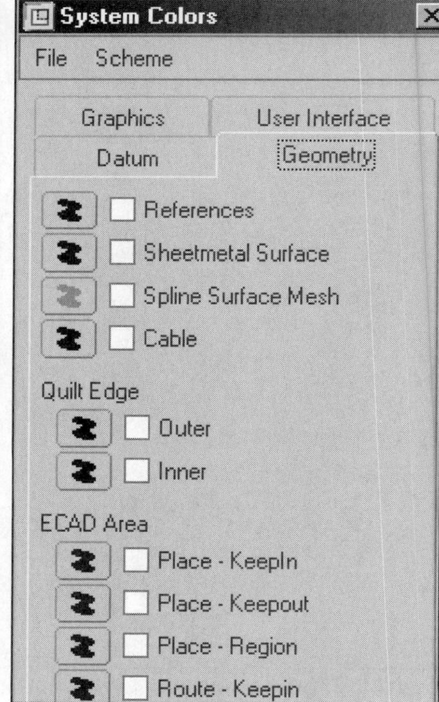

Figure 1.23(c) Geometry

Information Tools

At anytime during the design process you can request model, feature or other information. Clicking on a feature in the graphics window and then **RMB ⇒ Info** (Fig. 1.24) will provide information about that feature in the Browser [Fig. 1.25(a-b)]. This can also be accomplished by clicking on the feature name in the Model Tree and then **RMB**. Both feature and model information can be obtained using this method (Fig. 1.26). A variety of information can also be extracted using the **Info** tool from the menu bar.

Click once on the revolved protrusion (Fig. 1.24) ⇒ **RMB ⇒ Info ⇒ Feature** [Fig. 1.25(a-b)] ⇒ click on the "quick sash" to collapse the Browser ⇒ in the graphics window **RMB** (Fig. 1.26) ⇒ **Info ⇒ Model** (Fig. 1.27) ⇒ collapse the Browser

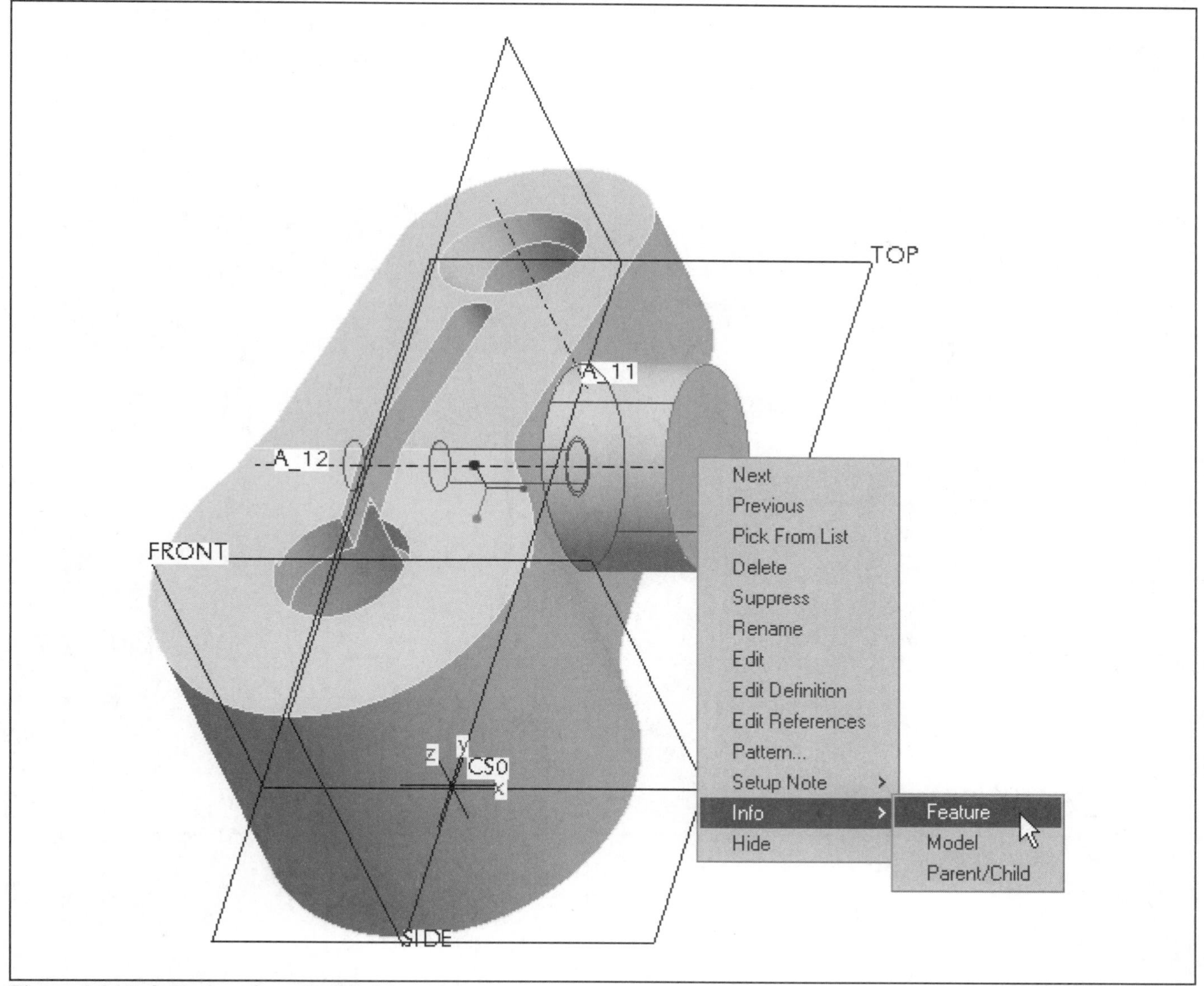

Figure 1.24 Info Feature from the Model

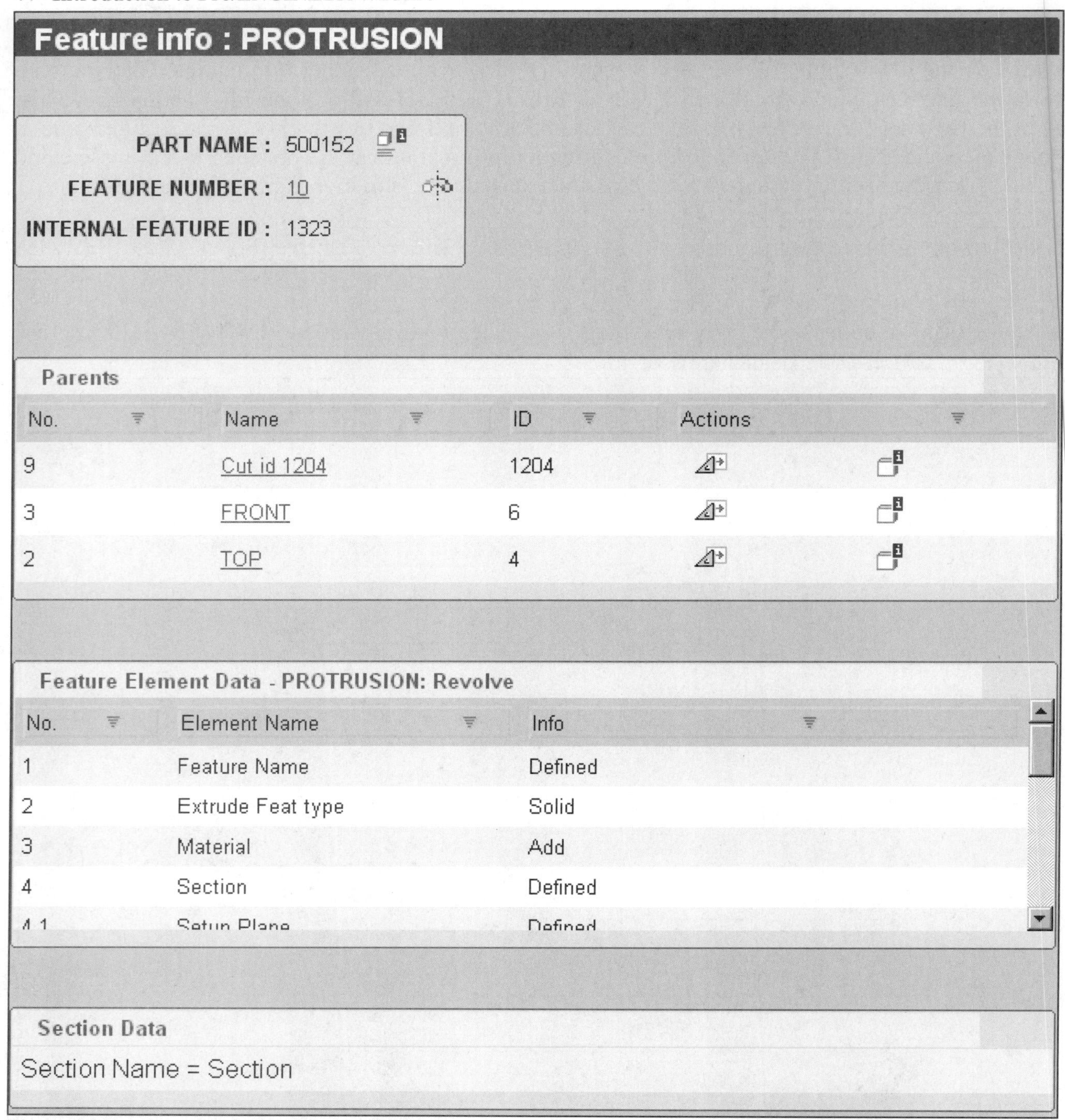

Figure 1.25(a) Feature Info Displayed in the Browser

FEATURE Is In Layer(s) :

Layer	Status
LAYER_FEATURE	OPERATION = SHOWN
LAYER_GEOM_FEAT	OPERATION = SHOWN
LAYER_AXIS	OPERATION = SHOWN
LAYER_PROTRUSION_FEAT	OPERATION = SHOWN

Figure 1.25(b) Layer Information

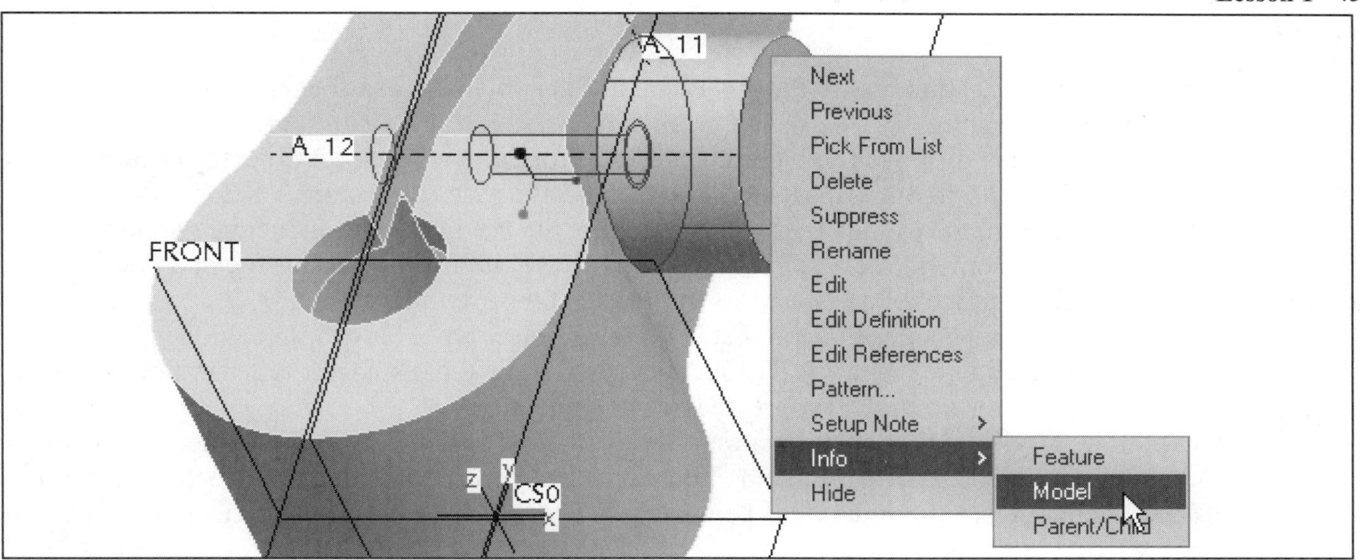

Figure 1.26 Info Model

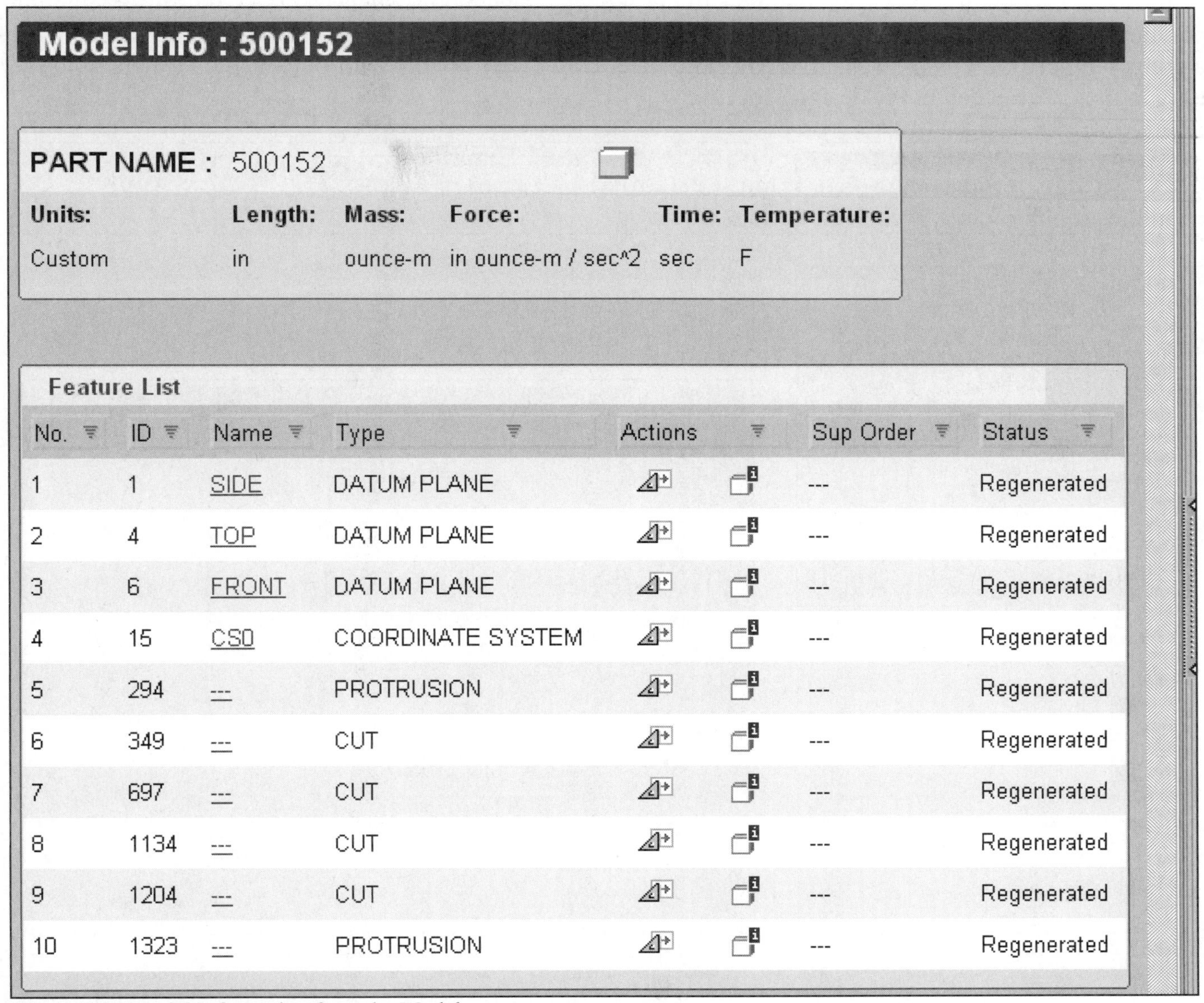

Model Info : 500152

PART NAME : 500152

Units:	Length:	Mass:	Force:	Time:	Temperature:
Custom	in	ounce-m	in ounce-m / sec^2	sec	F

Feature List

No.	ID	Name	Type	Actions		Sup Order	Status
1	1	SIDE	DATUM PLANE			---	Regenerated
2	4	TOP	DATUM PLANE			---	Regenerated
3	6	FRONT	DATUM PLANE			---	Regenerated
4	15	CS0	COORDINATE SYSTEM			---	Regenerated
5	294	---	PROTRUSION			---	Regenerated
6	349	---	CUT			---	Regenerated
7	697	---	CUT			---	Regenerated
8	1134	---	CUT			---	Regenerated
9	1204	---	CUT			---	Regenerated
10	1323	---	PROTRUSION			---	Regenerated

Figure 1.27 Model Information from the Model

The Model Tree

The **Model Tree** is a tabbed feature on the Pro/E navigator that contains a list of every feature or part in a current part, assembly, or drawing.

The model structure is displayed in hierarchical (tree) format with the root object (the current part or assembly) at the top of its tree and the subordinate objects (parts or features) below. If you have multiple Pro/E windows open, the Model Tree contents reflect the file in the current active window.

The Model Tree lists only the related feature and part level items in a current file and does not list the entities (such as edges, surfaces, curves, and so forth) that comprise the features.

Each Model Tree item contains an icon that reflects its object type, for example, hidden, assembly, part, feature, or datum plane (also a feature). The icon can also show the display or regeneration status for a feature, part, or assembly, for example, suppressed or unregenerated.

Selection in the Model Tree is object-action oriented; you select objects in the Model Tree without first specifying what you intend to do with them. You can select components, parts, or features using the Model Tree. You cannot select the individual geometry that makes up a feature (entities). To select an entity, you must select it in the graphics window.

With the **Settings** tab you can control what is displayed in the Model Tree.

You can add informational columns to the Model Tree window, such as **Tree Columns** containing parameters and values, assigned layers, or feature name for each item. You can use the cells in the columns to perform context-sensitive edits and deletions. These options will be covered elsewhere in the text, as they are needed in the design process.

Click: **Settings** tab ⇒ Tree Filters... [Fig. 1.28(a-b)]⇒ Notes ⇒ Suppressed Objects ⇒ **Apply** ⇒ **OK** ⇒ in the graphics window click **LMB** to deselect

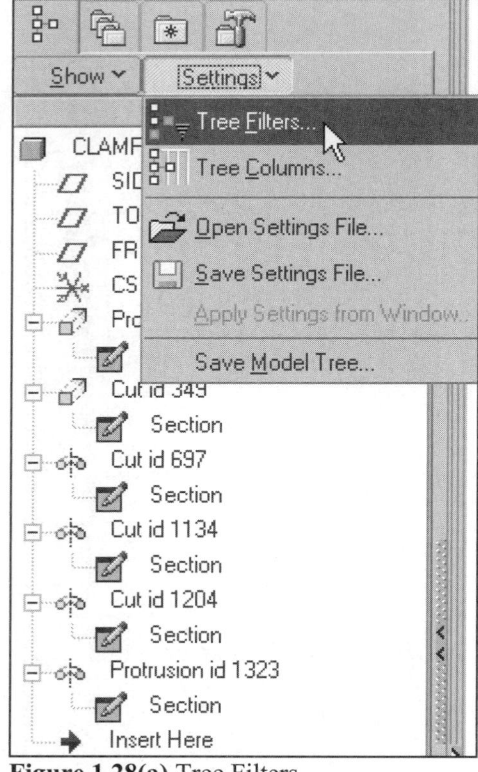

Figure 1.28(a) Tree Filters

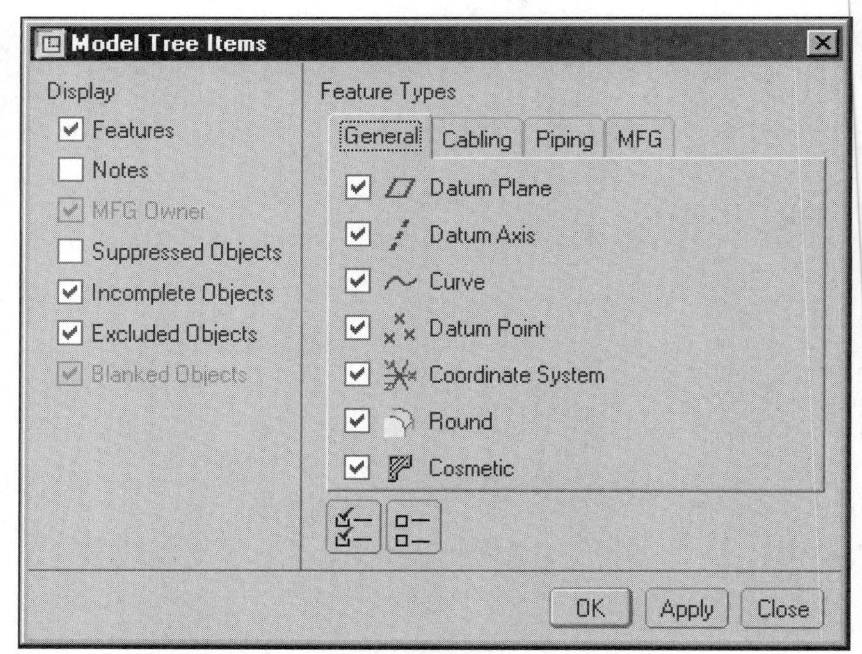

Figure 1.28(b) Model Tree Items Dialog Box

Working on the Model

In Pro/E, you can select objects to work on from within the graphics window or in the Model Tree by using the mouse or the keyboard. The object types that are available for selection vary depending on whether you select an object from within the graphics window or in the Model Tree. You can select any type of object, including features, 3-D notes, parts, datum objects (planes, axes, curves, points, and coordinate systems), and geometry (edges and surfaces) from within the graphics window. Additionally, since the Model Tree displays only parts, components, and features, you can select only those object types from within the Model Tree.

Selection in both the graphics window and the Model Tree can be action-object or object-action oriented, depending on the process you choose within Pro/E to build your model. You can specify the action you want to perform on an object before you select the object, or you can select the object before you specify the action.

You can *dynamically* modify certain features from within the graphics window as you work. Features can be edited by selecting them from the graphics window directly, or from the Model Tree. Dimensions of the following features can be modified dynamically:

- **Protrusions** Extruded protrusions with variable depth, and revolved protrusions of variable angle
- **Cuts** Extruded cuts with variable depth, and revolved cuts with variable angle
- **Surfaces** Extruded surfaces with variable depth, and revolved surfaces with variable angle
- **Rounds** Simple, constant, and edge chain

Click on the cylindrical protrusion ⇒ **RMB** [Fig. 1.29(a)] ⇒ **Delete** ⇒ **OK** [Fig. 1.29(b)]

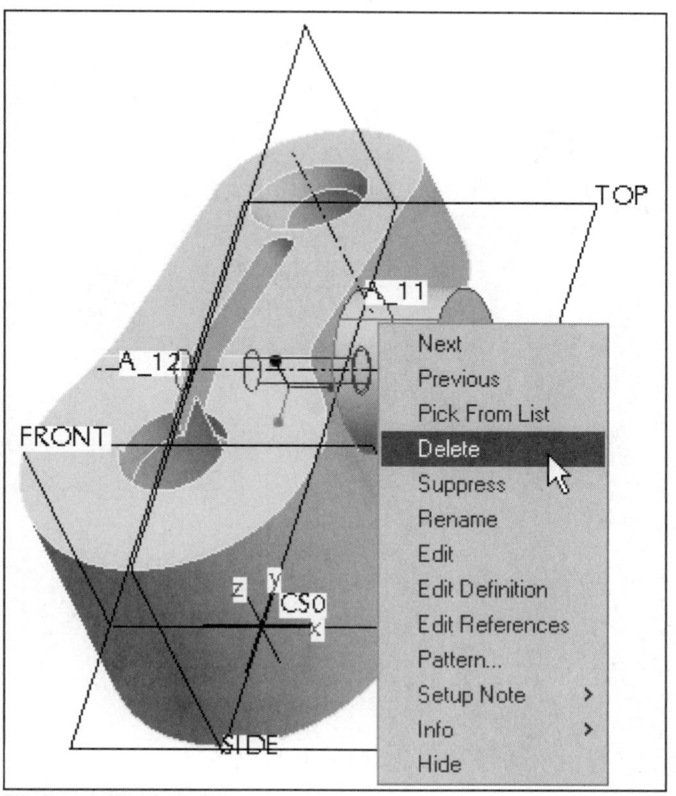

Figure 1.29(a) Delete

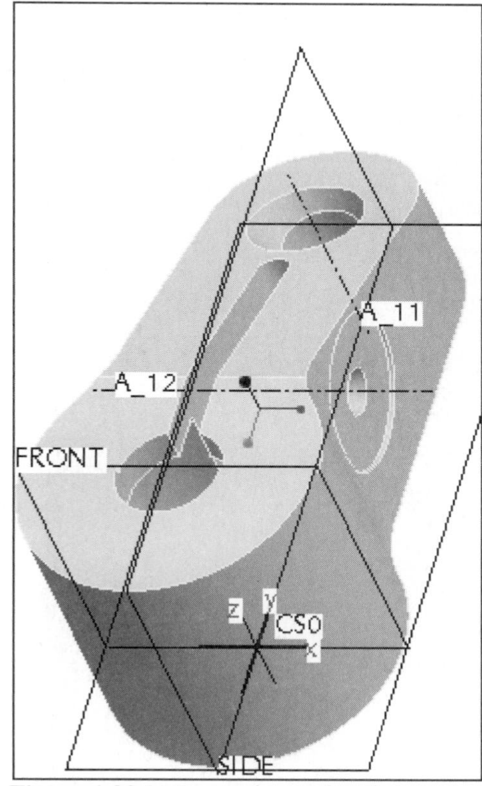

Figure 1.29(b) Protrusion Deleted

Click on the protrusion in the Model Tree (Fig. 1.30) ⇒ **RMB** ⇒ **Edit Definition** (Fig. 1.31)

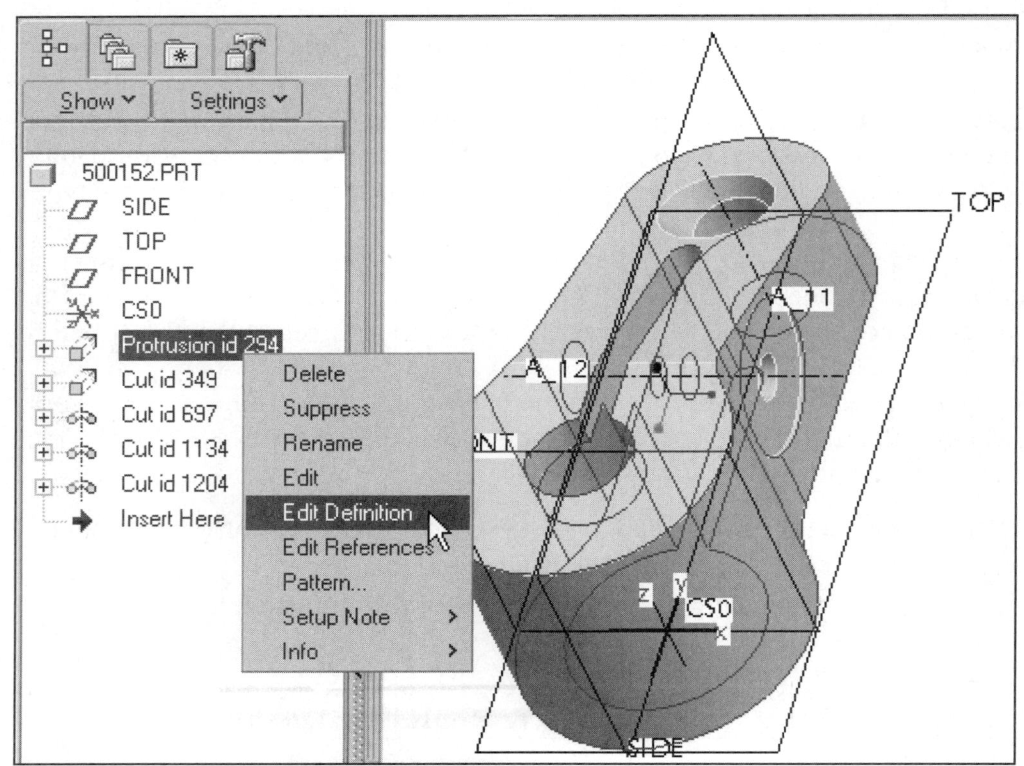

Figure 1.30 Redefining a Feature

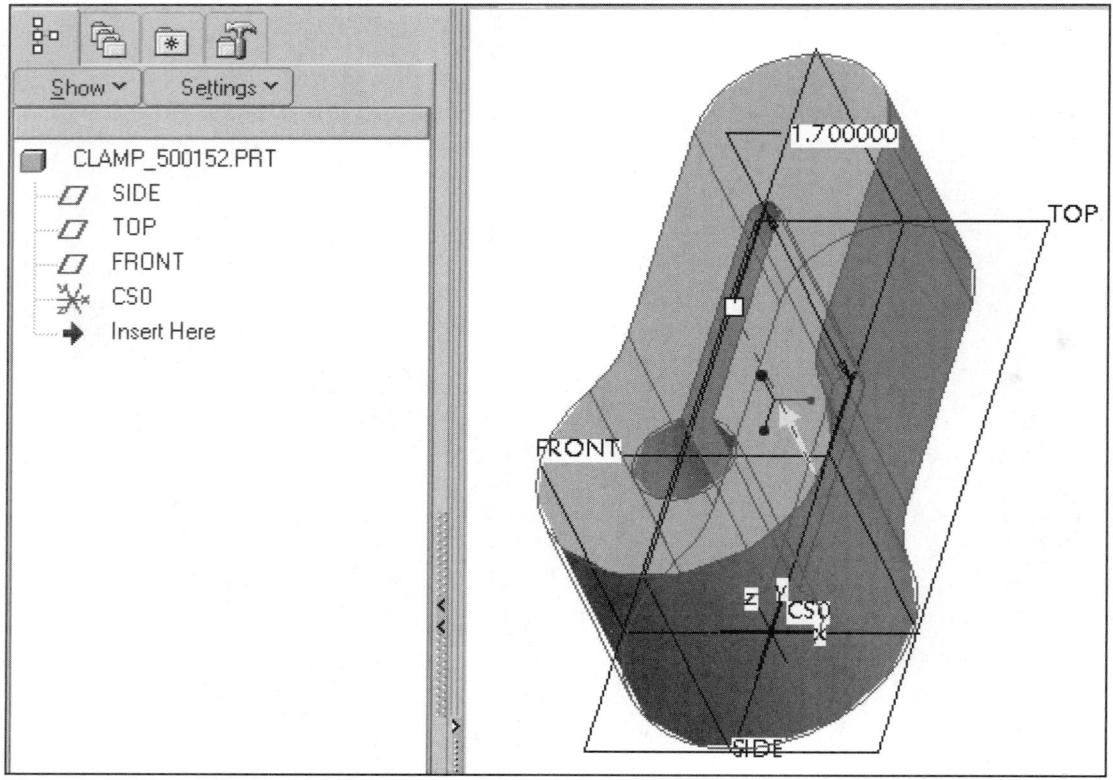

Figure 1.31 Dynamically Redefining a Feature

Click on and drag the white nodal handle to change the protrusion's height from **1.70** to **2.00** (Fig. 1.32) ⇒ **MMB** to regenerate the model (Fig. 1.33) ⇒ **LMB**

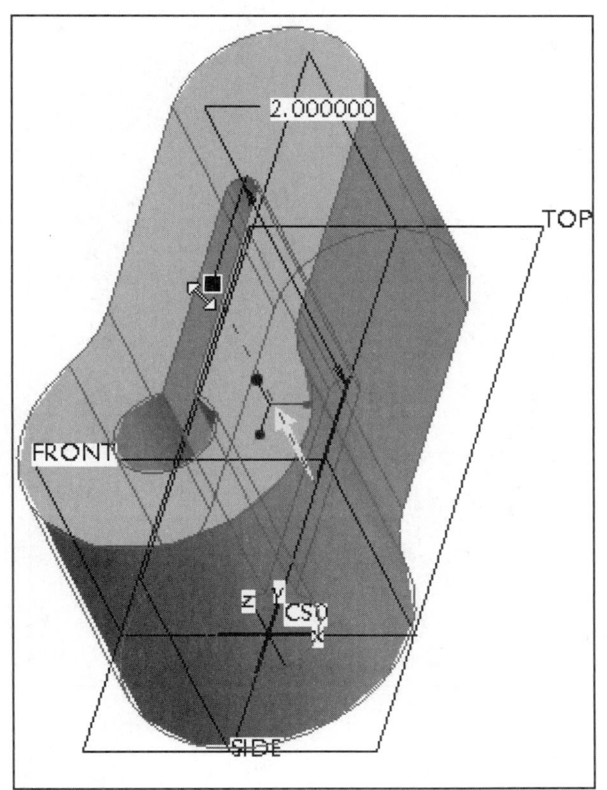

Figure 1.32 Dynamically Redefining a Feature

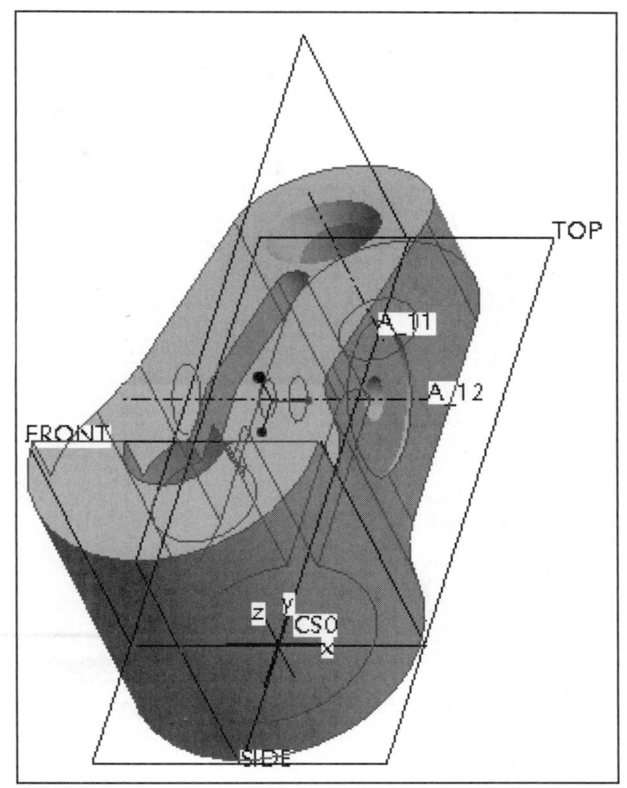

Figure 1.33 Redefined Feature

Although you may have not noticed it, there was other information and editing available during the redefine process. Repeat the command: Click on the protrusion in the Model Tree ⇒ **RMB** ⇒ **Edit Definition** ⇒ look at the "dashboard" on the lower part of the screen [Fig. 1.34(a)] ⇒ **Options** tab [Fig. 1.34(b)] ⇒ change the **2.00** to **2.25** ⇒ **Enter** ⇒ **MMB** (or ✓) [Fig. 1.35(a-b)]

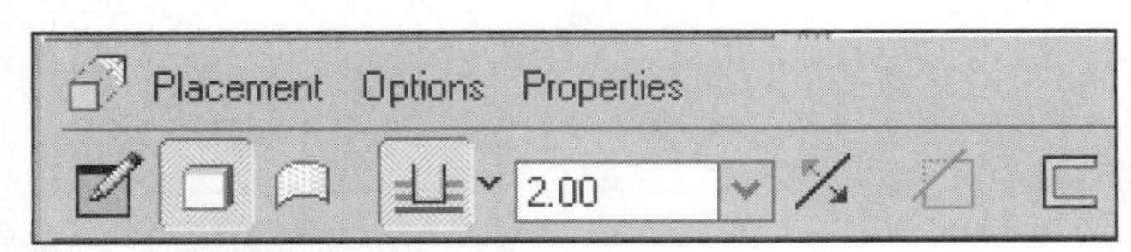

Figure 1.34(a) Dashboard

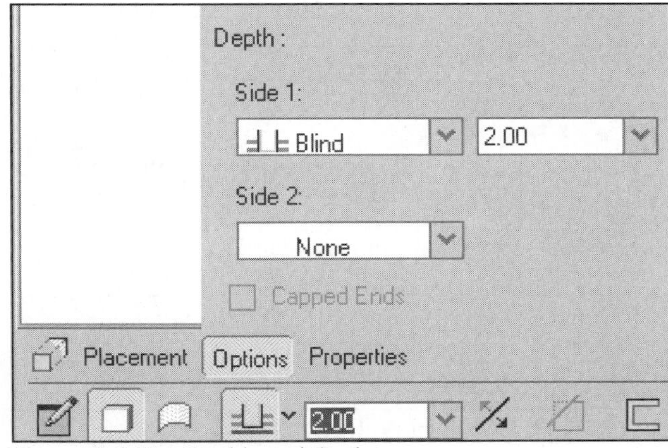

Figure 1.34(b) Dashboard Options

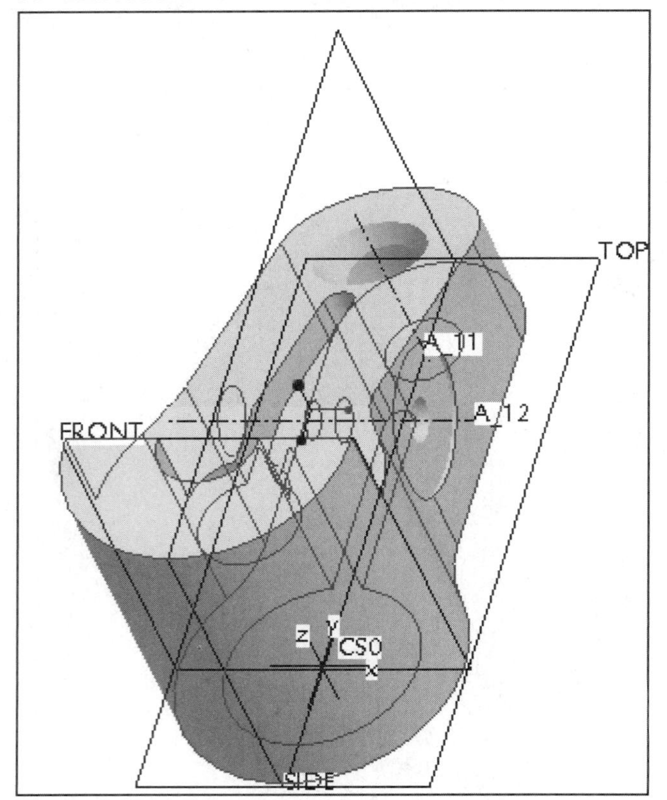

Figure 1.35(a) Edited Model

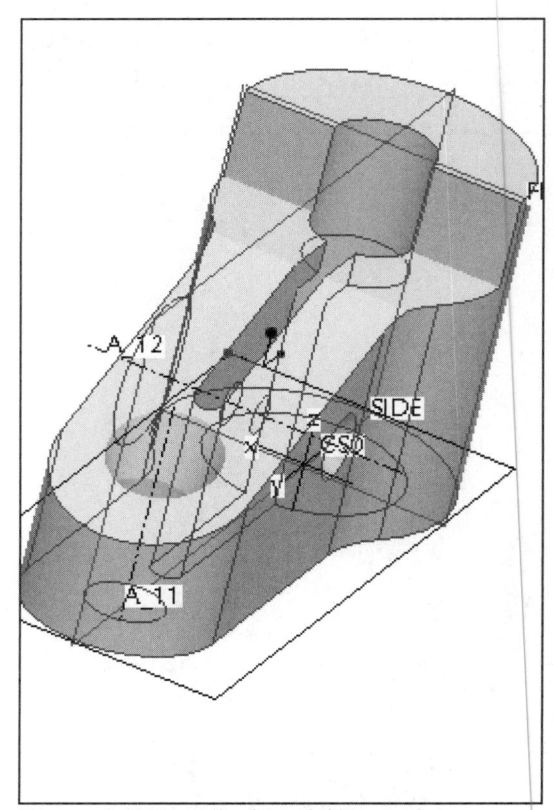

Figure 1.35(b) Edited Model Rotated

About the Dashboard

As you create and modify your models using direct graphical manipulation in the graphics window, the **Dashboard** guides you throughout the modeling process. This context sensitive interface monitors your actions in the current tool and provides you with basic design requirements that need to be satisfied to complete your feature. As you select individual geometry, the Dashboard narrows the available options enabling you to make only targeted modeling decisions. For advanced modeling, separate slide-up panels provide all relevant advanced options for your current modeling action. These Dashboard panels remain hidden until needed. This enables you to remain focused on successfully capturing the design intent of your model. The Dashboard for the cut feature is different from the protrusion feature [Fig. 1.36(a-c)].

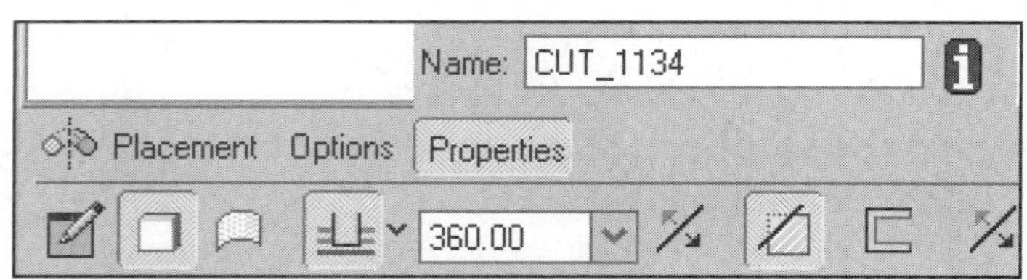

Figure 1.36(a) Revolved Cut Dashboard

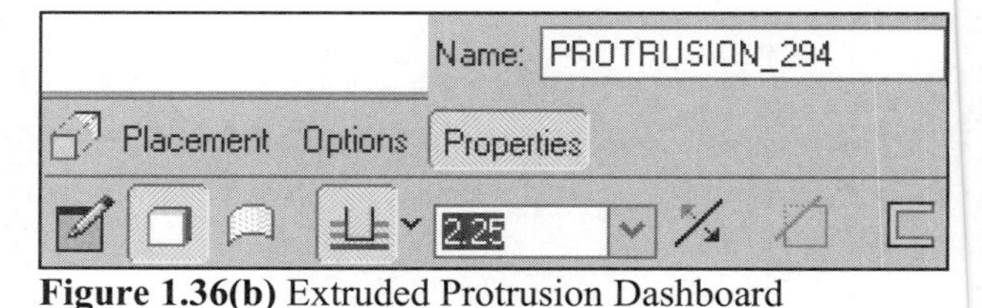

Figure 1.36(b) Extruded Protrusion Dashboard

Figure 1.36 (c) Revolved Protrusion Dashboard

This concludes the basic tour of Pro/E, Lesson 1. A Lesson Project is not provided here, at the end of Lesson 1. Instead, you may download another model from the PTC Catalog and practice navigating the user interface and using commands and tools introduced previously. If you do not have Internet access, ask your instructor to provide another model.

Lesson 2 Extrusions

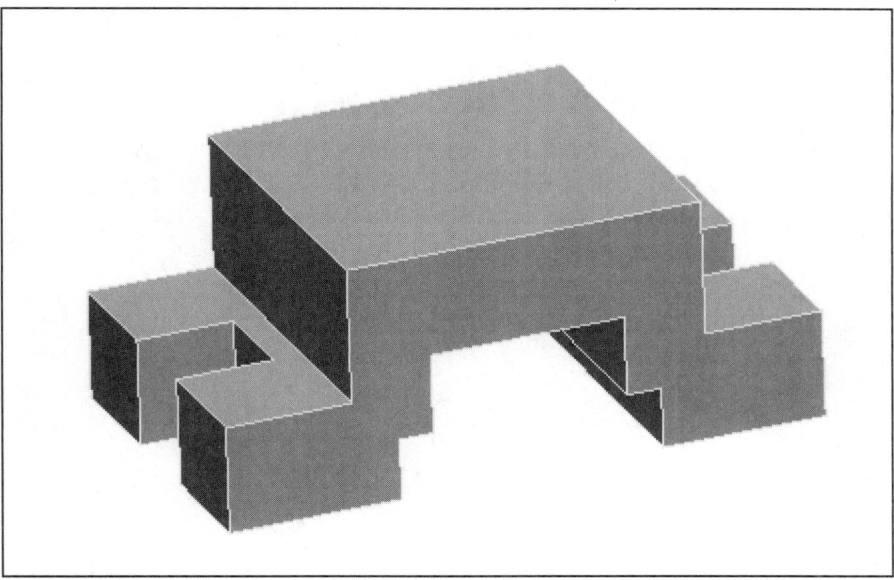

Figure 2.1 Clamp

OBJECTIVES

- Create a feature using an **Extruded** protrusion
- Understand **Setup** and **Environment settings**
- Define and set a **Material** type
- Create and use **Datum** features
- Sketch **protrusion** and **cut** feature geometry using the **Sketcher**
- Understand the feature **Dashboard**
- **Copy** a feature
- **Save** and **Delete Old Versions** of a part file

Extrusions

The design of a part using Pro/E starts with the creation of base features (normally datum planes), and a solid protrusion. Other protrusions and cuts are then added in sequence as required by the design. You can use various types of Pro/E features as building blocks in the progressive creation of solid parts (Fig. 2.1). Certain features, by necessity, precede other more dependent features in the design process. Those dependent features rely on the previously defined features for dimensional and geometric references.

The progressive design of features creates these dependent feature relationships known as *parent-child relationships*. The actual sequential history of the design is displayed in the Model Tree. The parent-child relationship is one of the most powerful aspects of Pro/E and parametric modeling in general. It is also very important after you modify a part. After a parent feature in a part is modified, all children are automatically modified to reflect the changes in the parent feature. It is therefore essential to reference feature dimensions so that Pro/E can correctly propagate design modifications throughout the model.

An **extrusion** is a part feature that adds or removes material. A protrusion is *always the first solid feature created*. This is usually the first feature created after a base feature of datum planes. The **Extrude Tool** is used to create both protrusions and cuts. A toolchest button is available for this command or it can be initiated using Insert ⇒ Extrude from the menu bar. Figure 2.2 shows four different types of basic protrusions.

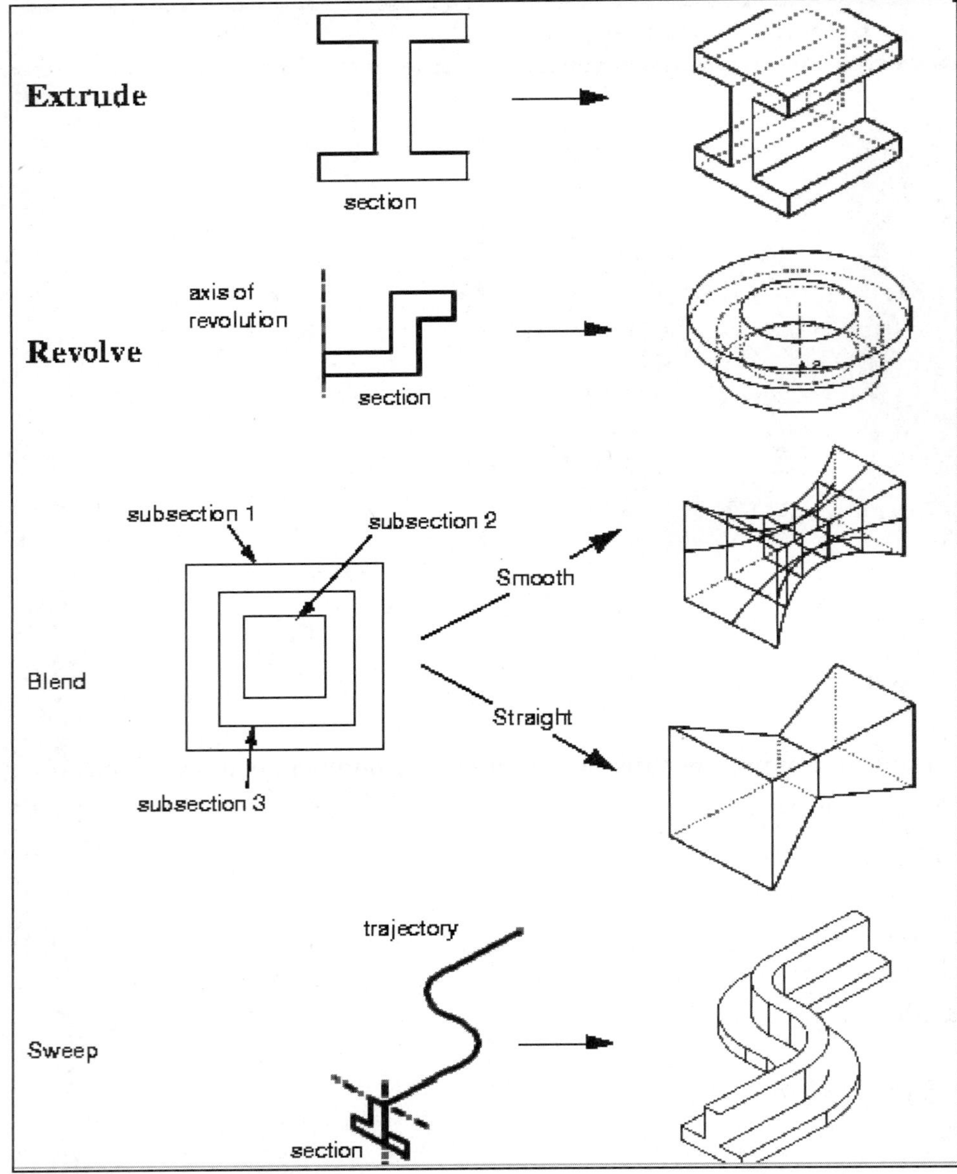

Figure 2.2 Basic Protrusions

The Design Process

You can immediately begin modeling in Pro/E. It is tempting to directly start creating models. Nevertheless, in order to build value into a design, you need to create a product that can keep up with the constant design changes associated with the design-through-manufacturing process. Flexibility must be "built in" to the design. Flexibility is the key to a friendly and robust product design while maintaining design intent, and you can accomplish it through planning.

To plan a design, you need to have a basic understanding of your model from a broad perspective. In other words, understand the overall function, form, and fit of the product. This understanding includes the following points:

- Overall size of the part
- Basic part characteristics
- The way in which the part can be assembled
- Approximate number of assembly components
- The manufacturing processes required to produce the part

Lesson 2 STEPS

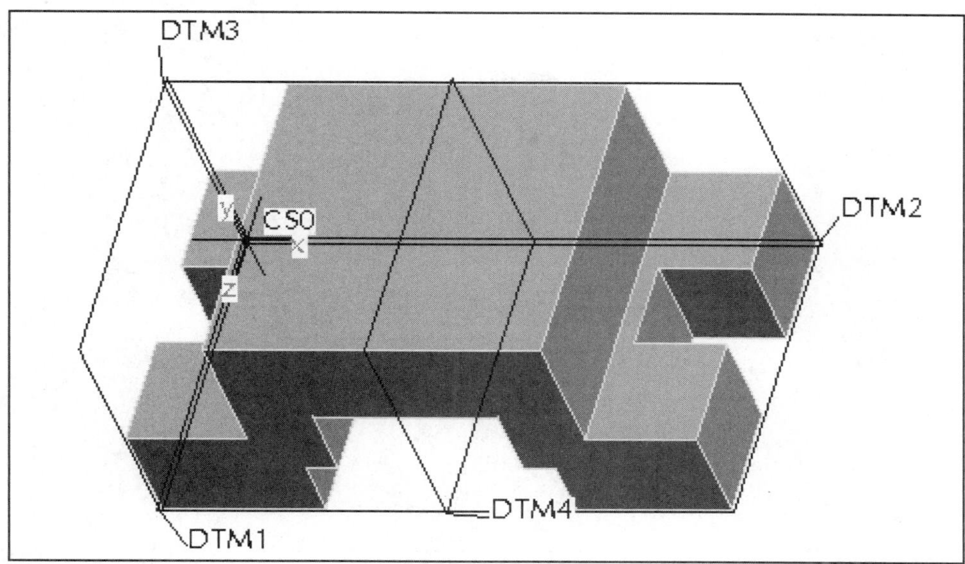

Figure 2.3 Clamp and Datum Planes

Clamp

The clamp in Figure 2.3 is composed of a simple protrusion and a cut. A number of things need to be established before you actually start modeling. These include setting up the *environment*, selecting the *units*, and establishing the *material* for the part.

Before you begin any part using Pro/E, you must plan the design. The **design intent** will depend on a number of things that are out of your control and on a number that you can establish. Asking yourself a few questions will clear up the design intent you will follow: Is the part a component of an assembly? If so, what surfaces or features are used to connect one part to another? Will geometric tolerancing be used on the part and assembly? What units are being used in the design, SI or decimal inch? What is the part's material? What is the primary part feature? How should I model the part, and what features are best used for the primary protrusion (the first solid mass)? On what datum plane should I sketch to model the first protrusion? These and many other questions will be answered as you follow the systematic lesson part. However, you must answer many of the questions on your own when completing the *lesson project*, which does not come with systematic instructions.

Start a new part, click: **File** ⇒ **Set Working Directory** ⇒ select the working directory ⇒ **OK** ⇒ **Create a new object** ⇒ ●**Part** ⇒ Name **CLAMP** ⇒ Use default template ⇒ **OK** ⇒ **Edit** ⇒ **Setup** ⇒ **Units** ⇒ Units Manager **millimeter Newton Second** ⇒ **Set** ⇒ ●**Convert Existing Numbers (Same Size)** [Fig. 2.4(a-b)] ⇒ **OK** ⇒ **Close** ⇒ **Material** ⇒ **Define** ⇒ type **STEEL** at the prompt ⇒ **MMB** [Fig. 2.4(c)] ⇒ **File** from material table ⇒ **Save** ⇒ **File** ⇒ **Exit** ⇒ **Assign** ⇒ pick **STEEL** ⇒ **Accept** ⇒ **MMB** ⇒ ⇒ **MMB** or

The material file, STEEL, is without any file information [Fig. 2.4(c)]. As an option, if your instructor provides you with the specifications, or you are familiar with setting up material specs, you can edit the file using: *Edit ⇒ Setup ⇒ Material ⇒ Edit ⇒ Steel ⇒ Accept ⇒ fill in the information ⇒ File ⇒ Save ⇒ File ⇒ Exit ⇒ Done.*

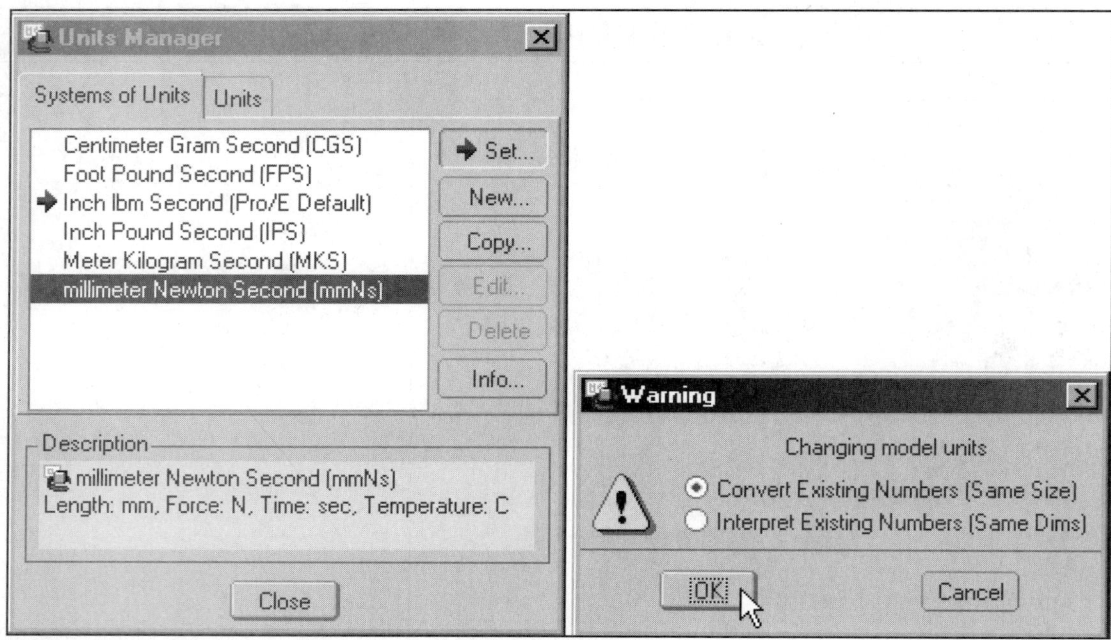

Figure 2.4(a-b) Units Manager and Warning Dialog Boxes

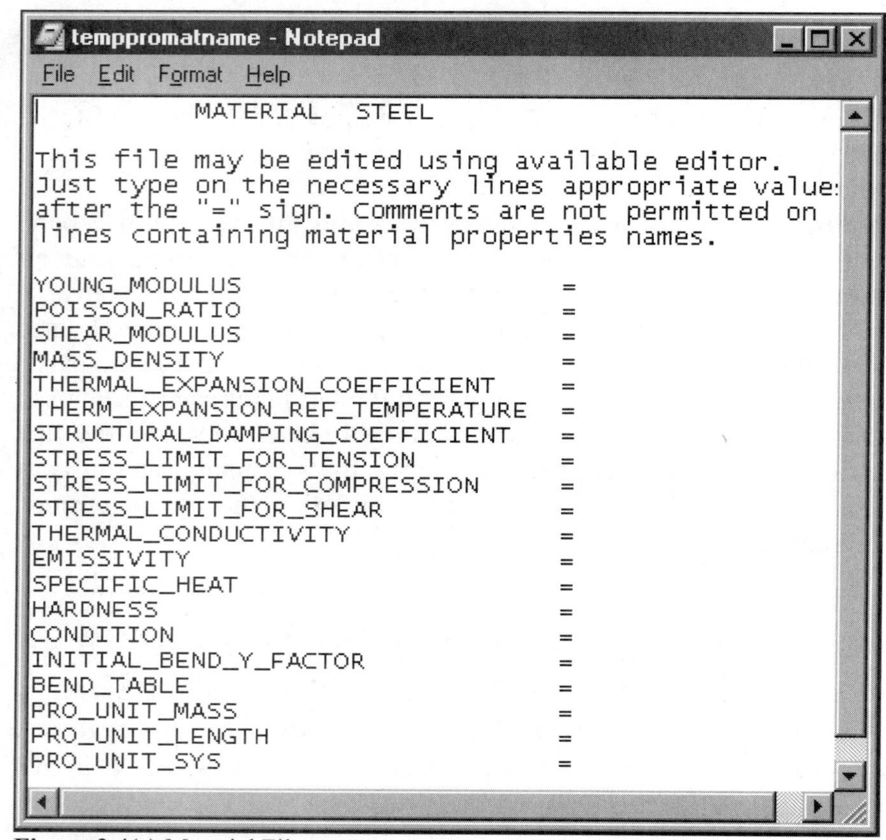

Figure 2.4(c) Material File

Since Use default template was selected, the default datum planes and the default coordinate system are displayed in the graphics window and in the Model Tree (Fig. 2.5). *The **default datum planes** and the **default coordinate system** will be the first features on all parts and assemblies.* The datum planes are used to sketch on and to orient the part's features. Having datum planes as the first features of a part, instead of the first protrusion, gives the designer more flexibility during the design process. Clicking on items in the Model Tree will highlight that item in red on the model (Fig. 2.5).

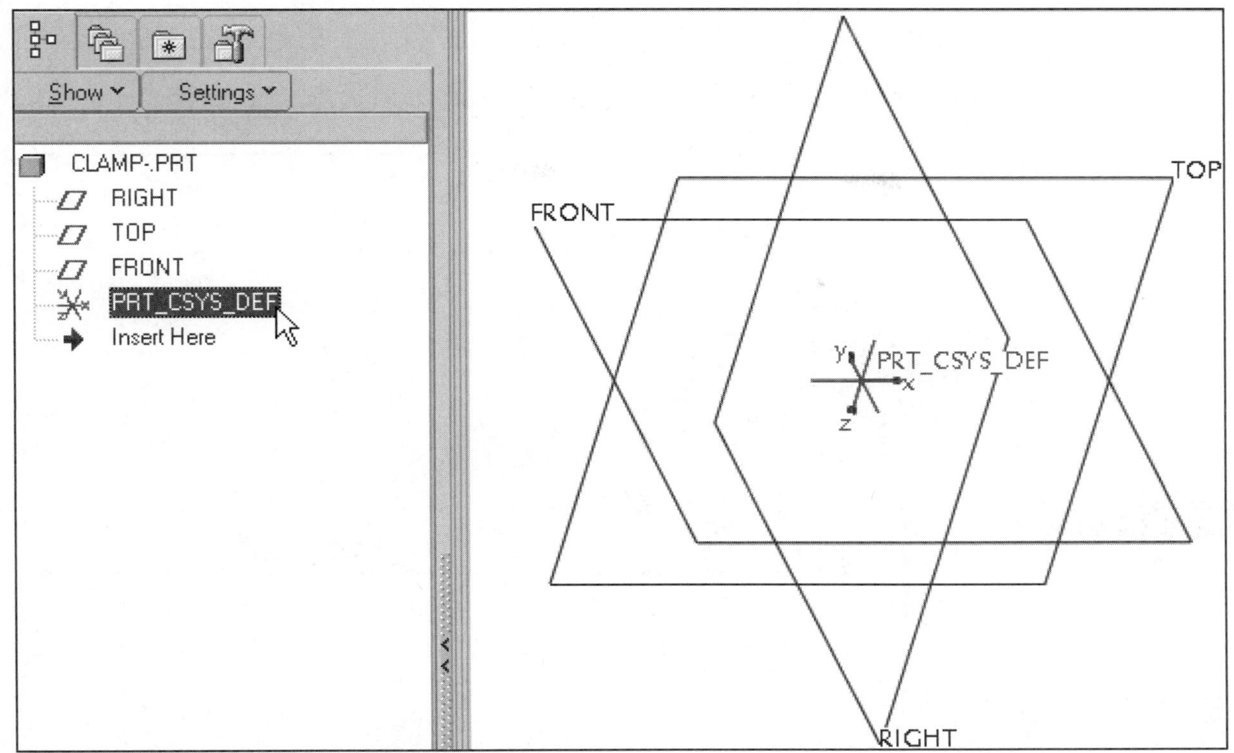

Figure 2.5 Default Datum Planes and Default Coordinate System

Create the first protrusion, click: **Extrude Tool** from the right Toolbar ⇒ **Create a section or redefine an existing section** from the dashboard ⇒ Section dialog box opens (Fig. 2.6) ⇒ Sketch Plane--- Plane: select **FRONT** datum from the model as the sketch plane ⇒ Sketch Orientation--- click inside Reference: box and select **TOP** datum from the model ⇒ Orientation: ⇒ **Top** (makes TOP datum plane face up) ⇒ click **Sketch** button

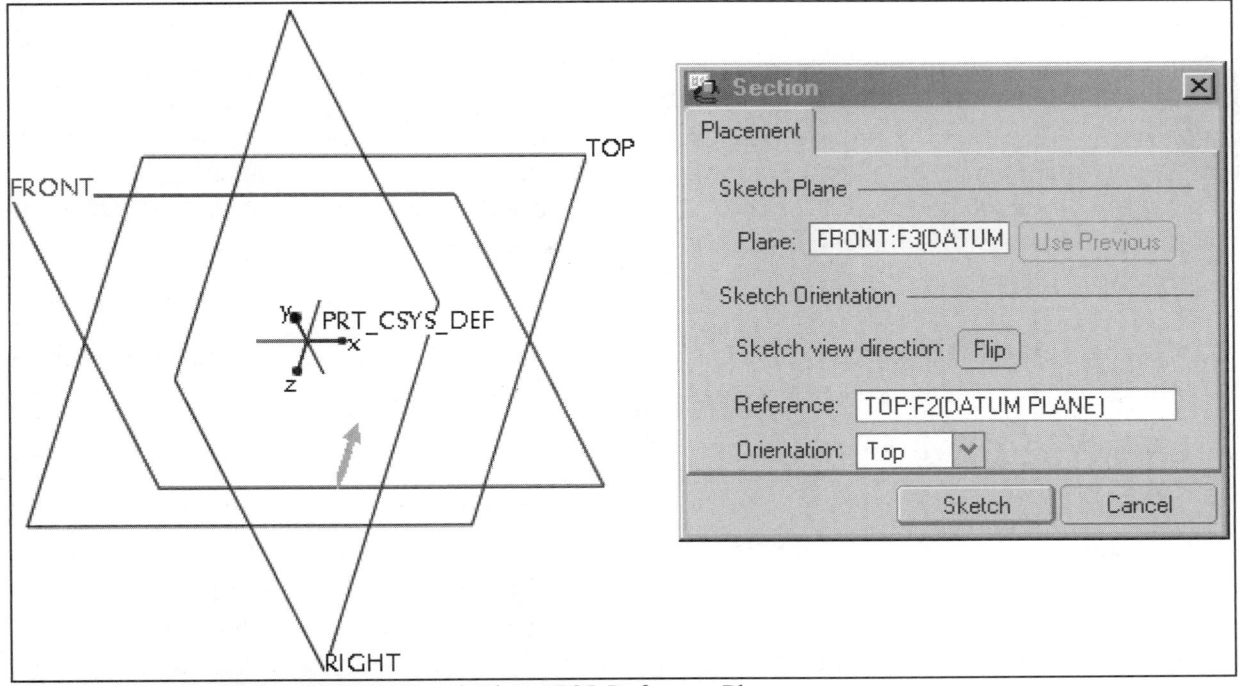

Figure 2.6 Section Dialog, FRONT Sketch Plane, TOP Reference Plane

Click: **Close** to accept the References (Fig. 2.7) ⇒ **Toggle the grid on** from the top toolbar (Fig. 2.8)

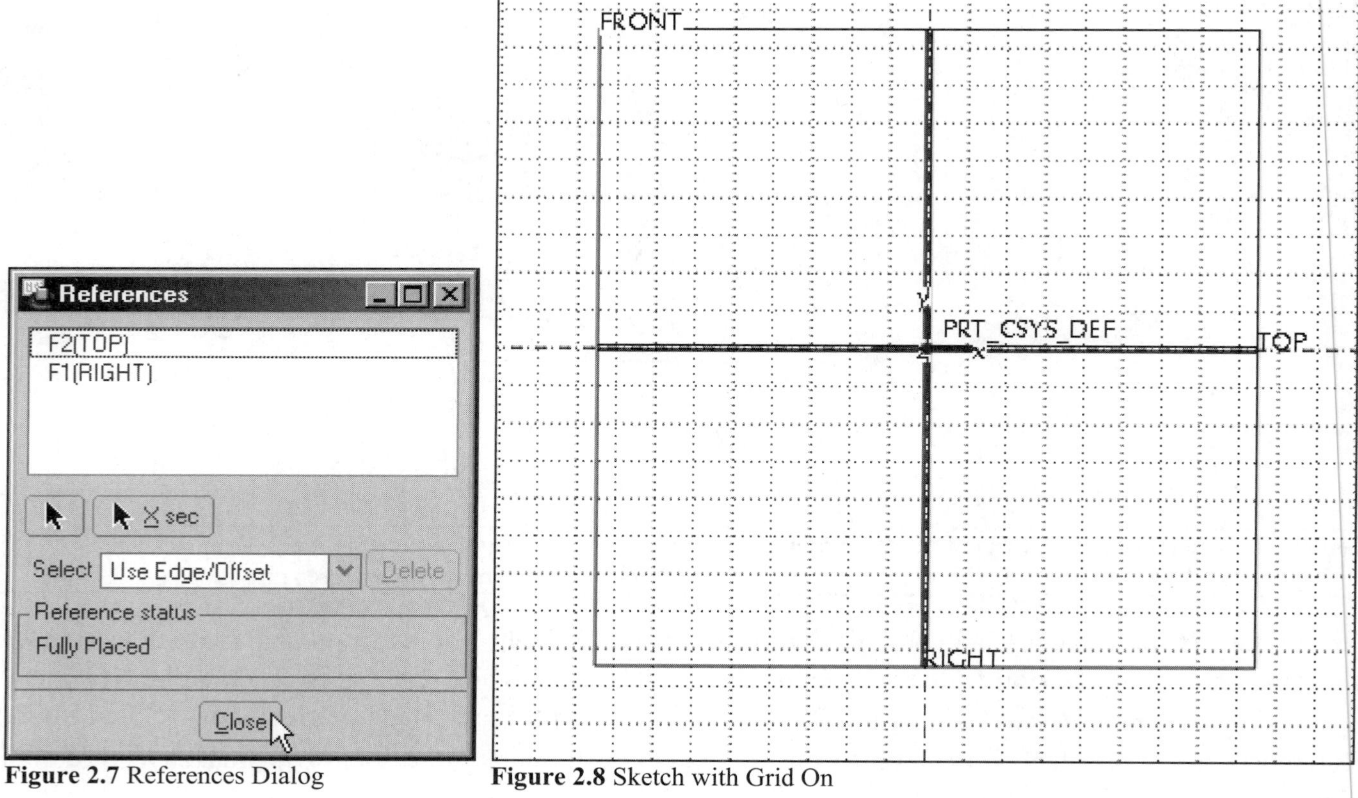

Figure 2.7 References Dialog

Figure 2.8 Sketch with Grid On

The sketch is now displayed and oriented in 2D (Fig. 2.8). The coordinate system is at the middle of the sketch, where datum RIGHT and datum TOP intersect. The X coordinate arrow points to the right and the Y coordinate arrow points up. The Z arrow is pointing toward you (out from the screen). The square box you see is the limited display of datum FRONT, which is like a piece of graph paper you will be sketching on when you create the protrusion's geometry. Pro/E is not a coordinate-based software, so you need not enter geometry with X, Y, and Z coordinates as with many other CAD systems.

In general, many of the part base protrusion features you will be modeling start with sketching in the first quadrant. To make this more convenient, and to have a greater sketching area, change the position and size of the sketch.

Use **Shift MMB** and **Ctrl MMB** to reposition and resize the sketch so that the first quadrant is more centrally located. In addition, since you now have a visible grid, it is a good idea to have your sketch picks snap to that grid. Click: **Tools** from menu bar ⇒ **Environment** ⇒ Snap to Grid (Fig. 2.9) ⇒ **Apply** ⇒ **OK**

You can control many aspects of the environment in which Pro/E runs with the Environment dialog box. To open the Environment dialog box, click Tools > Environment on the menu bar or click the appropriate icon in the toolbar. When you make a change in the Environment dialog box, it takes effect for the current session only. When you start Pro/E, the environment settings are defined by your configuration file, if any; otherwise, by Pro/E configuration defaults.

The following lists the options available in the **Environment** dialog box:

Display:

Dimension Tolerances Display model dimensions with tolerances
Datum Planes Display the datum planes and their names
Datum Axes Display the datum axes and their names
Point Symbols Display the datum points and their names
Coordinate Systems Display the coordinate systems and their names
Spin Center Display the spin center for the model
3D Notes Display model notes
Notes as Names Display the note as a name, not the full note
Reference Designators Display reference designation of cabling connections and ECAD components
Thick Cables Display a cable with 3-D thickness
Centerline Cables Display the centerline of a cable with location points
Internal Cable Portions Display cable portions that are hidden from view
Colors Display the models in colors
Textures Display textures on shaded models
Levels of Detail Controls levels of detail available in a shaded model during dynamic orientation

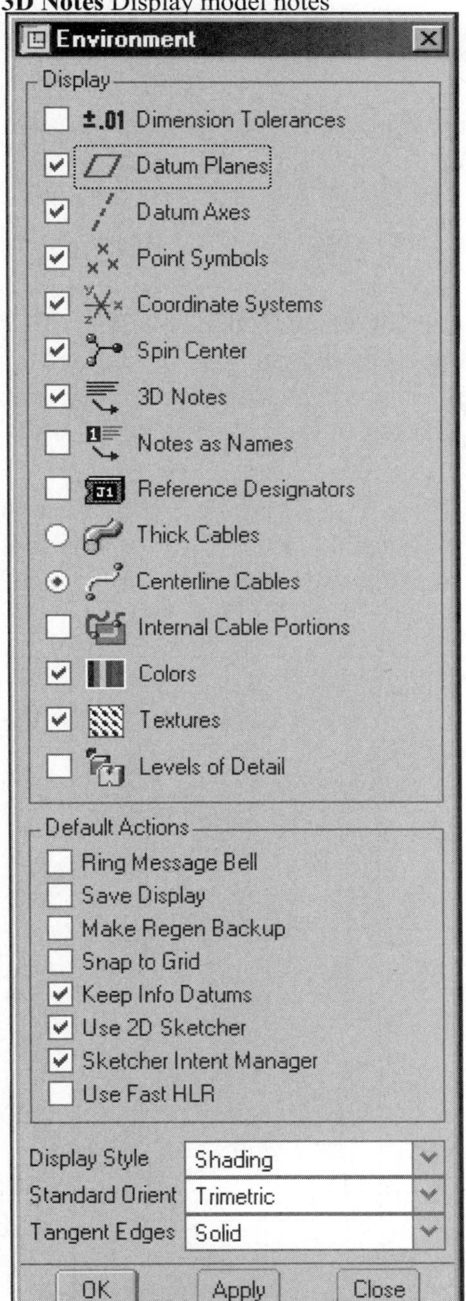

Figure 2.9 Environment Dialog Box

Default Actions:

Ring Message Bell Ring bell (beep) after each prompt or system message
Save Display Save objects with their most recent screen display
Make Regen Backup Backs up the current model before every regeneration
Snap to Grid Make points you select on the screen snap to a grid
Keep Info Datums Control how Pro/E treats datum planes, datum points, datum axes, and coordinate systems created on the fly under the Info functionality
Use 2D Sketcher Control the initial model orientation in Sketcher mode
Sketcher Intent Manager Use the Intent Manager when in Sketcher
Use Fast HLR Make possible the hardware acceleration of dynamic spinning with hidden lines, datums, and axes

Display Style:

- **Wireframe** Model is displayed with no distinction between visible and hidden lines
- **Hidden Line** Hidden lines are shown in gray
- **No Hidden** Hidden lines are not shown
- **Shading** All surfaces and solids are displayed as shaded

Standard Orient:

- **Isometric** Standard isometric orientation
- **Trimetric** Standard trimetric orientation
- **User Defined** User-defined orientation

Tangent Edges:

- **Solid** Display tangent edges as solid lines
- **No Display** Blank tangent edges
- **Phantom** Display tangent edges in phantom font
- **Centerline** Display tangent edges in centerline font
- **Dimmed** Display tangent edges in the Dimmed Menu system

Because you checked [✓ Snap to Grid], you can now sketch by simply picking grid points representing the part's geometry (outline). Because this is a sketch in the true sense of the word, you need only create geometry that *approximates* the shape of the feature; the sketch does not have to be accurate as far as size or dimensions are concerned. Even with the grid snap off Pro/E, constrains the geometry according to rules, which include but are not limited to the following:

- **RULE:** Symmetry
- **DESCRIPTION:** Entities sketched symmetrically about a centerline are assigned equal values with respect to the centerline
- **RULE:** Horizontal and vertical lines
- **DESCRIPTION:** Lines that are approximately horizontal or vertical are considered exactly horizontal or vertical.
- **RULE:** Parallel and perpendicular lines
- **DESCRIPTION:** Lines that are sketched approximately parallel or perpendicular are considered exactly parallel or perpendicular
- **RULE:** Tangency
- **DESCRIPTION:** Entities sketched approximately tangent to arcs or circles are assumed to be exactly tangent

The outline of the part's primary feature is sketched using a set of connected lines. The part's dimensions and general shape are provided in Figure 2.10. Because much of the part can be defined by sketching an outline similar to its front view, you can complete most of the part's geometry with one protrusion. The cuts on the sides will be the second feature created. Sketch only one series of lines (16 lines in this sketch). Do not sketch lines on top of lines.

It is important not to create any unintended constraints while sketching. Therefore, remember to exaggerate the sketch geometry and not to align edges that have no relationship. Pro/E is very smart: If you draw two lines at the same horizontal level, Pro/E thinks they are horizontally aligned.

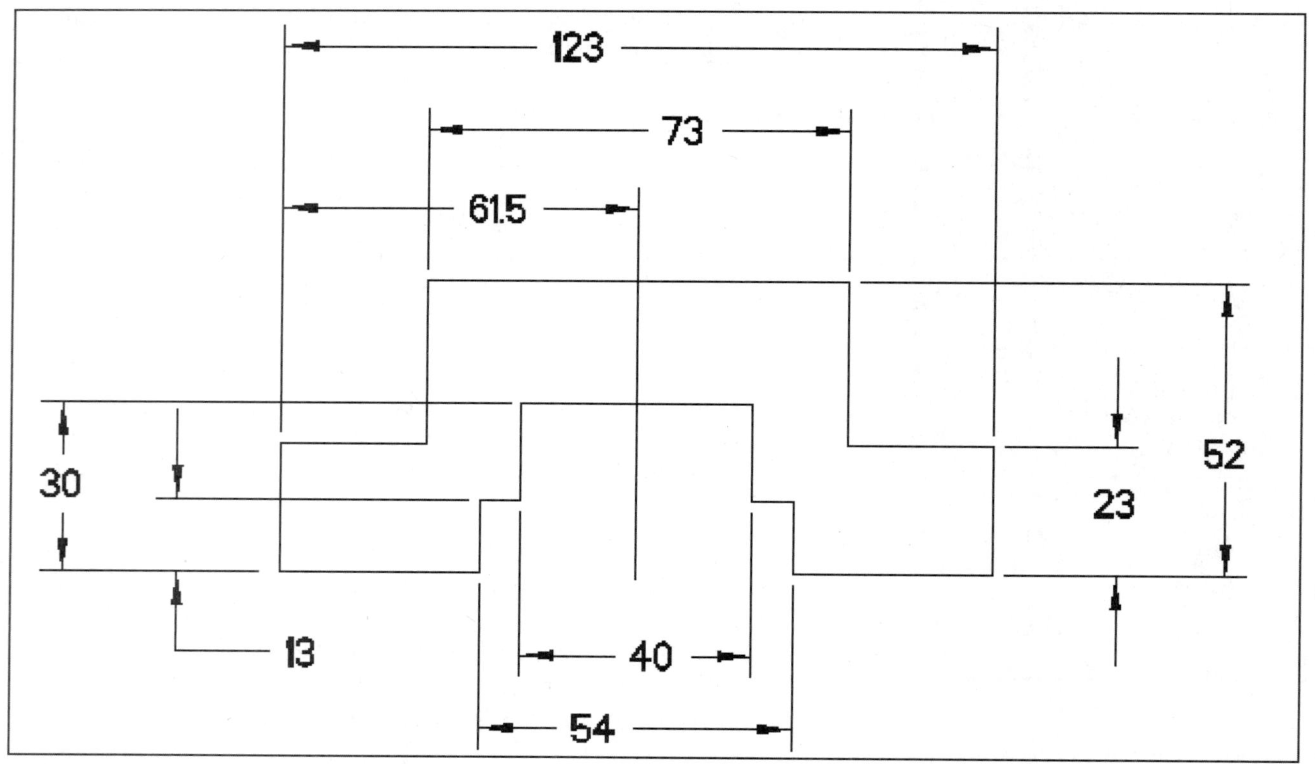

Figure 2.10 Dimensions for Clamp

Click: **Create 2 point lines** from the right vertical toolbar ⇒ Place the mouse near the center of the coordinate system at the intersection of the **RIGHT** and **TOP** datum planes and click. Continue picking until you have sketched an outline approximating Figure 2.11(a) ⇒ **MMB** to end the line sequence

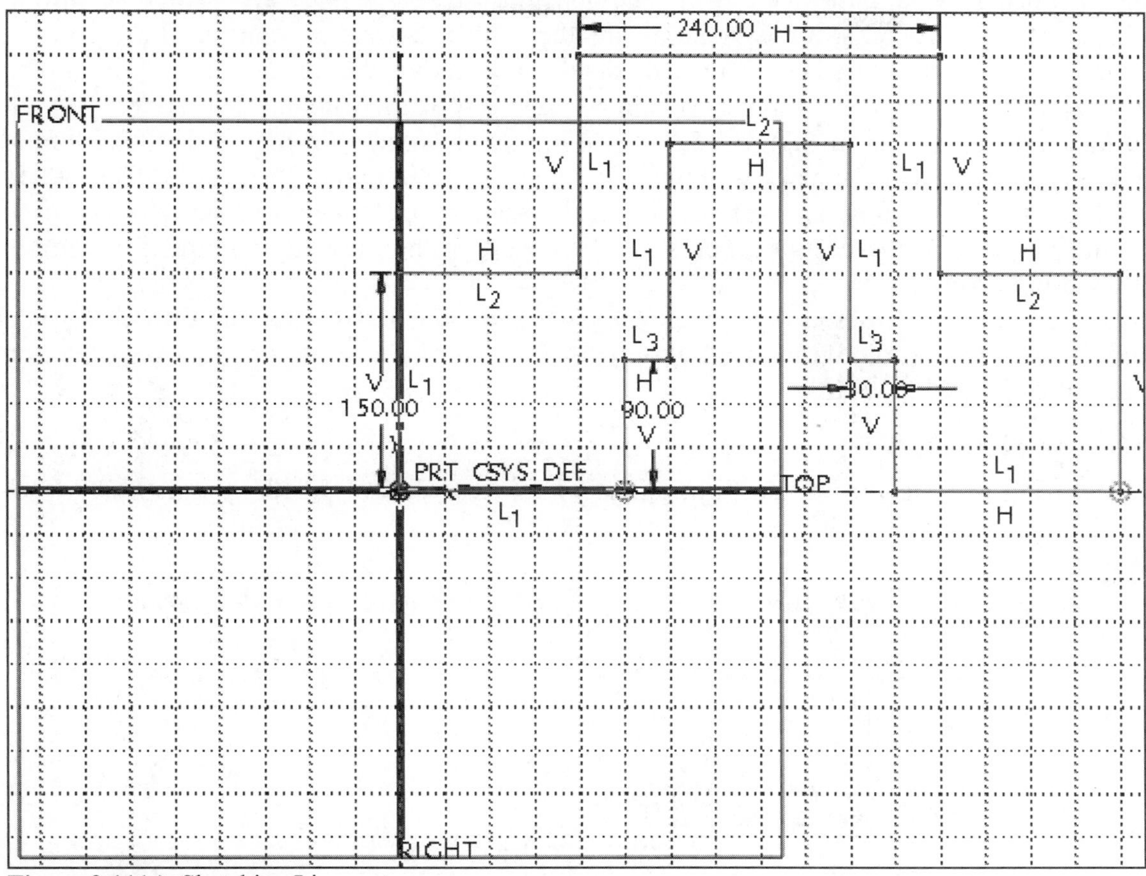

Figure 2.11(a) Sketching Lines

Click: **View** ⇒ **Repaint** ⇒ add a centerline, click: ⇒ **Create 2 point centerlines** [Fig. 2.11(b)] ⇒ pick two positions to create the vertical centerline

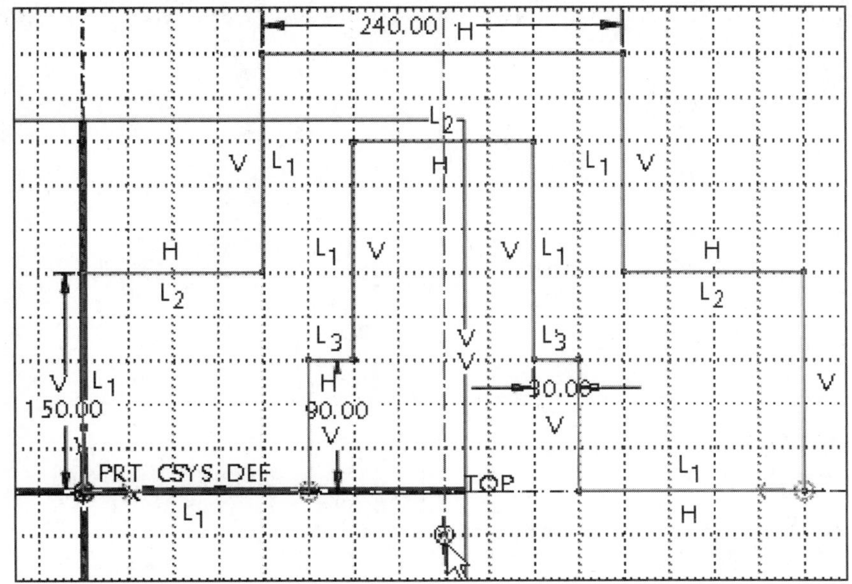

Figure 2.11(b) Sketching a Centerline

Dimensions, Constraints, Grid, and Vertices can be toggled on and off as needed using the toolbar buttons. A sketcher *constraint symbol* appears next to the entity that is controlled by that constraint. Sketcher constraints can be turned on or off (enabled or disabled) while you are sketching. An **H** next to a line means horizontal; a **T** means tangent. Dimensions display as they are needed according to the references selected and the constraints. Seldom are they the exact same required dimensioning scheme needed to manufacture the part. You can add, delete, and move dimensions as required. The dimensioning scheme is important, not the dimension value, which can be modified now or later.

Place the dimensions as shown in Figure 2.12. Do not be concerned with the perfect positioning of the dimensions, but try to follow the spacing and positioning standards found in the **ASME Geometric Tolerancing and Dimensioning** standards. This saves you time when you create a drawing of the part. Dimensions placed at this stage of the design process are displayed on the drawing document by simply asking to show all the dimensions.

To dimension between two lines, simply pick the lines with the left mouse button (LMB) and place the dimension value with the middle mouse button (MMB). To dimension a line, click on the line (LMB) and then place the dimension with MMB.

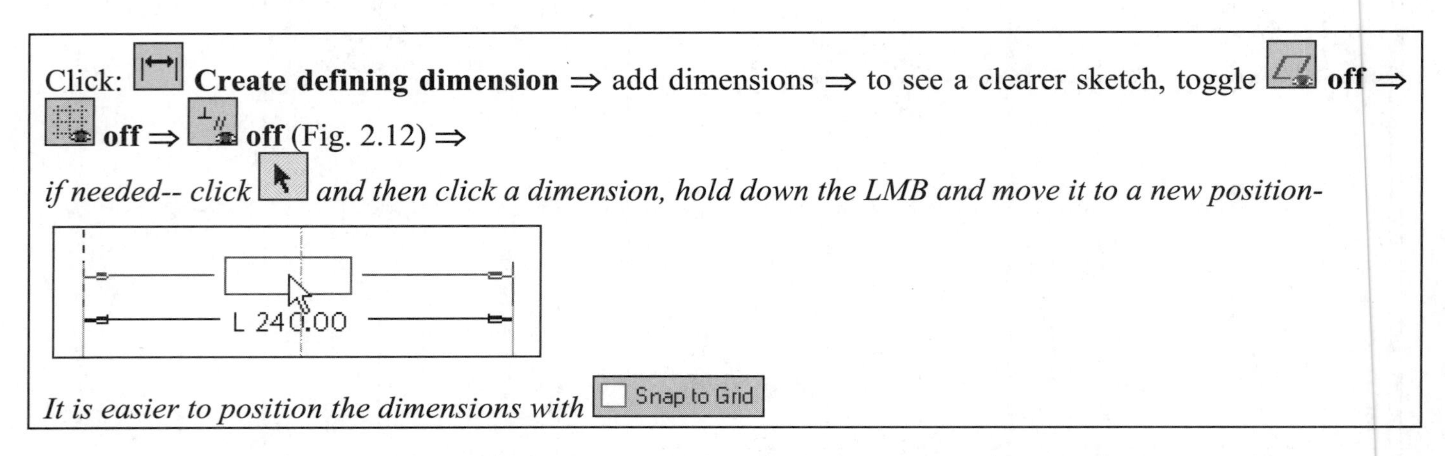

Click: **Create defining dimension** ⇒ add dimensions ⇒ to see a clearer sketch, toggle **off** ⇒ **off** ⇒ **off** (Fig. 2.12) ⇒

if needed-- click and then click a dimension, hold down the LMB and move it to a new position-

L 240.00

It is easier to position the dimensions with Snap to Grid

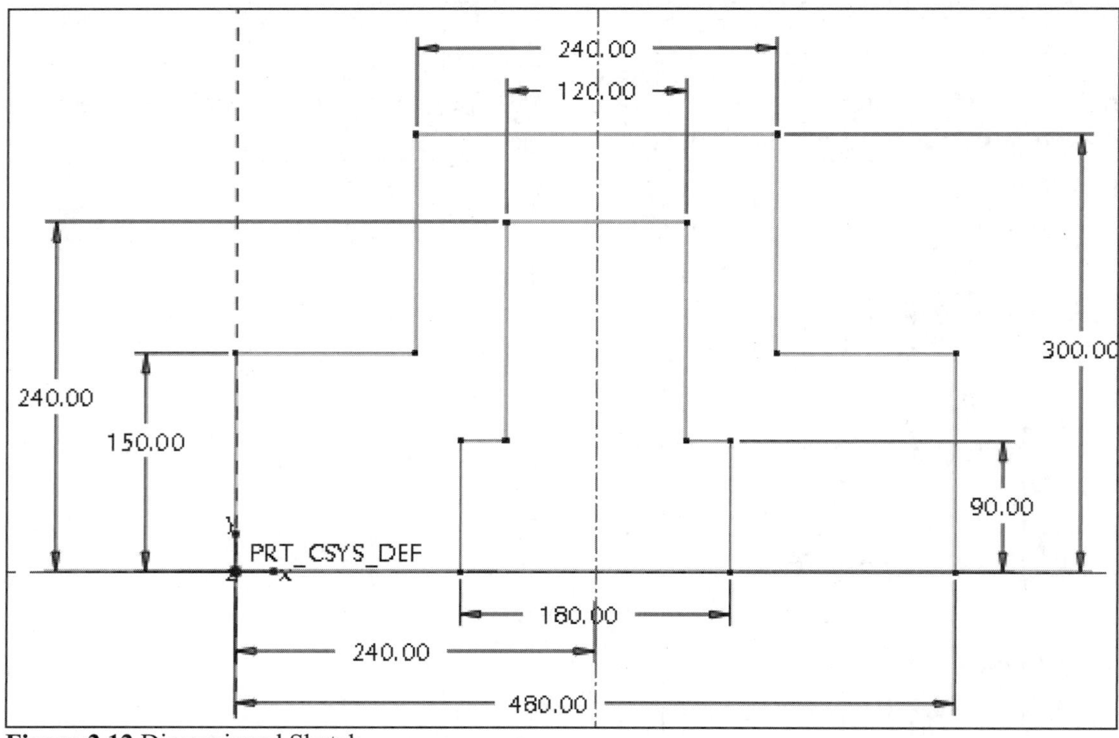

Figure 2.12 Dimensioned Sketch

You can now modify the dimensions to the *design sizes*, click: ⇒ Window-in the sketch (place the cursor at one corner of the window with the **LMB** depressed, drag the cursor to the opposite corner of the window and release the **LMB**) to capture all dimensions. They will turn red. ⇒ **Modify the values** (Fig. 2.13) ⇒ Regenerate ⇒ click on each dimension individually in the Modify Dimensions dialog box and type the design value (Fig. 2.10) at the prompt ⇒ Regenerate ⇒ **Regenerate the section and close the dialog** (Fig. 2.14)

Alternately, double click on any dimension individually and modify the value directly on the sketch

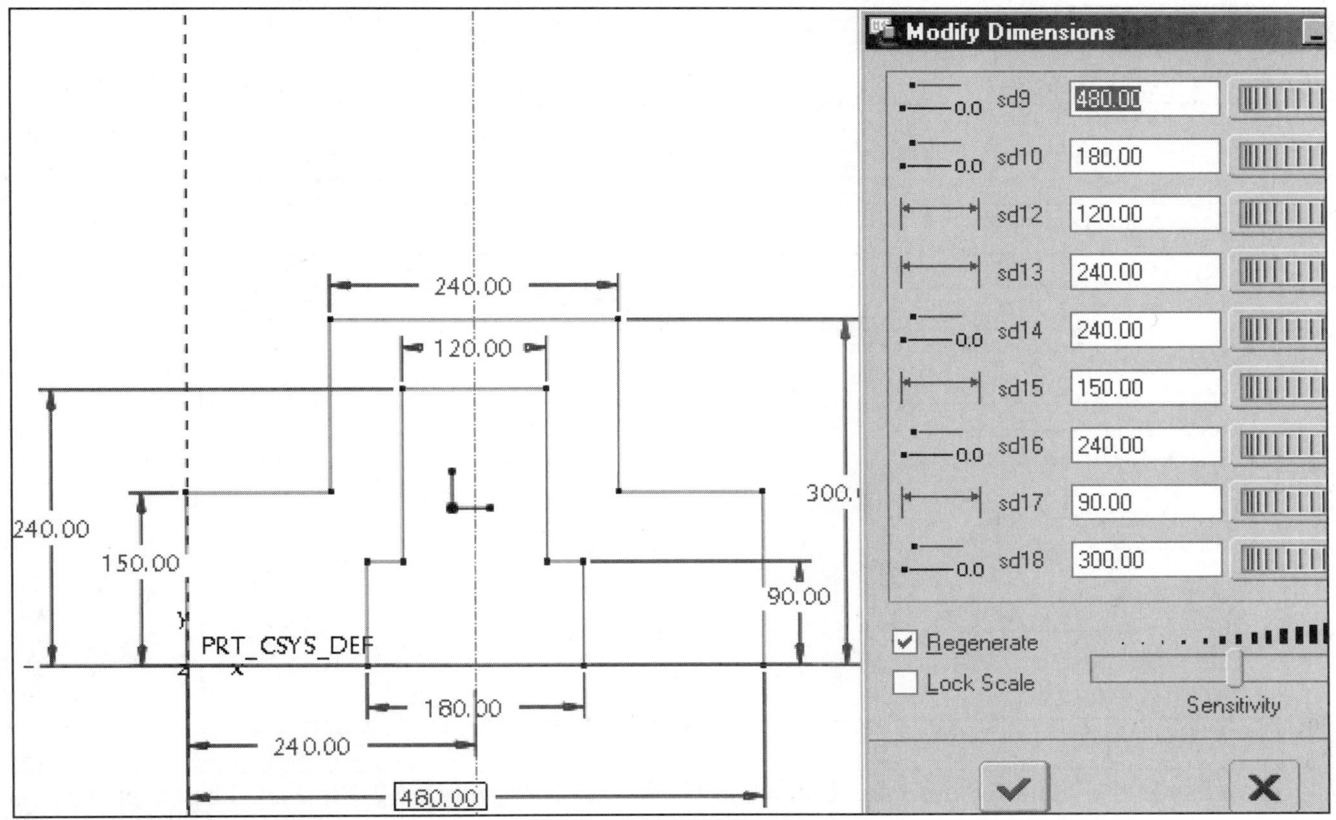

Figure 2.13 Modify

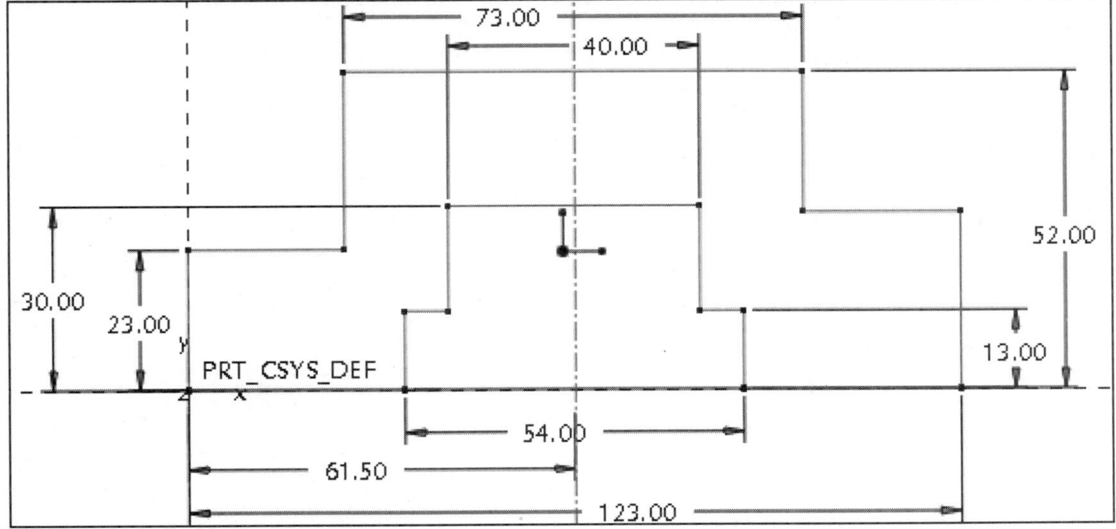

Figure 2.14 Modified and Regenerated Sketch

Click: ⇒ **Standard Orientation** (Fig. 2.15) ⇒ **on** [dimensions are toggled off for viewing (Fig. 2.16)] ⇒ **Continue with the current section** ⇒ note the yellow direction arrow ⇒ **Zoom Out** as needed to see the whole object

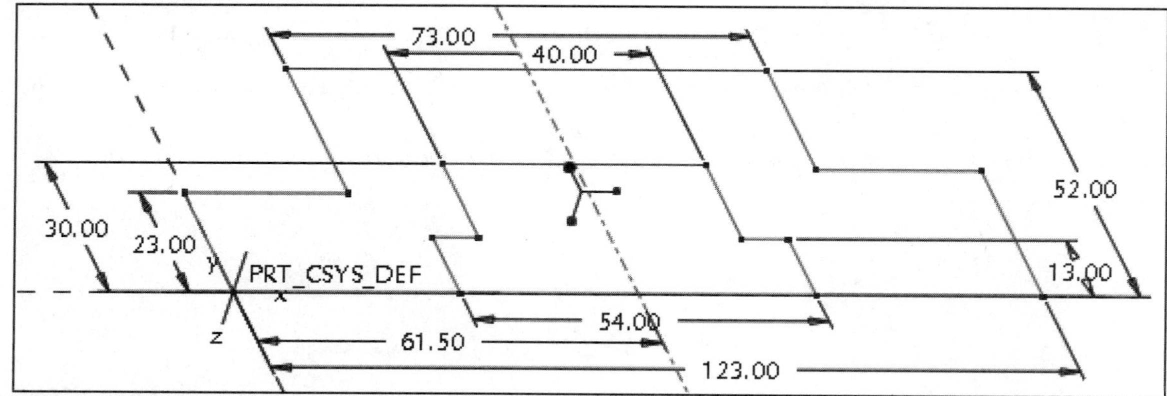

Figure 2.15 Regenerated Dimensions

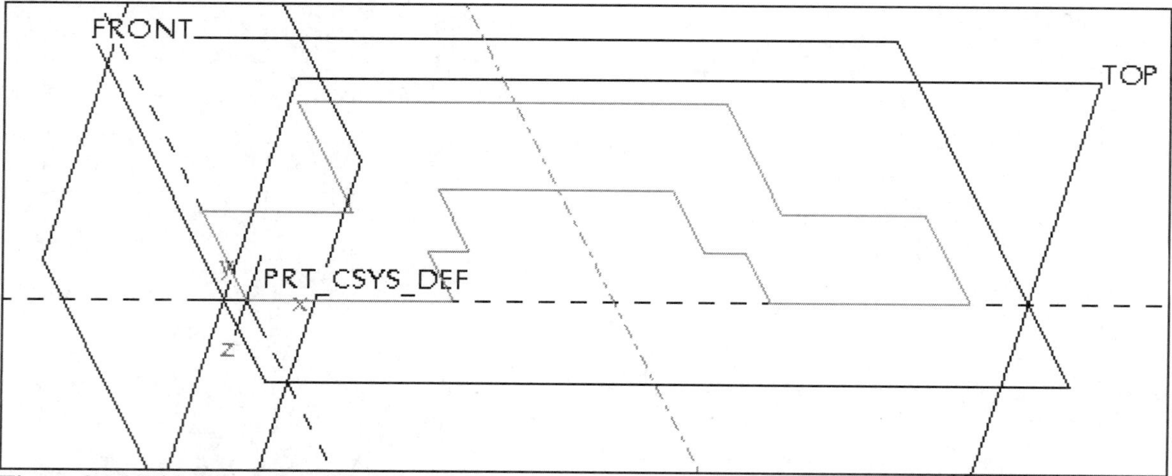

Figure 2.16 With Dimensions Not Displayed

Click on and slide the "handle" until the dimension is **70** [Fig. 2.17 and Fig. 2.18(a-b)]. The exact value is not needed since the depth dimension can be modified later. Release the **LMB** to end ⇒ (Fig. 2.19)

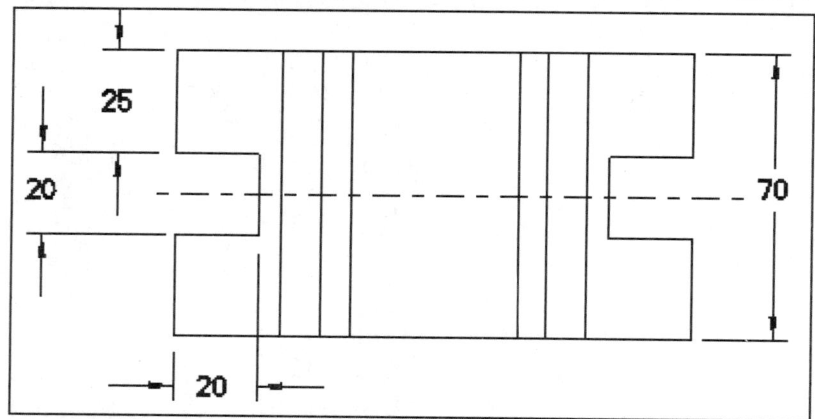

Figure 2.17 Depth and Cut Dimensions

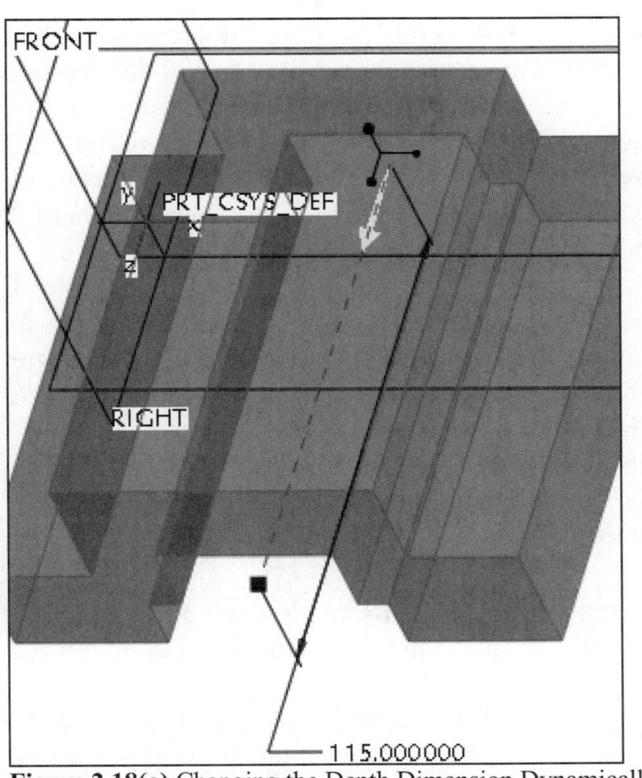

Figure 2.18(a) Changing the Depth Dimension Dynamically

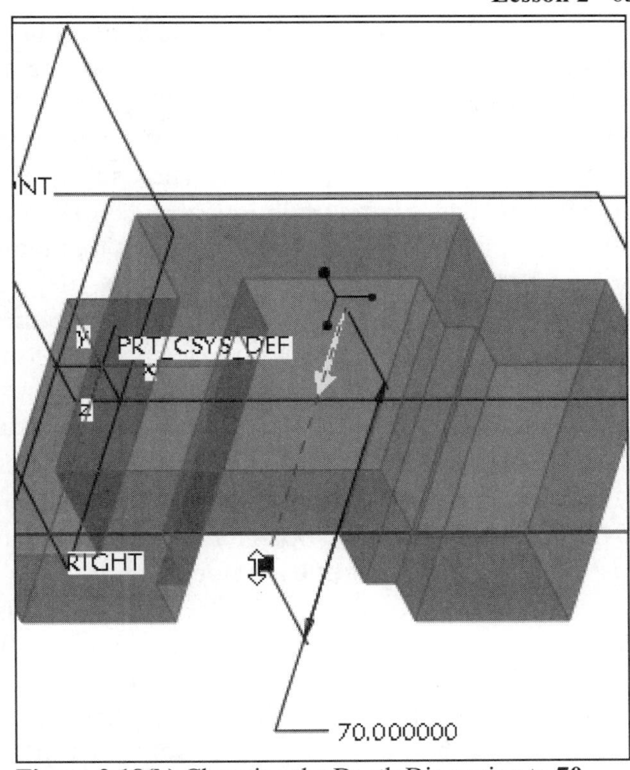

Figure 2.18(b) Changing the Depth Dimension to **70**

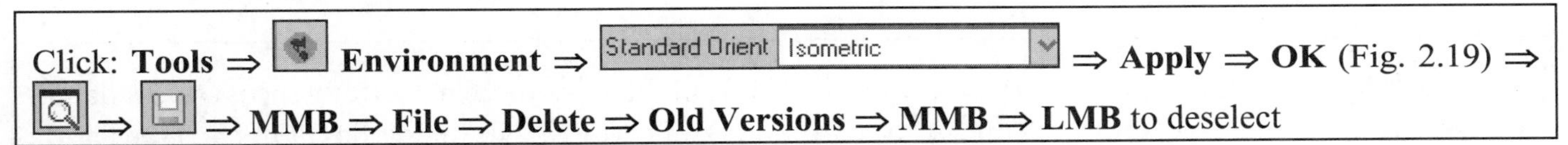

Click: **Tools** ⇒ **Environment** ⇒ Standard Orient Isometric ⇒ **Apply** ⇒ **OK** (Fig. 2.19) ⇒ ⇒ ⇒ **MMB** ⇒ **File** ⇒ **Delete** ⇒ **Old Versions** ⇒ **MMB** ⇒ **LMB** to deselect

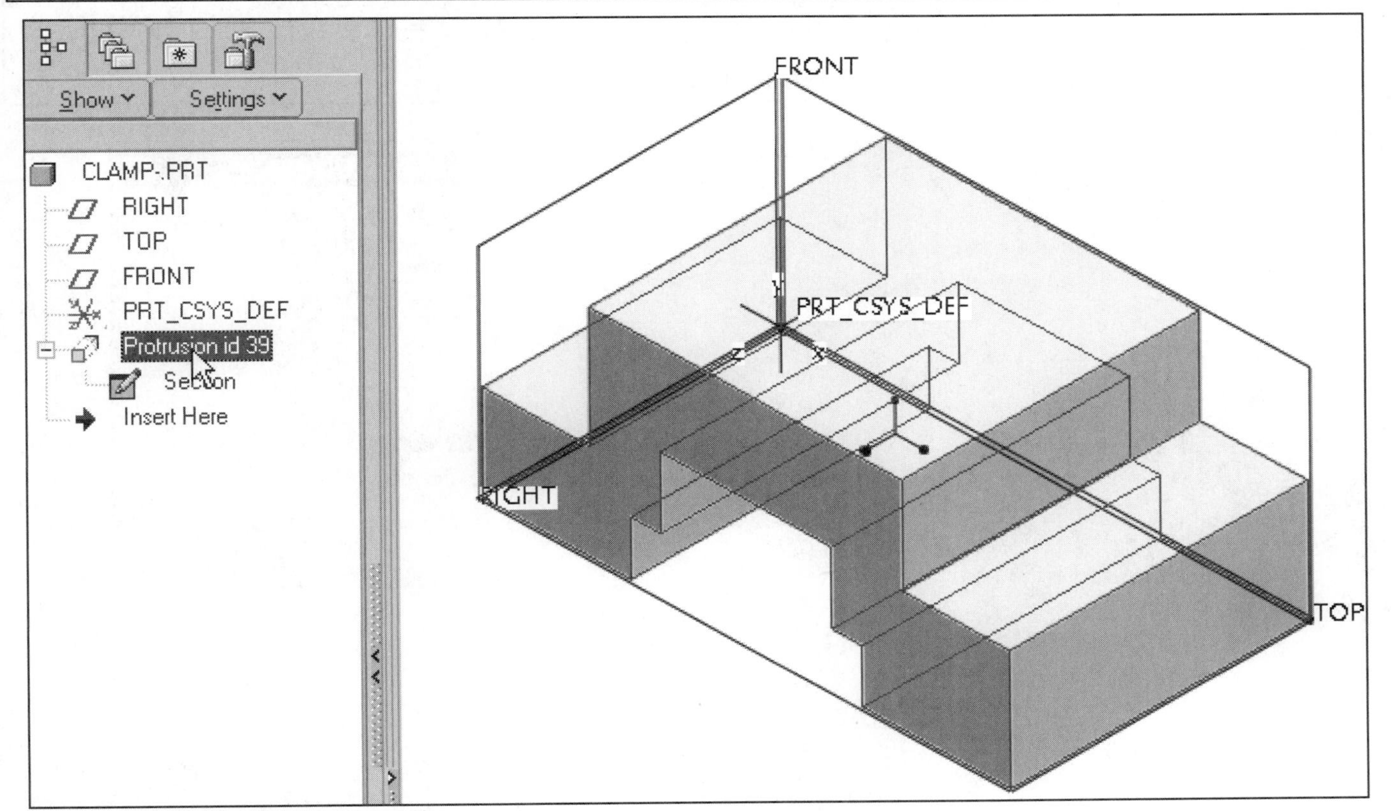

Figure 2.19 Completed First Protrusion

Storing an object on the disk does not overwrite an existing object file. To preserve earlier versions, Pro/E saves the object to a new file with the same object name but with an updated version number. Every time you store an object using Save, you create a new version of the object in memory, and write the previous version to disk. Pro/E numbers each version of an object storage file consecutively (for example, box.sec.1, box.sec.2, box.sec.3). If you save 25 times you have 25 versions of the object, all at different stages of completion. You can use *File ⇒ Delete ⇒ Old Versions* after the *Save* command to eliminate previous versions of the object that may have been stored.

When opening an existing object file, you can open any version that is saved. Although Pro/E automatically retrieves the latest saved version of an object, you can retrieve any previous version by entering the full file name with extension and version number (for example, **bracket.prt.5**). If you do not know the specific version number, you can enter a number relative to the latest version. For example, to retrieve a part from two versions ago enter **partname.prt.-2**.

You use *File ⇒ Erase* to remove the object and its associated objects from memory. If you close a window before erasing it, the object is still in memory. In this case, you use *File ⇒ Erase ⇒ Not Displayed* to remove the object and its associated objects from memory. This does not delete the object. It just removes it from active memory.

File ⇒ Delete ⇒ All Versions does just that, removes the file from memory and from disk completely. You are prompted with a Delete All Confirm dialog box when choosing this command. Be careful not to delete needed files.

The next feature will be a **20 X 20** centered cut (Fig. 2.17). Because the cut feature is identical on both sides of the part, you can mirror and copy the cut after it has been created. Start by creating a new datum plane that is offset from datum RIGHT.

Click: **Spin Center off** ⇒ **Datum Plane Tool** from the right toolbar ⇒ References: select datum **RIGHT** as the plane to offset from (Fig. 2.20) ⇒ In the DATUM PLANE dialog box, Offset: Translation type the distance **123/2** ***(123 divided by 2; Pro/E will do the math: 123/2 = 61.5)*** (Fig. 2.21) ⇒ **Enter** ⇒ **OK** DTM1 is then created (Fig. 2.22) ⇒ **LMB** to deselect ⇒

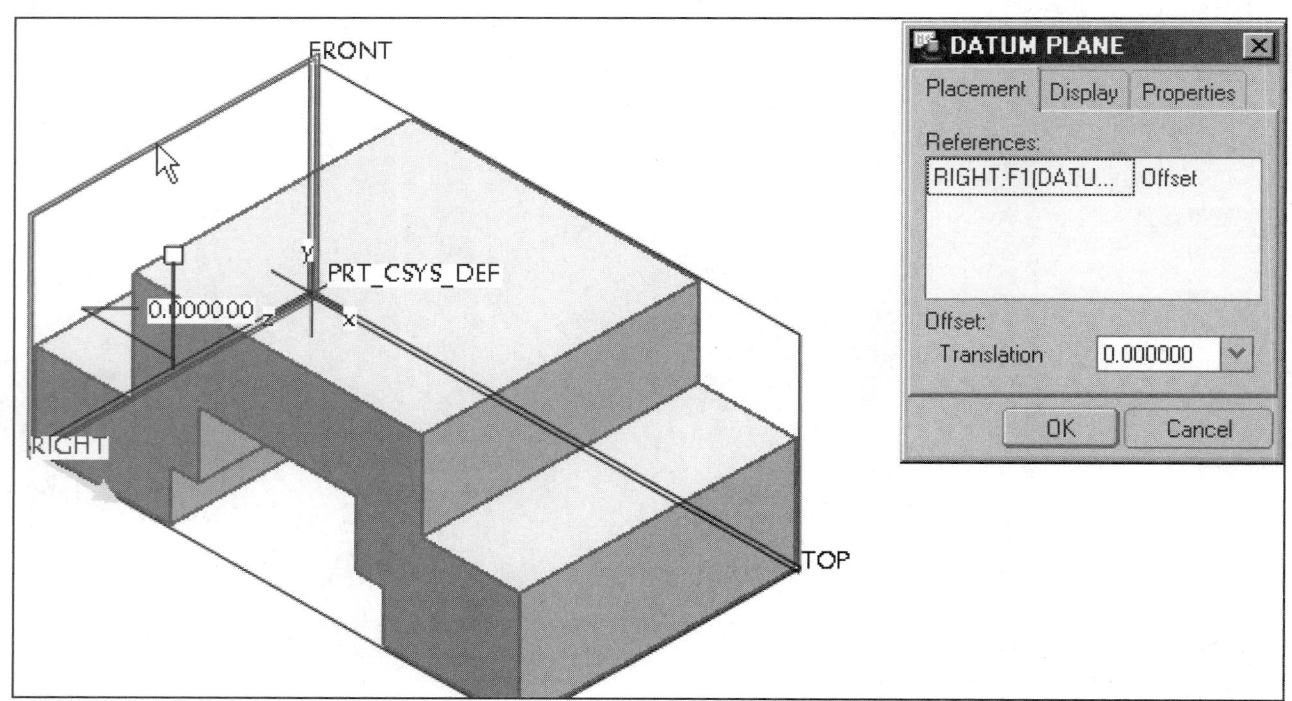

Figure 2.20 Creating an Offset Datum Plane

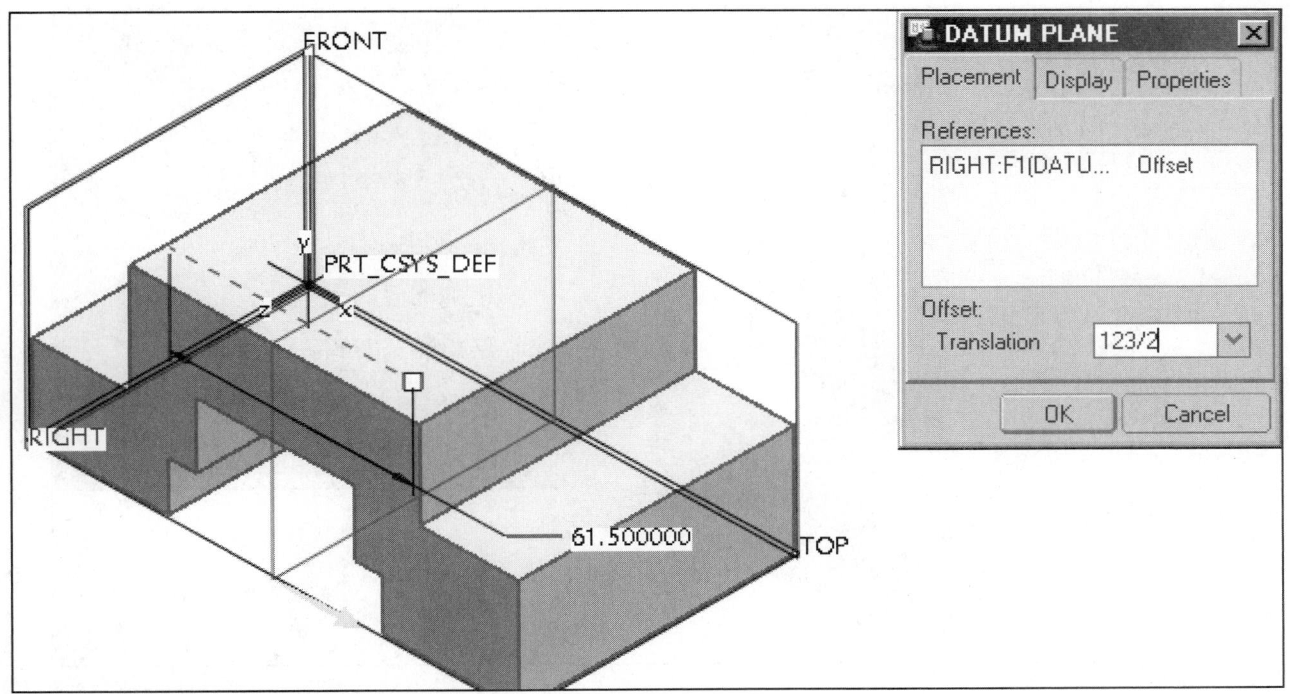

Figure 2.21 Datum Plane Offset One Half of the Part's Length (**123/2**)

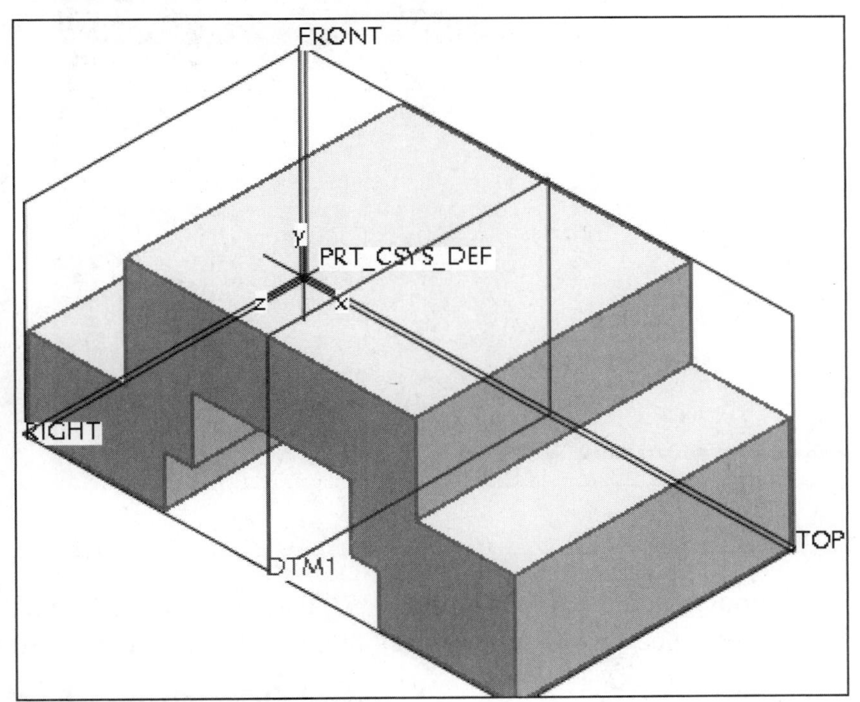

Figure 2.22 Datum Plane DTM1

Create the cut, click: **Extrude Tool** ⇒ ⇒ Section dialog box opens (Fig. 2.23) ⇒ Sketch Plane--- Plane: select **TOP** datum from the model as the sketch plane ⇒ Sketch Orientation--- Sketch view direction: **Flip** ⇒ select **RIGHT** datum from the model ⇒ Orientation: **Right** ⇒ click **Sketch** button (Fig. 2.23) ⇒ click **Close** to accept the References (Fig. 2.24) ⇒ **Tools** ⇒ **Environment** ⇒ Snap to Grid ⇒ **OK**

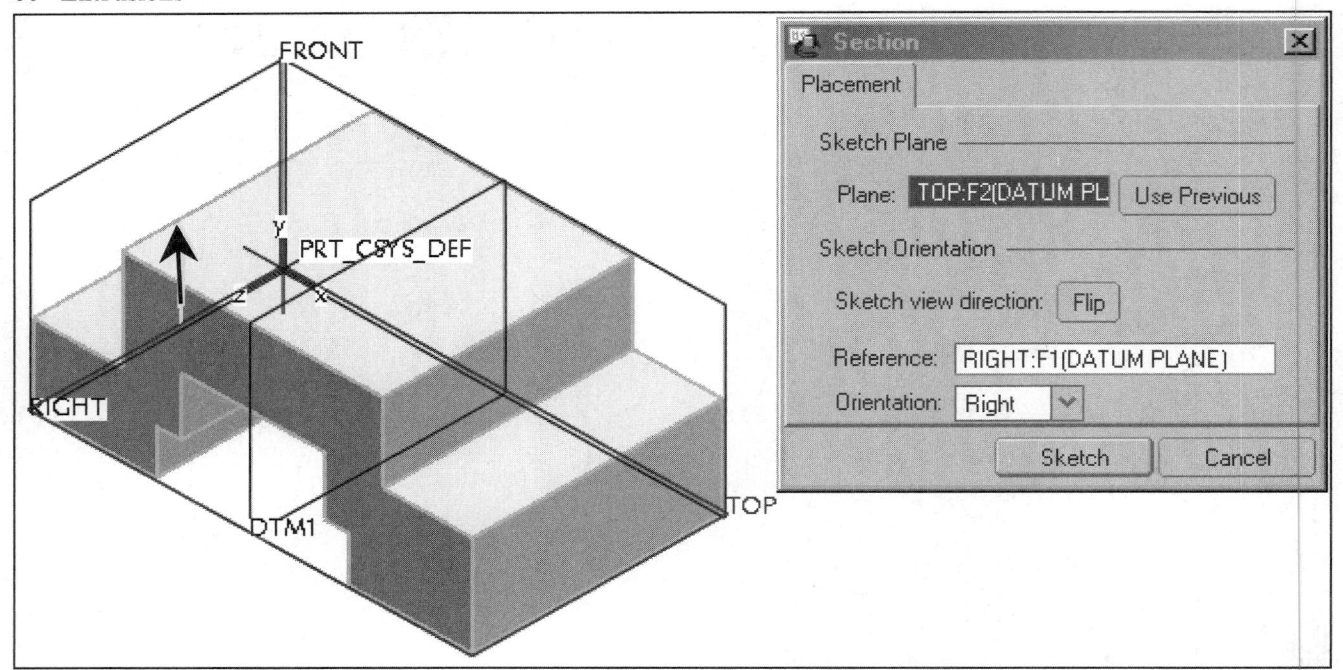

Figure 2.23 Section Dialog Box

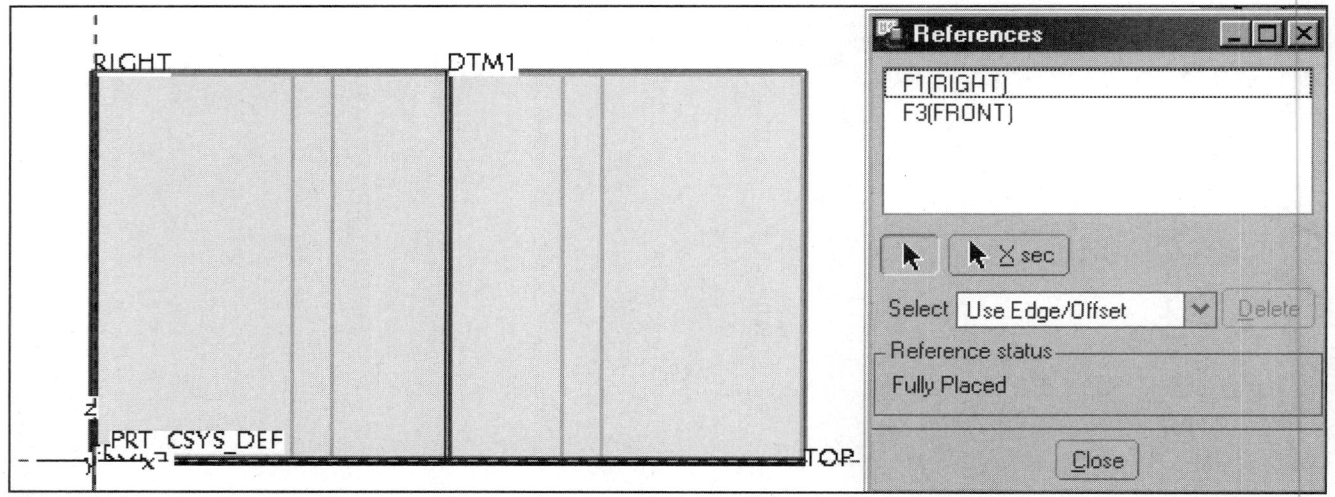

Figure 2.24 References Dialog Box

Click: **Create 2 point lines** ⇒ place the mouse on the edge of the **RIGHT** datum plane and create three lines (Fig. 2.25) ⇒ **MMB** to end the line sequence ⇒ add the required dimensions and modify the values for the three dimensions as shown (Fig. 2.26) ⇒ ⇒ ⇒ **Standard Orientation** ⇒ from the dashboard **Remove Material** (Fig. 2.27) ⇒ **Change depth direction of extrude to other side of sketch** ⇒ **Options** tab (Fig. 2.28) ⇒ **Through All (Extrude in first direction to interest with all surfaces)** (Fig. 2.29) ⇒ (Fig. 2.30) ⇒ **LMB** to deselect ⇒ ⇒ rotate your model using **MMB** to see the cut clearly ⇒ ⇒ **MMB**

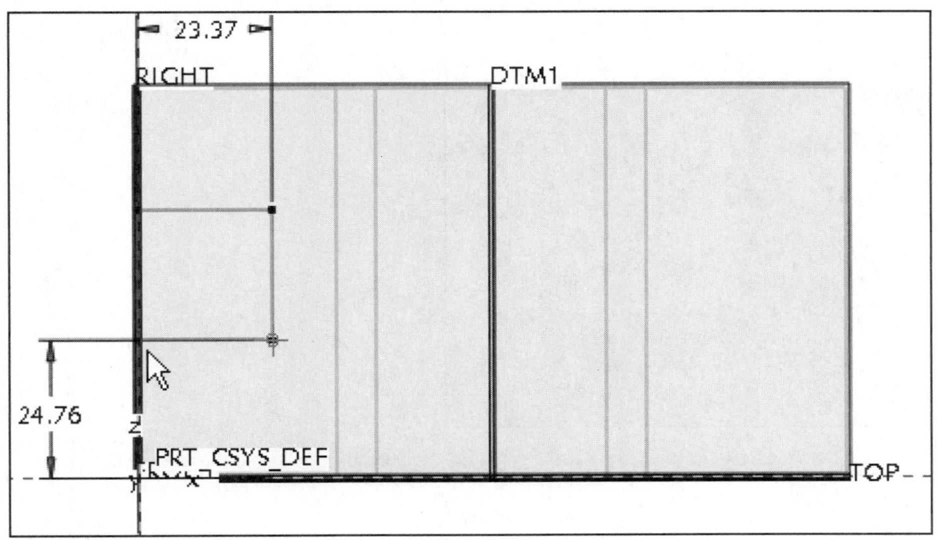

Figure 2.25 Three Line Sketch

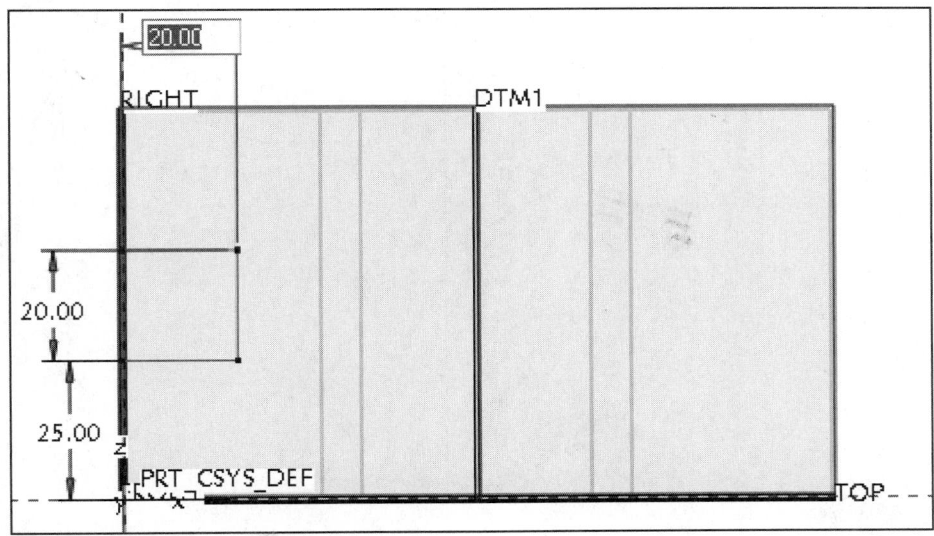

Figure 2.26 Add and Modify Dimensions

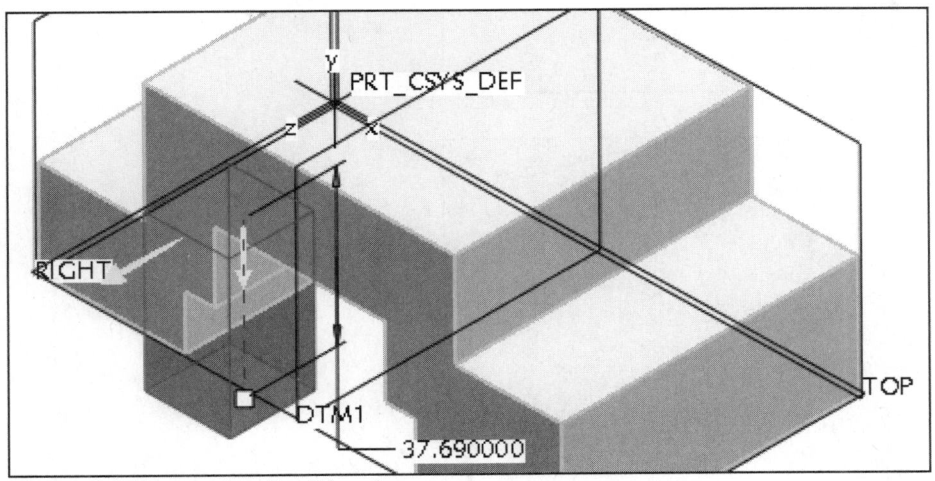

Figure 2.27 Determine Cut Direction

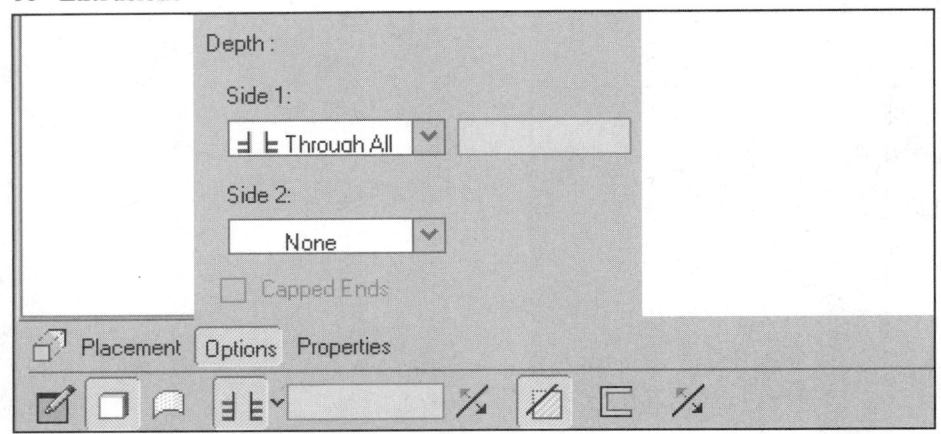

Figure 2.28 Dashboard Options

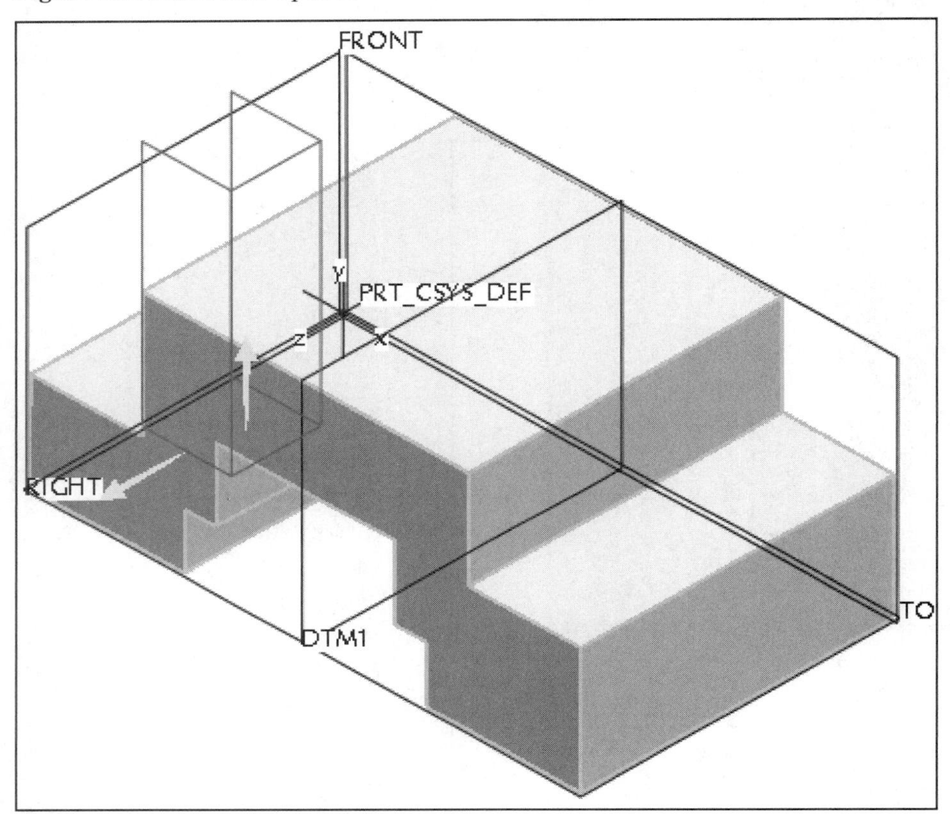

Figure 2.29 Cut Option Through All

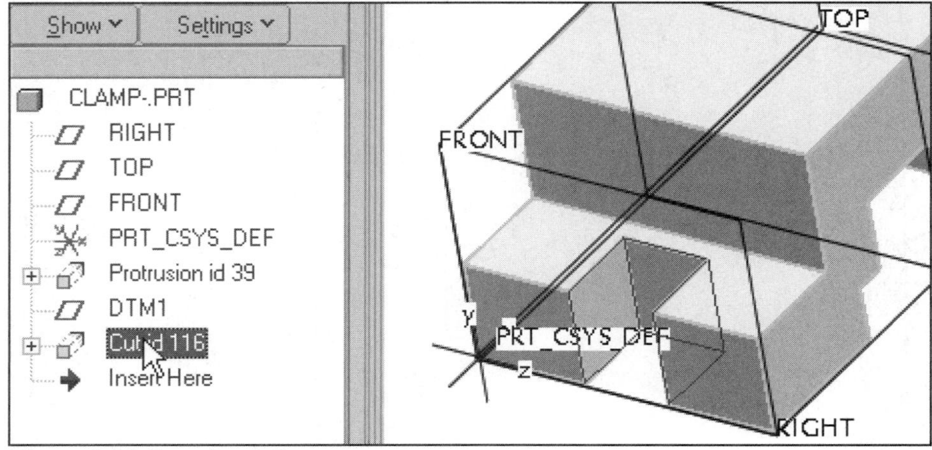

Figure 2.30 Completed Cut

The last step in the modeling of the part is to copy and mirror the cut to the other side, click: **Edit ⇒ Feature Operations ⇒ Copy ⇒ Mirror ⇒ Dependent ⇒ MMB ⇒** pick the cut from the model (Fig. 2.31) ⇒ **MMB ⇒ MMB ⇒** Select a plane or create a datum to mirror about**:** pick on **DTM1** (Fig. 2.32) ⇒ **MMB ⇒** [icon] ⇒ **Standard Orientation ⇒** [icon] ⇒ **MMB** (Fig. 2.33)

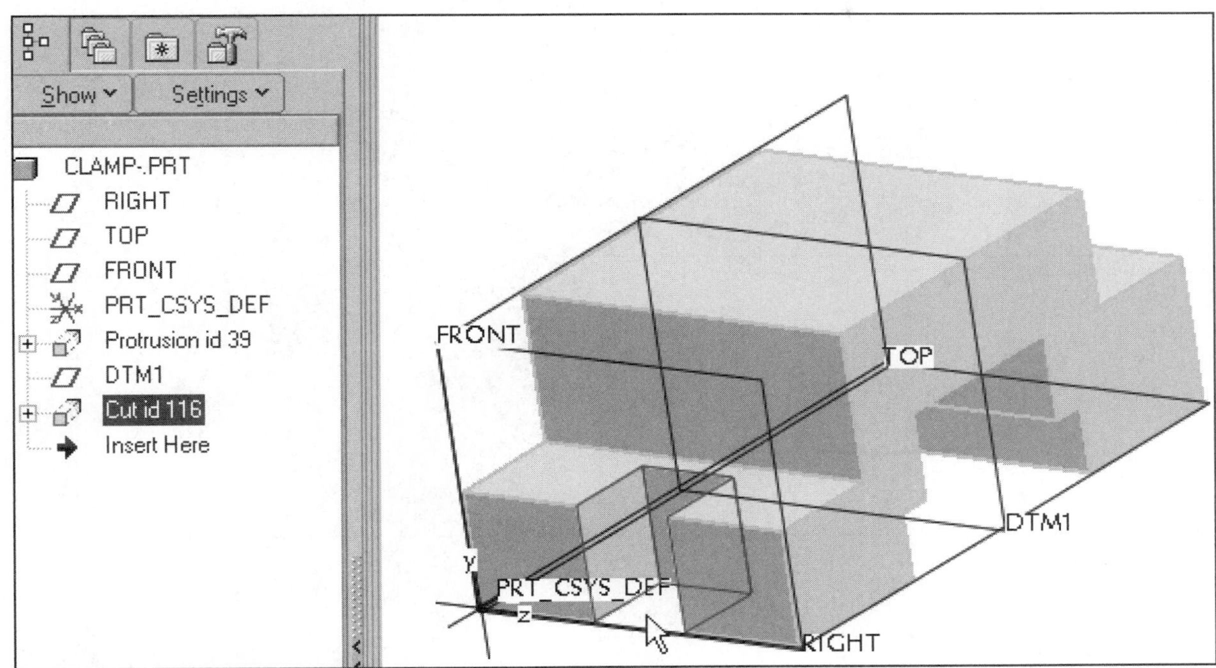

Figure 2.31 Selecting the Cut to be Mirrored and Copied

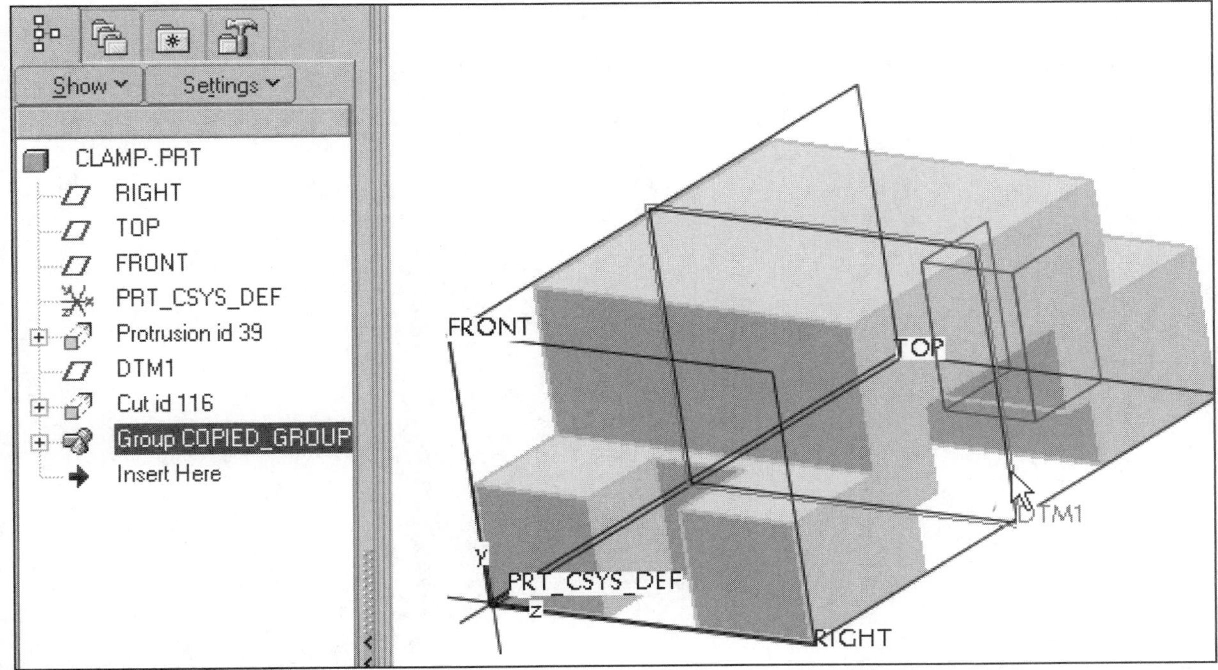

Figure 2.32 Mirrored Cut

Double click on the cut to see its dimensions ⇒ Double click on a dimension to modify the value (Fig. 2.34) ⇒ **MMB** to end the process without changing the value ⇒ [icon] ⇒ **File ⇒ Close Window**

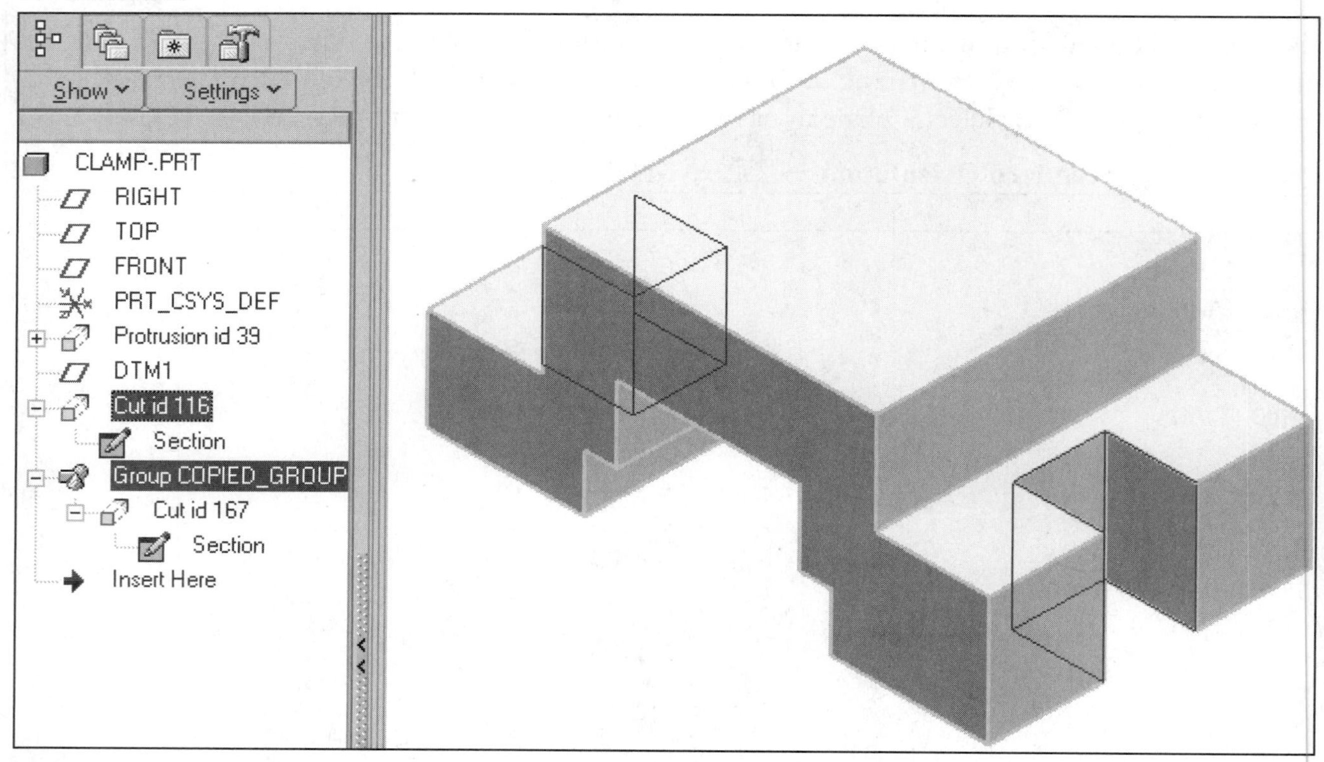

Figure 2.33 Completed Part

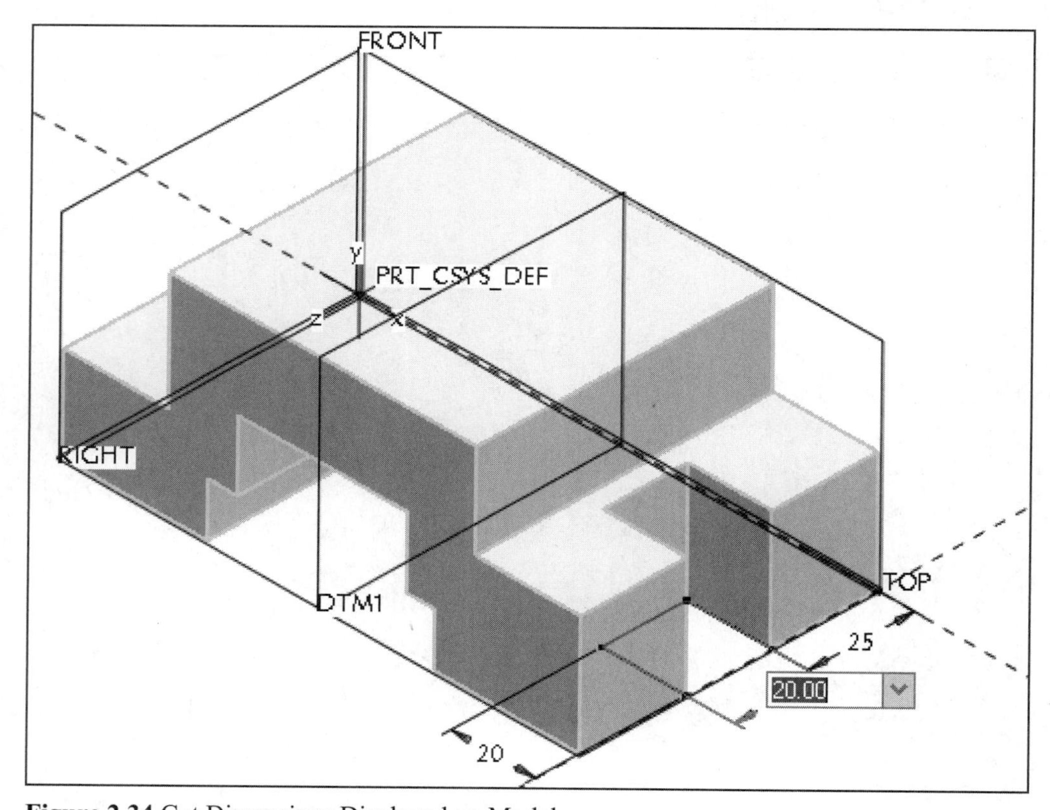

Figure 2.34 Cut Dimensions Displayed on Model

Lesson 2 is now complete. Continue with Lesson 2 Project. You will be required to plan your design and use most of the tools and commands presented previously.

Lesson 2 Project

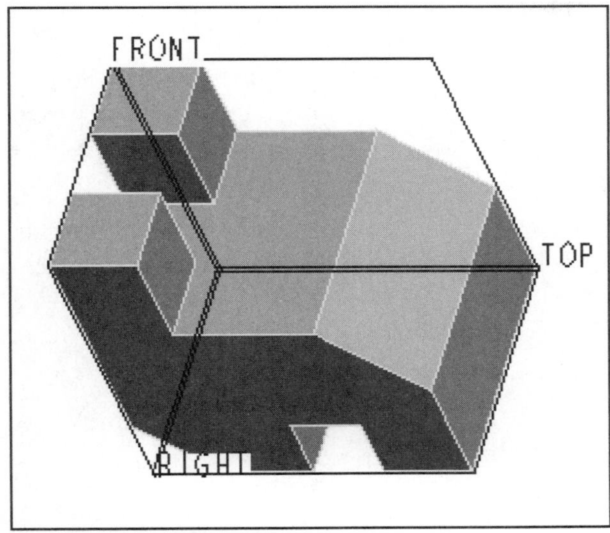

Figure 2.35 Angle Block

Angle Block

This lesson project is a simple block that requires many of the same commands as the Clamp. Using the default datums, create the part shown in Figures 2.35 through 2.39. Sketch the protrusion on datum FRONT. Sketch the cut on datum RIGHT and align it to the upper surface/plane of the protrusion, as shown in Figure 2.36.

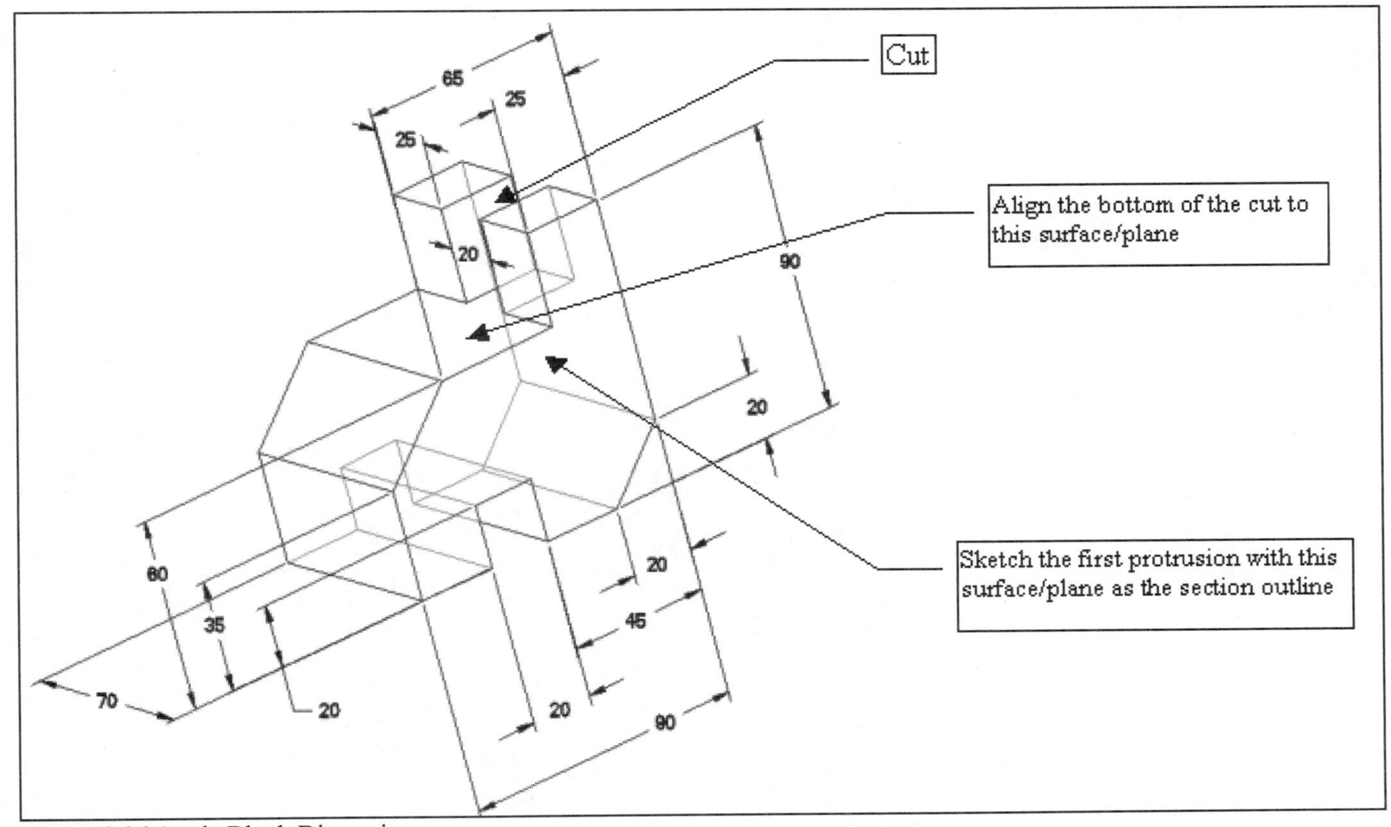

Figure 2.36 Angle Block Dimensions

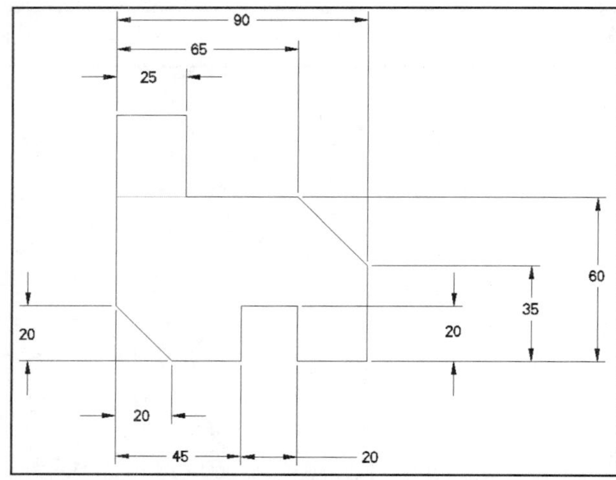

Figure 2.37 Angle Block Protrusion Sketch Dimensions

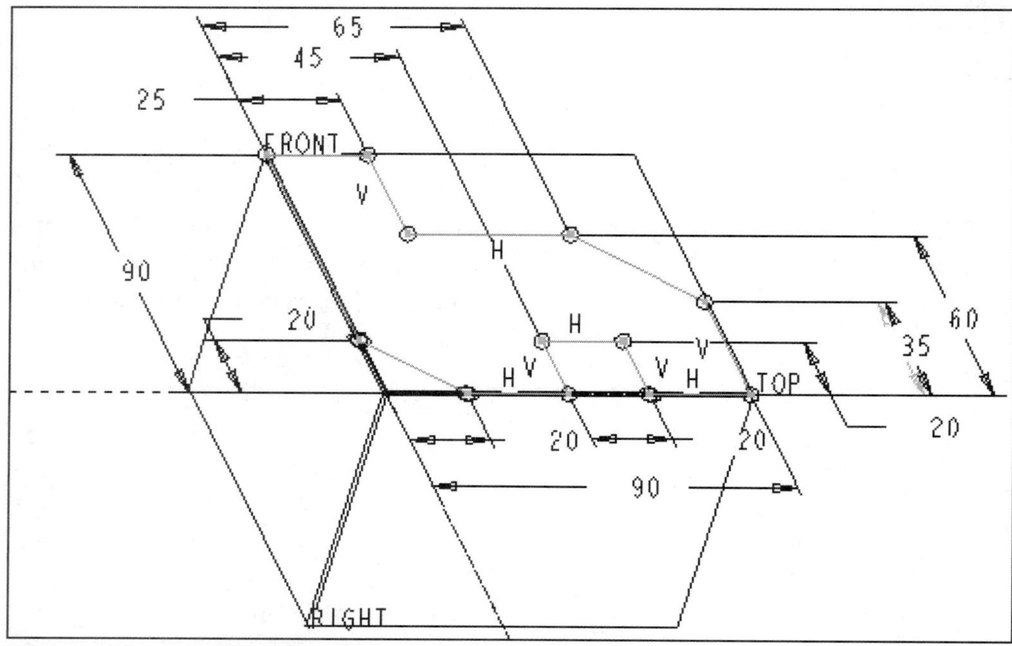

Figure 2.38 Angle Block Sketch

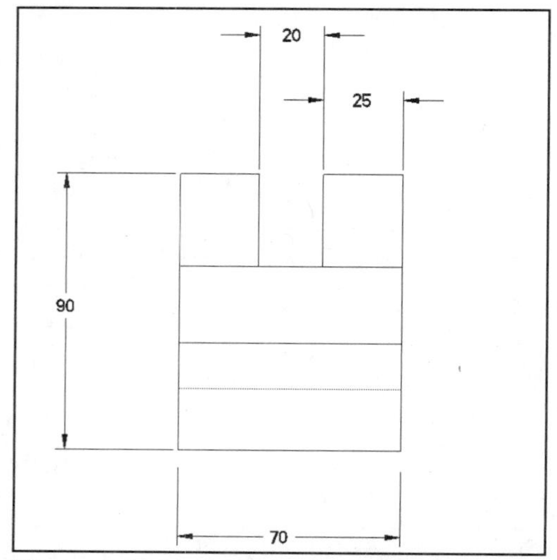

Figure 2.39(a) Cut Drawing Dimensions

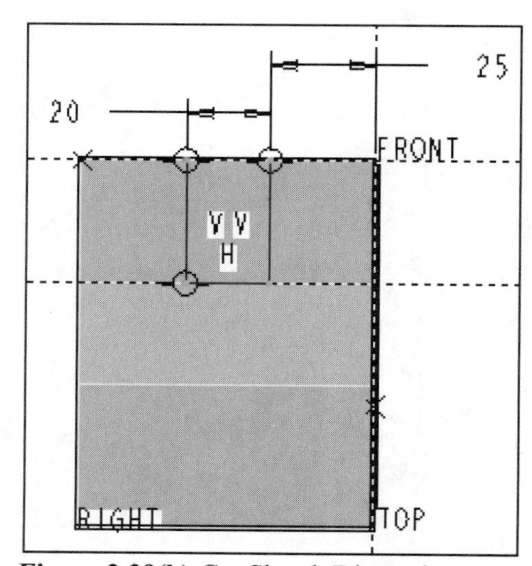

Figure 2.39(b) Cut Sketch Dimensions

Lesson 3 Editing

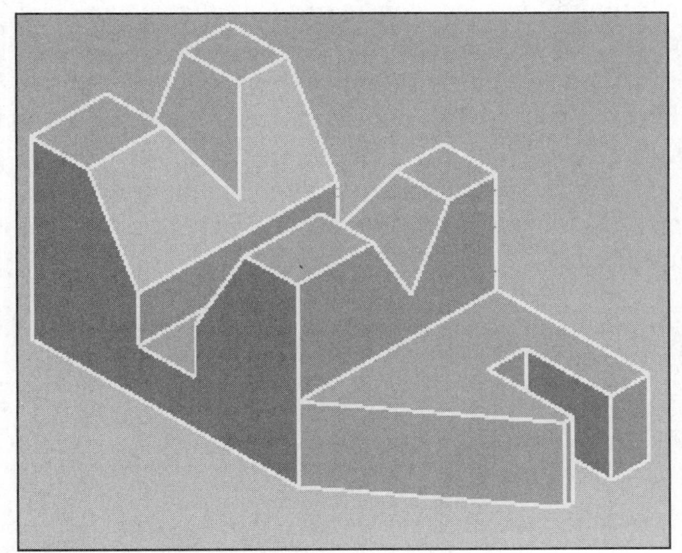

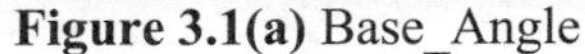

Figure 3.1(a) Base_Angle

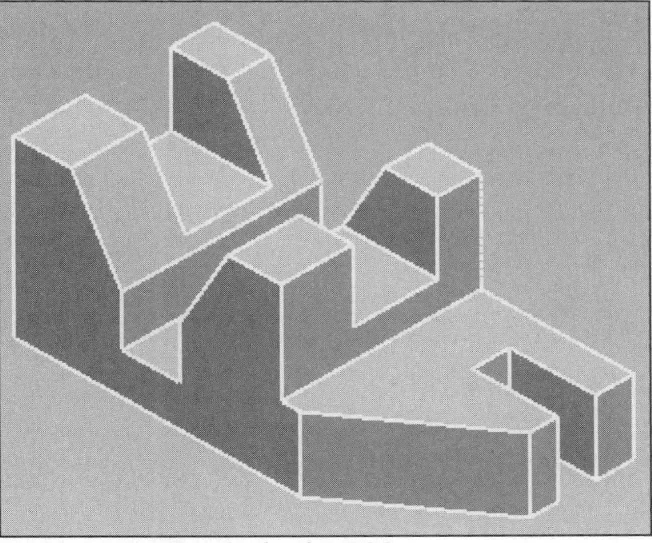

Figure 3.1(b) Base_Angle after ECO

OBJECTIVES

- Understand **Setup** and **Environment** settings
- **Edit** design dimensions
- Using **Edit Definition** to redefine a part's features
- Understand **Configuration** files
- Input ***config.pro*** changes using **Options**
- Use the **Model Tree** to edit and redefine
- Use **Sketcher Constraints**

EDITING

Both **Edit** and **Edit Definition** are important capabilities in the design of parts. You have learned to change dimensions in the Sketcher with **Modify**. Edit is used for changing the same sketch dimensions after the feature has been regenerated (outside the Sketcher), directly on the model. With Edit Definition, the feature's attribute, the placement plane, the placement/orientation references, the size and configuration of the feature can be redone. Edit is for simple dimensional changes; Edit Definition is implemented for comprehensive changes to a model's feature configuration.

You will be creating the Base_Angle [Fig. 3.1(a)] and then edit and redefine part features using specifications from an ECO [Fig. 3.1(b)].

Changing Dimension Values

When you modify or edit the value of a dimension, you can enter a new number or a **Relation**. Pro/E also supports the use of negative dimensions. The value entered depends on the displayed sign of the dimension. By default, Pro/E displays all dimensions as positive values, and entering a negative value tells Pro/E to create the section geometry on the opposite side, but the *direction* of a feature creation cannot be changed by entering a negative number. Use the Edit Definition option to redefine the direction of the feature.

Configuration Files

You can customize the look and feel of Pro/E and the way in which Pro/E runs by setting options in a configuration file (Fig. 3.2). Pro/E uses two important configuration files: ***config.pro*** and ***config.win***. The *config.pro* file is a text file that stores all of the settings that define the way in which Pro/E handles operations. The *config.win* file is a binary file that stores window configuration settings, such as toolbar visibility settings and Model Tree location settings. Each setting in the configuration files is called a configuration option. Pro/E provides default values for each option. You can set or change the configuration options. ***Config.sup*** is a protected system configuration file. Your system administrator uses this file to set configuration options that are used on a company-wide basis. Any values that are set in this file cannot be overridden by other (more local) *config.pro* files.

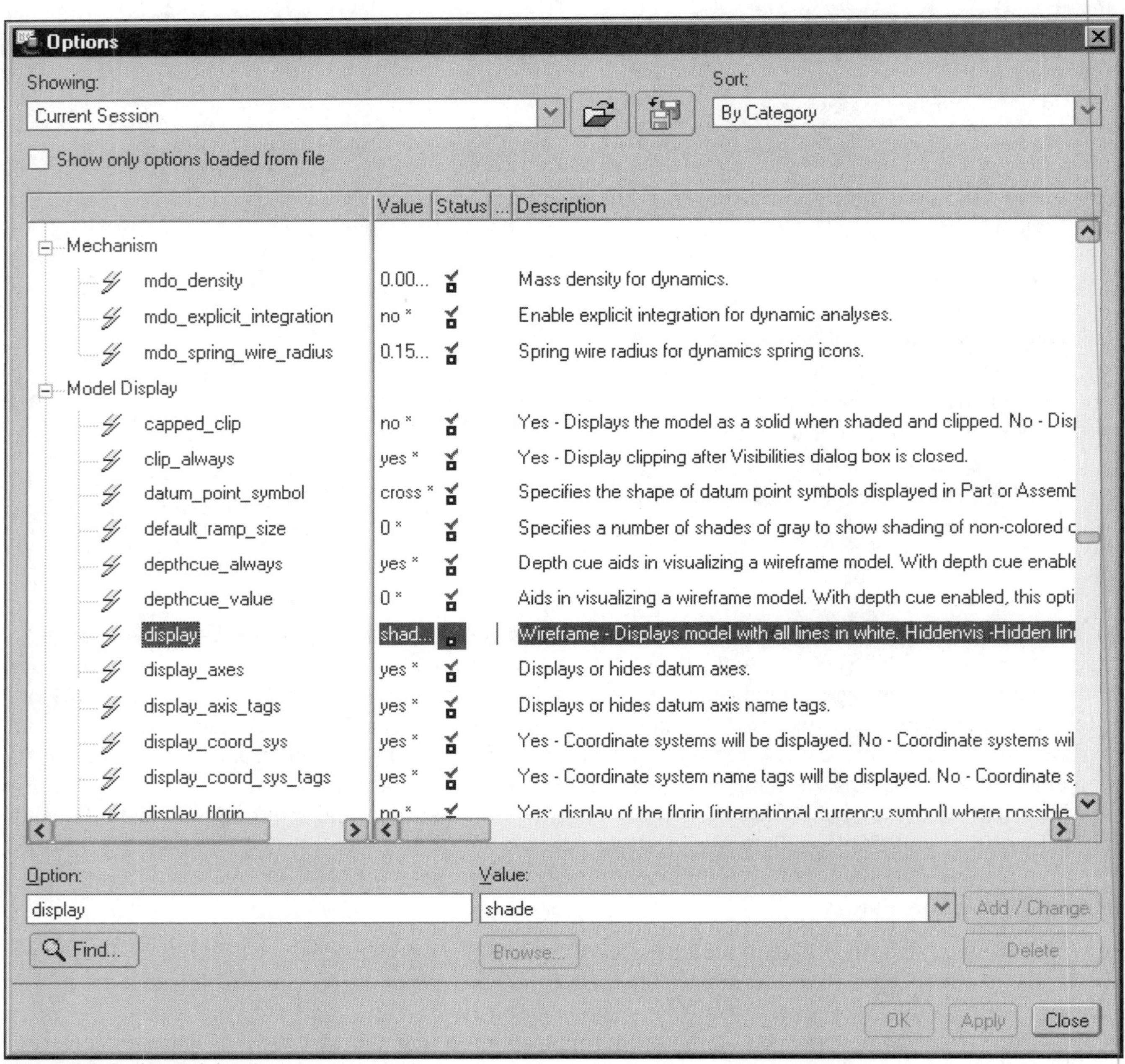

Figure 3.2 Options Dialog Box

Lesson 3 STEPS

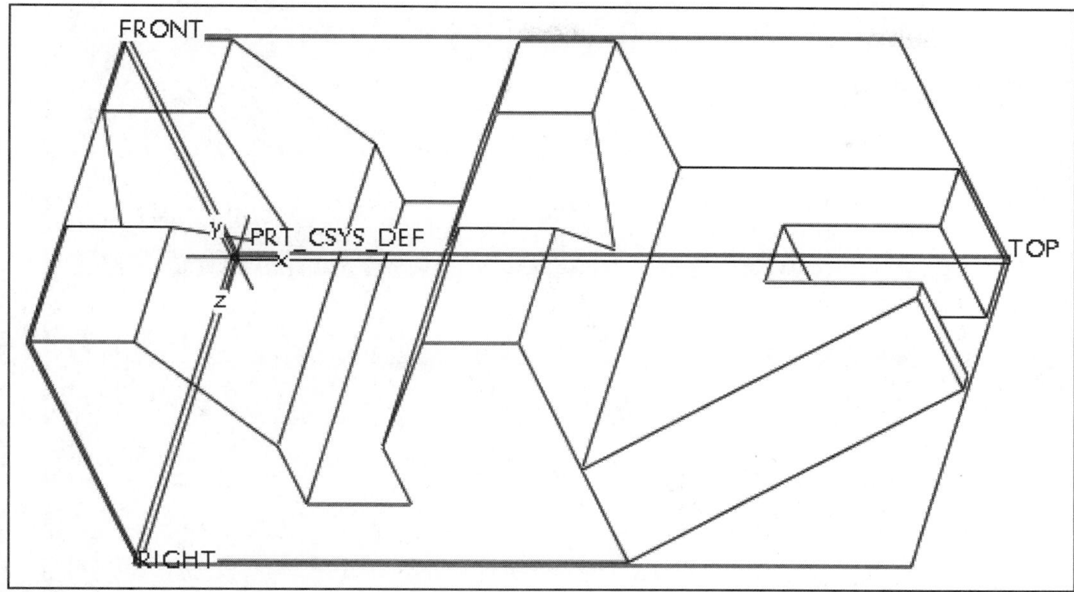

Figure 3.3 Base_Angle

Base Angle

The base angle (Fig. 3.3) is composed of one protrusion and three cuts. Along with creating a new part, you will use *Edit* and *Edit Definition* to change the design values of the part.

Include an underline character or dash when creating file names that have a space. BASE ANGLE, for example needs to be typed as BASE_ANGLE or BASE-ANGLE. No spaces are allowed in object (part, assembly, or drawing) names. A maximum number of 31 characters are allowed per name.

Start a new part, click: **File** ⇒ **Set Working Directory** ⇒ select the working directory ⇒ **OK** ⇒ **Create a new object** ⇒ ●**Part** ⇒ Name **BASE_ANGLE** ⇒ ☑ Use default template ⇒ **OK** ⇒ **Edit** ⇒ **Setup** ⇒ **Units** ⇒ Units Manager **Inch lbm Second** ⇒ **Close** ⇒ **Material** ⇒ **Define** ⇒ type **ALUMINUM** ⇒ **MMB** ⇒ **File** from the material table ⇒ **Save** ⇒ **File** ⇒ **Exit** ⇒ **Assign** ⇒ pick **ALUMINUM** ⇒ **Accept** ⇒ **MMB** ⇒ ⇒ **MMB** (or)

Next, you will change the ***default_dec_places*** and ***sketcher_dec_places*** configuration options:

- ***default_dec_places*** sets the default number of decimal places to be displayed in all model modes for nonangular dimensions; it does not affect the *previously displayed* number of digits as modified using NUM DIGITS.
- ***sketcher_dec_places*** controls the number of digits displayed when you are in the Sketcher.

Click: **Tools** ⇒ **Options** ⇒ Showing: **Current Session** (Fig. 3.4) ⇒ ❑ Show only options loaded from file ⇒ slide the bar down to the option or type, Option: ***default_dec_places*** (Fig. 3.5) ⇒ Value: **3** ⇒ **Add/Change** ⇒ slide the bar down to the option or type, Option: ***sketcher_dec_places*** ⇒ Value: **3** (Fig. 3.6) ⇒ **Add/Change** ⇒ **Apply** ⇒ **Save a copy of the currently displayed configuration file** ⇒ Name **base_angle.pro** ⇒ **Ok** ⇒ **Close** ⇒ **Folder Browser** ⇒ select your working directory and the Browser **Browser** will open (Fig. 3.7) ⇒ click on the quick sash to close the Browser

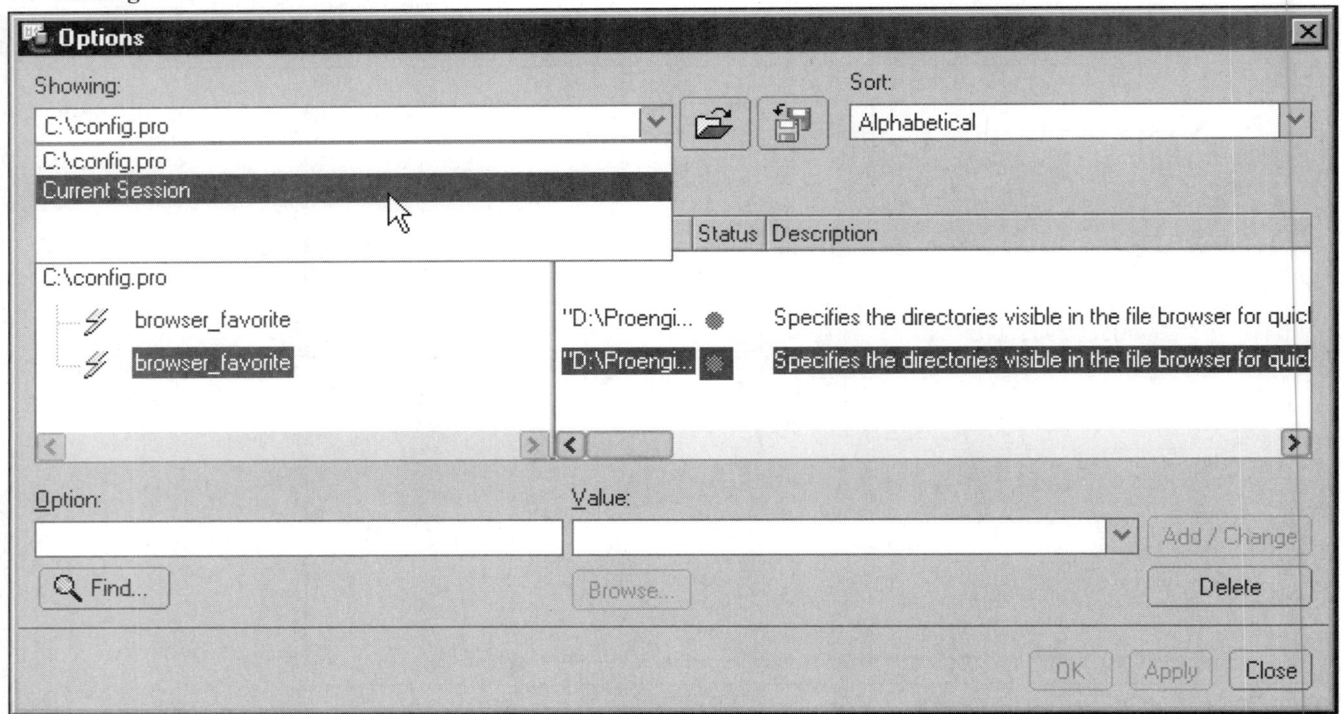

Figure 3.4 Current Session

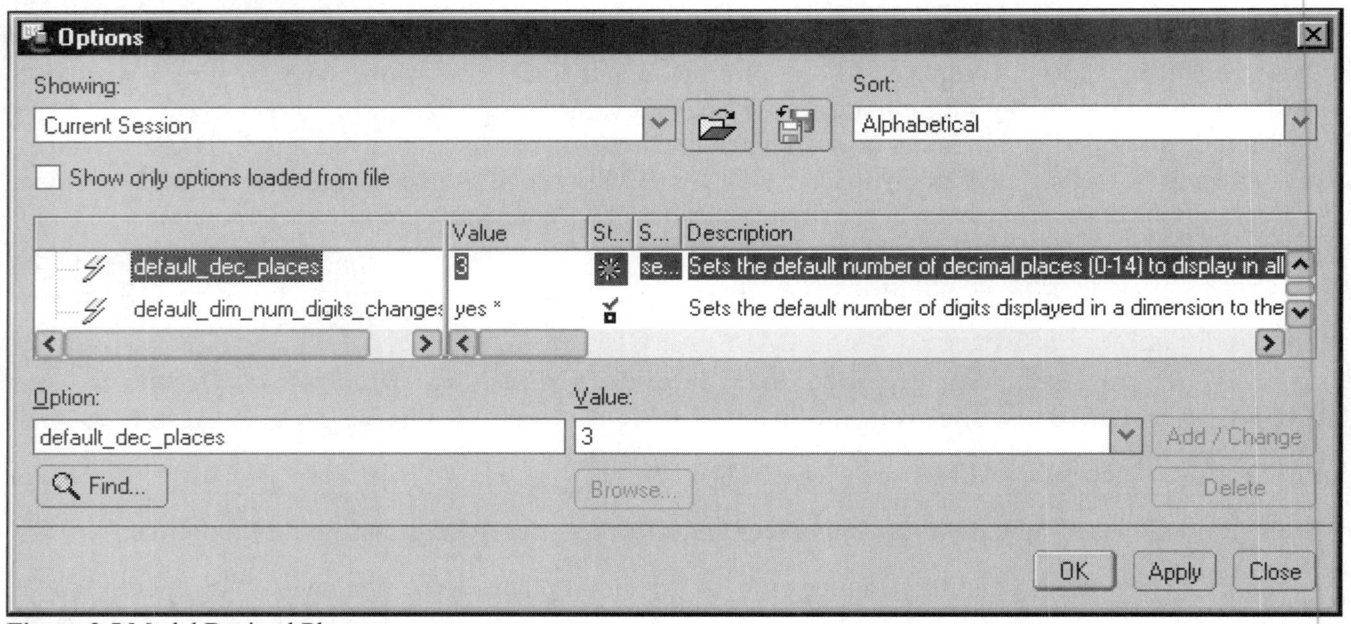

Figure 3.5 Model Decimal Places

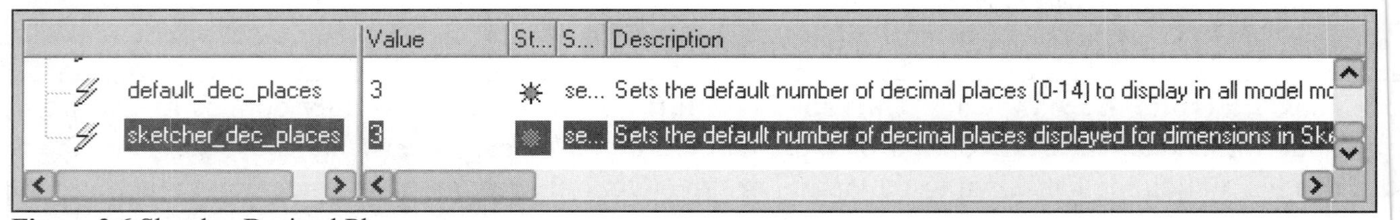

Figure 3.6 Sketcher Decimal Places

	base_angle.pro	25 KB	17-Sep-02 03:53:52 PM
	base_angle.prt	58 KB	17-Sep-02 01:12:36 PM

Figure 3.7 *Config.pro* File *(base_angle.pro)* in the Browser Window

The modeling process will be similar to Lesson 2 Steps.

Create the first protrusion, click: **Extrude Tool** ⇒ from the dashboard ⇒ Section dialog box opens (Fig. 3.8) ⇒ Sketch Plane--- Plane: select **FRONT** datum from the model ⇒ Sketch Orientation--- Reference: select **TOP** datum from the model ⇒ Orientation: **Top** (TOP datum to face up) ⇒ **Sketch** ⇒ **Close** to accept the References ⇒ **Toggle the grid on**

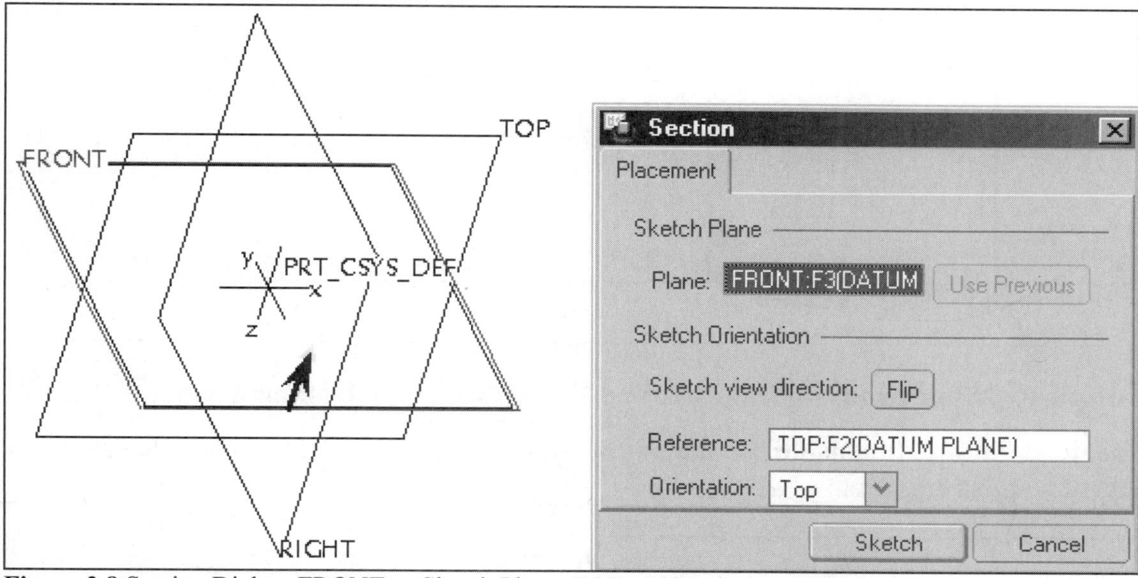

Figure 3.8 Section Dialog, FRONT as Sketch Plane, TOP as Sketch Orientation

Most of the part can be defined by sketching an outline similar to its front view. Therefore, you can complete most of the part's geometry with one protrusion. The cuts on the top and side will be created later. As you sketch, try not to create any constraints that you do not want as part of the design. Lines not intended to be at the same level (horizontal) are not sketched in line. Features of differing sizes are sketched with different lengths. Make sure that lines at the same angle are sketched that way, that symmetrical features are drawn equally about the centerline, and so on. Lines of differing dimensional lengths are sketched with different grid spacing. Though it would be considered poor practice in industry, it is OK for you to count the grid spacing for this feature's sketch. It is normal for different people to sketch different versions of the sketch. As long as the assumptions are the same, the sketch will be valid. Only the outline and proper constraints are important, not the size or distances.

As you sketch, Pro/E creates "weak" dimensions- dimensions that Pro/E needs to solve (has sufficient alignments and dimensions to build the feature) the sketch/section. As you add dimensions, which are called "strong" dimensions, Pro/E will remove weak dimensions that are no longer needed to solve the sketch/section.

Use **Shift MMB** and **Ctrl MMB** to reposition and resize the sketch. Click: **Tools** ⇒ **Environment** ⇒ Snap to Grid ⇒ **OK** ⇒ **Toggle display of dimensions off** ⇒ **Create 2 point lines** ⇒ sketch an outline as shown in Figure 3.9 ⇒ **MMB** ⇒ **Create 2 point centerlines** ⇒ add a centerline to locate the **V** shape

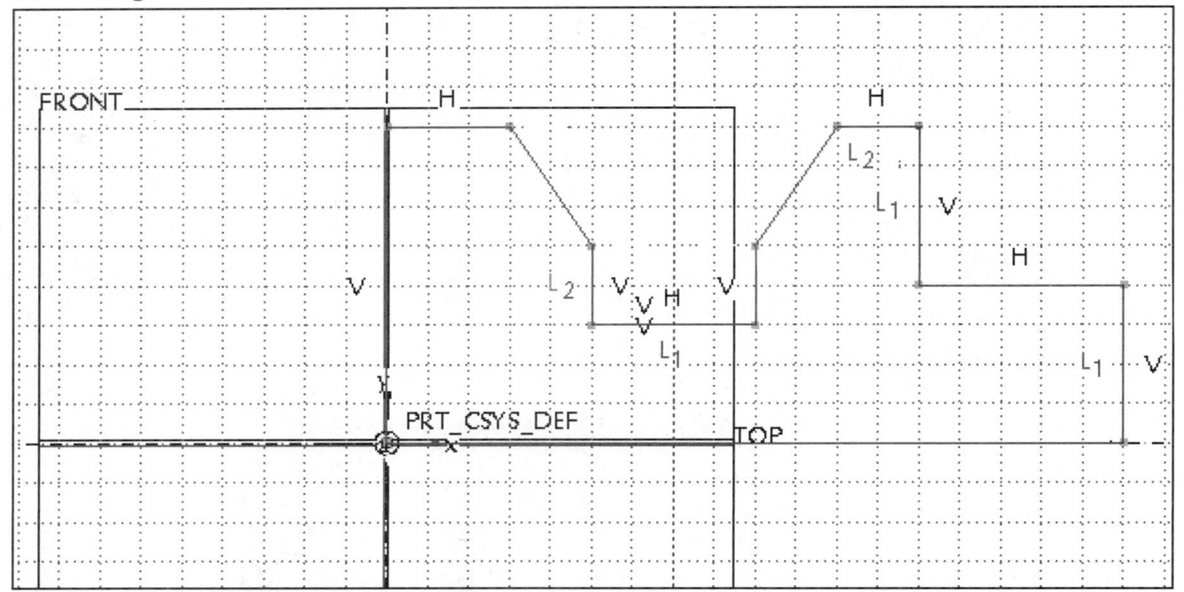

Figure 3.9 Sketch Twelve Connected Lines and a Centerline

Click: **Toggle display of dimensions on** (Fig. 3.10) ⇒ **Tools** ⇒ **Environment** ⇒ Snap to Grid ⇒ **OK** ⇒ **Create defining dimension** ⇒ add dimensions ⇒ ⇒ move the dimensions to conform with ASME standards (Fig. 3.11)

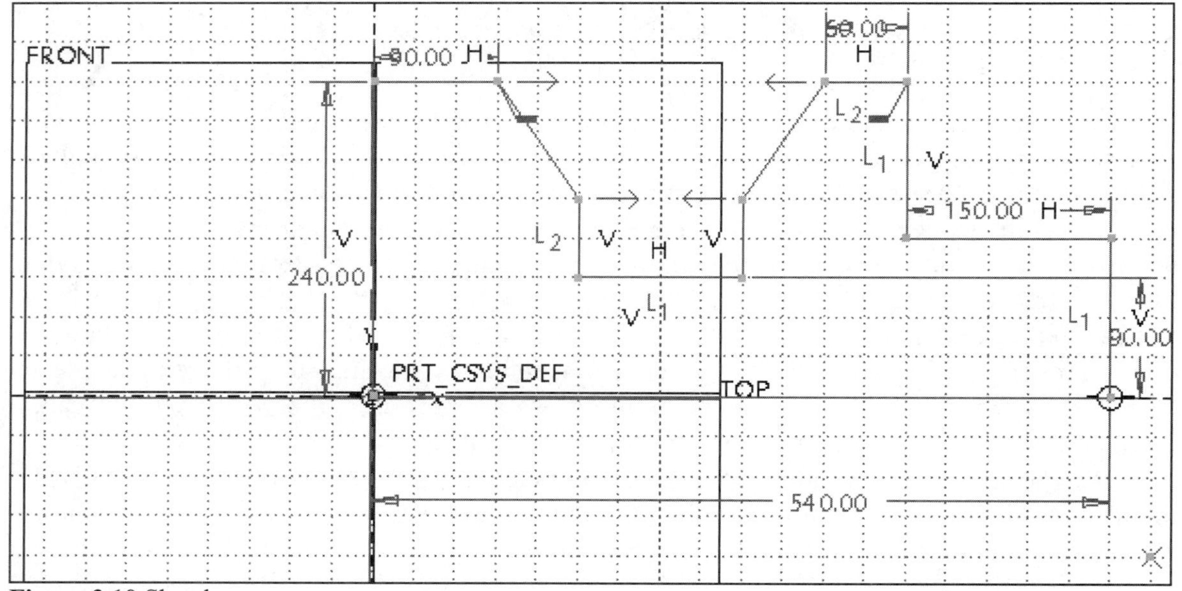

Figure 3.10 Sketch

Modify the dimensions to the *design sizes*, click: ⇒ window-in the sketch to capture all dimensions ⇒ **Modify the values of dimensions, geometry of splines, or text entities** ⇒ Regenerate (Fig. 3.12) ⇒ modify dimensions to the design values at the prompt ⇒ Regenerate ⇒ (Fig. 3.13)

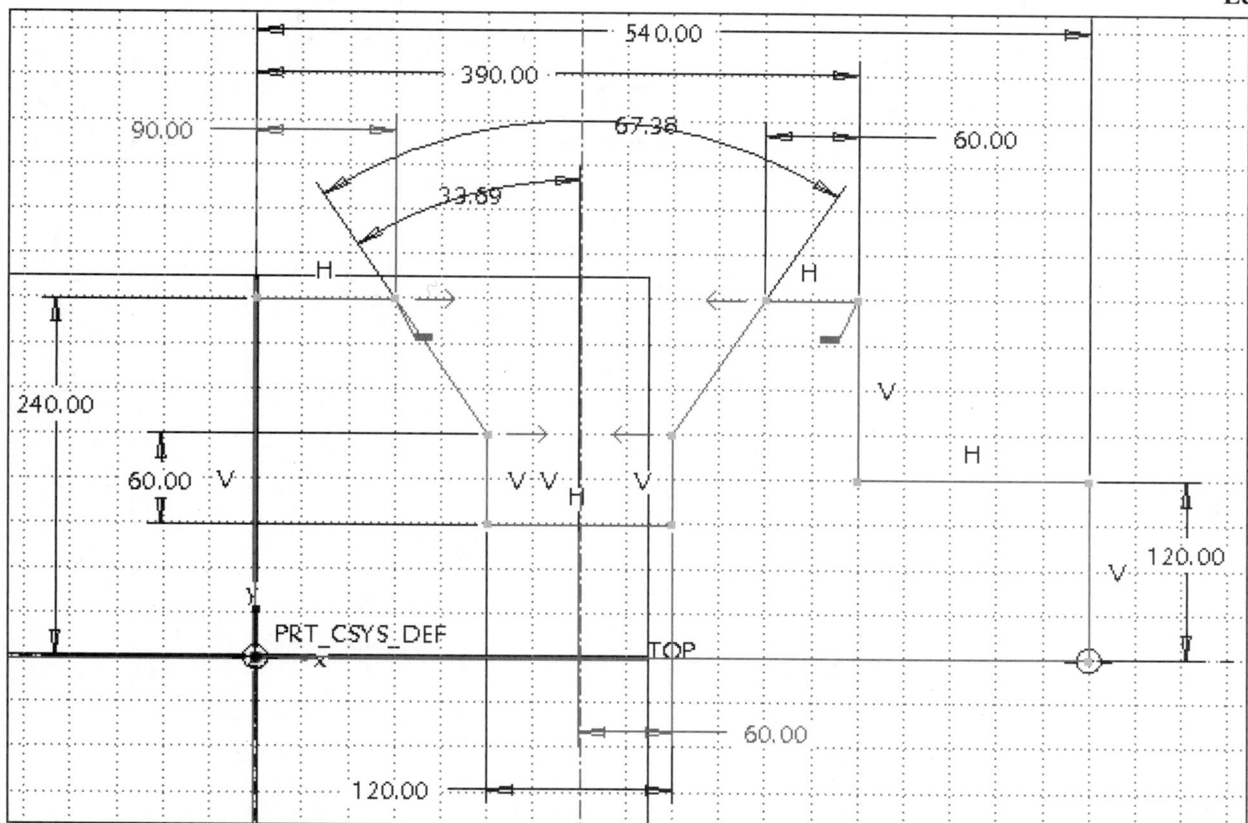

Figure 3.11 Dimensioned Sketch

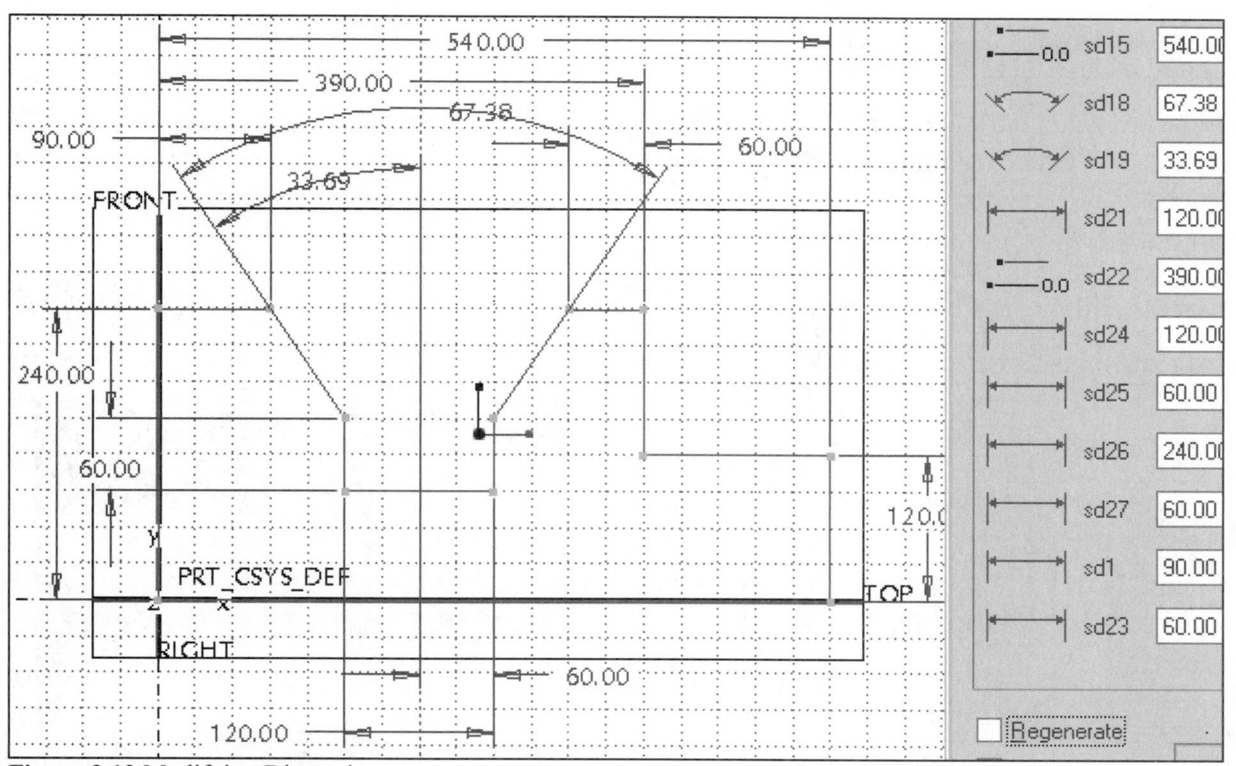

Figure 3.12 Modifying Dimensions

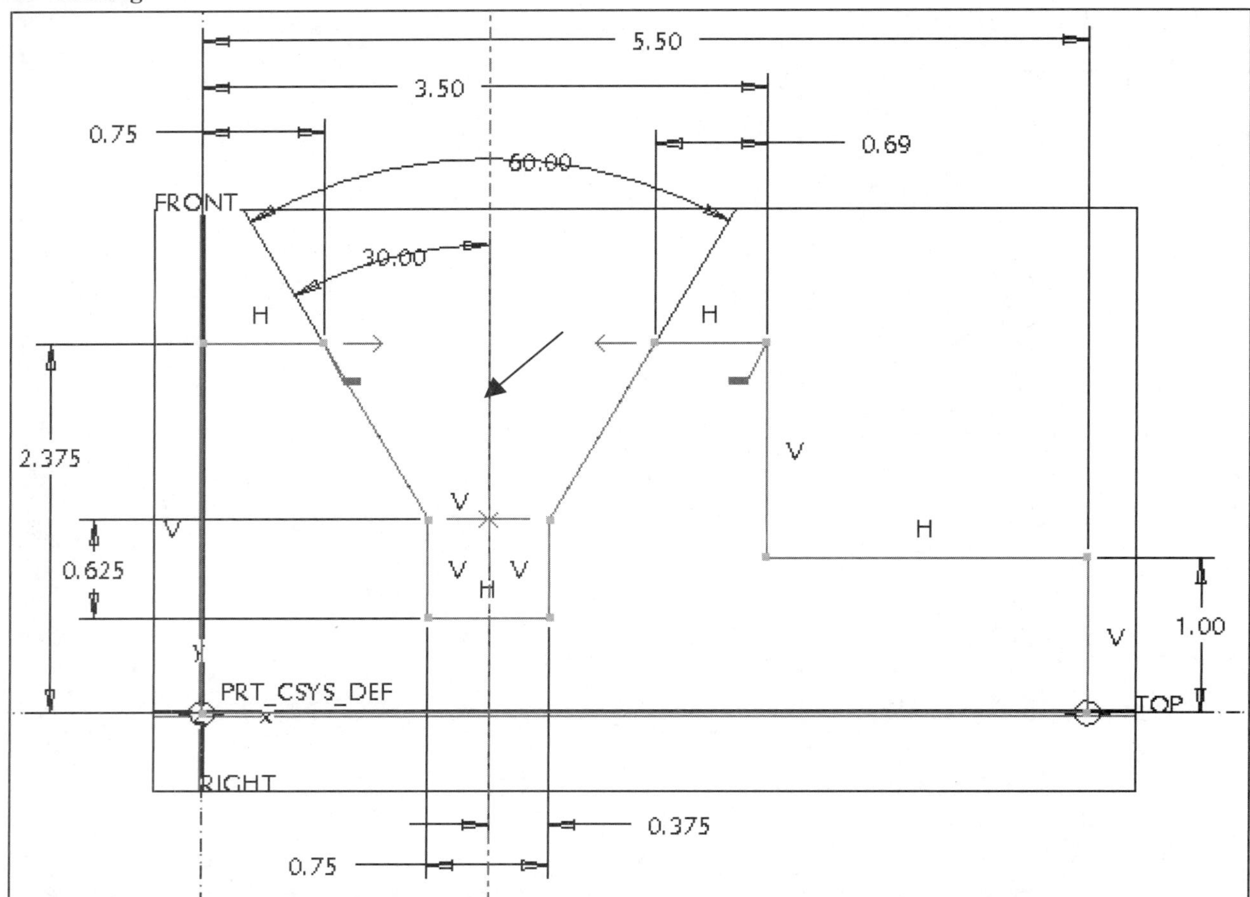

Figure 3.13 Modified and Repositioned Dimensions

Next, we will remove some dimensions, click: **Impose sketcher constraints on the section** ⇒ **Make two points or vertices symmetric about a centerline** (Fig. 3.14) ⇒ pick the centerline and then pick two vertices to be symmetric (Fig. 3.15) ⇒ the Resolve Sketch dialog box displays (Fig. 3.16) ⇒ pick **Dimension sd# = 0.375** ⇒ **Delete** ⇒ **Close** ⇒

Repeat the process and make the four vertices of the angled lines symmetrical (Fig. 3.17). You must use the Symmetric constraint twice. ⇒ ⇒ **Standard Orientation** (Fig. 3.18)

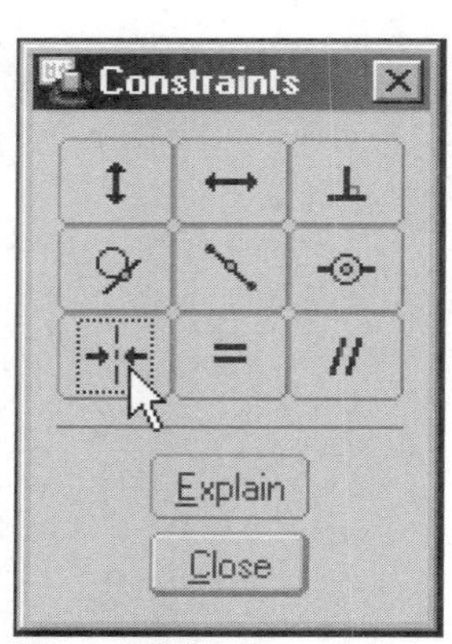

Figure 3.14 Constraints

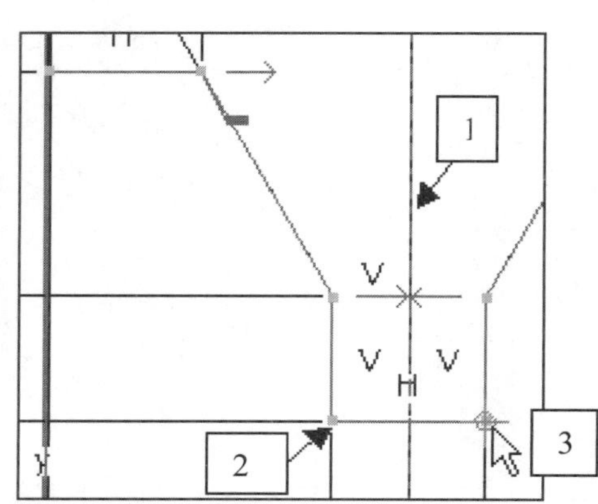

Figure 3.15 Picks

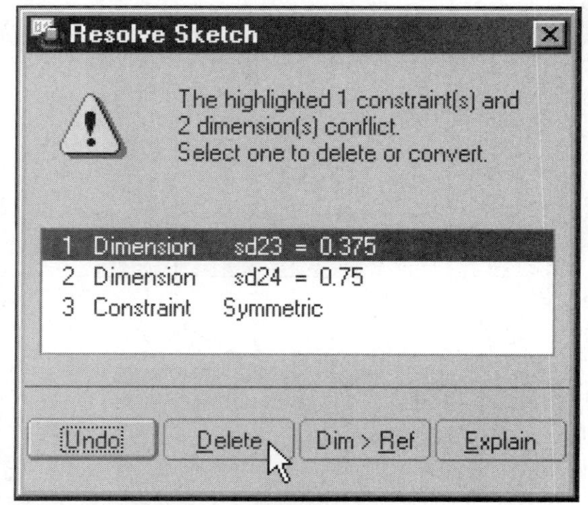

Figure 3.16 Resolve Sketch Dialog

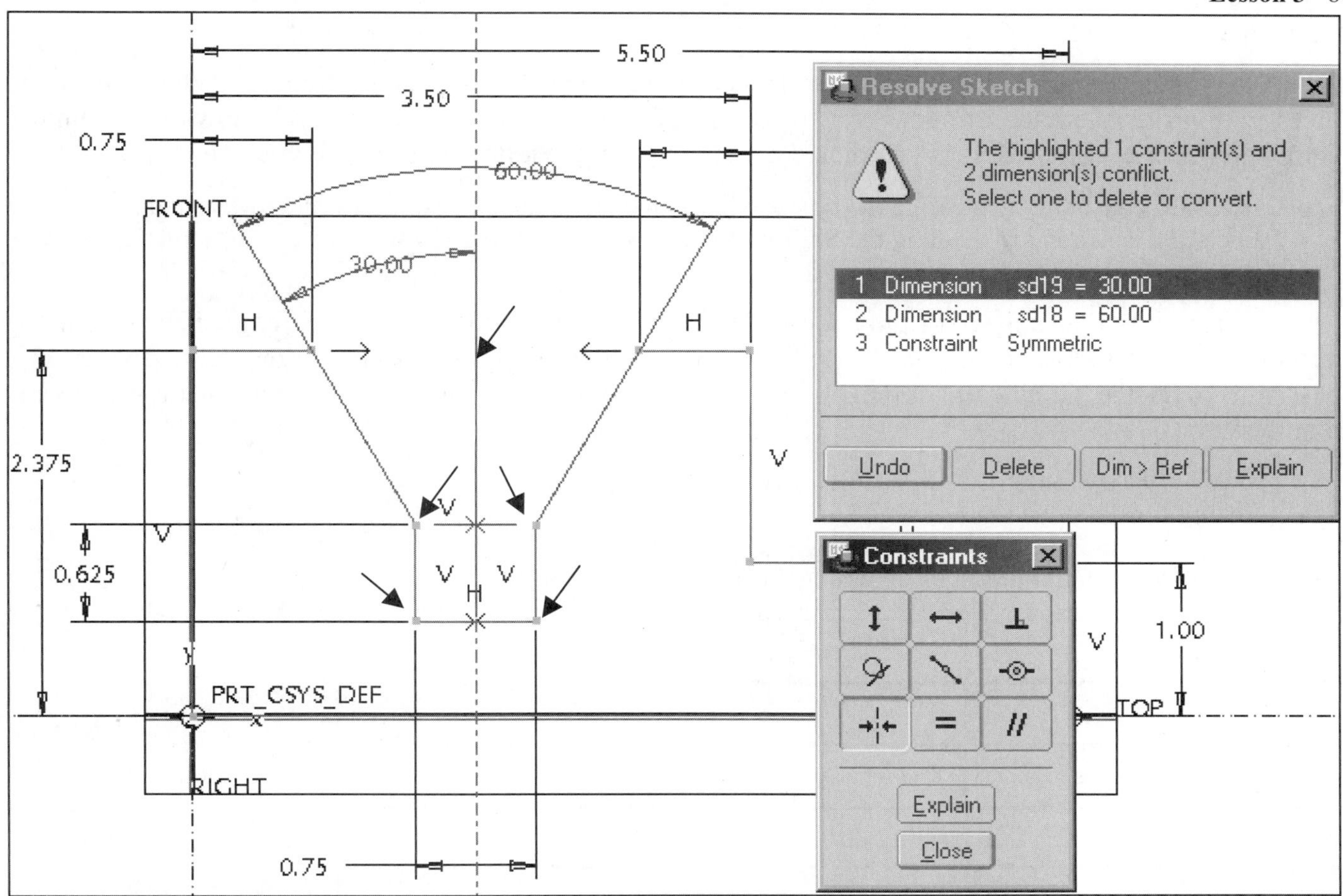

Figure 3.17 Removing the **30** degree Dimension

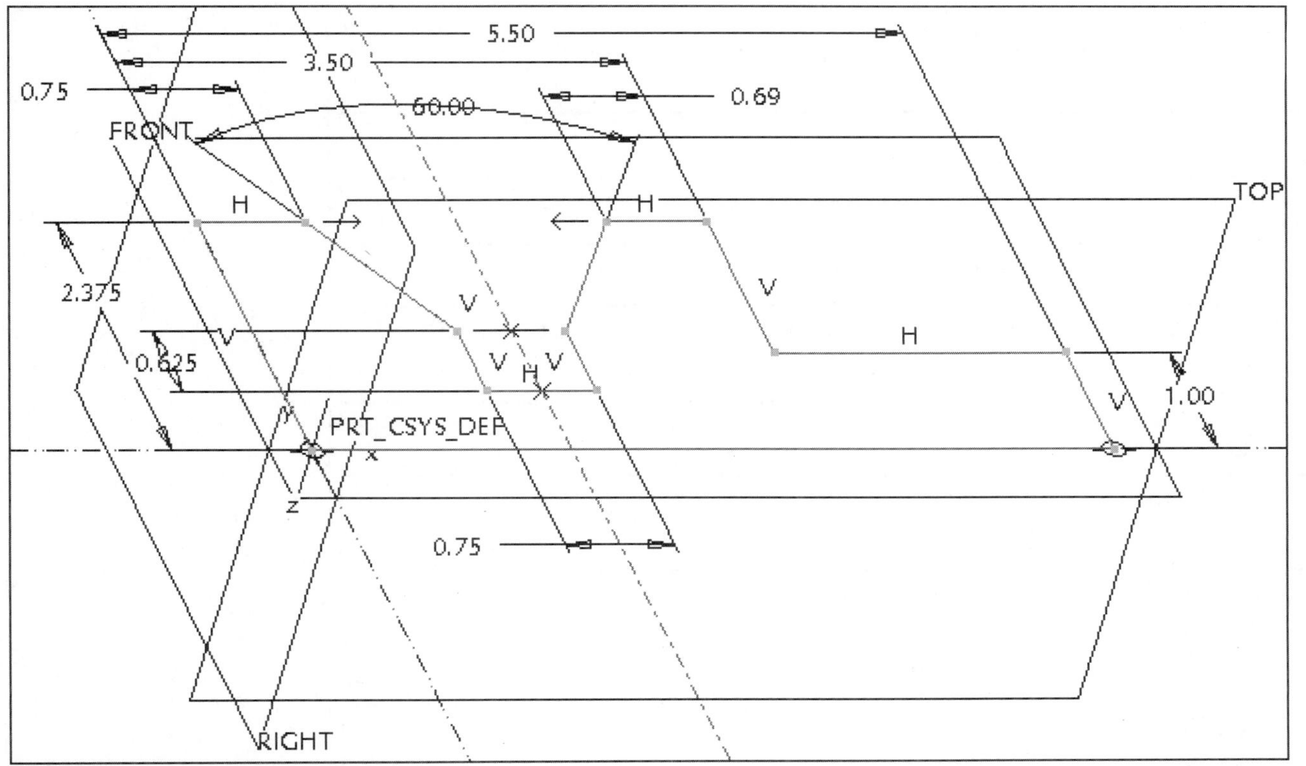

Figure 3.18 Standard Orientation of Sketch

Sketcher Constraints

When you sketch geometry, Pro/E uses certain assumptions to help you locate geometry. When the cursor comes within the tolerance of some constraints, Pro/E snaps to that constraint and shows its graphical symbol next to that entity. Before you pick the location with the left mouse button, you can:

- Disable a constraint by pressing the right mouse button. To enable it again, press the RMB again.
- Lock in a constraint by holding the Shift key and pressing the RMB. To unlock the constraint, repeat your actions.

Constraints that appear in gray are called "weak" constraints. Pro/E can remove them without warning. Figure 3.19 lists constraints with the corresponding graphical symbols. You can add your own constraints with the **Constrain** option in the Sketch menu. Pro/E shows constraints in the following default colors:

- **Weak** constraint gray
- **Strong** constraint yellow
- **Locked** constraint enclosed in a circle
- **Disabled** constraint with a line crossing the constraint symbol

Constraint	Symbol
Midpoint	M
Same points	o
Horizontal entities	H
Vertical entities	V
Point on entity	-O- - -
Tangent entities	T
Perpendicular entities	⊥
Parallel lines	$//_1$
Equal radii	R with an index in subscript
Line segments with equal lengths	L with an index in subscript (for example, L1)
Symmetry	→←
Entities are lined up horizontally or vertically	-- ¦
Collinear	═
Alignment	Symbol for the appropriate alignment type.
Use Edge/Offset Edge	— o

Figure 3.19 Constraints and their Symbols

Click: **Continue with the current section** ⇒ note the direction arrow (Fig. 3.20) ⇒ ⇒ slide the depth box until the dimension is approximately **2.500** (Fig. 3.21) ⇒ **MMB** (Fig. 3.22) ⇒ **LMB** to deselect ⇒ in the **Model Tree** click on **Protrusion** (Fig. 3.23) ⇒ **RMB** ⇒ **Edit** ⇒ double click on **2.500** and change to **2.625** (Fig. 3.24) ⇒ **Enter** ⇒ **Edit** from the menu bar ⇒ **Regenerate** ⇒ **Tools** ⇒ **Environment** ⇒ Standard Orient Isometric ⇒ **Apply** ⇒ **OK** (Fig. 3.25) ⇒ ⇒ ⇒ **MMB**

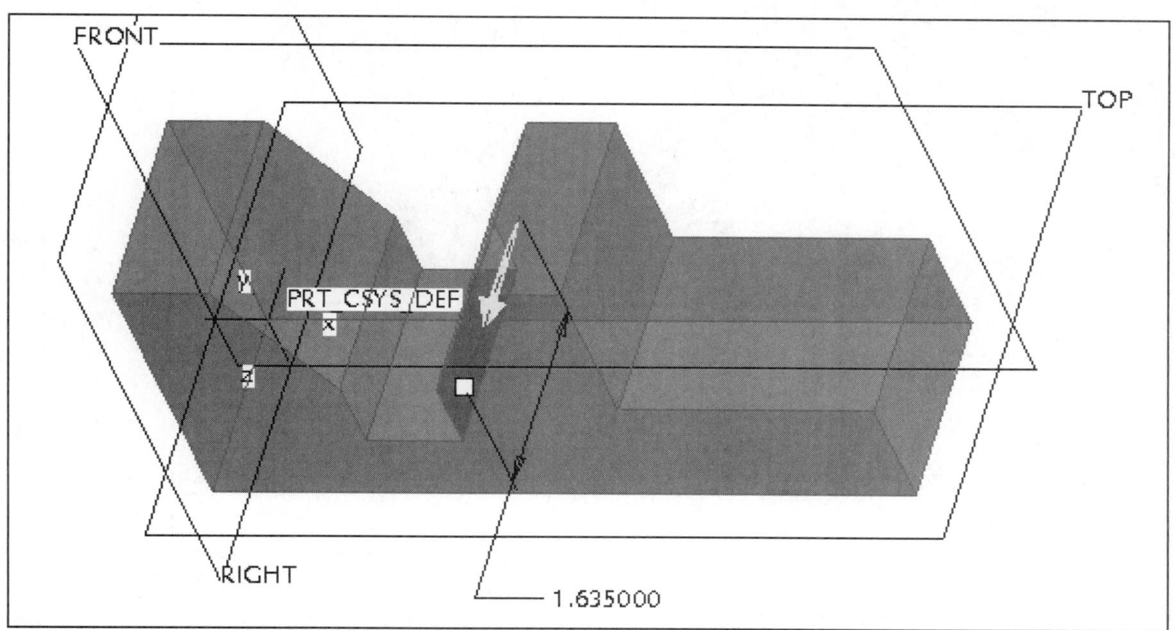

Figure 3.20 Direction Arrow

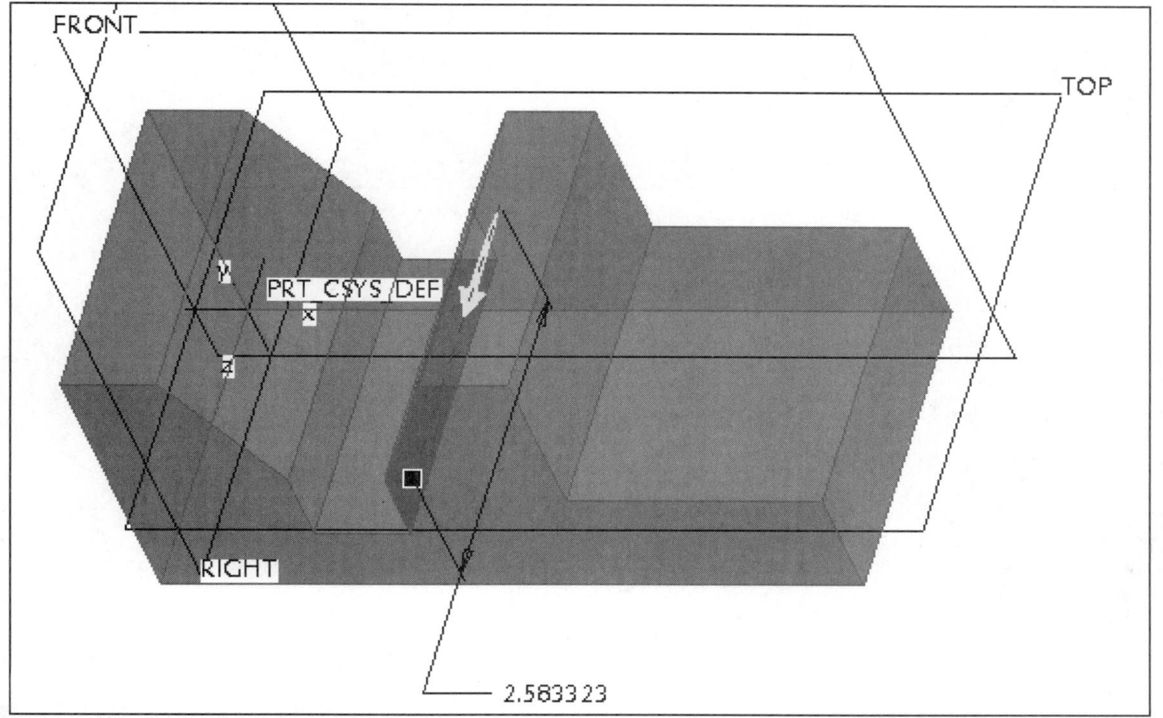

Figure 3.21 Dynamically Modify the Depth

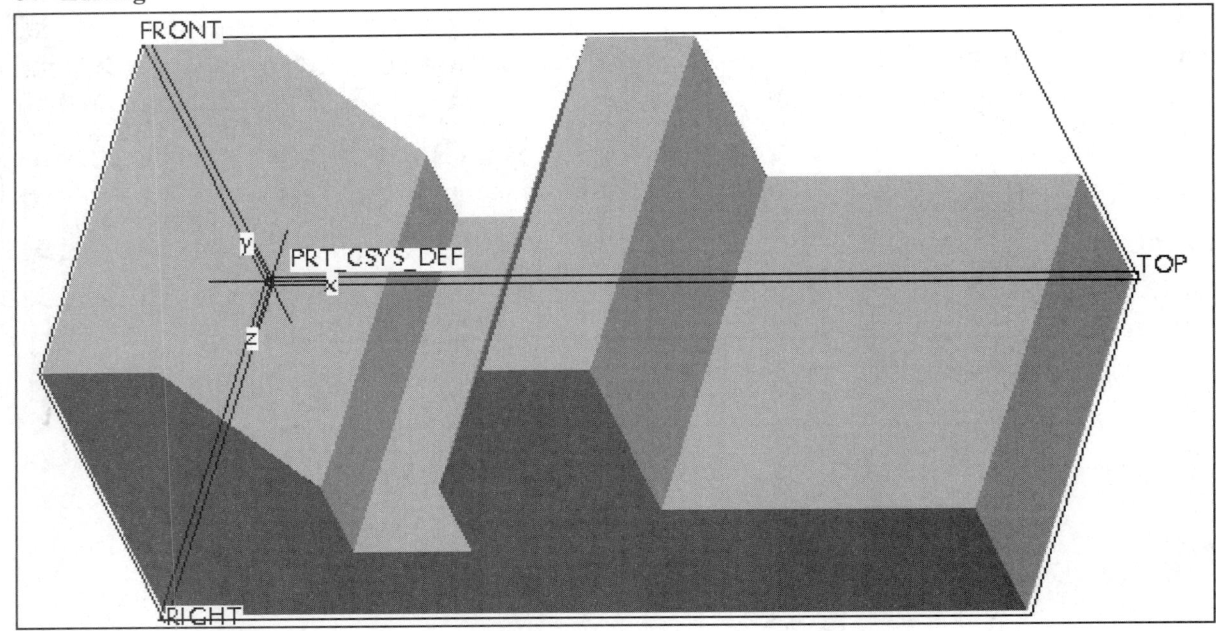

Figure 3.22 First Protrusion

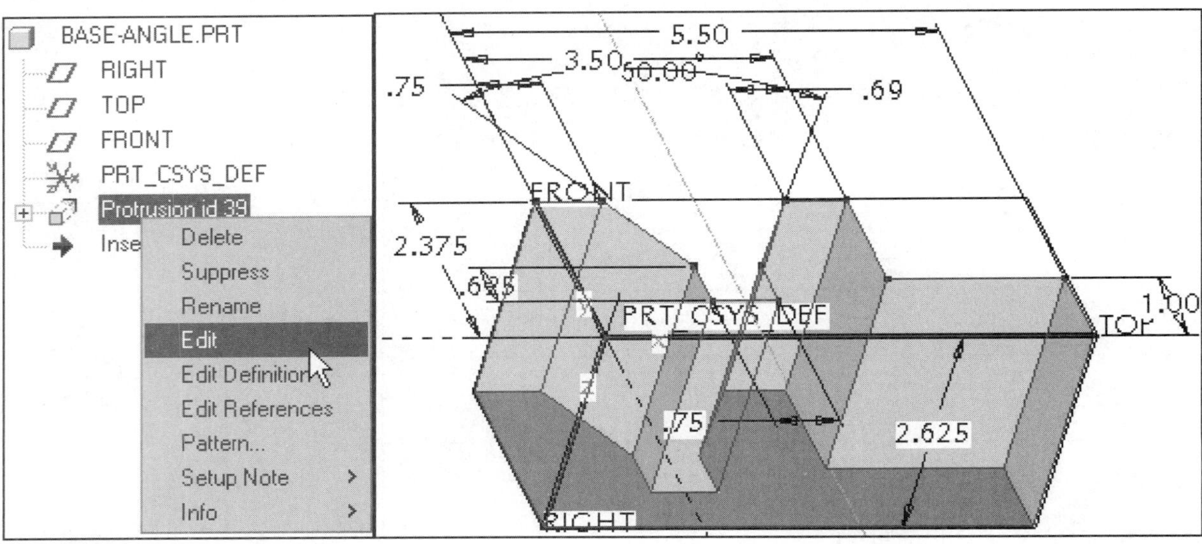

Figure 3.23 Edit

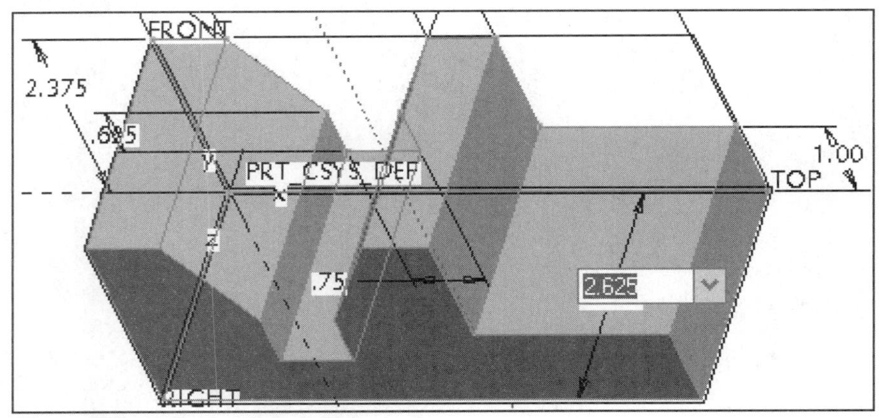

Figure 3.24 Edit Depth to **2.625**

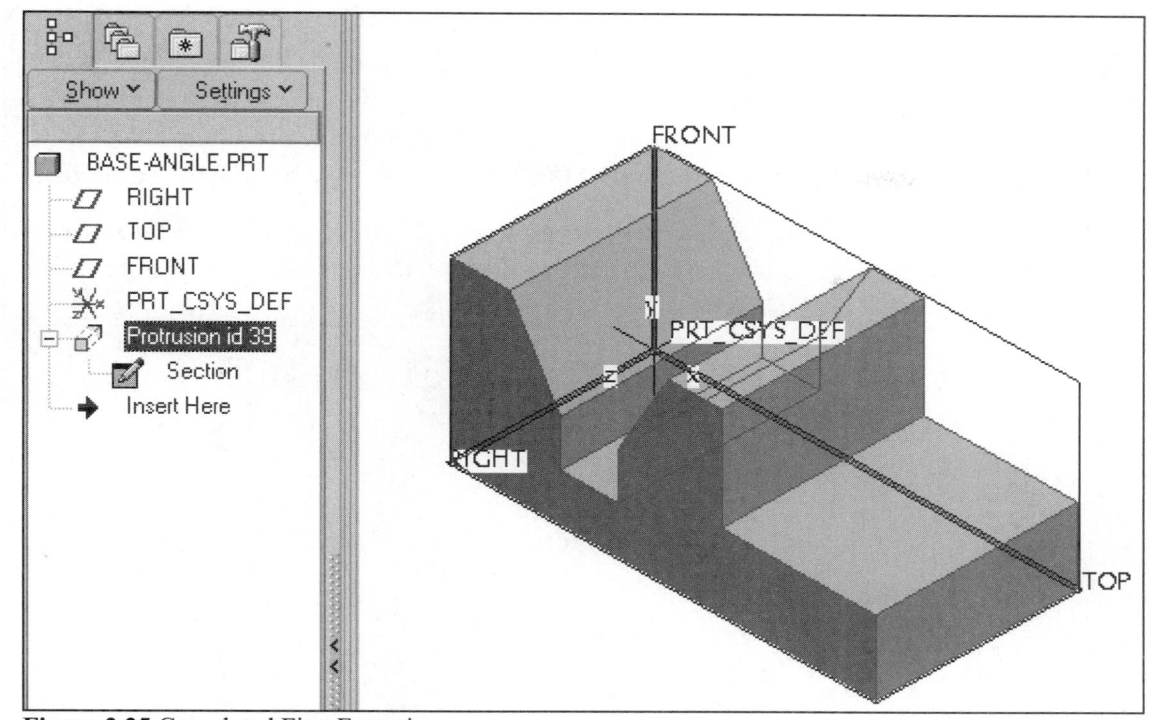

Figure 3.25 Completed First Extrusion

A sketch can be extruded to add material or remove material (cut). It is better *design intent* to make each cut extrusion separately. A *closed section* is a sketched set of entities that start and end at the same position, like a square. An *open section* does not form a closed figure; an example would be a **V** or **U**-shape. In general, try to use open sections for a cut extrusion when possible. The next feature that will be created is the V-shaped cut. The dimensions for the cut are shown in Figure 3.26(a-b). The **V**-cut is sketched similar to the first extrusion except that it removes material.

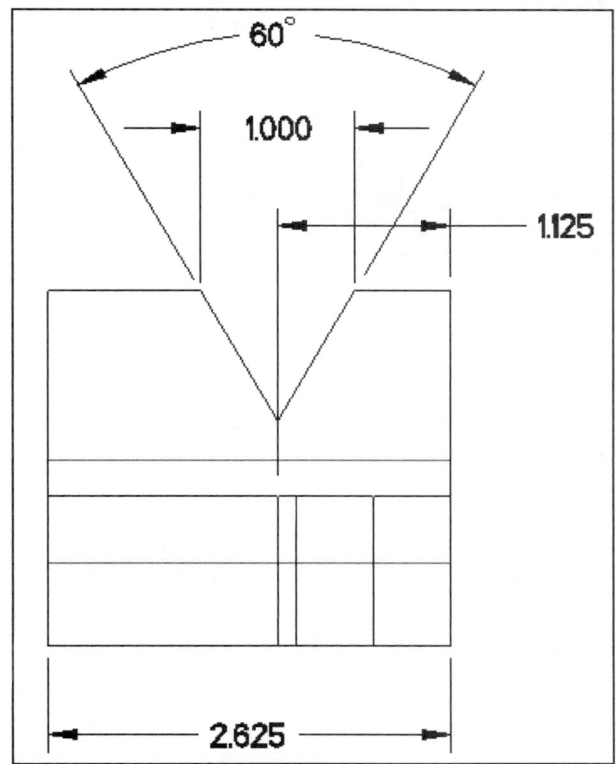

Figure 3.26(a) V-Cut Shown in Drawing

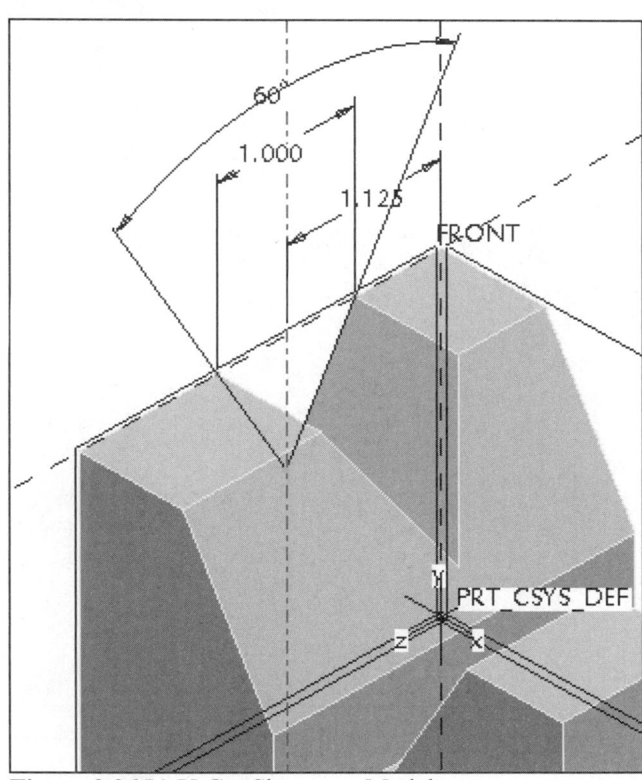

Figure 3.26(b) V-Cut Shown on Model

Click: **Extrude Tool** ⇒ ⇒ Section dialog box opens (Fig. 3.27) ⇒ Sketch Plane--- Plane: select **RIGHT** datum from the model as the sketch plane ⇒ Sketch Orientation--- Sketch view direction: **Flip** ⇒ click inside Reference: box (if the box is yellow, it is active, and you do not have to click inside of it) and then select **TOP** datum from the model ⇒ Orientation: **TOP** ⇒ click **Sketch** button ⇒ **Tools** from the menu bar ⇒ **Environment** ⇒ Snap to Grid ⇒ **OK** ⇒ **Toggle the grid off** ⇒ **Redraw the current view** (Fig. 3.28) ⇒ References: click on **F2(TOP)** ⇒ **Delete** ⇒ click on the top edge of the part to add a new reference **Edge:F5(PROTRUSION)** (Fig. 3.28) ⇒ **Close**

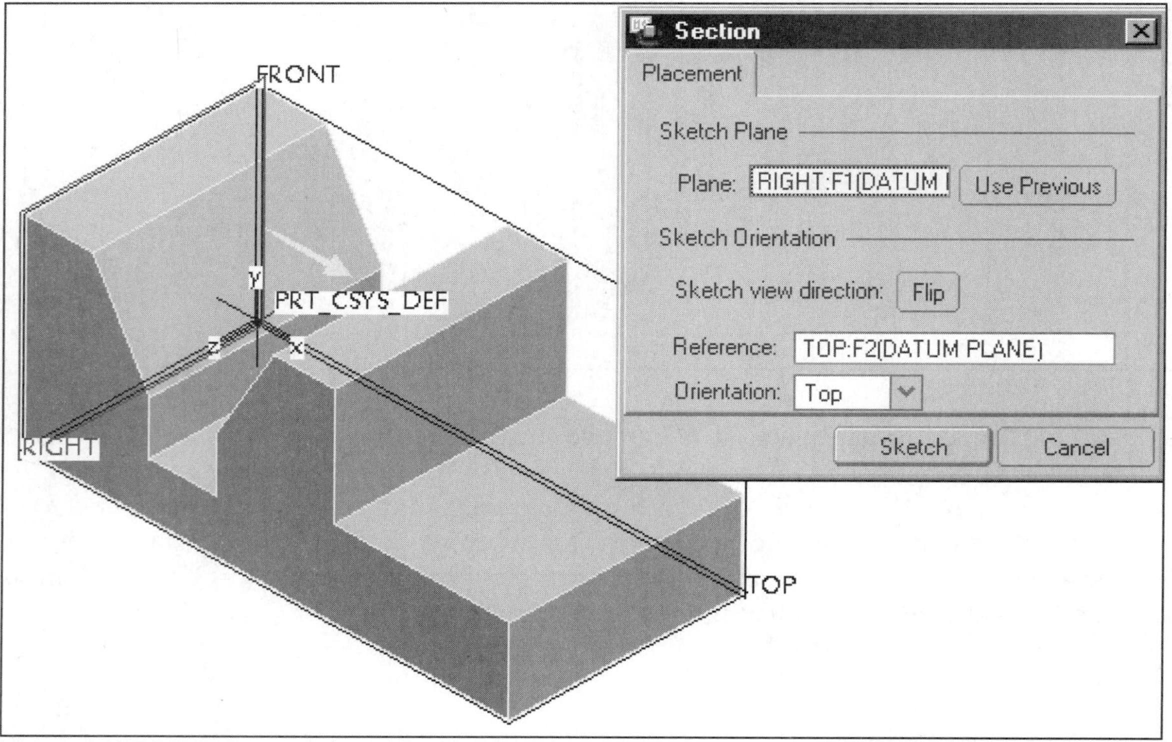

Figure 3.27 RIGHT Datum Selected for Sketch Plane, TOP Datum Selected for Sketch Orientation

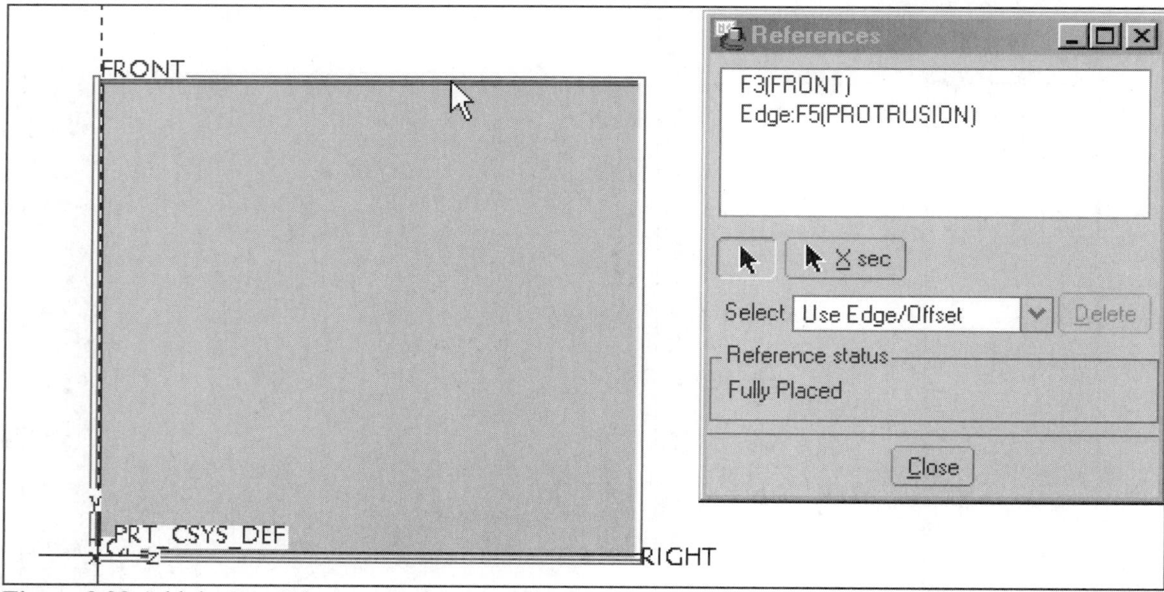

Figure 3.28 Add the Top Edge as a Reference

Add a centerline, click: ⇒ **Create 2 point centerlines** (Fig. 3.29) ⇒ sketch the vertical centerline ⇒ **Create 2 point lines** ⇒ sketch the two lines (Fig. 3.30) ⇒ **MMB** to end the line sequence ⇒ ⇒ reposition the weak dimensions (Fig. 3.31) ⇒ add and move the dimensions (Fig. 3.32) ⇒ modify the values (Fig. 3.33) ⇒ ⇒ **Continue with the current section** (Fig. 3.34) ⇒ ⇒ **Standard Orientation** (Fig. 3.35)

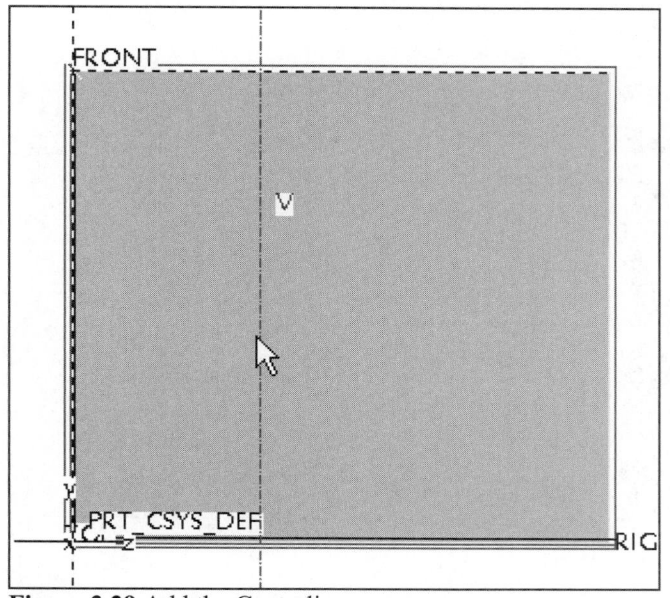

Figure 3.29 Add the Centerline

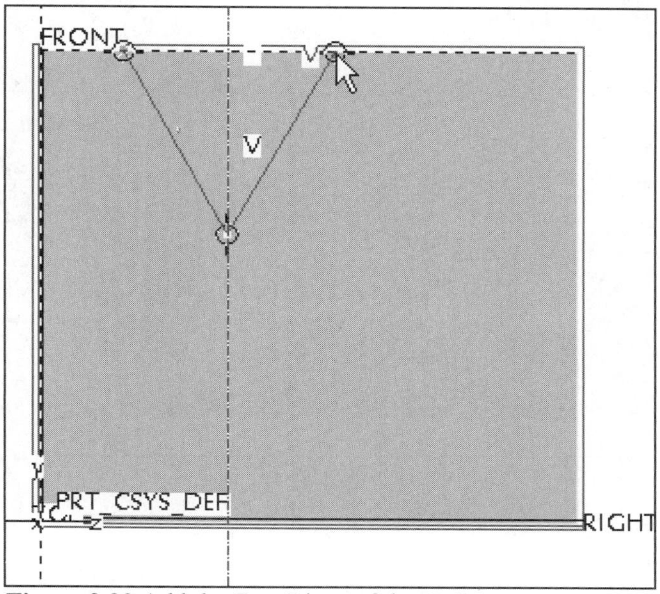

Figure 3.30 Add the Two Lines of the V-Cut

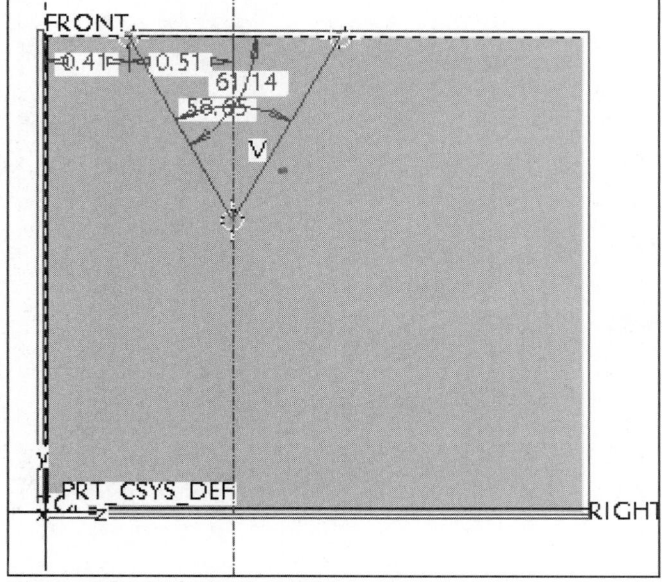

Figure 3.31 Default Weak Dimensions

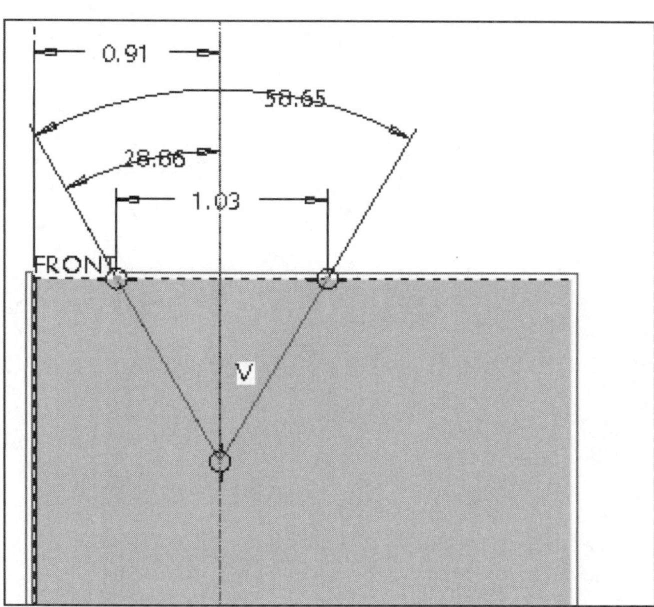

Figure 3.32 Added and Moved Dimensions

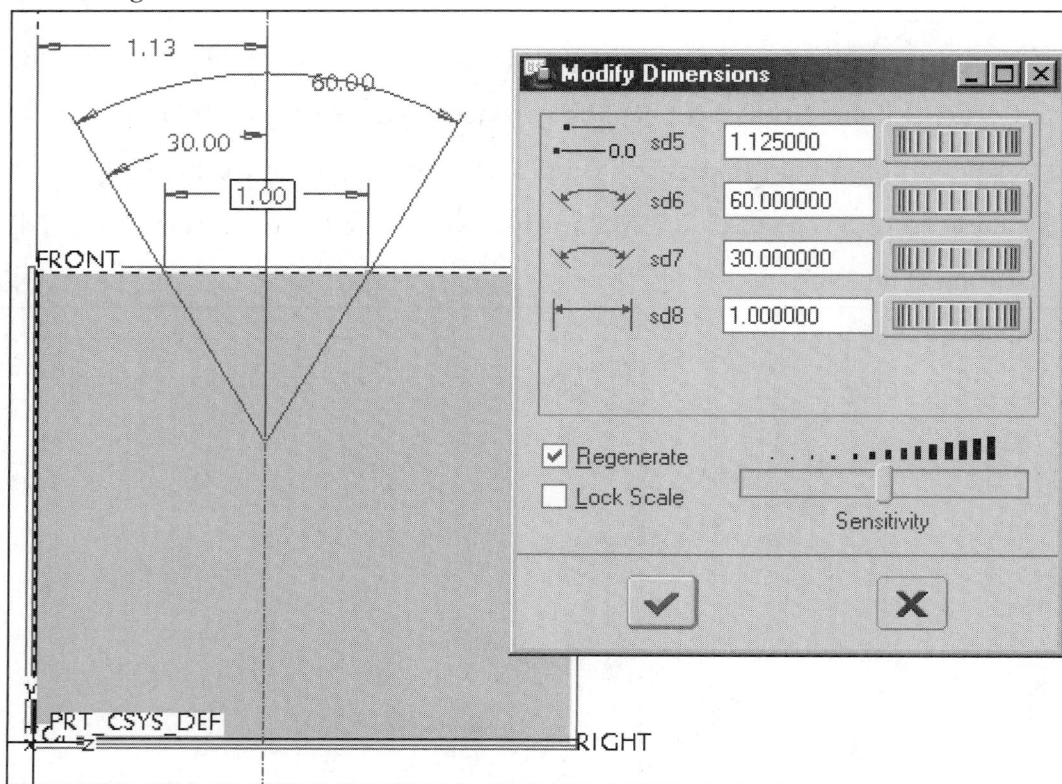

Figure 3.33 Modify the Dimensions

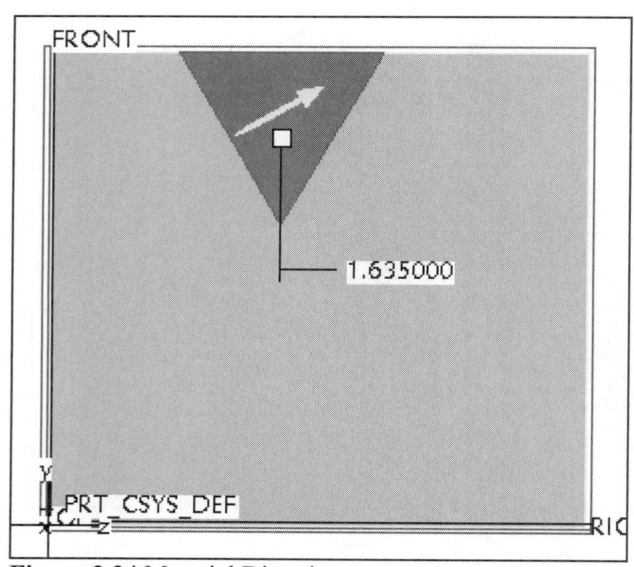

Figure 3.34 Material Direction

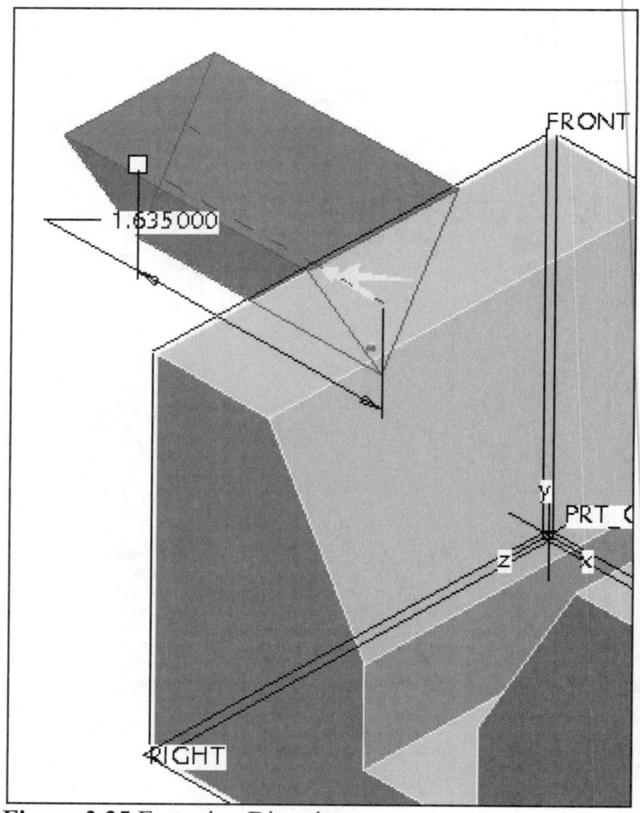

Figure 3.35 Extrusion Direction

From the dashboard, click: **Remove Material** ⇒ **Change depth direction of extrude to other side of sketch** (Fig. 3.36) ⇒ **Options** tab ⇒ **To Selected** ⇒ pick the surface as shown (Fig. 3.37) ⇒ (Fig. 3.38) ⇒ ⇒ **MMB** ⇒ **LMB** to deselect

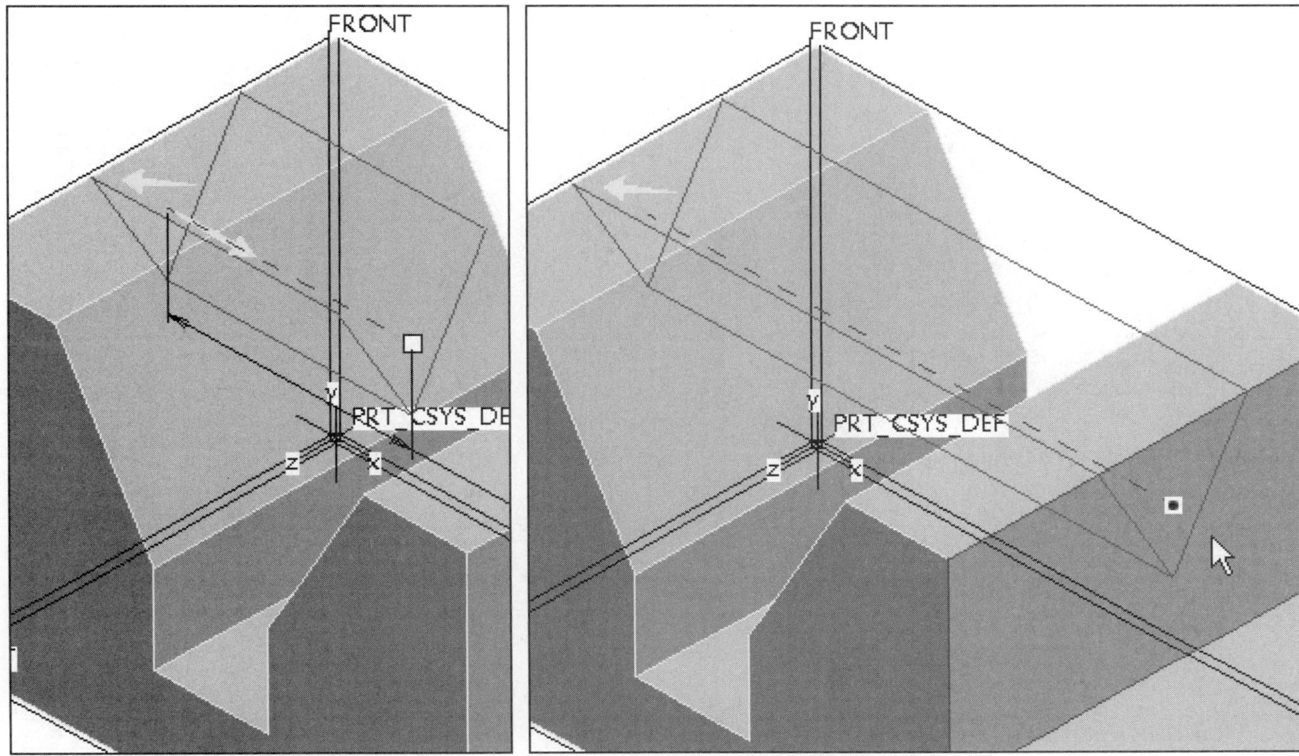

Figure 3.36 Extrusion Direction Reversed

Figure 3.37 Extrude To Selected Surface

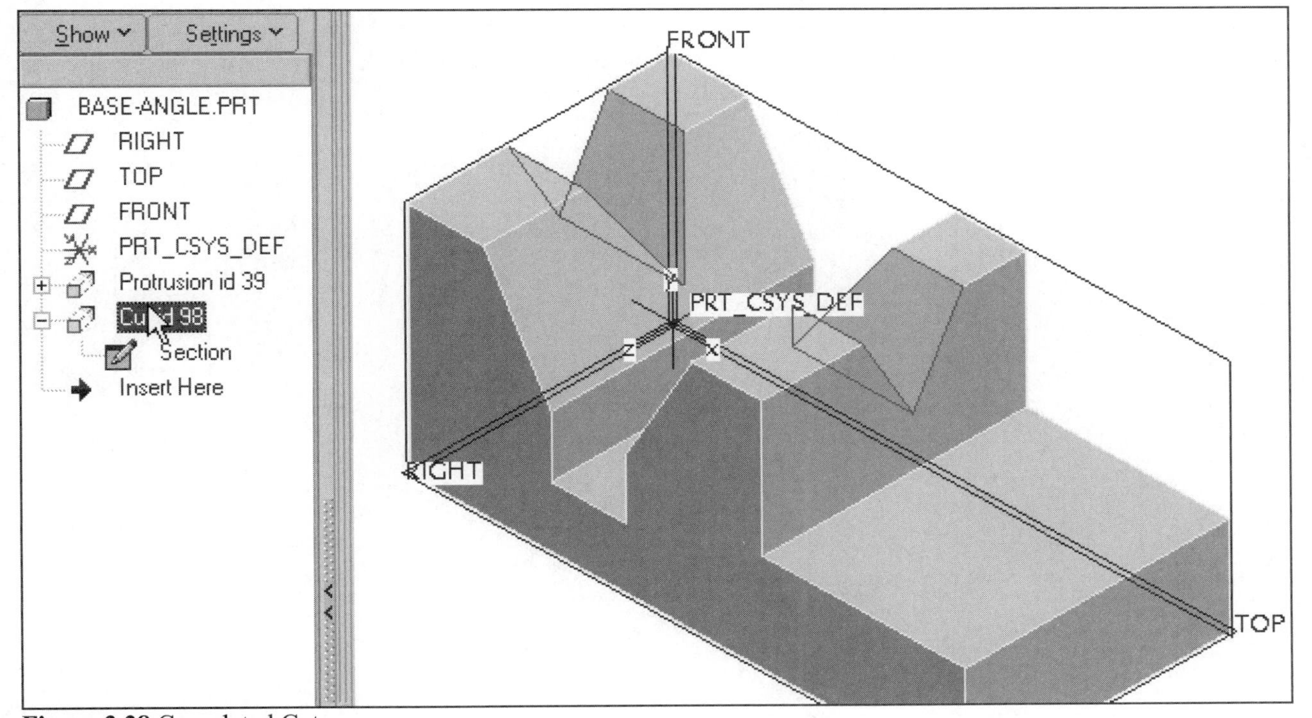

Figure 3.38 Completed Cut

Click on the cut feature in the Model Tree (Fig. 3.39) ⇒ **RMB** ⇒ **Edit Definition** (Fig. 3.40)

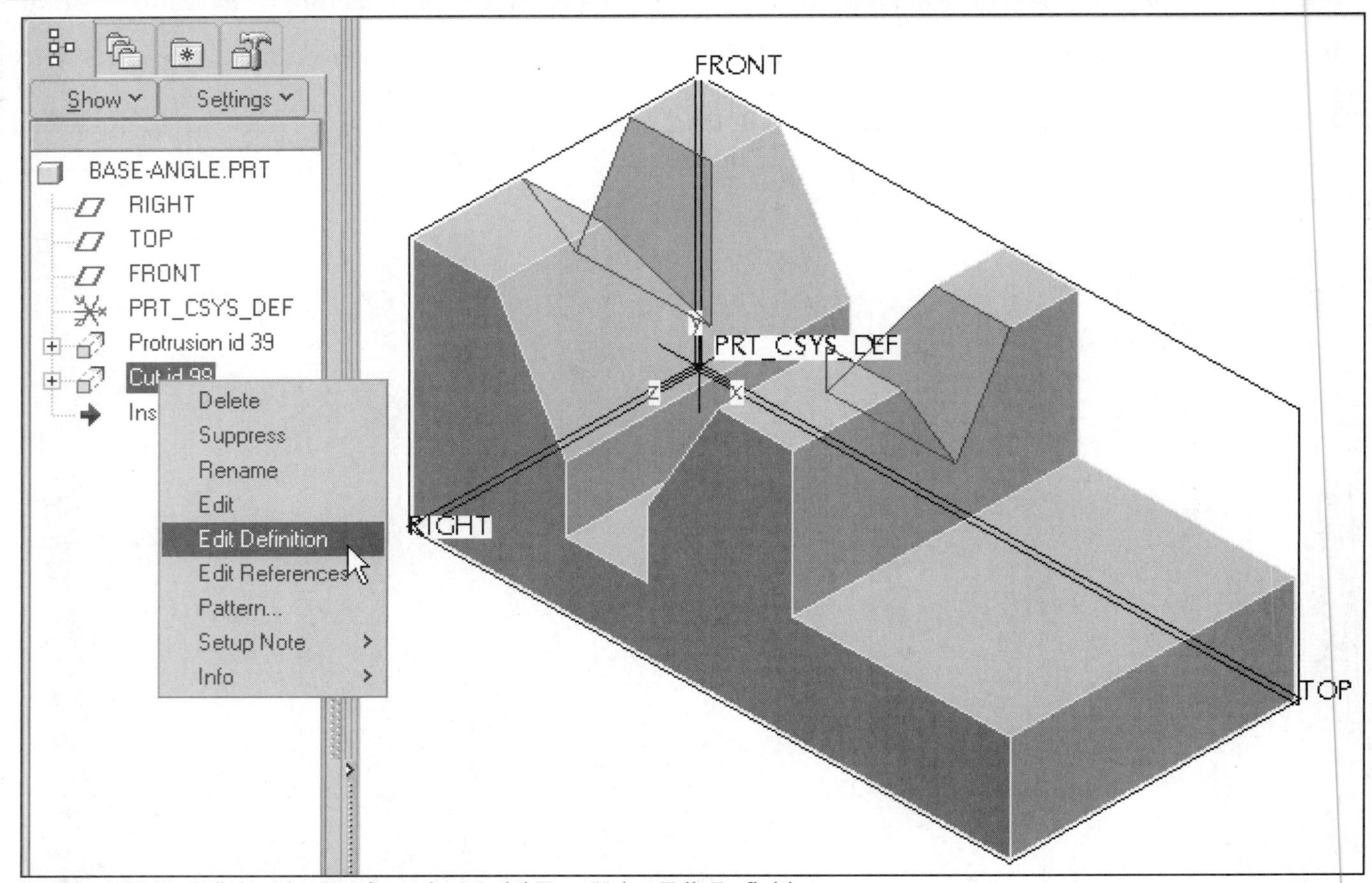

Figure 3.39 Redefining the Cut from the Model Tree Using Edit Definition

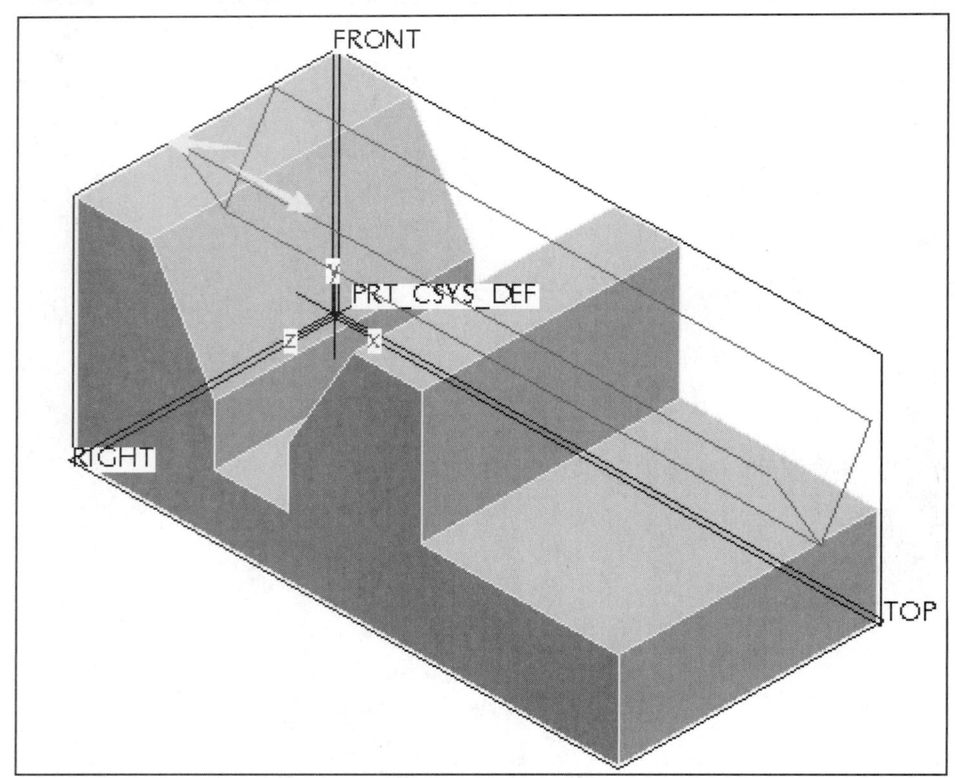

Figure 3.40 Cut Direction Arrow and Material Direction Arrow

Click: [icon] from the dashboard ⇒ **Sketch** from the Section dialog box ⇒ [icon] **Impose sketcher constraints** ⇒ [icon] **Make two points or vertices symmetric about a centerline** ⇒ pick the centerline and then pick the two upper vertices to be symmetric (Fig. 3.41) ⇒ Resolve Sketch dialog box displays (Fig. 3.41) ⇒ pick **Dimension 30.00** (your identifier may be different) ⇒ **Delete** ⇒ **Close** ⇒ [icon] ⇒ **Standard Orientation** ⇒ [icon] **Continue** ⇒ **OK** ⇒ **Options** tab (Fig. 3.42) ⇒ [icon] **Through Until** ⇒ pick a different surface than before (Fig. 3.43) ⇒ [icon] **Toggle display of dynamic display geometry** (Fig. 3.44) ⇒ [icon] **Resumes the previously paused tool** ⇒ **Options** tab ⇒ [icon] **Through All** ⇒ [icon] ⇒ [icon] ⇒ **MMB** [Fig. 3.45(a-b)] ⇒ **Tools** ⇒ **Environment** ⇒ Standard Orient | Trimetric ⇒ **OK**

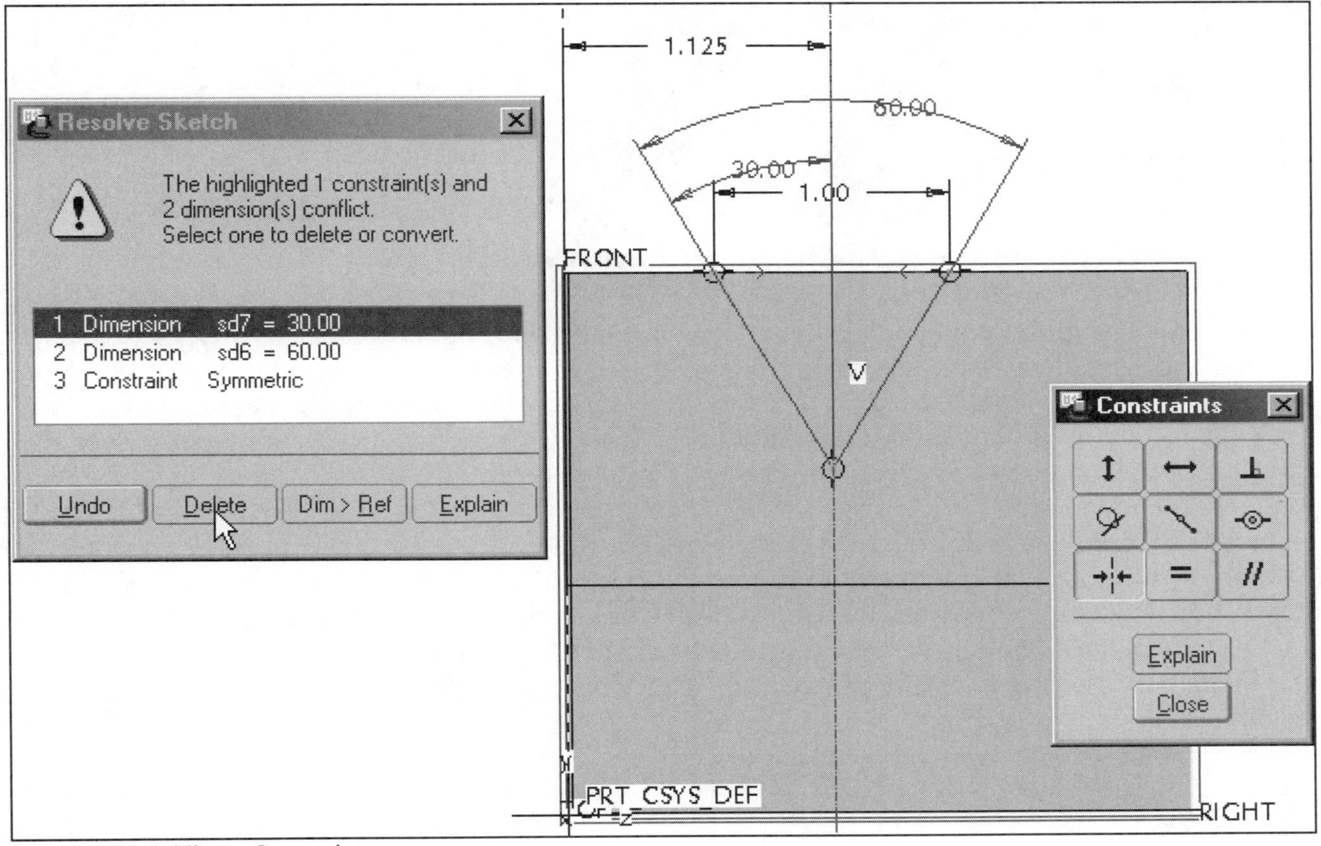

Figure 3.41 Adding a Constraint

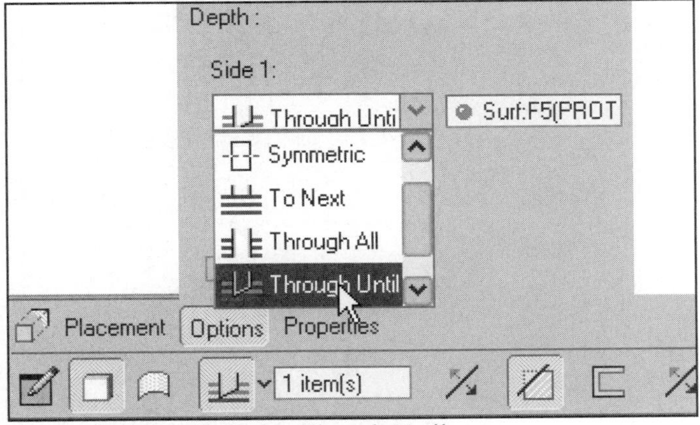

Figure 3.42 Depth Option Though Until

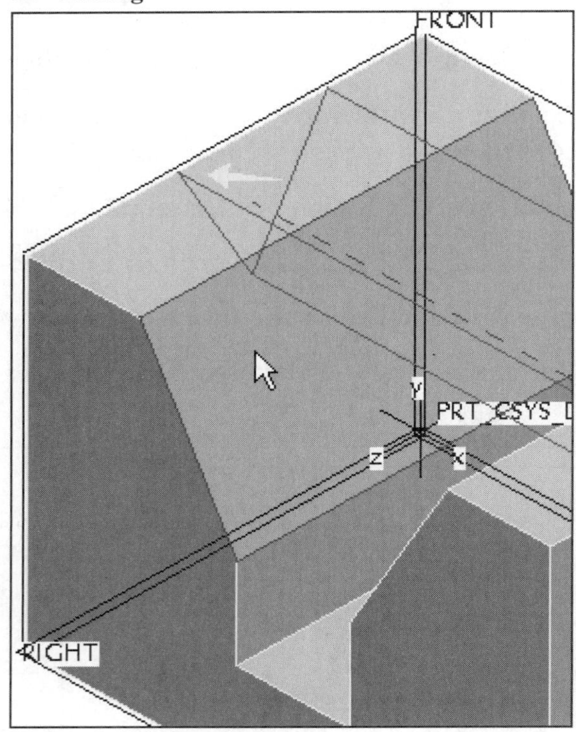

Figure 3.43 Select the Plane

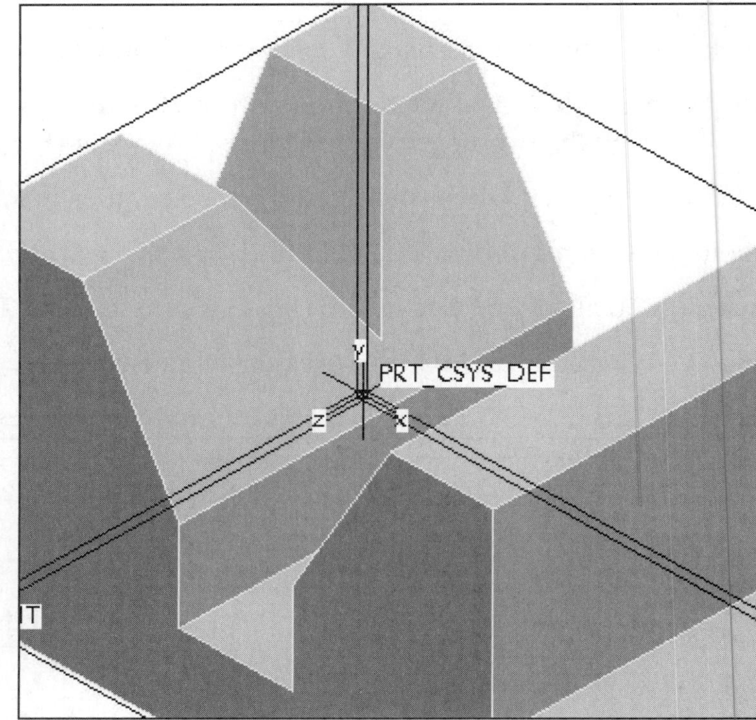

Figure 3.44 Feature Preview

The **Edit Definition** option allows you to change how a feature is created. The types of changes you can make depend on the selected feature. For example, if a feature was created using a section, you can redefine (Edit Definition) the section, feature references, dimensioning scheme and so on. Pro/E recreates the feature using the new feature definitions. When you redefine the feature sections, you may need to Edit Definition or reroute any child feature whose reference edge or surface was replaced. When you preview the redefinition, Pro/E removes the feature geometry and creates temporary geometry for your changes. When you exit from the user interface, Pro/E regenerates the part. If you quit the redefinition of a feature, Pro/E attempts to restore the part to its original state, without regenerating the geometry of the model. After you have redefined (Edit Definition) certain part features, and if you quit the redefinition, Pro/E must still regenerate the geometry of later features, of the feature just redefined. You eliminated an extra dimension by adding a symmetric constraint and made the cut through all instead of to a selected surface.

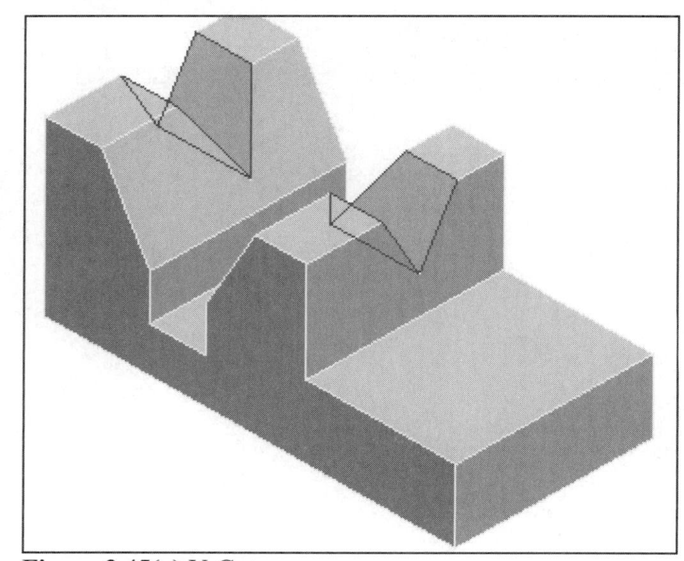

Figure 3.45(a) V-Cut

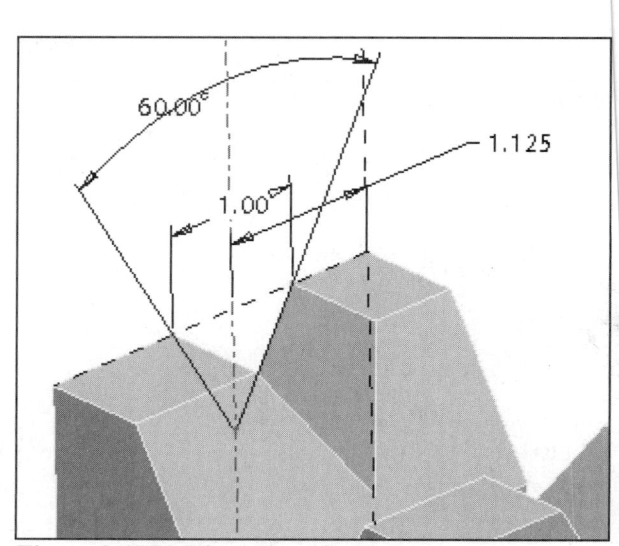

Figure 3.45(b) Dimensions for V-Cut

Two cuts are still required for the completion of the part; the angle cut and the **U**-shaped cut (Fig. 3.46). They can be created together using "closed sections", or separately, each with its own open section. It is better *design intent* to make the cuts separately. They do not have any particular relationship, except that they cut the same direction, and start on the same surface/plane.

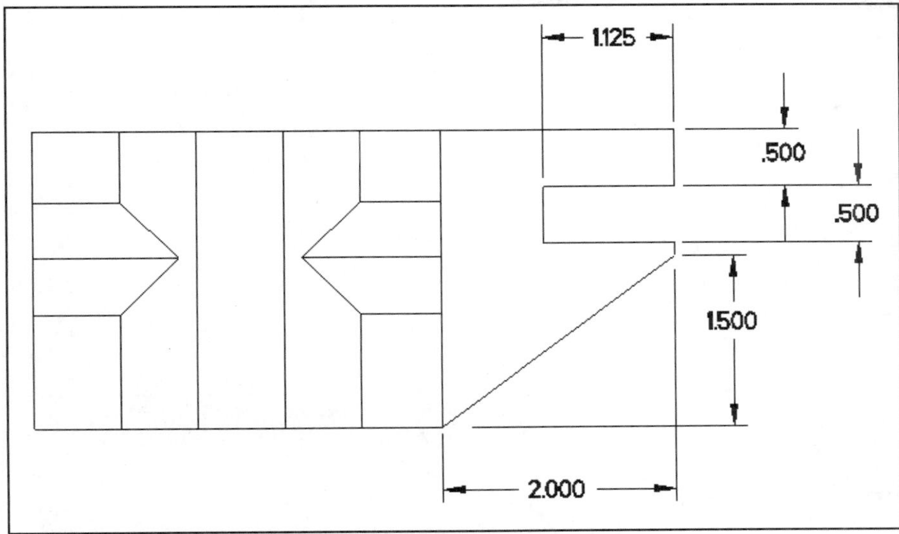

Figure 3.46 Top View Showing Dimensions for Two Cuts

Click: **Extrude Tool** ⇒ ⇒ Sketch Plane--- Plane: select **TOP** datum from the model as the sketch plane ⇒ Sketch Orientation--- Sketch view direction: **Flip** ⇒ Reference: select **RIGHT** datum from the model ⇒ Orientation: **Right** (Fig. 3.47) ⇒ click **Sketch** button ⇒ **Hidden Line** ⇒ **Tools** from the menu bar ⇒ **Environment** ⇒ Snap to Grid ⇒ **OK** ⇒ **Toggle the grid off** ⇒ ⇒ References: hold **Ctrl** key down and click on **RIGHT** and **FRONT** ⇒ **Delete** (Fig. 3.48) ⇒ click on the top and right side edges of the part to add these references **Edge:F5** (Fig. 3.49) ⇒ **Close** to accept the References ⇒ ⇒ sketch one line ⇒ **MMB** ⇒ add dimensions from each endpoint of the line to the corresponding part reference edge (Fig. 3.50) ⇒ ⇒ double click on each dimension and modify to the design values [Fig. 3.51(a-b)] ⇒ ⇒ **Standard Orientation** (Fig. 3.52) ⇒ (Fig. 3.53)

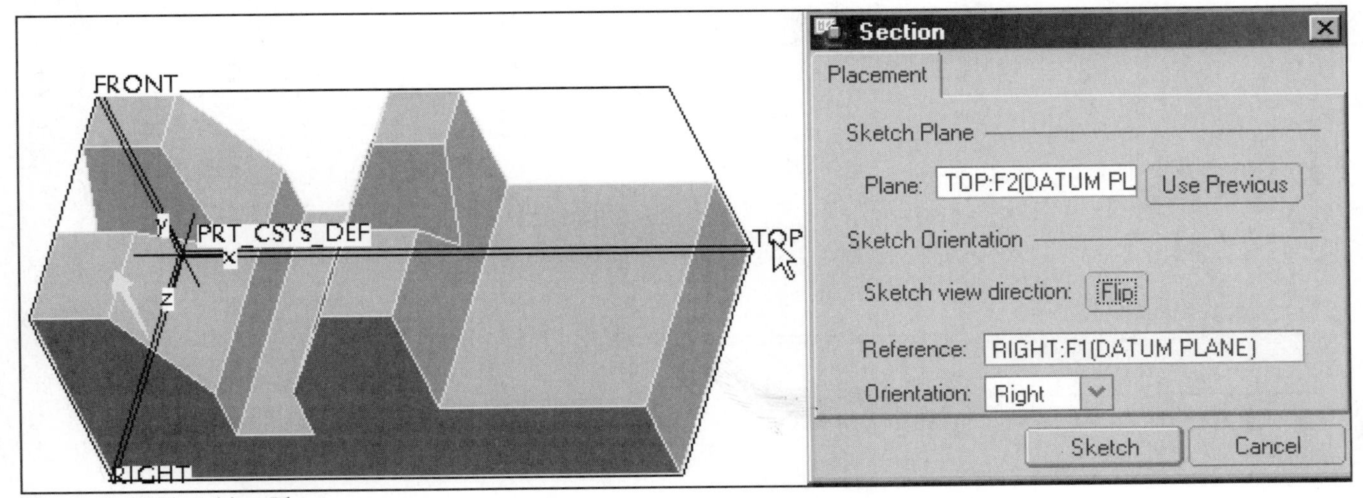

Figure 3.47 Sketching Plane

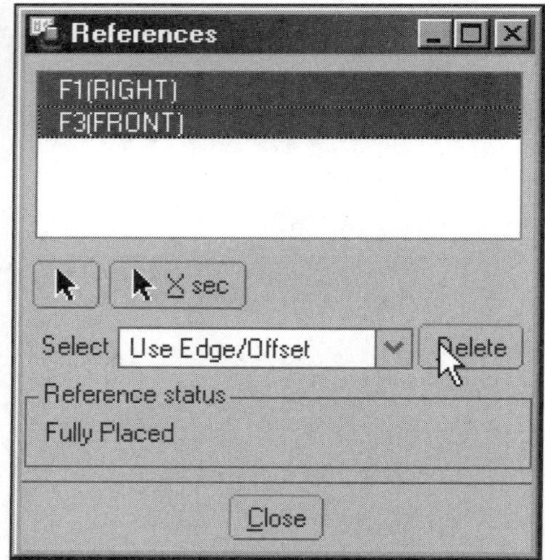

Figure 3.48 Delete References

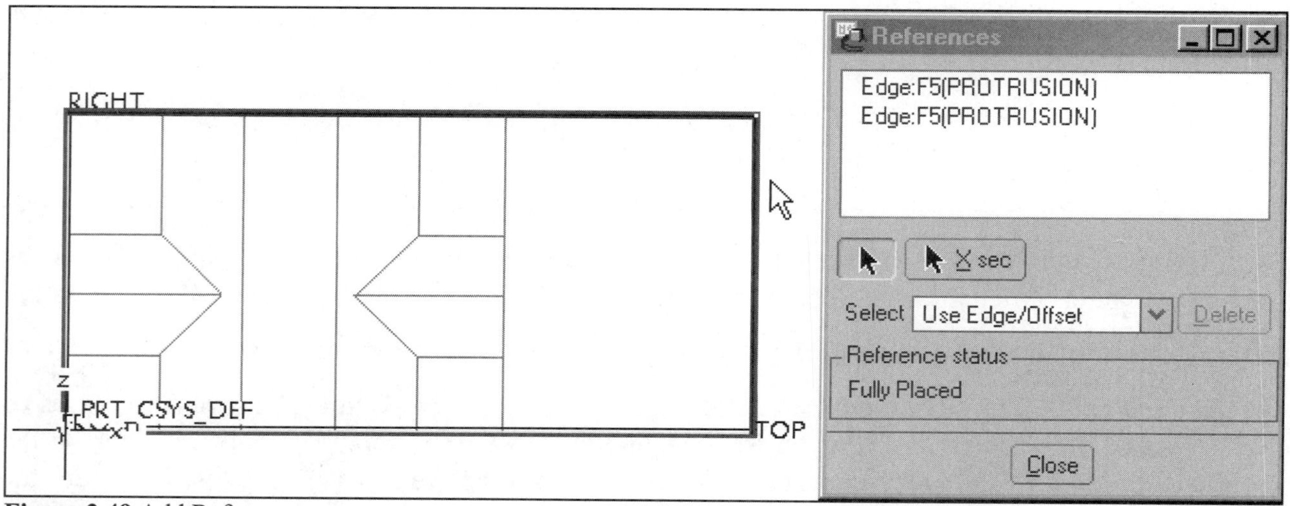

Figure 3.49 Add References

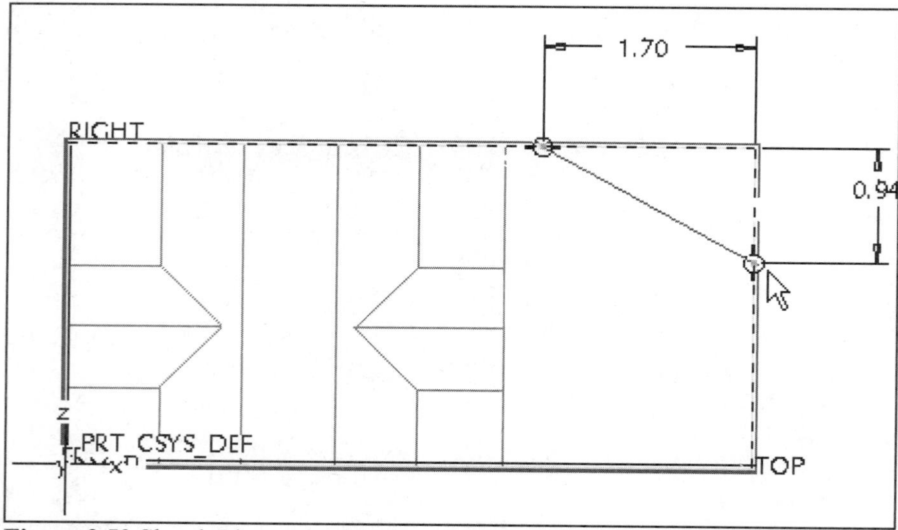

Figure 3.50 Sketch Line and Add Dimensions

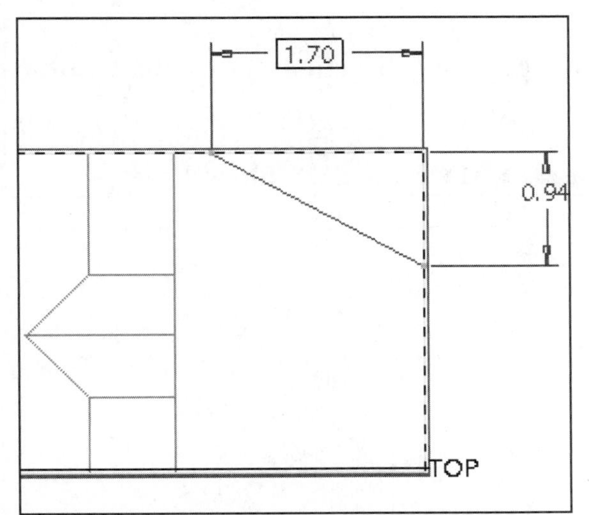

Figure 3.51(a) Modify to the Design Values

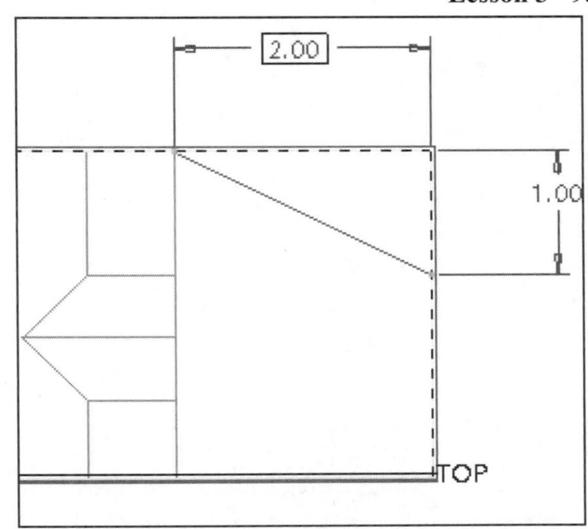

Figure 3.51(b) Design Values

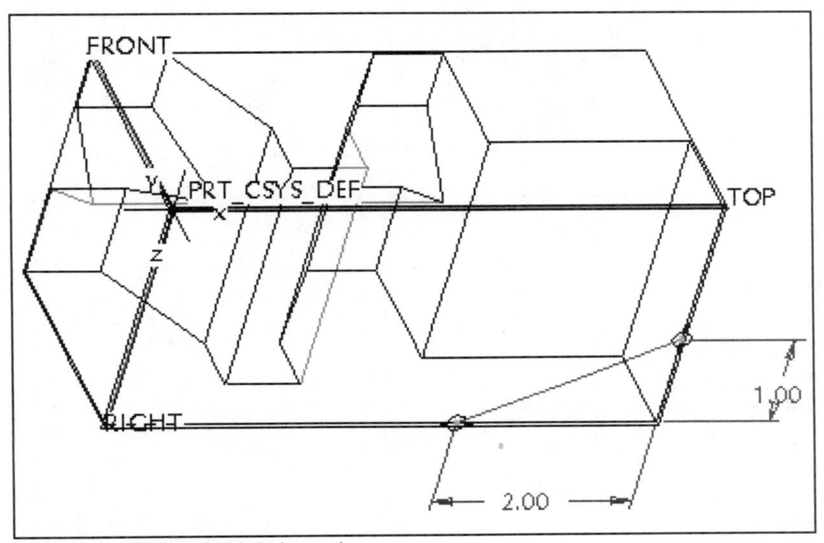

Figure 3.52 Standard Orientation

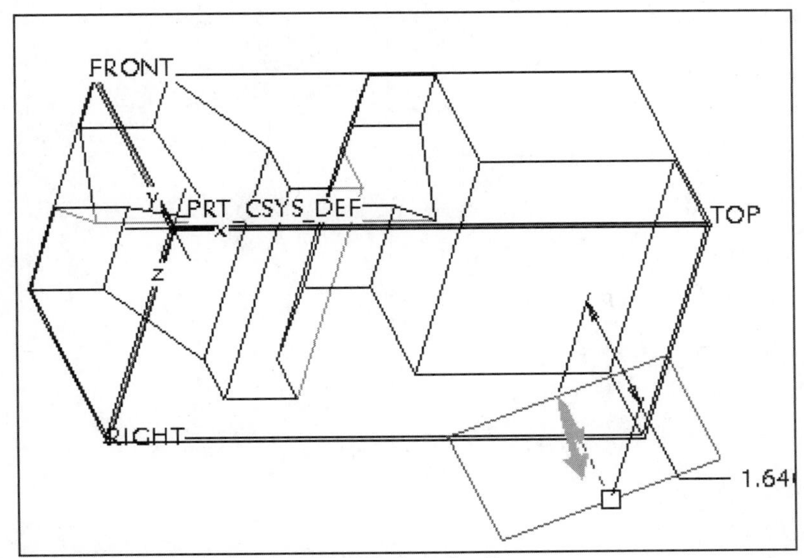

Figure 3.53 Cut Side Extending Below the Part

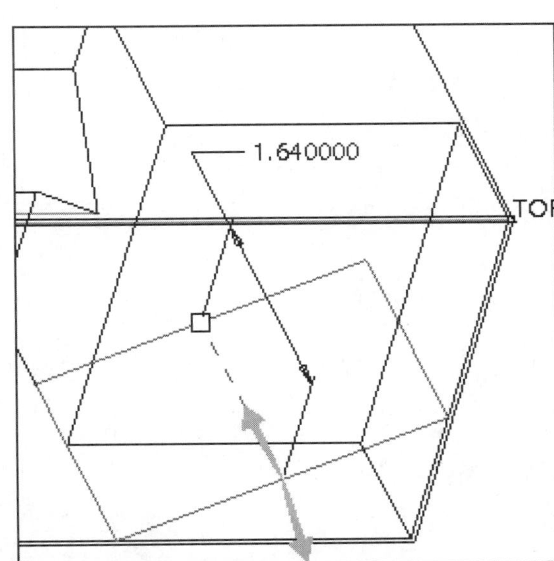

Figure 3.54 Cut Side Flipped

From the dashboard, click: **Remove Material** ⇒ **Change depth direction of extrude to other side of sketch** (see Fig. 3.54 previous page) ⇒ **Options** tab ⇒ To Selected (Fig. 3.55) ⇒ pick the surface as shown (Fig. 3.56) ⇒ **MMB** ⇒ (Fig. 3.57) ⇒ ⇒ **MMB** ⇒ **Hidden Line**

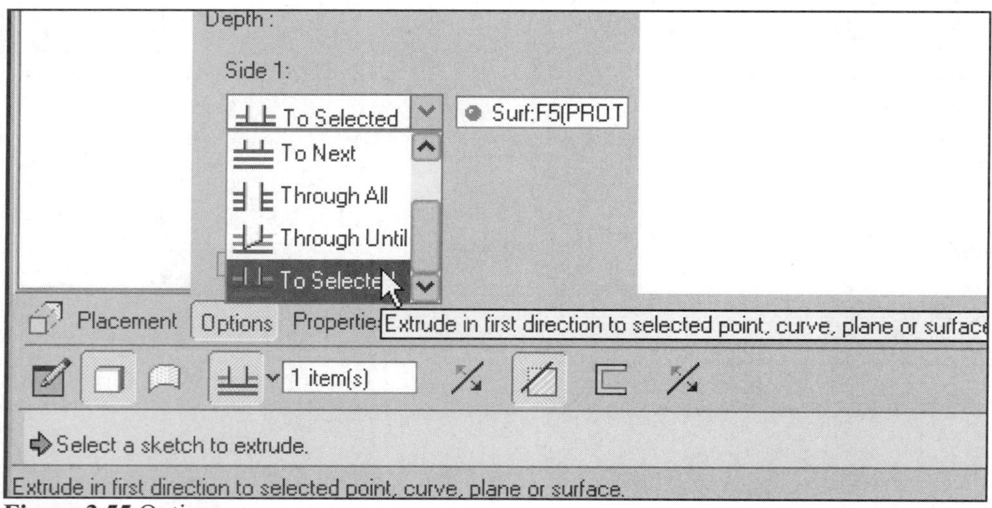

Figure 3.55 Options

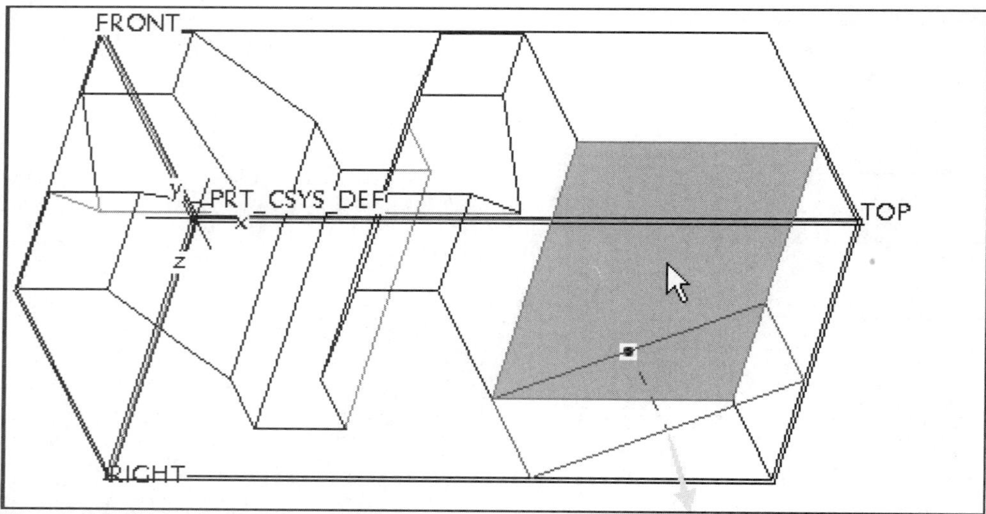

Figure 3.56 Extrude to Select Surface

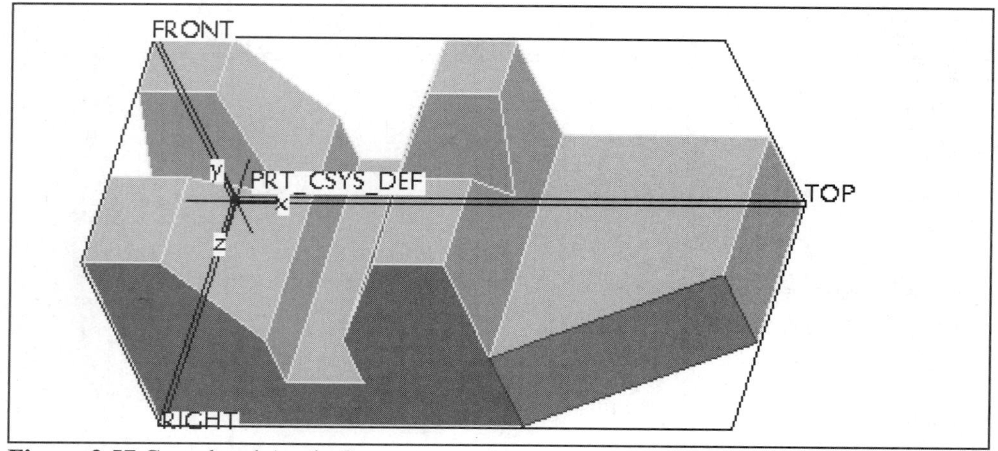

Figure 3.57 Completed Angle Cut

The last feature of the part is the **U**-shaped cut. Use the previous cut's placement plane and orientation.

Click: **Extrude Tool** ⇒ ⇒ **Use Previous** ⇒ **Sketch** ⇒ References: click on **F1(RIGHT)** ⇒ **Delete** ⇒ click on the right side edge of the part to add a new reference **Edge:F5(PRTRUSION)** (Fig. 3.58) ⇒ **Close** to accept the References ⇒ **Create 2 point lines** ⇒ sketch a **U**-shaped, three-line open section ⇒ **MMB** ⇒ add dimensions and modify the dimensions [Fig. 3.59(a-b)] ⇒ ⇒ ⇒ **Standard Orientation** ⇒ from the dashboard click: **Change depth direction of extrude to other side of sketch** (Fig. 3.60) ⇒ **Options** tab ⇒ To Selected ⇒ pick the surface (Fig. 3.61) ⇒ **Remove Material** ⇒ **MMB** ⇒ ⇒ ⇒ **MMB** (Fig. 3.62)

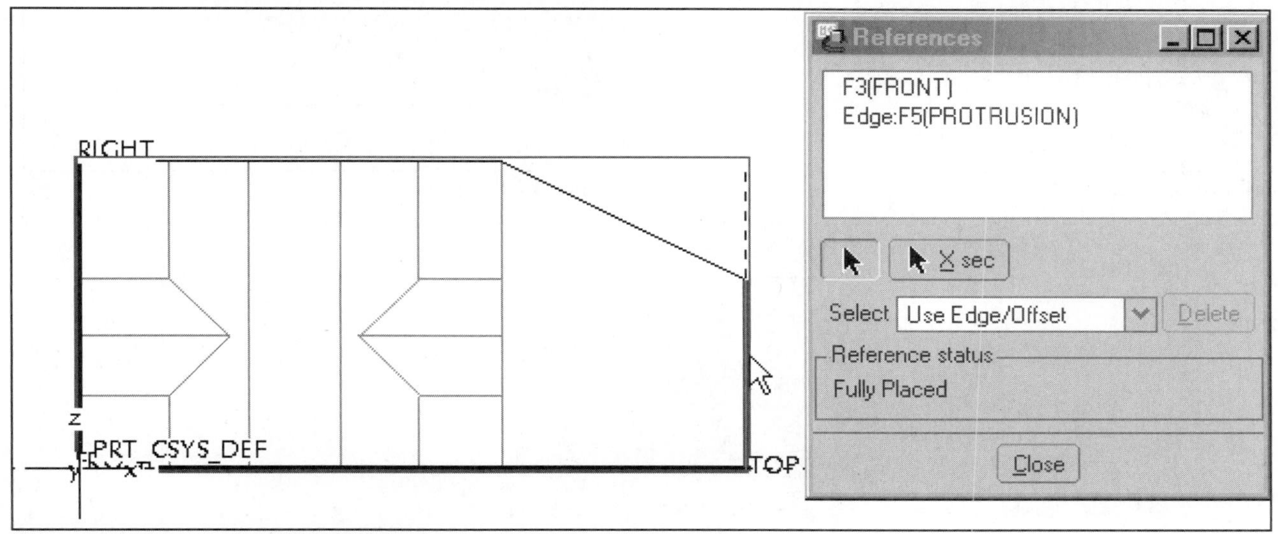

Figure 3.58 Delete the RIGHT Reference and Add the Edge

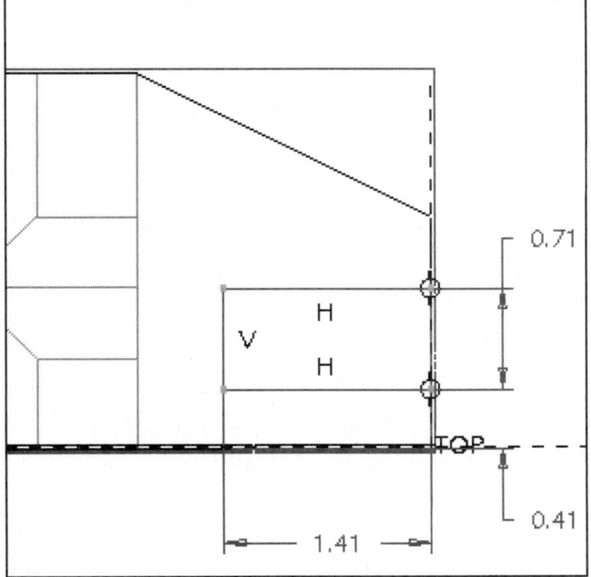

Figure 3.59(a) Modify the Dimensions

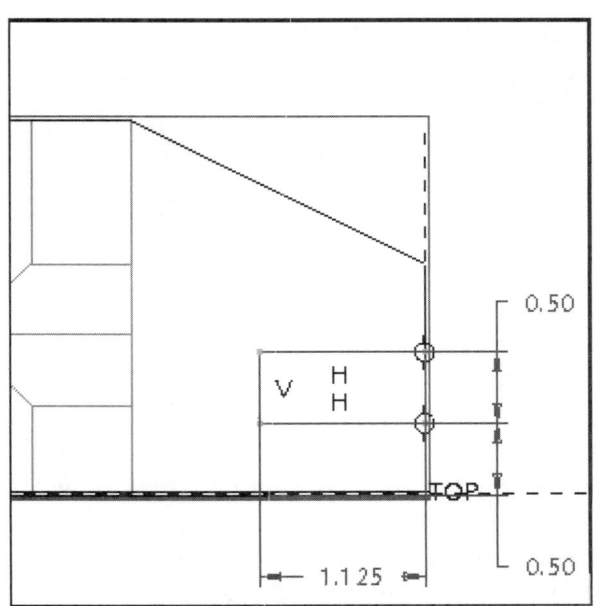

Figure 3.59(b) Modified Dimensions

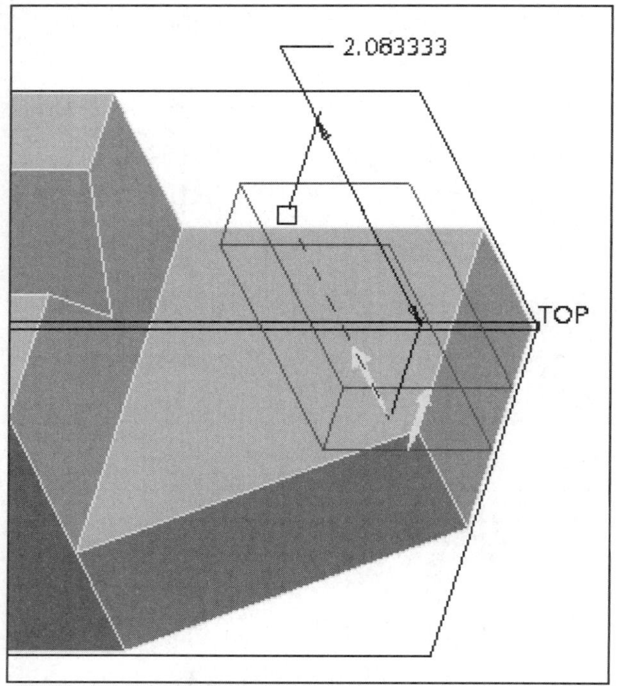

Figure 3.60 Flipped Feature Creation Direction

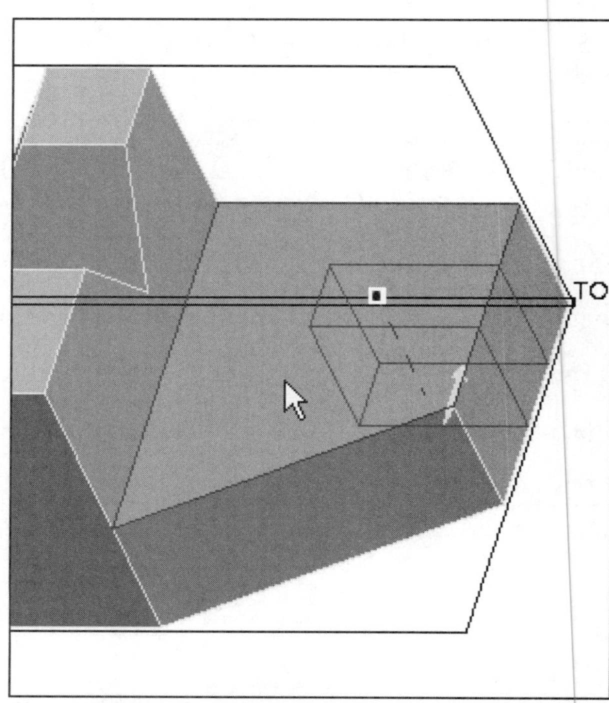

Figure 3.61 Surface Selection

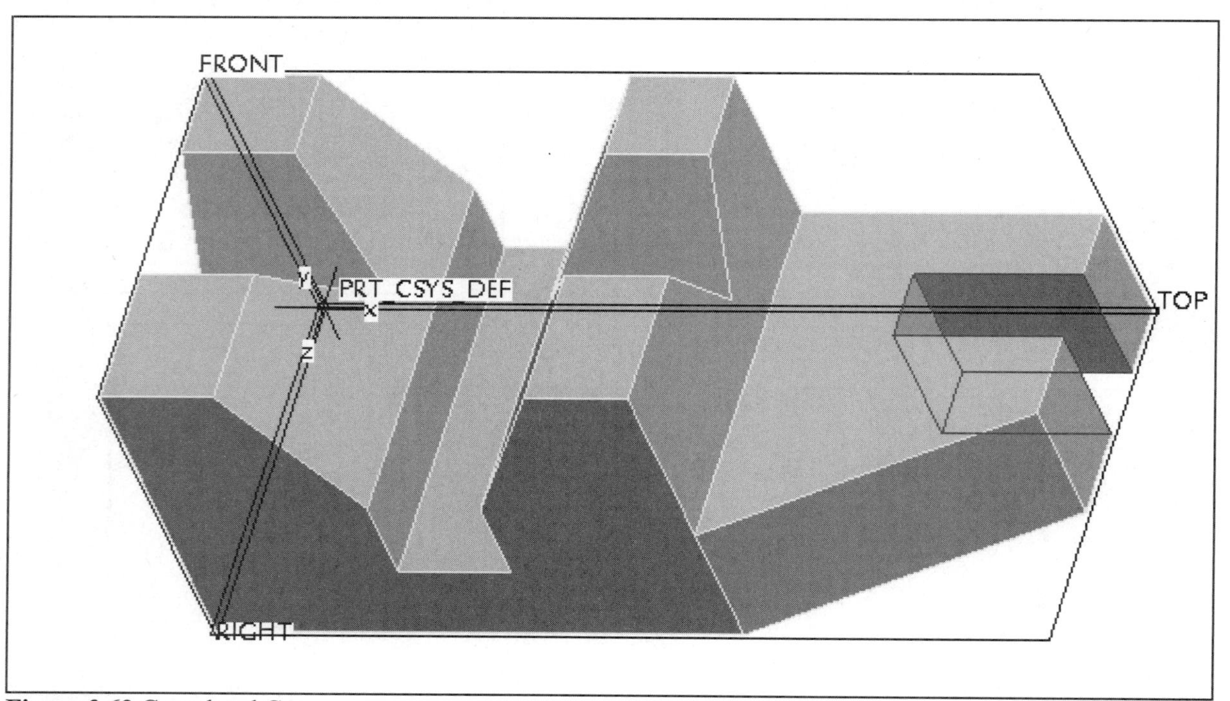

Figure 3.62 Completed Cut

You may be asking yourself, did the authors create the angled cut with the wrong dimension, or was this a teaching technique where the **Edit** tool could be introduced to modify the dimension to its correct shape? (We will be claiming that it was intentional). Regardless, if you followed the exact instructions and figures, the values for the cut were not completely correct.

Any dimension created on a sketch can be modified directly on the model, without having to redefine the feature and redo the sketch dimension. Simply by double-clicking on the feature in the graphics window, the design dimensions will display and they can be modified. The model must be regenerated after the value is changed for the feature to be updated.

Click on the second cut in the **Model Tree** ⇒ **RMB** ⇒ **Edit** (Fig. 3.63) ⇒ double click on the **1.000** dimension ⇒ change the dimension to **1.50** (Fig. 3.64) ⇒ **Enter** ⇒ **Edit** from menu bar ⇒ **Regenerate** (Fig. 3.65) ⇒ [save icon] ⇒ **MMB** ⇒ **File** ⇒ **Save a Copy** ⇒ New Name **BASE_ANGLE_Rev1** ⇒ **MMB** ⇒ **File** ⇒ **Close Window** ⇒ **File** ⇒ **Erase** ⇒ **Not Displayed** ⇒ **MMB**

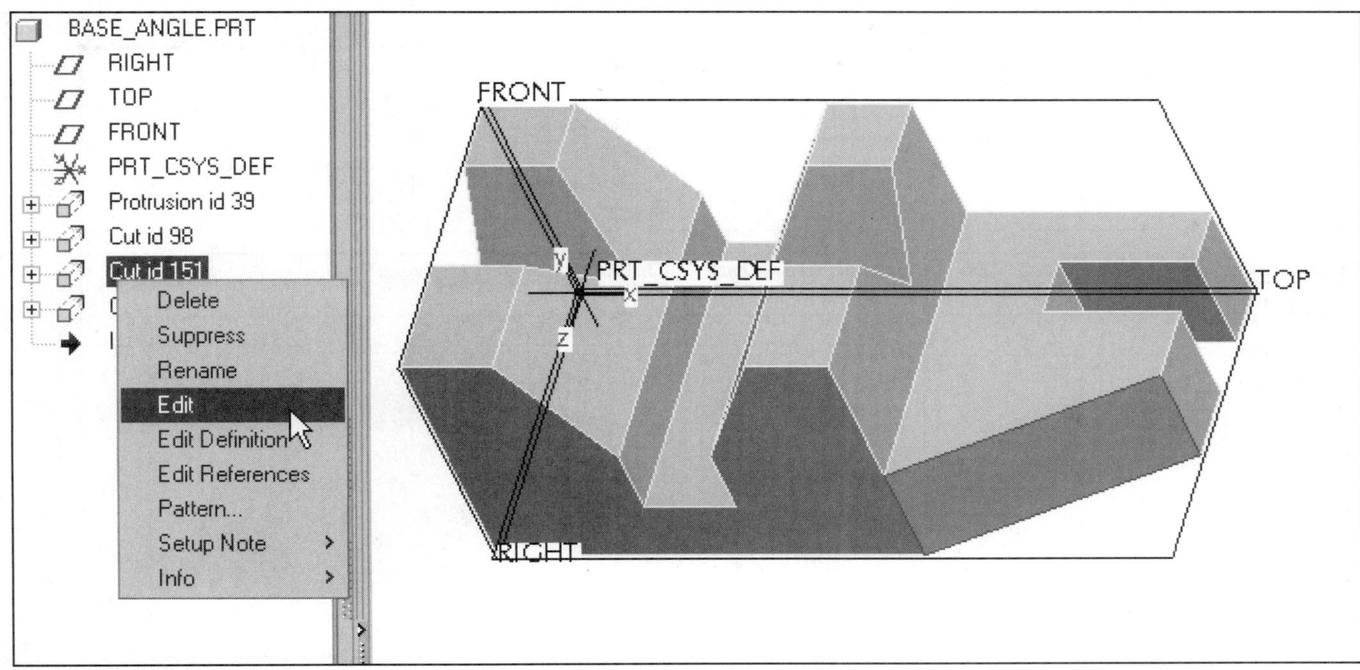

Figure 3.63 Editing

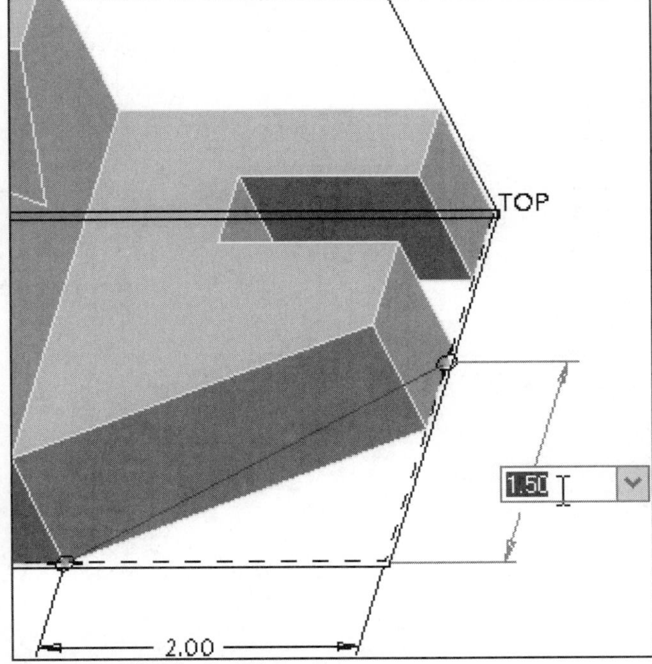

Figure 3.64 Edit the Dimension

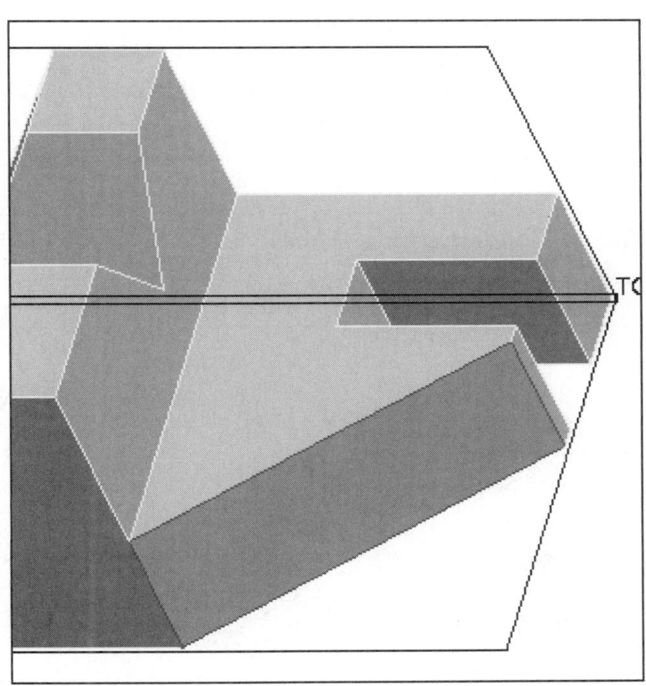
Figure 3.65 Modified Part

Engineering Change Orders (ECOs)

Few projects go through the design, engineering, and manufacturing phases without changes. Modifications can be simple dimensional changes in the part's size or more extreme changes in the part's configuration. For simple dimensional changes, the Edit tool is used; when configuration changes are needed, the Edit Definition tool is called upon. Engineering or Manufacturing departments normally release an **ECO** (Fig. 3.66).

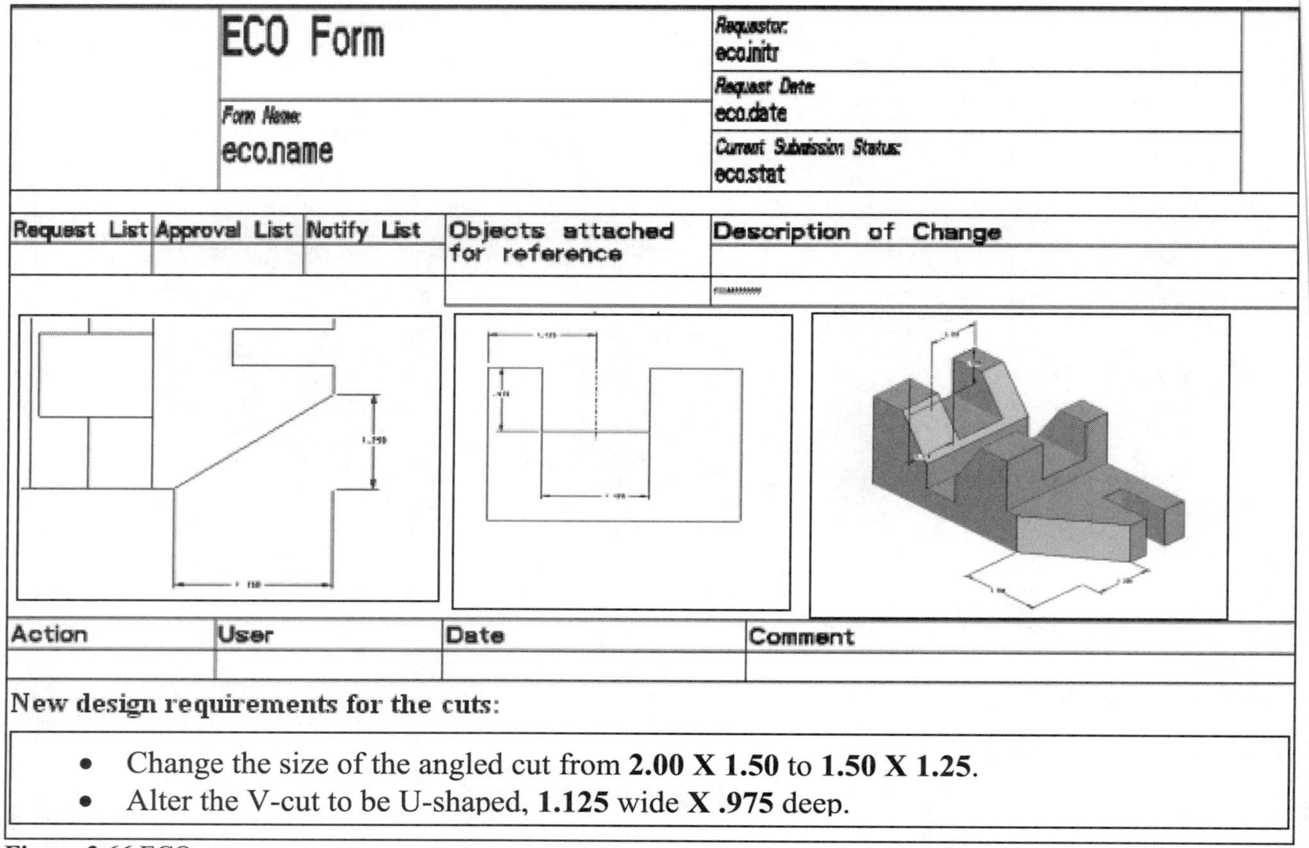

ECO Form

Form Name:
eco.name

Requestor:
eco.initr

Request Date:
eco.date

Current Submission Status:
eco.stat

Request List	Approval List	Notify List	Objects attached for reference	Description of Change

Action	User	Date	Comment

New design requirements for the cuts:

- Change the size of the angled cut from **2.00 X 1.50** to **1.50 X 1.25**.
- Alter the V-cut to be U-shaped, **1.125** wide **X .975** deep.

Figure 3.66 ECO

File ⇒ Open ⇒ (pick **BASE_ANGLE_REV1) ⇒ Preview** >>> spin the object with the **MMB ⇒ Open ⇒ Tools ⇒ Environment ⇒** Default Orient Isometric **⇒ OK** (Fig. 3.67)

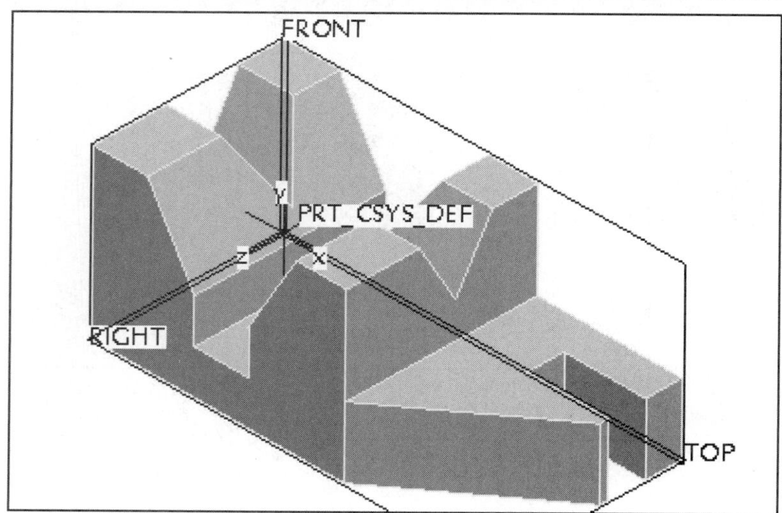

Figure 3.67 Isometric View

Double-click on the angle cut (Fig. 3.68) ⇒ Double-click on a dimension (Fig. 3.69) and edit its value. Change **2.00** to **1.50** in one direction. Repeat and change the **1.50** to **1.25** [Fig. 3.70(a)] in the other direction ⇒ **Regenerates Model** [Fig. 3.70(b)] ⇒ ⇒ **MMB**

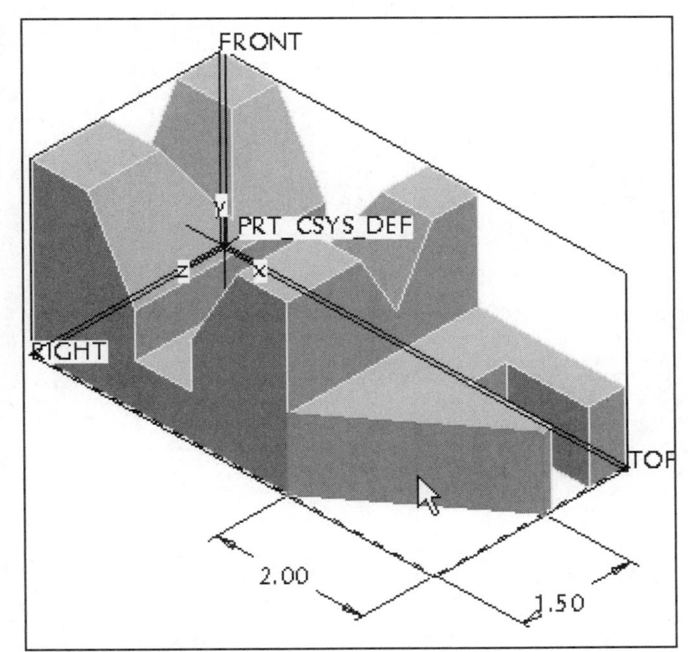

Figure 3.68 Double-click Directly on the Feature

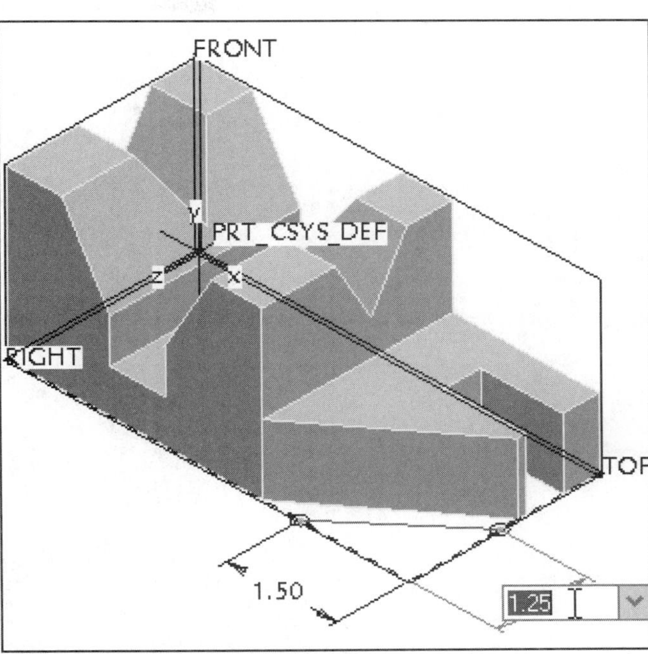

Figure 3.69 Double-click each Dimension and Change Value

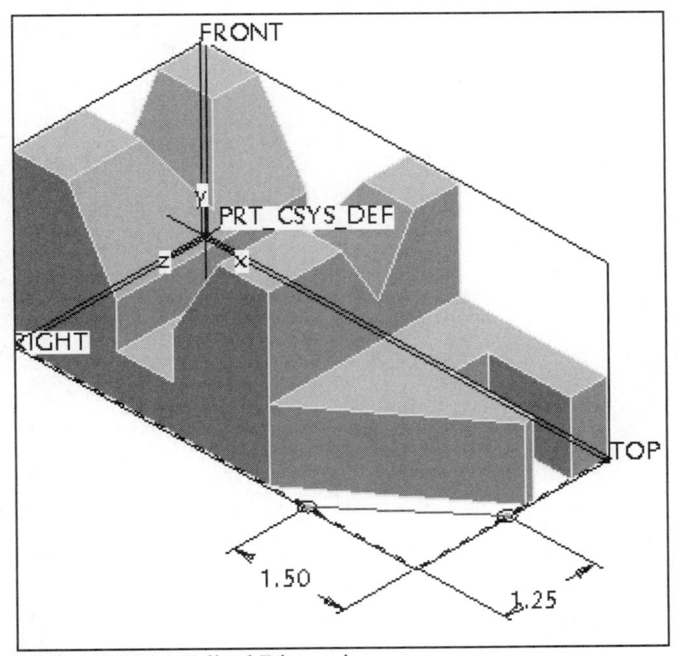

Figure 3.70(a) Edited Dimensions

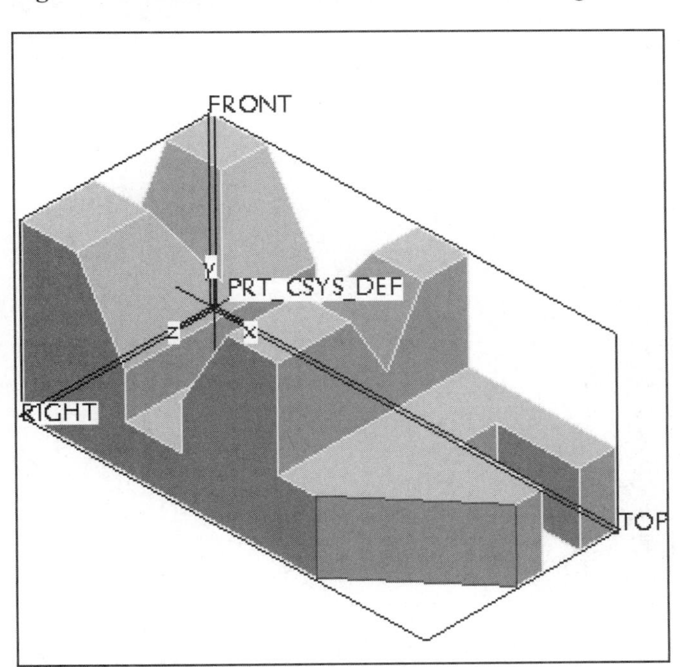

Figure 3.70(b) Regenerated Part

The simple dimensional changes have been made. The other requirement from the ECO was to change the shape and the size of the **V**-cut to be a **U**-shaped slot **1.125** wide by **.975** deep. This requires that you use **Edit Definition** to redefine the feature, enter the sketch mode, and rework the sketch as needed.

Directly on the model, click on the **V**-cut ⇒ **RMB** [Fig. 3.71(a)] ⇒ **Edit Definition** [Fig. 3.71(b)] ⇒ ⇒ **Sketch** *(if the Select dialog box appears, click Cancel to close)*

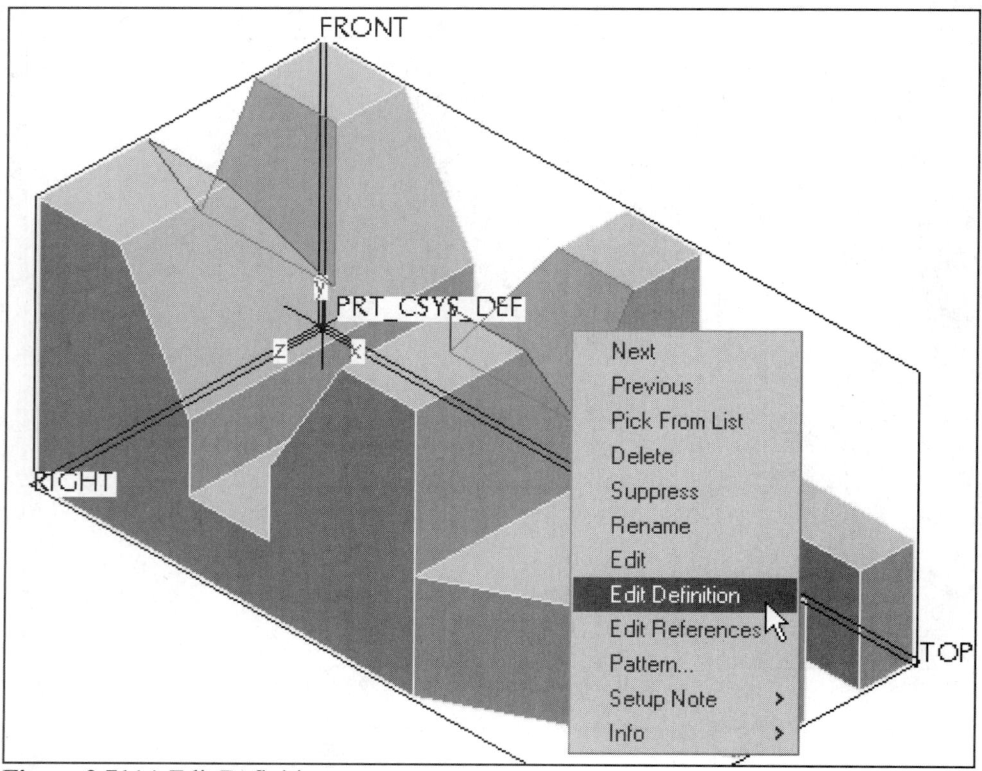

Figure 3.71(a) Edit Definition

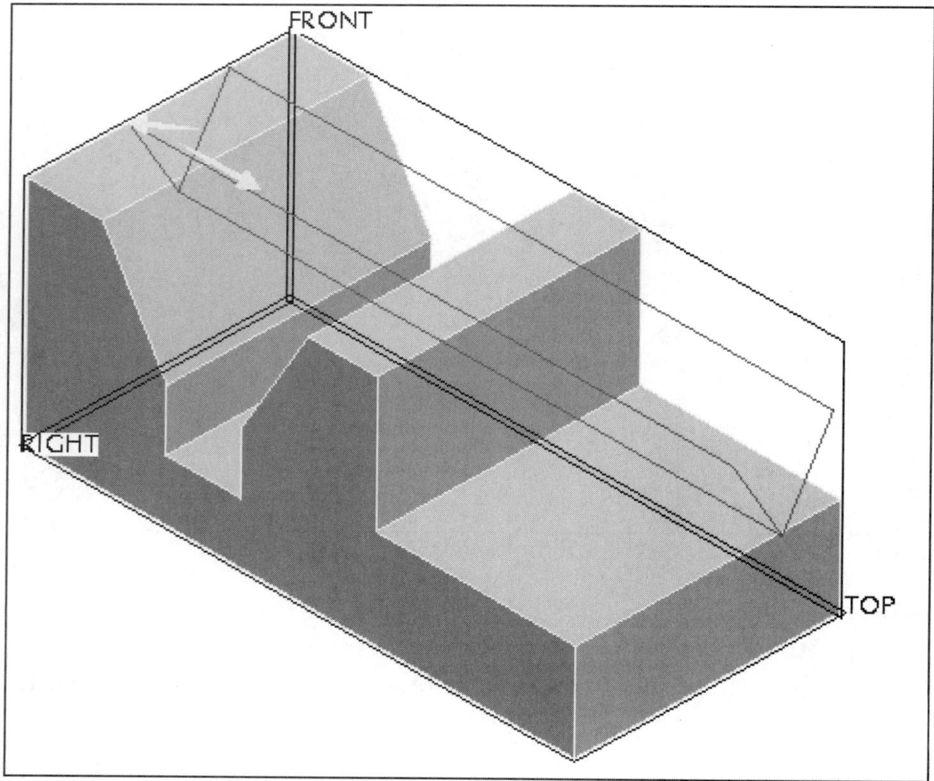

Figure 3.71(b) Redefining the V-Cut

Use the same centerline for the cut section (Fig. 3.72). Delete the two lines forming the **V** by holding down the **Ctrl** key and select both lines (they will turn *red*) (Fig. 3.73) ⇒ **RMB** to display the pop-up menu options ⇒ **Delete** (Fig. 3.74)

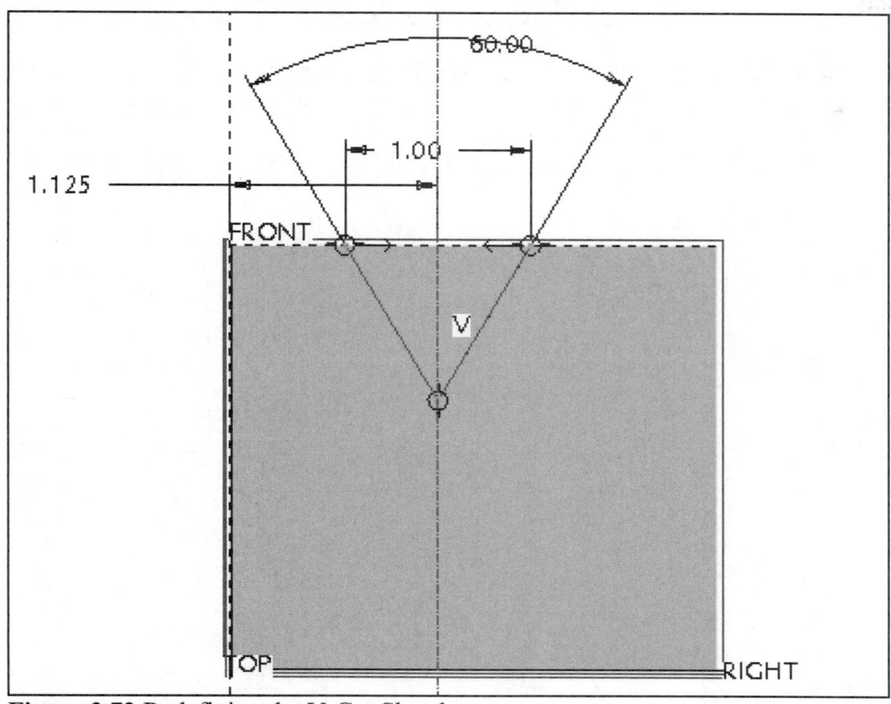

Figure 3.72 Redefining the V-Cut Sketch

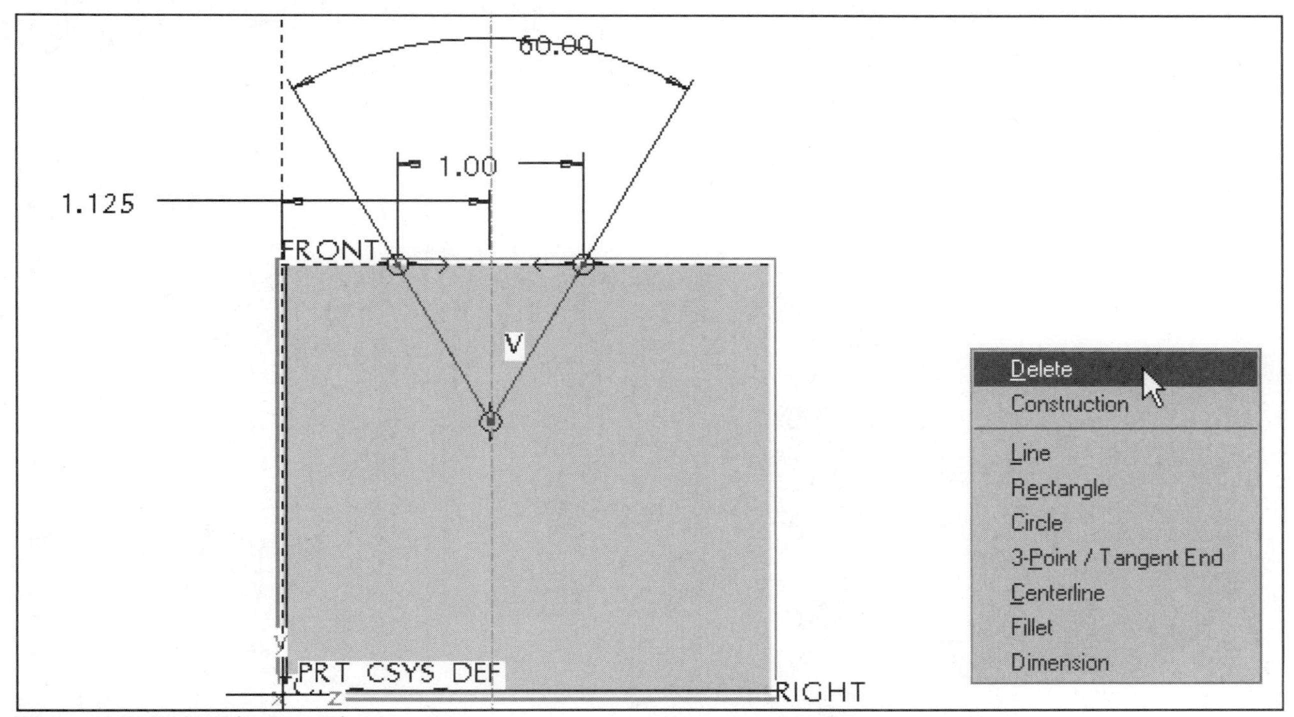

Figure 3.73 Delete the Two Lines

Click: **Create 2 point lines** ⇒ sketch three new lines for the slot (Fig. 3.75) ⇒ **Impose sketcher constraints on the section** ⇒ **Make two points or vertices symmetric about a centerline** (Fig. 3.76) ⇒ pick the centerline and then pick the two vertices to be symmetric ⇒ ⇒ window-in the sketch to capture all dimensions ⇒ **Modify the values of dimensions, geometry of splines, or text entities** ⇒ modify dimensions to the design values using the **ECO** sizes of **1.125** wide by **.975** deep (Fig. 3.77) ⇒ ⇒ ⇒ **Standard Orientation** (Fig. 3.78) ⇒ ⇒ **OK** (Fig. 3.79) ⇒ **Options** tab ⇒ **To Selected** ⇒ pick the surface [Fig. 3.80(a-b)] ⇒ **MMB**

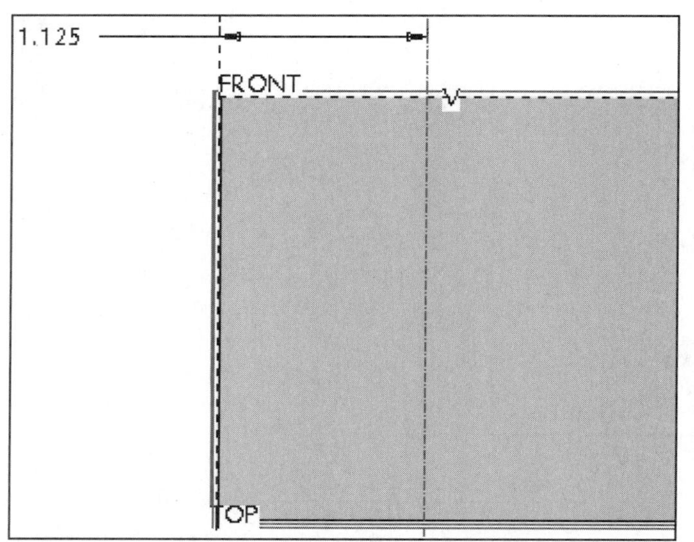

Figure 3.74 Sketch has only a Centerline and Dimension

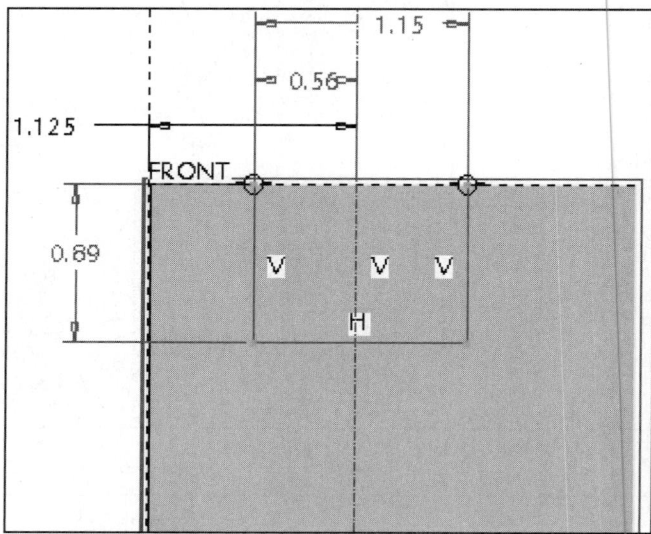

Figure 3.75 Sketched Lines

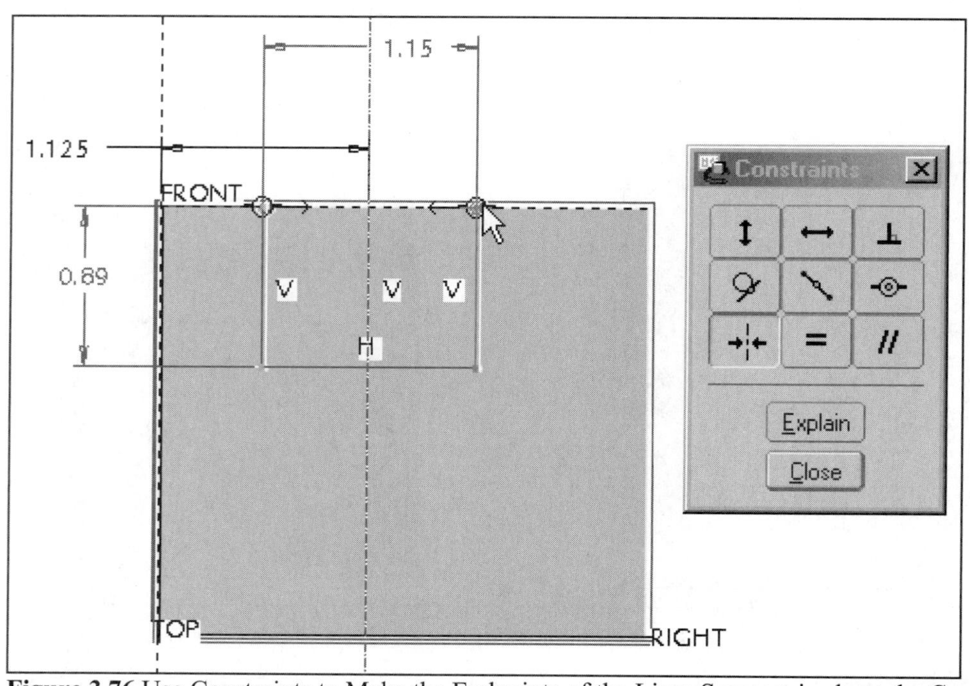

Figure 3.76 Use Constraints to Make the Endpoints of the Lines Symmetric about the Centerline

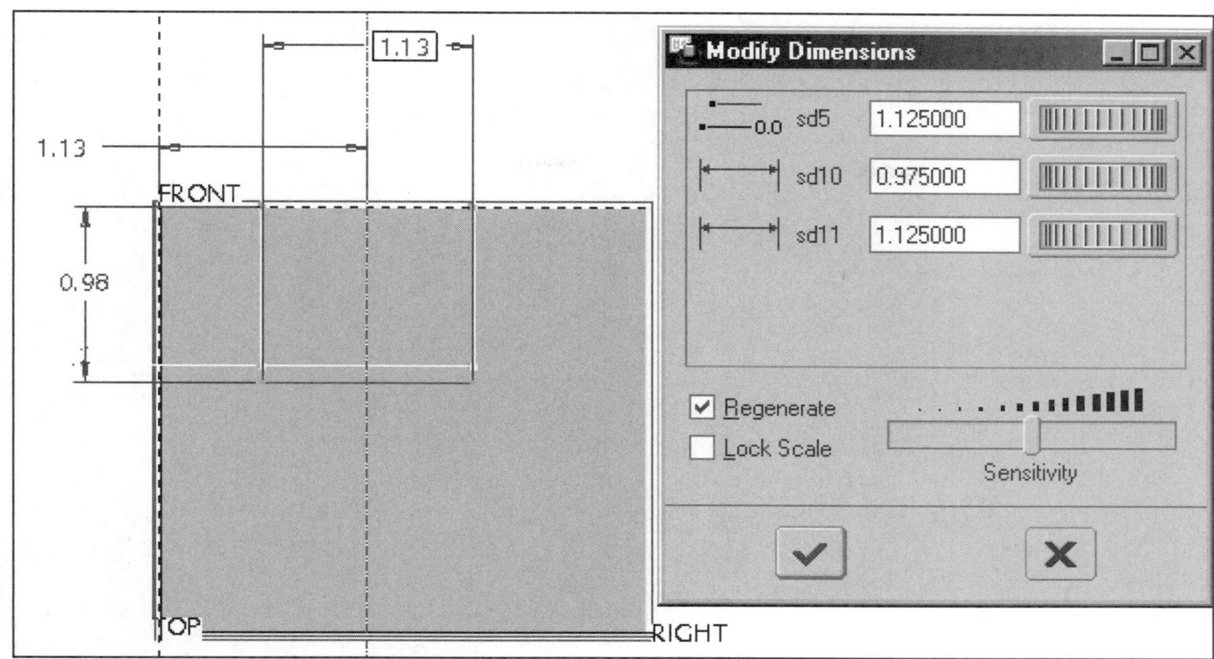

Figure 3.77 Modify the Dimensions

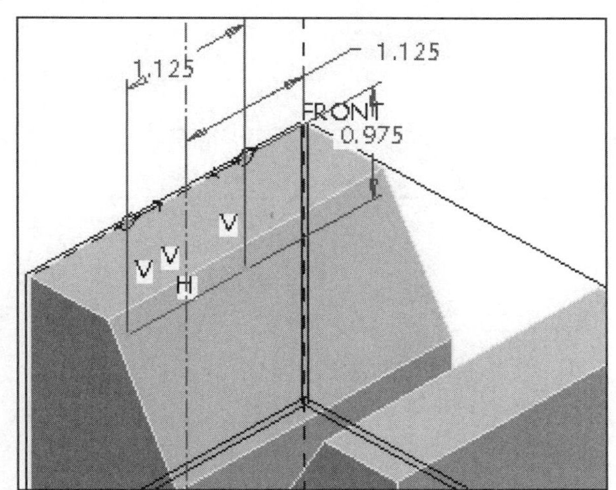

Figure 3.78 Standard Orientation

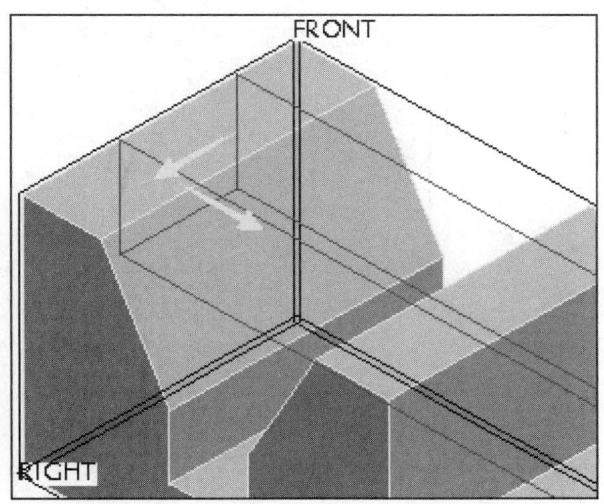

Figure 3.79 Material Removal and Feature Directions

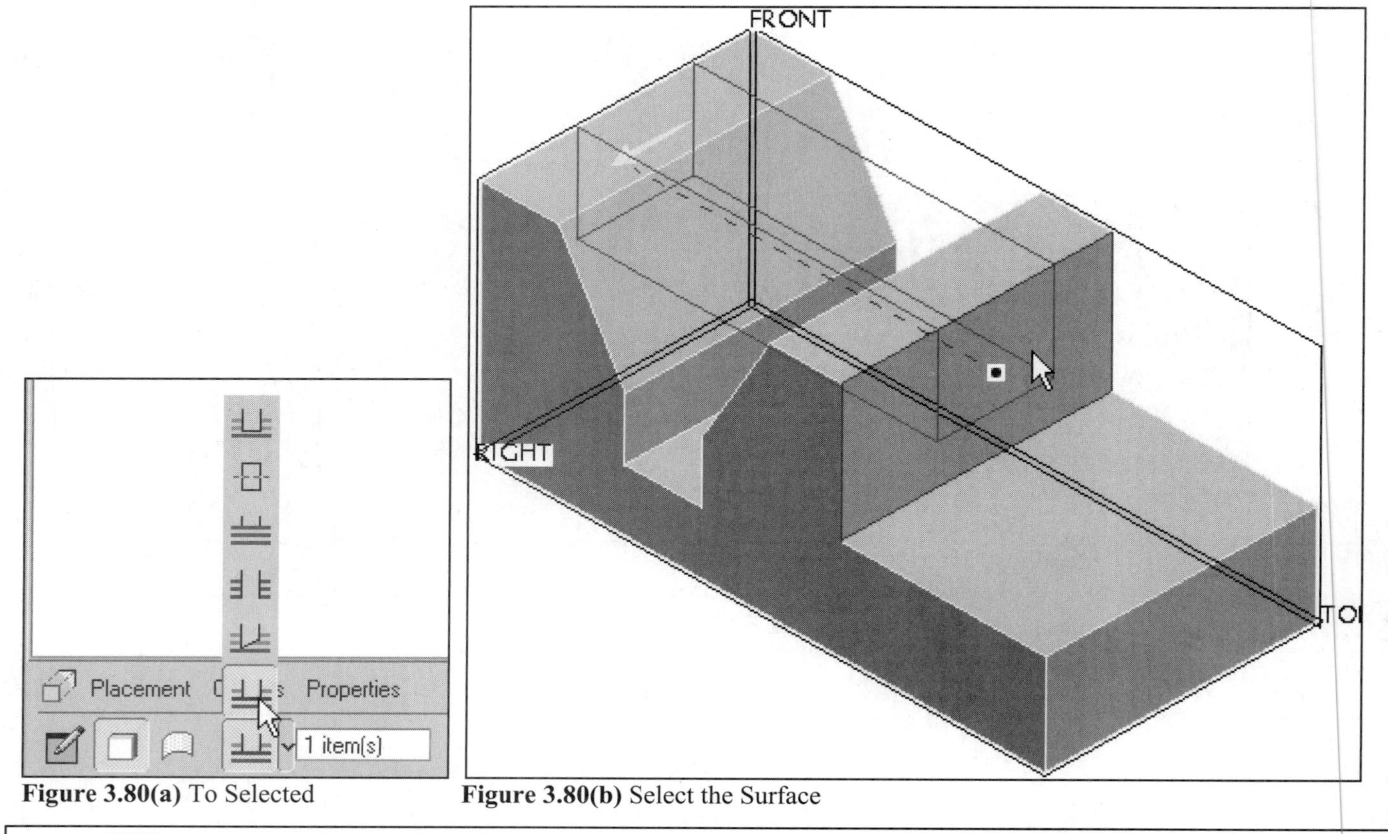

Figure 3.80(a) To Selected

Figure 3.80(b) Select the Surface

Click: **Save the active object ⇒ MMB** [Fig. 3.81(a-b)] **⇒ Window ⇒ Close**

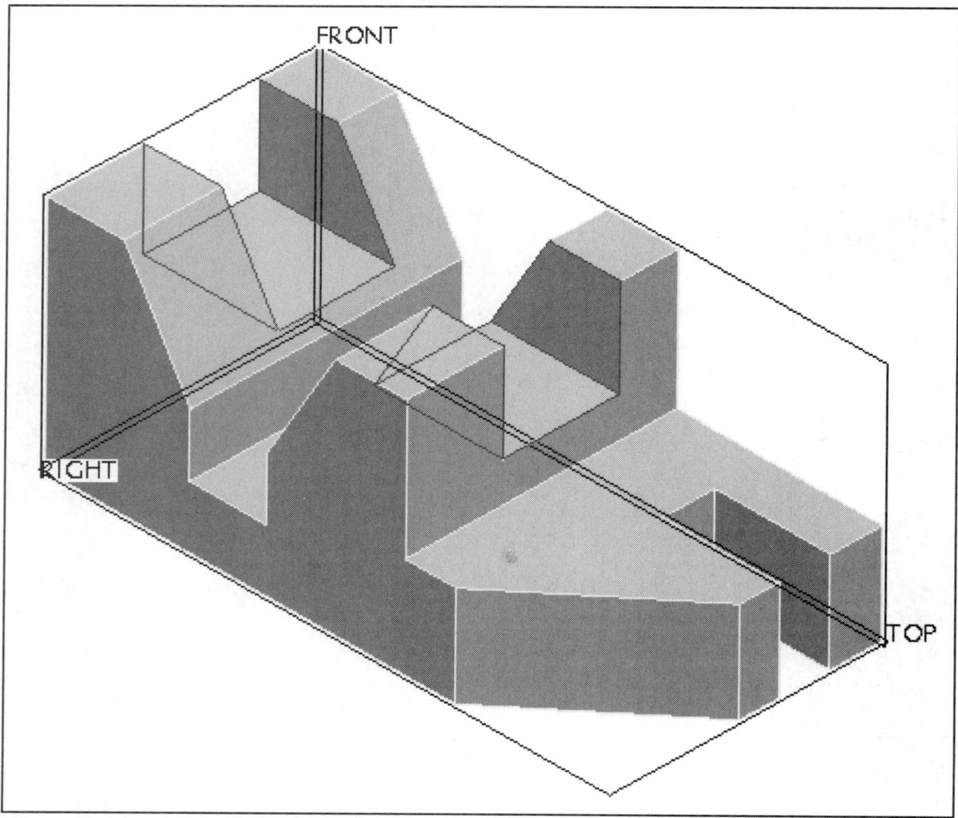

Figure 3.81(a) Completed Revision

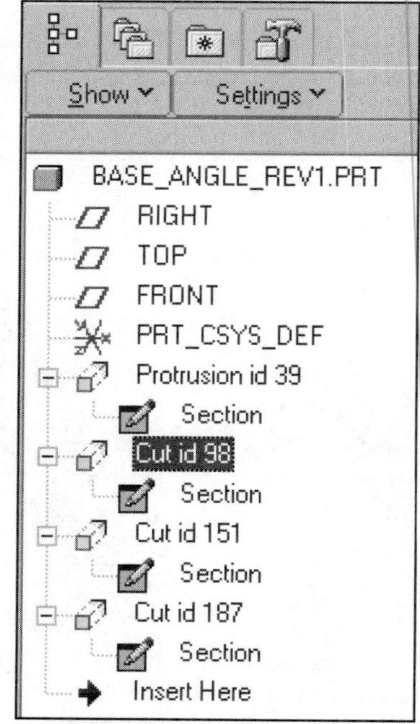

Figure 3.81(b) Model Tree

Lesson 3 Project

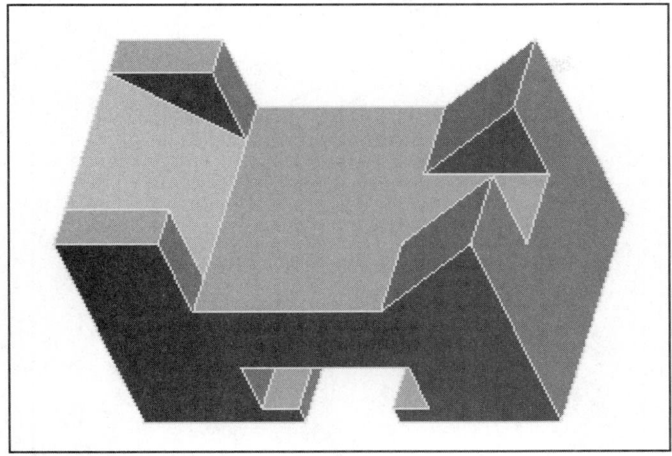

Figure 3.82 T-Block

Figure 3.83 T-Block ECO

T-Block

This lesson project is a simple block (Figs. 3.82 through 3.87) that is created with commands similar to those for the BASE_ANGLE. Sketch the protrusion on datum FRONT.

After the T-Block is modeled, you will be prompted to edit and redefine a number of features from an ECO (T-BLOCK_ECO) (Figs. 3.88 through 3.92).

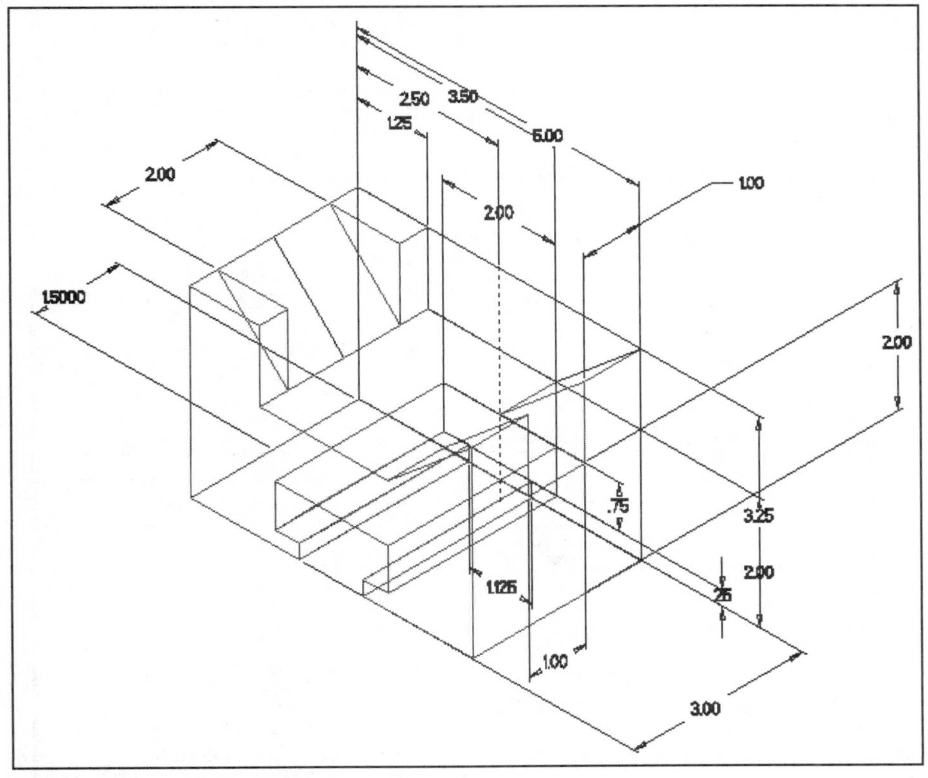

Figure 3.84 T-Block Dimensions

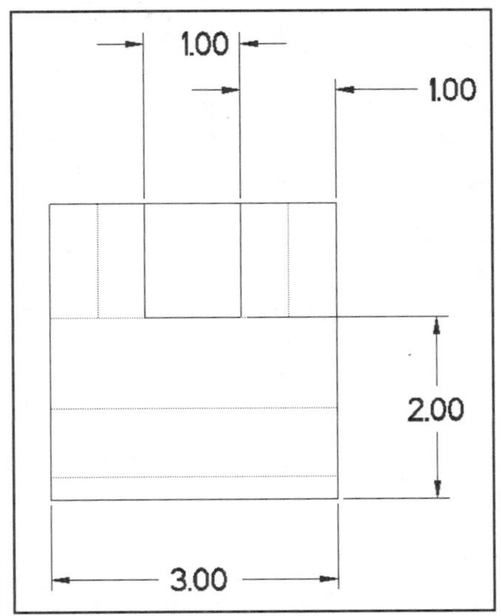

Figure 3.85 Right Side View

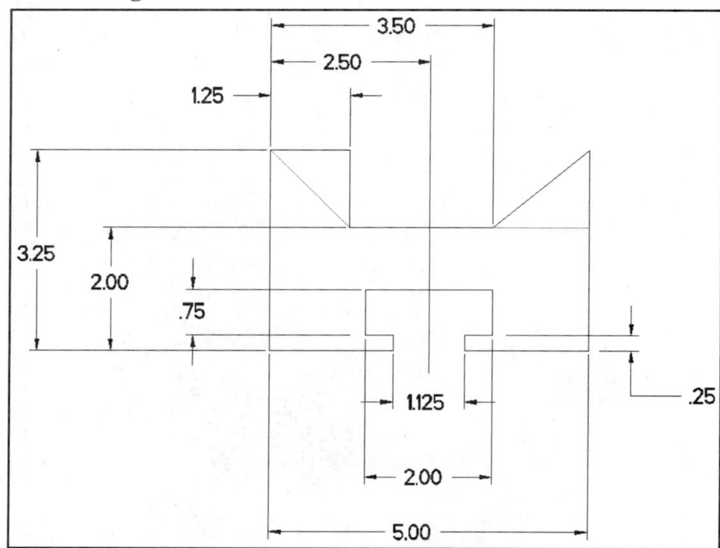

Figure 3.86 Front View

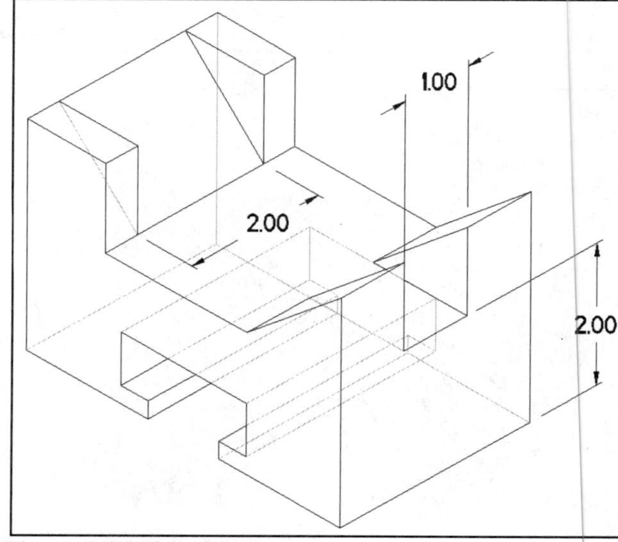

Figure 3.87 Pictorial View

Use **Edit Definition** and **Edit** to complete the ECO after you save it to another name. Figures 3.88 though 3.92 provide the ECO and feature redefinition requirements. The T-shaped slot is symmetrical and is located at the center of the part. When completing the ECO, be careful not to assume that this condition still applies.

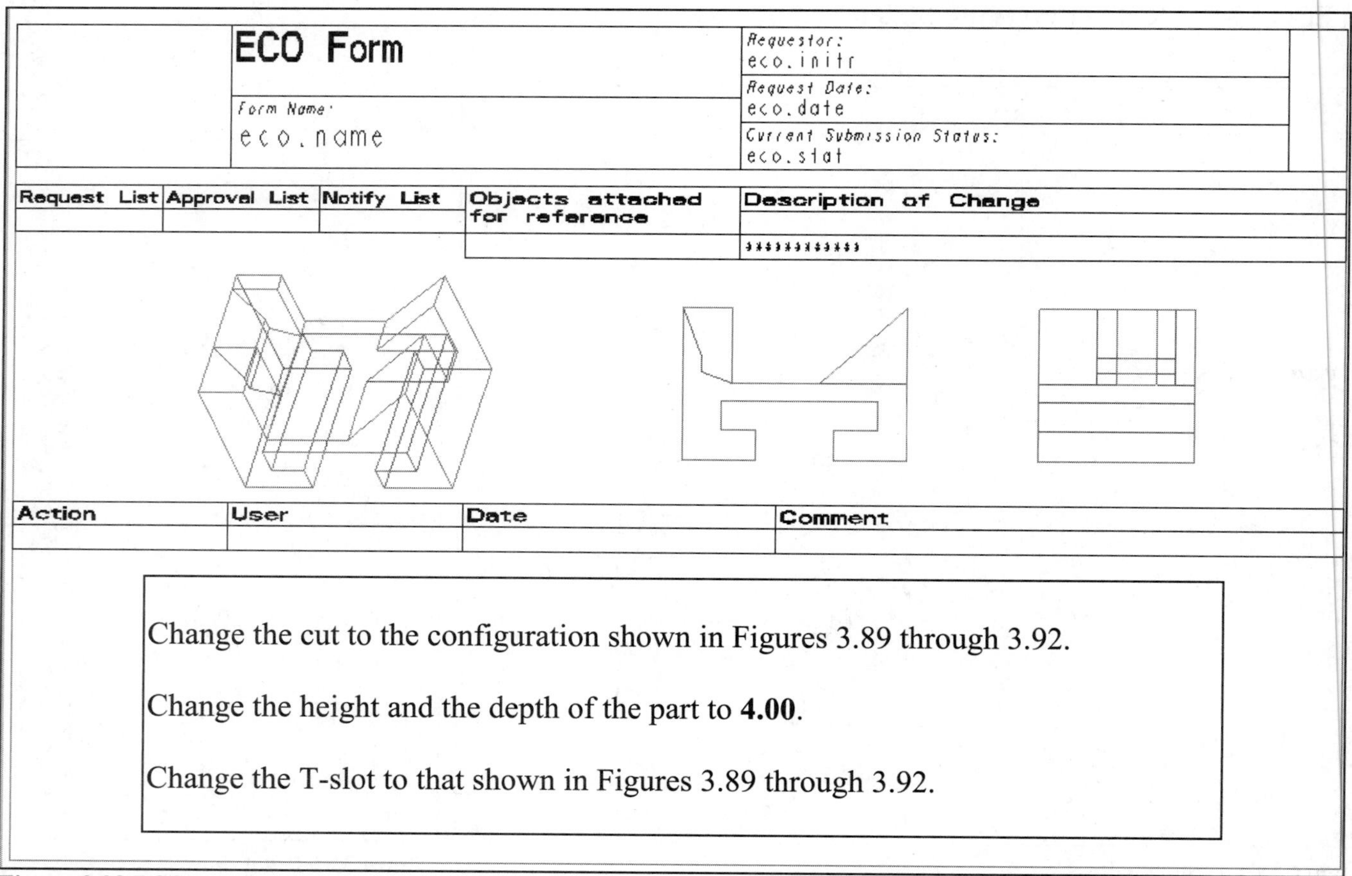

ECO Form

Form Name:
eco.name

Requestor:
eco.initr

Request Date:
eco.date

Current Submission Status:
eco.stat

Request List	Approval List	Notify List	Objects attached for reference	Description of Change
				************

Action	User	Date	Comment

Change the cut to the configuration shown in Figures 3.89 through 3.92.

Change the height and the depth of the part to **4.00**.

Change the T-slot to that shown in Figures 3.89 through 3.92.

Figure 3.88 ECO

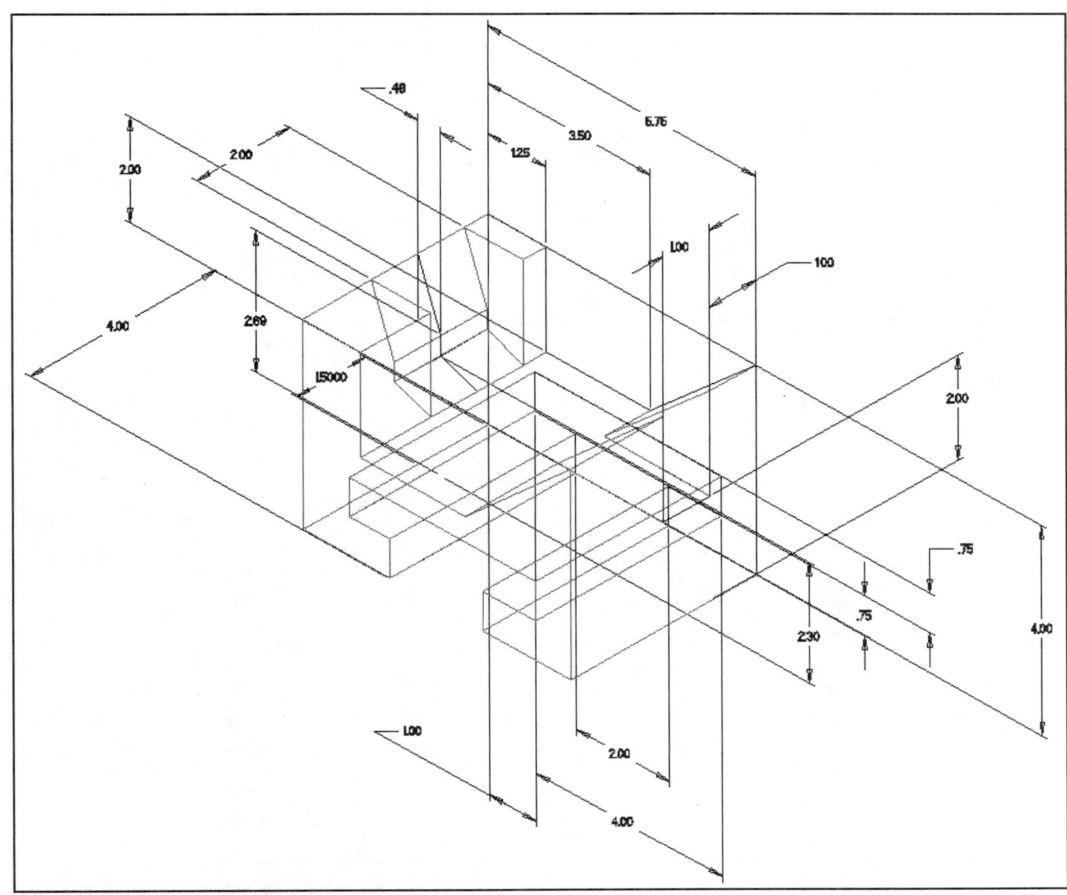

Figure 3.89 Modified Dimensions for T-Block

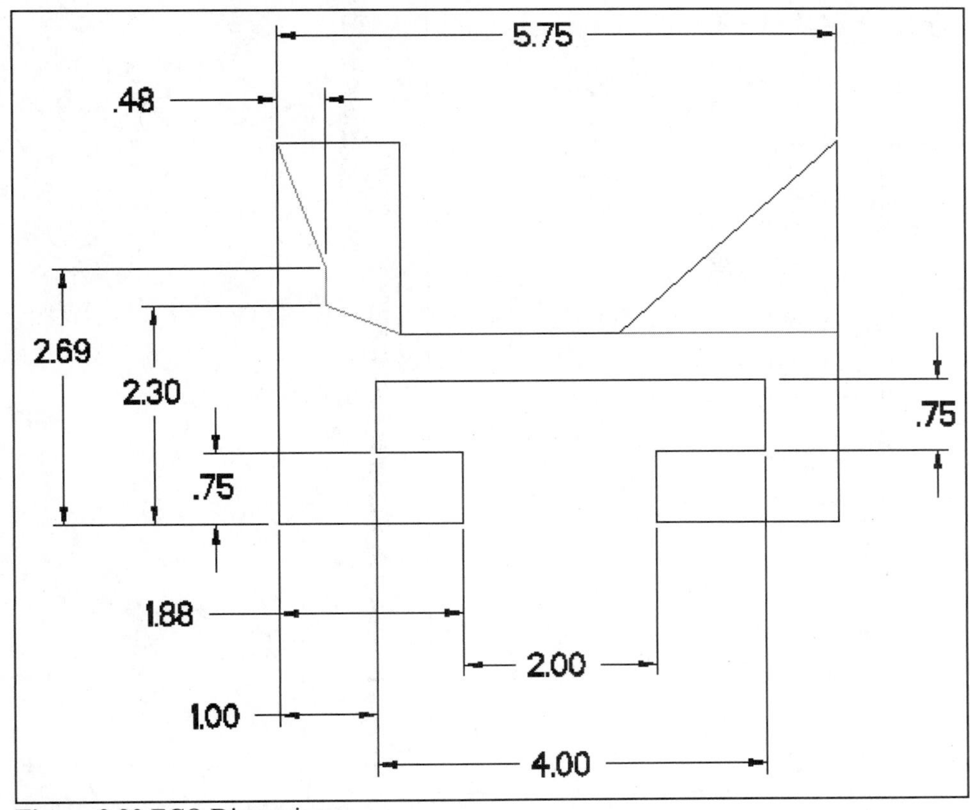

Figure 3.90 ECO Dimensions

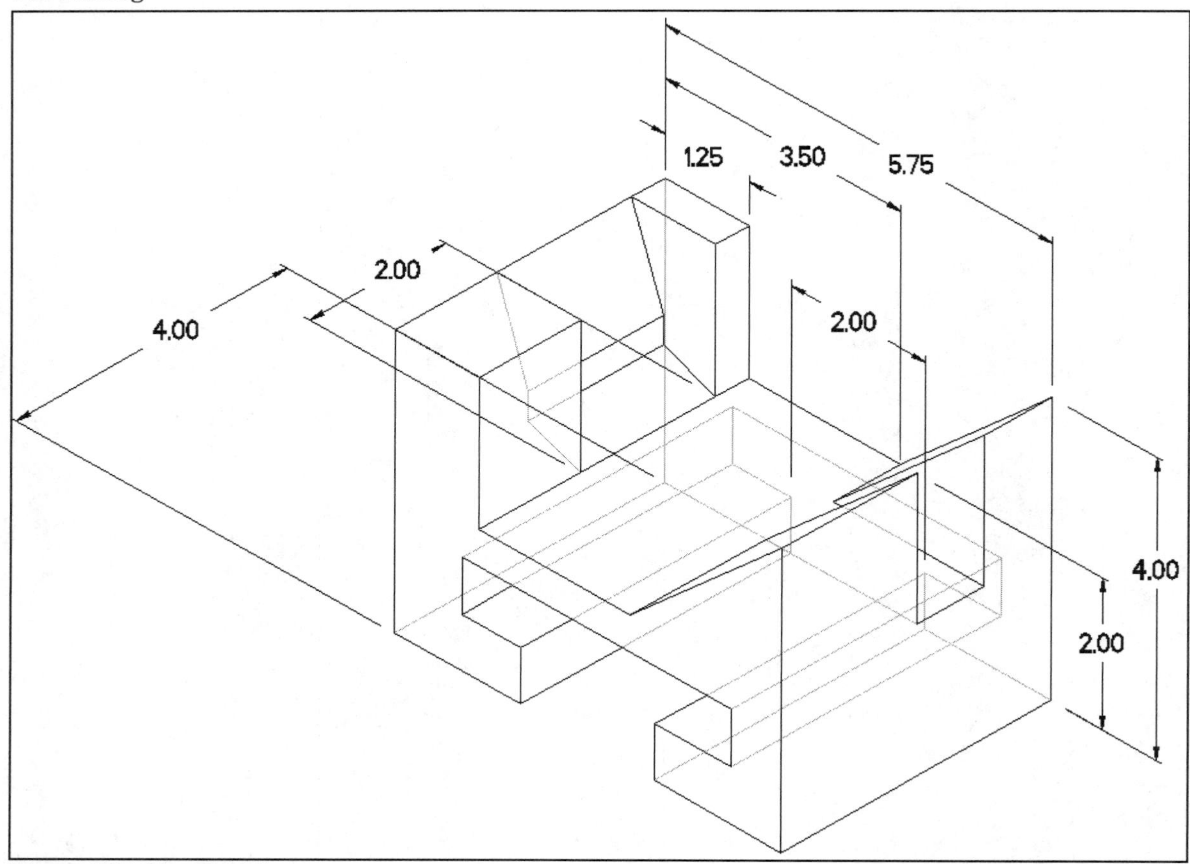

Figure 3.91 Pictorial View

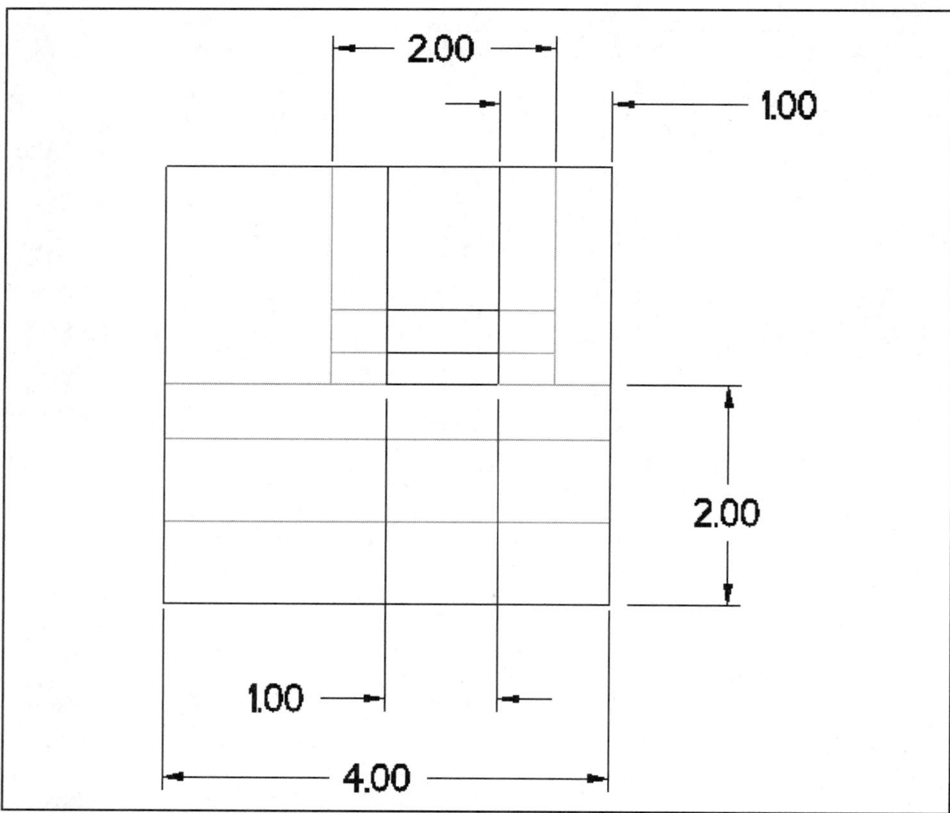

Figure 3.92 Cut Dimensions

Lesson 4 Holes and Rounds

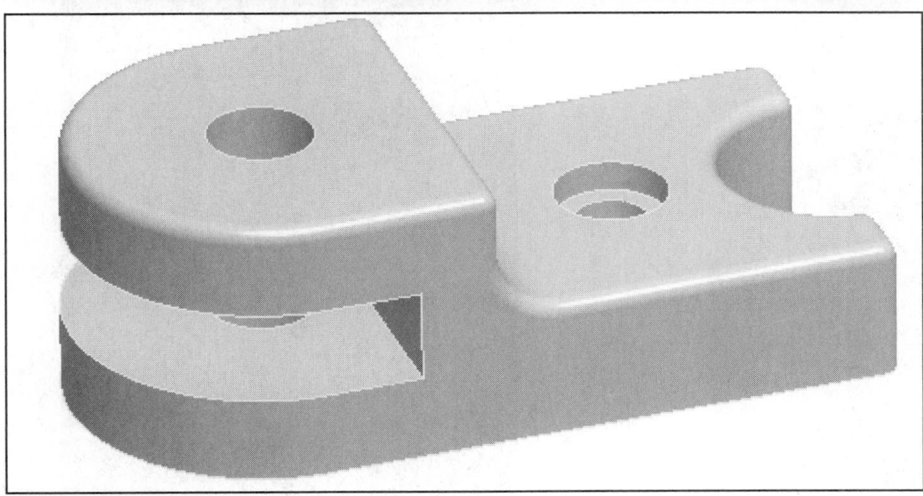

Figure 4.1 Breaker

OBJECTIVES

- Sketch **arcs** in sections
- Create a **straight hole** through a part
- Complete a **Sketched hole**
- Understand the **Hole Tool**
- Use **Info** to extract information about the features and the model
- **Set** and **Save Views**
- Create simple **Rounds** along model edges using **direct modeling**
- Understand the **Round Tool**

HOLES AND ROUNDS

A variety of geometric shapes and constructions are accomplished automatically with Pro/E, including *holes* and *rounds* (Fig. 4.1). These features are called *pick-and-place* features, because they are created automatically from your input and then placed according to prompts by Pro/E. A hole can also be created using the **Extrude Tool** and removing material, but it must be sketched. In general, pick-and-place features are not sketched (except for the **Sketched** option when you are creating a complex hole shape, such as a non-standard countersink or counterbore). The **Round Tool** creates a fillet, or a round on an edge, that is a smooth transition with a circular profile between two adjacent surfaces.

Holes

The **Hole Tool** creates a variety of holes. Types of hole geometry include:

- **Straight hole** An extruded slot with a circular section
- **Sketched hole** A revolved feature defined by a sketched section
- **Standard hole** A revolved feature created with UNC, UNF, or ISO standards

All *straight holes* are created with a constant diameter. A *sketched hole* is created by sketching a section for revolution and then placing the hole on the part [Fig. 4.2(a)]. Sketched holes are always blind and one-sided. Sketched holes must have a vertical centerline (*axis of revolution*), with at least one of the entities sketched normal to the axis centerline [Fig. 4.2(b)]. Pro/E aligns the normal entity with the placement plane. The remainder of the sketched feature is cut from the part, as with a revolved cut.

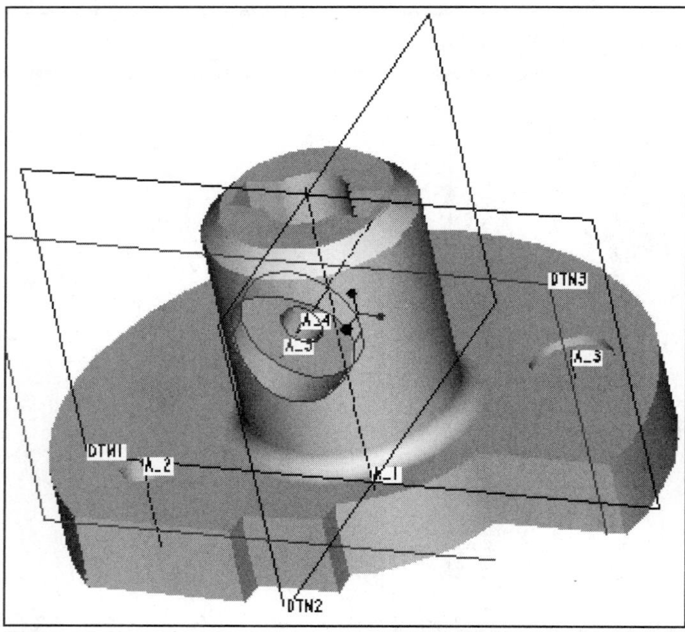

Figure 4.2(a) Sketched Holes (CADTRAIN, COAch for Pro/E)

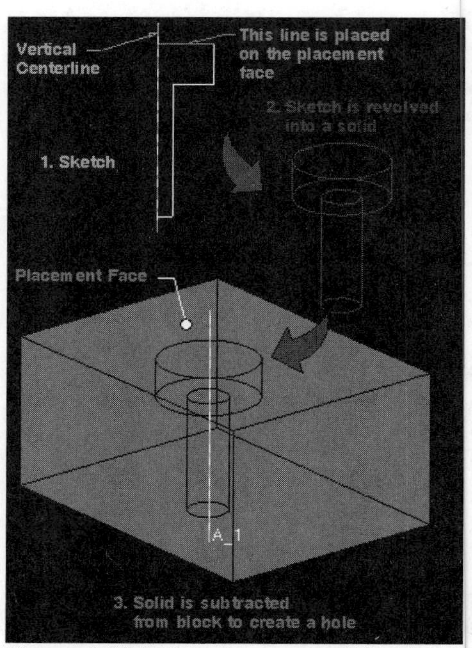

Figure 4.2(b) Hole Placement (CADTRAIN)

Rounds

Rounds (Fig. 4.3) are created at selected edges of the part. Tangent arcs are introduced as rounds between two adjacent surfaces of the solid model. There are cases in which rounds should be added early, but in general, wait until later in the design process to add the rounds. Introducing rounds into a complex design early in the project can cause a series of failures later.

Two categories of rounds are available: simple and advanced. Much of the time, you will create *simple* rounds. These rounds smooth the hard edges between two adjacent surfaces.

Reference Type	Original Geometry	Rounded Geometry
(a) Edge Chain/ One By One	Select these edges.	Resulting round
(b) Edge Chain/ Tangnt Chain	An edge from a tangent chain	Resulting round
(c) Surf-Surf	Select these two surfaces.	Resulting round

Figure 4.3 Rounds

Lesson 4 STEPS

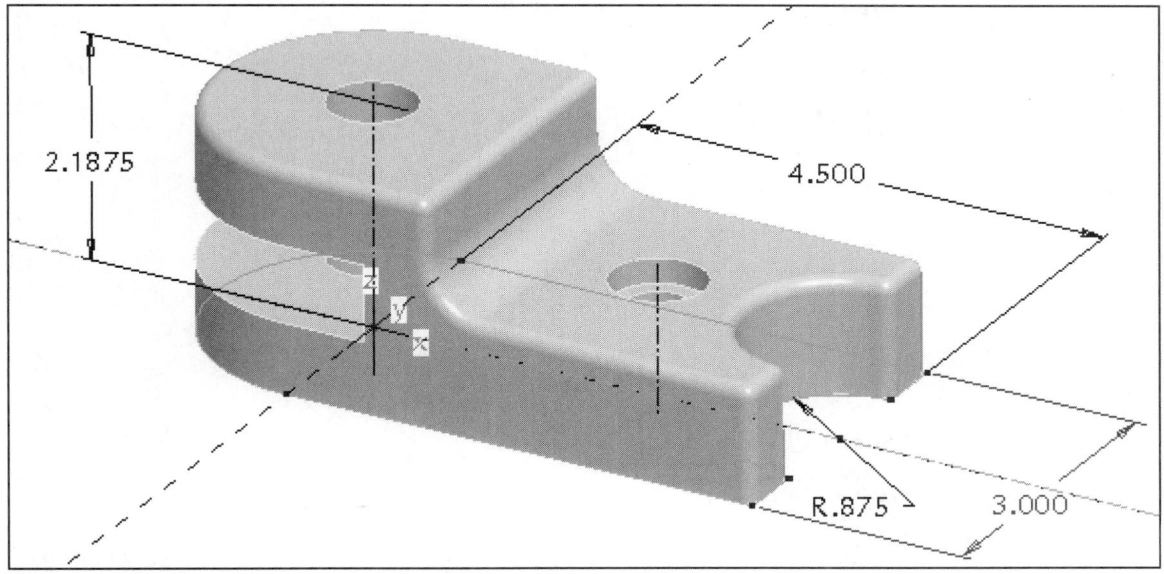

Figure 4.4 Breaker

Breaker

The Breaker (Fig. 4.4) introduces two new features, *holes* and *rounds*. The first protrusion is created using the Extrude Tool. In the Sketcher, the **Arc** command will be used to create the rounded end and the half-circle cut.

Start a new part, click: **File** ⇒ **Set Working Directory** ⇒ select the working directory ⇒ **OK** ⇒ **Create a new object** ⇒ ●**Part** ⇒ Name **Breaker** ⇒ ☑ Use default template ⇒ **OK** ⇒ **Edit** ⇒ **Setup** ⇒ **Units** ⇒ Units Manager **Inch lbm Second** ⇒ **Close** ⇒ **Material** ⇒ **Define** ⇒ type **aluminum** ⇒ **MMB** ⇒ **File** from material table ⇒ **Save** ⇒ **File** ⇒ **Exit** ⇒ **Assign** ⇒ pick **ALUMINUM** ⇒ **Accept** ⇒ **MMB** ⇒ **Tools** ⇒ **Options** ⇒ Showing: **Current Session** ⇒ ☐ Show only options loaded from file ⇒ Option: ***default_dec_places*** ⇒ Value: **3** ⇒ **Add/Change** ⇒ Option: ***sketcher_dec_places*** ⇒ Value: **3** ⇒ **Add/Change** ⇒ **Apply** (Fig. 4.5) ⇒ **Close**

	Value	Status	Description
default_dec_places	3	●	Sets the default number of decimal places (0-1
sketcher_dec_places	3	●	Sets the default number of decimal places disp

Figure 4.5 Options

Click: **Info** from menu bar ⇒ **Model** model information displays in the **Browser** window (MATERIAL FILENAME: ALUMINUM should show (Fig. 4.6)] ⇒ click on the quick sash to collapse the **Browser** ⇒ **Tools** ⇒ **Environment** ⇒ ☑ Snap to Grid, Display Style Hidden Line, Tangent Edges Dimmed ⇒ **Apply** ⇒ **OK** ⇒ **Save** ⇒ **MMB**

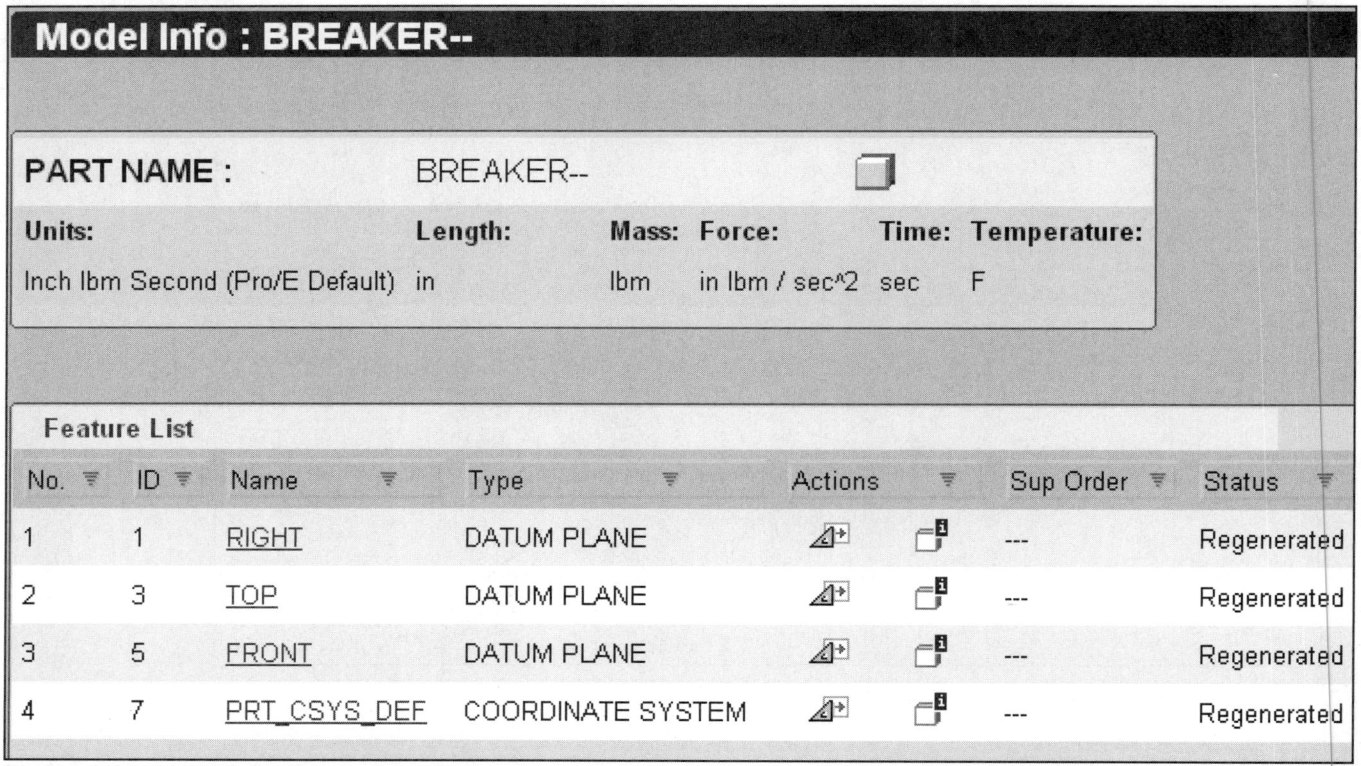

Model Info : BREAKER--

PART NAME :	BREAKER--				
Units:	**Length:**	**Mass:**	**Force:**	**Time:**	**Temperature:**
Inch lbm Second (Pro/E Default)	in	lbm	in lbm / sec^2	sec	F

Feature List

No.	ID	Name	Type	Actions		Sup Order	Status
1	1	RIGHT	DATUM PLANE			---	Regenerated
2	3	TOP	DATUM PLANE			---	Regenerated
3	5	FRONT	DATUM PLANE			---	Regenerated
4	7	PRT_CSYS_DEF	COORDINATE SYSTEM			---	Regenerated

Figure 4.6 Model Information shown in Browser

Create the first protrusion, click: **Extrude Tool** ⇒ ⇒ Sketch Plane--- Plane: select **FRONT** datum from the model as the sketch plane ⇒ Sketch Orientation--- Reference: select **TOP** datum from the model ⇒ Orientation: **Top** (makes TOP datum face up) ⇒ click **Sketch** button ⇒ References dialog box opens ⇒ **Close** ⇒ **Toggle the grid on**

Though it is not necessary for the sketching of this section, sometimes the grid spacing needs to be altered to a different size. Grid spacing defaults at **30** units (**30** inches or **30** millimeters depending on the units selected). You can change the grid size at any point in the sketching process.

Change the size of the grid spacing by choosing the following commands: **Sketch** from the menu bar ⇒ **Options** ⇒ Sketcher Preferences dialog box displays with **Display** tab active [Fig. 4.7(a)] ⇒ click **Constraints** tab to see options [Fig. 4.7(b)] ⇒ **Parameters** tab ⇒ Grid Spacing **Manual** ⇒ activate **Equal Spacing** ⇒ Values **X** ⇒ type **15** ⇒ **Enter** [Fig. 4.7(c)] ⇒

Notice that you can also change the number of digits displayed with this dialog instead of setting them in the *config.pro* settings. After the sketch is regenerated, the **15.00**-inch grid zooms out of view beyond the **6.00**-inch long part model. Change the grid **X** and **Y** spacing to **.25**.

Use the values and dimensioning scheme provided in Figure 4.8. Only three dimensions are required for the first extrusion.

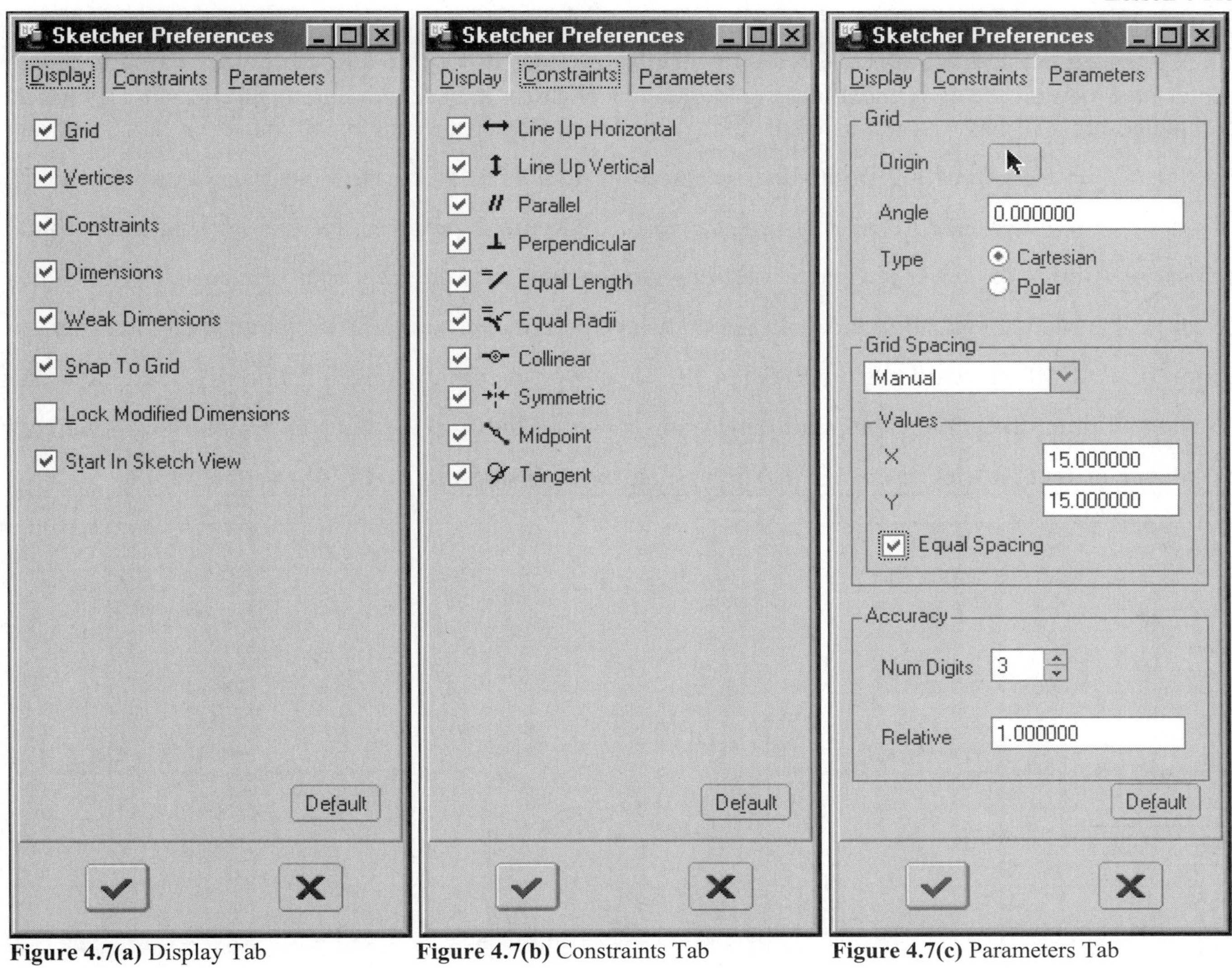

Figure 4.7(a) Display Tab

Figure 4.7(b) Constraints Tab

Figure 4.7(c) Parameters Tab

4.500

3.00

R.875

Figure 4.8 Top View

Click: **Create an arc by picking its center and endpoints** ⇒ click on the center and then on the first endpoint [Fig. 4.9(a)] ⇒ click on the second endpoint [Fig. 4.9(b)] ⇒ The completed arc will have its radius displayed (Fig. 4.10). Repeat the command and create a second arc (Fig. 4.11). ⇒ **Create 2 point lines** ⇒ Add the four lines to create a closed section (Fig. 4.12). *(Create two lines ⇒ MMB and then repeat).* ⇒ **Create 2 point centerlines** ⇒ add a horizontal centerline (Fig. 4.12) ⇒ **Create defining dimension** ⇒ Add the height dimension. The radius (weak) dimension for the first arc will automatically be removed. ⇒ ⇒ move the dimensions to appropriate ASME standard positions (Fig. 4.13) ⇒ modify the dimensions to the *design sizes*: click ⇒ window-in the sketch to capture all dimensions ⇒ **Modify the values of dimensions, geometry of splines, or text entities** ⇒ change the dimensions to the design values (Fig. 4.14) ⇒

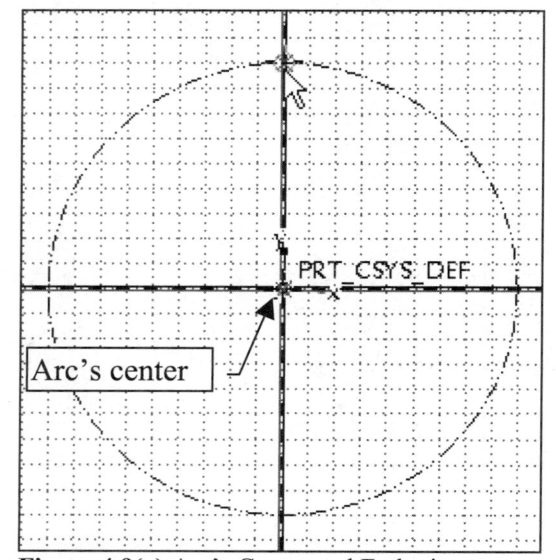

Figure 4.9(a) Arc's Center and Endpoint

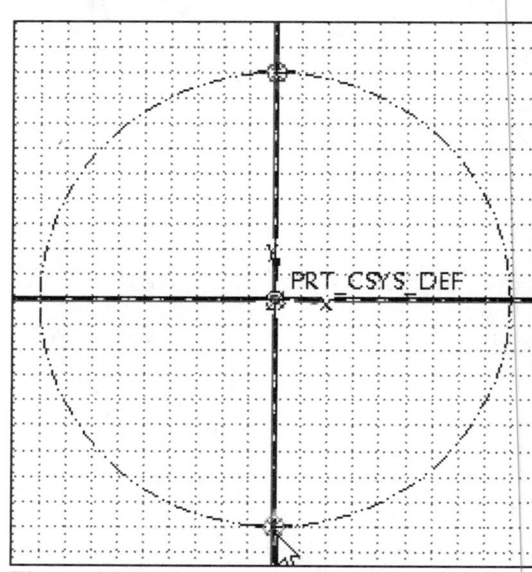

Figure 4.9(b) Arc's Second Endpoint

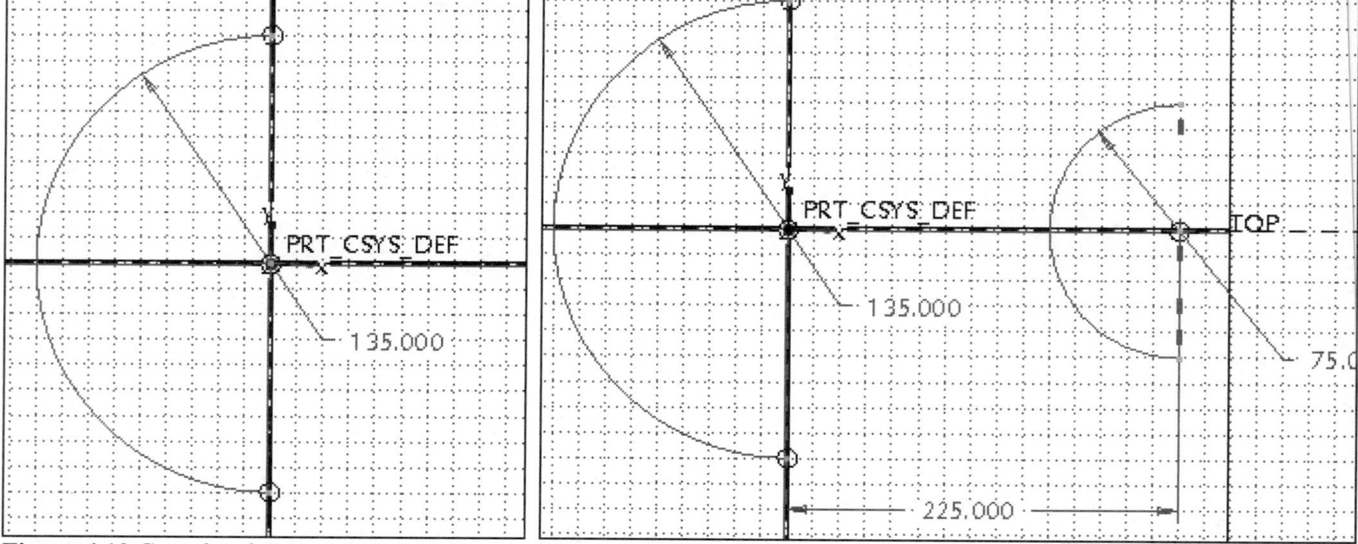

Figure 4.10 Completed Arc

Figure 4.11 Second Arc

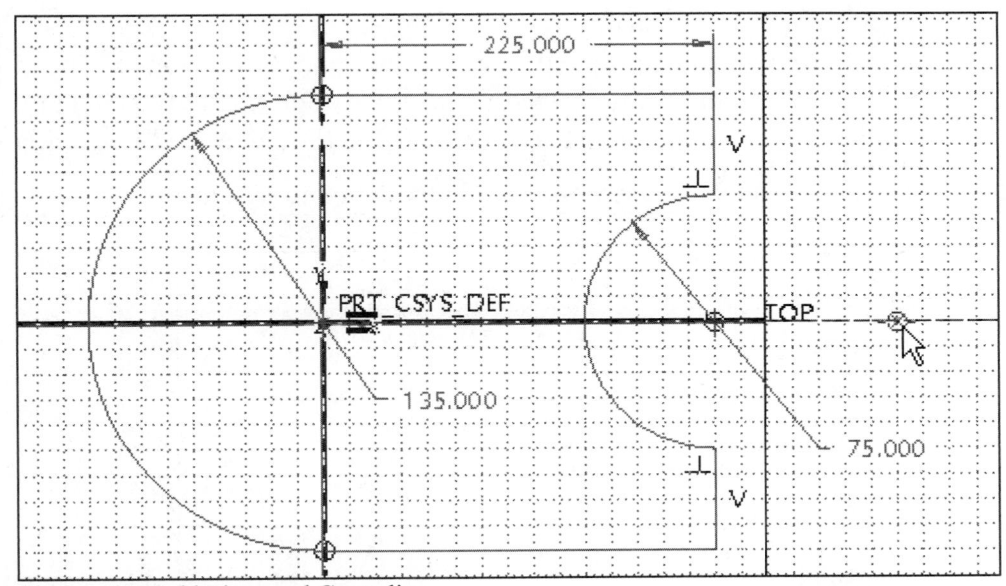

Figure 4.12 Add Lines and Centerline

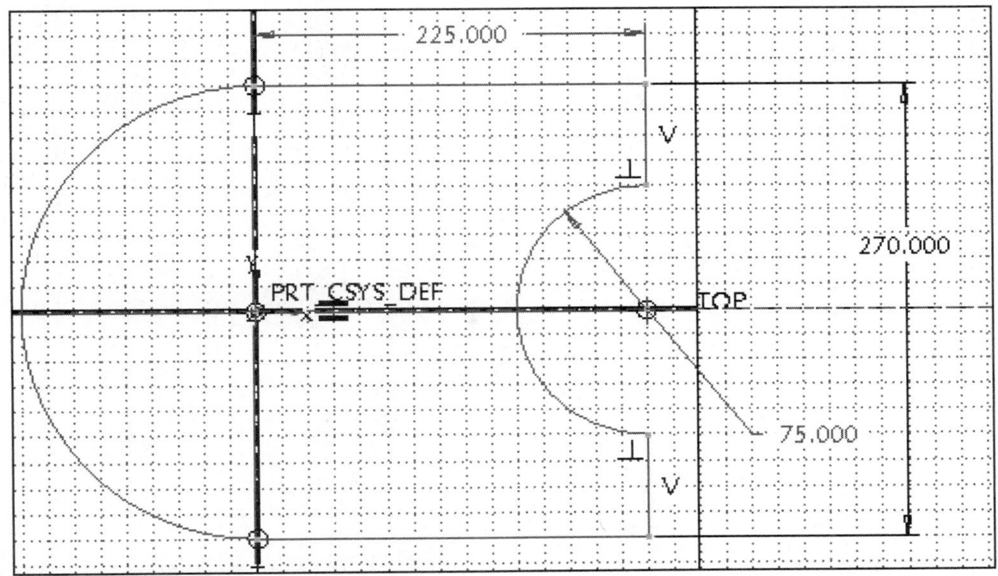

Figure 4.13 Add and Move Dimensions

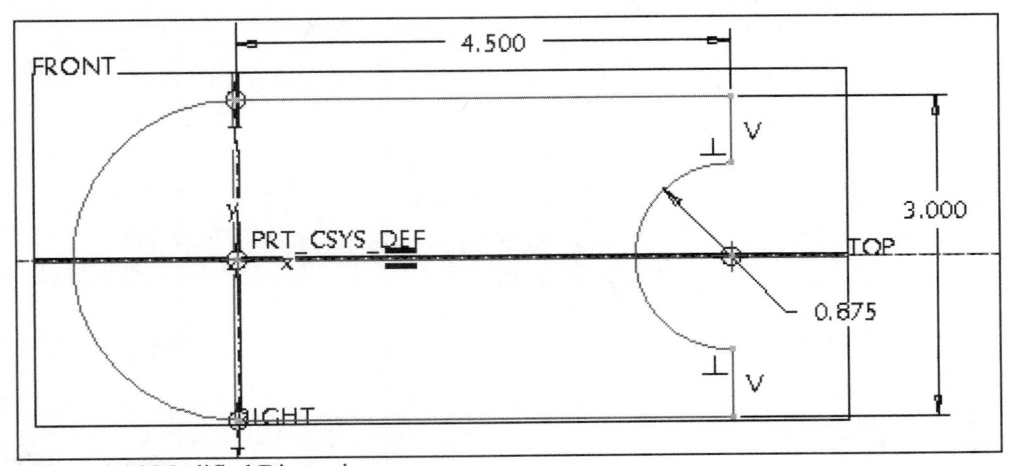

Figure 4.14 Modified Dimensions

Click: ✔ **Continue** ⇒ ⇒ **Standard Orientation** ⇒ note the yellow direction arrow (you may need to zoom out to see the complete part) ⇒ slide the handle until the dimension is approximately **2.00** (Fig. 4.15) ⇒ double click on the value and input the design value **2.188** (or **2.1875**) (Fig. 4.16 and Fig. 4.17) ⇒ **Enter** ⇒ ⇒ **Resumes the previously paused tool** ⇒ ✔ ⇒ **LMB** to deselect

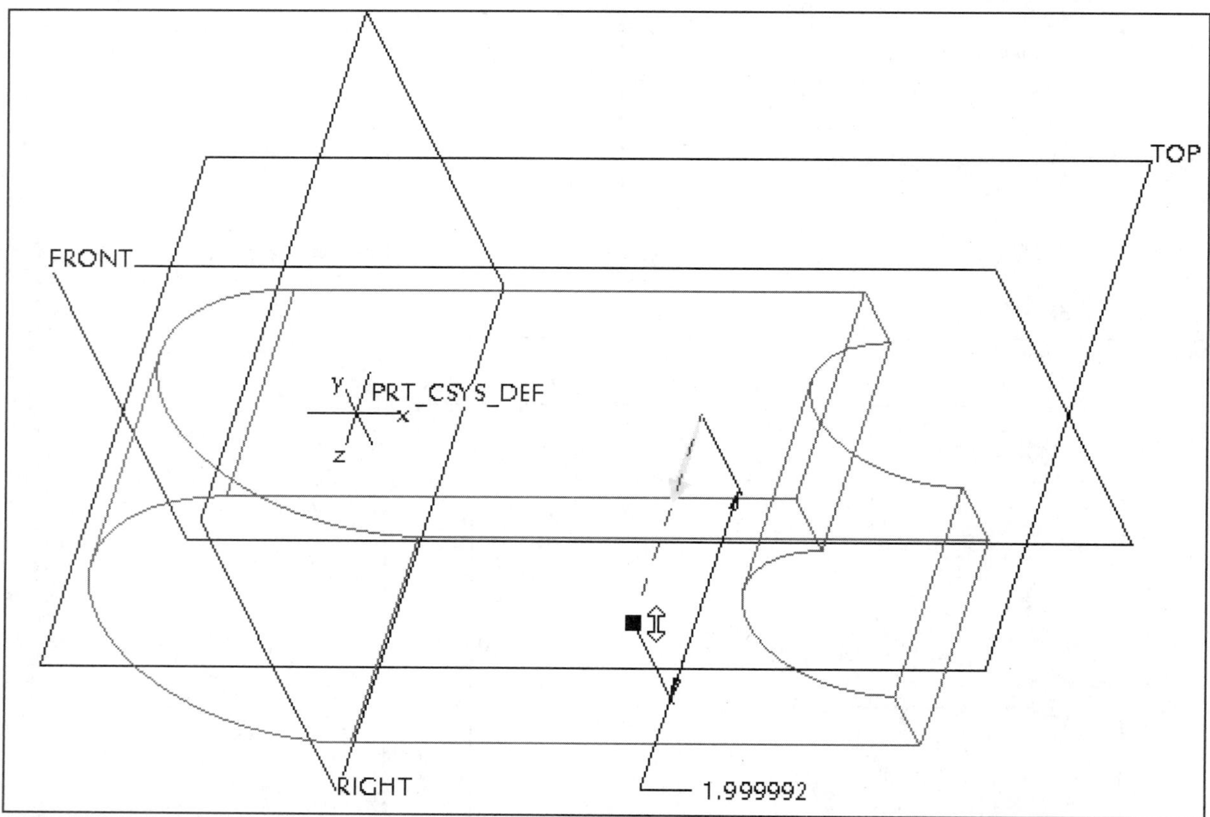

Figure 4.15 Slide the Handle Until the Value is Approximately **2.00**

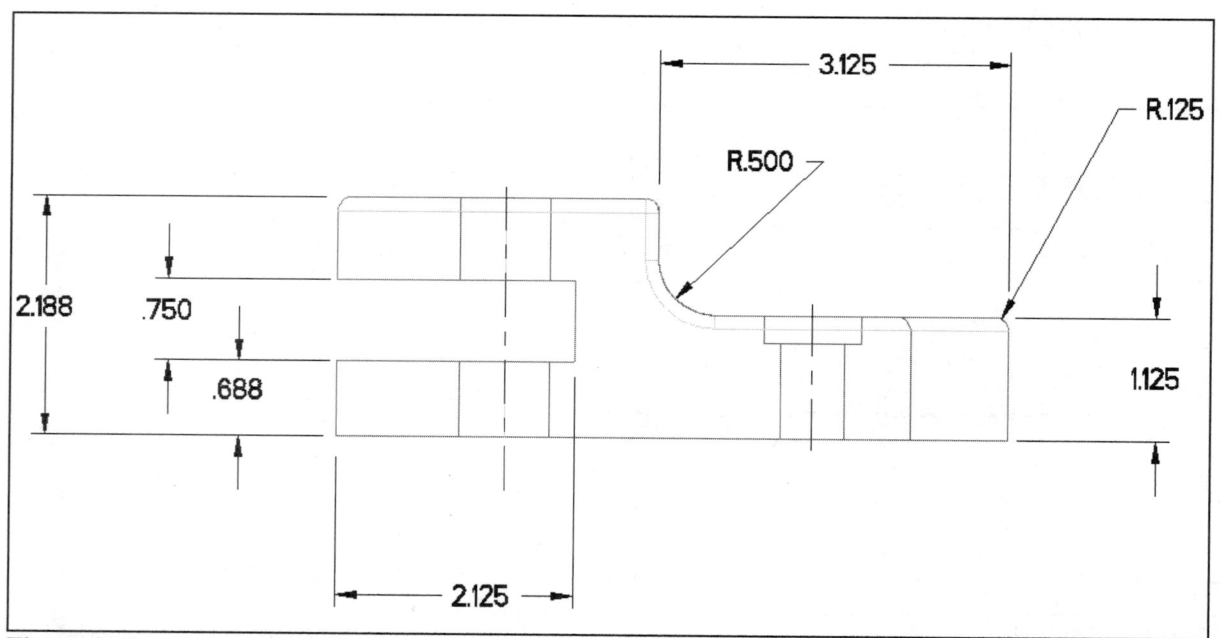

Figure 4.16 Front View

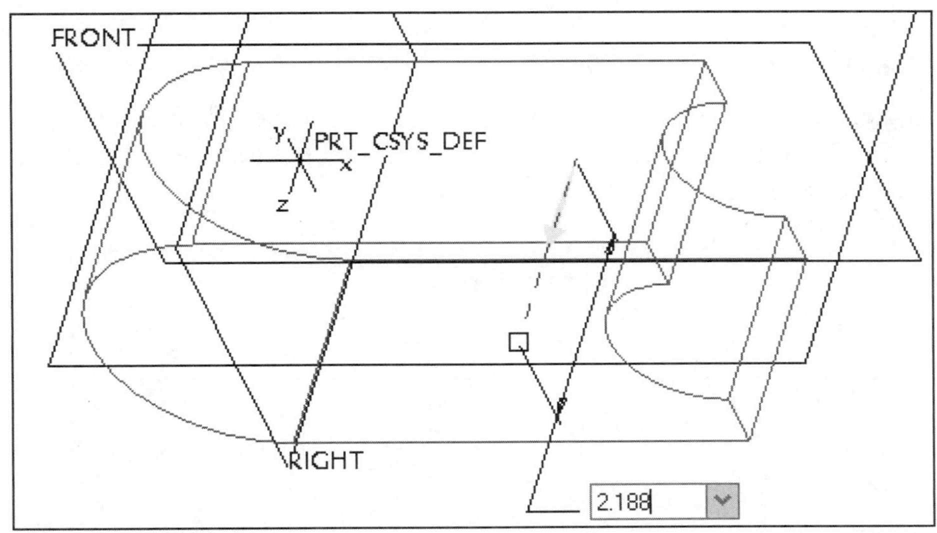

Figure 4.17 Modify Value

Click: ⇒ click on **Protrusion** in the **Model Tree** ⇒ **RMB** ⇒ **Edit** to show dimensions (Fig. 4.18) ⇒ ⇒ ⇒ **MMB** ⇒ **File** ⇒ **Delete** ⇒ **Old Versions** ⇒ **MMB**

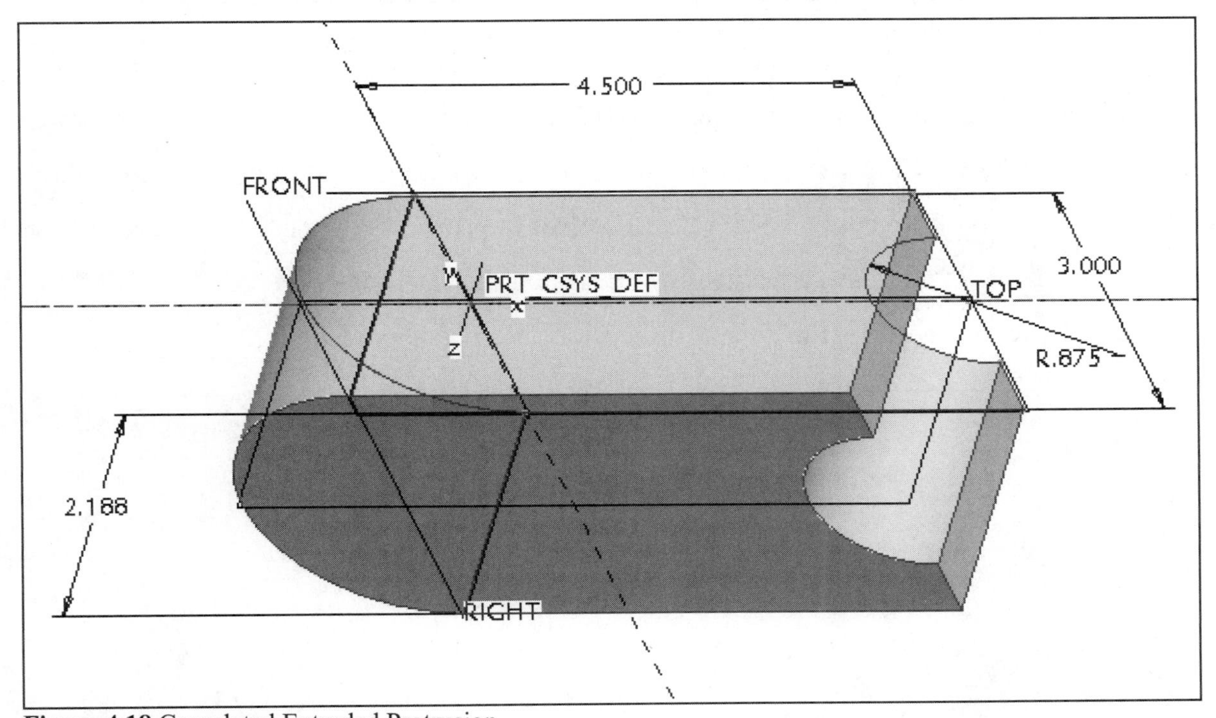

Figure 4.18 Completed Extruded Protrusion

The next features will be the cuts created to remove portions of the protrusion. The cuts will complete the primary features of the part.

In general, leave *holes* and *rounds* as the final features of the part. A majority of holes are *pick-and-place* features that are added to the model at a similar step, such as when they are drilled, reamed, or bored during actual manufacturing. In most cases, this means after most of the machining has been completed. Rounds are the very last features created. A good many model failures occur when a set of rounds is being created. Leaving them as the final features reduces the effort needed to resolve modeling problems.

Create the first cut, click: **Extrude Tool** ⇒ from dashboard click **Remove Material** ⇒ **Options** ⇒ **Symmetric** ⇒ ⇒ Section dialog box displays, Sketch Plane--- Plane: select **TOP** datum from the model as the sketch plane ⇒ **RIGHT** datum displays as default Sketch Orientation Reference (Fig. 4.19) ⇒ Sketch view direction: **Flip** ⇒ **Sketch** button

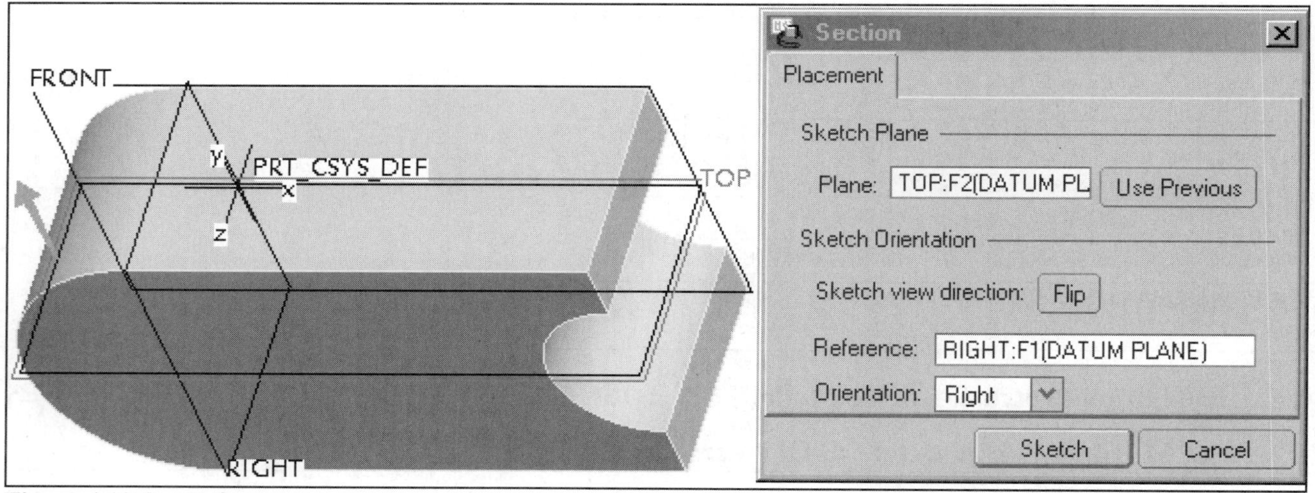

Figure 4.19 Cut References

Click: ⇒ **Standard Orientation** ⇒ delete the **RIGHT** Reference ⇒ ⇒ click on the three surfaces shown in Figure 4.20 ⇒ **Close** to accept the References ⇒ **Orient the sketching plane parallel to the screen** ⇒ **Tools** from menu bar ⇒ **Environment** ⇒ Snap to Grid ⇒ **OK** ⇒ **Toggle the grid off** ⇒ ⇒ **Create 2 point lines** ⇒ sketch the two lines (Fig. 4.21) ⇒ **MMB** to end the line sequence ⇒ ⇒ reposition the default dimensions as necessary (Fig. 4.22)

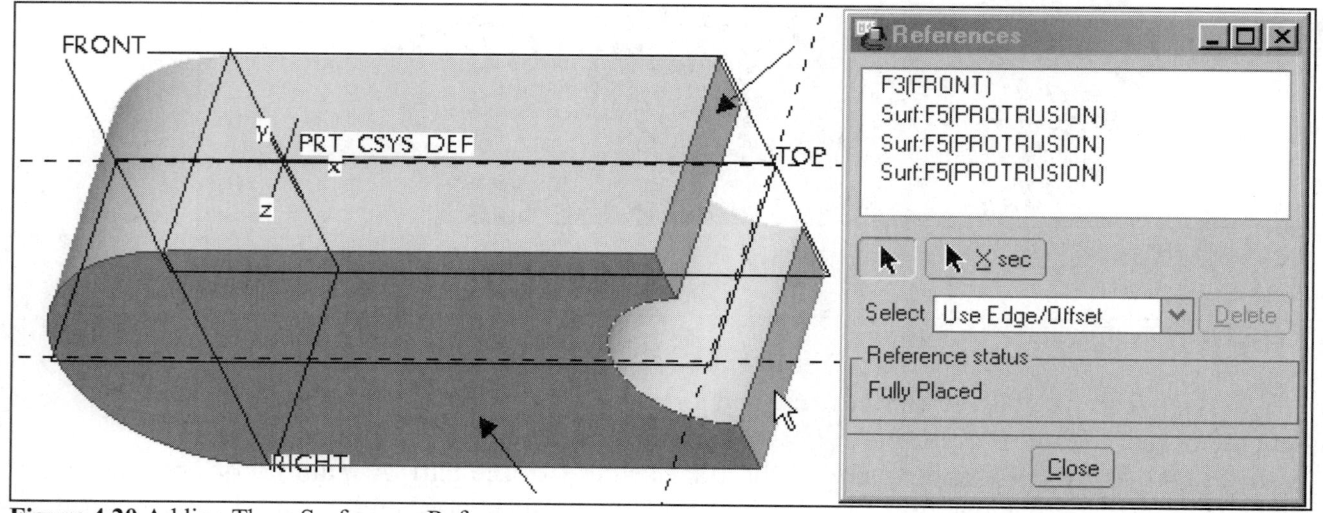

Figure 4.20 Adding Three Surfaces as References

Modify the values for the dimensions by double clicking on a dimension and typing a new value (Fig. 4.23) ⇒ ⇒ ⇒ **Standard Orientation** ⇒ **Coordinate systems off** ⇒ slide one of the depth handles to **3.00** (Fig. 4.24) ⇒ ⇒ ⇒ **MMB** (Fig. 4.25) ⇒ ⇒ **MMB**

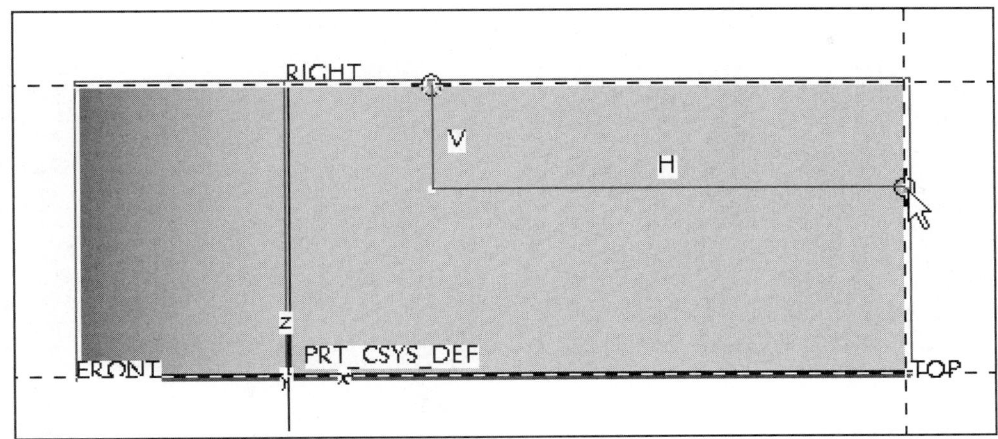

Figure 4.21 Sketch Two Lines from the References

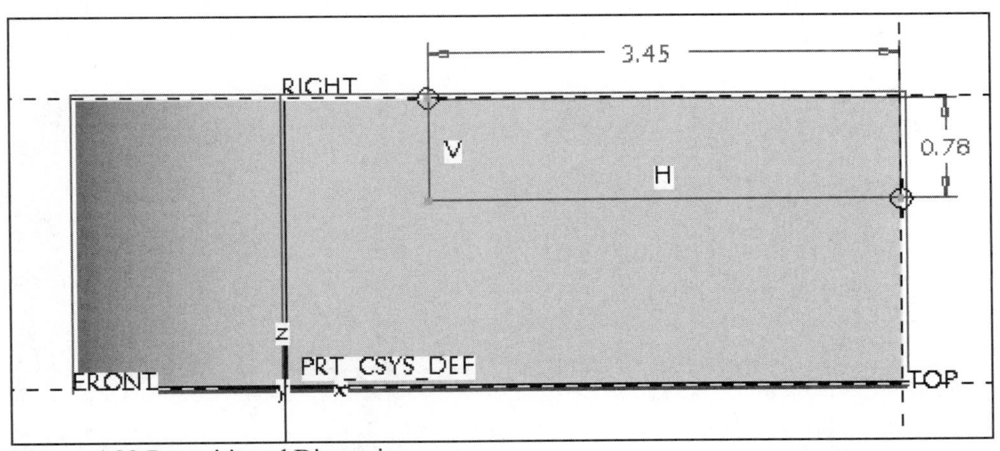

Figure 4.22 Repositioned Dimensions

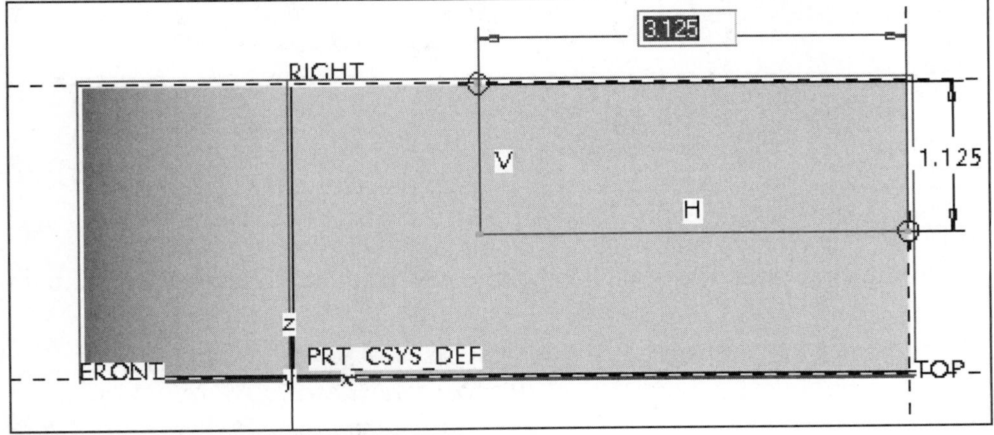

Figure 4.23 Modify Dimensions

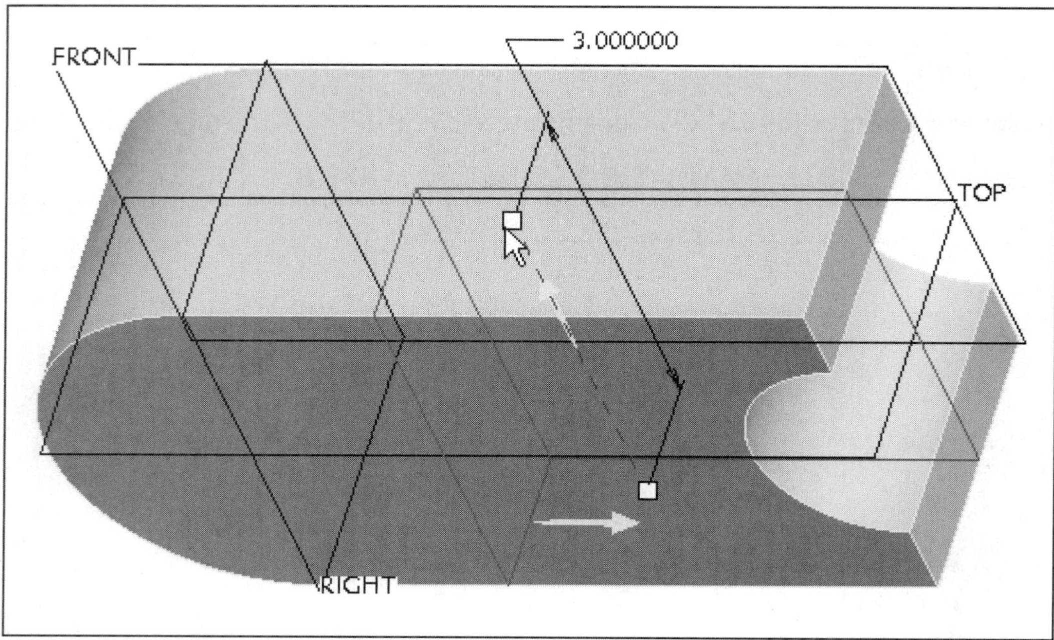

Figure 4.24 Slide the Cut Depth Handles to **3.00**

You can use drag handles on certain features to change their dimensions. As you dynamically drag the handles, the features get larger or smaller, depending upon the direction you drag.

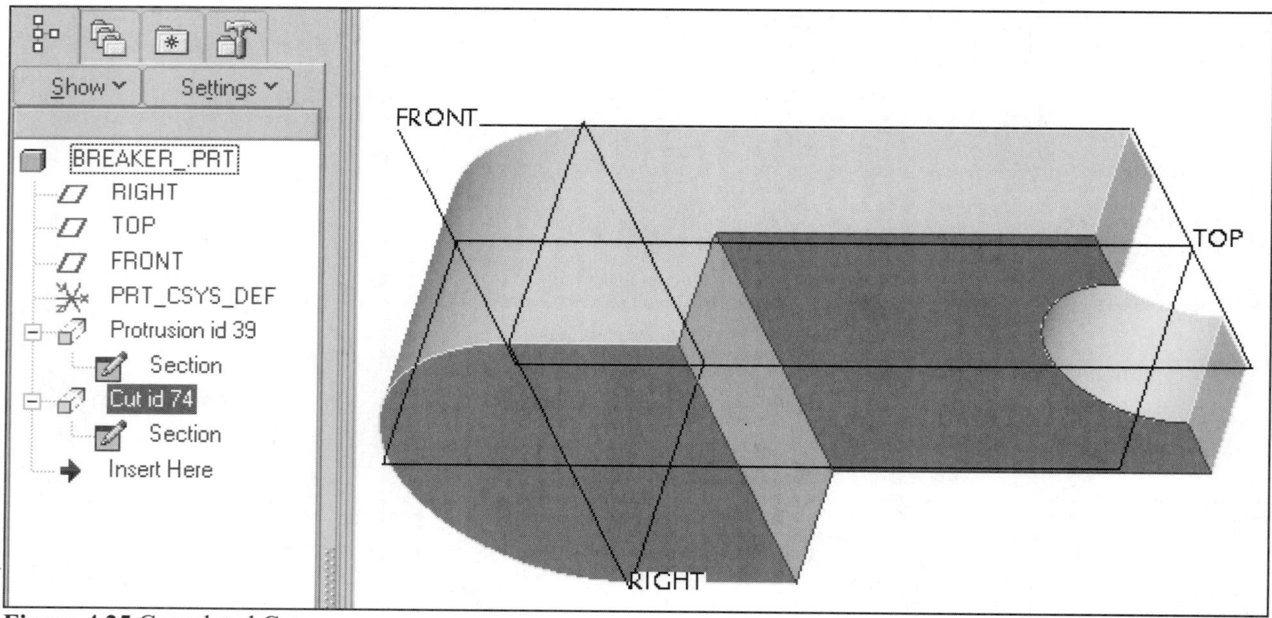

Figure 4.25 Completed Cut

The second cut is similar to the first one. The sketching plane and primary reference will remain the same.

Command sequences may not show, as many explanations, tool tips, or other descriptive information for tools and commands, which are now familiar. New tools, icons, and commands will have the tool description and tip provided. This practice will remain in effect for the remainder of the text.

Click: **Extrude Tool** ⇒ **Remove Material** ⇒ **Options** ⇒ ⇒ ⇒ **Use Previous** ⇒ **Sketch** ⇒ ⇒ click on the left edge surface shown in Figure 4.26 ⇒ **Close** to accept the References ⇒ ⇒ sketch the three lines (Fig. 4.27) ⇒ **MMB** to end the line sequence ⇒ ⇒ reposition the default dimensions as necessary (Fig. 4.28) ⇒ modify the values for the three dimensions by double clicking on a dimension and typing a new value (Fig. 4.29)

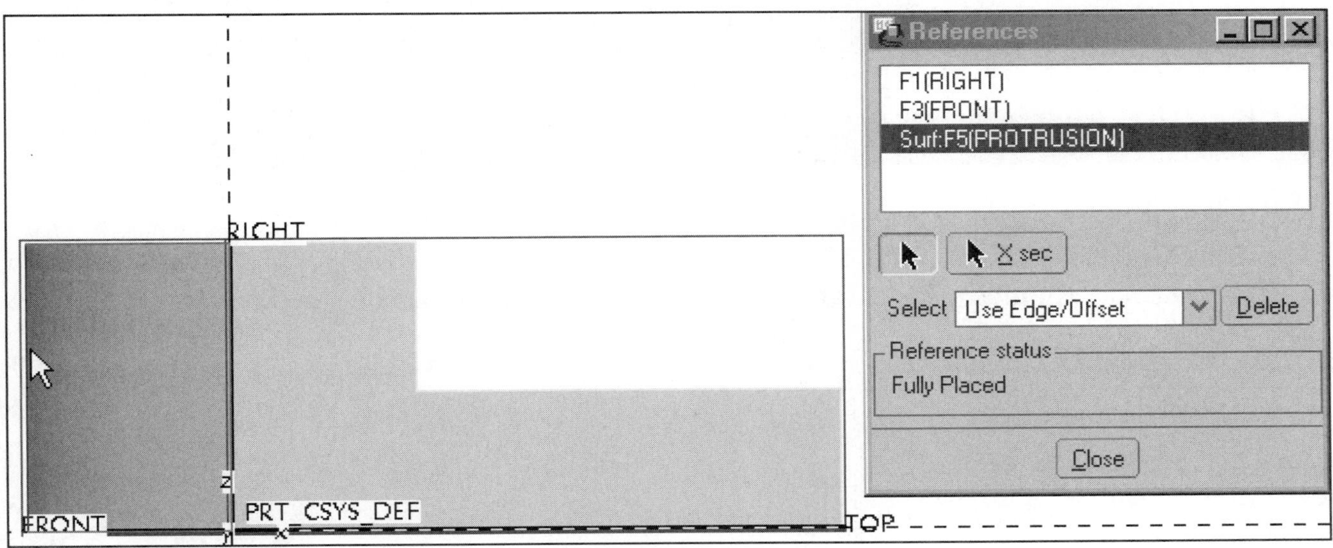

Figure 4.26 References

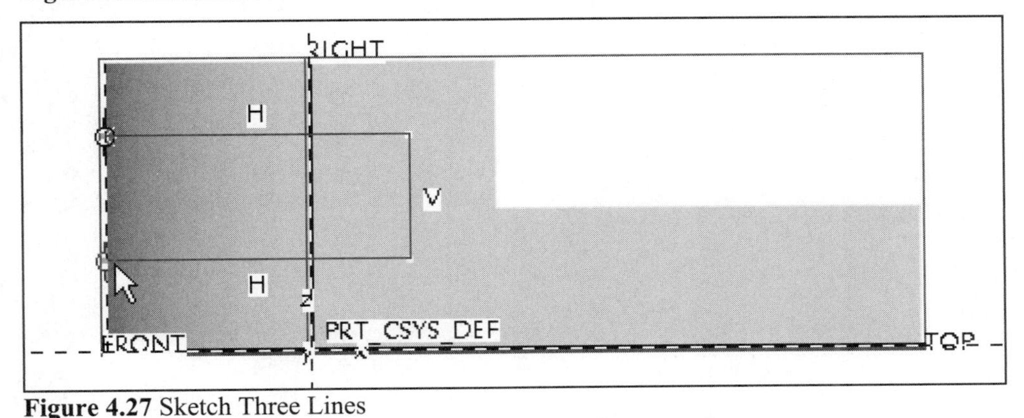

Figure 4.27 Sketch Three Lines

Figure 4.28 Move Dimensions as Required

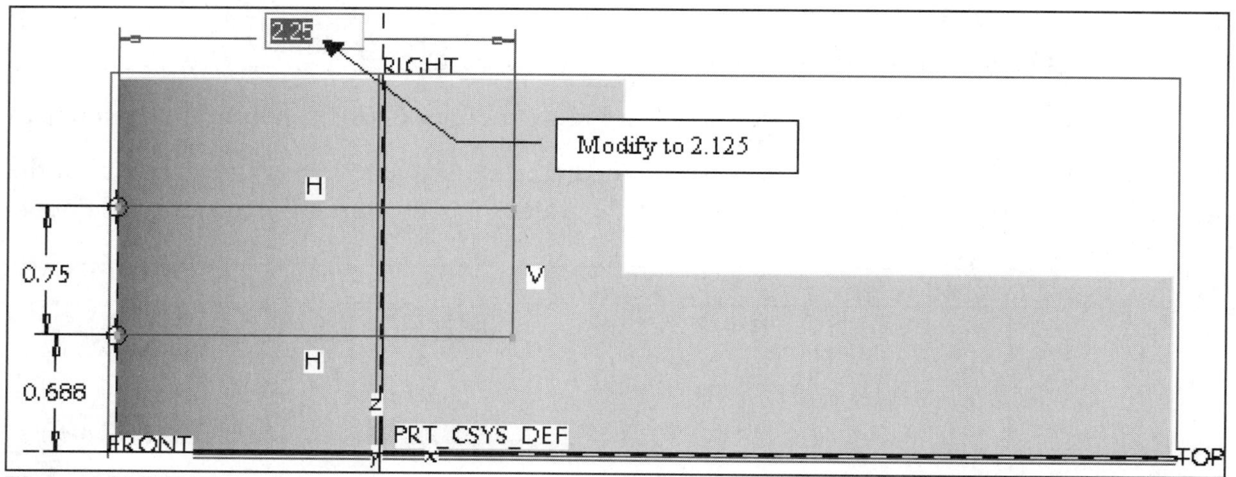

Figure 4.29 Edit Dimensions

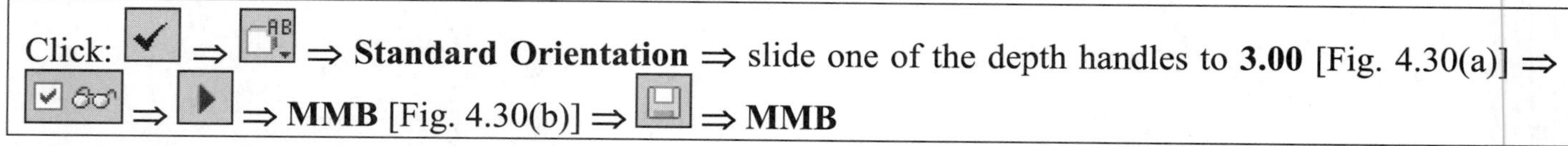

Click: ⇒ ⇒ **Standard Orientation** ⇒ slide one of the depth handles to **3.00** [Fig. 4.30(a)] ⇒ ⇒ ⇒ **MMB** [Fig. 4.30(b)] ⇒ ⇒ **MMB**

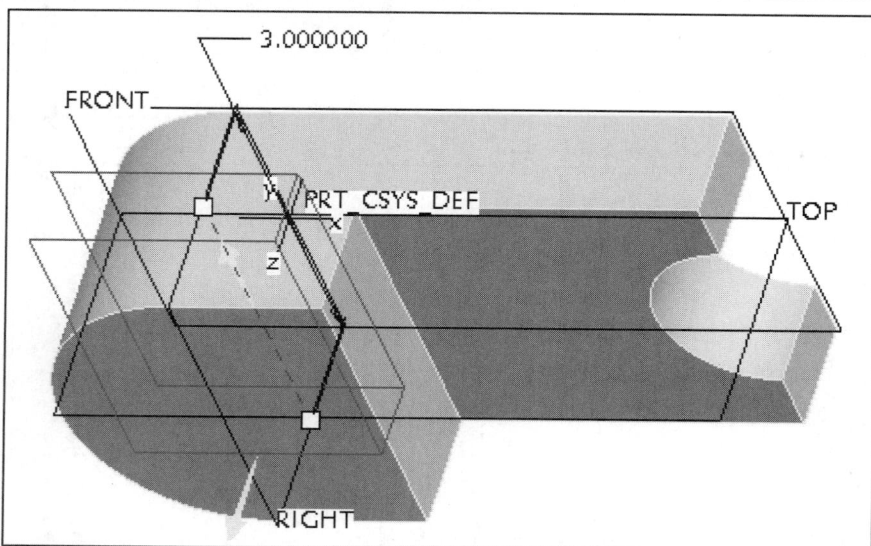

Figure 4.30(a) Slide Depth Handles to **3.00**

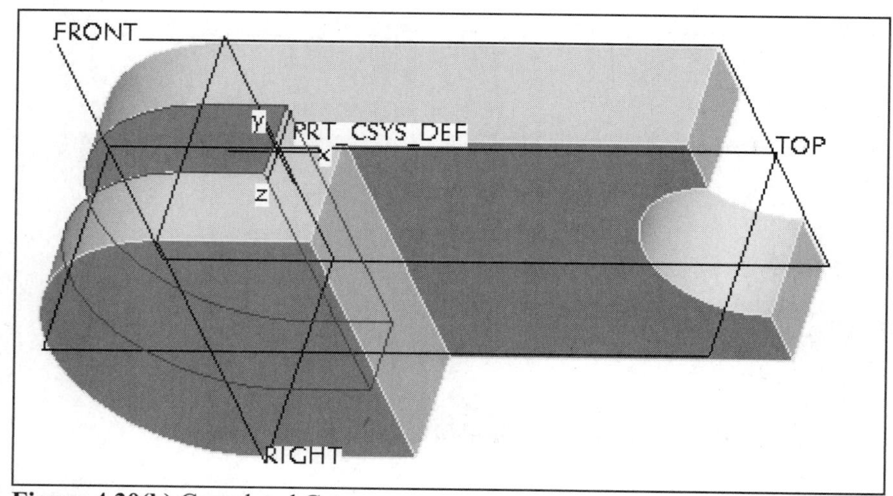

Figure 4.30(b) Completed Cut

Redefine the first cut and change the dimensioning scheme as per the design (Fig. 4.16).

Click on **Cut id** in the **Model Tree** (Fig. 4.31) ⇒ **RMB** ⇒ **Edit Definition** ⇒ [icon] ⇒ **Sketch** ⇒ add the dimension ⇒ **Delete** the unneeded dimension (Fig. 4.32) ⇒ [icon] ⇒ modify the new dimension (Fig. 4.33) ⇒ [icon] ⇒ **OK** ⇒ [icon] (Fig. 4.34) ⇒ [icon] ⇒ **Standard Orientation** (Fig. 4.35) ⇒ [icon] ⇒ **MMB**

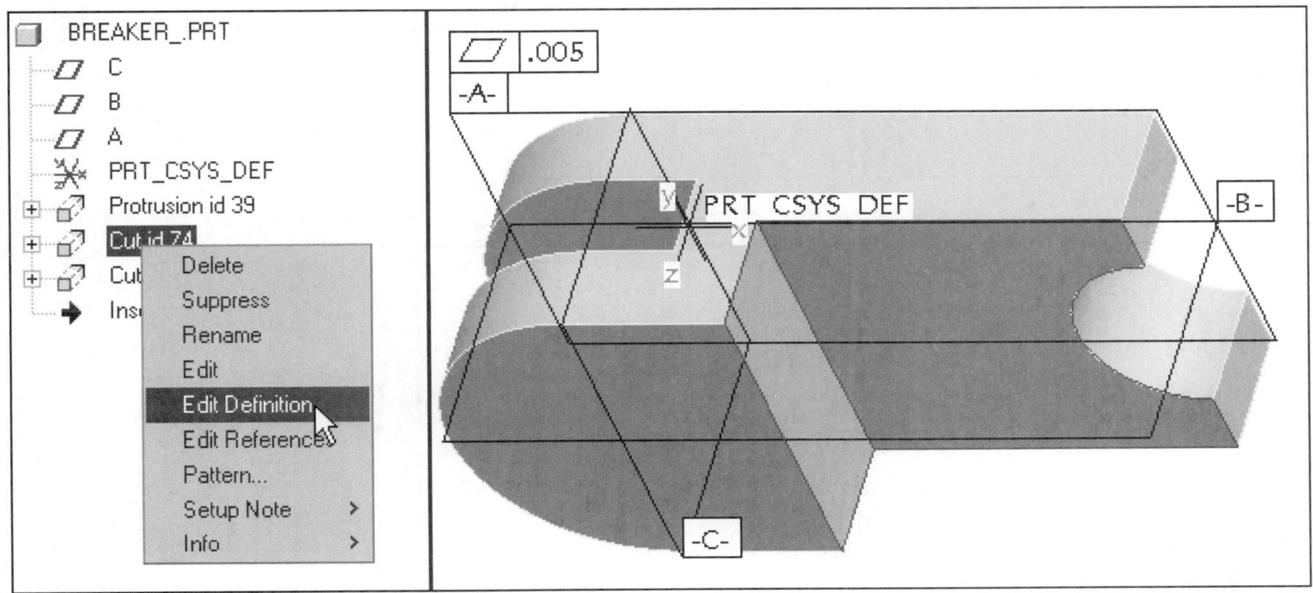

Figure 4.31 Redefine the First Cut using Edit Definition

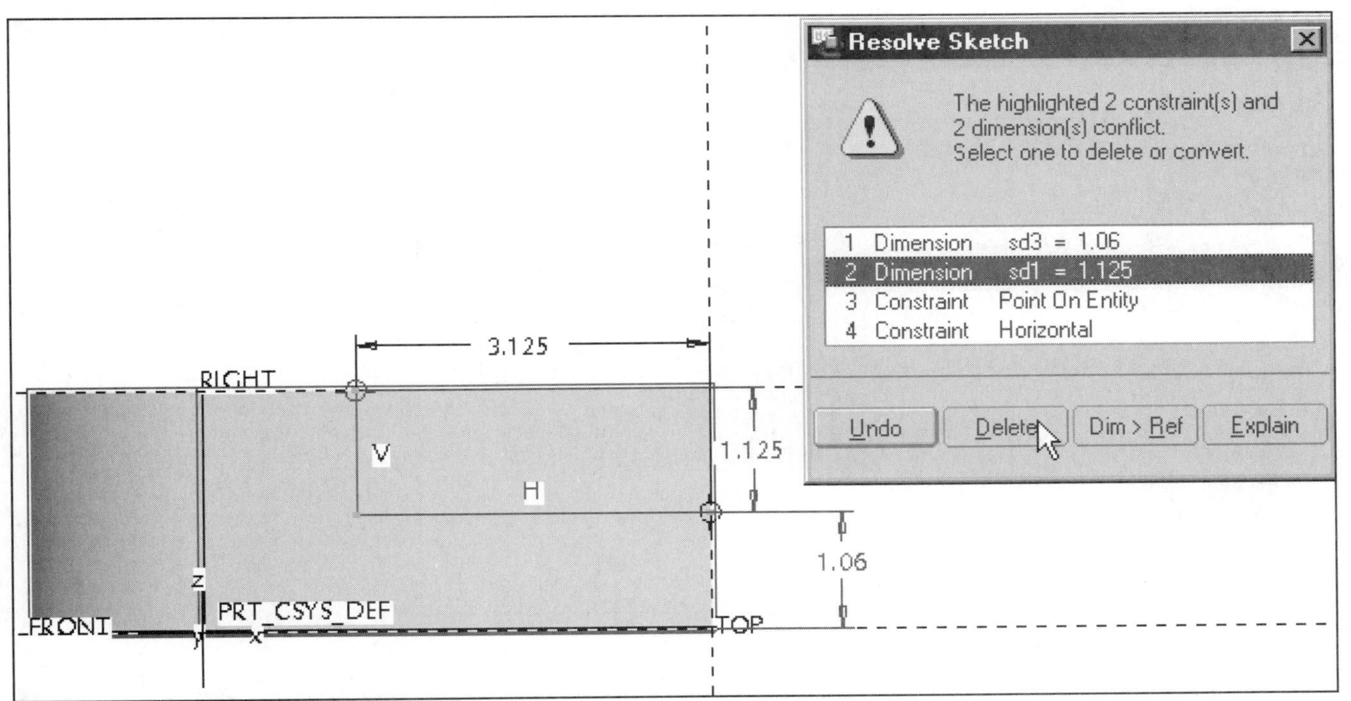

Figure 4.32 Add the new Dimension and Delete the **1.125** Dimension

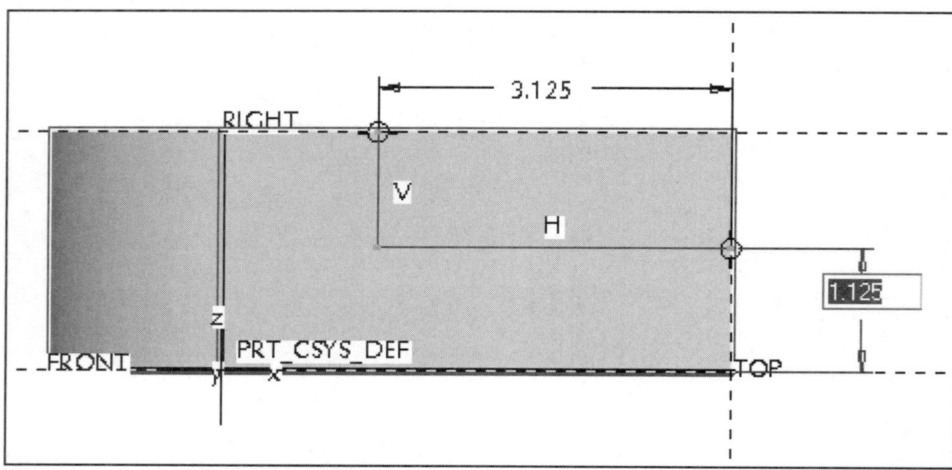

Figure 4.33 Modify the Dimension

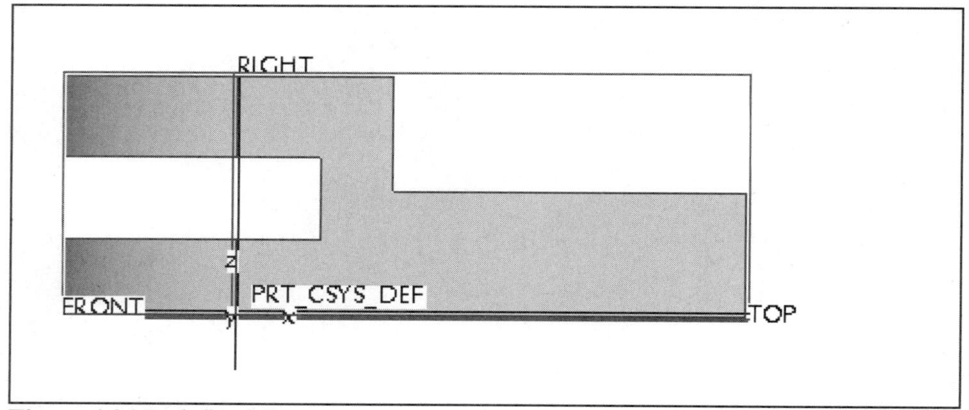

Figure 4.34 Redefined Cut

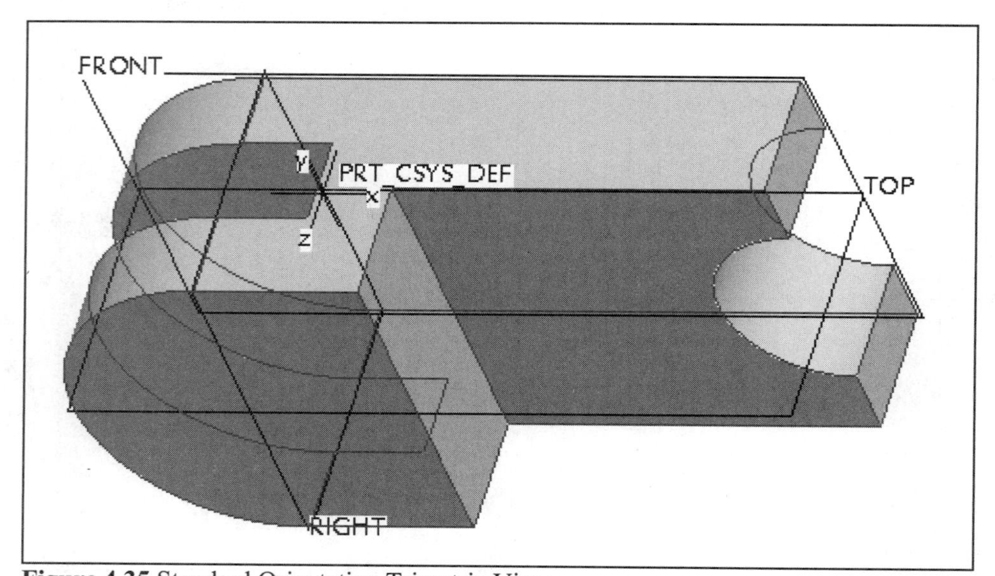

Figure 4.35 Standard Orientation Trimetric View

The next feature to be created is a hole. This *pick-and-place* (direct) feature does not require a sketch. Start by creating a datum axis through the cylindrical surface of the part. The top surface will be the second reference for the coaxial hole.

Spin your part to clearly see the cylindrical surface. Click: **Datum Axis Tool** ⇒ pick on the cylindrical surface (Fig. 4.36) ⇒ **OK** ⇒ **Hole Tool** from the right tool bar (status displays- Loading Hole Charts)

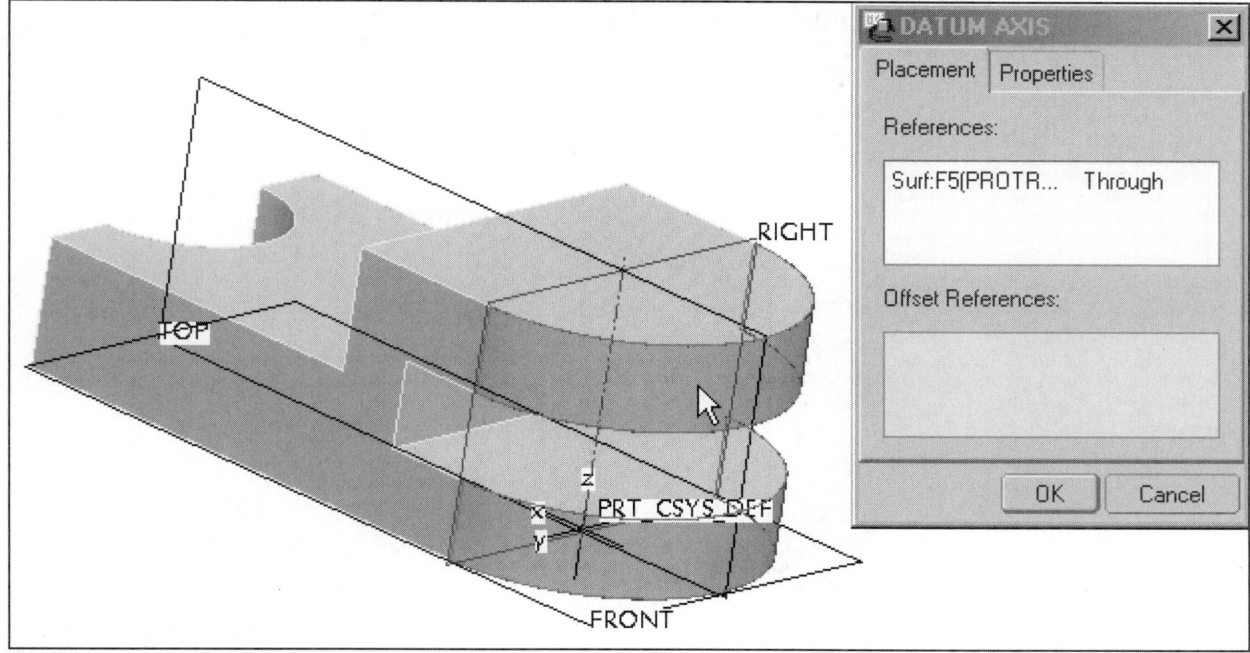

Figure 4.36 Creating a Datum Axis

Since a datum axis was created prior to the Hole Tool being selected [the datum axis is still selected (highlighted)], Pro/E will assume that the hole is to be coaxial (Fig. 4.37).

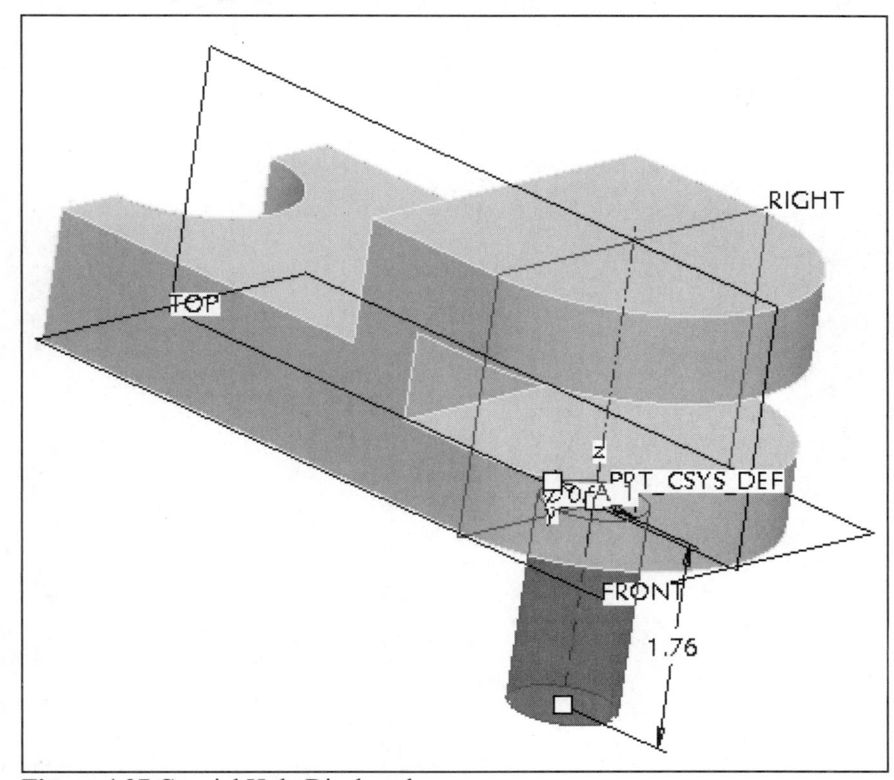

Figure 4.37 Coaxial Hole Displayed

From the dashboard, click: **Drill to intersect with all surfaces** ⇒ **Placement** tab (Fig. 4.38) ⇒ click in (No Items) box under- Secondary references: ⇒ pick the top surface (Fig. 4.38) as the Secondary reference ⇒ change the diameter to **.8125** ⇒ **Enter** ⇒ ⇒ ⇒ (Fig. 4.39) ⇒ ⇒ **Standard Orientation** ⇒ ⇒ **MMB**

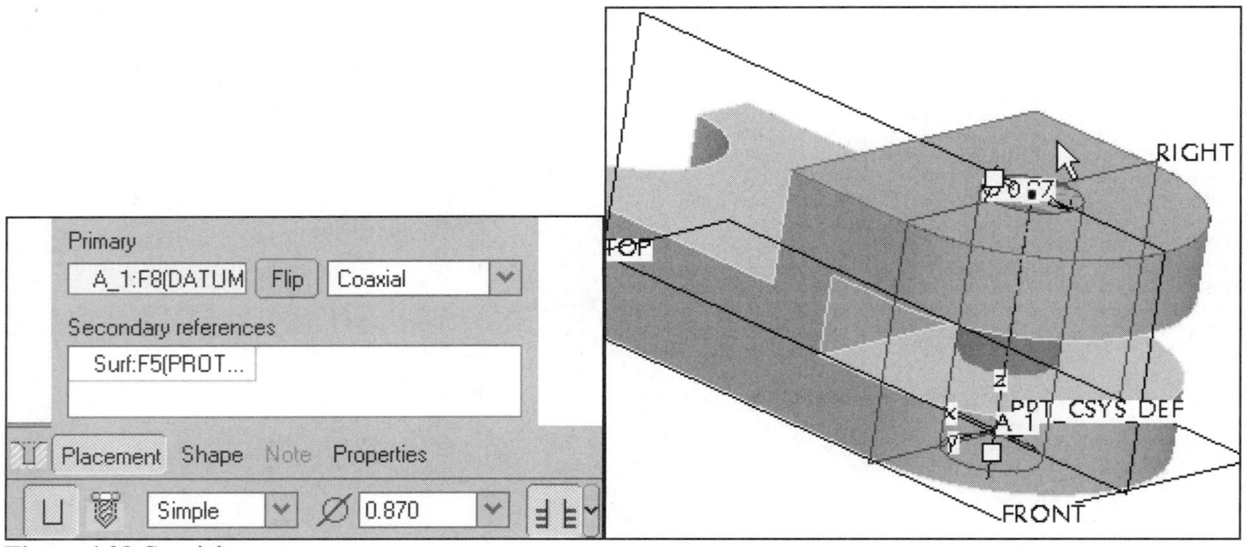

Figure 4.38 Coaxial

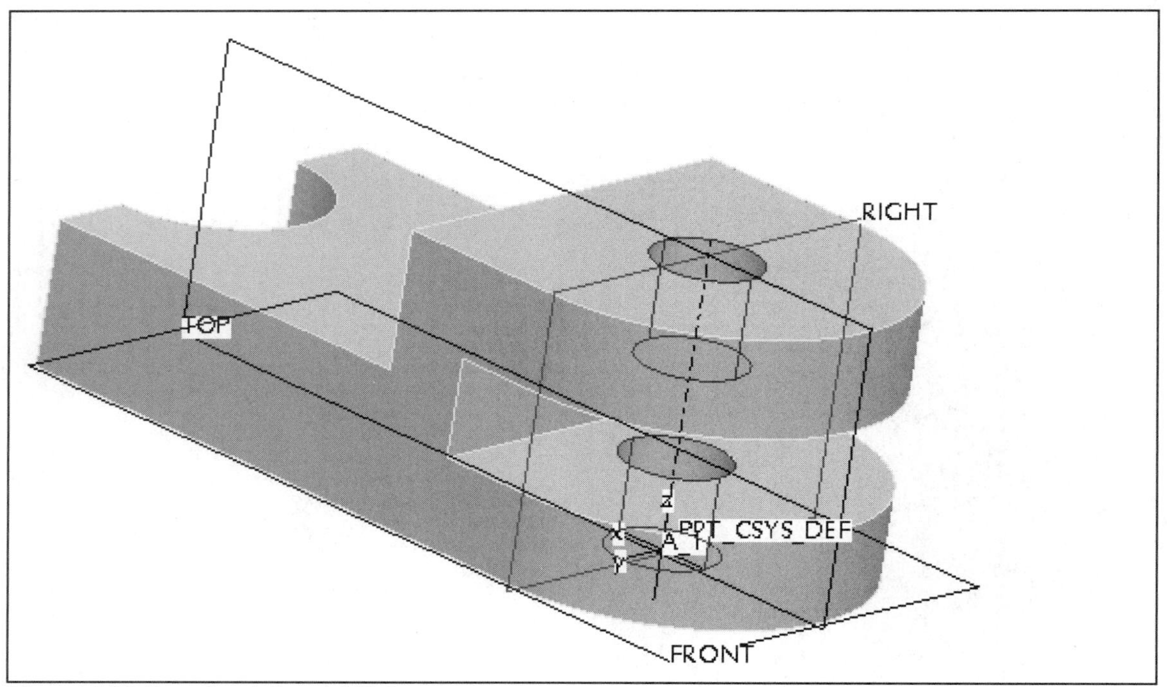

Figure 4.39 Completed Coaxial Hole

The second hole will be a non-standard counterbore hole. Instead of using the **Hole** command, we could also create this hole with a *revolved cut*. Sketched holes are really nothing more than revolved cuts.

Orient the part as shown (Fig. 4.40) ⇒ **Hole Tool** ⇒ pick on the horizontal surface of the first cut and the hole will display with handles for; hole position, diameter adjustment, depth adjustment, and two reference handles for establishing the dimensioning scheme (Fig. 4.40) ⇒ drag one handle to the right side surface and the other handle to the **TOP** datum plane (Fig. 4.41) ⇒ click on the **Placement** tab on dashboard to see **Secondary references**

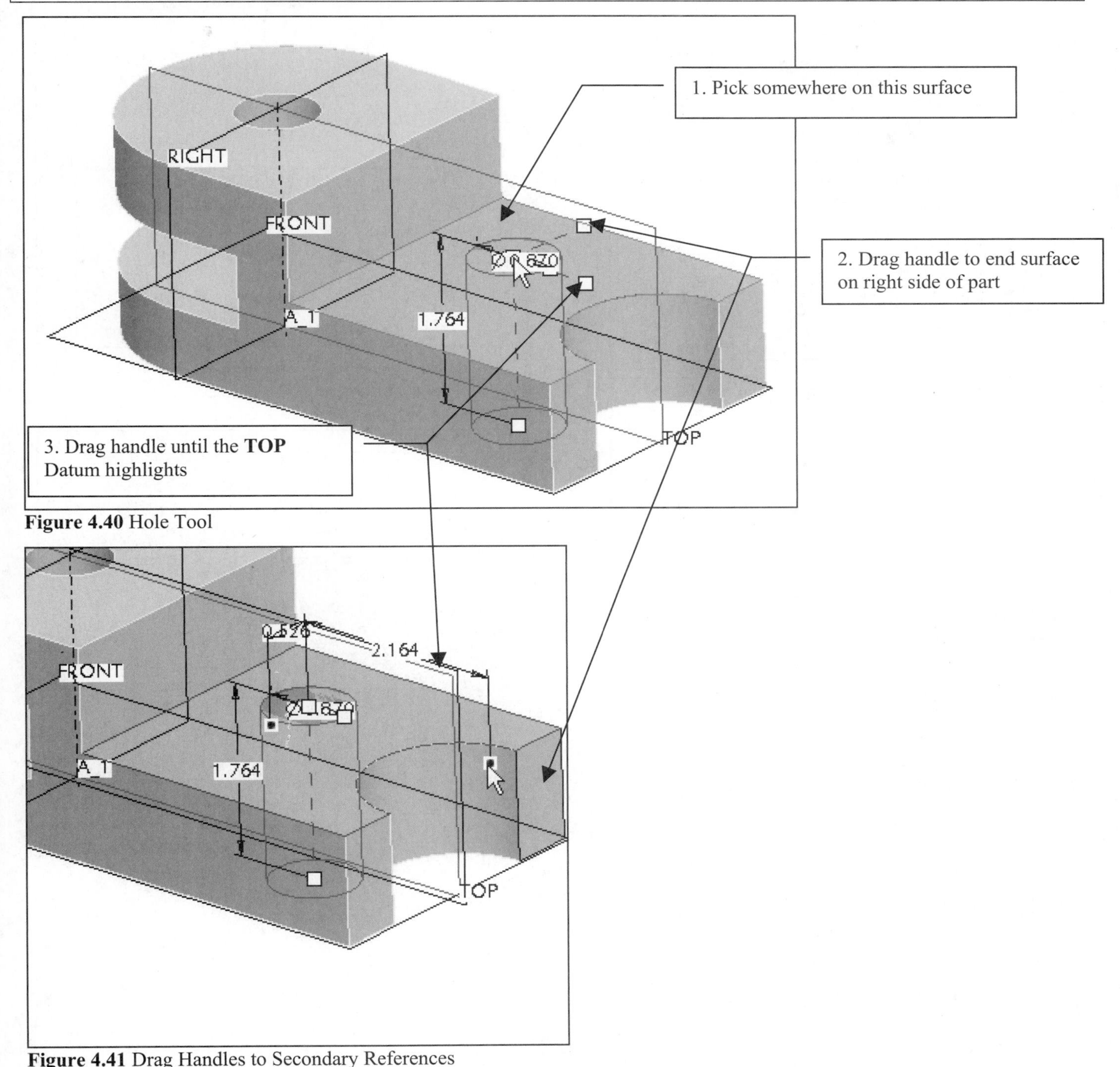

Figure 4.40 Hole Tool

Figure 4.41 Drag Handles to Secondary References

Click: ⇒ **Front** ⇒ click and hold the handle at the center of the hole [Fig. 4.42(a)] and move it about the surface [Fig. 4.42(b)] ⇒ double click on the dimension from the holes center to the **TOP** datum plane and modify the value to **0.00** [Fig. 4.43(a)] ⇒ **Enter** ⇒ ⇒ **Standard Orientation** [Fig. 4.43(b)] ⇒ Sketched (Fig. 4.44)

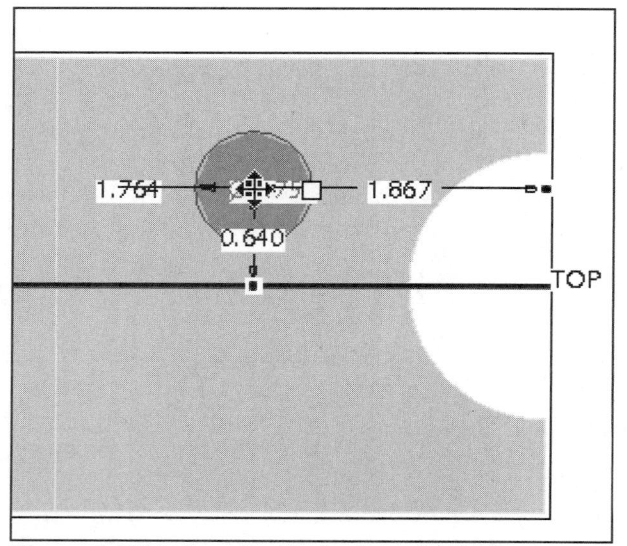

Figure 4.42(a) Drag the Circle about the Surface

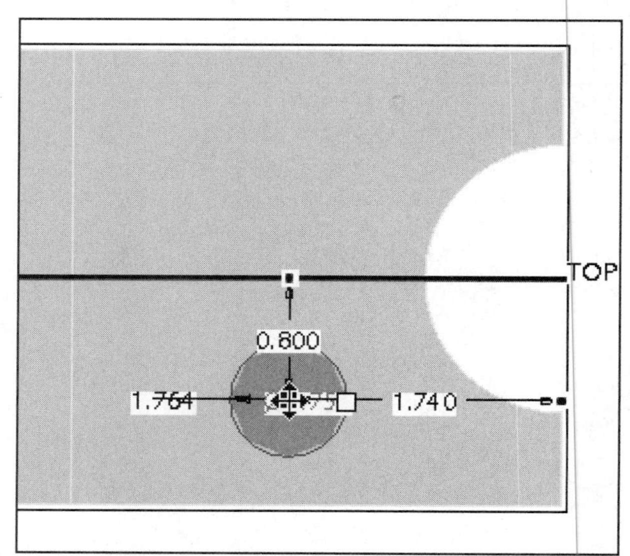

Figure 4.42(b) Dragging the Circle

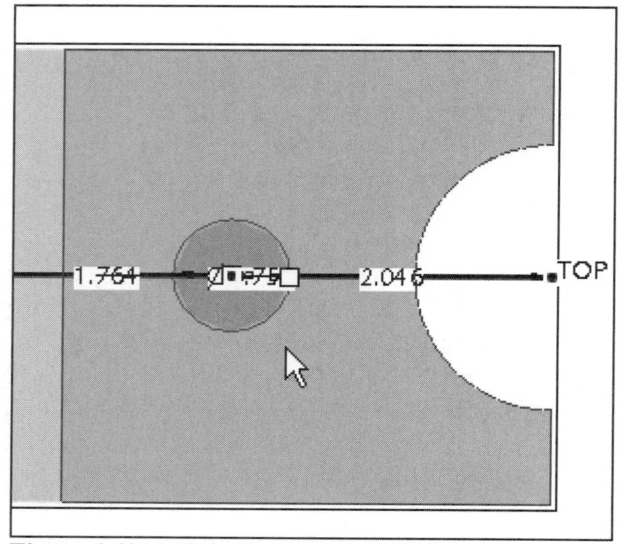

Figure 4.43(a) Hole Centered on TOP Datum

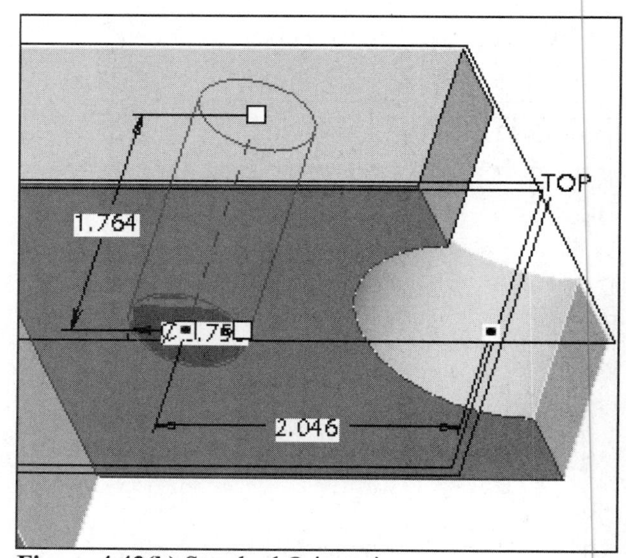

Figure 4.43(b) Standard Orientation

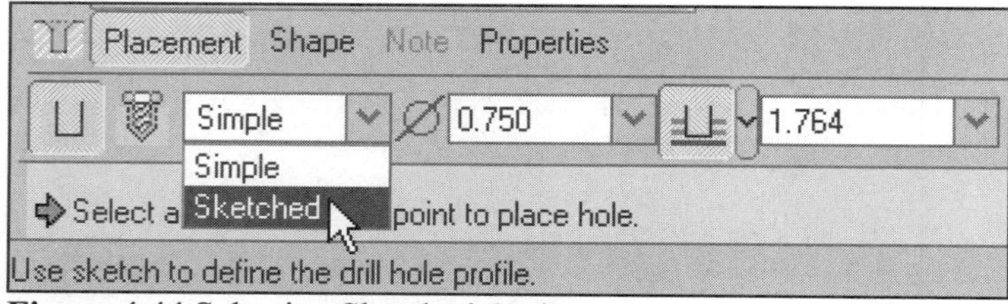

Figure 4.44 Selecting Sketched Option

Sketched holes and revolved cuts are created with a section sketch. *The section must be closed, and have a vertical centerline. All entities must be on one side of that centerline.*

Always use diameter dimensions. There is no such thing as a radius hole or a radius shaft. The counterbore *diameter* is **.875**. The thru hole *diameter* is **.5625** and a depth the same as the part (**1.125**). The depth of the counterbore is **.250**.

⌀.5625
⌴⌀.875
↧.250

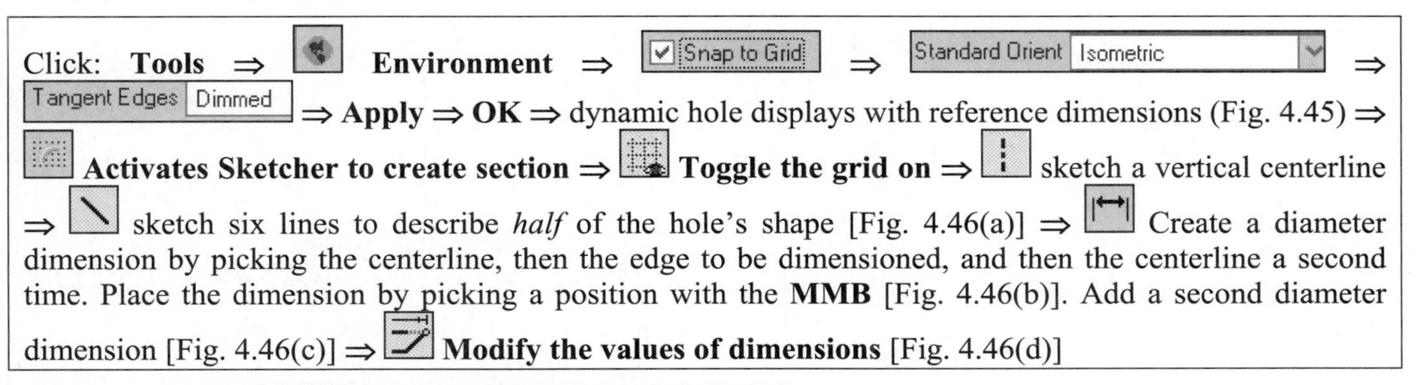

Click: **Tools** ⇒ **Environment** ⇒ Snap to Grid ⇒ Standard Orient Isometric ⇒ Tangent Edges Dimmed ⇒ **Apply** ⇒ **OK** ⇒ dynamic hole displays with reference dimensions (Fig. 4.45) ⇒ **Activates Sketcher to create section** ⇒ **Toggle the grid on** ⇒ sketch a vertical centerline ⇒ sketch six lines to describe *half* of the hole's shape [Fig. 4.46(a)] ⇒ Create a diameter dimension by picking the centerline, then the edge to be dimensioned, and then the centerline a second time. Place the dimension by picking a position with the **MMB** [Fig. 4.46(b)]. Add a second diameter dimension [Fig. 4.46(c)] ⇒ **Modify the values of dimensions** [Fig. 4.46(d)]

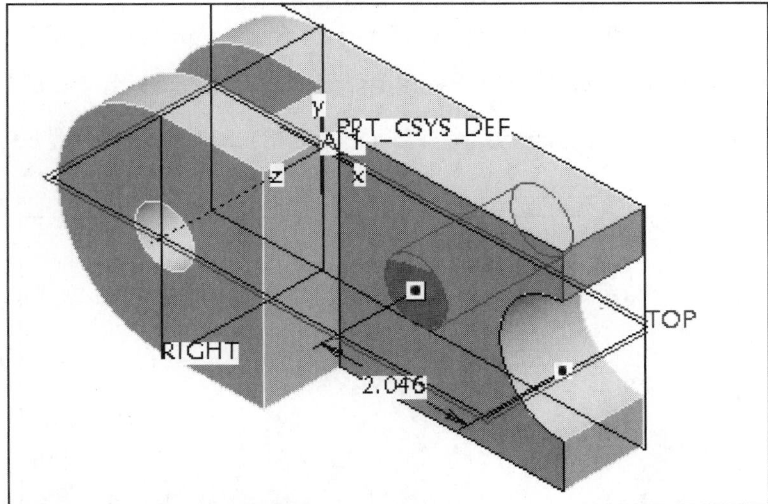

Figure 4.45 Hole Placement

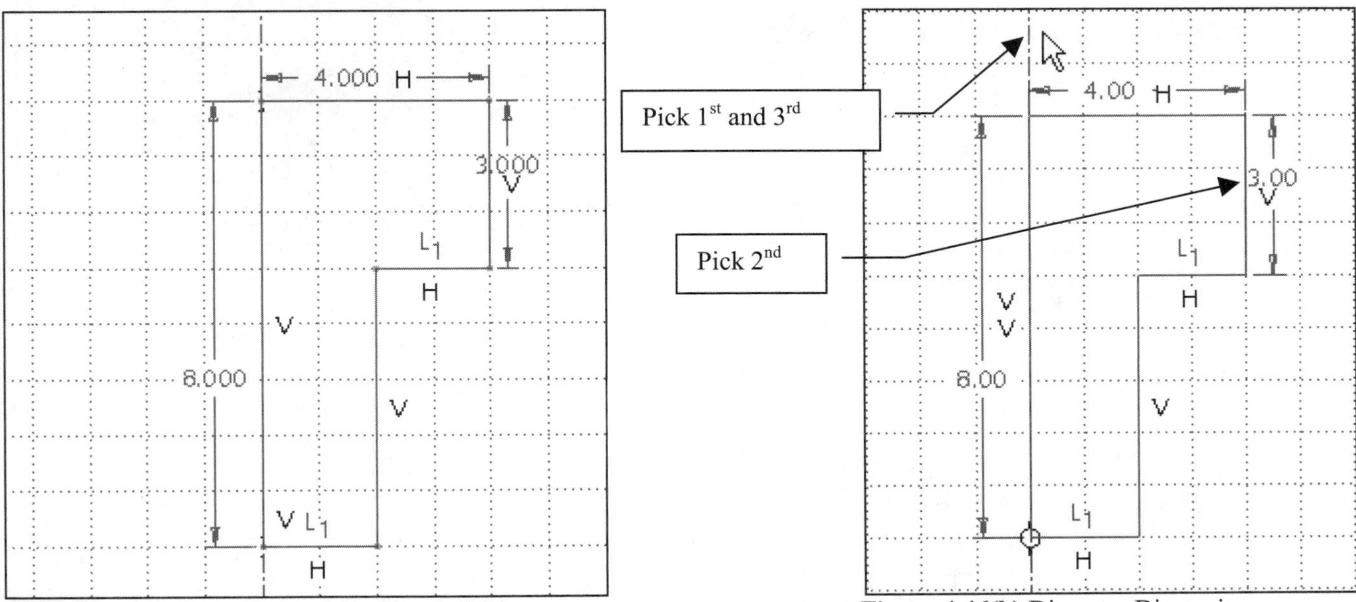

Figure 4.46(a) Centerline and Six Sketched Lines

Figure 4.46(b) Diameter Dimension

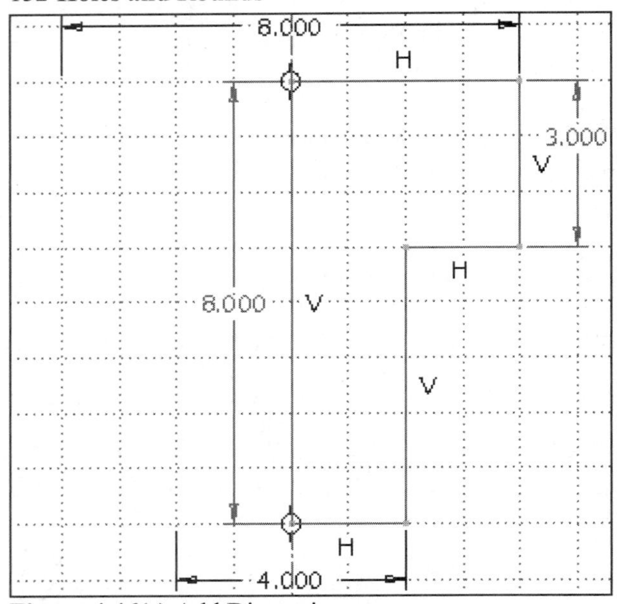

Figure 4.46(c) Add Dimensions

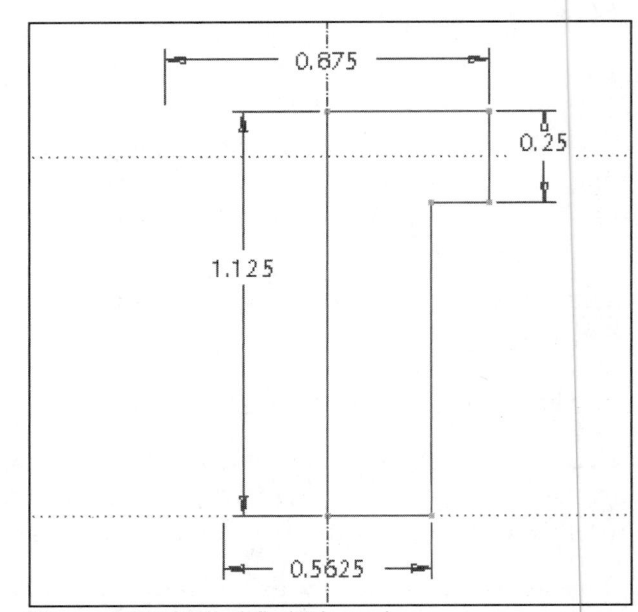

Figure 4.46(d) Modified Dimensions

Click: ⇒ ⇒ **Standard Orientation** ⇒ **Coordinate Systems off** ⇒ rotate with **MMB** (Fig.4.47) ⇒ double click on the distance to edge value and change to **1.75** (Fig.4.48) ⇒ ⇒ ⇒ **MMB** (Fig.4.49) ⇒ **LMB** to deselect

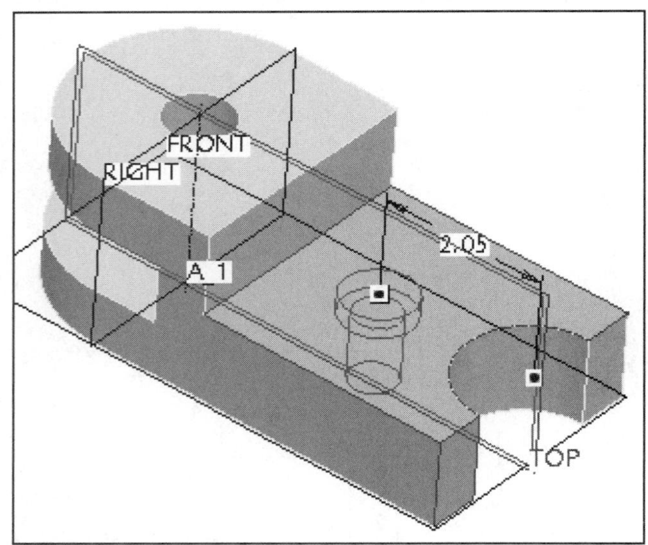

Figure 4.47 Hole

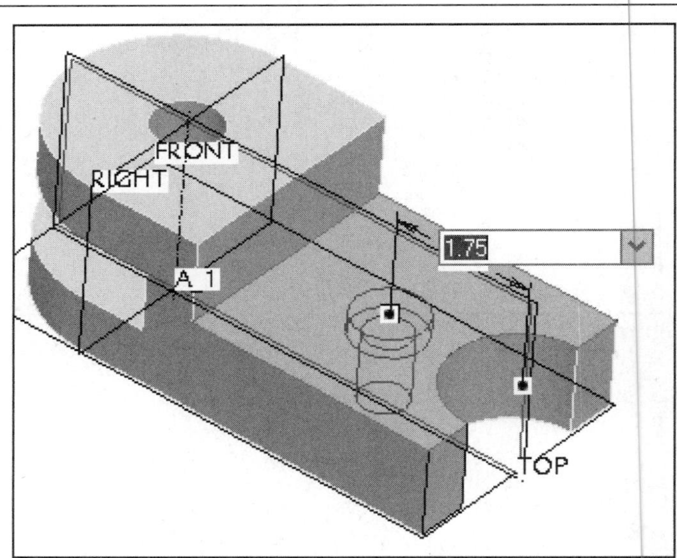

Figure 4.48 Modify Distance to Edge (**1.75**)

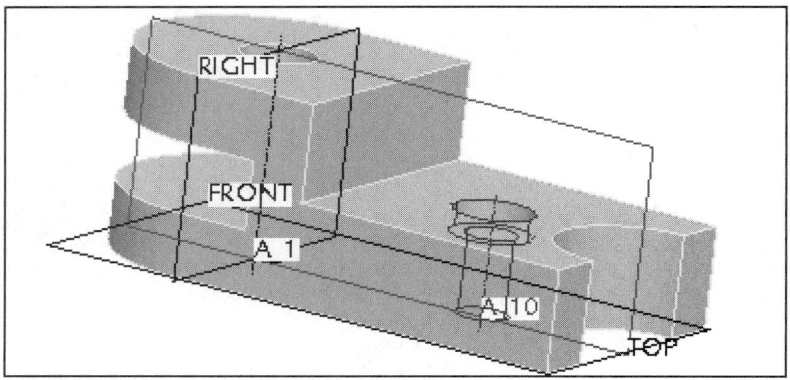

Figure 4.49 Completed Sketched Hole

Click on the counterbore hole in the **Model Tree** ⇒ **RMB** ⇒ **Info** ⇒ **Feature** ⇒ **Feature info: HOLE** displays in the **Browser** (Fig.4.50) ⇒ Take some time to view all the information available. Click on a Dimension ID in the Browser, the dimension will display on the model. ⇒ click on the quick sash to collapse the Browser ⇒ ⇒ **MMB** ⇒ **File** ⇒ **Delete** ⇒ **Old Versions** ⇒ **MMB**

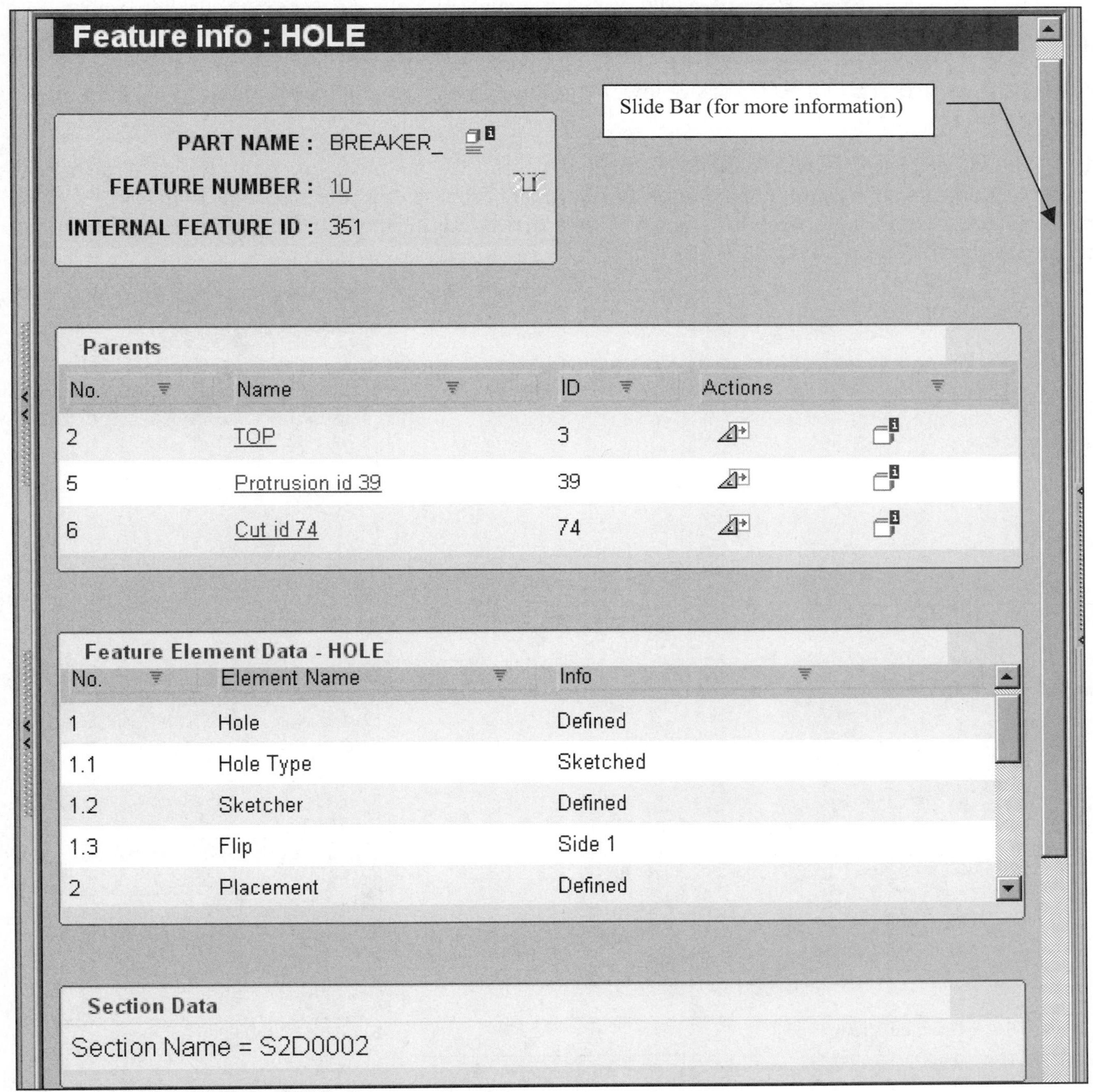

Figure 4.50 FEATURE info: HOLE

Both holes are now complete. In Lesson 18, you will detail this part in a drawing. If the counterbore were for a standard fastener, you could have created it with the *Create standard hole* option. This option allows the varying of the counterbore diameter and depth but does not permit the thru hole diameter to be altered for the screw shaft size.

To complete the part, a number of rounds need to be created. The first round is an edge round between the vertical and horizontal faces of the first cut. Before you start, create a user specified view and save it to be used later.

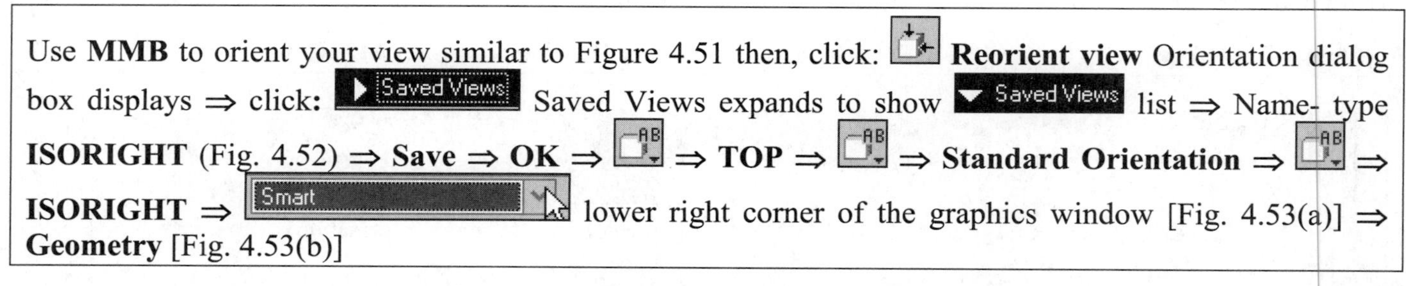

Use **MMB** to orient your view similar to Figure 4.51 then, click: **Reorient view** Orientation dialog box displays ⇒ click: Saved Views Saved Views expands to show Saved Views list ⇒ Name- type **ISORIGHT** (Fig. 4.52) ⇒ **Save** ⇒ **OK** ⇒ ⇒ **TOP** ⇒ ⇒ **Standard Orientation** ⇒ ⇒ **ISORIGHT** ⇒ Smart lower right corner of the graphics window [Fig. 4.53(a)] ⇒ **Geometry** [Fig. 4.53(b)]

The Smart filter provides context-sensitive access to the most common types of selectable geometry in any given situation. In addition to the Smart filter setting, the selection filter can be set to limit the scope of selectable items to a specific type depending on the situation and need. Here we are setting the filter to **Geometry**.

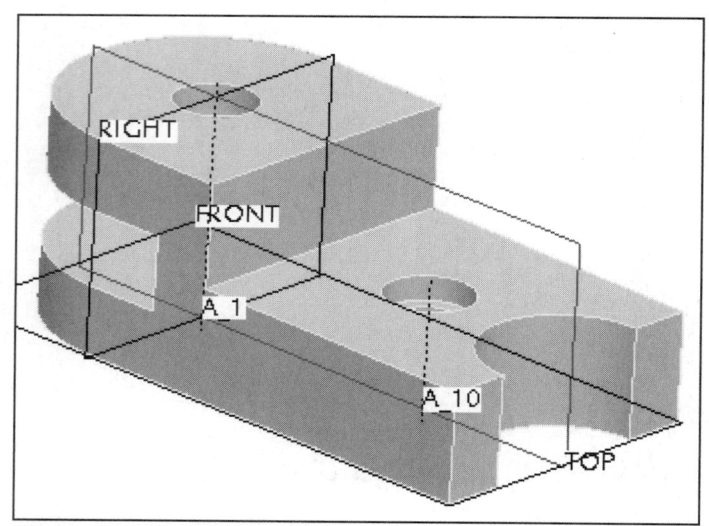

Figure 4.51 User Oriented View

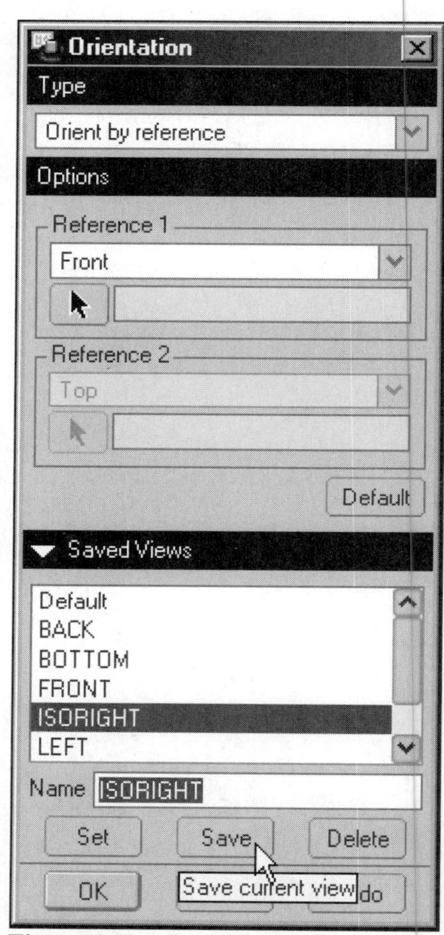

Figure 4.52 Orientation Dialog

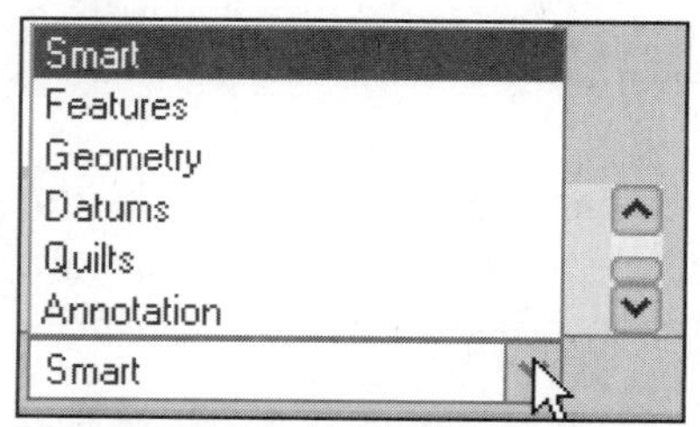

Figure 4.53(a) Selection Filter Smart

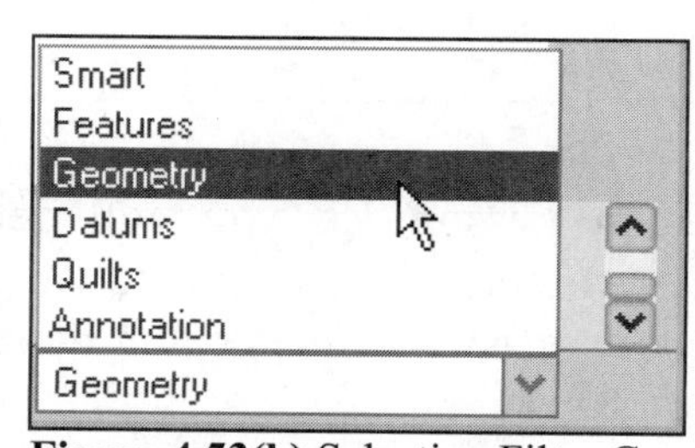

Figure 4.53(b) Selection Filter Geometry

Pick on the edge between the horizontal and vertical surfaces of the part ⇒ **RMB** ⇒ **Round Edges** (Fig. 4.54) ⇒ slide a drag handle until the radius is **.50** [Fig. 4.55(a-b)] ⇒ **MMB** (Fig. 4.56)

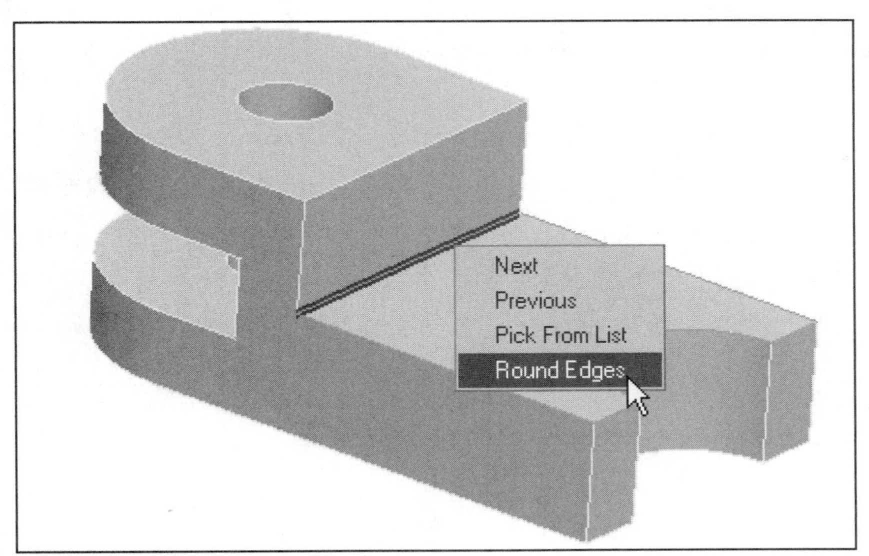

Figure 4.54 Pick the Edge then RMB

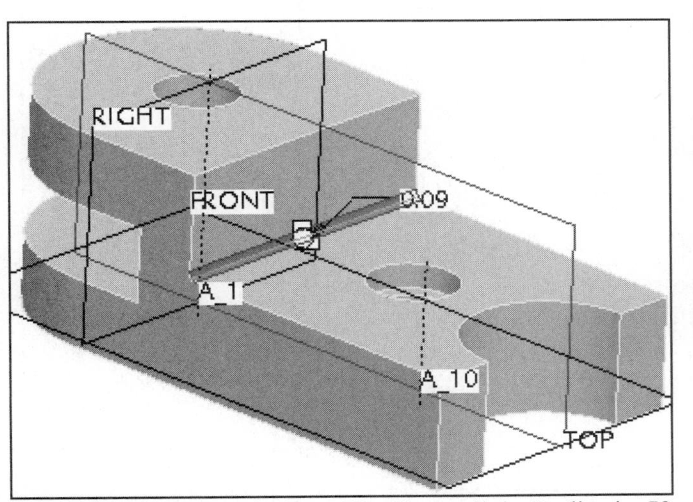

Figure 4.55(a) Move the Drag Handles until the Radius is **.50** **Figure 4.55(b)** Radius is **.50**

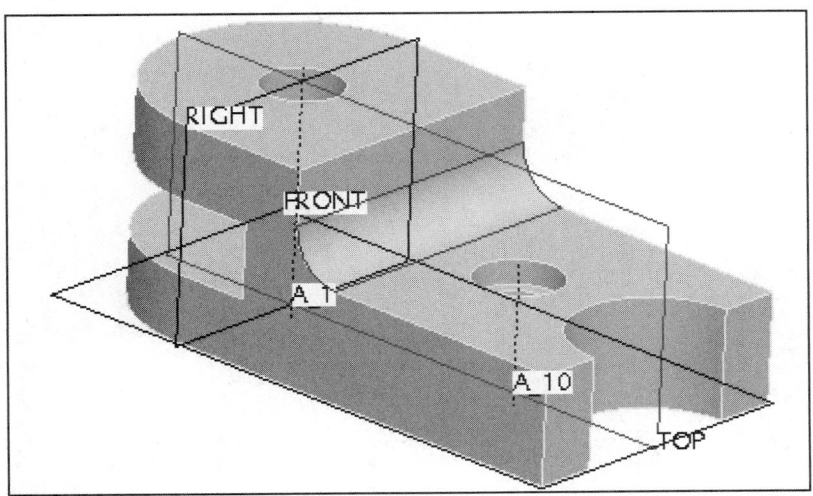

Figure 4.56 Completed Round

In general, consider these recommendations for creating rounds:

- Try to add rounds as late in the design as possible (but before machining features)
- Place all the rounds on a layer and then suppress that layer to speed up your working session
- To avoid creating children dependent on the round features, do *not* dimension to edges or tangent edges created by rounds

Click: **Round Tool** ⇒ ⇒ **Datum planes off** ⇒ **Datum axes off** ⇒ **Datum points off** ⇒ **Coordinate systems off** ⇒ (**All** is the default Selection Filter) ⇒ 0.125 type **.125** for the radius value ⇒ **Enter** ⇒ hold down the **Ctrl** key and select the edges [Fig. 4.57(a-e)] ⇒ **Sets** from the dashboard (Fig. 4.58) ⇒ **MMB** (Fig. 4.59) ⇒ ⇒ **MMB** ⇒ **File** ⇒ **Delete** ⇒ **Old Versions** ⇒ **MMB**

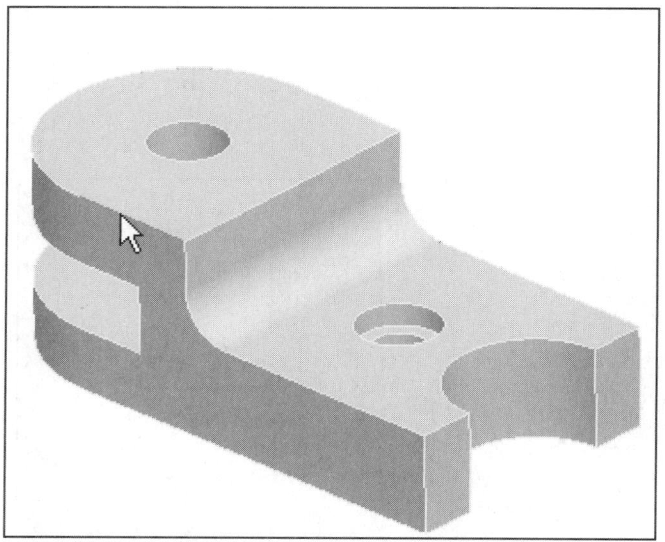

Figure 4.57(a) Datum Features Turned Off

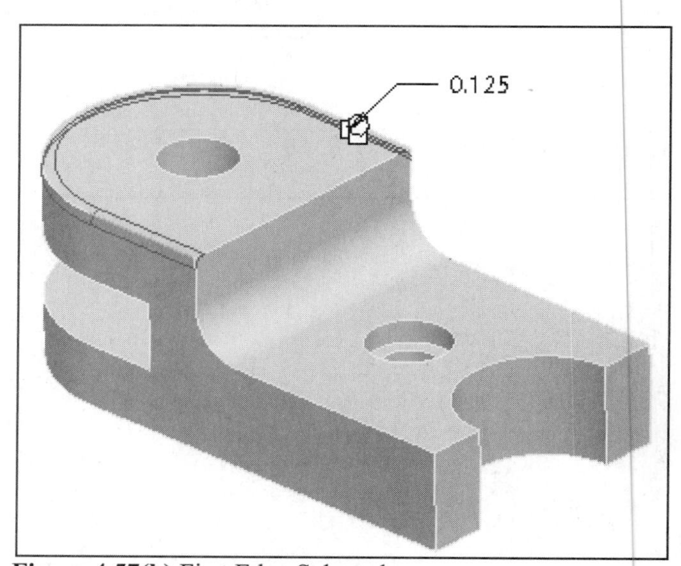

Figure 4.57(b) First Edge Selected

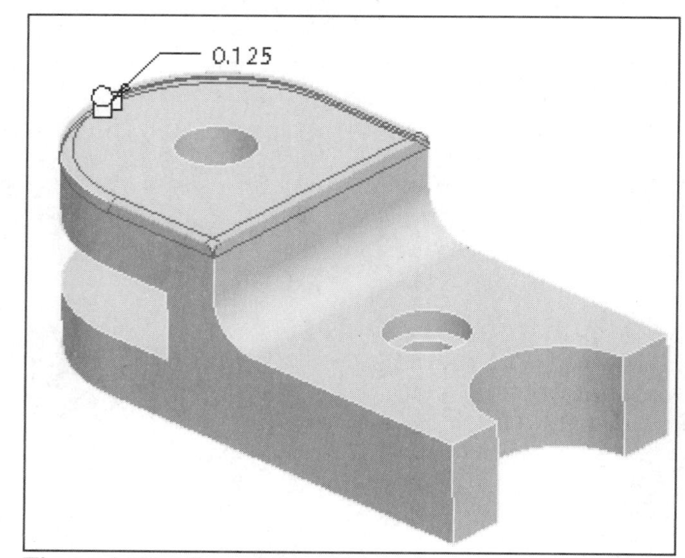

Figure 4.57(c) Second Edge Selected

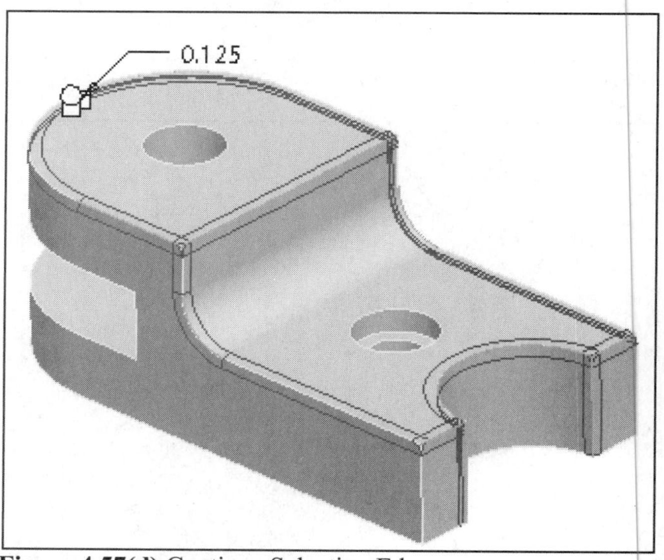

Figure 4.57(d) Continue Selecting Edges

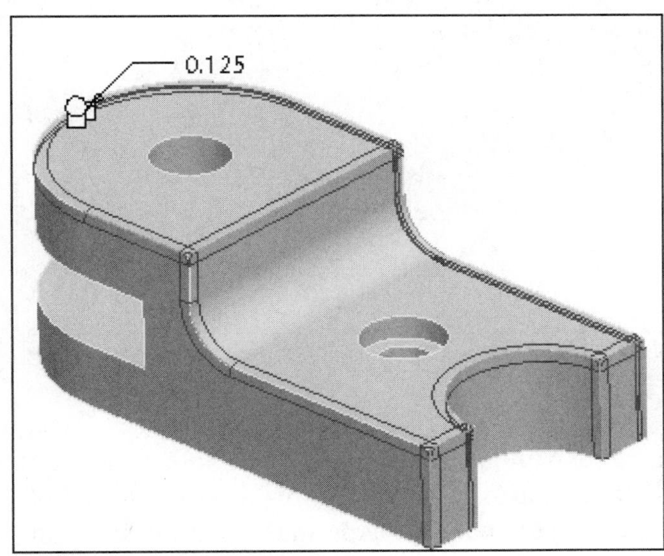

Figure 4.57(e) Selected Edges

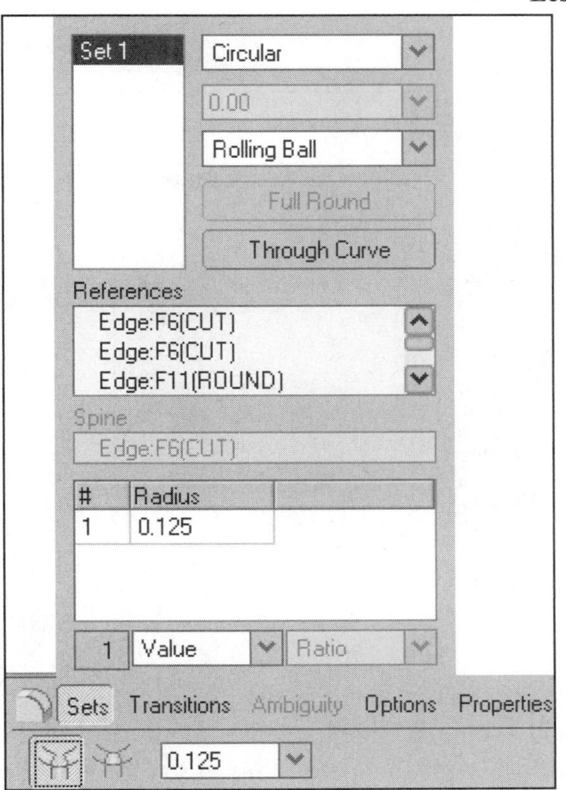

Figure 4.58 Sets

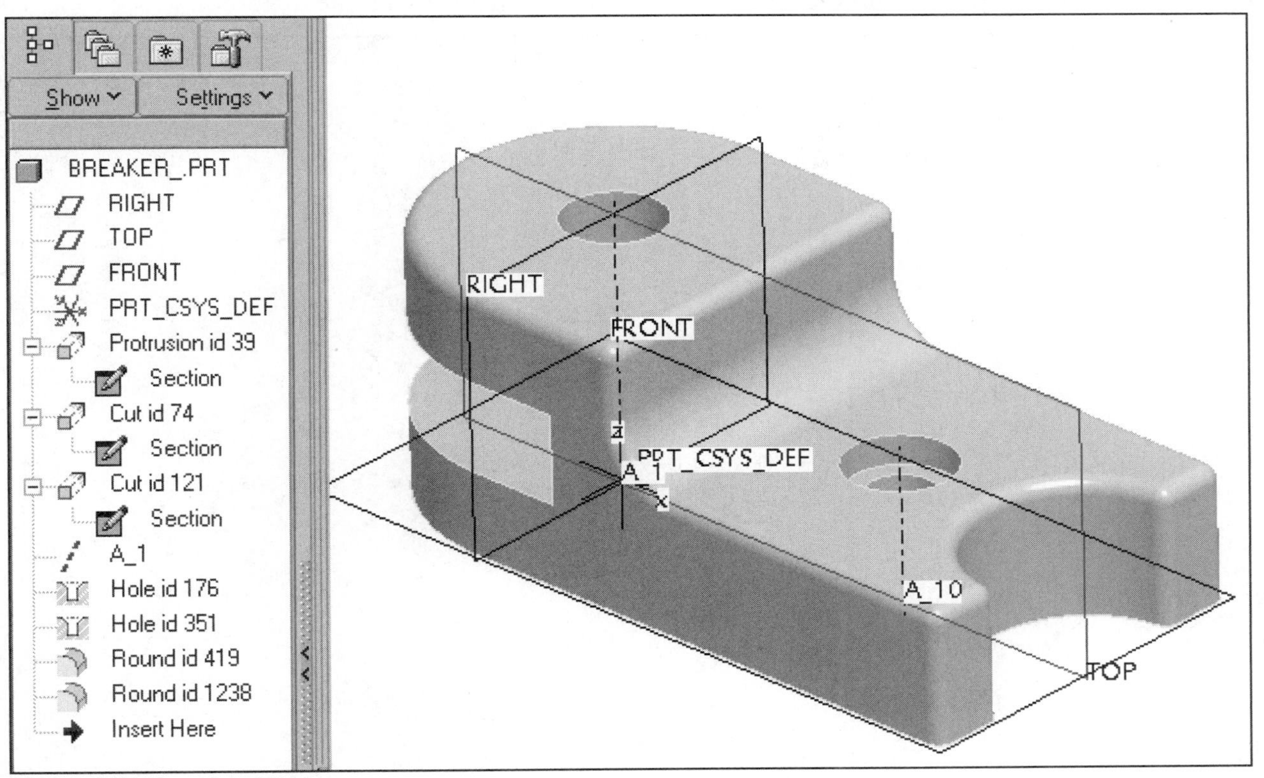

Figure 4.59 Completed Part

Lesson 4 is now complete; continue on to the lesson project.

Lesson 4 Project

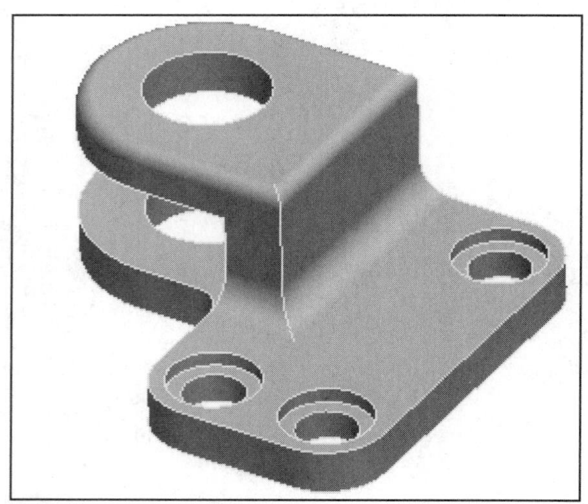

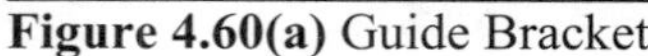

Figure 4.60(a) Guide Bracket

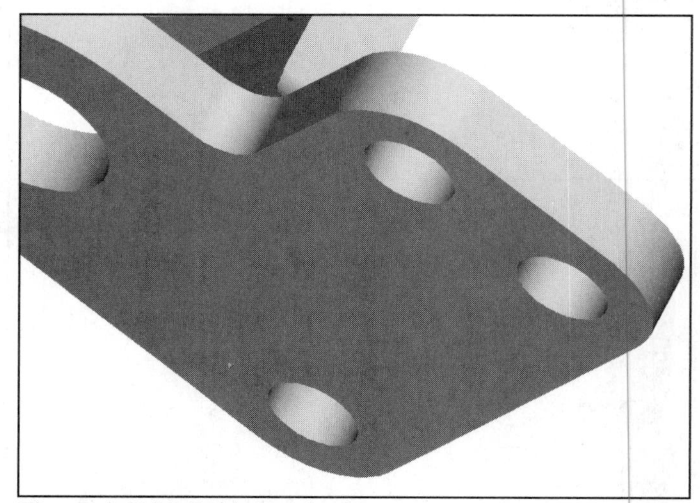

Figure 4.60(b) Guide Bracket Bottom

Guide Bracket

The Guide Bracket is a machined part that requires commands similar to the Breaker. Simple rounds and straight and sketched holes are part of the exercise. Create the part shown in Figures 4.60 through 4.65. At this stage in your understanding of Pro/E, you should be able to analyze the part and plan the steps and features required to model it. You must use the same dimensions and dimensioning scheme, but the choice and quantity of datum planes and the sequence of modeling features can be different.

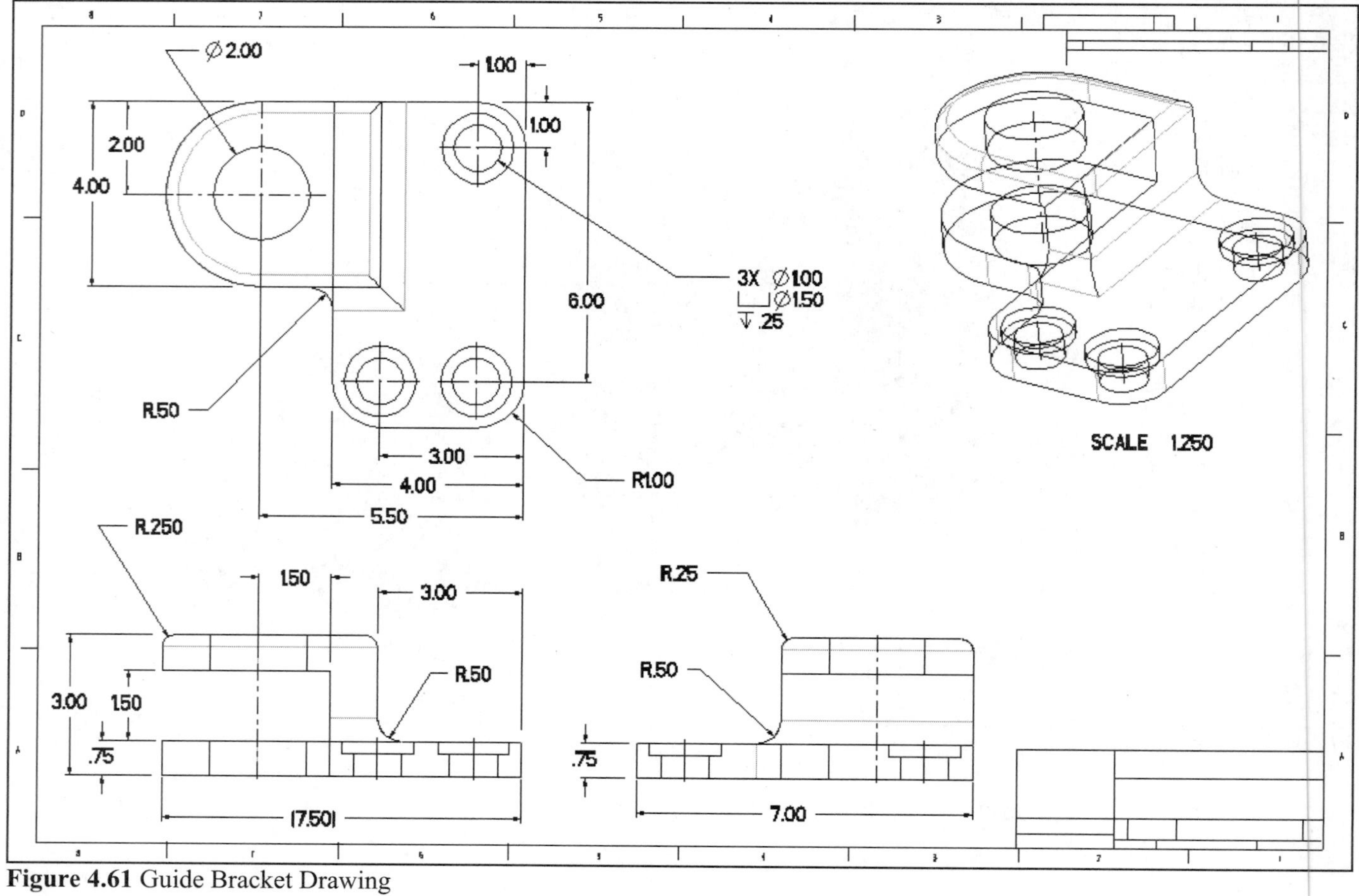

Figure 4.61 Guide Bracket Drawing

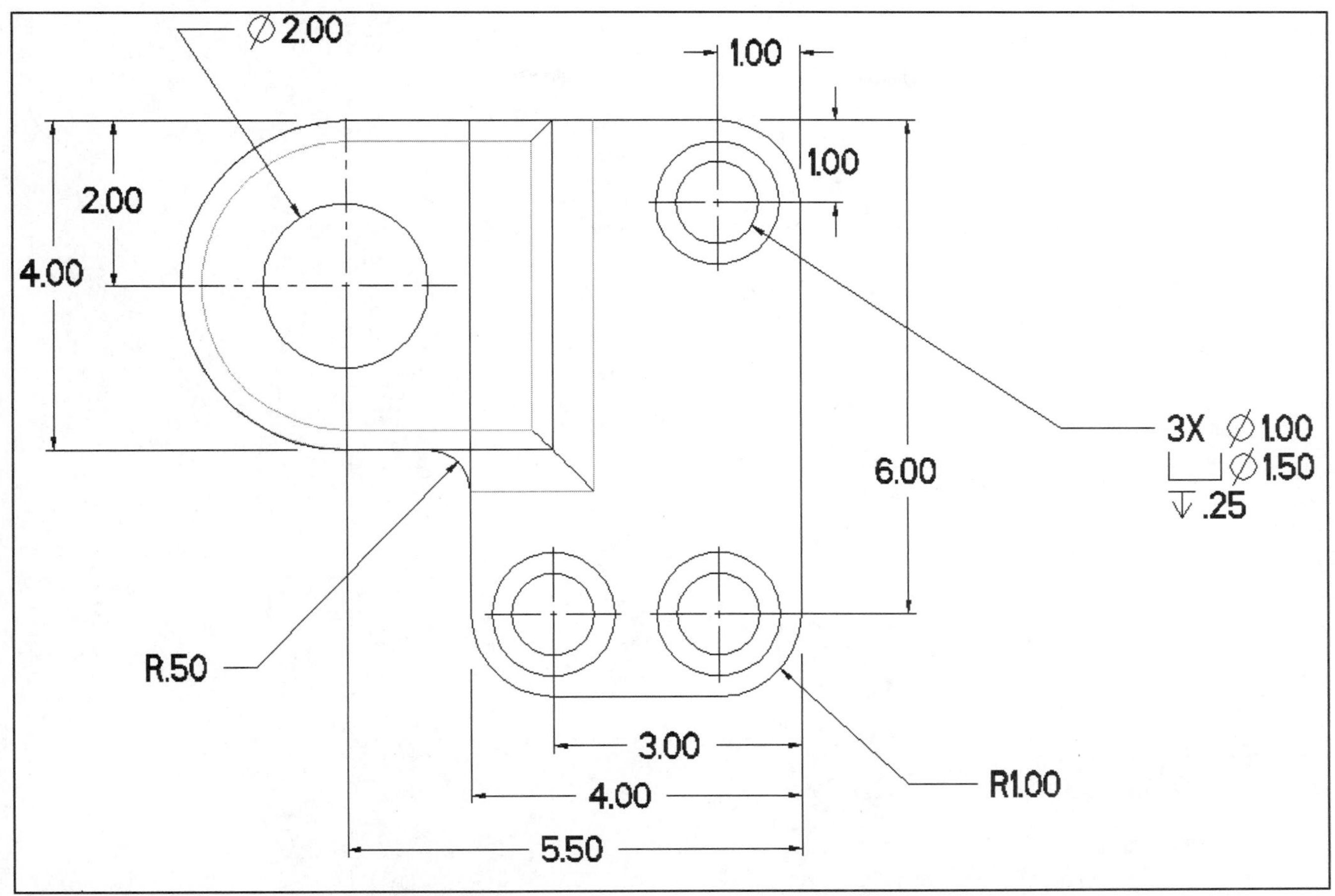

Figure 4.62 Guide Bracket Drawing, Top View

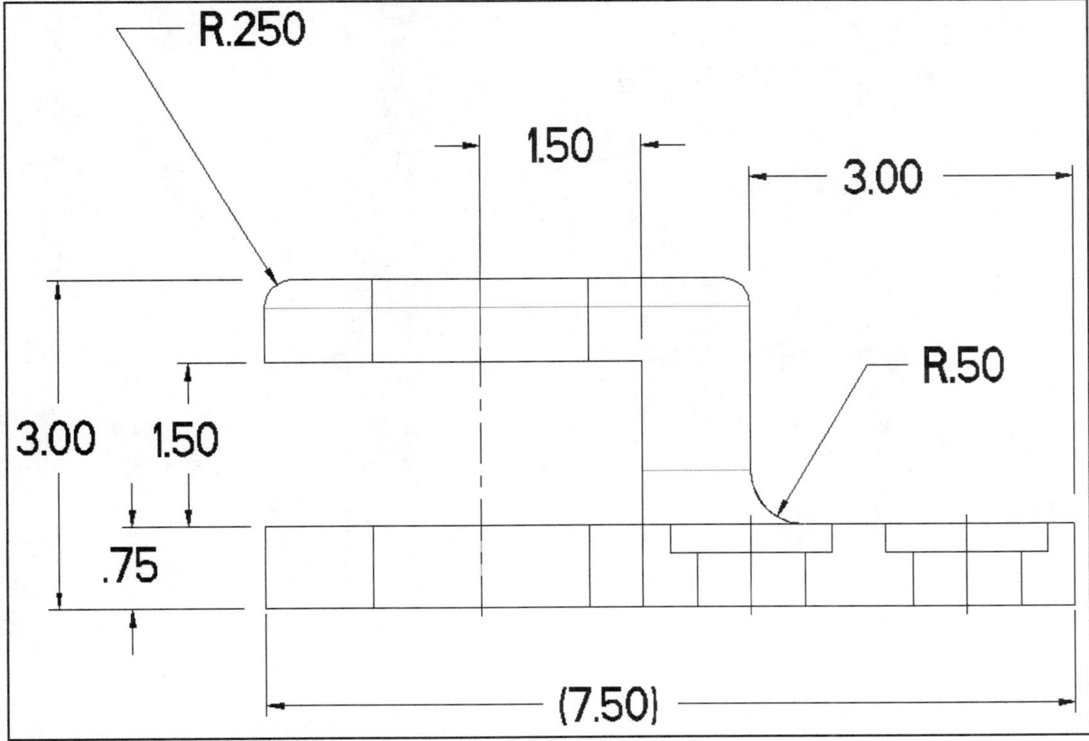

Figure 4.63 Guide Bracket Drawing, Front View

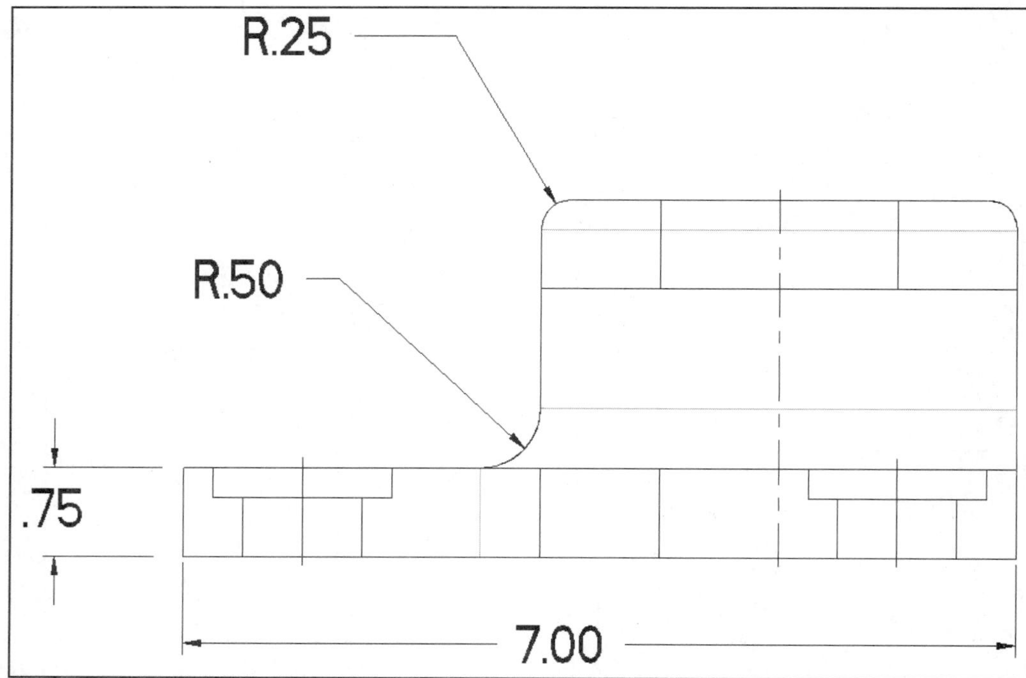

Figure 4.64 Guide Bracket Drawing, Right Side View

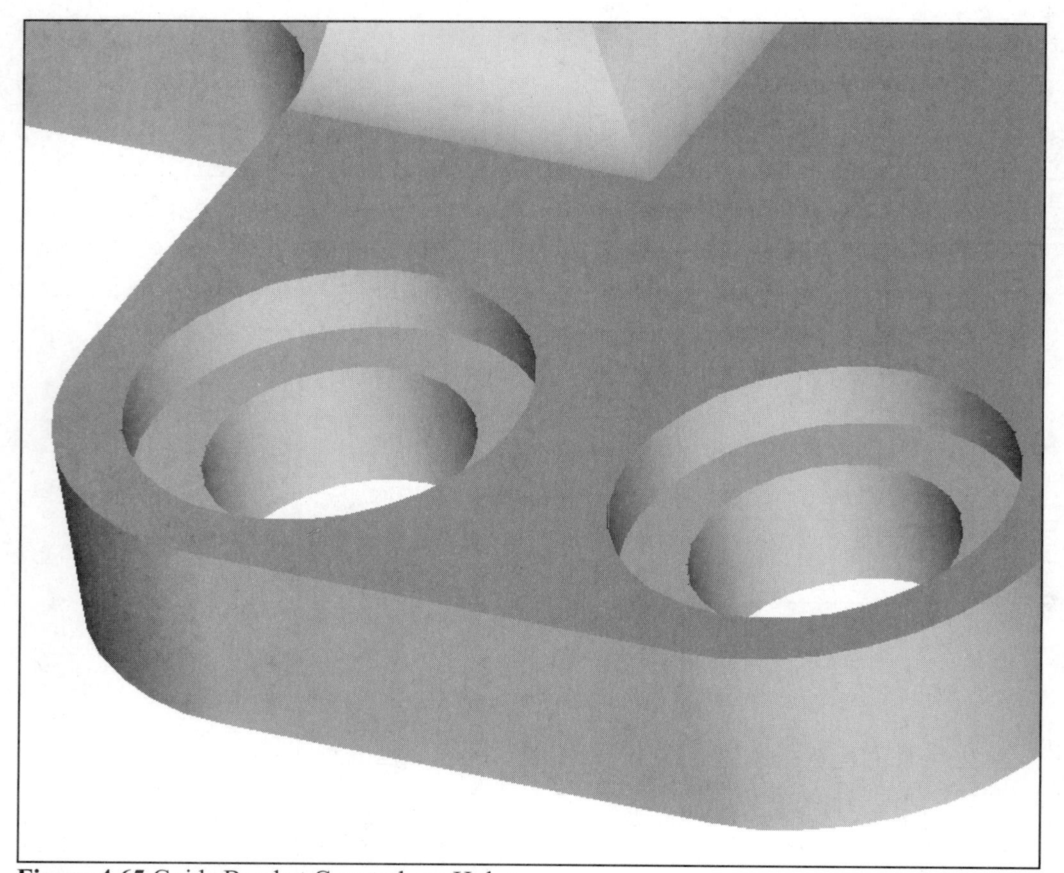

Figure 4.65 Guide Bracket Counterbore Holes

Lesson 5 Datums, Layers and Sections

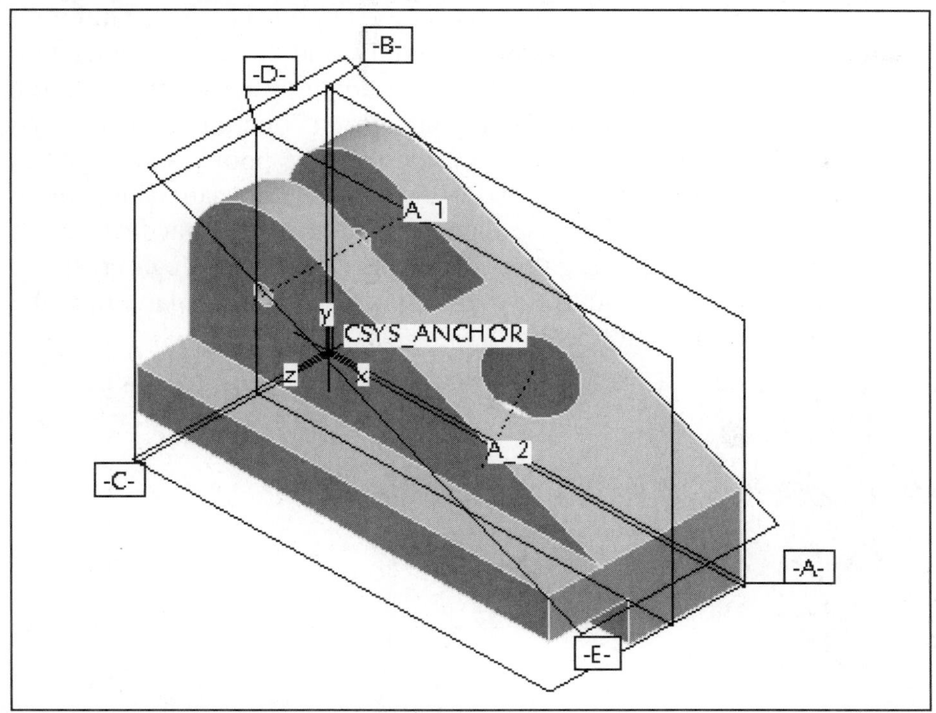

Figure 5.1 Anchor Model with Datum Features

OBJECTIVES

- Create **Datums** to locate features
- Set datum planes for **geometric tolerancing**
- Learn how to change the **Color** and **Shading** of models
- Use **Layers** to organize part features
- Use datum planes to establish **Model Sectioning**
- Add a simple **Relation** to control a feature
- Use **Info** command to extract **Relations** information

DATUMS AND LAYERS

Datums and **layers** are two of the most useful mechanisms for creating and organizing your design (Fig. 5.1). Features such as *datum planes* and *datum axes* are essential for the creation of all parts, assemblies, and drawings using Pro/E.

Layers are an essential tool for grouping items and performing operations on them, such as selecting, displaying or blanking, plotting, and suppressing. Any number of layers can be created. User-defined names are available, so layer names can be easily recognized.

Most companies have a layering scheme that serves as a *default standard* so that all projects follow the same naming conventions and objects/items are easily located by anyone with access. Layer information, such as display status, is stored with each individual part, assembly, or drawing.

Datum Features

You can create all datum feature types, including points, axes, planes, coordinate systems, curves, graphs, and analyses, by clicking Insert ⇒ Model Datum (Fig. 5.2) or clicking on the appropriate icon button on the tool bar. During feature creation, model investigation, component assembly, and other operations, you are required to select one or more datum points, axes, planes, and/or coordinate systems, but in many cases, the desired datum does not exist. It is unnecessary to abort your current operation in order to navigate the menus and create the desired datum feature. You can create datum points, axes, planes (including offset planes), curves, ribbons, and coordinate systems at any time. You can create these features even during the creation of another feature, and then use them as references of that feature. At any point in feature creation, measurement, or analysis, if a datum feature is needed, you can create it. This applies even for nested datum features, such as a condition where a point, axis, and plane must be created simply to get a reference for an offset or angled datum (Fig. 5.3).

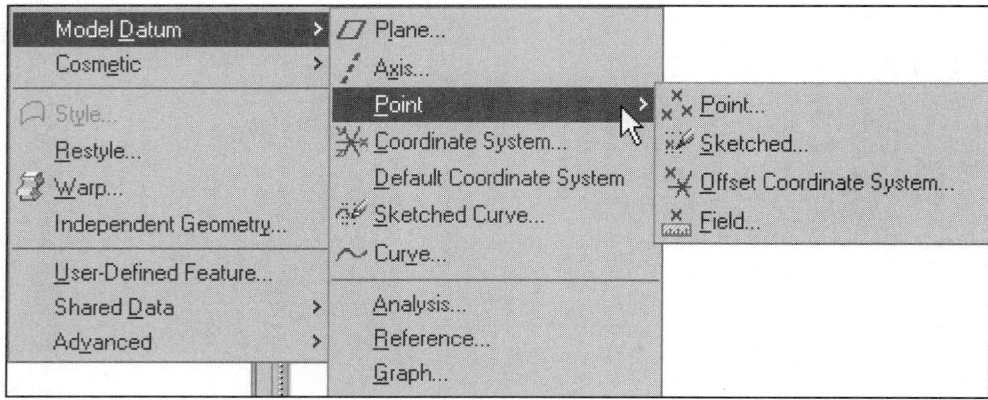

Figure 5.2 Datum Menu

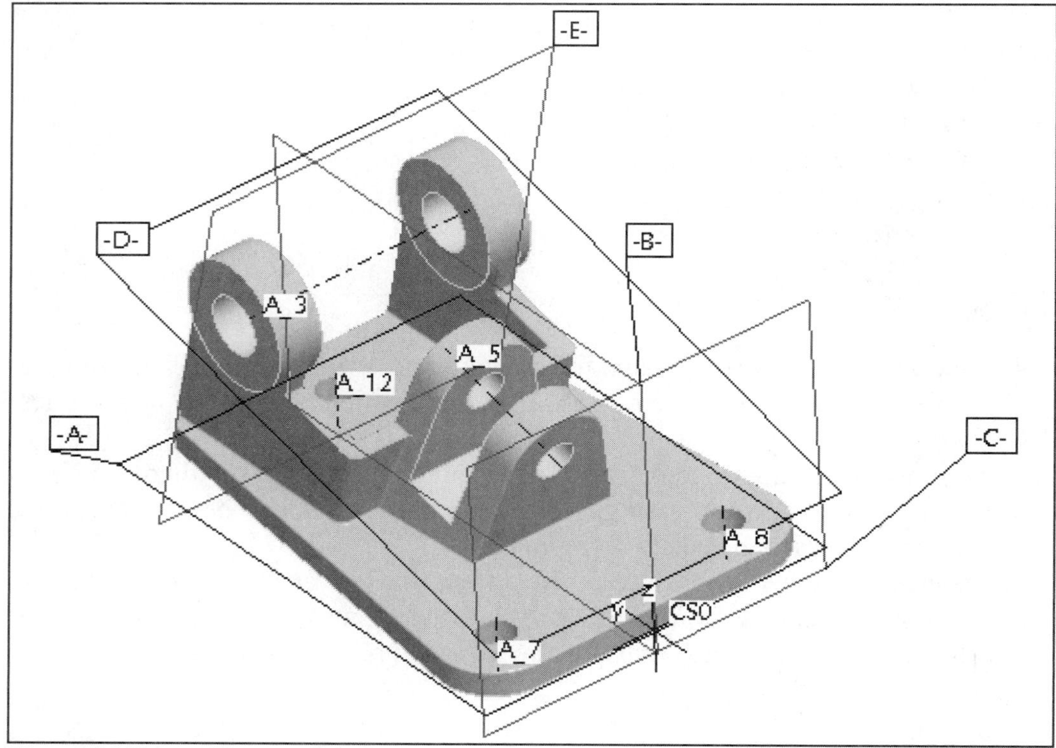

Figure 5.3 Datum Planes and Axes

Datum Planes

Datum planes are used to create a reference for a part that does not already exist. For example, you can sketch or place features on a datum plane when there is no appropriate planar surface; you can also dimension to a datum plane as though it were an edge. When you are constructing an assembly, you can use datums with assembly commands. Datum planes have a positive *tan* side and a negative *gray* side so that you know on which side you are working. You can see the default colors by picking: View ⇒ Display Settings ⇒ Systems Colors ⇒ Datum tab (Fig. 5.4). The color of datum will change if you change the Pro/E color scheme (Fig. 5.5).

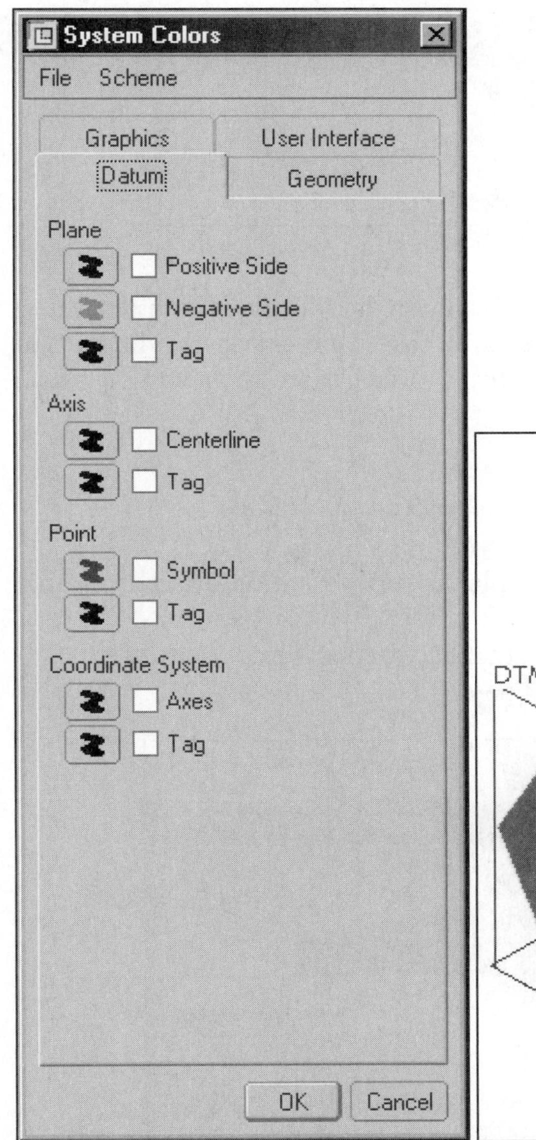

Figure 5.4 System Colors Datum

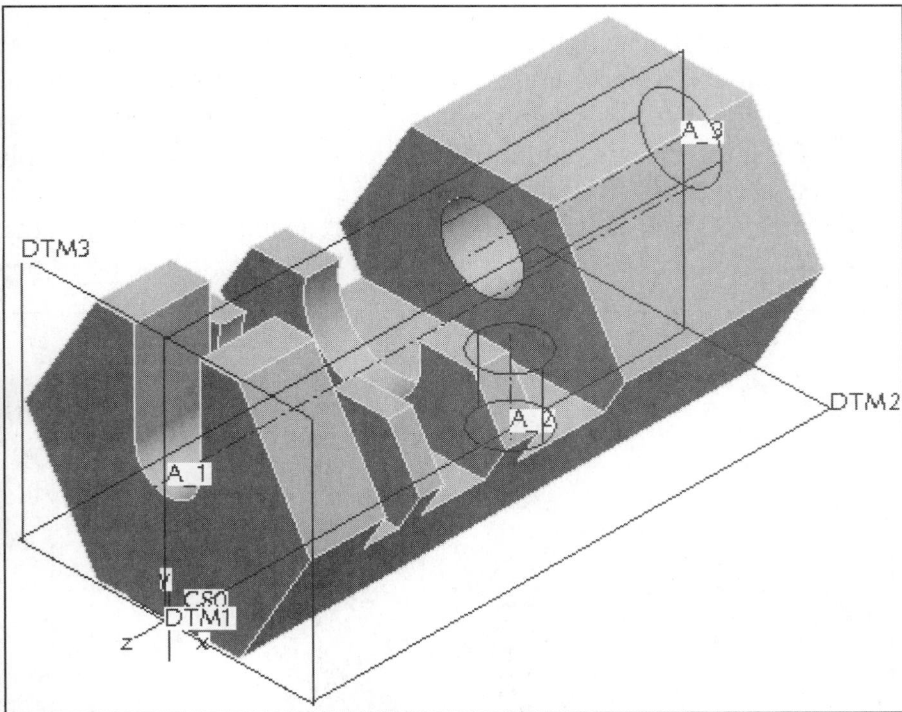

Figure 5.5 System Colors Black on White

In all models, the first features created are the three default datum planes. When you use [✓ Use default template], the default datum are named **FRONT, TOP,** and **RIGHT.** These three datums and the coordinate system are the parents of all other features. The base construction feature (a protrusion) is created using these datums. Datum features can be created as needed to complete the design. Datum planes can be used as references, as sketching planes, and as parent features for a variety of non-sketched part features. Assemblies (Fig. 5.6) also require that the first features be datum planes.

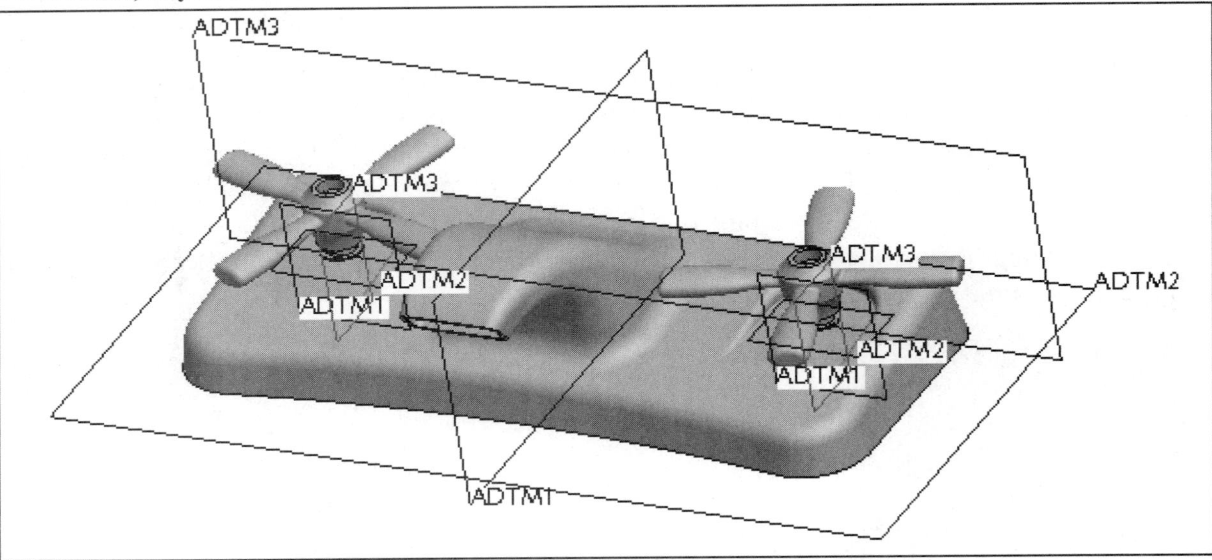

Figure 5.6 Assembly with Datum Features (CADTRAIN, COAch for Pro/ENGINEER)

A non-default datum is created by specifying one or more constraints that locate it with respect to existing geometry. As an example, a datum plane might be created to pass **Tangent** to a cylinder and **Parallel** to a planar surface. In Figure 5.7, a datum is being introduced offset from an existing surface. The following list provides some datum plane constraints:

- **Through** Creates a datum plane coincident with a planar surface
- **Normal** Creates a datum plane that is normal to an object that has been selected
- **Parallel** Creates a datum plane that is parallel to a selected object
- **Offset** Creates a datum plane that is parallel to a selected planar surface and is offset from that surface by a specified distance
- **Angle** Creates a datum plane that is at an angle to an object that has been selected
- **Tangent** Creates a datum plane that is tangent to a selected object

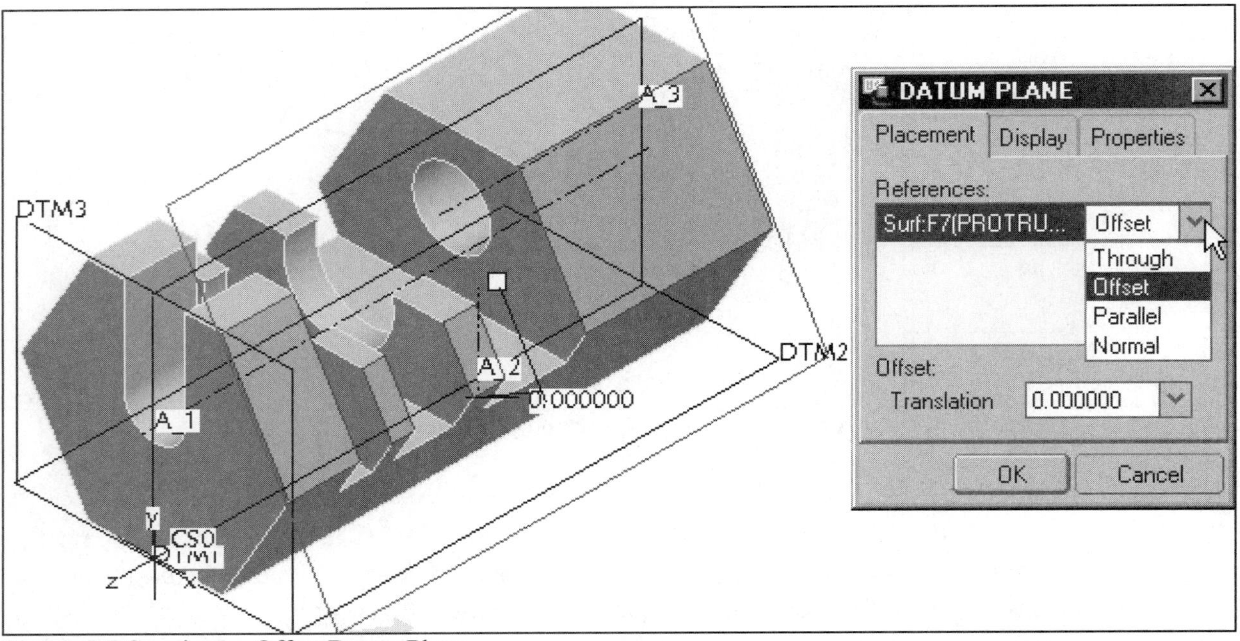

Figure 5.7 Creating an Offset Datum Plane

The size of a displayed datum plane changes with the dimensions of a part. Datum planes can be sized to specific geometry using Edit Definition (Fig. 5.8). You can make a datum plane as big as the model or as small as an edge or surface on the model. Radius sizes the datum plane to fit a specified radius, centering itself within the constraints of the model.

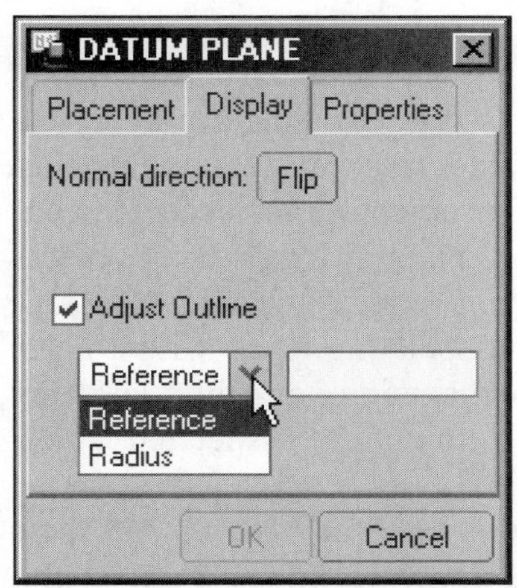

Figure 5.8 DATUM PLANE Dialog Box, Display

Datum Axes

Datum Axes (Fig. 5.9) can be used as references for feature creation, such as the coaxial placement of a hole. They are particularly useful for making datum planes, for placing items concentrically, and for creating radial patterns.

Axes can be used to measure from, place coordinate systems, and place specific features. The angle between a feature and an axis, the distance between an axis and a feature, and so on, can be determined using the **Info** command. Axes (appearing as centerlines) can be automatically created for:

- **Revolved features** All features whose geometry is revolved, including revolved base features, holes (Fig 5.9), shafts, revolved slots, cuts, and circular protrusions
- **Extruded circles** An axis is created for every extruded circle in any extruded feature
- **Extruded arcs** An axis can be created automatically for extruded arcs only when you set the configuration option *(show_axes_for_extr_arcs yes)*

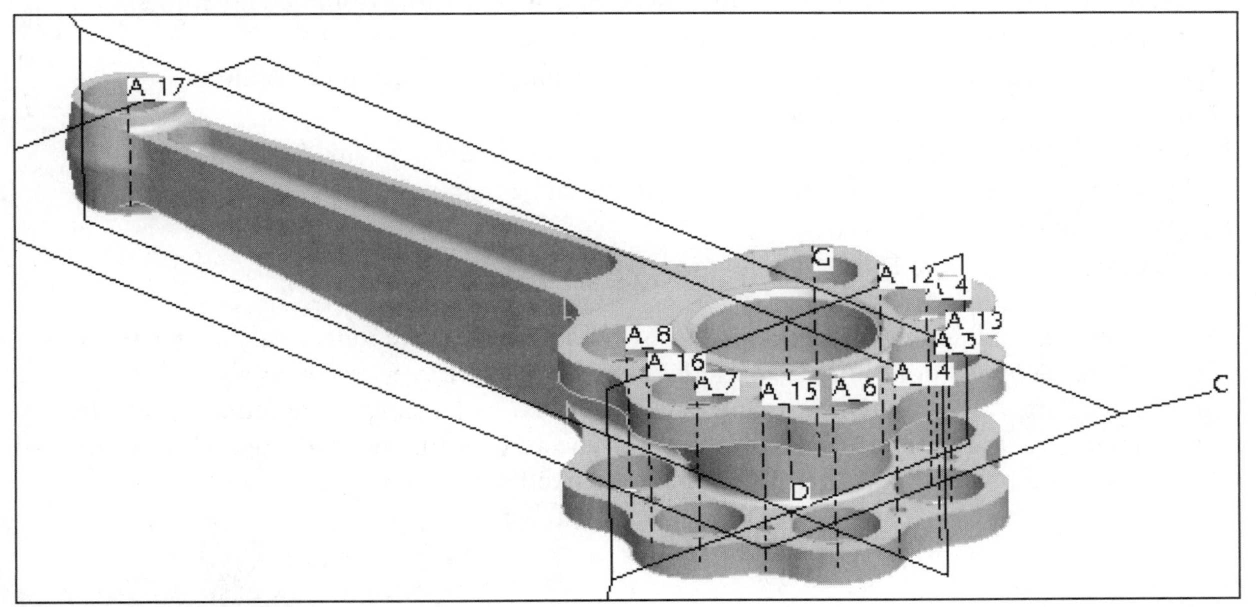

Figure 5.9 Anchor Model

Layers

Layers provide a means of organizing model items, such as features, datum planes, parts (in an assembly), and even other layers, so that you can perform operations on those items collectively. These operations primarily include ways of showing the items in the model, such as displaying or blanking, selecting, and suppressing.

You can have as many layers as you need in a model or a layout. You can also have items associated with more than one layer. For example, you could have an axis associated with several layers. Layer display status is stored locally with its object. This means that the changes made to the layer display status of one object do not affect similarly named layers present in any other object active in the current session of Pro/E. However, changes to the layers in assemblies may affect layers in lower-level objects (subassemblies or parts). Layer functionality is the same in Part mode as it is in Assembly mode.

You use layers in a model or a layout as an organizational tool, by associating items with a layer so that you can collectively manipulate them. One of the main reasons for organizing items into layers is to make it possible to show or blank them selectively. These functions affect how items associated with a layer are displayed in the Pro/E window.

Using the Layer Tree, you can manipulate layers, their items, and their display status. To access the Layer Tree, click: Show ⇒ Layer Tree in the Model Tree navigator window (Fig. 5.10). The Layers command is available while you are working below the highest-level functional menu of any mode (Drawing mode, Assembly mode, Part mode, and so forth), so that you can manipulate layer display status or layer membership as required without having to return to the Part, Assembly, or top-level menu.

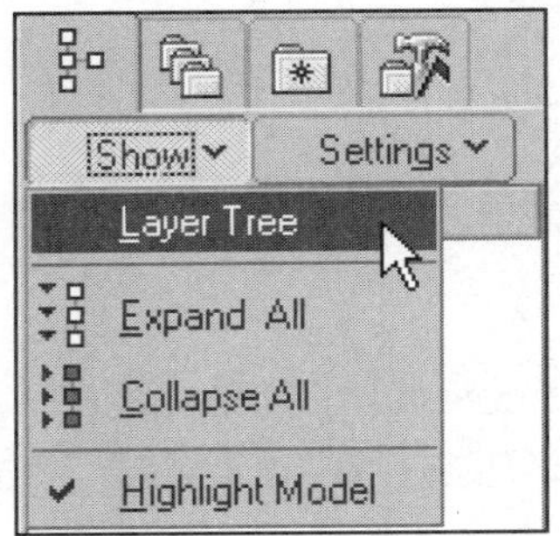

Figure 5.10 Show Layer Tree in Navigator Window

Layers are identified by name. Layer names can be expressed in numeric or alphanumeric form, with a maximum of 31 characters per name. When layers are displayed, numeric layer names are sorted first, then alphanumeric layer names. Layer names in alphabetic form are sorted alphabetically.

Using layers functionality, you can perform the following operations:

- Create, delete, and rename layers
- Copy all items of one layer to a different layer
- Add items to a layer, remove items from a layer, and switch items from one layer to another
- View and change the display status of layers
- Define global default layers using Pro/TABLE or using configuration file options
- Define local (model specific) layers
- Nest layers by including layers in other layers and show items in nested layers
- Automatically create layers in sub models with the same name
- Integrate with the selection tool
- Add a description note to layers

You can organize models into layers and then show or blank them selectively using the Layer Tree. Showing or blanking layers does not affect model geometry. The Layers are available while you are working in any mode (Part, Assembly, or Drawing, and so on), so that you can manipulate layer display status or layer membership as required. Figure 5.11 shows the Layers for the mouse part. Note that the SURFACES layer has been blanked, all other layers are unblanked.

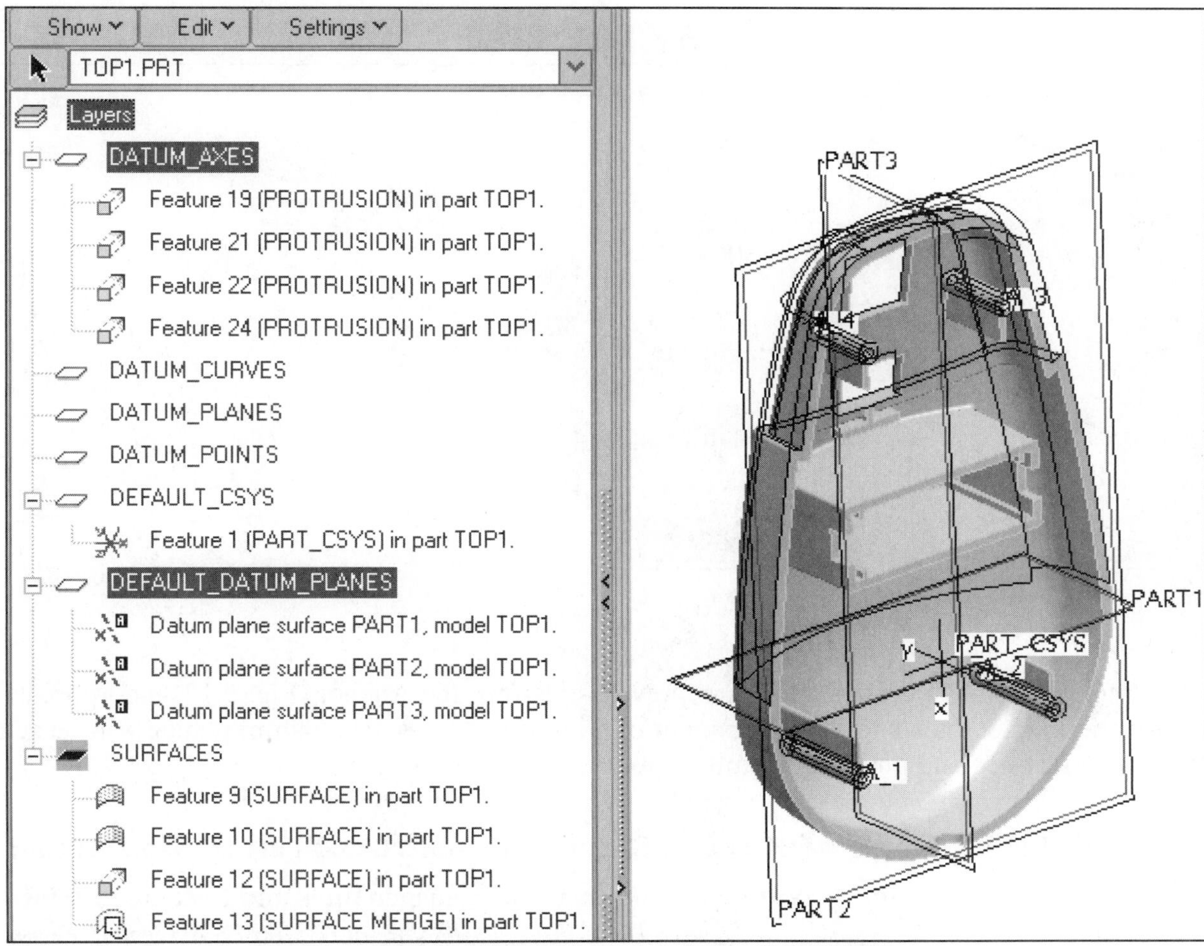

Figure 5.11 Layer Tree for Mouse Part (CADTRAIN, COAch for Pro/ENGINEER

To use layers:

1. Set up the layers to which items will be added:
   - Layer names cannot exceed 31 characters
   - No blank spaces allowed in the name

2. Add items to the specified layers:
   - Using configuration file options, items can be automatically added as they are created
   - An item can be added by selecting its type and then picking the items themselves
   - Features are also added by selecting an option that adds all features of a particular type
   - A range of features can be added to a layer
   - One layer can be added onto another layer
   - Items can be copied from one layer to an existing layer or to a new layer

3. Set the display status of the layers:
   - A layer's name can be picked from an Object name list box
   - A layer status file can be retrieved that contains the desired layer status, which automatically sets each layer to the status specified in the file
   - The current layer file can be edited

Lesson 5 STEPS

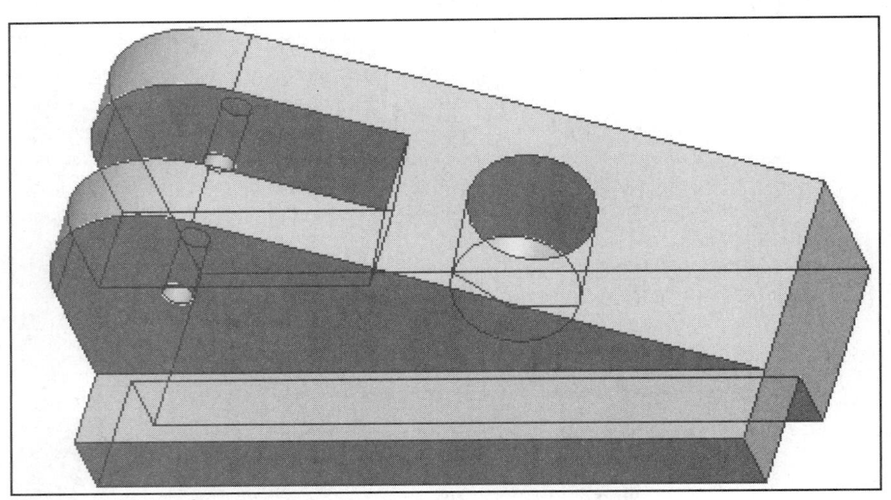

Figure 5.12 Anchor

Anchor

Though default datum planes have been sufficient in previous lessons, the Anchor (Fig. 5.12) incorporates the creation of user-defined datums and the assignment of datums to layers. The datum planes will be set as geometric tolerance features and put on a separate layer.

Click: **File** ⇒ **Set Working Directory** ⇒ **OK** ⇒ **Create a new object** ⇒ ●**Part** ⇒ Name **Anchor** ⇒ Use default template ⇒ **OK** ⇒ **Edit** ⇒ **Setup** ⇒ **Units** ⇒ Units Manager **Inch lbm Second** ⇒ **Close** ⇒ **Material** ⇒ **Define** *or if you have previously saved a material file you may simply* ***Assign*** *it* ***From File*** ⇒ type **Steel** ⇒ **Enter** ⇒ **File** ⇒ **Save** ⇒ **File** ⇒ **Exit** ⇒ **Assign** ⇒ pick **STEEL** ⇒ **Accept** ⇒ **MMB** ⇒ click on the default coordinate system name on the model-- **PRT_CSYS_DEF** ⇒ **RMB** (Fig. 5.13) ⇒ **Rename** ⇒ type **CSYS_ANCHOR** in the **Model Tree** (Fig. 5.14) ⇒ **Enter**

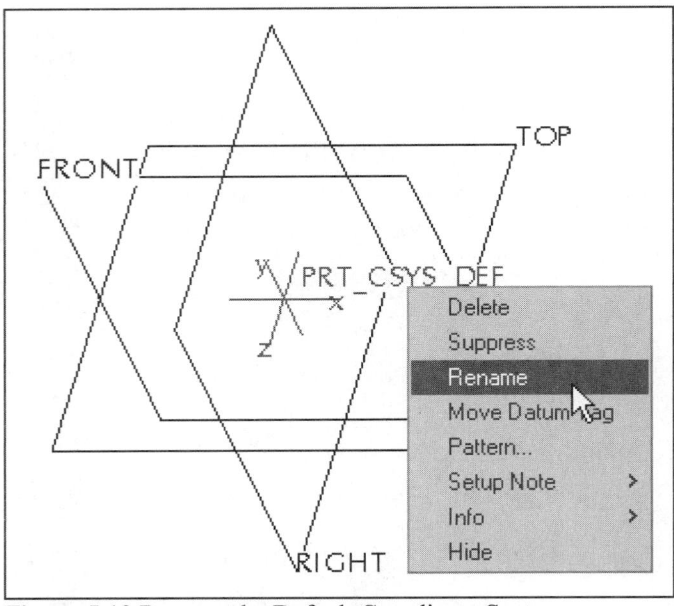

Figure 5.13 Rename the Default Coordinate System

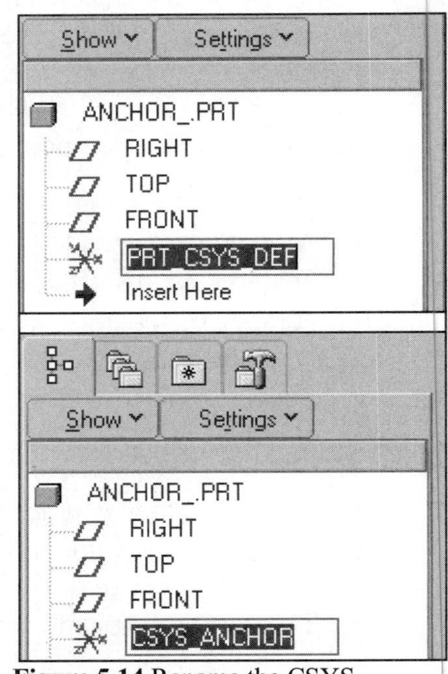

Figure 5.14 Rename the CSYS

It is considered good practice to rename the default coordinate system to something similar to the components name. If an assembly has 25 components, it will have 25 default coordinate systems. Renaming the coordinate system to the component name will make it easier to identify. In the next set of steps, you will show the Layer Tree and see the default layering that automatically takes place as you model. The default datum planes and the default coordinate system are automatically placed on two layers each. For the datum planes; the *part default datum plane layer* and the *part all datum plane layer*. The coordinate system will also be layered in a similar fashion. In Figure 5.16, the *part all datum layer* for the coordinate system displays the new name created for the coordinate system.

Click: **Show** in the **Navigator** [Fig. 5.15(a)] ⇒ **Show** [Fig. 5.15(b)] ⇒ **LMB** to deselect --Layer Tree displays in place of the Model Tree (Fig. 5.16) ⇒ **Save** ⇒ **MMB** ⇒ **Show** in the **Navigator** ⇒ **Model Tree**

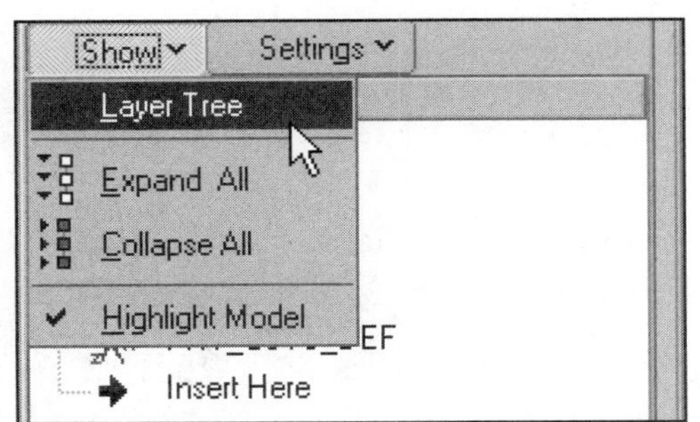

Figure 5.15(a) Show the Layer Tree in the Navigator

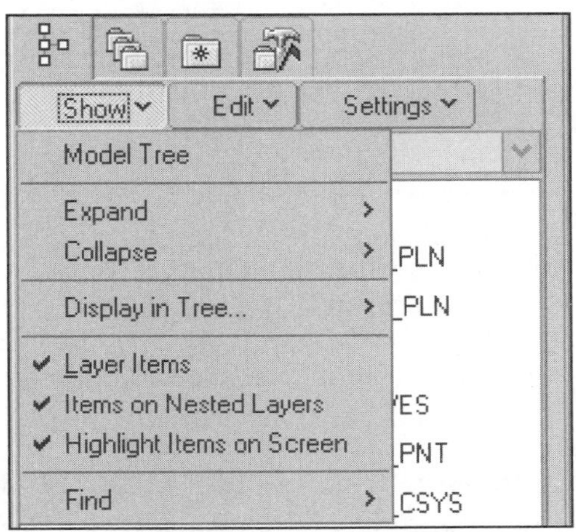

Figure 5.15(b) Show the Layer Tree

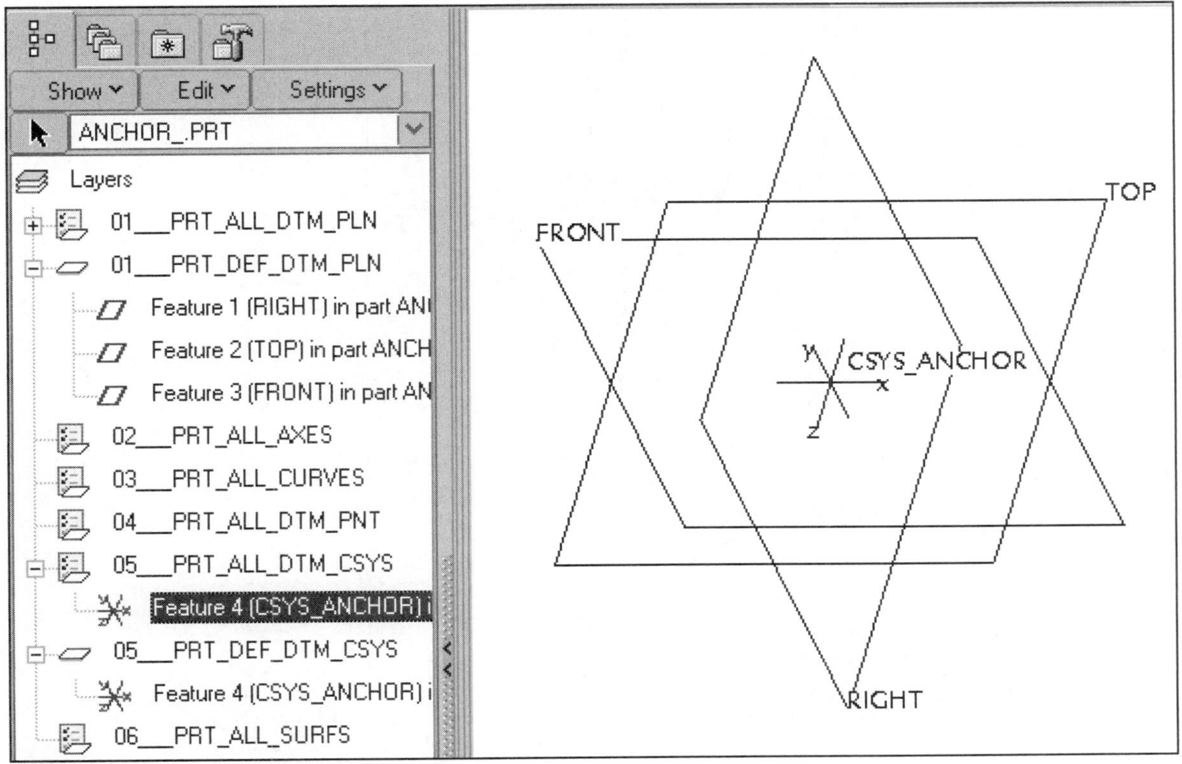

Figure 5.16 Layers Displayed in Layer Tree

As in previous lessons, the first protrusion will be sketched on **FRONT**. Use Figure 5.17 for the protrusion dimensions. Use only the **5.50**, **R1.00**, **1.125**, and **25°** dimensions.

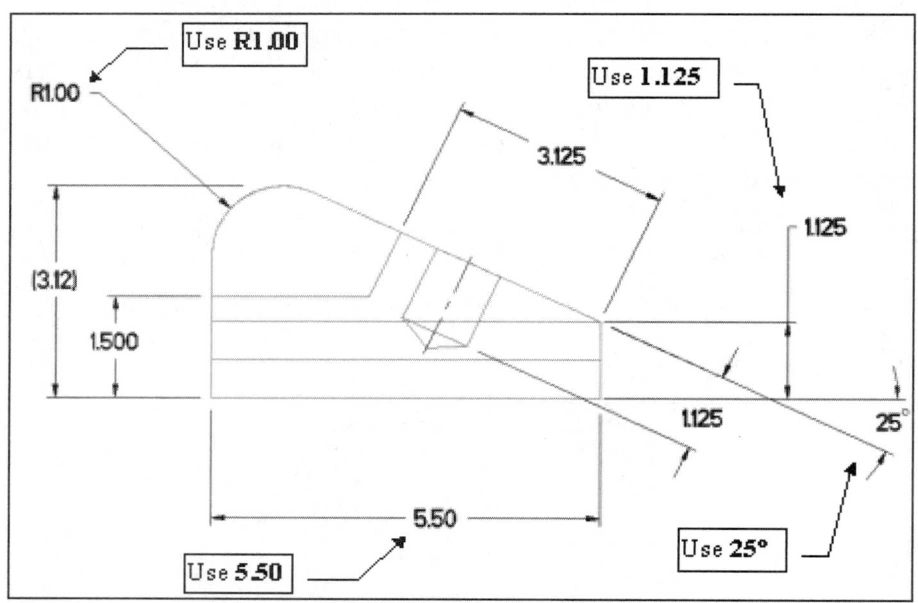

Figure 5.17 Anchor Drawing Front View

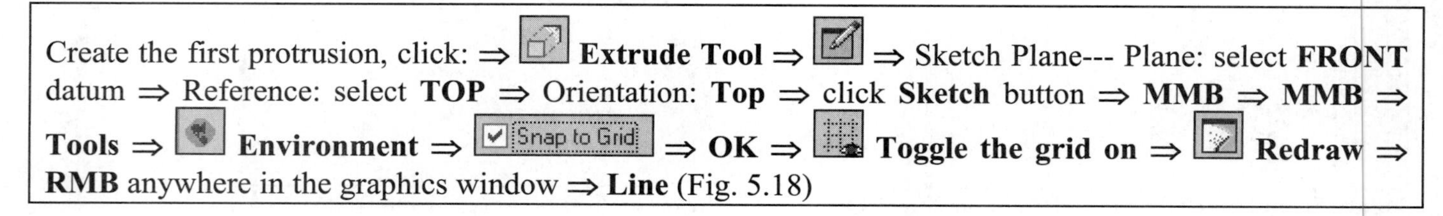
Create the first protrusion, click: ⇒ **Extrude Tool** ⇒ ⇒ Sketch Plane--- Plane: select **FRONT** datum ⇒ Reference: select **TOP** ⇒ Orientation: **Top** ⇒ click **Sketch** button ⇒ **MMB** ⇒ **MMB** ⇒ **Tools** ⇒ **Environment** ⇒ Snap to Grid ⇒ **OK** ⇒ **Toggle the grid on** ⇒ **Redraw** ⇒ **RMB** anywhere in the graphics window ⇒ **Line** (Fig. 5.18)

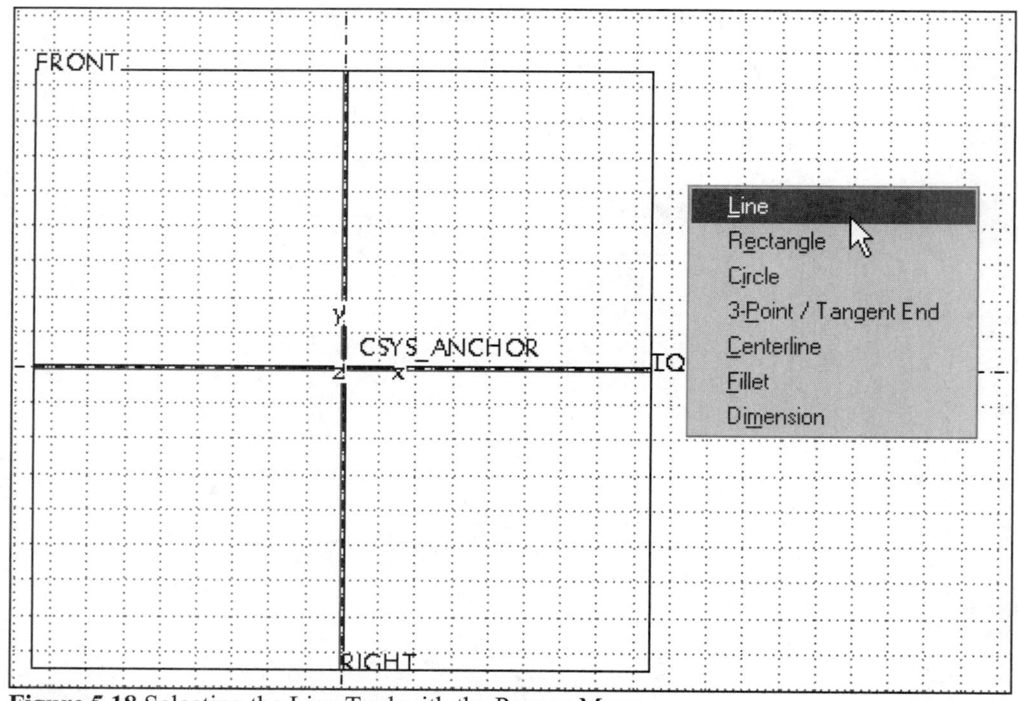

Figure 5.18 Selecting the Line Tool with the Pop-up Menu

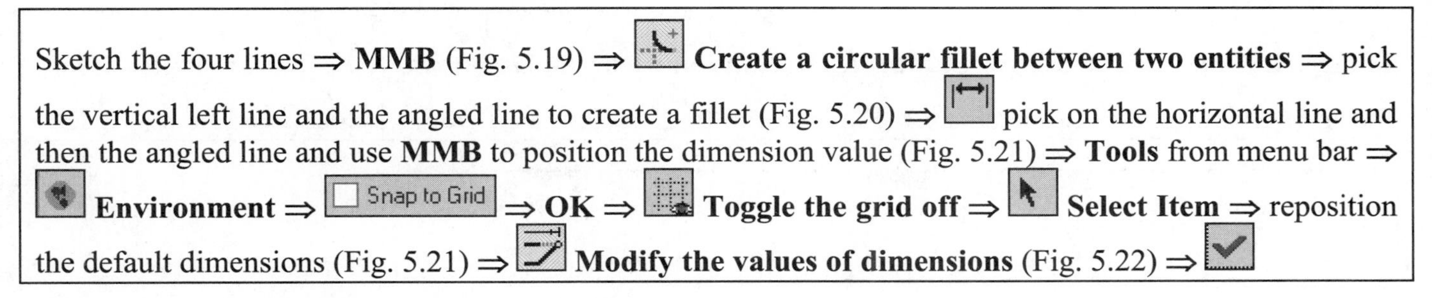
Sketch the four lines ⇒ **MMB** (Fig. 5.19) ⇒ **Create a circular fillet between two entities** ⇒ pick the vertical left line and the angled line to create a fillet (Fig. 5.20) ⇒ pick on the horizontal line and then the angled line and use **MMB** to position the dimension value (Fig. 5.21) ⇒ **Tools** from menu bar ⇒ **Environment** ⇒ Snap to Grid ⇒ **OK** ⇒ **Toggle the grid off** ⇒ **Select Item** ⇒ reposition the default dimensions (Fig. 5.21) ⇒ **Modify the values of dimensions** (Fig. 5.22) ⇒

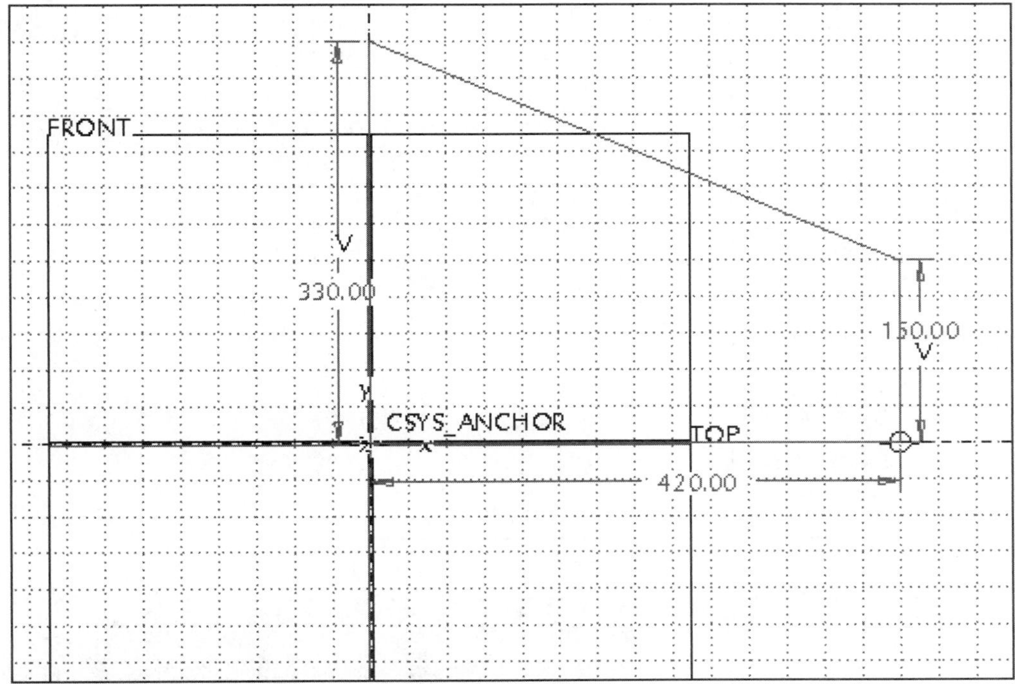

Figure 5.19 Sketch Four Lines to Create a Closed Section

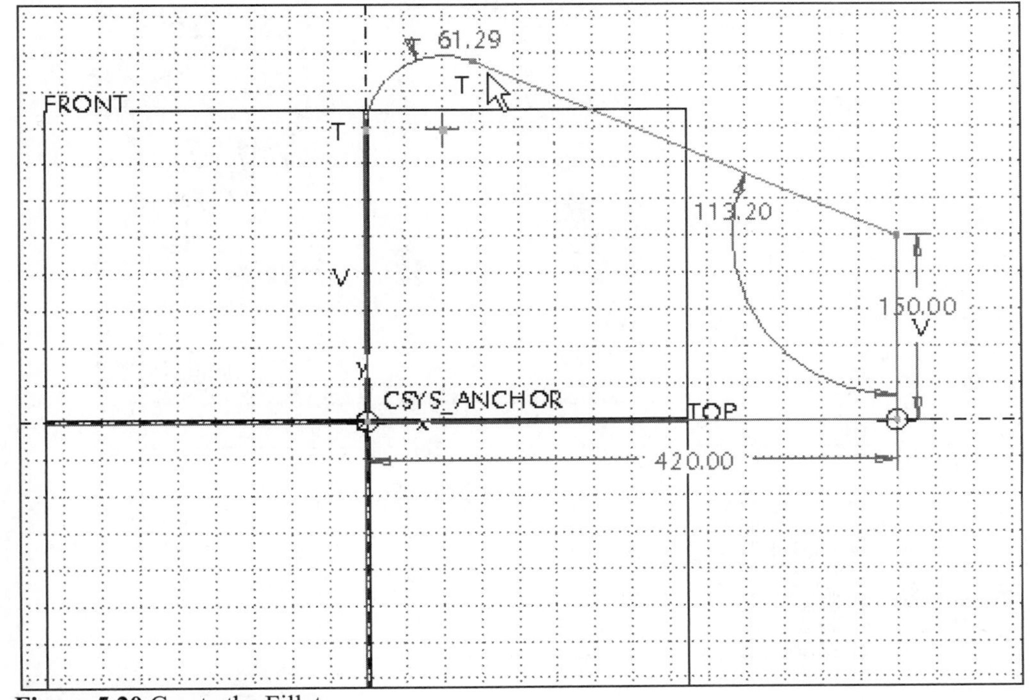

Figure 5.20 Create the Fillet

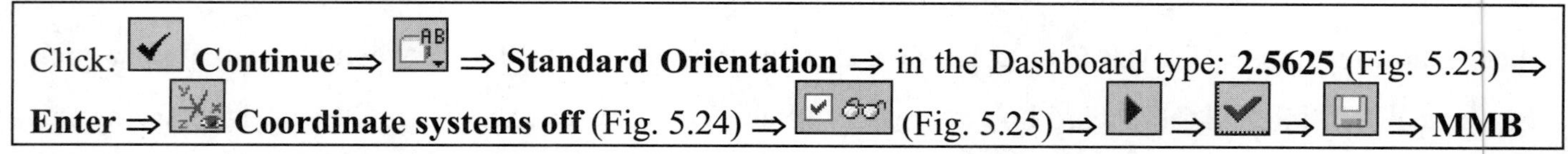

Click: **Continue** ⇒ ⇒ **Standard Orientation** ⇒ in the Dashboard type: **2.5625** (Fig. 5.23) ⇒ **Enter** ⇒ **Coordinate systems off** (Fig. 5.24) ⇒ (Fig. 5.25) ⇒ ⇒ ⇒ ⇒ **MMB**

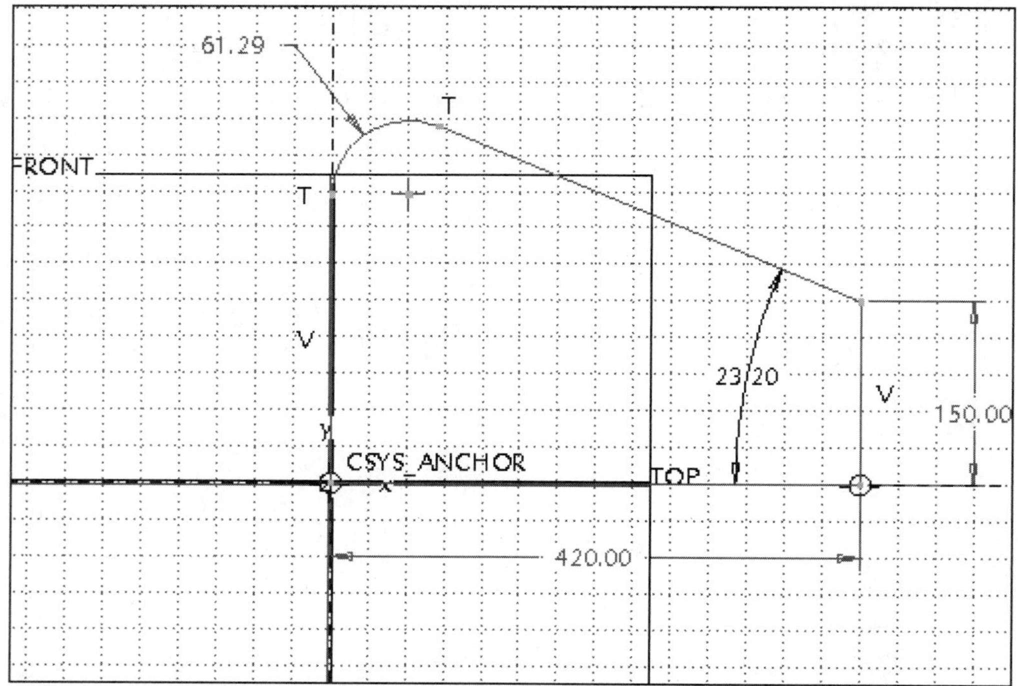

Figure 5.21 Create the Angle Dimension and Reposition the other Dimensions

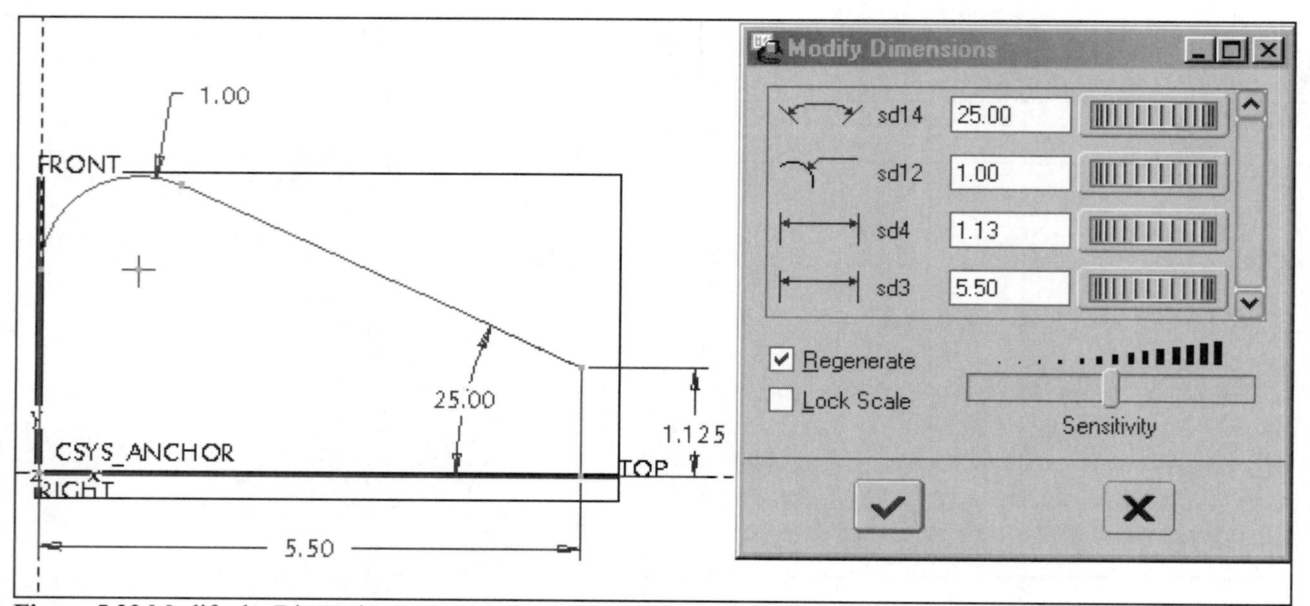

Figure 5.22 Modify the Dimension Values

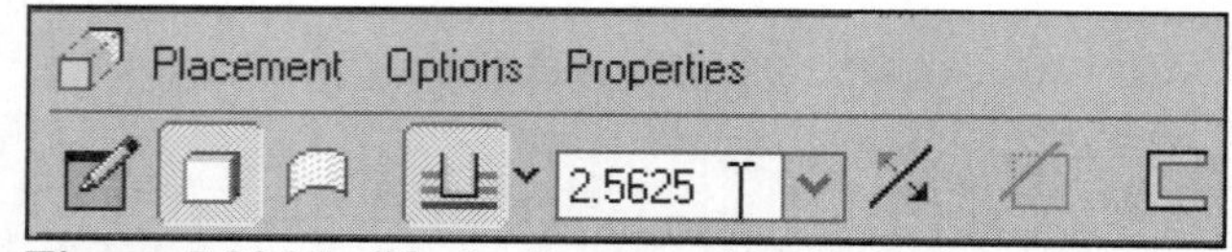

Figure 5.23 Modify the Depth to **2.5625** in the Dashboard

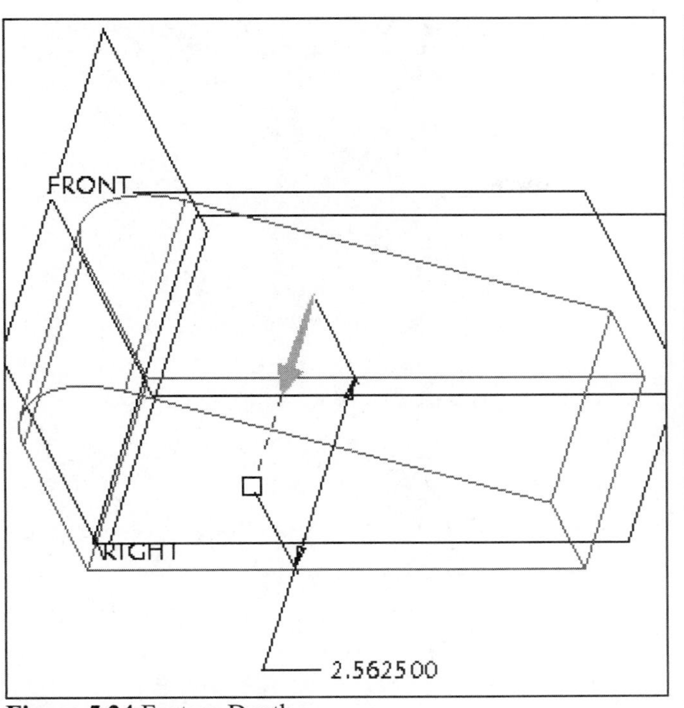

Figure 5.24 Feature Depth

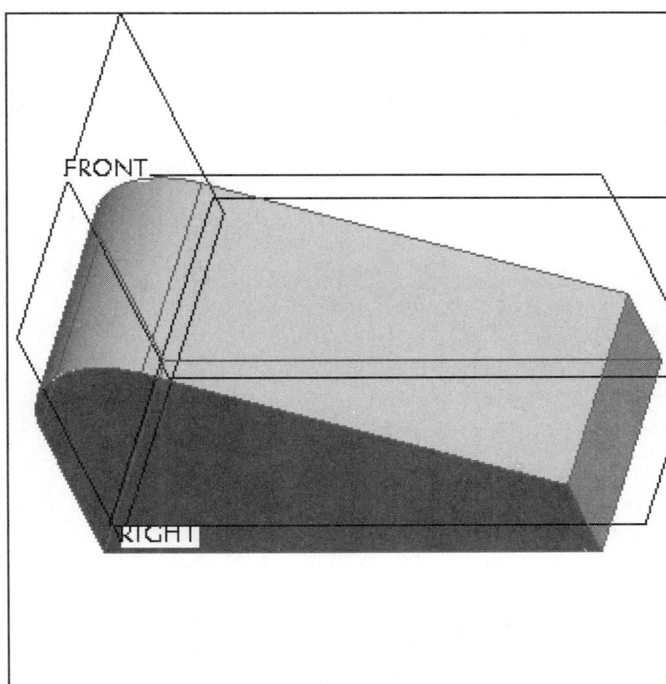

Figure 5.25 Preview

Colors

Colors are used to define material and light properties. Use the **Color Editor** to create and modify colors and to customize the display of geometric items such as curves and surfaces and interface items such as fonts and background. You can access the Color Editor by clicking a color swatch to set the **Appearance Color** or the **Highlight Color**.

Next, you will change the color of the model from the default. In general, try to avoid the colors that are used as Pro/E defaults. Because feature and entity highlighting is defaulted to *red*, colors similar to *red* should be avoided. Datum planes have *tan* and *gray* sides; therefore, they should be avoided. Shades of *blue* and *green* work well. It is really up to you to choose colors that are pleasant to look at and work well with the type of project you are modeling. Remember, if you will be using the parts in an assembly, each component should have a unique color scheme.

Click: **View** ⇒ **Color and Appearance** (Fig. 5.26) ⇒ **Properties** if needed (Fig. 5.27) ⇒ **Basic** tab ⇒ click [+] in the Appearance Editor dialog box

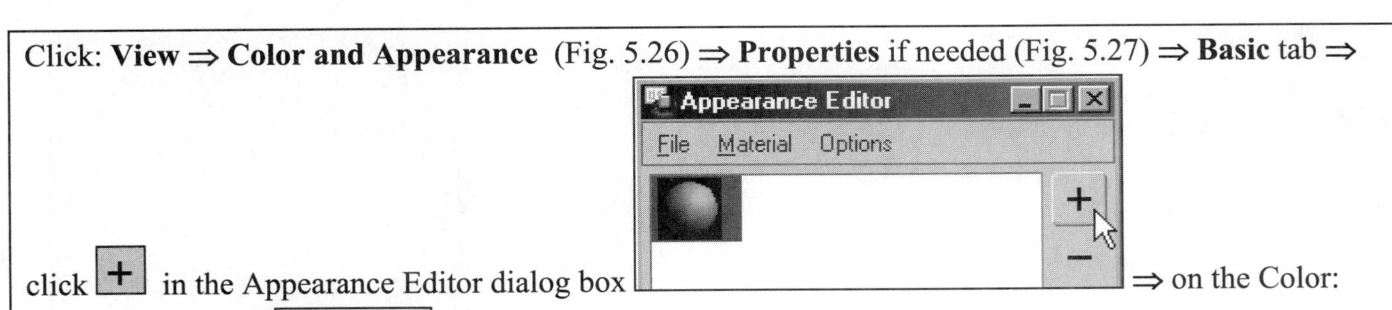

⇒ on the Color: color swatch click [Color] to open the **Color Editor** dialog box (Fig. 5.28) ⇒ slide the RGB/HSV bars to create a new color (Fig. 5.29) ⇒ **Close** ⇒ repeat to make a few more colors, adding each to the palette (Fig. 5.29) ⇒ also, add some Highlights to the colors using the same basic method (Fig. 5.30) ⇒ also, try creating colors with the **Color Wheel** or the **Blending Palette** [Fig. 5.31(a-b)] ⇒ **File** ⇒ **Save** ⇒ type a name for your color file (you can open and use this file for other projects instead of recreating new colors) ⇒ **OK** (Fig. 5.32) ⇒ **Apply** ⇒ **Close** from the Appearance Editor dialog box

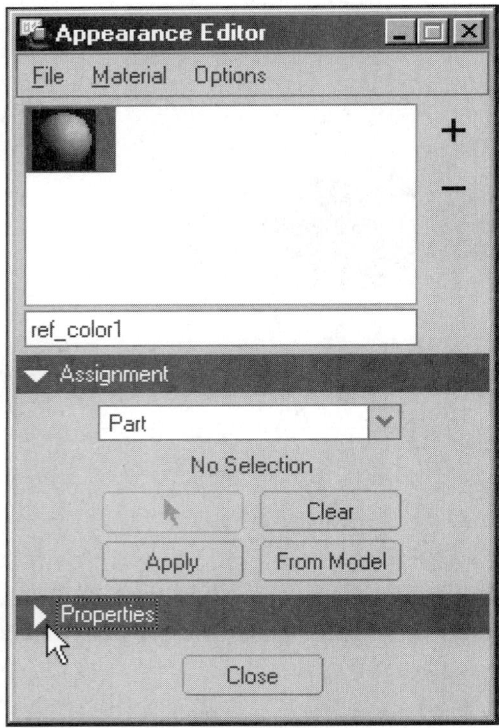

Figure 5.26 Appearance Editor

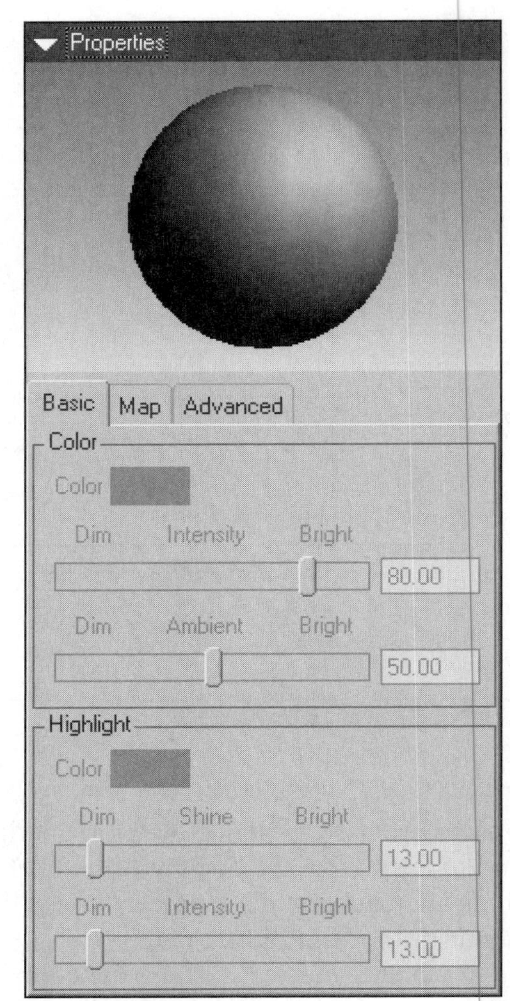

Figure 5.27 Properties

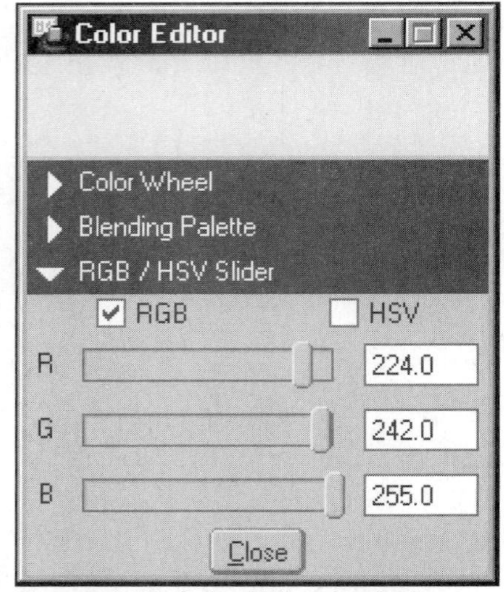

Figure 5.28 Color Editor Default

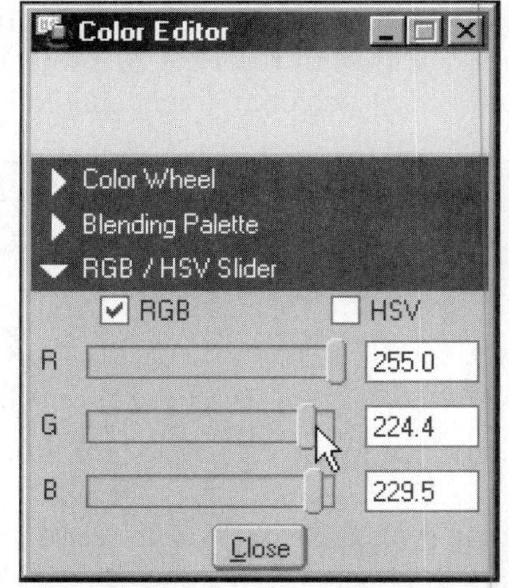

Figure 5.29 Color Editor New Color

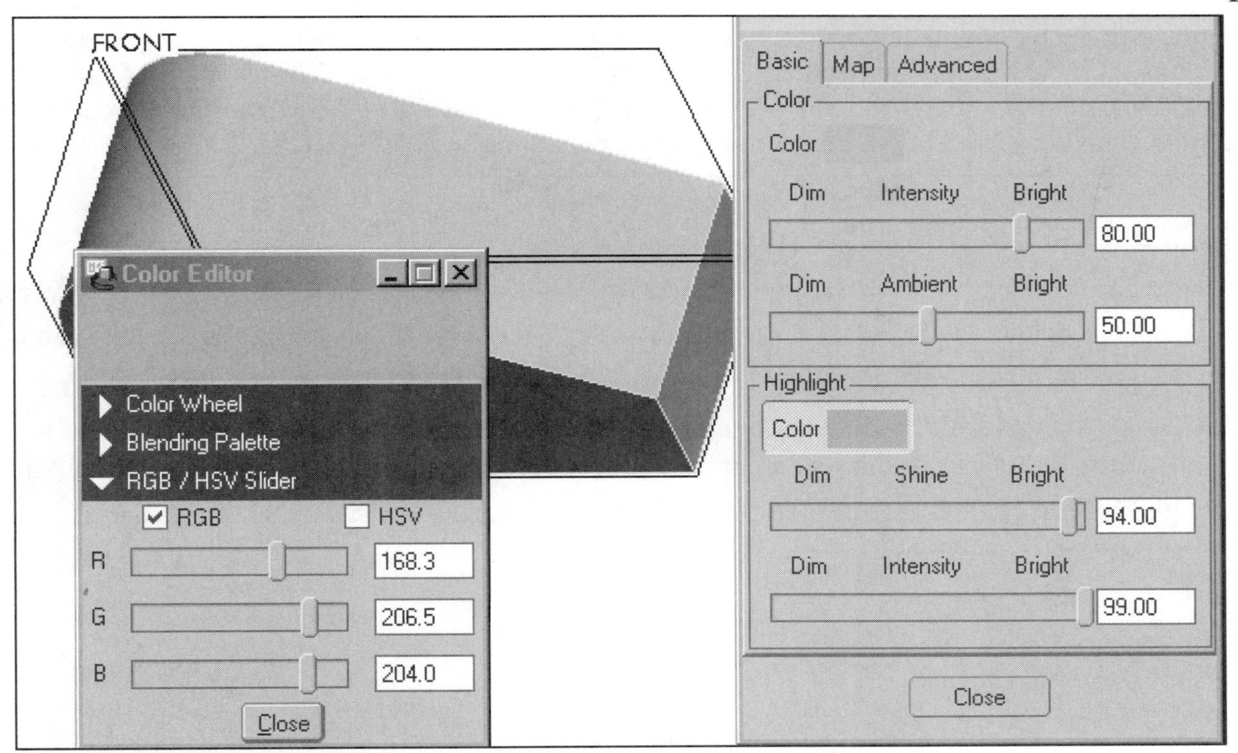

Figure 5.30 Highlight Color Editor

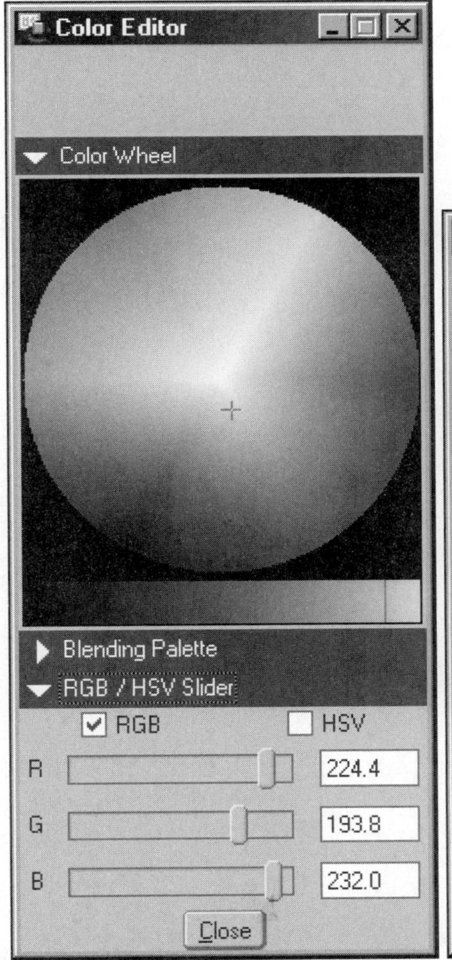

Figure 5.31(a) Color Wheel

Figure 5.31(b) Blending Palette

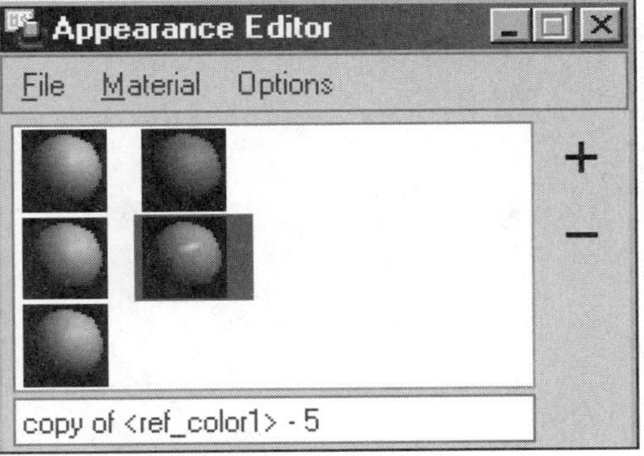

Figure 5.32 Completed Color Palette

The next two features will remove material. For both of the cuts, use RIGHT as the sketching plane, use TOP as the reference plane. Each cut requires just two lines, and two dimensions. Use the first cut's sketching/placement plane and reference/orientation (**Use Previous**) for the second cut. Create the cuts separately, each as an open section. Figure 5.33 shows the dimensions for each cut.

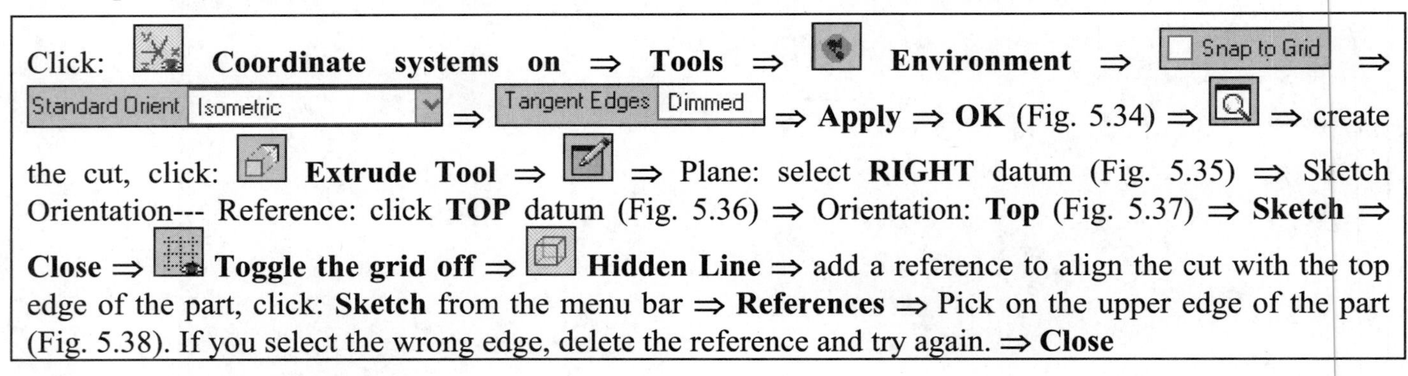

Click: **Coordinate systems on** ⇒ **Tools** ⇒ **Environment** ⇒ Snap to Grid ⇒ Standard Orient Isometric ⇒ Tangent Edges Dimmed ⇒ **Apply** ⇒ **OK** (Fig. 5.34) ⇒ ⇒ create the cut, click: **Extrude Tool** ⇒ ⇒ Plane: select **RIGHT** datum (Fig. 5.35) ⇒ Sketch Orientation--- Reference: click **TOP** datum (Fig. 5.36) ⇒ Orientation: **Top** (Fig. 5.37) ⇒ **Sketch** ⇒ **Close** ⇒ **Toggle the grid off** ⇒ **Hidden Line** ⇒ add a reference to align the cut with the top edge of the part, click: **Sketch** from the menu bar ⇒ **References** ⇒ Pick on the upper edge of the part (Fig. 5.38). If you select the wrong edge, delete the reference and try again. ⇒ **Close**

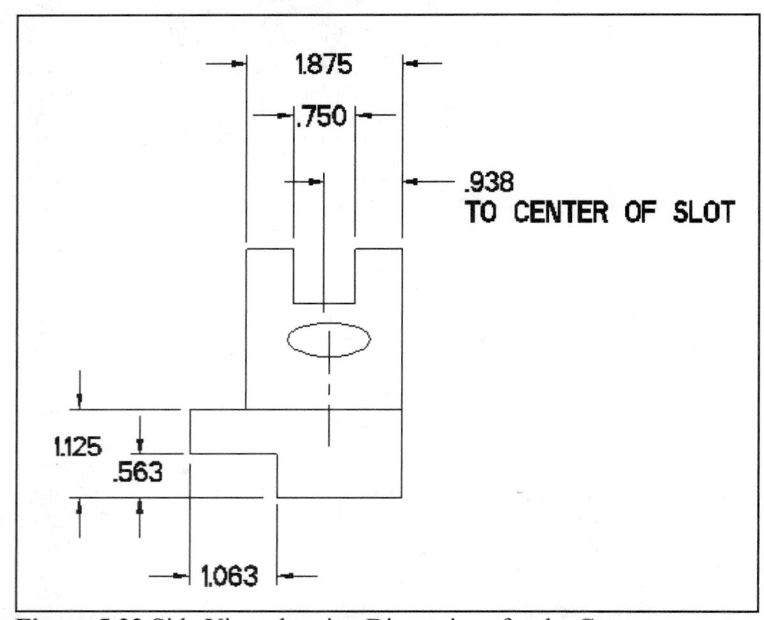

Figure 5.33 Side View showing Dimensions for the Cuts

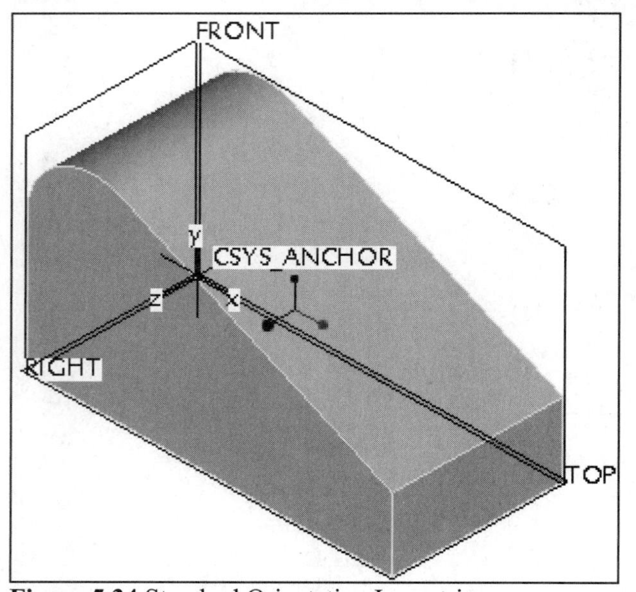

Figure 5.34 Standard Orientation Isometric

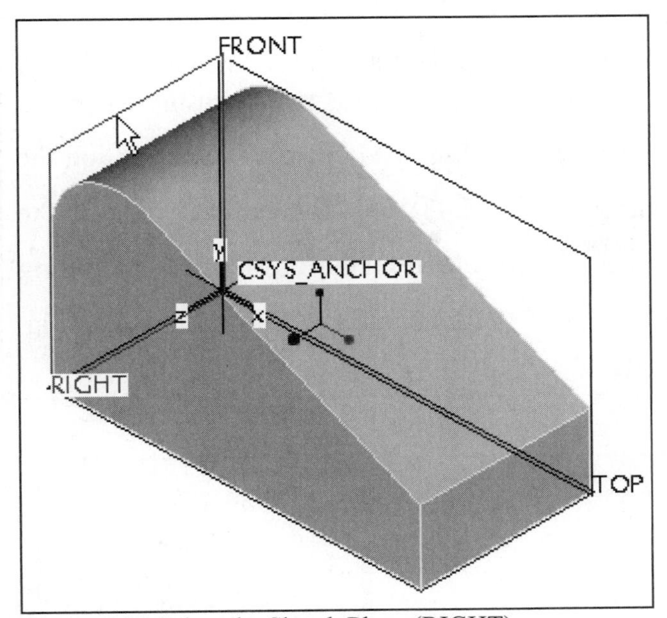

Figure 5.35 Select the Sketch Plane (RIGHT)

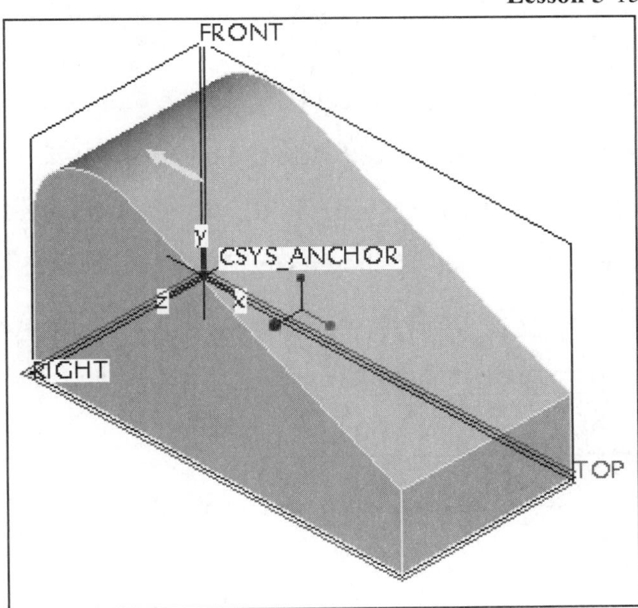

Figure 5.36 Select the Orientation Plane (TOP)

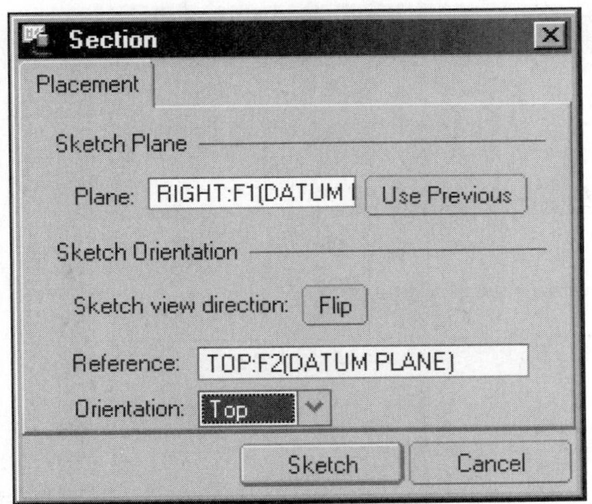

Figure 5.37 Section Dialog Box

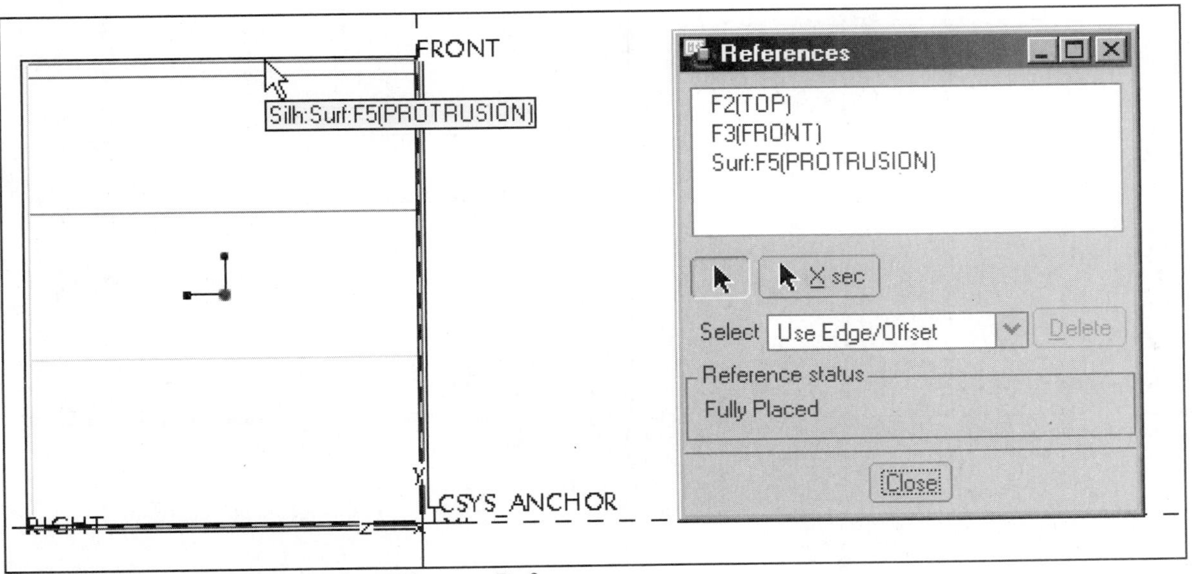

Figure 5.38 Add the upper Surface Edge as a Reference

Click: **Create 2 point lines** ⇒ sketch the lines- start from the top with a vertical line (Fig. 5.39) ⇒ **MMB** to end the line sequence ⇒ the **Confirm** dialog box will display asking if you want the end point of the line to be aligned with the edge of the part (Fig. 5.40) ⇒ **Yes** (Fig. 5.41) ⇒ ⇒ reposition the default dimensions (Fig. 5.42) ⇒ add the two dimensions (Fig. 5.43) ⇒ ⇒ modify the dimension values as shown (Fig. 5.44) ⇒ ⇒ ⇒ ⇒ **Standard Orientation** ⇒ **Shading** (Fig. 5.45) ⇒ from the dashboard click: **Remove Material** ⇒ **Options** tab ⇒ **Through All** (Fig. 5.46) ⇒ (Fig. 5.47) ⇒ ⇒ ⇒ ⇒ **MMB**

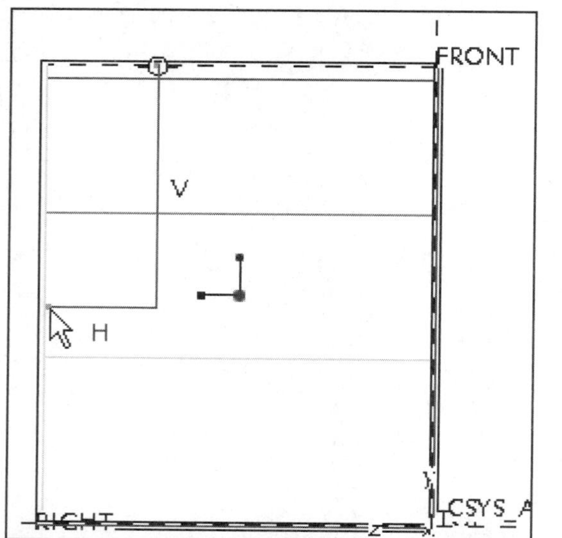

Figure 5.39 Sketch a Vertical and a Horizontal Line

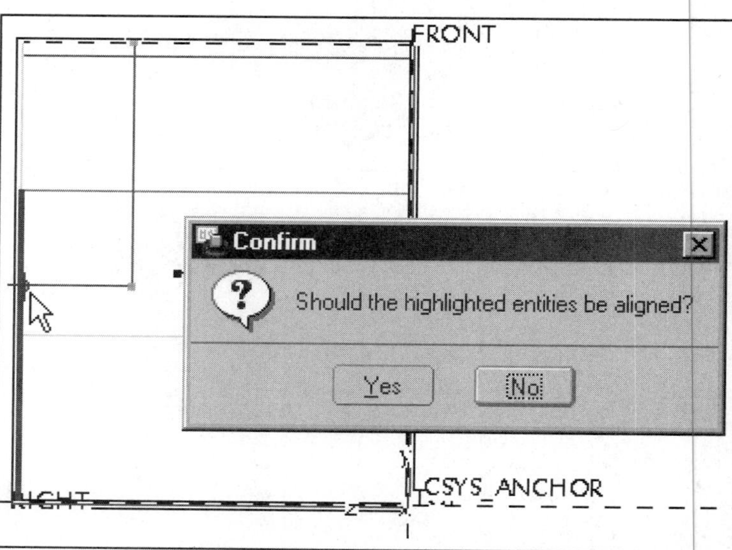

Figure 5.40 Align the End of the Horizontal Line

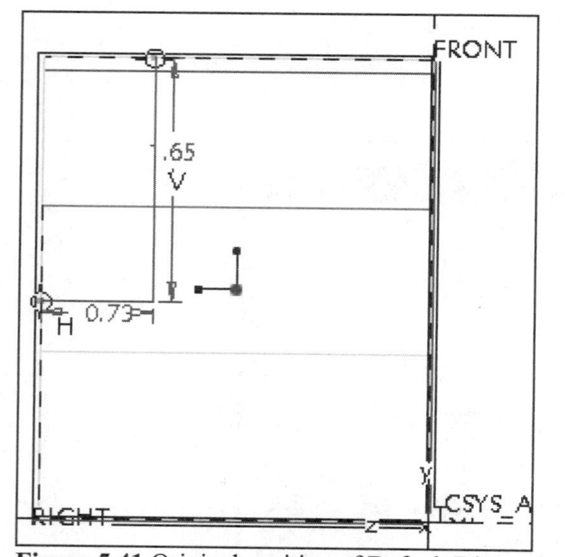

Figure 5.41 Original position of Default Dimensions

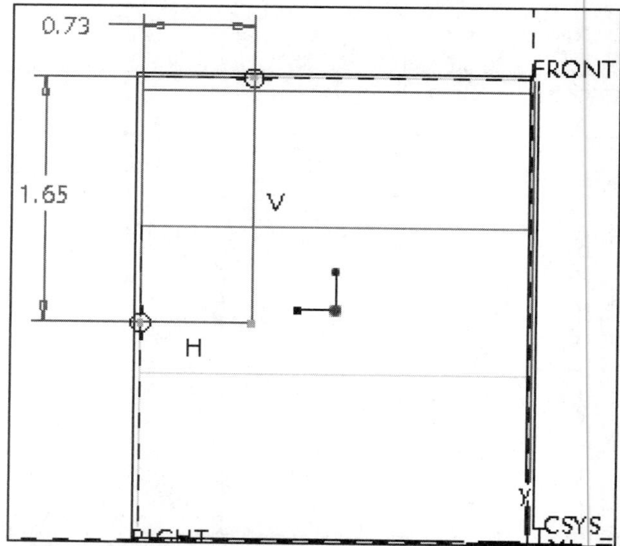

Figure 5.42 Moved Default Dimensions

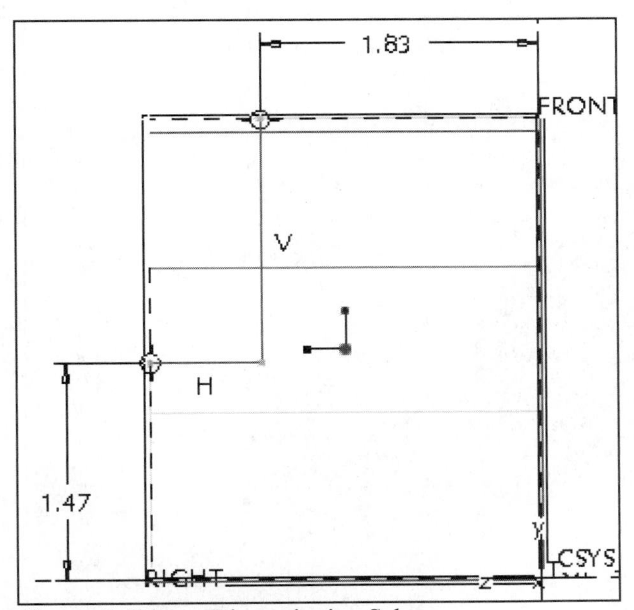

Figure 5.43 New Dimensioning Scheme

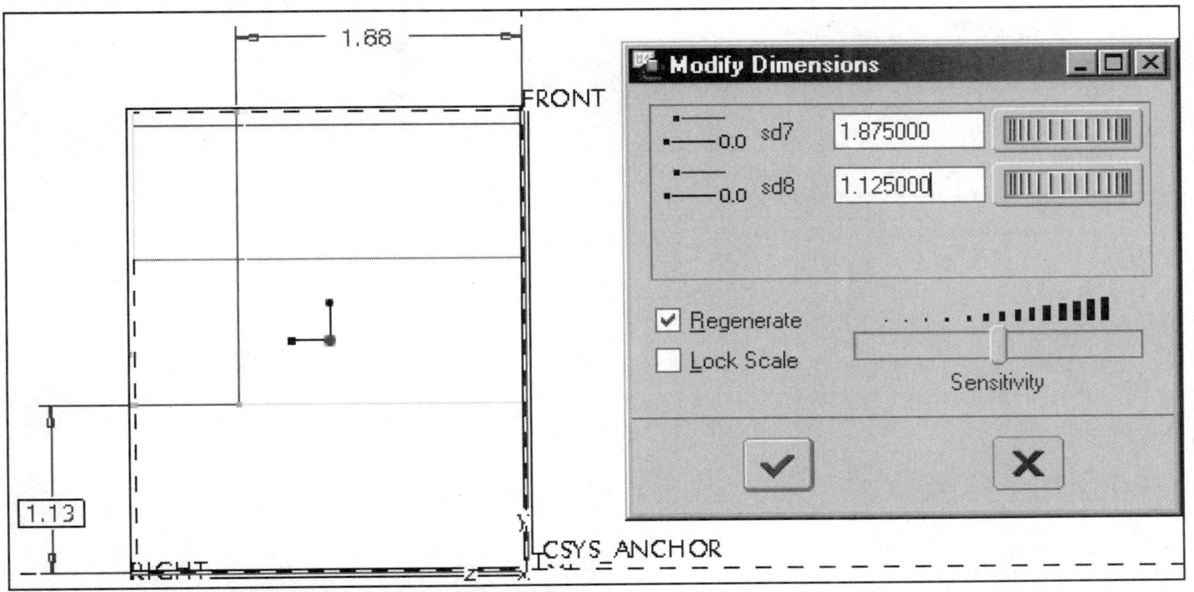

Figure 5.44 Modified Dimensions **1.875** and **1.125**

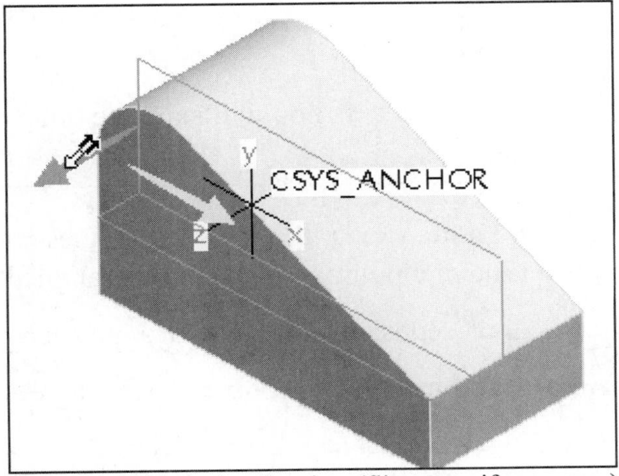

Figure 5.45 Extrusion Preview (flip arrow if necessary)

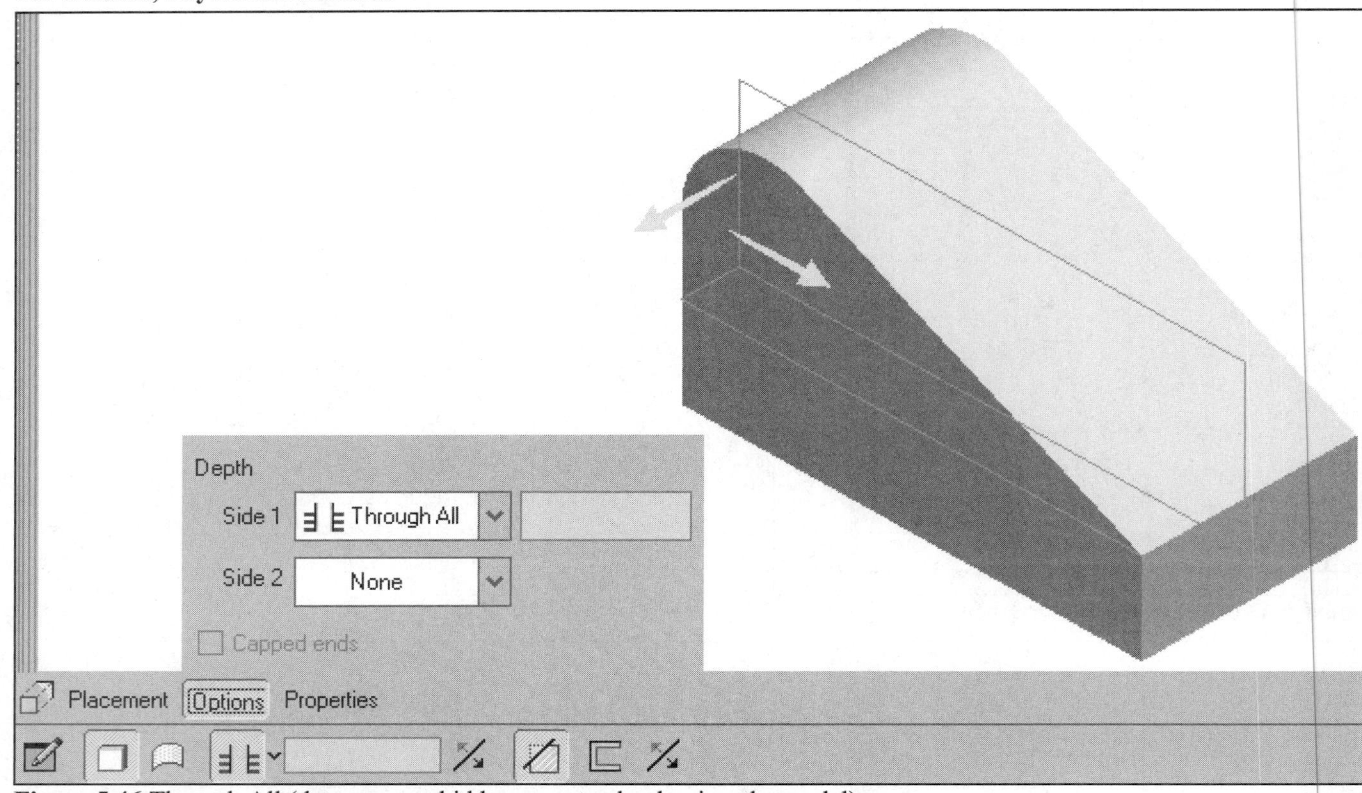

Figure 5.46 Through All (datums were hidden so as to clearly view the model)

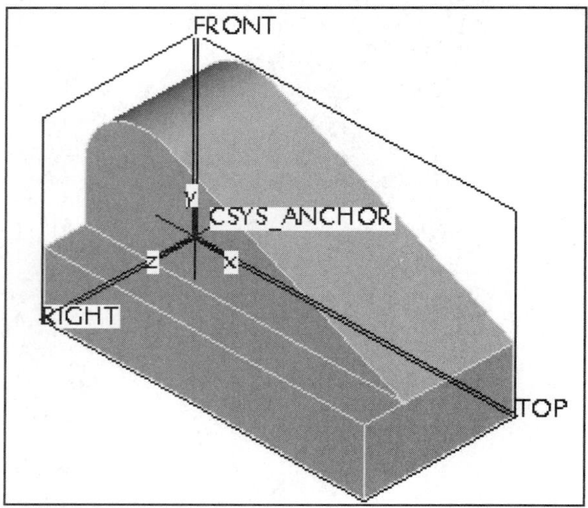

Figure 5.47 Preview

Complete the second cut. The dimensions are **.563** and **1.063** for this cut. You will expedite things by selecting Use Previous for the sketching and orientation planes.

Click: ⇒ ⇒ **Use Previous** ⇒ **Sketch** ⇒ **Close** ⇒ **No Hidden** ⇒ ⇒ sketch the lines ⇒ **MMB** ⇒ the **Confirm** dialog box will display asking if you want the end point of the line to be aligned with the edge of the part (Fig. 5.48) ⇒ **Yes** [Fig. 5.49(a)] ⇒ ⇒ reposition the default dimensions [Fig. 5.49(b)] ⇒ [Fig. 5.49(c)] ⇒ ⇒ ⇒ ⇒ **Standard Orientation** ⇒ ⇒ **Remove Material** ⇒ **Options** tab ⇒ **Through All** (Fig. 5.50) ⇒ ⇒ ⇒ ⇒ ⇒ **MMB** (Fig. 5.51) ⇒ **LMB** to deselect

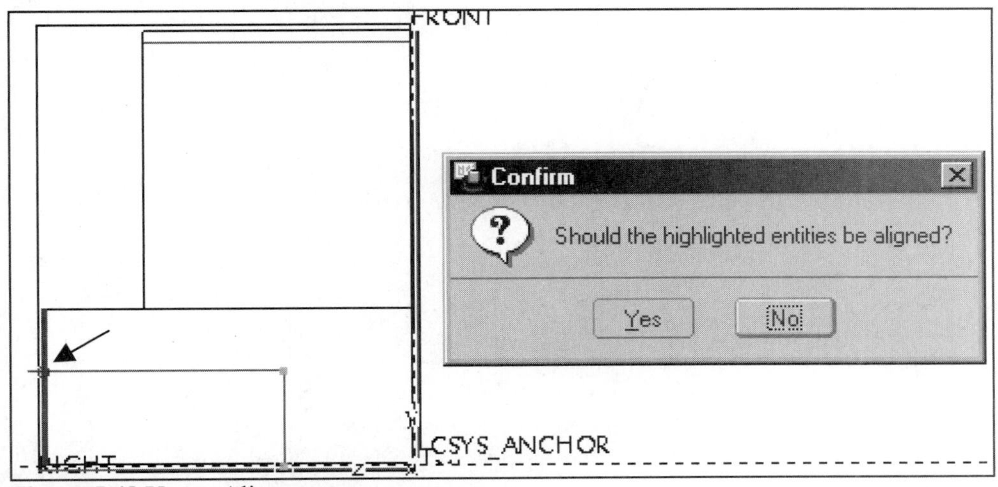

Figure 5.48 Yes to Alignment

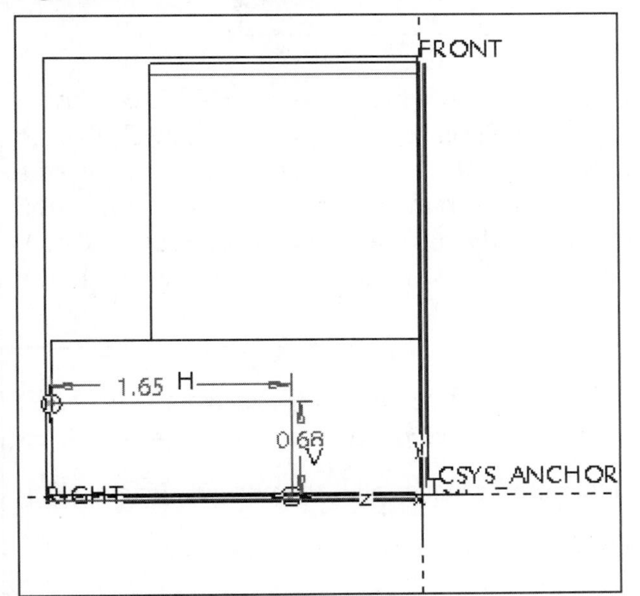

Figure 5.49(a) Default Dimensions

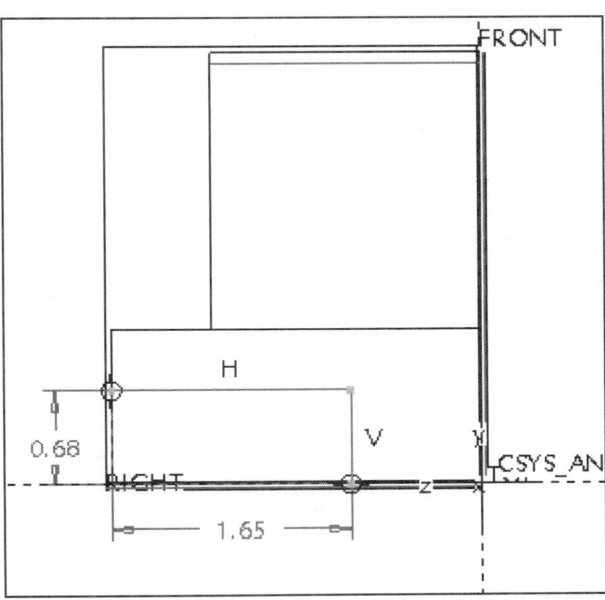

Figure 5.49(b) Reposition Dimensions

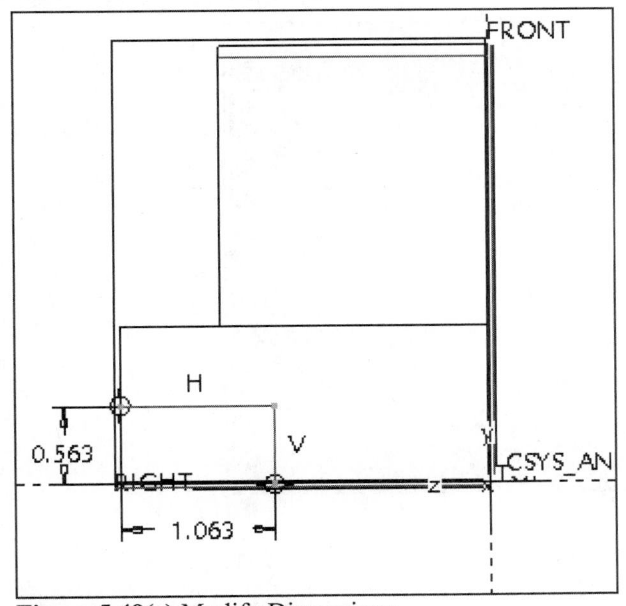

Figure 5.49(c) Modify Dimensions

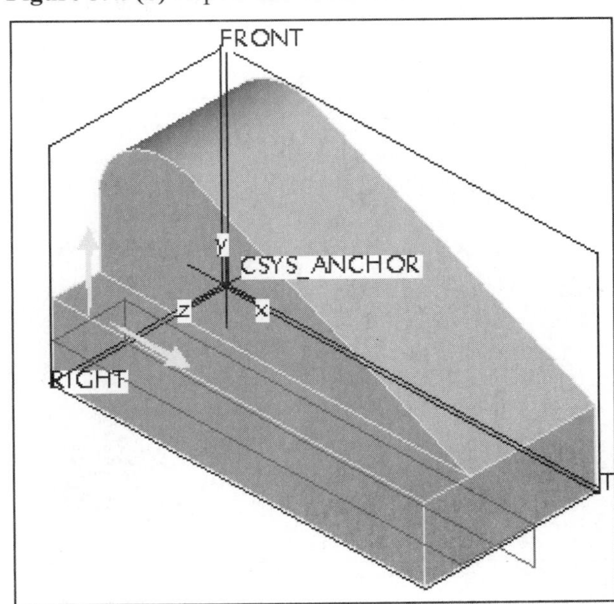

Figure 5.50 Standard Orientation

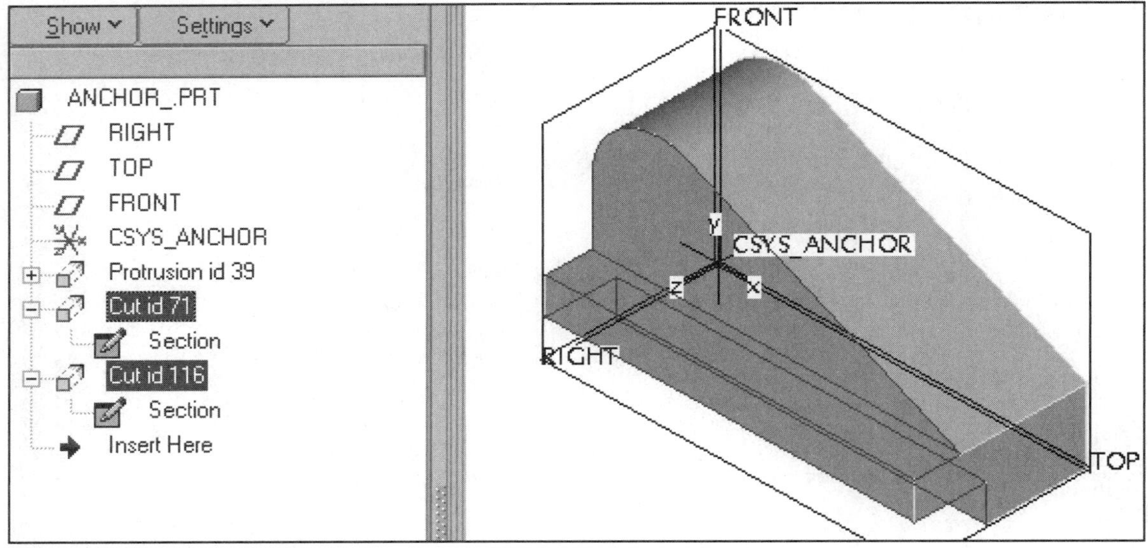

Figure 5.51 Completed Cuts

For the next feature, we will need to create a new datum plane on which to sketch. The new datum plane is DTM1. New datum planes are by default numbered sequentially. Another datum will also be constructed at this time. A datum axis will also be created at this time, to be used later for the creation of a hole. The A_1 datum axis will be created through the curved top of the part. A datum axis can be inserted through any curved feature. Holes and circular features automatically have axes when they are created. Features created with arcs, fillets, and so on need to have axes added afterward, unless set in the *configuration file* as the default.

Click: **Datum Plane Tool** ⇒ References: pick datum **FRONT** as the plane to offset from (Fig. 5.52) ⇒ in the DATUM PLANE dialog box, Offset: Translation, type the distance **1.875/2** ***(1.875/2 = .9375)*** ⇒ **Enter** ⇒ **OK** ⇒ **LMB** to deselect ⇒ **Datum Axis Tool** ⇒ References: pick the curved surface (Fig. 5.53) ⇒ **OK** ⇒ **LMB** to deselect ⇒ **Datum Plane Tool** ⇒ References: pick the angled surface ⇒ Translation- type **0** ⇒ **Enter** (Fig. 5.54) ⇒ **OK** ⇒ ⇒ **MMB** (Fig. 5.55) ⇒ **LMB** to deselect

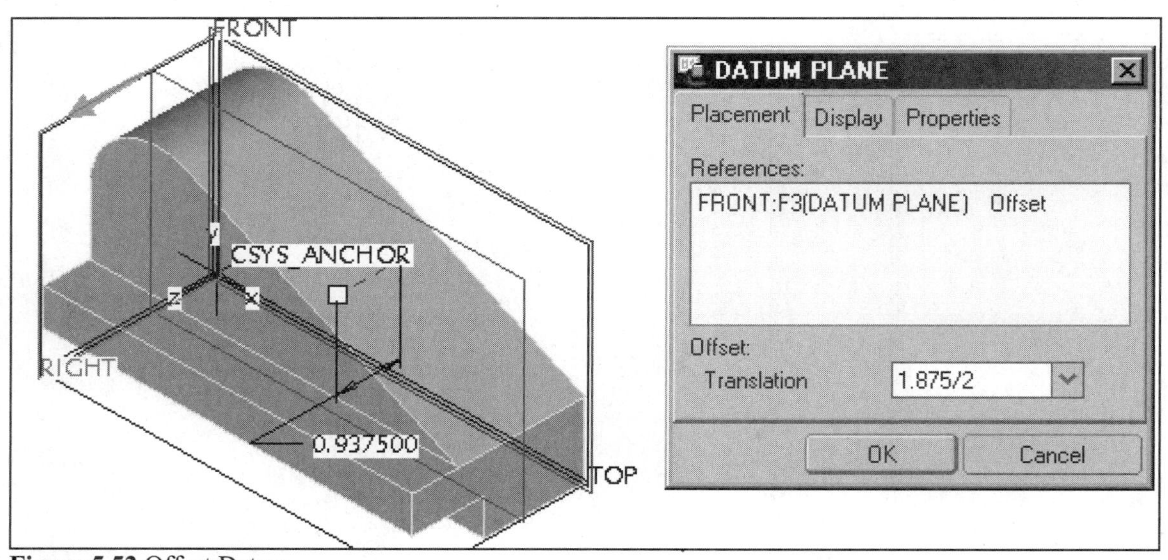

Figure 5.52 Offset Datum

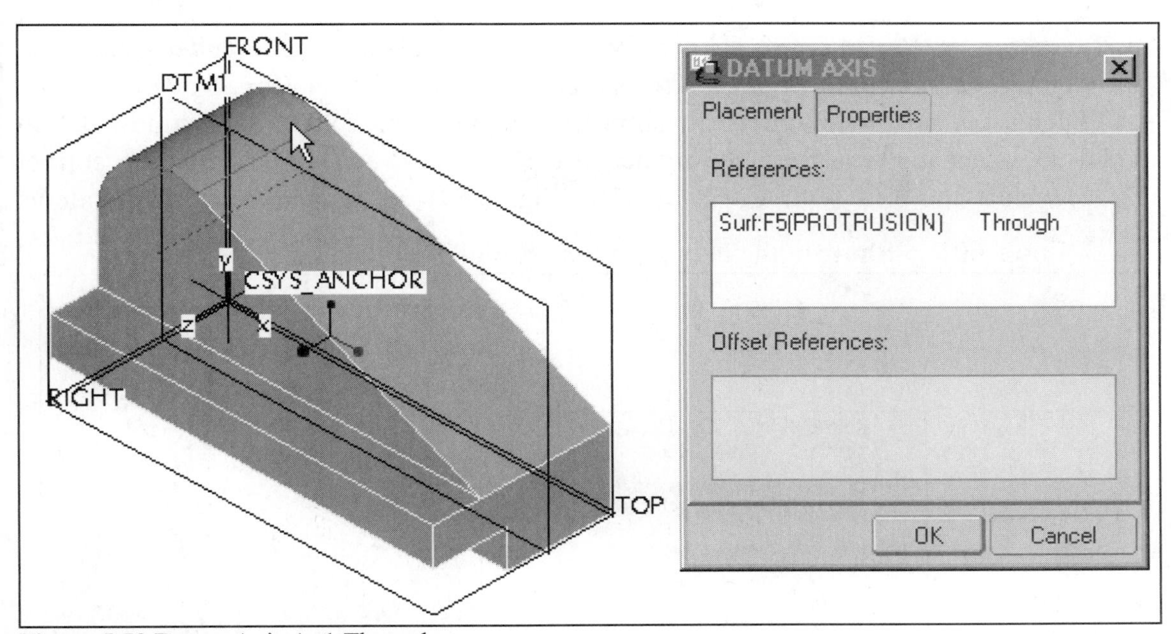

Figure 5.53 Datum Axis A_1 Through

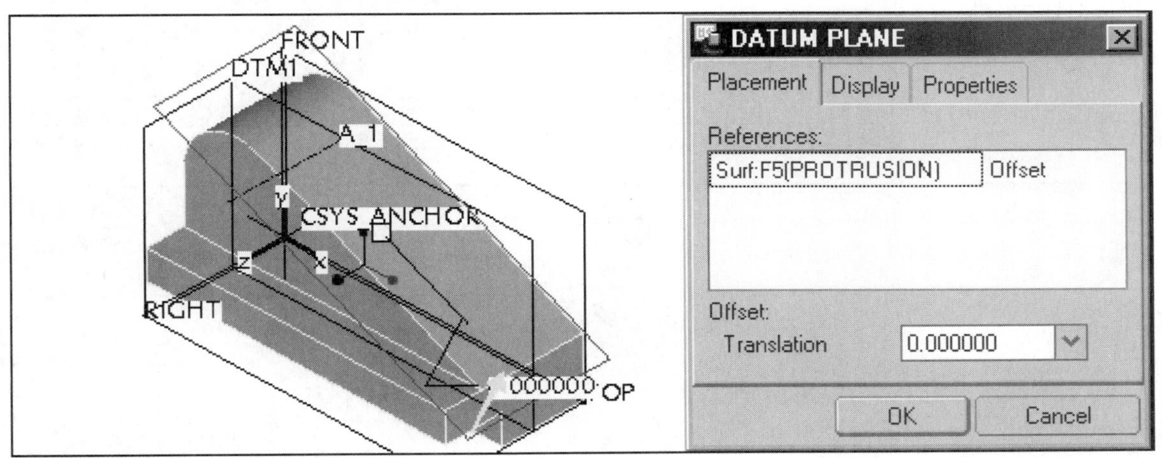

Figure 5.54 Datum Plane DTM2

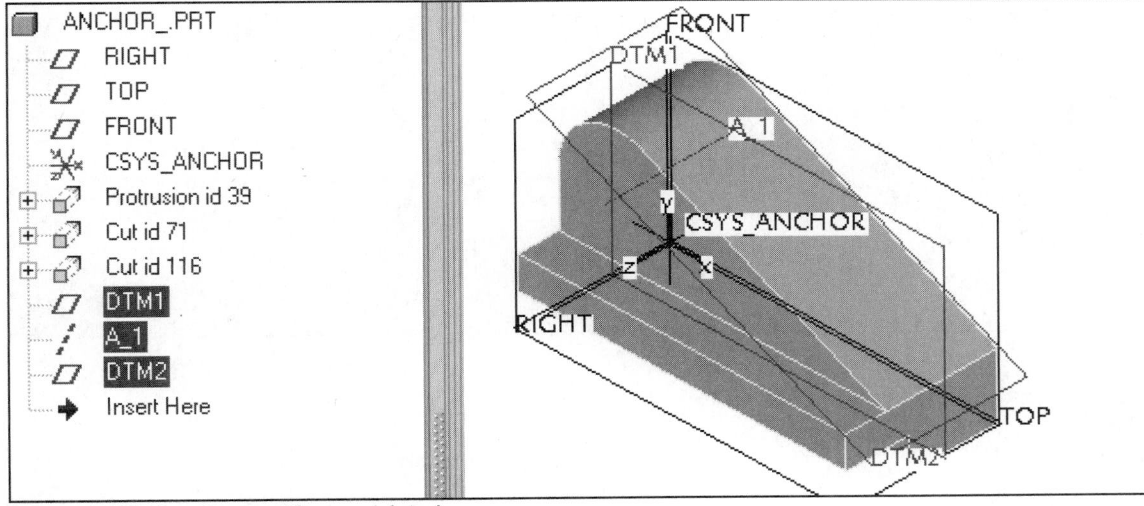

Figure 5.55 New Datum Planes and Axis

Create a new layer and add the new datum planes and axis to it.

Click: **Show** in the **Navigator** ⇒ **Layer Tree** --Layer Tree displays in place of the **Model Tree** (Fig. 5.56) ⇒ click on and highlight **Layers** in the Layer Tree ⇒ **RMB** ⇒ **New Layer** (Fig. 5.57) ⇒ Name: type **DATUM_FEATURES** *(do not press Enter)* as name for new layer ⇒ select the two new datum planes from the model ⇒ select the axis from the model, use **RMB** ⇒ **Next** to toggle to the correct selection (Fig. 5.58) ⇒ select the axis with **LMB** ⇒ **OK** (Fig. 5.59) ⇒ expand layer to see items DATUM_FEATURES (Fig. 5.60) ⇒ **Show** in the **Navigator** ⇒ **Model Tree** ⇒ ⇒ **MMB**

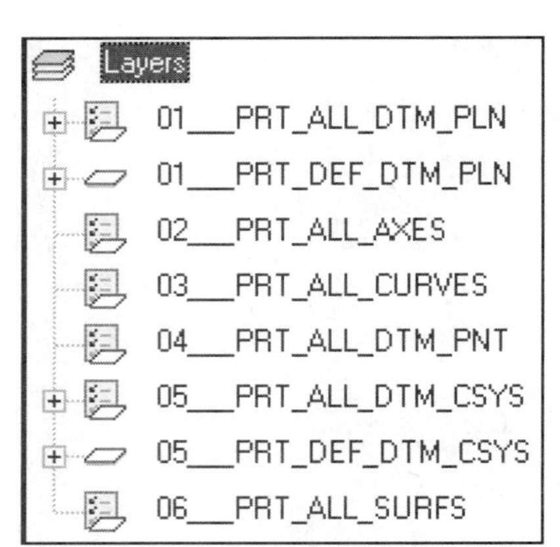

Figure 5.56 Layer Tree

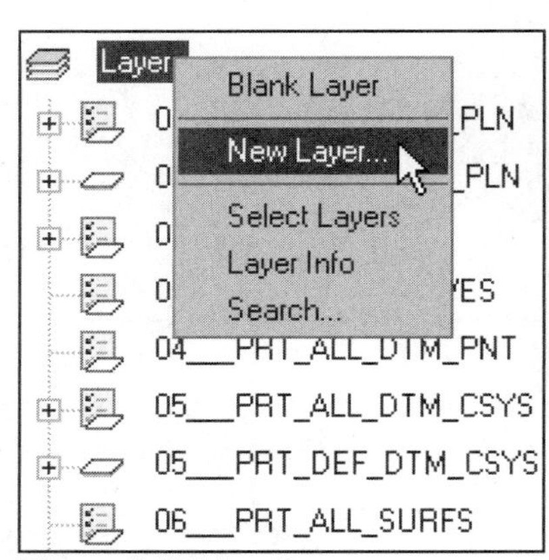

Figure 5.57 New Layer

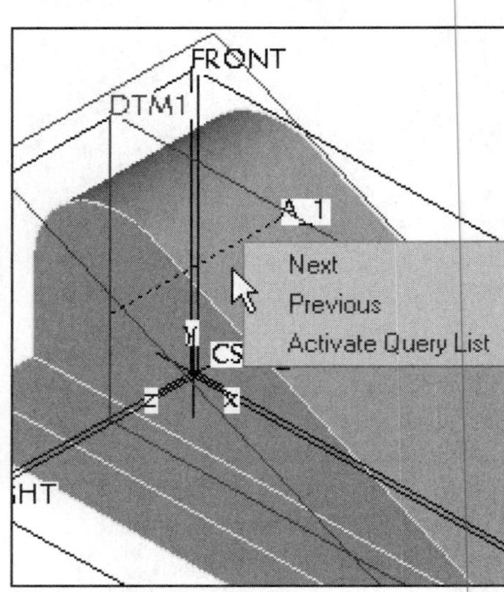

Figure 5.58 Highlighting Axis using Next

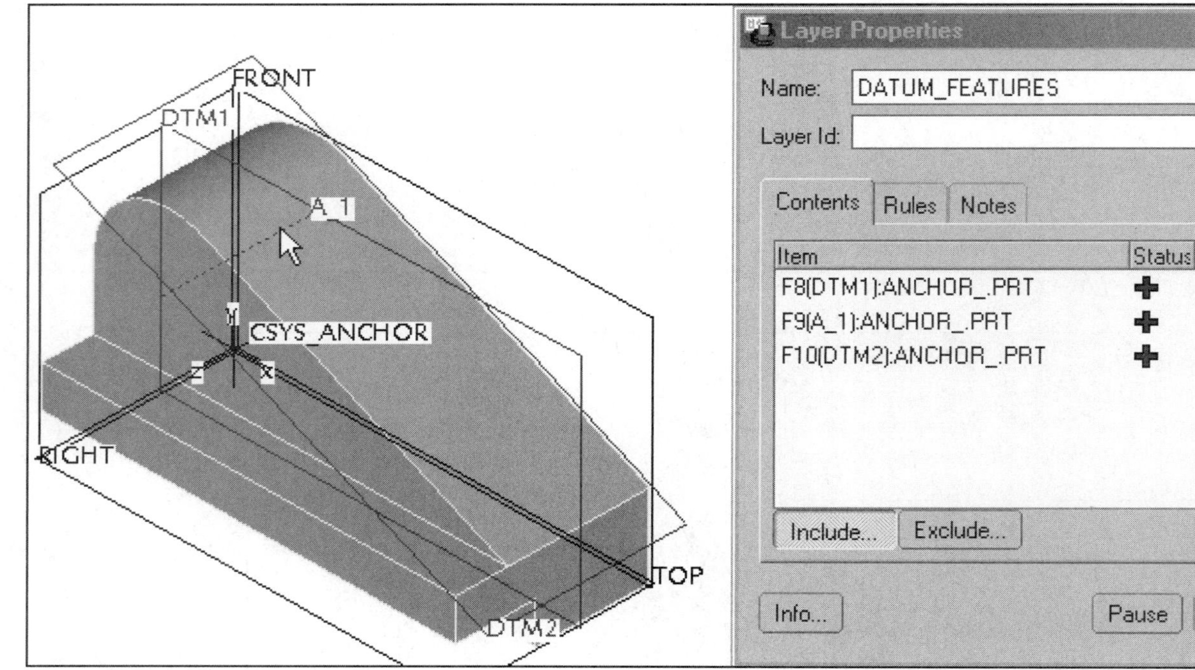

Figure 5.59 Adding Items to a Layer

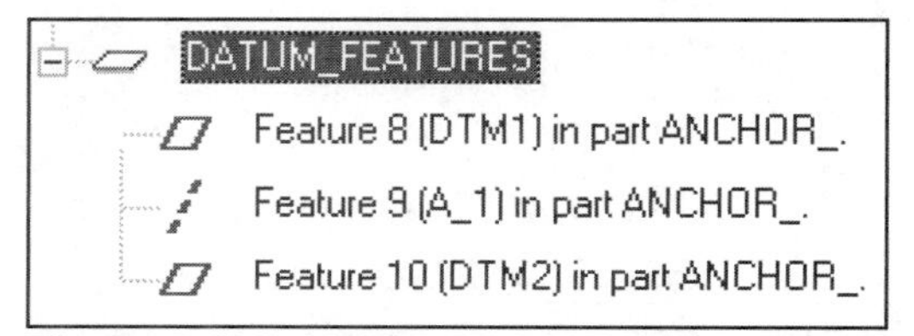

Figure 5.60 DATUM_FEATURES in Layer Tree

Geometric Tolerances

Before continuing with the modeling of the part, the datums used and created thus far will be "set" for geometric tolerancing. Geometric tolerances (GTOLs) provide a comprehensive method of specifying where on a part the critical surfaces are, how they relate to one another, and how the part must be inspected to determine if it is acceptable. They provide a method for controlling the location, form, profile, orientation, and run out of features. When you store a Pro/E GTOL in a solid model, it contains parametric references to the geometry or feature it controls—its reference entity—and parametric references to referenced datums and axes. As a result, Pro/E updates a GTOL's display when you rename a referenced datum.

In Assembly mode, you can create a GTOL in a subassembly or a part. A GTOL that you create in Part or Assembly mode automatically belongs to the part or assembly that occupies the window; however, it can refer only to set datums belonging to that model itself, or to components within it. It cannot refer to datums outside of its model in some encompassing assembly, unlike assembly created features. You can add GTOLs in Part or Drawing mode, but they are reflected in all other modes. Pro/E treats them as annotations, and they are always associated with the model. Unlike dimensional tolerances, though, GTOLs do not affect part geometry.

Before you can reference a datum plane or axis in a GTOL, you must set it as a reference. Pro/E encloses its name using the set datum symbol. After you have set a datum, you can use it in the usual way to create features and assemble parts. You enter the set datum command by clicking Edit ⇒ Setup ⇒ Geom Tol ⇒ Set Datum, and then select the datum plane or axis. Pro/E encloses the datum name in a feature control frame. If needed, type a new name in the Name box of the Datum dialog box. Most datums will follow the alphabet, A, B, C, D, and so on. You can blank a reference datum plane only by placing it on a layer and then blanking the layer, or using hide.

Click: **Edit ⇒ Setup ⇒ Geom Tol ⇒ Set Datum ⇒** pick **TOP** from the model [Fig. 5.61(a)] ⇒ Name- type **A ⇒ OK ⇒** pick **FRONT ⇒** Name- **B** [Fig. 5.61(b)] ⇒ **OK ⇒** pick **RIGHT ⇒** Name- type **C** [Fig. 5.61(c)] ⇒ **OK ⇒** pick **DTM1 ⇒** Name- type **D** [Fig. 5.61(d)] ⇒ **OK ⇒** pick **DTM2 ⇒** Name- type **E ⇒ OK ⇒ MMB ⇒ MMB ⇒ MMB** (Fig. 5.62) ⇒ ⇒ ⇒ **MMB ⇒ File ⇒ Delete ⇒ Old Versions ⇒ MMB**

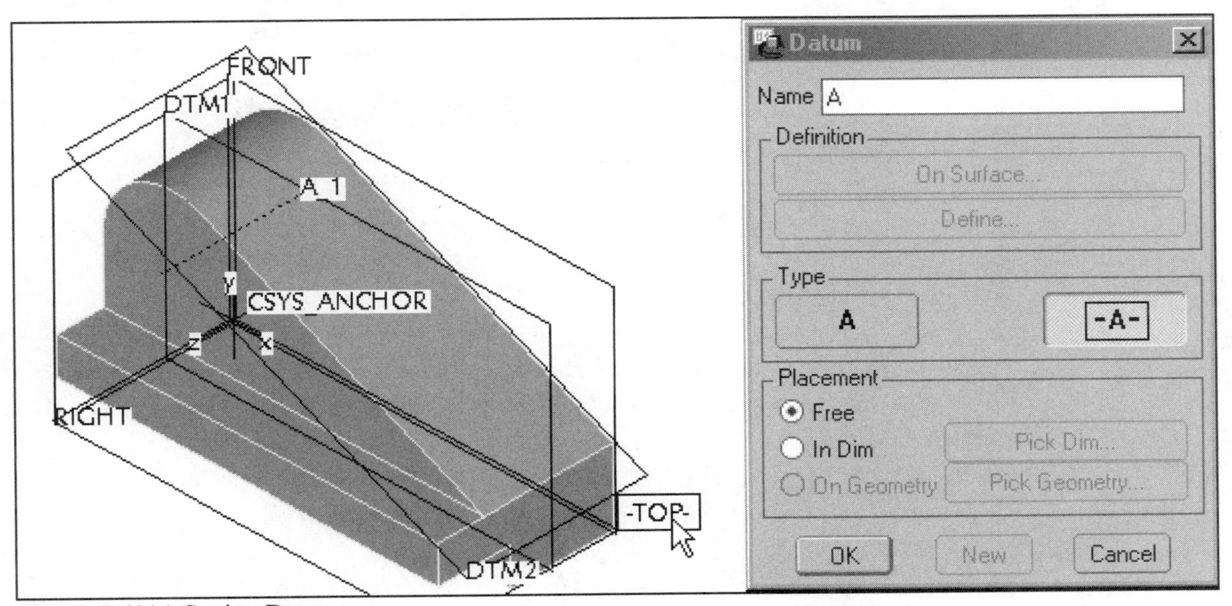

Figure 5.61(a) Setting Datums

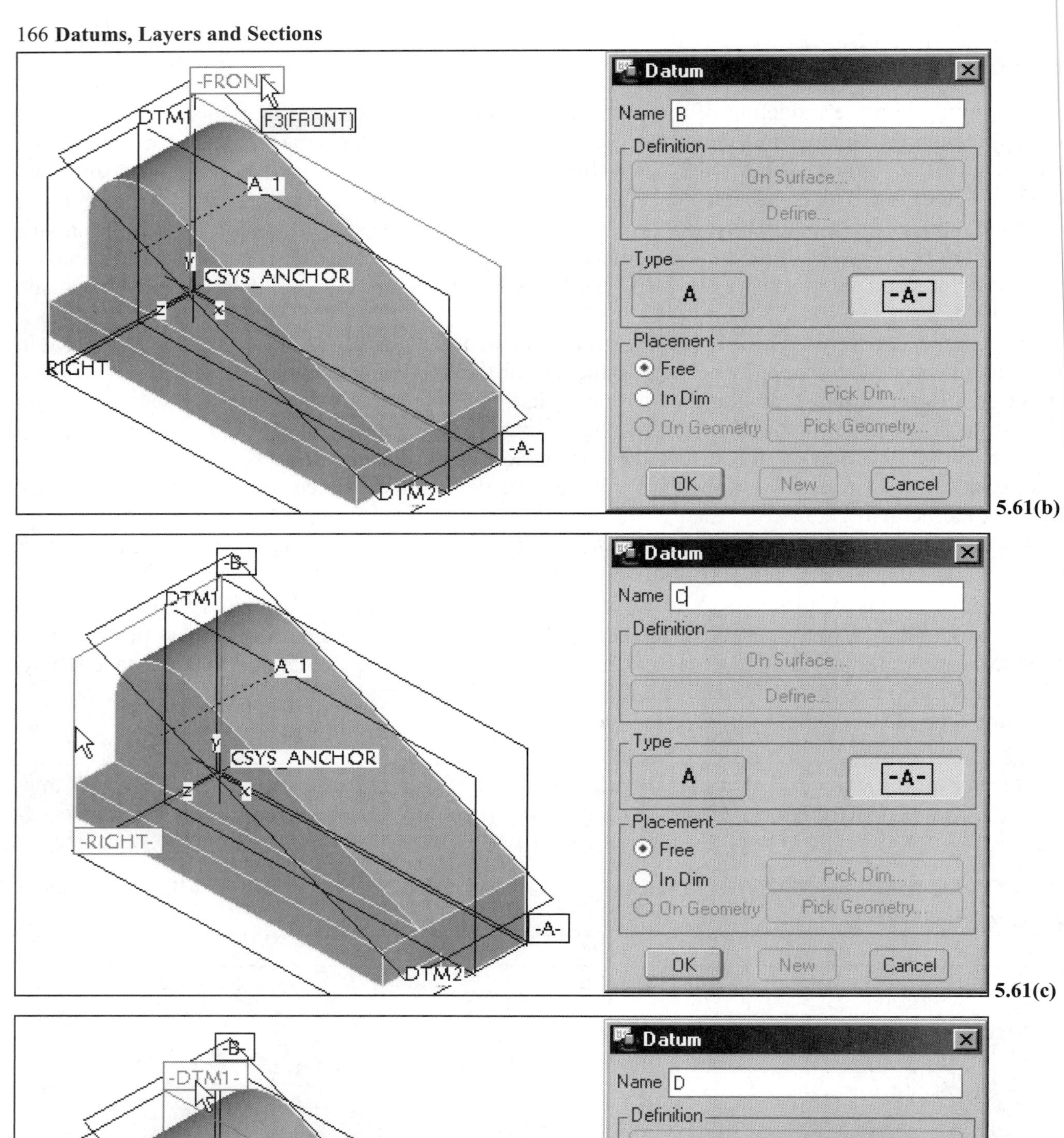

5.61(b)

5.61(c)

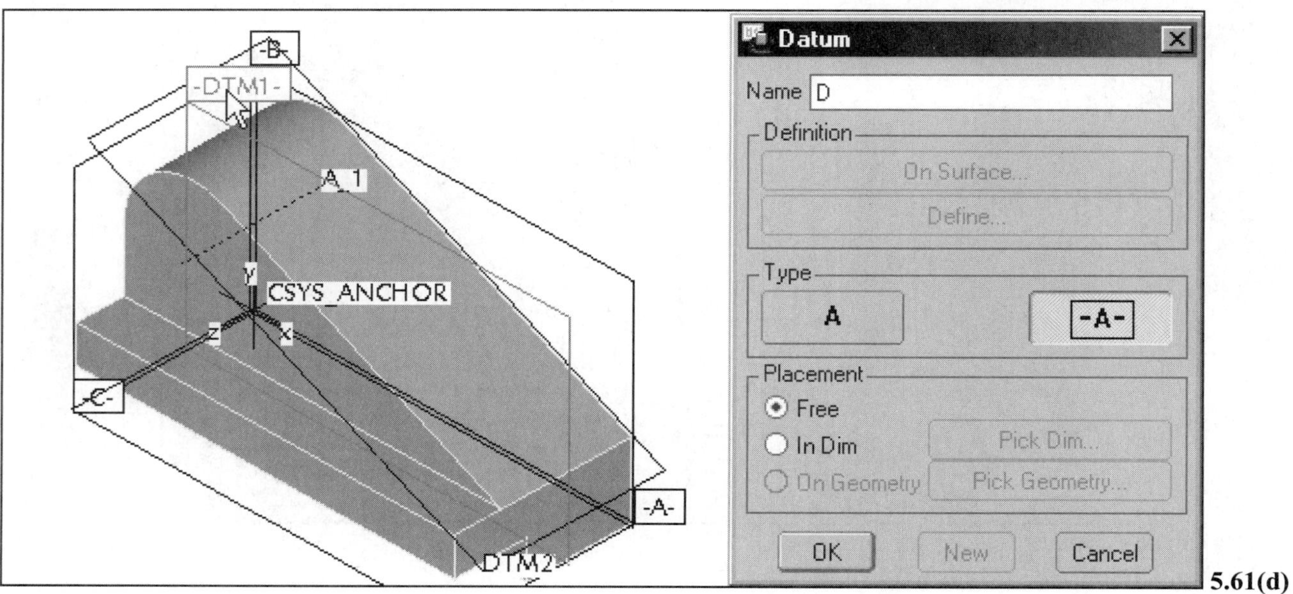

5.61(d)

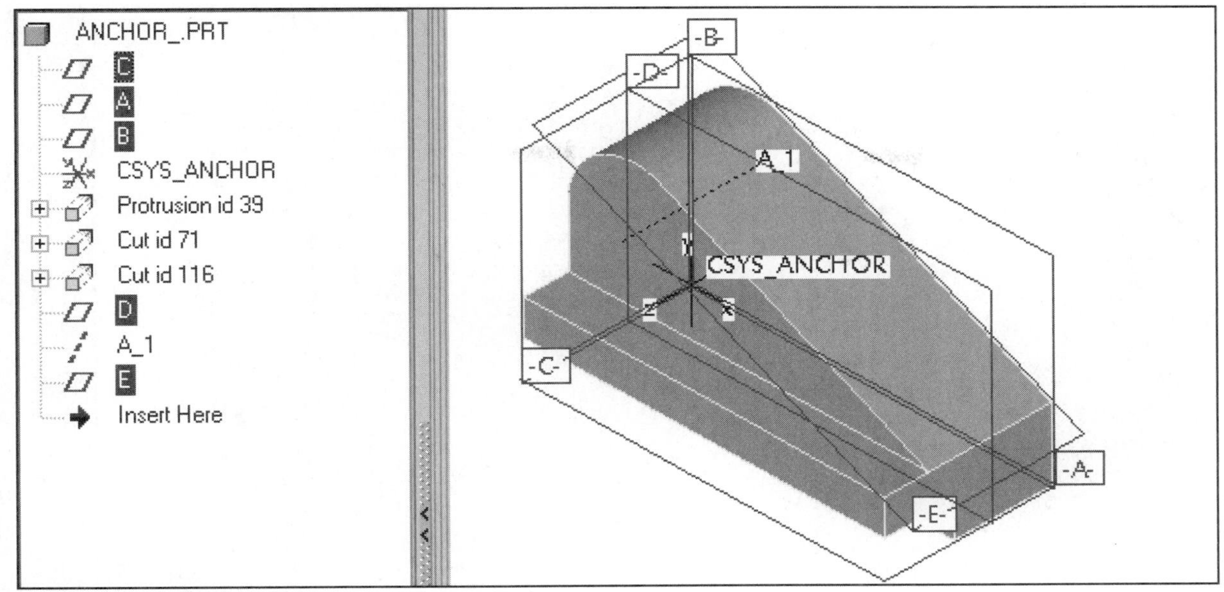

Figure 5.62 Set Datums

The slot on the top of the part is created by sketching on datum D and projecting it toward both sides. Use the Model Tree to select the appropriate datum planes.

Click: ⇒ ⇒ Plane: select datum **D** (Fig. 5.63) ⇒ Sketch Orientation--- Sketch view direction: select datum **C** ⇒ Orientation: **Right** ⇒ **Sketch** ⇒ **Close** ⇒ **Hidden Line** ⇒ add a reference to align the cut with the angled edge of the part, click: **Sketch** from the menu bar ⇒ **References** ⇒ pick on datum **E** ⇒ pick the small vertical surface on the right side of the part as the fourth reference (Fig. 5.64) ⇒ **Close**

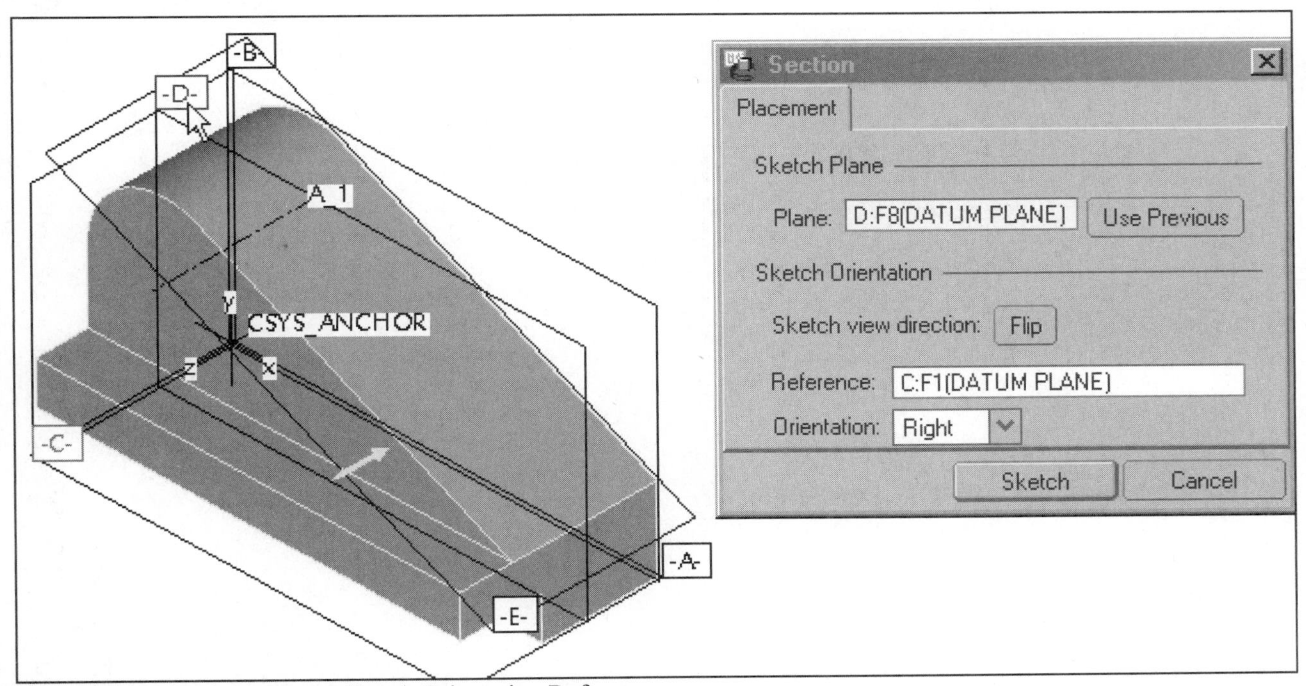

Figure 5.63 Sketch Plane and Sketch Orientation Reference

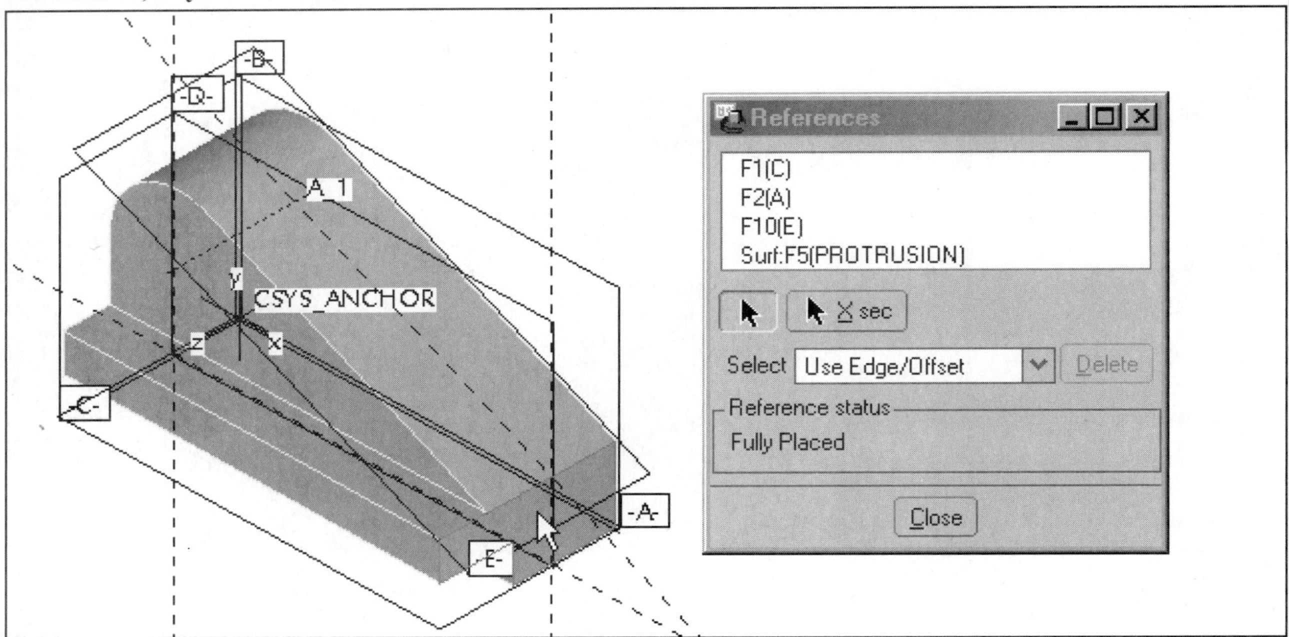

Figure 5.64 New References (Datum E and Right Vertical Surface)

Click: **Orient the sketching plane parallel to the screen** ⇒ **Create points** pick at the corner of the angled datum **E** and the right vertical reference (Fig. 5.65) ⇒ **MMB**

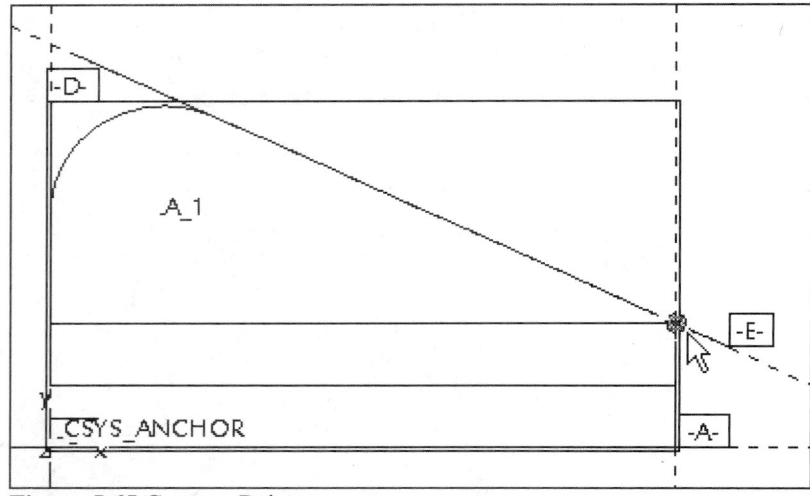

Figure 5.65 Create a Point

Click: **Create 2 point lines** ⇒ sketch the first line from and perpendicular to the angled edge (Fig. 5.66) ⇒ continue to sketch the second line horizontal (Fig. 5.67) ⇒ **MMB** (Fig. 5.68)

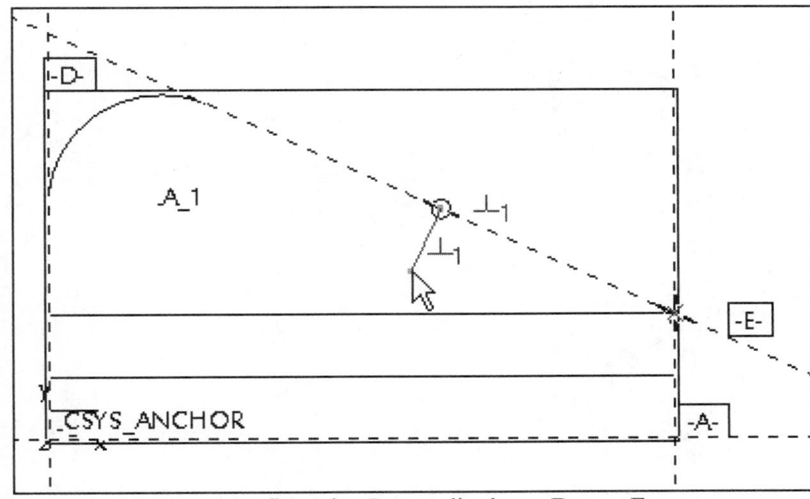

Figure 5.66 Create the first Line Perpendicular to Datum E

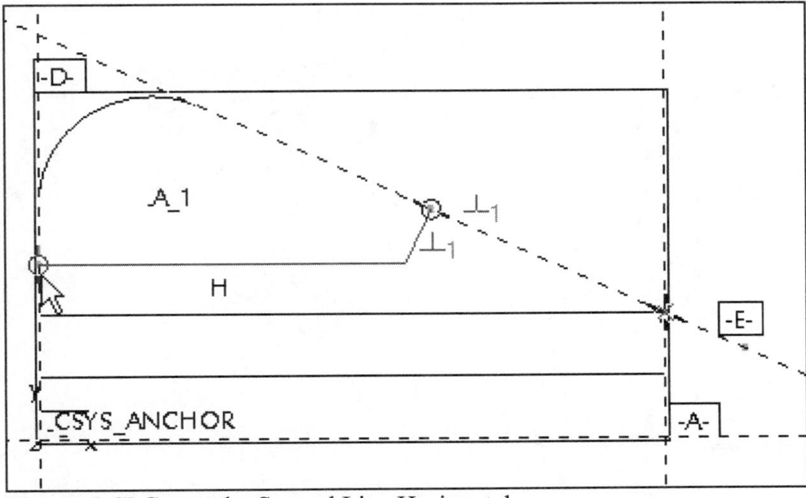

Figure 5.67 Create the Second Line Horizontal

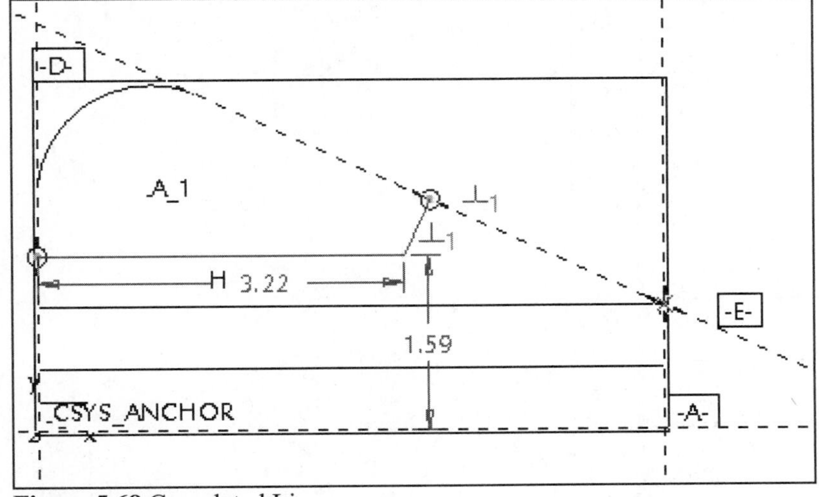

Figure 5.68 Completed Lines

Click: add a dimension by picking the first line (Fig. 5.69) and then the point (do not pick endpoint to point) ⇒ **MMB** to place the dimension (Fig. 5.70) ⇒ add a vertical dimension (Fig. 5.71)

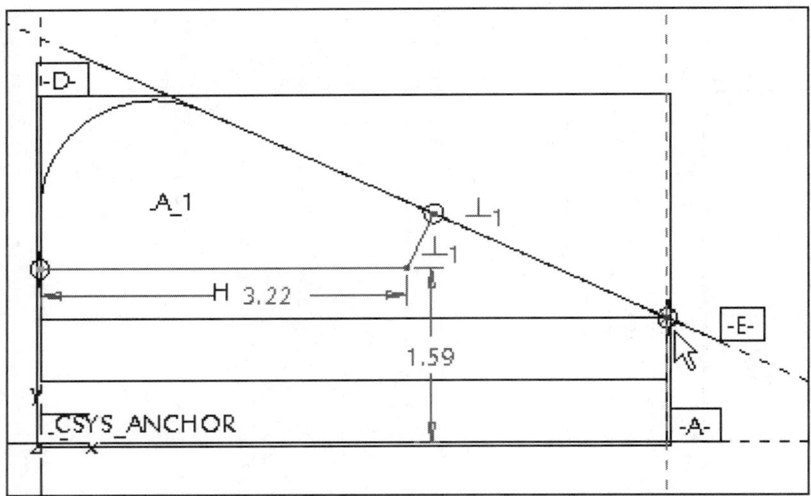

Figure 5.69 Create Dimension from First Line to Point (not point to point!)

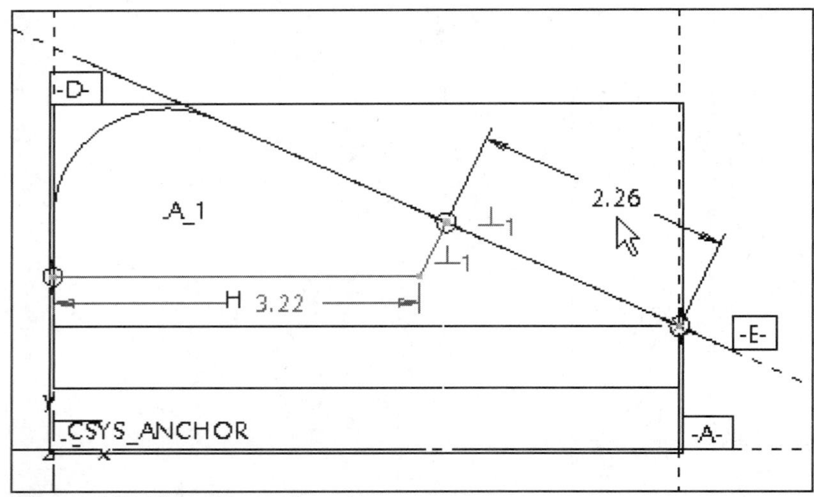

Figure 5.70 Place Dimension

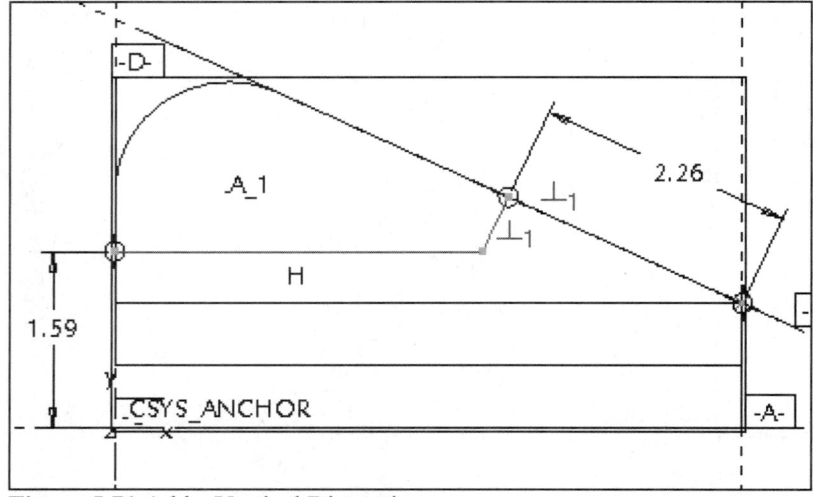

Figure 5.71 Add a Vertical Dimension

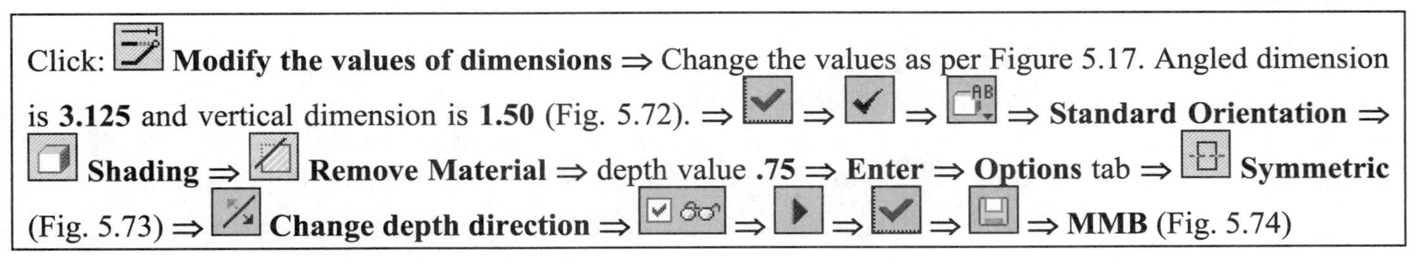

Click: **Modify the values of dimensions** ⇒ Change the values as per Figure 5.17. Angled dimension is **3.125** and vertical dimension is **1.50** (Fig. 5.72). ⇒ ⇒ ⇒ ⇒ **Standard Orientation** ⇒ **Shading** ⇒ **Remove Material** ⇒ depth value **.75** ⇒ **Enter** ⇒ **Options** tab ⇒ **Symmetric** (Fig. 5.73) ⇒ **Change depth direction** ⇒ ⇒ ⇒ ⇒ ⇒ **MMB** (Fig. 5.74)

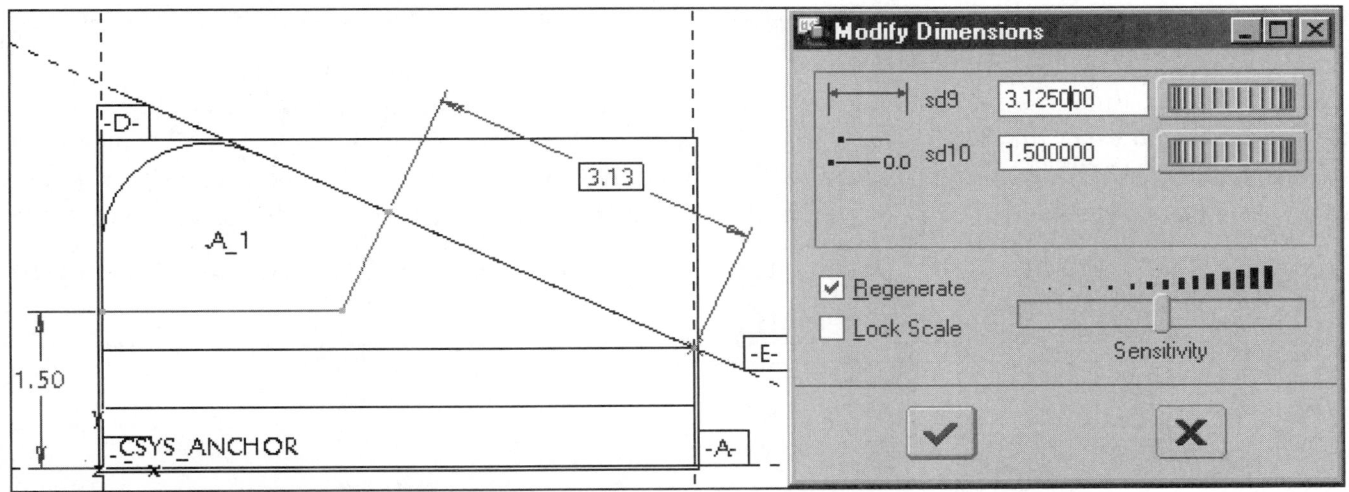

Figure 5.72 Modify Dimensions to **3.125** and **1.50**

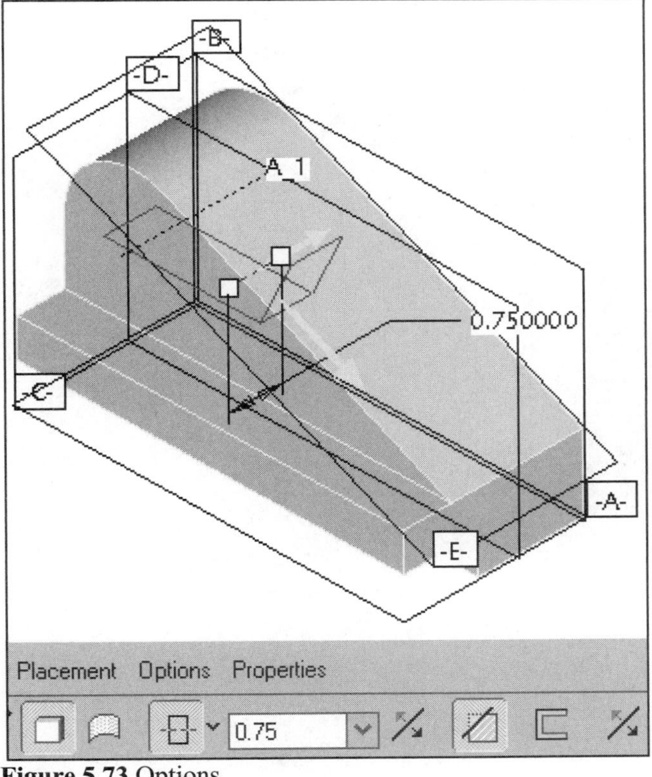

Figure 5.73 Options

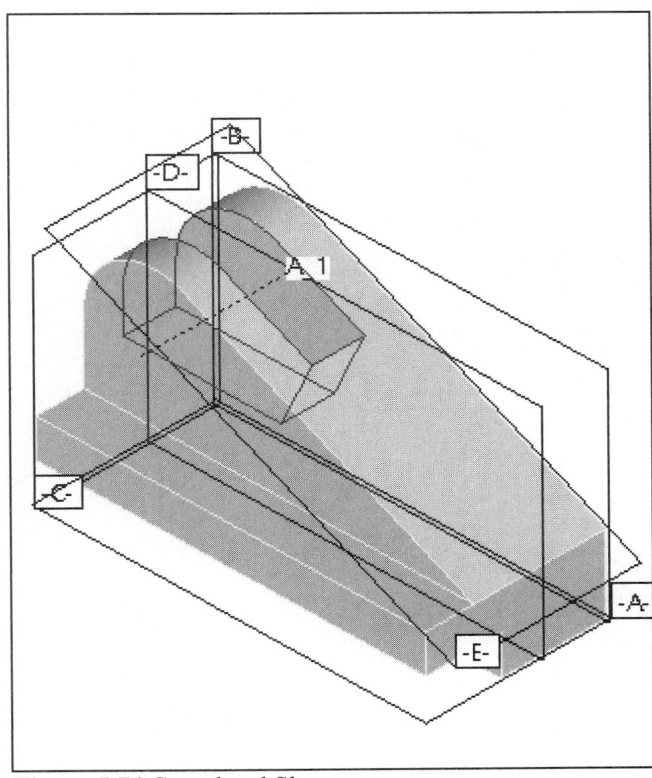

Figure 5.74 Completed Slot

The hole drilled in the angled surface appears to be aligned with datum D. Upon closer inspection (Fig. 5.33), it can be seen that the hole is at a different distance (**.875** from B) and is not in line with the slot and datum plane *(D was offset **.9375** from B)*. Create the feature using a sketched hole. The drill tip (**118** degrees) at the bottom of the hole needs to be modeled so the hole is created as a sketched hole.

Click: **Hole Tool** ⇒ Sketched ⇒ pick on the angled face and the hole will display with handles ⇒ drag one handle to datum **B** and the other to the edge between the angled surface and the right vertical face ⇒ double click on the dimension from the hole's center to datum **B** and modify the value to **.875** ⇒ **Enter** ⇒ double click on the dimension from the hole's center to the edge and modify the value to **2.0625** ⇒ **Enter** ⇒ **Placement** tab [(Fig. 5.75(a-b)] ⇒ ⇒ **Tools** ⇒ **Environment** ⇒ Snap to Grid ⇒ **Apply** ⇒ **OK** ⇒ **Toggle the grid on** ⇒ sketch a vertical centerline ⇒ sketch the four lines required to describe half of the hole's shape (Fig. 5.76) (default dimensions show in Fig. 5.77) ⇒ Create a diameter dimension by picking the centerline with the **LMB**, then the edge to be dimensioned with the **LMB**, and then the centerline a second time with the **LMB**. Place the dimension by picking a position with the **MMB**. ⇒ add the angle and the depth dimension [Fig. 5.78(a)] ⇒ add centerlines, point, and symmetric constraint [Fig. 5.78(b)] ⇒ add the full angle dimension ⇒ delete the half-angle dimension ⇒ **Modify** to the design values (Fig. 5.79) ⇒ ⇒ ⇒ ⇒ **Standard Orientation** ⇒ (Fig. 5.80) ⇒ ⇒ **MMB**

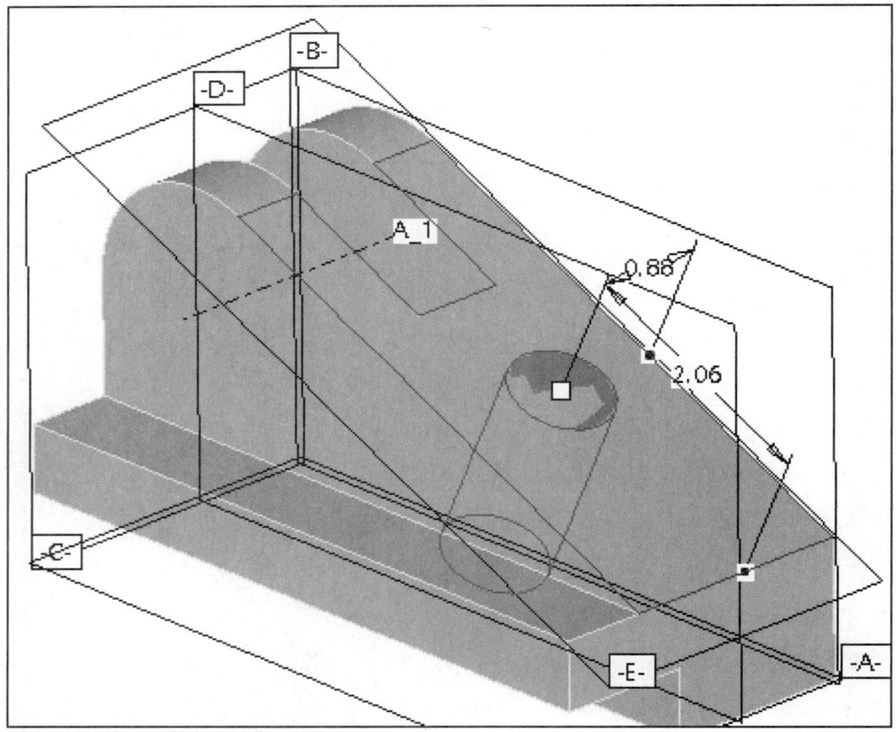

Figure 5.75(a) Hole Placement

Primary

Surf:F5(PROTR	Flip	Linear

Secondary references

B:F3(DATUM ...	Offset	0.875
Edge:F6(CUT)	Offset	2.0625

Placement Shape Note Properties

Sketched

Figure 5.75(b) Hole Placement

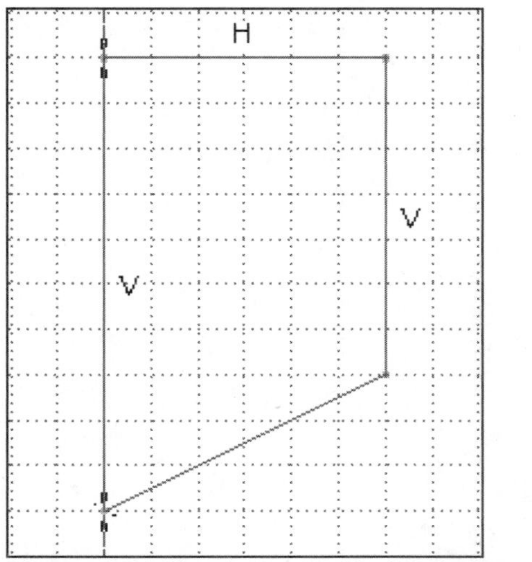

Figure 5.76 Sketch One Half of Hole Geometry

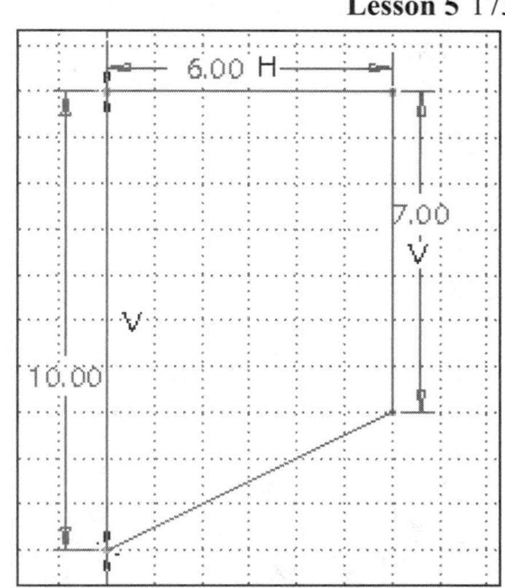

Figure 5.77 Default Dimensions

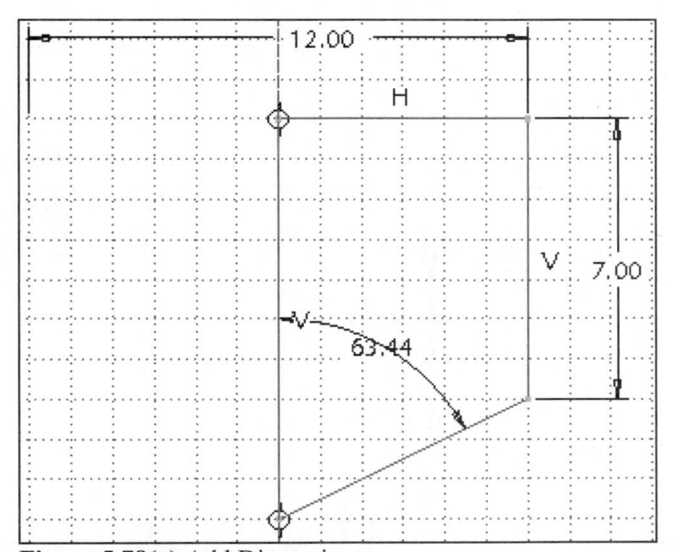

Figure 5.78(a) Add Dimensions

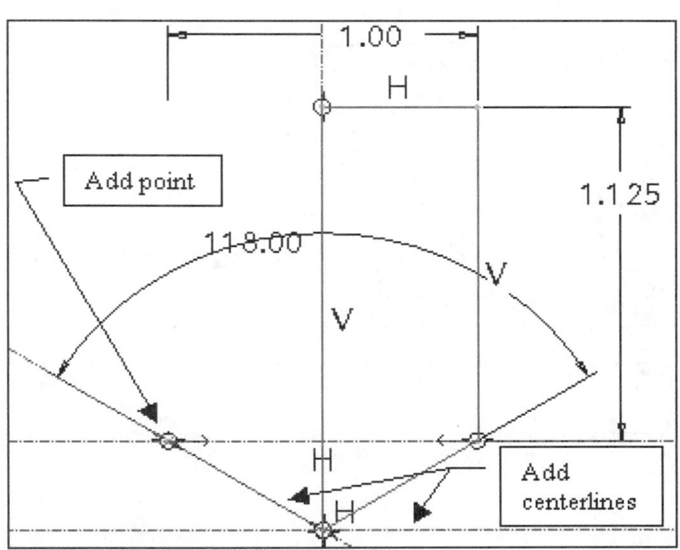

Figure 5.78(b) Add Centerlines, Point, and Dimension

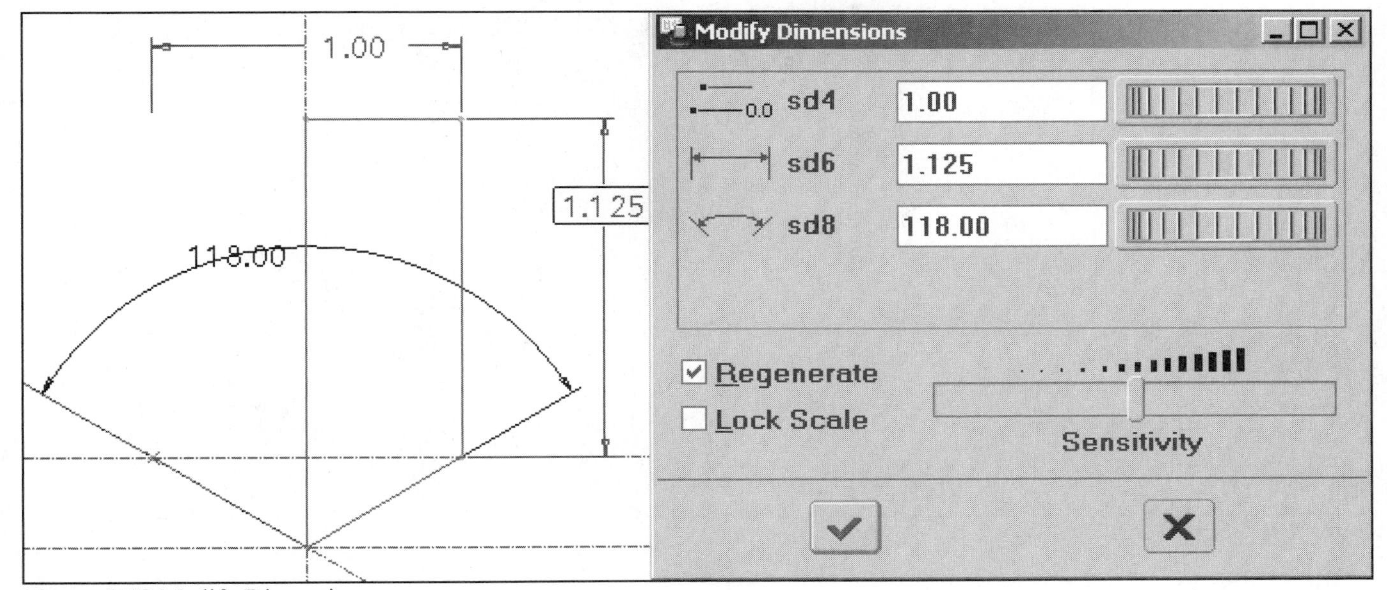

Figure 5.79 Modify Dimensions

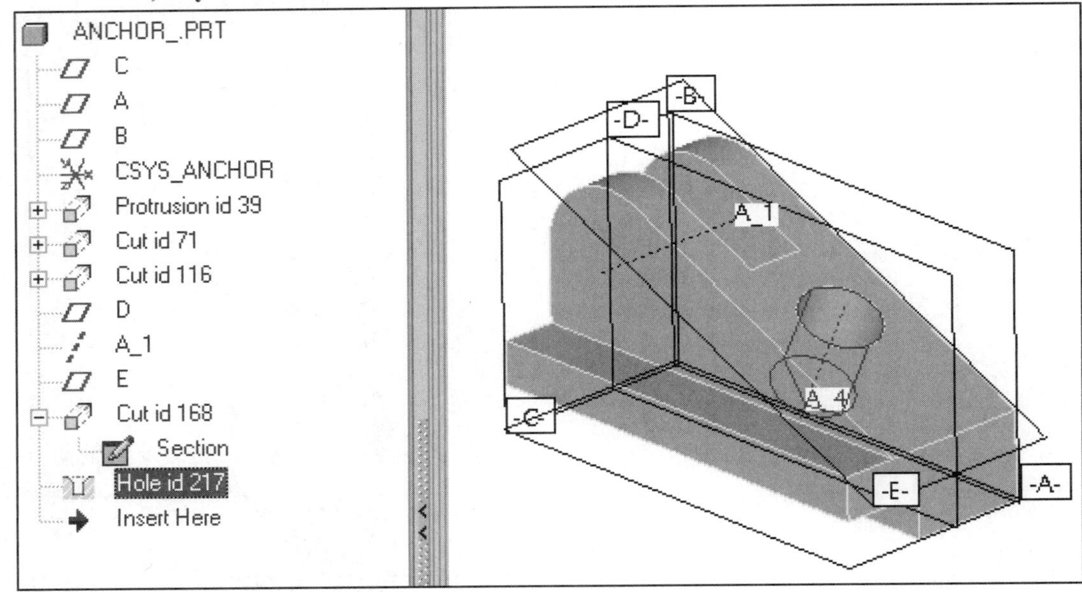

Figure 5.80 Completed Sketched Hole

The last feature to create is a **∅.250** hole. Click: **Hole Tool** ⇒ pick **A_1** from model ⇒ **Placement** tab ⇒ **Coaxial** ⇒ click in (No Items) box under- Secondary references: pick datum **D** from the model ⇒ diameter **.250** ⇒ **Enter** ⇒ **Symmetric** ⇒ depth **1.875** (Fig. 5.81) ⇒ **Enter** ⇒ ⇒ ⇒ ⇒ ⇒ **MMB** ⇒ **LMB** to deselect

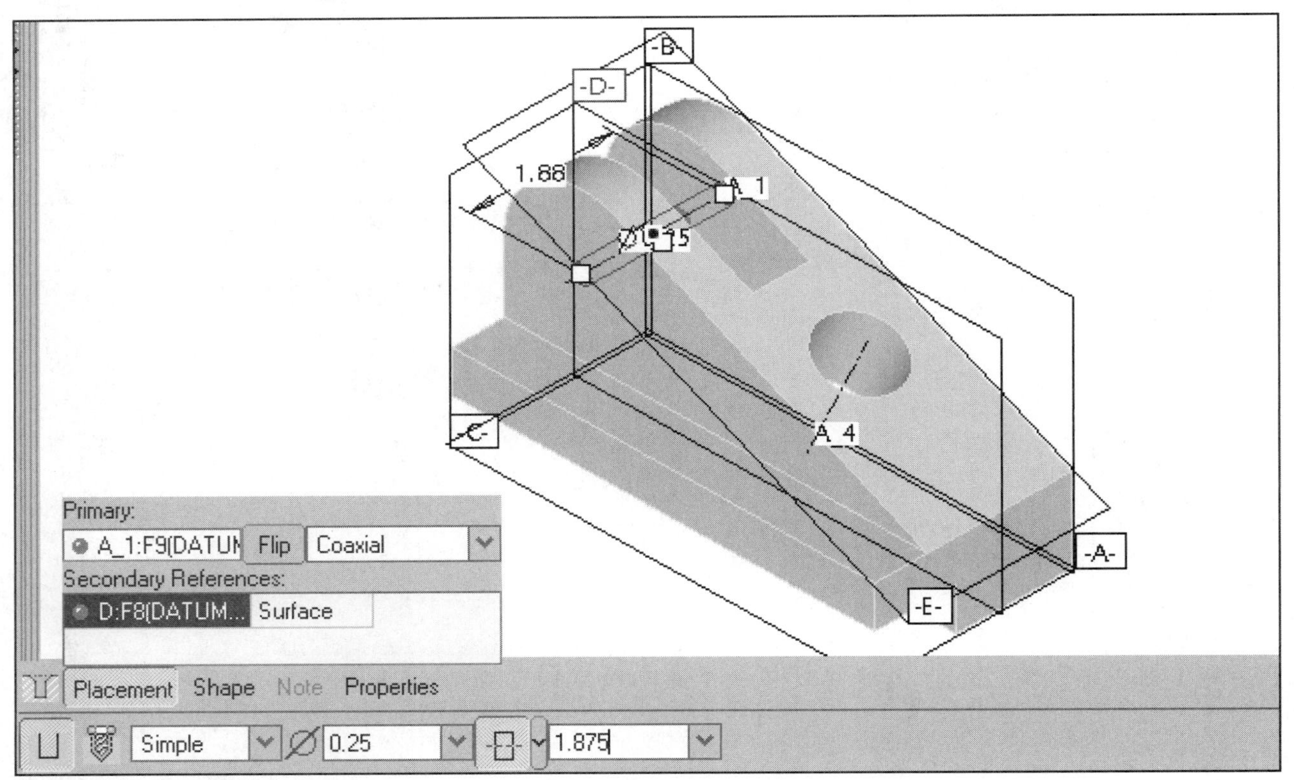

Figure 5.81 Coaxial Hole

Suppressing and Resuming Features

Next, you will create a new layer and add the two holes to it. The holes will then be selected in the Layer Tree and suppressed. Suppressed features are temporarily removed from the model along with their children (if any). In the example, you will notice that the holes and one axis (A_4) will be suppressed. Since axis A_1 is the parent of the small hole, it will not be suppressed. Suppressing features is like removing them from regeneration temporarily. However, you can "unsuppress" (resume) suppressed features at any time.

You can suppress features on a part to simplify the part model and decrease regeneration time. For example, while you work on one end of a shaft, it may be desirable to suppress features on the other end of the shaft. Similarly, while working on a complex assembly, you can suppress some of the features and components for which the detail is not essential to the current assembly process. Suppress features to do the following:

- Concentrate on the current working area by suppressing other areas
- Speed up a modification process because there is less to update
- Speed up the display process because there is less to display
- Temporarily remove features to try different design iterations

Unlike other features, the base feature cannot be suppressed. If you are not satisfied with your base feature, you can redefine the section of the feature, or start another part.

Click: **Show** in the **Navigator** ⇒ **Layer Tree** --Layer Tree displays in place of the **Model Tree** ⇒ **Show** ⇒ activate **Layer Items** ⇒ click on and highlight **Layers** in the Layer Tree ⇒ **RMB** ⇒ **New Layer** ⇒ Name: type **HOLES** as the name for the new layer ⇒ Selection Filter (bottom right-hand side below graphics window) click: **Feature** ⇒ select the two holes from the model (Fig. 5.82) ⇒ **OK** ⇒ expand the HOLES layer ⇒ press and hold down the **Ctrl** key and click on the two layer items (Fig. 5.83) ⇒ **Edit** from the menu bar ⇒ **Suppress** ⇒ **OK** from the Suppress dialog box (Fig. 5.84) ⇒ **Edit** from the menu bar ⇒ **Resume** ⇒ **All** ⇒ **Show** ⇒ **Model Tree** ⇒ [save icon] ⇒ **MMB**

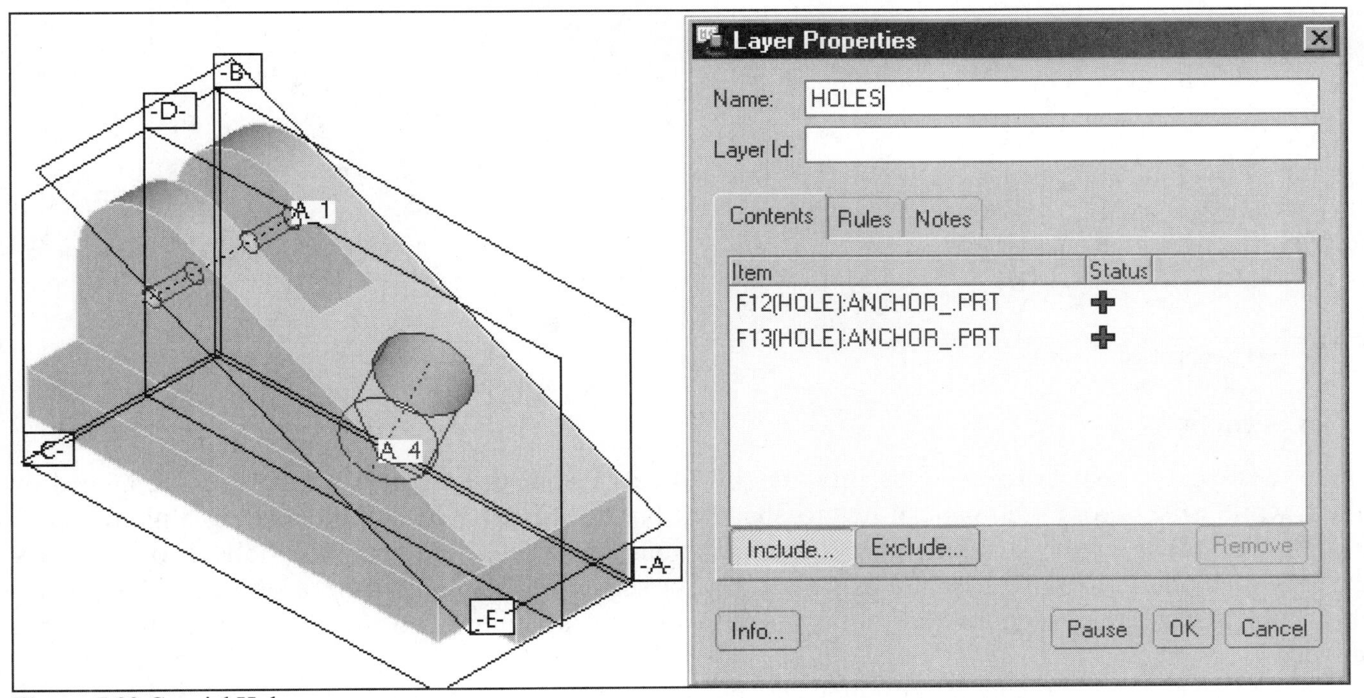

Figure 5.82 Coaxial Hole

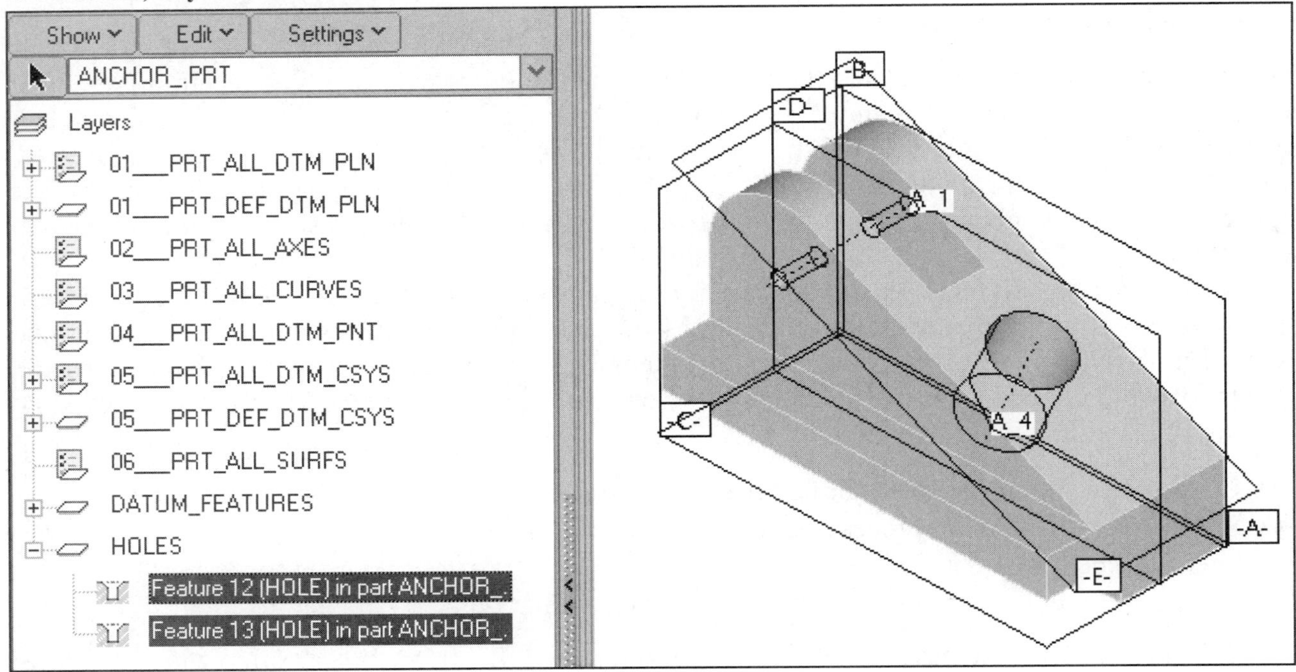

Figure 5.83 HOLES Layer

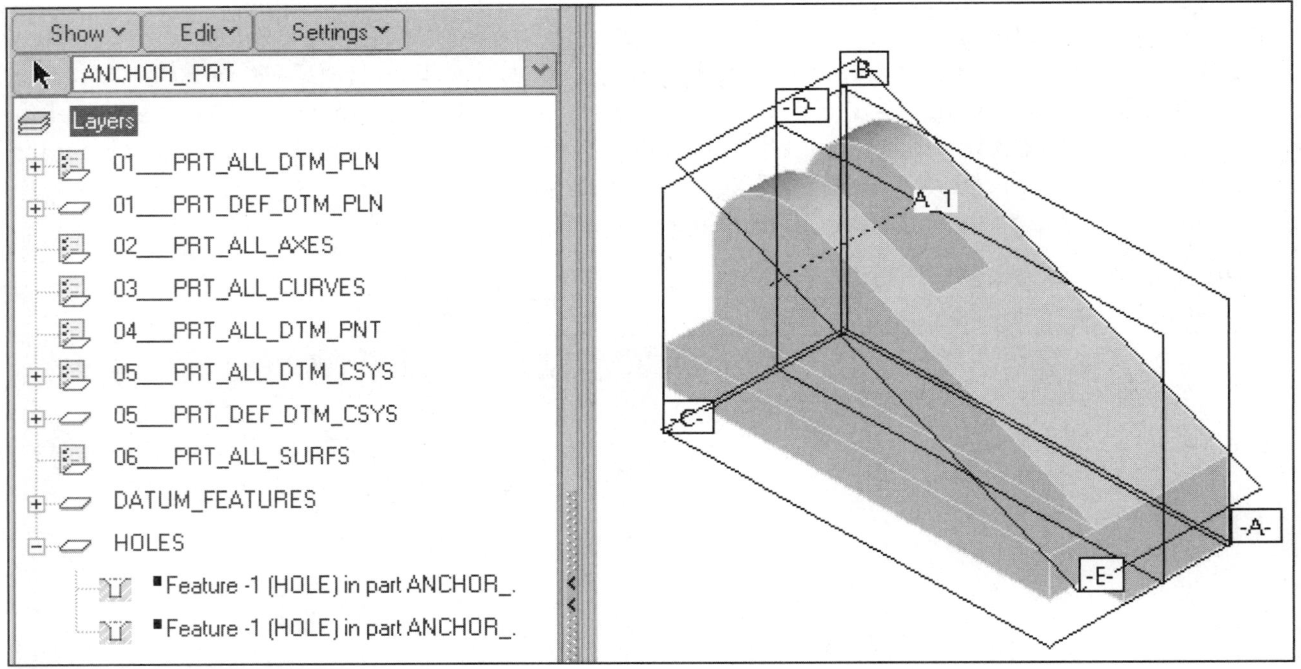

Figure 5.84 Suppressed Holes

Cross Sections

There are two types of cross sections: **planar** and **offset**. Planar cross sections can be crosshatched or filled, while offset cross sections can be crosshatched but not filled. You will be creating a planar cross section. Pro/E can create standard planar cross sections of models (parts or assemblies), offset cross sections of models (parts or assemblies), planar cross sections of datum surfaces or quilts (Part mode only) and planar cross sections that automatically intersect all quilts and all geometry in the current model.

Cross section cut planes do not intersect cosmetic features in a model.

Click: **Tools** ⇒ Model Sectioning... ⇒ Cross Section dialog box displays ⇒ **New** [Fig. 5.85(a)] ⇒ type name **A** [Fig. 5.85(b)] ⇒ **Enter** ⇒ **Planar** ⇒ **Single** ⇒ **MMB** ⇒ **Plane** ⇒ pick datum **D** ⇒ **Display** ⇒ **Show X-Section** (Fig. 5.86) ⇒ **Edit** ⇒ **Redefine** ⇒ **Hatching** ⇒ **Fill** (Fig. 5.87) ⇒ **Hatch** ⇒ **Spacing** ⇒ **Half** (Fig. 5.88) ⇒ **MMB** ⇒ **MMB** ⇒ **Close** ⇒ ⇒ **MMB**

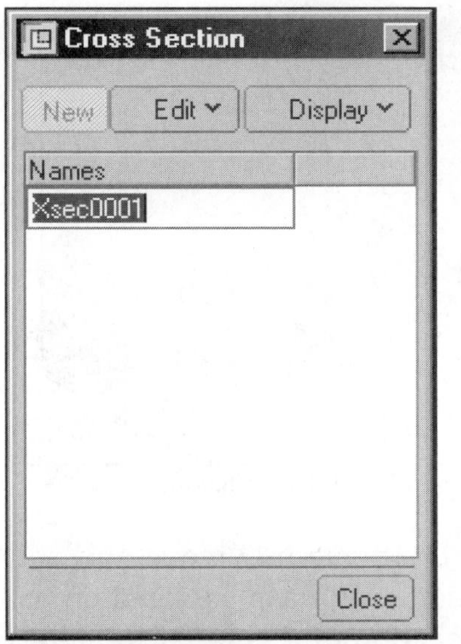

Figure 5.85(a) Default X-Section Name

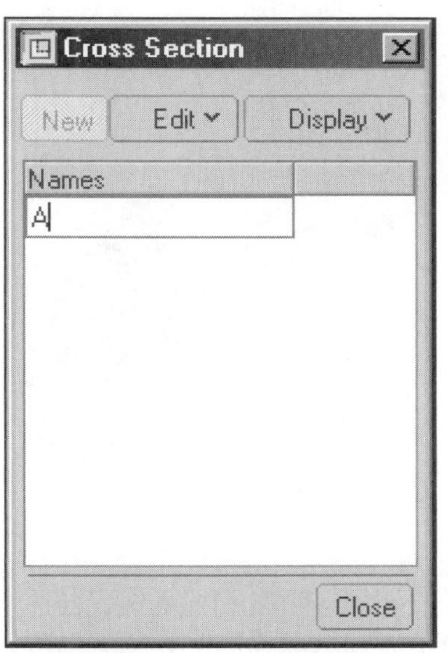

Figure 5.85(b) A as New X-Section Name

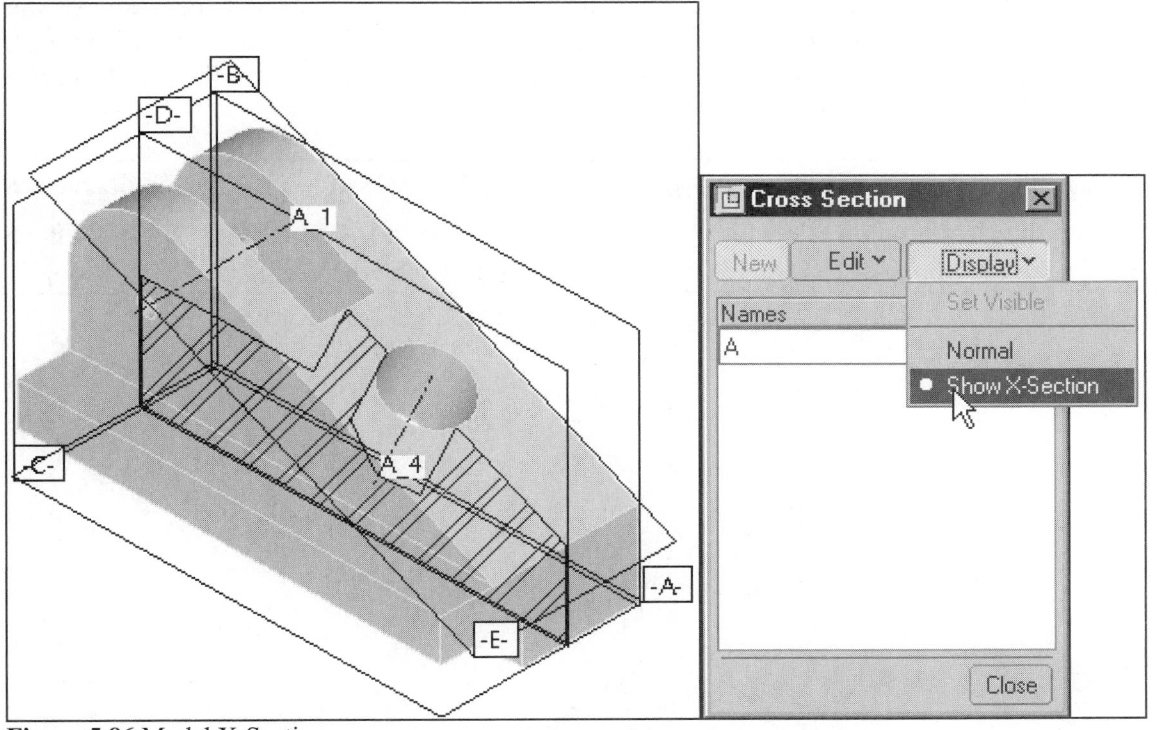

Figure 5.86 Model X-Section

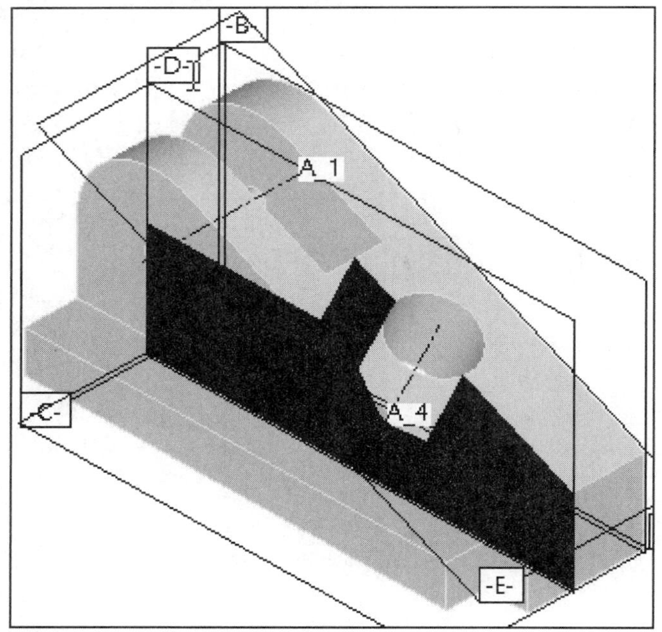

Figure 5.87 Model X-Section Fill

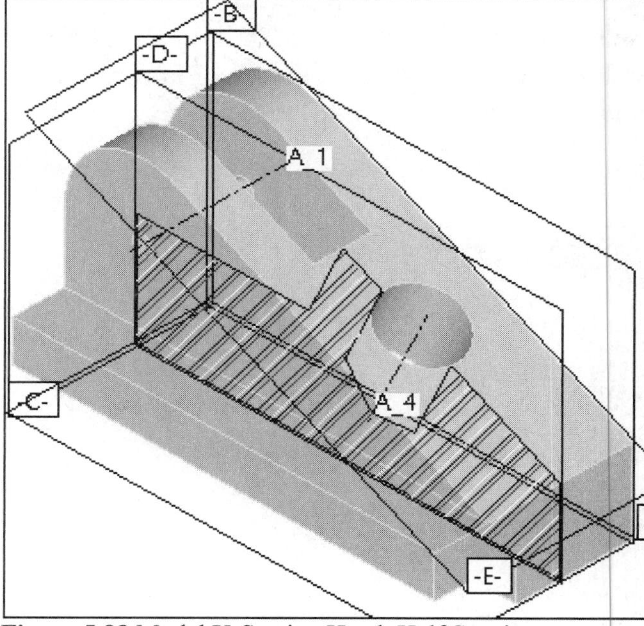

Figure 5.88 Model X-Section Hatch Half Spacing

The section passes through datum D. The slot is the *child* of datum D. If datum D moves, so will the slot and the small hole (and the X-Section). In order to ensure that the datum D stays centered on the upper portion of the part, you will need to create a relation to control the location of datum D. Relations will be covered in-depth in Lesson 9. The first cut used a dimension from datum B for location (**1.875**).

The relation states that the distance from datum B to datum D will be one-half the value of the distance from datum B to the cut surface. If the thickness of the upper portion of the part (**1.875**) changes, datum D will remain centered, as will the slot, and the X-Section. To start, you must first find out the feature dimension symbols (**d#**) required for the relation.

Click on the first cut in the **Model Tree** ⇒ **RMB** ⇒ **Edit** (Fig. 5.89) ⇒ **Info** from the menu bar ⇒ Switch Dimensions - note the **d** symbol **d6** (Fig. 5.91) ⇒ click on datum **D** in the **Model Tree** ⇒ **RMB** ⇒ **Edit** (Fig. 5.90) ⇒ note the **d** symbol **d12** (Fig. 5.92) (your "**d**" value may be different)

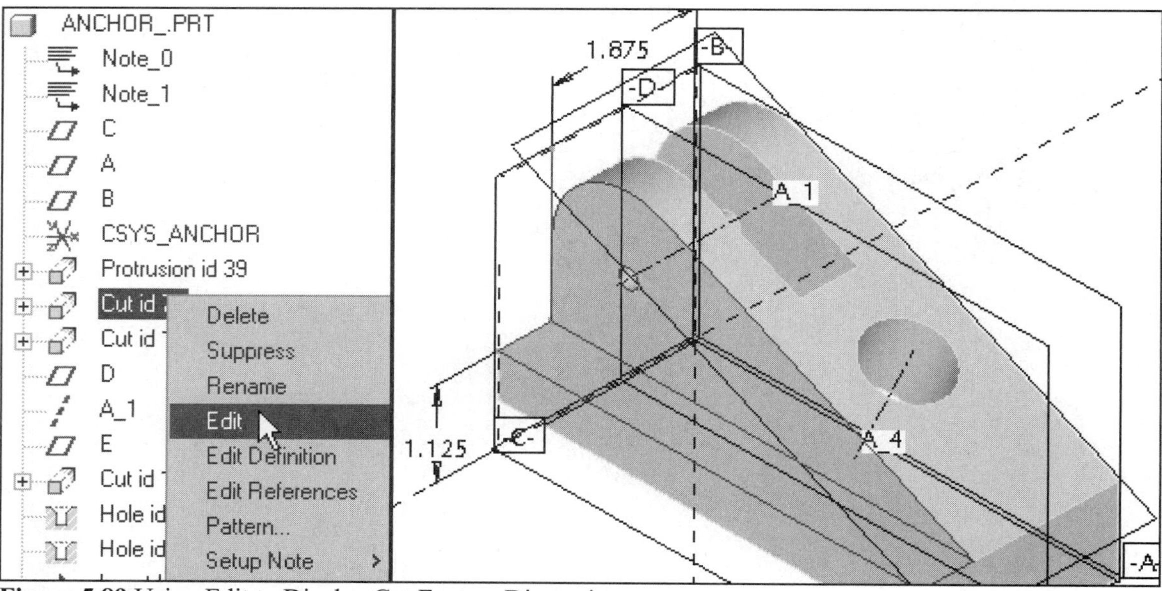

Figure 5.89 Using Edit to Display Cut Feature Dimensions

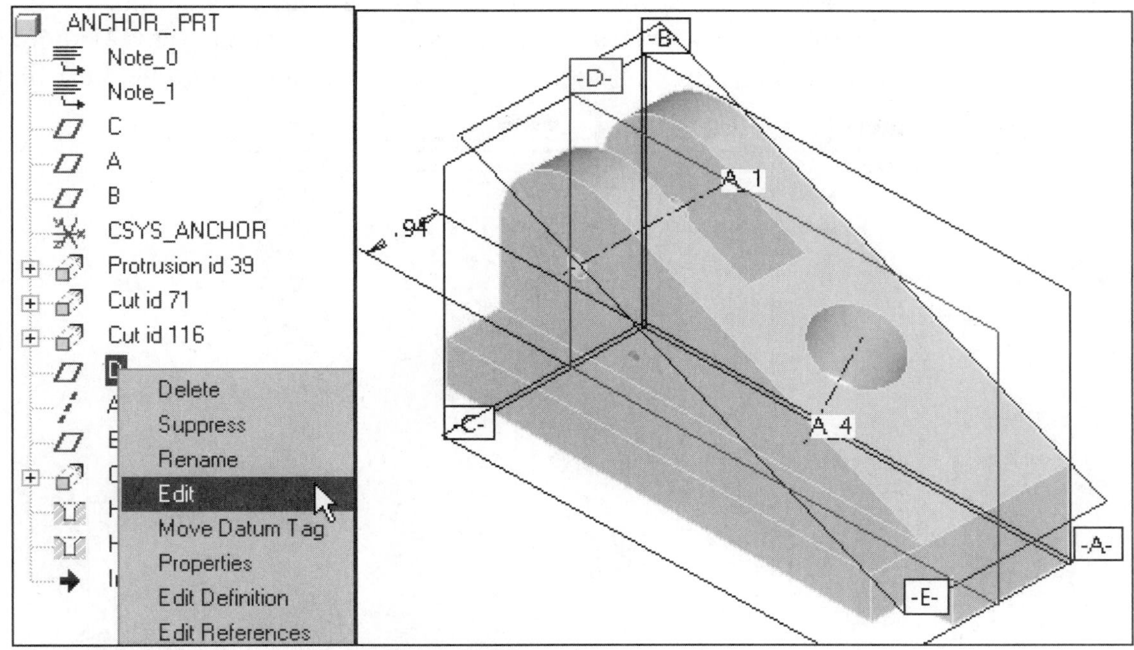

Figure 5.90 Using Edit to Display Datum D Dimension

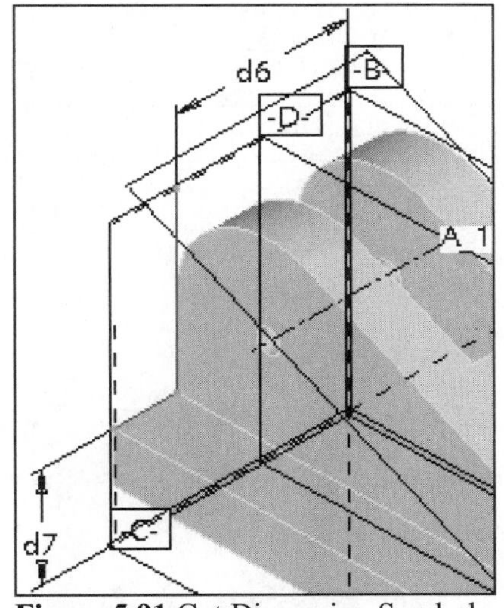

Figure 5.91 Cut Dimension Symbols

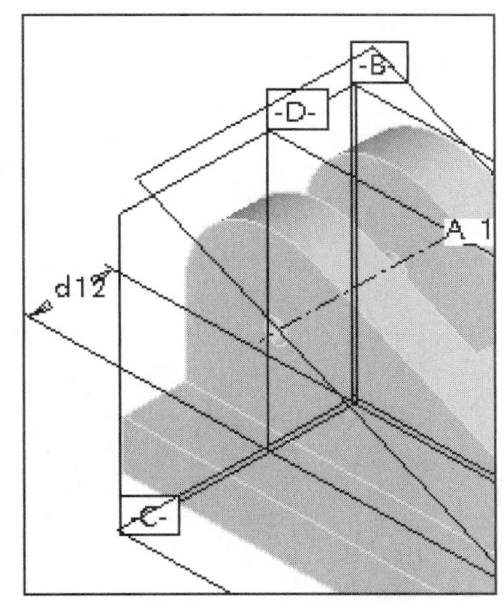

Figure 5.92 Datum D Dimension Symbols

Click: **Tools** from the menu bar ⇒ **Relations** relations dialog box displays ⇒ type **d12=d6/2** (Fig. 5.93) *your "d" values may be different* ⇒ **Ok** ⇒ **Info** ⇒ **Switch Dimensions** ⇒ [save icon] ⇒ **MMB**

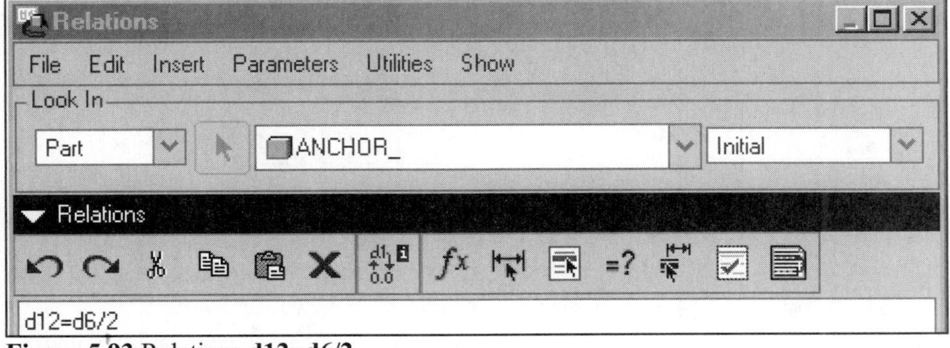

Figure 5.93 Relations **d12=d6/2**

Double-click on the large hole [Fig. 5.94(a)] ⇒ **Info** ⇒ **Switch Dimensions**- note the **d** symbol **d21** [Fig. 5.94(b)] (hole dimension from datum B) ⇒ **Tools** ⇒ **Relations** ⇒ below the first relation type **d21=d6/2** ⇒ **Ok** ⇒ **Info** ⇒ **Switch Dimensions** ⇒ double-click on the first cut on the model ⇒ double-click on **1.875** [Fig. 5.94(c)] and modify to **2.00** [Fig. 5.94(d)] ⇒ **Enter** ⇒ ⇒ ⇒ **TOP** (Fig. 5.95)

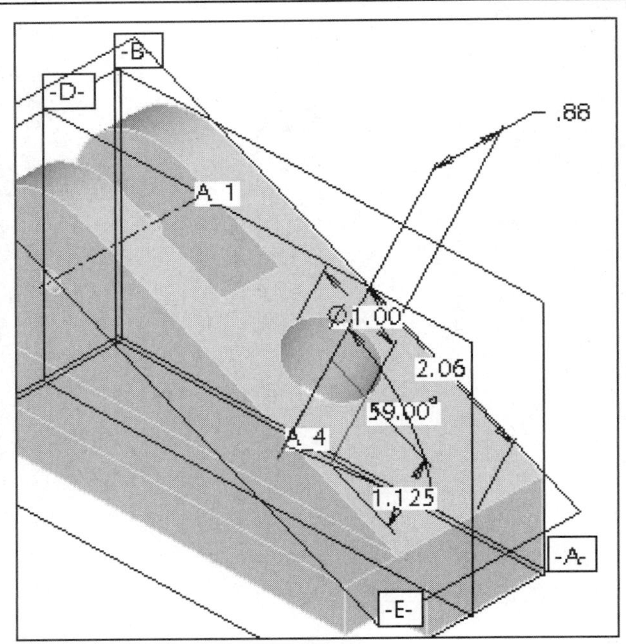

Figure 5.94(a) Hole Dimensions

Figure 5.94(b) Hole Dimension Symbols

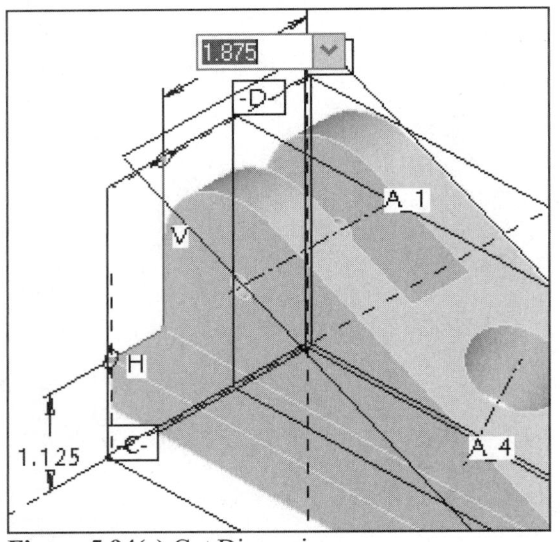

Figure 5.94(c) Cut Dimensions

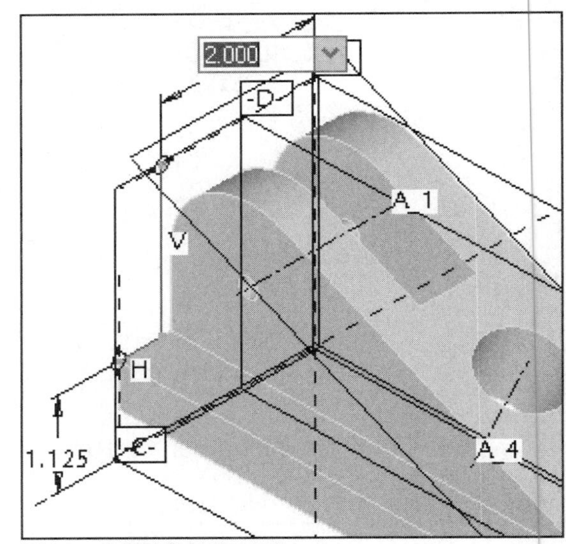

Figure 5.94(d) Edited Cut

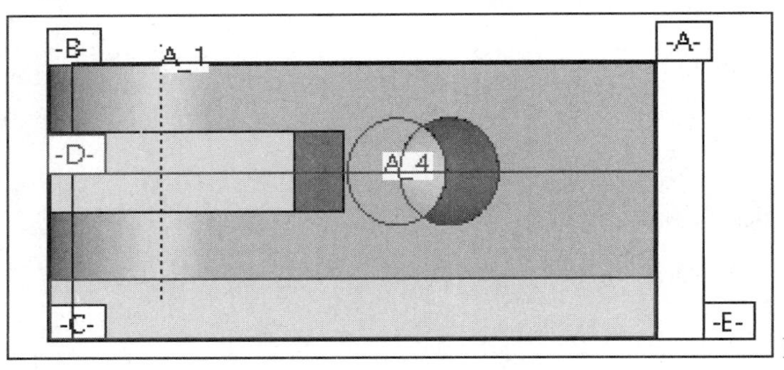

Figure 5.95 Top View, Slot, Hole, and Datum are Aligned

Click: ⇒ **Standard orientation** ⇒ change the **2.00** dimension back to **1.875** and regenerate, then click: **Info** ⇒ **Relations and Parameters** ⇒ click on **d6**, **d12**, and **d21** in the Browser in order to display them in the graphics window (Fig. 5.96) ⇒ close the Browser with the quick sash ⇒ ⇒ **MMB** ⇒ **File** ⇒ **Delete** ⇒ **Old Versions** ⇒ **MMB** ⇒ **File** ⇒ **Close Window**

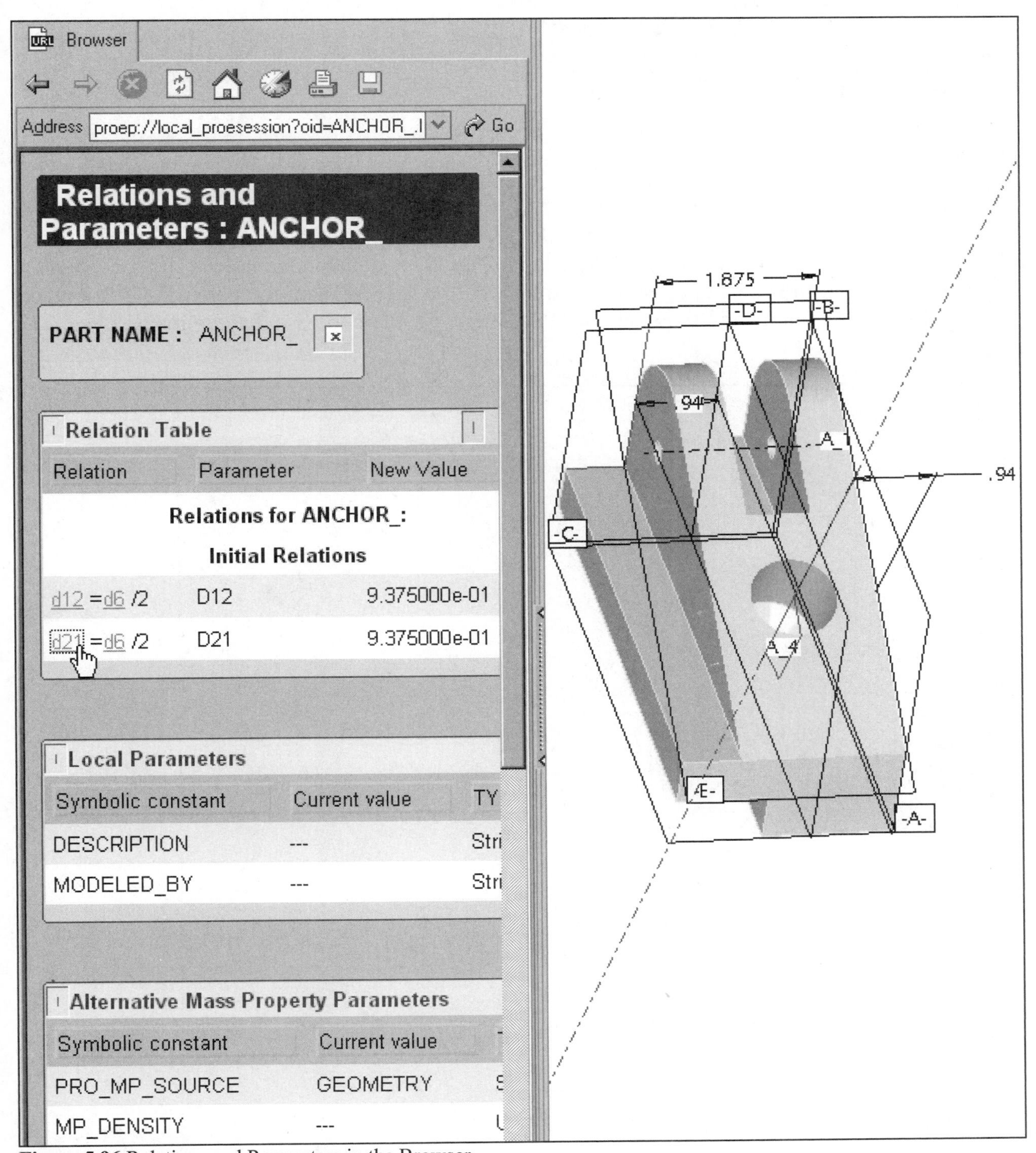

Figure 5.96 Relations and Parameters in the Browser

Lesson 5 is now complete. Continue with the Lesson Project.

Lesson 5 Project

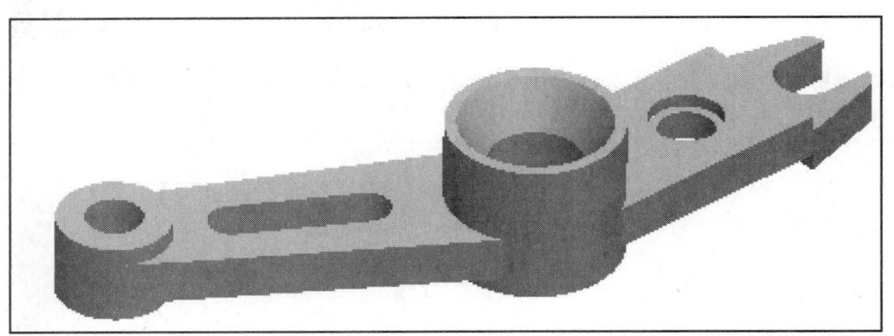

Figure 5.97 Angle Frame

Angle Frame

The Angle Frame is a machined part that requires the use of a variety of datum planes and a layering scheme. You will also add a relation to control the depth of the large countersink hole at the part's center. Analyze the part and plan the steps and features required to model it. Create the part shown in Figures 5.97 through 5.107. Set the units and the material (aluminum) for the part. Set the datums as Basic, and rename to A, B, C, and so on.

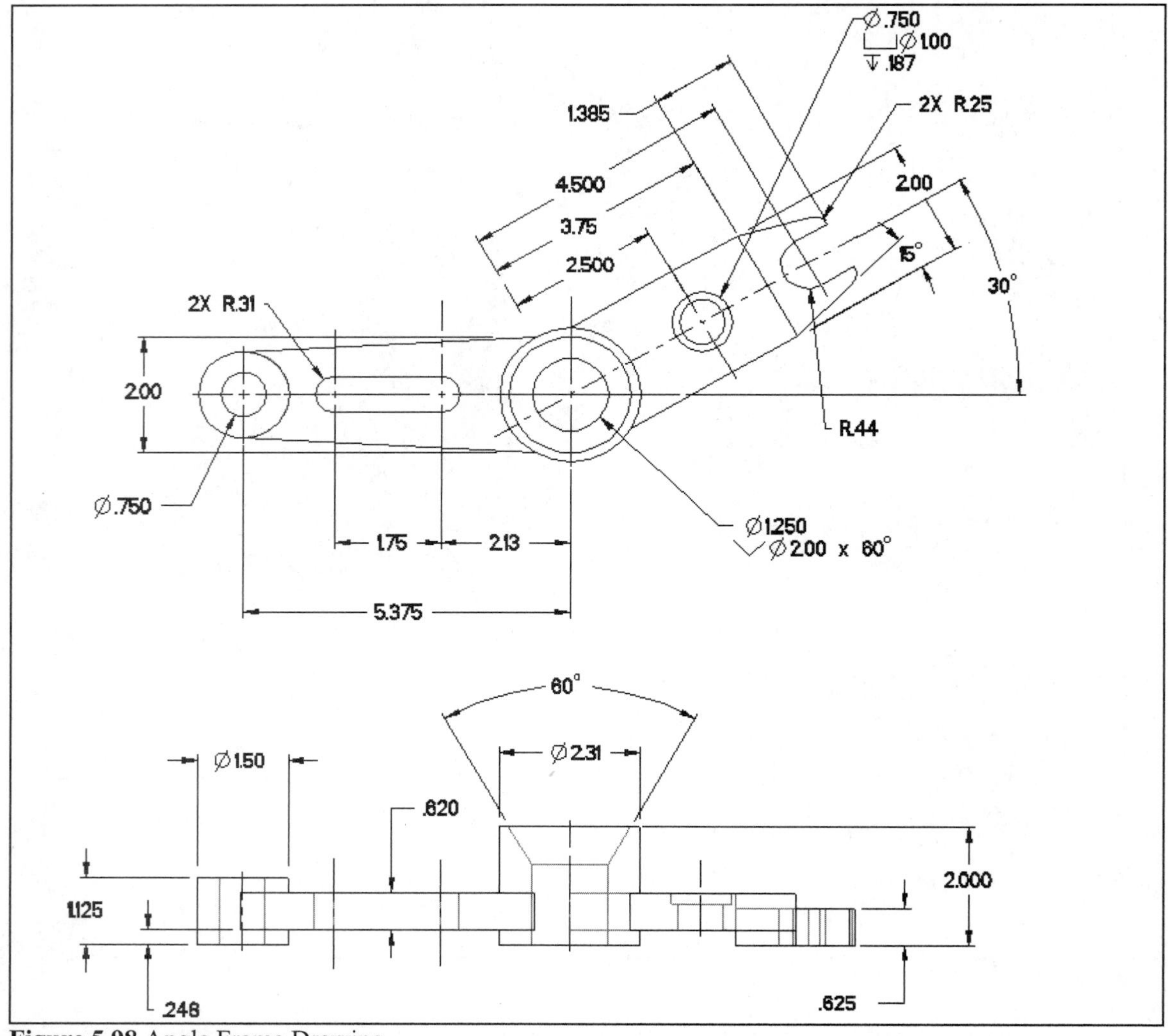

Figure 5.98 Angle Frame Drawing

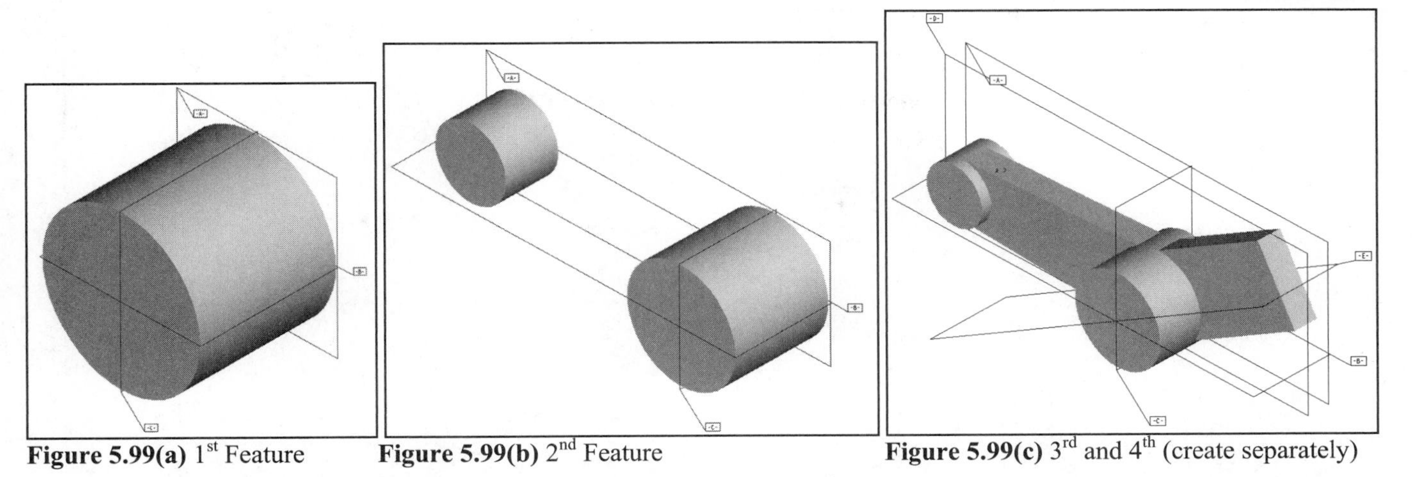

Figure 5.99(a) 1st Feature **Figure 5.99(b)** 2nd Feature **Figure 5.99(c)** 3rd and 4th (create separately)

Set the datums with the appropriate geometric tolerance names: A, B, C, and so on. Create two *sections* through the Angle Frame to be used later in a Drawing Lesson. For the sections, use datum planes B and E, which pass vertically through the center of the part. Name the cross sections A (SECTION A-A) and B (SECTION B-B).

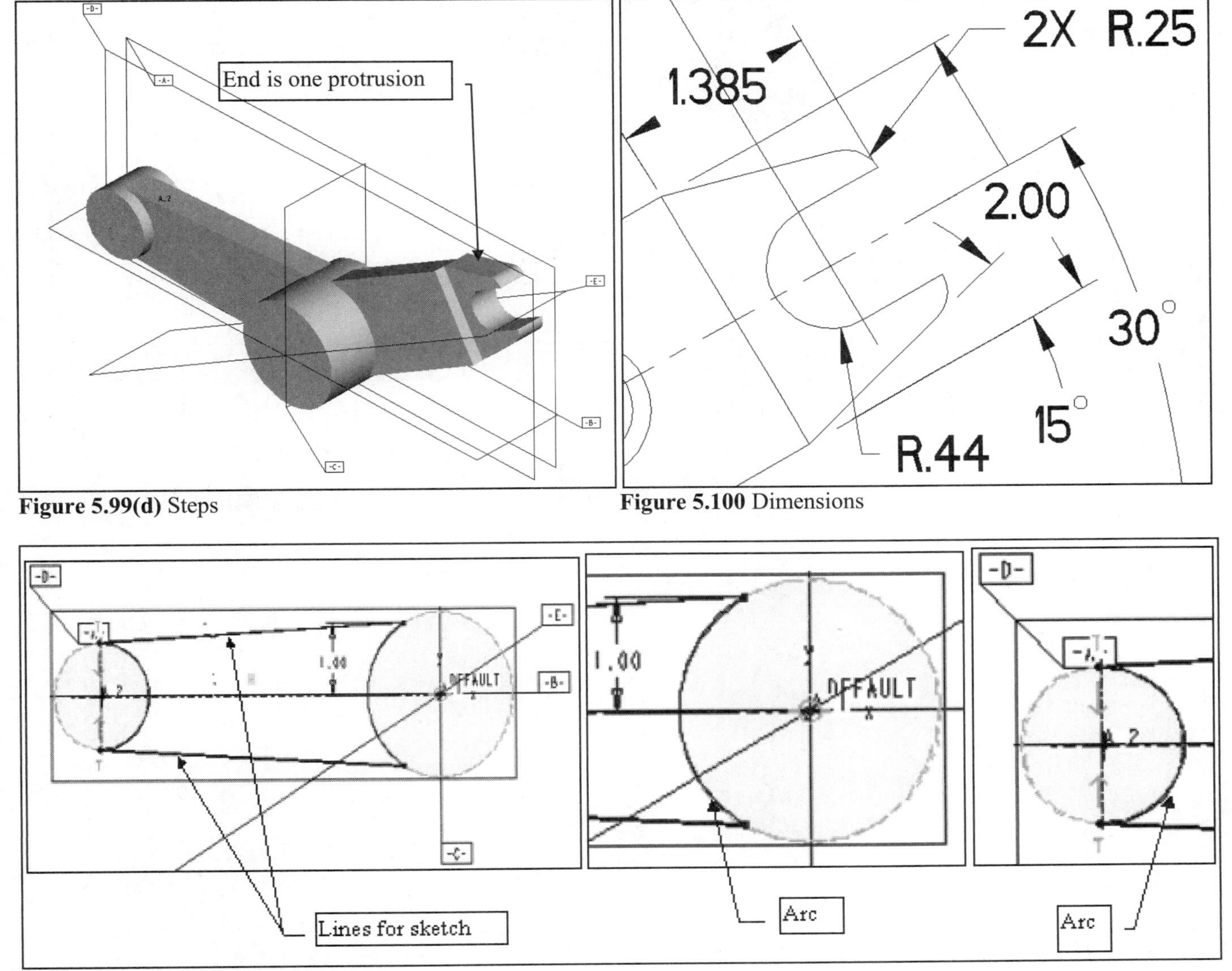

Figure 5.99(d) Steps **Figure 5.100** Dimensions

Figure 5.101 Sketching the Third Feature

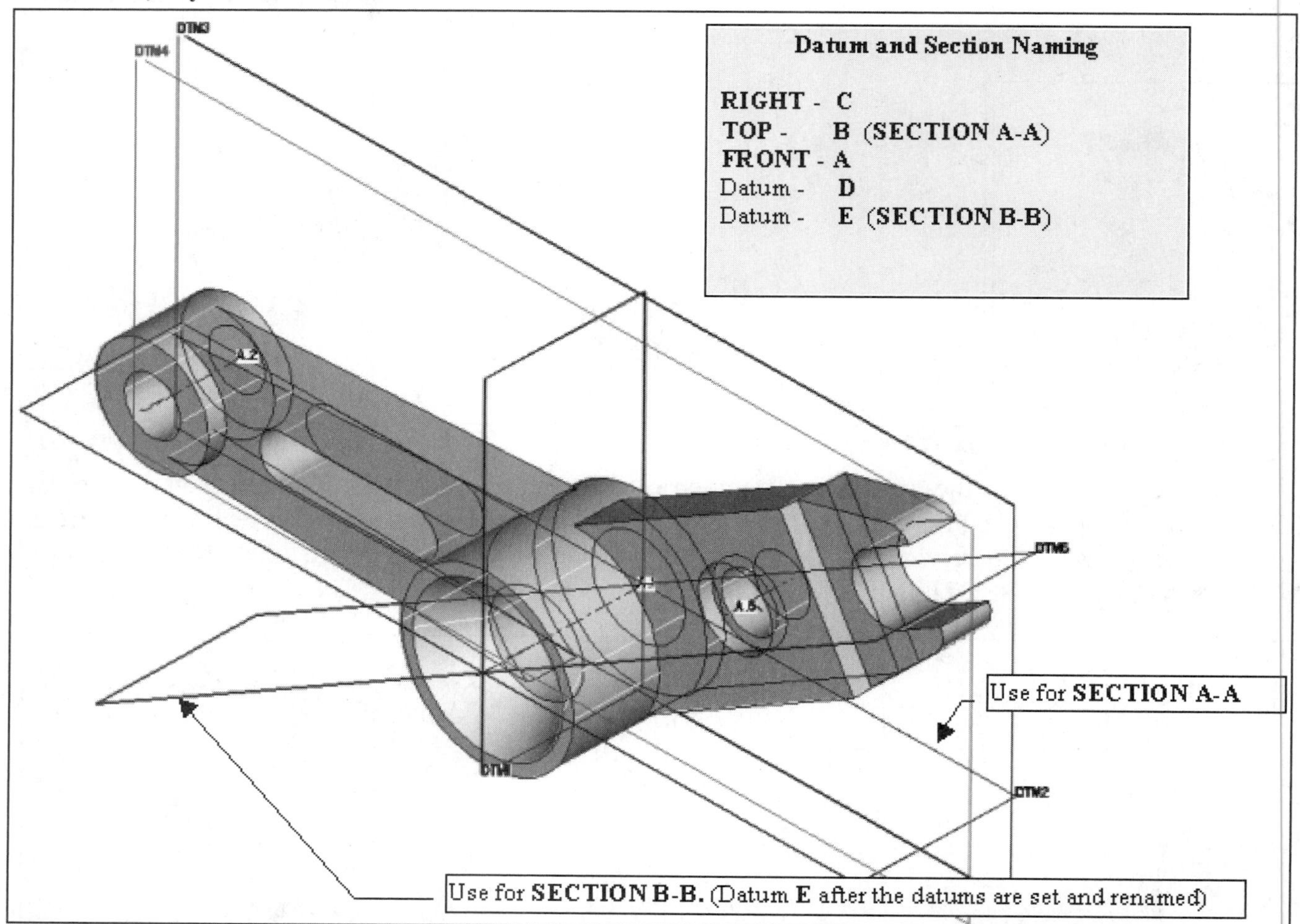

Figure 5.102 Datums. Datum E is through the axis of the first protrusion, and at an angle to datum B.

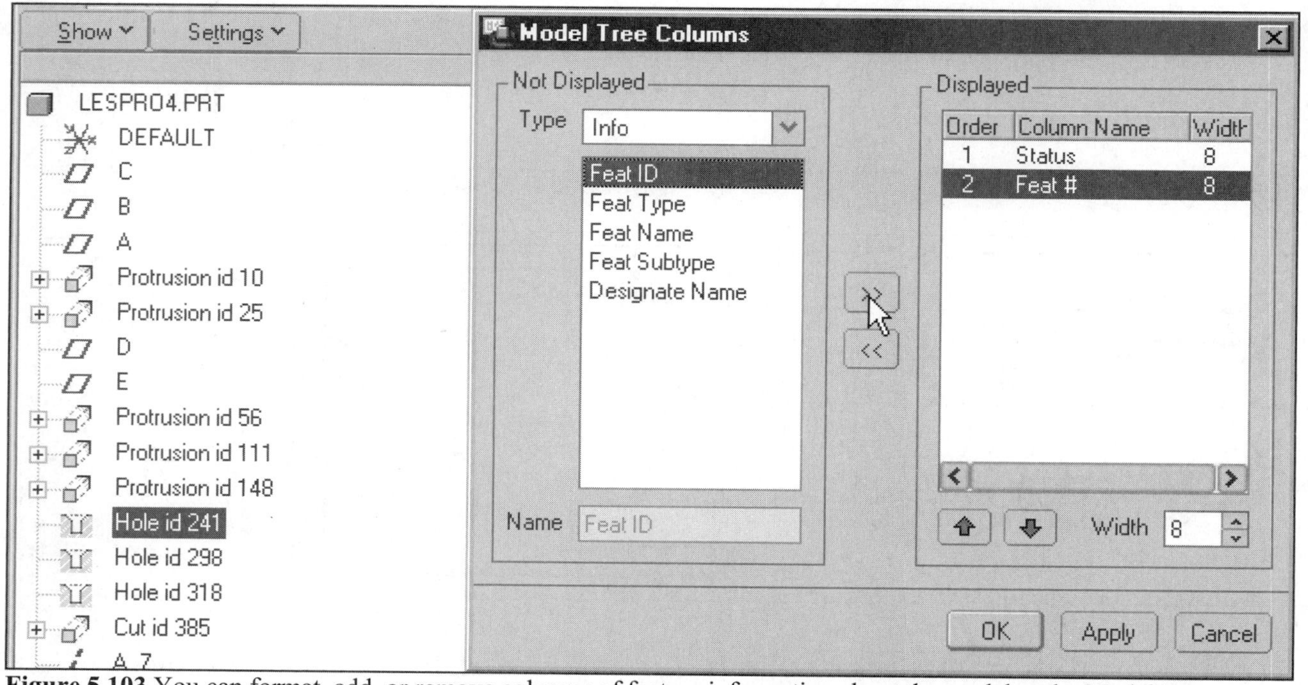

Figure 5.103 You can format, add, or remove columns of feature information about the model to the Model Tree

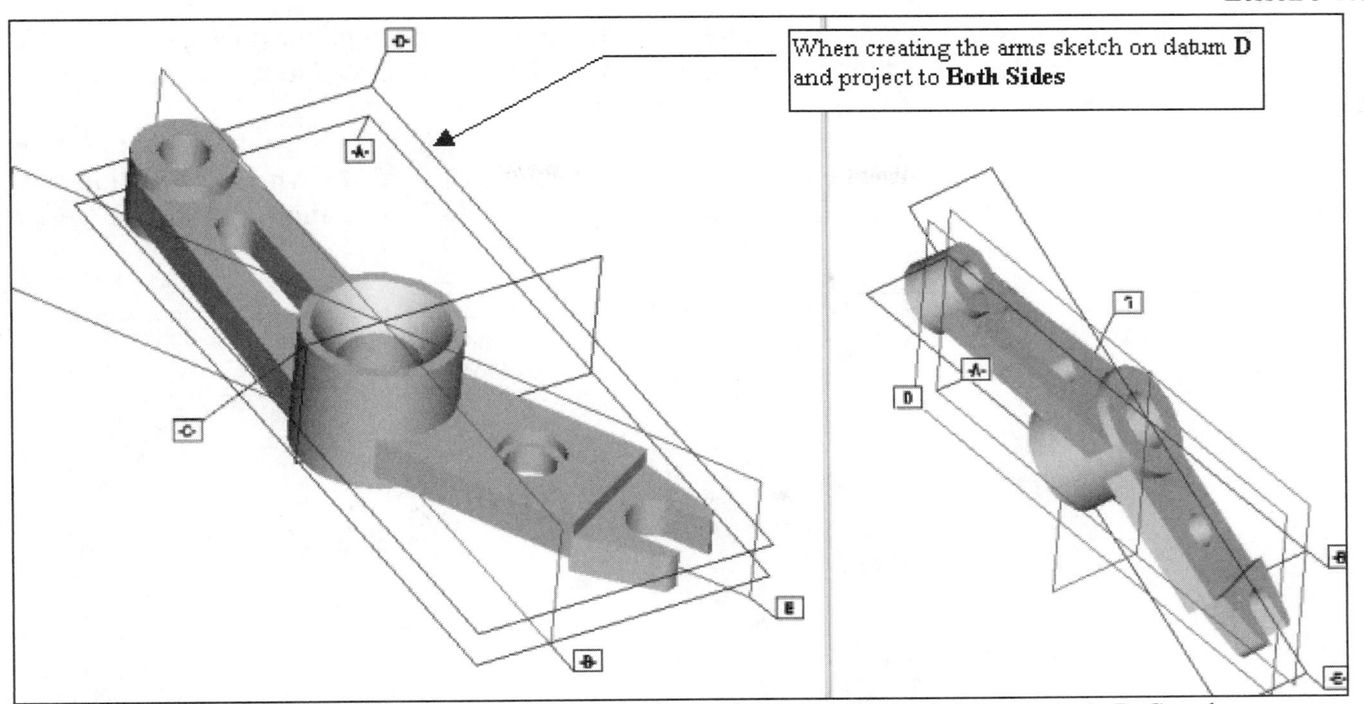

Figure 5.104 Using the Datums to Create Features. Set the datum planes and change their names to A, B, C, and so on.

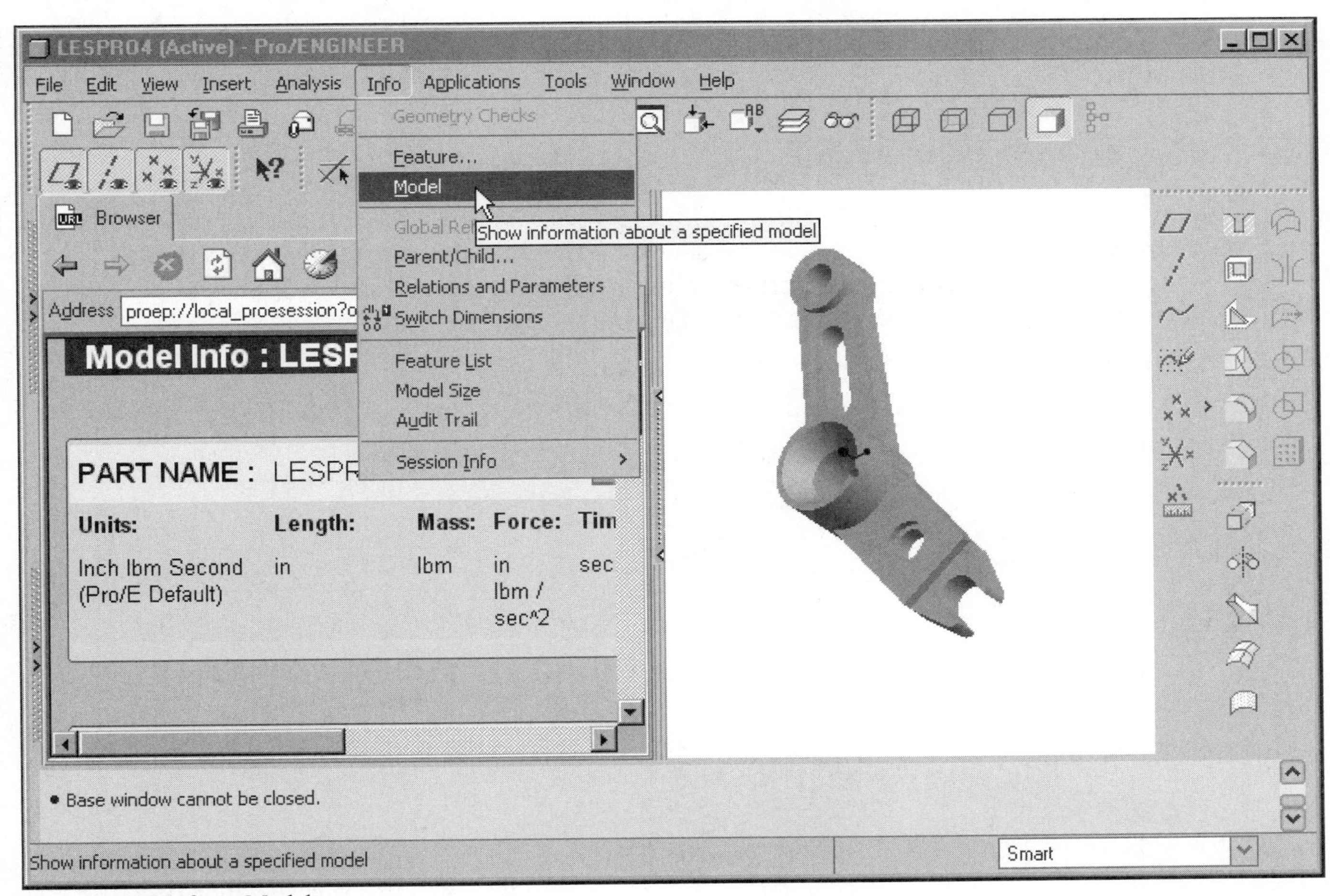

Figure 5.105 Info ⇒ Model

Modify the thickness of the boss from **2.00** to **2.50**. Note that the hole does not go through the part. Modify the boss back to the original design dimension of **2.00**. Add a relation to say that the depth of the hole should be equal to the thickness of the boss **(d43=d6)**.

Your **d#** symbols will probably be different from the ones shown here. Change the thickness of the boss (original protrusion) to see that the hole still goes through the part. No matter what the boss thickness dimension changes to, the hole will always go completely through it. This relation controls the *design intent* of the hole.

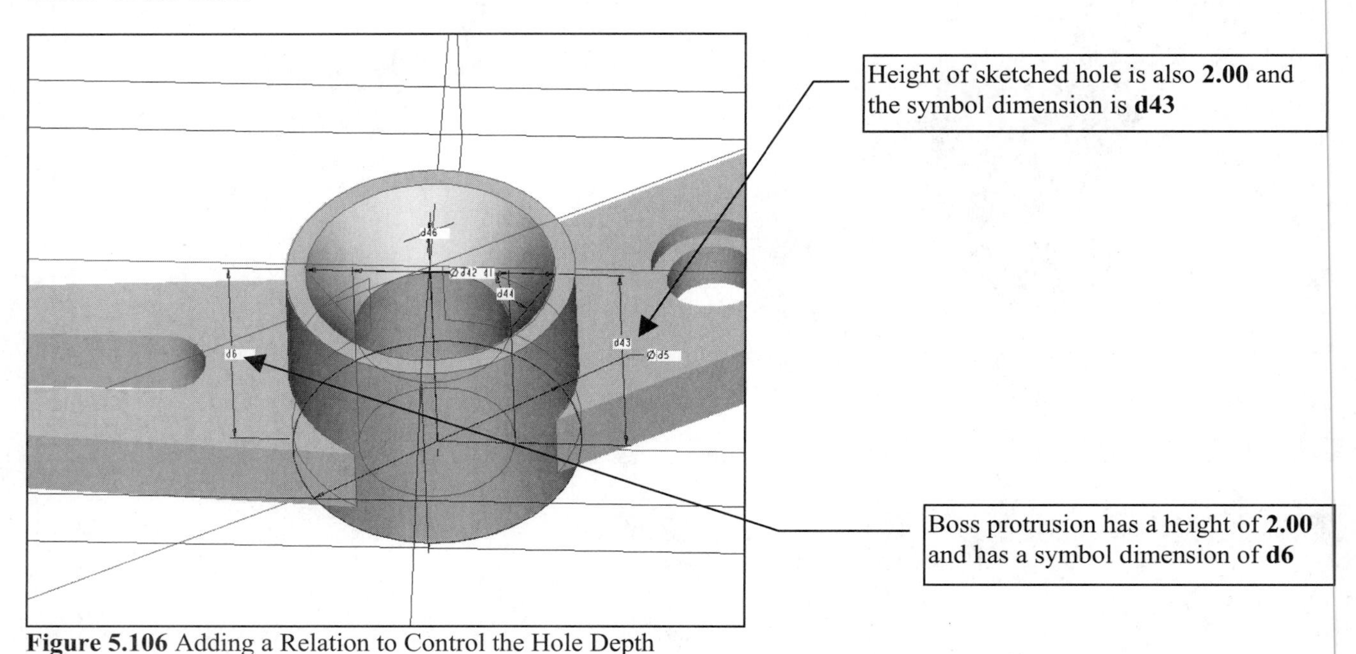

Figure 5.106 Adding a Relation to Control the Hole Depth

Relations are used to control features and preserve the *design intent* of the part. Lesson 9 will cover relations in more detail. Add a relation that says **d43=d6** (your **d#** symbols may be different).

Create the sections required to describe the part while you are in Part Mode so that they will be available for use when you are detailing the part in Drawing Mode.

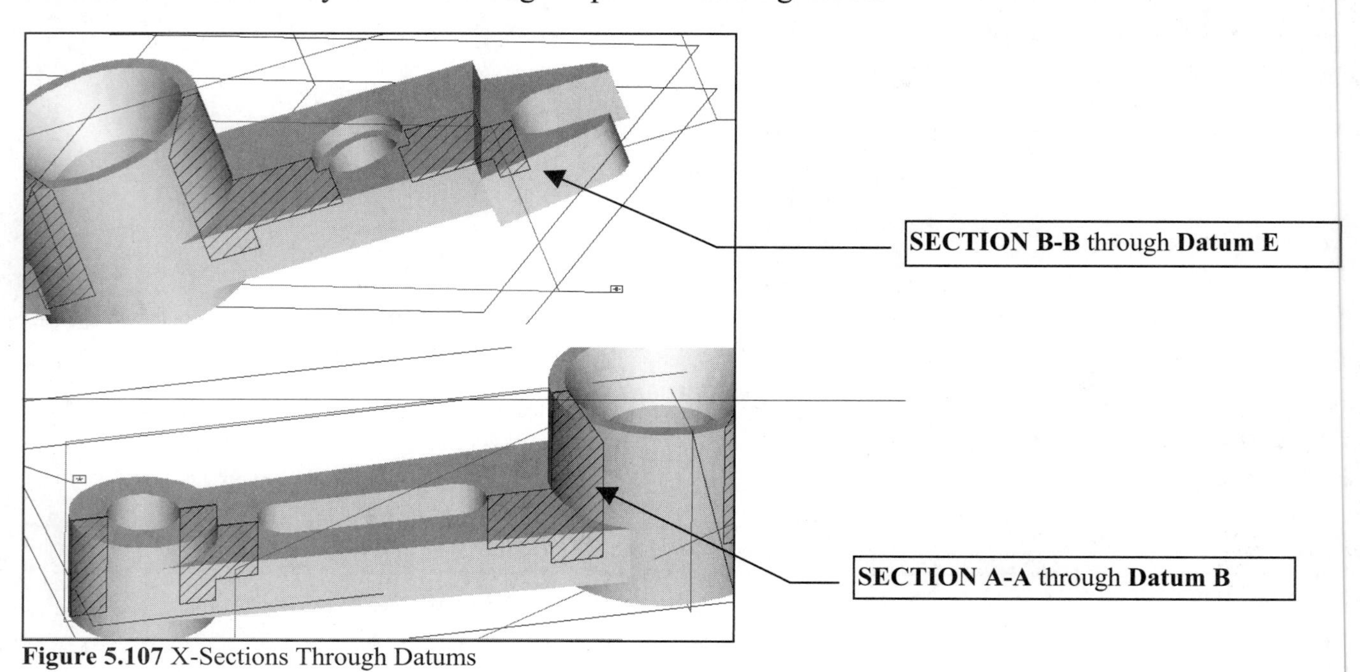

Figure 5.107 X-Sections Through Datums

Lesson 6 Revolved Protrusions and Revolved Cuts

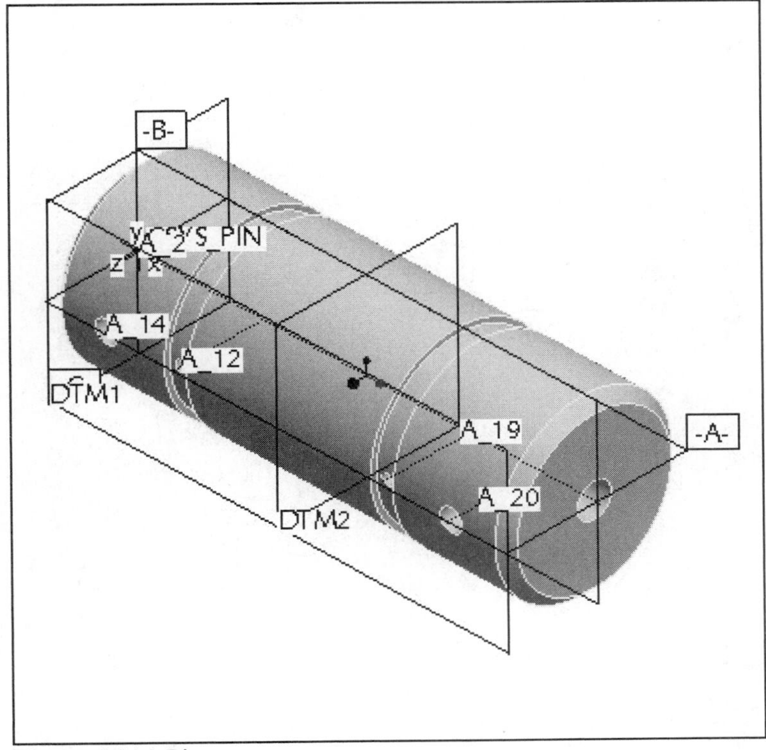

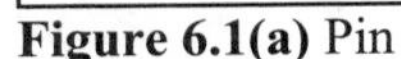
Figure 6.1(a) Pin

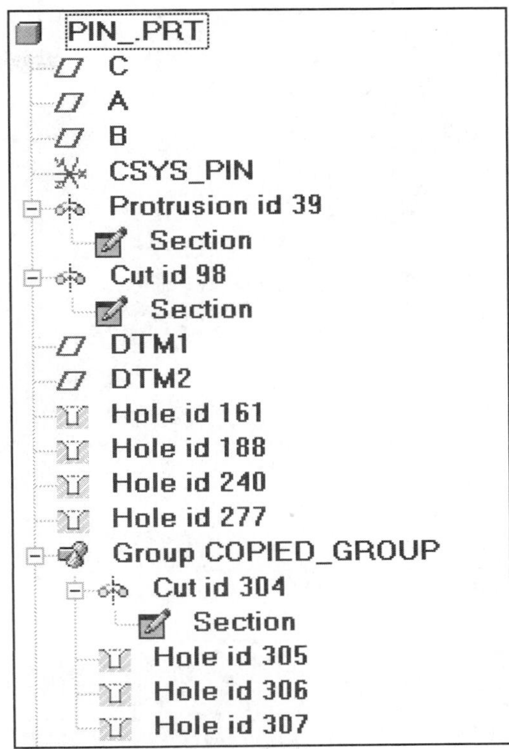

Figure 6.1(b) Pin's Model Tree

OBJECTIVES

- Alter and set the **Items** and **Columns** displayed in the **Model Tree**
- Create a **Revolved Protrusion**
- Use **Datums** to **locate holes**
- Create a **conical revolved cut**
- Edit **Dimension Properties**
- Use the **Model Player** to extract information and dimensions
- Get a hard copy using the **Print** command

REVOLVED PROTRUSIONS AND REVOLVED CUTS

The **Revolve Tool** creates a *revolved solid* or a *revolved cut* by revolving a sketched section around a centerline from the sketching plane [Fig. 6.1(a-b)]. You can have any number of centerlines in your sketch/section, but the first centerline will be the one used to rotate your section geometry [Fig. 6.2(a-b)]. To create a revolved section, create a centerline and the geometry that will be revolved about that centerline. Rules for sketching a revolved feature include:

- The revolved section must have a centerline [Fig. 6.3(a)]
- If you use more than one centerline in the sketch, Pro/E uses the first centerline sketched as the *axis of revolution* (you may select a different axis)
- The geometry must be sketched on only one side of the *axis of revolution*
- The section must be closed for a protrusion [Fig. 6.3(a-b)] but can be open for a cut

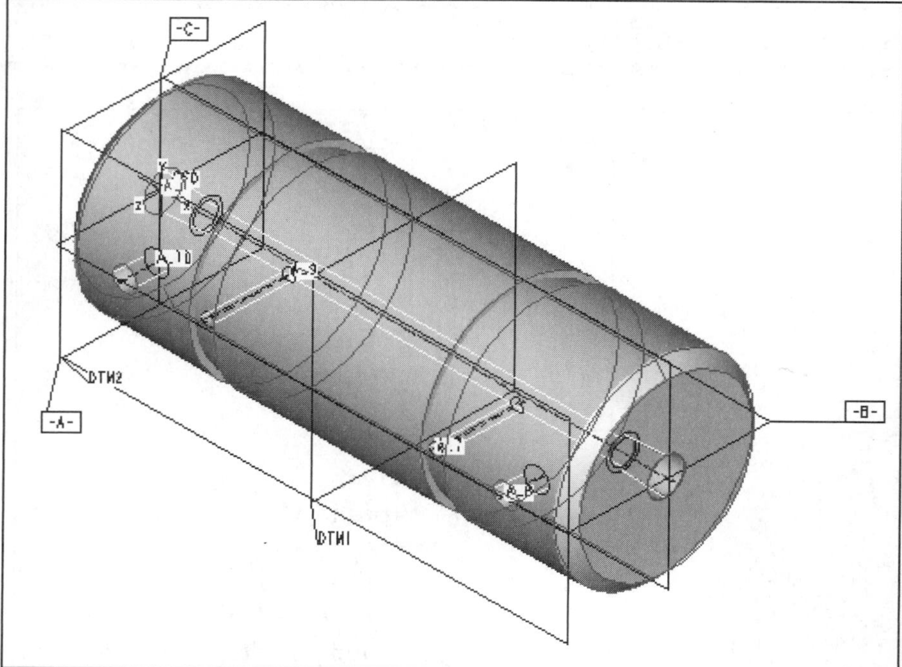

Figure 6.2(a) Pin and Datum Features

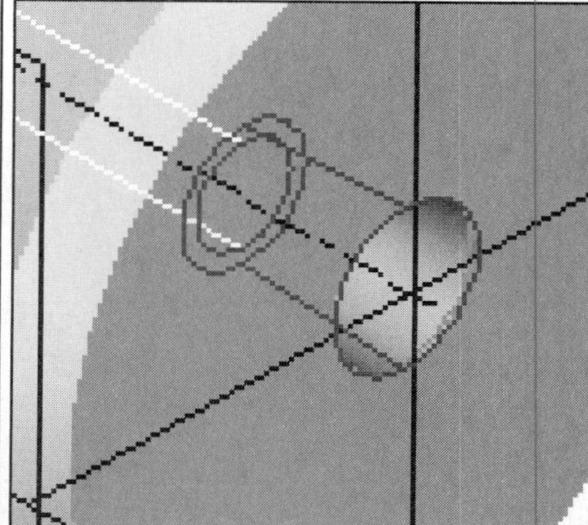

Figure 6.2(b) Pipe Tap

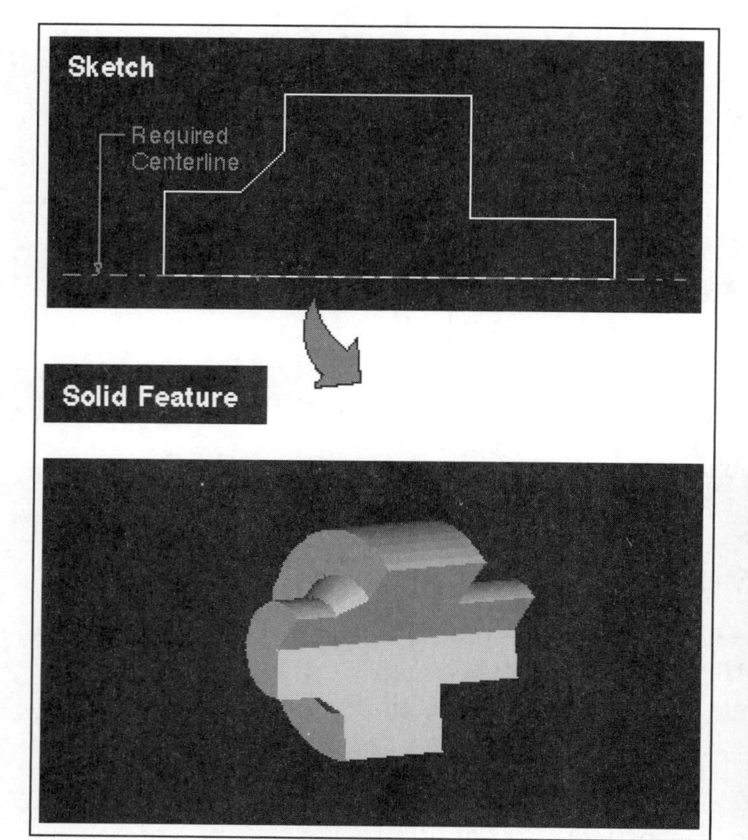

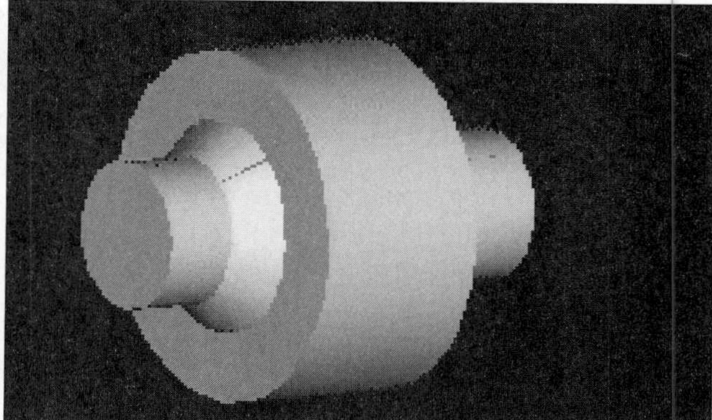

Figure 6.3(a-b) Revolved Protrusion (CADTRAIN, COAch for Pro/ENGINEER)

Lesson 6 STEPS

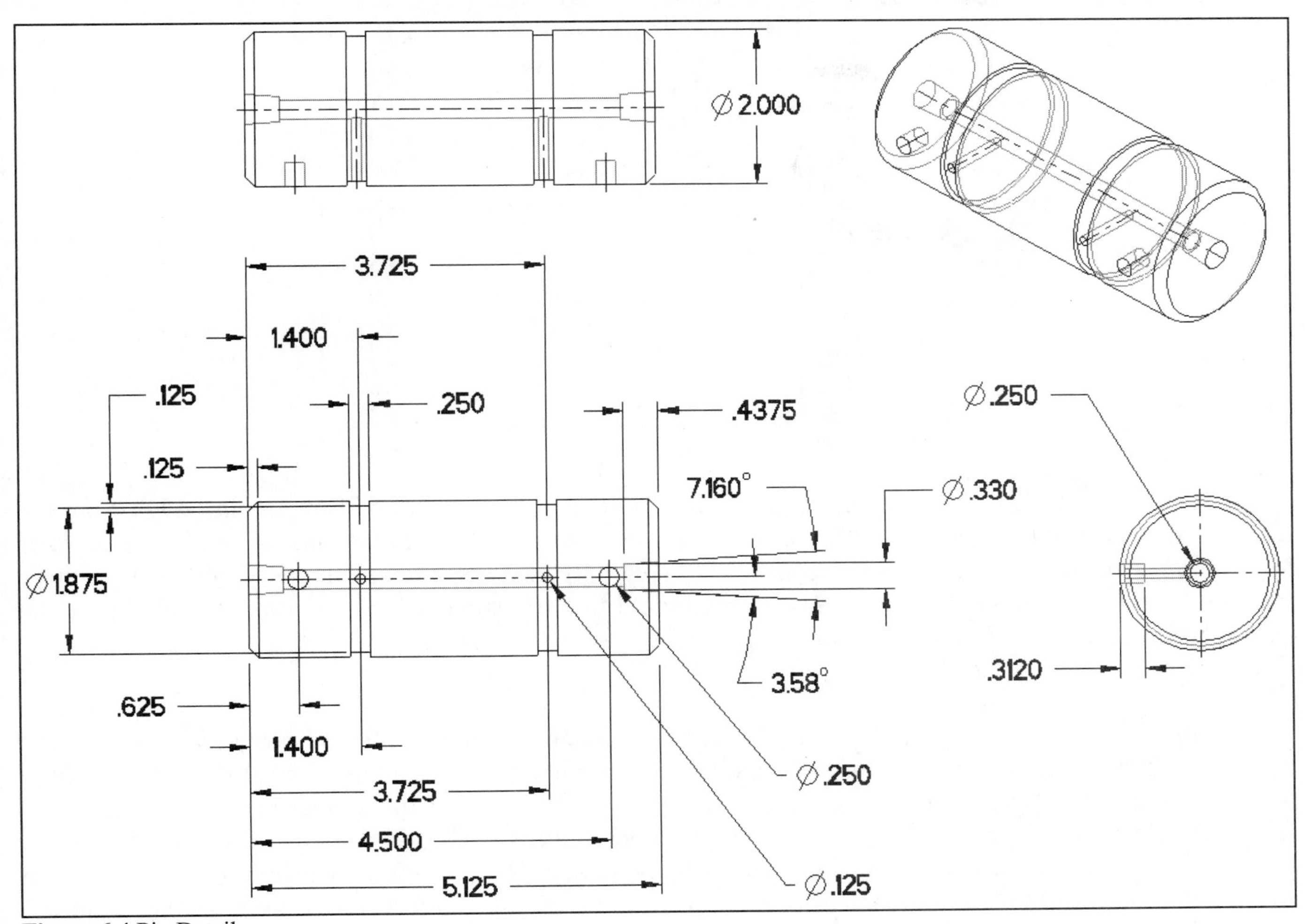

Figure 6.4 Pin Detail

Pin

The Pin is an example of a part created by revolving one section about a centerline (Fig. 6.4). The pin was created as a **revolved protrusion**. The chamfers are created on the first revolved protrusion. The grooves were created with revolved cuts. The holes were added using datum axes and a new datum plane.

For most parts, the basic shape should be the first protrusion, followed by the most important secondary features (such as secondary protrusions and cuts). The holes required for the part are then created.

Click: **File ⇒ Set Working Directory** ⇒ select the working directory ⇒ **OK** ⇒ [icon] ⇒ ●**Part** ⇒ Name **Pin** ⇒ [☑ Use default template] ⇒ **OK** ⇒ **Edit** ⇒ **Setup** ⇒ **Units** ⇒ Units Manager **Inch lbm Second** ⇒ **Close** ⇒ **Material** ⇒ **Define** *(or if you have previously saved a material file you may click **Assign** ⇒ **From File** ⇒ **Open**)* ⇒ type **Steel** ⇒ **MMB** ⇒ **File** from material table ⇒ **Save As** ⇒ type **Steel.mat** ⇒ **Ok** ⇒ **File** ⇒ **Exit** ⇒ **Assign** ⇒ pick **STEEL** ⇒ **Accept** ⇒ **MMB** ⇒ click on **PRT_CSYS_DEF** in the Model Tree ⇒ **RMB** ⇒ **Rename** [Fig. 6.5(a)] ⇒ type **CSYS_PIN** [Fig. 6.5(b-c)] ⇒ **Enter** ⇒ [icon] ⇒ **MMB**

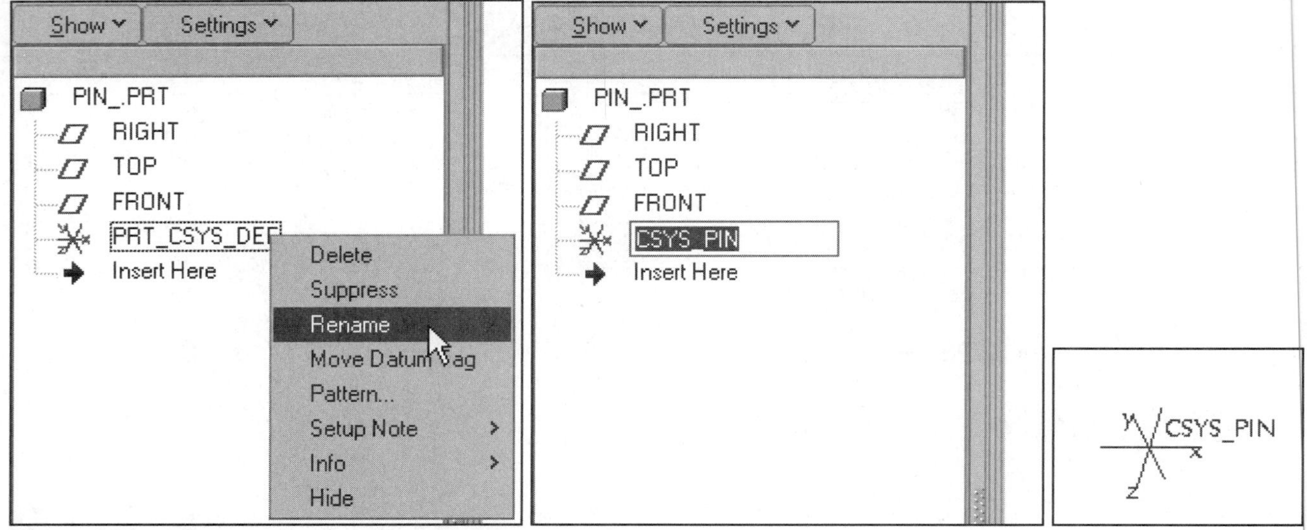

Figure 6.5(a-c) Rename the Coordinate System

The first protrusion is a revolved protrusion. Create the chamfers with the main body of the pin with the first protrusion. *In Lesson 7, you will see that chamfers, like rounds, are normally added near the end of the modeling sequence.* The commands, sketch, references, and so on, are very similar to those used in the previous lessons. Start by setting the model tree to display notes and suppressed objects and adding information columns to the Model Tree display.

Model Tree

The Model Tree is a tabbed feature on the Pro/E navigator that contains a list of every feature or part in the current part, assembly, or drawing. The model structure is displayed in hierarchical (tree) format with the root object (the current part or assembly) at the top of its tree and the subordinate objects (parts or features) below. If you have multiple Pro/E windows open, the Model Tree contents reflect the file in the current "active" window. The Model Tree lists only the related feature- and part-level objects in a current file and does not list the entities (such as edges, surfaces, curves, and so forth) that comprise the features.

Each Model Tree item contains an icon that reflects its object type, for example, hidden, assembly, part, feature, or datum plane (also a feature). The icon can also show the display or regeneration status for a feature, part, or assembly, for example, suppressed or unregenerated. The information in the Model Tree can be saved as a text (.txt) file.

Selection in the Model Tree is object-action oriented; you select objects in the Model Tree without first specifying what you intend to do with them. You can select components, parts, or features using the Model Tree. You cannot select the individual geometry that makes up a feature (entities). To select an entity, you must select it in the graphics window.

You can add or remove items from the Model Tree column display using Settings in the Navigator:

- Select features, parts, or assemblies, and perform object-specific operations on them using the shortcut menu
- Filter the display by item type or status- showing or hiding datum features, or showing or hiding suppressed features
- Open a part within an assembly file by right-clicking the part in the Model Tree
- Create or modify features and perform other operations such as deleting or redefining parts or features, and rerouting parts or features using the shortcut menu
- Search the Model Tree for model properties or other feature information
- Show the display or regeneration status for an object, for example, suppressed or unregenerated

Click: **Settings** in the Navigator (Fig. 6.6) ⇒ **Tree Filters** ⇒ ☑ Notes ⇒ ☑ Suppressed Objects ⇒ **OK**

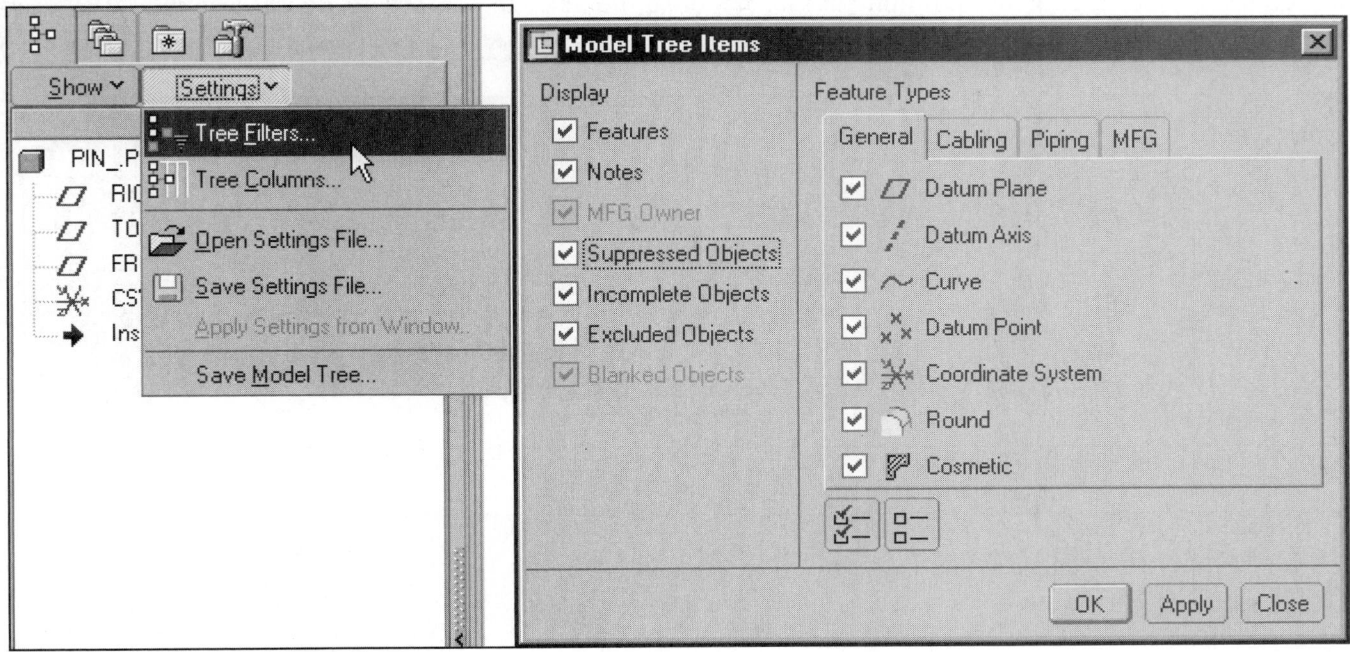

Figure 6.6 Setting Model Tree Filters

Click: **Settings** in the Navigator ⇒ **Tree Columns** ⇒ Model Tree Columns dialog box opens: Type **Info** (Fig. 6.7) ⇒ add types to the Displayed list, click: **Feat #** ⇒ >> ⇒ **Feat ID** ⇒ >> ⇒ Type ⌄ ⇒ select **Layer** ⇒ **Layer Names** ⇒ >> ⇒ **Layer Status** ⇒ >> [Fig. 6.7(a-b)] ⇒ **Apply** ⇒ **OK**

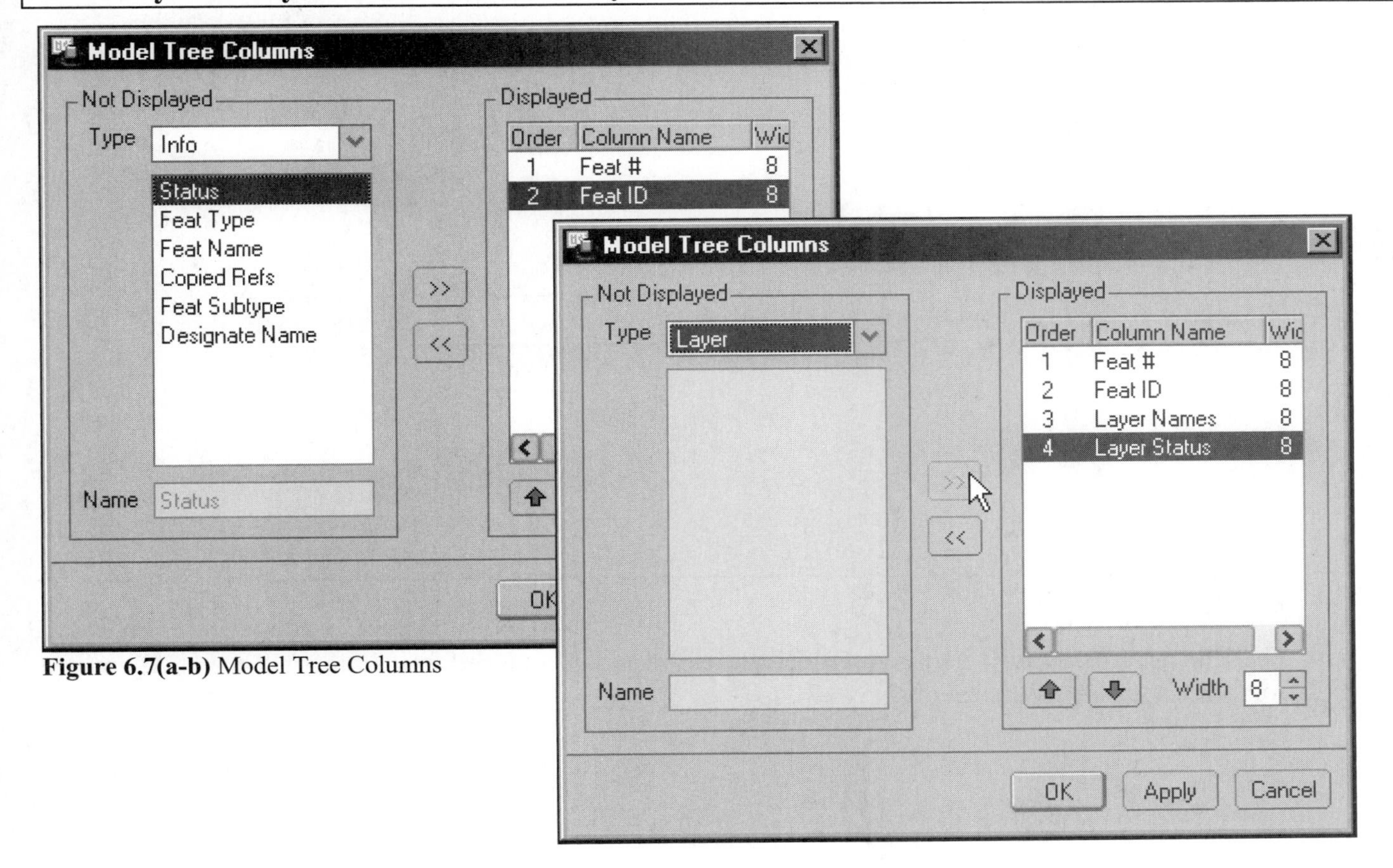

Figure 6.7(a-b) Model Tree Columns

Click on the sash ⇒ drag sash to the right to expand window (Fig. 6.8) ⇒ adjust the width of each column by dragging the column divider ⇒ drag the sash to the left to decrease window size as shown in Figure 6.9

Show Settings

LMB and drag to adjust columns

	Feat #	Feat ID	Layer Names	Layer Status
PIN_.PRT				
RIGHT	1	1	01___PRT_ALL_DTM_PLN, 01___PRT_DEF_DTM_PLN	Displayed
TOP	2	3	01___PRT_ALL_DTM_PLN, 01___PRT_DEF_DTM_PLN	Displayed
FRONT	3	5	01___PRT_ALL_DTM_PLN, 01___PRT_DEF_DTM_PLN	Displayed
CSYS_PIN	4	7	05___PRT_ALL_DTM_CSYS, 05___PRT_DEF_DTM_CSYS	Displayed
Insert Here				

Figure 6.8 Expand the Navigator Window

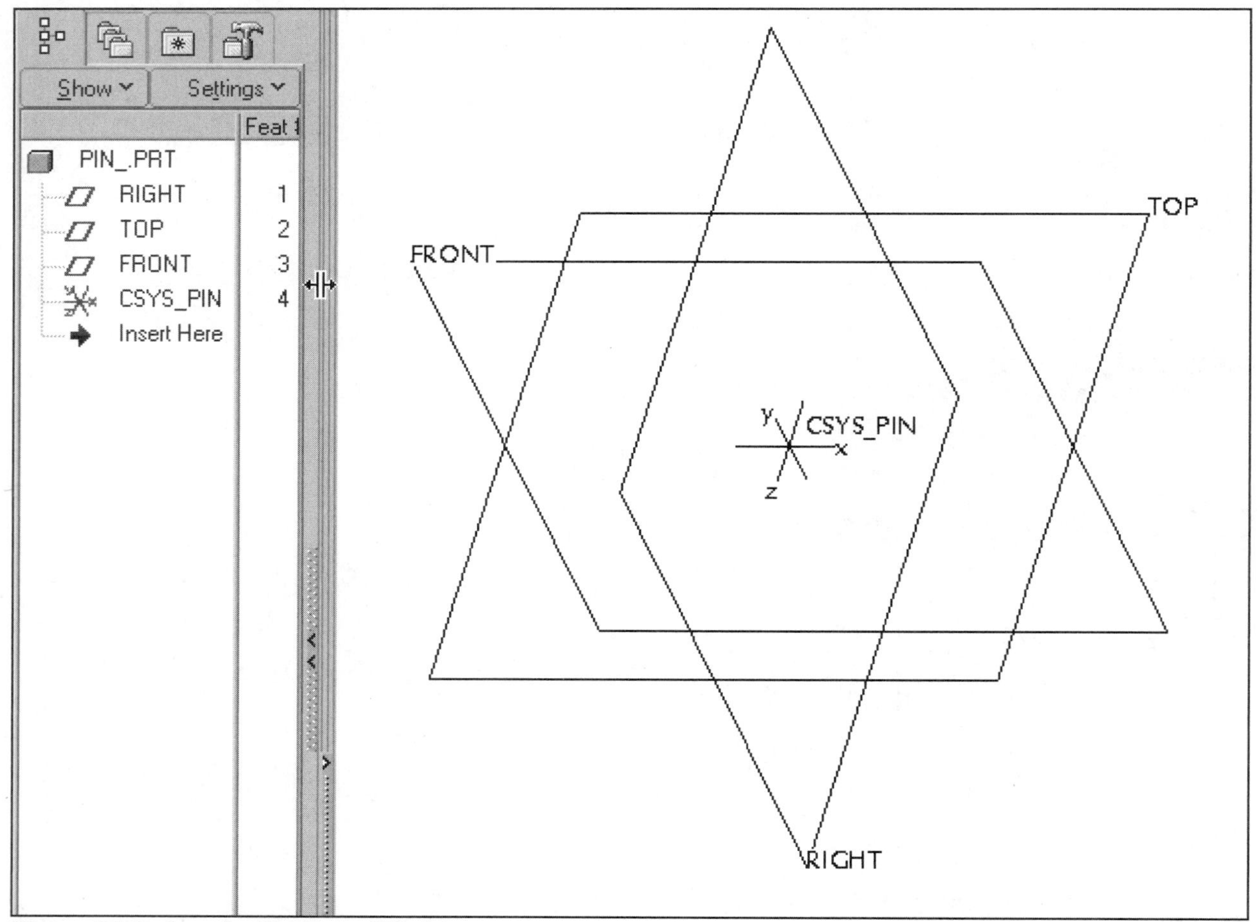

Figure 6.9 Reduce the Navigator Window

The first protrusion will be a revolved feature and using a closed boundary. All sketched entities must lie on only one side of the revolving axis.

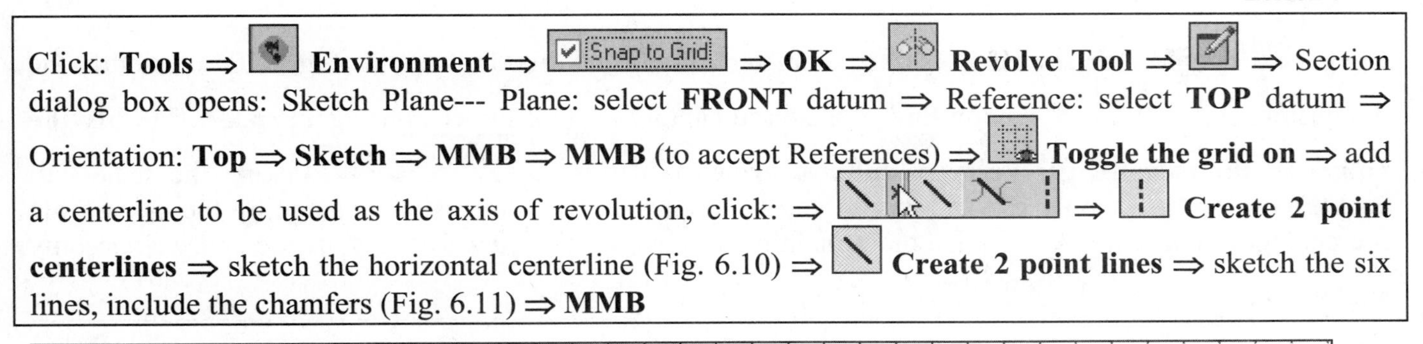

Click: **Tools** ⇒ **Environment** ⇒ Snap to Grid ⇒ **OK** ⇒ **Revolve Tool** ⇒ ⇒ Section dialog box opens: Sketch Plane--- Plane: select **FRONT** datum ⇒ Reference: select **TOP** datum ⇒ Orientation: **Top** ⇒ **Sketch** ⇒ **MMB** ⇒ **MMB** (to accept References) ⇒ **Toggle the grid on** ⇒ add a centerline to be used as the axis of revolution, click: ⇒ ⇒ **Create 2 point centerlines** ⇒ sketch the horizontal centerline (Fig. 6.10) ⇒ **Create 2 point lines** ⇒ sketch the six lines, include the chamfers (Fig. 6.11) ⇒ **MMB**

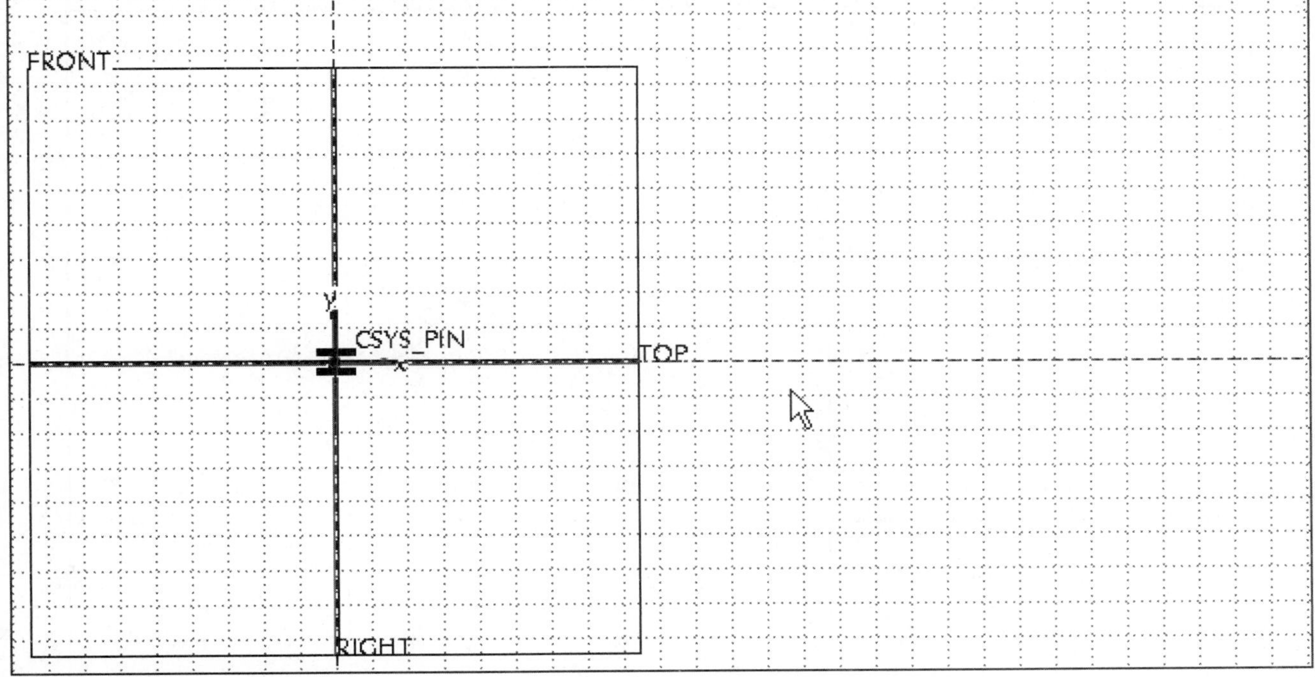

Figure 6.10 Sketch a Centerline

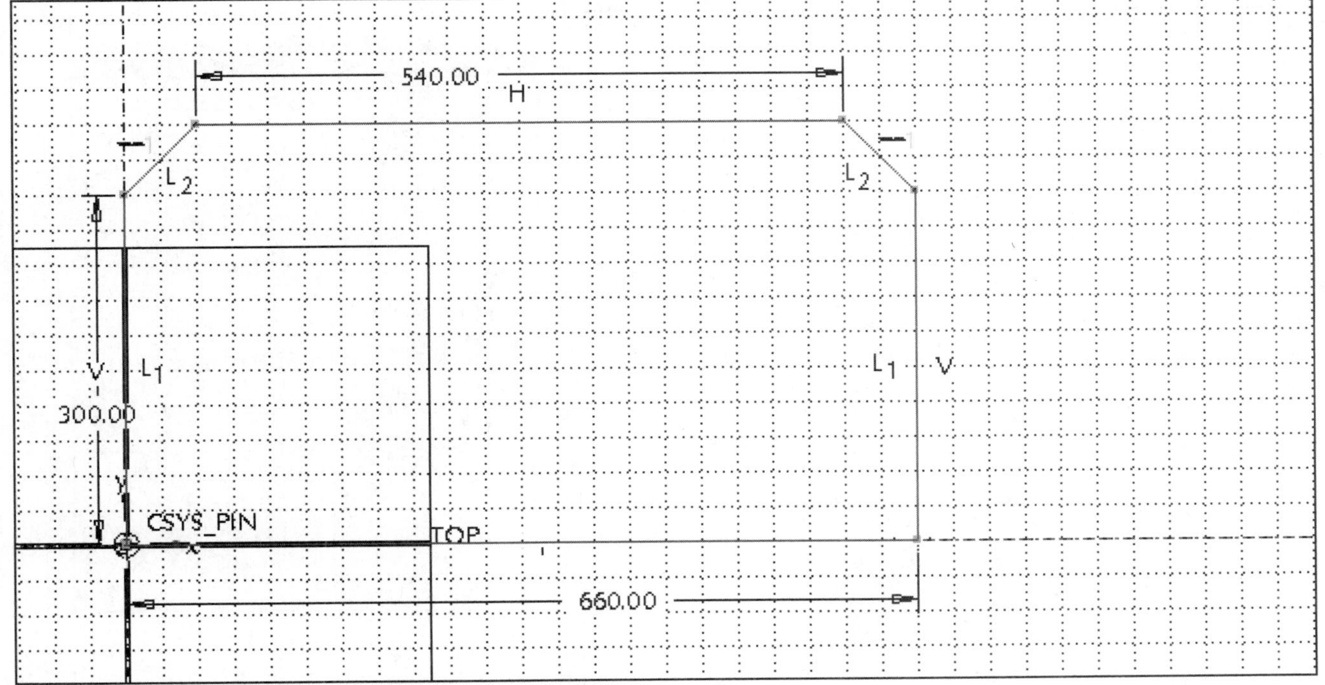

Figure 6.11 Sketch the Six Lines of One Half of the Pin

Click: **Toggle the grid off** ⇒ add the diameter and the chamfer dimensions (Fig. 6.12) ⇒ modify the dimensions, click: ⇒ window-in the sketch to capture all four dimensions again ⇒ **Modify the values of dimensions, geometry of splines, or text entities** ⇒ Lock Scale [Locking the scale will enable you to change only one dimension and have all the other dimensions change by the same scale (percentage). You can then modify the remaining values, but the view will regenerate and change only slightly each time] ⇒ Regenerate (Fig. 6.13)

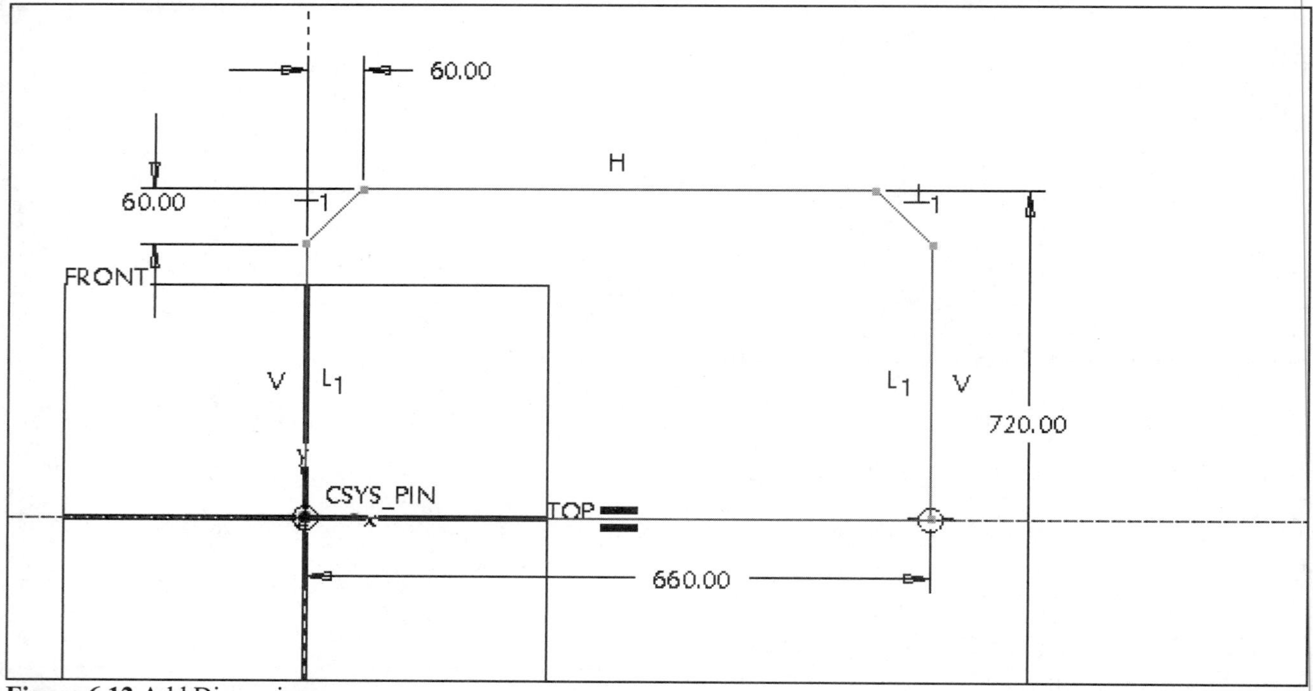

Figure 6.12 Add Dimensions

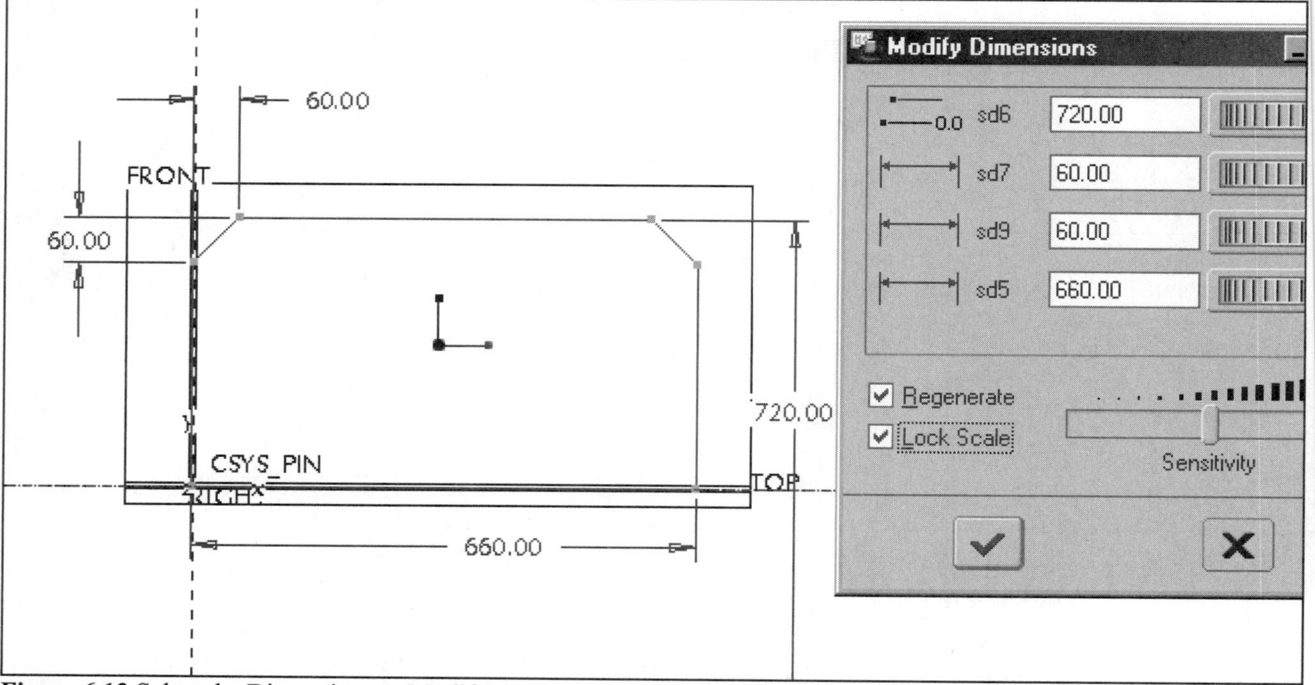

Figure 6.13 Select the Dimensions to Modify

Modify only the diameter dimension (Fig. 6.14) to the design value of **2.00** (see Figure 6.4) ⇒ **Enter** (Fig. 6.15) ⇒ ✓

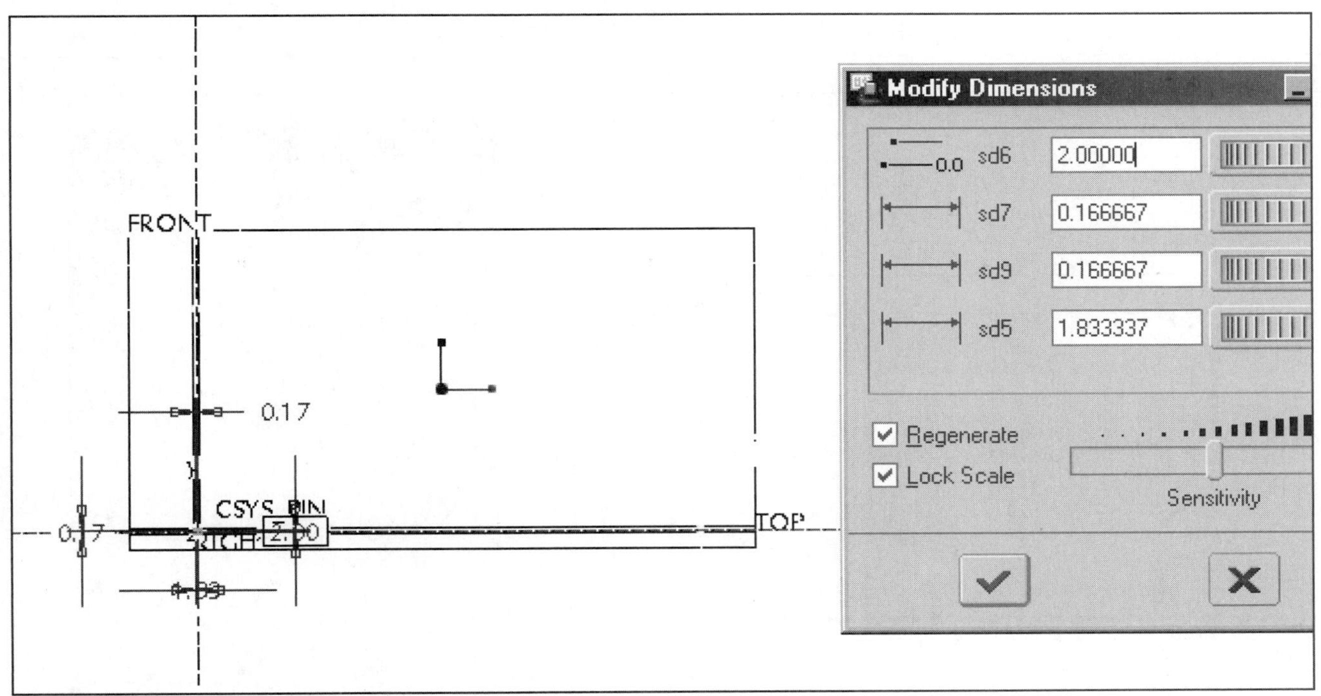

Figure 6.14 Modify only One Dimension

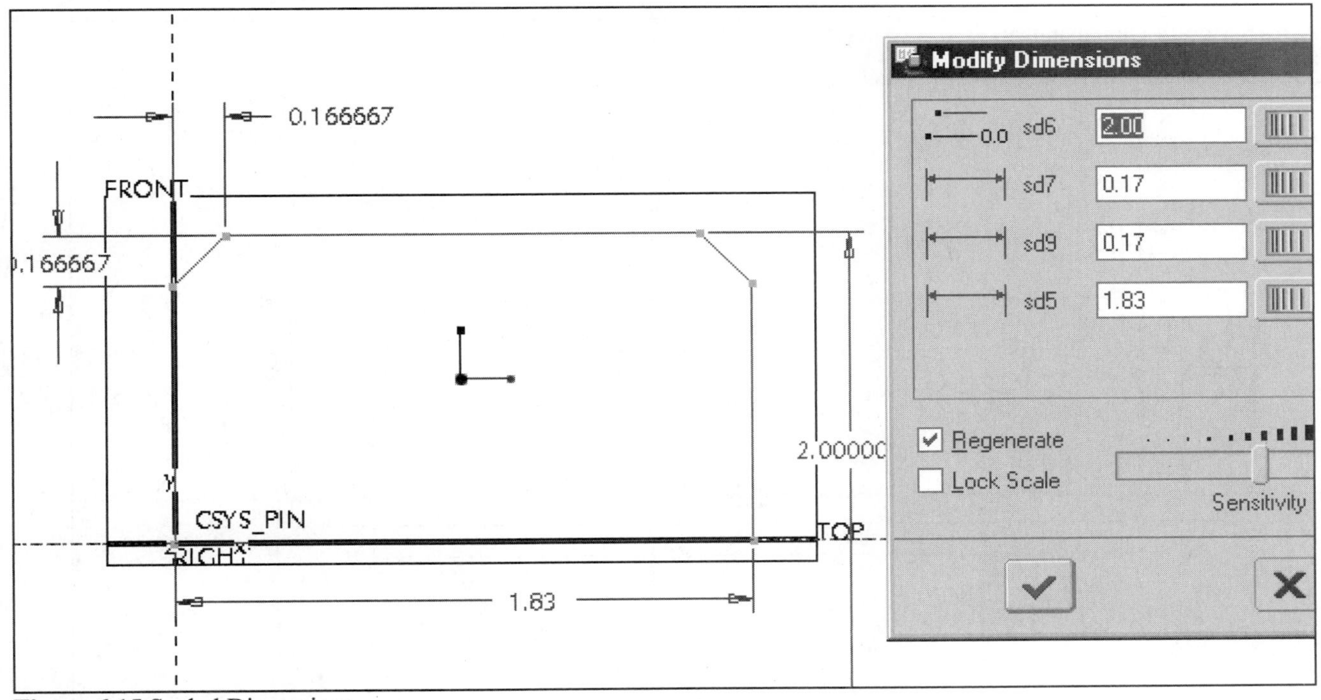

Figure 6.15 Scaled Dimensions

Modify the remaining dimensions, click: [arrow icon] ⇒ window-in the sketch to capture all dimensions ⇒ [modify icon] **Modify the values of dimensions** ⇒ [Regenerate] ⇒ modify the dimensions (Fig. 6.16) ⇒ [check icon] (Fig. 6.17) ⇒ [view icon] ⇒ **Standard Orientation**

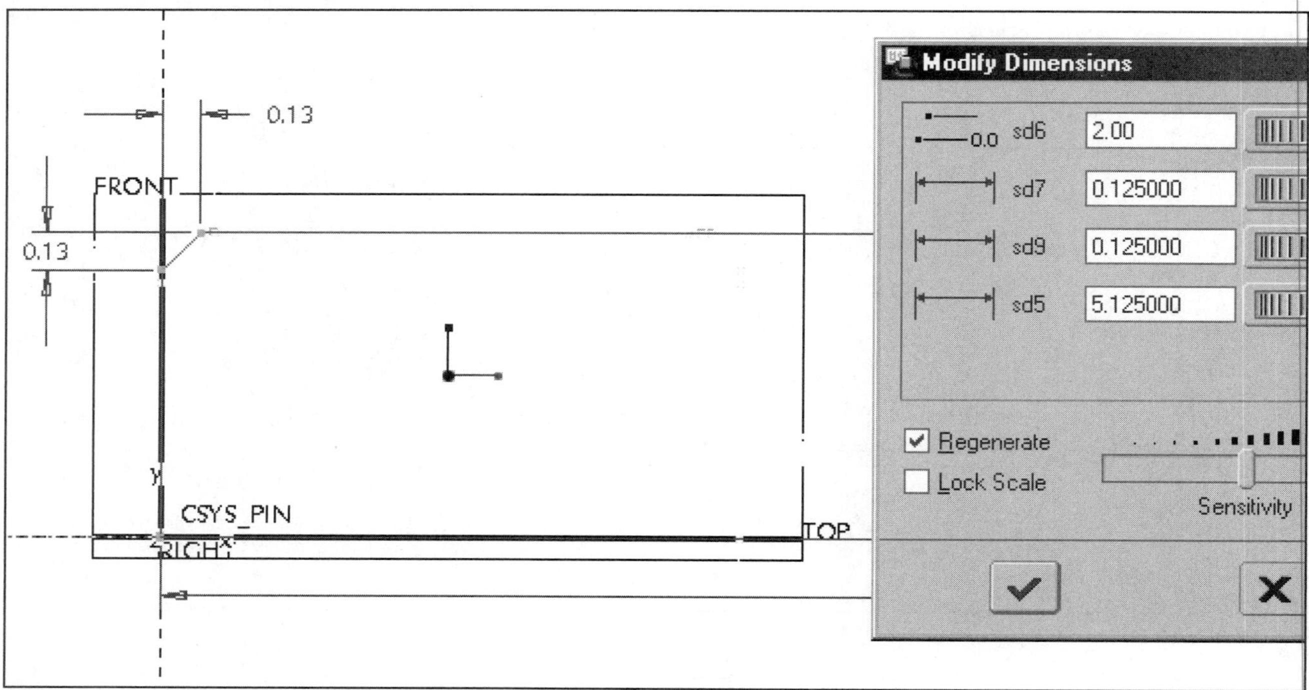

Figure 6.16 Modify Dimensions to Design Values (see Figure 6.4)

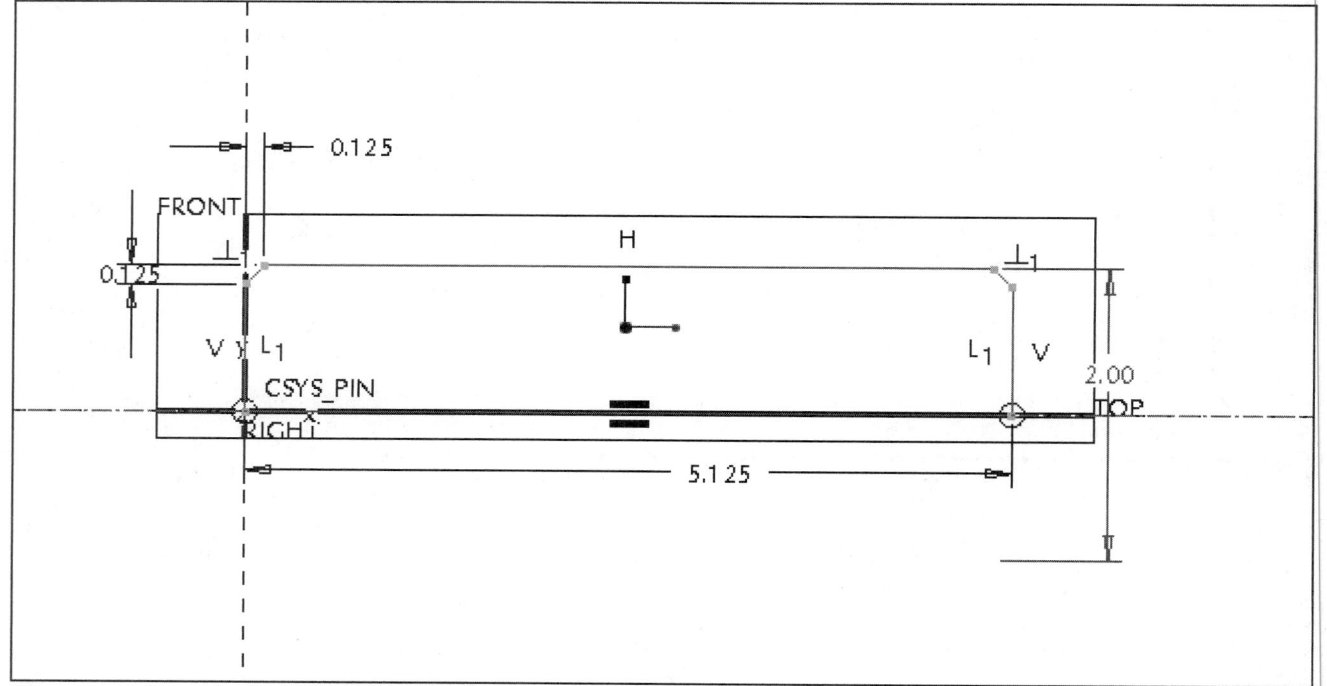

Figure 6.17 Regenerated Sketch

Click: [✓] (Fig. 6.18) ⇒ move the drag handle from **360** to about **90** degrees (Fig. 6.19) and then back again to **360** ⇒ **MMB** (Fig. 6.20) ⇒ [icon] ⇒ [icon] ⇒ [icon] ⇒ **MMB**

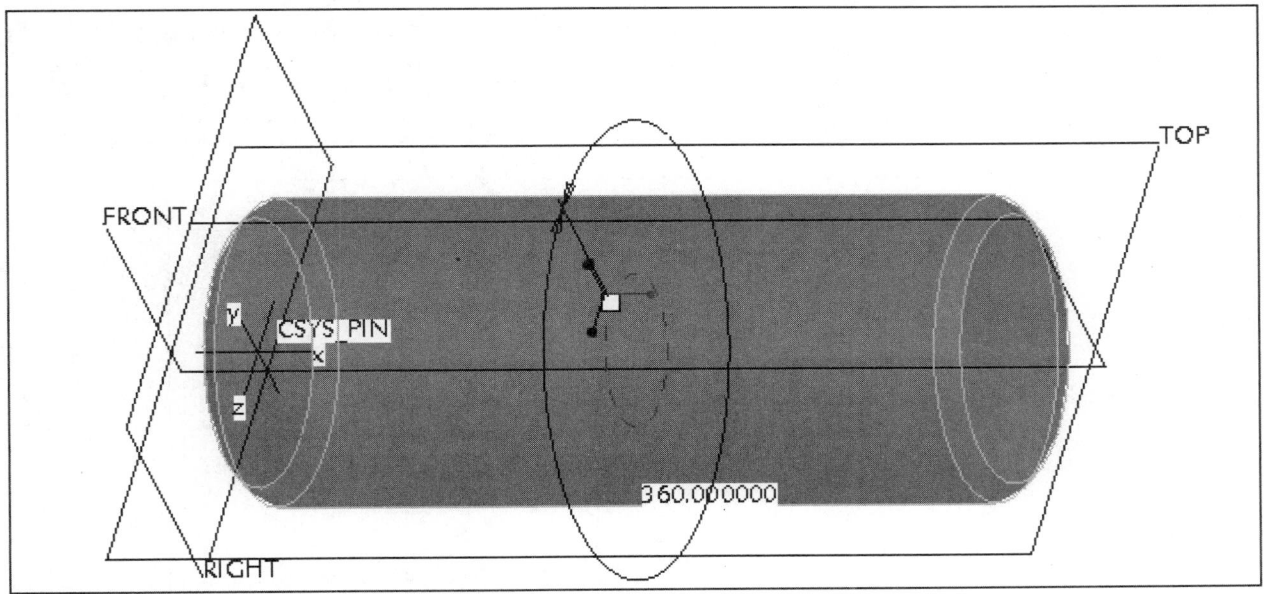

Figure 6.18 Revolve Angle

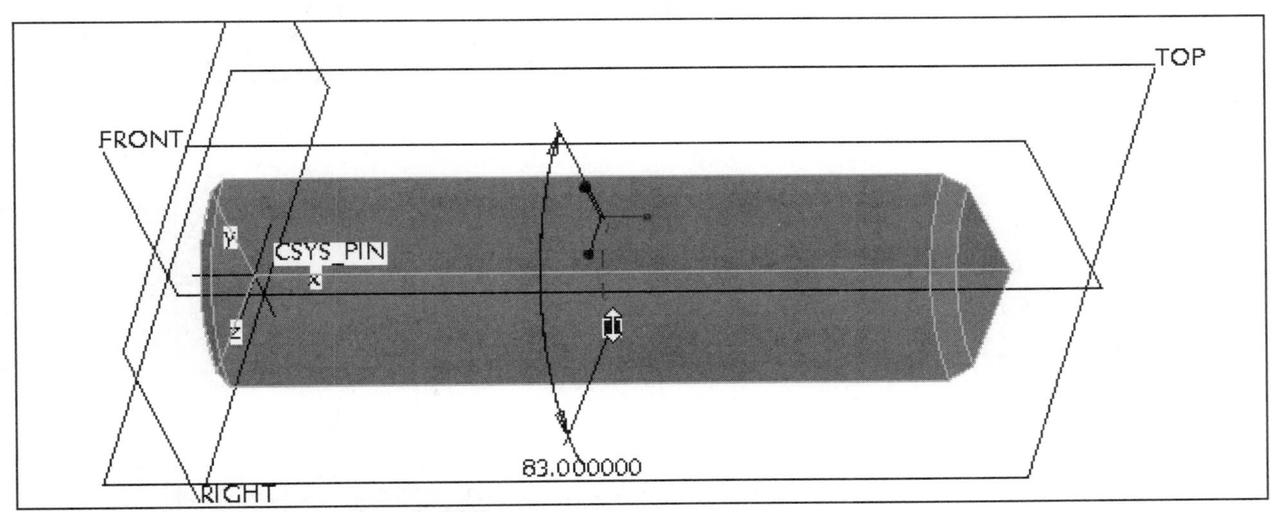

Figure 6.19 Revolve to Approximately **90** Degrees

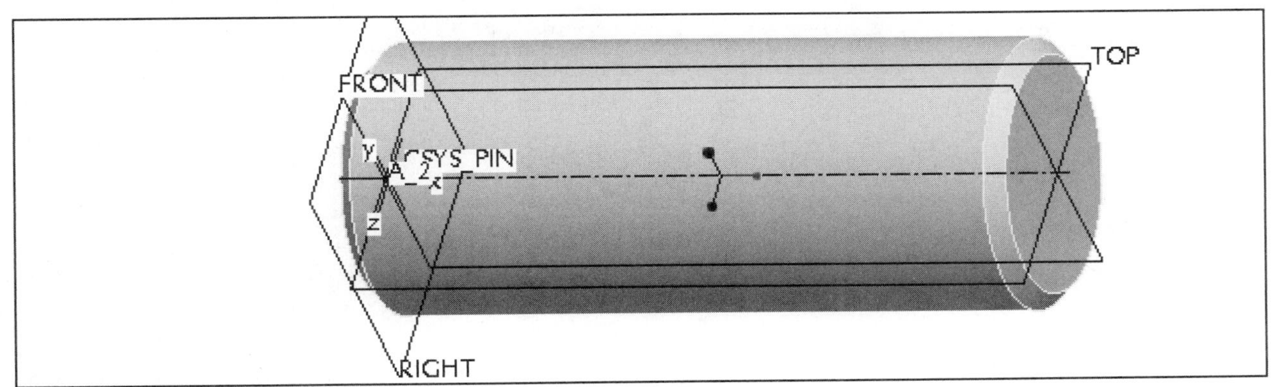

Figure 6.20 Revolved Extrusion

The next feature will be the cut to create the groove in the pin.

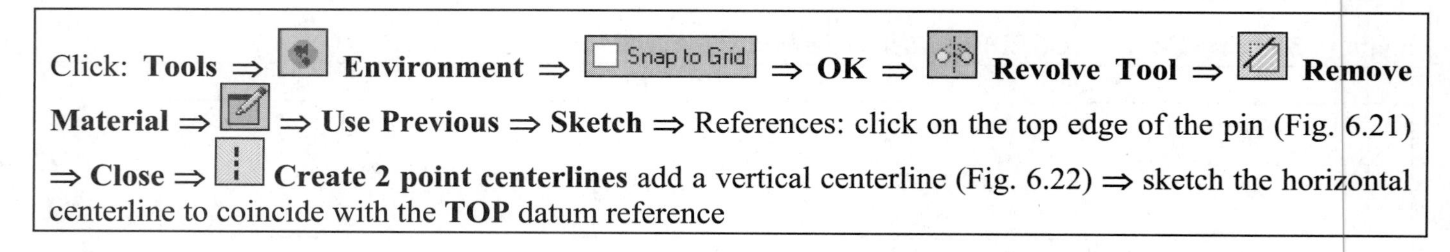

Click: **Tools** ⇒ **Environment** ⇒ Snap to Grid ⇒ **OK** ⇒ **Revolve Tool** ⇒ **Remove Material** ⇒ ⇒ **Use Previous** ⇒ **Sketch** ⇒ References: click on the top edge of the pin (Fig. 6.21) ⇒ **Close** ⇒ **Create 2 point centerlines** add a vertical centerline (Fig. 6.22) ⇒ sketch the horizontal centerline to coincide with the **TOP** datum reference

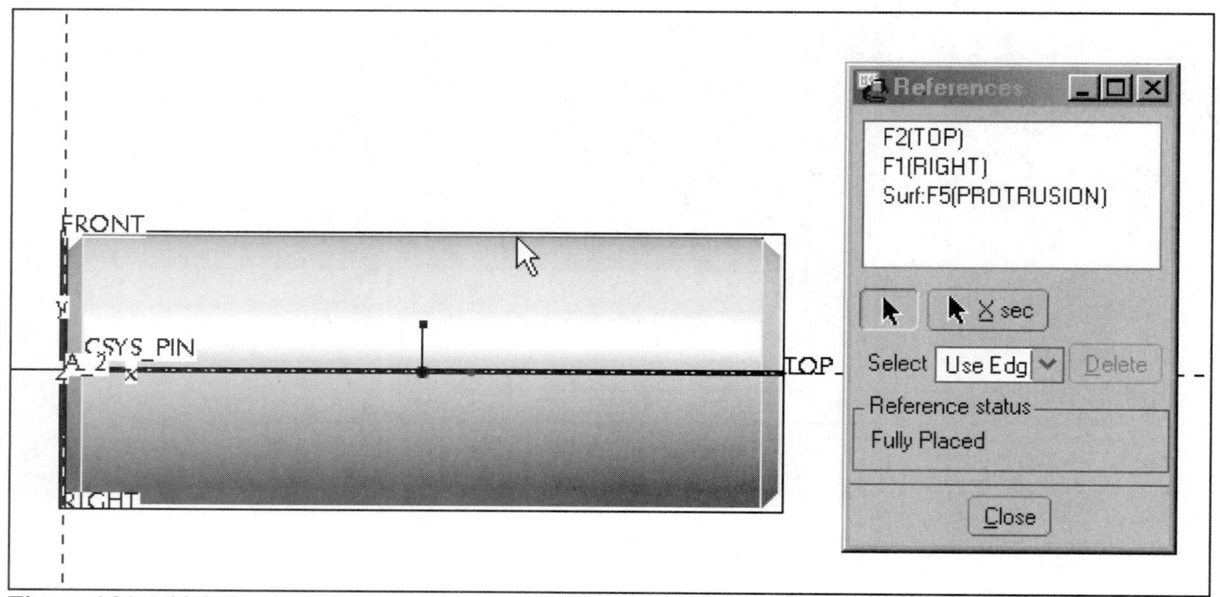

Figure 6.21 Add the Upper Edge as a Reference

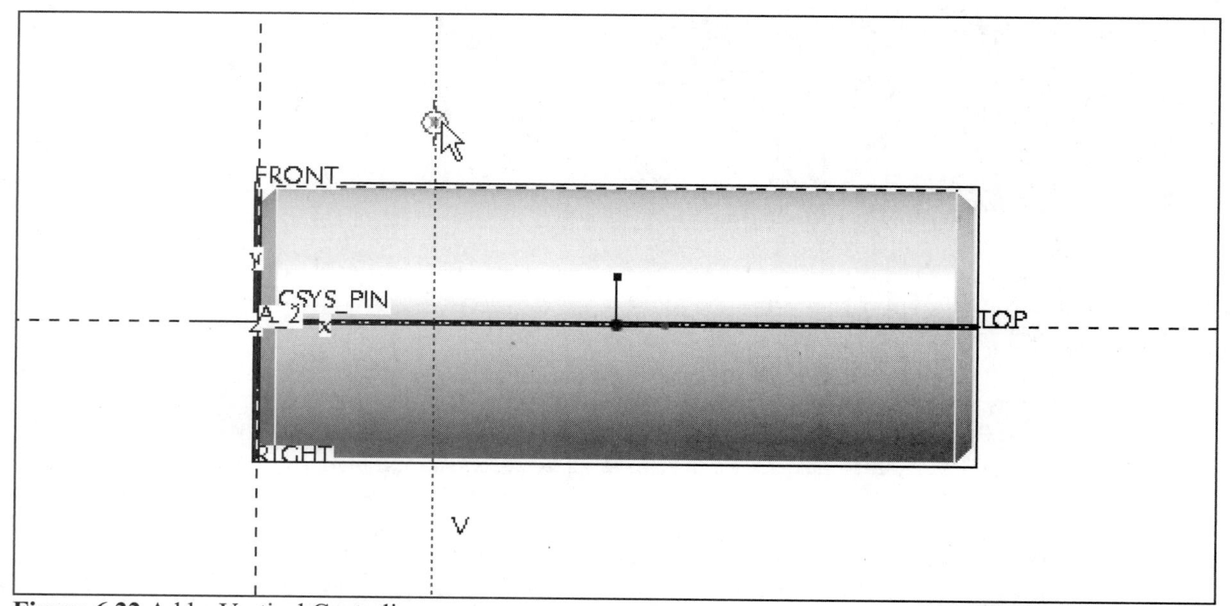

Figure 6.22 Add a Vertical Centerline

Click: [icon] **Create 2 point lines** ⇒ sketch the three lines ⇒ **MMB** ⇒ [icon] ⇒ click on and move the dimensions (Fig. 6.23) ⇒ add the dimension to locate the vertical centerline from the left edge (datum **RIGHT**) ⇒ add the diameter dimension by clicking on the horizontal centerline and then the horizontal sketch line of the groove and then the horizontal centerline a second time, place the dimension with the **MMB** (Fig. 6.24) ⇒ [icon] ⇒ pick the horizontal axis ⇒ **Sketch** from the menu bar ⇒ **Feature Tools** ⇒ **Axis of Revolution** (Fig. 6.25)

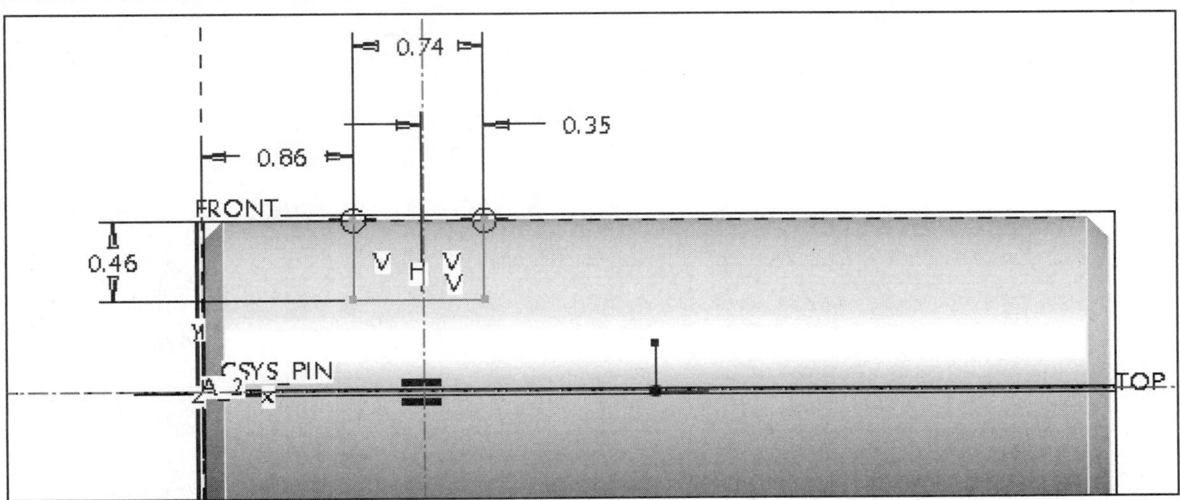

Figure 6.23 Sketch Three Lines and Move the Dimensions

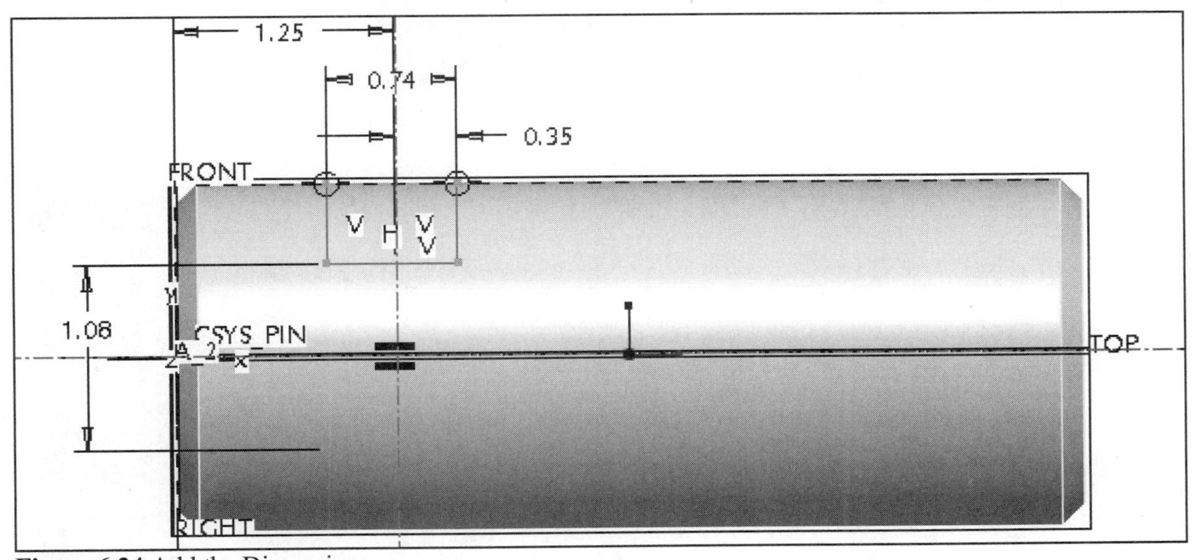

Figure 6.24 Add the Dimensions

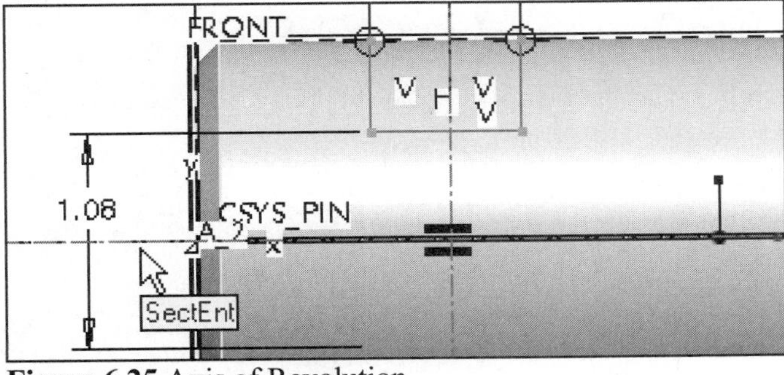

Figure 6.25 Axis of Revolution

Click: **Impose sketcher constraints on the section** ⇒ **Make two points or vertices symmetric about a centerline** ⇒ pick the vertical centerline ⇒ pick the endpoints of the two vertical lines (Fig. 6.26) ⇒ **Close** (Fig. 6.27) ⇒ modify the dimensions, click: ⇒ window-in the sketch to capture all three dimensions ⇒ **Modify the values of dimensions** (**1.400** from datum **RIGHT**, **.250** width x **1.875** groove diameter, also see Figure 6.4) ⇒ Regenerate (Fig. 6.28) ⇒

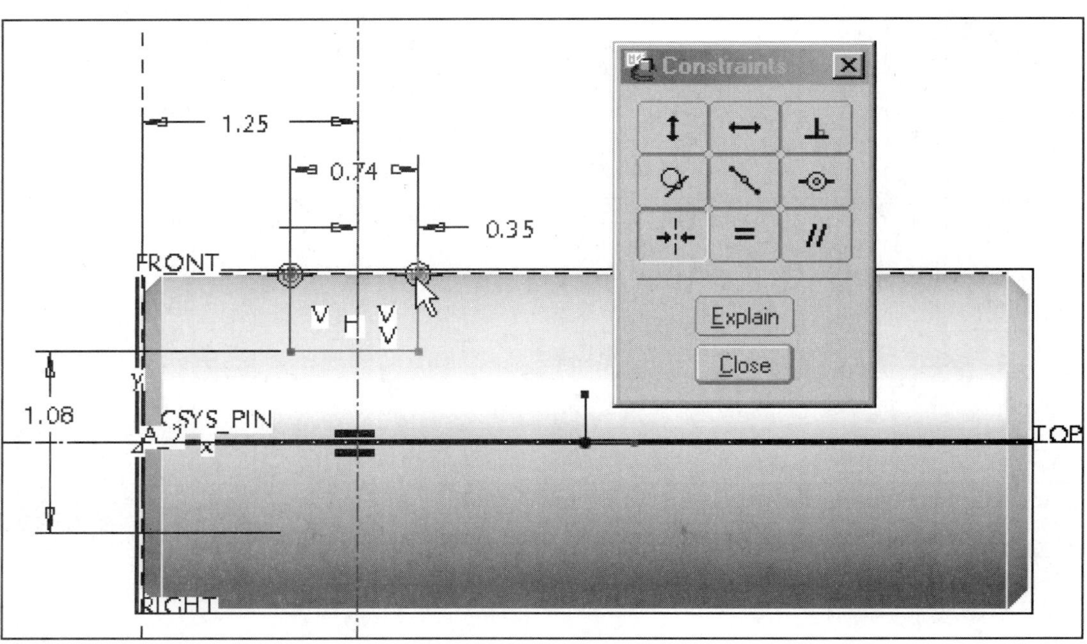

Figure 6.26 Add Symmetric Constraint to Eliminate Half Distance Measurement

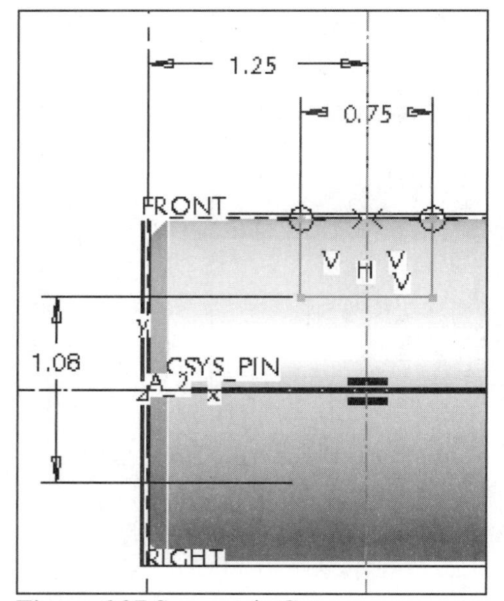

Figure 6.27 Symmetric Cut

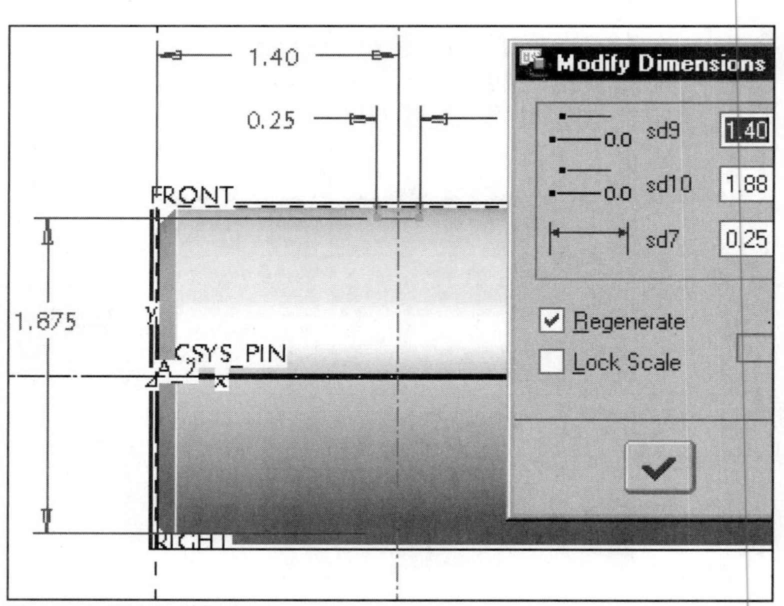

Figure 6.28 Modified Dimensions

Click: ⇒ ⇒ **Standard Orientation** (Fig. 6.29) ⇒ (Fig. 6.30) ⇒ ⇒ (Fig. 6.31) ⇒ **LMB** to deselect ⇒ click on the **Cut** in the Model Tree (Fig. 6.32) ⇒ **RMB** ⇒ **Edit** feature dimensions display (Fig. 6.32) ⇒ click on the **1.40** dimension ⇒ **RMB** (Fig. 6.33) ⇒ **Properties** Dimension Properties dialog box displays (Fig. 6.34) ⇒ **Move** ⇒ select a new position for the **1.40** dimension (Fig. 6.35) ⇒ **OK** ⇒ repeat to move the other dimensions as required (Fig. 6.36) ⇒ ⇒ **MMB** ⇒ **File** ⇒ **Delete** ⇒ **Old Versions** ⇒ **MMB** ⇒ **LMB** to deselect

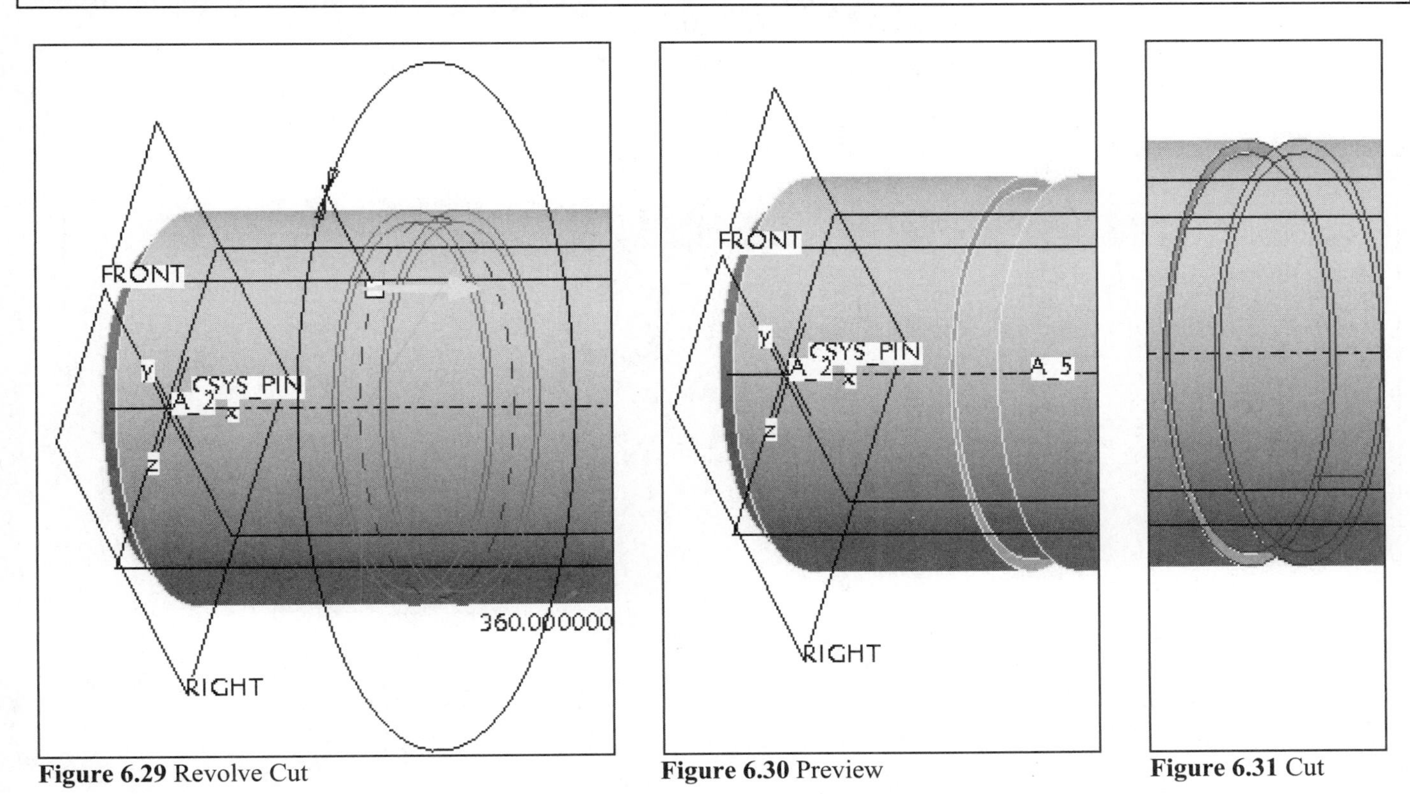

Figure 6.29 Revolve Cut

Figure 6.30 Preview

Figure 6.31 Cut

Figure 6.32 Using Edit To Display Feature Dimensions

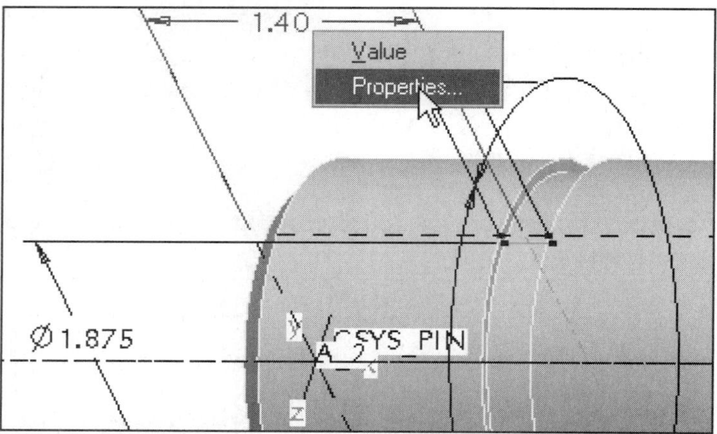

Figure 6.33 Properties

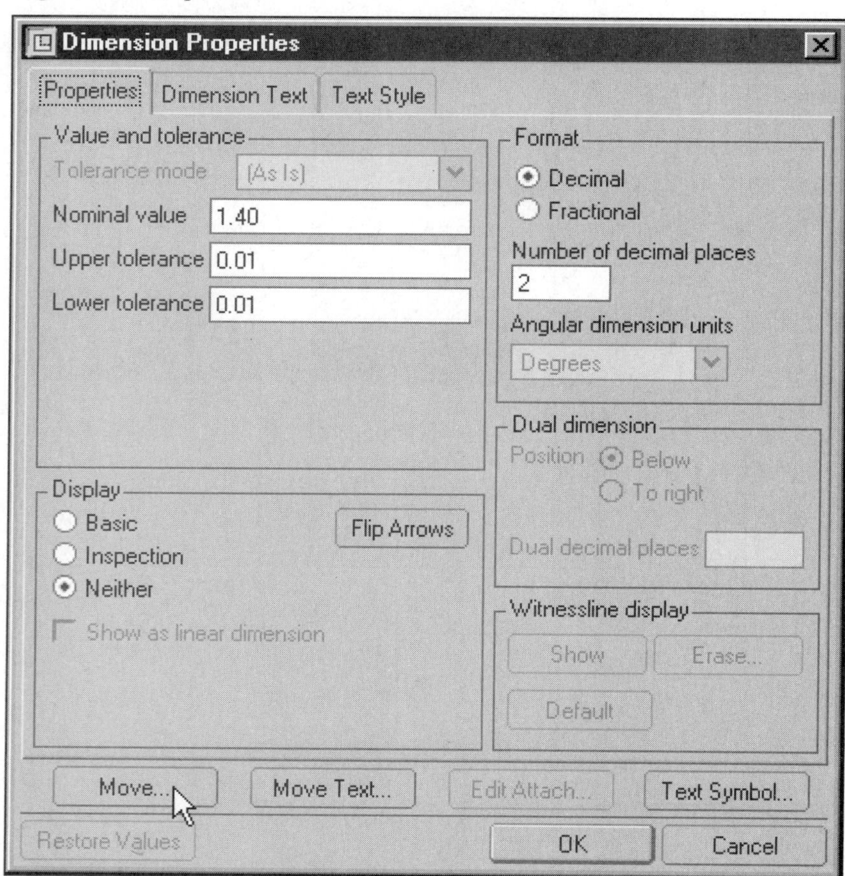

Figure 6.34 Dimension Properties Dialog Box

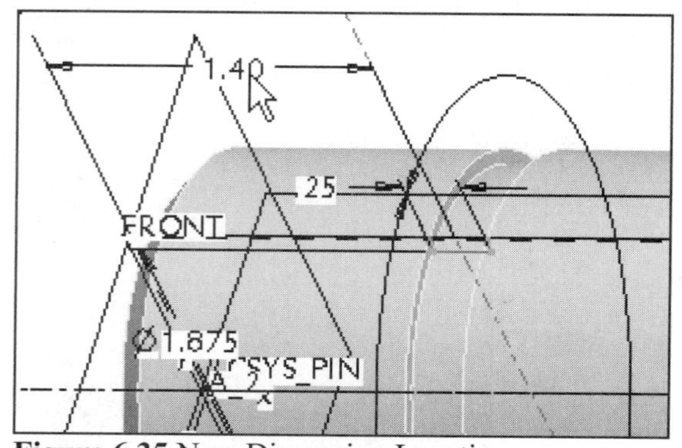

Figure 6.35 New Dimension Location

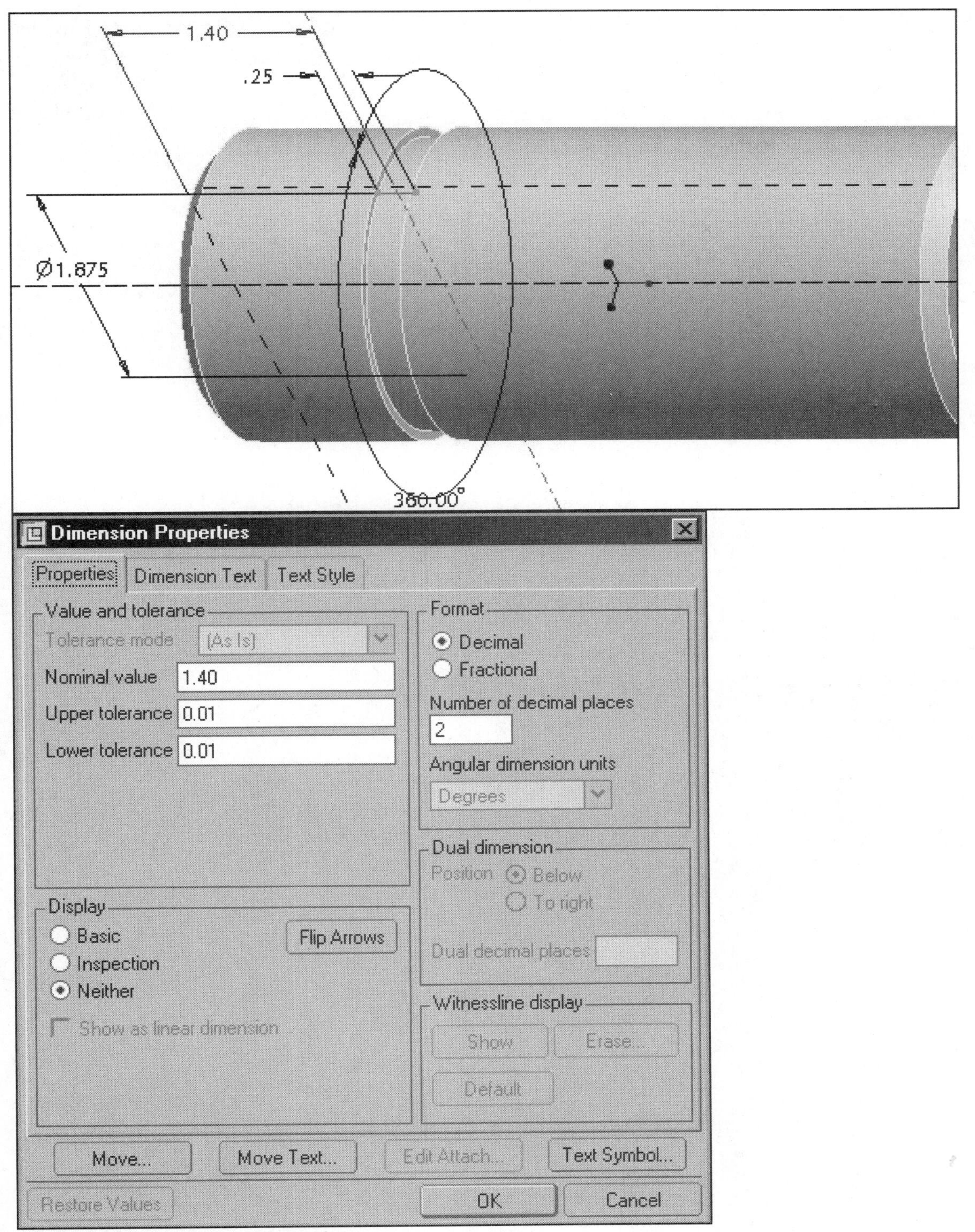

Figure 6.36 Repositioned Dimensions

Dimension Properties

Besides the value of a dimension and its position, there are several attributes that you can change using the Dimension Properties dialog box. The Properties tab (Fig. 6.36) provides options for Value and tolerance modification, Format options, and Display variables including Flip Arrows, which allows the toggling of dimension arrows inside or outside of its extension lines.

Move, which is available on all three tabs, moves the dimension itself, and the associated leader lines, to a new location. Move Text only moves the text associated with the dimension to a new location. Text Symbol provides a dialog box with symbology.

The Dimension Text tab (Fig. 6.37) shows the parametric dimension symbol. The Text Style tab (Fig. 6.38) provides options for Character, and Note/Dimension variations. Take some time and explore the options provided in the Dimension Properties dialog box.

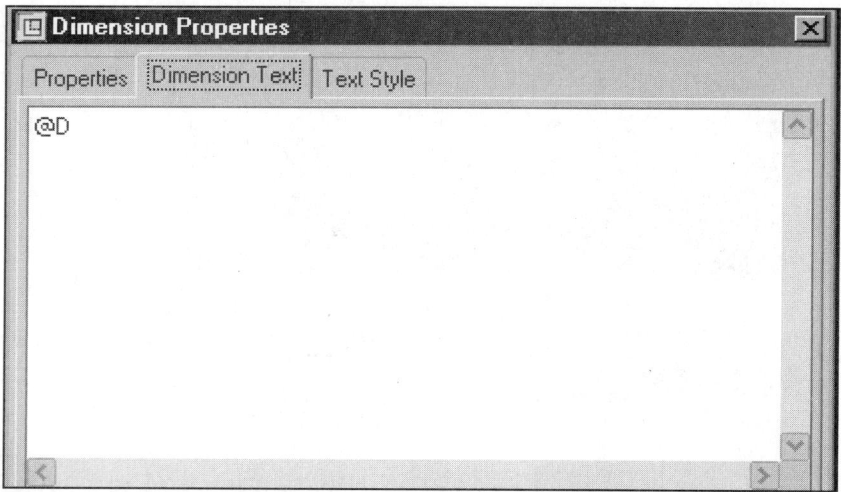

Figure 6.37 Dimension Properties, Dimension Text Tab

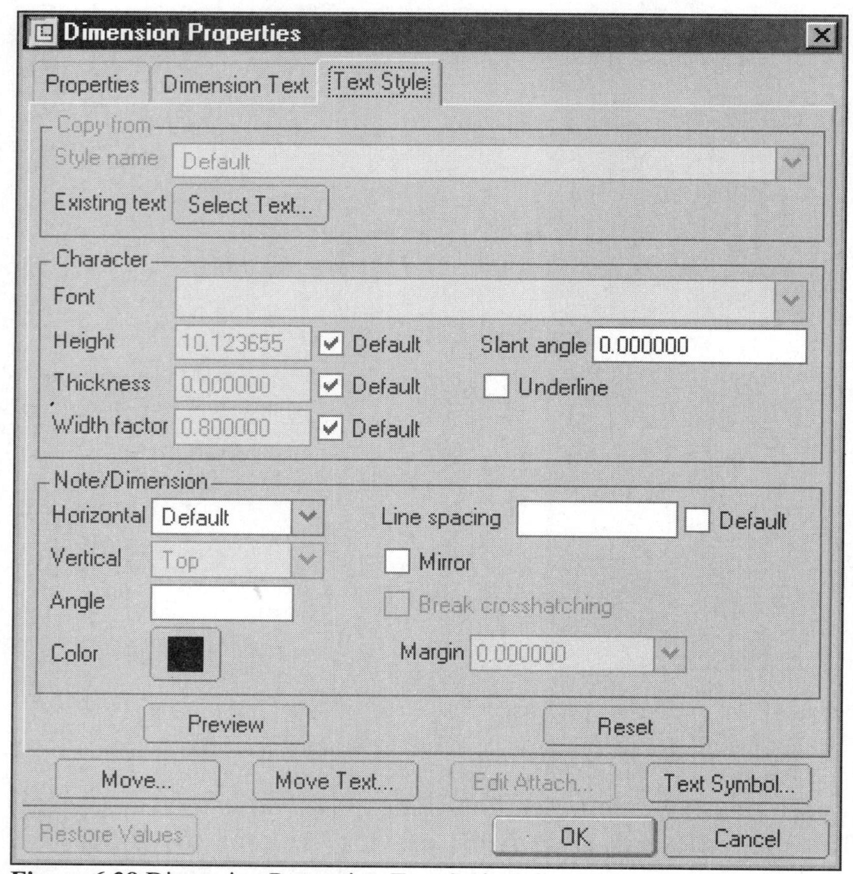

Figure 6.38 Dimension Properties, Text Style Tab

Next, you will create a couple of datum planes to be used when constructing the holes.

Click: **Tools** ⇒ **Environment** ⇒ Standard Orient Isometric ⇒ **Apply** ⇒ **OK** ⇒ **Datum Plane Tool** ⇒ References: pick the parts cylindrical surface (Fig. 6.39) ⇒ **Tangent** in the DATUM PLANE dialog box ⇒ press and hold the **Ctrl** key down and add the **FRONT** datum plane as the second reference ⇒ **Parallel** (Fig. 6.40) ⇒ **OK** ⇒ ⇒ **MMB** ⇒ **LMB** to deselect

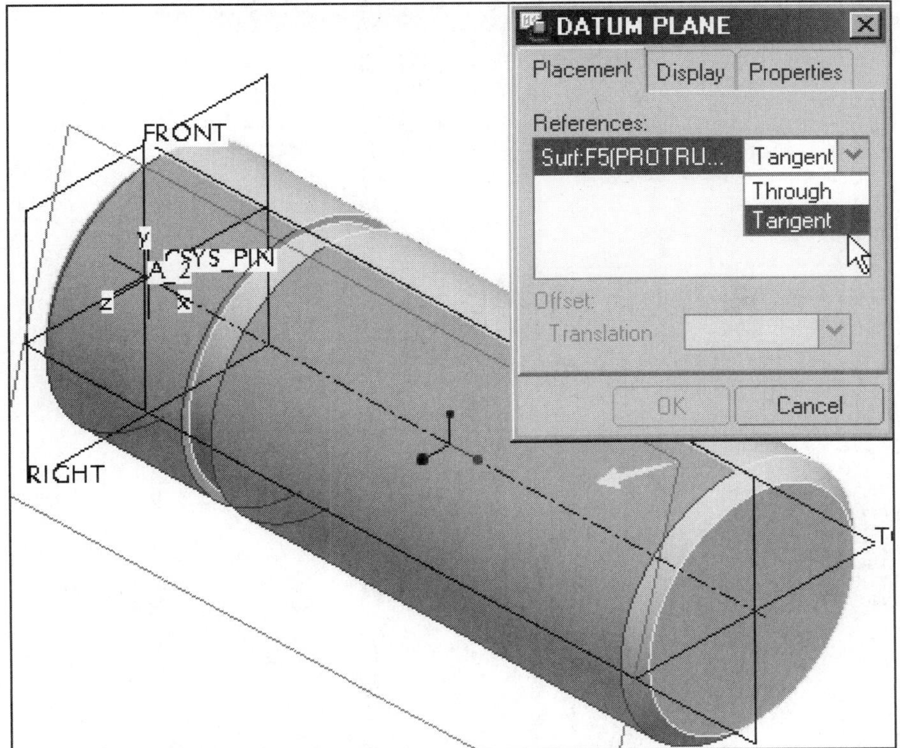

Figure 6.39 Surface Tangent

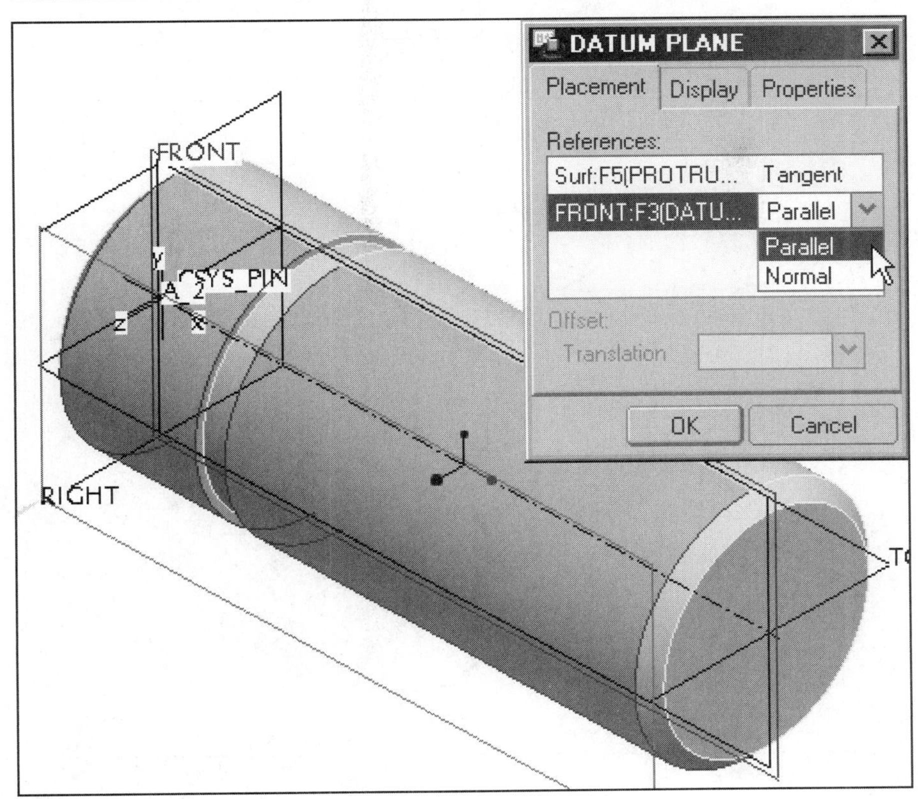

Figure 6.40 Datum Parallel

Click: **Datum Plane Tool** ⇒ References: pick datum **RIGHT** (Fig. 6.41) ⇒ Offset: Translation-type **5.125/2** (length of the part divided by two) ⇒ **Enter** ⇒ **OK** ⇒ ⇒ **MMB** (Fig. 6.42) ⇒ **LMB**

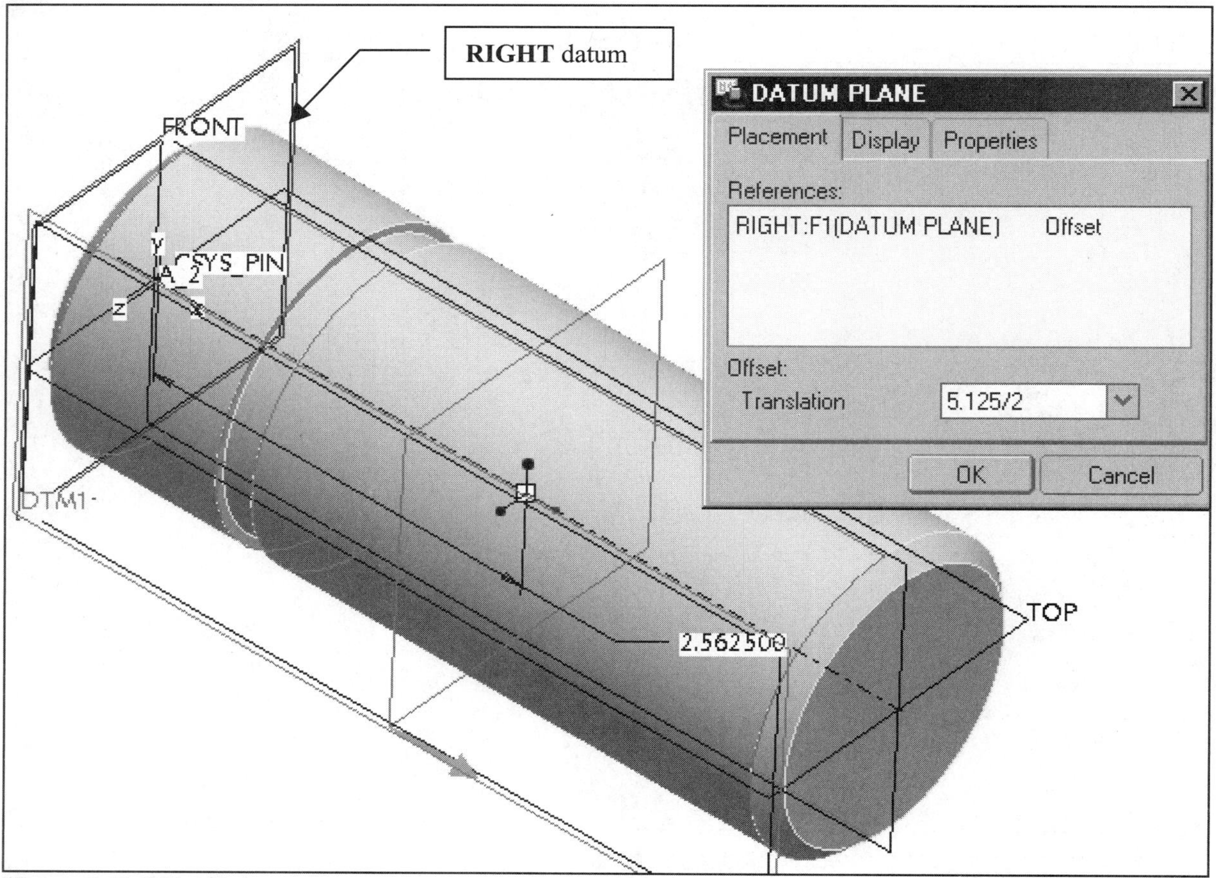

Figure 6.41 Offset Datum

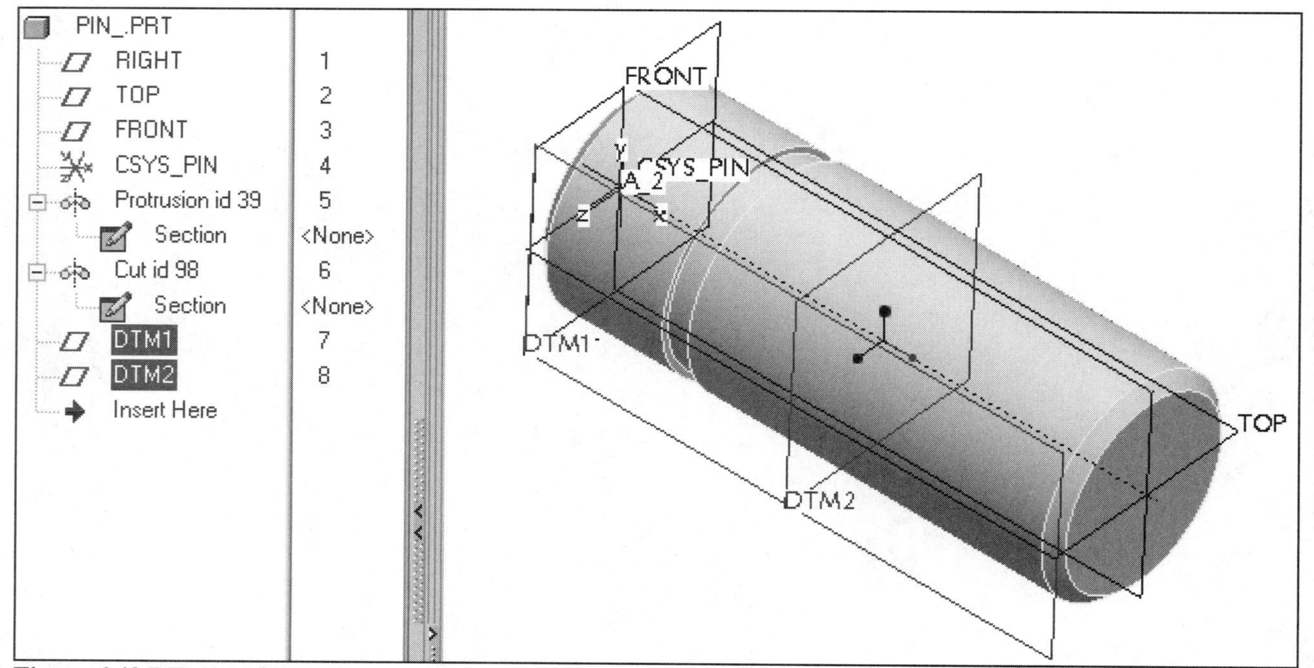

Figure 6.42 DTM1 and DTM2

The first hole to create is the **Ø.250** hole through the center (coaxial) of the pin, click: **Coordinate systems off** ⇒ **Hole Tool** ⇒ **Drill to intersect with all surfaces** ⇒ **Placement** tab (Fig. 6.43) ⇒ Primary: pick axis **A_2** ⇒ **Coaxial** ⇒ **Flip** ⇒ Secondary references: click on [No Items] ⇒ pick the **RIGHT** datum plane ⇒ **Ø.250** ⇒ **Enter** ⇒ (Fig. 6.44) ⇒ ⇒ **MMB** ⇒ **LMB** to deselect

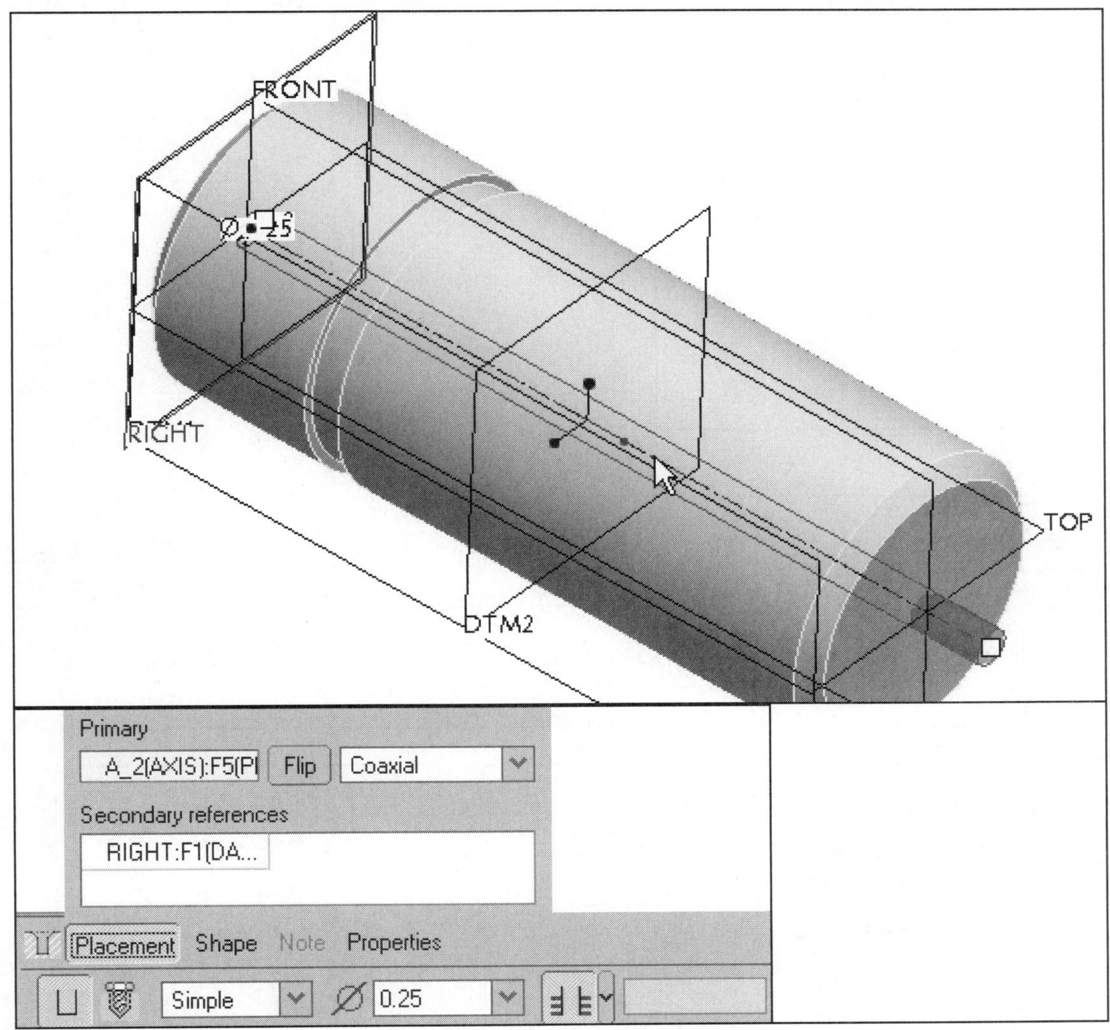

Figure 6.43 Hole Options

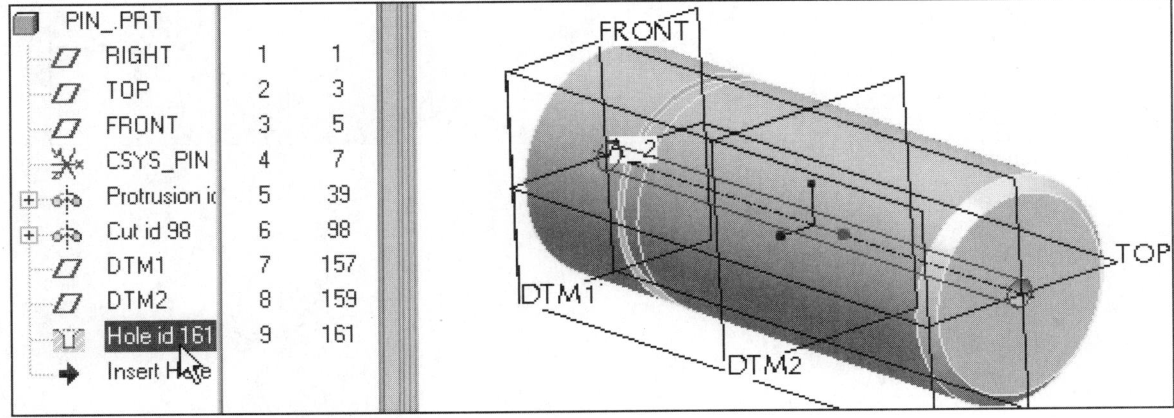

Figure 6.44 Completed Coaxial Hole

The pin has a conical cut (hole) at both ends that is coaxial with the **∅.250** hole and axis A_2. Make the conical feature with a *sketched hole*, for a pipe thread tap drill. Sketched holes and revolved cuts are created with a section sketch. The section must be closed, and have a *vertical centerline*. All entities must be on one side of that centerline. You are creating a drill tip and that the drill will be held vertical to the surface being drilled. The counterbore *diameter* is **∅.330** at the small end. The conical hole depth is **.4375** with an *angle* of **7.160** degrees (Fig. 6.45)

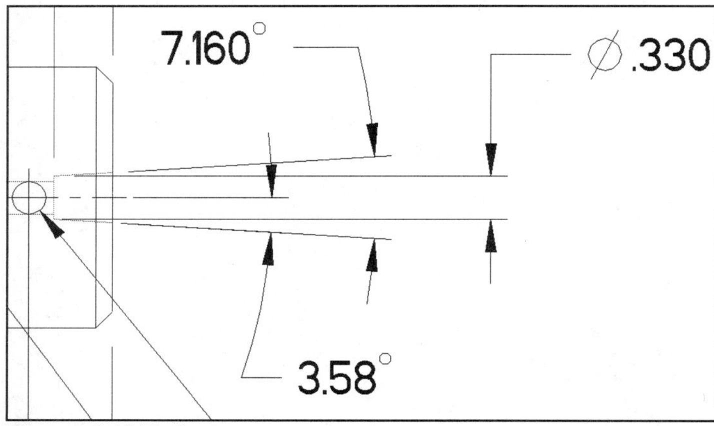

Figure 6.45 References

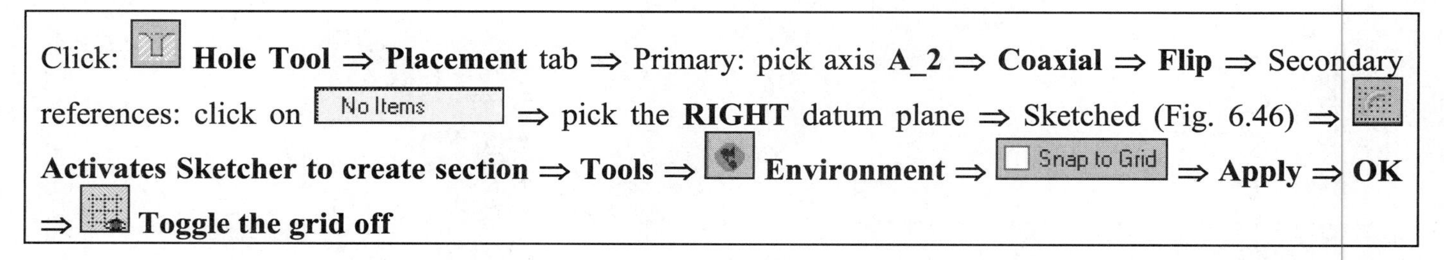
Click: **Hole Tool** ⇒ **Placement** tab ⇒ Primary: pick axis **A_2** ⇒ **Coaxial** ⇒ **Flip** ⇒ Secondary references: click on No Items ⇒ pick the **RIGHT** datum plane ⇒ Sketched (Fig. 6.46) ⇒ **Activates Sketcher to create section** ⇒ **Tools** ⇒ **Environment** ⇒ Snap to Grid ⇒ **Apply** ⇒ **OK** ⇒ **Toggle the grid off**

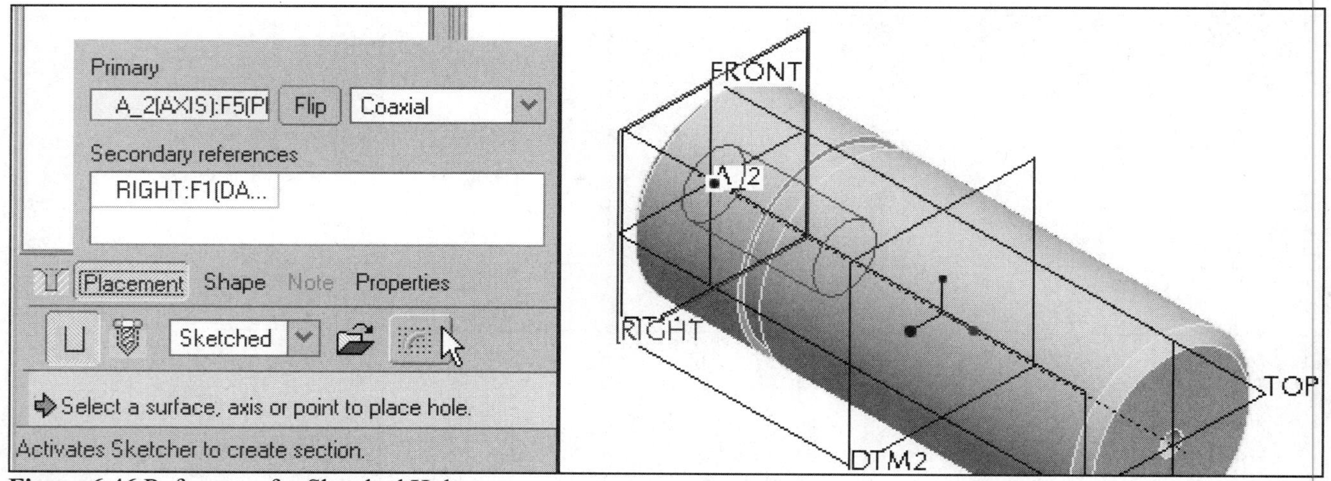

Figure 6.46 References for Sketched Hole

Click: sketch a vertical centerline ⇒ sketch the four lines required to describe half of the conical hole's shape [Fig. 6.47(a)] ⇒ Create the dimension for the lower conical diameter (**∅.330**) by picking the vertical centerline with the **LMB**, then the endpoint to be dimensioned with the **LMB**, and then the centerline a second time with the **LMB**. Place the dimension by picking a position with the **MMB**. [Fig. 6.47(b)] ⇒ add the angle dimension (**7.160/2**) ⇒ reposition the dimensions as needed ⇒ **Modify** dimensions to design values [Fig. 6.47(c)] ⇒ **Continue** [Fig. 6.48(a)] ⇒ ⇒ ⇒ [Fig. 6.48(b)] ⇒ ⇒ **MMB** ⇒ **File** ⇒ **Delete** ⇒ **Old Versions** ⇒ **MMB**

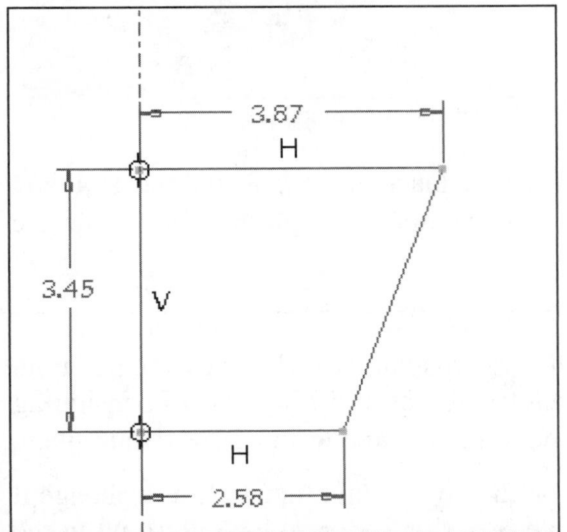

Figure 6.47(a) Centerline and Four Sketched Lines

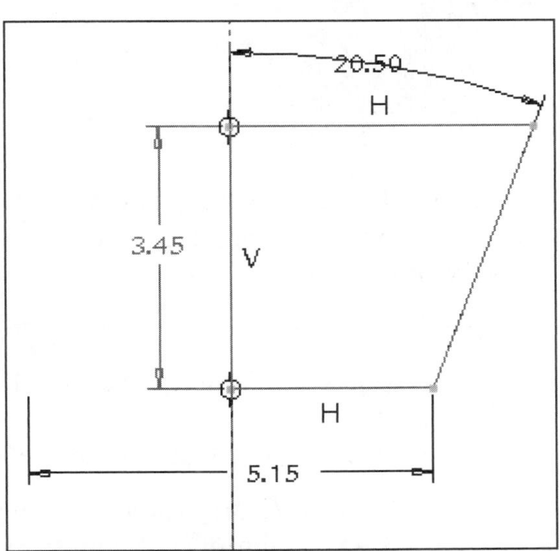

Figure 6.47(b) Strong Dimensions Added

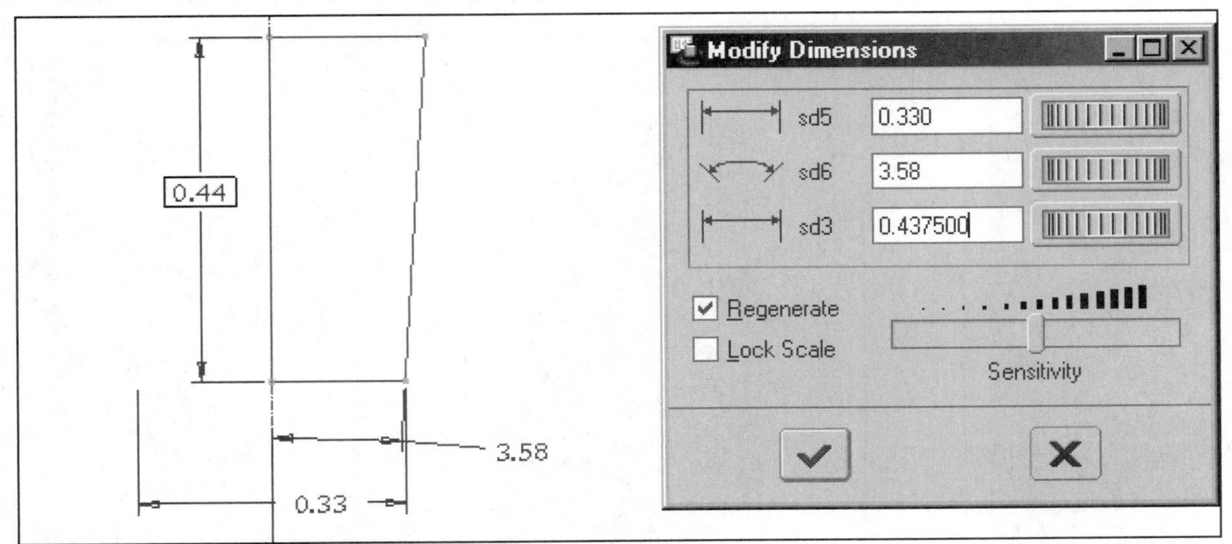

Figure 6.47(c) Modified Dimensions

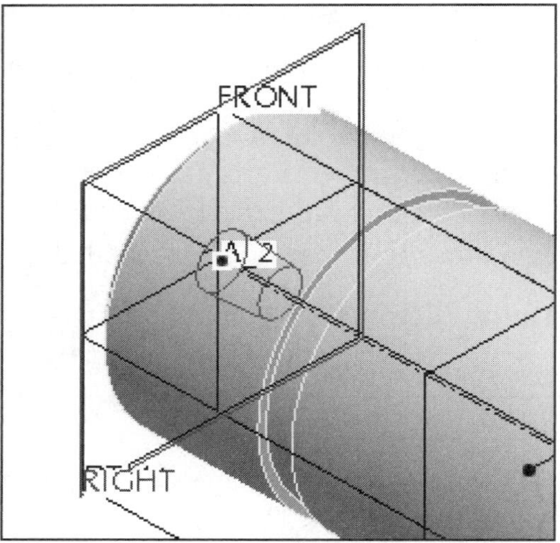

Figure 6.48(a) Conical Hole Positioned

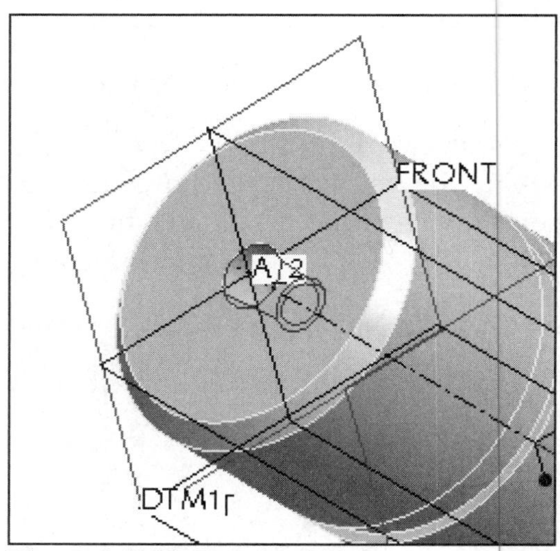

Figure 6.48(b) Completed Hole

Create the next hole using the tangent datum plane DTM1 as the placement plane and the TOP and RIGHT datum as the secondary references. The **∅.250** coaxial hole running through the pin will be the ending surface for the holes depth.

Click: **Hole Tool** ⇒ pick on datum plane **DTM1**, the hole will display with handles for position, diameter adjustment, depth adjustment, and two reference handles for establishing the dimensioning scheme [Fig 6.49(a)] ⇒ move the drag handles to the **RIGHT** datum plane and to the **TOP** datum plane respectively [Fig 6.49(b)] ⇒ **Hidden Line** ⇒ double click on the horizontal dimension and change it to **1.40** (**1.40** from the **RIGHT** datum plane), and then edit the vertical dimension to be **0.00** (**0.00** to the **TOP** datum) [Fig 6.49(c)] ⇒ **∅.250** ⇒ **Enter**

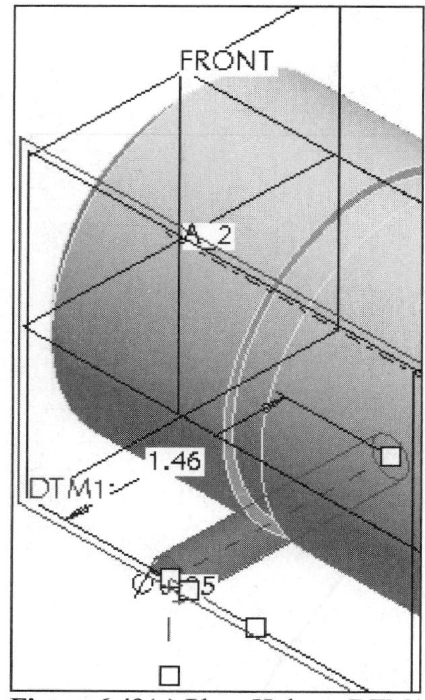

Figure 6.49(a) Place Hole on DTM1

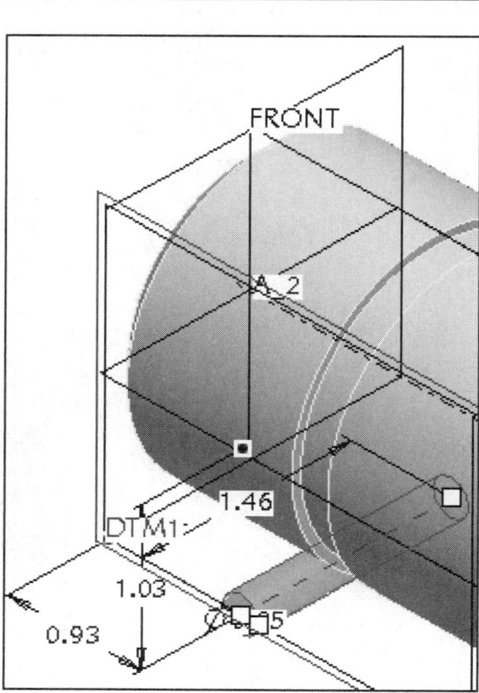

Figure 6.49(b) Move Drag Handles

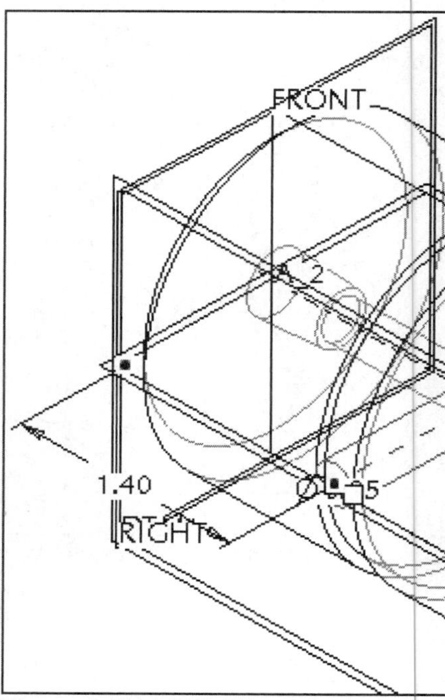

Figure 6.49(c) Edit Dimensions

Click: **Drill to selected surface** ⇒ place the cursor over the coaxial hole ⇒ click **RMB** until the hole is highlighted Showing surface created by feature 9 (HOLE), model PIN_. Confirm selection. (Fig. 6.50) ⇒ **LBM** to accept (Fig. 6.51) ⇒ (Fig. 6.52) ⇒ ⇒ ⇒ (Fig. 6.53) ⇒ **LMB** to deselect

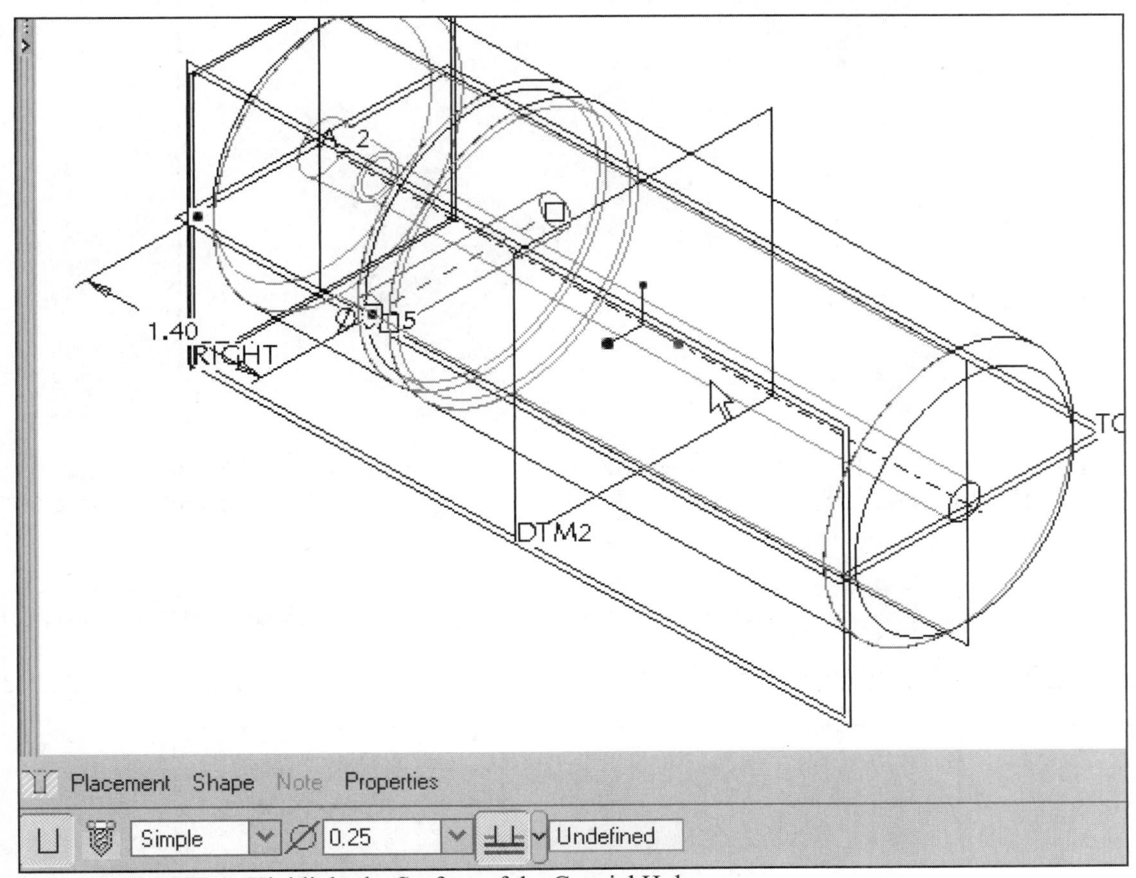

Figure 6.50 RMB to Highlight the Surface of the Coaxial Hole

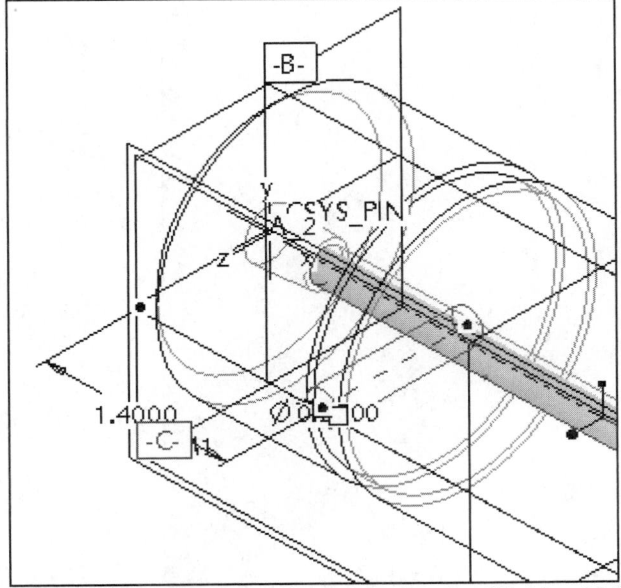

Figure 6.51 Accept Highlighted Surface by Clicking LMB

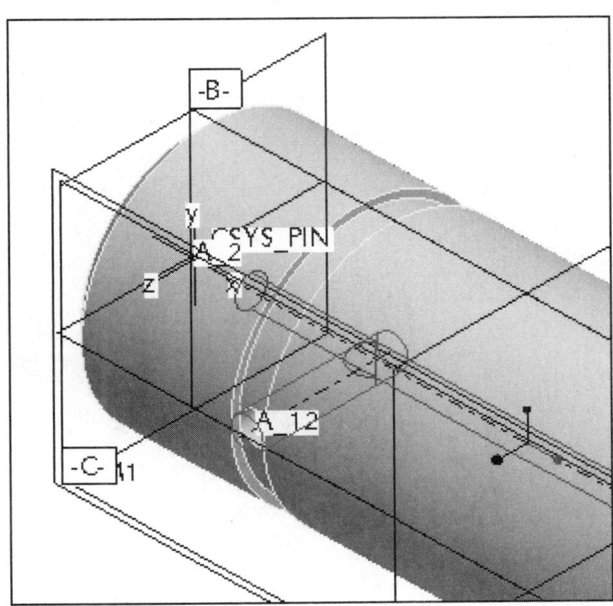

Figure 6.52 Hole Preview

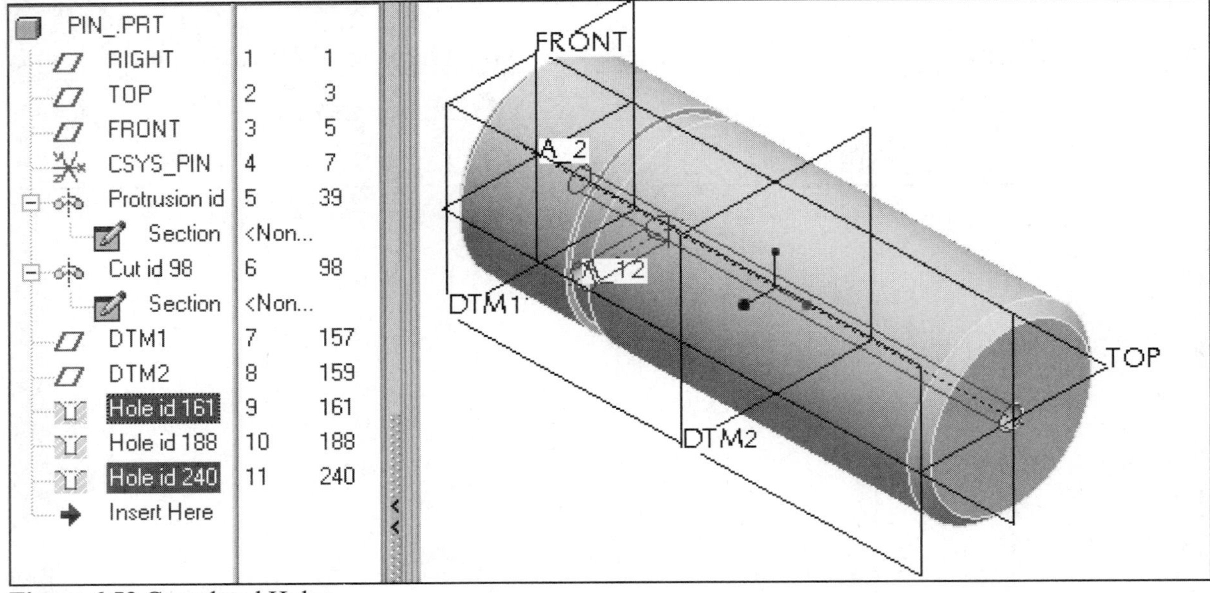

Figure 6.53 Completed Holes

Change the diameter of the hole to the proper design value. The diameter for the hole was supposed to be **.125**, not **.250** (no such thing as making a mistake, just use Edit or Edit Definition and regenerate).

Click on the last hole in the **Model Tree** ⇒ **RMB** ⇒ **Edit** ⇒ double click on the **.250** dimension and type **.125** (Fig. 6.54) ⇒ **Enter** ⇒ **Regenerates Model** ⇒ ⇒ **MMB**

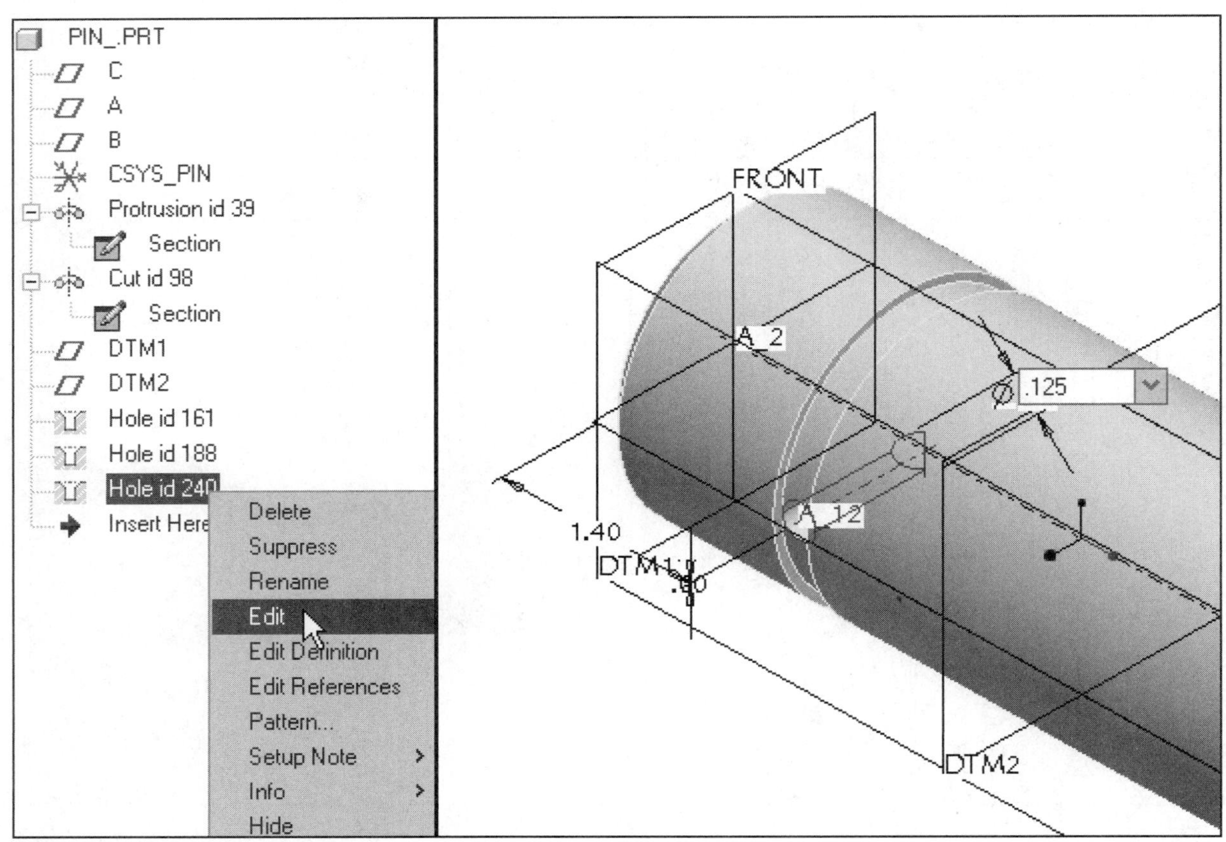

Figure 6.54 Edit the Hole Diameter

Now create the last hole, also on the part's side. Use the same steps as outlined for the previous hole, with the same placement plane and references; except that the diameter is **.250**, its distance from datum **RIGHT** is **.625**, and **0.00** from the **TOP** datum. It is a *"blind"* hole with a depth of **.3120** [Fig. 6.55(a)] ⇒ **LMB** to deselect ⇒ click on the last hole in the Model Tree ⇒ **RMB** ⇒ **Edit** to see dimensions [Fig. 6.55(b)] ⇒ [save icon] ⇒ **MMB**

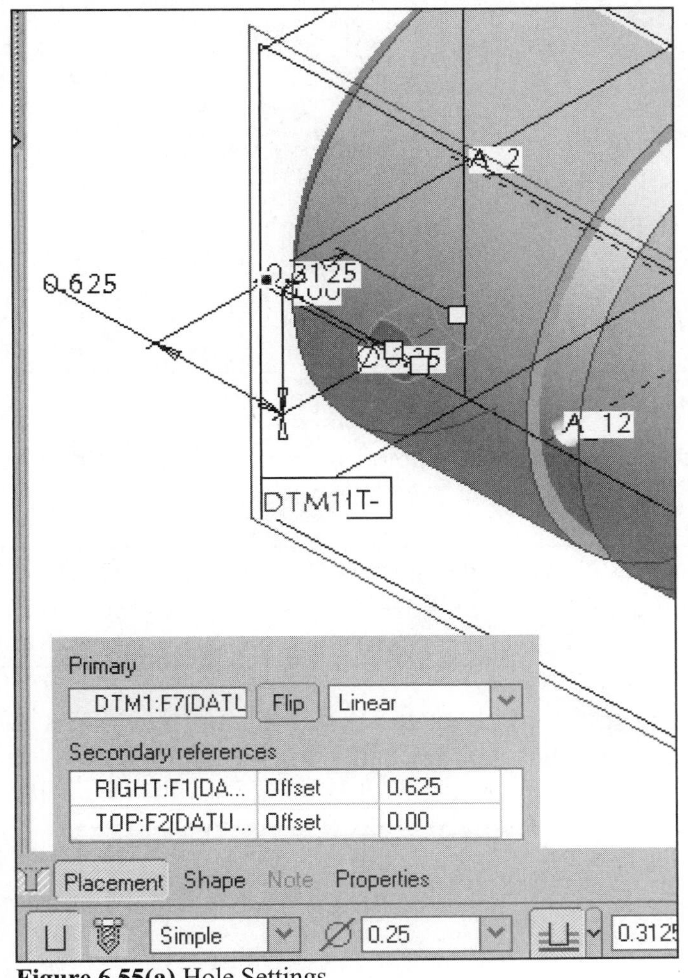

Figure 6.55(a) Hole Settings

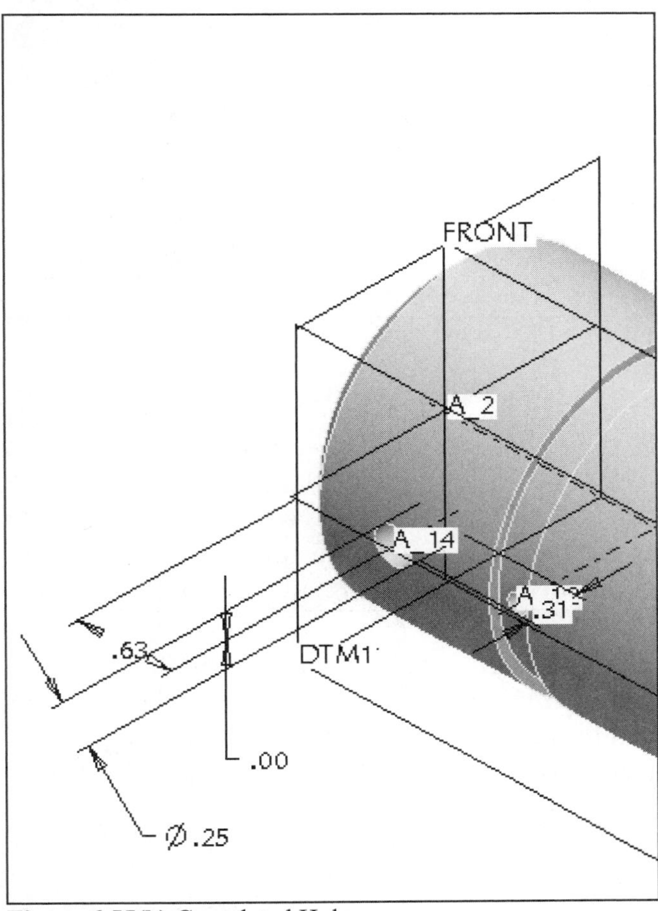

Figure 6.55(b) Completed Hole

Rename and set the datum planes, click: **TOP** in the Model Tree ⇒ **RMB** ⇒ **Properties** ⇒ [-A-] ⇒ Name- type **A** (Fig. 6.56) ⇒ **Enter** ⇒ **OK** ⇒ click: **FRONT** ⇒ **RMB** ⇒ **Properties** ⇒ [-A-] ⇒ **B** ⇒ **Enter** ⇒ **OK** ⇒ click: **RIGHT** ⇒ **RMB** ⇒ **Properties** ⇒ [-A-] ⇒ **C** ⇒ **Enter** ⇒ **OK** (Fig. 6.57)

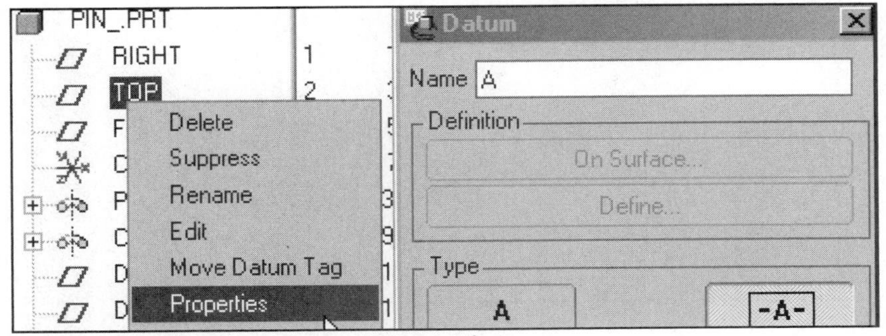

Figure 6.56 Changing a Datum's Properties

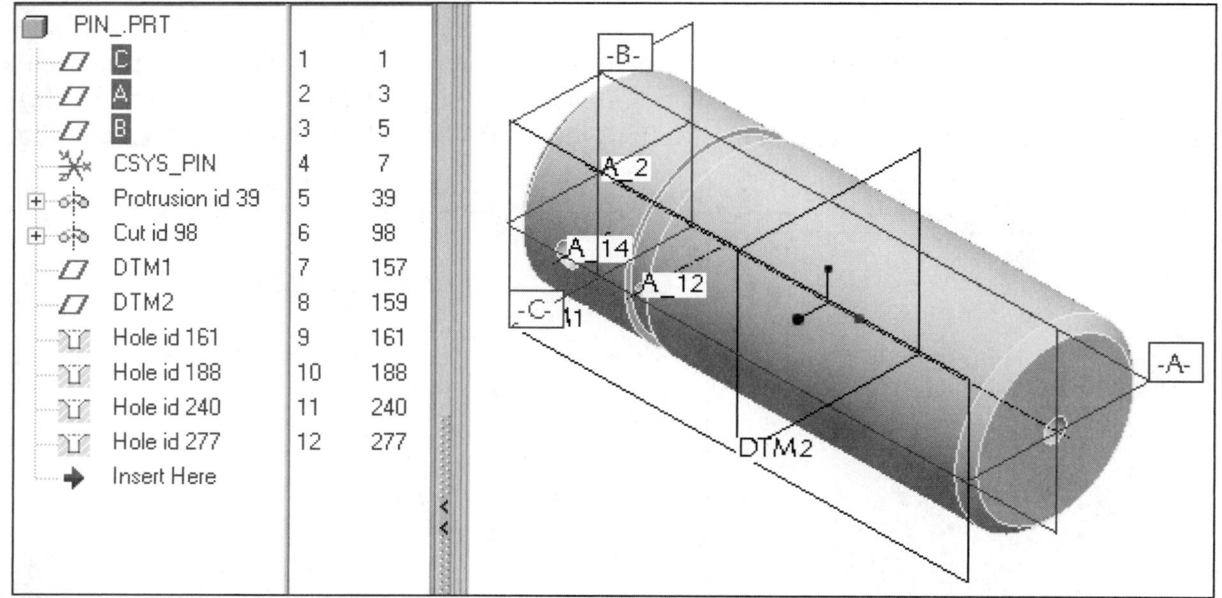

Figure 6.57 Renamed and Set Datums

Three holes (not the coaxial hole) and the groove can now be copied by mirroring about datum plane DTM2.

Click: **Edit** from the menu bar ⇒ **Feature Operations** ⇒ **Copy** ⇒ **Mirror** ⇒ **Dependent** ⇒ **MMB** ⇒ ⇒ Select features to be mirrored. Press and hold down the **Ctrl** key and click on the cut, and three holes in the Model Tree (Fig. 6.58). ⇒ **MMB** ⇒ **MMB** ⇒ **Plane** ⇒ click on **DTM2** (Figs. 6.59) ⇒ **MMB**

The copied features (Fig. 6.60) are grouped and mirrored about the datum plane (Fig. 6.61). A relation must be added to control the position of DTM2. Regardless of the change in length of the part, the datum must always be in the middle, otherwise the copied conical hole will become "swallowed" (the feature will remain but will be inside the protrusion) by the base revolved protrusion, as it gets longer.

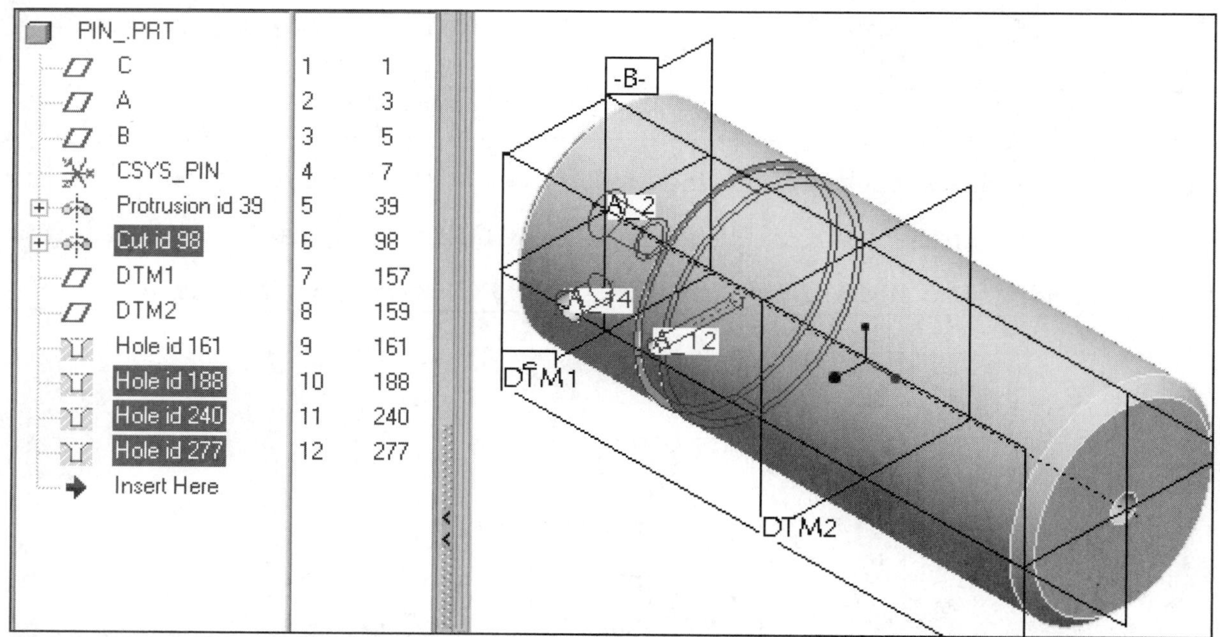

Figure 6.58 Select the Cut and Holes to be Copied

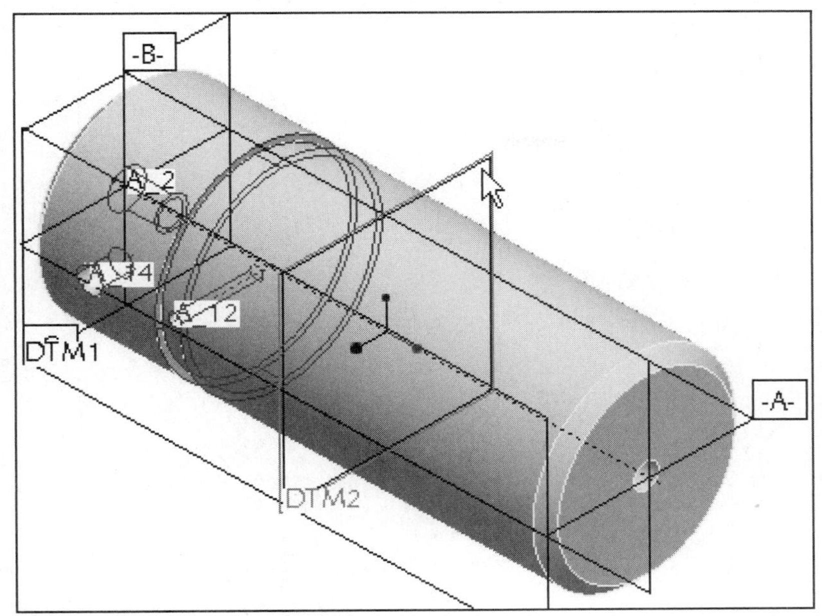

Figure 6.59 Pick Datum DTM2

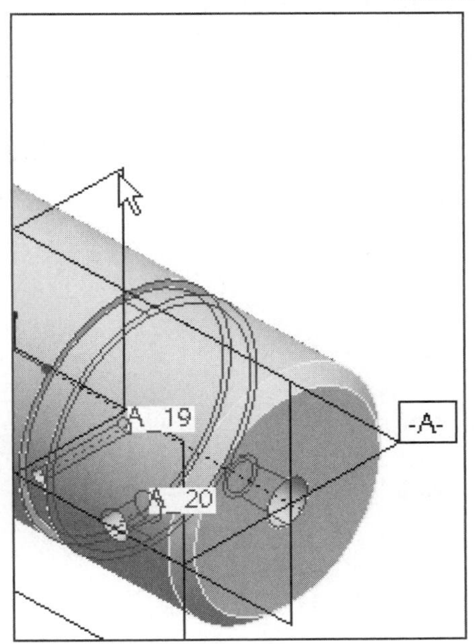

Figure 6.60 Copied features

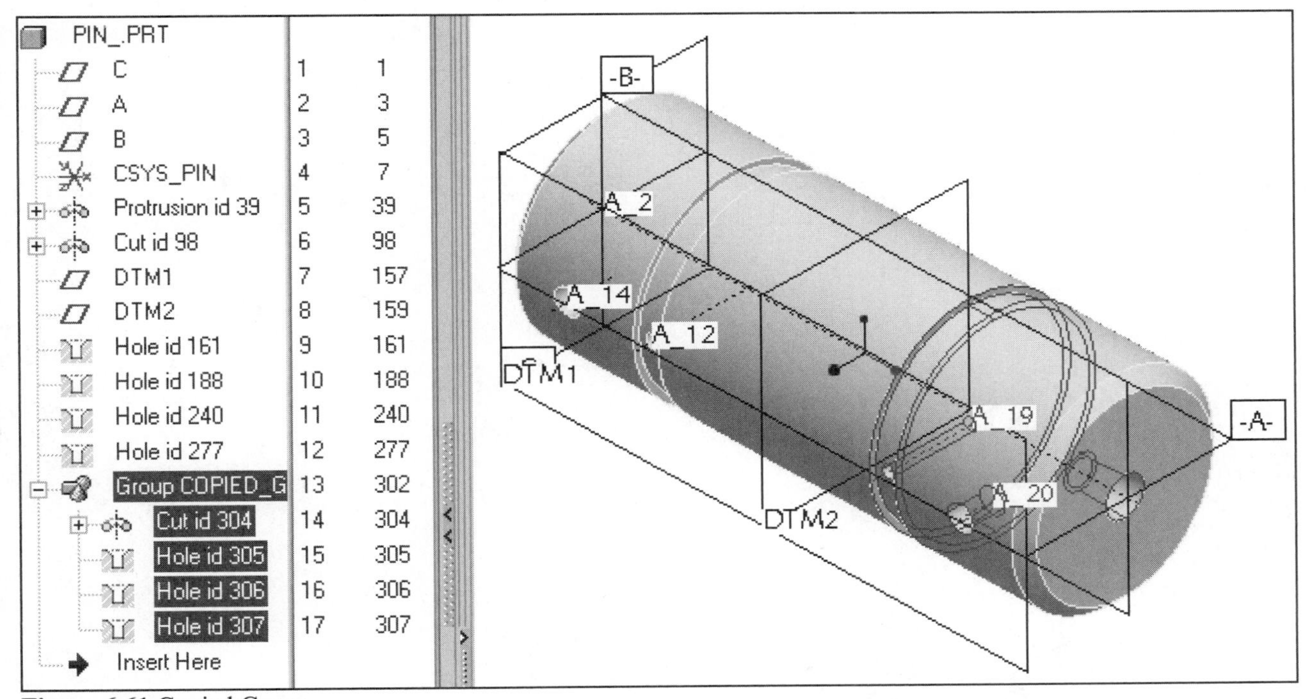

Figure 6.61 Copied Group

Click on **Protrusion** in the **Model Tree** ⇒ **RMB** ⇒ **Edit** (Fig. 6.62) ⇒ **Info** from the menu bar ⇒ Switch Dimensions ⇒ note the **d#** symbol for the length (**5.125** dimension), here it is **d4** (Fig. 6.63) ⇒ click on **DTM2** in the **Model Tree** ⇒ **RMB** ⇒ **Edit** ⇒ note the **d#** symbol for the length (**5.125/2=2.5625**) dimension, here it is **d11** (your **d** symbols may differ) (Fig. 6.64) ⇒ **Tools** ⇒ **Relations** ⇒ **d11=d4/2** (remember, your **d** symbols may differ) (Fig. 6.65) ⇒ **Ok** ⇒ ⇒ **Info** from the menu bar ⇒ Switch Dimensions ⇒ ⇒ **MMB** ⇒ **File** ⇒ **Delete** ⇒ **Old Versions** ⇒ **MMB**

*If you get a failure, ⇒ Undo Changes ⇒ Confirm ⇒ Tools ⇒ Relations ⇒ remove any relations ⇒ Ok ⇒ pick DTM2 ⇒ Edit ⇒ change dimension to **5.125/2** ⇒ Regenerate. Then redo the relations.*

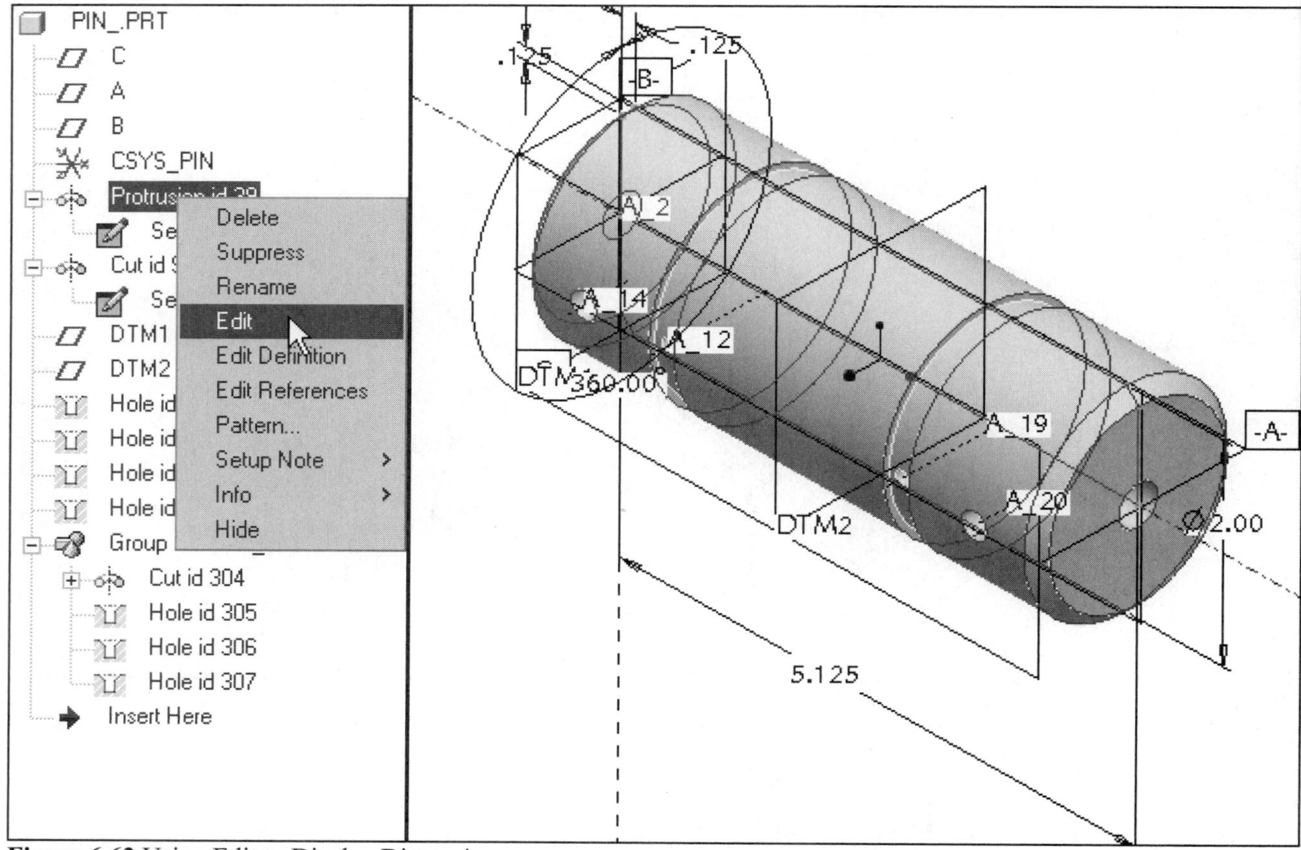

Figure 6.62 Using Edit to Display Dimensions

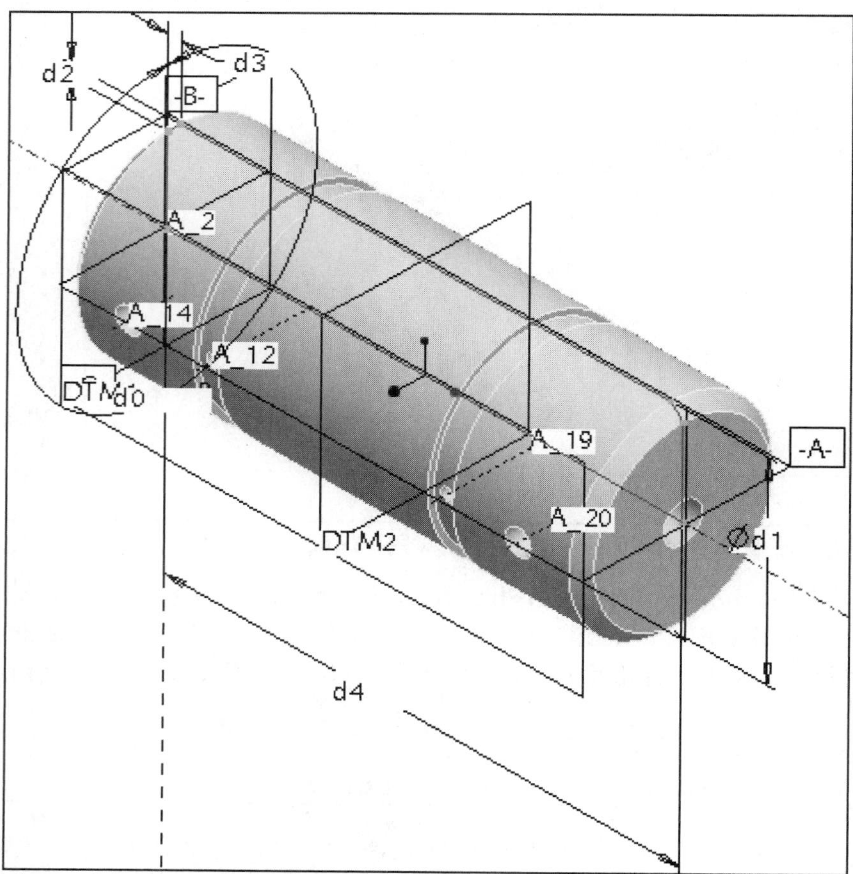

Figure 6.63 Dimension Symbols (**d4**)

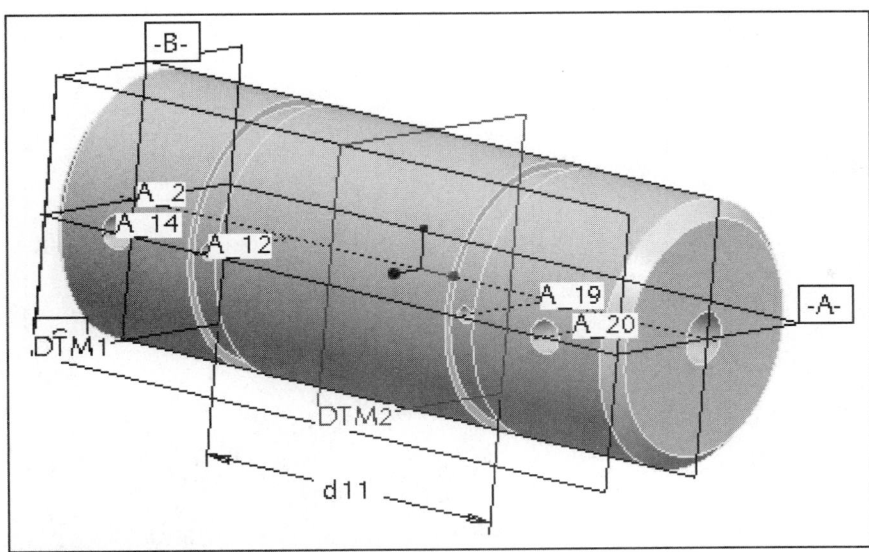

Figure 6.64 Dimension Symbol for DTM2 (**d11**)

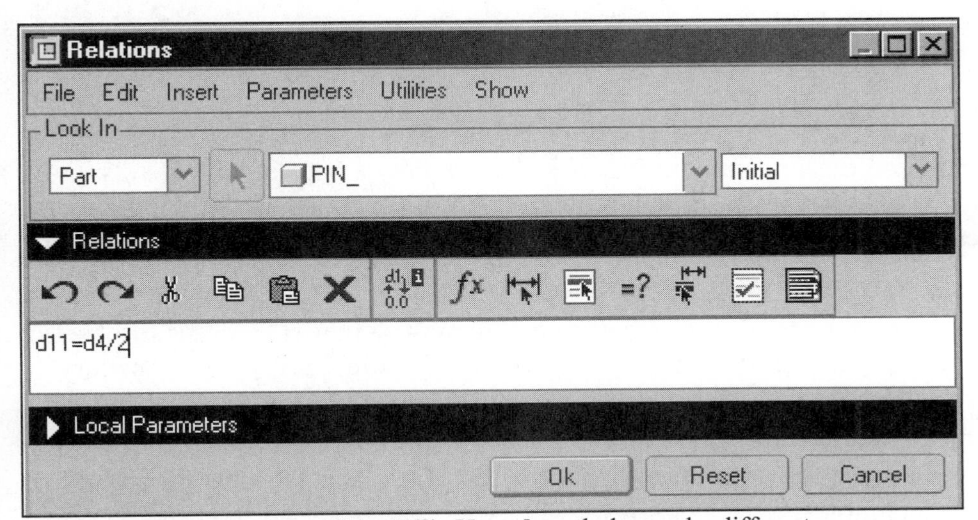

Figure 6.65 Add Relations (**d11=d4/2**). Your **d** symbols may be different.

Add another relation (Fig. 6.66) to control the location of the small **.125** hole to always be centered on the revolved cut. Remember, your **d** symbols will probably be different. See Figures 6.67(a-b) for the **d** symbols. ⇒ **Regenerates Model** ⇒ ⇒ ⇒ ⇒ **MMB**

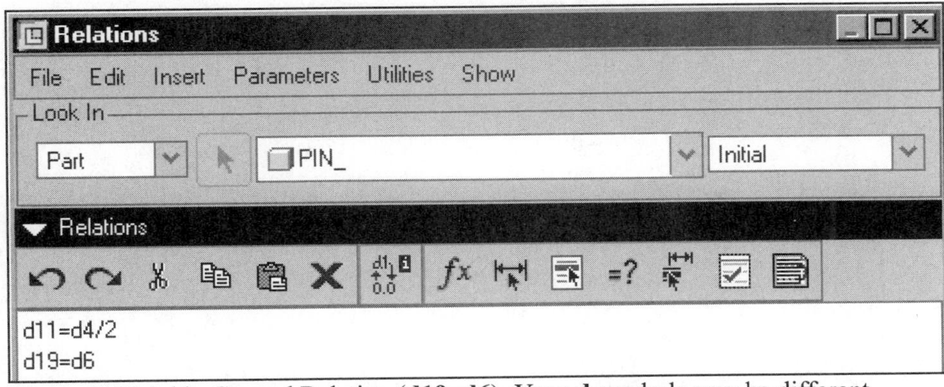

Figure 6.66 Add a Second Relation (**d19=d6**). Your **d** symbols may be different.

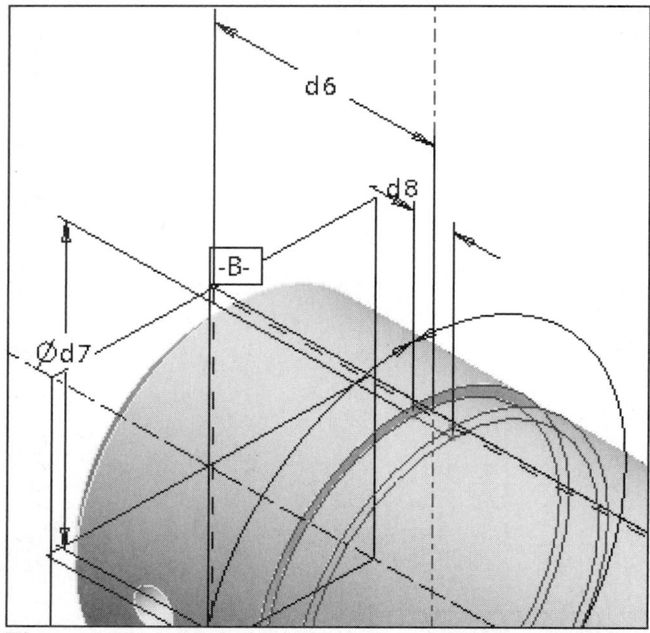

Figure 6.67(a) Symbol **d6** Dimension to Center of Groove

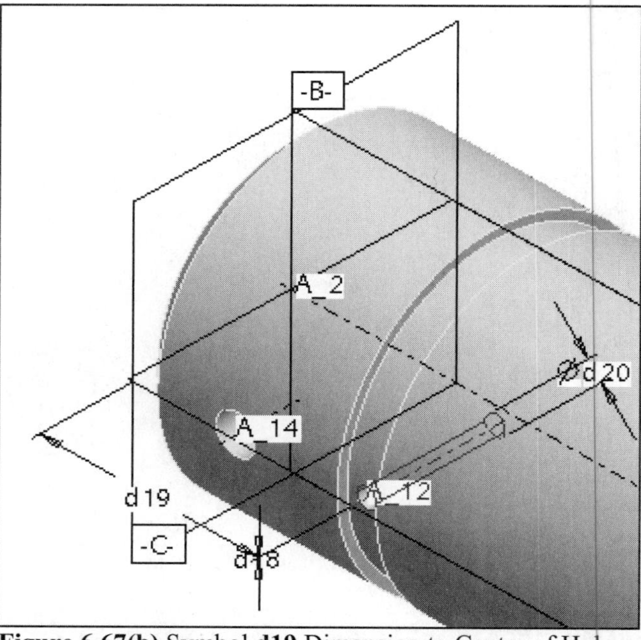

Figure 6.67(b) Symbol **d19** Dimension to Center of Hole

Viewing Model Creation History

The **Model Player** (Fig. 6.68) option on the **Tools** menu lets you observe how a part is built. It aids in the diagnosis of bad features in a part by allowing you to view the creation history of a model. You can use this option at any time, or instead of regenerating a model after you modify it. It allows you to do the following:

- Move backward or forward through the feature-creation history in the model in order to observe how the model was created. You can start the model playback at any point in its creation history
- Regenerate each feature in sequence, starting from the specified feature
- Display each feature as it is regenerated or rolled forward
- Update (regenerate all the features in) the entire display when you reach the desired feature or when the playback process is complete
- Obtain information about the current feature (the feature that was currently regenerated when you stopped the model playback process) (you can show dimensions, obtain regular feature information, investigate geometry errors, and enter Fix Model mode)

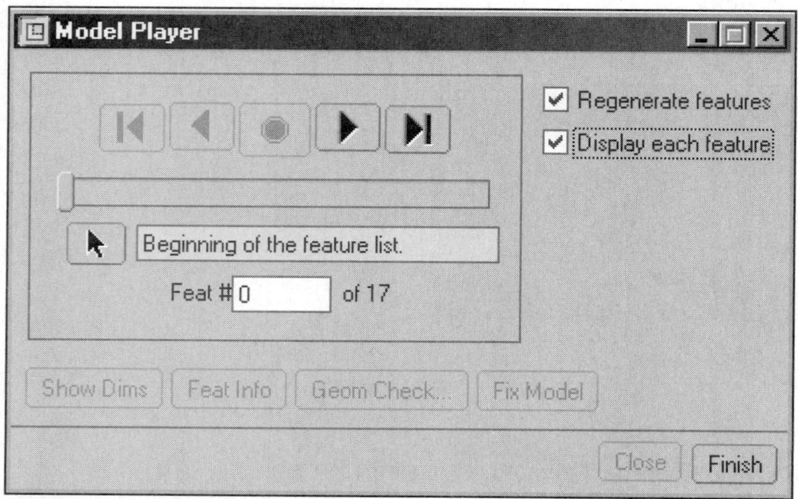

Figure 6.68 Model Player

Using the Model Player

To open the **Model Player** dialog box, click Tools ⇒ Model Player. You can select one of the following commands for controlling feature regeneration and display:

- **Regenerate features** Regenerates each feature in sequence, starting from the specified feature, as the model moves forward. If this box is cleared, features will be rolled forward without regenerating. If you have made changes to the model without regenerating, the command is checked and unavailable.
- **Display each feature** Displays each feature in the graphics window as it is being regenerated or rolled forward. Future features are not displayed until they are regenerated. If this box is cleared, the entire display is updated only when the desired feature is reached and the model playback is complete, or when you stop model playback.
- **Compute CL** (Available in Manufacturing mode only) When selected, the CL data is recalculated for each NC sequence during regeneration.

Select one of the following commands to select the place (feature) in the model creation history at which to start the regeneration process:

Moves immediately to the beginning of the model (suppresses all features)

Steps backward through the model one feature at a time and regenerates the preceding feature

Moves immediately to the end of the model (resume all features)

Steps forward through the model one feature at a time and regenerates the next feature

Slider Bar Drags the slider handle to the first feature at which you want model playback to begin. The features are highlighted in the graphics window as you move through their position with the slider handle. The feature number and type are displayed in the selection panel [such as #16 (HOLE)], and the feature number is displayed in the **Feat #** box. When you release the slider, the model immediately rolls or regenerates to that feature.

Lets you select a starting feature from the graphics window or the Model Tree. Opens the **SELECT FEAT** and **SELECT** menus. After you select a starting feature, its number and ID are displayed in the selection panel, and the feature number is displayed in the **Feat #** box.
Feat # Lets you specify a starting feature by typing the feature number in the box. After you enter the feature number, the model immediately rolls or regenerates to that feature.

To stop playback, click the **Stop** button. You can use the following commands to obtain information about the current feature (the feature that was currently regenerated when you stopped the model playback process):

- **Show Dims** Displays the dimensions of the current feature.
- **Feat Info** Provides regular feature information about the current feature in an Information window.
- **Geom Check** Investigates the geometry error for the current feature. This command is accessible only when Pro/E encounters a geometry error.
- **Fix Model** Activates Resolve mode by forcing the current feature to abort regeneration. When you exit Resolve mode, Pro/E returns to the Model Player at the current feature. You can then continue to move backward or forward through the model.
- **Close** Closes the Model Player and enters Insert mode at the current feature (the last regenerated feature). You can choose this command anytime during the model playback.
- **Finish** Closes the Model Player and returns to the last feature in the model. Pro/E restores all features. You can choose this command anytime during the model playback.

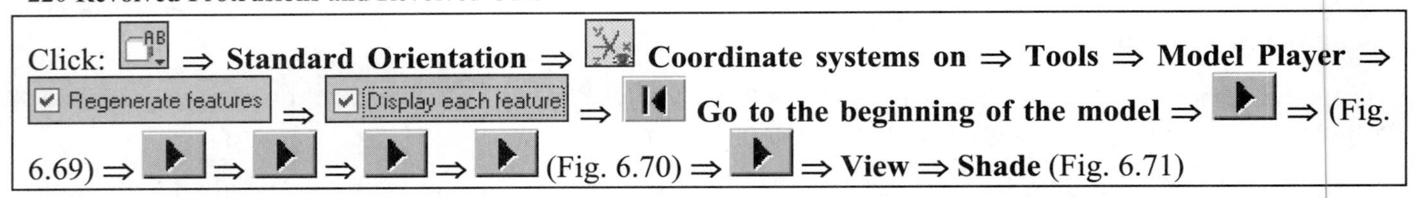
Click: ⇒ **Standard Orientation** ⇒ **Coordinate systems on** ⇒ **Tools** ⇒ **Model Player** ⇒ Regenerate features ⇒ Display each feature ⇒ **Go to the beginning of the model** ⇒ ⇒ (Fig. 6.69) ⇒ ⇒ ⇒ ⇒ (Fig. 6.70) ⇒ ⇒ **View** ⇒ **Shade** (Fig. 6.71)

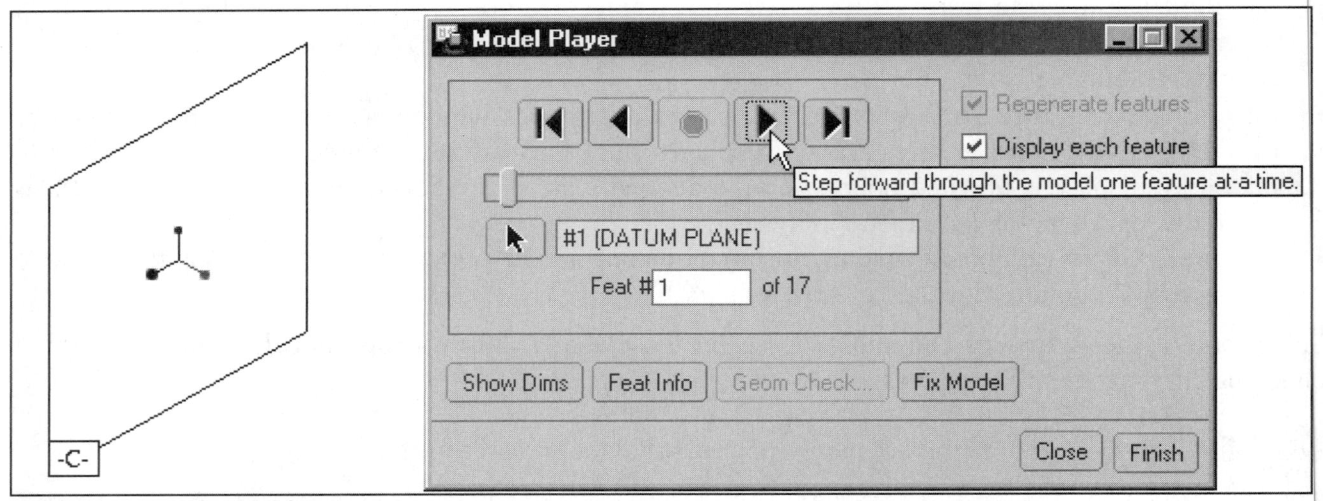

Figure 6.69 Regenerate First Feature

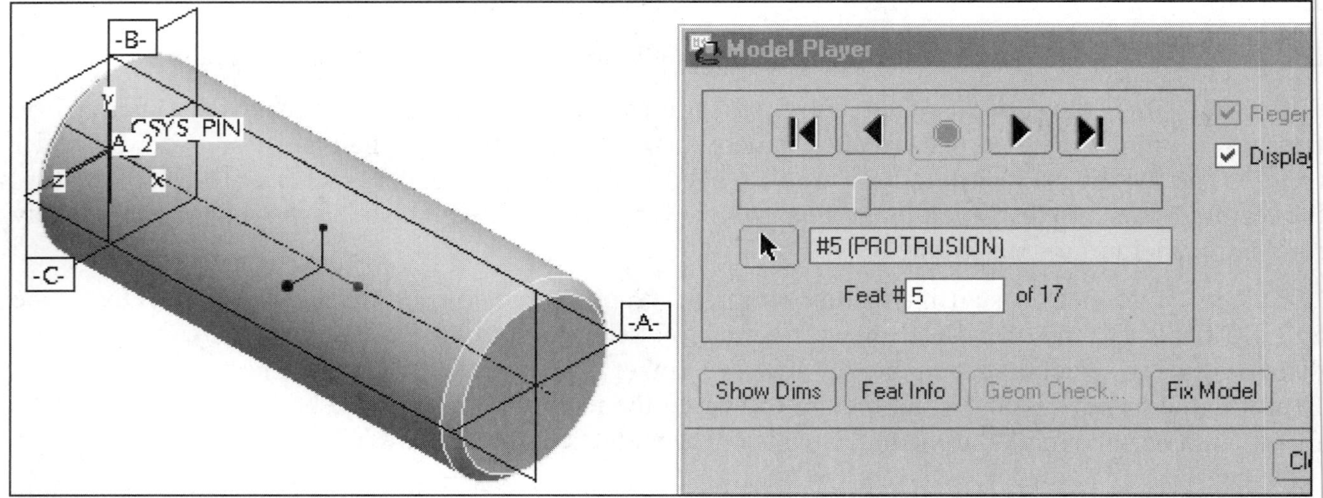

Figure 6.70 Regenerate First Protrusion

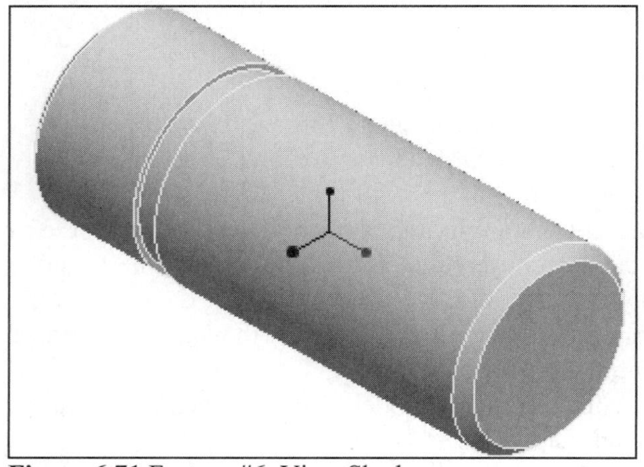

Figure 6.71 Feature #6, View Shade

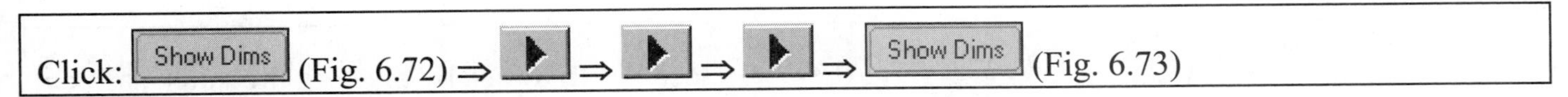

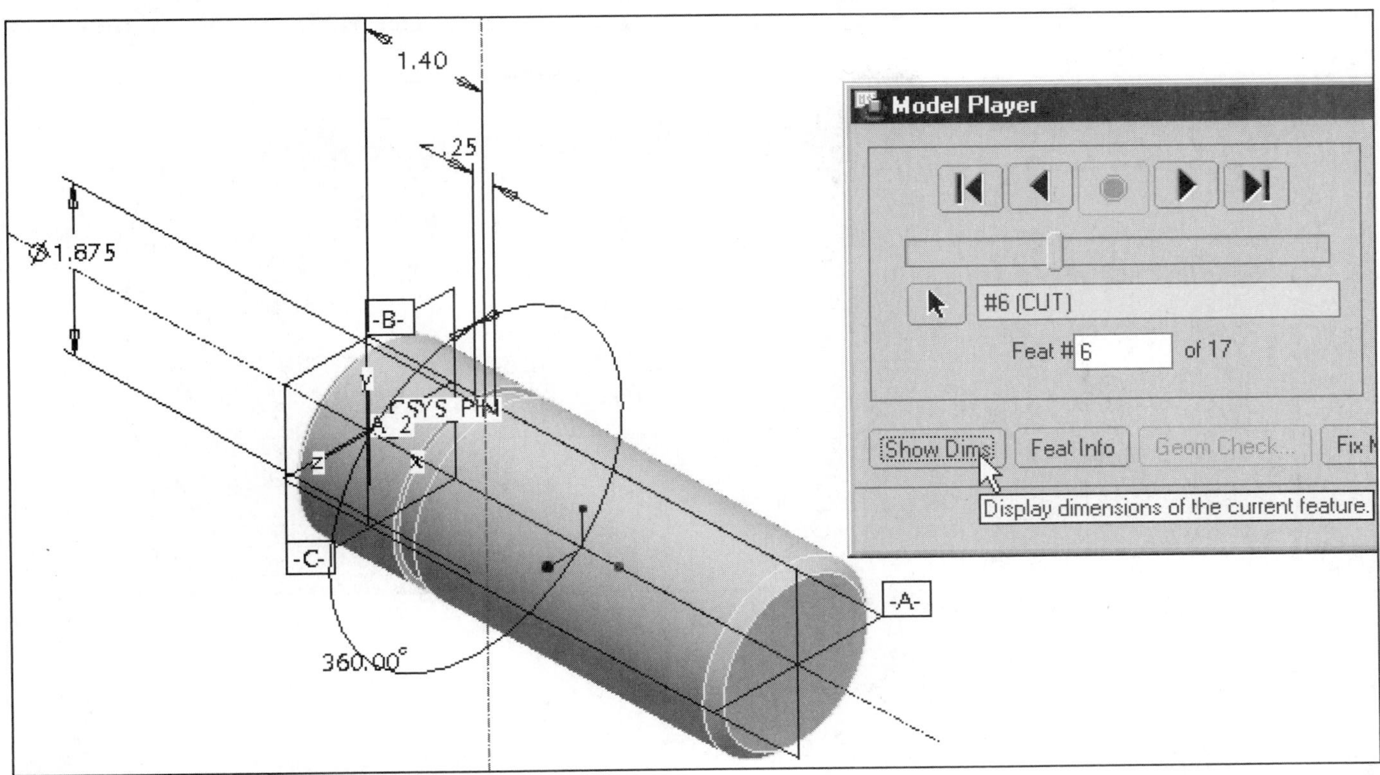

Figure 6.72 Feature **#6**, Cut Dimensions

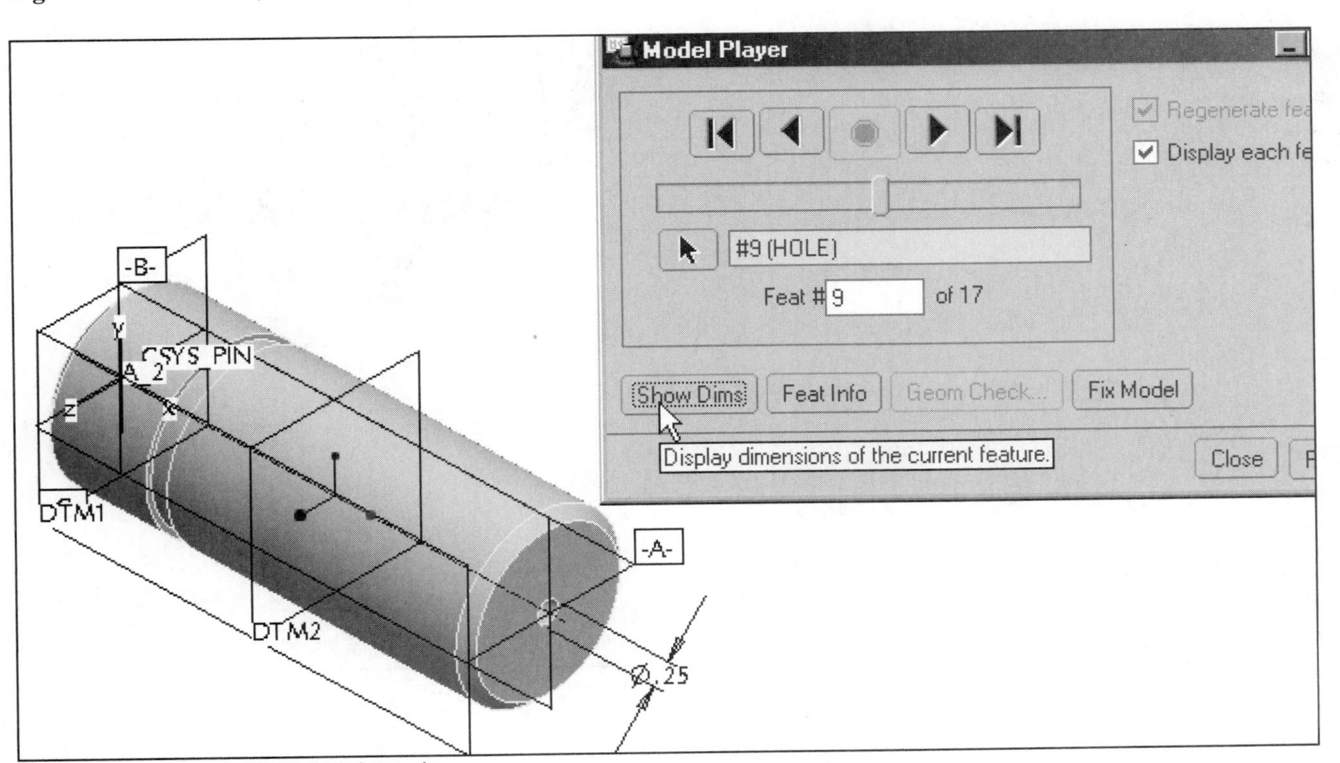

Figure 6.73 Feature **#9**, Hole Dimensions

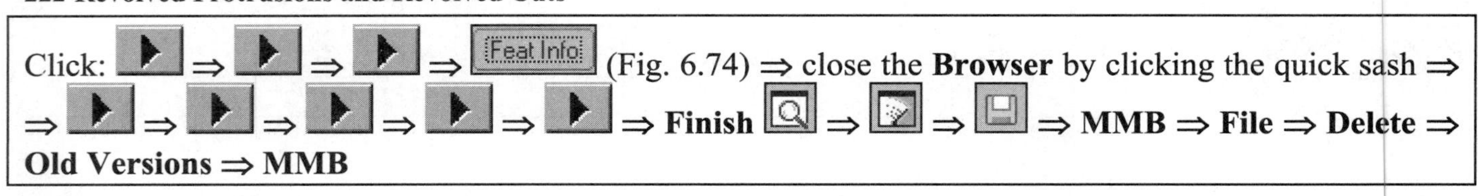

Click: ⇒ ⇒ ⇒ Feat Info (Fig. 6.74) ⇒ close the **Browser** by clicking the quick sash ⇒ ⇒ ⇒ ⇒ ⇒ ⇒ ⇒ **Finish** ⇒ ⇒ ⇒ **MMB** ⇒ **File** ⇒ **Delete** ⇒ **Old Versions** ⇒ **MMB**

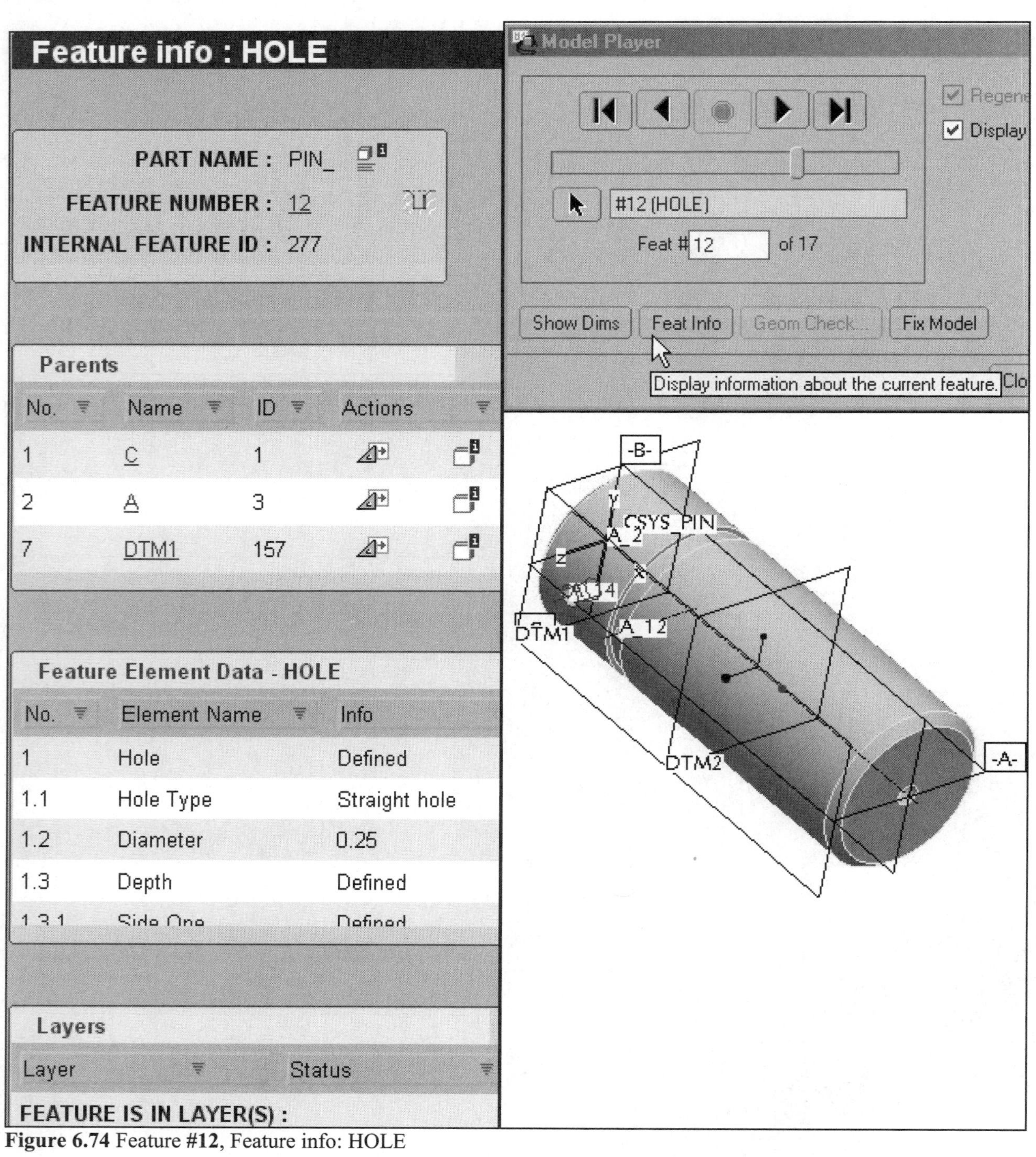

Figure 6.74 Feature **#12**, Feature info: HOLE

Printing and Plotting

From the File menu, you can print with the following options: scaling, clipping, displaying the plot on the screen, or sending the plot directly to the printer. Shaded images can also be printed from this menu. You can create plot files of the current object (sketch, part, assembly, drawing, or layout) and send them to the print queue of a plotter. The plotting interface to HPGL and PostScript formats is standard.

You can configure your printer using the Printer Configuration dialog box, available from the Print dialog box. If you are printing a shaded image, the Shaded Image Configuration dialog box opens instead of the Printer Configuration dialog box.

The following applies for plotting:

- Hidden lines appear as gray for a screen plot, but as dashed lines on paper.
- When Pro/E plots Pro/E line fonts, it scales them to the size of a sheet. It does not scale the user-defined line fonts, which do not plot as defined.
- You can use the configuration file option *use_software_linefonts* to make sure that the plotter plots a user-defined font exactly as it appears in Pro/E.
- You can plot a cross section from Part or Assembly mode.
- With a Pro/PLOT license, you can write plot files in Calcomp, Gerber, HPGL2, and Versatec format.

Print your object, click: **File ⇒** Print... **⇒ OK** (Fig. 6.75) **⇒ OK** (Fig. 6.76) **⇒** [Save icon] **⇒ MMB ⇒ File ⇒ Delete ⇒ Old Versions ⇒ MMB**

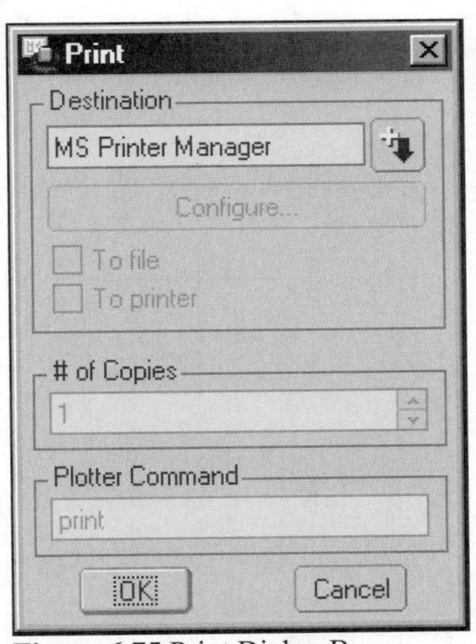

Figure 6.75 Print Dialog Box

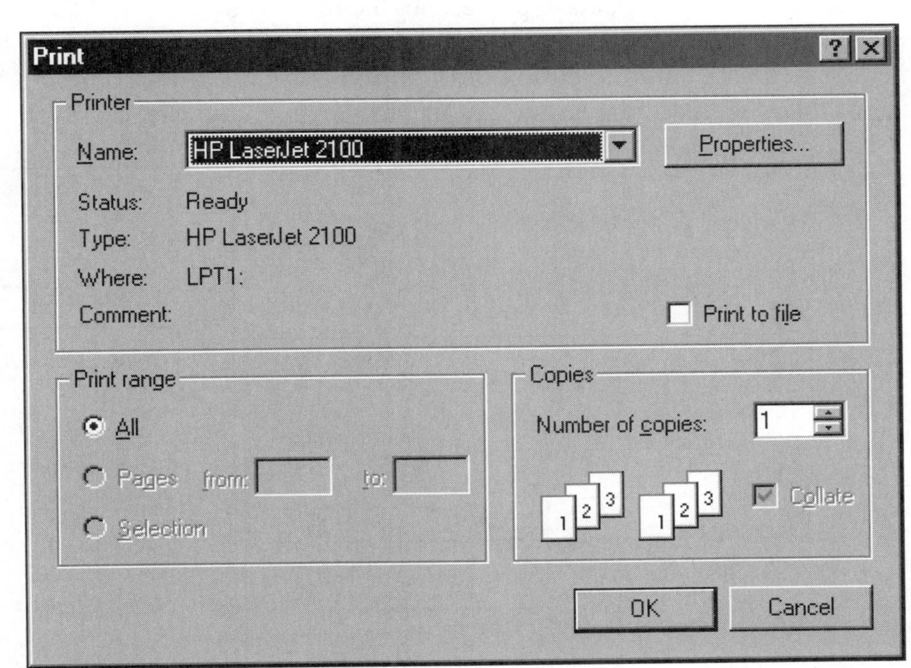

Figure 6.76 Windows Print Dialog Box

You have completed the lesson. Continue with the Lesson Project. Be sure to apply the methods and commands used in this and previous lessons.

Lesson 6 Project

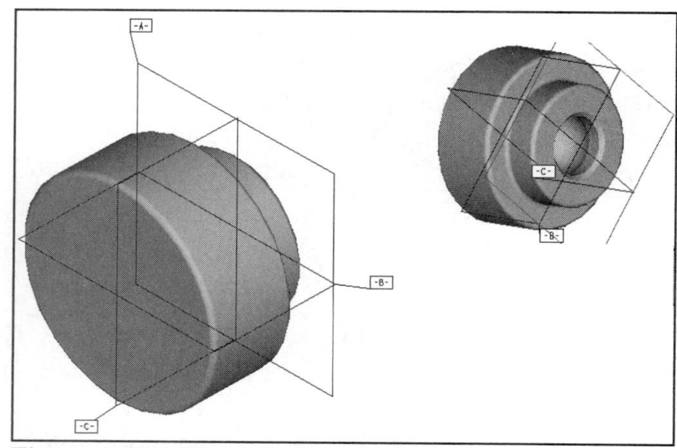

Figure 6.77(a) Clamp Foot

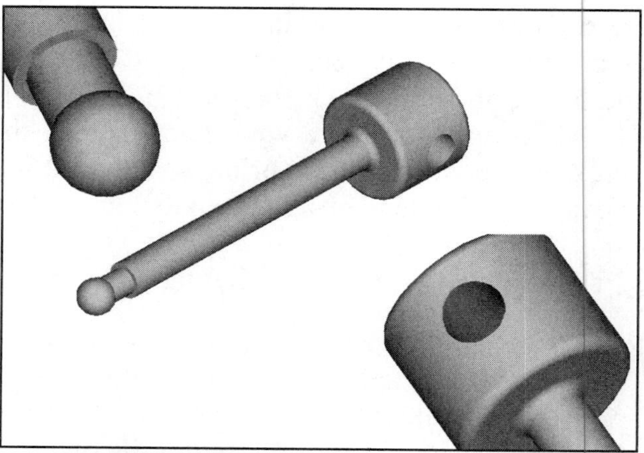

Figure 6.77(b) Clamp Swivel

Clamp Foot and Clamp Swivel

Two separate projects are provided in this lesson. You will use both of these parts in Lessons 15 and 16 when creating an assembly. Both the Clamp Foot and the Clamp Swivel are simple revolved protrusions (Figs. 6.77 through 6.83). The Clamp Foot is nylon and the Clamp Swivel is steel. The Clamp Swivel fits inside the Clamp Foot.

Analyze each part and plan the steps and features required to model it. Remember to set up the environment, establish datum planes, set the planes as basic, and add them to layers.

Create the two parts with revolved protrusions and cuts using datum C (RIGHT) as the sketching plane. Create the internal cut on the Clamp Foot with a *revolved cut*. Add the rounds on both parts at the end of the modeling process; do not include them on the first revolved protrusions.

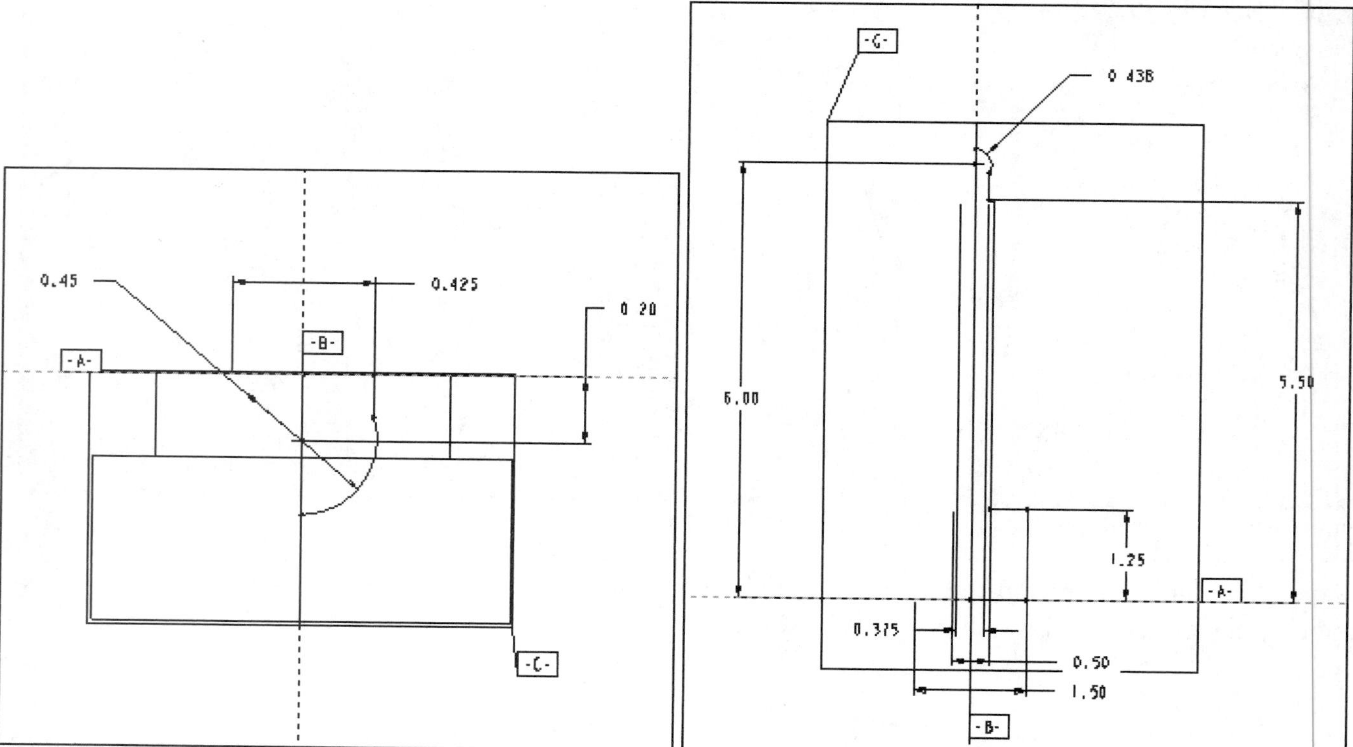

Figure 6.78 Revolved Cut sketch

Figure 6.79 Revolved Protrusion sketch

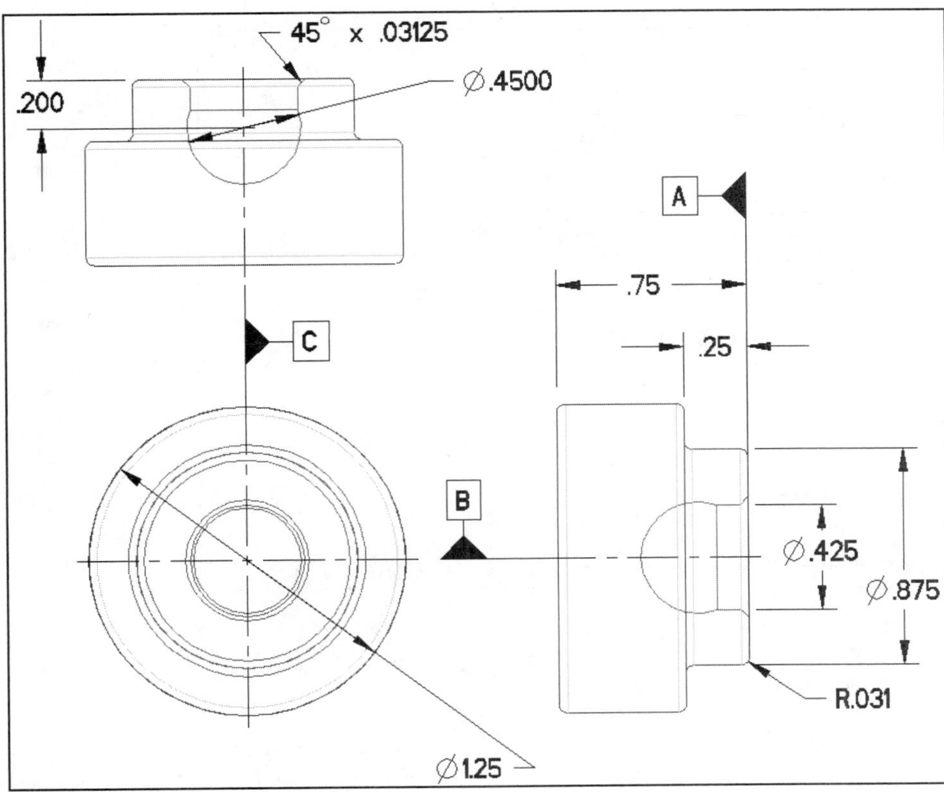

Figure 6.80 Clamp Foot Dimensions

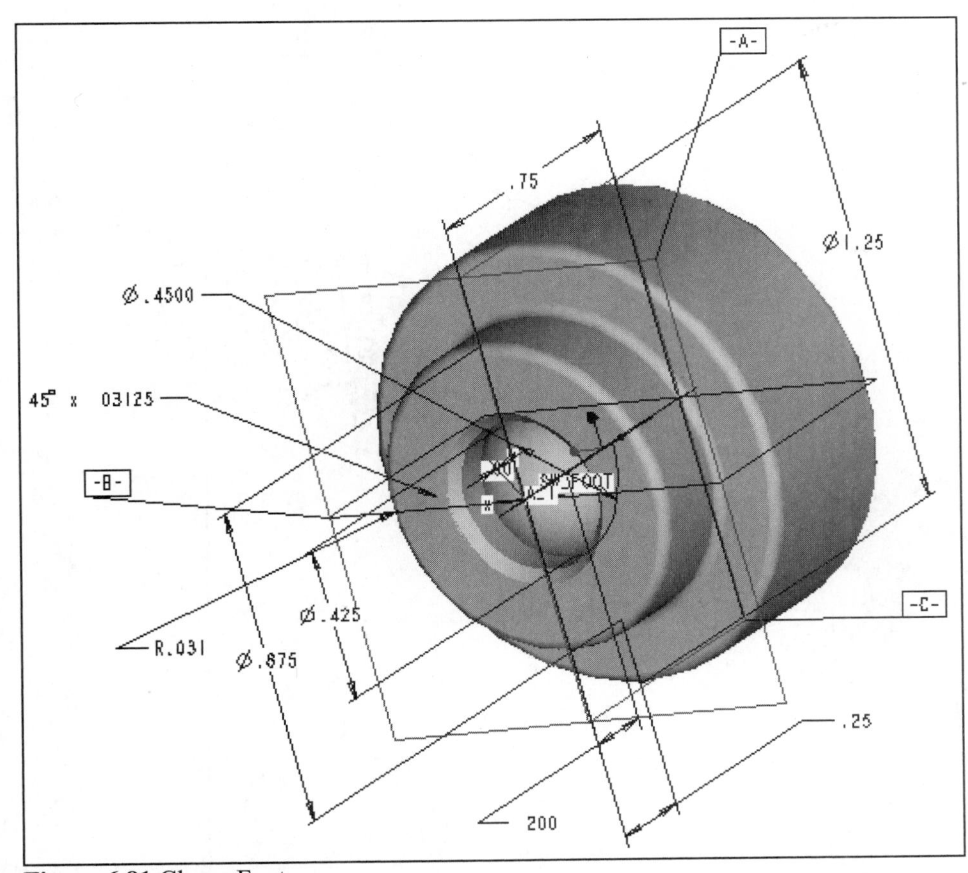

Figure 6.81 Clamp Foot

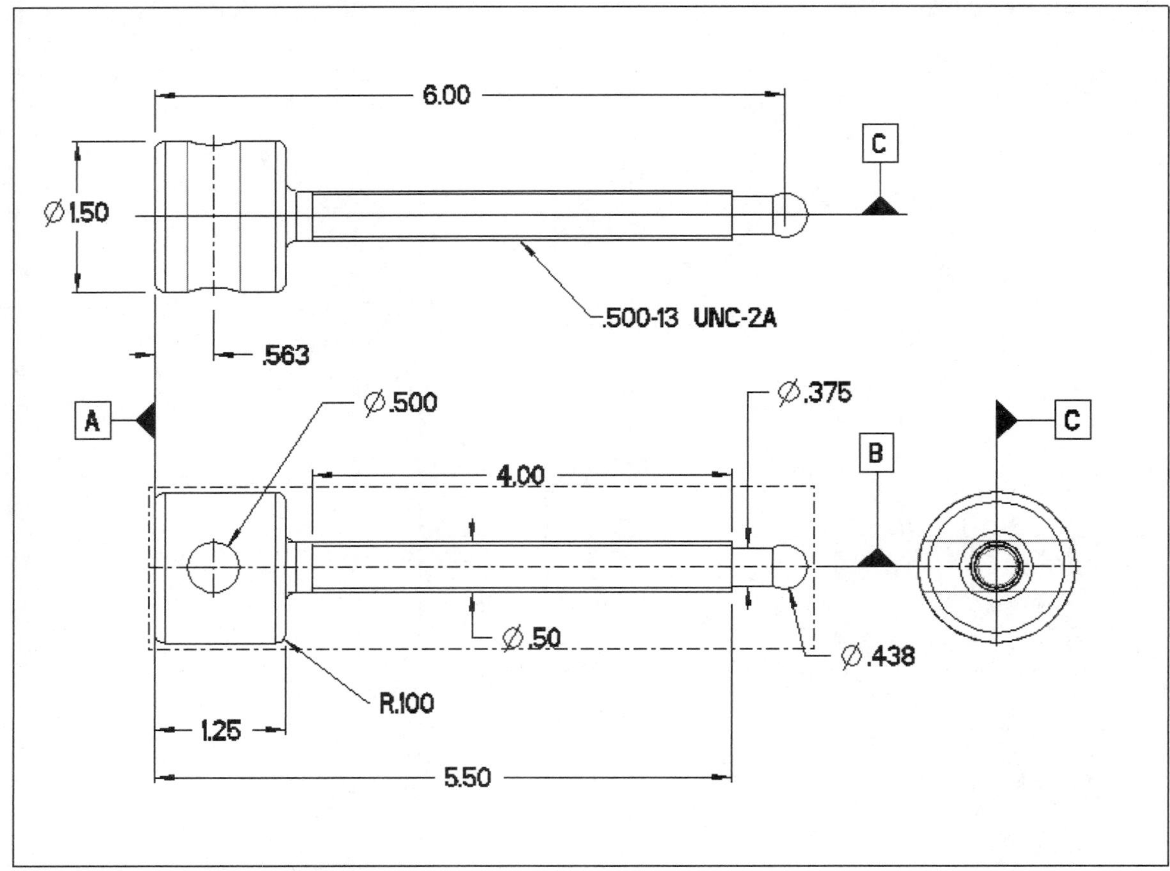

Figure 6.82 Clamp Swivel Dimensions

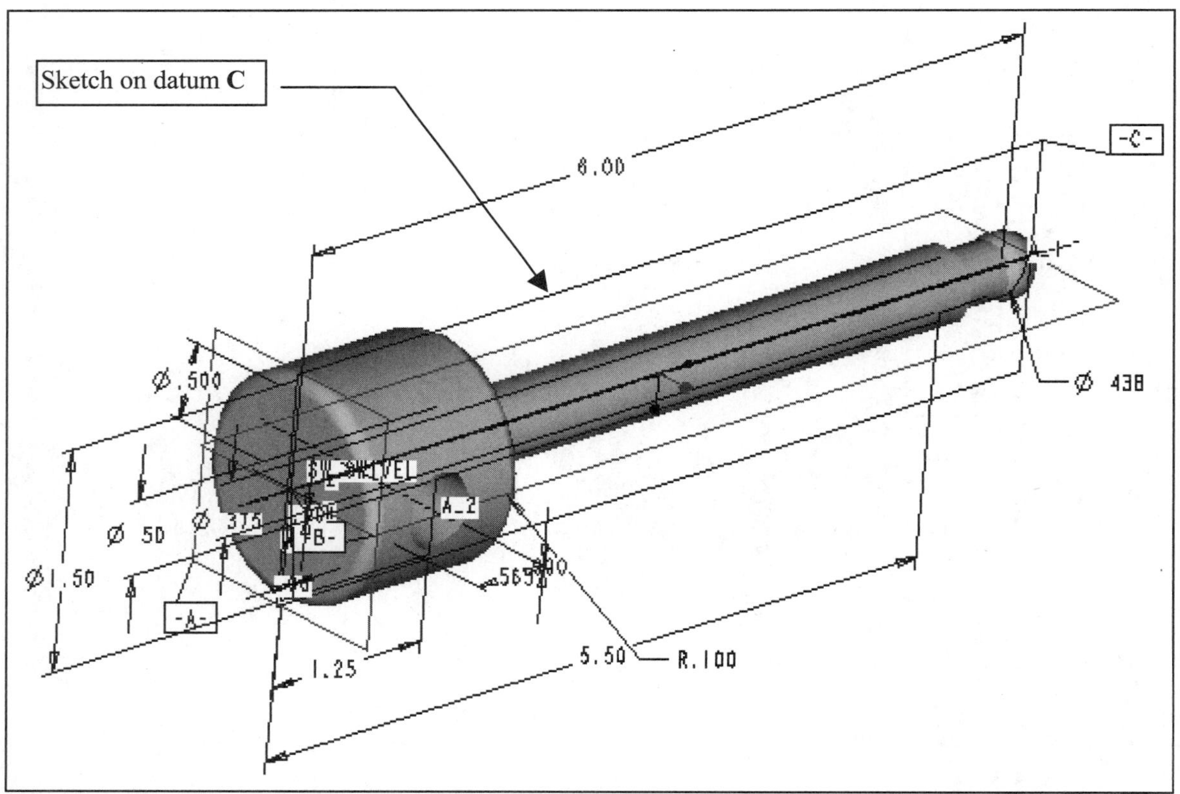

Figure 6.83 Clamp Swivel

Lesson 7 Chamfers and Threads

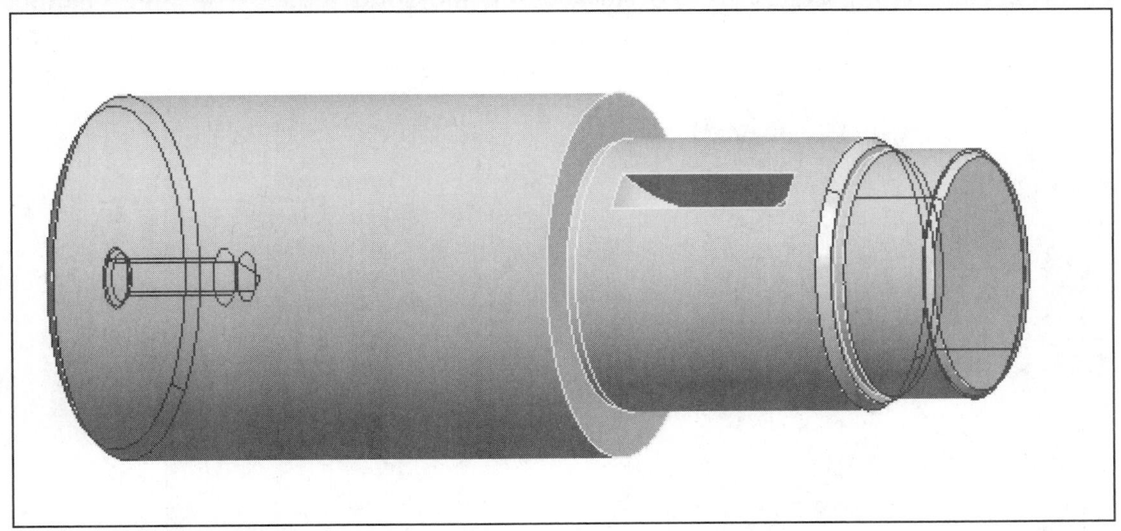

Figure 7.1 Cylinder Rod

OBJECTIVES

- Create **Chamfers** along part edges
- Learn how to **Sketch in 3D**
- Understand and use the **Navigation browser**
- Create standard **Tapped Holes**
- Create **Cosmetic Threads**
- Complete **tabular information** for threads

CHAMFERS AND THREADS

A variety of geometric shapes and constructions are accomplished using parametric modeling. For instance, **chamfers** are created at selected edges of the part (Figs. 7.1). Chamfers are *pick-and-place* features [Fig. 7.2(a-b)]. **Threads** can be a *cosmetic feature* [Fig. 7.3(a-b)] representing the *nominal diameter* or the *root diameter* of the thread. Information can be embedded in the feature. Threads show as a unique color *(magenta)*. By putting cosmetic threads on a separate layer, you can display, blank, or suppress them as required.

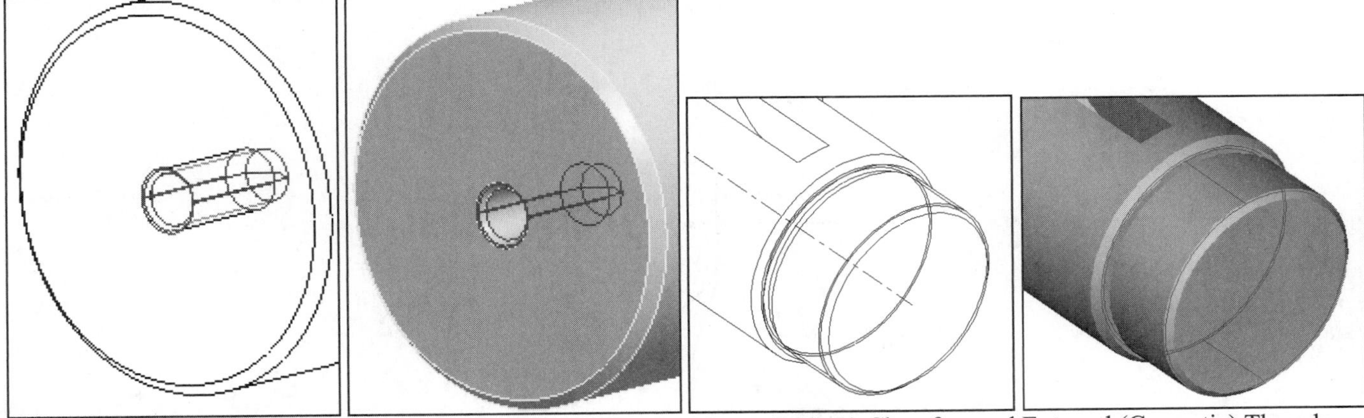

Figure 7.2(a-b) Chamfered Standard Tapped Hole **Figure 7.3(a-b)** Chamfers and External (Cosmetic) Threads

Chamfers

Chamfers are created between abutting edges of two surfaces on the solid model. An edge chamfer removes a flat section of material from a selected edge to create a beveled surface between the two original surfaces common to that edge (Fig. 7.4). Multiple edges can be selected.

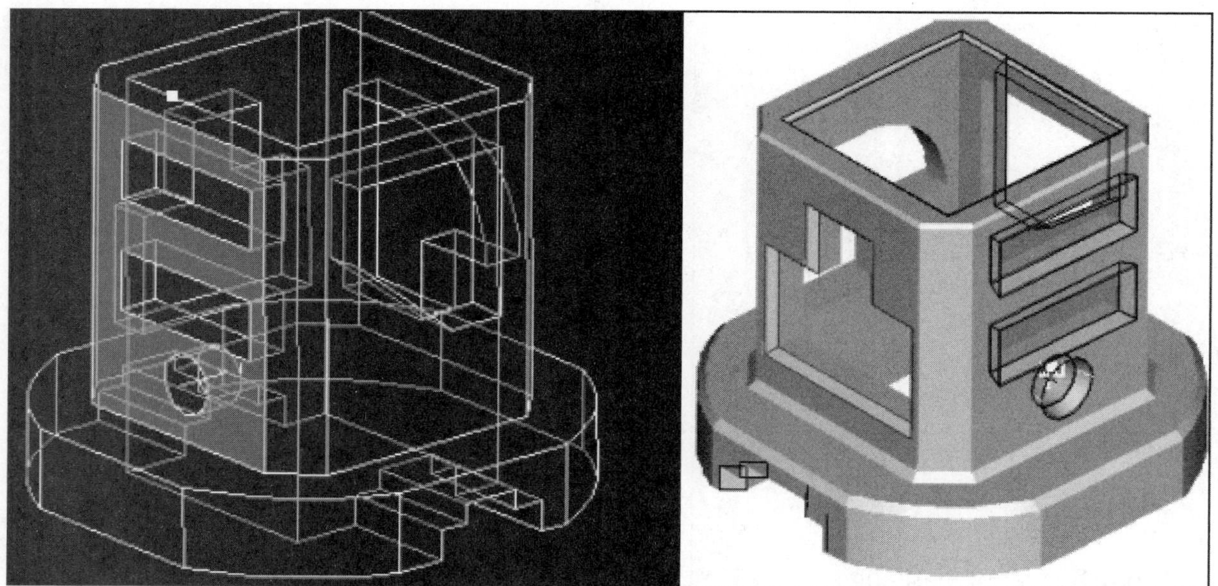

Figure 7.4 Chamfers (CADTRAIN, COAch for Pro/ENGINEER)

There are four basic dimensioning schemes for edge chamfers (Fig. 7.5):

- **45 x d** Creates a chamfer that is at an angle of **45°** to both surfaces and a distance **d** from the edge along each surface. The distance is the only dimension to appear when edited. **45 x d** chamfers can be created only on an edge formed by the intersection of two *perpendicular* surfaces.
- **d x d** Creates a chamfer that is a distance **d** from the edge along each surface. The distance is the only dimension to appear when edited.
- **d1 x d2** Creates a chamfer at a distance **d1** from the selected edge along one surface and a distance **d2** from the selected edge along the other surface. Both distances appear along their respective surfaces.
- **Ang x d** Creates a chamfer at a distance **d** from the selected edge along one adjacent surface at an **Angle** to that surface.

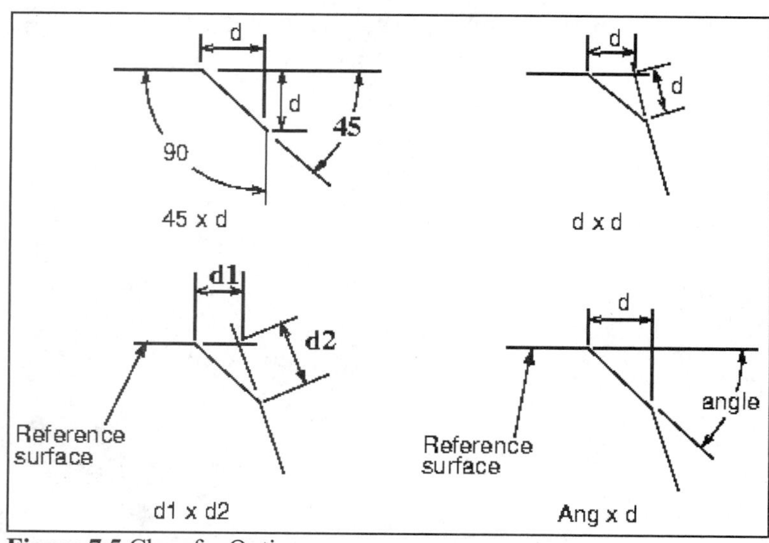

Figure 7.5 Chamfer Options

Threads

Cosmetic threads are displayed with *magenta* lines and circles. Cosmetic threads can be external or internal, blind or through. In the cylinder rod part, one end has external blind threads and the opposite end has internal blind threads. A thread has a set of supported parameters that can be defined at its creation or later, when the thread is added.

Standard Holes

Standard holes are a combination of sketched and extruded geometry. It is based on industry-standard fastener tables. You can calculate either the tapped or clearance diameter appropriate to the selected fastener. You can use Pro/E supplied standard lookup tables for these diameters or create your own. Besides threads, standard holes can be created with chamfers (Fig. 7.6).

Cosmetic Threads

A **cosmetic thread** is a feature that "represents" the diameter of a thread (Fig. 7.6) without having to show the actual threaded surfaces. Since a threaded feature is memory intensive, using cosmetic threads can save an enormous amount of visual memory on your computer. It is displayed in a unique default color. Unlike other cosmetic features, you cannot modify the line style of a cosmetic thread, nor are threads affected by hidden line display settings in the ENVIRONMENT menu.

Threads are created with the default tolerance setting of limits. Cosmetic threads can be external or internal, and blind or through. You create cosmetic threads by specifying the minor or major diameter (for external and internal threads, respectively), starting surface, and thread length or ending edge.

For a starting surface, you can select a quilt surface, regular Pro/E surface, or split surface (such as a surface that belongs to a revolved feature, chamfer, round, or swept feature). For an "up to" surface, you can select any solid surface or a datum plane. A cosmetic thread that uses a depth parameter (a blind thread) cannot be defined from a non-planar surface.

The following table lists the parameters that can be defined for a cosmetic thread at its creation or later when the cosmetic thread is added. In this table, "pitch" is the distance between two threads.

PARAMETER NAME	PARAMETER VALUE	PARAMETER DESCRIPTION
MAJOR_DIAMETER	Number	Thread major diameter
THREADS_PER_INCH	Number	Threads per inch (1/pitch)
THREAD FORM	String	Thread form
CLASS	Number	Thread class
PLACEMENT	Character	Thread placement (A-external, B-internal)
METRIC	TRUE/FALSE	Thread is metric

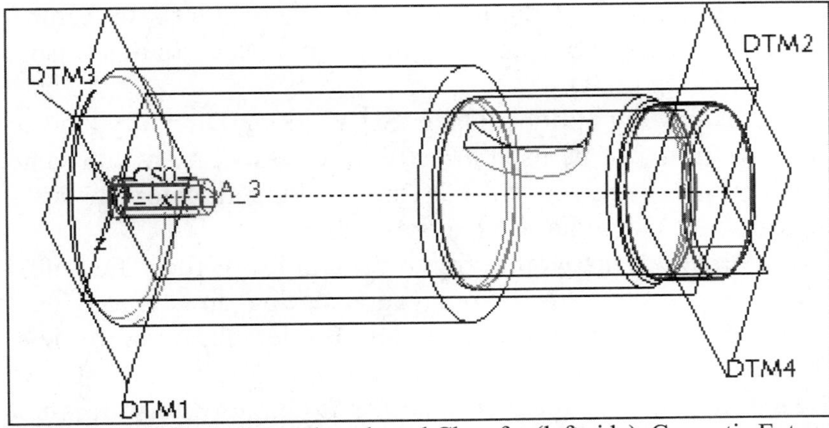

Figure 7.6 Standard Hole, Threads and Chamfer (left side), Cosmetic External Threads (right side)

Lesson 7 STEPS

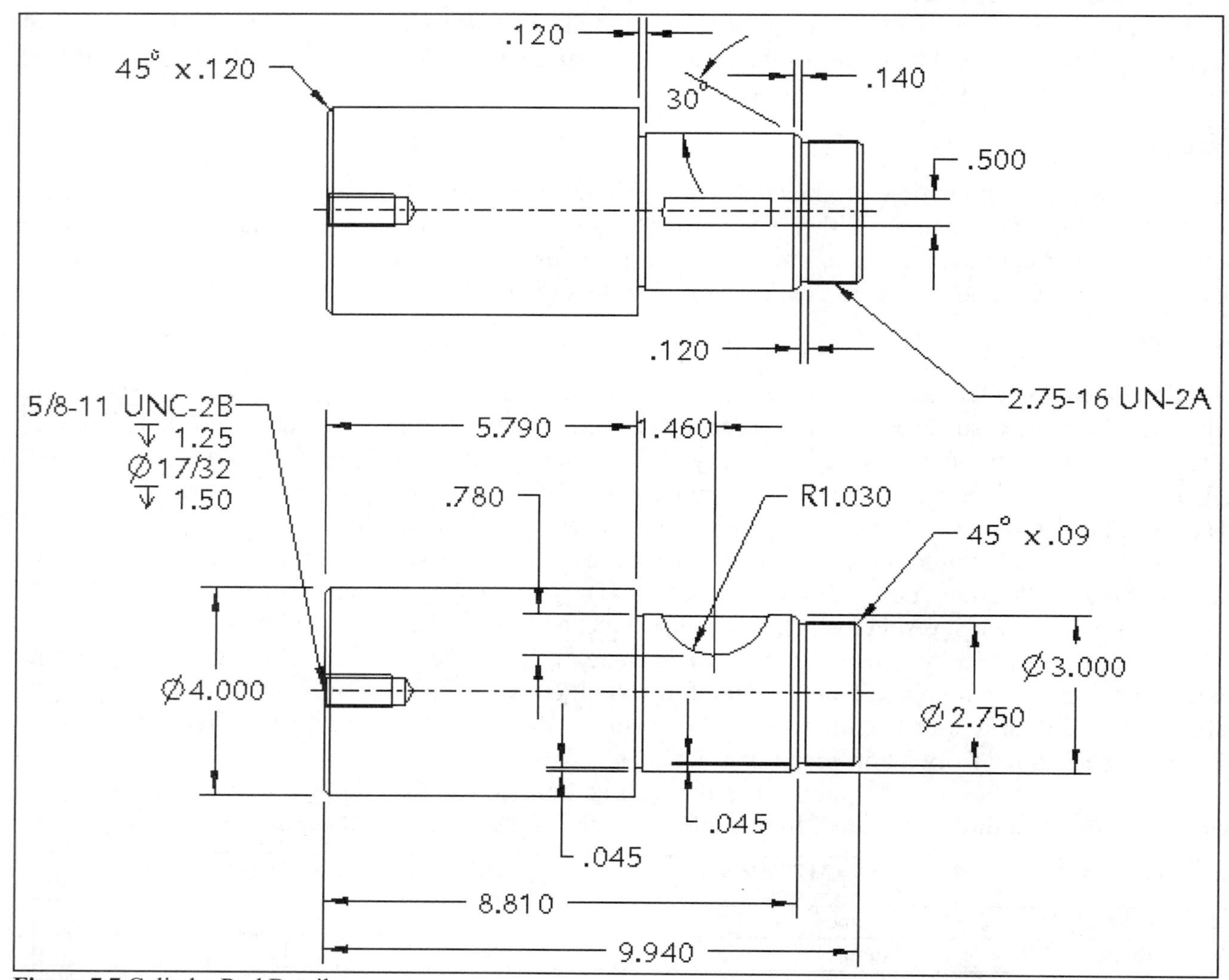

Figure 7.7 Cylinder Rod Detail

Cylinder Rod

The Cylinder Rod is modeled by creating a revolved protrusion, similar to the pin in the previous lesson. The geometry of the revolved feature is shown in Figure 7.7. After the revolved protrusion (base feature) is created, the necks (reliefs), the chamfers, the key seat, and the tap drill hole are modeled. In this lesson, we will create our first revolved protrusion by sketching in 3D.

Besides using the File command and corresponding options (Open, Set Working Directory, and so forth), the *Navigation area* of the Pro/E main window can be used directly to access many of the same functions. Before you start the part and begin modeling, you will set the working directory using a different method, and explore some of the possibilities using the Navigator browser.

As you know, the working directory is a directory that you set up to contain Pro/E files. You must have read/write access to this directory. You usually start Pro/E from your working directory. A new working directory setting is not saved when you exit the current Pro/E session. By default, if you retrieve a file from a non-working directory, rename the file and then save it, the renamed file is saved to the directory from which it was originally retrieved, if you have read/write access to that directory. It is not saved in the current working directory, unless the config.pro option *save_object_in_current* is set to *yes*.

Navigation Area

As introduced in Lesson 1, the navigation area is located on the left side of the Pro/E main window. It includes tabs for the Model Tree and Layer Tree, Folder Browser, Favorites, and Connections:

- **Model Tree** (default)
- **Layer Tree (Show ⇒ Layer Tree)**
- **Folder Browser**
- **Favorites**
- **Connections**

Folder Browser

The Folder browser (**Folder Browser**) is an expandable tree that lets you browse the file systems and other locations that are accessible from your computer. As you navigate the folder, the contents of the selected folder appear in the Pro/E browser as the Contents page. The Folder browser contains top-level nodes for accessing file systems and other locations that are known to Pro/E:

- **In Session** Pro/E objects that have been retrieved into local memory.
- **Shared Spaces** This is a shared file location accessed through the PTC Conference Center.
- **All registered servers** The browser lists all servers that you have registered with the Server Registry dialog box. These registered servers may include a Windchill server, a Pro/INTRALINK server, and an FTP server.
- **Node for the local file system** When you open the Folder browser, the local file system appears in the browser with the startup directory node expanded and highlighted.
- **Network Neighborhood** *(only for Windows)* The navigator shows computers on the networks to which you have access. The operations you can perform depend on your permissions to the remote computers.

Manipulating Folders

To work with folders, you can use the Folder browser toolbar or the shortcut menu. You can perform the following tasks with the toolbar:

- **Create a new folder**
- **Delete selected folders**
- **Open the Windchill Cabinets** This icon is available only when you are working with an active Windchill server.
- **Open the Windchill Workspace** This icon is available only when you are working with an active Windchill server.
- **Working directory**

Using the Shortcut Menu in the Folder browser

To open a shortcut menu, **RMB** click on an item in the Folder browser. The commands on the shortcut menu vary depending on the task and your permissions. The shortcut menu lists the following commands:

- **New Folder** Add a subfolder to the selected folder.
- **Open** Open a folder in the Pro/ENGINEER browser.
- **Expand** Expand a node (if not open). **Collapse** a node (if open).
- **Server Registry** Access the Server Registry dialog box.
- **Make Working Directory** Designate the selected directory as the working directory.
- **Rename** Rename a selected folder.
- **Delete** Delete a selected folder and all subfolders.

Click: **Folder Browser** in the Navigator (Fig. 7.8) ⇒ click on the directory you wish to set as the working directory ⇒ **RMB** ⇒ **Make Working Directory** (Fig. 7.9)

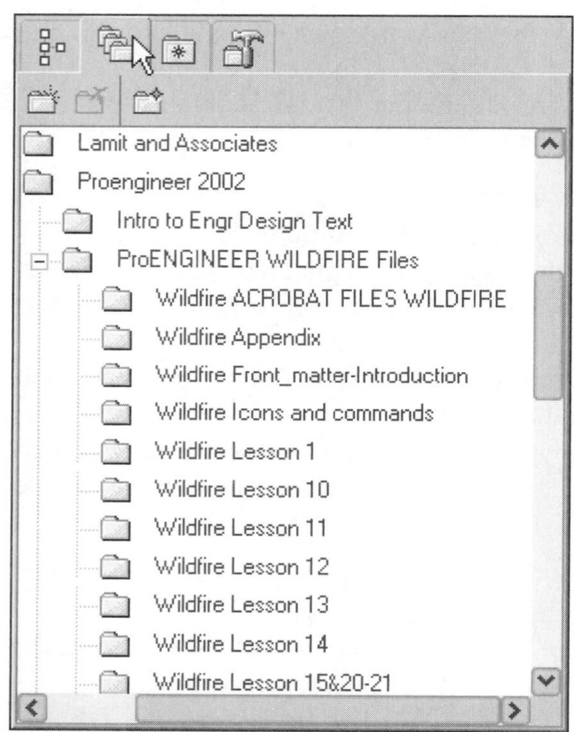

Figure 7.8 Folder Browser

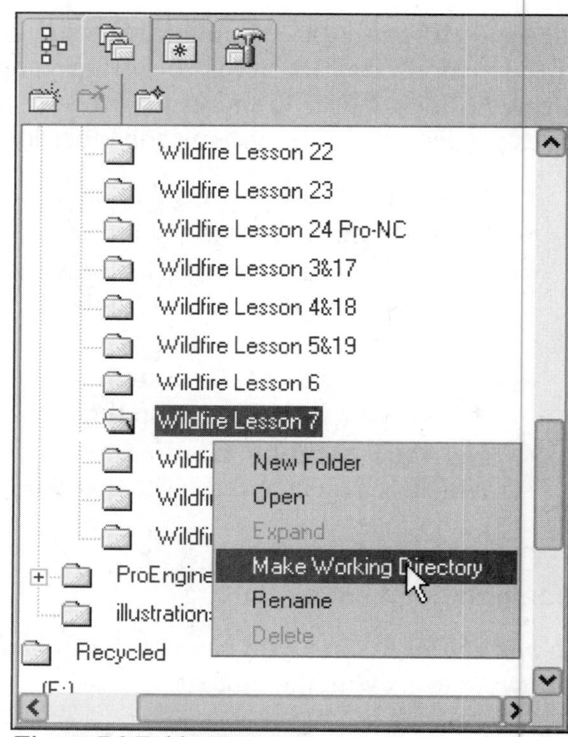

Figure 7.9 Folder Browser

Click: **Tools** ⇒ **Options** ⇒ Showing: **Current Session** ⇒ Show only options loaded from file ⇒ Slide the bar down to the option or type. Option: ***default_dec_places*** ⇒ Value: **3** ⇒ **Enter** ⇒ Slide the bar down to the option or type. Option: ***sketcher_dec_places*** ⇒ Value: **3** ⇒ **Enter** ⇒ Option: ***sketcher_starts_in_2d*** ⇒ Value: **no** ⇒ **Enter** ⇒ Option: (type) ***def_layer*** ⇒ Value: ⇒ **layer_axis datum_axes** (Fig. 7.10) ⇒ **Add/Change** ⇒ **Apply** ⇒ ☑ Show only options loaded from file (Fig. 7.11) ⇒ **Save a copy of the currently displayed configuration file** (Fig. 7.12) ⇒ Name **cylinder_rod.pro** (Fig. 7.13) ⇒ **Ok** ⇒ **Close** ⇒ **Create a new object** ⇒ ●**Part** ⇒ Name **CYLINDER_ROD** ⇒ ☑ Use default template ⇒ **OK** ⇒ **Edit** ⇒ **Setup** ⇒ **Units** ⇒ Units Manager **Inch lbm Second** ⇒ **Close** ⇒ **MMB**

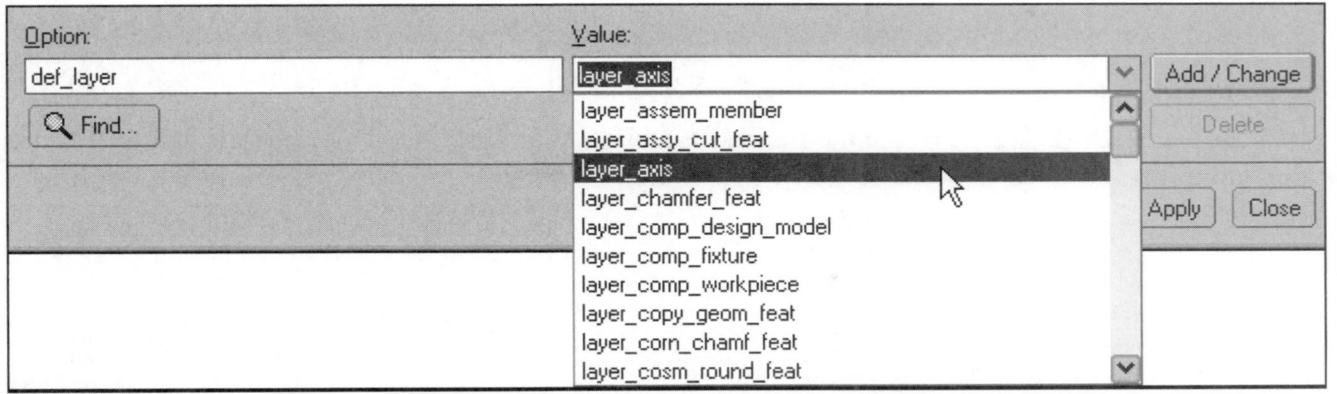

Figure 7.10 Setting *def_layer* with Value of *layer_axis datum_axes* (datum_axes is the name of the new layer)

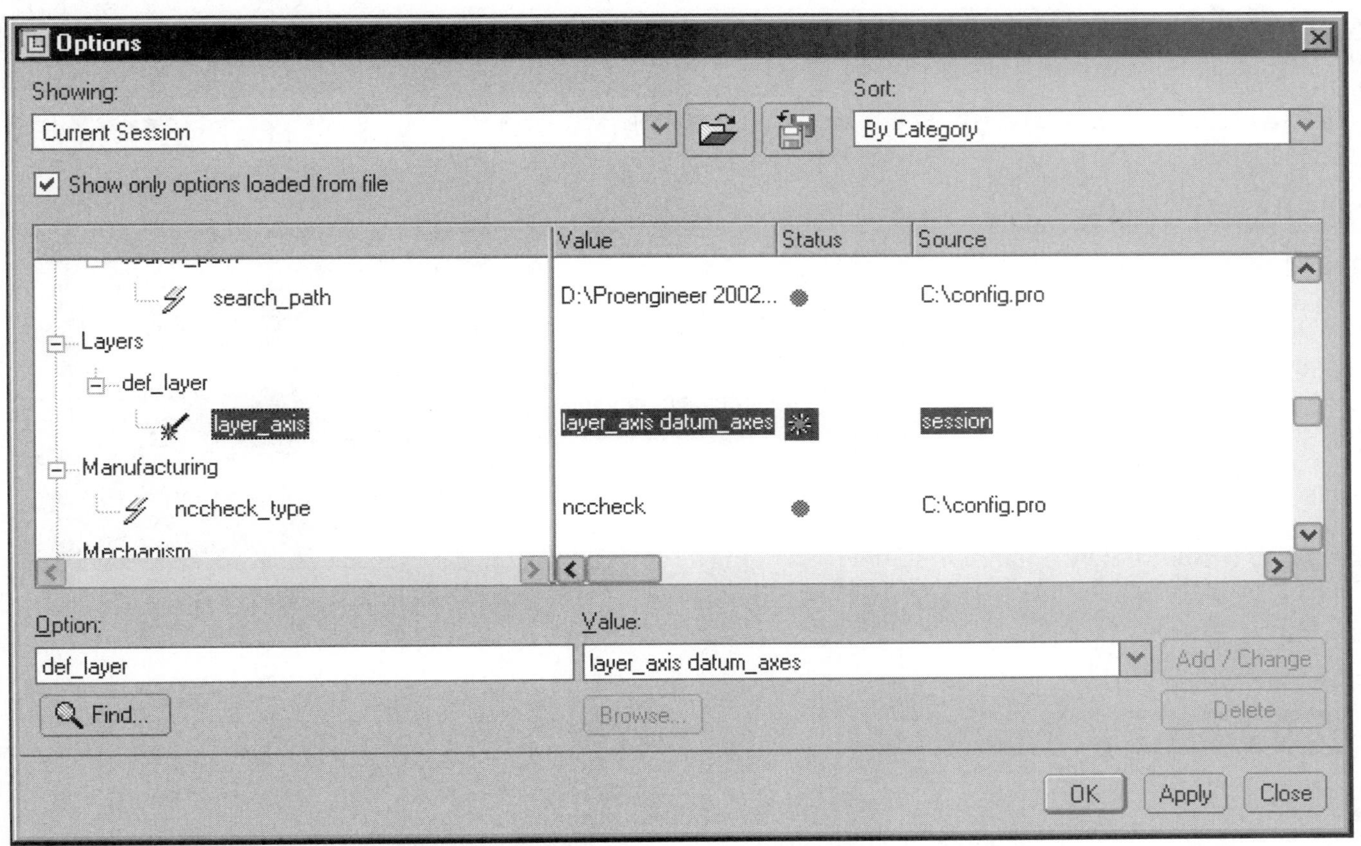

Figure 7.11 Showing only Options Loaded From File

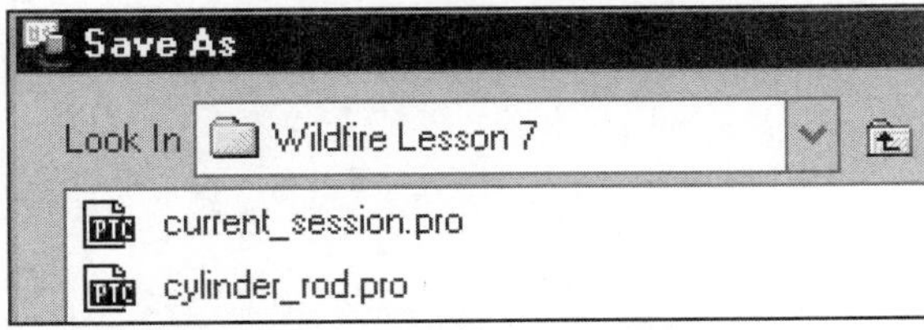

Figure 7.12 Config.pro Files

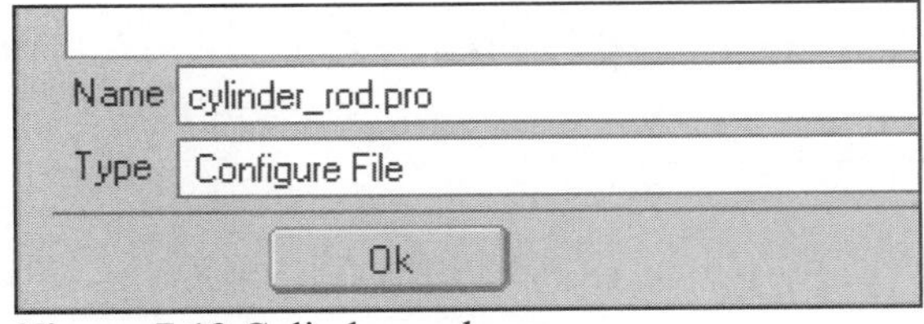

Figure 7.13 Cylinder_rod.pro

Open the **Browser** by clicking on the quick sash ⇒ Address: type **www.matweb.com** ⇒ **Enter** ⇒ type **aluminum 2024** in the upper right-hand corner box (Fig. 7.14) ⇒ click **SEARCH** aluminum 2024 SEARCH ⇒ pick 1 Aluminum 2024-O ⇒ use the material specifications (Fig. 7.15) to complete as much of the material table (Fig. 7.16) ⇒ **Edit** ⇒ **Setup** ⇒ **Material** ⇒ **Define** ⇒ type **ALUMINUM-2024** ⇒ **Enter** ⇒ **File** ⇒ **Save** ⇒ **File** ⇒ **Exit** ⇒ **Assign** ⇒ pick **ALUMINUM-2024** ⇒ **Accept** ⇒ **MMB**

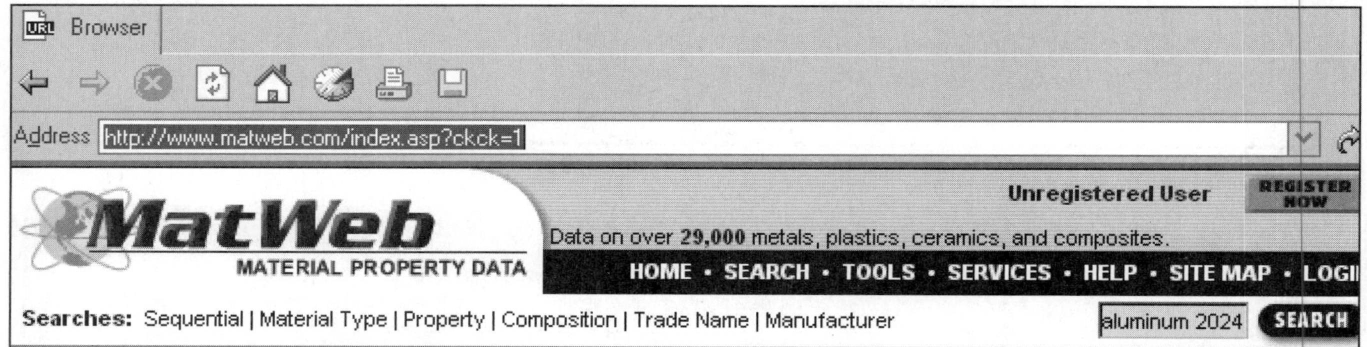

MatWeb, Your Source for Materials Information

What is MatWeb?

MatWeb's database of material properties includes thermoplastic and thermoset polymers such as ABS, nylon, polycarbonate, polyester, and polyolefins; metals such as aluminum, cobalt, copper, lead, magnesium, nickel, steel, superalloys, titanium and zinc alloys; ceramics; plus a growing list of semiconductors, fibers, and other engineering materials.

MatWeb is still freely available and does not require registration. You can still access all of the features previously available on the old site. However, advanced features are only available to our Registered and Premium users.

New Look, Better Features:

- MatWeb's New, More Powerful Search Pages
- Side-By-Side Material Comparisons
- New Data Export Tools
- More Utilities, Including Calculators, Converters, and Definitions

- Free -
  - Everything Previously Available on MatWeb.com!
- Registered -
  - Free Limited Access to Premium Features!

Figure 7.14 www.matweb.com

Aluminum 2024-O

Printer friendly version

Subcategory: 2000 Series Aluminum Alloy; Aluminum Alloy; Metal; Nonferrous Metal

Key Words: Aluminium 2024-O; UNS A92024; ISO AlCu4Mg1; NF A-U4G1 (France); DIN AlCuMg2; AA2024-O, ASME SB211; CSA CG42 (Canada)

Component	Wt. %	Component	Wt. %	Component	Wt. %
Al	93.5	Fe	Max 0.5	Si	Max 0.5
Cr	Max 0.1	Mg	1.2 - 1.8	Ti	Max 0.15
Cu	3.8 - 4.9	Mn	0.3 - 0.9	Zn	Max 0.25

Material Notes:
General 2024 characteristics and uses (from Alcoa): Good machinability and surface finish capabilities. A high strength material of adequate workability. Has largely superceded 2017 for structural applications. Use of 2024-O not recommended unless subsequently heat treated.

Uses: Aircraft fittings, gears and shafts, bolts, clock parts, computer parts, couplings, fuse parts, hydraulic valve bodies, missile parts, munitions, nuts, pistons, rectifier parts, worm gears, fastening devices, veterinary and orthopedic equipment, structures.

Physical Properties	Metric	English	Commen
Density	2.78 g/cc	0.1 lb/in³	
Mechanical Properties			
Hardness, Brinell	47	47	500 kg load with 10 mm b
Tensile Strength, Ultimate	185 MPa	26800 psi	
Tensile Strength, Yield	75 MPa	10900 psi	
Elongation @ break	20 %	20 %	In 5 cm; Sample 1.6 mm thic
Modulus of Elasticity	72.4 GPa	10500 ksi	Average of Tension and Compression. Aluminum alloys, the compressive modulus typically 2% greater than the tensile modulu
Ultimate Bearing Strength	345 MPa	50000 psi	Edge distance/pin diameter = 2
Bearing Yield Strength	131 MPa	19000 psi	Edge distance/pin diameter = 2
Poisson's Ratio	0.33	0.33	
Fatigue Strength	95 MPa	13800 psi	500,000,000 Cycle
Machinability	30 %	30 %	0-100 Scale of Aluminum Alloy
Shear Modulus	28 GPa	4060 ksi	
Shear Strength	125 MPa	18100 psi	

Electrical Properties

Figure 7.15 Aluminum 2024-O Properties

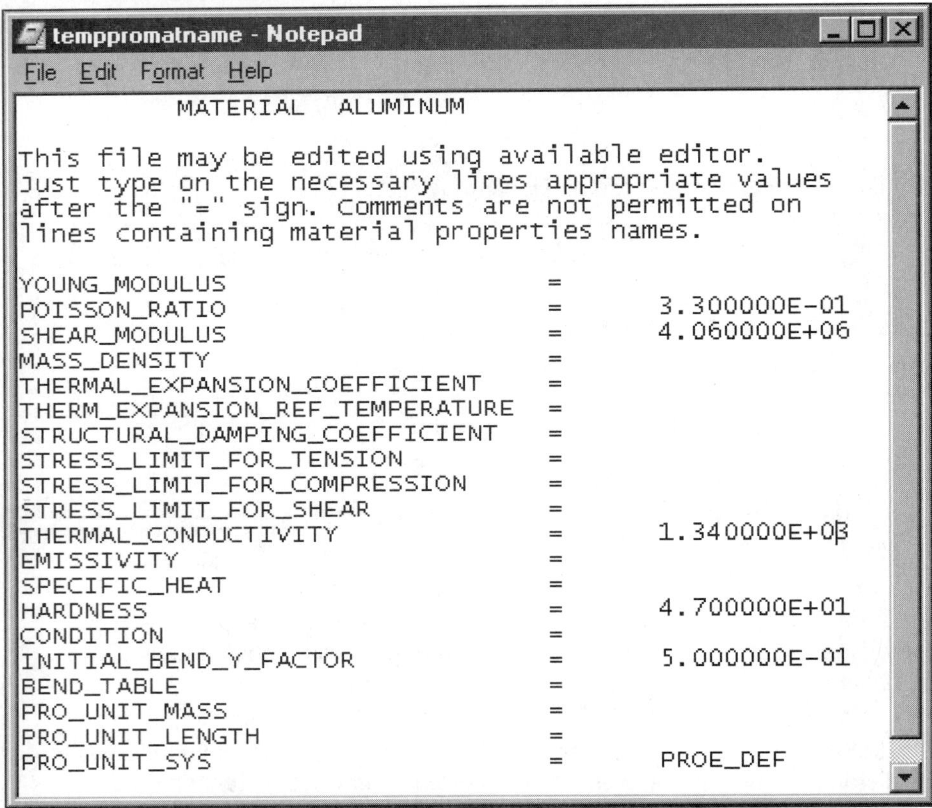

```
temppromatname - Notepad
File  Edit  Format  Help
           MATERIAL   ALUMINUM

This file may be edited using available editor.
Just type on the necessary lines appropriate values
after the "=" sign. Comments are not permitted on
lines containing material properties names.

YOUNG_MODULUS                          =
POISSON_RATIO                          =       3.300000E-01
SHEAR_MODULUS                          =       4.060000E+06
MASS_DENSITY                           =
THERMAL_EXPANSION_COEFFICIENT          =
THERM_EXPANSION_REF_TEMPERATURE        =
STRUCTURAL_DAMPING_COEFFICIENT         =
STRESS_LIMIT_FOR_TENSION               =
STRESS_LIMIT_FOR_COMPRESSION           =
STRESS_LIMIT_FOR_SHEAR                 =
THERMAL_CONDUCTIVITY                   =       1.340000E+03
EMISSIVITY                             =
SPECIFIC_HEAT                          =
HARDNESS                               =       4.700000E+01
CONDITION                              =
INITIAL_BEND_Y_FACTOR                  =       5.000000E-01
BEND_TABLE                             =
PRO_UNIT_MASS                          =
PRO_UNIT_LENGTH                        =
PRO_UNIT_SYS                           =       PROE_DEF
```

Figure 7.16 Sample Material File

Click: ⇒ **MMB** or ⇒ **Info** ⇒ **Model** (Fig. 7.17) ⇒ click the quick sash to close **Browser**

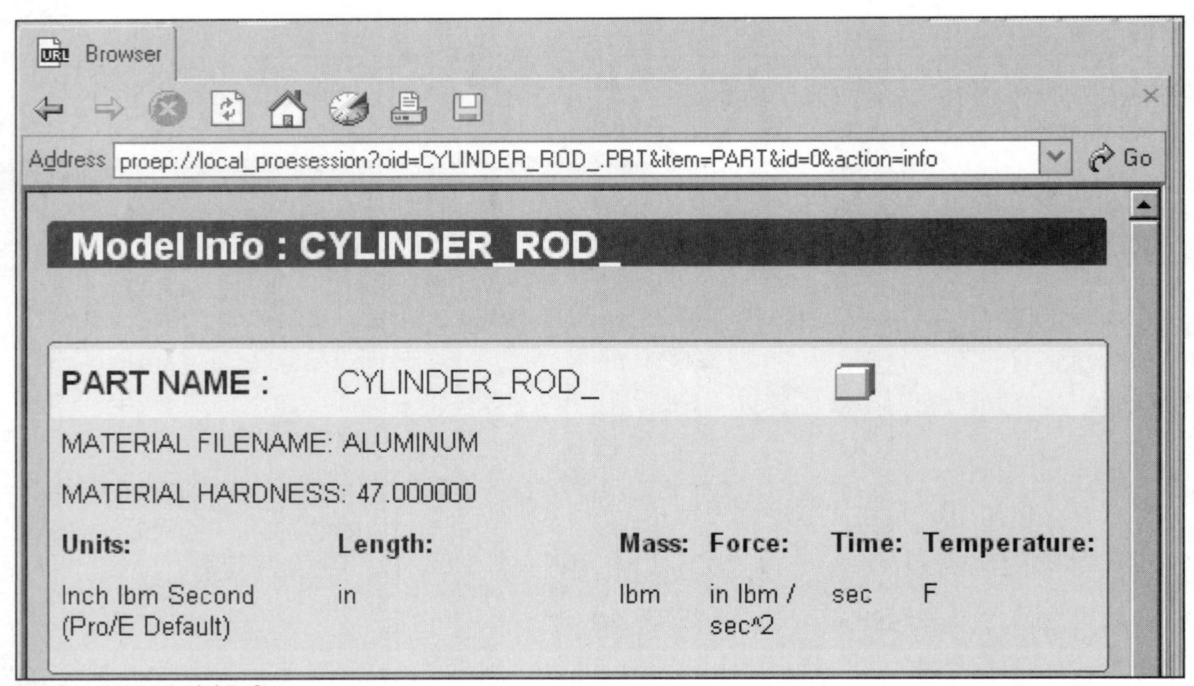

Figure 7.17 Model Info

The first protrusion is a revolved protrusion created with the Revolve Tool, you will be sketching the section in 3D since the configuration option; *sketcher_starts_in_2d,* was previously set to *no.* In the Environment dialog box, you can see that ☐ Use 2D Sketcher is deactivated (Fig. 7.18).

Click: **Tools** ⇒ **Environment** ⇒ ☑ Snap to Grid (Fig. 7.18) ⇒ **OK** ⇒ **Revolve Tool** ⇒ Section dialog box opens ⇒ Sketch Plane--- Plane: select **FRONT** datum ⇒ click **Sketch** button to accept the default orientation and start the sketcher (Fig. 7.19) ⇒ **MMB** ⇒ **MMB** (to accept References) ⇒ **Toggle the grid on** ⇒ sketch a horizontal centerline through the default coordinate system to be used as the axis of revolution ⇒ sketch the eight lines ⇒ **MMB** (Fig. 7.20)

Figure 7.18 Activate Snap to Grid. Use 2D Sketcher is Deactivated. Keep Standard Orient as Trimetric.

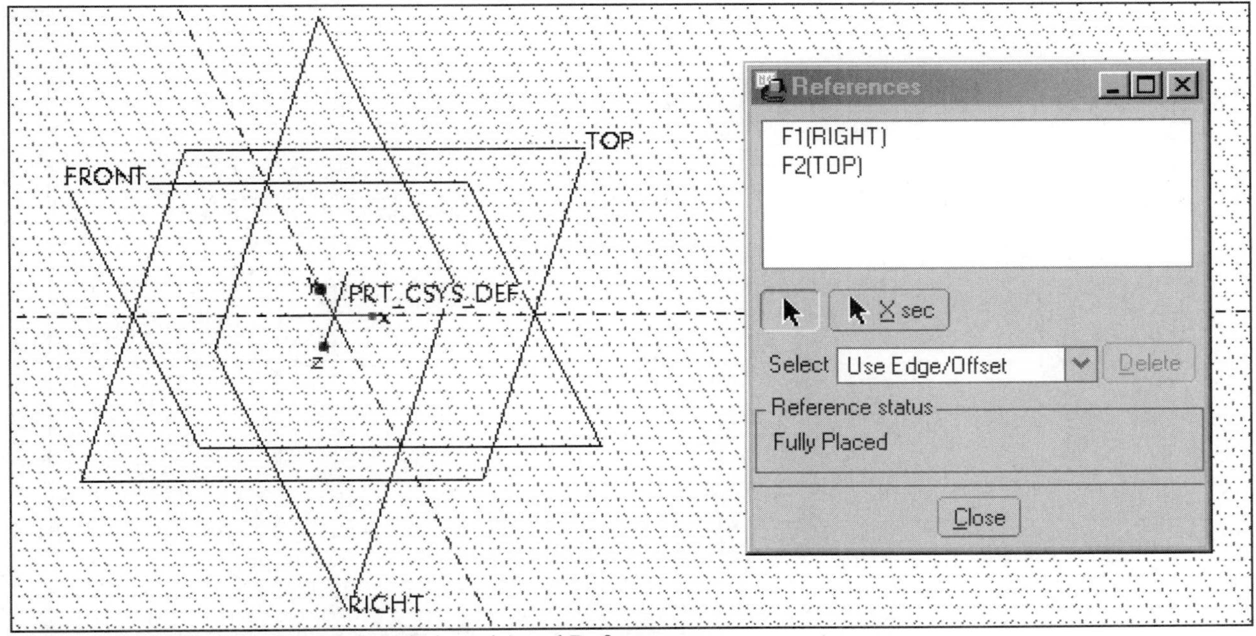

Figure 7.19 3D Default Orientation (Trimetric) and References

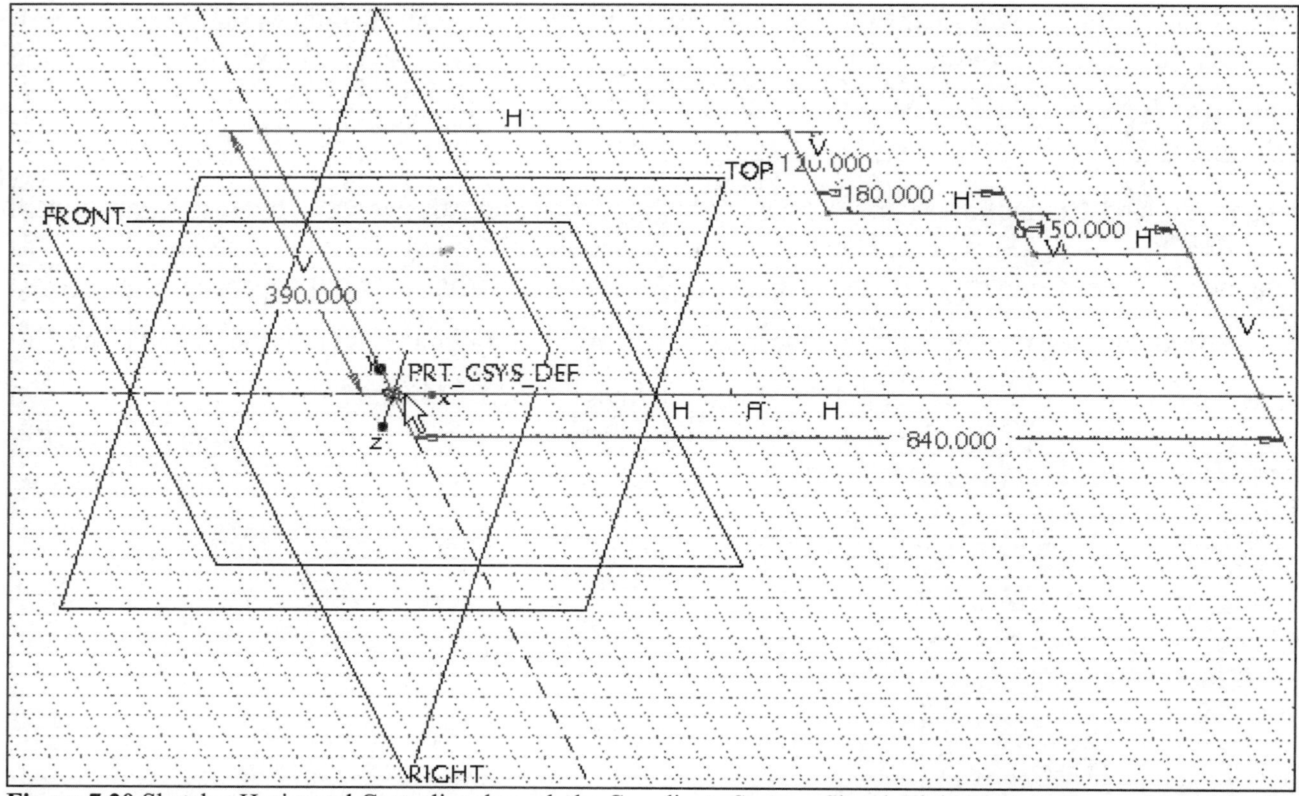

Figure 7.20 Sketch a Horizontal Centerline through the Coordinate System. Sketch Eight Lines to Form a Closed Section

Click: **Sketch** ⇒ **Options** ⇒ Display tab ⇒ Snap to Grid ⇒ ✔ ⇒ **Toggle the grid off** ⇒ add the three strong diameter dimensions (Fig. 7.21) ⇒ Add the two strong horizontal dimensions and move the overall dimension (Fig. 7.22) ⇒ **off** (Fig. 7.23) ⇒ **on**

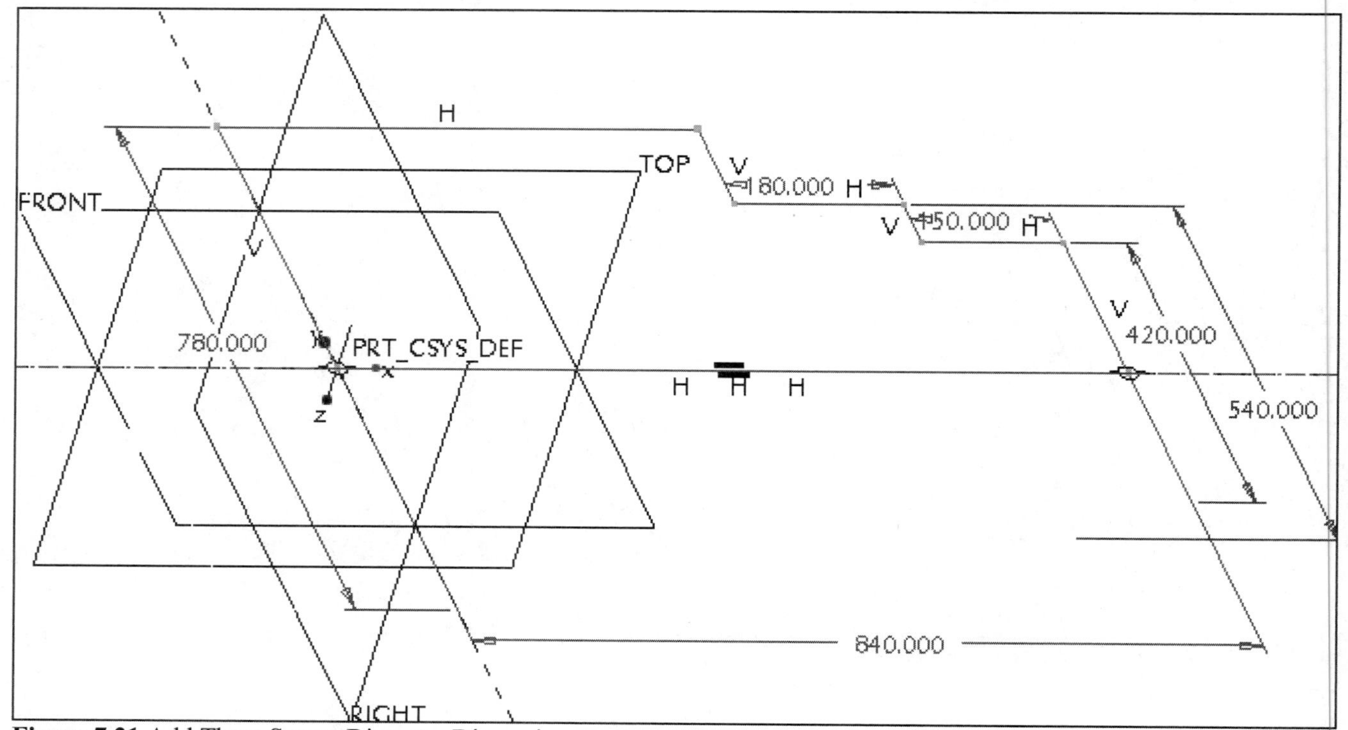

Figure 7.21 Add Three Strong Diameter Dimensions

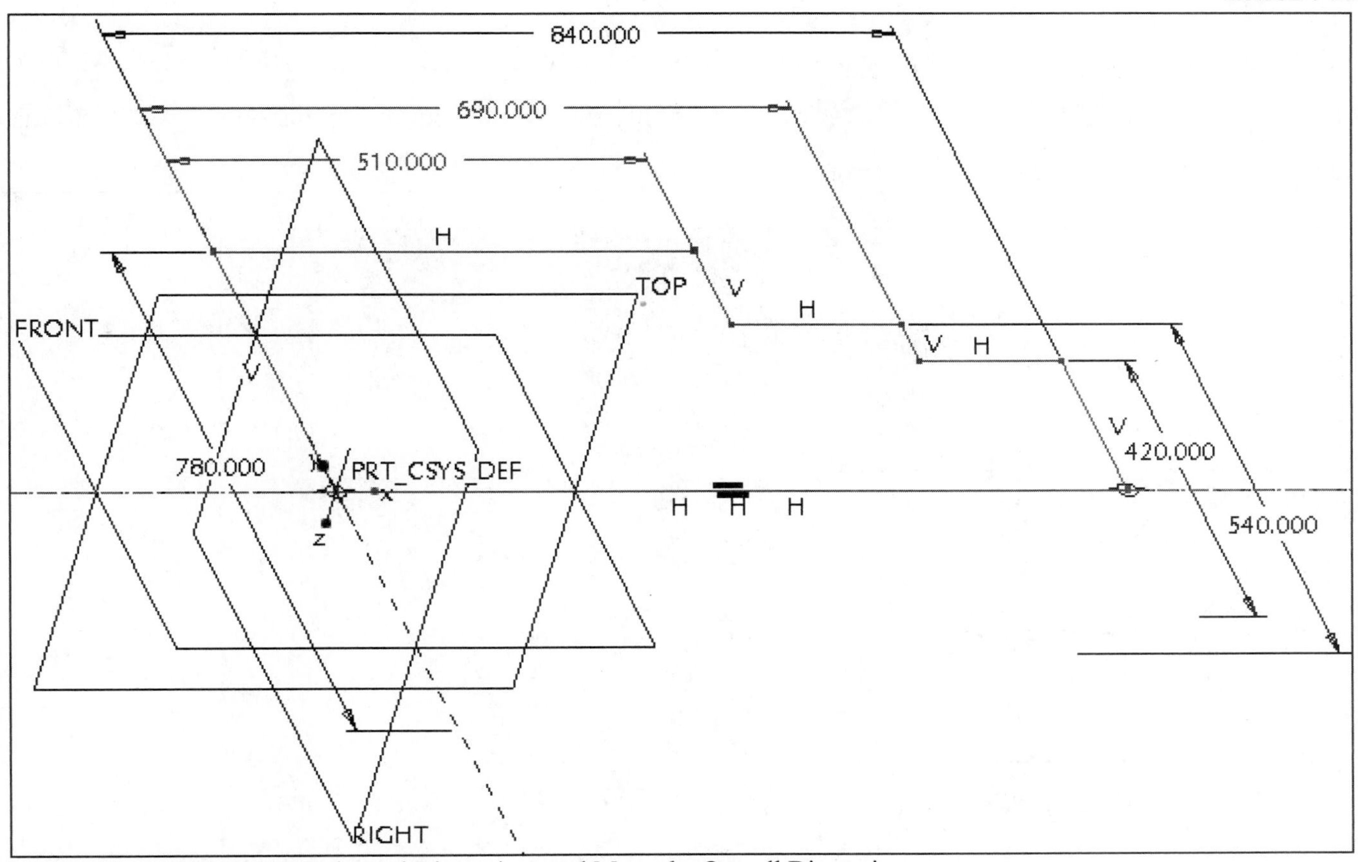

Figure 7.22 Add the Two Horizontal Dimensions and Move the Overall Dimension

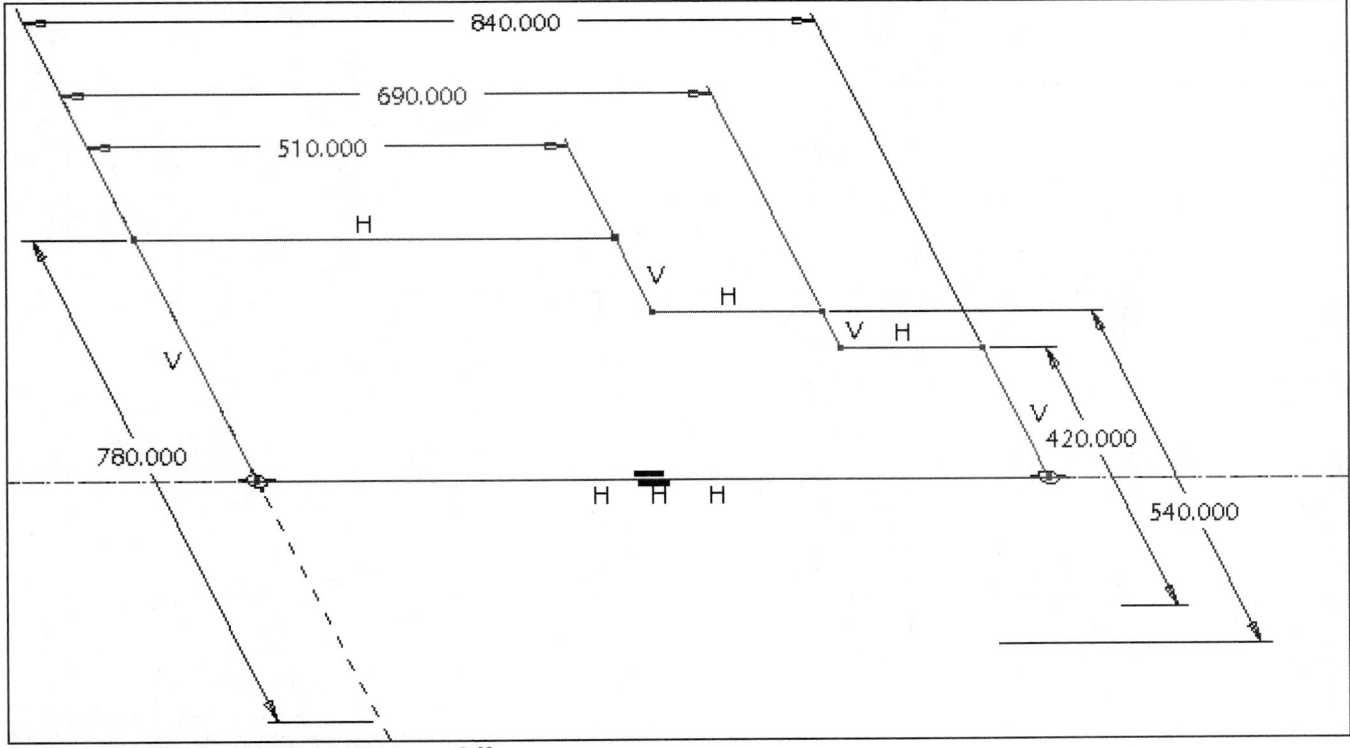

Figure 7.23 Sketch with Datum Features Off

Modify the dimensions, click: [select icon] ⇒ window-in the sketch to capture all six dimensions ⇒ [modify icon] **Modify the values** ⇒ [✓ Lock Scale] ⇒ [✓ Regenerate] ⇒ modify only the overall length dimension (Fig. 7.24) to the design value of **9.94** ⇒ **Enter** ⇒ [✓] (Fig. 7.25)

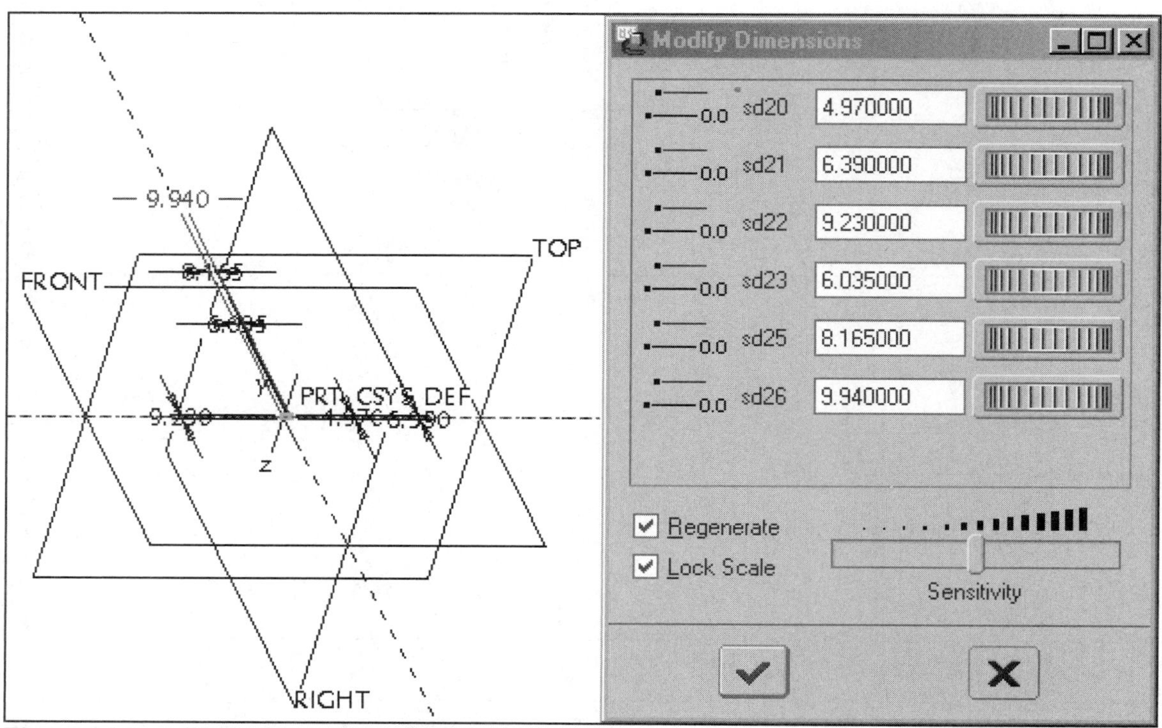

Figure 7.24 Modify the Overall Length to **9.94**

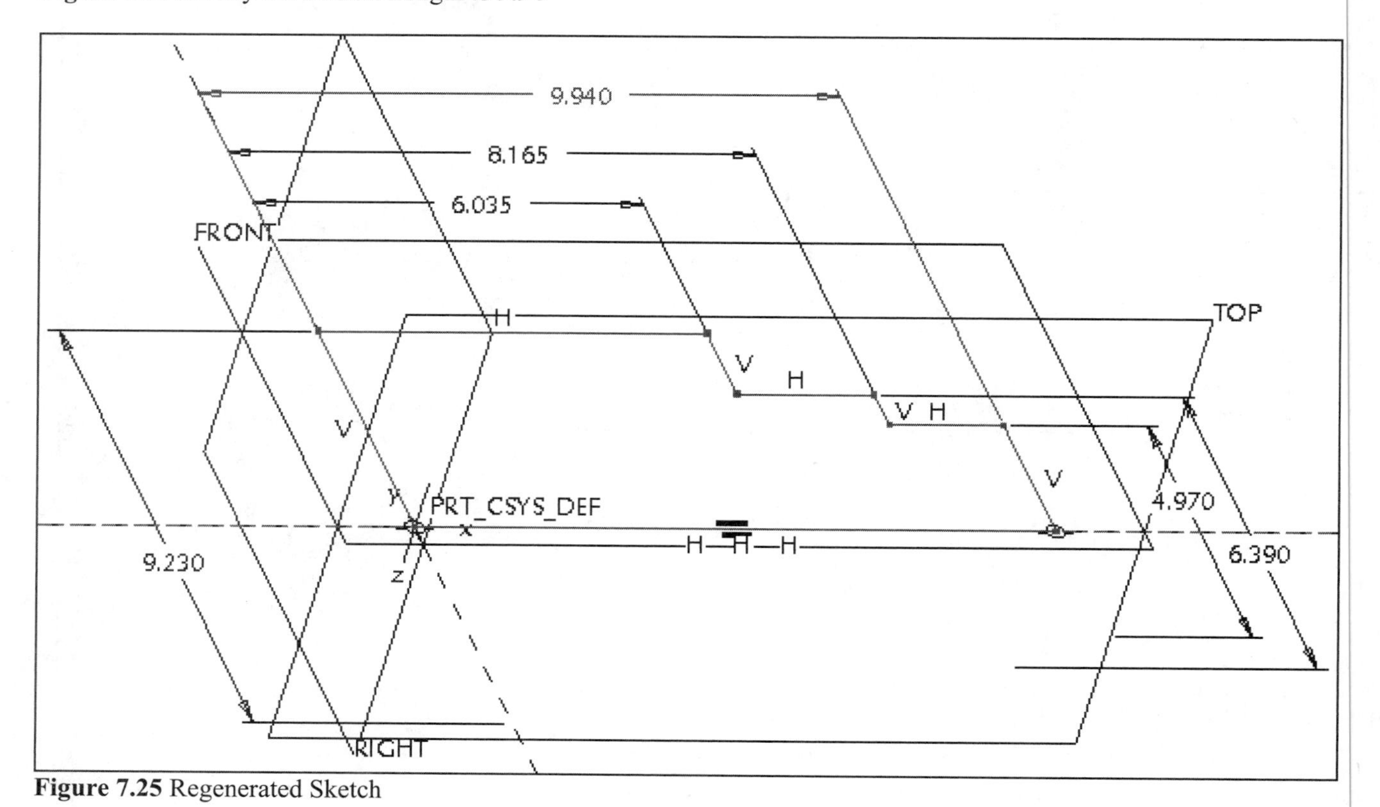

Figure 7.25 Regenerated Sketch

Click: [arrow icon] ⇒ click on a horizontal line, it will highlight in red, click again and hold (Fig. 7.27) ⇒ drag the line to a value closer to the design value ⇒ release the **LMB** ⇒ Repeat until all of the diameter dimensions are somewhat near the design value as provided in Figure 7.26. This process can also be used when the sketch is in 2D. You can also use the thumbwheels in the Modify Dimensions dialog box .

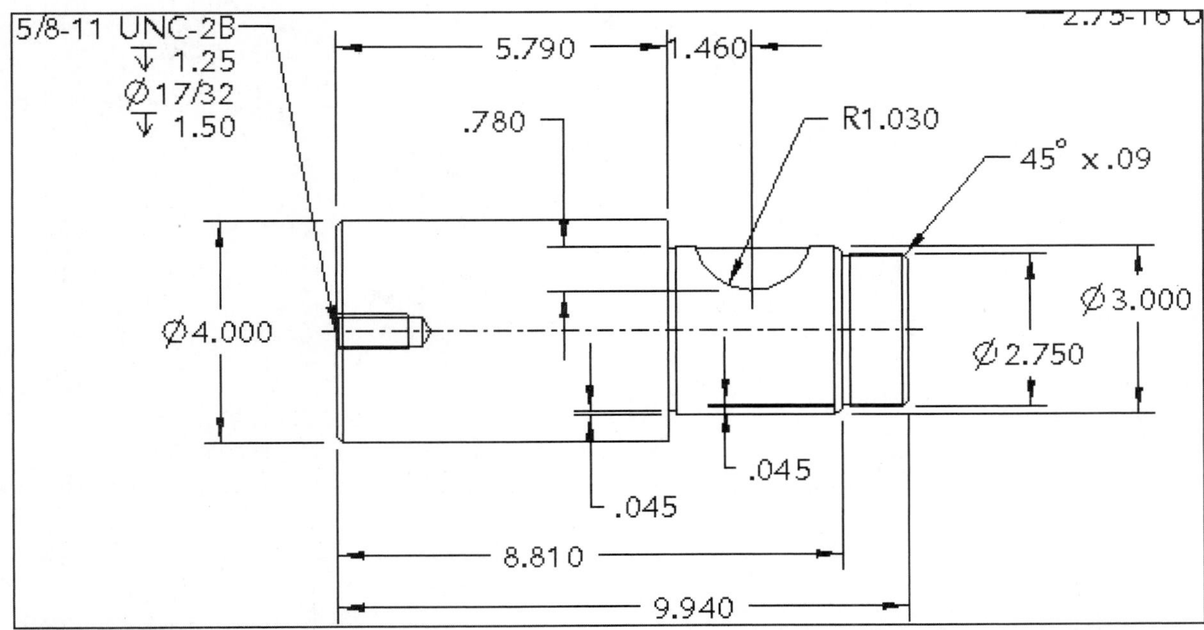

Figure 7.26 Design Dimensions

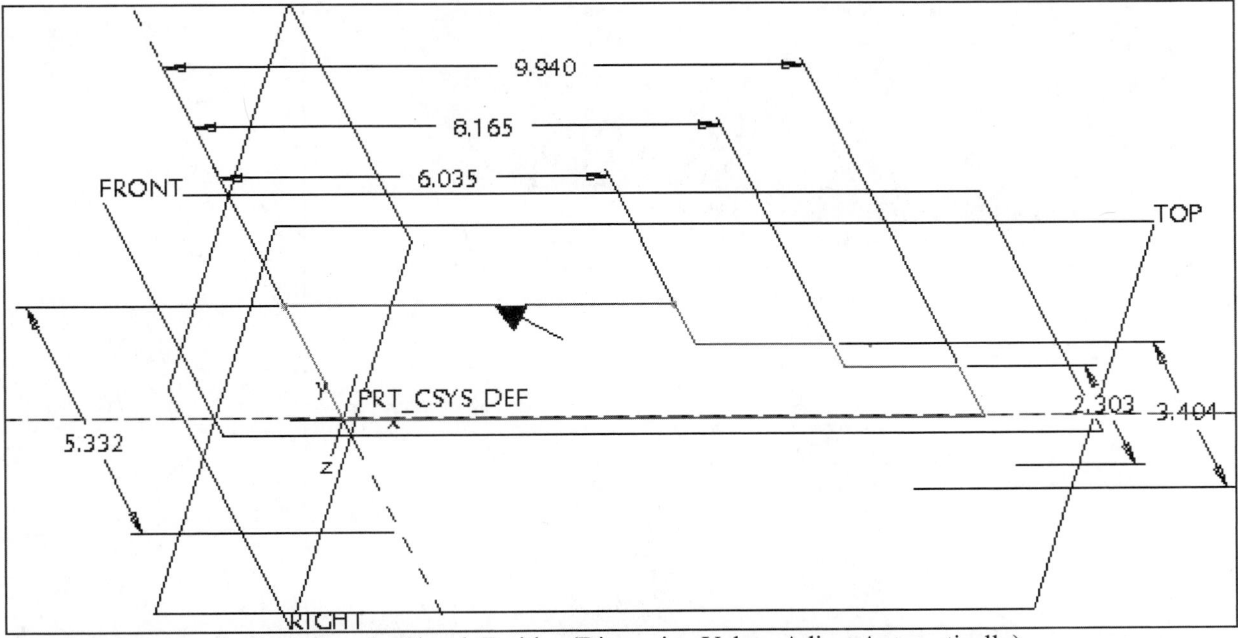

Figure 7.27 Dynamically Adjust the Sketch Entities (Dimension Values Adjust Automatically)

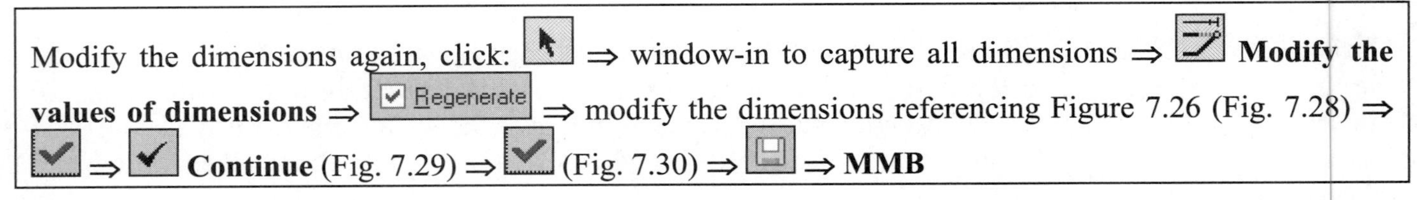

Modify the dimensions again, click: ⇒ window-in to capture all dimensions ⇒ **Modify the values of dimensions** ⇒ Regenerate ⇒ modify the dimensions referencing Figure 7.26 (Fig. 7.28) ⇒ ⇒ **Continue** (Fig. 7.29) ⇒ (Fig. 7.30) ⇒ ⇒ **MMB**

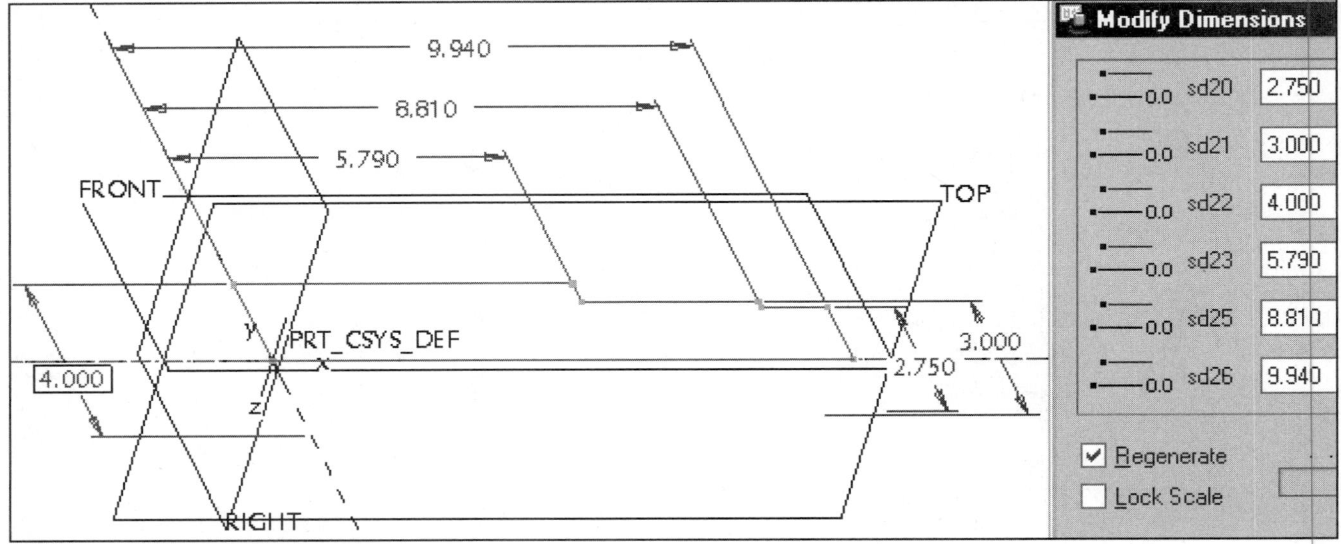

Figure 7.28 Modify the Dimensions

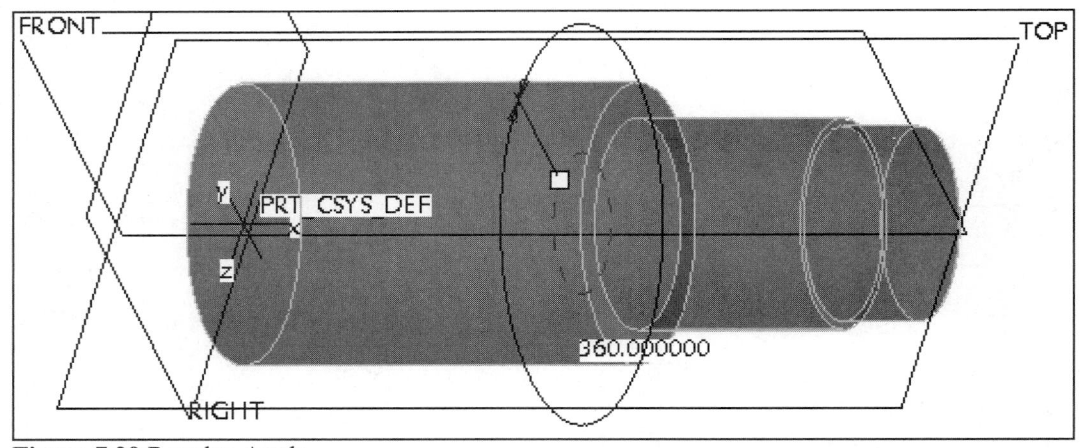

Figure 7.29 Revolve Angle

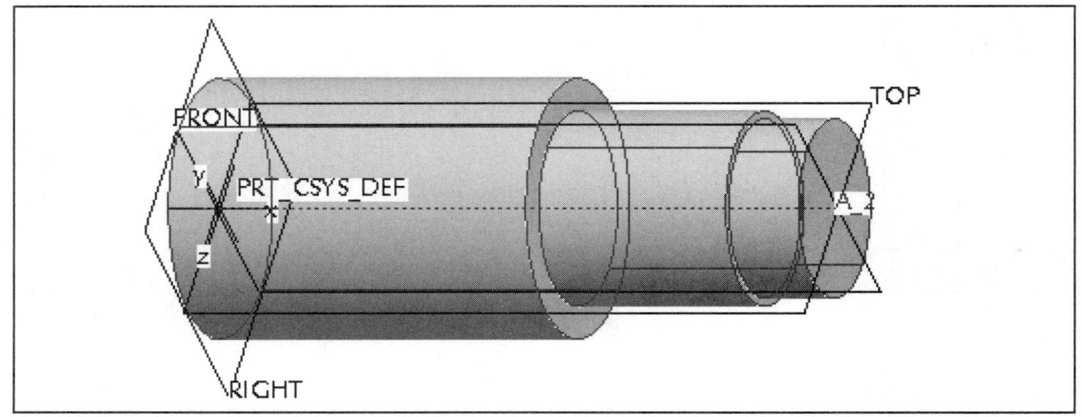

Figure 7.30 Completed Cylinder Rod Base Feature

Click on the default coordinate system name in the Model Tree-- **PRT_CSYS_DEF** ⇒ **RMB** ⇒ **Rename** ⇒ type **CSYS_CYL_ROD** ⇒ **Enter** ⇒ pick **RIGHT** in the Model Tree ⇒ **RMB** ⇒ **Properties** ⇒ -A- ⇒ Name- type **A** ⇒ **MMB** ⇒ pick **FRONT** in the Model Tree ⇒ **RMB** ⇒ **Properties** ⇒ -A- ⇒ Name- type **B** ⇒ **MMB** ⇒ pick **TOP** in the Model Tree ⇒ **RMB** ⇒ **Properties** ⇒ -A- ⇒ Name- type **C** ⇒ **MMB** (Fig. 7.31) ⇒

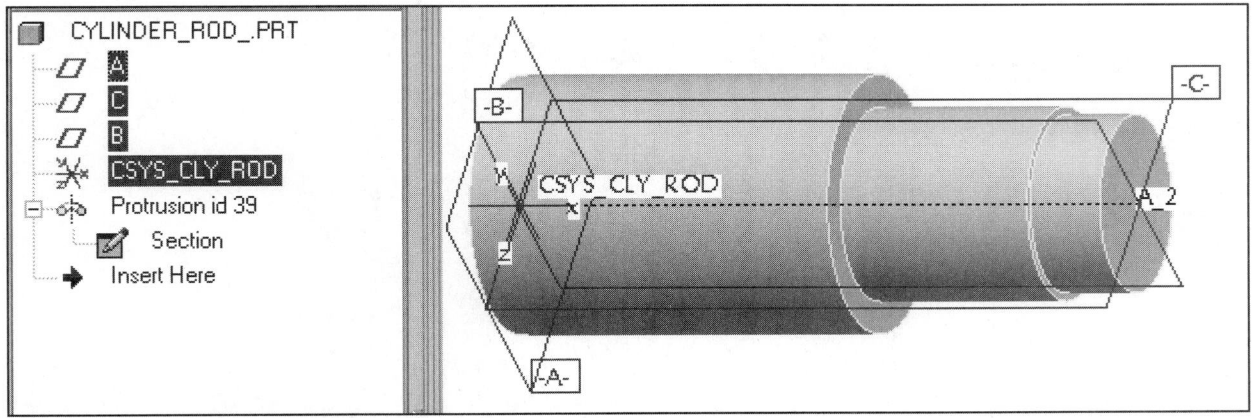

Figure 7.31 Highlighted Datum features

Click: **Tools** ⇒ **Environment** ⇒ Standard Orient Isometric ⇒ **Apply** ⇒ **OK** ⇒ **Show** in Navigator ⇒ **Layer Tree** ⇒ **Show** ⇒ ✔ Layer Items ⇒ **Expand** ⇒ **All** ⇒ press and hold **Ctrl** key and click on the coordinate system and the three datum planes in the Layer Tree (Fig. 7.32) ⇒ **LMB** to deselect ⇒ ⇒ ⇒ ⇒ **MMB** ⇒ **File** ⇒ **Delete** ⇒ **Old Versions** ⇒ **MMB** ⇒ **Show** ⇒ **Model Tree**

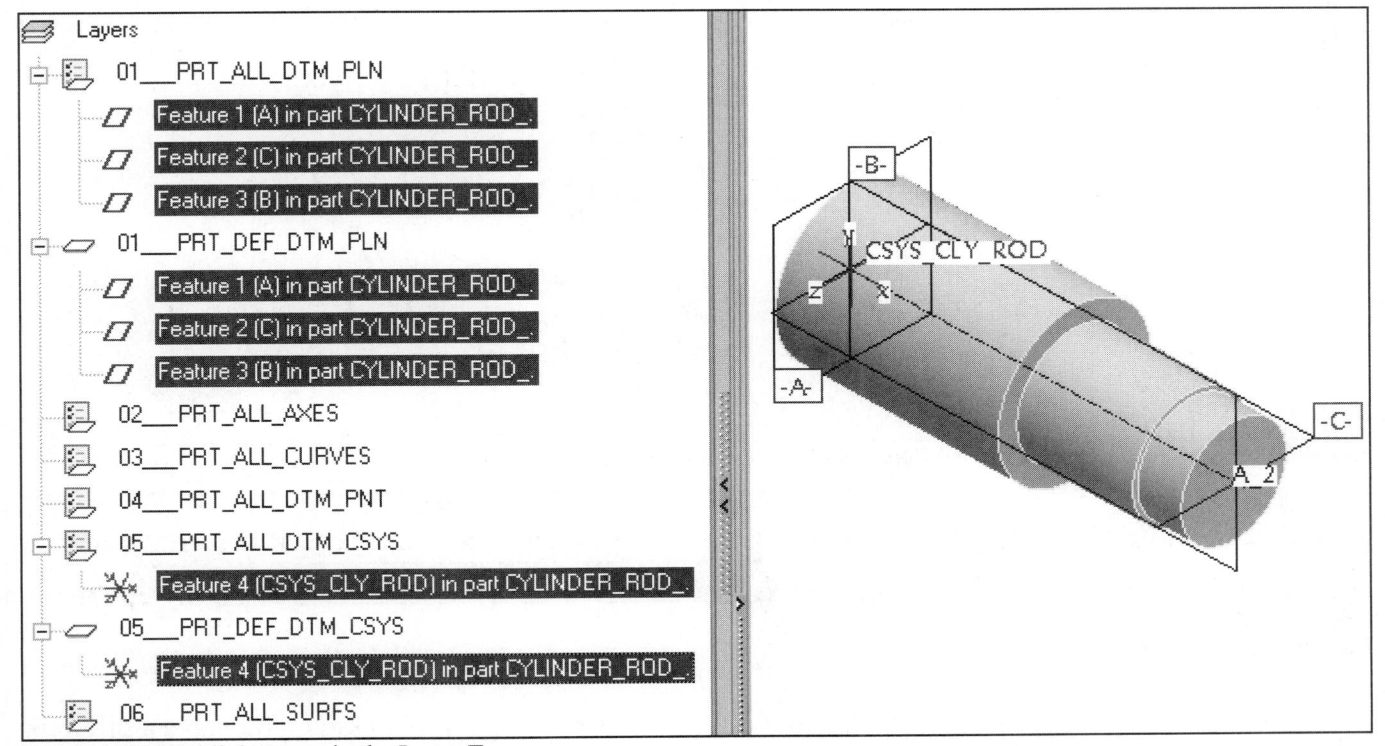

Figure 7.32 Highlight Items in the Layer Tree

Click: **Settings** ⇒ **Tree Columns** ⇒ [>>] move **Feat #**, **Feat Type**, and **Feat Name** to the Displayed column (Fig. 7.33) ⇒ **Apply** ⇒ **OK** ⇒ adjust the column widths (Fig. 7.34)

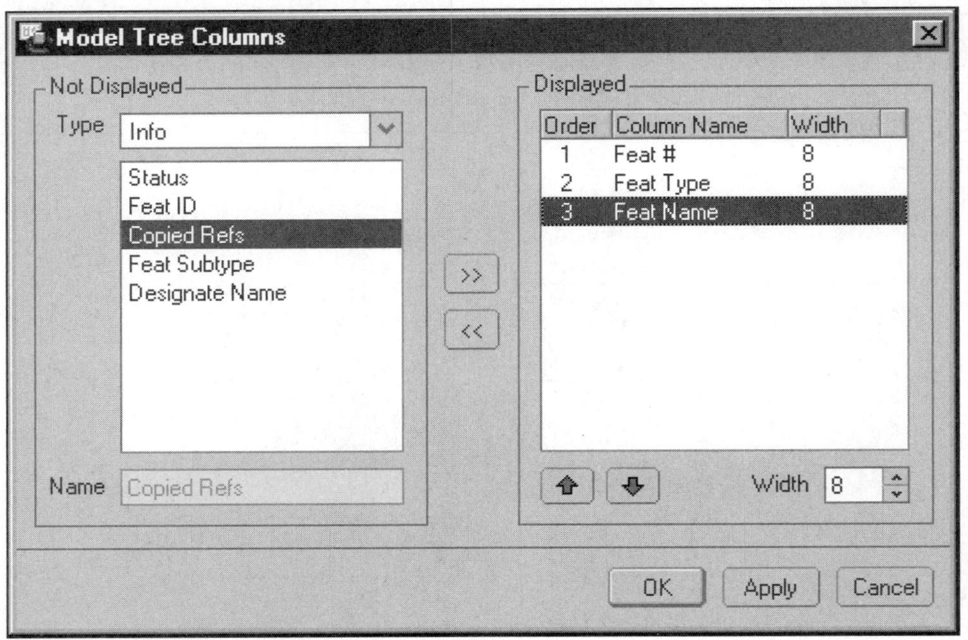

Figure 7.33 Model Tree Columns Dialog Box

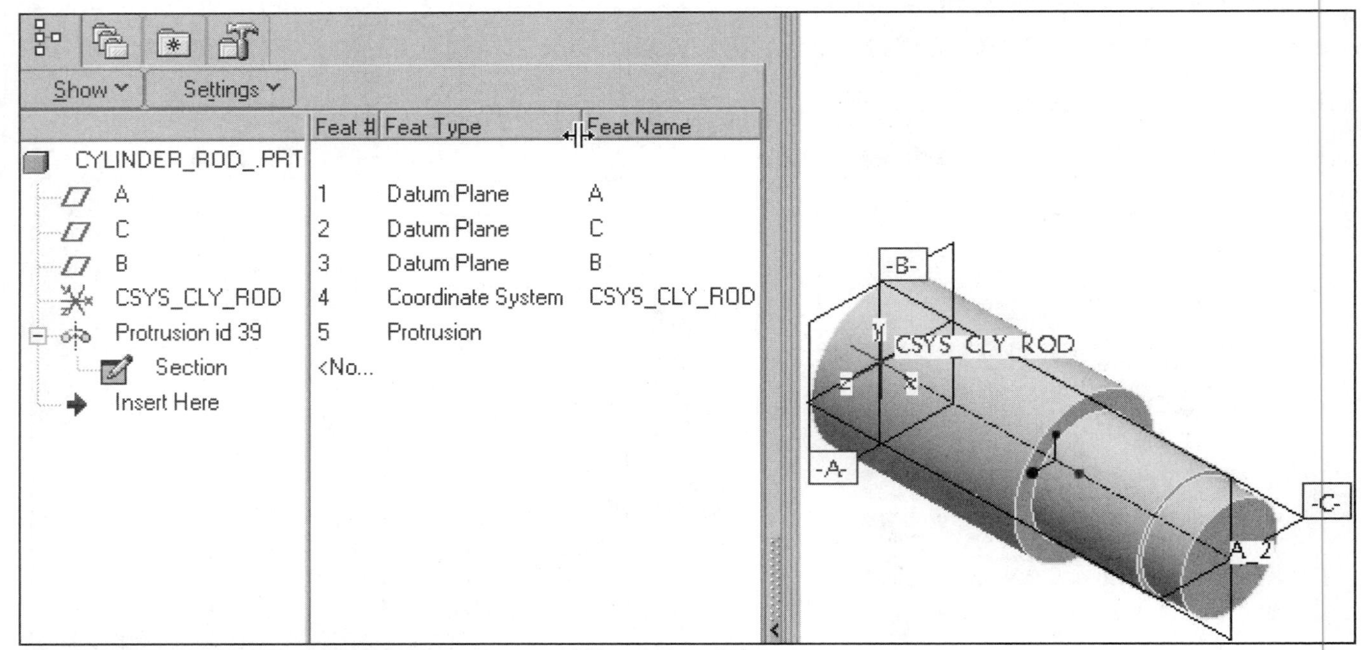

Figure 7.34 Model Tree with Feat #, Feat Type, and Feat Name Columns

Click: [icon] ⇒ close the Model Tree by clicking on the quick sash ⇒ [icon] ⇒ check your Environment settings, click: **Tools** ⇒ [icon] **Environment** ⇒ [Snap to Grid] ⇒ [Tangent Edges Dimmed] ⇒ **Apply** ⇒ **OK** ⇒ [icon] ⇒ [icon] ⇒ **MMB**

The next two features are revolved cuts similar to those for the Pin in Lesson 6 but have a different dimensioning scheme. The "necks" are each **.120 X .045 DEEP**.

Click: **Revolve Tool** ⇒ **Remove Material** ⇒ ⇒ Section dialog box opens ⇒ **Use Previous** ⇒ click **Sketch** button ⇒ add (by clicking on) the edge and the vertical circular surface to the References (Fig. 7.35) ⇒ **Close** ⇒ **Orient the sketching plane parallel to the screen** ⇒ **Sketch** ⇒ **References** add the circular surface/horizontal edge to the References (Fig. 7.36) ⇒ **Close** ⇒ **Create 2 point centerlines** sketch the horizontal centerline (Fig. 7.37) *(you **cannot** use a centerline from a previous sketch/section)* ⇒ **Hidden Line**

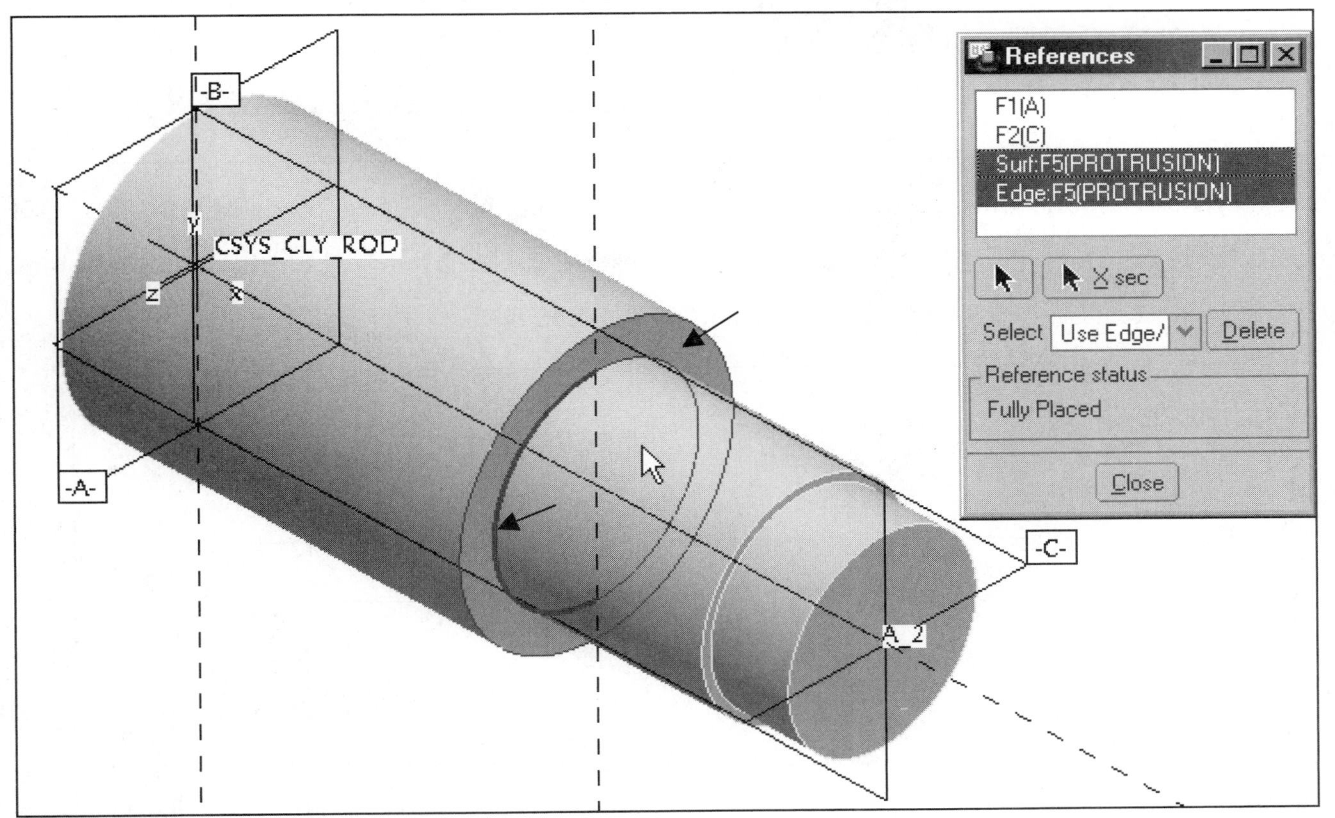

Figure 7.35 Add References

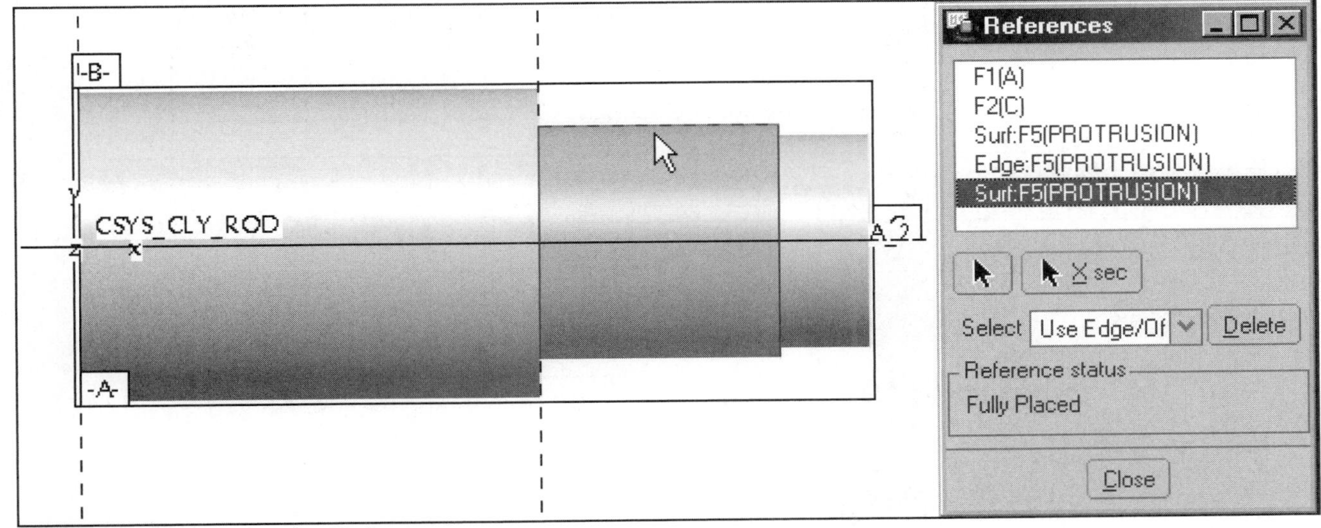

Figure 7.36 Adding Another Reference

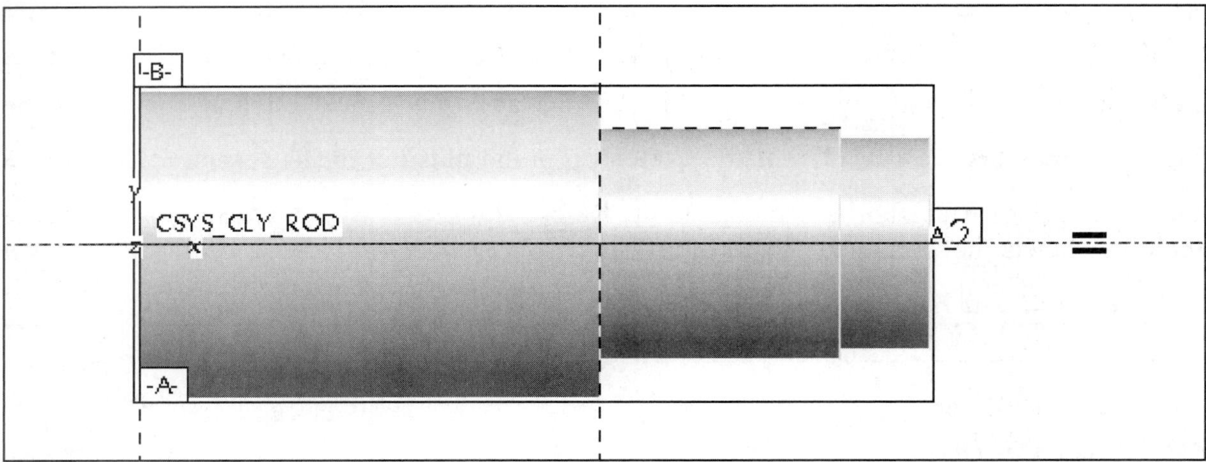

Figure 7.37 Sketch a Centerline

Click: **Create 2 point lines** ⇒ sketch the three lines defining the open section ⇒ **MMB** (Fig. 7.38) ⇒ modify the dimensions, click: ⇒ select both dimensions ⇒ **Modify the values of dimensions** (Fig. 7.39) ⇒ Regenerate ⇒ modify the dimensions (**.12 X .045 DEEP**) (Fig. 7.40) ⇒

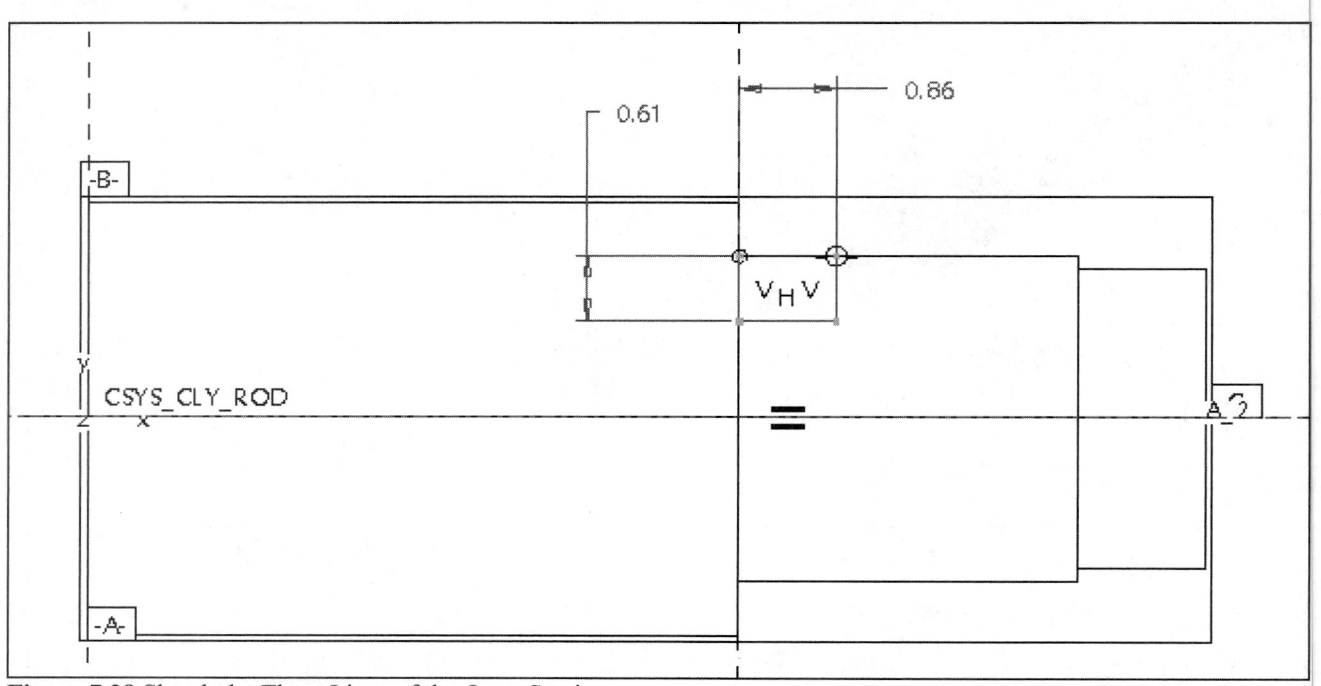

Figure 7.38 Sketch the Three Lines of the Open Section

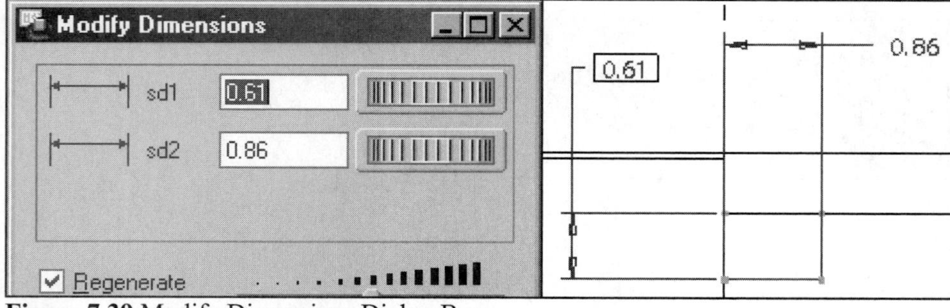

Figure 7.39 Modify Dimensions Dialog Box

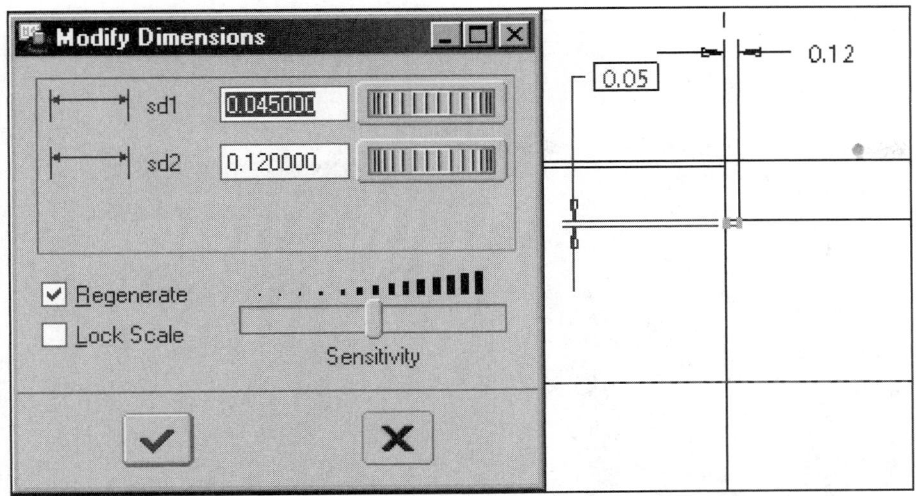

Figure 7.40 Modify to **.12 X .045**

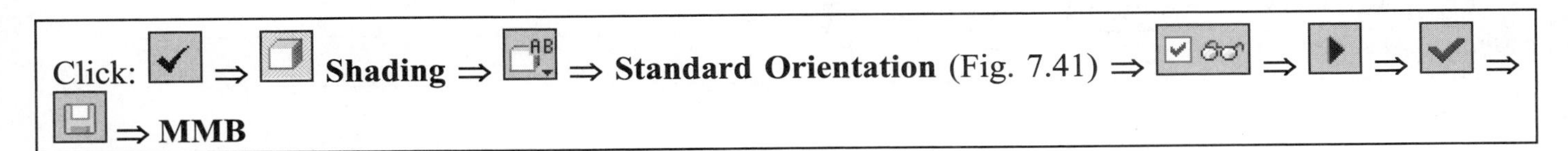

Click: ⇒ **Shading** ⇒ ⇒ **Standard Orientation** (Fig. 7.41) ⇒ ⇒ ⇒ ⇒ ⇒ **MMB**

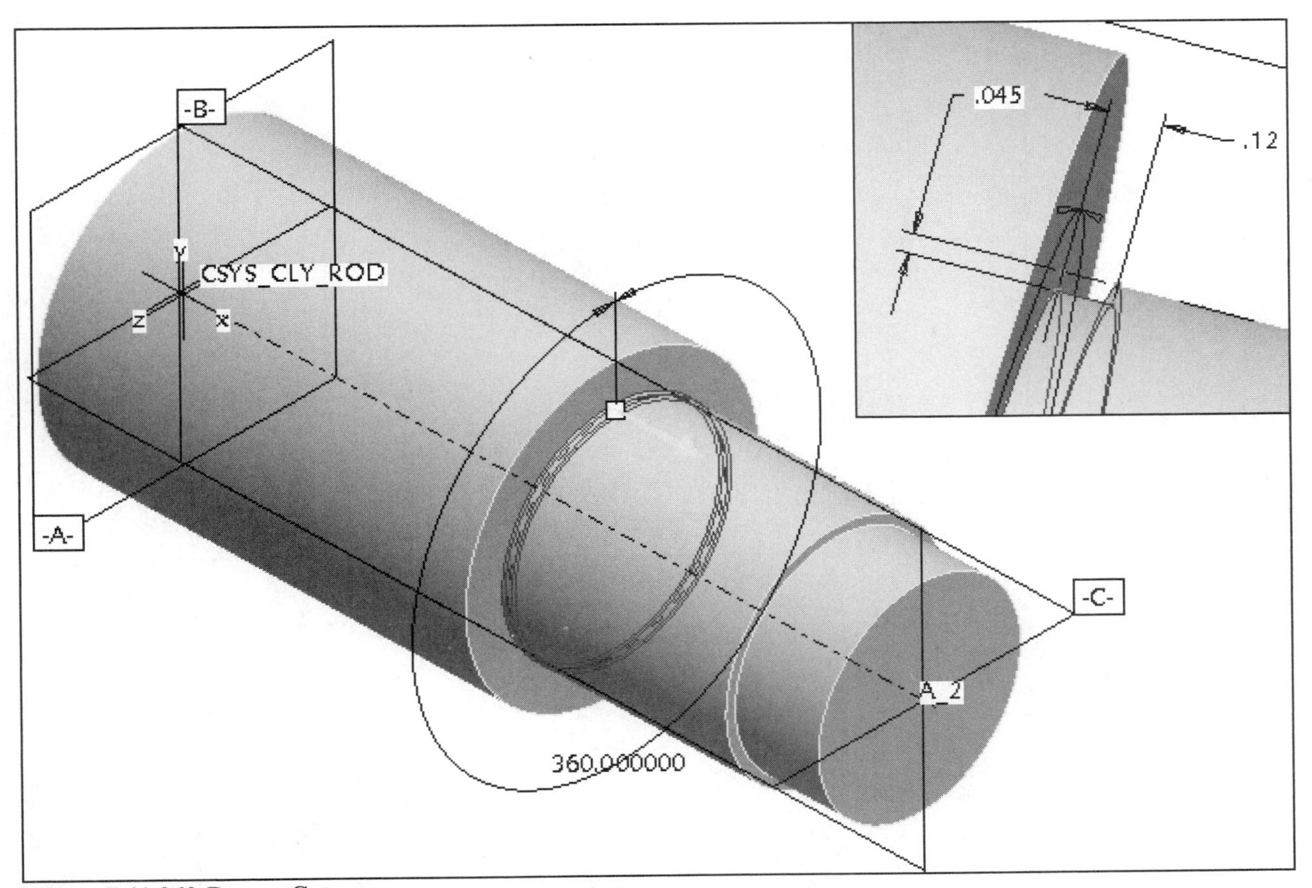

Figure 7.41 360-Degree Cut

Use the same commands and technique to create the second "neck" cut on the cylinder rod. Use similar references (Fig. 7.42). Sketch the section using three lines to form an open section (Fig. 7.43). Add the horizontal centerline at any time in the sketching process (Fig. 7.44). If you forget your centerline, the section will not regenerate. The completed cut is shown in Figure 7.45.

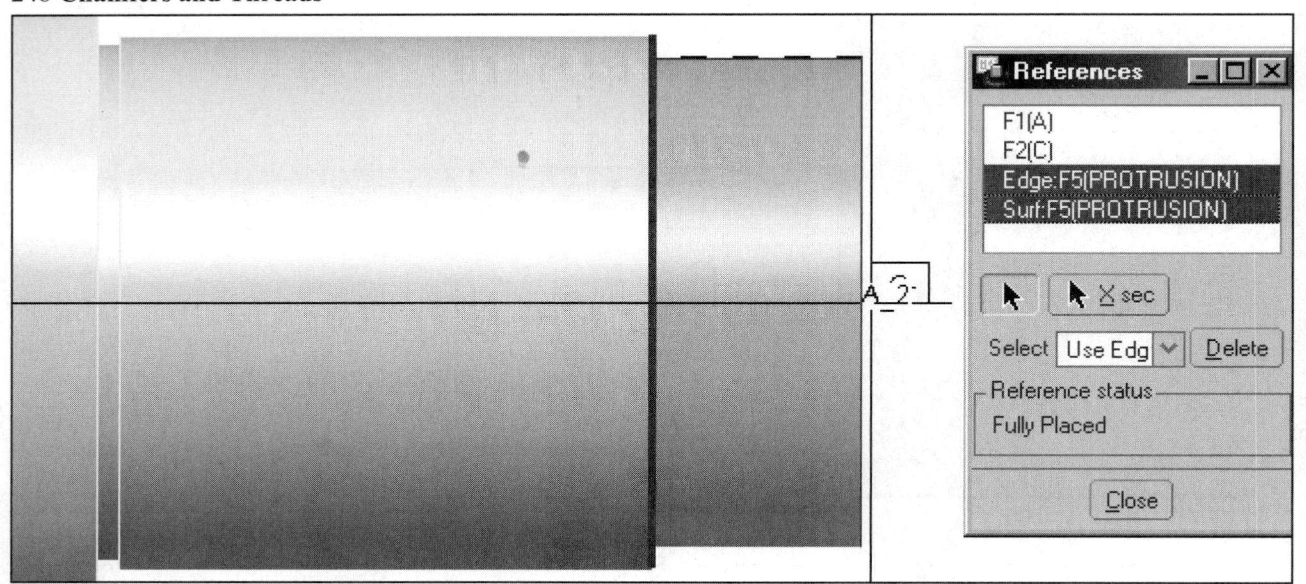

Figure 7.42 Add References

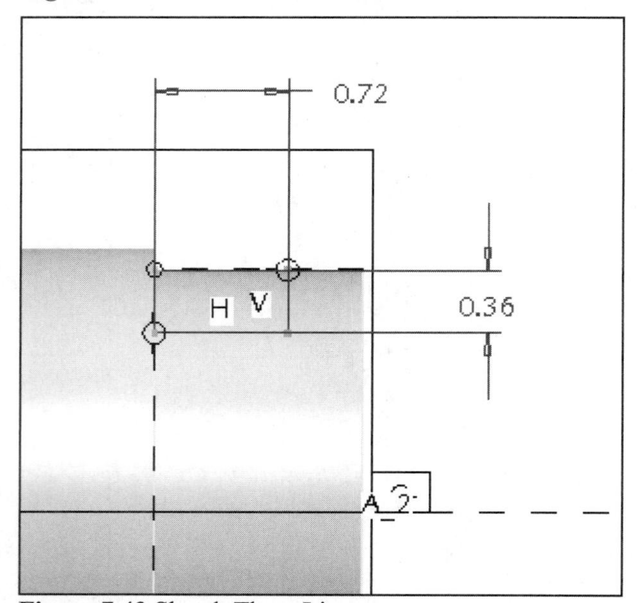

Figure 7.43 Sketch Three Lines

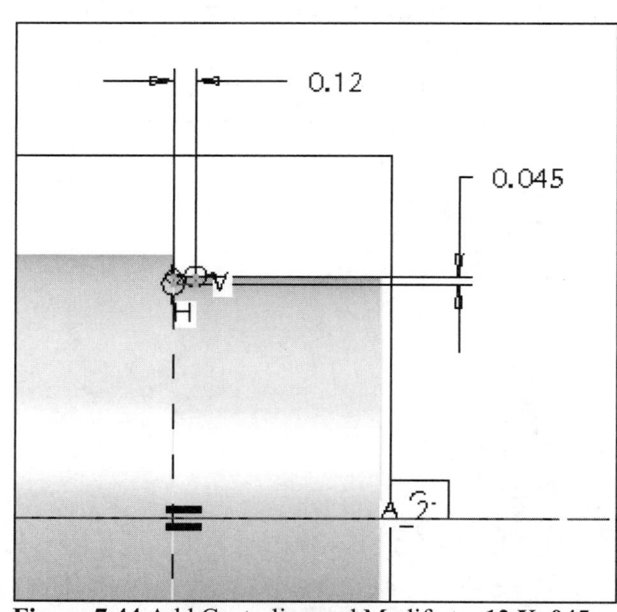

Figure 7.44 Add Centerline and Modify to **.12 X .045**

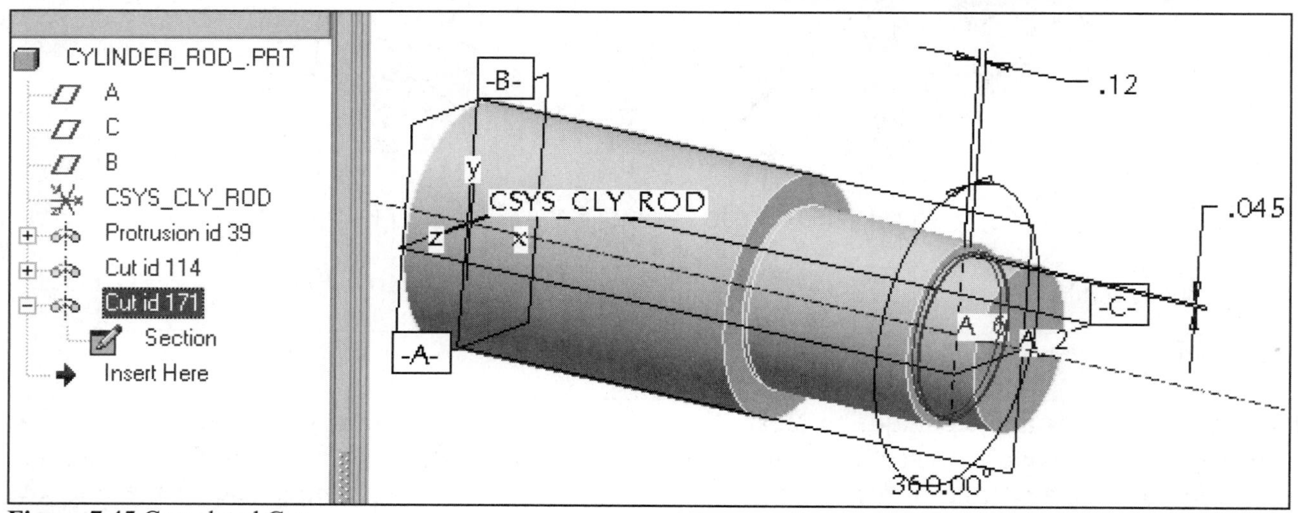

Figure 7.45 Completed Cut

Create the next feature (keyseat) using the design values of **.780** deep, **R1.030**, **1.460** from the edge and **.500** wide [Fig. 7.46(a-b)].

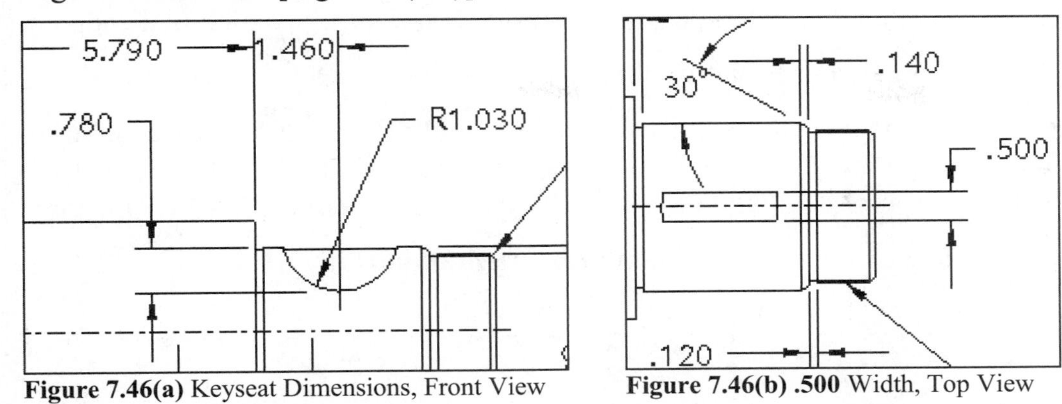

Figure 7.46(a) Keyseat Dimensions, Front View **Figure 7.46(b)** **.500** Width, Top View

Click: **Extrude Tool** ⇒ **Remove Material** ⇒ **Extrude on both sides of sketch plane by half the specified depth value in each direction** ⇒ type **.500** ⇒ **Enter** .500 ⇒ ⇒ **Use Previous** ⇒ click **Sketch** button ⇒ References: click on the surface shown in Figure 7.47 ⇒ **Orient the sketching plane parallel to the screen** ⇒ add (by clicking on) the horizontal edge to the References (Fig. 7.48) ⇒ **Close**

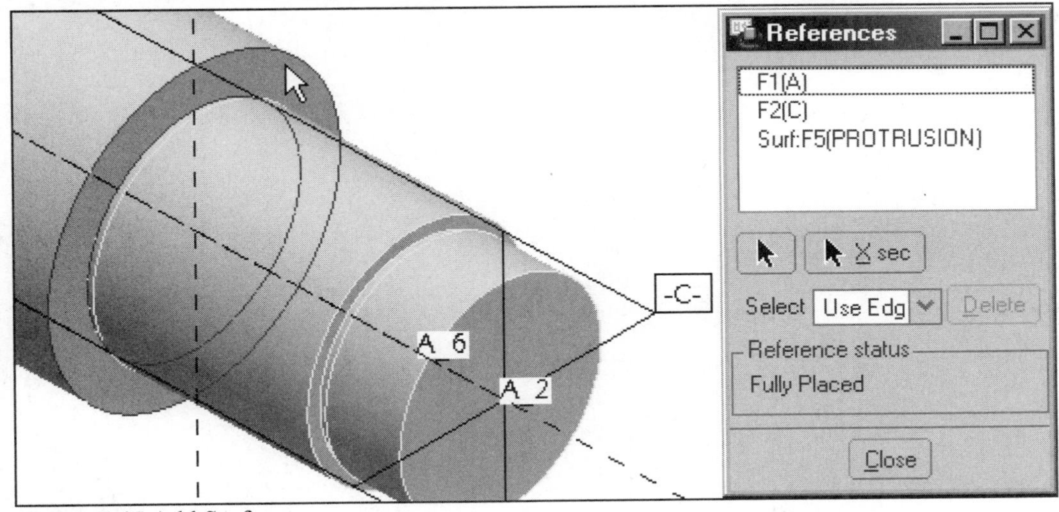

Figure 7.47 Add Surface

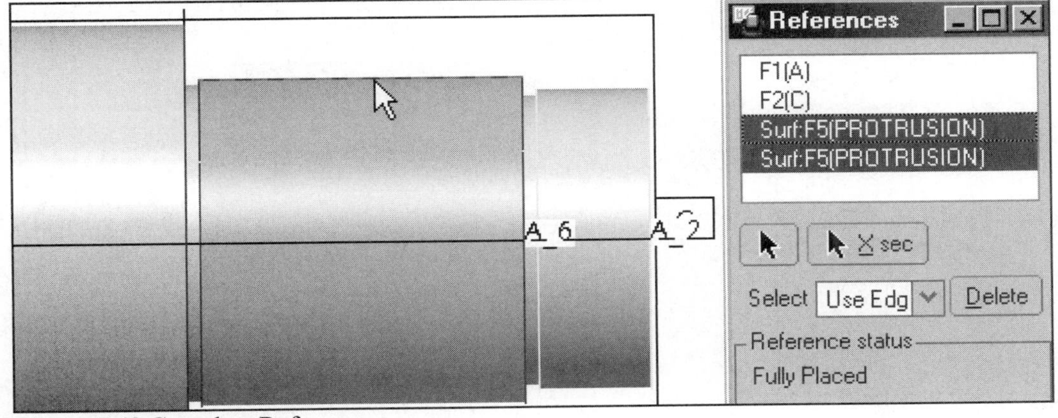

Figure 7.48 Complete References

Click: **Create 2 point centerlines** sketch the vertical centerline ⇒ **Create an arc by picking its center and endpoints** pick the center on the centerline above the part [Fig. 7.49(a)] ⇒ pick first endpoint on the edge [Fig. 7.49(b)] ⇒ pick second endpoint on the edge [Fig. 7.49(c)] ⇒ **Create defining dimension** add arc dimension and move other dimensions [Fig. 7.49(d)] ⇒ **Create defining dimension** ⇒ pick the bottom of the arc [Fig. 7.50(a)] then pick the horizontal edge of the part [Fig. 7.50(b)] ⇒ **MMB** to place the dimension [Fig. 7.50(c)] ⇒ modify the values (see Fig. 7.46) for the three dimensions [Fig. 7.50(d)] ⇒ ⇒ ⇒ ⇒ **Standard Orientation** (Fig. 7.51) ⇒ (Fig. 7.52) ⇒ ⇒ **MMB** (Fig. 7.53) ⇒ ⇒ **MMB**

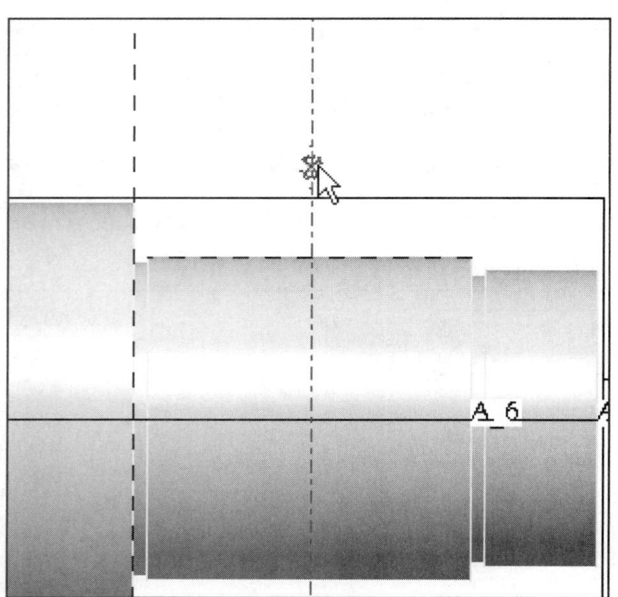

Figure 7.49(a) Pick Center for Arc

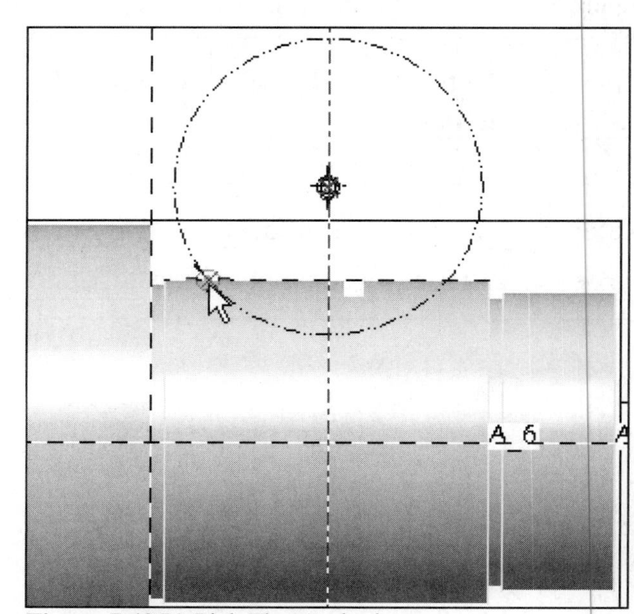

Figure 7.49(b) Pick First Endpoint

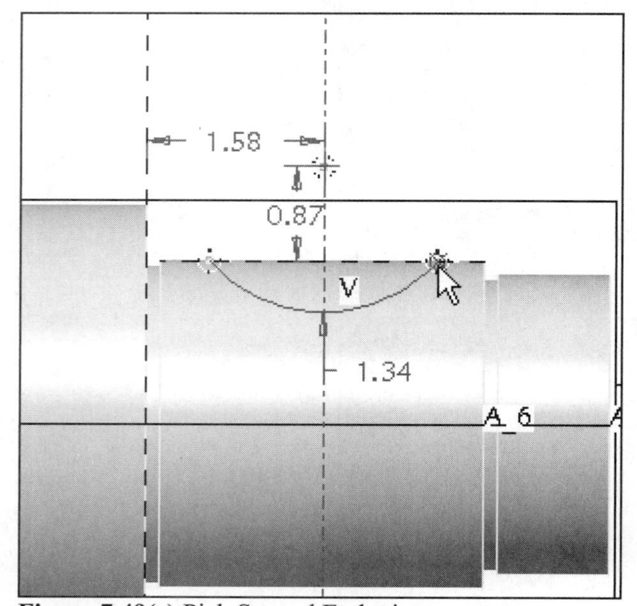

Figure 7.49(c) Pick Second Endpoint

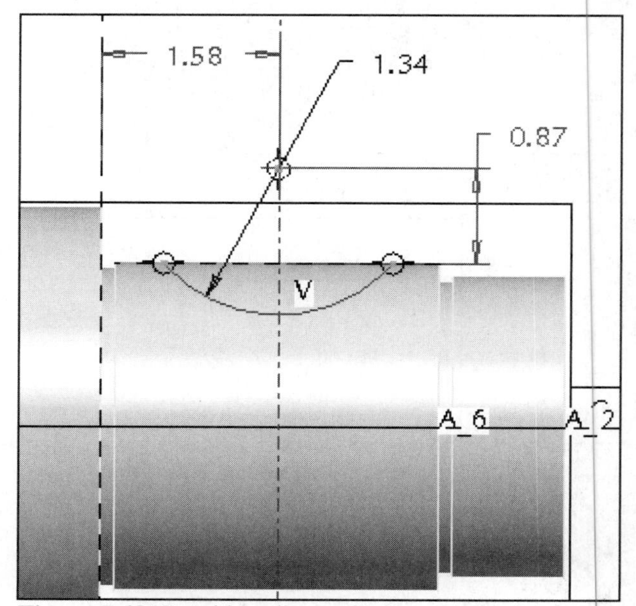

Figure 7.49(d) Add Radius and Move Dimensions

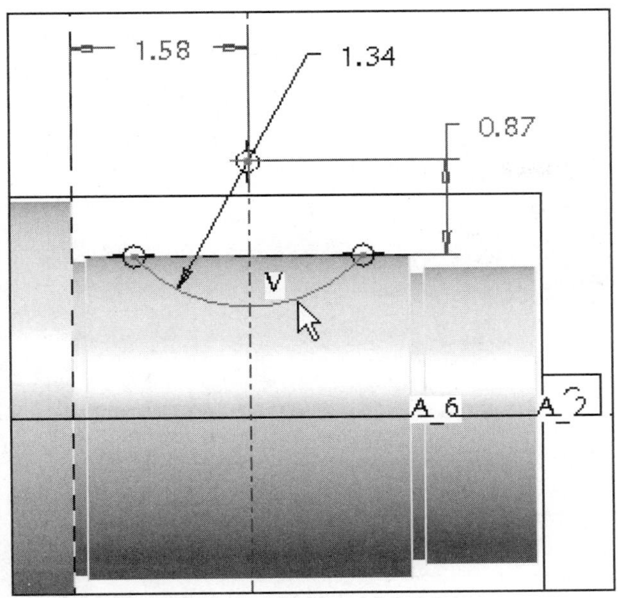

Figure 7.50(a) Pick the Bottom of the Arc

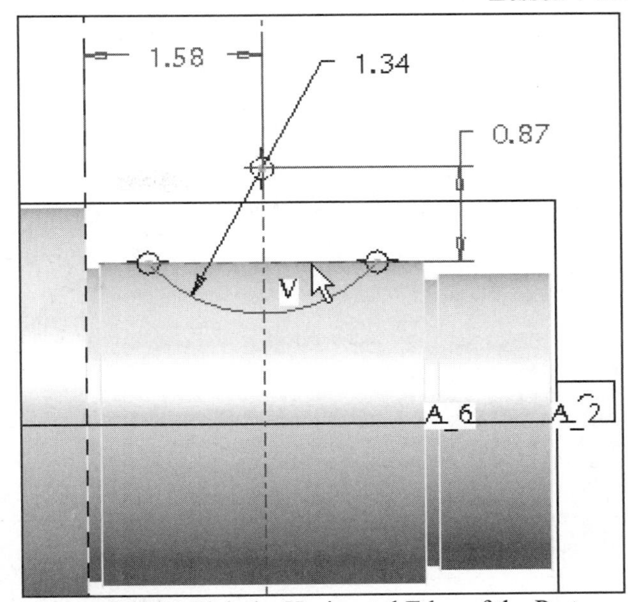

Figure 7.50(b) Pick the Horizontal Edge of the Part

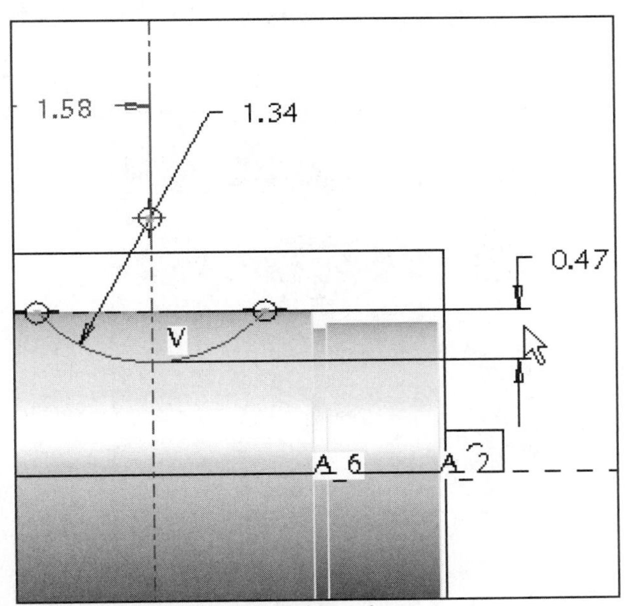

Figure 7.50(c) Place the Dimension

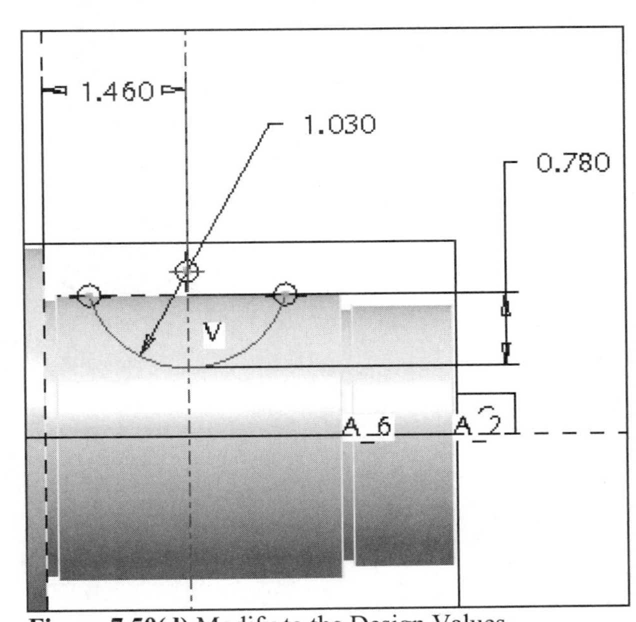

Figure 7.50(d) Modify to the Design Values

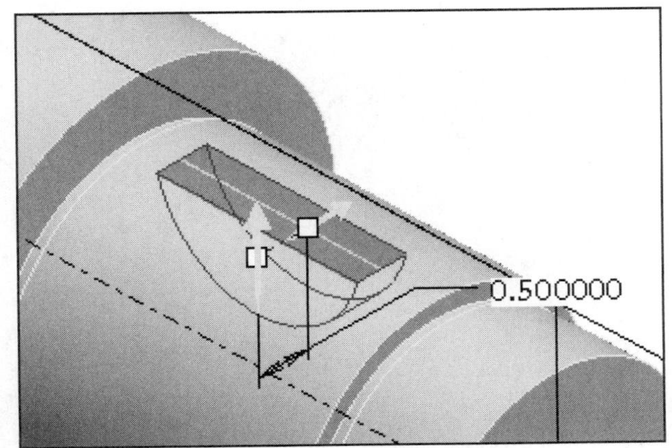

Figure 7.51 Material Removal Direction

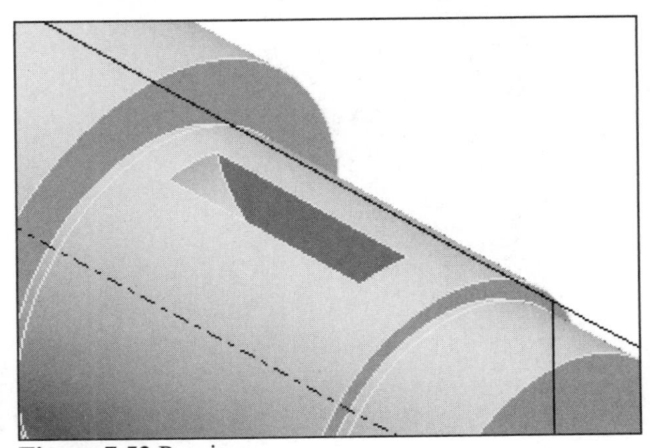
Figure 7.52 Preview

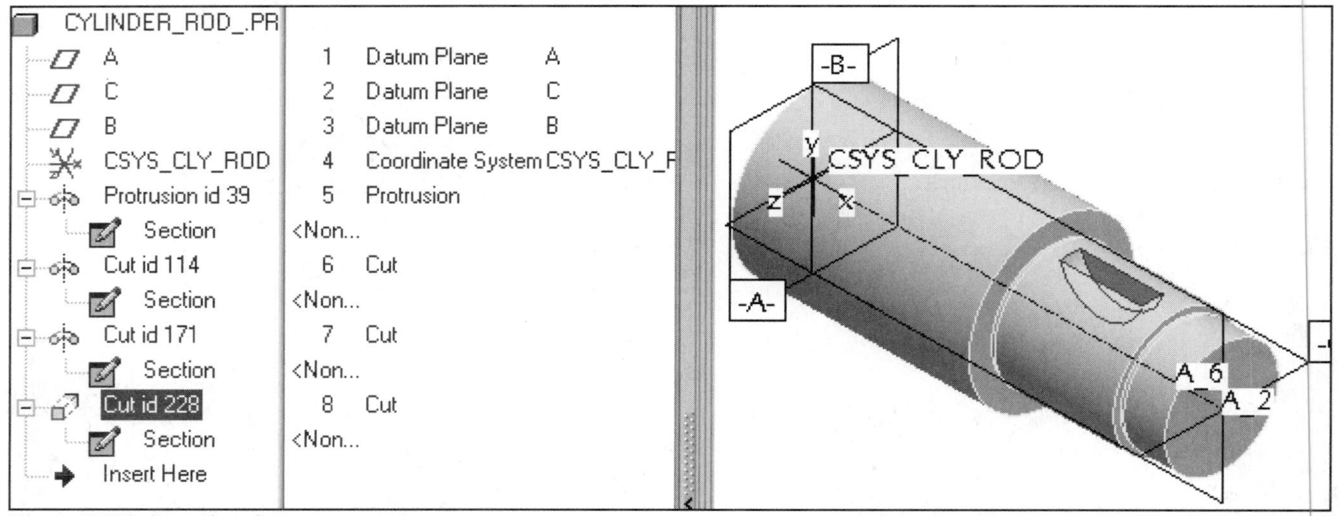

Figure 7.53 Completed Keyseat

Spin the part to see the large diameter end, click: **View** ⇒ **Orientation** ⇒ **Reorient** ⇒ Type **Dynamic orient** ⇒ Spin **Spin using spin center axis** ⇒ ⇒ slide bar to about **108** degrees (Fig. 7.54) ⇒ **OK**

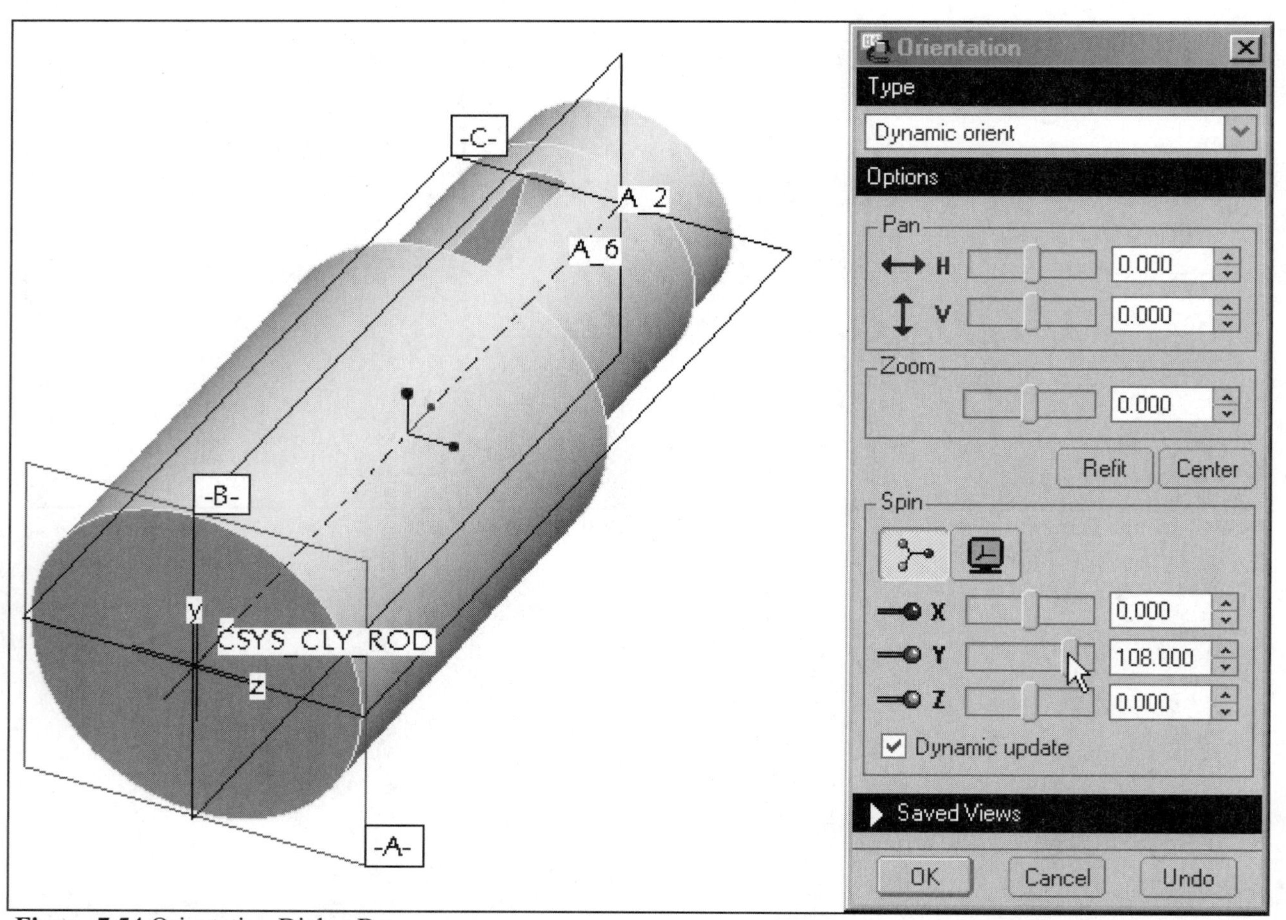

Figure 7.54 Orientation Dialog Box

Add a coaxial *standard hole* at the large-diameter end of the part. A *standard hole* is based on industry-standard fastener tables.

Hole charts are used to lookup diameters for a given fastener size. You can create custom hole charts and specify their directory location with the configuration file option *hole_parameter_file_path.* UNC, UNF and ISO hole charts are supplied with Pro/E. Charts are located in the loadpoint and *should not be changed.* Create a standard **5/8-11 UNC-2B** hole, **1.25** thread depth, **1.50** tap drill. Include a standard chamfer.

5/8-11 UNC-2B
↧ 1.25
⌀17/32
↧ 1.50

Click: **Hole Tool** from the tool bar ⇒ **Create standard hole** ⇒ **UNC** ⇒ **5/8-11** ⇒ **Add countersink drilling** ⇒ **Shape** tab ⇒ Variable **1.25** ⇒ **1.50** depth of tap drill (Fig. 7.55) ⇒ **Placement** tab ⇒ click on axis **A_2** ⇒ click in the Secondary references: No Items ⇒ pick on the end surface of the cylinder rod [Fig. 7.56(a)] ⇒ [Fig. 7.56(b)] ⇒ [Fig. 7.57(a)] ⇒ [Fig. 7.57(b)]

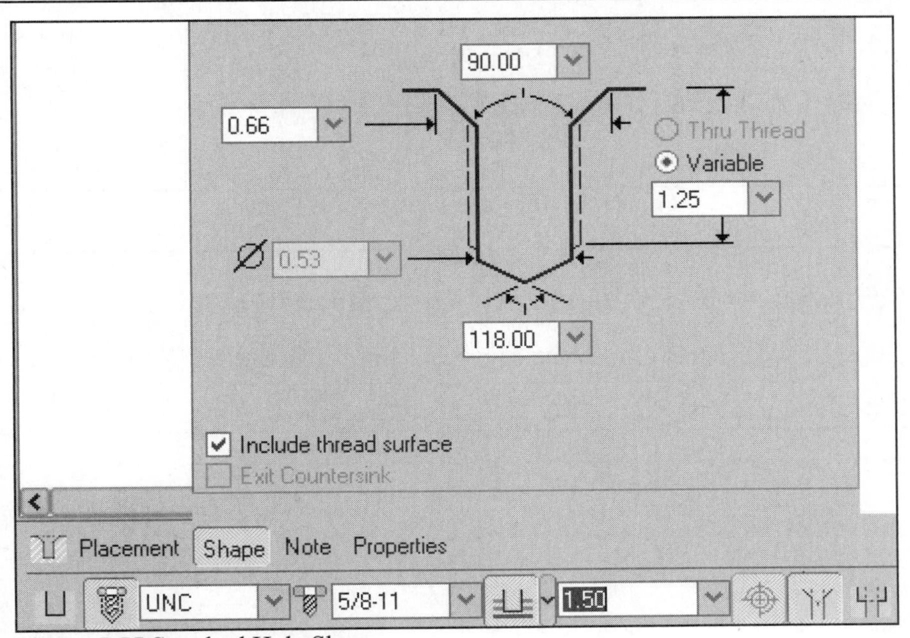

Figure 7.55 Standard Hole Shape

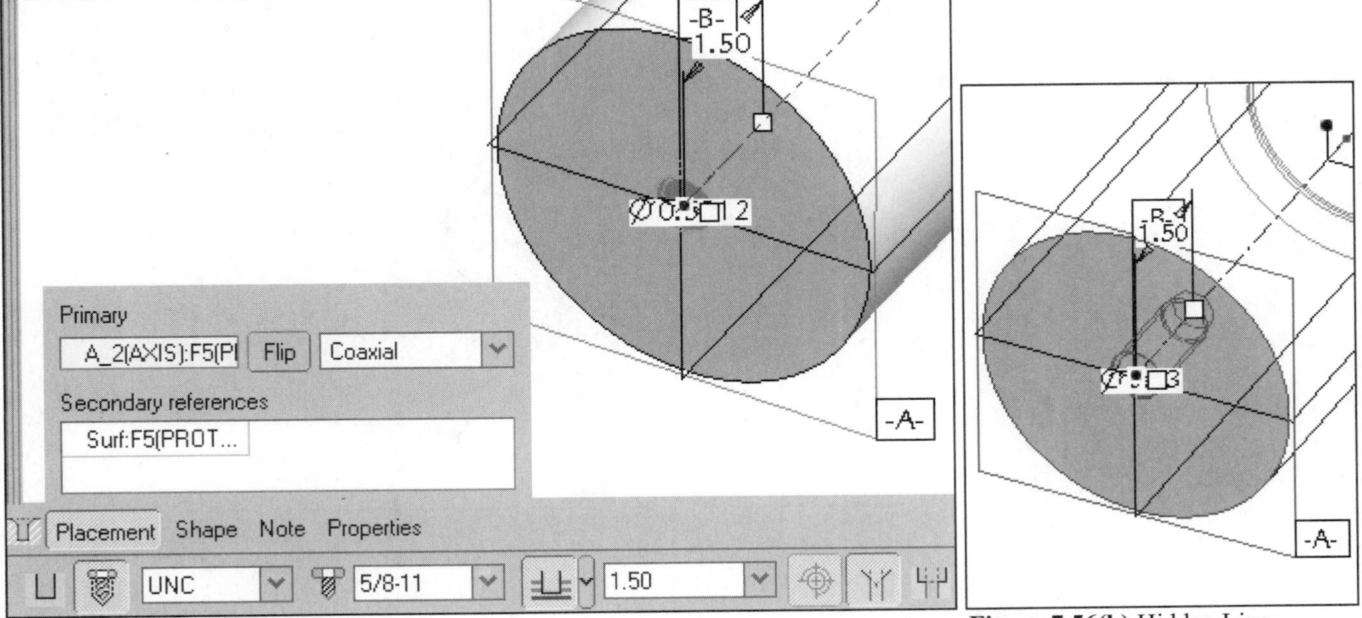

Figure 7.56(a) Placement, Shaded

Figure 7.56(b) Hidden Line

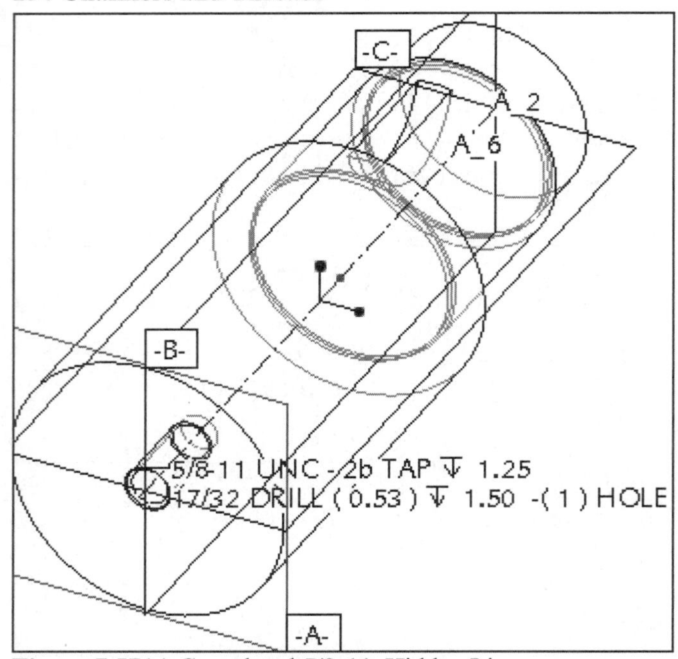

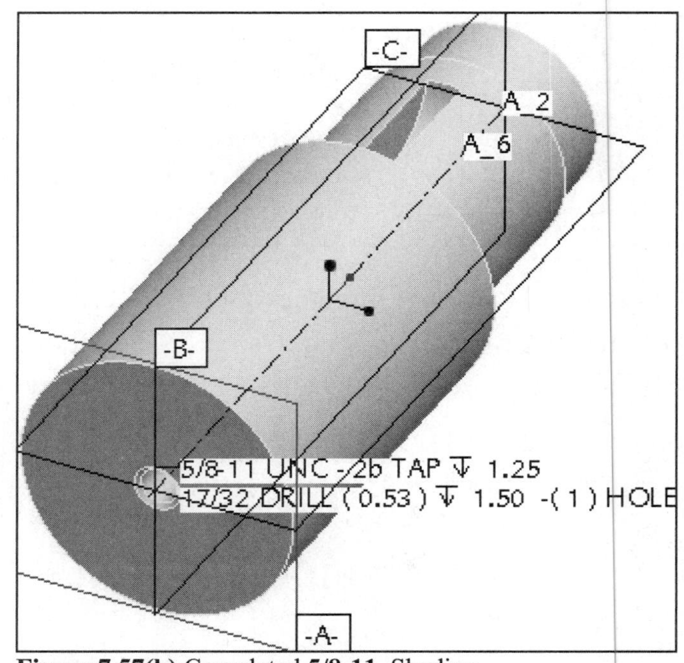

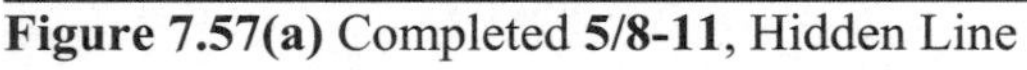

Figure 7.57(a) Completed **5/8-11**, Hidden Line

Figure 7.57(b) Completed **5/8-11**, Shading

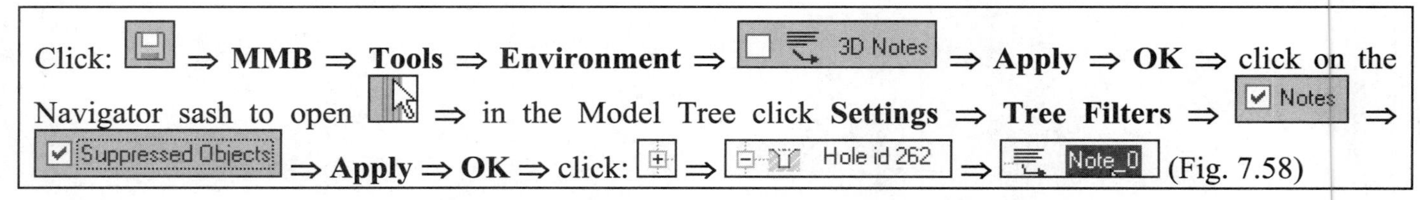

Click: ⇒ **MMB** ⇒ **Tools** ⇒ **Environment** ⇒ 3D Notes ⇒ **Apply** ⇒ **OK** ⇒ click on the Navigator sash to open ⇒ in the Model Tree click **Settings** ⇒ **Tree Filters** ⇒ Notes ⇒ Suppressed Objects ⇒ **Apply** ⇒ **OK** ⇒ click: ⇒ Hole id 262 ⇒ Note_0 (Fig. 7.58)

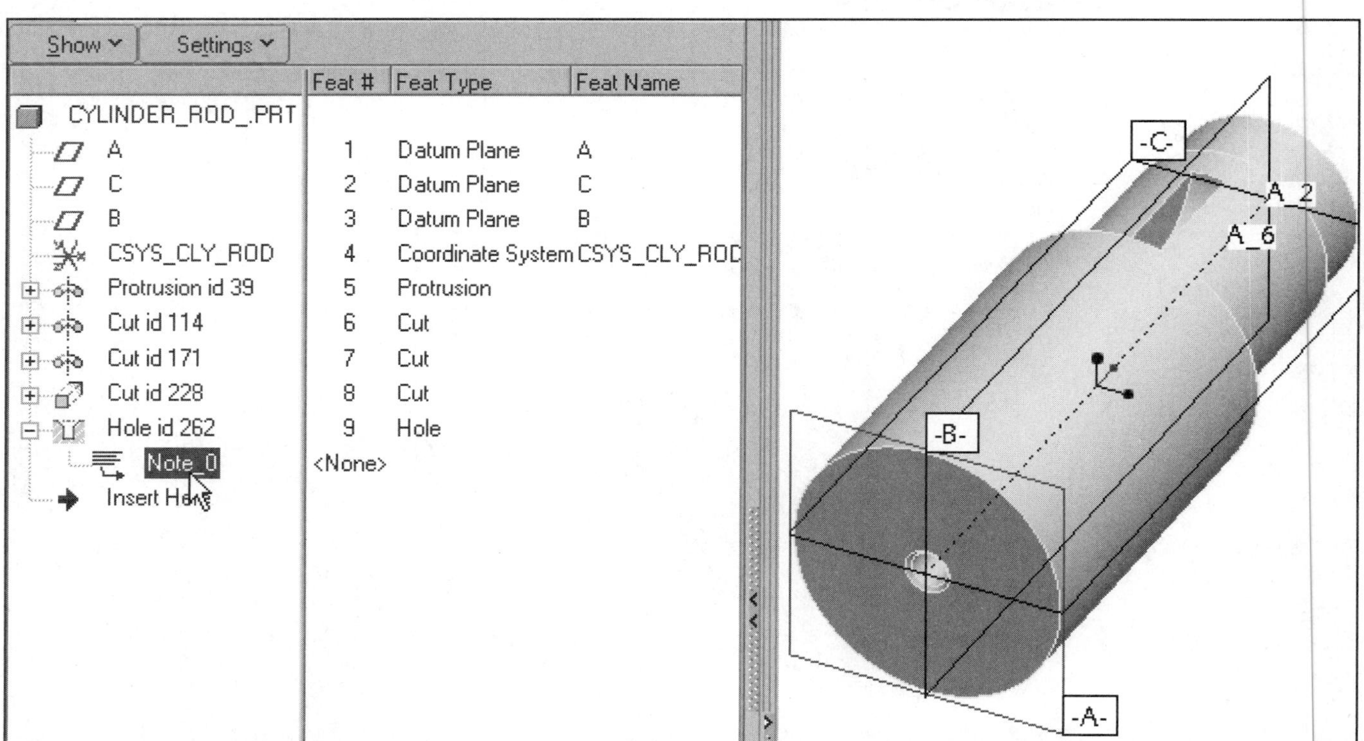

Figure 7.58 Note Item Displayed in Model Tree

Create the **45° X .120** chamfer on the large diameter end of the part 45° x.120.

Click: **Chamfer Tool** ⇒ **45 x D** ⇒ D **.120** [Fig. 7.59(a)] ⇒ pick **Edge:F5** ⇒ **Sets** tab [Fig. 7.59(b)] ⇒ **Options** tab ⇒ Attachment Solid ⇒ ⇒ ⇒ [Fig. 7.59(c)] ⇒ ⇒ **MMB**

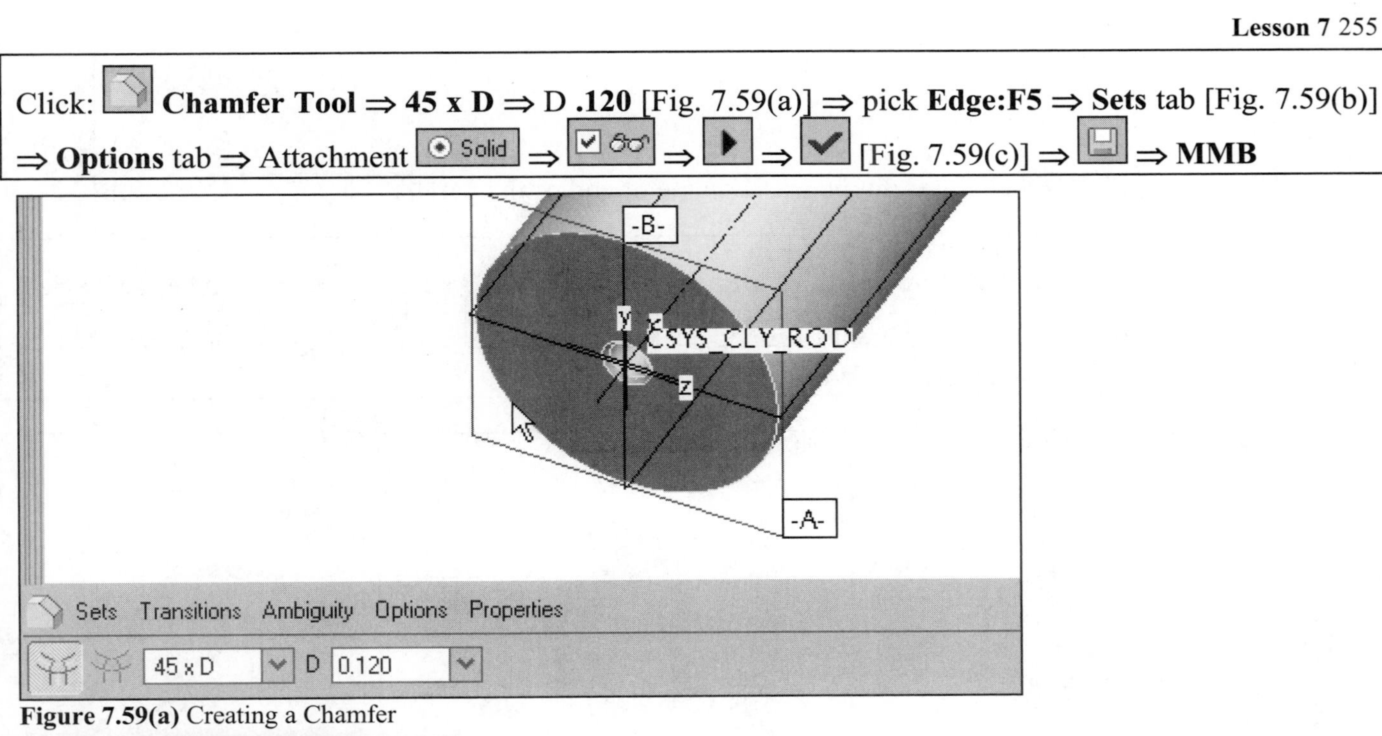

Figure 7.59(a) Creating a Chamfer

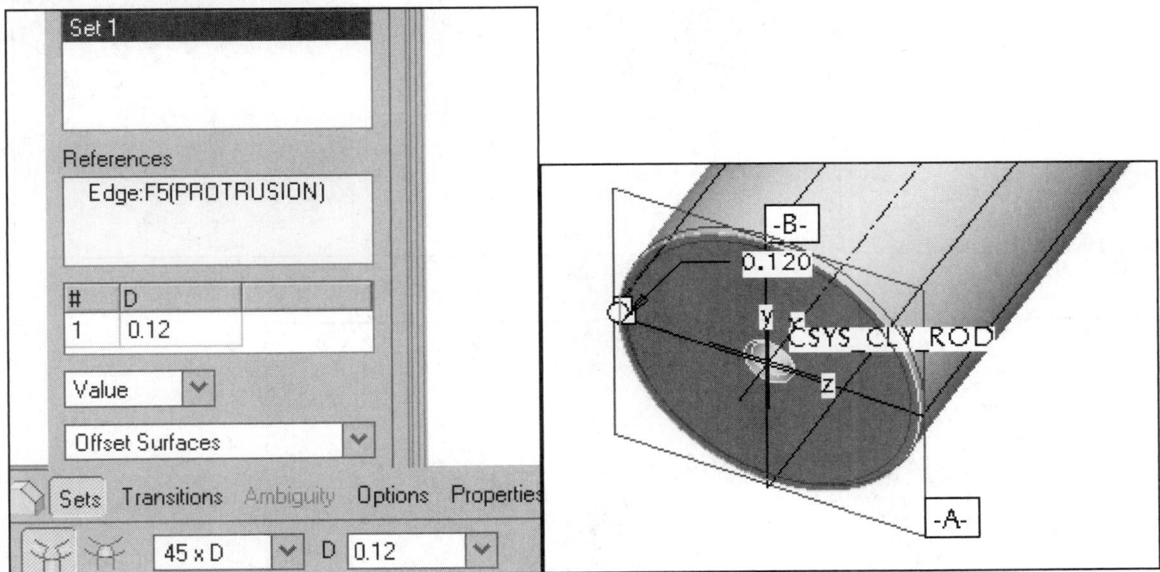

Figure 7.59(b) Set 1 Chamfer

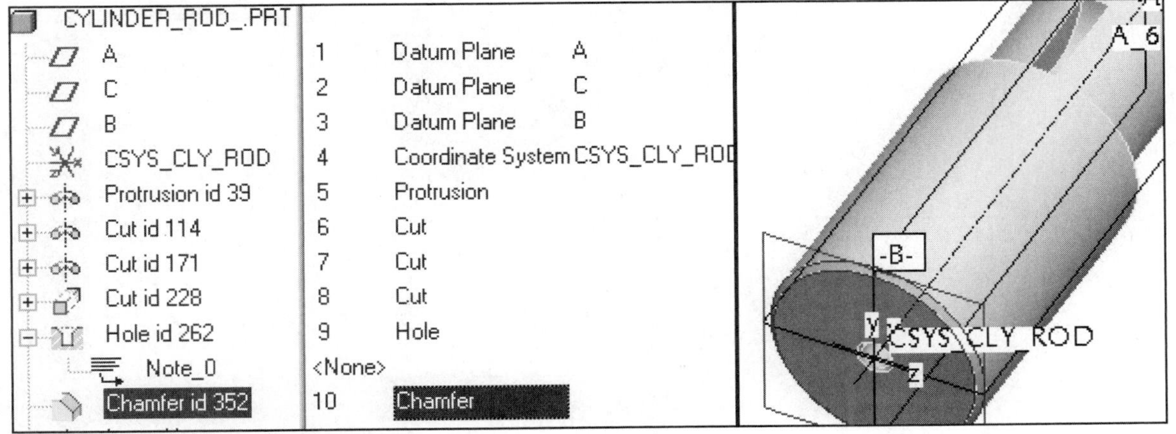

Figure 7.59(c) Completed Chamfer

Create the **45° X .09** chamfer on the small diameter end of the part

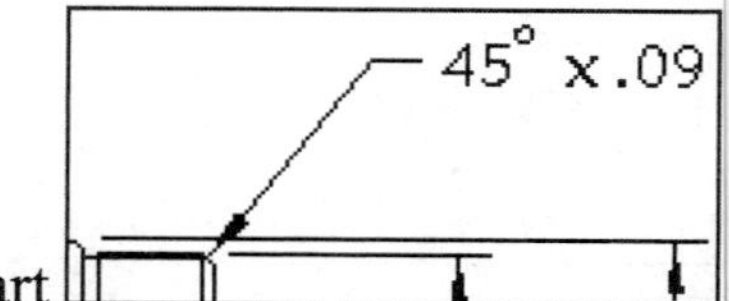

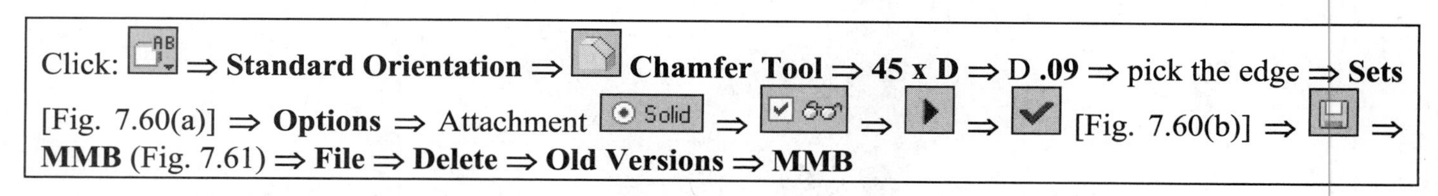

Click: ⇒ **Standard Orientation** ⇒ **Chamfer Tool** ⇒ **45 x D** ⇒ D **.09** ⇒ pick the edge ⇒ **Sets** [Fig. 7.60(a)] ⇒ **Options** ⇒ Attachment Solid ⇒ ⇒ ⇒ [Fig. 7.60(b)] ⇒ ⇒ **MMB** (Fig. 7.61) ⇒ **File** ⇒ **Delete** ⇒ **Old Versions** ⇒ **MMB**

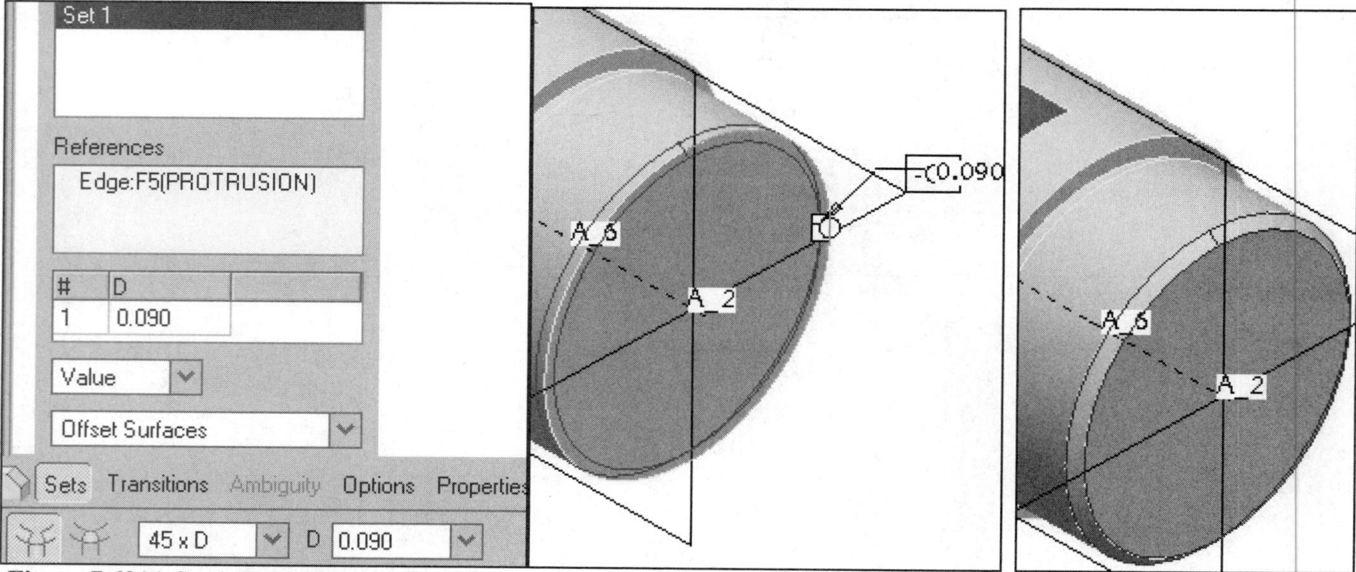

Figure 7.60(a) Set 1 **45 X .09** Chamfer

Figure 7.60(b) Chamfer

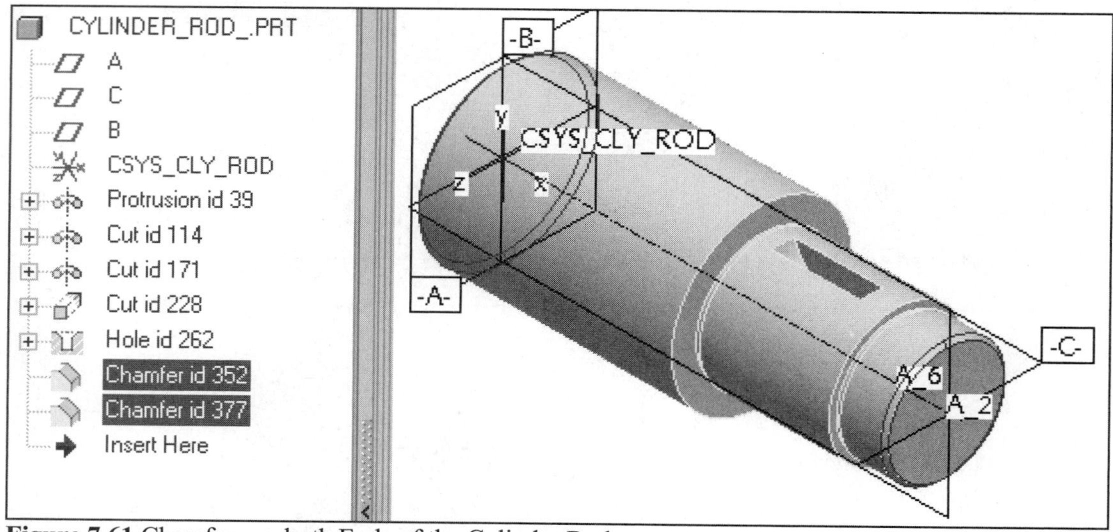

Figure 7.61 Chamfers on both Ends of the Cylinder Rod

Create the **30° X .140** chamfer 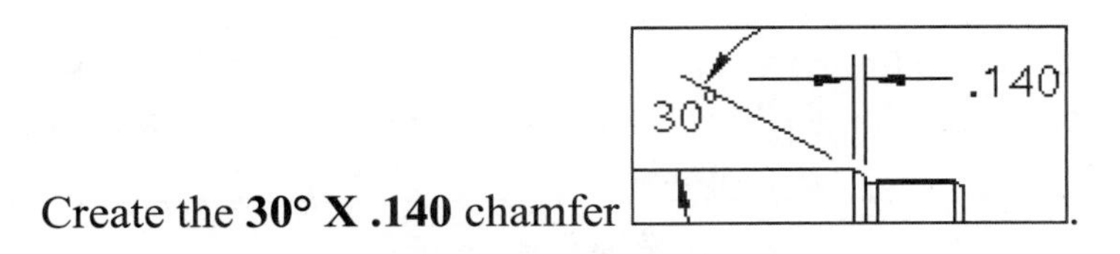

.

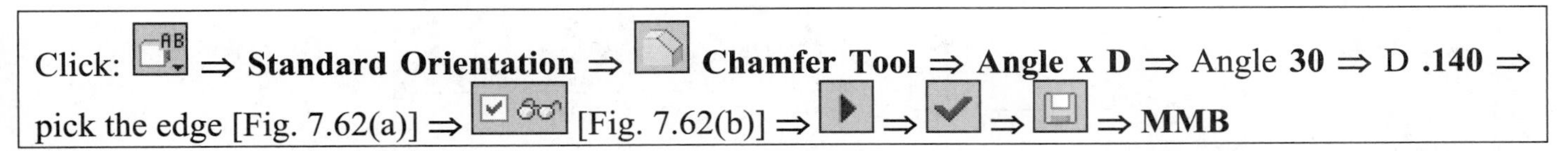
Click: ⇒ **Standard Orientation** ⇒ **Chamfer Tool** ⇒ **Angle x D** ⇒ Angle **30** ⇒ D **.140** ⇒ pick the edge [Fig. 7.62(a)] ⇒ [Fig. 7.62(b)] ⇒ ⇒ ⇒ ⇒ **MMB**

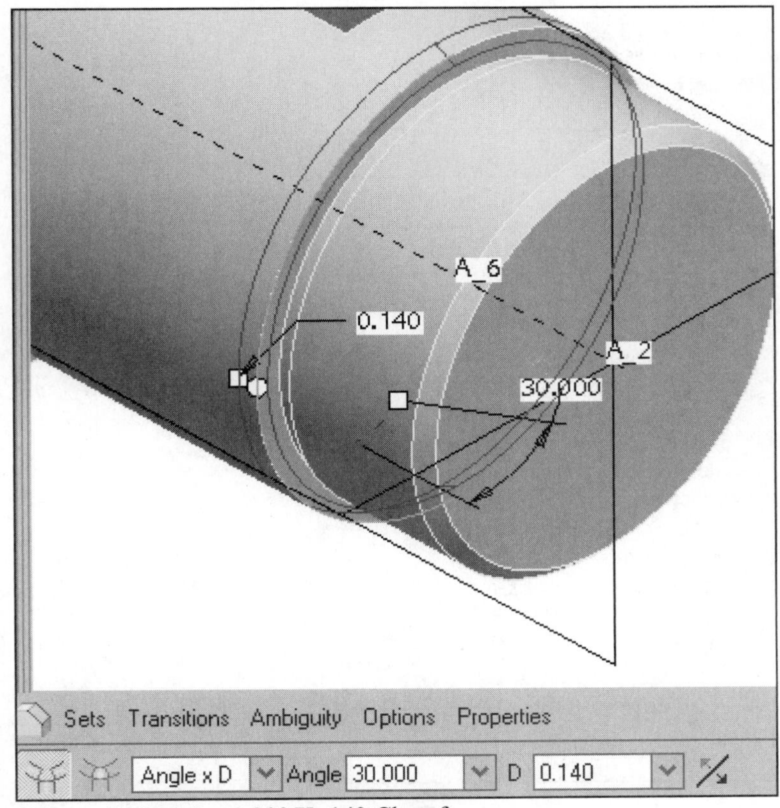

Figure 7.62(a) Set 1 **30° X .140** Chamfer

Figure 7.62(b) Chamfer

A **cosmetic thread** "represents" the diameter of a thread. You cannot modify the line style of a cosmetic thread, nor are threads affected by hidden line display settings.

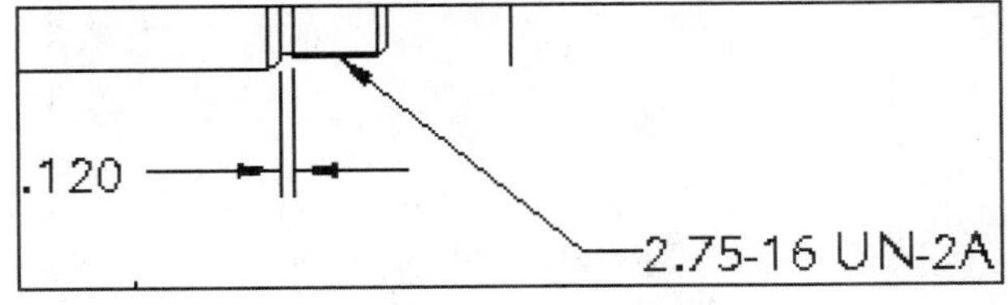

Threads are created with the default tolerance setting of limits. Here, the external cosmetic thread is specified by the minor diameter (external threads), starting surface, and ending edge.

- ***Internal cosmetic*** threads were created automatically when you created the Standard (Tapped) Hole. For threaded shafts you must create the cosmetic threads.
- ***External cosmetic*** threads represent the *root diameter.*

After creating the cosmetic thread, edit the thread table. The thread size of an external thread must be changed to the nominal size from the root diameter defaulted on the Table. Create an external cosmetic thread (**2.75-16 UN-2A**) using the **∅2.75** surface. The thread starts at the "neck cut" and goes to the edge of the chamfer.

Click: **Insert** ⇒ **Cosmetic** ⇒ **Thread** ⇒ pick the cylindrical surface [Fig. 7.63(a)] ⇒ pick the Start Surf--the edge lip surface created by the "neck" cut [Fig. 7.63(b)] ⇒ **Okay** arrow must point toward the small diameter end of the part [Fig. 7.63(c)]

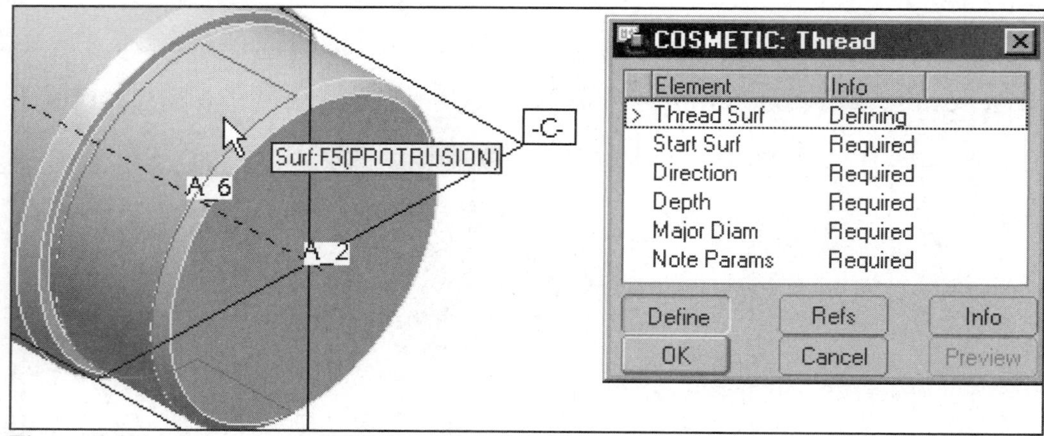

Figure 7.63(a) Select the Thread Surface

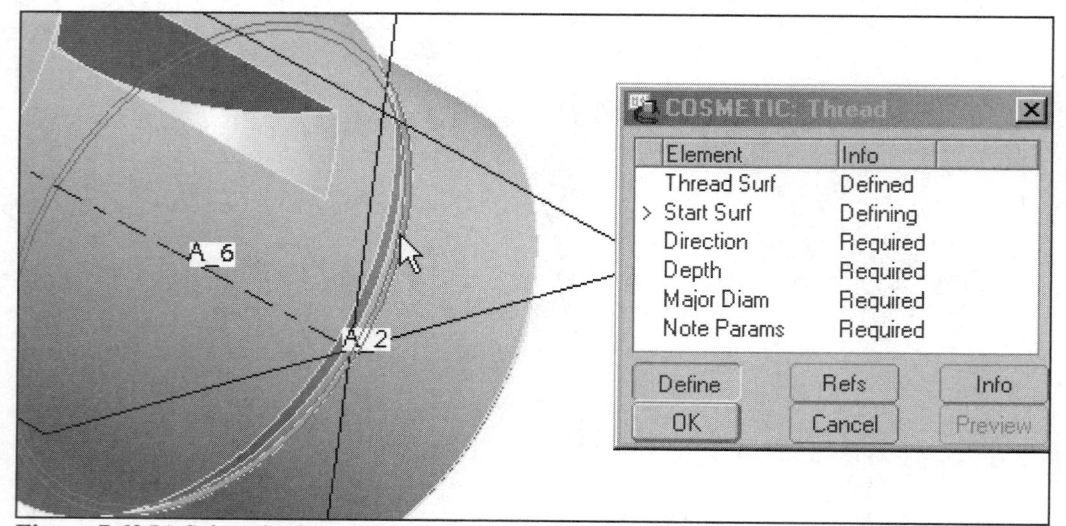

Figure 7.63(b) Select the Start Surface

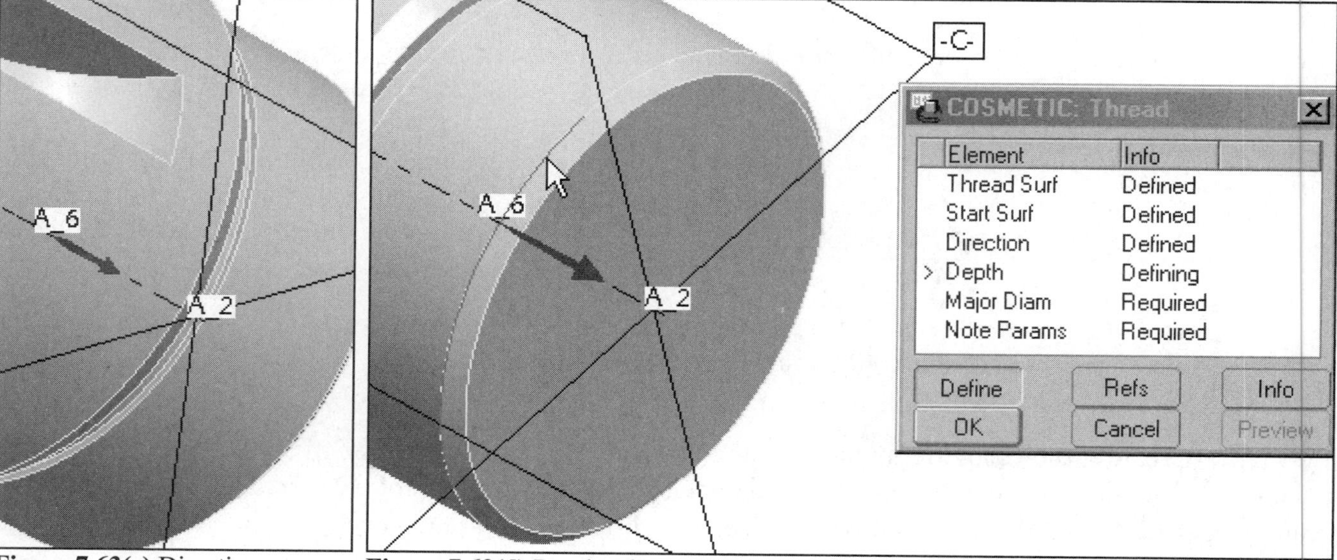

Figure 7.63(c) Direction **Figure 7.63(d)** Depth using UpTo Curve

Click: **UpTo Curve** ⇒ **MMB** ⇒ pick on the edge created by the chamfer [Fig. 7.63(d)] ⇒ type the diameter of the cosmetic thread root diameter: **2.6875** [Fig. 7.63(e)] ⇒ **MMB** ⇒ **Mod Params** ⇒ edit the table

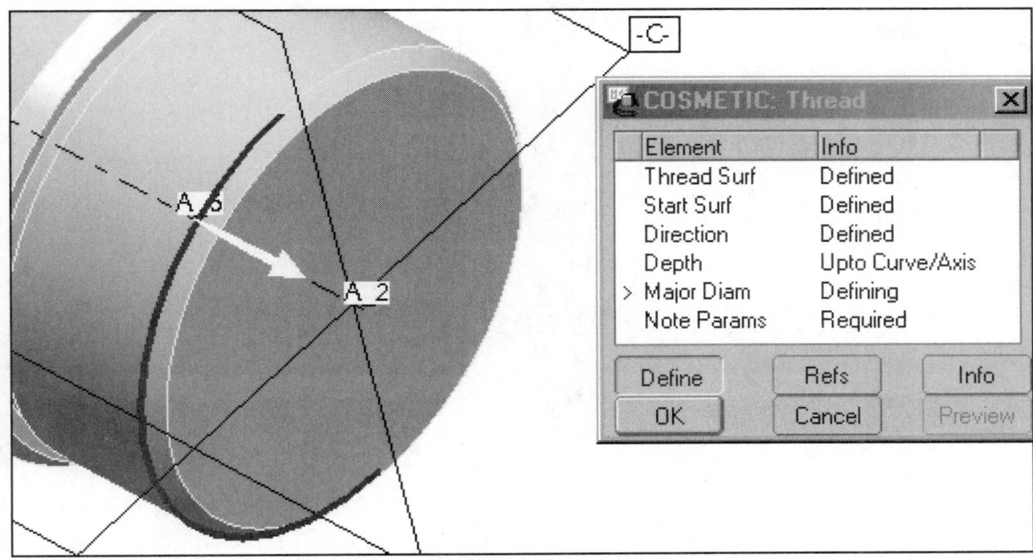

Figure 7.63(e) Enter the Minor Diameter (Root Diameter)

The Pro/TABLE information [Fig. 7.64(a)] shows the diameter as **2.6875** (**2.688** rounded), because the *cosmetic* cylinder representing the thread on your screen will be **∅2.688** (rounded). Since you are cosmetically representing the *root diameter* of the thread on the model, the *thread diameter* is *smaller* than the *nominal* thread size.

Change the **2.6875** dimension to **2.75** on the Thread Table [Fig. 7.64(a-b)] ⇒ **16** ⇒ **UN** ⇒ **2** ⇒ **A** ⇒ **False** ⇒ **File** ⇒ **Save** ⇒ **File** ⇒ **Exit** ⇒ **Show** (Fig. 7.65) ⇒ **Close** ⇒ **MMB** ⇒ **OK** (Fig. 7.66) ⇒ ⇒ **MMB** ⇒ **LMB** to deselect

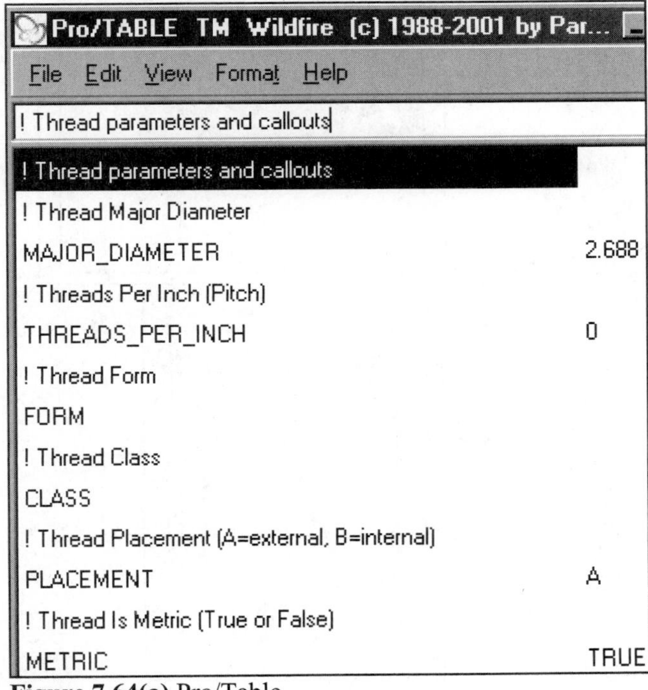

Figure 7.64(a) Pro/Table

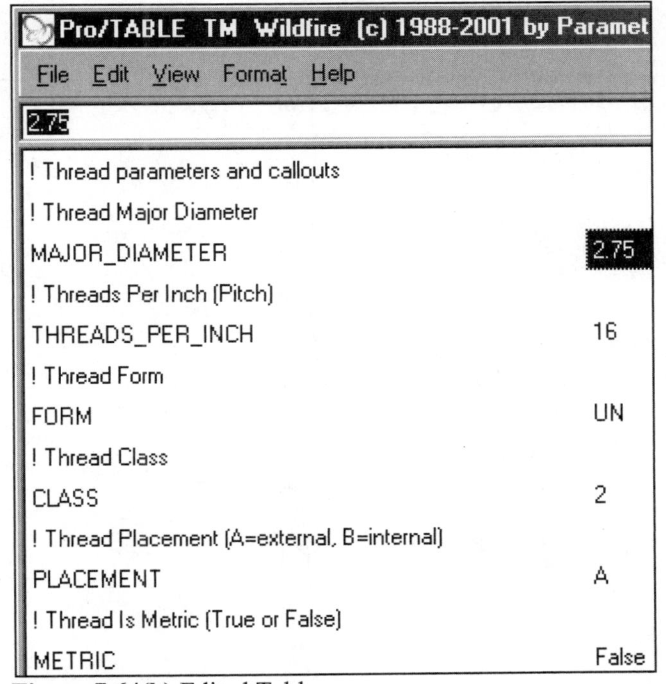

Figure 7.64(b) Edited Table

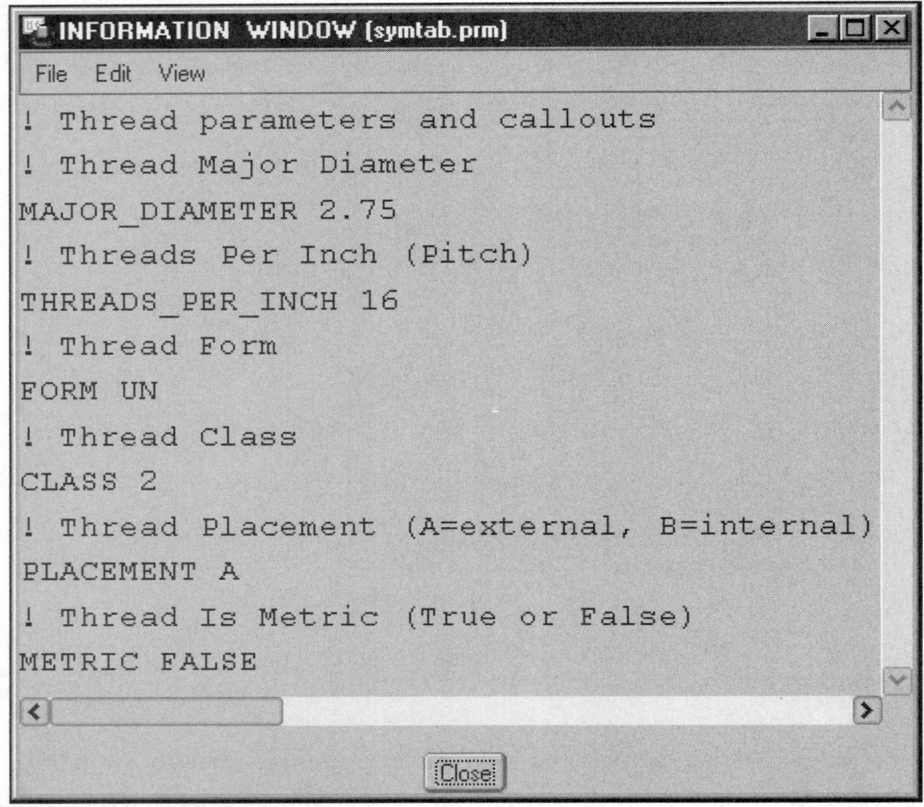

Figure 7.65 Thread Information

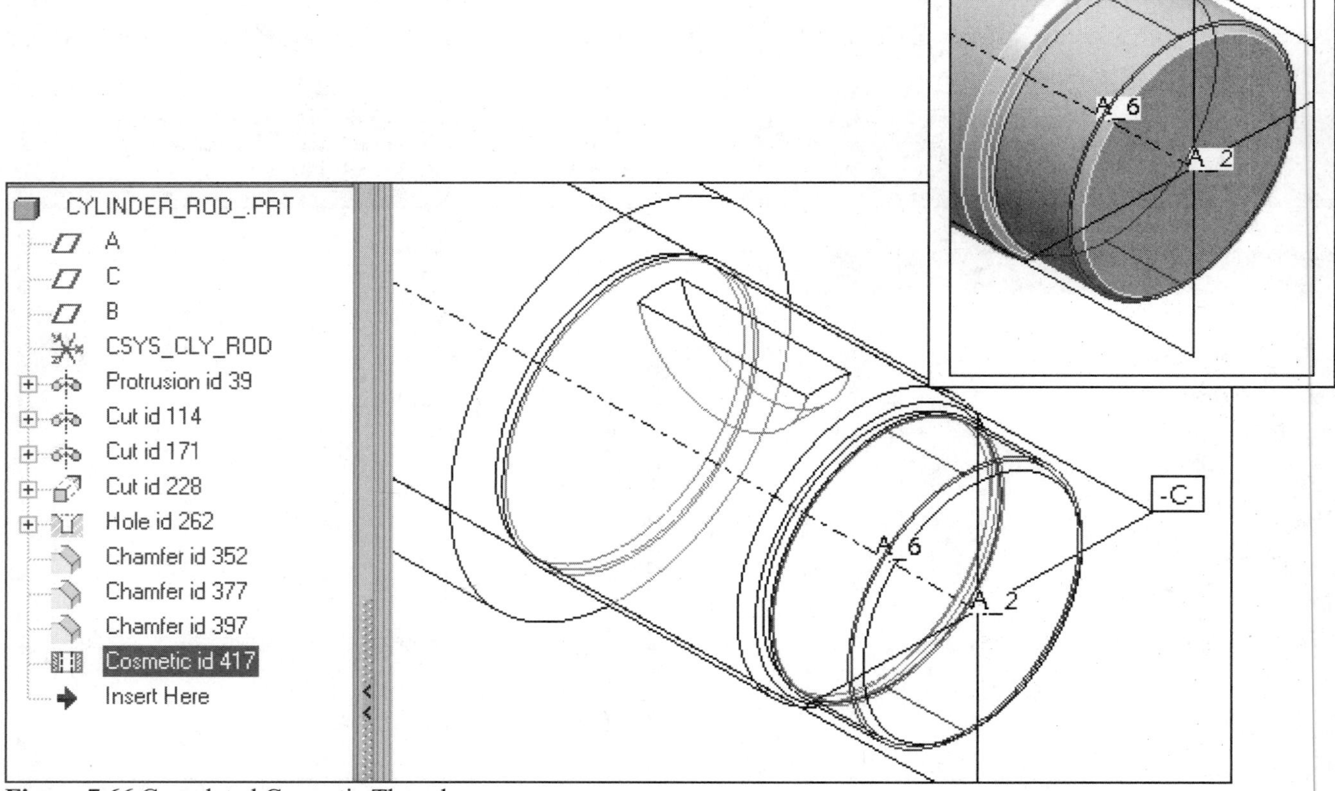

Figure 7.66 Completed Cosmetic Thread

Click on **Cosmetic id** in the Model Tree (Fig. 6.67) ⇒ **RMB** ⇒ **Info** ⇒ **Feature** ⇒ click on the quick sash to close the **Browser** ⇒ ⇒ **MMB**

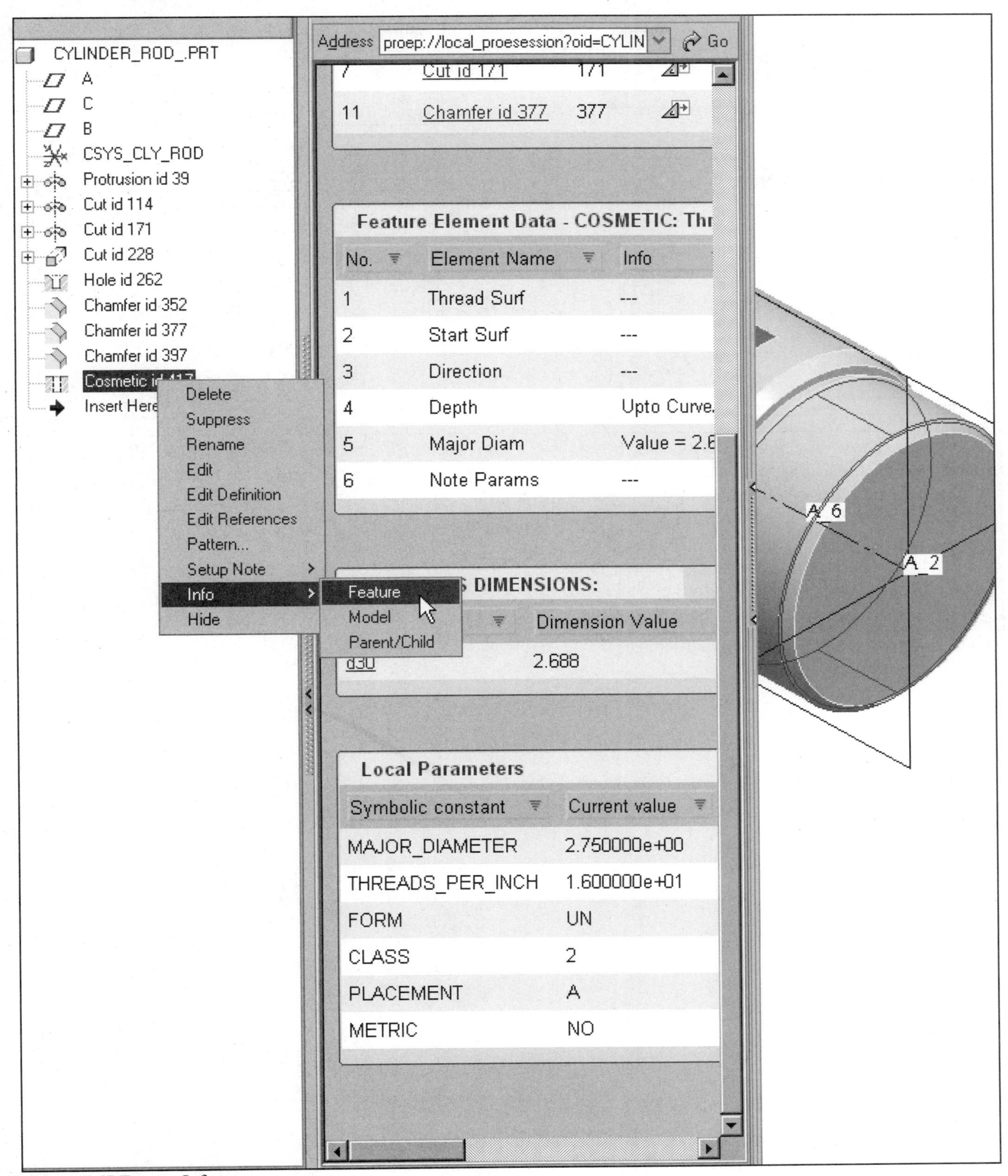

Figure 7.67 Feature Info

Before going on to the Lesson 7 Project, complete the **ECO** shown in Figure 7.68 for the Pin part from Lesson 6.

Look up the pipe thread in your **Machinery's Handbook** or a drafting/engineering drawing text to check the geometry sizes.

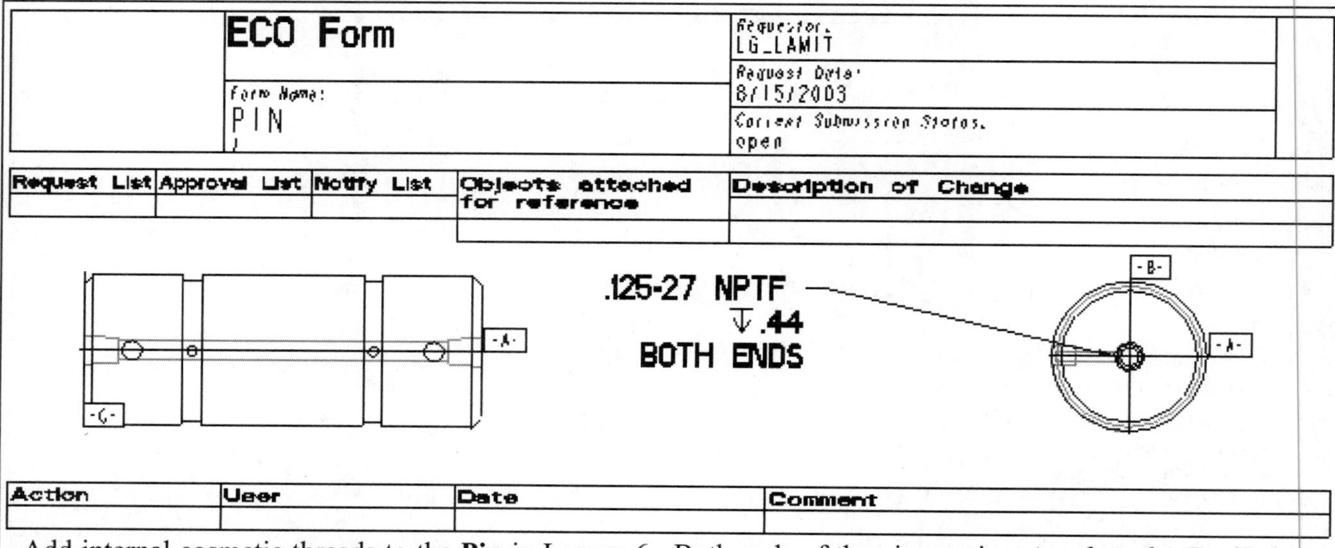

ECO Form

Form Name:
PIN

Requestor:
LG_LAMIT

Request Date:
8/15/2003

Current Submission Status:
open

Request List	Approval List	Notify List	Objects attached for reference	Description of Change

Action	User	Date	Comment

Add internal cosmetic threads to the **Pin** in Lesson 6. Both ends of the pin require *pipe threads*. Pro/E does not create a *conical* pipe cosmetic thread. Therefore, the cosmetic thread created will not be correct as far as the visual geometry is concerned--it will appear *cylindrical*.

The pipe thread for both ends is: **.125-27 NPTF**
.44 DEEP

The diameter at the small end of the tapered thread is **.339** and the angle of taper is **1.48°** (taper of thread on diameter is *3/4 inch per foot*) from the centerline.

Modify the existing conical cut before you create the cosmetic threads.

Figure 7.68 ECO for the Pin in Lesson 6

Lesson 7 Project

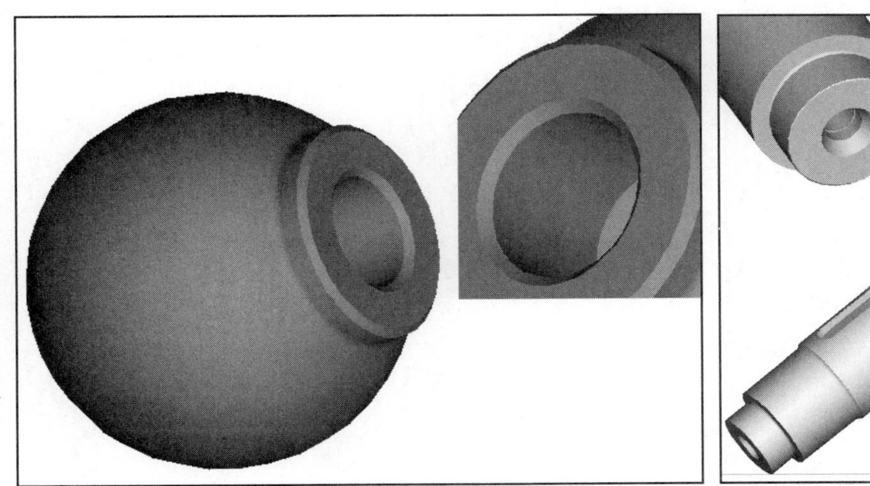

Figure 7.69 Clamp Ball

Figure 7.70 Coupling Shaft

Clamp Ball and Coupling Shaft

As with the previous lesson project, two lesson projects are provided here in Lesson 7 (Figs. 7.69 through 7.84). You will use both parts in Lessons 15 and 16 when creating different assemblies. Both the Clamp **Ball** (decimal inch) and the Coupling Shaft (SI units) are revolved protrusions. The Clamp Ball is simpler and easier to complete. The Clamp Ball is *black plastic* and the Coupling Shaft is *steel*. Create all cosmetic threads required on each part. The two parts are used on different assemblies. Analyze the parts and plan the steps and features required to model them. Remember to set up the environment, set datum planes, and add layers to the project parts.

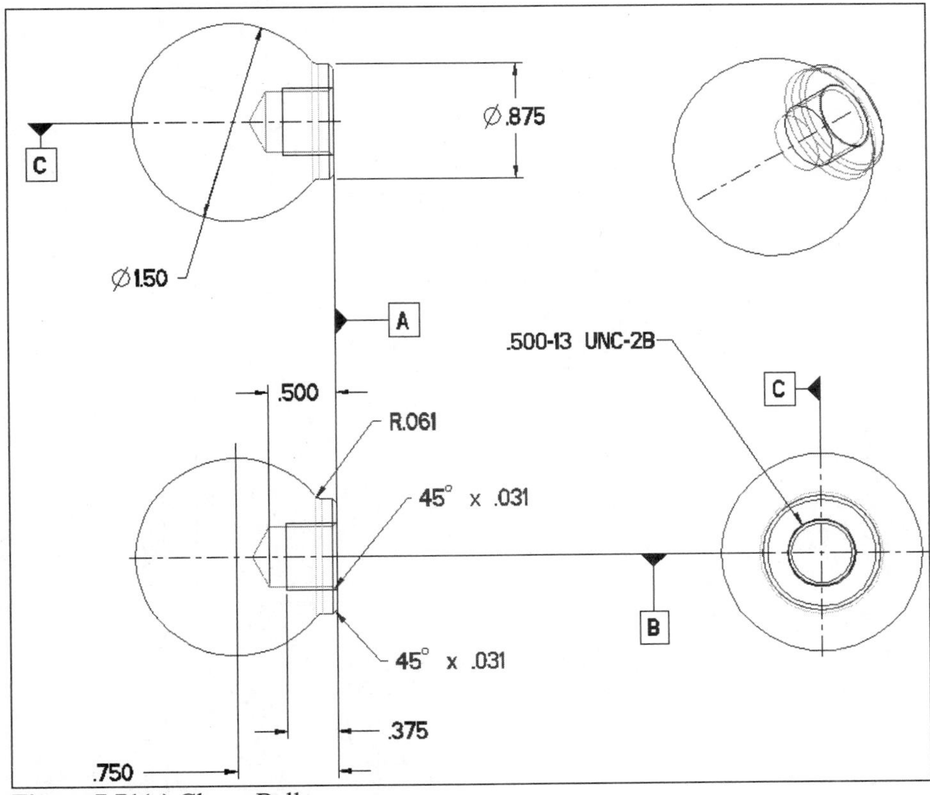

Figure 7.71(a) Clamp Ball

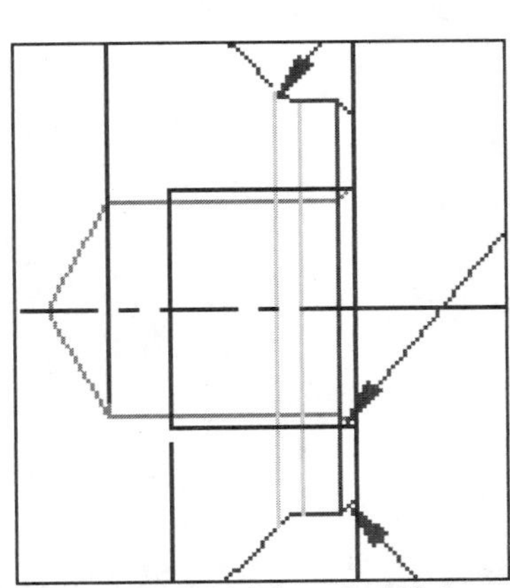

Figure 7.71(b) Threaded Hole

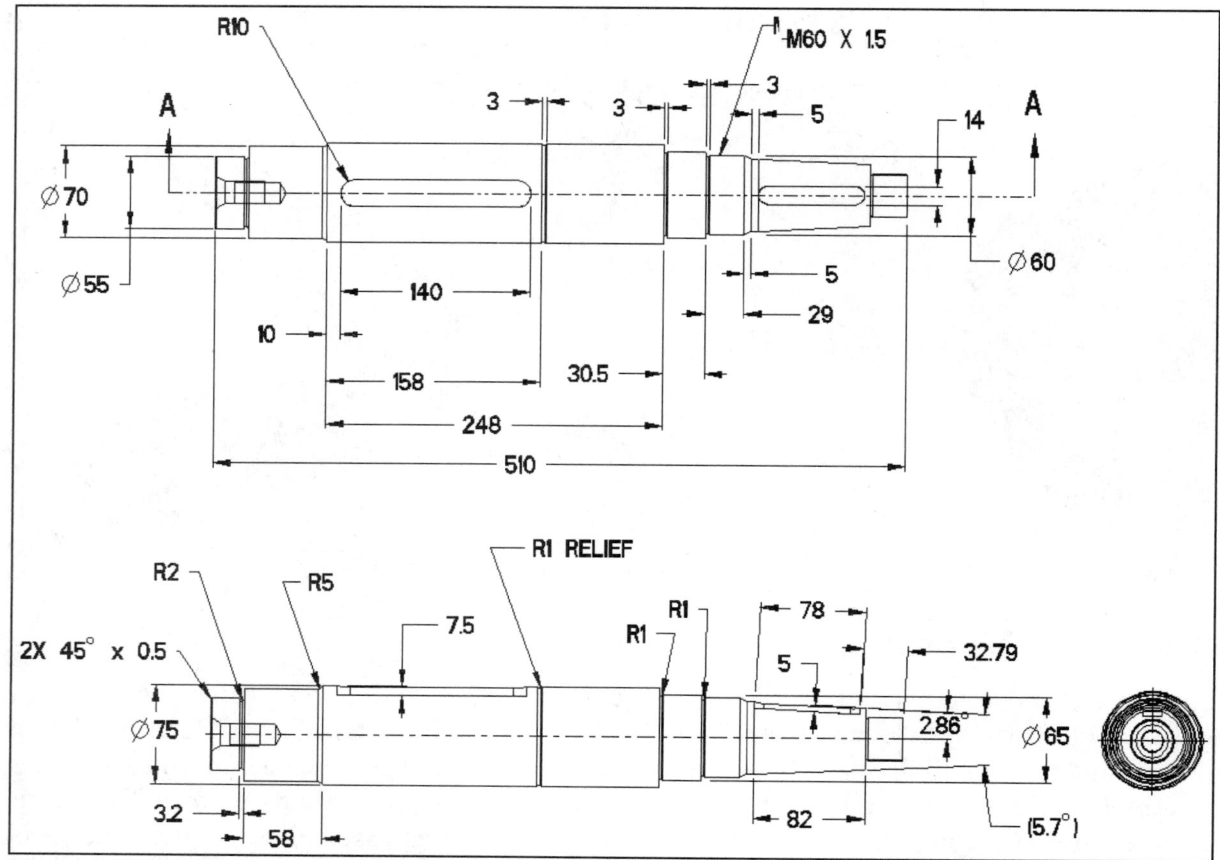

Figure 7.72 Coupling Shaft Drawing, Sheet One

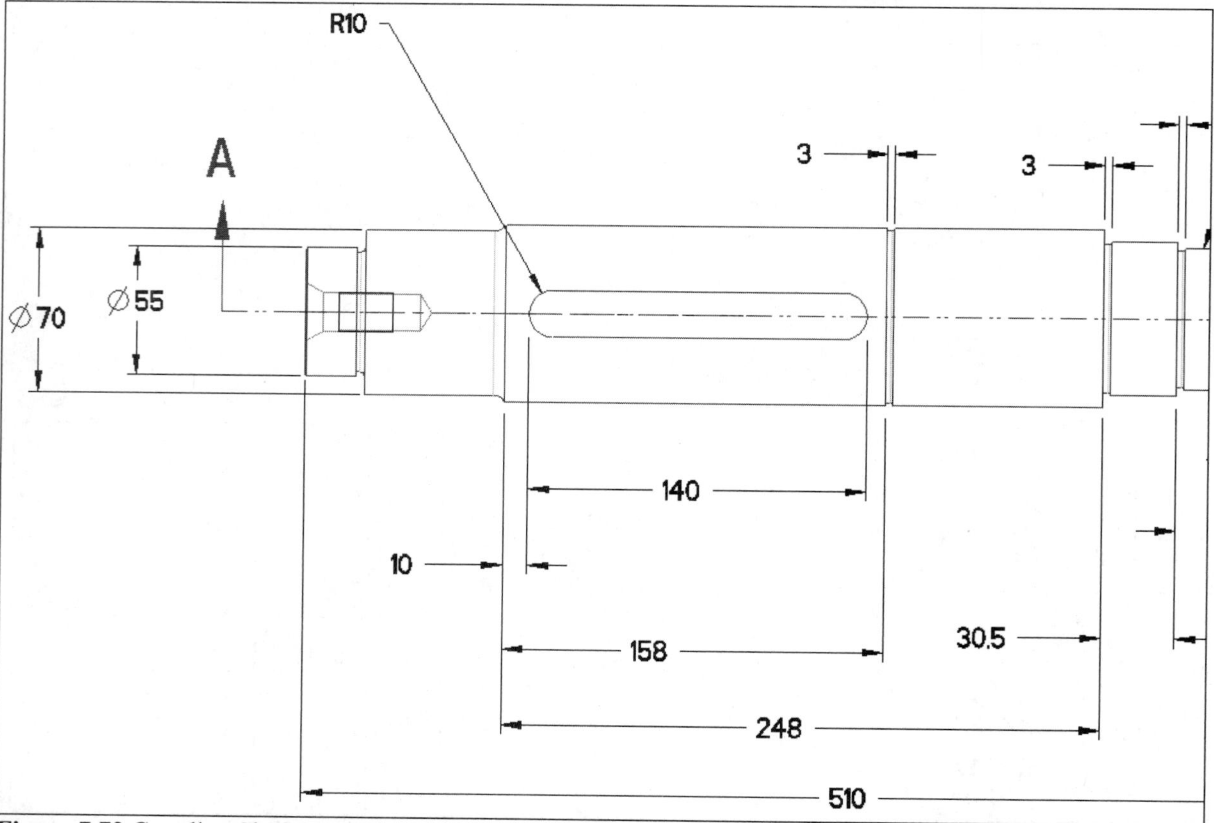

Figure 7.73 Coupling Shaft Drawing, Top View, Left Side

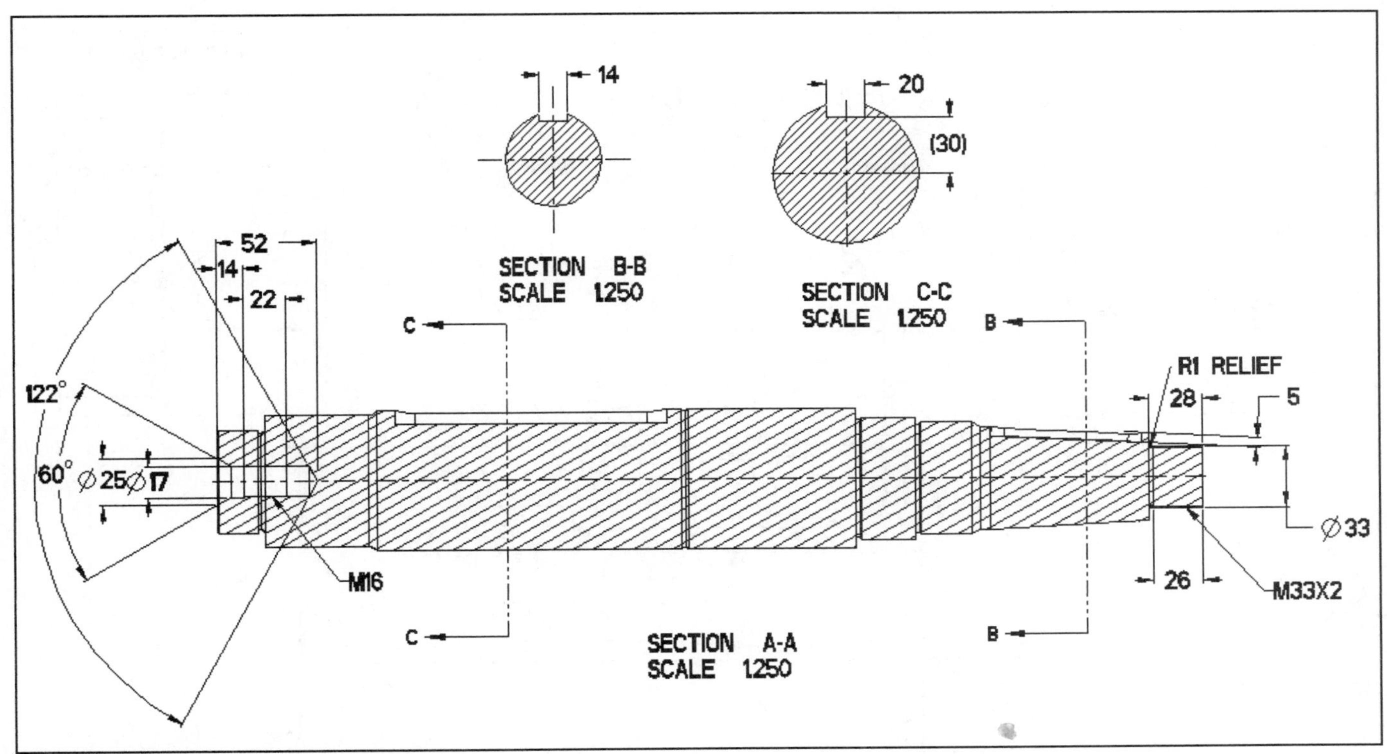

Figure 7.74 Coupling Shaft Drawing, Sheet Two

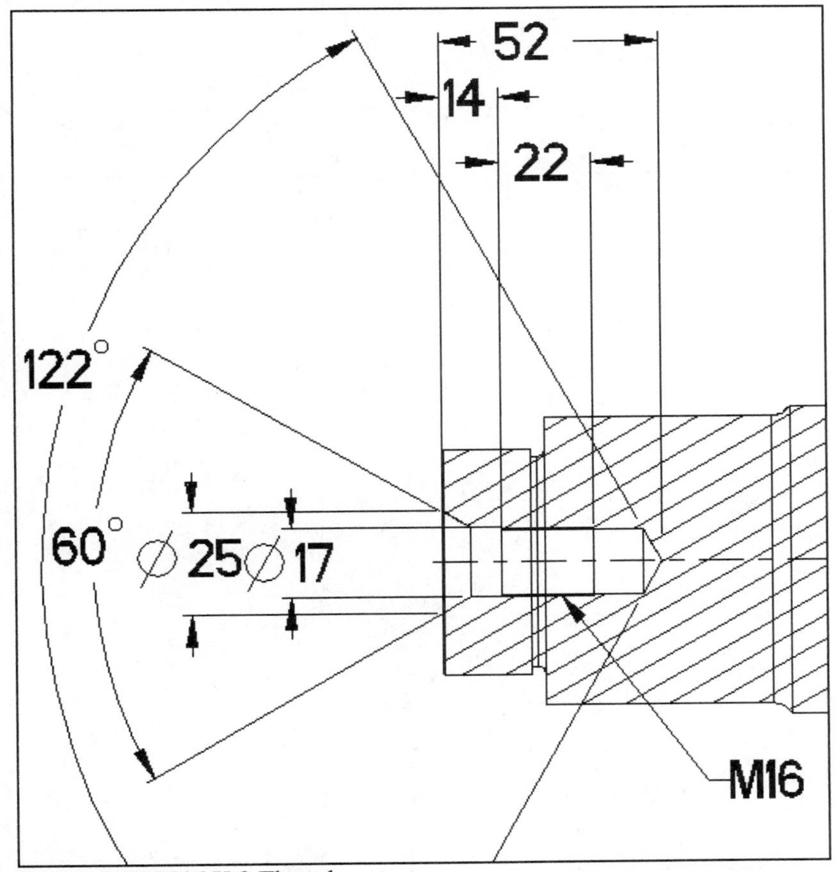

Figure 7.75 M16 X 2 Thread

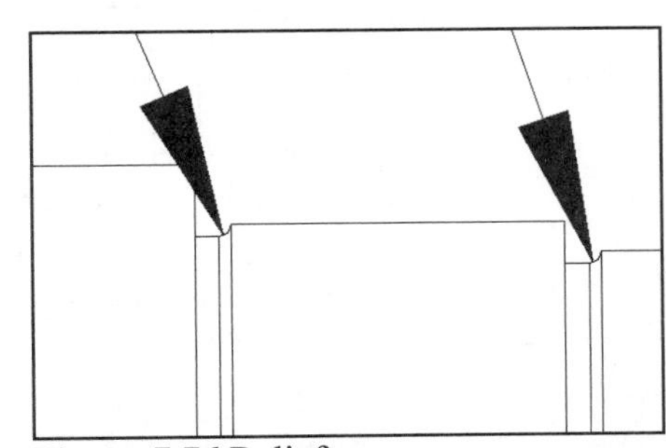

Figure 7.76 Reliefs

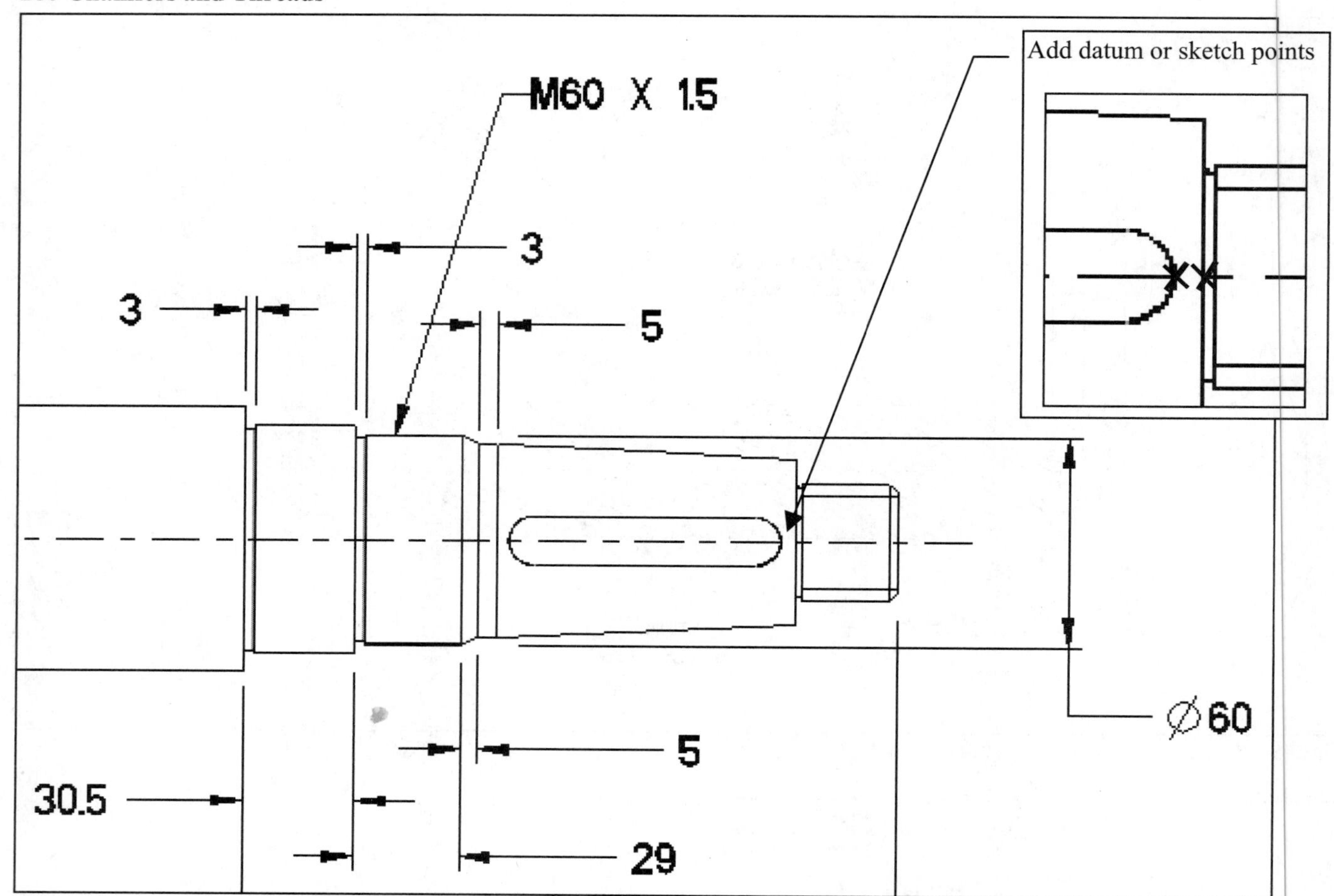

Figure 7.77 Coupling Shaft Drawing, Top View, Right Side

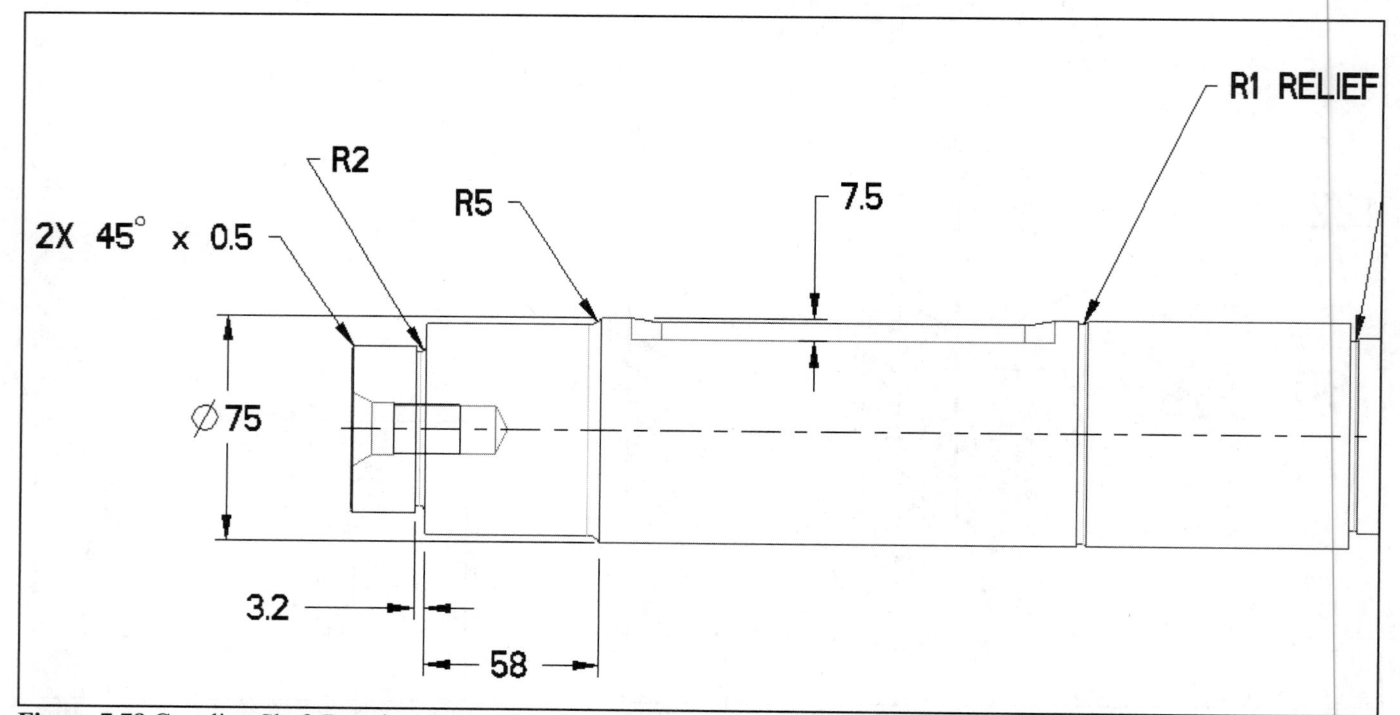

Figure 7.78 Coupling Shaft Drawing, Front View, Left Side

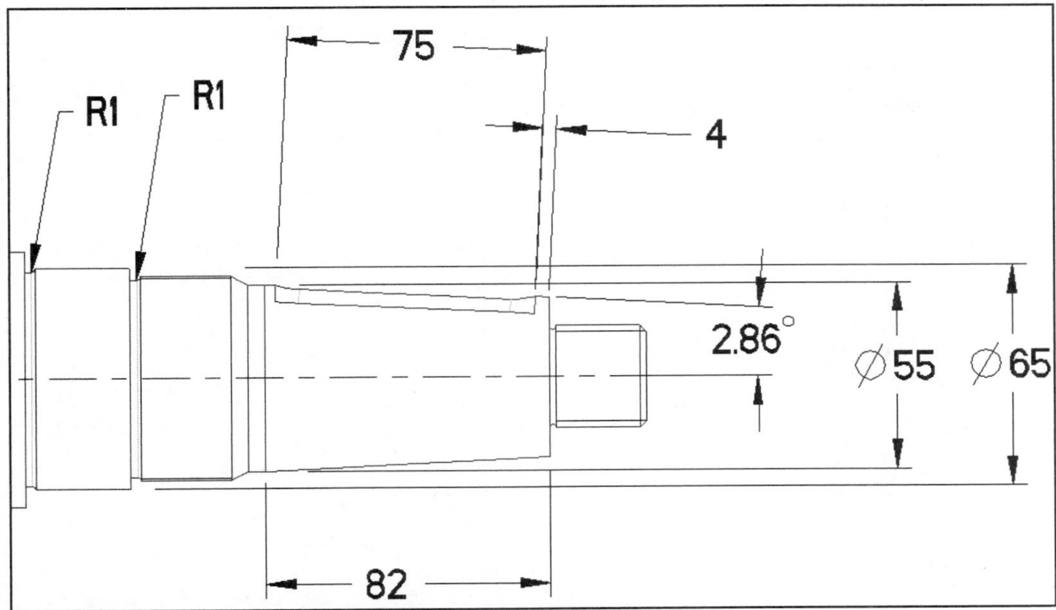

Figure 7.79 Coupling Shaft Drawing, Front View, Right Side

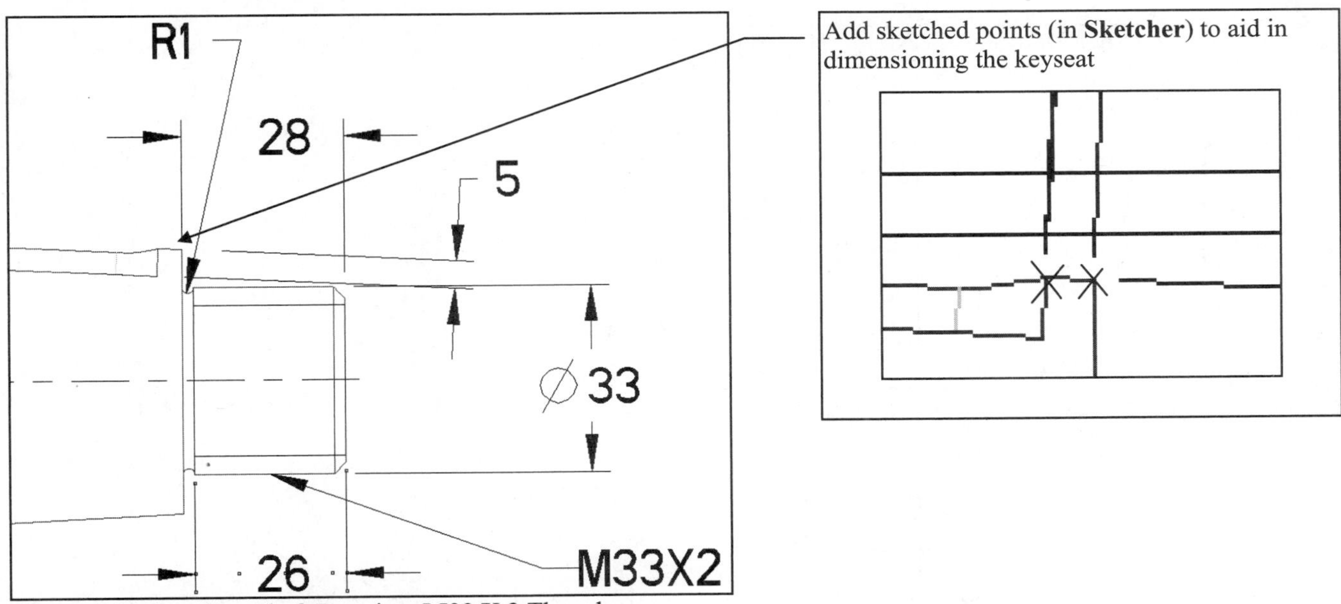

Figure 7.80 Coupling Shaft Drawing, **M33 X 2** Threads

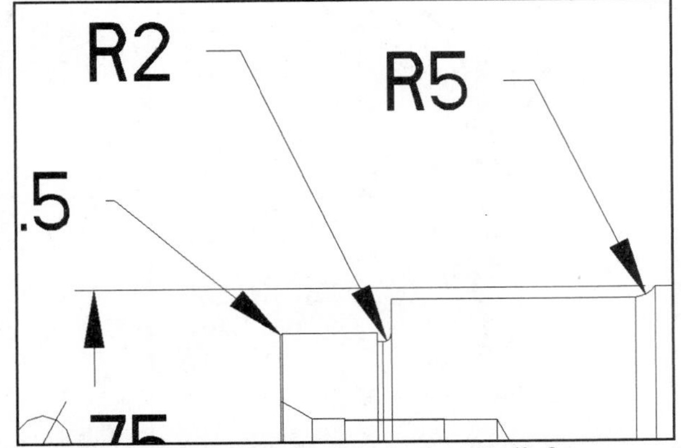

Figure 7.81 Coupling Shaft Drawing, Reliefs

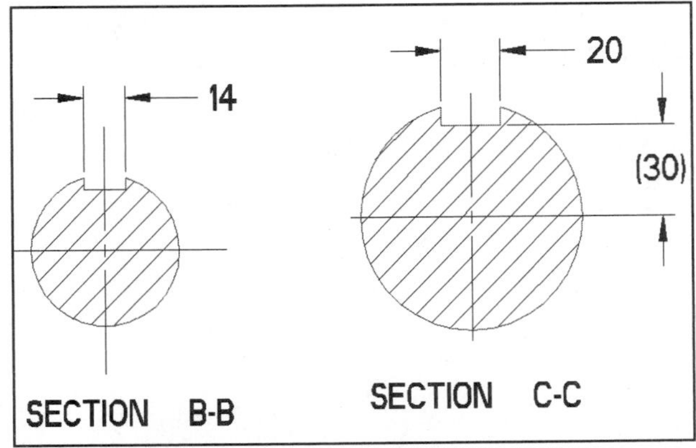

Figure 7.82 SECTION B-B and SECTION C-C

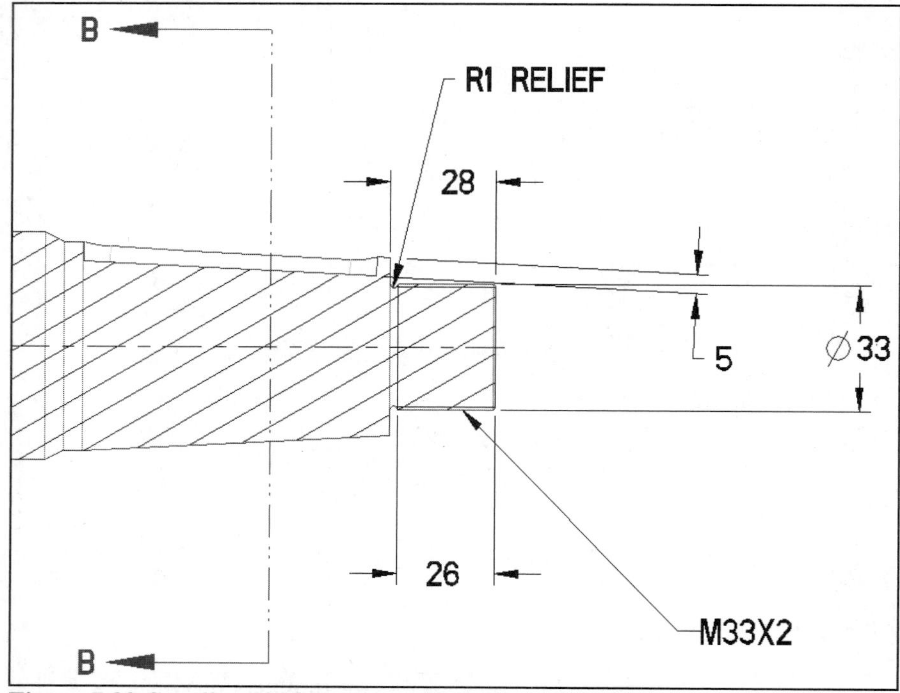

Figure 7.83 Coupling Shaft Drawing, Sheet Two, SECTION A-A Right Side

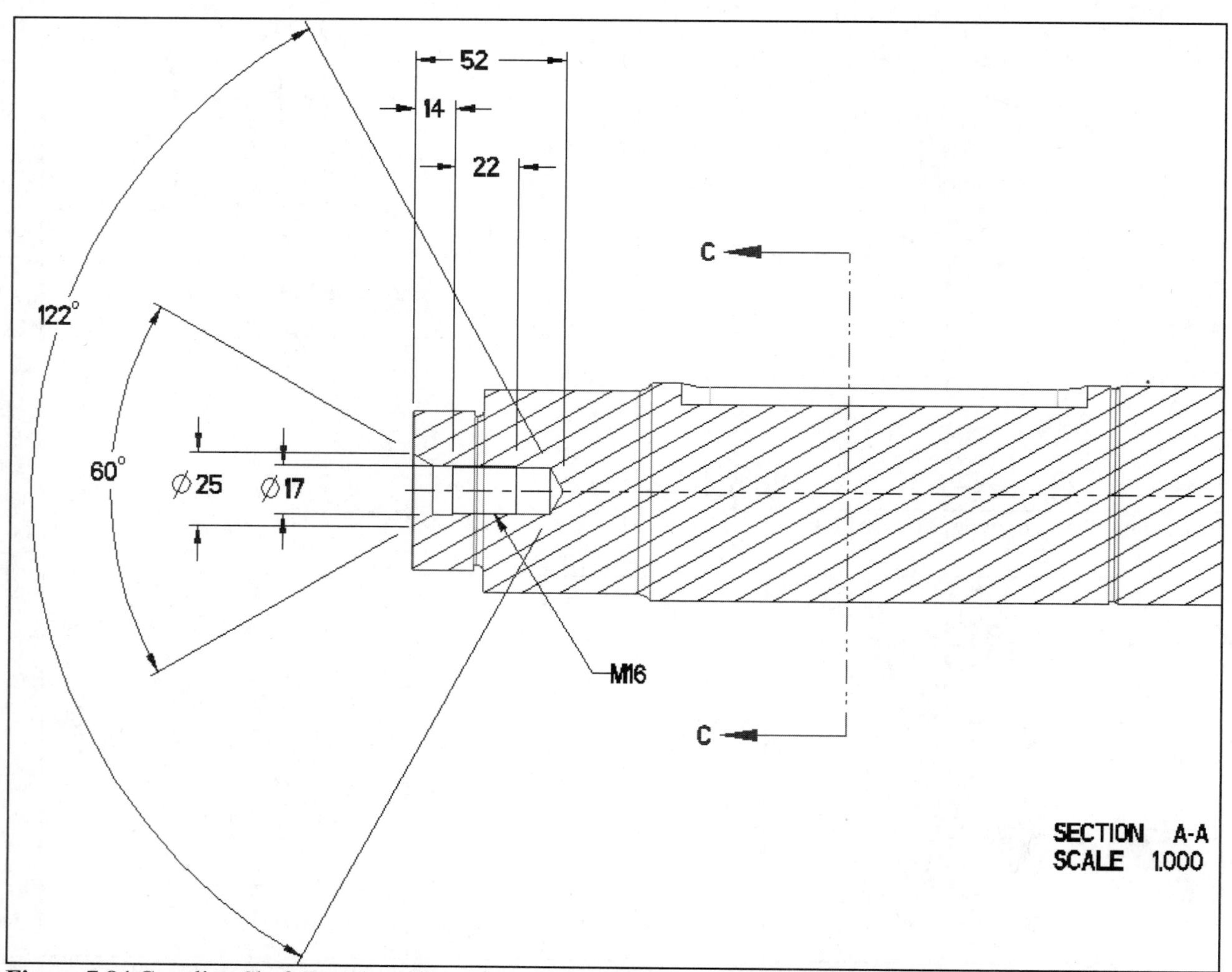

Figure 7.84 Coupling Shaft Drawing, Sheet Two, SECTION A-A Left Side

Lesson 8 Groups and Patterns

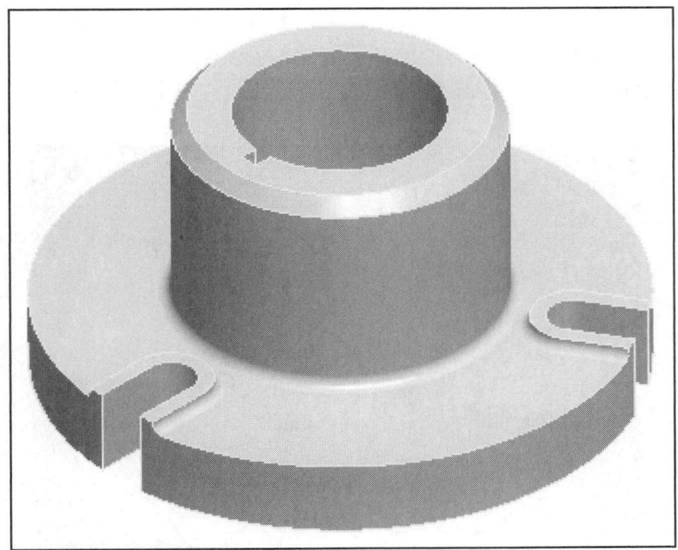

Figure 8.1(a) Post Reel

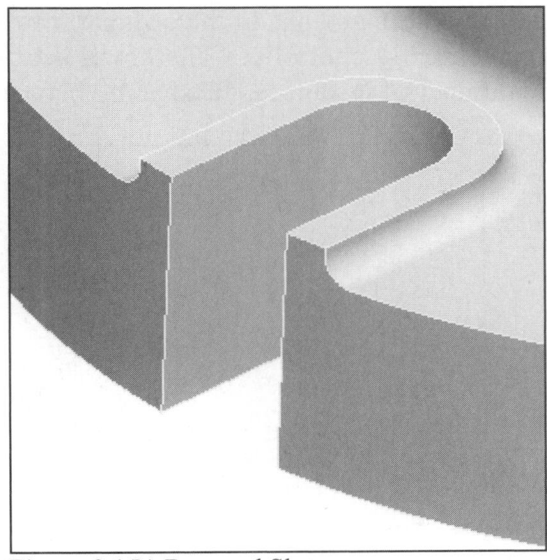

Figure 8.1(b) Boss and Slot

OBJECTIVES

- Create **Mapkeys**
- Personalize the Pro/E interface using **Customize Screen**
- Create and manipulate **Groups**
- **Pattern** features
- Understand how to **Pattern Groups**

Groups and Patterns

The **Group** option unites a series of features. To create multiple features from a single feature (or group of features), the **Pattern** option can be used. After it is created, a pattern behaves as if it were a single feature. When you create a pattern, you create instances (copies) of the selected feature (or group of features). Creating a pattern is a quick way to reproduce a feature or a set of features that are related and grouped for easy manipulation. Manipulating a pattern may be more advantageous for you than operating on individual features. The post reel [Fig. 8.1(a-b)] has a protrusion, slot, and rounds grouped and patterned.

Group

When you create a *local group*, you must select the features in the consecutive (sequential) order of the regeneration list. A quick way to do this is to select the intended group by range. If there are features between the specified features in the regeneration list, Pro/E asks whether you want to group all the features in between. Features that are already in other groups cannot be grouped again.

Pattern

When you create a pattern, Pro/E assumes it is a “single” feature. Creating a pattern is a quick way to reproduce a feature. Patterning is an easier and more effective way to perform a single operation on the multiple features contained in a pattern, rather than on the individual features [Fig. 8.2(a-b)]. For example, you can easily suppress a pattern or add it to a layer.

You can pattern a feature with dimensions (**Dimension**), using a table (**Table**), referencing (**Reference**) an existing pattern and by filling (**Fill**) in a sketched boundary.

Modifying patterns is more efficient than modifying individual features. In a pattern, when you change the dimensions of the original feature, the whole pattern can be updated automatically. A pattern is parametrically controlled. Therefore, a pattern can be modified by changing pattern parameters, such as the number of instances, the spacing between instances, and the original feature dimensions (Fig. 8.3).

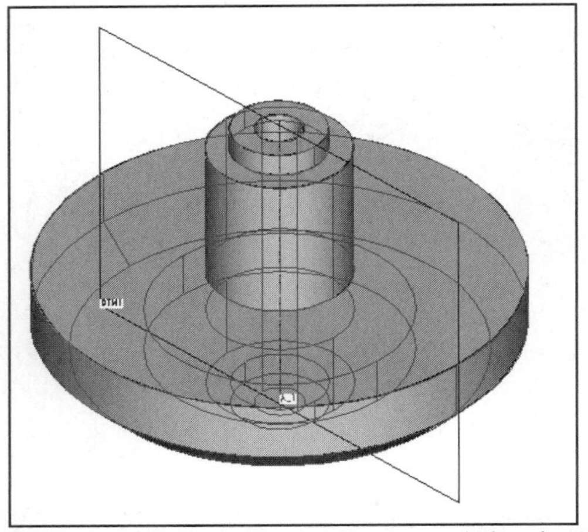

Figure 8.2(a) Duplicating Features (Circular Pattern)

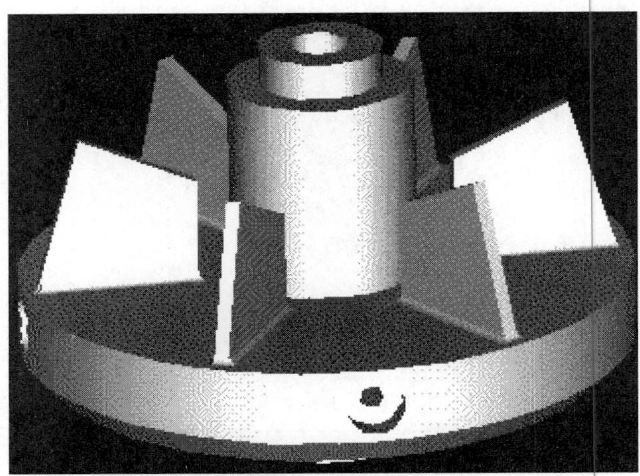

Figure 8.2(b) Ribs (COAch for Pro/ENGINEER)

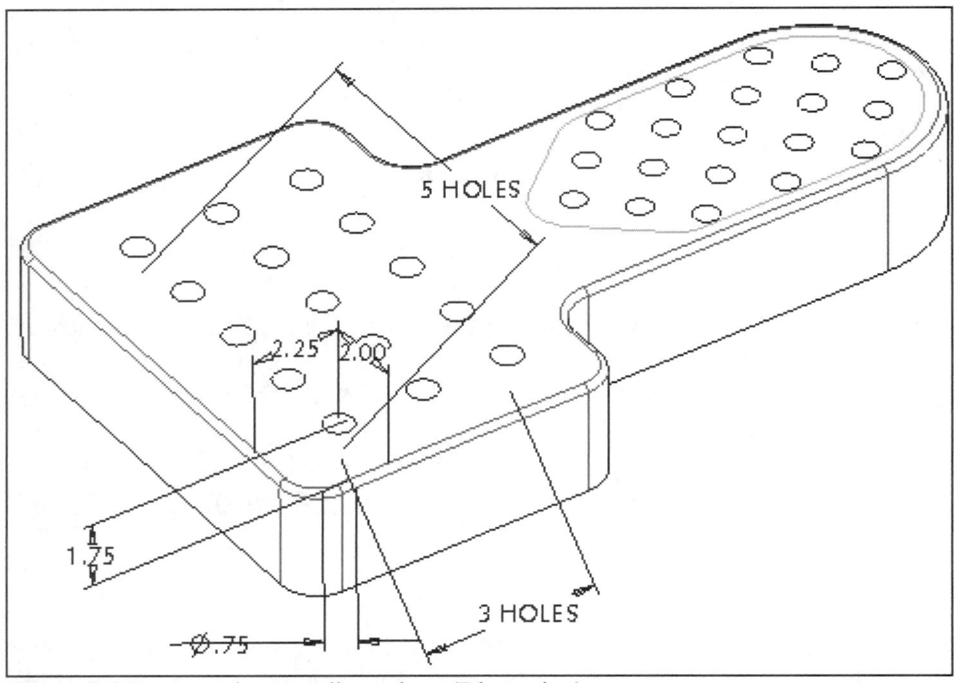

Figure 8.3 Pattern using two dimensions (Dimension)

Pattern Tables

You can pattern features using pattern tables. Pattern tables allow you to create complicated or irregular patterns of features or groups by letting you specify unique dimensions for each instance in the pattern through an editable table. Multiple tables can be established for a pattern, so you can change the pattern by switching the table that drives it.

You can modify a pattern table at any time after you create the pattern. Suppressing or deleting a table-driven pattern suppresses or deletes the pattern leader, as well. You can use pattern tables in Assembly mode to pattern assembly features and components.

Pattern tables are *not* family tables. Pattern tables (Fig. 8.4) can only drive pattern dimensions, and unless they are unpatterned, pattern instances cannot be made independent. You can also include pattern tables in family tables so a particular family table instance can use a specified pattern table.

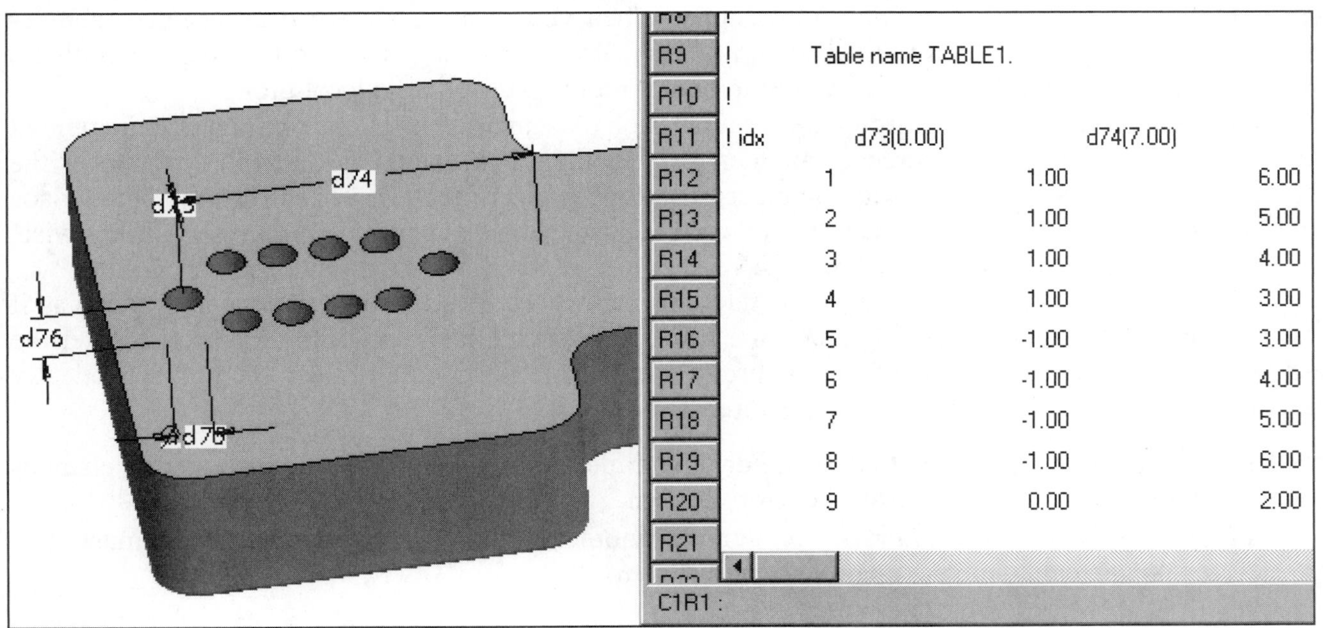

Row	! idx	d73(0.00)	d74(7.00)
R12	1	1.00	6.00
R13	2	1.00	5.00
R14	3	1.00	4.00
R15	4	1.00	3.00
R16	5	-1.00	3.00
R17	6	-1.00	4.00
R18	7	-1.00	5.00
R19	8	-1.00	6.00
R20	9	0.00	2.00
R21			

Figure 8.4 Pattern using a Table (Table)

Pattern Fill

Another option is to fill a specified closed boundary with a pattern. The spacing, angle, style, and edge distance from the sketched curve can all be specified to meet specific design requirements.

Square, **Diamond** (Fig. 8.5), **Triangle** and **Circle** are the fill space options.

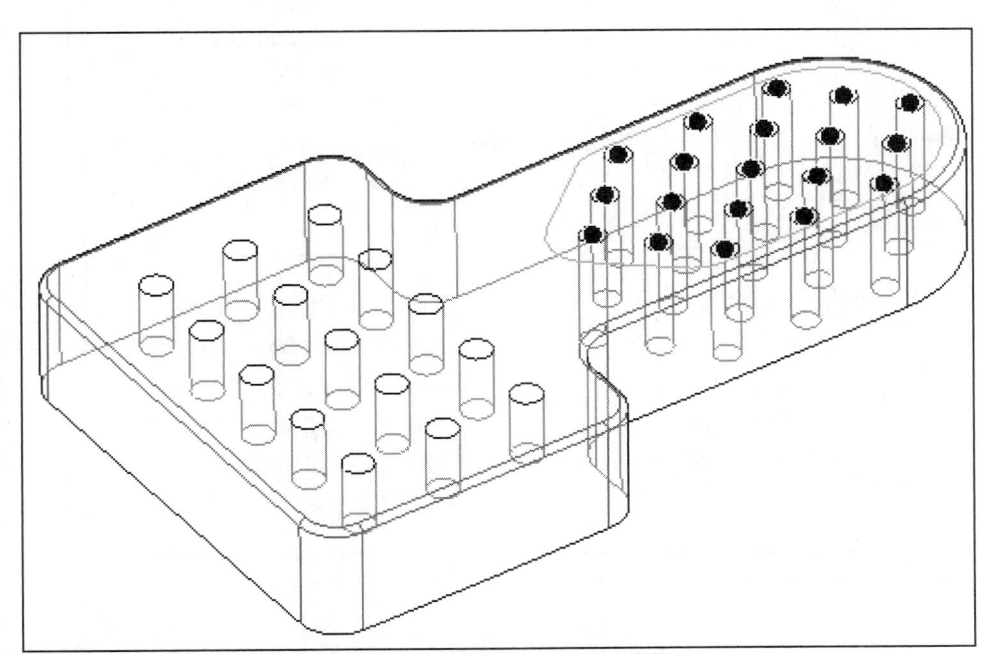

Figure 8.5 Pattern using a Sketched Section Boundary (Fill)

Patterning Groups

You can pattern groups created from UDFs (user-defined features) and local groups using the GROUP menu option Pattern. This option differs from the FEAT menu option Pattern in that the GROUP menu option treats an entire group as a single entity. The FEAT menu Pattern option is used to pattern one feature at a time.

You can select all the dimensions in the selected group, except those used to create a feature pattern within the group, as incremental dimensions. When you create a patterned group, one member represents the whole group. When regenerating, however, Pro/E regenerates all the features individually. To pattern a group, you must first name and group two or more features into a local group.

When you pattern or copy a group, be careful which placement dimensions you select to increment or vary. If a feature in a group references another for placement (for example, a chamfer references the edge of a hole), you need to change only the placement dimensions of the referenced feature. If you place features in a group separately, you must change the placement dimensions of *each member*. Otherwise, features with unchanged dimensions will have several copies superimposed onto one another.

Pro/E allows you to pattern a single feature only. However, you can pattern several features as if they were a single feature by arranging them in a **Local Group** and then patterning the group. After the pattern is created, you can unpattern and ungroup the instances and then make them independently modifiable. There are two ways to pattern a feature:

- **Dim Pattern** Controls the pattern using driving dimensions to determine the incremental changes to the pattern. This is the case for the entire pattern.
- **Ref Pattern** Controls the pattern by referencing another pattern. The dimension pattern must exist before you can create the next pattern type (Fig. 8.6).

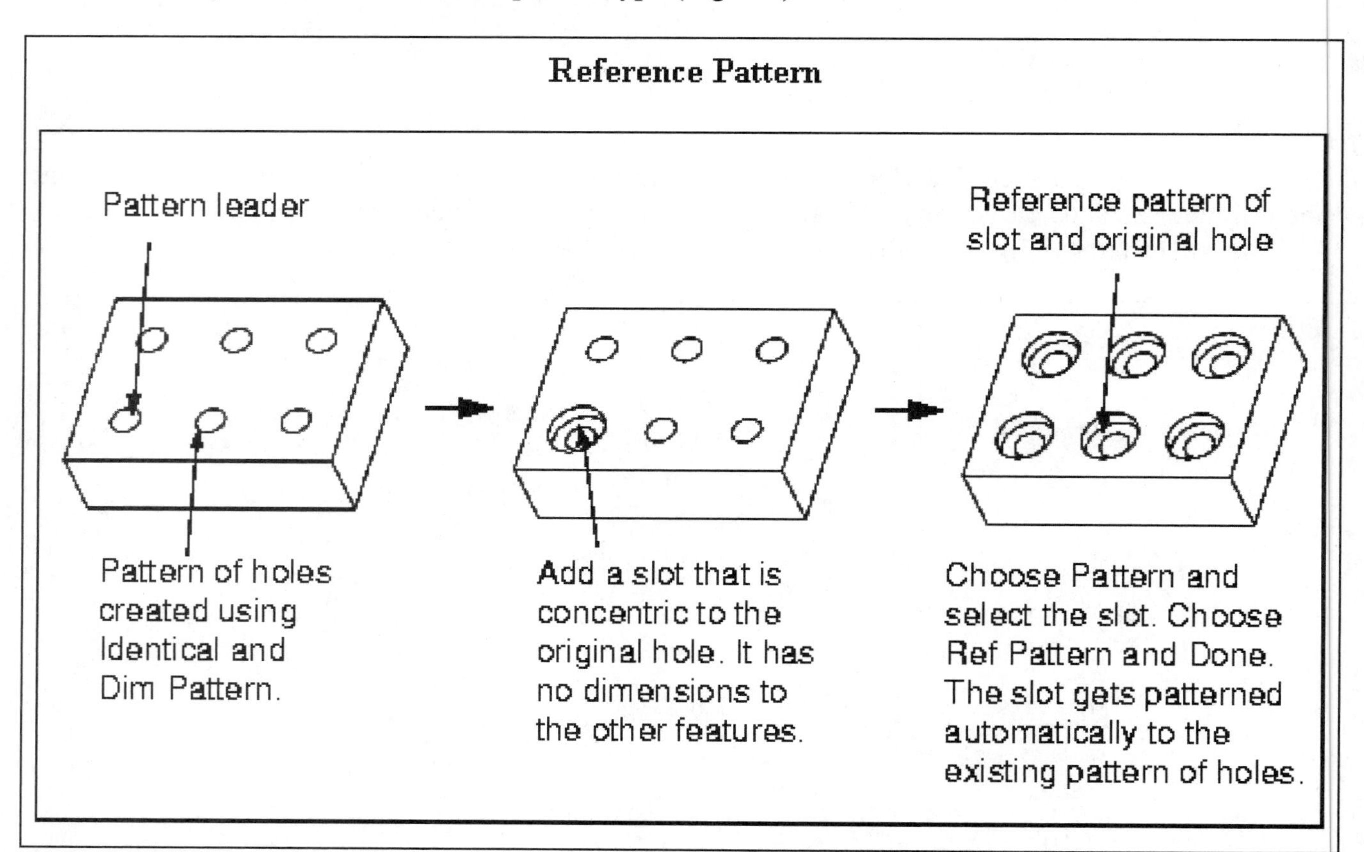

Figure 8.6 Reference Pattern

Lesson 8 STEPS

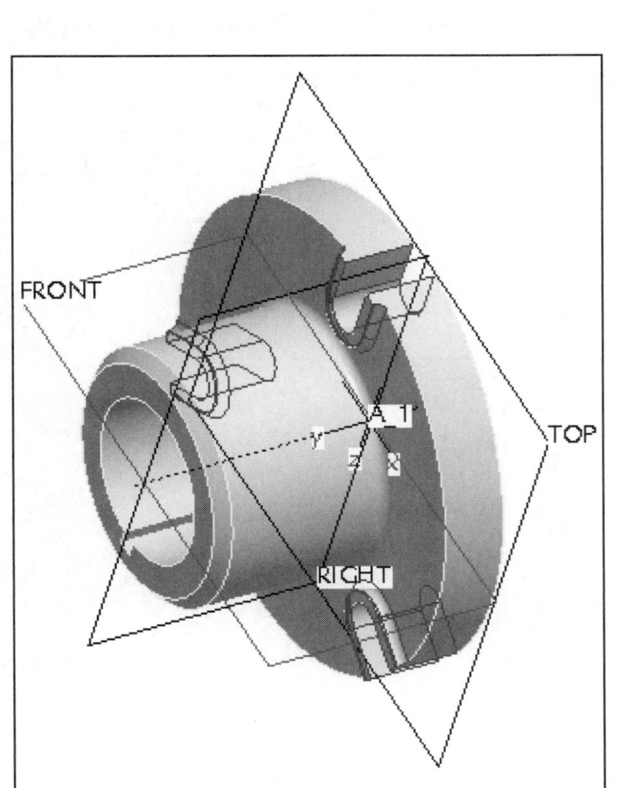

Figure 8.7(a) Post Reel Model

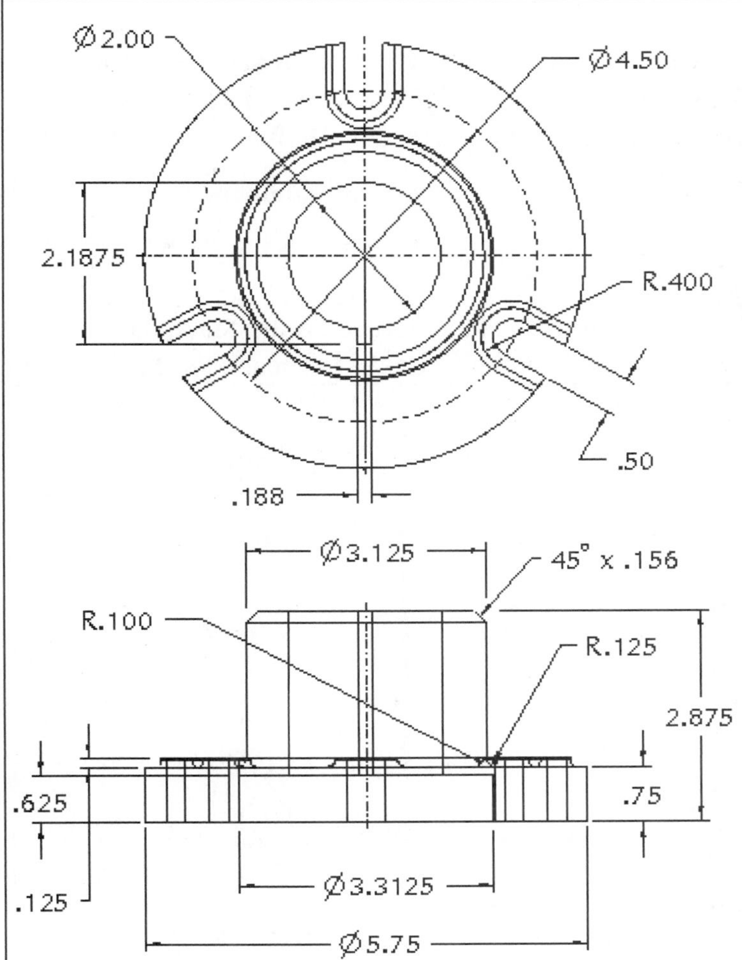

Figure 8.7(b) Post Reel Detail

Post Reel

The Post Reel [Fig. 8.7(a-b)] is created with a *revolved protrusion*, as in Lesson 6 and Lesson 7. The internal geometry of the Post Reel can be created with a *revolved cut* instead of two holes of differing diameters or a sketched hole. The chamfers and the rounds are simple pick-and-place features.

One boss and one slot are created using a non-default datum plane, then grouped and patterned to complete the part. A detailed set of instructions will be supplied only for the boss and slot, because the other geometry is similar to that presented in previous lessons.

The dimensions for the part are provided in Figure 8.7(a-b).

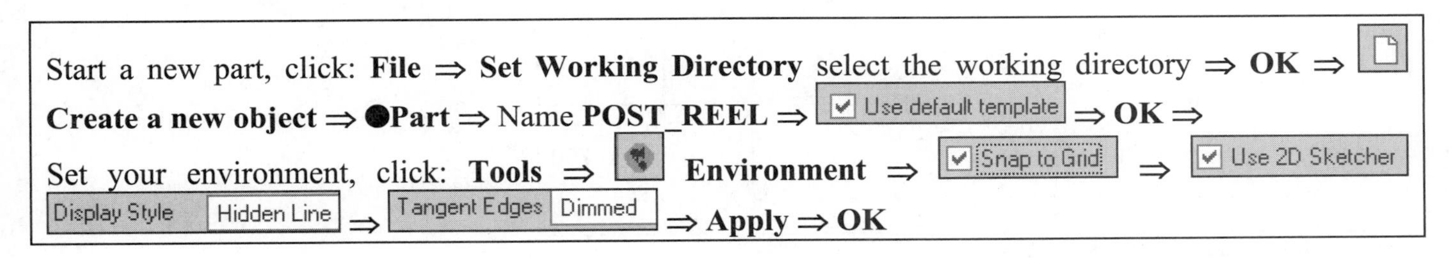
Start a new part, click: **File** ⇒ **Set Working Directory** select the working directory ⇒ **OK** ⇒ **Create a new object** ⇒ ●**Part** ⇒ Name **POST_REEL** ⇒ ☑ Use default template ⇒ **OK** ⇒
Set your environment, click: **Tools** ⇒ **Environment** ⇒ ☑ Snap to Grid ⇒ ☑ Use 2D Sketcher
Display Style Hidden Line ⇒ Tangent Edges Dimmed ⇒ **Apply** ⇒ **OK**

Set your configuration options, click: **Tools** ⇒ **Options** ⇒ Showing: **Current Session** ⇒ Option: ***default_dec_places***⇒ Value: **3** ⇒ **Enter** ⇒ Option: ***sketcher_dec_places*** ⇒ Value: **3** ⇒ **Enter** ⇒ **Apply** ⇒ **Close**

Set your material, click: **Edit** ⇒ **Setup** ⇒ **Units** ⇒ Units Manager **Inch lbm Second** ⇒ **Close** ⇒ **Material** ⇒ **Define** ⇒ type **STEEL_1020** ⇒ **MMB** ⇒ **File** (from material table) ⇒ **Save** ⇒ **File** ⇒ **Exit** ⇒ **Assign** ⇒ pick **STEEL_1020** ⇒ **Accept** ⇒ **MMB**

Change the default coordinate system name ⇒ double-click on the default coordinate system in the Model Tree-- **PRT_CSYS_DEF** ⇒ type **CSYS_POST_REEL** ⇒ **Enter** ⇒ [save icon] ⇒ **MMB**

The first protrusion is a revolved protrusion and consists of flange-like geometry. Sketch on **FRONT**, and revolve the section **360°** about a vertical axis (Figures 8.8 through 8.11).

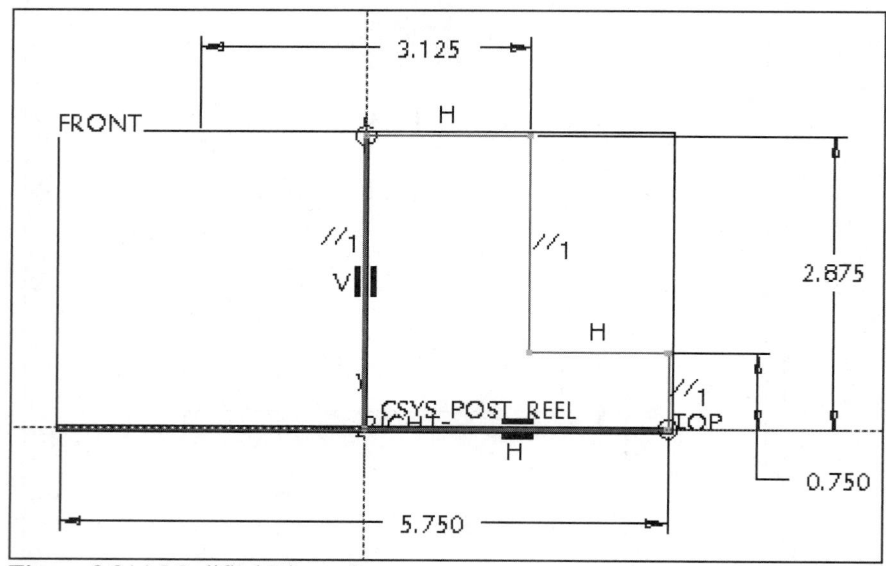

Figure 8.8(a) Modified Dimensions

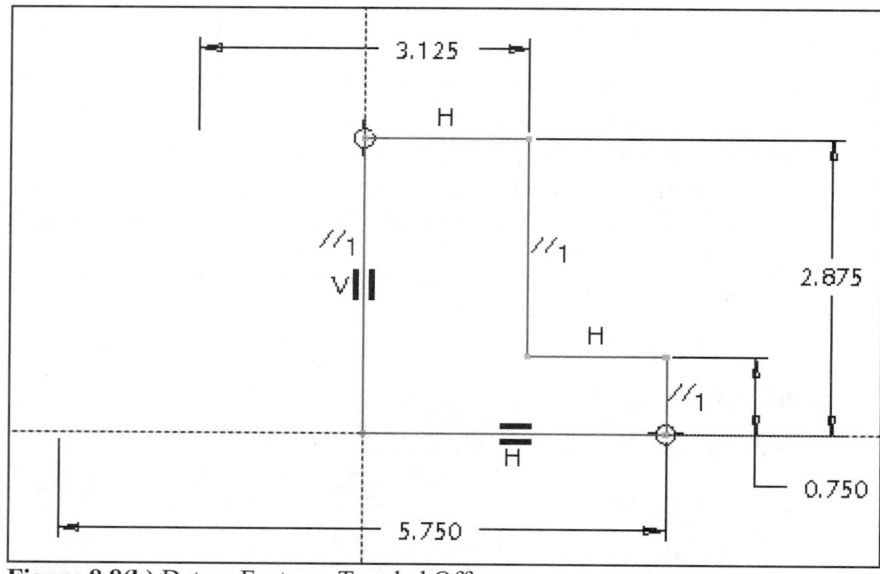

Figure 8.8(b) Datum Features Toggled Off

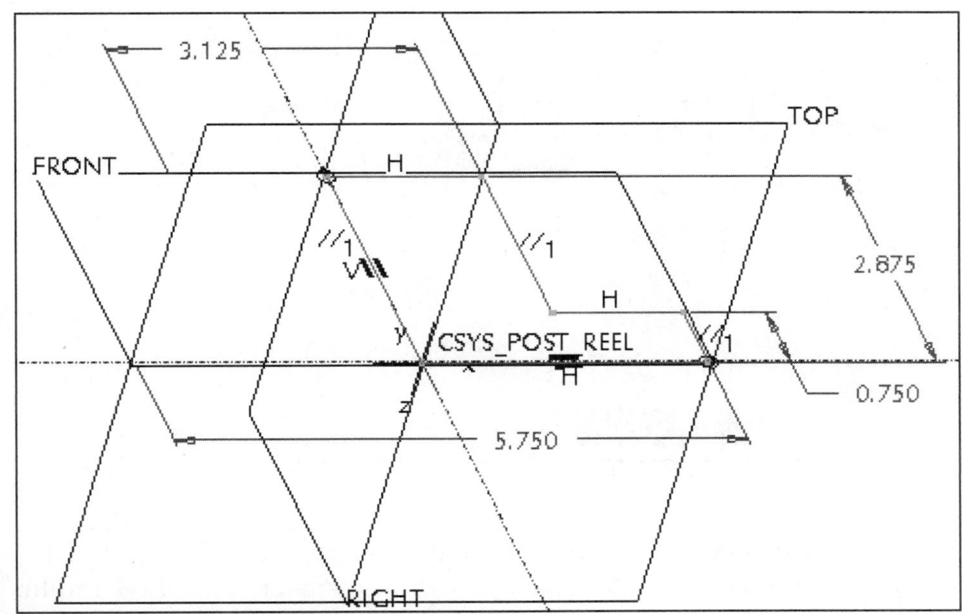

Figure 8.9 Closed Section Sketch shown in Trimetric

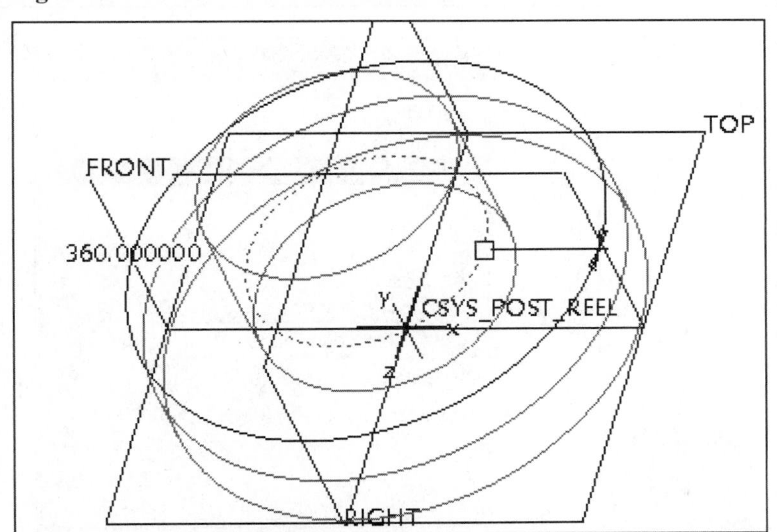

Figure 8.10(a) Angle **360** Degrees

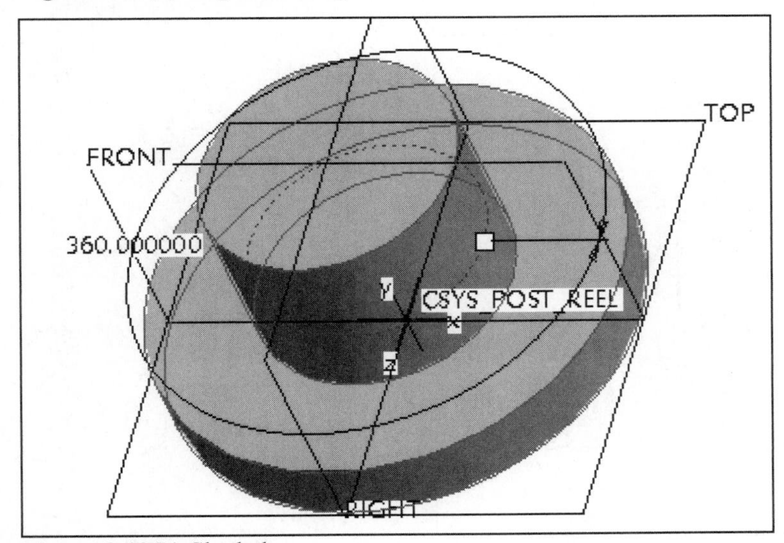

Figure 8.10(b) Shaded

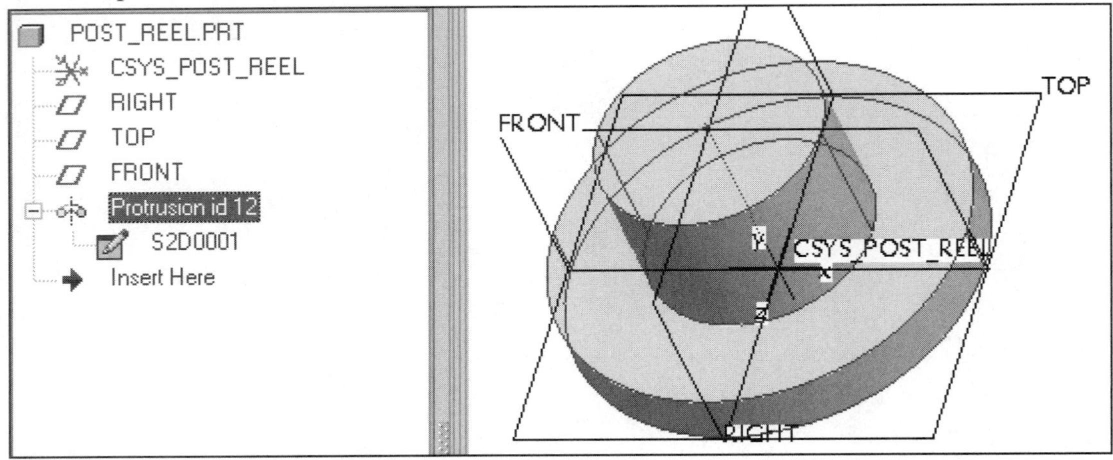

Figure 8.11 Revolved Protrusion

Create the internal revolved cut. Use the previous sketching plane and add the edge, shown in Figure 8.12, as a reference. Use the illustrations provided in Figures 8.12 through 8.16 to complete the cut. Turn the grid snap off in the environment.

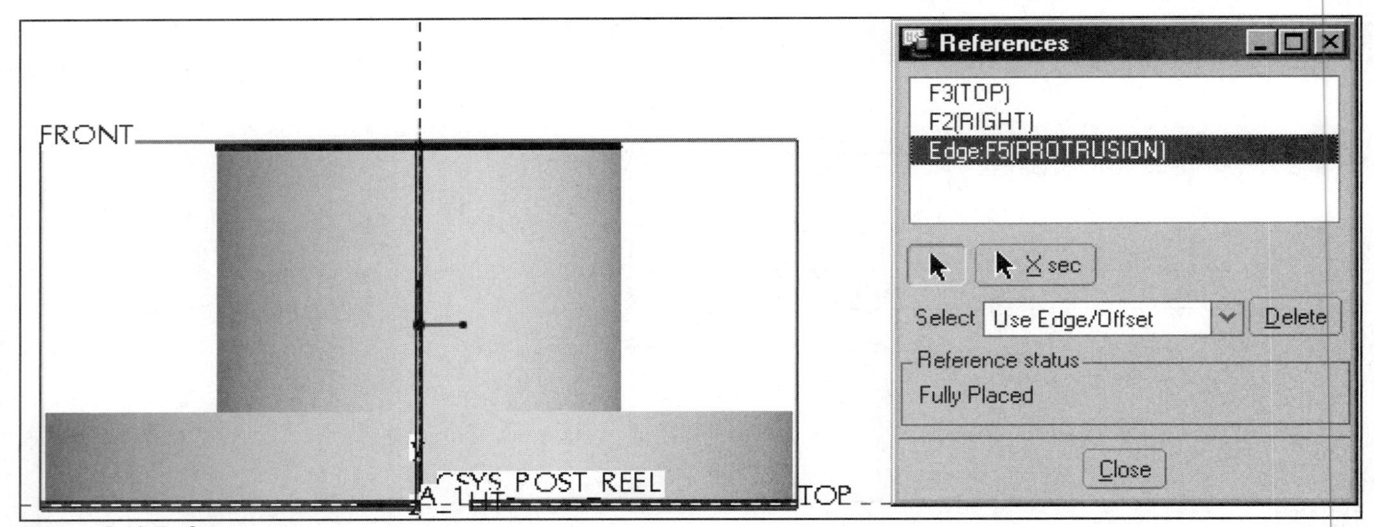

Figure 8.12 References

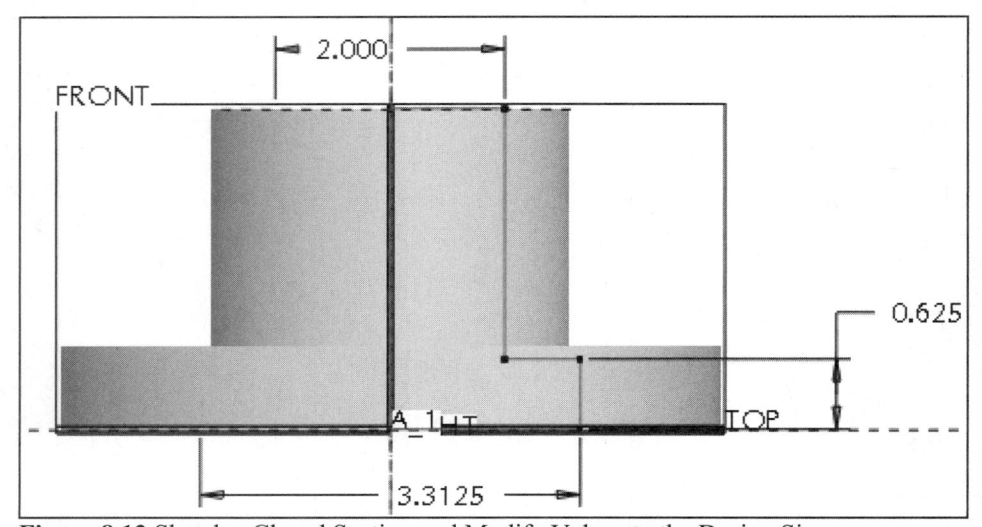

Figure 8.13 Sketch a Closed Section and Modify Values to the Design Sizes

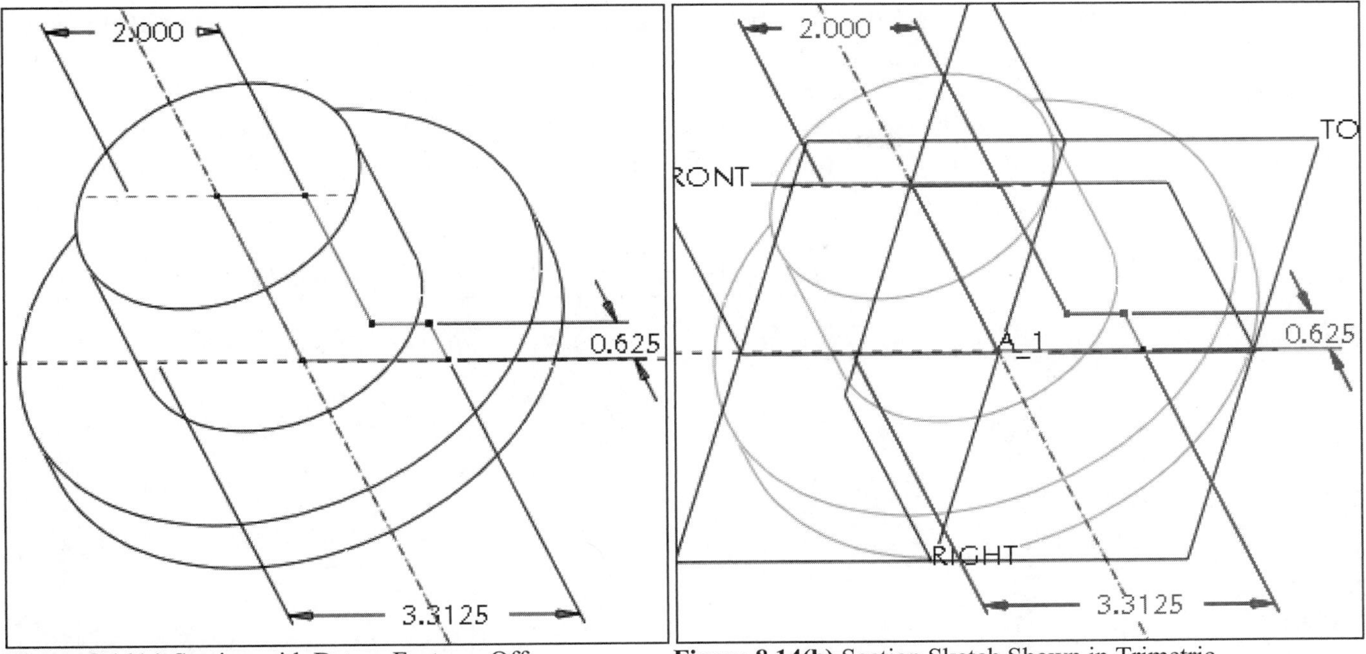

Figure 8.14(a) Section with Datum Features Off

Figure 8.14(b) Section Sketch Shown in Trimetric

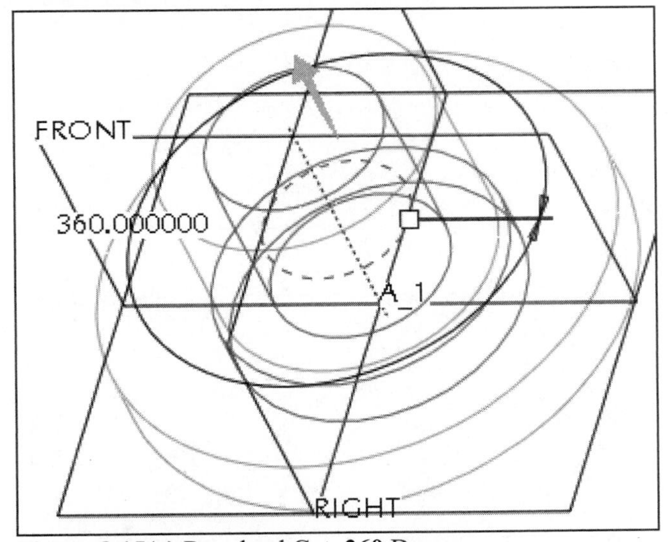

Figure 8.15(a) Revolved Cut, **360** Degrees

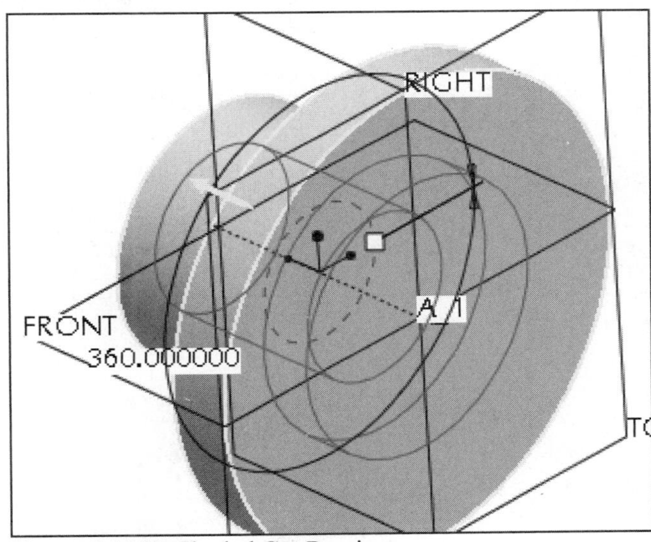

Figure 8.15(b) Shaded Cut Preview

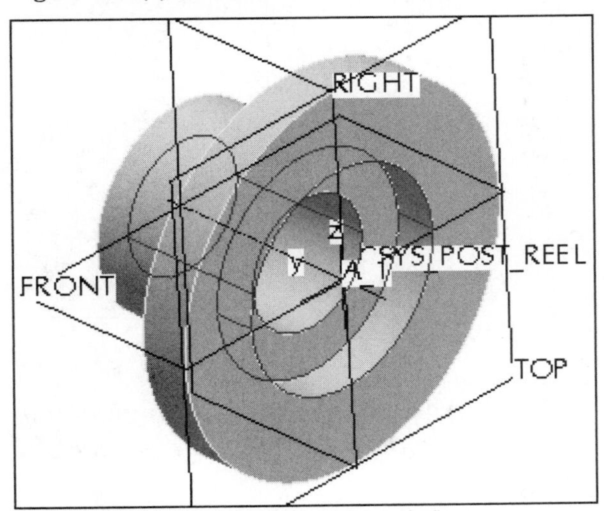

Figure 8.16(a) Completed Cut (Shaded)

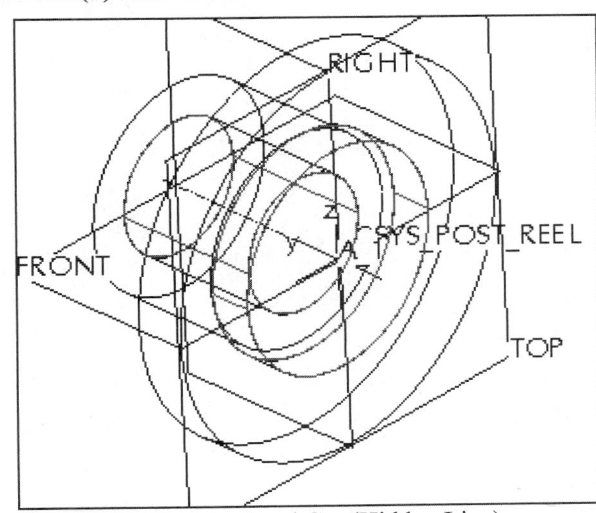

Figure 8.16(b) Completed Cut (Hidden Line)

The keyseat can be created with a cut from the end of the Post Reel using three lines sketched on **TOP** (or the top face), or it can be created with one line from the side, sketching on **RIGHT** and projecting to both sides. Be careful to use the proper dimensioning scheme.

Keyseats are normally dimensioned by giving the width of the key slot and the distance between the top of the keyseat cut and the tangent edge of the shaft hole. The size of the keyseat is driven by the shaft size.

Create the keyseat cut. Use the illustrations provided in Figures 8.17 and 8.18 to create the cut.

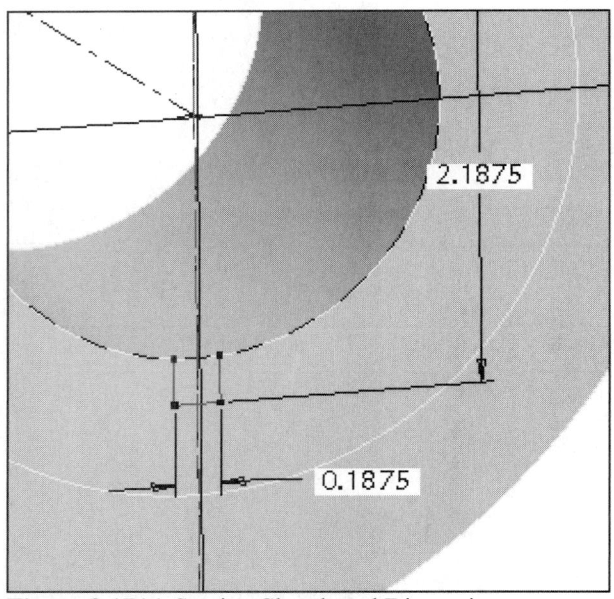

Figure 8.17(a) Section Sketch and Dimensions

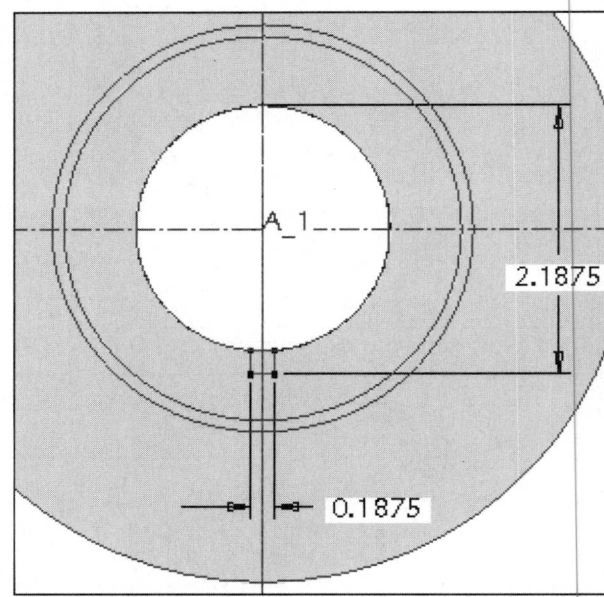

Figure 8.17(b) Three Lines

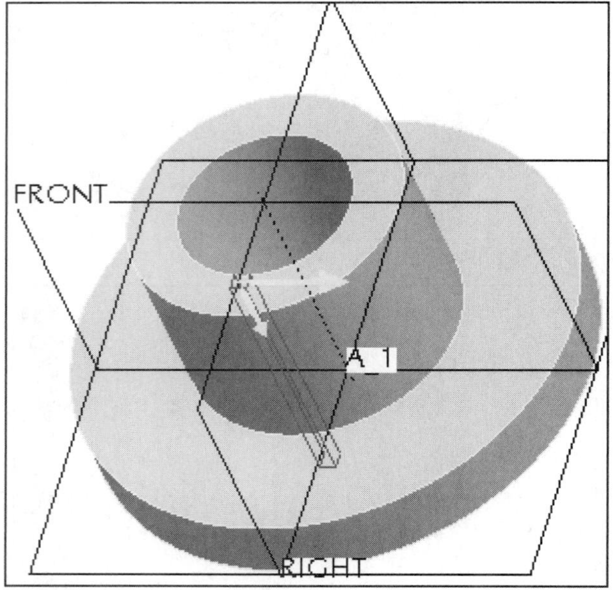

Figure 8.18(a) Cut Direction

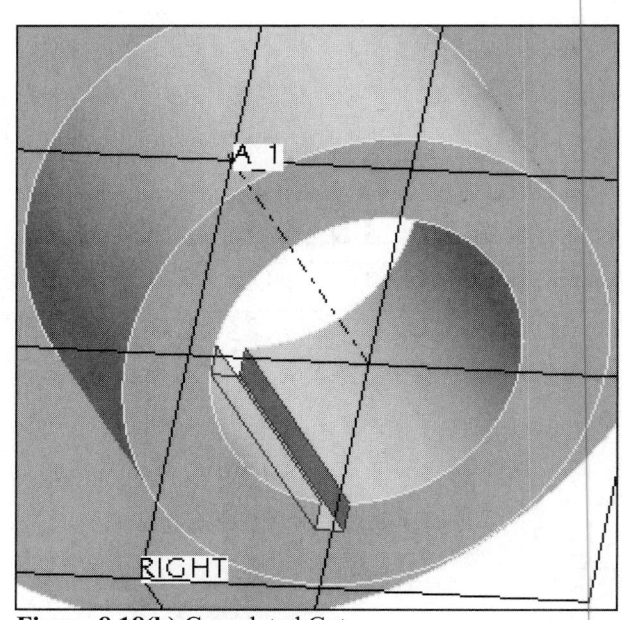

Figure 8.18(b) Completed Cut

*Time permitting recalculate the keyseat size using your **Machinery's Handbook** or **Engineering Graphics** textbook. Then, create a relation that controls the relationship of the keyseat size as per any modified shaft size.*

Add the **45° X .156** chamfer (Fig. 8.19) to the top of the Post Reel and the **R.125** round (Fig. 8.20). Figure 8.21 shows the chamfer and round in the Model Tree.

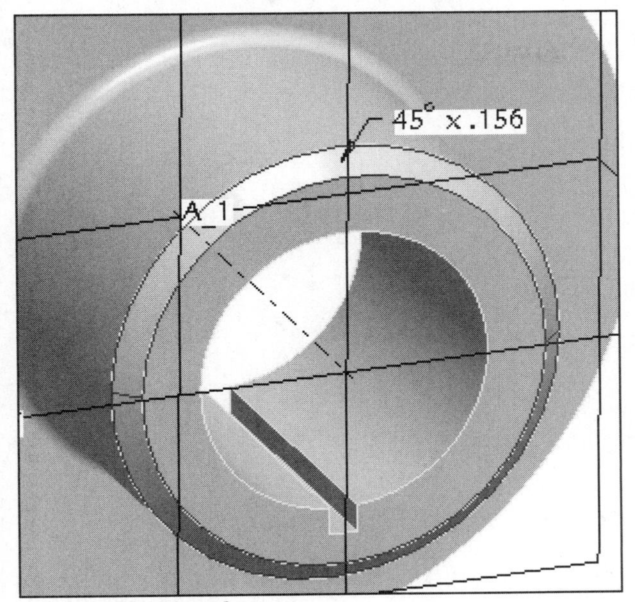

Figure 8.19 Chamfer

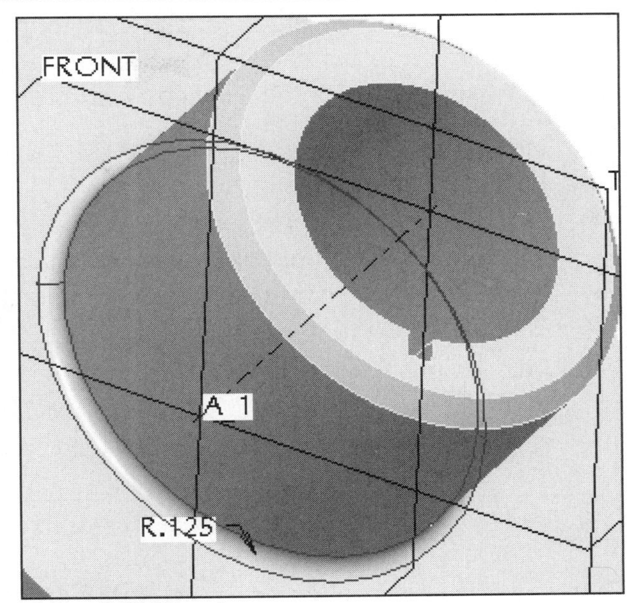

Figure 8.20 Round

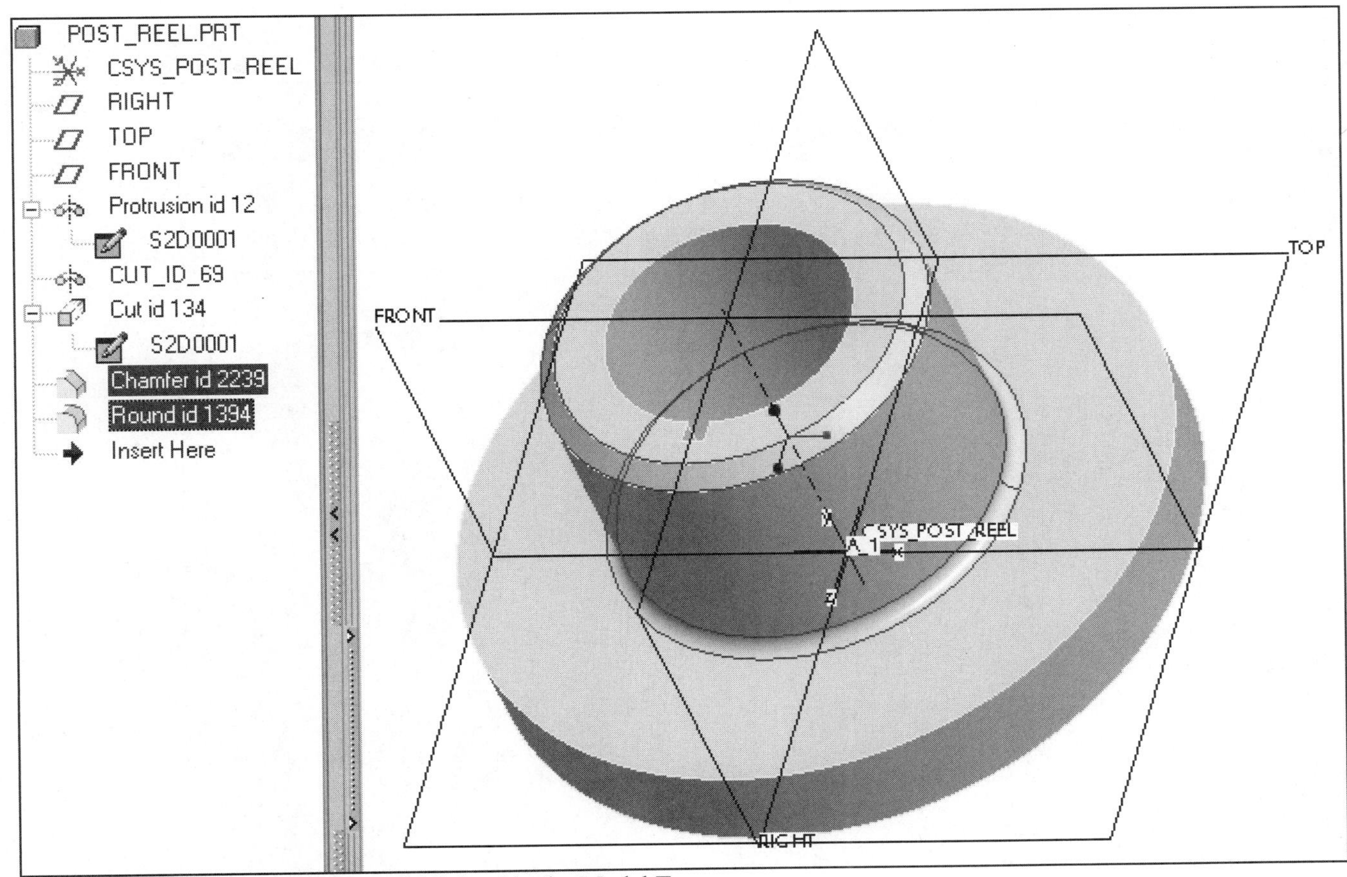

Figure 8.21 Chamfer and Round Highlighted in the Model Tree

Customizing the User Interface (UI)

The final feature creation sequence consists of a boss protrusion, a slot cut, and a series of rounds. The three types of features are grouped, and the group is then patterned. Before completing the part, customize your user interface to increase your efficiency in modeling. You can customize the Pro/E user interface, according to your needs or the needs of your group or company, to include the following:

- Create keyboard macros, called *mapkeys*, and add them to the menus and toolbars
- Add or remove existing toolbars
- Add split buttons to the toolbars (split buttons contain multiple closely-related commands and save space by hiding all but the first command)
- Move or remove commands from the menus or toolbars
- Change the location of the message area
- Add options to the Menu Manager
- Blank (make unavailable) options in the Menu Manager
- Set default command choices for Menu Manager menus

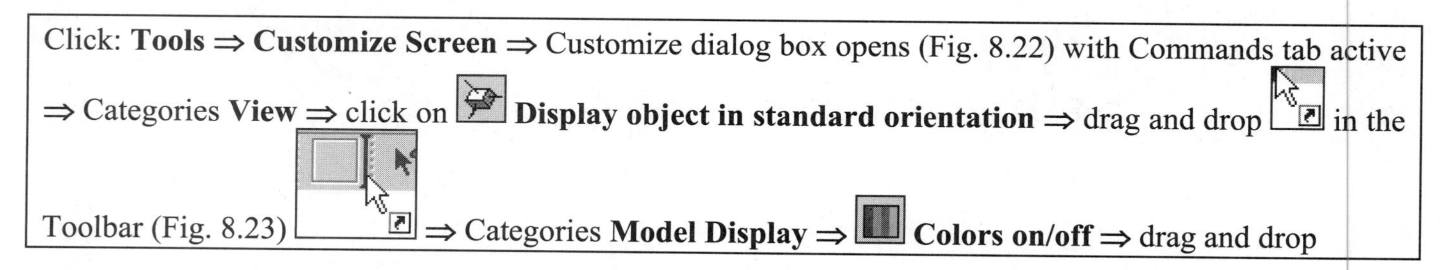

Click: **Tools** ⇒ **Customize Screen** ⇒ Customize dialog box opens (Fig. 8.22) with Commands tab active ⇒ Categories **View** ⇒ click on **Display object in standard orientation** ⇒ drag and drop in the Toolbar (Fig. 8.23) ⇒ Categories **Model Display** ⇒ **Colors on/off** ⇒ drag and drop

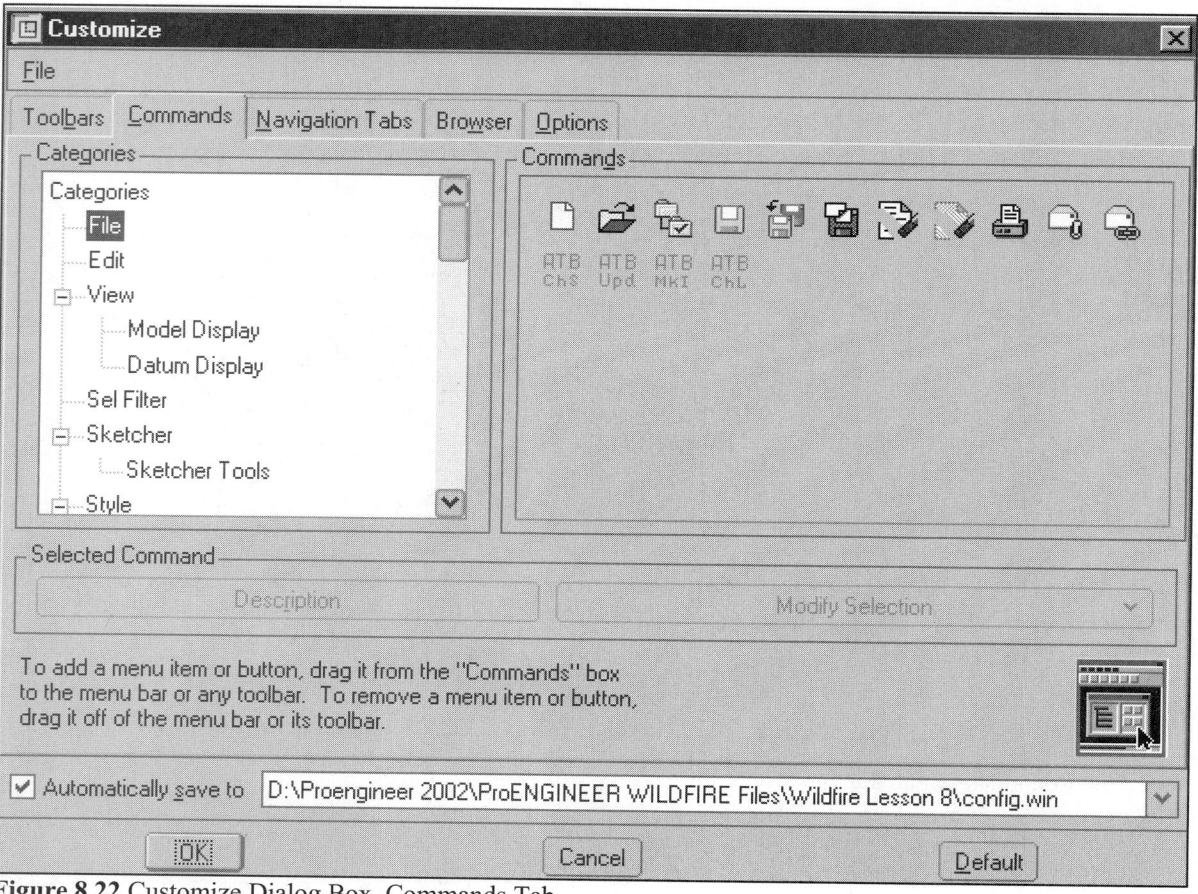

Figure 8.22 Customize Dialog Box, Commands Tab

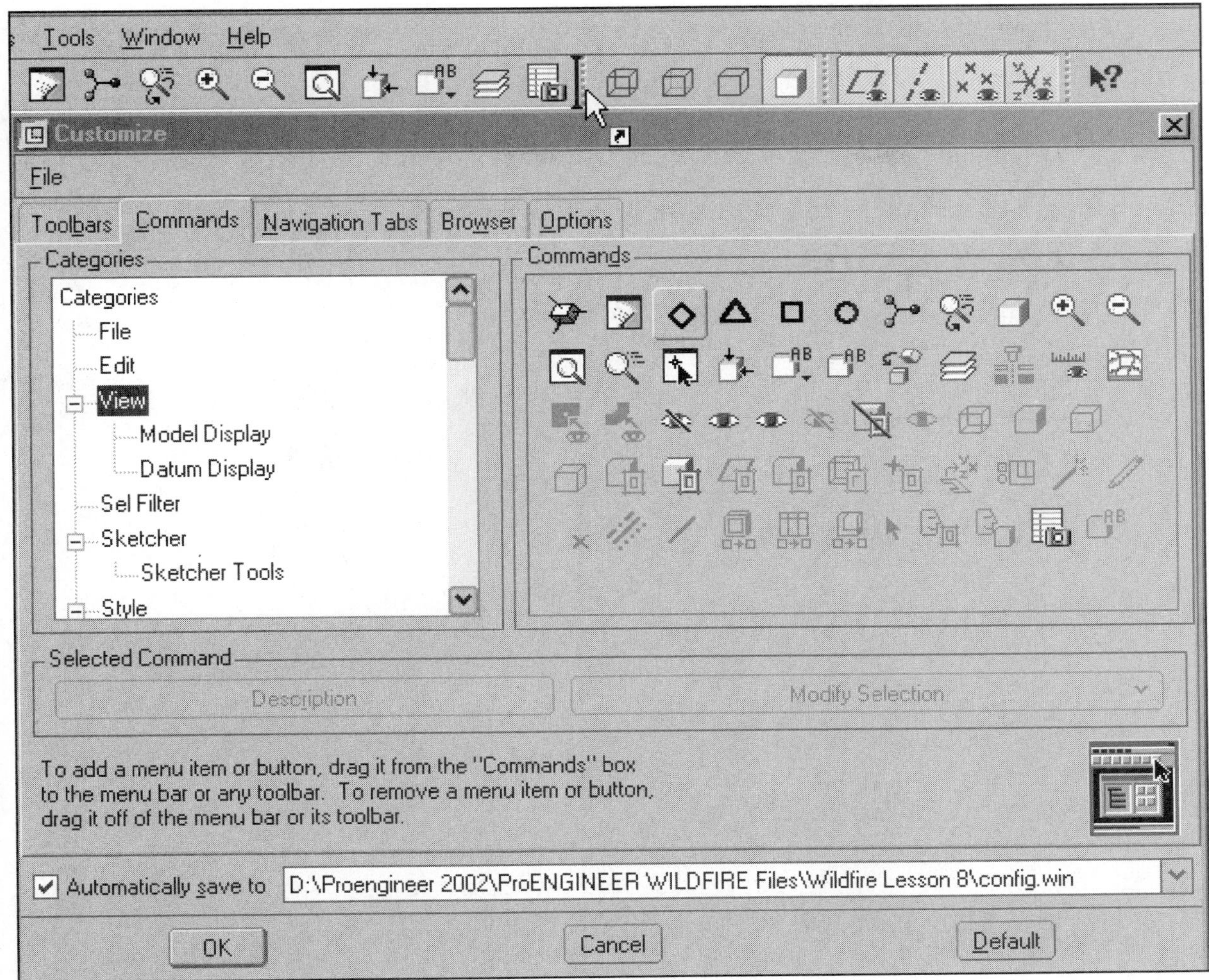

Figure 8.23 Dragging and Dropping a Command Button into the Toolbar

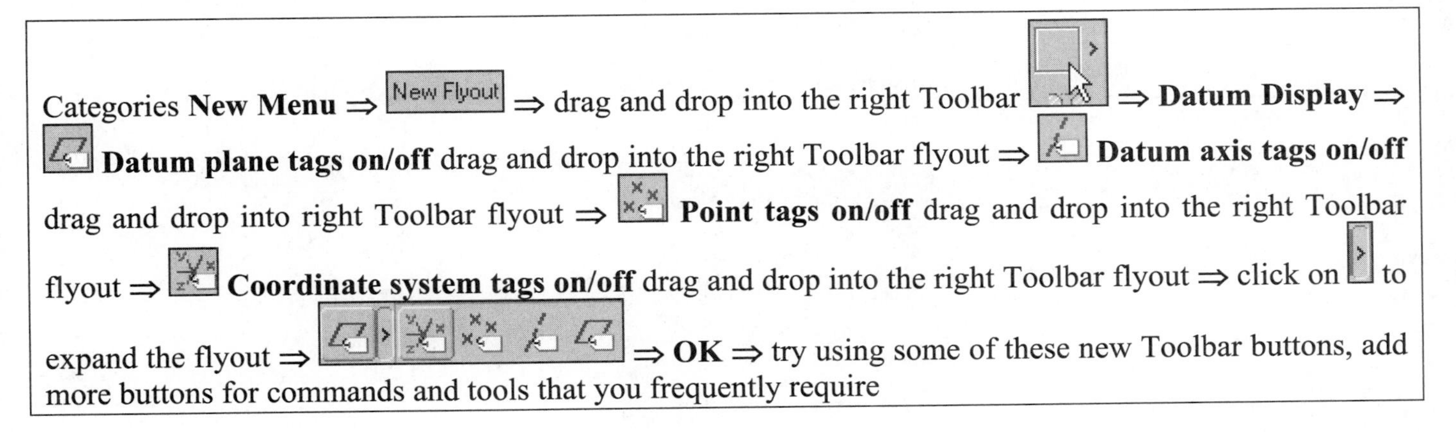

Categories **New Menu** ⇒ New Flyout ⇒ drag and drop into the right Toolbar ⇒ **Datum Display** ⇒ **Datum plane tags on/off** drag and drop into the right Toolbar flyout ⇒ **Datum axis tags on/off** drag and drop into right Toolbar flyout ⇒ **Point tags on/off** drag and drop into the right Toolbar flyout ⇒ **Coordinate system tags on/off** drag and drop into the right Toolbar flyout ⇒ click on > to expand the flyout ⇒ ⇒ **OK** ⇒ try using some of these new Toolbar buttons, add more buttons for commands and tools that you frequently require

Your new Toolbar buttons are automatically saved in the current working directory: Automatically save to D:\Proengineer 2002\ProENGINEER WILDFIRE Files\textbook. You can recall saved settings by clicking: Tools ⇒ Customize Screen ⇒ File ⇒ Open Settings ⇒ select the file ⇒ Open ⇒ OK. Buttons can be removed from the Toolbar using the exact same method, except, drag them away from the Toolbar and release the mouse button.

Next, add a Toolbar to the left side of the Navigator Browser, click: **Tools ⇒ Customize Screen ⇒** click Toolbars tab [Fig. 8.24(a)] ⇒ Tools [Fig. 8.24(b)] ⇒ **Left ⇒ ⇒ OK** (Fig. 8.25)

The Tools toolbar includes: **Set various environment options**, **Run trail or training file**, **Create macros**, and **Select hosts for distributed computing**.

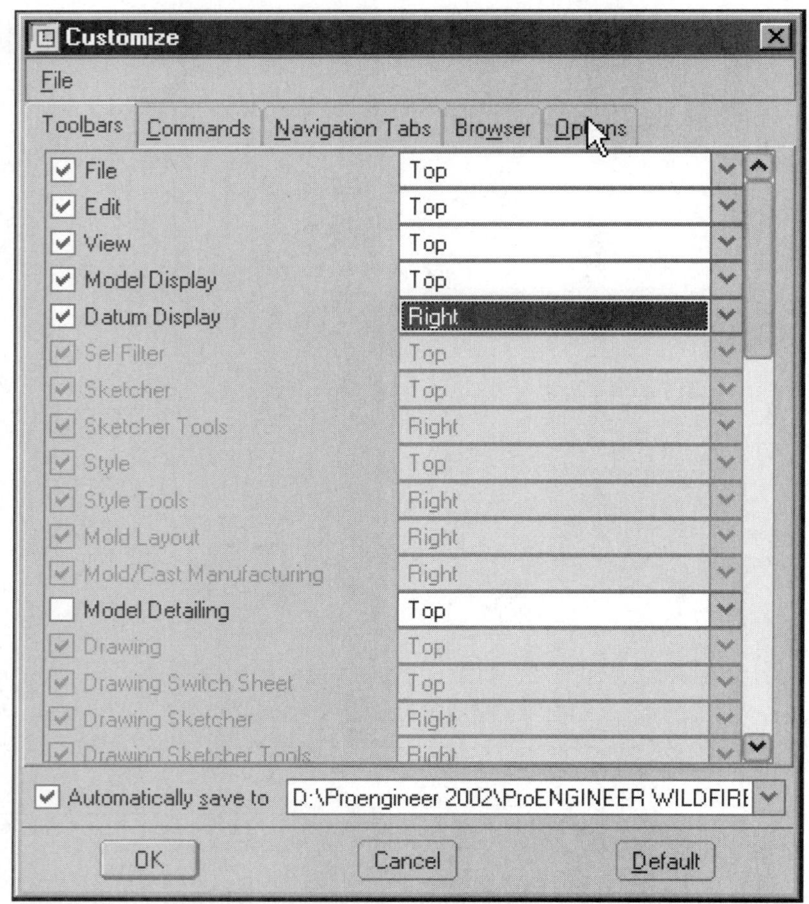

Figure 8.24(a) Customize Dialog Box, Toolbars Tab

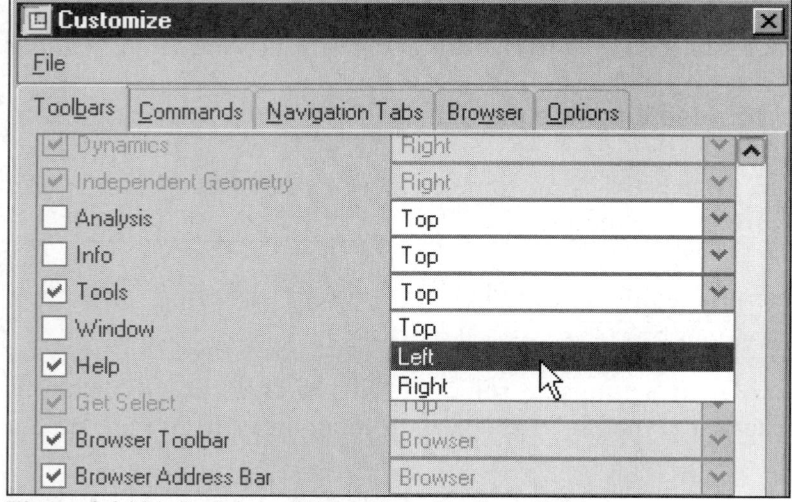

Figure 8.24(b) Customize Dialog Box, Toolbars Tab

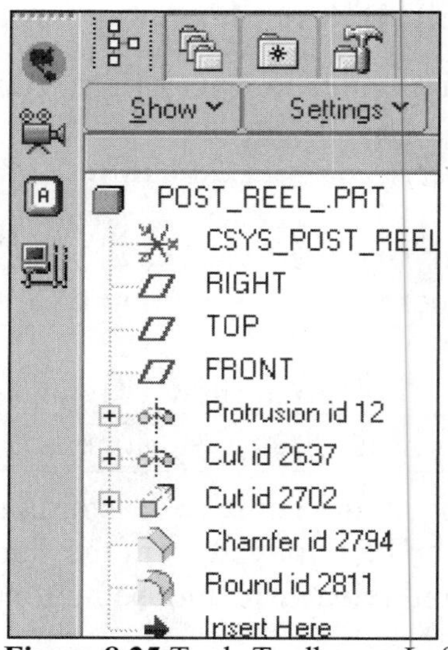

Figure 8.25 Tools Toolbar on Left

The Navigation Tabs tab provides options for controlling the location of the Navigator (left or right), its width setting, and its placement in relation to the Model Tree settings. The Browser tab is used to control its window width and animation option. The Options tab provides settings to locate the Dashboard, Secondary Window size, and Menu display.

Click: **Tools** ⇒ **Customize Screen** ⇒ click **Navigation Tabs** tab [Fig. 8.26(a)] and explore the settings ⇒ click **Browser** tab [Fig. 8.26(b)] and explore the options ⇒ click **Options** tab [Fig. 8.26(c)] and explore its capabilities ⇒  ⇒ **OK**

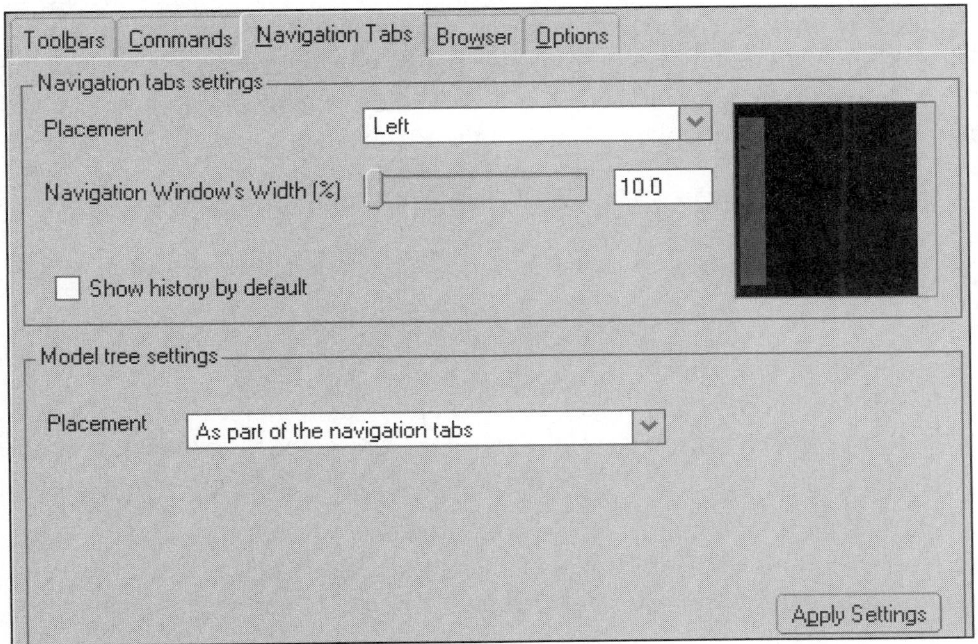

Figure 8.26(a) Customize Dialog Box, Navigation Tabs

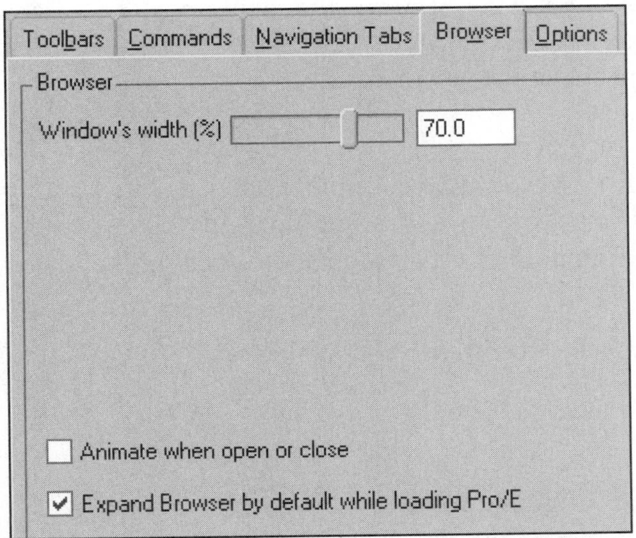

Figure 8.26(b) Customize Dialog Box, Browser Tab

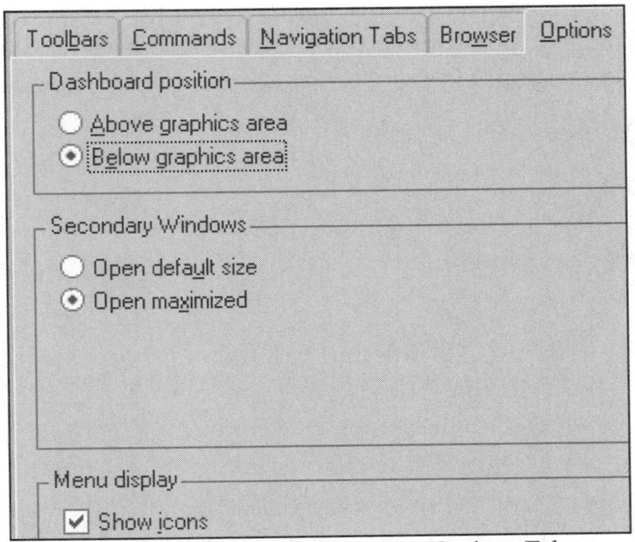

Figure 8.26(c) Customize Dialog Box, Options Tab

Mapkeys

In Pro/E, a **Mapkey** is a keyboard macro that maps frequently used command sequences to certain keyboard keys or sets of keys. Mapkeys are saved in the configuration file, and are identified with the option *mapkey*, followed by the identifier and then the macro. You can define a unique key or combination of keys which, when pressed, executes the mapkey macro (for example, F6). You can create a mapkey for virtually any task you perform frequently within Pro/E.

By adding custom mapkeys to your toolbar or menu bar, you can use mapkeys with a single mouse click or menu command and thus streamline your workflow in a visible way.

To create a mapkey, you can use the configuration file option *mapkey*, or, on the Pro/E menu bar, click Tools ⇒ Mapkeys, and then in the Mapkeys dialog box, you click New and record your mapkey in the Record Mapkey dialog box. Pro/E records your mapkey as you step through the sequence of keystrokes or command executions to define it. After you define the mapkey, Pro/E creates a corresponding icon and places it in the Customize dialog box under the Mapkeys category. To open the Customize dialog box, click: Tools ⇒ Customize Screen. On the Toolbars tabbed page, select the Mapkeys category. You can then drag the visible mapkey icon onto the Pro/E main toolbar. You can also create a label for the new mapkey.

You can also nest one mapkey within another, so that one mapkey initiates another. To do so, you include the mapkey name in the sequence of commands of the new mapkey you are defining.

Mapkeys include the ability to do the following:

- Pause for user interaction.
- Handle message window input more flexibly.
- Run operating system scripts and commands. The Record Mapkey dialog box contains the OS Script tabbed page, whose options allow you to run OS commands instead of Pro/E commands.

When you define a mapkey, Pro/E automatically records a pause when you make screen selections, so that you can make new selections while the mapkey is running. In addition, you can record a pause at any place in the mapkey along with a user-specified dialog prompt, which will appear at the corresponding point while the mapkey is running.

If you create a new mapkey that contains actions that open and make selections from dialog boxes, then when you run the mapkey, it does not pause for user input when it opens the dialog box. To set the mapkey to pause for user input when opening dialog boxes, you must select *Pause for keyboard input* on the Pro/E tab in the Record Mapkey dialog box before you create the new mapkey.

Mapkeys Dialog Box

You use the Mapkeys dialog box [Fig. 8.27(a)] to define new mapkeys, modify, and delete existing mapkeys, run a mapkey chosen from the list, and save mapkeys to a configuration file. To open the Mapkeys dialog box, click Tools ⇒ Mapkeys. The following defines each command option on the dialog box:

- **New** Allows you to define a new mapkey and opens the Record Mapkey dialog box
- **Modify** Allows you to modify the selected mapkey
- **Run** Allows you to run the selected mapkey
- **Delete** Allows you to delete the selected mapkey
- **Save** Allows you to save the selected mapkey to a configuration file
- **Changed** Allows you to save only the mapkeys changed in the current session
- **All** Save all the mapkeys

In the Record Mapkey dialog box [Fig. 8.27(b)], you can type the key sequence that is to be used to execute the mapkey in the Key Sequence text box. To use a function key, precede its name with a dollar sign (**$**). For example, to map to **F7**, type **$F7**. Type the Name and Description of the mapkey in the appropriate text boxes. On the Pro/E tab, specify how the Pro/E will handle the prompts when running the mapkey by selecting one of the following commands:

- **Record keyboard input** (default selection) Record the keyboard input for prompts when defining the mapkey, and use it when running the macro
- **Accept system defaults** Accept the system defaults when running the macro
- **Pause for keyboard input** Pause for user input when running the macro

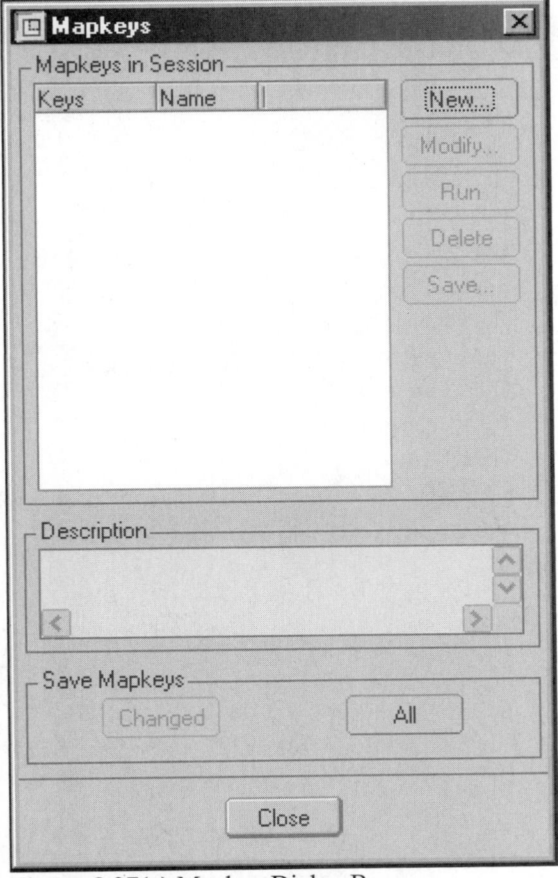

Figure 8.27(a) Mapkey Dialog Box

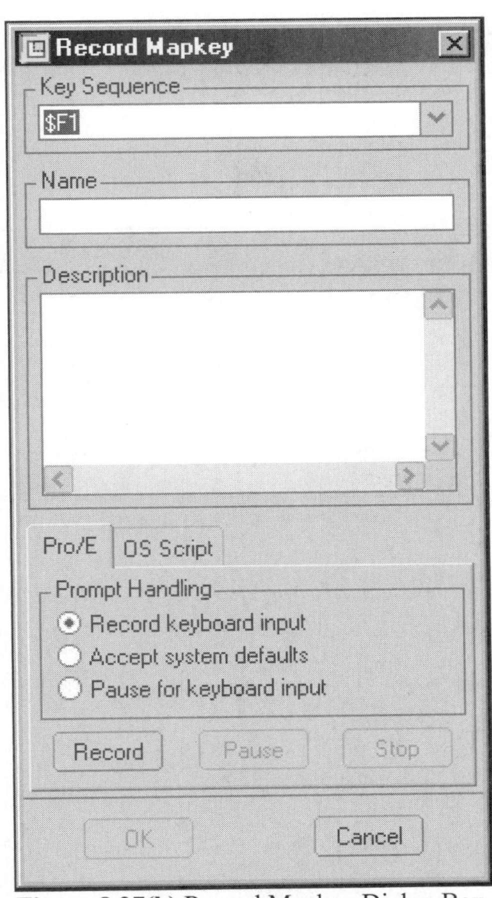

Figure 8.27(b) Record Mapkey Dialog Box

If you create a new mapkey that contains actions that open and make selections from dialog boxes, then when you run the mapkey, it does not pause for user input when it opens the dialog box. To set the mapkey to pause for user input when opening dialog boxes, you must select *Pause for keyboard* input. Use Record to start recording the macro by selecting menu commands in the appropriate order. Use Pause to indicate where to pause while running the mapkey. Type the prompt in the Resume Prompt dialog box. Use Resume and continue recording the mapkey. When you run the macro, Pro/E will pause, display the prompt you typed, and give you the options to Resume running the macro or Cancel. Use Stop when you are finished recording the macro.

After you define the mapkey, a corresponding button appears in the **Customize Toolbars** dialog box under the **Mapkeys** category. You can then drag the mapkey onto the toolbar just like the Pro/E-supplied buttons. Mapkeys include the ability to pause for user interaction, handle message window input flexibly, and run operating system commands.

Create a mapkey, click: **Tools** ⇒ **Mapkeys** Mapkeys dialog box opens ⇒ **New** Record Mapkey dialog box opens ⇒ Key Sequence- **$F5** ⇒ Name **COLORS** ⇒ Description **Opens Appearance Editor and Starts New Color Definition** (Fig. 8.28) ⇒ **Record** ⇒ **View** ⇒ **Color and Appearance** ⇒ [+] **Add new appearance** ⇒ Select Color: **Color** button ⇒ **Stop** ⇒ **OK** ⇒ Save Mapkeys **All** ⇒ **Ok** ⇒ **Close** ⇒ **Close** Color Editor ⇒ **Close** Appearance Editor ⇒ **Tools** ⇒ **Customize Screen** ⇒ **Commands** tab ⇒ **Mapkeys** from the Categories list ⇒ click on the new mapkey **COLORS** ⇒ press the **RMB** (Fig. 8.29) ⇒ pick **Choose Button Image** ⇒ select ◇ (Fig. 8.30) ⇒ **Modify Selection** (Fig. 8.31) ⇒ **Edit Button Image** Button Editor Opens (Fig. 8.32) ⇒ click on a color block and edit the picture as you wish (Fig. 8.33) ⇒ **OK**

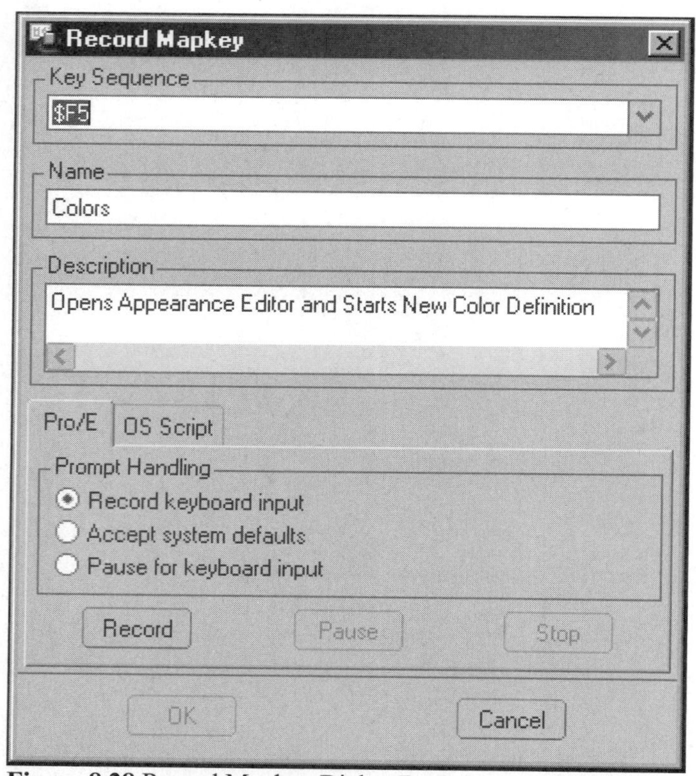

Figure 8.28 Record Mapkey Dialog Box

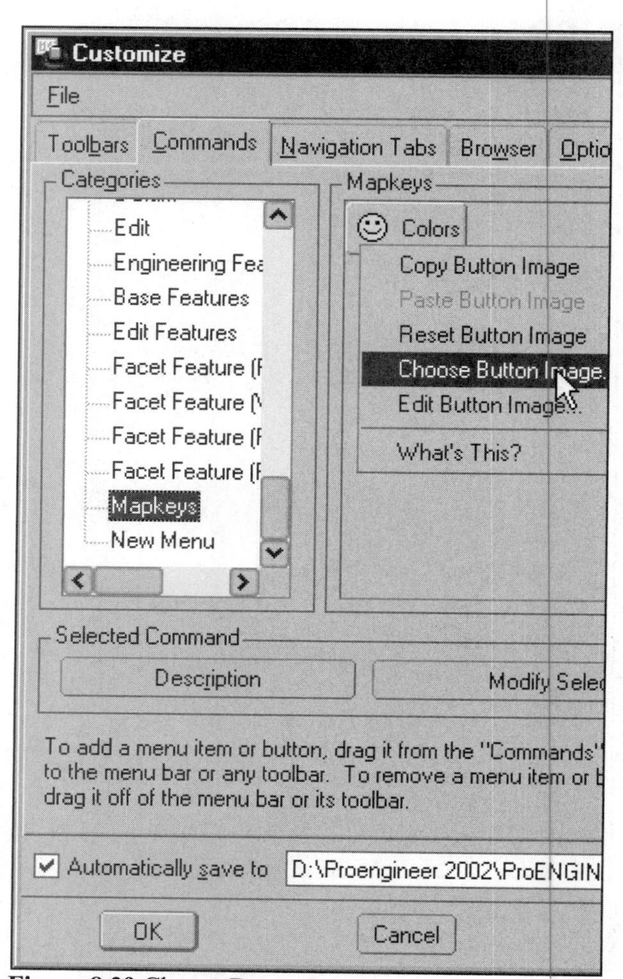

Figure 8.29 Choose Button Image

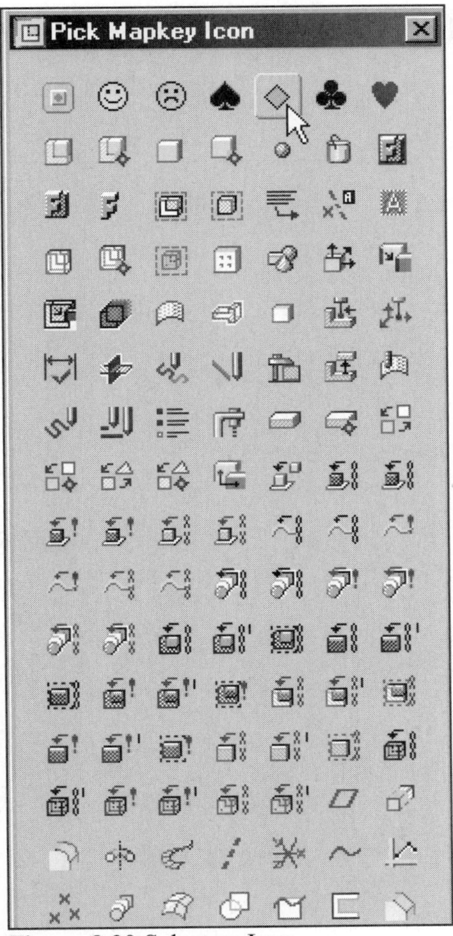

Figure 8.30 Select an Icon

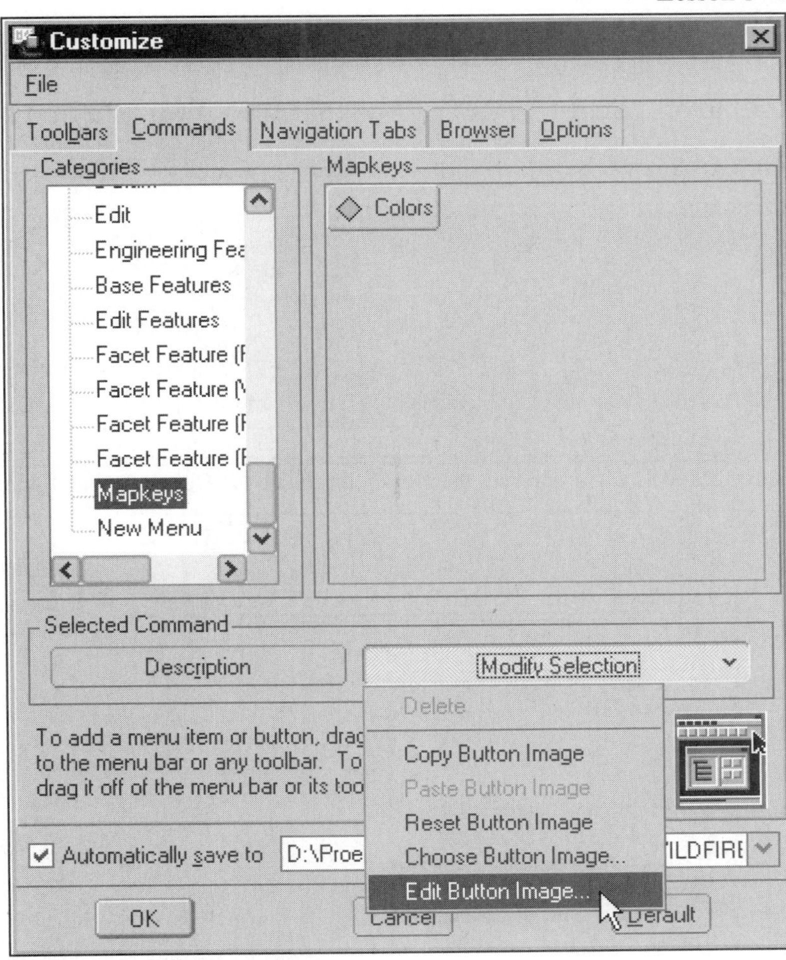

Figure 8.31 Edit Button Image

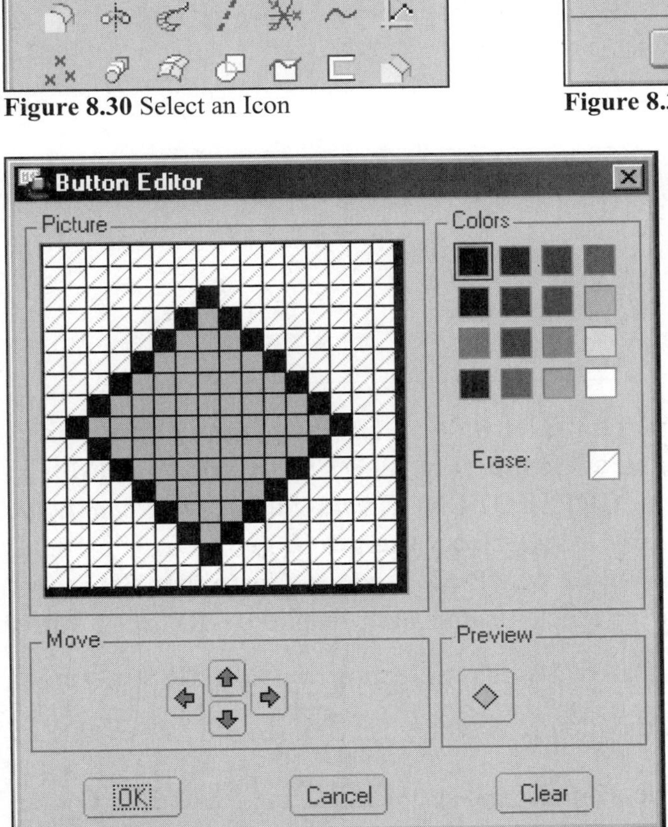

Figure 8.32 Button Editor

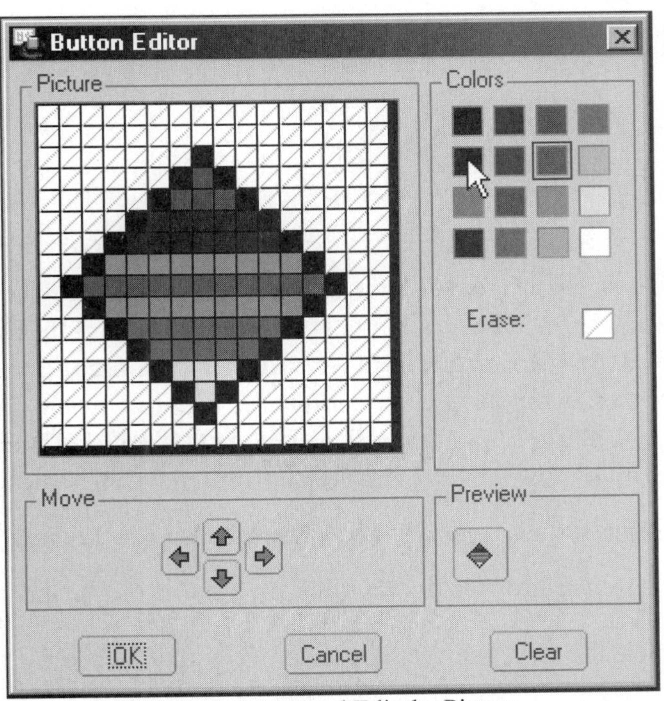

Figure 8.33 Pick on Colors and Edit the Picture

Pick the new mapkey: Colors (Fig. 8.34) ⇒ drag to the Toolbar and drop (Fig.8.35) ⇒ **OK** ⇒ test the new button, click: **Opens Appearance Editor and Starts New Color Definition** Color Editor dialog box opens (Fig. 8.36) ⇒ **Close** ⇒ **Close** ⇒ ⇒ **MMB**

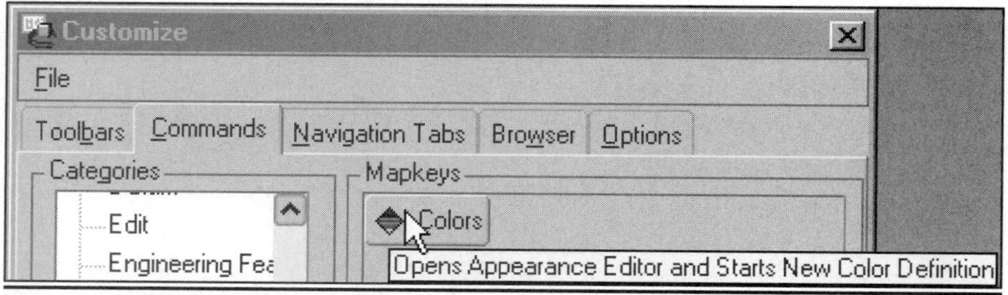

Figure 8.34 Place Cursor over the New Icon to see the Description. Click, Drag and Drop in Toolbar

Figure 8.35 Drag and Drop in Toolbar

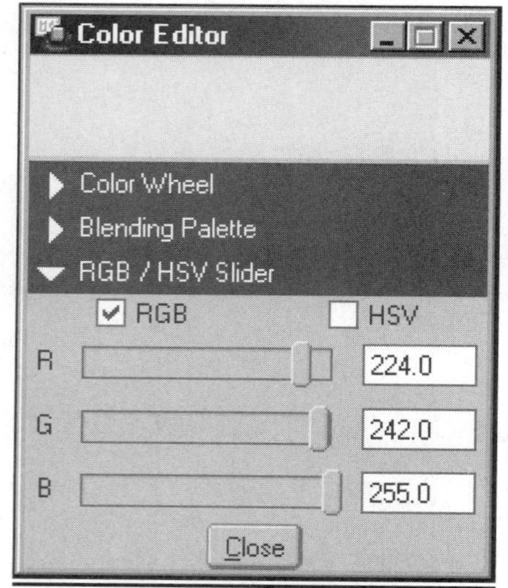

Figure 8.36 Color Editor Opens

Click: **Create macros** from your new Tools Toolbar on the left side of the Navigator ⇒ **New** ⇒ Key Sequence- **$F6** ⇒ Name **GRP** ⇒ Description **Creates a Group** (Fig. 8.37) ⇒ **Record** ⇒ **Edit** ⇒ **Feature Operations** ⇒ **Group** ⇒ **Local Group** ⇒ type **GRPSEQUENCE** *you will have to Rename the group in the Model Tree after it is created, when you use this button* ⇒ **Enter** ⇒ **Stop** ⇒ **OK** (Fig. 8.38) ⇒ **Changed** ⇒ **Ok** ⇒ **Close** ⇒ **MMB** ⇒ **MMB** ⇒ **MMB** ⇒ **MMB** ⇒ **Tools** ⇒ **Customize Screen** ⇒ Commands tab ⇒ **Mapkeys** from the Categories list ⇒ click on the new mapkey **GRP** ⇒ **Modify Selection** ⇒ pick **Choose Button Image** ⇒ select ⇒ **Modify Selection** ⇒ **Edit Button Image** Button Editor Opens ⇒ click on a color block and edit the picture as you wish (Fig. 8.39) ⇒ **OK** ⇒ pick the new mapkey GRP (Fig. 8.40) ⇒ drag to the Toolbar and drop ⇒ **OK** ⇒ test the new button, click **Creates a Group** ⇒ **MMB** ⇒ **MMB** ⇒ **MMB** ⇒ **MMB** ⇒ ⇒ **MMB** ⇒ **LMB** to deselect

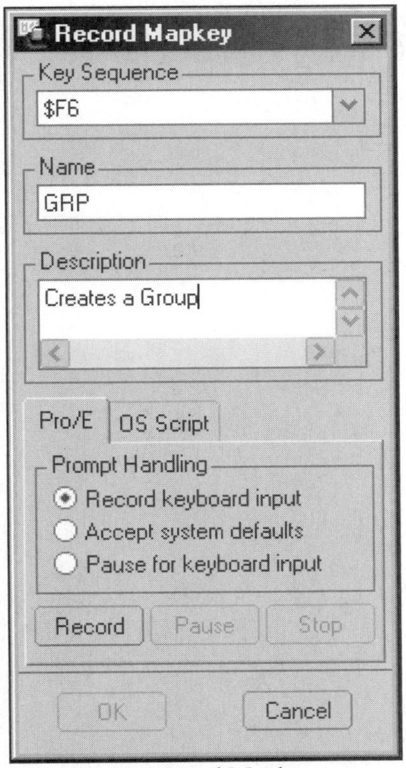

Figure 8.37 Record Mapkey

Figure 8.38 Mapkeys

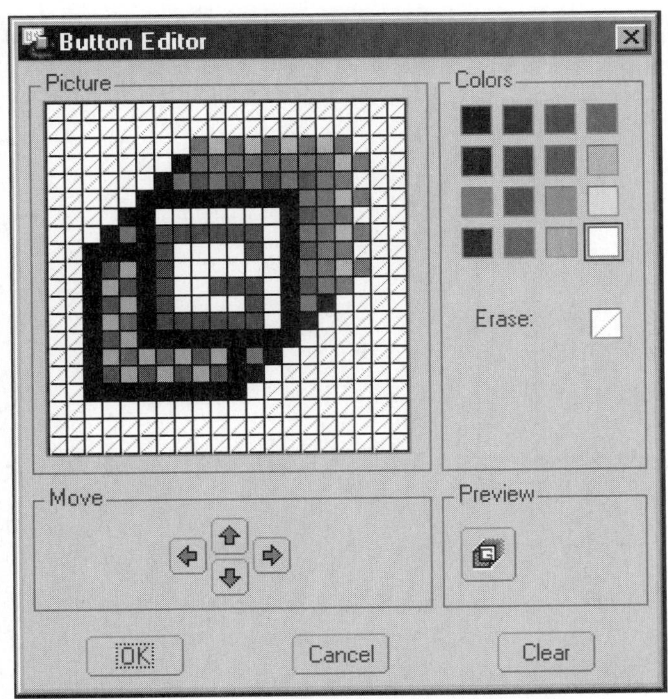

Figure 8.39 Button Editor (yes, you can do better!)

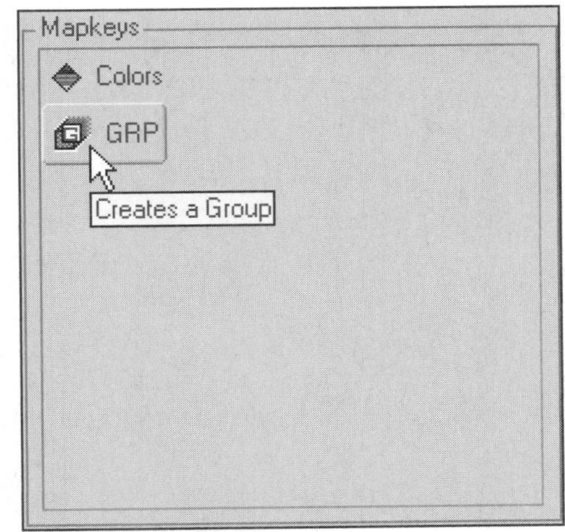

Figure 8.40 New Mapkey

The final feature creation sequence consists of a boss protrusion, a slot cut, and a series of rounds (Fig. 8.41). The three types of features are grouped, and the group is then patterned. The protrusion and the slot must be created with a *non-default datum*, which is created through the parts axis and at an angle to an existing datum plane. The new datum plane is used to provide the direction of rotation and is to orient the sketch plane for the feature sketch. The three equally spaced (**120** degrees) grouped features are on a **∅4.50** bolt circle [Fig. 8.42(a-b)].

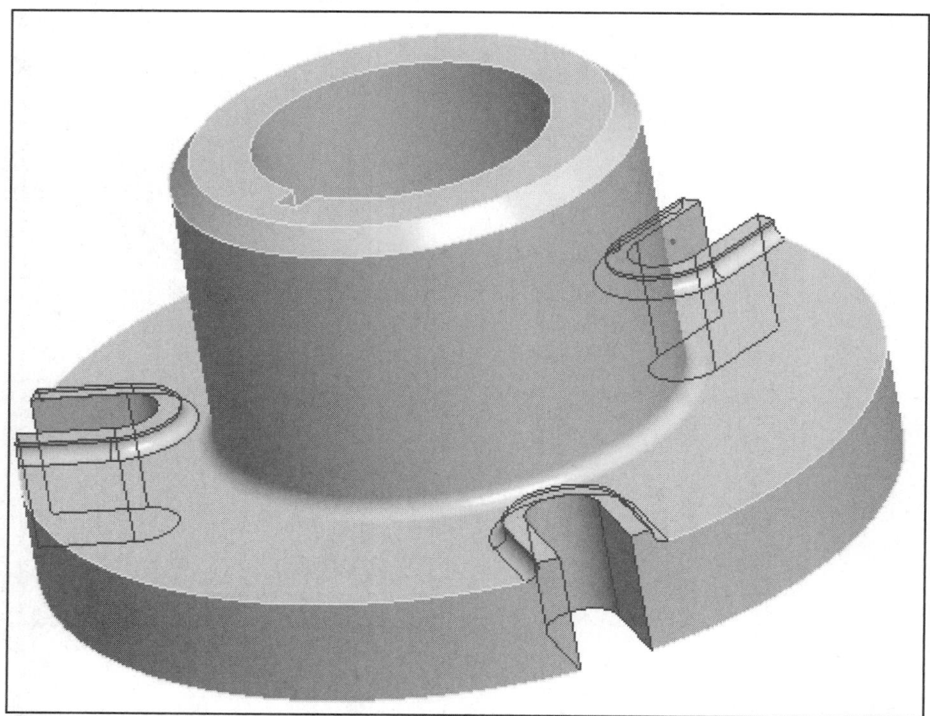

Figure 8.41 Boss, Slot, and Rounds, Grouped and Patterned

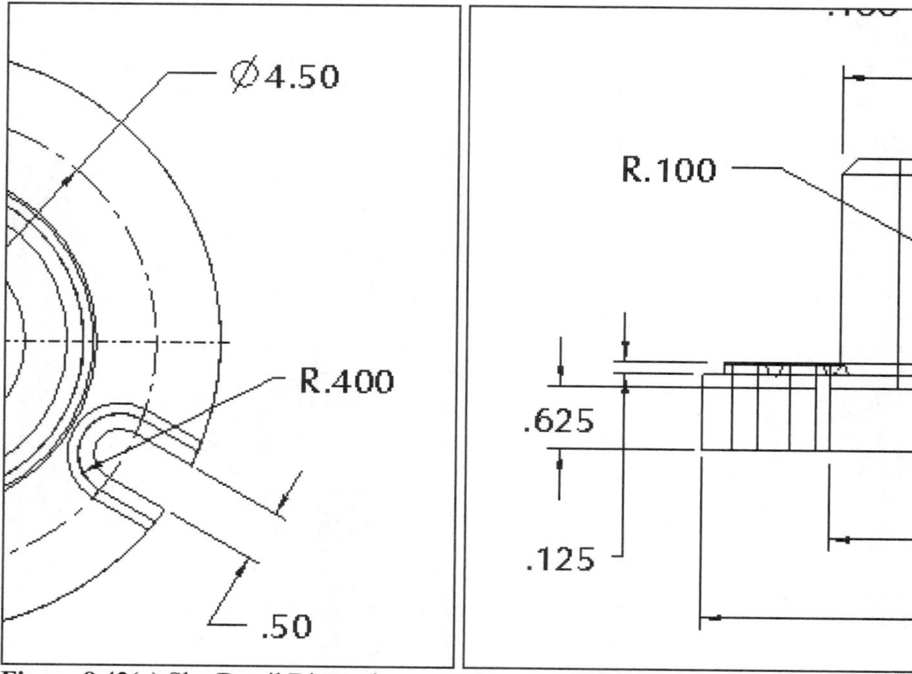

Figure 8.42(a) Slot Detail Dimensions **Figure 8.42(b)** Boss Dimensions

Create the boss protrusion, click: **Datum Plane Tool** ⇒ References: pick axis **A_1** ⇒ press and hold the **Ctrl** key ⇒ pick on datum **FRONT** as the second reference (Fig. 8.43) ⇒ Rotation: type **30** *(depending on the side of the plane you select, you may need to use 330 degrees)* ⇒ **Enter** ⇒ **OK** ⇒ **Coordinate system tags off** ⇒ ⇒ **MMB**

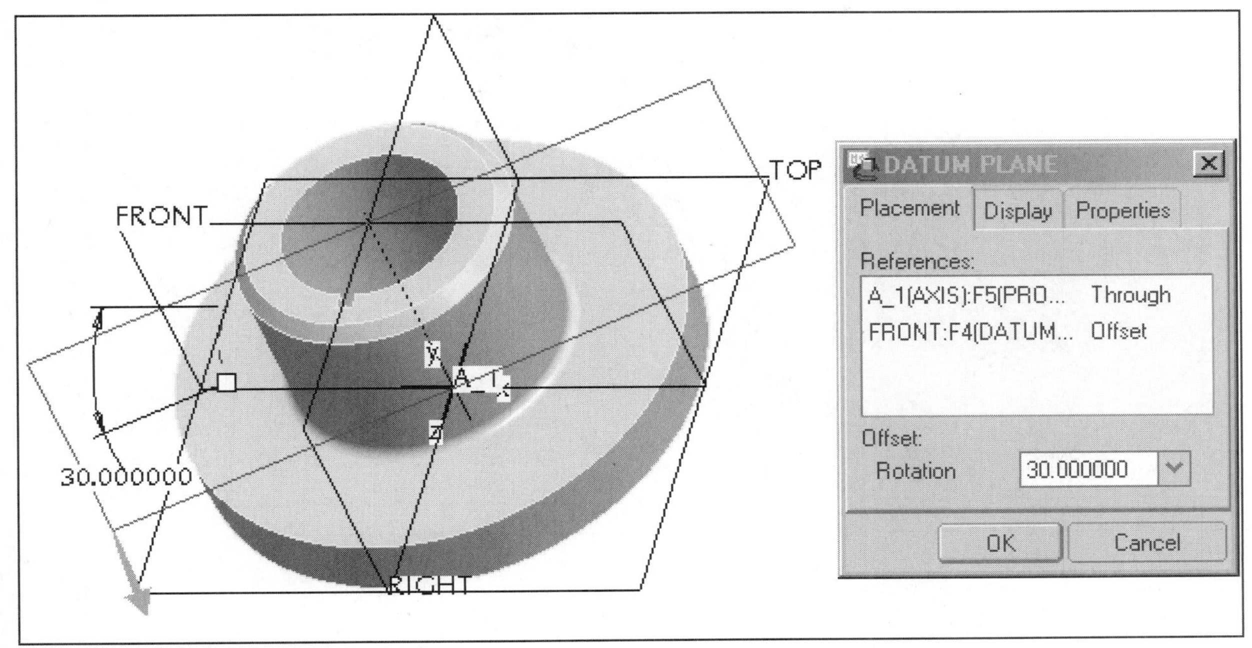

Figure 8.43 Datum Creation

Click: **Extrude Tool** ⇒ ⇒ Section dialog box opens ⇒ Sketch Plane--- Plane: select the top face of the flange ⇒ Reference: **DTM1** ⇒ Orientation: **Bottom** (Fig. 8.44) ⇒ **Sketch** ⇒ ⇒ **Standard Orientation**

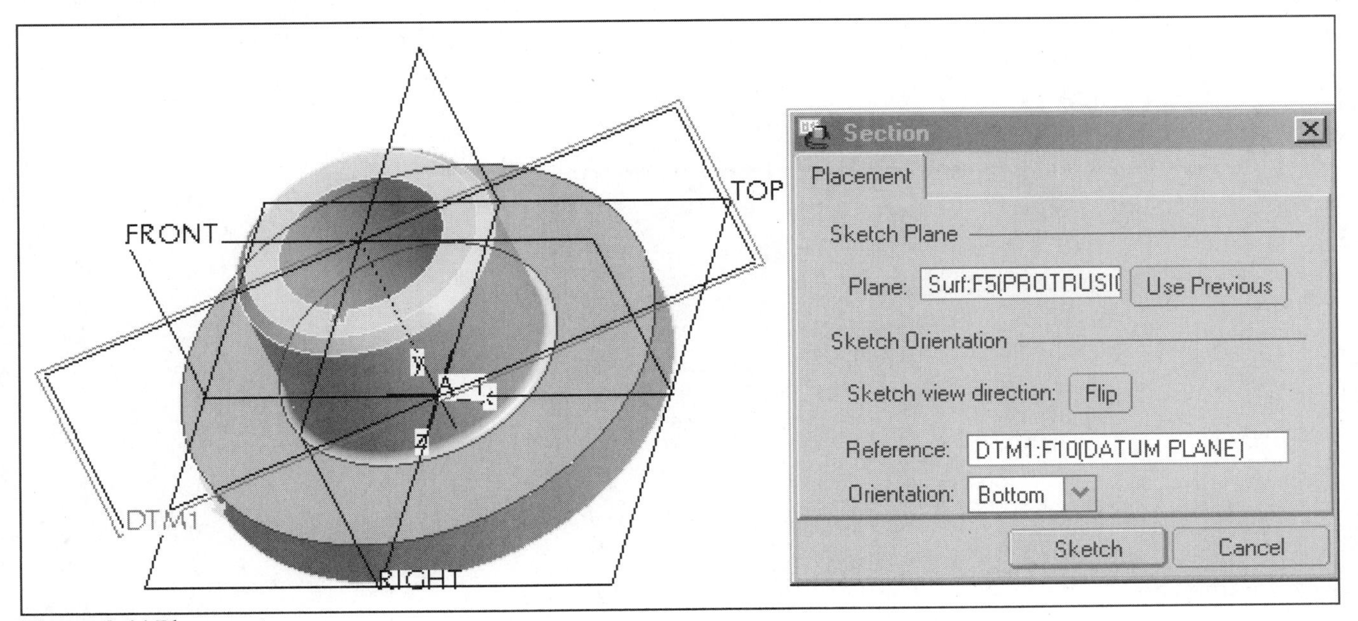

Figure 8.44 Placement

Pick on the outer flange surface for the first reference ⇒ pick on datum plane **DTM1** as the second reference (Fig. 8.45) ⇒ **Close** the References dialog box ⇒ **Orient the sketching plane parallel to the screen** ⇒ **Create 2 point centerlines** ⇒ sketch a horizontal centerline on **DTM1** (Fig. 8.46)

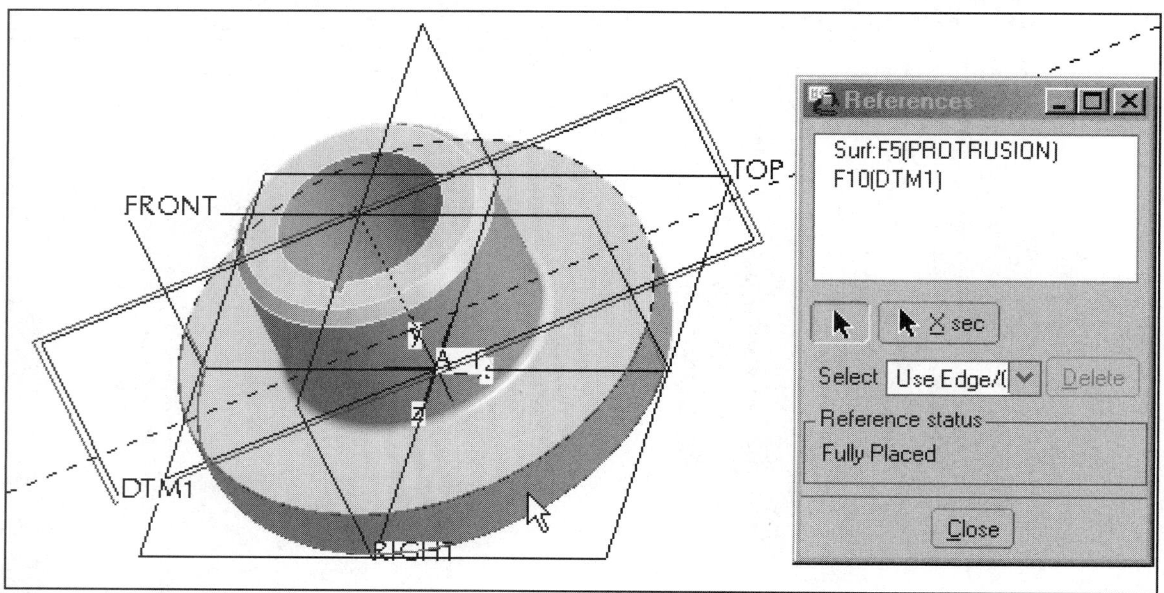

Figure 8.45 References

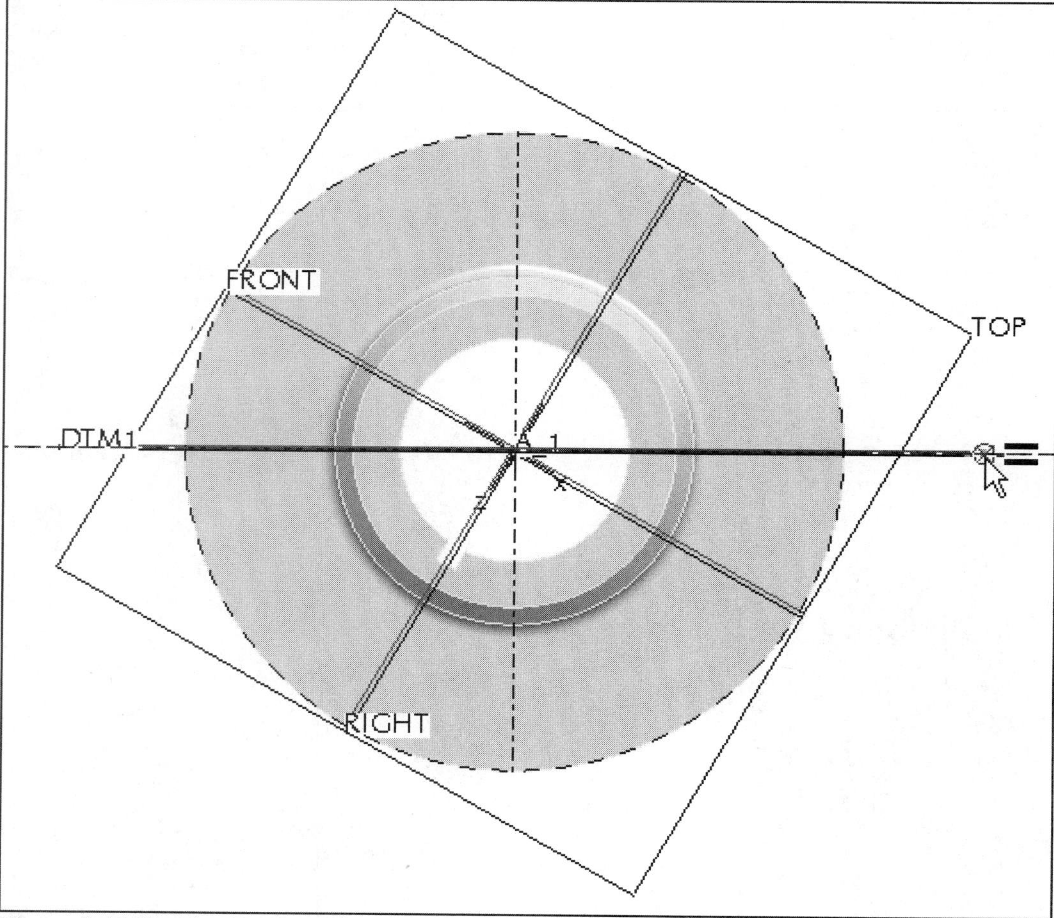

Figure 8.46 Sketch a Horizontal Centerline

Check your Environment settings, click: **Tools** ⇒ **Set various environment options** ⇒ Snap to Grid ⇒ **OK** ⇒ ⇒ **Create an arc by picking its center and endpoints** ⇒ pick on the centerline (and **DTM1**) as the starting point [Fig. 8.47(a)] ⇒ pick the start point of the arc ⇒ pick the ending point for the arc [Fig. 8.47(b)] ⇒ **Create 2 point lines** ⇒ sketch the first line from the end of the arc to the edge of the part [Fig. 8.47(c)] ⇒ **MMB** ⇒ sketch the second line from the end of the arc to the edge of the part [Fig. 8.47(d)] ⇒ **MMB** [Fig. 8.47(e)]

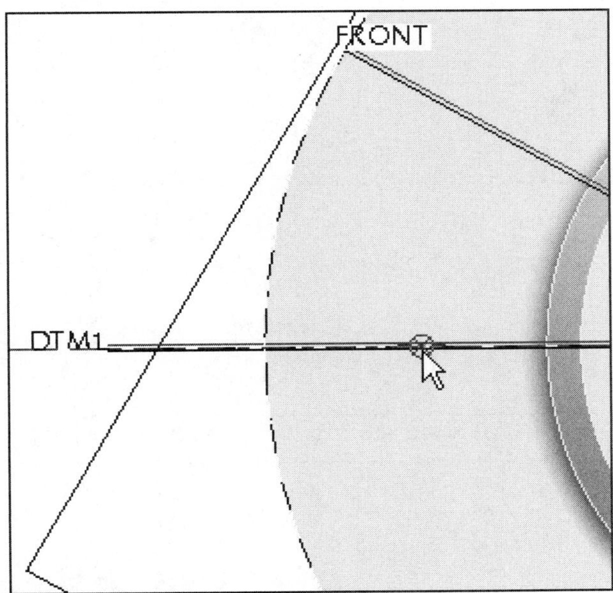

Figure 8.47(a) Arc Center Point

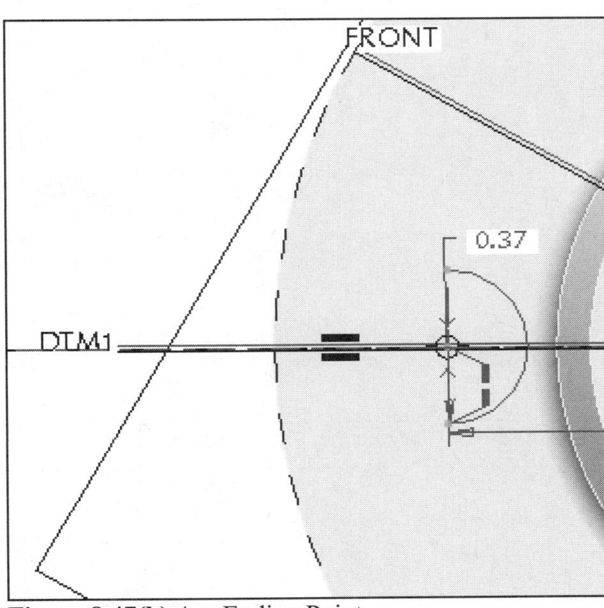

Figure 8.47(b) Arc Ending Point

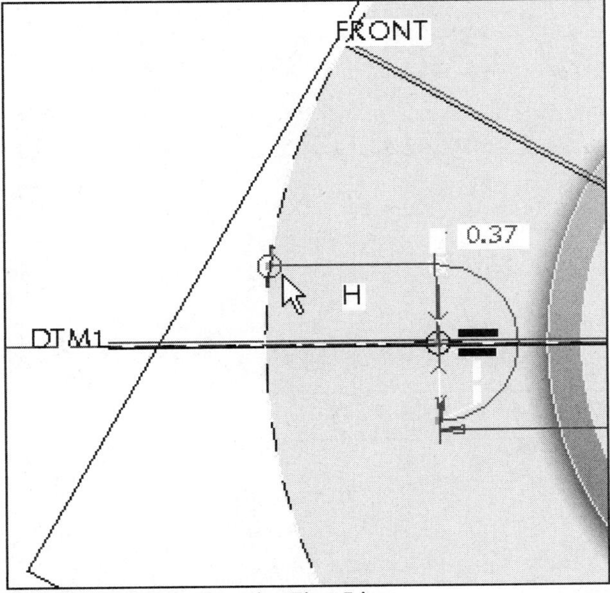

Figure 8.47(c) Create the First Line

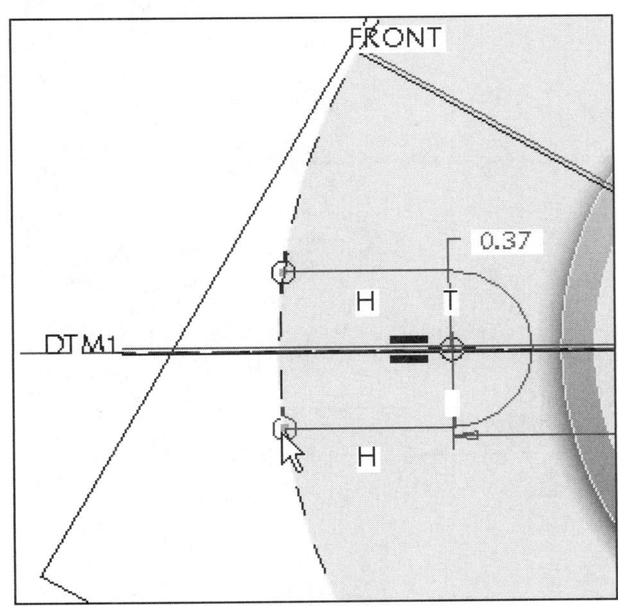

Figure 8.47(d) Create the Second Line

Click: ⇒ **Create concentric arc** ⇒ pick on the circular edge of the center hole [Fig. 8.47(f)] ⇒ pick the start point of the arc at the end of the line [Fig. 8.47(g)] ⇒ pick the end point of the arc at the end of the other line [Fig. 8.47(h)], completed closed section [Fig. 8.47(i)] ⇒ **MMB**

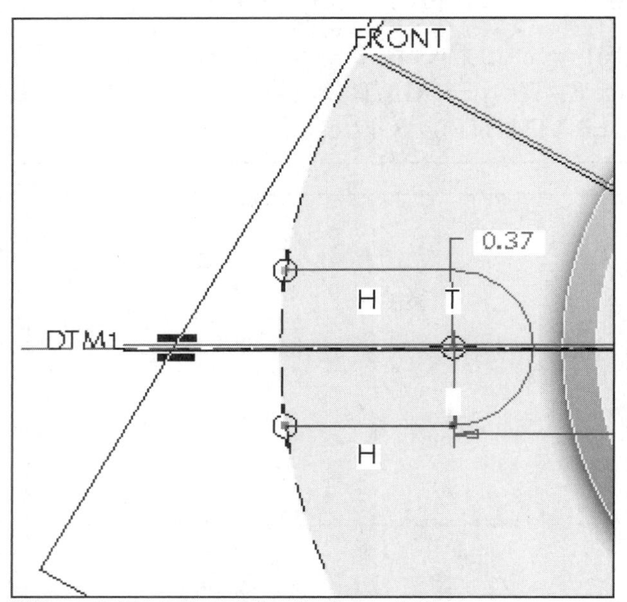

Figure 8.47(e) Complete the Second Line

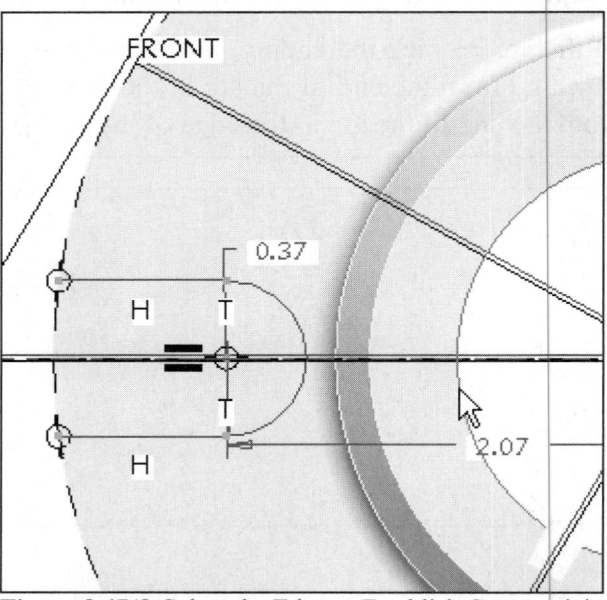

Figure 8.47(f) Select the Edge to Establish Concentricity

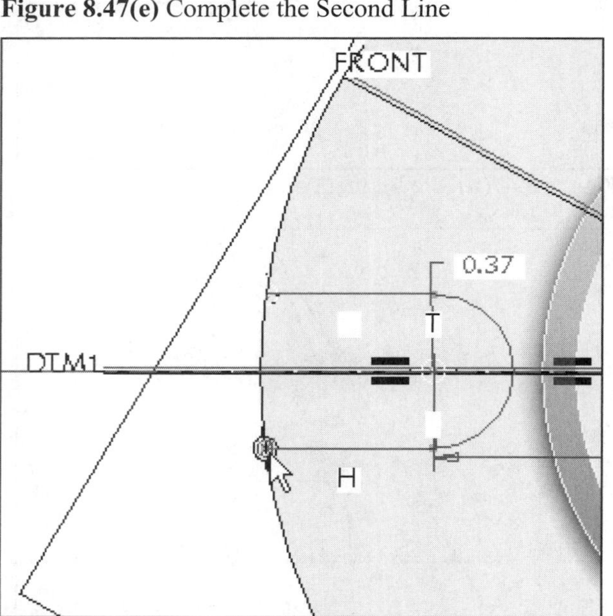

Figure 8.47(g) Start the Arc

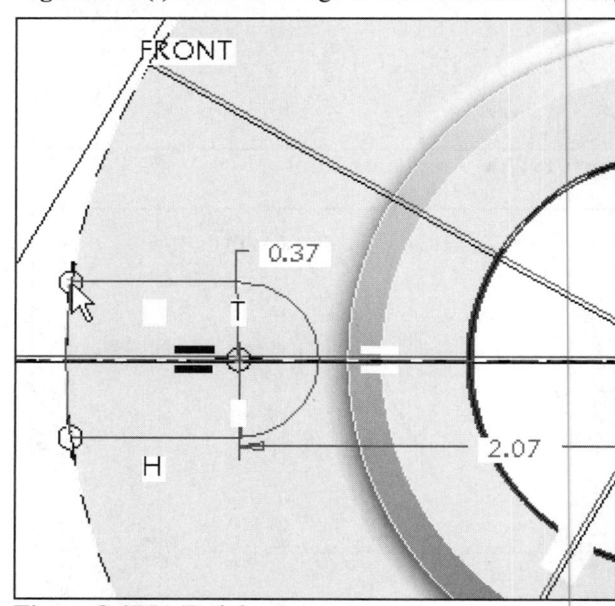

Figure 8.47(h) End the Arc

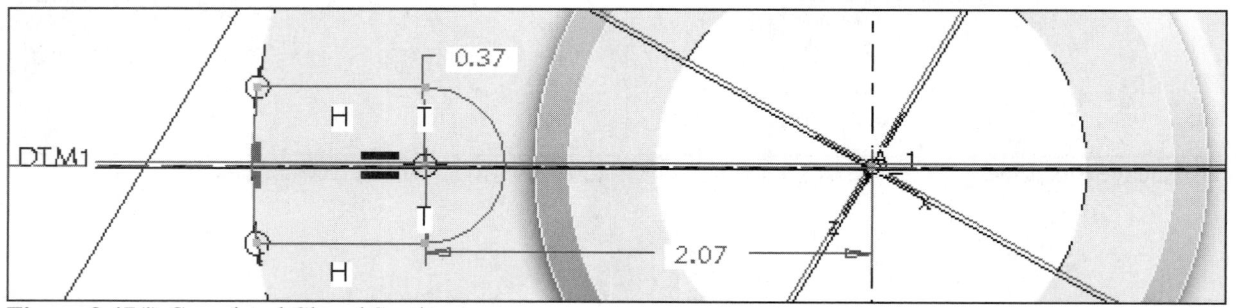

Figure 8.47(i) Completed Closed Section

Click: [toolbar icons] ⇒ [icon] **Create concentric circle** ⇒ pick on the circular edge of the center hole ⇒ pick the center point of the first arc [Fig. 8.47(j)] ⇒ **MMB** ⇒ [icon] ⇒ pick on the new circle ⇒ **RMB** ⇒ **Construction** [Fig. 8.47(k)]

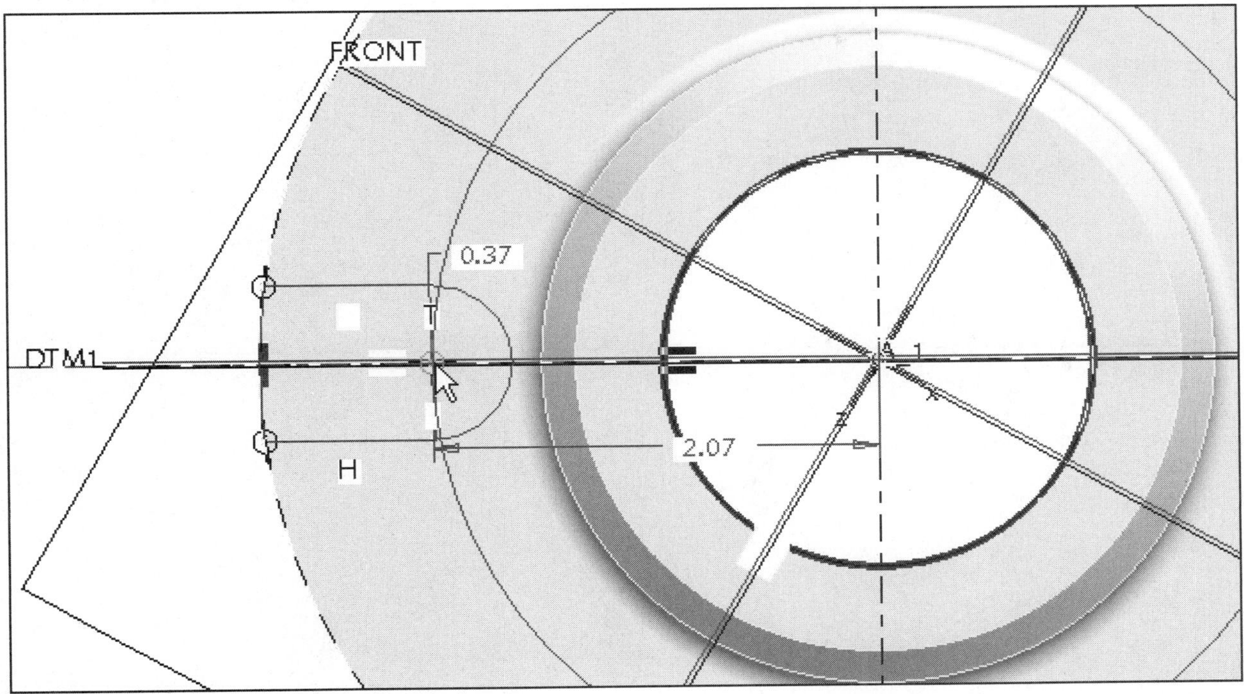

Figure 8.47(j) Create the Concentric Circle

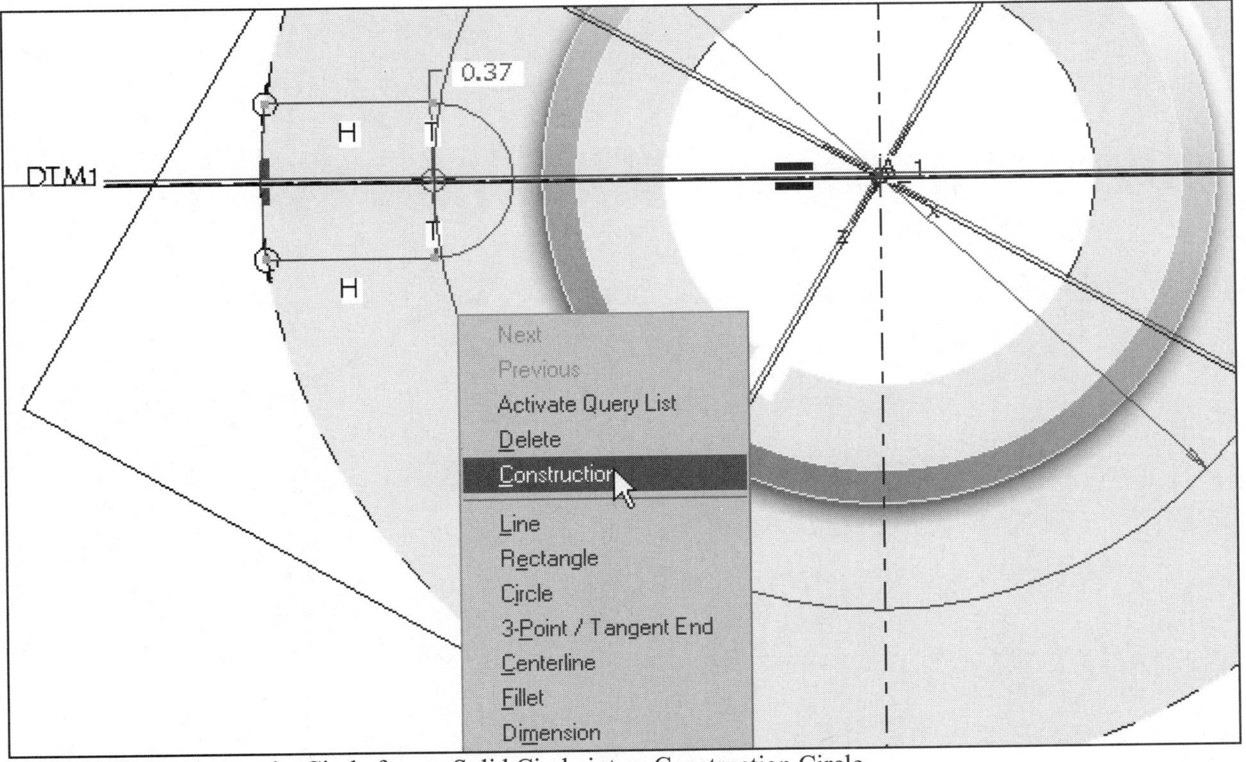

Figure 8.47(k) Change the Circle from a Solid Circle into a Construction Circle

Modify the dimensions: ⇒ **Ø4.50** for the construction bolt circle and **R.40** for the protrusion size [Fig. 8.47(l)] ⇒ move the dimensions to a more appropriate positions ⇒ ✔ **Continue with the current section** [Fig. 8.47(m)]

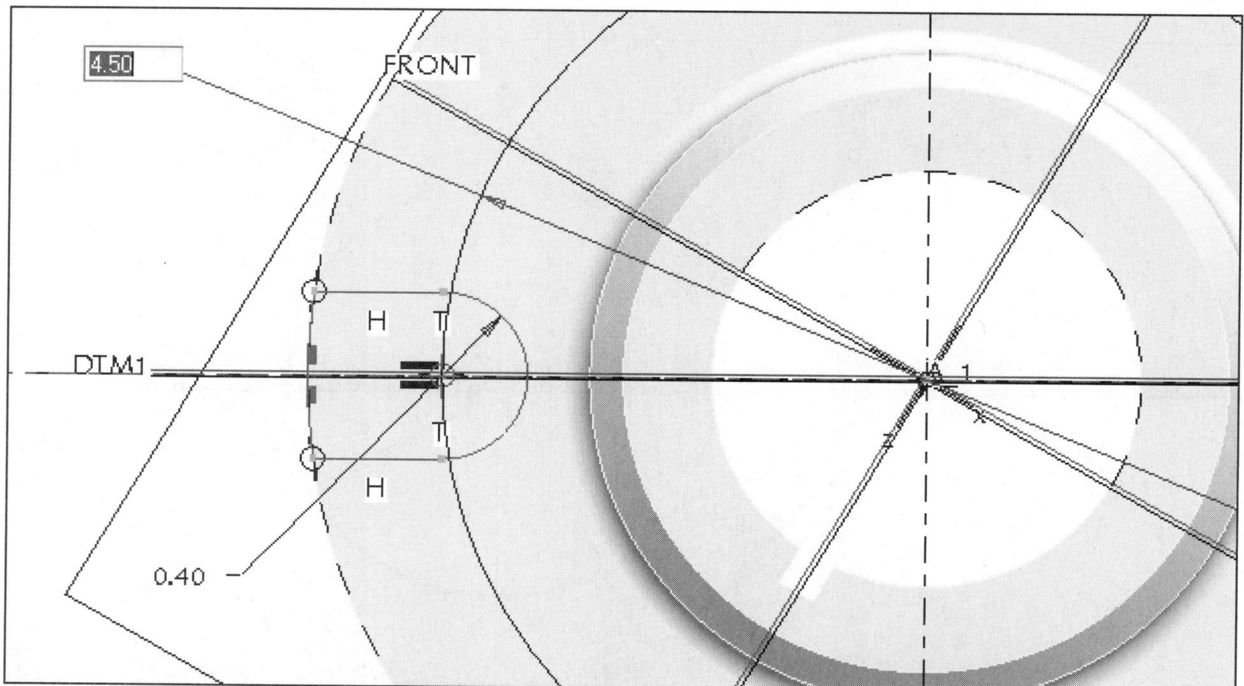

Figure 8.47(l) Move and Modify the Dimensions

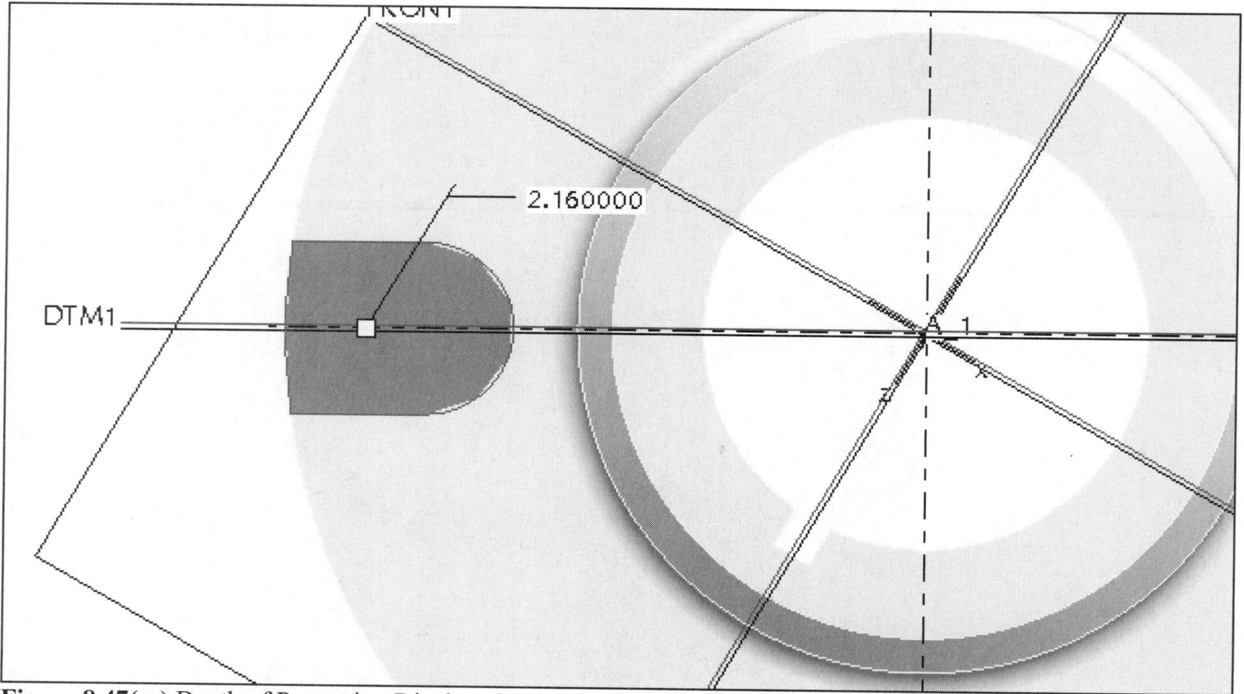

Figure 8.47(m) Depth of Protrusion Displayed

Click: **Display object in standard orientation** [Fig. 8.47(n)] ⇒ move the drag handle to approximately **.10** and then modify the dimension to the design size of **.125** [Fig. 8.47(o)] ⇒ ⇒ ⇒ **MMB**

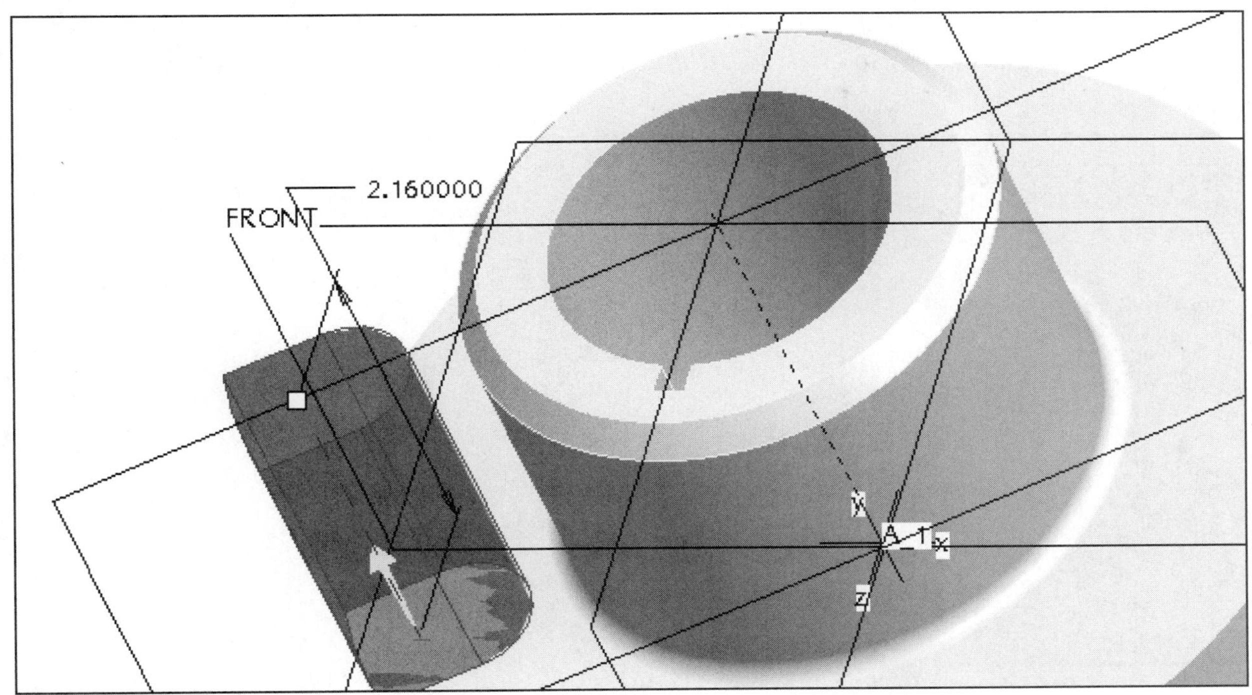

Figure 8.47(n) Standard Orientation

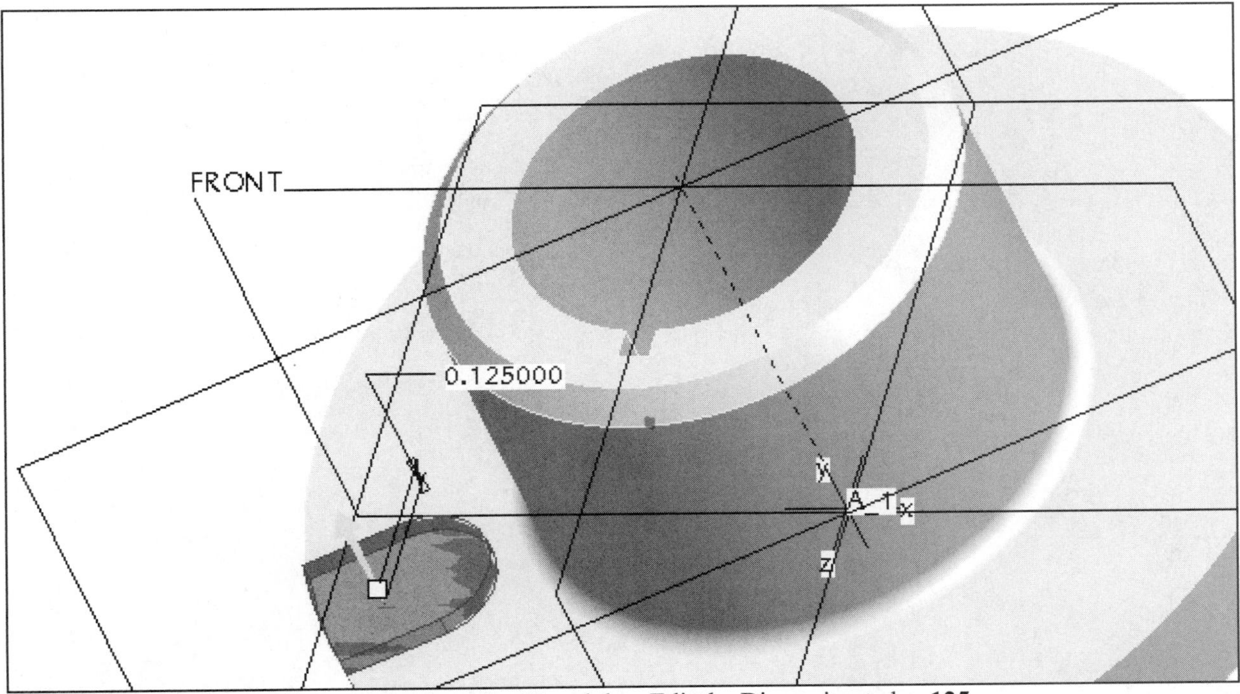

Figure 8.47(o) Drag Handle to Approximately **.10** and then Edit the Dimension to be **.125**

Click: **Opens Appearance Editor and Starts New Color Definition** the Color Editor will open (Fig. 8.48) ⇒ create a new color ⇒ **Close** ⇒ click on the new color patch in the Appearance Editor dialog box ⇒ Assignment Surfaces ⇒ ⇒ pick on the surface of the new protrusion (Fig. 8.49) ⇒ **OK** from the Select dialog box (you may need to move some of your windows in order to see the Select dialog box) *(you can also click MMB to end the selection process)* ⇒ **Apply** ⇒ **Close** ⇒ ⇒ **MMB**

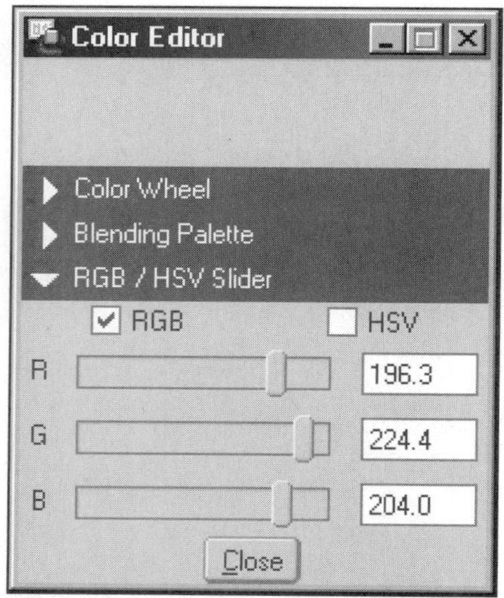

Figure 8.48 Create a New Color Using the Color Editor

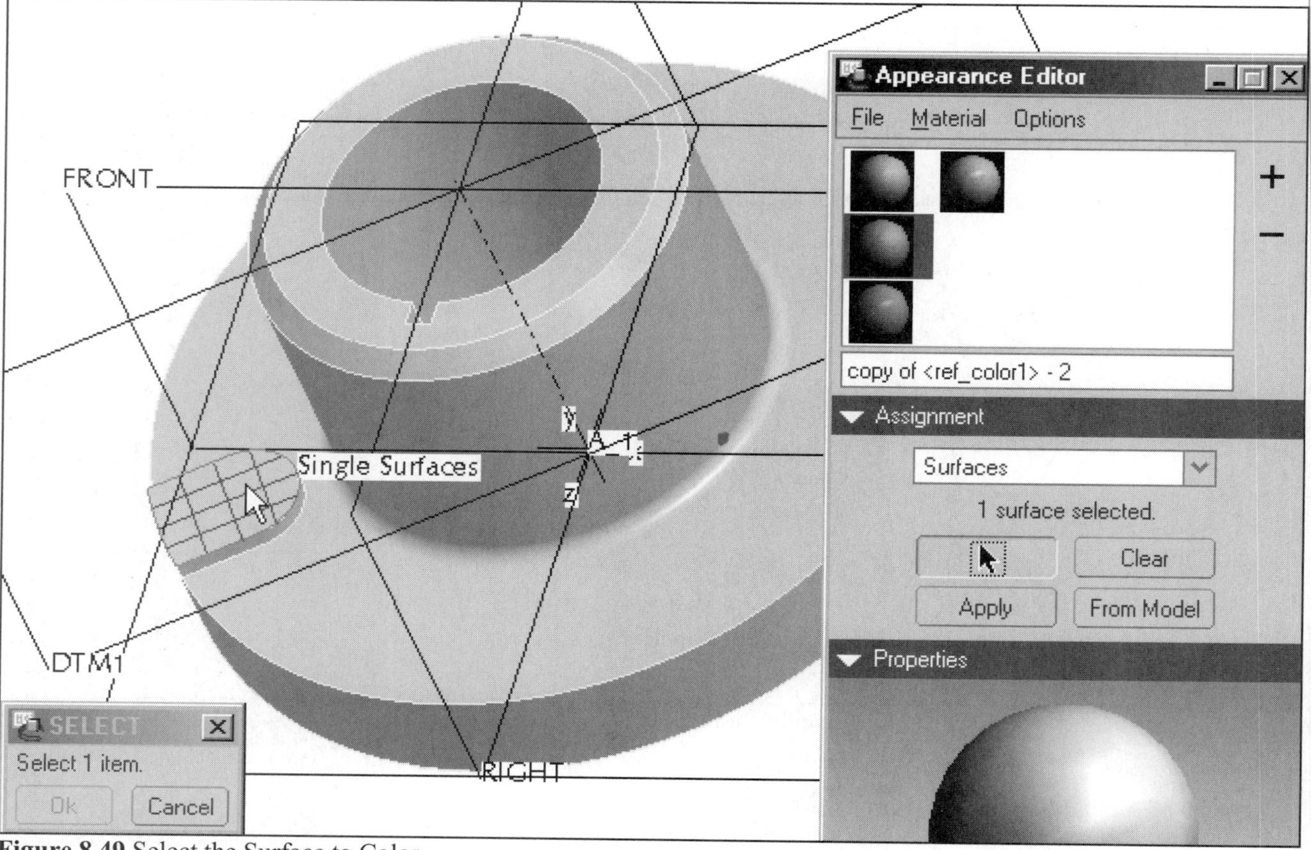

Figure 8.49 Select the Surface to Color

Create the cut; click ⇒ **Extrude Tool** ⇒ **Remove Material** ⇒ **Extrude to intersect all surface** ⇒ ⇒ pick on the top face of the boss protrusion as the sketch plane ⇒ Reference: **DTM1** ⇒ Orientation: **Bottom** [Fig. 8.50(a)] ⇒ **Sketch** ⇒ **Display object in standard orientation** ⇒ ⇒ pick on the outer flange surface for the first reference ⇒ pick the two arcs and two lines of the boss protrusion [Fig. 8.50(b)] ⇒ **Close** the References dialog box

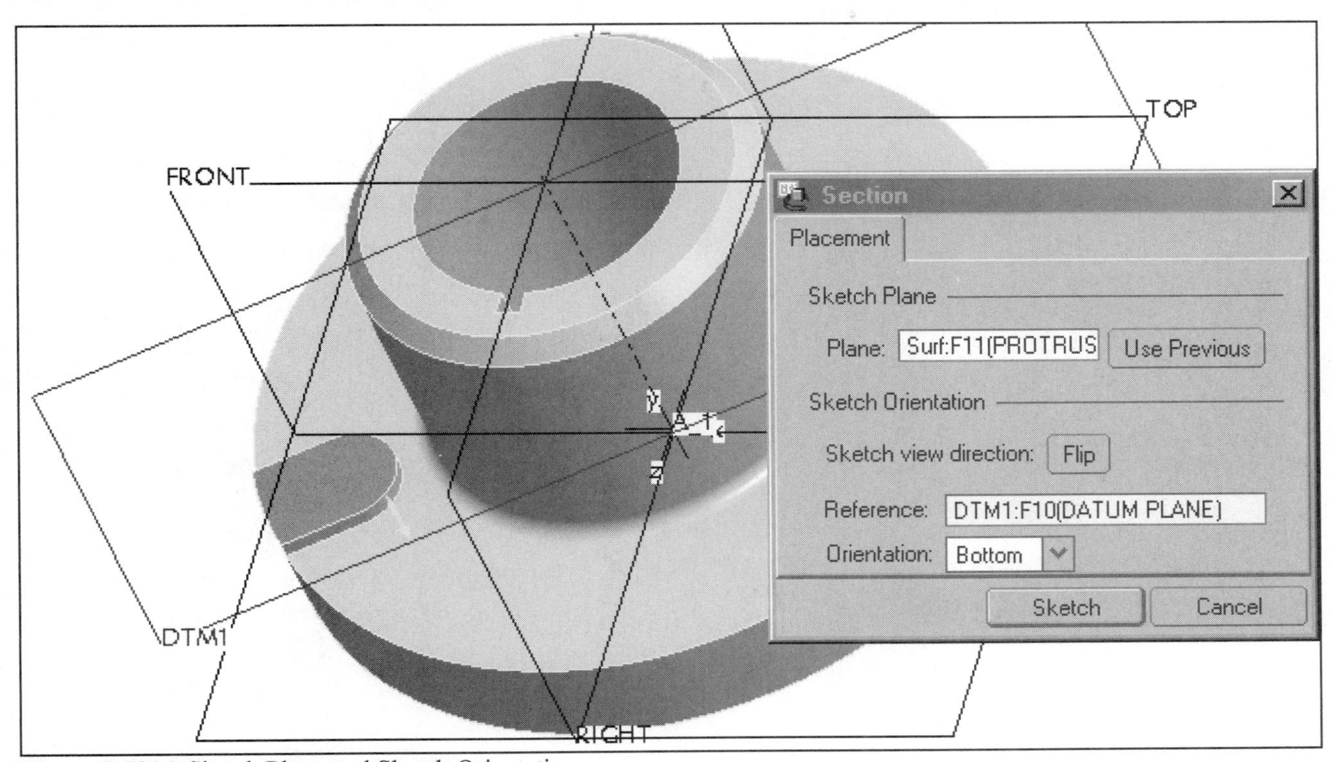

Figure 8.50(a) Sketch Plane and Sketch Orientation

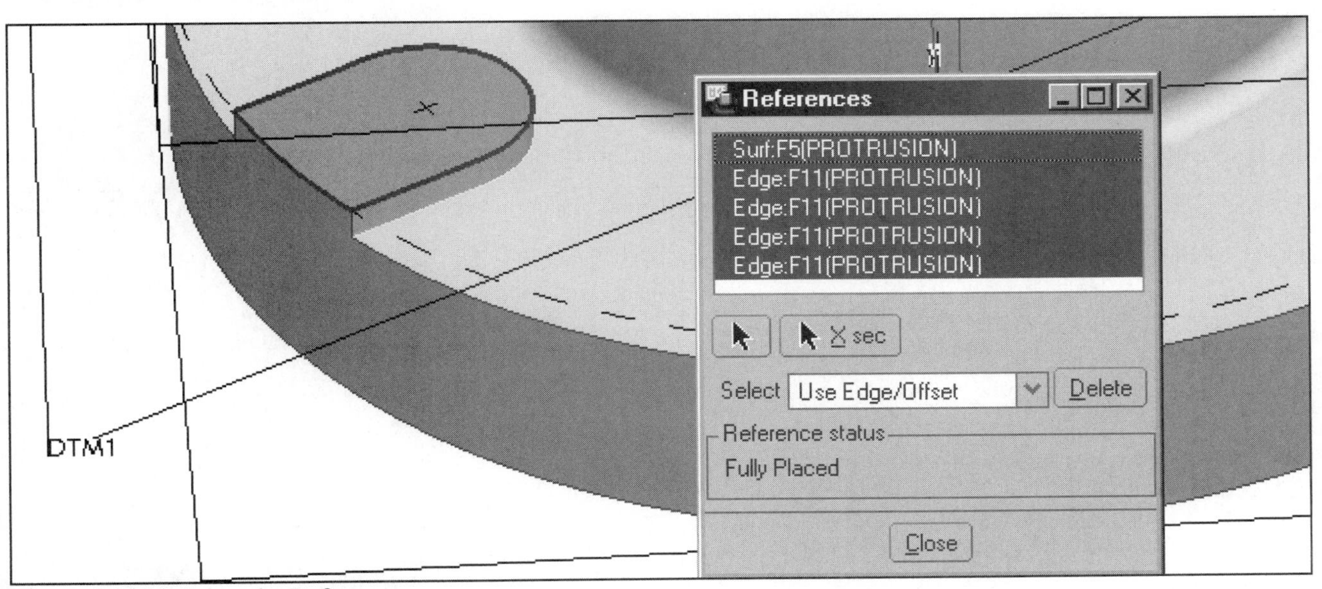

Figure 8.50(b) Select the References

Click: **Orient the sketching plane parallel to the screen** [Fig. 8.50(c)]⇒ **Create 2 point centerlines** ⇒ sketch a horizontal centerline on **DTM1** [Fig. 8.50(d)] ⇒ ⇒ **Create an entity by offsetting an edge** ⇒ Single [Fig. 8.50(e] ⇒ pick on the edge of the arc [Fig. 8.50(f)] ⇒ type **-.15** (*negative .15*) at the prompt ⇒ **Enter** [Fig. 8.50(g)] ⇒ **Close** Type dialog box

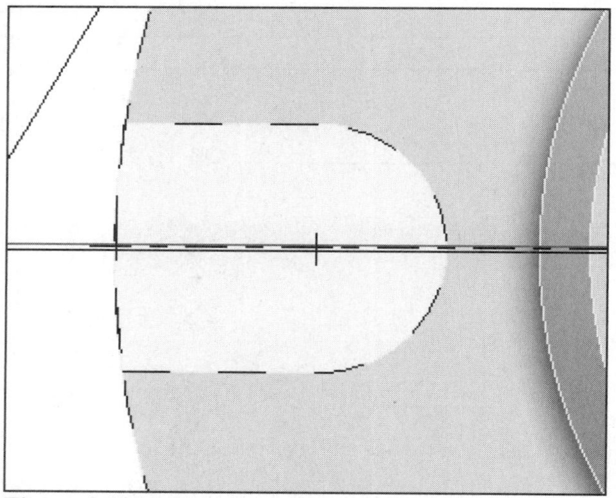

Figure 8.50(c) Sketching Plane Parallel to the Screen

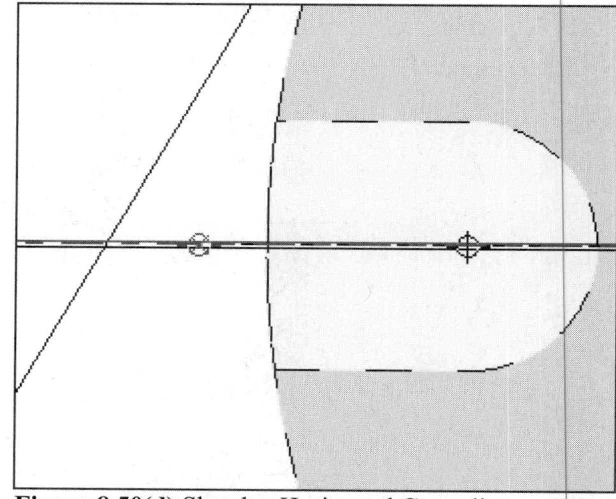

Figure 8.50(d) Sketch a Horizontal Centerline on DTM1

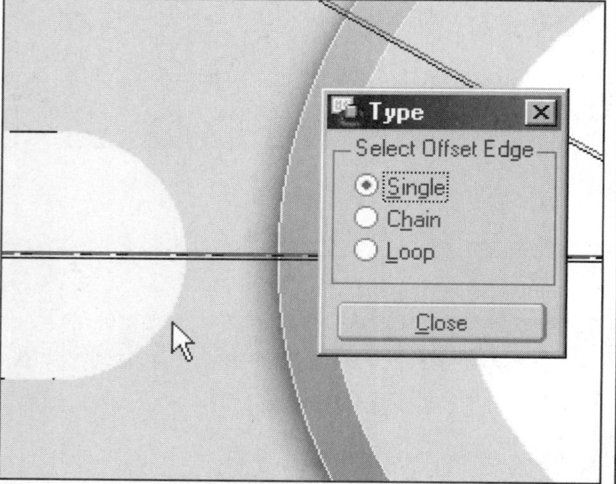

Figure 8.50(e) Type Dialog Box

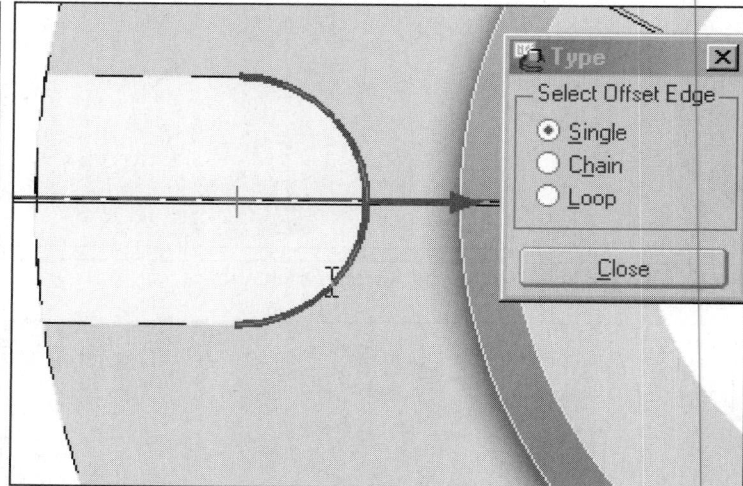

Figure 8.50(f) Select the Arc's Edge

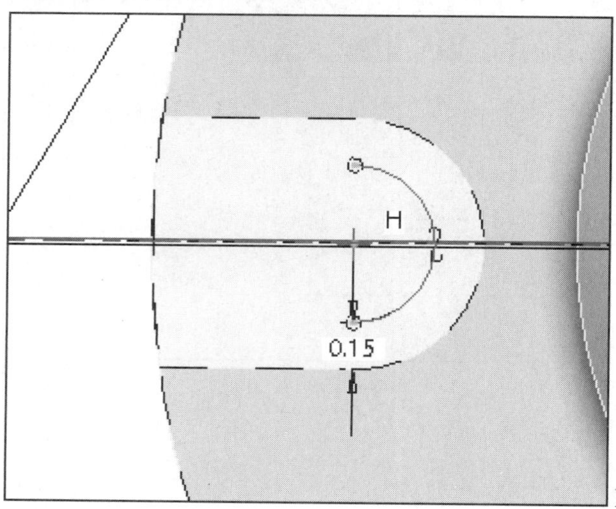

Figure 8.50(g) Offset Arc

Click: **Create 2 point lines** ⇒ sketch the first line from the end of the arc to the edge of the part [Fig. 8.50(h)] ⇒ **MMB** ⇒ sketch the second line [Fig. 8.50(i)] from the end of the arc to the edge of the part [Fig. 8.50(j)] ⇒ **MMB** ⇒ **MMB** ⇒ click on the **.15** dimension ⇒ **RMB** ⇒ **Delete** [Fig. 8.50(k)] ⇒ **.25** dimension displays automatically [Fig. 8.50(l)] ⇒ reposition the **.25** dimension [Fig. 8.50(m)]

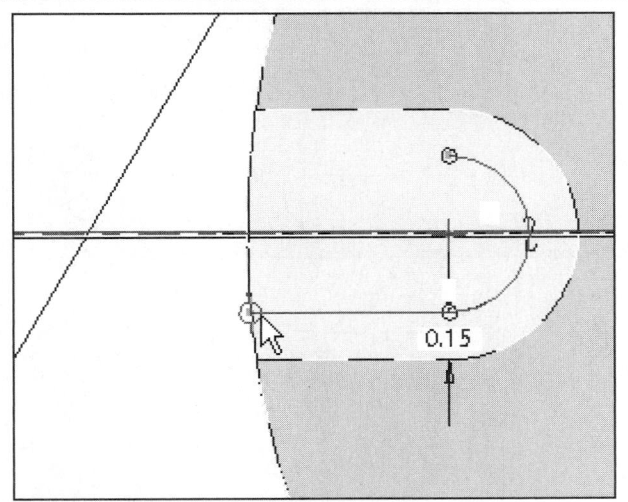

Figure 8.50(h) Sketch First Line

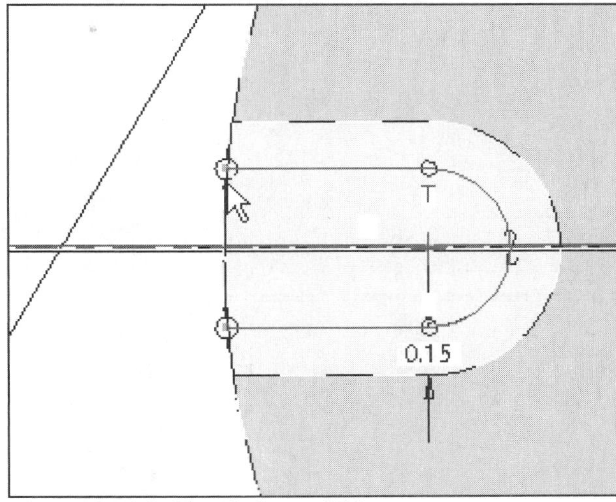

Figure 8.50(i) Sketch Second Line

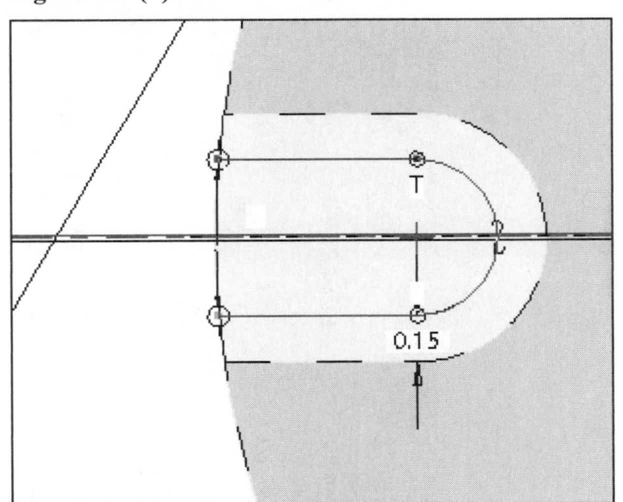

Figure 8.50(j) Completed Open Section

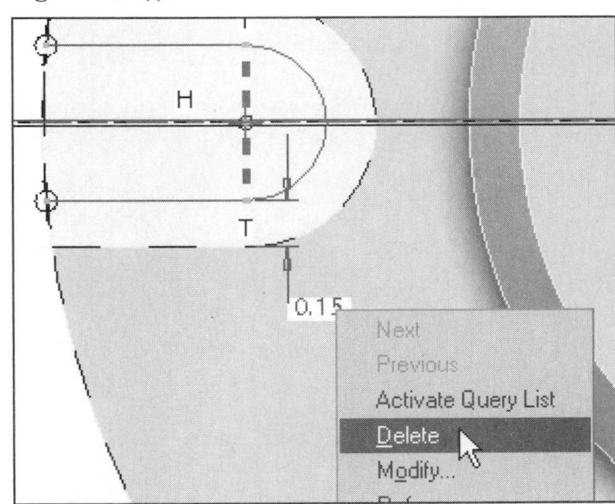

Figure 8.50(k) Delete Default Dimension

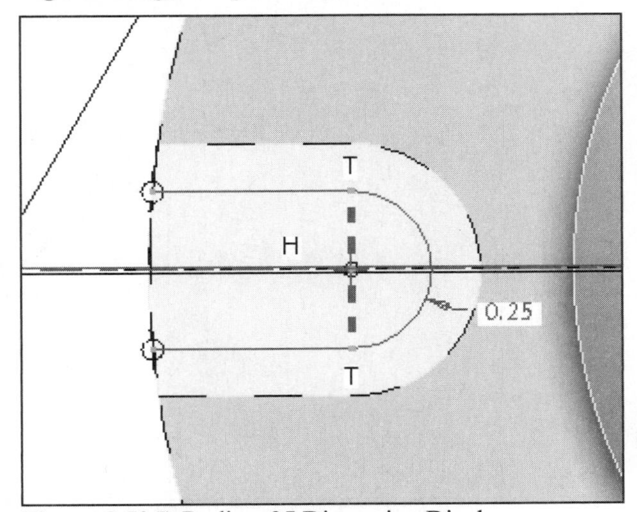

Figure 8.50(l) Radius **.25** Dimension Displays

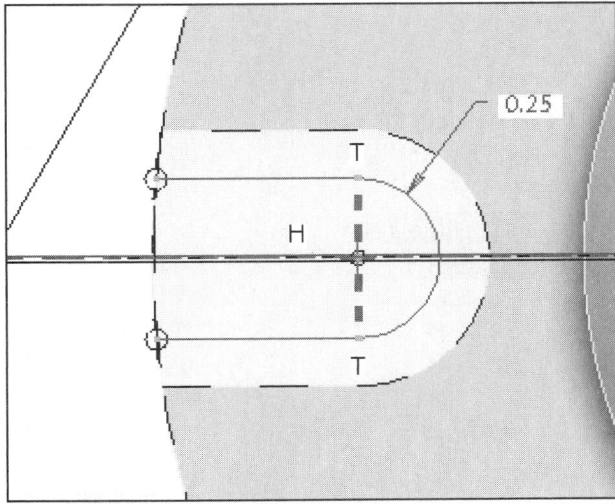

Figure 8.50(m) Move the Dimension

Click on the **.25** dimension ⇒ **RMB** [Fig. 8.50(n)] ⇒ **Strong** [Fig. 8.50(o)] ⇒ ⇒ **Display object in standard orientation** [Fig. 8.50(p)] (*change the depth direction if necessary*) ⇒ ⇒ ⇒ [Fig. 8.50(q)] ⇒ highlight new features in the Model Tree [Fig. 8.50(r)] ⇒ ⇒ ⇒ **MMB**

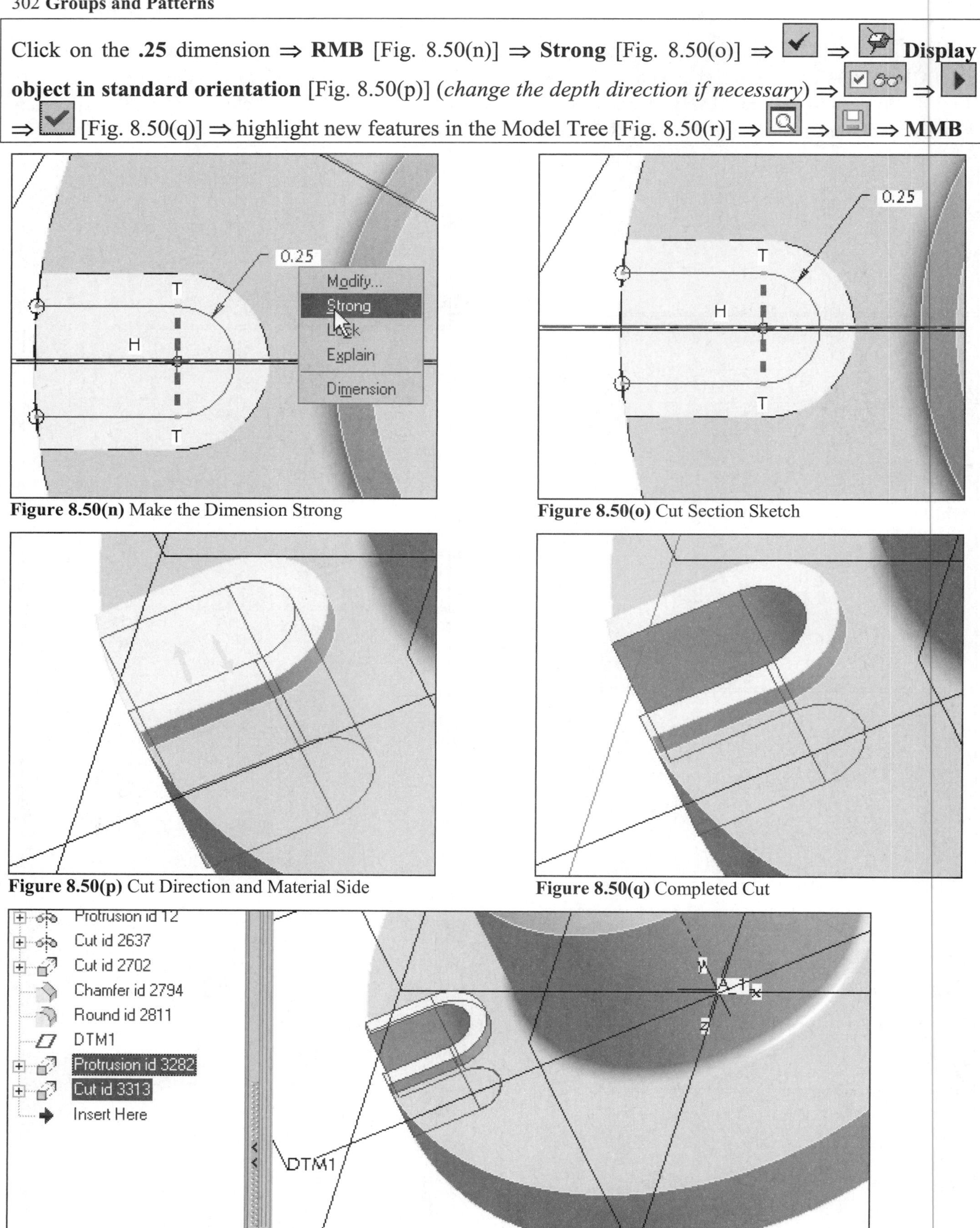

Figure 8.50(n) Make the Dimension Strong

Figure 8.50(o) Cut Section Sketch

Figure 8.50(p) Cut Direction and Material Side

Figure 8.50(q) Completed Cut

Figure 8.50(r) Protrusion and Cut

Create a set of rounds around the boss, click: **Round Tool** ⇒ type **.100** for the radius value ⇒ **Enter** ⇒ **Options** ⇒ **Solid** ⇒ **Sets** ⇒ hold down the **Ctrl** key and select the horizontal face of the flange and the vertical face of the protrusion [Fig. 8.51(a-b)] ⇒ **MMB** ⇒ ⇒ **MMB**

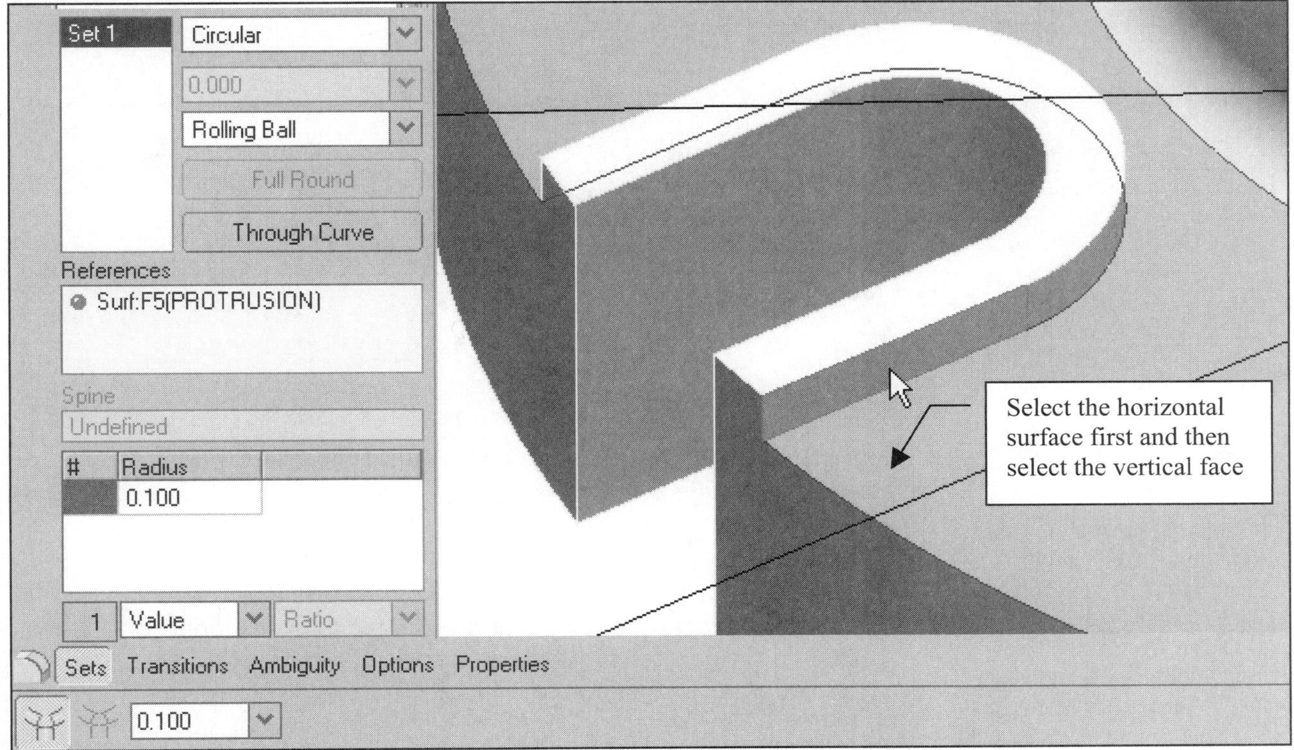

Figure 8.51(a) Select the Horizontal Surface of the Flange and then the Vertical Face of the Protrusion

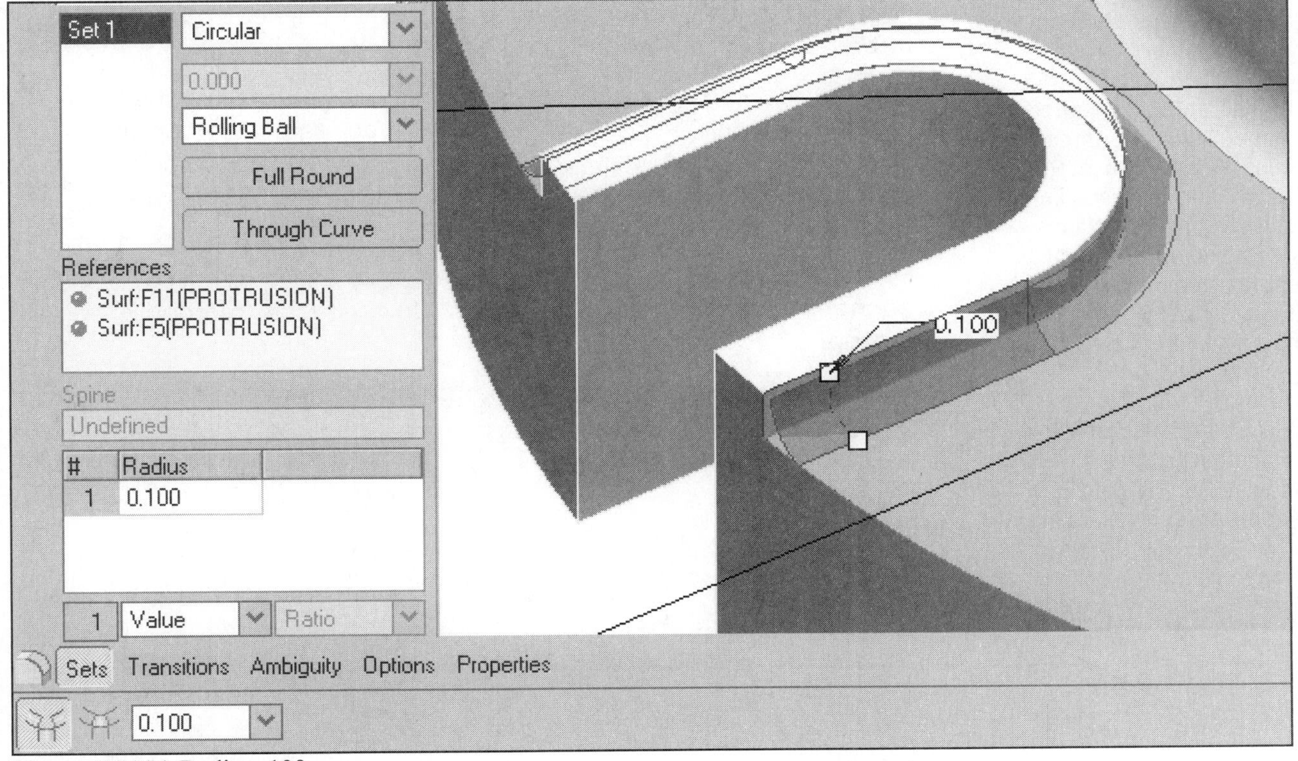

Figure 8.51(b) Radius **.100**

Before creating the local group, click: **Help** ⇒ **Help Center** ⇒ **Part Modeling** ⇒ **Part** ⇒ **Search** tab ⇒ type **Group** ⇒ **Enter** ⇒ **About Local Groups** (Fig. 8.52) ⇒ **Close**

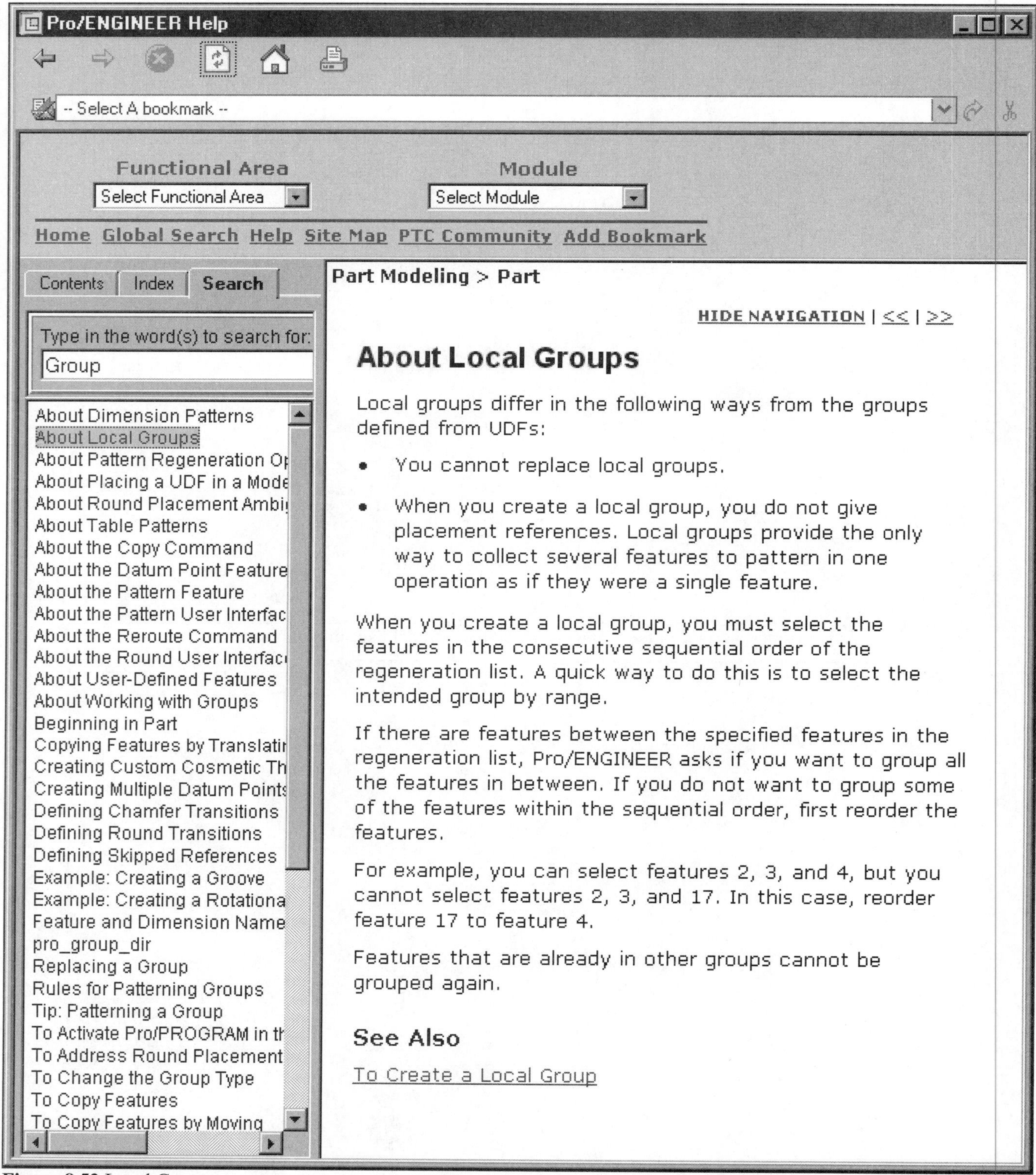

Figure 8.52 Local Groups

For creating the group, use the new mapkey button you created. *(If you did not create a group mapkey then use: Edit ⇒ Feature Operations ⇒ Group ⇒ Local Group ⇒ name ⇒)*

Zoom and spin the model slightly so as to see the features better ⇒ **Creates a Group** (you will have to Rename the group in the Model Tree after it is created, when you use this mapkey button) ⇒ press and hold down the Ctrl key and select **DTM1**, the protrusion, cut and round from the model or Model Tree (Fig. 8.53) ⇒ **MMB** ⇒ **MMB** ⇒ **MMB** ⇒ **MMB** ⇒ **GRPSEQUENCE** should still be highlighted in the Model Tree ⇒ **RMB** ⇒ **Rename** (Fig. 8.54) ⇒ change the name to **BOSS** (Fig. 8.55) ⇒ **MMB**

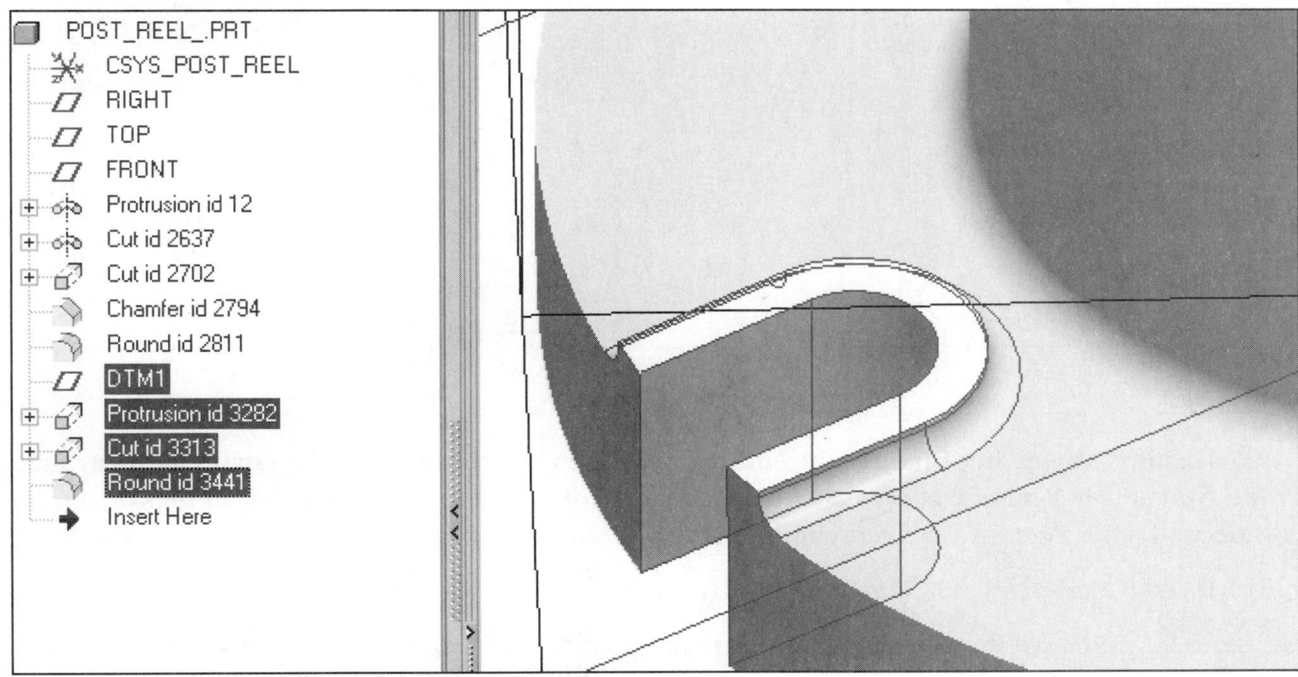

Figure 8.53 Grouping DTM1, Protrusion, Cut, and Round

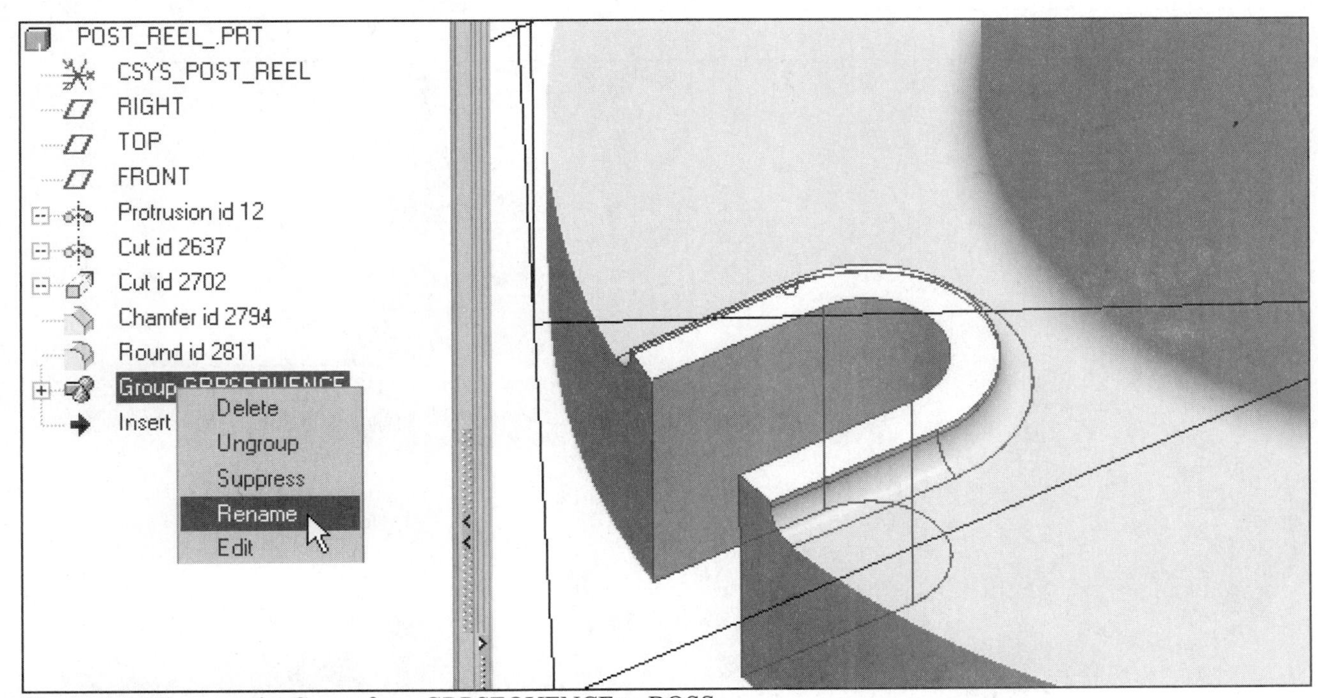

Figure 8.54 Rename the Group from GRPSEQUENCE to BOSS

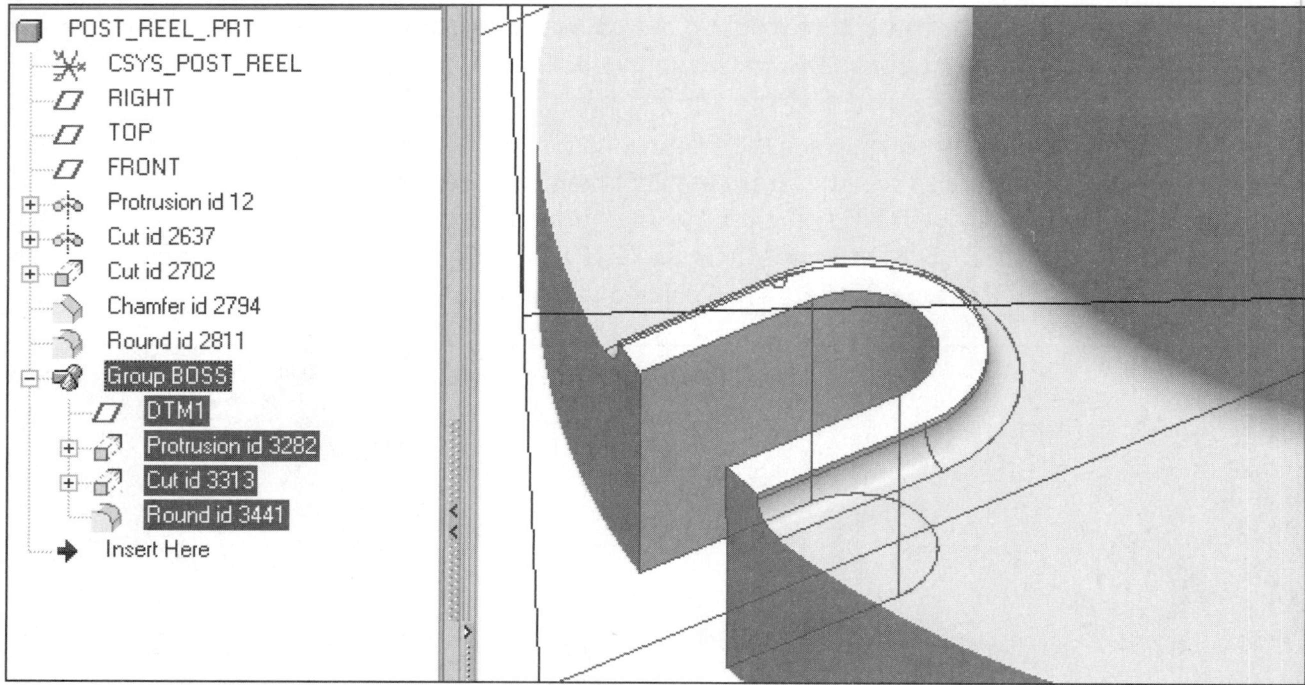

Figure 8.55 Group BOSS

Click: **Display object in standard orientation** ⇒ **View** ⇒ **Shade** ⇒ **Opens Appearance Editor and Starts New Color Definition** ⇒ define a new color ⇒ **Close** ⇒ **Surfaces** ⇒ press the **Ctrl** key and select all ten surfaces associated with the group to have the same new color (Fig. 8.56) ⇒ **MMB** ⇒ **Apply** ⇒ **Close** ⇒ ⇒ **MMB** ⇒ **LMB** to deselect ⇒ Before creating the pattern, click Group BOSS in the Model Tree ⇒ **Context sensitive help** ⇒ click on the **Pattern Tool** button on the right Toolbar ⇒ **About the Pattern** (Fig. 8.57) ⇒ **Close**

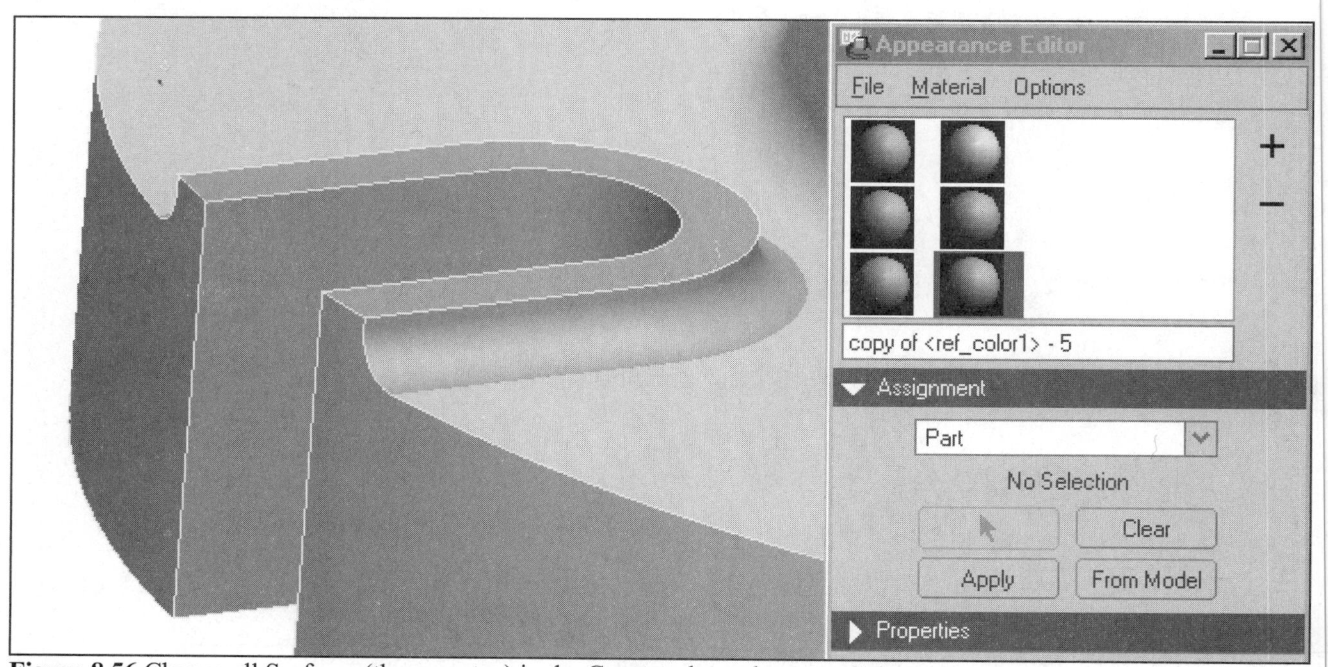

Figure 8.56 Change all Surfaces (there are ten) in the Group to have the same Color

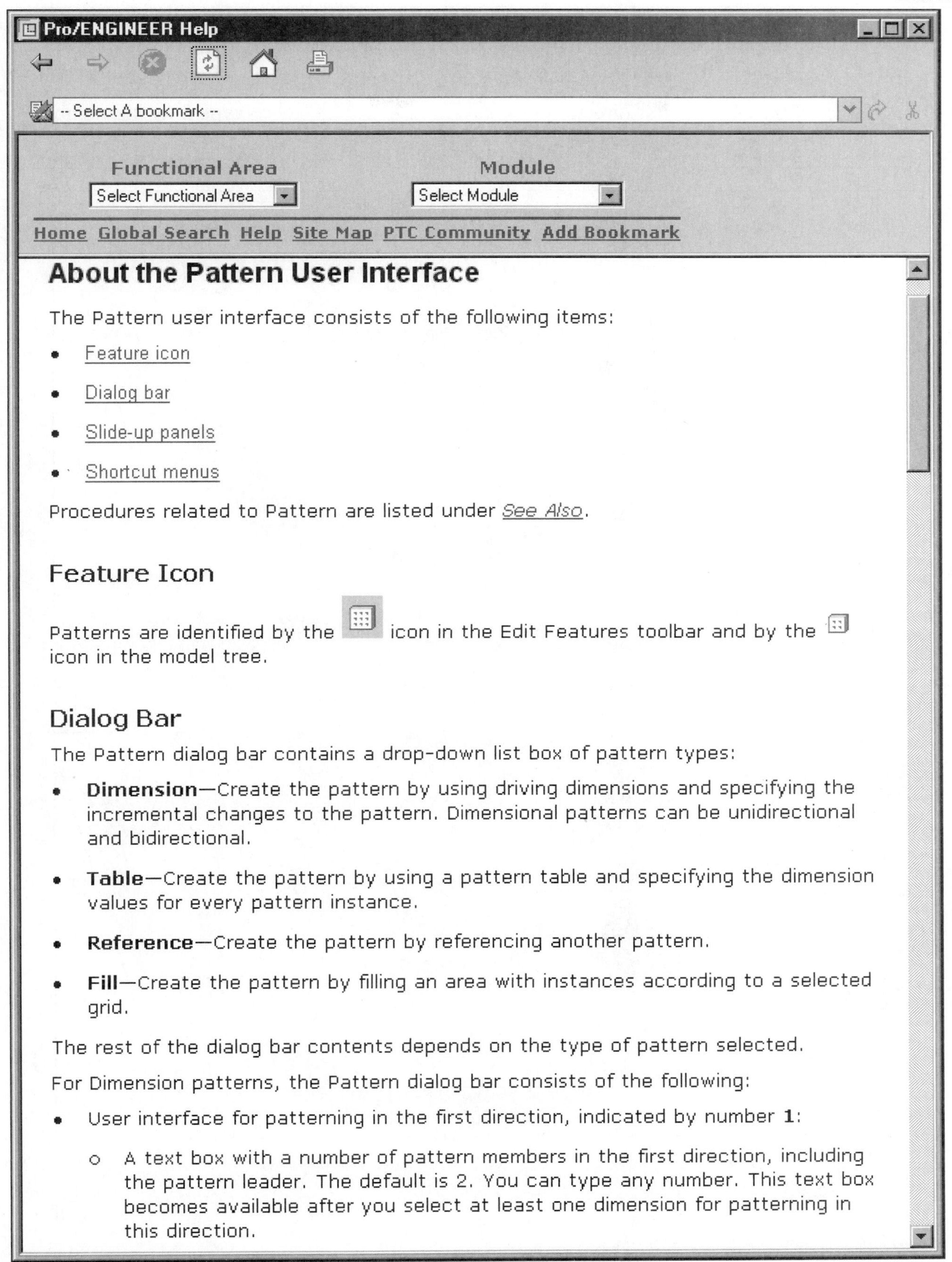

Figure 8.57 Pattern Help

Click: ⇒ ⇒ Before creating the pattern, click on Group BOSS in the Model Tree ⇒ **Pattern Tool** [Fig. 8.58(a)] ⇒ **Dimensions** tab [Fig. 8.58(b)] ⇒ click on the **30.00** degree dimension ⇒ type **120** ⇒ **Enter** ⇒ pick in the first direction box 1: 3 1 item(s) ⇒ type **3** as the number of members required ⇒ **Enter** [Fig. 8.58(c)] ⇒ **MMB** [Fig. 8.58(d)] ⇒ ⇒ **MMB** ⇒ **LMB** to deselect

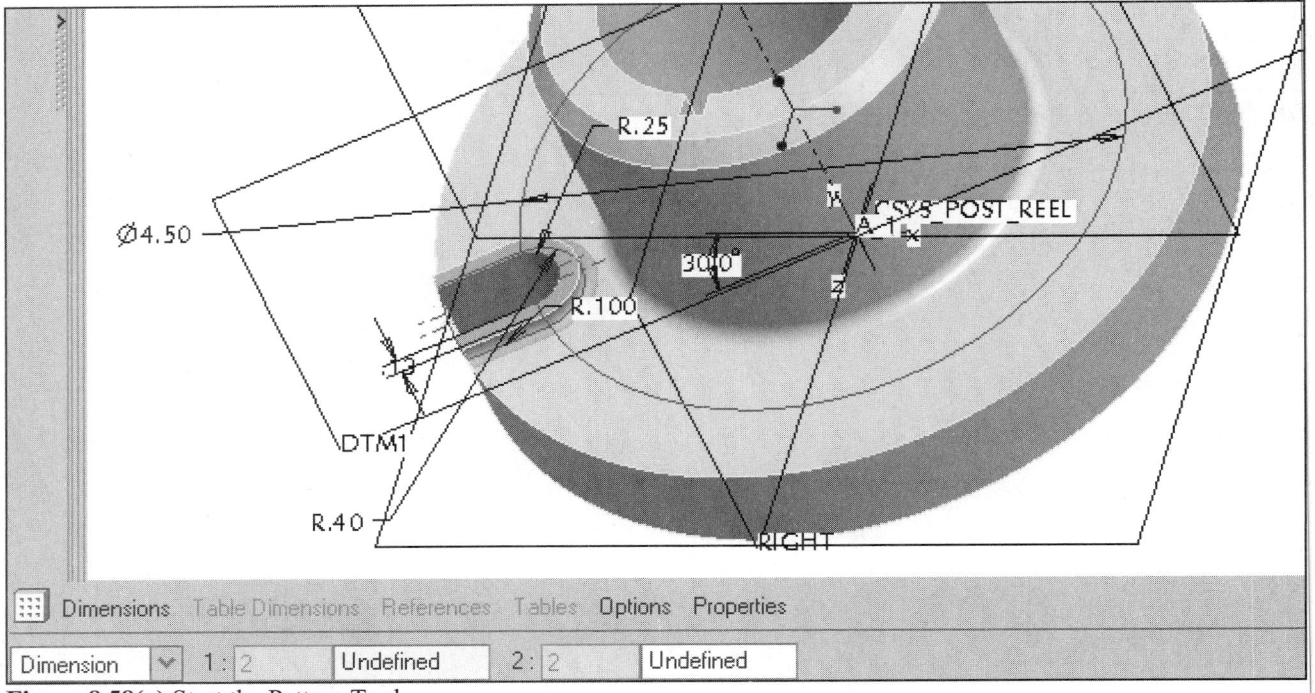

Figure 8.58(a) Start the Pattern Tool

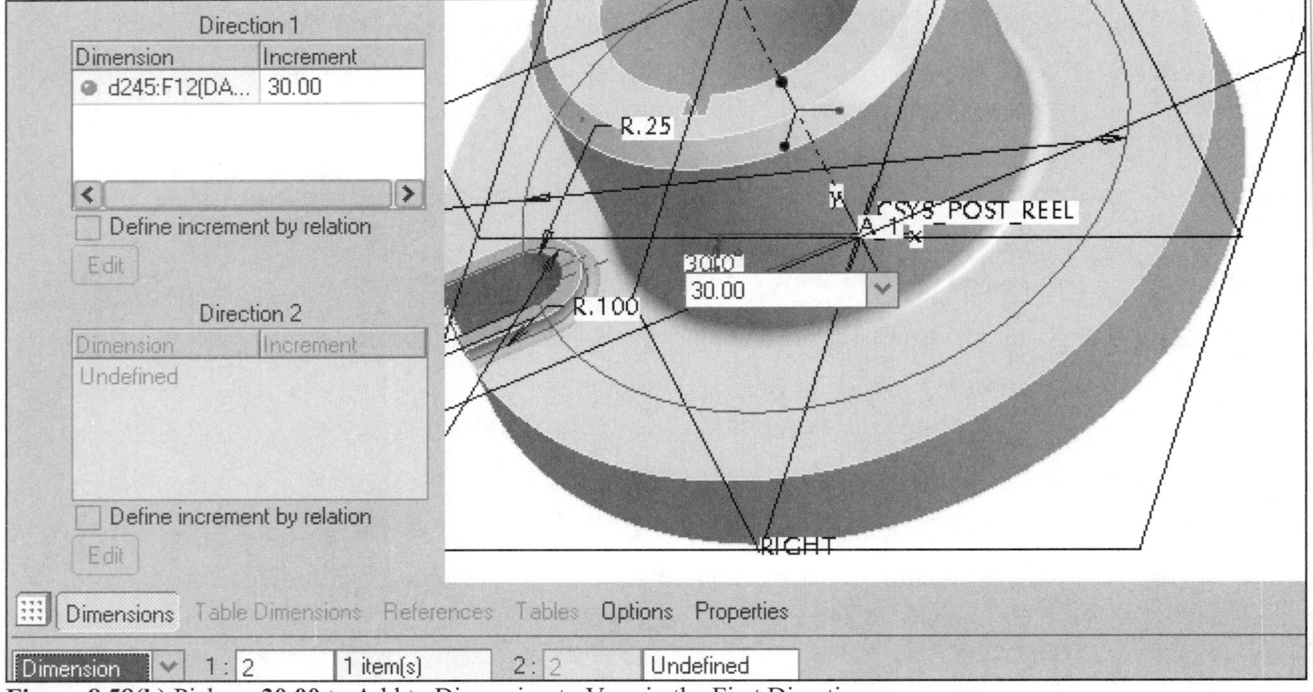

Figure 8.58(b) Pick on **30.00** to Add to Dimension to Vary in the First Direction

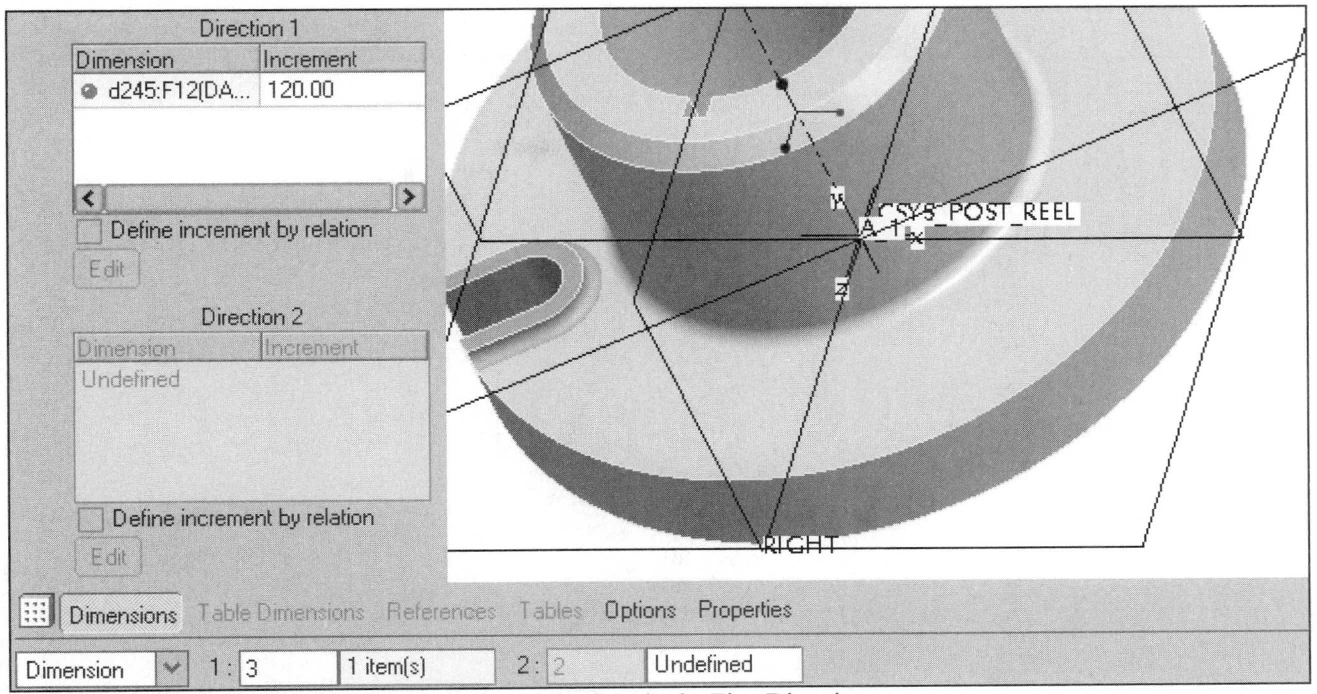

Figure 8.58(c) Type **3** for the Number of Pattern Members in the First Direction

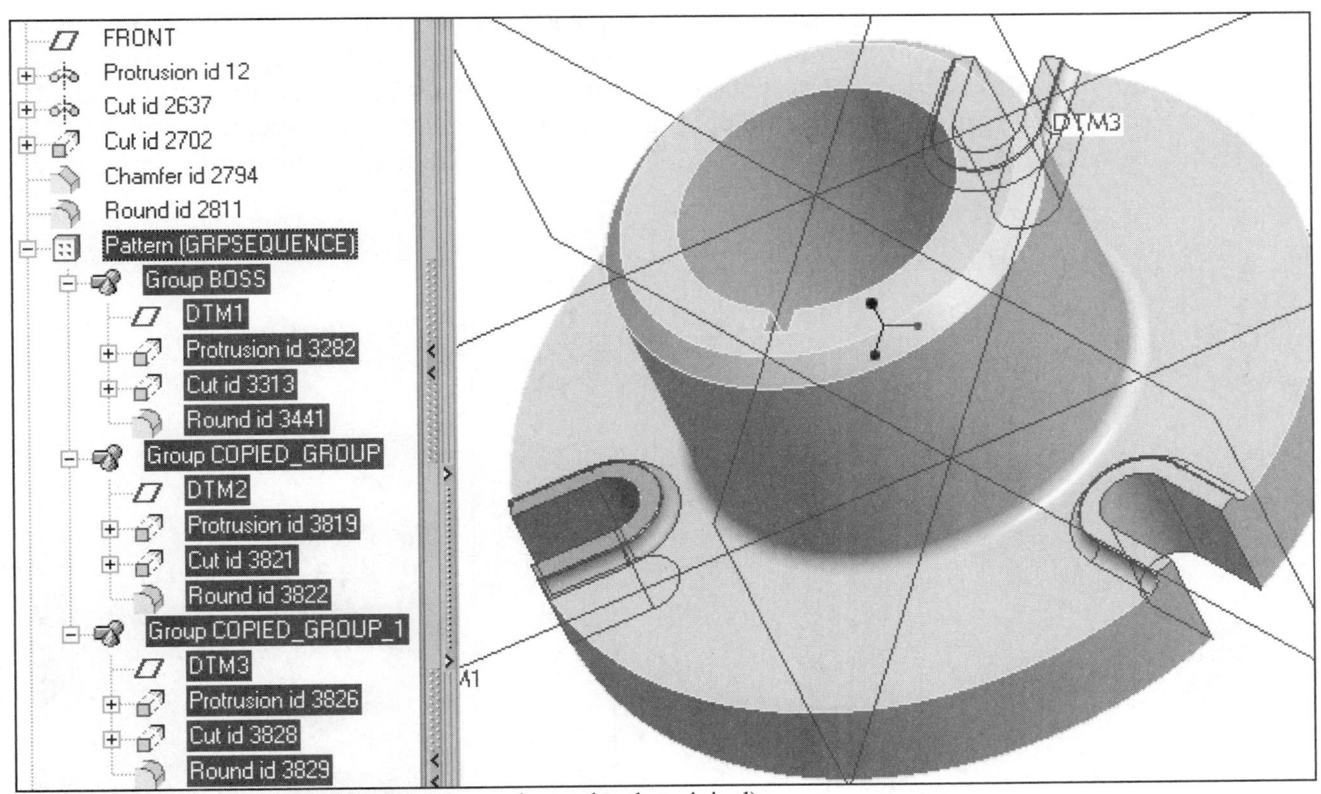

Figure 8.58(d) Three Members (two new members, plus the original)

You are now finished with Lesson 8. A radial pattern was created in this lesson and will be required for the Lesson Project. Rectangular patterns and patterns created with tables will be required in other lessons.

Complete the Lesson 8 Project (Taper Coupling).

Lesson 8 Project

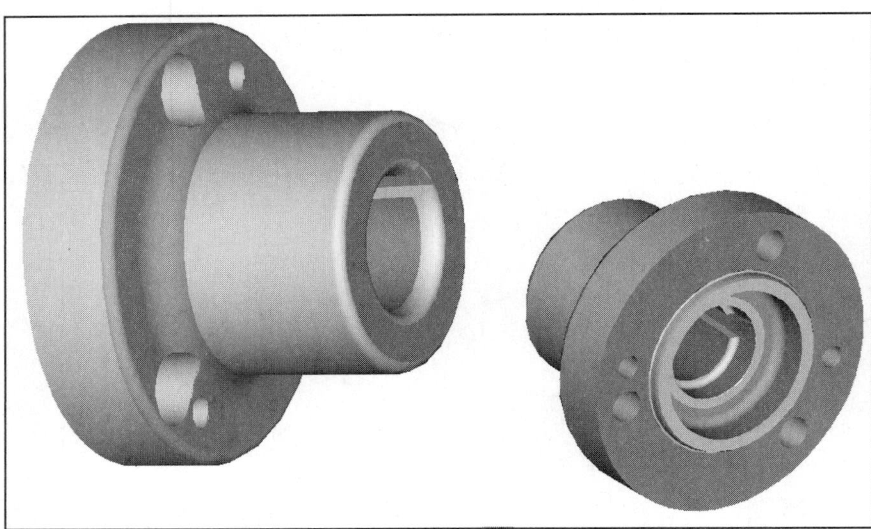

Figure 8.59 Taper Coupling

Taper Coupling

The Taper Coupling is a machined part that requires commands similar to those used in the Post Reel. Create the part shown in Figures 8.59 through 8.79. Plan the feature creation and the parent-child relationships for the part. The taper coupling will be used in an assembly in Lesson Projects 15 and 16. The machined face of the coupling mates with and is fastened to a similar surface when assembled. Plan your geometric tolerancing requirements accordingly. Set the datums to anticipate the mating surfaces.

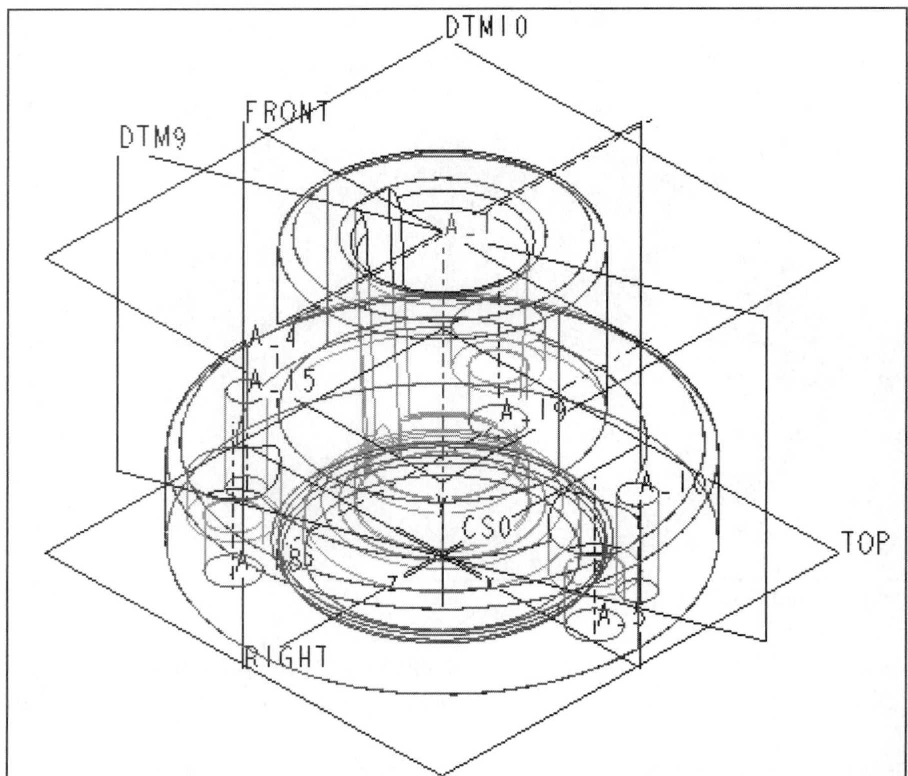

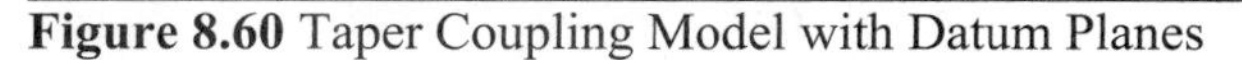

Figure 8.60 Taper Coupling Model with Datum Planes

Figure 8.61 Counterbore

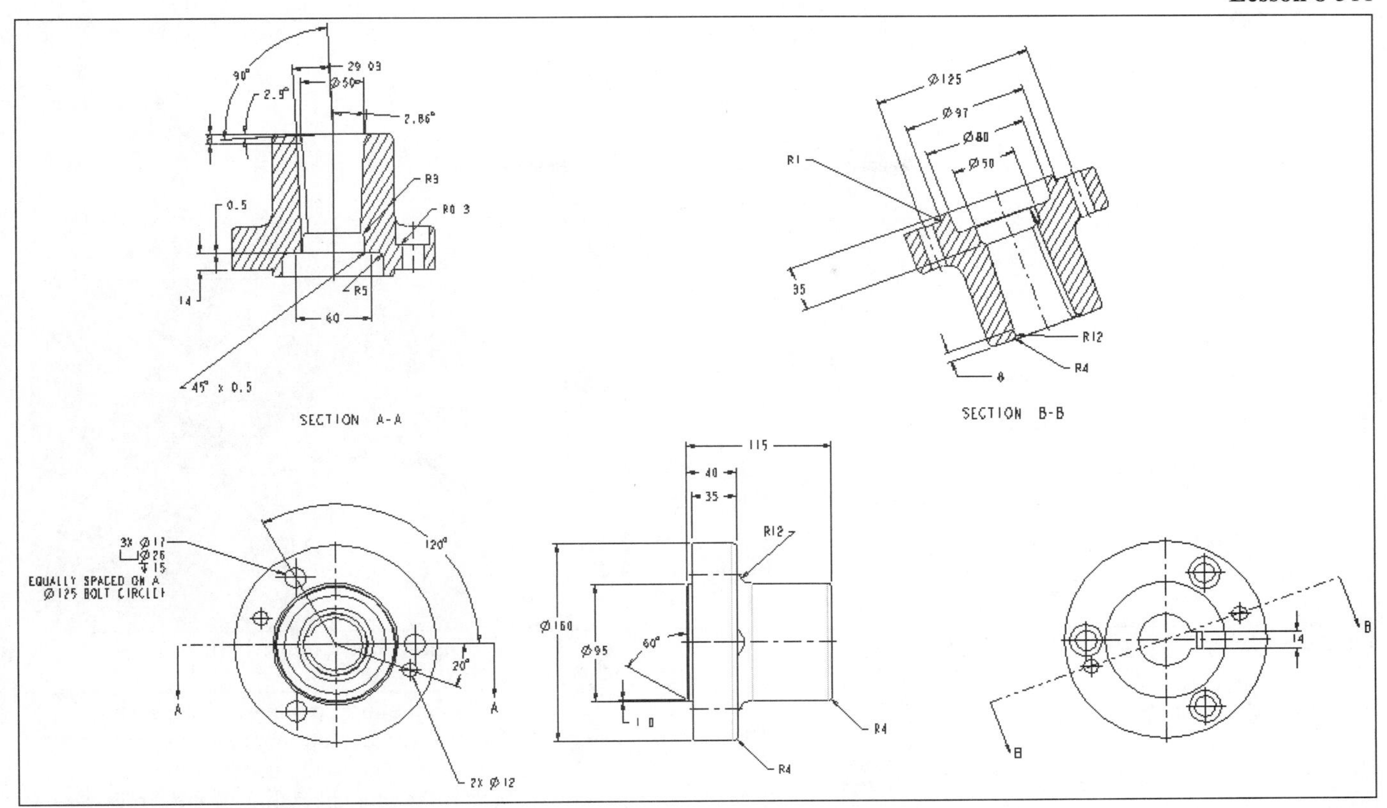

Figure 8.62 Taper Coupling Drawing

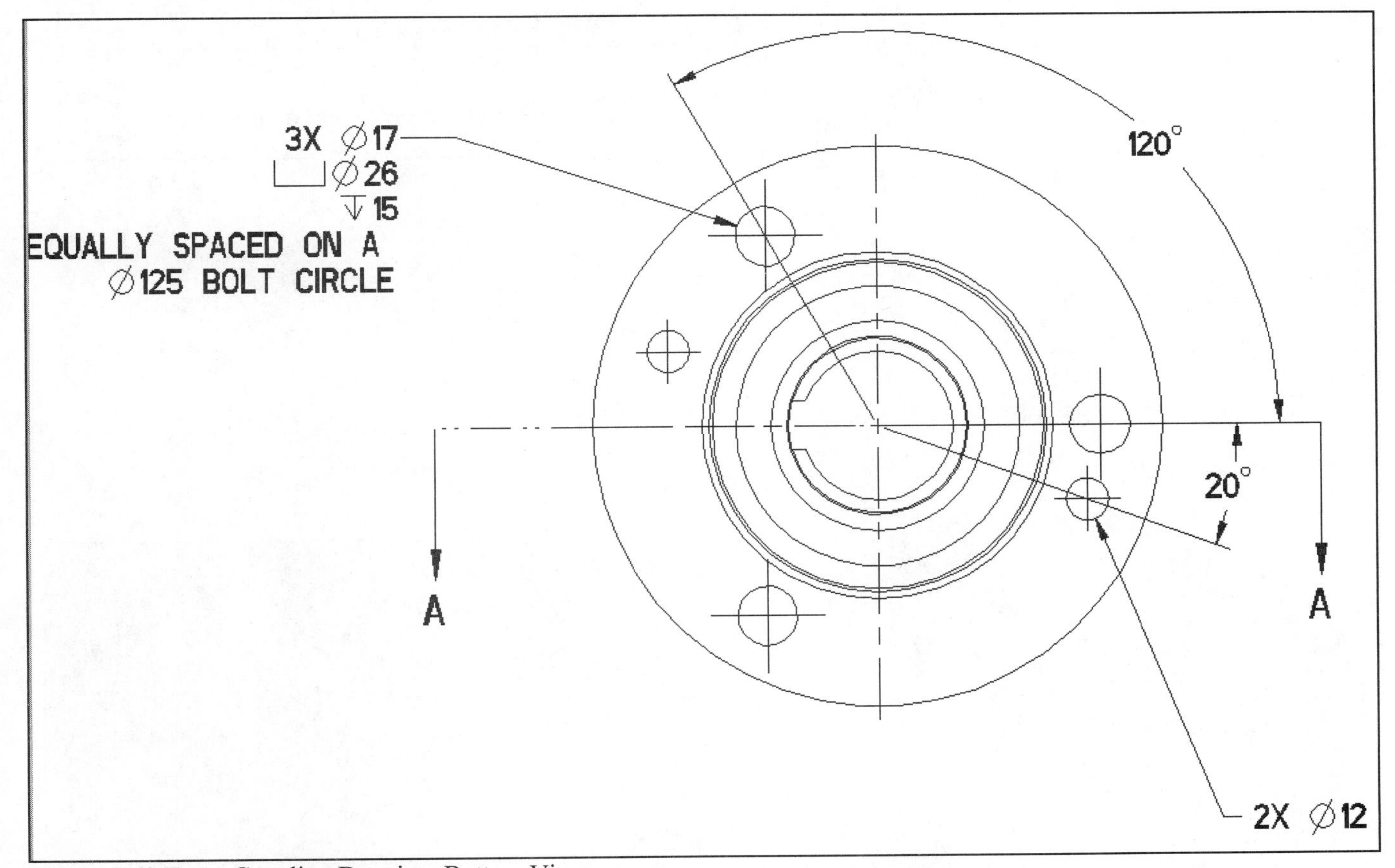

Figure 8.63 Taper Coupling Drawing, Bottom View

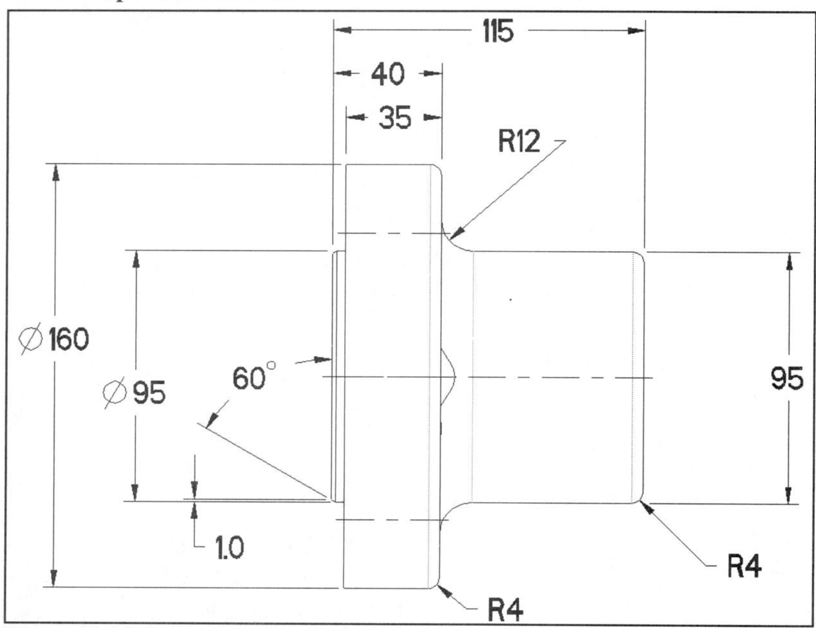

Figure 8.64 Taper Coupling Drawing, Side View

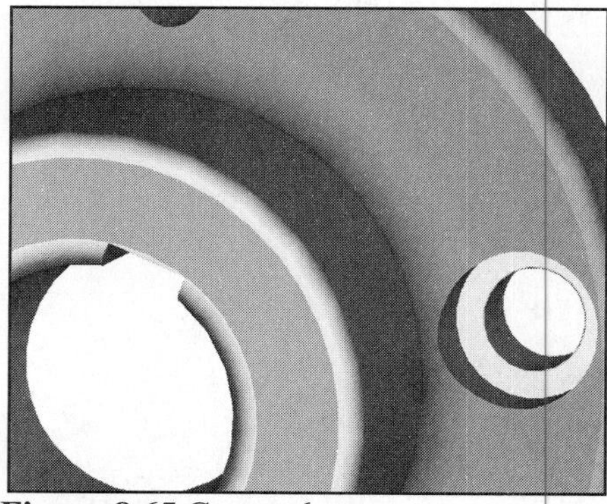

Figure 8.65 Counterbore

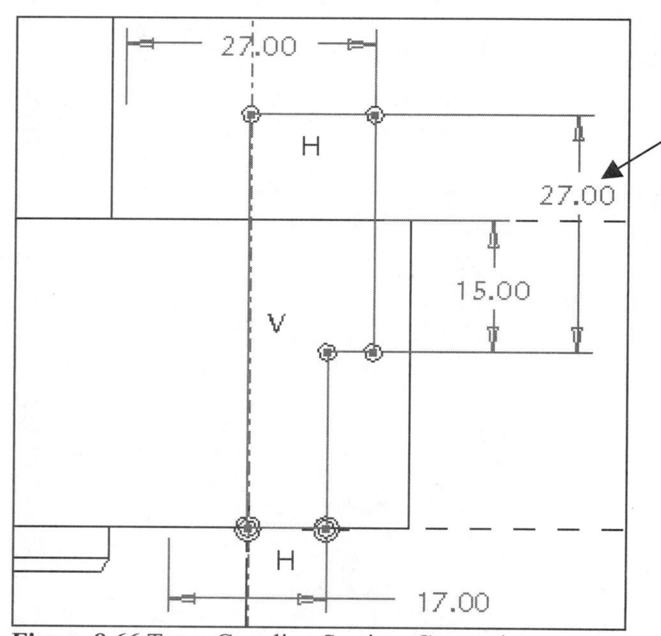

Figure 8.66 Taper Coupling Section, Counterbore

After completing Lesson 9, return to this project. Write a relation that will keep this dimension equal to the depth of the counterbore plus the radius of the large round (**R12**).

Dim=15+Radius

d18=d9+d6

Your dim values (**d#'s**) may differ.

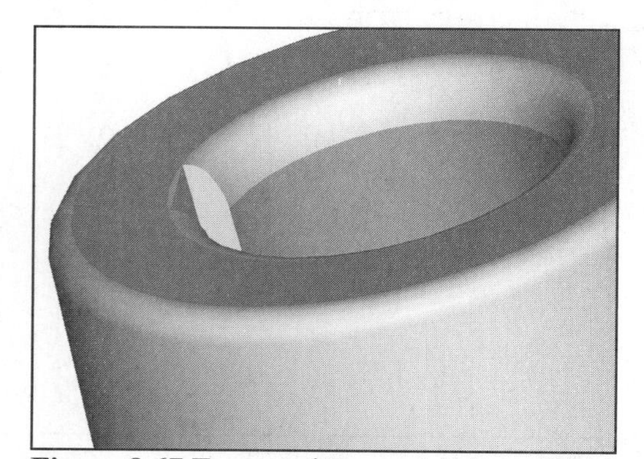

Figure 8.67 Taper and Keyseat (**14mm** Wide)

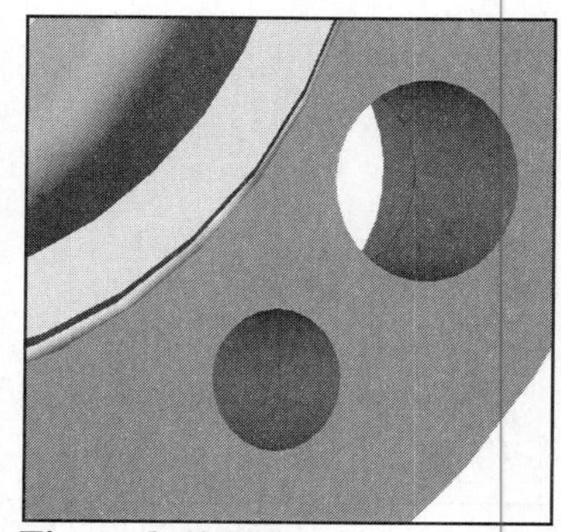

Figure 8.68 Holes

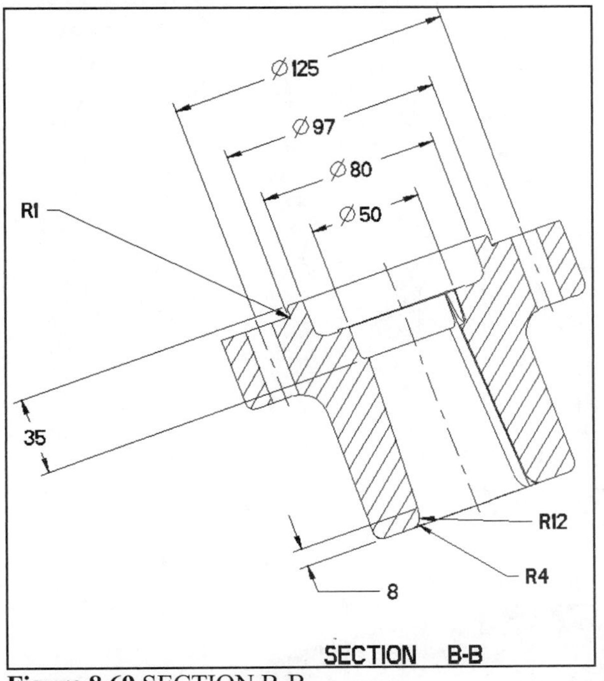

Figure 8.69 SECTION B-B

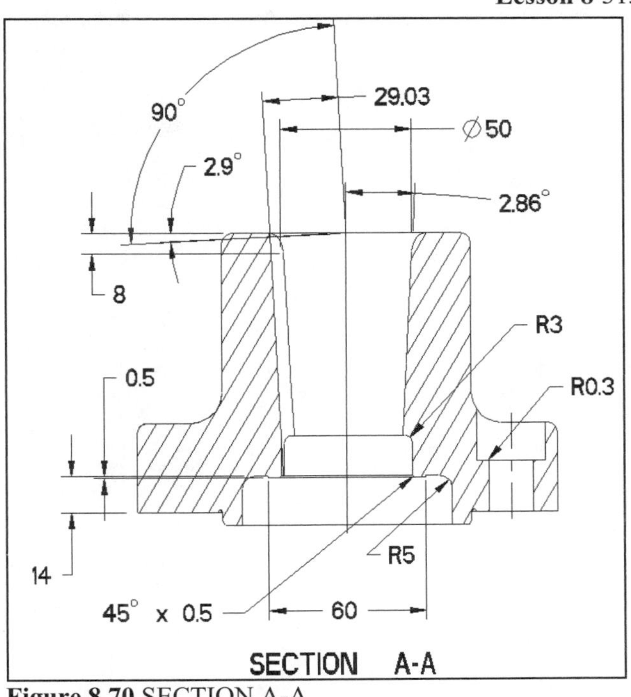

Figure 8.70 SECTION A-A

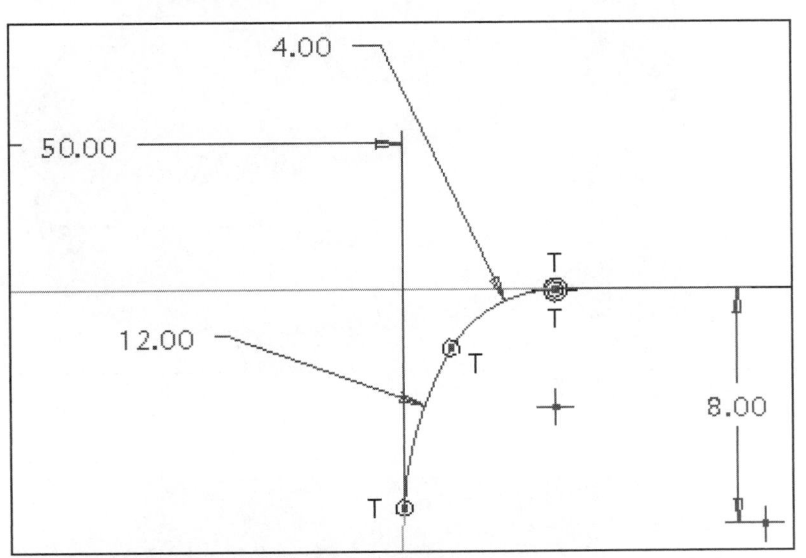

Figure 8.71 Section Sketch Rounds

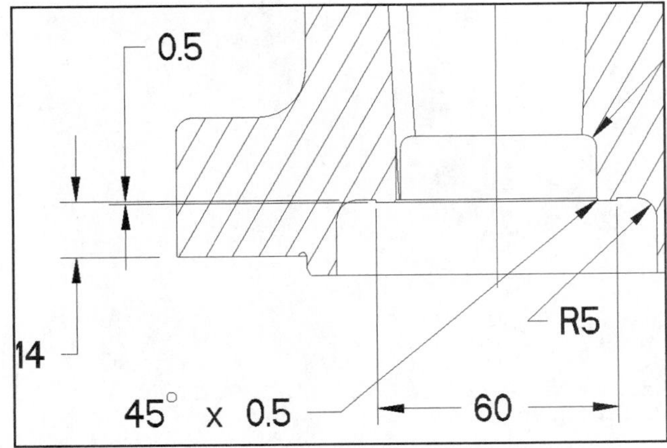

Figure 8.72 Taper Coupling Drawing, SECTION A-A

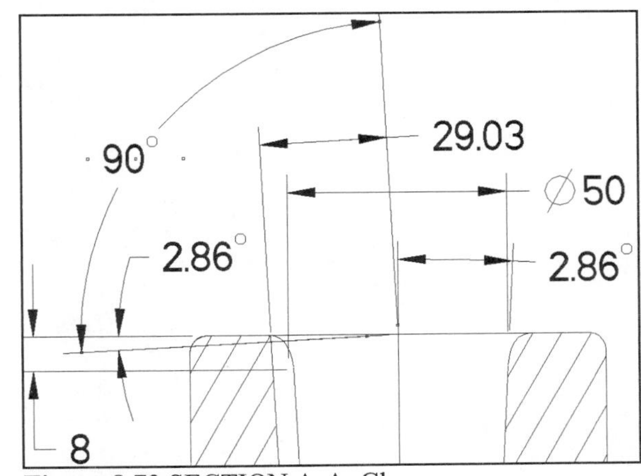

Figure 8.73 SECTION A-A, Close-up

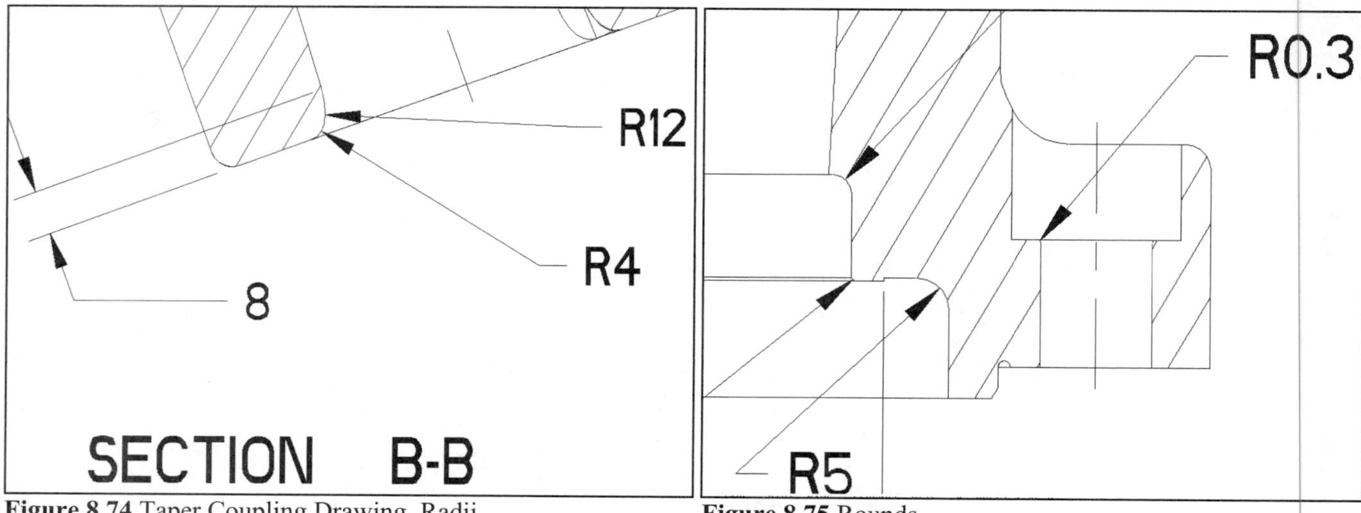

Figure 8.74 Taper Coupling Drawing, Radii

Figure 8.75 Rounds

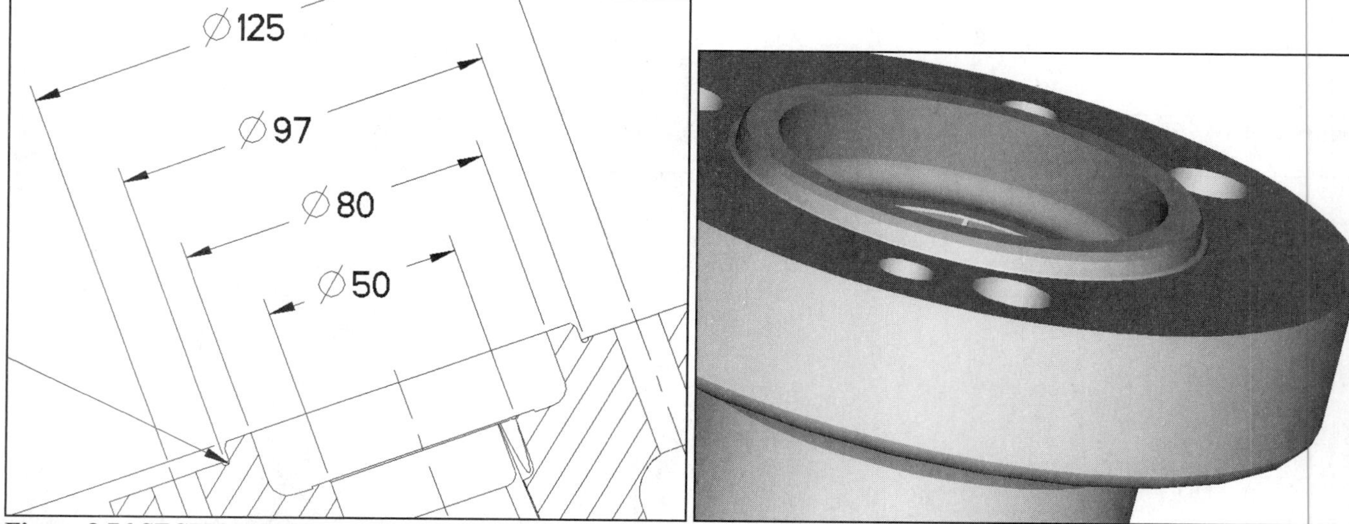

Figure 8.76 SECTION B-B, Mating Diameters

Figure 8.77 Mating Surface

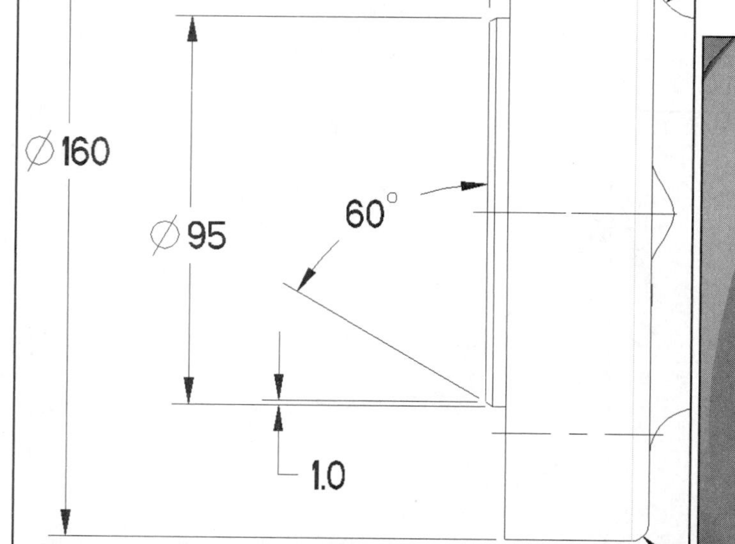

Figure 8.78 Side View, Close-up

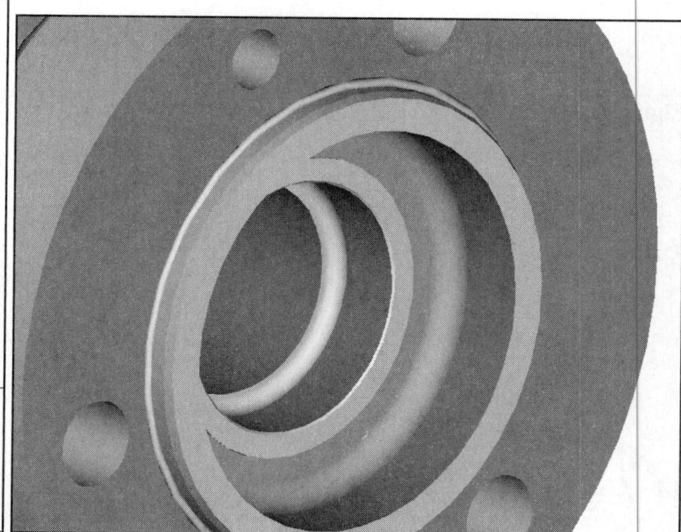

Figure 8.79 Internal View

Lesson 9 Ribs, Relations, Failures, and Family Tables

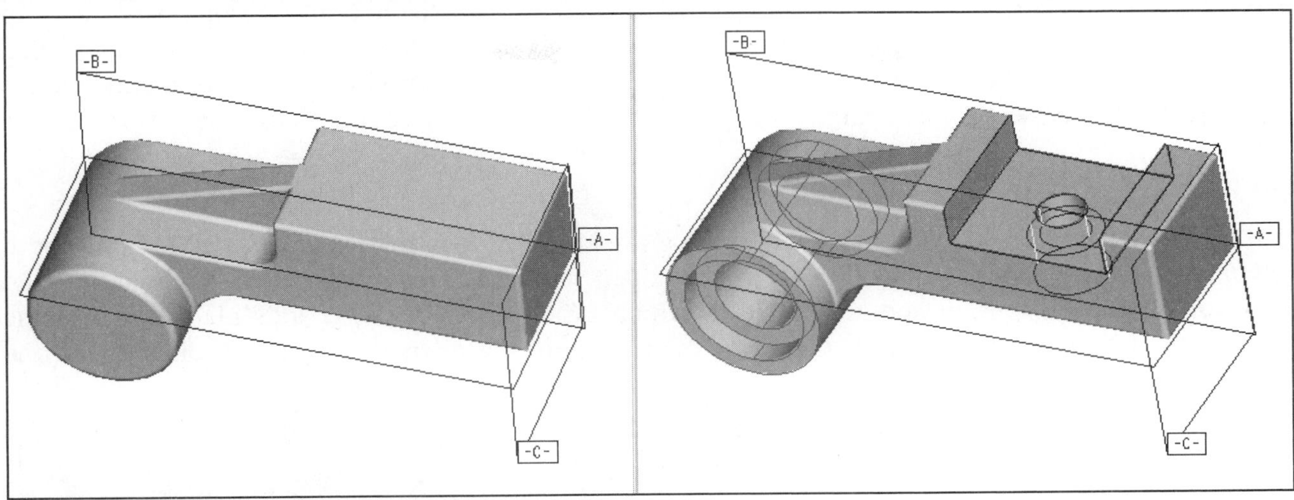

Figure 9.1 Adjustable Guide, Casting and Machine Part

OBJECTIVES

- Create **Ribs**
- Understand **Parameters**
- Write **Relations** to control features
- Understand and solve **Failures**
- Edit standard hole **Notes**
- Create a workpiece using a **Family Table**

RIBS, RELATIONS, FAILURES, AND FAMILY TABLES

A **Rib** is a feature (Figures 9.1 and 9.2) designed to create a thin fin or web that is attached to a part.

Relations are equations written between symbolic dimensions and parameters. By writing relations between dimensions in a part or an assembly, we can control the effects of modifications.

Failures happen when the model cannot be regenerated. You need to know how to avoid and how to resolve part and assembly failures.

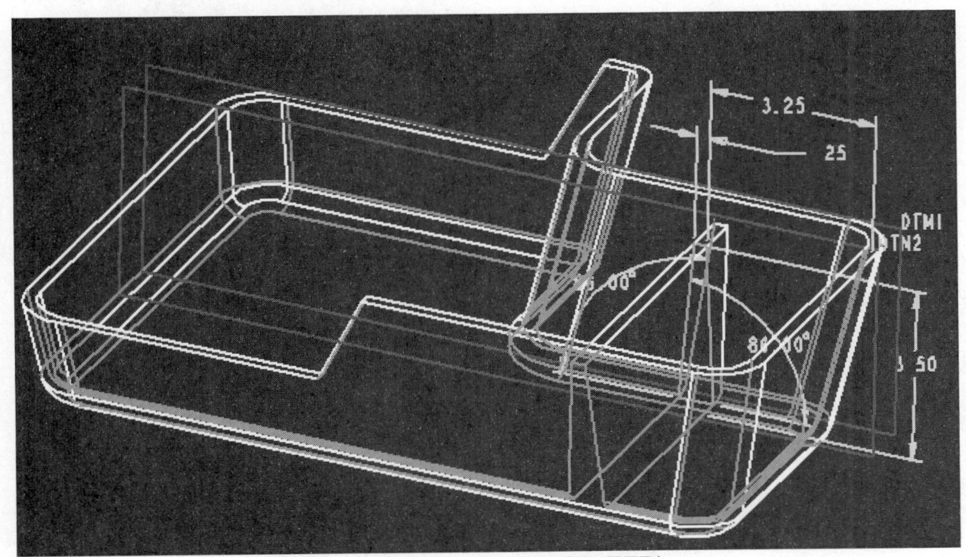

Figure 9.2 Ribs (CADTRAIN, COAch for Pro/ENGINEER)

Ribs

A rib is a special type of protrusion [Fig. 9.3(a-b)] designed to create a thin fin or web that is attached to a part. You always sketch a rib from a side view, and it grows about the sketching plane symmetrically or to either side. Because of the way ribs are attached to the parent geometry, they are always sketched as open sections. A rib must "see" material everywhere it attaches to the part; otherwise, it becomes an unattached feature. There are two types of ribs: straight and rotational. The following sections describe ribs in detail.

Straight Ribs

Ribs that are not created on through axis datum planes are extruded symmetrically or to either side about the sketching plane. You must still sketch the ribs as open sections. Because you are sketching an open section, Pro/E may be uncertain about the side to which to add the rib. Pro/E adds all material in the direction of the arrow. If the incorrect choice is made, toggle the arrow direction by clicking on the direction arrow on the screen.

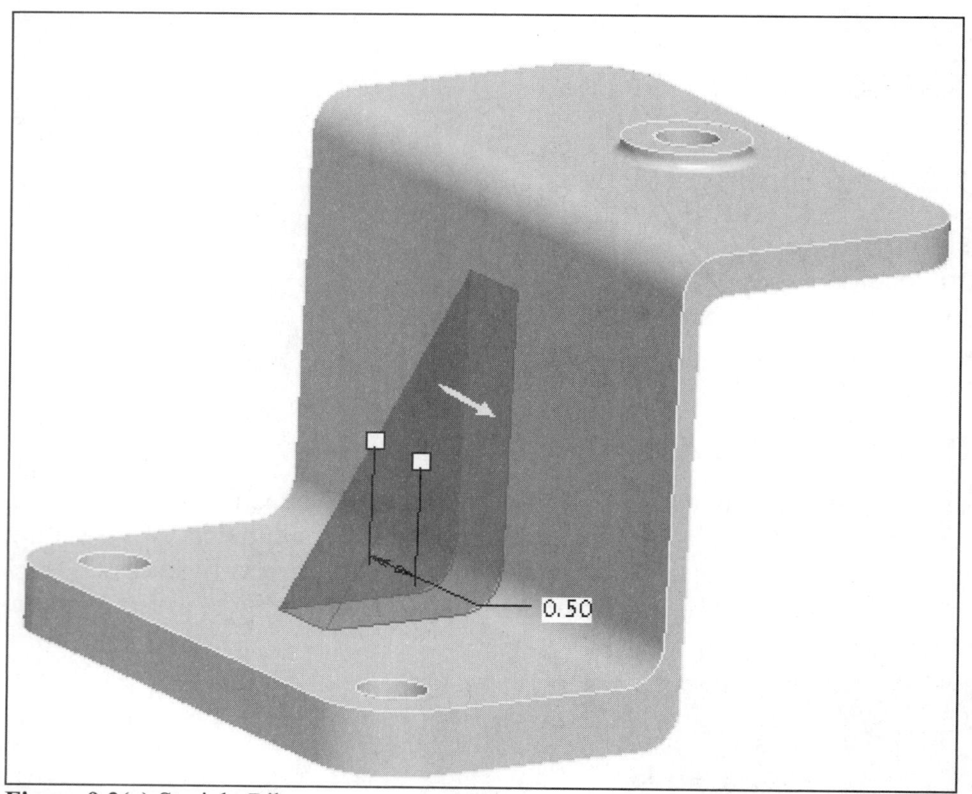

Figure 9.3(a) Straight Rib

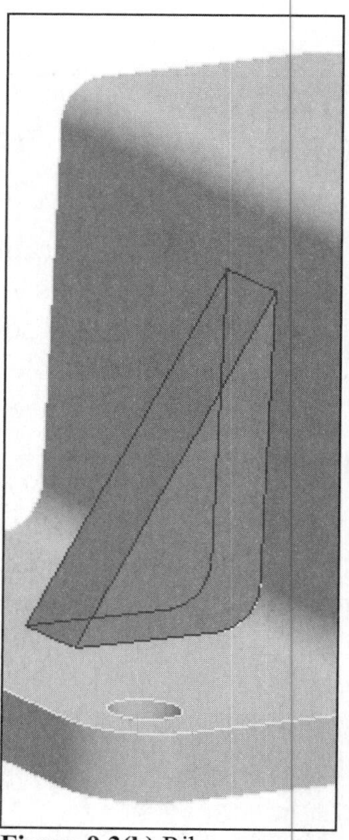

Figure 9.3(b) Rib

Rotational Ribs

You create rotational ribs on through axis datum planes. You sketch the rib to the silhouette of the parent feature. To create the solid geometry, Pro/E revolves the section about the axis of the parent, making a wedge that is symmetrical about the sketching plane. Pro/E then trims the wedge with two planes parallel to the sketching surface; the distance between these planes corresponds to the thickness of the rib. You can place a rotational rib on any surface of revolution. Note that the angled surface of the rib is conical, not planar.

Relations

Relations (also known as parametric relations) are user-defined equations written between symbolic dimensions and parameters. Relations capture design relationships within features or parts, or among assembly components, thereby allowing users to control the effects of modifications on models (Fig. 9.4).

Relations are a way of capturing design knowledge and intent. Like parameters, they are used to drive models: change the relation and you change the model.

Relations can be used to control the effects of modifications on models, to define values for dimensions in parts and assemblies, and to act as constraints for design conditions (for example, specifying the location of a hole in relation to the edge of a part). They are used in the design process to describe conditional relationships between different features of a part or an assembly. Relations can be simple values (for example, d1=4) or complex conditional branching statements.

Relations can be used to provide a value for a dimension. However, they can also be used to notify you when a condition has been violated, such as when a dimension exceeds a certain value. There are two basic types of relations, *equality* and *comparison*. An *equality relation* equates a parameter on the left side of the equation to an expression on the right side. This type of relation is used for assigning values to dimensions and parameters. The following are a few examples of equality relations:

d2 = 25.500   **d8 = d4/2**   **d7 = d1+d6/2**   **d6 = d2*(sqrt(d7/3.0+d4))**

A *comparison relation* compares an expression on the left side of the equation to an expression on the right side. This type of relation is commonly used as a constraint or as a conditional statement for logical branching. The following are examples of comparison relations:

d1 + d2 > (d3 + 5.5)	Used as a constraint
IF (d1 + 5.5) > = d7	Used in a conditional statement

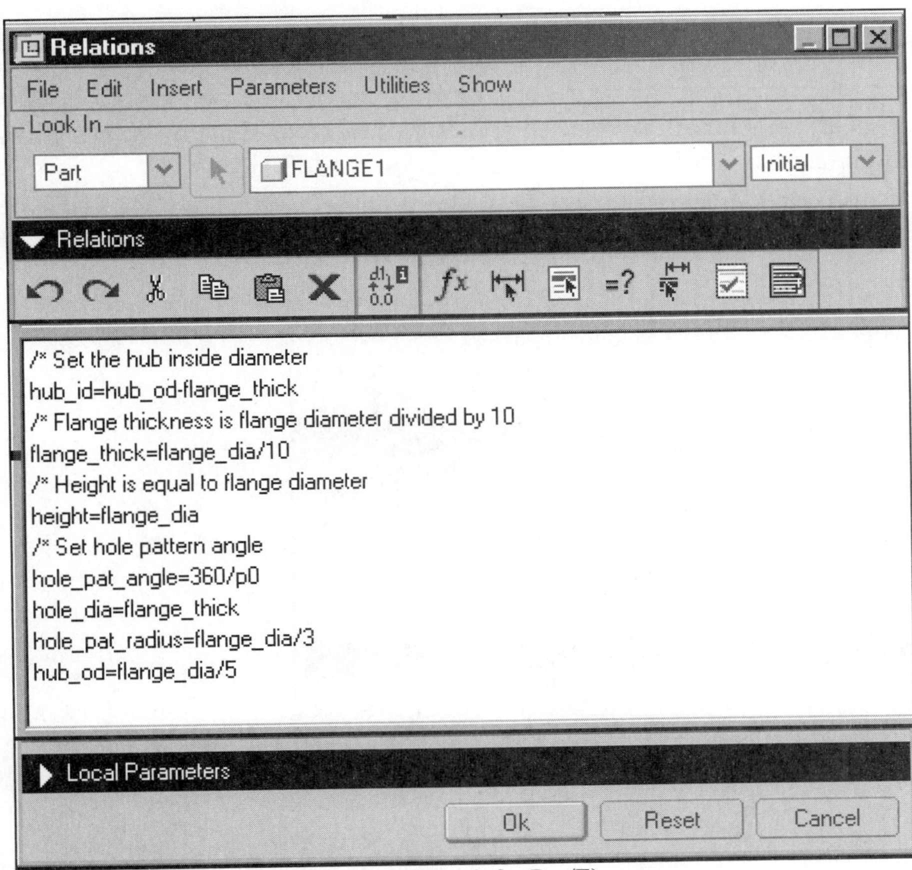

Figure 9.4 Relations (CADTRAIN, COAch for Pro/E)

Parameter Symbols

Four types of parameter symbols are used in relations:

- **Dimensions** These are dimension symbols, such as **d8**, **d12**.
- **Tolerances** These are parameters associated with ± symmetrical and plus-minus tolerance formats. These symbols appear when dimensions are switched from numeric to symbolic.
- **Number of Instances** These are integer parameters for the number of instances in a direction of a pattern.
- **User Parameter** These can be parameters defined by adding a parameter or a relation (e.g., **Volume = d3 * d4 * d5**).

Operators and Functions

The following operators and functions can be used in equations and conditional statements:

Arithmetic Operators

+	**Addition**
–	**Subtraction**
/	**Division**
*	**Multiplication**
^	**Exponentiation**
()	**Parentheses for grouping** [for example, **(d0 = (d1–d2)*d3)**]

Assignment Operator

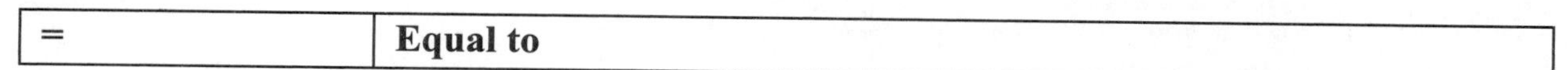

=	**Equal to**

The = (equals) sign (Fig. 9.5) is an assignment operator that equates the two sides of an equation or relation. When it is used, the equation can have only a single parameter on the left side.

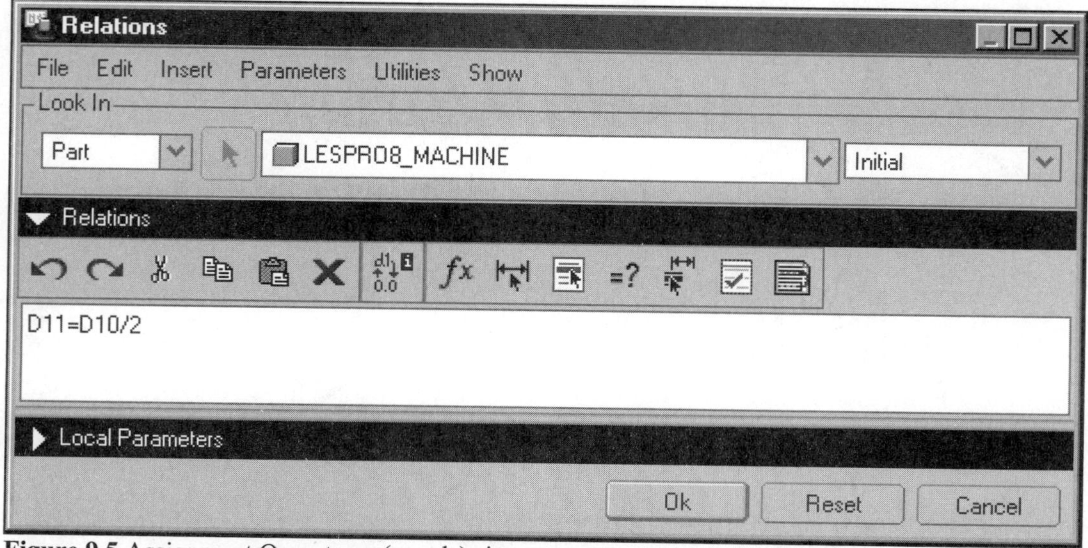

Figure 9.5 Assignment Operator = (equals) sign

Comparison Operators

Comparison operators are used whenever a TRUE/FALSE value can be returned. For example, the relation **d1 >= 3.5** returns TRUE whenever d1 is greater than or equal to **3.5**. It returns FALSE whenever **d1** is less than **3.5**. The following comparison operators are supported:

==	**Equal to**
>	**Greater than**
>=	**Greater than or equal to**
!=, <>,~=	**Not equal to**
<	**Less than**
<=	**Less than or equal to**
		Or
&	**And**
~, !	**Not**

Mathematical Functions

The following operators can be used in relations, both in equations and in conditional statements. Relations may include the following mathematical functions:

cos ()	**cosine**
tan ()	**tangent**
sin ()	**sine**
sqrt ()	**square root**
asin ()	**arc sine**
acos ()	**arc cosine**
atan ()	**arc tangent**
sinh ()	**hyperbolic sine**
cosh ()	**hyperbolic cosine**
tanh ()	**hyperbolic tangent**

Note: All trigonometric functions use degrees.

log()	**base 10 logarithm**
ln()	**natural logarithm**
exp()	**e to an exponential degree**
abs()	**absolute value**
ceil()	**the smallest integer not less than the real value**
floor()	**the largest integer not greater than the real value**

Failures

Sometimes model geometry cannot be constructed because features that have been modified or created conflict with or invalidate other features (Fig. 9.6). For example, this can happen when the following occurs:

- A protrusion is created that is unattached and has a one-sided edge
- New features are created that are unattached and have one-sided edges
- A feature is resumed that now conflicts with another feature (such as having two chamfers on the same edge)
- The intersection of features is no longer valid because dimensional changes have moved the intersecting surfaces
- A relation constraint has been violated

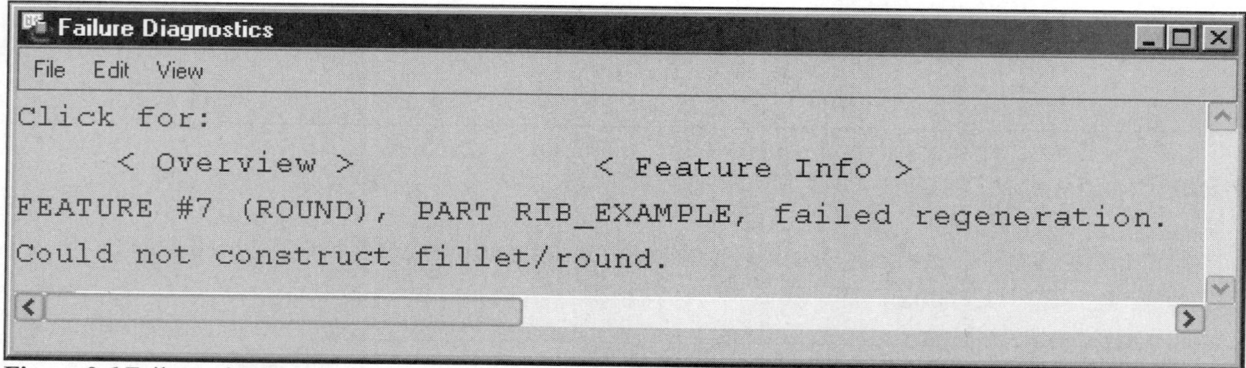

Figure 9.6 Failure of a Feature

Resolve Feature

After the feature fails, Pro/E enters Resolve Feature mode (Fig. 9.7). Use the commands in the RESOLVE FEAT menu to fix the failed feature:

- **Undo Changes** Undo the changes that caused the failed regeneration attempt, and return to the last successfully regenerated model
- **Investigate** Investigate the cause of the regeneration failure using the Investigate submenu
- **Fix Model** Roll the model back to the state before failure and select commands to fix the problem
- **Quick Fix** Choose an option from the **QUICK FIX** menu, the options are as follows:
    - **Redefine** Redefine the failed feature
    - **Reroute** Reroute the failed feature
    - **Suppress** Suppress the failed feature and its children
    - **Clip Supp** Suppress the failed feature and all the features after it
    - **Delete** Delete the failed feature

Figure 9.7 Resolving a Failure

Failed Features

If a feature fails during creation and it does not use the dialog box interface, Pro/E displays the FEAT FAILED menu with the following options:

- **Redefine** Redefine the feature.
- **Show Ref** Display the SHOW REF menu so you can see the references of the failed feature. Pro/E displays the reference number in the Message Window.
- **Geom Check** Check for problems with overlapping geometry, misalignment, and so on. This command may be dimmed. If a shell, offset surface, or thickened surface fails, Pro/E stores information about the surfaces that could not be offset. The GEOM CHECK menu displays a list of features with failed geometry and a Restore command.
- **Feat Info** Get information about the feature.

If a feature fails, you can redisplay the part with all failed geometry highlighted in different colors. Pro/E displays the corresponding error messages in an Information Window. Features can fail during creation for the following reasons:

- **Overlapping geometry** A surface intersects itself. If Pro/E finds a self-intersecting surface, it does not perform any further surface checks. Pro/E highlights the overlapping geometry in red and the corresponding points of intersection in white, and displays an error message.
- **Surface has edges that coincide** The surface has no area. Pro/E highlights the surface in red and displays an error message.
- **Inverted geometry** Pro/E highlights the inverted geometry in purple and displays an error message.
- **Bad edges** Pro/E highlights bad edges in blue and displays an error message.
- **Sheetmetal form** Pro/E highlights sheetmetal form features that fail in red.

Fixing the Model

If you choose the **Fix Model** option, Pro/E displays the FIX MODEL menu. Choose either Current Modl or Backup Modl and an appropriate option from the FIX MODEL menu.

The FIX MODEL menu has the following options:

- **Current Modl** Perform operations on the current active (failed) model.
- **Backup Modl** Perform operations on the backup model, displayed in a separate window from the current model in the active window.
- **Feature** Perform feature operations on the model using the standard FEAT menu. Pro/E displays the CONFIRMATION menu so you can confirm or cancel the request only if the Undo Changes option is not possible. However, the Undo Changes option is always possible if you used the Regen Backup option in the ENVIRONMENT menu. If you confirm the request, choose Redefine and the **SELECT FEAT** menu opens. The Failed Feat command selects the latest failed feature. Pro/E displays a message in the Message Window when you successfully redefine the feature and automatically regenerates the model. When you choose Done from the FEAT menu, Pro/E displays the Information Window with a message stating that the model has been successfully regenerated. Pro/E also displays instructions so you can either exit the Resolve environment or continue to make changes before you exit.

- **Modify** Modify dimensions using the standard MODIFY menu
- **Regenerate** Regenerate the model
- **Switch Dim** Switch the dimension display from symbols to values or vice versa
- **Restore** Display the RESTORE menu so you can restore dimensions, parameters, relations, or all of these to their values prior to the failure, the RESTORE menu options are as follows:
    - **All Changes** Restore all the changed items
    - **Dimensions** Restore the dimensions
    - **Parameters** Restore the parameters
    - **Relations** Restore the relations

- **Relations** Add, delete, or modify relations, as necessary, to be able to regenerate the model, using the Relations dialog box (for more information, see Introduction to Pro/E)
- **Set Up** Display the standard PART SETUP menu to perform additional part setup procedures
- **X-Section** Create, modify, or delete a cross-sectional view using the Cross Section dialog box (for more information, see Introduction to Pro/E)
- **Program** Access Pro/PROGRAM capabilities

The Investigate Function

The **INVESTIGATE** menu (Fig. 9.8). lists the following options:

- **Current Modl** Perform operations on the current active (failed) model (Fig. 9.9).
- **Backup Modl** Perform operations on the backup model, displayed in a separate window (Pro/E displays the current model in the active window).
- **Diagnostics** Toggle on or off the display of the failed feature diagnostic window.
- **List Changes** Show the modified dimensions in the Main Window and in a pre-regenerated model window (Review Window), if available. Also, display a table that lists all the modifications and changes.

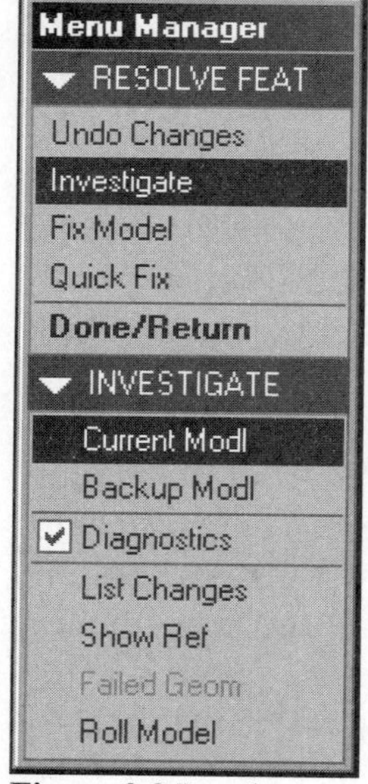

Figure 9.8 Resolve

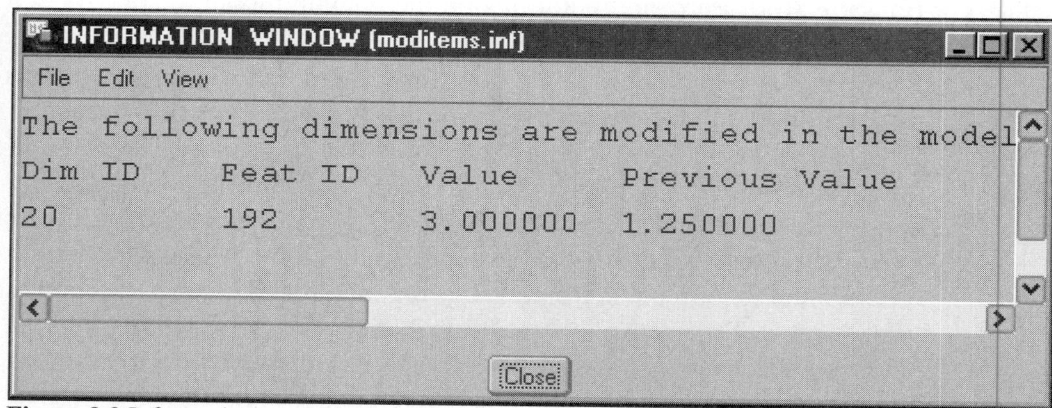

Figure 9.9 Information

- **Show Ref** Display the Reference Information dialog box [Fig. 9.10(a)] to show all the references [Fig. 9.10(b)] for the failed feature in the models, in both the Review Window and the Main Window. Pro/E highlights the selected reference in the reference color.
- **Failed Geom** Displays invalid geometry of the failed feature. This command may be unavailable. The **FAILED GEOM** menu displays a list of features with failed geometry and a restore command.
- **Roll Model** Roll the model back to the option selected in the submenu. The options are as follows:
    - **Failed Feat** Roll the model back to the failed feature (for the backup model only)
    - **Before Fail** Roll the model back to the feature just before the failed feature
    - **Last Success** Roll the model back to the state it was in at the end of the last successful feature regeneration
    - **Specify** Roll the model back to the specified feature

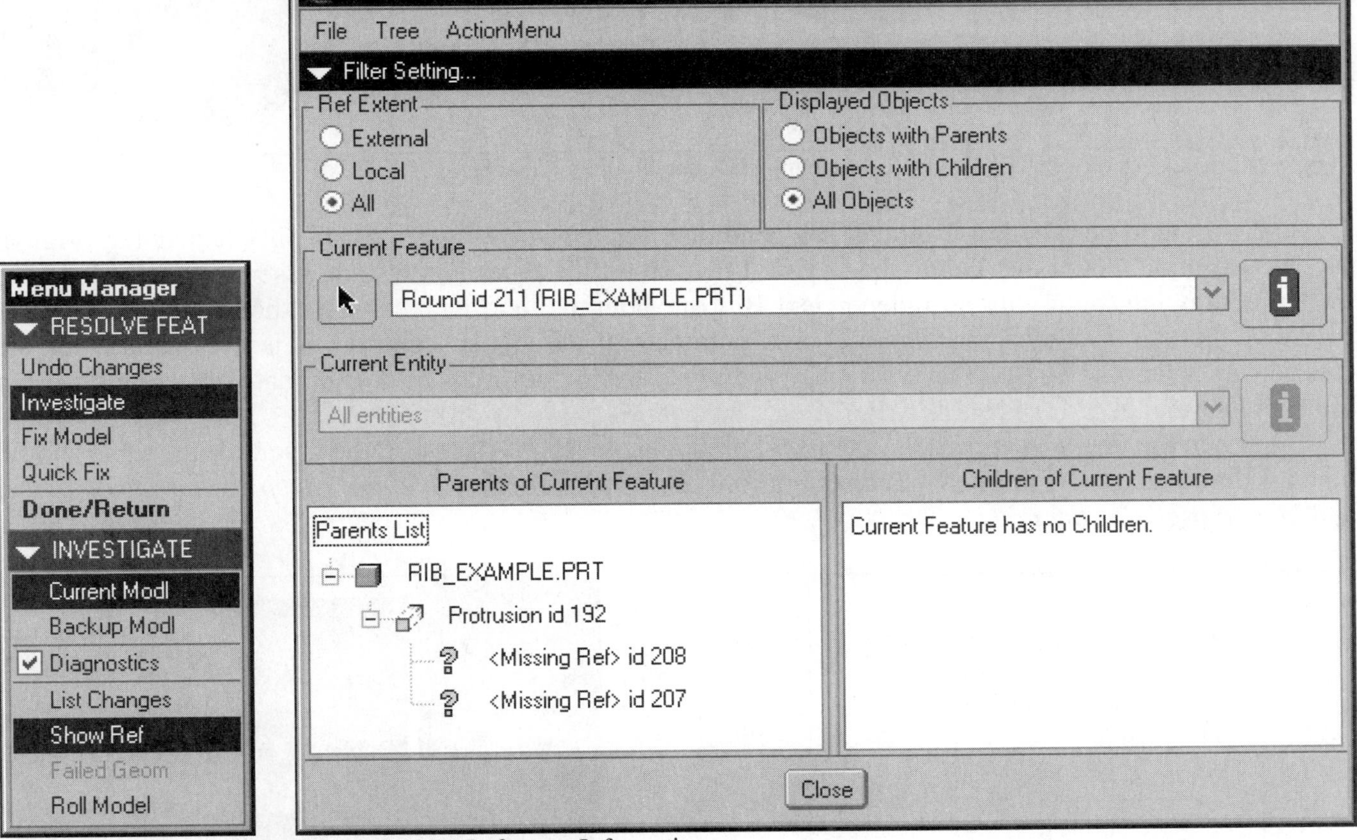

Figure 9.10(a) Investigate **Figure 9.10(b)** Reference Information

Family Tables

Family Tables are effective for two main reasons: they provide a beneficial tool, and they are easy to use. You need to understand the functionality of Family Tables, and you must understand when a Family Table is required and what circumstances should promote its use.

Family Tables are used any time a part or assembly [Fig. 9.11(a)] has several unique iterations developed from the original model. The iterations must be considered separate parts or assemblies, not just iterations of the same model.

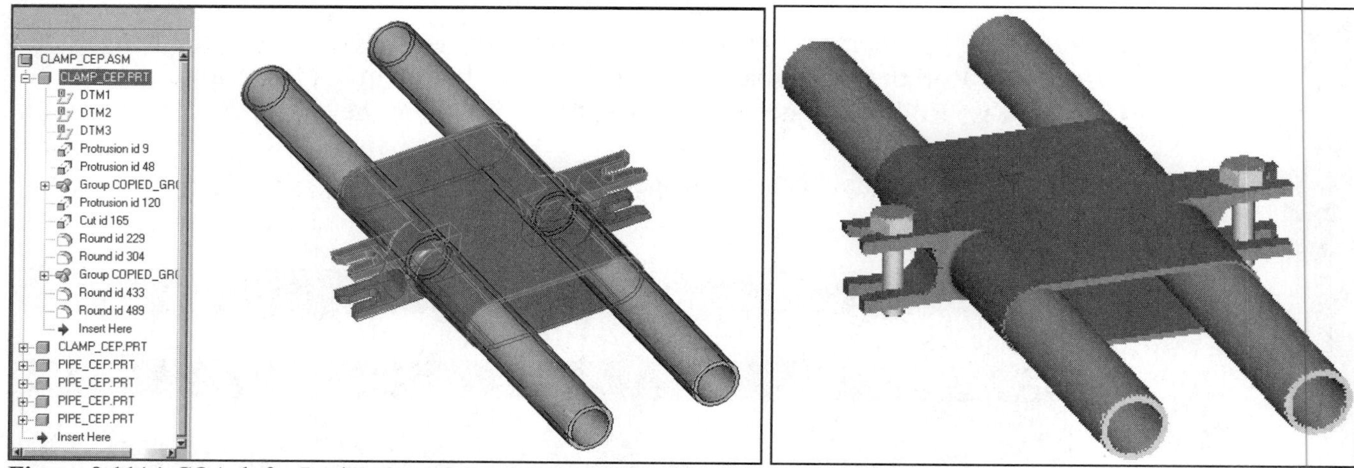

Figure 9.11(a) COAch for Pro/ENGINEER, Modeling II (Assembly Family Tables) (CADTRAIN)

To determine whether a model is a candidate for a Family Table: establish whether the original and the variation would ever have to co-exist at the same time (both in the same assembly, both shown in the same drawing, both with an independent Bill of Materials) and whether they should be tied together (most of the same dimensions, features, and parameters). If so, the component is a candidate for the creation of a Family Table [Fig. 9.11(b)], otherwise, the model may be a candidate for copying to an independent model or for Pro/PROGRAM.

When the decision is made to create a Family Table [Fig. 9.11(c-e)], the layout of the table should be considered next. The three main considerations are *what is in the table, how deep the table should be,* and *how to organize the table.*

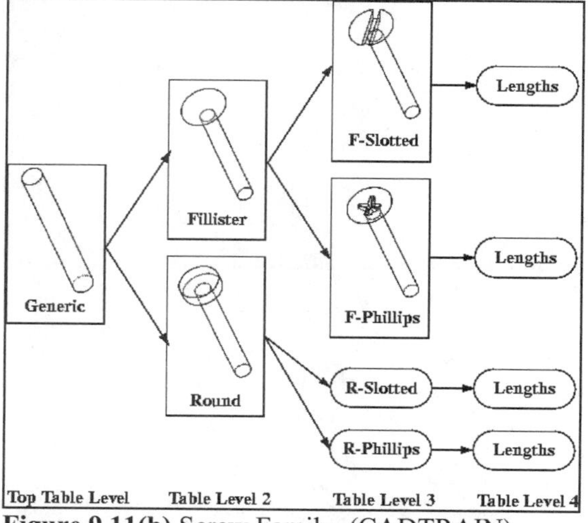

Figure 9.11(b) Screw Family (CADTRAIN)

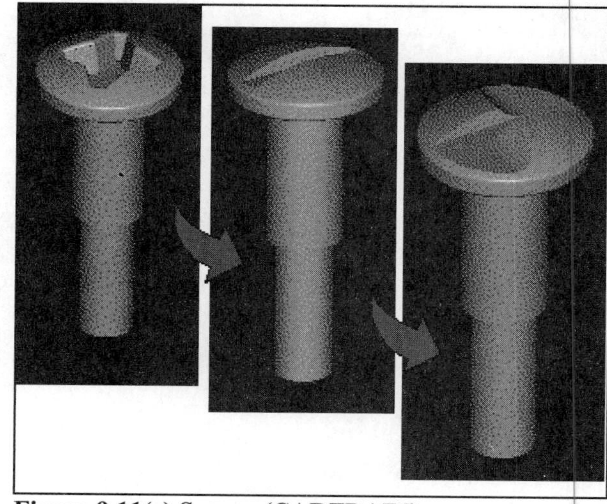

Figure 9.11(c) Screws (CADTRAIN)

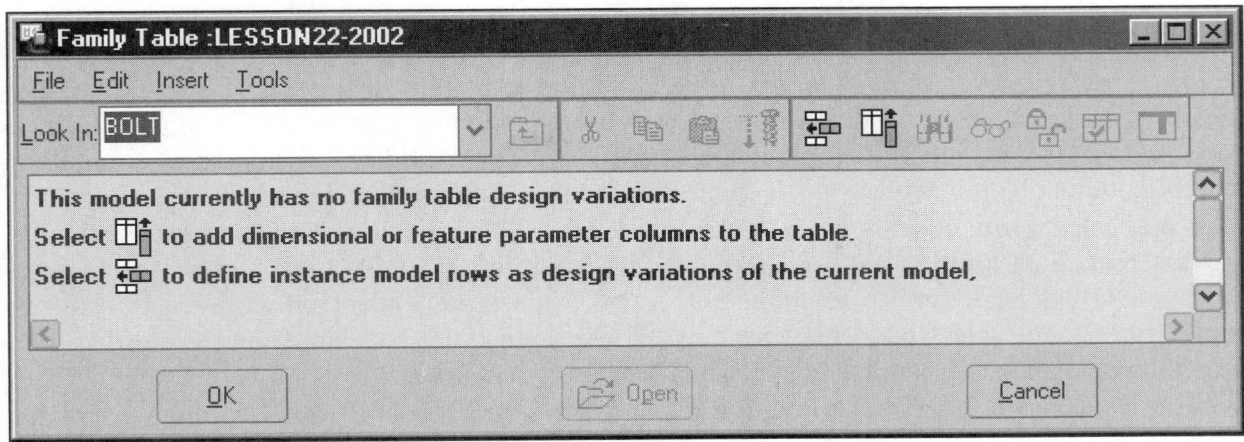

Figure 9.11(d) Family Table Dialog Box

Cut selected cells

Copy selected cells

Paste previously copied cells

Copy the selected instance with increments (patternize)

Insert a new instance at the selected row

Add/delete the table columns

Find instance in the currently edited table

Preview the selected instance

Lock/Unlock the selected instance

Verify instances of the family

Edit the current table using Excel

Open **Retrieve the selected instance in a separate window**

Figure 9.11(e) Family Table Icons

Table Considerations

- The *first consideration*, *what is in the table*, is defined by how the instances vary from the original. A table can include dimensions, features, components, parameters, groups, reference models, pattern tables, and system parameters. Each of these should be considered as a variation possibility, although most tables vary in their dimensions, parameters, and features.
- The *second consideration* is *how deep the table should be.* The model can have a single-level table (e.g., different-sized Phillips flat-head screws) or a multiple-level table (e.g., screws to flat-head screws to Phillips-head screws to individual sizes). This is dependent on a trade-off between complexity (a single-level table may be unwieldy if it becomes too large) and ease of use (a multiple-level table can make instance retrieval more cumbersome).
- The *third consideration* is *how to organize the table.* Pro/ENGINEER will prompt users for information to identify an instance by the order in which it appears in the table [Fig. 9.11(f-g)]. Therefore, if length is the most important differentiating aspect of the instance, it should be the first item in the table.

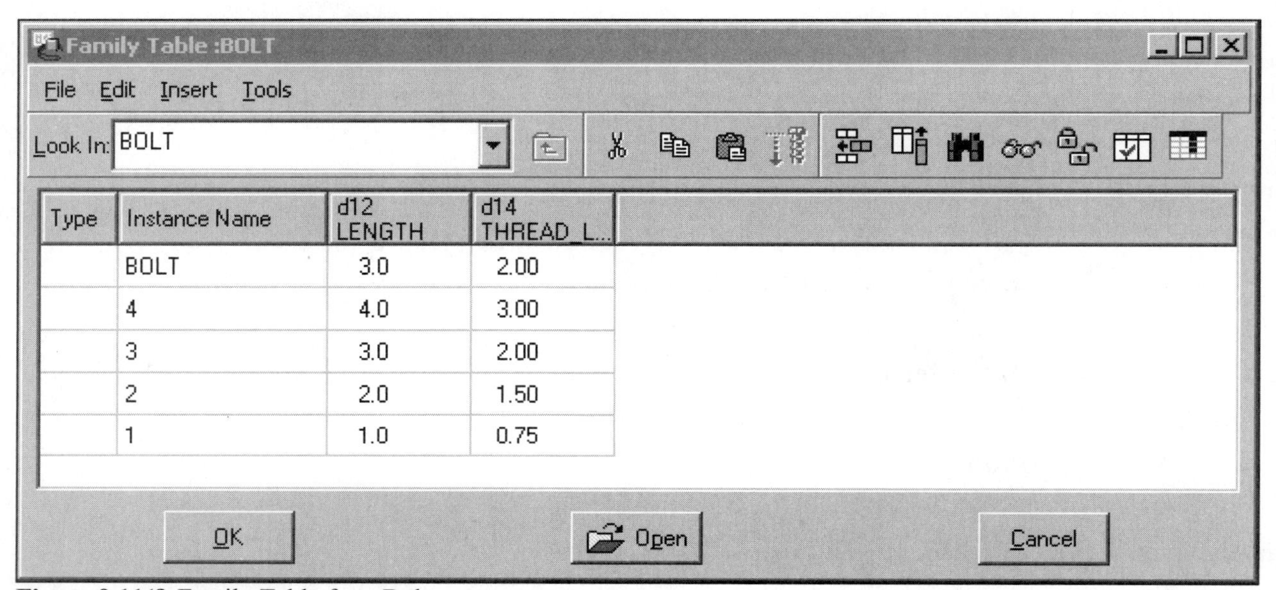

Type	Instance Name	d12 LENGTH	d14 THREAD_L...
	BOLT	3.0	2.00
	4	4.0	3.00
	3	3.0	2.00
	2	2.0	1.50
	1	1.0	0.75

Figure 9.11(f) Family Table for a Bolt

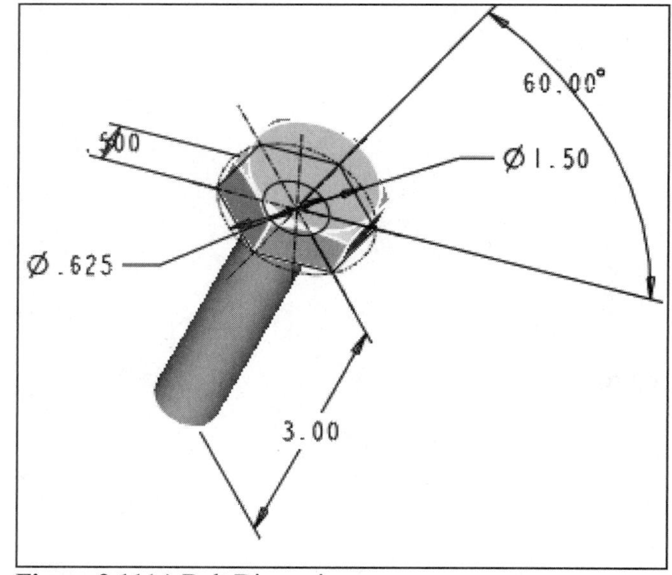

Figure 9.11(g) Bolt Dimensions

Lesson 9 STEPS

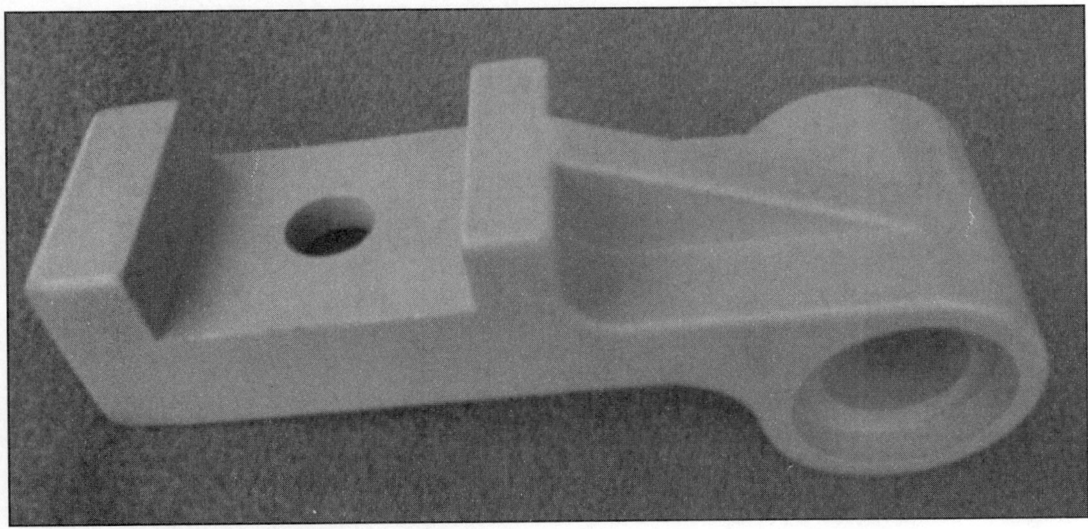

Figure 9.12 Adjustable Guide

Adjustable Guide

The Adjustable Guide (Fig. 9.12) is modeled in two stages. Model the casting using the casting detail, and then use the machine detail to complete the last (machined) features. The last step will be to create a Family Table with an instance that suppresses the machined features. By having a *casting part* (which is called a ***workpiece*** in **Pro/NC**) and a separate but almost identical *machined part* (which is called a ***design part*** in **Pro/NC**), you can create an operation for machining and an NC sequence. During the manufacturing process, *you merge the workpiece into the design part* and create a ***manufacturing model*** [Fig. 9.13(a) and (b)].

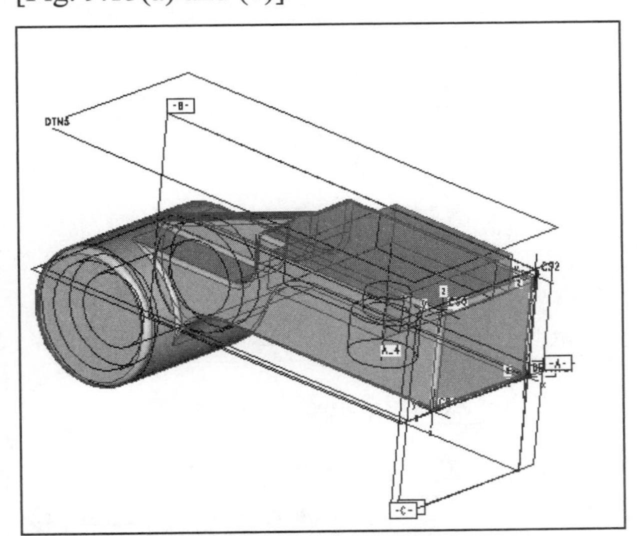

Figure 9.13(a) Manufacturing Model

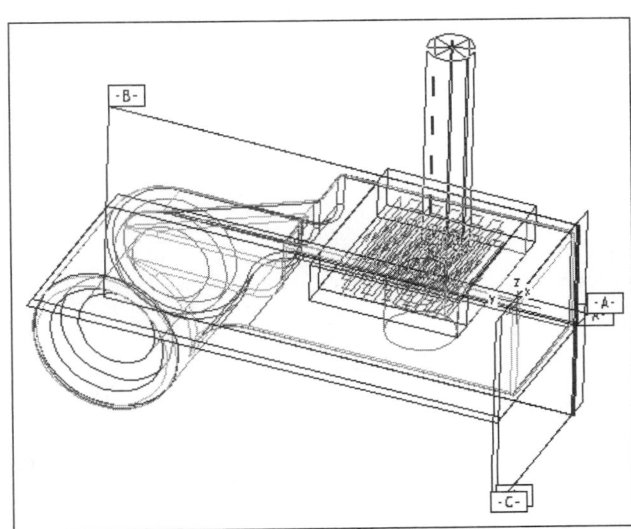

Figure 9.13(b) NC Sequence and a CL File

The difference between the two files is the difference between the volume of the casting part and the volume of the machined part. The removed volume can be seen as *material removal* when you are performing an **NC Check** operation on the manufacturing model. If the machining process gouges the part, the gouge will be displayed as *cyan*. The cutter location (CL) can also be displayed as an animated machining process [Fig. 9.13(b)].

The Adjustable Guide casting (Fig. 9.14) is a simple part, so the process of describing systematic commands and tools will start with the creation of the rib. The rib, created in the casting model, will have a relation added to it to control its location. The relation will keep the rib centered on the rectangular side of the part. The rounds are added late in the modeling process, for they can cause the model to fail in many cases. The process of fixing the part, so that the rounds do not make the regeneration fail, is also described.

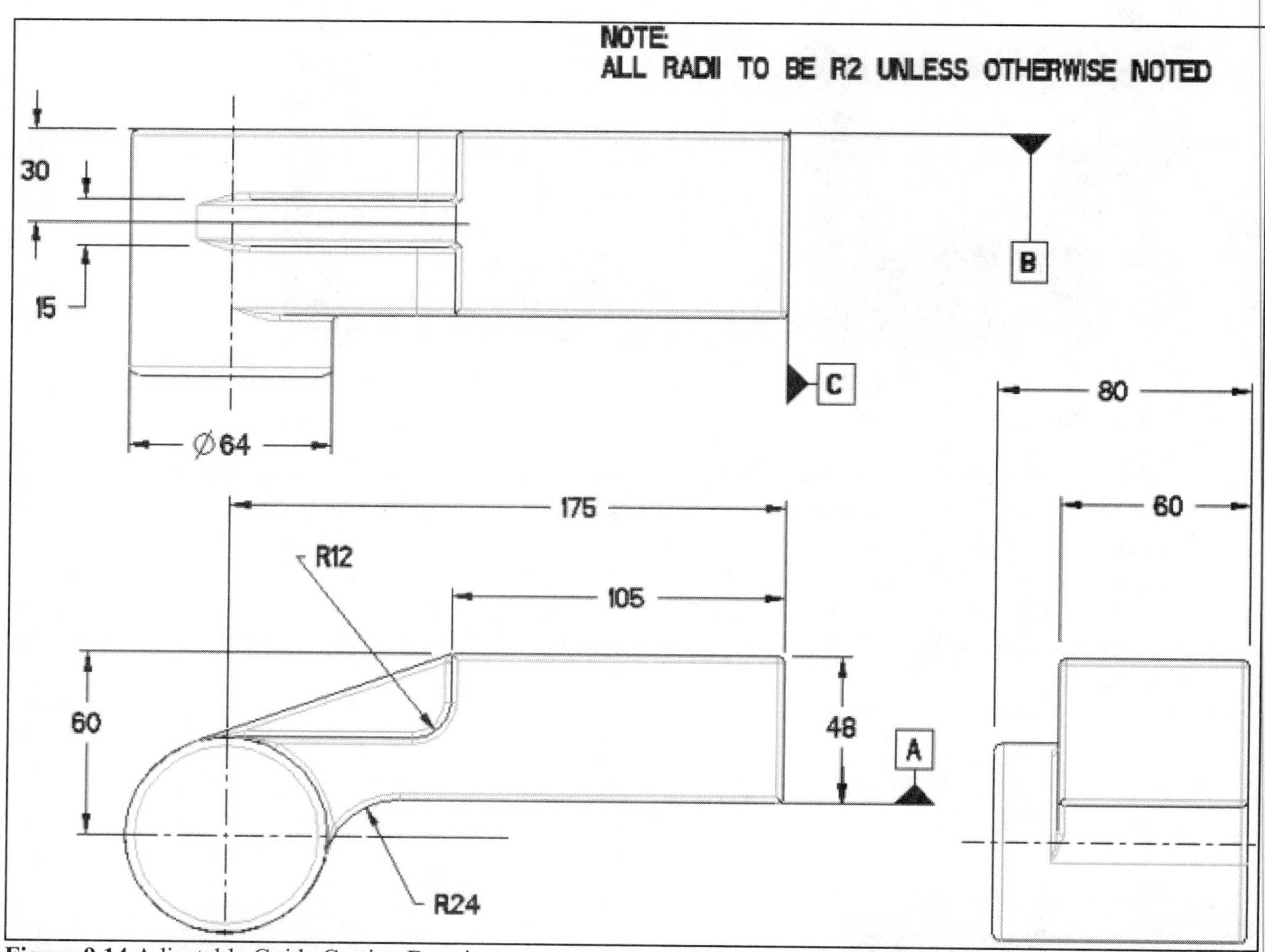

Figure 9.14 Adjustable Guide Casting Drawing

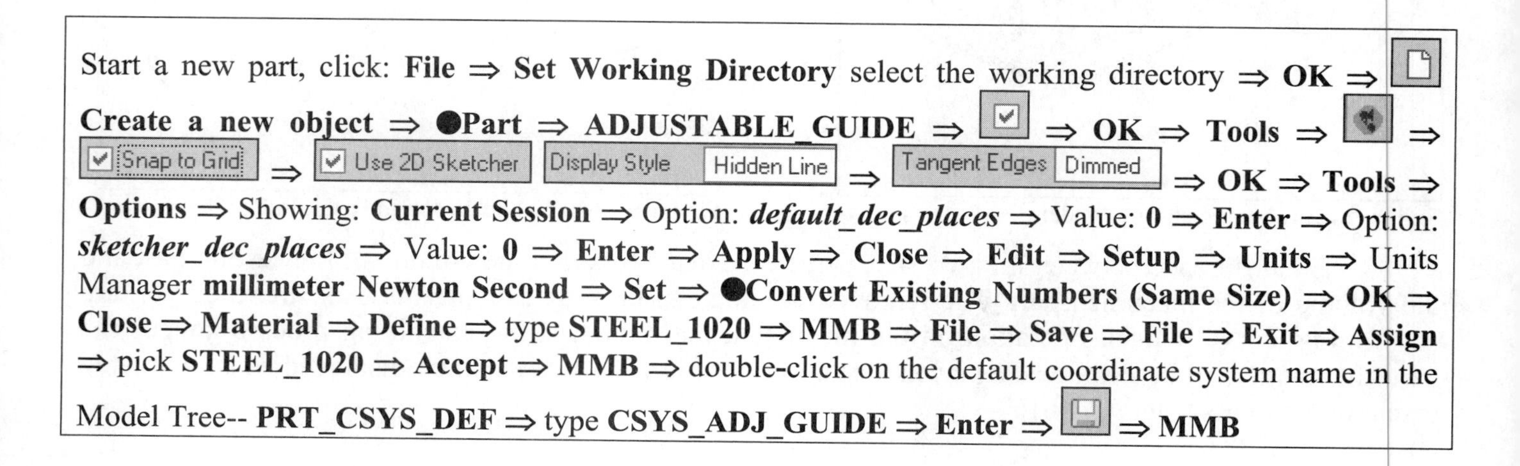
Start a new part, click: **File** ⇒ **Set Working Directory** select the working directory ⇒ **OK** ⇒ **Create a new object** ⇒ ●**Part** ⇒ **ADJUSTABLE_GUIDE** ⇒ ⇒ **OK** ⇒ **Tools** ⇒ ⇒ Snap to Grid ⇒ Use 2D Sketcher Display Style Hidden Line ⇒ Tangent Edges Dimmed ⇒ **OK** ⇒ **Tools** ⇒ **Options** ⇒ Showing: **Current Session** ⇒ Option: ***default_dec_places*** ⇒ Value: **0** ⇒ **Enter** ⇒ Option: ***sketcher_dec_places*** ⇒ Value: **0** ⇒ **Enter** ⇒ **Apply** ⇒ **Close** ⇒ **Edit** ⇒ **Setup** ⇒ **Units** ⇒ Units Manager **millimeter Newton Second** ⇒ **Set** ⇒ ●**Convert Existing Numbers (Same Size)** ⇒ **OK** ⇒ **Close** ⇒ **Material** ⇒ **Define** ⇒ type **STEEL_1020** ⇒ **MMB** ⇒ **File** ⇒ **Save** ⇒ **File** ⇒ **Exit** ⇒ **Assign** ⇒ pick **STEEL_1020** ⇒ **Accept** ⇒ **MMB** ⇒ double-click on the default coordinate system name in the Model Tree-- **PRT_CSYS_DEF** ⇒ type **CSYS_ADJ_GUIDE** ⇒ **Enter** ⇒ ⇒ **MMB**

Click: **Edit** ⇒ **Setup** ⇒ **Geom Tol** ⇒ **Set Datum** ⇒ pick **TOP** from the model ⇒ Name- type **A** ⇒ **MMB** ⇒ **Set Datum** ⇒ pick **FRONT** ⇒ Name- type **B** ⇒ **MMB** ⇒ **Set Datum** ⇒ pick **RIGHT** ⇒ Name- type **C** ⇒ **MMB** ⇒ **MMB** ⇒ **MMB** ⇒ **MMB** ⇒ ⇒ **MMB** (Fig. 9.15) ⇒ load your previously saved customization, click **Tools** ⇒ **Customize Screen** ⇒ **File** from the Customize dialog box ⇒ **Open Settings** ⇒ click on your saved file (*config.win*) ⇒ **Open** ⇒ **OK** ( )

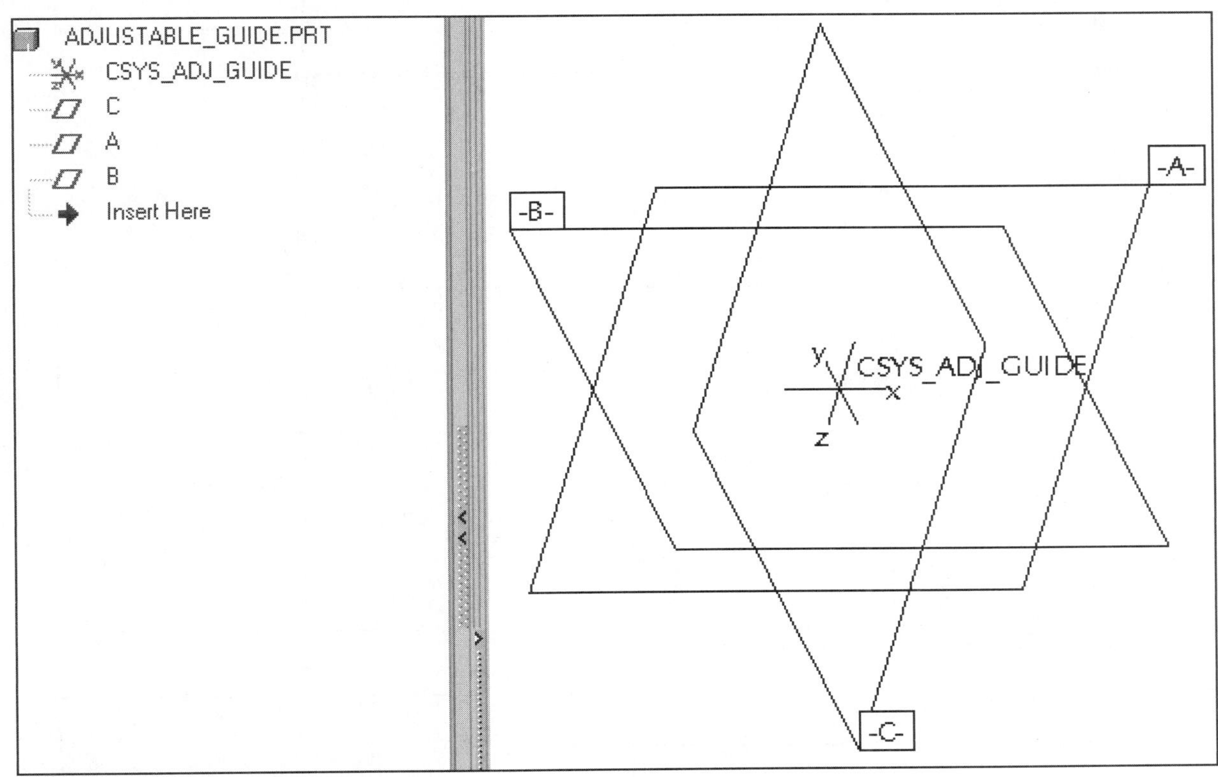

Figure 9.15 Renamed and Set Datums

Model the Adjustable_Guide casting using Figure 9.14 and Figures 9.16(a) though 9.16(g).

Figure 9.16(a) Section Dialog Box

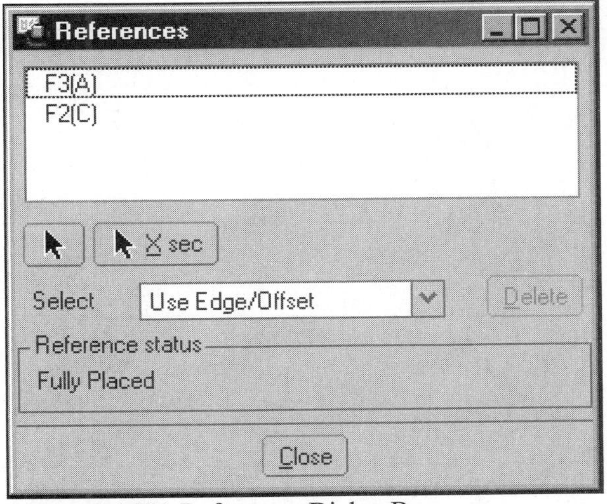

Figure 9.16(b) References Dialog Box

Sketch the section in the *second quadrant*, not in the first quadrant as was done for most of the other lessons. Note the location of the coordinate system. Pay particular attention to the tangencies.

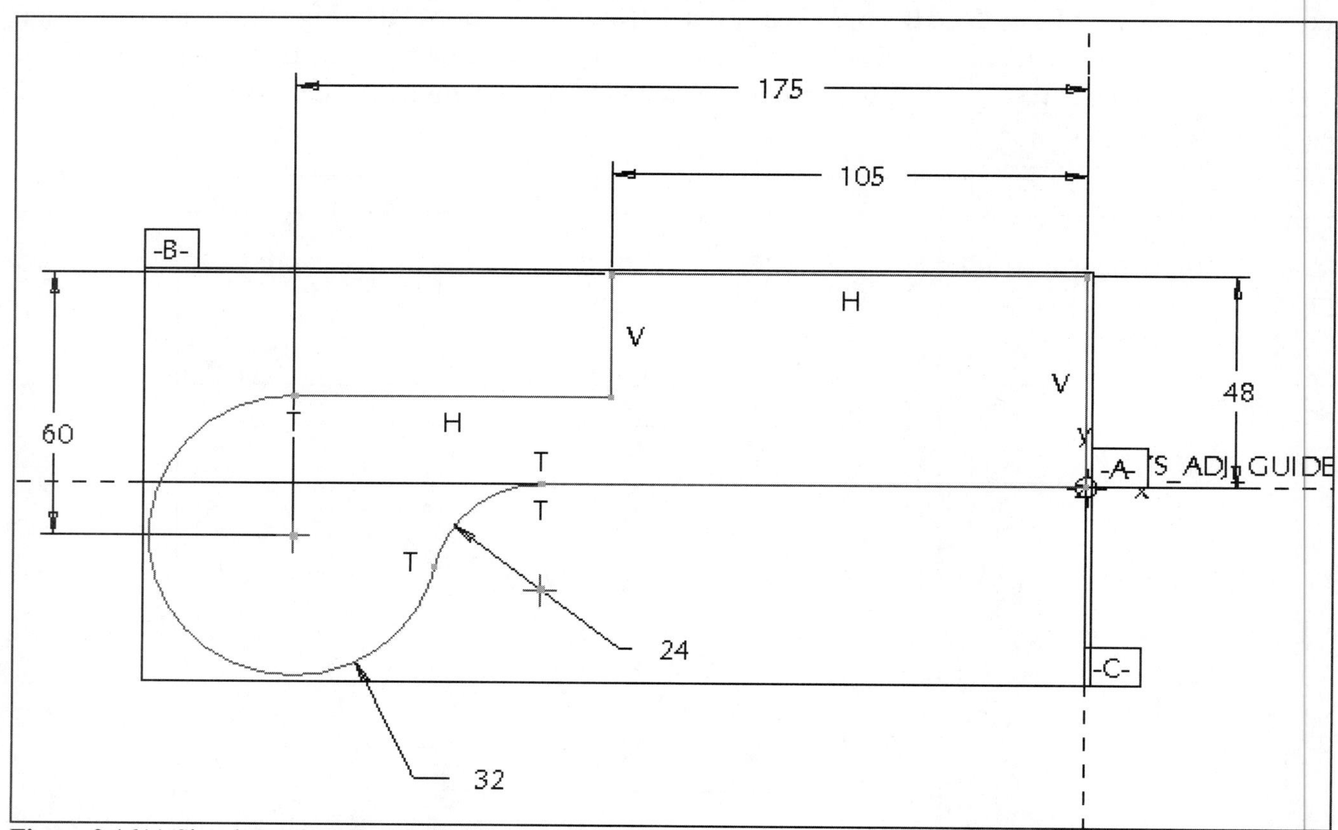

Figure 9.16(c) Sketch Section (Parallel to the Screen)

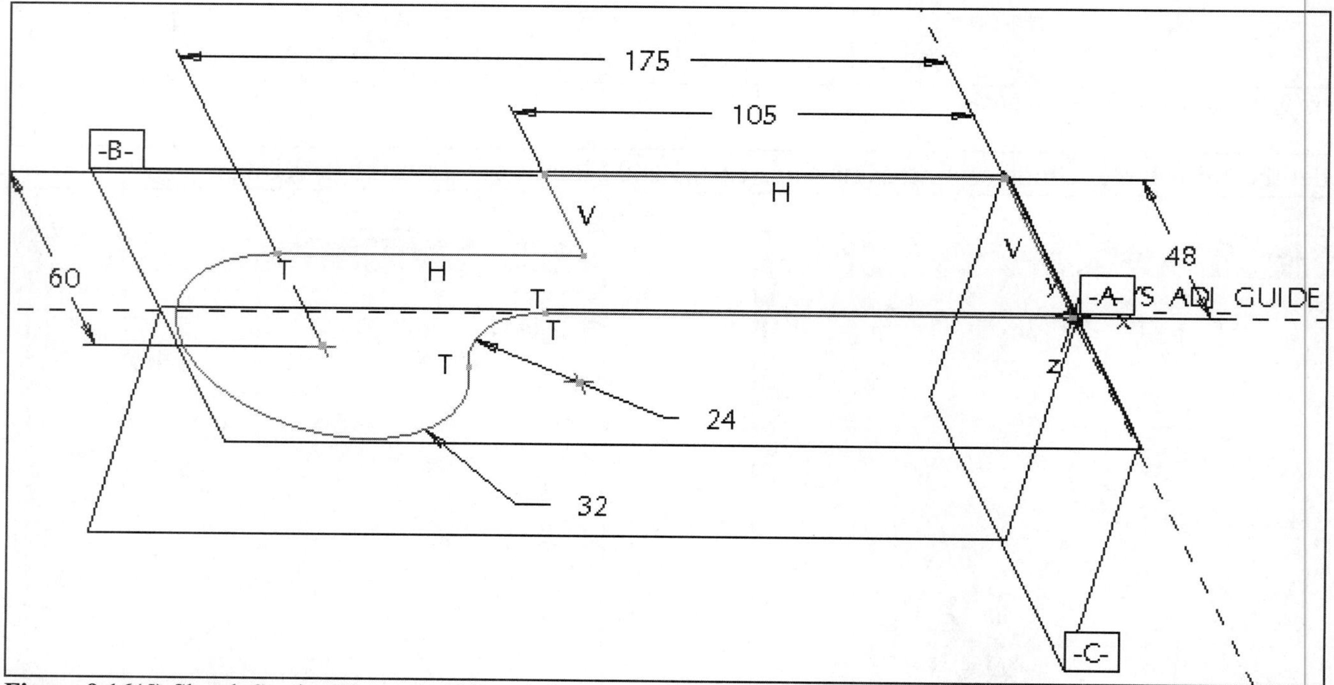

Figure 9.16(d) Sketch Section Shown in Trimetric

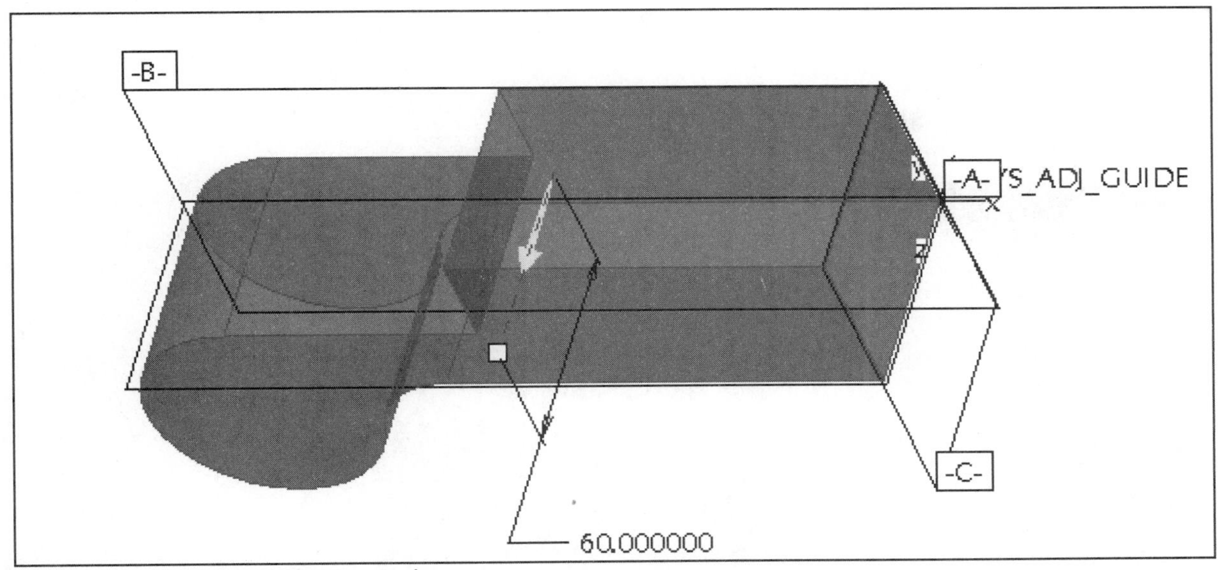

Figure 9.16(e) Depth of First Protrusion

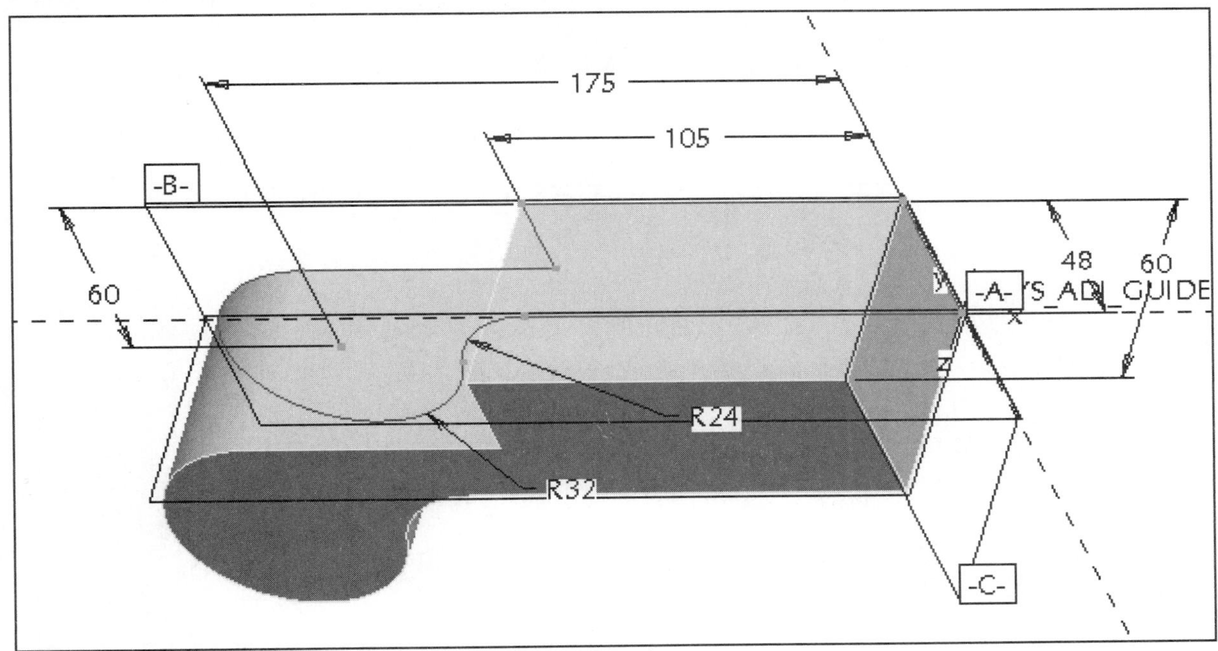

Figure 9.16(f) Dimensions Shown in the Standard Orientation

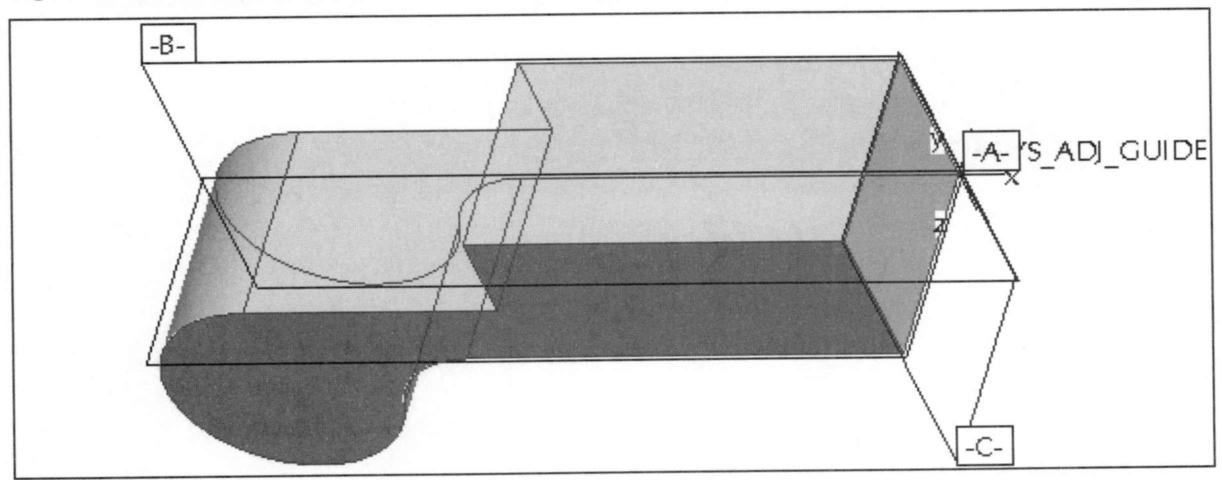

Figure 9.16(g) Completed Protrusion

Model the second protrusion using Figure 9.14 and Figures 9.17(a) though 9.17(d).

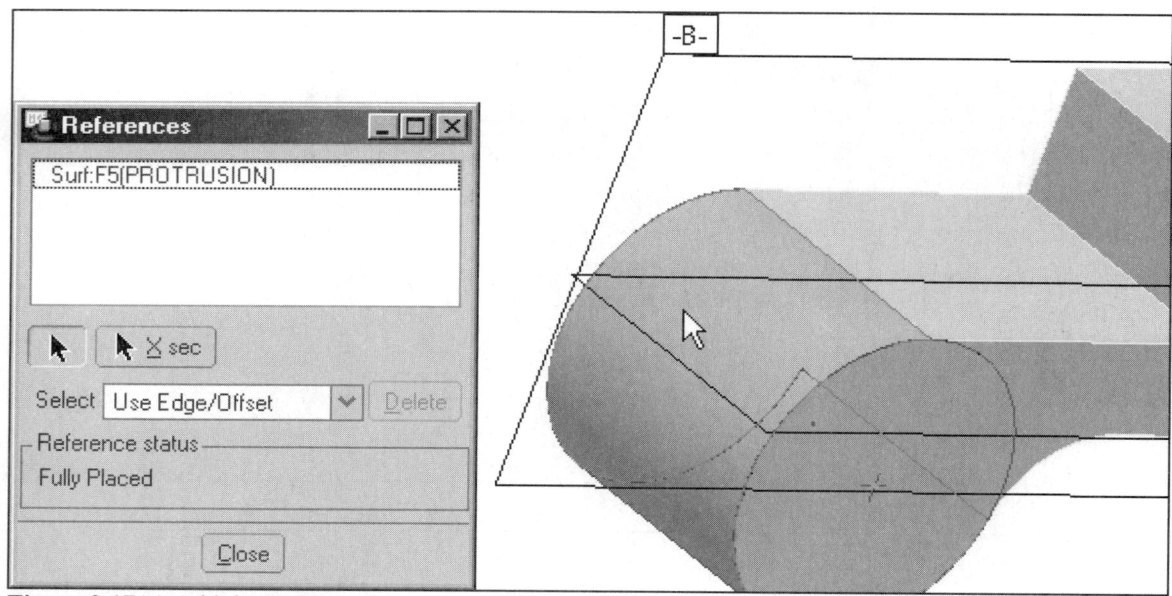

Figure 9.17(a) Add the Reference

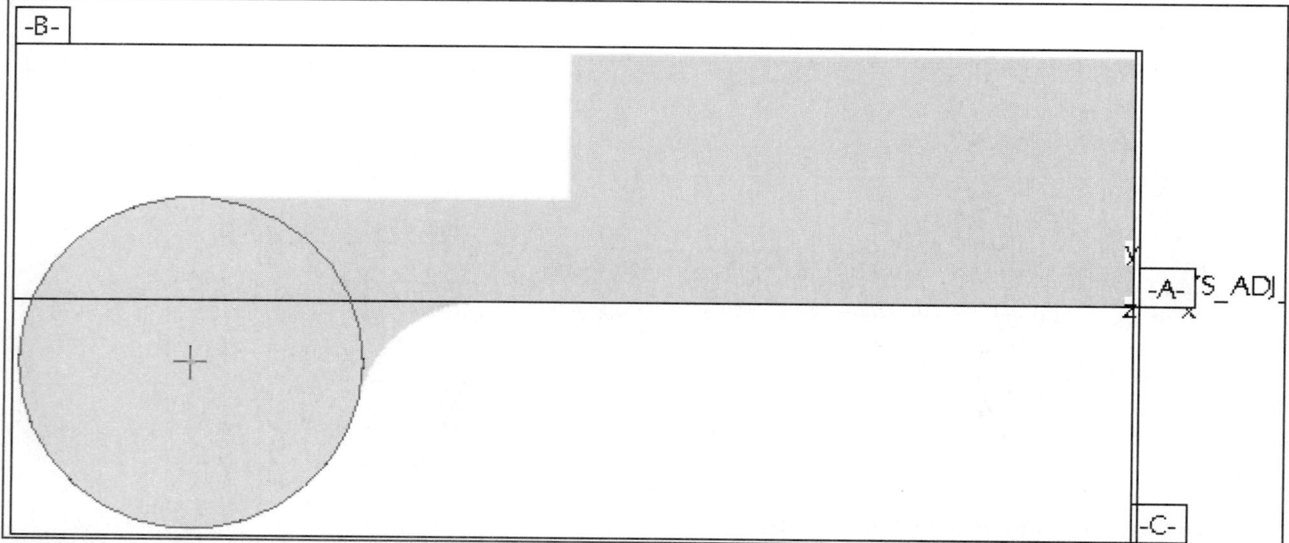

Figure 9.17(b) Sketch One Concentric Circle

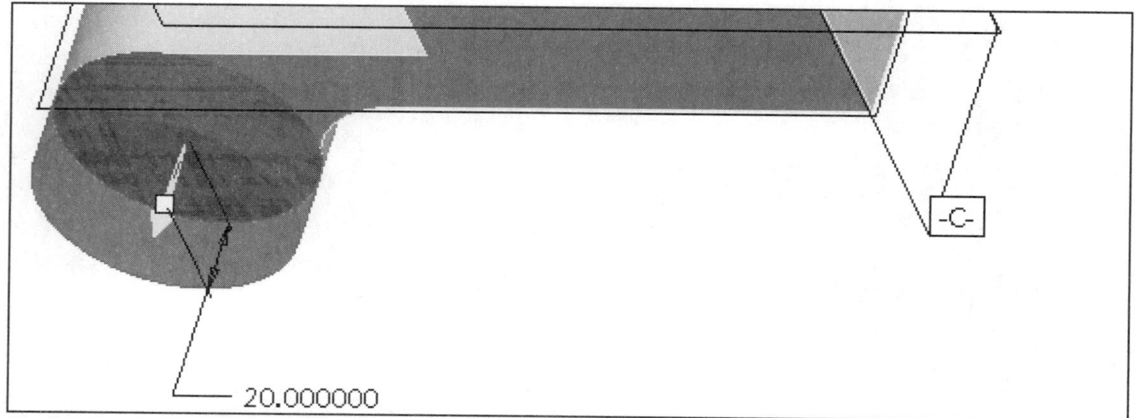

Figure 9.17(c) Depth of the Second Protrusion

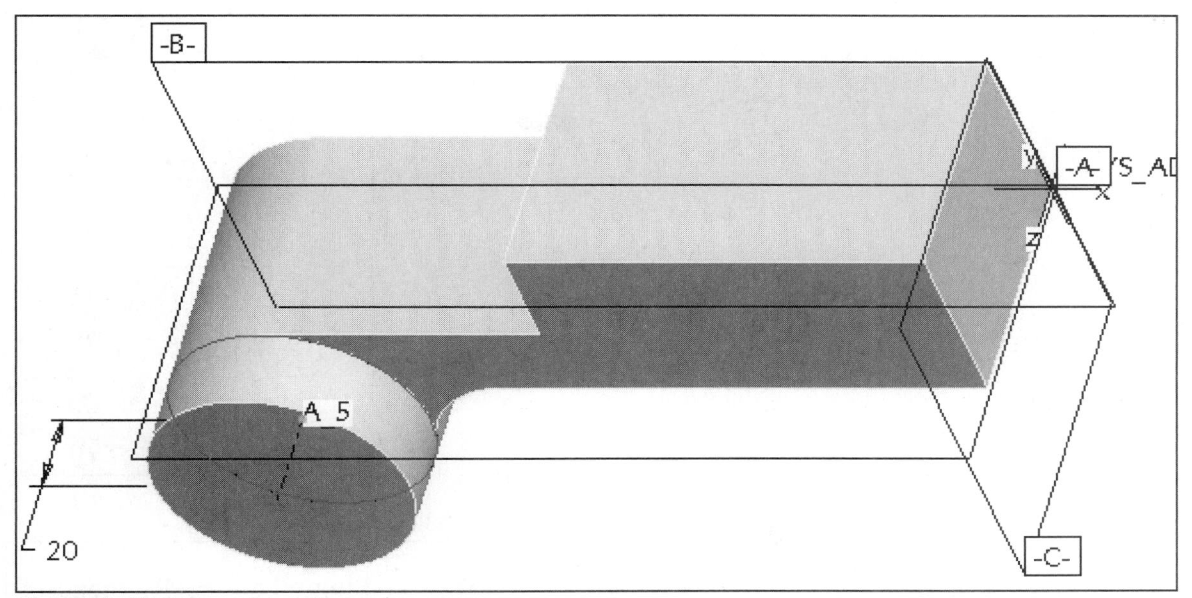

Figure 9.17(d) Completed Protrusion

Create a **R2** round along the top edge [Fig. 9.18(a)]. Select the first edge, then hold down the **Ctrl** key ⇒ click on the three remaining edges ⇒ **RMB** ⇒ **Round Edges** [Fig. 9.18(b)] ⇒ **MMB** (Fig. 9.19)

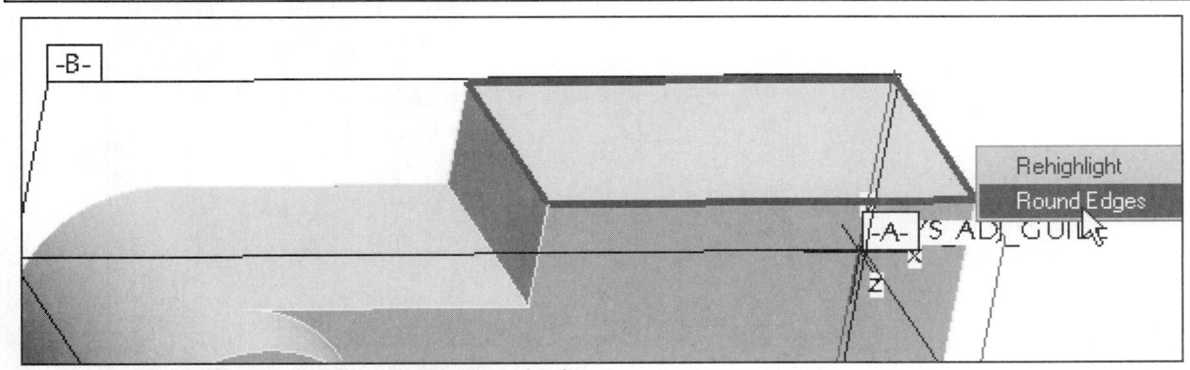

Figure 9.18(a) Select Each Edge, RMB, Round Edges

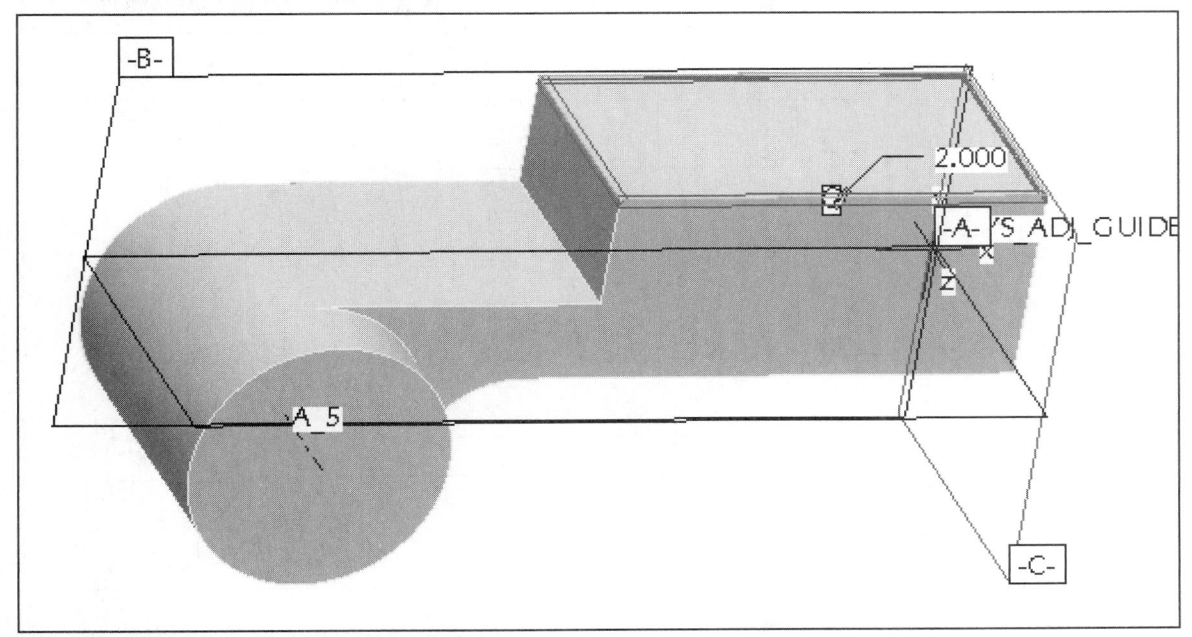

Figure 9.18(b) Move Drag Handle to **2.000**

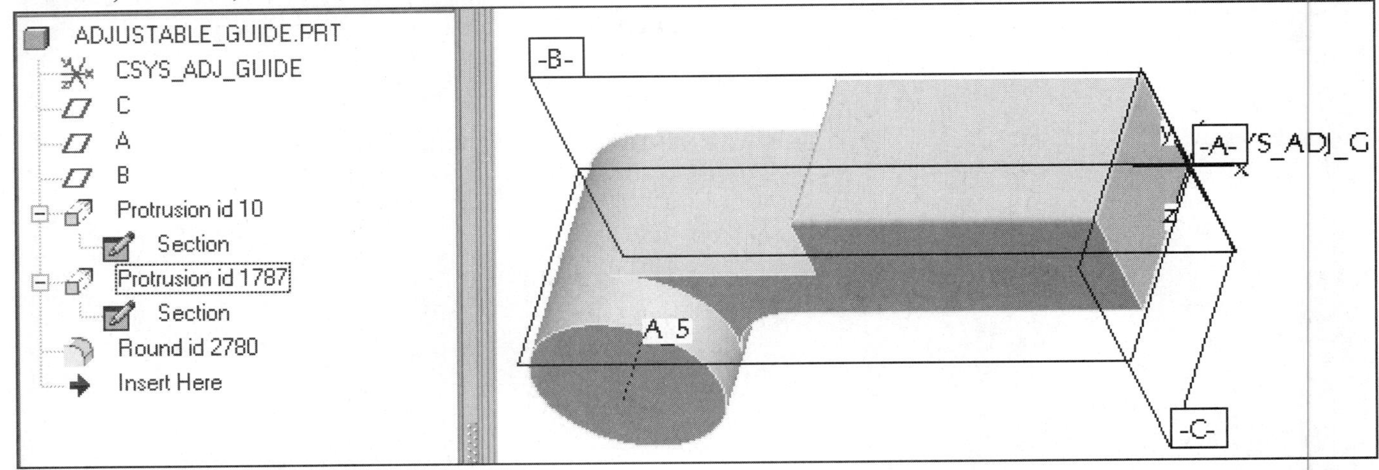

Figure 9.19 Model And Model Tree

Create a datum plane, offset from datum plane **B**, **30mm**, click: **Datum Plane Tool** ⇒ References: pick datum **B** as the plane to offset from ⇒ In the DATUM PLANE dialog, Offset: Translation **30** ⇒ **OK**

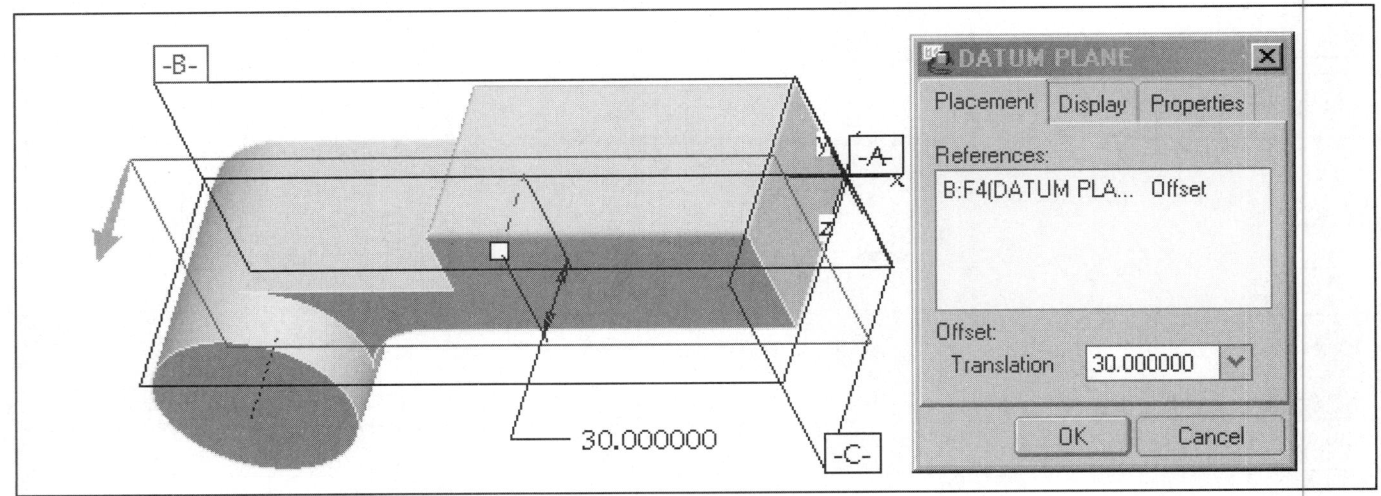

Figure 9.20(a) Create an Offset Datum Plane

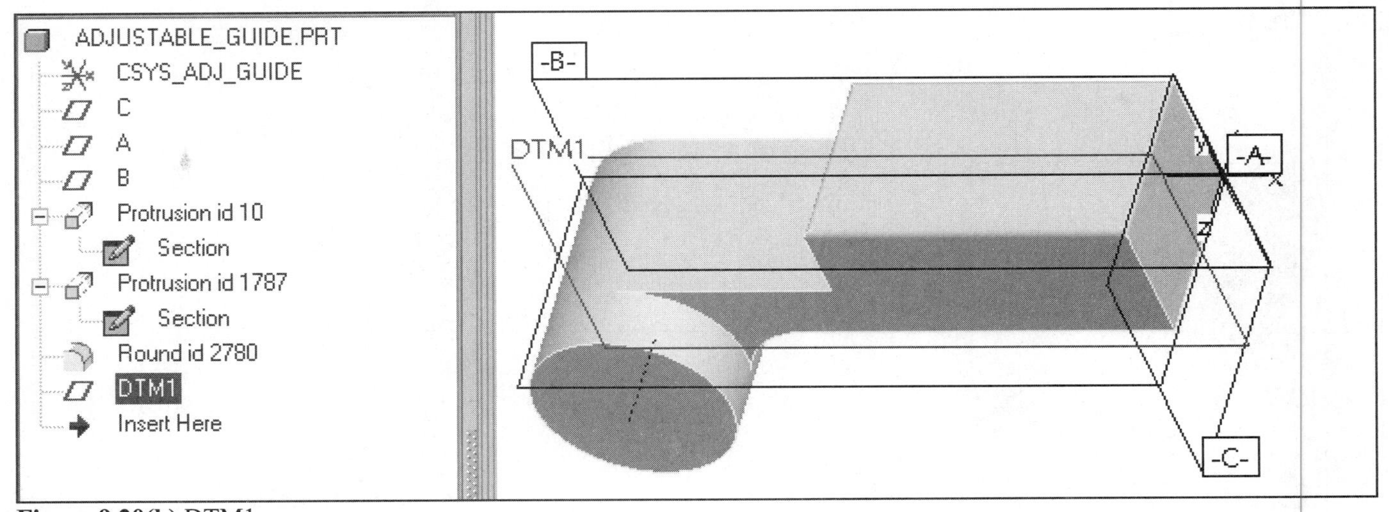

Figure 9.20(b) DTM1

Write a relation to control the position of the **DTM1**. The datum (and its children) must remain centered on the protrusion (**60mm**), regardless of modifications during an ECO change.

Click: **Tools ⇒ Relations ⇒ Insert Dimension symbol from Screen ⇒** click on **DTM1 ⇒** click on the offset **d** symbol dimension (**d153**) *(your symbols will be different)* [Fig. 9.21(a)] ⇒ type = *(equal sign)* ⇒ **Insert Dimension symbol from Screen ⇒** click on the protrusion ⇒ click on the depth **d** symbol dimension (**d11**) *(your symbols may be different)* ⇒ type **/2** *(divide by 2)* [Fig. 9.21(b)] ⇒ **Enter ⇒ Ok ⇒ Info ⇒ Relations and Parameters** (Fig. 9.22) ⇒ close quick sash ⇒ ⇒ **MMB**

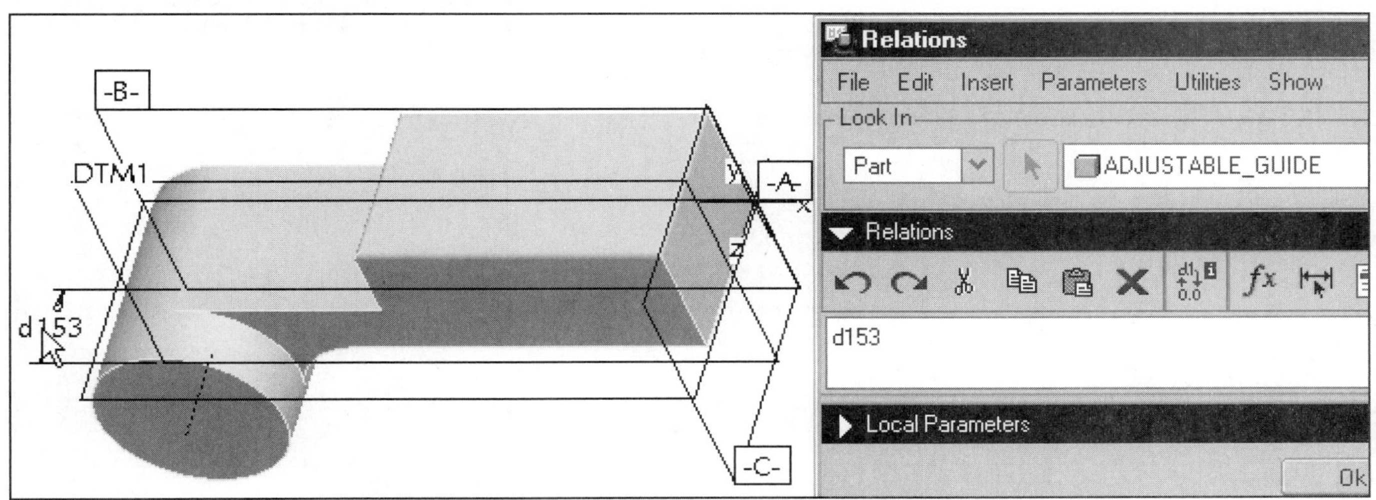

Figure 9.21(a) Select the First **d** Symbol Dimension

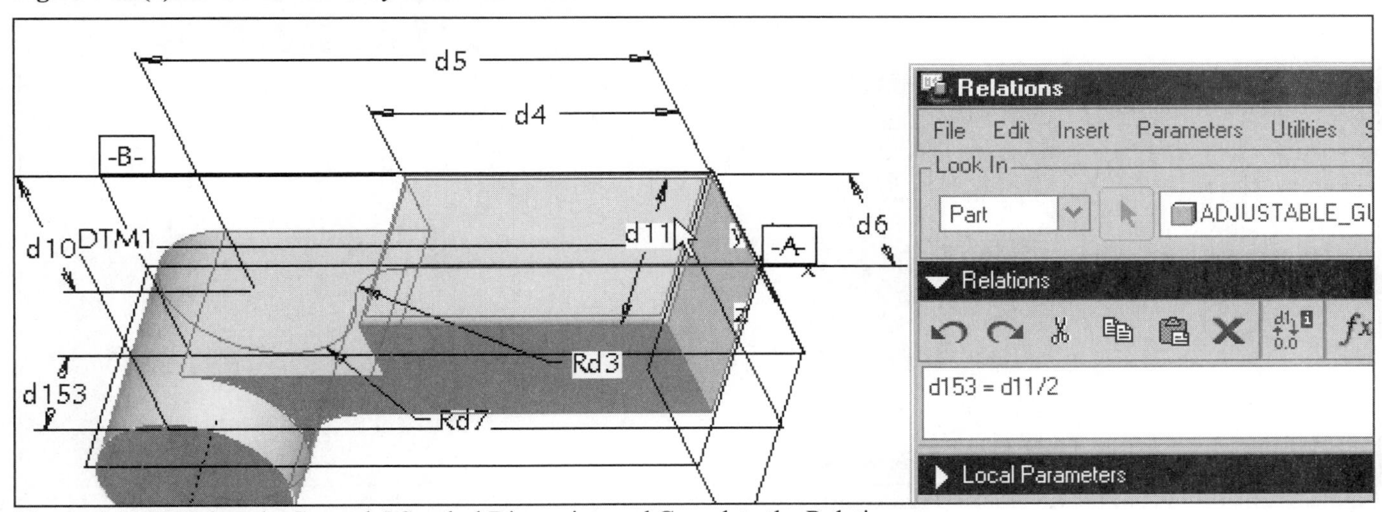

Figure 9.21(b) Select the Second **d** Symbol Dimension and Complete the Relation

Relation Table

Relation	Parameter	New Value
Relations for ADJUSTABLE_GUIDE:		
Initial Relations		
d153 = d11 /2	D153	3.000000e+01

Figure 9.22 Relation Table

The sketch of the rib will require just one tangent line between the large circular protrusion and the round along the part's upper edge. Zoom in to see the tangent position correctly.

Click: ⇒ ⇒ **Rib Tool** ⇒ ⇒ Sketch Plane- Plane: pick **DTM1** [Fig. 9.23(a)] ⇒ **Sketch** ⇒ **MMB** spin the model ⇒ delete the two existing references ⇒ click on the cylindrical surface as the first reference [Fig. 9.23(b)] ⇒ **Zoom In** ⇒ click on the round as the second reference [Fig. 9.23(c)] ⇒ **Close**

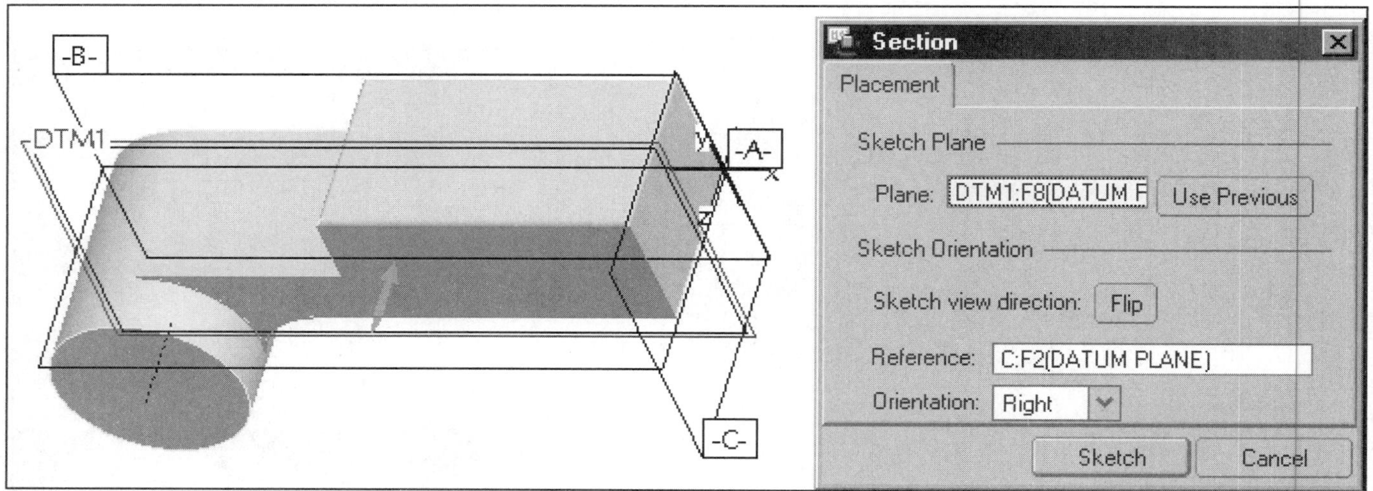

Figure 9.23(a) Sketch Plane DTM1

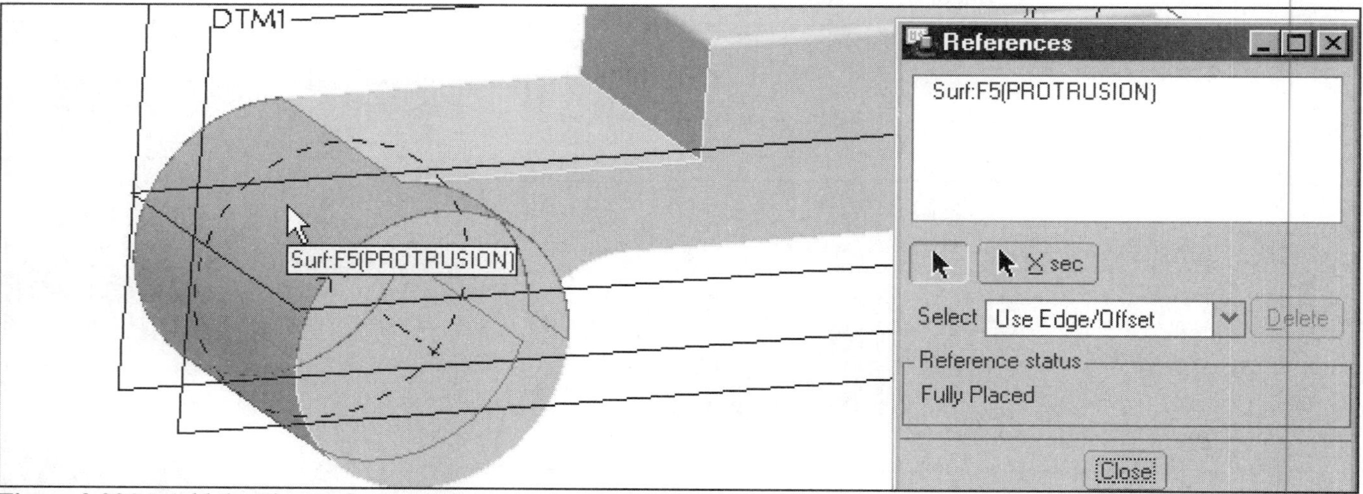

Figure 9.23(b) Add the First Reference (Protrusion)

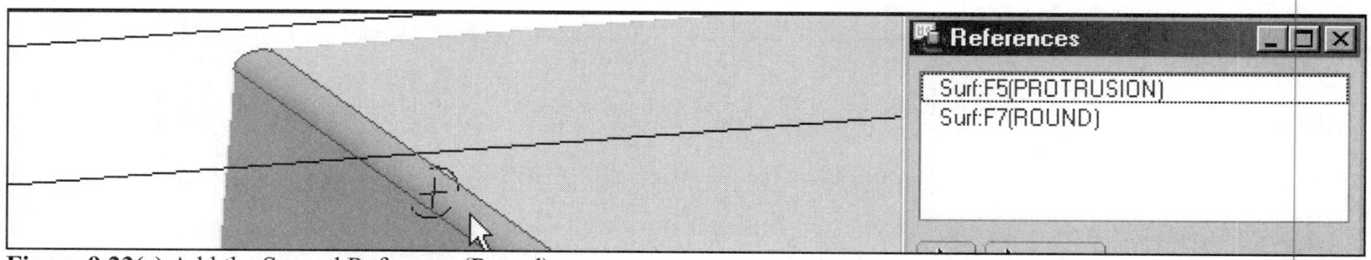

Figure 9.23(c) Add the Second Reference (Round)

Click: ⇒ ⇒ ⇒ Snap to Grid ⇒ **Create Lines** ⇒ pick the starting point of the line *near* the point of tangency [Fig. 9.23(d)] ⇒ **Zoom In** ⇒ **MMB** to end zoom ⇒ pick the ending point of the line *near* the point of tangency [Fig. 9.23(e)] ⇒ **MMB** [Fig. 9.23(f)]

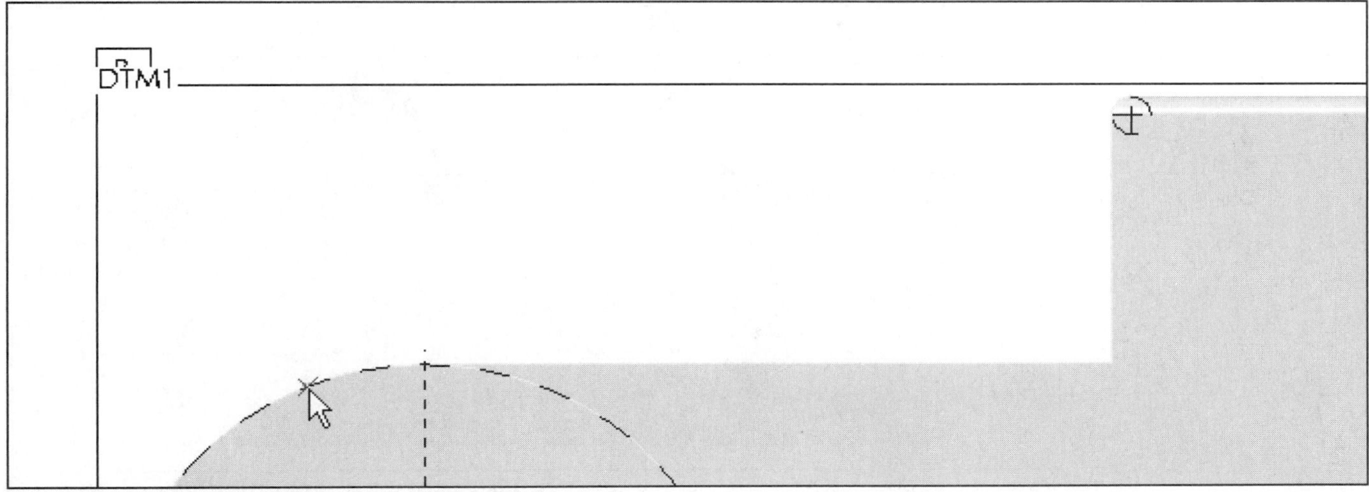

Figure 9.23(d) Pick Starting Point for Tangent Line

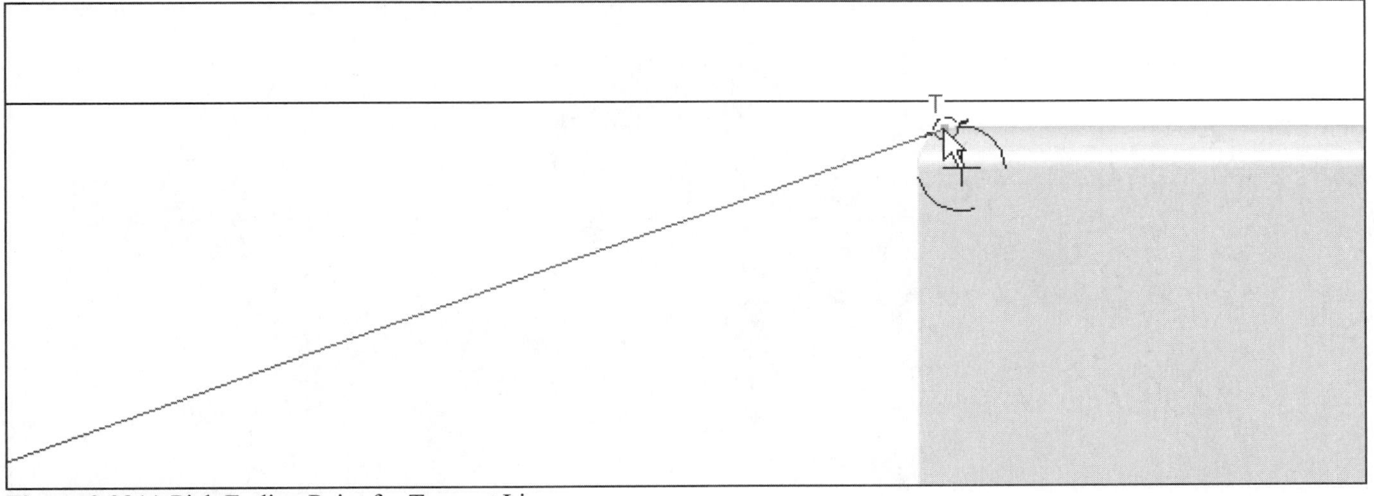

Figure 9.23(e) Pick Ending Point for Tangent Line

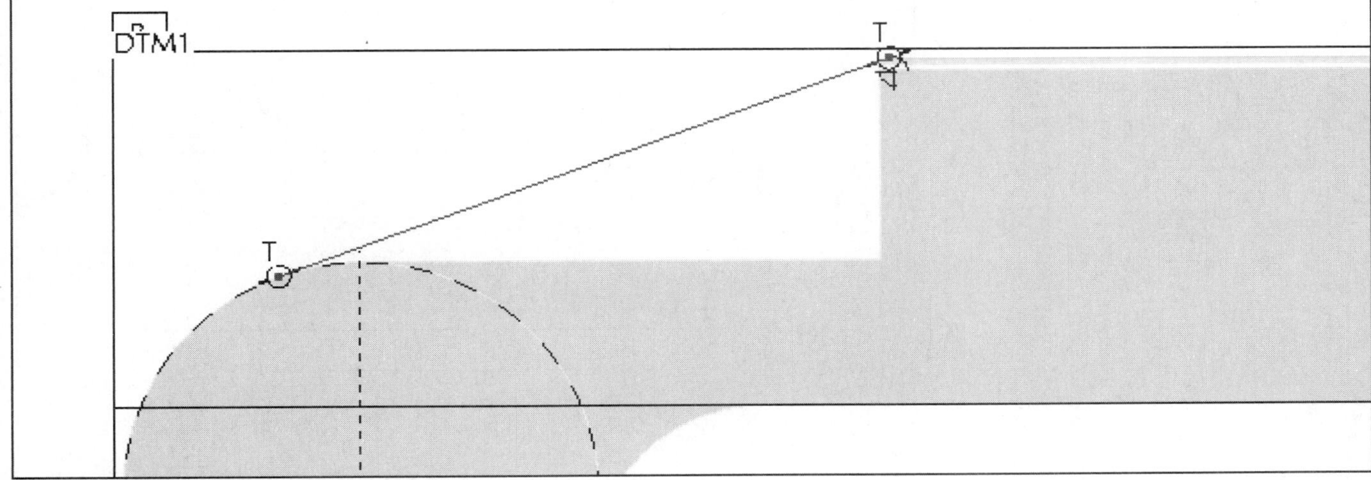

Figure 9.23(f) Completed Section Sketch

Click: **Display object in standard orientation** ⇒ [Fig. 9.23(g)] ⇒ click on the yellow arrow to flip the direction of the rib creation towards the part [Fig. 9.23(h)] ⇒ modify the rib thickness to **15** ⇒ **Enter** 15 ⇒ [Fig. 9.23(i)] ⇒ ⇒ **MMB**

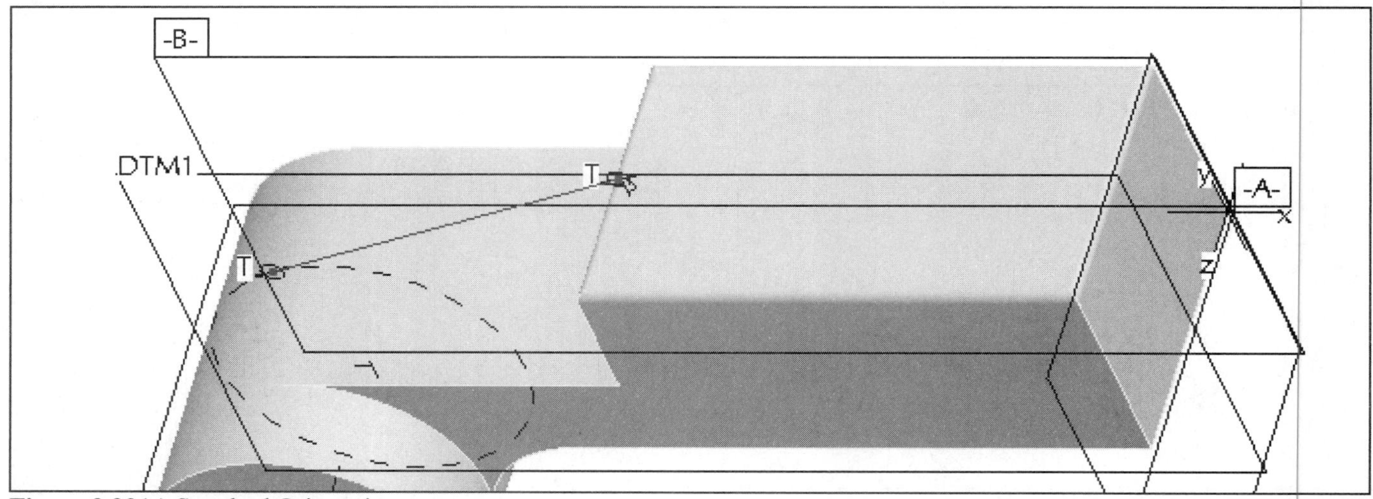

Figure 9.23(g) Standard Orientation

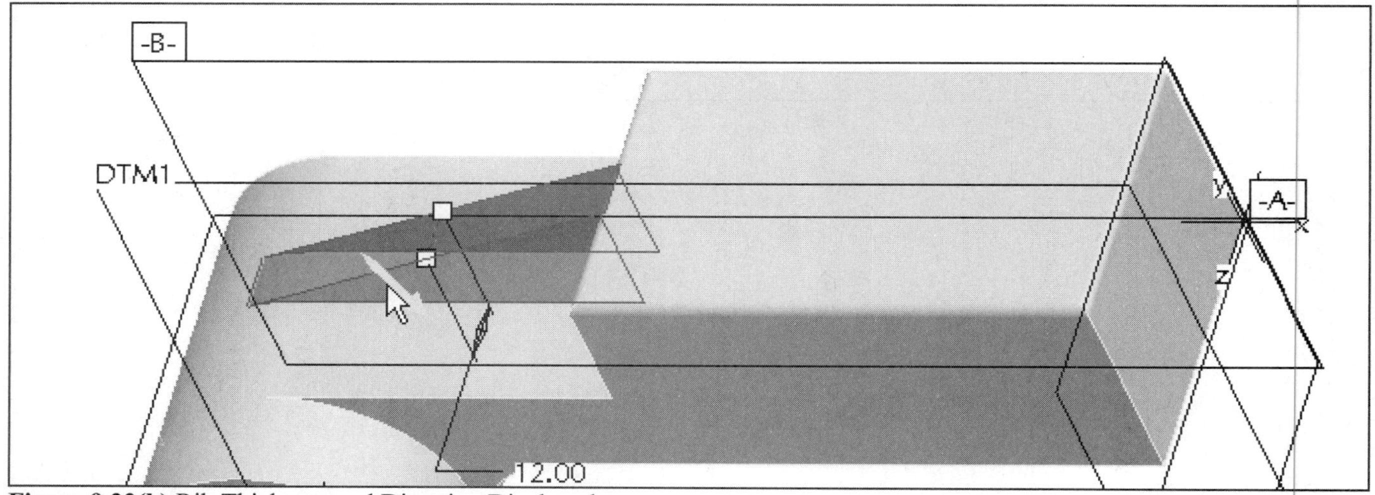

Figure 9.23(h) Rib Thickness and Direction Displayed

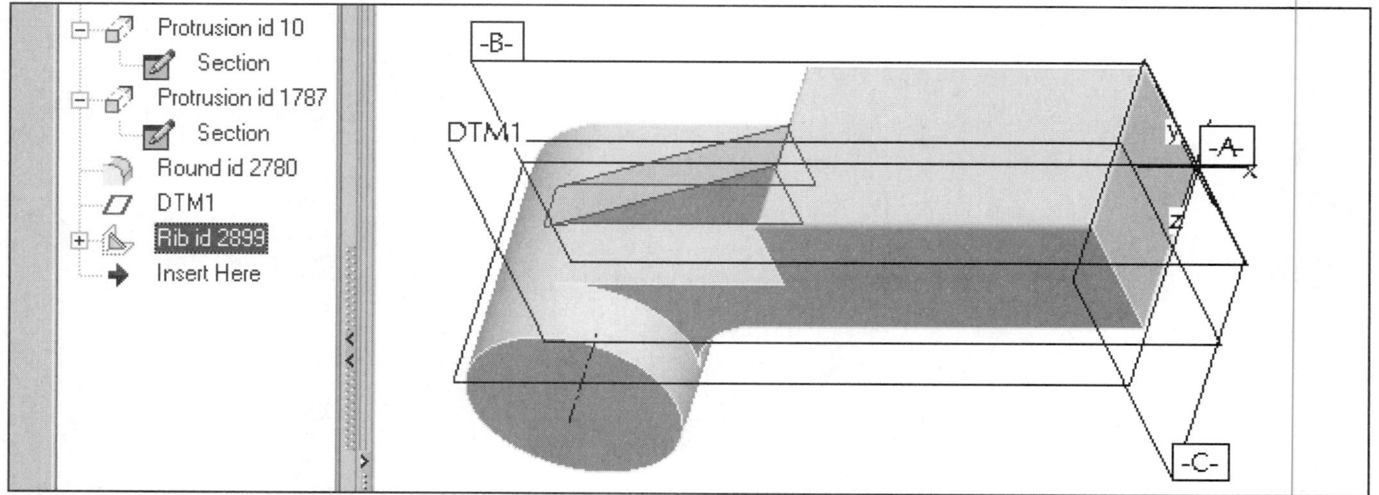

Figure 9.23(i) Completed Rib

(Note that clicking in the ***Dashboard*** *of the* ***Rib Tool*** *toggles the rib from centered, to the right, or to the left of the datum plane).*

Create the rounds, click on the edge shown in Figure 9.24(a) until it highlights in red ⇒ press and hold the **Ctrl** key and pick on the second edge until it highlights in red ⇒ **RMB** ⇒ **Round Edges** ⇒ modify radius to **2.000** [Fig. 9.24(b)] ⇒ **MMB** [Fig. 9.24(c)]

Figure 9.24(a) Round Edges

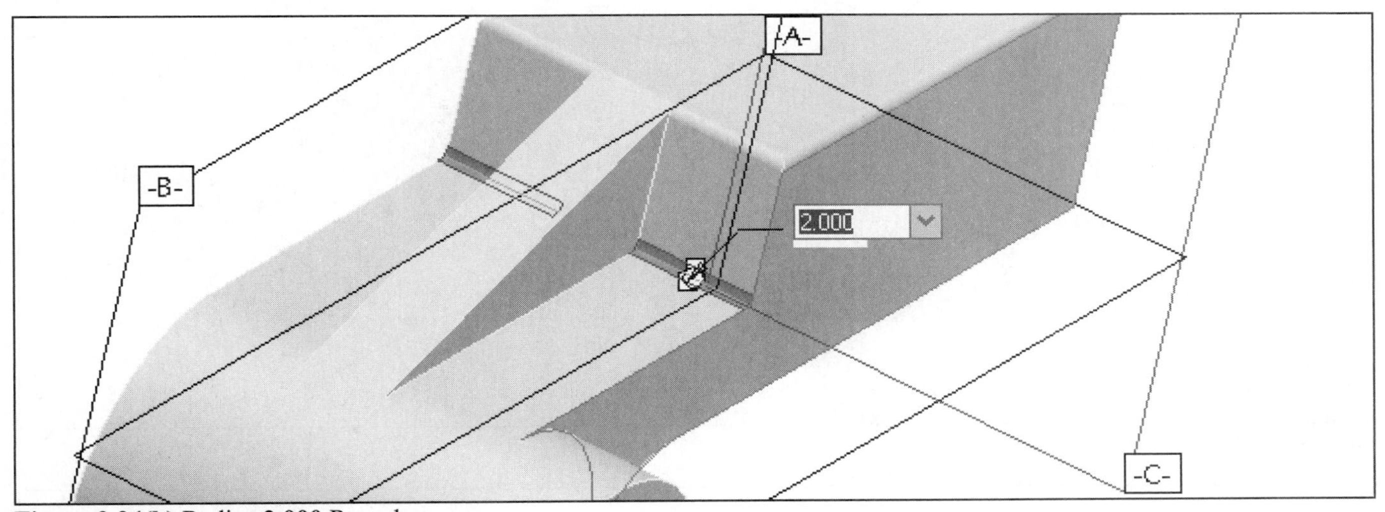

Figure 9.24(b) Radius **2.000** Round

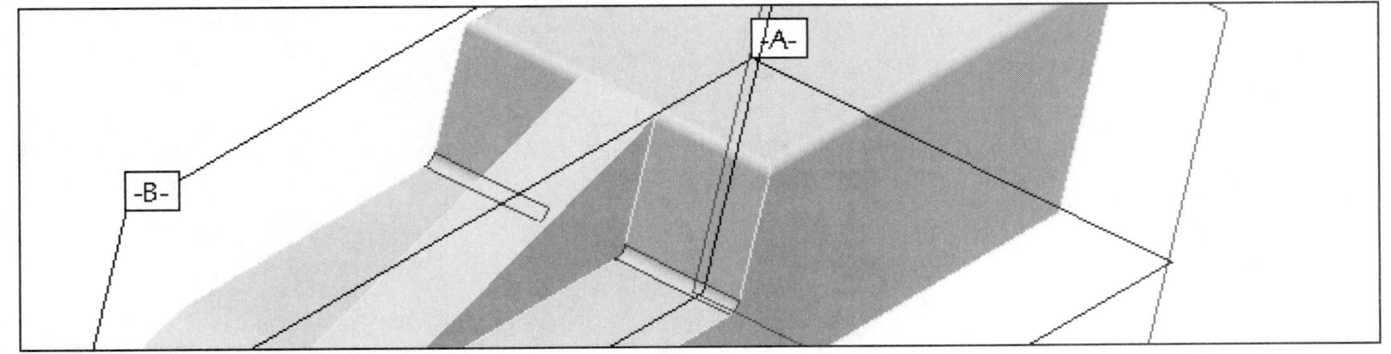

Figure 9.24(c) Completed Round

Click on the vertical edge shown in Figure 9.24(d) until it highlights in red ⇒ press and hold the **Ctrl** key and pick on the second edge until it highlights ⇒ **RMB** ⇒ **Round Edges** ⇒ modify radius to **2.00** [Fig. 9.24(e-f)] ⇒ **Sets** ⇒ click in the **Set 1** block [Fig. 9.24(g)] ⇒ **RMB** ⇒ **Add** [Fig. 9.24(h)] ⇒ **MMB** spin the part ⇒ click on the corresponding vertical edge [Fig. 9.24(i)] ⇒ **MMB**

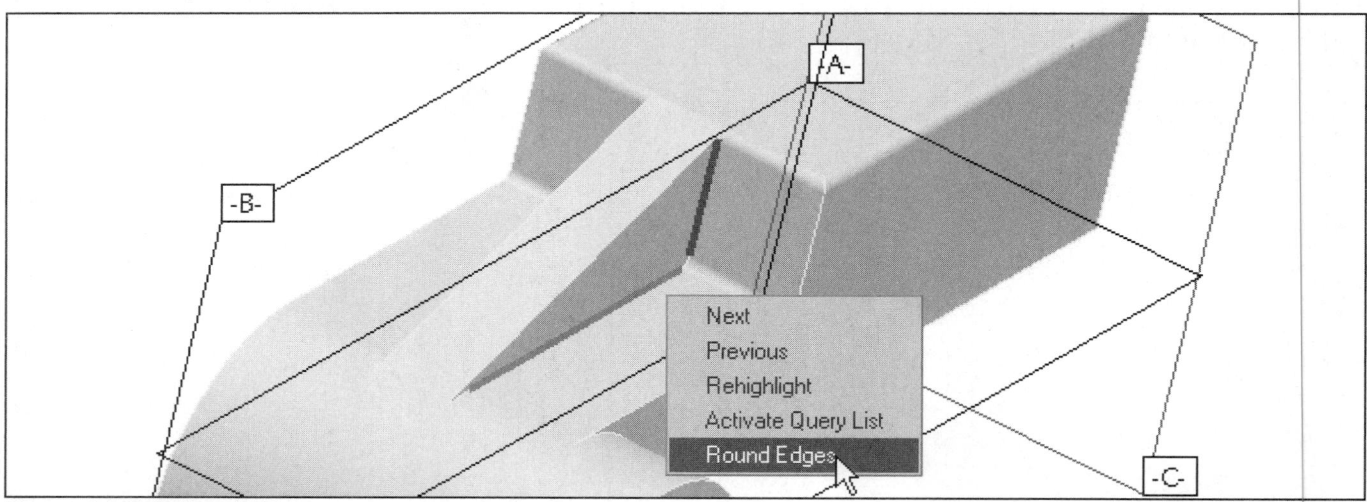

Figure 9.24(d) Round Edges

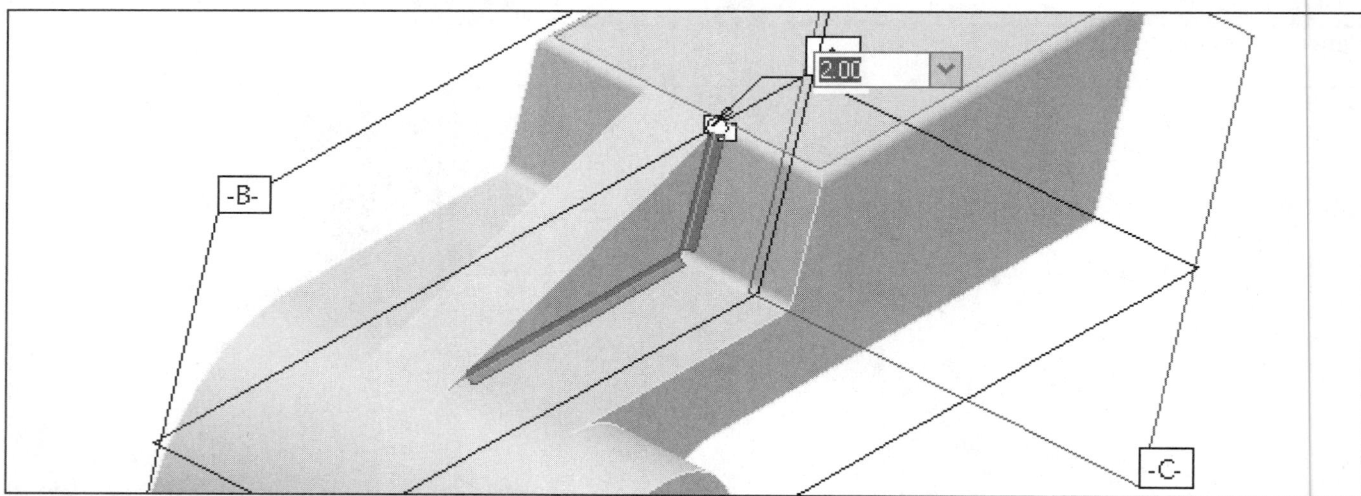

Figure 9.24(e) Round **R2.00**

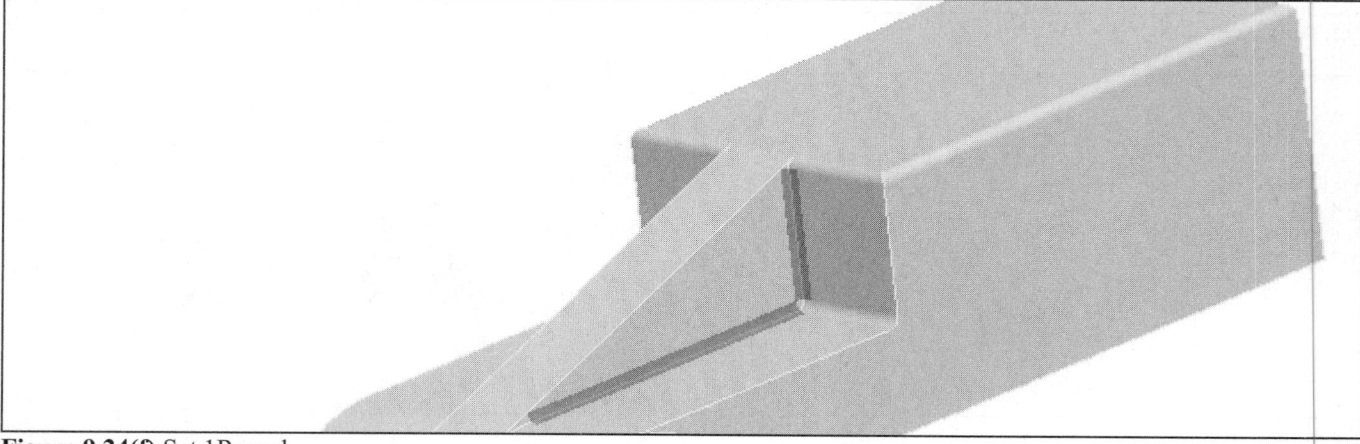

Figure 9.24(f) Set 1Round

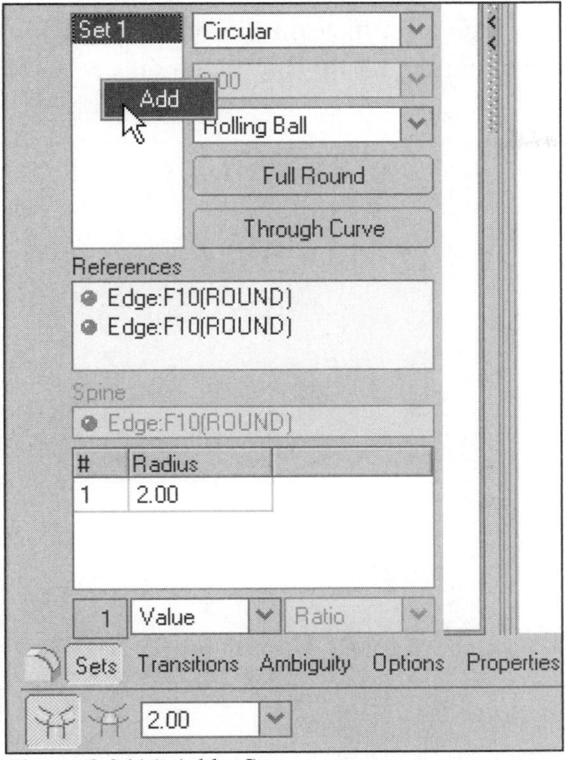

Figure 9.24(g) Add a Set

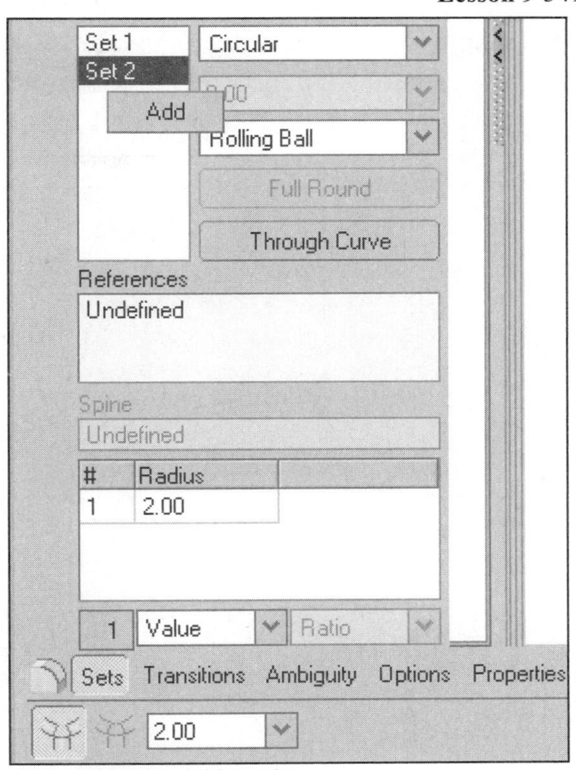

Figure 9.24(h) Add Set 2

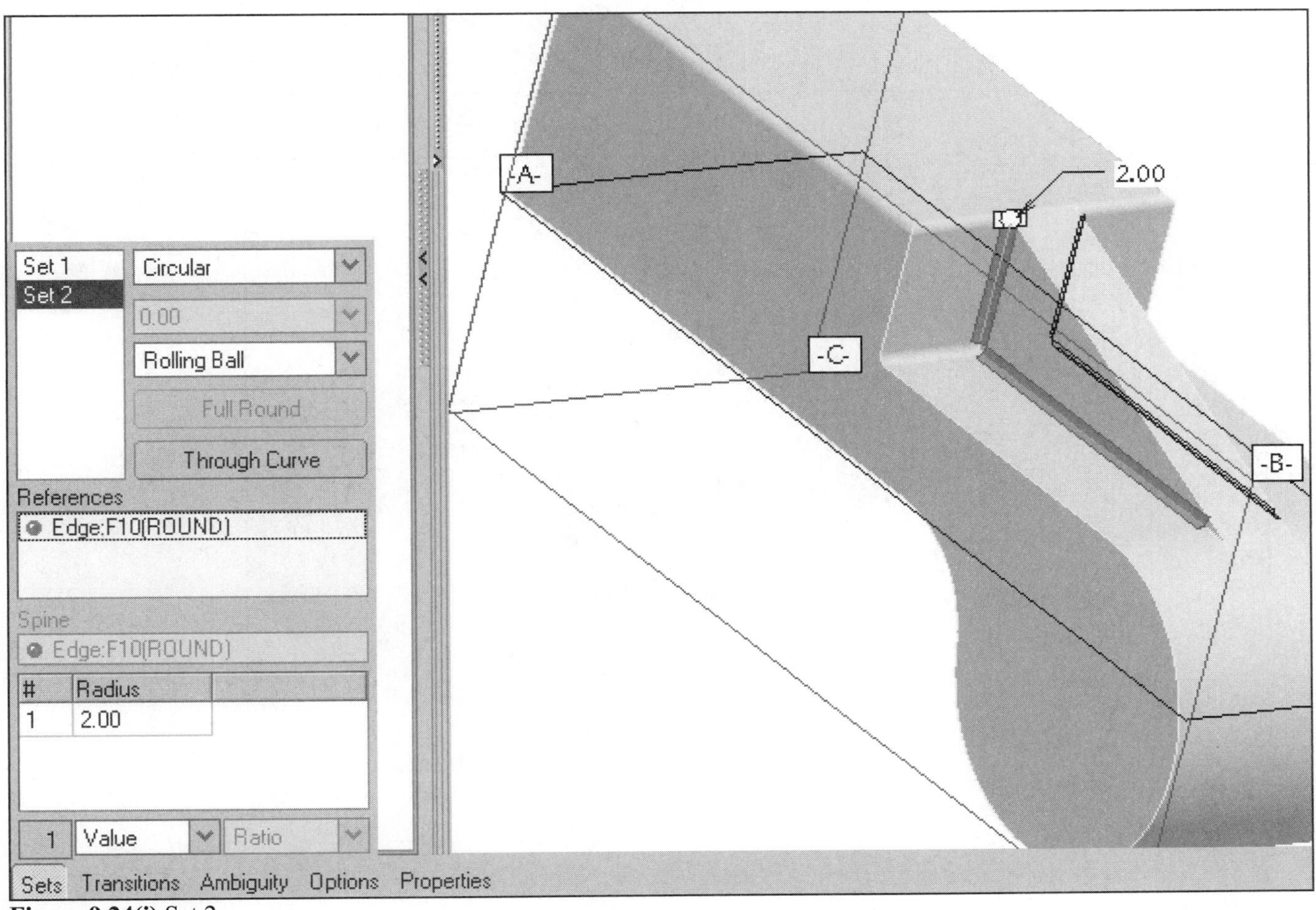

Figure 9.24(i) Set 2

MMB spin the part ⇒ select an edge ⇒ press and hold the **Ctrl** key and pick on the remaining edges shown in Figure 9.24(j) ⇒ **RMB** ⇒ **Round Edges** ⇒ modify the radius to **2.00** [Fig. 9.24(k)] ⇒ **MMB** ⇒ **MMB** spin the part ⇒ [save icon] ⇒ **MMB**

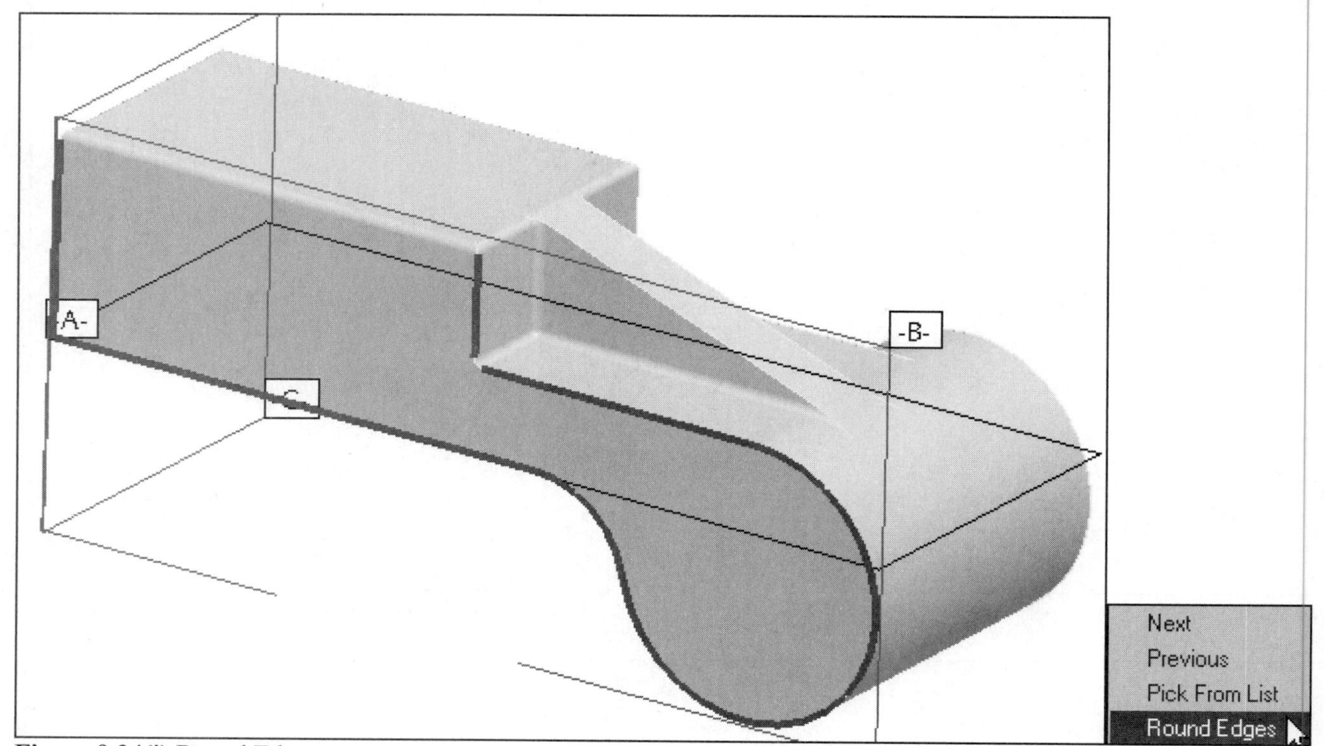

Figure 9.24(j) Round Edges

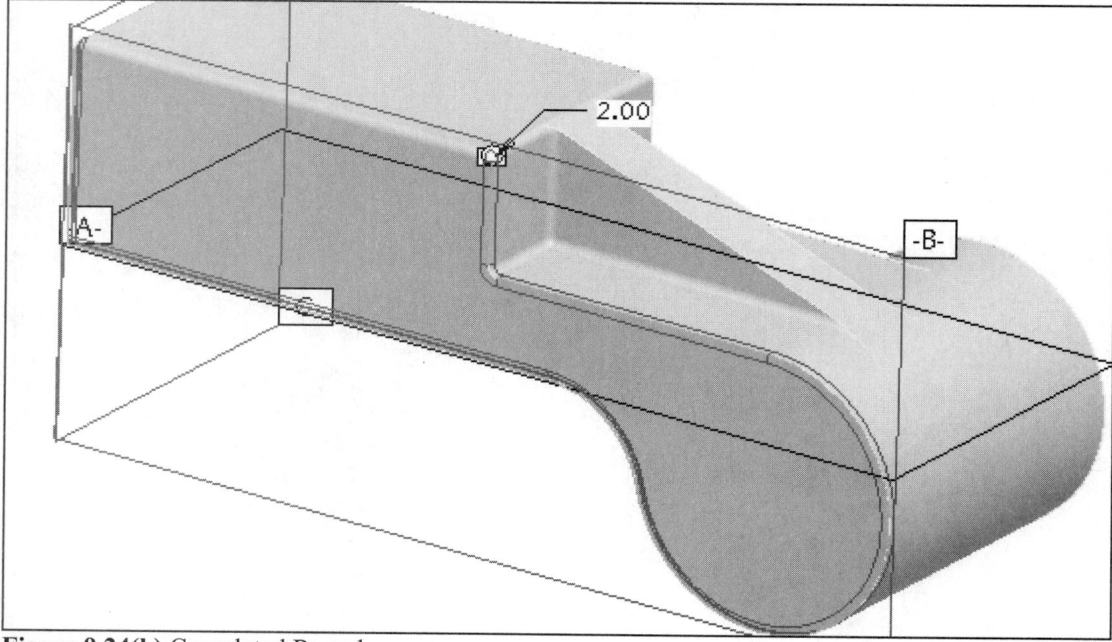

Figure 9.24(k) Completed Round

Use the previous sequence and create the **2** mm radius rounds shown in Figure 9.24(l-m) ⇒ **MMB** spin the part ⇒ use the previous sequence and create the **2** mm radius rounds shown in Figure 9.24(n-o) ⇒ **Sets**

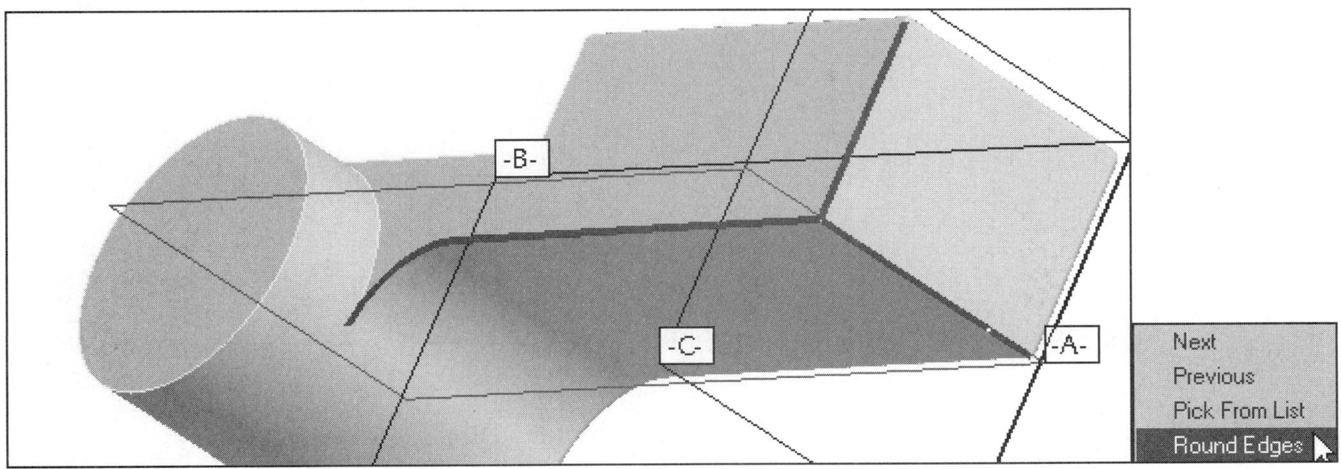

Figure 9.24(l) Round Edges

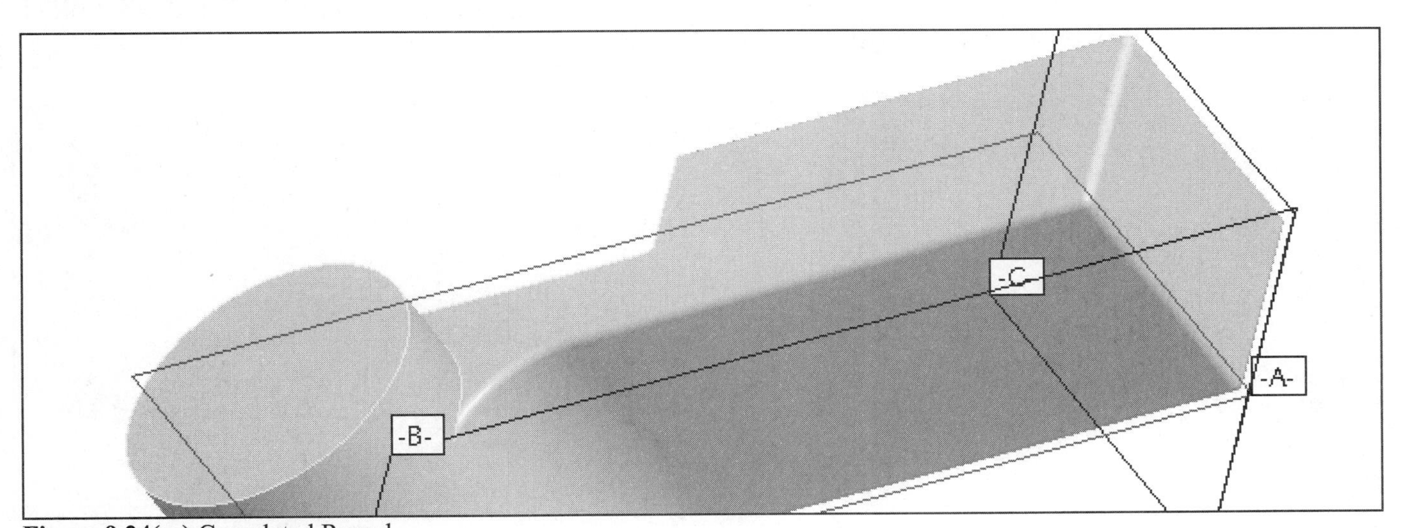

Figure 9.24(m) Completed Round

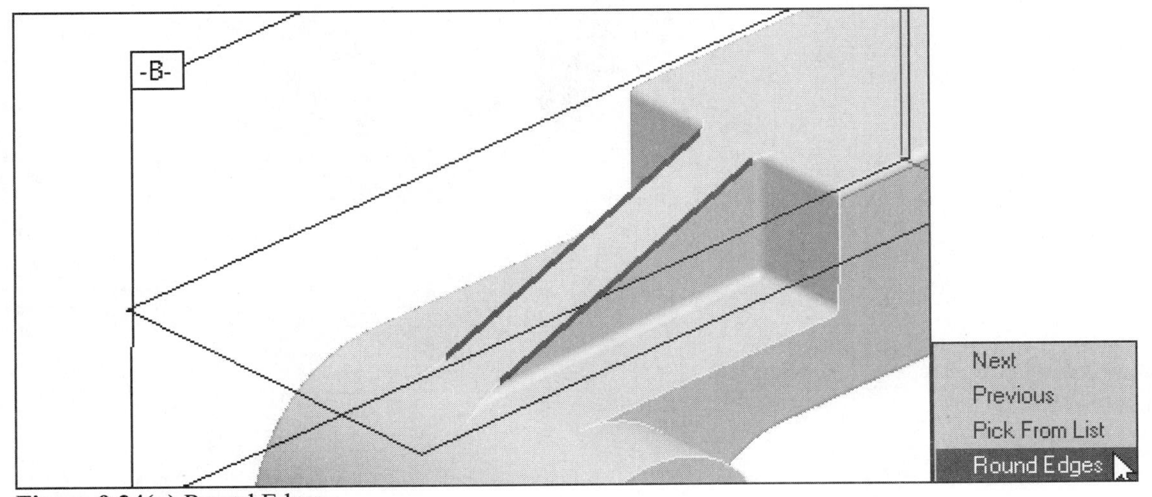

Figure 9.24(n) Round Edges

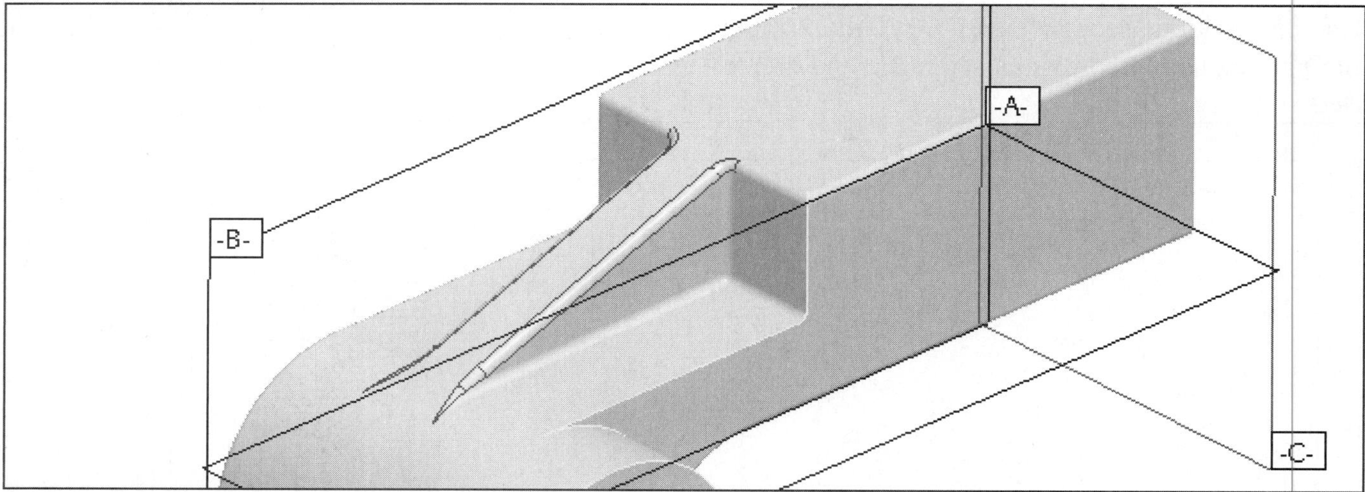

Figure 9.24(o) Set 1Round

Click in the **Set 1** block ⇒ **RMB** ⇒ **Add** ⇒ click on the circular edge [Fig. 9.24(p-q)] ⇒

Instead of using a second set, remove the **SET 2** and make the round later ⇒ click: **Set 2** in the Set block [Fig. 9.24(r)] ⇒ **RMB** ⇒ **Delete** ⇒ **MMB**

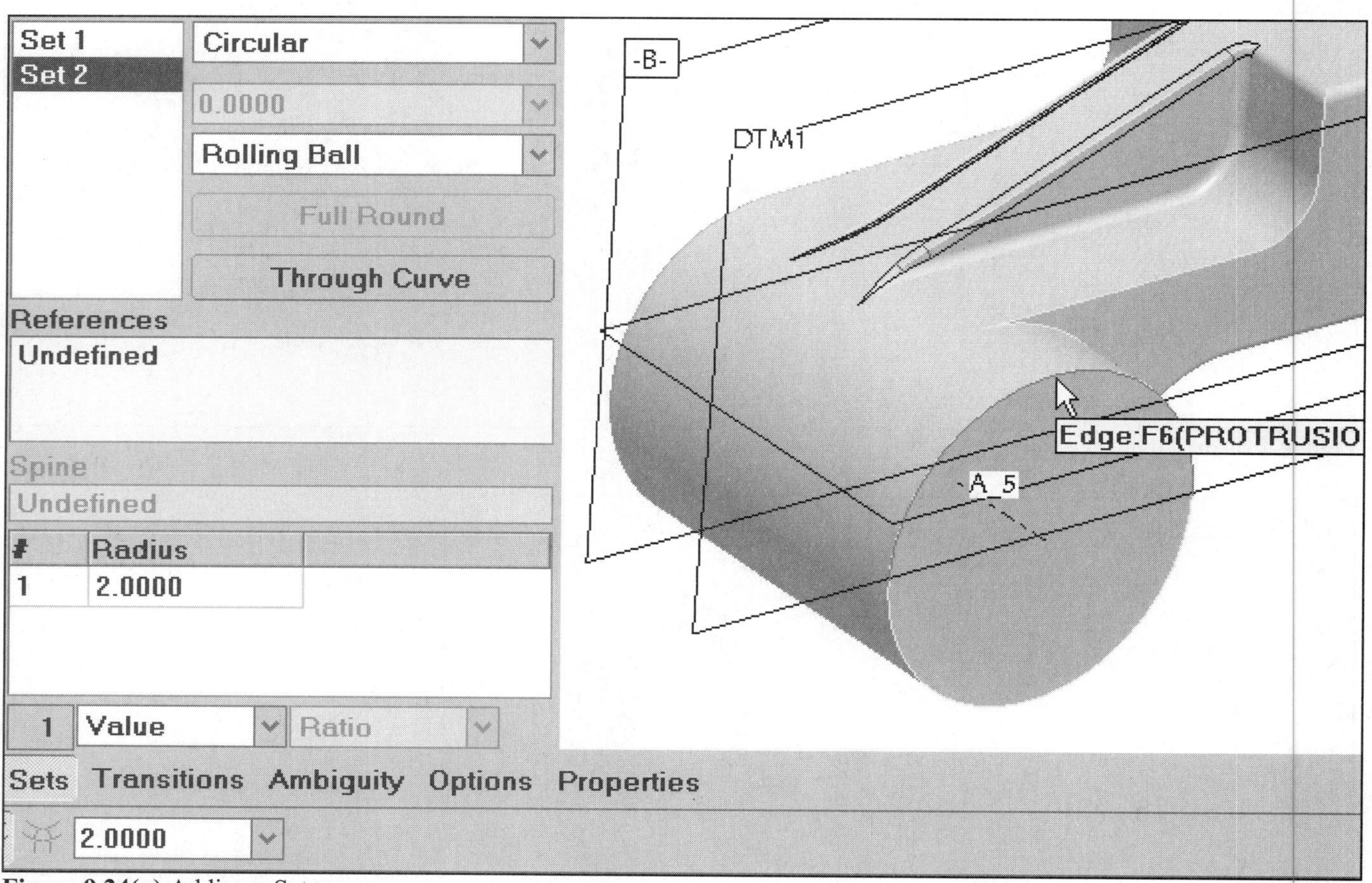

Figure 9.24(p) Adding a Set

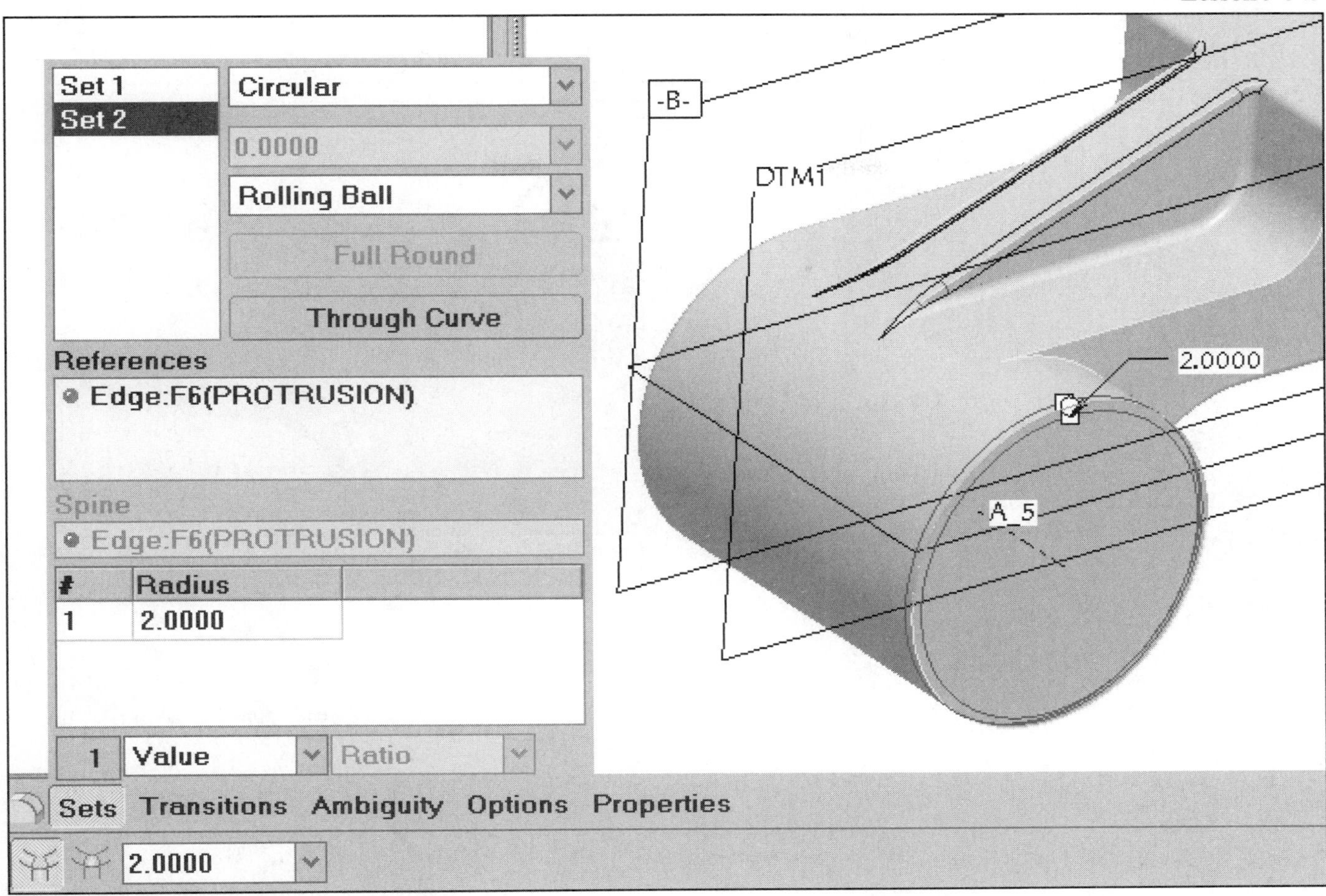

Figure 9.24(q) Set 2

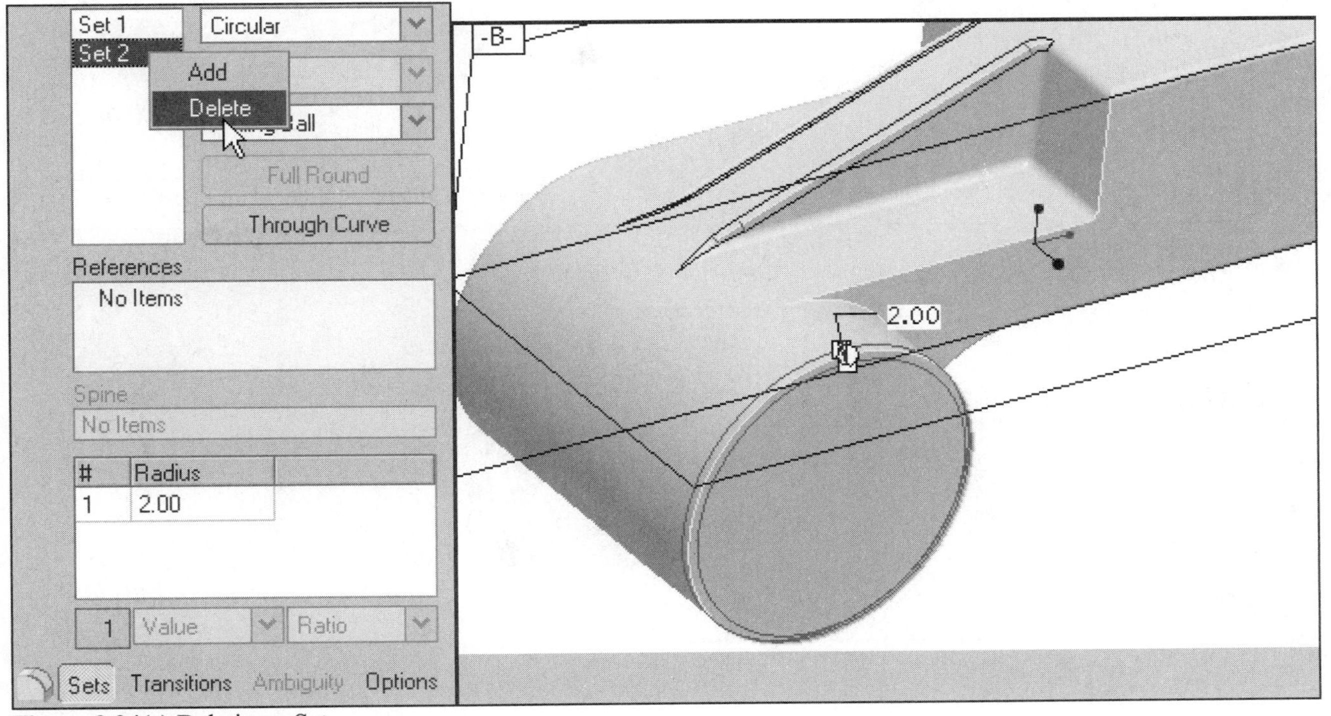

Figure 9.24(r) Deleting a Set

Create the remaining rounds separately [Fig. 9.24(s-v)] ⇒ [Save icon] ⇒ **MMB** (Fig. 9.25)

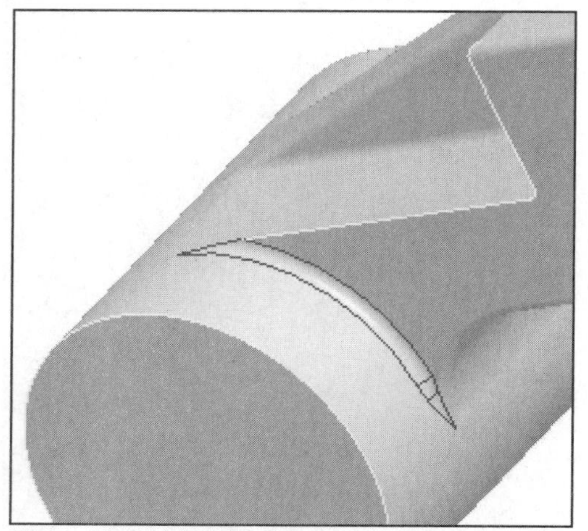

Figure 9.24(s) Add Corner Tangent Round

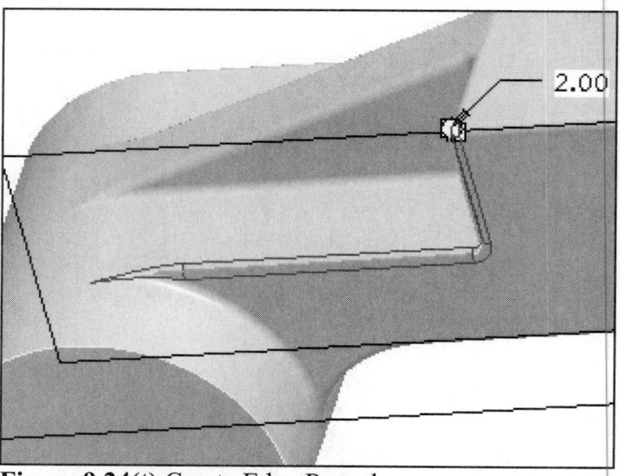

Figure 9.24(t) Create Edge Round

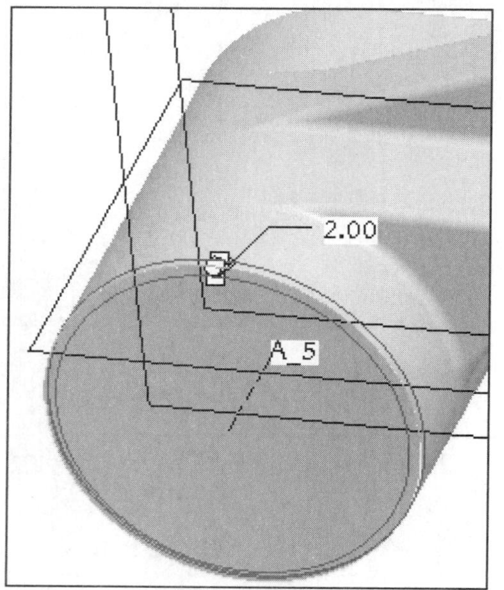

Figure 9.24(u) Circular Round

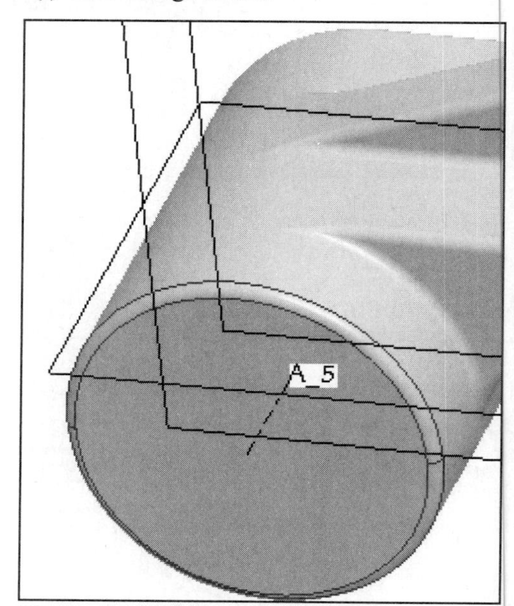

Figure 9.24(v) Completed Circular Round

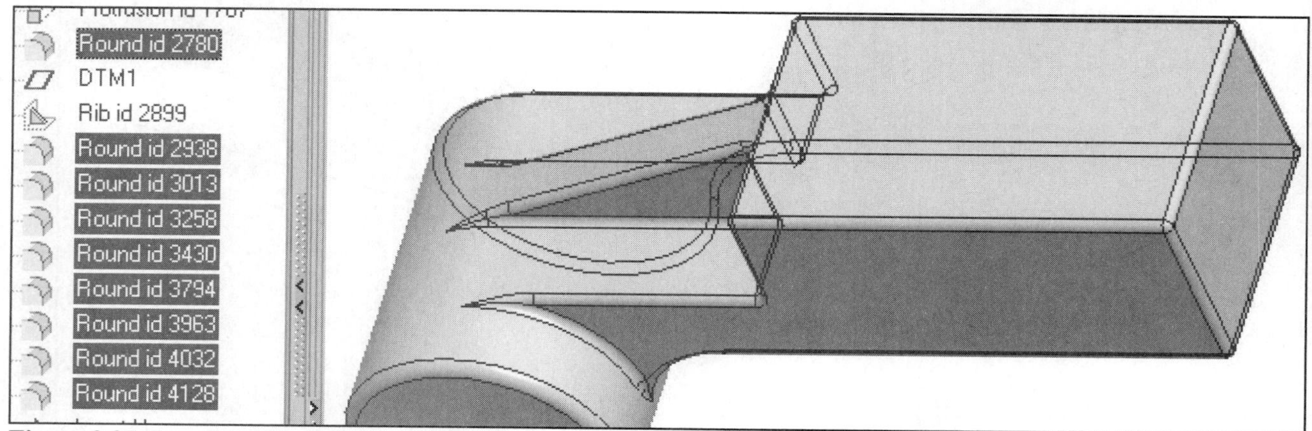

Figure 9.25 Completed Rounds *(Completed Casting-Workpiece)*

The "machined" features of the part can be created with a cut to "face" the end of the cylindrical protrusion, a cut on the top, a counterbored hole, and a thru hole with two counterbores (created with a revolved cut). Use the machine detail (Fig. 9.26) for the dimensions.

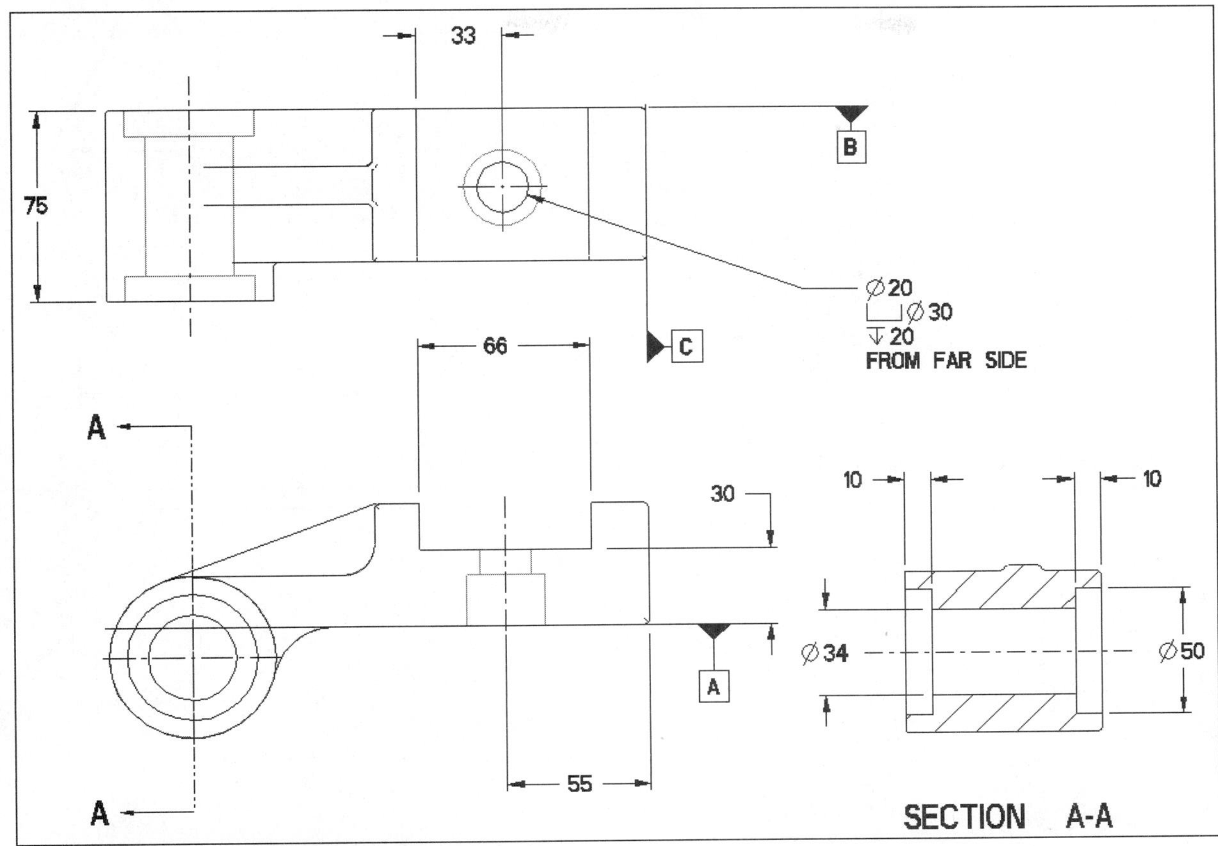

Figure 9.26 Adjustable Guide Machining Drawing

Cut the front face of the cylindrical protrusion from **80mm** to the design size of **75mm**. Create and use **DTM2** as the sketching plane [Fig. 9.27(a)]. Only one reference is required, the cylindrical surface. Sketch a concentric circle. No dimensions are necessary [Fig. 9.27(b)].

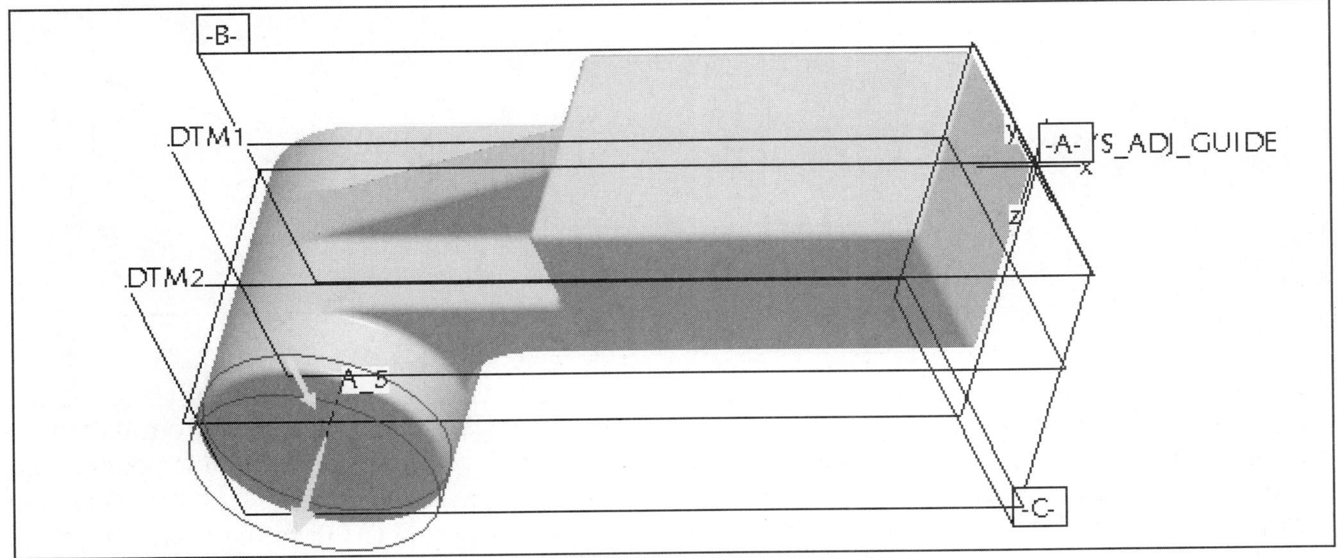

Figure 9.27(a) Offset a new Datum Plane (DTM2) from Datum B (**75mm**)

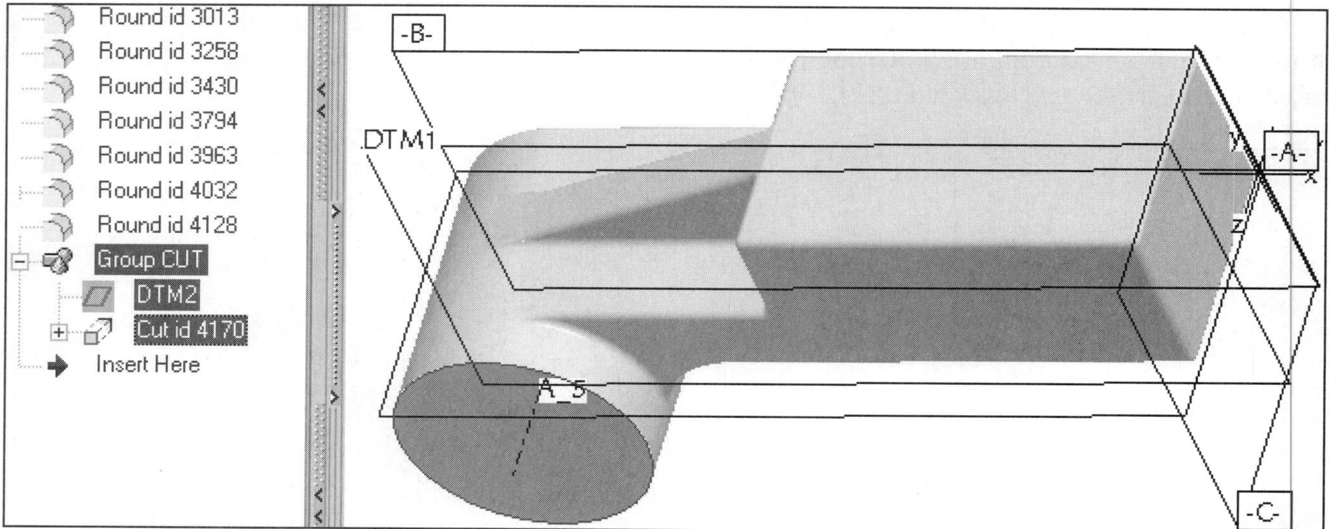

Figure 9.27(b) Completed Cut

Before continuing, edit the **R2** round [Fig. 9.28(a)] to the design size of **R12** [Fig. 9.28(b)].

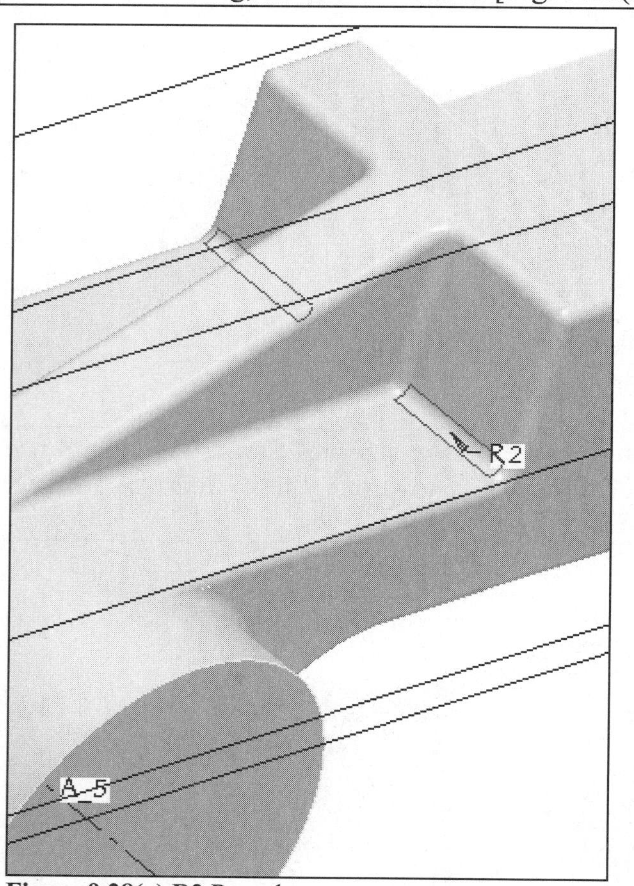

Figure 9.28(a) R2 Round

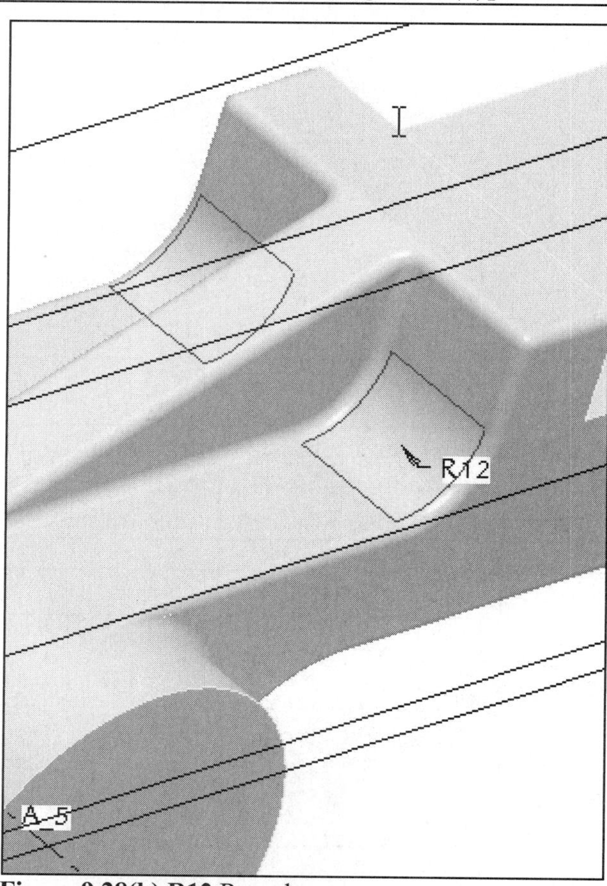

Figure 9.28(b) R12 Round

The slot/cut should be modeled next. The machining dimensions (Fig. 9.26) of **30mm** from datum **A**, **55mm** from datum **C** to the center of the slot, and the slot/cut width of **66mm** must be incorporated into the features dimensioning scheme so as to have the proper driving dimensions [Fig. 9.29(a-b)]. There are two ways of accomplishing this; using an internal datum (offset datum **DTM3)** as the sketching plane [Fig. 9.29(a)], or sketching on **DTM1** and projecting the cut through both sides. Create the slot/cut.

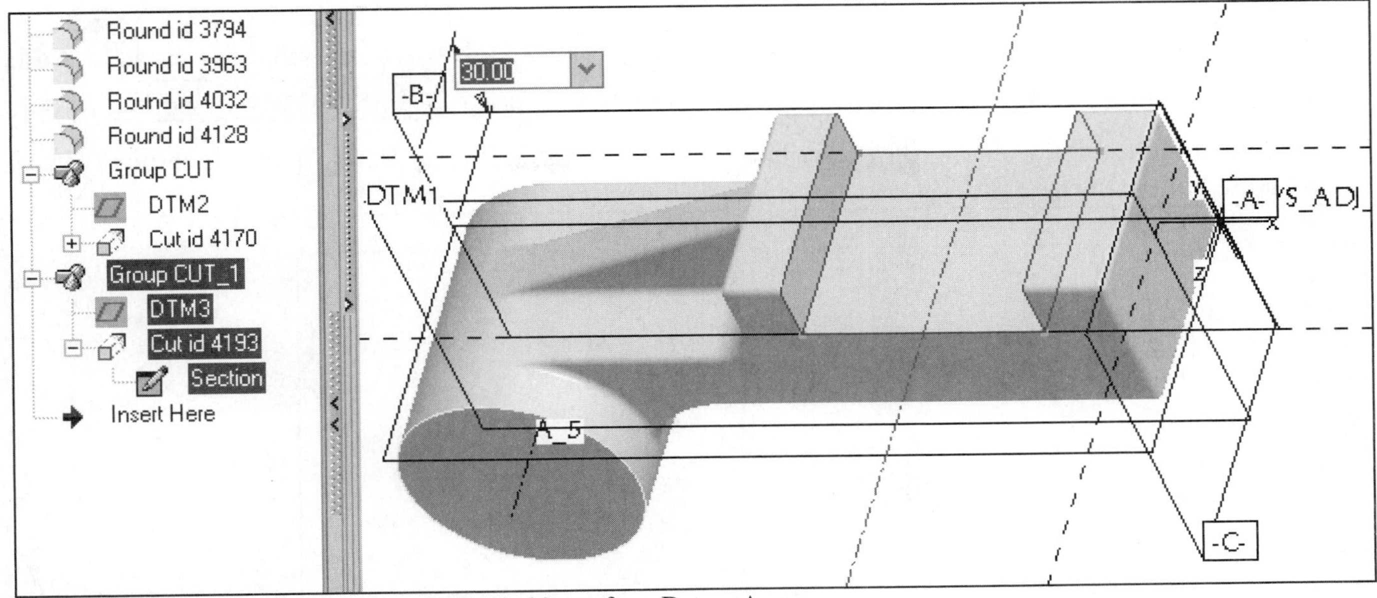

Figure 9.29(a) Slot/Cut Sketch Plane (DTM3), **30mm** from Datum A

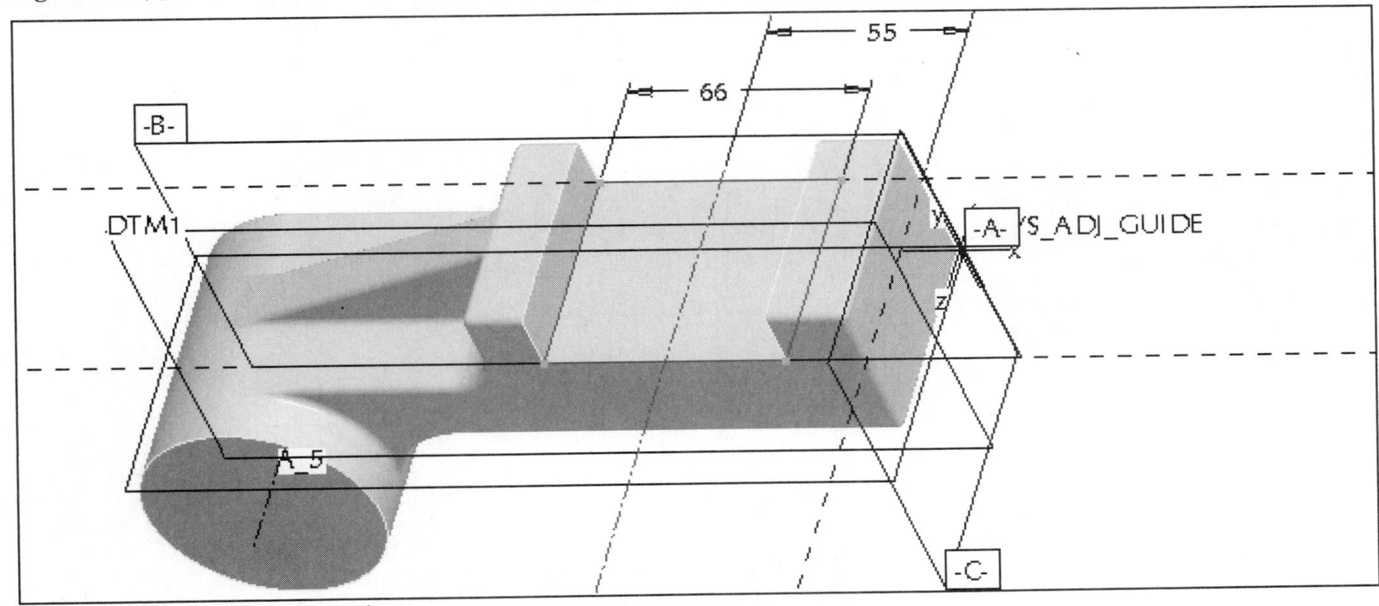

Figure 9.29(b) Slot/Cut Dimensions

MMB spin the model (Fig. 9.30) and save the view.

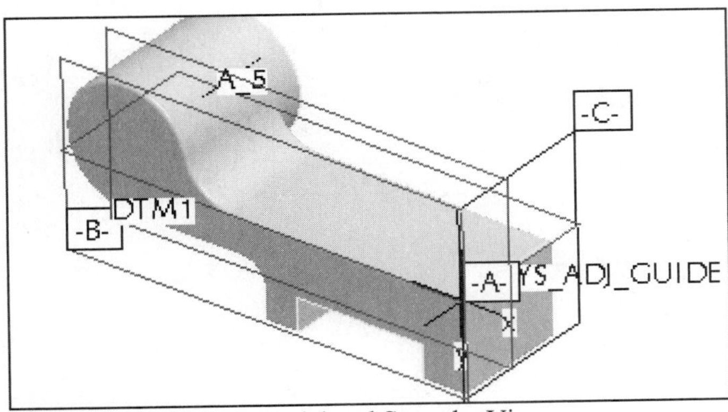

Figure 9.30 Spin the Model and Save the View

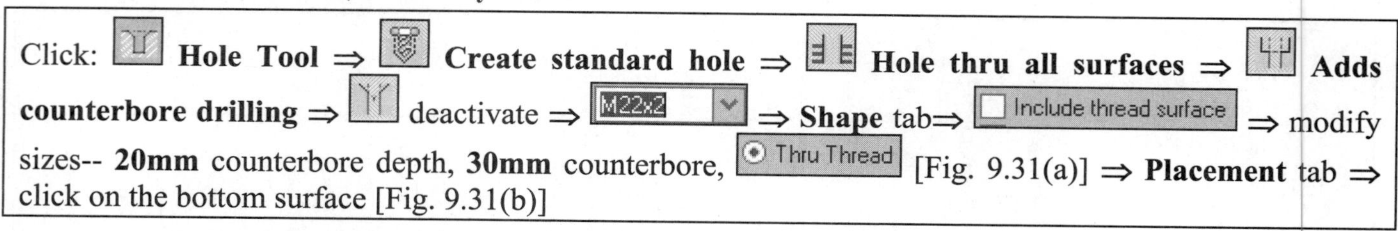

Click: **Hole Tool** ⇒ **Create standard hole** ⇒ **Hole thru all surfaces** ⇒ **Adds counterbore drilling** ⇒ deactivate ⇒ M22x2 ⇒ **Shape** tab⇒ Include thread surface ⇒ modify sizes-- **20mm** counterbore depth, **30mm** counterbore, Thru Thread [Fig. 9.31(a)] ⇒ **Placement** tab ⇒ click on the bottom surface [Fig. 9.31(b)]

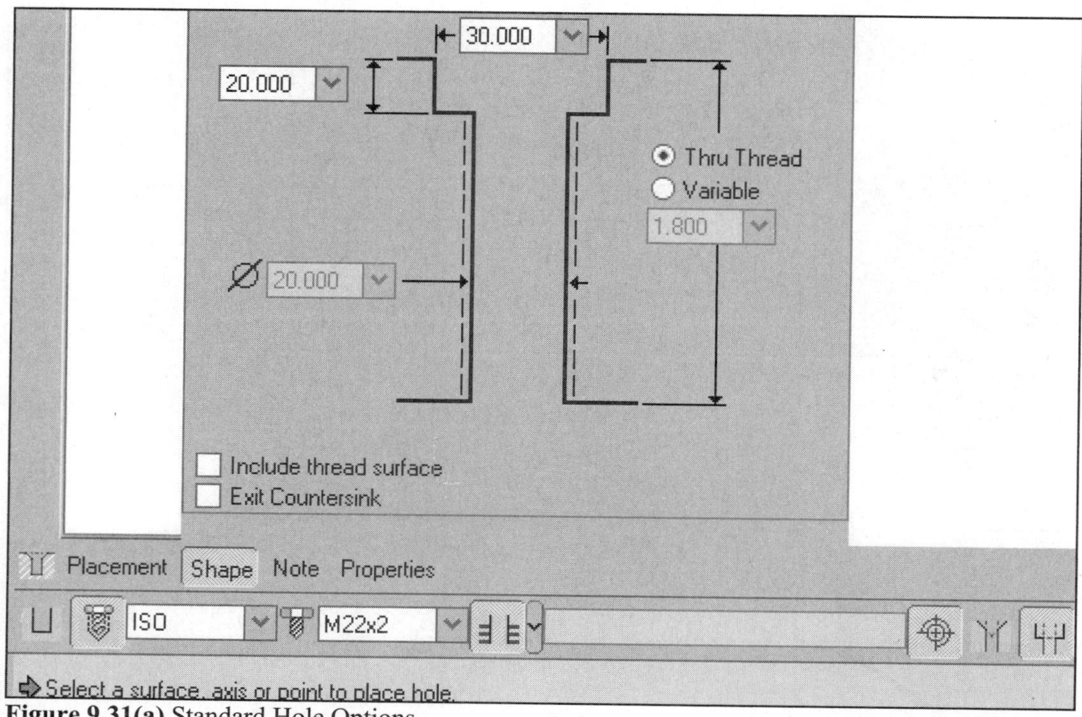

Figure 9.31(a) Standard Hole Options

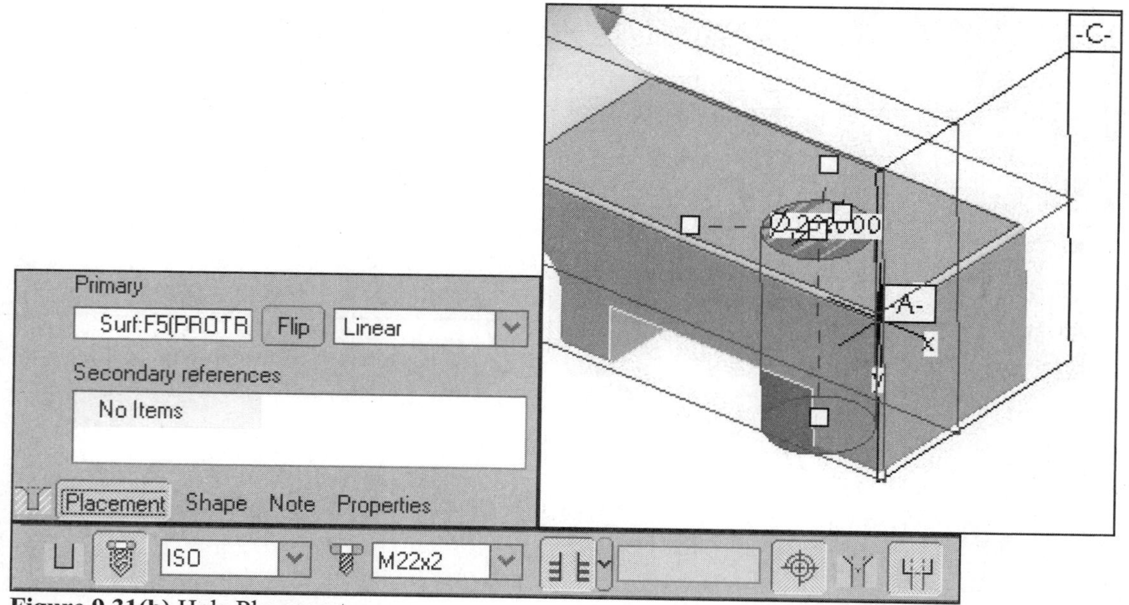

Figure 9.31(b) Hole Placement

Move the drag handles to **DTM1** and datum **C** respectively [Fig. 9.31(c)] ⇒ edit the distance to **DTM1** to **0** ⇒ edit the distance to datum **C** to **55** (if you edit the value on the model type **55** [Fig. 9.31(d)], if you edit the value in the dashboard Secondary references, type **–55**) [Fig. 9.31(e)] ⇒ **MMB** [Fig. 9.31(f)]

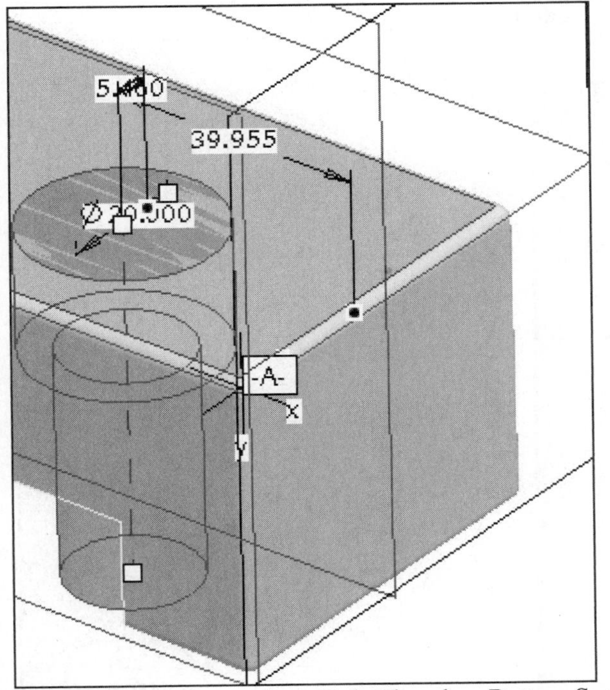

Figure 9.31(c) Counterbore Hole Placed on Bottom Surface

Figure 9.31(d) Hole Positioned to Design Values

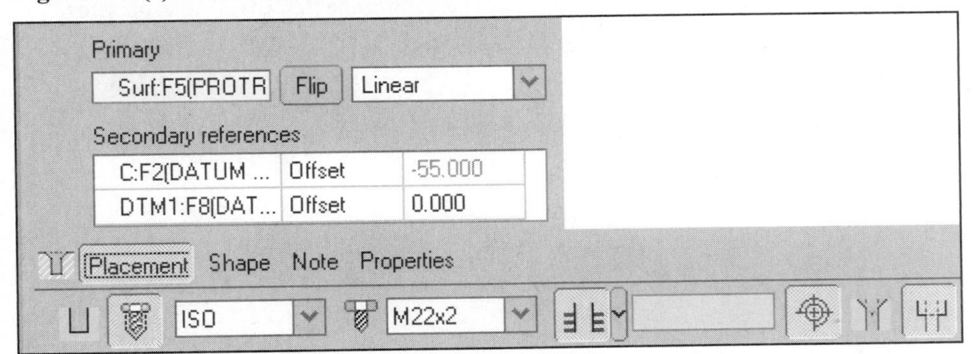

Figure 9.31(e) Hole Dashboard

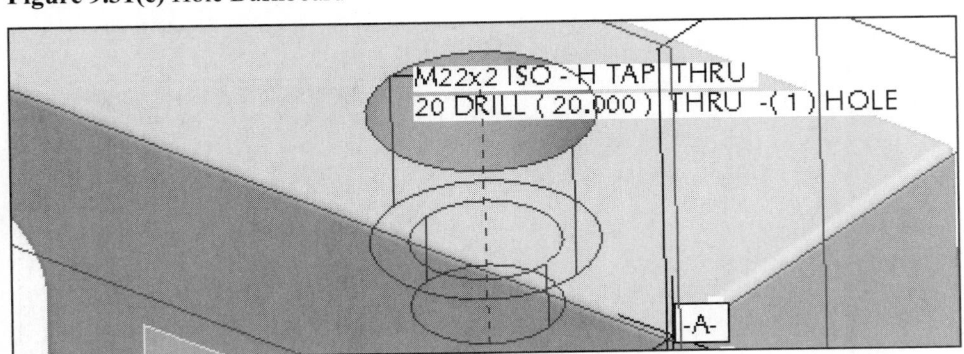

Figure 9.31(f) Completed Hole (with incorrect note)

Usually you would simply turn off the threads by toggling off the taped hole option: ***Taps the drilled hole****. The method used here was used so that manual editing of the note/callout and the hole's shape could be introduced. In Lesson 15, the holes for the Clamp Plate are created with the proper method.*

In Model Tree, click: **Settings** ⇒ **Tree Filters** ⇒ Display [Notes] and [Suppressed Objects] [Fig. 9.32(a)] ⇒ **Apply** ⇒ **OK** ⇒ in the Model Tree click **Hole id** and expand the tree ⇒ click on **Note_3** ⇒ **RMB** ⇒ **Properties** [Fig. 9.32(b)] ⇒ Note dialog box displays [Fig. 9.32(c)], click **Symbols** tab [Fig. 9.32(d)] ⇒ highlight all of the text in the Text area ⇒ **RMB** ⇒ **Delete**

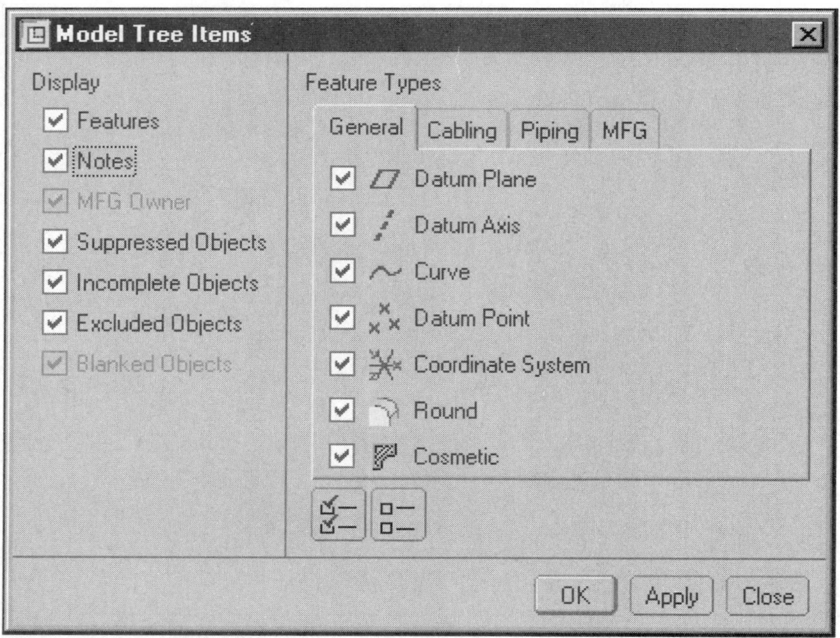

Figure 9.32(a) Model Tree Items Dialog Box

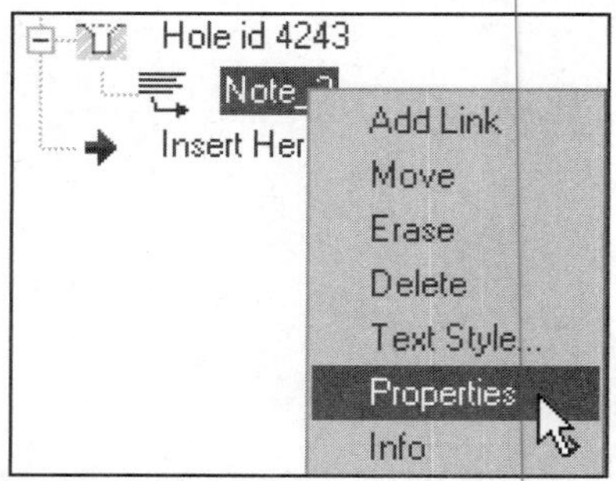

Figure 9.32(b) Note Properties

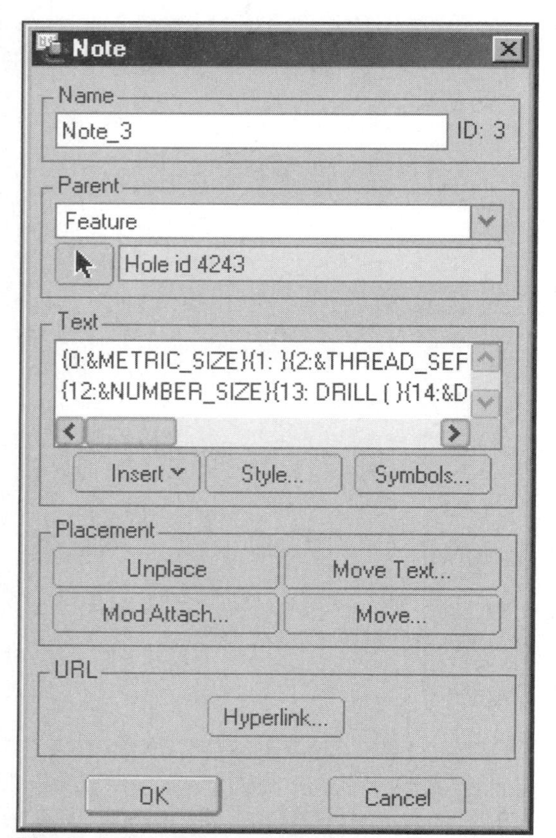

Figure 9.32(c) Note Dialog Box

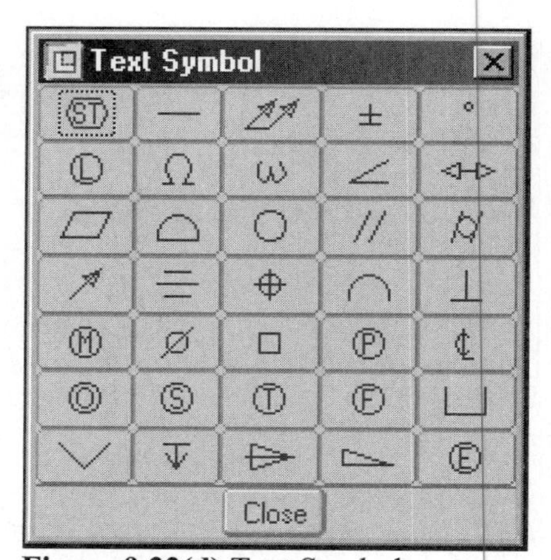

Figure 9.32(d) Text Symbol

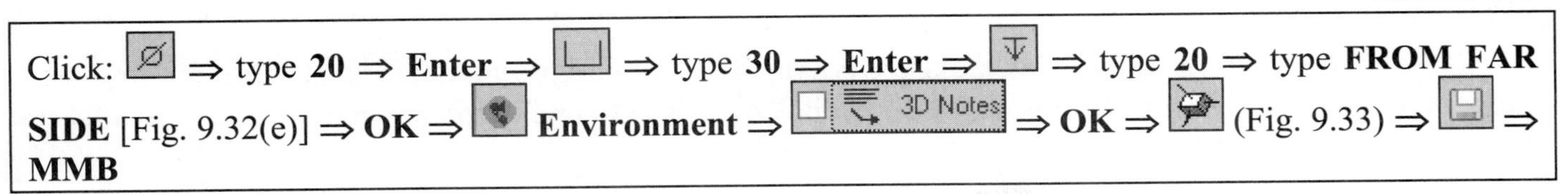

Click: [icon] ⇒ type **20** ⇒ **Enter** ⇒ [icon] ⇒ type **30** ⇒ **Enter** ⇒ [icon] ⇒ type **20** ⇒ type **FROM FAR SIDE** [Fig. 9.32(e)] ⇒ **OK** ⇒ [icon] **Environment** ⇒ 3D Notes ⇒ **OK** ⇒ [icon] (Fig. 9.33) ⇒ [icon] ⇒ **MMB**

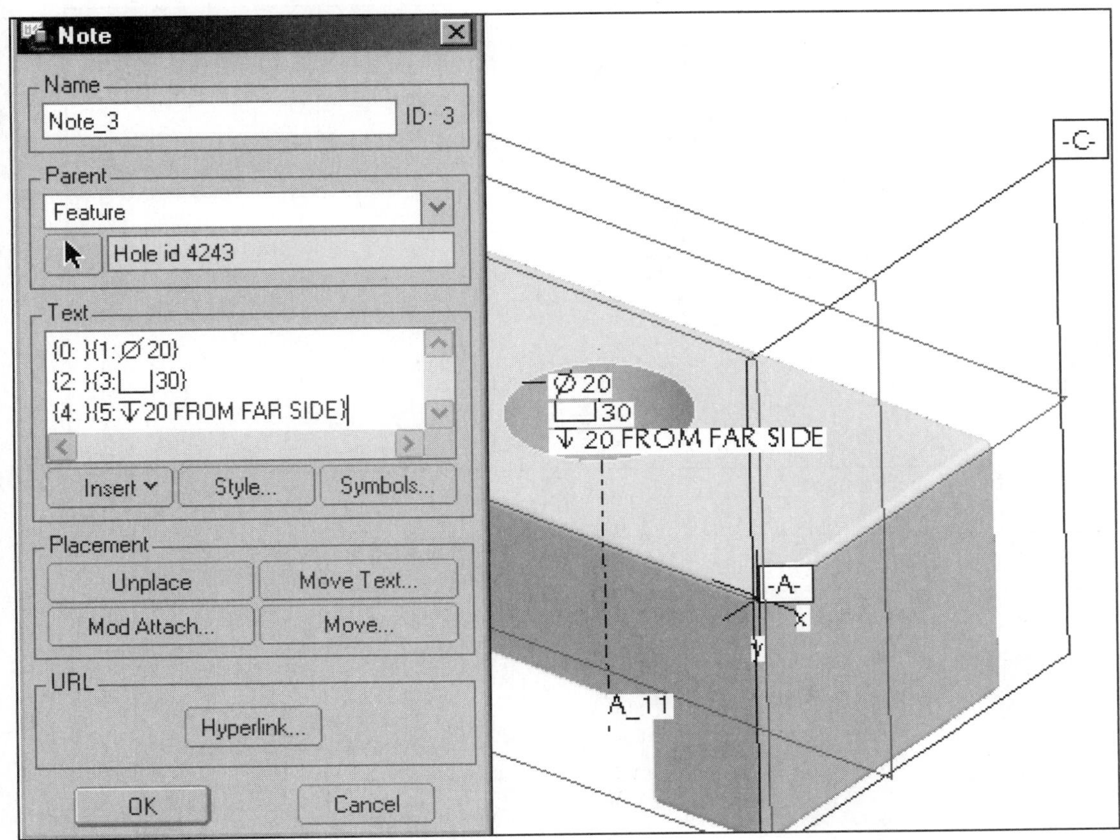

Figure 9.32(e) Edit Note

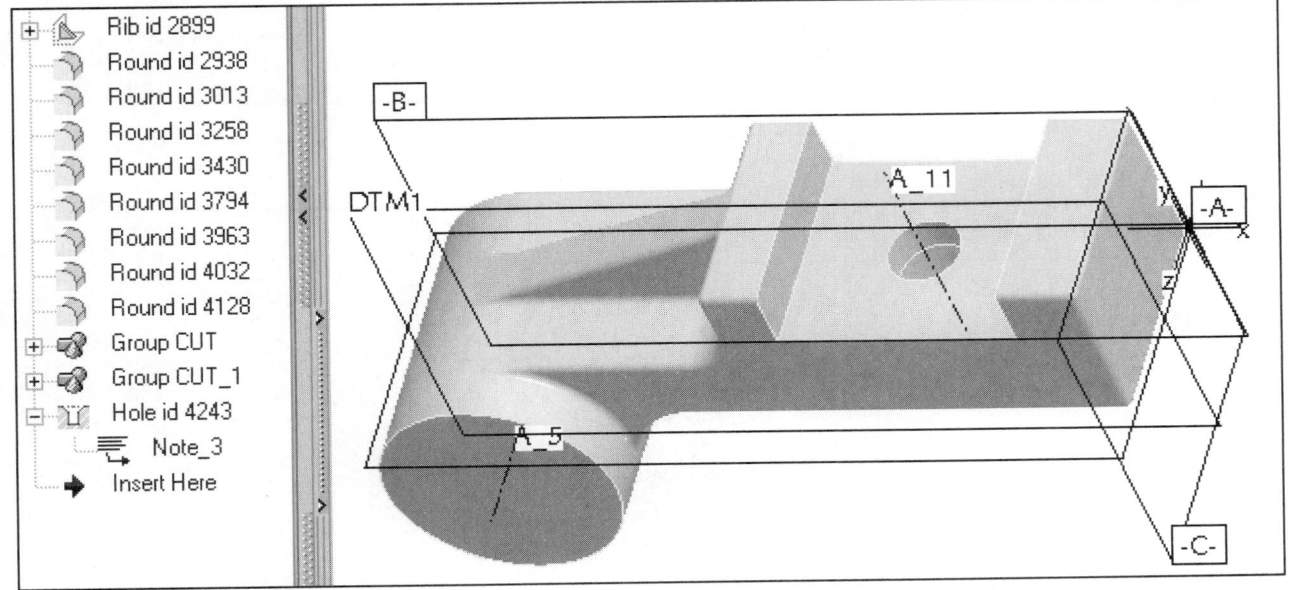

Figure 9.33 Completed Counterbore Hole

For the last feature, you will need to create a datum plane through the axis of the cylinder (A_5) and parallel to datum **C** for the sketching plane.

Click: **Datum Plane Tool** ⇒ pick axis **A_5** axis of the second protrusion [Fig. 9.34(a)] ⇒ press and hold **Ctrl** key ⇒ pick datum **C** [Fig. 9.34(b)] ⇒ click on Offset to open drop down options ⇒ **Parallel** ⇒ **OK** [Fig. 9.34(c)] ⇒ **LMB** to deselect

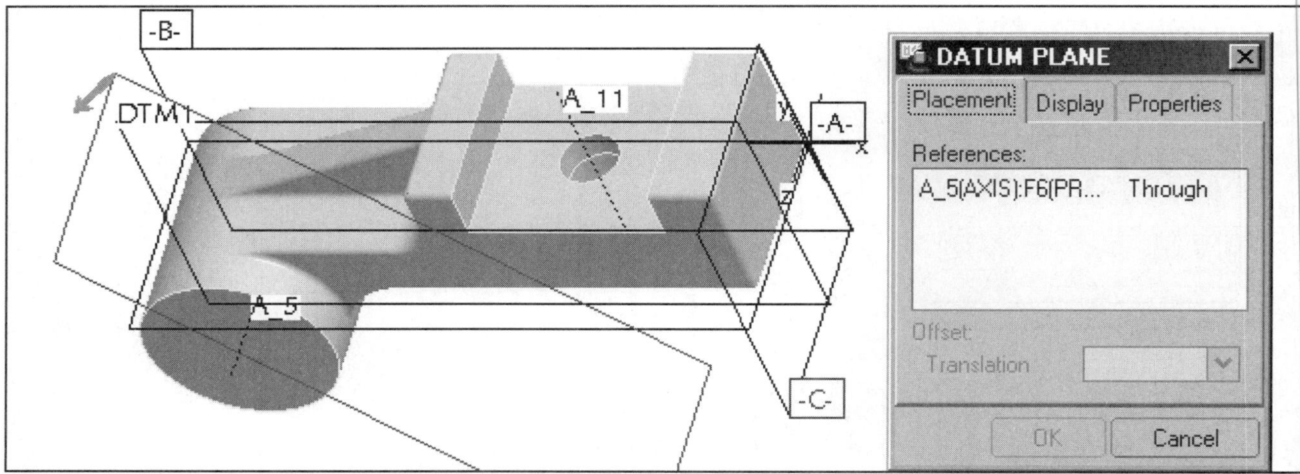

Figure 9.34(a) New Datum Plane Set to be Through A_5

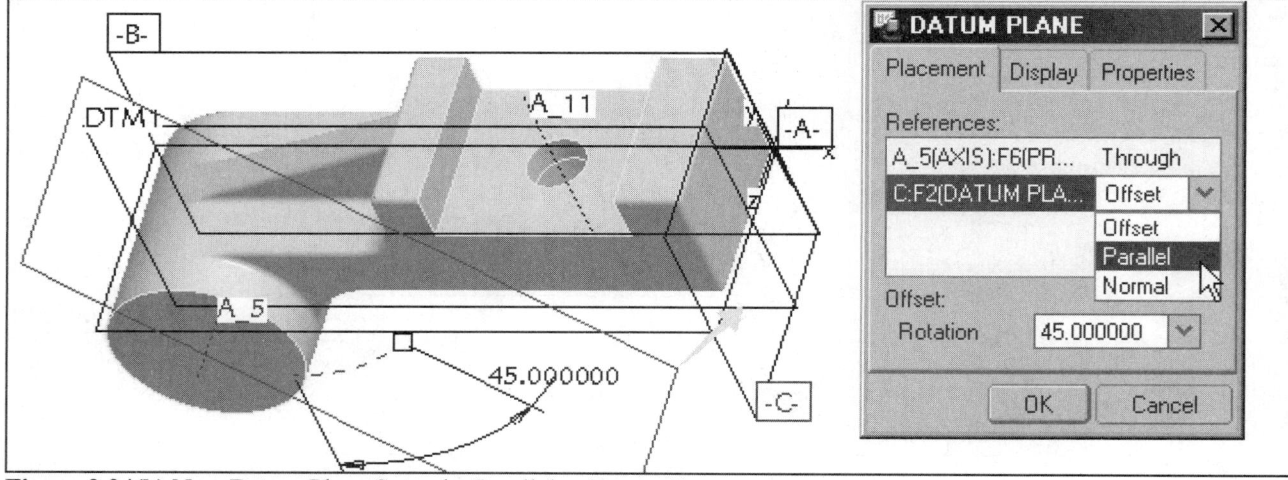

Figure 9.34(b) New Datum Plane Set to be Parallel to Datum C

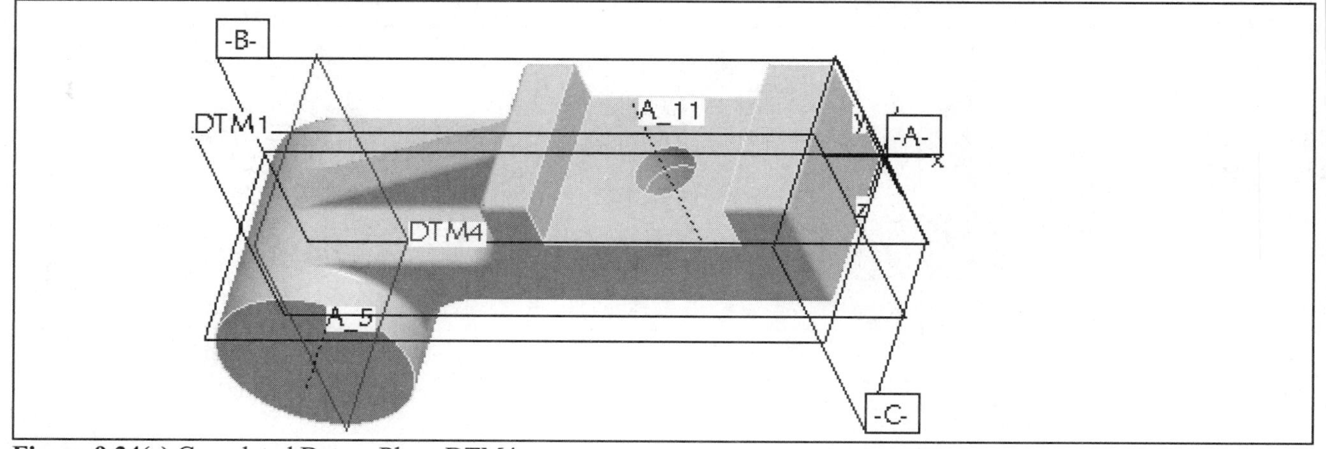

Figure 9.34(c) Completed Datum Plane DTM4

Click: **Tools** ⇒ **Environment** ⇒ Snap to Grid ⇒ **OK** ⇒ **Revolve Tool** ⇒ **Remove Material** ⇒ ⇒ Sketch Plane--- Plane: **DTM4** ⇒ **Flip** [Fig. 9.35(a)] ⇒ Reference: datum **A** ⇒ Orientation: **Top** ⇒ **Sketch** References dialog box displays ⇒ delete datum **A** as a reference ⇒ spin the model and add **Surf:F20(CUT)** and **A_5(AXIS)** as references [Fig. 9.35(b)] ⇒ **Close** ⇒ **Orient the sketching plane parallel to the screen**

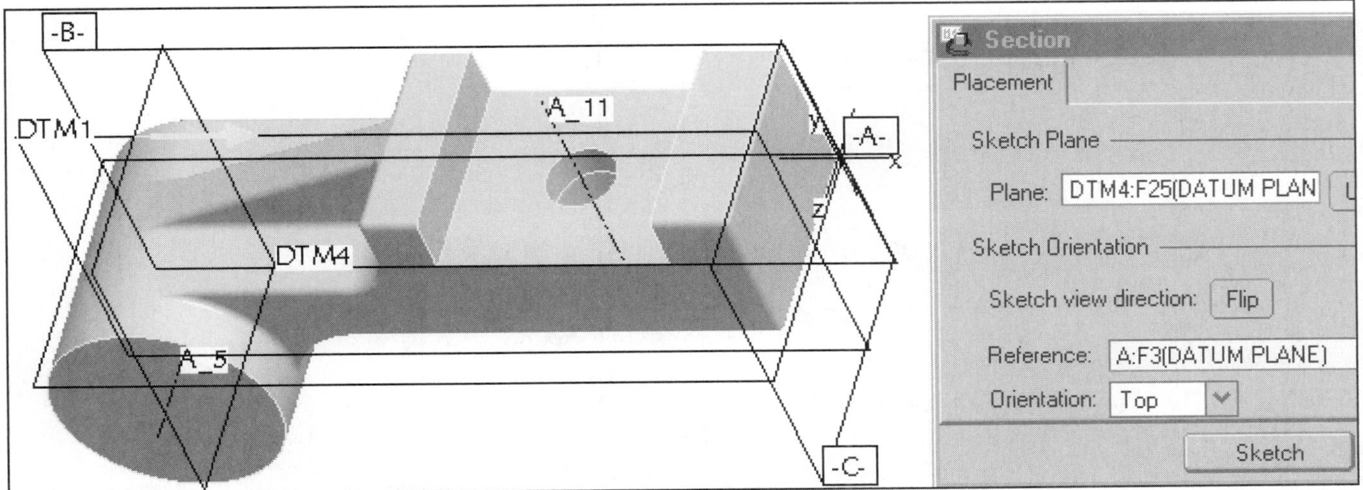

Figure 9.35(a) Cut Sketch Plane DTM4, Sketch Orientation- Datum A

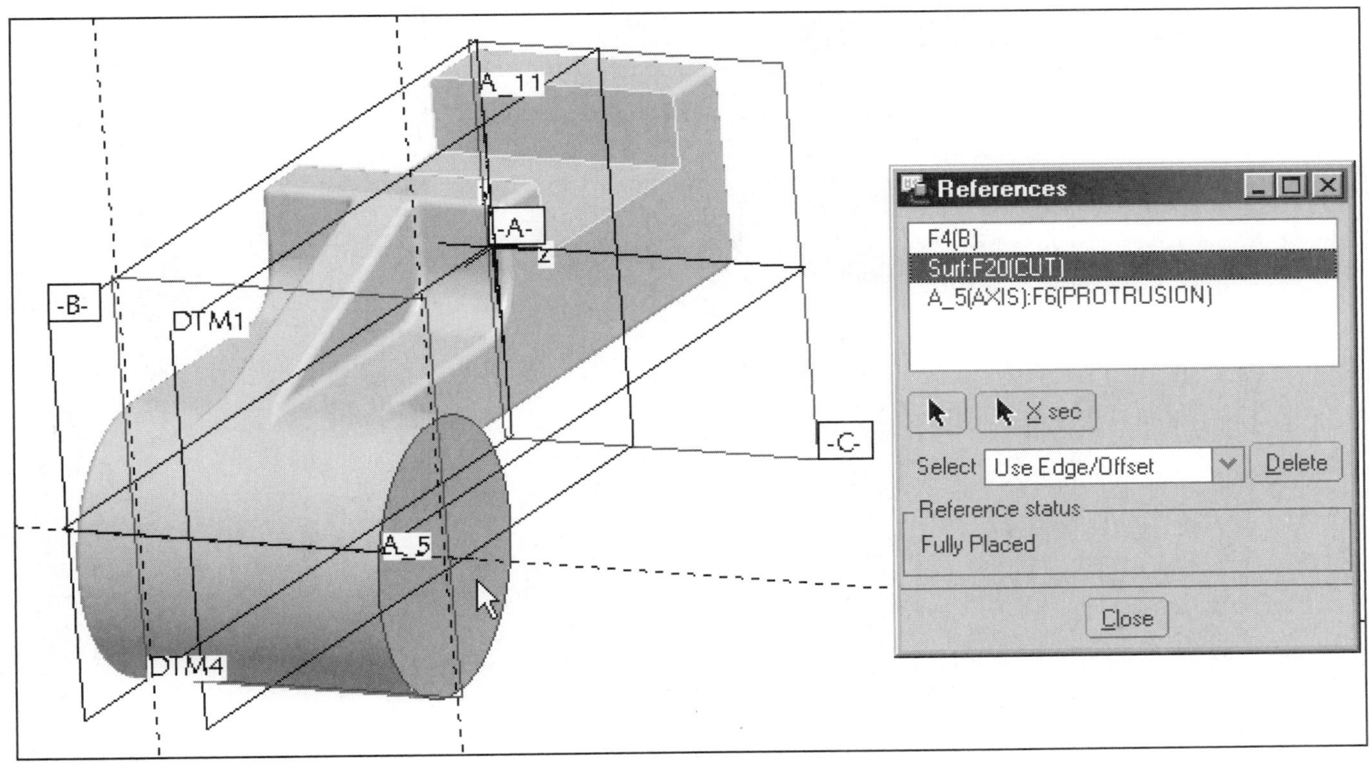

Figure 9.35(b) References

Sketch a horizontal centerline ⇒ sketch eight lines to form a *closed* section ⇒ add the required dimensions ⇒ reposition the dimensions ⇒ modify dimensions to the design values [Fig. 9.34(c-d)] ⇒ ✓ [Fig. 9.34(e)] ⇒ ✓ [Fig. 9.34(f)] ⇒ ⇒ **Standard Orientation** ⇒ ⇒ **MMB**

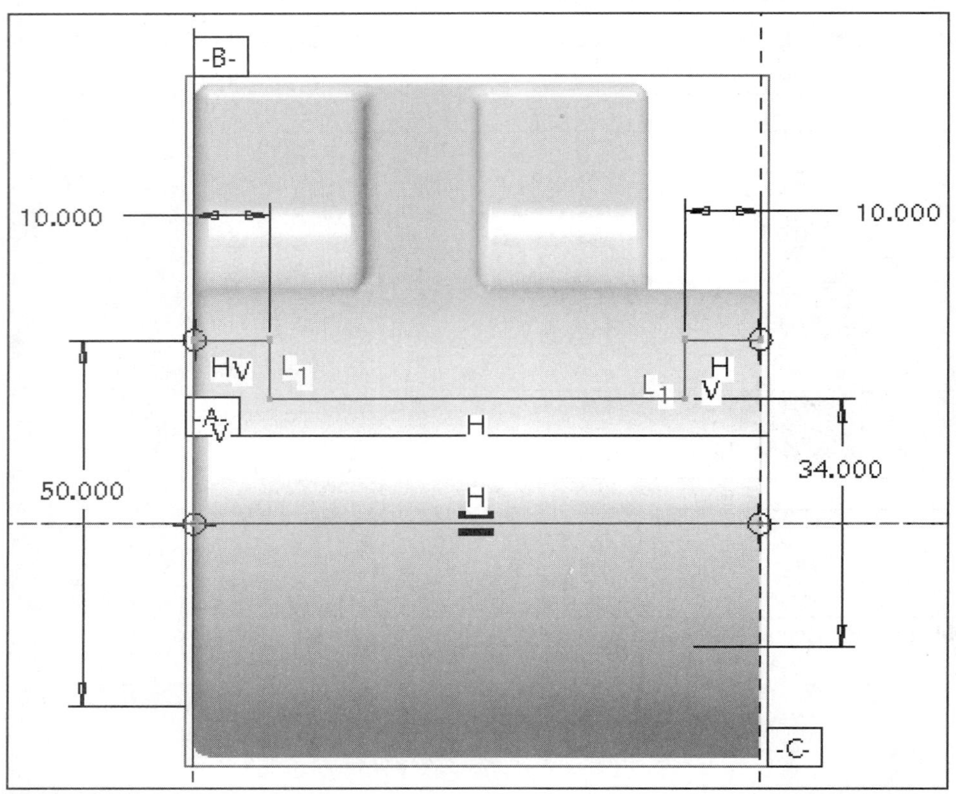

Figure 9.35(c) Section Sketch

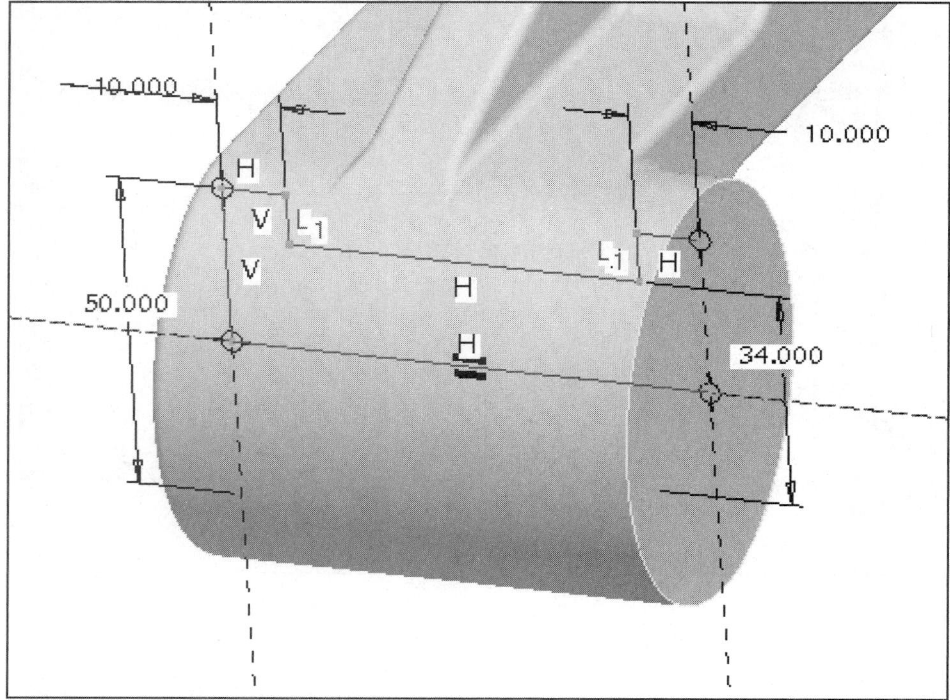

Figure 9.35(d) Section Sketch Rotated

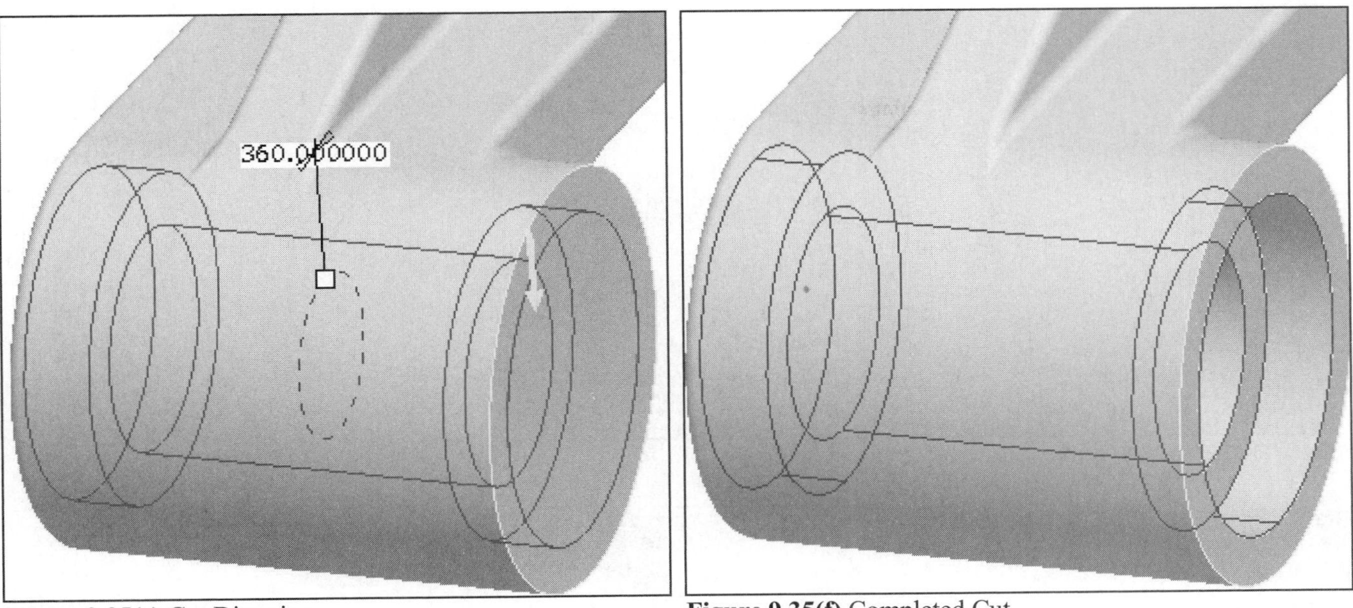

Figure 9.35(e) Cut Direction   **Figure 9.35(f)** Completed Cut

Write a relation to control the position of the counterbore hole. The hole must remain centered on the cut. You may add other relations that will control the cut position, and any other features you feel need to be controlled.

Click on **Group CUT_1** in the Model Tree ⇒ **RMB** ⇒ **Edit** ⇒ **Info** from the menu bar ⇒ **Switch Dimensions** ⇒ dimension **d201** *(your symbols will be different)* locates the center of the slot/cut from datum **C** [Fig. 9.36(a)] ⇒ click on **Hole id** in the Model Tree ⇒ **RMB** ⇒ **Edit** [Fig. 9.36(b)] ⇒ dimension **d206** *(your symbols will be different)* locates the center of the hole from datum **C** ⇒ **Tools** ⇒ **Relations** ⇒ type **d206=d201** [Fig. 9.36(c)] ⇒ **Ok** ⇒ **Info** ⇒ **Relations and Parameters** [Fig. 9.36(d)] ⇒ close quick sash ⇒ **Display object in standard orientation** ⇒ ⇒ **MMB** ⇒ **Info** ⇒ Switch Dimensions ⇒

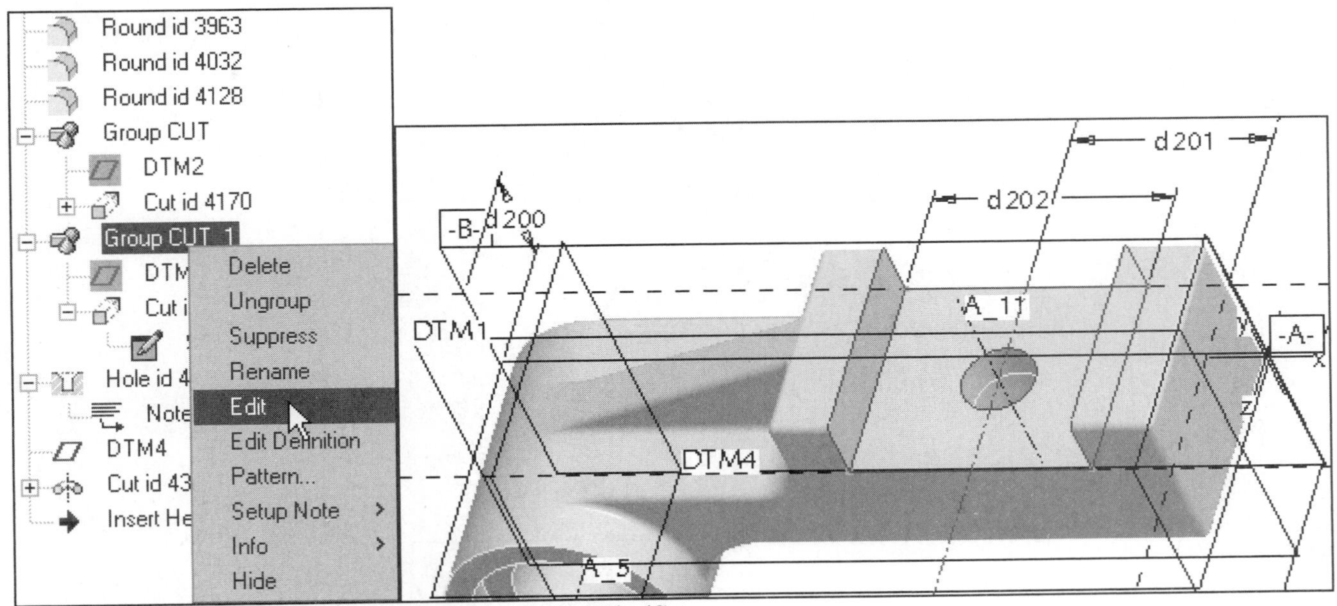

Figure 9.36(a) Switch Dimensions to Symbols for the Slot/Cut

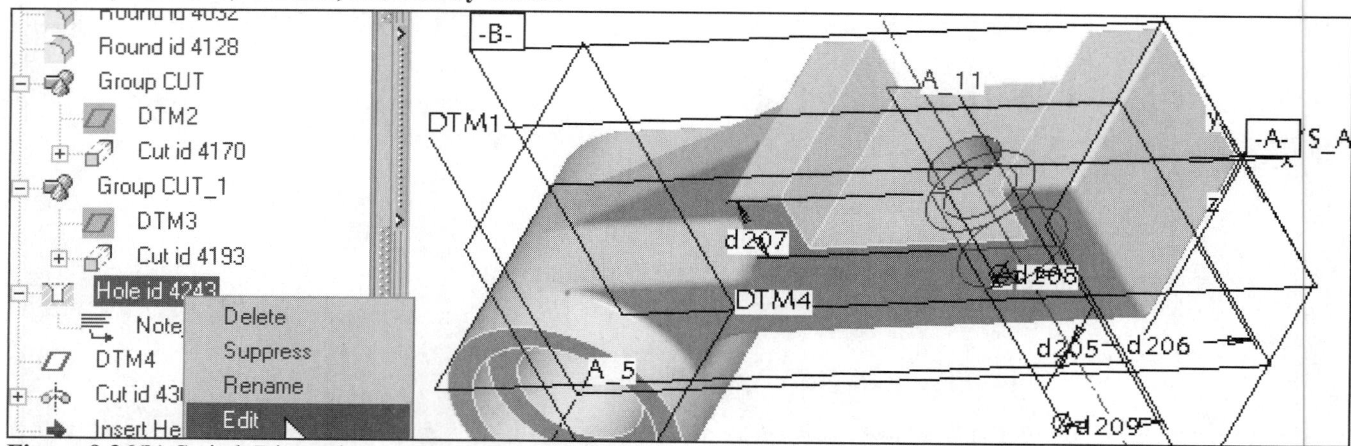

Figure 9.36(b) Switch Dimensions to Symbols for the Hole

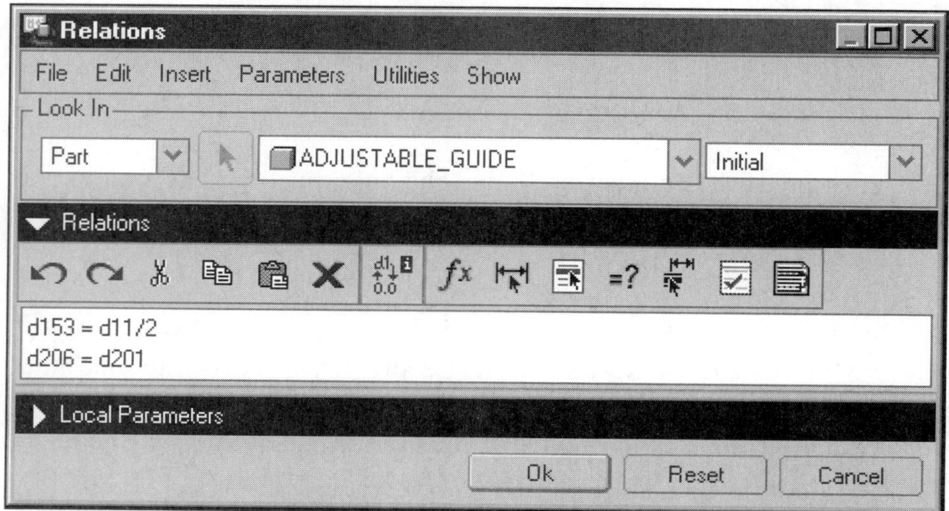

Figure 9.36(c) Add New Relation

Features Containing Relations/Parameters

ID	Name	Type	Actions
4243	---	HOLE	

Relation Table

Relation	Parameter	New Value
Relations for ADJUSTABLE_GUIDE:		
Initial Relations		
d153 = d11 /2	D153	3.000000e+01
d206 = d201	D206	5.500000e+01

Figure 9.36(d) Relations and Parameters Information

Family Tables are used any time a part or assembly has several unique iterations developed from the original model. The iterations must be considered separate parts, not just iterations of the same model. In this lesson, we will use an instance of a family table to create a casting of the Adjustable Guide, a version without the face cut, slot/cut, counterbore, or double-ended counterbore hole.

You will be creating a Family Table from the adjustable guide model. The (base) model is the **Generic**. Each variation is referred to as an **Instance**. When you create a Family Table, Pro/E allows you to *select dimensions,* which can vary between instances. You can also *select features* to add to the Family Table. Features can vary by being suppressed or resumed in an instance. When you are finished selecting items (e.g., dimensions, features, and parameters), the Family Table is automatically generated.

When adding features to the table, enter an **N** to suppress the feature, or a **Y** to resume the feature. Each instance must have a unique name. Here we will use ADJ_GUIDE_CASTING.

Family tables are spreadsheets, consisting of columns and rows. *Rows* contain instances and their corresponding values; *columns* are used for items. The column headings include the *instance name* and the names of all of the *dimensions, parameters, features, members,* and *groups* that were selected to be in the table. You use the Family Table dialog box to create and modify family tables.

Family tables include:

- The base object (generic object or *generic*) on which all members of the family are based.
- Dimensions, parameters, feature numbers, user-defined feature names, and assembly member names that are selected to be table-driven (*items*).
    - **Dimensions** are listed by name (for example, **d125**) with the associated symbol name (if any) on the line below it (for example, depth).
    - **Parameters** are listed by name (dim symbol).
    - **Features** are listed by feature number with the associated feature type (for example, [cut]) or feature name on the line below it. The generic model is the first row in the table. The table entries belonging to the generic can be changed only by modifying the actual part, suppressing, or resuming features; *you cannot change the generic model by editing its entries in the family table.*
- Names of all family members (*instances*) created in the table and the corresponding values for each of the table-driven items

Use the following commands to create a family table, click: **Tools ⇒ Family Table**--the Family Table: dialog box opens [Fig. 9.37(a)]

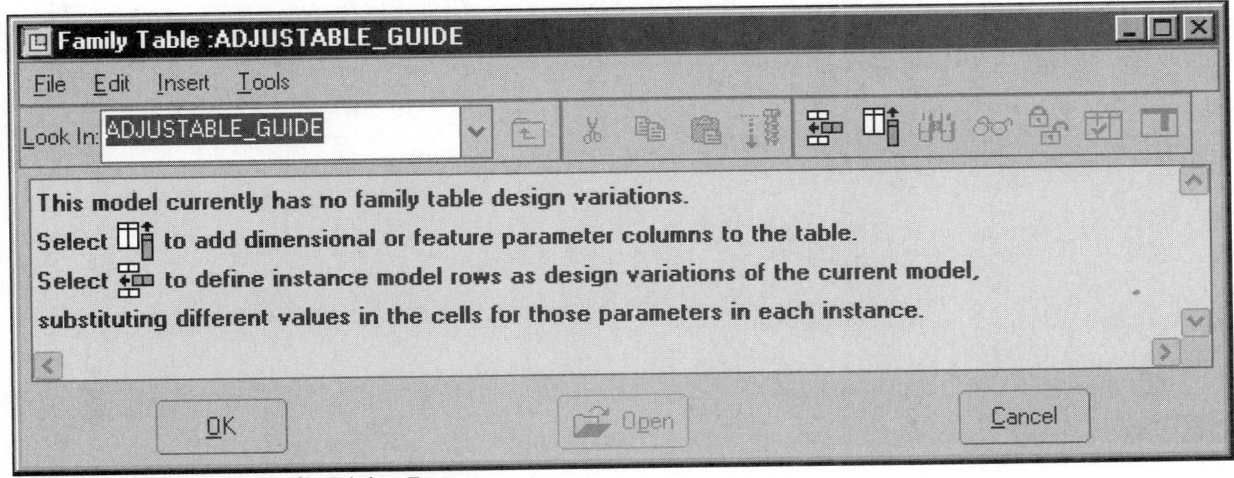

Figure 9.37(a) Family Table Dialog Box

Click: **Add/delete the table columns** ⇒ Family Items, Generic: dialog box displays ⇒ Feature (from the Add Item list) ⇒ select the three cuts and the hole from the model or the Model Tree [Fig. 9.37(b)] ⇒ **OK** from the Select dialog box ⇒ **MMB** ⇒ **MMB**

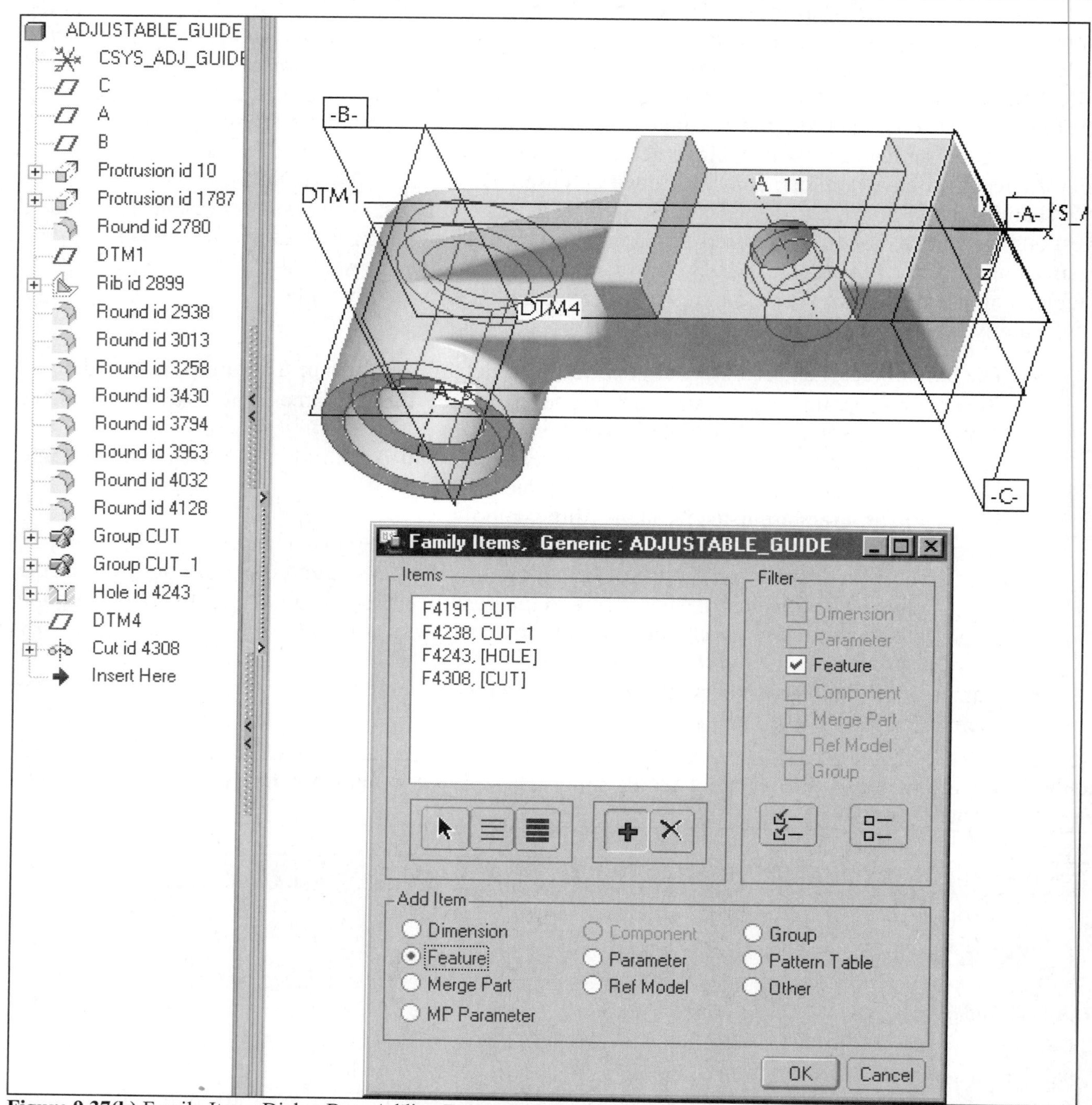

Figure 9.37(b) Family Items Dialog Box, Adding Features

Add one instance, click: **Insert a new instance at the selected row** [Fig. 9.37(c)] ⇒ click on the name of the first instance **ADJUSTABLE_GUIDE_INST** [Fig. 9.37(d)] ⇒ type **ADJ_GUIDE_CASTING** ⇒ make all features **N** [Fig. 9.37(e)] ⇒ **Verify** ⇒ **Verify** ⇒ **Close** ⇒ **OK** ⇒ ⇒ **MMB**

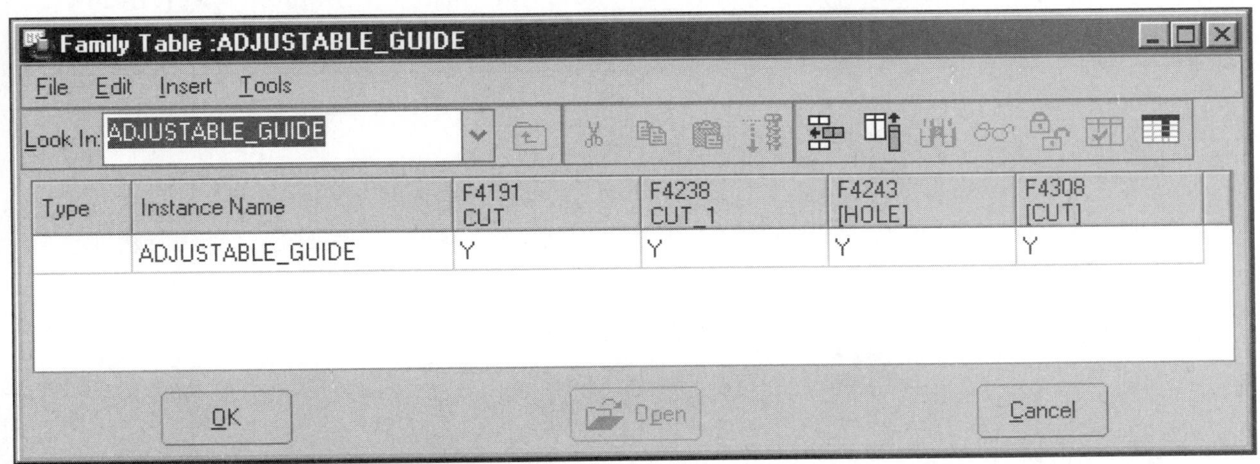

Figure 9.37(c) Family Table with Family Items

Family Table :ADJUSTABLE_GUIDE

File Edit Insert Tools

Look In: ADJUSTABLE_GUIDE

Type	Instance Name	F4191 CUT	F4238 CUT_1	F4243 [HOLE]	F4308 [CUT]
	ADJUSTABLE_GUIDE	Y	Y	Y	Y
	ADJUSTABLE_GUIDE_INST	*	*	*	*

OK Open Cancel

Figure 9.37(d) New Row and Instance. Rename the Instance

Family Table :ADJUSTABLE_GUIDE

File Edit Insert Tools

Look In: ADJUSTABLE_GUIDE

Type	Instance Name	F4191 CUT	F4238 CUT_1	F4243 [HOLE]	F4308 [CUT]
	ADJUSTABLE_GUIDE	Y	Y	Y	Y
	ADJ_GUIDE_CASTING	N	N	N	N

Y

N

*

OK Open Cancel

Figure 9.37(e) Rename the Instance to ADJ_GUIDE_CASTING and choose **N** for all of its Family Items

A Family Table controls whether a feature is present or not for a given design instance, not whether a feature is displayed, as with the layer display function.

Click: **Tools** ⇒ **Family Table** ⇒ click on **ADJ_GUIDE_CASTING** ⇒ **RMB** ⇒ Open ⇒ adjust your windows to view both models [Fig. 9.37(f)] ⇒ **Window** from the menu bar of ADJ_GUIDE_CASTING ⇒ **Activate** ⇒ **Window** ⇒ **Close** *the Instance window* ⇒ **Window** ⇒ **Activate** INSTANCE: GENERIC ⇒ [save icon] ⇒ **MMB**

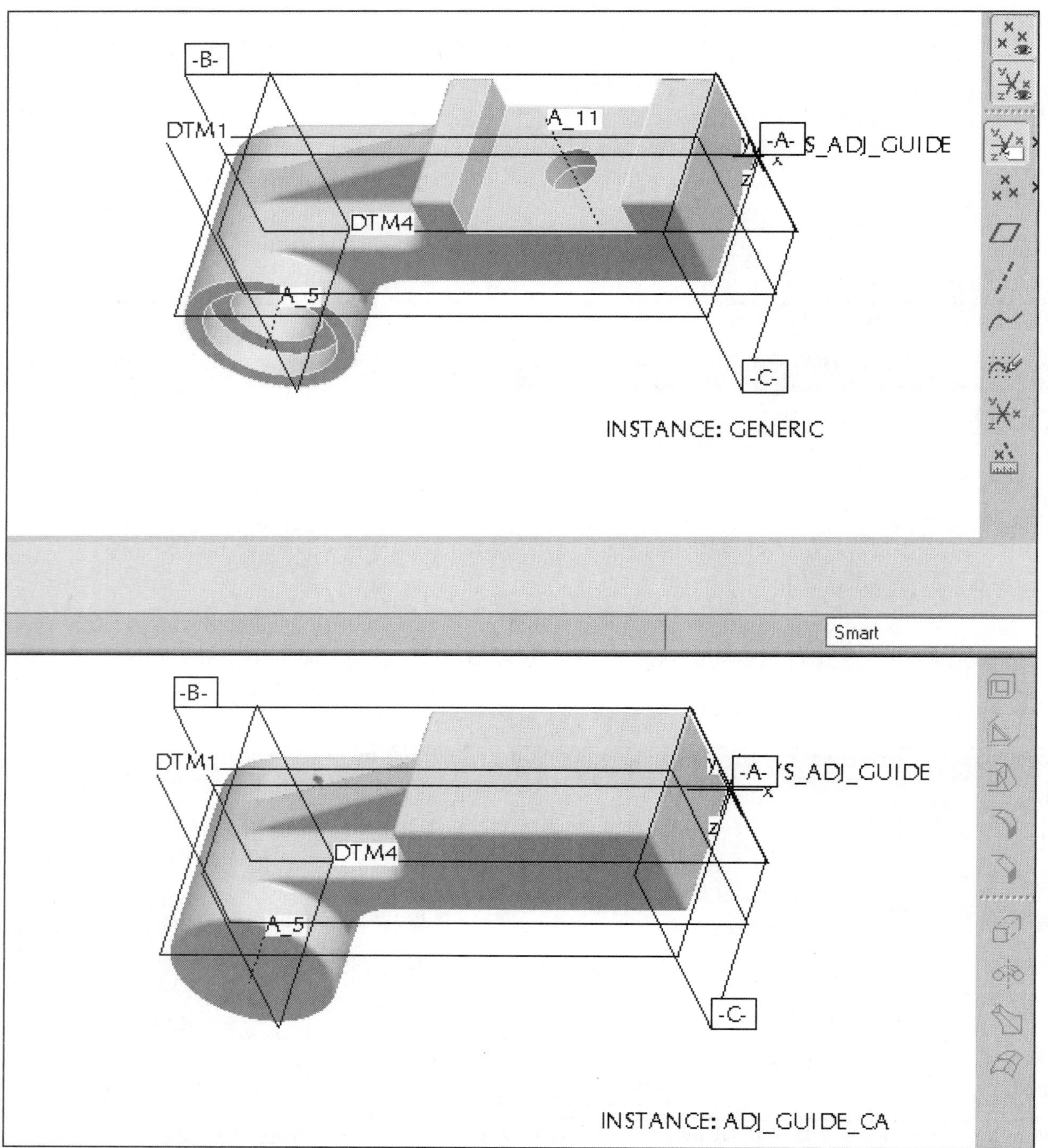

Figure 9.37(f) Design (Machined) Part (Instance: GENERIC), Workpiece (Casting) Part (Instance: ADJ_GUIDE_CASTING)

You now have two separate models, a casting (*workpiece*) and a machined part (*design part*). During the manufacturing process, the workpiece is merged (assembled) into the design part thereby creating a *manufacturing model*. The difference between the two files is the difference between the volume of the casting and the volume of the machined part. During the design of a typical part there are many modifications made to the design. The ability to make changes without causing failures is important. "Flexing" the model, changing and editing dimension values to see if the model integrity withstands these modifications, establishes how robust your design is.

Click: ⇒ double-click on the models first protrusion to see the dimensions ⇒ double-click on the **60** (depth) value ⇒ type **80** *(if it does not fail, try 100)* [Fig. 9.38(a)] ⇒ **Enter** ⇒ **Regenerates Model** ⇒ Failure Diagnostics Dialog box opens [Fig. 9.38(b)] ⇒ FEATURE #13 (ROUND), failed because the geometry was overlapping ⇒

Menu Manager
RESOLVE FEAT
Undo Changes
Investigate
Fix Model
Quick Fix
Done/Return

Click: **Undo Changes** ⇒ **Confirm**

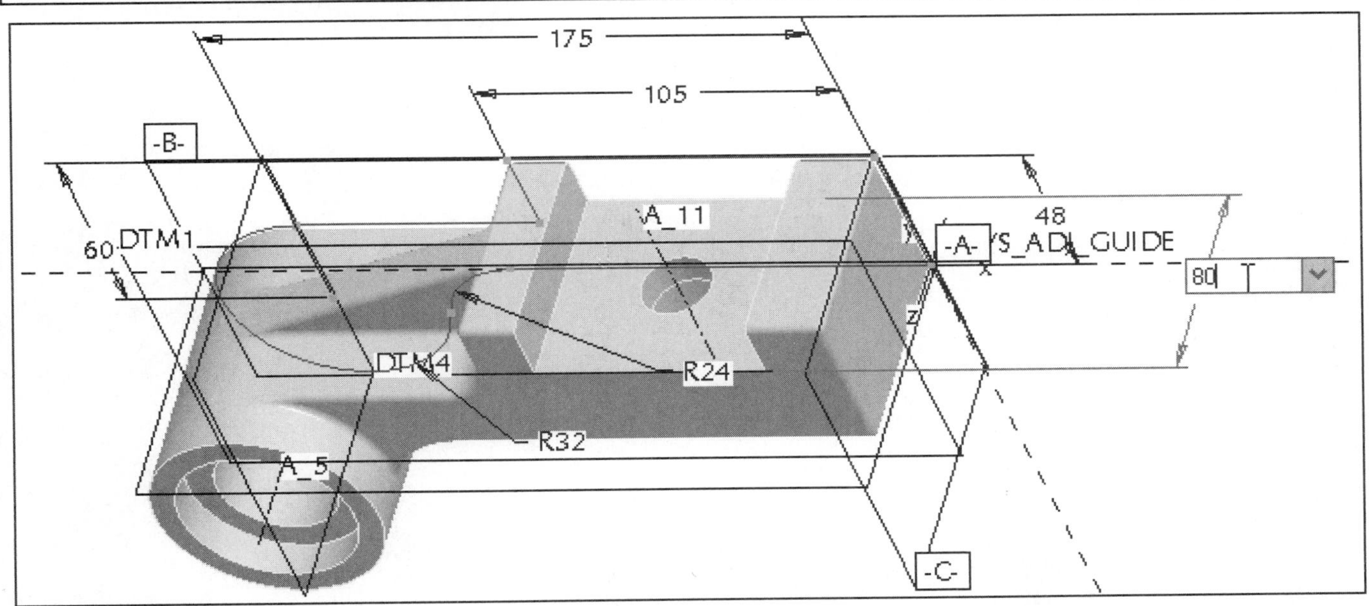

Figure 9.38(a) Edit the Dimension

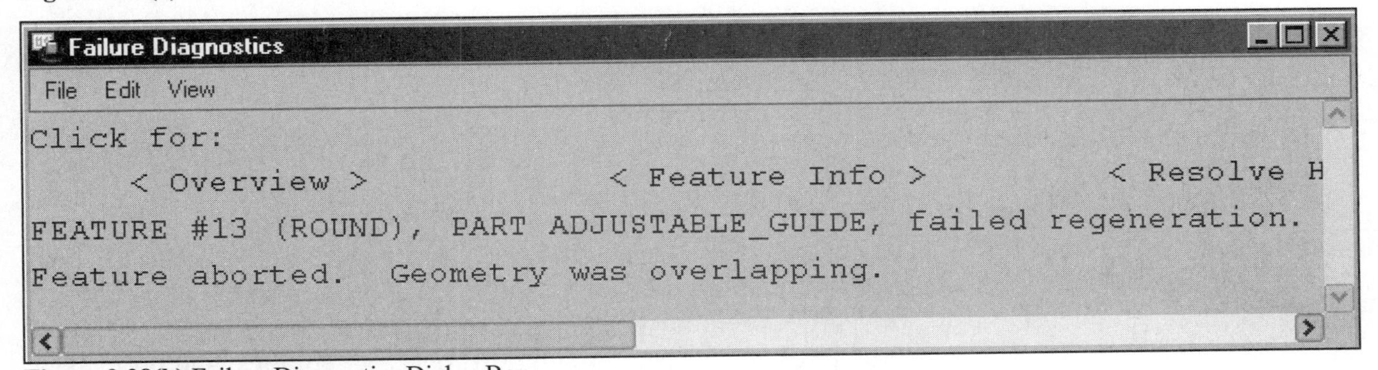

Figure 9.38(b) Failure Diagnostics Dialog Box

Double-click on the **60** value again and type **70** ⇒ **Enter** ⇒ **Regenerates Model** ⇒ **View** ⇒ **Shade** [Fig. 9.38(c)] ⇒ ⇒ **TOP** note that the rib and the counterbore hole are still centered [Fig. 9.38(d)] ⇒ ⇒ return the model to the original design size, click on the first protrusion in the Model Tree ⇒ **RMB** ⇒ **Edit** ⇒ double-click on the **70** value and type **60** ⇒ **Enter** ⇒ [Fig. 9.38(e)]

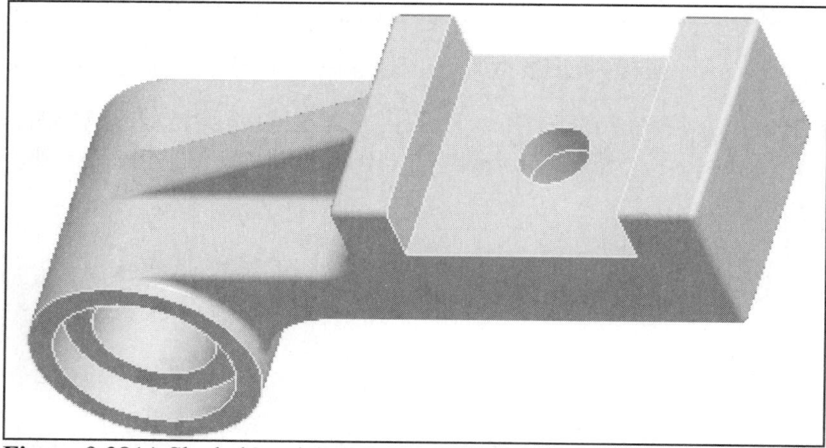

Figure 9.38(c) Shaded Model with new Width Value

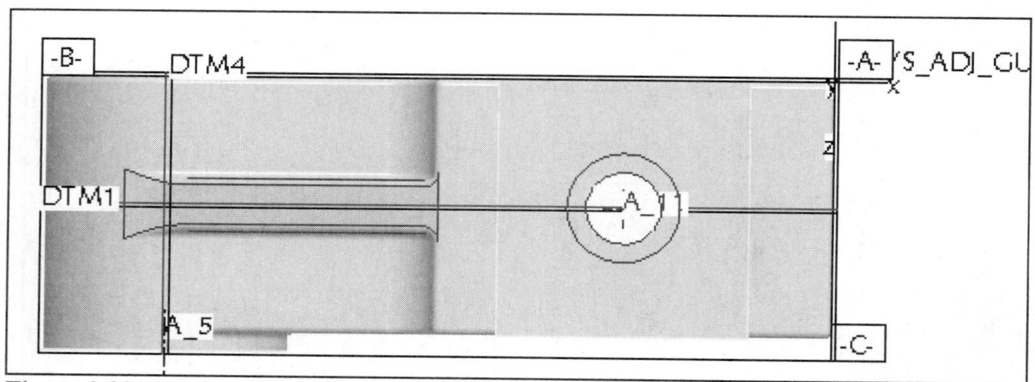

Figure 9.38(d) Orient the Model to View from the Top

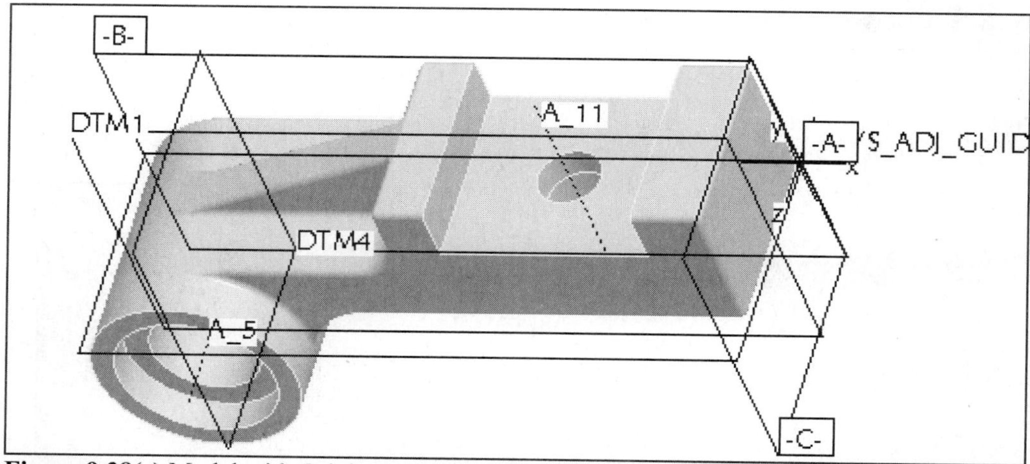

Figure 9.38(e) Model with Original Design Dimensions

You have completed the lesson part. Continue with the Lesson Project. A variety of different failures and responses will be encountered throughout the remaining text parts, assemblies, and projects.

Lesson 9 Project

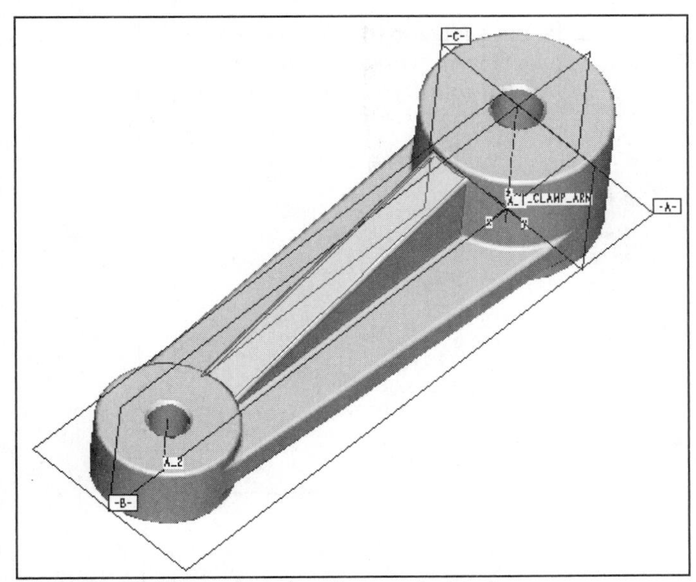

Figure 9.39 Clamp Arm (casting-workpiece)

Figure 9.40 Purchased Swing Clamp Assembly

Clamp Arm

The Clamp Arm is a cast part that requires commands similar to those used for the Adjustable Guide. Create the part shown in Figures 9.39 through 9.48. The Clamp Arm is used in the assembly of Lessons 15 and 16. Create two versions of the Clamp Arm; one with all cast surfaces and the other with machined ends (use a Family Table). Write a relation to control the position of the horizontal ribs. They must remain at the center of the smaller circular end. Include cosmetic threads where required. Machine the top and bottom surfaces of both circular protrusions (**.10** off either end).

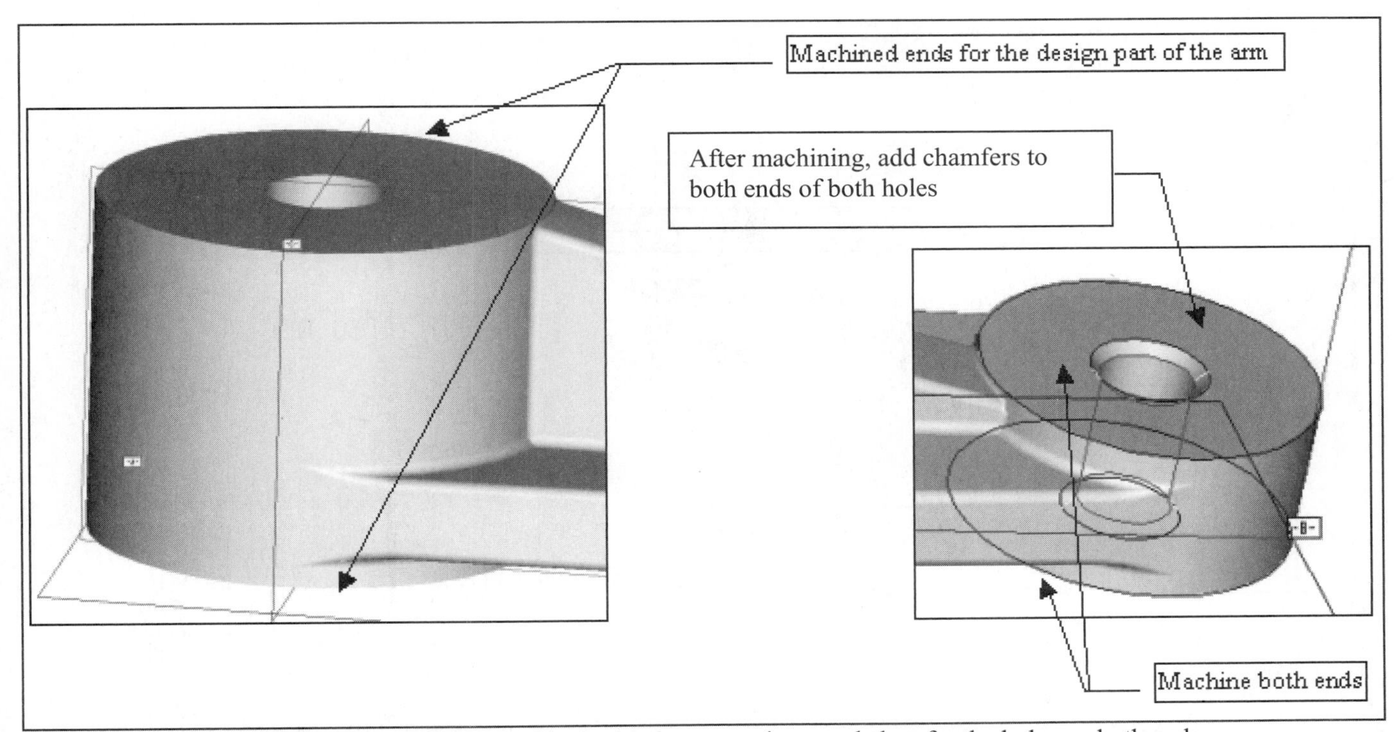

Figure 9.41(a-b) Machine the top and bottom of both circular protrusions, and chamfer the holes on both ends.

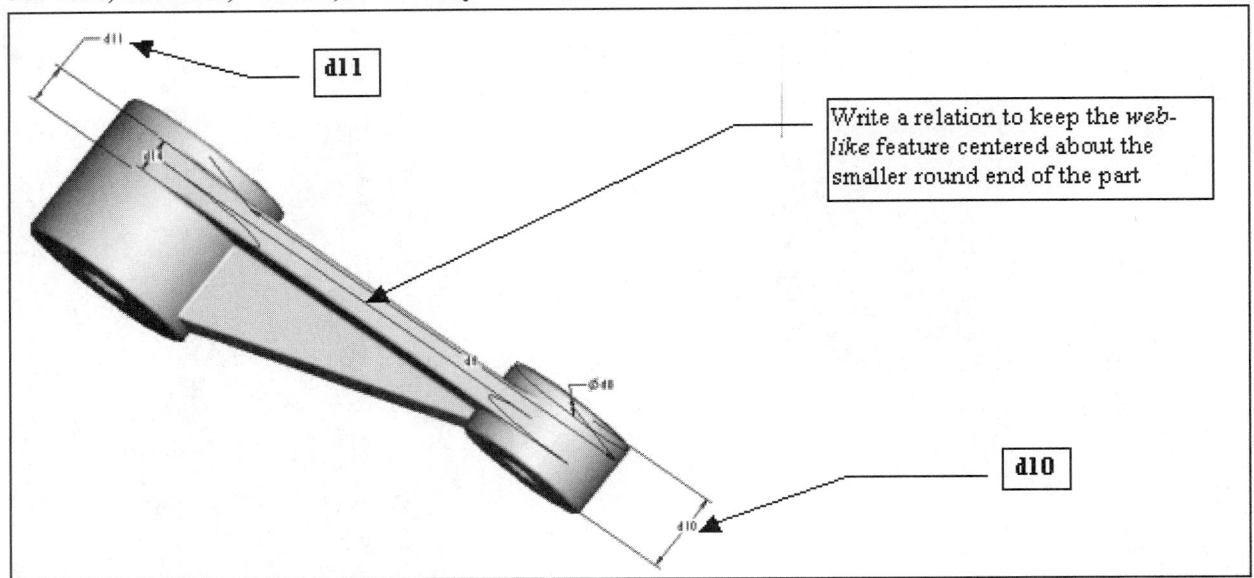

Figure 9.42 Relation **D11=D10/2**

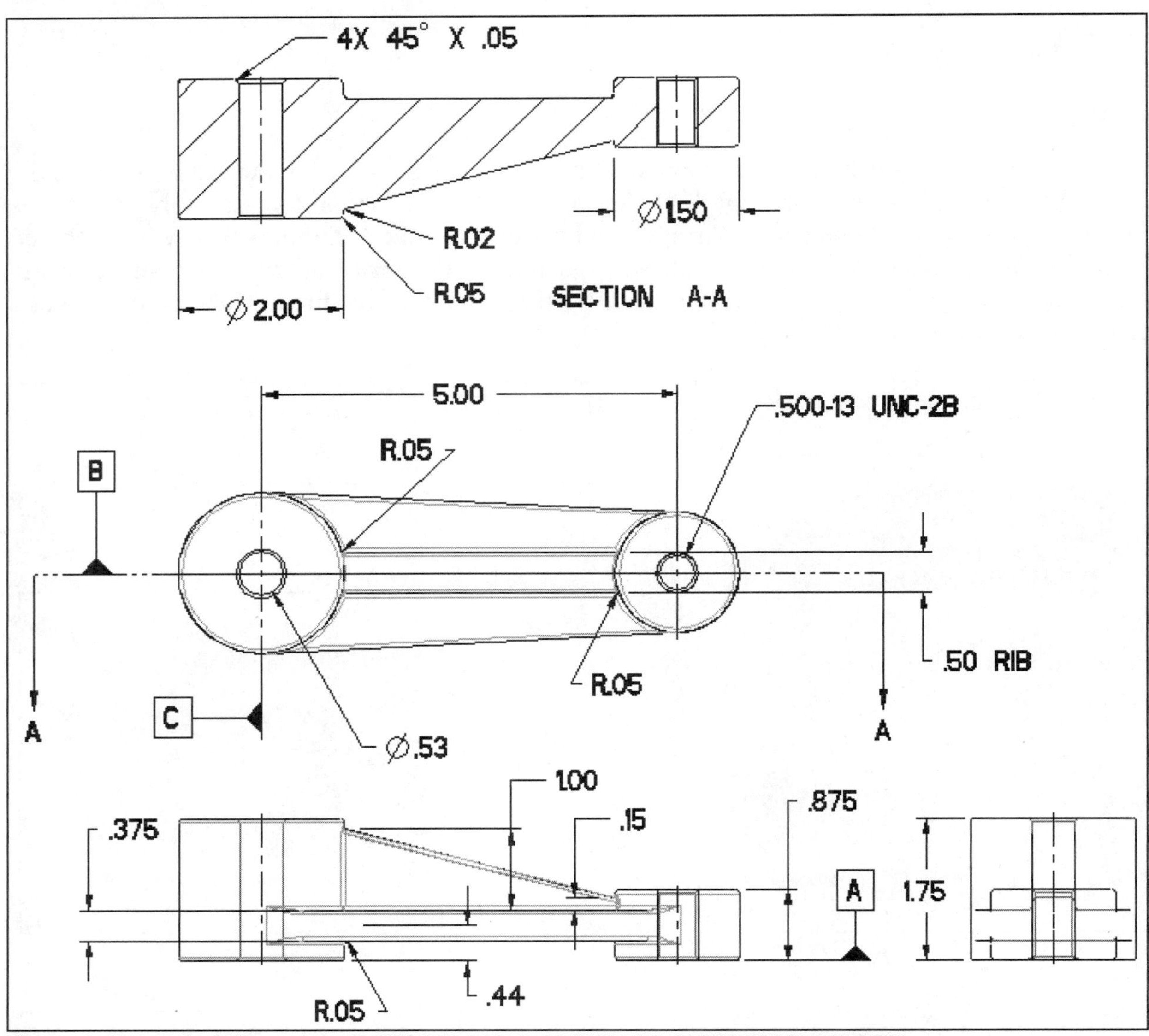

Figure 9.43 Detail of Clamp Arm (ROUNDS ARE R.02 UNLESS OTHERWISE NOTED)

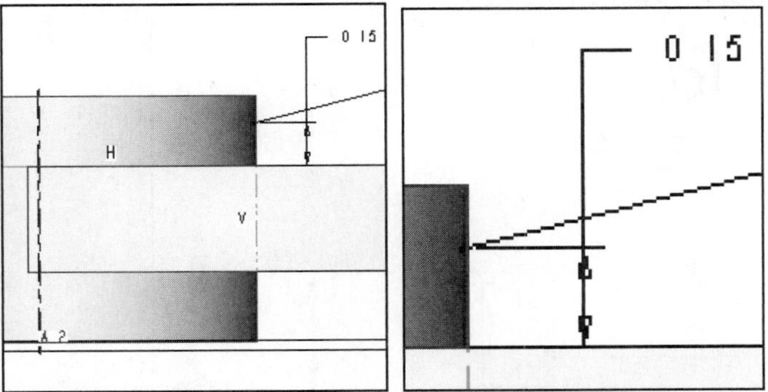

Figure 9.44(a-b) Rib Section Sketch, **.15** to rib *before* the round is added

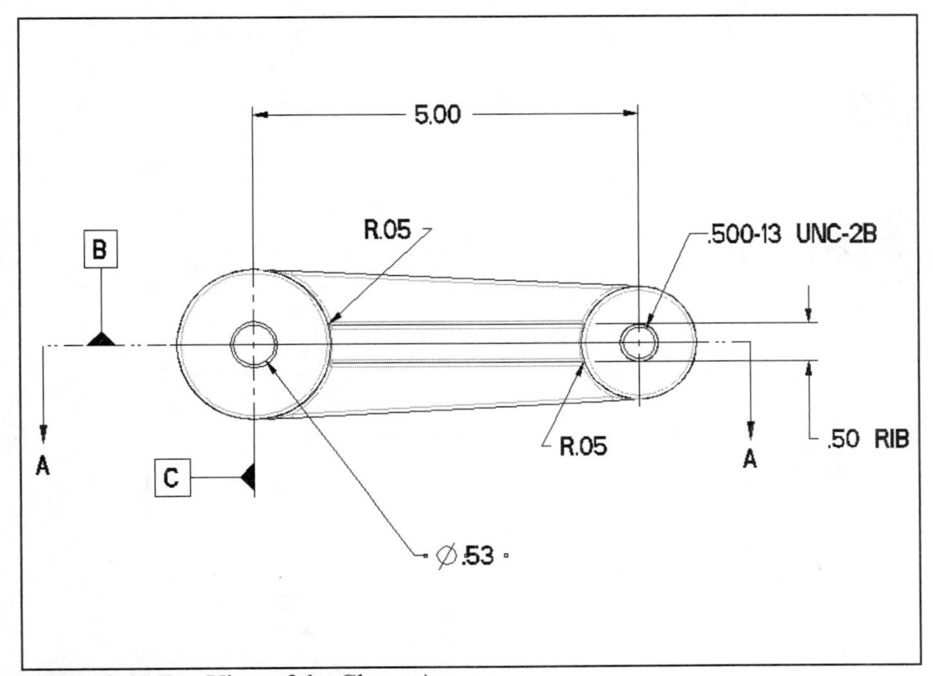

Figure 9.45 Top View of the Clamp Arm

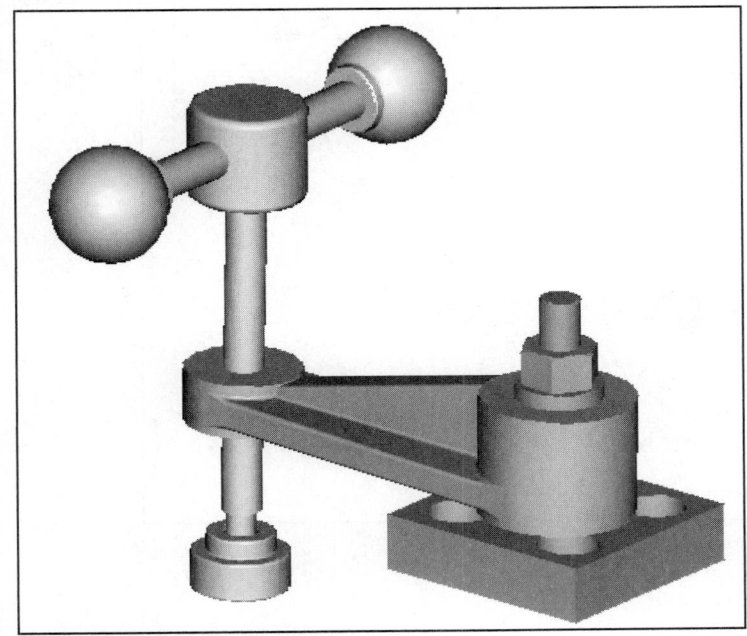

Figure 9.46 Swing Clamp Assembly

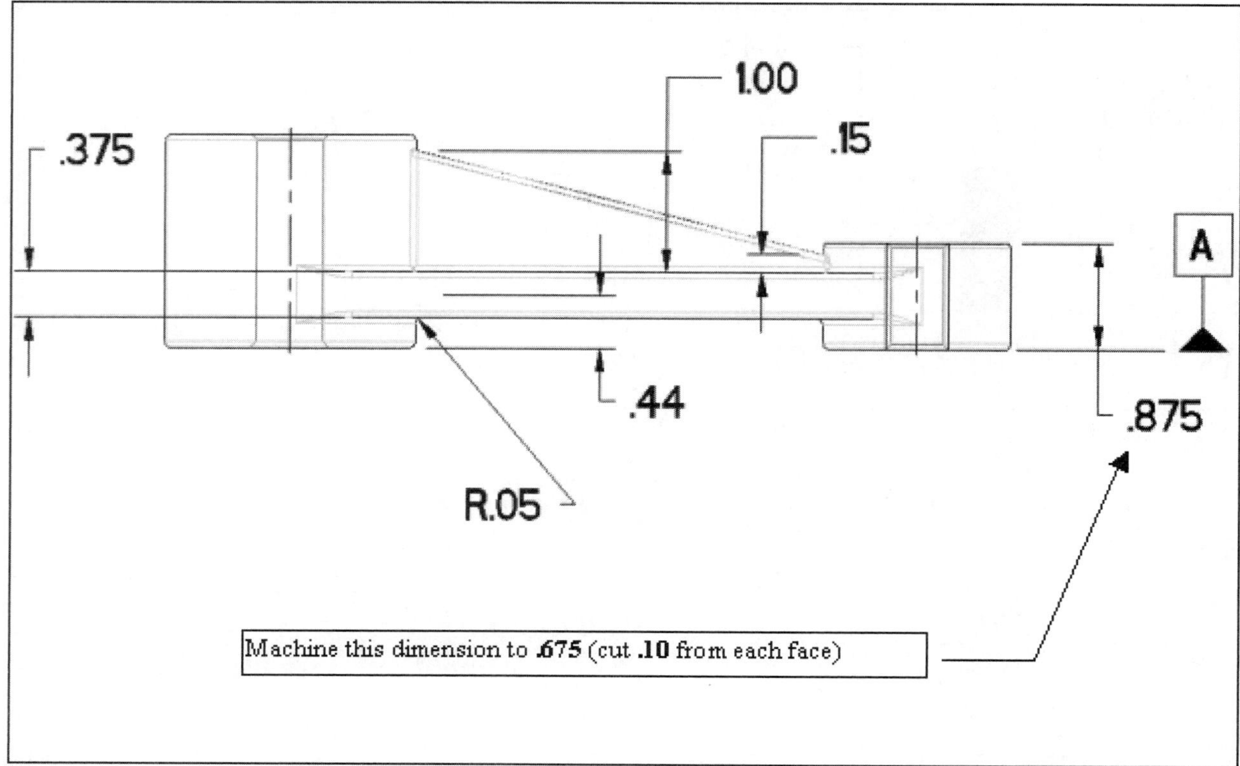

Figure 9.47 Front View of the Clamp Arm

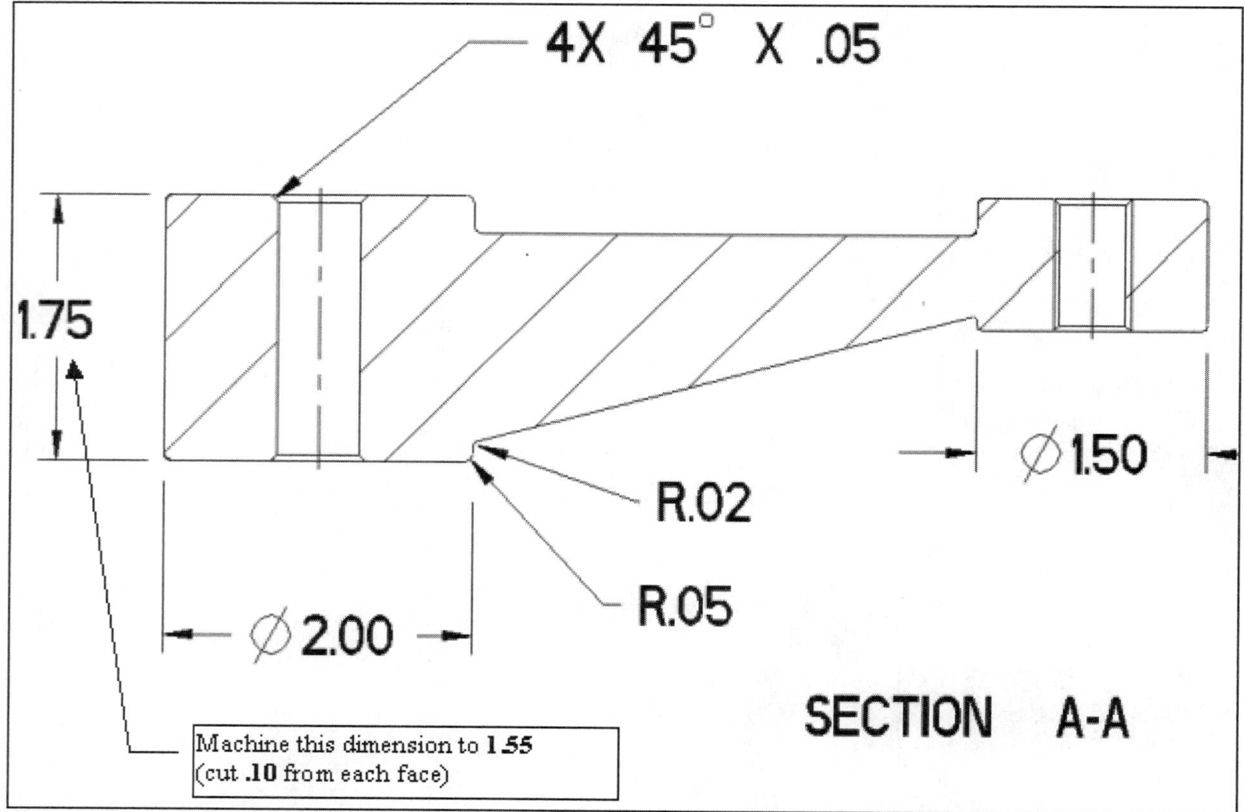

Figure 9.48 SECTION A-A of the Clamp Arm

Lesson 10 Drafts, Suppress, and Text Protrusions

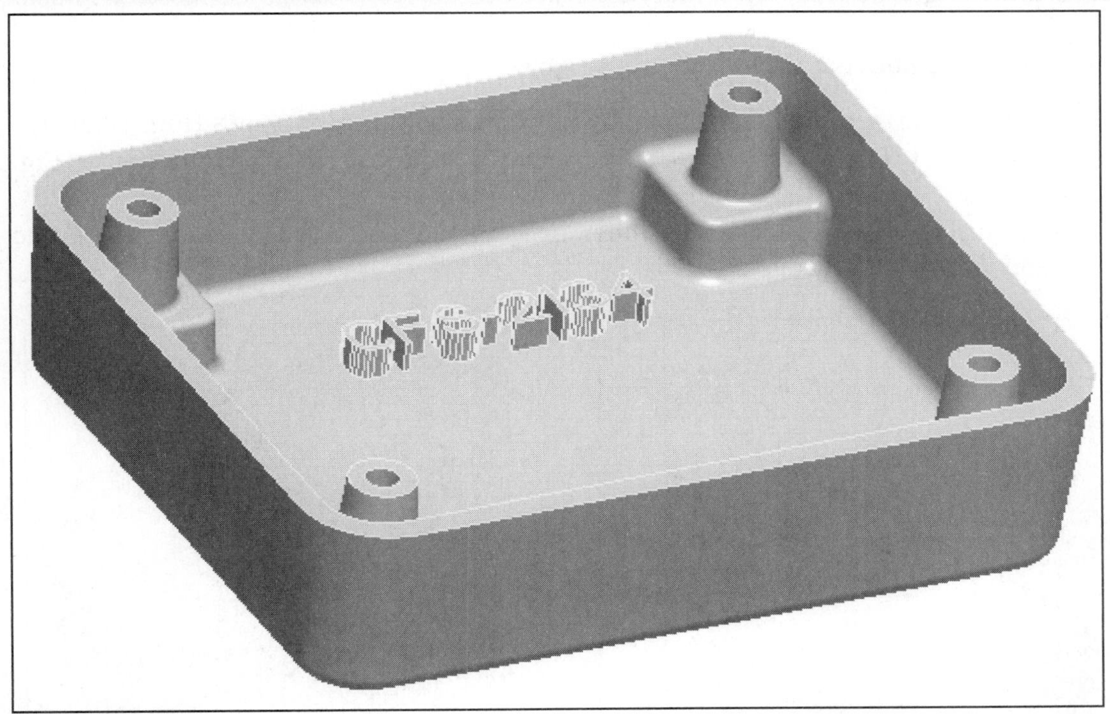

Figure 10.1 Enclosure

OBJECTIVES

- Create **Draft** features
- **Shell** a part
- **Suppress** features to decrease regeneration time
- **Resume** a set of suppressed features
- Create **Text** features on parts

DRAFTS, SUPPRESS, AND TEXT PROTRUSIONS

The **Draft** feature adds a draft angle between surfaces. A wide range of parts incorporate drafts into their design. Casting, injection mold, and die parts normally have drafted surfaces. The ENCLOSURE in Figure 10.1 and the object in Figure 10.2 are plastic injection-molded parts.

Suppressing features by using the **Suppress** command temporarily removes them from regeneration. Suppressed features can be "unsuppressed" (**Resume**) at any time. It is sometimes convenient to suppress text protrusions and rounds to speed up regeneration of the model. Suppressing removes the item from regeneration and requires you to resume the item later.

Hide is another option. Pro/E allows you to hide and unhide some types of model entities. When you hide an item, Pro/E removes the item from the graphics window. The hidden item remains in the Model Tree list, and its icon dims to reveal its hidden status. When you unhide an item, its icon returns to normal display (undimmed) and the item is redisplayed in the graphics window. The hidden status of items is saved with the model. Unlike the suppression of items, hidden items are regenerated.

Text can be included in a sketch for extruded protrusions and cuts, trimming surfaces, and cosmetic features. To decrease regeneration time of the model, text can be suppressed after it has been created. Text can also be drafted.

Drafts

The **Draft Tool** adds a draft angle between two individual surfaces or to a series of selected planar surfaces [Fig. 10.2(a)].

During draft creation, remember the following:

- You can draft only the surfaces that are formed by tabulated cylinders or planes [Fig. 10.2(b)].
- The draft direction must be normal to the neutral plane if a draft surface is cylindrical, [Fig. 10.2(c-d)].
- You cannot draft surfaces with fillets around the edge boundary. However, you can draft the surfaces first, and then fillet the edges [Fig. 10.2(e)].

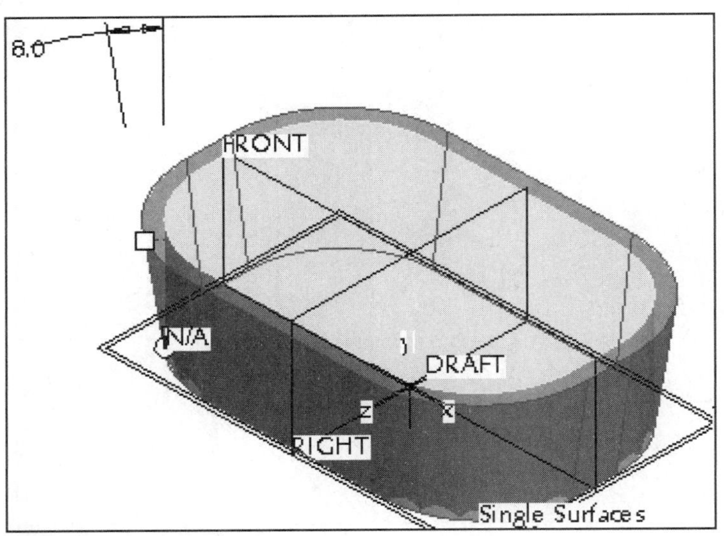

Figure 10.2(a) Draft

Figure 10.2(b) Drafted Protrusion

The following table lists the terminology used in drafts.

TERM	DEFINITION
Draft surfaces	Model surfaces selected for drafting.
Draft Hinges	Draft surfaces are pivoted about the intersection of the neutral plane with the draft surfaces.
Neutral curve	The curve on the draft surfaces that is used as an axis of rotation for draft surfaces; draft surfaces are rotated about the neutral curve.
Pull direction	Direction that is used to measure the draft angle. It is defined as normal to the reference plane.
Draft angle	Angle between the draft direction and the resulting drafted surfaces. If the draft surfaces are split, you can define two independent angles for each portion of the draft.
Direction of rotation	Direction that defines how draft surfaces are rotated with respect to the neutral plane or neutral curve.
Split areas	Areas of the draft surfaces to which you can apply different draft angles. Split object is also a choice.

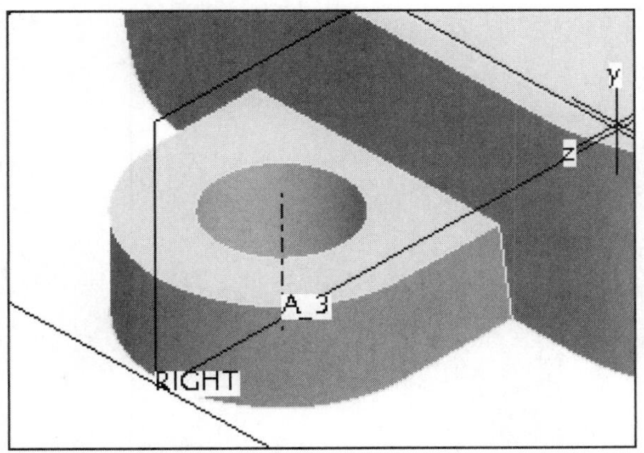

Figure 10.2(c) Second Protrusion

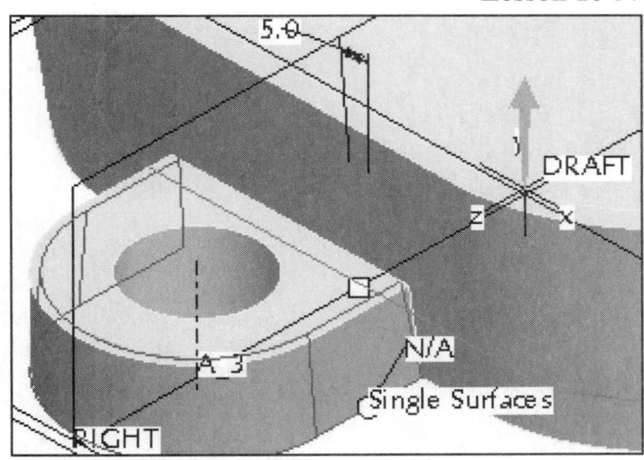

Figure 10.2(d) Drafted Protrusion

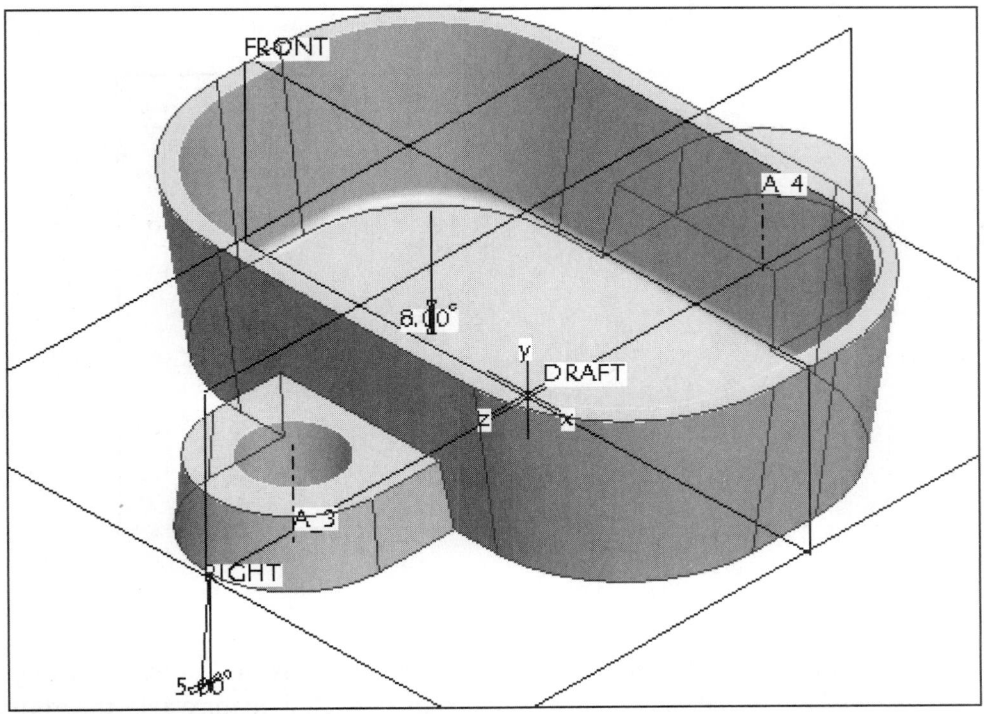

Figure 10.2(e) Drafts

Suppressing and Resuming Features

Suppressing a feature is similar to removing the feature from regeneration temporarily. You can "unsuppress" (**Resume**) suppressed features at any time. Features on a part can be suppressed to simplify the part model and decrease regeneration time. For example, while you work on one end of a shaft, it may be desirable to suppress features on the other end of the shaft. Similarly, while working on a complex assembly, you can suppress some of the features and components for which the detail is not essential to the current assembly process.

Unlike other features, the base feature cannot be suppressed. If you are not satisfied with your base feature, you can redefine the section of the feature, or you can delete it and start over again. Select a feature(s) to suppress by: picking on it, selecting from the Model Tree [Fig. 10.3(a-b)], specifying a *range*, entering its *feature number* or *identifier*, or using *layers*.

You can use **Suppress** and **Resume** to simplify the part before inserting features such as text protrusions. In addition, you may wish to suppress the text protrusion if there is other work to be done on the part. Text protrusions take time to regenerate, and increase the file size considerably.

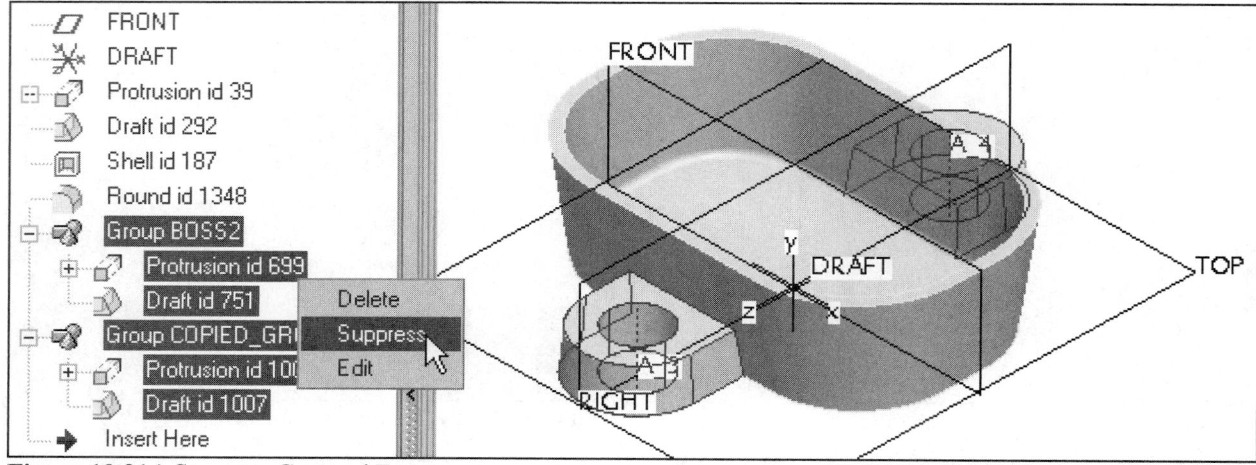

Figure 10.3(a) Suppress Grouped Features

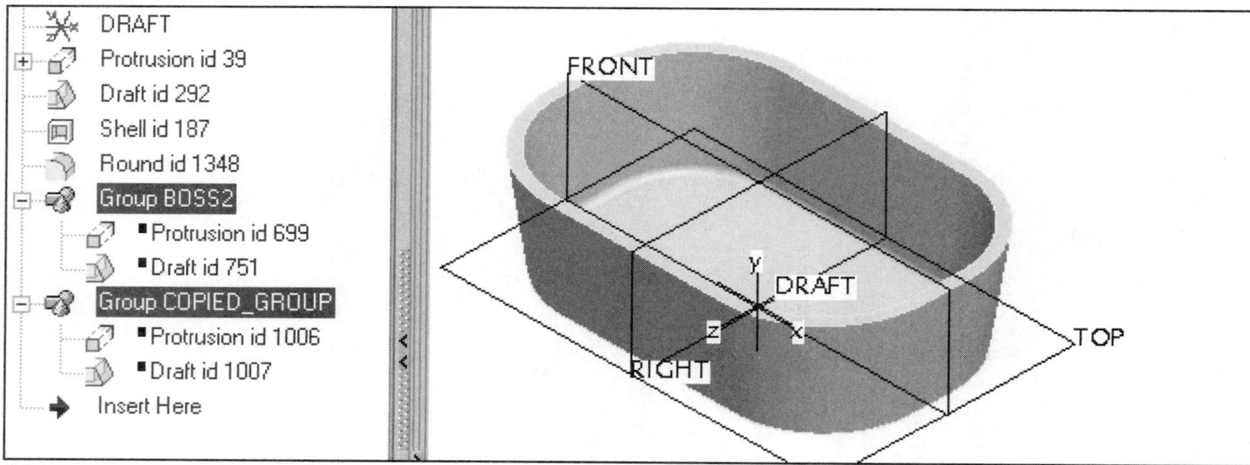

Figure 10.3(b) Suppressed Features

Text Extrusions

When you are modeling, **Text** can be included in a sketch for extruded protrusions and cuts, trimming surfaces, and cosmetic features (Fig. 10.4). The characters that are in an extruded feature use the font **font3d** as the default. Other fonts are available.

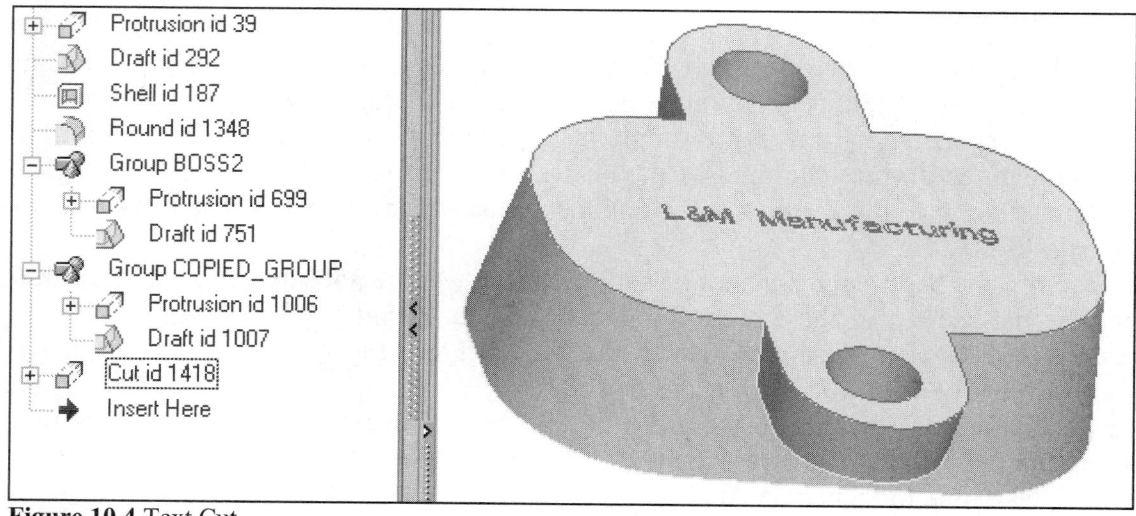

Figure 10.4 Text Cut

Lesson 10 STEPS

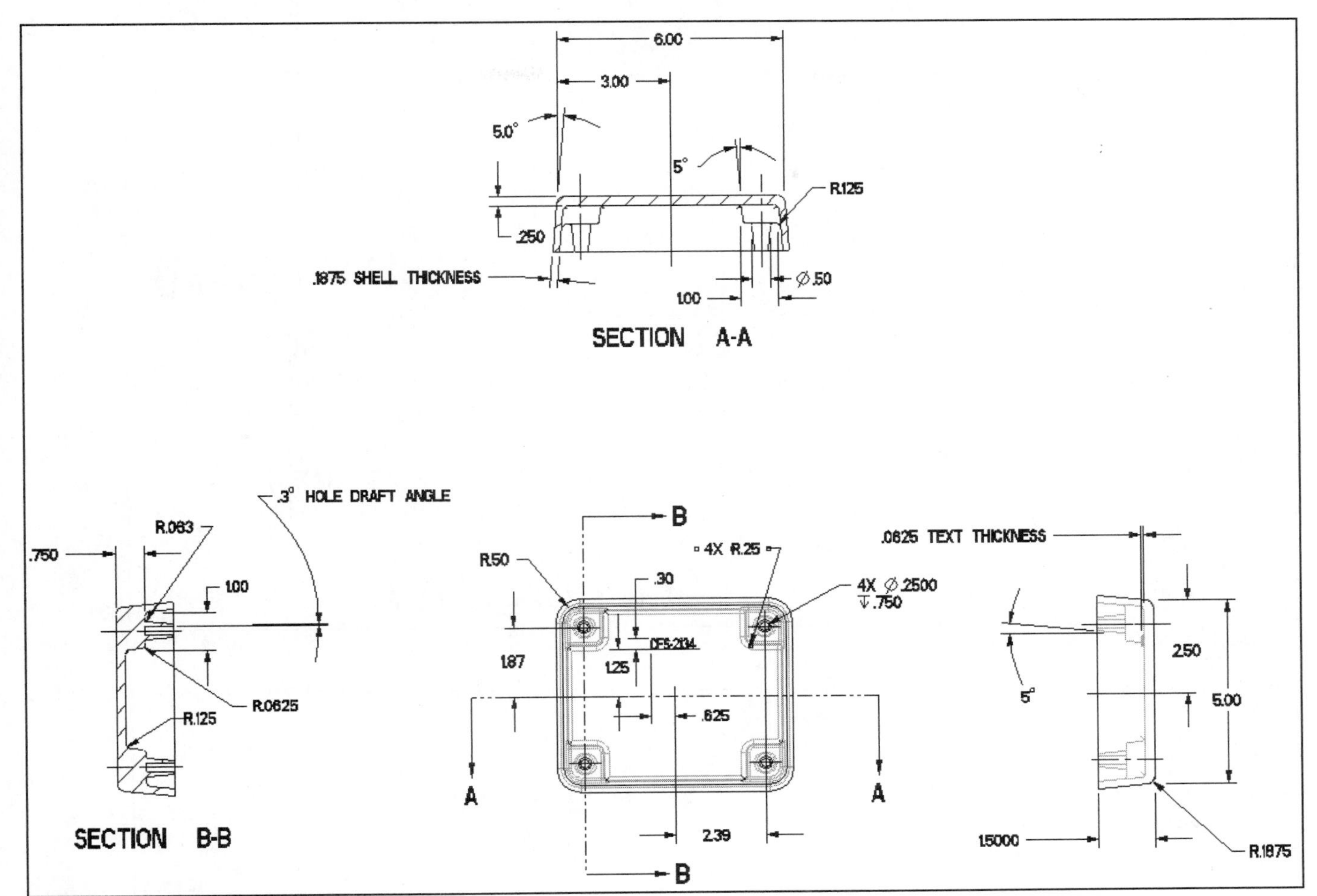

Figure 10.5 Enclosure

Enclosure

The Enclosure is a plastic injection-molded part. A variety of drafts will be used in the design of this part. A *raised text protrusion* will be modeled on the inside of the Enclosure, as shown in Figure 10.5. The dimensions for the part are provided in Figures 10.6 through 10.9.

Start a new part, click: **File** ⇒ **Set Working Directory** select the working directory ⇒ **OK** ⇒ **Create a new object** ⇒ ●**Part** ⇒ **ENCLOSURE** ⇒ ⇒ **OK** ⇒ **Tools** ⇒ ⇒ Snap to Grid ⇒ Use 2D Sketcher Display Style Hidden Line ⇒ Tangent Edges Dimmed ⇒ **OK** ⇒ **Tools** ⇒ **Options** ⇒ Showing: **Current Session** ⇒ Option: ***default_dec_places*** ⇒ Value: **3** ⇒ **Enter** ⇒ Option: ***sketcher_dec_places*** ⇒ Value: **3** ⇒ **Enter** ⇒ **Apply** ⇒ **Close** ⇒ **Edit** ⇒ **Setup** ⇒ **Units** ⇒ Units Manager **Inch lbm Second** ⇒ **Close** ⇒ **Material** ⇒ **Define** ⇒ type **PLASTIC** ⇒ **MMB** ⇒ **File** ⇒ **Save** ⇒ **File** ⇒ **Exit** ⇒ **Assign** ⇒ pick **PLASTIC** ⇒ **Accept** ⇒ **MMB** ⇒ double-click on the default coordinate system name in the Model Tree-- **PRT_CSYS_DEF** ⇒ type **CSYS_ENCLOSURE** ⇒ **Enter** ⇒ ⇒ **MMB** ⇒ load your customization by clicking **Tools** ⇒ **Customize Screen** ⇒ **File** ⇒ **Open Settings** ⇒ click on your saved file ⇒ **Open** ⇒ **OK**

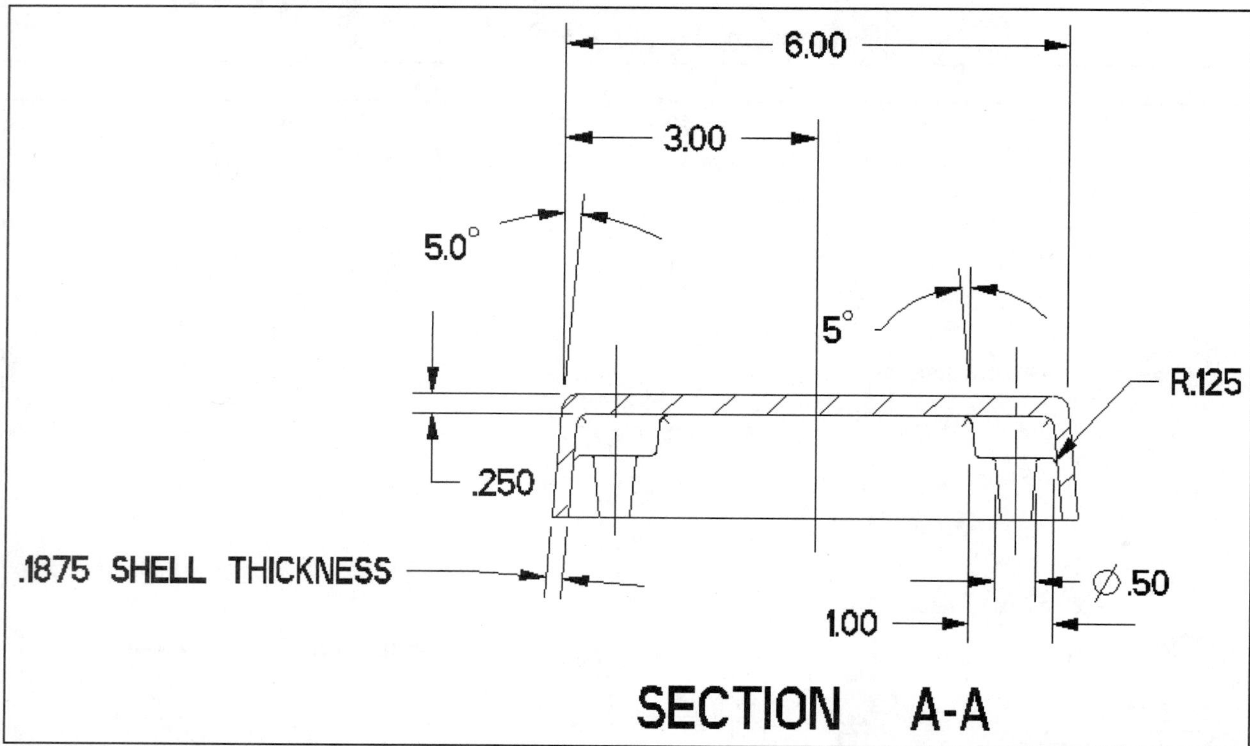

Figure 10.6 SECTION A-A (Top View)

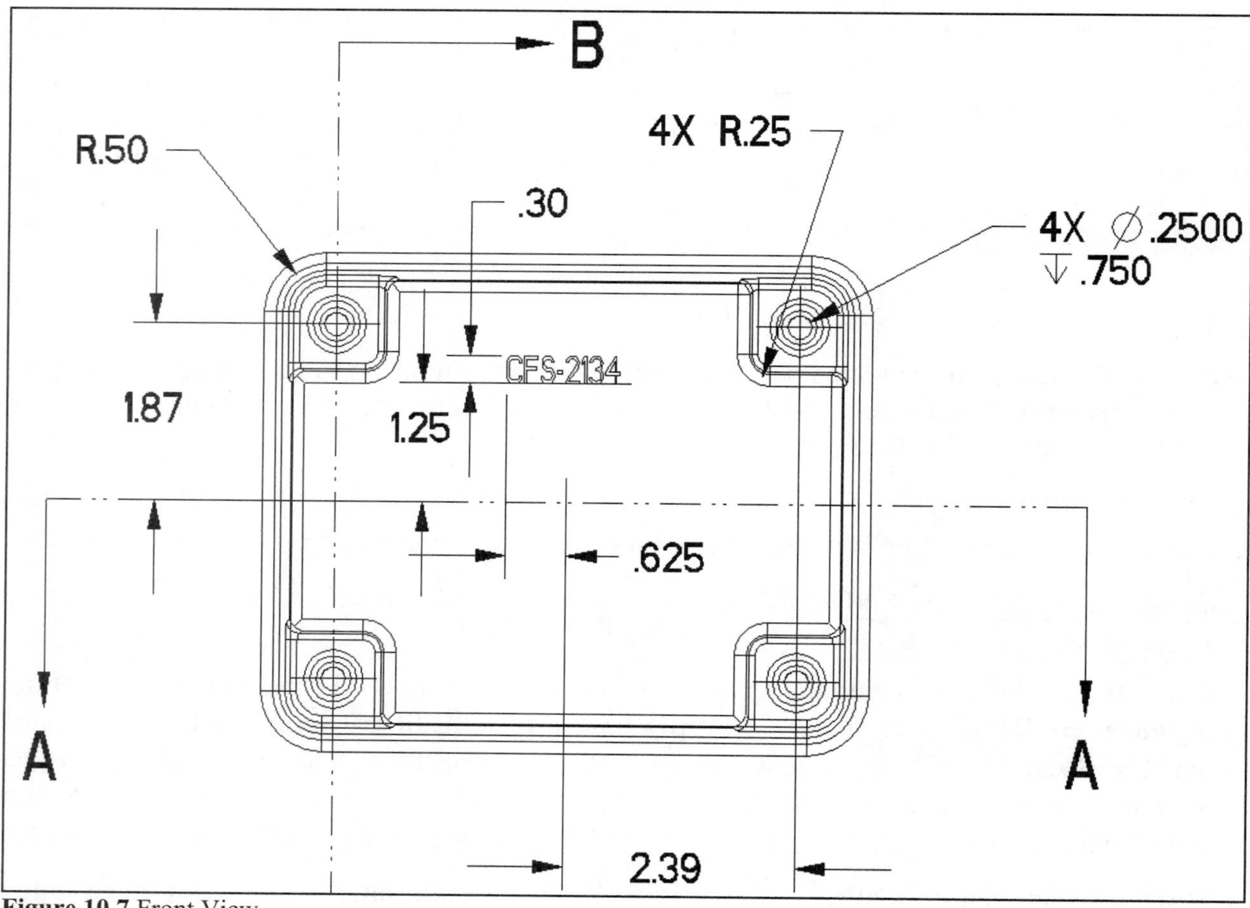

Figure 10.7 Front View

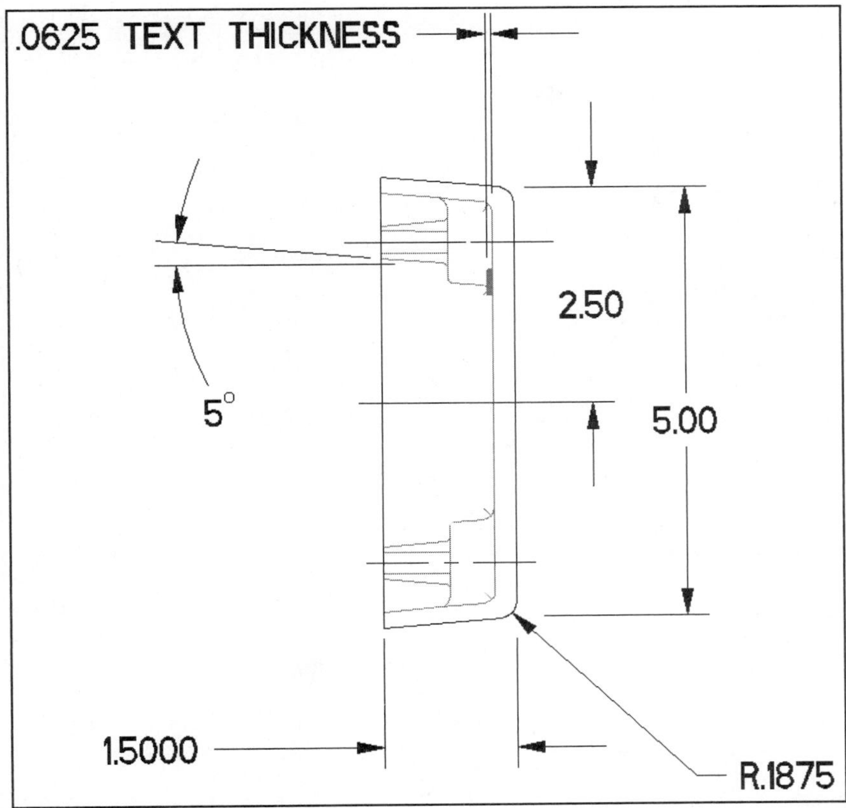

Figure 10.8 Right Side View

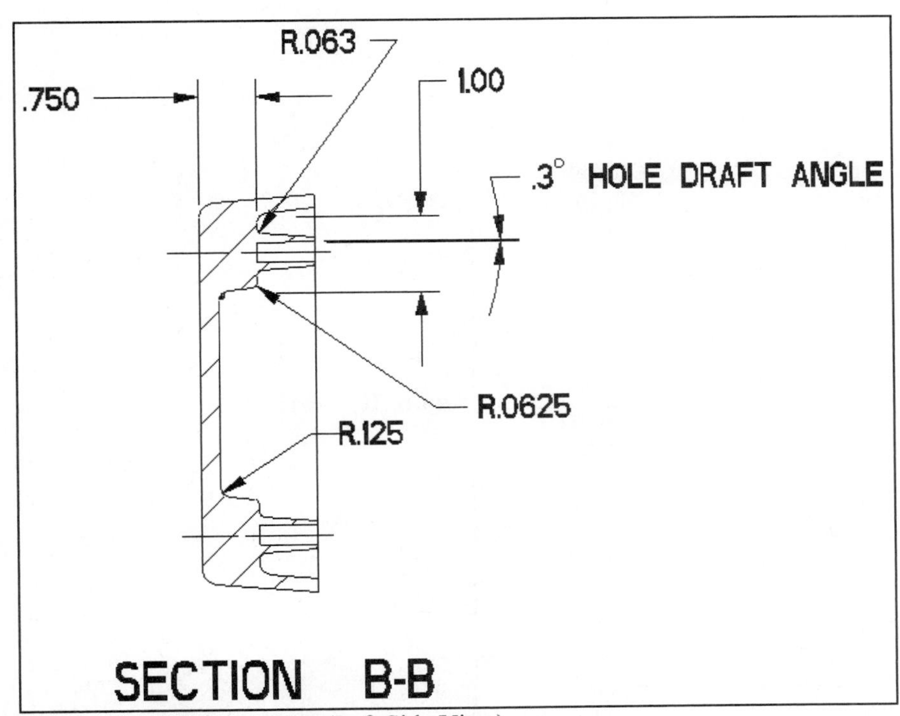

Figure 10.9 SECTION B-B (Left Side View)

Make first protrusion **6.00** (width) **X 5.00** (height) **X 1.50** (depth), with **R.50** rounds (add the fillets to the sketch rather than after the first protrusion is complete) [Fig. 10.10(a-b)]. Sketch on datum **FRONT** [Fig. 10.10(c)]. Center the first protrusion horizontally on datum **TOP** and vertically on datum **RIGHT** [Fig. 10.10(d)]. Add constraints if needed to control your sketch geometry.

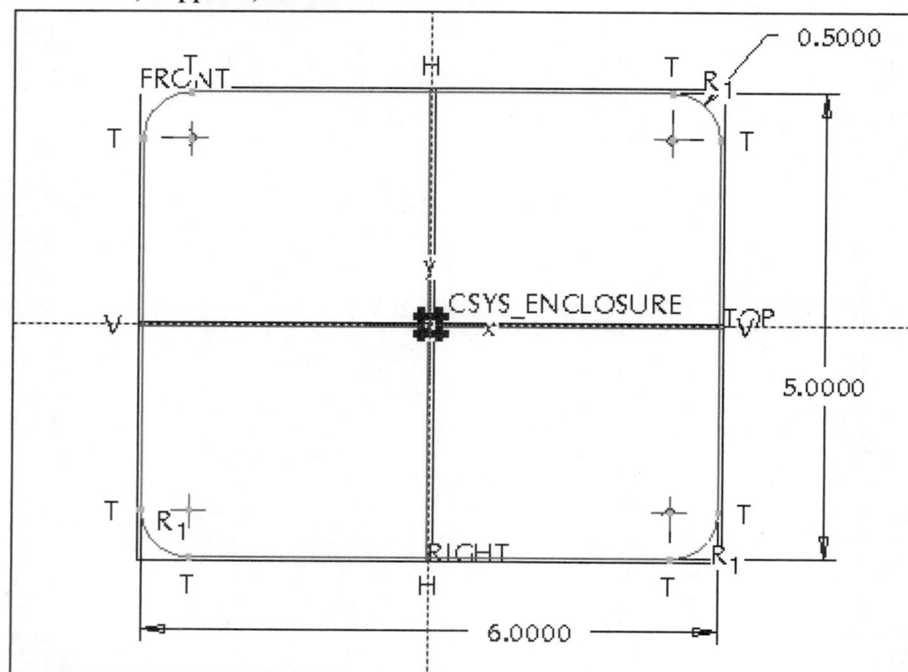

Figure 10.10(a) Sketch

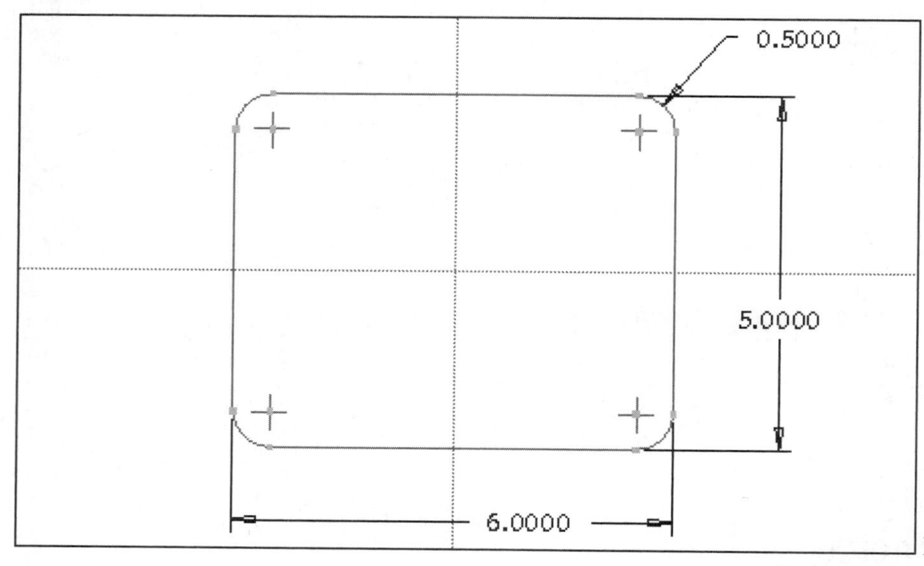

Figure 10.10(b) Dimensions

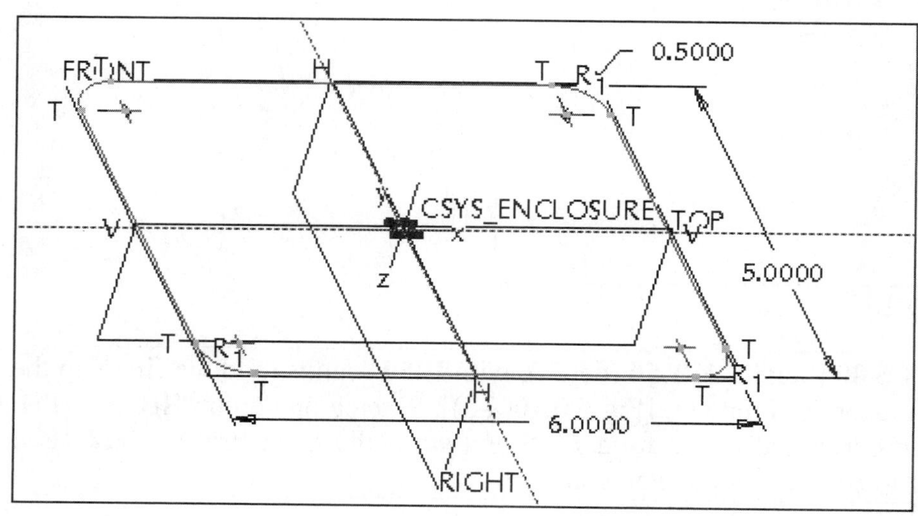

Figure 10.10(c) Standard Orientation

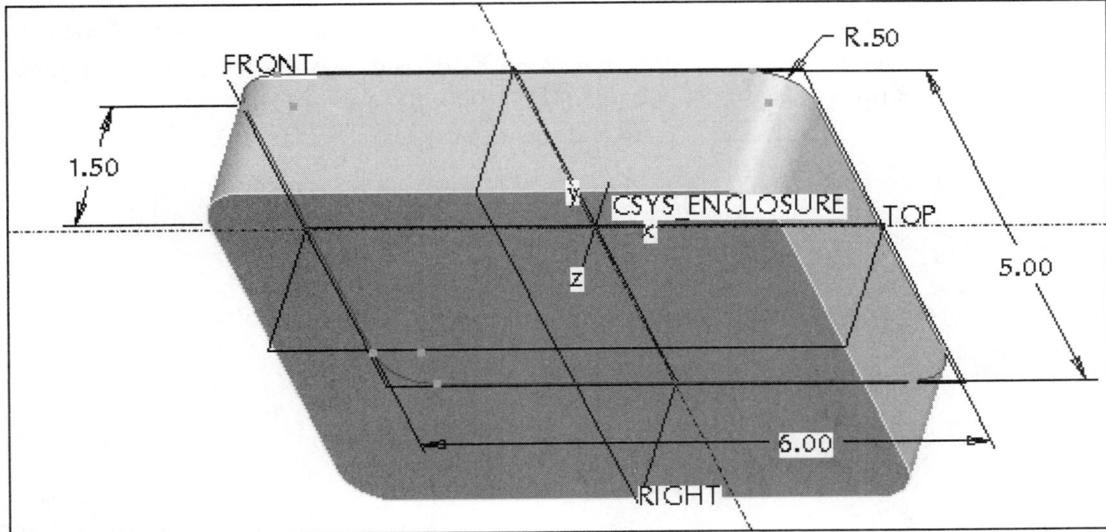

Figure 10.10(d) First Protrusion, **6.00 X 5.00 X 1.50**, and **R.50** rounds

Create the draft for the lateral surfaces of the protrusion, click: **Draft Tool** ⇒ **Options** tab ⇒ Draft tangent surfaces ⇒ Intersect ⇒ **References** tab ⇒ select one of the lateral surfaces [Fig. 10.11(a)]

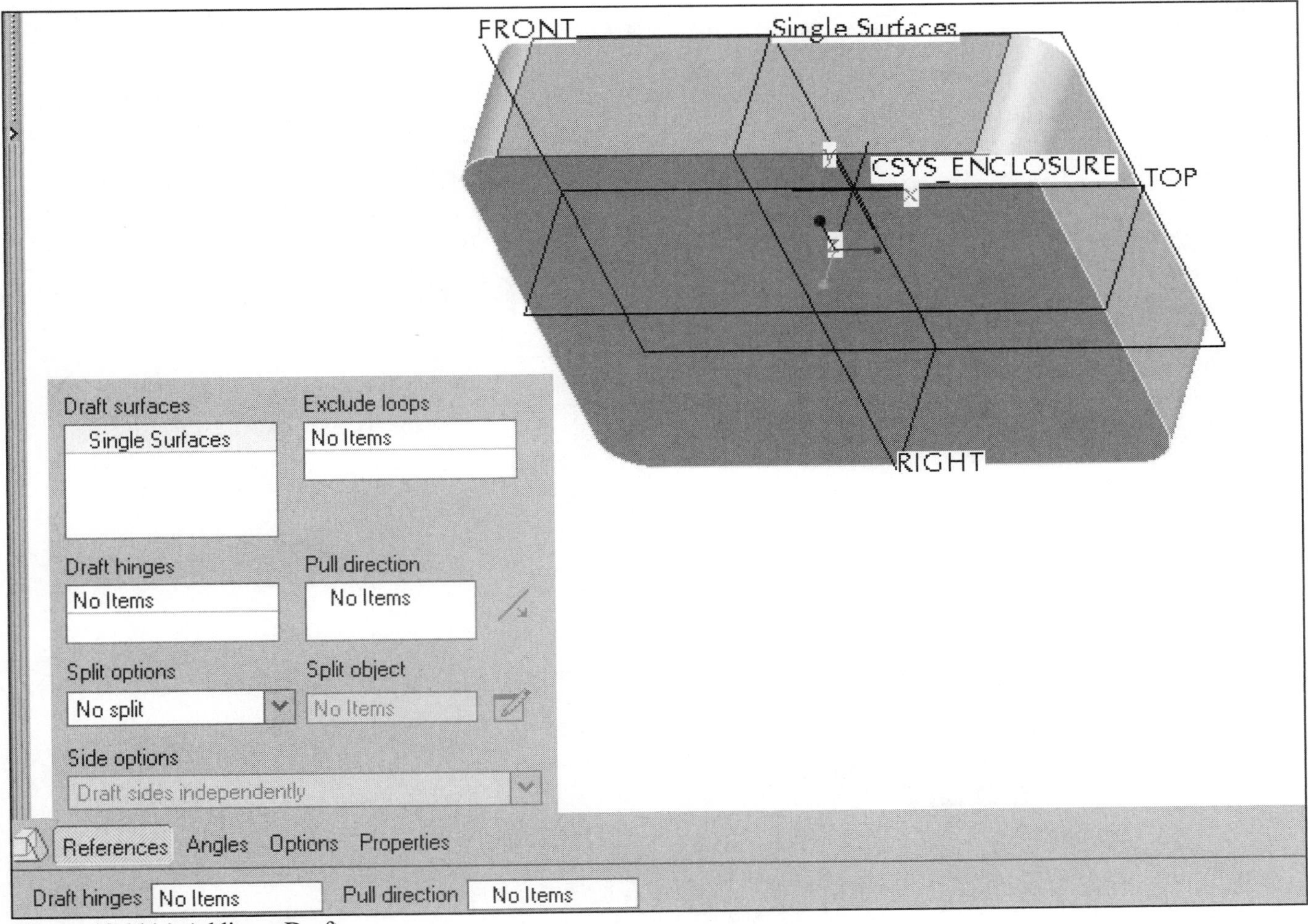

Figure 10.11(a) Adding a Draft

Click on **No Items** in the Draft hinges collector ⇒ pick **FRONT** ⇒ type **5** [Fig. 10.11(b)] ⇒ **Enter** ⇒ **MMB** ⇒ with Draft highlighted in the Model Tree, click **RMB** ⇒ **Edit** [Fig. 10.11(c)] ⇒ click on **5.000** degrees ⇒ **RMB** ⇒ **Properties** ⇒ Flip Arrows ⇒ Number of decimal places **0** ⇒ **OK**

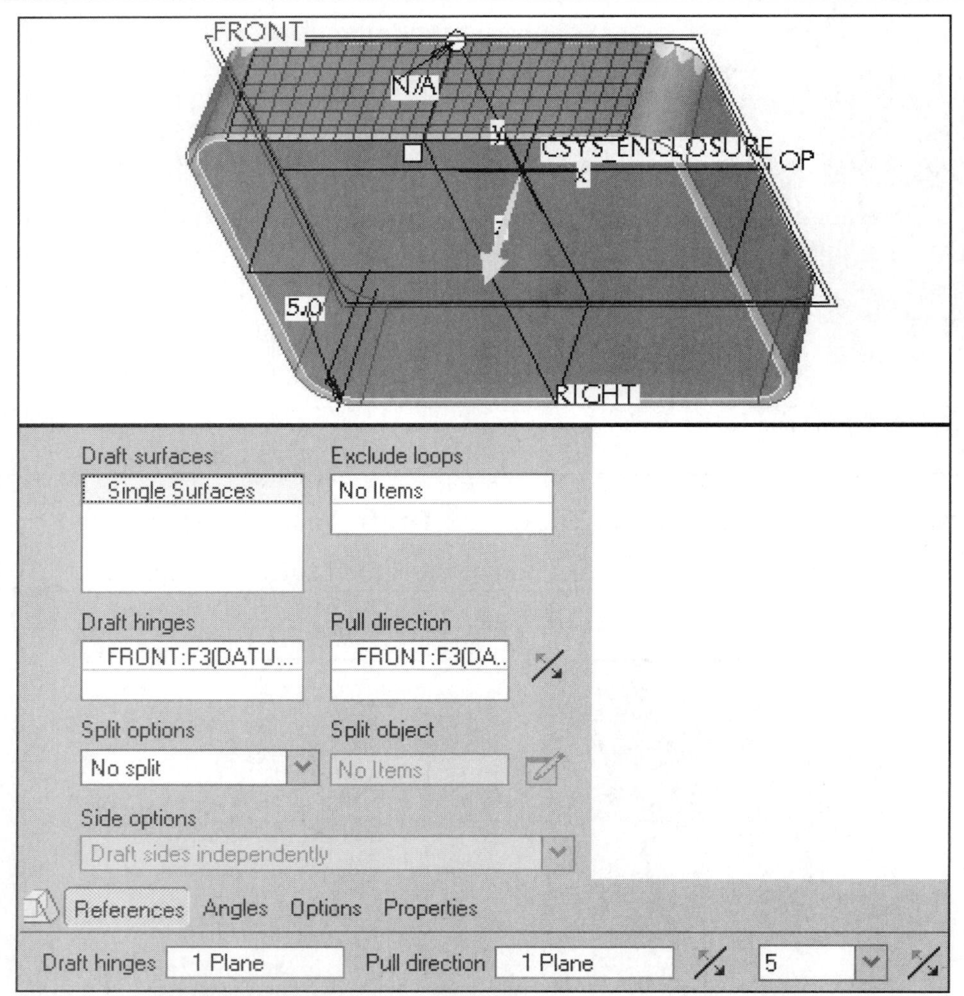

Figure 10.11(b) Draft Dialog Boa and Collectors

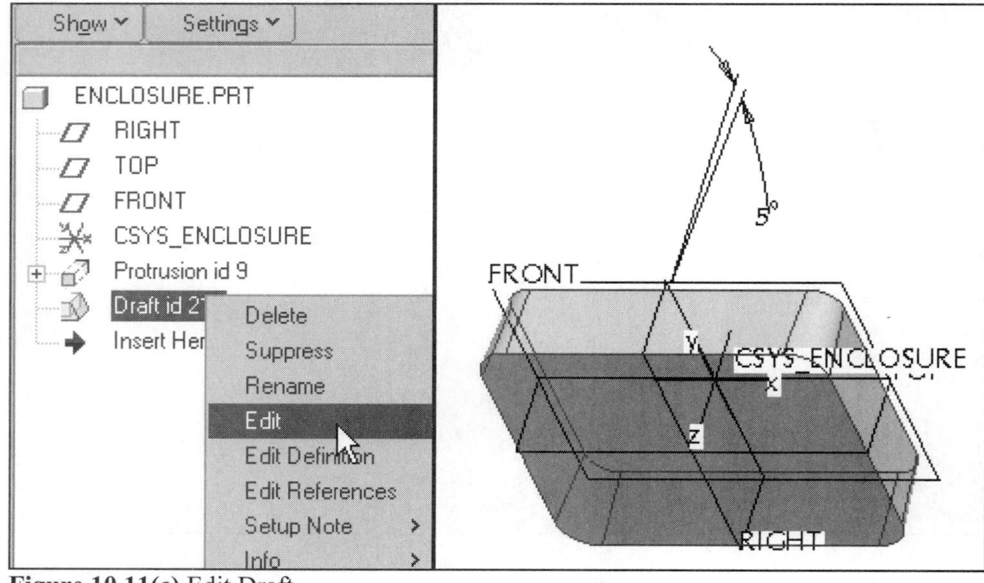

Figure 10.11(c) Edit Draft

Next, use the **Shell Tool** to create a part with a specific thickness.

Click: **Shell Tool** ⇒ type **.1875** in Dimension field ⇒ **References** ⇒ pick the face to be removed [Fig. 10.12(a)] ⇒ [Fig. 10.12(b)] ⇒

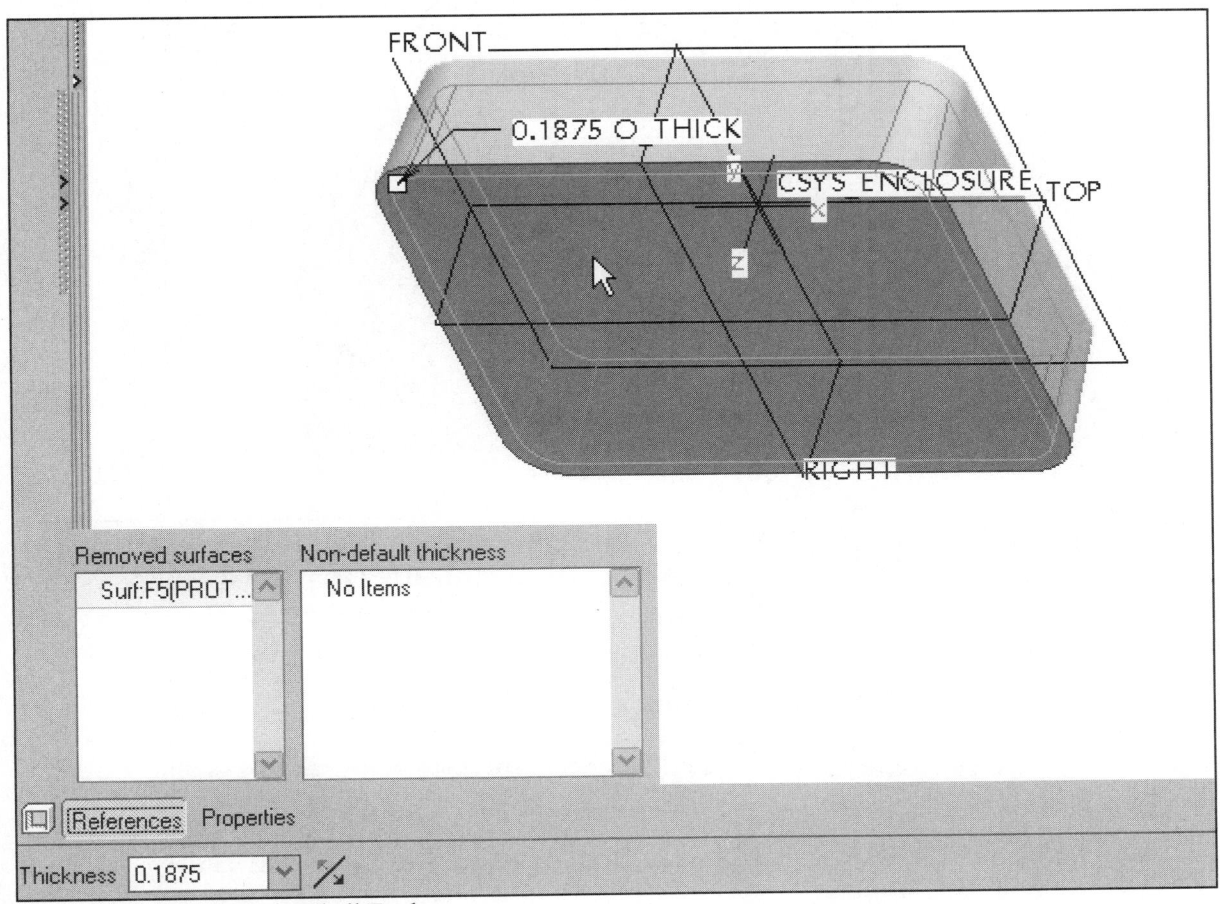

Figure 10.12(a) Using the Shell Tool

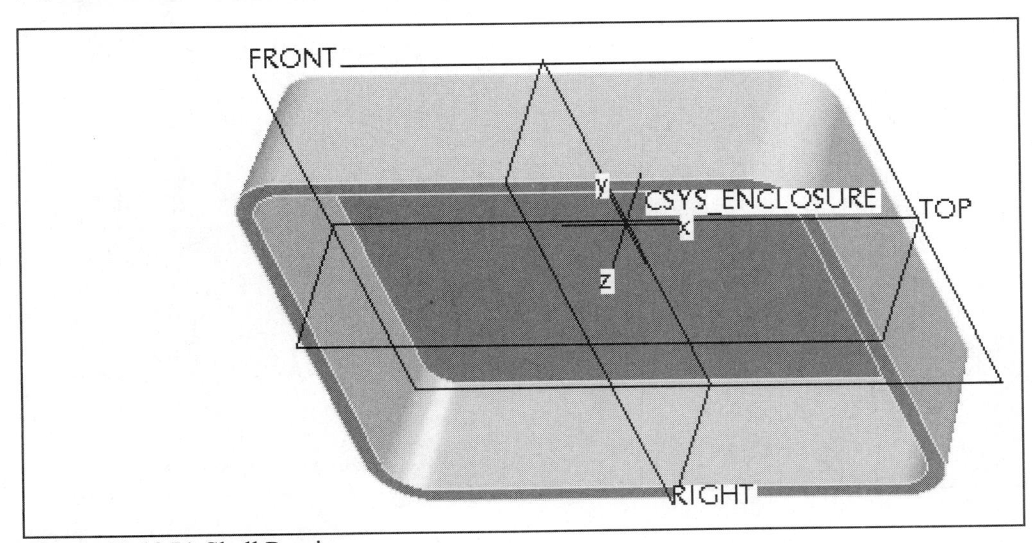

Figure 10.12(b) Shell Preview

A complete description of the Shell Tool appears in Lesson 11.

Change the thickness of the enclosure to be **.25**. The walls will remain **.1875**, click: **References** ⇒ click **No Items** in the Non-default thickness collector ⇒ pick on the face (*highlights*) [Fig. 10.12(c)] ⇒ type **.250** in the Dimension field [Surf:F5(PROT... 0.250] ⇒ **Enter** ⇒ **MMB** [Fig. 10.12(d)]

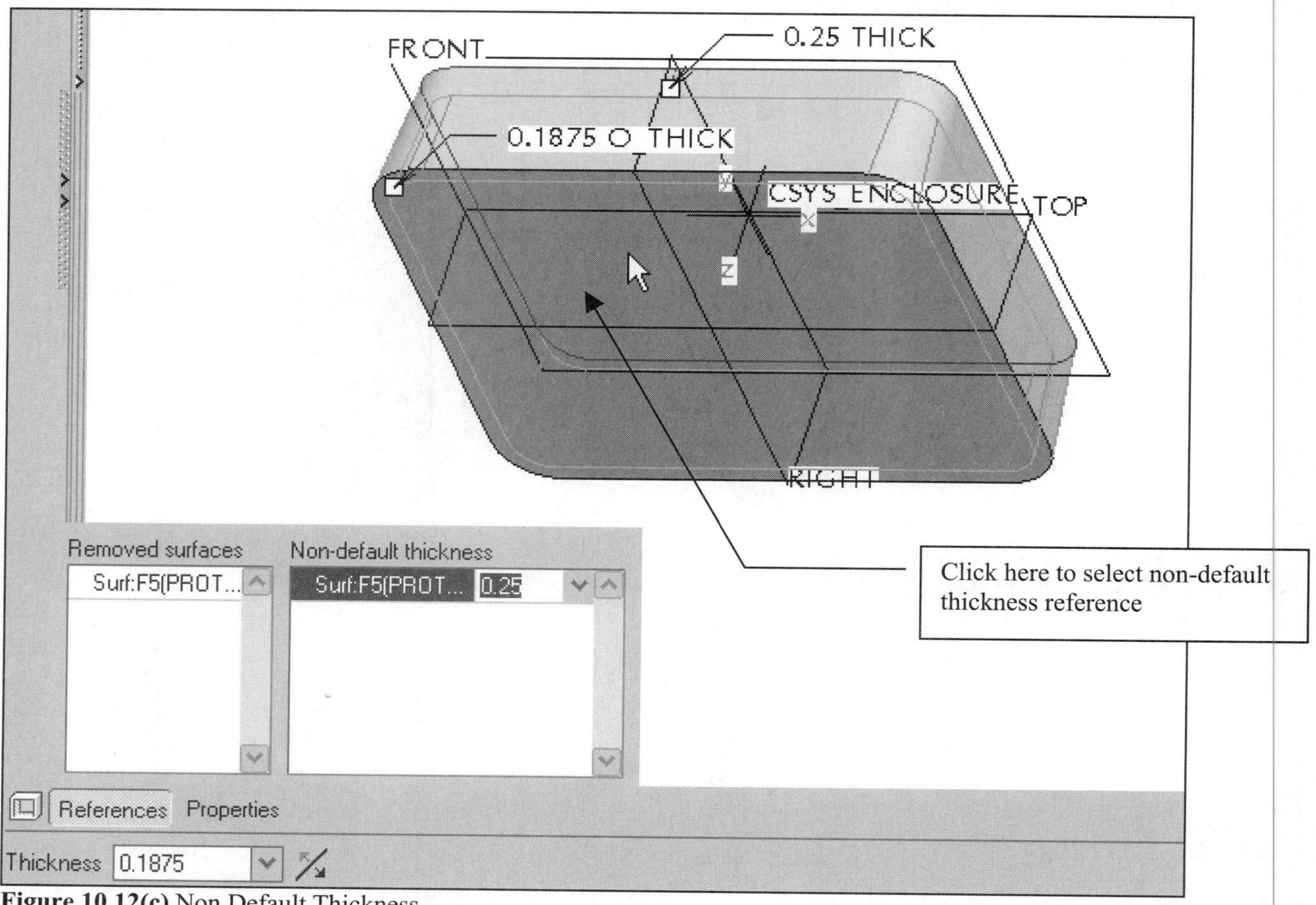

Figure 10.12(c) Non Default Thickness

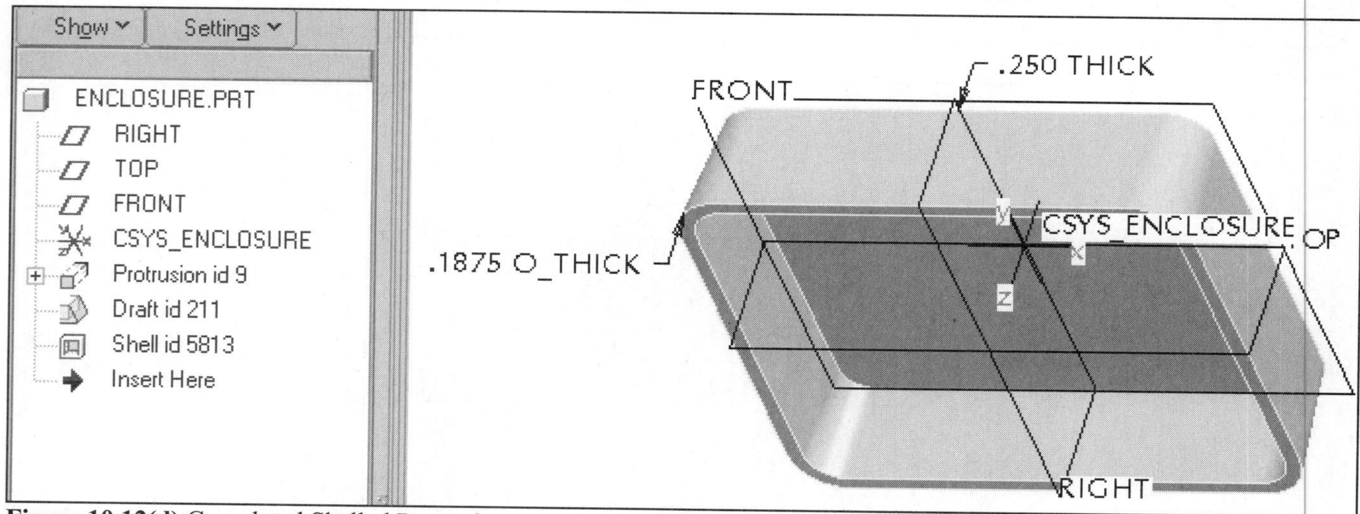

Figure 10.12(d) Completed Shelled Protrusion

Create a raised pedestal-like extrusion on the inside surface. Start by modeling a datum plane offset from datum **FRONT** by **.75** [Fig. 10.13(a)]. **DTM1** will be used to control the height of the pedestal. ⇒ select only the References shown in Figure 10.13(b) ⇒ use an existing *internal_shelled_edge* to start the section geometry [Fig. 10.13(c)] ⇒ add four lines and a fillet to complete the section [Fig. 10.13(d)] ⇒ use the dimensioning scheme shown in Figure 10.13(e) ⇒ ✓ ⇒ [Fig. 10.13(f)]

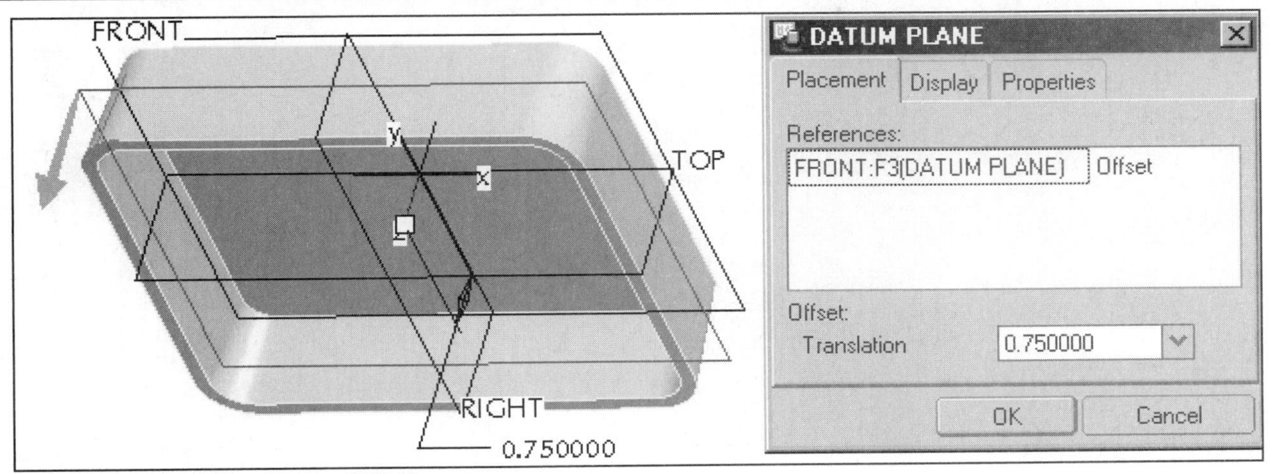

Figure 10.13(a) Offset Datum DTM1

Figure 10.13(b) Extrusion References

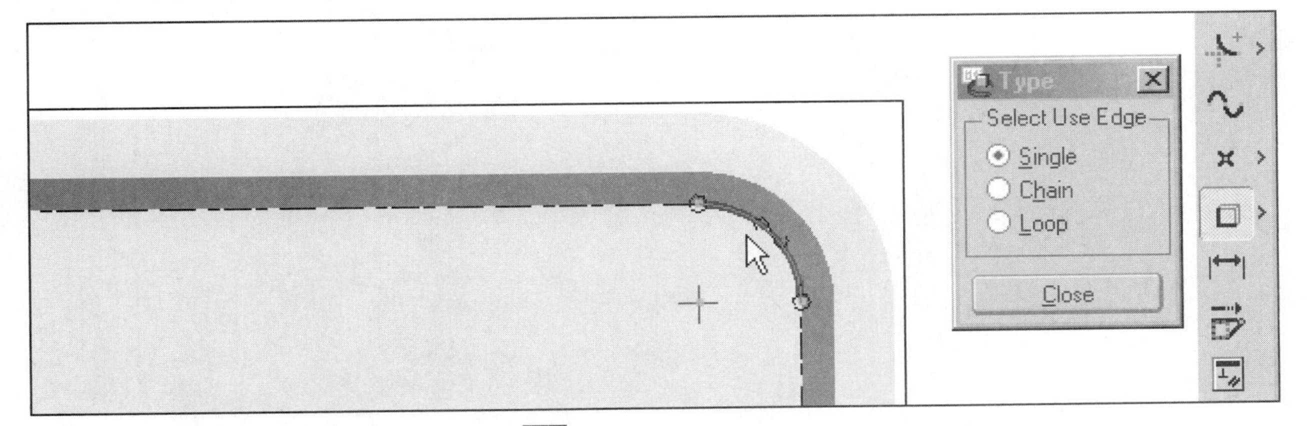

Figure 10.13(c) Create the First Entity using: **Create an entity from an edge**

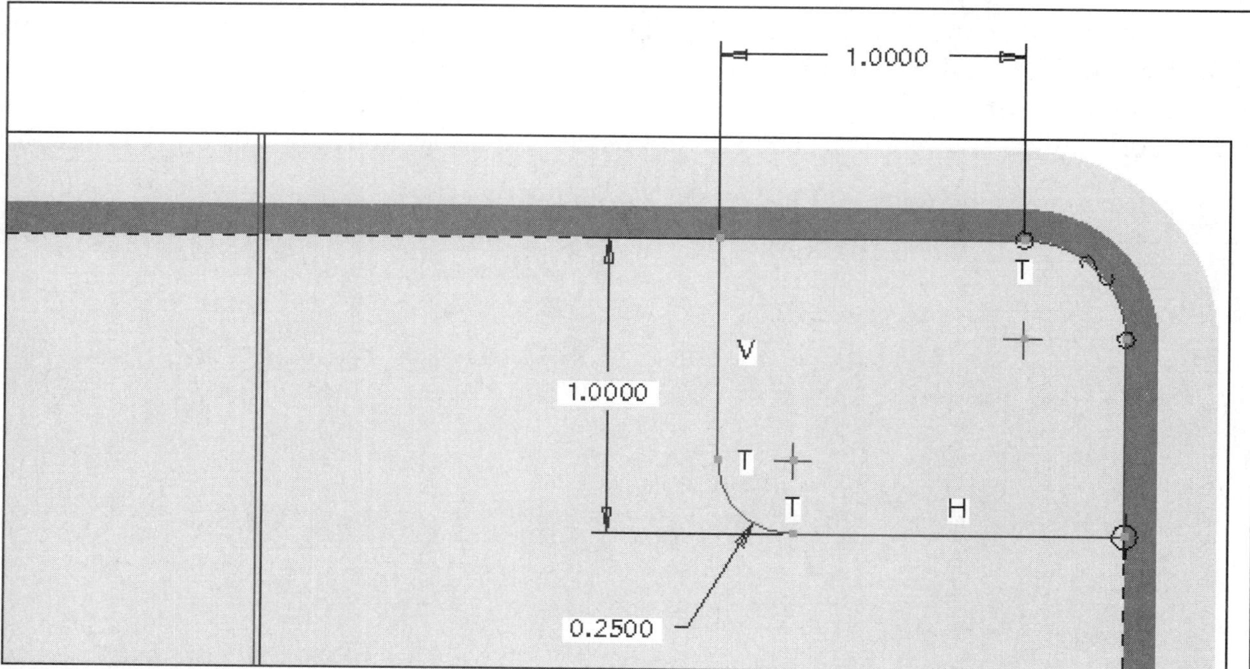

Figure 10.13(d) Section Sketch

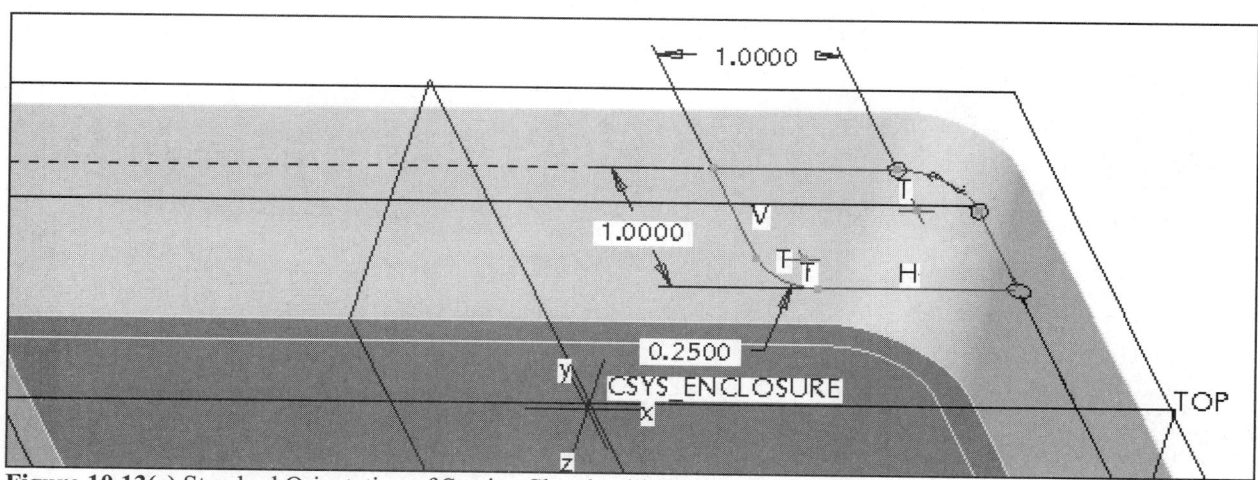

Figure 10.13(e) Standard Orientation of Section Sketch with Design Dimensions

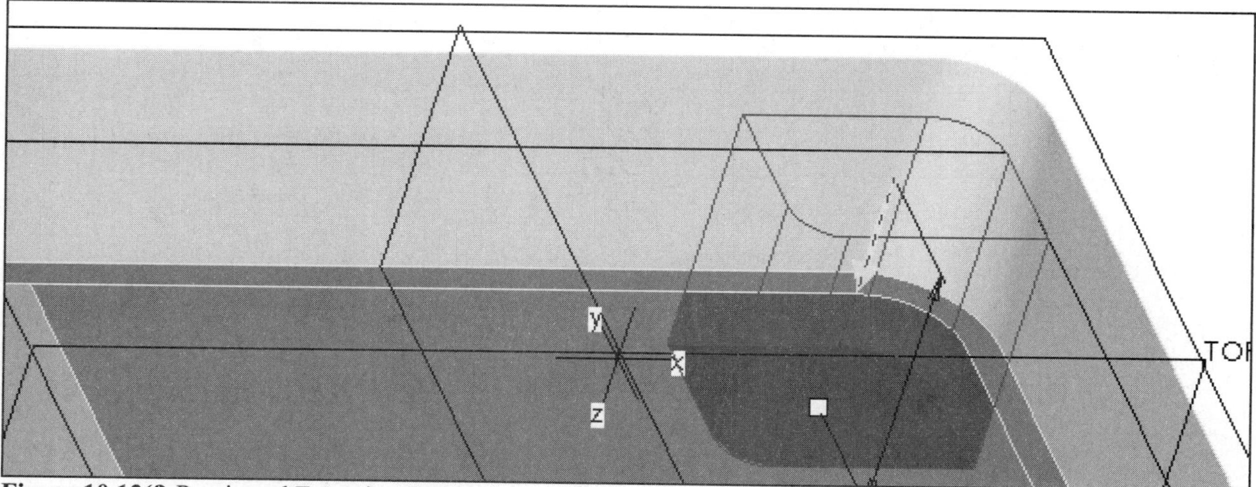

Figure 10.13(f) Previewed Extrusion Depth

Click on the depth dimension ⇒ **RMB** ⇒ **To Selected** [Fig. 10.13(g)] ⇒ pick **DTM1** [Fig. 10.13(h)] ⇒ **MMB** ⇒ spin the model [Fig. 10.13(i)] ⇒ [save icon] ⇒ **MMB**

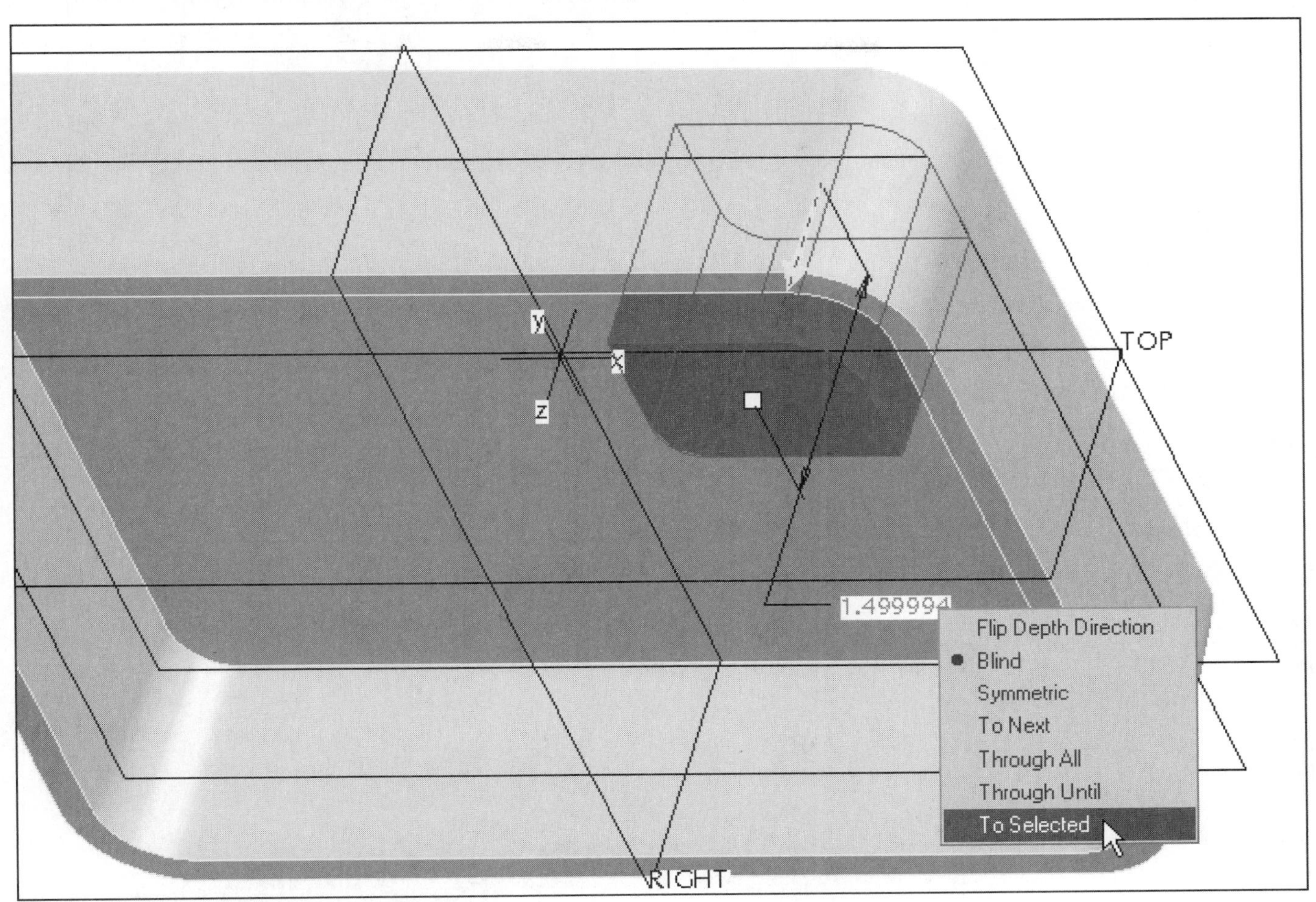

Figure 10.13(g) Determining the Depth Option

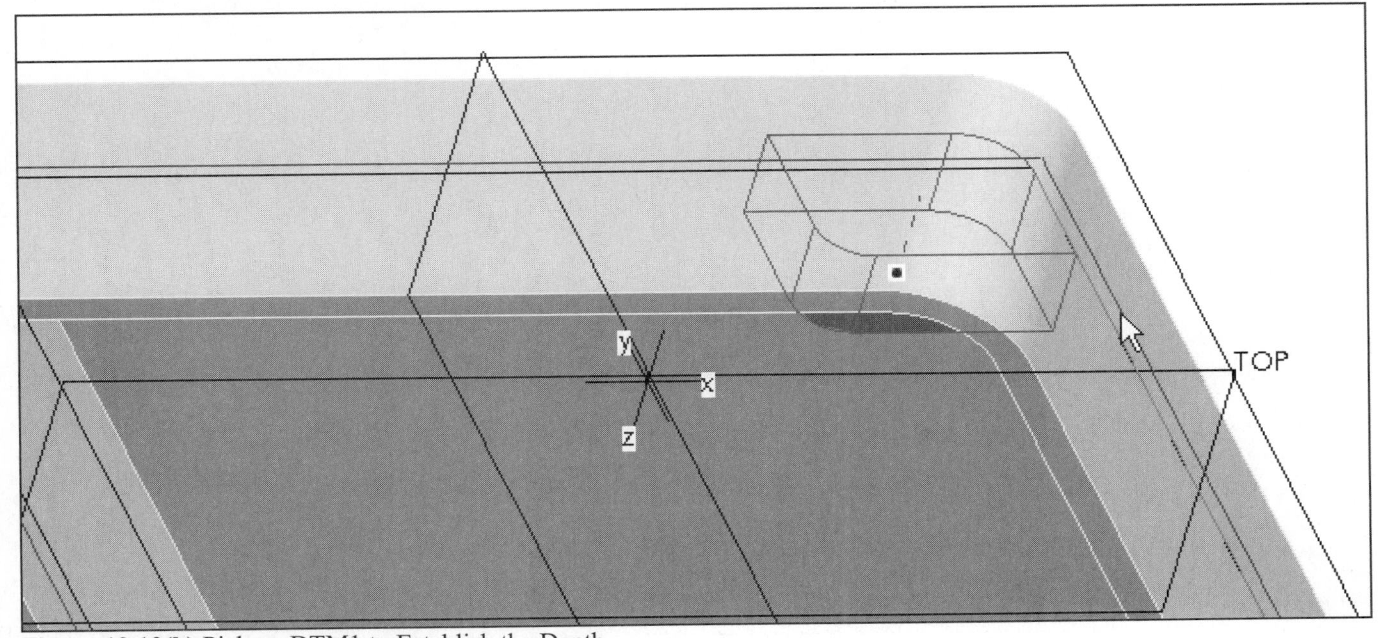

Figure 10.13(h) Pick on DTM1 to Establish the Depth

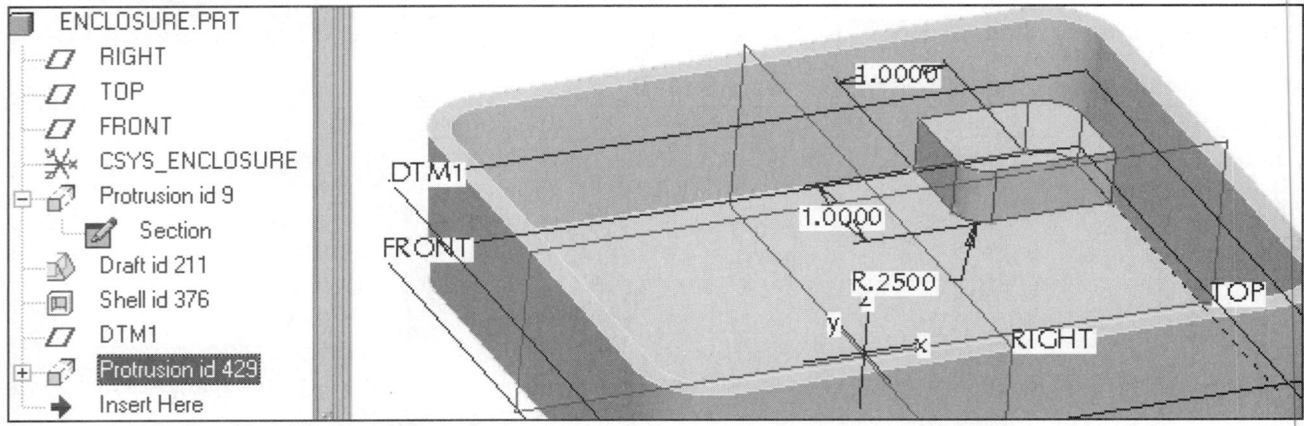

Figure 10.13(i) Completed Pedestal Extrusion

Click: **Draft Tool** ⇒ **References** tab ⇒ select one vertical surface of the pedestal ⇒ click **No Items** in the Draft hinges collector ⇒ pick **FRONT** ⇒ type **5** in Dimension field ⇒ **Enter** ⇒ **Reverse pull direction** (Fig. 10.14) ⇒ **MMB**

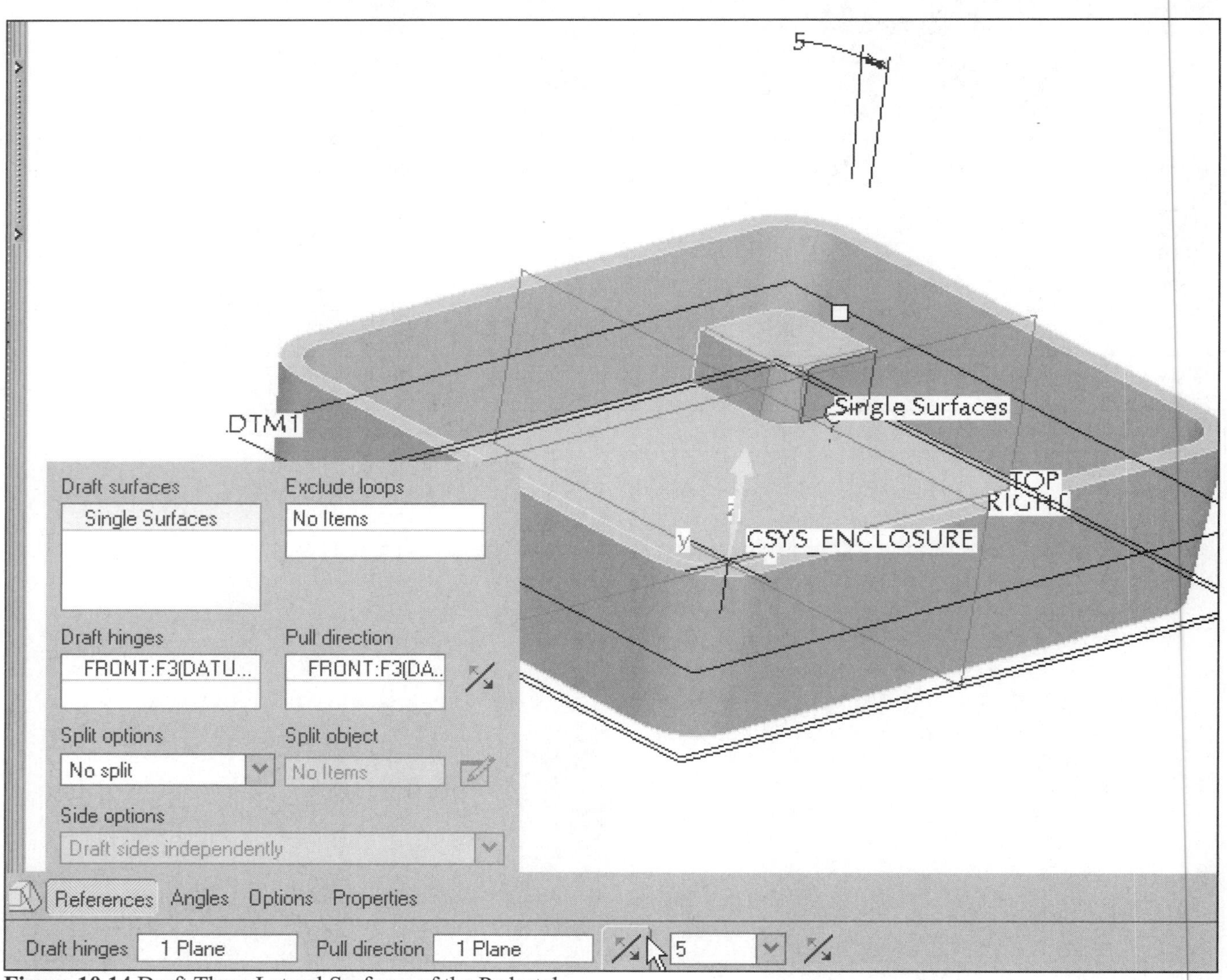

Figure 10.14 Draft Three Lateral Surfaces of the Pedestal

Model the circular protrusion ⇒ use the top surface of the pedestal as the sketching plane ⇒ Keep the default references. The section consists of one circle [Fig. 10.15(a)]. ⇒ ✓ ⇒ rotate the model [Fig. 10.15(b)] ⇒ click on the drag handle ⇒ **RMB** ⇒ **To Selected** [Fig. 10.15(c)] ⇒ pick on the surface [Fig. 10.15(d)] ⇒ **MMB** [Fig. 10.15(e)]

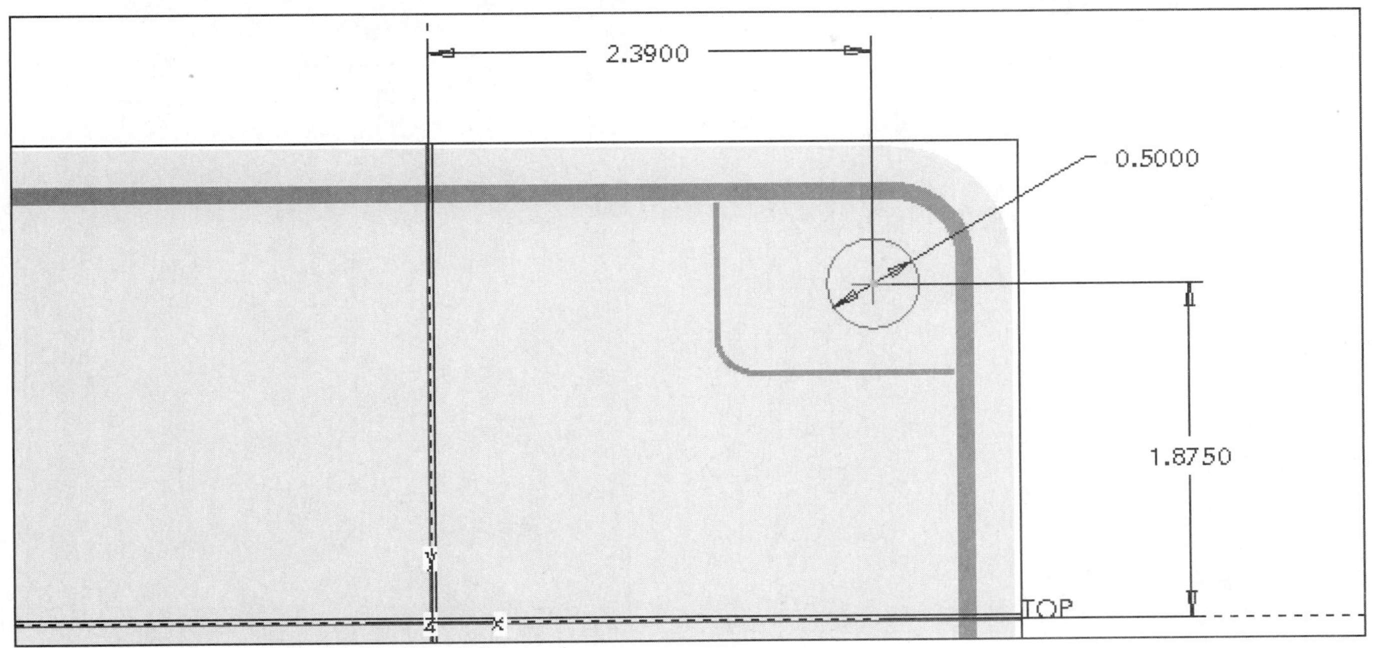

Figure 10.15(a) Section Sketch for Circular Extrusion

Figure 10.15(b) Extrusion Depth

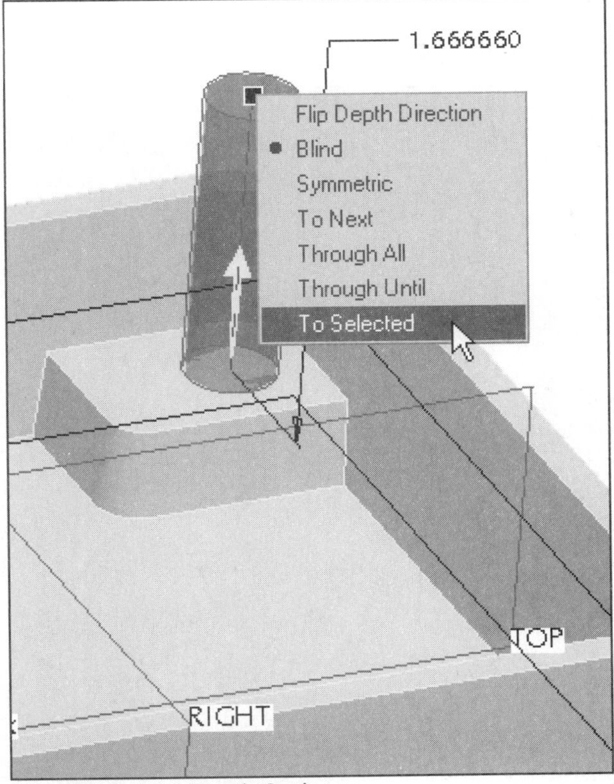

Figure 10.15(c) Depth Options

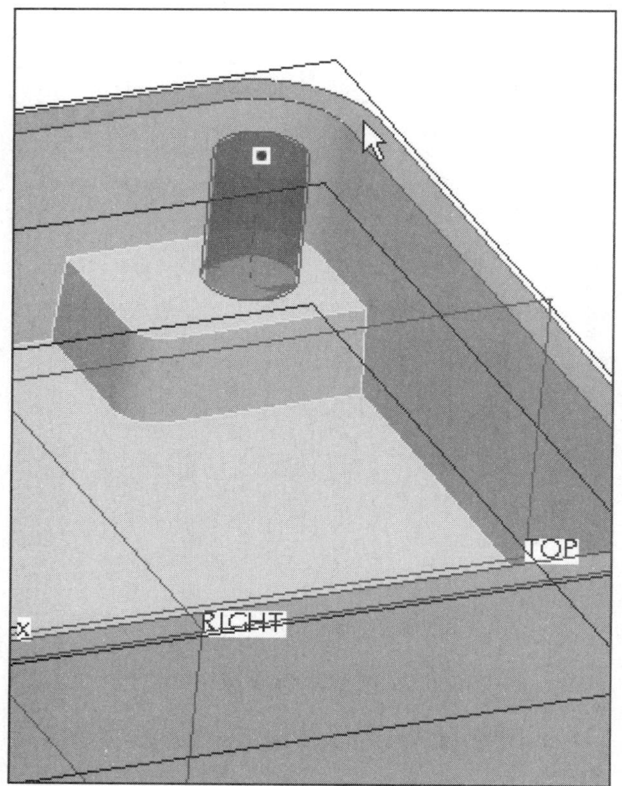

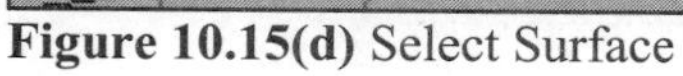
Figure 10.15(d) Select Surface

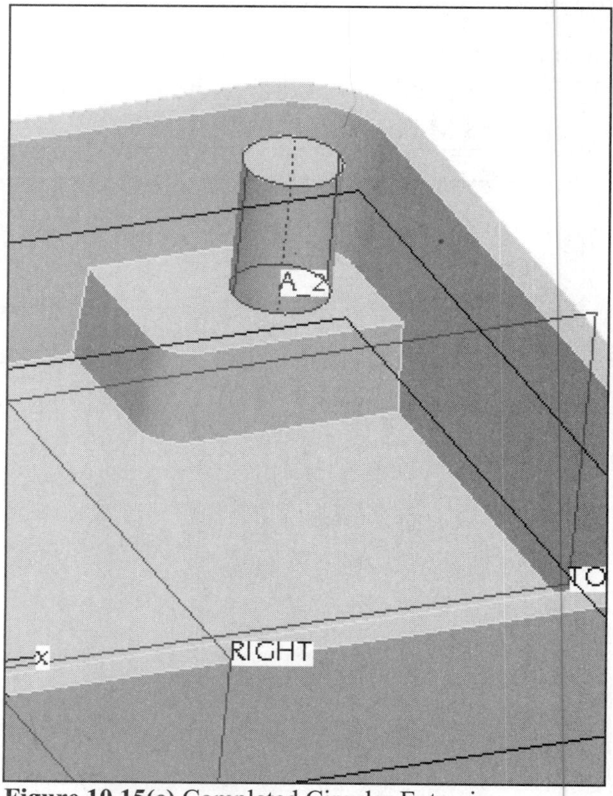

Figure 10.15(e) Completed Circular Extrusion

The circular feature looks correct, but there seems to be a problem with the pedestal.

Click on the pedestal protrusion in the graphics window ⇒ **RMB** ⇒ **Edit Definition** [Fig. 10.16(a)]

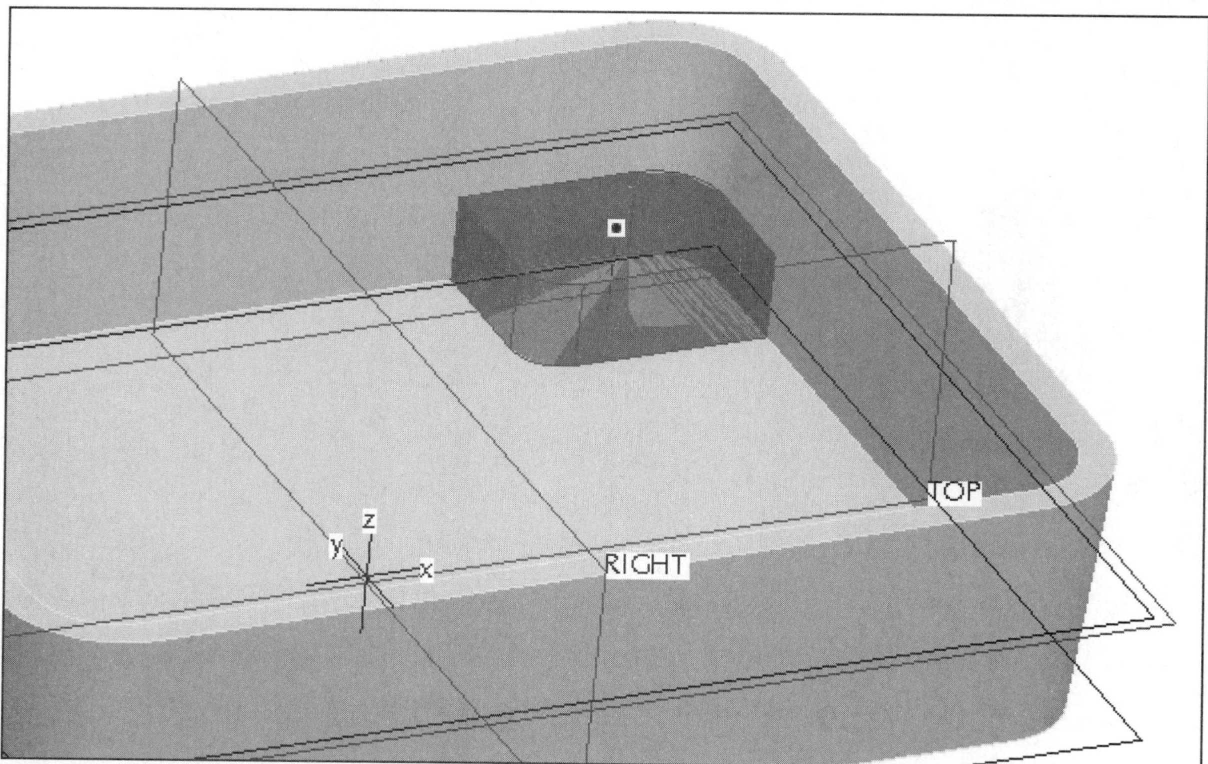

Figure 10.16(a) Redefining the Pedestal

Click: [icon] ⇒ **Sketch** [Fig. 10.16(b)] ⇒ The dimension is referencing the end of the arc instead of the edge. Create a new defining dimension and modify the new dimension to **1.000** [Fig. 10.16(c)] ⇒ [icon] ⇒ **OK** ⇒ [icon] ⇒ [Fig. 10.16(d-e)]

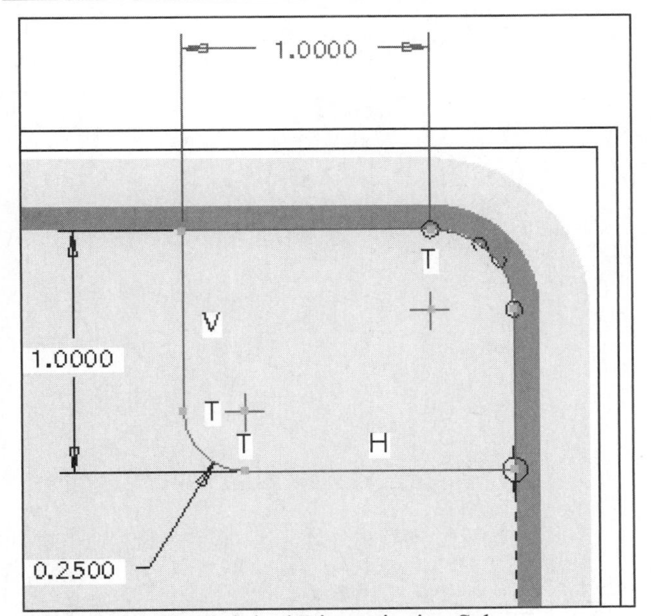

Figure 10.16(b) Original Dimensioning Scheme

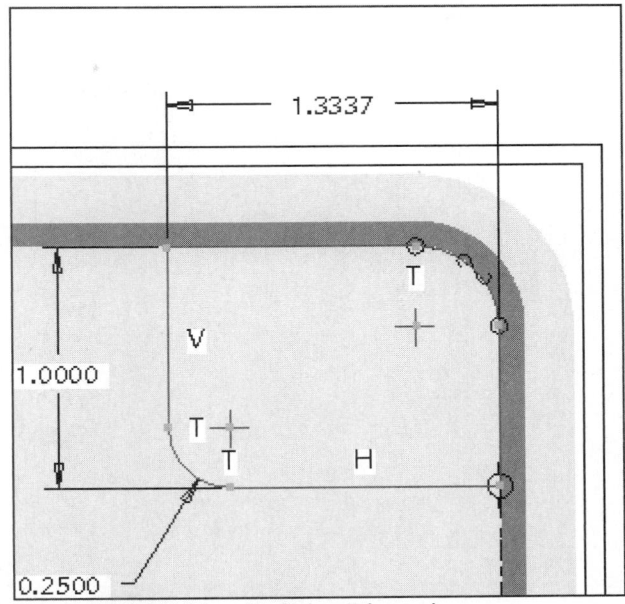

Figure 10.16(c) New Defining Dimension

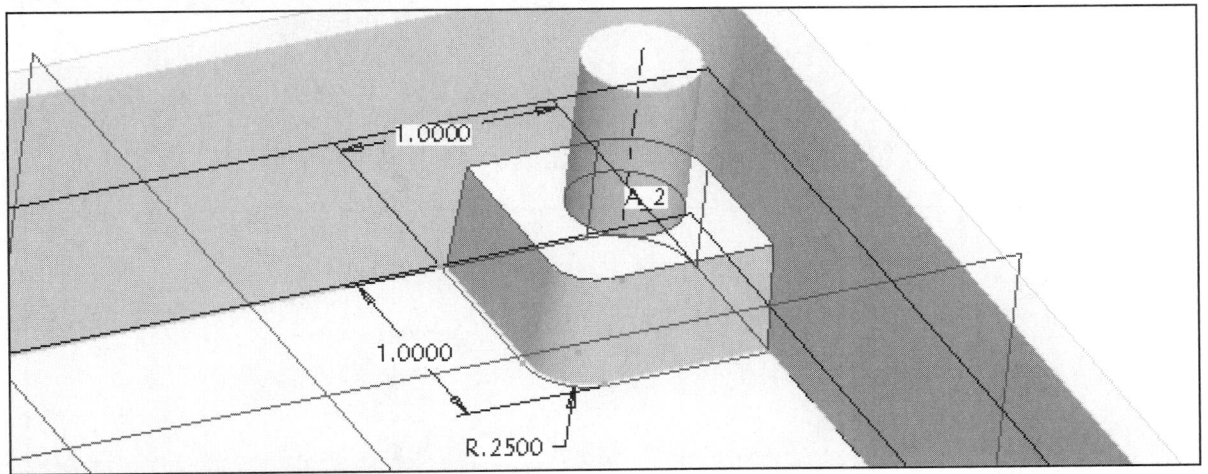

Figure 10.16(d) Redefined Pedestal

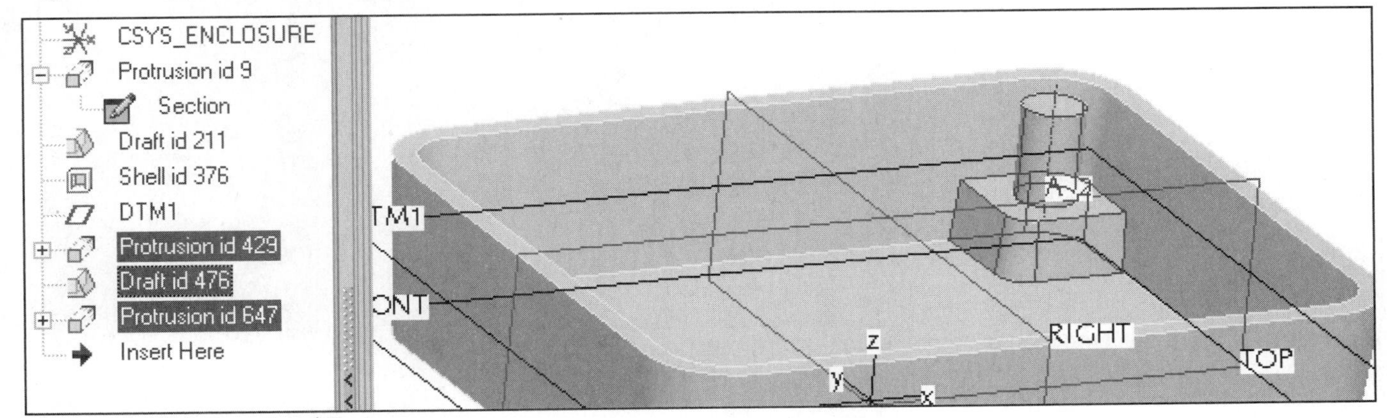

Figure 10.16(e) Completed Features

Draft the circular protrusion at **5°**. Use the upper surface of the circular protrusion as the Draft Hinge [Fig. 10.17(a)]. Select the drag handle. ⇒ **RMB** ⇒ **Flip Angle** ⇒ **MMB** ⇒ **RMB** ⇒ change the dimensions properties to Flip Arrows [Fig. 10.17(b)].

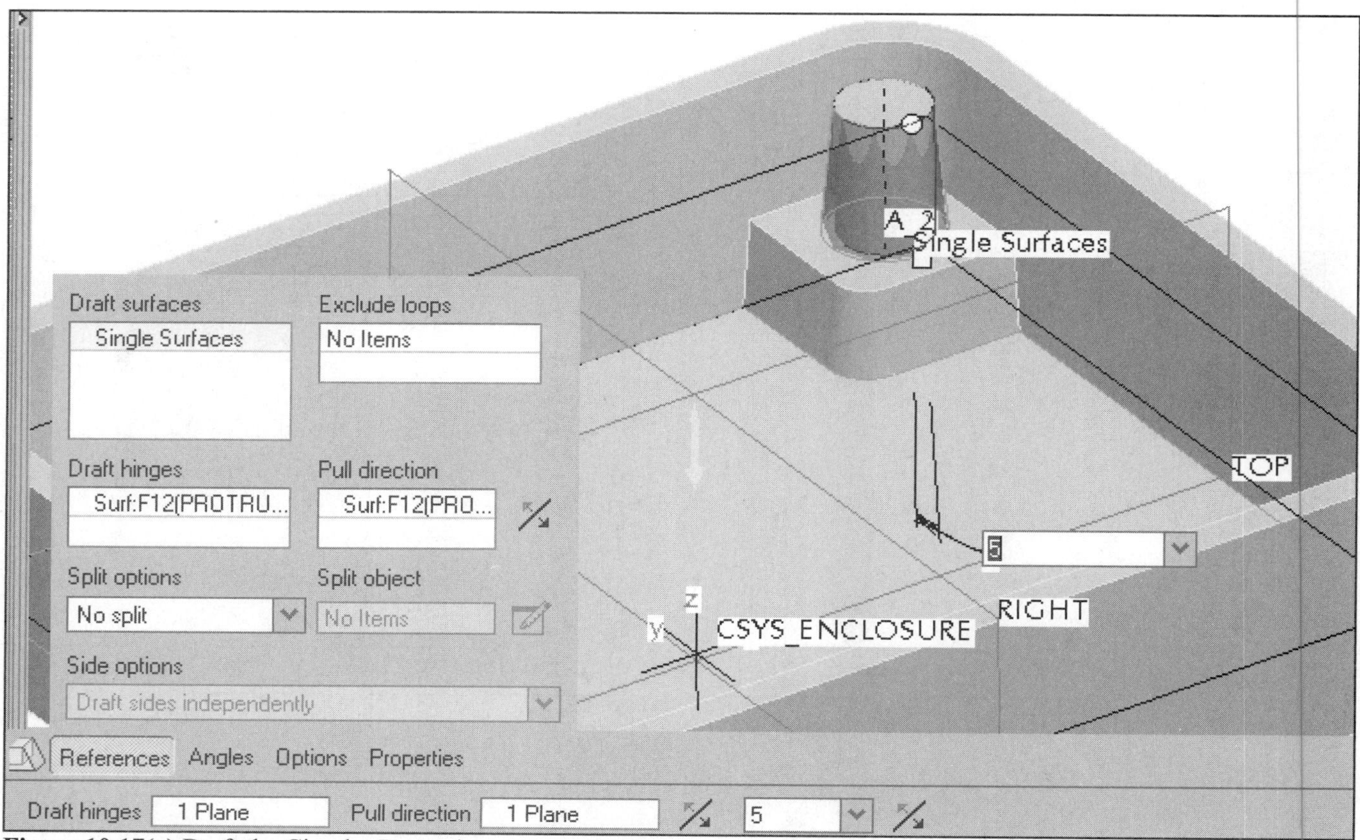

Figure 10.17(a) Draft the Circular Protrusion

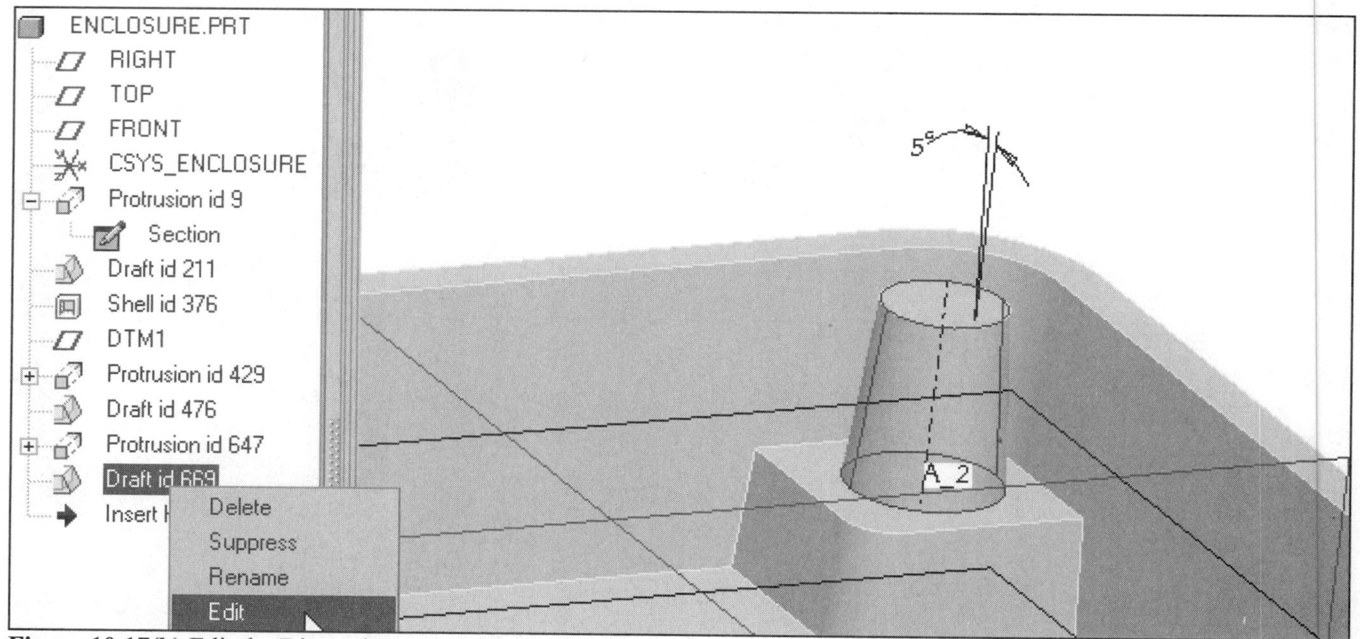

Figure 10.17(b) Edit the Dimensions Properties to Flip Arrows

Create a **.250** diameter coaxial hole on the upper surface of the circular protrusion [Fig. 10.18(a)]. Use "To Selected" to establish the hole's depth [Fig. 10.18(b)]. Complete the feature and save the part.

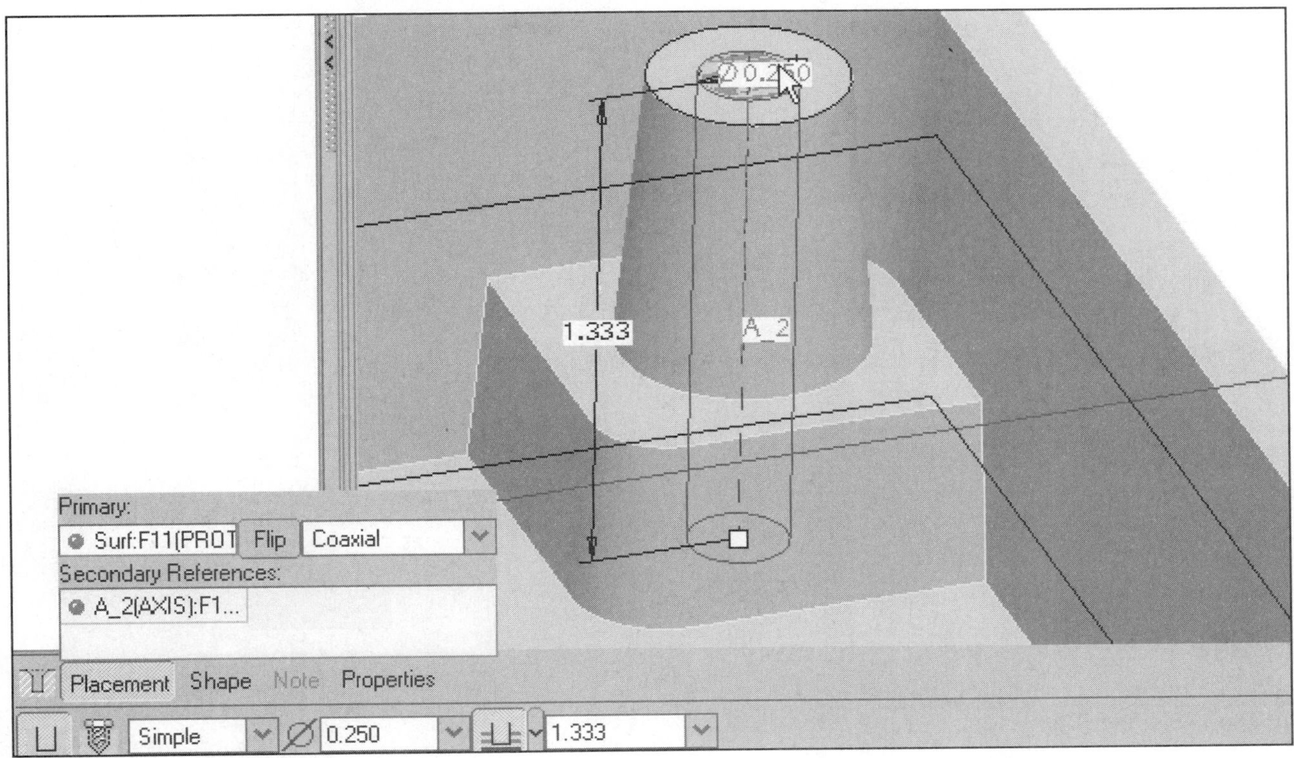

Figure 10.18(a) Coaxial Hole

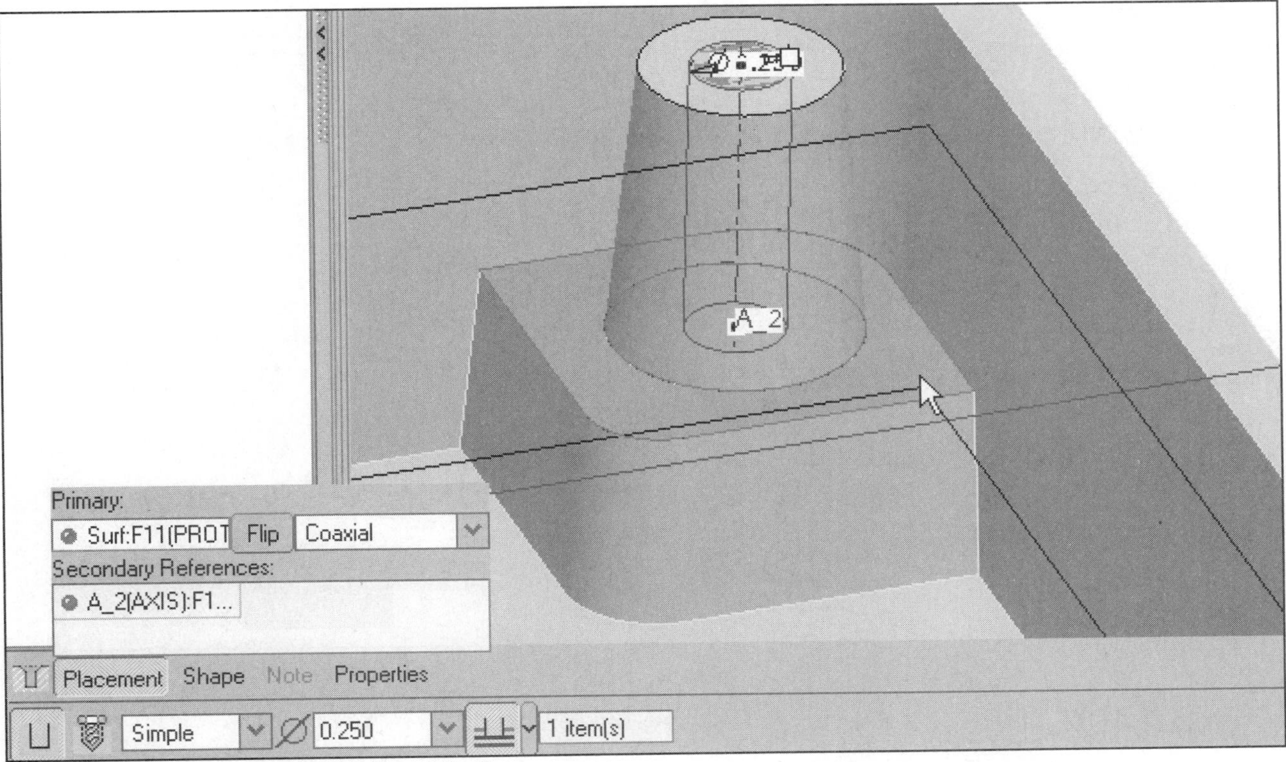

Figure 10.18(b) Hole Depth to Selected Surface

Add an internal draft of **.3°** to the coaxial hole [Fig. 10.18(c)].

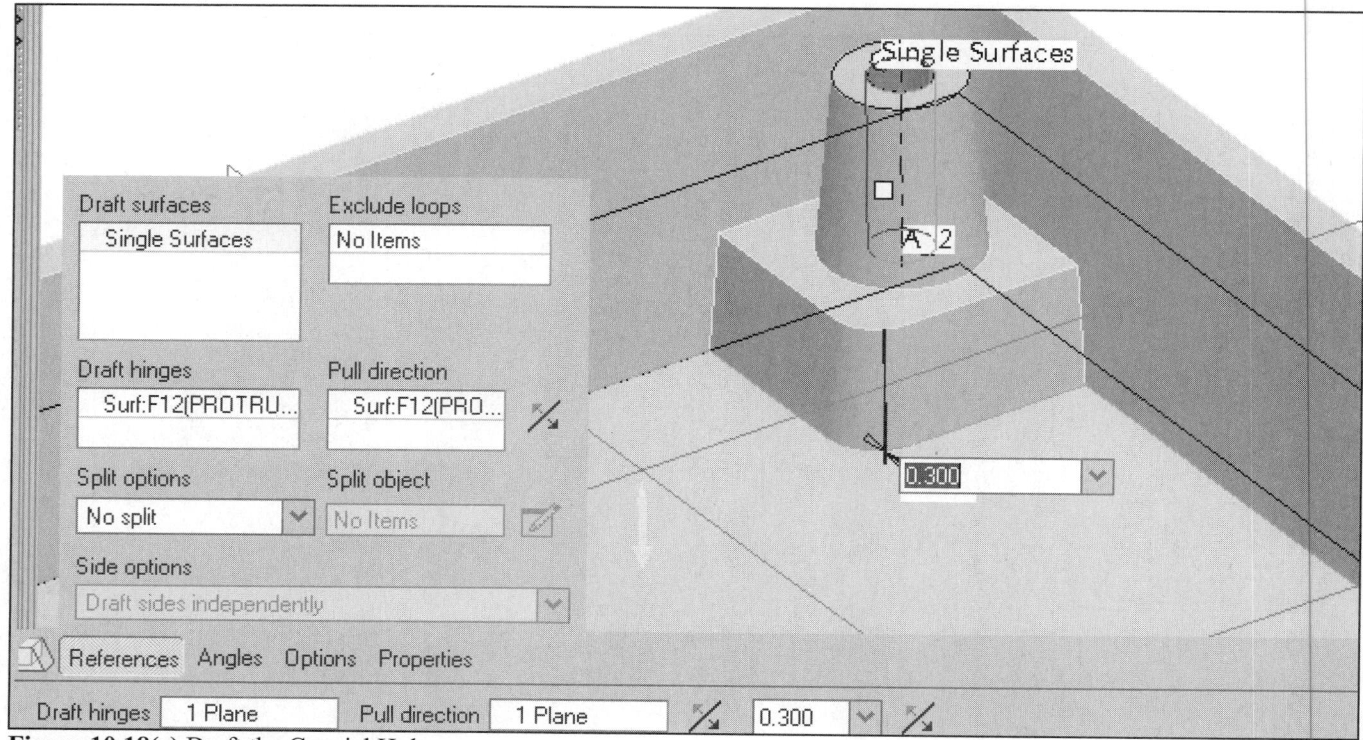

Figure 10.18(c) Draft the Coaxial Hole

Create the **.0625** and **.125** rounds Figures 10.19 (a-c).

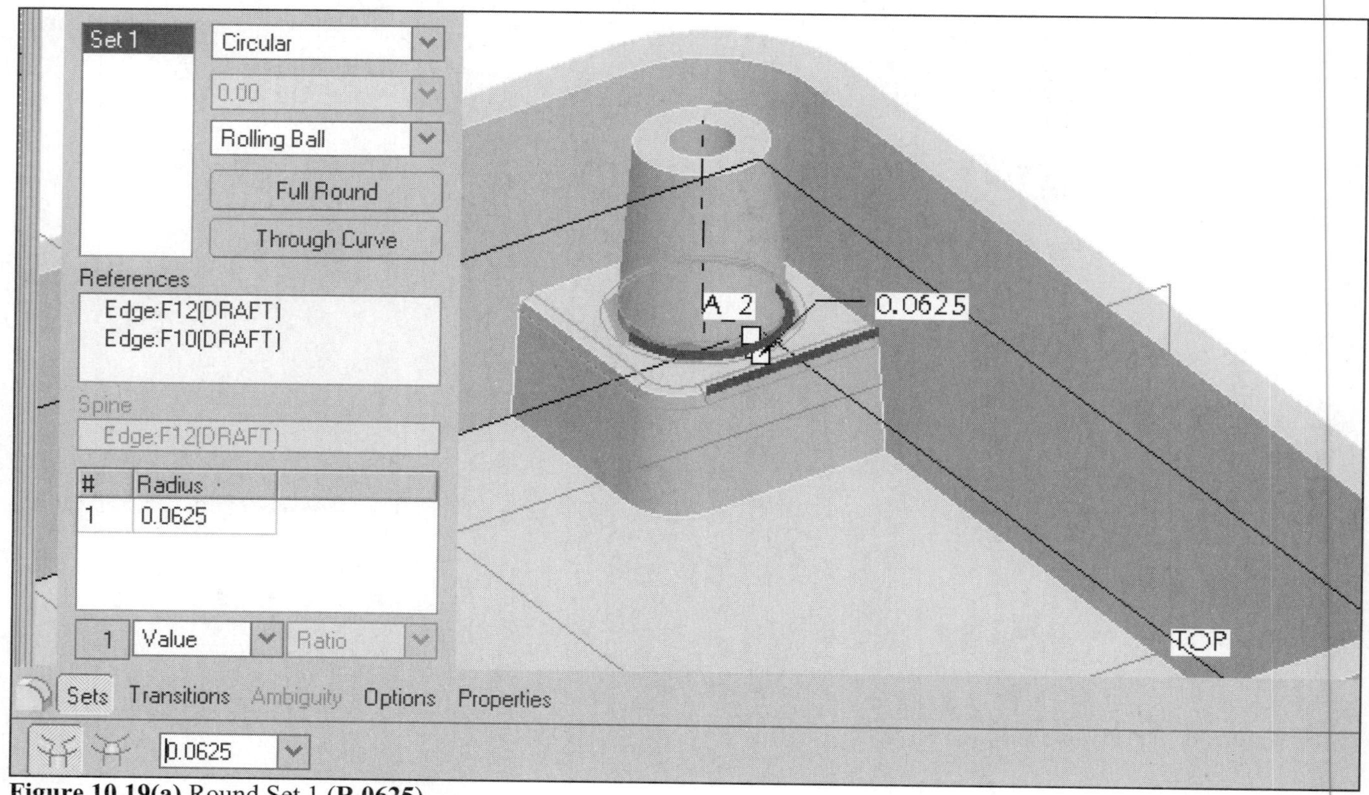

Figure 10.19(a) Round Set 1 (**R.0625**)

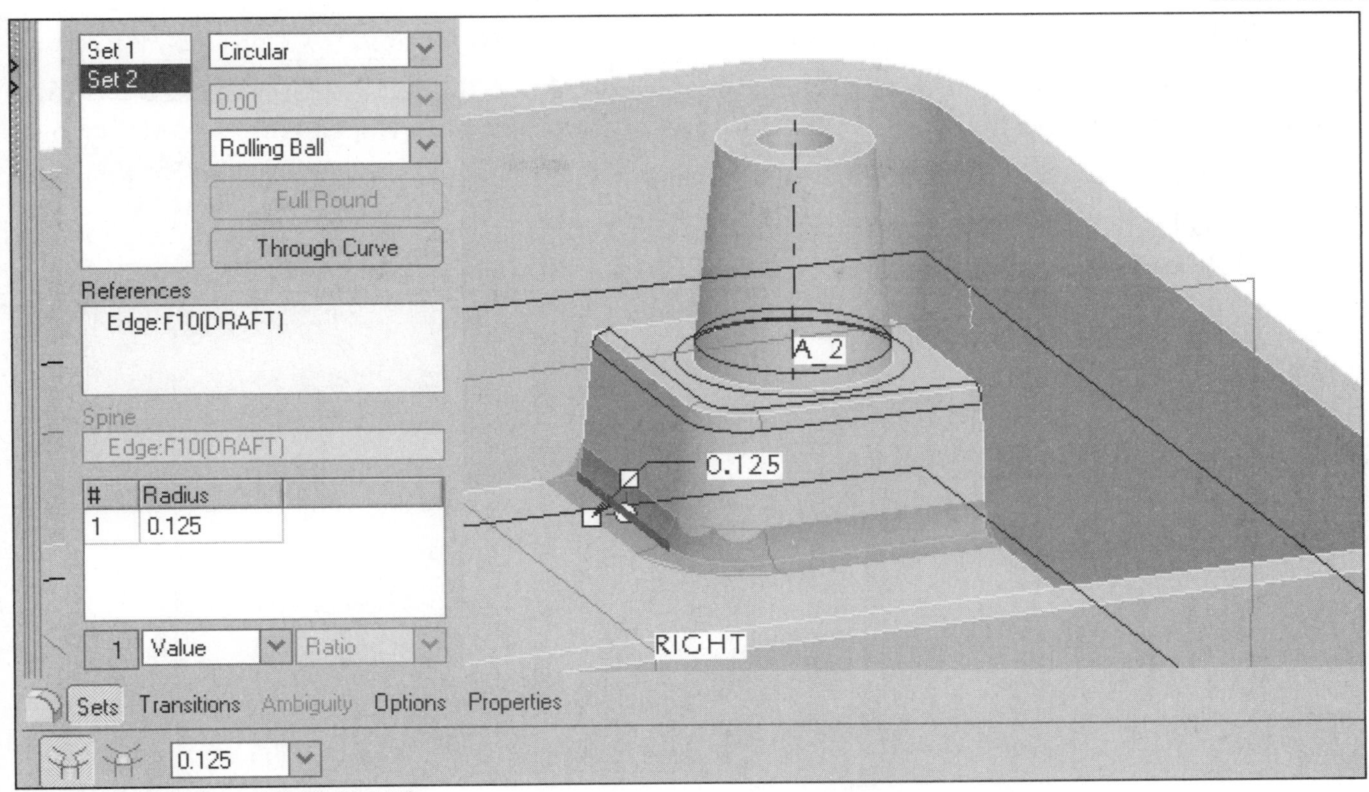

Figure 10.19(b) Round Set 2 (**R.125**)

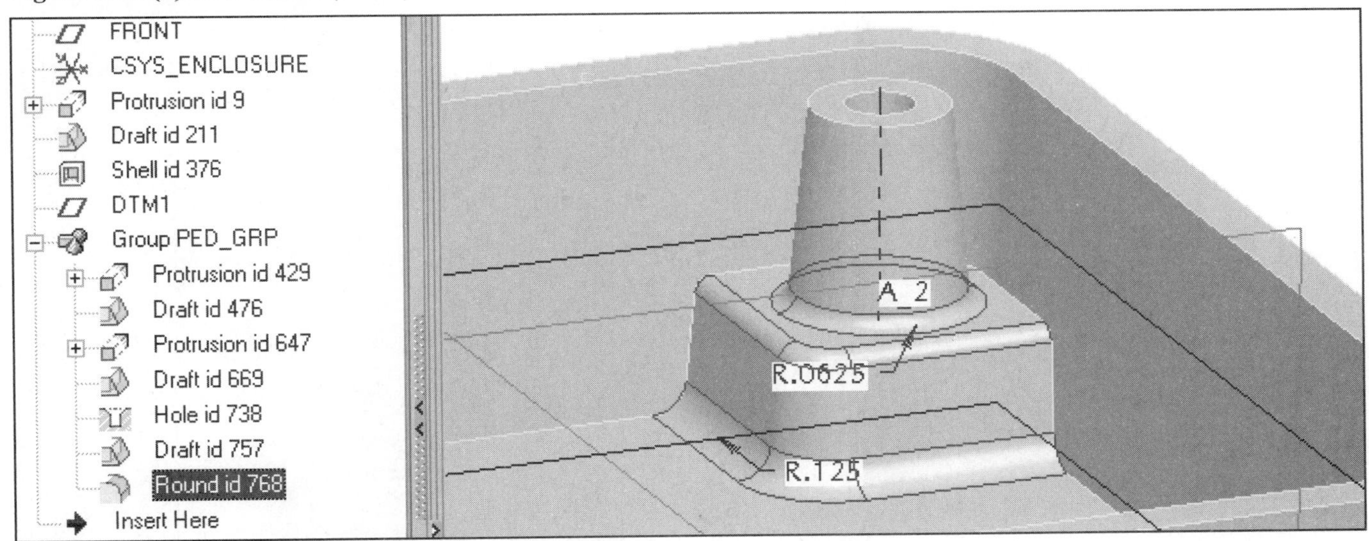

Figure 10.19(c) Completed Rounds (**R.125** and **R.0625**)

Group the protrusions, hole and rounds, click: **Edit** ⇒ **Feature Operations** ⇒ **Group** ⇒ **Local Group** ⇒ type **PED_GRP** ⇒ **MMB** ⇒ select features from the Model Tree (Fig. 10.20) ⇒ **MMB** ⇒ **MMB** ⇒ **MMB** ⇒ **MMB** ⇒ **LMB** to deselect

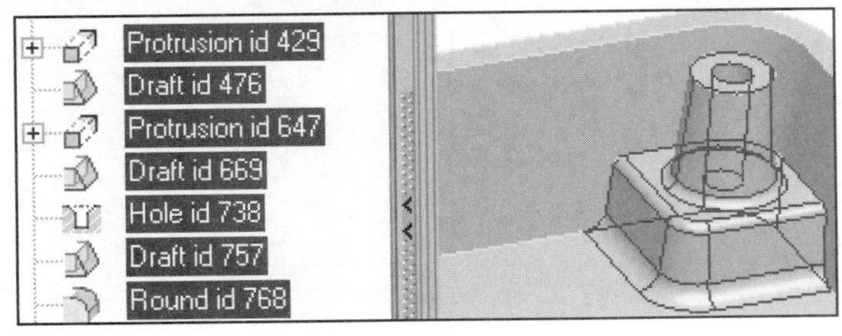

Figure 10.20 Create a Group

Create three identical grouped features, click: **Edit** ⇒ **Feature Operations** ⇒ **Copy** ⇒ **Mirror** ⇒ **Dependent** ⇒ **MMB** ⇒ from the Model Tree, pick: **Group PED_GRP** Group PED_GRP ⇒ **MMB** ⇒ **MMB** ⇒ select datum **RIGHT** [Fig. 10.21(a)] ⇒ **Copy** ⇒ **Mirror** ⇒ **Dependent** ⇒ **MMB** ⇒ pick **Group PED_GRP** Group PED_GRP and **Group COPIED_GROUP** Group COPIED_GROUP from the Model Tree ⇒ **MMB** ⇒ **MMB** ⇒ select datum **TOP** [Fig. 10.21(b)] ⇒ **MMB** ⇒ **Save** ⇒ **MMB** ⇒ **File** ⇒ **Delete** ⇒ **Old Versions** ⇒ **MMB**

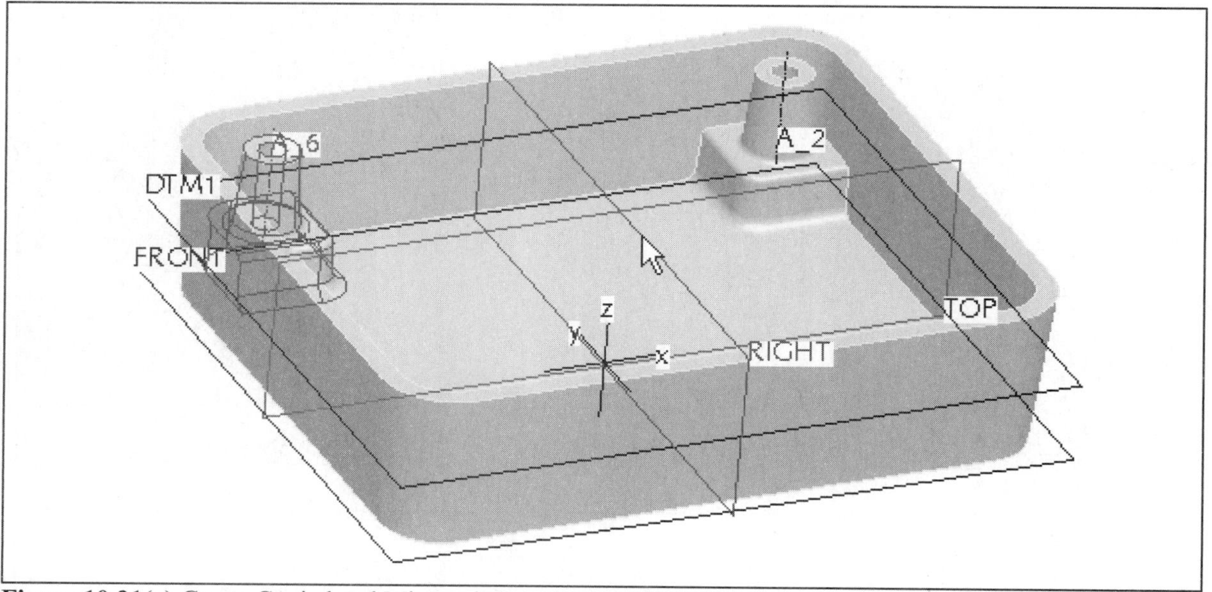

Figure 10.21(a) Group Copied and Mirrored about Datum RIGHT

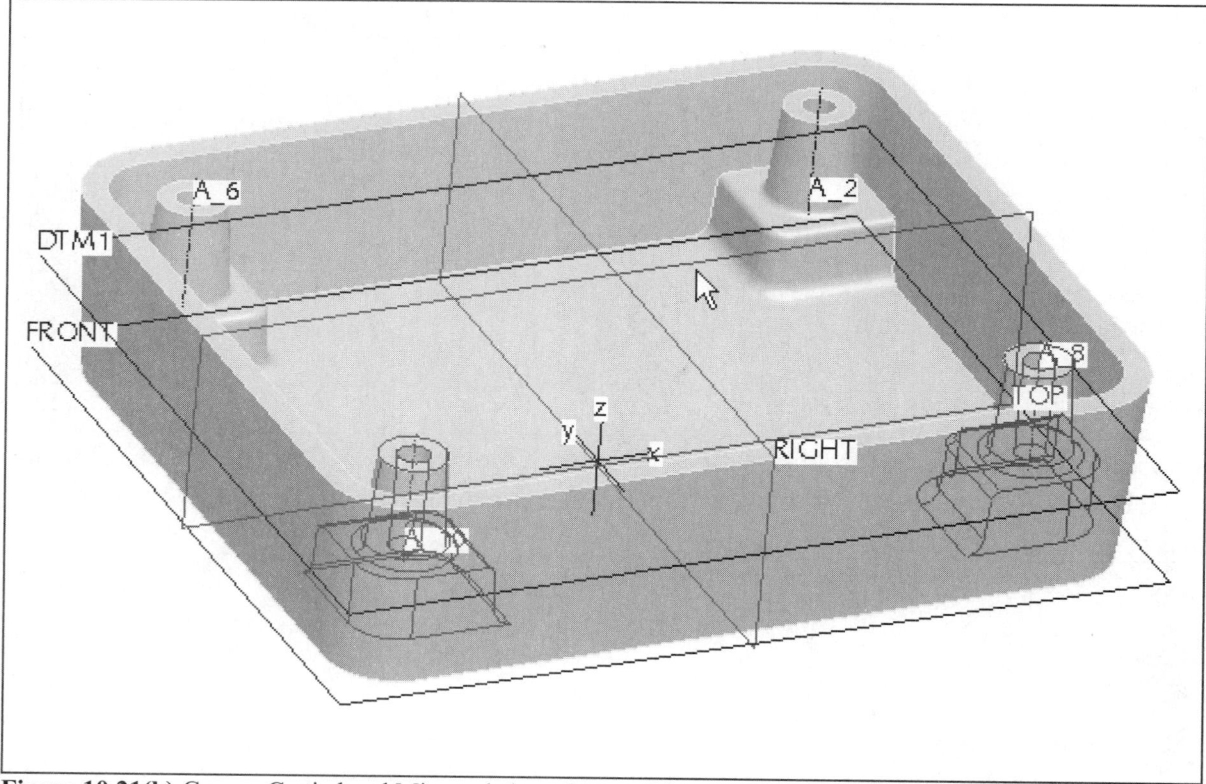

Figure 10.21(b) Groups Copied and Mirrored about Datum TOP

Create the internal round, click: **Round Tool** ⇒ type **.125** ⇒ select the inside of the shelled wall as the first reference [Fig. 10.22(a)] ⇒ **Sets** tab ⇒ press and hold the **Ctrl** key ⇒ select the top surface of the pedestal as the second reference [Fig. 10.22(b)]

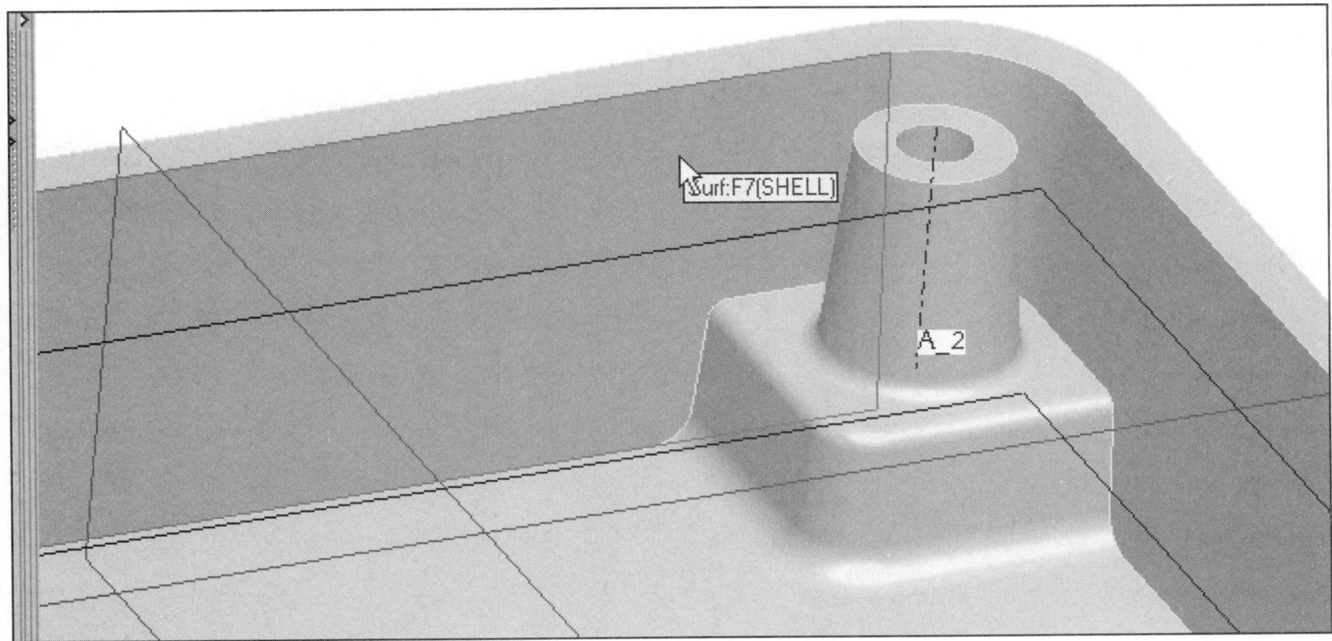

Figure 10.22(a) Select First Reference

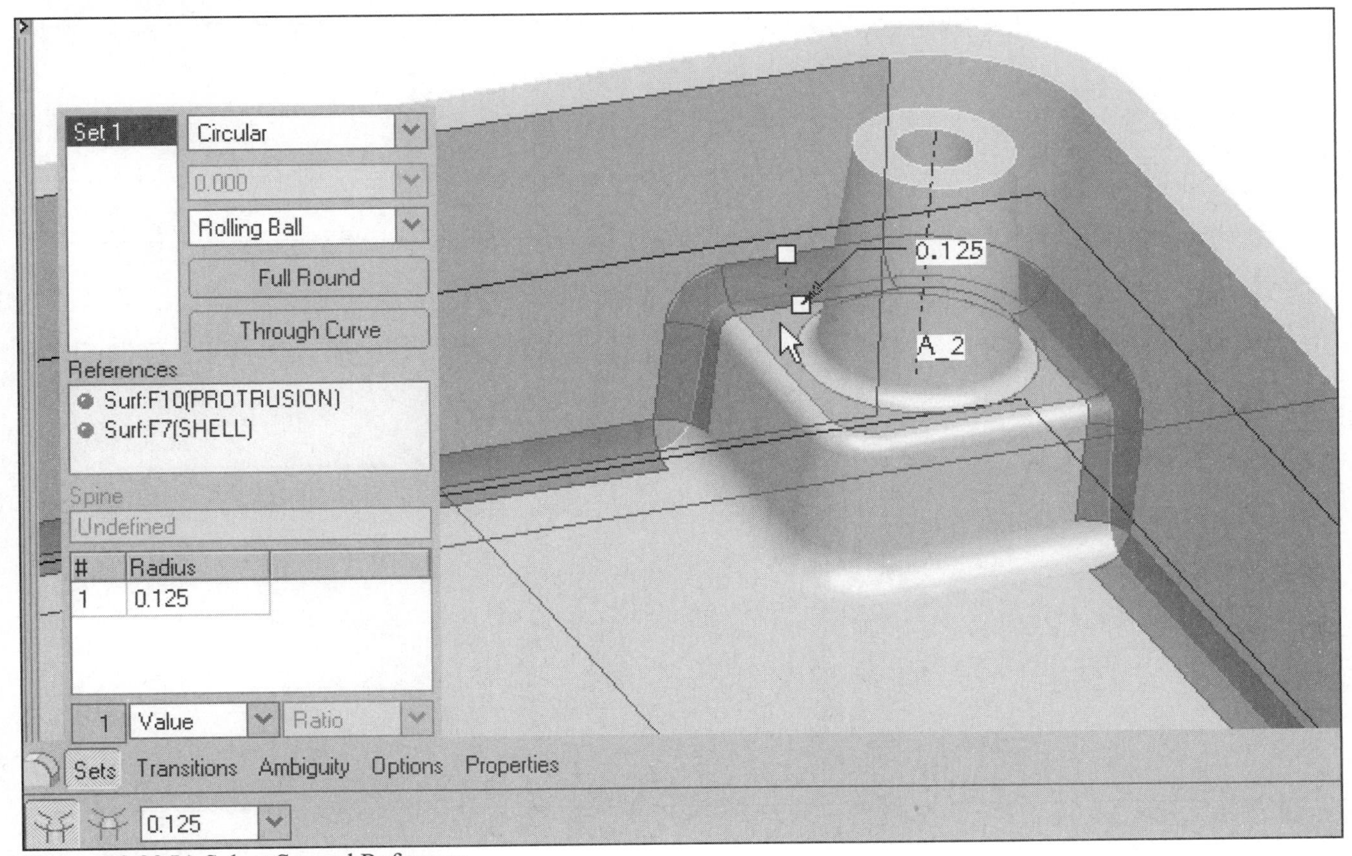

Figure 10.22(b) Select Second Reference

Click: [icon] ⇒ SHOW ERRORS menu **MMB** ⇒ [icon] from Dashboard ⇒ RESOLVE FEAT menu **Quick Fix** ⇒ **Redefine** ⇒ **Confirm** ⇒ **Options** tab ⇒ **Surface** [Fig. 10.22(c)] ⇒ [icon] [Fig. 10.22(d)] ⇒ **Yes** ⇒ [icon] ⇒ **MMB**

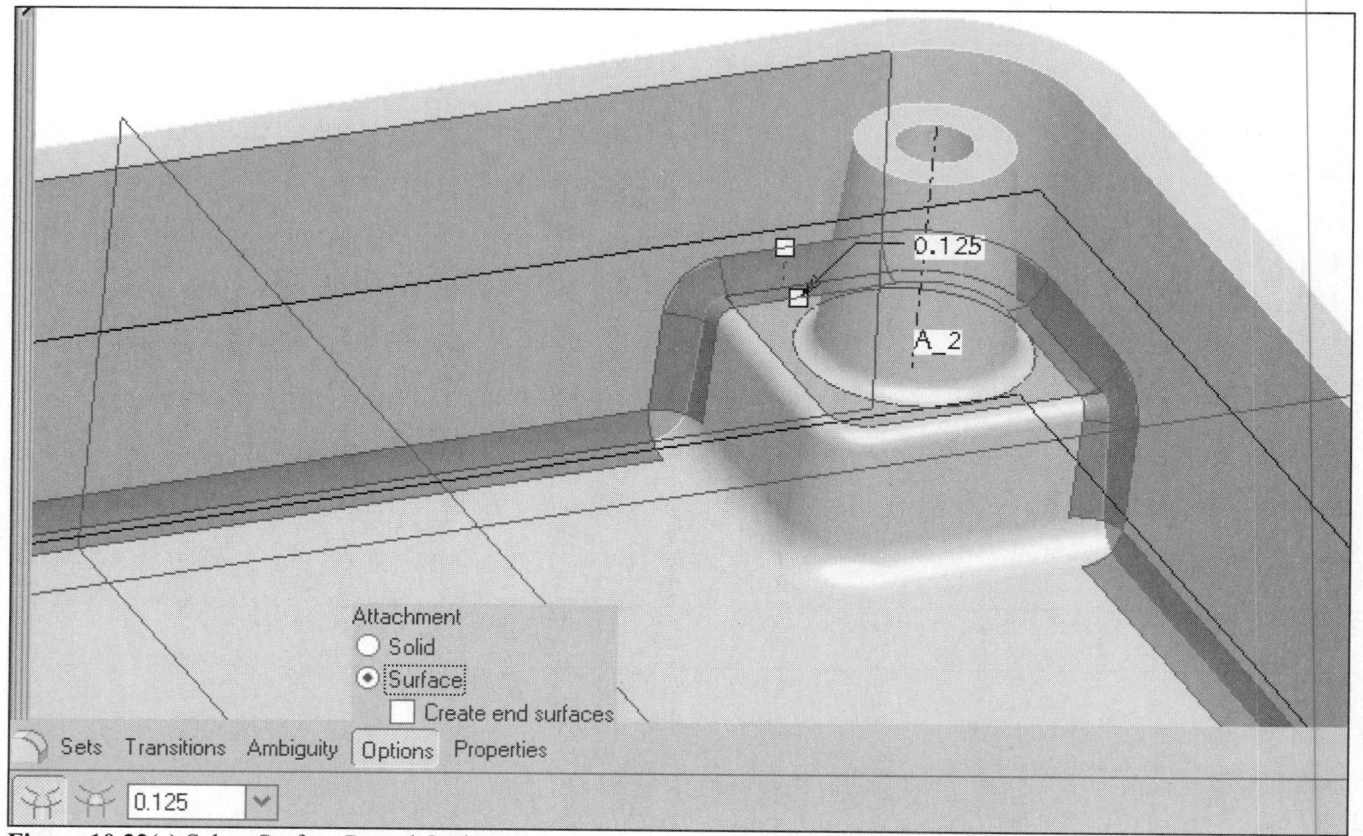

Figure 10.22(c) Select Surface Round Option

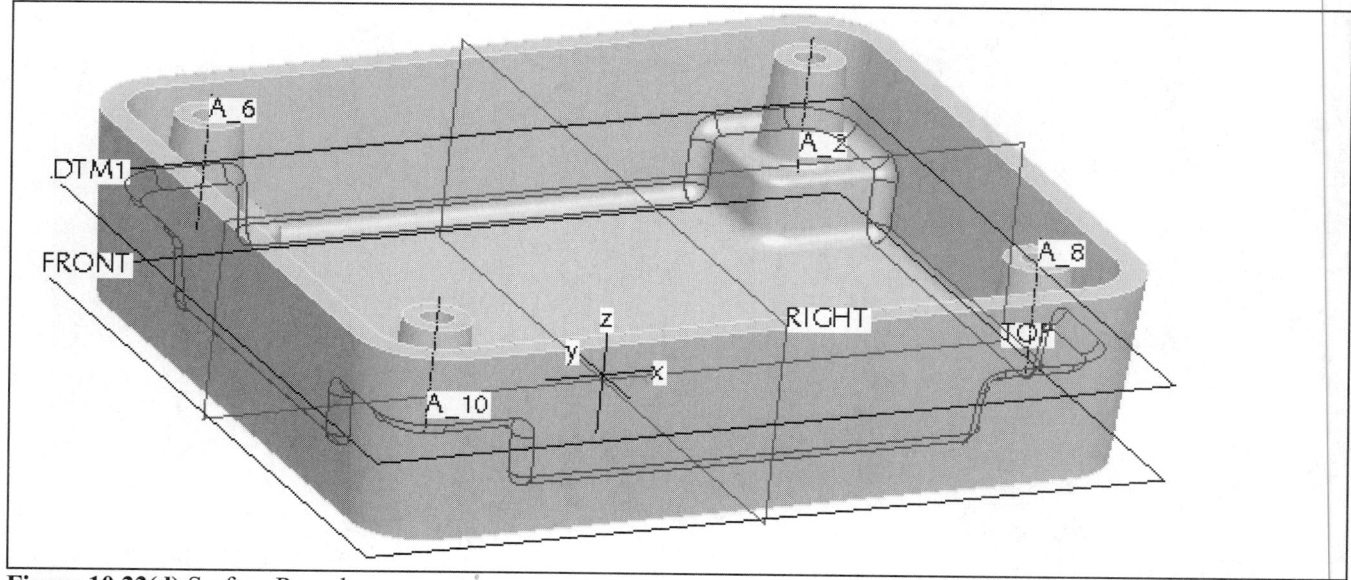

Figure 10.22(d) Surface Round

The part's internal round may look correct, but it is not a solid round. Next, model a datum plane through the axis of two pedestals and use it to create a cross section.

Click: **Datum Plane Tool** ⇒ References: pick axis **A_2** [Fig. 10.23(a)] ⇒ press and hold the **Ctrl** key ⇒ References: pick axis **A_6** *(your id's may be different)* [Fig. 10.23(b)] ⇒ **OK** (creates **DTM2**)

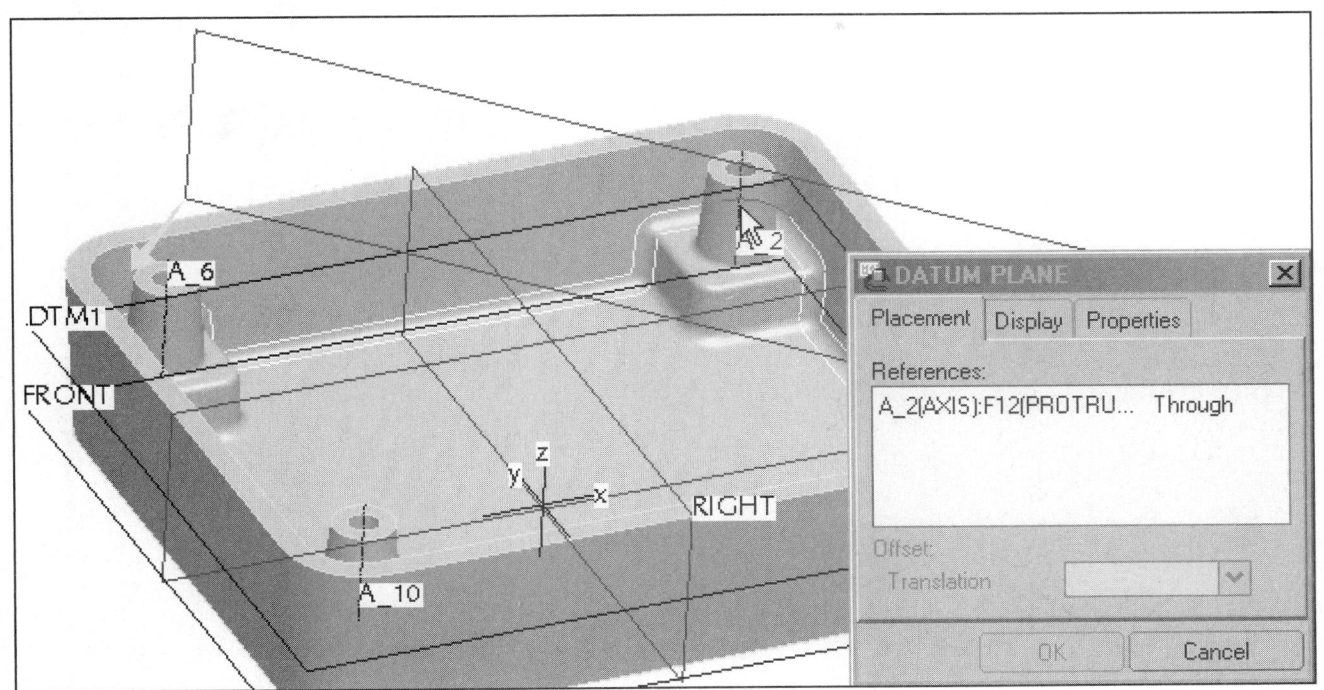

Figure 10.23(a) Select Axis A_2

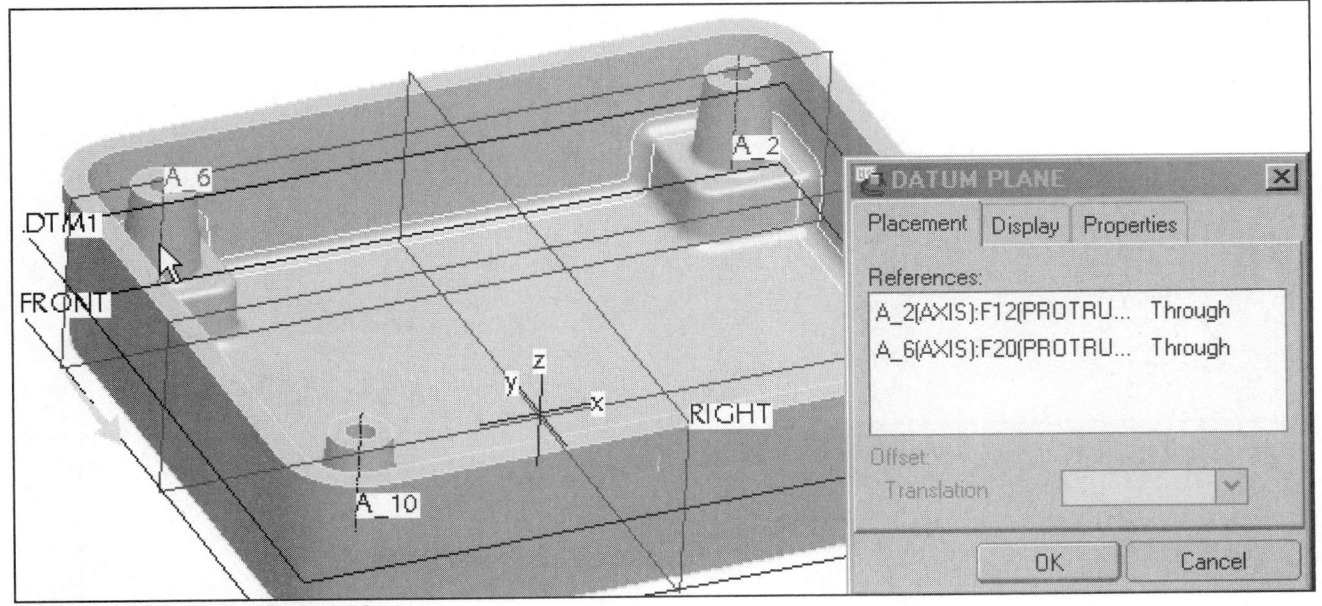

Figure 10.23(b) Select Axis A_6

Create a cross section through the part, using datum **DTM2**.

Click: **Tools** ⇒ Model Sectioning [Fig. 10.24(a)] ⇒ **New** ⇒ type **A** ⇒ **MMB** ⇒ **Model** ⇒ **Planar** ⇒ **Single** ⇒ **MMB** ⇒ **Plane** ⇒ pick **DTM2** ⇒ **Display** ⇒ **Show X-Section** ⇒ click on section **A** in the Cross Section dialog box ⇒ **RMB** ⇒ **Redefine** [Fig. 10.24(b)] ⇒ **Hatching** ⇒ **Fill** ⇒ **MMB** ⇒ **MMB** [Fig. 10.24(c)] ⇒ **Close**

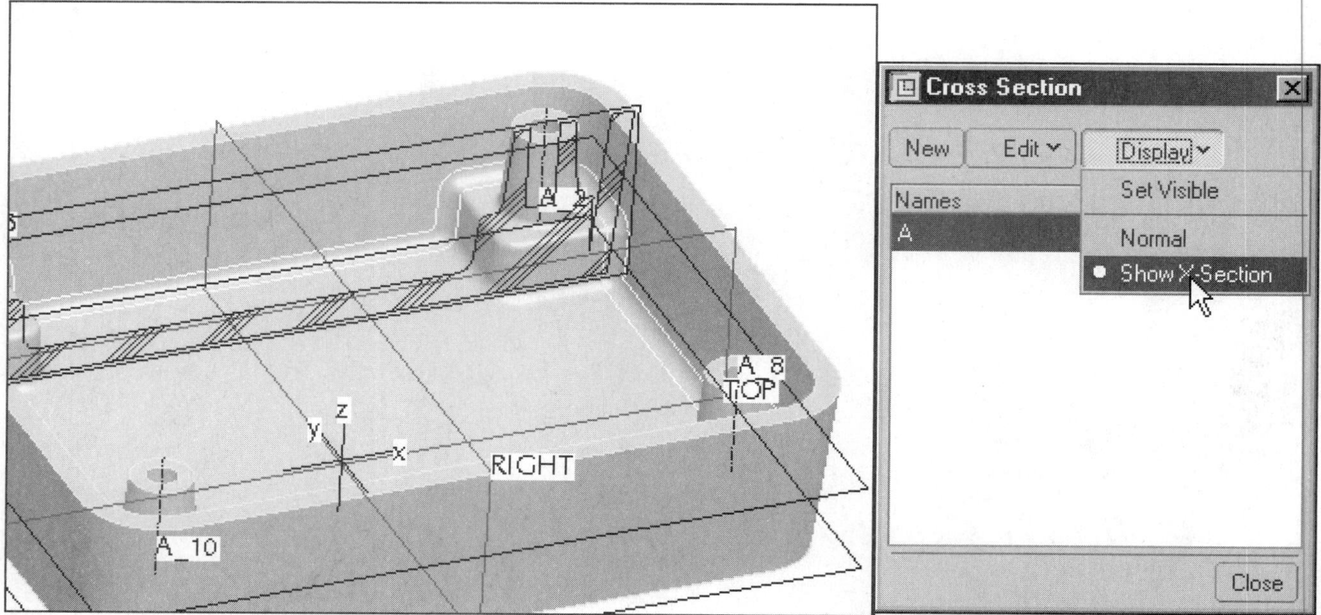

Figure 10.24(a) Show X-Section

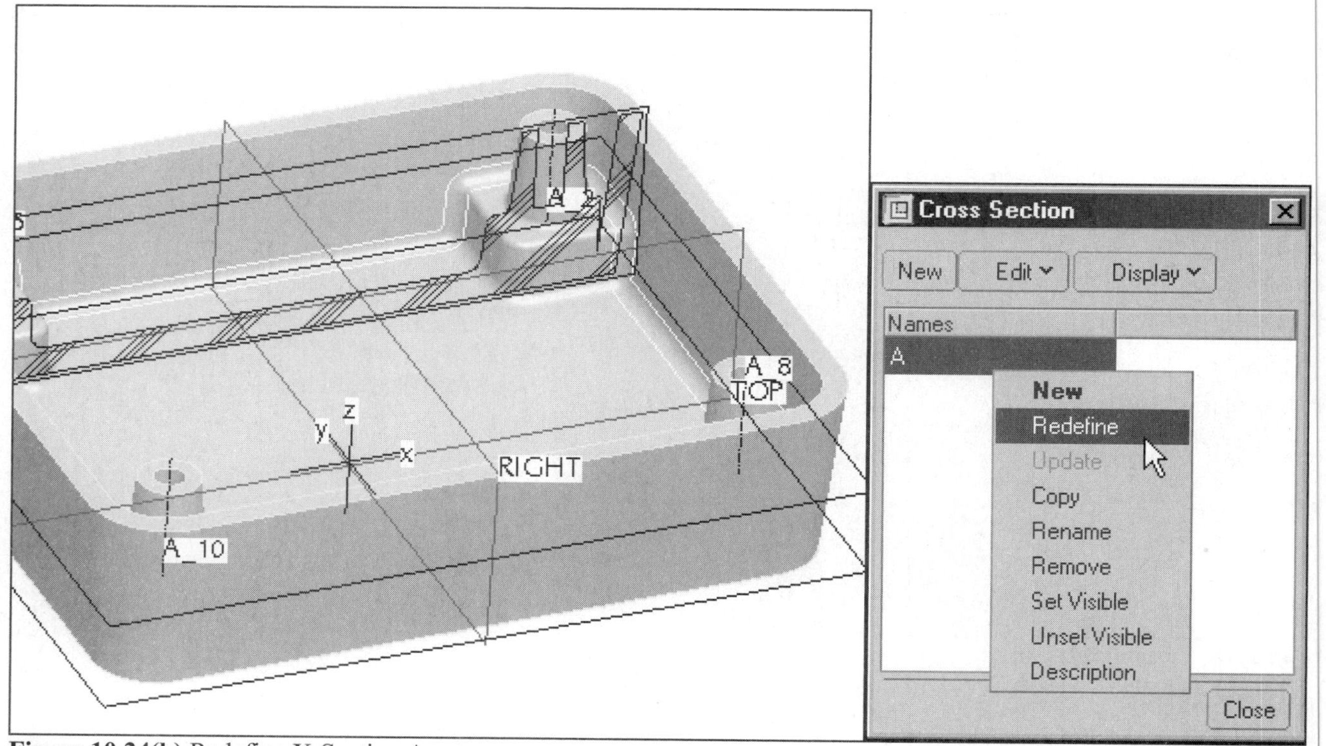

Figure 10.24(b) Redefine X-Section A

The gaps between the pedestals and the shelled walls were not filled by the Round Tool because the round was a surface round.

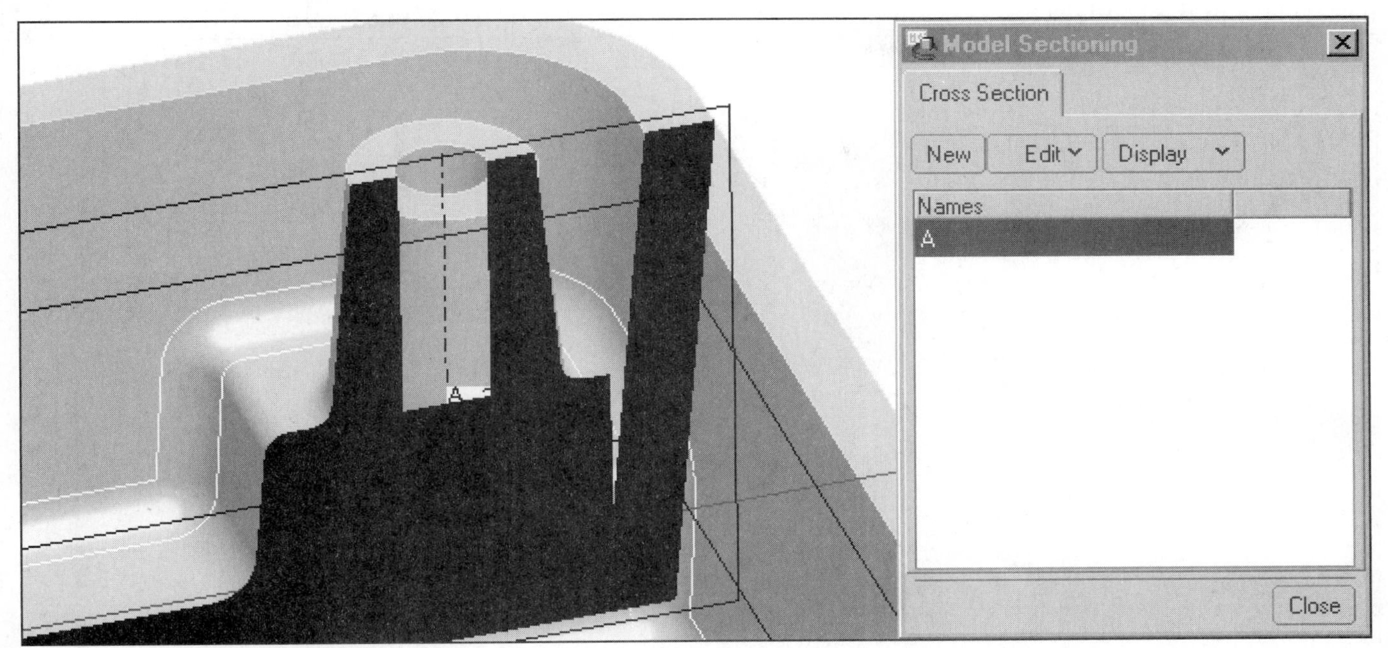

Figure 10.24(c) Fill the X-Section

Click on the surface round in the Model Tree (it will highlight on the model) ⇒ **Edit** from the menu bar ⇒ **Solidify** ⇒ **References** tab (Fig. 10.25) ⇒ **MMB**

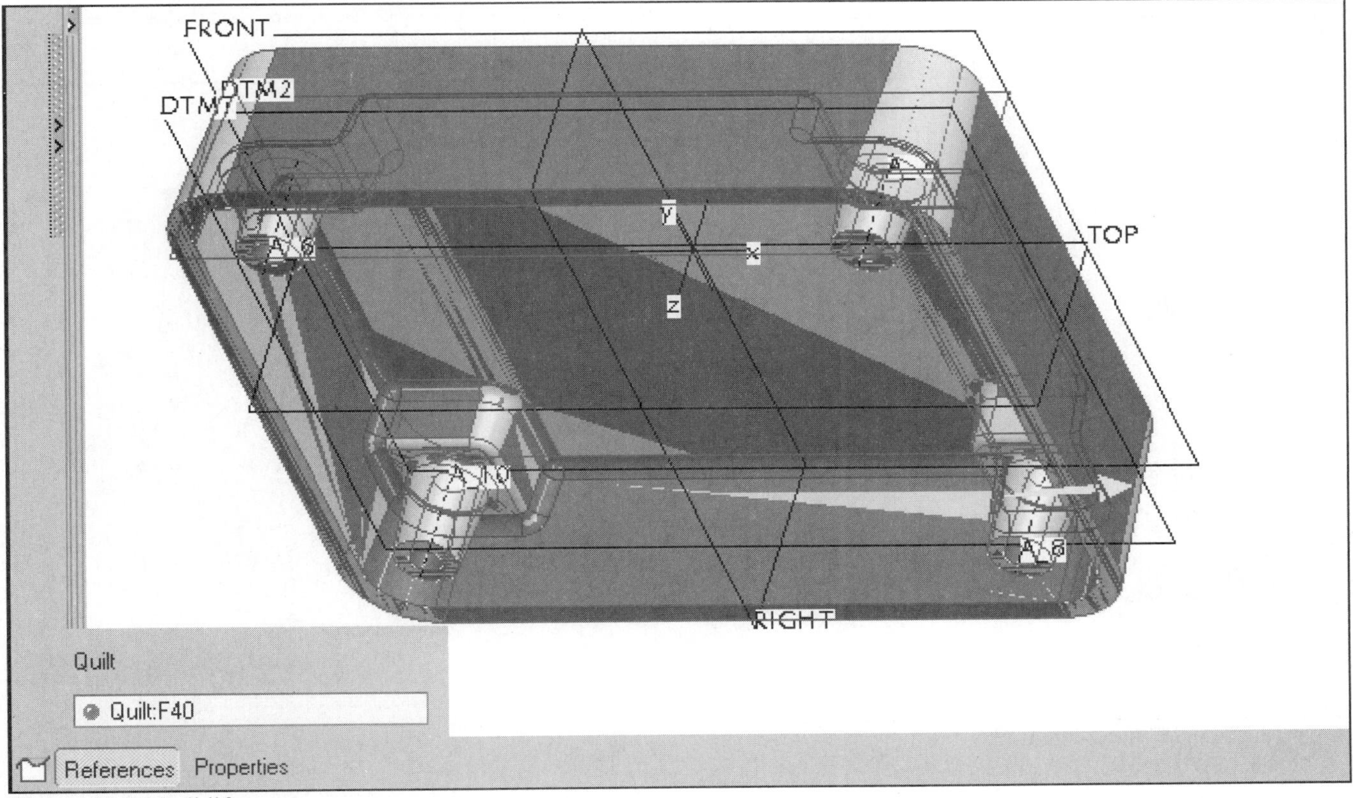

Figure 10.25 Solidify

Click: **Tools** ⇒ Model Sectioning ⇒ **Display** ⇒ **Show X-Section** (Fig. 10.26) ⇒ **Close** ⇒ click on the last feature in the Model Tree to see it highlighted on the model (Fig. 10.27) ⇒ ⇒ **MMB**

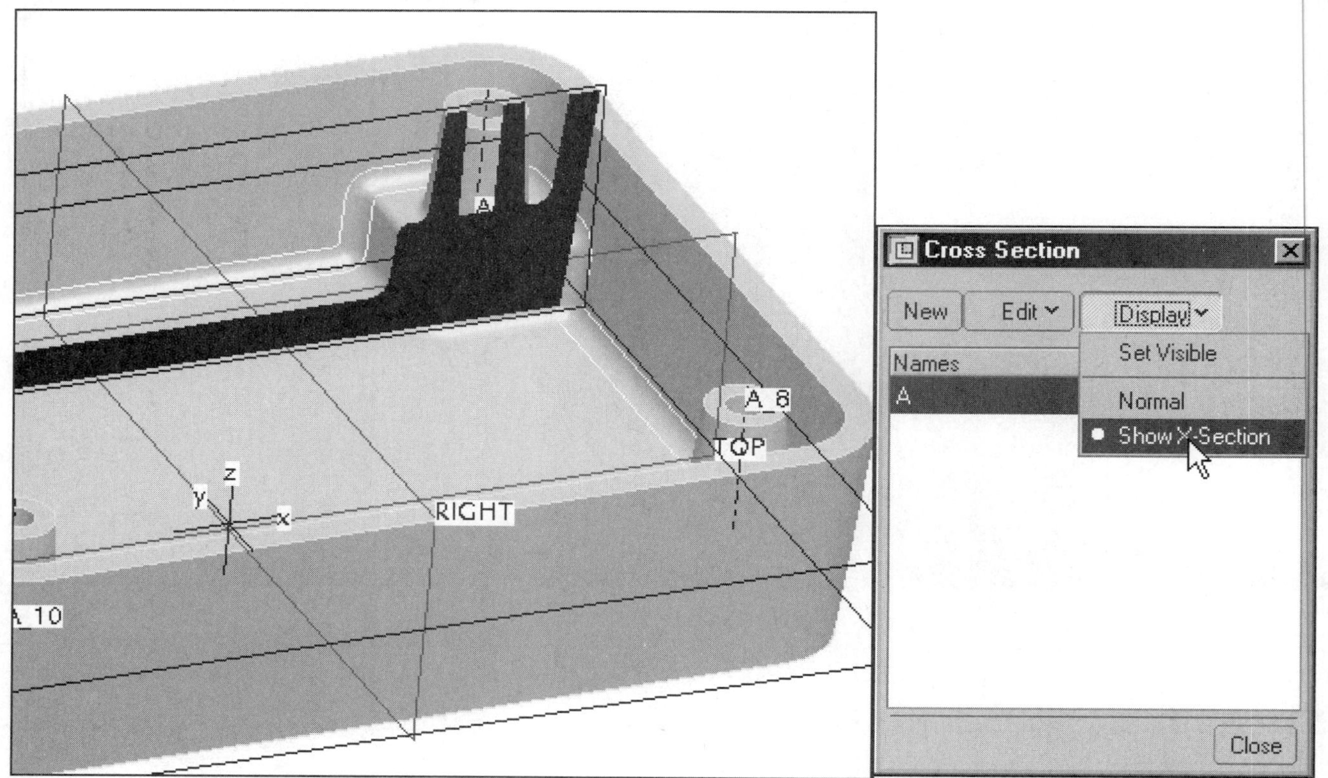

Figure 10.26 Gap has been Solidified

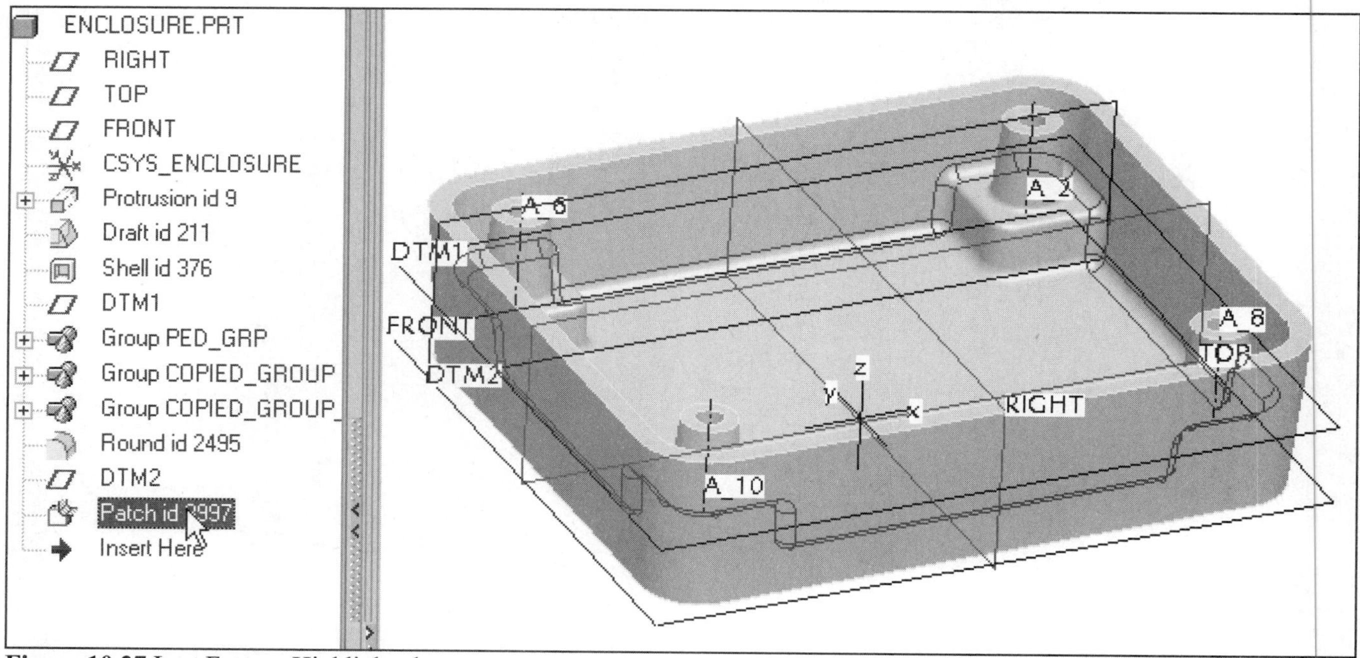

Figure 10.27 Last Feature Highlighted

Before creating the text protrusion, **Suppress** all the features after the shell command. Expand the Model Tree to include the feature number and status.

Click: **Settings** ⇒ **Tree Filters** ⇒ ☑ Notes ⇒ ☑ Suppressed Objects (Fig. 10.28) ⇒ **Apply** ⇒ **OK**

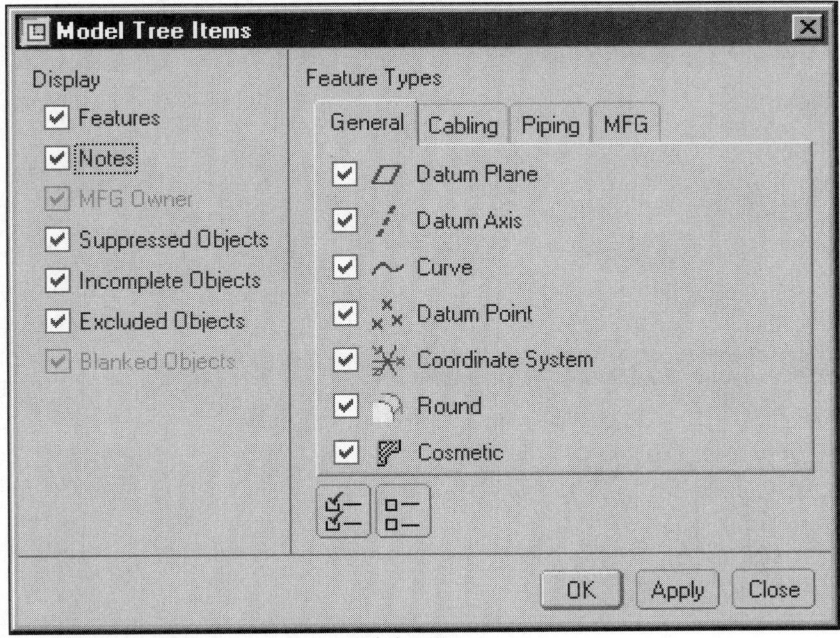

Figure 10.28 Model Tree Items Dialog Box

Click on **Group PED_GRP** in the model tree ⇒ press and hold the **Shift** key ⇒ click on the last feature in the Model Tree [Fig. 10.29(a)] ⇒ **RMB** ⇒ **Suppress** [Fig. 10.29(b)] ⇒ **OK** [Fig. 10.29(c)]

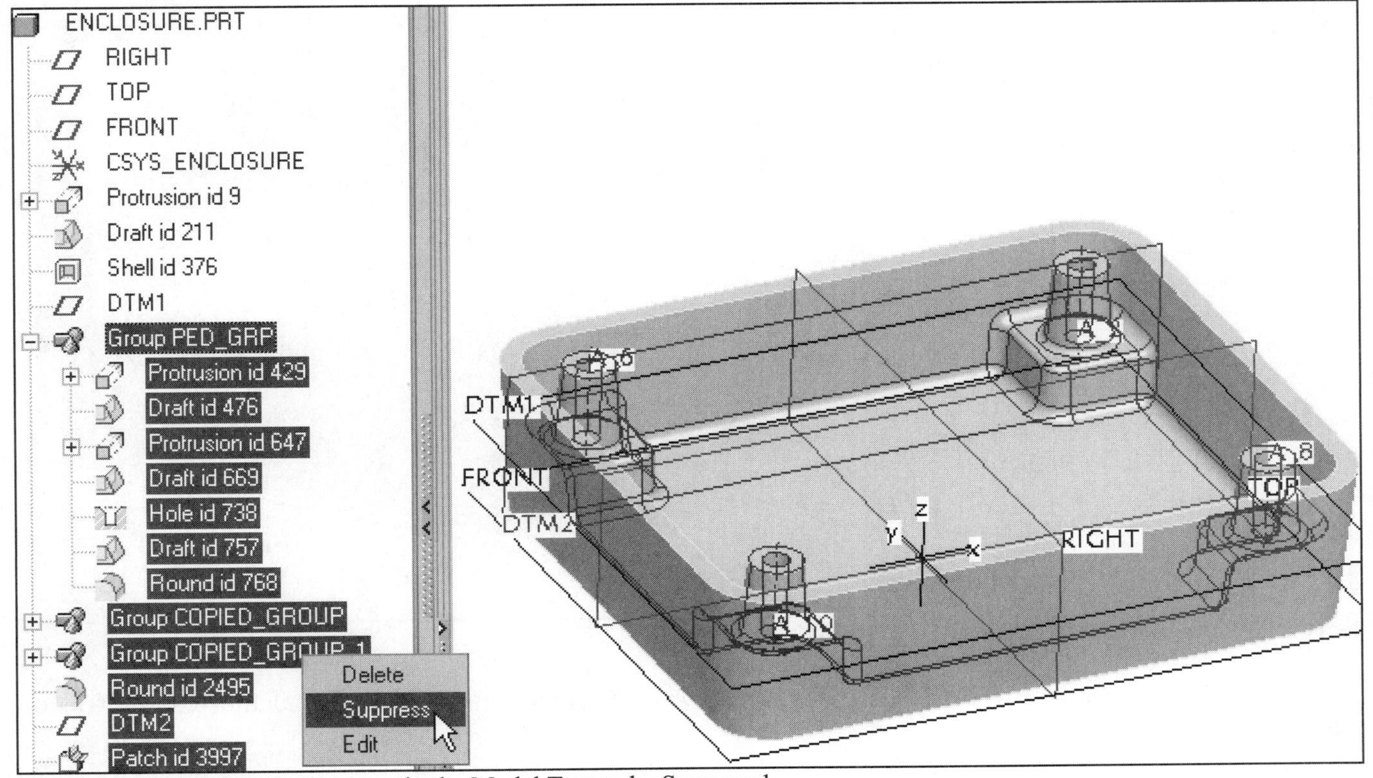

Figure 10.29(a) Select the Features in the Model Tree to be Supressed

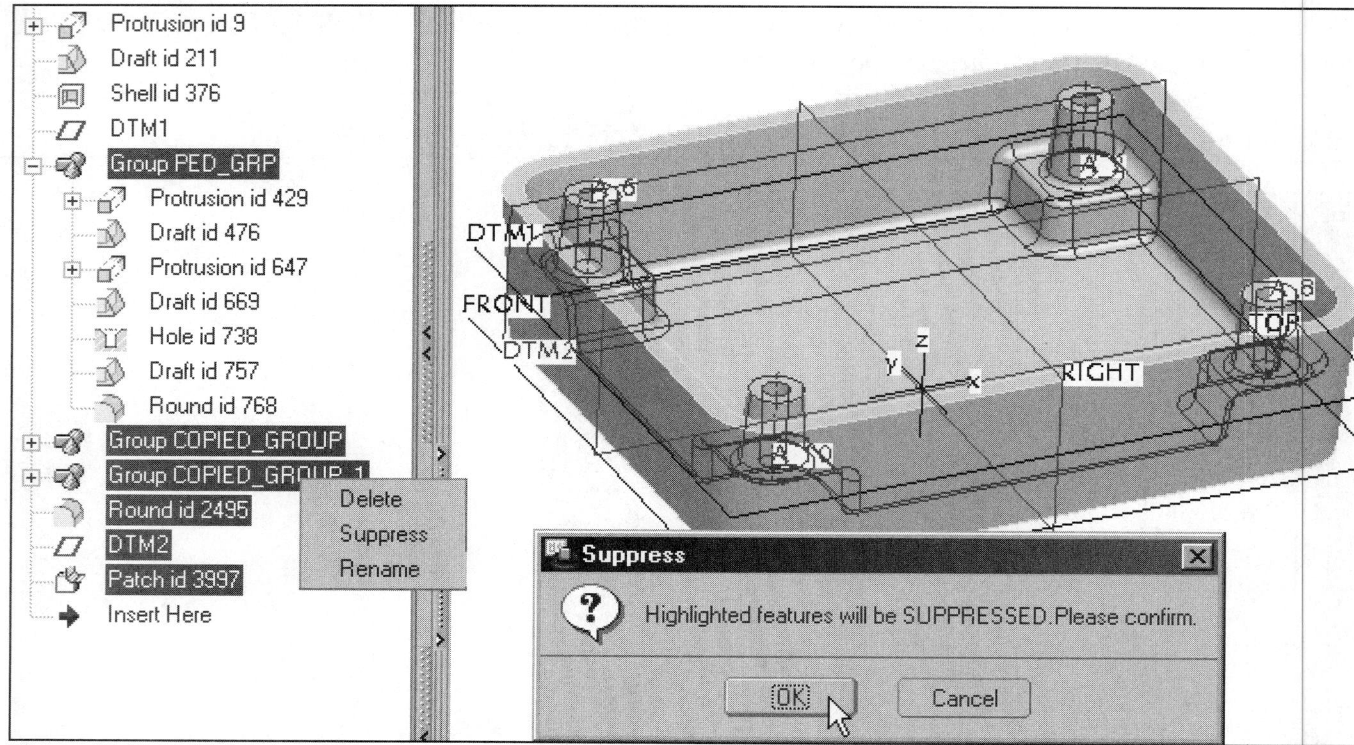

Figure 10.29(b) Highlighted Features

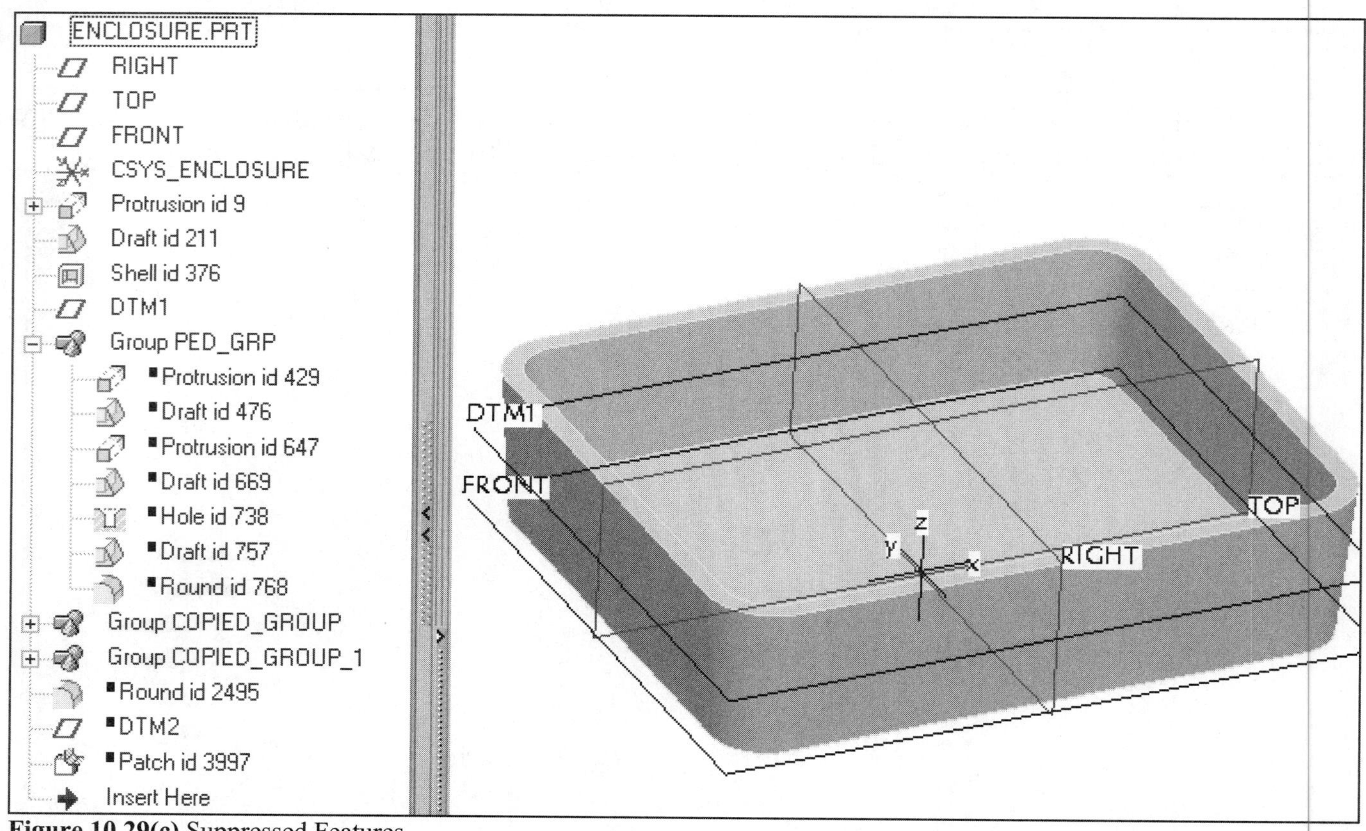

Figure 10.29(c) Suppressed Features

The regeneration time for your model will now be much shorter. Next, add the text protrusion (**CFS_2134**).

Click: **Display object in standard orientation** ⇒ **Extrude Tool** ⇒ ⇒ Sketch Plane---Plane: pick the inside of the enclosure for the sketching plane ⇒ Reference: **TOP** ⇒ Orientation: **Top** [Fig. 10.30(a)] ⇒ **Sketch** ⇒ **Close** ⇒ **Set various environment options** ⇒ Snap to Grid ⇒ **Apply** ⇒ **OK** ⇒ **Toggle the grid on** ⇒ **Create text as a part of a section** ⇒ select *start point* of line to determine text height and orientation [Fig. 10.30(b)] ⇒ select *second point* of line to determine text height and orientation [Fig. 10.30(c)]

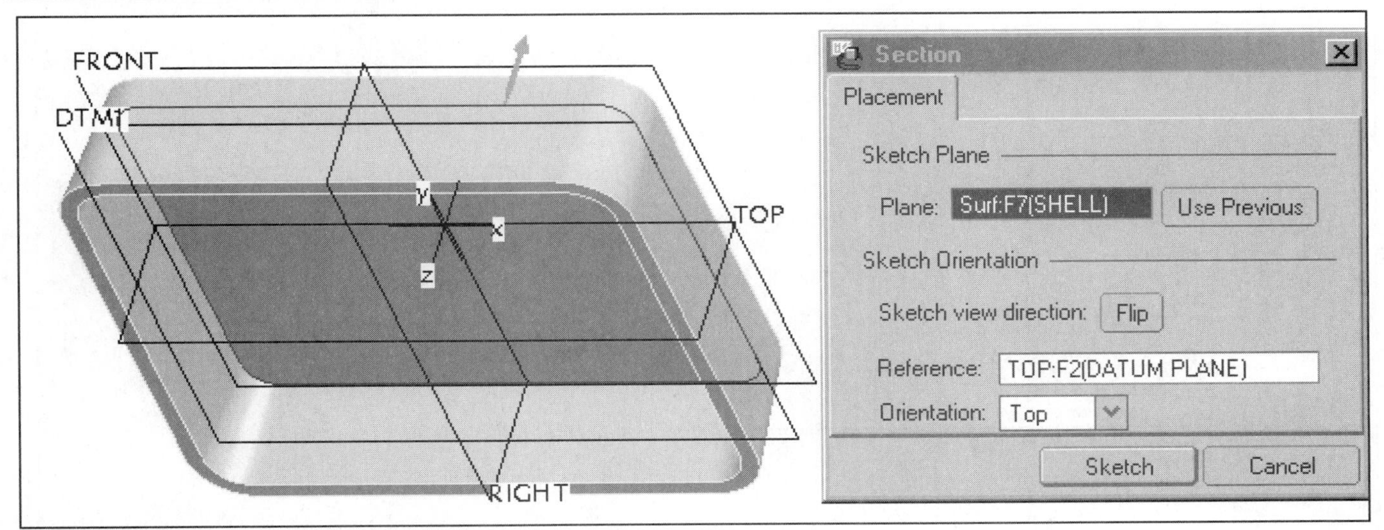

Figure 10.30(a) Sketching Plane, Inside Surface

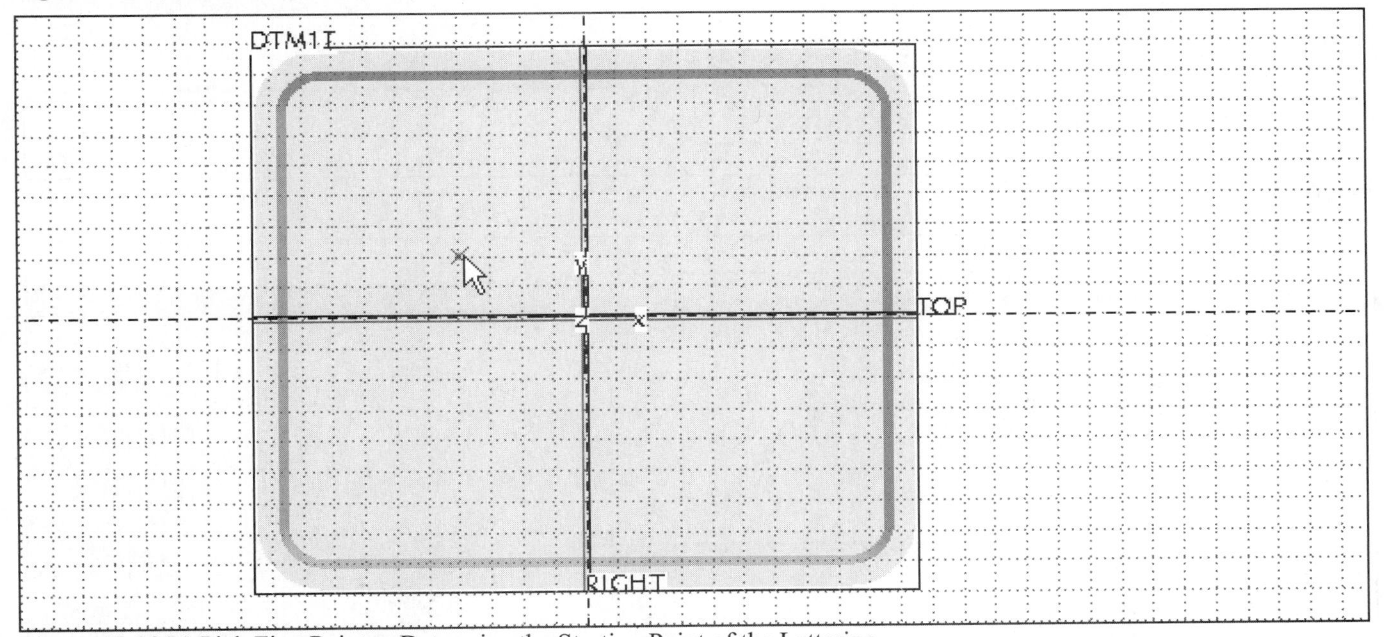

Figure 10.30(b) Pick First Point to Determine the Starting Point of the Lettering

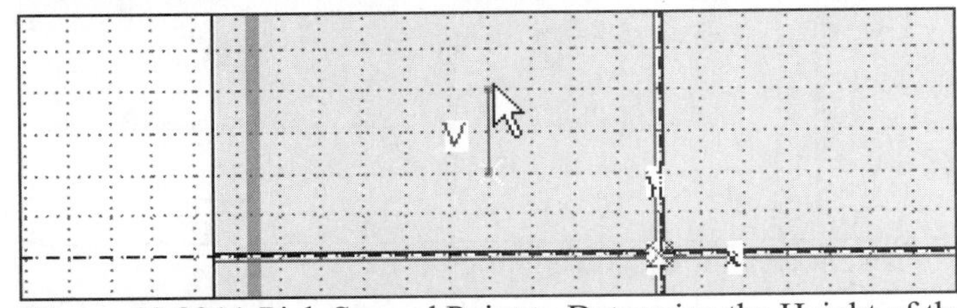

Figure 10.30(c) Pick Second Point to Determine the Height of the Lettering

Text Line- type **CFS-2134** [Fig. 10.30(d)] ⇒ ✓ ⇒ ↖ ⇒ window-in the sketch to capture all dimensions ⇒ **Modify the values of dimensions, geometry of splines, or text entities** ⇒ ☑ Regenerate ⇒ modify the dimensions [Fig. 10.30(e)] ⇒ ✓ ⇒ ✓ **Continue** ⇒ press and hold **MMB** to spin the model [Fig. 10.30(f)] ⇒ double-click on the height dimension and modify it to **.0625** ⇒ **Enter** [Fig. 10.30(g)] ⇒ **MMB** [Fig. 10.30(h)] ⇒ 💾 ⇒ **MMB**

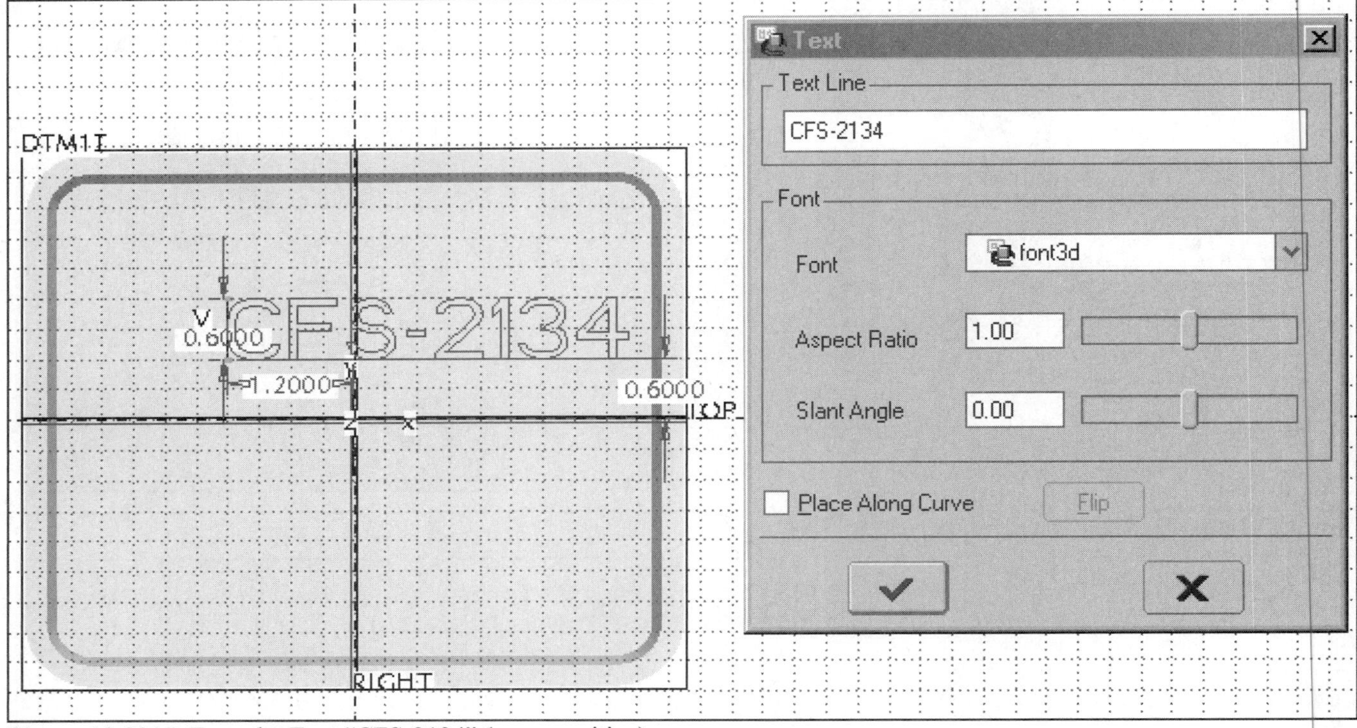

Figure 10.30(d) Type the Text "CFS-2134" (case sensitive)

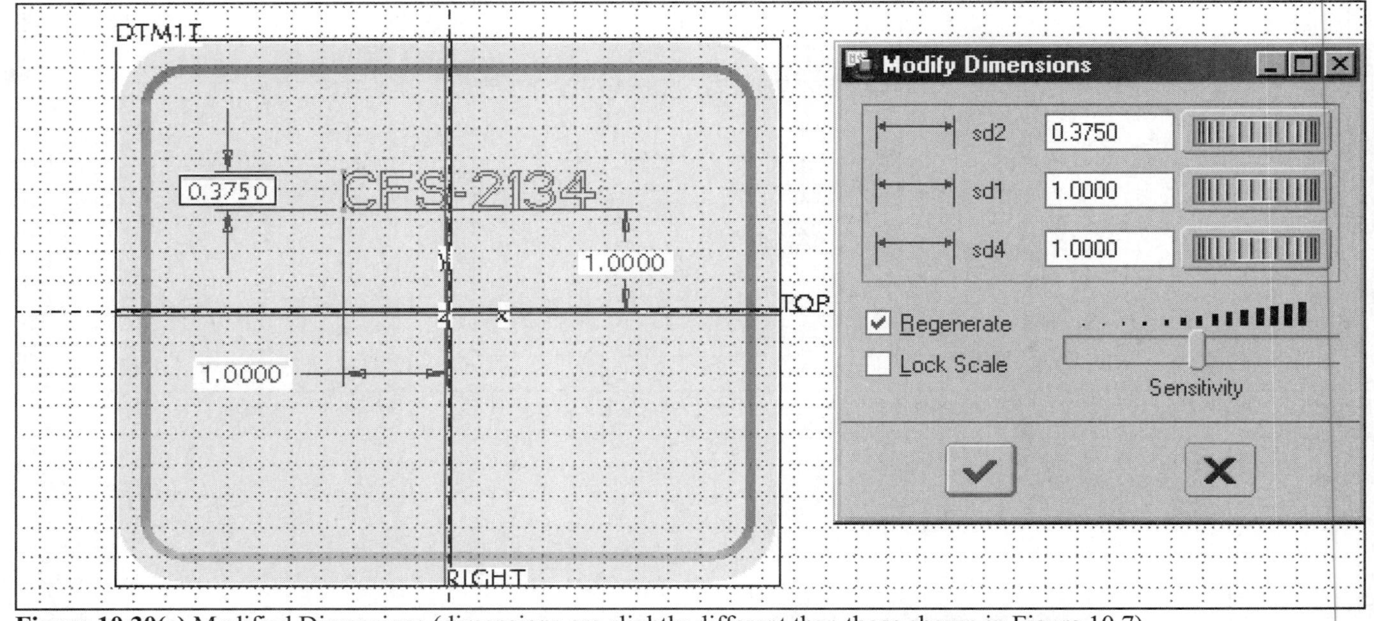

Figure 10.30(e) Modified Dimensions (dimensions are slightly different than those shown in Figure 10.7)

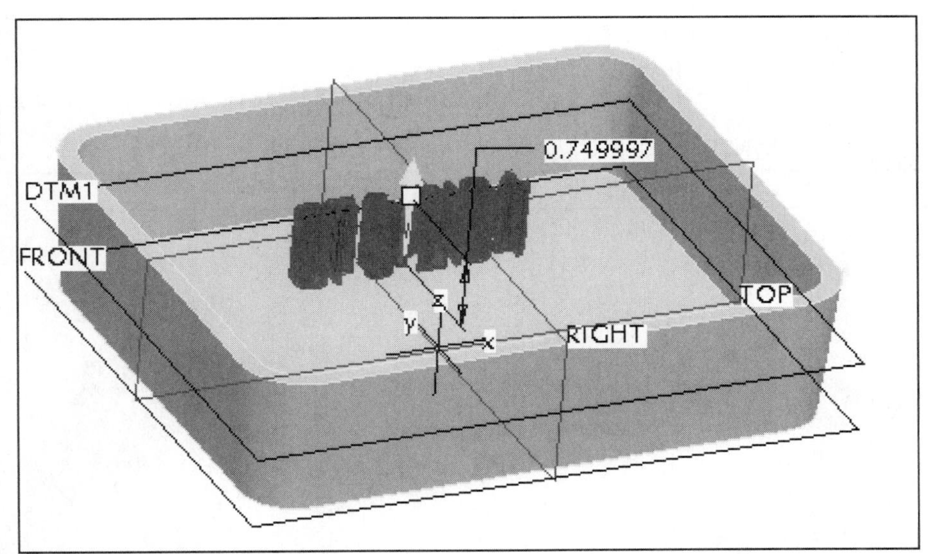

Figure 10.30(f) Depth Preview

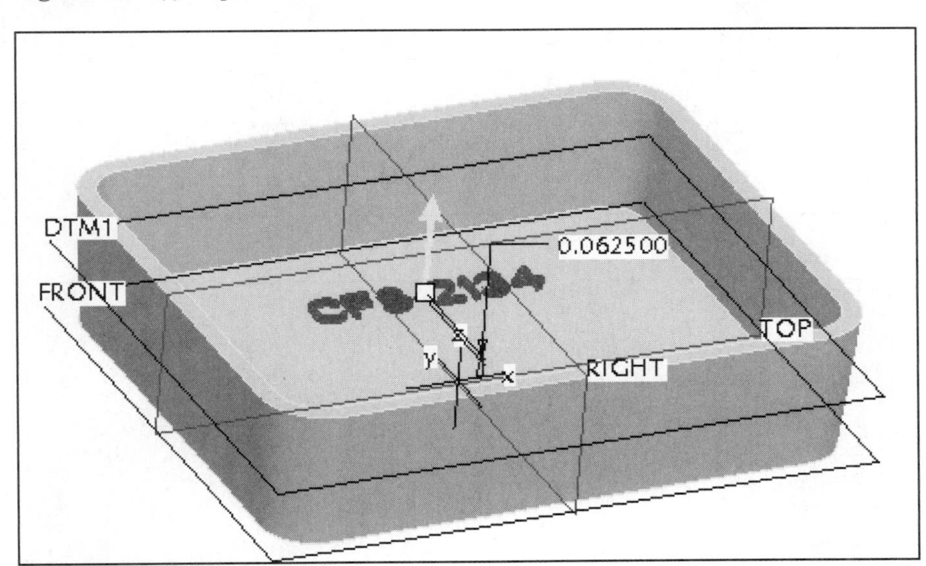

Figure 10.30(g) Modified Depth

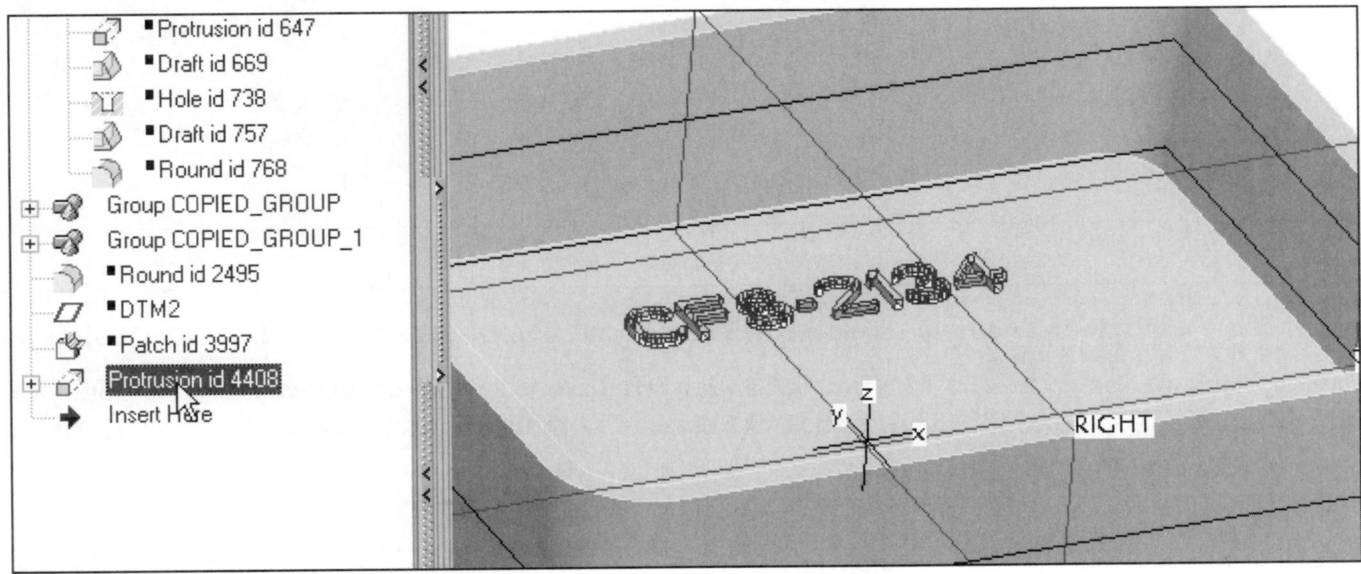

Figure 10.30(h) Completed Text Protrusion

Click on **Group PED_GRP** in the Model Tree ⇒ **RMB** [Fig. 10.31(a)] ⇒ **Resume** [Fig. 10.31(b)] ⇒ **Edit** from menu bar ⇒ **Resume** ⇒ **All** ⇒ ⇒ ⇒ ⇒ **Standard Orientation** ⇒ **File** ⇒ **Delete** ⇒ **Old Versions** ⇒ **MMB** ⇒ ⇒ **MMB**

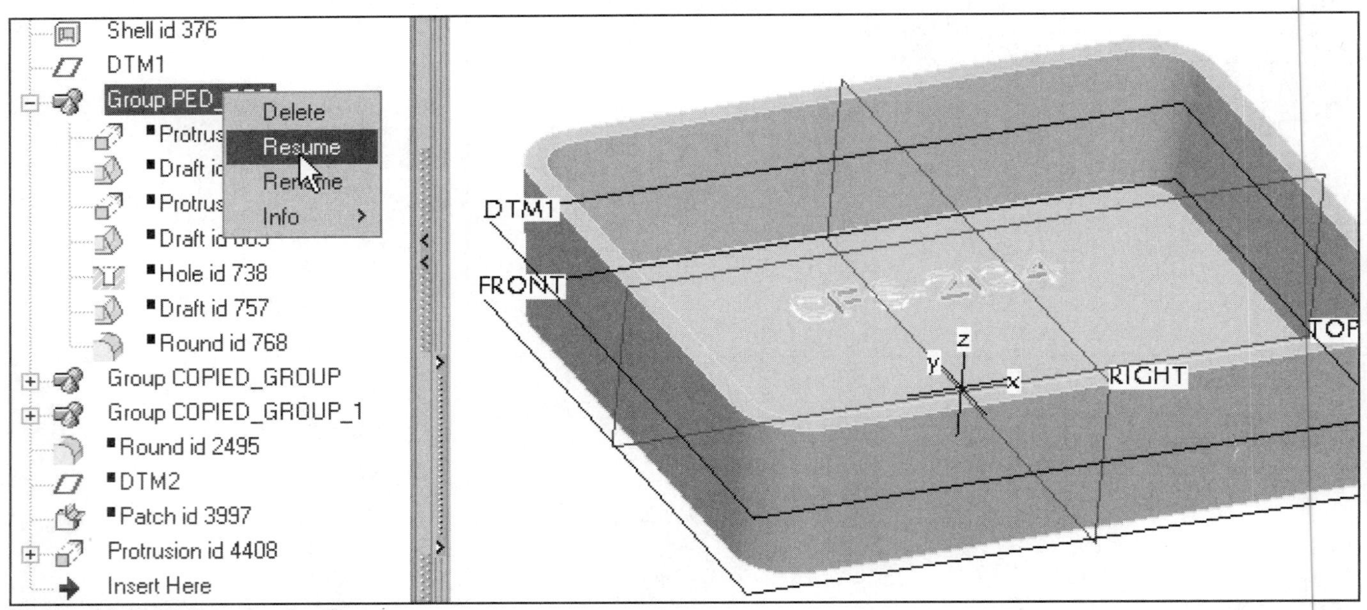

Figure 10.31(a) Resume the Group

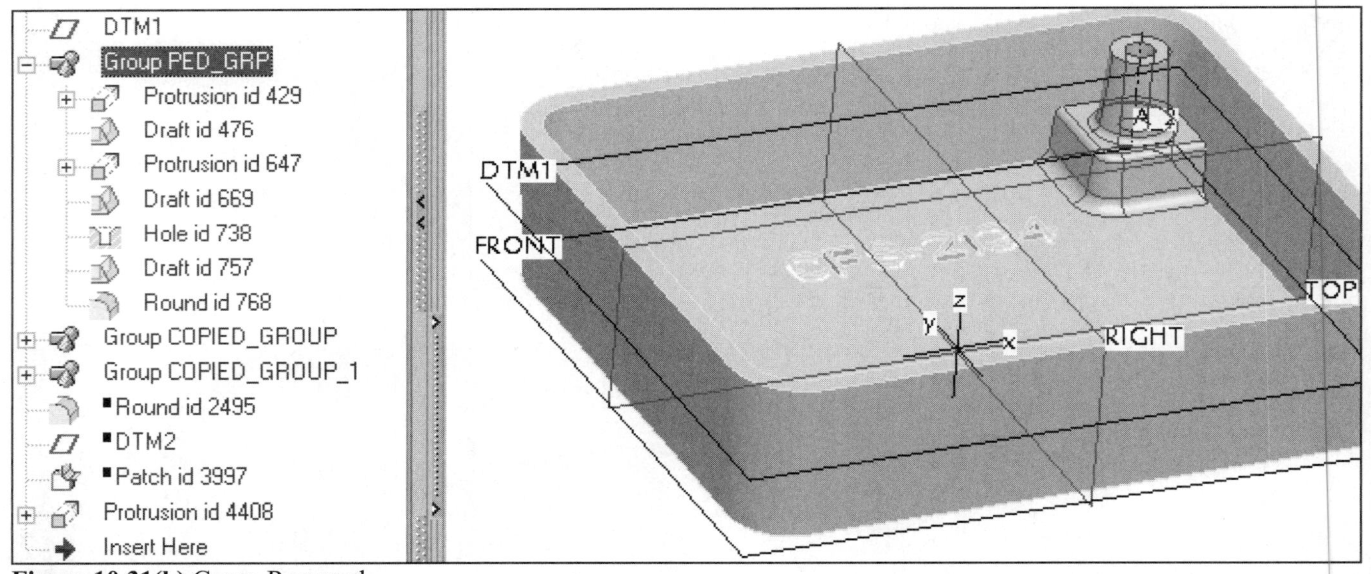

Figure 10.31(b) Group Resumed

Spin the part ⇒ **View** ⇒ **Shade** ⇒ create a **.1875** edge round [Fig. 10.32(a-b)] ⇒ **Display object in standard orientation** ⇒ ⇒ **MMB** ⇒ **View Mode on** ⇒ **RMB** ⇒ **Velocity** [Fig. 10.33(a)] ⇒ hold down **MMB** and move the cursor about the screen to orbit the model [Fig. 10.33(b)] ⇒ **RMB** ⇒ **Exit Viewing** ⇒ ⇒ **Standard Orientation** ⇒ ⇒ ⇒ **MMB**

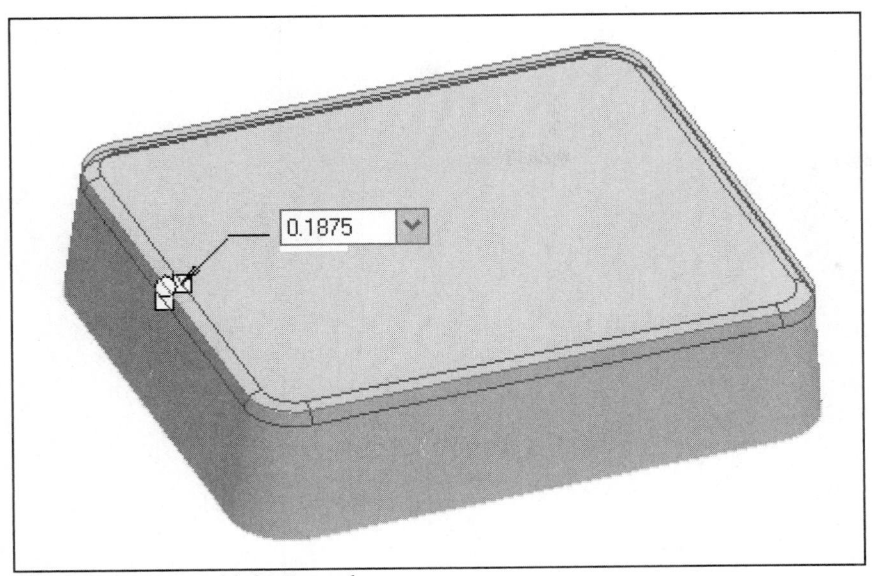

Figure 10.32(a) Add the Round

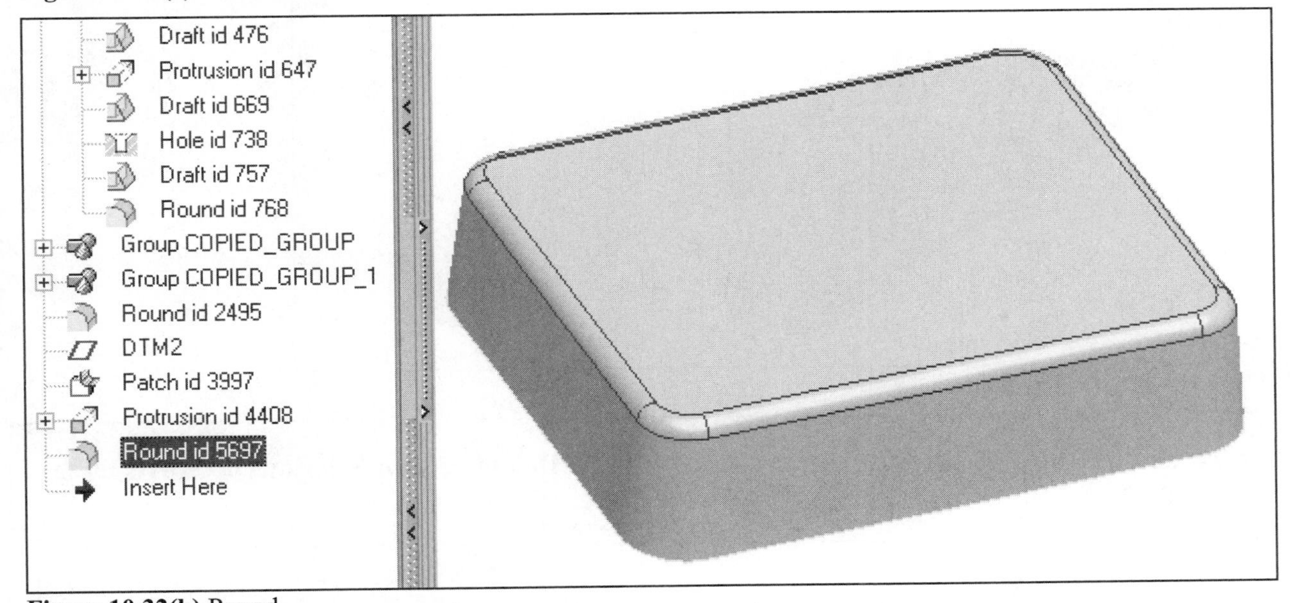

Figure 10.32(b) Round

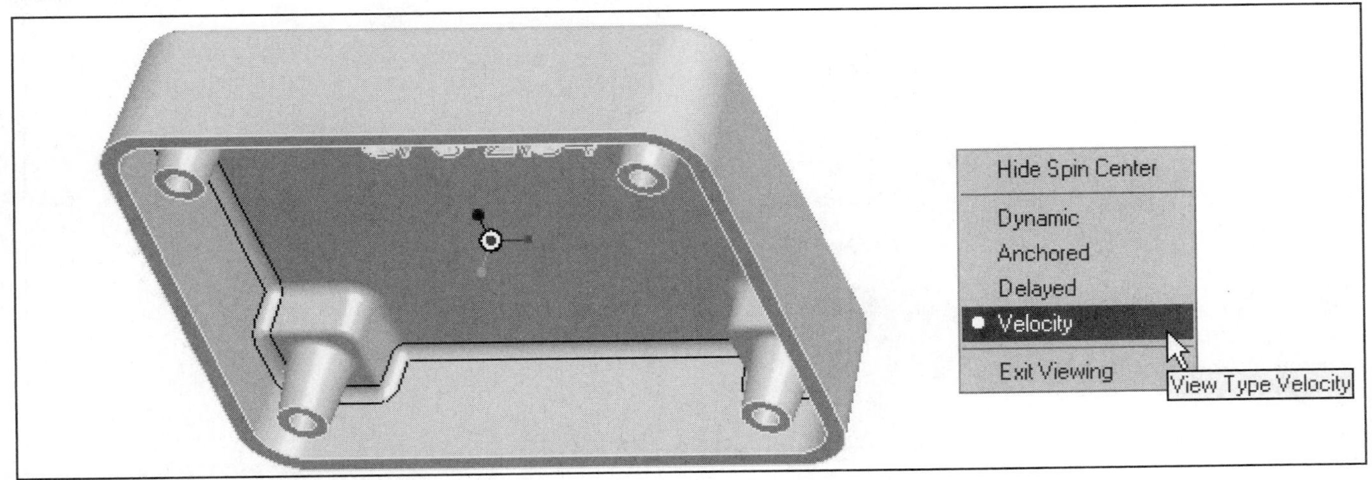

Figure 10.33(a) View Mode, Velocity

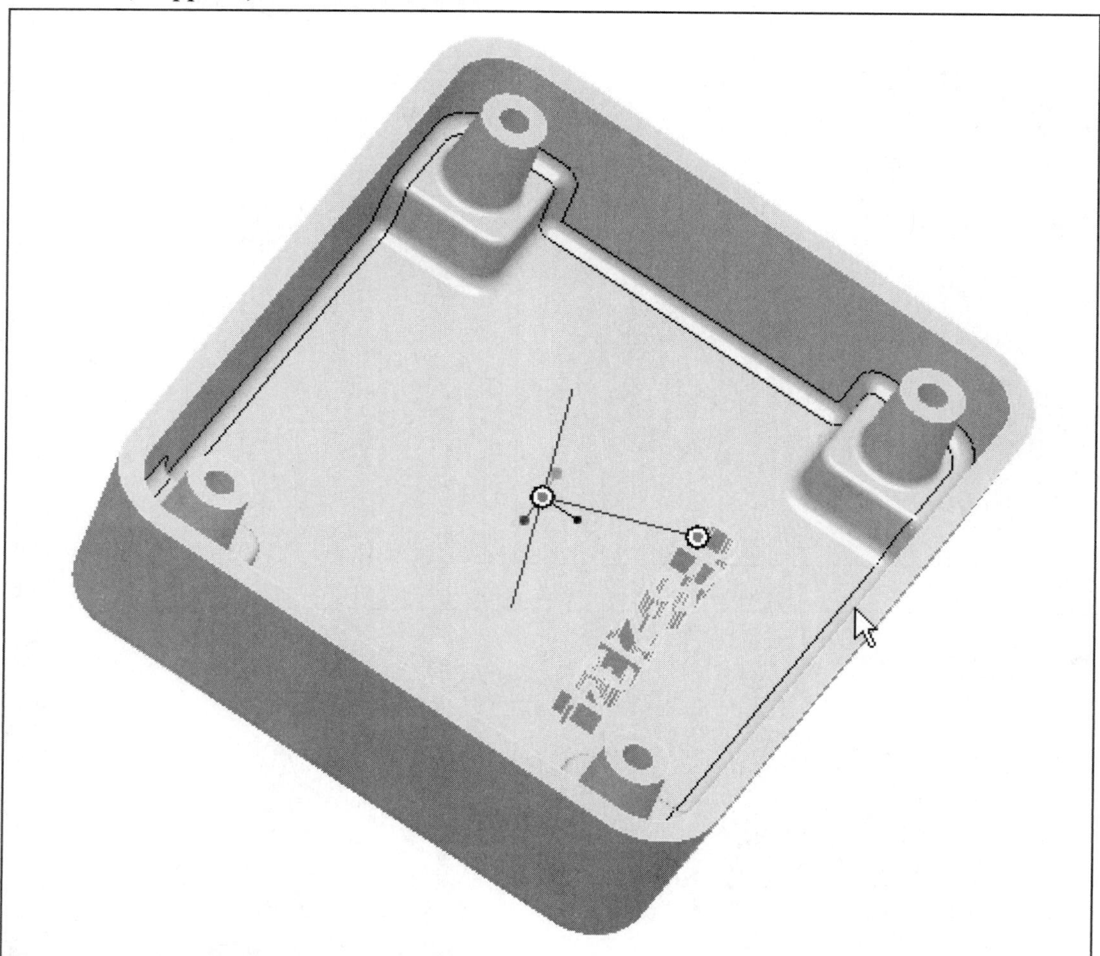

Figure 10.33(b) Spinning the Model in Real Time

Modify the text depth to be **.250** (Fig. 10.34) ⇒ [icon] ⇒ **File** ⇒ **Delete** ⇒ **Old Versions**⇒ **MMB** ⇒ [icon] ⇒ **MMB** (Fig. 10.35)

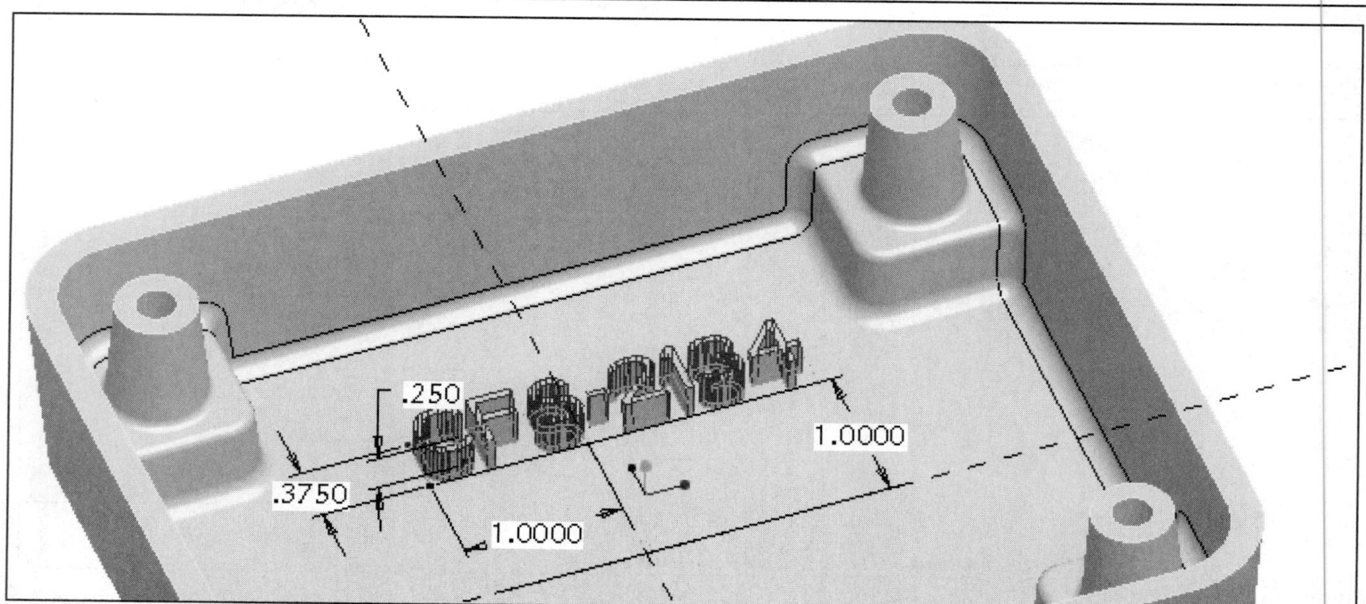

Figure 10.34 Modify the Text Protrusion to have a **.250** Depth (dimension is different than shown in Figure 10.8)

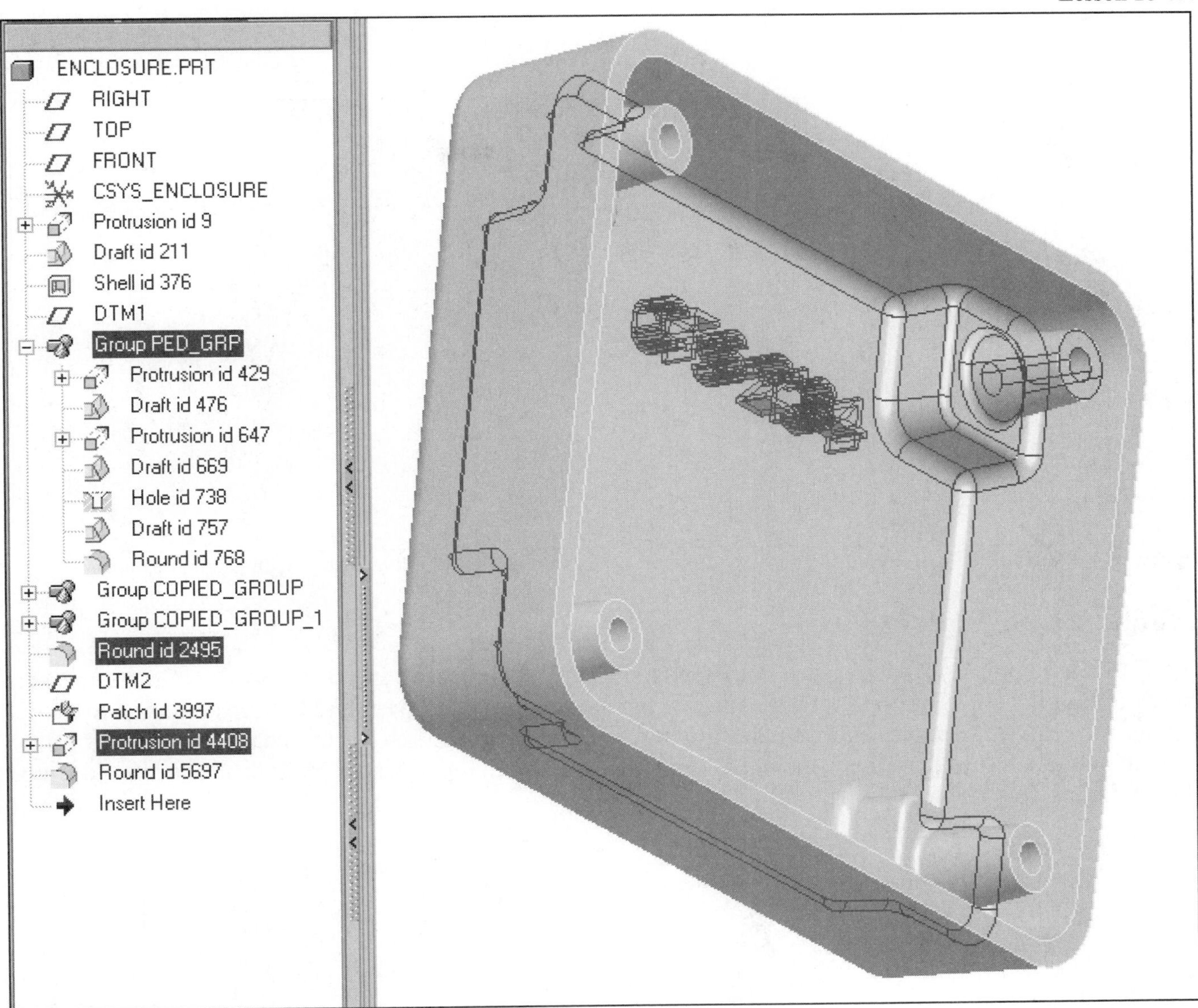

Figure 10.35 Completed Enclosure

You have now completed Lesson 10. Continue and model the Cellular Phone Bottom in the Lesson Project.

Lesson 10 Project

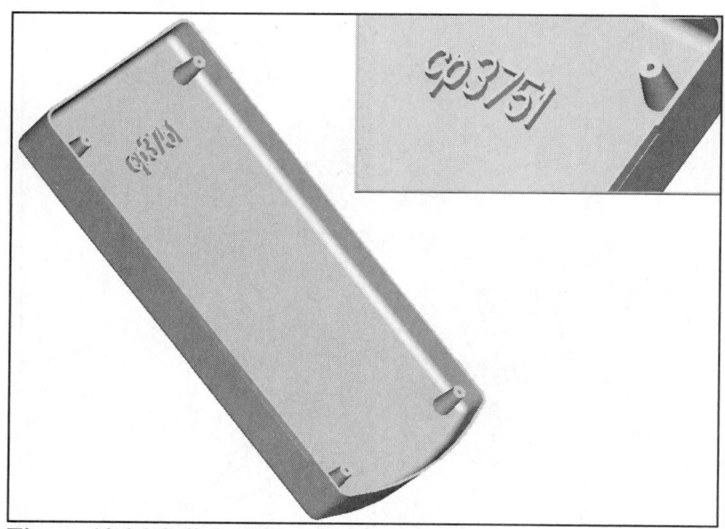

Figure 10.36 Cellular Phone Bottom

Figure 10.37 Cellular Phone with Datum Features

Cellular Phone Bottom

The Cellular Phone Bottom requires commands similar to those used in the Enclosure. Create the part shown in Figures 10.36 through 10.47. Analyze the part and plan the steps and features required to model it. Plan the feature creation sequence and the parent-child relationships for the part. The top half of the cellular phone is created in the Lesson 11 Project. Shell the Cellular Phone Bottom **.04** on its sides and **.500** on its inside bottom.

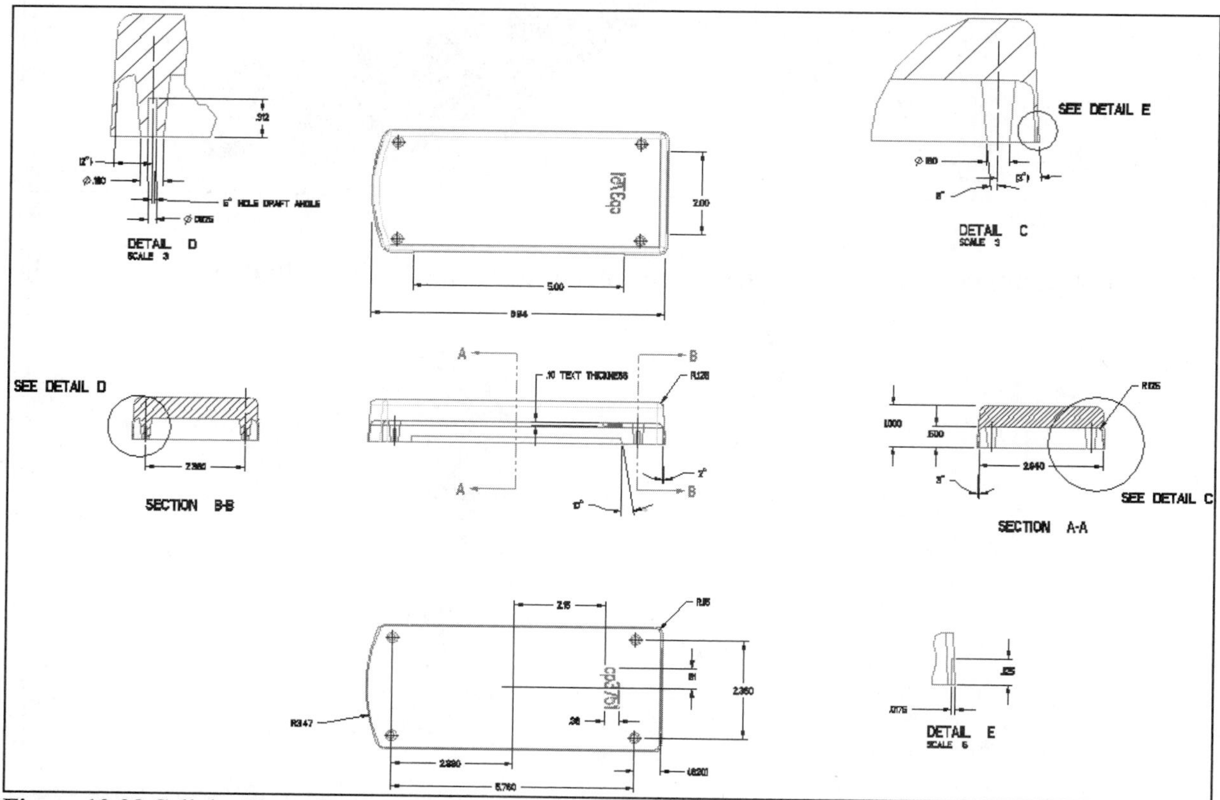

Figure 10.38 Cellular Phone Bottom Drawing

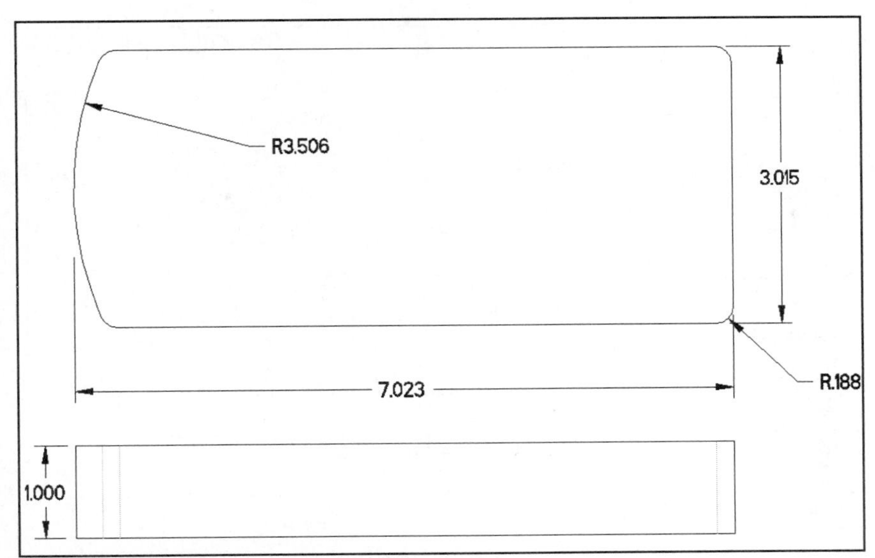

Figure 10.39 Dimensions for First Protrusion

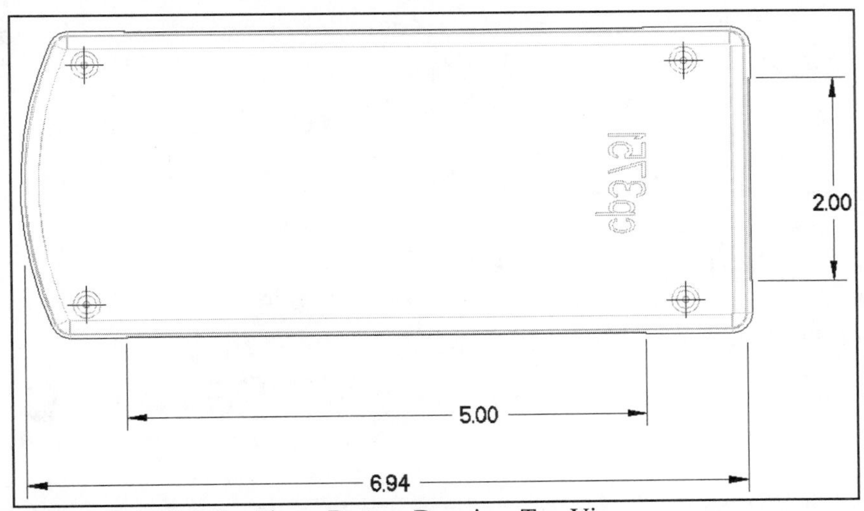

Figure 10.40 Cellular Phone Bottom Drawing, Top View

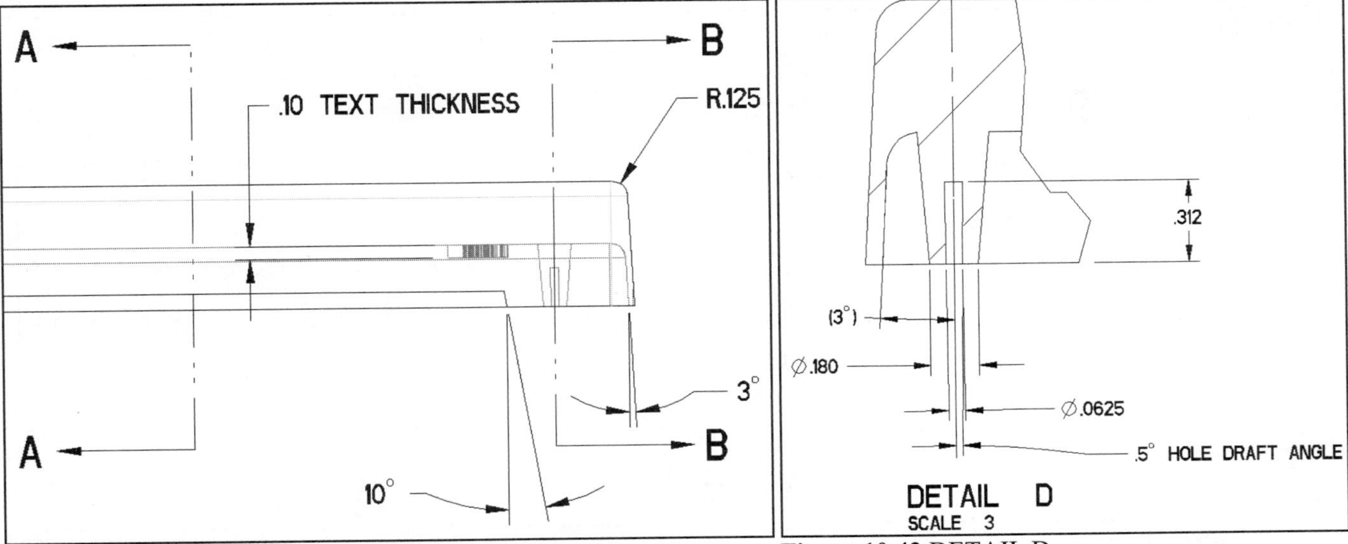

Figure 10.41 Cellular Phone Bottom Drawing, Front View

Figure 10.42 DETAIL D

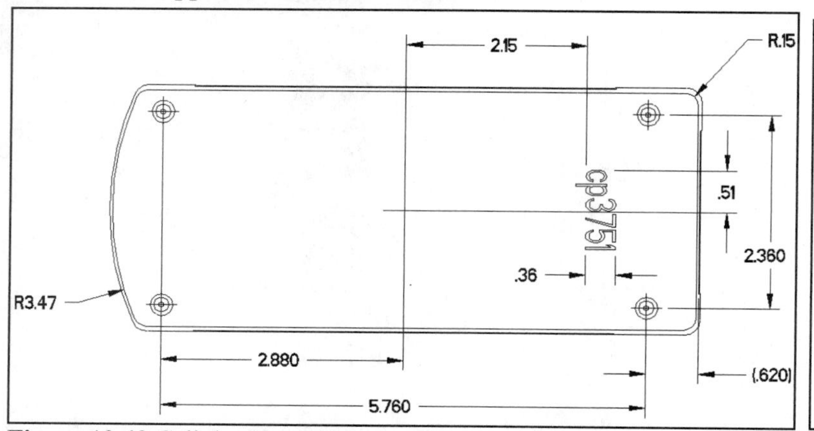

Figure 10.43 Cellular Phone Drawing, Bottom View

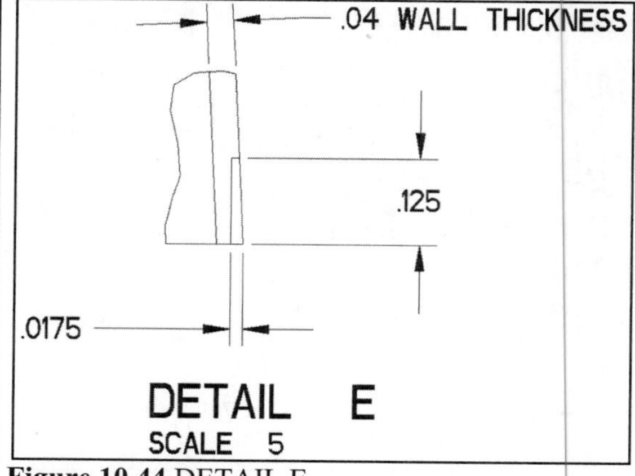

Figure 10.44 DETAIL E

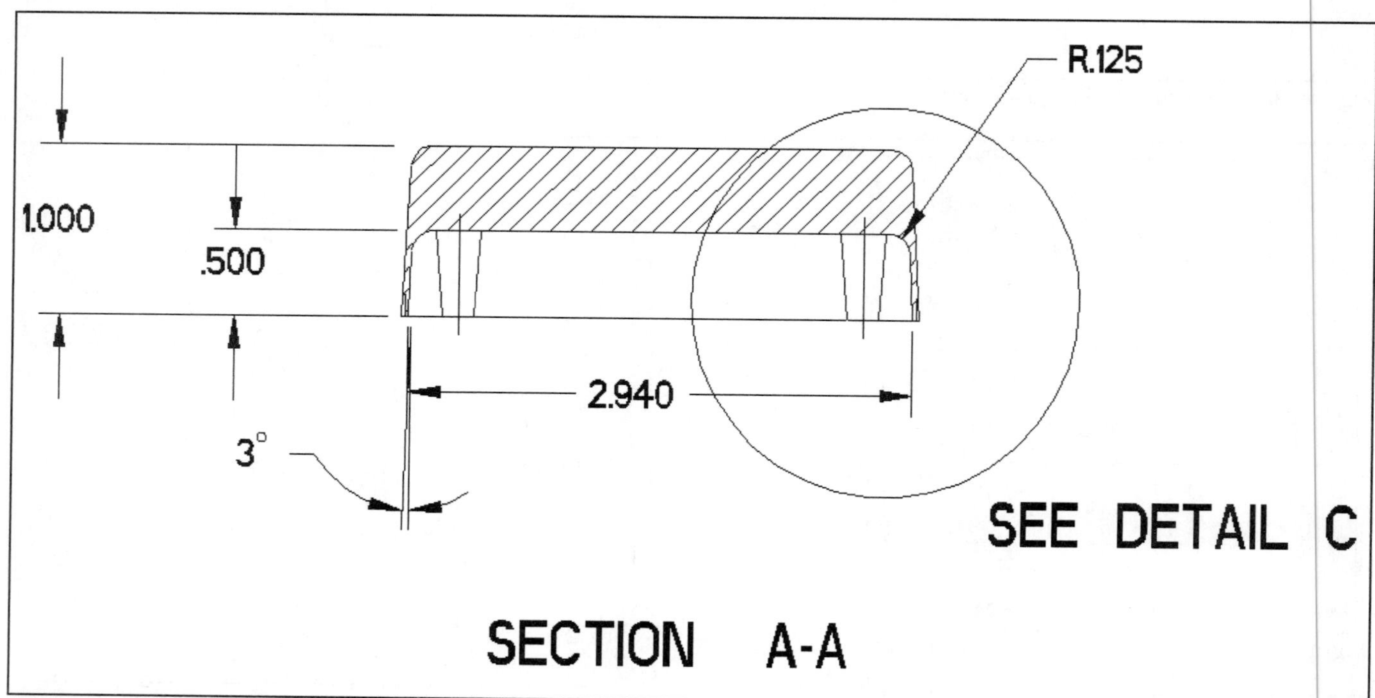

Figure 10.45 SECTION A-A

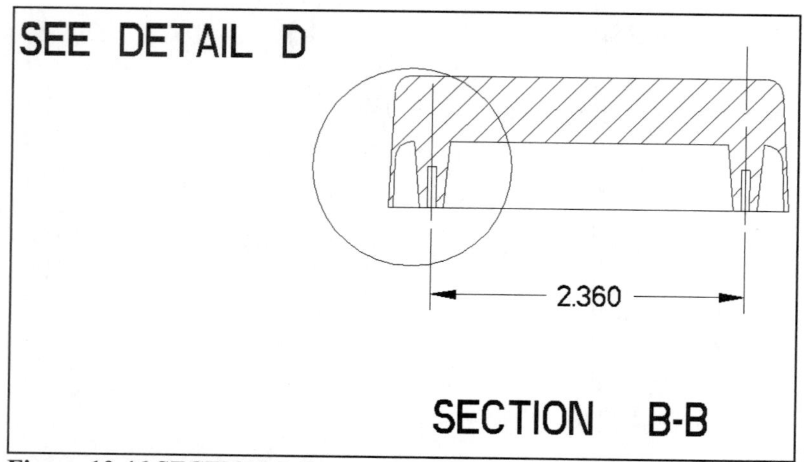

Figure 10.46 SECTION B-B

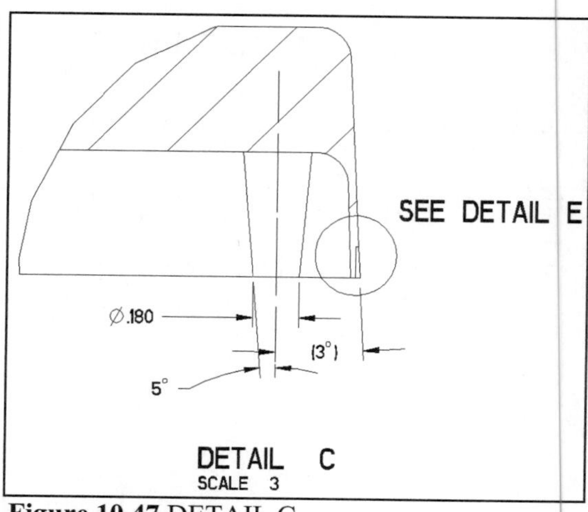

Figure 10.47 DETAIL C

Lesson 11 Shell, Reorder, and Insert Mode

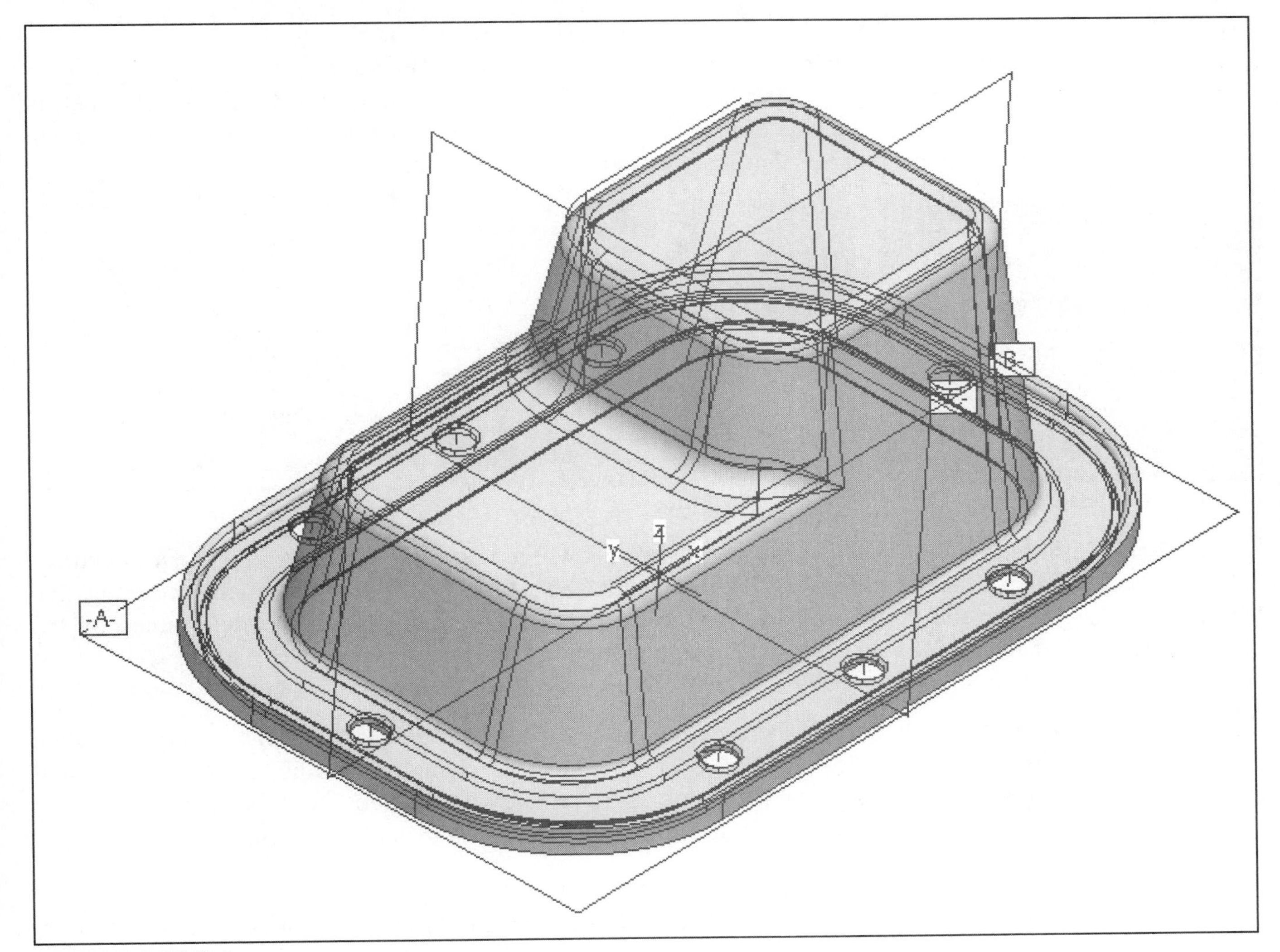

Figure 11.1 Oil Sink

OBJECTIVES

- Master the use of the **Shell Tool**
- Alter the creation sequence with **Reorder**
- **Insert** a feature at a specific point in the design order
- Create a **Hole Pattern** using a **Table**

SHELL, REORDER, AND INSERT MODE

The **Shell Tool** removes a surface or surfaces from the solid and then hollows out the inside of the solid, leaving a shell of a specified wall thickness, as in the Oil Sink (Fig. 11.1). When Pro/E makes the shell, all the features that were added to the solid before you chose the Shell Tool are hollowed out. Therefore, the *order of feature creation* is very important when you use the Shell Tool. You can alter the feature creation order by using the **Reorder** option. Another method of placing a feature at a specific place in the feature/design creation order is to use the **Insert Mode** option.

Creating Shells

The Shell Tool [Fig. 11.2(a-c)] enables you to remove a surface or surfaces from the solid, then hollows out the inside of the solid, leaving a shell of a specified wall thickness. If you flip the thickness side by entering a negative value, dragging a handle, or using the **Change thickness direction** icon, the shell thickness is added to the outside of the part. If you do not select a surface to remove, a "closed" shell is created, with the whole inside of the part hollowed out and no access to the hollow. In this case, you can add the necessary cuts or holes to achieve proper geometry at a later time.

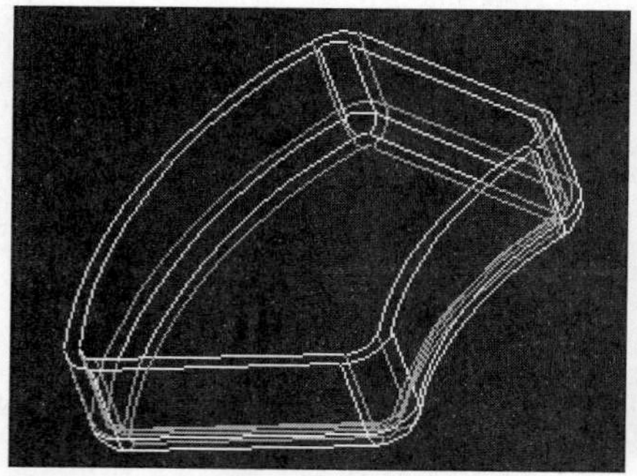
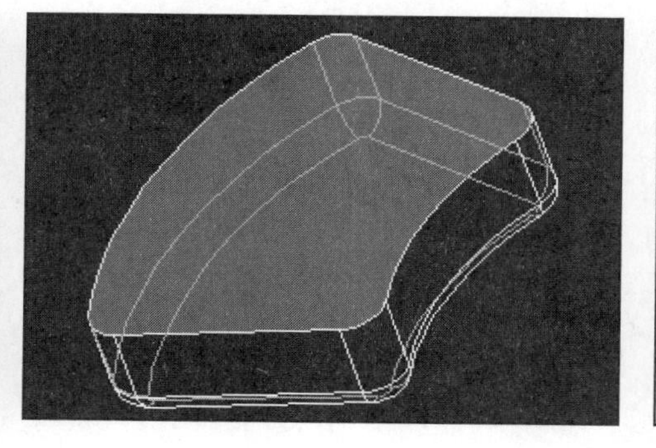
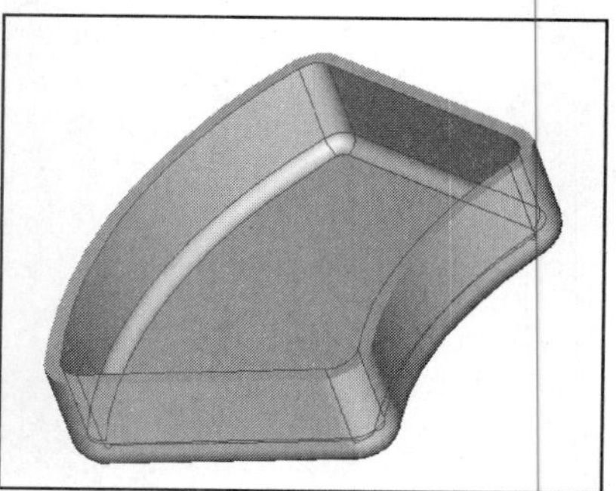

Figure 11.2(a-c) Shell

When defining a shell, you can also select surfaces where you want to assign a different thickness. You can specify independent thickness values for each such surface. However, you cannot enter negative thickness values, or flip the thickness side, for these surfaces. The thickness side is determined by the default thickness of the shell. When Pro/E makes the shell, all the features that were added to the solid before you started the Shell Tool are hollowed out. Therefore, the order of feature creation is very important when you use the Shell Tool. To access the Shell Tool, click icon in the Toolbar, or click Insert ⇒ Shell on the top menu bar. The Thickness box lets you change the value for the default shell thickness. You can type the new value, or select one of the recently used values from the drop-down list.

In the graphics window, you can use the shortcut (**RMB**) menu to access the following options:

- **Remove Surfaces** Activates the collector of surfaces. You can select any number of surfaces
- **Non Default Thickness** Activates the collector of surfaces with a different thickness
- **Clear** Remove all references from the collector that is currently active
- **Flip** Change the shell side direction

The Shell Dashboard displays the following slide-up/down panels:

- **References** Contains the collectors of references used in the Shell feature
- **Properties** Contains the feature name and an icon to access feature information

The References slide-up/down panel contains the following elements:

- The **Removed surfaces** collector lets you select the surfaces to be removed. If you do not select any surfaces, a "closed" shell is created.
- The **Non-default thickness** collector lets you select surfaces where you want to assign a different thickness. For each surface included in this collector, you can specify an individual thickness value.

The Properties panel contains the Name text box Name SHELL_ID_4989, where you can type a custom name for the shell feature, to replace the automatically generated name. It also contains the icon that you can click to display information about this feature in the Browser.

Reordering Features

You can move features forward or backward in the feature creation (regeneration) order list, thus changing the order in which they are regenerated [Fig. 11.3(a-b)]. Use Edit ⇒ Feature Operations ⇒ Reorder to active the command.

You can reorder multiple features in one operation, as long as these features appear in *consecutive* order. Feature reorder *cannot* occur under the following conditions:

- **Parents** Cannot be moved so that their regeneration occurs after the regeneration of their children
- **Children** Cannot be moved so that their regeneration occurs before the regeneration of their parents

You can select the features to be reordered by choosing an option:

- **Select** Select features to reorder by picking on the screen and/or from the Model Tree
- **Layer** Select all features from a layer by selecting the layer
- **Range** Specify the range of features by entering the regeneration numbers of the starting and ending features

You can reorder features in the Model Tree by dragging one or more features to a new location in the feature list. If you try to move a child feature to a higher position than its parent feature, the parent feature moves with the child feature in context, so the parent/child relationship is maintained.

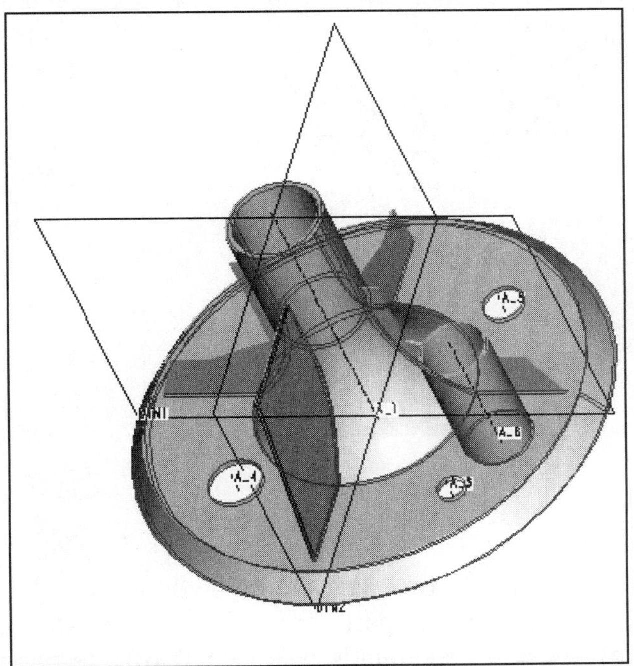

Figure 11.3(a) Reorder (CADTRAIN, COAch for Pro/ENGINEER)

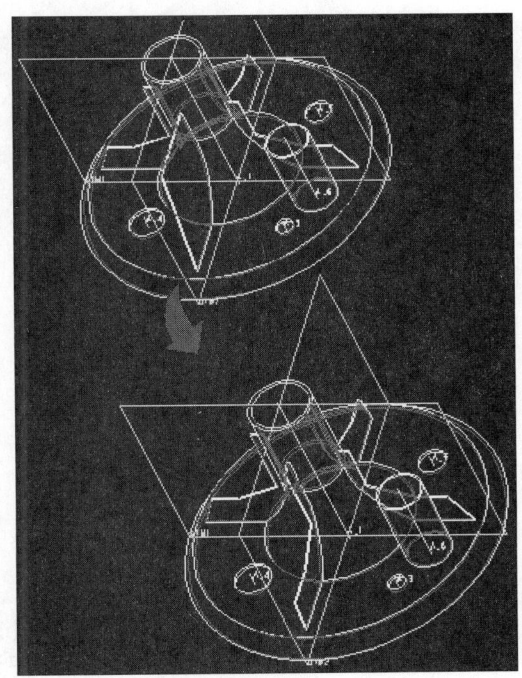

Figure 11.3(b) Reorder

Inserting Features

Normally, Pro/E adds a new feature after the last existing feature in the part, including suppressed features. Insert Mode allows you to add new features at any point in the feature sequence, except before the base feature or after the last feature. You can also Insert features using the Model Tree. There is an arrow-shaped icon on the Model Tree that indicates where features will be inserted upon creation. By default, it is always at the end of the Model Tree. You may drag the location of the *arrow* higher or lower in the tree to insert features at a different point. When the *arrow* is dropped at a new location, the model is rolled backward or forward in response to the insertion *arrow* being moved higher or lower in the tree.

Lesson 11 STEPS

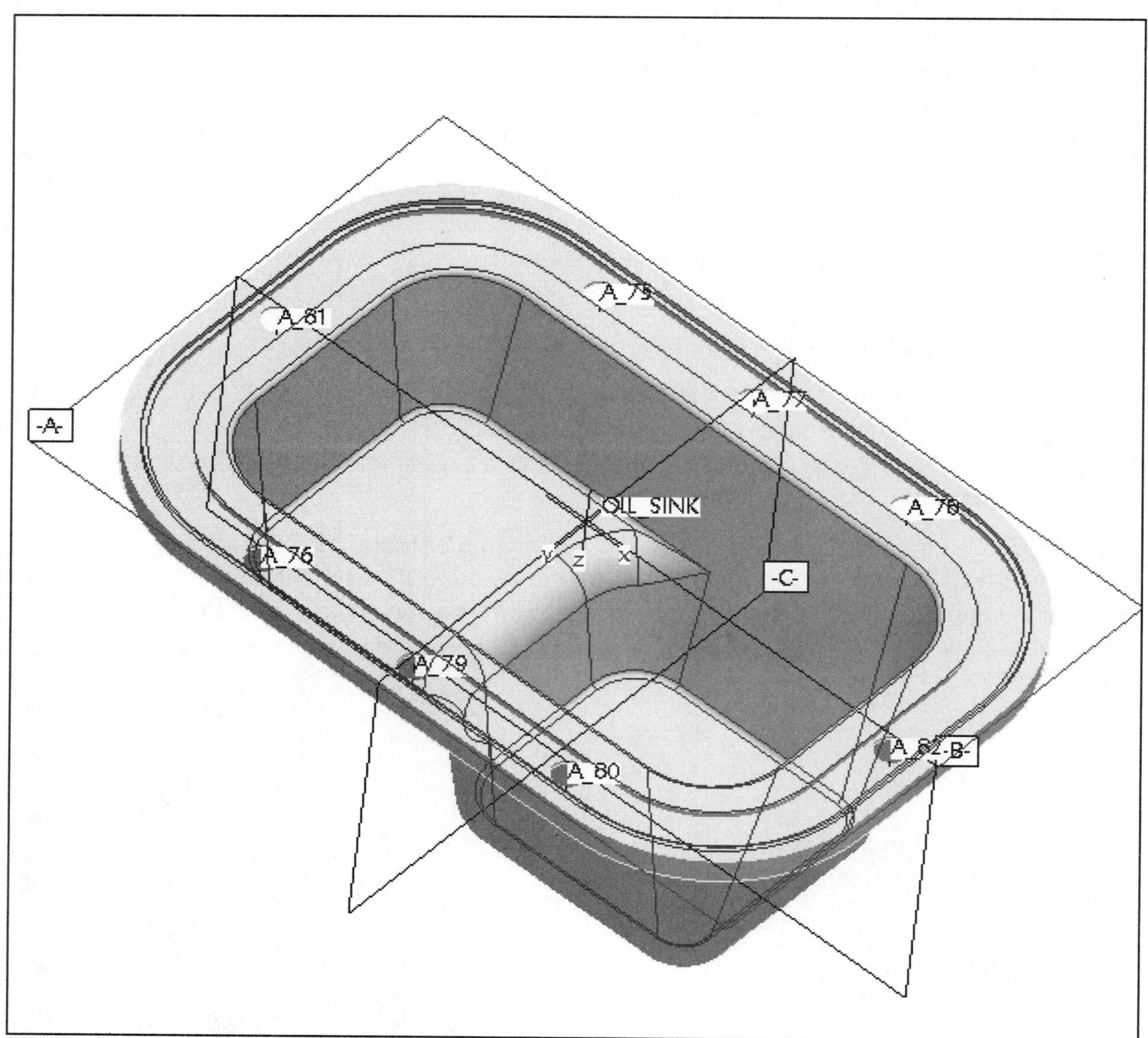

Figure 11.4 Oil Sink

Oil Sink

The Oil Sink (Fig. 11.4) requires the use of the **Shell Tool**. The shelling of a part should be done after the desired protrusions and most rounds have been modeled. This lesson part will have you create a protrusion, a cut, and a set of rounds (Fig. 11.5). Some of the required rounds will be left off the part model on purpose. Pro/E's **Insert Mode** option enables you to insert a set of features at an earlier stage in the design of the part. In other words, you can create a feature after or before a selected existing feature even if the whole model has been completed. You can also *move the order in which a feature was created* and therefore have subsequent features affect the reordered feature. A round created after a shell operation can be reordered to appear before the shell, to have the shell be affected by the round.

In this lesson, you will also insert a round or two before the existing shell feature using Insert Mode. The rounds will be shelled after the **Resume** option is picked, because the rounds now appear before the shell feature. The details shown in Figures 11.5(a) through (m) provide the design dimensions.

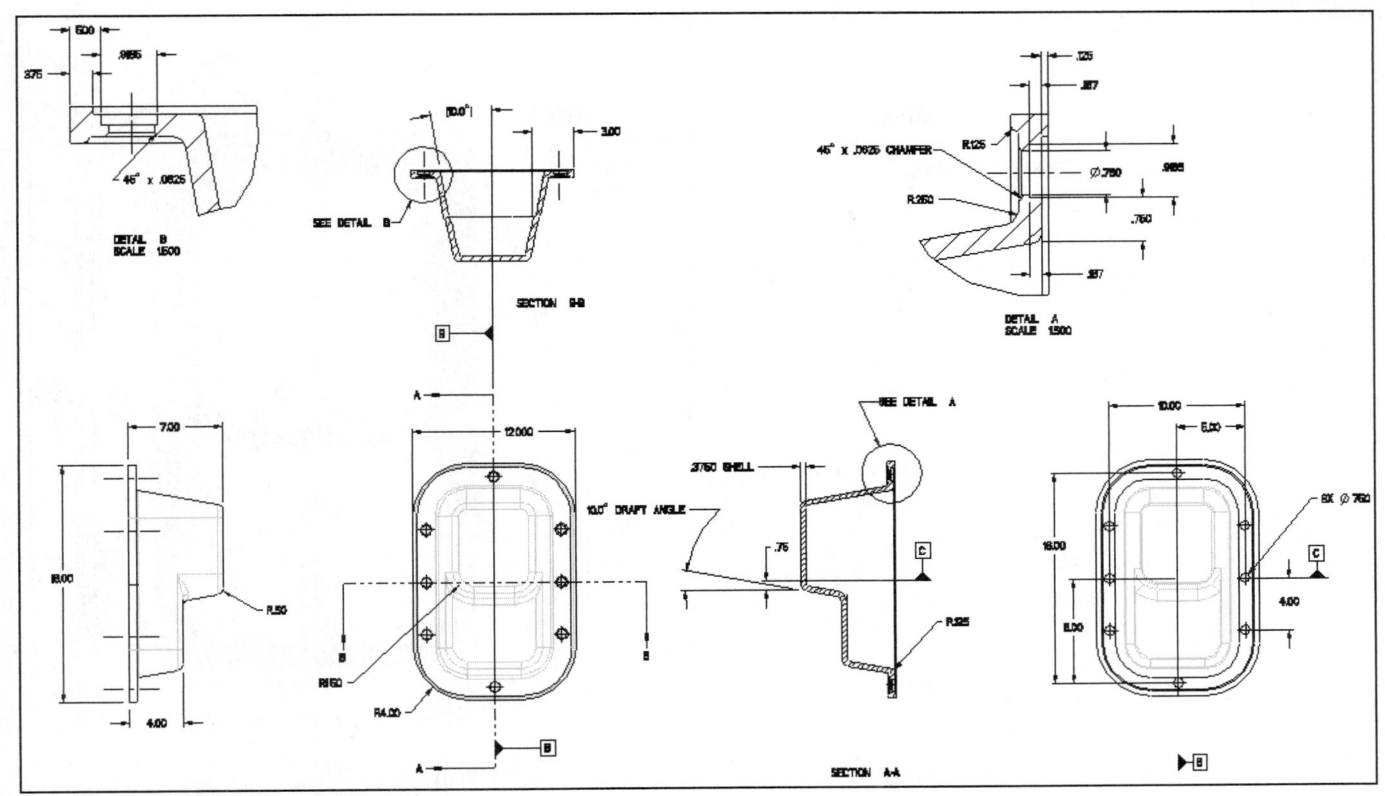

Figure 11.5(a) Oil Sink Detail Drawing

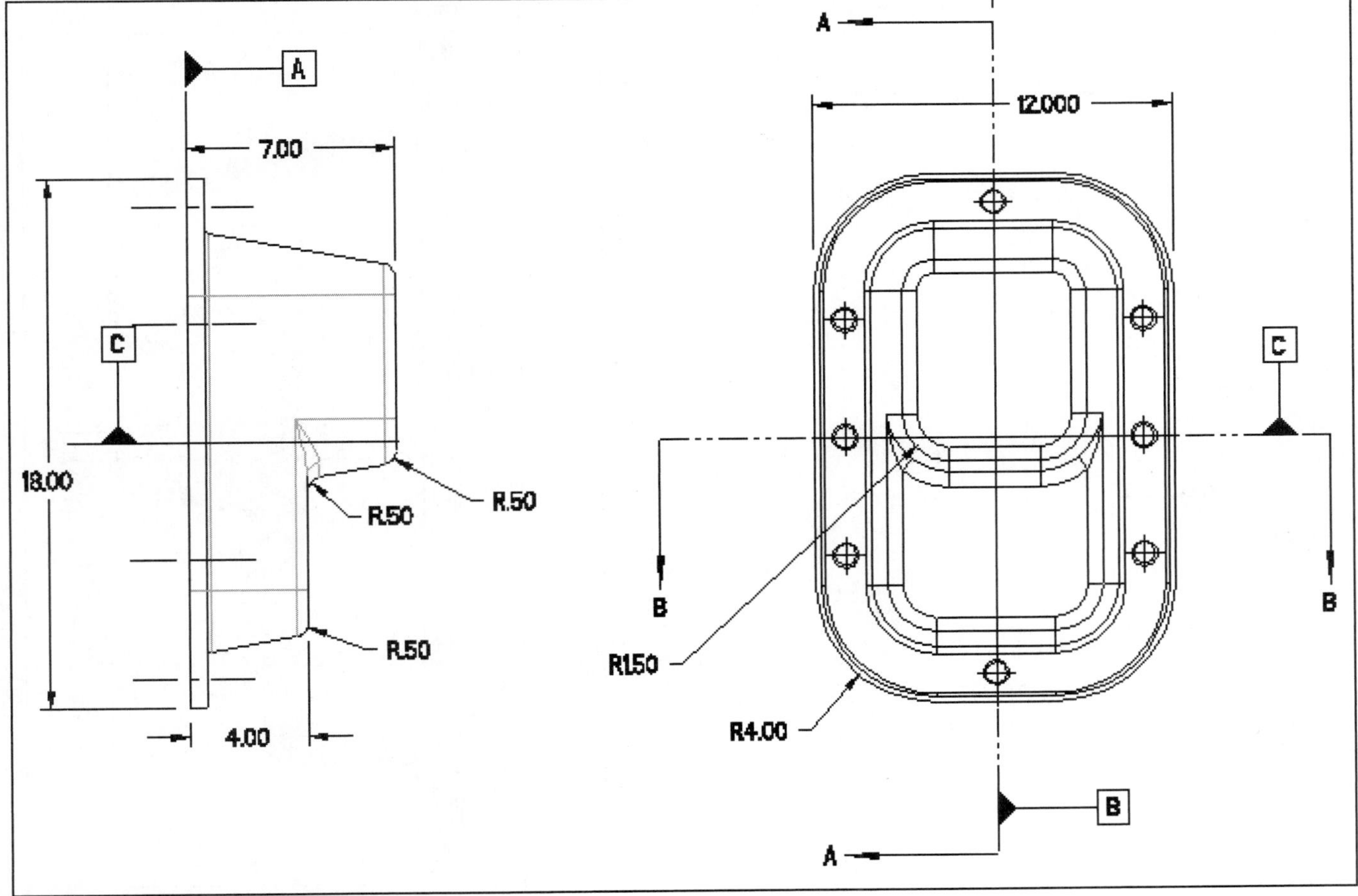

Figure 11.5(b) Oil Sink Left and Top Views

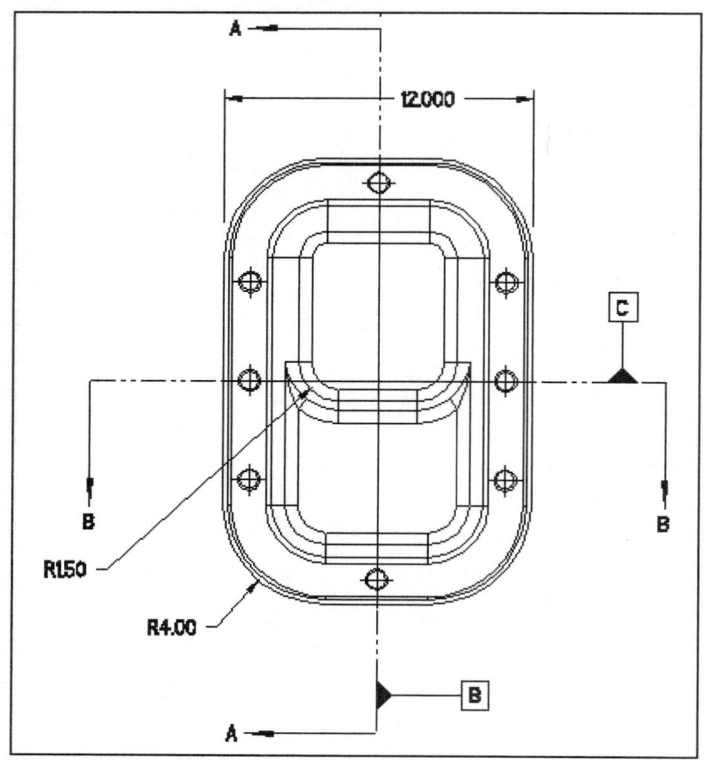

Figure 11.5(c) Oil Sink Top View Dimensions

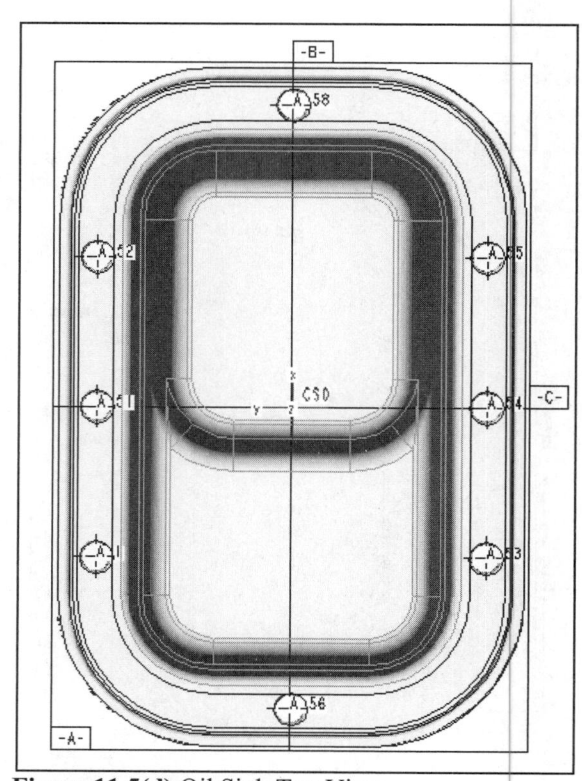

Figure 11.5(d) Oil Sink Top View

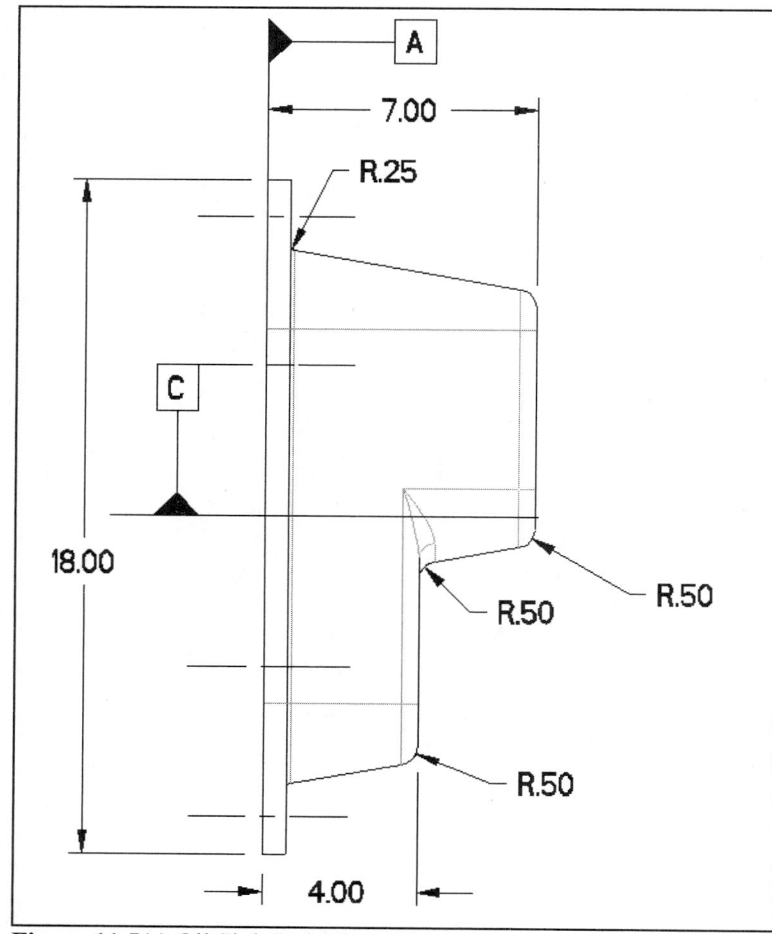

Figure 11.5(e) Oil Sink Left View Dimensions

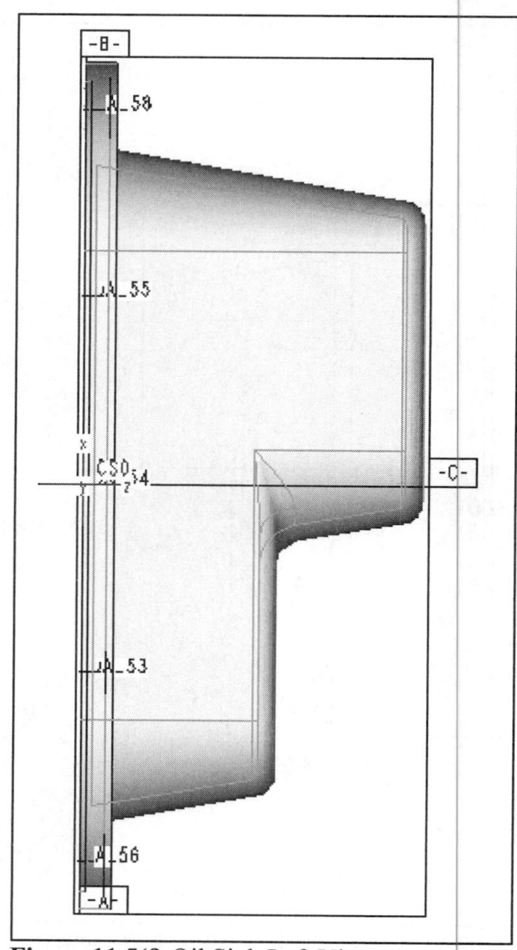

Figure 11.5(f) Oil Sink Left View

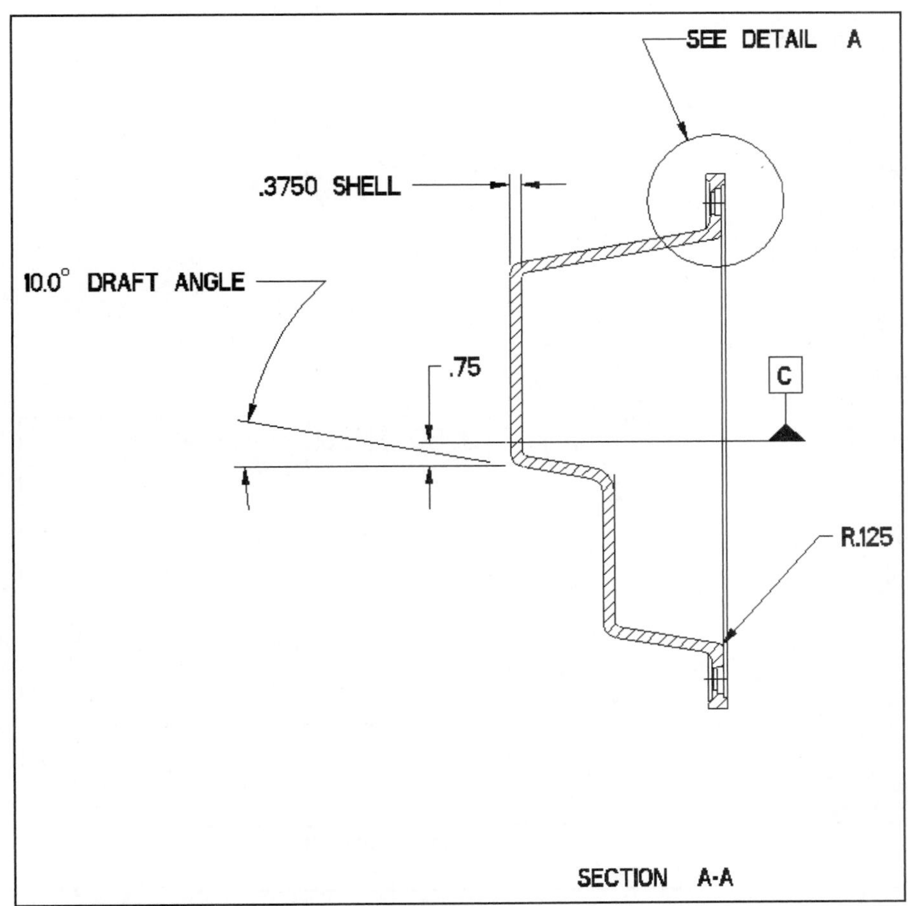

Figure 11.5(g) Oil Sink SECTION A-A

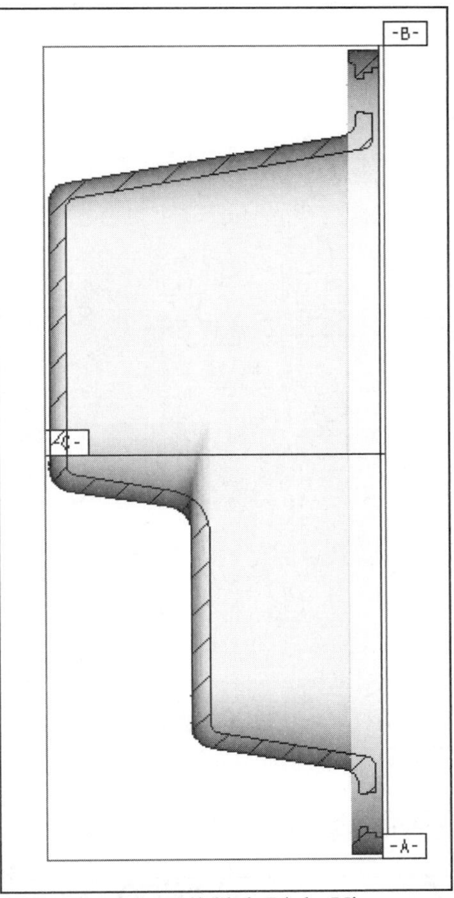

Figure 11.5(h) Oil Sink Right View

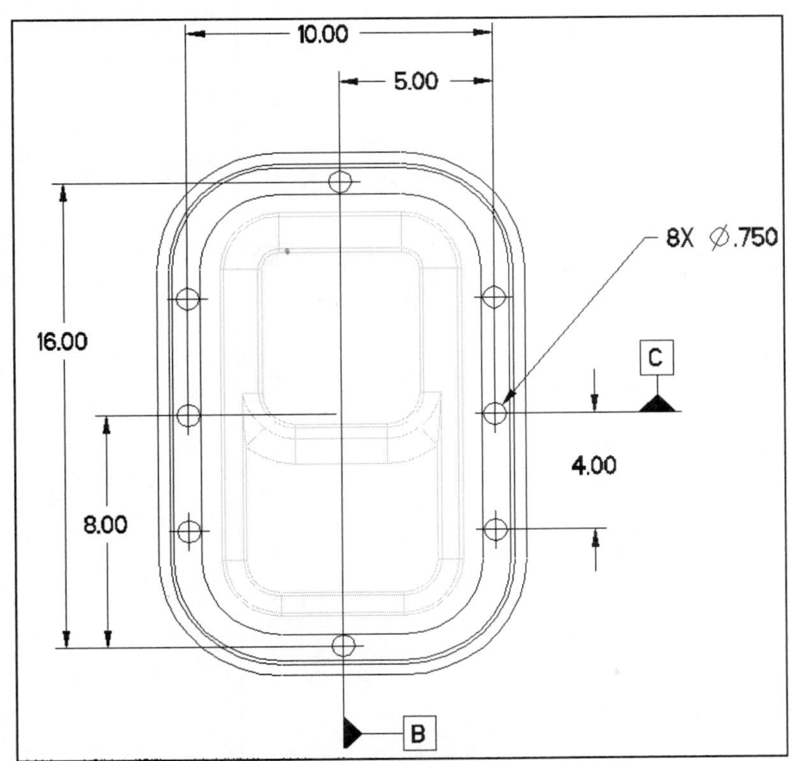

Figure 11.5(i) Oil Sink Bottom View Dimensions

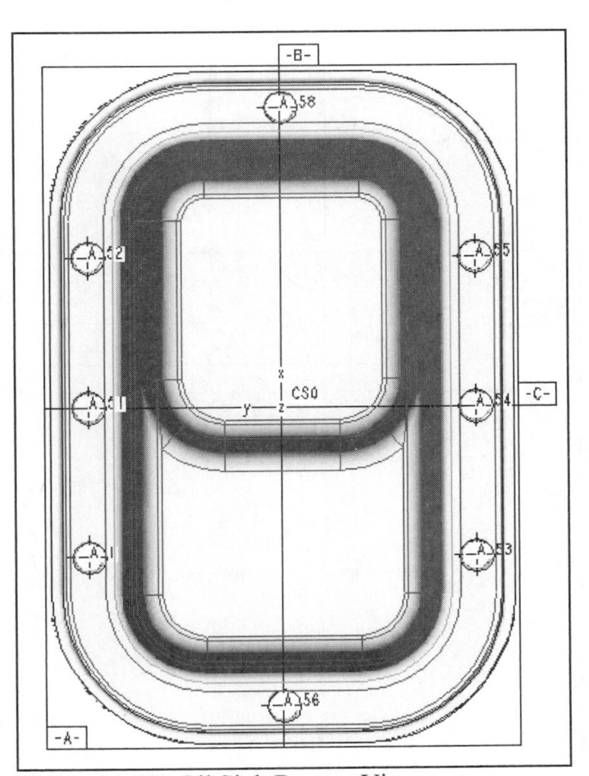

Figure 11.5(j) Oil Sink Bottom View

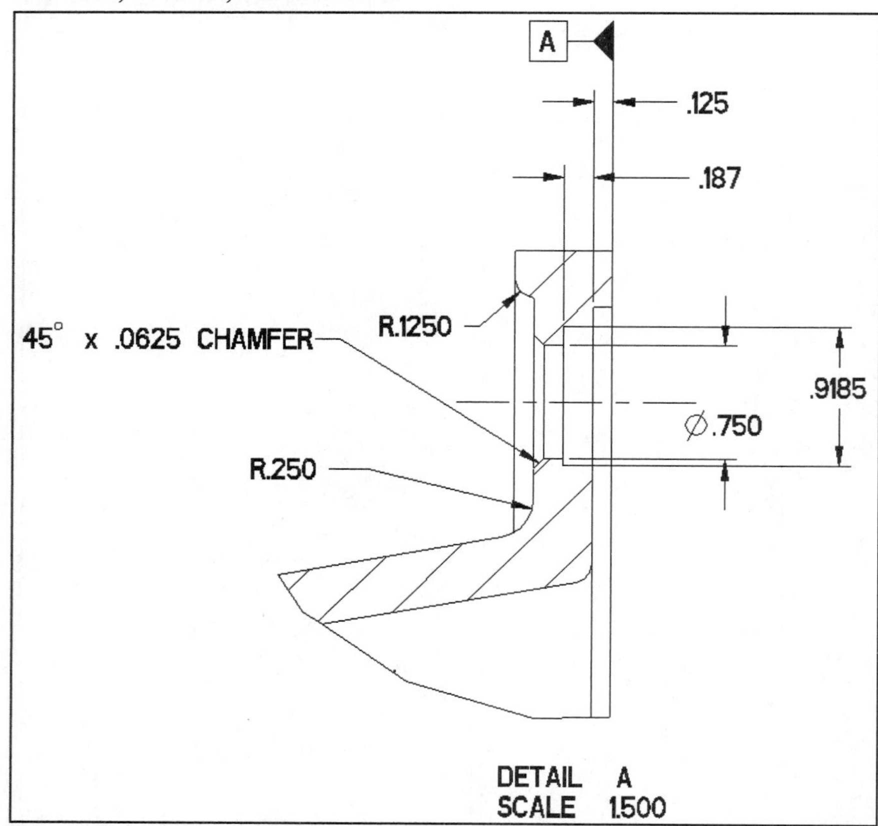

Figure 11.5(k) Oil Sink DETAIL A

.3750 SHELL
.75

Figure 11.5(l) Oil Sink Shell

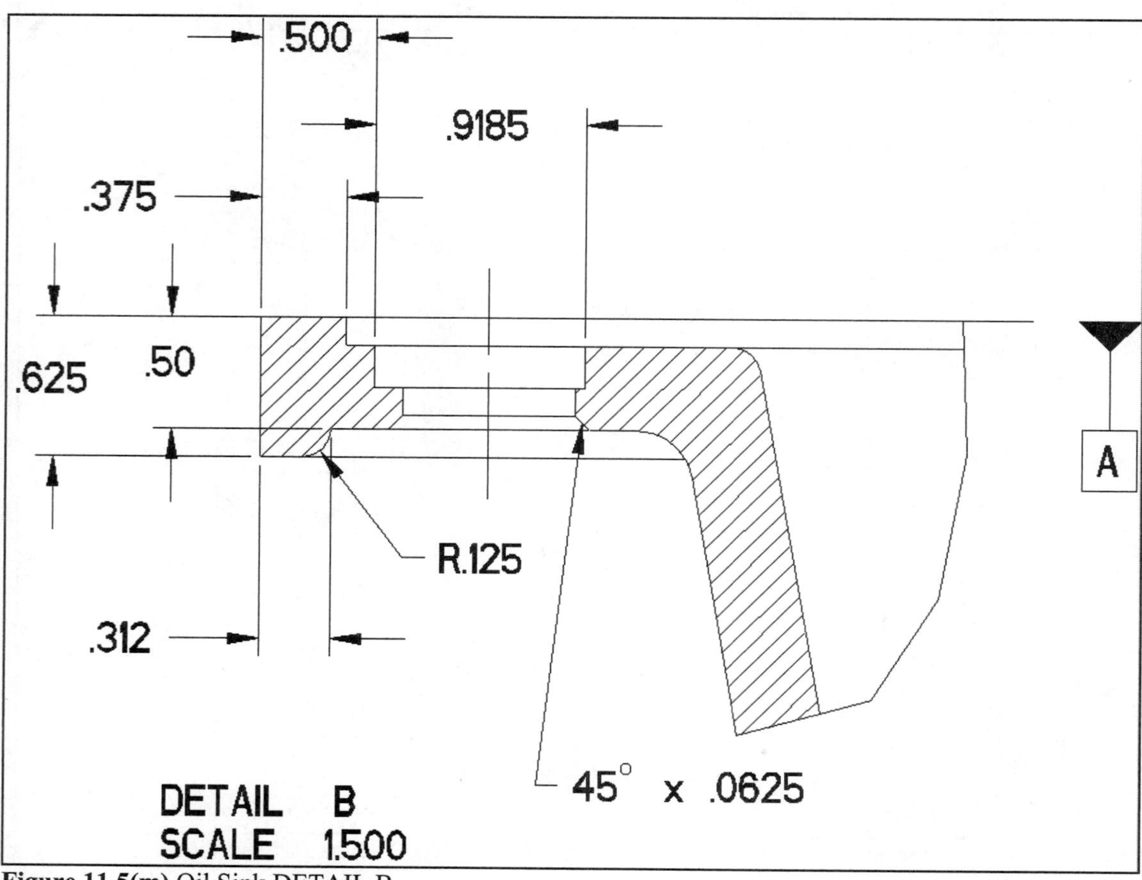

Figure 11.5(m) Oil Sink DETAIL B

Start a new part: **File** ⇒ **Set Working Directory** select the working directory ⇒ **OK** ⇒ [icon] ⇒ ●**Part** ⇒ **OIL_SINK** ⇒ [☑] ⇒ **OK** ⇒ **Tools** ⇒ [icon] ⇒ [☑ Snap to Grid] ⇒ [☑ Use 2D Sketcher] ⇒ [Display Style | Hidden Line] ⇒ [Tangent Edges | Dimmed] ⇒ **OK**

Set up the working environment and defaults: **Tools** ⇒ **Options** ⇒ Showing: **Current Session** ⇒ Option: ***default_dec_places*** ⇒ Value: **3** ⇒ **Enter** ⇒ Option: ***sketcher_dec_places*** ⇒ Value: **3** ⇒ **Enter** ⇒ **Apply** ⇒ **Close** ⇒ load your saved customization file ⇒ **Tools** ⇒ **Customize Screen** ⇒ **File** ⇒ **Open Settings** ⇒ click on your saved file ⇒ **Open** ⇒ **OK**

Set and Assign the material: **Edit** ⇒ **Setup** ⇒ **Units** ⇒ Units Manager **Inch lbm Second** ⇒ **Close** ⇒ **Material** ⇒ **Define** ⇒ type **STEEL** ⇒ **MMB** ⇒ **File** ⇒ **Save** ⇒ **File** ⇒ **Exit** ⇒ **Assign** ⇒ pick **STEEL** ⇒ **Accept** ⇒ **MMB**

Change the coordinate system name and set the datums: double-click on the default coordinate system name in the Model Tree-- **PRT_CSYS_DEF** ⇒ **OIL_SINK** ⇒ **MMB** ⇒ **Edit** ⇒ **Setup** ⇒ **Geom Tol** ⇒ **Set Datum** ⇒ pick **TOP** from the model ⇒ Name- **B** ⇒ **OK** ⇒ pick **FRONT** ⇒ Name- **A** ⇒ **OK** ⇒ pick **RIGHT** ⇒ Name- **C** ⇒ **MMB** ⇒ **MMB** ⇒ **MMB** ⇒ **MMB** ⇒ [icon] ⇒ **MMB** (Fig. 11.6)

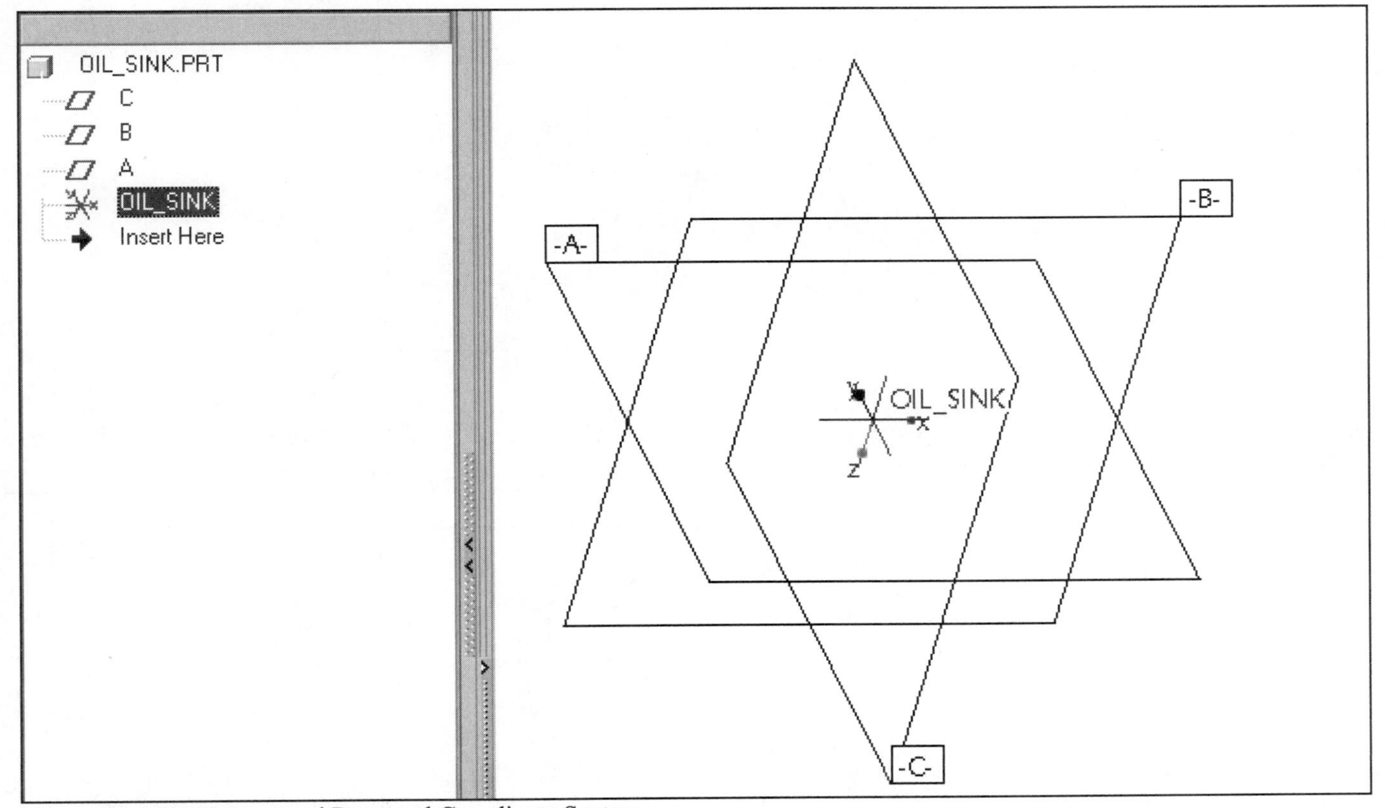

Figure 11.6 Set Datums and Renamed Coordinate System

Make the first protrusion **.50** (thickness) **X 12.00** (height) **X 18.00** (length), with **R4.00** rounds (add the fillets to the sketch). Sketch on datum plane **A**, and center the first protrusion horizontally on datum **B** and vertically on datum **C** [Fig. 11.7(a-b)].

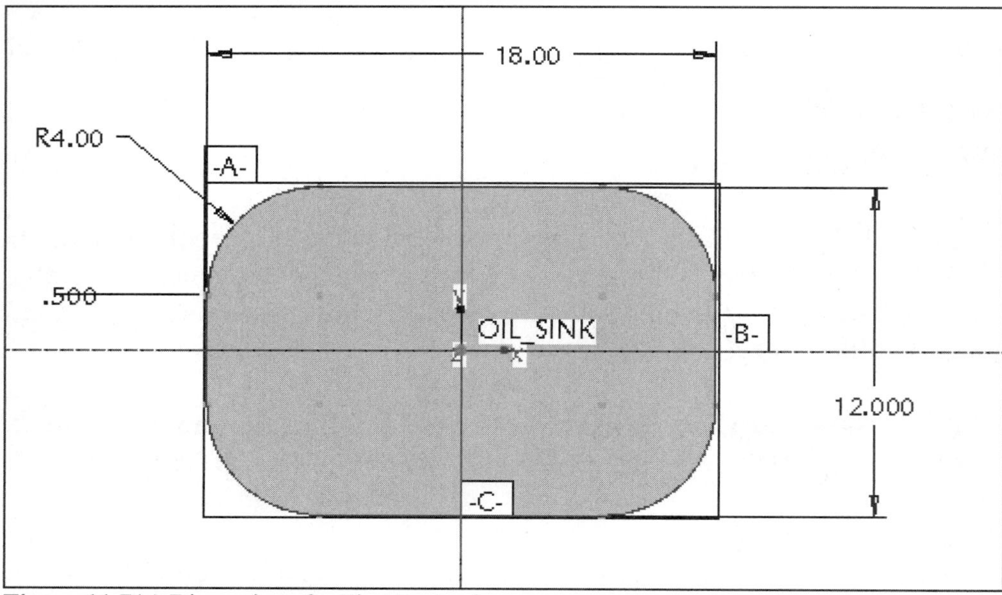

Figure 11.7(a) Dimensions for First Protrusion

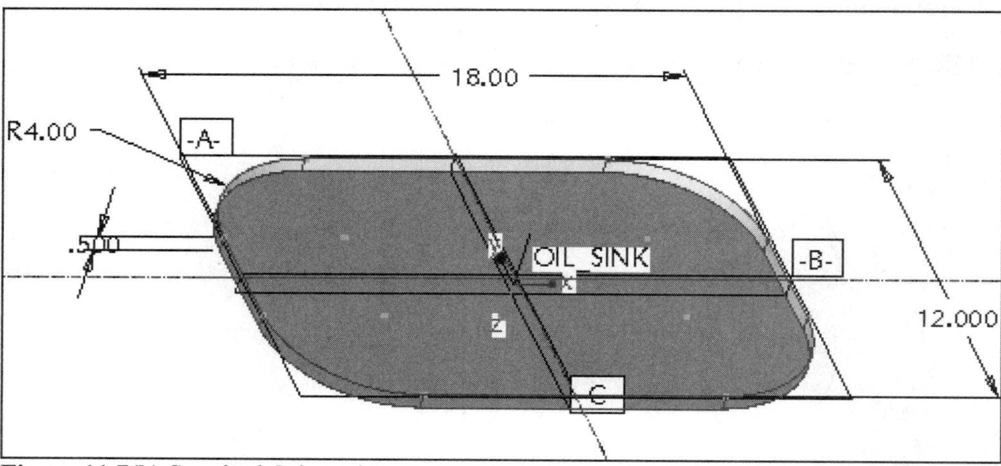

Figure 11.7(b) Standard Orientation

Make the second protrusion offset from the edge of the first protrusion **3.00**, with a height of **7.00** [Fig. 11.8(a-b)]. Sketch on top surface of the first protrusion. Then, create the cut [Fig. 11.9(a-b)].

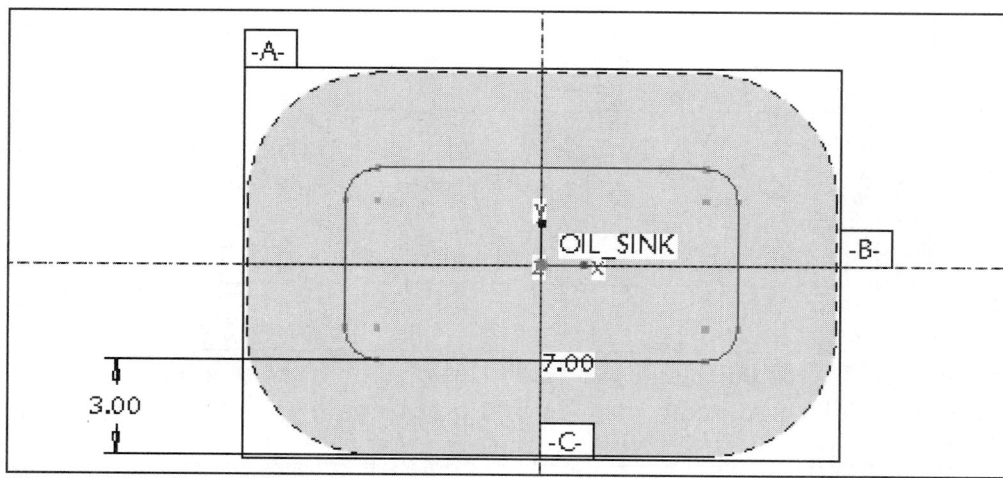

Figure 11.8(a) Second Protrusion is Offset from the Edge of the First Protrusion

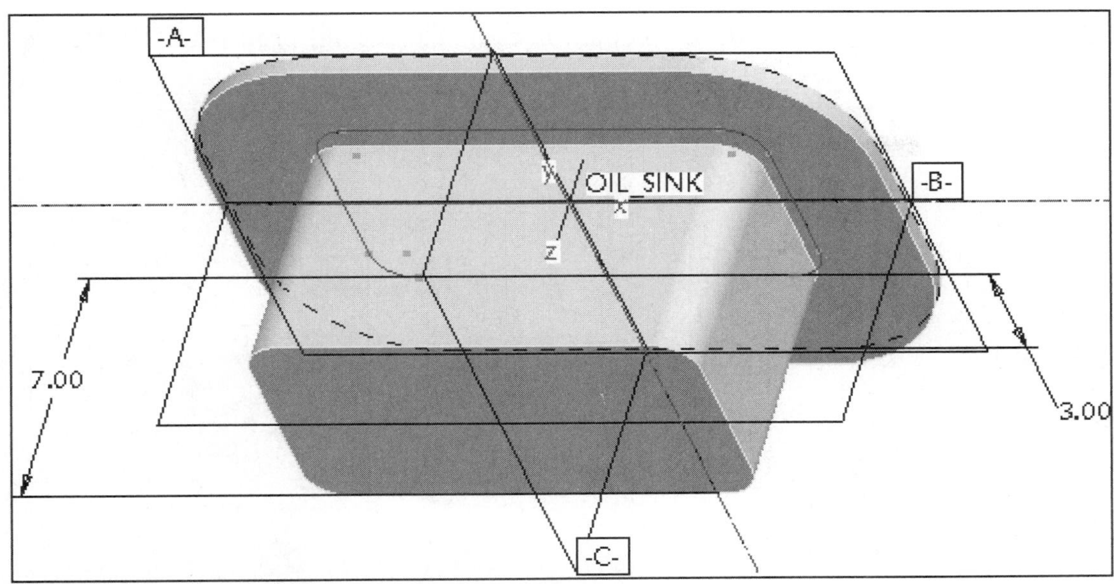

Figure 11.8(b) Second Protrusion

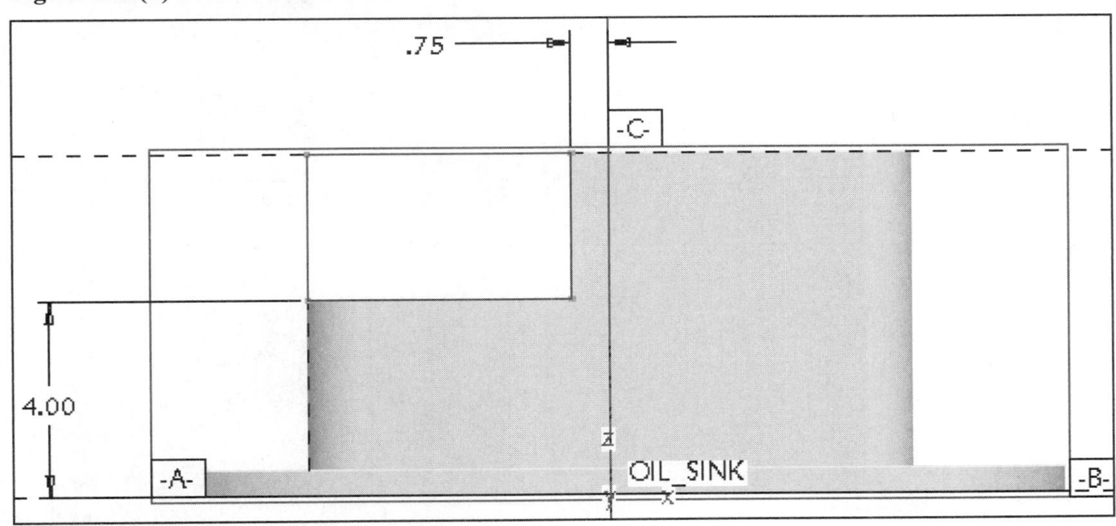

Figure 11.9(a) Cut

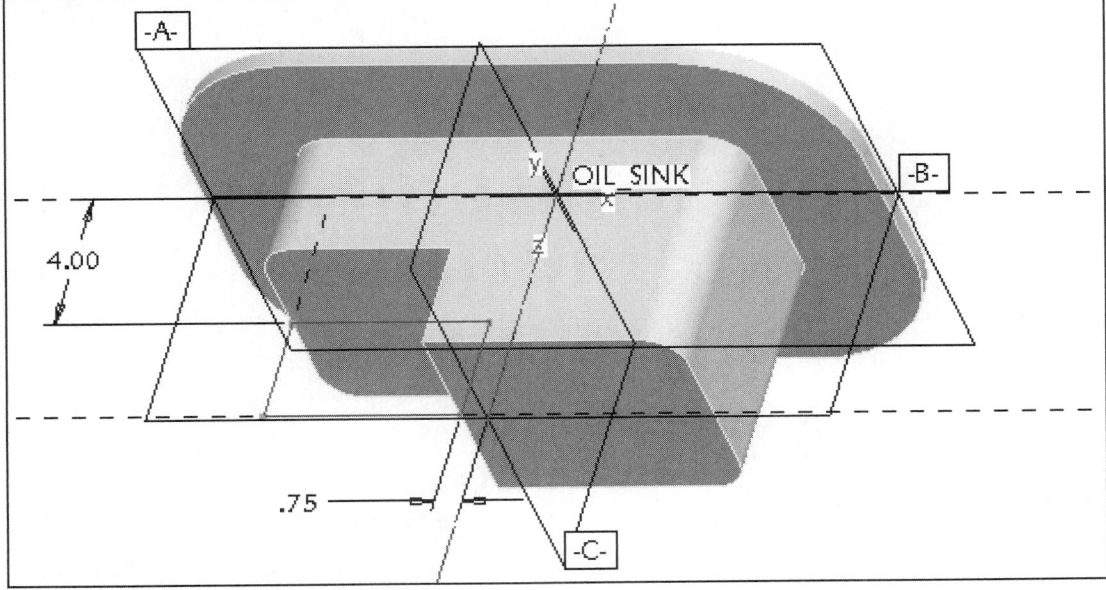

Figure 11.9(b) Standard Orientation of the Cut

Add the **R1.50** rounds [Fig. 11.10(a-b)]. Draft all vertical surfaces of the second protrusion **10** degrees. Use the top surface as the Draft hinge [Fig. 11.11(a-b)].

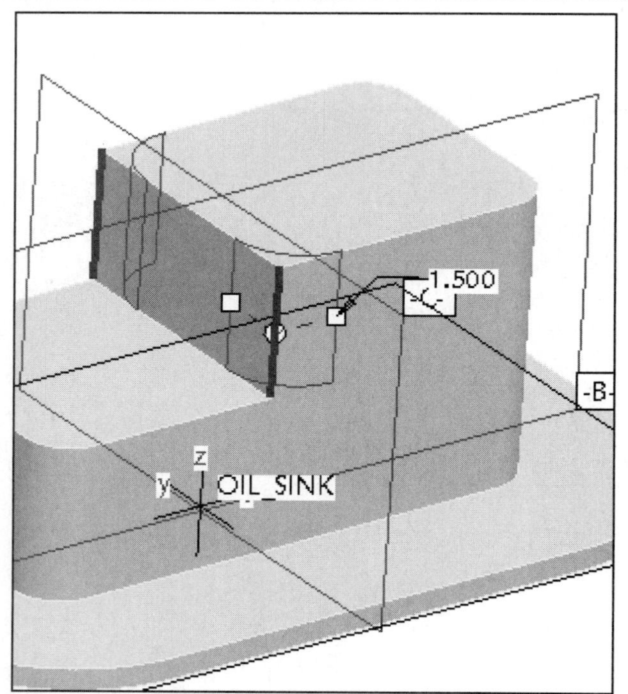

Figure 11.10(a) Create the **R1.50** Rounds

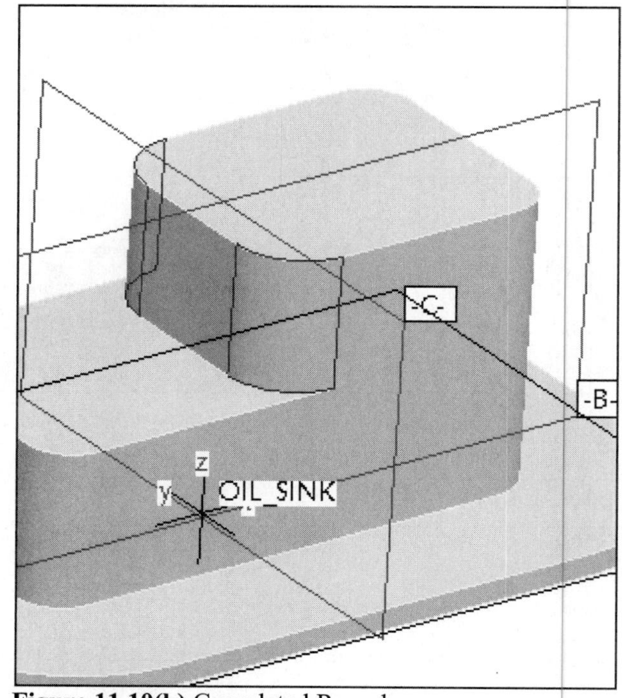

Figure 11.10(b) Completed Rounds

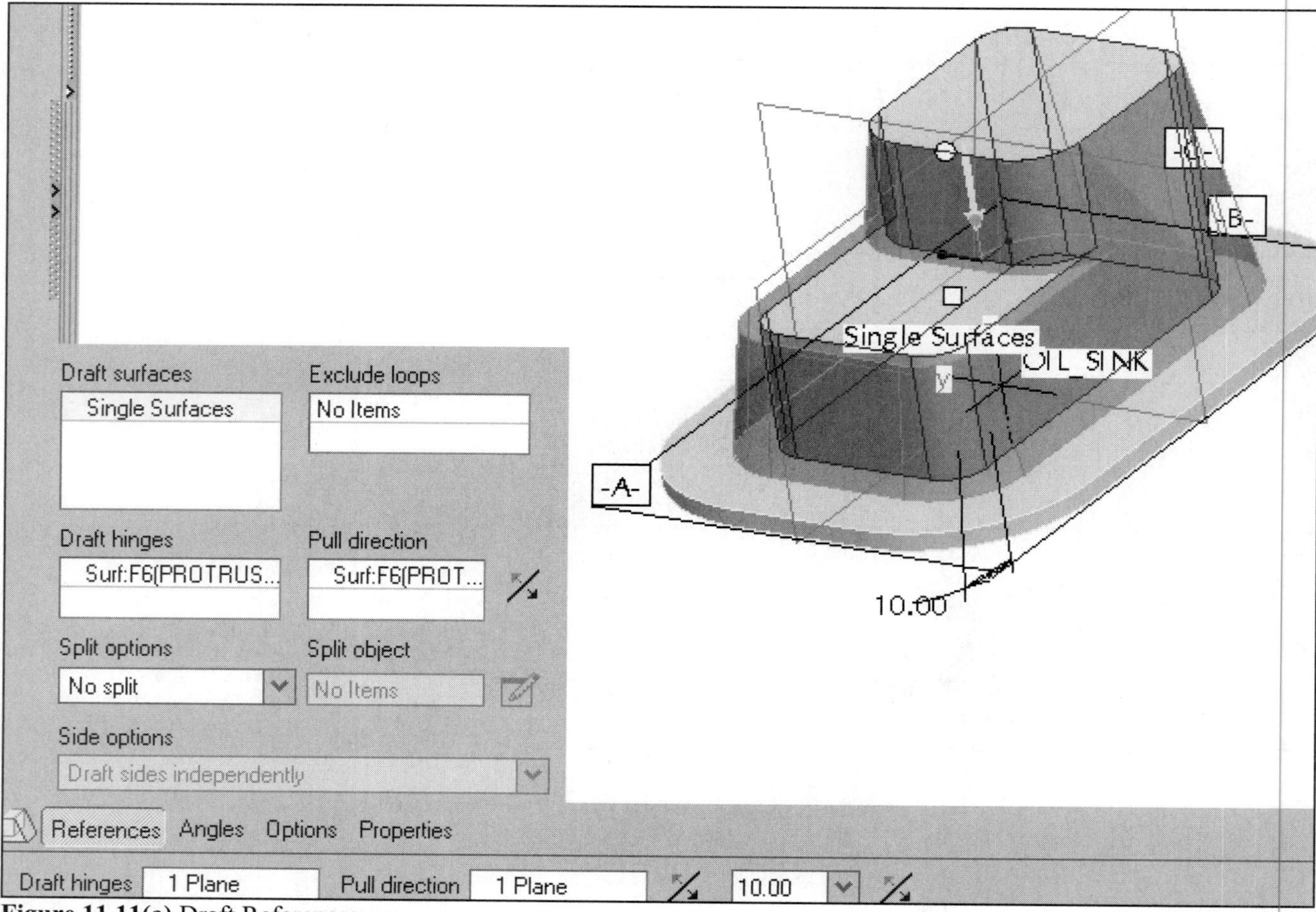

Figure 11.11(a) Draft References

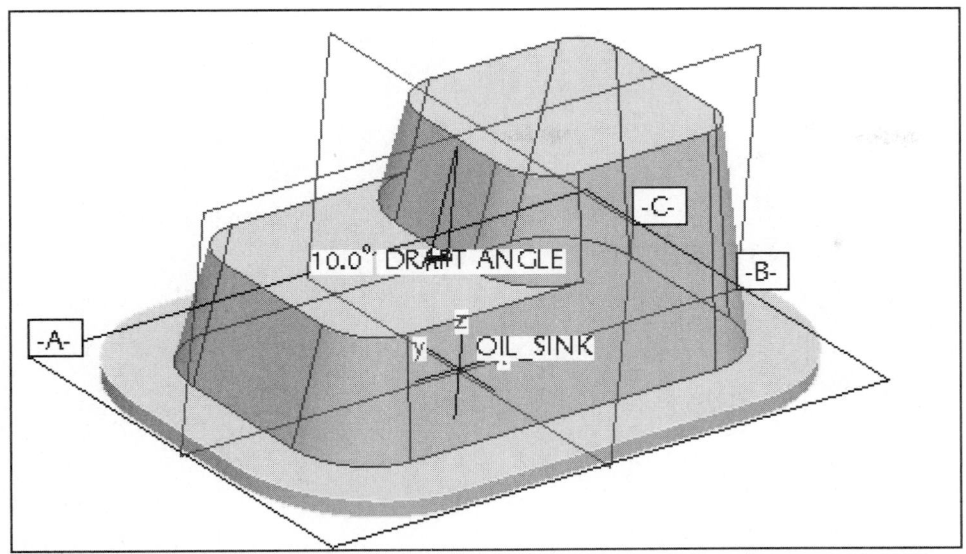

Figure 11.11(b) Drafted Sides

Click: **Shell Tool** ⇒ Thickness **.375** ⇒ spin the model ⇒ **References** tab ⇒ Removed surfaces--select the bottom surface of the part [Fig. 11.12(a)] ⇒ ⇒ ⇒ **MMB** ⇒ ⇒ ⇒ ⇒ **MMB** ⇒ **File** ⇒ **Delete** ⇒ **Old Versions** ⇒ **MMB** [Fig. 11.12(b)]

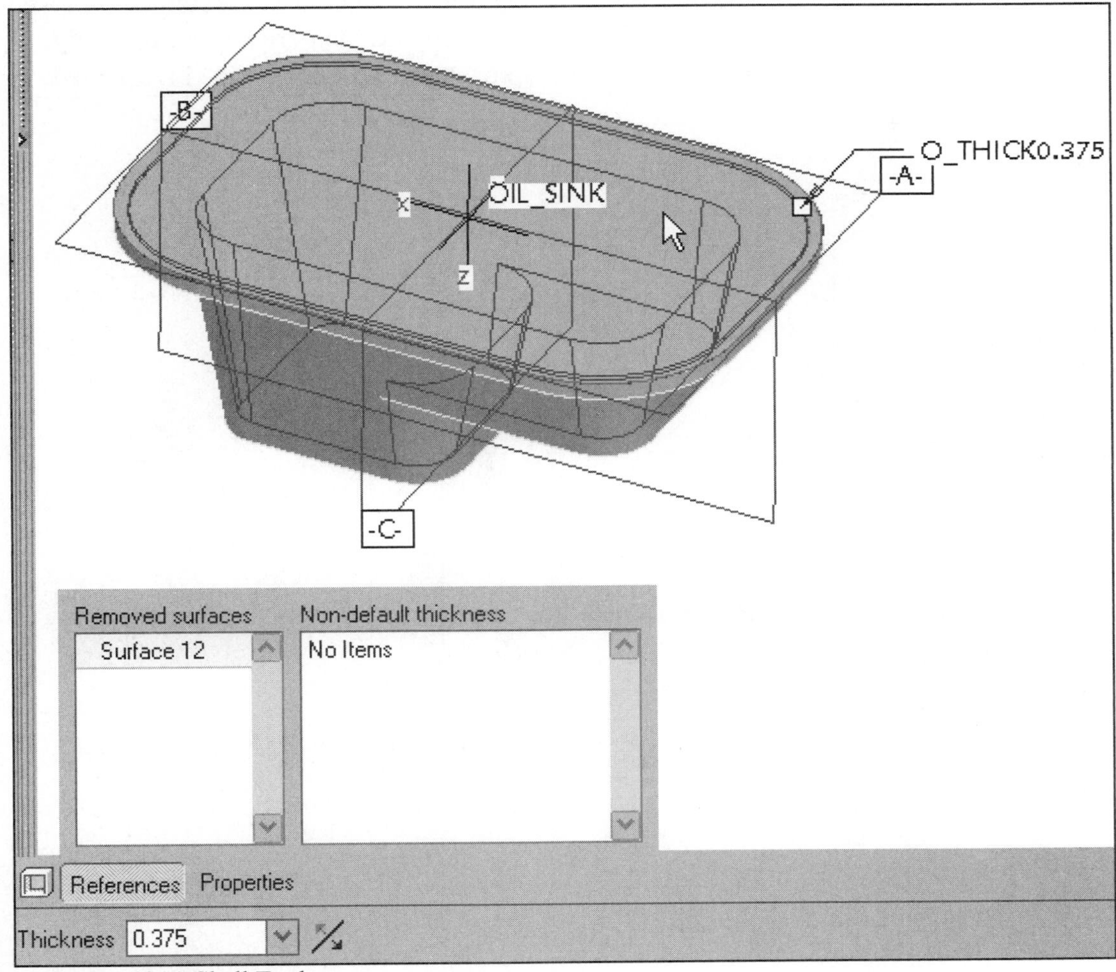

Figure 11.12(a) Shell Tool

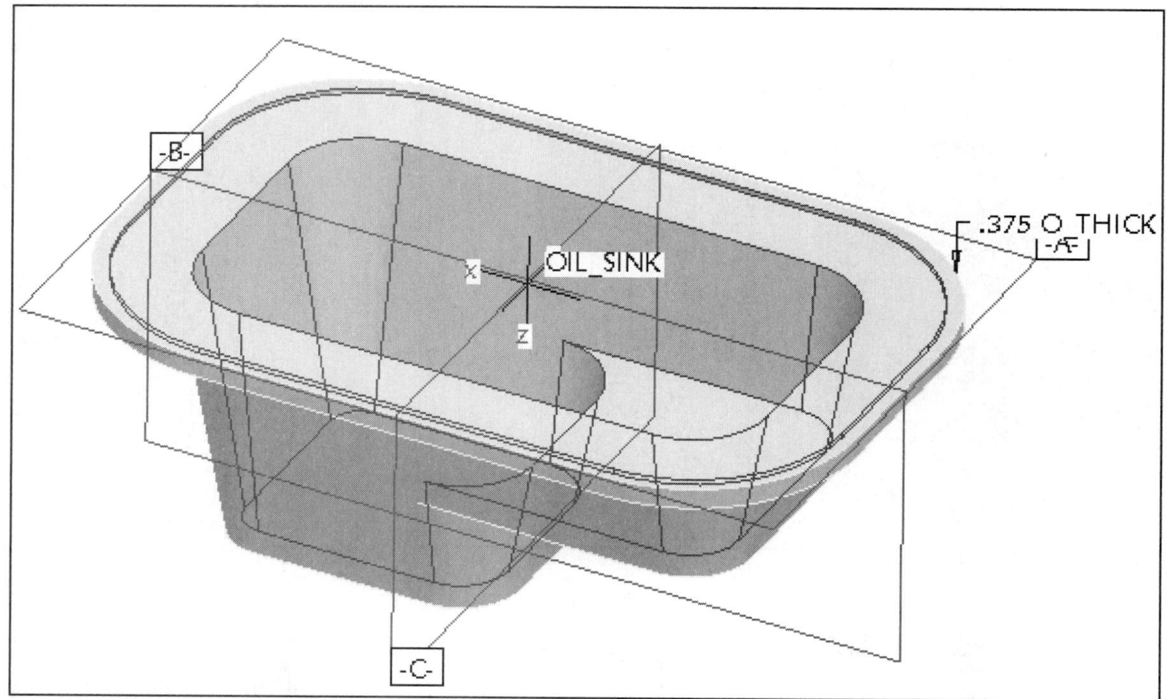

Figure 11.12(b) Shelled Part

The next feature you need to create is a *"lip"* around the part using a protrusion ⇒ Sketch two closed loops [Fig. 11.13(a)]. Use the edge of the first protrusion for the first loop and then create an offset edge (**-.3125**) for the second loop [Fig. 11.13(b)]. ⇒ The depth of the lip protrusion is **.125** [Fig. 11.13(c-d)].

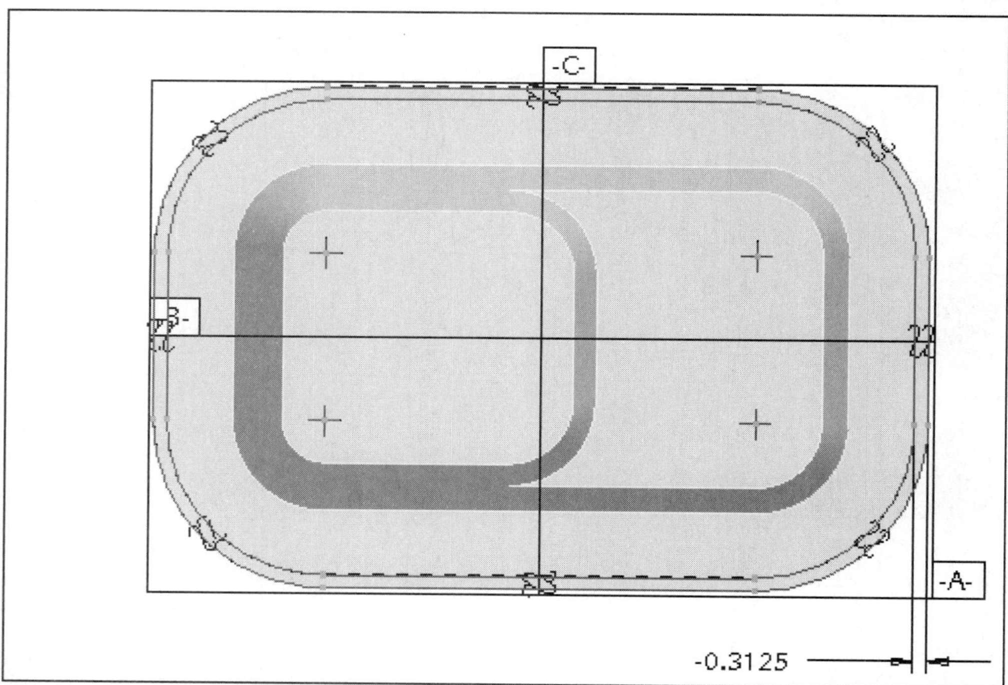

Figure 11.13(a) Sketch

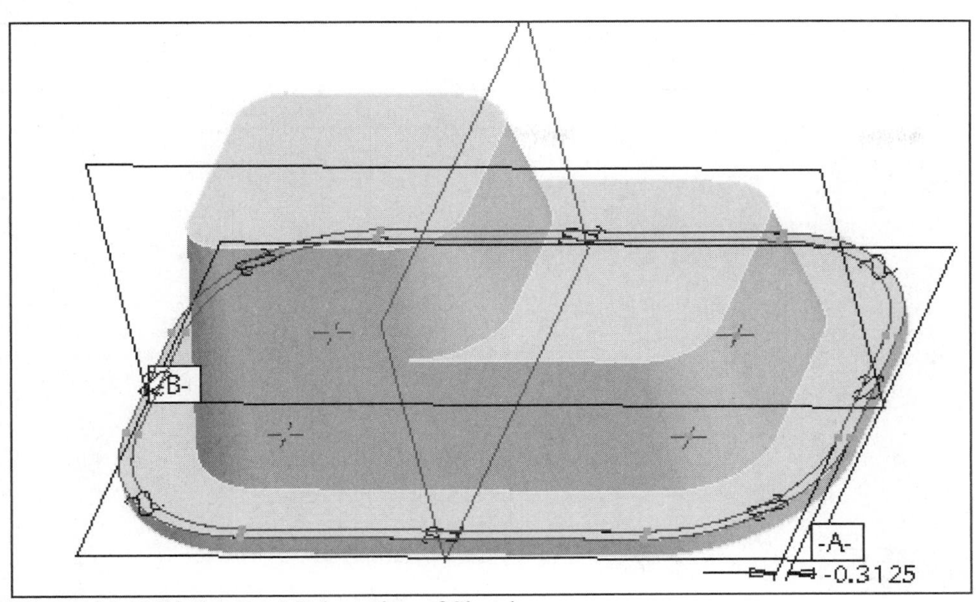

Figure 11.13(b) Standard Orientation of Sketch

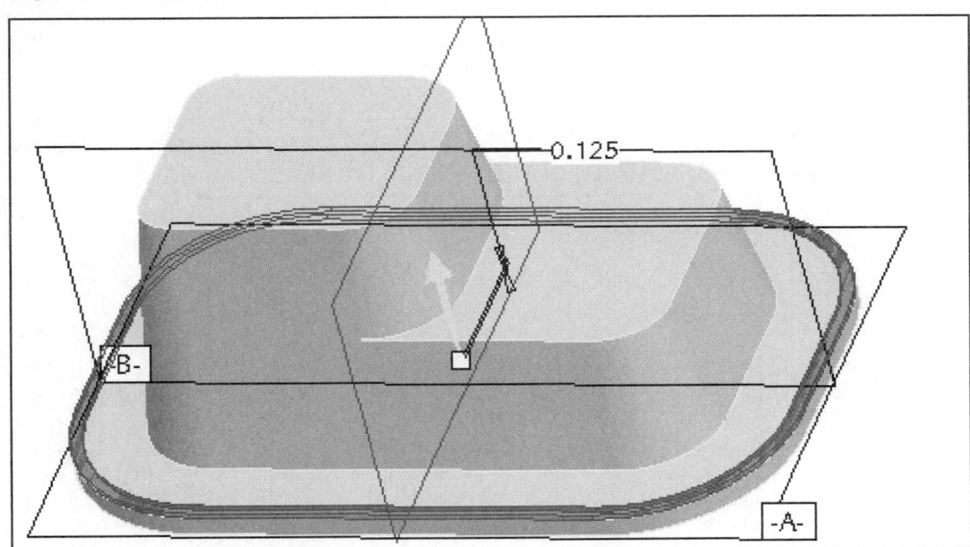

Figure 11.13(c) Depth **.125**

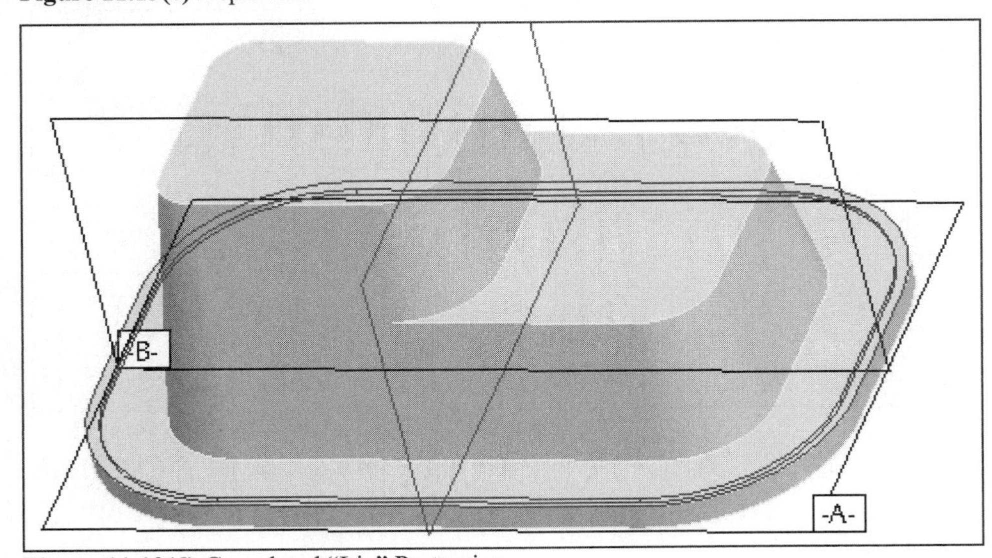

Figure 11.13(d) Completed "Lip" Protrusion

Add the **R.125** round to the inside of the *"lip"* (Fig. 11.14) ⇒ [save icon] ⇒ **MMB**

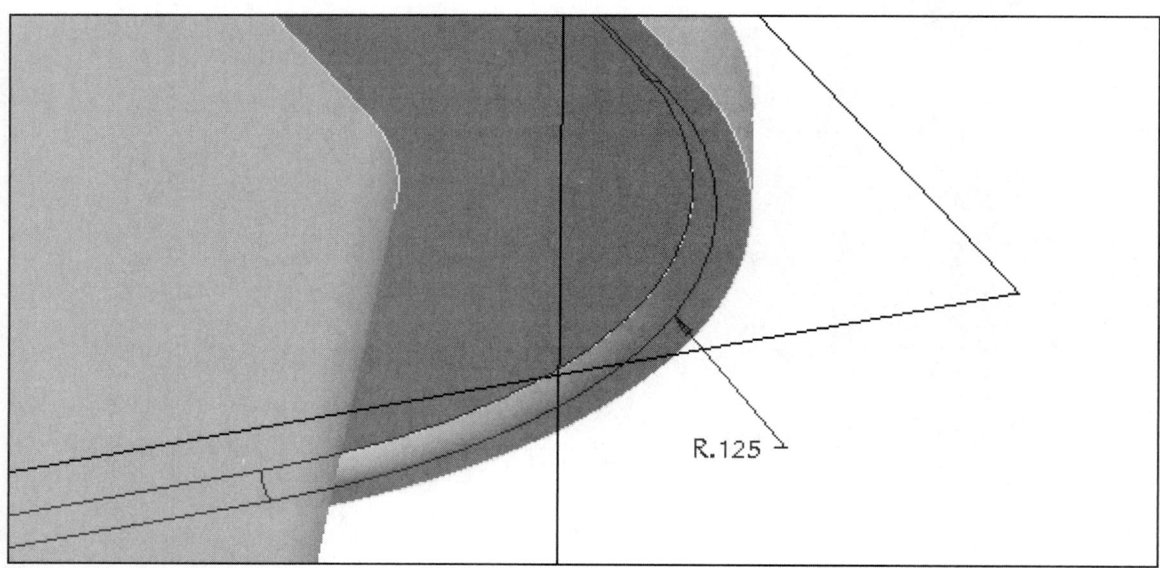

Figure 11.14 Round **R.125**

The next feature is a cut measuring **.9185** wide by **.187** deep [Fig. 11.15(a-b)]. The sections sketch will be composed of two closed loops as with the lip-like protrusion created previously [Fig. 11.16(a-c)].

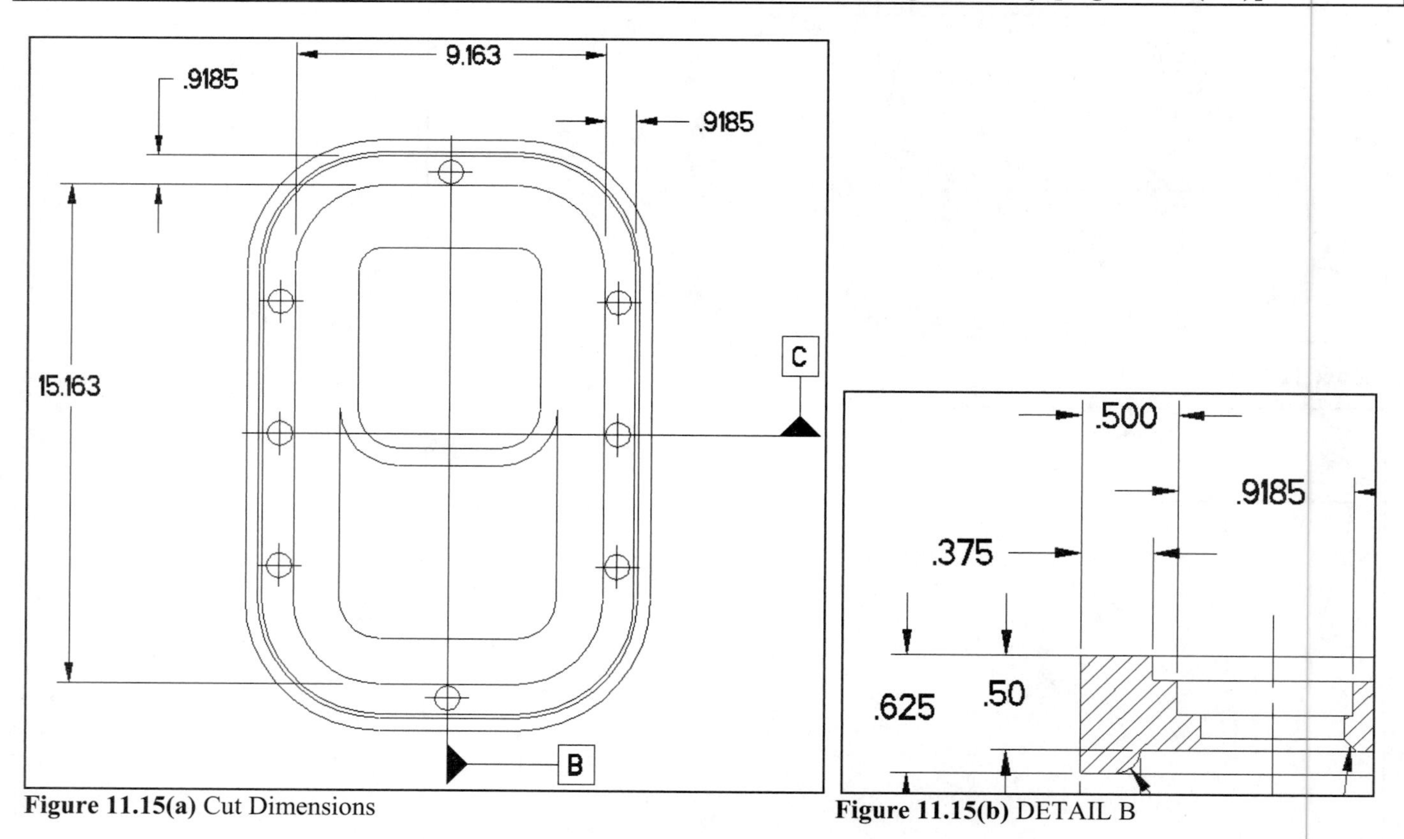

Figure 11.15(a) Cut Dimensions

Figure 11.15(b) DETAIL B

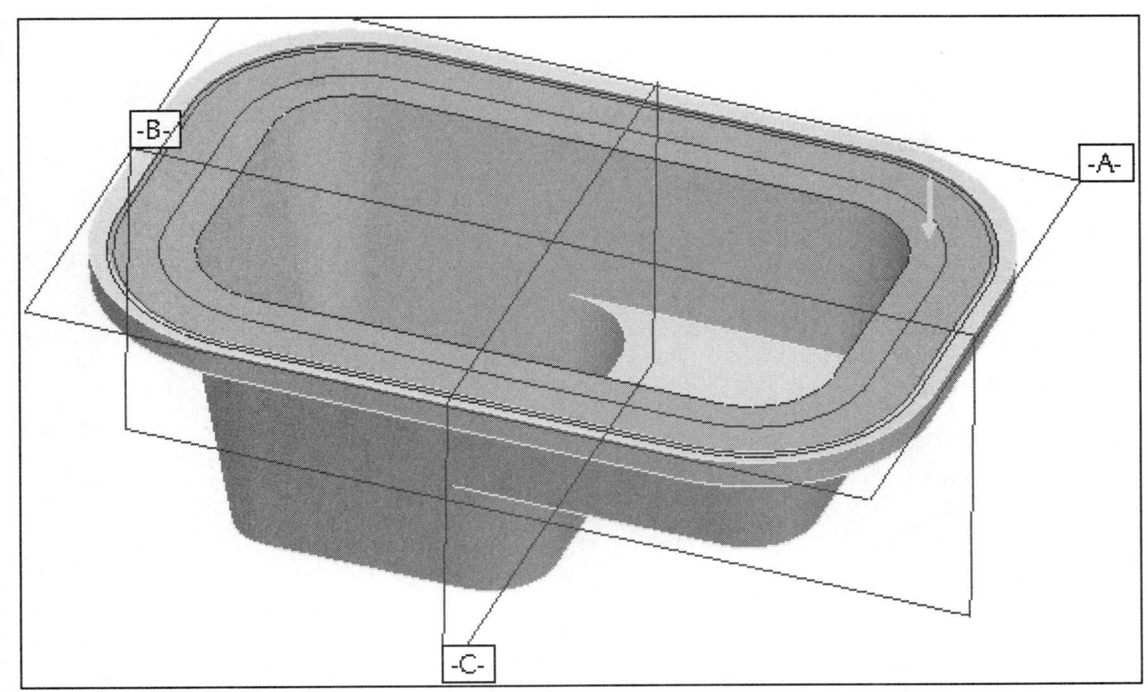

Figure 11.16(a) Cut Surface

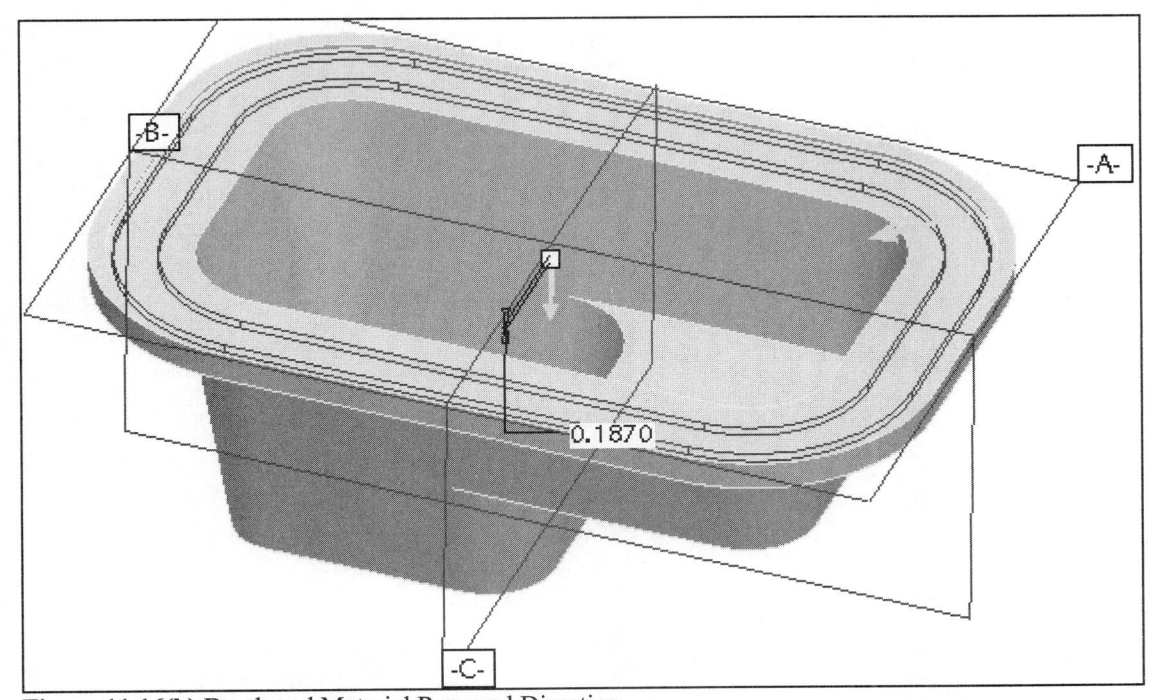

Figure 11.16(b) Depth and Material Removal Direction

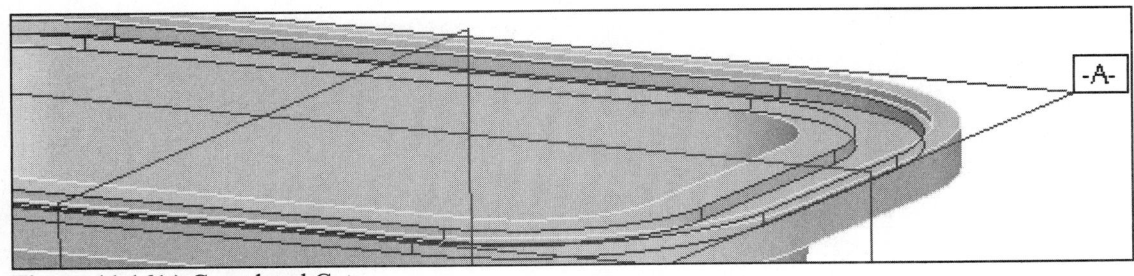

Figure 11.16(c) Completed Cut

Add another **R.125** round to the inside edge [Fig. 11.17(a-b)] ⇒ [save icon] ⇒ **MMB**

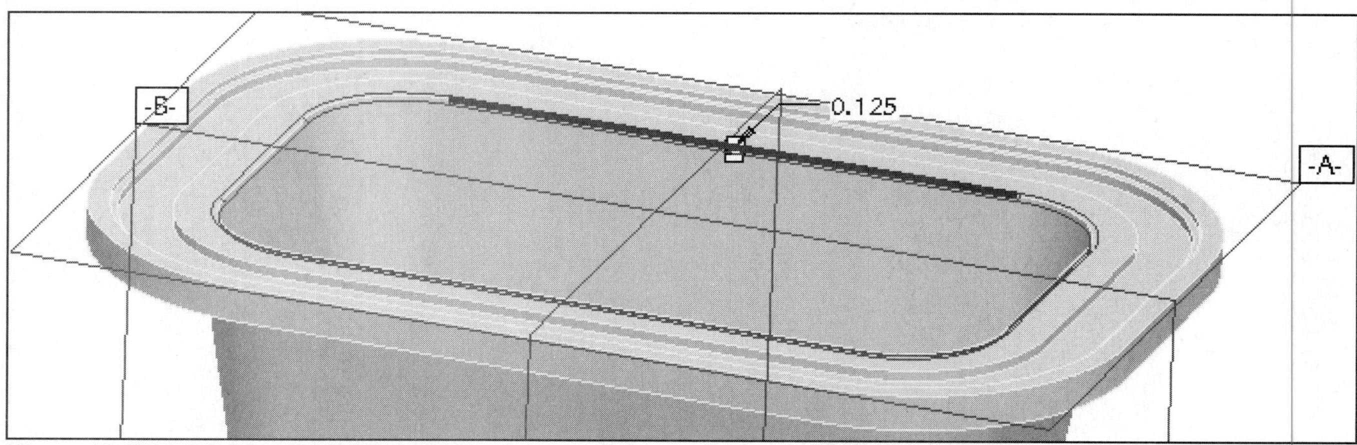

Figure 11.17(a) Edge Round **R.125**

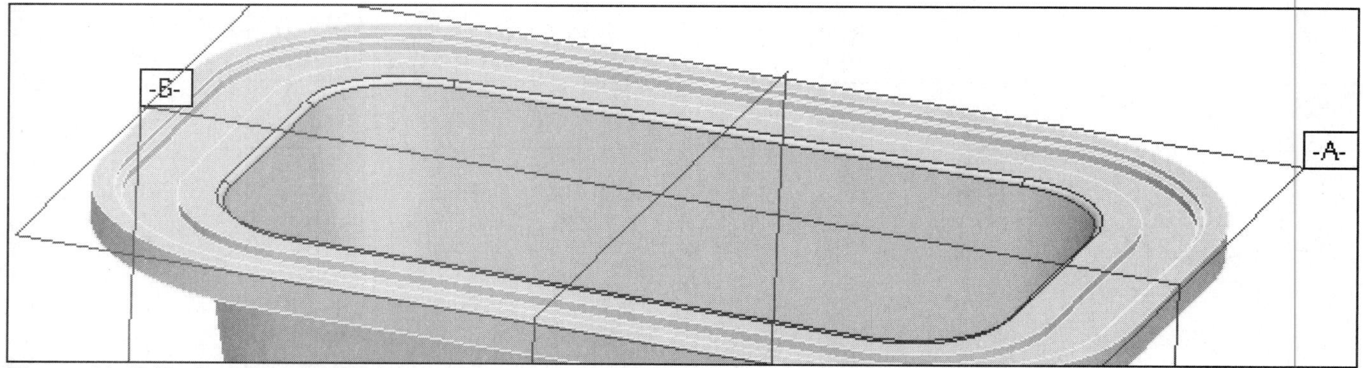

Figure 11.17(b) Completed Round

Create a **R.250** round as shown in Figure 11.18 ⇒ [save icon] ⇒ **MMB**

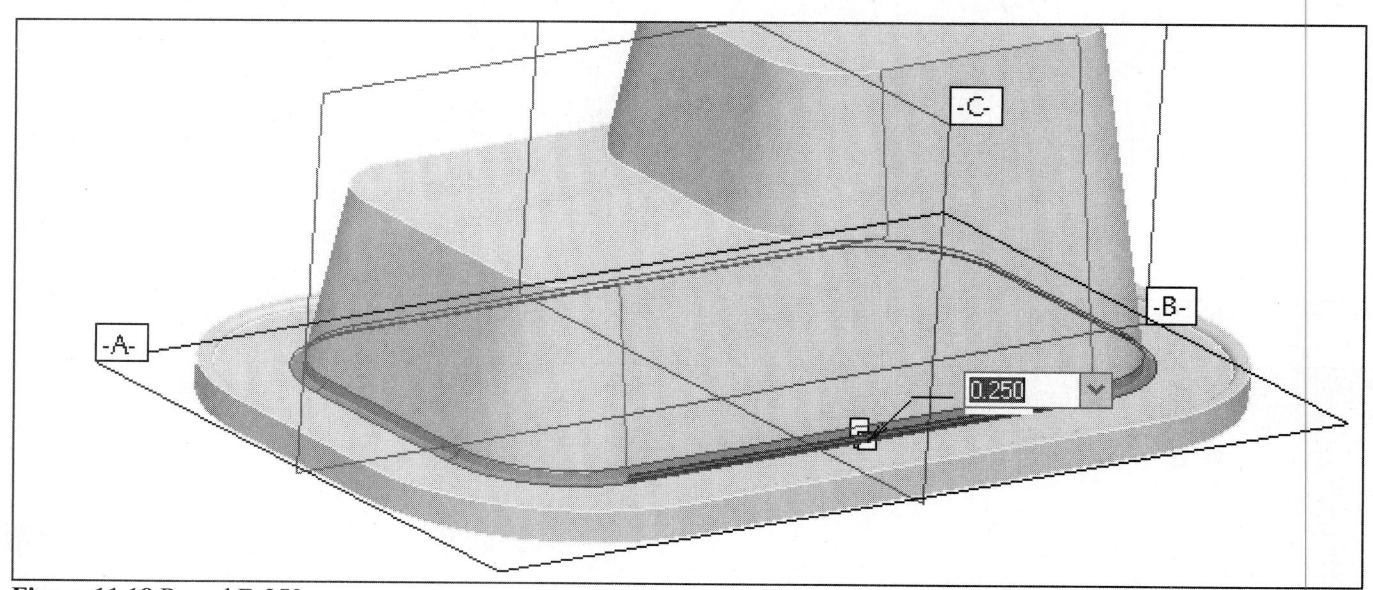

Figure 11.18 Round **R.250**

The countersunk holes will be added next [Fig. 11.19(a-b)].

Click: **Hole Tool** from the tool bar ⇒ spin the part ⇒ **Drill to intersect with all surfaces** ⇒ change the diameter to **.750** ⇒ ⇒ **Placement** tab ⇒ select the surface for placement [Fig. 11.19(c)]

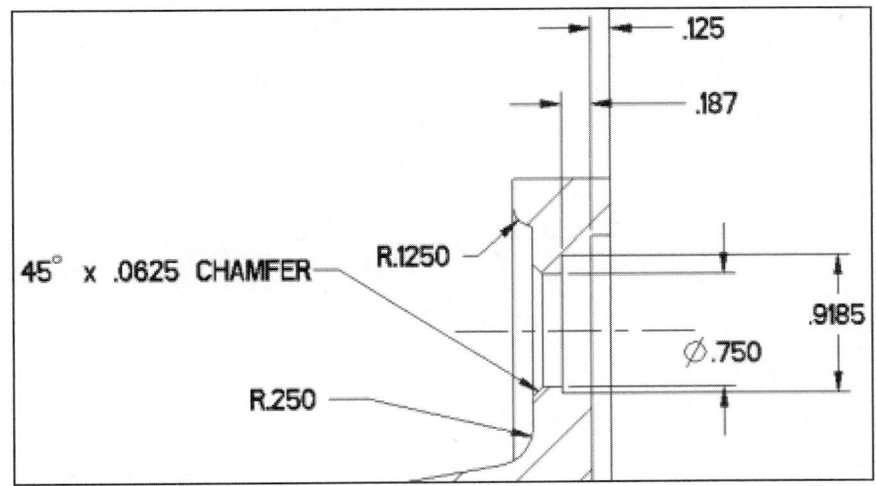

Figure 11.19(a) Hole Dimensions

Figure 11.19(b) X-Section of Hole

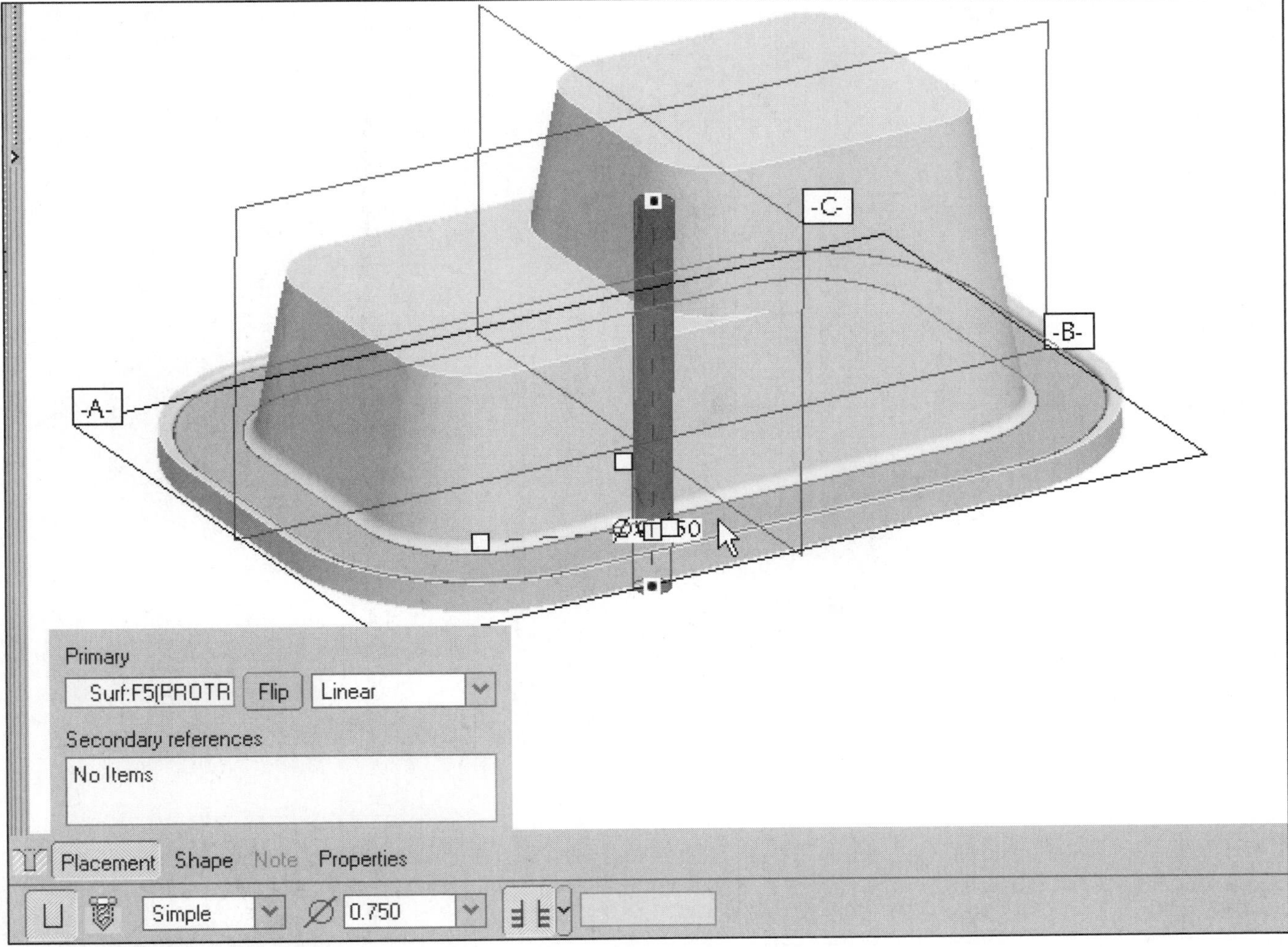

Figure 11.19(c) Hole Placement

Move one drag handle to reference datum **B** and the other to reference datum **C** [Fig. 11.19(d)] ⇒ modify the values to be **-4.00** *(negative 4.00)* from datum **C** and **-5.00** *(negative 5.00)* from datum **B** [Fig. 11.19(e)] ⇒ ⇒ ⇒ ⇒ ⇒ **MMB**

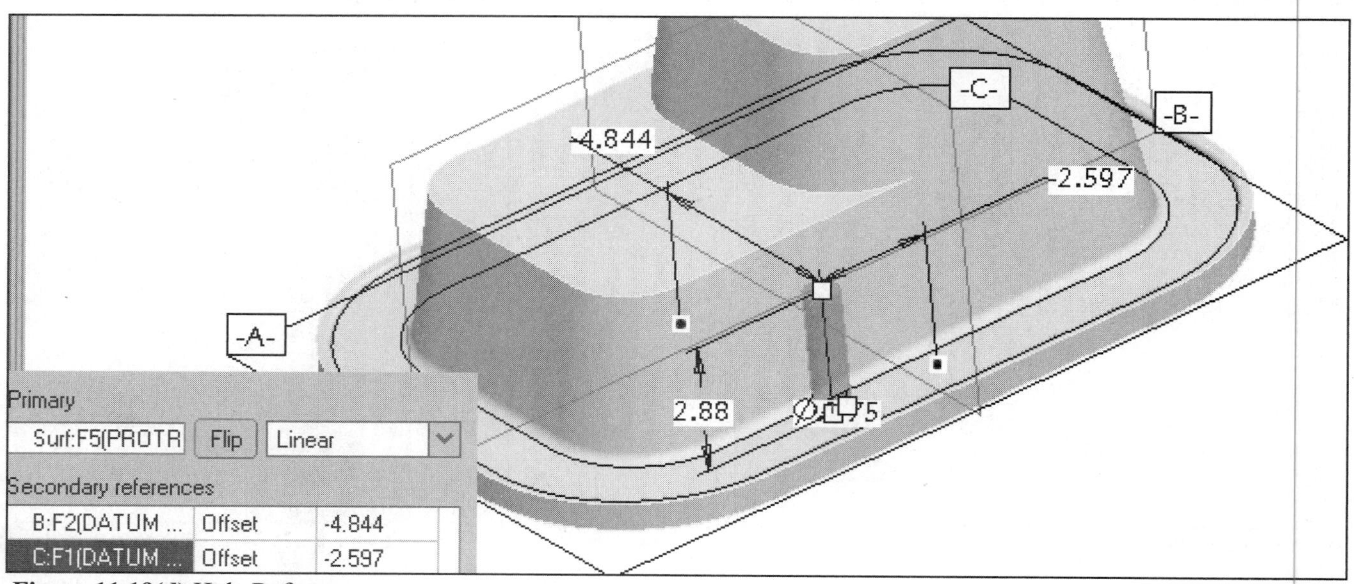

Figure 11.19(d) Hole References

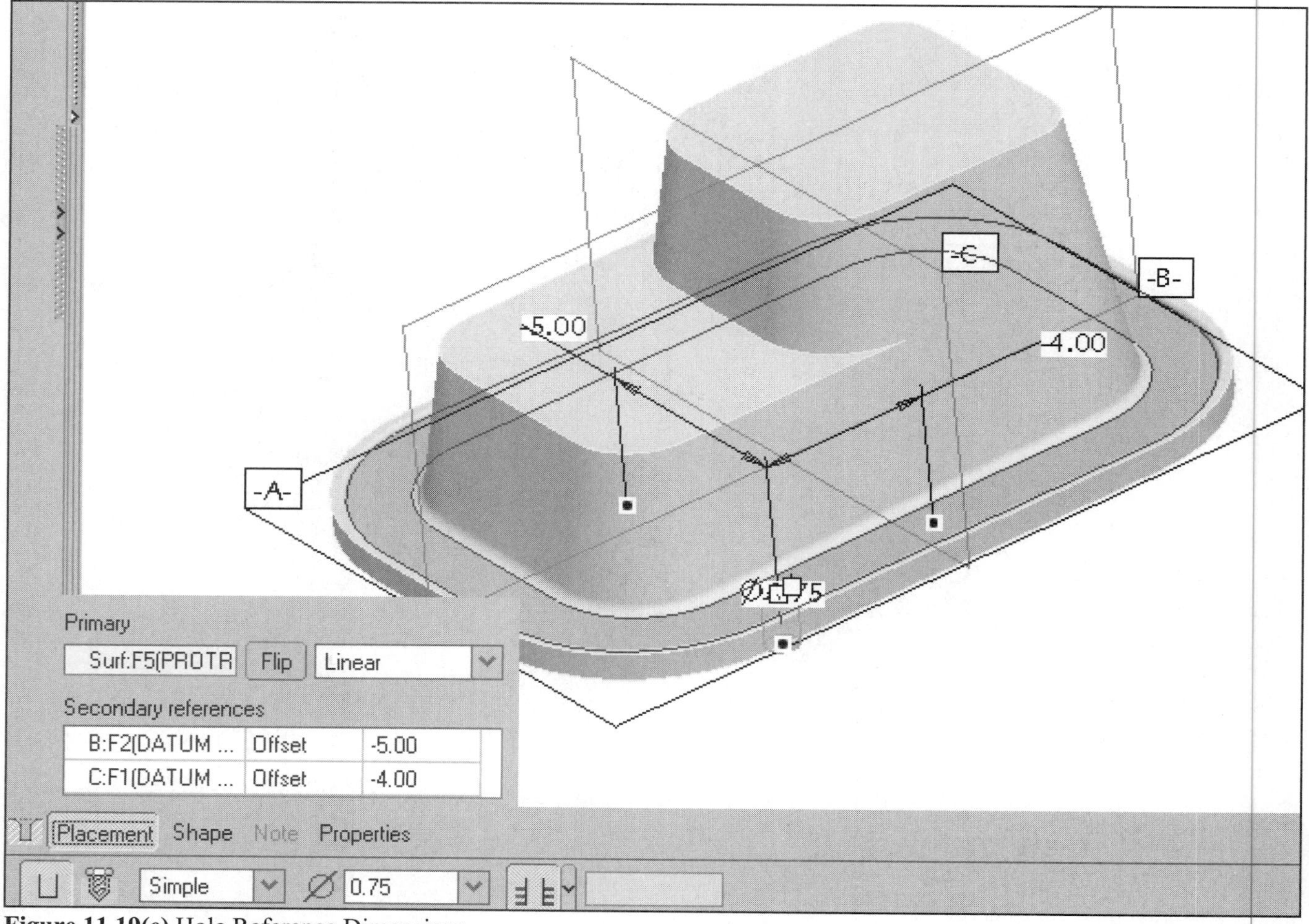

Figure 11.19(e) Hole Reference Dimensions

Create a **45 X .0625** chamfer on the hole [Fig. 11.20(a-b)] ⇒ group the hole and the chamfer, click: **Edit** ⇒ **Feature Operations** ⇒ **Group** ⇒ **Local Group** ⇒ type a name for the group: **CH_HOLE** ⇒ **MMB** ⇒ click on the hole and the chamfer in the Model Tree ⇒ **MMB** ⇒ **MMB** ⇒ **MMB** ⇒ **MMB** [Fig. 11.20(c)] ⇒ [icon] ⇒ **MMB** ⇒ **LMB** to deselect ⇒ [icon]

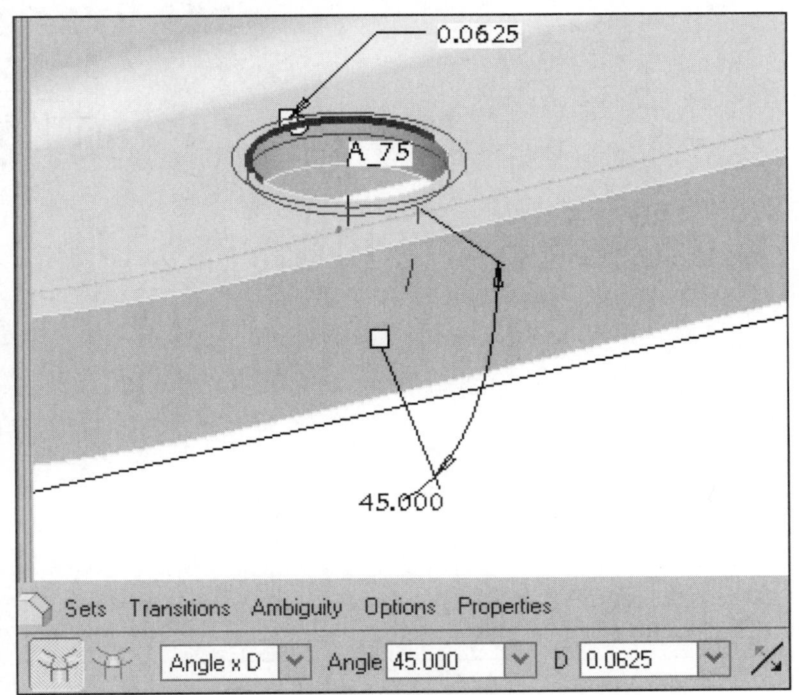

Figure 11.20(a) Chamfer **45 X .0625**

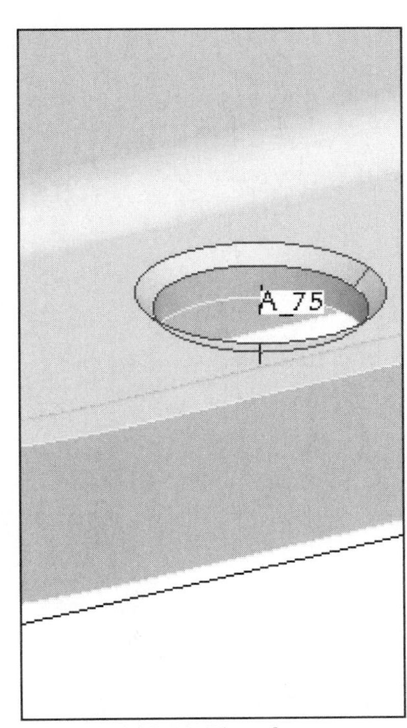

Figure 11.20(b) Chamfer

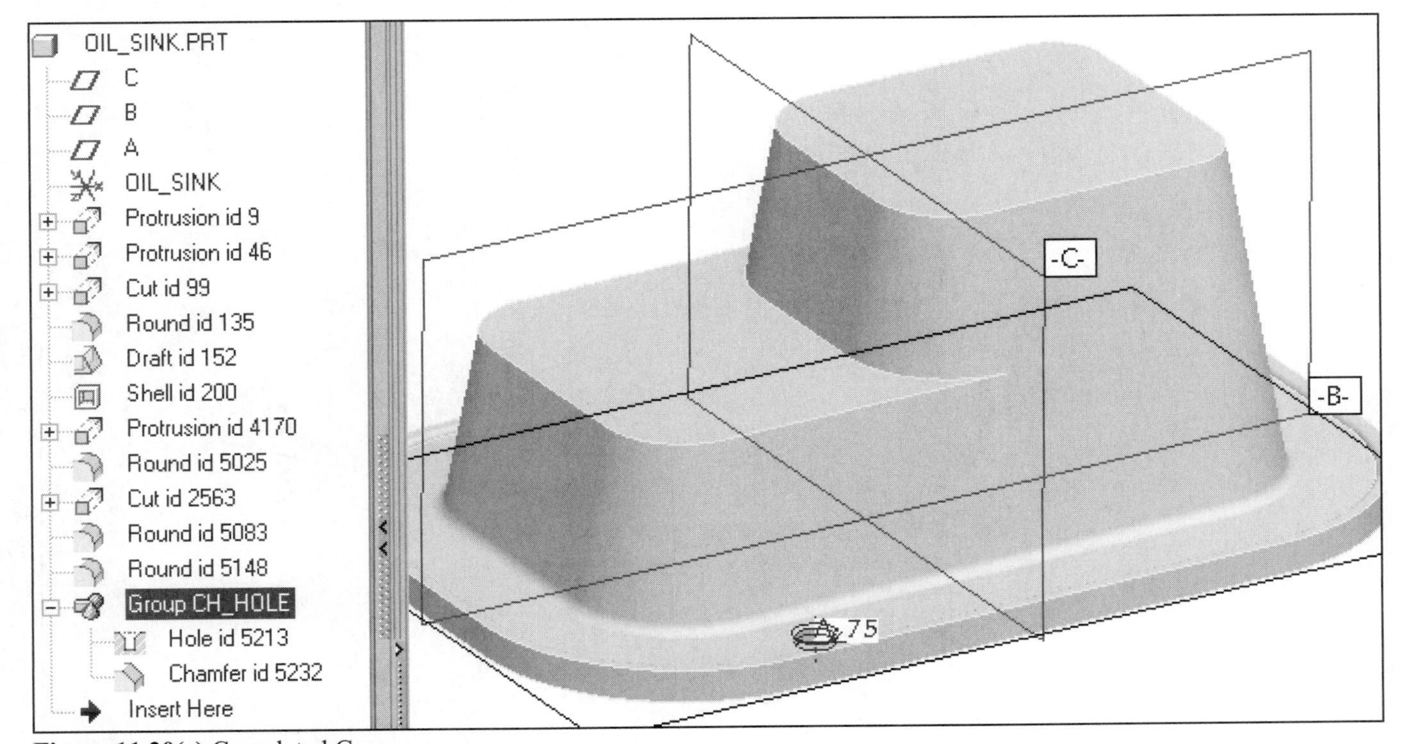

Figure 11.20(c) Completed Group

With the next set of commands, you will pattern the Hole.

Click on **Group CH_HOLE** in the Model Tree [Fig. 11.20(d)] ⇒ **RMB** ⇒ **Pattern** Dimension ⇒ ⇒ Table ⇒ with the **Ctrl** key pressed, click on **4.000** and then on **5.000** [Fig. 11.20(e)]

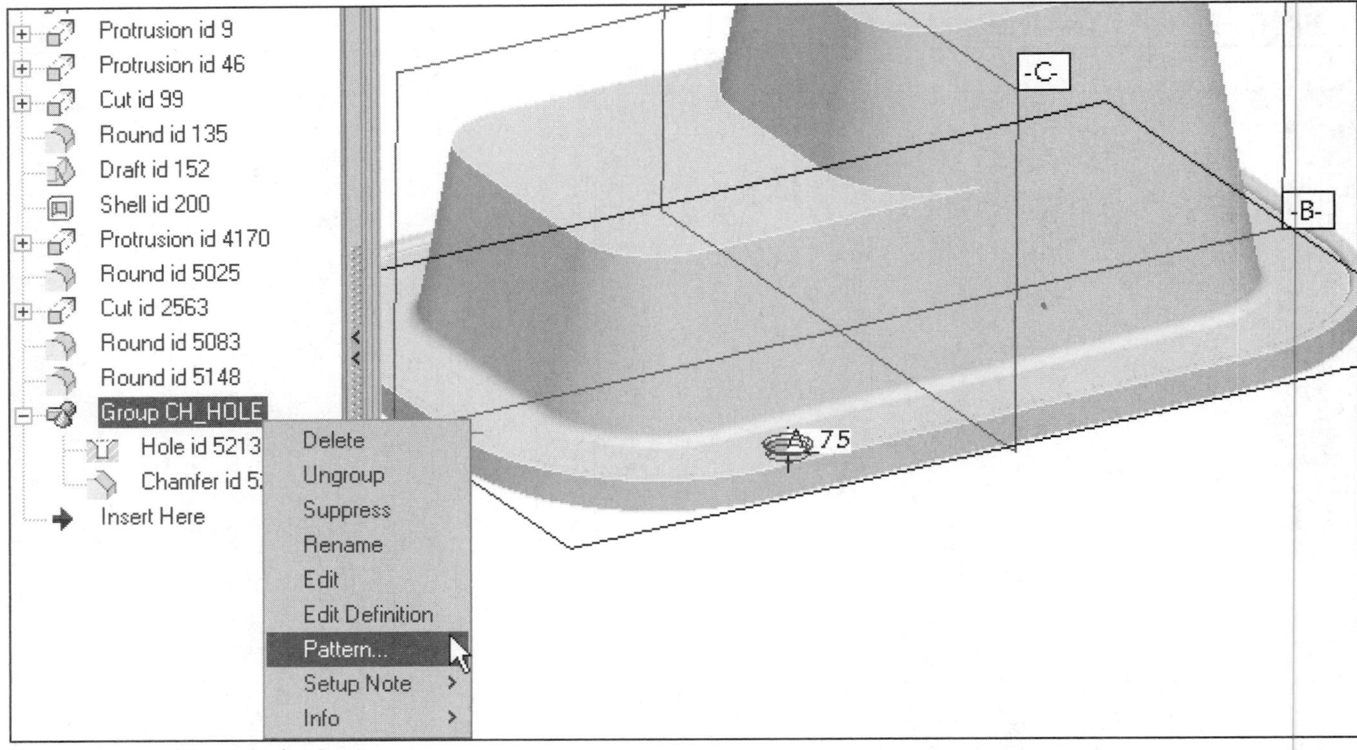

Figure 11.20(d) Pattern the Group

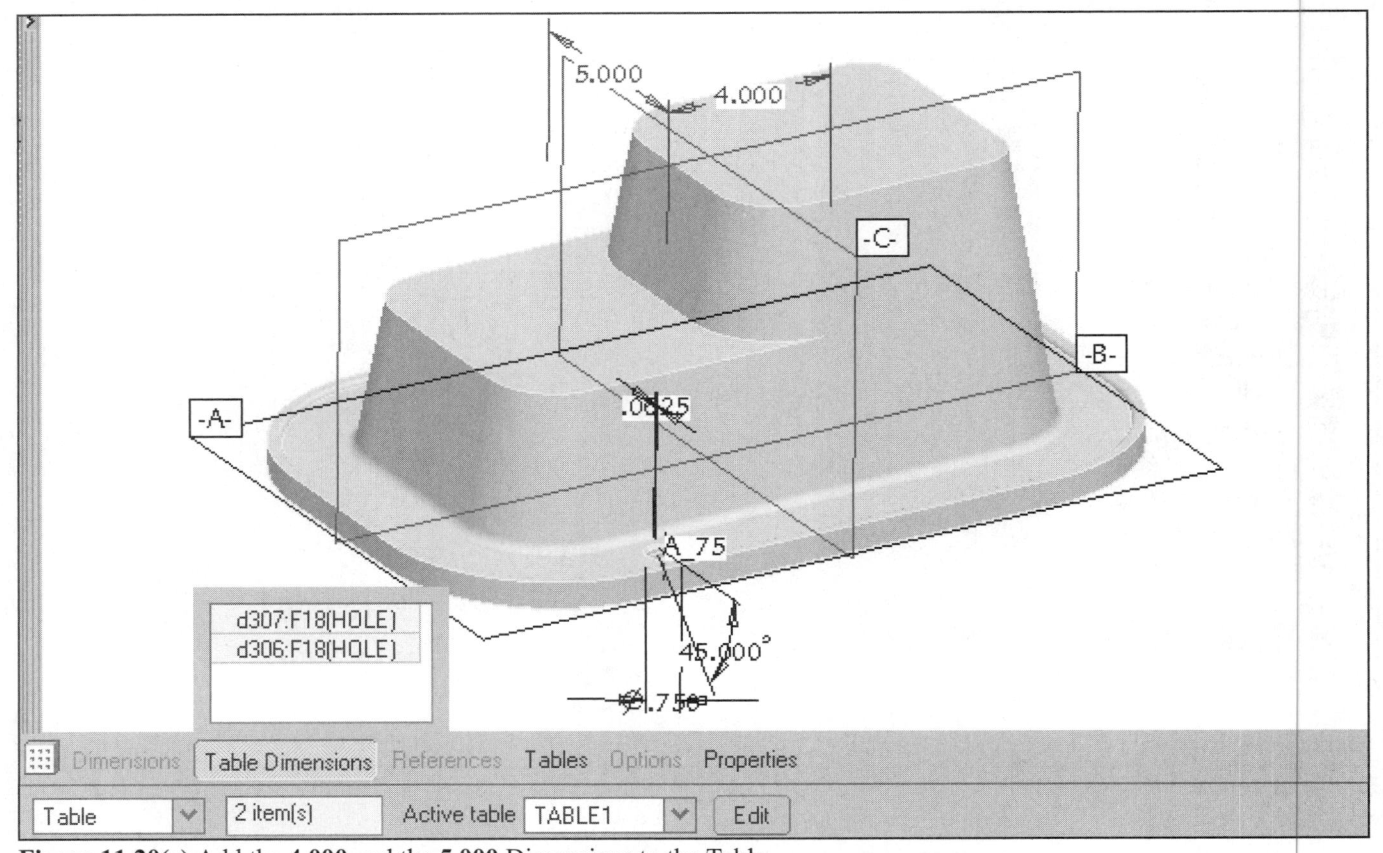

Figure 11.20(e) Add the **4.000** and the **5.000** Dimensions to the Table

Click: Edit [Fig. 11.20(f)] ⇒ add the column and row information [Fig. 11.20(g-j)] ⇒ **File** ⇒ **Exit**

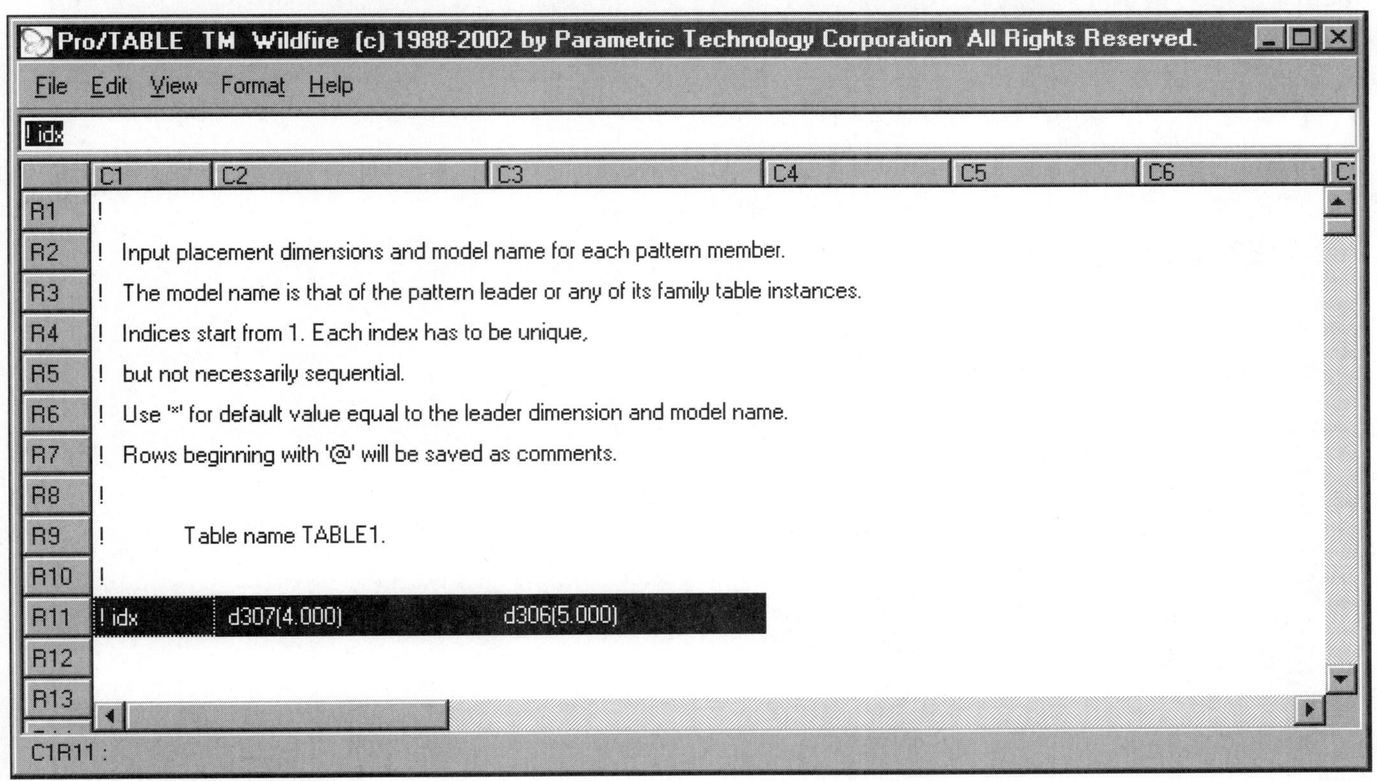

Figure 11.20(f) Pattern Table

Row	C1
R9	! Tal
R10	!
R11	! idx
R12	1
R13	2
R14	3
R15	4
R16	5
R17	6
R18	7

Figure 11.20(g) Add numbers 1-7

R9	! Table name TABLE1.	
R10	!	
R11	! idx	d307(4.000)
R12	1	*
R13	2	0.00
R14	3	-4.00
R15	4	0.00
R16	5	-4.00
R17	6	8.00
R18	7	-8.00

Figure 11.20(h) Add Values in the Second Column (* means identical value)

R9	! Table name TABLE1.		
R10	!		
R11	! idx	d307(4.000)	d306(5.000)
R12	1	*	-5.00
R13	2	0.00	*
R14	3	-4.00	*
R15	4	0.00	-5.00
R16	5	-4.00	-5.00
R17	6	8.00	0.00
R18	7	-8.00	0.00

Figure 11.20(i) Add Values in the Third Column

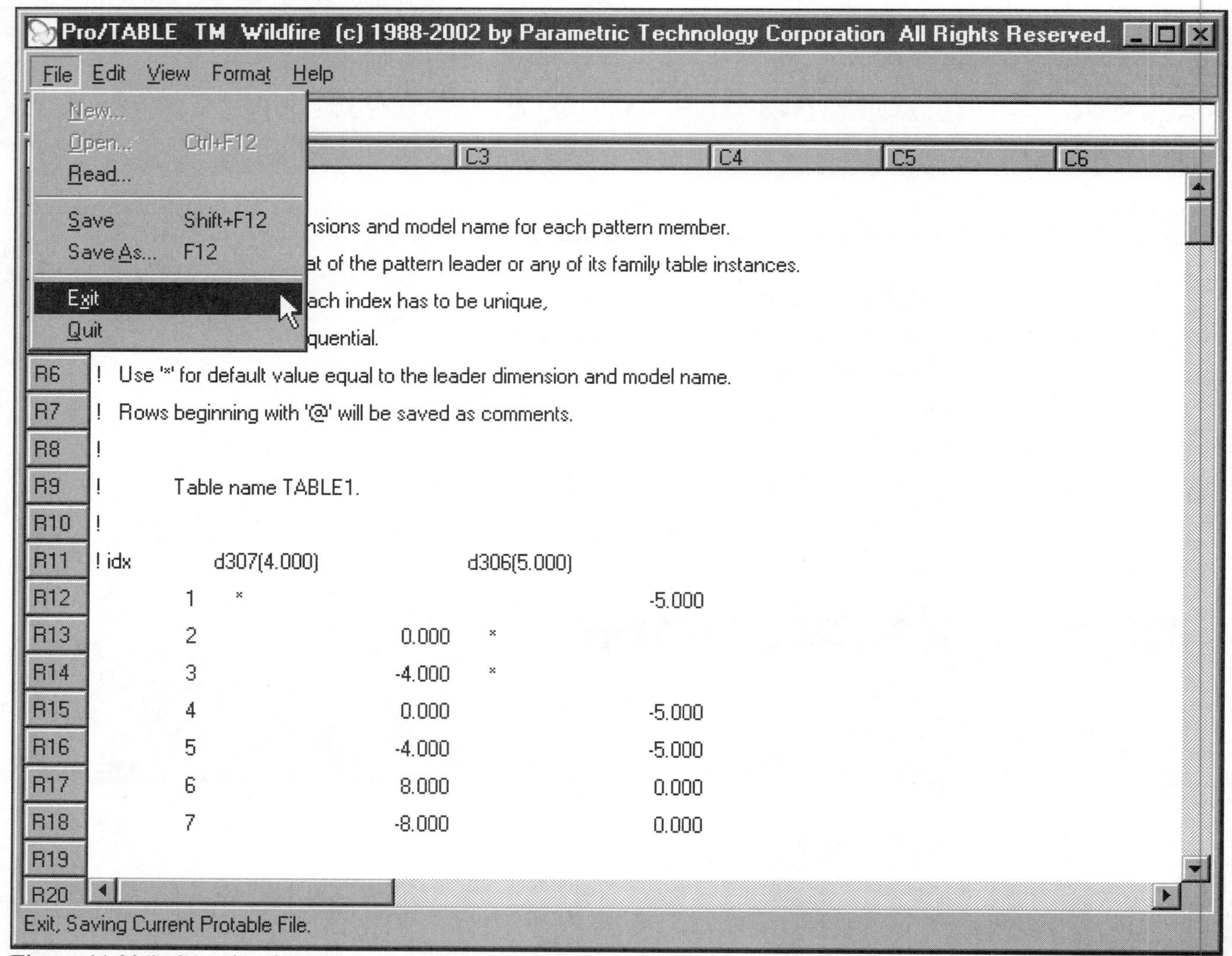

Figure 11.20(j) Completed Table

Click: ✓ [Fig. 11.20(k)] ⇒ 💾 ⇒ **MMB**

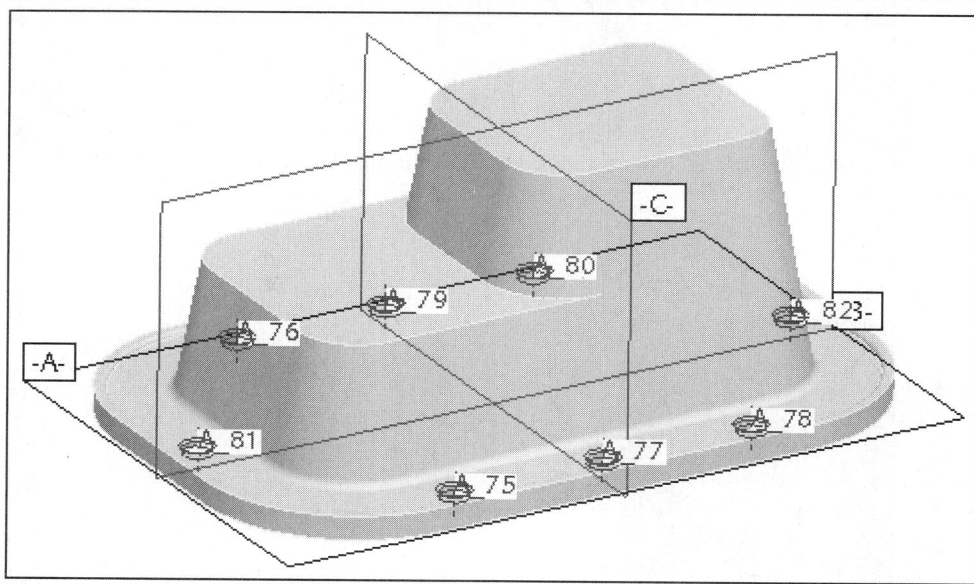

Figure 11.20(k) Completed Pattern

Click: **Tools** ⇒ [icon] **Environment** ⇒ Standard Orient Isometric ⇒ **Apply** ⇒ **OK** ⇒ expand the pattern in the Model Tree ⇒ click on the first instance **Group CH_HOLE** ⇒ **RMB** ⇒ **Edit** [Fig. 11.20(l)]

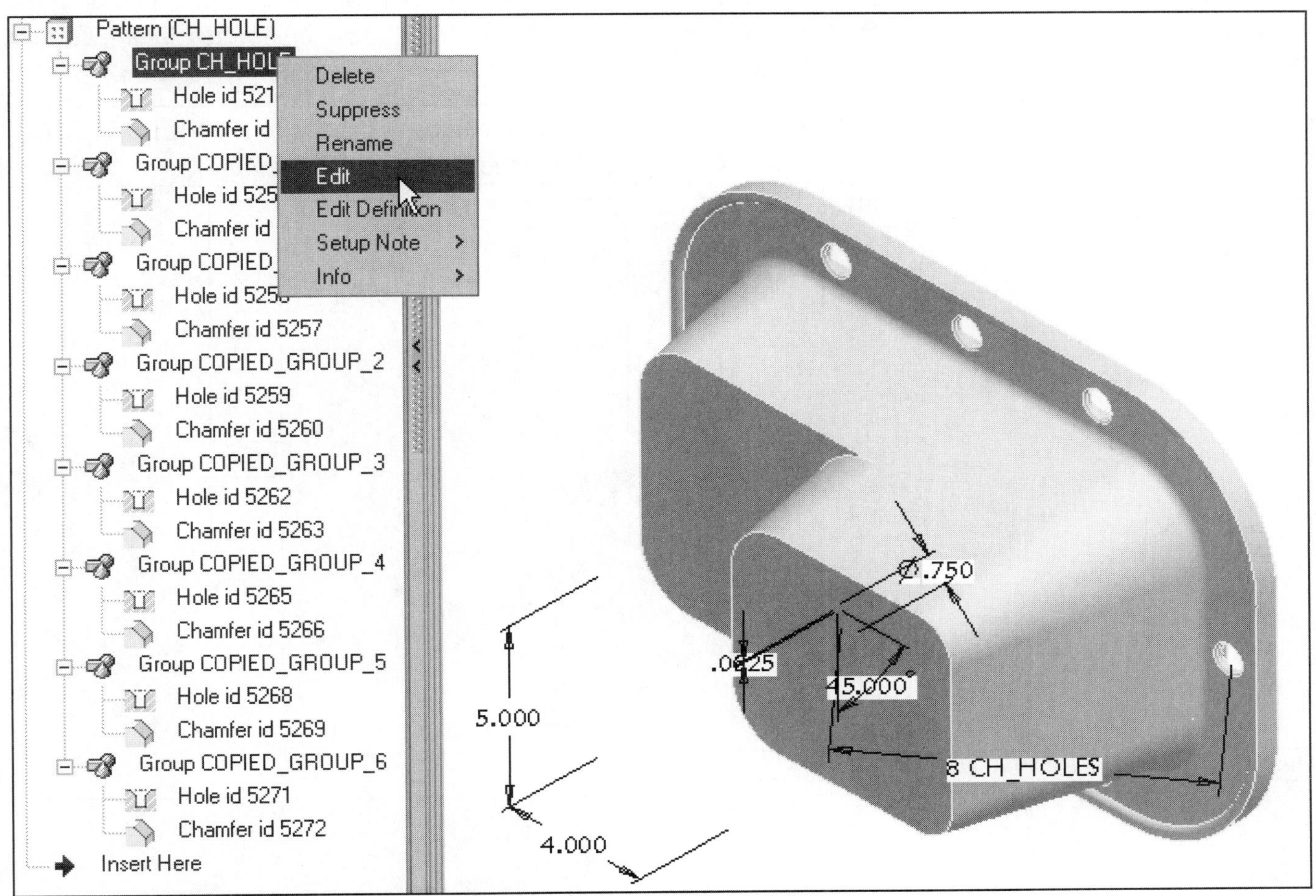

Figure 11.20(l) Feature, Group, and Pattern Values Displayed on the Model

Click: **Settings** in the Navigator ⇒ **Tree Filters** ⇒ ☑ Suppressed Objects (Fig. 11.21) ⇒ **Apply** ⇒ **OK**

Display
Features
Notes
MFG Owner
Suppressed Objects
Incomplete Objects
Excluded Objects
Blanked Objects

Feature Types
General | Cabling | Piping | MFG
Datum Plane
Datum Axis
Curve
Datum Point
Coordinate System
Round
Cosmetic

Figure 11.21 Display Suppressed Objects

The next series of features will purposely be created at the wrong stage of this project. You will now create the **R.50** round [Fig. 11.22(a)]. Because the design intent is to have a constant thickness for the part, the round should have been created before the shell. The Reorder capability will be used to change the position of this round in the design sequence. Using the Model Tree, you can pick and drag the round to a new location in the feature list. *The Reorder capability can also be completed using Edit ⇒ Feature Operations ⇒ Reorder ⇒ pick the feature to reorder ⇒ Done ⇒ select the new position.*

Click on the top edge of the part ⇒ **RMB** ⇒ **Round Edges** ⇒ move drag handle to **.500** [Fig. 11.22(a)] ⇒ **MMB** [Fig. 11.22(b)] ⇒ spin the part, and then click on the inner surface [Fig. 11.22(c)]

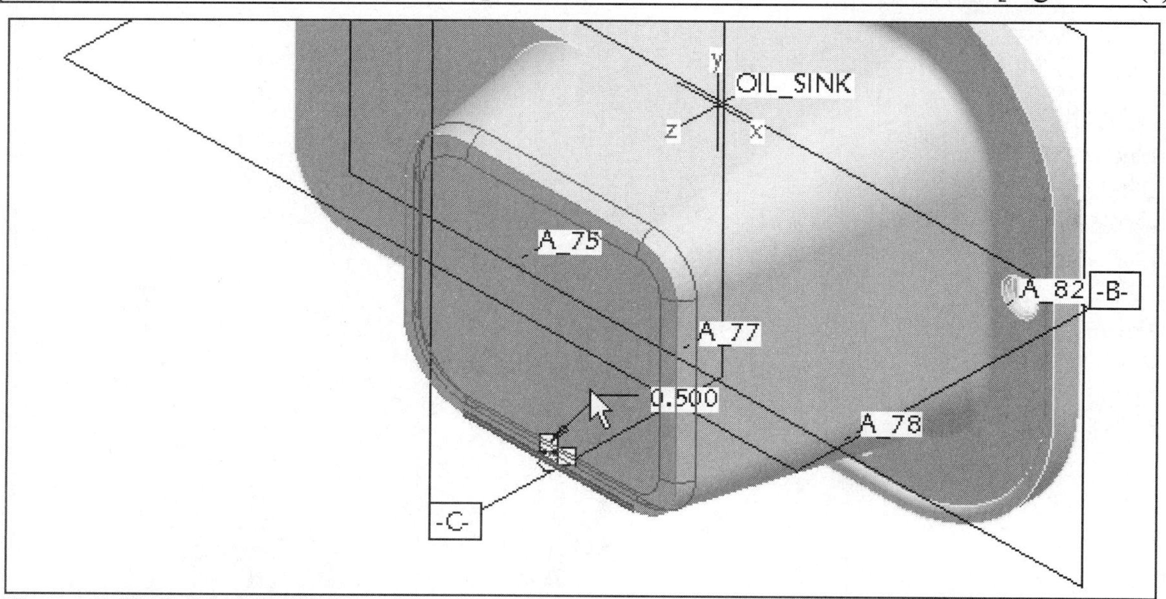

Figure 11.22(a) Add a **R.500** Round

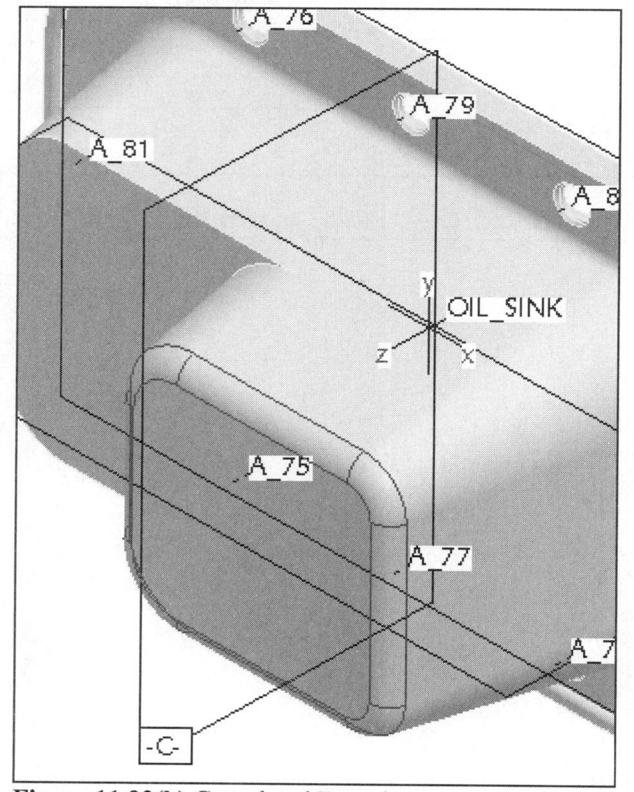

Figure 11.22(b) Completed Round

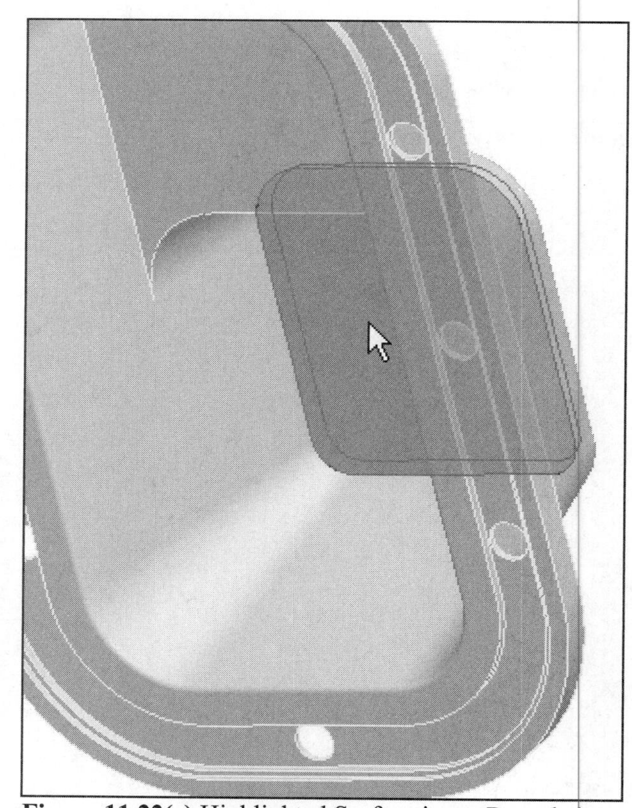

Figure 11.22(c) Highlighted Surface is not Rounded

Reorder the round to appear before the Shell: click on the **Round** [Fig. 11.23(a)] and drag Round id 5518 to a position before/above Shell id 200 [Fig. 11.23(b)] and drop [Fig. 11.23(c)] ⇒ ⇒ **MMB**

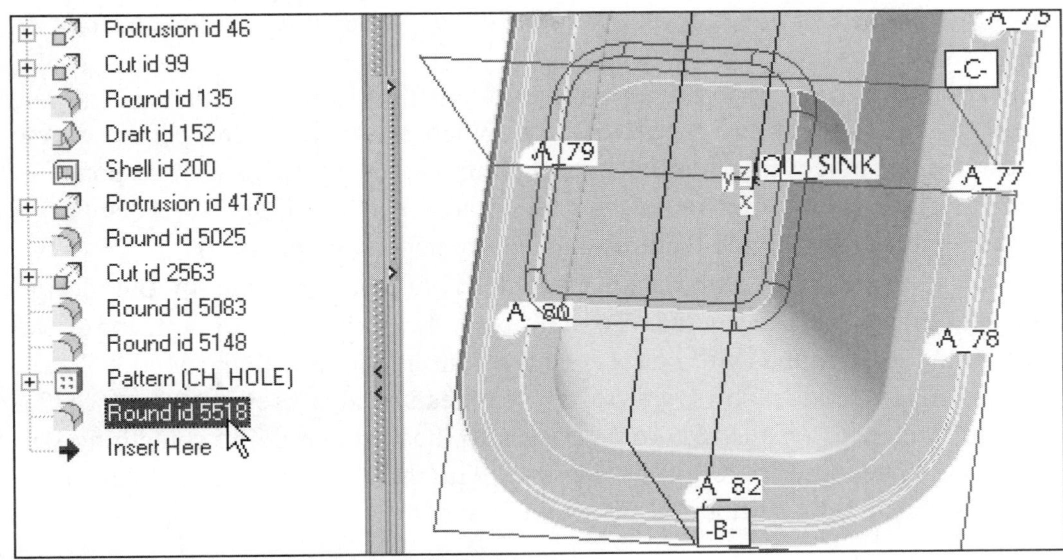

Figure 11.23(a) Click on Round in the Model Tree

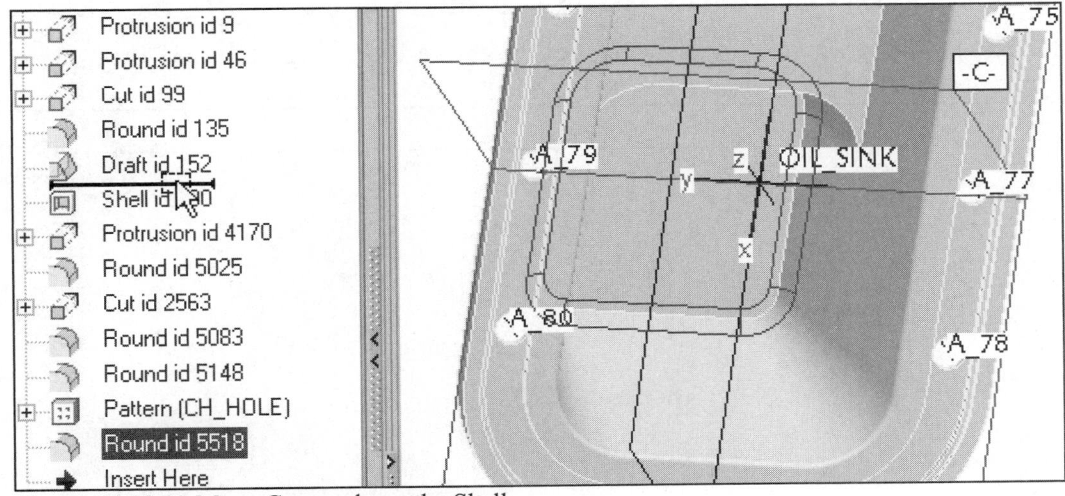

Figure 11.23(b) Move Cursor above the Shell

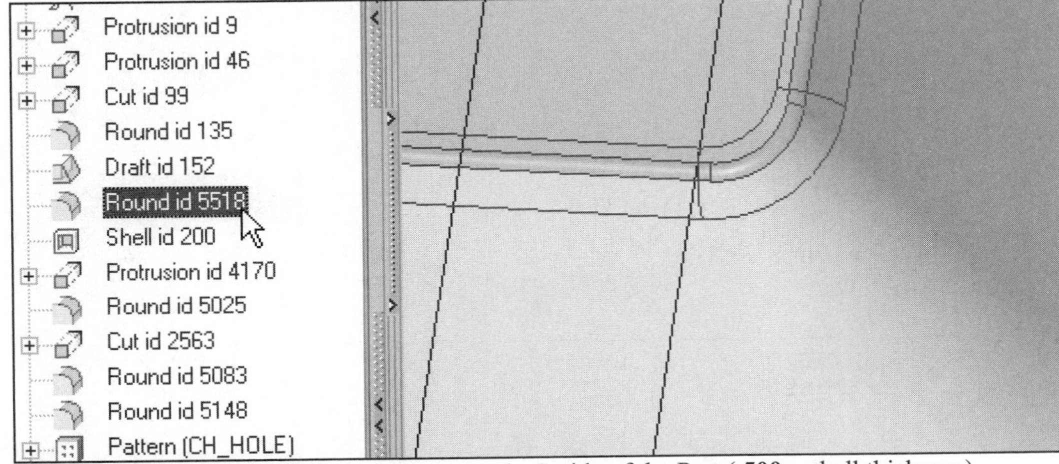

Figure 11.23(c) Reordered Round Shows on the Inside of the Part (**.500** – shell thickness)

You can also Insert new features using the Model Tree. The arrow-shaped icon [Insert Here] in the Model Tree indicates where features will be inserted upon creation and is by default at the end of the Model Tree.

The Insert capability can also be completed using Edit ⇒ Feature Operations ⇒ Insert Mode ⇒ Activate ⇒ Select a feature to insert after ⇒ etc.

By dragging the location of the insert node higher so that its position is before existing features, you can insert a new feature at that stage of the model history. When the *node* is dropped at a new location, the model is rolled backward (suppressed) or forward in response to the insertion node being moved higher or lower. The Model Tree displays a small square ■ next to the features that are not active (suppressed).

Round id 5518
Insert Here
■Shell id 200
■Protrusion id 4170
■Round id 5025
■Cut id 2563
■Round id 5083
■Round id 5148
■Pattern (CH_HOLE)

The previous round was created at the wrong stage in the design sequence and then reordered. To eliminate the reordering of a feature, the remaining **R.50** rounds will be created using Insert Mode with the Model Tree.

Insert Mode allows you to insert a feature at a previous stage of the design sequence. This is like going back into the past and doing something you wish you had done before--not possible with life, but with Pro/E less of a problem. Add the additional **R.50** rounds.

Click: **Datum axis tags off** ⇒ **Coordinate system tags off** ⇒ in the Model Tree click on [Insert Here] [Fig. 11.24(a)] and drag it to a position before/above [Shell id 200] and drop [Fig. 11.24(b)]

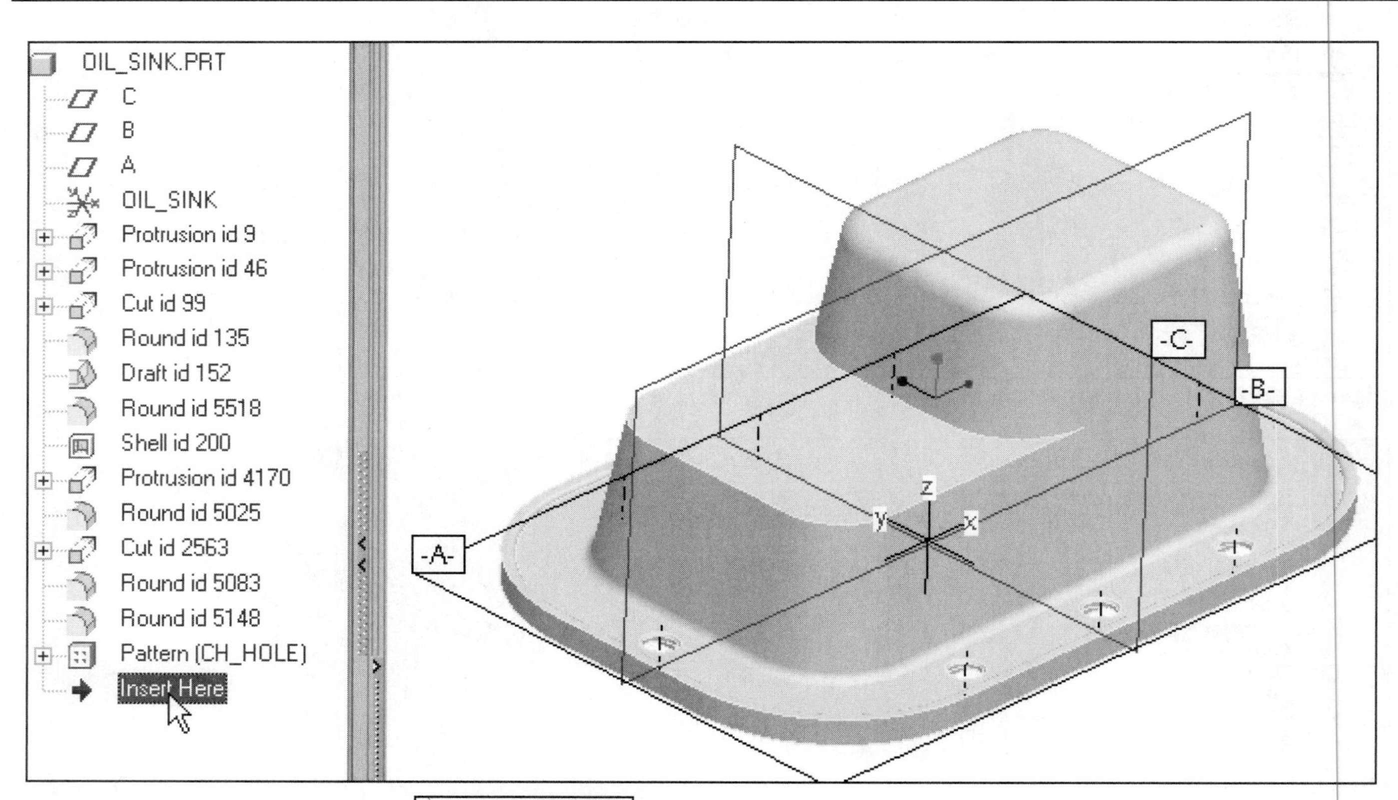

Figure 11.24(a) Insert Here Pointer [Insert Here]

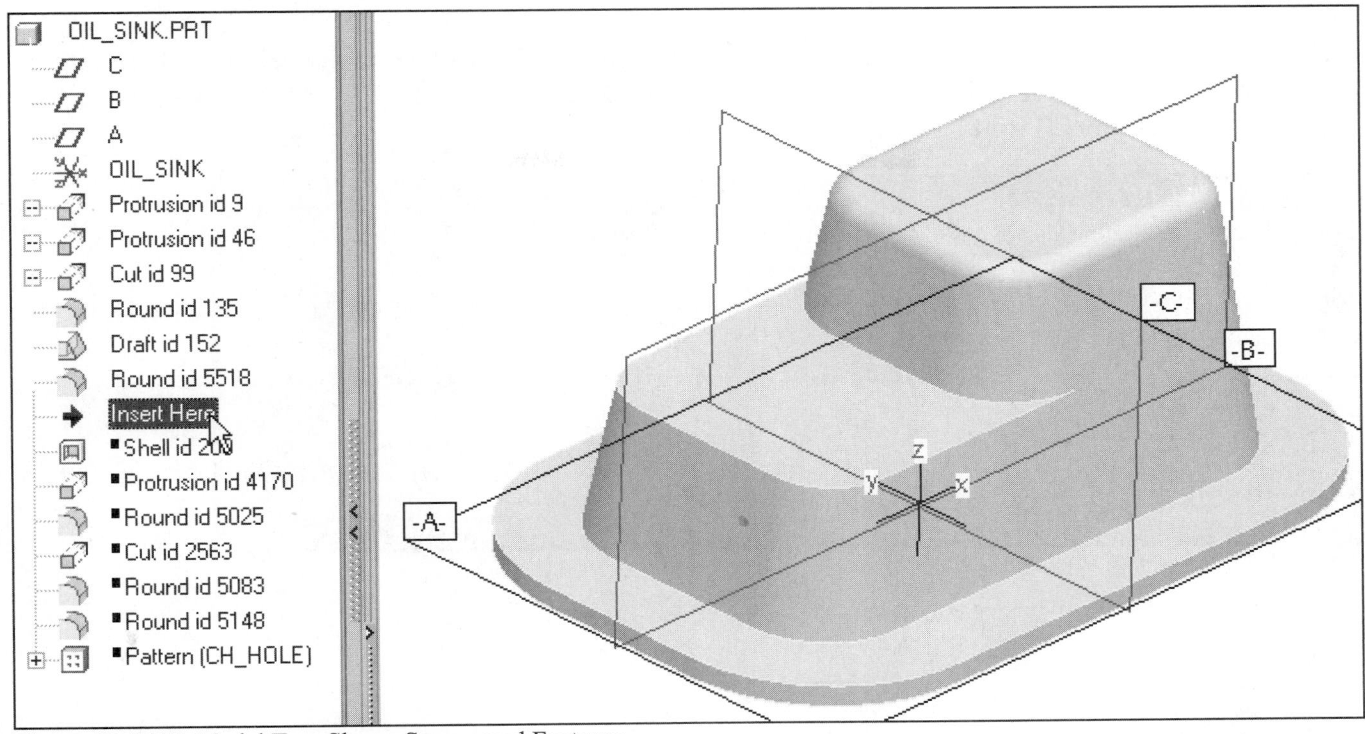

Figure 11.24(b) Model Tree Shows Suppressed Features

Create two sets of rounds [Fig. 11.25(a)]

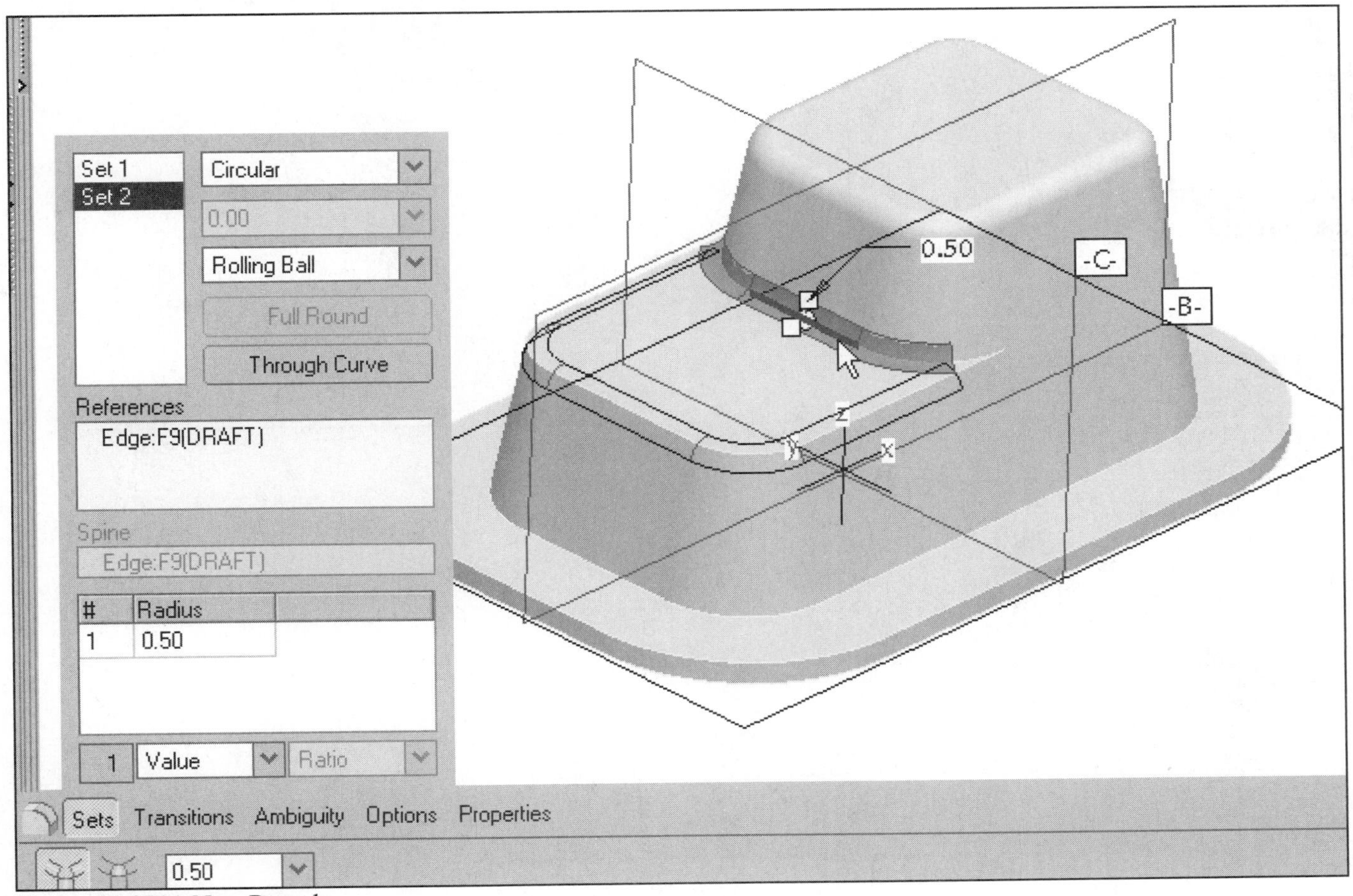

Figure 11.25(a) New Rounds

Click: ⇒ **Enter fix model mode** [Fig. 11.25(b)] ⇒ **Quick Fix** ⇒ **Redefine** ⇒ **Confirm** ⇒ **Sets** tab ⇒ click on **Set 2** [Fig. 11.25(c)] ⇒ **RMB** ⇒ **Delete** ⇒ **MMB** [Fig. 11.27(d)] ⇒ create the second round separately [Fig. 11.27(e)]

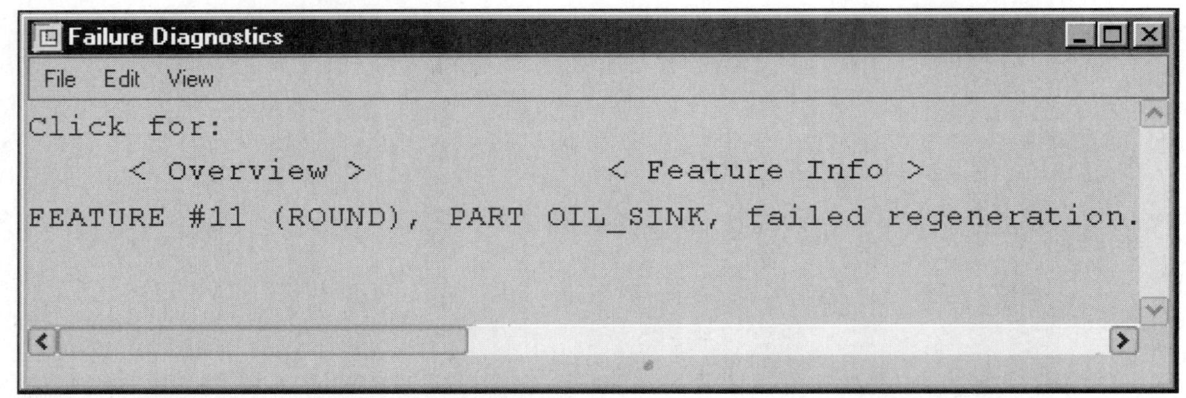

Figure 11.25(b) Failure Diagnostics Dialog Box

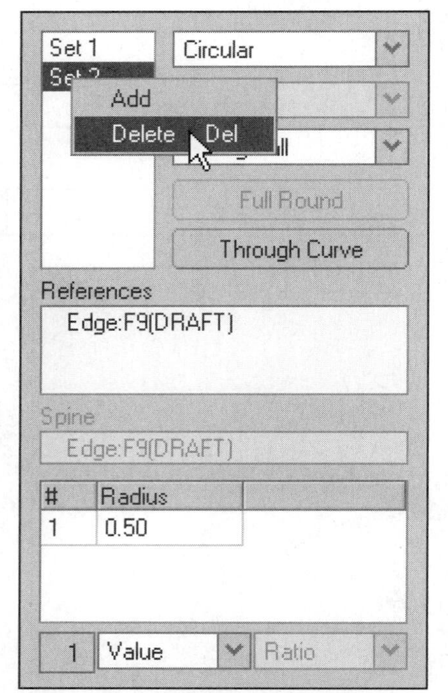

Figure 11.25(c) Delete Set 2

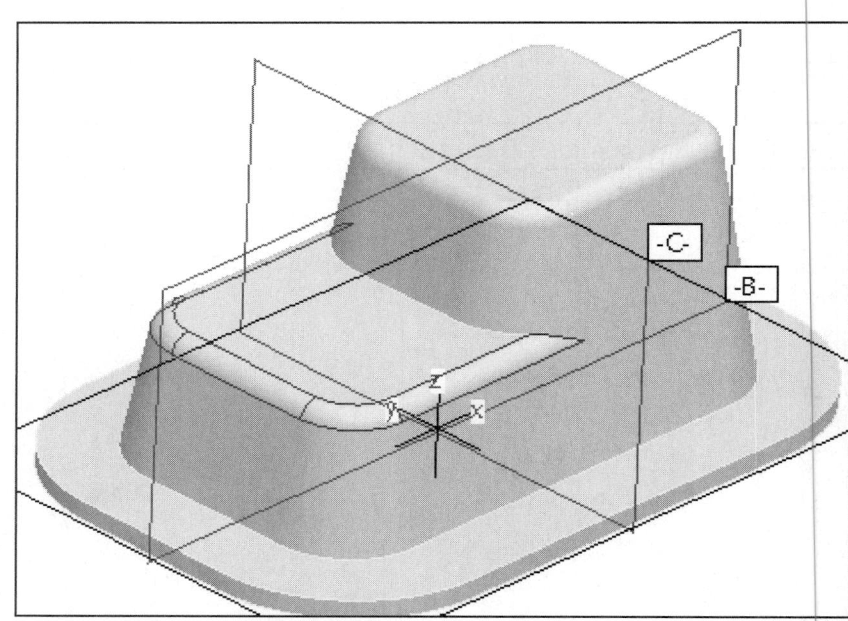

Figure 11.25(d) Completed Round

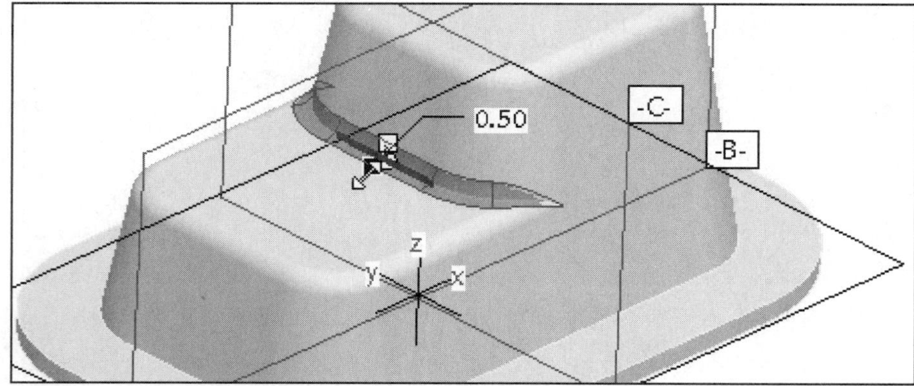

Figure 11.25(e) Second Round Added Separately

Click on and hold Insert Here [Fig. 11.26(a)] ⇒ drag it to the bottom of the Model Tree list and drop [Fig. 11.26(b)] ⇒ ⇒ **MMB** [Fig. 11.26(c)] ⇒ **File** ⇒ **Delete** ⇒ **Old Versions** ⇒ **MMB**

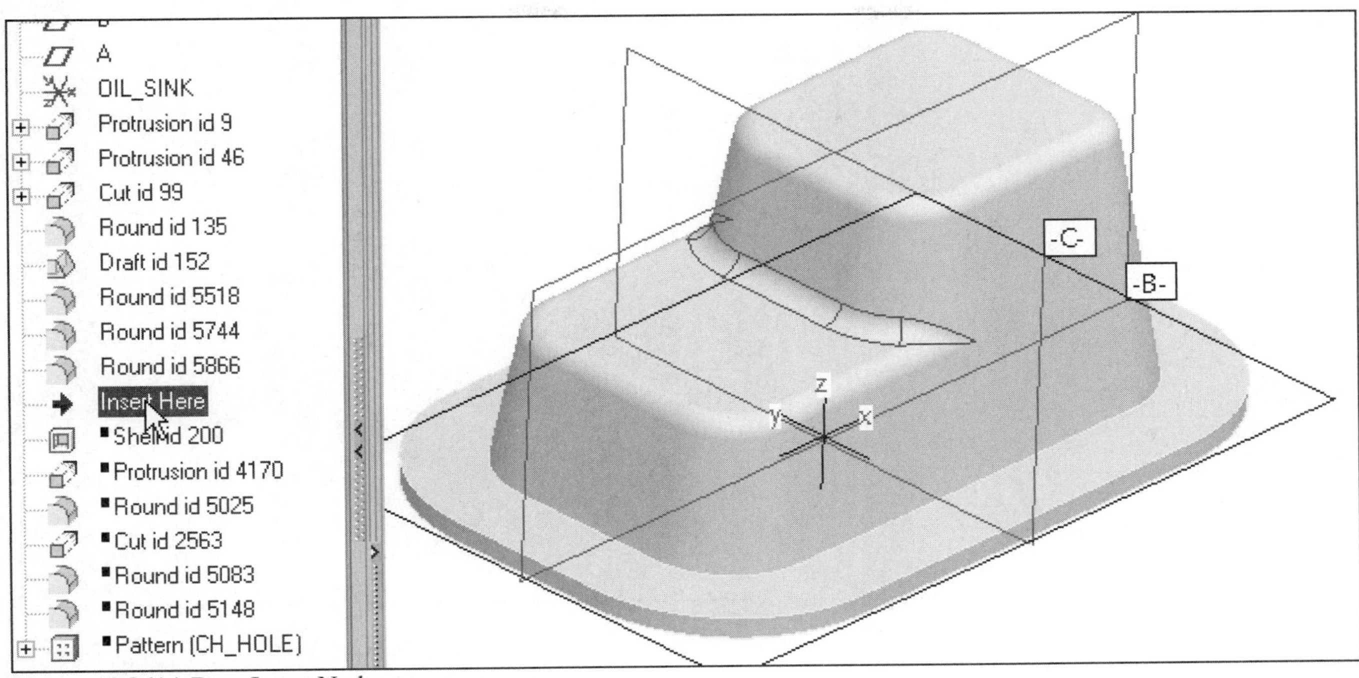

Figure 11.26(a) Drag Insert Node

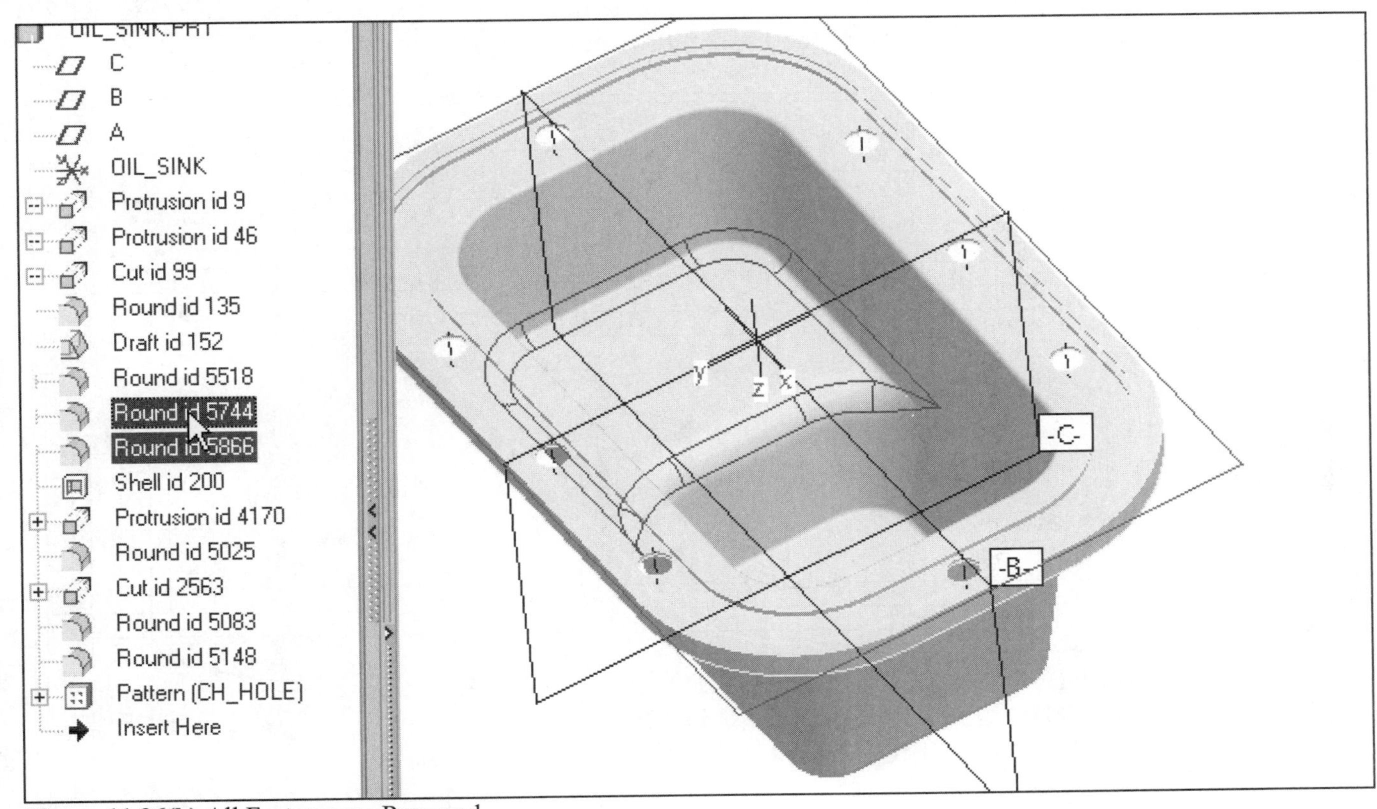

Figure 11.26(b) All Features are Resumed

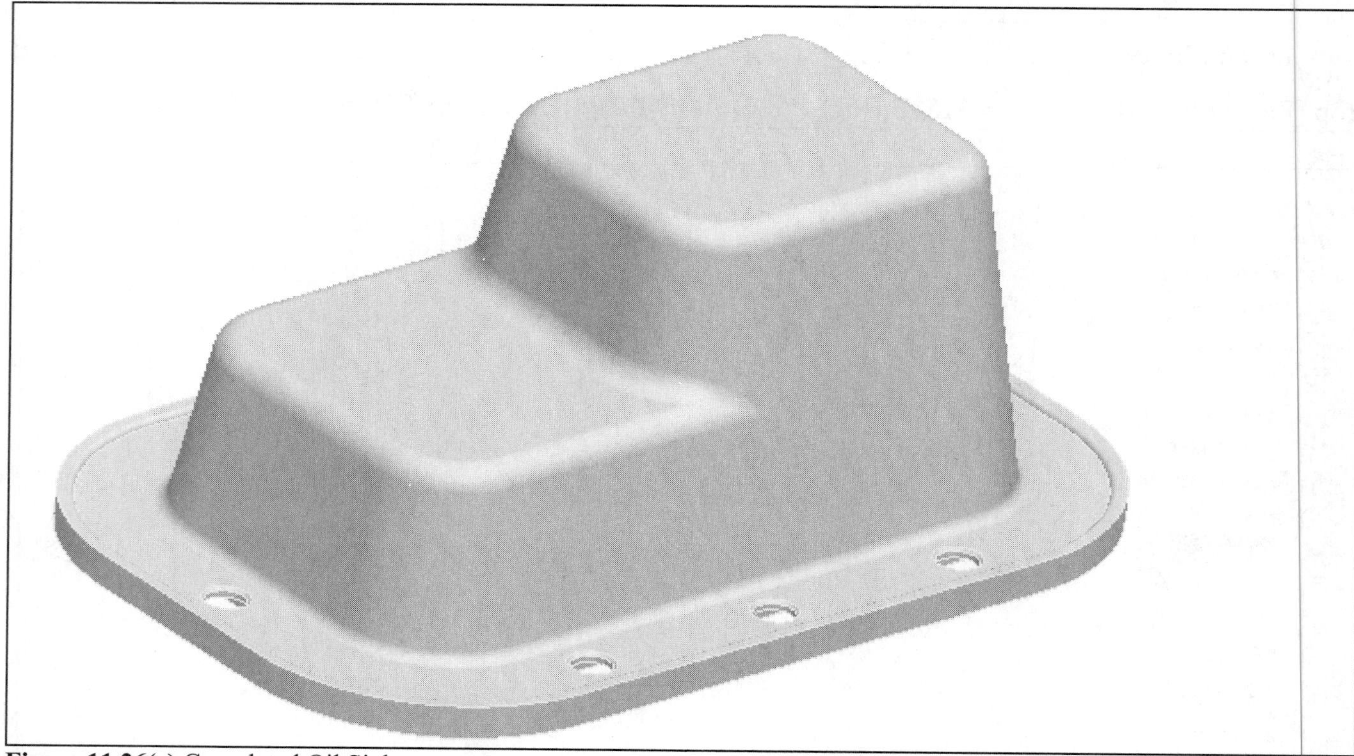

Figure 11.26(c) Completed Oil Sink

The rounds are now in the proper design sequence for the shell feature to have a constant shell thickness. After the part is finished, save it under a new name and then complete the **ECO** (Fig. 11.27) using the current file name. Continue on to the Lesson Project afterwards.

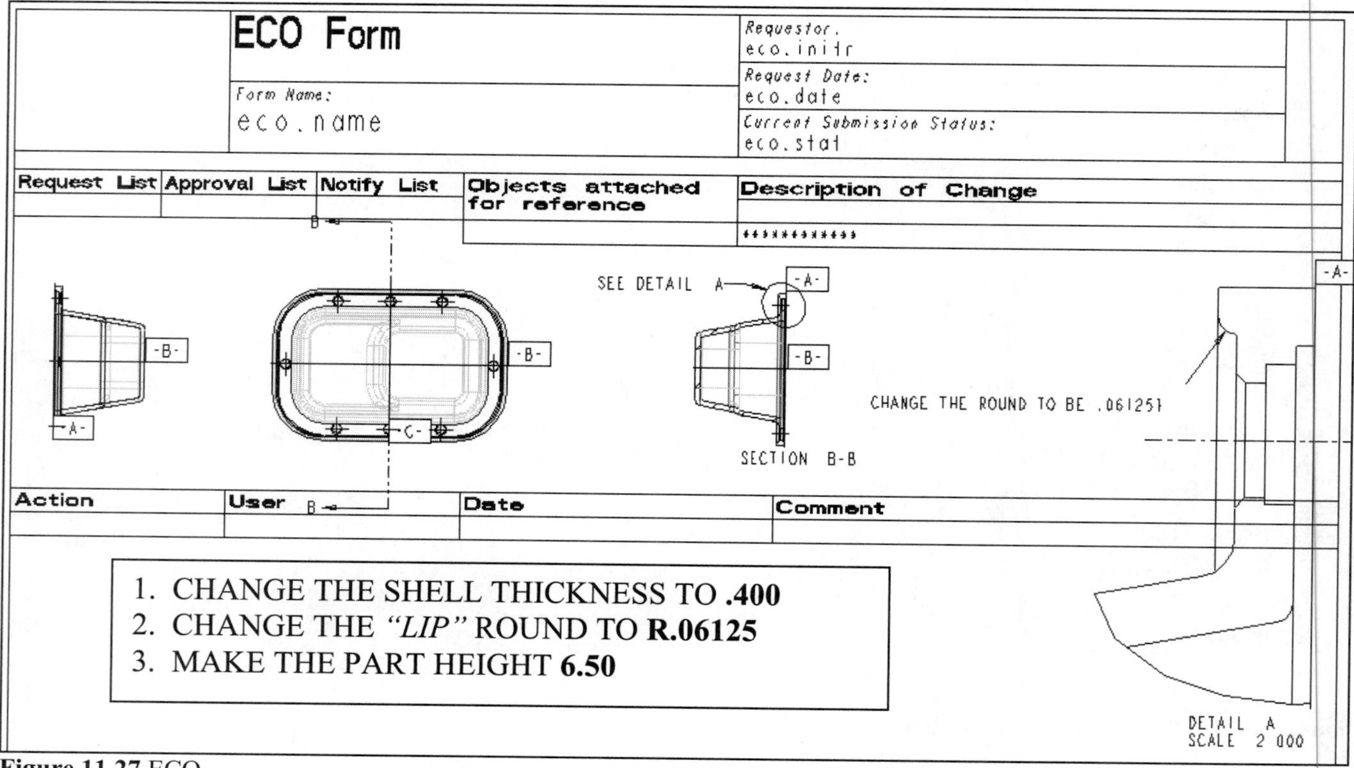

ECO Form

Form Name:
eco.name

Requestor.
eco.initr

Request Date:
eco.date

Current Submission Status:
eco.stat

Request List	Approval List	Notify List	Objects attached for reference	Description of Change
				************

SEE DETAIL A
-A-
-B-
-C-
B
CHANGE THE ROUND TO BE .06125
SECTION B-B
DETAIL A
SCALE 2.000

Action	User	Date	Comment

1. CHANGE THE SHELL THICKNESS TO **.400**
2. CHANGE THE *"LIP"* ROUND TO **R.06125**
3. MAKE THE PART HEIGHT **6.50**

Figure 11.27 ECO

Lesson 11 Project

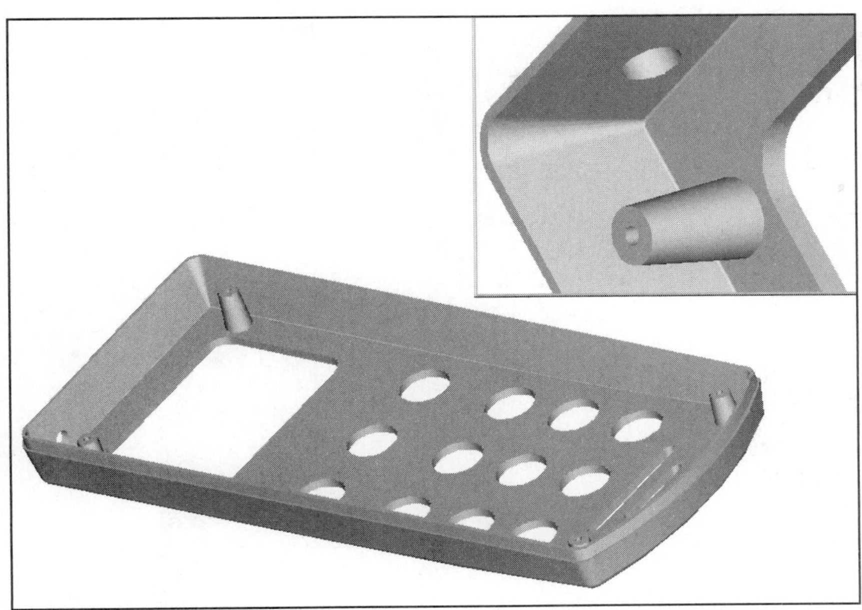

Figure 11.28 Cellular Phone Top

Cellular Phone Top

The Cellular Phone Top (Figures 11.28 through 11.46) is one of two major components for a cellular phone. You created the other as a project in Lesson 10. The part is made of the same plastic as the Cellular Phone Bottom. If time permits, try to assemble the two pieces after completing Lesson 15 (you may need to modify these parts to attain a correct fit). Analyze the part and plan the steps and features required to model it.

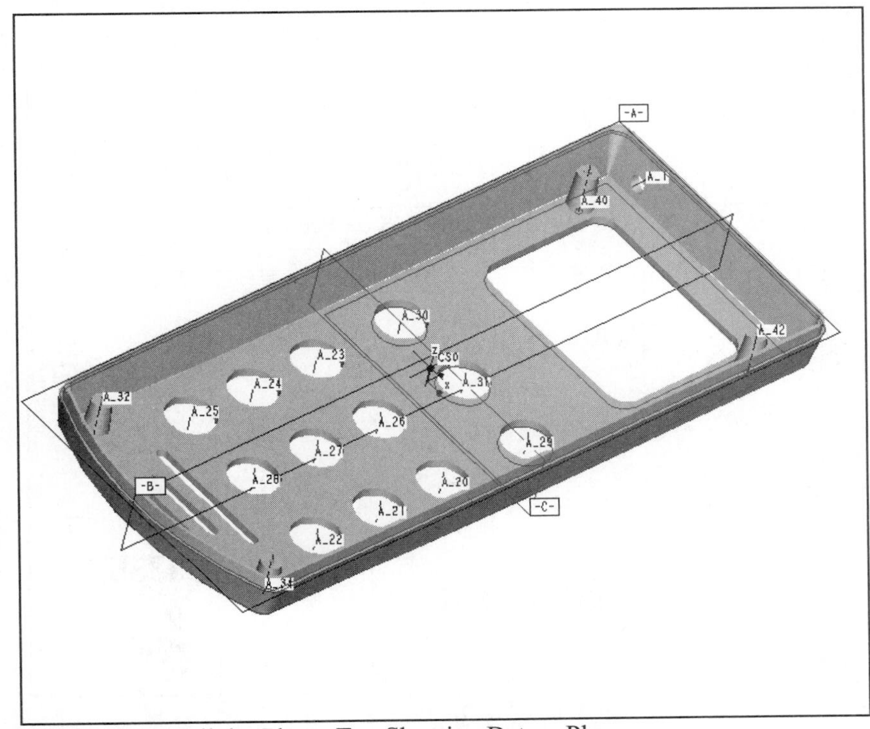

Figure 11.29 Cellular Phone Top Showing Datum Planes

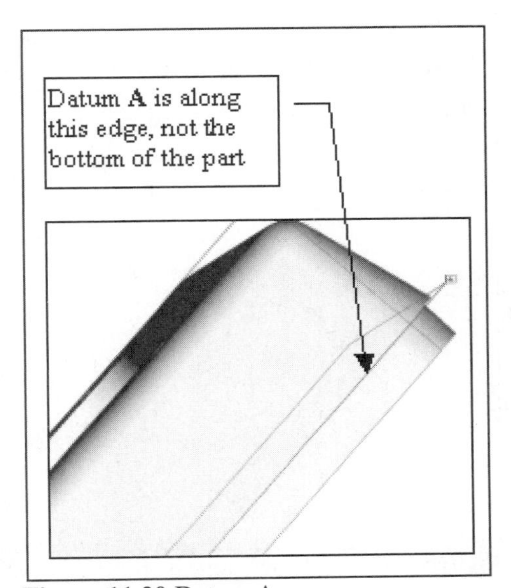

Figure 11.30 Datum A

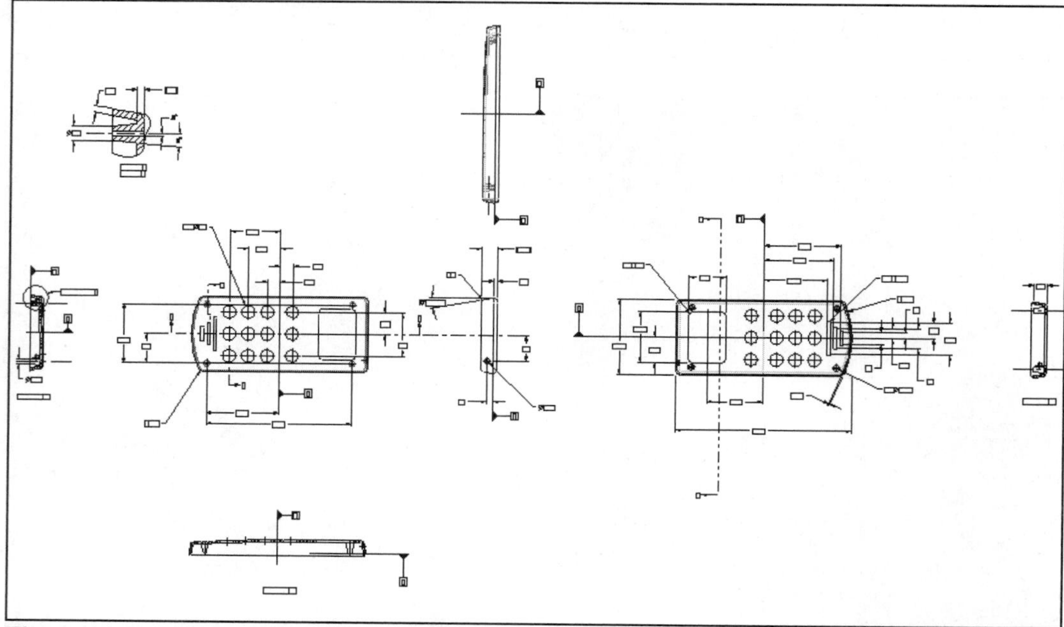

Figure 11.31 Cellular Phone Top, Detail Drawing

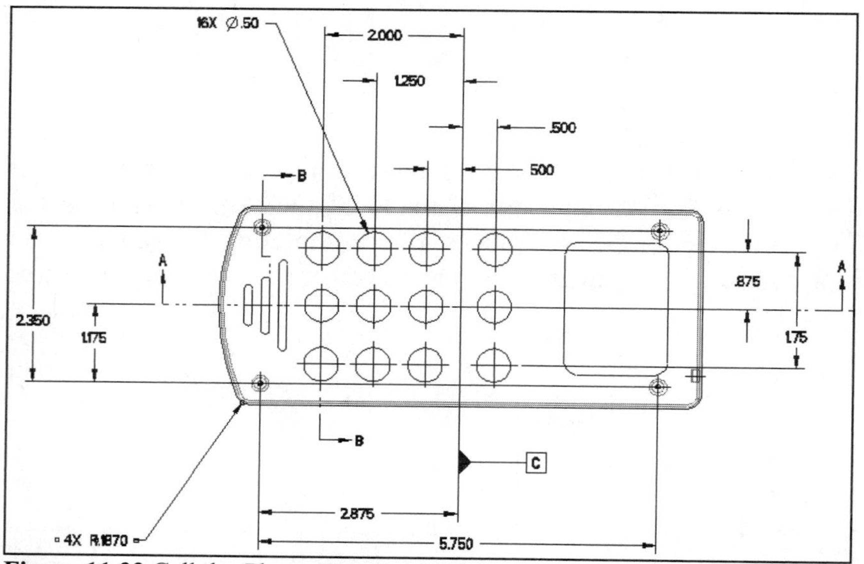

Figure 11.32 Cellular Phone Top, Front View

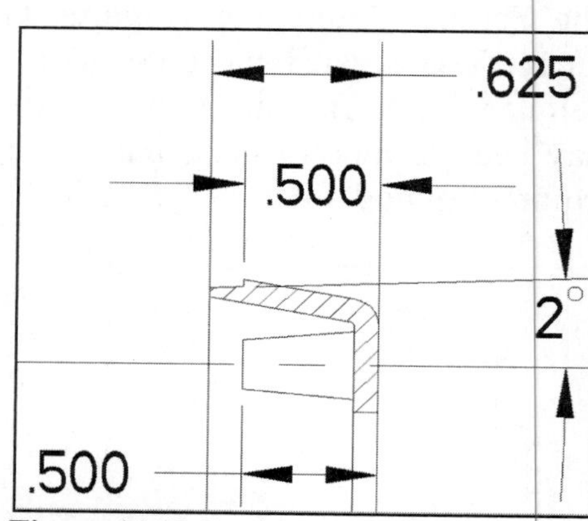

Figure 11.33 Section

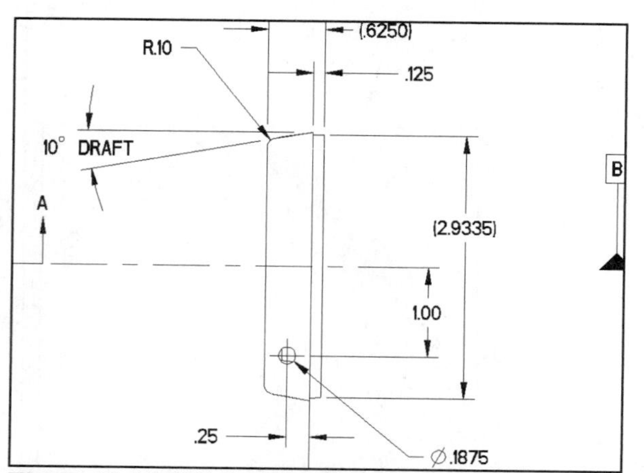

Figure 11.34 Cellular Phone Top, Right Side View

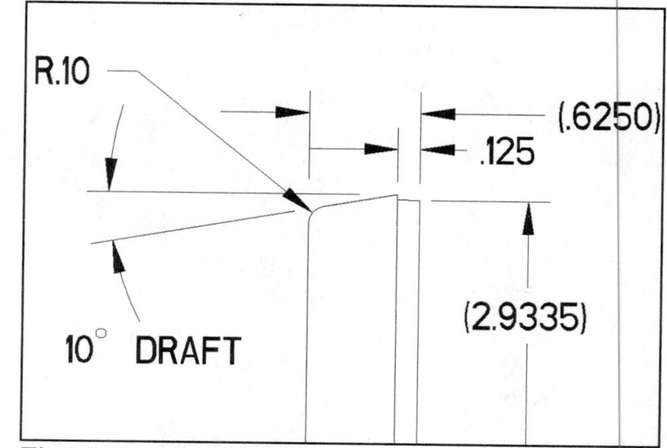

Figure 11.35 Section

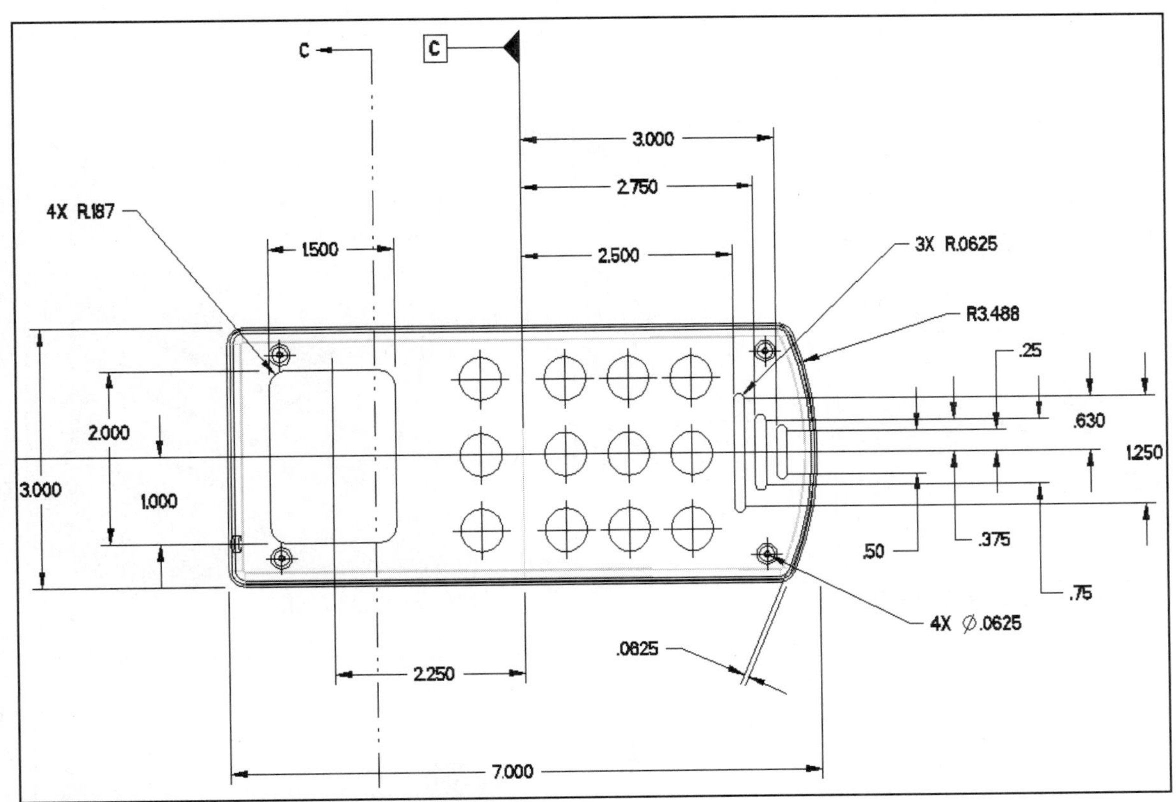

Figure 11.36 Cellular Phone Top, Bottom View

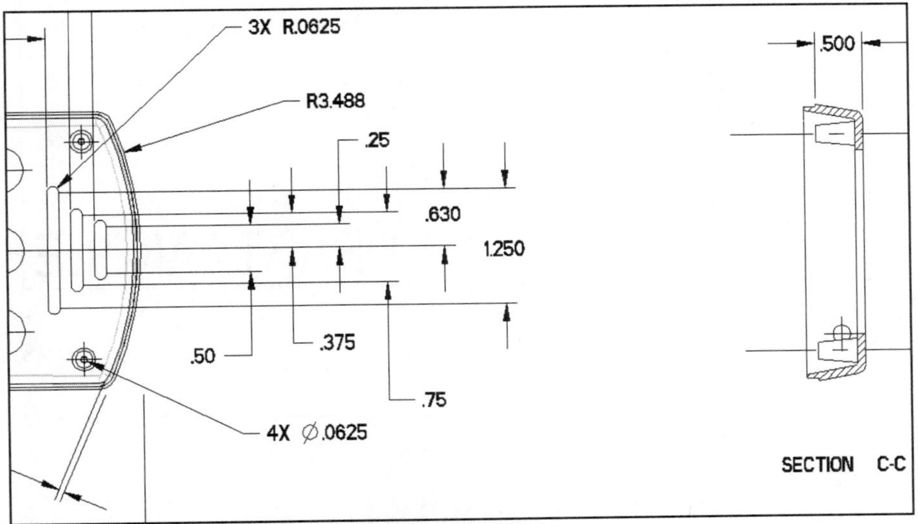

Figure 11.37 Cellular Phone Top, SECTION C-C

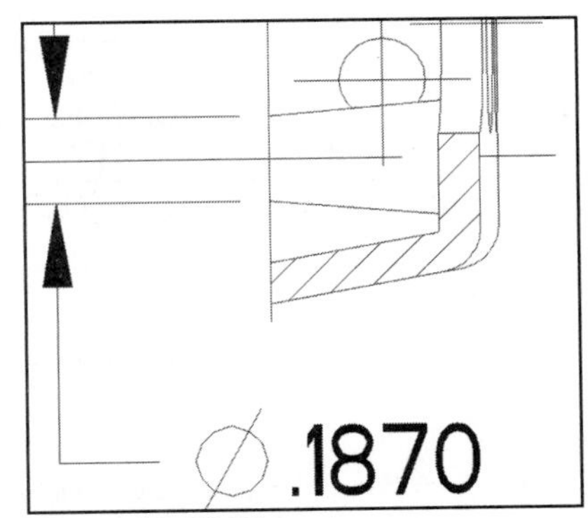

Figure 11.38 Diameter

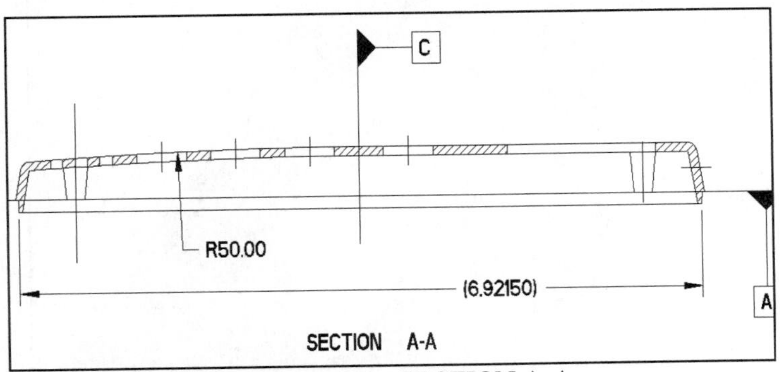

Figure 11.39 Cellular Phone Top, SECTION A-A

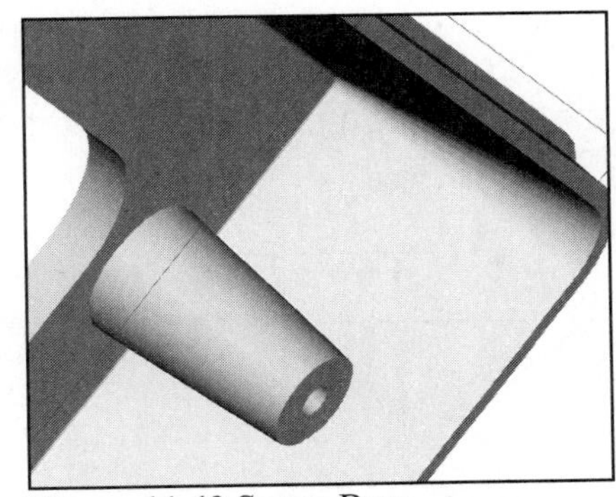

Figure 11.40 Screw Boss

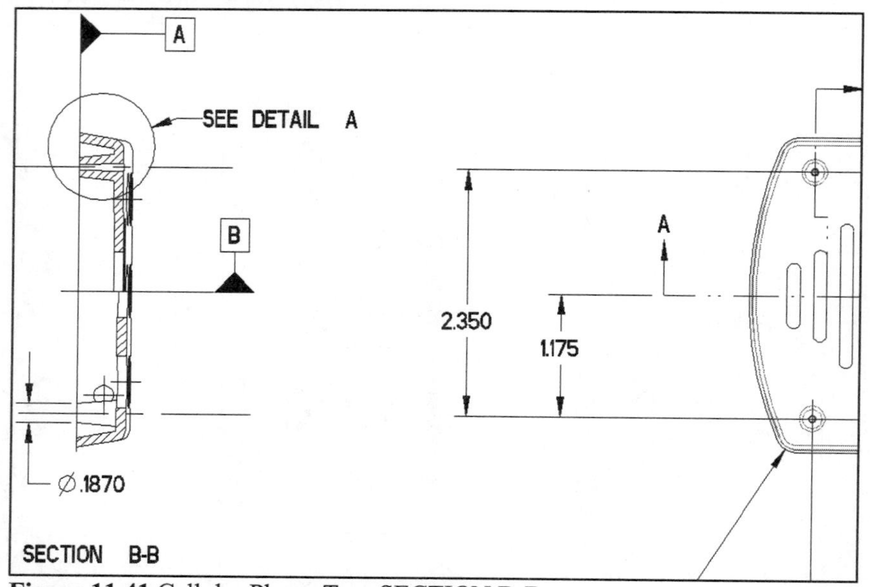

Figure 11.41 Cellular Phone Top, SECTION B-B

Figure 11.42 Grill

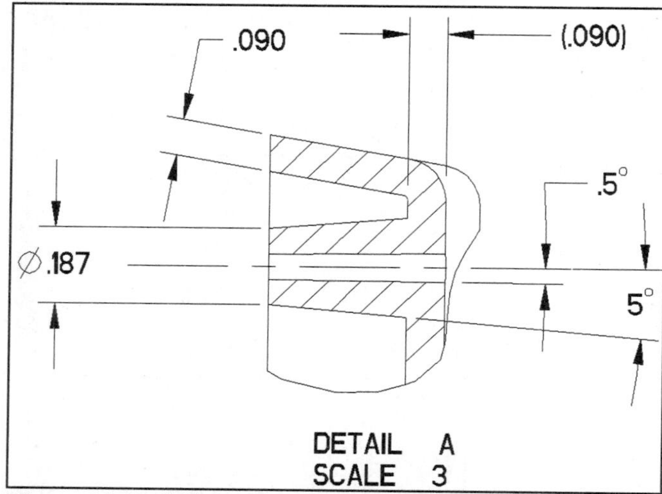

Figure 11.43 Cellular Phone Top, DETAIL A

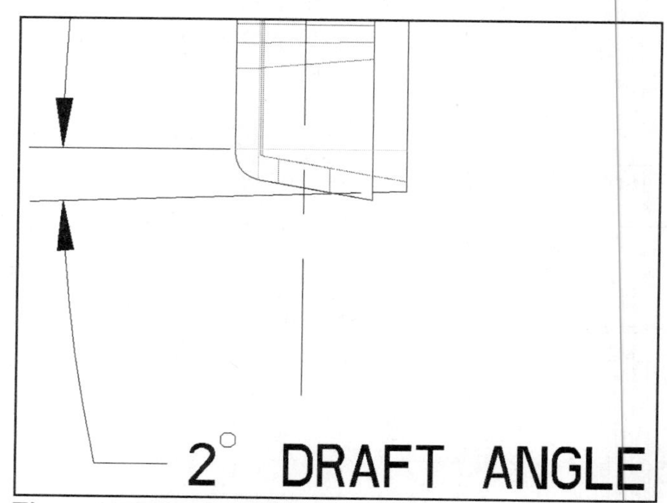

Figure 11.44 Draft Angle

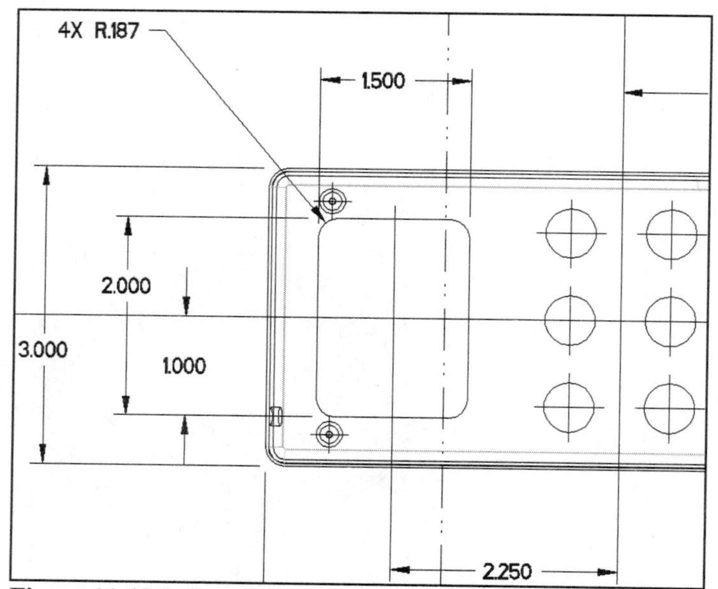

Figure 11.45 Cellular Phone Top, Opening

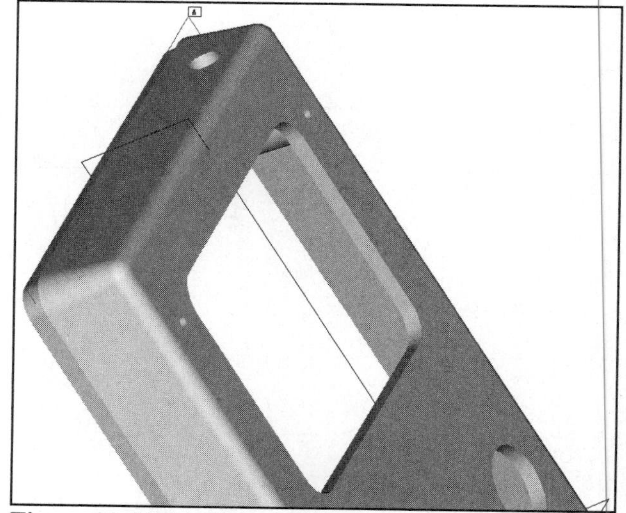

Figure 11.46 Opening

Lesson 12 Sweeps

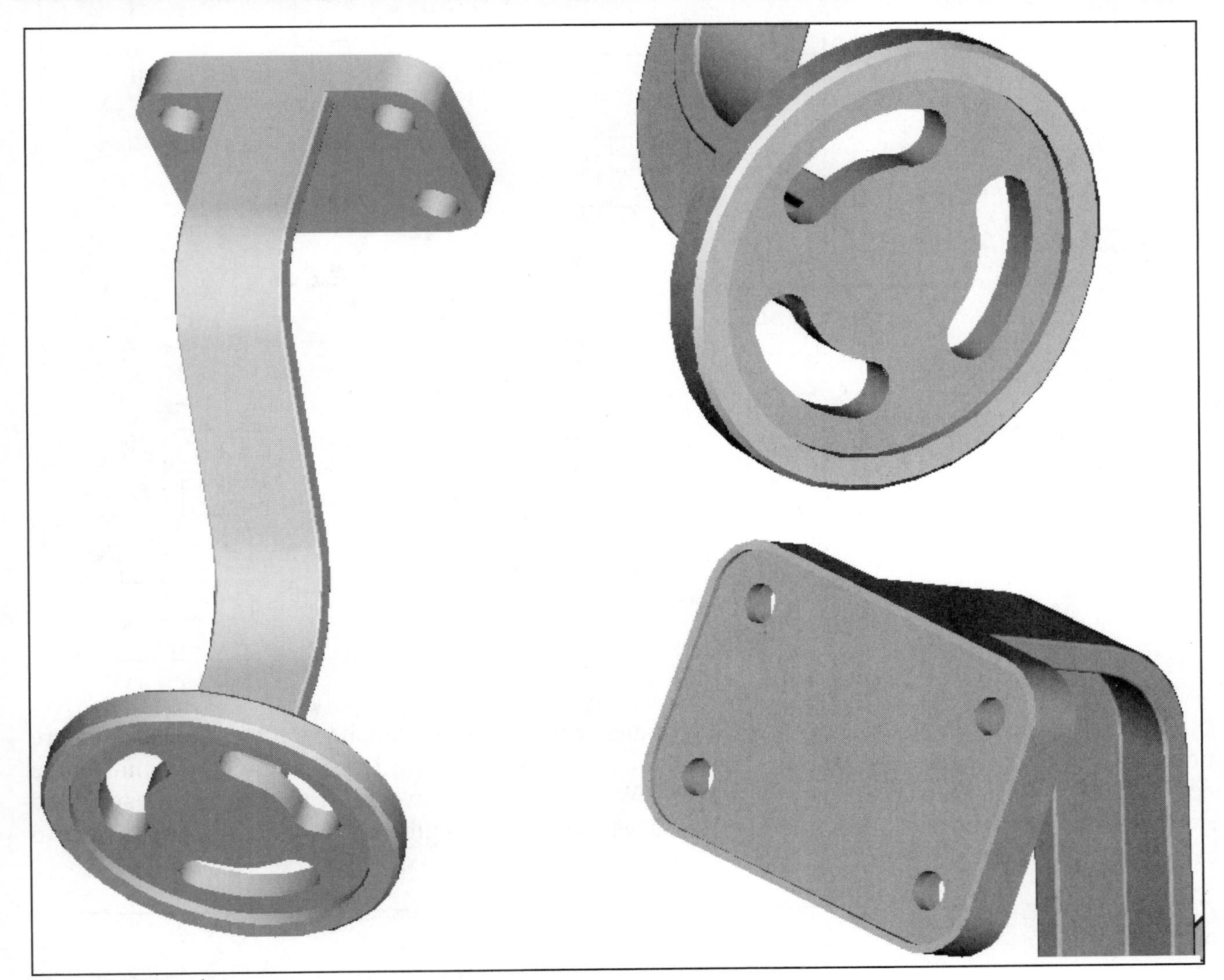

Figure 12.1 Bracket

OBJECTIVES

- Create a **constant-section sweep** feature
- Sketch a **Trajectory** for a sweep
- Sketch and locate a **Sweep section**
- Understand the difference between adding and not adding **Inner Faces**
- Be able to **Edit** a sweep
- Understand the difference between a **Sketched Trajectory** and a **Selected Trajectory**
- Create **Variable Sweeps**

SWEEPS

A Sweep is created by sketching or selecting a *trajectory* and then sketching a *section* to follow along it. The Bracket, shown in Figures 12.1 and 12.2, uses a simple sweep in its design.

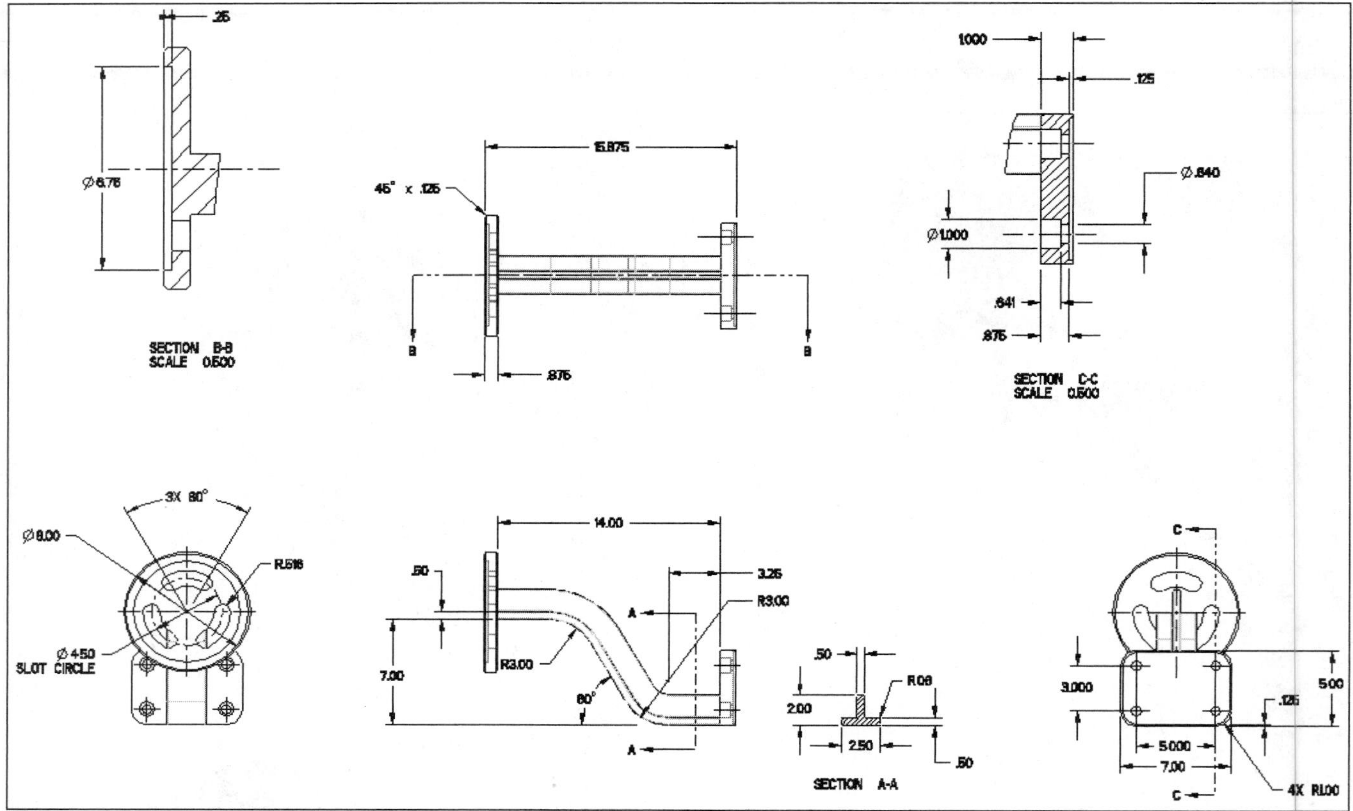

Figure 12.2 Bracket Detail

A *constant-section sweep* can use either trajectory geometry sketched at the time of feature creation or a trajectory made up of selected datum curves or edges (Fig. 12.3). The trajectory must have adjacent reference surfaces or be planar (Fig. 12.4). When defining a sweep, Pro/E checks the specified trajectory for validity and establishes normal surfaces. When ambiguity exists, Pro/E prompts you to select a normal surface.

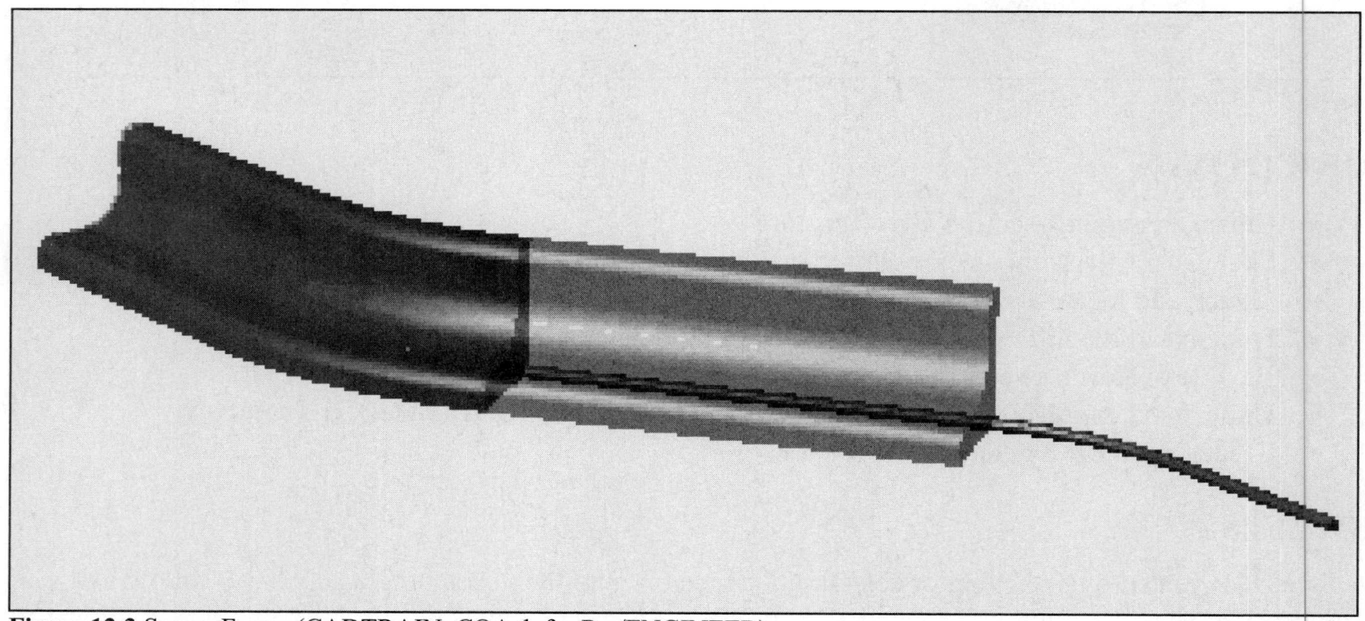

Figure 12.3 Sweep Forms (CADTRAIN, COAch for Pro/ENGINEER)

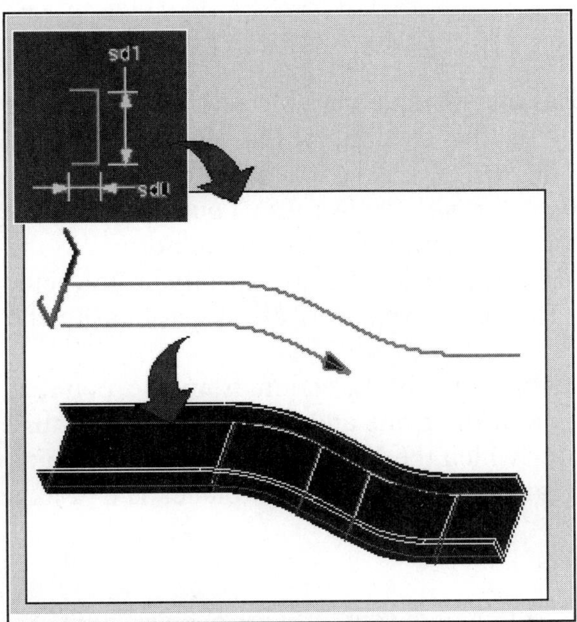

Figure 12.4 Sweep Trajectory and Section (CADTRAIN, COAch for Pro/ENGINEER)

The following options are available for sweeps:

- **Sketch Traj** Sketch the sweep trajectory using Sketcher mode
- **Select Traj** Select a chain of existing curves or edges as the sweep trajectory
- **Merge Ends** Merge the ends of the sweep, if possible, into the adjacent solid
- **Free Ends** Do not attach the sweep end to adjacent geometry
- **Add Inn Fcs** For open sections, add top and bottom faces to close the swept solid (planar, closed trajectory, and open section)
- **No Inn Fcs** Do not add top and bottom faces (Fig. 12.5)

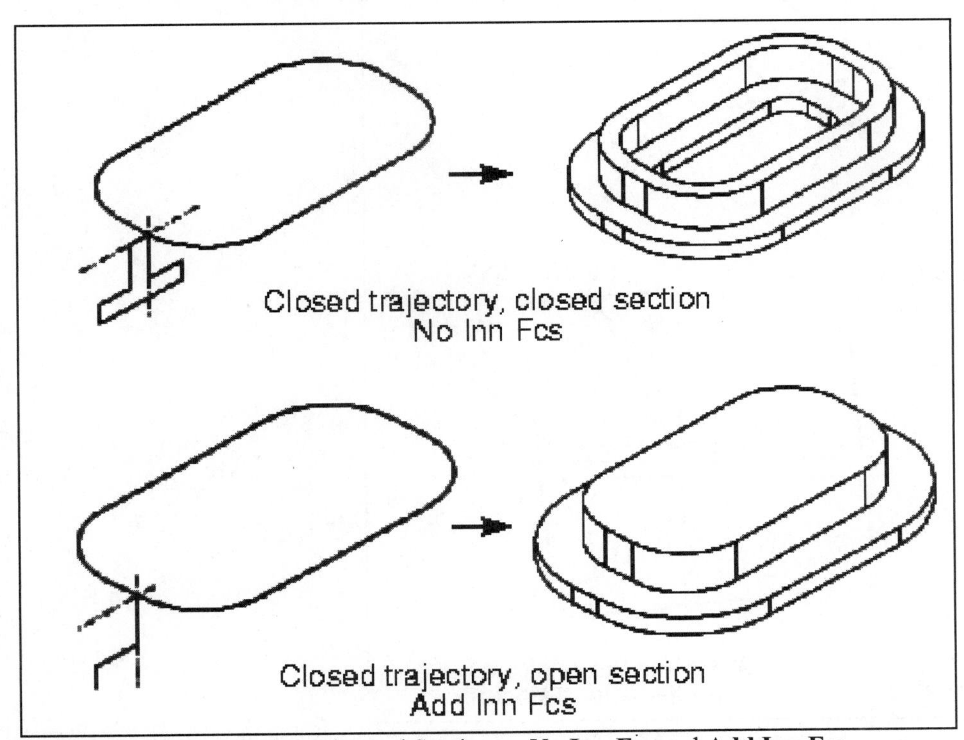

Figure 12.5 Sweep Trajectories and Sections—**No Inn Fcs** and **Add Inn Fcs**

Variable-Section Sweeps

A solid sweep feature using one or more longitudinal trajectories and a single variable section can also be created (Fig. 12.6). The parameters of the section can vary as the section moves along the sweep trajectories. These sweeps are called variable-section sweeps.

Every variable-section sweep requires one *longitudinal "spine" trajectory*. You can define a sweep for which the **X**-axis of the section follows the **X**-vector trajectory while remaining normal to the spine at all times as it sweeps along the spine (Nrm to Spine). To do so, you must also specify an **X**-vector trajectory to orient the section as it sweeps along the spine. The section plane is always normal to the spine trajectory at the point of their intersection.

The **X**-axis of each section's coordinate system is defined by the direction from the point of intersection of the plane and the spine to the point of intersection of the plane and the **X**-vector trajectory for that section. You can also define a variable-section sweep for which the **Y**-axis of the section remains constant. The section will follow the spine such that it is normal to the selected pivot plane. The **X**-axis and **Z**-axis will still follow the spine and **X**-vector trajectories.

Variable-section sweeps need the following trajectories:

- **Spine trajectory** The trajectory along which the section is swept. If you choose **Nrm To Spine**, the origin of the section (crosshairs) is always located on the spine trajectory, with the **X** axis pointing toward the **X**-vector trajectory.
- **X-vector trajectory** Sweeps created using **Nrm To Spine** need this additional trajectory. It defines the orientation of the **X**-axis of the section coordinate system. *The X-vector and spine trajectories cannot intersect.*

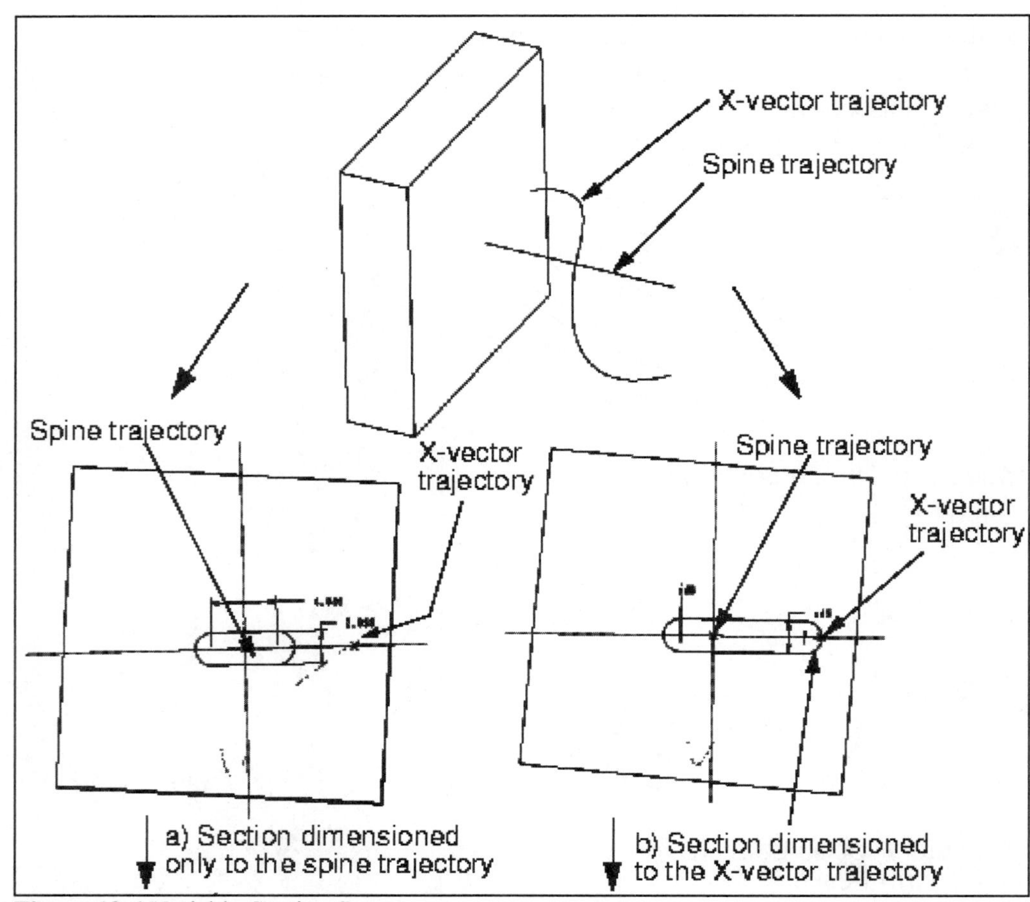

Figure 12.6 Variable-Section Sweeps

Lesson 12 STEPS

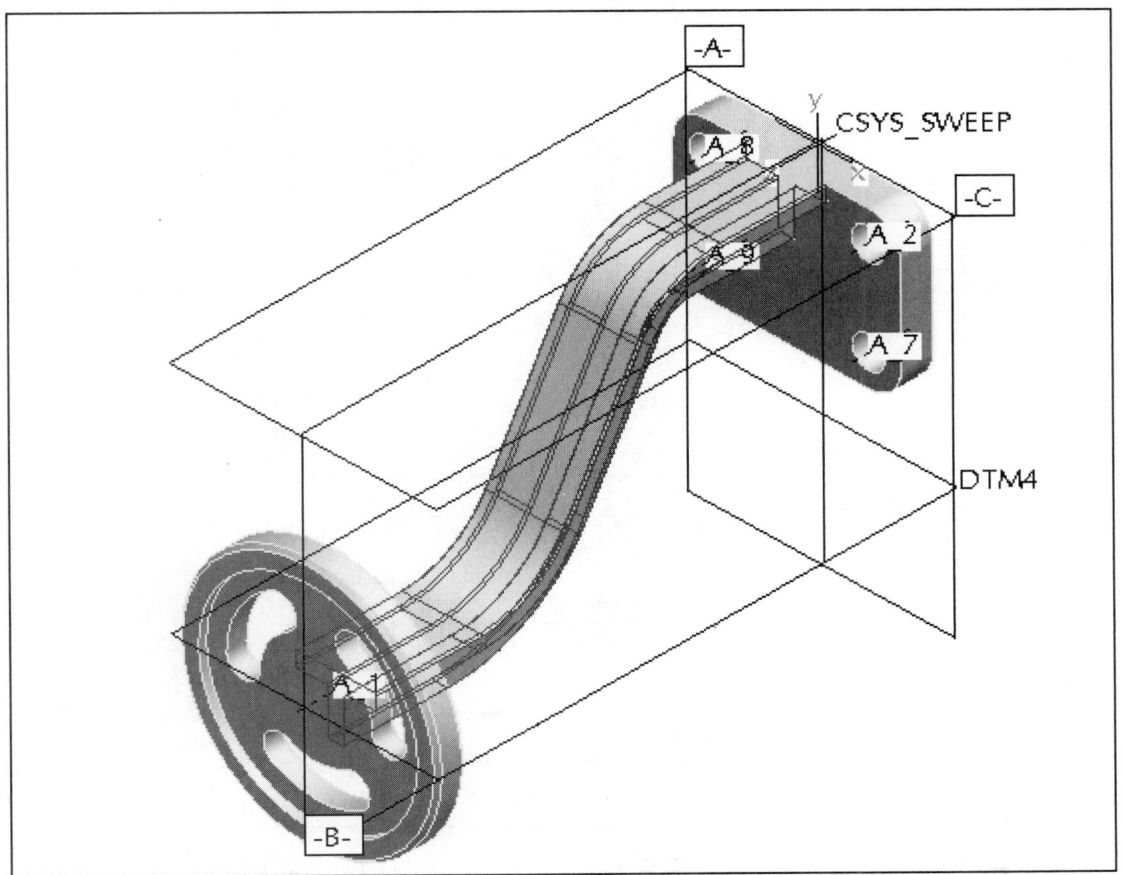

Figure 12.7 Bracket Showing Set Datum Planes and Coordinate System

Bracket

The Bracket (Fig. 12.7) requires the use of the Sweep command. The T-shaped section is swept along the sketched *trajectory*. The protrusions on both sides of the swept feature are to be created with the dimensions shown in Figures 12.8 through 12.14. Systematic commands are provided only for the sweep trajectory and its cross section.

Start a new part. Click: [icon] ⇒ ●**Part** ⇒ Name **Bracket** ⇒ [☑ Use default template] ⇒ **OK** ⇒ **Edit** ⇒ **Setup**

- **Material** = CAST_IRON
- **Units** = Inch lbm Second

Set Datum and **Rename** the default datum planes and coordinate system:

- Datum **TOP** = **C**
- Datum **FRONT** = **A**
- Datum **RIGHT** = **B**
- Coordinate System = **CSYS_SWEEP**

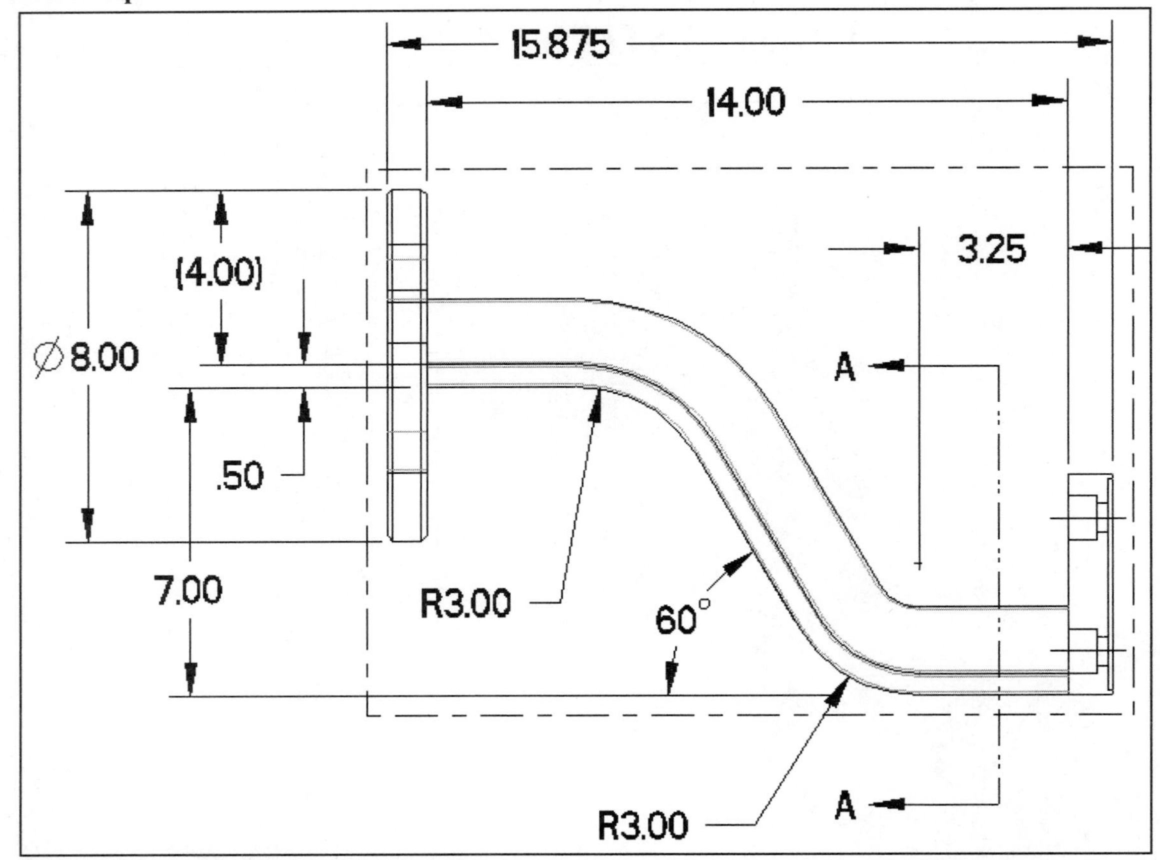

Figure 12.8 Bracket Drawing, Front View

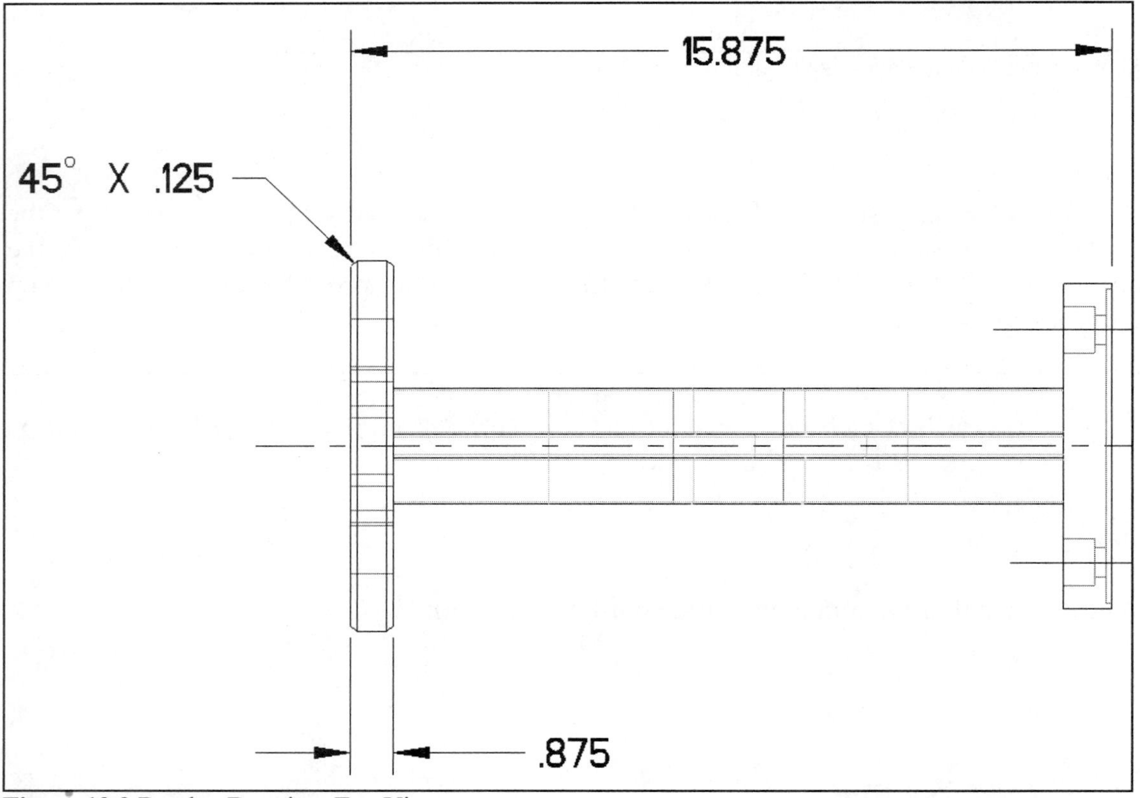

Figure 12.9 Bracket Drawing, Top View

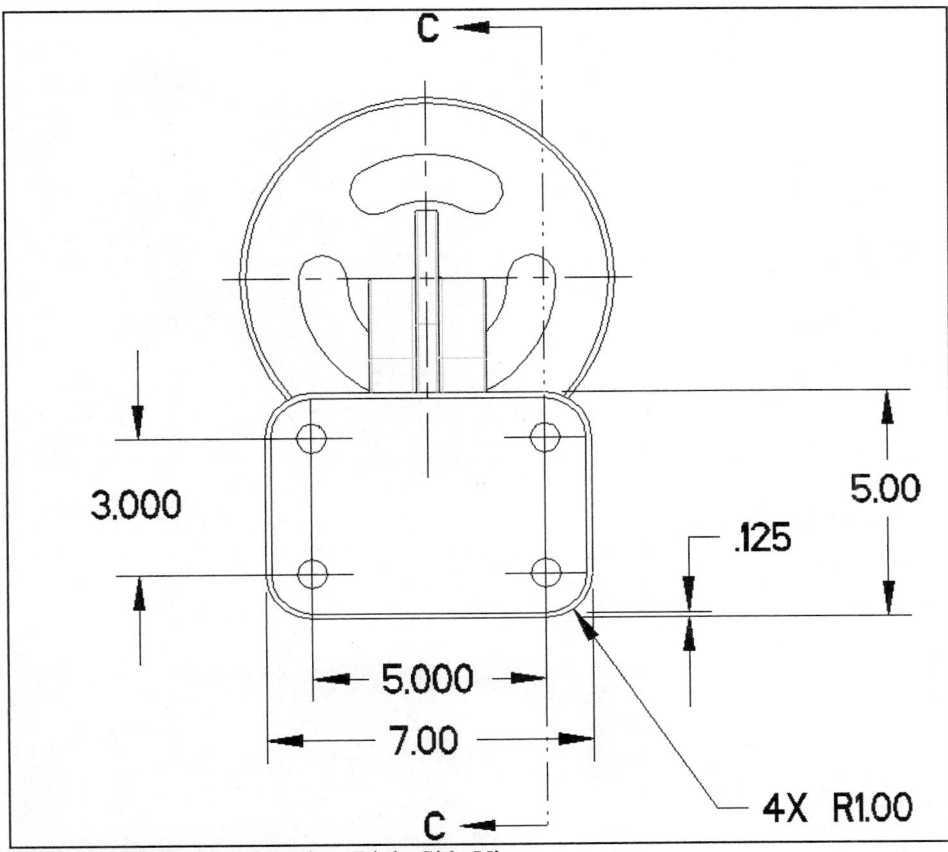

Figure 12.10 Bracket Drawing, Right Side View

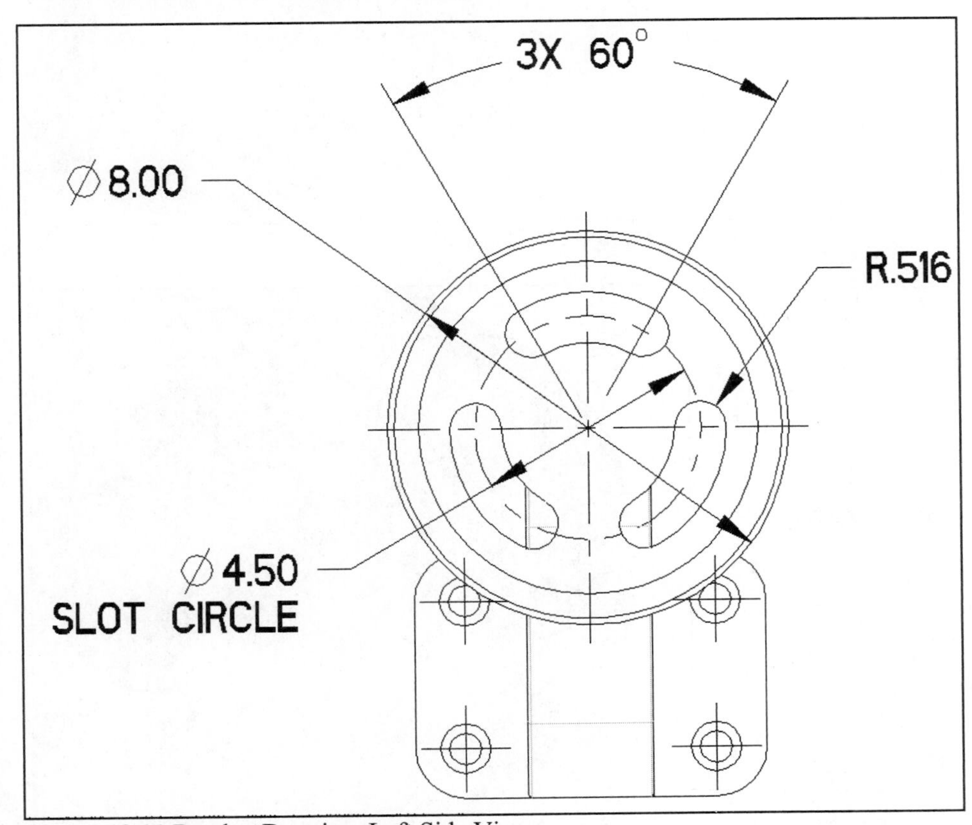

Figure 12.11 Bracket Drawing, Left Side View

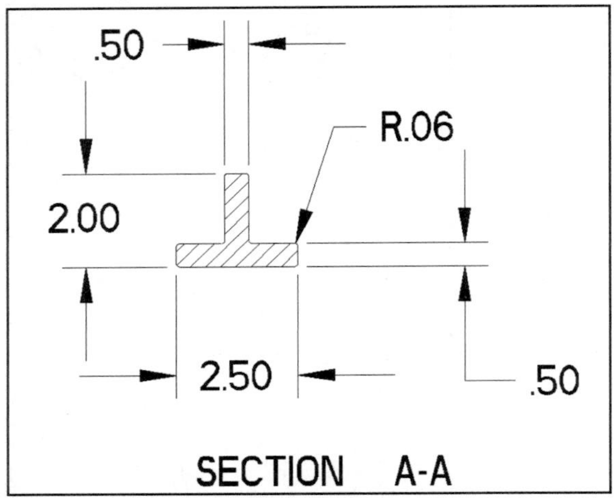

Figure 12.12(a) SECTION A-A

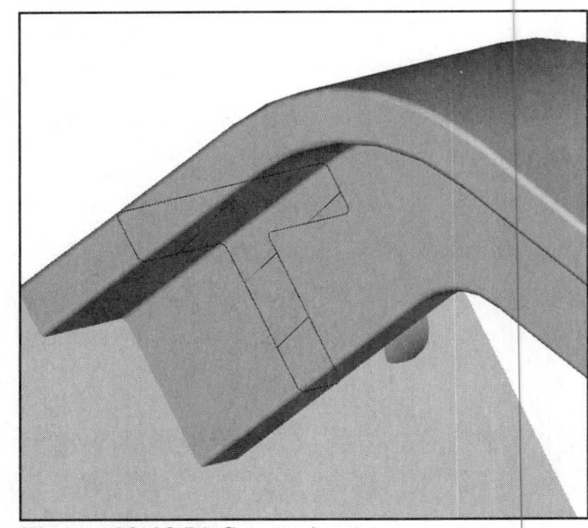

Figure 12.12(b) Swept Arm

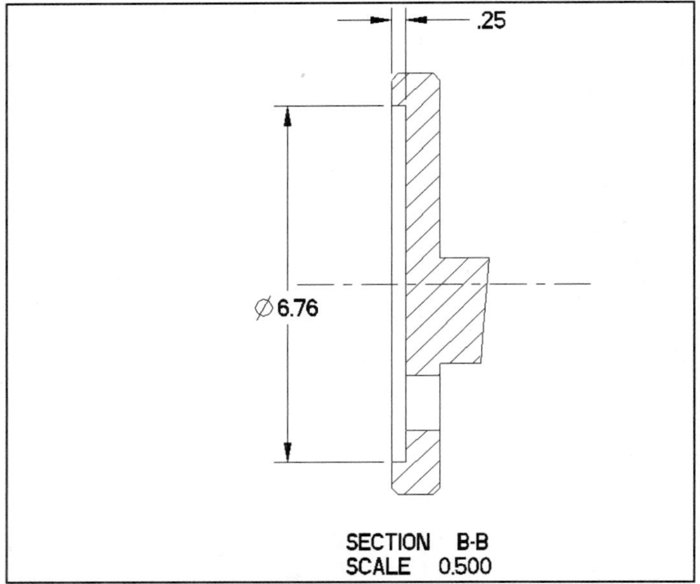

Figure 12.13(a) SECTION B-B

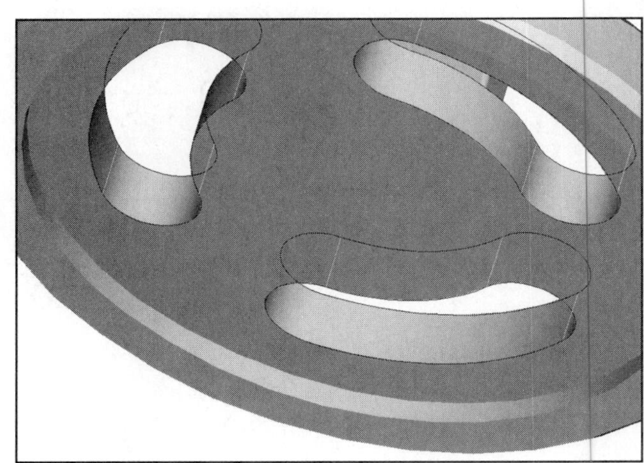

Figure 12.13(b) Cut

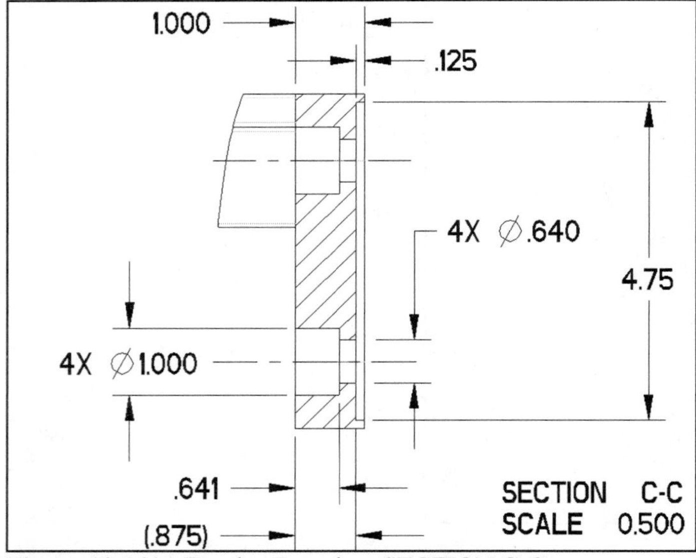

Figure 12.14(a) Bracket Drawing, SECTION C-C

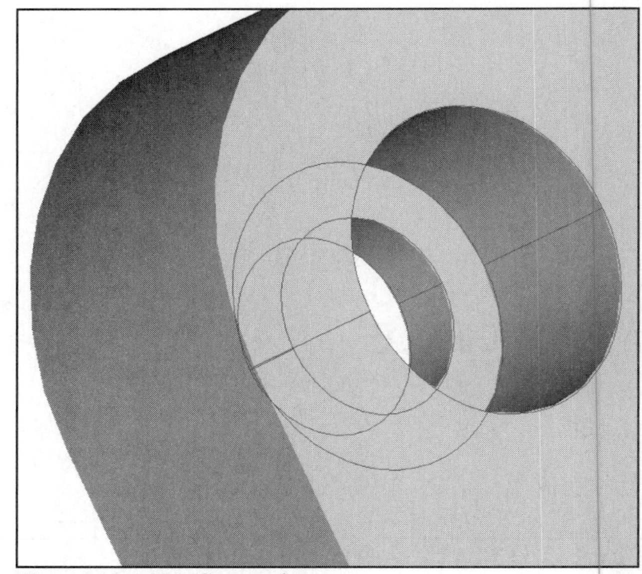

Figure 12.14(b) Counterbore

Start the Bracket by modeling the protrusion shown in Figure 12.15. This protrusion will be used to establish the sweep's position in space. Sketch the protrusion on datum **A**. The second protrusion is a sweep and is shown in Figure 12.16.

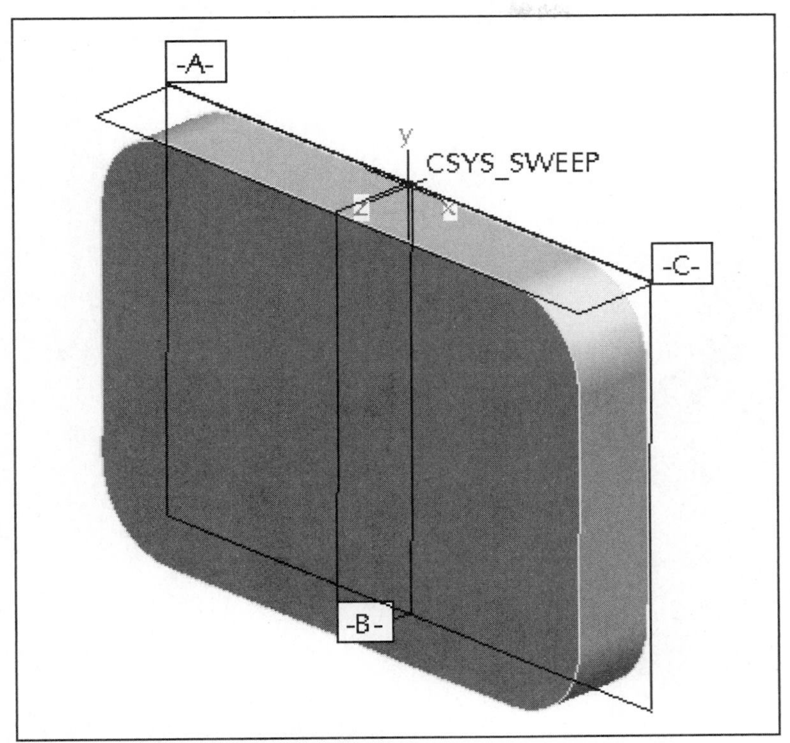

Figure 12.15 Bracket's First Protrusion

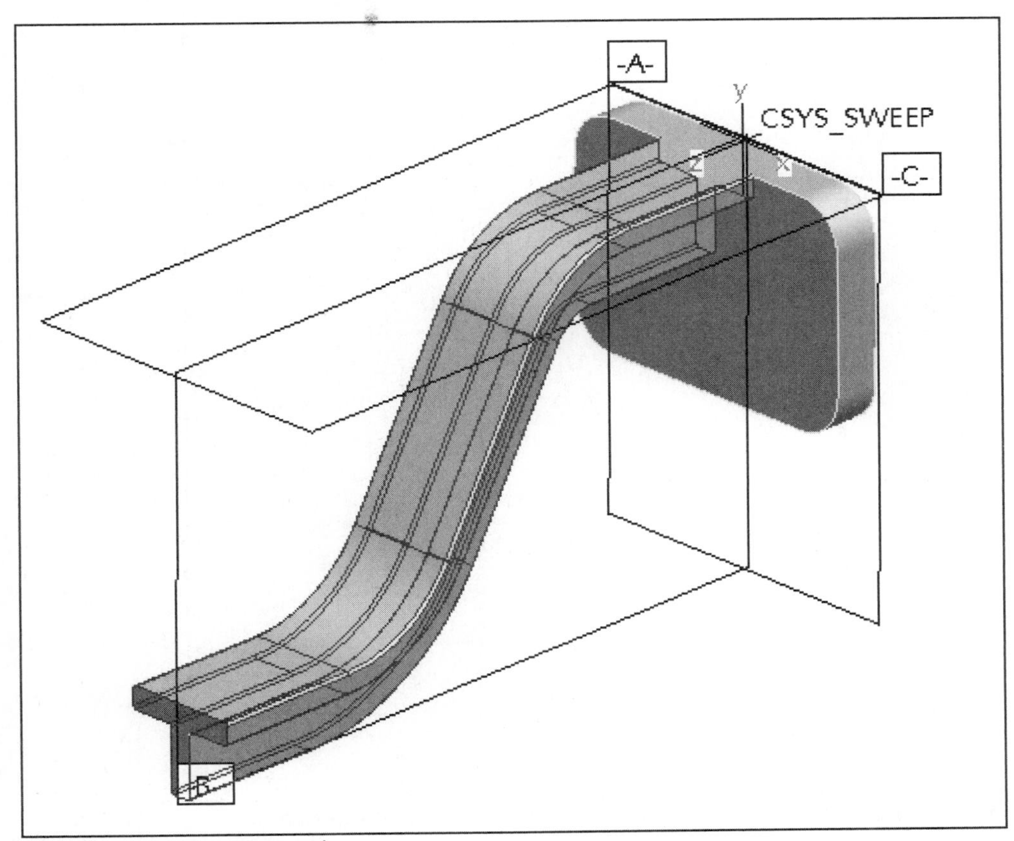

Figure 12.16 Swept Protrusion

Create the sweep, from the Menu Bar click: **Insert** ⇒ **Sweep** ⇒ **Protrusion** ⇒ **Sketch Traj** ⇒ pick datum plane **B** as the sketching plane for the trajectory (Fig. 12.17) ⇒ **MMB** ⇒ **Top** ⇒ pick datum plane **C** as the orientation plane ⇒ from the References dialog box, delete datum **A** and add the front face of the protrusion ⇒ **Close** ⇒ sketch, dimension, and modify the trajectory (Figs. 12.18 through 12.22) ⇒ ✓

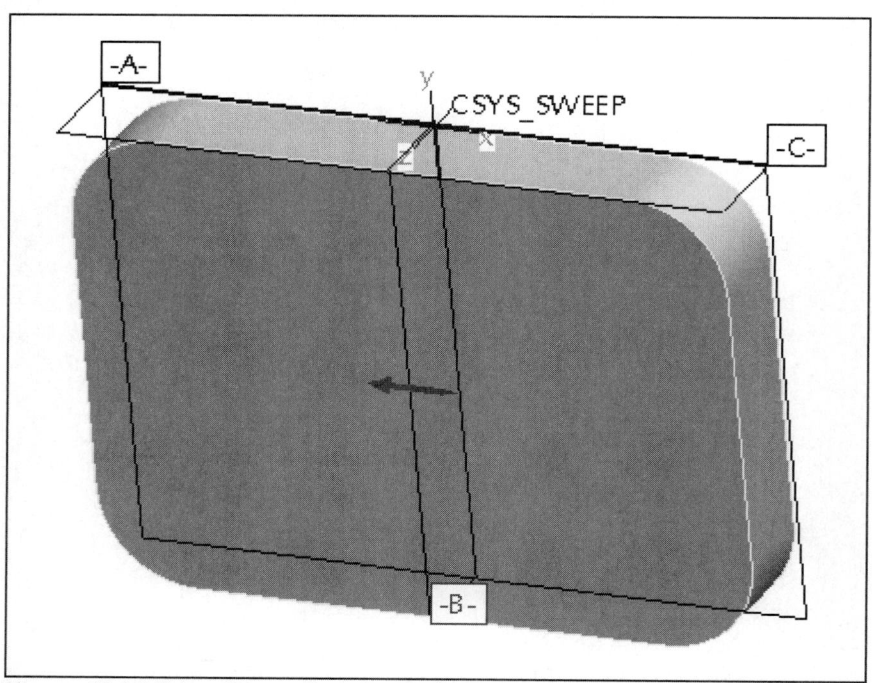

Figure 12.17 Select Datum B as the Trajectory Sketching Plane

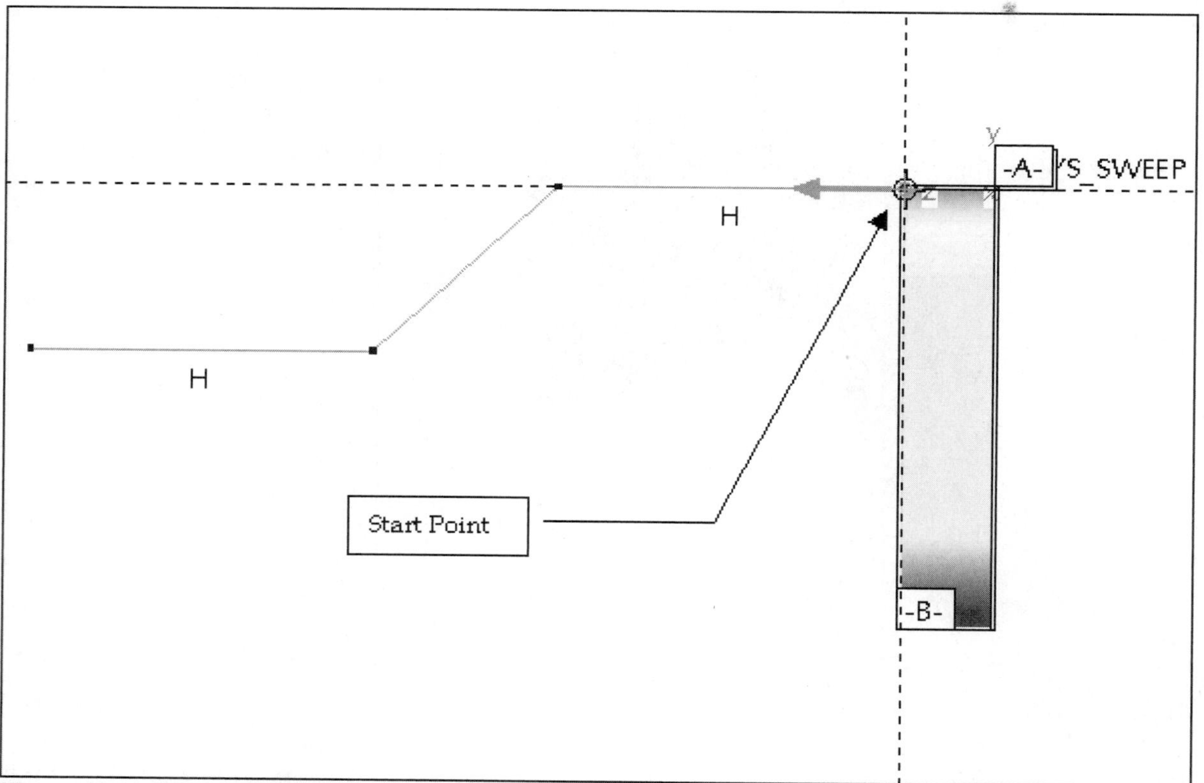

Figure 12.18 Sketch the Three Lines. Start the trajectory by sketching a horizontal line from this position.

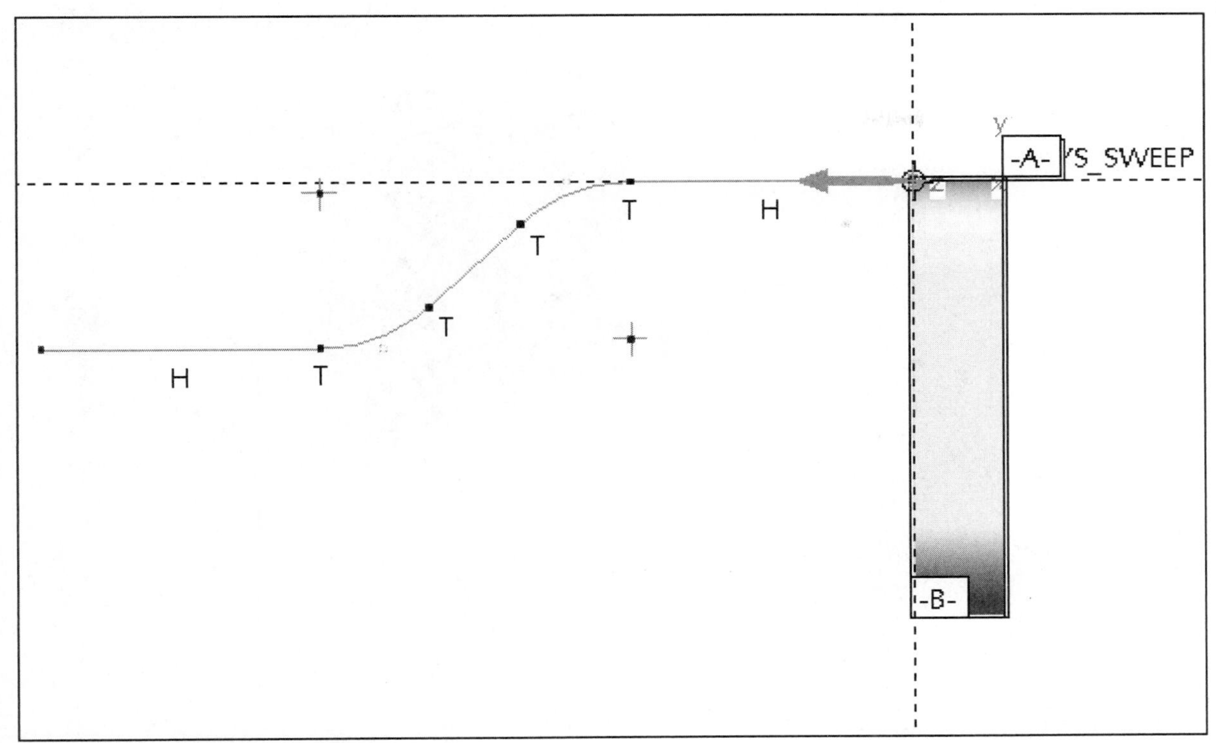

Figure 12.19 Add the Arc Fillets

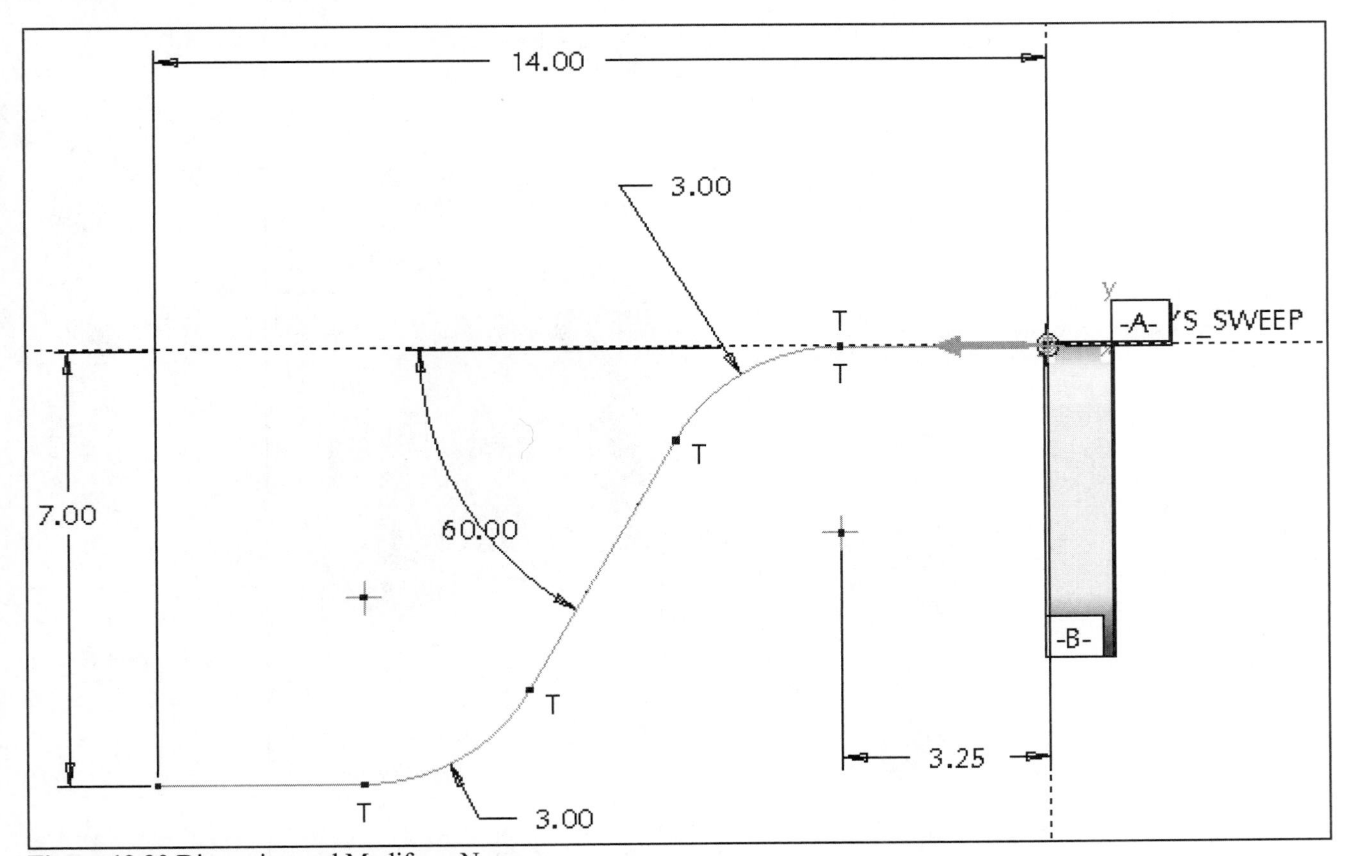

Figure 12.20 Dimension and Modify as Necessary

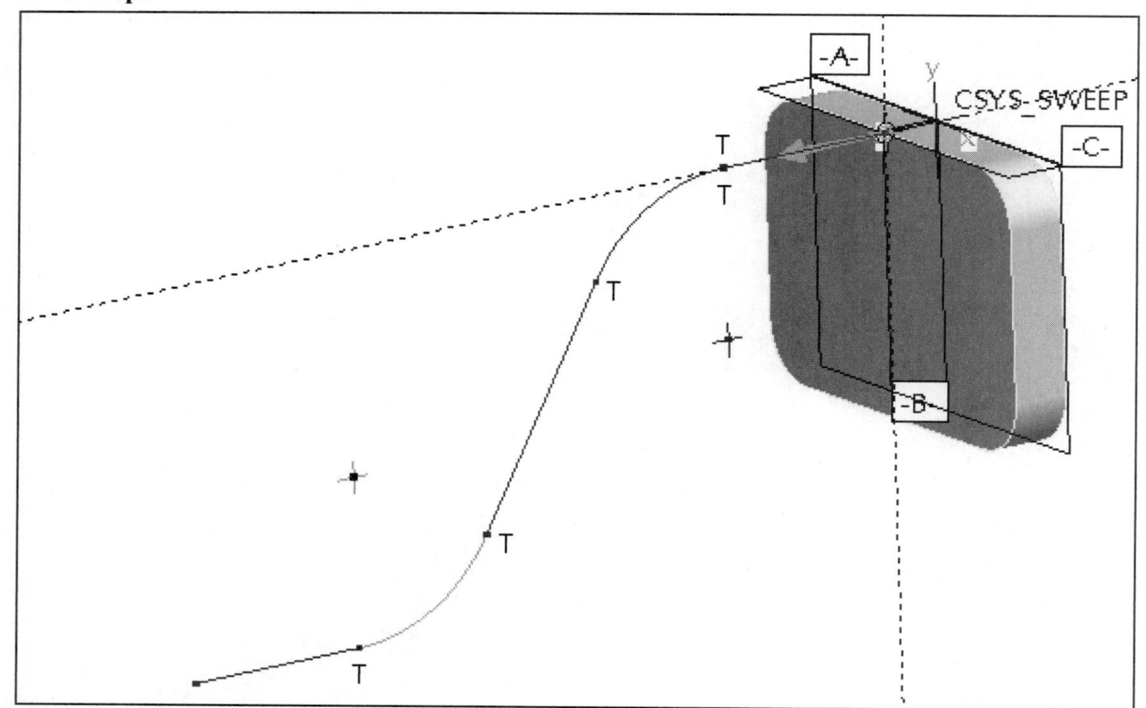

Figure 12.21 Trajectory Sketch

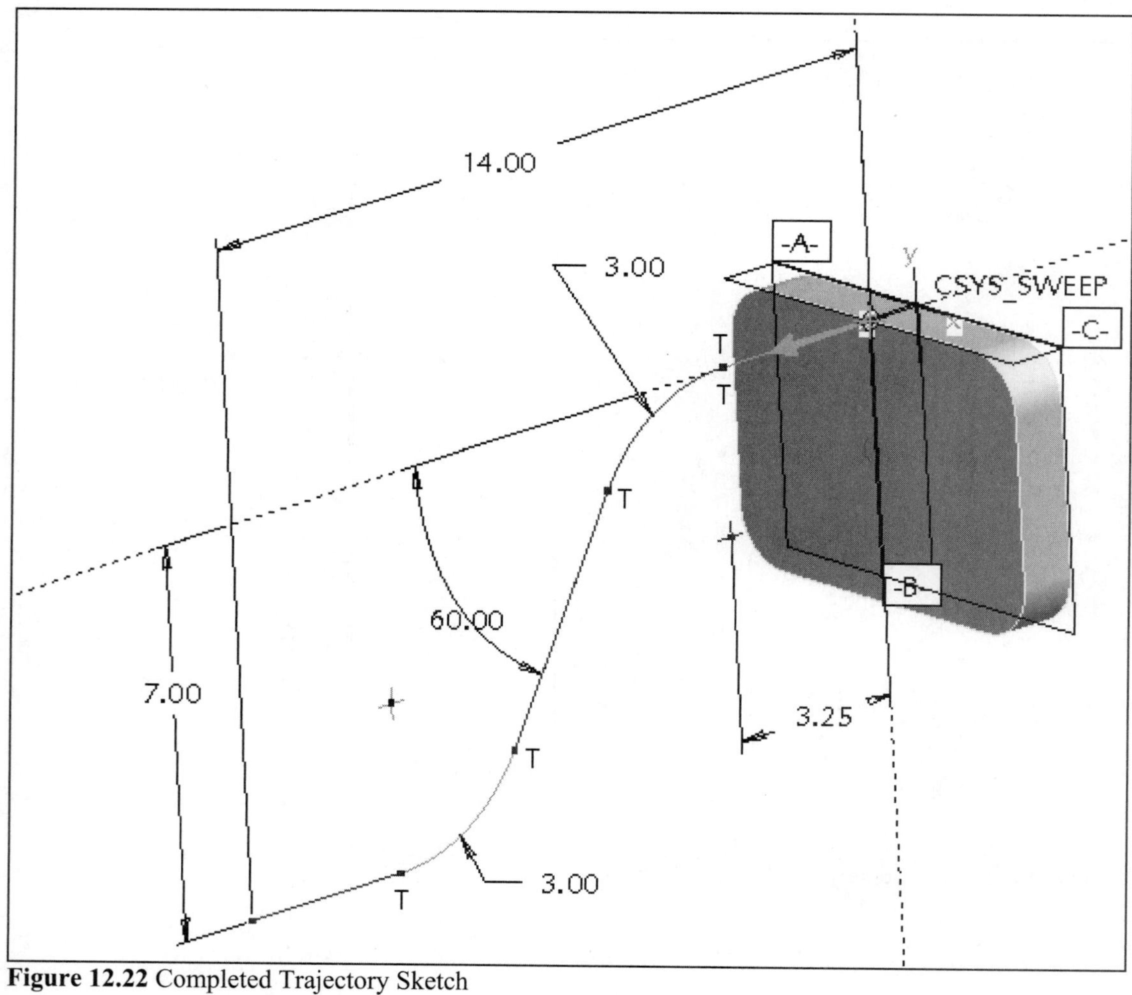

Figure 12.22 Completed Trajectory Sketch

Complete the sweep, click: **Free Ends** ⇒ **MMB** ⇒ **Sketch** (from menu bar) ⇒ **Options** ⇒ **Display** tab ⇒ Grid ⇒ Snap To Grid ⇒ **Parameters** tab ⇒ Grid Spacing: **Manual** ⇒ Equal Spacing ⇒ X = **.25** ⇒ ✔ ⇒ sketch the *eight* lines of the section [Fig. 12.23(a-b)] ⇒ add a horizontal centerline and sketch *eight* fillets (Fig. 12.24) ⇒ add and reposition dimensions as needed (Fig. 12.25 and Fig. 12.26) ⇒ modify the section [Figs. 12.27(a-b) and 12.28] ⇒ ✔ ⇒ **Preview** (Fig. 12.29) ⇒ **OK** ⇒ 💾 ⇒ **MMB**

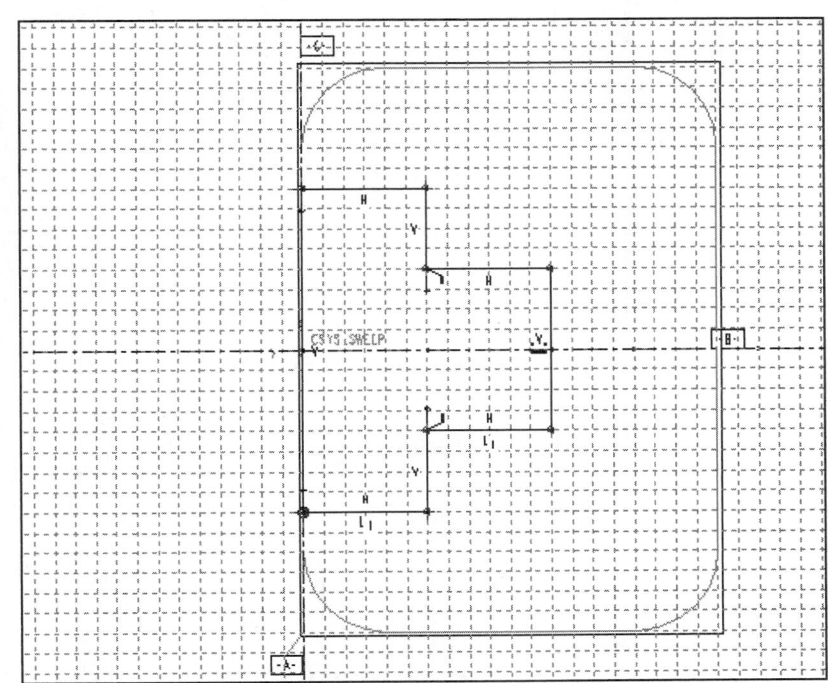

Figure 12.23(a) Sketch the Eight Lines

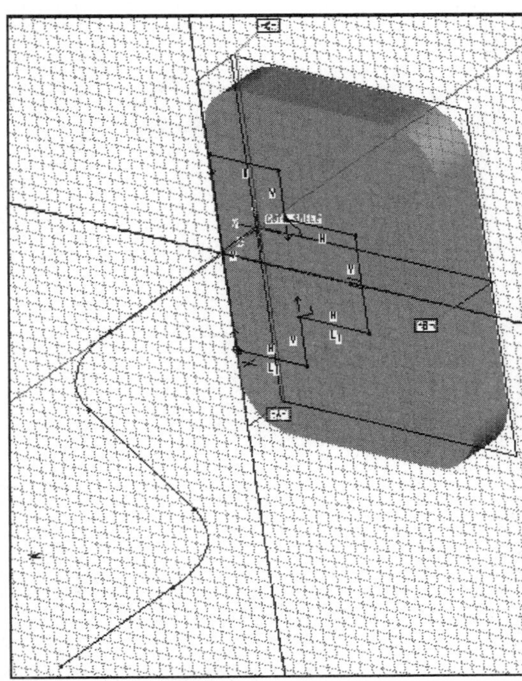

Figure 12.23(b) Sketch

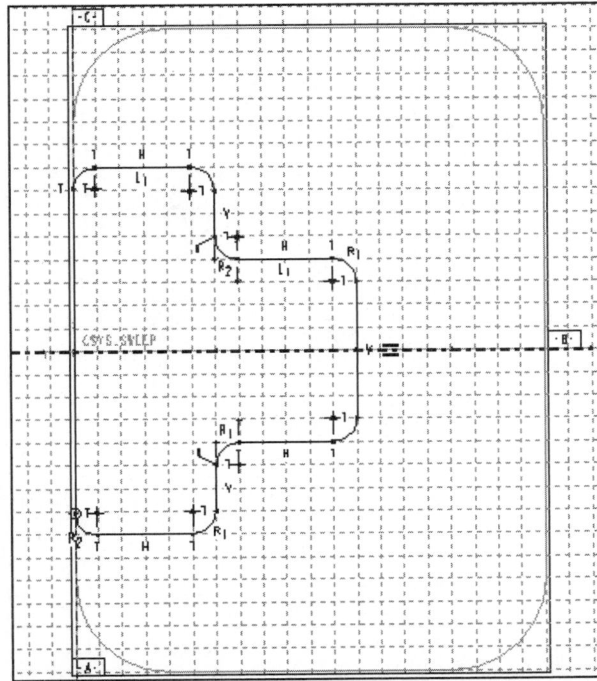

Figure 12.24 Sketch the Centerline and Arc Fillets

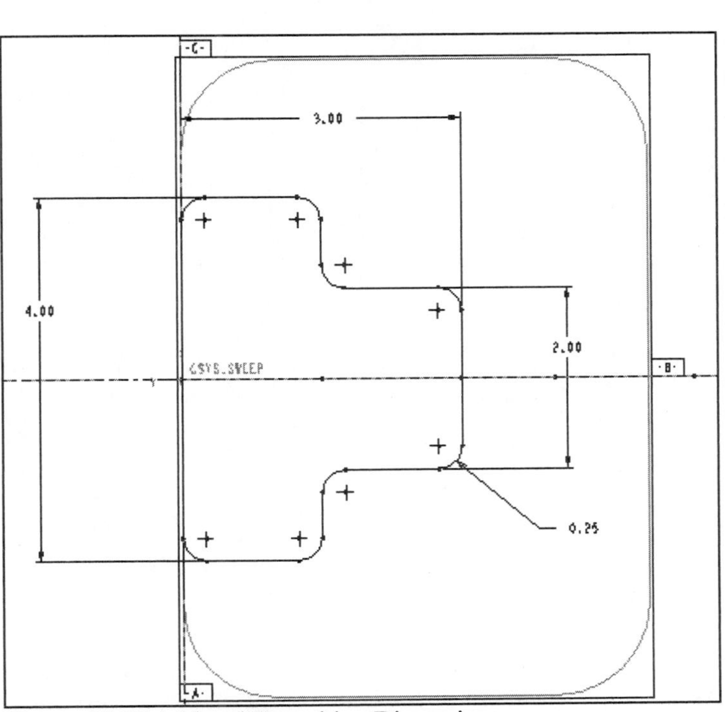

Figure 12.25 Add and Reposition Dimensions

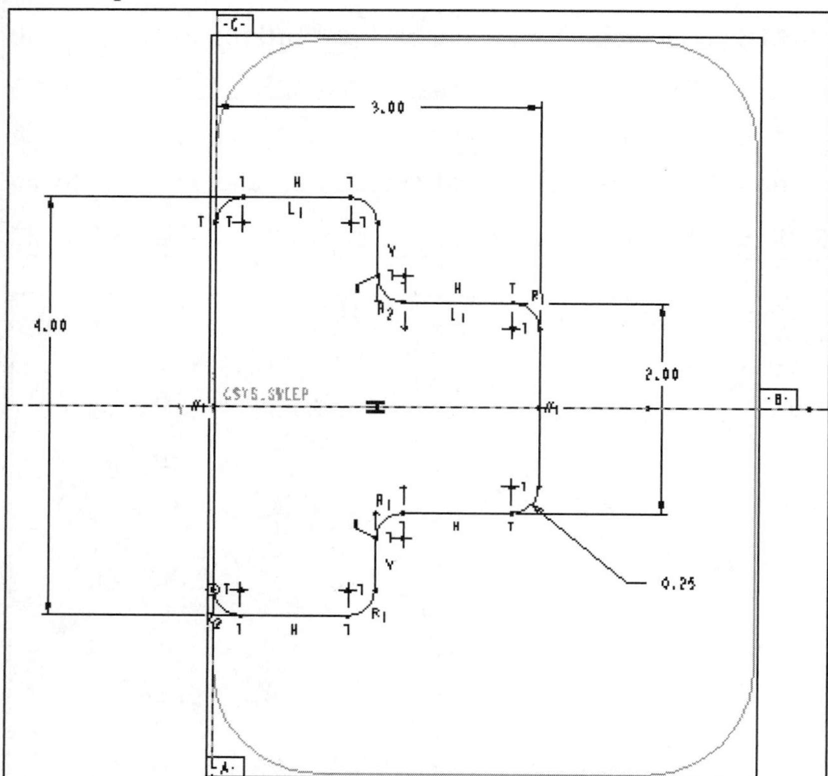

Figure 12.26 Constraints On

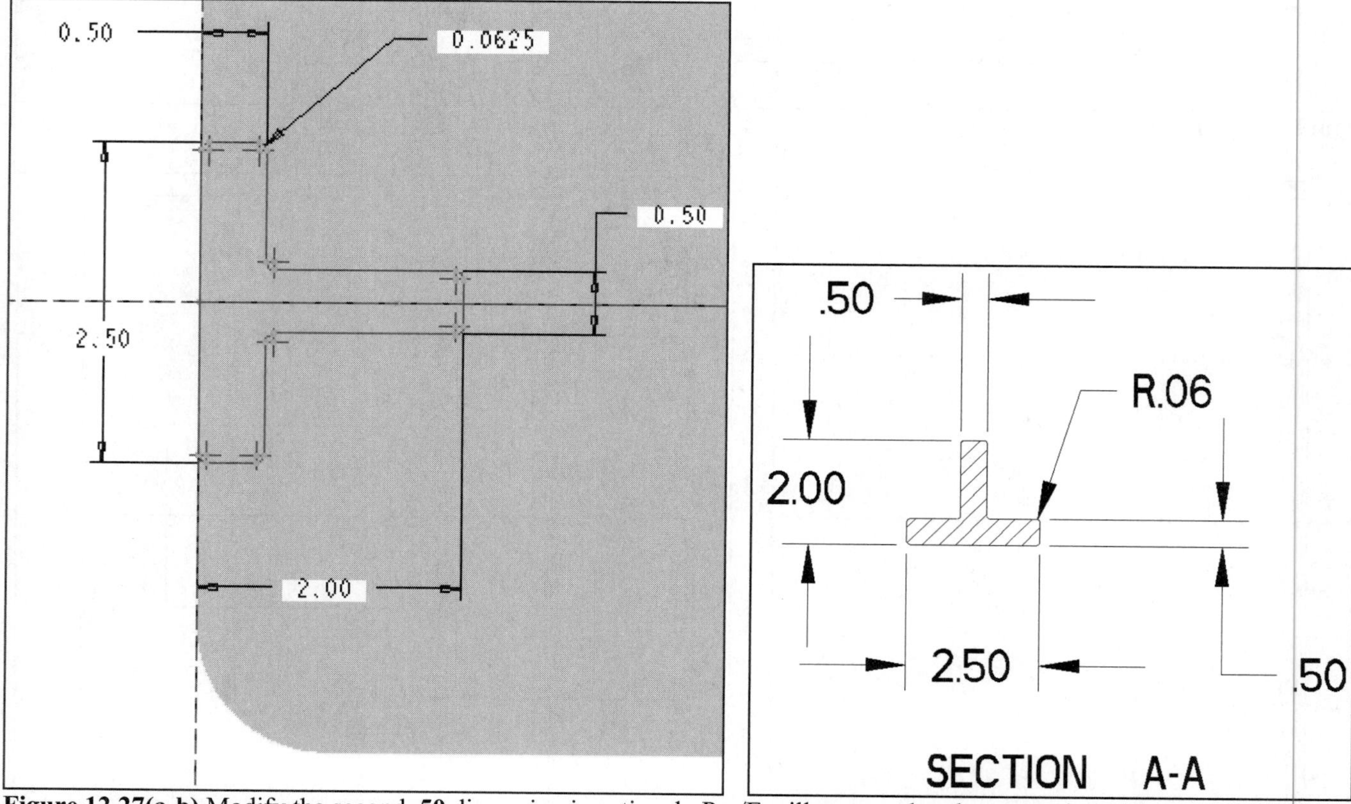

Figure 12.27(a-b) Modify the second **.50** dimension is optional. Pro/E will assume that the two web thicknesses are equal. (As per design intent, if they are both to be displayed on a drawing, then both should be on the section sketch.)

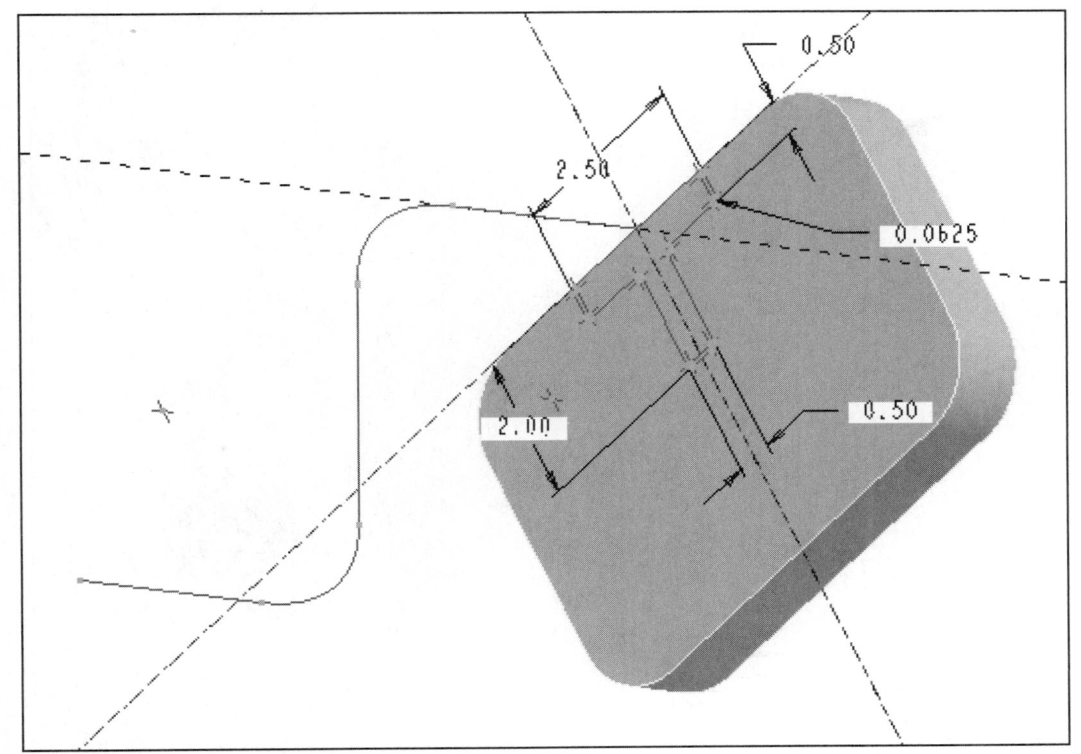

Figure 12.28 Section Rotated

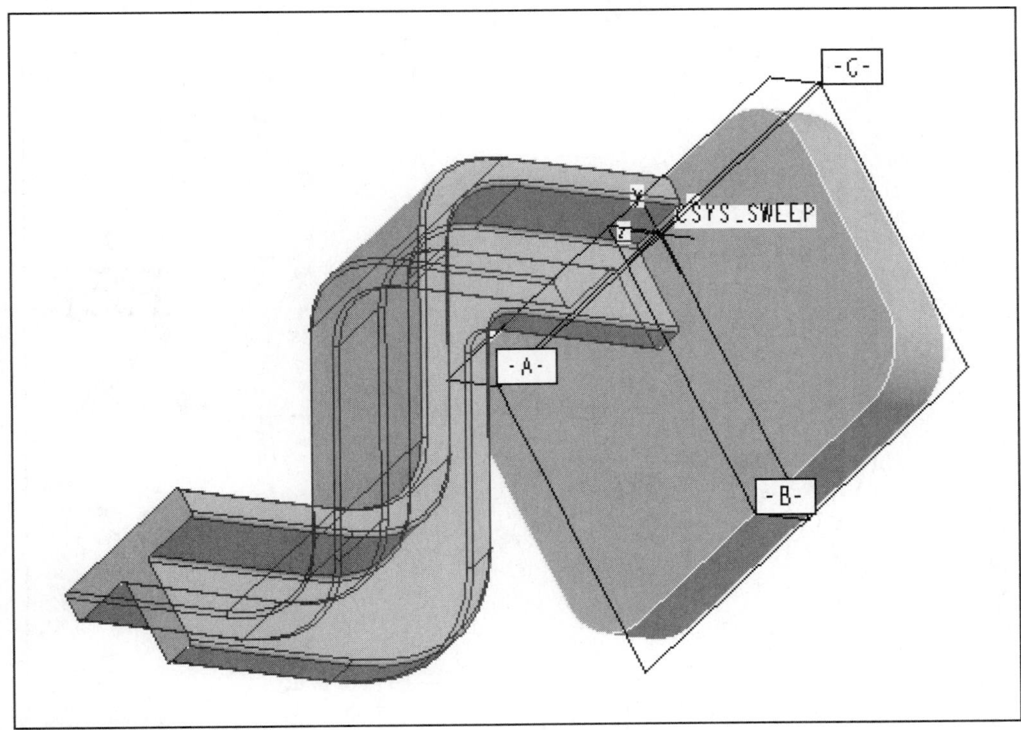

Figure 12.29 Preview Sweep

Complete the part by modeling the remaining features (Figs. 12.30 through 12.34).

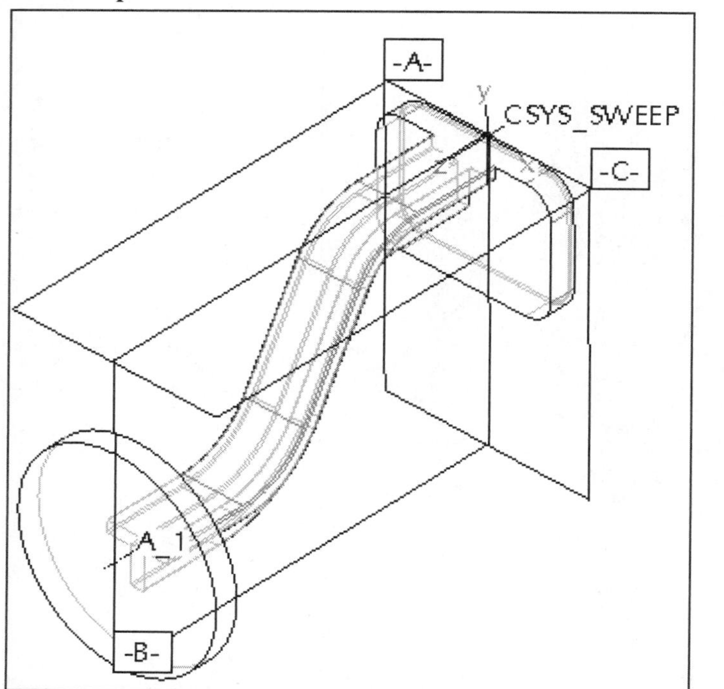

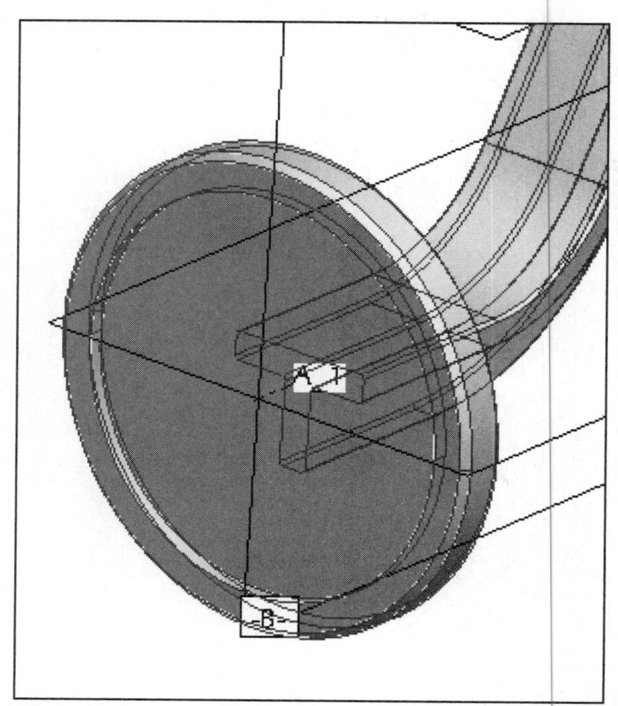

Figure 12.30(a-b) Add the Third Protrusion; Add the Cut (**Ø6.76** by **.250** deep) and the Chamfers (**45° X .125**)

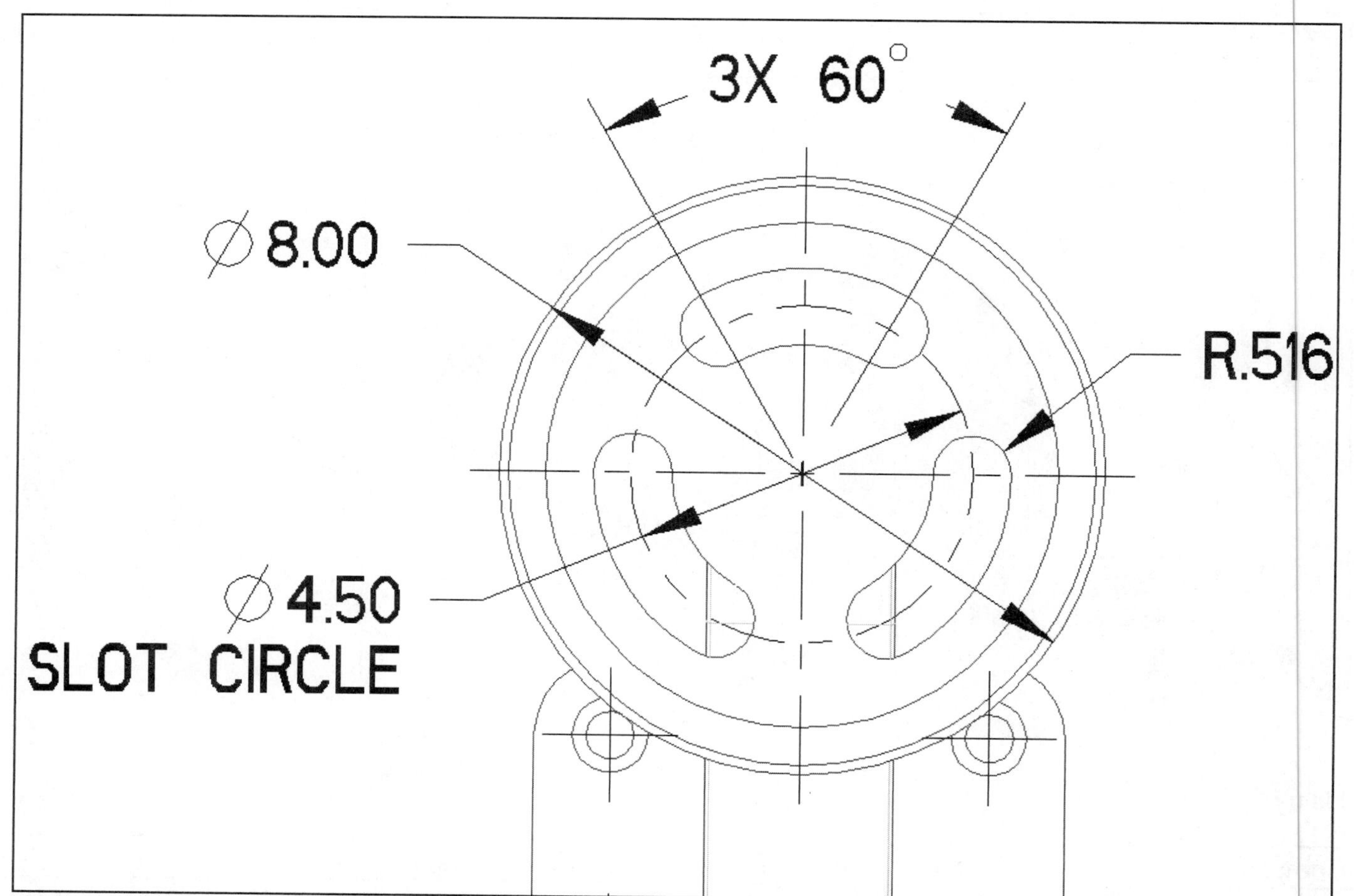

Figure 12.31(a) In order to pattern the slot feature, create a new datum through A_1 and at an angle, to use as the orientation (reference) plane.

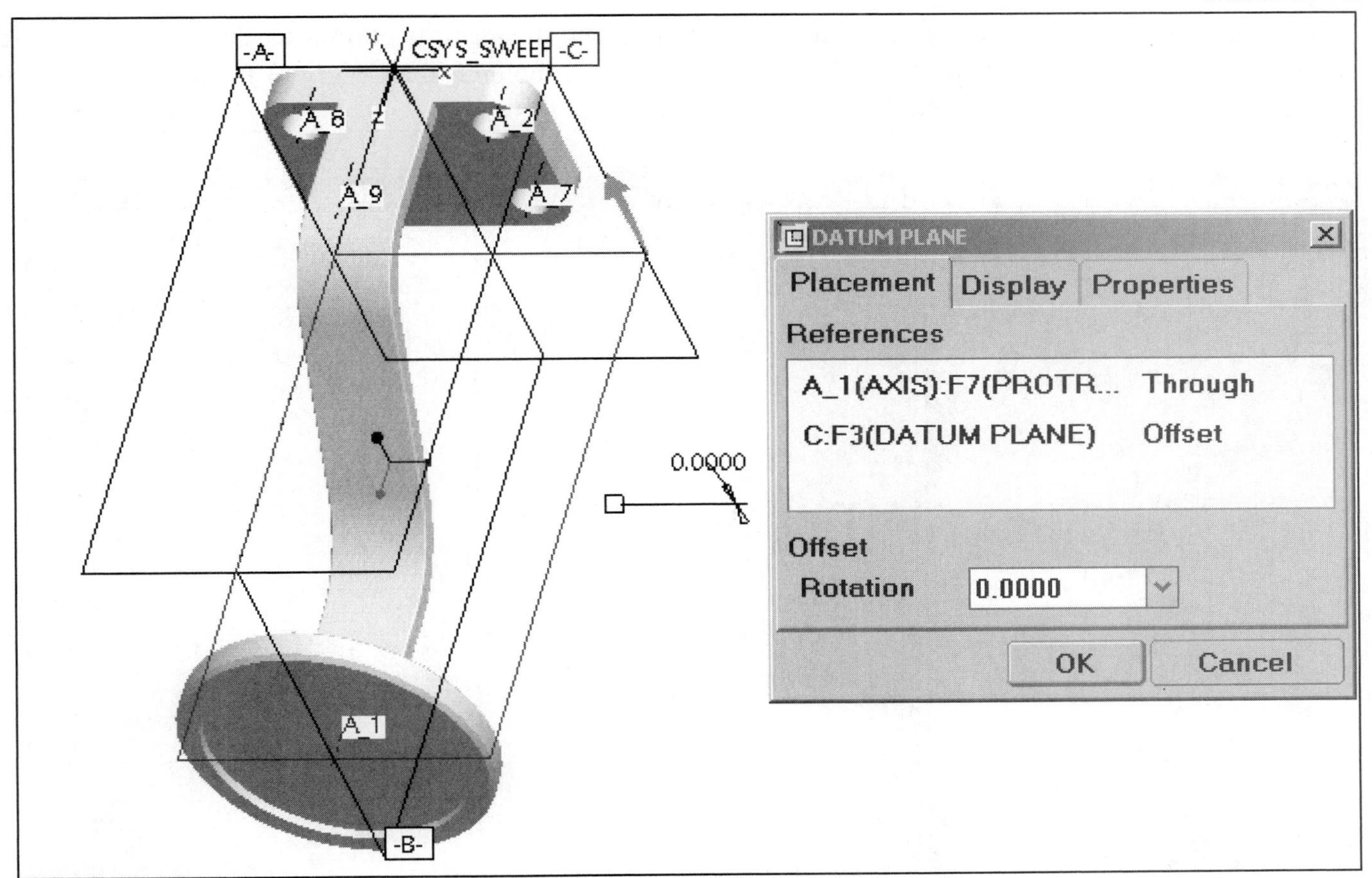

Figure 12.31(b) Datum through A_1 and at an angle of **0** degrees to Datum C

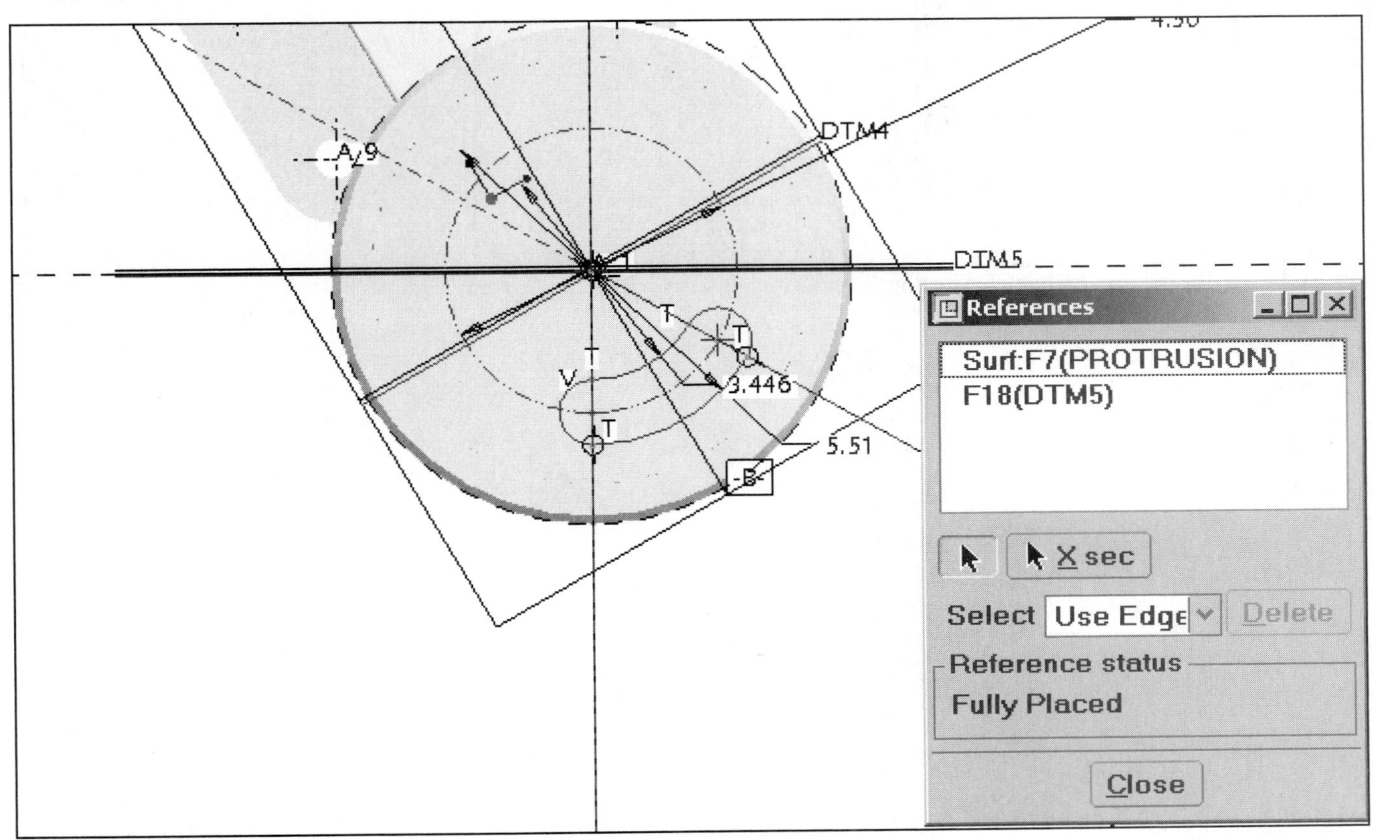

Figure 12.31(c) References

Add a Sketcher Point at the center of the round protrusion and then set the options as shown [Fig. 12.31(d)] (**Sketcher** ⇒ **Options** ⇒ **Parameters** tab) ⇒ Use the sketcher point as the **Origin** of the grid ⇒ Sketch the section [Fig. 12.31(e)] ⇒ Complete the feature [Fig. 12.31(f-g)]

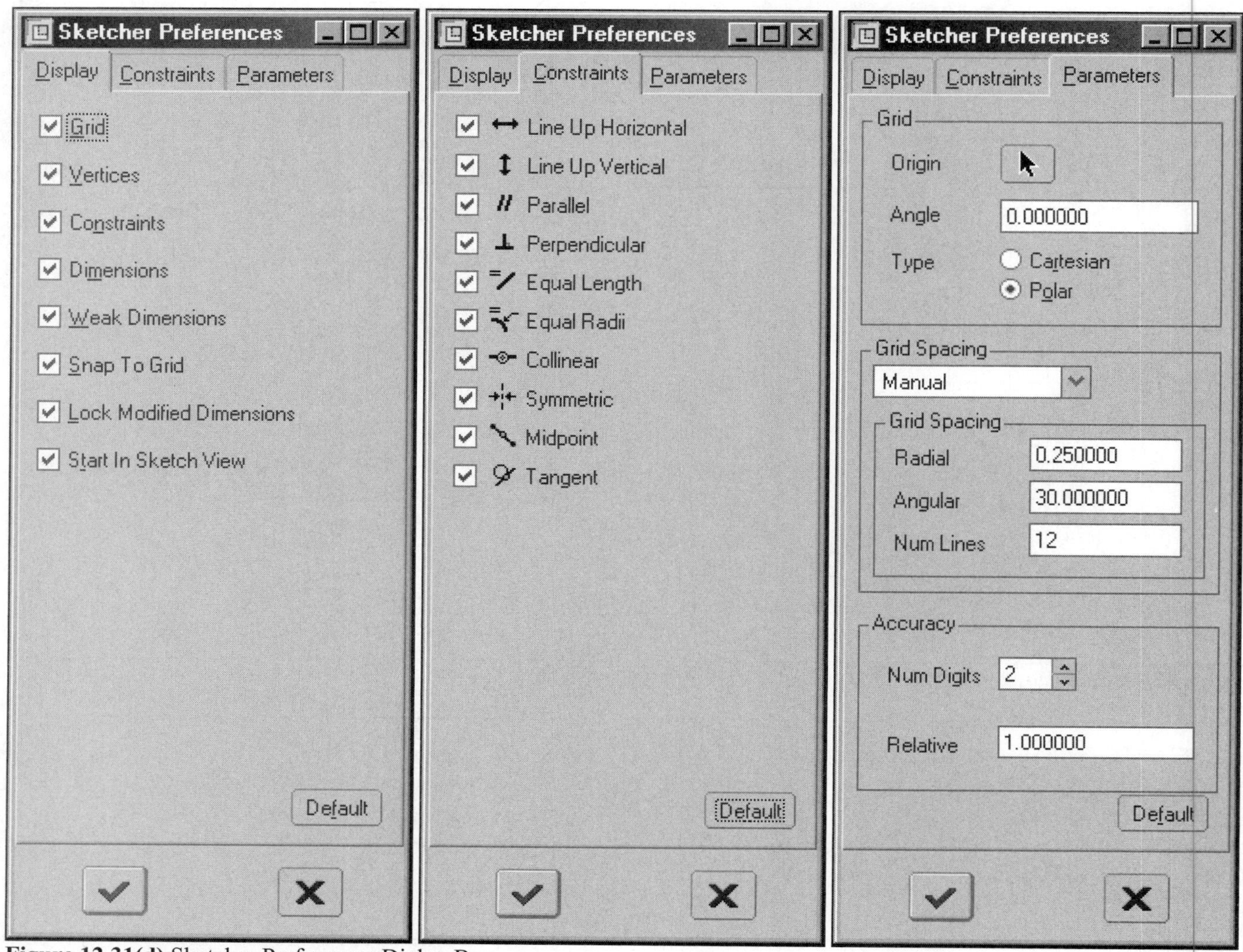

Figure 12.31(d) Sketcher Preferences Dialog Box

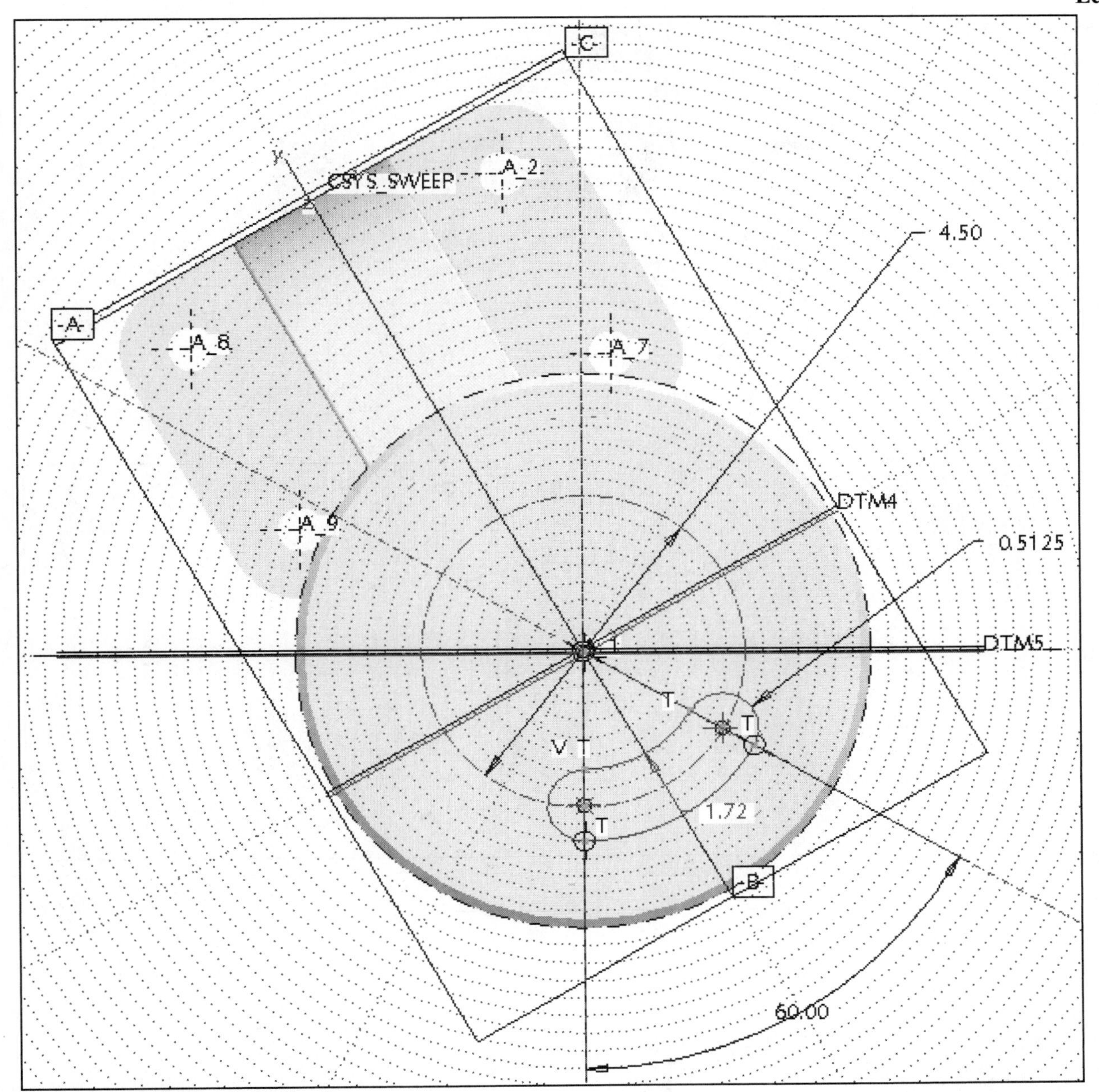

Figure 12.31(e) Sketch the Section

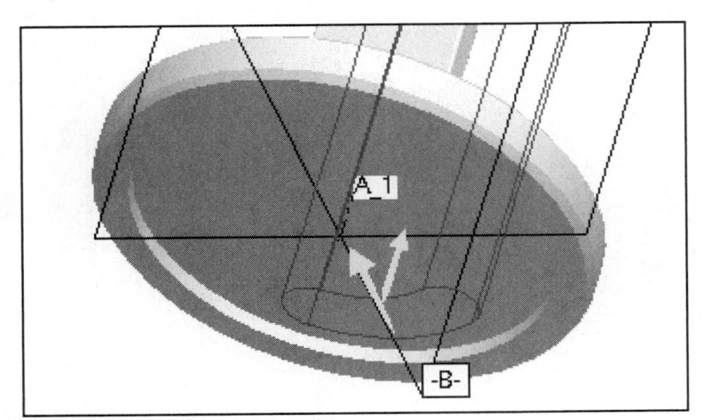

Figure 12.31(f) Cut Direction

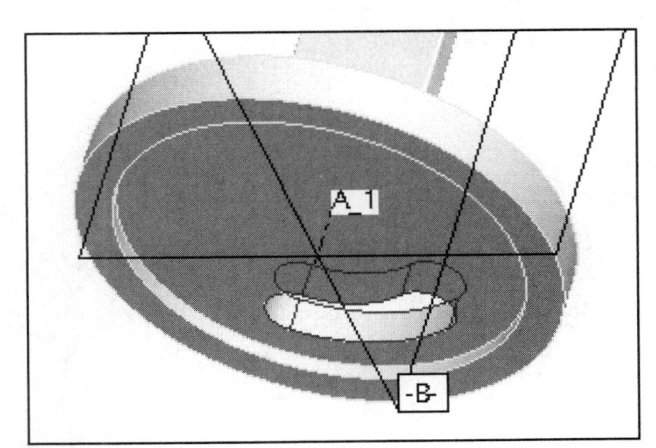

Figure 12.31(g) Completed Cut

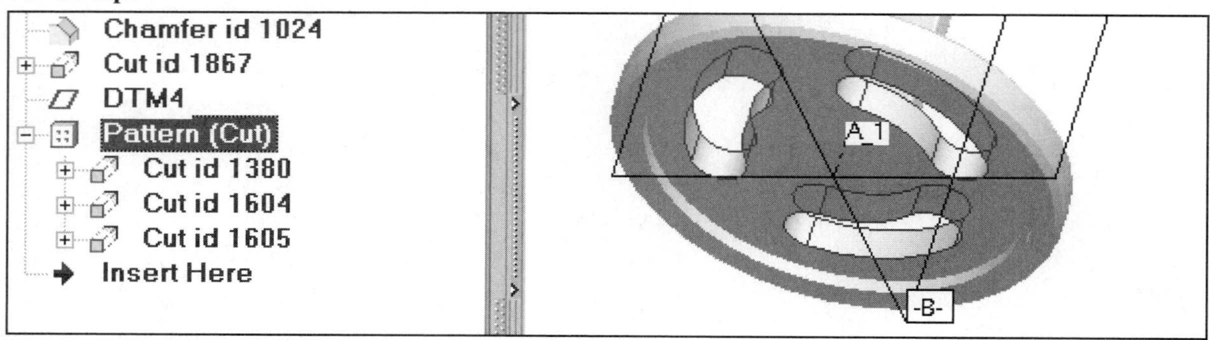

Figure 12.32 Pattern the Slots

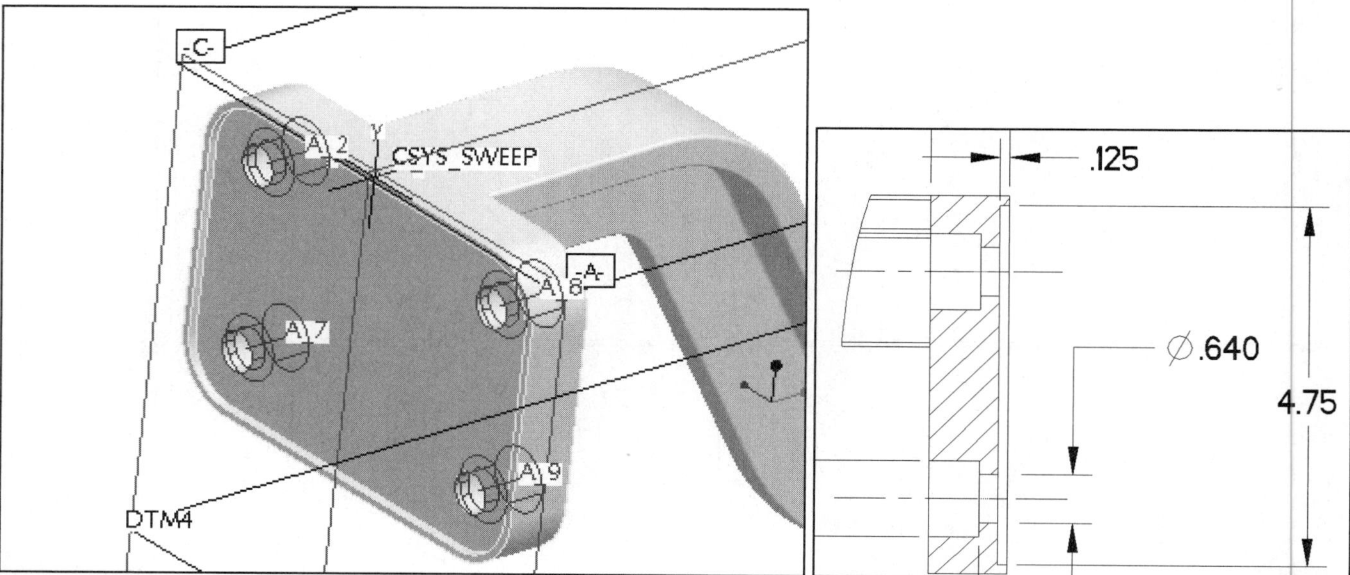

Figure 12.33 Create the Cut and Pattern the Counterbore Holes

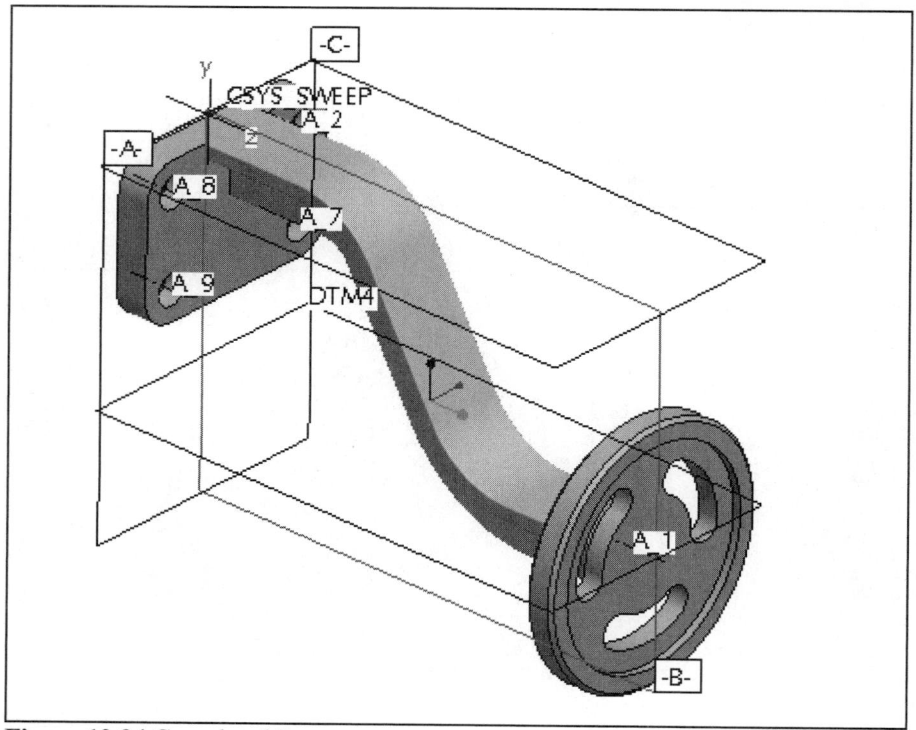

Figure 12.34 Completed Part

Lesson 12 Project

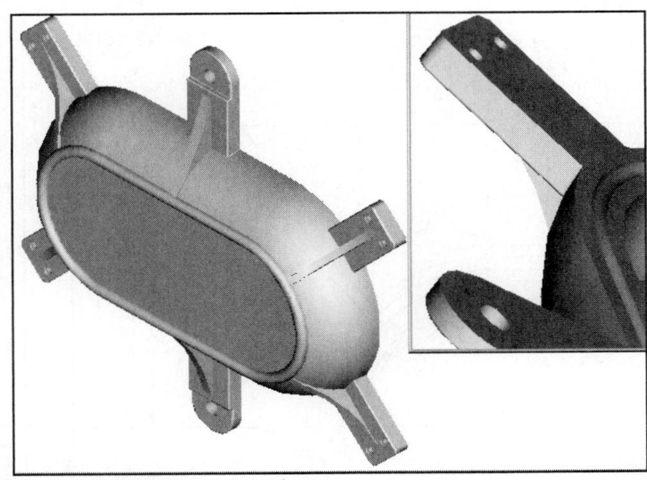

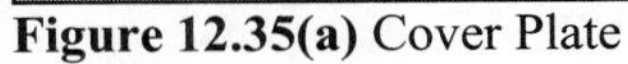

Figure 12.35(a) Cover Plate

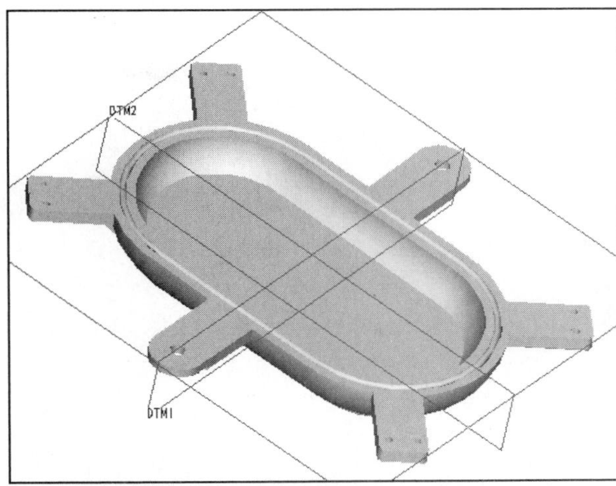

Figure 12.35(b) Cover Plate Bottom View

Cover Plate

The Cover Plate is a cast-iron part. Create the part shown in Figures 12.35 through 12.51. The sweep will have a *closed trajectory* with *inner faces included*. Analyze the part and plan the steps and features required to model it. Establish the feature creation sequence and the parent-child relationships for the part. Add rounds on all non-machined edges.

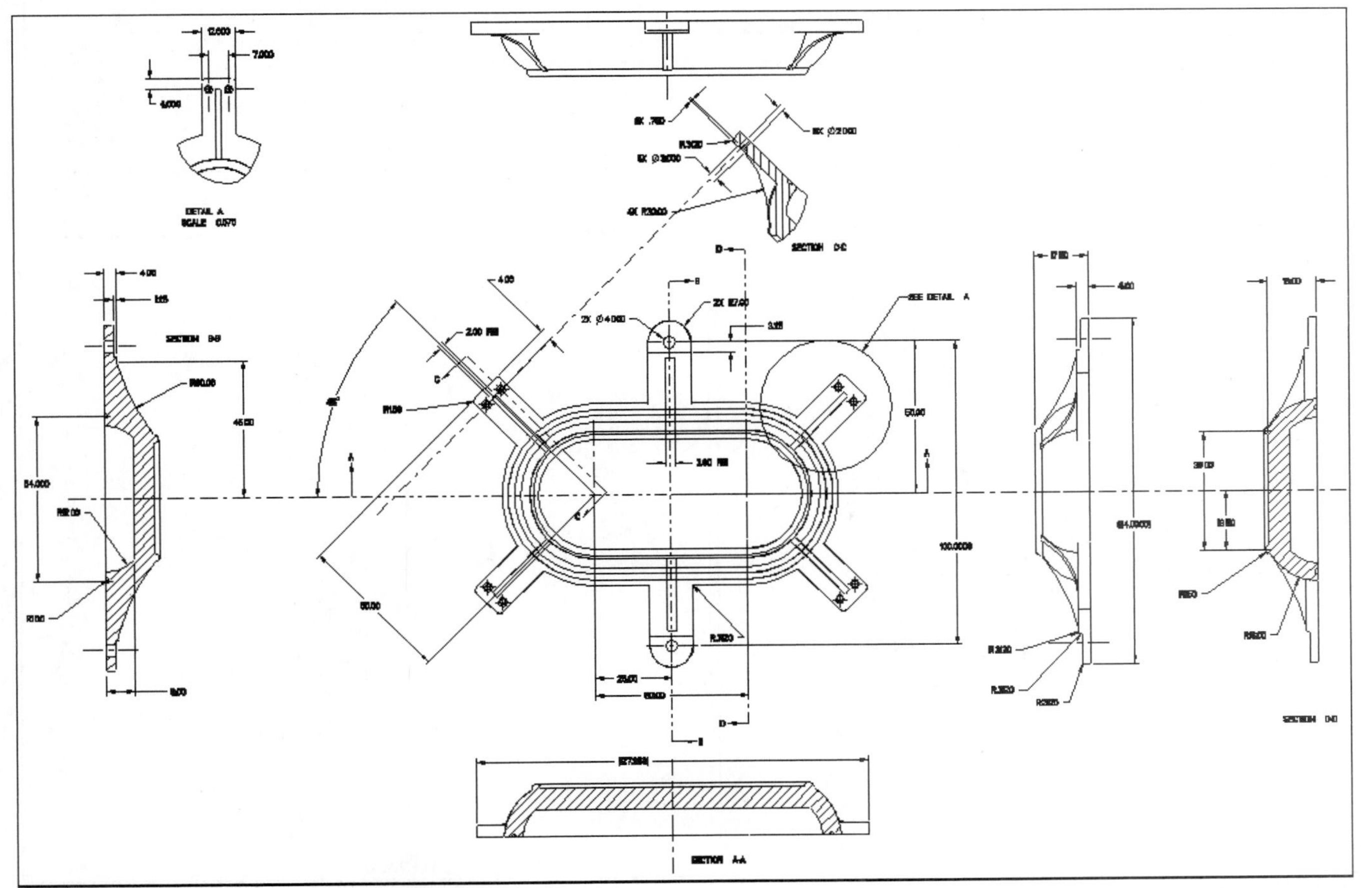

Figure 12.36 Cover Plate Detail Drawing, Sheet One

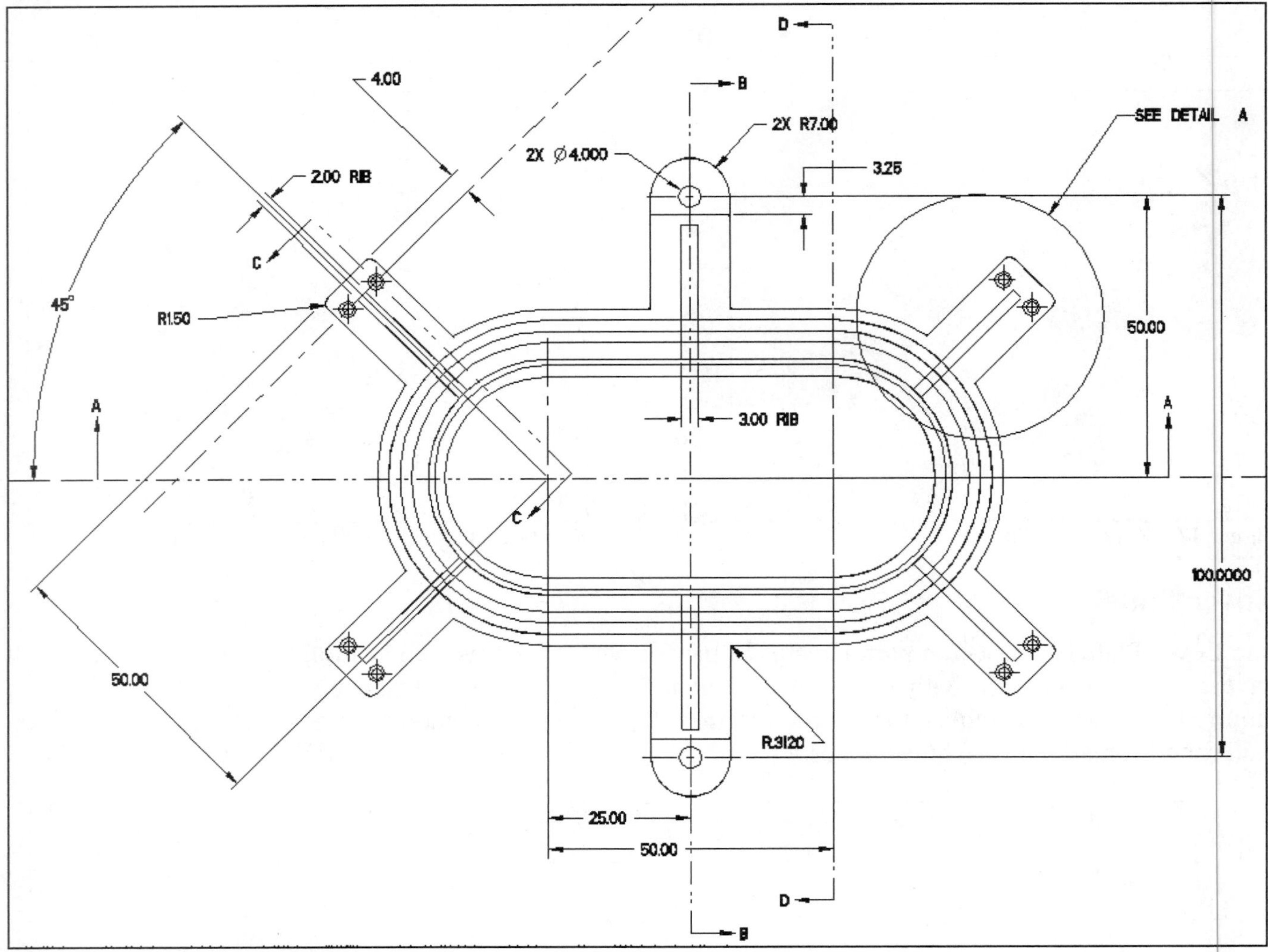

Figure 12.37 Cover Plate Drawing, Top View

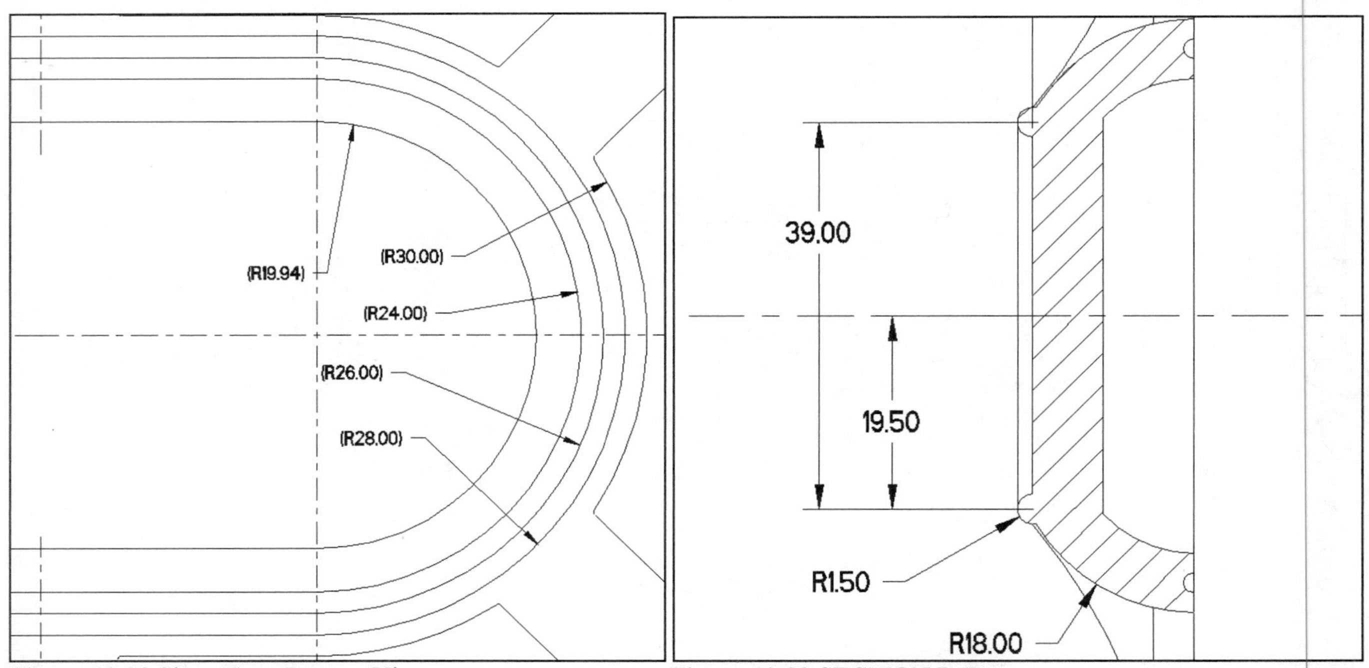

Figure 12.38 Sheet Two, Bottom View

Figure 12.39 SECTION D-D

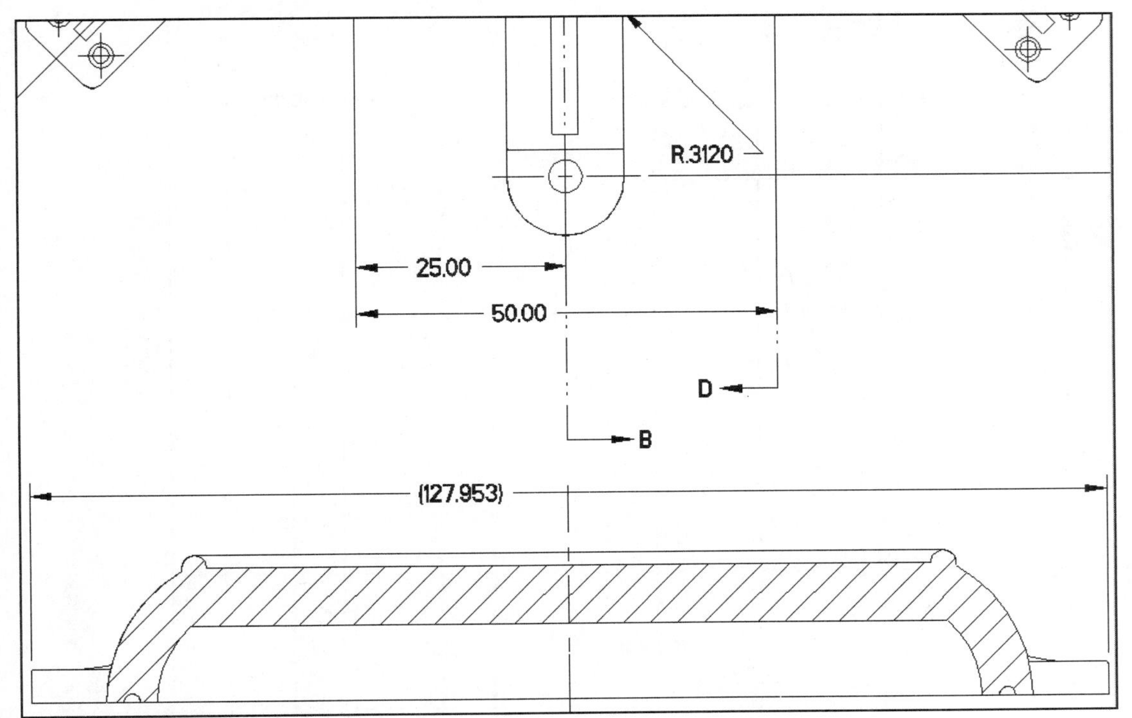

Figure 12.40 SECTION A-A

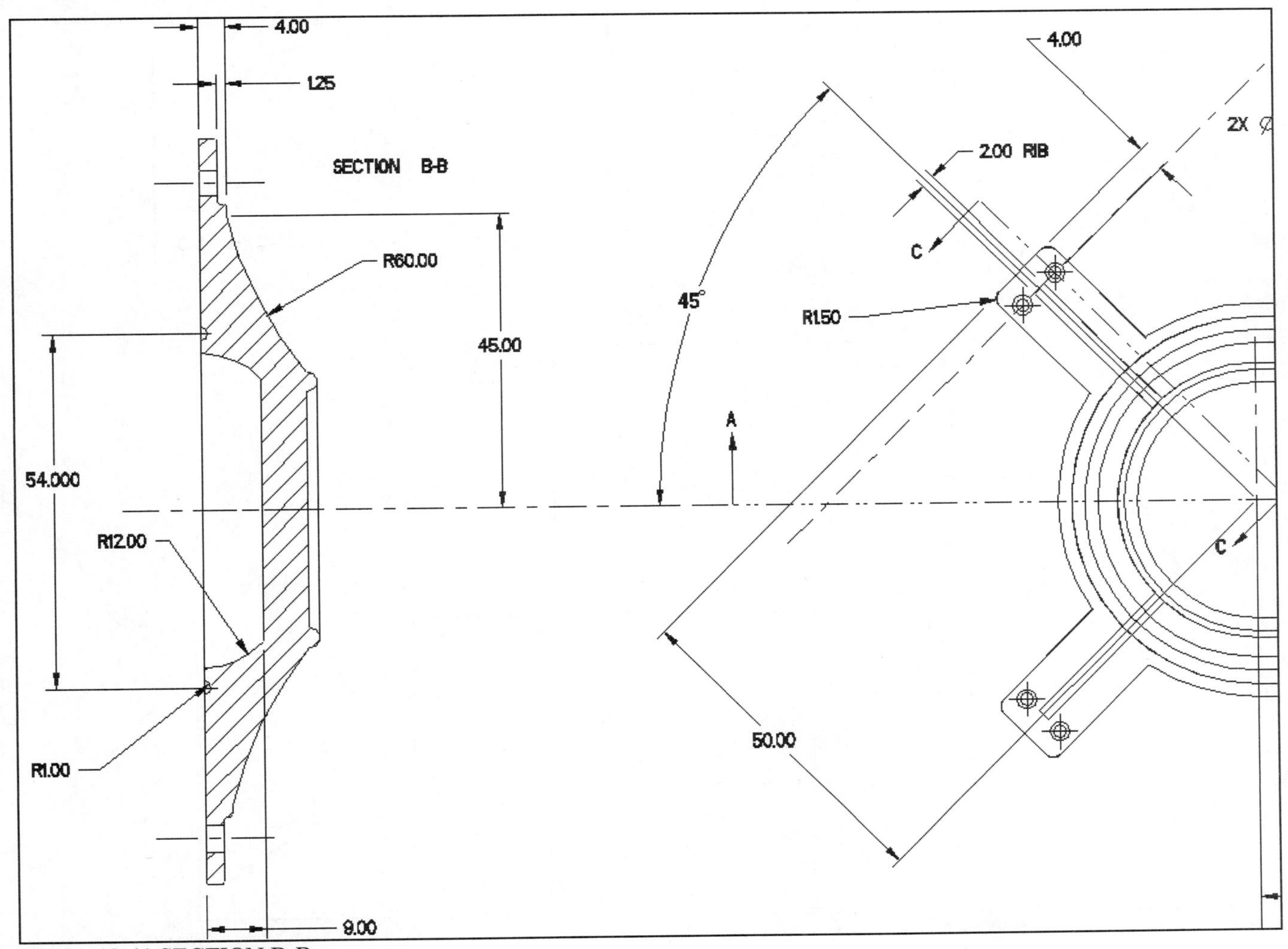

Figure 12.41 SECTION B-B

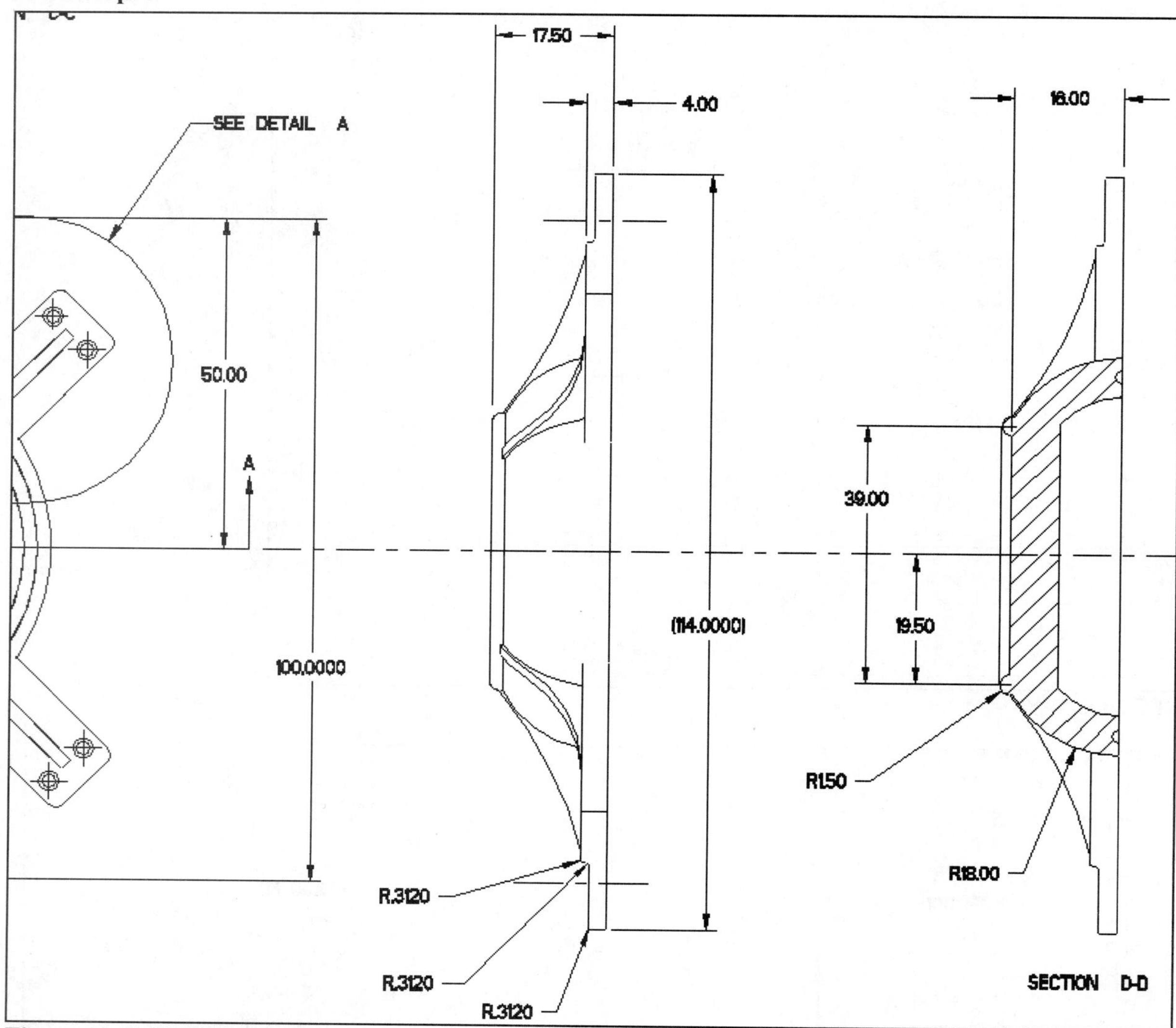

Figure 12.42 SECTION D-D

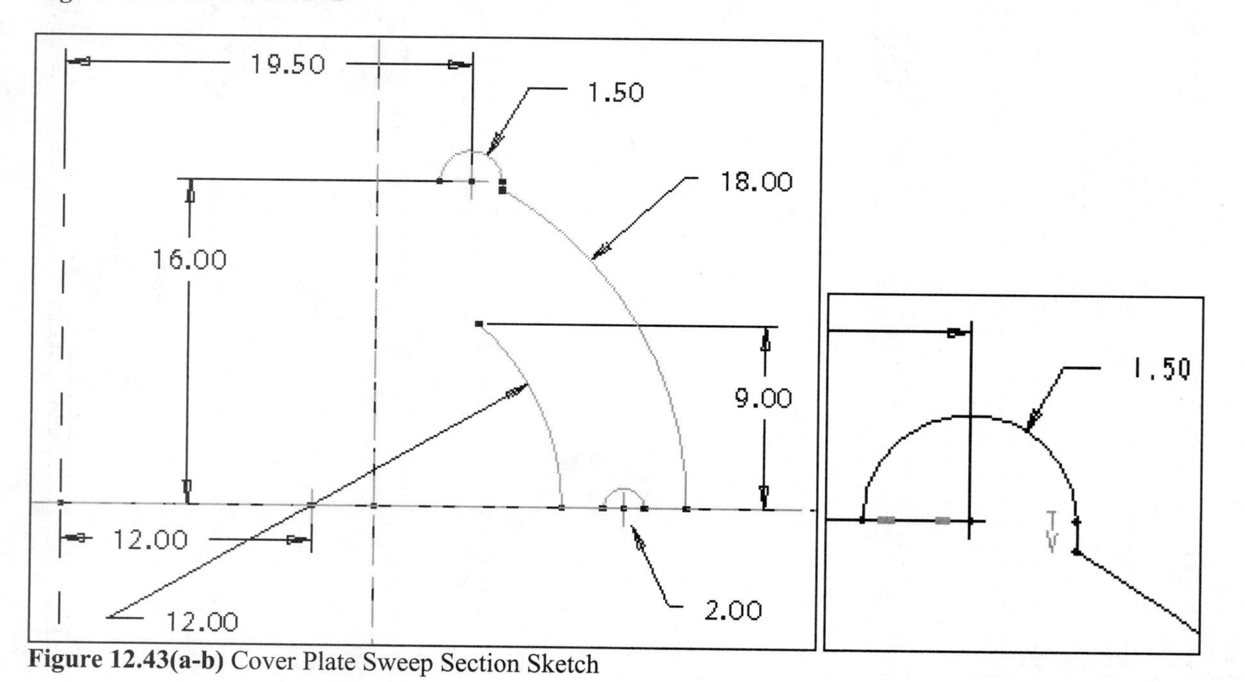

Figure 12.43(a-b) Cover Plate Sweep Section Sketch

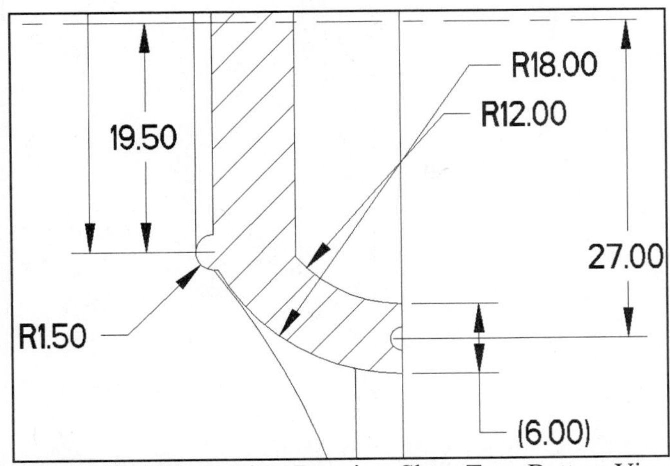

Figure 12.44 Cover Plate Drawing, Sheet Two, Bottom View

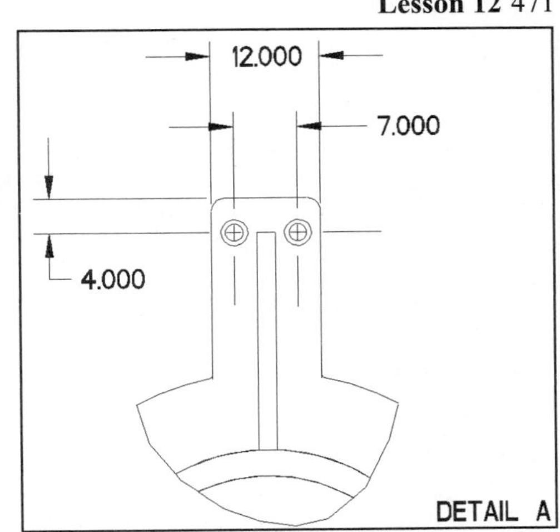

Figure 12.45 DETAIL A

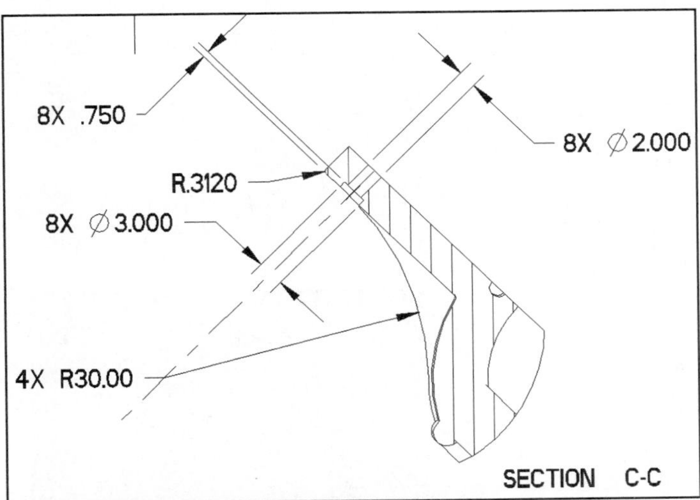

Figure 12.46 Cover Plate Drawing, SECTION C-C

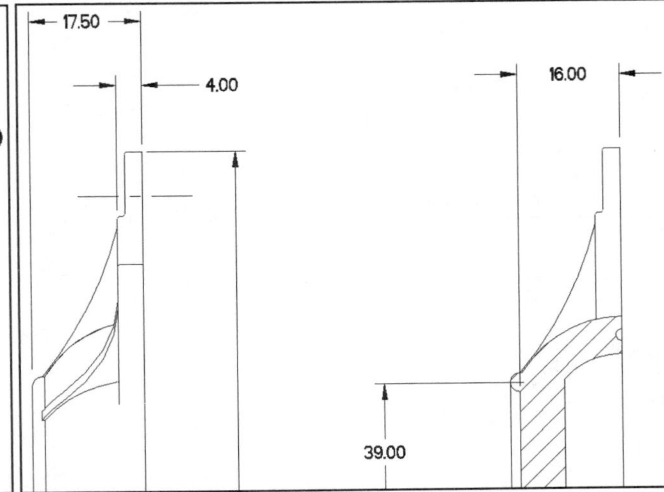

Figure 12.47 Cover Plate Drawing, Dimensions

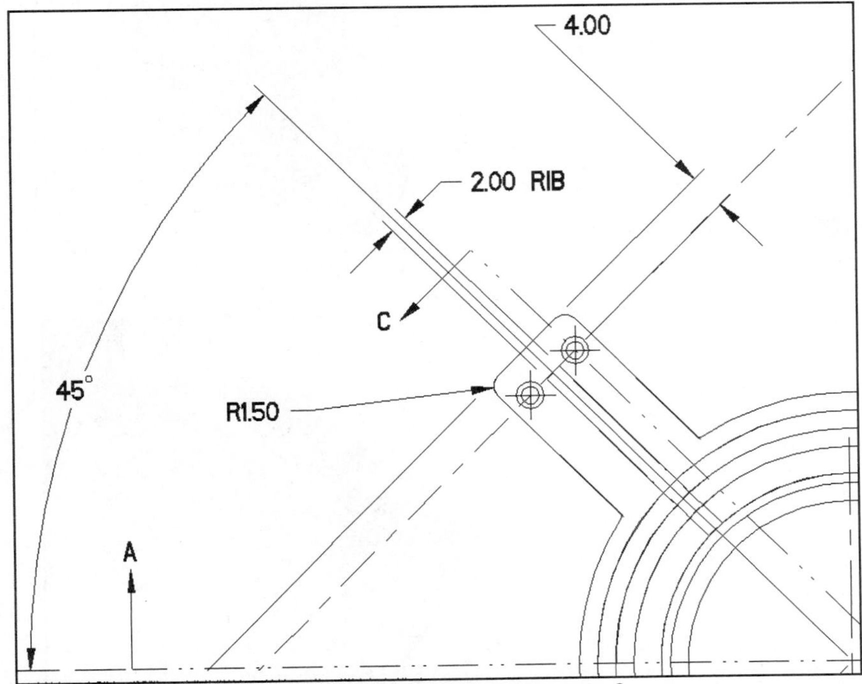

Figure 12.48 Cover Plate Drawing, Close-up of Top Left

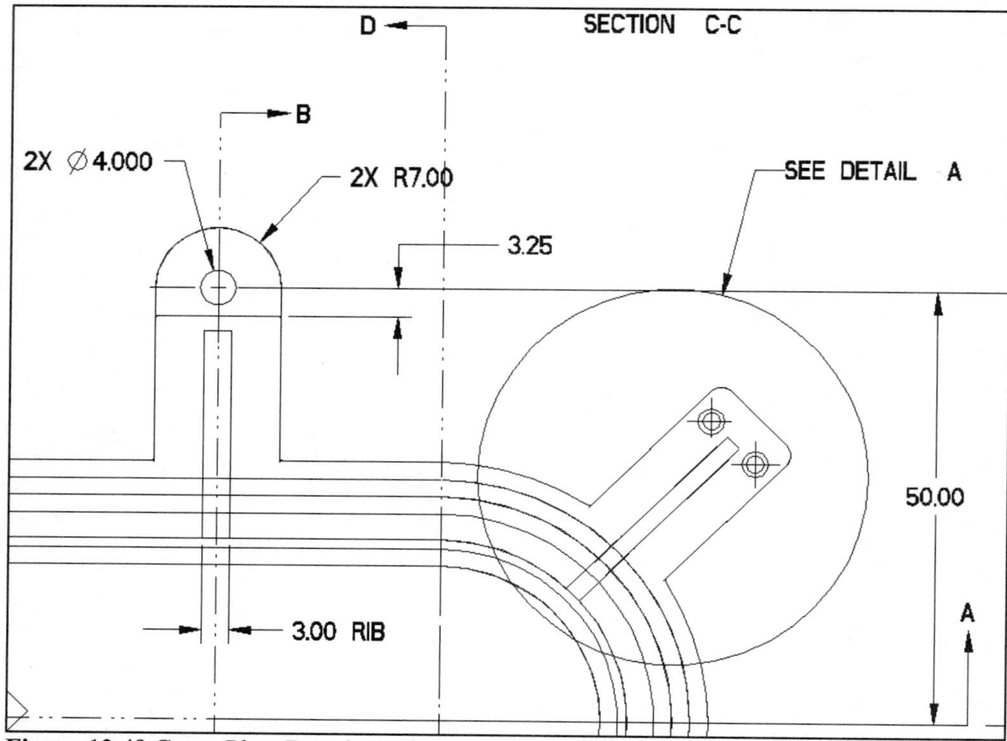

Figure 12.49 Cover Plate Drawing, Close-up of Top Right

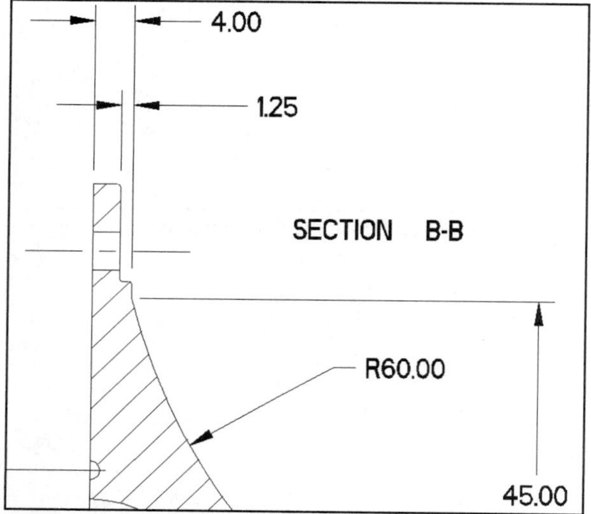

Figure 12.50(a) SECTION B-B

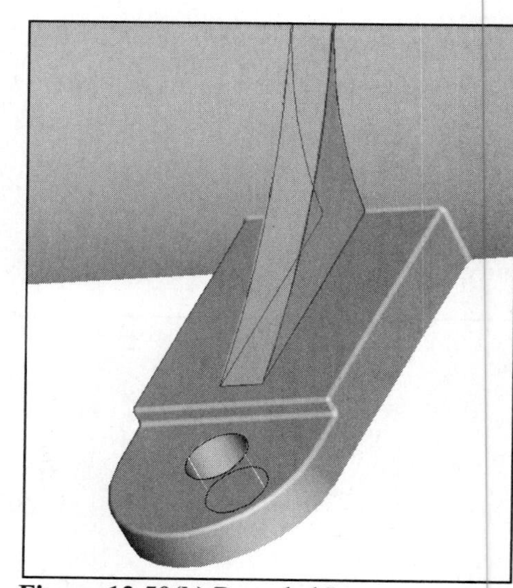

Figure 12.50(b) Rounded Leg

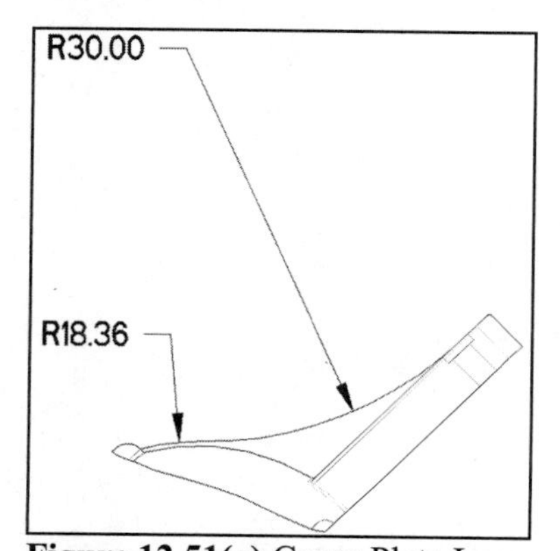

Figure 12.51(a) Cover Plate Legs

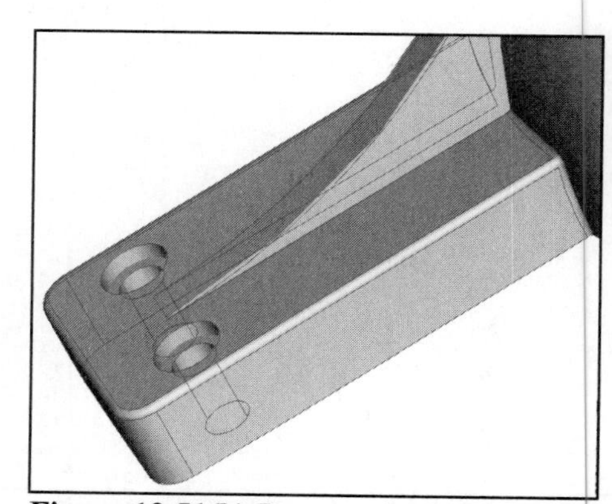

Figure 12.51(b) Rectangular Leg

Lesson 13 Blends and Splines

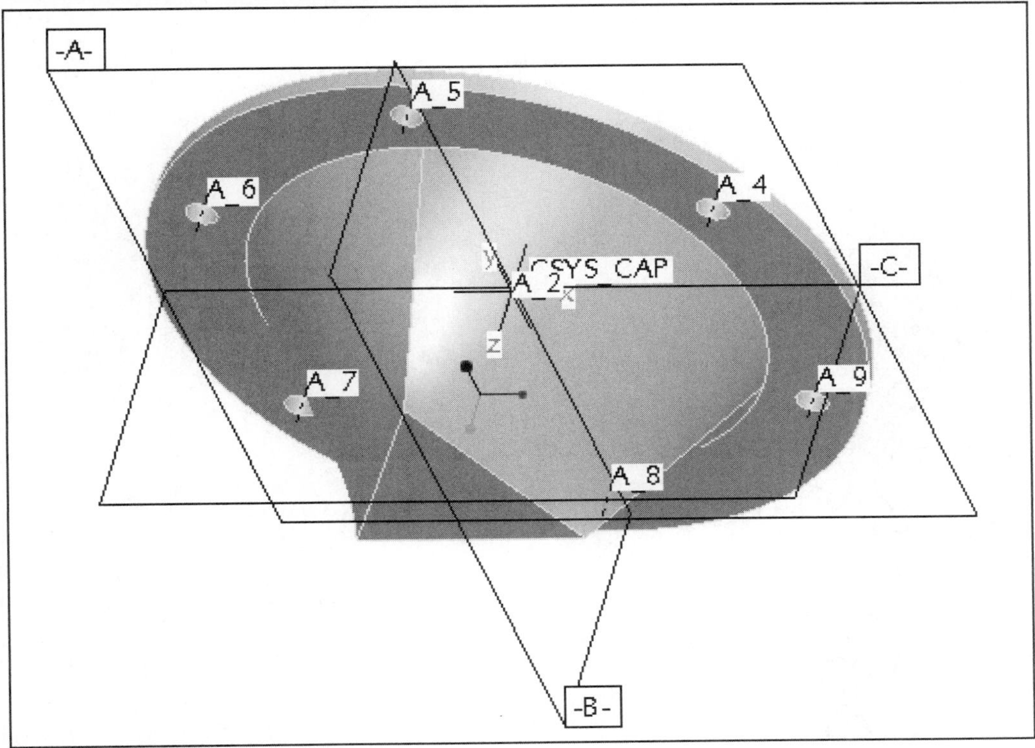

Figure 13.1 Cap

OBJECTIVES

- Create a **Parallel Blend** feature
- Use the **Shell Tool**
- Create a **Swept Blend**
- Create a **Spline** and use it as a **Trajectory** in a swept blend

BLENDS AND SPLINES

A blended feature consists of a series of at least two planar sections that are joined together at their edges with transitional surfaces to form a continuous feature. The Cap (Figs. 13.1 and 13.2) uses a blend feature in its design. A Blend can be created as a **Parallel Blend**, or you can construct a **Swept Blend**.

A spline is similar to an irregular curve and is used in a variety of designs.

Blend Sections

Blended surfaces are created between the corresponding sections. Figure 13.3 shows a parallel blend for which the *section* consists of four *subsections*. Each segment in the subsection is matched with a segment in the following subsection; to create the transitional surfaces, Pro/E connects the *starting points* of the sections and continues to connect the vertices of the sections in a clockwise manner. By changing the starting point of a blend section, you can create blended surfaces that twist between the sections. The default starting point is the first point sketched in the subsection. You can position the starting point to the endpoint of another segment by choosing the option Start Point and selecting the new position.

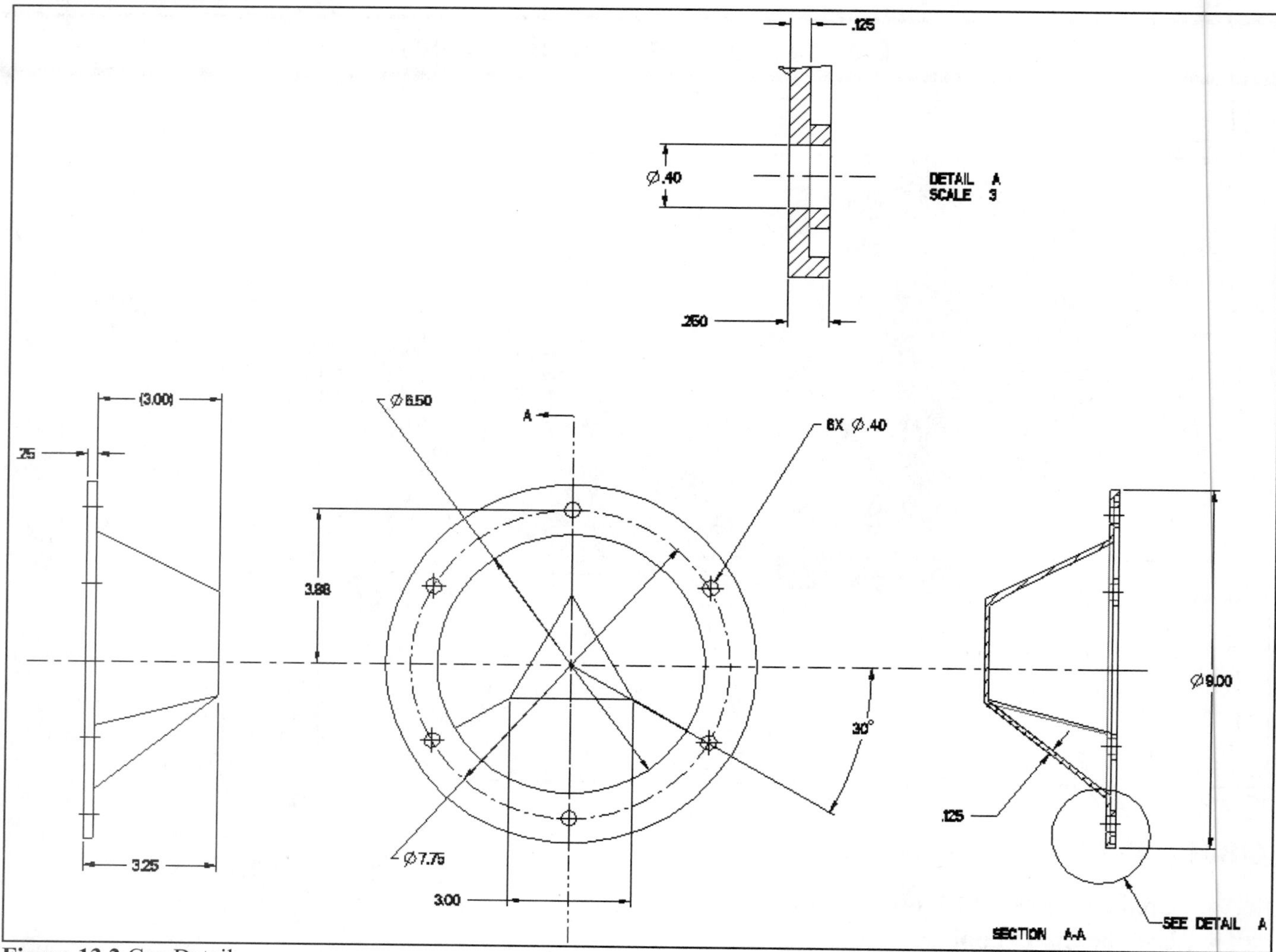

Figure 13.2 Cap Detail

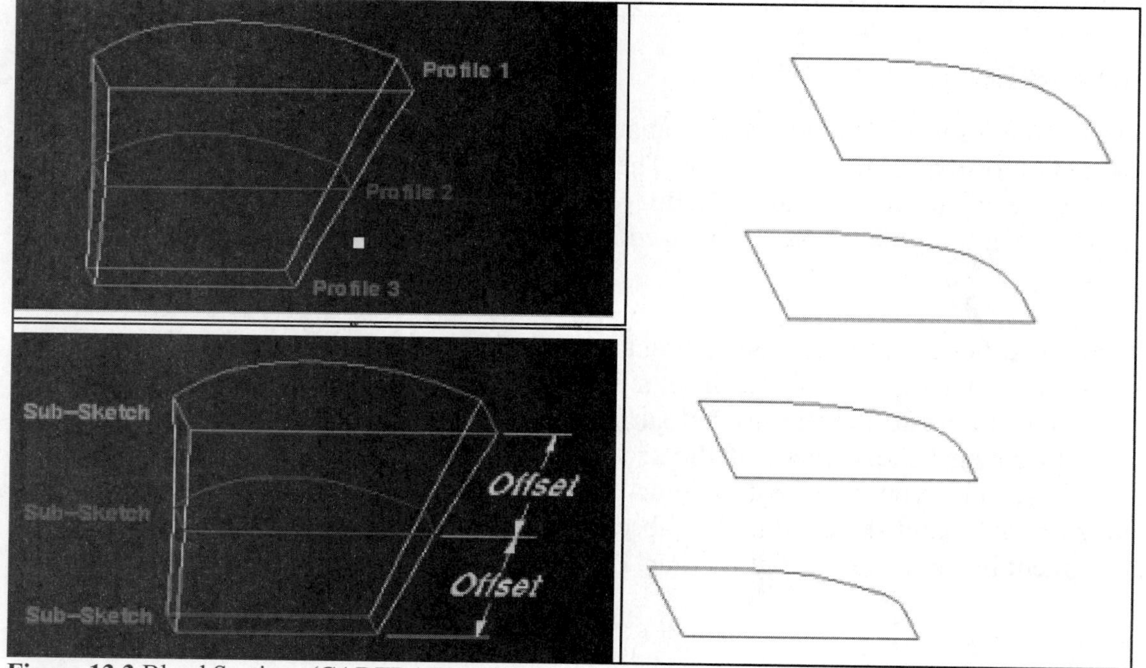

Figure 13.3 Blend Sections (CADTRAIN, COAch for Pro/ENGINEER)

Options

Blends use one of the following transitional surface options:

- **Straight** Create a straight blend by connecting vertices of different subsections with straight lines. Edges of the sections are connected with ruled surfaces.
- **Smooth** Create a smooth blend by connecting vertices of different subsections with smooth curves. Edges of the sections are connected with ruled (spline) surfaces.
- **Parallel** All blend sections lie on parallel planes in one section sketch.
- **Rotational** The blend sections are rotated about the **Y** axis, up to a maximum of **120°**. Each section is sketched individually and aligned using the coordinate system of the section.
- **General** The sections of a general blend can be rotated about and translated along the **X**, **Y**, and **Z** axes. Sections are sketched individually and aligned using the coordinate system of the section.
- **Regular Sec** The feature will use the regular sketching plane.
- **Project Sec** The feature will use the projection of the section on the selected surface. This is used for parallel blends only.
- **Select Sec** Select section entities (not available for parallel blends).
- **Sketch Sec** Sketch section entities.

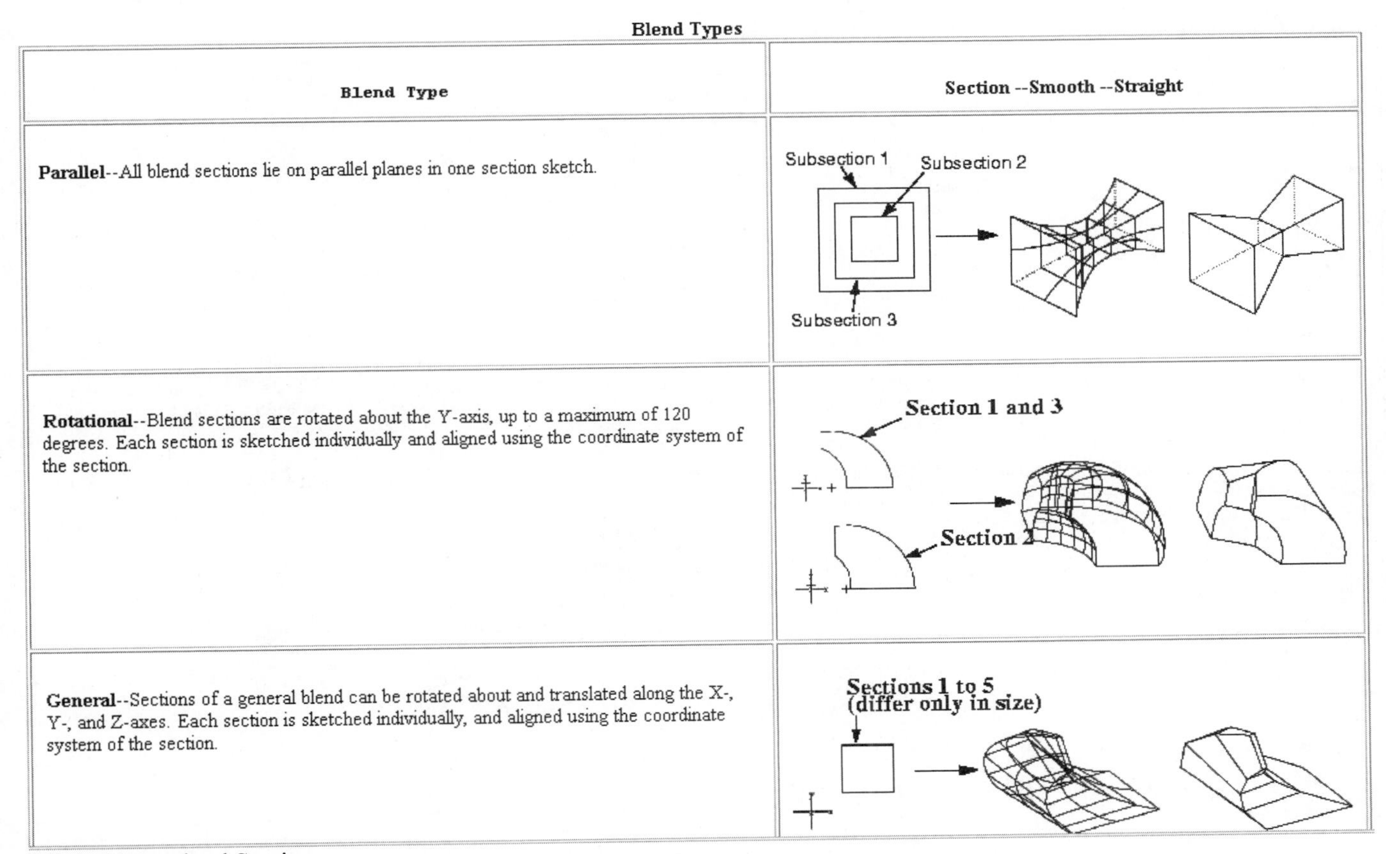

Blend Types

Blend Type	Section --Smooth --Straight
Parallel--All blend sections lie on parallel planes in one section sketch.	Subsection 1 Subsection 2 Subsection 3
Rotational--Blend sections are rotated about the Y-axis, up to a maximum of 120 degrees. Each section is sketched individually and aligned using the coordinate system of the section.	Section 1 and 3 Section 2
General--Sections of a general blend can be rotated about and translated along the X-, Y-, and Z-axes. Each section is sketched individually, and aligned using the coordinate system of the section.	Sections 1 to 5 (differ only in size)

Figure 13.4 Blend Sections

Parallel Blends

You create parallel blends using the Parallel option. A parallel blend is created from a single section that contains multiple sketches called *subsections*. A first or last subsection can be defined as a point or a blend vertex (Fig. 13.5). The starting point for each subsection must be carefully selected as per the design requirements (Fig. 13.6).

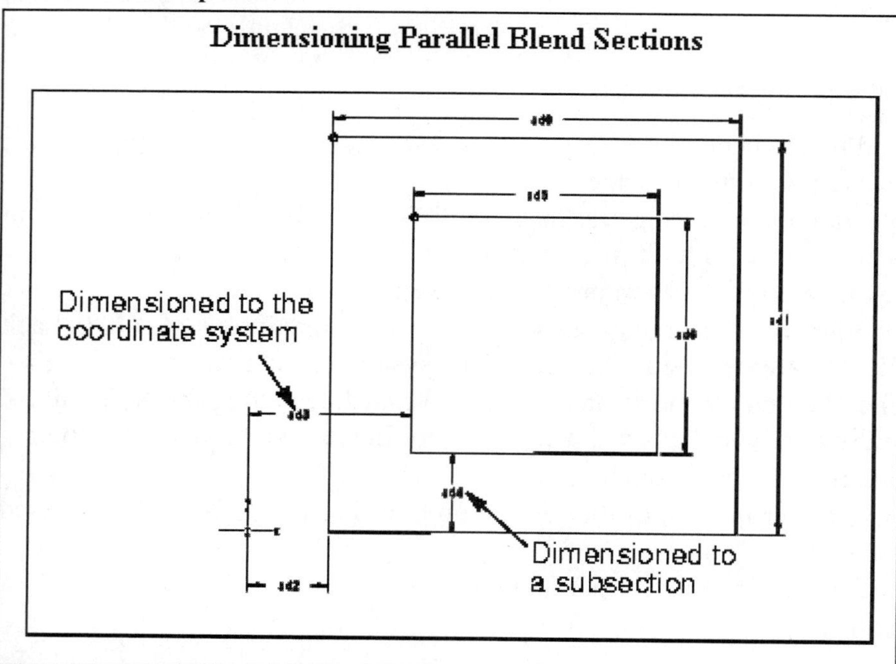

Figure 13.5 Dimensioning Parallel Blend Sections

Straight Parallel Blend

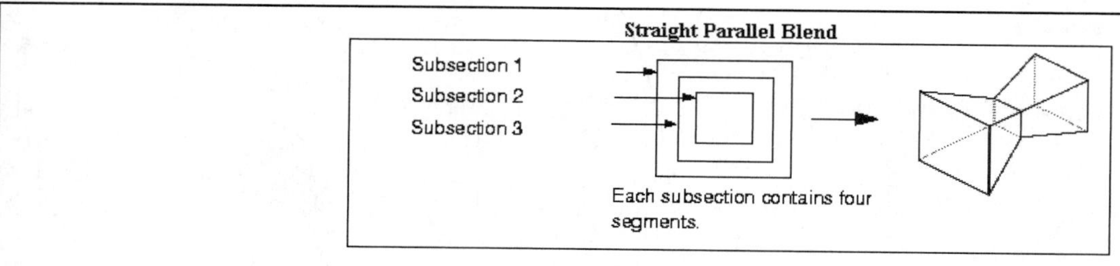

Starting Point of a Section

To create the transitional surfaces, Pro/ENGINEER connects the starting points of the sections and continues to connect the vertices of the sections in a clockwise manner. By changing the starting point of a blend subsection, you can create blended surfaces that twist between the sections (see the illustration Starting Points and Blend Shape).

The default starting point is the first point sketched in the subsection. You can place the starting point at the endpoint of another segment by choosing the option **Start Point** from the
SEC TOOLS menu and selecting the point.

Starting Points and Blend Shape

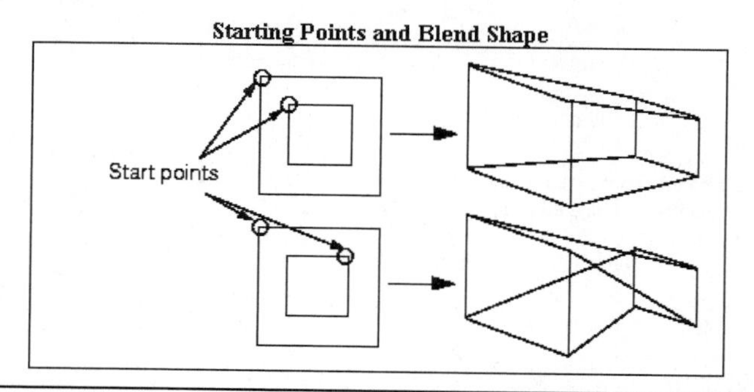

Figure 13.6 Starting Points

Swept Blends

A swept blend [Figs. 13.7(a-b)] is created using a single trajectory (a spine) and multiple sections. You create the spine of the swept blend by sketching or selecting a datum curve or an edge. Spines can be created with splines. You sketch the sections at specified segment vertices or datum points on the spine. Each section can be rotated about the **Z**-axis with respect to the section immediately preceding it.

The following restrictions apply:

- A section cannot be located at a sharp corner in the spine.
- For a closed trajectory profile, sections must be sketched at the start point and at least one other location. Pro/E uses the first section at the endpoint.
- For an open trajectory profile, you must create sections at the start point and endpoint. You cannot skip placement of a section at those points.
- Sections cannot be dimensioned to the model, because modifying the trajectory would invalidate those dimensions.
- A composite datum curve cannot be selected for defining sections of a swept blend (**Select Sec**). Instead, you must select one of the underlying datum curves or edges for which a composite curve is determined.
- If you choose **Pivot Dir** and **Select Sec**, all selected sections must lie in planes that are parallel to the pivot direction.
- You cannot use a nonplanar datum curve from an equation as a swept blend trajectory.

Creating a Swept Blend

To create a **Swept Blend**, you can define the trajectory by sketching a trajectory or by selecting existing curves and edges and extending or trimming the first and last entities in the trajectory.

The options include:

- **Select Sec** Select existing curves or edges to define each section.
- **Sketch Sec** Sketch new section entities to define each section.
- **NrmToOriginTraj** The section plane remains normal to the *Origin Trajectory* throughout its length. The generic **Sweep** behaves this way.
- **Pivot Dir** The section remains normal to the *Origin Trajectory* as it is viewed along the *Pivot direction.* The upward direction of the section remains parallel to the *Pivot Direction.*
- **Norm To Traj** Two trajectories must be selected to determine the location and the orientation of the section. The *Origin Trajectory* determines the origin of the section along the length of the feature. The section plane remains normal to the *Normal Trajectory* along the length of the feature.
- **Sketch Traj** Sketch a spine. The spine can have sharp corners (a discontinuous tangent to the curve), except at the endpoint of a closed curve. At nontangent vertices, Pro/E miters the geometry as in constant section sweeps.
- **Select Traj** Define the spine trajectory using existing curves and edges.

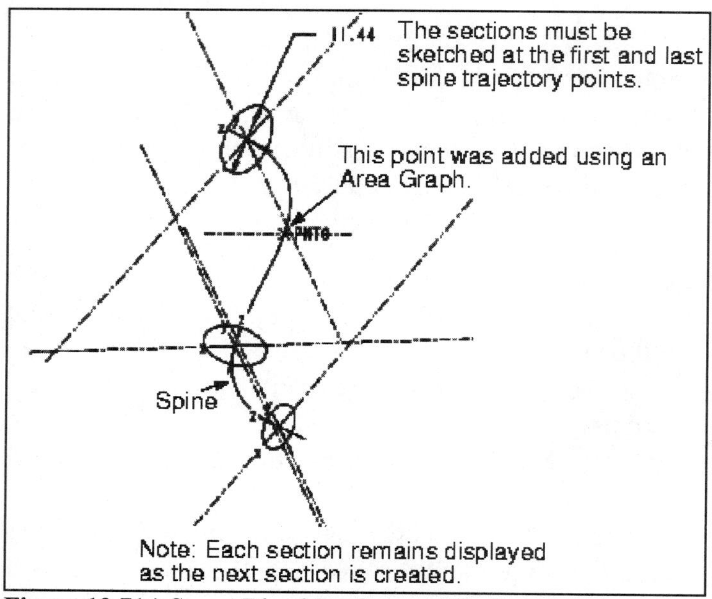

Figure 13.7(a) Swept Blend Sections and Trajectory

Figure 13.7(b) Swept Blend

Splines

Sketching a **Spline** is similar to drawing an irregular curve (Fig. 13.8). Splines are created by picking or sketching a series of specific points. Spline options include:

- **Sketch Points** Create a spline by picking screen points for the spline to pass through.
- **Select Points** Create a spline by selecting existing Sketcher points. Once the point has been selected, there is no further link between the point and the spline.

Splines

Splines are curves that smoothly pass through any number of intermediate points. The tangency angle and radius of curvature can be set at the ends of a spline to control its shape further.

Spline Curve

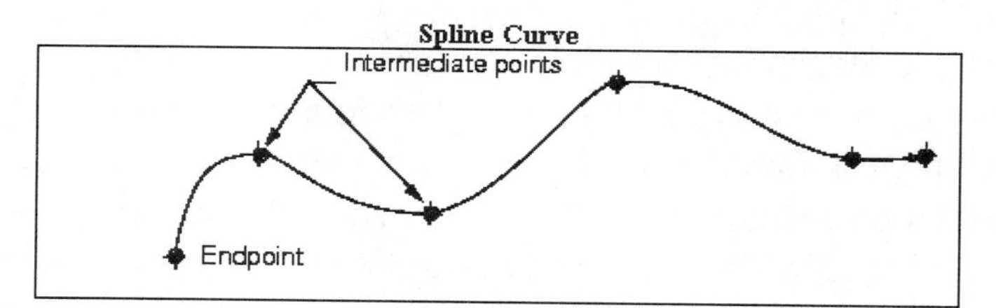

You can define spline tangency for the endpoints before you create the spline using the Tangency menu options, or by modifying the spline after it is sketched.

If the spline is to be tangent to other geometry, the sketched geometry does not have to be present when you first sketch the spline. However, when the section is regenerated, either adjacent entity or angular dimensions must exist. The tangency to the sketched entities will not actually be displayed until the section is regenerated.

Conversely, if a spline endpoint is dimensioned with an angular dimension and the endpoint has not been defined with a tangency, you must add the tangency, or remove the dimension. You must also set tangency if you are controlling curvature of the spline at its endpoints with curvature dimensions.

You can modify tangency conditions after the spline has been created (see Modifying the Tangency of a Spline).

A closed spline must have a tangency condition of **None** and will be made tangent at its endpoints. Closed splines that are non-tangent at their endpoints cannot be created.

In a spline with one or more tangent endpoints, if you move the first or last interior point on the spline, the spline immediately adjusts to its new shape.

Creating a Spline

The following procedure explains how to create a spline.

➤ **How to create a spline**

Figure 13.8 Splines

Lesson 13 STEPS

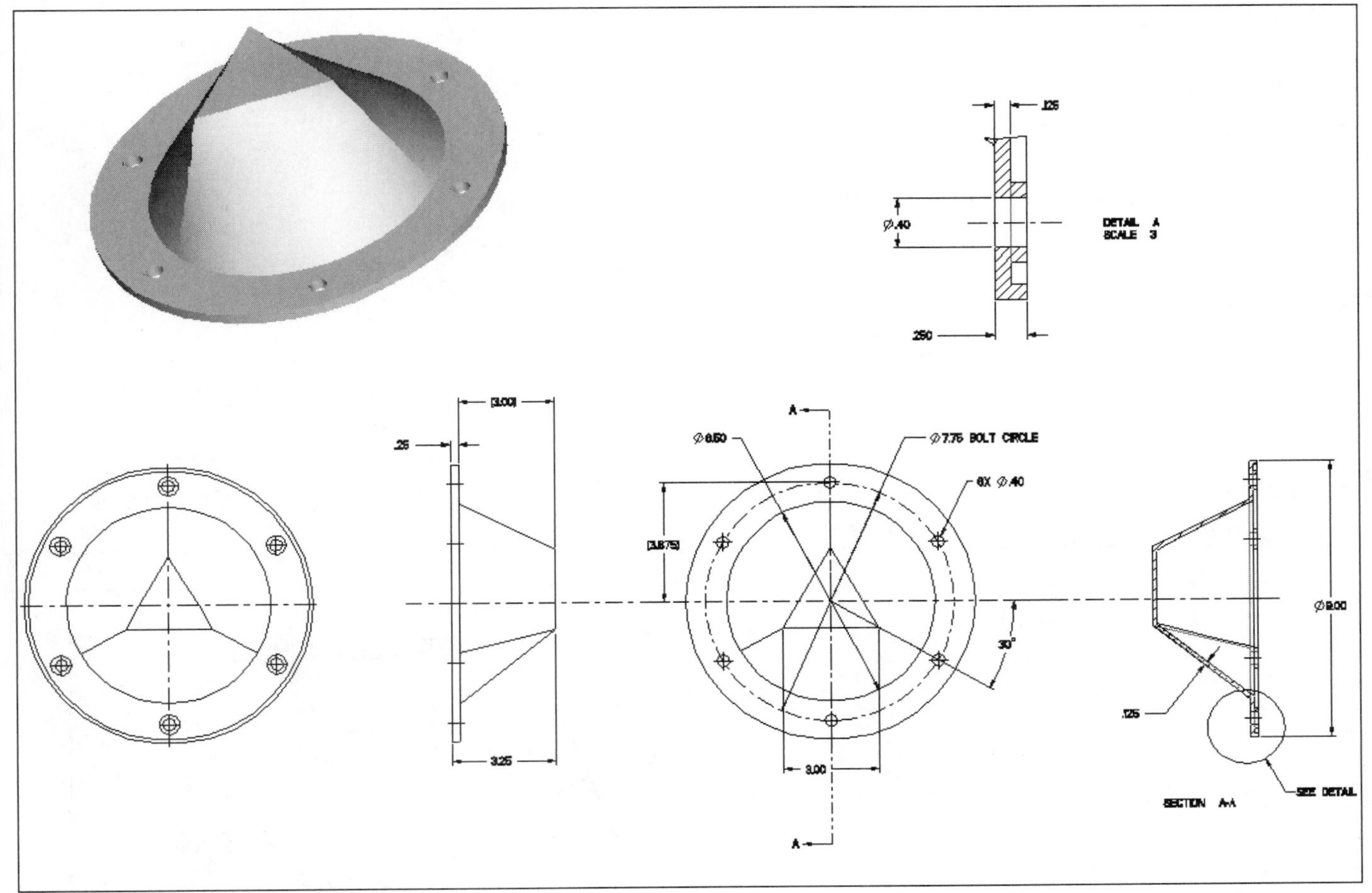

Figure 13.9 Cap Drawing

Cap

The Cap is a part created with a Parallel Blend (Figs. 13.9 through 13.16). The blend sections are a circle and a triangle. Because *the sections of a blend must have equal segments*, the "circle" is actually three equal arcs. The part is shelled as the last feature in its creation. The Shell Tool will create *bosses* around each hole as it hollows out the part. For this part, a cross section will be created in the Part mode to be used when you are detailing the Cap in the Drawing mode.

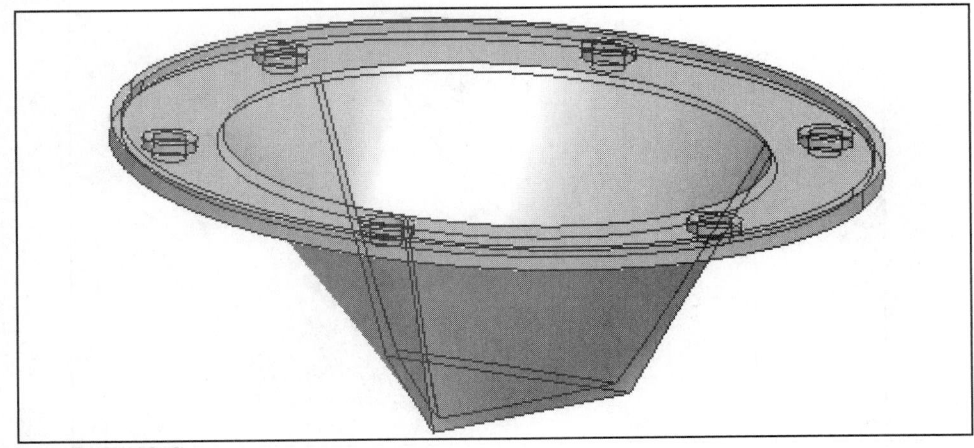

Figure 13.10 Cap

Start a new part. Click: **Create a new object** ⇒ ●**Part** ⇒ Name **CAP** ⇒ Use default template ⇒ **MMB** ⇒ **Edit** ⇒ **Setup**

- **Material** = PLASTIC
- **Units** = Inch lbm Second

Set Datum and **Rename** the default datum planes and coordinate system:
- Datum TOP = **C**
- Datum FRONT = **A**
- Datum RIGHT = **B**
- Coordinate System = **CSYS_CAP**

Tools ⇒ **Options**
- ***sketcher_dec_places*** ***3***
- ***default_dec_places*** ***3***
- ***tol_mode*** ***plusminus***

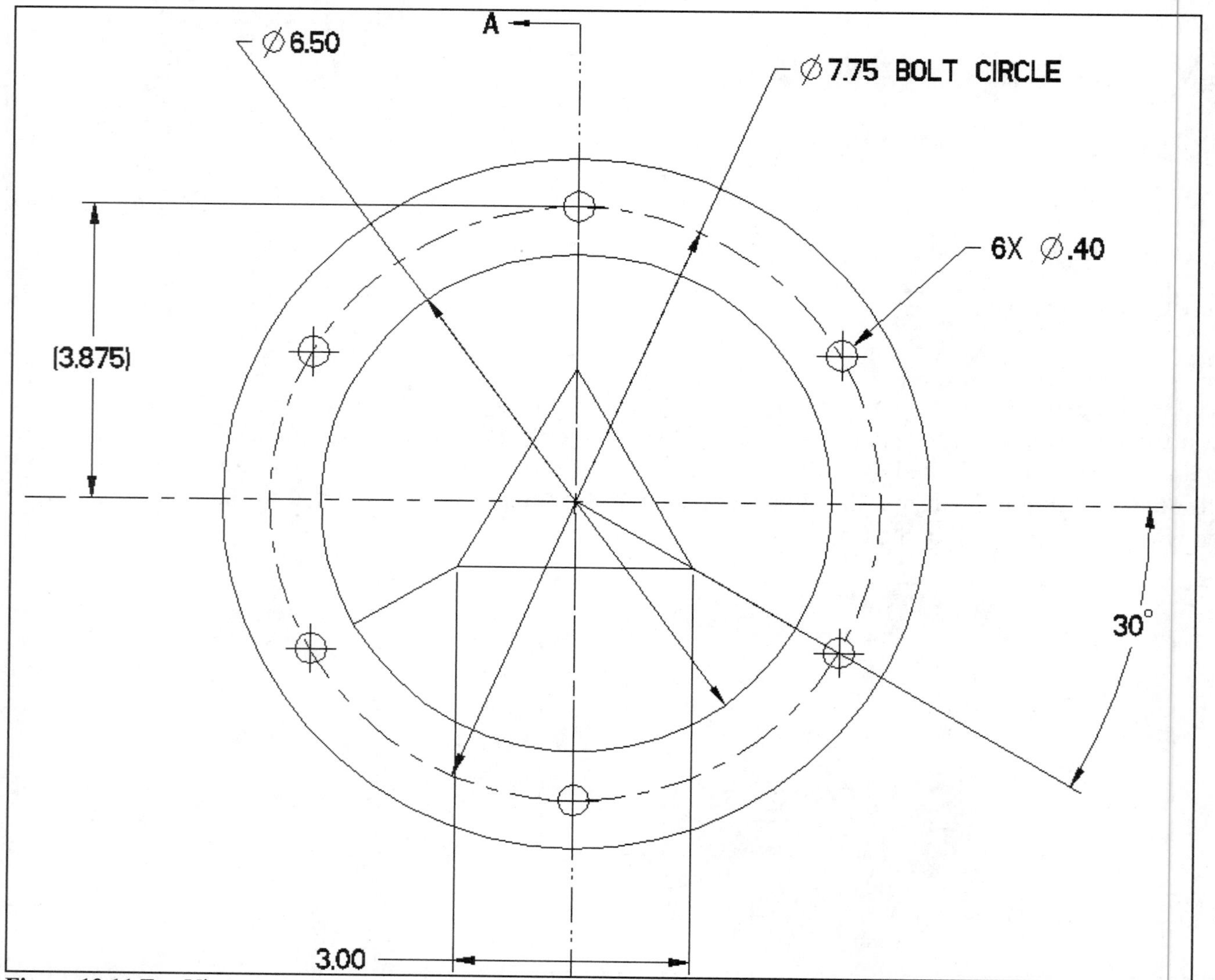

Figure 13.11 Top View

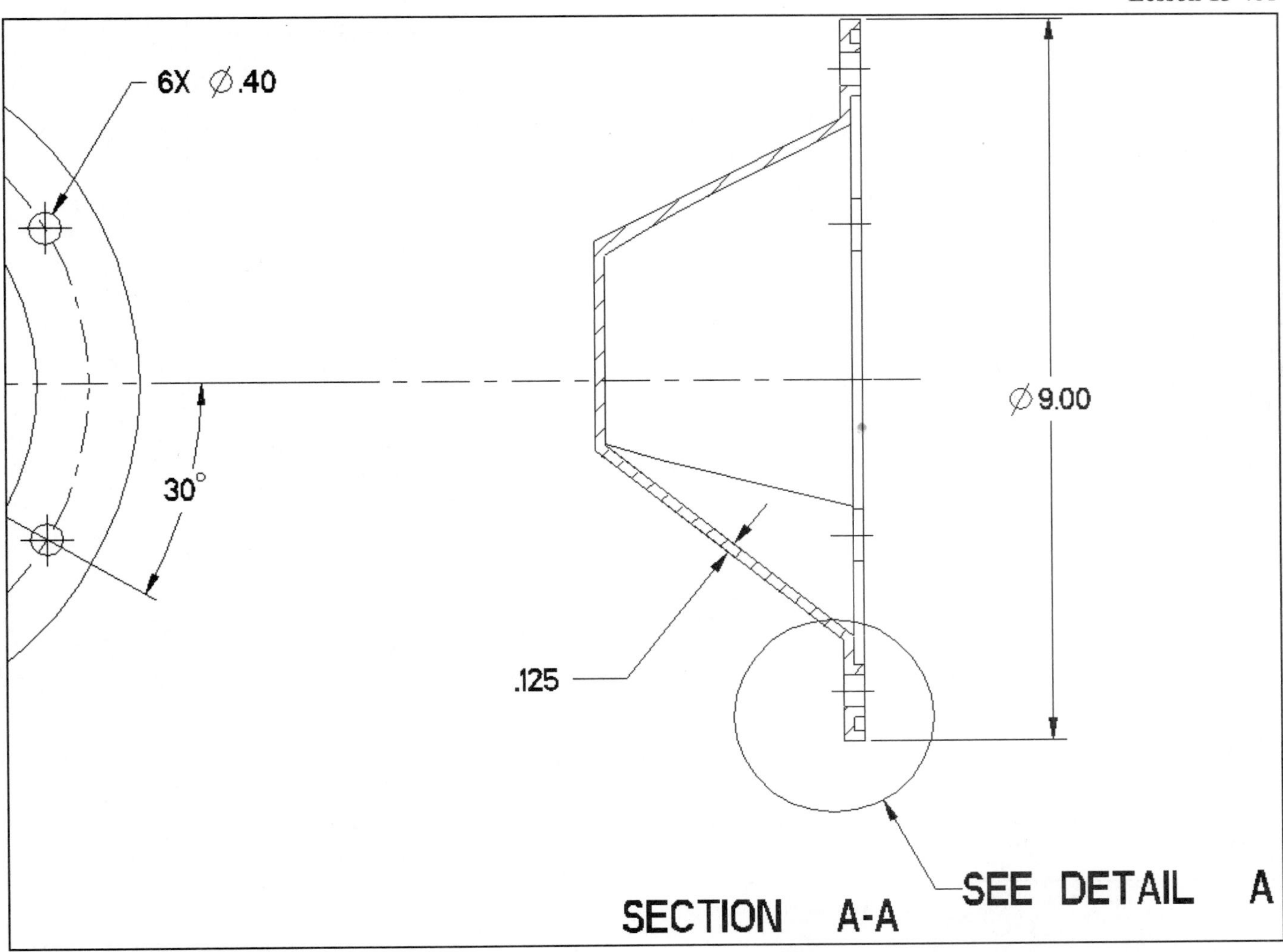

Figure 13.12 SECTION A-A

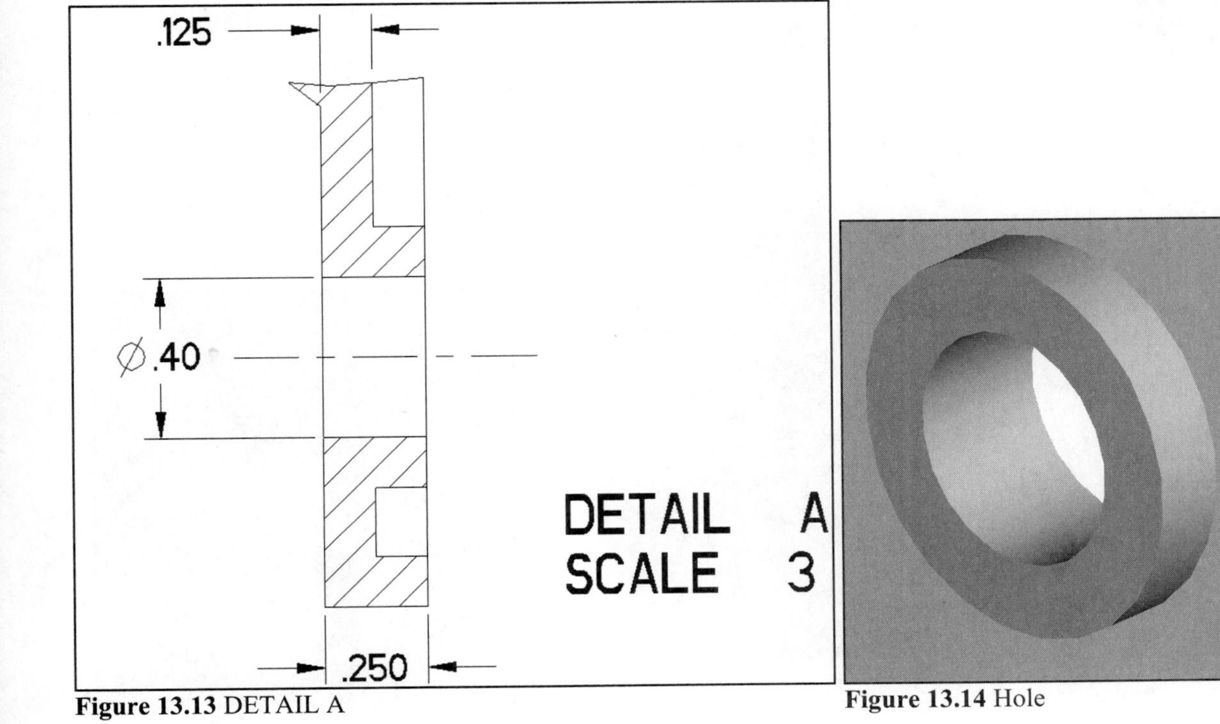

Figure 13.13 DETAIL A

Figure 13.14 Hole

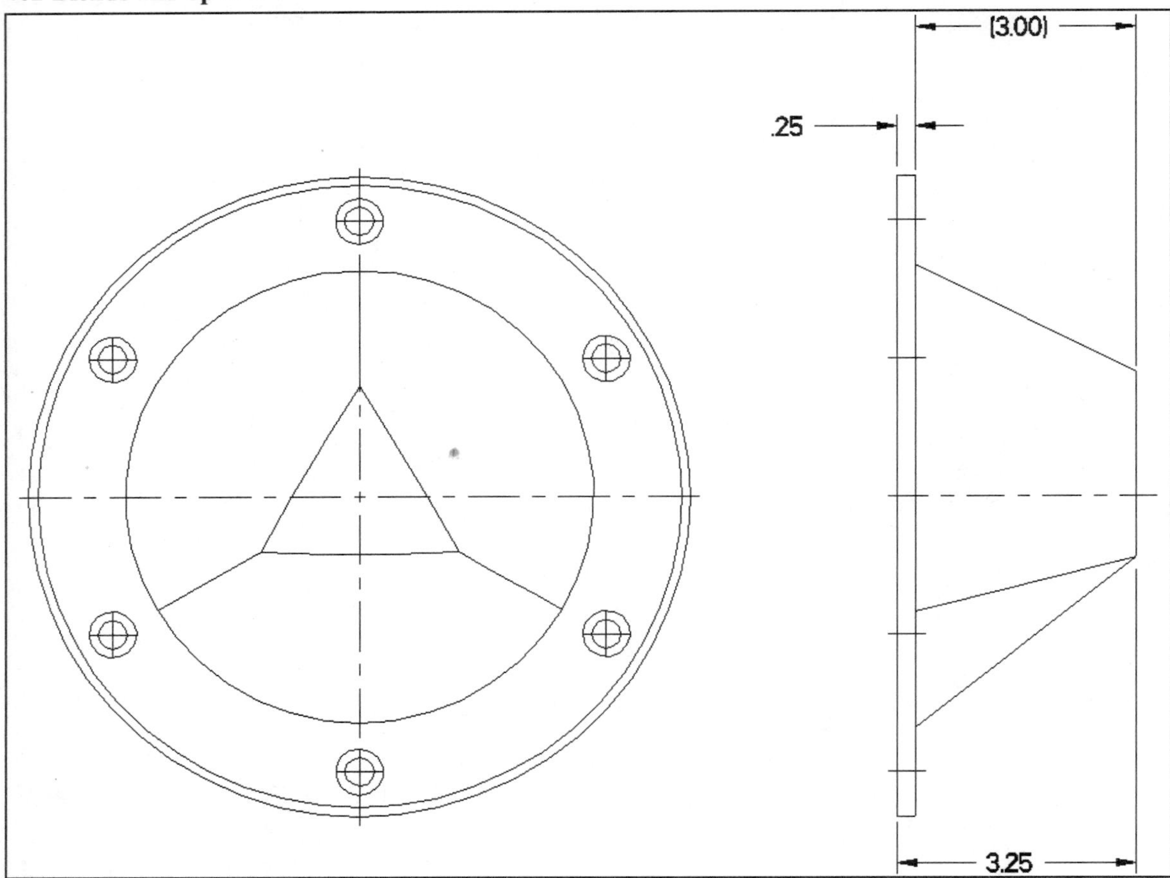

Figure 13.15 Bottom View

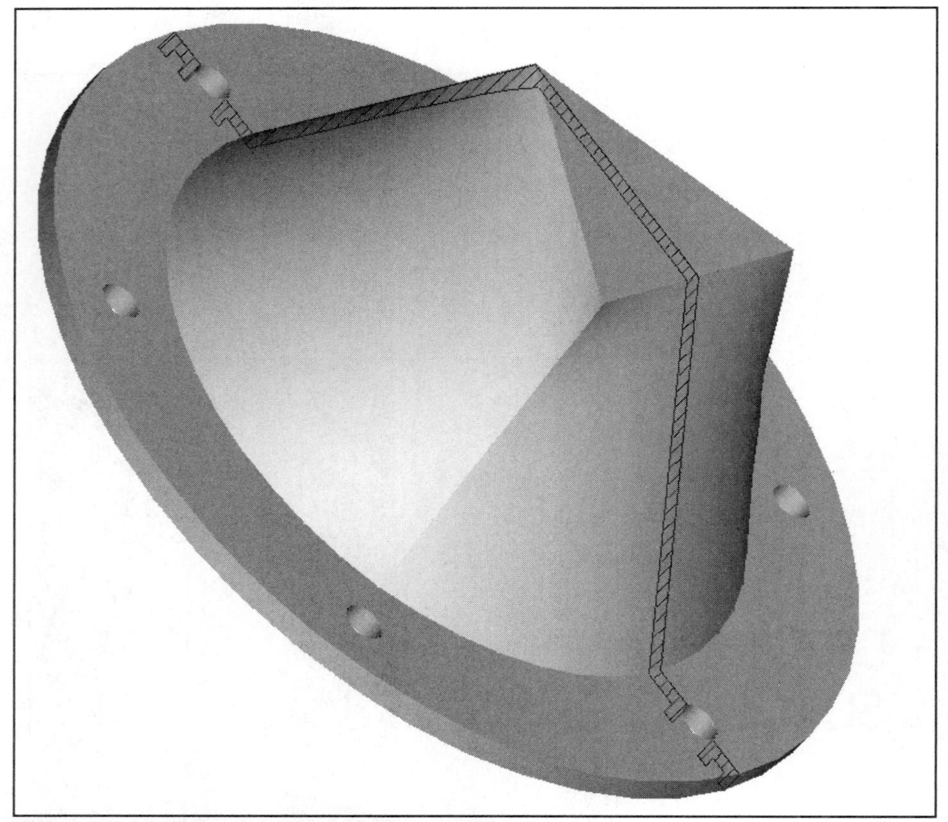

Figure 13.16(a) Section Through Hole and Boss

Figure 13.16(b) Hole and Boss

Model the circular protrusion that is **∅9.00** by **.25** thick shown in Figure 13.17. Sketch the first protrusion on datum **A** (**FRONT**) and centered on **B** (**RIGHT**) and **C** (**TOP**). The **Blend** feature is modeled next (Fig. 13.18).

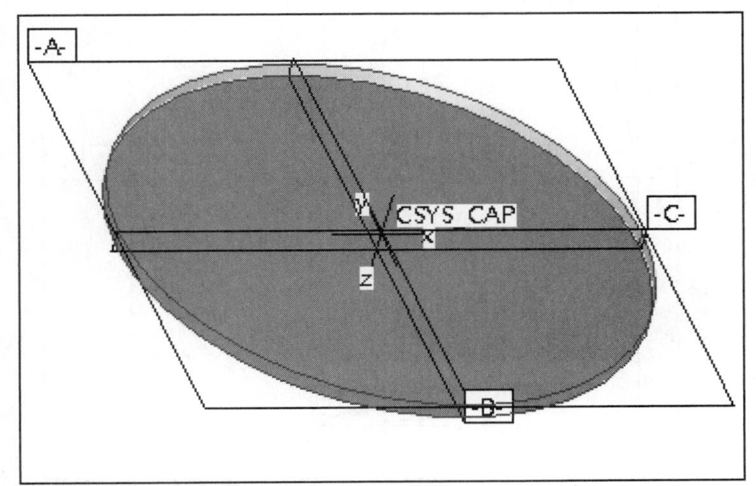

Figure 13.17 First Protrusion

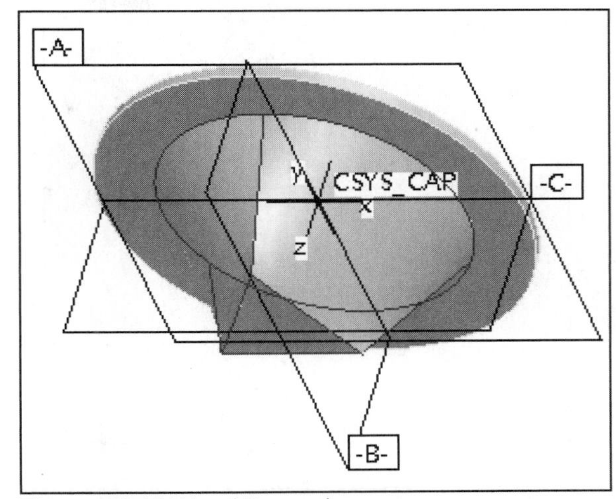

Figure 13.18 Blend Protrusion

Create the blend protrusion, click: **Insert ⇒ Blend ⇒ Protrusion ⇒ Parallel ⇒ Regular Sec ⇒ Sketch Sec ⇒ MMB ⇒ Straight ⇒ MMB ⇒** pick the top surface of the first protrusion (Fig. 13.19) **⇒ MMB** to confirm direction of feature creation **⇒ MMB** default orientation **⇒ MMB ⇒ MMB** closes References dialog box ⇒ **Toggle the grid on** (Fig. 13.20)

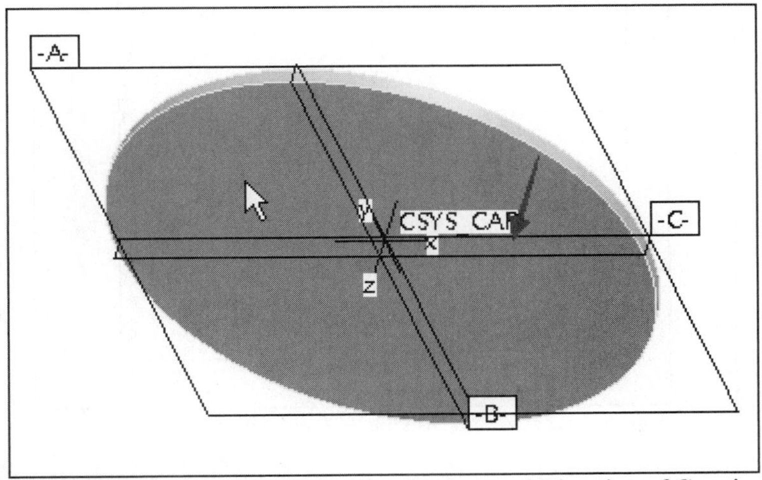

Figure 13.19 Blend Feature Starting Surface and Direction of Creation

The section grid would be better utilized if it were a Polar grid rather than a Cartesian grid (Fig. 13.20). Change the grid type and size, click: **Sketch ⇒ Options ⇒ Display** tab ⇒ Snap to Grid ⇒ **Parameters** tab ⇒ Polar ⇒ Grid Spacing **Manual** ⇒ Radial **.50** ⇒ Angle **30°** ⇒ ✔ ⇒ **Hidden Line** (Fig. 13.21) ⇒ add three centerlines (Figs. 13.22 to 13.25)

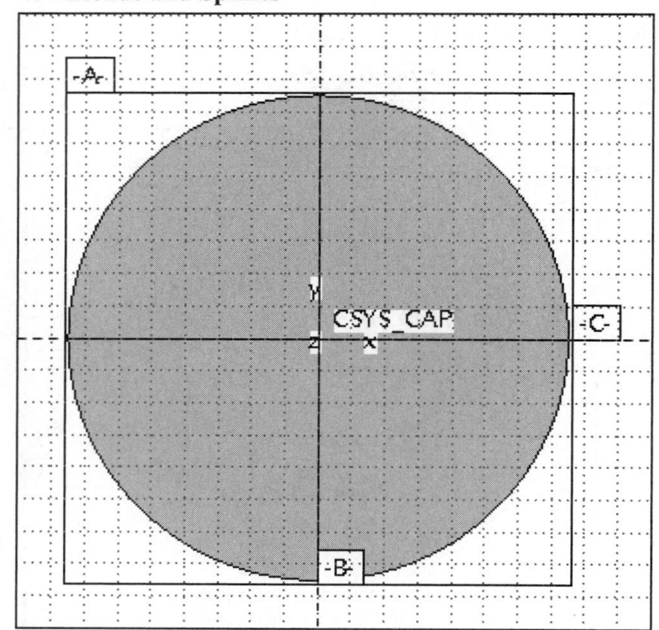

Figure 13.20 Sketcher Showing Cartesian Grid

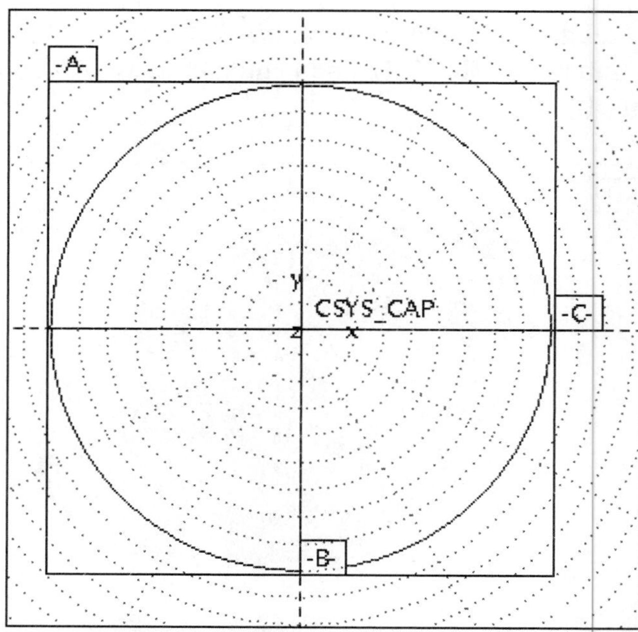

Figure 13.21 Sketcher Showing Polar Grid

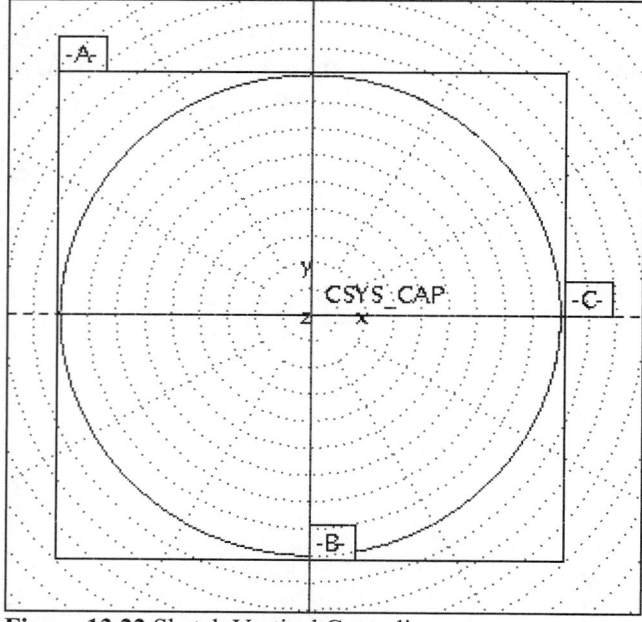

Figure 13.22 Sketch Vertical Centerline

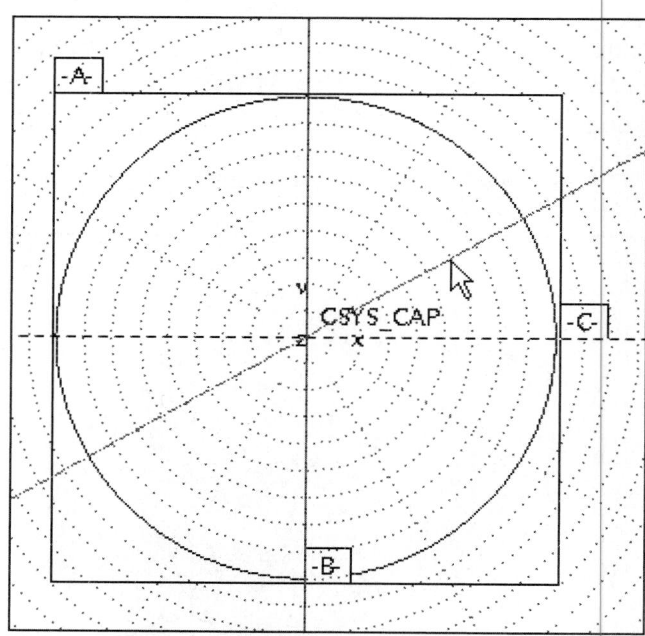

Figure 13.23 Sketch **30** Degree Centerline

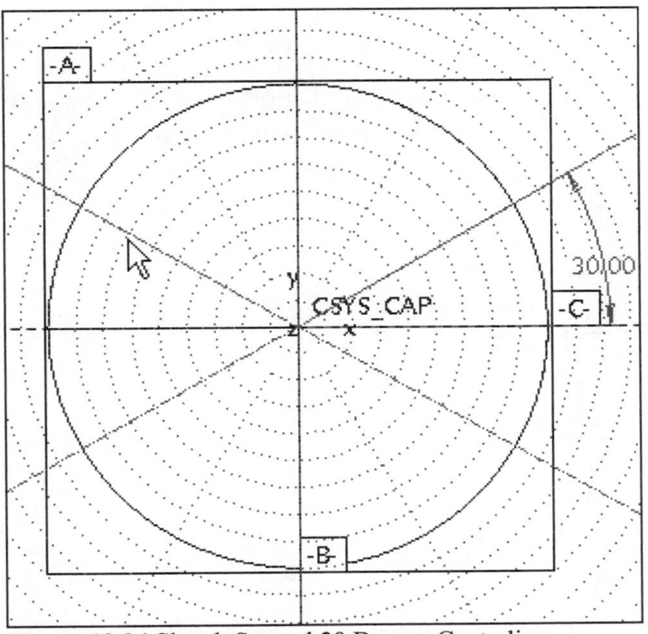

Figure 13.24 Sketch Second **30** Degree Centerline

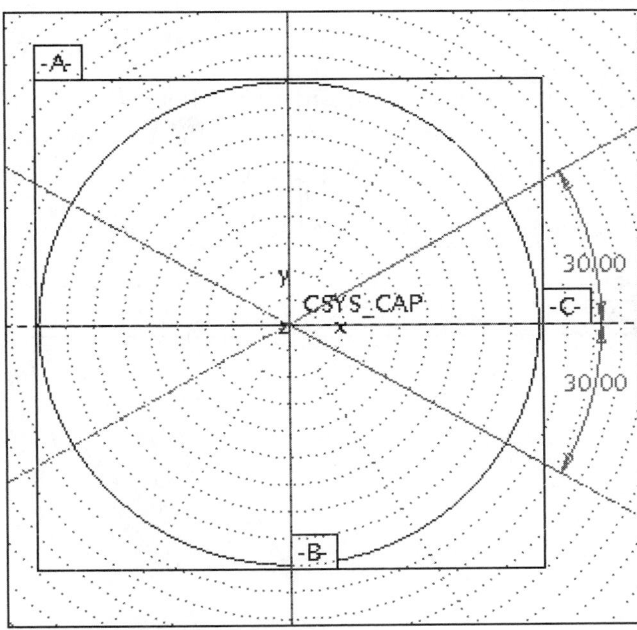

Figure 13.25 Three Centerlines

Sketch the *first section* of the blend by creating *three* equal **120°** arcs, click: **Create an arc by picking its center and endpoints** sketch each arc in a counterclockwise direction. Note the location of the start point for this section. (Fig.13.26) ⇒ add a diameter dimension (Fig. 13.26)

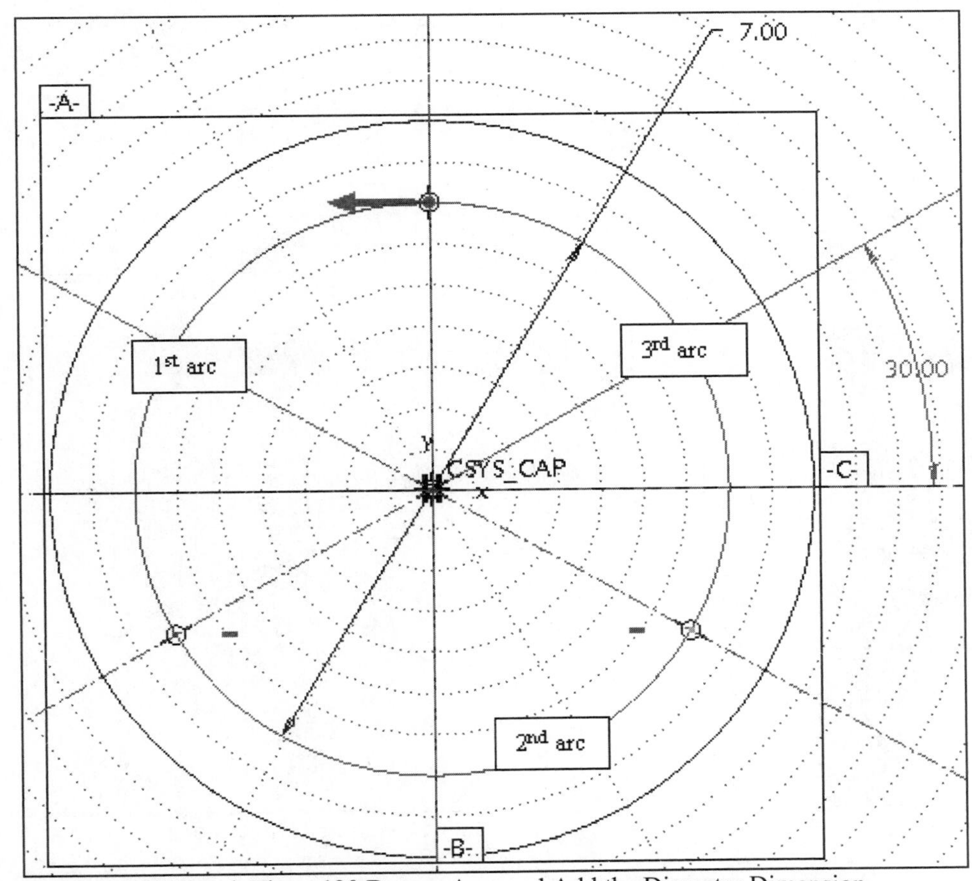

Figure 13.26 Sketch Three **120** Degree Arcs and Add the Diameter Dimension

Click: **RMB** ⇒ **Toggle Section** to sketch the second parallel section (the first section is *grayed out* ⇒ [icon] ⇒ [icon] **Create 2 point lines** sketch the three lines of the triangle starting at the top so that the start point is the same as the first section and picking points *in the same direction* in which the arcs were created [Fig. 13.27(a-b)] ⇒ dimension and modify [Fig. 13.28(a-b)] ⇒ add the two **120** degree dimensions (Fig. 13.29) ⇒ [icon]

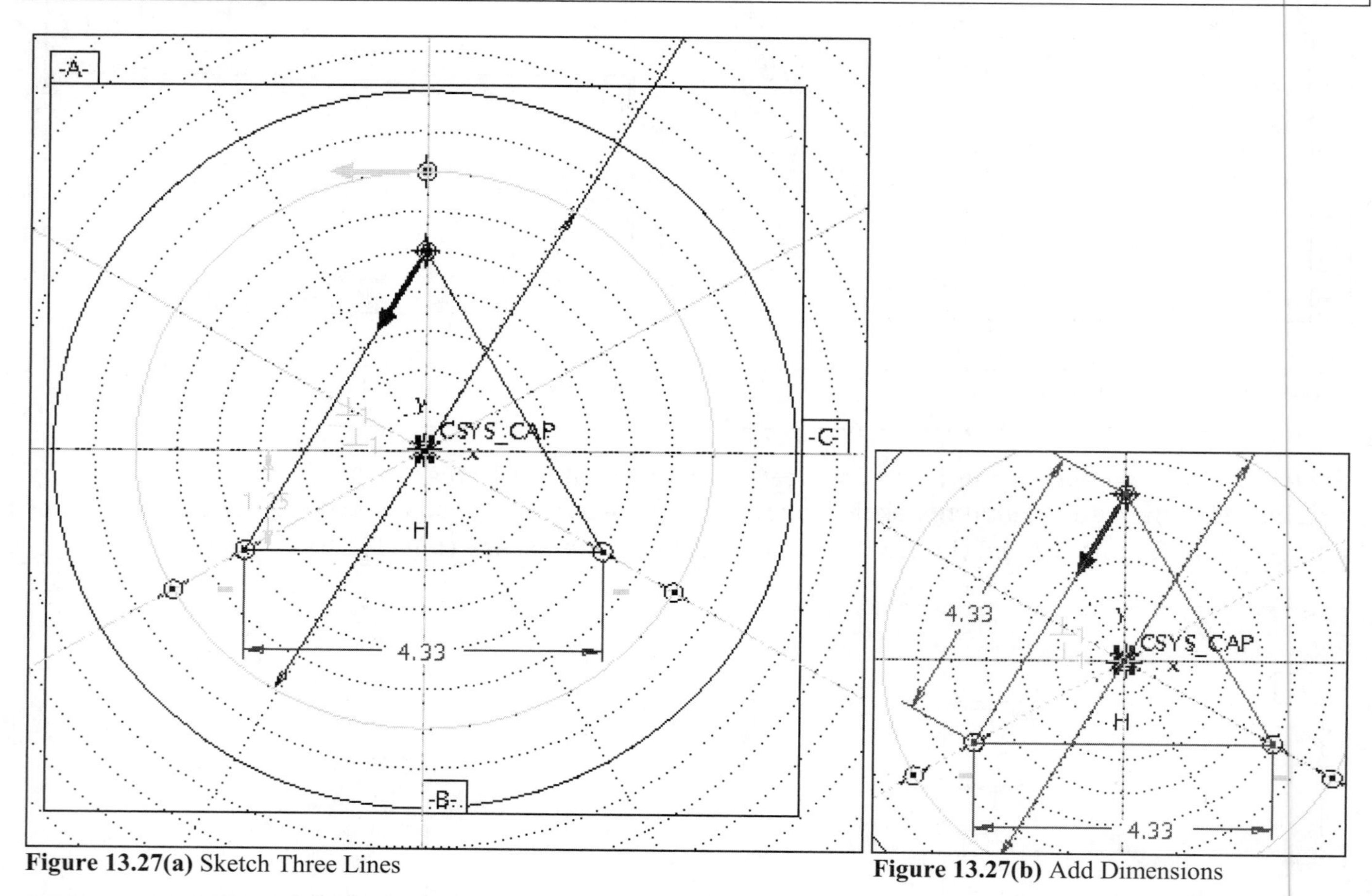

Figure 13.27(a) Sketch Three Lines

Figure 13.27(b) Add Dimensions

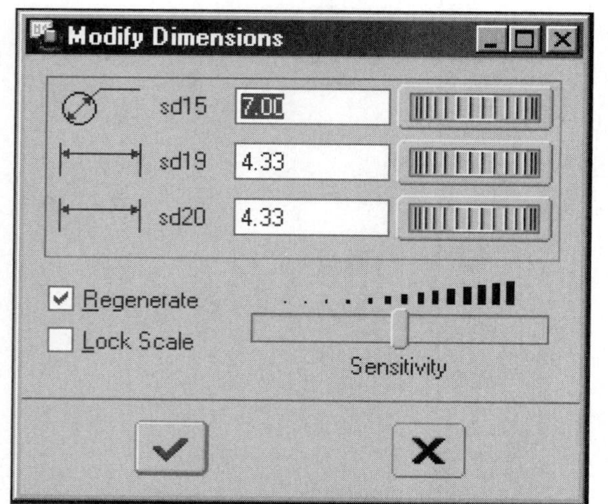

Figure 13.28(a) Modify the Dimensions

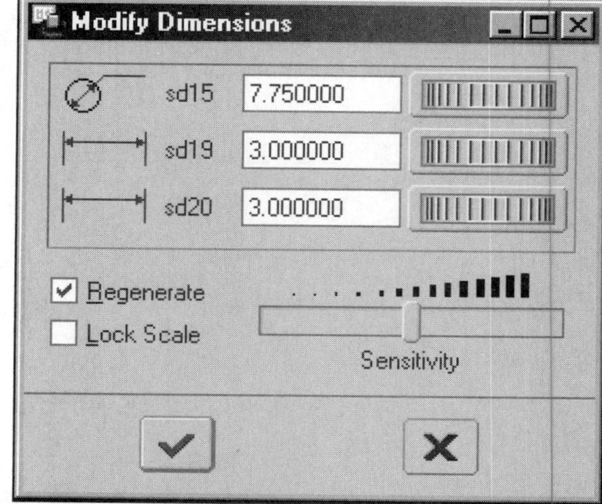

Figure 13.28(b) Modified Dimensions

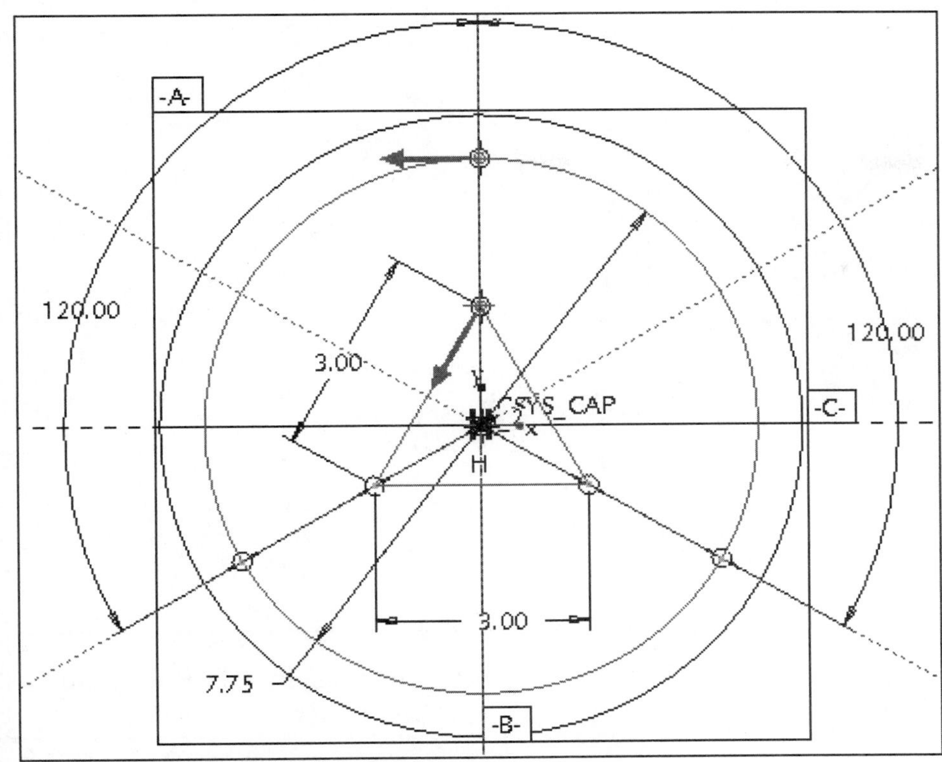

Figure 13.29 Add the **120** Degree Dimensions

Blind ⇒ **MMB** ⇒ type **3.00** at the prompt ⇒ **MMB** ⇒ **Preview** ⇒ **MMB** ⇒ with the blend highlighted on the model and Model Tree, click **RMB** ⇒ **Edit** ⇒ double-click on the **7.75** dimension and change the diameter to **6.50** (Fig. 13.30) ⇒ **Enter** ⇒ **Regenerates Model** ⇒ ⇒ **MMB** (Fig. 13.31)

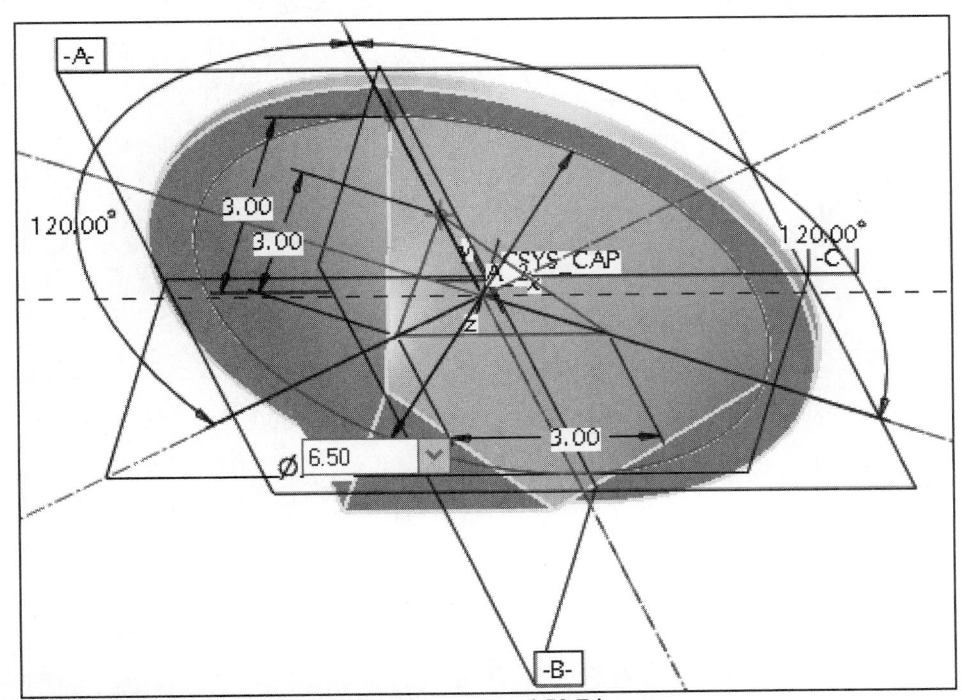

Figure 13.30 Edit the Blend, **7.75** Diameter to **6.50** Diameter

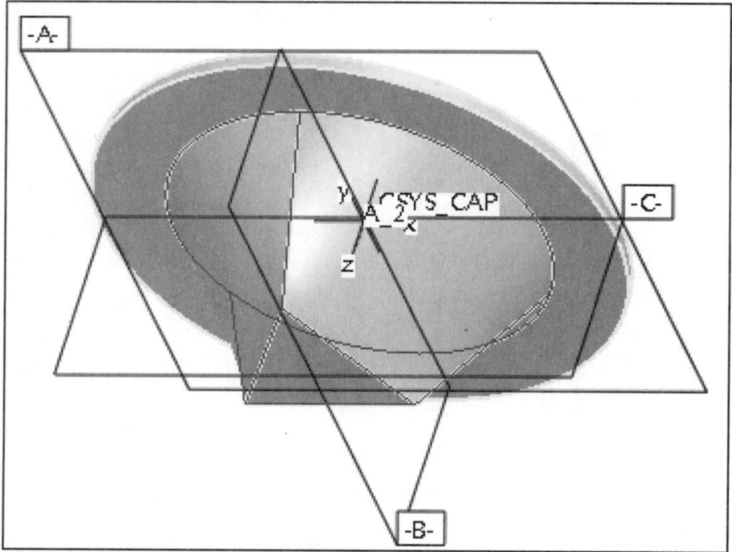

Figure 13.31 Completed Blend

Create and pattern the six equally spaced holes **∅.400** on a **∅7.75** bolt circle, click: **Hole Tool** ⇒ complete the hole with the options and references provided in Figure 13.32 ⇒ **LMB** to deselect

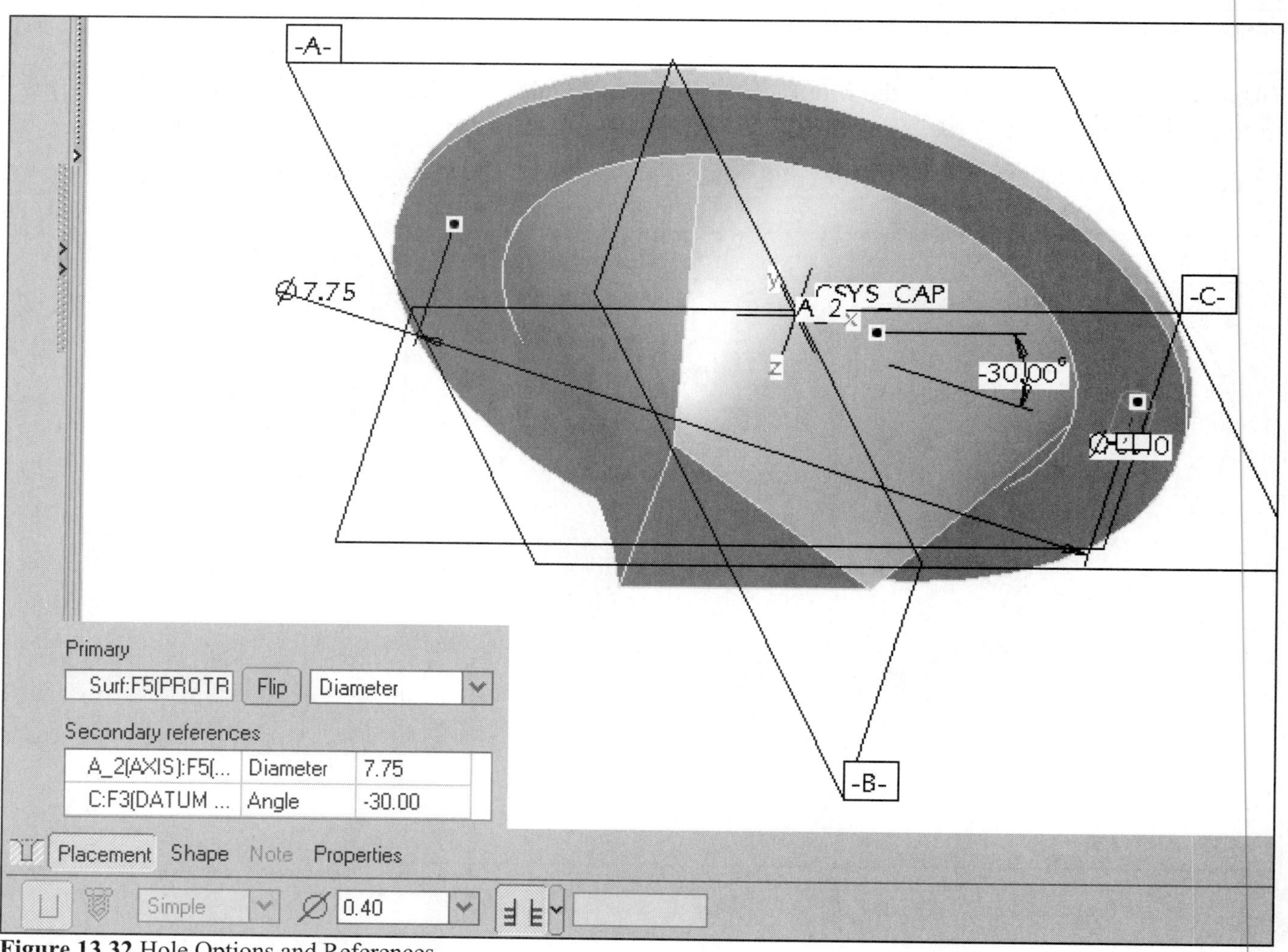

Figure 13.32 Hole Options and References

Click on the hole on the model or in the Model Tree ⇒ **RMB** ⇒ **Pattern** ⇒ Complete the pattern with the options and references provided in Figure 13.33 and Figure 13.34. (Hint: select the **30** degrees dimension to pattern).

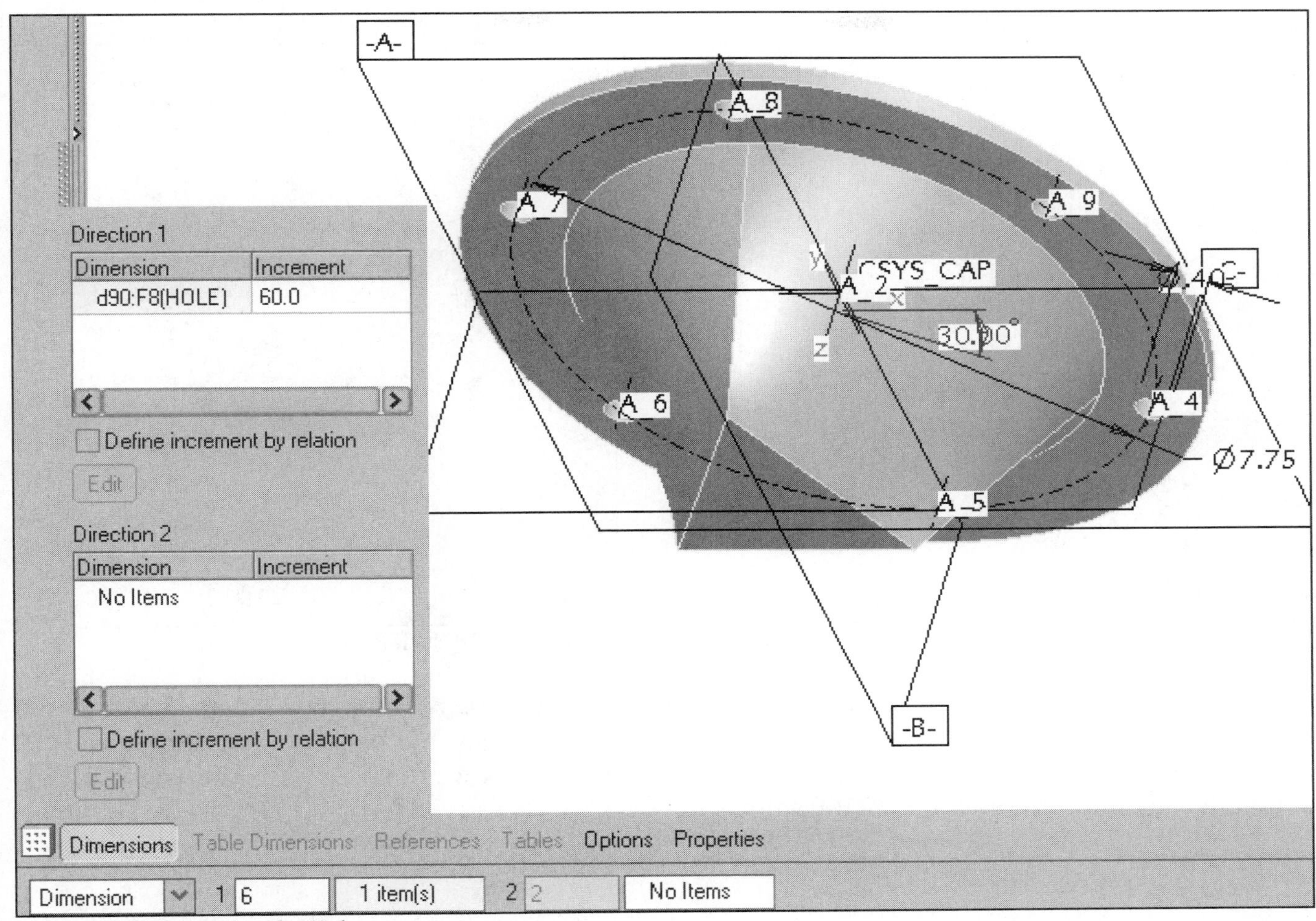

Figure 13.33 Patterning the Hole

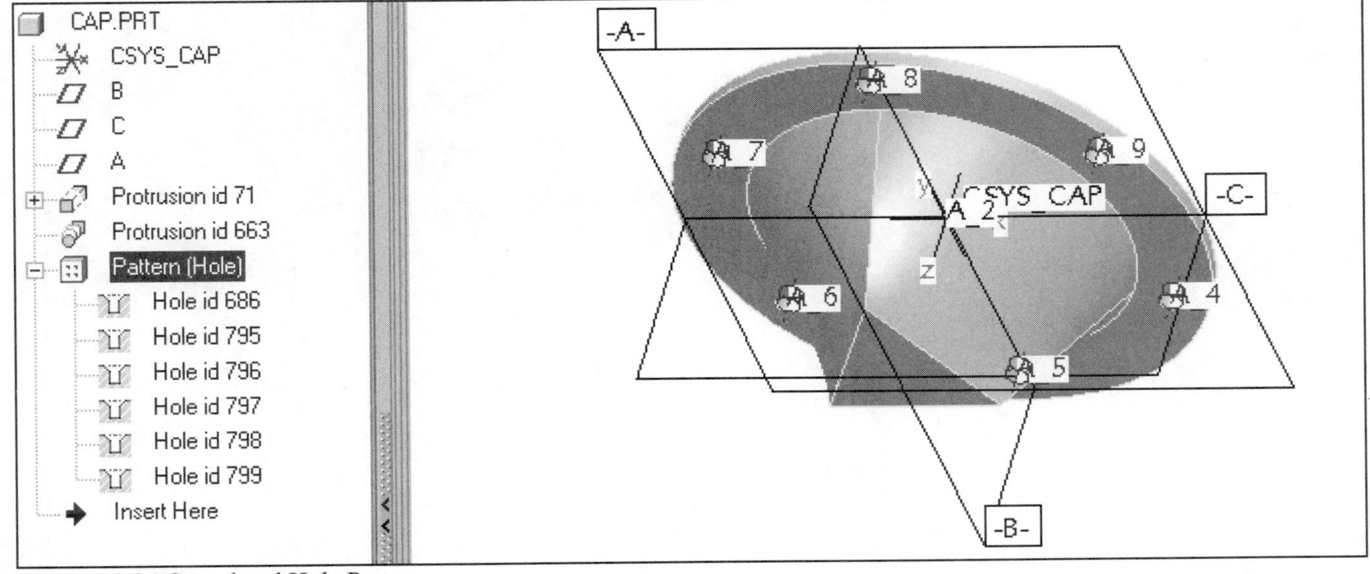

Figure 13.34 Completed Hole Pattern

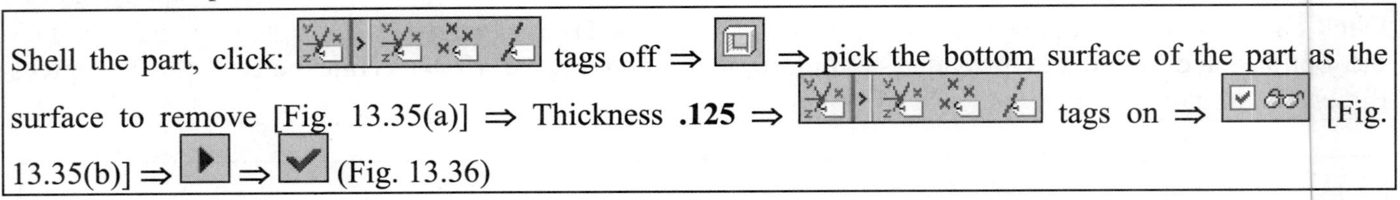

Shell the part, click: [icons] tags off ⇒ [icon] ⇒ pick the bottom surface of the part as the surface to remove [Fig. 13.35(a)] ⇒ Thickness **.125** ⇒ [icons] tags on ⇒ [icon] [Fig. 13.35(b)] ⇒ [icon] ⇒ [icon] (Fig. 13.36)

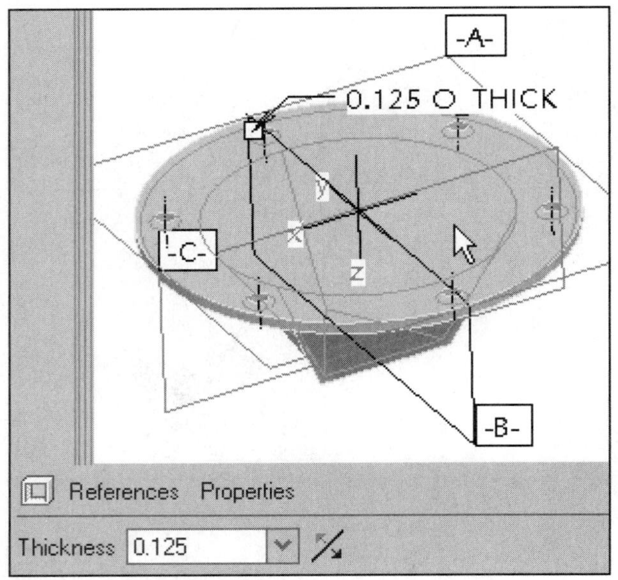

Figure 13.35(a) Shell Tool

-A-
-C-
-B-
References Properties
Thickness 0.125

Figure 13.35(b) Shell Preview

The shell will automatically create the bosses (Fig. 13.36), because the **.125** thickness is left around all previously created features. The *bosses* around the holes are created automatically at **.250** larger than the holes: (**.125 + .125 + .400 = ∅.650**). If the bosses were not desired, you would simply reorder the holes to come after the shell feature.

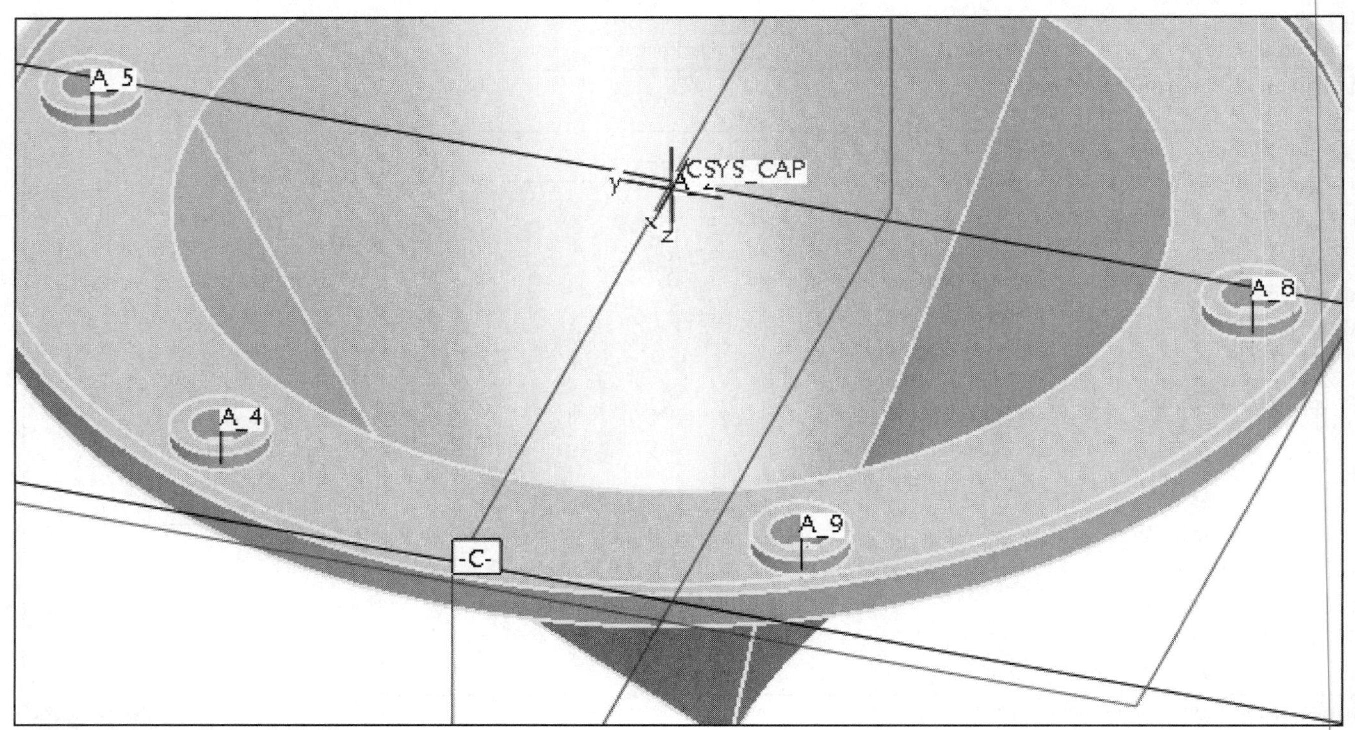

Figure 13.36 Completed Shell

Measure the size of the boss: **Analysis** ⇒ **Measure** ⇒ click [icon] to open the Type drop down selections ⇒ **Diameter** ⇒ pick the vertical surface of the boss (Fig. 13.37) ⇒ **Close**

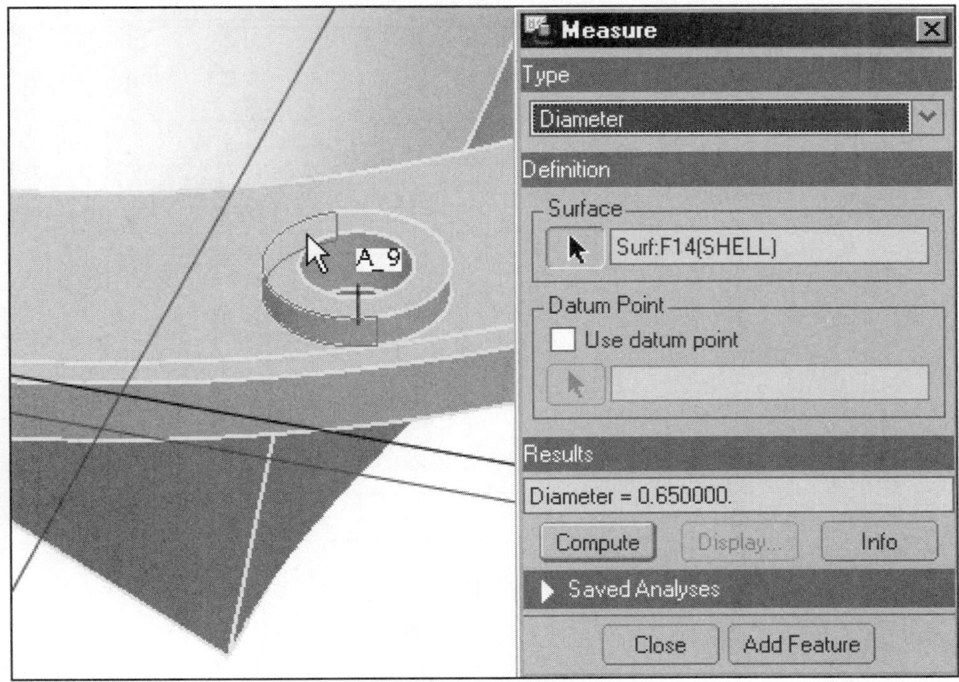

Figure 13.37 Measure (Results **Diameter = 0.650000**)

Now create a cross section to be used in the Drawing mode when detailing the Cap. Choose the following commands: **Tools** ⇒ [Model Sectioning] ⇒ **New** ⇒ type **A** (Fig. 13.38) ⇒ **Enter** ⇒ **Planar** ⇒ **Single** ⇒ **MMB** ⇒ pick datum **B** ⇒ **Display** ⇒ **Show X-Section** (Fig. 13.38) ⇒ **Close** ⇒ [icon] ⇒ **Standard Orientation** ⇒ [icon] ⇒ [icon] ⇒ **MMB**

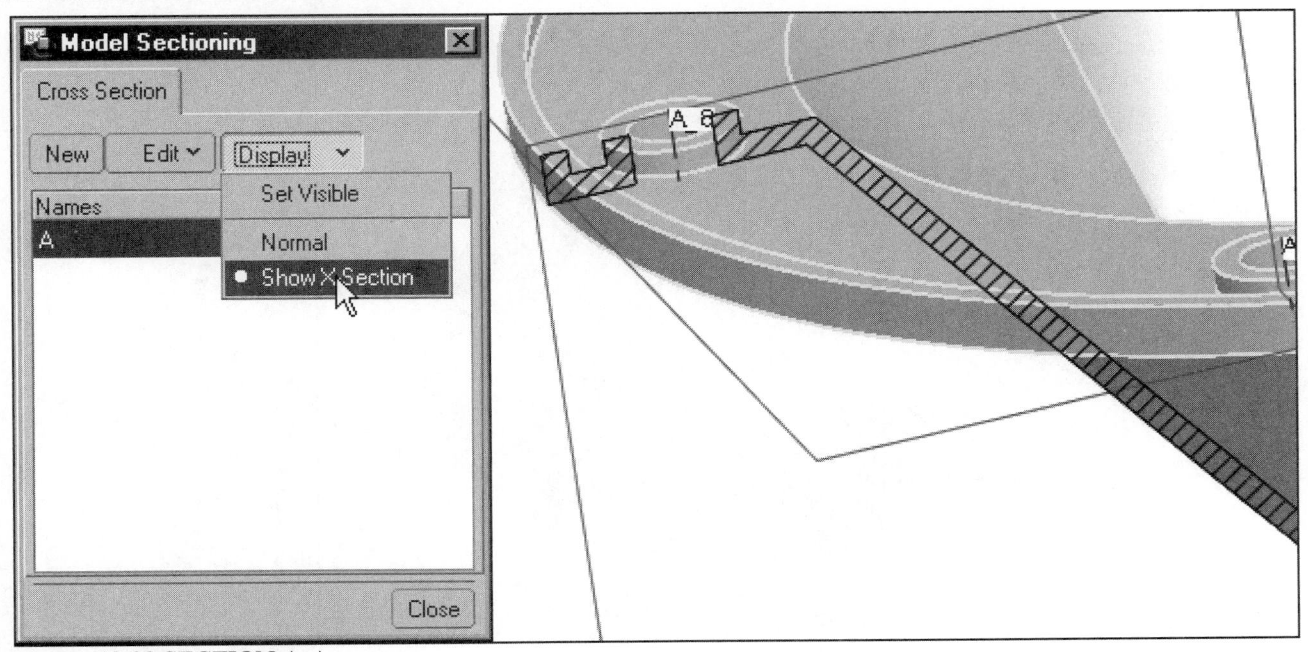

Figure 13.38 SECTION A-A

Save the Cap under a new name, and complete the **ECO** [Fig. 13.39(a-b)] to establish a second part with different dimensional sizes and features. Add **R.20** rounds to the edges. Insert the rounds to be affected by the shell.

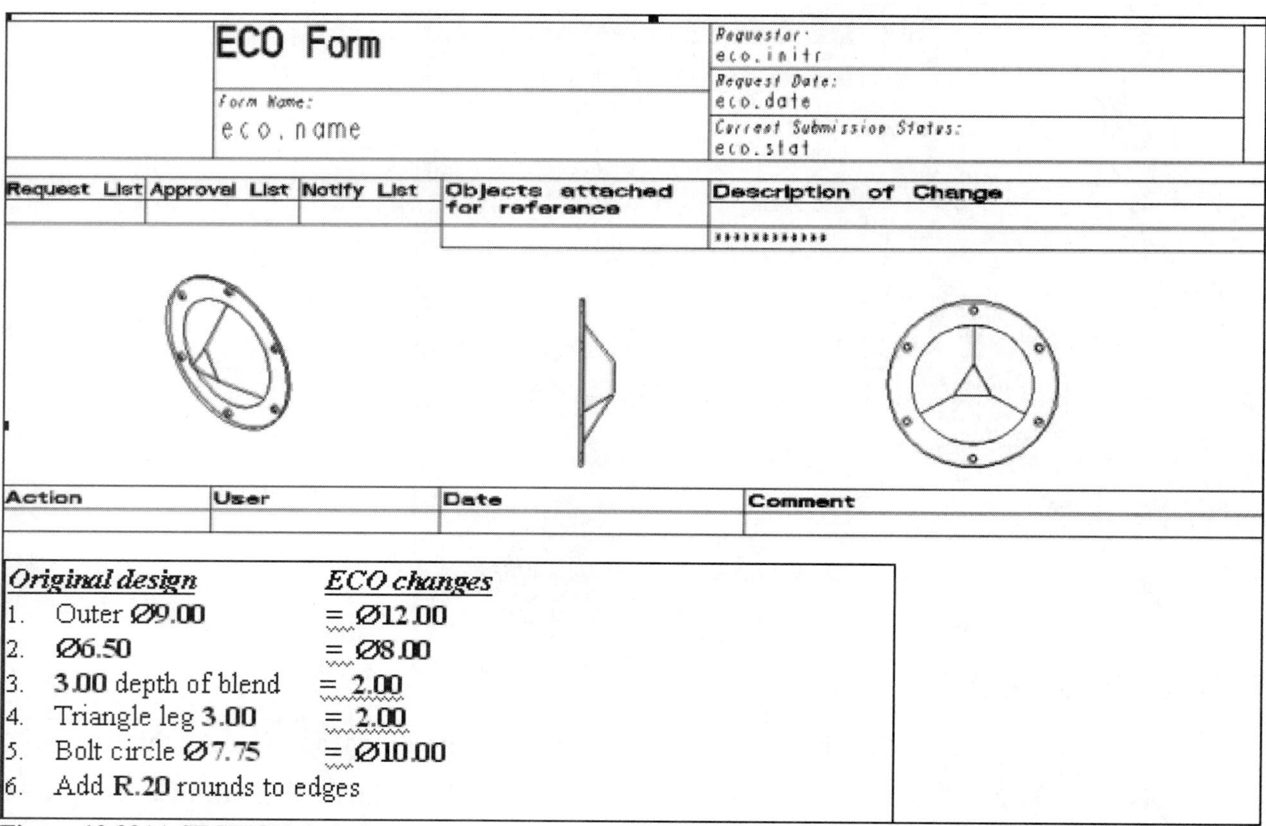

ECO Form

Form Name:
eco.name

Requestor:
eco.initr

Request Date:
eco.date

Current Submission Status:
eco.stat

Request List	Approval List	Notify List	Objects attached for reference	Description of Change
				************

Action	User	Date	Comment

Original design	*ECO changes*
1. Outer **Ø9.00**	= **Ø12.00**
2. **Ø6.50**	= **Ø8.00**
3. **3.00** depth of blend	= **2.00**
4. Triangle leg **3.00**	= **2.00**
5. Bolt circle **Ø7.75**	= **Ø10.00**
6. Add **R.20** rounds to edges	

Figure 13.39(a) SECTION A-A

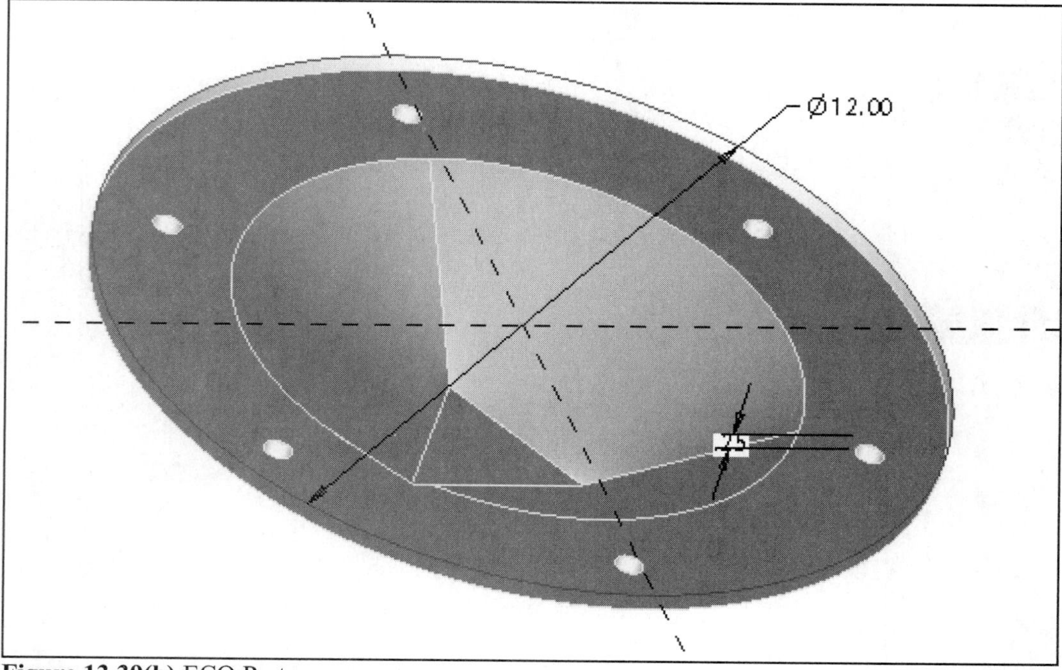

Figure 13.39(b) ECO Part

Lesson 13 Project

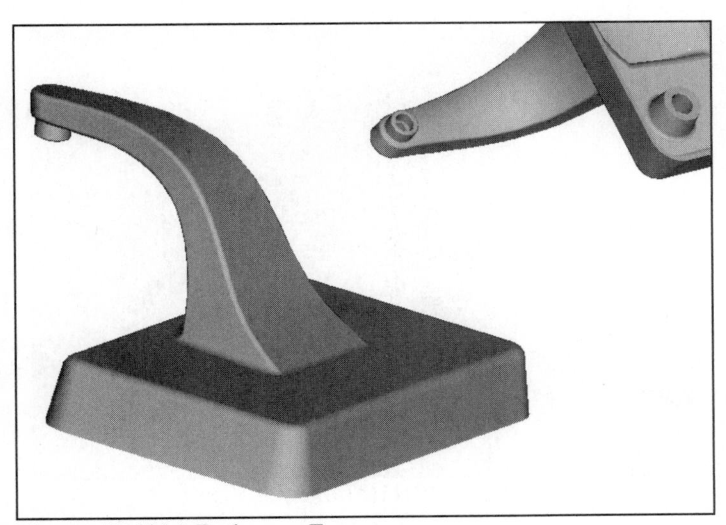

Figure 13.40(a) Bathroom Faucet

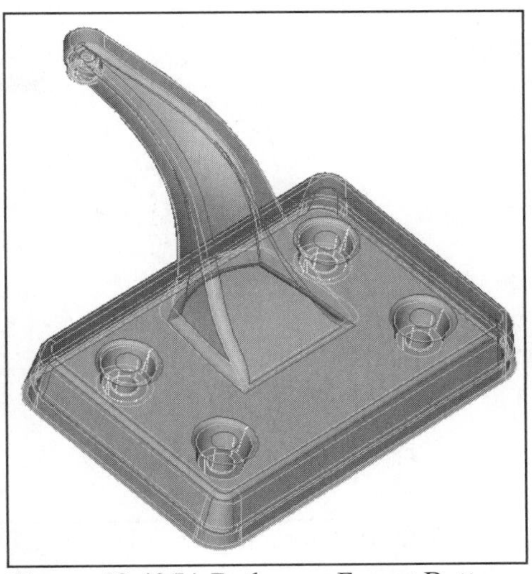

Figure 13.40(b) Bathroom Faucet Bottom

Bathroom Faucet

This is an advanced lesson project. Because you have created over twenty parts, you should be able to use that knowledge to model the Swept Blend required to create the Bathroom Faucet (Figs. 13.40 through 13.81). Some instructions accompany this lesson project, but you will be required to research documentation (Pro/HELP) and learn about Splines and Swept Blends. After the model is complete, create a number of sections that can be used in Drawing mode when you are detailing the Faucet.

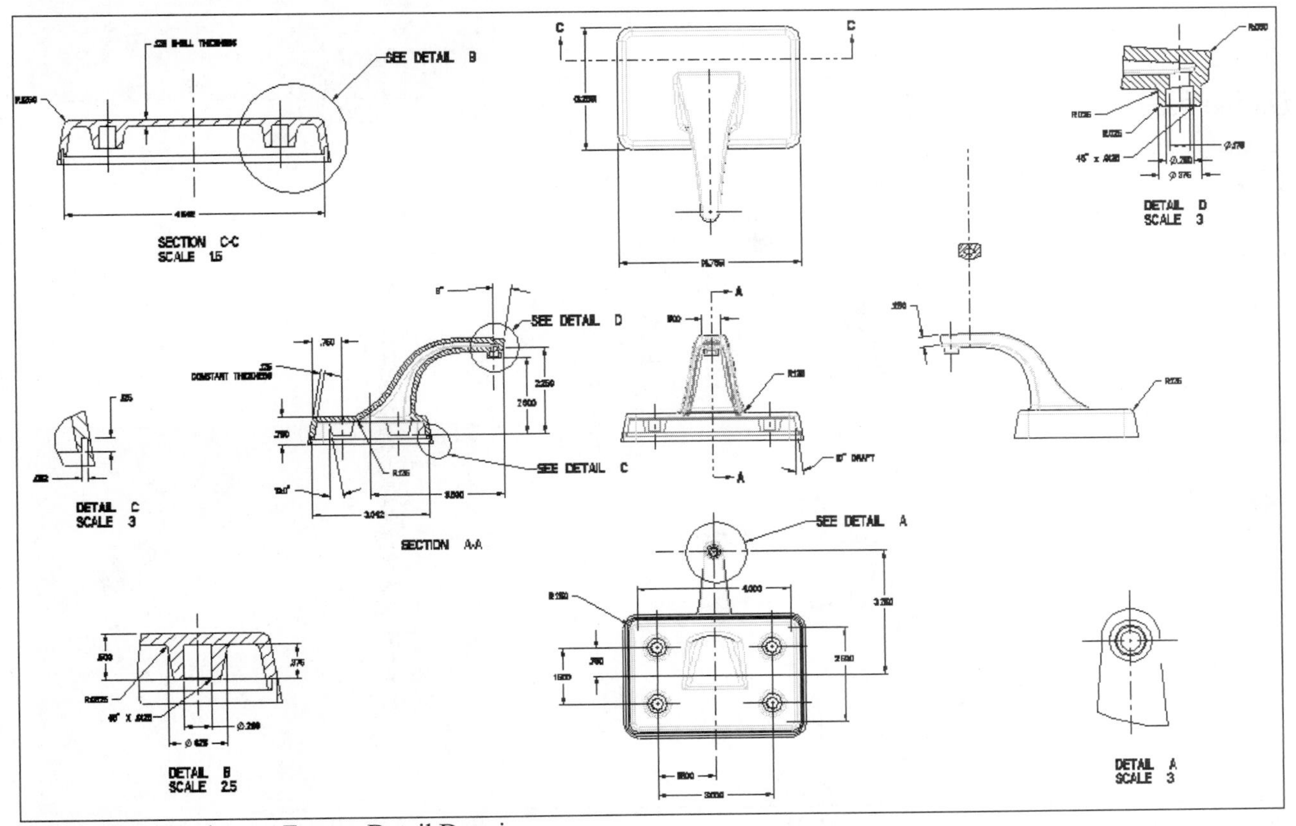

Figure 13.41 Bathroom Faucet, Detail Drawing

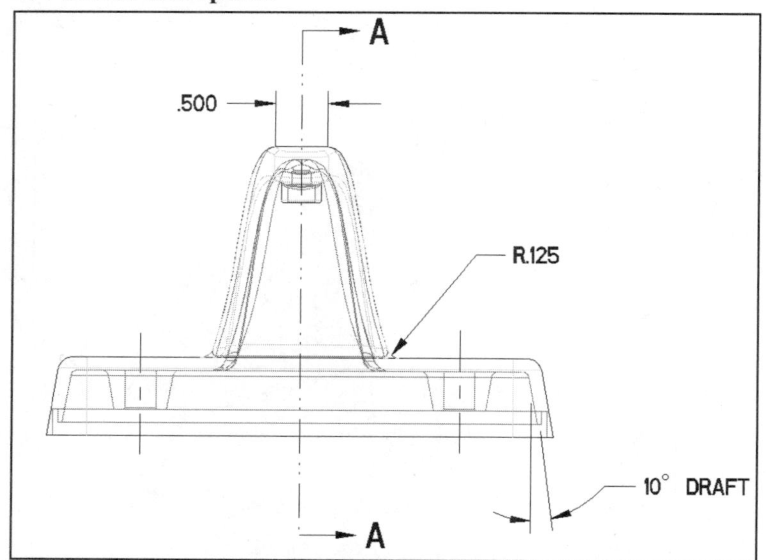

Figure 13.42(a) Bathroom Faucet Drawing, Front View

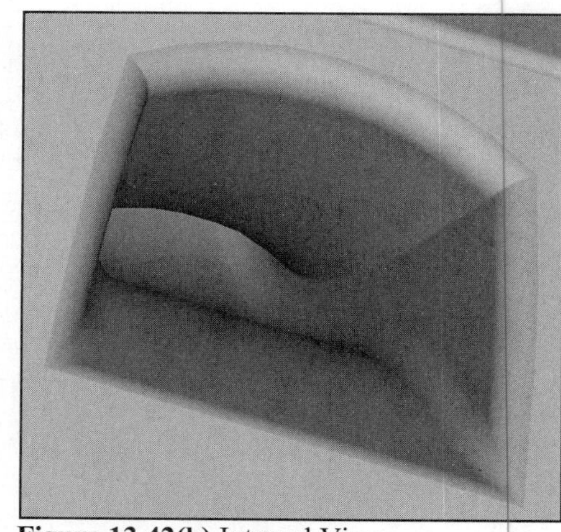

Figure 13.42(b) Internal View

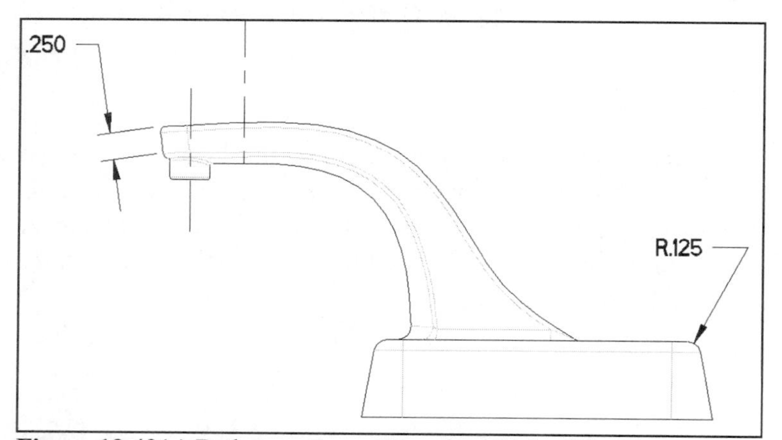

Figure 13.43(a) Bathroom Faucet Drawing, Right Side View

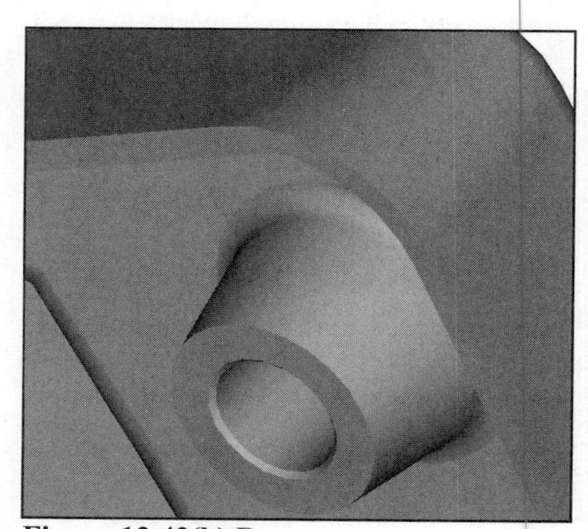

Figure 13.43(b) Boss

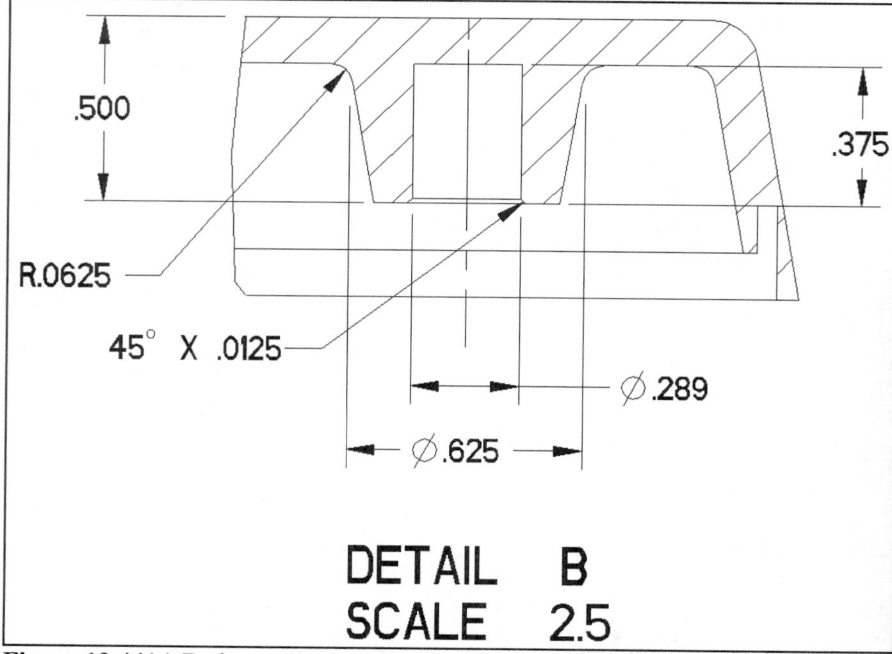

Figure 13.44(a) Bathroom Faucet Drawing, DETAIL B

Figure 13.44(b) Draft

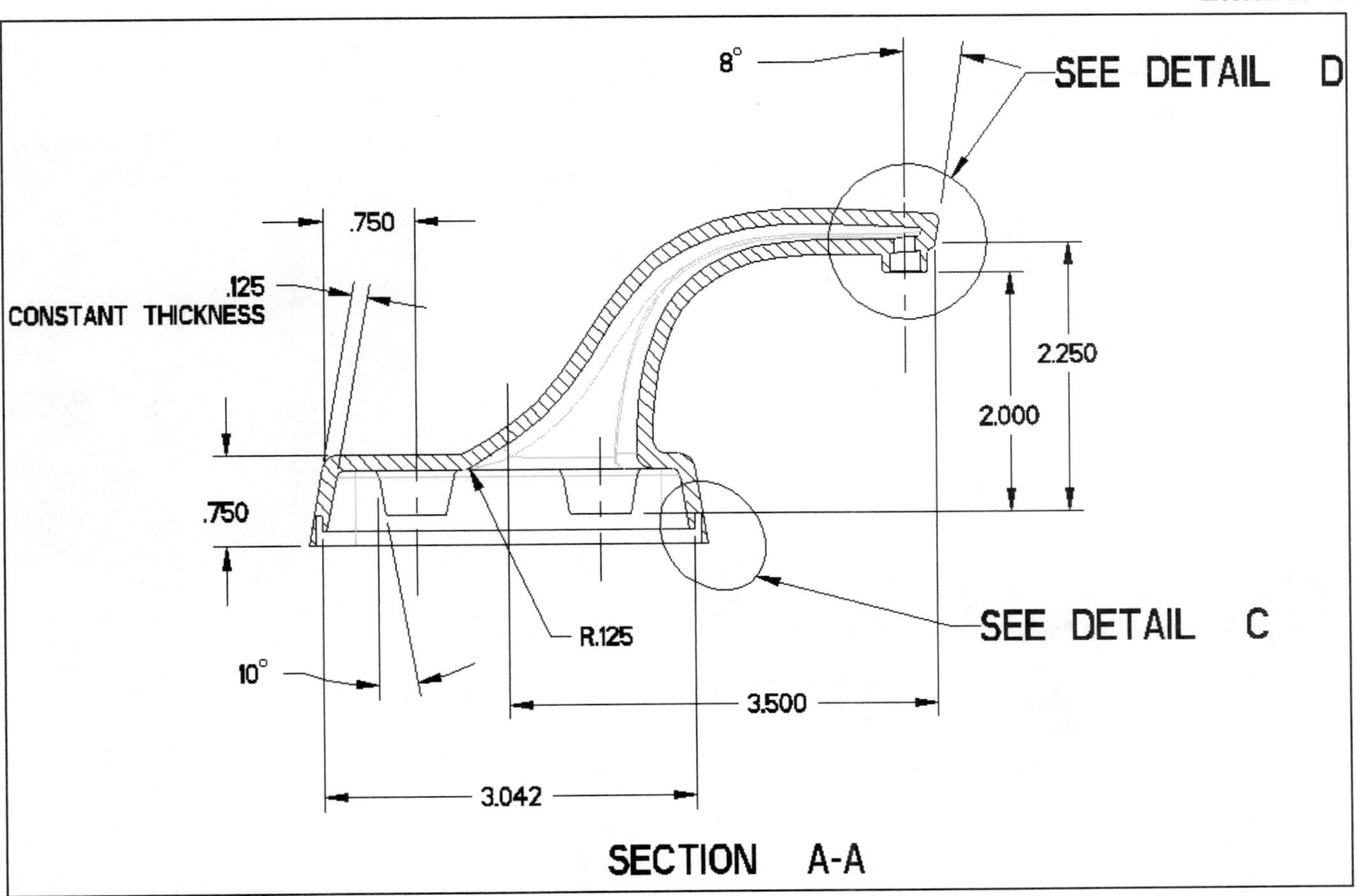

Figure 13.45 SECTION A-A

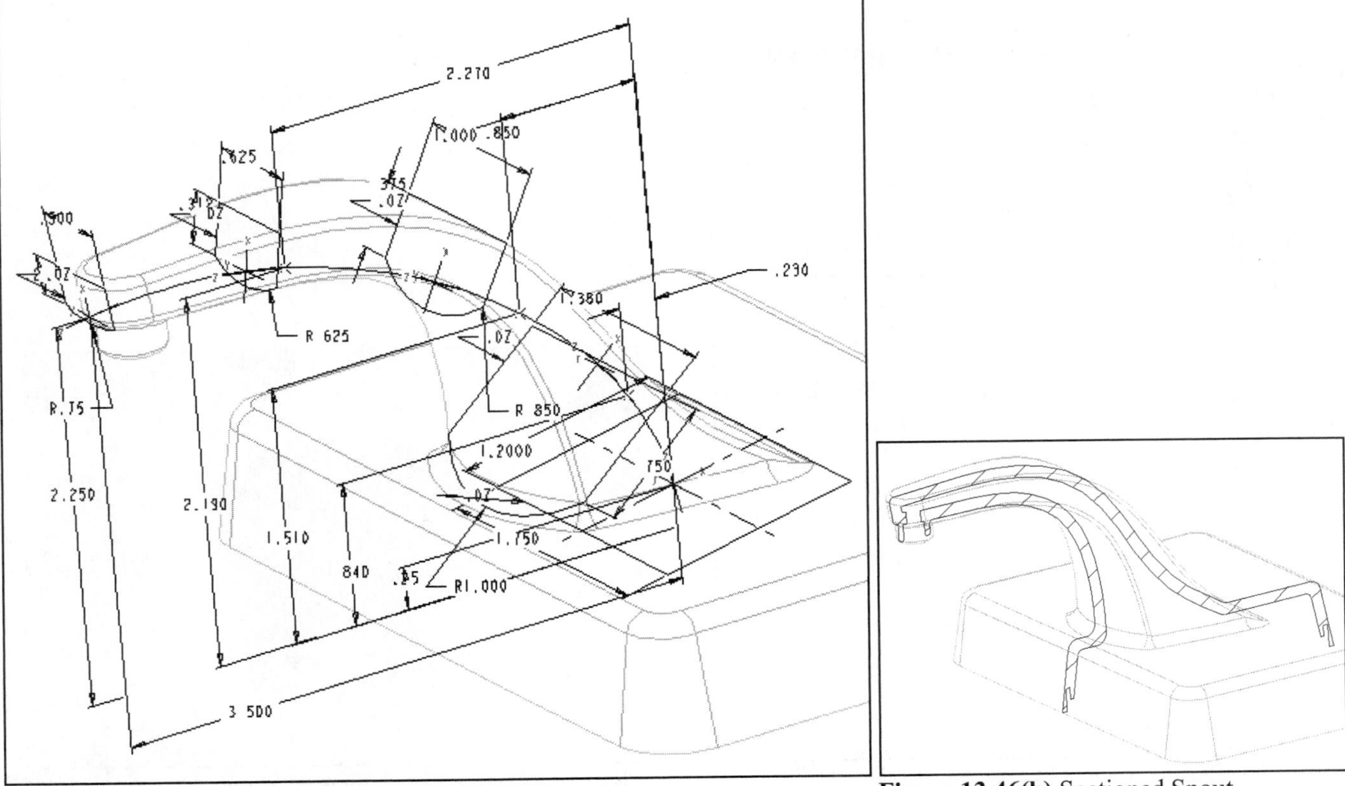

Figure 13.46(a) Sweep Sections

Figure 13.46(b) Sectioned Spout

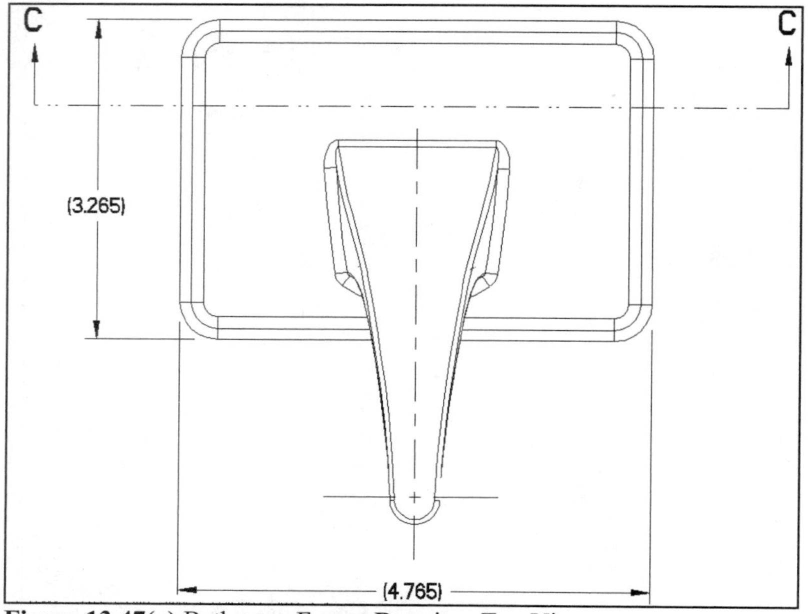

Figure 13.47(a) Bathroom Faucet Drawing, Top View

Figure 13.47(b) End Section

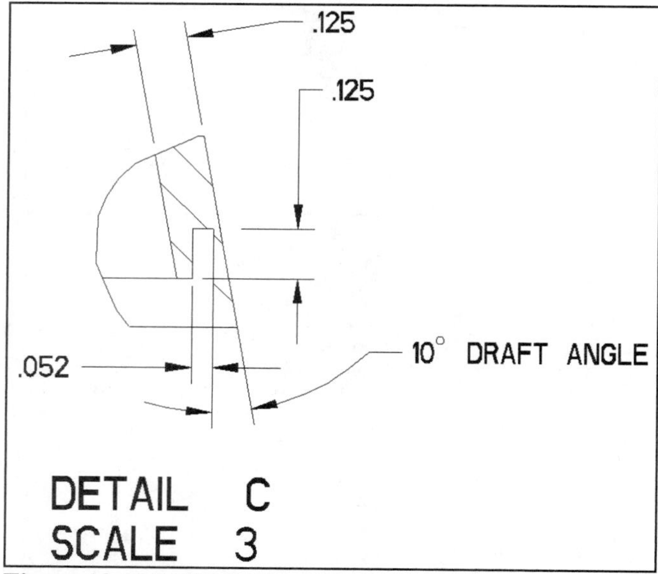

Figure 13.48 DETAIL C

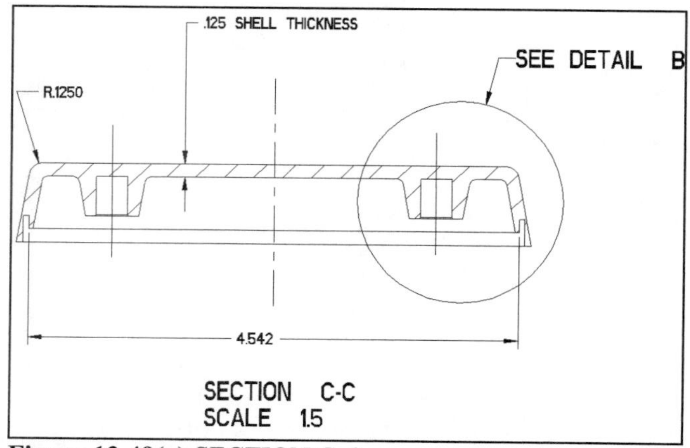

Figure 13.49(a) SECTION C-C

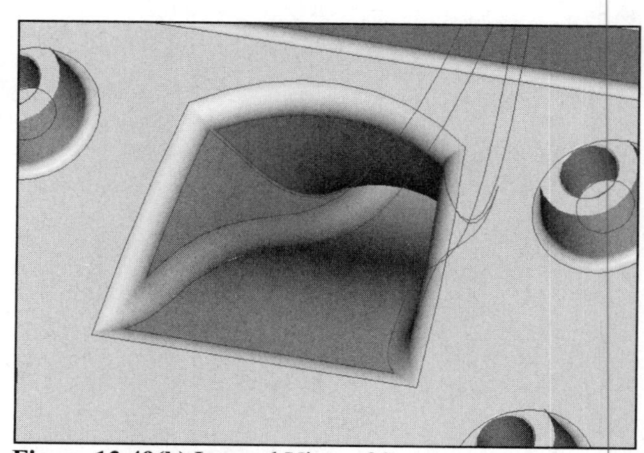

Figure 13.49(b) Internal View of Spout

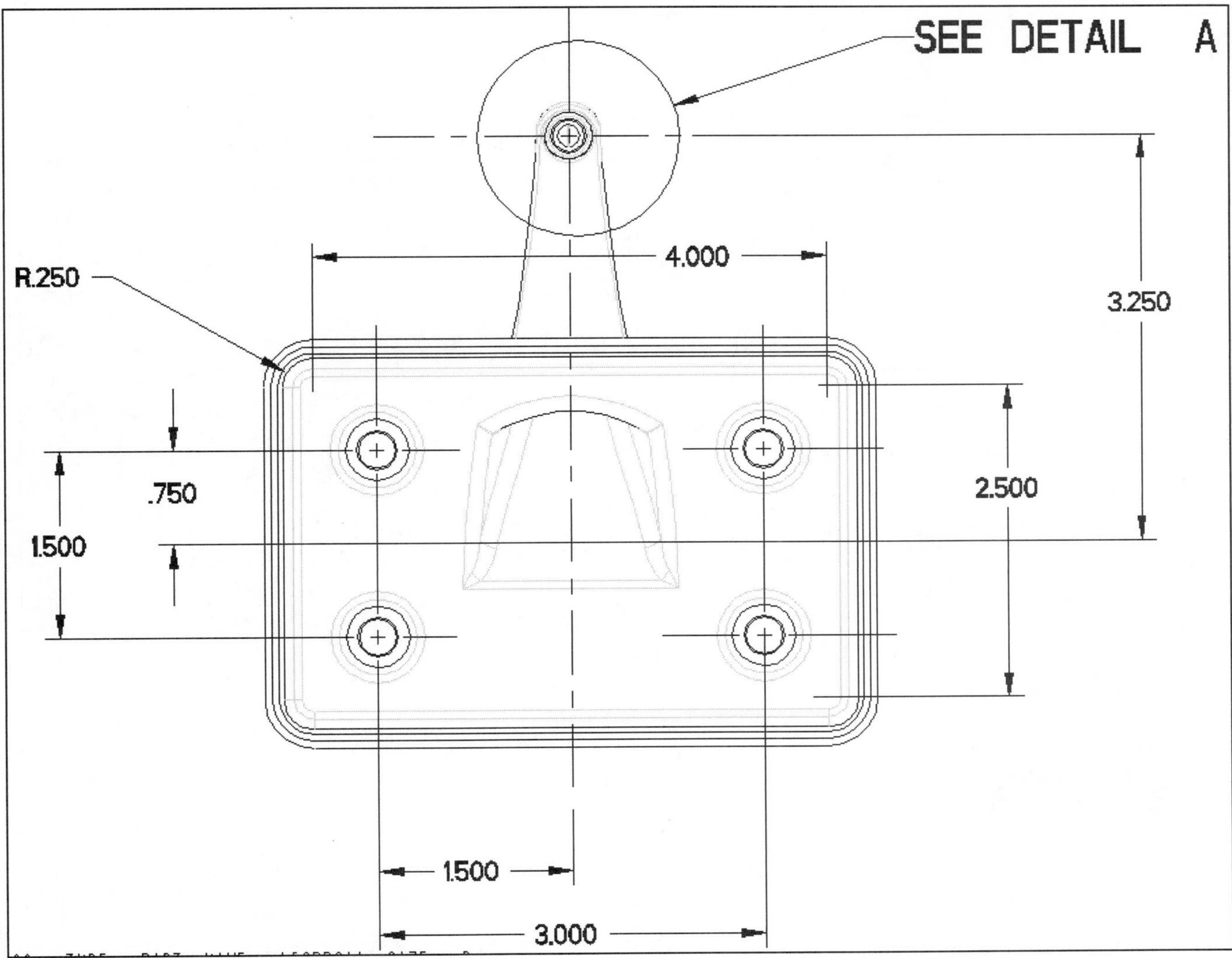

Figure 13.50 Bathroom Faucet Drawing, Bottom View

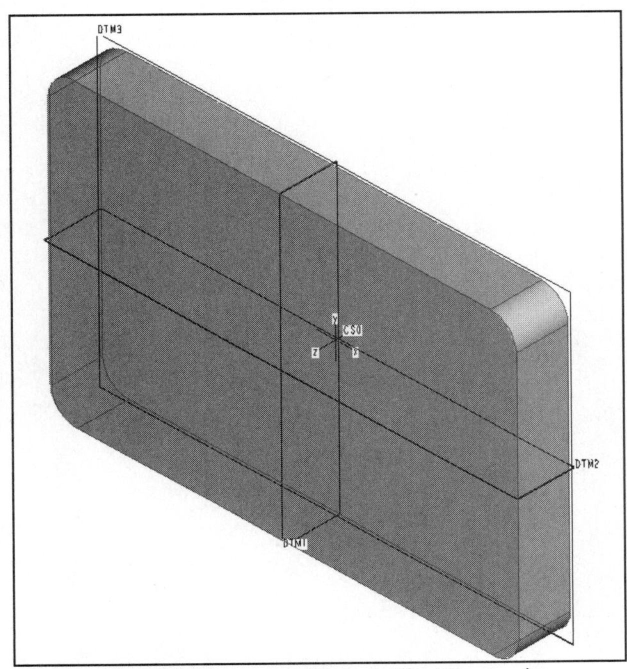

Figure 13.51(a) Bathroom Faucet, First Protrusion

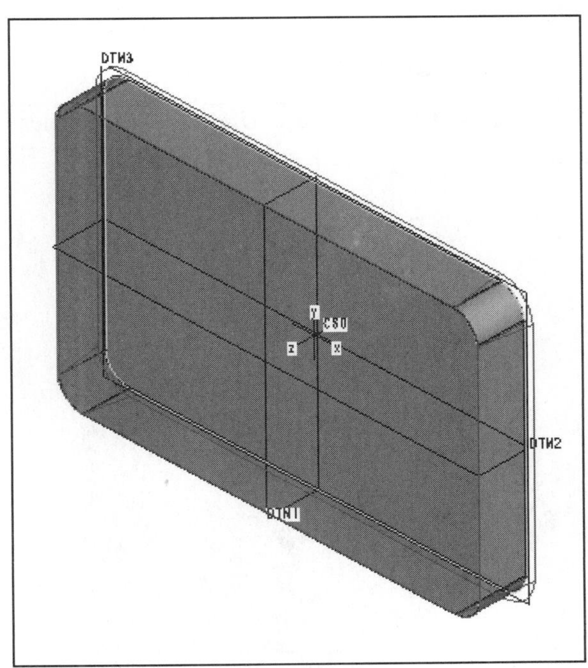

Figure 13.51(b) Previewed Draft

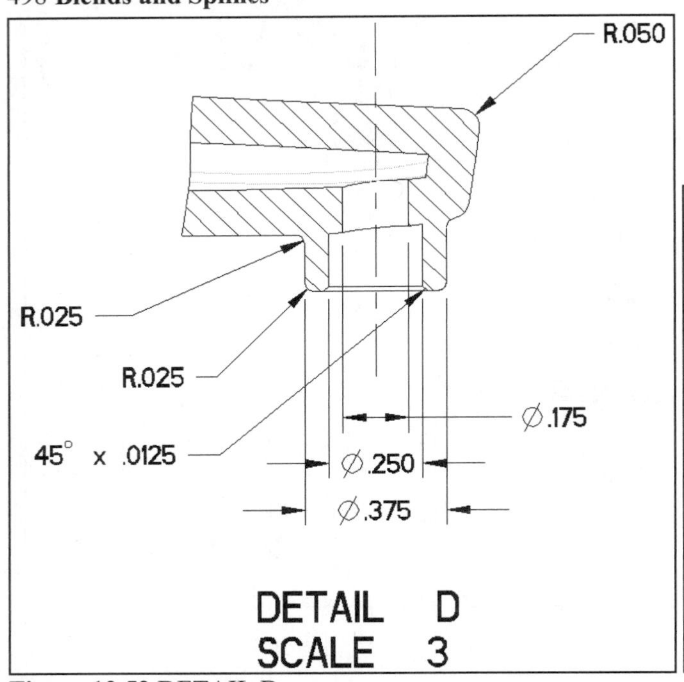

Figure 13.52 DETAIL D

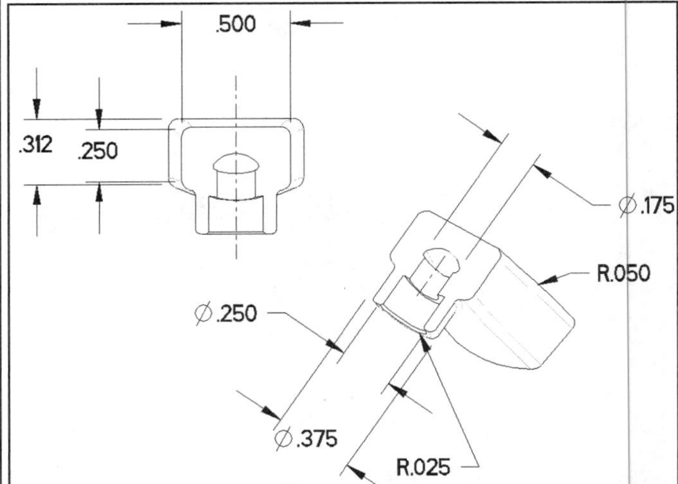

Figure 13.53 Bathroom Faucet Drawing, Spout Section

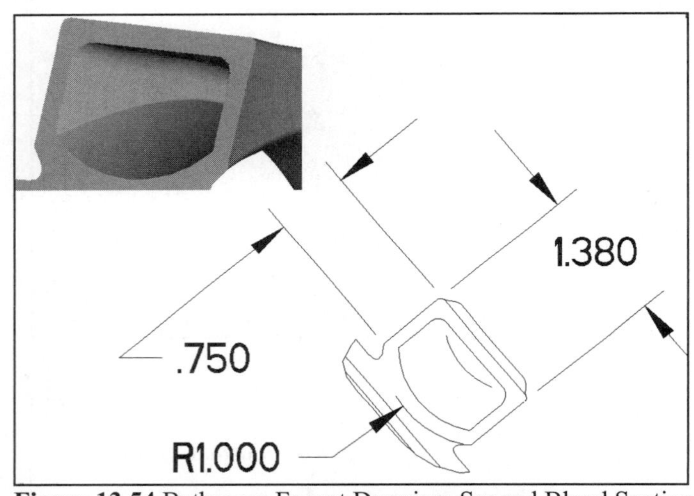

Figure 13.54 Bathroom Faucet Drawing, Second Blend Section

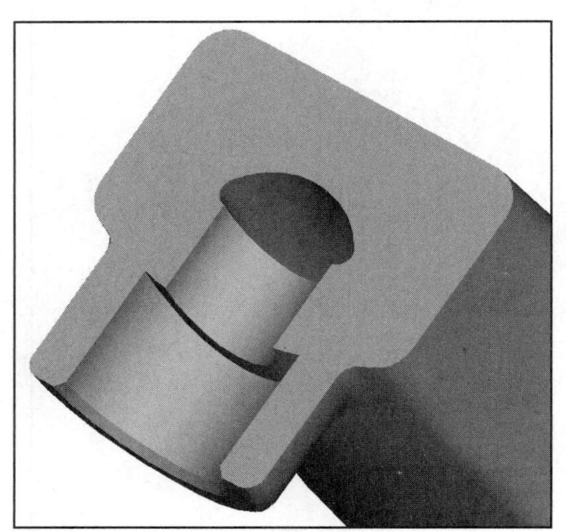
Figure 13.55 Sectioned End

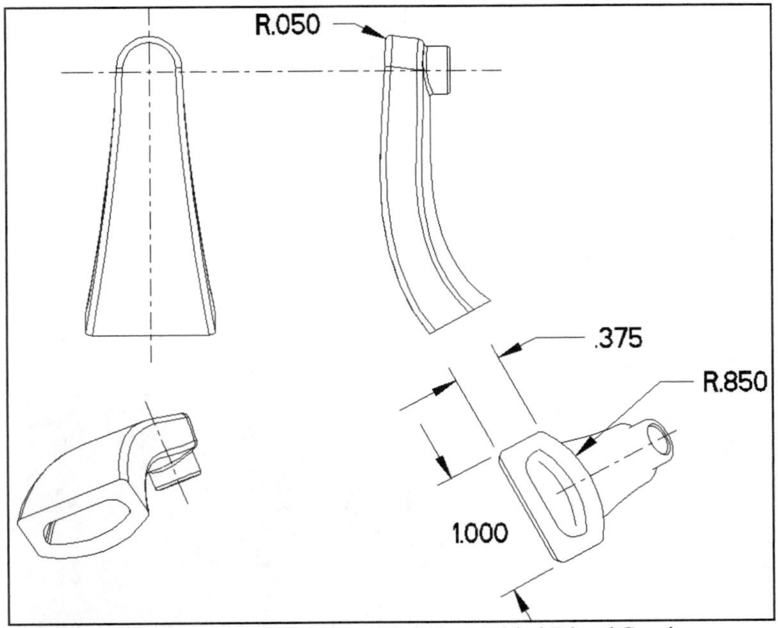

Figure 13.56(a) Bathroom Faucet Drawing, Third Blend Section

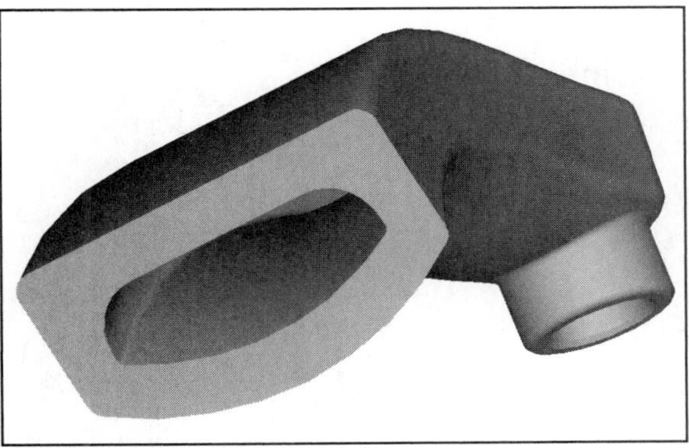

Figure 13.56(b) Third Section

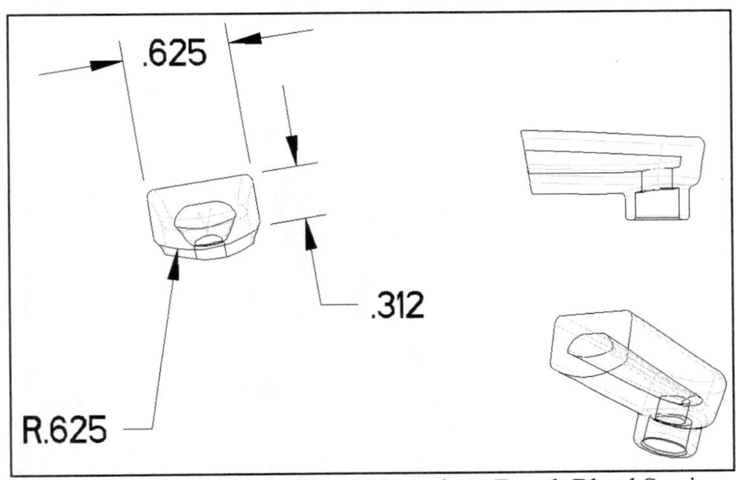

Figure 13.57(a) Bathroom Faucet Drawing, Fourth Blend Section

Figure 13.57(b) Fourth Section

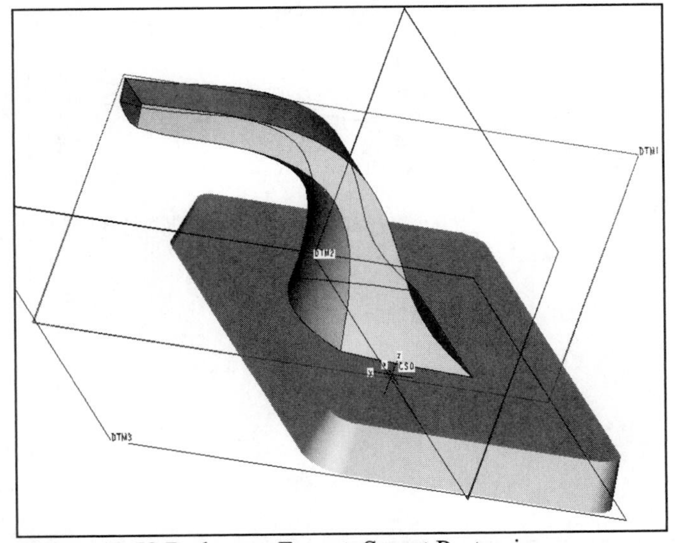

Figure 13.58 Bathroom Faucet, Swept Protrusion

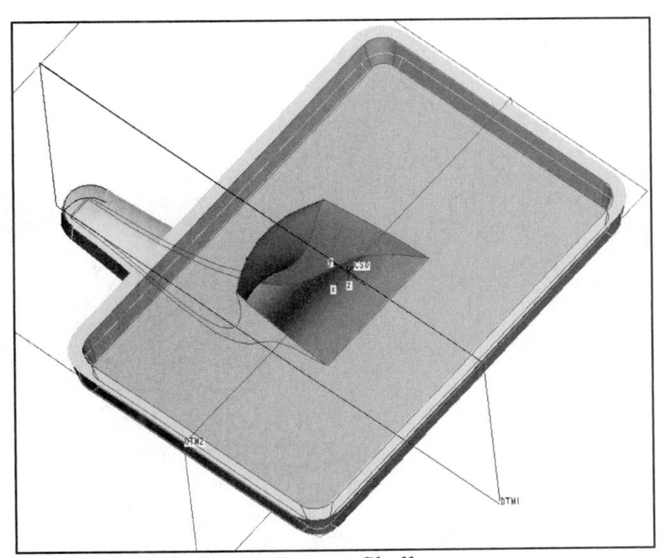

Figure 13.59 Bathroom Faucet, Shell

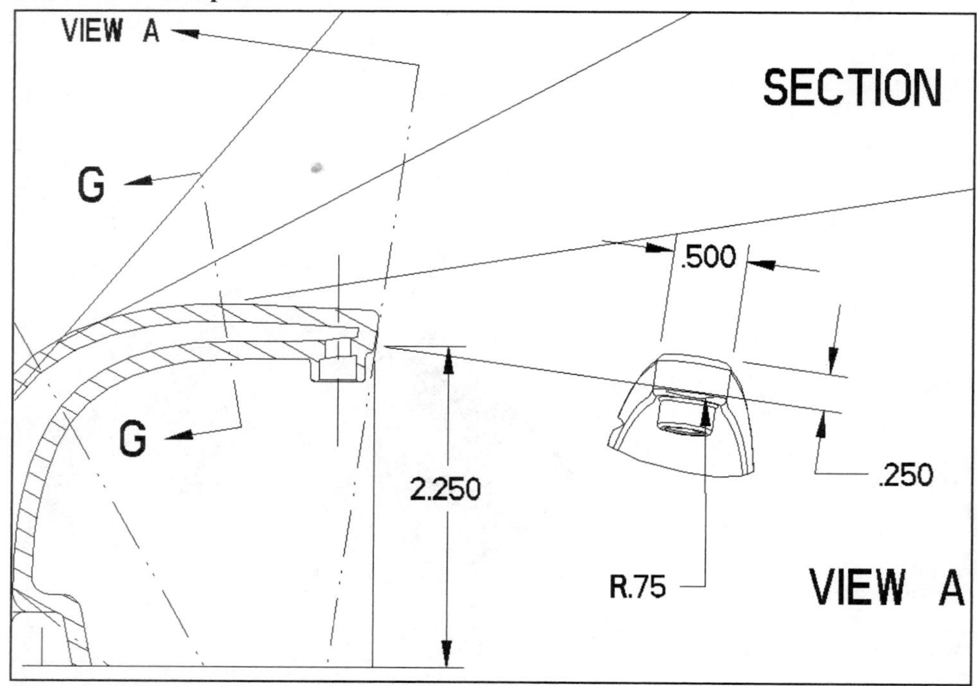

Figure 13.60(a) Bathroom Faucet Drawing, Fifth Blend Section

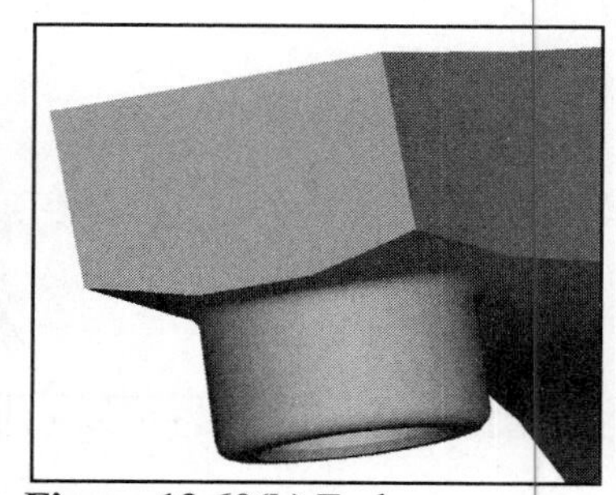

Figure 13.60(b) End

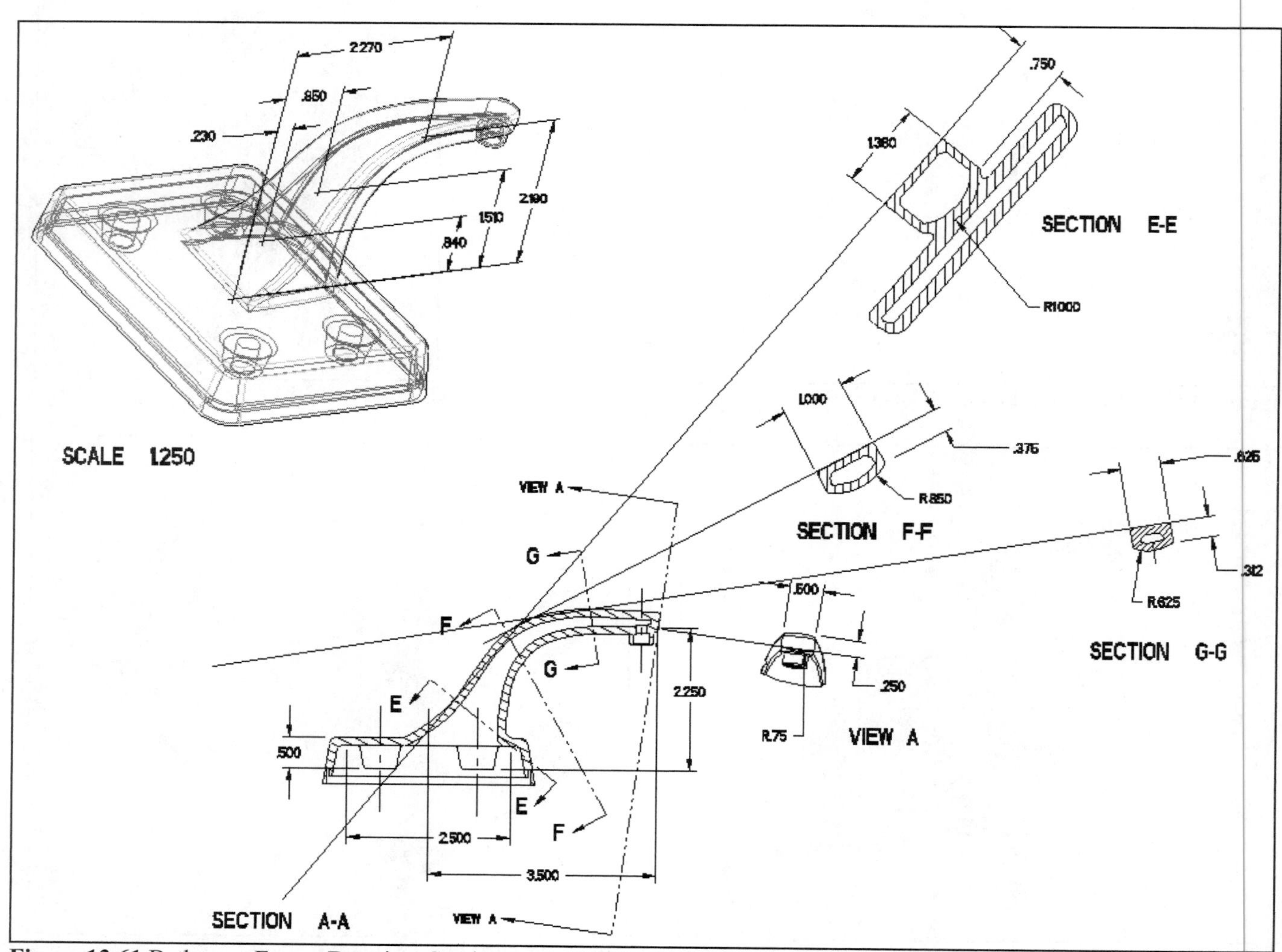

Figure 13.61 Bathroom Faucet Drawing, Sections

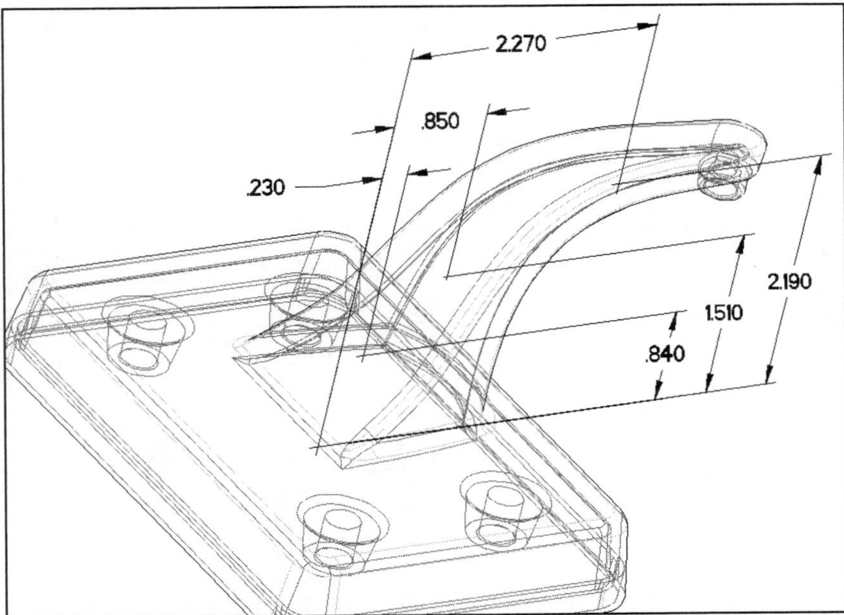

Figure 13.62 Bathroom Faucet Drawing, First Three Blend Section Locations

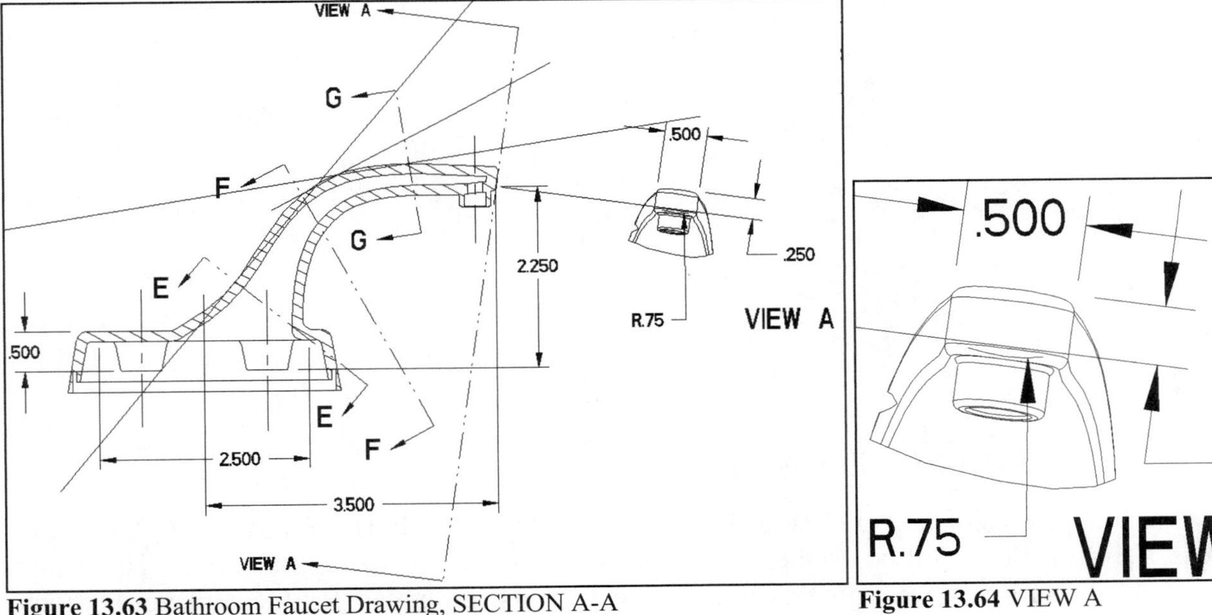

Figure 13.63 Bathroom Faucet Drawing, SECTION A-A

Figure 13.64 VIEW A

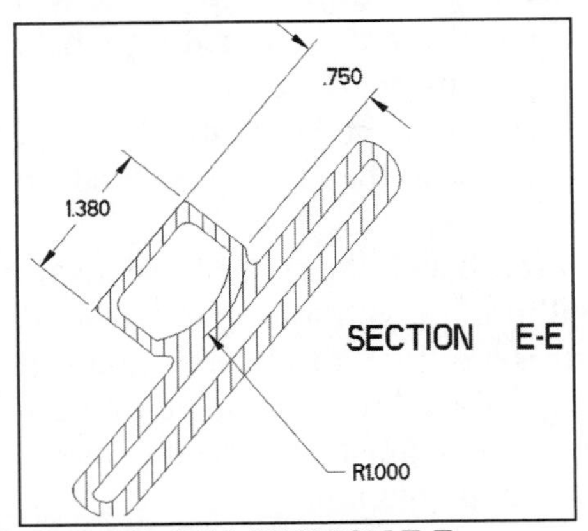

Figure 13.65(a) SECTION E-E

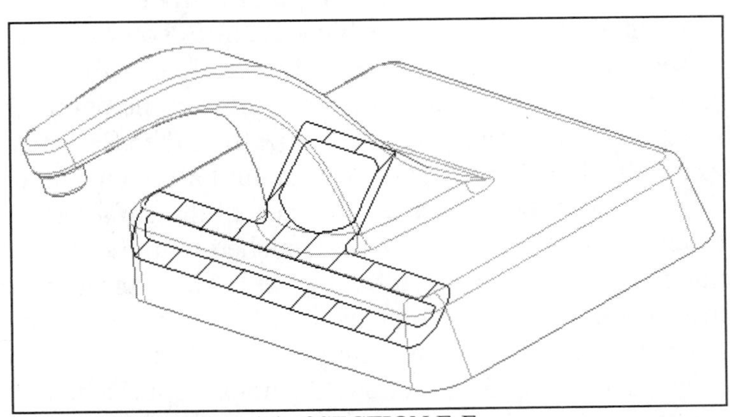

Figure 13.65(b) Pictorial of SECTION E-E

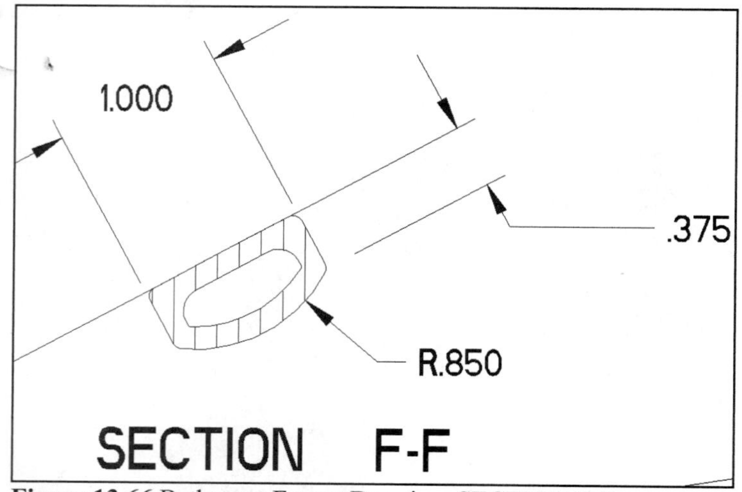

Figure 13.66 Bathroom Faucet Drawing, SECTION F-F

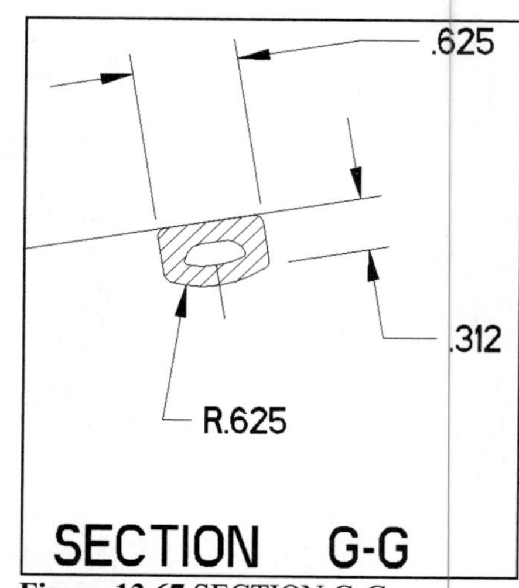

Figure 13.67 SECTION G-G

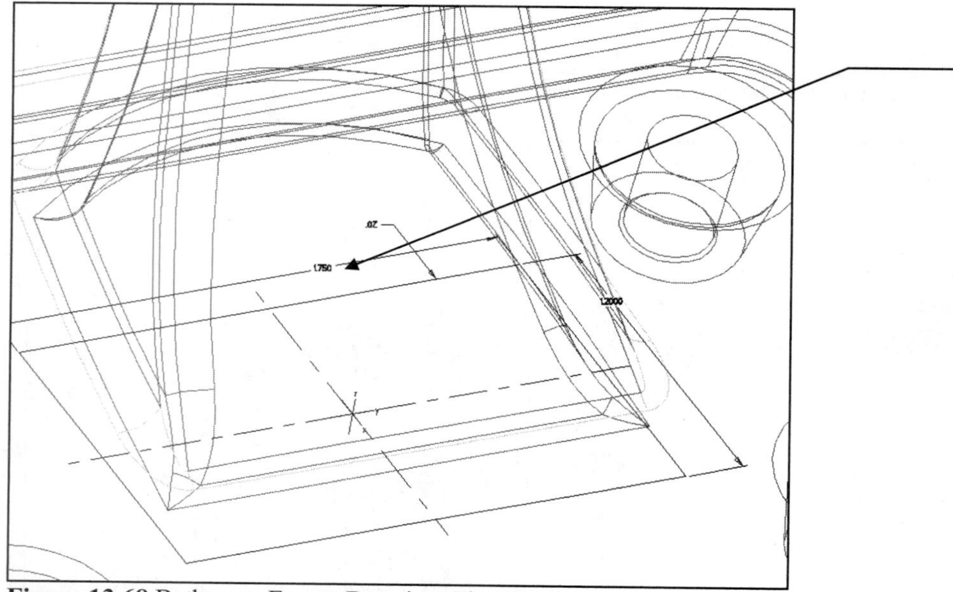

Figure 13.68 Bathroom Faucet Drawing, First Section

Create the first protrusion and draft (Fig. 13.69). (Note: DTM1 = RIGHT, DTM2 = TOP, DTM3 = FRONT). Next, a Swept Blend will be used to create the geometry for the Bathroom Faucet. The default options of Sketch Sec and NrmToOriginTraj will be used. When prompted for the trajectory, choose Sketch Traj. Sketch a trajectory (spine) with the Spline. Create five points along the trajectory. These will locate the five sections of the blend. Before you begin to sketch any sections, you are prompted for where the sections are to be located. Five sections will be sketched for this protrusion. Two will be at the endpoints (mandatory) and the other three at the datum points. The first location to be highlighted will be the second point of the sketched trajectory, then Select Accept. The next point will then be highlighted, so Accept this location. Choose Accept until all points have been accepted.

To sketch the first section, accept the default **Z**-axis rotation of zero. Sketch the section. Be aware of the *start point* location. When finished with the first section, Pro/E will prompt you for the next **Z**-axis rotation. Accept the default value and sketch the second section. Again, keep track of the start point; it must match up with the first section. Sketch the third section, again using a **Z**-axis rotation value of zero. Sketch the fourth section, again using a **Z**-axis rotation value of zero. When finished with the fourth section, sketch the fifth section. Again, accept the default of zero for the **Z**-axis rotation.

Preview your feature elements to check for twist in the swept blend because of start points that do not line up. If there is a problem, use the Section element from the dialog box and choose Define to change the start point(s) so that they are aligned.

Each section in the Swept Blend has four entities. Keep the start point direction arrow of each section in the same region and facing the same direction.

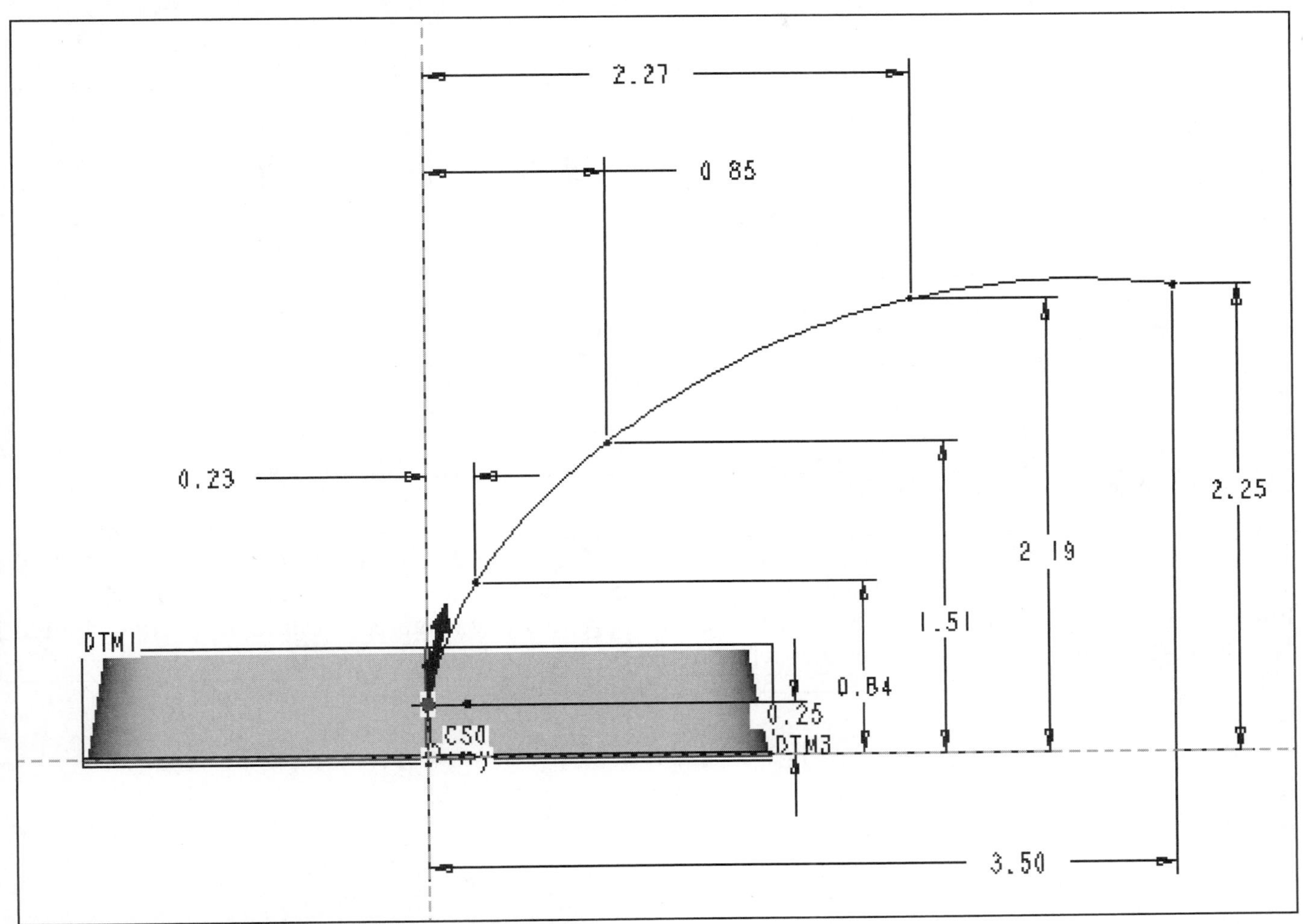

Figure 13.69 Bathroom Faucet, Trajectory Dimensions

Create the Swept Blend with the following commands, click: **Insert** ⇒ **Swept Blend** ⇒ **Protrusion** ⇒ **Sketch Sec** ⇒ **NrmToOriginTraj** ⇒ **MMB** ⇒ **Sketch Traj** ⇒ **Setup New** ⇒ **Plane** ⇒ pick **DTM1-RIGHT** ⇒ **Okay** ⇒ **Right** ⇒ pick **DTM2-FRONT** ⇒ **MMB** ⇒ **MMB** accept the References ⇒ **Create a spline curve** ⇒ sketch the ***five*** points of the trajectory (Fig. 13.69) ⇒ dimension and modify (Fig.13.70) ⇒ ⇒ **Accept** (All three middle points will be highlighted one at a time as you accept them. The first and last points are automatically accepted.) ⇒ **MMB** accept the default of **0°** ⇒ sketch the first section centered about the crosshairs, be aware of the starting point, each section will have one and must face the same direction and start at the same corner ⇒ ⇒ continue creating the sections ⇒ **Preview** ⇒ **MMB**

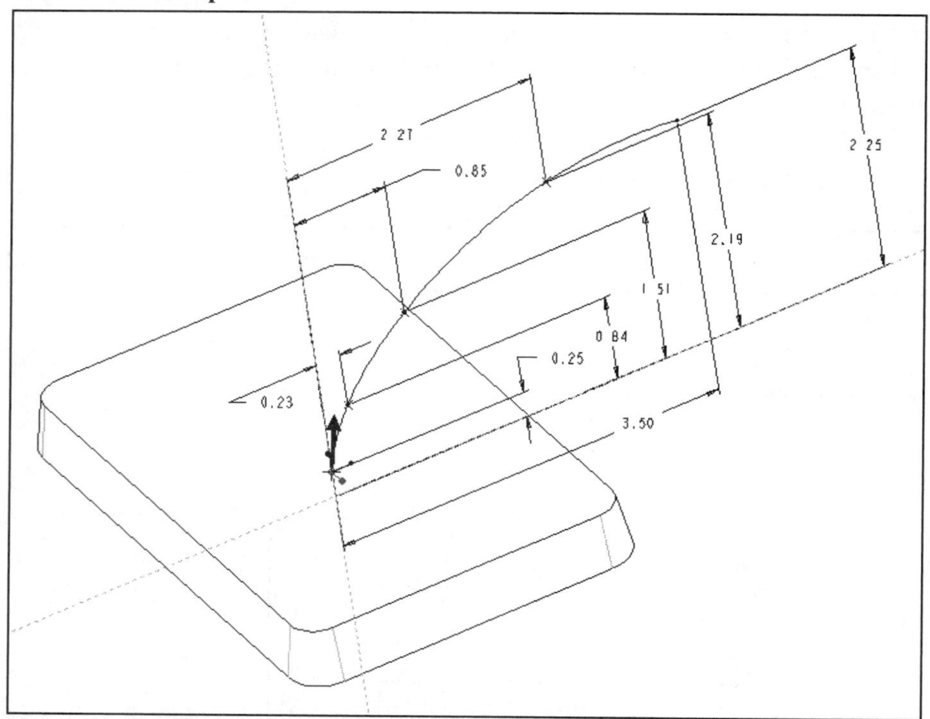

Figure 13.70 Bathroom Faucet, First Section: Pictorial

The *first section* is **1.75 X 1.20**. The *second section* is **1.380 X .75 X R1.00**. The *third section* is **1.00 X .375 X R.850**. The *fourth section* is **.625 X .312 X R.625**. The *fifth section* is **.500 X .25 X R.75**, shown in Figures 13.68 through 13.81. Create the first section approximately centered about its respective colored crosshairs (starting point). Use centerlines when possible.

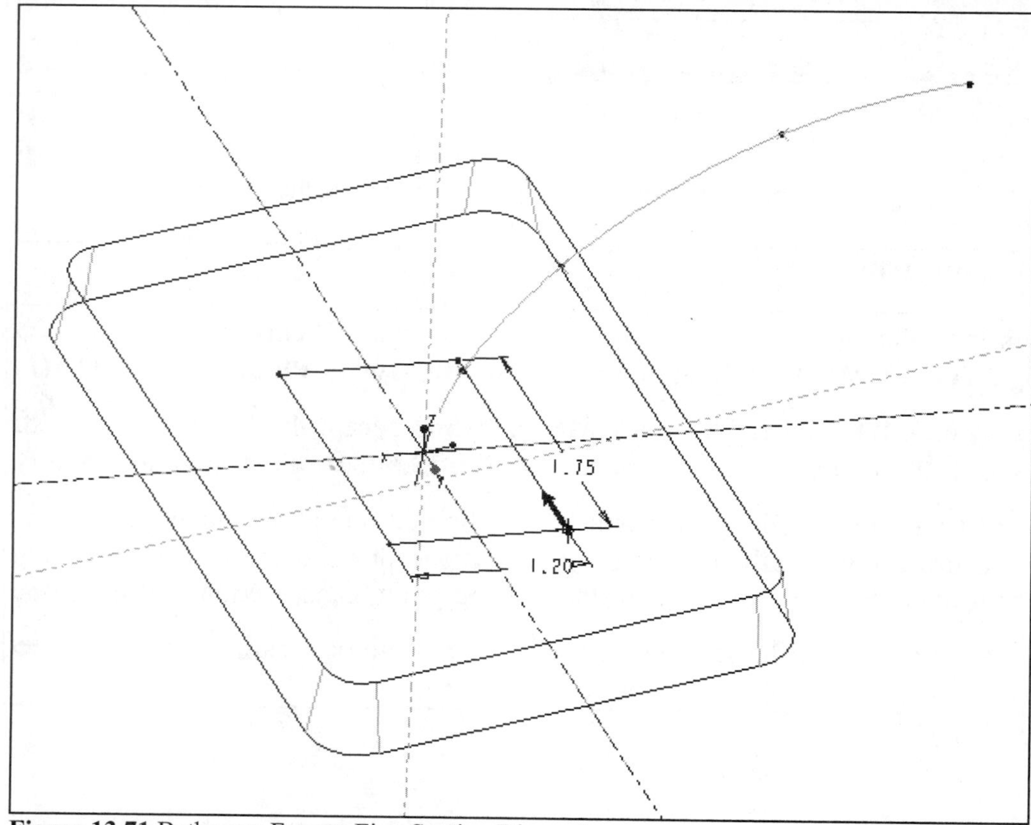

Figure 13.71 Bathroom Faucet, First Section Dimensions

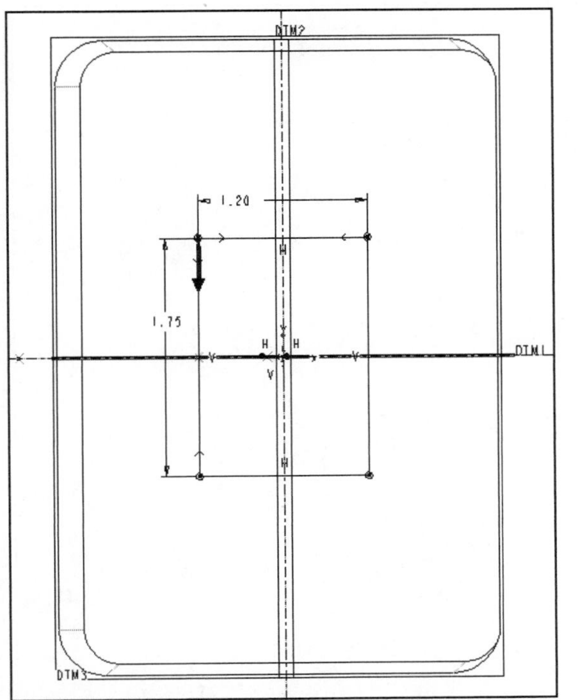

Figure 13.72 Bathroom Faucet, First Section: **1.75 X 1.20**

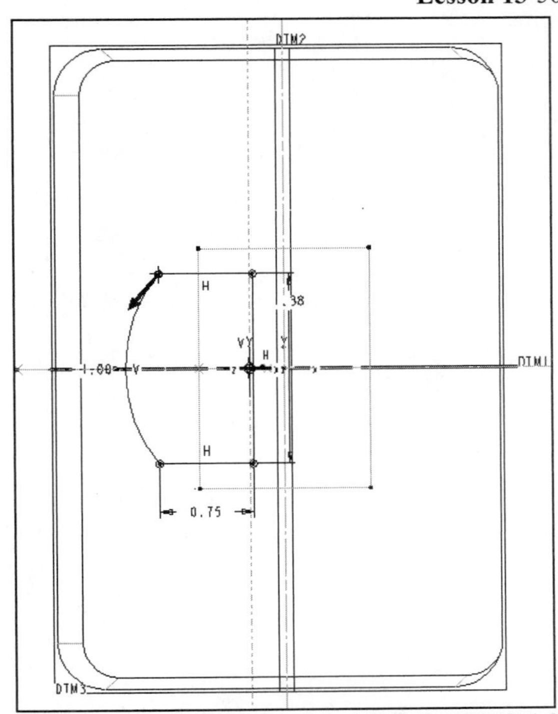

Figure 13.73 Second Section: **1.380 X .75 X R1.00**

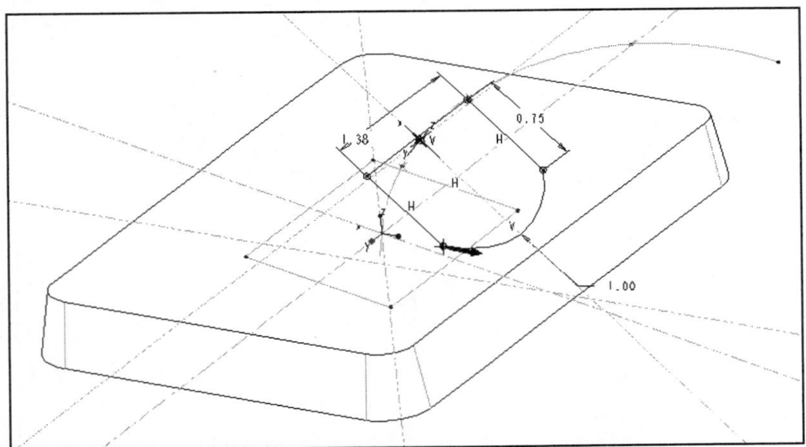

Figure 13.74 Second Section: Pictorial For the second section, align the arc center to the colored crosshairs (starting point).

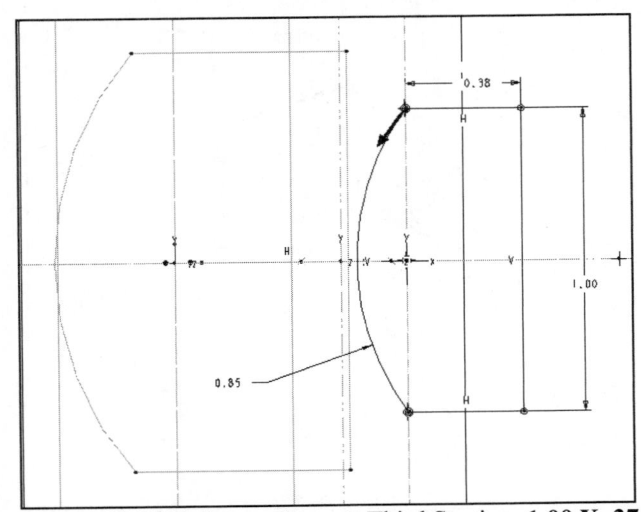

Figure 13.75 Bathroom Faucet, Third Section: **1.00 X .375 X R.850**.

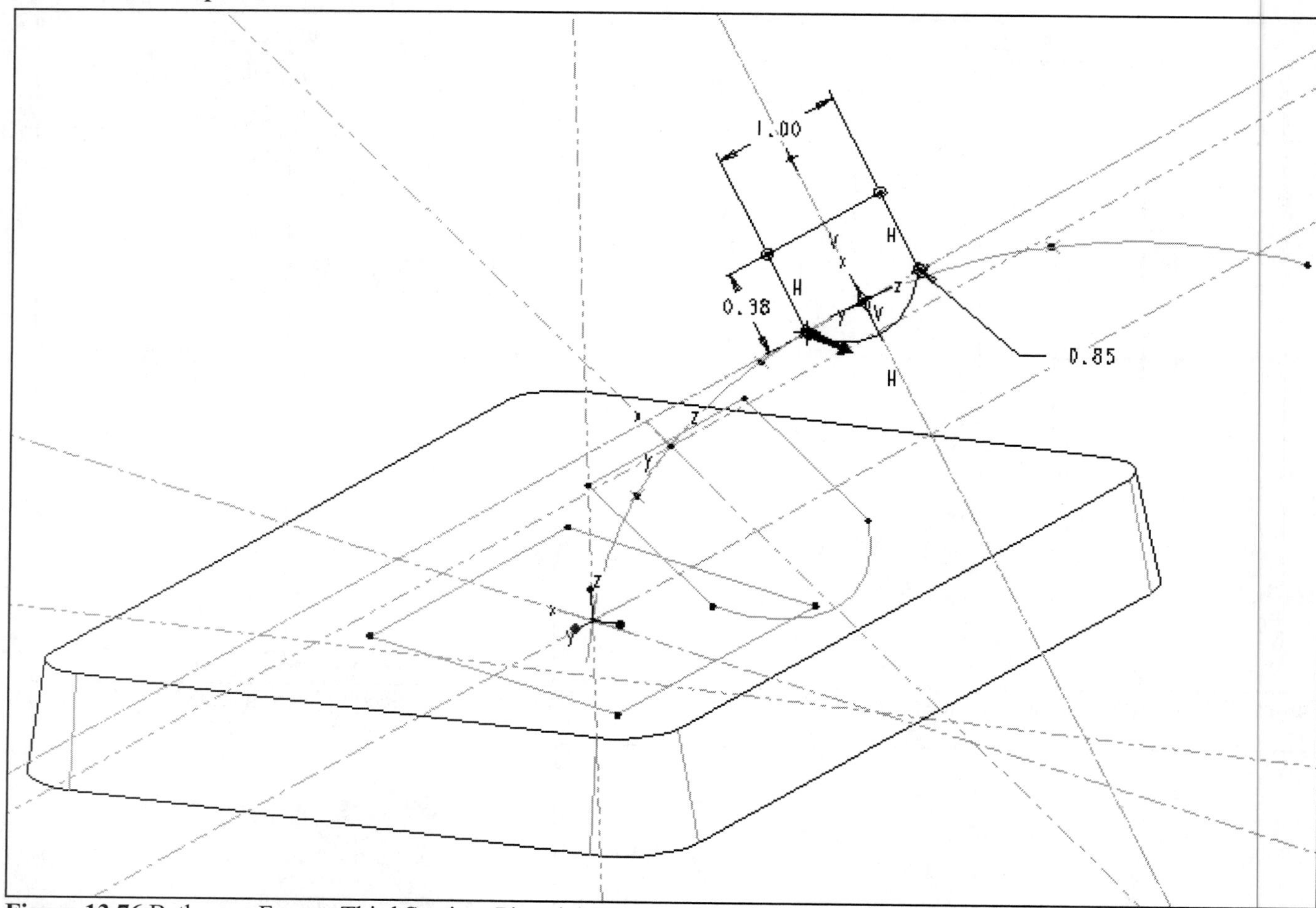

Figure 13.76 Bathroom Faucet, Third Section: Pictorial

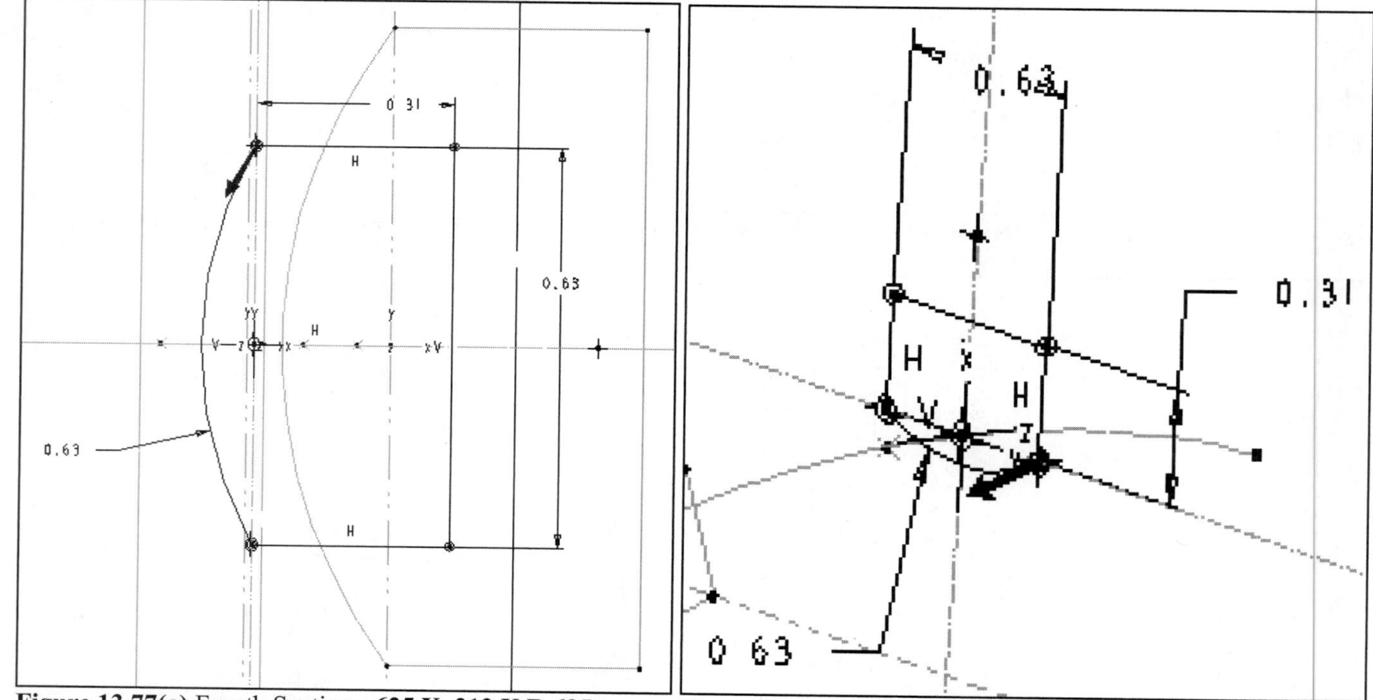

Figure 13.77(a) Fourth Section**: .625 X .312 X R.625**

Figure 13.77(b) Fourth Section

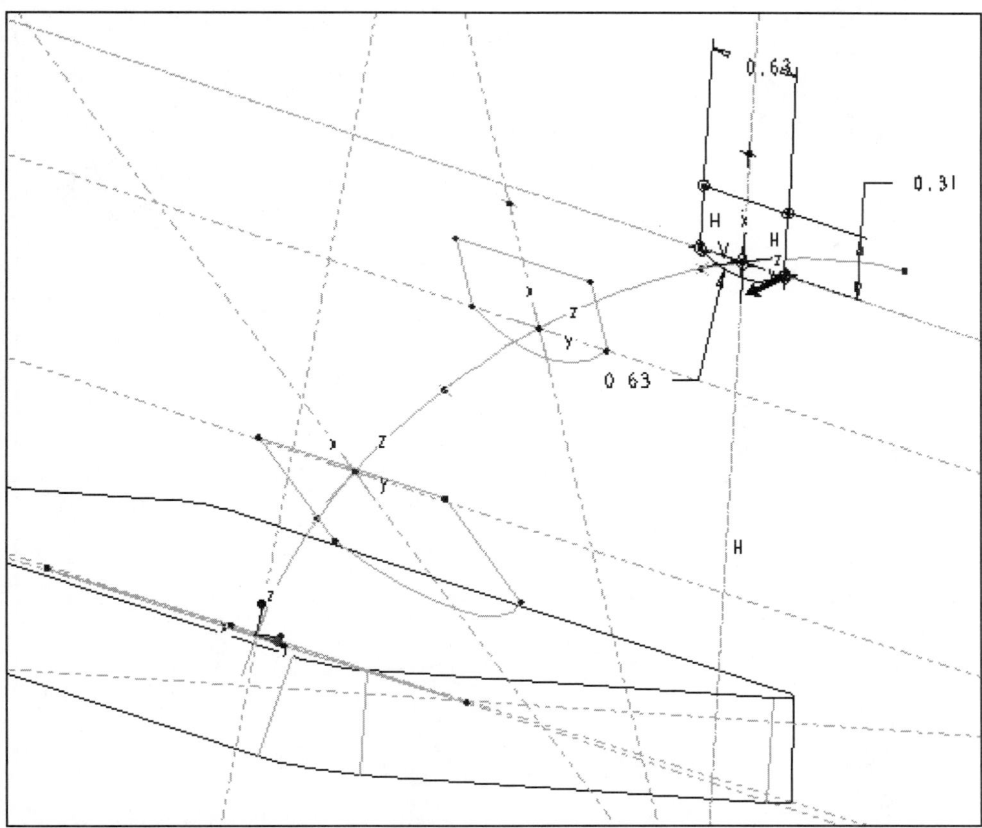

Figure 13.78 Bathroom Faucet, Fourth Section: Pictorial

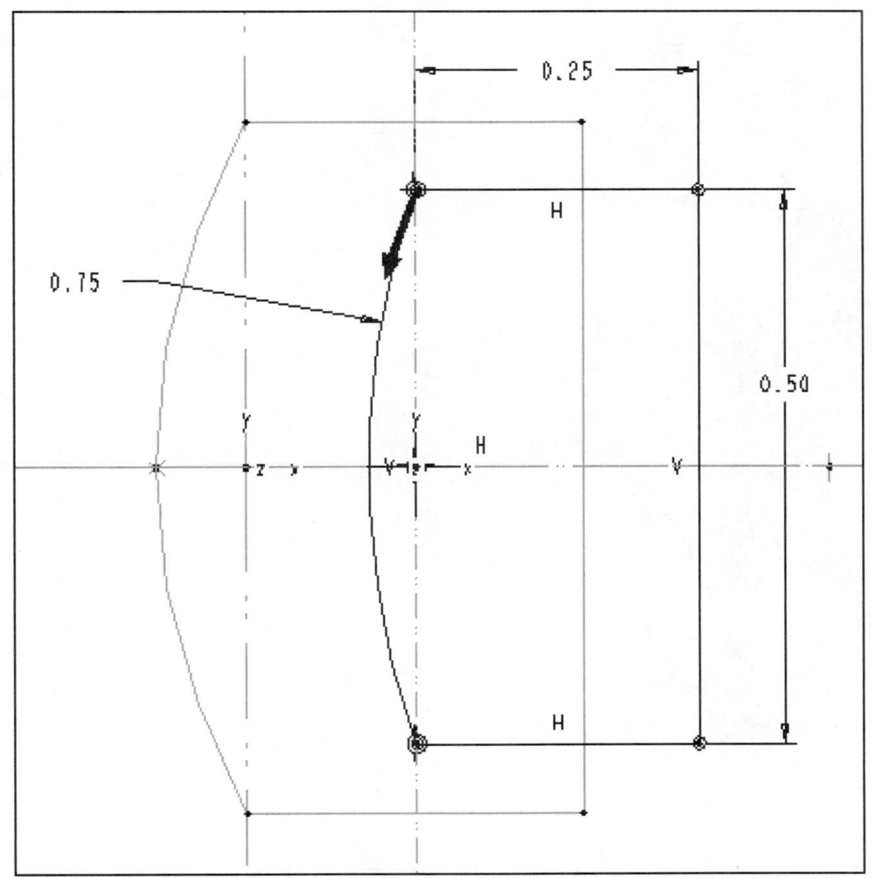

Figure 13.79 Bathroom Faucet, Fifth Section: **.500 X .25 X R.75**

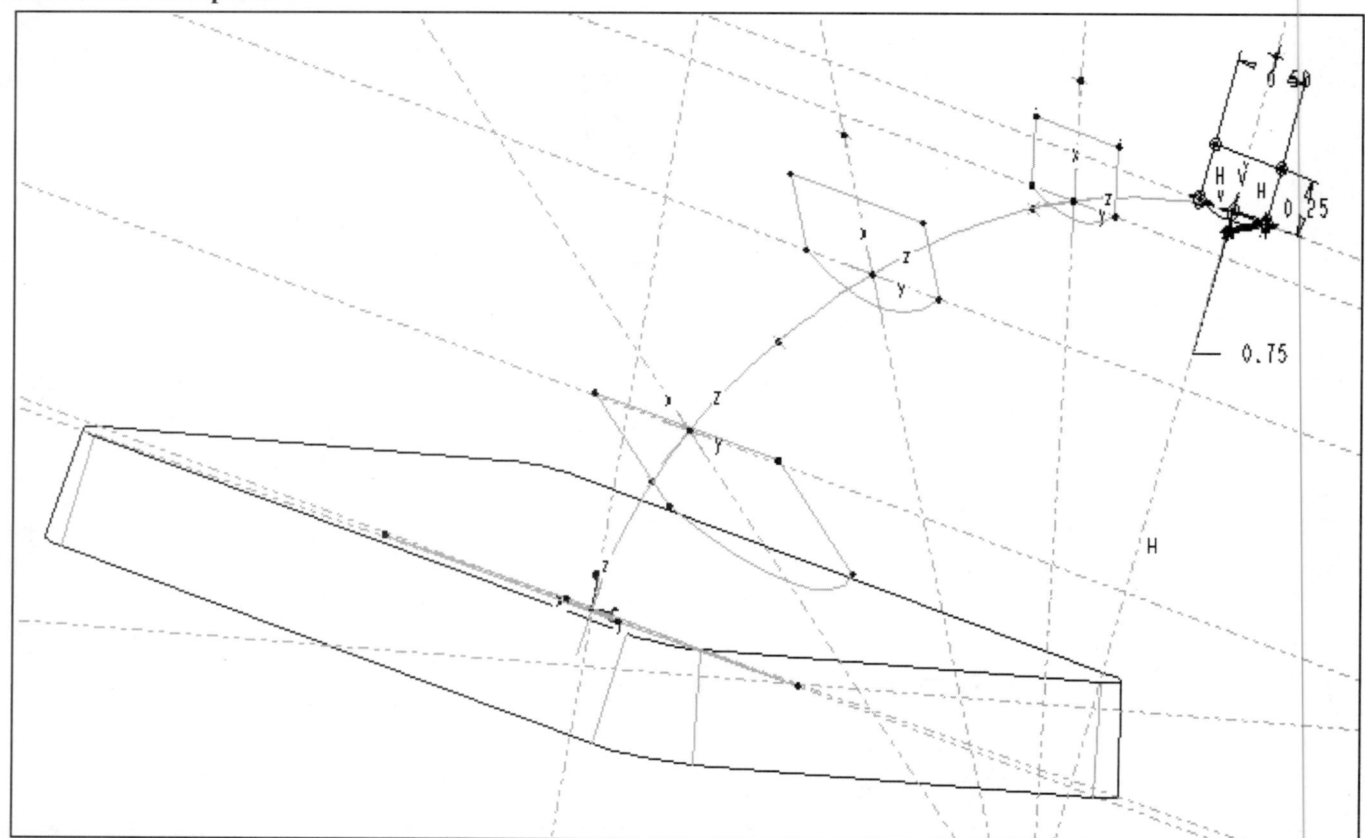

Figure 13.80 Bathroom Faucet, Fifth Section: Pictorial

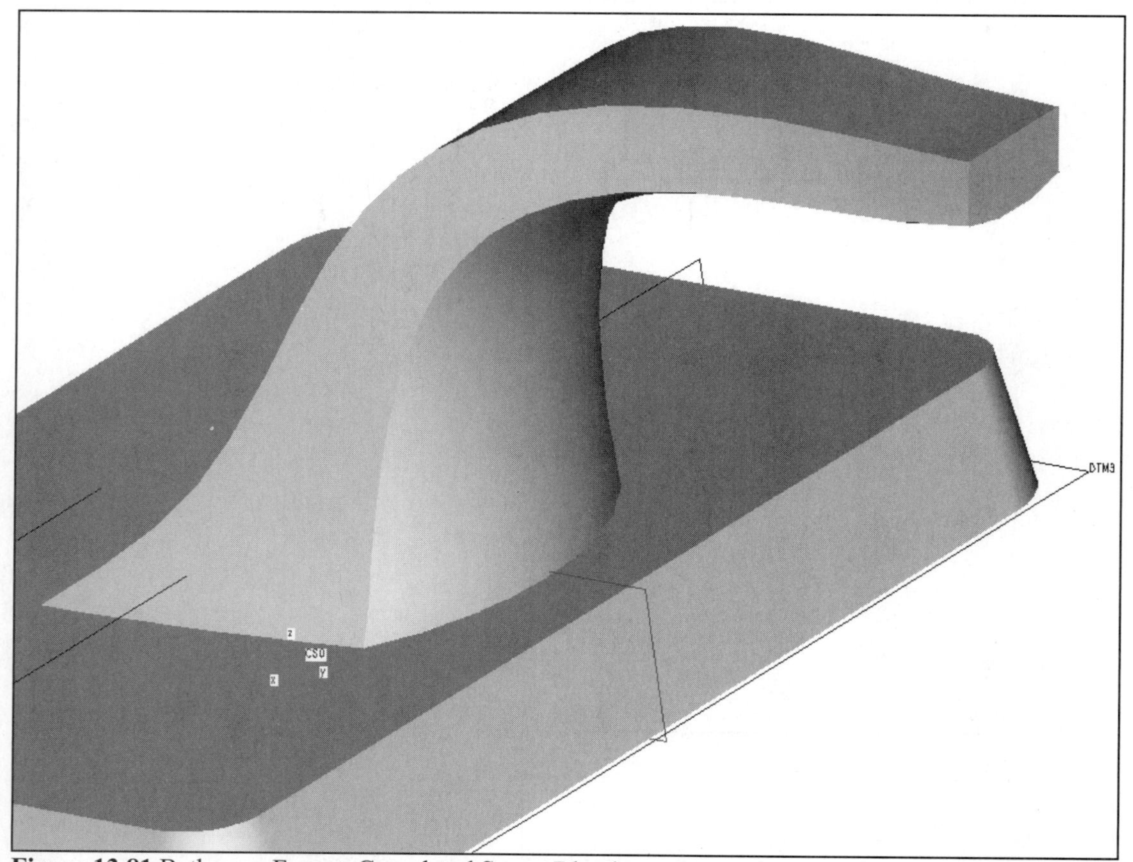

Figure 13.81 Bathroom Faucet, Completed Swept Blend

Lesson 14 Helical Sweeps and 3D Model Notes

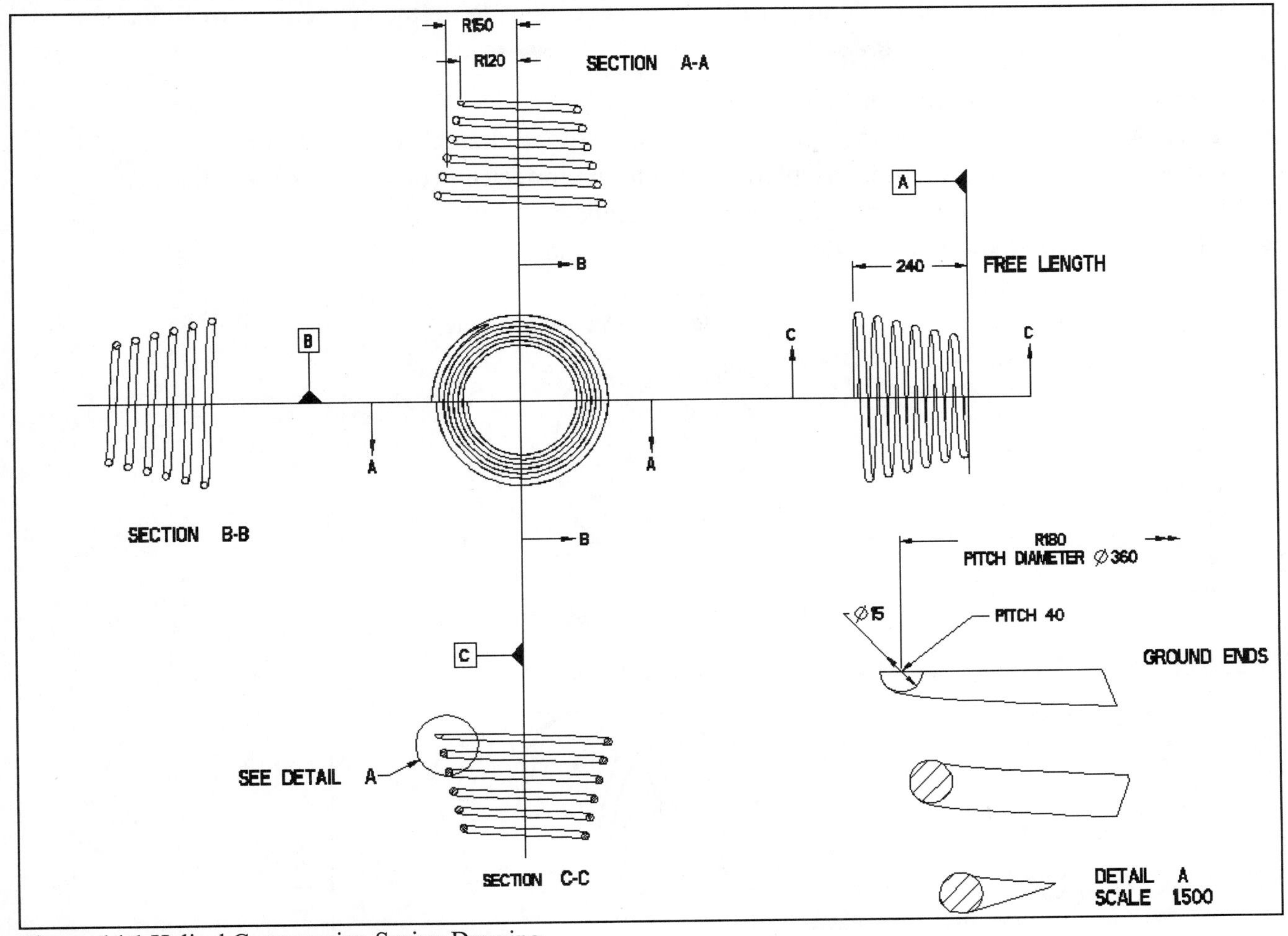

Figure 14.1 Helical Compression Spring Drawing

OBJECTIVES

- Create springs with a **Helical Sweep**
- Model a **helical compression spring**
- Use sweeps to create **hooks** on **extension springs**
- Design an **extension spring** with a **machine hook**
- Create **plain ground ends** on a spring
- Model a **convex spring**
- Create **3D Notes**

HELICAL SWEEPS AND 3D MODEL NOTES

A **helical sweep** (Fig. 14.1) is created by sweeping a section along a helical *trajectory*. The trajectory is defined by both the *profile* of the *surface of revolution* (which defines the distance from the section origin of the helical feature to its *axis of revolution*) and the *pitch* (the distance between coils). The trajectory and the surface of revolution are construction tools and do not appear in the resulting geometry. **3D Model Notes** are pieces of text, which can contain links (URL's) to World Wide Web pages, which you can attach to objects in Pro/E. Model notes, increase the amount of information that you can attach to any object in your model.

Helical Sweeps

The Helical Sweep command is available (Fig. 14.2) for both solid and surface features. Use the following ATTRIBUTES menu options, presented in mutually exclusive pairs, to define the helical sweep feature:

- **Constant** The pitch is constant
- **Variable** The pitch is variable and defined by a graph
- **Thru Axis** The section lies in a plane that passes through the axis of revolution
- **Norm To Traj** The section is oriented normal to the trajectory (or surface of revolution)
- **Right Handed** The trajectory is defined by the right-hand rule
- **Left Handed** The trajectory is defined by the left-hand rule

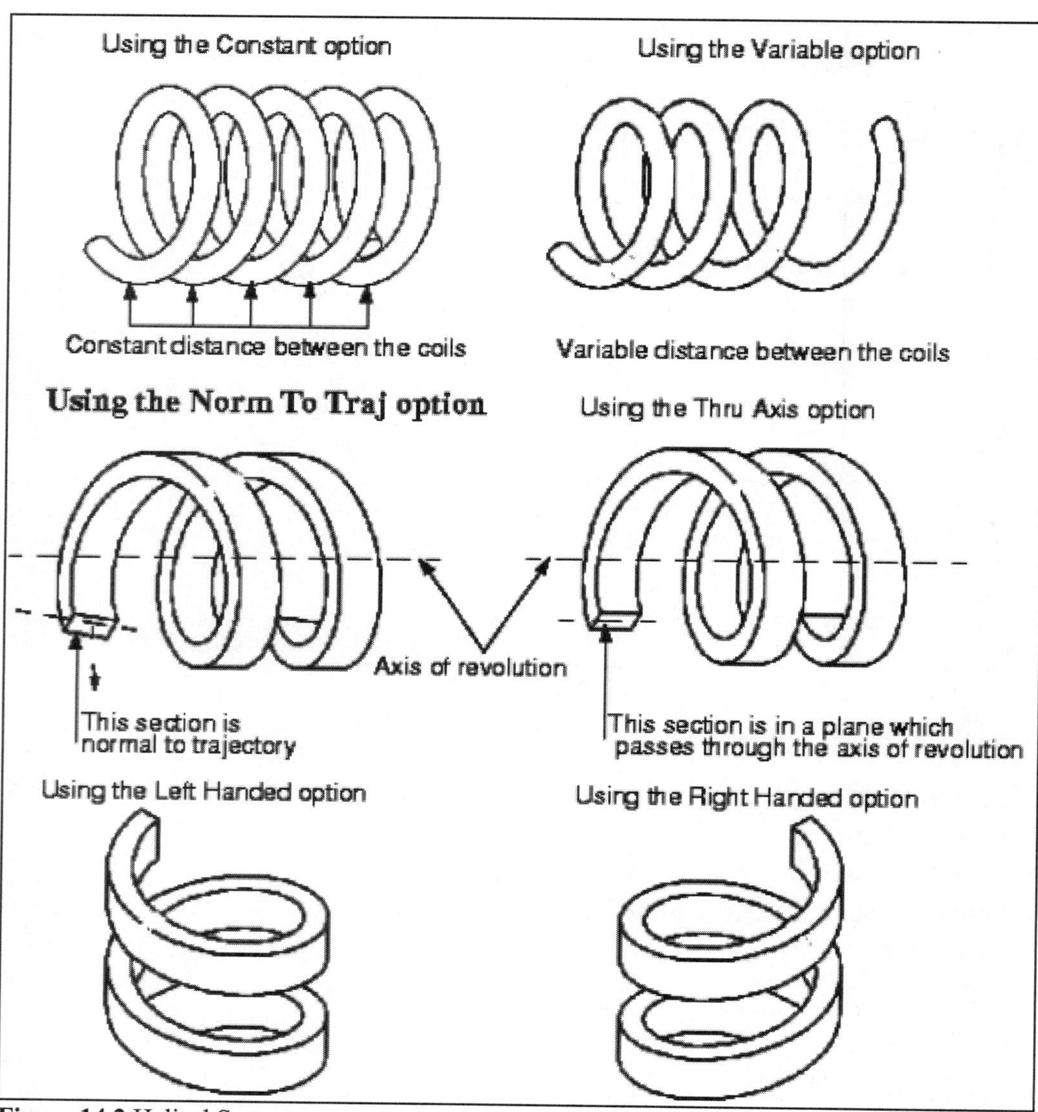

Figure 14.2 Helical Sweeps

3D Model Notes

Model notes are text strings that you can attach to objects [Fig. 14.3(a)]. You can attach any number of notes to any object in your model. When you attach a note to an object, the object is considered the parent of the note. When you delete the parent object, all child notes are deleted with it. The attachment is at the end of the model note leader line. Notes do not have to be attached to a parent. You can also allocate a URL to each model note [Fig. 14.3(b)]. You can use model notes to:

- Communicate with members of your workgroup as to how to review or use a model
- Explain how you approached or solved a design problem when modeling
- Explain changes that you have made to the features of a model over time

Modifying Note Text Style

You can modify the following text style attributes of an existing model note:

- Style name
- Font, width factor, and slant angle of text
- Line spacing and angle
- Horizontal and vertical justification
- Note orientation. That is, you can make the note a mirror image of itself
- System and user-defined colors for note text

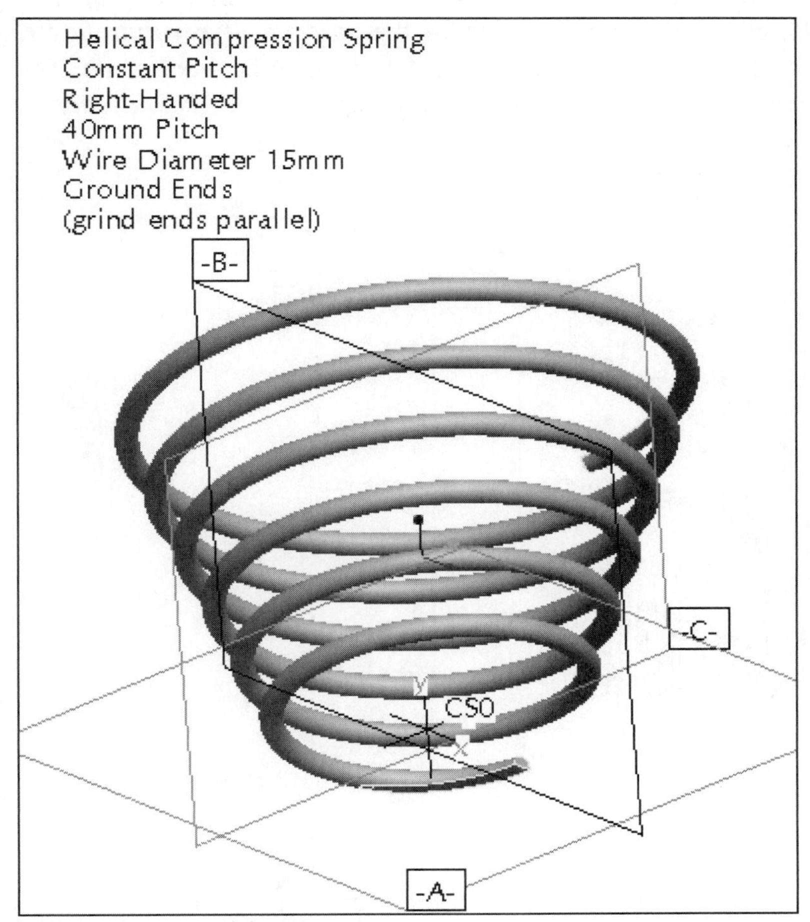

Figure 14.3(a) Model Notes

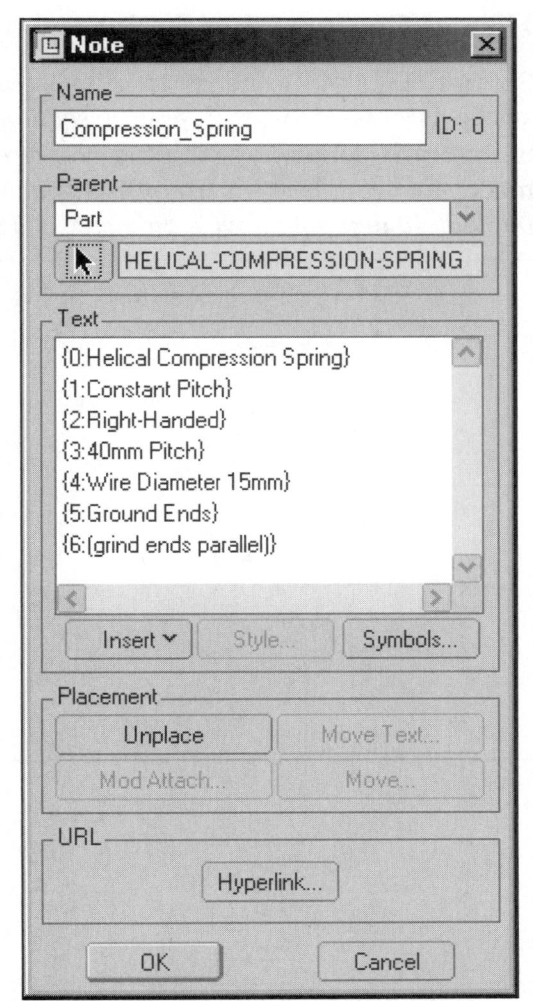

Figure 14.3(b) Note Dialog Box

Lesson 14 STEPS

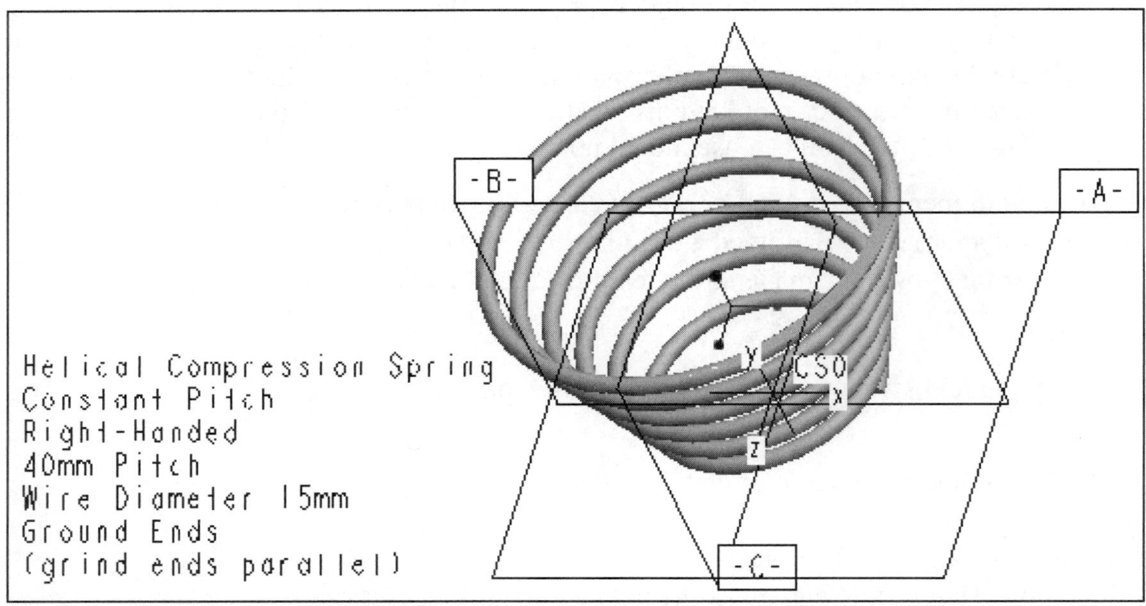

Figure 14.4 Helical Compression Spring with Datum Planes and Model Note

Helical Compression Spring

Springs (Fig. 14.4) and other helical features are created with the Helical Sweep command. A helical sweep is created by sweeping a *section* along a *trajectory* that lies in the *surface of revolution:* the trajectory is defined by both the *profile* of the surface of revolution and the distance between coils. The model for this lesson is a *constant-pitch right-handed helical compression spring with ground ends, a pitch of **40 mm**, and a wire diameter of **15 mm*** (Figs. 14.4 through 14.8).

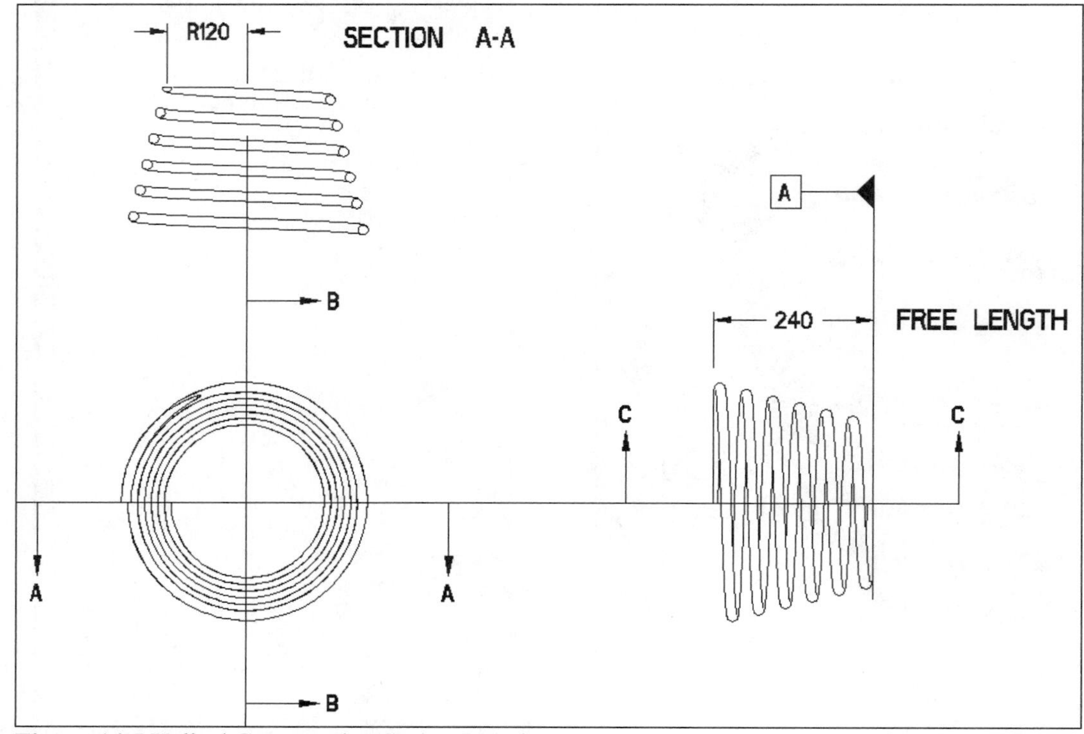

Figure 14.5 Helical Compression Spring Drawing

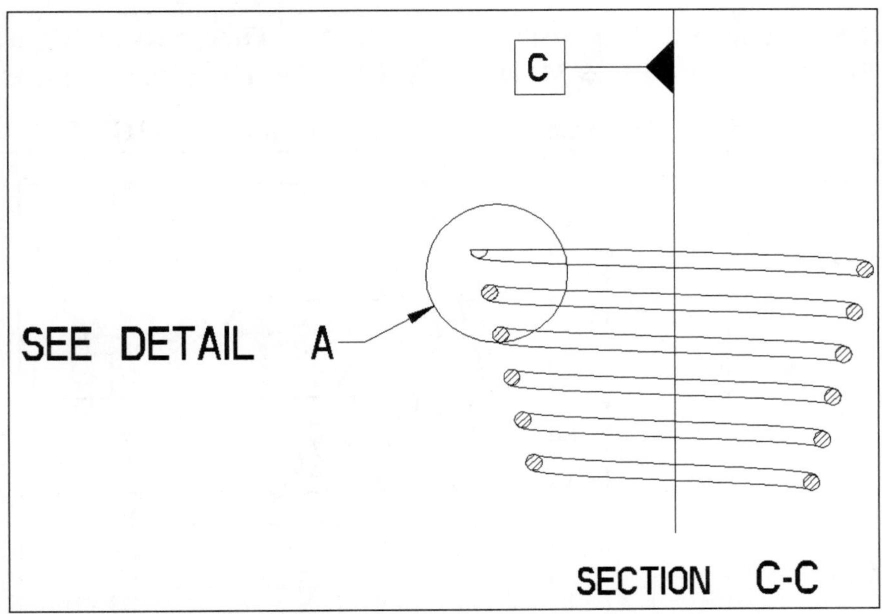

Figure 14.6 Helical Compression Spring Drawing, SECTION C-C

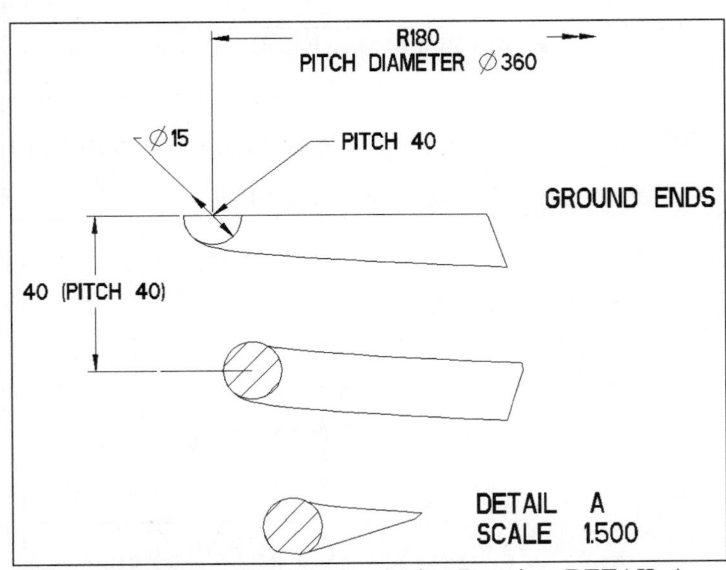

Figure 14.7 Helical Compression Spring Drawing, DETAIL A

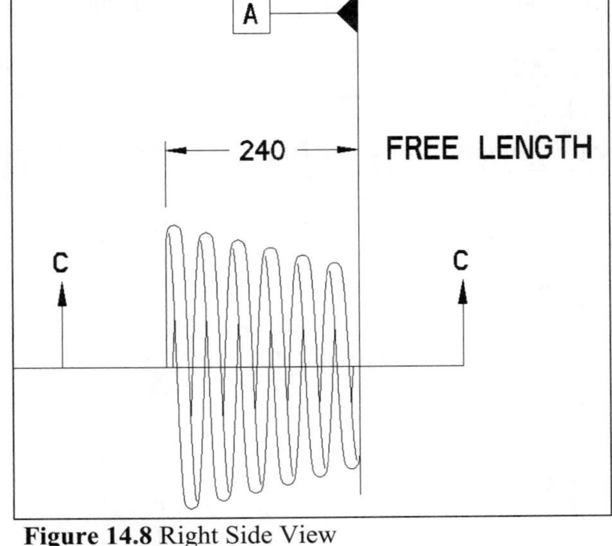

Figure 14.8 Right Side View

Start a new part. Click: **Create a new object** ⇒ ●**Part** ⇒ Name **Helical_Compression_Spring** ⇒ Use default template ⇒ **MMB** ⇒ **Edit** ⇒ **Setup** ⇒ setup up your project as follows:

- **Material** = SPRING_STEEL
- **Units** = millimeters

Set Datum and **Rename** the default datum planes and coordinate system:

- Datum TOP = **A**
- Datum FRONT = **B**
- Datum RIGHT = **C**
- Coordinate System = **CS0**

Create the first protrusion: **Insert ⇒ Helical Sweep ⇒ Protrusion ⇒ Constant ⇒ Thru Axis ⇒ Right Handed ⇒ MMB ⇒** pick datum **B (FRONT) ⇒ MMB ⇒ MMB ⇒ MMB ⇒ MMB ⇒** sketch a line [(Fig. 14.9(a-b)] ⇒ **Create 2 point centerlines** add a vertical centerline along datum **C (RIGHT)**

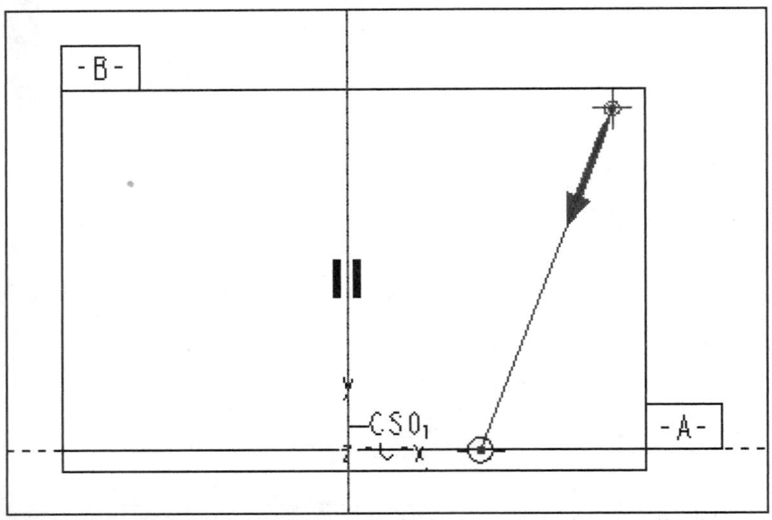

Figure 14.9(a) Sketch the Profile Line and a Vertical Centerline

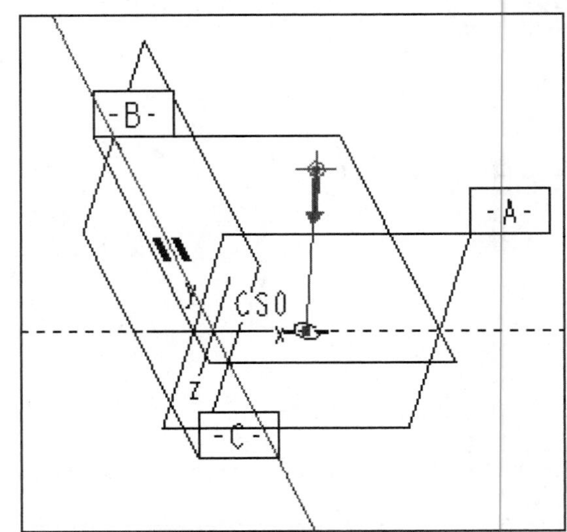

Figure 14.9(b) Sketch in 3D

Click: create all required dimensions ⇒ **Modify the values of dimensions** change the values to the design sizes [Fig. 14.10(a-b)] ⇒ ⇒ enter the pitch value **40** at the prompt ⇒ **MMB**

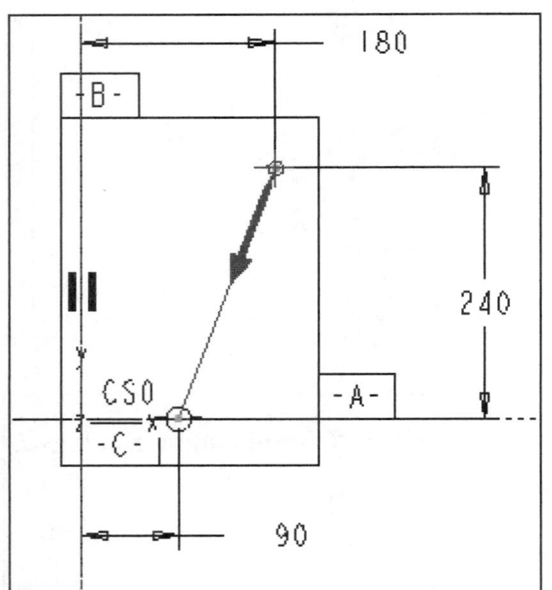

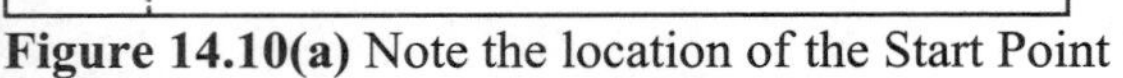

Figure 14.10(a) Note the location of the Start Point

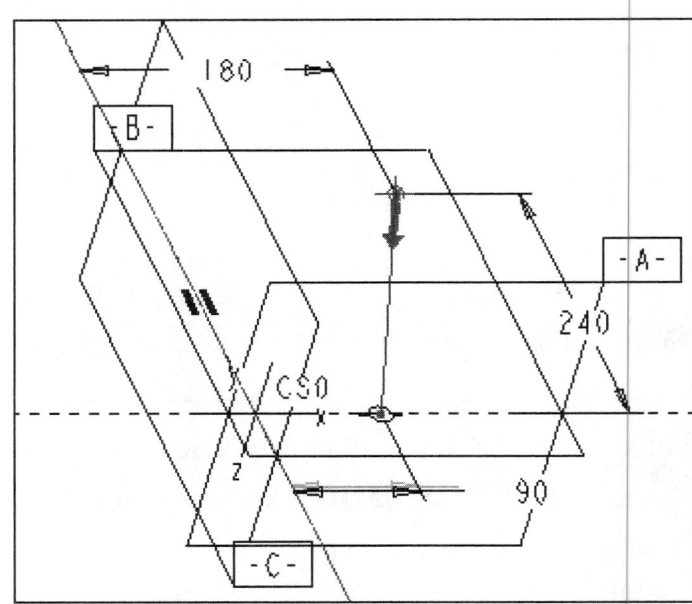

Figure 14.10(b) Start point in 3D

Sketch the section geometry of the spring at the crosshairs (a circle), click: **Create circle** (Fig. 14.11) ⇒ **Modify the values of dimensions (15)** (Fig. 14.12) ⇒ **MMB** ⇒ ⇒ **MMB** ⇒ **Shading** ⇒ ⇒ **MMB** ⇒ ⇒ **Standard Orientation** (Fig. 14.13)

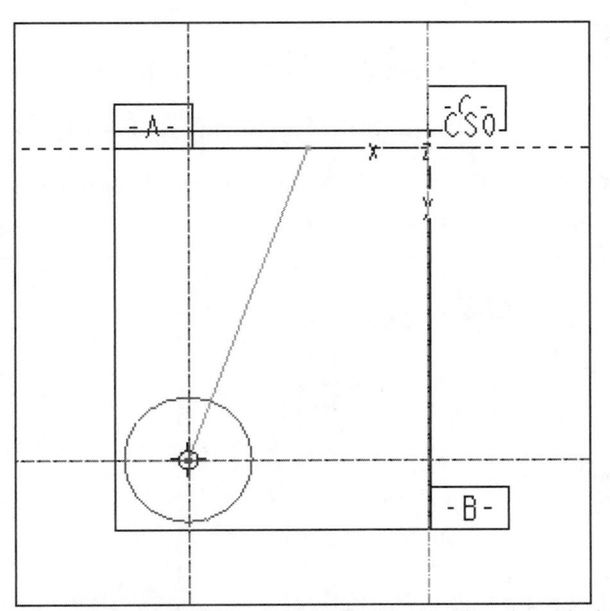

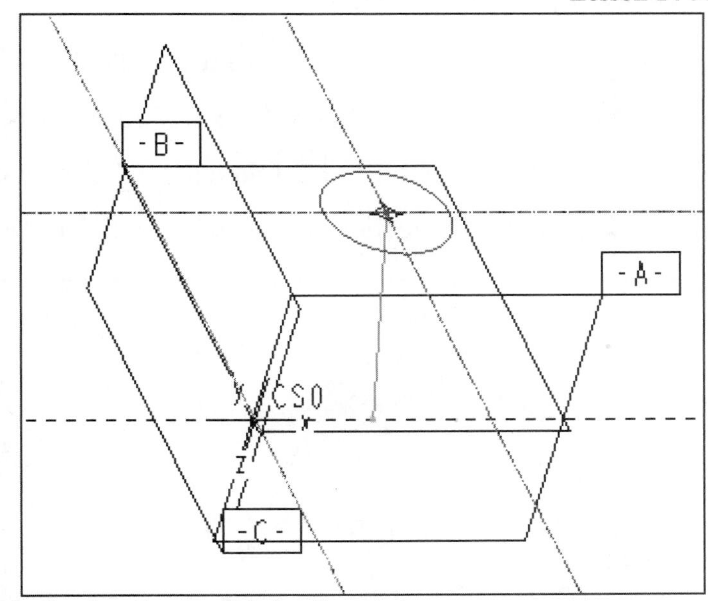

Figure 14.11(a-b) Sketching the Circle, Sketch a Circle as the Section Geometry (wire diameter)

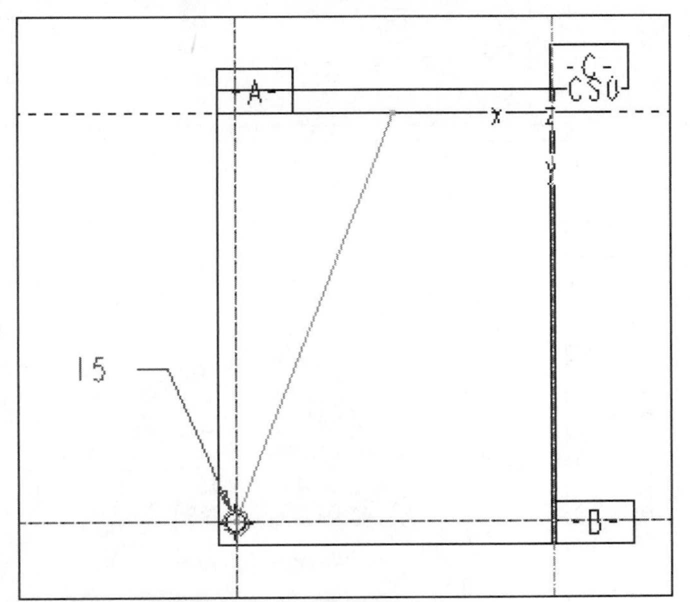

Figure 14.12(a) Wire Diameter **15**

Figure 14.12(b) Wire Diameter Sketch

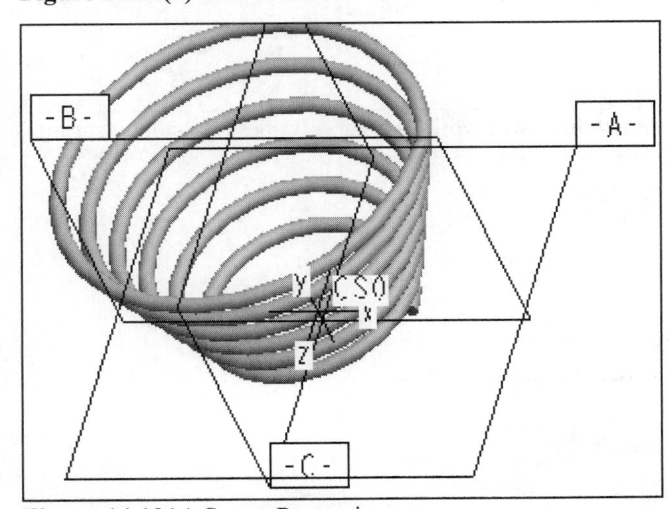

Figure 14.13(a) Swept Protrusion

Figure 14.13(b) Shaded Spring

Create the *ground ends*, click: **Extrude Tool** ⇒ **Remove Material** ⇒ **Extrude on both sides** ⇒ ⇒ Plane**:** pick datum **C** ⇒ Reference: pick datum **A** ⇒ Orientation: **Bottom** (Fig. 14.14) ⇒ **MMB** ⇒ **MMB** ⇒ **Sketch** ⇒ sketch one horizontal line [Fig. 14.15(a-b)] ⇒ modify the dimensions (Fig. 14.16) ⇒ spin the model as needed ⇒ (Fig. 14.17)

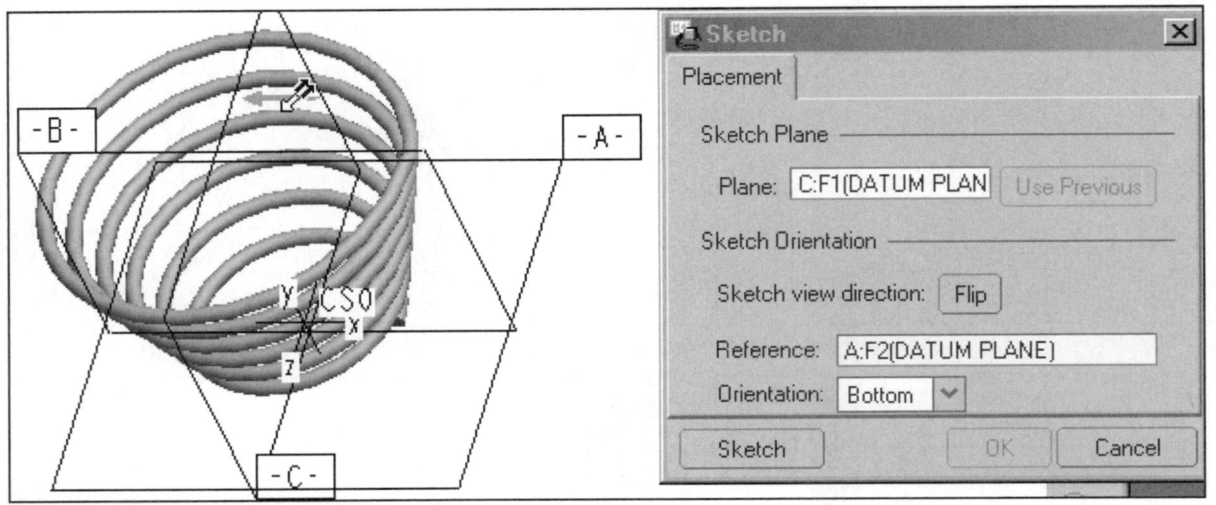

Figure 14.14 Cut Sketch Orientation

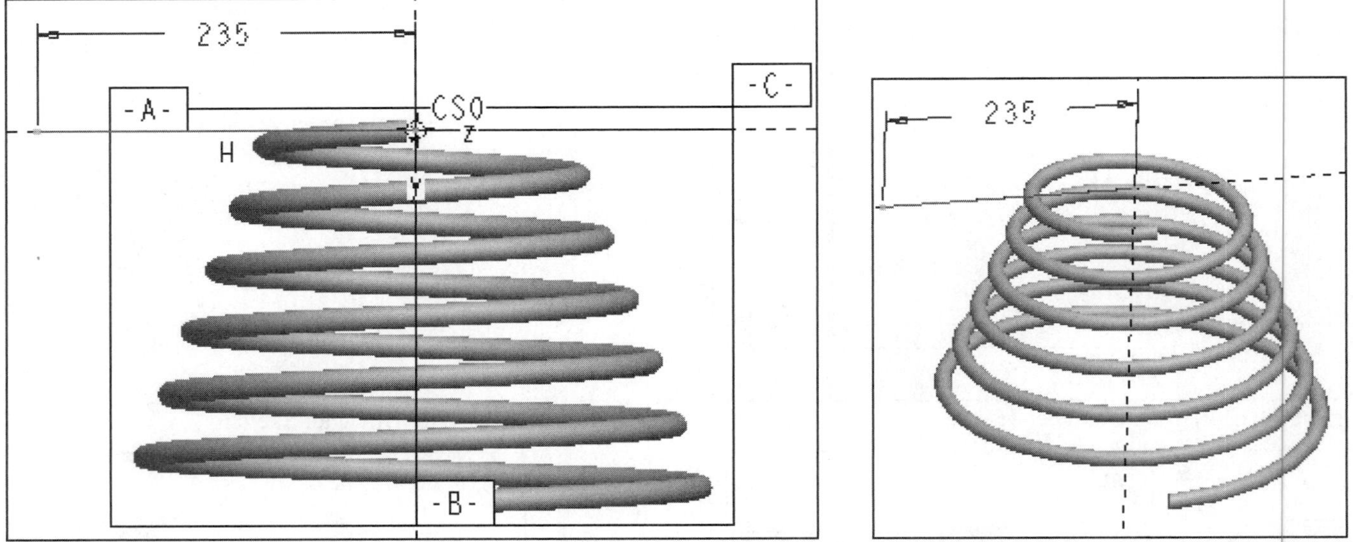

Figure 14.15(a-b) Creating Ground Ends (One dimension required; any length will work as long as it goes beyond the spring)

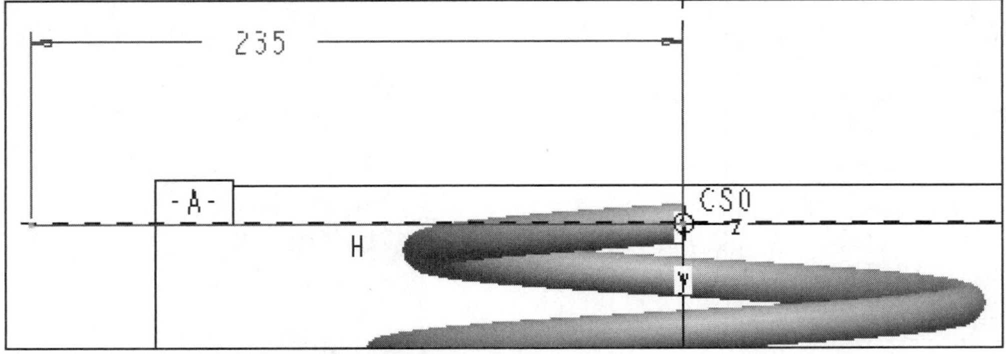

Figure 14.16 Ground End Cut Dimension

Extend the depth handle to include the spring (Fig. 14.17 and Fig. 14.18) ⇒ [icon] (Fig. 14.19) ⇒ verify material direction ⇒ [icon] ⇒ [icon] ⇒ [icon] ⇒ [icon] ⇒ [icon] ⇒ **MMB** ⇒ **File** ⇒ **Delete** ⇒ **Old Versions** ⇒ **MMB**

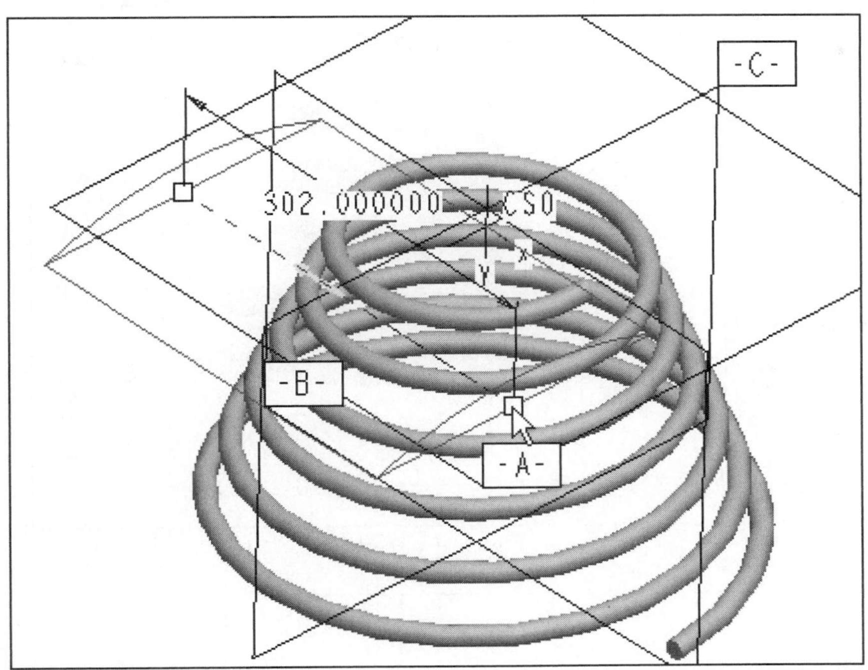

Figure 14.17 Depth Handle

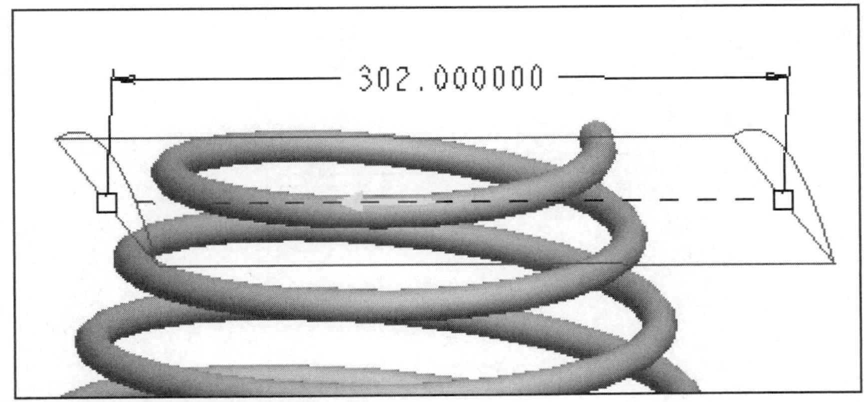

Figure 14.18 Ground End Depth

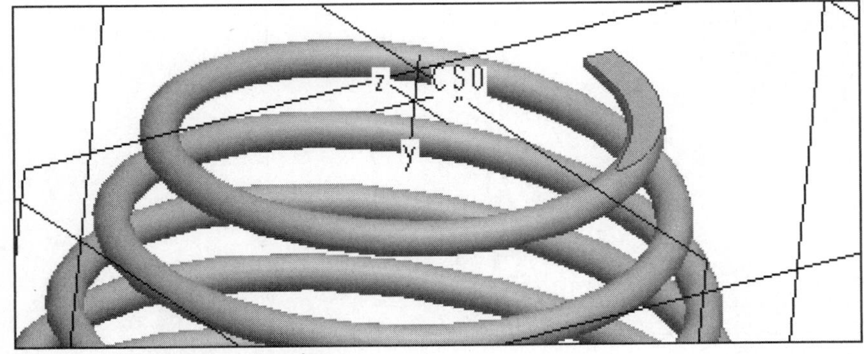

Figure 14.19 Feature Preview

The second ground end (Figs. 14.20 and 14.21) is created using similar commands. The completed spring is shown in Figure 14.22.

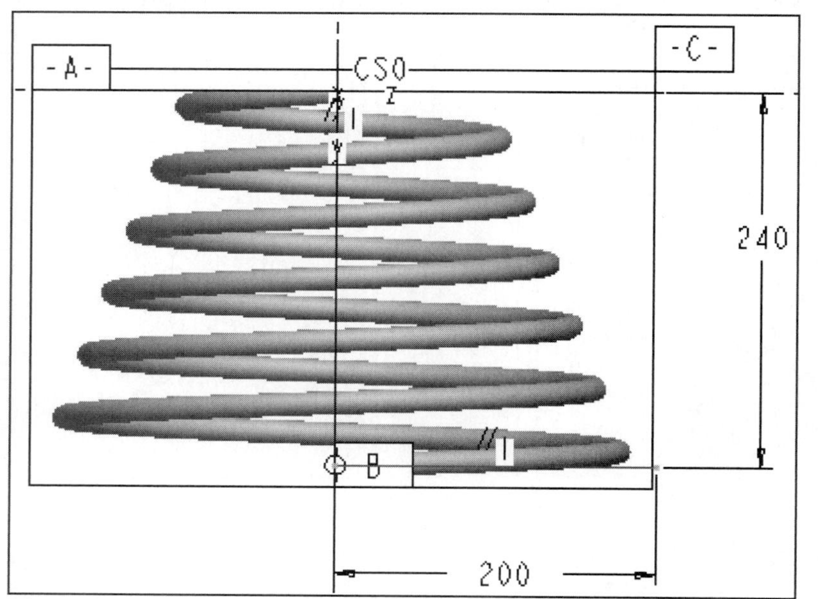

Figure 14.20 Creating the Second Ground End

Figure 14.21 Shaded Sketch

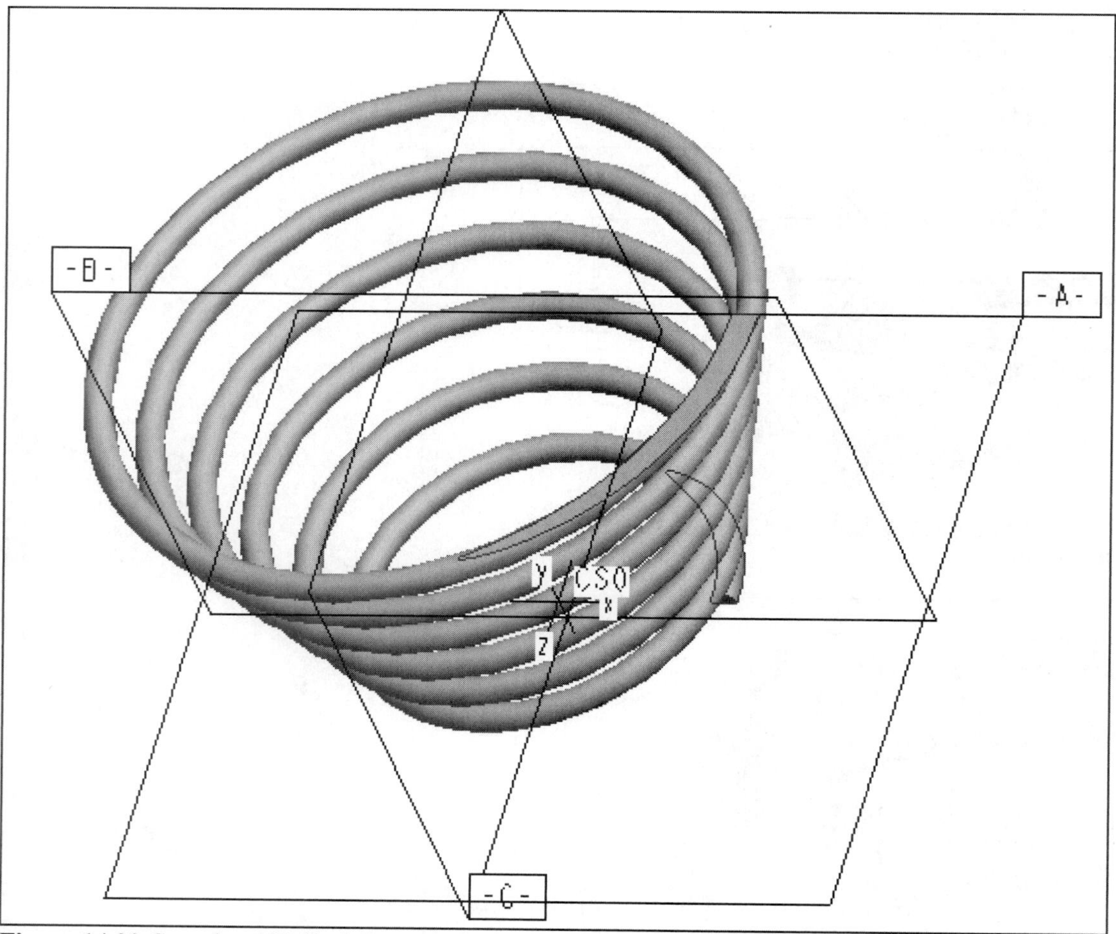

Figure 14.22 Completed Helical Compression Spring

Save a copy the Helical Compression Spring by clicking: **File ⇒ Save a Copy ⇒** Type a different name--**HEL_COMP_SPR_GRND_ENDS**. Figure 14.23 provides an **ECO** for the new spring. Rename the file you are working on by clicking: **File ⇒ Rename ⇒** provide a unique name such as **HEL_EXT_SPR_MACH_ENDS** ⇒ delete the existing ground ends ⇒ modify the pitch to **10 mm** ⇒ change the wire diameter to **7.5 mm** ⇒ Complete the extension spring (Figs. 14.24 through Fig. 14.33). The free length is to be **120 mm**. The large radius will now be **90 mm** (*was* ***180mm***), and the small radius will be **60 mm** (*was* ***90mm***).

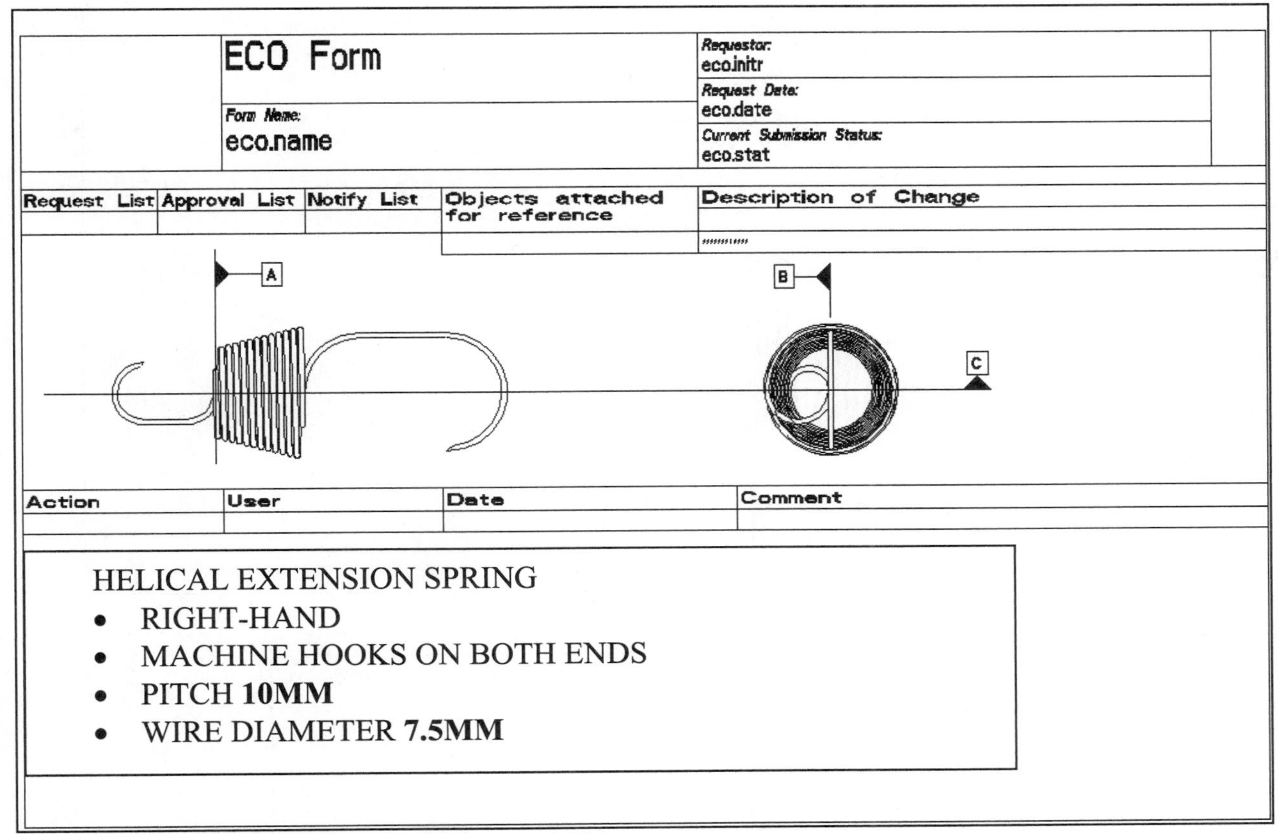

ECO Form

Form Name:
eco.name

Requestor:
eco.initr

Request Date:
eco.date

Current Submission Status:
eco.stat

Request List	Approval List	Notify List	Objects attached for reference	Description of Change

Action	User	Date	Comment

HELICAL EXTENSION SPRING

- RIGHT-HAND
- MACHINE HOOKS ON BOTH ENDS
- PITCH **10MM**
- WIRE DIAMETER **7.5MM**

Figure 14.23 ECO to Create a Helical Extension Spring

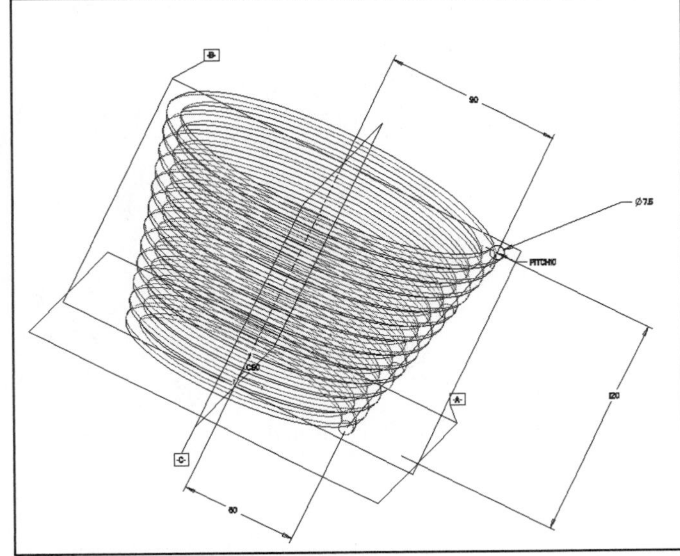

Figure 14.24 ECO Changes

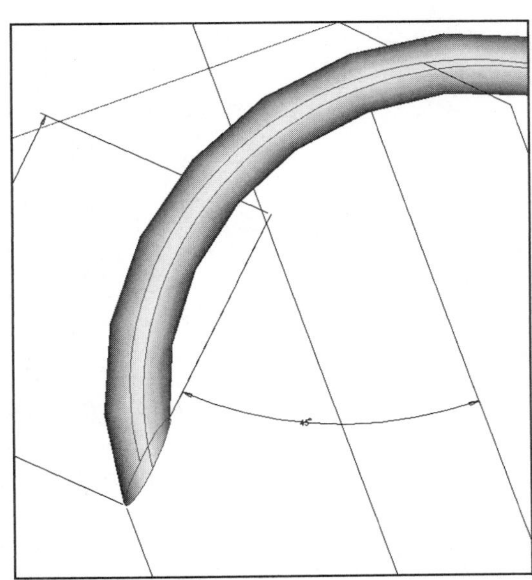

Figure 14.25 Ground End

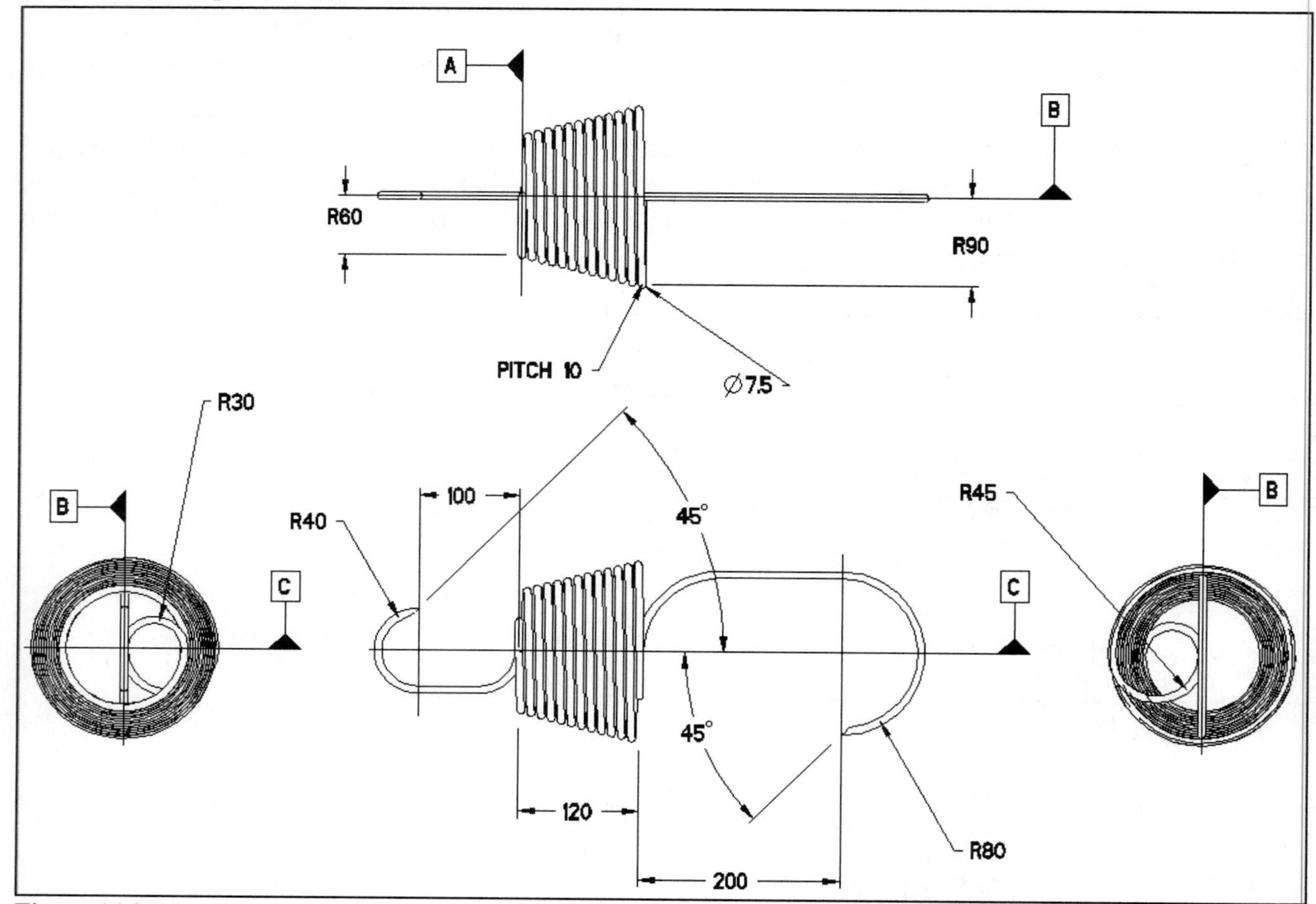

Figure 14.26 Detail Drawing of Helical Extension Spring with Machine Hook Ends

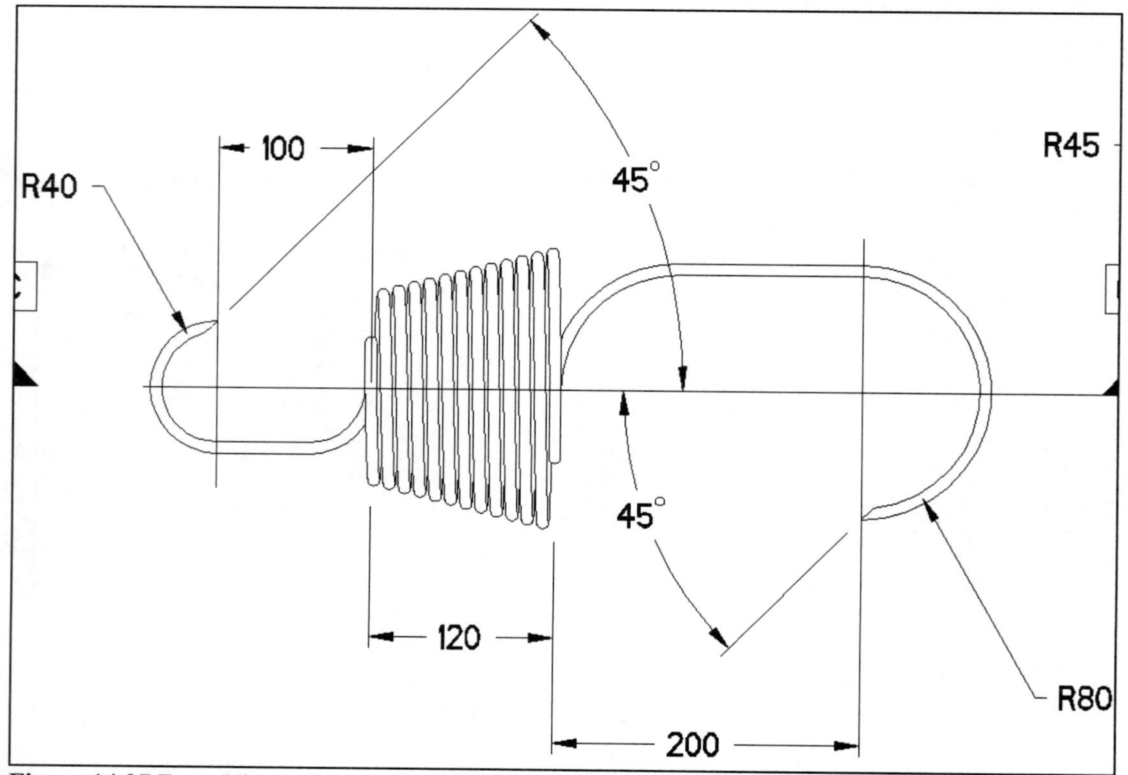

Figure 14.27 Front View

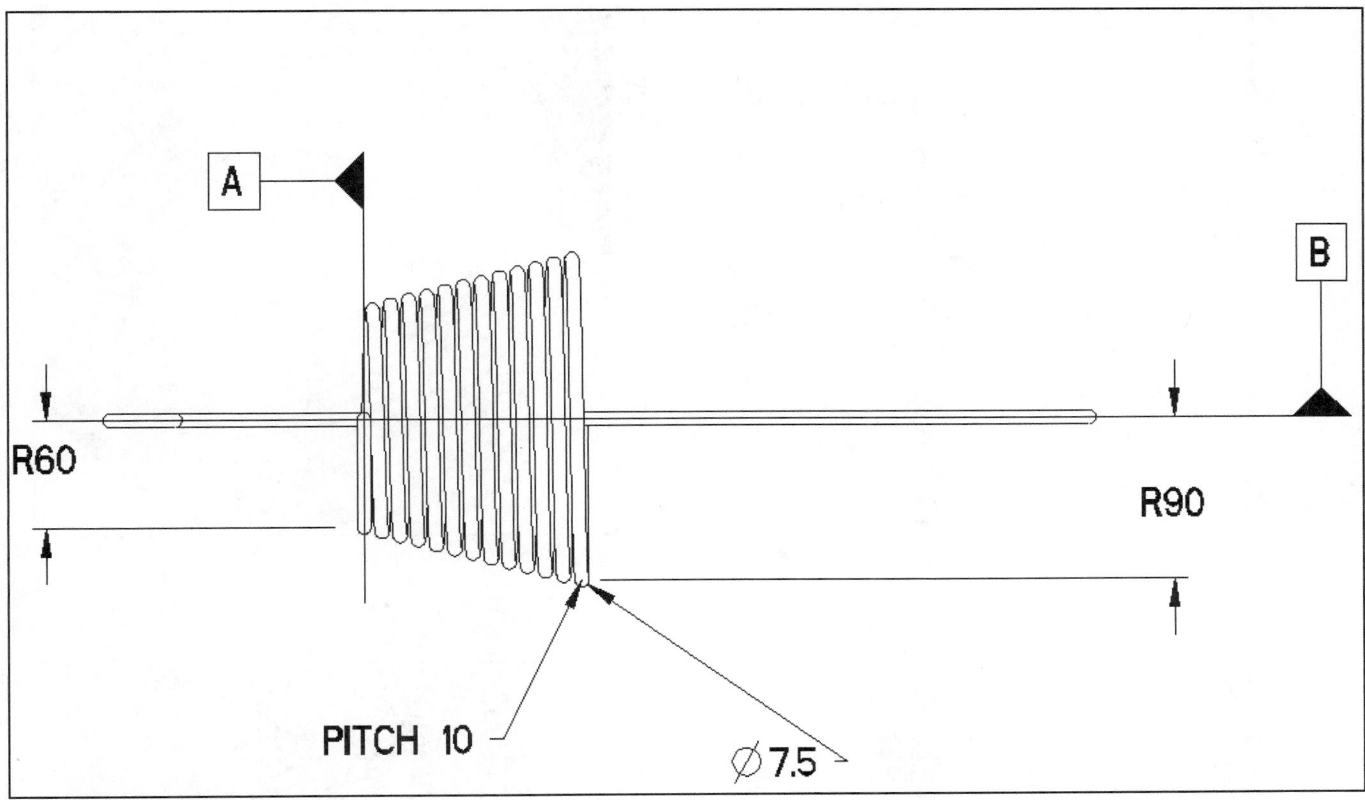

Figure 14.28 Top View

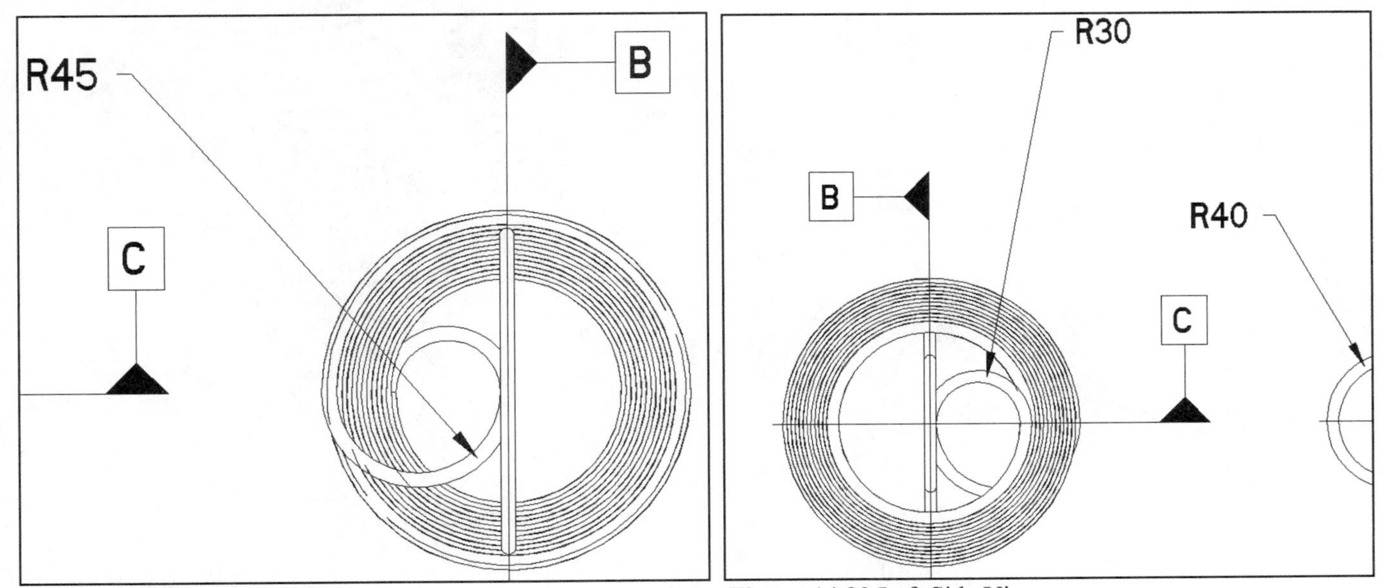

Figure 14.29 Right Side View

Figure 14.30 Left Side View

Create the machine hooks using simple sweeps and cuts, as shown in Figures 14.31 through 14.33.

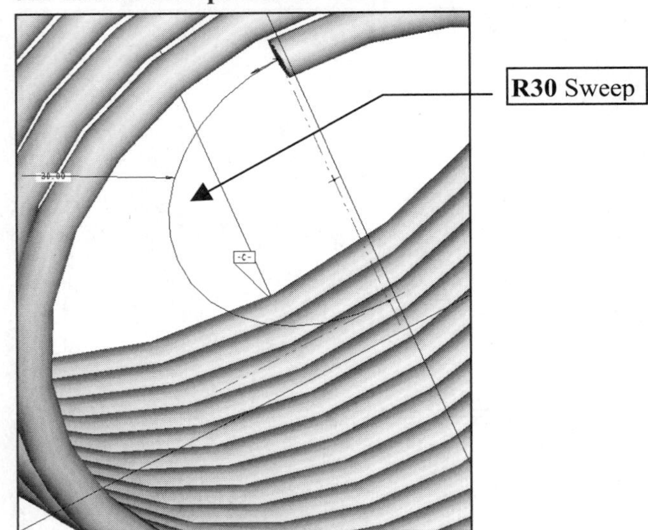

Figure 14.31(a) Sweep **R30**

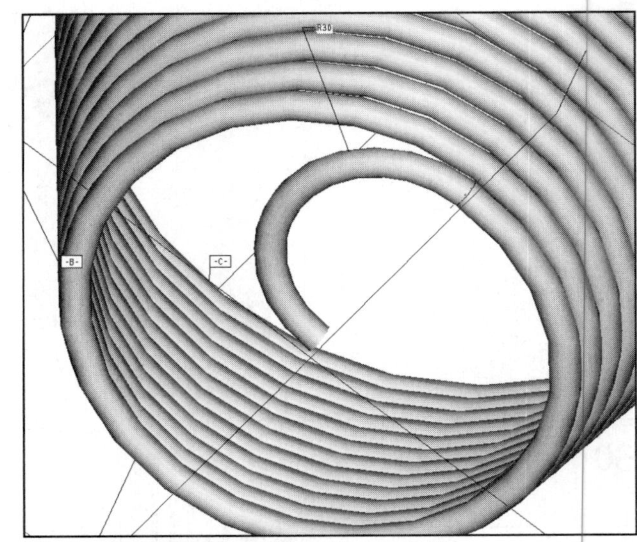
Figure 14.31(b) Completed Sweep **R30**

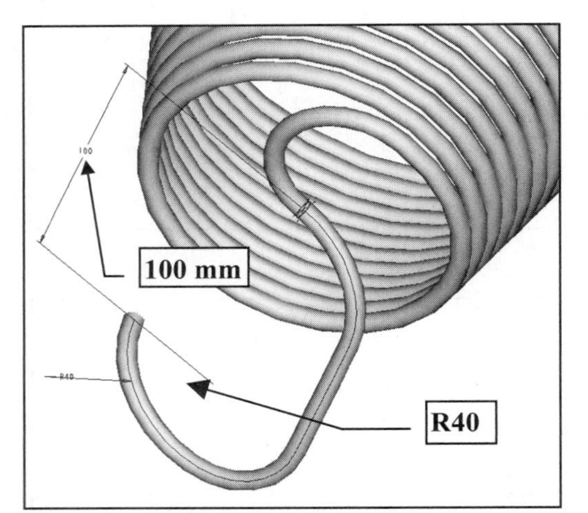

Figure 14.32(a) Small Hook End Sweep

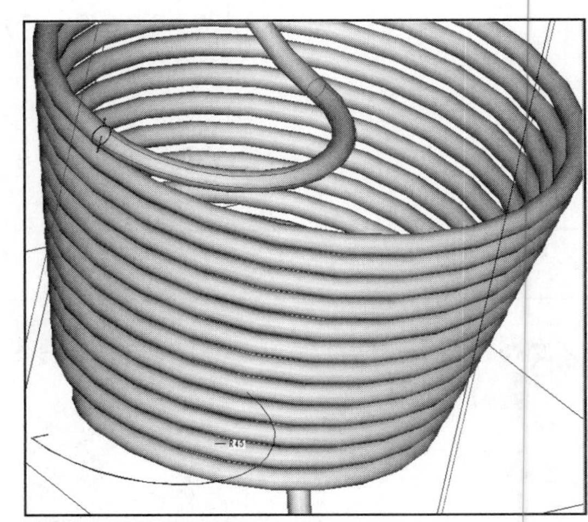
Figure 14.32(b) Hook Sweep

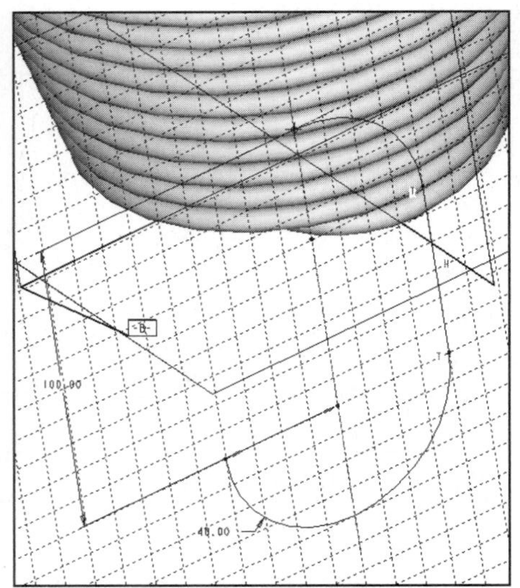
Figure 14.33(a) Large Hook End Sweep

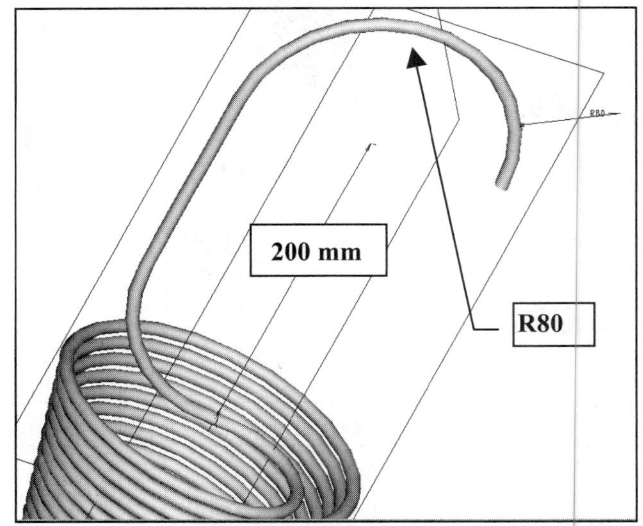

Figure 14.33(b) Large Hook Dimensions

Model Notes

When you attach a note to an object, the object is considered the "parent" of the note. Deleting the parent deletes all of the notes of the parent. You can attach model notes anywhere in the model; they do not have to be attached to a parent. Here we will add a note to the part and describe the spring.

Open the saved spring file that has the ground ends. Choose the following commands, click: **Edit** ⇒ **Setup** ⇒ **Notes** ⇒ **New** ⇒ type **Compression_Spring** as the name of the note; no spaces are allowed in the name ⇒ pick in the Text area and type the note (Fig. 14.34):

Helical Compression Spring
Constant Pitch
Right-handed
40 mm Pitch
Wire Diameter 15 mm
Ground Ends
(grind ends parallel)

Click: **Place** ⇒ **No Leader** ⇒ **Standard** ⇒ **MMB** ⇒ pick a place on the screen to place the note [Fig. 14.35(a-b)] ⇒ **OK** ⇒ **Done/Return** ⇒ **Done**

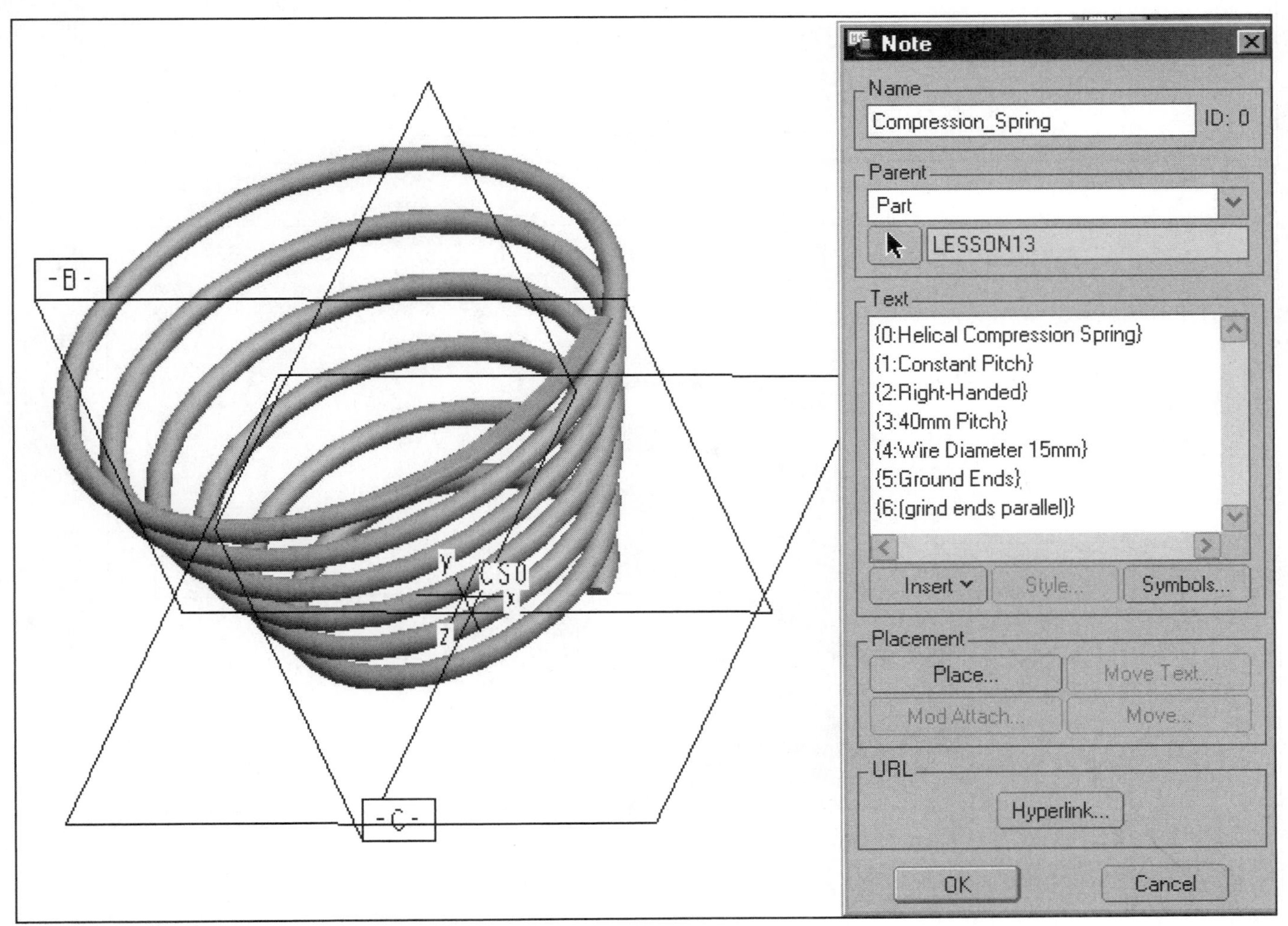

Figure 14.34 3D Notes

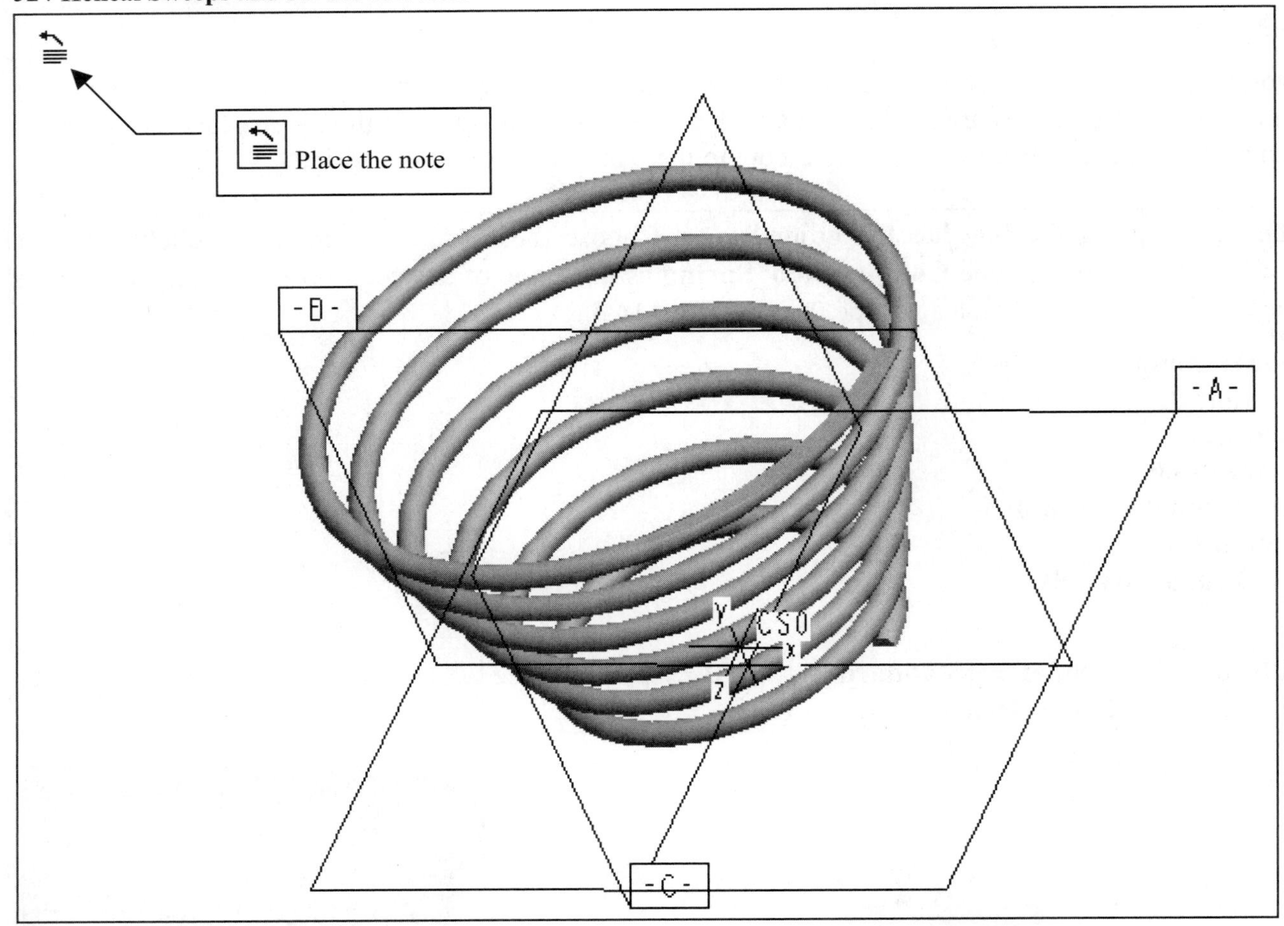

Figure 14.35(a) Placing the Note

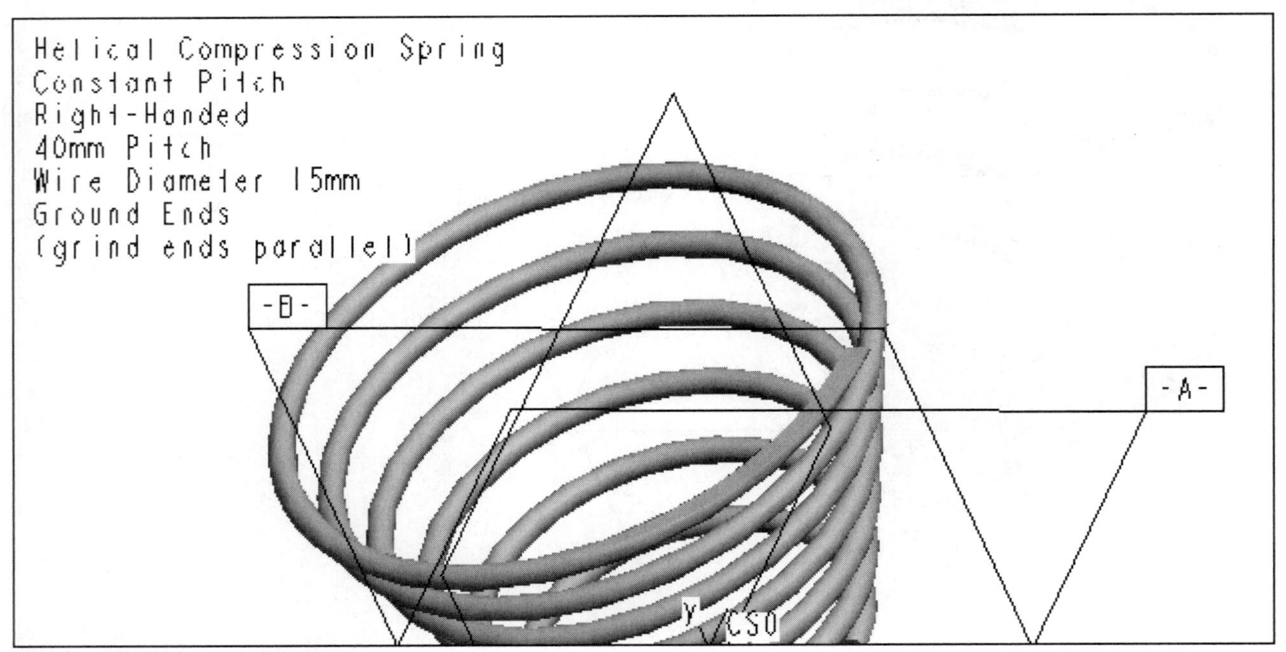

Figure 14.35(b) Completed Note

You can toggle the note 3D Notes and 3D Notes in the Environment dialog box (Fig. 14.36). Display the note in the Model Tree by clicking: **Settings ⇒ Tree Filters ⇒** Notes (Fig. 14.37) ⇒ **Apply ⇒ OK**

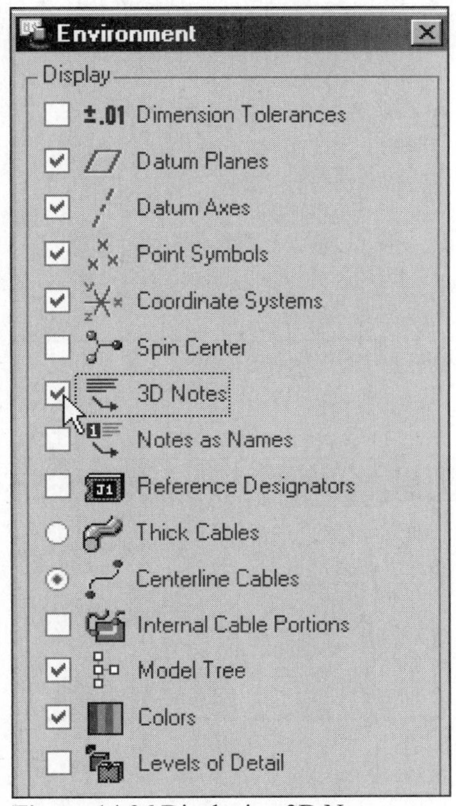

Figure 14.36 Displaying 3D Notes

Figure 14.37 Displaying Notes in the Model Tree

You can also perform a variety of functions directly from the Model Tree. Click on Compression_Spring in the Model Tree ⇒ **RMB** [Fig. 14.38(a)] ⇒ **Properties** [Fig. 14.38(b)] ⇒ URL click: **Hyperlink** (Fig. 14.39) ⇒ type the URL or internal link: **www.americanprecspring.com** (Fig. 14.40)

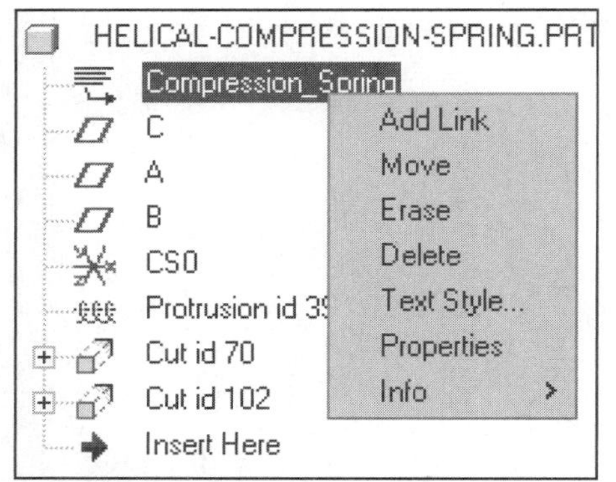

Figure 14.38(a) Model Tree (RMB)

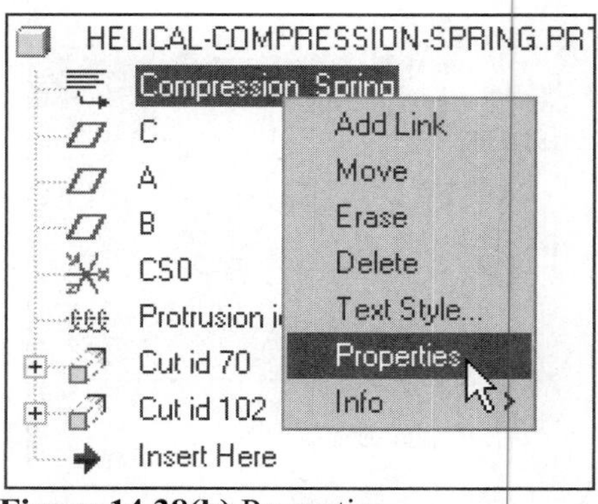

Figure 14.38(b) Properties

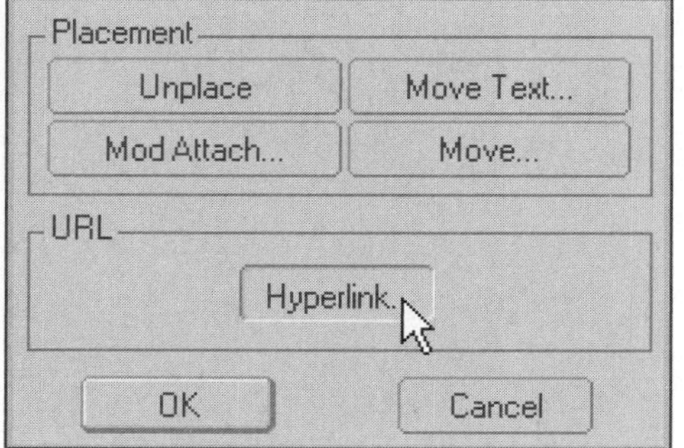

Figure 14.39 Hyperlink

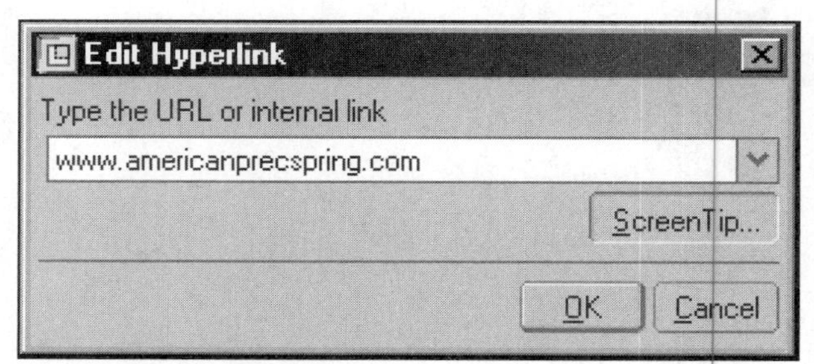

Figure 14.40 URL

Create a screen tip that will display as your cursor passes over the note, click: ScreenTip... ⇒ type **SPRING COMPANY** [Fig. 14.41(a-b)] ⇒ **OK** ⇒ **MMB** ⇒ **MMB** ⇒ ⇒ ⇒ ⇒ **MMB** ⇒ **File** ⇒ **Delete** ⇒ **Old Versions** ⇒ **MMB**

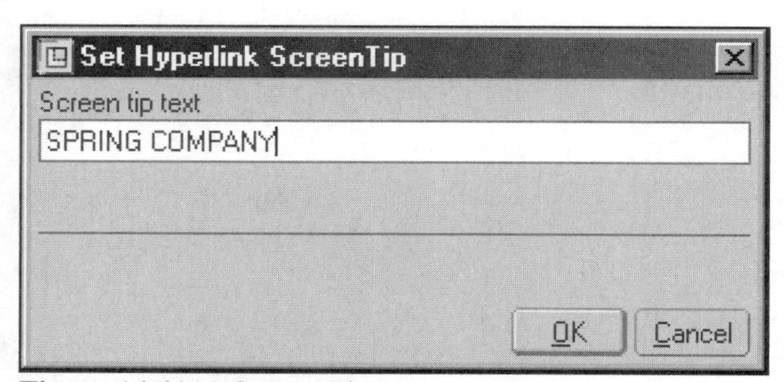

Figure 14.41(a) Screen Tip

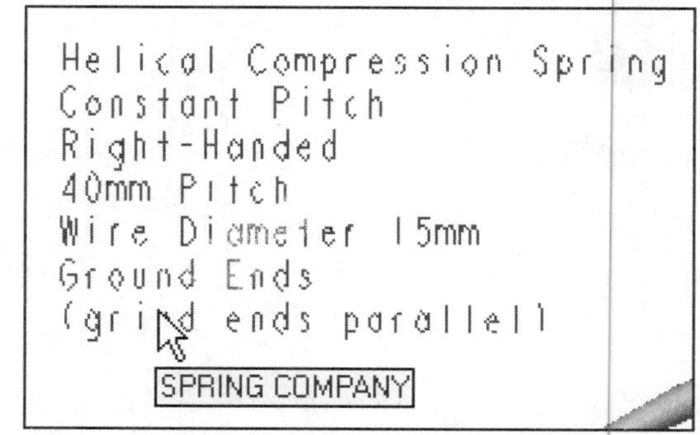

Figure 14.41(b) Screen Tip Displayed

Open the URL, click: Compression_Spring ⇒ **RMB** ⇒ Open URL (Fig. 14.42) ⇒ pick in the browser window (Fig. 14.43)

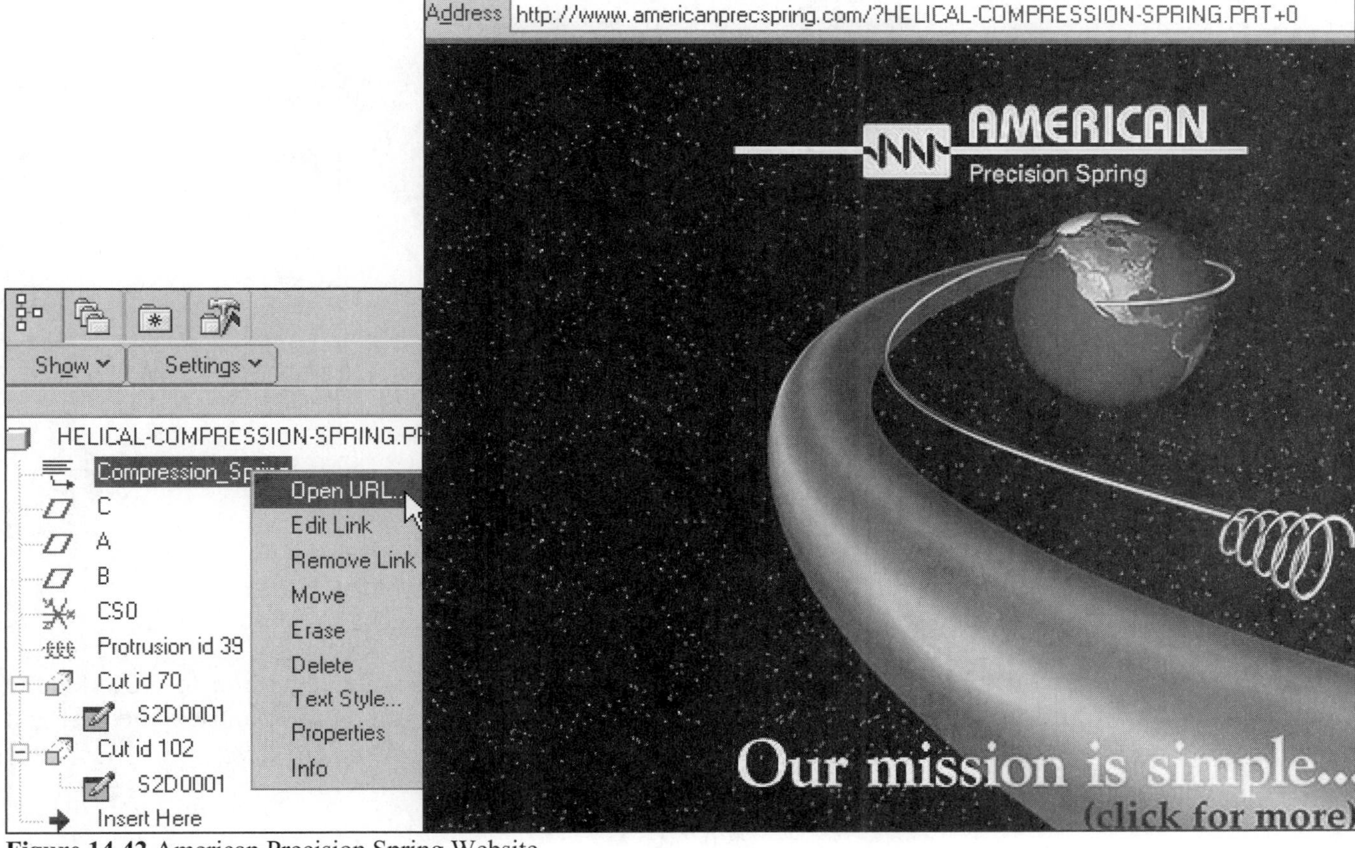

Figure 14.42 American Precision Spring Website

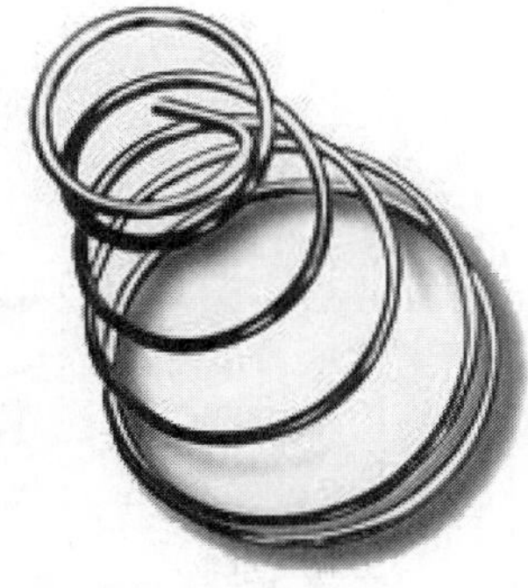

Figure 14.43 Spring Website

Lesson 14 Project

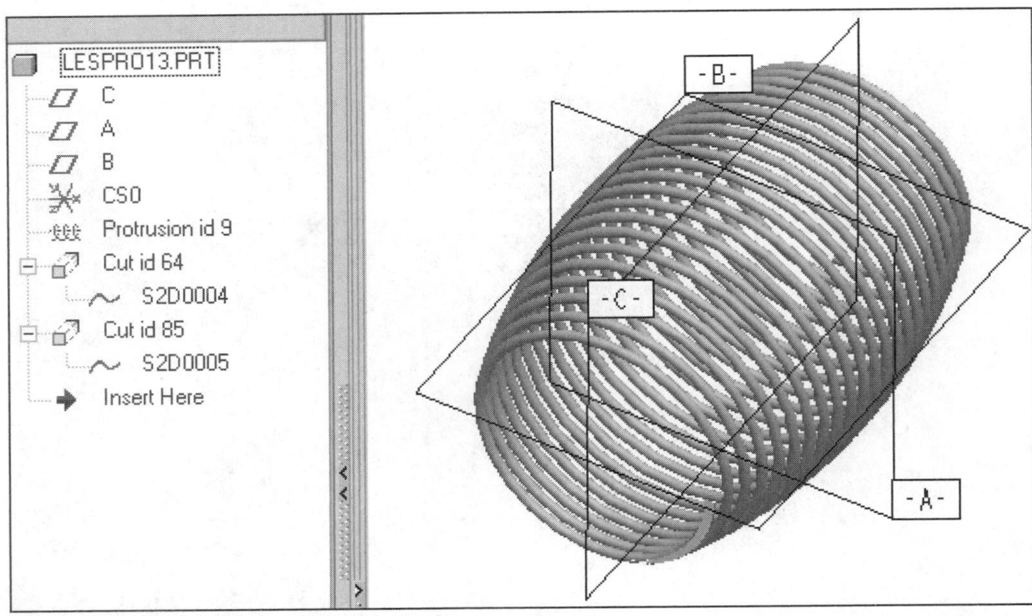

Figure 14.44 Convex Compression Spring

Convex Compression Spring

This lesson project is a Convex Compression Spring. This project uses commands similar to those for the Helical Compression Spring. Create the part shown in Figures 14.44 through 14.49. Analyze the part and plan the steps and features required to model it. The spring is made of *spring steel*. Add 3D Notes describing the spring.

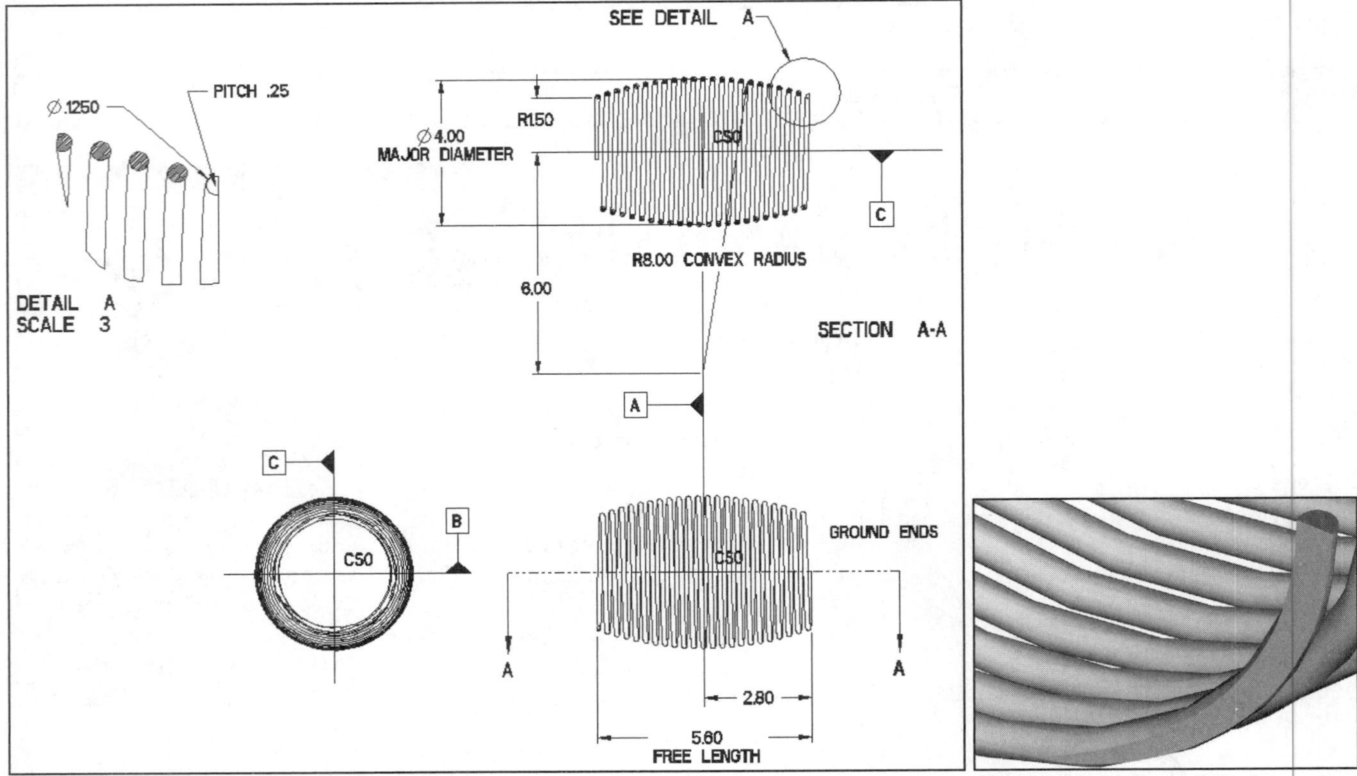

Figure 14.45 Convex Compression Spring, Detail Drawing (ground ends)

Figure 14.46 Ground Ends

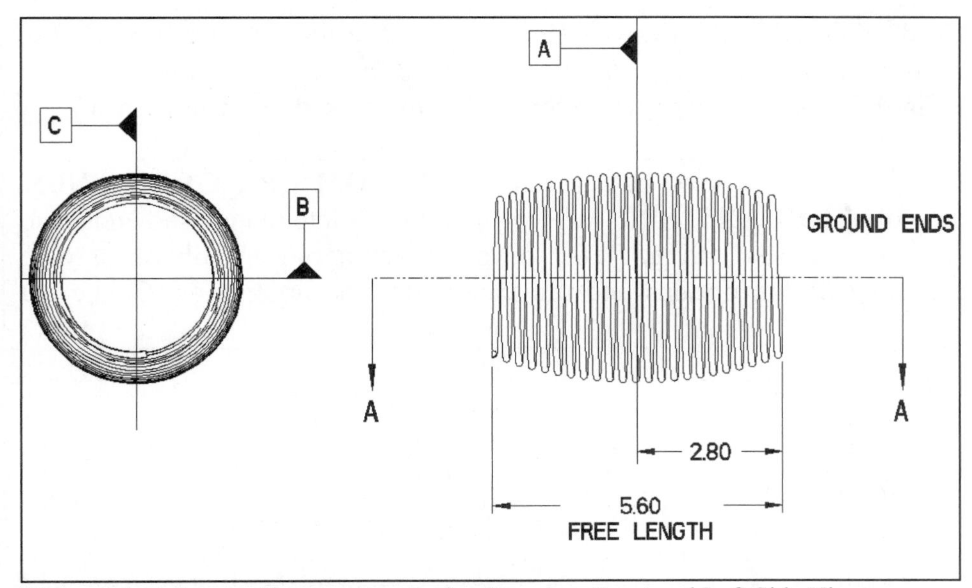

Figure 14.47 Convex Compression Spring Drawing, Front and Left Side Views

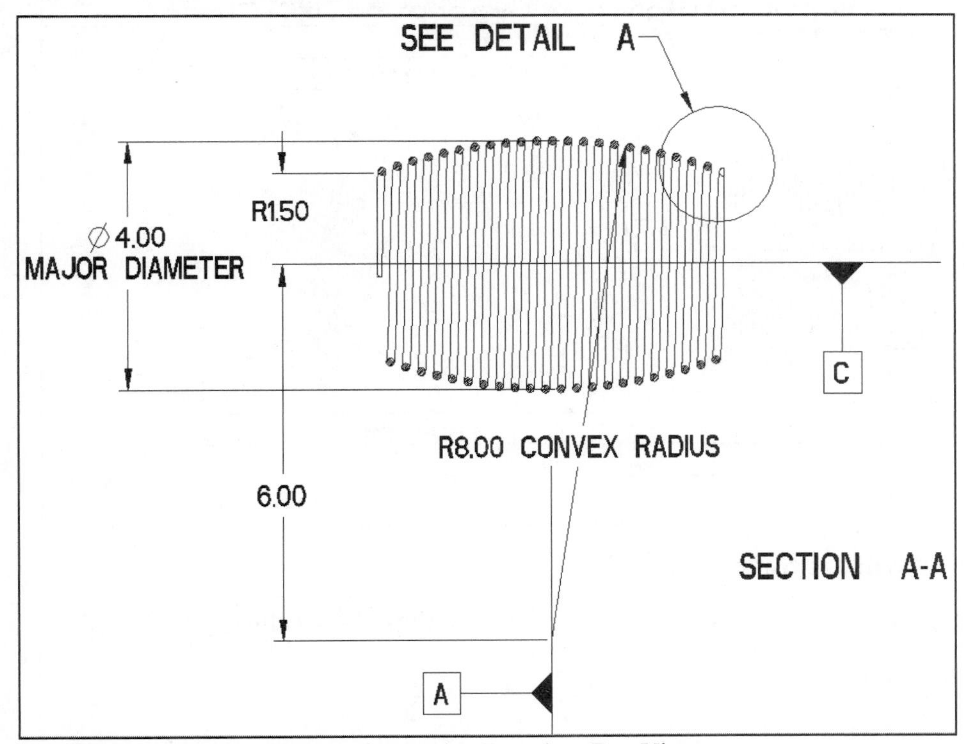

Figure 14.48 Convex Compression Spring Drawing, Top View

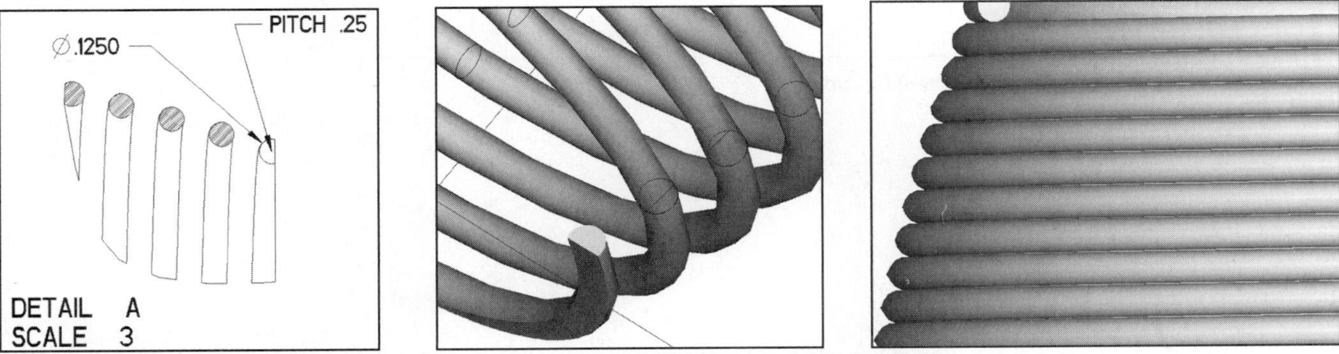

Figure 14.49(a-c) Convex Compression Spring Drawing, DETAIL A, and Shaded Views

The ECO in Figure 14.50 does *not* alter the Convex Compression Spring you just created. *The ECO requests that a different extension spring be designed with hook ends instead of ground ends.* The same size and dimensions required in the *Convex Compression Spring* are to be used in the new spring.

Save the Convex Compression Spring under a new name-- **CONVEX_COM_SPR_GRND_ENDS**. Rename the active part to something like **CON_EXT_SPR_HOOK_ENDS**. Delete the ground ends, and design machine hooks for both ends of the new spring. Refer to your Machinery's Handbook or your engineering graphics text for acceptable design options and dimensions for the hook ends. Create **3D Notes** to describe the spring.

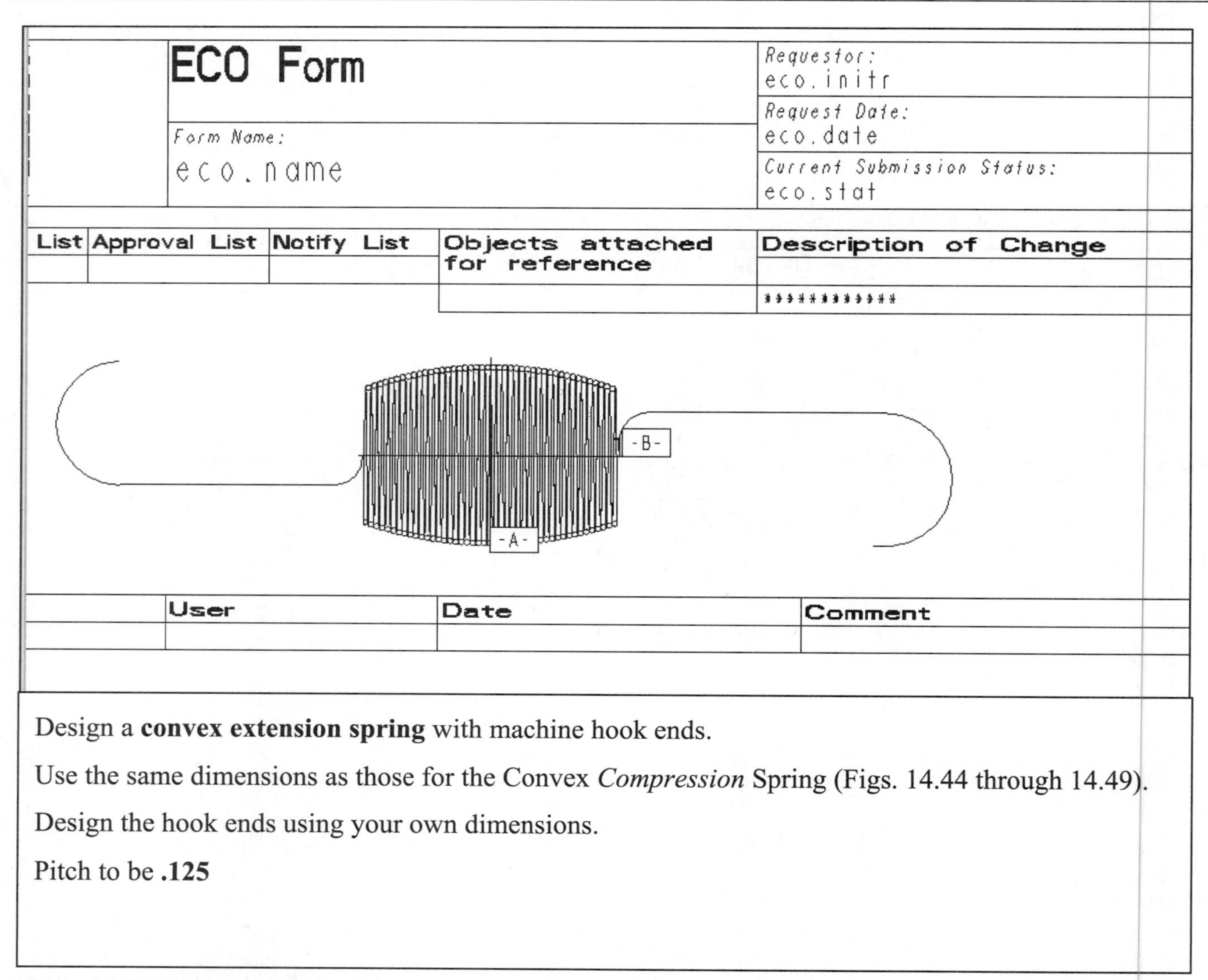

ECO Form

Requestor:
eco.initr

Request Date:
eco.date

Form Name:
eco.name

Current Submission Status:
eco.stat

List	Approval List	Notify List	Objects attached for reference	Description of Change
				************

	User	Date	Comment

Design a **convex extension spring** with machine hook ends.

Use the same dimensions as those for the Convex *Compression* Spring (Figs. 14.44 through 14.49).

Design the hook ends using your own dimensions.

Pitch to be **.125**

Figure 14.50 ECO for New Convex *Extension* Spring

Lesson 15 Assembly Constraints

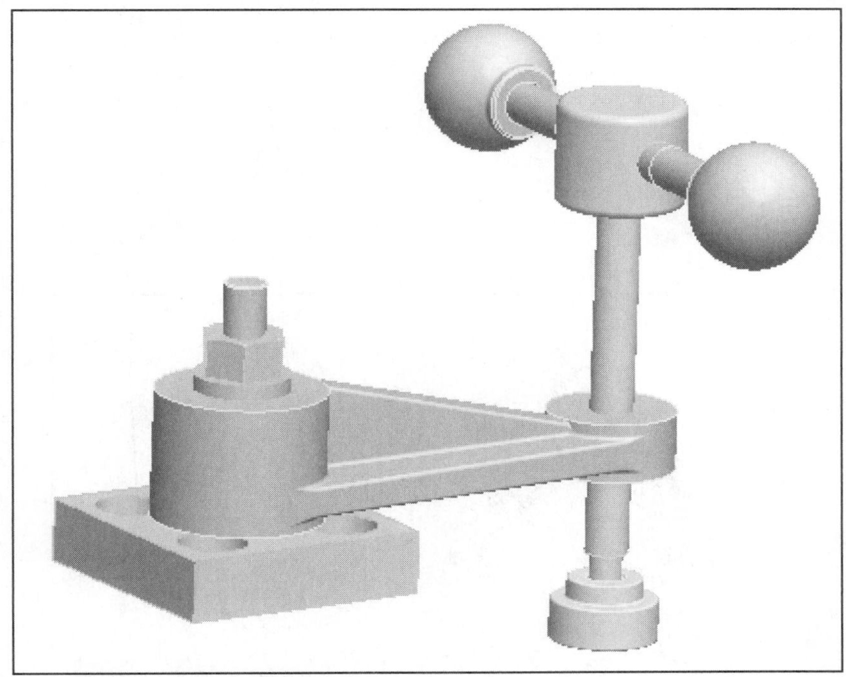

Figure 15.1 Swing Clamp Assembly

OBJECTIVES

- **Assemble** components to form an assembly
- Create a **subassembly**
- Understand and use a variety of **Assembly Constraints**
- **Modify** a component constraint
- **Edit** a constraint value
- Check for **clearance** and **interference**

ASSEMBLY CONSTRAINTS

Assembly mode allows you to place together components and subassemblies to form an assembly (Fig. 15.1). Assemblies (Fig. 15.2) can be modified, reoriented, documented, or analyzed. An assembly can be assembled into another assembly, thereby becoming a subassembly.

Figure 15.2 Swing Clamp (**CARRLANE** at **www.carrlane.com**)

Assembly mode is used for the following functions:

- Placing components (Fig. 15.3)
- Exploding views
- Part and feature modification
- Analysis

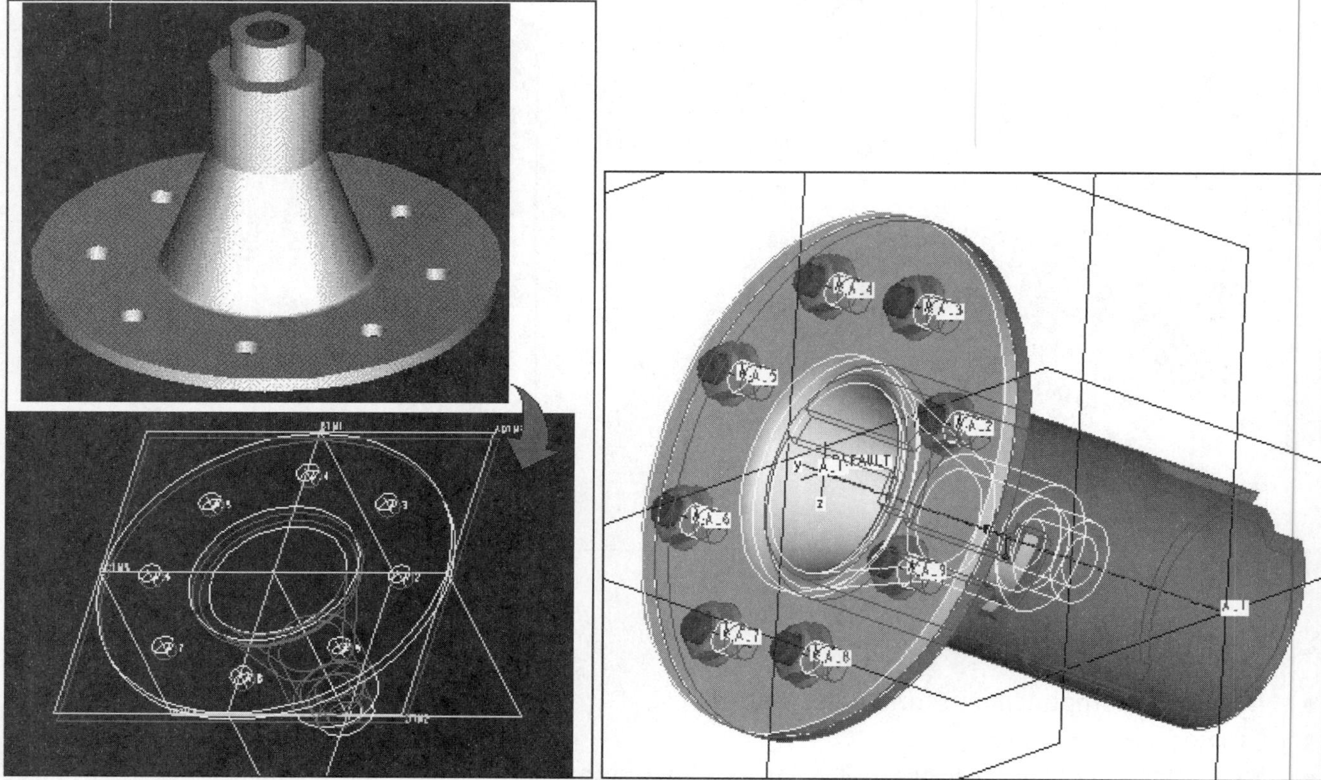

Figure 15.3 Assemblies, Bottom-Up Design (CADTRAIN, COAch for Pro/ENGINEER)

In Pro/Assembly you can:

- Place together component parts and subassemblies to form assemblies [Fig. 15.4(a-b)].
- Remove or replace assembly components.
- Modify assembly placement offsets, and create and modify assembly datum planes, coordinate systems, and cross sections.
- Edit parts directly in Assembly mode.
- Get assembly engineering information, perform viewing and layer operations, create reference dimensions, and work with interfaces.
- Create new parts in Assembly mode.
- Create sheet metal parts in Assembly mode (using Pro/SHEETMETAL).
- Mirror parts in Assembly mode (create a new part).
- Replace components automatically by creating interchangeable groups. Create assembly features, existing only in Assembly mode, that intersect several components.
- Create families of assemblies, using the family table.
- Simplify the assembly representation.
- Use Move and Copy for assembly components.
- Use Pro/PROGRAM to create design programs that allow users to respond to program prompts to alter the design model.

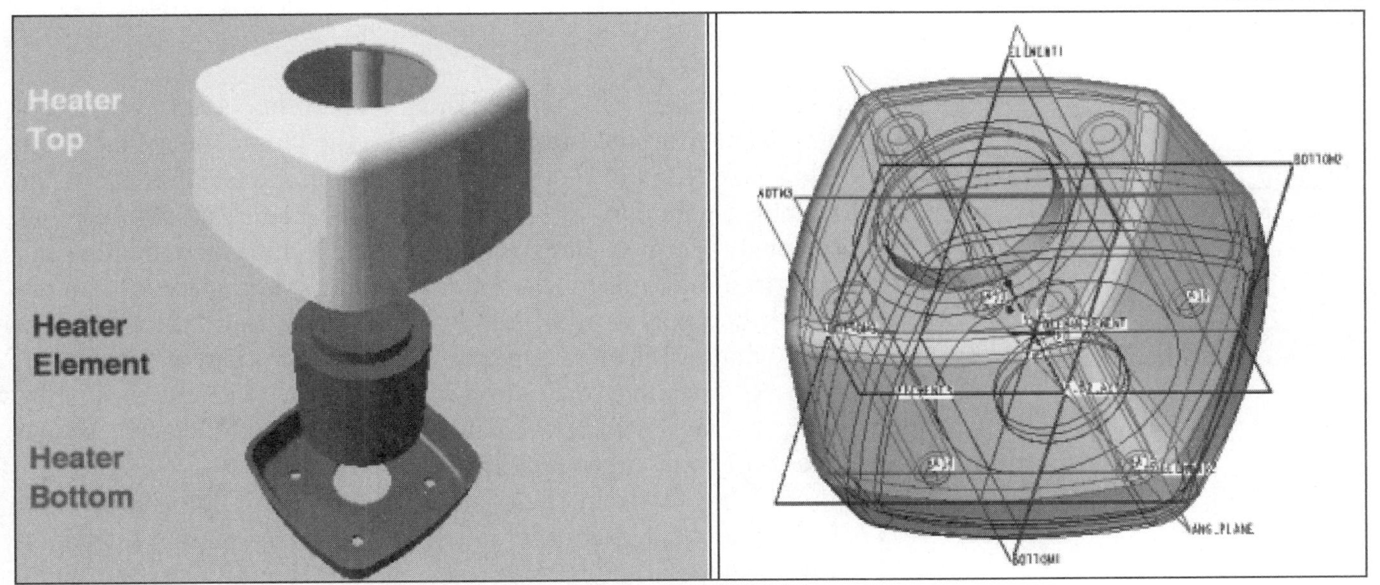

Figure 15.4(a-b) Assemblies (CADTRAIN, COAch for Pro/ENGINEER)

Assembling Components

The process of creating an assembly involves adding components (parts/subassemblies) to a base component (parent part/subassembly) using a variety of constraints (Fig. 15.5). Components can also be created in Assembly mode using existing components as references.

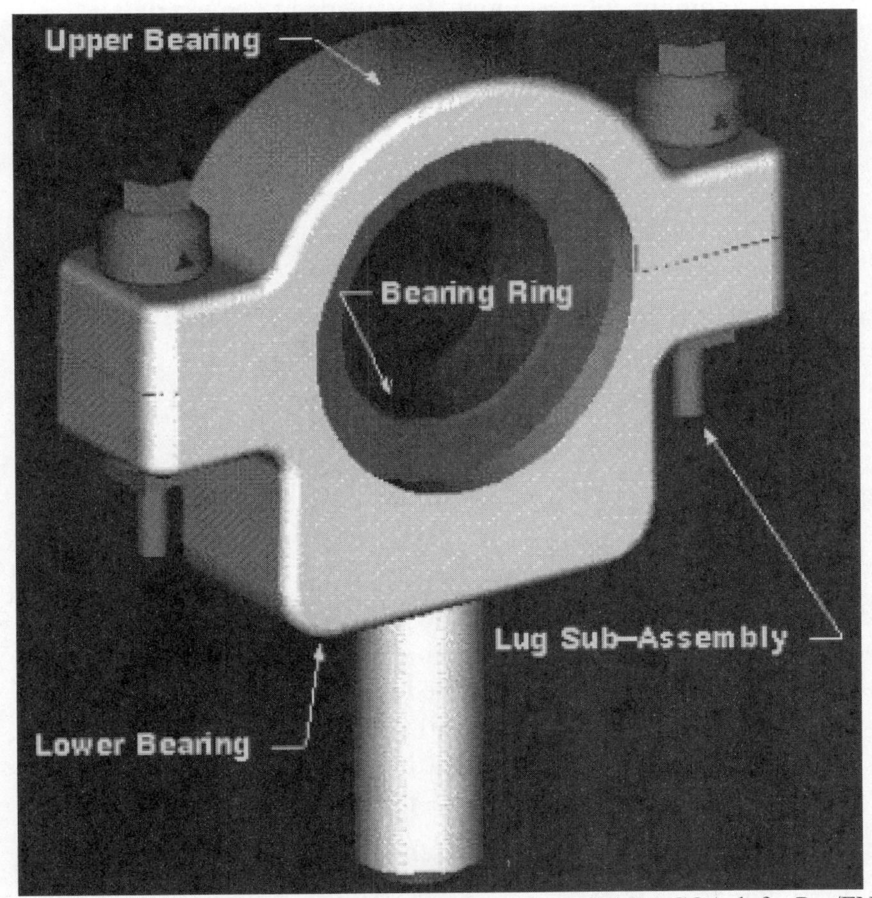

Figure 15.5 Assemblies and Sub-Assemblies (CADTRAIN, COAch for Pro/ENGINEER)

Placing Components

To assemble components you simply use or Insert ⇒ Component ⇒ Assemble. After selecting a component from the Open dialog box, the Component Placement dialog box (Fig. 15.6) opens and the component appears in the assembly window. If the component being assembled has previously defined interfaces, the Select Interface dialog box opens to allow you to select an interface. Alternatively, you can select a component from a browser window and drag it into the Pro/E window. If there is an assembly in the window, Pro/E will begin to assemble the component into the current assembly. Either the Component Placement or the Select Interface dialog boxes will come up as previously described.

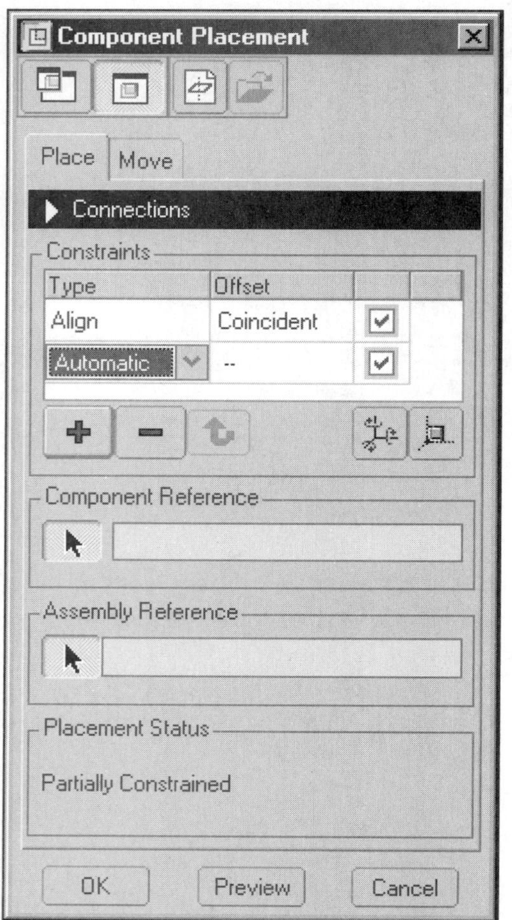

Figure 15.6 Component Placement Dialog Box

Using icons in the toolbar at the top of the dialog box, specify the screen window in which the component is displayed while you position it. You can change windows at any time using:

- **Separate Window** Displays the component in its own window while you specify its constraints.
- **Assembly Window** Displays the component in the main assembly graphics window and updates component placement as you specify constraints.

The Automatic placement constraint is selected by default when a new component is introduced into an assembly for placement. Do one of the following:

- Select a reference on the component Component Reference SW_CLAMP_STUD35: Surface and a reference on the assembly Assembly Reference SW_CLAMP_PLATE: Surface, in either order, to define a placement constraint. After you select a pair of valid references from the assembly and the component, Pro/E automatically selects a constraint type appropriate to the specified references.
- Before selecting references, change the type of constraint by selecting a type from the Constraints Type list (Fig. 15.6). Clicking the current constraint in the Type box shows the list. You may also select an offset type from the Offset list. Coincident is the default offset, but you may also choose **Oriented** or 0.0 from the list. If you choose **0.0**, type in the value of the offset in the cell 5.00 and then press Enter.

After you define a constraint, **Specify a new constraint** is automatically selected, and you can define another constraint. You can define as many constraints as you want (up to Pro/E limit of 50 constraints). As you define constraints, each constraint is listed under the Constraints area and the status of the component is reported in the Placement Status area as you select constraint references.

You can select one of the constraints listed in the Constraints area at any time and change the constraint type, flip between Mate and Align with **Change orientation of constraint**, modify the offset value, reset and Align constraint to forced or unforced, or switch between allowing and disallowing Pro/E assumptions.

Use **Assemble component at default location** to align the default Pro/E -created coordinate system of the component to the default Pro/E -created coordinate system of the assembly. Pro/E places the component at the assembly origin. Use **Fix component to current position** to fix the current location of the component that was moved or packaged.

You can select **Remove the selected constraint** or Preview **Preview component placement** at any time.

- To delete a placement constraint for the component, select one of the constraints listed in the Constraints area, and then click .
- Click Preview to show the location of the component, as it would be with the current placement constraints.

Click **OK** when the Placement Status of the component is shown as Placement Status Fully Constrained, Placement Status Partially Constrained, or Placement Status No Constraints. Pro/E places the component with the current constraints.

If the Placement Status is Placement Status Constraints Invalid, you should correct or complete the constraint definition.

If constraints are incomplete, you can leave the component as packaged. A packaged component is one that is included in the assembly but not fully placed. Packaged components follow the behavior dictated by the configuration file option *package_constraints*.

The warning "You are leaving this component as packaged. • COMPONENT has been created successfully." will display if you leave the component partially constrained, and the model tree (ASM0002.ASM, SW_CLAMP_PLATE.PRT, ▫SW_CLAMP_STUD35.PRT) will display the component with a small box in front of its name ▫SW.

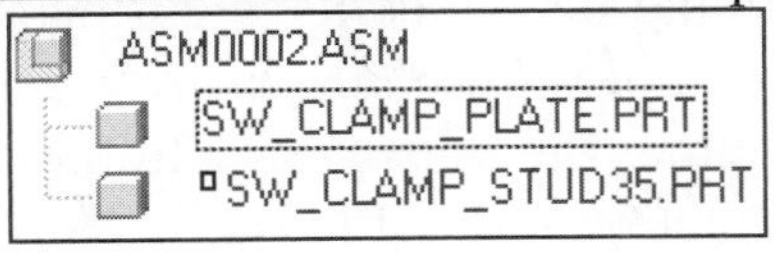

If constraints are conflicting, you can restart or continue placing the component. Restarting erases all previously defined constraints for the component. You can also uncheck a constraint in the Constraints table to make it inactive.

Lesson 15 STEPS

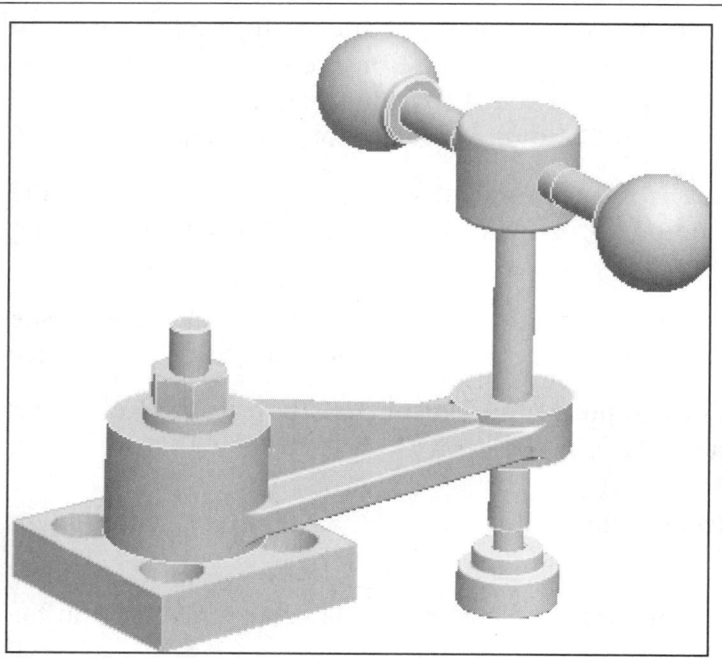

Figure 15.7(a) Swing Clamp Main Assembly

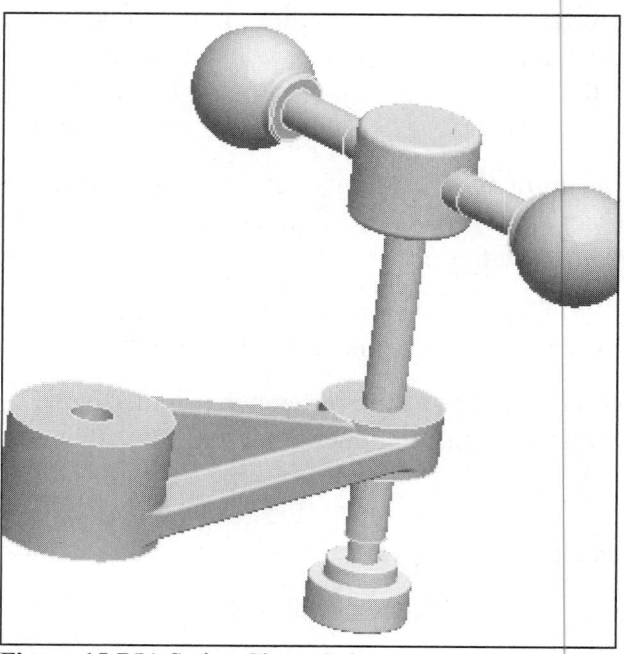

Figure 15.7(b) Swing Clamp Sub-Assembly

Swing Clamp Assembly

Most of the parts required in this lesson are lesson projects from this text. The **Clamp Foot** and **Clamp Swivel** are from Lesson 6. The **Clamp Ball** is from Lesson 7. The **Clamp Arm** was created in Lesson 9. *If you have not modeled these parts previously, please do so before you start the following systematic instructions.* The other components required for the assembly are standard *off-the-shelf* hardware items that you can get from Pro/E by accessing the library. If your system does not have a Pro/LIBRARY license for the Basic and Manufacturing libraries, model the parts using the detail drawings provided here. The **Flange Nut**, the **3.50 Double-ended Stud**, and the **5.00 Double-ended Stud** are standard. The **Clamp Plate** component is the first component of the main assembly and will be modeled later when completing the main assembly [Fig. 15.7(a)] and using the *top-down design* approach.

Because you will be creating the sub-assembly [Fig. 15.7(b)] using the *bottom-up design* approach, all the components must be available before the assembling starts. *Bottom-up design* means that existing parts are assembled one by one until the assembly is complete. The assembly starts with a set of datum planes and a coordinate system. The parts are constrained to the datum features of the assembly. The sequence of assembly will determine the parent-child relationships between components.

Top-down design is the design of an assembly where one or more component parts are created in Assembly mode as the design unfolds. Some existing parts are available, such as standard components and a few modeled parts. The remaining design evolves during the assembly process. The main assembly will involve creating one part using the *top-down design* approach.

Regardless of the design method, the assembly datum planes and coordinate system should be on their own separate *assembly layer*. Each part should also be placed on separate assembly layers; the part's datum features should already be on *part layers*.

Before starting the assembly, you will be modeling each part or retrieving *standard parts* from the library and saving them under unique names in *your* directory. *Unless instructed to do so, do not use the library parts directly in the assembly.* Start this process by retrieving the standard parts from the library [Fig. 15.8(a-c)].

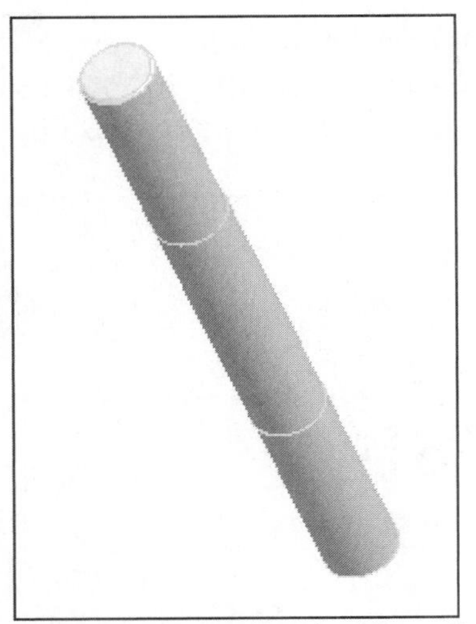
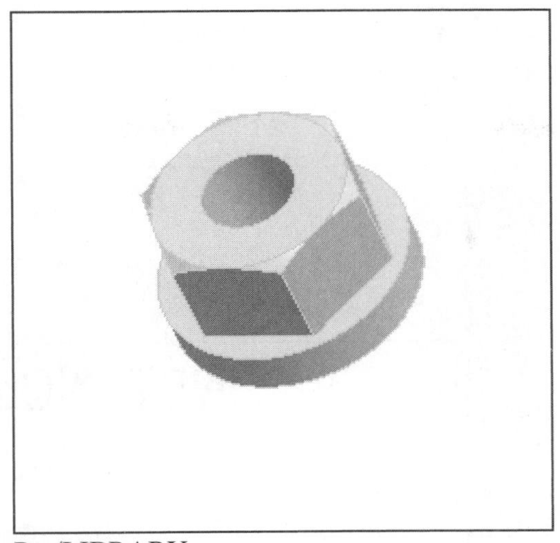
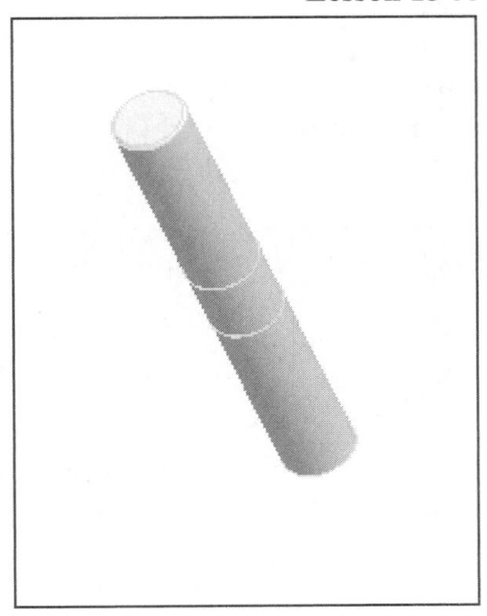

Figure 15.8(a-c) Standard Parts from Pro/LIBRARY

File ⇒ Set Working Directory ⇒ select your working directory **⇒ OK ⇒ File ⇒ Open ⇒** navigate to **prolibrary ⇒ mfglib ⇒ Open ⇒ fixture_lib ⇒ Open ⇒ nuts_bolts_screws ⇒ Open ⇒ st.prt ⇒ Open ⇒ By Parameter** (from Select Instance dialog box) **⇒ d0,thread_dia ⇒ .500 ⇒ d8,stud_length ⇒ 3.500** [Fig. 15.9(a-b)] **⇒ Open ⇒ File ⇒ Save a Copy ⇒ CLAMP_STUD35 ⇒ OK ⇒ File ⇒ Erase ⇒ Current ⇒ Yes**

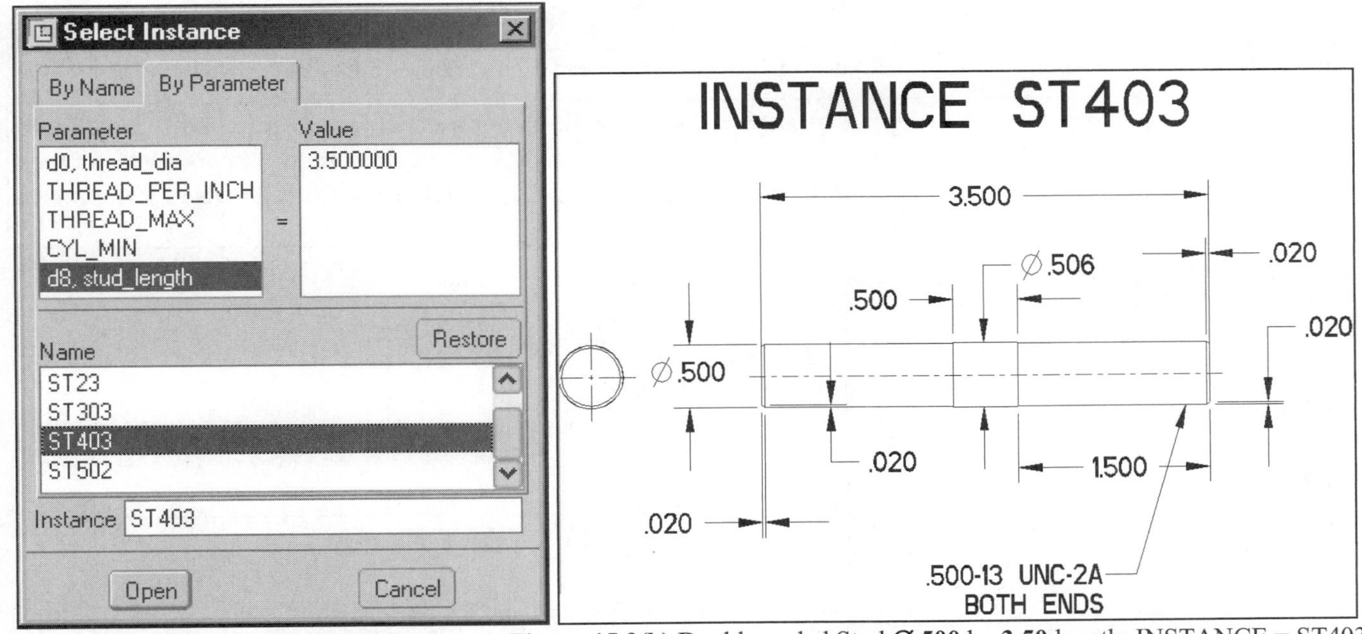

Figure 15.9(a) Select Instance Dialog Box **Figure 15.9(b)** Double-ended Stud **Ø.500** by **3.50** length: INSTANCE = ST403

File ⇒ **Open** ⇒ navigate to **prolibrary** ⇒ **mfglib** ⇒ **Open** ⇒ **fixture_lib** ⇒ **Open** ⇒ **nuts_bolts_screws** ⇒ **Open** ⇒ **st.prt** ⇒ **Open** ⇒ **By Parameter** (from Select Instance dialog box) ⇒ **d0,thread_dia** ⇒ **.500** ⇒ **d8,stud_length** ⇒ **5.00** [Fig. 15.10(a-b)] ⇒ **Open** ⇒ **File** ⇒ **Save a Copy** ⇒ **CLAMP_STUD5** ⇒ **OK** ⇒ **File** ⇒ **Erase** ⇒ **Current** ⇒ **Yes**

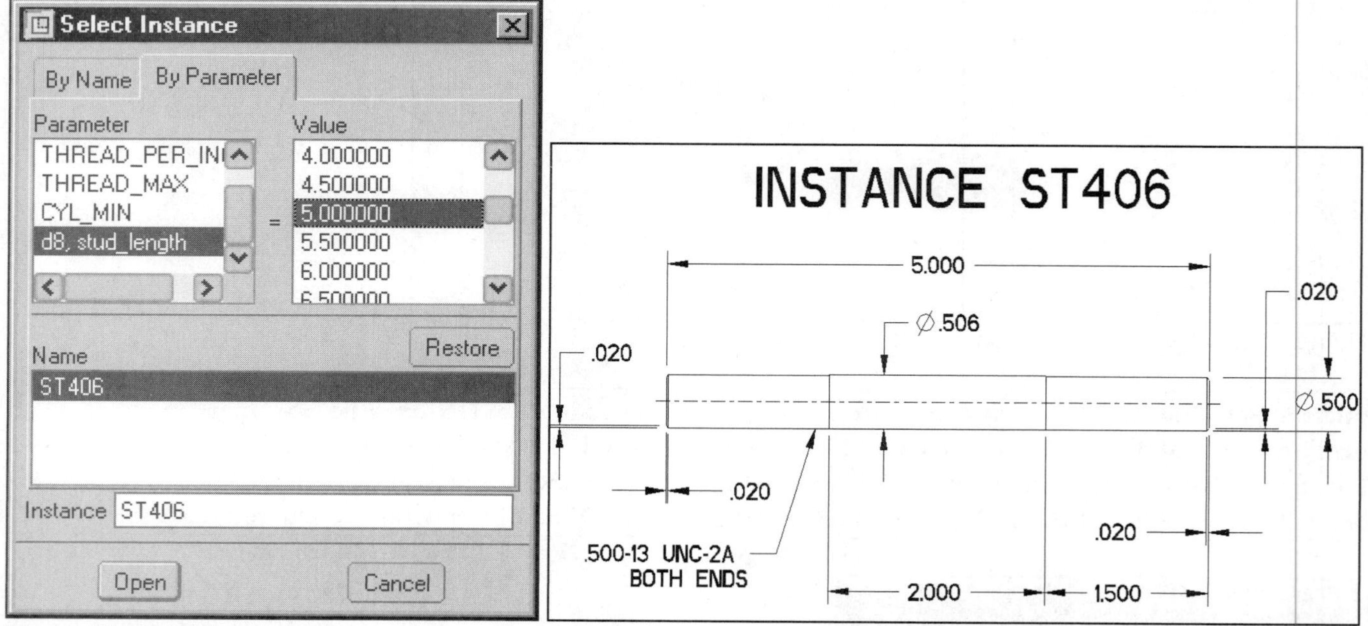

Figure 15.10(a) Select Instance

Figure 15.10(b) Double-ended Stud **Ø.500** by **5.00** length: INSTANCE = ST406

File ⇒ **Open** ⇒ navigate to **prolibrary** ⇒ **mfglib** ⇒ **Open** ⇒ **fixture_lib** ⇒ **Open** ⇒ **nuts_bolts_screws** ⇒ **Open** ⇒ **fn.prt** ⇒ **Open** ⇒ **By Parameter** (from Select Instance dialog box) ⇒ **d4,thread_dia** ⇒ **.500** ⇒ **Open** ⇒ **File** ⇒ **Save a Copy** ⇒ **CLAMP_FLANGE_NUT** [Fig. 15.11(a-b)] ⇒ **OK** ⇒ **File** ⇒ **Erase** ⇒ **Current** ⇒ **Yes**

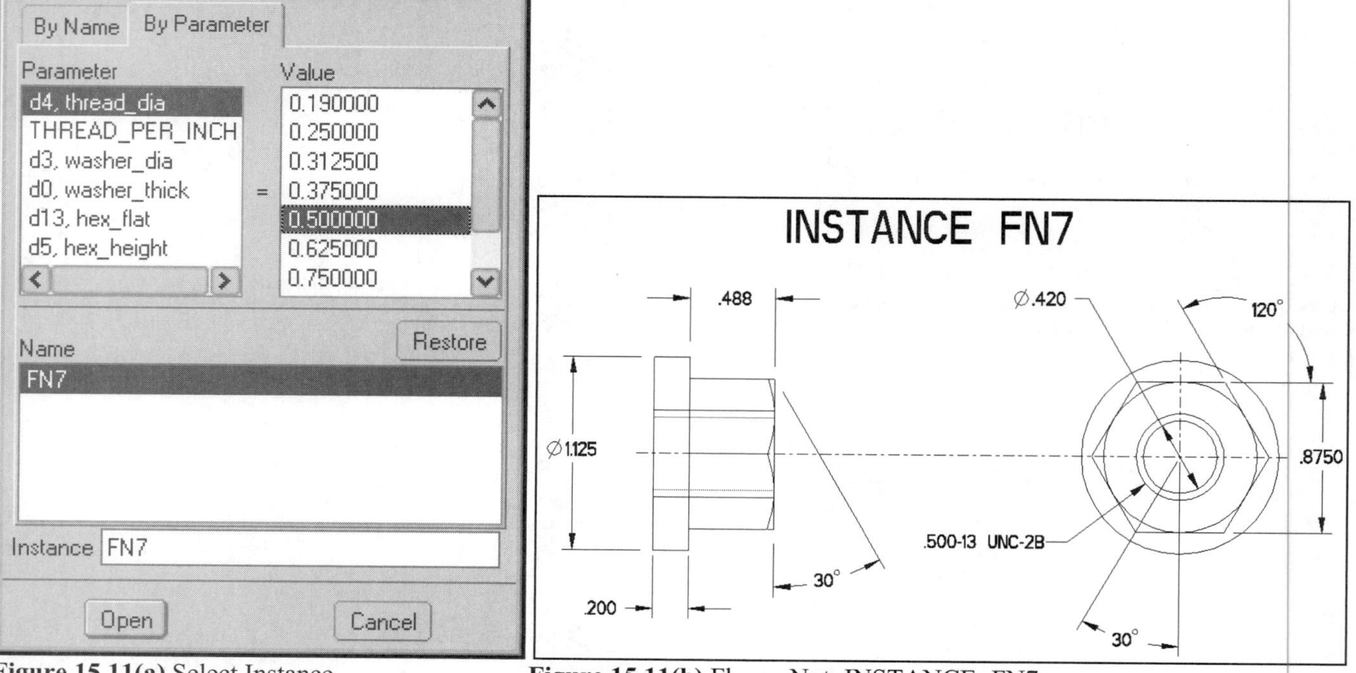

Figure 15.11(a) Select Instance

Figure 15.11(b) Flange Nut: INSTANCE=FN7

Open the Clamp_Arm (Fig. 15.12), the Clamp_Swivel (Fig. 15.13), the Clamp_Ball (Fig. 15.14) and the Clamp_Foot (Fig. 5.15) ⇒ review the components and standard parts for correct color, layering, coordinate system naming, and datum planes ⇒ **Save a Copy** of each to your working directory.

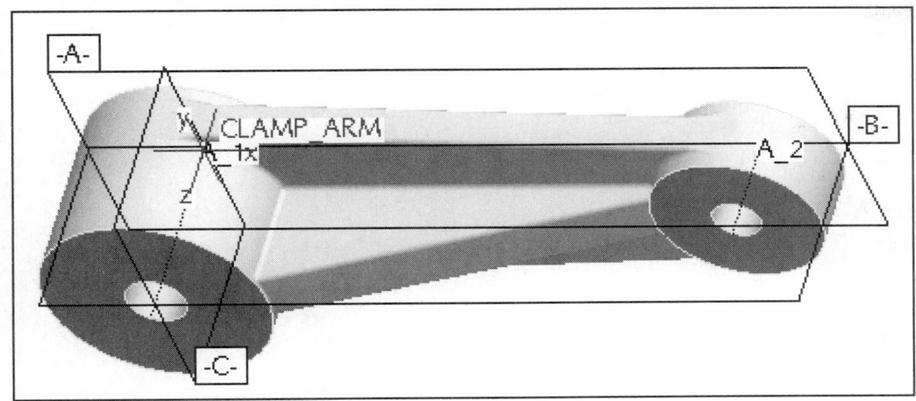

Figure 15.12 Clamp_Arm

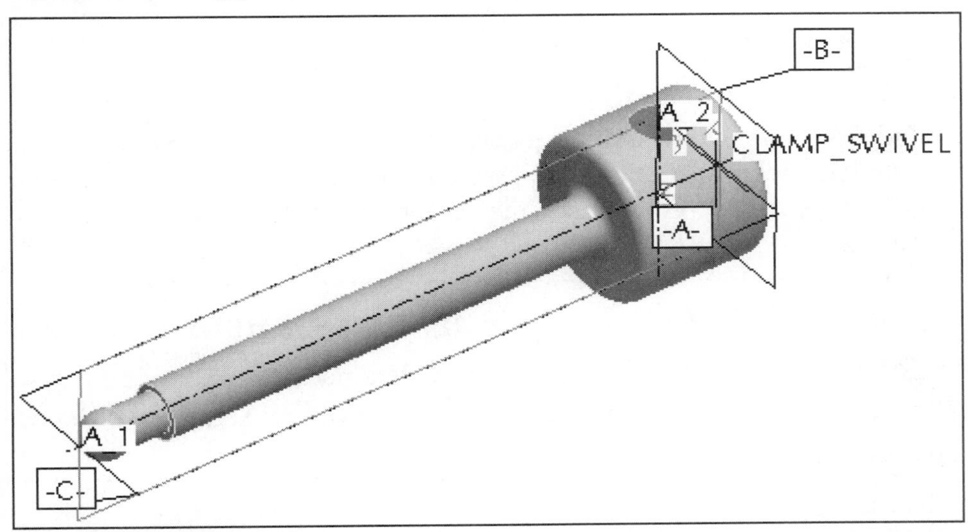

Figure 15.13 Clamp_Swivel

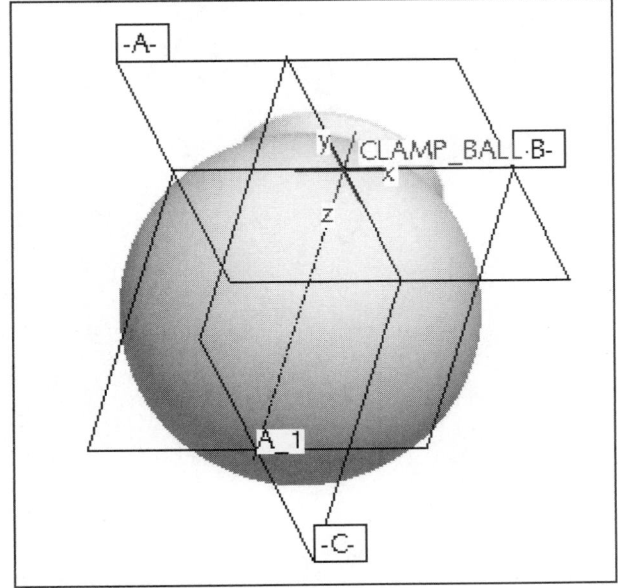

Figure 15.14 Clamp_Ball

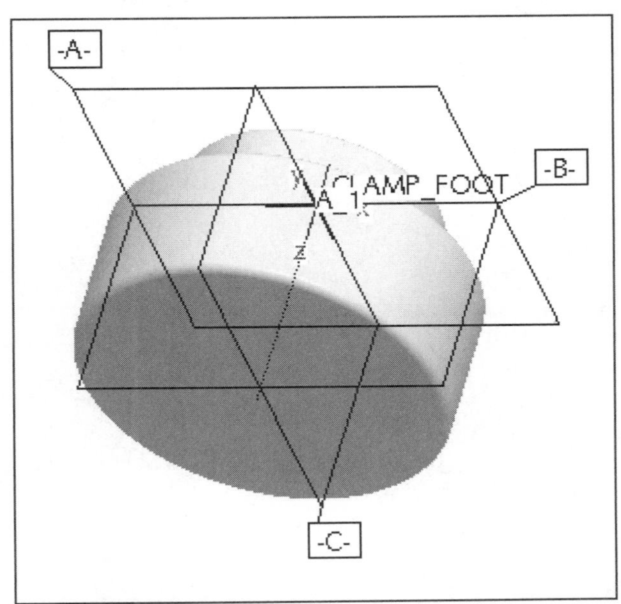

Figure 15.15 Clamp_Foot

You now have eight components (two identical Clamp_Ball components are used) required for the assembly. The ninth component- the Clamp_Plate (Fig. 15.16), will be created using *top-down design* procedures when you start the main assembly. *All parts must be in the same working directory used for the assembly.*

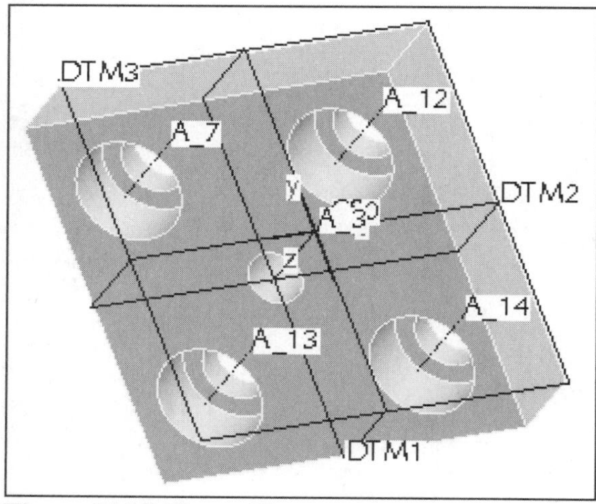

Figure 15.16 Clamp_Plate

A subassembly will be created first. The main assembly is created second. The subassembly will be added to the main assembly to complete the project. Note: the assembly and the components can have different units. Therefore, you must check and correctly set the assembly units before creating or assembling components or sub-assemblies.

Start the subassembly, click: [New] ⇒ [Assembly] ⇒ Name **CLAMP_SUBASSEMBLY** ⇒ Sub-type [Design] ⇒ [Use default template] ⇒ **OK** ⇒ Template [inlbs_asm_design] ⇒ Parameters [DESCRIPTION] type a simple description ⇒ [MODELED_BY] type your name [Fig. 15.17(a)] ⇒ **OK** ⇒ **Edit** ⇒ **Setup** ⇒ **Units** ⇒ [Inch lbm Second (Pro/E Default)] ⇒ **Close** ⇒ **MMB**

New File Options
Template
inlbs_asm_design
Browse...
Empty
inlbs_asm_design
inlbs_asm_ecad
inlbs_mfg_cast
Parameters
DESCRIPTION sub-assembly for swing clamp
MODELED_BY Louis Gary Lamit
Copy associated drawings
OK
Cancel

Figure 15.17(a) New File Options Dialog Box

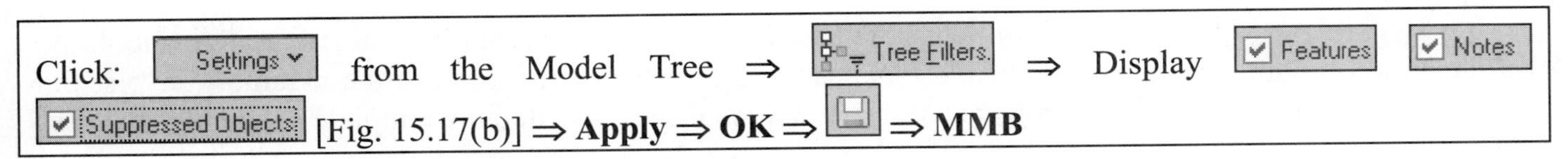

Click: Settings from the Model Tree ⇒ Tree Filters... ⇒ Display ☑ Features ☑ Notes ☑ Suppressed Objects [Fig. 15.17(b)] ⇒ **Apply** ⇒ **OK** ⇒ ⇒ **MMB**

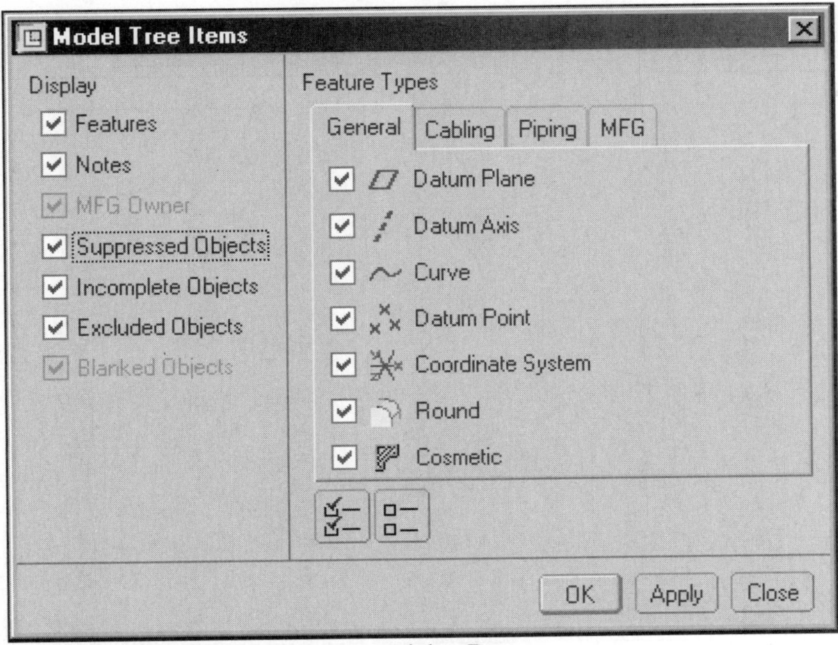

Figure 15.17(b) Model Tree Items Dialog Box

Datum planes and the coordinate system are created per the template provided by Pro/E. The datum planes will have the names, **ASM_RIGHT**, **ASM_TOP**, and **ASM_FRONT**.

Change the coordinate system name: slowly double-click on ASM_DEF_CSYS in the Model Tree ⇒ type new name **SUB_ASM_CSYS** SUB_ASM_CSYS ⇒ click in the graphics window [Fig. 15.17(c)].

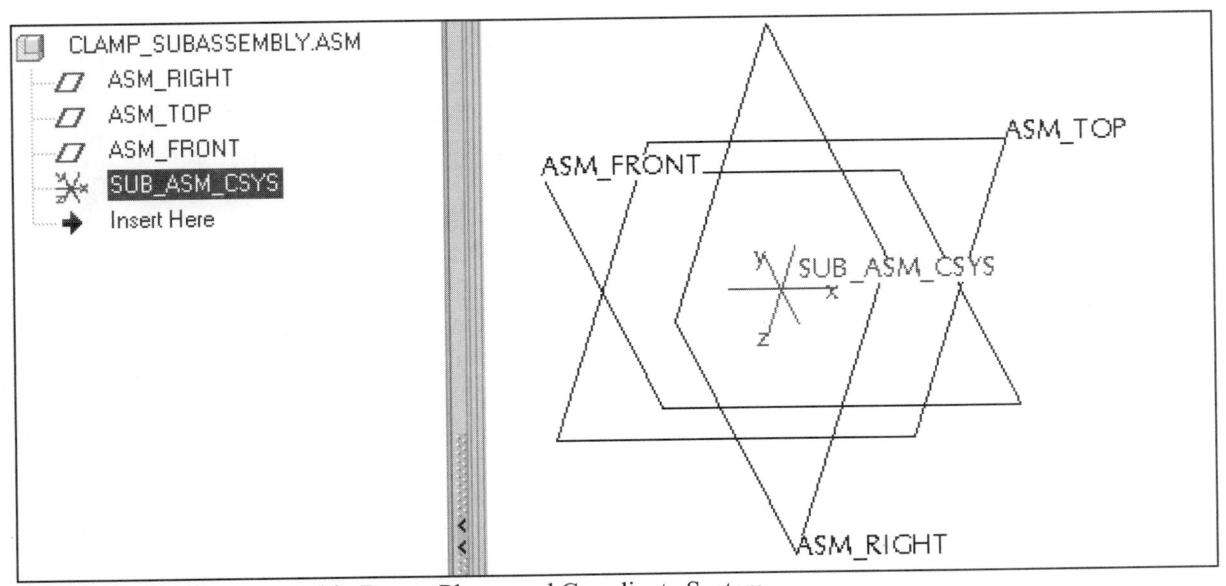

Figure 15.17(c) Sub-Assembly Datum Planes and Coordinate System

Regardless of the design mythology, the assembly datum planes and coordinate system should be on their own separate *assembly layer*. Each part should also be placed on separate assembly layers; the part's datum features should already be on *part layers*. Check the default template assembly layering system.

Click: Show ⇒ Layer Tree ⇒ expand the branches [Fig. 15.17(d)] ⇒ Show ⇒ Model Tree

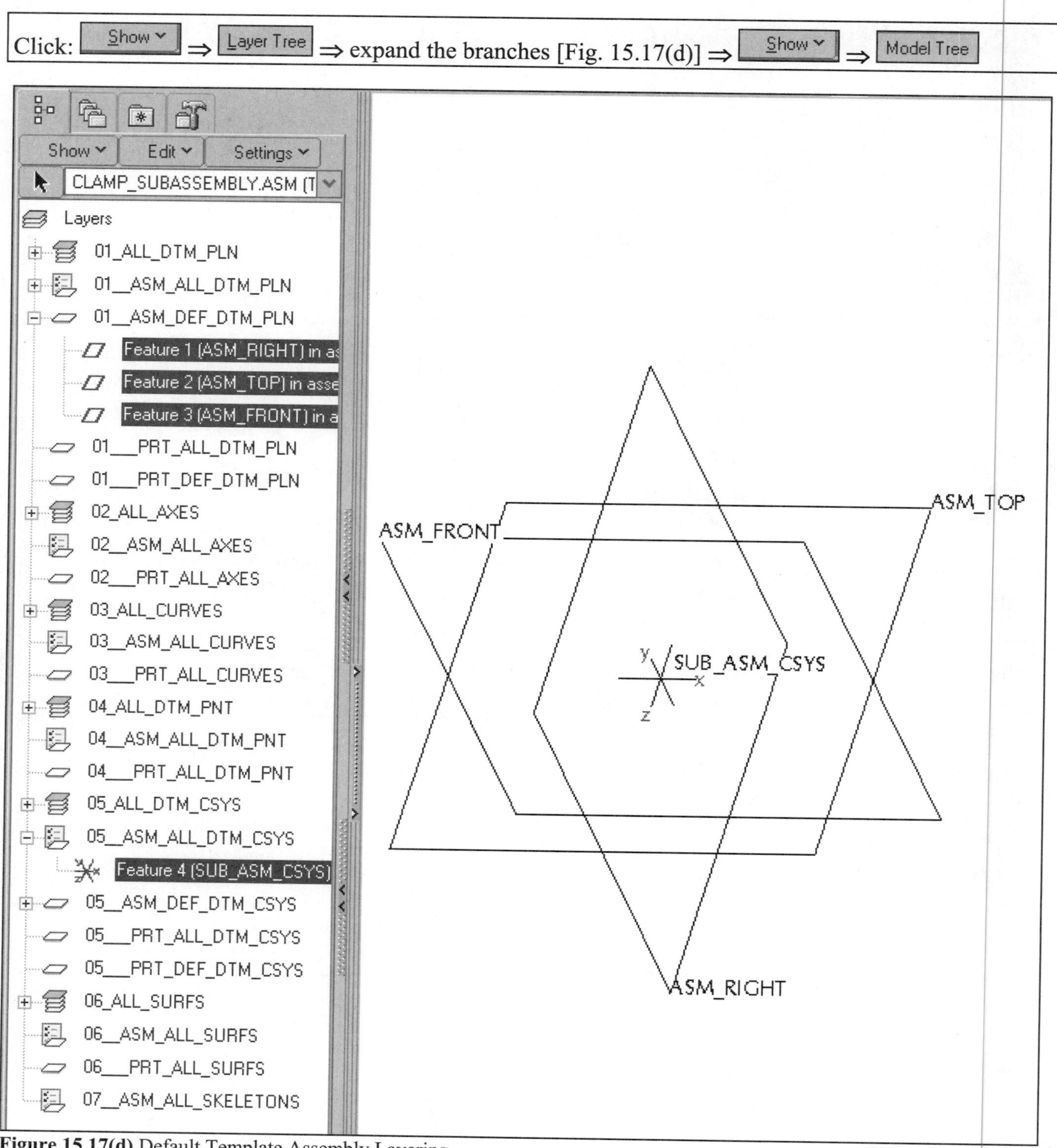

Figure 15.17(d) Default Template Assembly Layering

The first component to be assembled to the subassembly is the Clamp_Arm (Fig. 15.13). The simplest and quickest method of adding a component to an assembly is to match the coordinate systems. The first component assembled is usually where this *constraint* is used, because after the first component is established, few if any of the remaining components are assembled to the assembly coordinate system (with the exception of *top-down design*) or, for that matter, other parts' coordinate systems. Make sure all your models are in the same working directory before you start the assembly process.

If you have named your models to something that is not listed here, pick the appropriate part model as requested.

Click: **Add component to the assembly** ⇒ pick the **clamp_arm.prt** from the Open dialog box [the machined part (generic), not the casting (instance)] ⇒ **Preview** [Fig. 15.18(a)] ⇒ **Open** ⇒ **Show component in a separate window while specifying constraints** (you may need to resize your windows)

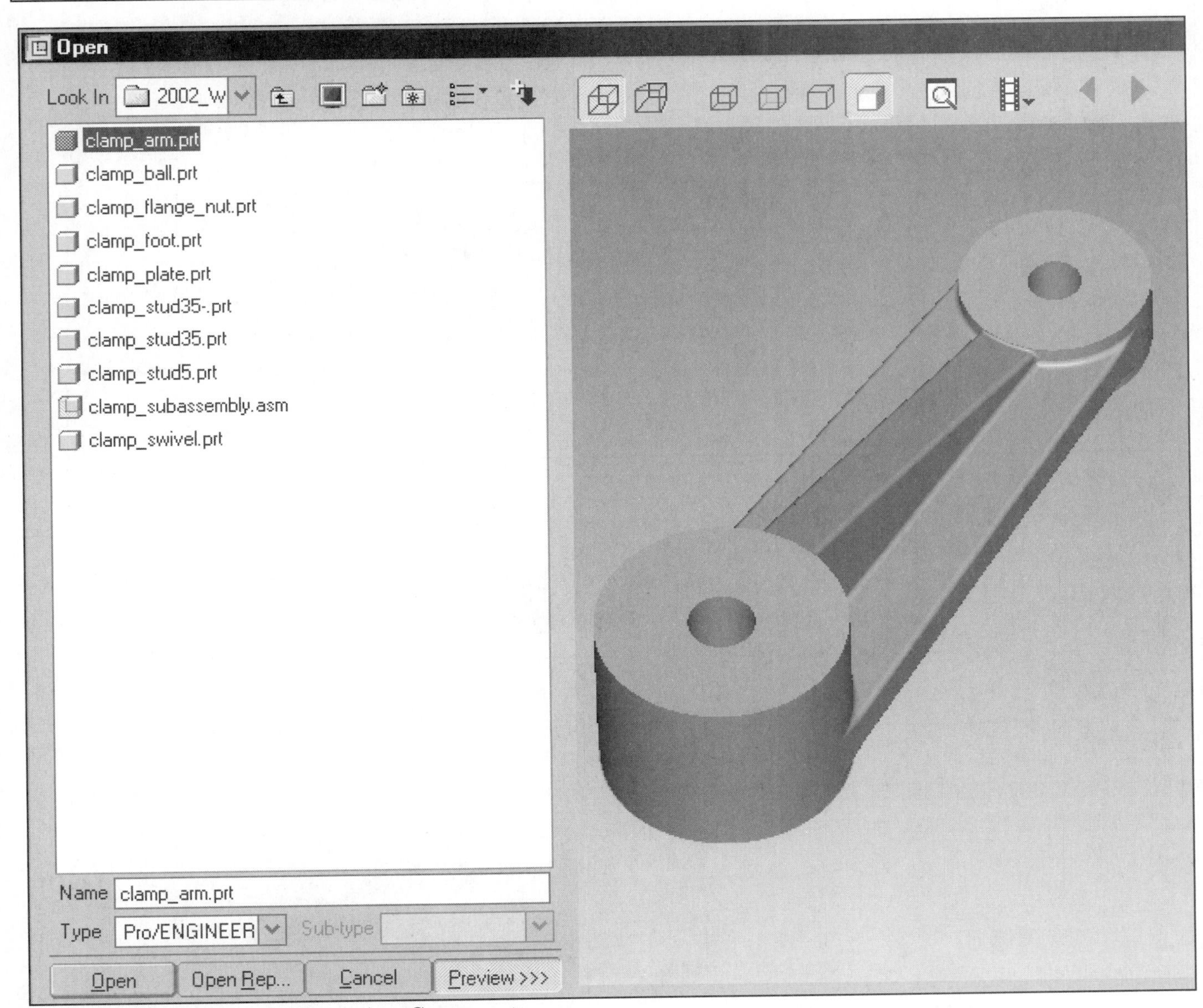

Figure 15.18(a) Previewed Clamp_Arm Component

Click: Constraints **Assemble component at default location** [Fig. 15.18(b)] ⇒ **OK** [Fig. 15.18(c)]

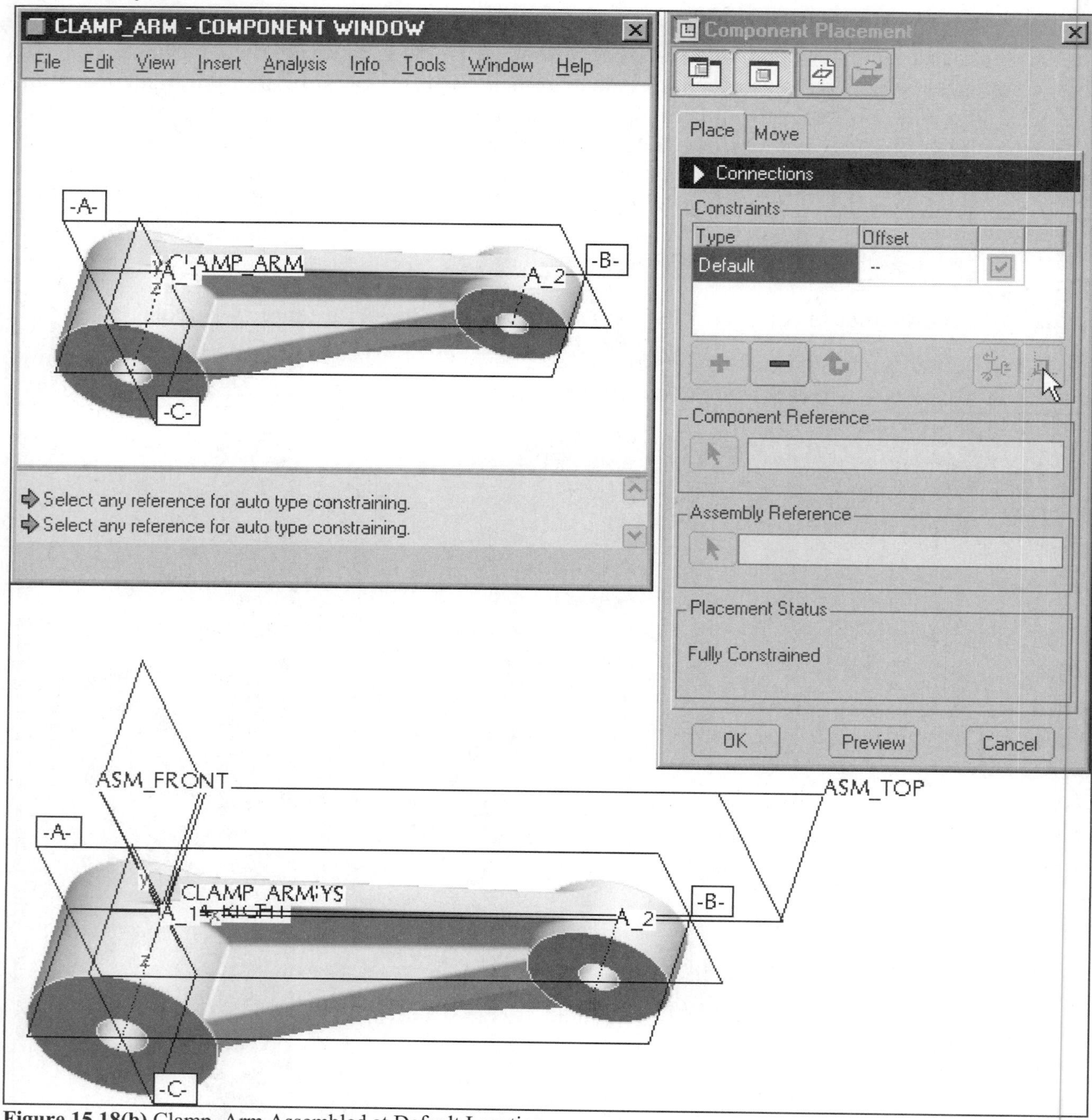

Figure 15.18(b) Clamp_Arm Assembled at Default Location

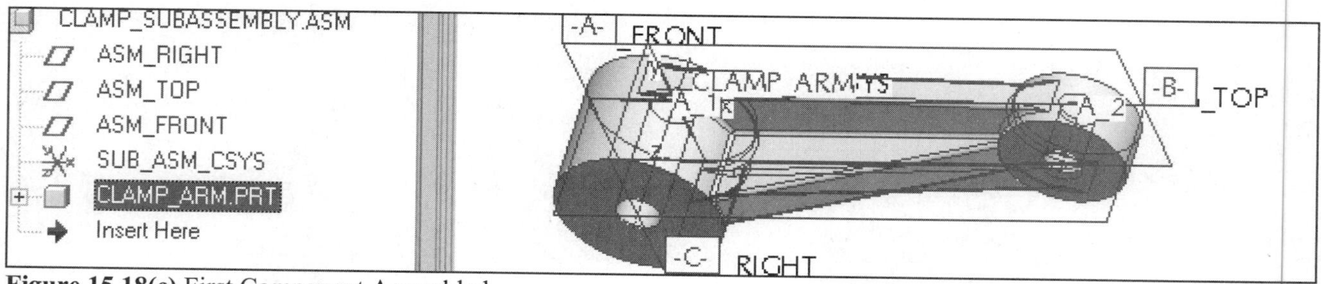

Figure 15.18(c) First Component Assembled

The Clamp_Arm has been assembled to the default location. This means to align the default Pro/E-created coordinate system of the component to the default Pro/E-created coordinate system of the assembly. Pro/E places the component at the assembly origin. By using the constraint, Coord Sys and selecting the assembly and then the component's coordinate systems would have accomplished the same thing, but with more picks. There will be situations where this constraint is useful, as when one or more of the coordinate systems used in the assembly is not the default coordinate system created when the default template is selected.

The next component to be assembled is the Clamp_Swivel. Two constraints will be used with this component: Insert and Mate (Offset). *Placement constraints* are used to specify the relative position of a *pair of surfaces/references* between two components. The Mate, Align, Insert, commands are placement constraints. The two surfaces/references must be of the same type. When using a datum plane as a placement constraint, specify Mate or Align. When using Mate (Offset) or Align (Offset), enter the offset distance. The *offset direction* is displayed with an arrow. If you need an offset in the opposite direction, enter a negative value.

Click: **Add component to the assembly** ⇒ pick the **clamp_swivel.prt** from the Open dialog box ⇒ **Preview** [Fig. 15.19(a)] ⇒ **Open** ⇒ clicking on the active radio button, toggles off the component window ⇒ **Tools** ⇒ **Environment** ⇒ Standard Orient Isometric ⇒ **Apply** ⇒ **OK**

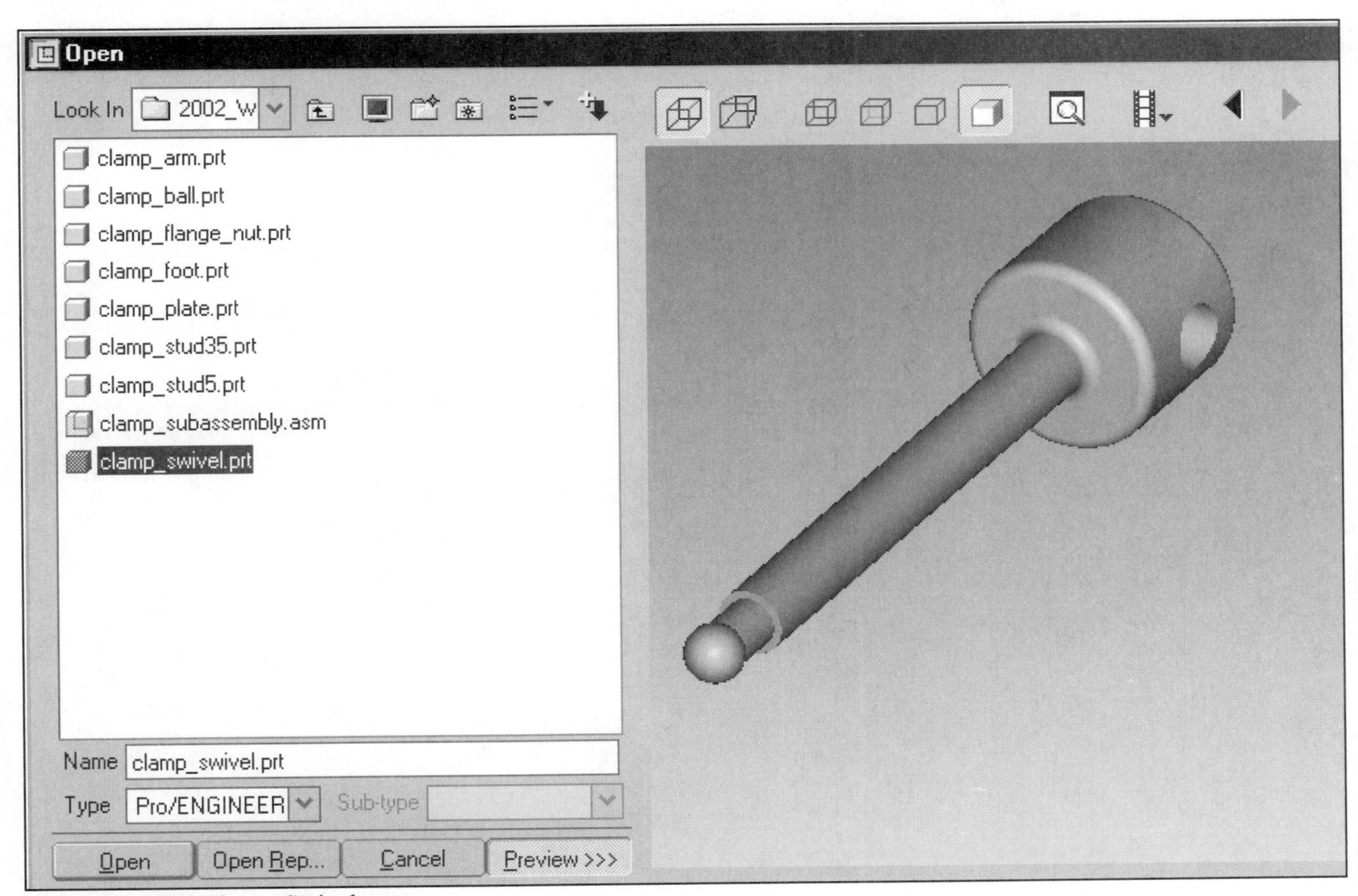

Figure 15.19(a) Clamp_Swivel

Click: ✚ (**Automatic**) ⇒ pick the cylindrical surface of the Clamp_Swivel [Fig. 15.19(b)] ⇒ pick the hole surface of the Clamp_Arm [Fig. 15.19(c)] *constraint type becomes **Insert*** [Fig. 15.19(d)]

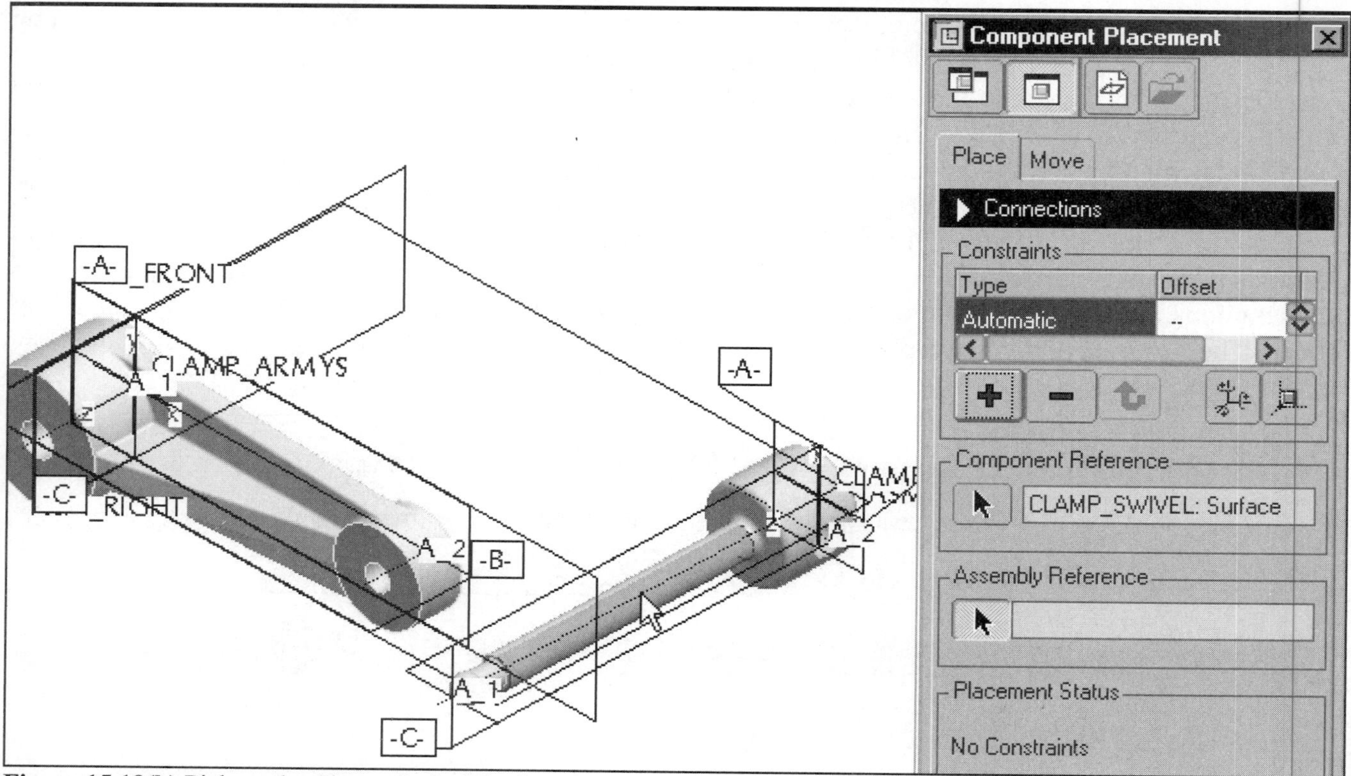

Figure 15.19(b) Pick on the Clamp_Swivel Surface

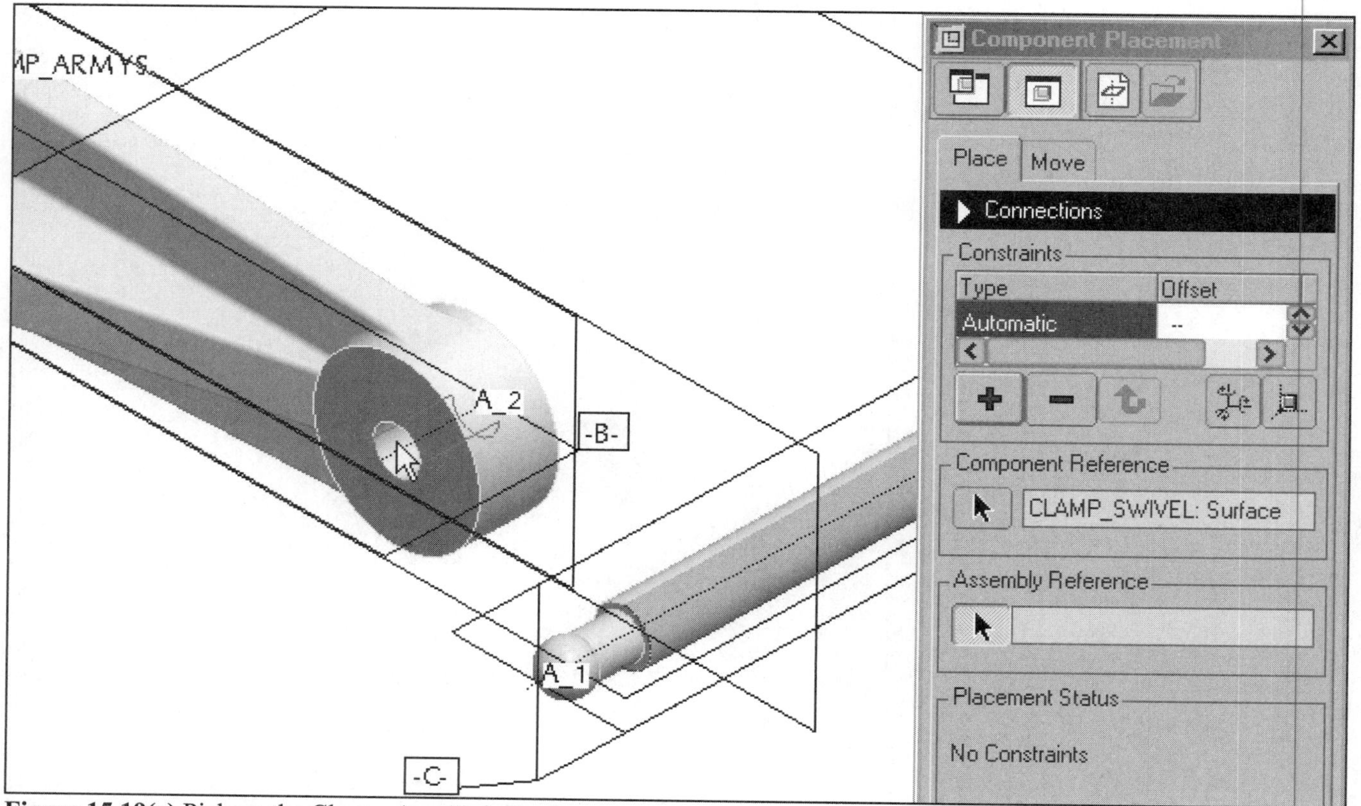

Figure 15.19(c) Pick on the Clamp_Arm Hole Surface

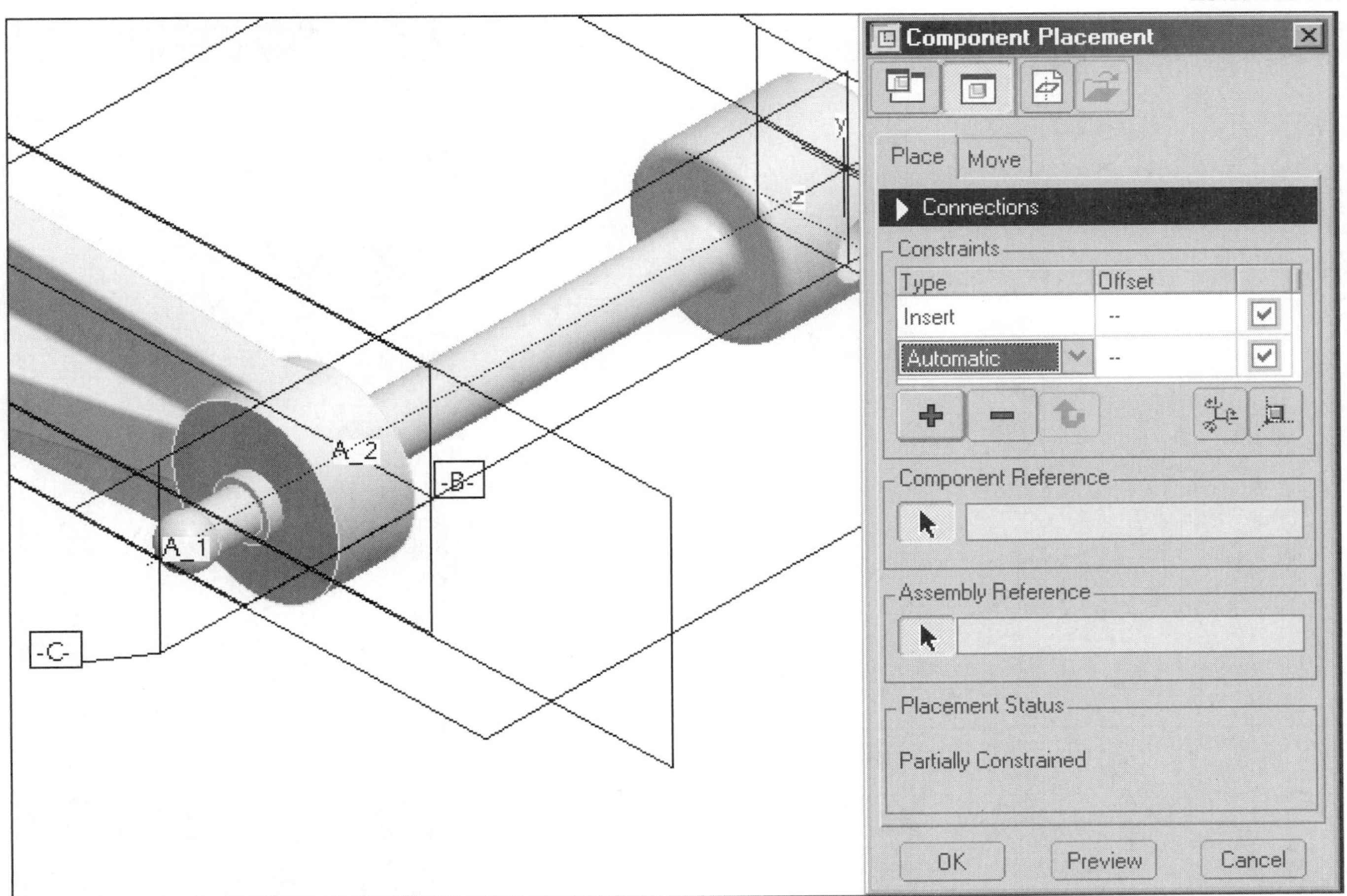

Figure 15.19(d) Insert Constraint Completed

Constraints [Automatic] pick the underside surface of the Clamp_Swivel [Fig. 15.19(e)] ⇒ pick the top surface of the Clamp_Arm [Fig. 15.19(f)] ⇒ Offset (ins) in indicated direction: type **1.50** ⇒ [✓] ⇒ [Align] ⇒ [▾] ⇒ [Mate] [Fig. 15.19(g)] *the Clamp_Swivel reverses* [Fig. 15.19(h)]

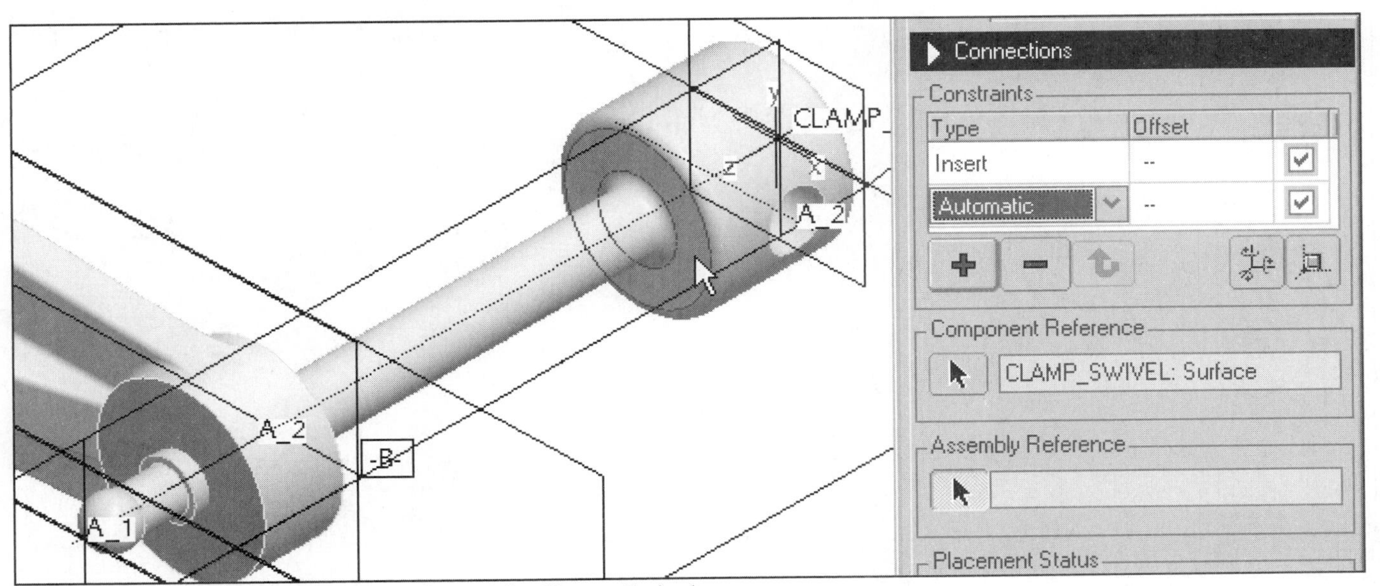

Figure 15.19(e) Select Underside Surface of the Clamp_Swivel

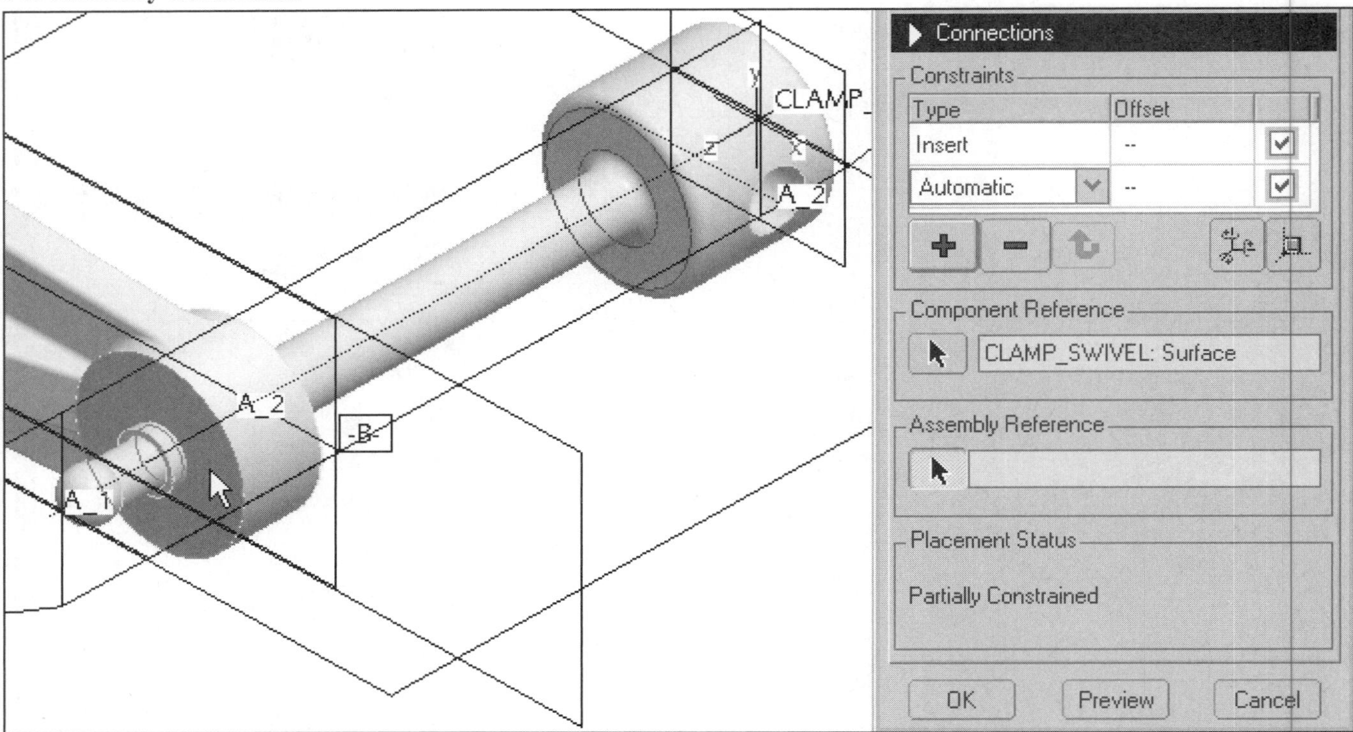

Figure 15.19(f) Select Top Surface of the Small Circular Protrusion

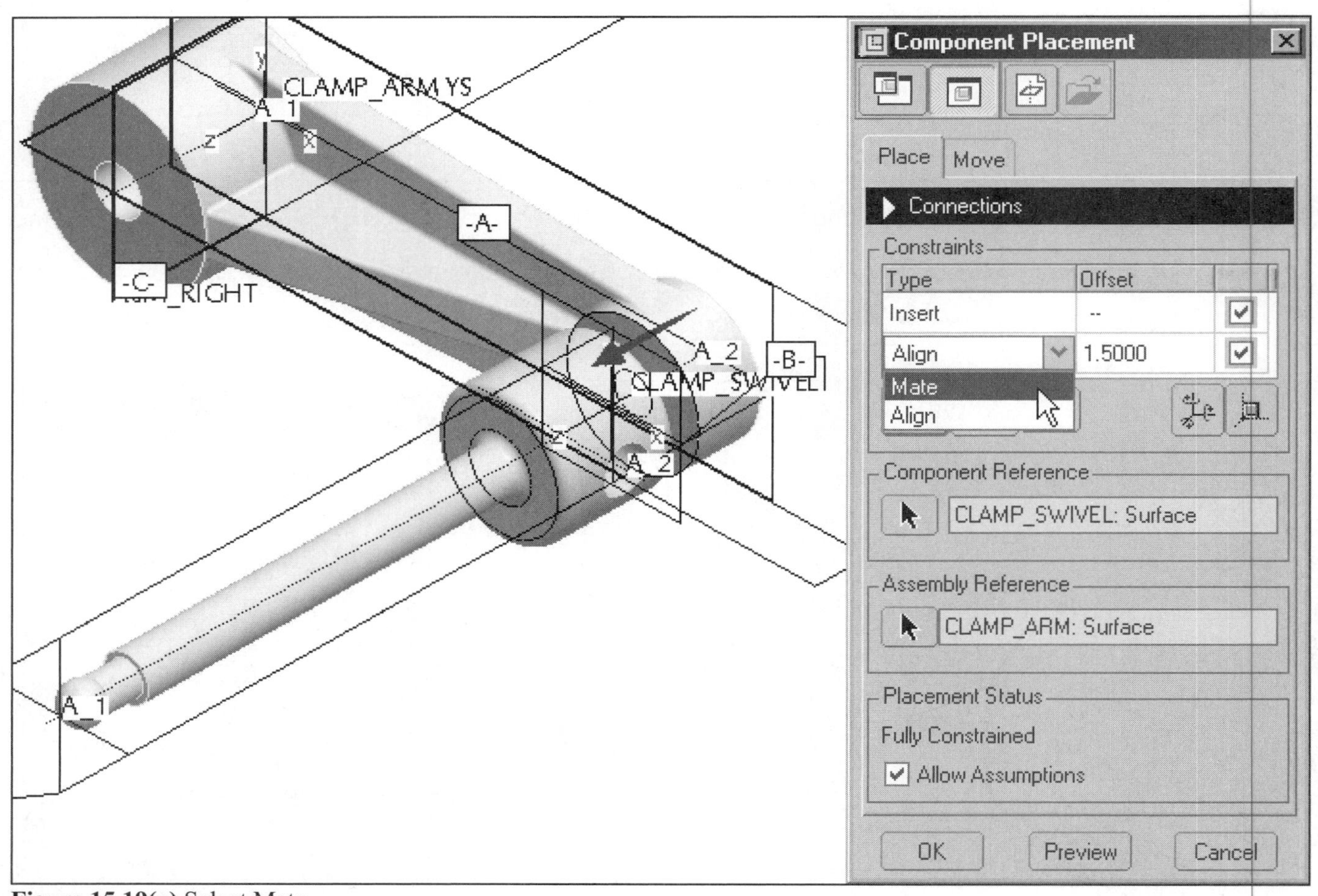

Figure 15.19(g) Select Mate

The Placement Status shows the component is Fully Constrained since Allow Assumptions is checked by default [Fig. 15.19(h)] ⇒ **OK** *the second component is now assembled* (Fig. 15.20)

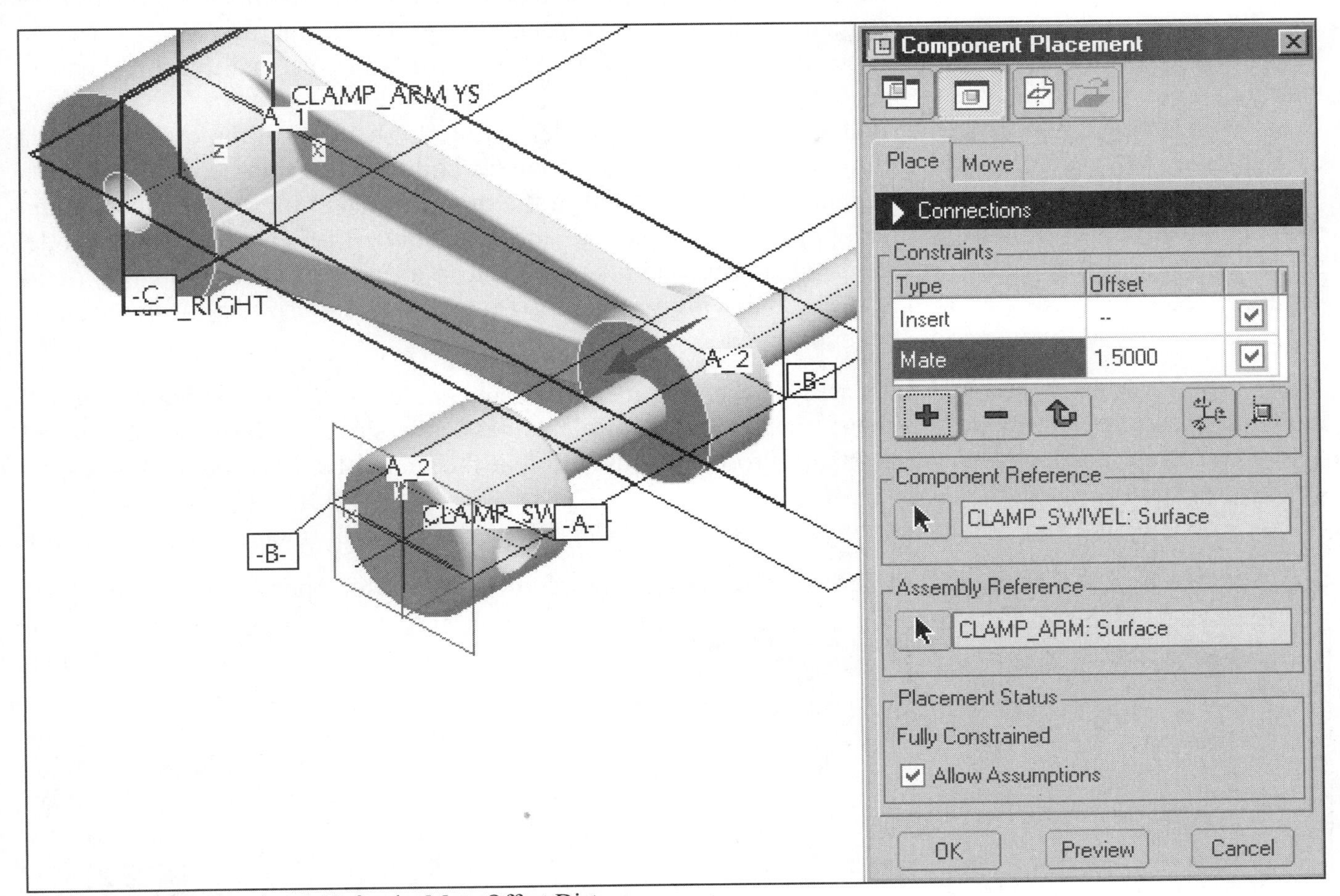

Figure 15.19(h) Showing **1.50** for the Mate Offset Distance

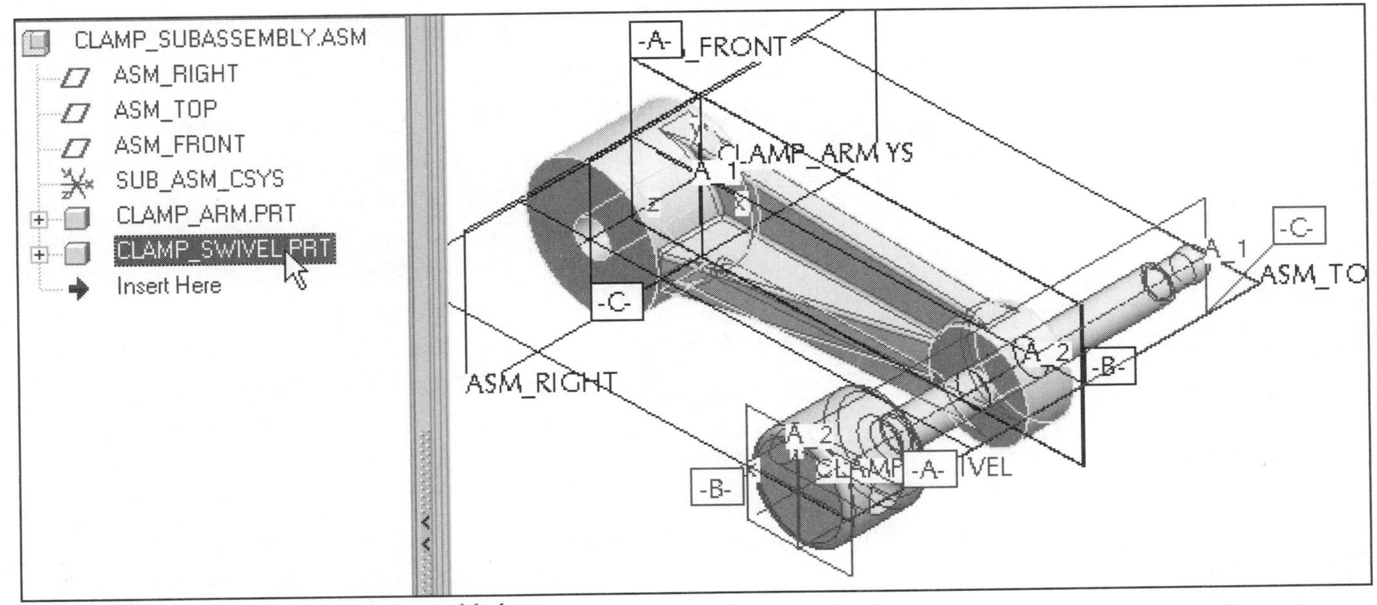

Figure 15.20 Second Component Assembled

Regenerating Models

You can use Regenerate to find bad geometry, broken parent-child relationships, or any other problem with a part feature or assembly component. In general, it is a good idea to regenerate the model every time you make a change, so that you can see the effects of each change in the graphics window as you build the model. By regenerating often, it helps you stay on course with your original design intent by helping you to resolve failures as they happen.

When Pro/E regenerates a model, it recreates the model feature by feature, in the order in which each feature was created, and according to the hierarchy of the parent-child relationship between features.

In an assembly, component features are regenerated in the order in which they were created, and then in the order in which each component was added to the assembly. Pro/E regenerates a model automatically in many cases, including when you open, save, or close a part or assembly or one of its instances, or when you open an instance from within a Family Table. You can also use the Regenerate command to manually regenerate the model.

The Regenerate command, located on the Edit menu or using the icon **Regenerates Model**, lets you recalculate the model geometry, incorporating any changes made since the last time the model was saved. If no changes have been made, Pro/E informs you that the model has not changed since the last regeneration.

The Custom Regenerate command (Assembly Mode), located on the Edit menu or using the icon **Specify the list of modified features or components to regenerate**, opens the Regeneration Manager dialog box if there are features or components that have been changed that require regeneration (Fig. 15.21).

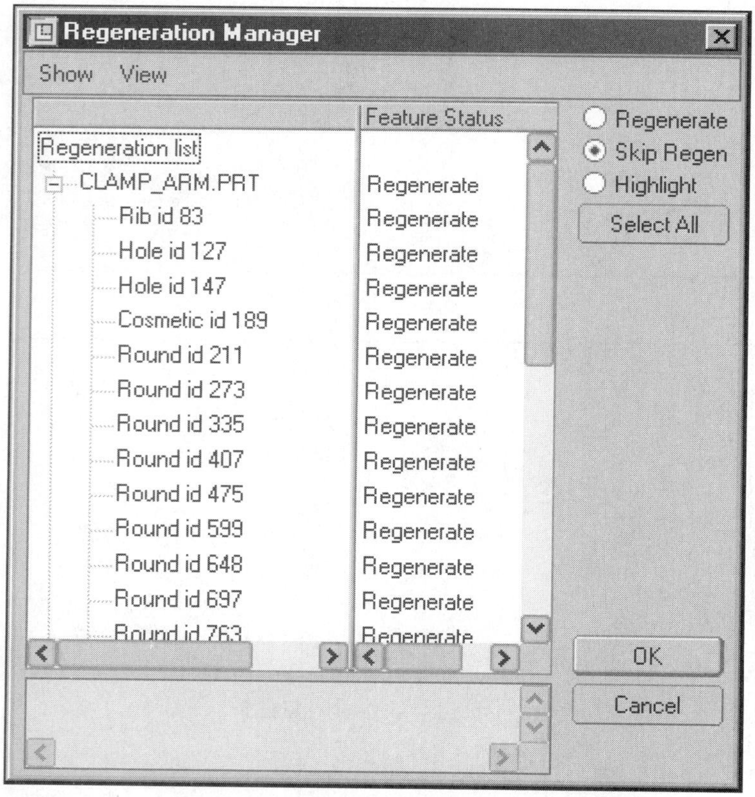

Initially, Pro/E expands the tree to the first level in the Regeneration list column. To display features, choose Show from the menu bar. To expand or collapse the tree to any level, choose View from the menu bar. Information appears below the Regeneration list.

In the dialog box do one of the following:

- To select all components for regeneration, select Regenerate and then Select All.
- To omit all components from regeneration, select Skip Regen and then Select All.
- To determine the reason an object requires regeneration, select Highlight and select an entry in the dialog box.

Figure 15.21 Regeneration Manager Dialog Box

If you select a subassembly, Pro/E selects all components and features within that subassembly even if it is broken up in the regeneration list. A column next to the regeneration list indicates whether you have selected a component or feature to skip regeneration or to be regenerated.

Click: **Specify the list of modified features or components to regenerate** ⇒ Regen Manager Info dialog box opens (Fig. 15.22) ⇒ click **OK** since there are no objects to be regenerated at this time

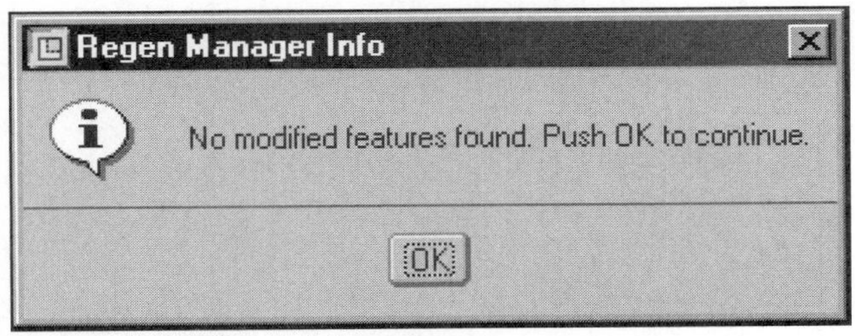

Figure 15.22 Second Component

The next component to be assembled is the Clamp_Foot.

Click: **Add component to the assembly** ⇒ pick the **clamp_foot.prt** from the Open dialog box ⇒ **Preview** [Fig. 15.23(a)] ⇒ **Open** ⇒ **Show component in a separate window while specifying constraints** ⇒ spin and zoom in on your model and resize your windows as desired ⇒ Constraints Automatic pick the *internal cylindrical surface* of the Clamp_Foot [Fig. 15.23(b)] ⇒ pick the *external cylindrical surface* of the Clamp_Swivel [Fig. 15.23(c)] *constraint turns to* ***Insert*** ⇒ Constraints Automatic pick the spherical hole of the Clamp_Foot [Fig. 15.23(d)] ⇒ pick the spherical end of the Clamp_Swivel [Fig. 15.23(e)] *constraint turns to* ***Mate*** ⇒ **Preview** ⇒ **View** ⇒ **Shade** [Fig. 15.23(f)] ⇒ **OK** ⇒ ⇒ **MMB**

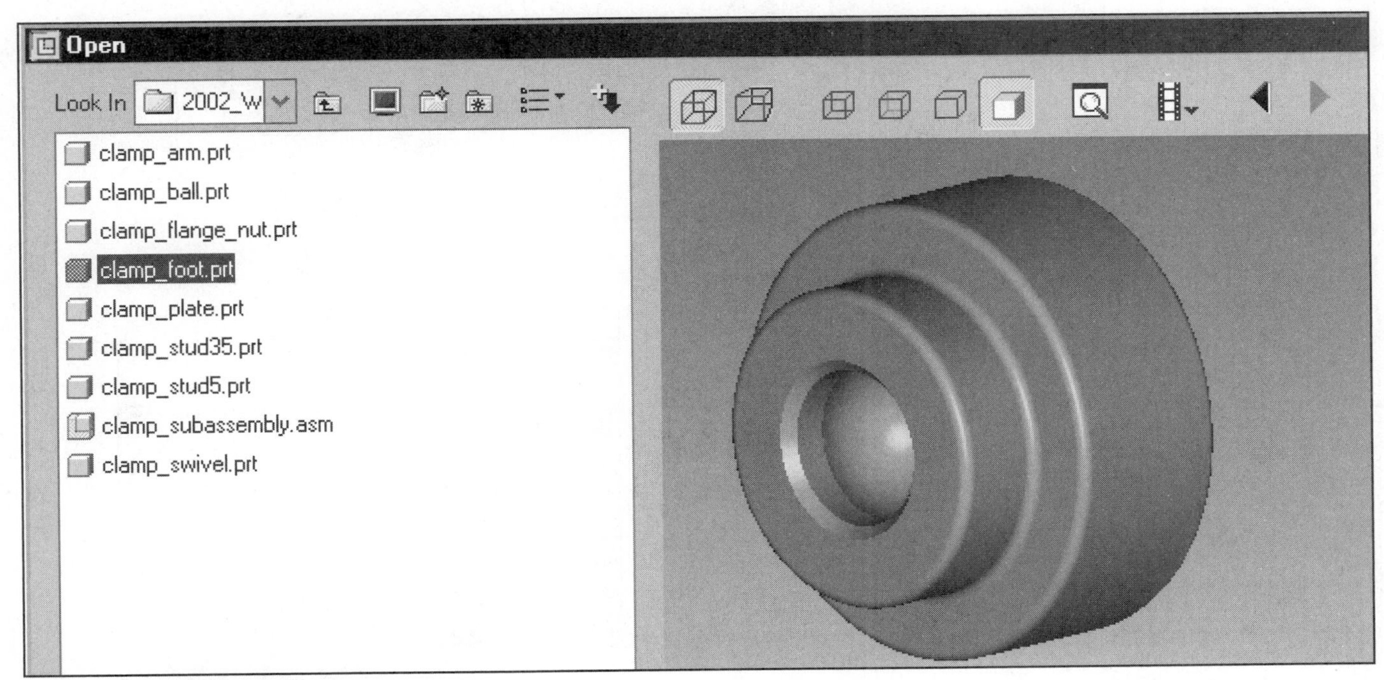

Figure 15.23(a) Clamp_Foot

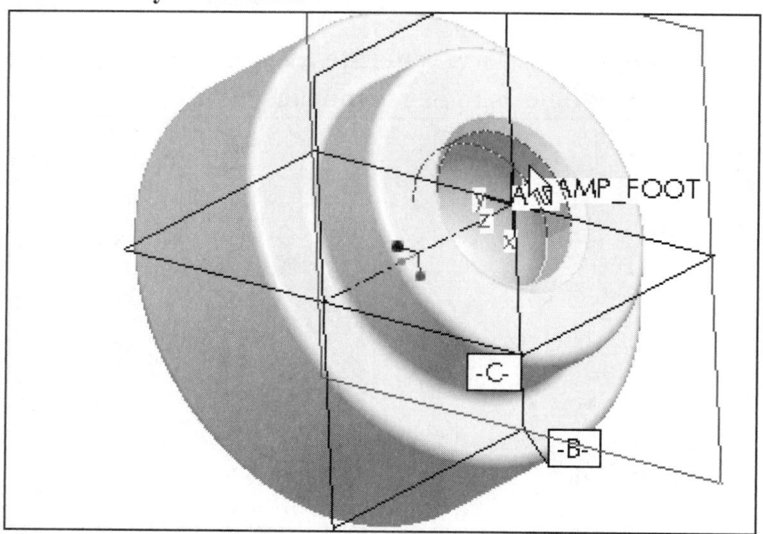

Figure 15.23(b) Pick on the Internal Cylindrical Surface

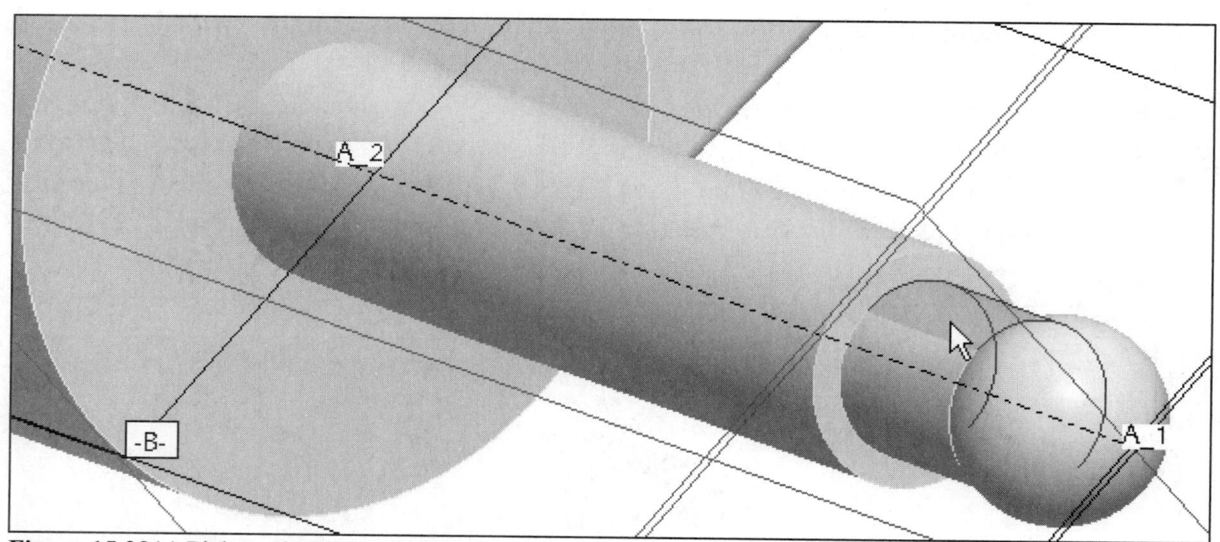

Figure 15.23(c) Pick on the External Cylindrical Surface

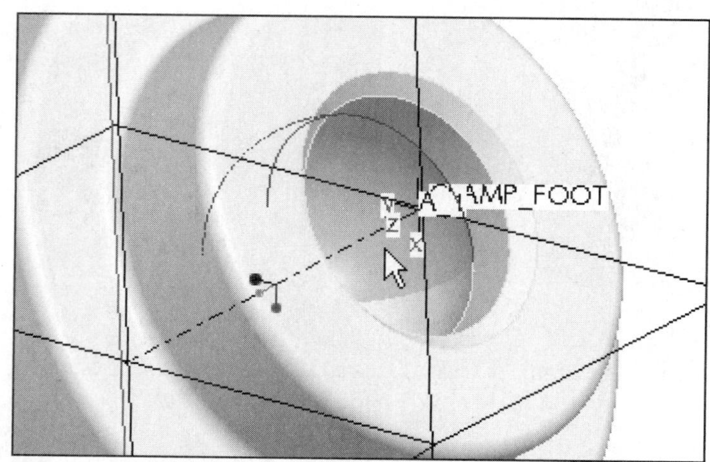

Figure 15.23(d) Pick on the Internal Spherical Surface

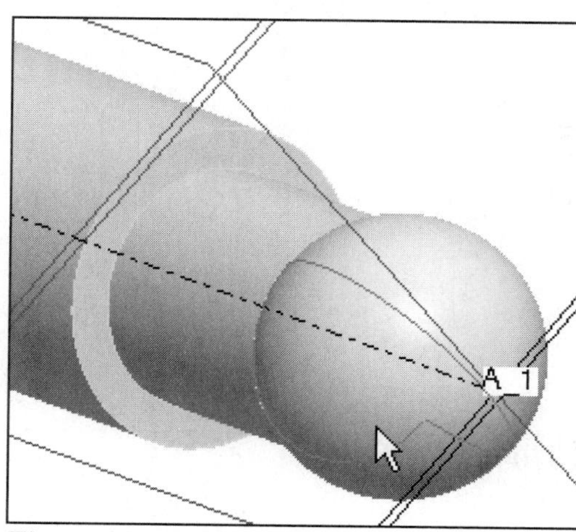

Figure 15.23(e) Pick on the External Spherical Surface

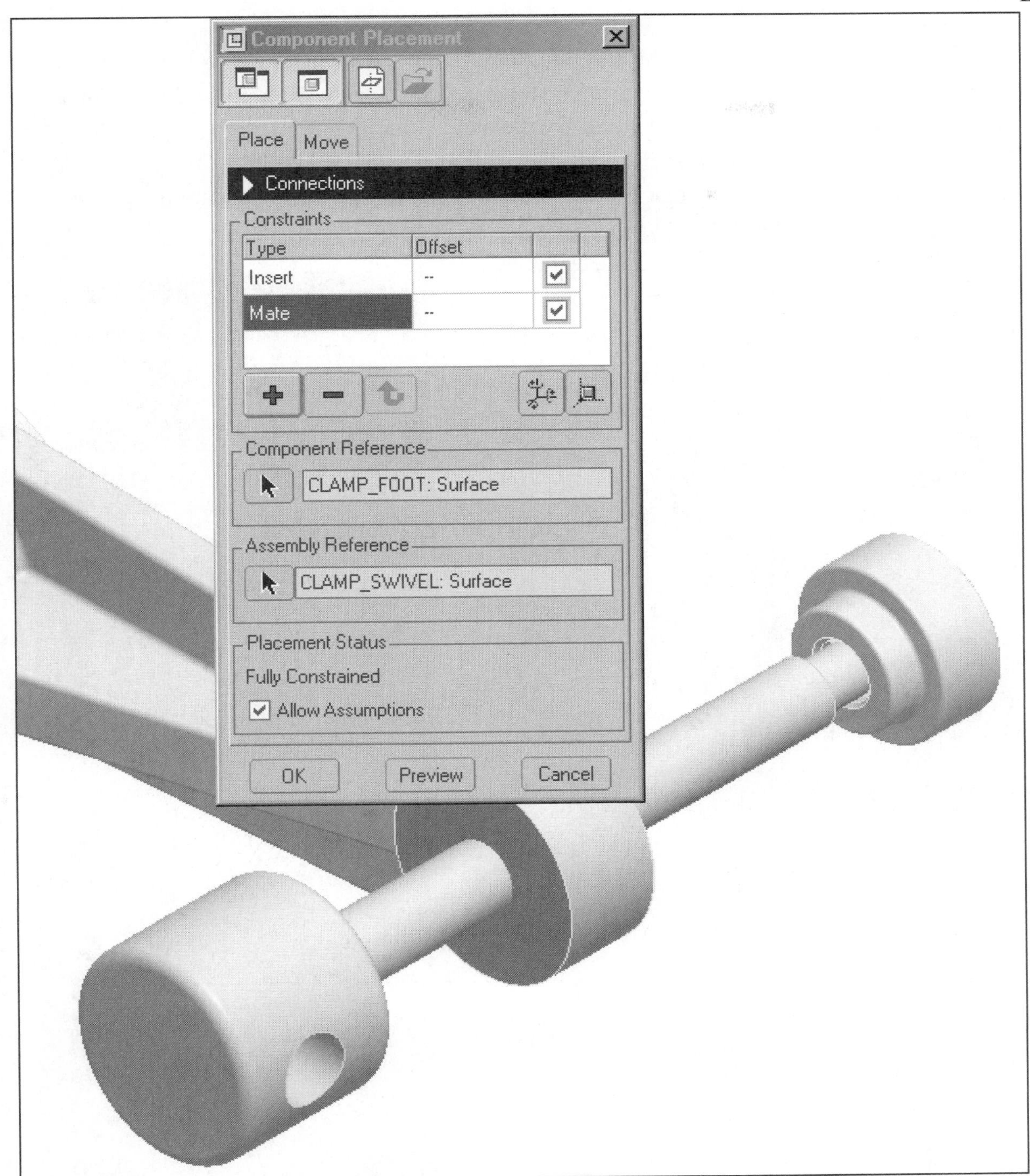

Figure 15.23(f) Fully Constrained Clamp_Foot

Assemble the **5.00** double-ended stud. Click: **Add component to the assembly** ⇒ pick **clamp_stud5.prt** ⇒ **Preview** [Fig. 15.24(a)]

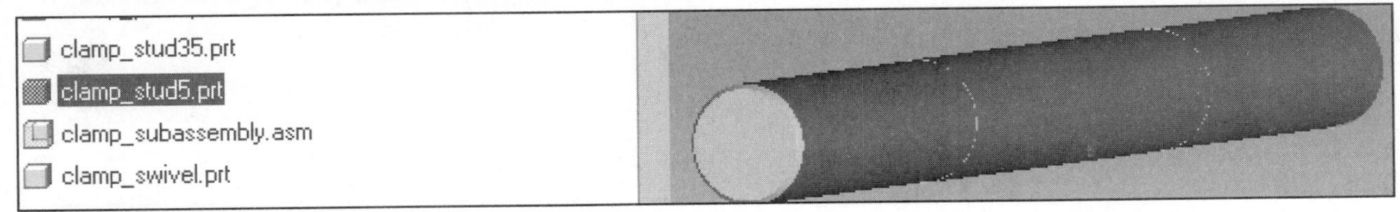

Figure 15.24(a) Clamp_Stud5

Click: **Open** ⇒ [icon] clicking on the active button, toggles off the component window ⇒ Constraints [Automatic] pick the *external cylindrical surface* of the Clamp_Stud5 [Fig. 15.25(b)] ⇒ pick the *internal cylindrical surface of the hole* of the Clamp_Swivel [Fig. 15.25(b)] *constraint turns to* ***Insert*** ⇒ Constraints [Automatic] pick the end surface of the Clamp_Stud5 [Fig. 15.25(c)] ⇒ to pick the appropriate datum of the Clamp_Swivel, place the cursor on/near the appropriate datum reference ⇒ **RMB** ⇒ **Pick From List** ⇒ click [C:F2(DATUM PLANE):CLAMP_SWIVEL] [Fig. 15.25(d)] ⇒ **MMB**

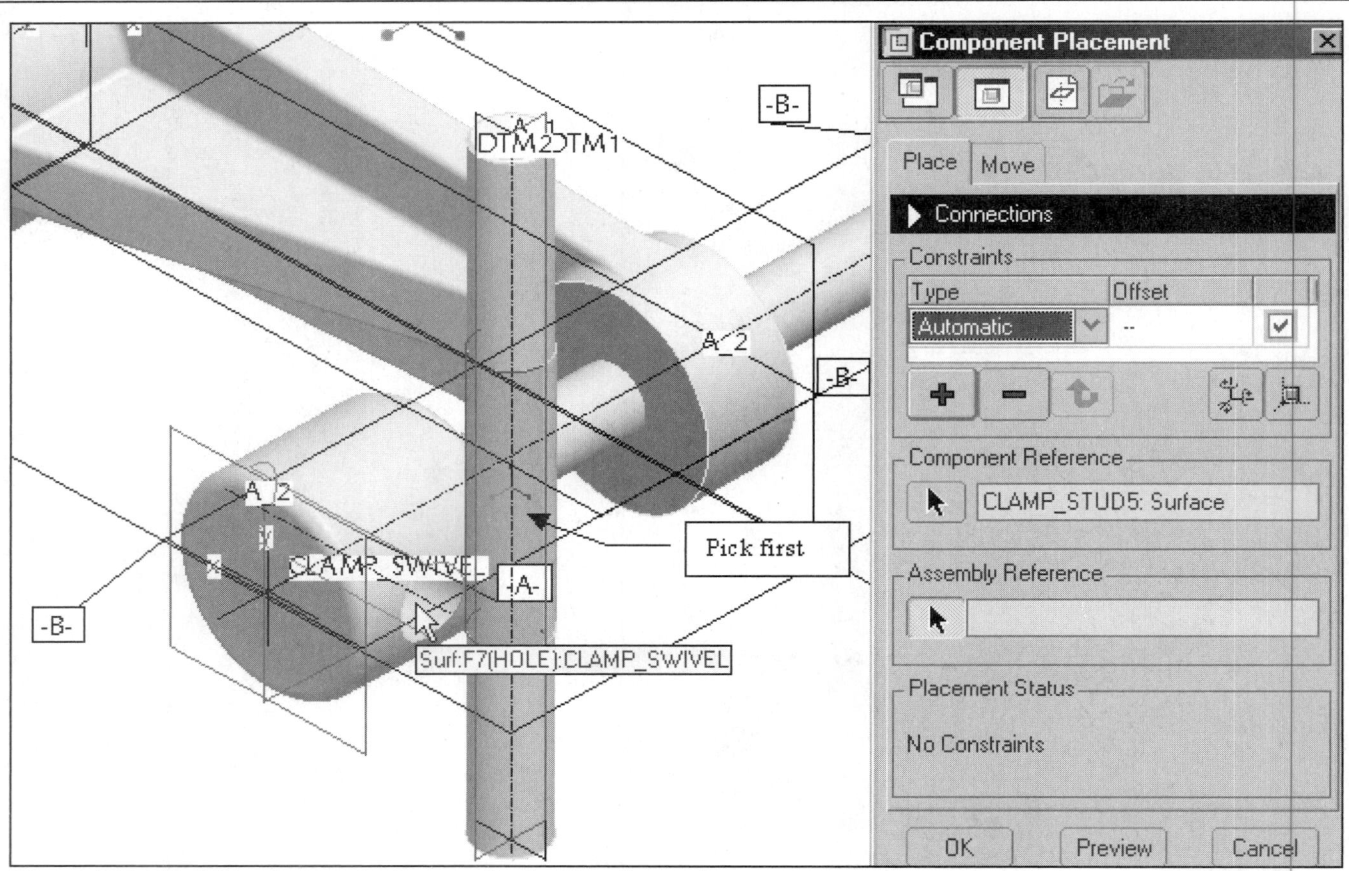

Figure 15.24(b) Select the Two References for the First Constraint

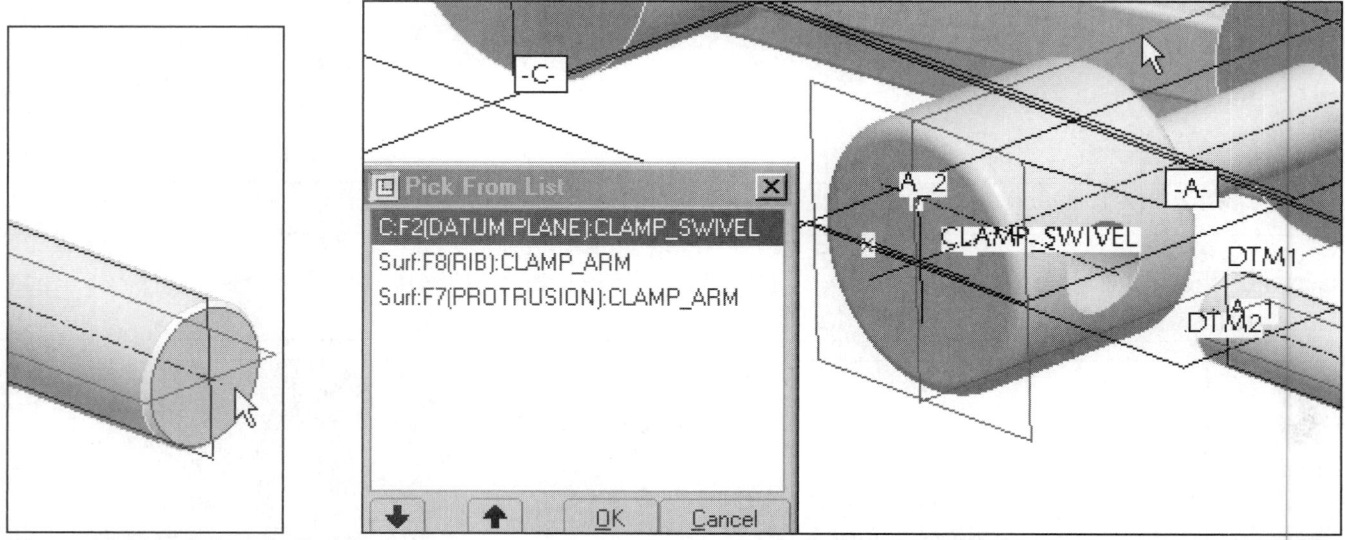

Figure 15.24(c) Select the End Surface

Figure 15.24(d) Select Datum C of the Clamp_Swivel

Offset (ins) in indicated direction: type **–2.50** ⇒ **Enter** [Mate -2.5000] *constraint turns to* ***Mate*** [Fig. 15.25(a)] ⇒ **OK** ⇒ [save icon] ⇒ **MMB** ⇒ click **Settings** from the Model Tree ⇒ **Tree Columns** ⇒ use [>>] to add the item Types shown in Figure 15.25(b) ⇒ **Apply** ⇒ **OK** ⇒ adjust the columns in the Model Tree [Fig. 15.25(c)]

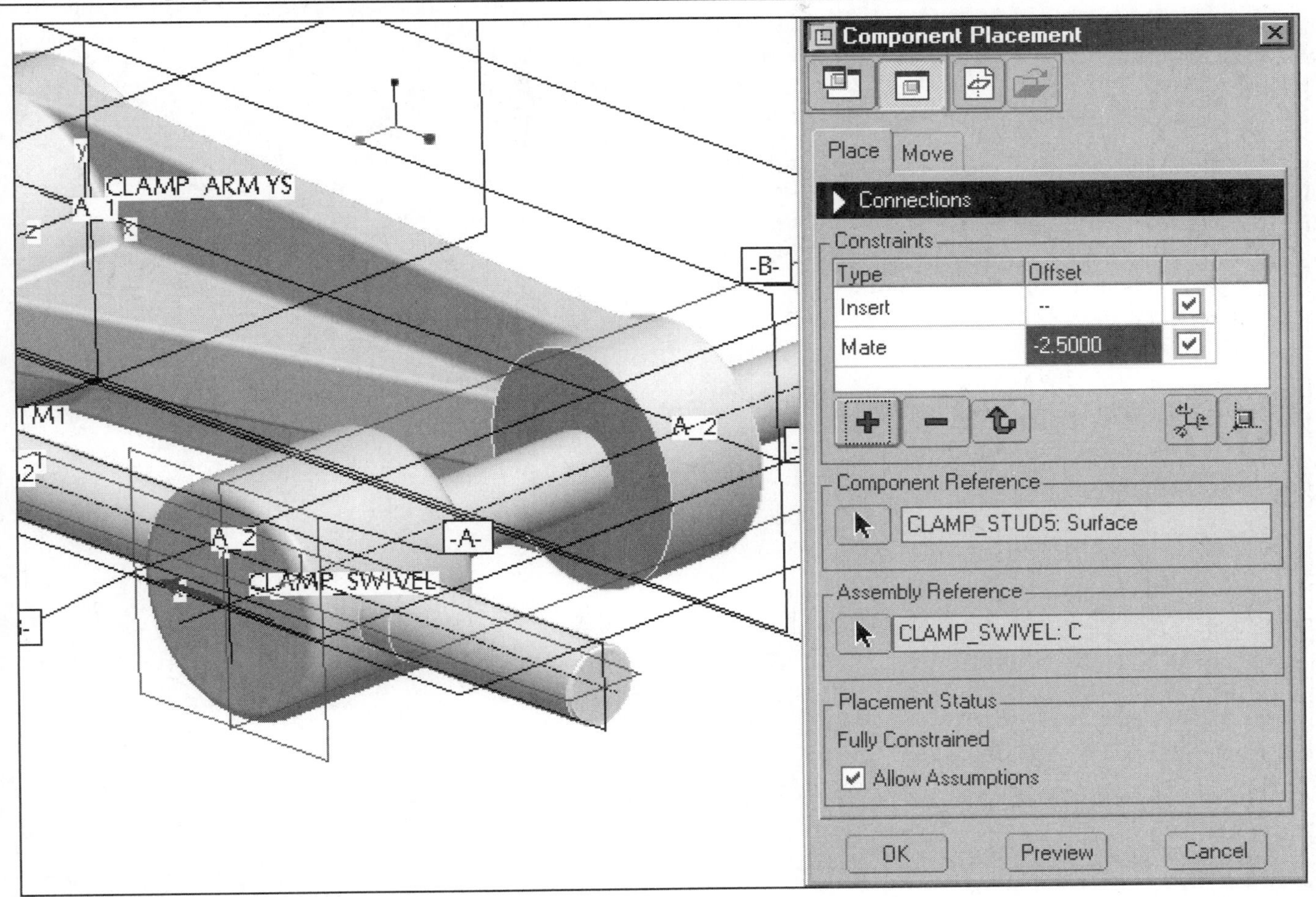

Figure 15.25(a) Mate Offset Constraint

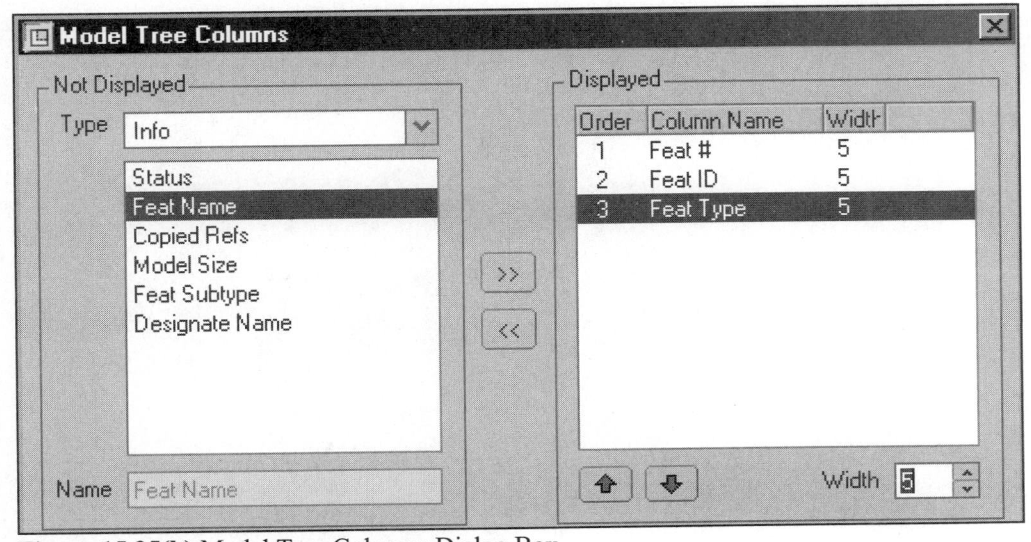

Figure 15.25(b) Model Tree Columns Dialog Box

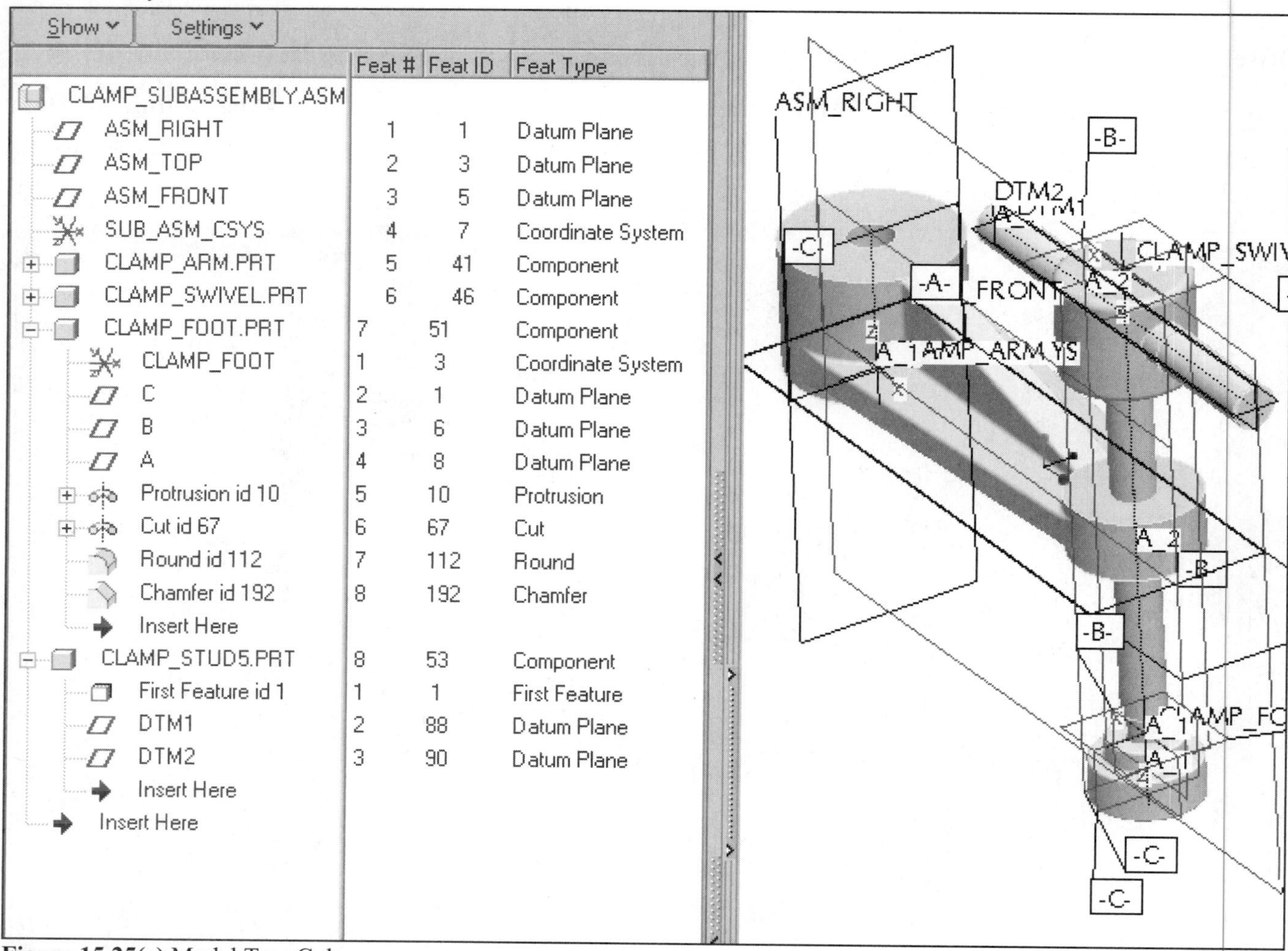

Figure 15.25(c) Model Tree Columns

The Clamp_Ball handles are the last components of the Clamp_Subassembly. Instructions are provided for assembling one Clamp_Ball; you must assemble the other on your own.

Click: **Add component to the assembly** ⇒ pick **clamp_ball.prt** ⇒ **Preview** ⇒ ⇒ **MMB** spin the model to see the hole [Fig. 15.26(a)]

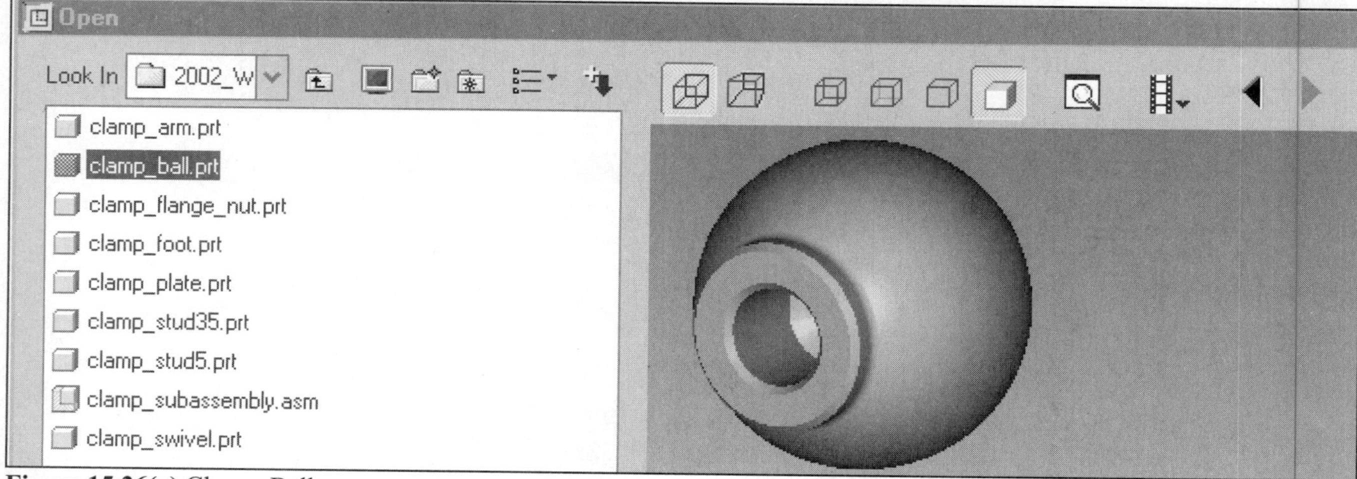

Figure 15.26(a) Clamp_Ball

Click: **Open** [Fig. 15.26(b)] ⇒ spin the model ⇒ Constraints Automatic pick the *internal cylindrical surface of the hole* of the Clamp_Ball [Fig. 15.26(c)] ⇒ pick the *external cylindrical surface* of the Clamp_Stud5 [Fig. 15.26(c)] *constraint turns to **Insert***

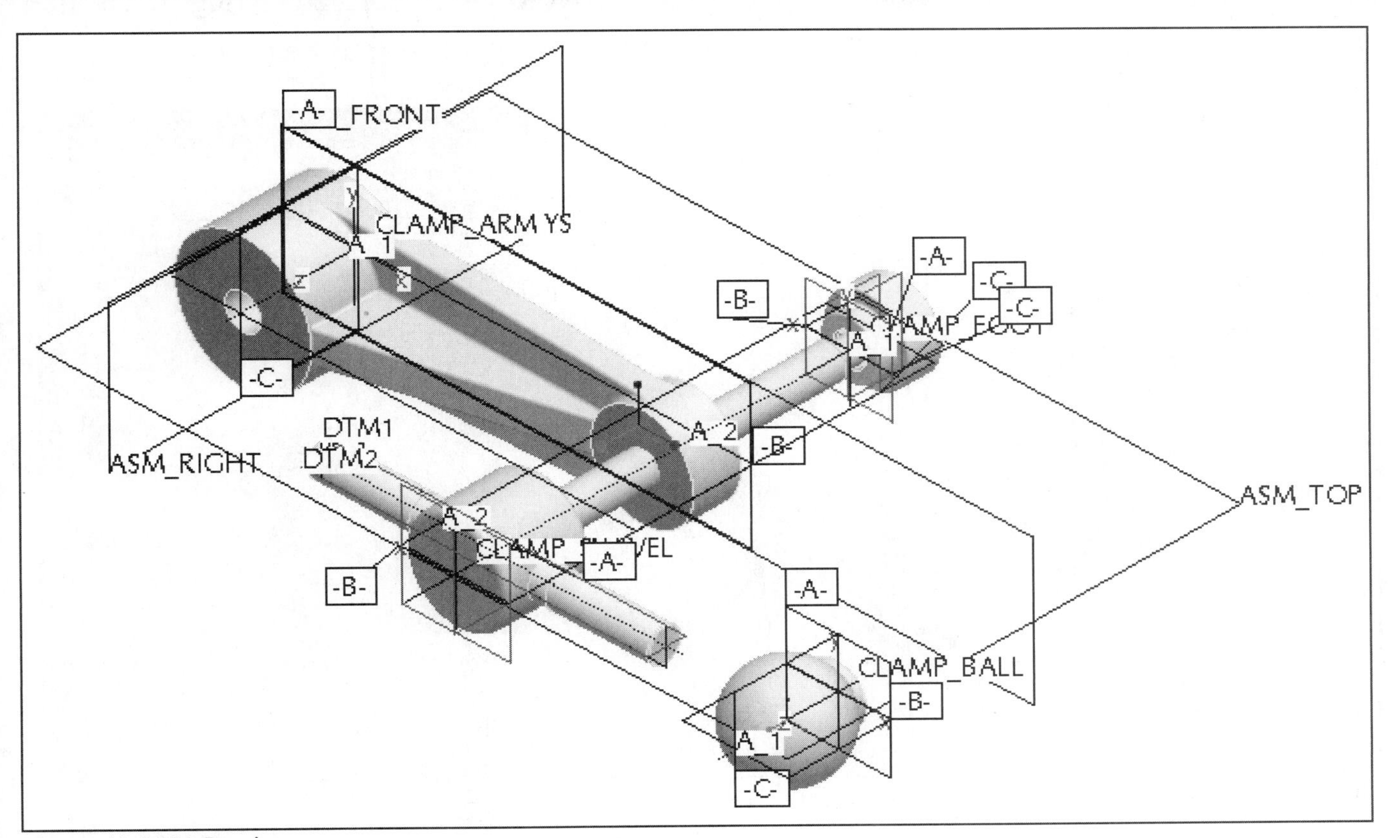

Figure 15.26(b) Preview

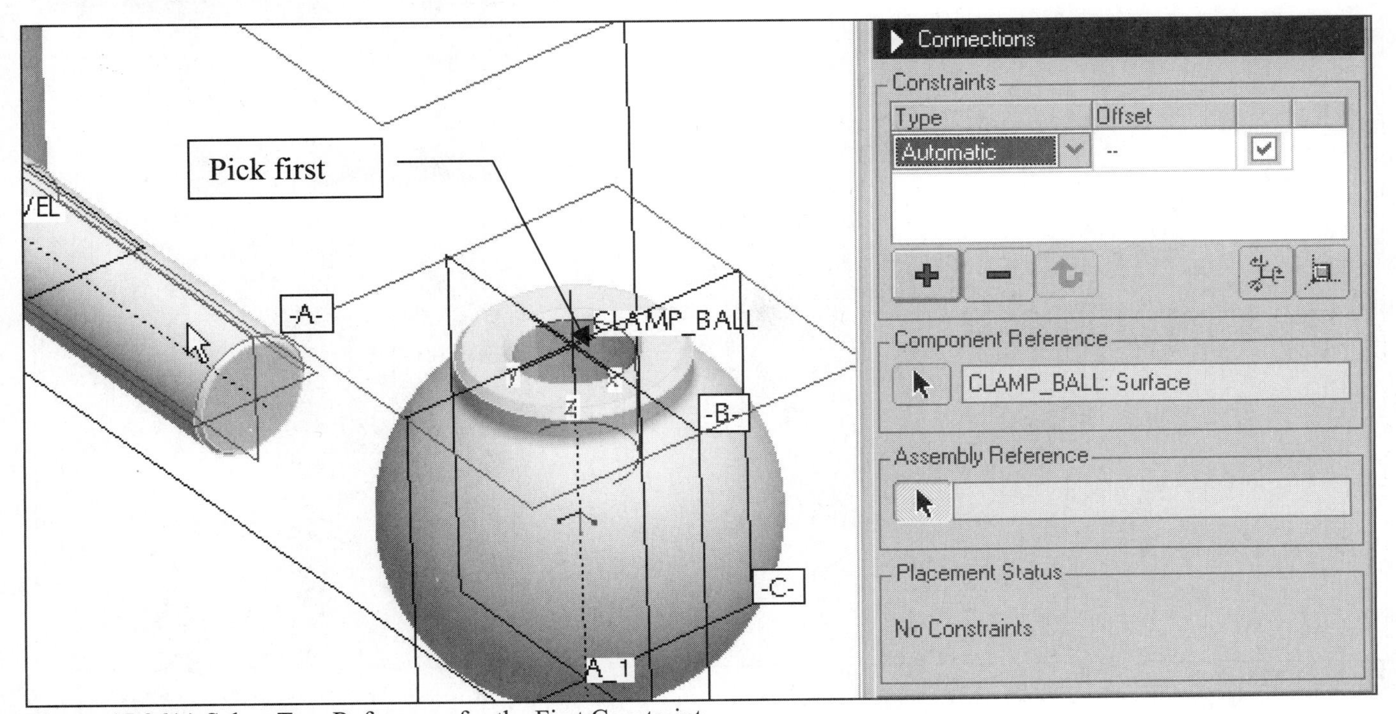

Figure 15.26(c) Select Two References for the First Constraint

Constraints [Automatic] pick the flat end of the Clamp_Ball [Fig. 15.26(d)] ⇒ pick the end surface of the Clamp_Stud5 [Fig. 15.26(d-e)] ⇒ type **-.500** as Offset distance ⇒ **Enter** ⇒ click [Align] in the Constraints box ⇒ click [▾] for drop down choices ⇒ click **Mate** [Mate] [Fig. 15.26(f)] ⇒ **OK** ⇒ [Save] ⇒ **MMB** ⇒ assemble the second Clamp Ball [Fig. 15.26(g)] ⇒ [Save] ⇒ **MMB**

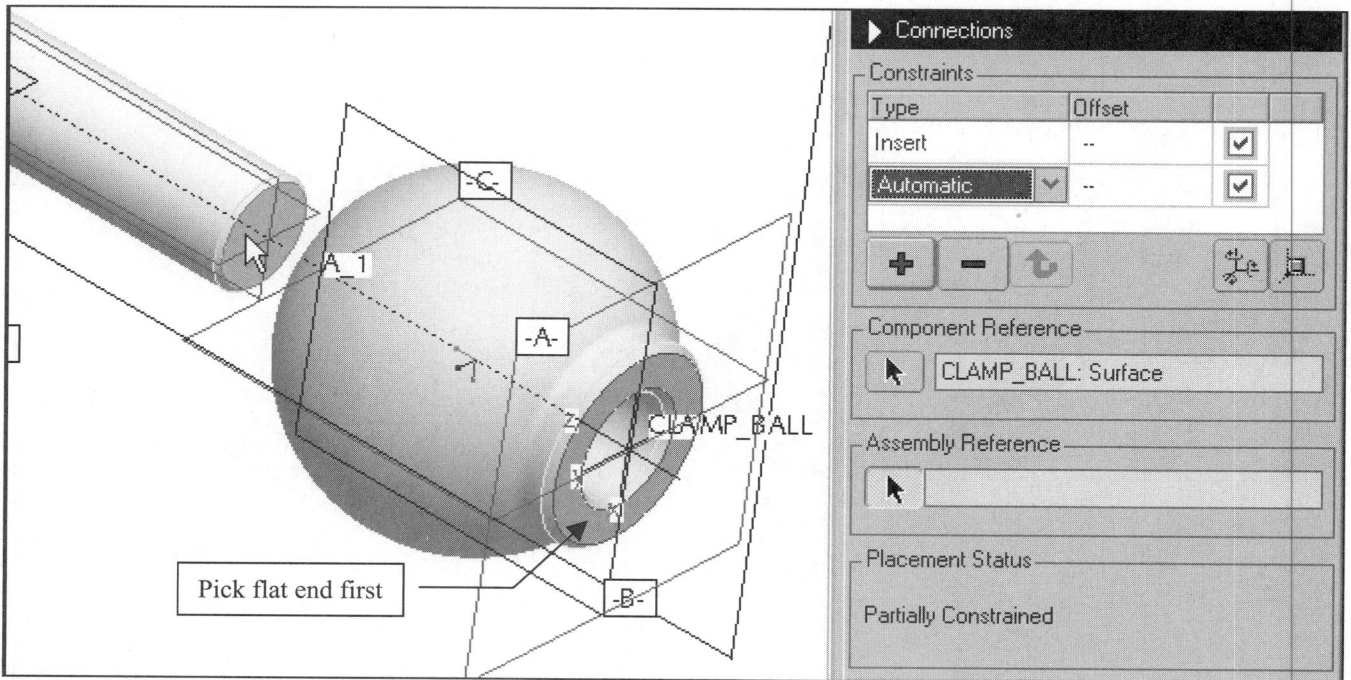

Figure 15.26(d) Select Two References for the Second Constraint

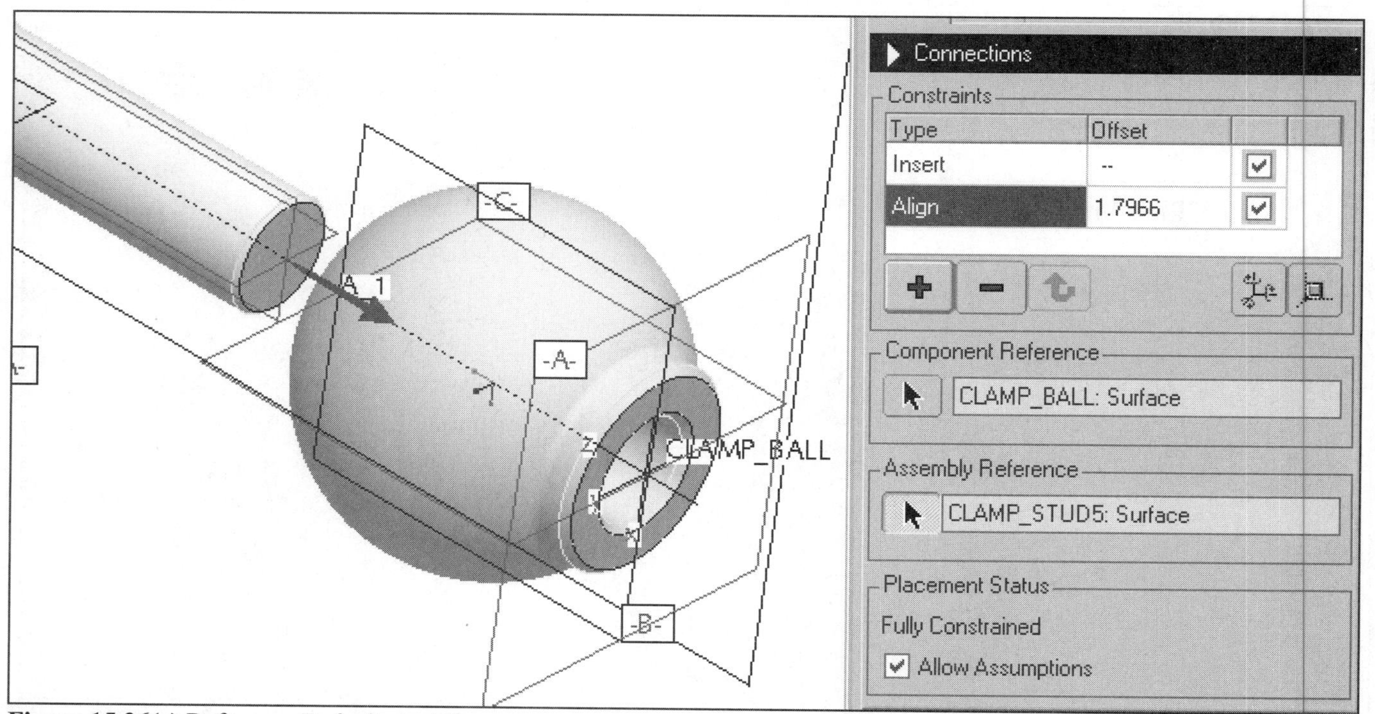

Figure 15.26(e) Reference Defaults to Align

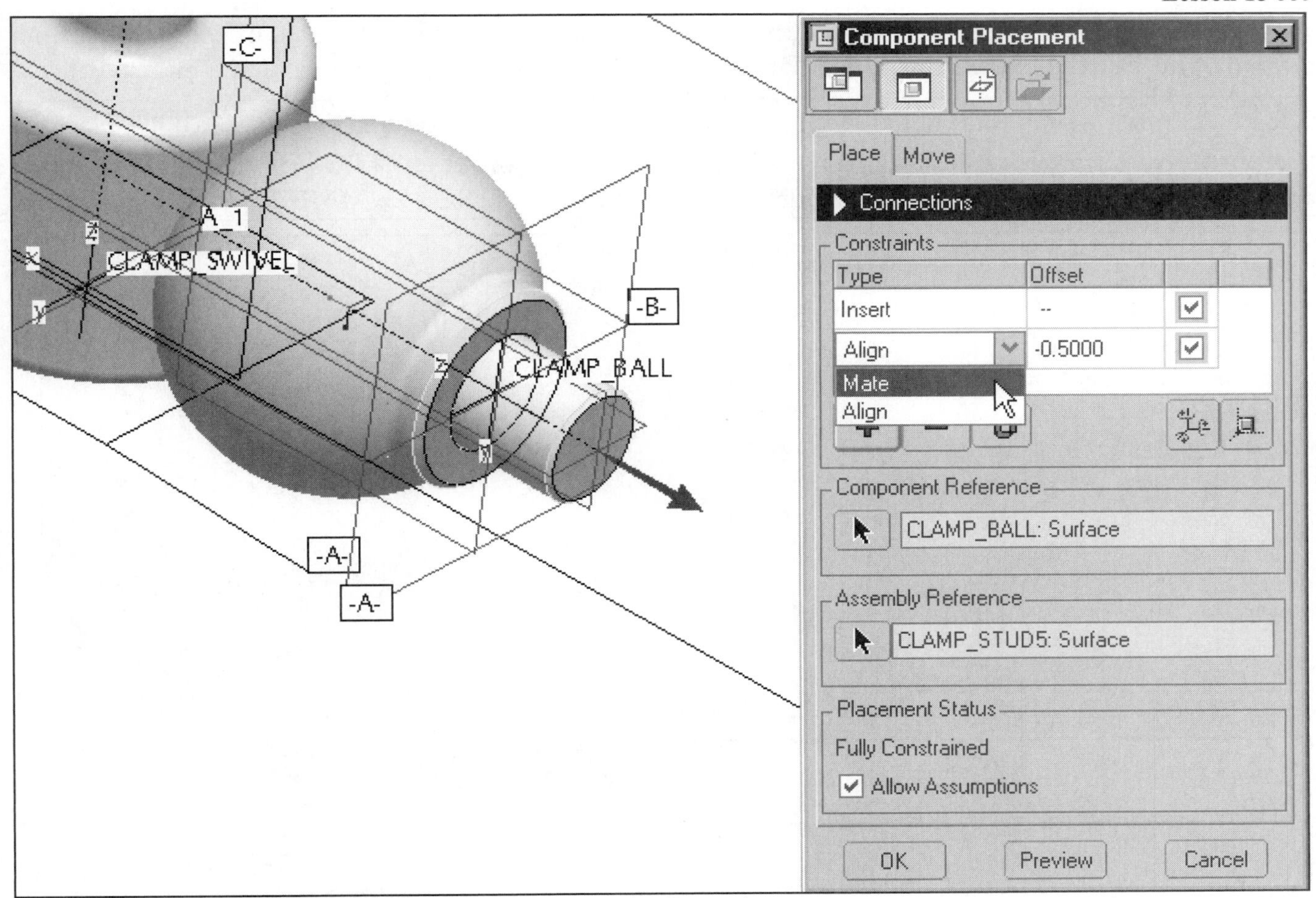

Figure 15.26(f) Change Offset to **-.500** and Select Mate as the Constraint

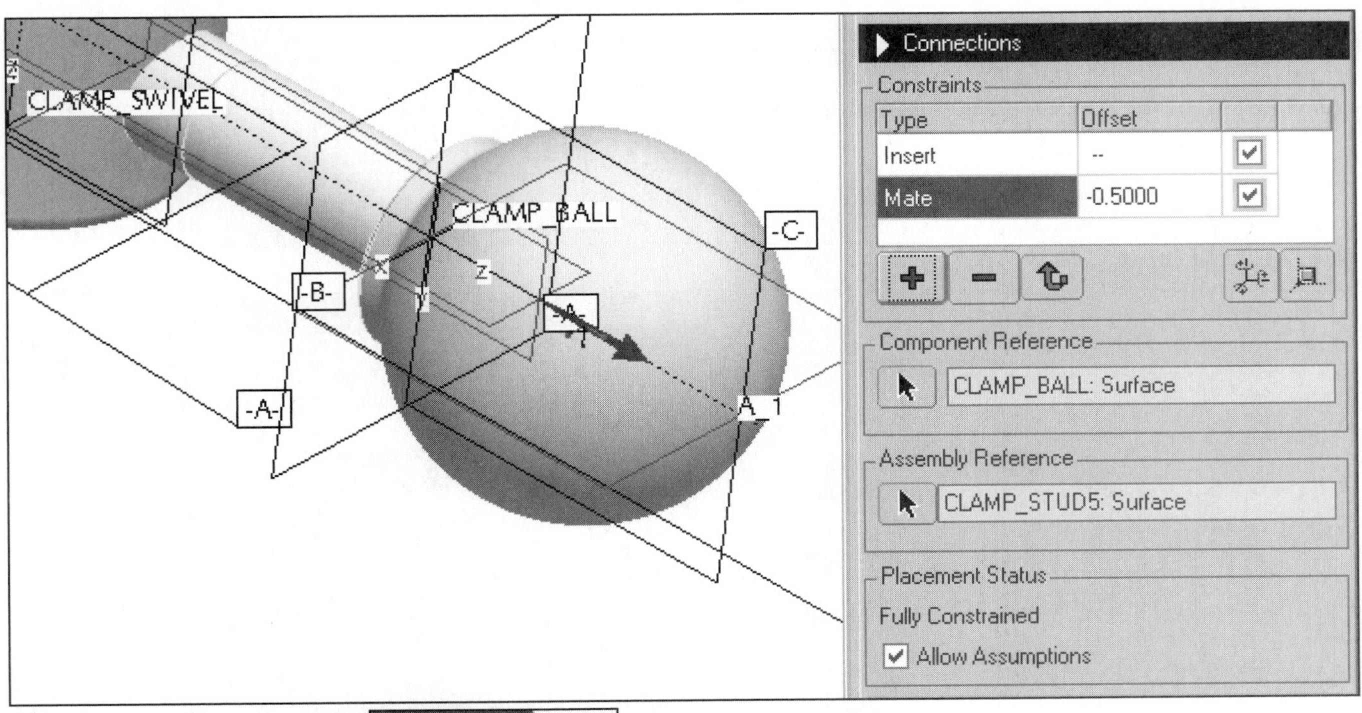

Figure 15.26(g) Mate Constraint Mate -0.5000 is now Active for the Second Clamp_Ball

As a minor **ECO**, redefine or modify the Clamp_Ball component to have an offset of **.4375** from the end of the shaft so that Clamp_Stud5 does not *bottom-out* in the Clamp_Ball's hole.

Click on the first Clamp_Ball in the Model Tree ⇒ **RMB** ⇒ **Edit Definition** [Fig. 15.27(a)] ⇒ select the appropriate view ⇒ click -0.5000 [Fig. 15.27(b)] ⇒ change the offset to **-.4375** [Fig. 15.27(c)] ⇒ **Enter** ⇒ **OK** ⇒ **View** ⇒ **Shade** ⇒ spin the model (Fig. 15.28) ⇒ **File** ⇒ **Save** ⇒ **MMB**

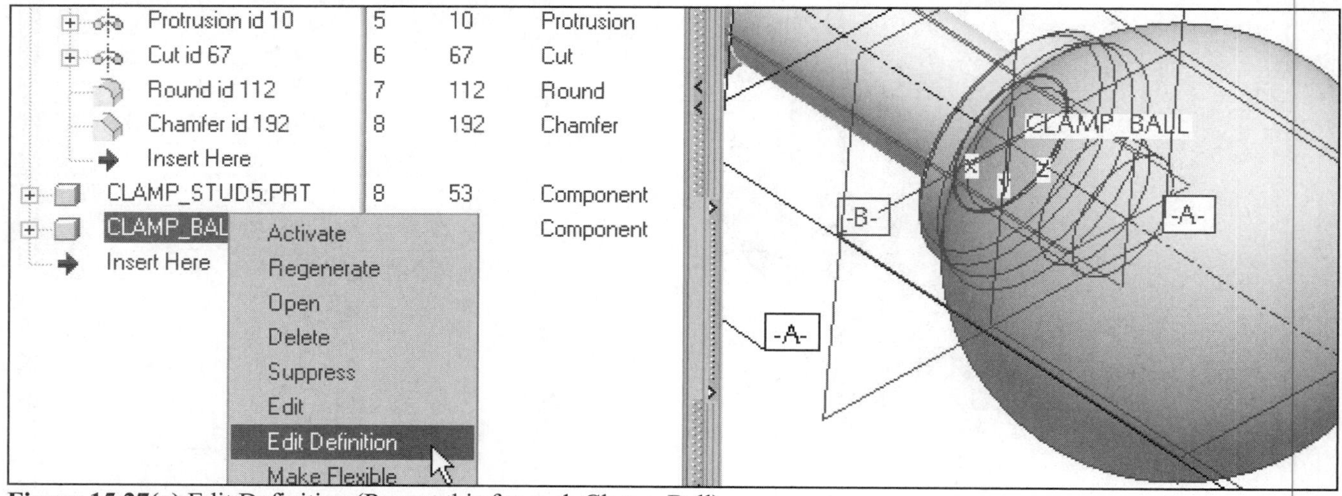

Figure 15.27(a) Edit Definition (Repeat this for each Clamp_Ball)

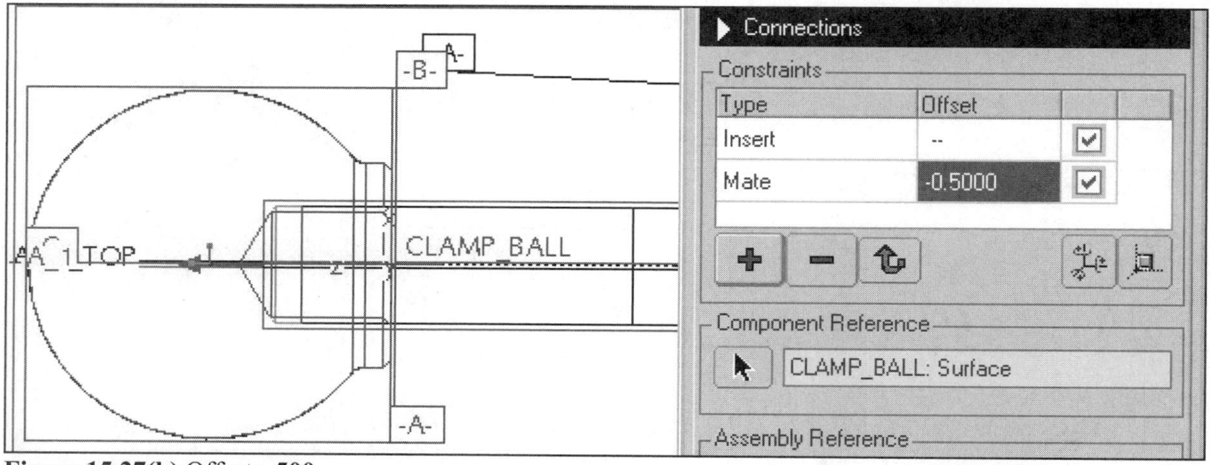

Figure 15.27(b) Offset **-.500**

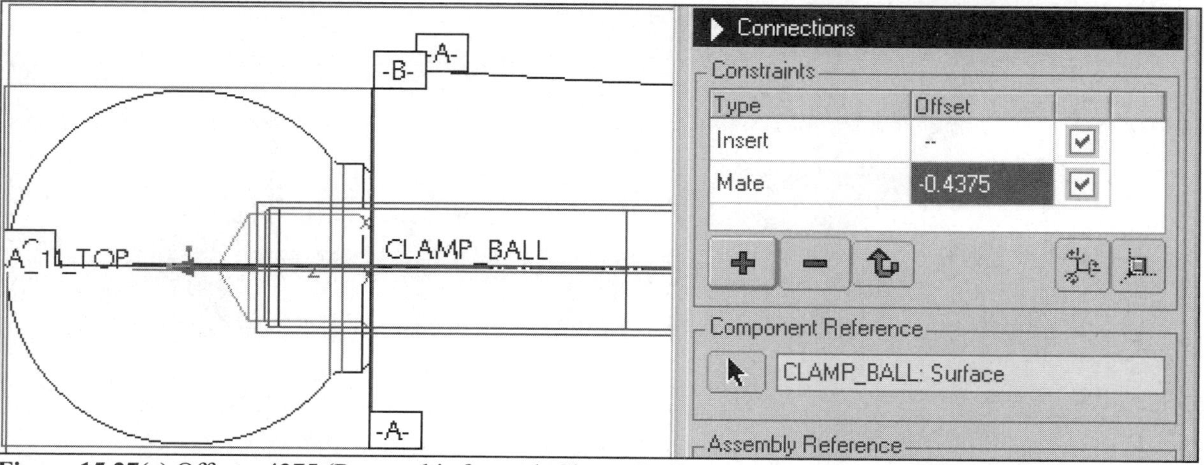

Figure 15.27(c) Offset **-.4375** (Repeat this for each Clamp_Ball)

Click: **Info** from menu bar ⇒ **Bill of Materials** ⇒ Top Level [Fig. 15.29(a-b)] ⇒ **OK** ⇒ (save icon) ⇒ **MMB**

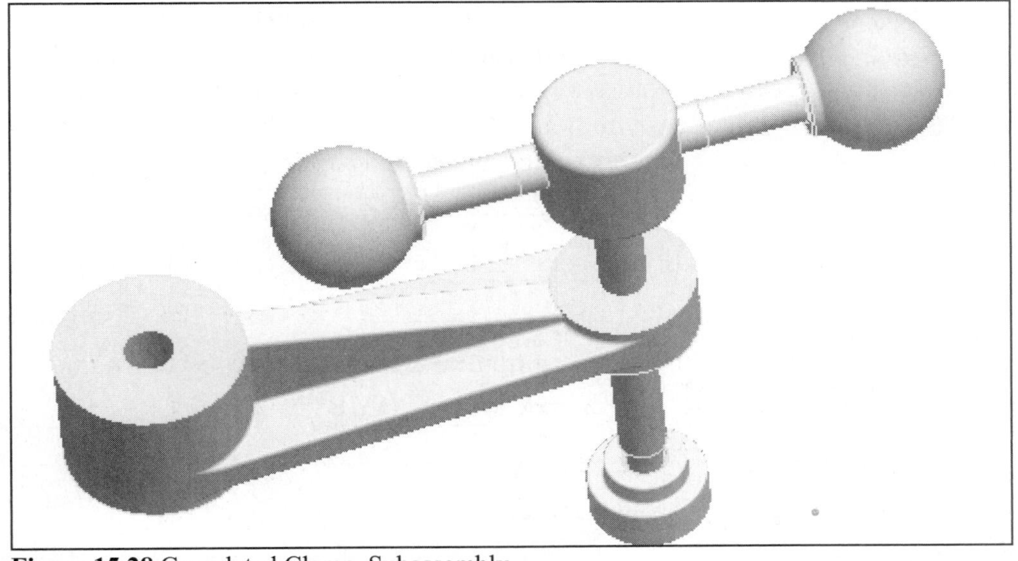

Figure 15.28 Completed Clamp_Subassembly

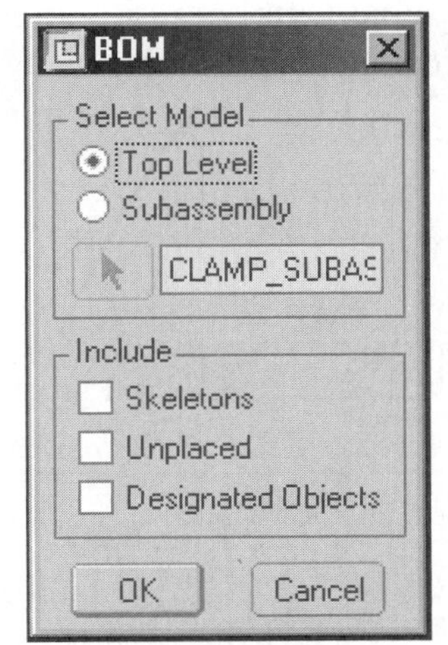

Figure 15.29(a) BOM

Bom Report : CLAMP_SUBASSEMBLY

Assembly CLAMP_SUBASSEMBLY contains:

Quantity	Type	Name	Actions
1	Part	CLAMP_ARM	
1	Part	CLAMP_SWIVEL	
1	Part	CLAMP_FOOT	
1	Part	CLAMP_STUD5	
2	Part	CLAMP_BALL	

Summary of parts for assembly CLAMP_SUBASSEMBLY:

Quantity	Type	Name	Actions
1	Part	CLAMP_ARM	
1	Part	CLAMP_SWIVEL	
1	Part	CLAMP_FOOT	
1	Part	CLAMP_STUD5	
2	Part	CLAMP_BALL	

Figure 15.29(b) Bill of Materials (BOM)

Swing Clamp Assembly

The first features for the main assembly will be the default datum planes and coordinate system. The first part assembled on the main assembly will be the Clamp_Plate, which will be created using *top-down design*; where the assembly is active, and the component is created within the assembly mode. The subassembly is still "in session-in memory" even though it will not show on the screen after its window is closed. For the two standard parts of the main assembly, you will use specific types of selected constraints instead of using Automatic, which allows Pro/E to default to an appropriate constraint.

Click: **Window** ⇒ **Close** ⇒ start the assembly, click: [new] ⇒ Assembly ⇒ Name **CLAMP_ASSEMBLY** ⇒ Sub-type Design ⇒ Use default template ⇒ **OK** ⇒ Template inlbs_asm_design ⇒ Parameters DESCRIPTION type a simple description ⇒ MODELED_BY type your name ⇒ **OK** ⇒ **Edit** ⇒ **Setup** ⇒ **Units** ⇒ Inch lbm Second (Pro/E Default) ⇒ **Close** ⇒ **MMB**

Click: Settings from the Model Tree ⇒ Tree Filters ⇒ Display Features Notes Suppressed Objects ⇒ **Apply** ⇒ **OK** ⇒ [save] ⇒ **MMB**

Change the coordinate system name: slowly double-click on ASM_DEF_CSYS in the Model Tree ⇒ type new name **ASM_CSYS** ⇒ click anywhere in the graphics window ⇒ slowly double-click on each of the datum identifiers in the Model Tree and add **CL_** as a prefix for each (i.e. **CL_ASM_TOP**) (Fig. 15.30) ⇒ **Tools** ⇒ [environment] **Set various environment options** ⇒ Standard Orient Trimetric ⇒ **Apply** ⇒ **OK**

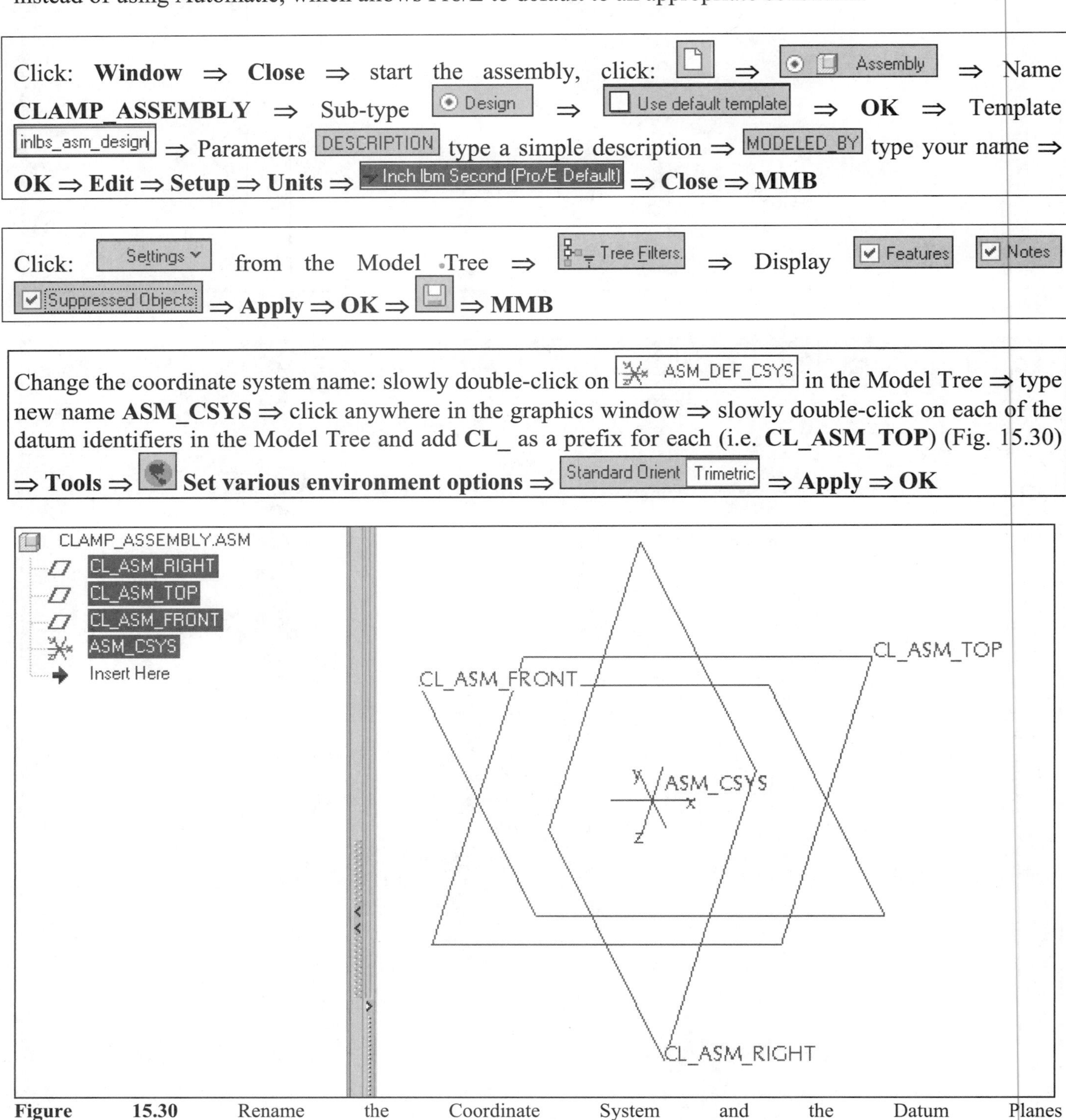

Figure 15.30 Rename the Coordinate System and the Datum Planes

Creating Components in the Assembly Mode

The Clamp_Subassembly is now complete. The subassembly will be added to the assembly after the Clamp_Plate is created and assembled. Using the Component Create dialog box (Fig. 15.31), you can create different types of components: parts, subassemblies, skeleton models, and bulk items. You cannot reroute components created in Assembly mode.

The following methods allow component creation in the context of an assembly without requiring external dependencies on the assembly geometry:

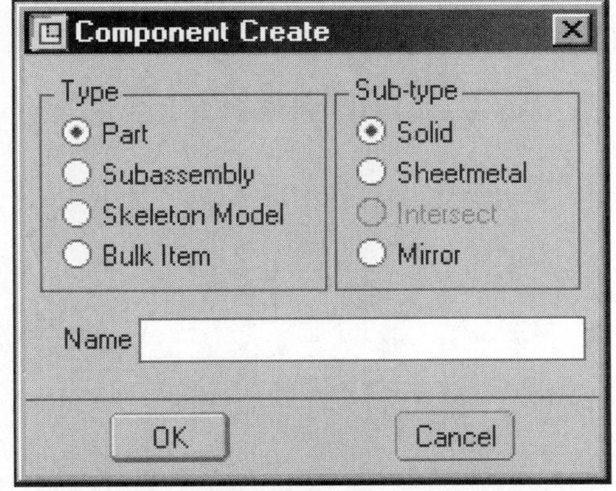

Figure 15.31 Component Create Dialog Box

- Create a component by copying another component or existing start part or start assembly
- Create a component with default datums
- Create an empty component
- Create the first feature of a new part; this initial feature is dependent on the assembly
- Create a part from an intersection of existing components
- Create a mirror copy of an existing part or subassembly
- Create Solid or Sheetmetal components
- Mirror Components

Main Assembly, Top-Down Design

The Clamp_Plate component is the first component of the main assembly. The Clamp_Plate is a new part. You will be modeling the plate "inside" the assembly using *top-down design*. Figure 15.32 provides the dimensions necessary to model the Clamp_Plate.

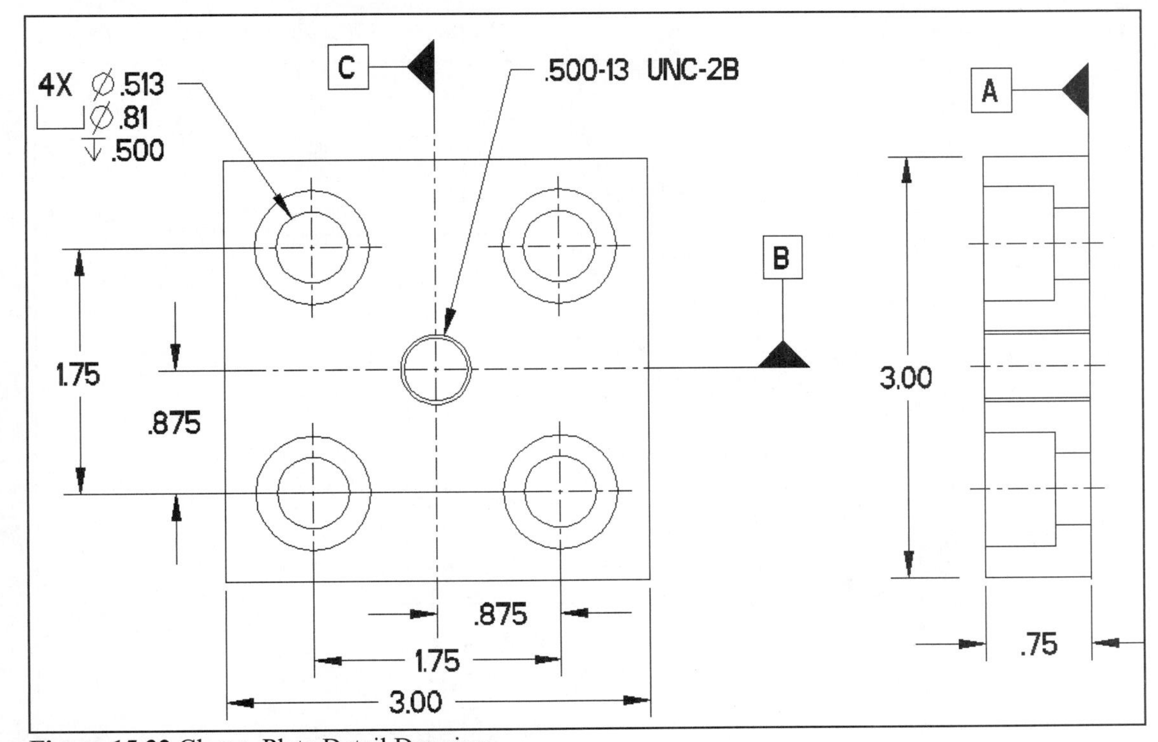

Figure 15.32 Clamp_Plate Detail Drawing

Click: **Create a component in assembly mode** Component Create dialog box displays [Fig. 15.33(a)] ⇒ Type Part ⇒ Sub-type Solid ⇒ Name **CLAMP_PLATE** ⇒ **OK** Creation Options dialog box displays [Fig. 15.33(b)] ⇒ Locate Default Datums ⇒ Align Csys To Csys ⇒ **OK** ⇒ pick **ASM_CSYS** from the graphics window and CLAMP_PLATE.PRT displays in the Model Tree [Fig. 15.33(c)], *the small green symbol/icon indicates that this component is now active* ⇒ ⇒ **MMB** *(if you cannot save, RMB on the CLAMP_ASSEMBLY.ASM in the Model Tree ⇒ Activate ⇒ RMB on the CLAMP_PLATE.PRT ⇒ Activate ⇒ File ⇒ Save⇒ MMB)*

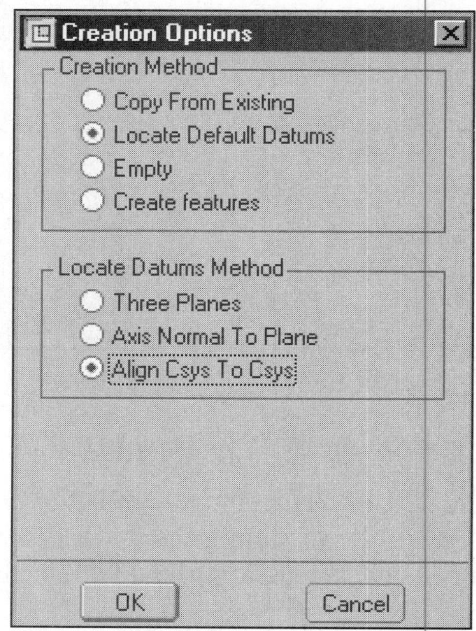

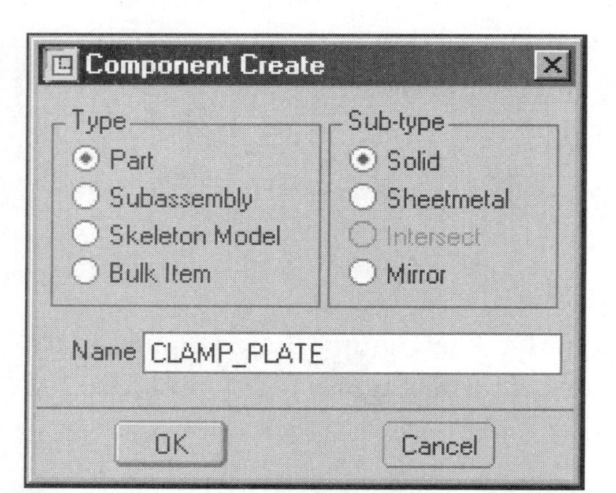

Figure 15.33(a) Component Create Dialog Box

Figure 15.33(b) Creation Options Dialog

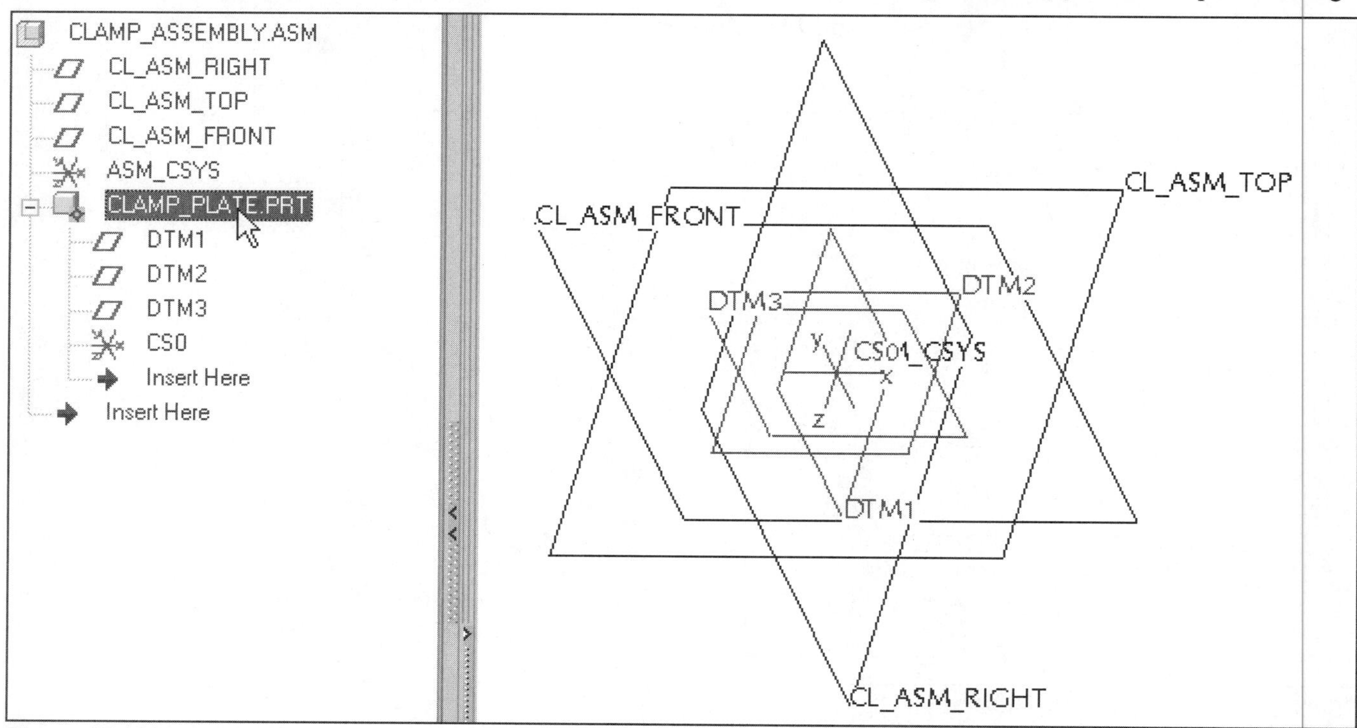

Figure 15.33(c) New Component Default Datums and Coordinate System Created and Assembled

Assembly Tools are now unavailable in the right column Toolbar. The current component is the Clamp_Plate. You are now effectively in the Part mode, except that you can see the assembly features and components. Be sure to reference only part features as you model (part datum planes and part coordinate system), otherwise you will create unwanted external references. There are many situations where external references are needed and desired, but in this case, you are simply modeling a new part.

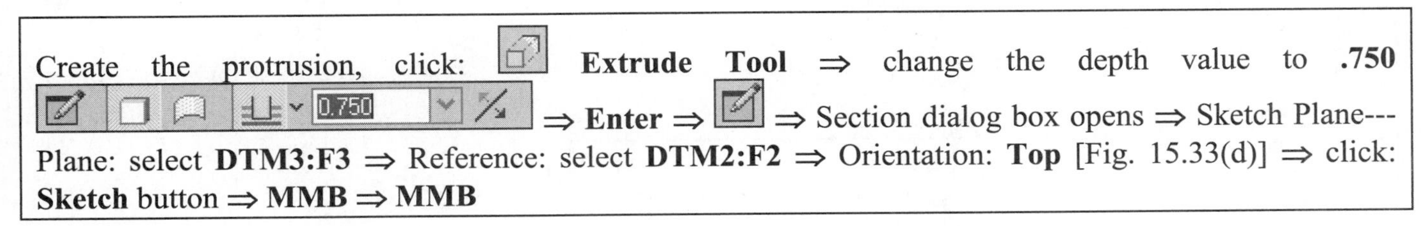

Create the protrusion, click: **Extrude Tool** ⇒ change the depth value to **.750** ⇒ **Enter** ⇒ ⇒ Section dialog box opens ⇒ Sketch Plane---Plane: select **DTM3:F3** ⇒ Reference: select **DTM2:F2** ⇒ Orientation: **Top** [Fig. 15.33(d)] ⇒ click: **Sketch** button ⇒ **MMB** ⇒ **MMB**

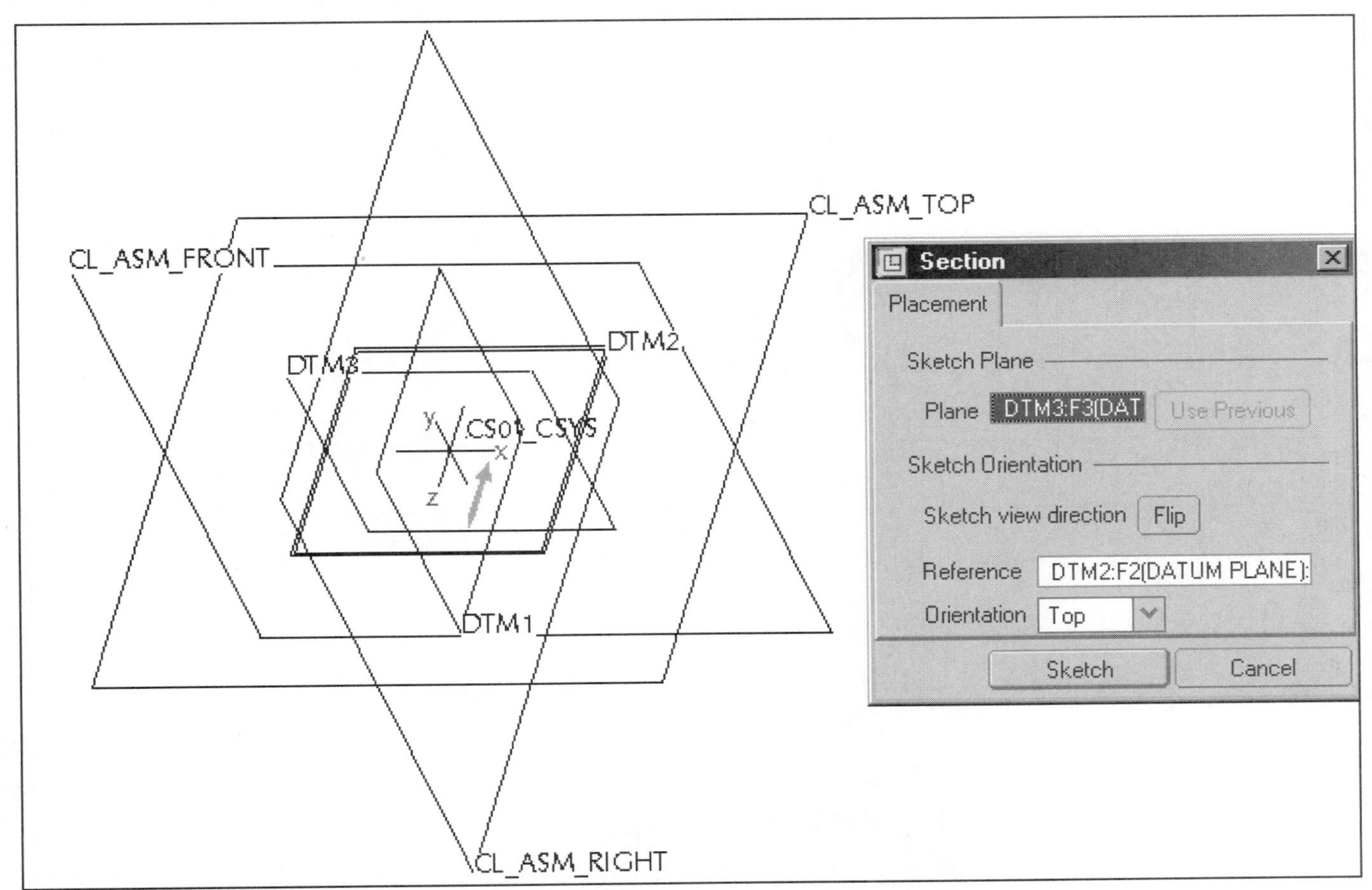

Figure 15.33(d) Sketch Plane and Reference Orientation

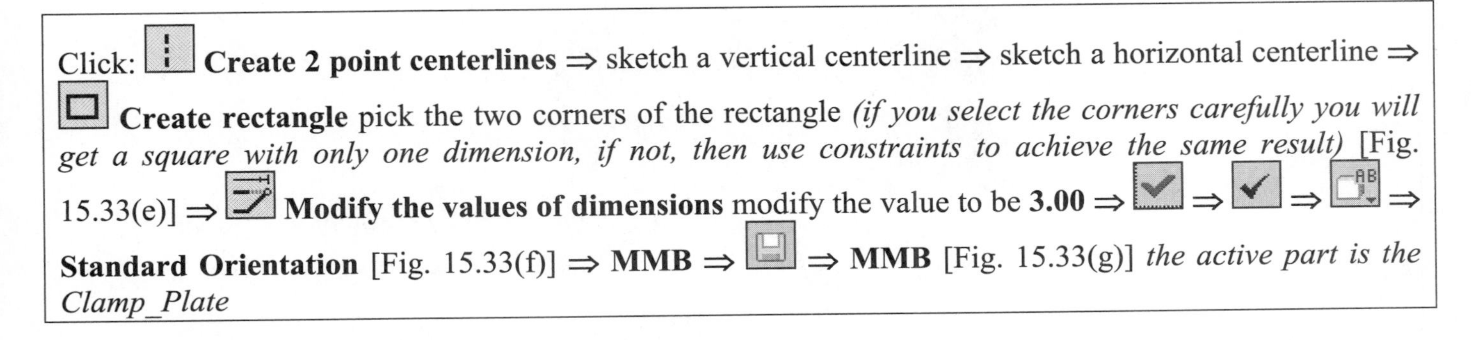

Click: **Create 2 point centerlines** ⇒ sketch a vertical centerline ⇒ sketch a horizontal centerline ⇒ **Create rectangle** pick the two corners of the rectangle *(if you select the corners carefully you will get a square with only one dimension, if not, then use constraints to achieve the same result)* [Fig. 15.33(e)] ⇒ **Modify the values of dimensions** modify the value to be **3.00** ⇒ ⇒ ⇒ ⇒ **Standard Orientation** [Fig. 15.33(f)] ⇒ **MMB** ⇒ ⇒ **MMB** [Fig. 15.33(g)] *the active part is the Clamp_Plate*

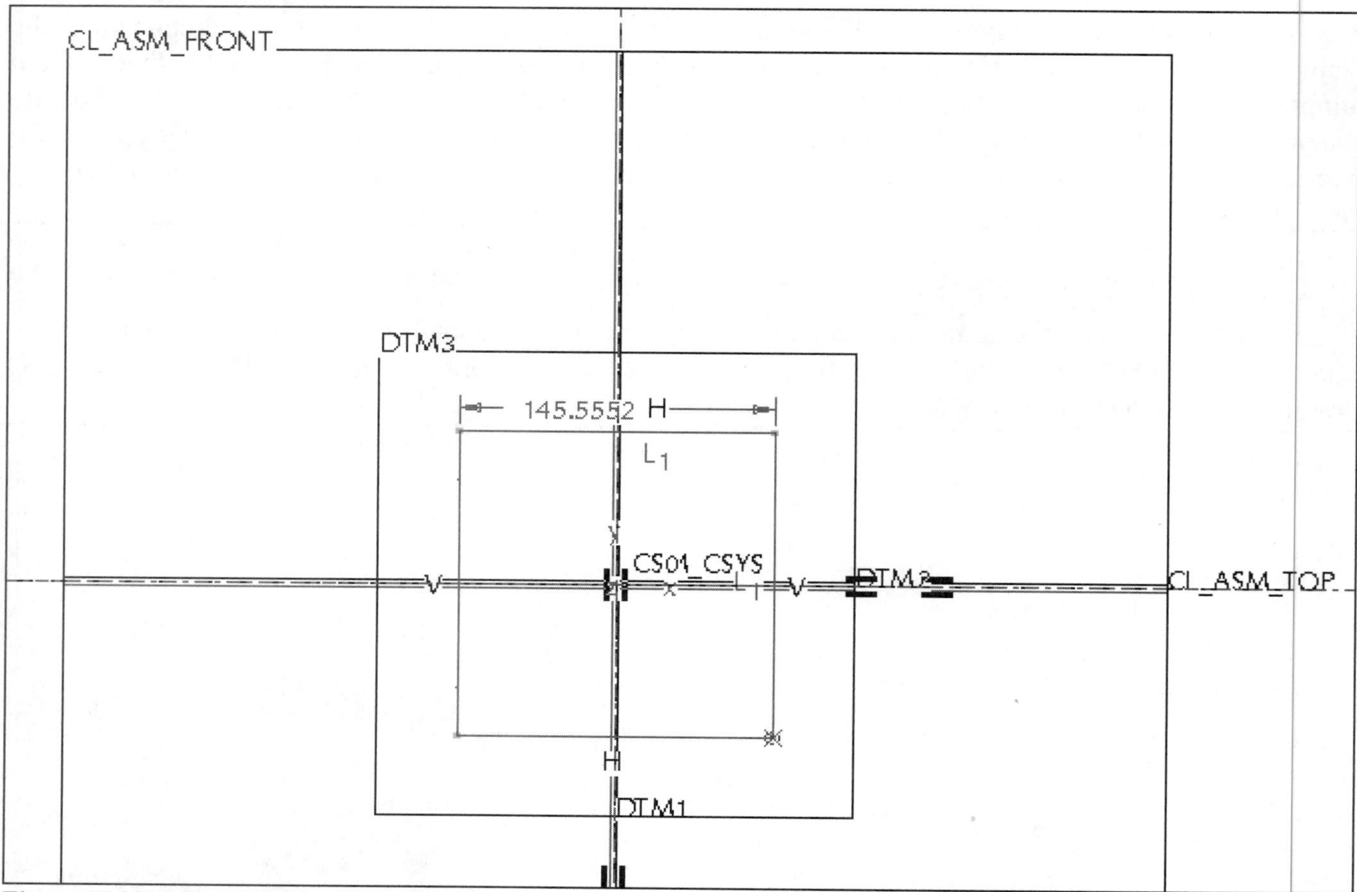

Figure 15.33(e) Sketch a Square Section

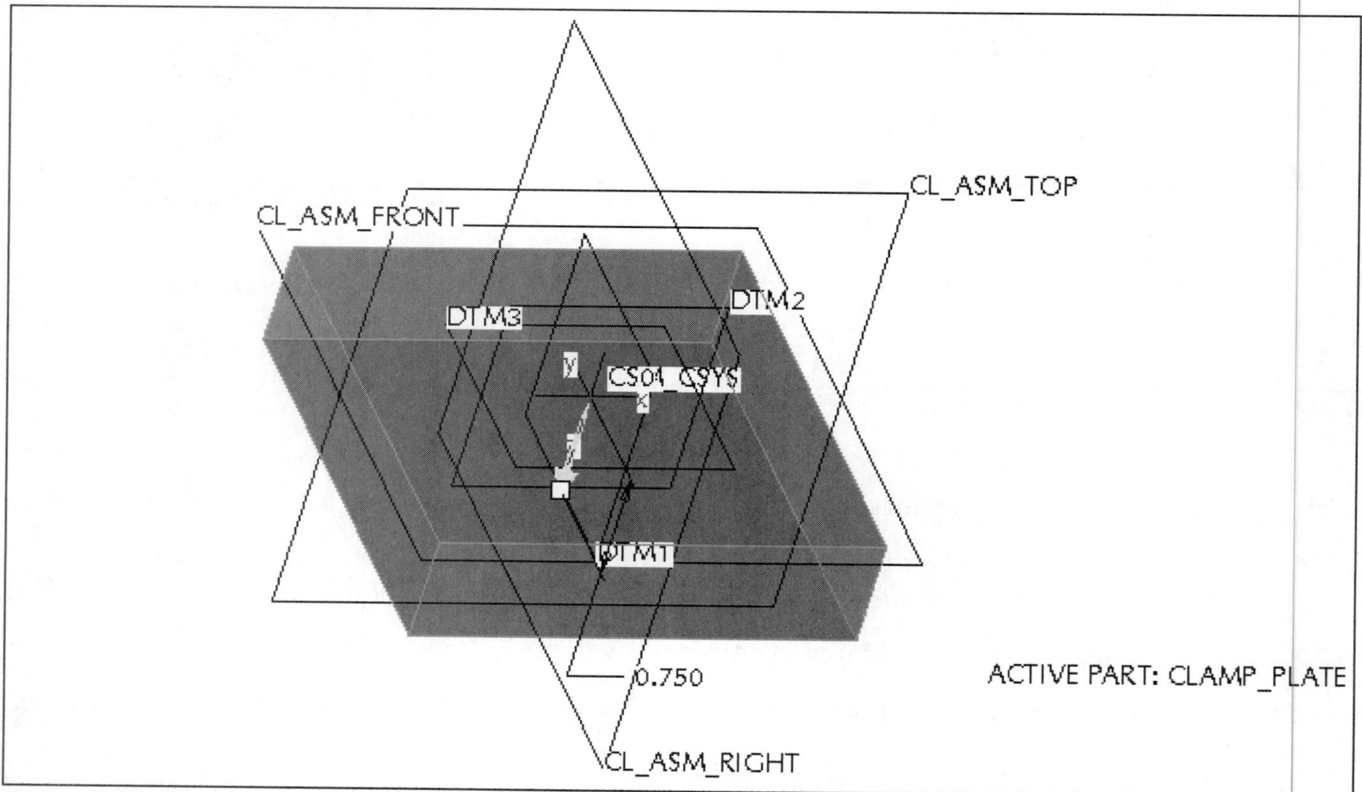

Figure 15.33(f) Depth and Direction Preview

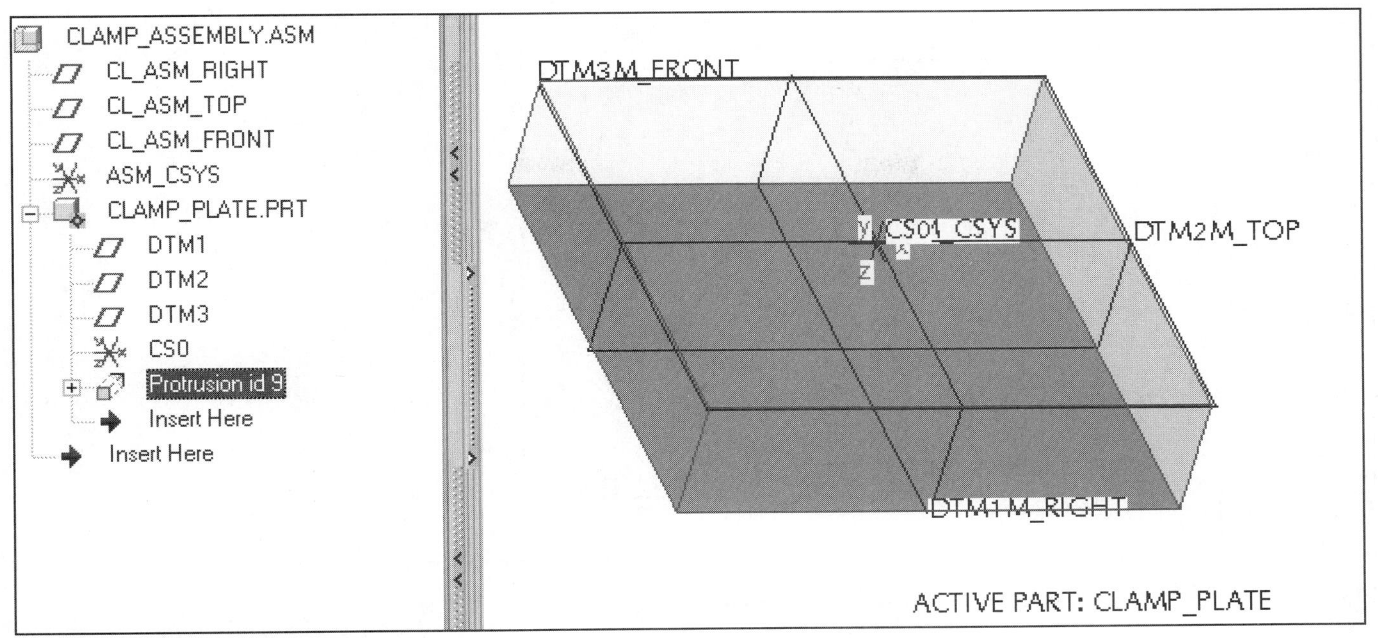

Figure 15.33(g) Completed Protrusion

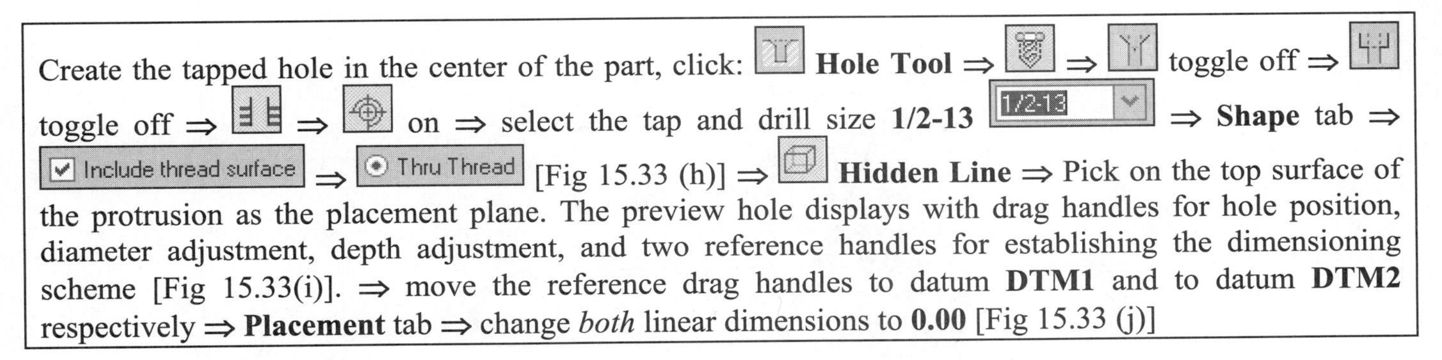

Create the tapped hole in the center of the part, click: **Hole Tool** ⇒ ⇒ toggle off ⇒ toggle off ⇒ ⇒ on ⇒ select the tap and drill size **1/2-13** ⇒ **Shape** tab ⇒ Include thread surface ⇒ Thru Thread [Fig 15.33 (h)] ⇒ **Hidden Line** ⇒ Pick on the top surface of the protrusion as the placement plane. The preview hole displays with drag handles for hole position, diameter adjustment, depth adjustment, and two reference handles for establishing the dimensioning scheme [Fig 15.33(i)]. ⇒ move the reference drag handles to datum **DTM1** and to datum **DTM2** respectively ⇒ **Placement** tab ⇒ change *both* linear dimensions to **0.00** [Fig 15.33 (j)]

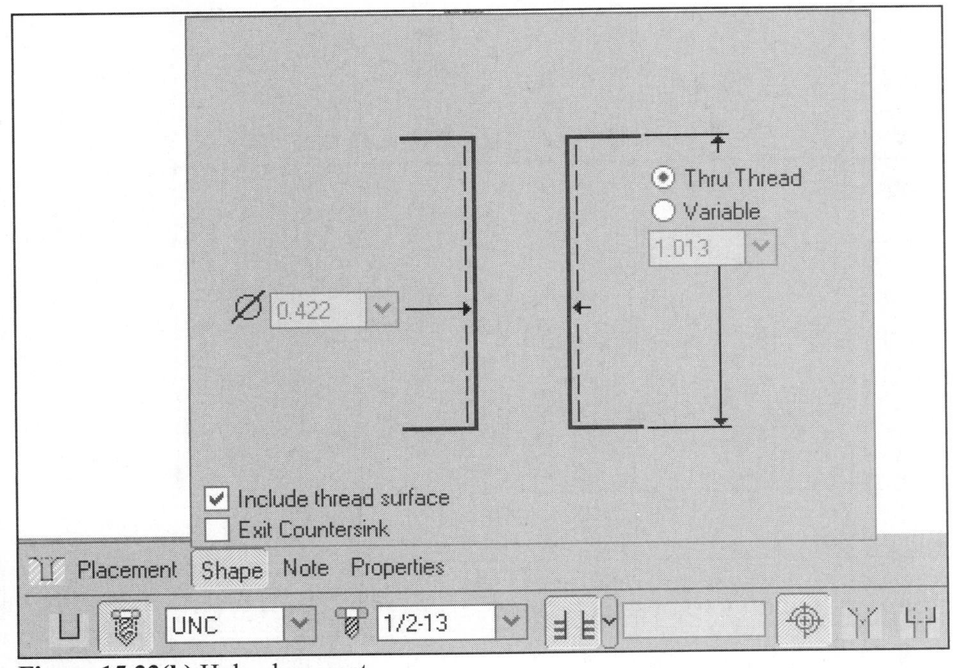

Figure 15.33(h) Hole placement

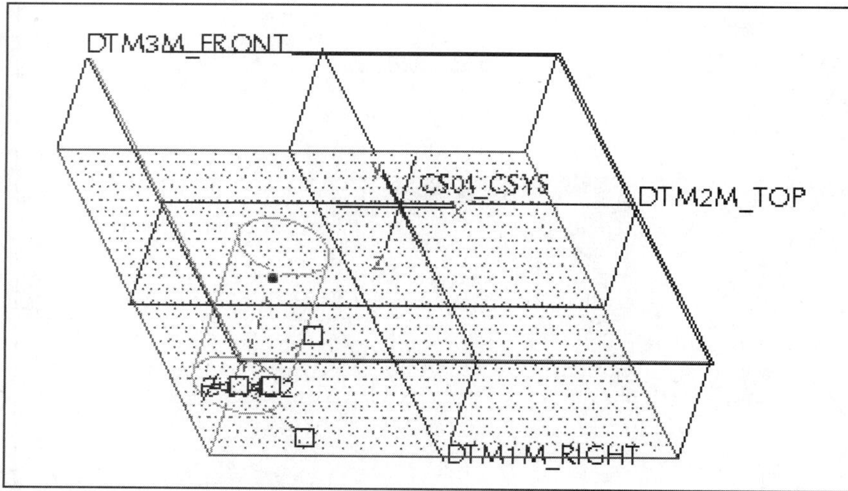

Figure 15.33(i) Initial Hole Placement

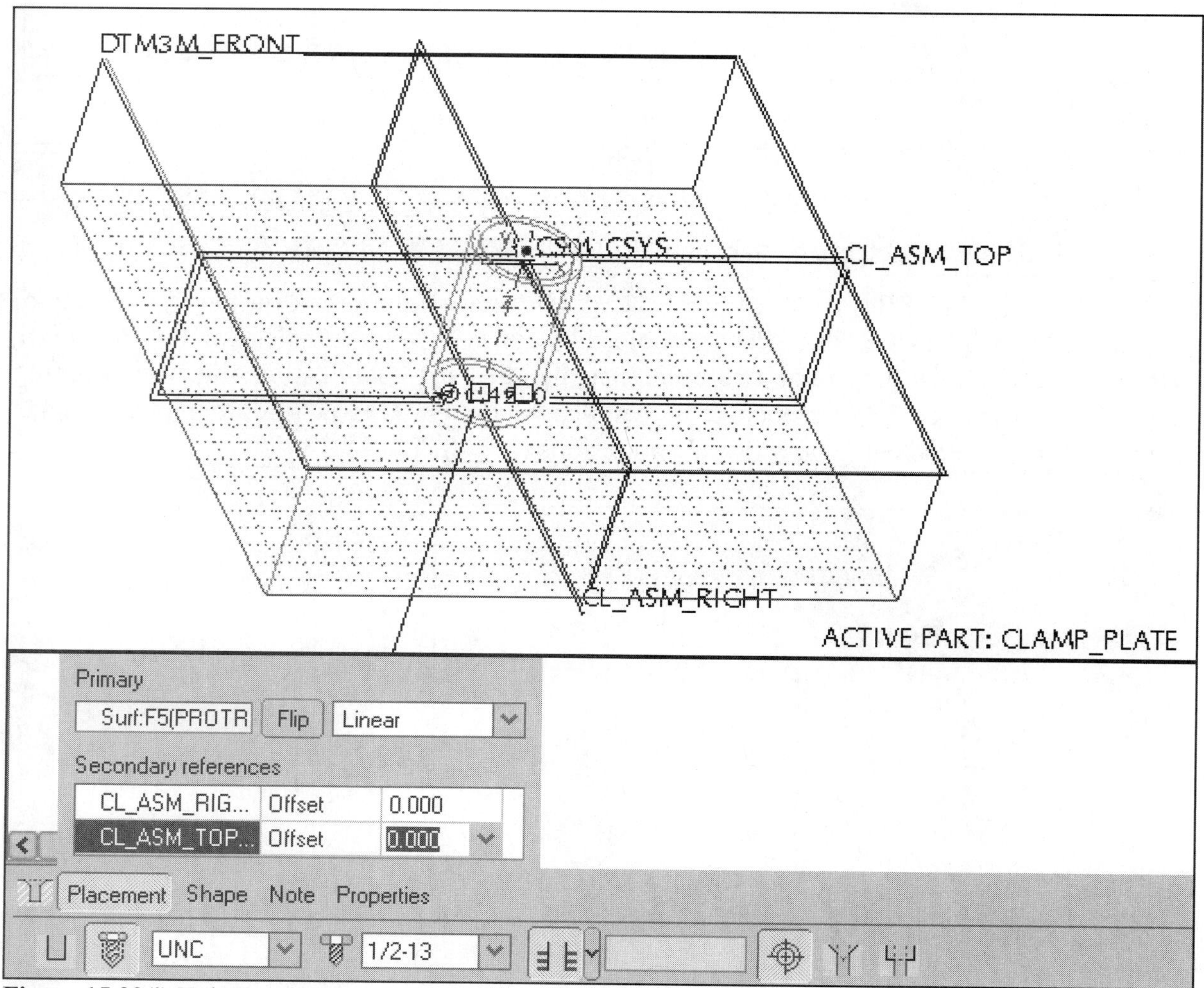

Figure 15.33(j) Hole Preview

Click: ⇒ ⇒ **LMB** in the graphics window to deselect ⇒ ⇒ **MMB** [Fig. 15.33(k)]

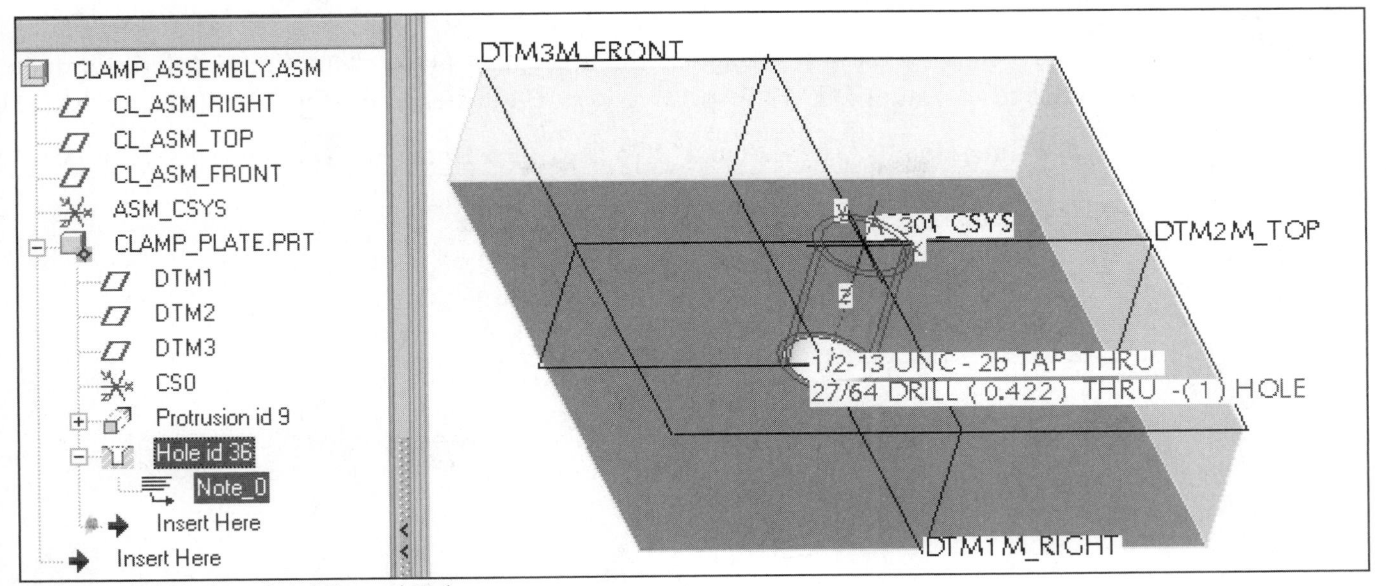

Figure 15.33(k) Completed Tapped Hole

Create a counterbore hole, click: **Hole Tool** ⇒ **Drill to intersect with all surfaces** ⇒ ⇒ 1/2-13 ⇒ ⇒ toggle off ⇒ toggle off ⇒ **Adds counterbore drilling** ⇒ **Shape** tab ⇒ leave the defaults for the counterbores sizes [Fig. 15.33(l)] *note that the detail drawing callouts are slightly different* [Fig. 15.33(m)]

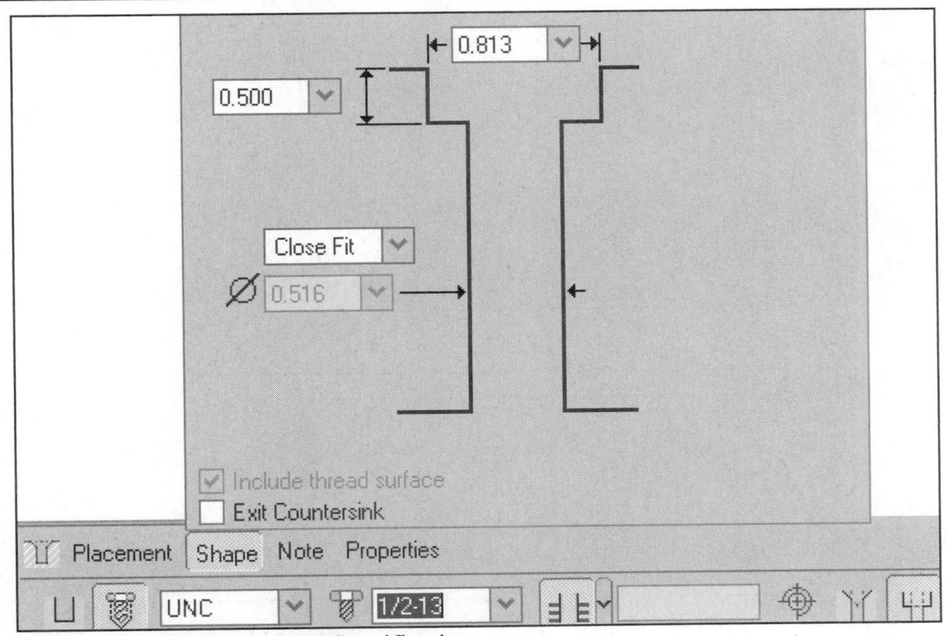

Figure 15.33(l) Counterbore Specifications

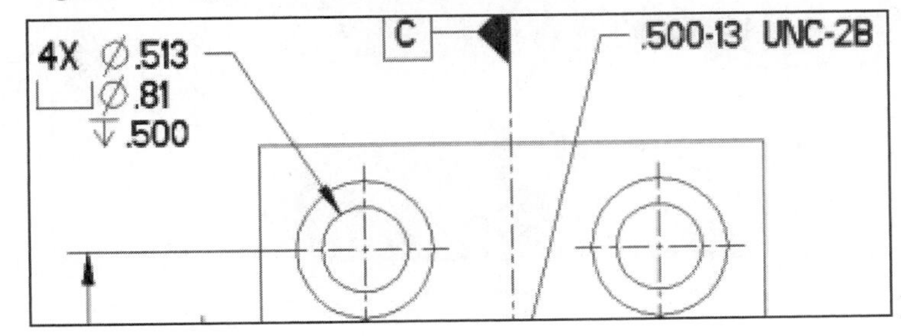

Figure 15.33(m) Hole Callout on Detail Drawing

Click: **Hidden Line** ⇒ pick on the top surface as the placement plane ⇒ move the reference drag handles to datum **DTM1** and to datum **DTM2** respectively ⇒ **Placement** tab [Fig 15.33(n)] ⇒ change linear dimensions to **-.875** *(negative .875* -0.875*)* and **.875** (0.875) ⇒ **Enter** ⇒ [Fig. 15.33(o)]

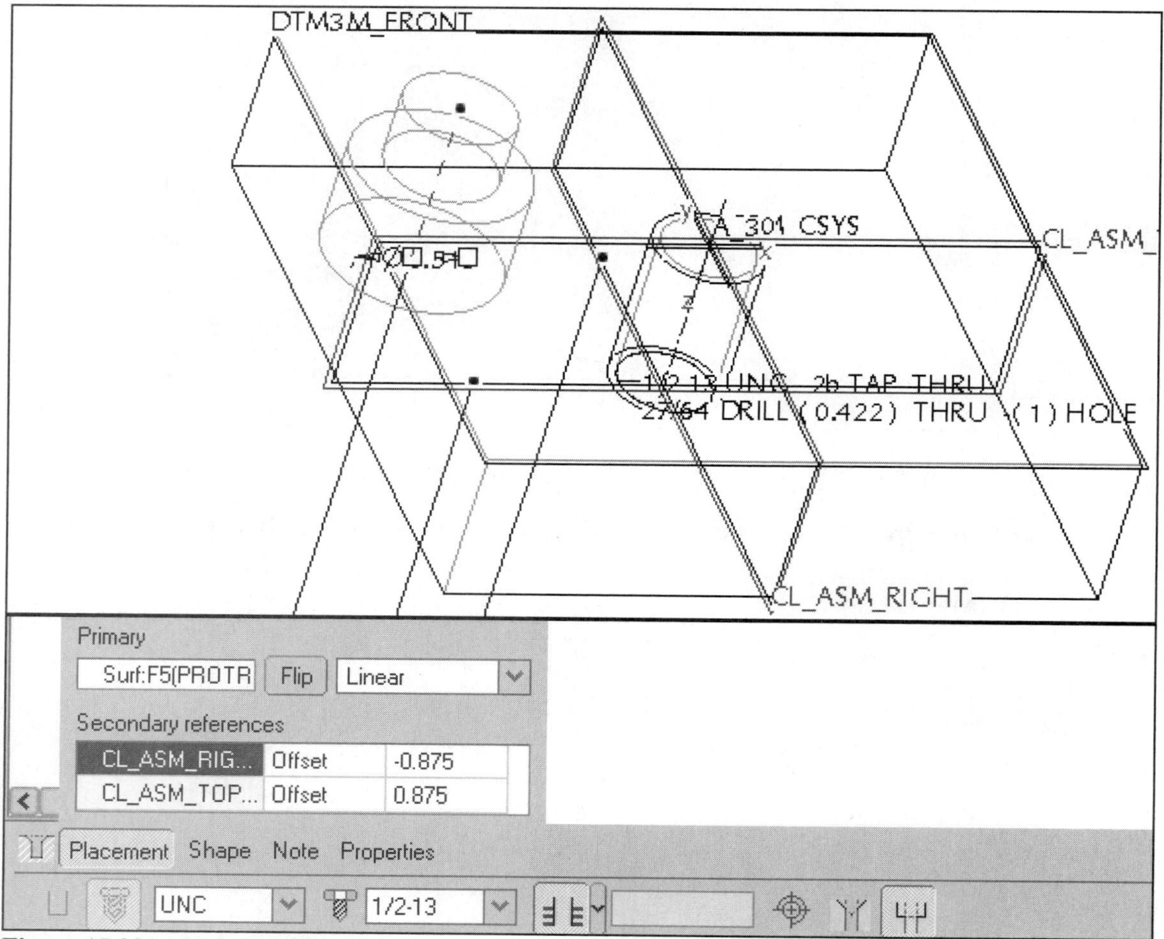

Figure 15.33(n) Hole Specifications

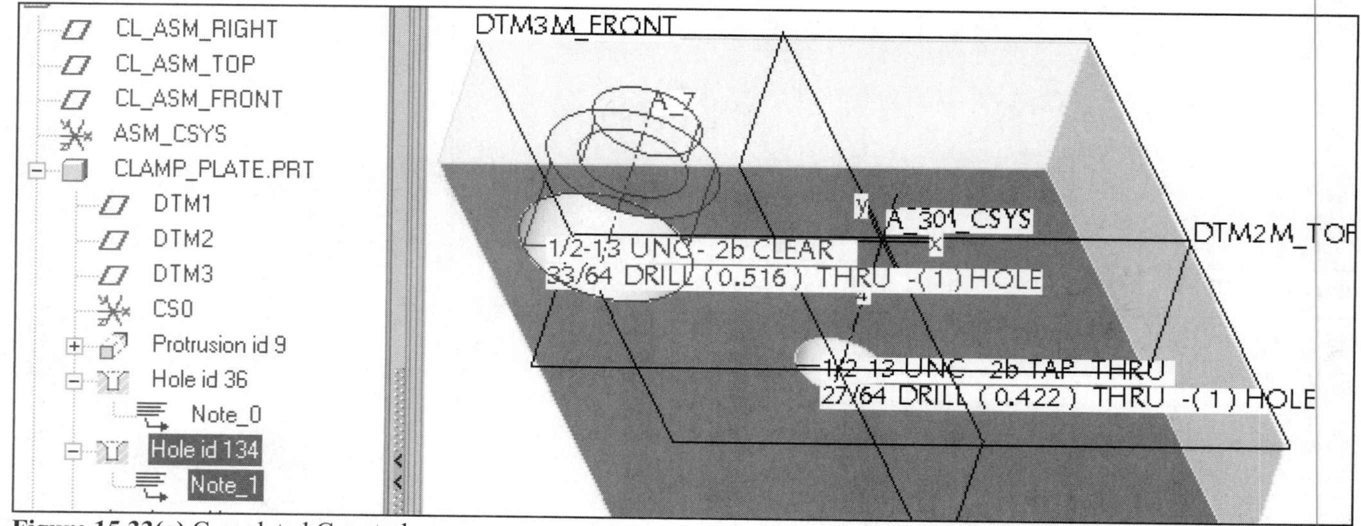

Figure 15.33(o) Completed Counterbore

The counterbore "seems" correct, but still check for external references.

Click: **Info** ⇒ **Global Reference Viewer** [Fig 15.33(p)] ⇒ Filter Setting... ⇒ expand to show CLAMP_PLATE.PRT [Fig. 15.33(q)] ⇒ **Close**

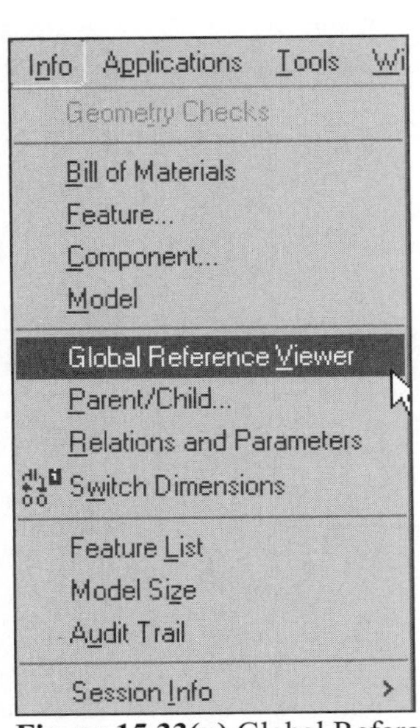

Figure 15.33(p) Global Reference Viewer

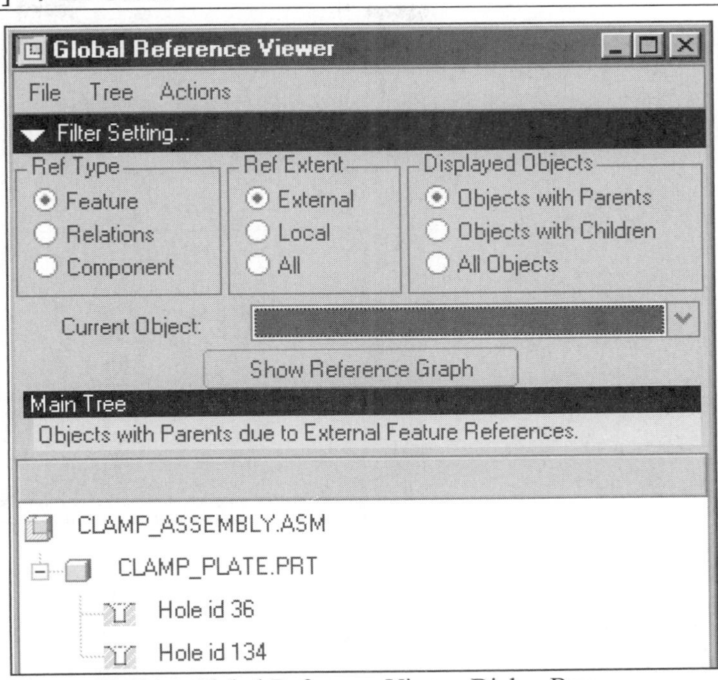

Figure 15.33(q) Global Reference Viewer Dialog Box

If you look back at Figure 15.33(j) and Figure 15.33(n) you will see that the assembly datums were used to place the holes instead of DTM1 and DTM2. The following commands will show you how to edit the references of one of the holes. Redo the references of the other hole using similar steps.

Click: **RMB** on the hole in the Model Tree ⇒ **Edit Definition** [Fig. 15.33(r)]

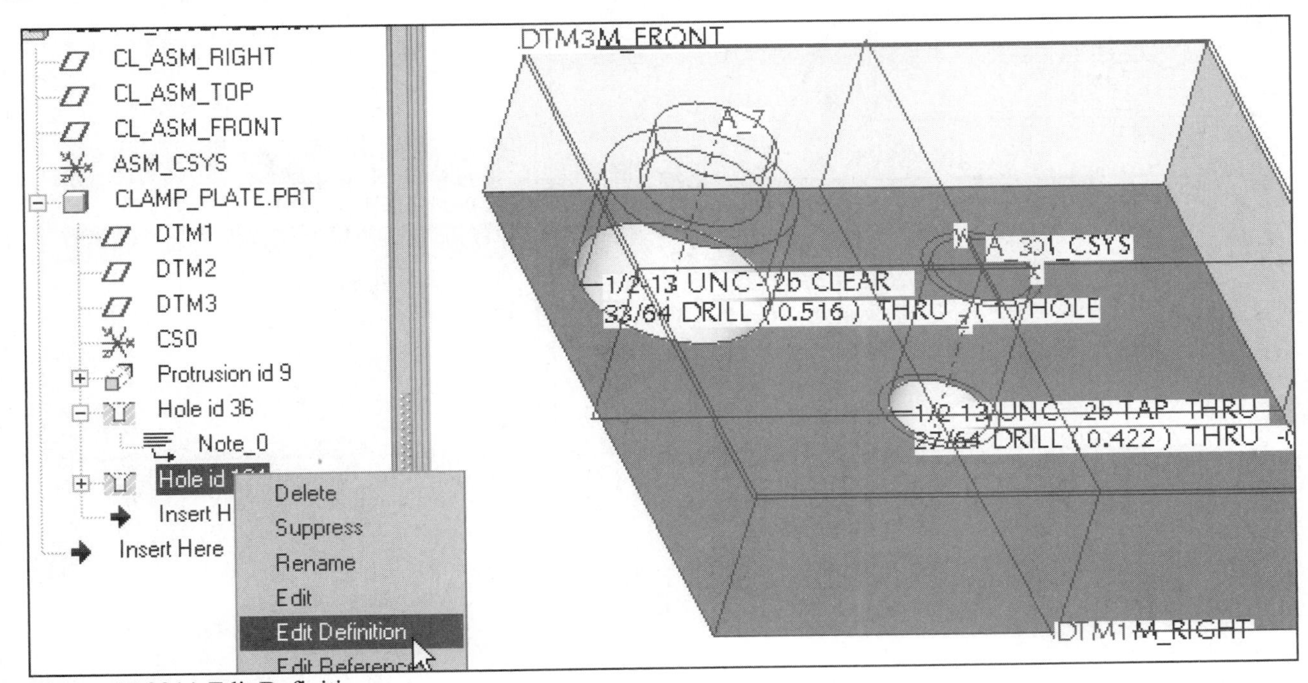

Figure 15.33(r) Edit Definition

Click: **Placement** tab ⇒ click on the first Secondary reference ⇒ **RMB** ⇒ **Remove** [Fig. 15.33(s)] ⇒ **RMB** ⇒ **Remove** removes the second reference ⇒ move one of the drag handles near **DTM1** and drop it ⇒ **RMB** ⇒ **Pick From List** [Fig. 15.33(t)] ⇒ click on **DTM1** from the list [Fig. 15.33(u)] ⇒ **OK** ⇒ press and hold **Ctrl** key

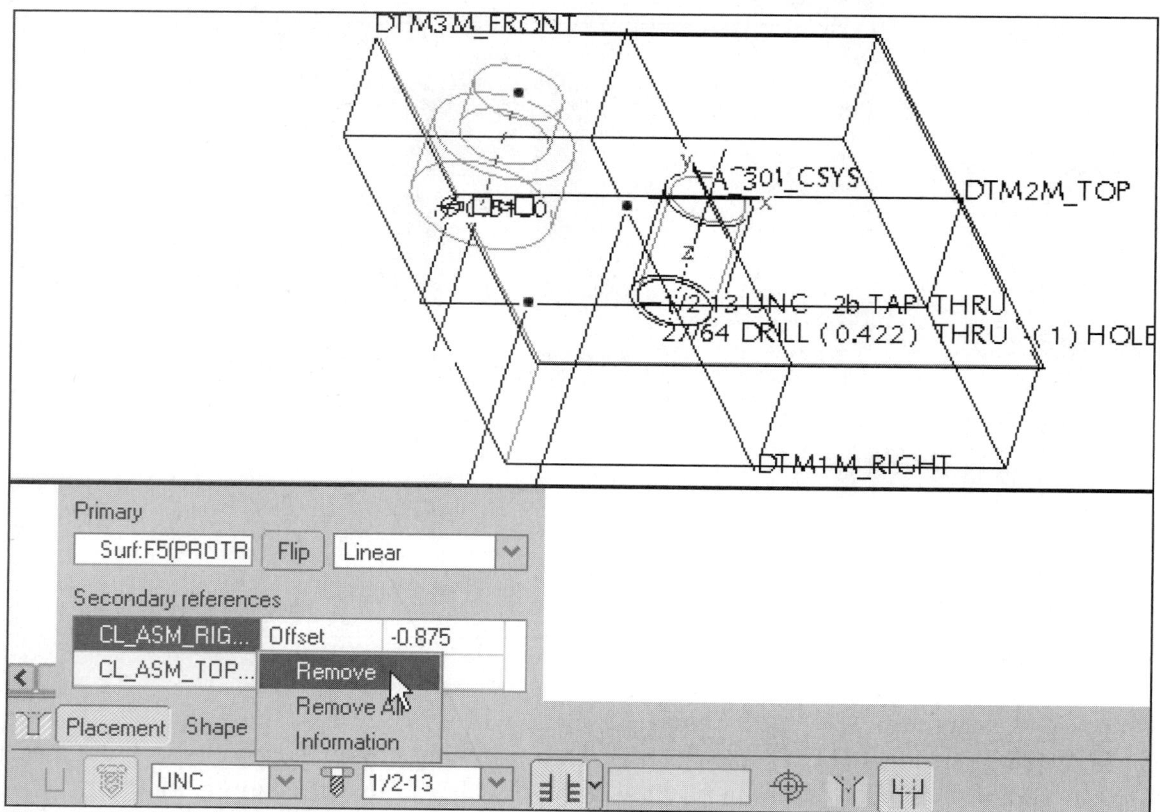

Figure 15.33(s) Remove Secondary References

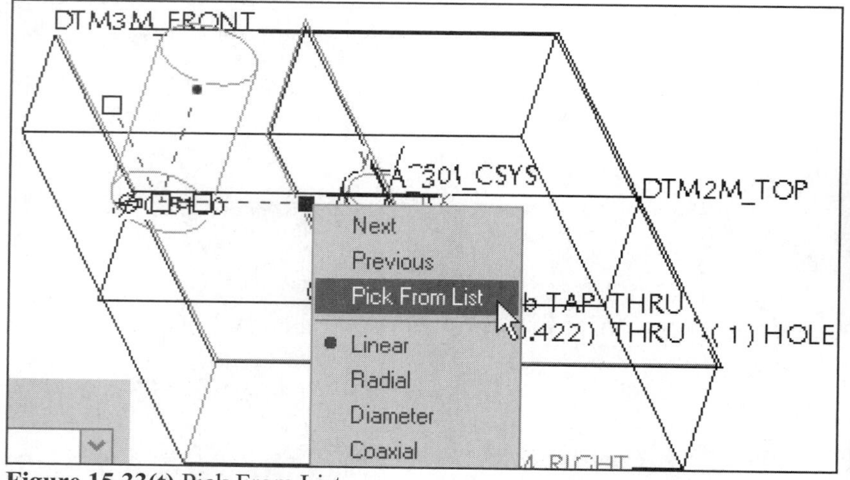

Figure 15.33(t) Pick From List

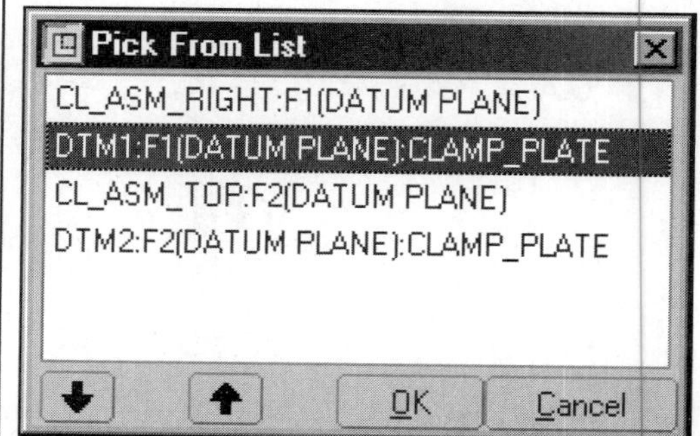

Figure 15.33(u) Pick From List Dialog Box

Move the other drag handle near **DTM2** and drop it ⇒ **RMB** ⇒ **Pick From List** [Fig. 15.33(t)] ⇒ click on **DTM2** from the list [Fig. 15.33(v)] ⇒ release the **Ctrl** key ⇒ **OK** [Fig. 15.33(w)] ⇒

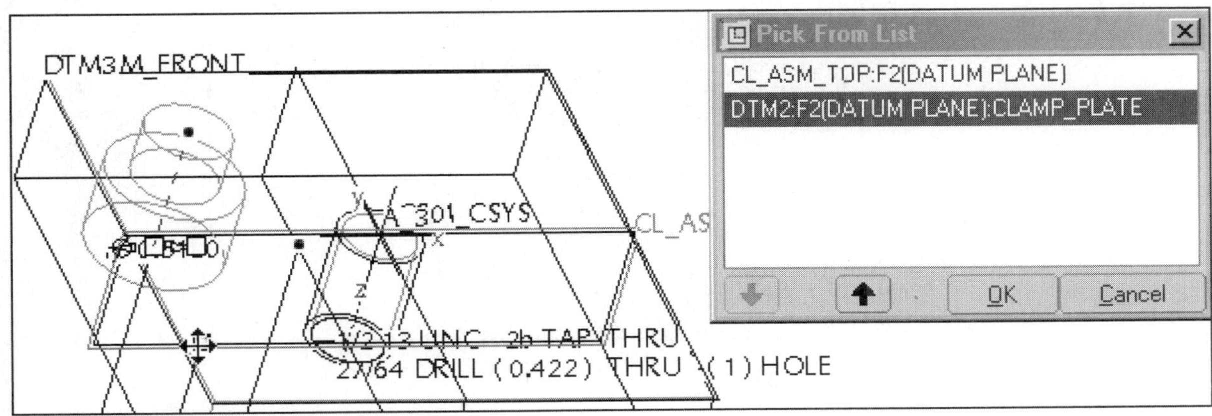

Figure 15.33(v) Redo Next Secondary Reference

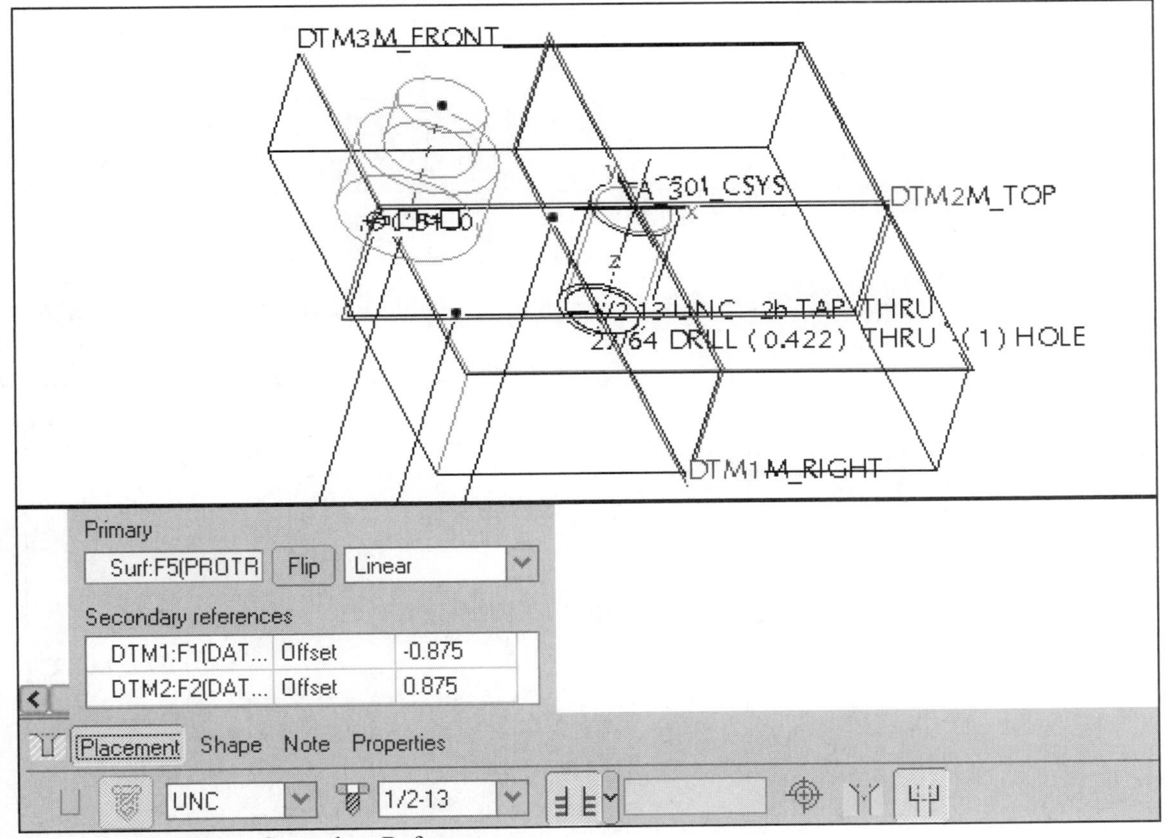

Figure 15.33(w) New Secondary References

Click: **Info** ⇒ **Global Reference Viewer** [Fig. 15.33(x)] *only one hole shows as having an external reference* ⇒ ⇒ repeat this process for the tapped hole in the center of the part [Fig. 15.33 (y)]

Figure 15.33(x) One Hole Left with External References

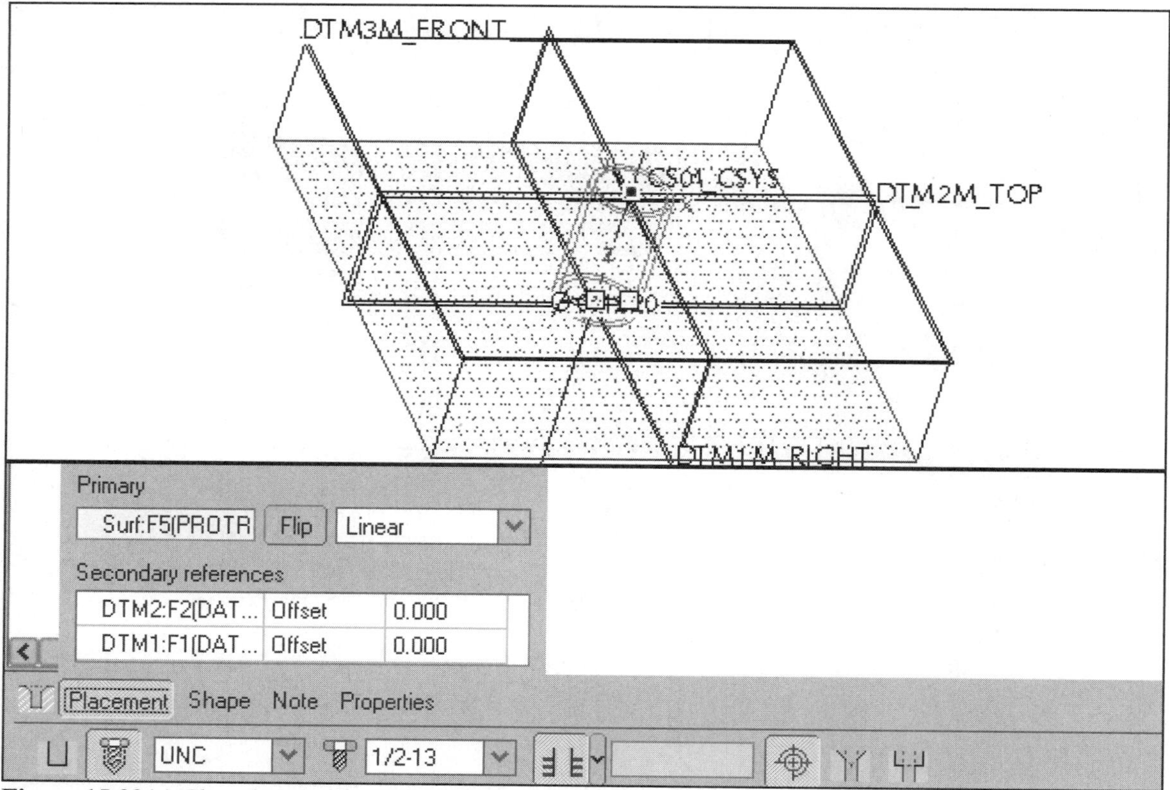

Figure 15.33(y) Changing the Secondary References for the Tapped Hole

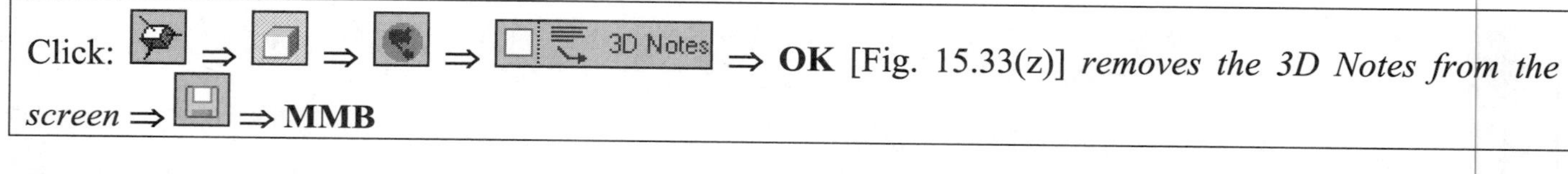

Click: ⇒ ⇒ ⇒ 3D Notes ⇒ **OK** [Fig. 15.33(z)] *removes the 3D Notes from the screen* ⇒ ⇒ **MMB**

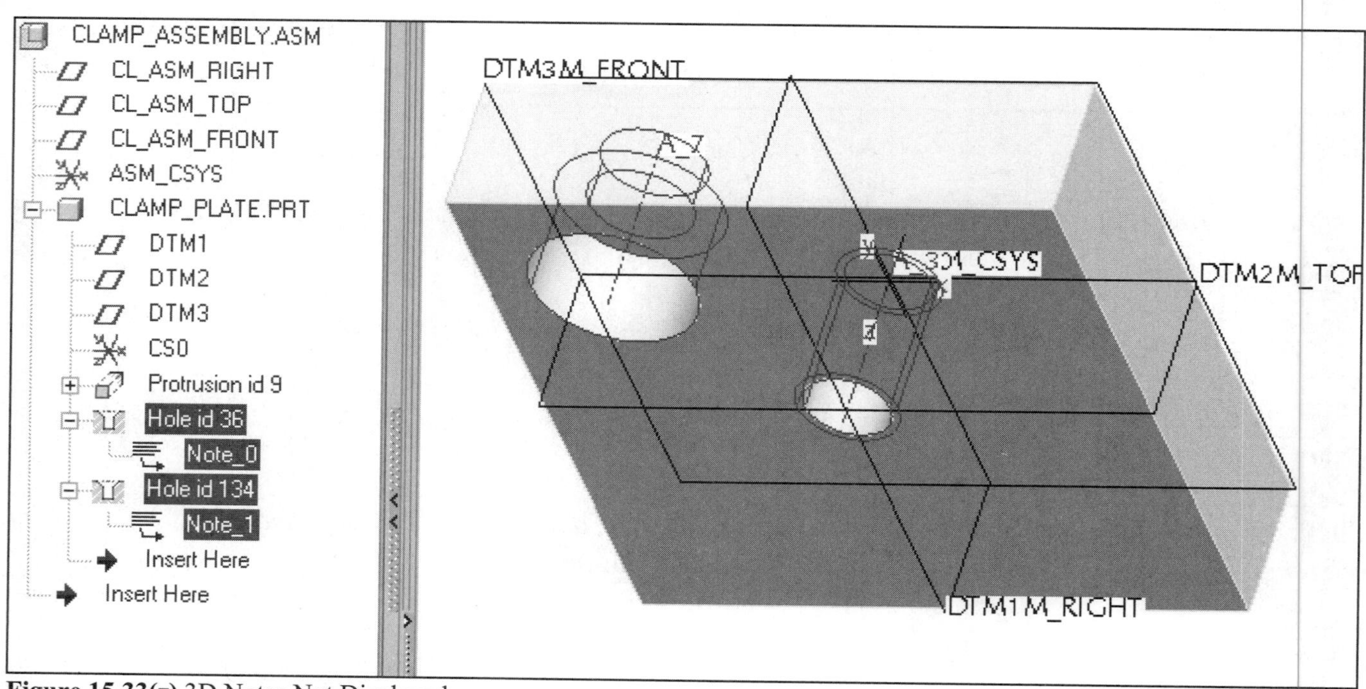

Figure 15.33(z) 3D Notes Not Displayed

Pattern the counterbore holes, click on the last hole in the Model Tree [Hole id] ⇒ **RMB** ⇒ **Pattern** [Fig. 15.34(a)] ⇒ **Dimensions** tab ⇒ click on the horizontal **.875** dimension [Fig. 15.34(b)] ⇒ highlight value and type **–1.75** ⇒ **Enter** ⇒ click **No Items** in Direction 2 collector box ⇒ click on the vertical **.875** dimension [Fig. 15.34(c)] ⇒ highlight value and type **–1.75** ⇒ **Enter** ⇒ **MMB** ⇒ **Tools** ⇒ [icon] **Environment** ⇒ [Standard Orient | Isometric] ⇒ **Apply** ⇒ **OK** [Fig. 15.34(d)] ⇒ [icon] ⇒ [icon] ⇒ **MMB**

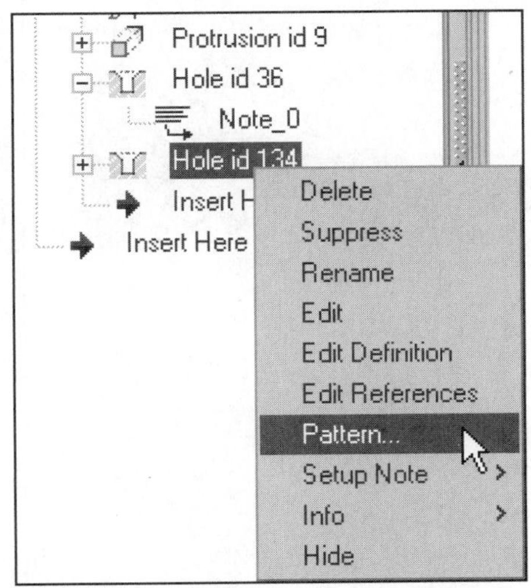

Figure 15.34(a) Pattern the Counterbore Hole

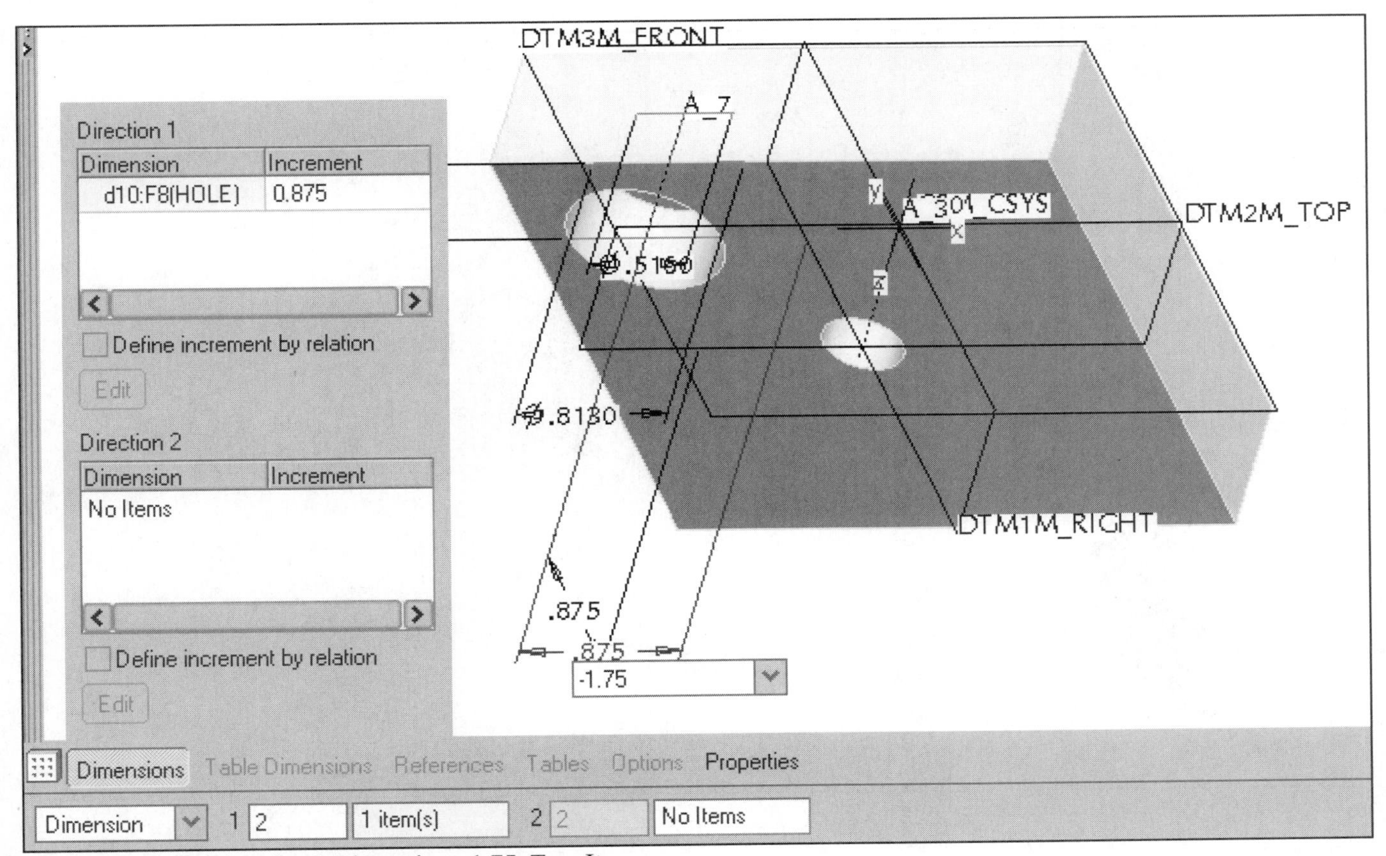

Figure 15.34(b) Direction 1 Dimension **–1.75**, Two Items

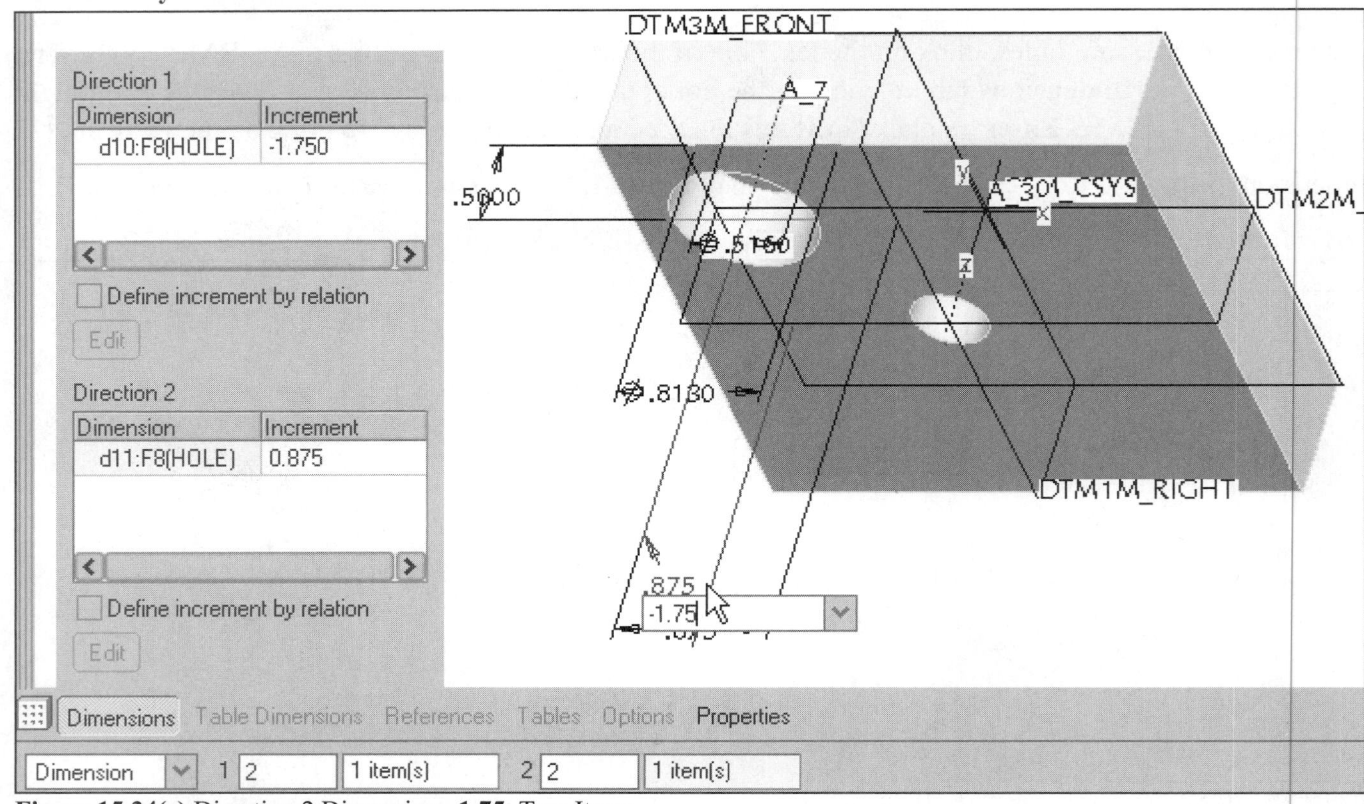

Figure 15.34(c) Direction 2 Dimension **–1.75**, Two Items

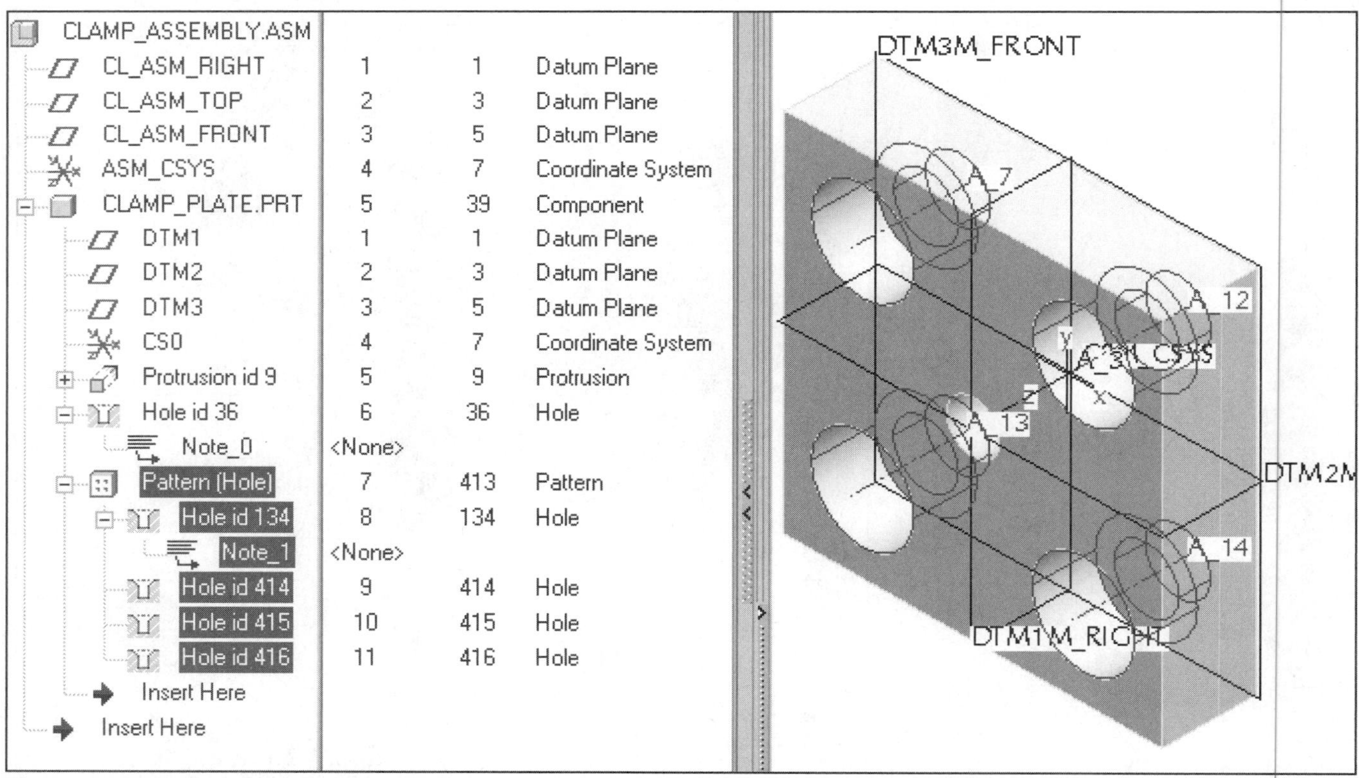

Figure 15.34(d) Patterned Counterbore Holes

Click CLAMP_ASSEMBLY.ASM in the Model Tree ⇒ **RMB** ⇒ **Activate** ⇒ click on the datum tag CL_ASM_FRONT in the graphics window [Fig. 15.35(a)] ⇒ Move Datum Tag ⇒ pick a new position [Fig. 15.35(b)] ⇒ repeat this process and move the other assembly datum tags ⇒ **Add component to the assembly** ⇒ pick **clamp_subamssembly.asm** ⇒ **Preview** [Fig. 15.36(a)] ⇒ **Open**

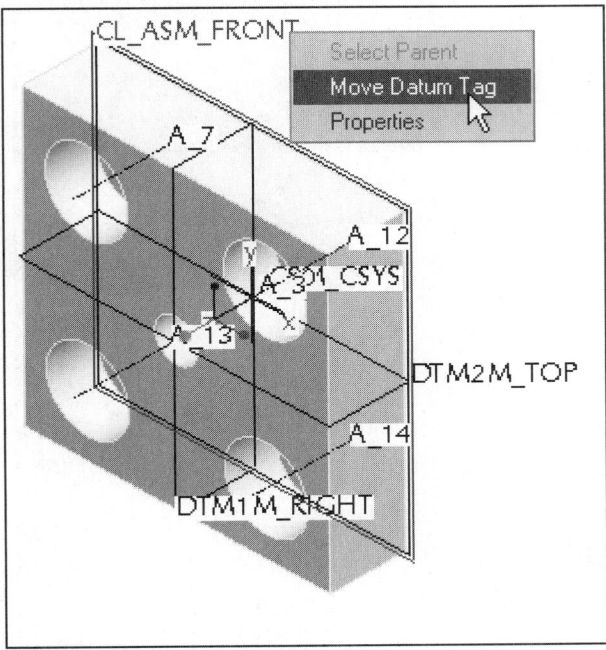

Figure 15.35(a) Move Datum Tag

Figure 15.35(b) Repositioned Datum Tag

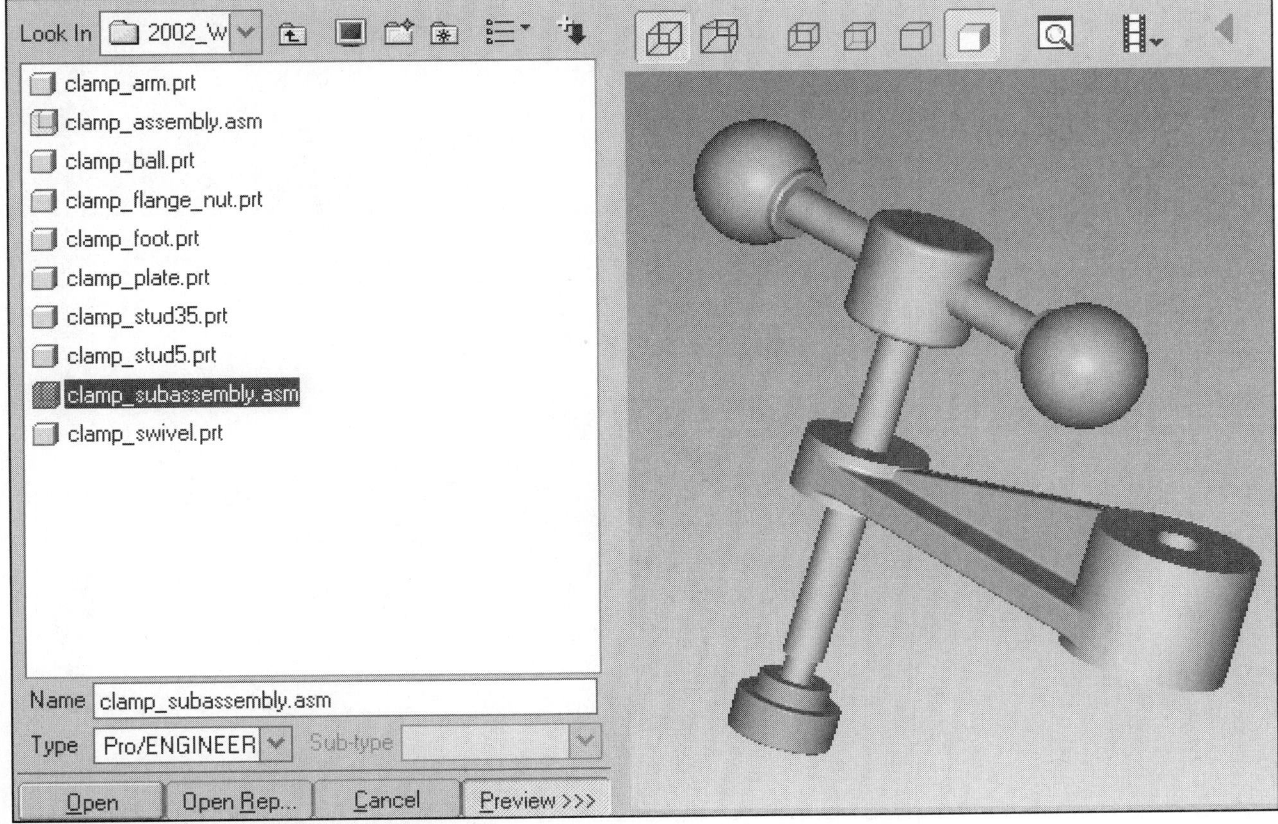

Figure 15.36(a) Preview Clamp_Subassembly

Click: [icon] toggle off component window ⇒ Constraints **Mate** [Fig. 15.36(b)] ⇒ spin the model ⇒ **View** ⇒ **Shade** ⇒ pick the *bottom surface* of the Clamp_Arm (of the Clamp_Subassembly) [Fig. 15.36(c)]

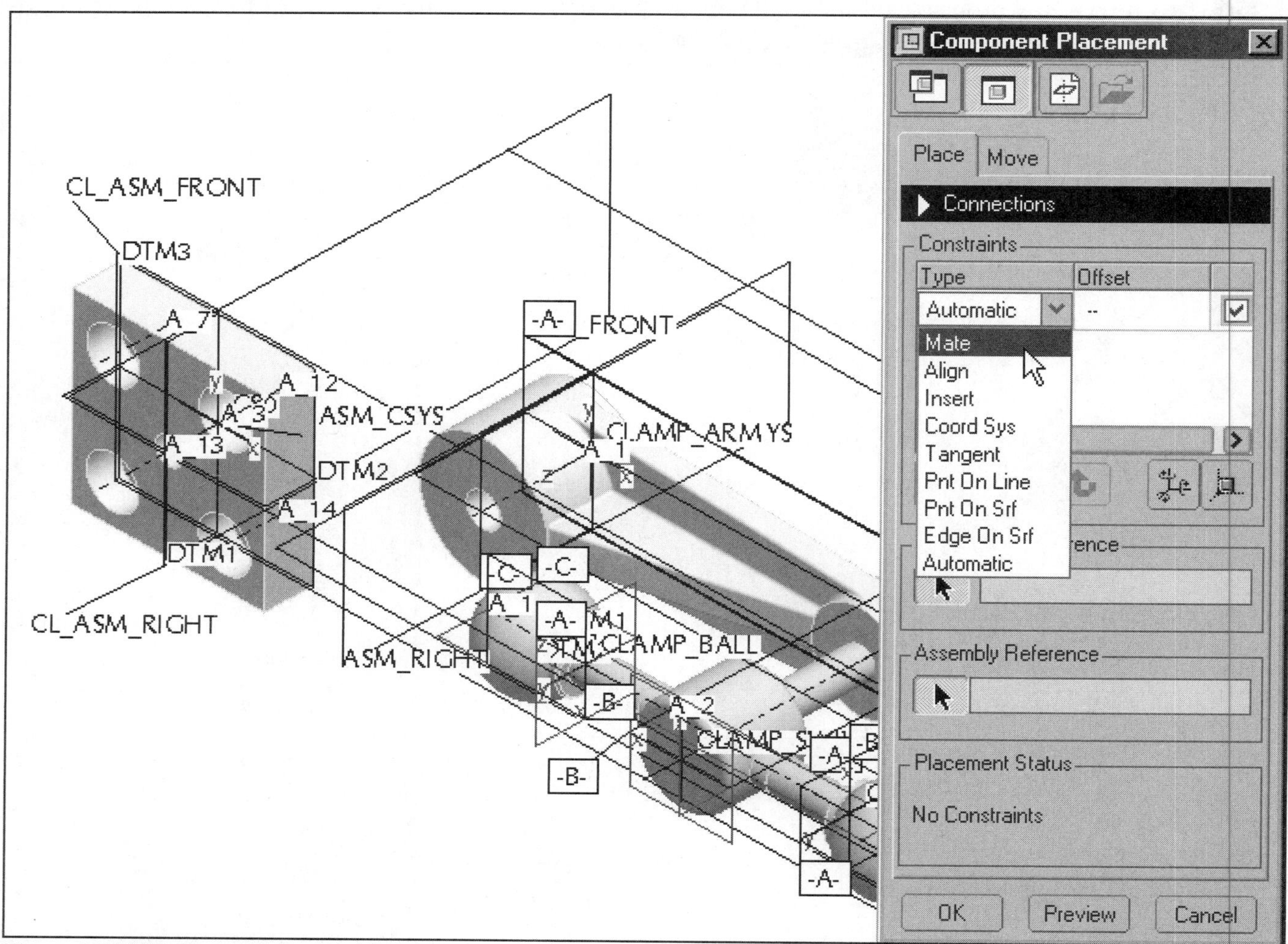

Figure 15.36(b) Clamp_Subassembly

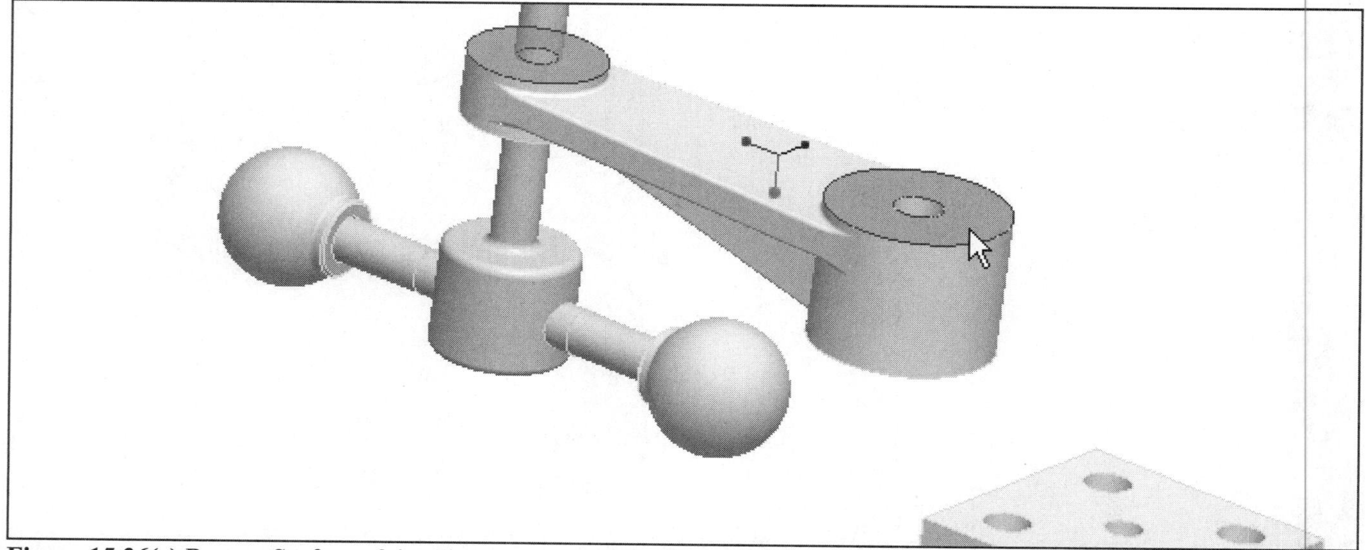

Figure 15.36(c) Bottom Surface of the Clamp_Arm is selected as the Component Reference

Spin the model ⇒ **View** ⇒ **Shade** ⇒ pick the *top surface* of the Clamp_Plate [Fig. 15.36(d)] ⇒ ✚ **Specify a new constraint** ⇒ Automatic ⇒ ⌄ ⇒ **Align** ⇒ pick the *hole axis* of the Clamp_Arm [Fig. 15.36(e)] ⇒ pick the *center tapped hole axis* of the Clamp_Plate [Fig. 15.36(f)] *note that you could have picked the two respective hole surfaces and achieve the same results*

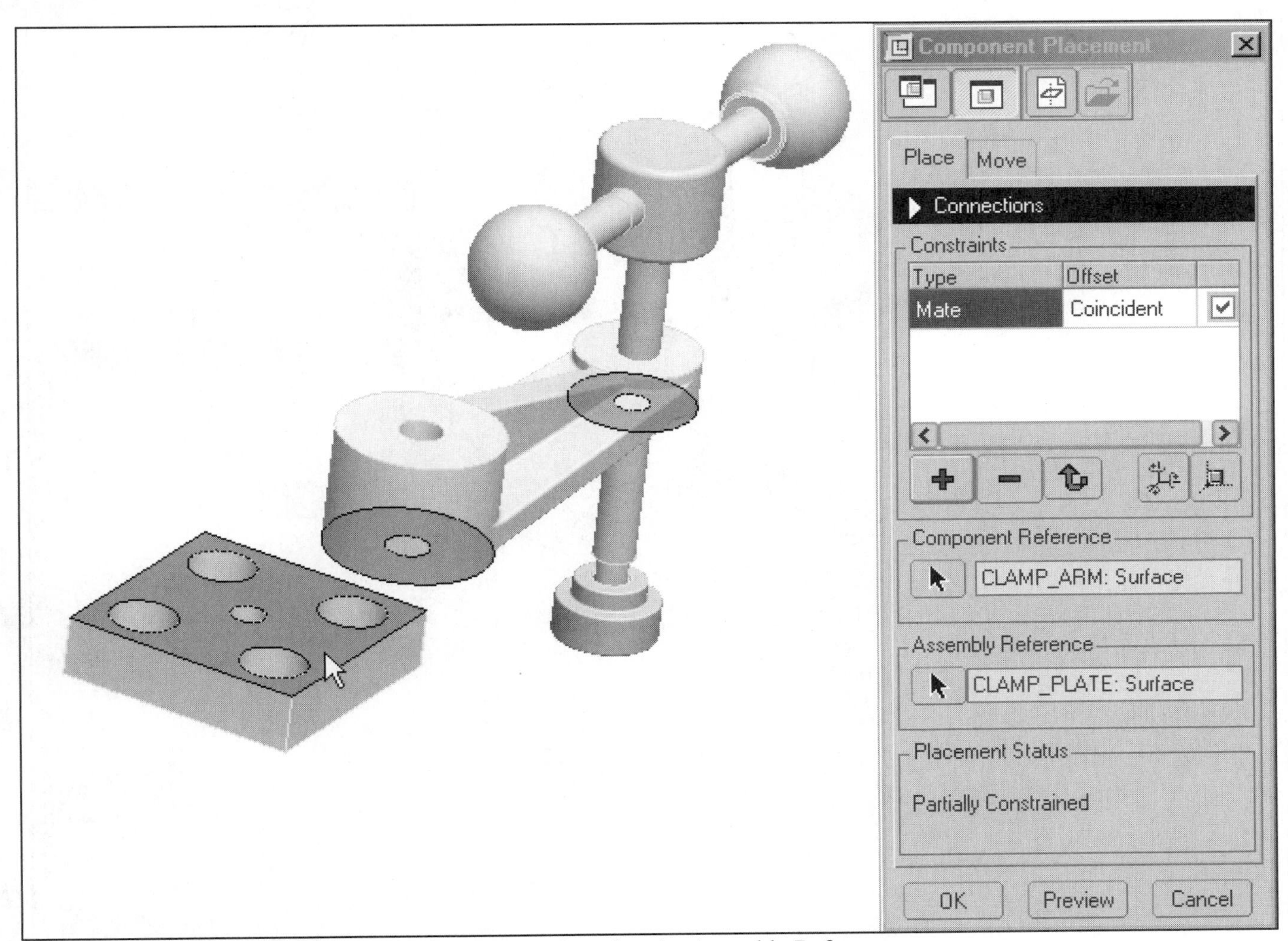

Figure 15.36(d) Top Surface of the Clamp_Plate is selected as the Assembly Reference

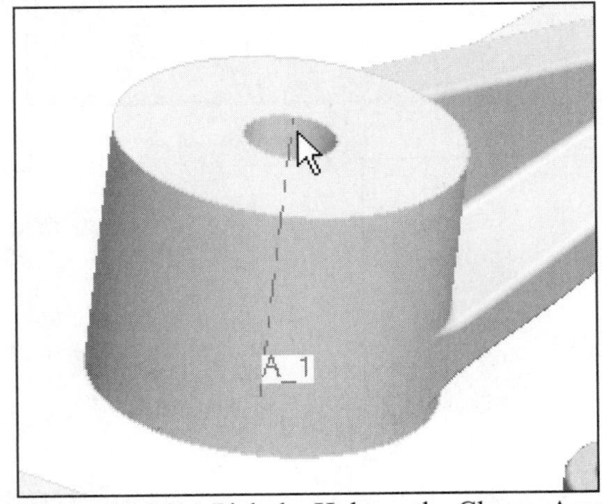

Figure 15.36(e) Pick the Hole on the Clamp_Arm

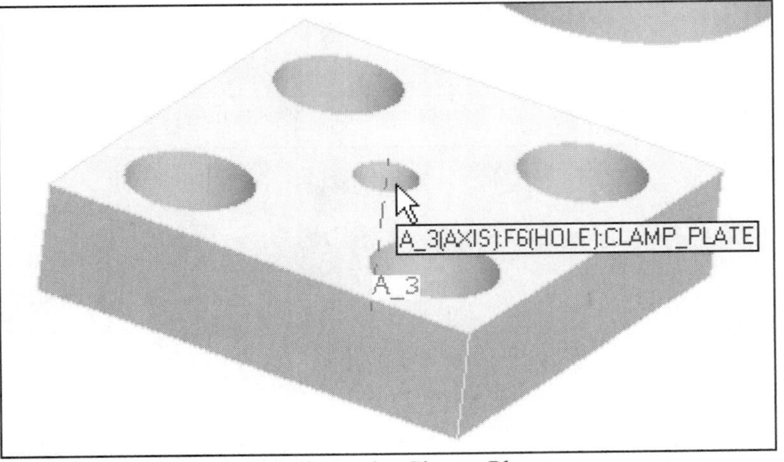

Figure 15.36(f) Pick the Hole on the Clamp_Plate

Click: **OK** [Fig. 15.36(g)] ⇒ [icon] ⇒ **FRONT** (Fig. 15.37) ⇒ **View** ⇒ **Shade** ⇒ [icon] ⇒ [icon] ⇒ **MMB**

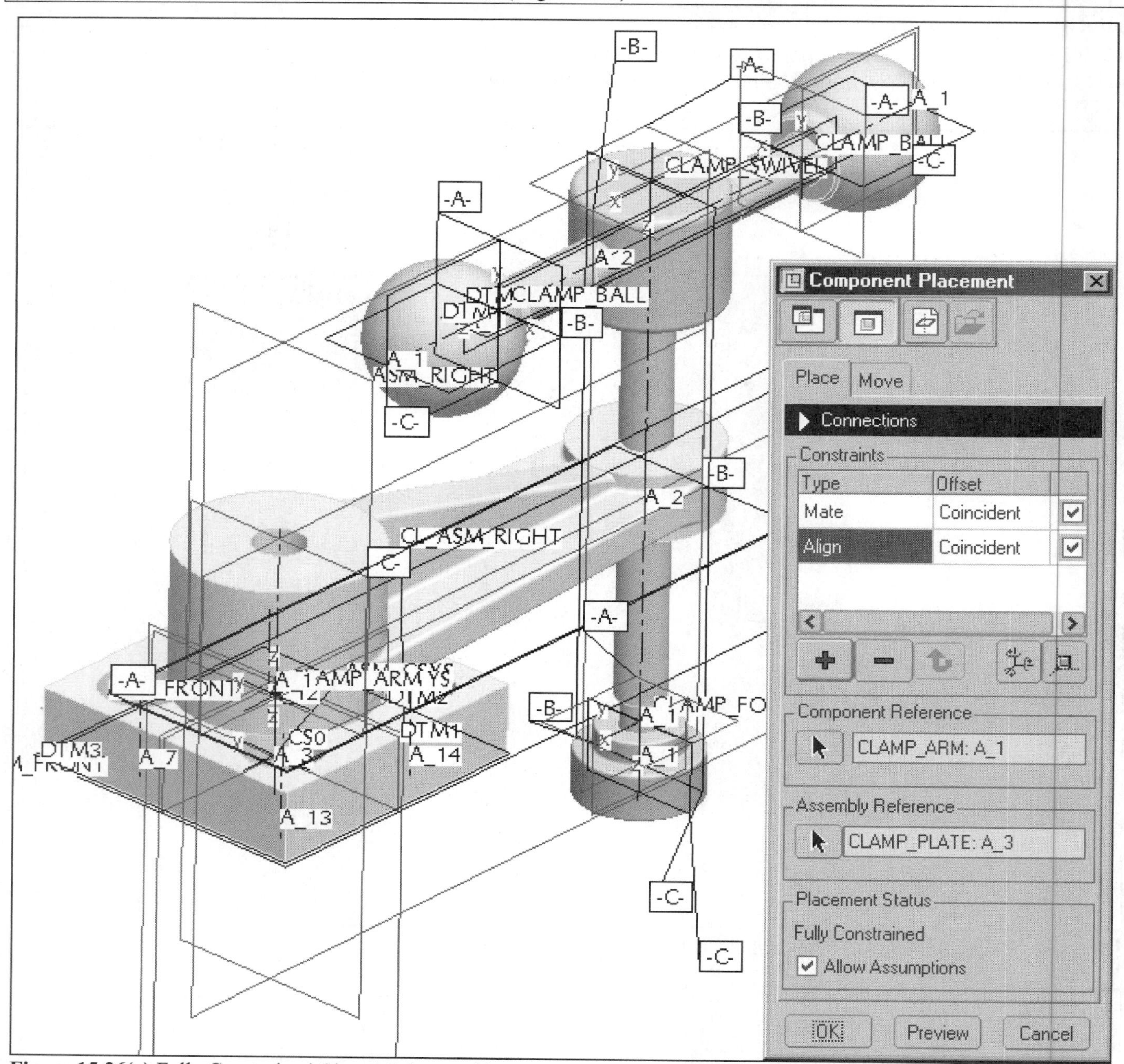

Figure 15.36(g) Fully Constrained Clamp_Subassembly

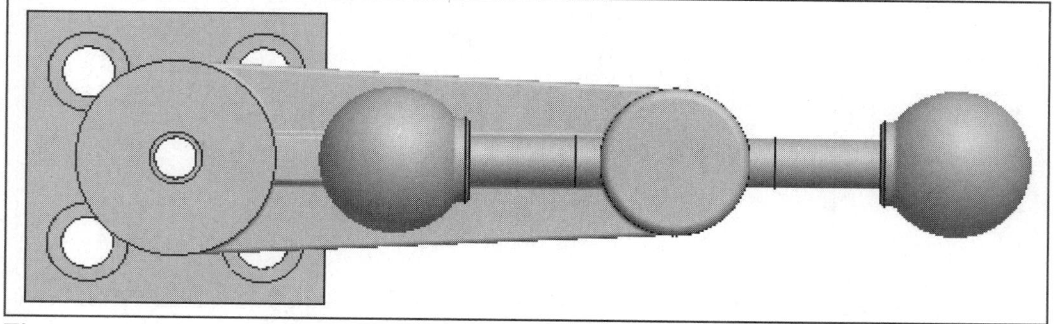
Figure 15.37 Front View

Freeform Mouse-Driven Component Manipulation

The last two components of the Clamp_Assembly will now be added. In both cases, you will use Freeform Mouse-Driven Component Manipulation. Whenever the Component Placement dialog box (Fig. 15.38) is available for placing a component or redefining placement constraints, a spin center for the active component is always visible, and you can manipulate the active component using a combination of mouse and keyboard commands *(Ctrl+Alt pressed at the same time)*. Manipulating a component is easier than moving to a separate tab in the Component Placement dialog box to package-move a component around on the screen. You can switch between full view navigation and component manipulation easily with the Ctrl+Alt keys. You can perform translation and rotation adjustments while you establish constraints. The component motion respects any constraints as they are established, as is the case with *Move tab* functionality. The spin icon appears during the entire component placement operation and defaults to the bounding box center. You can modify this location, using the Preferences dialog box, accessed from the Orientation dialog box.

Regardless of the selected tab- Place or Move: you can translate or rotate the component dynamically using:

- **Translate** Press and hold: **Ctrl+Alt** with **RMB**
- **Rotate** Press and hold: **Ctrl+Alt** with **MMB**
- **Z-axis** Press and hold: **Ctrl+Alt** with **LMB**

Because of the orthographic projection used by Pro/E, motion in the Z-axis, or screen normal, is noticeable only if objects intersect. Thus, while a component is moving in the Z-axis, the camera angle is adjusted to provide a noticeable effect.

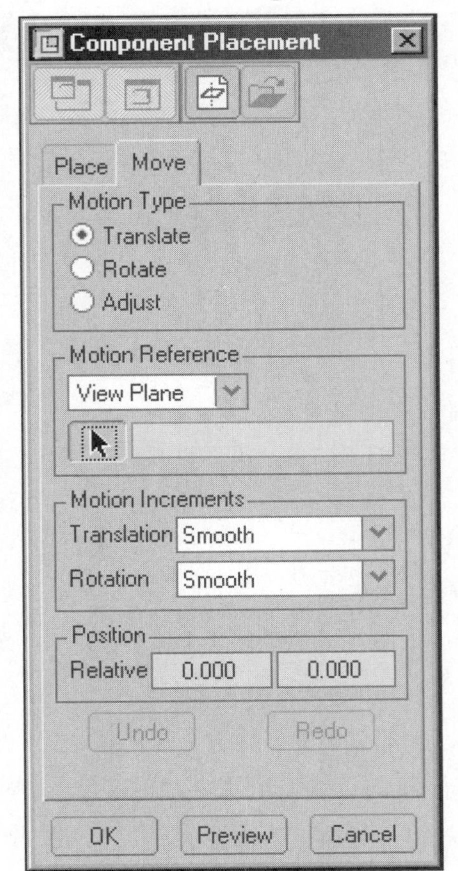

Figure 15.38 Component Placement Dialog Box (Move Tab Active)

Click: **Add component to the assembly** ⇒ pick **clamp_stud35.prt** ⇒ **Preview** [Fig. 15.39(a)] ⇒ **Open**

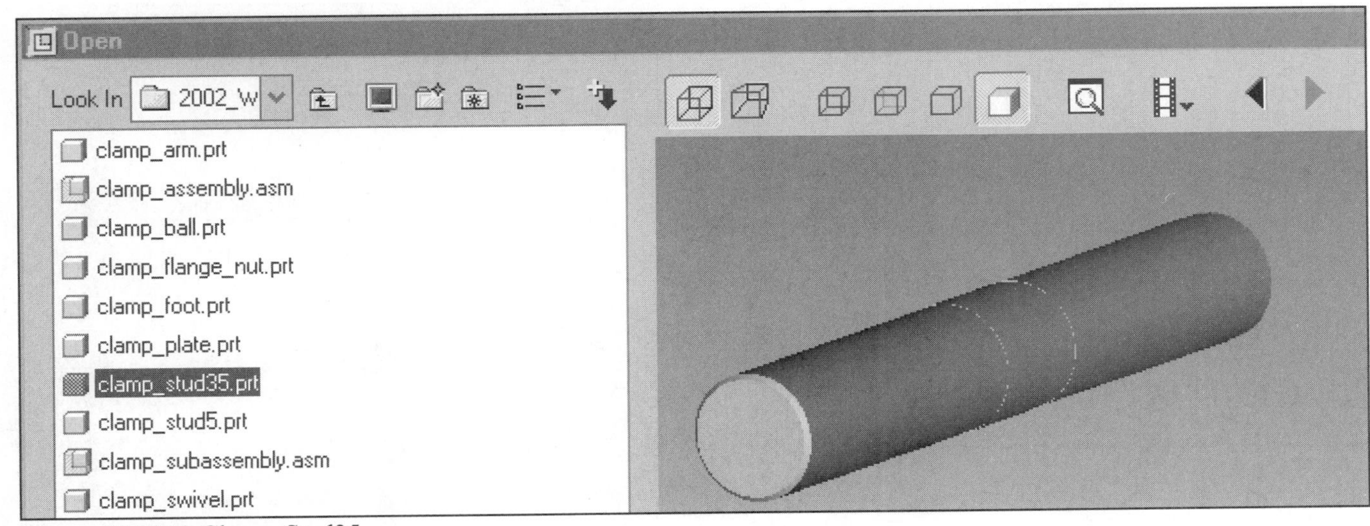

Figure 15.39(a) Clamp_Stud35

Click: **Datum planes off** ⇒ Constraints Automatic, click ⇒ Insert ⇒ pick the *cylindrical surface* of the Clamp_Stud35 ⇒ pick the *hole surface* of the Clamp_Arm [Fig. 15.39(b-c)]

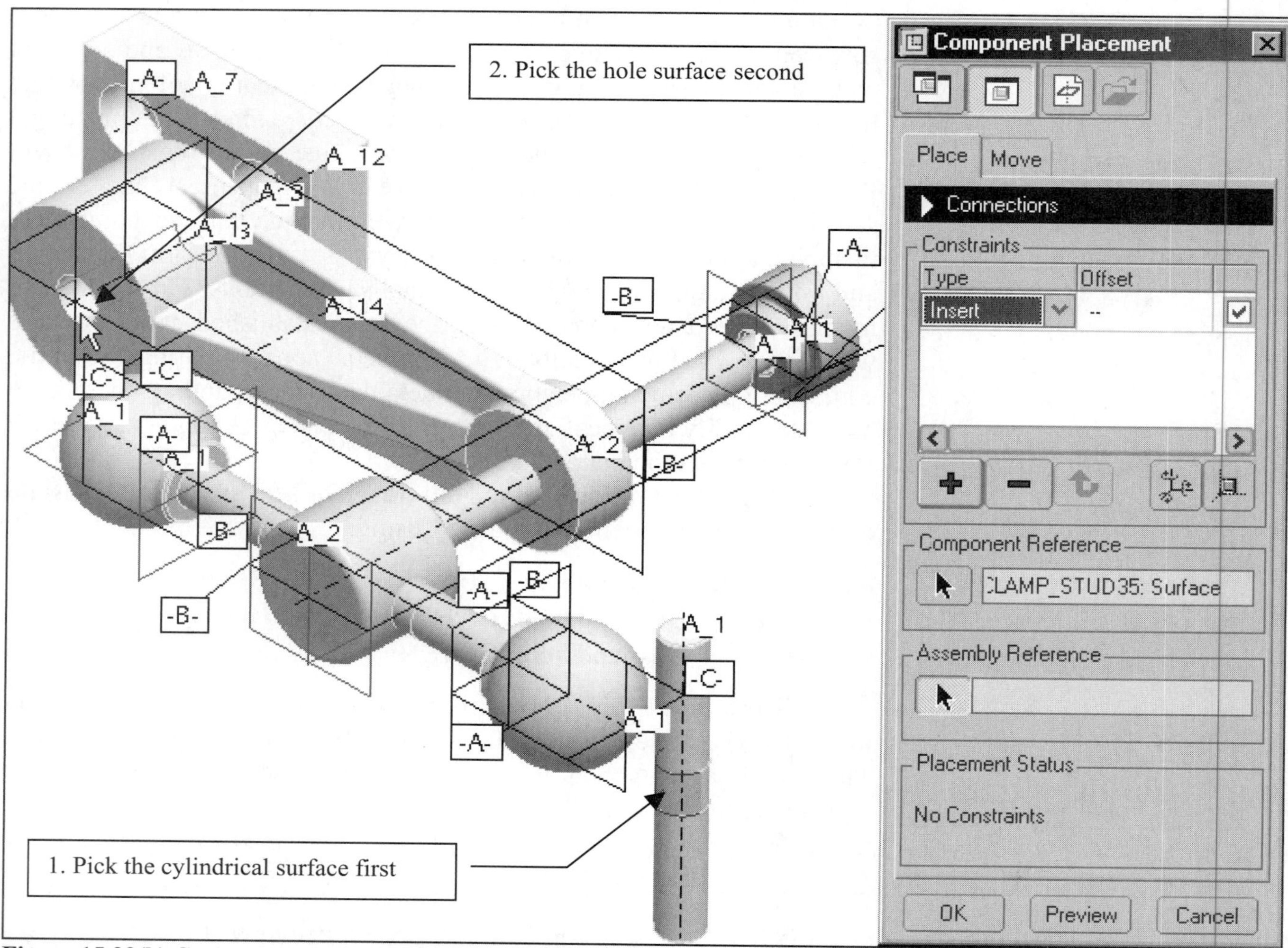

Figure 15.39(b) Component Placement Dialog Box, Place Tab

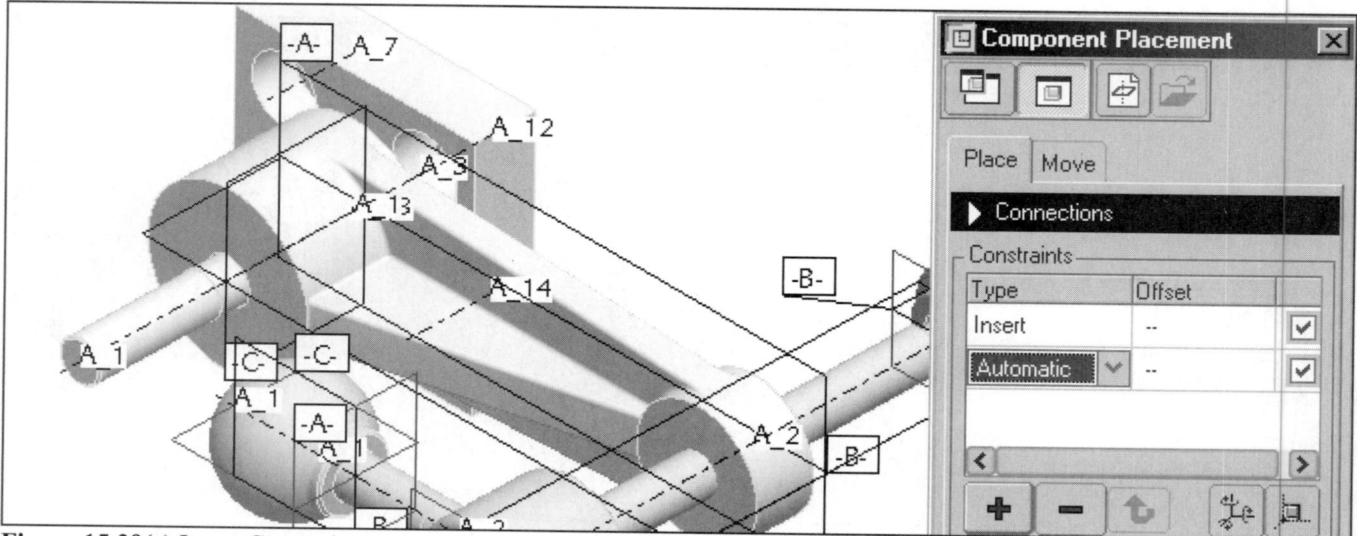

Figure 15.39(c) Insert Constraint

Click: ⇒ **LEFT** ⇒ **Move** tab ⇒ Motion Type Translate ⇒ ⇒ pick the Clamp_Stud35 and slide it deeper into the hole ⇒ **LMB** [Fig. 15.39(d)] ⇒ ⇒ **TOP** ⇒ **Hidden Line** ⇒ pick the Clamp_Stud35 and translate it until it is slightly inside the Clamp_Plate [Fig. 15.39(e)] ⇒ **LMB**

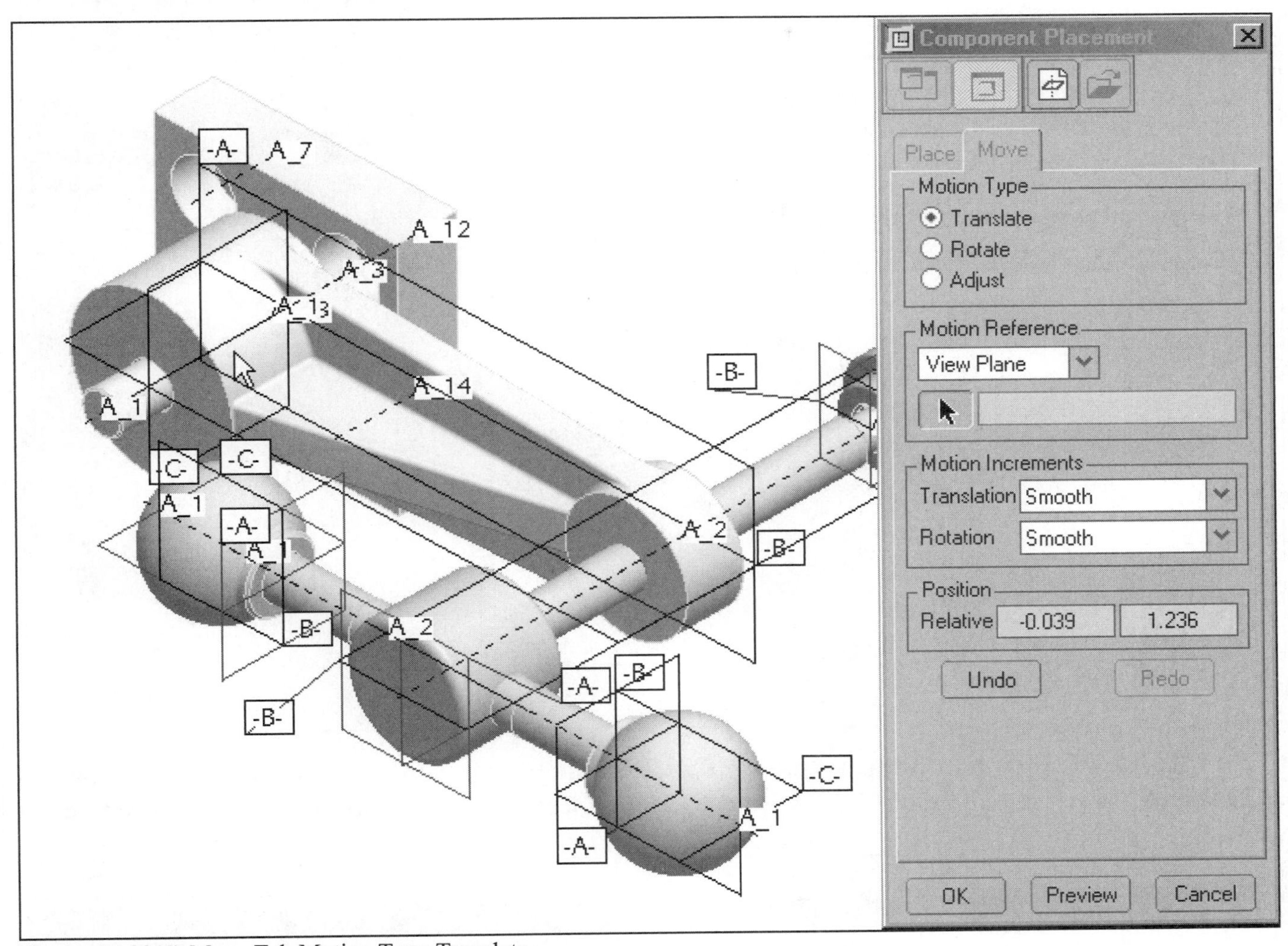

Figure 15.39(d) Move Tab Motion Type Translate

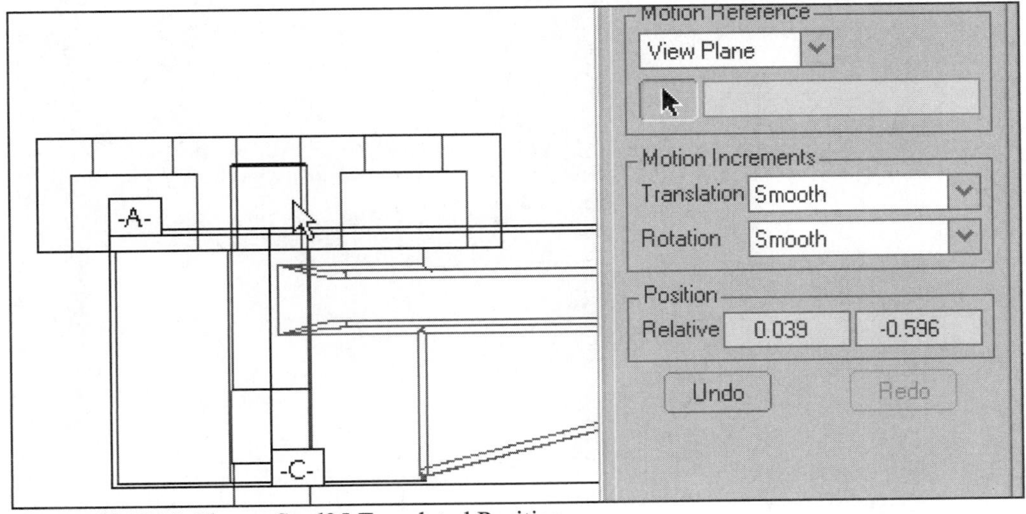

Figure 15.39(e) Clamp_Stud35 Translated Position

Click: **Place** tab ⇒ ⇒ **Standard Orientation** ⇒ **Shading** ⇒ **Fix component to current position** [Fig. 15.39(f)] ⇒ **OK** ⇒ ⇒ ⇒ **MMB**

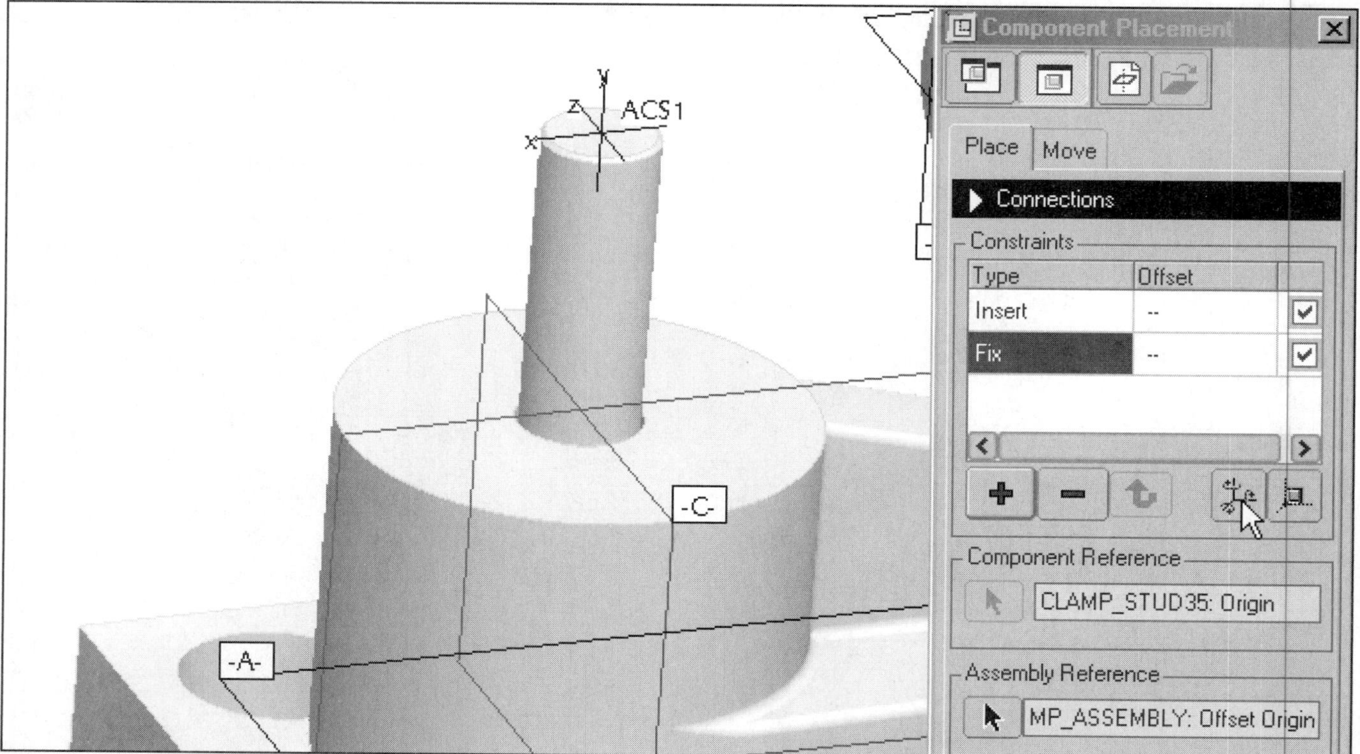

Figure 15.39(f) Add Fix Constraint

Click: **Add component to the assembly** ⇒ pick **clamp_flange_nut.prt** ⇒ **Preview** [Fig. 15.40(a)] ⇒ **Open** ⇒ **Move** tab ⇒ Motion Type **Adjust** ⇒ Motion Reference **Sel Plane** ⇒ pick the top surface of the Clamp_Arm [Fig. 15.40(b)]

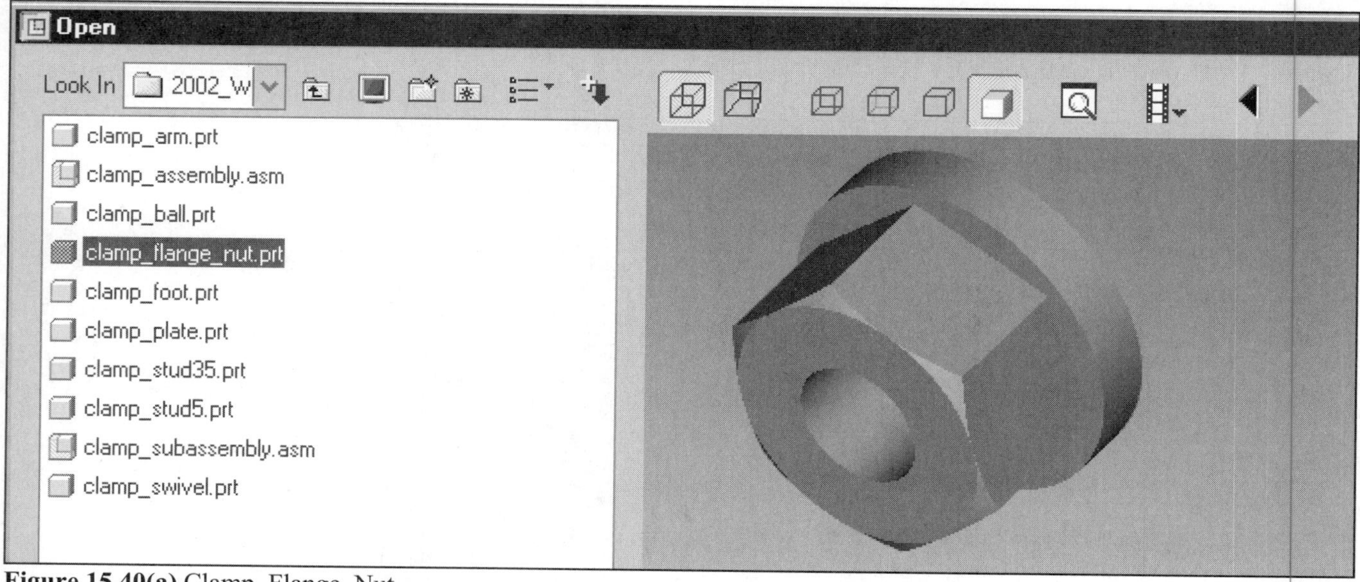

Figure 15.40(a) Clamp_Flange_Nut

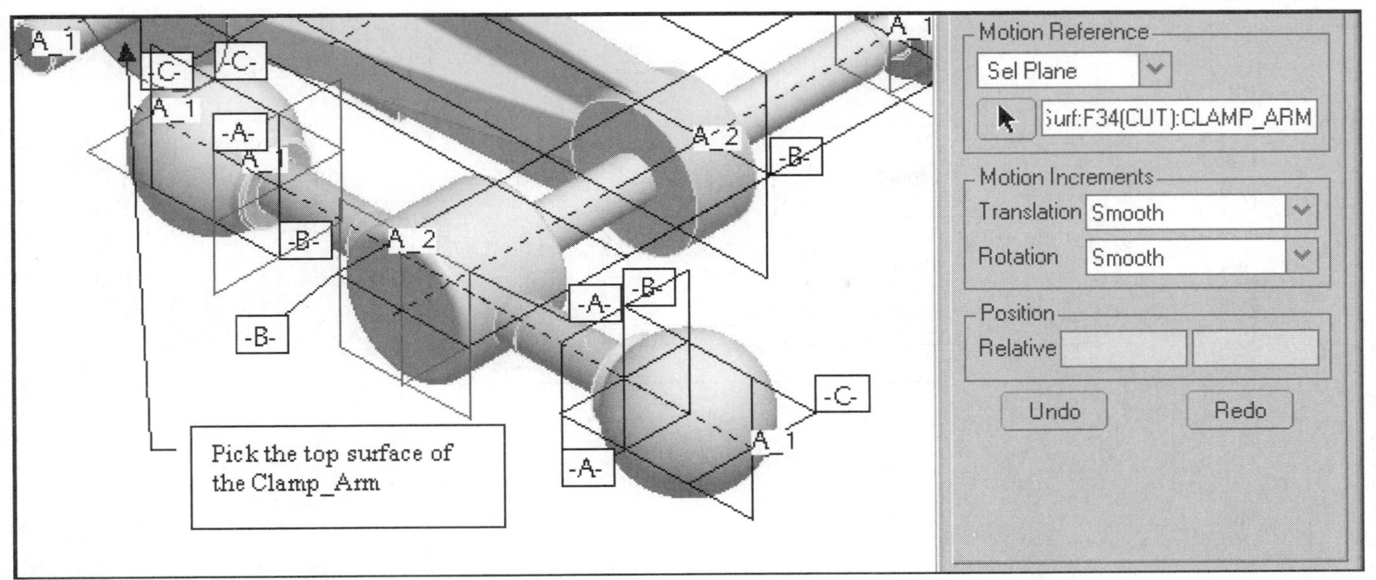

Figure 15.40(b) Sel Plane, Select the Top Surface of the Clamp_Arm

Pick the top surface of the Clamp_Flange_Nut [Fig. 15.40(c)] ⇒ **Align Offset** ⇒ type **2.00** ⇒ **Enter** ⇒ Motion Type **Translate** ⇒ Motion Reference **View Plane** ⇒ [icon] ⇒ **FRONT** ⇒ pick the Clamp_Flange_Nut and slide it near the Clamp_Stud35 [Fig. 15.40(d-e)] ⇒ **LMB** again to place [Fig. 15.40(f)] ⇒ [icon]

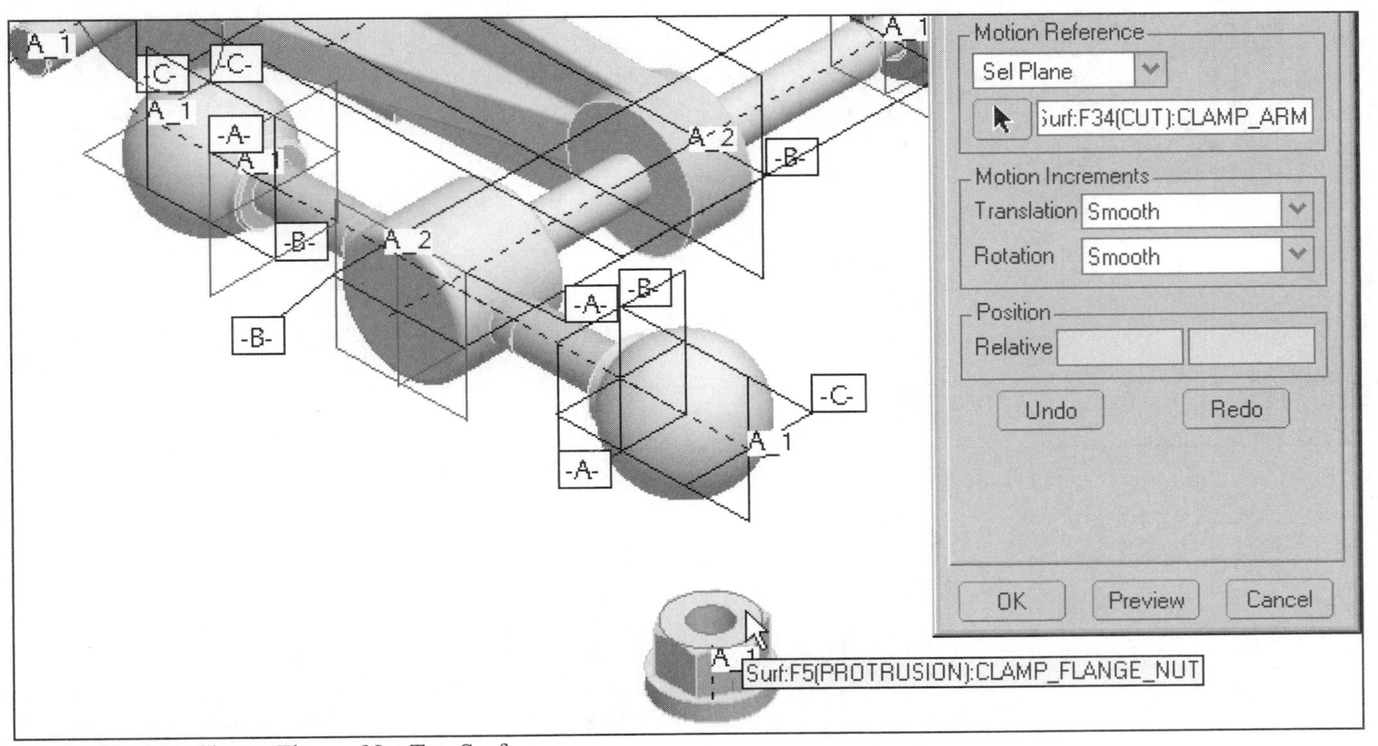

Figure 15.40(c) Clamp_Flange_Nut Top Surface

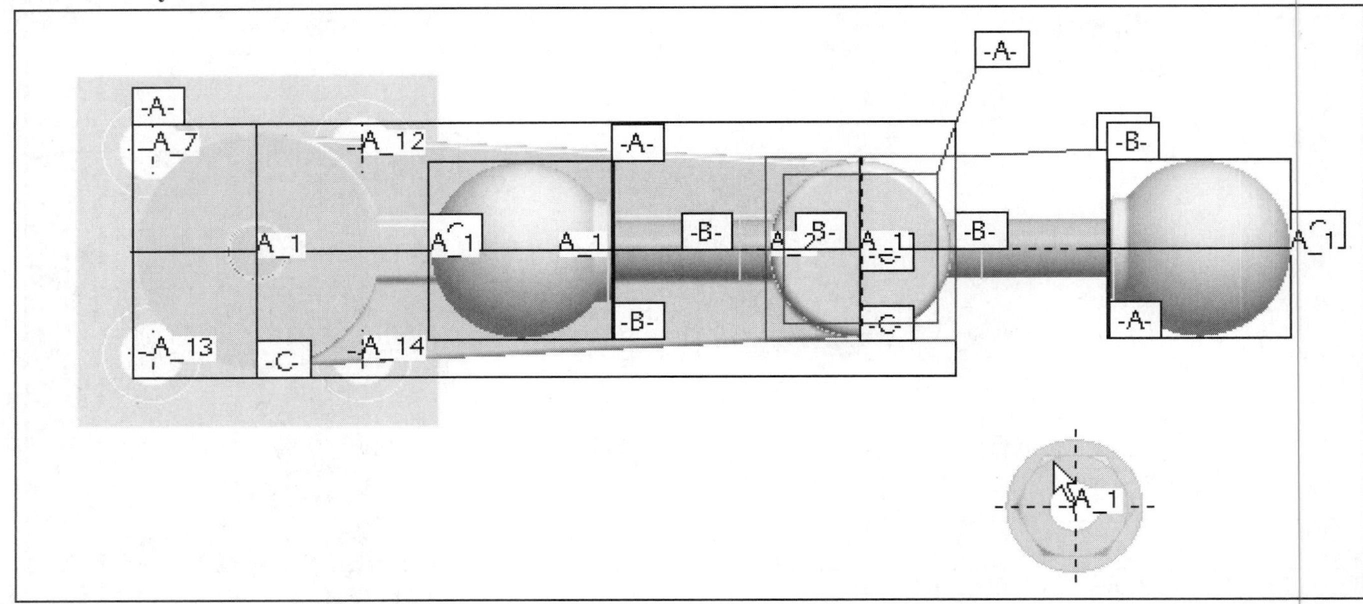

Figure 15.40(d) Move the Clamp_Flange_Nut

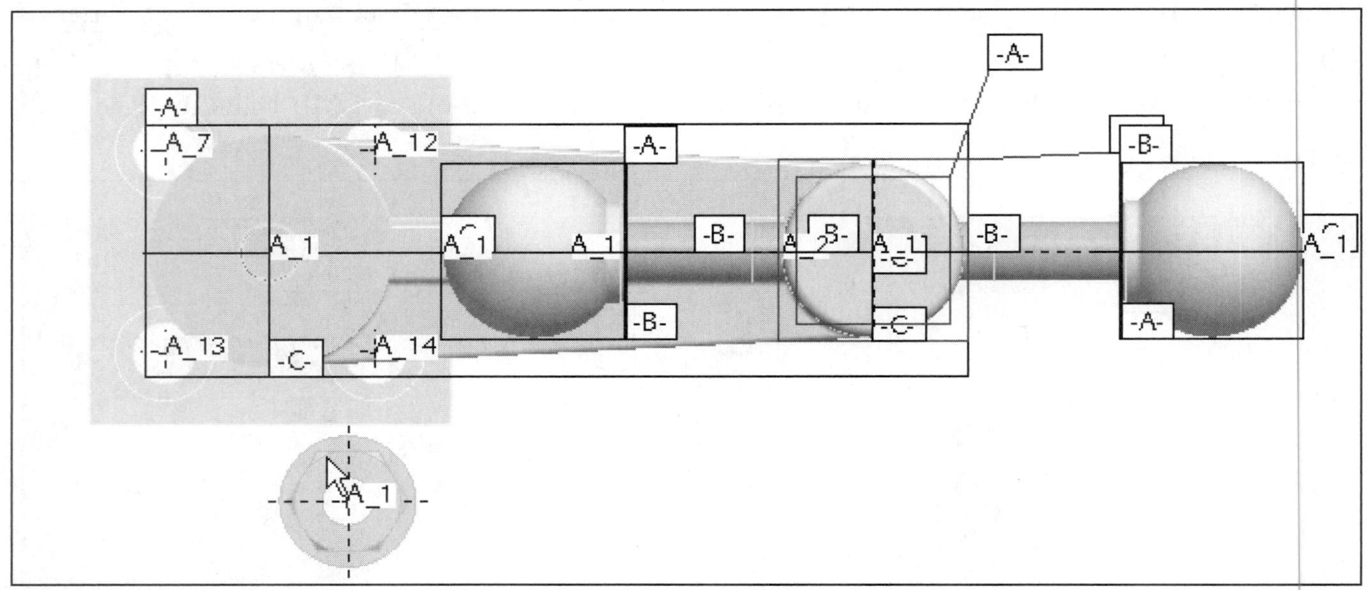

Figure 15.40(e) Moving

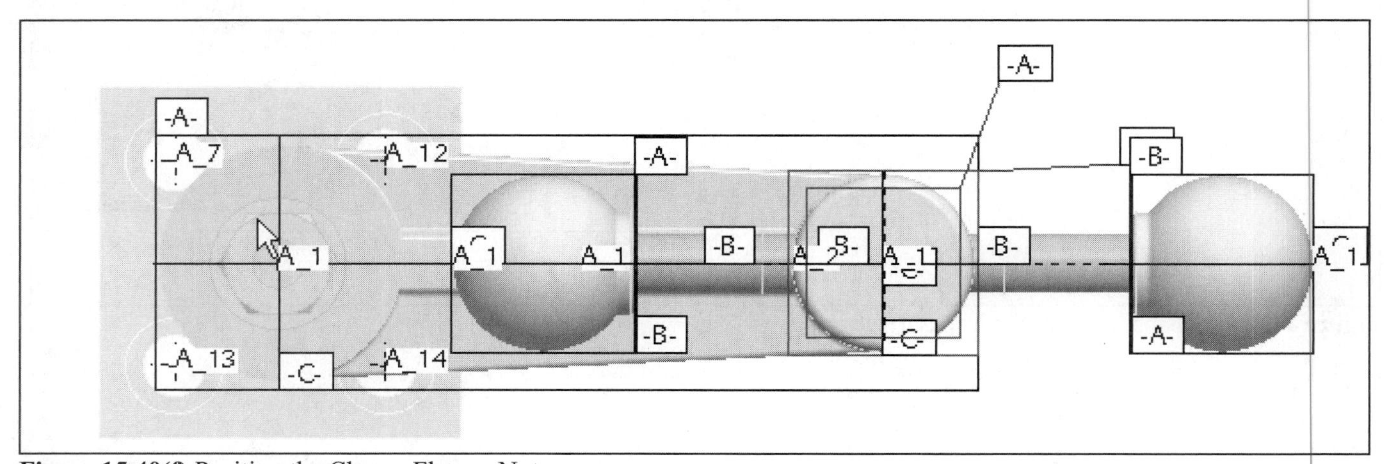

Figure 15.40(f) Position the Clamp_Flange_Nut

Click: **Place** tab [Automatic] ⇒ pick the hole's surface of the Clamp_Flange_Nut ⇒ pick on the shaft of the Clamp_Stud35 [Fig. 15.40(g)] *changes to* ***Insert*** ⇒ [Automatic] pick on the upper surface of the Clamp_Arm ⇒ place the cursor on the Clamp_Flange_Nut [Fig. 15.40(h)] ⇒ **RMB** ⇒ **Next** ⇒ **LMB** to accept [Fig. 15.40(i)] *changes to* ***Mate*** ⇒ type **0.00** ⇒ **Enter** [Fig. 15.40(j)] ⇒ **OK** ⇒ [icon] ⇒ **Standard Orientation** ⇒ [icon] ⇒ **MMB**

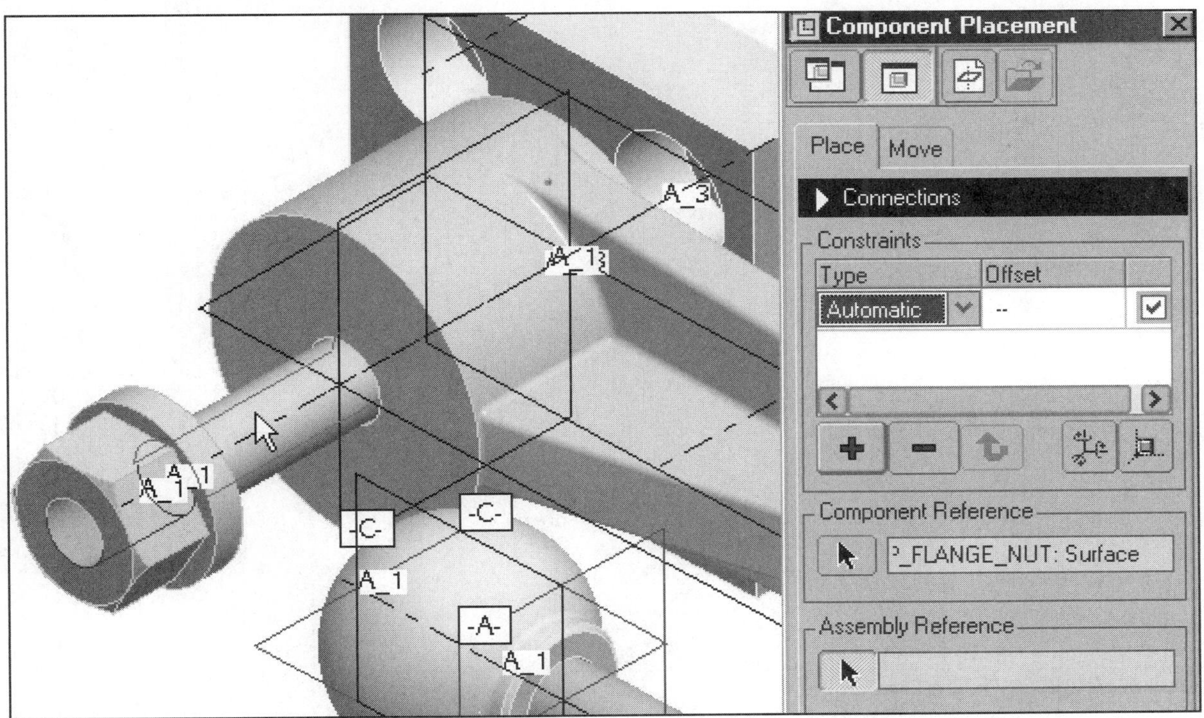

Figure 15.40(g) Select the Hole of the Clamp_Flange_Nut and the Shaft of the Clamp_Stud35

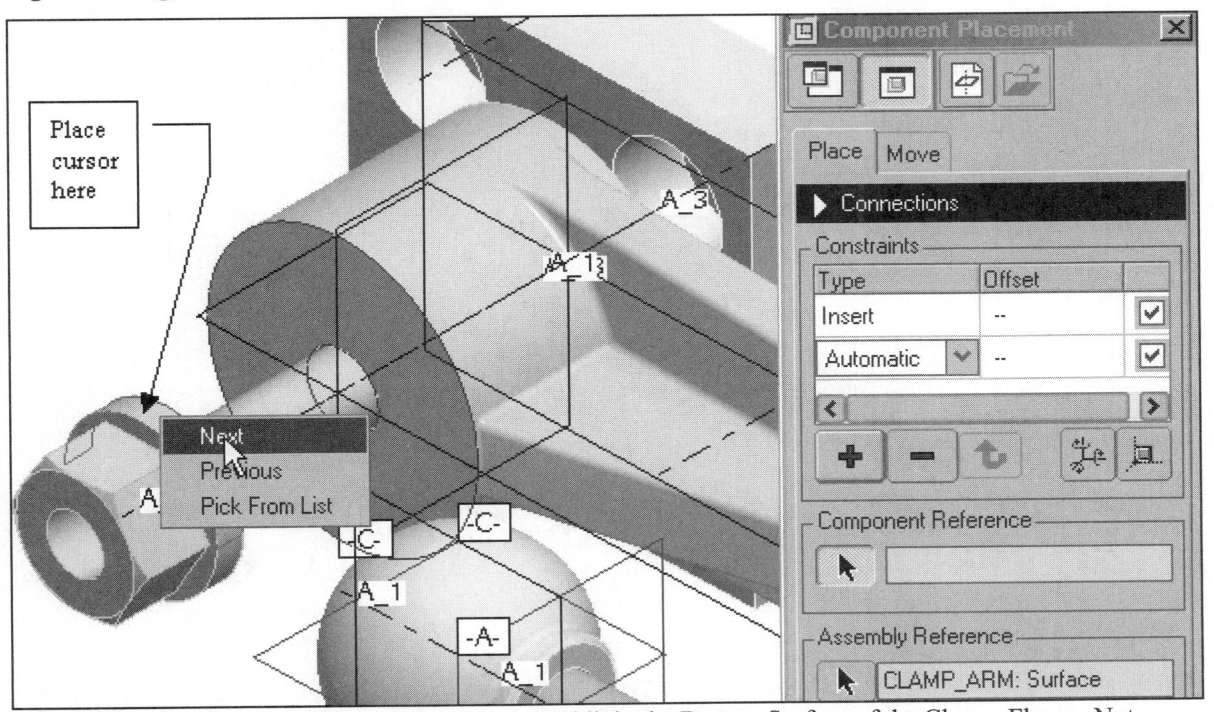

Figure 15.40(h) Filter through the Selections to Highlight the Bottom Surface of the Clamp_Flange_Nut

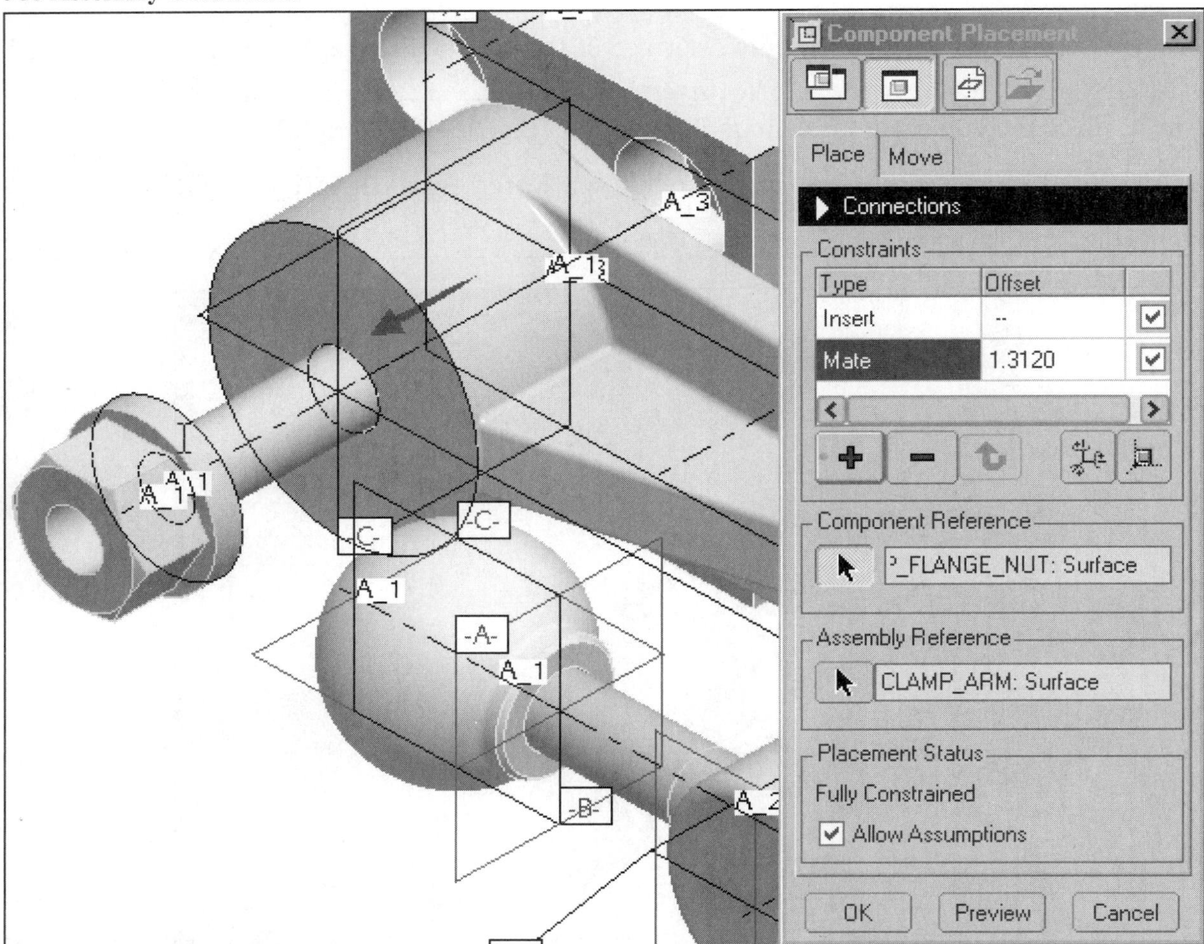

Figure 15.40(i) Accept Bottom Surface of the Clamp_Flange_Nut. Constraint changes to Mate

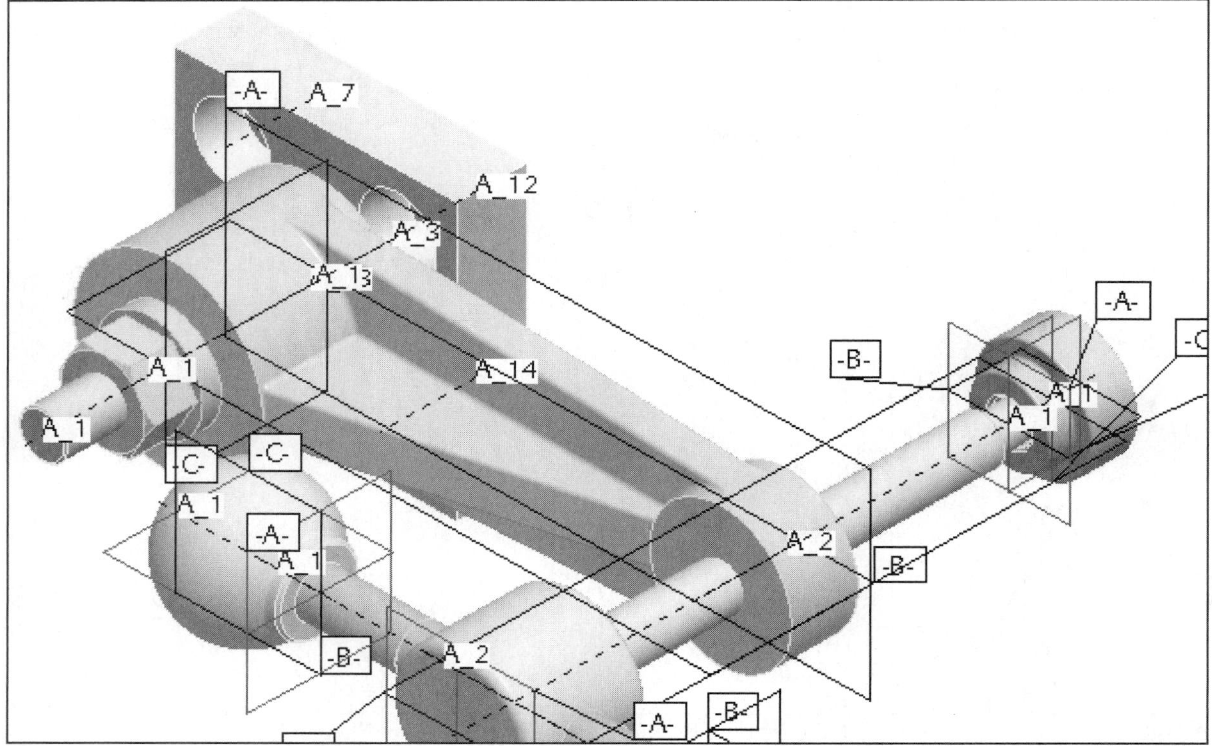

Figure 15.40(j) Mate Offset of **0.00**

The final procedure you will perform before proceeding is to check the assembly using the **Analysis** command. If you look at the long (**5.00**) stud's detail drawing [see Figure 15.10(b)], you will see that the shaft diameter is greater than **.500** at the center (**.506**) of the stud and that each end has a **.500-13 UNC** thread. The hole in the swivel is **.500**. This means that there should be a slight interference (Press Fit) between the two components. Check the clearance and interference between these components.

Click: **Analysis** ⇒ **Model Analysis** ⇒ Type **Pairs Clearance** ⇒ Definition- From **Whole Part** ⇒ pick the Clamp_Swivel ⇒ Definition- To **Whole Part** ⇒ pick the Clamp_Stud5 (Fig. 15.41) ⇒ Results Volume of interference is 0.00690137. ⇒ **Close**

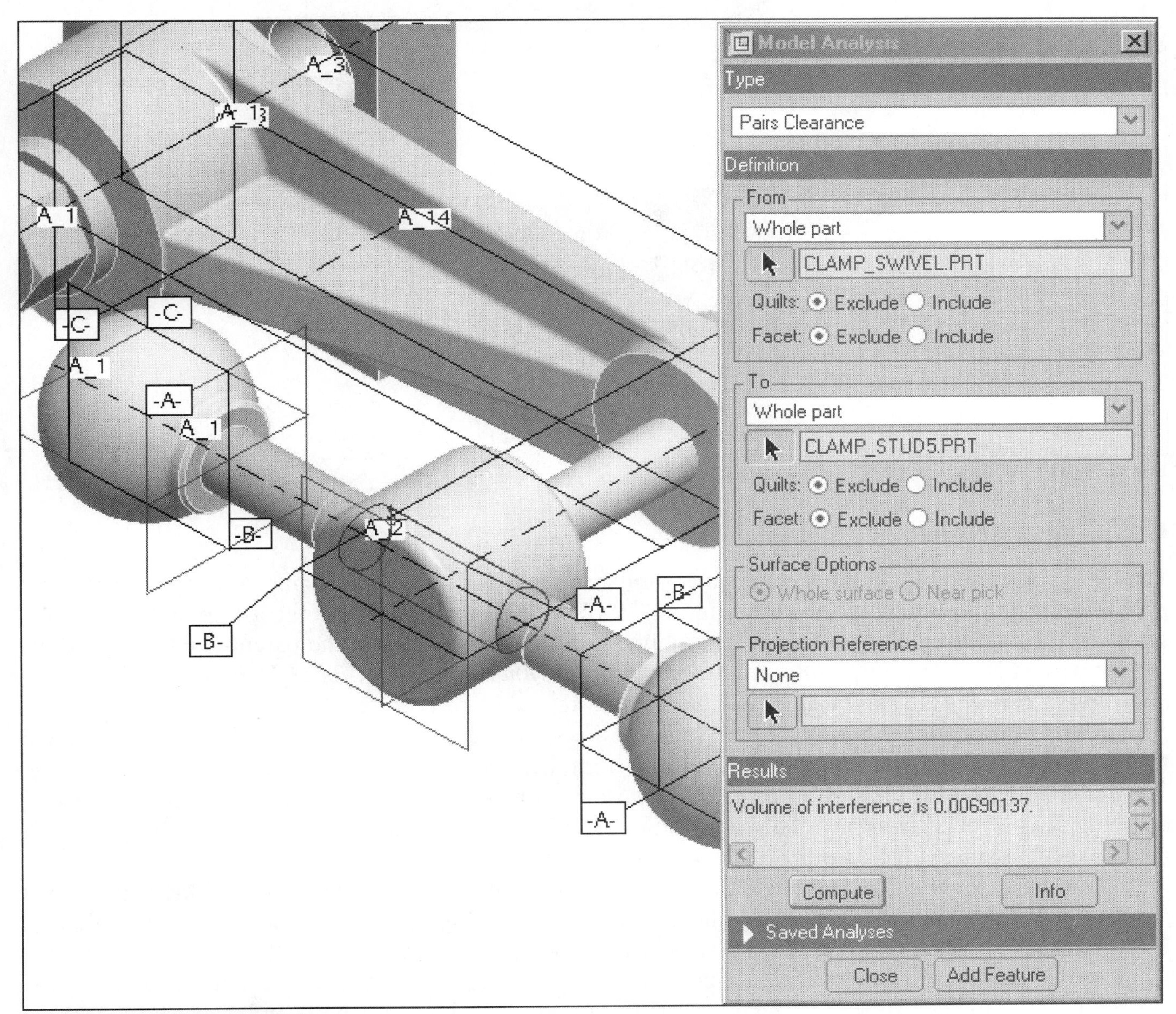

Figure 15.41 Volume of Interference

The assembly is now complete. In the next lesson, you will learn how to move and rotate components in the assembly, establish views for use in the Drawing mode, create exploded views of the assembly, generate a bill of materials, and change the component visibility. Complete the lesson project.

Lesson 15 Project

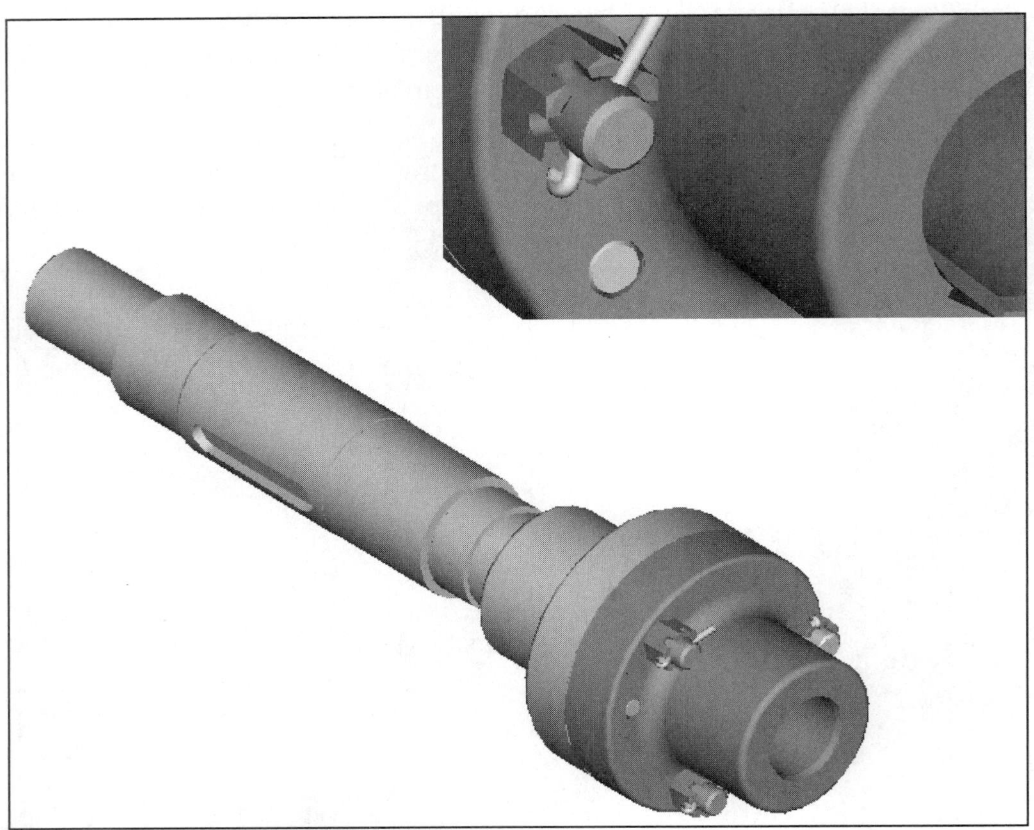

Figure 15.42 Coupling Assembly

Coupling Assembly

The Coupling Assembly requires commands similar to those for the Swing Clamp Assembly. Model the parts and create the assembly shown in Figures 15.42 through 15.73. Analyze the assembly and plan the steps required to assemble it. Plan the assembly component sequence and the parent-child relationships for the assembly. *After completing the assembly, do an Analysis using Global Interference. If there is interference between the shaft and the key, modify the key to the correct size.*

You will use the **Coupling Shaft** from the Lesson 7 Project. The Coupling Shaft should be the first component assembled. The **Taper Coupling** from the Lesson 8 Project is also used in the assembly. The detail drawings for the **second coupling** are provided here, in this lesson project. Model this second coupling *before* you start the assembly. Depending on the library parts available on your system, you may need to model the **Key**, the **Dowel**, and the **Washer**.

Because not all organizations purchase the libraries, details are provided for all the components required for the assembly, including the standard *off-the-shelf* parts available in Pro/E's library. Pro/LIBRARY commands to access the standard components are provided for those of you who have them loaded on your systems. The instance name is given for every standard component used in the assembly. The **Slotted Hex Nut**, **Socket Head Cap Screw**, **Hex Jam Nut**, and **Cotter Pin** are all standard parts from the library. The Cotter Pin is in *inch* units, and the remaining items are *metric*.

For this project, do not assemble the library parts directly from the library. Save each library part in your own directory with a new name, and then use the new part names in the assembly.

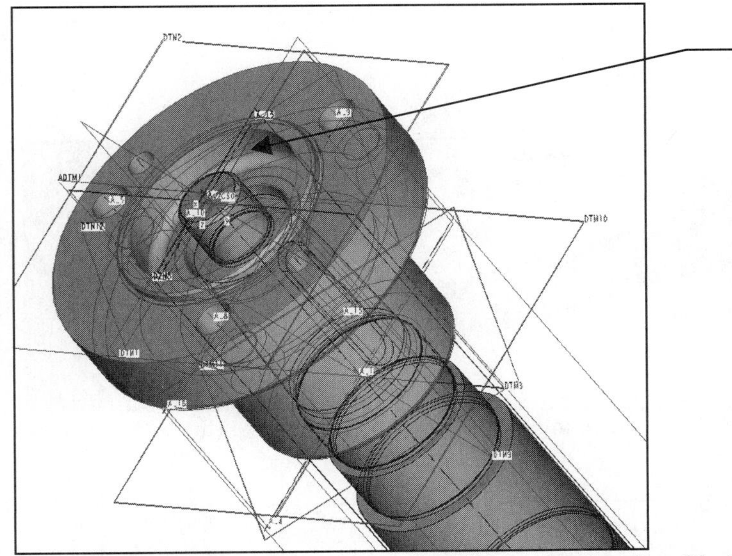
Figure 15.43 Assembling the Taper Coupling to the Coupling Shaft

Redesign the length of the threaded end to accommodate the washer and nut. *You* decide the new length based on the combined thickness of the two components. The other end also requires a modified length.

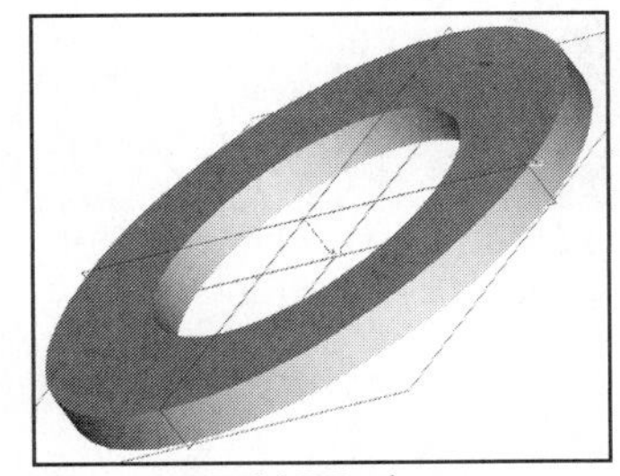
Figure 15.44 Washer

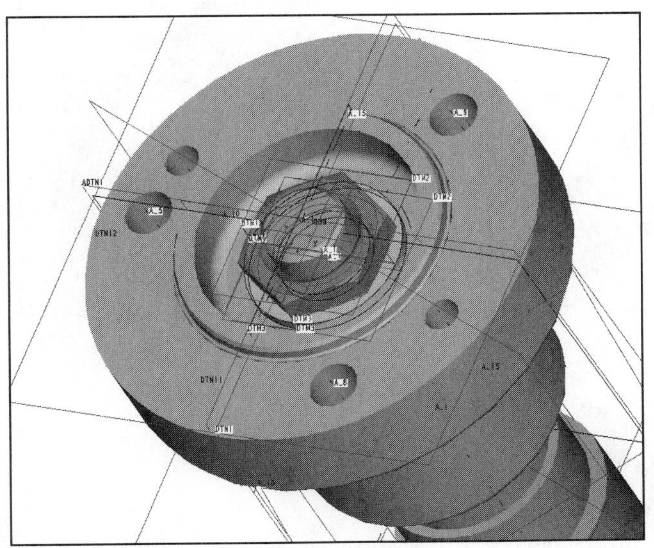
Figure 15.45 Hex Jam Nut and Washer

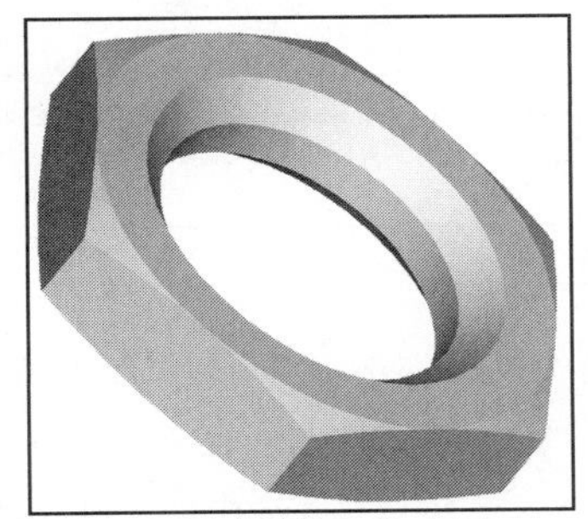
Figure 15.46 Hex Jam Nut

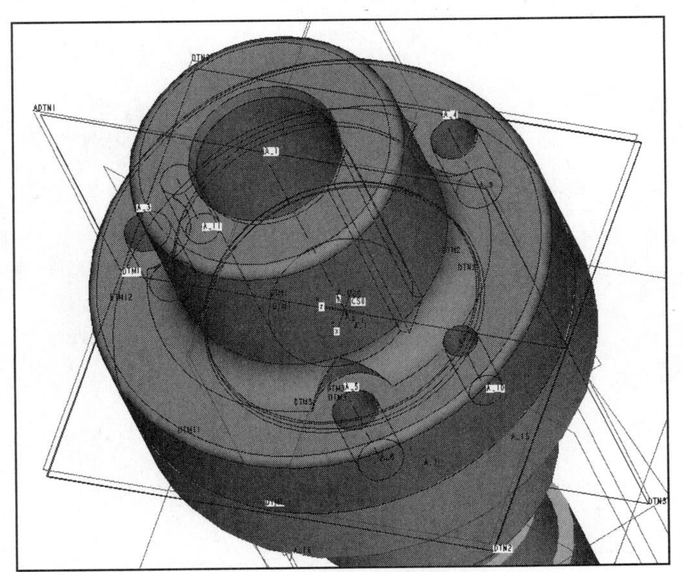
Figure 15.47 Second Coupling Assembled

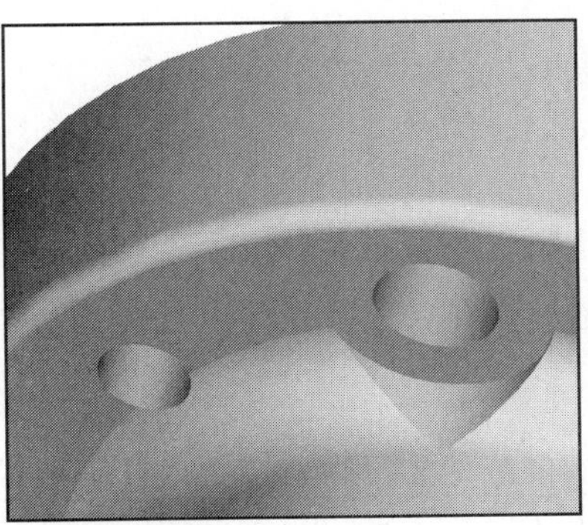
Figure 15.48 Holes

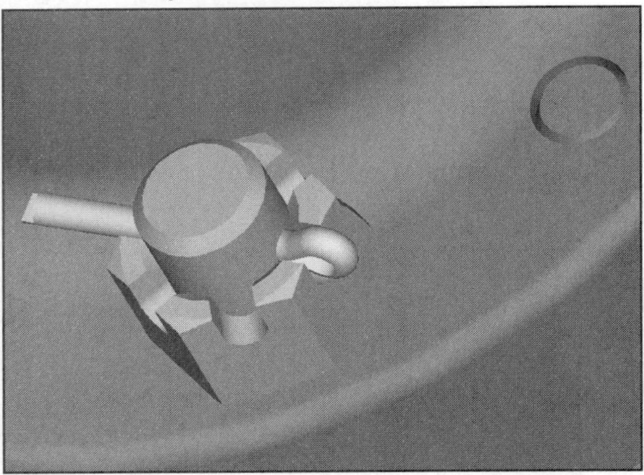

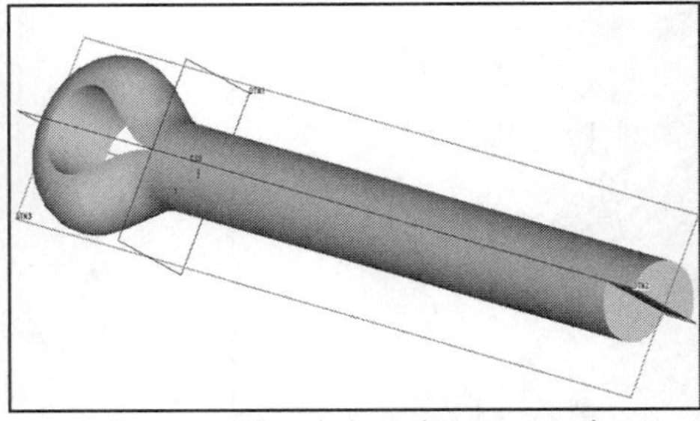

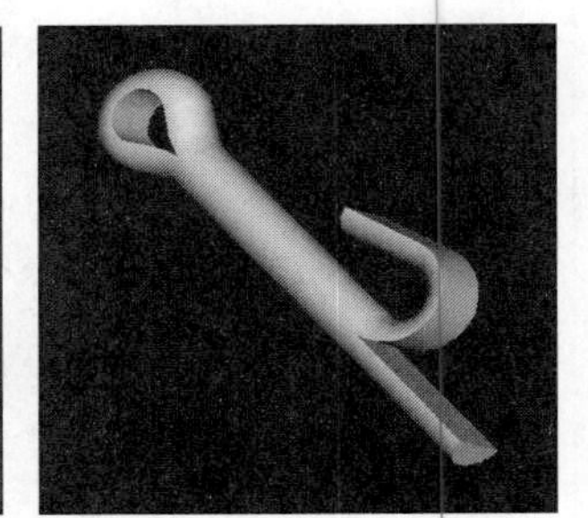

Figure 15.49(a-c) Dowel, Slotted Hex Nut, and Cotter Pin. After constraining the cotter pin, try redefining (the trajectory) it to bend one or both of its prongs.

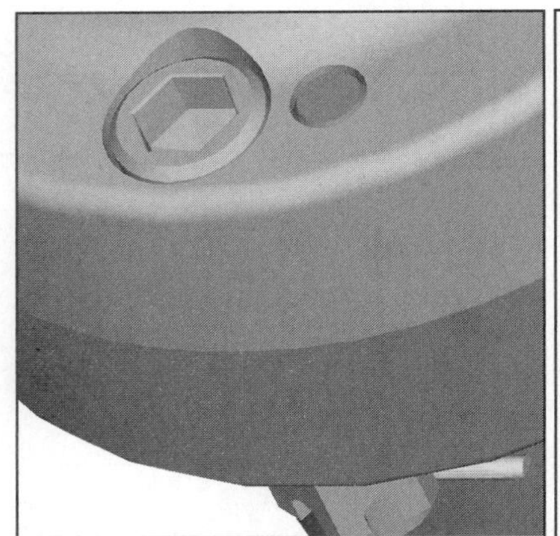

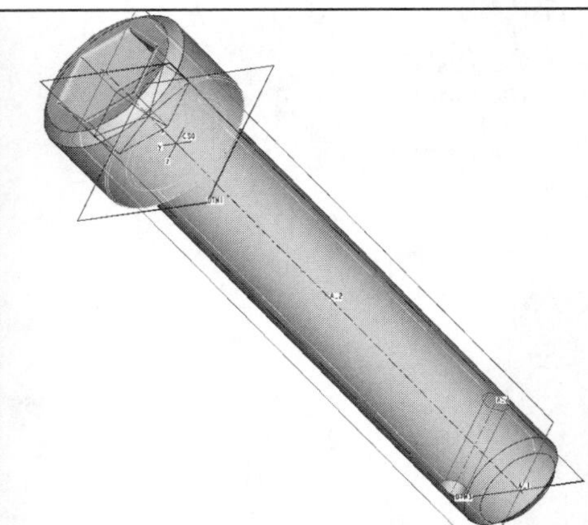

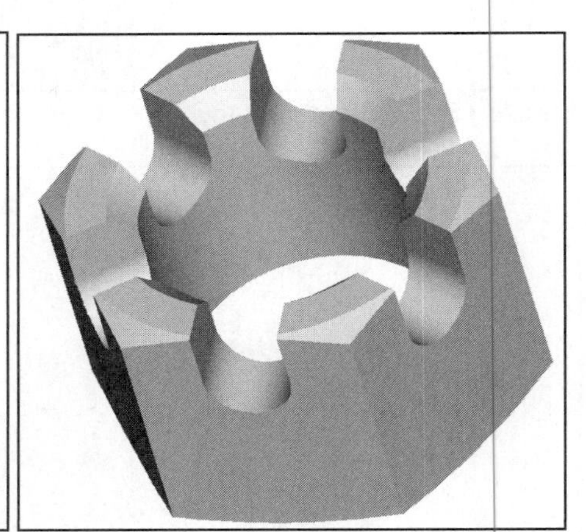

Figure 15.50(a-c) Socket Head Cap Screw and Slotted Hex Nut

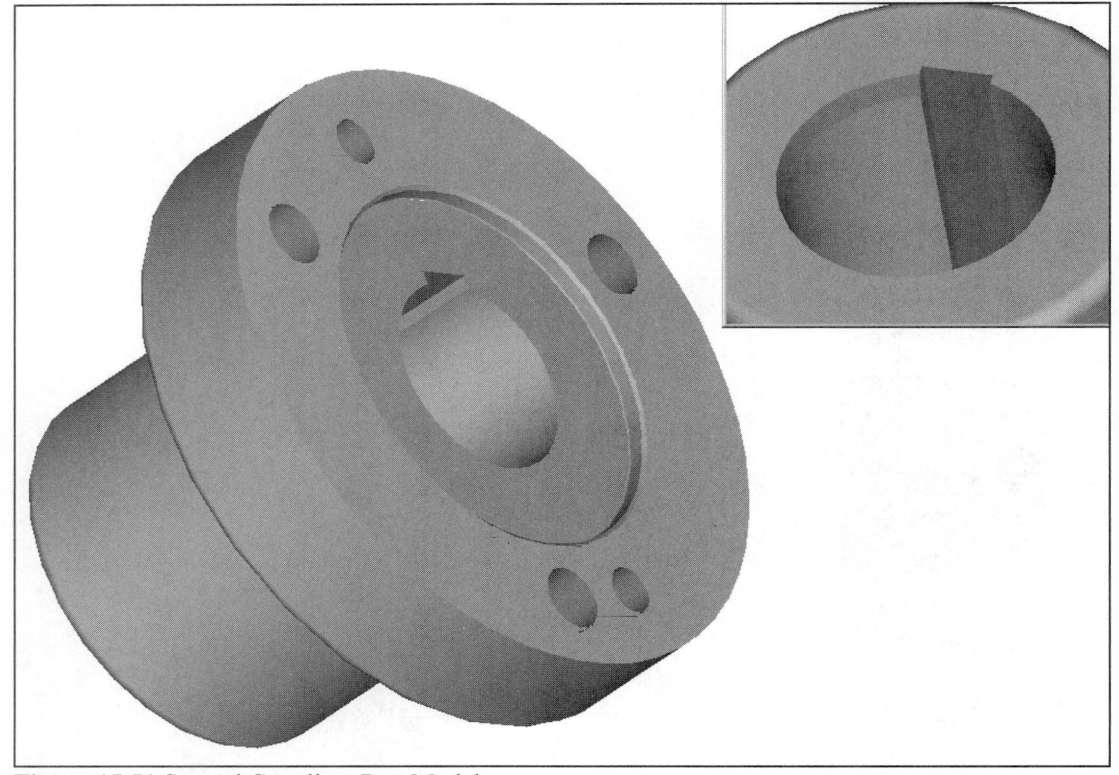

Figure 15.51 Second Coupling, Part Model

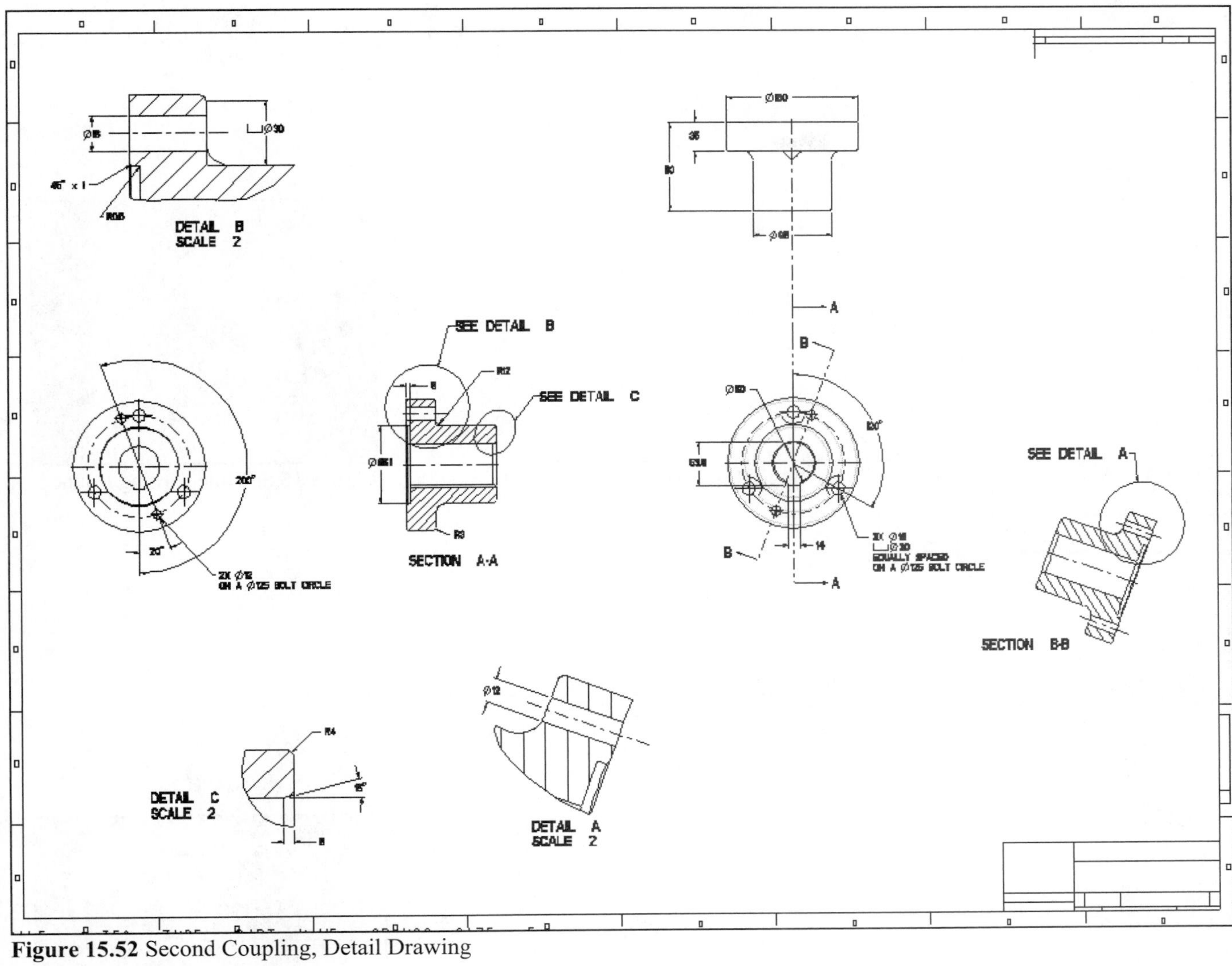

Figure 15.52 Second Coupling, Detail Drawing

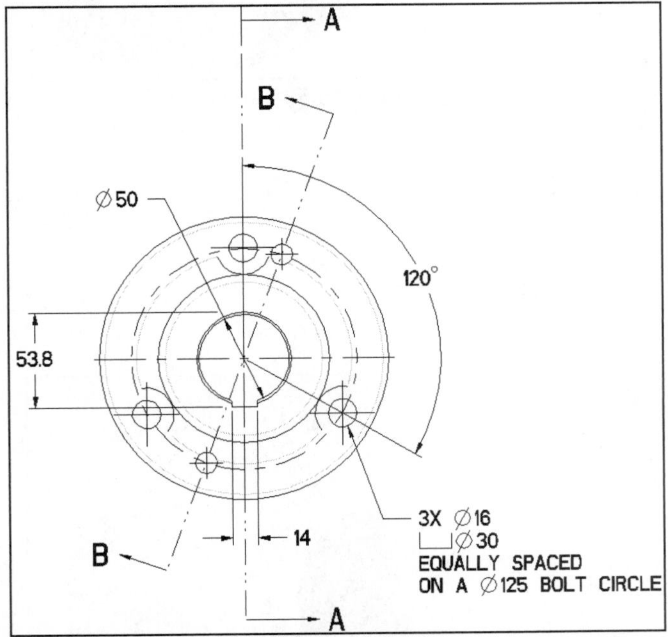

Figure 15.53 Second Coupling, Detail Drawing, Front View

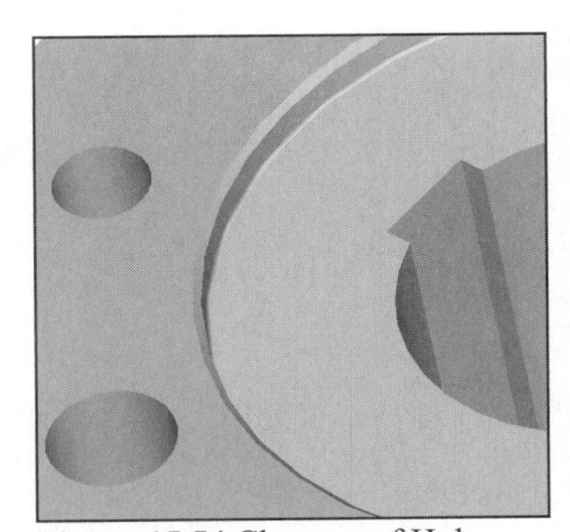

Figure 15.54 Close-up of Holes

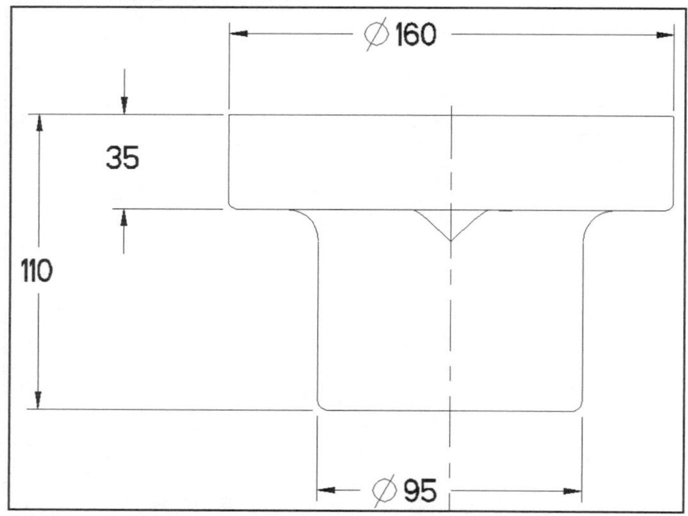

Figure 15.55 Second Coupling, Detail Drawing, Top View

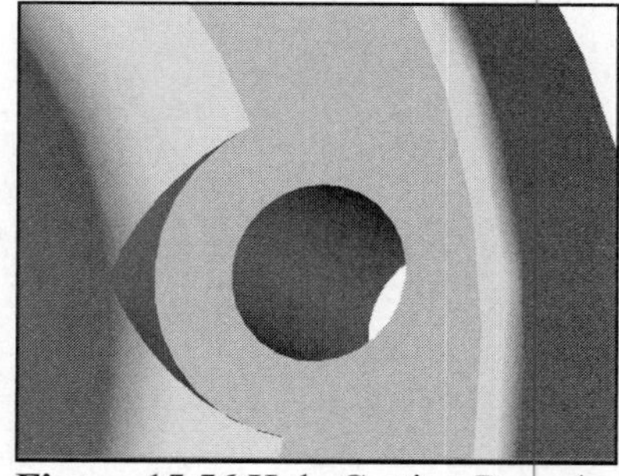
Figure 15.56 Hole Cutting Round

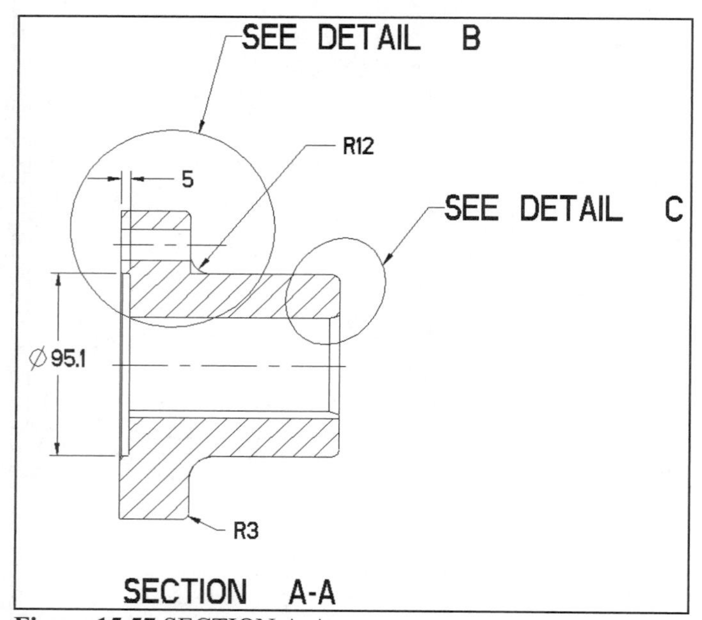

Figure 15.57 SECTION A-A

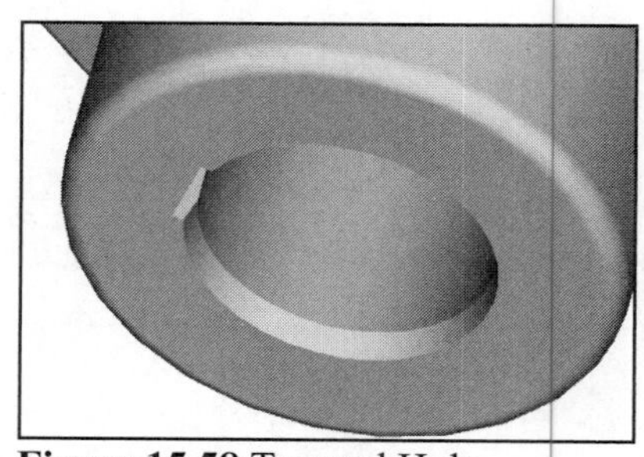
Figure 15.58 Tapered Hole

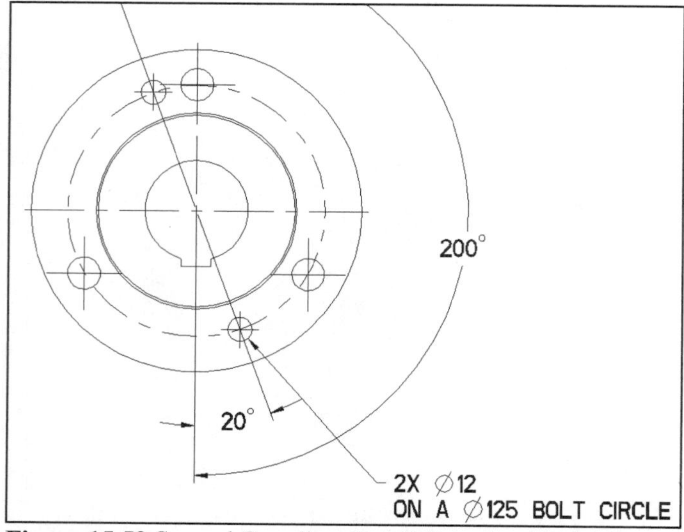

Figure 15.59 Second Coupling, Detail Drawing, Back View

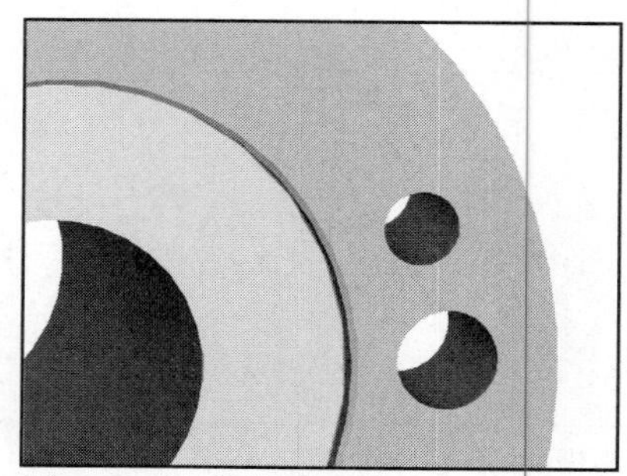
Figure 15.60 Holes from Bottom

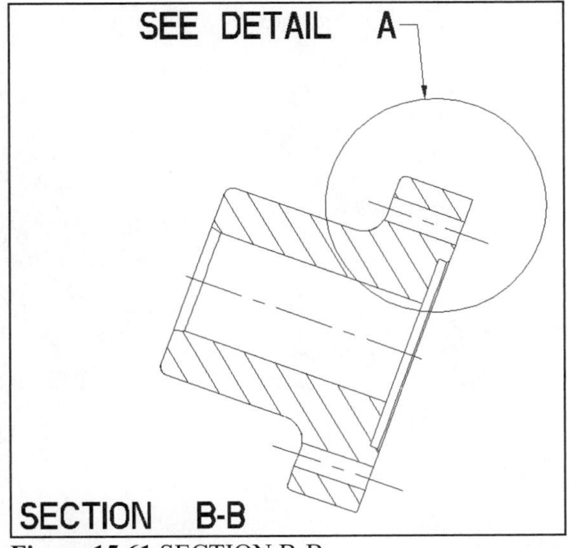

Figure 15.61 SECTION B-B

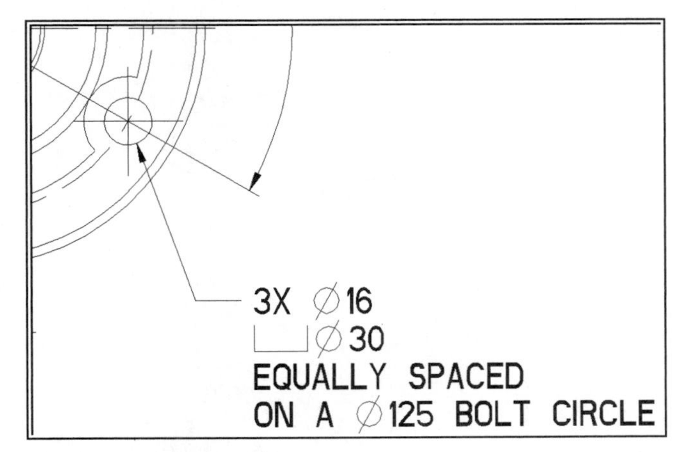

Figure 15.62 Hole Callout

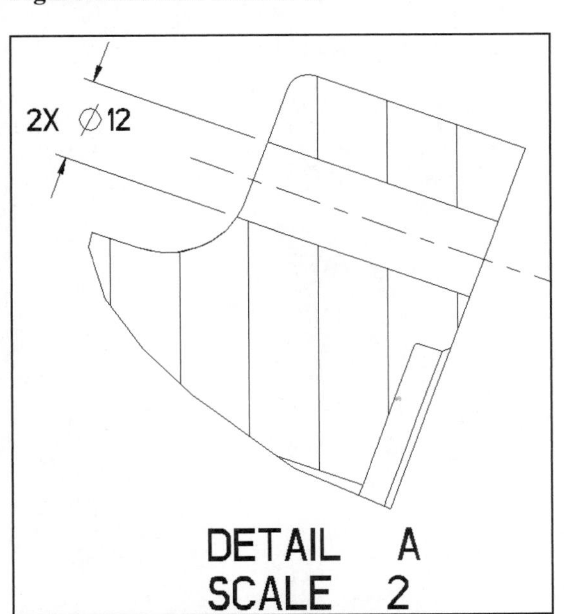

Figure 15.63 DETAIL A

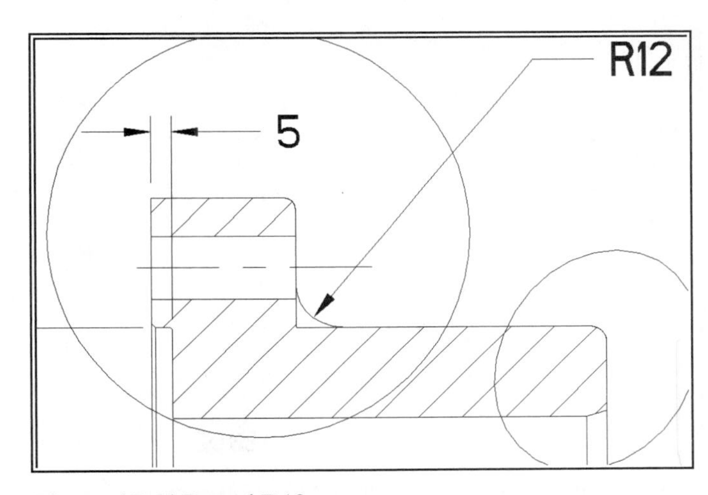

Figure 15.64 Round **R12**

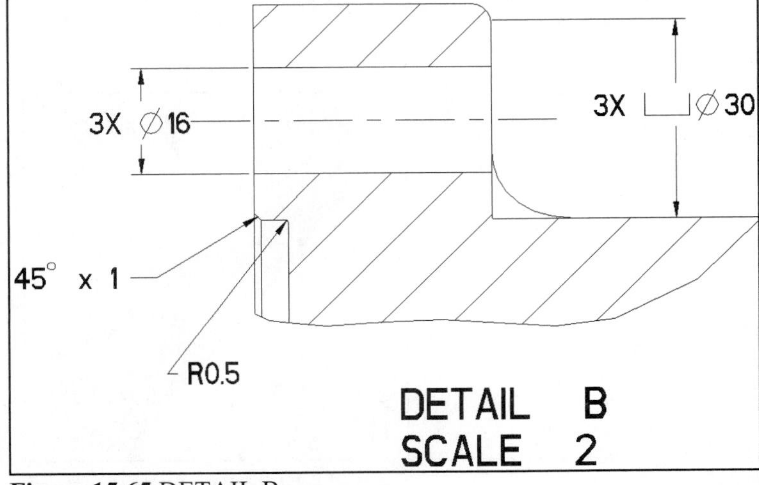

Figure 15.65 DETAIL B

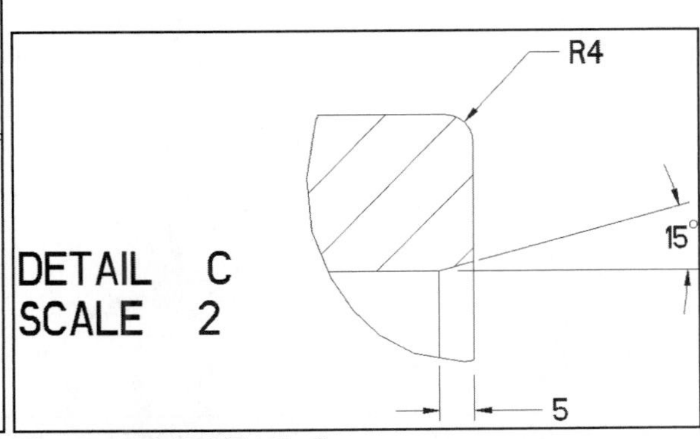

Figure 15.66 DETAIL C

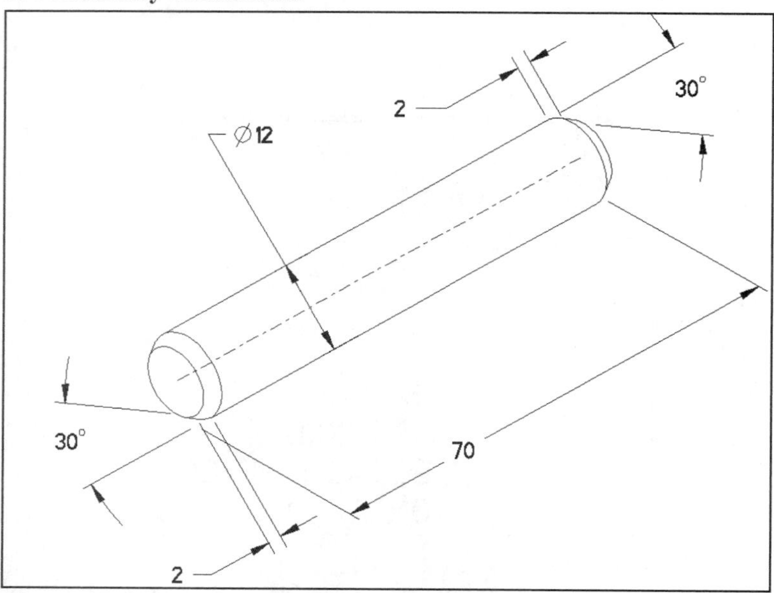

Figure 15.67 Dowel (model this component)

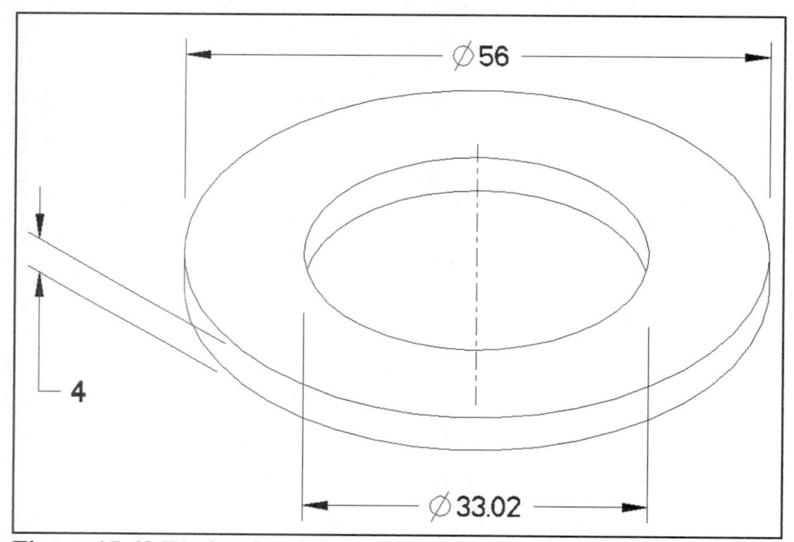

Figure 15.68 Washer (model this component)

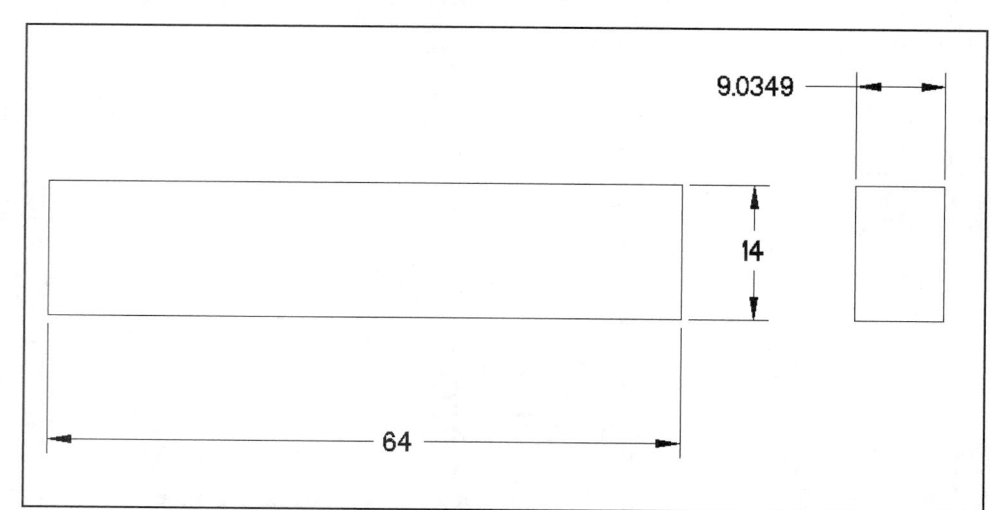

Figure 15.69(a) Key (model this component)

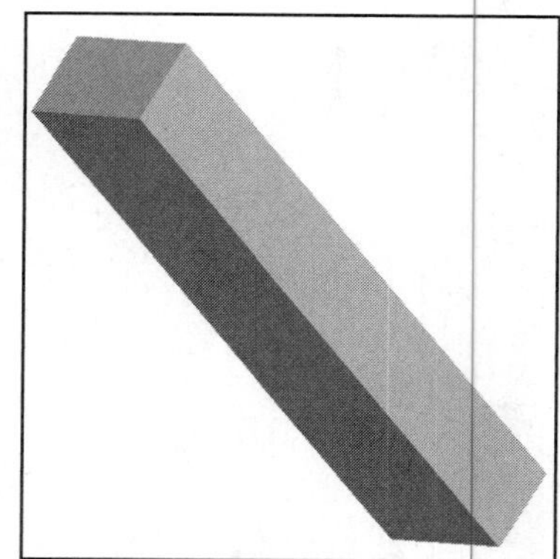

Figure 15.69(b) Key

Note: your directory structure may be slightly different. Consult your instructor or system administrator for the proper library path.

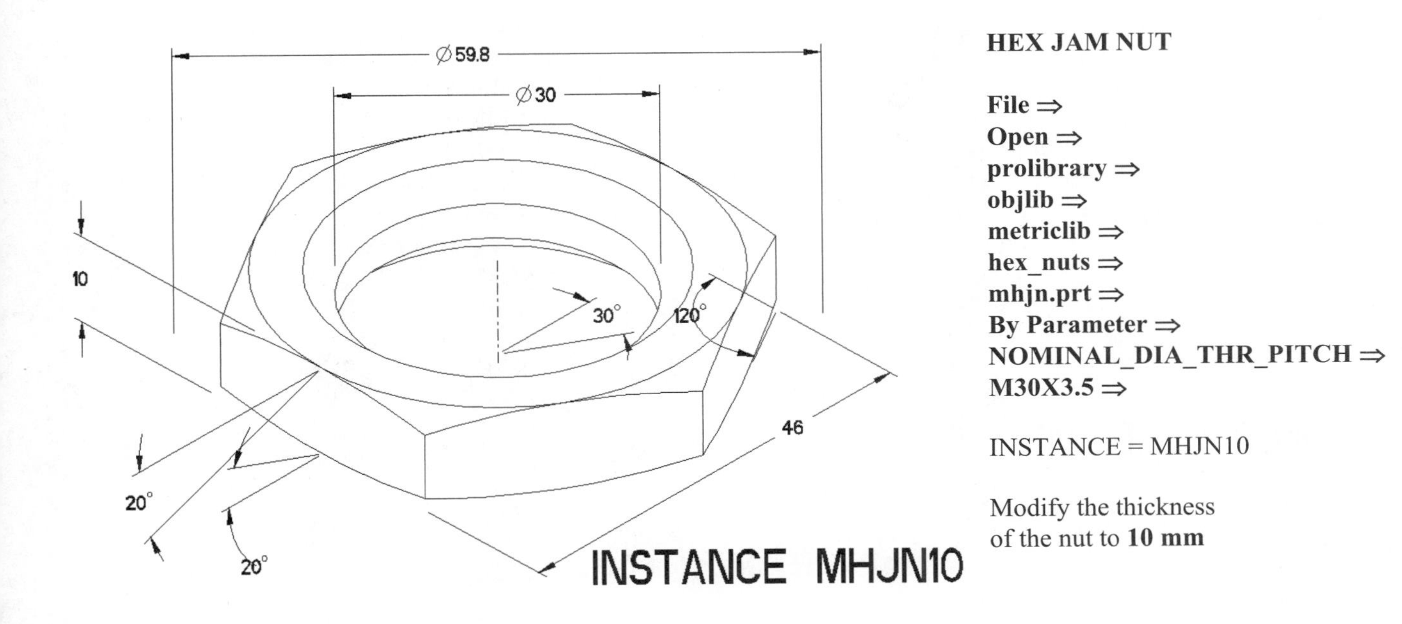

HEX JAM NUT

File ⇒
Open ⇒
prolibrary ⇒
objlib ⇒
metriclib ⇒
hex_nuts ⇒
mhjn.prt ⇒
By Parameter ⇒
NOMINAL_DIA_THR_PITCH ⇒
M30X3.5 ⇒

INSTANCE = MHJN10

Modify the thickness of the nut to **10 mm**

Figure 15.70 Hex Jam Nut

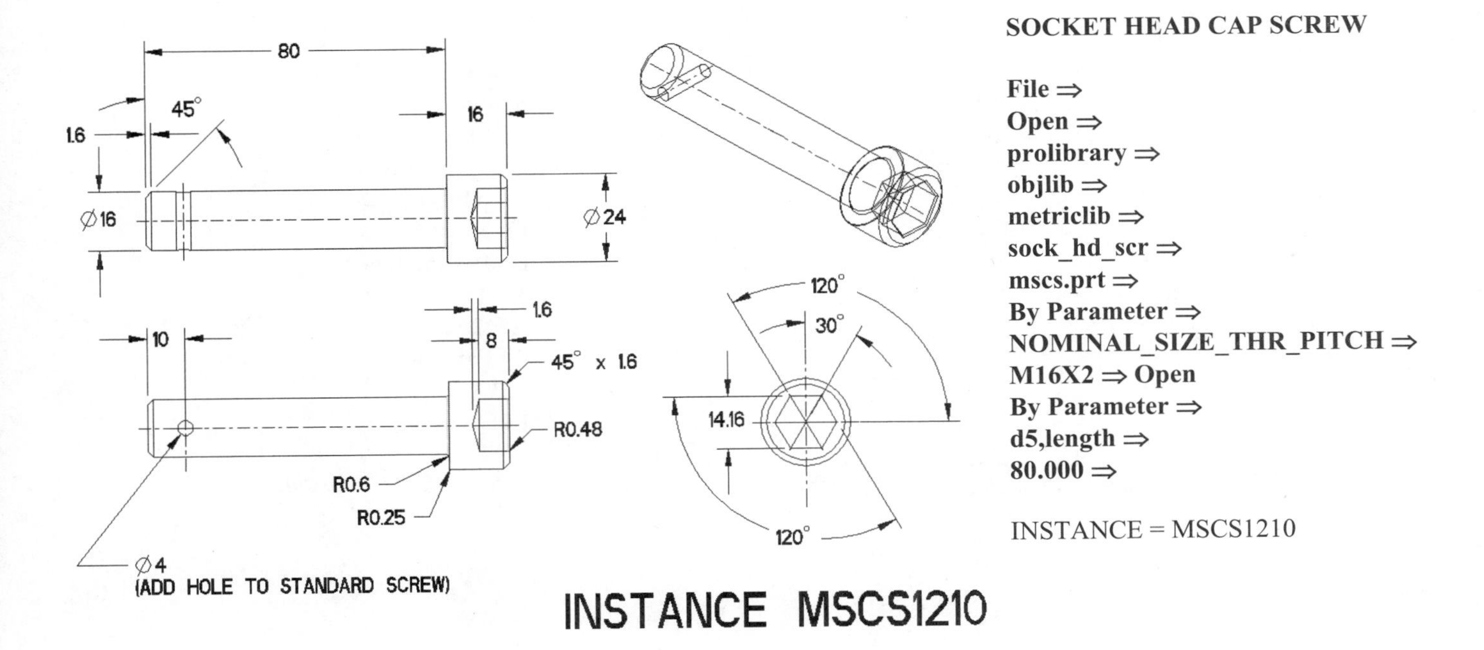

SOCKET HEAD CAP SCREW

File ⇒
Open ⇒
prolibrary ⇒
objlib ⇒
metriclib ⇒
sock_hd_scr ⇒
mscs.prt ⇒
By Parameter ⇒
NOMINAL_SIZE_THR_PITCH ⇒
M16X2 ⇒ Open
By Parameter ⇒
d5,length ⇒
80.000 ⇒

INSTANCE = MSCS1210

Figure 15.71 Socket Head Cap Screw

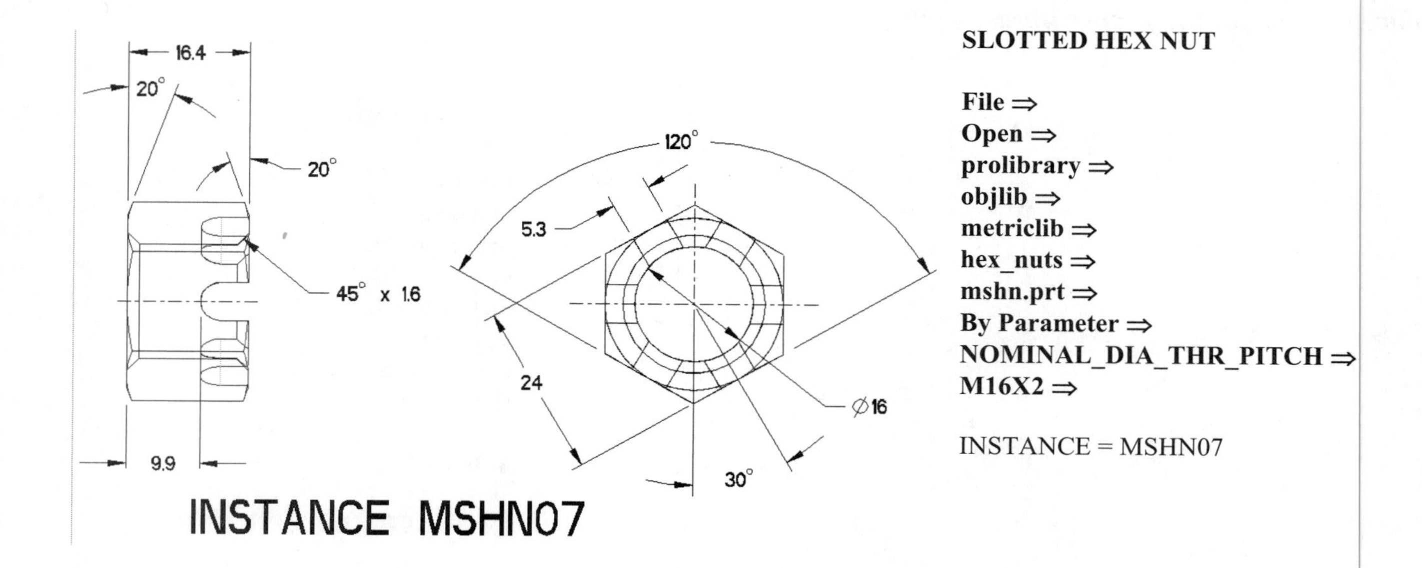

SLOTTED HEX NUT

File ⇒
Open ⇒
prolibrary ⇒
objlib ⇒
metriclib ⇒
hex_nuts ⇒
mshn.prt ⇒
By Parameter ⇒
NOMINAL_DIA_THR_PITCH ⇒
M16X2 ⇒

INSTANCE = MSHN07

Figure 15.72 Slotted Hex Nut

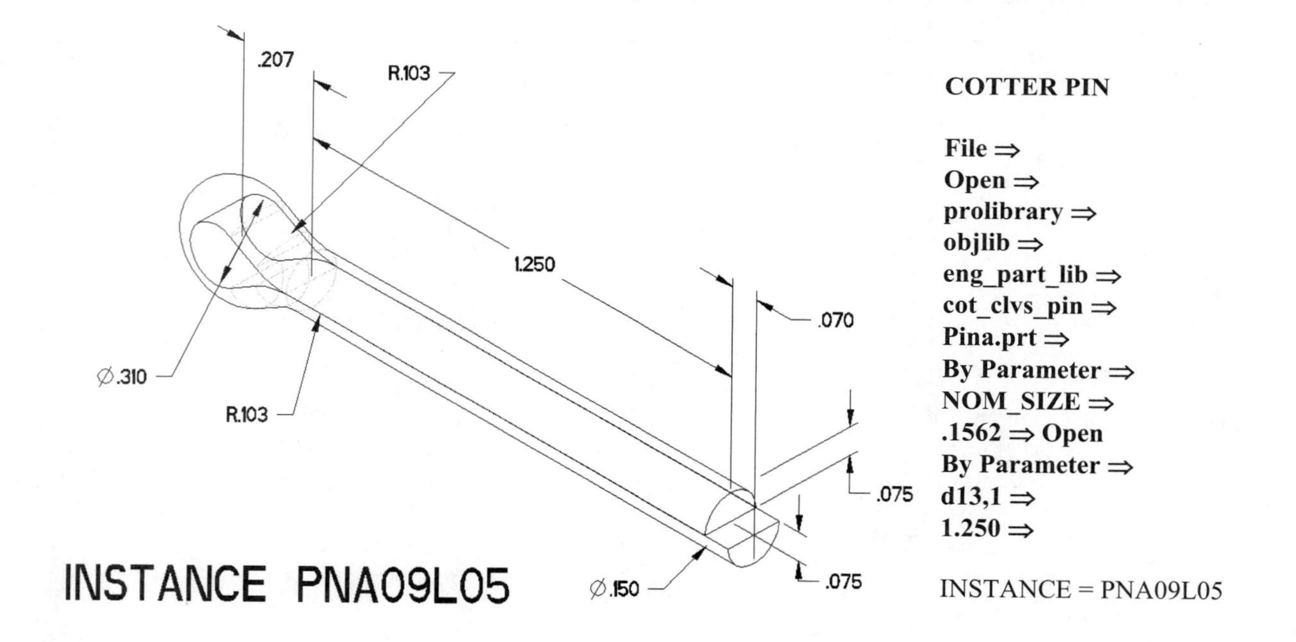

COTTER PIN

File ⇒
Open ⇒
prolibrary ⇒
objlib ⇒
eng_part_lib ⇒
cot_clvs_pin ⇒
Pina.prt ⇒
By Parameter ⇒
NOM_SIZE ⇒
.1562 ⇒ Open
By Parameter ⇒
d13,1 ⇒
1.250 ⇒

INSTANCE = PNA09L05

Figure 15.73 Cotter Pin

Lesson 16 Exploded Assemblies

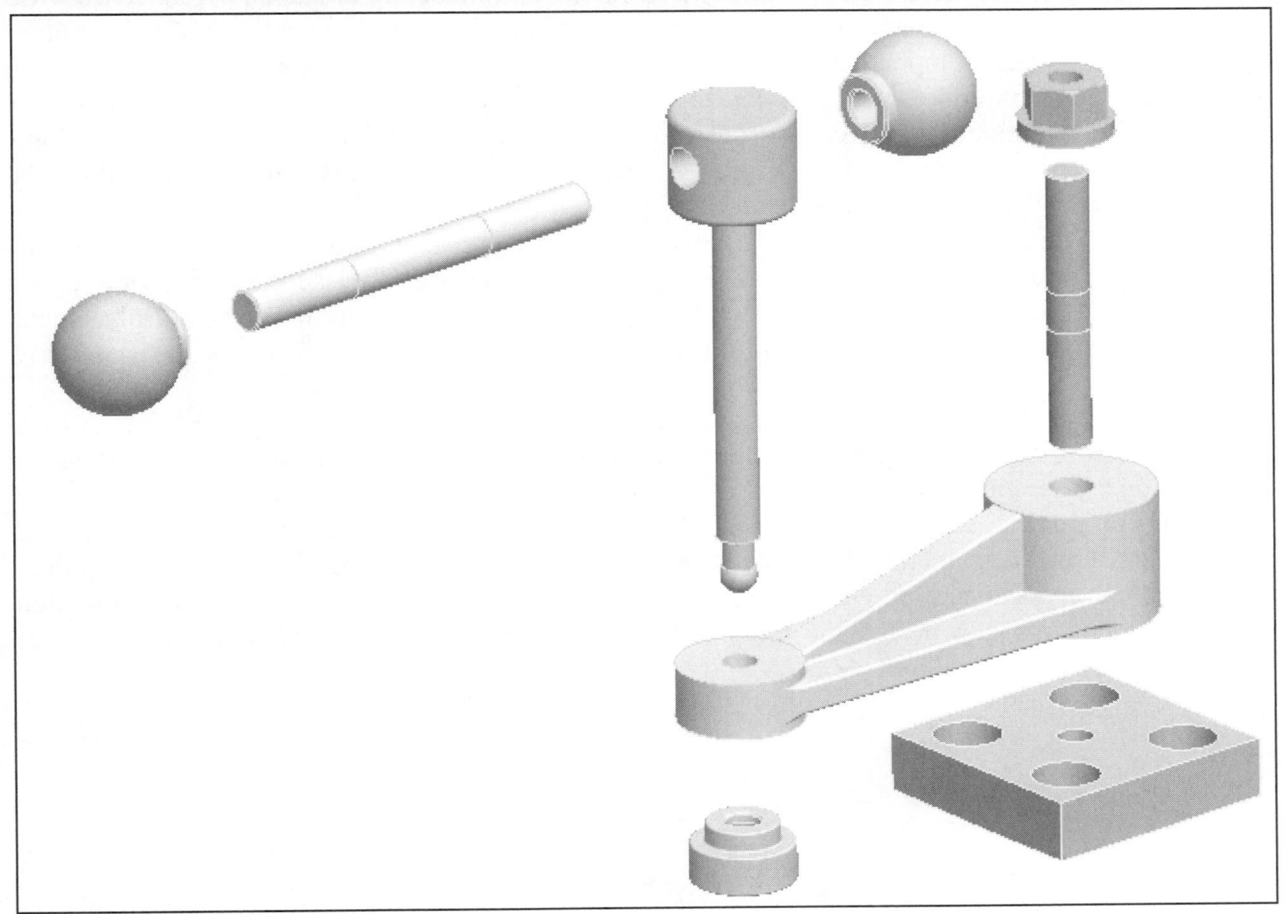

Figure 16.1 Exploded Swing Clamp Assembly

OBJECTIVES

- Create **Exploded Views**
- Create **Explode States**
- Utilize the **View Manager** to organize and control views states
- Create unique component **visibility settings (Style States)**
- **Move** and **Rotate** components in an assembly
- Add a **URL** to a **3D Note**
- Create **Perspective** views of the model
- View the **BOM** for an assembly

EXPLODED ASSEMBLIES

Pictorial illustrations, such as exploded views, are generated directly from the 3D model database (Fig. 16.1). The model can be displayed and oriented in any position. Each component in the assembly can have a different display type: wireframe, hidden line, no hidden, and shading. You can select and orient the component to provide the required view orientation to display the component from underneath or from any side or position. Perspective projections are made with selections from menus. The assembly can be spun around, reoriented, and even clipped to show the interior features. You have the choice of displaying all components and subassemblies or any combination of components in the design.

Creating Exploded Views

Using the **Explode State** option in the **View Manager**, you can automatically create an exploded view of an assembly (Fig. 16.2). Exploding an assembly affects only the display of the assembly; it does not alter true design distances between components. Explode states are created to define the exploded positions of all components. For each explode state, you can toggle the explode status of components, change the explode locations of components, and create and modify explode offset lines to show how explode components align when they are in their exploded positions. The Explode state Explode Position functionality is similar to the Package/Move functionality.

You can define multiple explode states for each assembly and then explode the assembly using any of these explode states at any time. You can also set an explode state for each drawing view of an assembly. Pro/E gives each component a default explode position determined by the placement constraints. By default, the reference component of the explode is the parent assembly (top-level assembly or subassembly).

To explode components, you use a drag-and-drop user interface similar to the Package/Move functionality. You select the motion reference and one or more components, and then drag the outlines to the desired positions. The component outlines drag along with the mouse cursor. You control the move options using a Preferences setting. Two types of explode instructions can be added to a set of components. The children components follow the parent component being exploded or they do not follow the parent component. Each explode instruction consists of a set of components, explode direction references, and dimensions that define the exploded position from the final (installed) position with respect to the explode direction references.

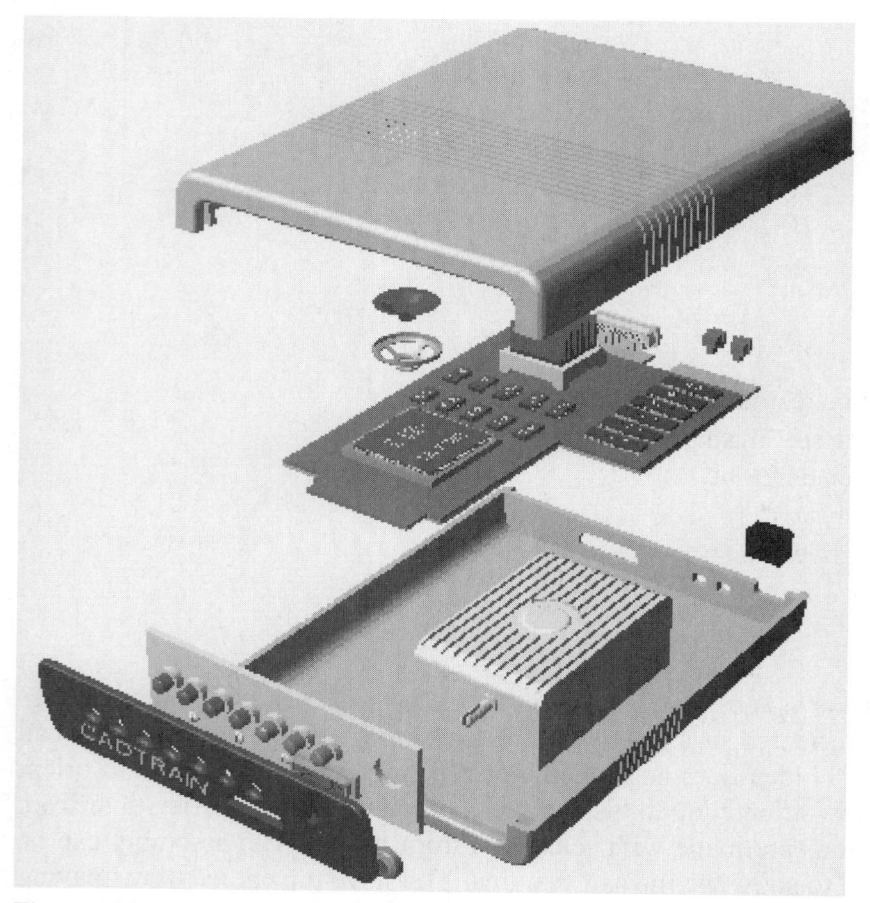

Figure 16.2 Exploded Assemblies (CADTRAIN, COAch for Pro/ENGINEER)

When using the explode functionality, keep in mind the following:

- You can select individual parts or entire subassemblies from the Model Tree or main window.
- If you explode a subassembly, in the context of a higher-level assembly, you can specify the explode state to use for each subassembly.
- You do not lose component explode information when you turn the status off. Pro/E retains the information so that the component has the same explode position if you turn the status back on.
- All assemblies have a default explode state called "Default Explode", which is the default explode state Pro/E creates from the component placement instructions.
- Multiple occurrences of the same subassembly can have different explode characteristics at a higher-level assembly.

Component Display (Style) States

Style accessed through using: View ⇒ View Manager [Fig. 16.3(a-b)], manages the display styles of an assembly. Wireframe, hidden line, no hidden, shaded, or blanked display styles can be assigned to each component. The components will be displayed according to their assigned display styles in the current style state (that is, blanked, shaded, drawn in hidden line color, and so on). The current setting, in the Environment menu, controls the display of unassigned components.

Figure 16.3(a) View Manager Style

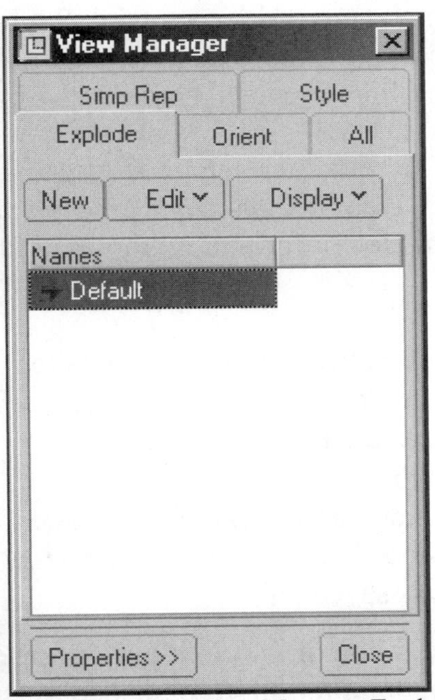

Figure 16.3(b) View Manager Explode

Components appear in the currently assigned style state (Fig. 16.4). The current setting is indicated in the display style column in the Model Tree window. Component display or style states can be modified without using the View Manager. You can select desired models from the graphics window, Model Tree, or search tool, and then use the View ⇒ Display Style commands to assign a display style (wireframe, shaded, and so forth) to the selected models. The Style representation is temporarily changed. These temporary changes can then be stored to a new style state, or updated to an existing style state. You can also define default style states. If the default style state is updated to reflect changes different from that of the master style state, then that default style state will be reflected each time the model is retrieved.

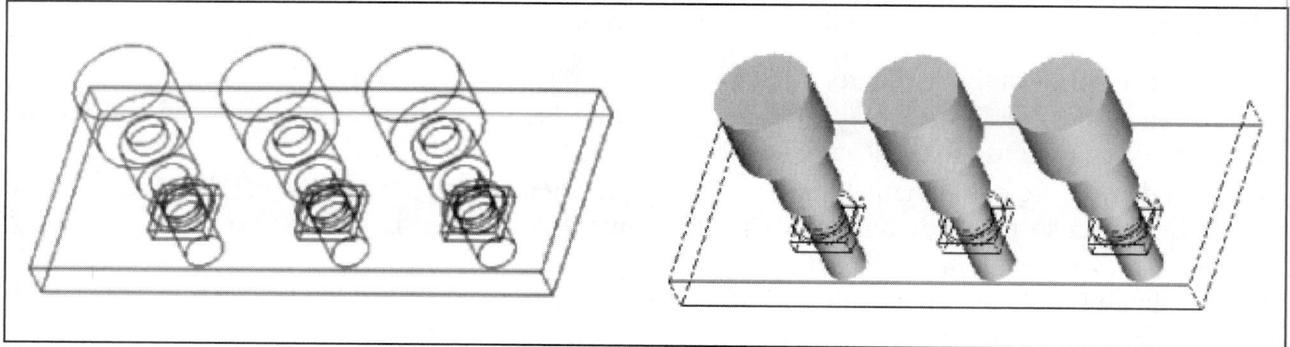

Figure 16.4 Style States

Types of Representations

The main types of simplified representations: **master**, **geometry**, **graphics**, and **symbolic**, designate which representation appears, using the commands in the View Manager dialog box.

Graphics and geometry representations speed up the retrieval process of large assemblies. All simplified representations provide access to components in the assembly and are based upon the Master Representation.

You cannot modify a feature in a graphics representation, but you can do so in a geometry representation.

Assembly features are displayed when you retrieve a model. Subtractive assembly features such as cuts and holes are represented in graphics and geometry representations, making it possible to use these simplified representations for performance improvement while still displaying on screen a completely accurate geometric model.

You can access model information for graphics and geometry representations of part models from the Information menu and from the Model Tree. Because part graphics and geometry representations do not contain feature history of the part model, information for individual features of the part is not accessible from these representations.

- The *Master Representation* always reflects the fully detailed assembly, including all of its members. The Model Tree lists all components in the Master Representation of the assembly, and indicates whether they are included, excluded, or substituted.
- The *Graphics Representation* contains information for display only and allows you to browse through a large assembly quickly. You cannot modify or reference graphics representations. The type of graphic display available depends on the setting of the *save_model_display* configuration option, the last time the assembly was saved:
    - ***wireframe*** (default) The wireframe of the components appear.
    - ***shading_low***, ***shading_med***, ***shading_high*** A shaded version of the components appears. The different levels indicate the density of the triangles used for shading.
    - ***shading_lod*** The level of detail depends on the setting in the View Performance dialog box. To access the View Performance dialog box, click View ⇒ Display Settings ⇒ Performance.

While in a simplified representation, Pro/E applies changes to an assembly, such as creation or assembly of new components, to the Master Representation. It reflects them in all of the simplified representations (Pro/PROGRAM processing also affects the Master Representation). It applies all suppressing and resuming of components to the Master Representation. However, it applies the actions of a simplified representation only to currently resumed members, that is, to members that are present in the BOM of the Master Representation.

Lesson 16 STEPS

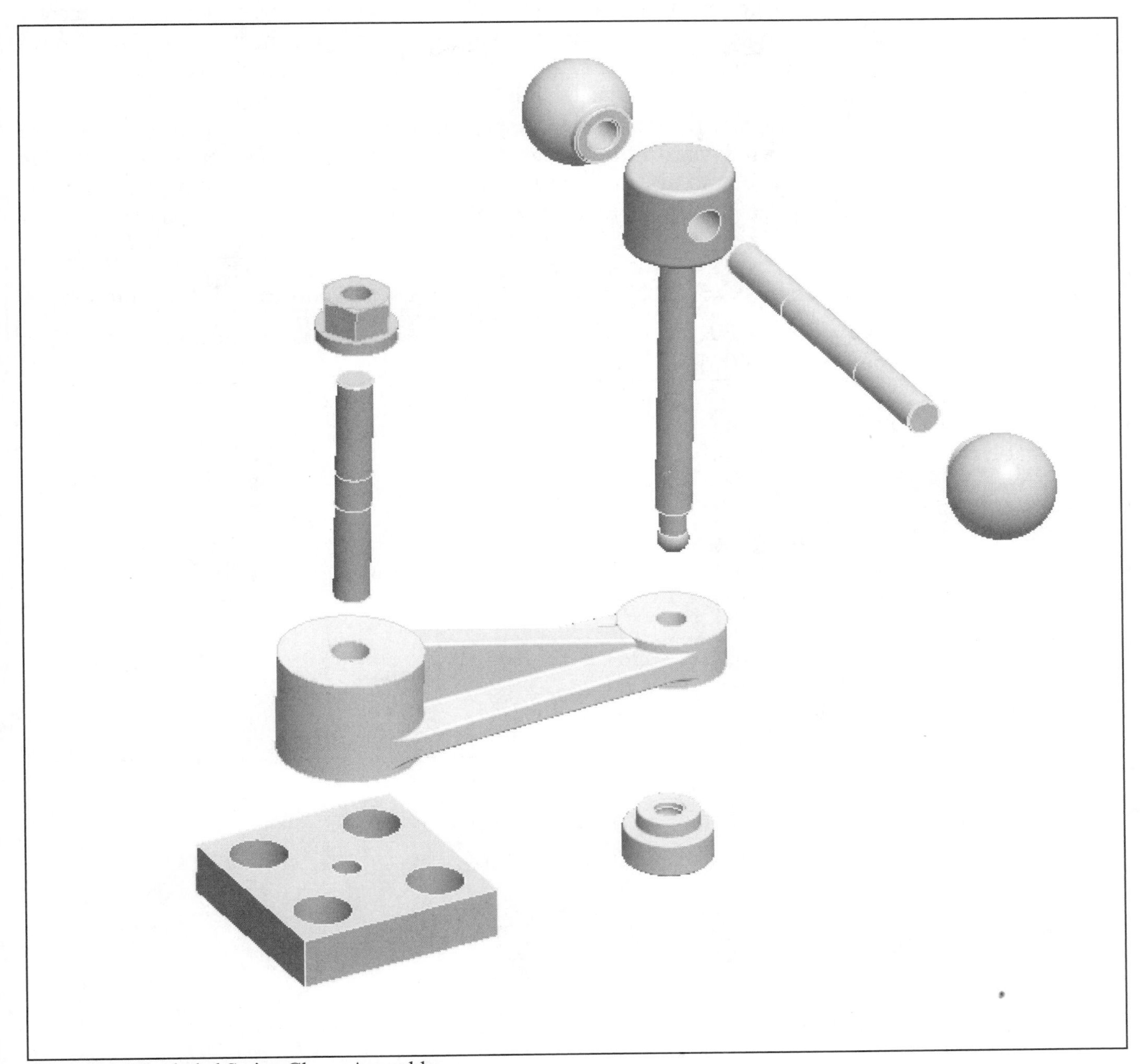

Figure 16.5 Exploded Swing Clamp Assembly

Exploded Swing Clamp Assembly

In this lesson, you will use the subassembly and assembly (Fig. 16.5) created in Lesson 15 to establish and save new views, exploded views, and views with component style states that differ from one another. The View Manager dialog box will be employed to control and organize a variety of different states.

You will also be required to move and rotate components of the assembly before cosmetically displaying the assembly in an exploded state. The creation and assembly of new components will not be required. A bill of materials will also be displayed using the Info command. A 3D Note with a URL will be added to the model as the last information feature.

Rotating Components of an Assembly

To rotate an existing component or set of components of an assembly (or subassembly), you select a coordinate system to use as a reference, pick one or more components, and give the rotation angle about a chosen axis of a coordinate system. The Clamp_Subassembly has a Clamp_Swivel, Clamp_Foot, Clamp_Stud5, and two Clamp_Ball components that can be rotated about the Clamp_Arm during normal operation of the assembly. You will rotate these components so that the Clamp_Stud5 and two Clamp_Ball components are perpendicular to the Clamp_Arm. This position looks better when you are displaying the assembly as exploded (and in its unexploded state). The components will be rotated in the subassembly, and the change will be propagated to the assembly.

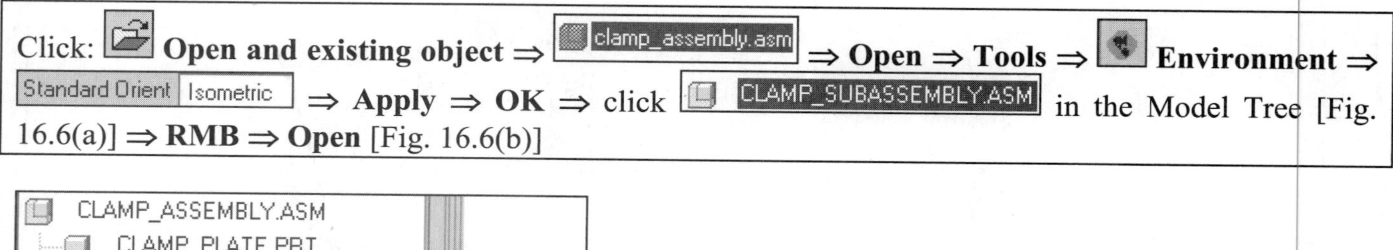

Click: **Open and existing object** ⇒ clamp_assembly.asm ⇒ **Open** ⇒ **Tools** ⇒ **Environment** ⇒ Standard Orient Isometric ⇒ **Apply** ⇒ **OK** ⇒ click CLAMP_SUBASSEMBLY.ASM in the Model Tree [Fig. 16.6(a)] ⇒ **RMB** ⇒ **Open** [Fig. 16.6(b)]

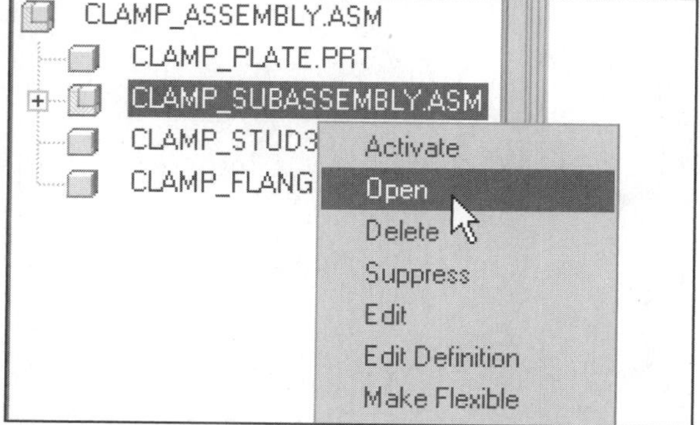

Figure 16.6(a) Open the Clamp Subassembly

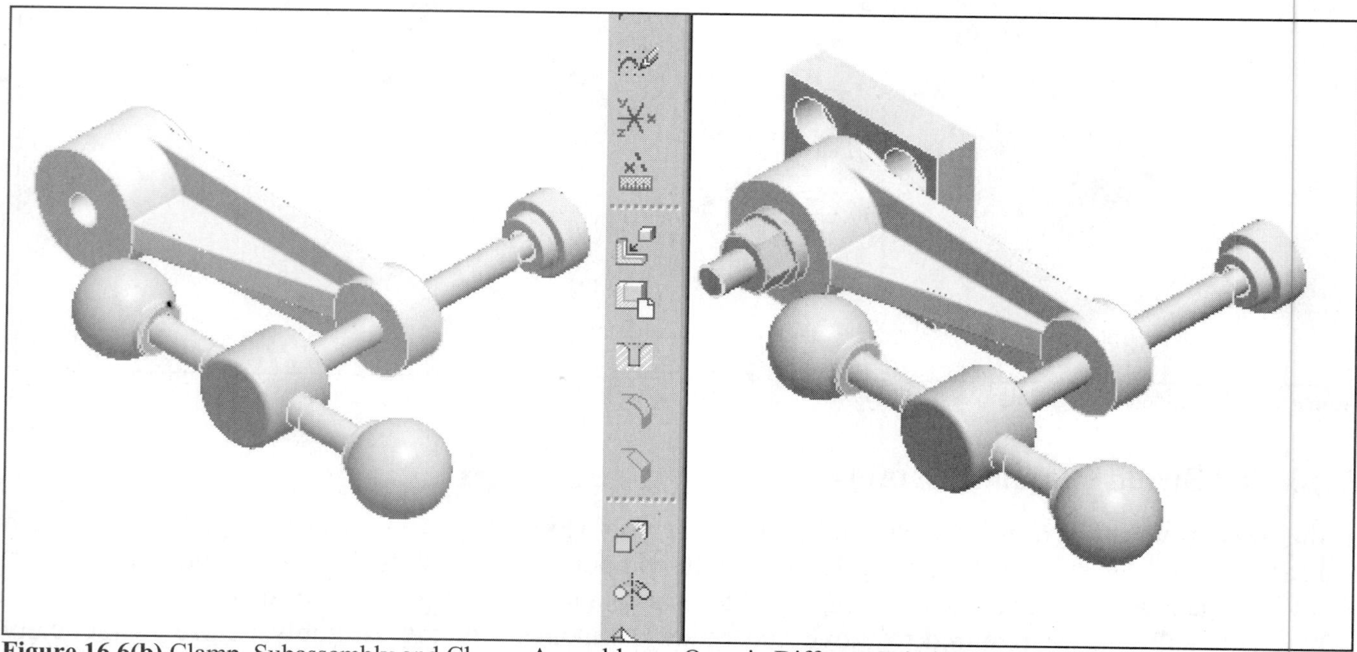

Figure 16.6(b) Clamp_Subassembly and Clamp_Assembly are Open in Different Windows

Click: **Window** in the Clamp_Subassembly menu bar ⇒ **Activate** to make sure you are working on the correct object ⇒ **Datum planes off** ⇒ **Redraw the current view** ⇒ **Edit** ⇒ **Component Operations** ⇒ ⇒ **Transform** ⇒ Select coordinate system: pick coordinate system **CLAMP_SWIVEL** [Fig. 16.7(a)] *or choose the coordinate system from the Model Tree* ⇒ Select components to move: pick the **CLAMP_SWIVEL.PRT** component [Fig. 16.7(b)] *you may also choose the component from the Model Tree* ⇒ **MMB** ⇒ **Rotate** ⇒ **Z Axis** *check your model, if you modeled it according to the instructions in Lesson Project 6, the Z axis of the coordinate system will run along the protrusion axis of the Clamp_Swivel.prt, if not, then select the appropriate axis on your component* ⇒ type **90** as the angle of rotation ⇒ **Enter** ⇒ **MMB** ⇒ **MMB** ⇒ **Regenerates Model** [Fig. 16.7(c)]

Only the Clamp_Swivel was selected, but the Clamp_Stud5, both Clamp_Balls, and the Clamp_Foot were rotated **90** degrees. These other components were children of the Clamp_Swivel and therefore rotated through the same **90** degree angle.

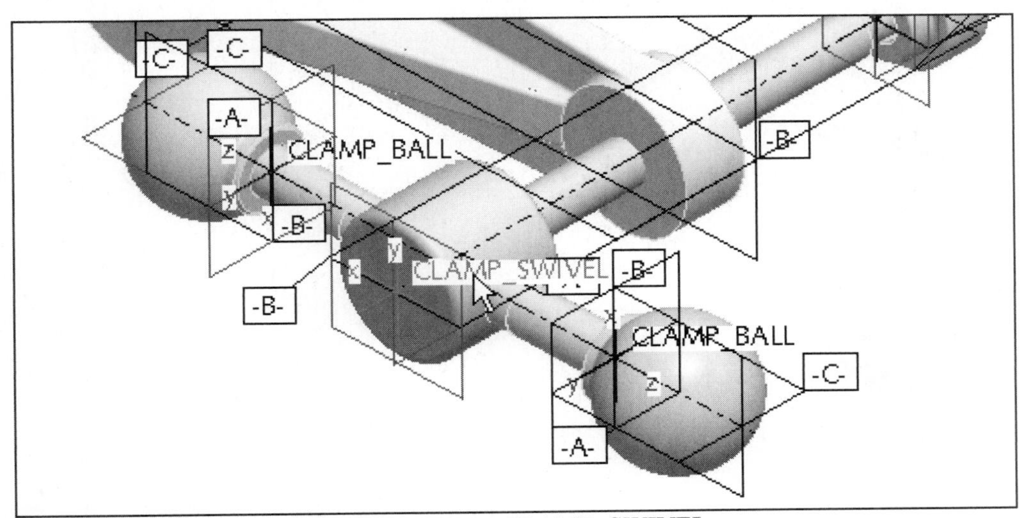

Figure 16.7(a) Select the Coordinate System CLAMP_SWIVEL

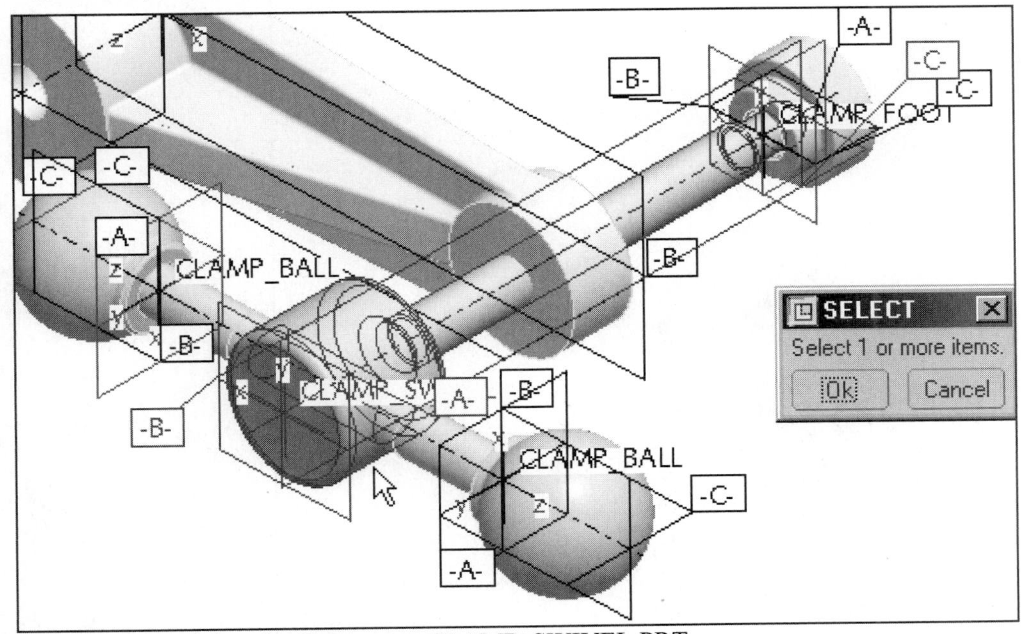

Figure 16.7(b) Select the Component CLAMP_SWIVEL.PRT

Click: **Window** ⇒ 1 CLAMP_ASSEMBLY.ASM ⇒ **Window** ⇒ Activate ⇒ **Refit object** [Fig. 16.7(d)]

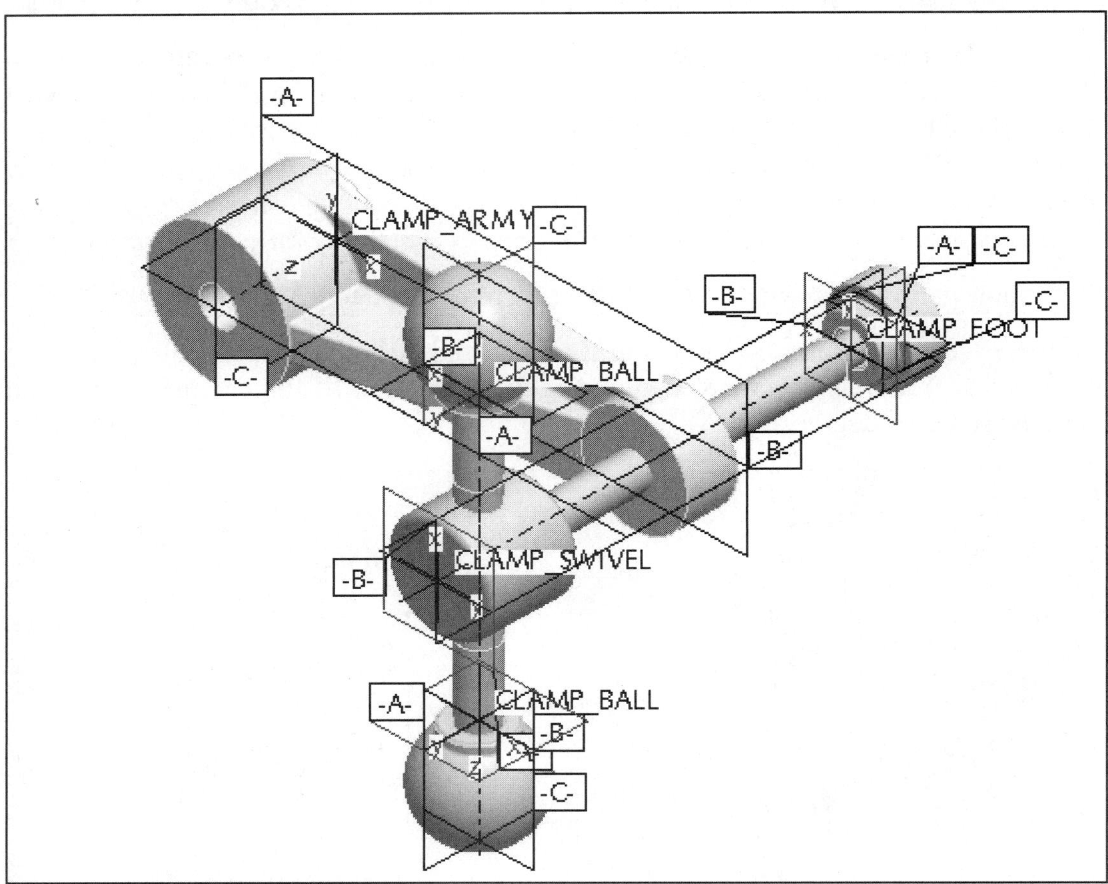

Figure 16.7(c) Rotated Swivel and Component Children

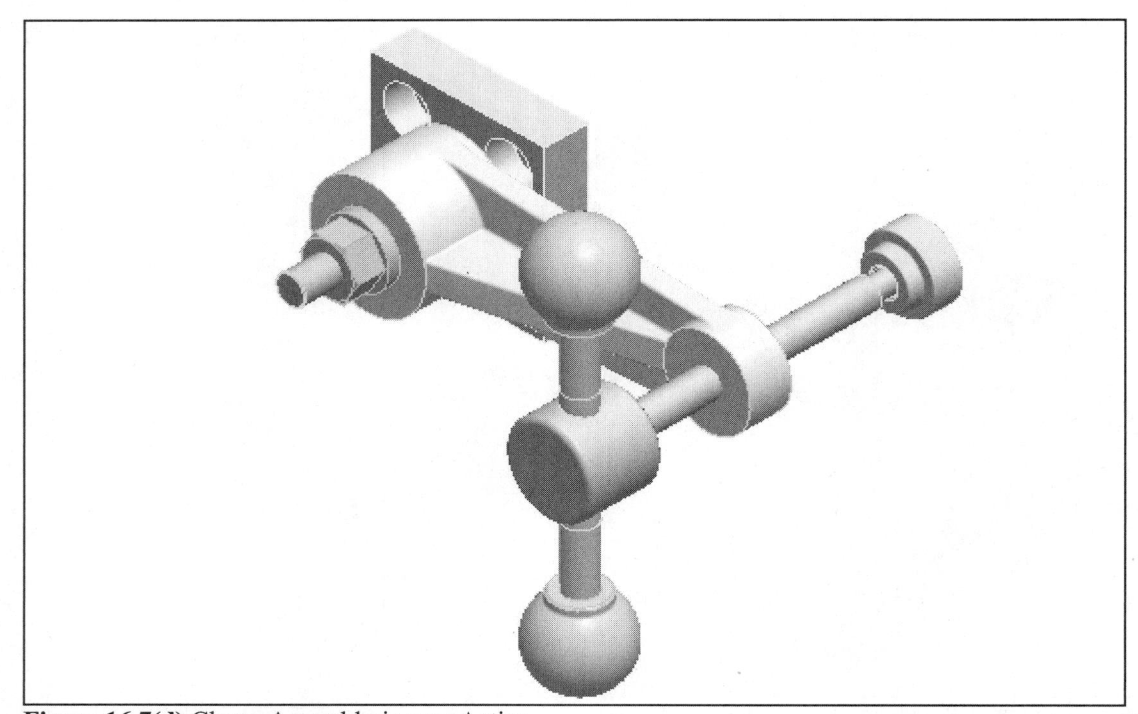
Figure 16.7(d) Clamp Assembly is now Active

Bill of Materials

The Bill of Materials (BOM) lists all parts and part parameters in the current assembly or assembly drawing. It can be displayed in HTML or text format and is separated into two parts: breakdown and summary. The Breakdown section lists what is contained in the current assembly or part. The Summary section lists the total quantity of each part included in the assembly, and is the list of all the parts needed to build the assembly from the part level. The default BOM HTML format provides hyperlinks to parts in the current assembly. It also provides hyperlinks that allow you to highlight, open or see additional information about the parts in the assembly. The BOM HTML breakdown section lists quantity, type, name (hyperlink), and three actions (highlight, information and open) about each member or sub-member of your assembly:

- **Quantity** Lists the number of components or drawings
- **Type** Lists the type of the assembly component (part or sub-assembly)
- **Name** Lists the assembly component and is hyperlinked to that item. Selecting this hyperlink highlights the component in the graphics window
- **Action** is divided into three areas:
    - **Highlight** Highlights the selected component in the assembly graphics window
    - **Information** Provides model information on the relevant component
    - **Open** Opens the component in another Pro/ENGINEER window

A bill of materials (BOM) can be seen by clicking: **Info** ⇒ **Bill of Materials** [Fig. 16.8(a)] ⇒ Top Level ⇒ check all Include options [Fig. 16.8(b)] ⇒ **OK** [Fig. 16.8(c)] ⇒ click quick sash

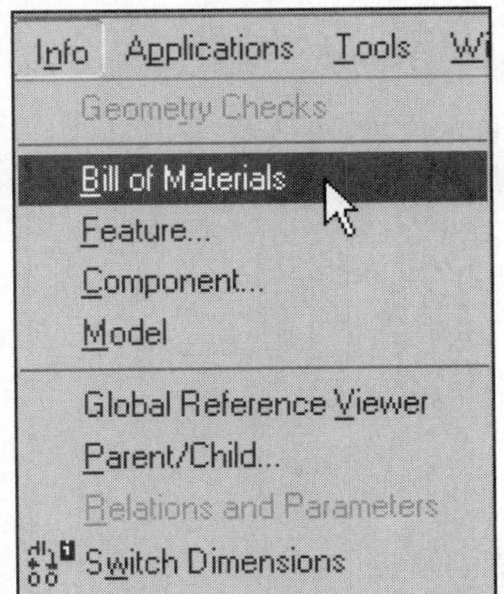

Figure 16.8(a) Info Bill of Materials

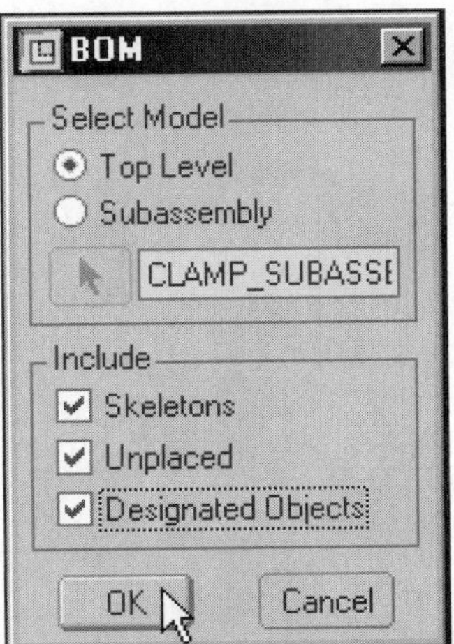

Figure 16.8(b) BOM Dialog Box

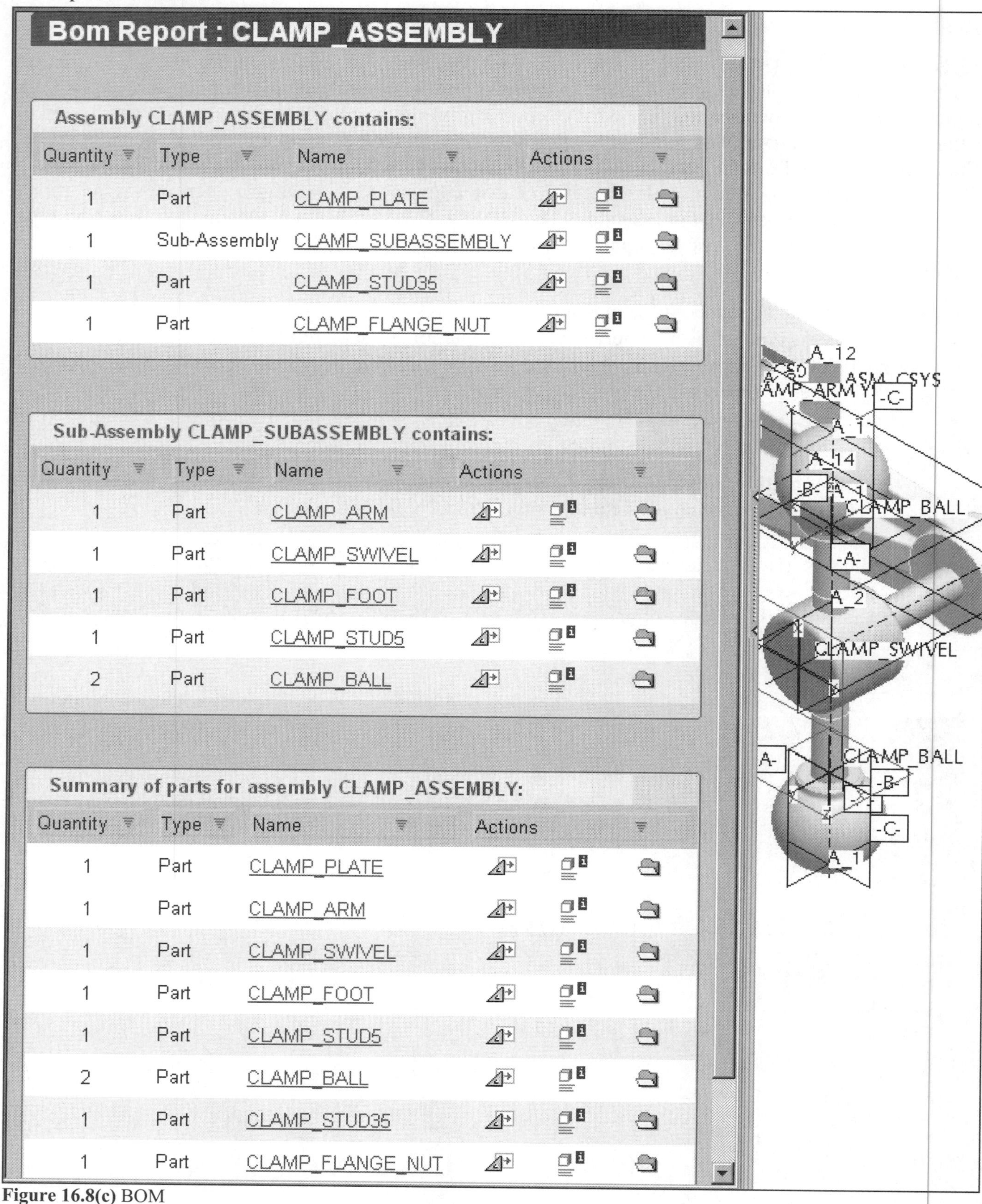

Bom Report : CLAMP_ASSEMBLY

Assembly CLAMP_ASSEMBLY contains:

Quantity	Type	Name	Actions
1	Part	CLAMP_PLATE	
1	Sub-Assembly	CLAMP_SUBASSEMBLY	
1	Part	CLAMP_STUD35	
1	Part	CLAMP_FLANGE_NUT	

Sub-Assembly CLAMP_SUBASSEMBLY contains:

Quantity	Type	Name	Actions
1	Part	CLAMP_ARM	
1	Part	CLAMP_SWIVEL	
1	Part	CLAMP_FOOT	
1	Part	CLAMP_STUD5	
2	Part	CLAMP_BALL	

Summary of parts for assembly CLAMP_ASSEMBLY:

Quantity	Type	Name	Actions
1	Part	CLAMP_PLATE	
1	Part	CLAMP_ARM	
1	Part	CLAMP_SWIVEL	
1	Part	CLAMP_FOOT	
1	Part	CLAMP_STUD5	
2	Part	CLAMP_BALL	
1	Part	CLAMP_STUD35	
1	Part	CLAMP_FLANGE_NUT	

Figure 16.8(c) BOM

URLs and Model Notes

Since the assembly you created is a standard clamp from CARR LANE Manufacturing Co.; you could go to CARR LANE's Website and see the assembly, order it (about $60.00 U.S.) and download a 3D IGES model. Create and attach a 3D Note to the assembly, identifying the manufacturer and Website.

Click: **Edit** ⇒ **Setup** ⇒ **Notes** ⇒ **New** ⇒ Name **CARR_LANE** ⇒ Text: type text lines [Fig. 16.9(a)] ⇒ URL **Hyperlink** ⇒ type **www.carrlane.com** [Fig. 16.9(b)] ⇒ **ScreenTip** ⇒ Screen tip text: **CARR LANE Manufacturing** [Fig. 16.9(c)] ⇒ **OK** ⇒ **OK** ⇒ **Place** ⇒ **No Leader** ⇒ **Standard** ⇒ **MMB** ⇒ Select LOCATION for note: [icon] pick near the assembly in graphics window ⇒ **OK** ⇒ **MMB** ⇒ **MMB** ⇒ **Settings** ⇒ **Tree Filters** ⇒ [Notes] (Fig. 16.10) ⇒ **OK** ⇒ **Tools** ⇒ [icon] **Environment** ⇒ [3D Notes] ⇒ **OK** (Fig. 16.11)

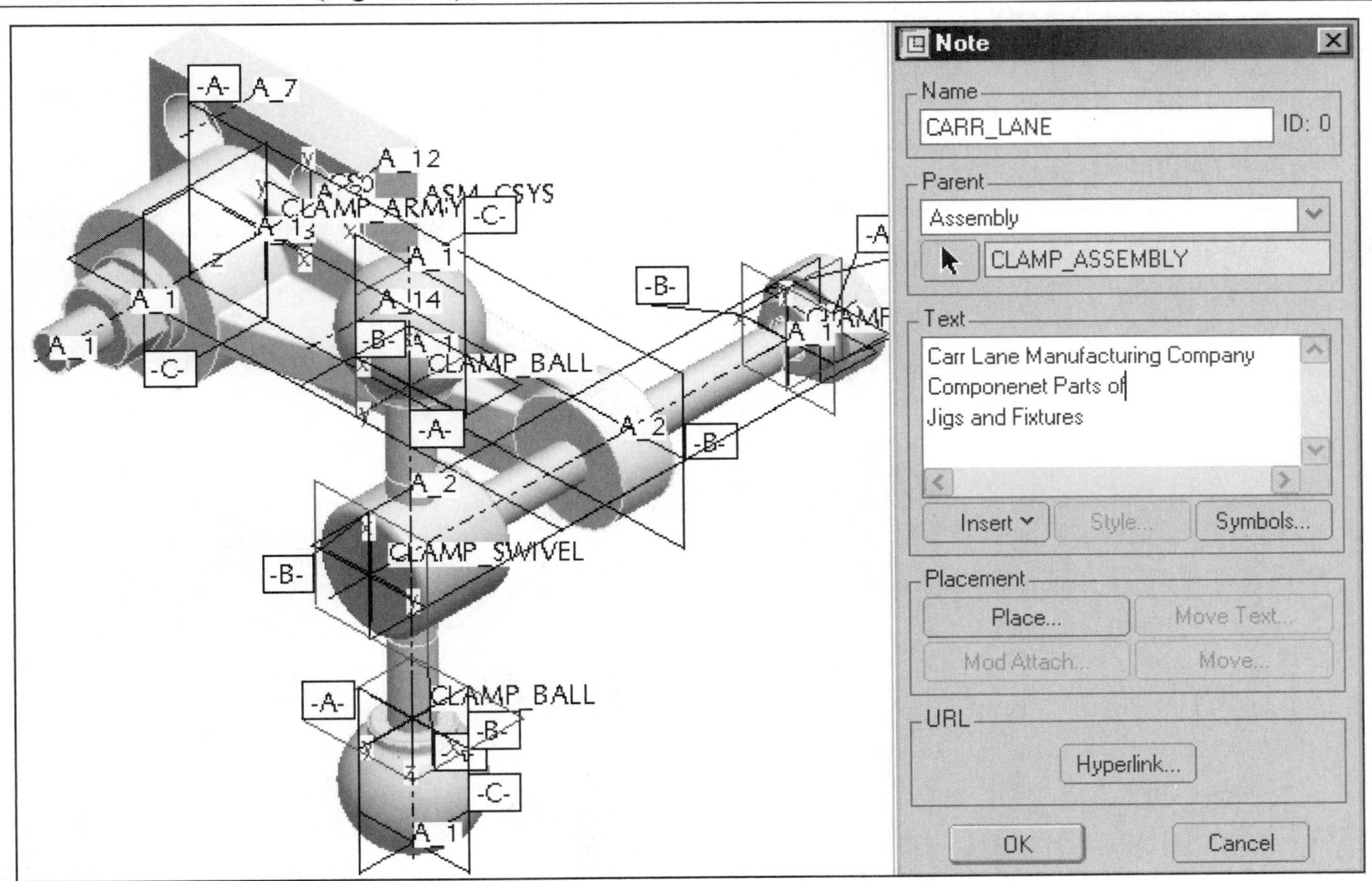

Figure 16.9(a) Note Dialog Box

Figure 16.9(b) Edit Hyperlink Dialog Box

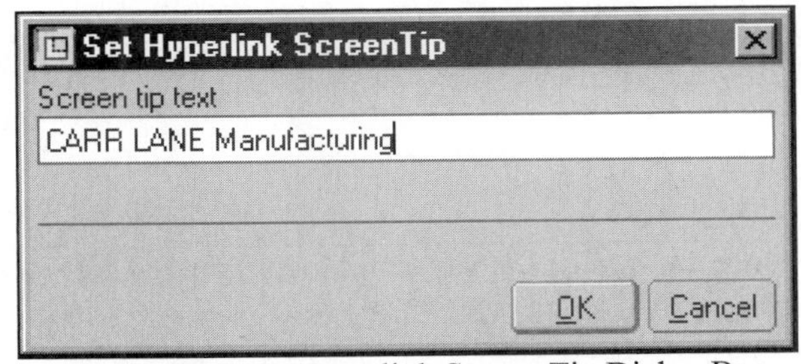

Figure 16.9(c) Set Hyperlink Screen Tip Dialog Box

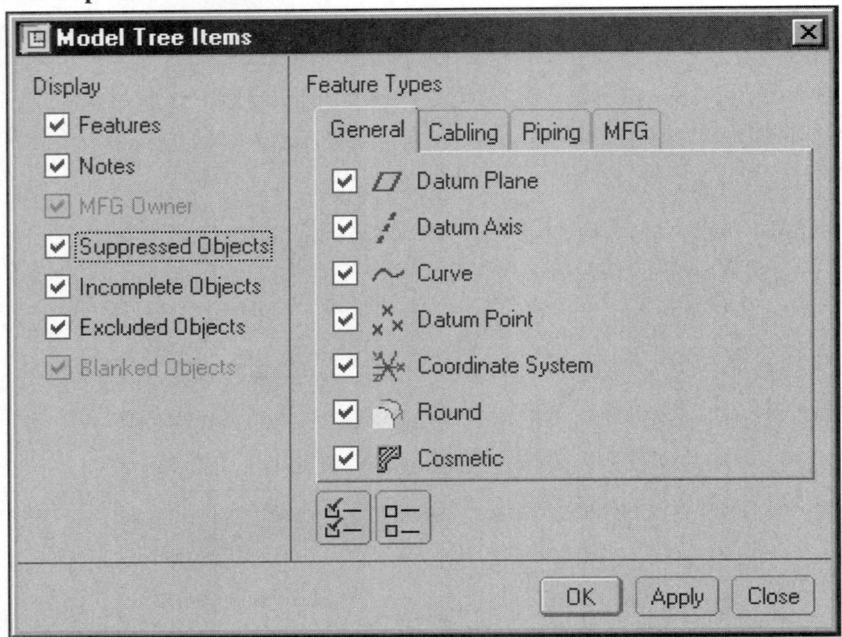

Figure 16.10 Notes Checked to Display

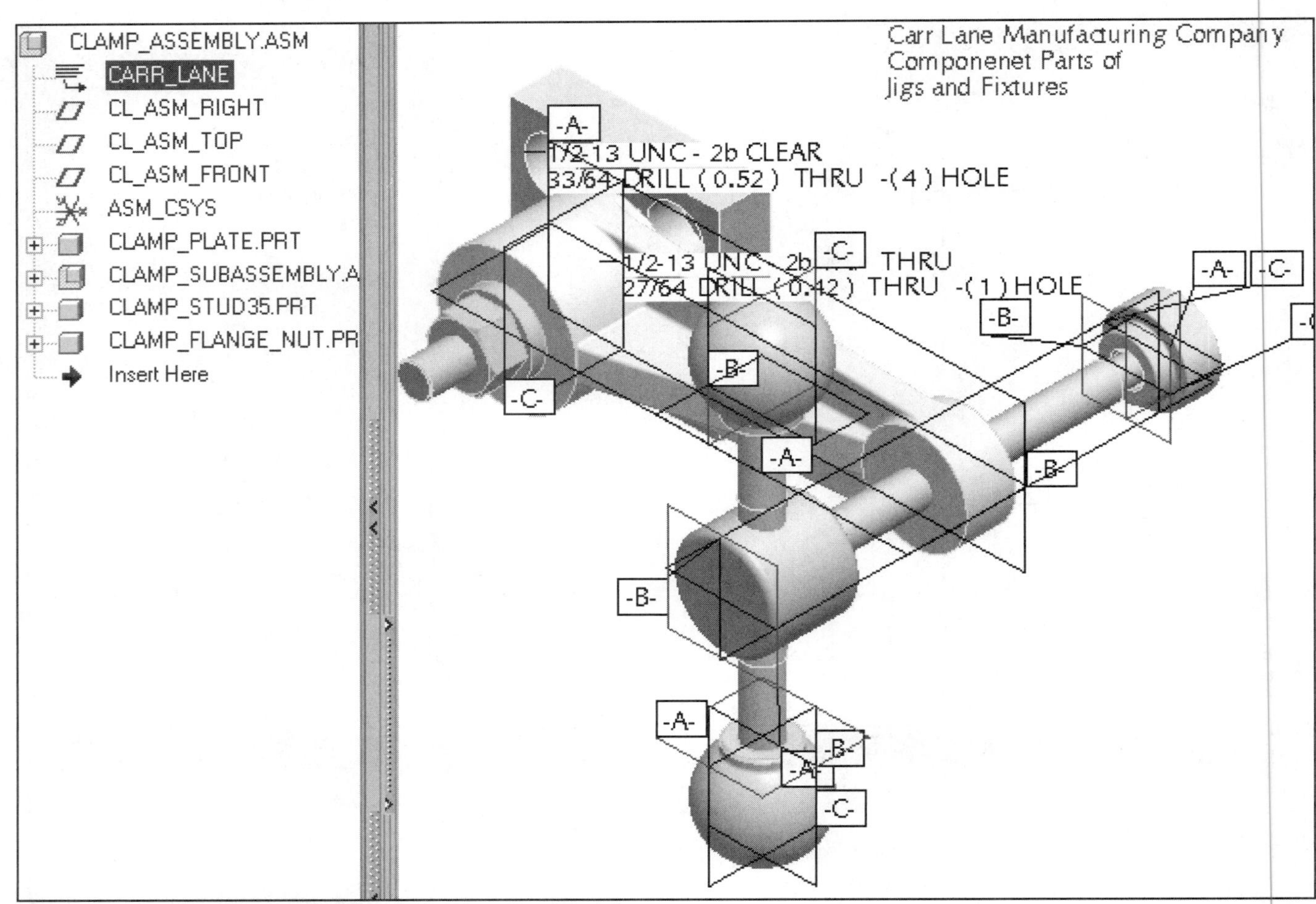

Figure 16.11 Note Displayed in Model Tree and on Model

To open a Hyperlink defined in a model note and to launch the embedded Web browser and go to a World Wide Web URL (Universal Resource Locator) associated with a model note, use one of three methods.

1. Click on the note in the graphics window or the Model Tree ⇒ **RMB** [Fig. 16.12(a)] the shortcut menu displays ⇒ **Open URL** [Fig. 16.12(b)] the associated URL opens in the Pro/E embedded browser ⇒ navigate the site to locate the Swing Clamp [Fig. 16.12(c-f)] ⇒ use the scroll bars to the see the drawing [Fig. 16.12(g-h)]) ⇒ close the Browser by clicking on the quick sash ⇒ **Save** ⇒ **MMB**
(Note that to pick the 3D Note in the graphics window, you may have to change the Selection Filter to ***Annotation****)*

or

2. Move the cursor over the hyperlinked note ⇒ press ***Ctrl*** *the cursor changes to a hand icon ⇒ click* ***LMB*** *the associated URL opens in the Pro/E embedded browser ⇒ navigate the site to the Swing Clamp*

or

3. Click: ***Edit*** *⇒* ***Setup*** *⇒* ***Notes*** *⇒* ***Modify*** *⇒ pick on the note in the Model Tree or from the screen to see the note before opening the URL ⇒* ***OK*** *to close the Note dialog box ⇒* ***Open URL*** *⇒ pick the 3D Note from the screen or the Model Tree ⇒ navigate the site to locate the Swing Clamp*

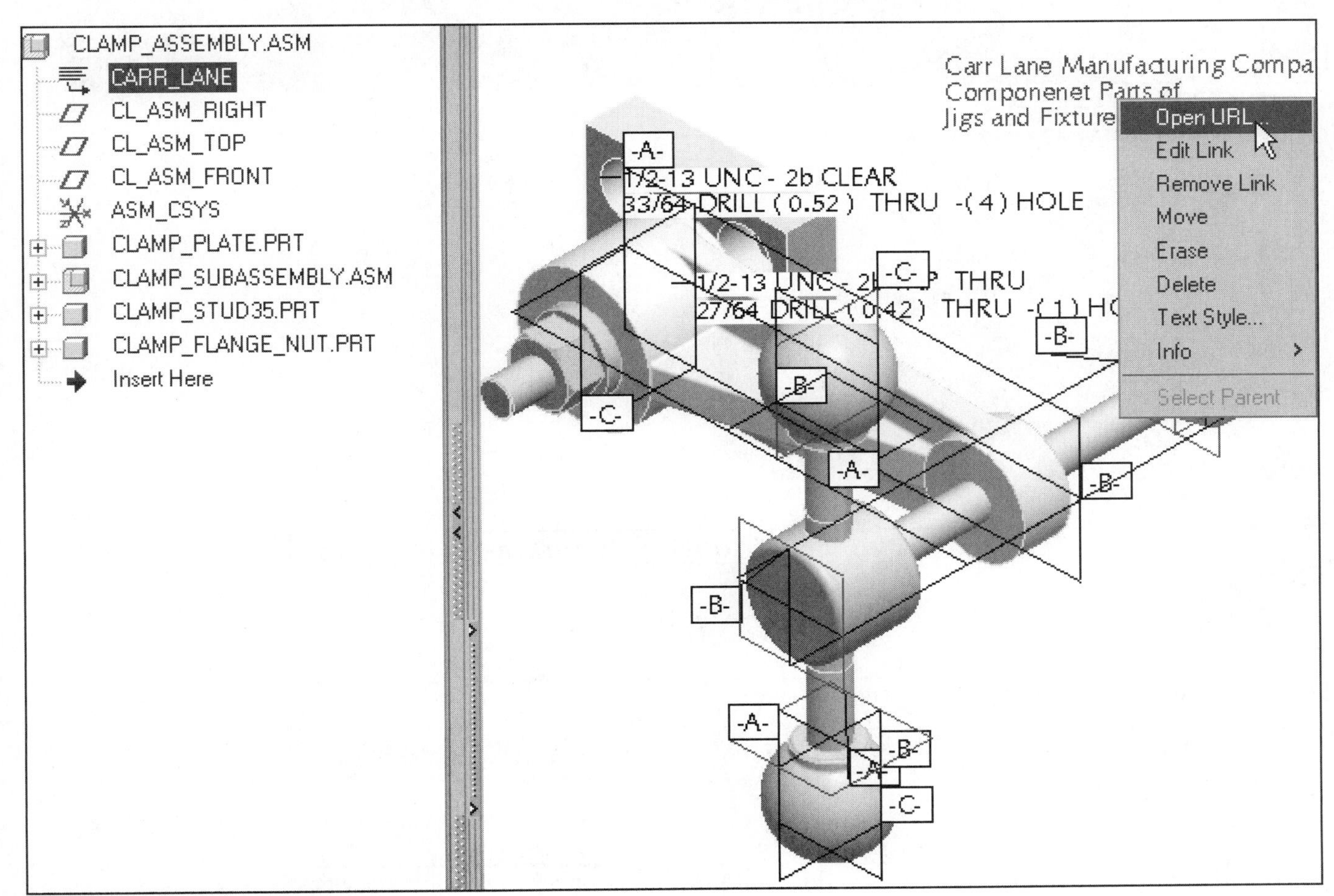

Figure 16.12(a) Open URL

Figure 16.12(b) Carr Lane Site

Figure 16.12(c) Click on Clamp Icon

Figure 16.12(d) Click on Tooling Components & Clamps

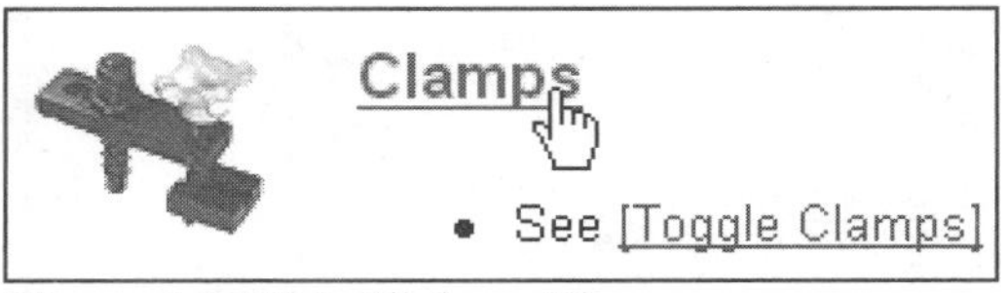

Figure 16.12(e) Click on Clamps

Figure 16.12(f) Click on Swing Clamp Assemblies

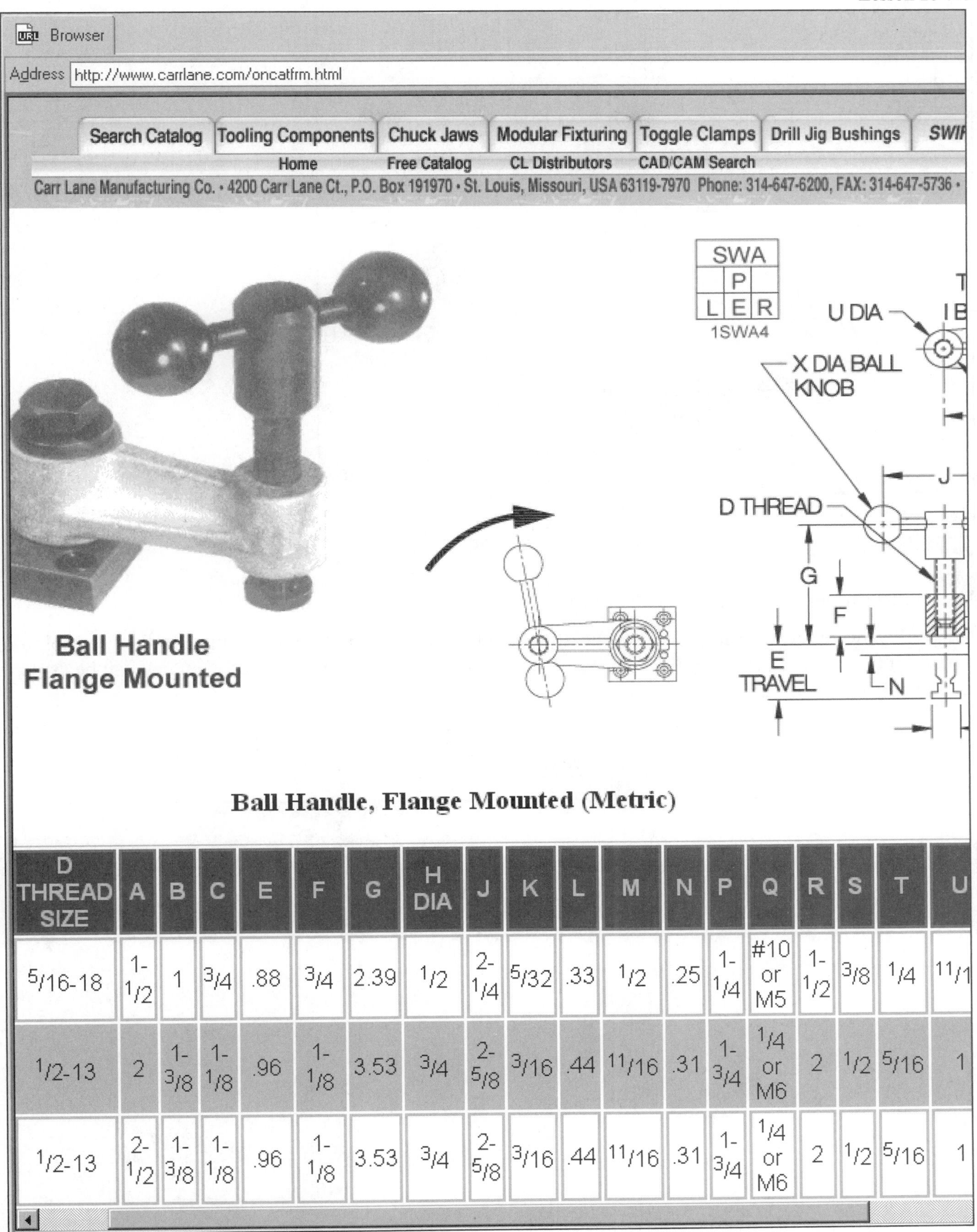

Ball Handle, Flange Mounted (Metric)

D THREAD SIZE	A	B	C	E	F	G	H DIA	J	K	L	M	N	P	Q	R	S	T	U
5/16-18	1-1/2	1	3/4	.88	3/4	2.39	1/2	2-1/4	5/32	.33	1/2	.25	1-1/4	#10 or M5	1-1/2	3/8	1/4	11/1
1/2-13	2	1-3/8	1-1/8	.96	1-1/8	3.53	3/4	2-5/8	3/16	.44	11/16	.31	1-3/4	1/4 or M6	2	1/2	5/16	1
1/2-13	2-1/2	1-3/8	1-1/8	.96	1-1/8	3.53	3/4	2-5/8	3/16	.44	11/16	.31	1-3/4	1/4 or M6	2	1/2	5/16	1

Figure 16.12(g) Ball Handle Flange Mounted Swing Clamp

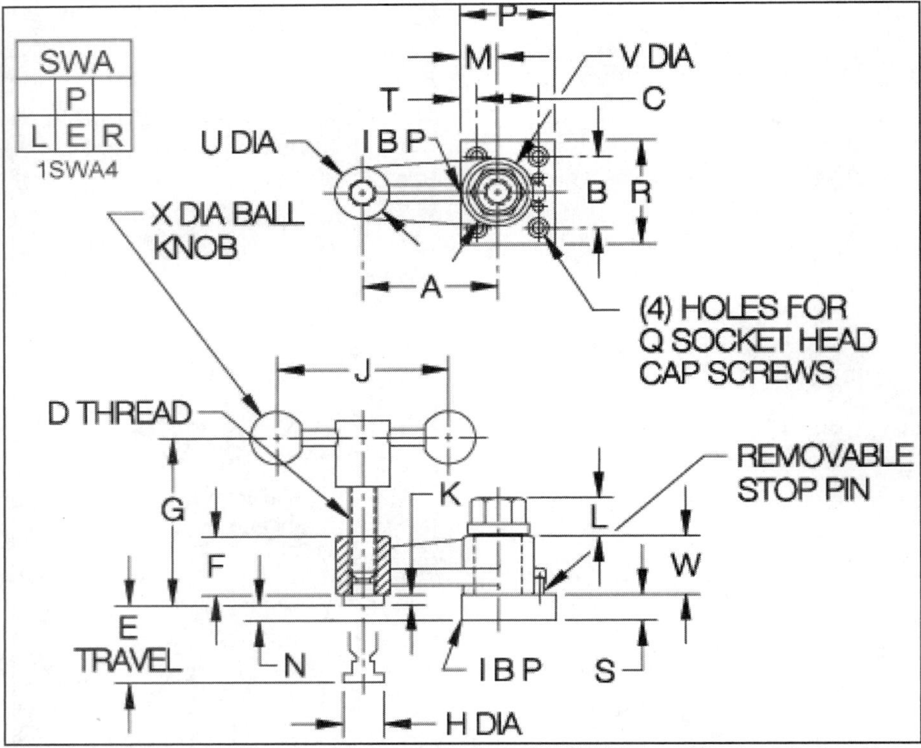

Figure 16.12(h) Specification Drawing for Ball Handle Flange Mounted Swing Clamp

Views: Perspective, Saved, and Exploded

You will now create a variety of cosmetically altered view states. Cosmetic changes to the assembly do not affect the model itself, only the way it is displayed on the screen. One type of view that can be created is the *perspective view*.

Click: **View** ⇒ **Model Setup** ⇒ **Perspective** (Fig. 16.13) ⇒ **Reset** ⇒ **OK** ⇒ ⇒

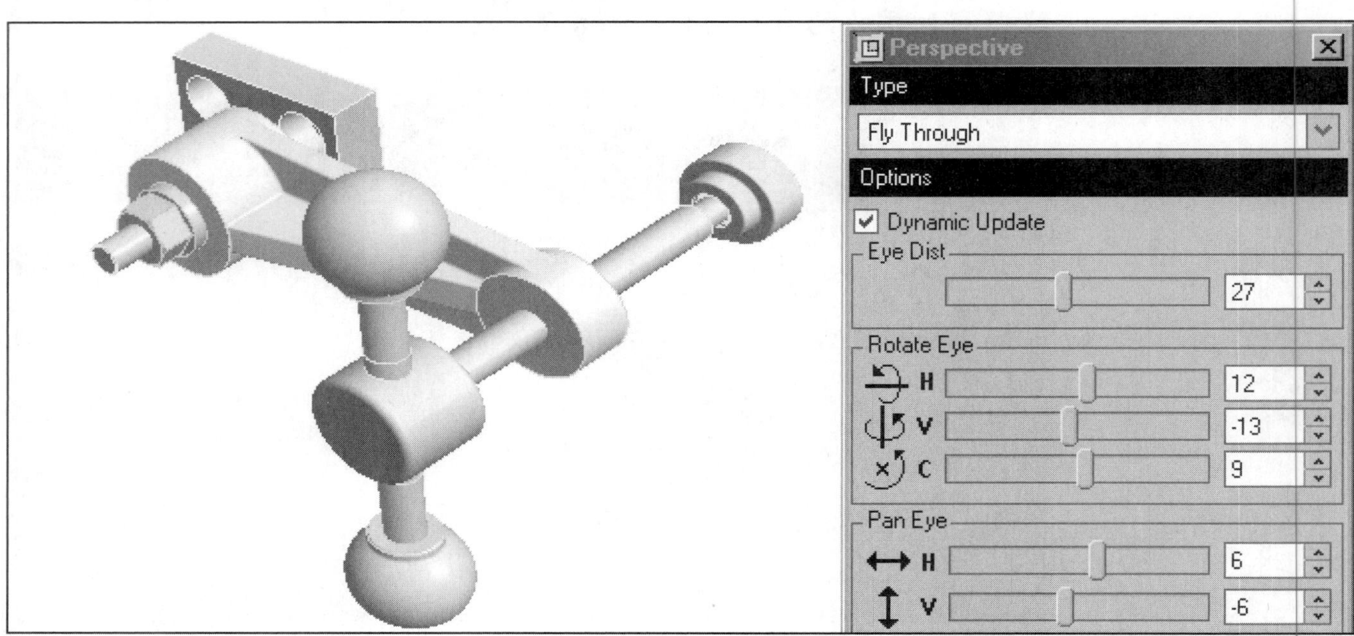

Figure 16.13 Perspective View

Saved Views

Before using the View Manager to set display and explode states, it is a good idea to create one or more saved views to be used later for exploding. The default trimetric and isometric views do not adequately represent the assembly in its functional position.

Using the **MMB**, rotate the model from its default position [Fig. 16.14(a)] to one where the Clamp_Swivel is vertical [Fig. 16.14(b)] ⇒ **Reorient view** [Fig. 16.14(c)] ⇒ **Saved Views** [Fig. 16.14(d)] ⇒ Name: type **EXPLODE1** ⇒ **Save** [Fig. 16.14(e)] ⇒ **OK**

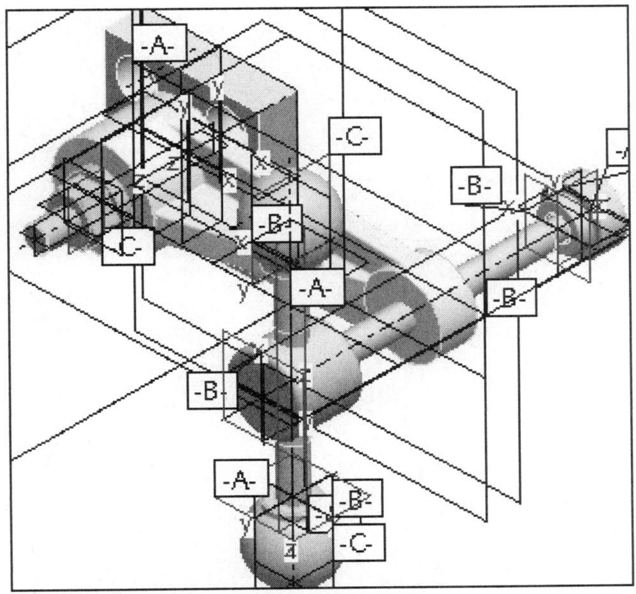

Figure 16.14(a) Isometric View

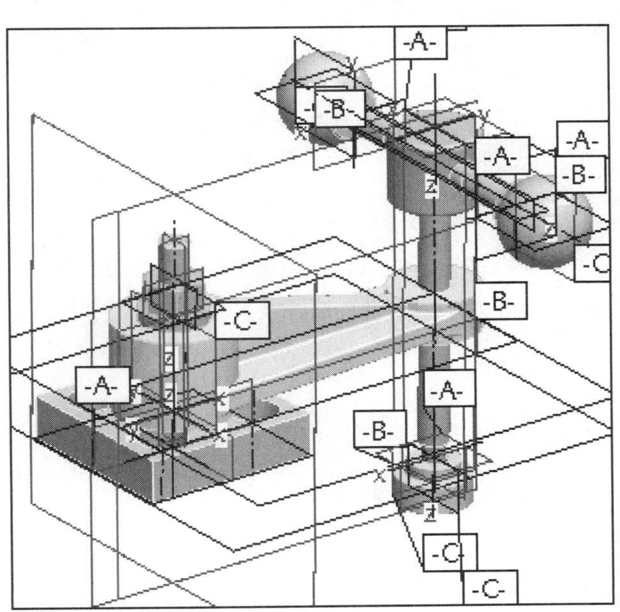

Figure 16.14(b) Reoriented View

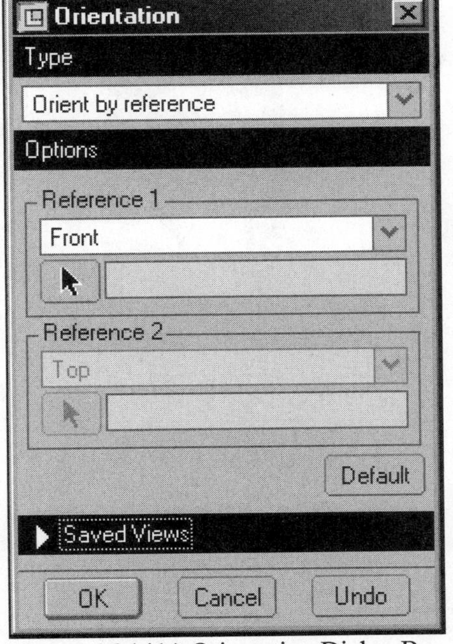

Figure 16.14(c) Orientation Dialog Box

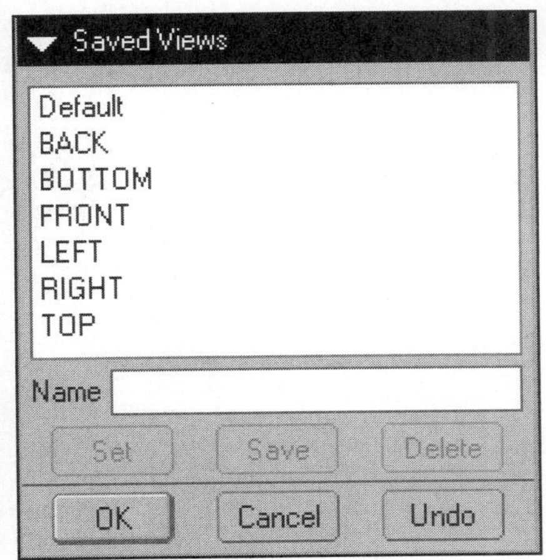

Figure 16.14(d) Saved Views

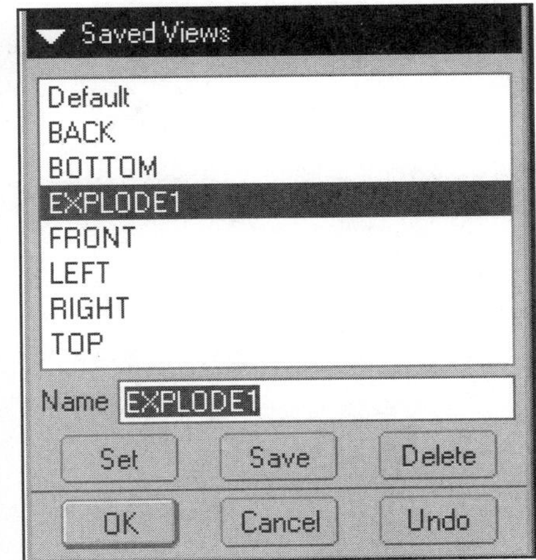

Figure 16.14(e) EXPLODE1 View

Default Exploded View

When you create an *exploded view*, Pro/E moves apart the components of an assembly to a set default distance. The default position is seldom the most desirable.

Click: **Display Datum** all **off** ⇒ **Datum tags** all **off** ⇒ **Datum axes on** ⇒ **View** ⇒ **Explode** [Fig. 16.15(a)] ⇒ **Explode View** [Fig. 16.15(b)] ⇒ **View** ⇒ **Shade** [Fig. 16.15(c)] ⇒ **View** ⇒ **Explode** ⇒ **Unexplode View** ⇒ ⇒ ⇒ **MMB**

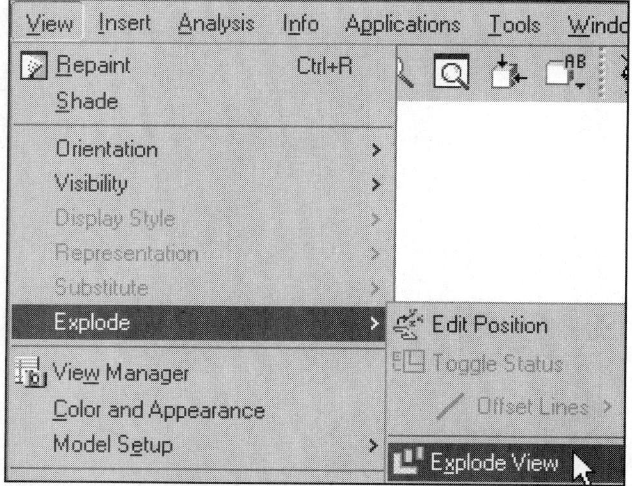

Figure 16.15(a) Explode Command

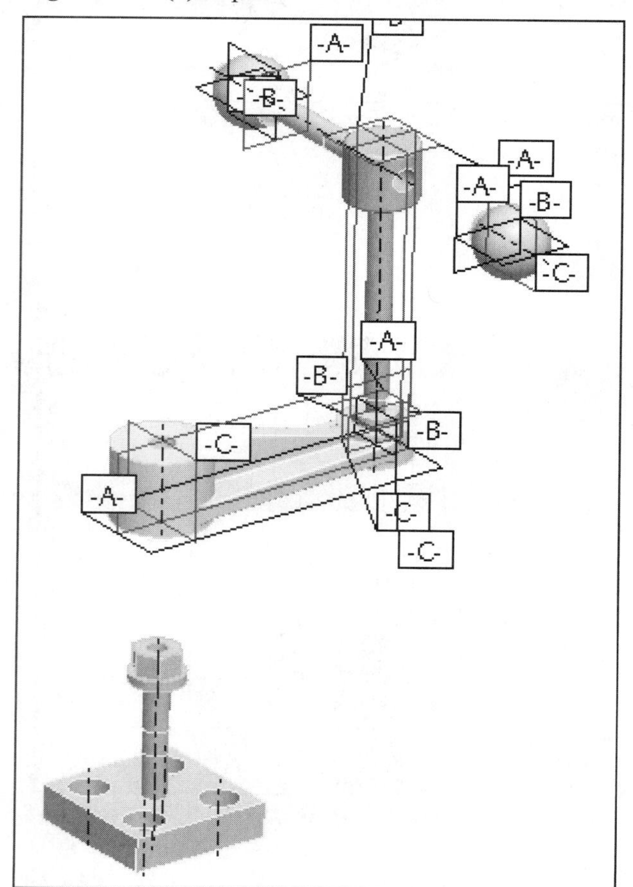

Figure 16.15(b) Default Exploded View

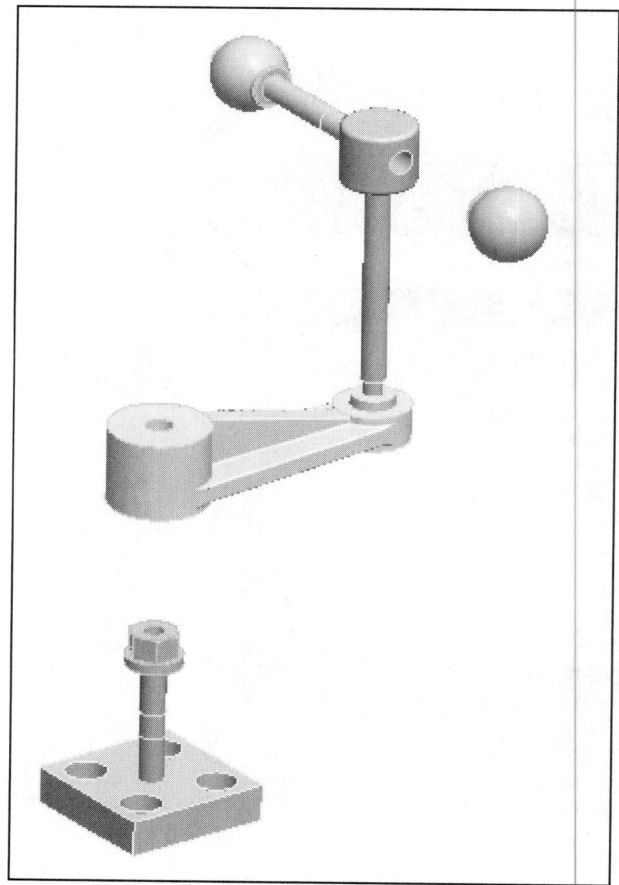

Figure 16.15(c) Shaded Default Exploded View

View Manager

The View Manager allows you to see and control the models representation [Fig. 16.16(a)], orientation [Fig. 16.16(b)], explode [Fig. 16.16(c)], style [Fig. 16.16(d)], and all (combined) states [Fig. 16.16(e)].

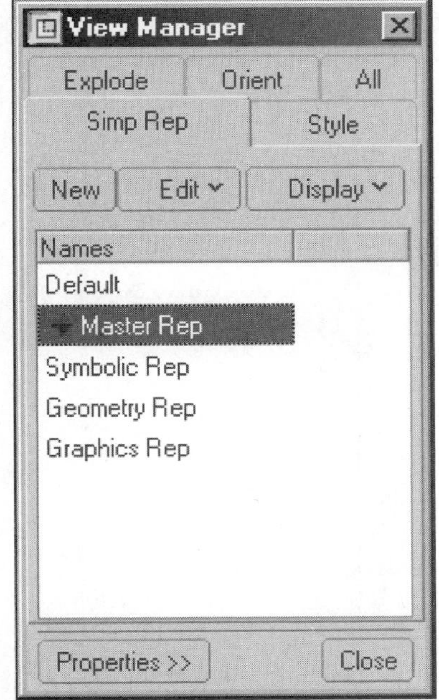

Figure 16.16(a) Representation

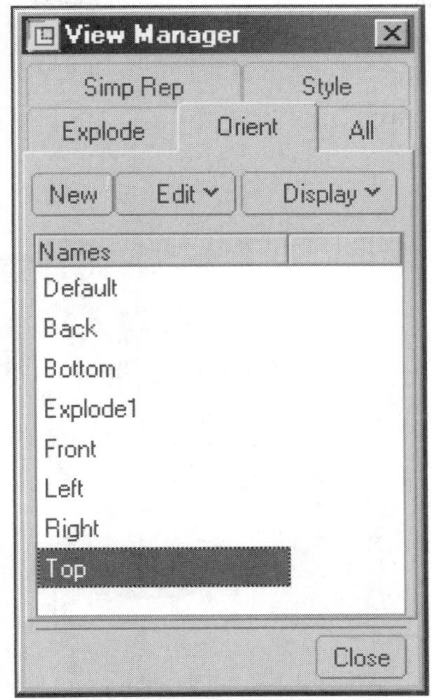

Figure 16.16(b) Orientation

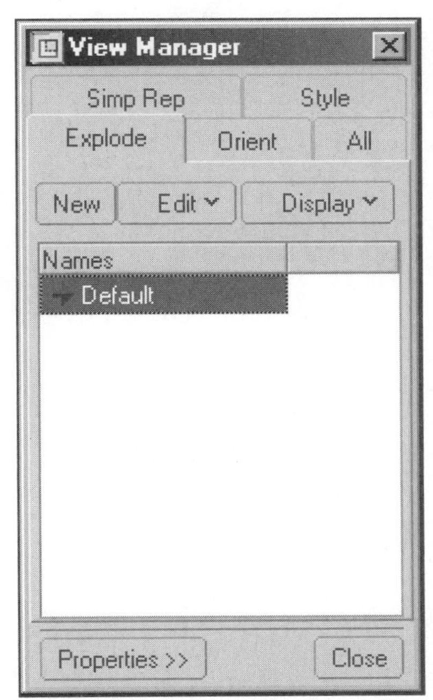

Figure 16.16(c) Explode State

Figure 16.16(d) Style State

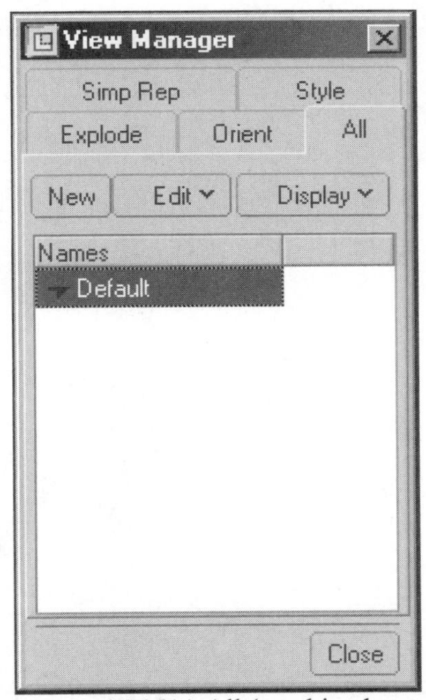

Figure 16.16(e) All (combined state)

Click: **View** ⇒ **View Manager** ⇒ **Simp Rep** tab ⇒ **Master Rep** ⇒ **Display** ⇒ **Set** ⇒ **Orient** tab ⇒ **Explode1** ⇒ **Display** ⇒ **Set** [Fig. 16.17(a)] ⇒ **Explode** ⇒ **New** ⇒ **Enter** to accept the default name: Exp0001 ⇒ **Edit** ⇒ **Redefine** [Fig. 16.17(b)]

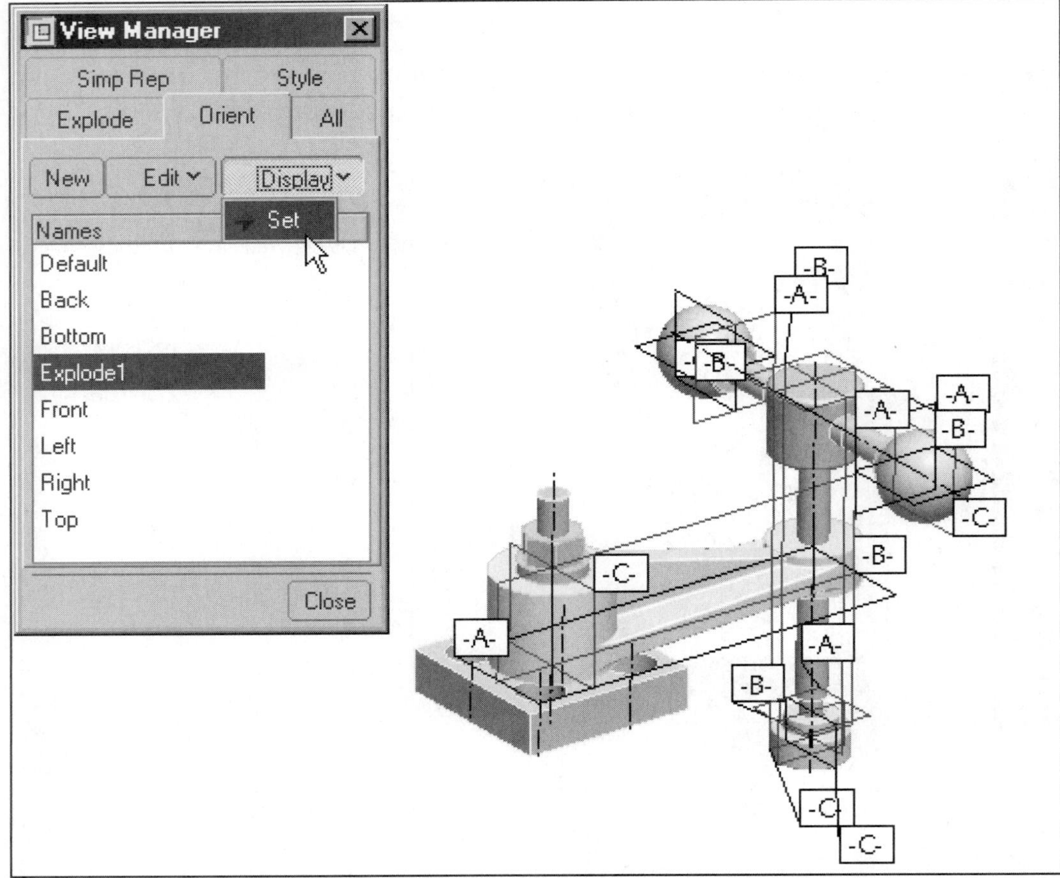

Figure 16.17(a) View Manager Orient

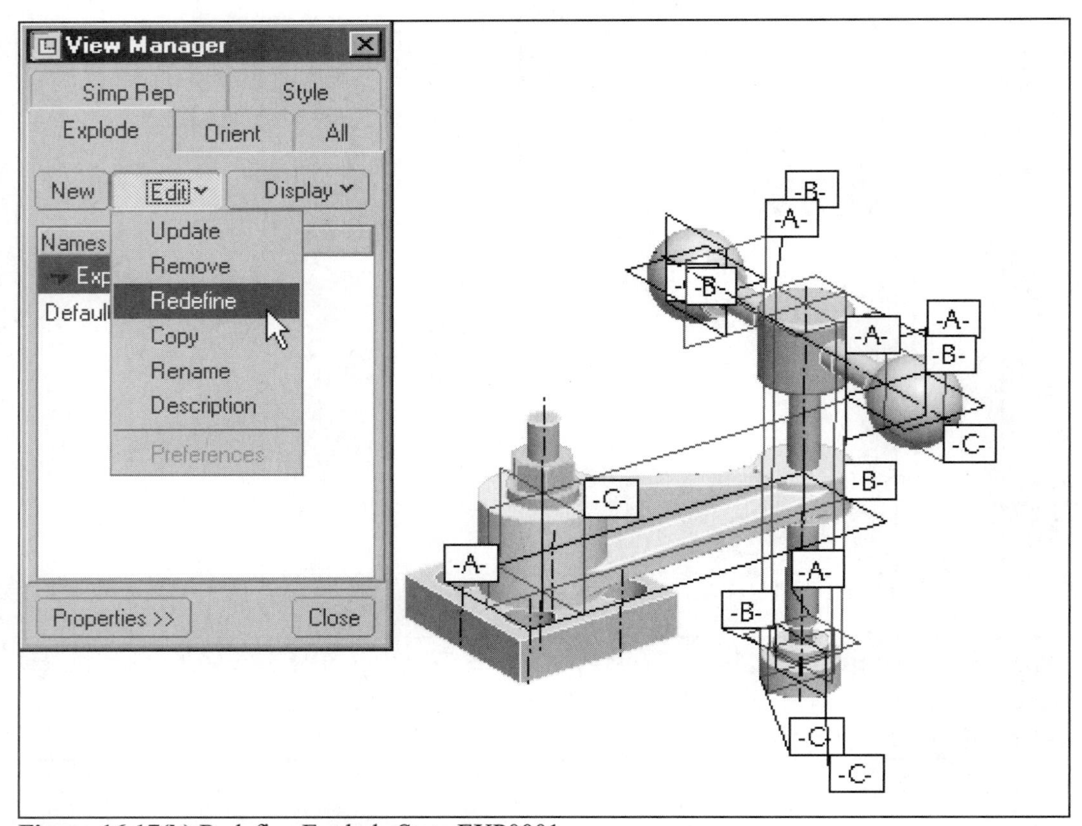

Figure 16.17(b) Redefine Explode State EXP0001

Click: **Position** [Fig. 16.17(c)] ⇒ Motion Reference ⇒ select the vertical axis of the CLAMP_STUD35 [Fig. 16.17(d)] *or any vertical edge of the CLAMP_PLATE* ⇒ **Preferences** ⇒ **Move Many** ⇒ **Close** ⇒ select the **CLAMP_STUD35.PRT** ⇒ press and hold the **Ctrl** key and select **CLAMP_FLANGE_NUT.PRT** [Fig. 16.17(e)] ⇒ **MMB** ⇒ click **LMB** in the graphics window and start dragging [Fig. 16.17(f)] ⇒ click **LMB** to place

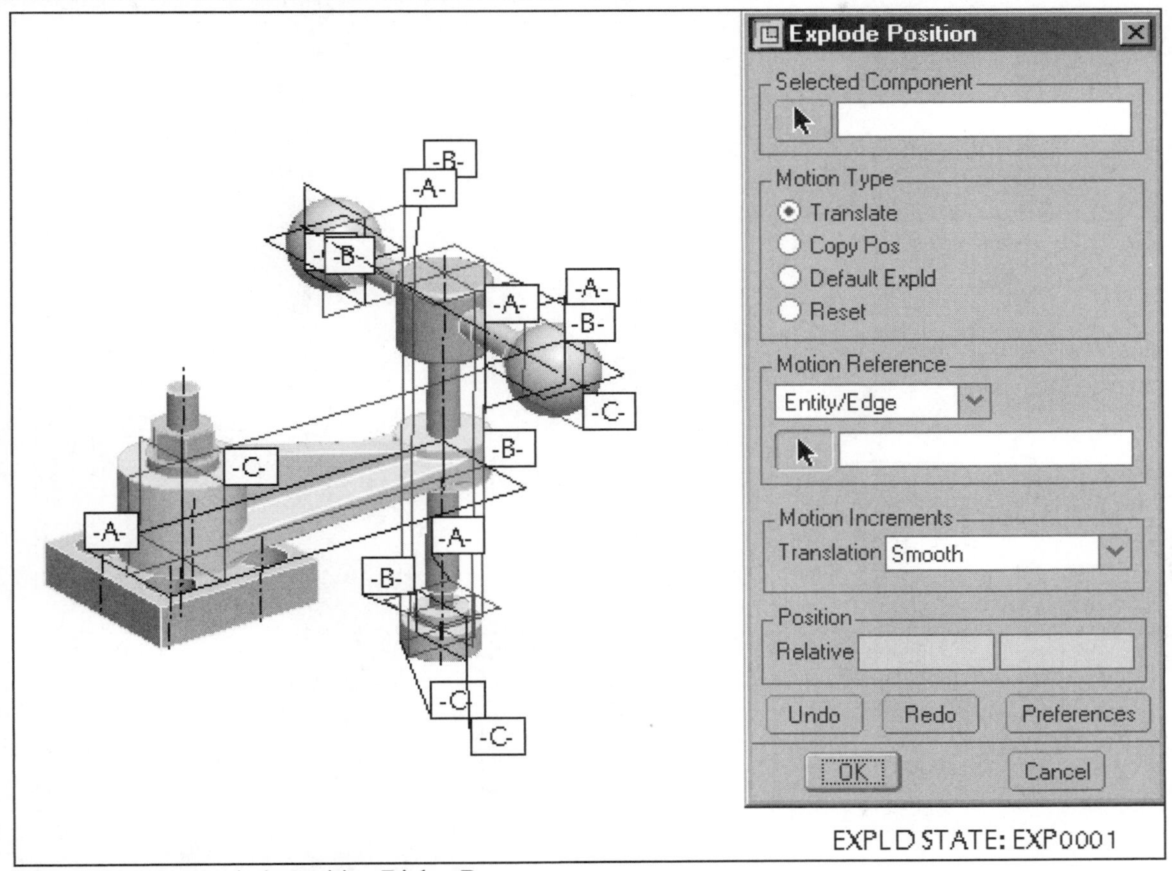

Figure 16.17(c) Explode Position Dialog Box

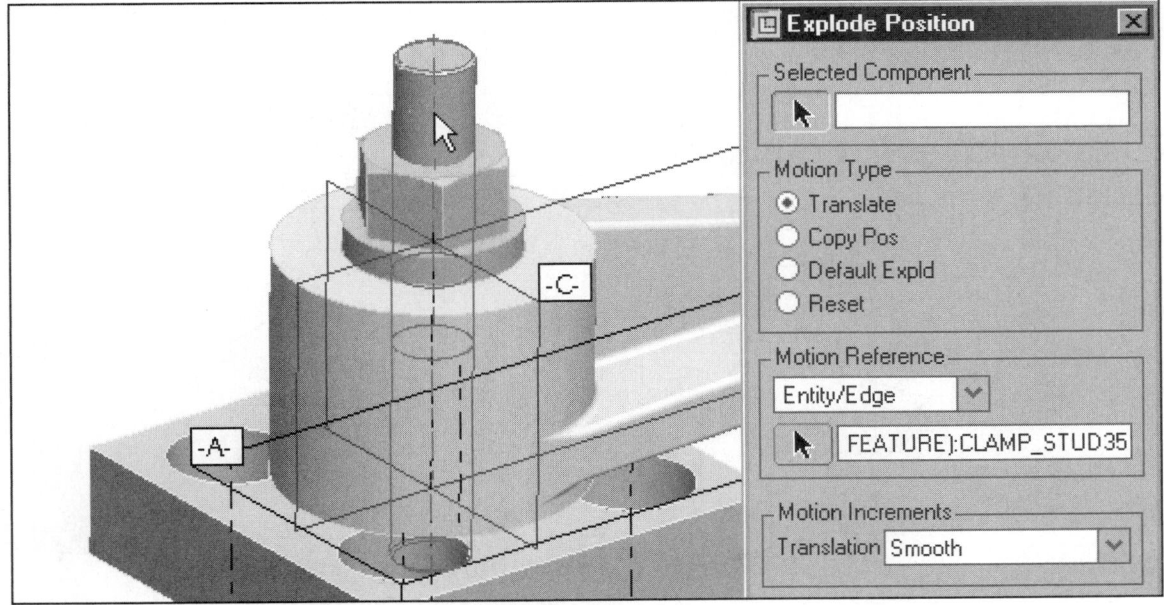

Figure 16.17(d) Select the Axis as the Motion Reference

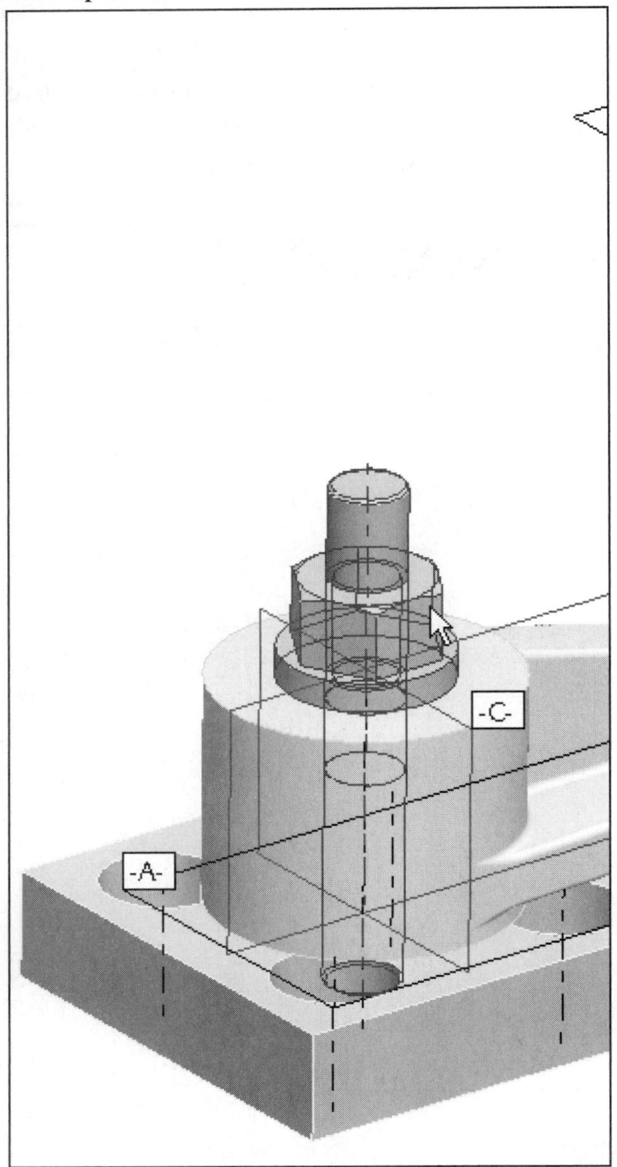

Figure 16.17(e) Select Components

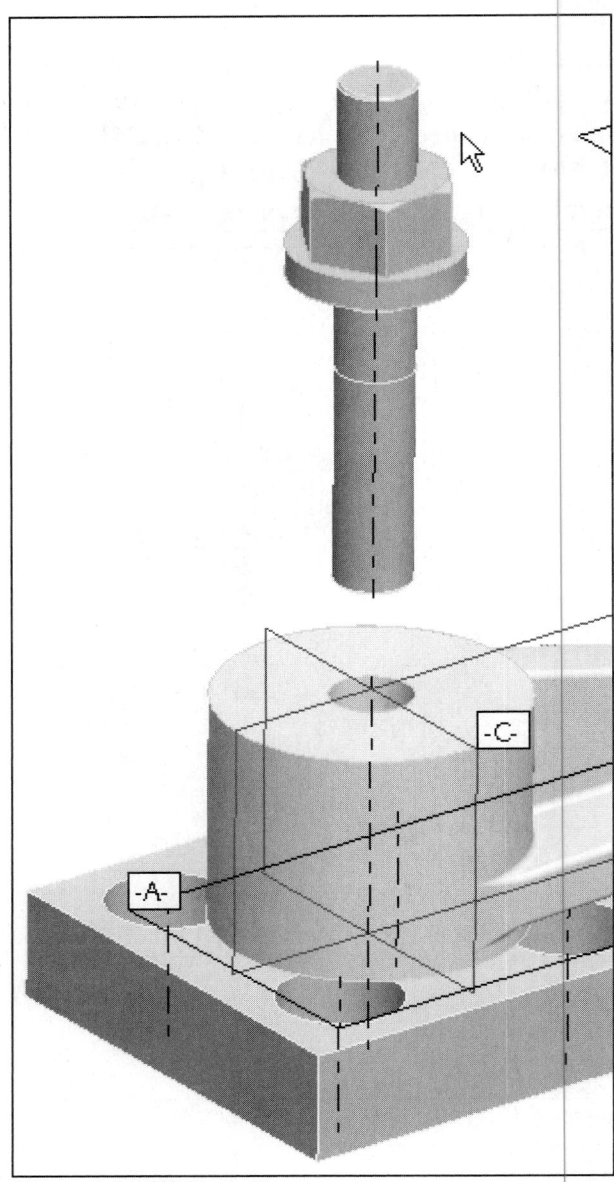

Figure 16.17(f) Drag Components

Select **CLAMP_FLANGE_NUT.PRT** [Fig. 16.17(g)] ⇒ **MMB** ⇒ click **LMB** in the graphics window and start dragging [Fig. 16.17(h)] ⇒ click **LMB** to place

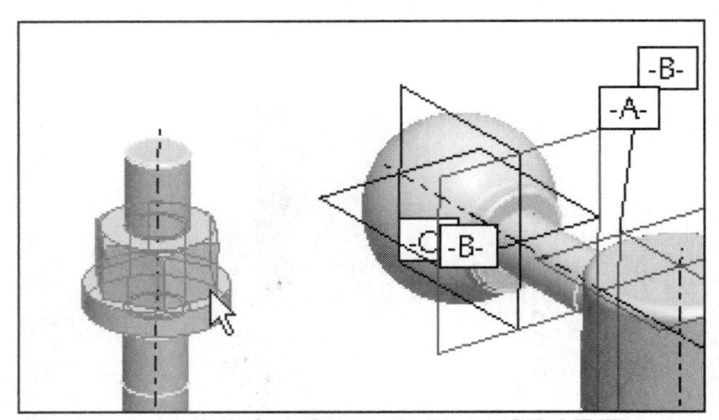

Figure 16.17(g) Select CLAMP_FLANGE_NUT

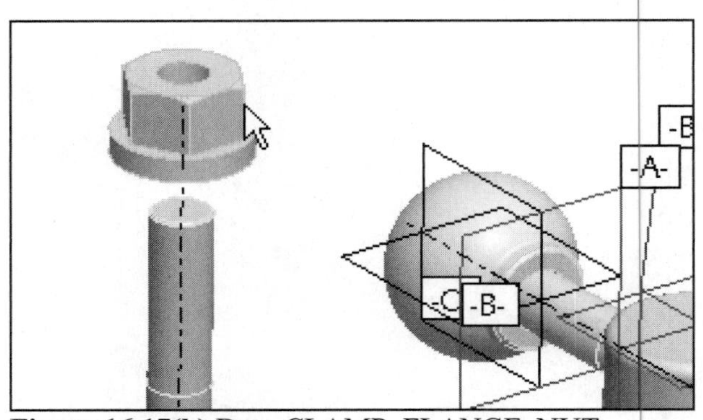

Figure 16.17(h) Drag CLAMP_FLANGE_NUT

Select **CLAMP_PLATE.PRT** [Fig. 16.17(i)] ⇒ **MMB** ⇒ click **LMB** in the graphics window and start dragging [Fig. 16.17(j)] ⇒ click **LMB** to place ⇒ **Preferences** ⇒ **Move With Children** ⇒ **Close** ⇒ select **CLAMP_SWIVEL.PRT** [Fig. 16.17(k)] ⇒ start dragging [Fig. 16.17(l)] ⇒ click **LMB** to place⇒ **Preferences** ⇒ **Move One** ⇒ **Close** ⇒ select **CLAMP_FOOT.PRT** [Fig. 16.17(m)] ⇒ start dragging [Fig. 16.17(n)] ⇒ click **LMB** to place

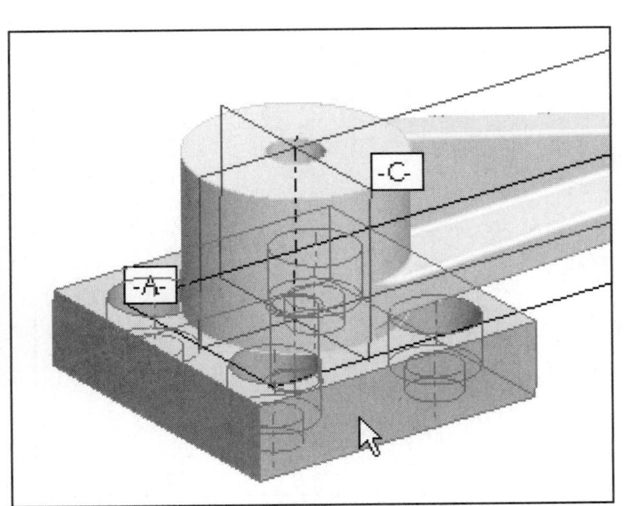

Figure 16.17(i) Select CLAMP_PLATE

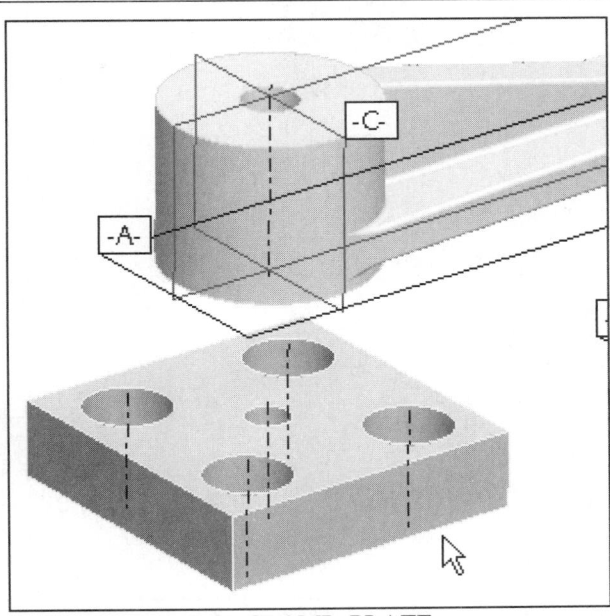

Figure 16.17(j) Drag CLAMP_PLATE

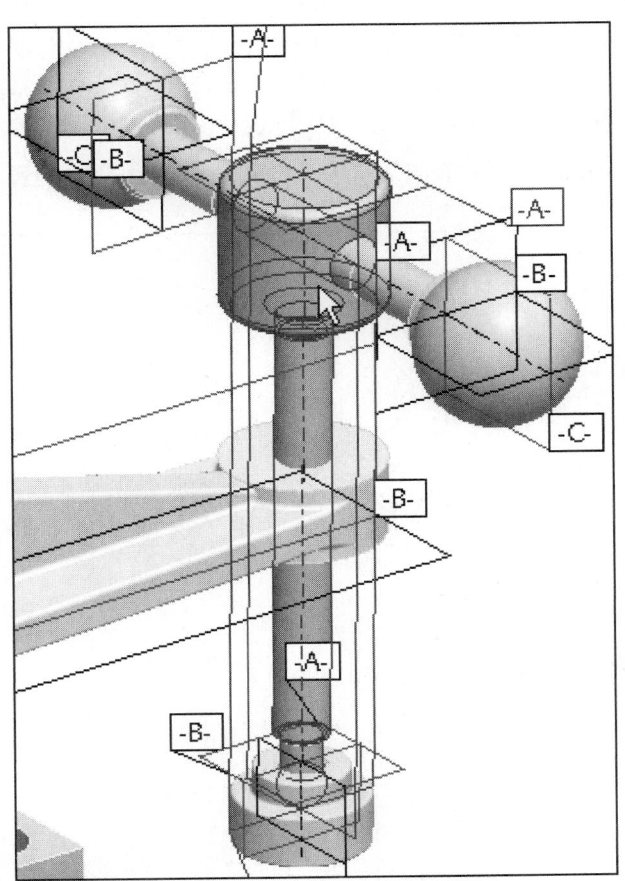

Figure 16.17(k) Select CLAMP_SWIVEL

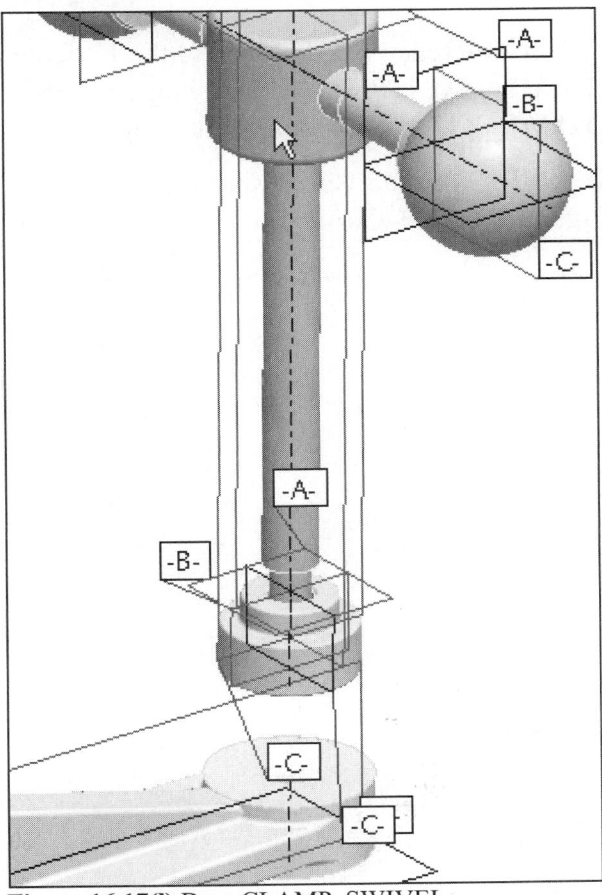

Figure 16.17(l) Drag CLAMP_SWIVEL

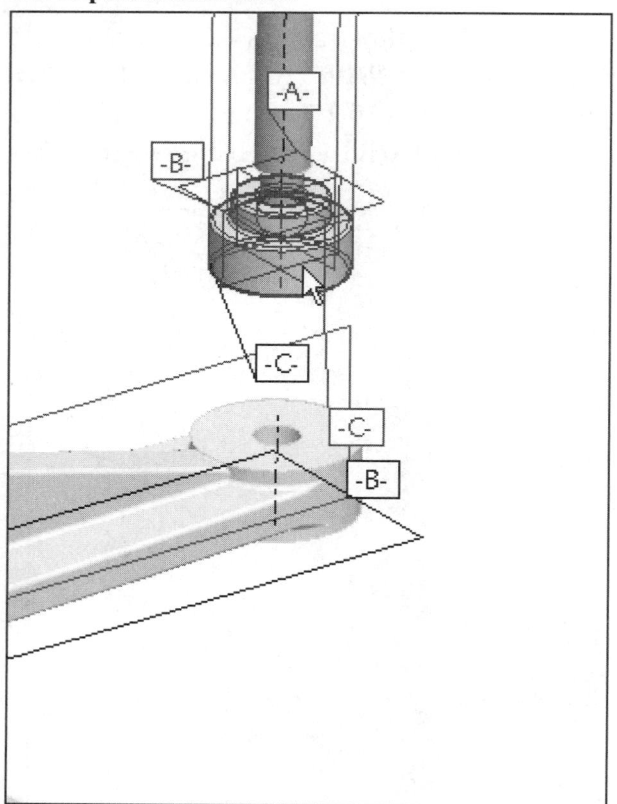

Figure 16.17(m) Select CLAMP_FOOT

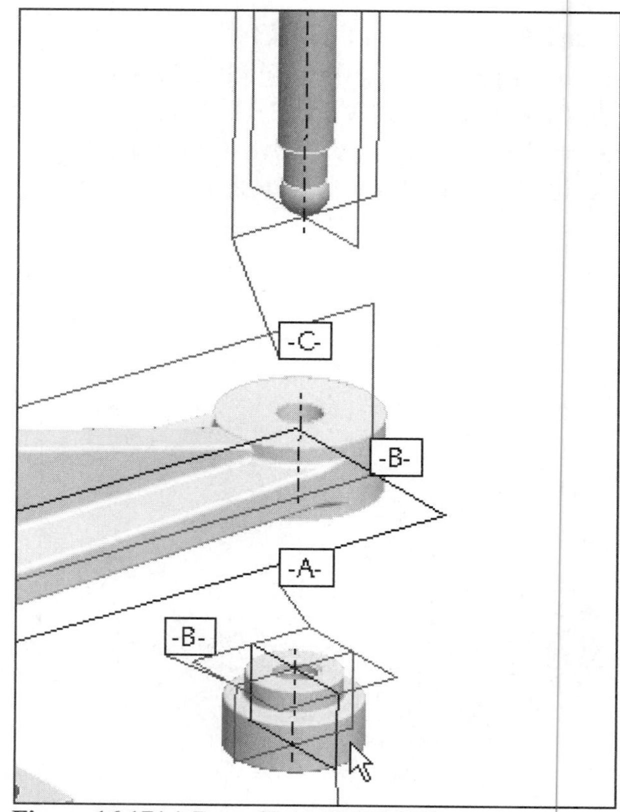

Figure 16.17(n) Drag CLAMP_FOOT

Motion Reference ⇒ select the horizontal edge of the CLAMP_PLATE [Fig. 16.17(o)] ⇒ select **CLAMP_BALL.PRT** ⇒ start dragging ⇒ click **LMB** to place [Fig. 16.17(p)] ⇒ select the second **CLAMP_BALL.PRT** ⇒ start dragging ⇒ click **LMB** to place [Fig. 16.17(q)] ⇒ select the **CLAMP_STUD5.PRT** ⇒ start dragging ⇒ click **LMB** to place [Fig. 16.17(r)] ⇒ **OK** ⇒ ⇒ **MMB** [Fig. 16.17(s)] ⇒ **Close**

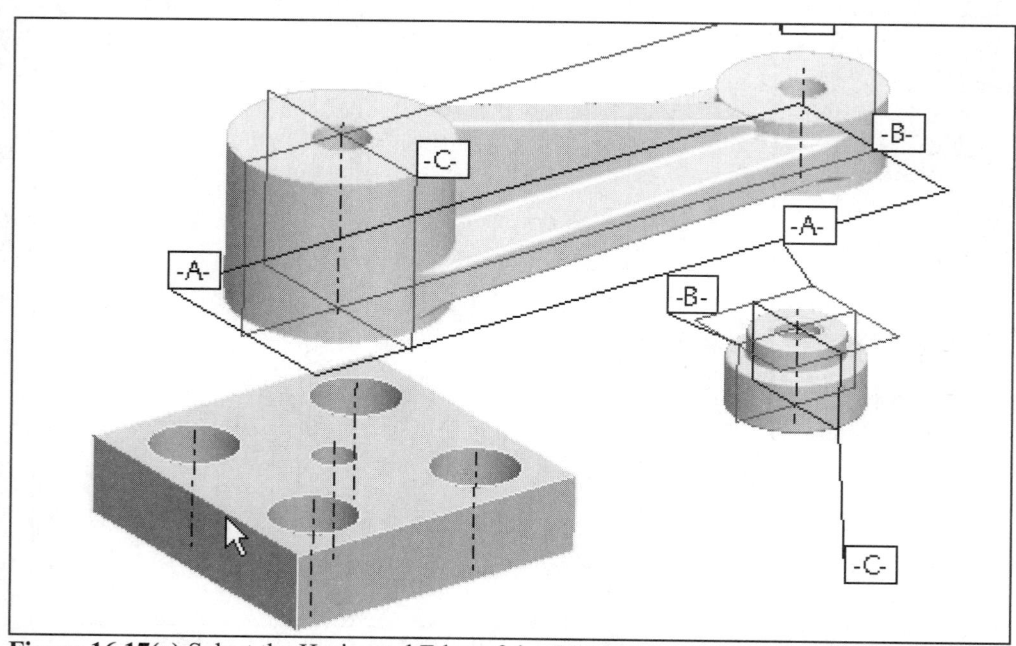

Figure 16.17(o) Select the Horizontal Edge of the CLAMP_PLATE

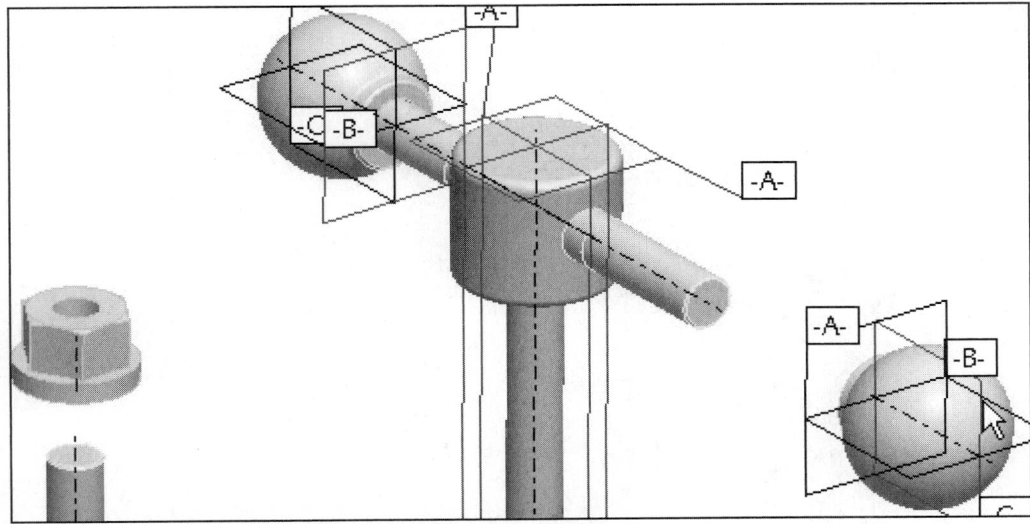

Figure 16.17(p) Select and Move the CLAMP_BALL

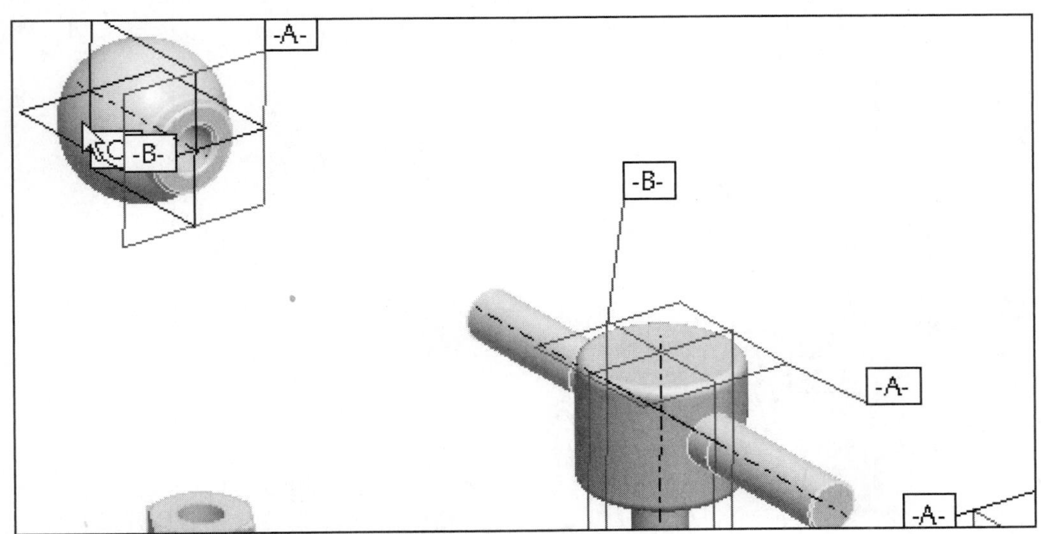

Figure 16.17(q) Select and Move the Second CLAMP_BALL

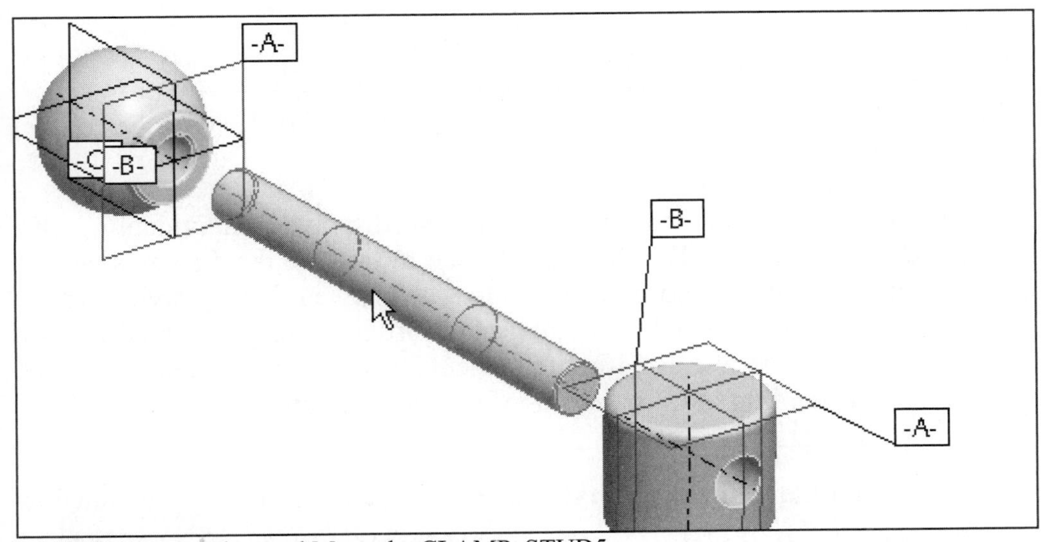

Figure 16.17(r) Select and Move the CLAMP_STUD5

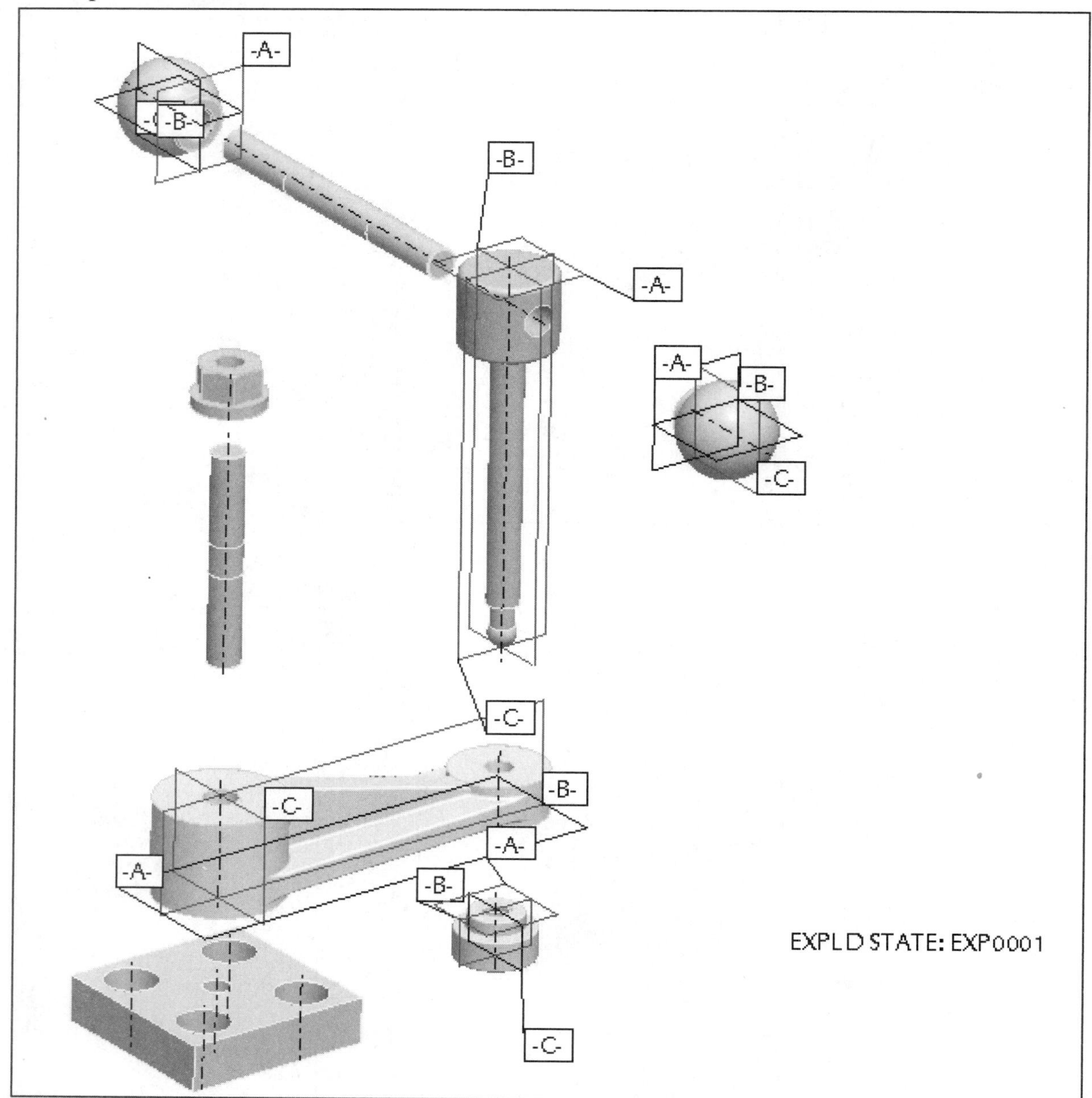

Figure 16.17(s) Completed Exploded State EXP0001

View Style

The components of an assembly (whether exploded or not) can be displayed individually with **Wireframe**, **Hidden Line**, **No Hidden**, **Shading**, or **Blank**. Style is used to manage the display styles of an assembly's components. Wireframe, hidden line, no hidden, shaded, or blanked display styles can be assigned to individual components. The components will be displayed according to their assigned display style in the current style state. The setting, in the Environment menu, controls the display style of unassigned components and they appear according to the current mode of the environment.

Click: **Start the view manager** from the tool bar ⇒ **Style** tab ⇒ **New** ⇒ **Enter** [Fig. 16.18(a)] to accept the default name: Style0001 ⇒ **Edit** ⇒ **Redefine** ⇒ **Show** tab [Fig. 16.18(b)] ⇒ Wireframe ⇒ select **CLAMP_SWIVEL.PRT** and **CLAMP_STUD35.PRT** ⇒ **Update**

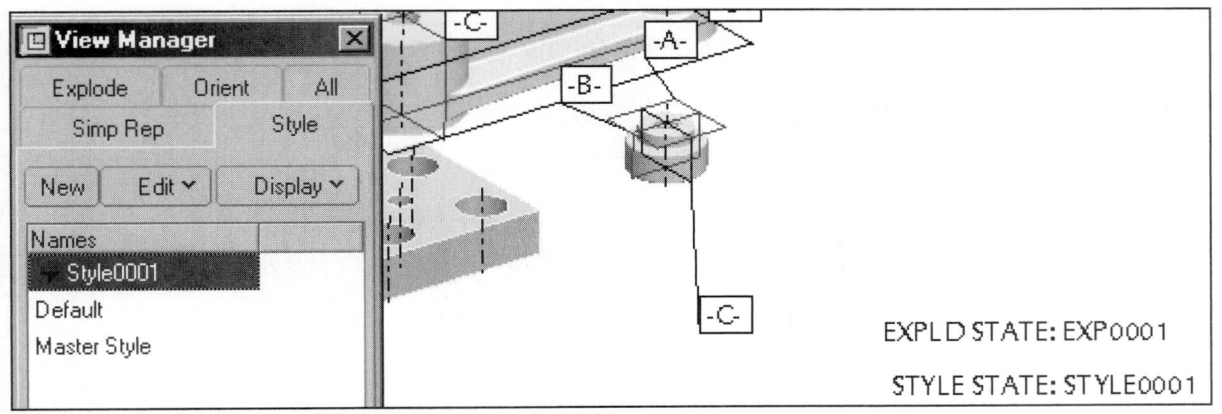

Figure 16.18(a) Style State STYLE0001

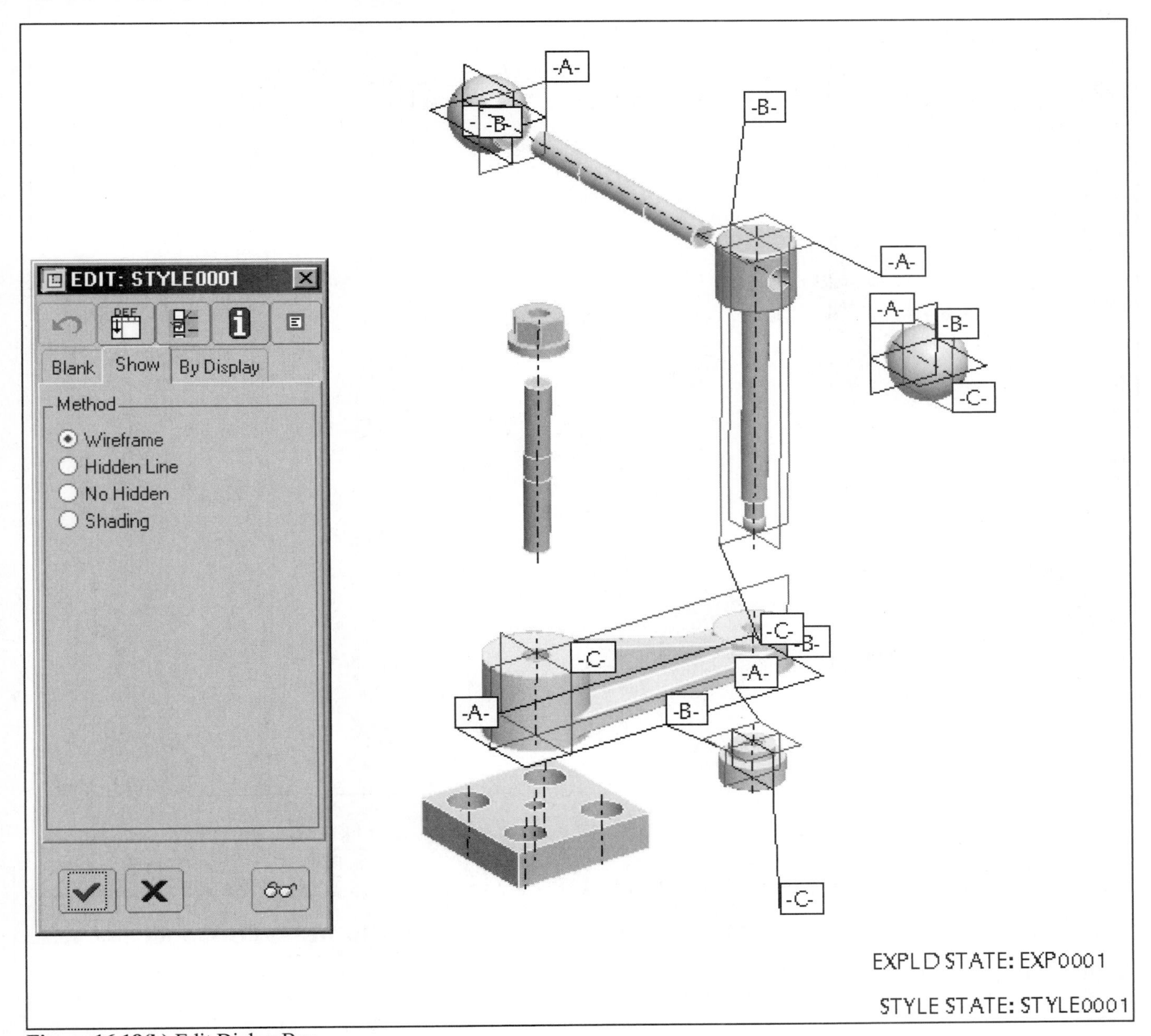

Figure 16.18(b) Edit Dialog Box

Click: [Hidden Line] ⇒ select the **CLAMP_PLATE.PRT** and the **CLAMP_FOOT.PRT** [Fig. 16.18(c)] ⇒ [glasses icon] **Update Model View According To Specified Changes**

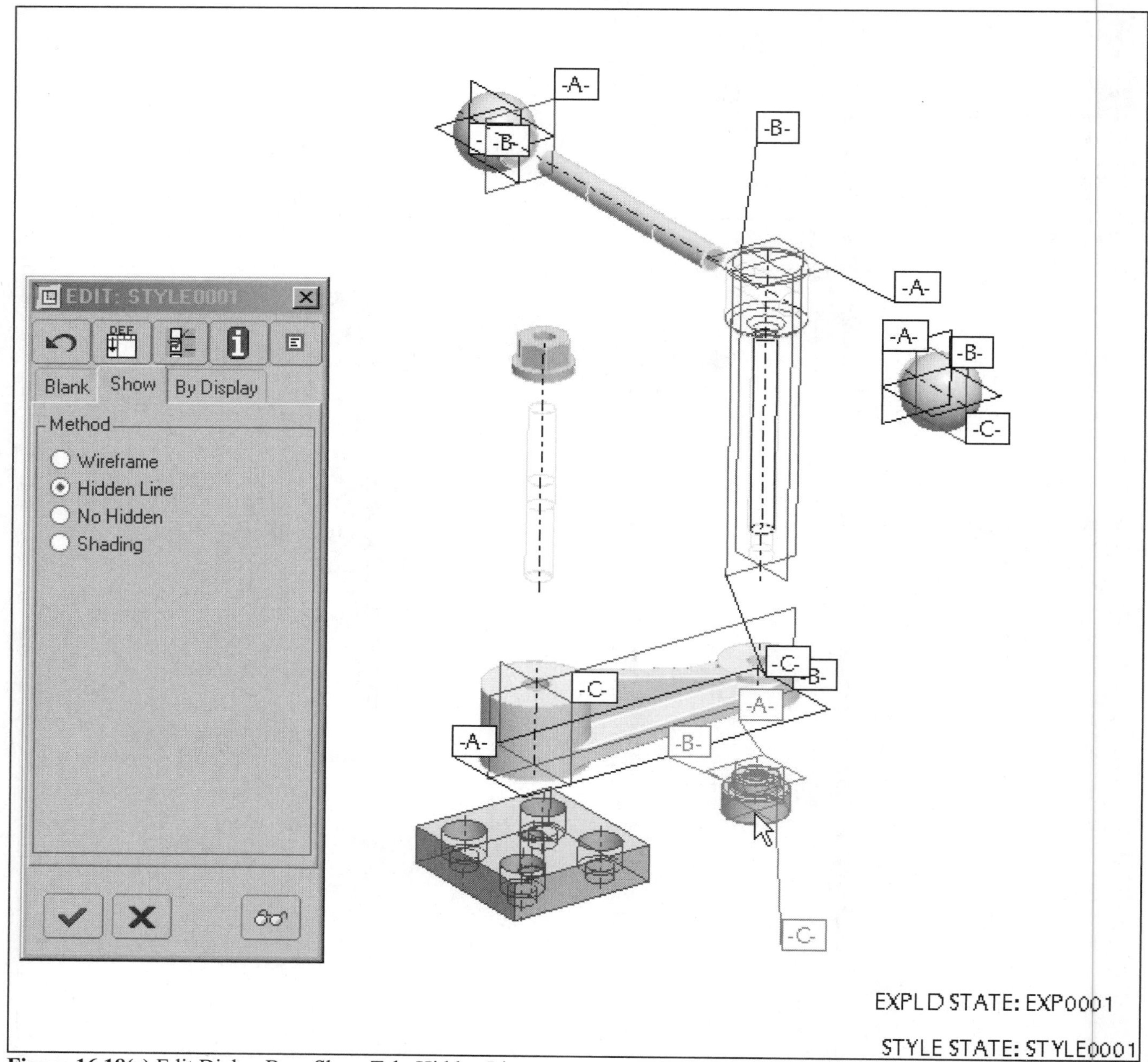

Figure 16.18(c) Edit Dialog Box, Show Tab, Hidden Line

Click: [No Hidden] ⇒ select the two **CLAMP_BALL.PRT** components ⇒ [glasses icon] **Update** ⇒ [Shading] ⇒ select **CLAMP_FLANGE_NUT.PRT**, **CLAMP_ARM.PRT**, and **CLAMP_STUD5.PRT** ⇒ [glasses icon] **Update Model View According To Specified Changes** [Fig. 16.18(d)] ⇒ [✔]

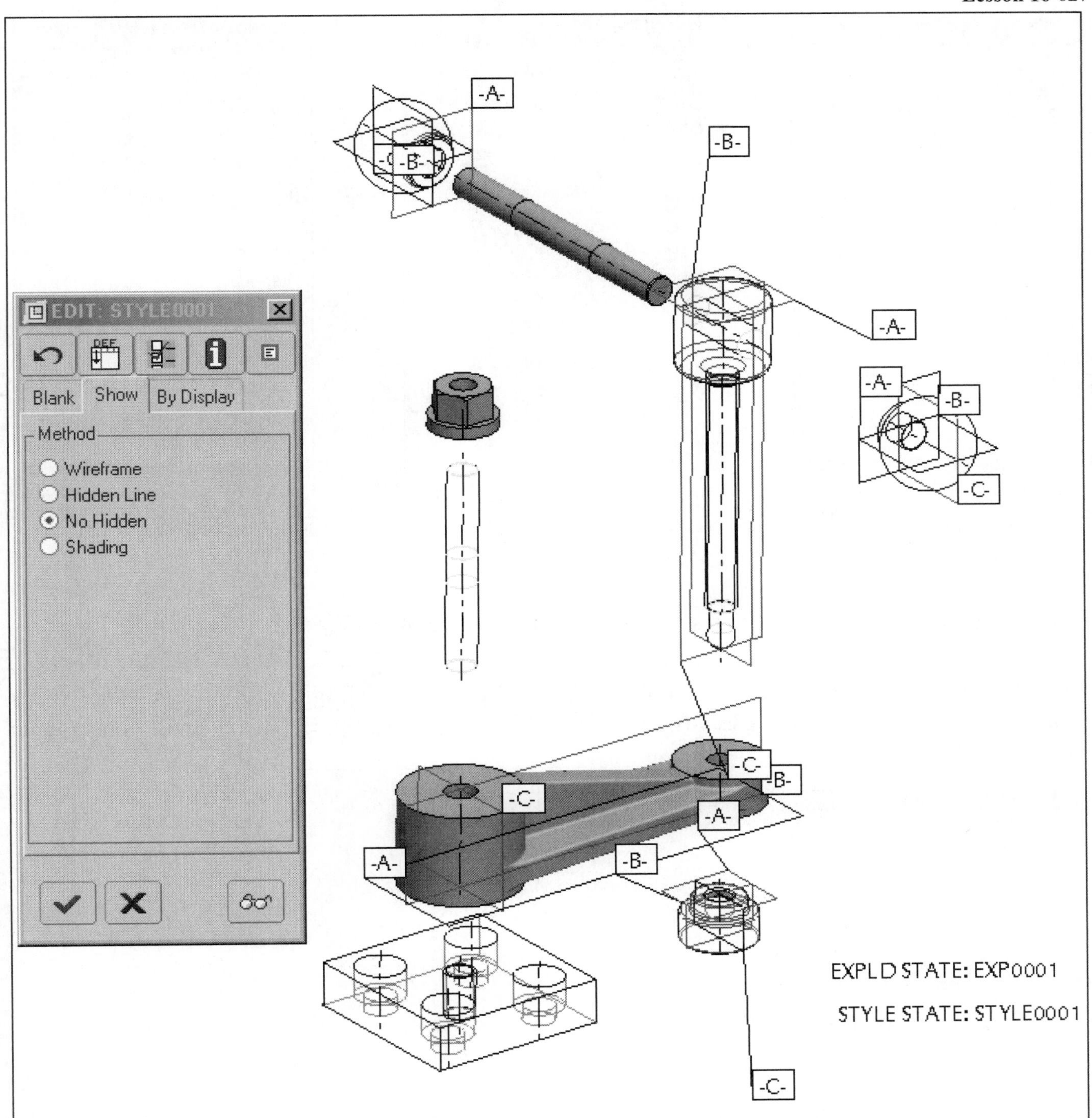

Figure 16.18(d) Completed Style State

Click: **All** tab ⇒ **New** ⇒ **Enter** to accept the default name: Comb0001 ⇒ **Reference Originals** [Fig. 16.18(e)] ⇒ **Edit** ⇒ **Update** [Fig. 16.18(f)] ⇒ **Orient** tab ⇒ **View0001** has been created [Fig. 16.18(g)] ⇒ **Close** ⇒ [save icon] ⇒ **MMB**

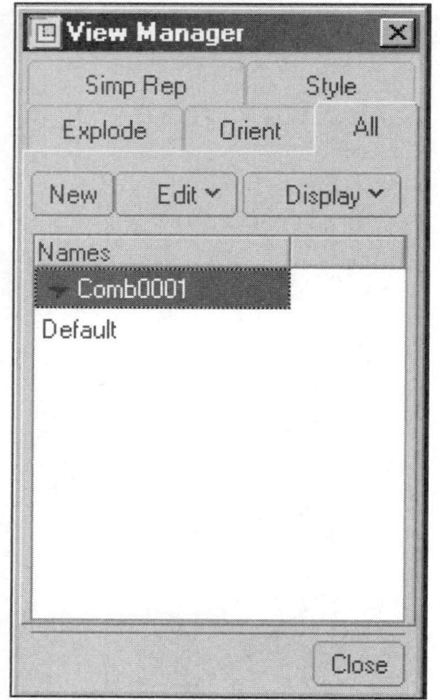

Figure 16.18(e) All Tab

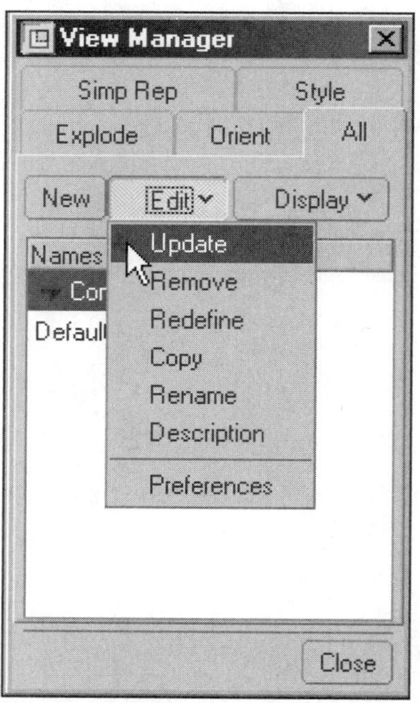

Figure 16.18(f) All Edit Update

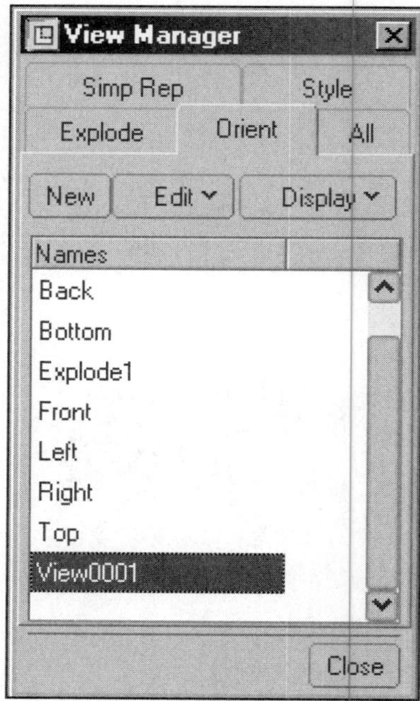

Figure 16.18(g) Orient Tab

Model Tree

You can display component style states, and explode states in the Model Tree.

Click: **Settings** from Model Tree ⇒ **Tree Columns** ⇒ Type: click [Fig. 16.19(a)] ⇒ **Display Styles** ⇒ **Display Style** ⇒ >> ⇒ **STYLE0001** ⇒ >> [Fig. 16.19(b)] ⇒ **Apply** ⇒ **OK** ⇒ adjust your column format [Fig. 16.19(c)]

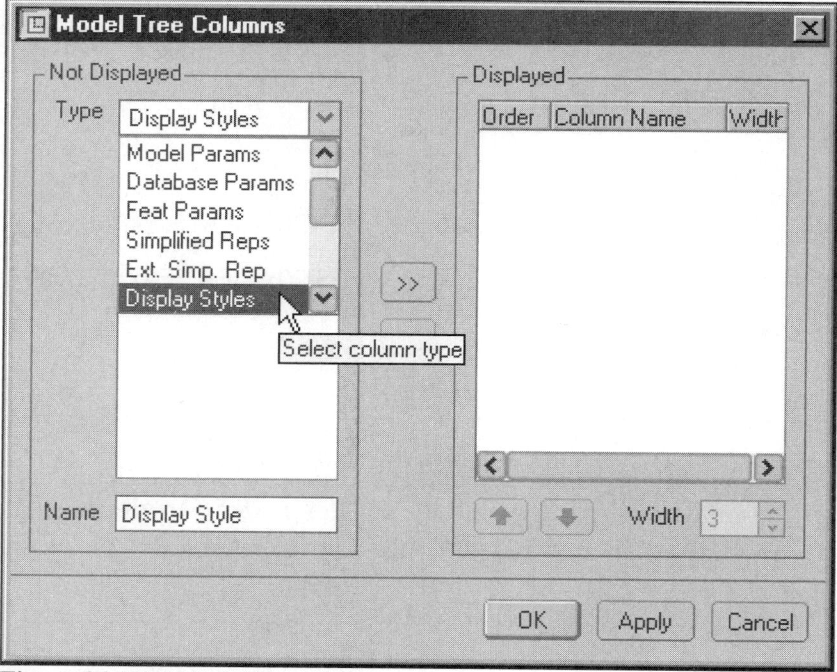

Figure 16.19(a) Model Tree Columns

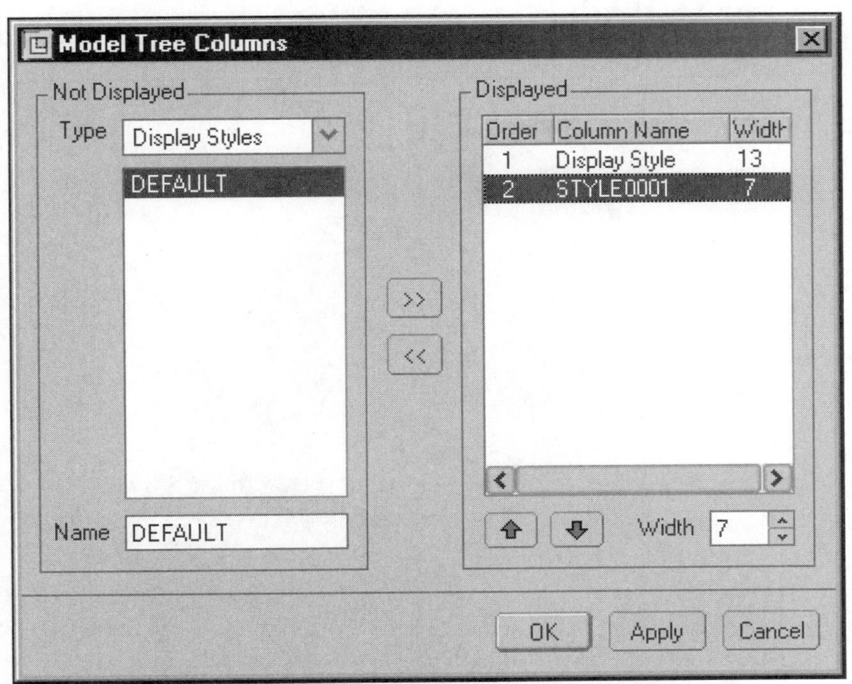

Figure 16.19(b) Adding Display Styles to the Model Tree Columns

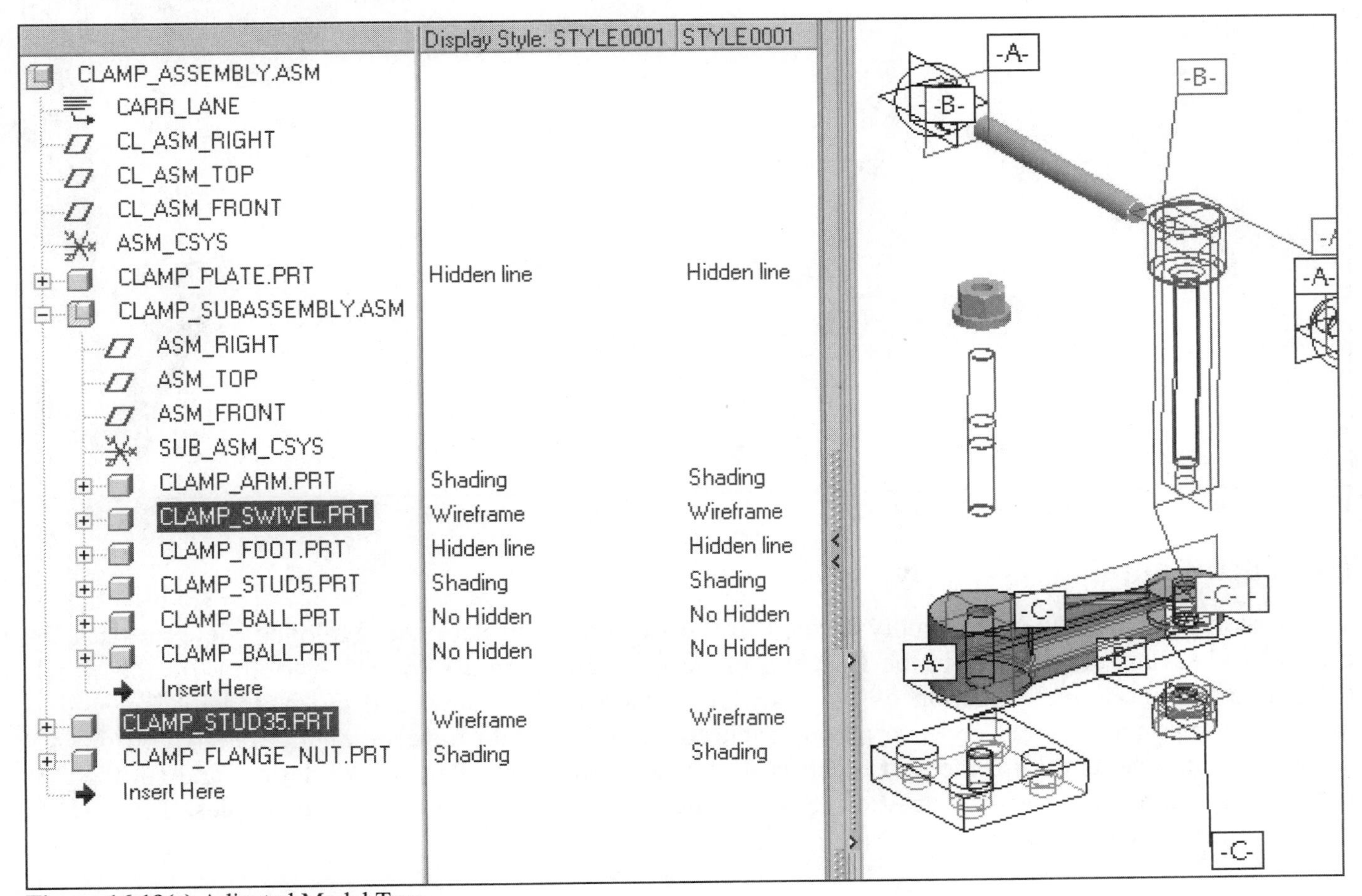

Figure 16.19(c) Adjusted Model Tree

You have now completed this lesson. Continue with the lesson project.

Lesson 16 Project

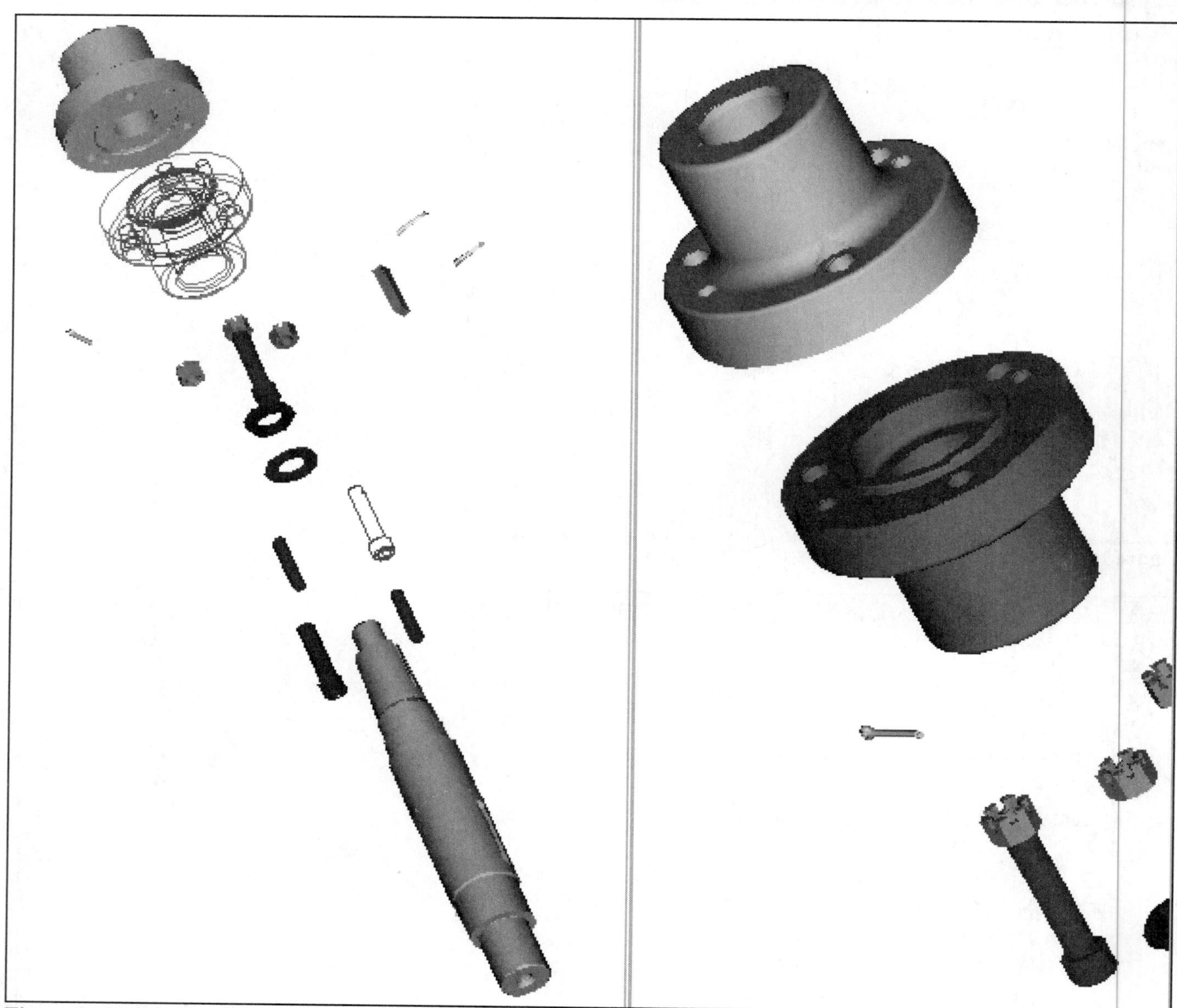

Figure 16.20 Exploded Coupling Assembly

Exploded Coupling Assembly

This lesson project uses the assembly created in the Lesson 15 Project. An exploded view needs to be created and saved for use later in the Drawing mode for the Lesson 20 Project. Varieties of other views are suggested, including a section of the assembly, a perspective view, and an exploded view with a different component display style variation. Each component should have its own color. If you did not color the components during the part creation, bring up each part in Part mode, define, and set the part with a color. Create three or four View States. You do not need to match the examples (Figs. 16.20 through 16.24) provided here.

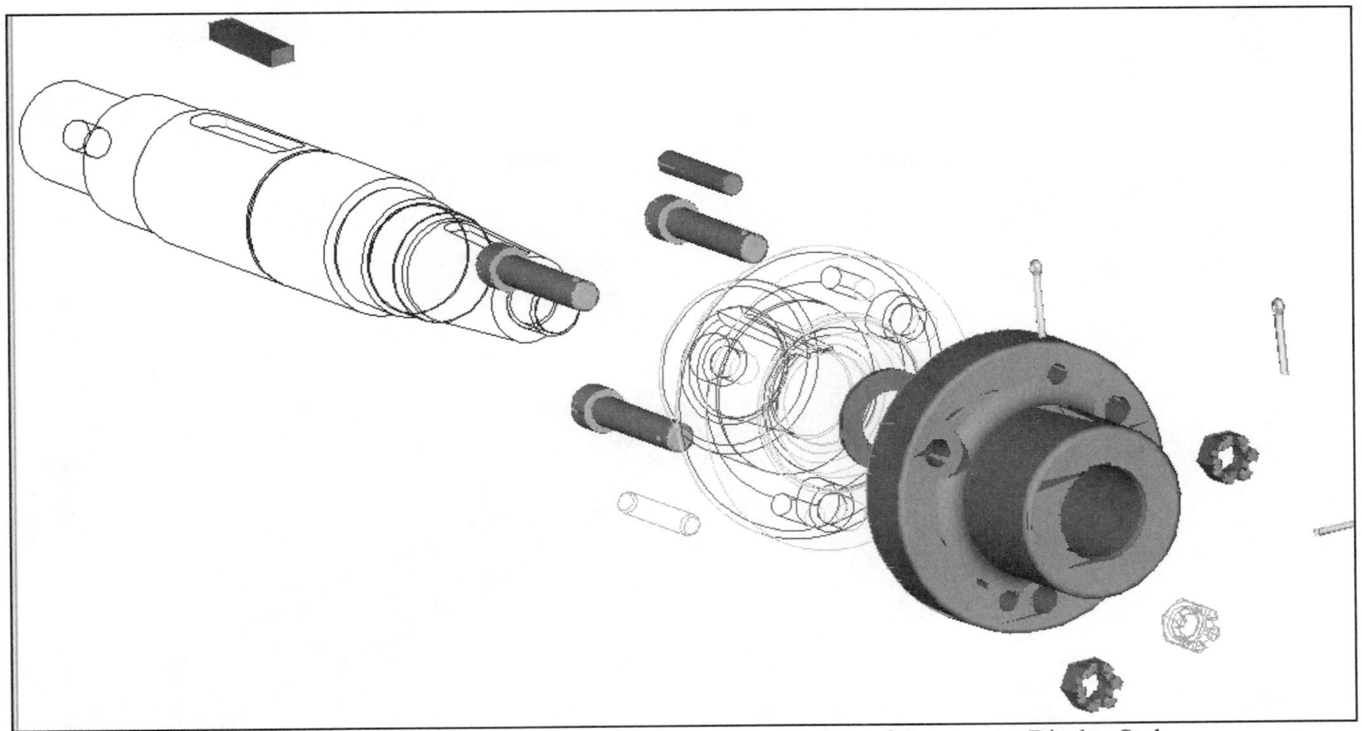

Figure 16.21 Perspective View of Exploded Coupling Assembly with a Variety of Component Display Styles

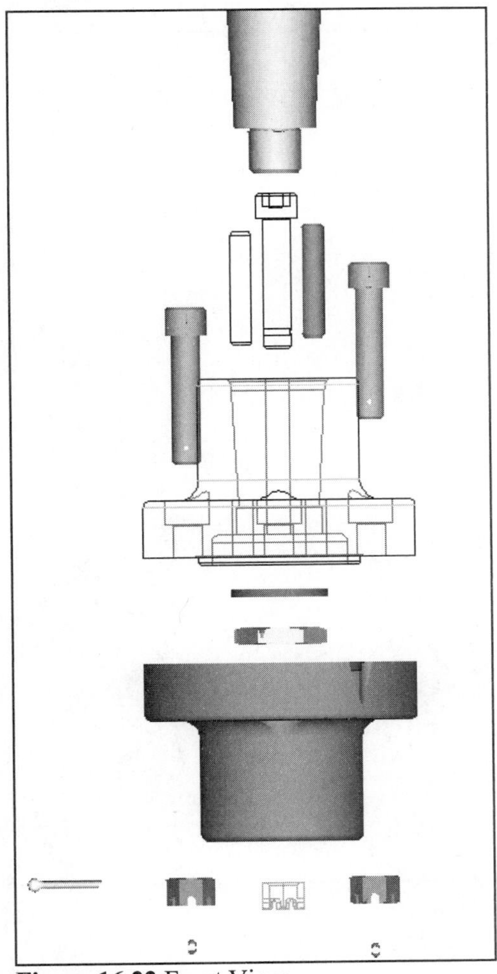

Figure 16.22 Front View

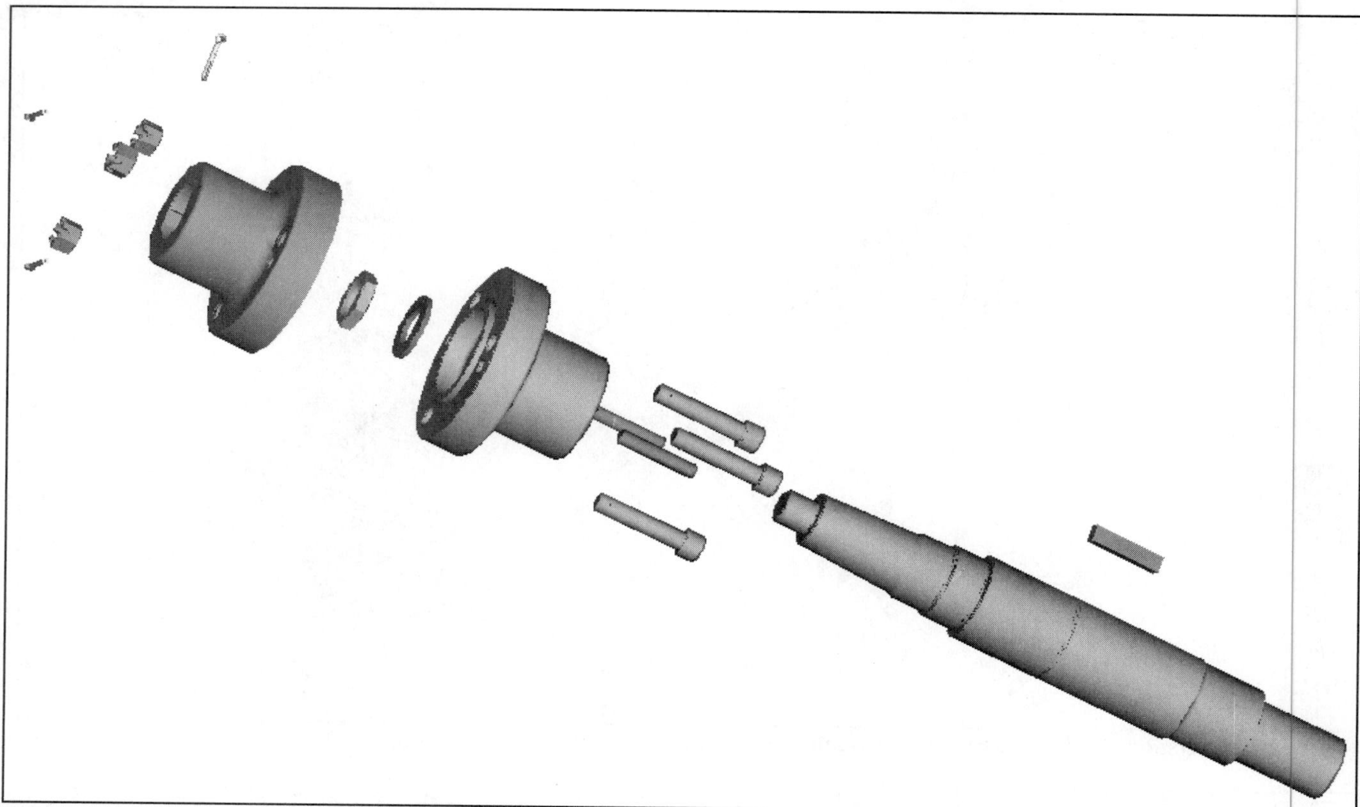

Figure 16.23 Shaded Exploded Coupling Assembly

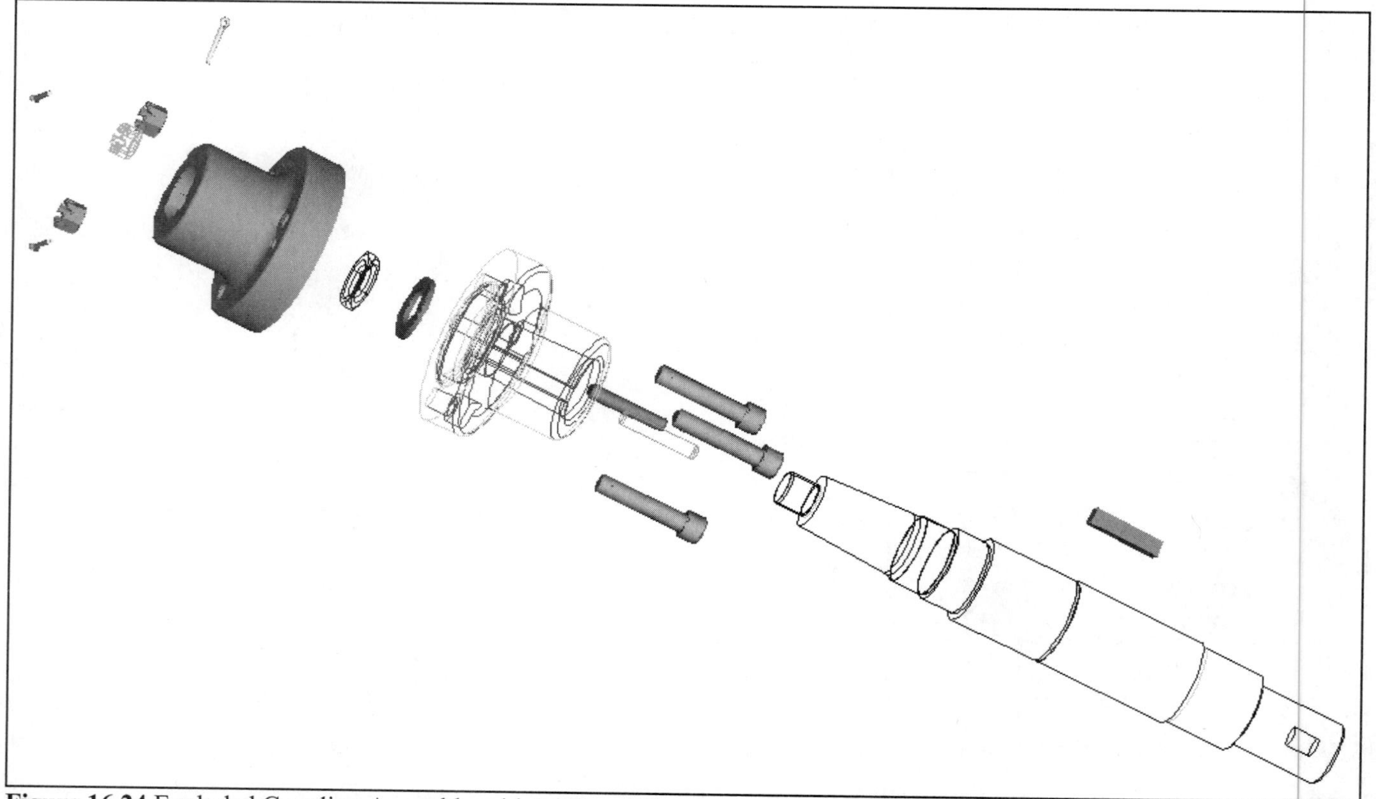

Figure 16.24 Exploded Coupling Assembly with Different Component Display Styles

Lesson 17 Formats, Title Blocks, and Views

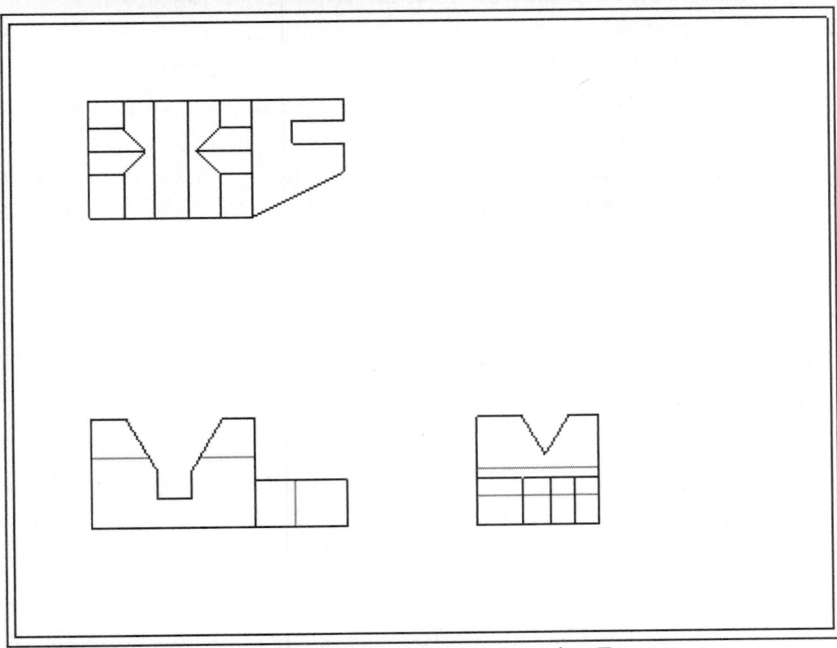

Figure 17.1(a) Base_Angle Drawing without Drawing Format

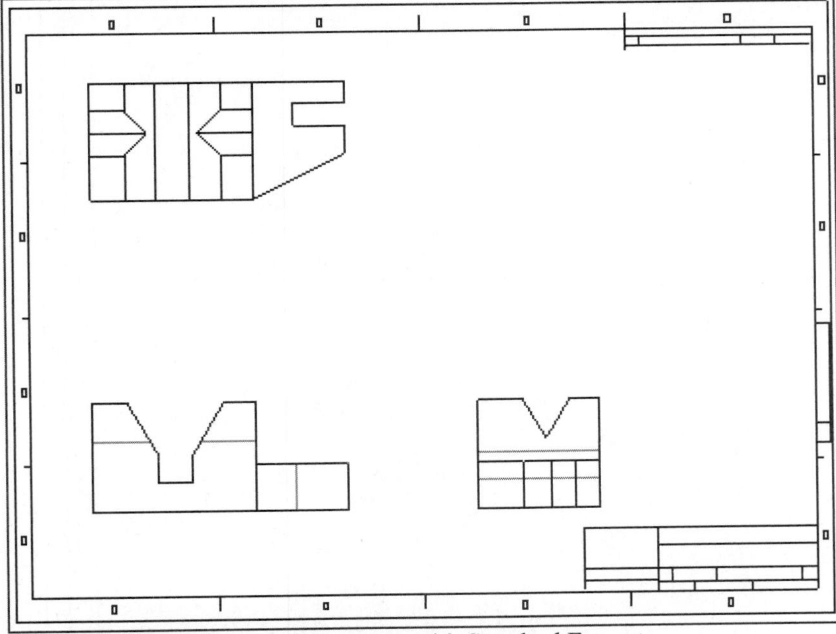

Figure 17.1(b) Base_Angle Drawing with Standard Format

OBJECTIVES

- Specify and retrieve standard **Format** paper size and units
- Create and save **title blocks** and **Formats**
- Create **Drawings** with **Views**
- Change the **Scale** of a view
- **Display Views** for detailing a project
- **Move**, **Erase**, and **Delete** views

FORMATS, TITLE BLOCKS, AND VIEWS

Formats are user-defined drawing sheet layouts. A drawing can be created with an empty format [Fig. 17.1(a)] and have a standard size format added as needed [Fig. 17.1(b)]. A format can be added to any number of drawings. The format can also be modified or replaced in a Pro/E drawing at any time.

Title Blocks are standard or sketched line entities that can contain parameters (object name, tolerances, scale, and so on) that will show when the format is added to the drawing (Fig. 17.3).

Views created by Pro/E are identical to views constructed manually by a designer on paper. The same rules of projection apply; the only difference is that you choose commands in Pro/E to create the views as needed.

Formats

Formats consist of draft entities, not model entities. There are two types of formats: standard and sketched. You can select from a list of **Standard Formats** (**A-F** and **A0-A4**) or create a new size by entering values for length and width.

Sketched Formats created in Sketcher mode (Fig. 17.2) may be parametrically modified, enabling you to create nonstandard-size formats or families of formats.

Formats can be altered to include note text, symbols, tables, and drafting geometry, including drafting cross sections and filled areas.

Figure 17.2 Sketched Format (CADTRAIN, COAch for Pro/ENGINEER)

With Pro/E, you can do the following in Format mode:

- Create draft geometry, notes
- Move, mirror, copy, group, translate, and intersect geometry
- Use and modify the draft grid
- Enter user attributes
- Create drawing tables
- Use interface tools to create plot, DXF, SET, and IGES files
- Import IGES, DXF, and SET files into the format
- Create user-defined line styles
- Create, use, and modify symbols
- Include drafting cross sections in a format

Whether you use a standard format or a sketched format, the format is added to a drawing that is created for a set of specified views of a parametric 3D model.

When you place a format on your drawing, the system automatically writes the appropriate notes based on information in the model you use.

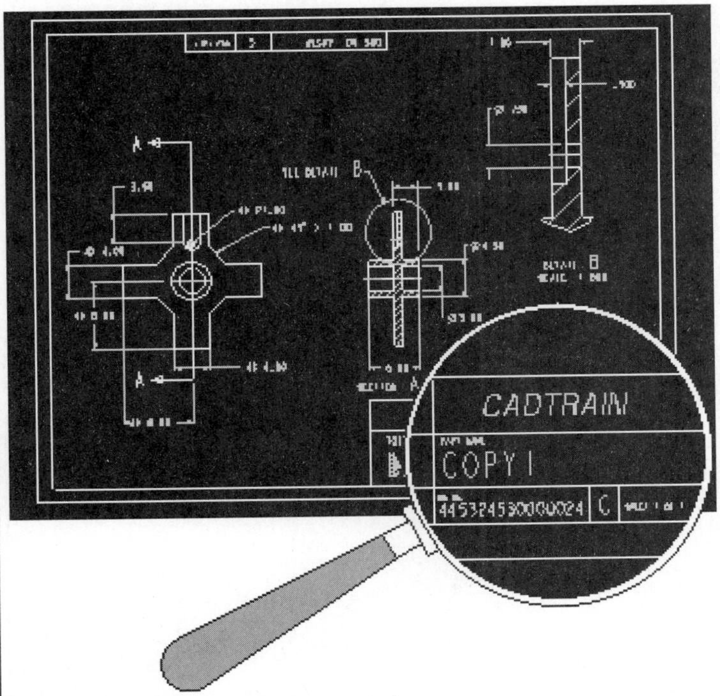

Figure 17.3 Format with Parametric Notes, Added to a Drawing (CADTRAIN, COAch for Pro/ENGINEER)

Specifying the Format Size when Creating a New Drawing

If you want to use an existing template, select the template listed [Fig. 17.4(a)], or select the template from the appropriate directory. If you want to use the existing format [Fig. 17.4(b)], using Browse will open Pro/E's System Formats folder [Fig. 17.4(c)]. Select from the list of standard formats or navigate to a directory with user created formats.

If you want to create your own variable size format [Figs. 17.4(d-e)], enter values for width and length. The main grid spacing and format text units depend on the units selected for a variable size format.

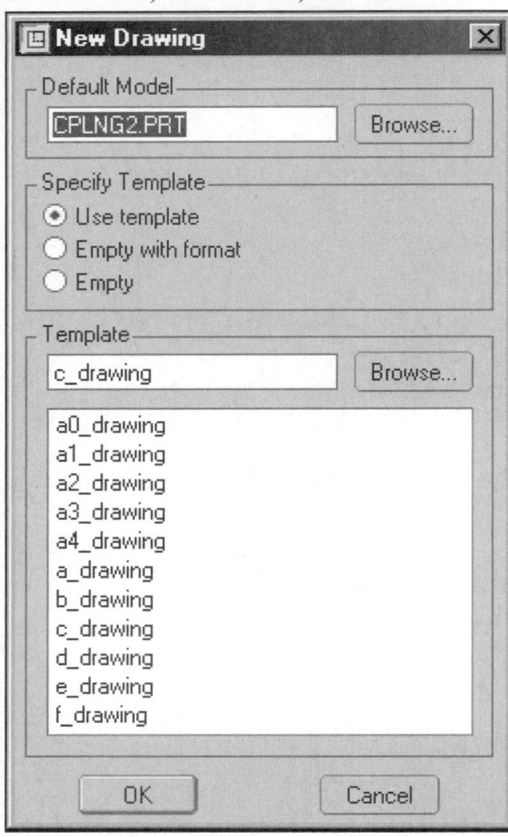

Figure 17.4(a) New Drawing Dialog Box (Use Template)

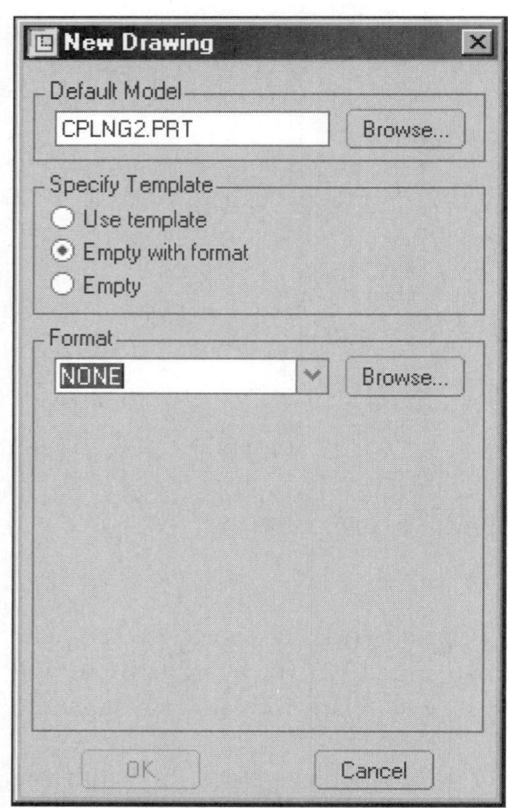

Figure 17.4(b) Empty with format

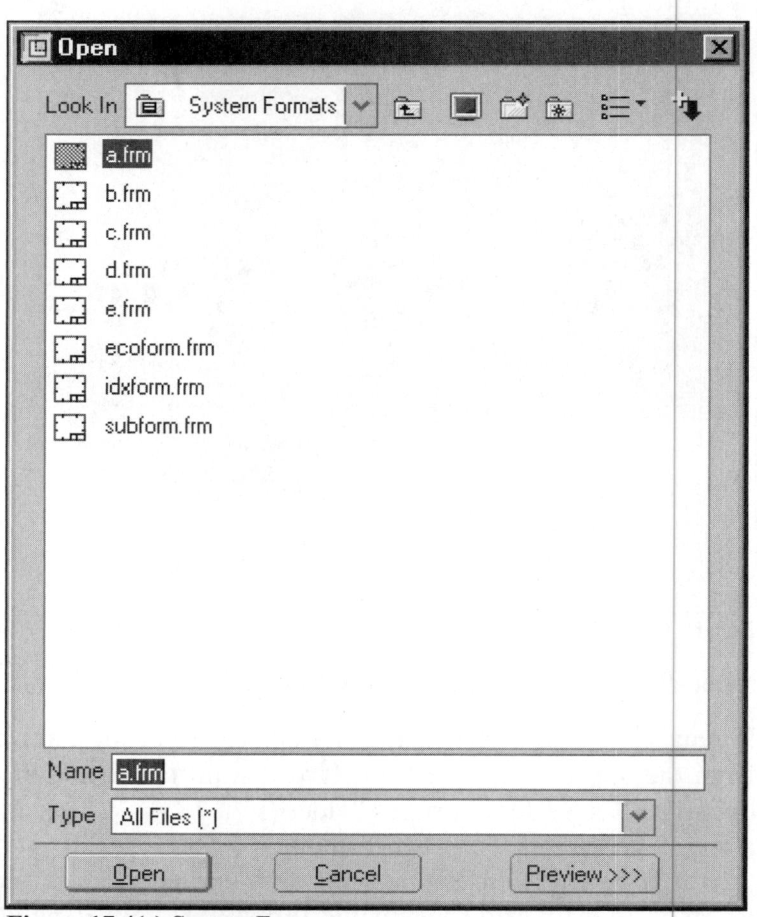

Figure 17.4(c) System Formats

The New Drawing dialog box also provides options for the orientation of the format sheet:

- **Portrait** Uses the larger of the dimensions of the sheet size for the format's height; uses the smaller for the format's width
- **Landscape** Uses the larger of the dimensions of the sheet size for the format's width; uses the smaller for the format's height [Fig. 17.4(d)]
- **Variable** Select the unit type, Inches or Millimeters, and then enter specific values for the Width and Height of the format [Fig. 17.4(e)]

A0	841 X 1189	mm	**A**	8.5 X 11	in.
A1	594 X 841	mm	**B**	11 X 17	in.
A2	420 X 594	mm	**C**	17 X 22	in.
A3	297 X 420	mm	**D**	22 X 34	in.
A4	210 X 297	mm	**E**	34 X 44	in.
			F	28 X 40	in.

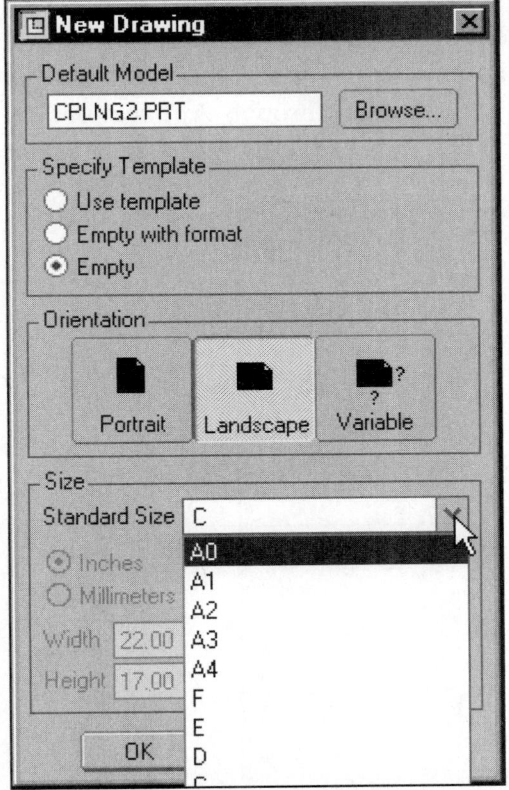

Figure 17.4(d) New Drawing Dialog Box (Empty- Landscape)

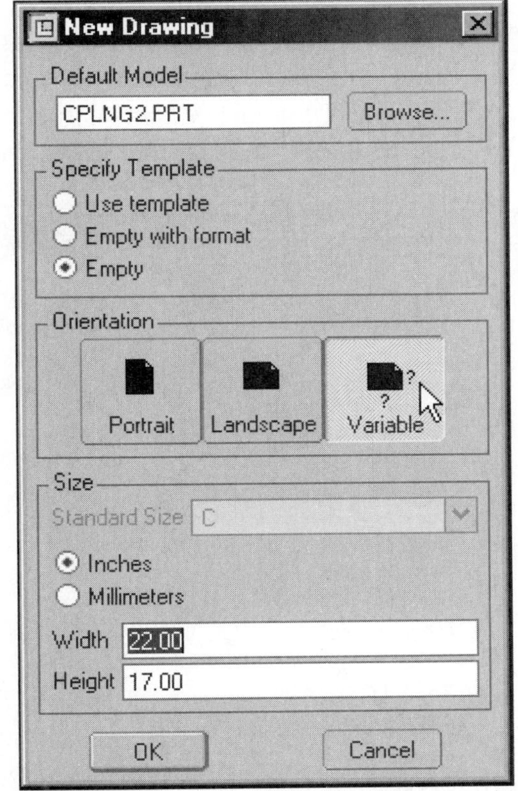

Figure 17.4(e) New Drawing (Empty- Variable)

Drawing Templates

Drawing templates may be referenced when creating a new drawing. Templates can automatically create the views, set the desired view display, create snap lines, and show model dimensions. Drawing templates contain three basic types of information for creating new drawings. The first type is basic information that makes up a drawing but is not dependent on the drawing model, such as notes, symbols, and so forth. This information is copied from the template into the new drawing. The second type is instructions used to configure drawing views and the actions that are performed on their views. The instructions are used to build a new drawing object. The third type is a parametric note. Parametric notes are notes that update to new drawing model parameters and dimension values. The notes are re-parsed or updated when the template is instantiated. Use the templates to:

- Define the layout of views
- Set view display
- Place notes
- Place symbols
- Define tables
- Create snap lines
- Show dimensions

Template View

You can also create customized drawing templates for the different types of drawings that you create. Creating a template allows you to create portions of drawings automatically, using the customizable template. The Template View Instructions dialog box (Fig. 17.5) is accessed through Applications ⇒ Template ⇒ Insert ⇒ Template View, when in the Drawing mode.

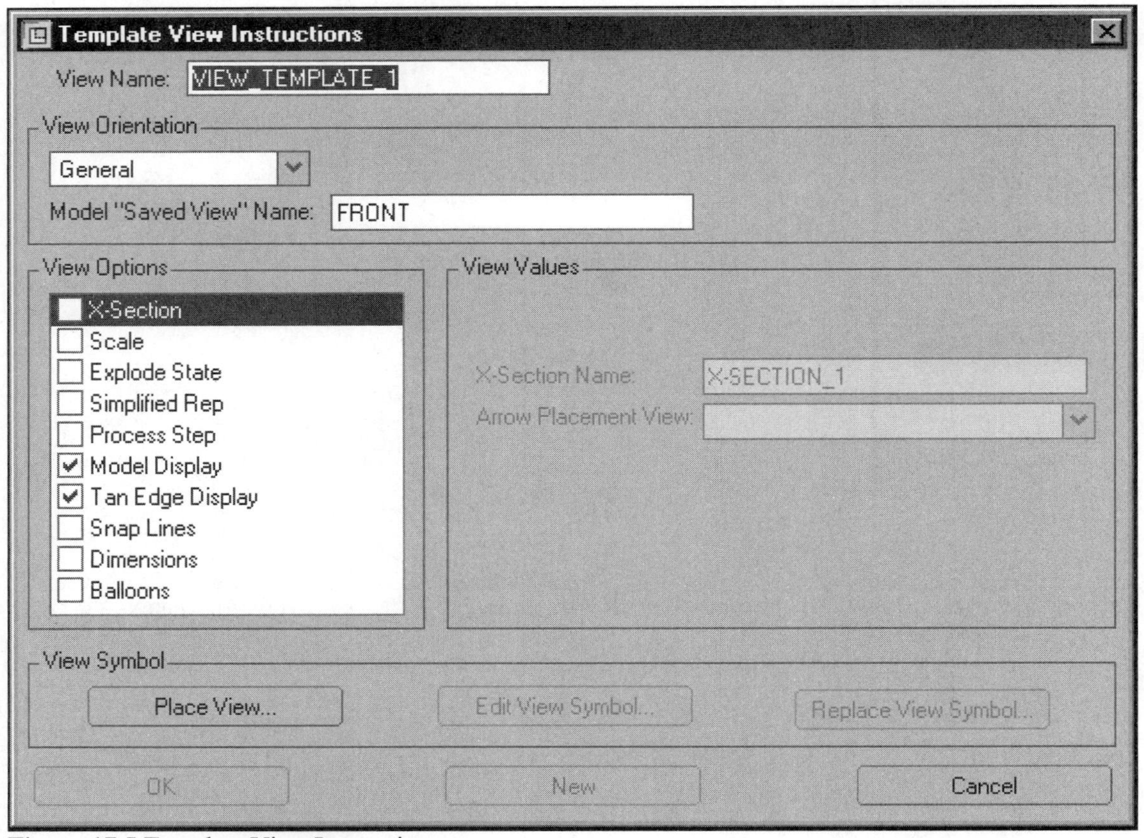

Figure 17.5 Template View Instructions

Use the following options in the Template View Instructions dialog box to customize your drawing templates:

- **View Name** Set the name of the drawing view that will be used as the view symbol label
- **View Orientation** Create a General view or a Projection view
- **Model "Saved View" Name** Orient the view based on a named view in the model
- **Place View** Places the view after you have set the appropriate options and values
- **Edit View Symbol** Allows you to edit the view symbol using the Symbol Instance dialog box
- **Replace View Symbol** Allows you to replace the view symbol using the Symbol Instance dialog box

View Options	View Values
X-Section Set cross-section. Only full, total cross-sections are supported	X-Section Name, Arrow Placement View
Scale Enter the scale value or no scale	View Scale
Explode State Set the exploded view	Explode Name
Simplified Rep Set the view as a simplified representation. If it is not a simplified representation, the default is a master representation	Simplified Rep Name
Process Step Set the process step for the view	Step Number
Model Display Set the view display for the drawing view	Wireframe, Hidden Line, No Hidden, Default
Tan Edge Display Set the tangent edge display	Tan Solid, No Disp Tan, Tan Ctrln, Tan Phantom, Tan Dimmed, Tan Default
Snap Lines Set the number, spacing and offset of the snap lines	Number, Incremental Spacing, Initial Offset
Dimensions Show dimensions on the view	Create Snap Lines, Incremental Spacing, Initial Offset
Balloons Show balloons on the view	

Sketching the Format

To create format geometry, you use draft geometry. The sheet outline is the border of the standard drawing format you selected. Because it is the actual border, it may not show up on "pen plots" unless you use a paper size larger than the drawing size.

Everything within the sheet outline border also plots, but you should make an allowance for the plotter's hold-down rollers. When you add a sketched format to a drawing, Pro/E aligns the lower left corner (the origin) of the format with the lower left corner (the origin) of the drawing, and centers all items in the drawing on the new sheet in the locations that correspond to their positions on the original sheet. If necessary, it adjusts the drawing scale, maintaining relative distances between items.

You can draw lines, arcs, splines, etc., on a drawing format to represent the border and title block. An easier method for making custom formats is to use a standard format, save a copy of it under a new name and edit the newly copied format as desired for your company or school.

Views

A wide variety of views (Fig. 17.6) can be derived from the parametric model. Among the most common are projection views. Pro/E creates projection views by looking to the left of, to the right of, above, and below the picked view location to determine the orientation of a projection view. When conflicting view orientations are found, you are prompted to select the view that will be the parent view. A view will then be constructed from the selected view.

At the time when they are created, projection, auxiliary, detailed, and revolved views have the same representation and explosion offsets, if any, as their parent views. From that time onward, each view can be simplified, be restored, and have its explosion distance modified without affecting the parent view. The only exception to this is detailed views, which will always be displayed with the same explosion distances and geometry as their parent views.

Once a model has been added to the drawing, you can place views of the model on a sheet. When a view is placed, you can determine how much of the model to show in a view, whether the view is of a single surface or shows cross sections, and how the view is scaled. You can then show the associative dimensions passed from the 3D model, or add reference dimensions as necessary.

Basic view types used by Pro/Engineer include general, projection, auxiliary, and detailed:

- **General** Creates a view with no particular orientation or relationship to other views in the drawing. The model must first be oriented to the desired view orientation established by you.
- **Projection** Creates a view that is developed from another view by projecting the geometry along a horizontal or vertical direction of viewing (orthographic projection). The projection type is specified by you in the drawing setup file and can be based on third-angle (default) or first-angle rules.
- **Auxiliary** Creates a view that is developed from another view by projecting the geometry at right angles to a selected surface or along an axis. The surface selected from the parent view must be perpendicular to the plane of the screen.
- **Detailed** Details a portion of the model appearing in another view. Its orientation is the same as that of the view it is created from, but its scale may be different so that the portion of the model being detailed can be better visualized.

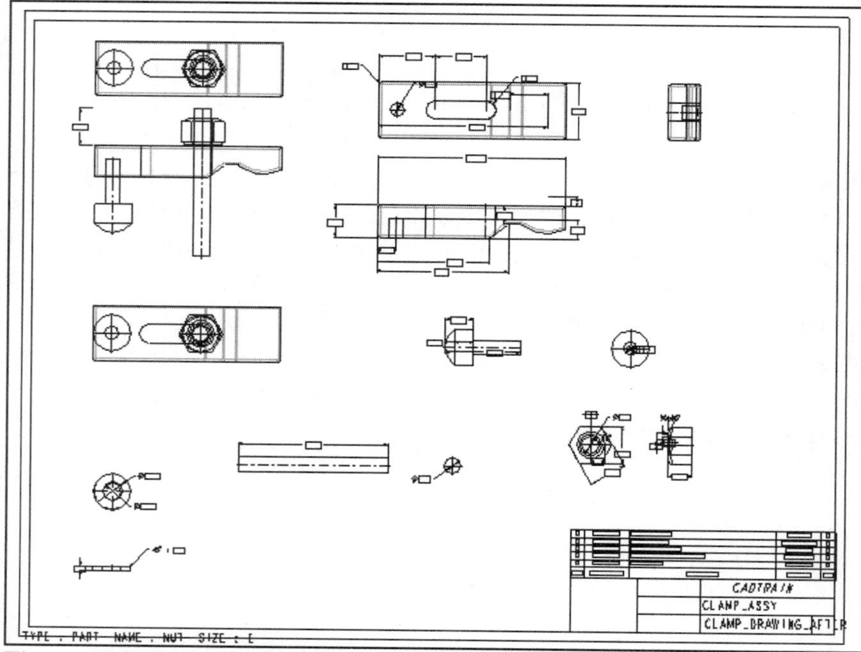

Figure 17.6 Views (CADTRAIN, COAch for Pro/ENGINEER)

The view options that determine how much of the model is visible in the view are:

- **Full View** Shows the model in its entirety.
- **Half View** Removes a portion of the model from the view on one side of a cutting plane.
- **Broken View** Removes a portion of the model from between two selected points and closes the remaining two portions together within a specified distance.
- **Partial View** Displays a portion of the model in a view within a closed boundary. The geometry appearing within the boundary is displayed; the geometry outside of it is removed.

The options that determine whether the view is of a single surface or has a cross section are:

- **Section** Displays an existing cross section of the view if the view orientation is such that the cross-sectional plane is parallel to the screen (Fig. 17.7)
- **No Xsec** Indicates that no cross section is to be displayed.
- **Of Surface** Displays a selected surface of a model in the view. The single-surface view can be of any view type except detailed.

The options that determine whether the view is scaled are:

- **Scale** Allows you to create a view with an individual scale shown under the view. When a view is being created, Pro/E will prompt you for the scale value. This value can be modified later. General and detailed views can be scaled.
- **No Scale** A view will be scaled automatically using a pre-defined scale value.
- **Perspective** Creates a perspective general view.

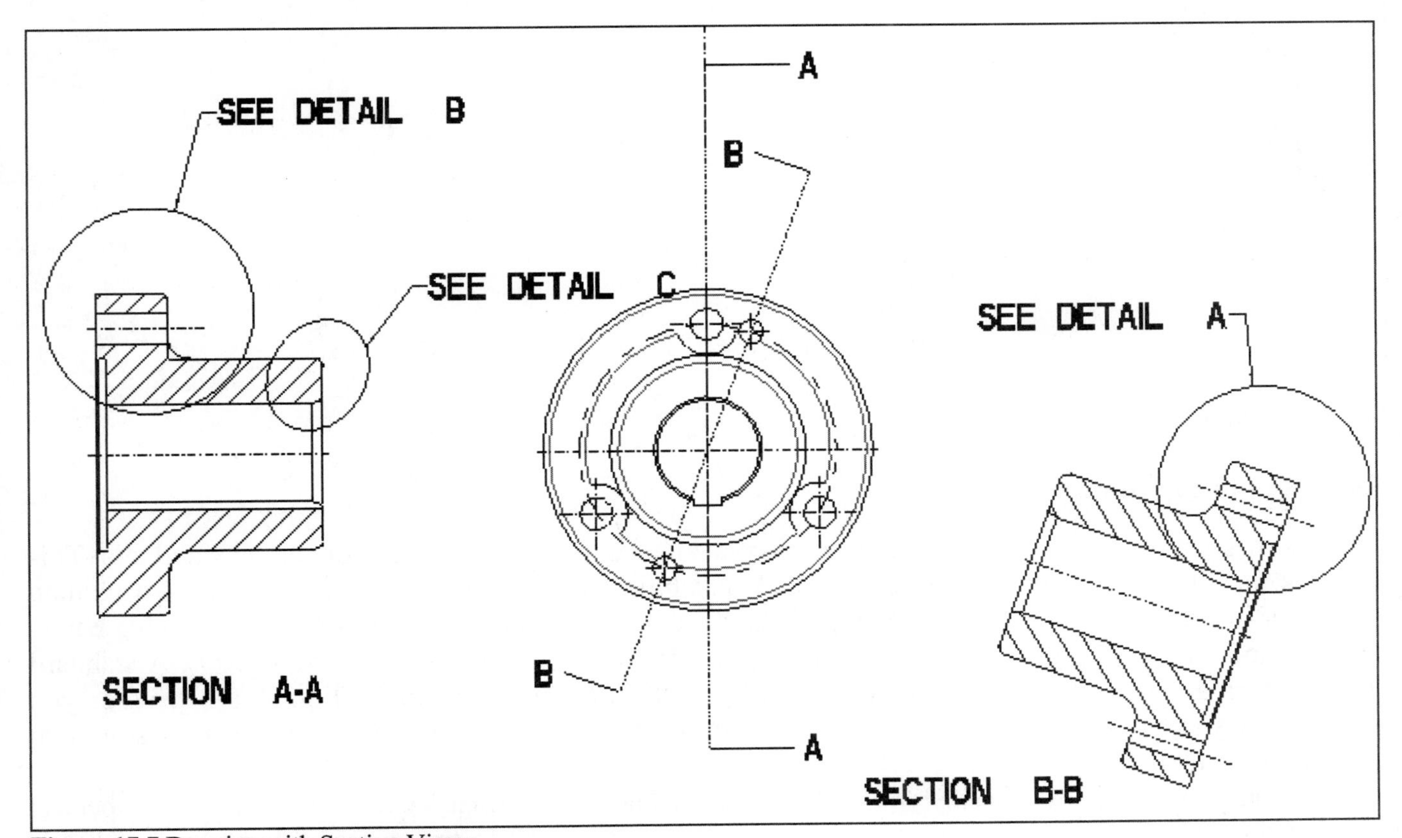

Figure 17.7 Drawing with Section Views

Lesson 17 STEPS

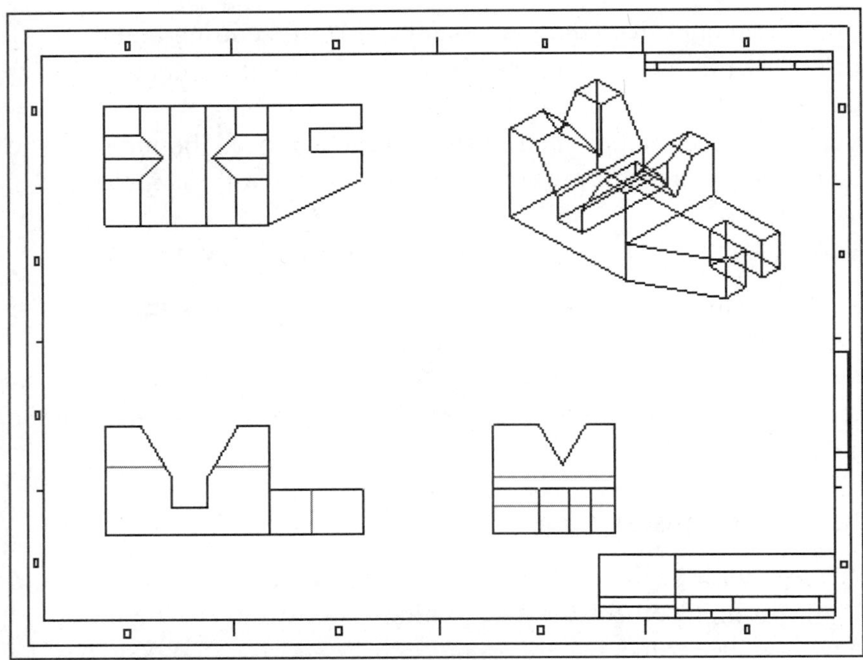

Figure 17.8(a) Base_Angle Drawing with Standard Views

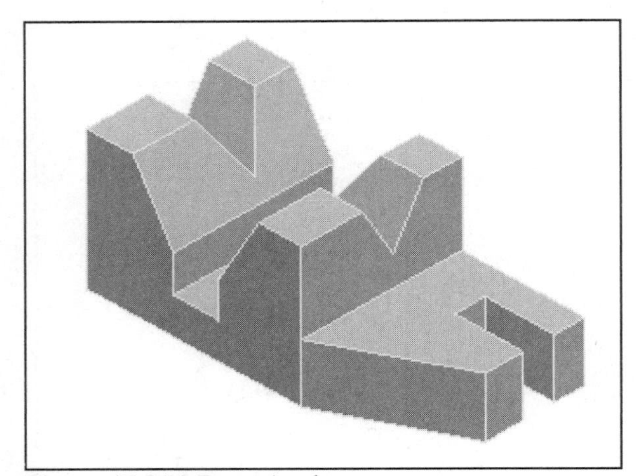

Figure 17.8(b) Base_Angle

Base Angle Drawing

Pro/E allows you to create your own drawing formats to support your company or school standards. This includes the lines that make up the border and title block, as well as text. The text can also contain parameter information that refers to Pro/E-generated values, such as the drawing scale, the drawing name, and the model name. In most cases, you will use the standard formats that come with Pro/E. A standard "**C**" size format has been added to the drawing [Fig. 17.8(a)] of the Base_Angle [Fig. 17.8(b)]. When you add a format to your drawing, Pro/E can write the appropriate notes based on information stored or defined in the model you use.

You can create two basic types of formats: a standard format, which is *locked* to a specific drawing size, or a sketched format, which is parametrically linked to the size of the drawing and thus changes as the drawing size changes.

Drawing Mode and Views

When you start a drawing, you can automatically have default views of the part or assembly displayed, or you can add the views of your liking to a blank drawing. In this lesson, you will use the [Use default template] setting and have the three primary views automatically displayed for you. *This will work only if the part or assembly has a Saved View named FRONT.*

You must be in the same working directory as your part. It is also a good practice to bring up the part in a separate Pro/E window.

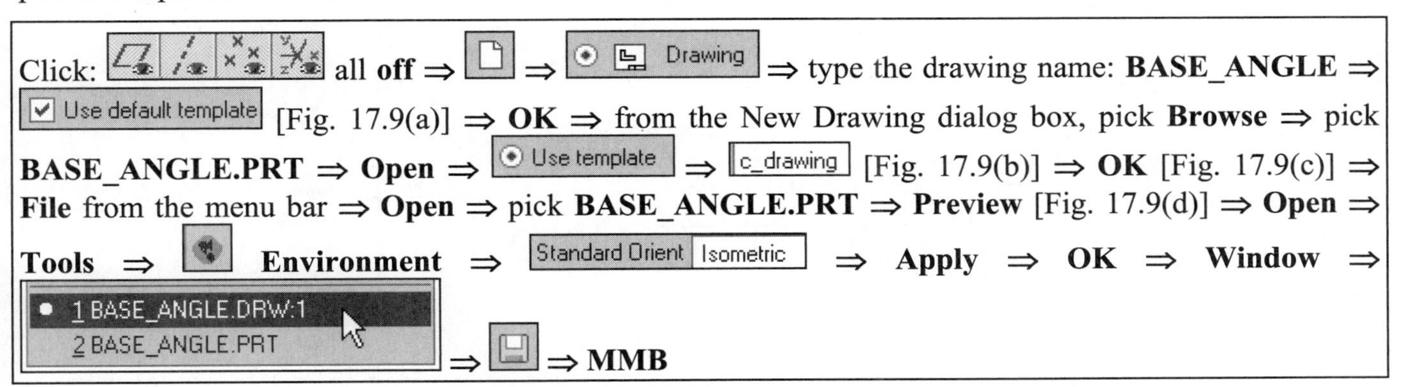

Click: all **off** ⇒ ⇒ Drawing ⇒ type the drawing name: **BASE_ANGLE** ⇒ Use default template [Fig. 17.9(a)] ⇒ **OK** ⇒ from the New Drawing dialog box, pick **Browse** ⇒ pick **BASE_ANGLE.PRT** ⇒ **Open** ⇒ Use template ⇒ c_drawing [Fig. 17.9(b)] ⇒ **OK** [Fig. 17.9(c)] ⇒ **File** from the menu bar ⇒ **Open** ⇒ pick **BASE_ANGLE.PRT** ⇒ **Preview** [Fig. 17.9(d)] ⇒ **Open** ⇒ **Tools** ⇒ **Environment** ⇒ Standard Orient Isometric ⇒ **Apply** ⇒ **OK** ⇒ **Window** ⇒ 1 BASE_ANGLE.DRW:1 / 2 BASE_ANGLE.PRT ⇒ ⇒ **MMB**

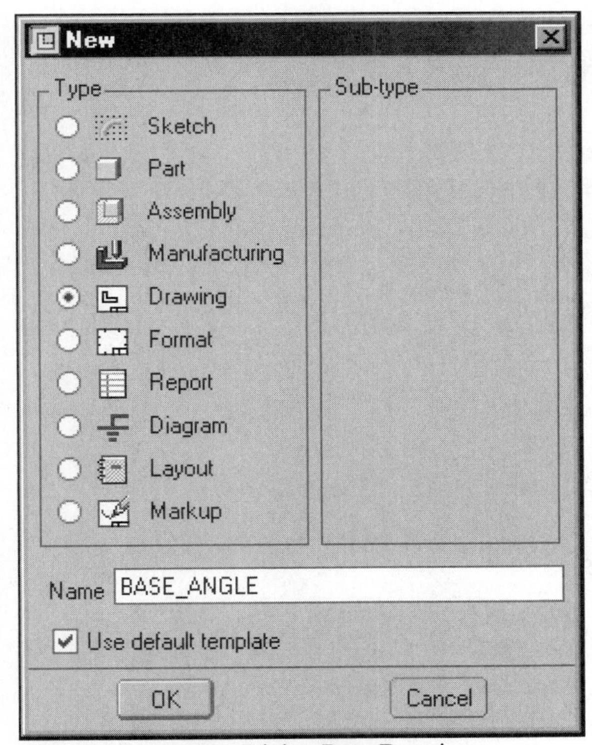

Figure 17.9(a) New Dialog Box, Drawing

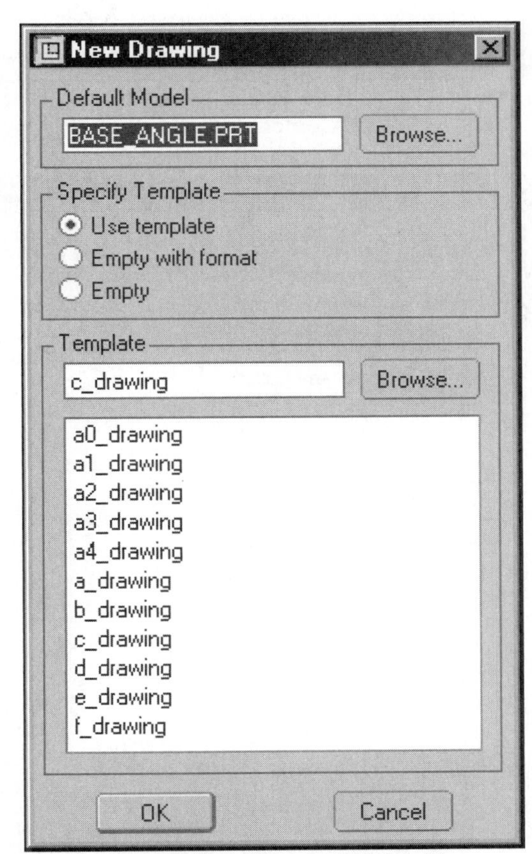

Figure 17.9(b) New Drawing Dialog Box

This will give you a drawing of the Base_Angle part (before ECO modifications), displayed with three standard views. Later, you may also wish to create a drawing for the ECO part. The Base_Angle part is in another window. Toggle between the drawing and the part as needed (Fig. 17.10).

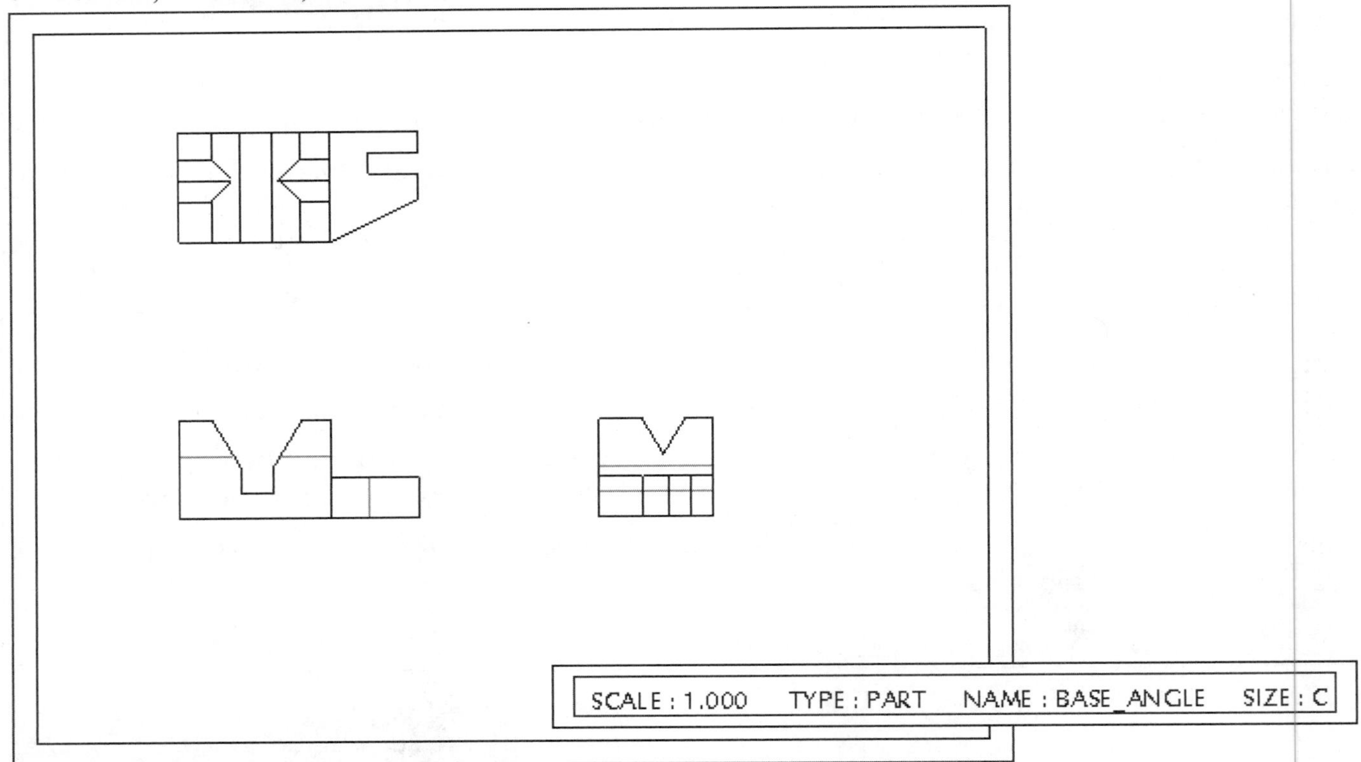

Figure 17.9(c) Base_Angle Drawing with Three Views of the Base_Angle Part. Scale, Type, Name, and Size show in the Lower Left-hand Corner of the Graphics Window.

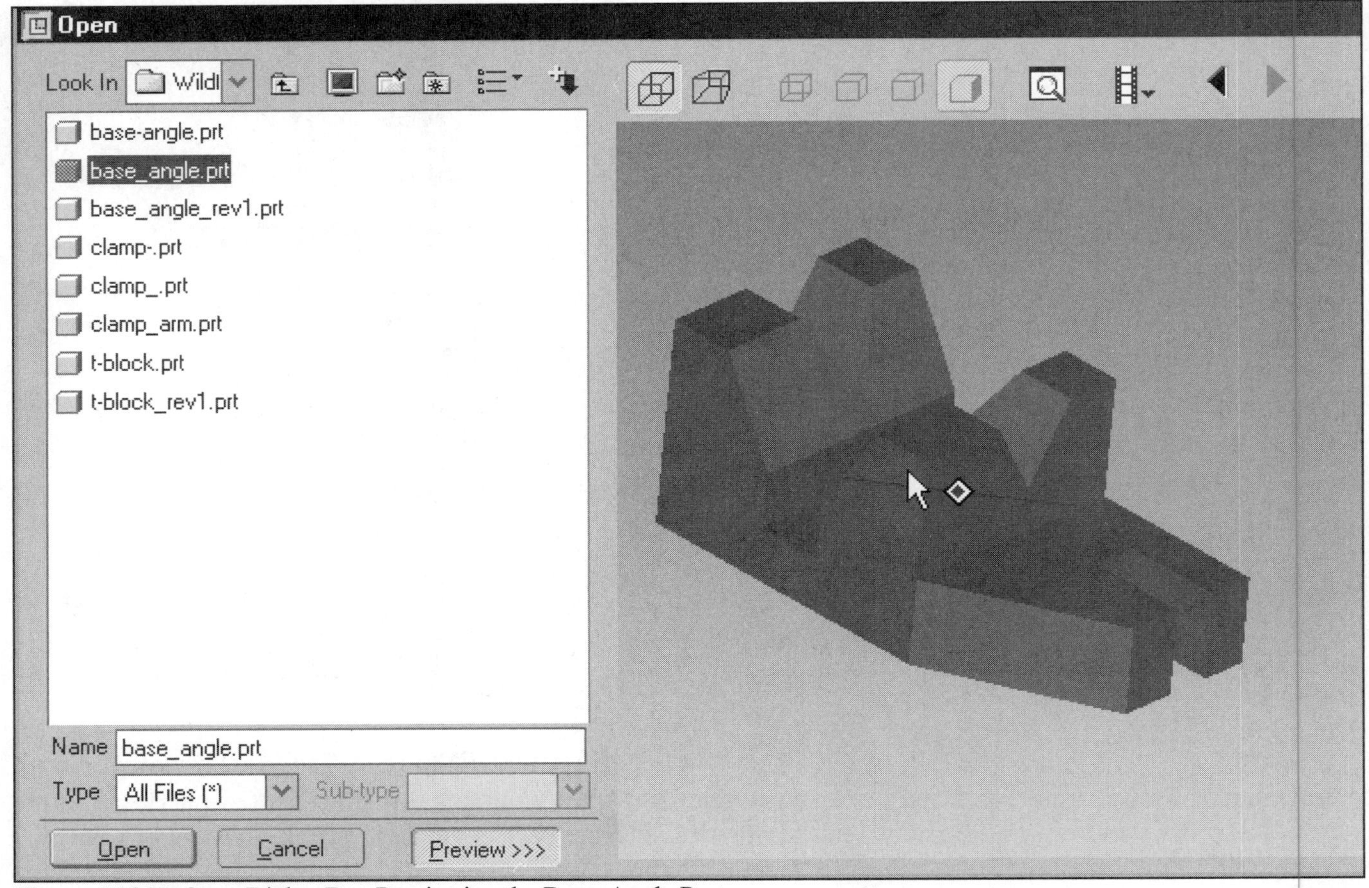

Figure 17.9(d) Open Dialog Box Previewing the Base_Angle Part

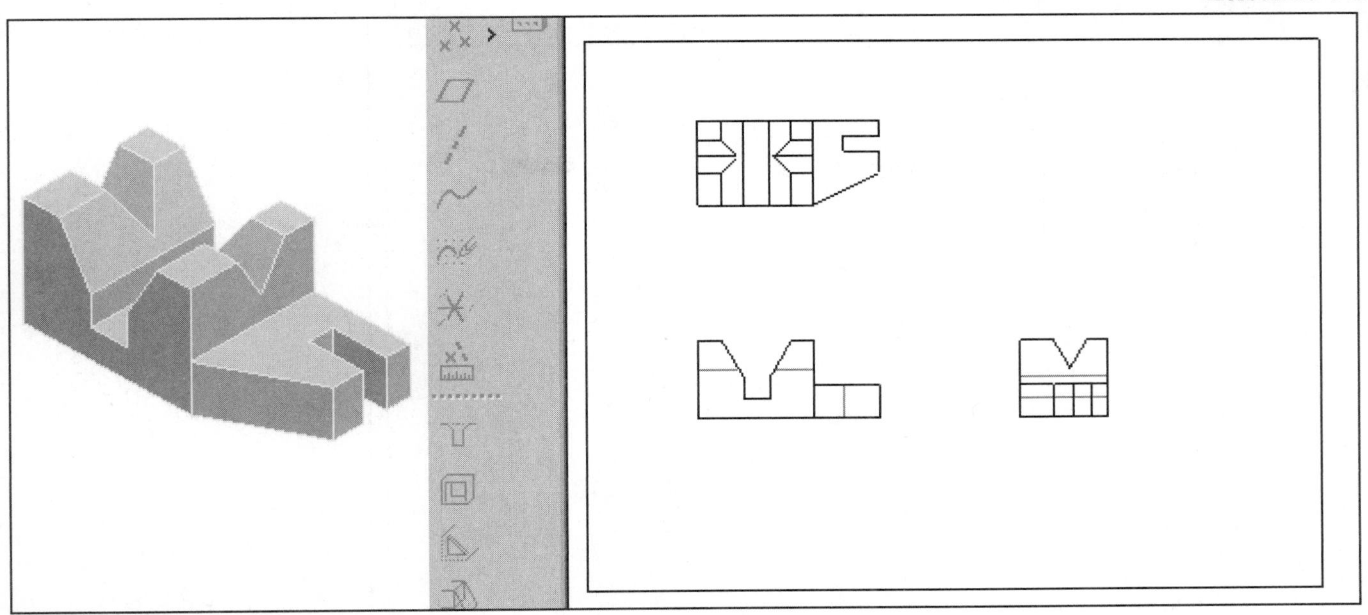
Figure 17.10 Part and Drawing

Add another projected view, click: **Insert a drawing view of the active model ⇒ Projection ⇒ Full View ⇒ No Xsec ⇒ No Scale ⇒ MMB ⇒** pick the position for the new view [Fig. 17.11(a)] *Pro/E prompts: Conflict in parent view exists. Select parent view for making the projection.* ⇒ click on the right side view [Fig. 17.11(b)] ⇒ **LMB** in the graphics window to deselect [Fig. 17.11(c)] ⇒ **Disallow the movement of drawing views with the mouse** off

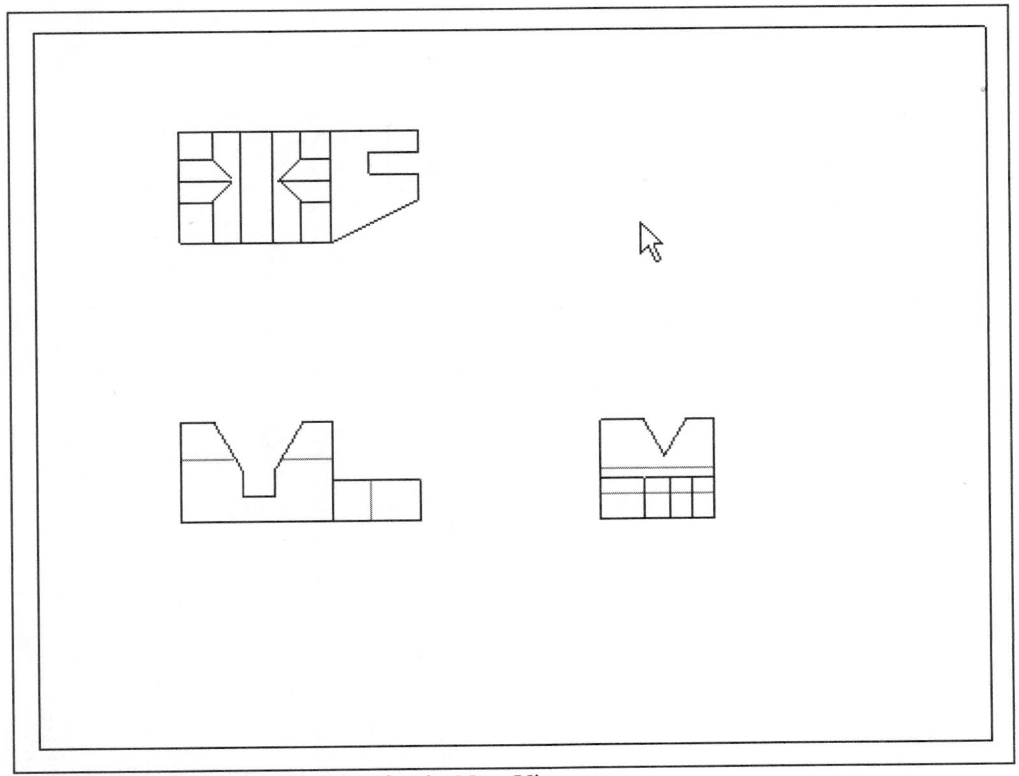
Figure 17.11(a) Pick the Position for the New View

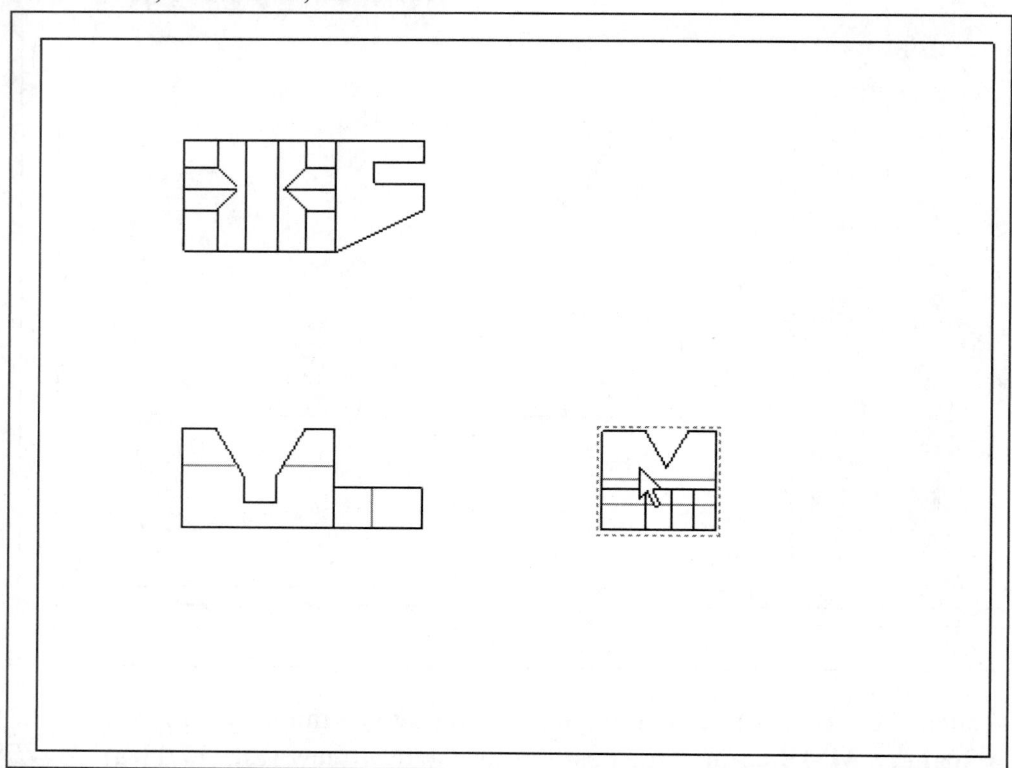

Figure 17.11(b) Pick the Parent View

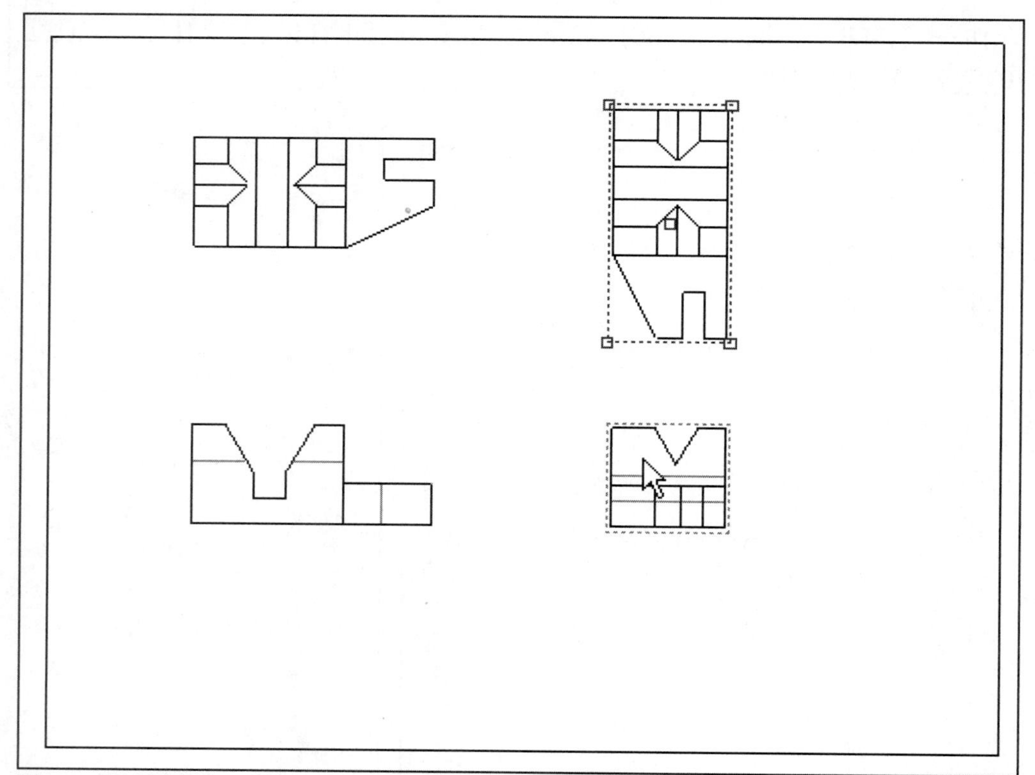

Figure 17.11(c) New View

Delete the new view by clicking on the view ⇒ **RMB** ⇒ **Delete** [Fig. 17.11(d)] ⇒ **Yes** ⇒ **LMB**

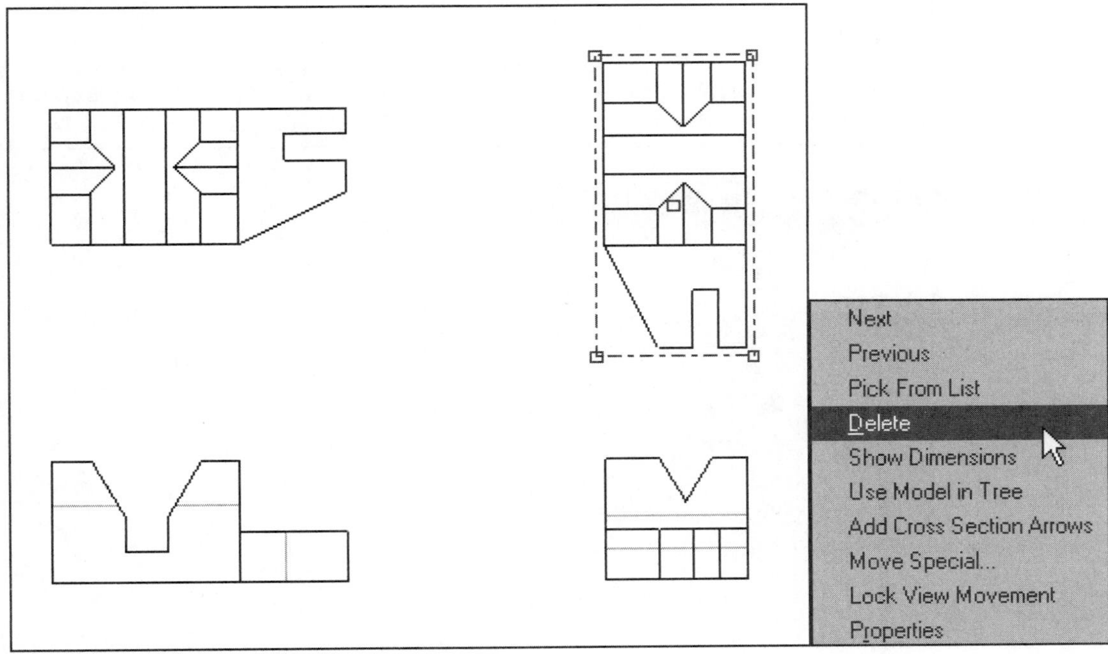

Figure 17.11(d) Delete the View

Add a general view (default display setting in the object's Environment: Standard Orient- Isometric) by clicking: **Insert a drawing view of the active model** ⇒ **General** ⇒ **Full View** ⇒ **No Xsec** ⇒ **No Scale** ⇒ **MMB** ⇒ pick the position for the new view (Fig. 17.12) ⇒ **OK** ⇒ **LMB** ⇒ ⇒ **MMB**

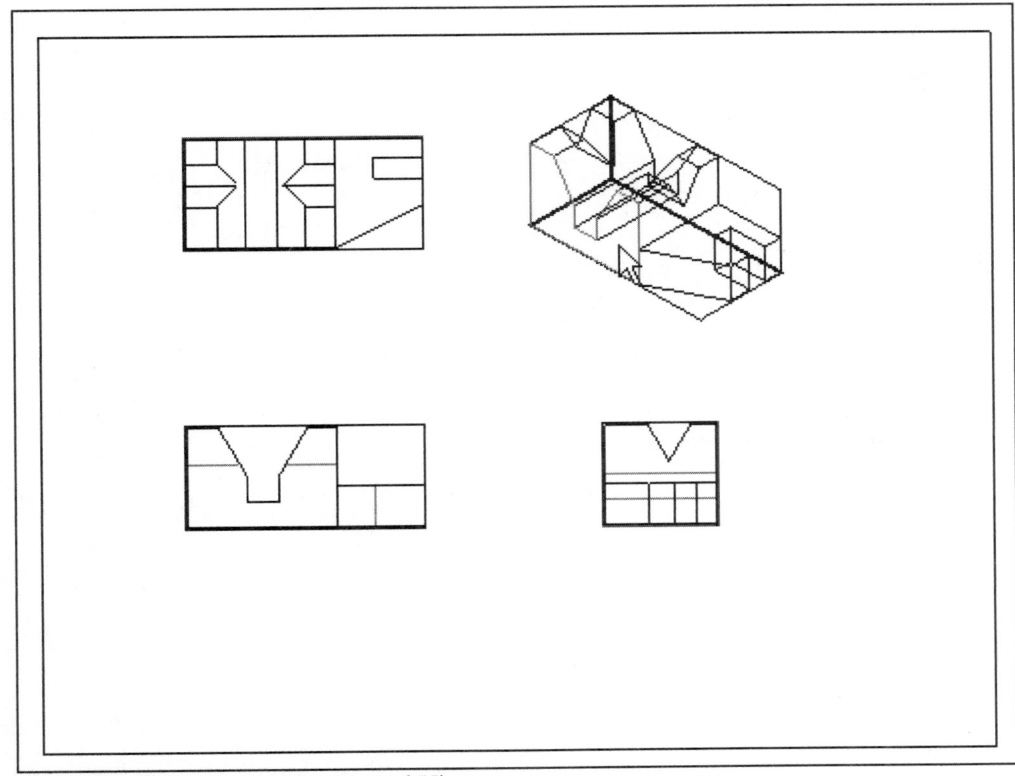

Figure 17.12 Position of the General View

The next step will be to add a standard ANSI "**C**" size format to the drawing sheet.

Click: **File** ⇒ **Page Setup** ⇒ click inside the Format field [Fig. 17.13(a)] ⇒ expand the drop down menu [Fig. 17.13(b)]

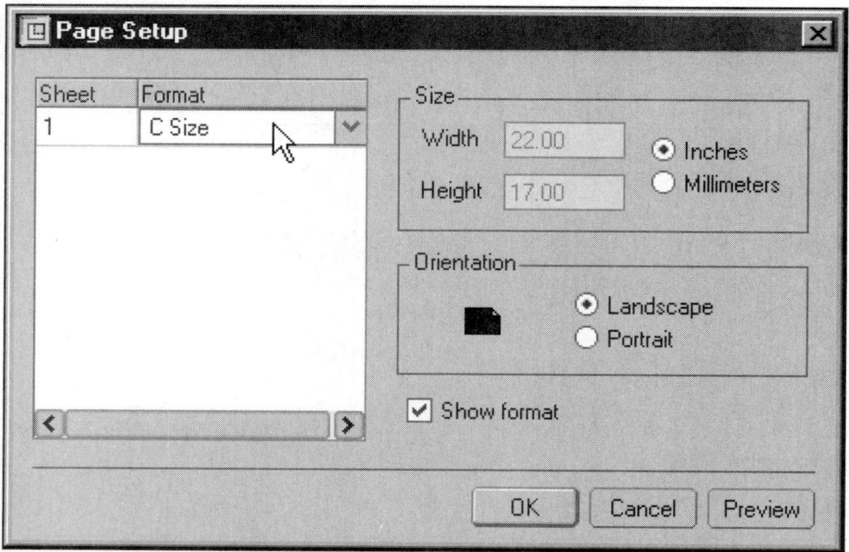

Figure 17.13(a) Page Setup Dialog Box, Format Field

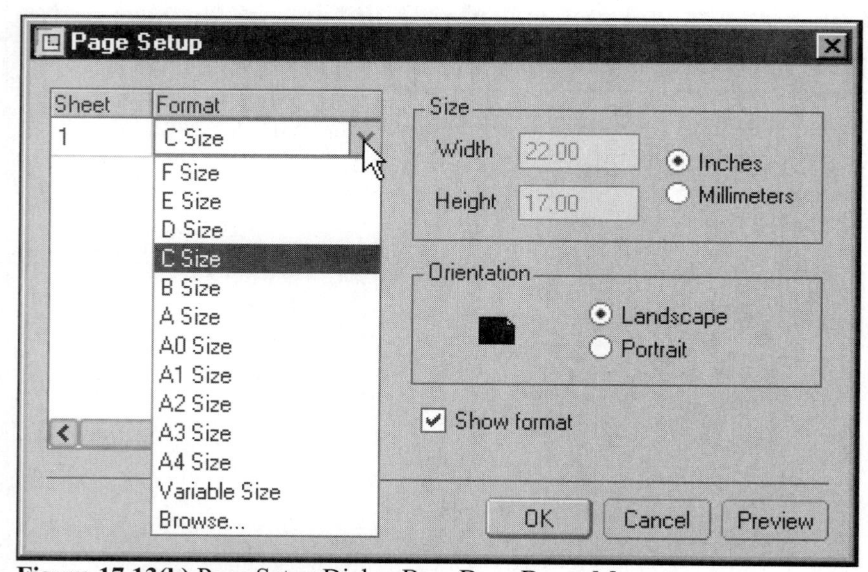

Figure 17.13(b) Page Setup Dialog Box, Drop Down Menu

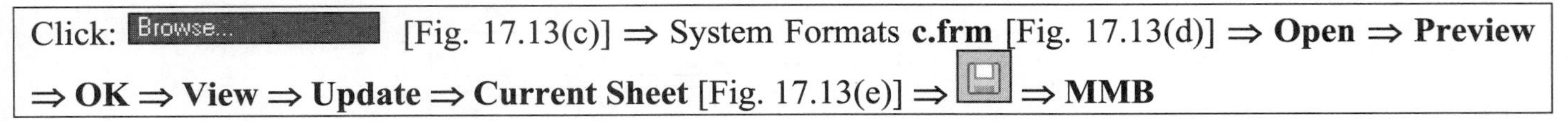
Click: Browse... [Fig. 17.13(c)] ⇒ System Formats **c.frm** [Fig. 17.13(d)] ⇒ **Open** ⇒ **Preview** ⇒ **OK** ⇒ **View** ⇒ **Update** ⇒ **Current Sheet** [Fig. 17.13(e)] ⇒ ⇒ **MMB**

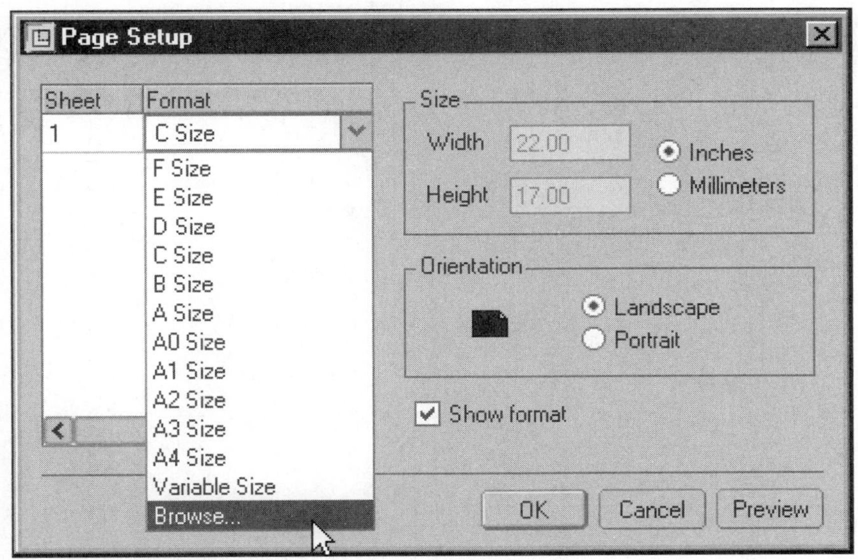

Figure 17.13(c) Click Browse

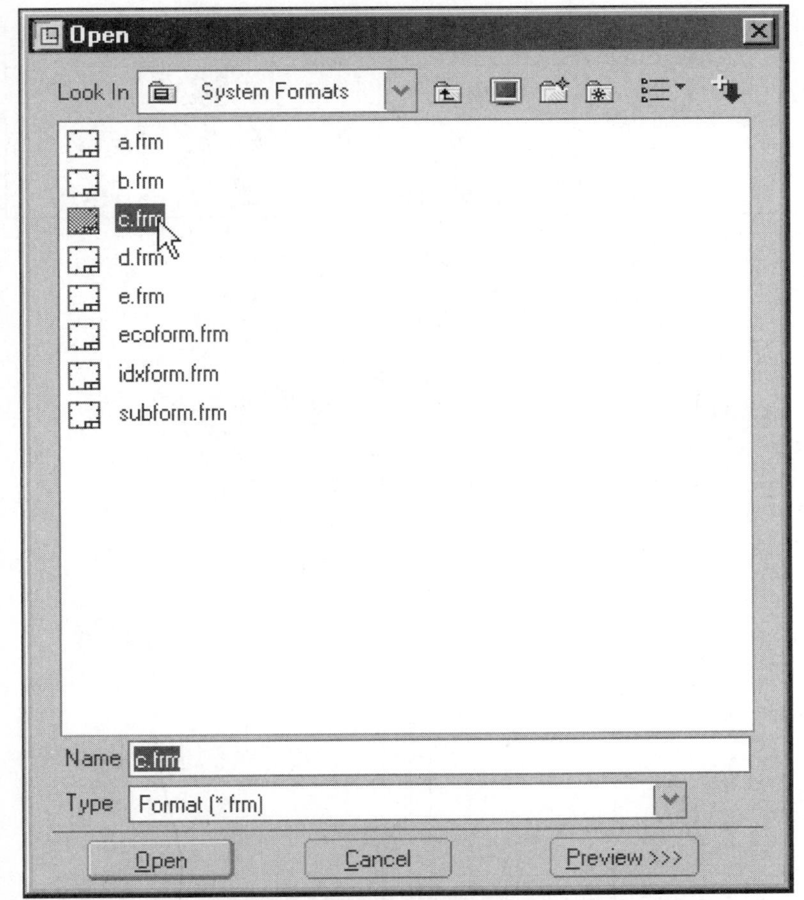

Figure 17.13(d) Open Dialog Box Showing System Formats

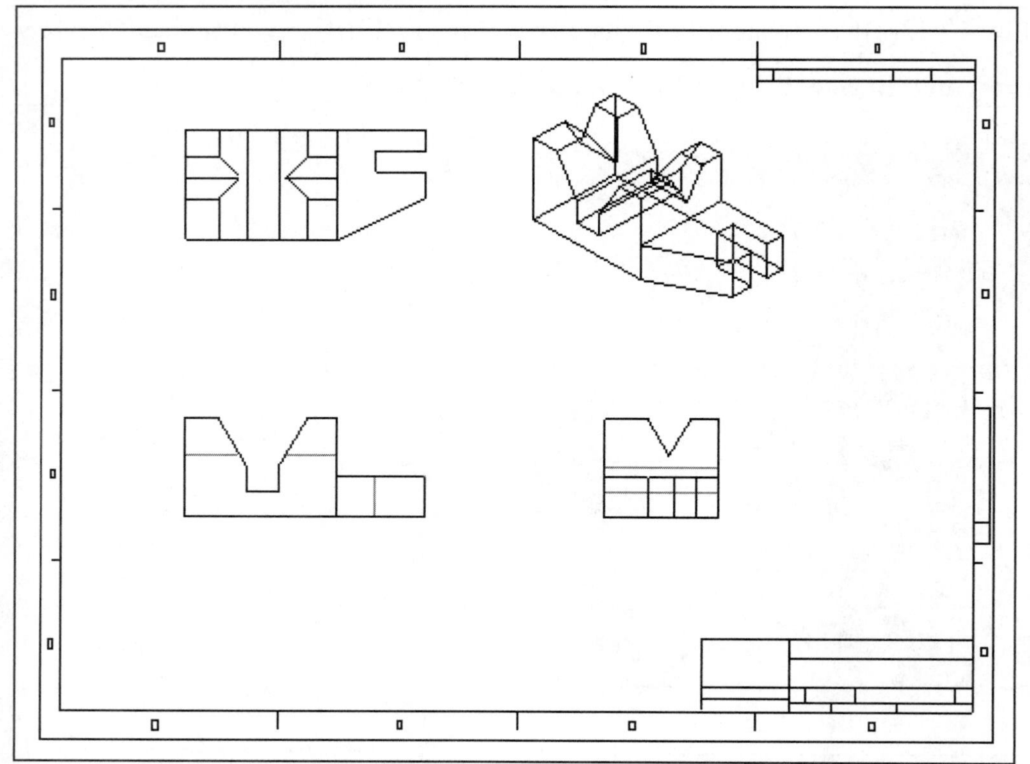

Figure 17.13(e) Drawing with "C" Format

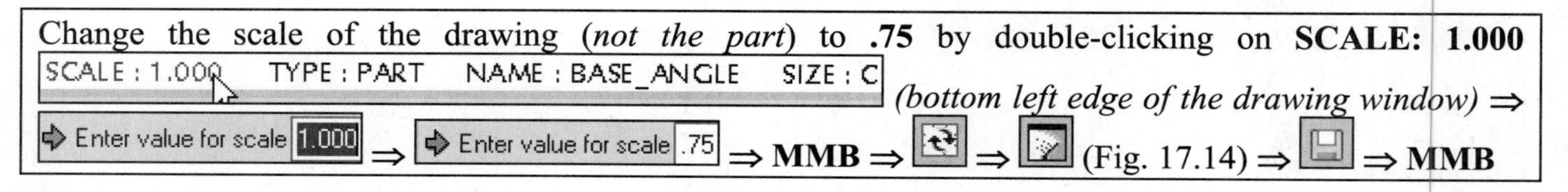

Change the scale of the drawing (*not the part*) to **.75** by double-clicking on **SCALE: 1.000** *(bottom left edge of the drawing window)* ⇒ ⇒ ⇒ **MMB** ⇒ ⇒ (Fig. 17.14) ⇒ ⇒ **MMB**

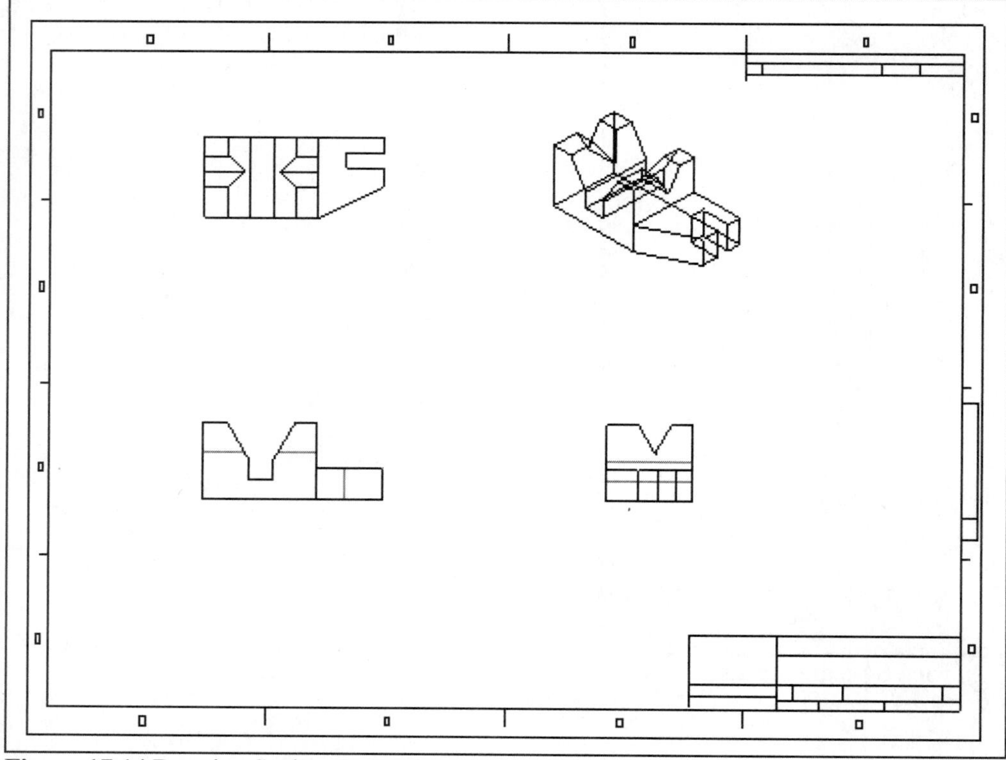

Figure 17.14 Drawing Scale now **.75**

Finally, reposition the views on the drawing. After you select the view to move, it will be highlighted with dashed lines. If the selected view is a parent of another view, that view will also be highlighted, and will move with the selected view.

Pick on the front view [Fig. 17.15(a)] ⇒ cursor will display when positioned over the view [Fig. 17.15(b)] ⇒ Press and hold the **LMB** and drag the front view down and to the left. Both, of the other views are also *highlighted*, being *children* of the front view. The top view follows the front view (right and left). The right side view follows the front view (up and down). [Fig. 17.15(c)] ⇒ release the **LMB** to drop the view ⇒ move the view to a new position, try a variety of positions [Fig. 17.15(d)] ⇒ release **LMB** to place the view ⇒ ⇒ **MMB**

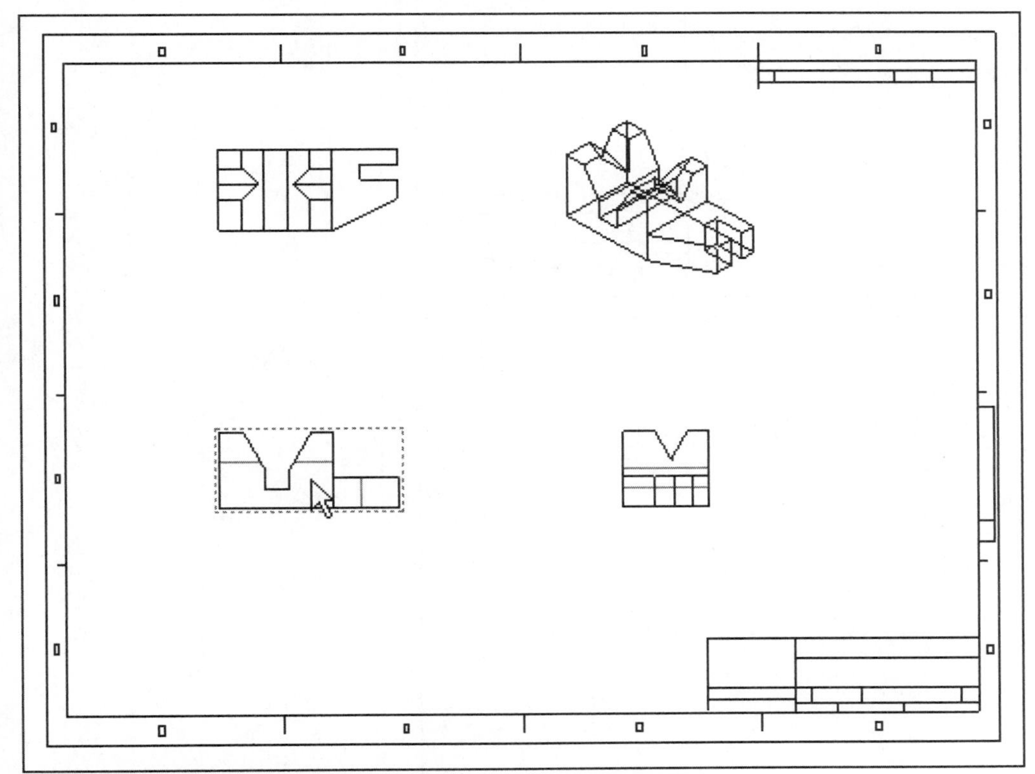

Figure 17.15(a) Click on the Front View

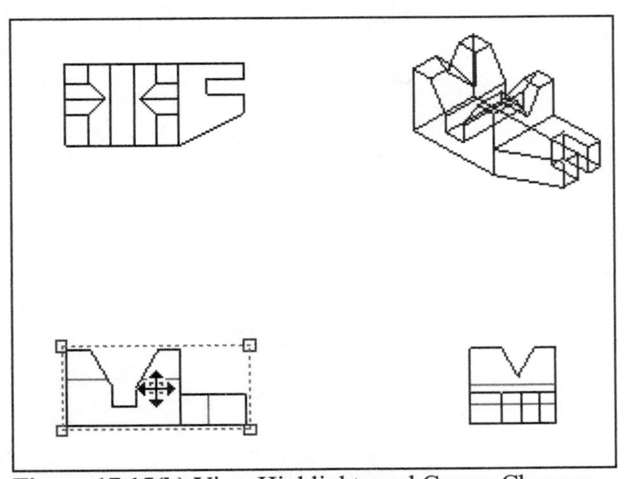

Figure 17.15(b) View Highlights and Cursor Changes

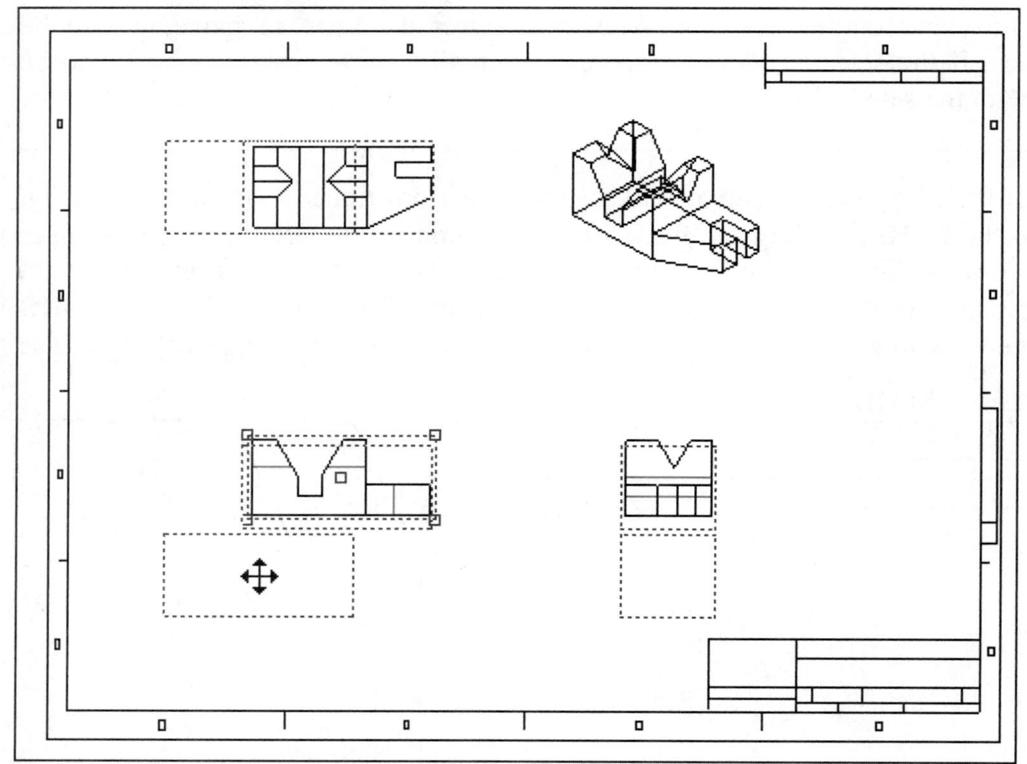

Figure 17.15(c) Drag the Front View

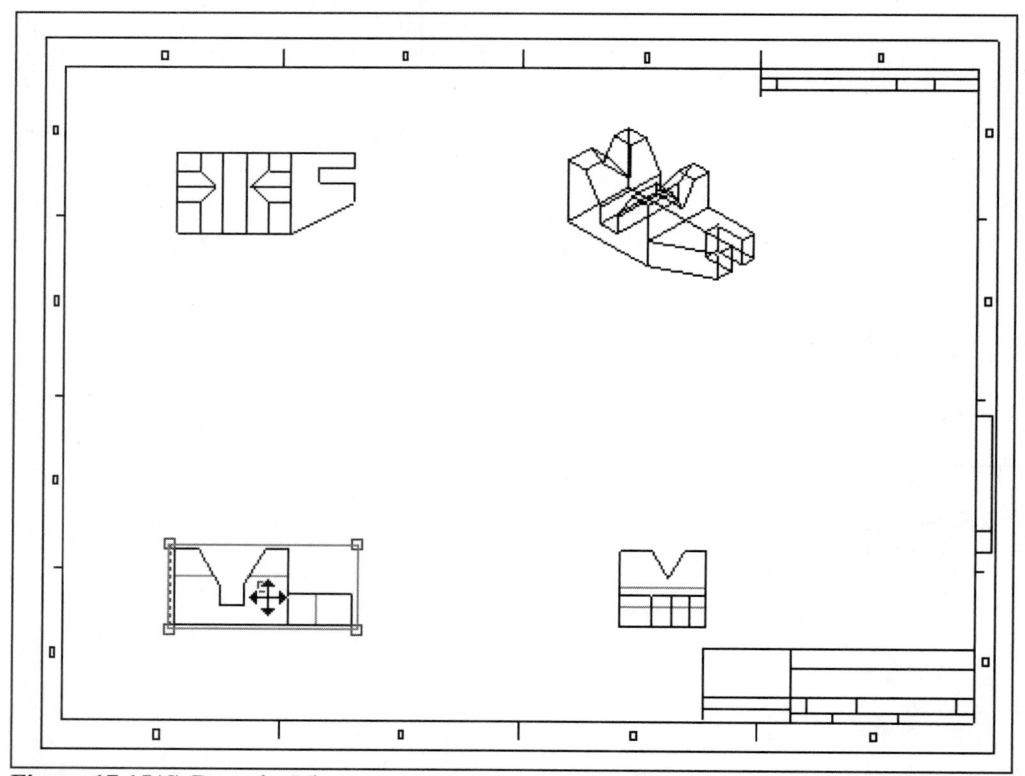

Figure 17.15(d) Drop the View

Move the right side view to the left ⇒ move the General (pictorial) view to the right [Fig. 17.15(e)] ⇒ continue until you are satisfied with the view placement [Fig. 17.15(f)] ⇒ ⇒ ⇒ ⇒ **MMB**

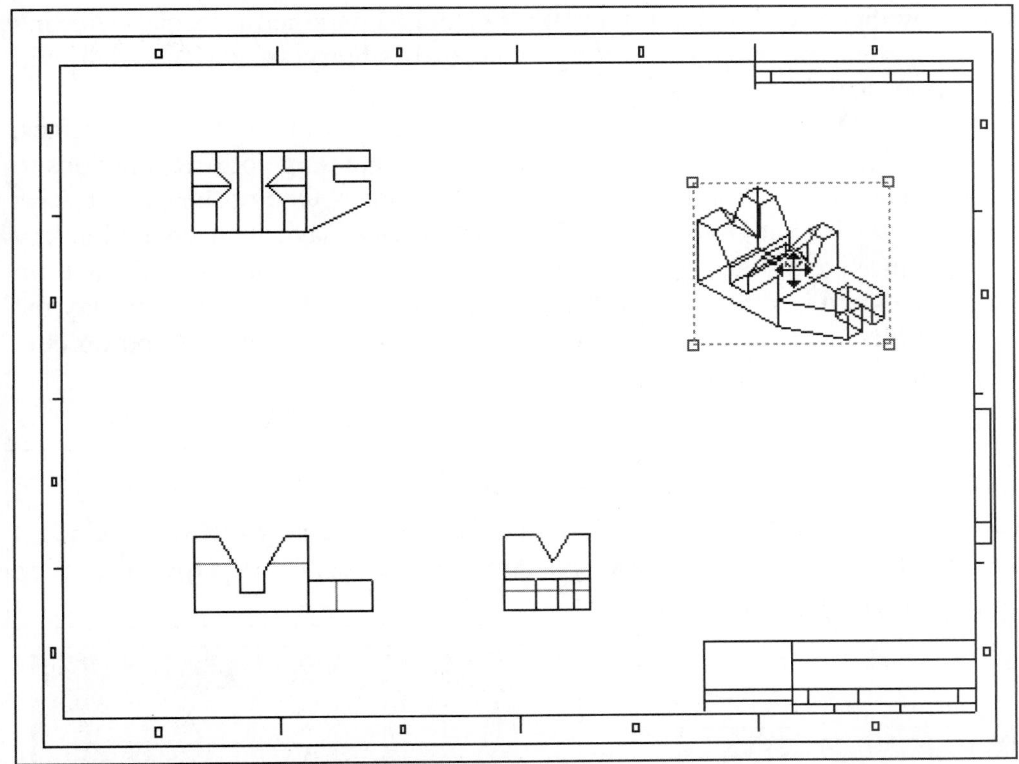

Figure 17.15(e) Repositioned General View

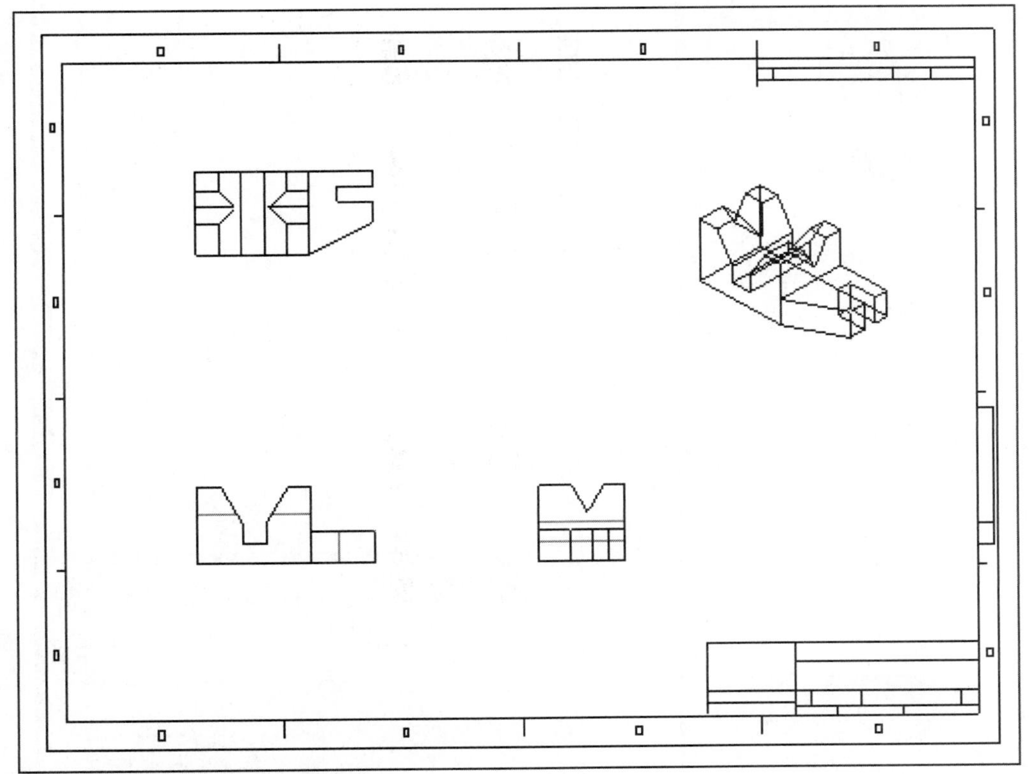

Figure 17.15(f) Repositioned Views

Creating a Format

This section deals with altering an existing standard format to match your format requirements. After you master the techniques required to make this type of format, you can easily extend these principles to using the Sketcher-like commands to draw the border outlines (to make a sketched parametric format). Adding parameters to the format sheet is the next step in creating your own library of formats that will automatically display the part's name, scale, tolerances, and so on.

Pro/E provides a subset of the 2D Drafting functionality to allow you to draw lines, arcs, splines, and so on to define a format border. The Format Mode also supports a complete range of text functions to create notes on the format that serve as title block information or standard notes. One of the easiest ways to create a format is to use an existing format from the Pro/E format library, save it in your directory under a different name, and alter it as required. This method insures that your format will have an ANSI/ASME standard size with a standard title block. You can add parameters to the sheet to display the appropriate information in the title block and at other locations on the format sheet where appropriate. Formats have an **.frm** extension and are read-only files.

Make sure you have the *BASE_ANGLE drawing* open, or at least in session, before starting.

Click: **Open an existing object** [Fig. 17.16(a)] ⇒ **In Session** *(or from Pro/E Format Directory)* ⇒ **c.frm** ⇒ **Open** ⇒ **File** ⇒ **Save a Copy** ⇒ type a unique name for your format: **FORMAT_C** [Fig. 17.16(b)] ⇒ **OK** ⇒ **Window** ⇒ **Close** ⇒ **Window** ⇒ **Activate** ⇒ **File** ⇒ **Open** ⇒ pick **FORMAT_C.FRM** ⇒ **Open** [Fig. 17.17(a)]

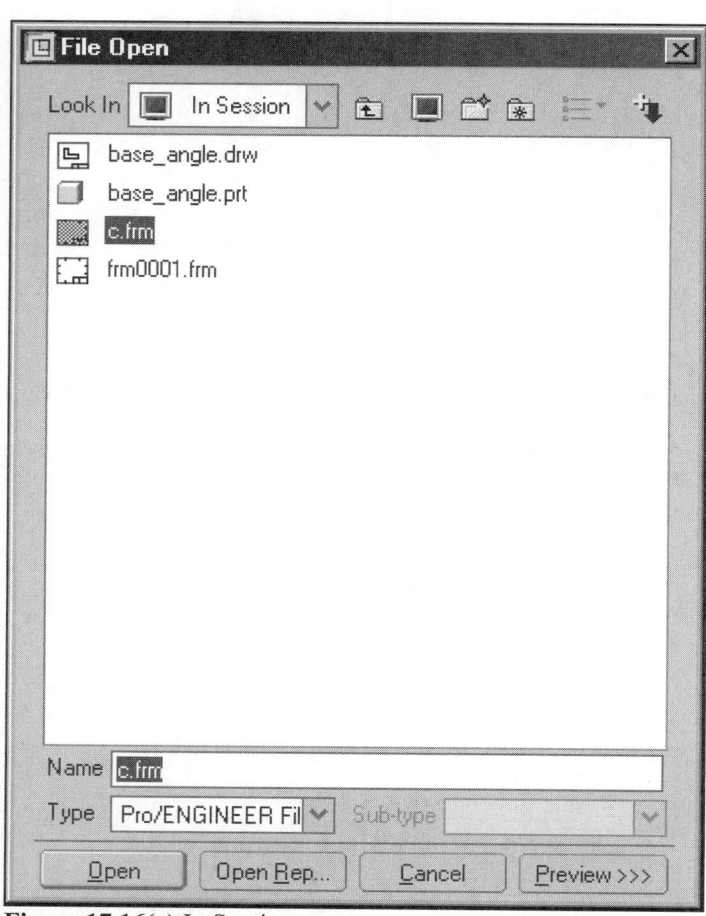

Figure 17.16(a) In Session

Figure 17.16(b) Save a Copy, FORMAT_C

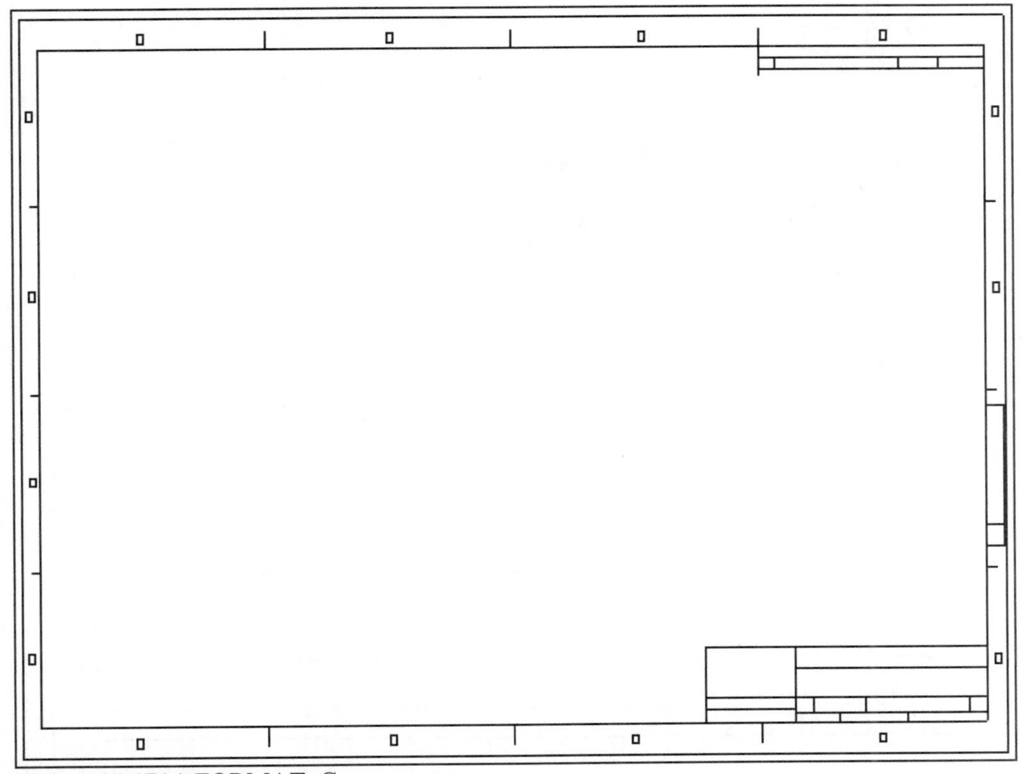

Figure 17.17(a) FORMAT_C

Sketching the Format Using Draft Geometry

This part of the lesson leads you through modifying a "**C**" size, inches format [Fig. 17.17(a)]. You can substitute metric values to create an equivalent-size metric format if desired.

To create the desired format (borders, title block, notes, and so forth) you can draw lines, arcs, splines, and so on using [toolbar]. Even though the option to draw geometry is called "sketch," you need to be aware that the Format Mode does not support the true Sketcher. Format drawing entities are 2D.

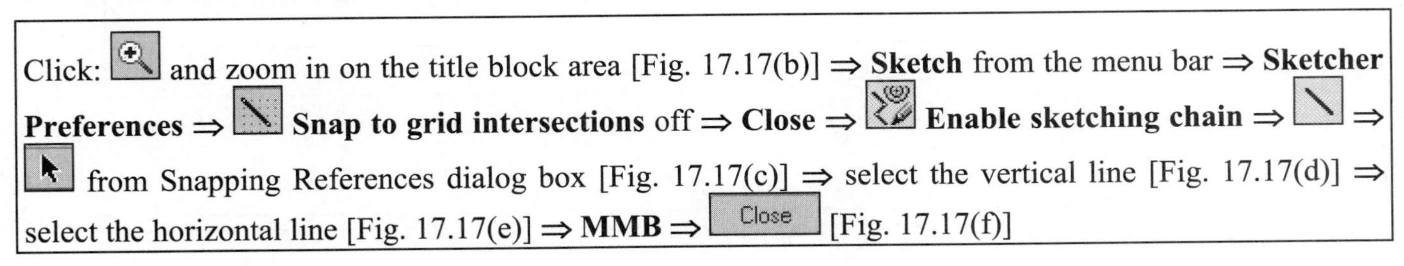

Click: [zoom icon] and zoom in on the title block area [Fig. 17.17(b)] ⇒ **Sketch** from the menu bar ⇒ **Sketcher Preferences** ⇒ [icon] **Snap to grid intersections** off ⇒ **Close** ⇒ [icon] **Enable sketching chain** ⇒ [line icon] ⇒ [arrow icon] from Snapping References dialog box [Fig. 17.17(c)] ⇒ select the vertical line [Fig. 17.17(d)] ⇒ select the horizontal line [Fig. 17.17(e)] ⇒ **MMB** ⇒ Close [Fig. 17.17(f)]

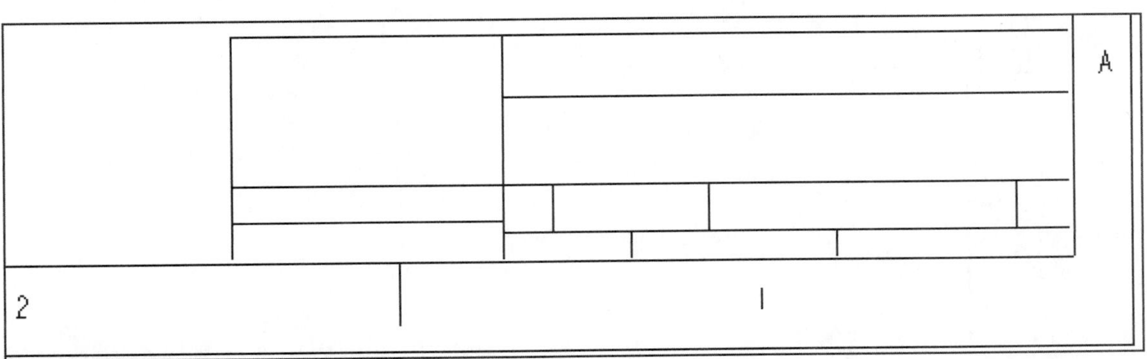

Figure 17.17(b) Title Block Area

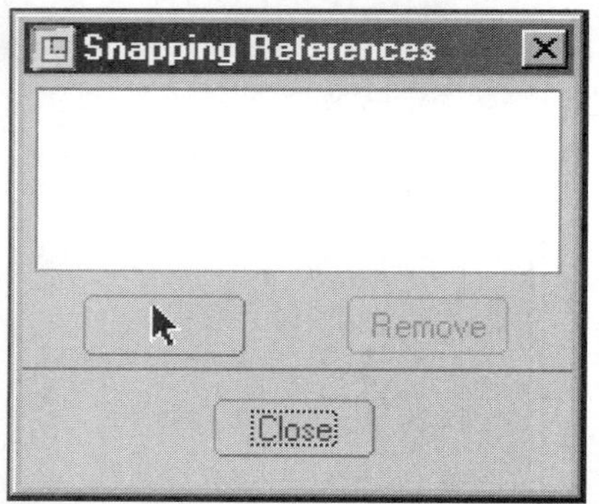

Figure 17.17(c) Snapping References Dialog Box

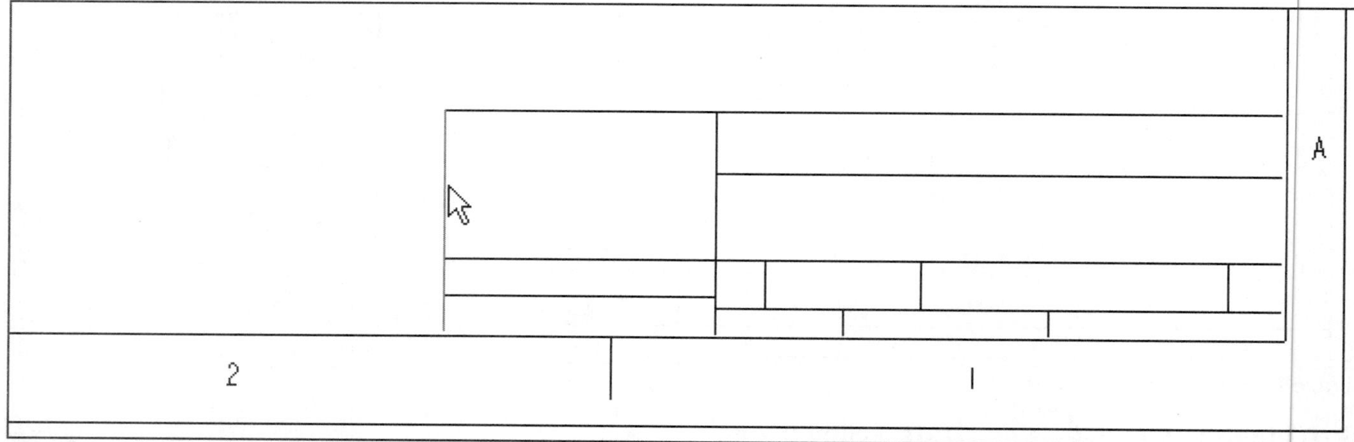

Figure 17.17(d) Vertical Line Snapping Reference

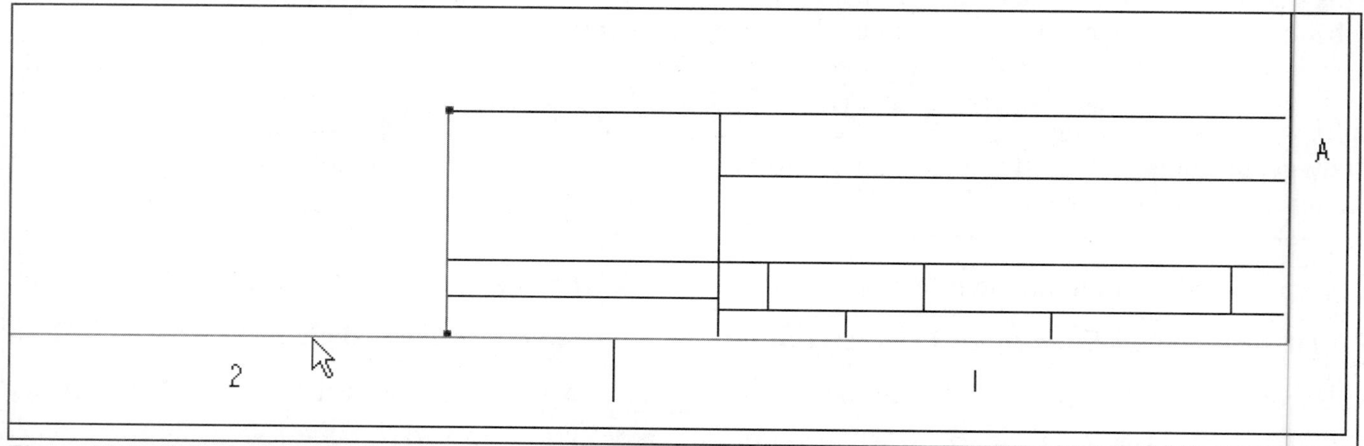

Figure 17.17(e) Horizontal Line Snapping Reference

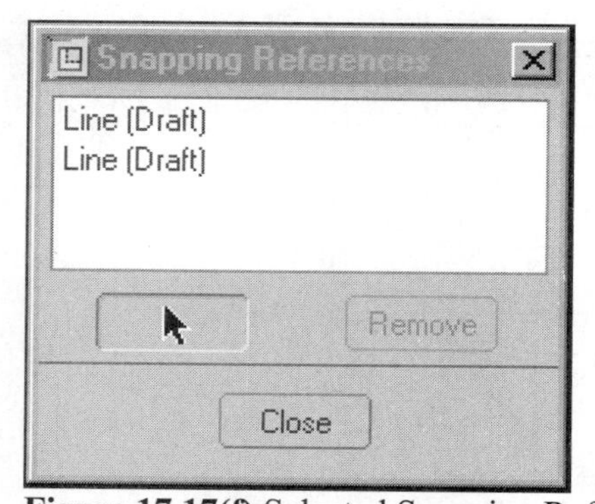

Figure 17.17(f) Selected Snapping References

Click: ⇒ start the line chain on the vertical line at its midpoint [Fig. 17.18(a)] ⇒ draw the horizontal line [Fig. 17.18(b)] ⇒ draw the vertical line [Fig. 17.18(c)] ⇒ **MMB** ⇒ **MMB** ⇒ **LMB** [Fig. 17.18(d)]

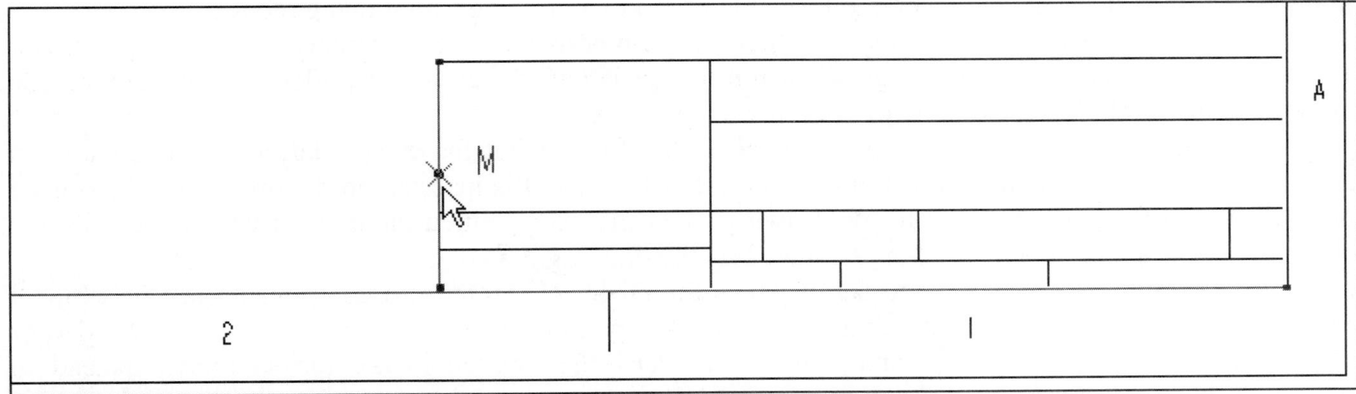

Figure 17.18(a) Select Midpoint of the Vertical Line as Starting Point

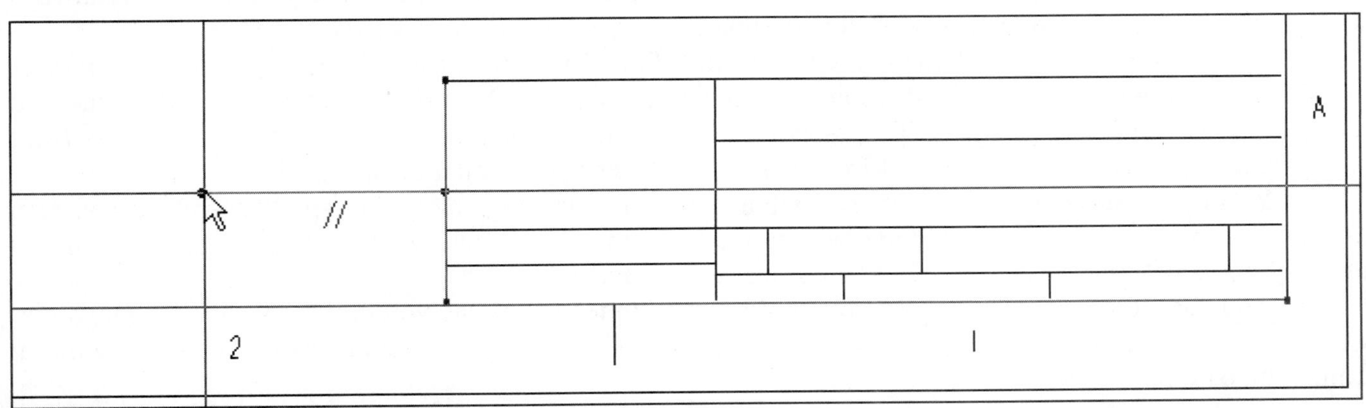

Figure 17.18(b) Pick Second Endpoint

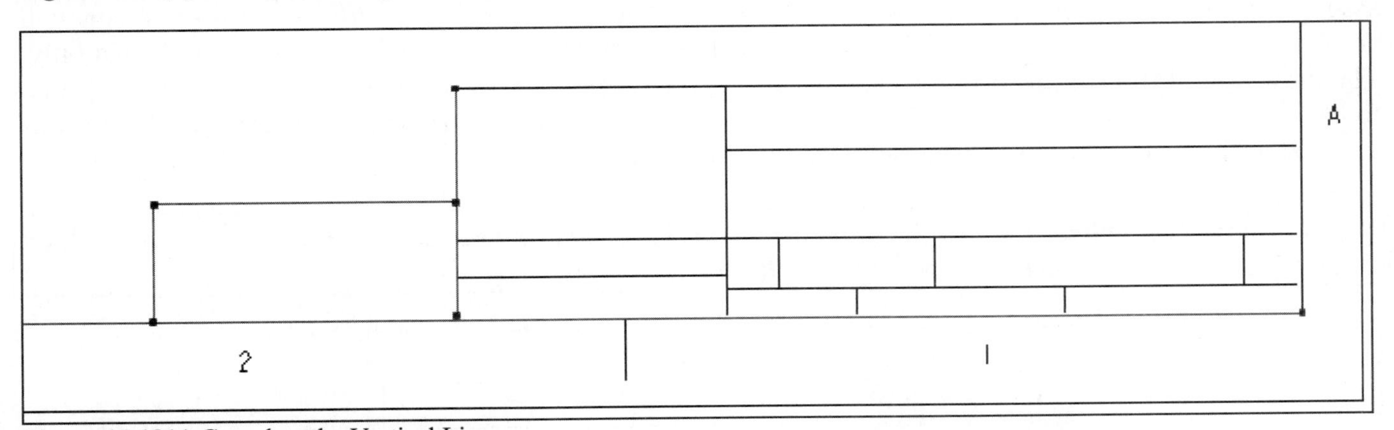

Figure 17.18(c) Complete the Vertical Line

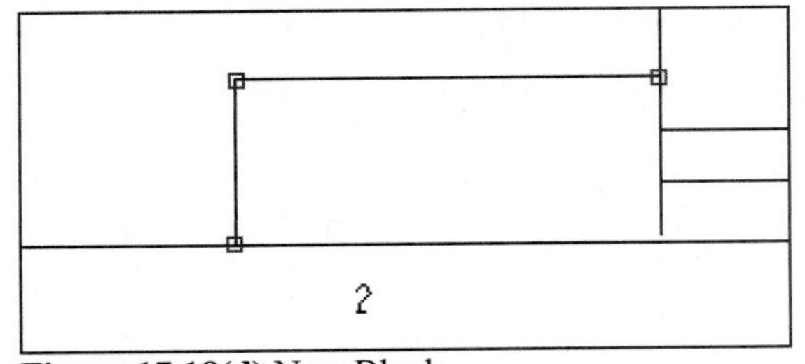

Figure 17.18(d) New Block

Adding Text

For every drawing, the title block is filled in with standard data or data unique to that design. Parameters for the title block information are created and saved in a format in Format mode. *Parameter text* will automatically reflect the design when the format is added to a drawing (in Drawing mode).

Examples of *parameter text* include tolerance standards and sheet numbering. Besides parameter text, *plain text,* which does not change when the format is added, can also be added in the format. The company name might be an example of *plain text* on a format.

User-defined parameters can be included in the format, but these user-defined parameters must also be embedded in the part or assembly model. After the model is inserted on the drawing, views made and placed, and the format added to the drawing, then the user-defined parameters will be read by the format parameters and the information displayed on the drawing.

An example of a user-defined parameter is the DESCRIPTION of a component used in a bill of material (BOM) on an assembly drawing. If user-defined parameters are not defined in the model (part or assembly), then the format will display the *parameter titles* created in the format mode, instead of converting the parameters to model values.

You can create and place the notes almost anywhere on the format, because you will nearly always want to alter and move the parameter and plain text after they have been created.

A format has its own Drawing Options **.dtl** file. The .dtl entries listed apply to a *format*. A *drawing* .dtl file will have these entries and more (see Lesson 19). You may want to establish a standard .dtl file for use only with formats. This is especially true if your standards call for text parameters (font size, arrow size, etc.) that are different from those used when you detail a drawing.

You may also want to create items such as your company logo and other special symbols you plan to use (e.g., projection angle and inspection symbols) before you create your formats. You can then add these symbols without having to redraw them for each format size (A, B, C, etc.).

You should also be aware that you could make copies of text as you create the format. However, when you make a copy, any parameters in the copy become simple text. This means that the copy will *not* utilize the parametric value.

Each *format* will have a .dtl file associated with it, and every *drawing* will have a different associated .dtl file. *These are separate .dtl files, and have nothing to do with each other.* When you activate a drawing and then add a format, the .dtl for the format controls the font etc., for the format only. The drawing .dtl file still needs to be established. As an example, if you changed the .dtl file in the format to have a filled font, the format will show as filled font when used on a drawing. The drawing dimensions and text will *not* show as filled because it uses a separate .dtl file.

Click: **File ⇒ Properties** *(or click RMB anywhere on the format and select Properties [Fig. 17.19(a)])* an Options dialog box will then display [Fig. 17.19(b)]

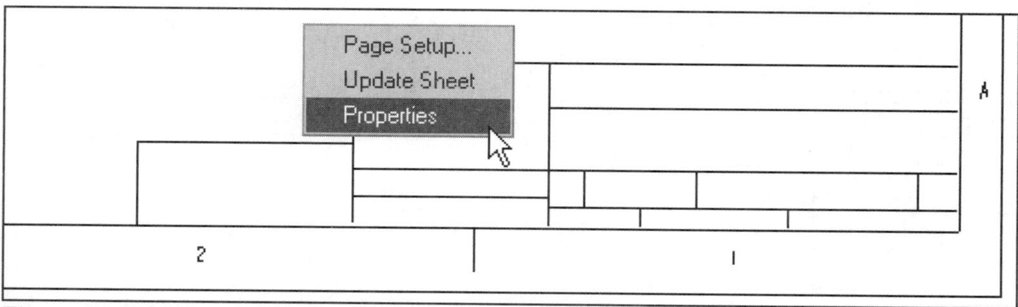

Figure 17.19(a) Format Properties

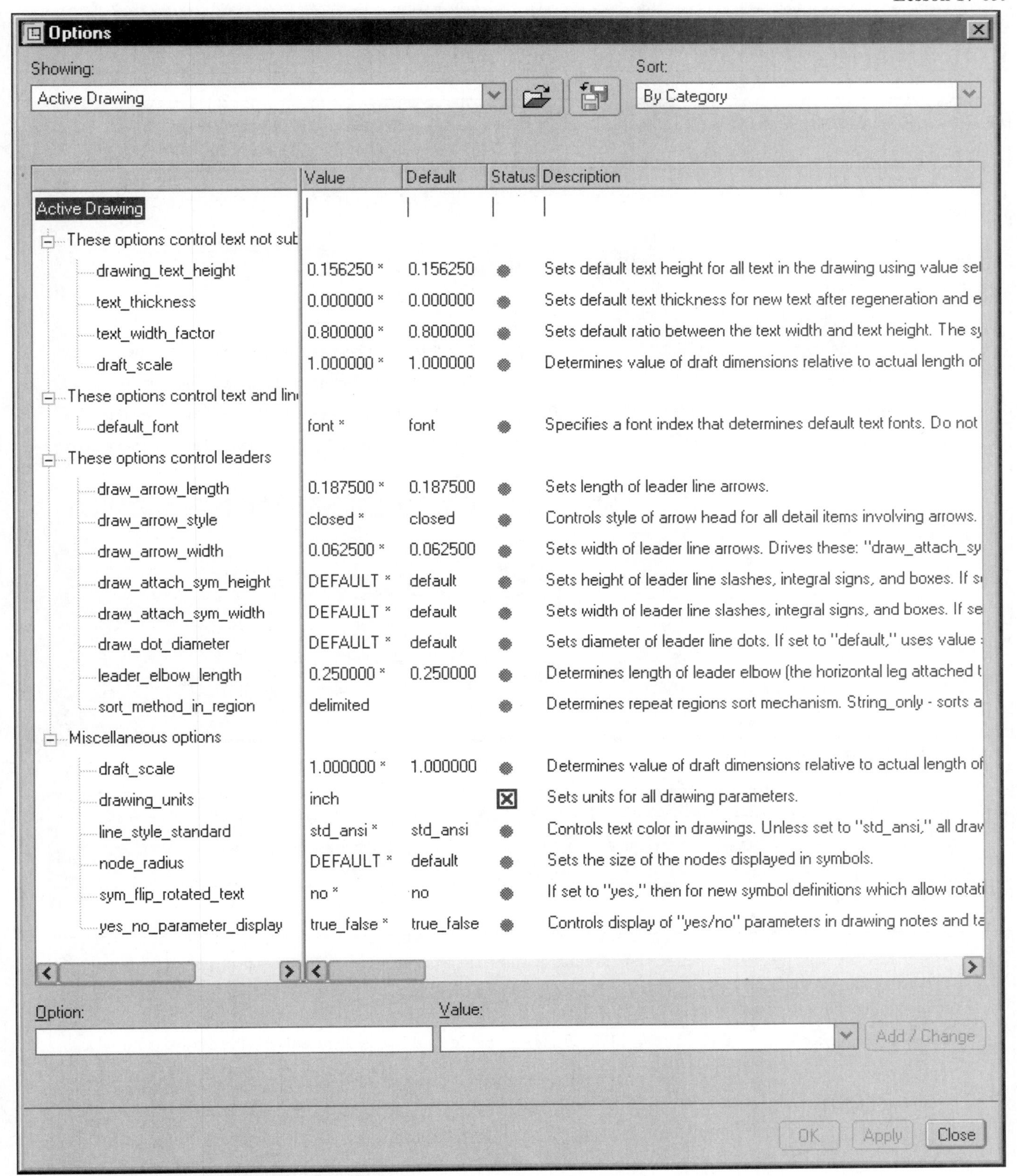

Figure 17.19(b) Options Dialog Box for a Format

Click on **default_font** [Fig. 17.19(c)] ⇒ Value: type **filled** [Fig. 17.19(d)] ⇒ **Add/Change** [Fig. 17.19(e)] ⇒ **Apply** ⇒ **Close** [Fig. 17.19(f)] text on format now displays filled-bold

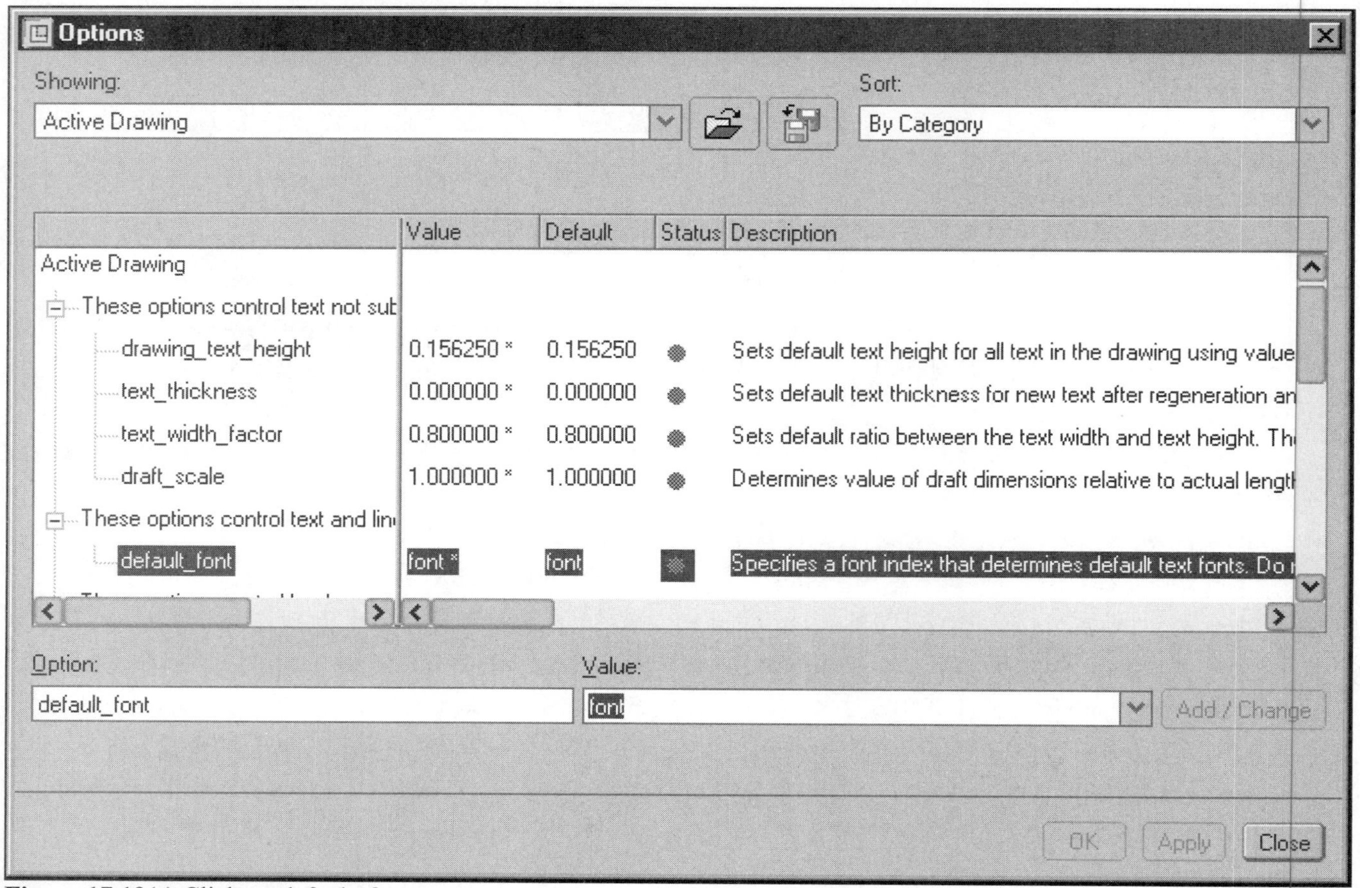

Figure 17.19(c) Click on default_font

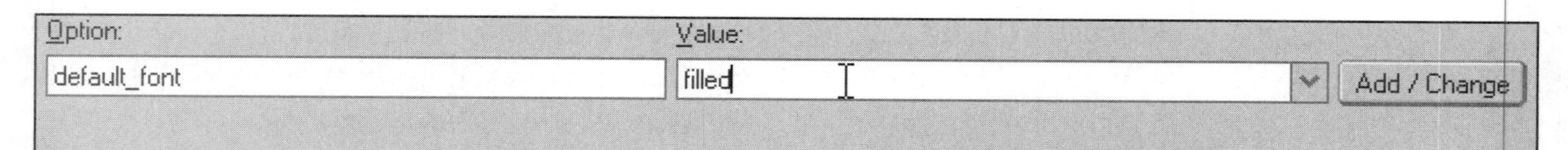

Figure 17.19(d) Change Value to filled

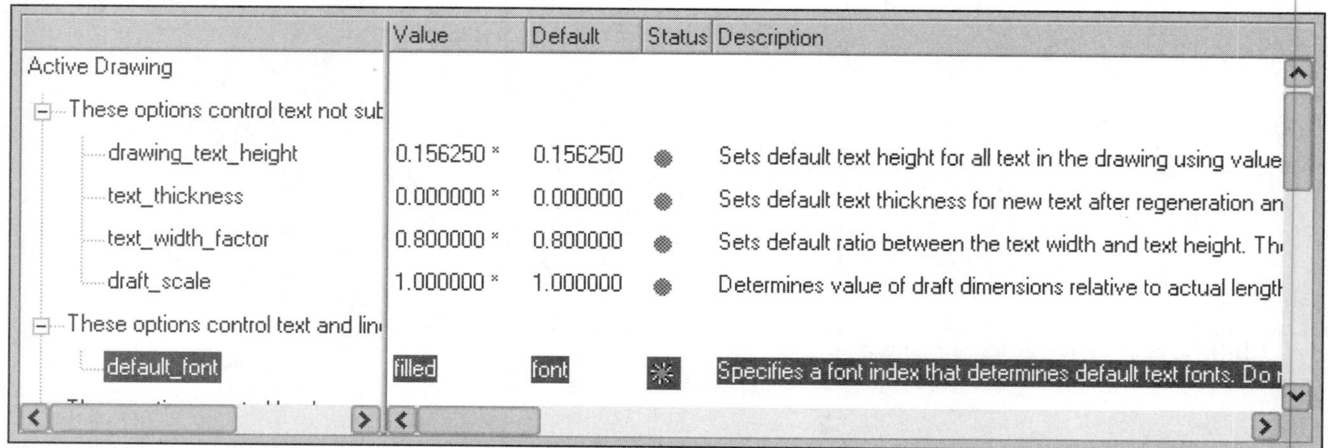

Figure 17.19(e) Add/Change

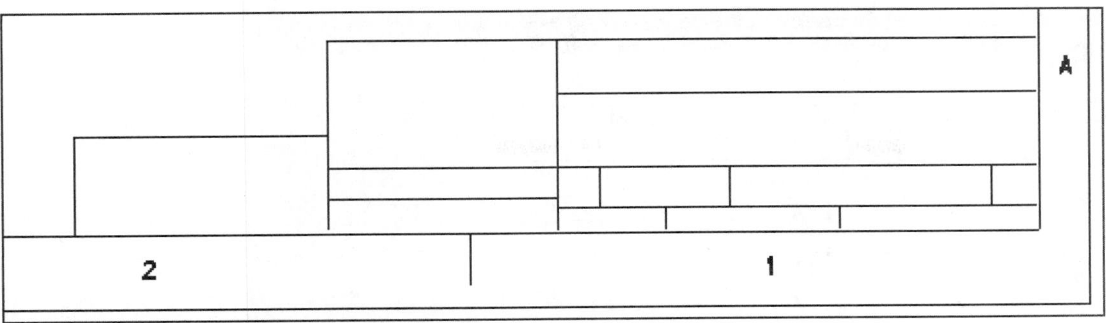

Figure 17.19(f) Text is now Filled (Bold)

Pro/ENGINEER Drawing Symbols

Before completing other areas of the title block by inserting plain text and parametric text, add a symbol to the new title block area. Pro/E symbols area contains libraries of Pro/E symbols that are available with the Pro/DETAIL module. This area is *read-only*. To retrieve a symbol in the Format or Drawing mode, click Insert ⇒ Drawing Symbol ⇒ From Palette. If you have other symbols that you want to use, then in the Symbol Instance Palette dialog box, click Open. The Open dialog box then opens, and you can retrieve different symbols from disk.

The Welding Symbols Library provides a collection of generic symbols according to the ANSI standard, and a collection of symbols according to the ISO standard. Using this library, you can create a variety of welding, brazing, and examination symbols in a drawing. Before you create an instance, familiarize yourself with the procedure for adding instances.

The Symbol Instance Palette dialog box is used to transfer symbols to and from the drawing and symbol instance palette. When you add a symbol to a drawing, you create an instance.

To find the name and directory path of a symbol corresponding to an instance, click Format ⇒ Symbol Gallery, and use the Show Name command in the SYM GALLERY menu. Using the Symbol Instance Palette dialog box, you can create new symbol instances and preview them during the creation process. You can specify symbol characteristics including: content, height, angle, position, and location.

Symbol instances exist *only* in the drawing or format. In Drawing mode, you cannot edit symbols that you created and added to formats in Format mode if you add the format to a drawing. To edit the symbol, retrieve the format in Format mode and make any of the necessary changes. Or you can add symbols to the drawing in the Drawing mode with the same commands used in the Format Mode.

For this format, we will add a symbol from the palette into the new title block area just created. The symbol will be from the library and will establish that the drawing is to Canadian Standards.

Click: **Insert ⇒ Drawing Symbol** [Fig. 17.20(a)] ⇒ **From Palette** the Symbol Instance Palette dialog box opens [Fig. 17.20(b)]

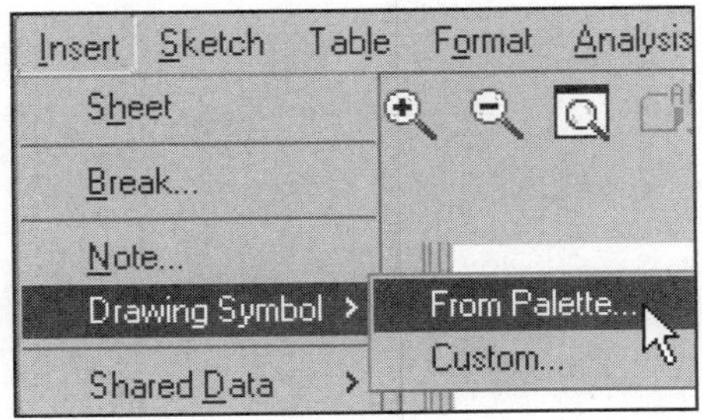

Figure 17.20(a) Drawing Symbol, From Palette (or use )

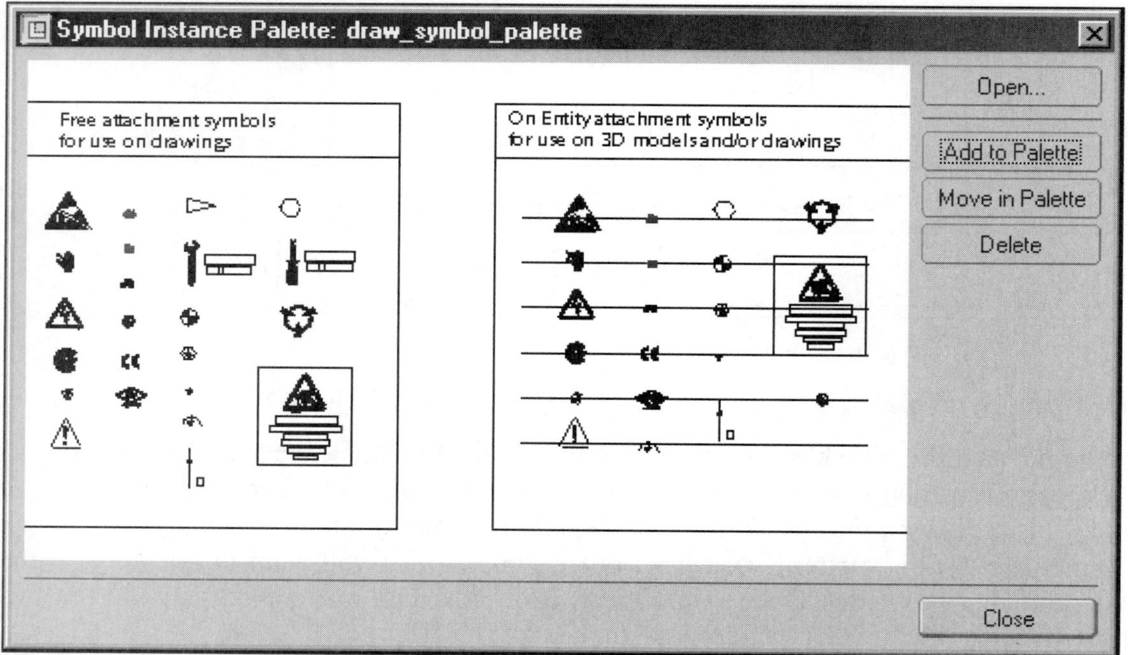

Figure 17.20(b) Symbol Instance Palette Dialog Box

Use **Ctrl+MMB** and **Shift+MMB** to zoom and pan on the symbols ⇒ pick on the Canadian Standard symbol [Fig. 17.20(c)]

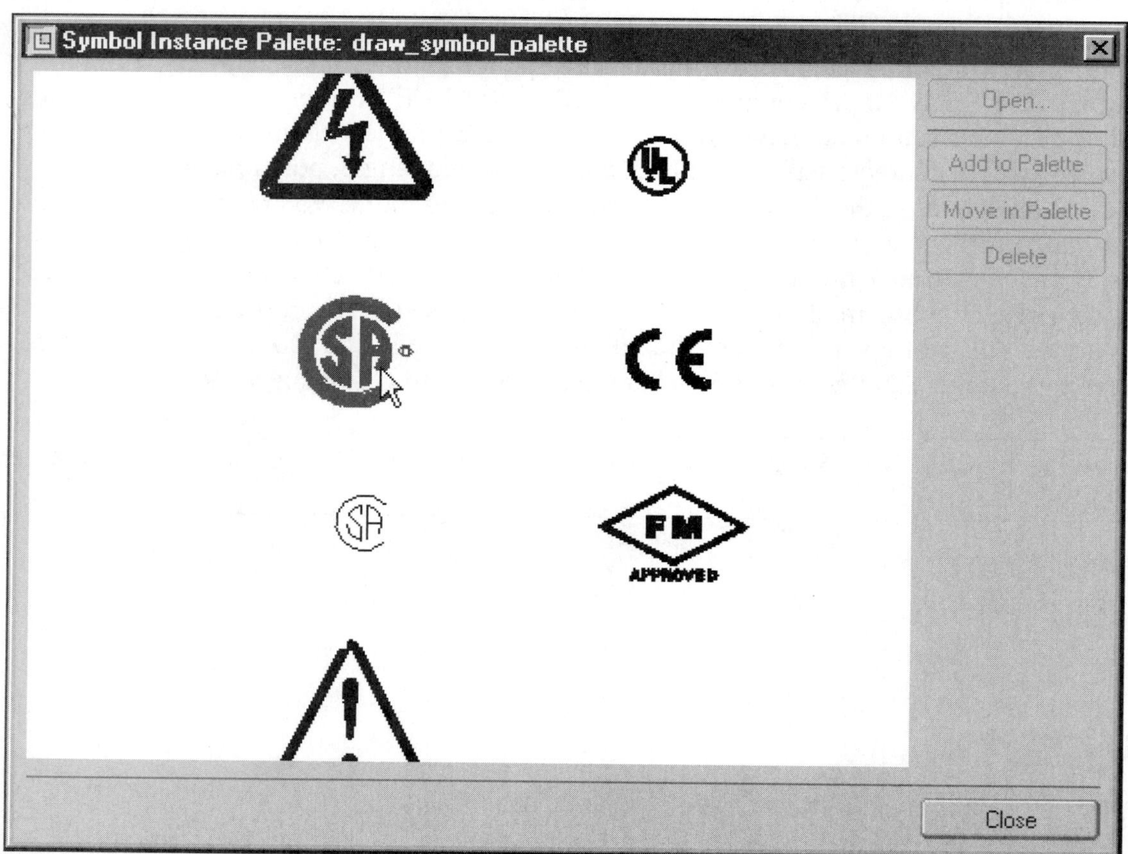

Figure 17.20(c) Pick on the Symbol

Drag and drop the symbol into the new title block area [Fig. 17.20(d)] ⇒ **LMB** to place ⇒ **RMB** ⇒ **Close** [Fig. 17.20(e)] ⇒ click on the symbol ⇒ **RMB** [Fig. 17.20(f)] ⇒ **Properties** [Fig. 17.20(g)] ⇒ **OK** ⇒ **LMB** ⇒

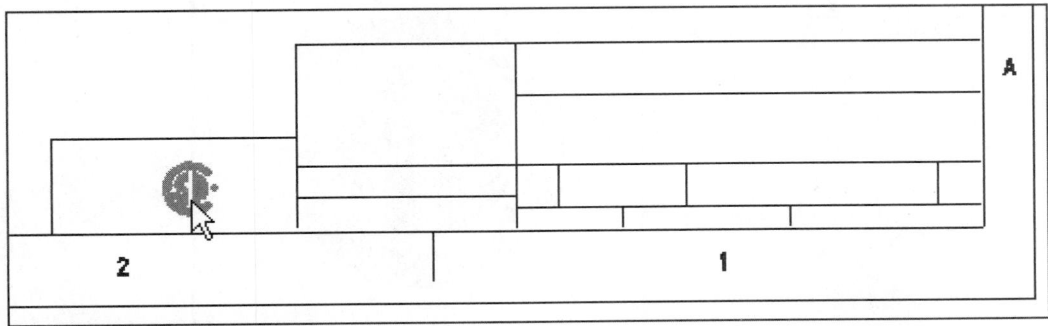

Figure 17.20(d) Drag and Drop the Symbol

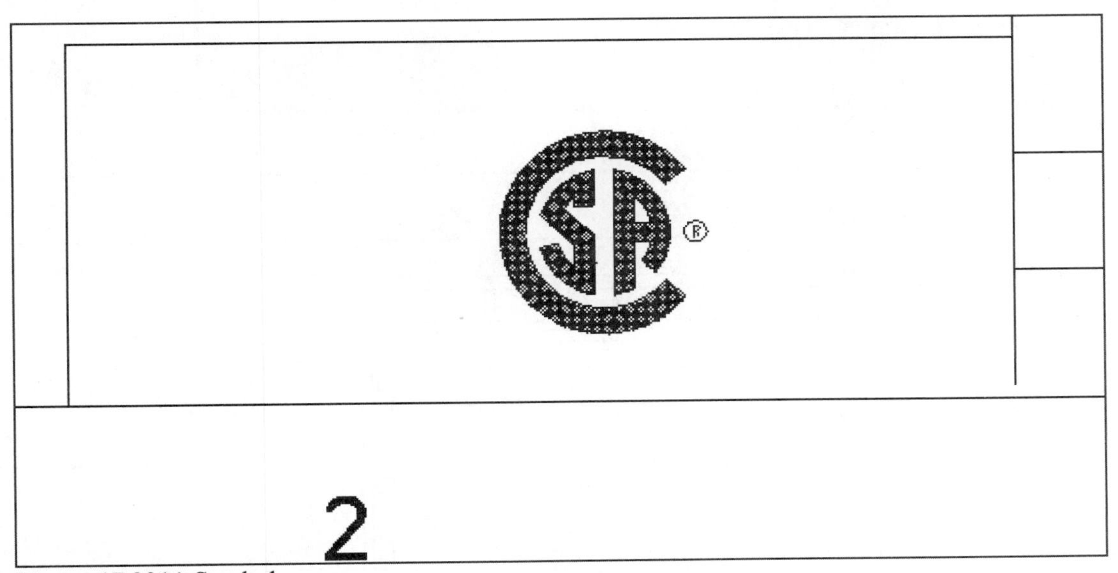

Figure 17.20(e) Symbol

Figure 17.20(f) Symbol Properties

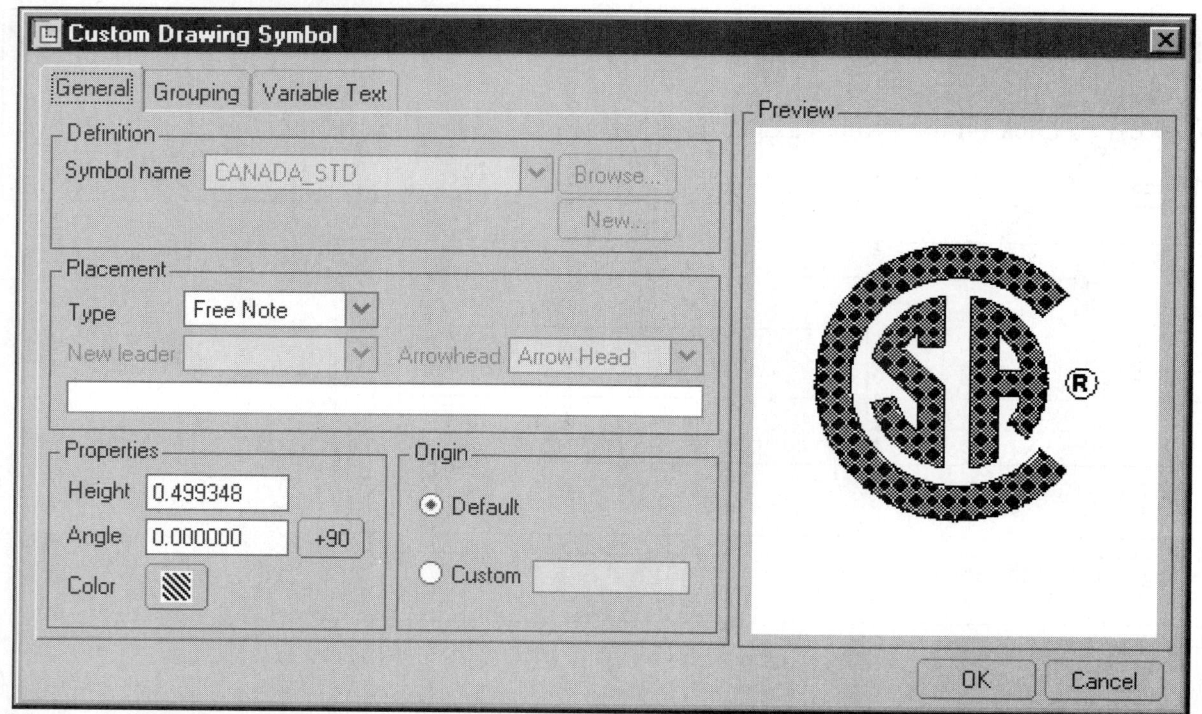

Figure 17.20(g) Custom Drawing Symbol Dialog Box

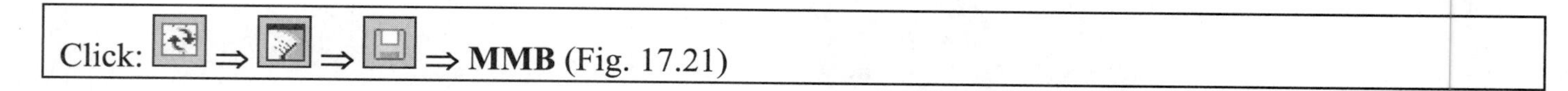

Click: ⇒ ⇒ ⇒ **MMB** (Fig. 17.21)

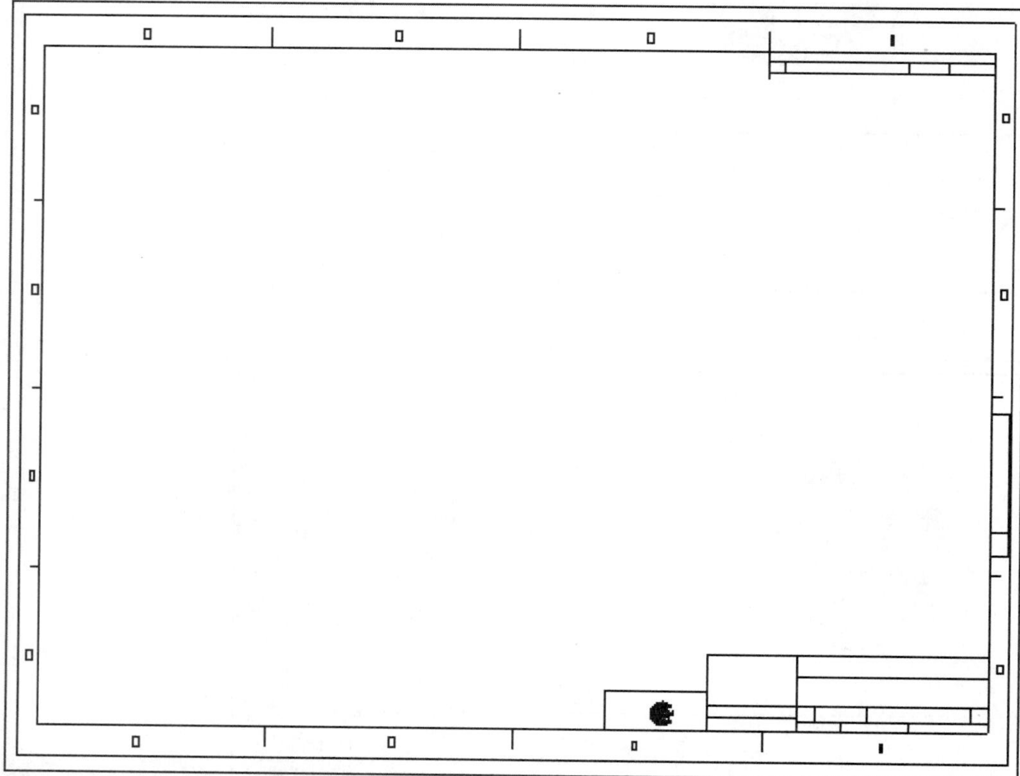

Figure 17.21 Format with Symbol

Creating Notes

You can use the grid to line up and position text. In the Format mode and the Drawing mode, the grid is 2D and is aligned with the lower left-hand corner of the sheet.

Click: and zoom in on the title block area ⇒ **Tools** ⇒ **Environment** ⇒ Snap to Grid ⇒ **OK** ⇒ **View** ⇒ **Draft Grid** ⇒ **Show Grid** ⇒ **Grid Params** ⇒ **X&Y Spacing** ⇒ type the new spacing value- **.100** ⇒ **MMB** ⇒ **MMB** ⇒ **MMB** [Fig. 17.22(a)] ⇒ **Create a note** ⇒ **Make Note** ⇒ pick a position for the note [Fig. 17.22(b)] ⇒ type **DESIGN ASSOCIATES** ⇒ **Enter** ⇒ **Enter** ⇒ [Fig. 17.22(c)]

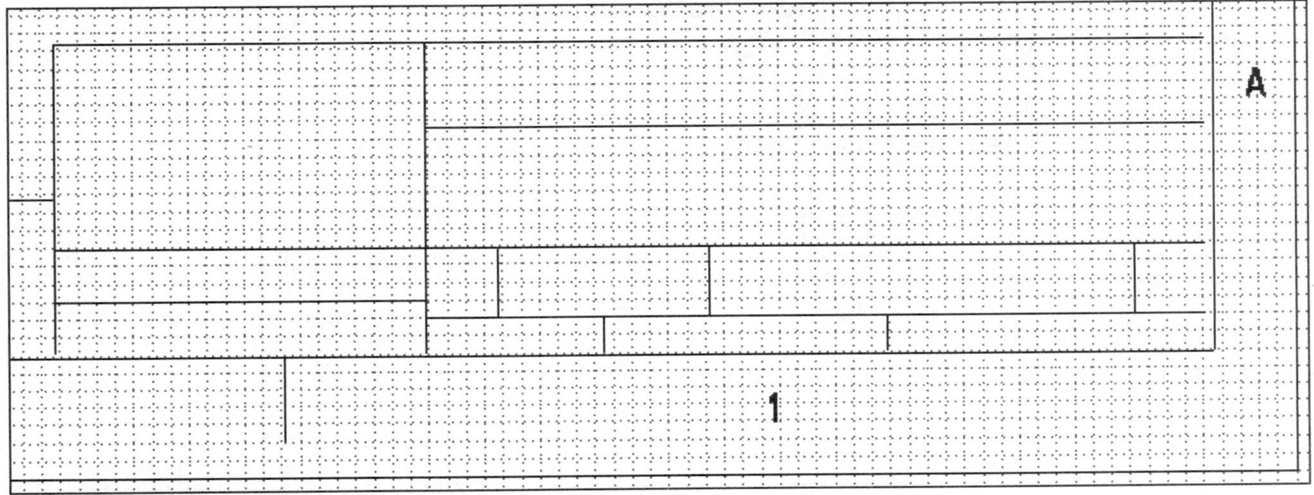

Figure 17.22(a) Draft Grid

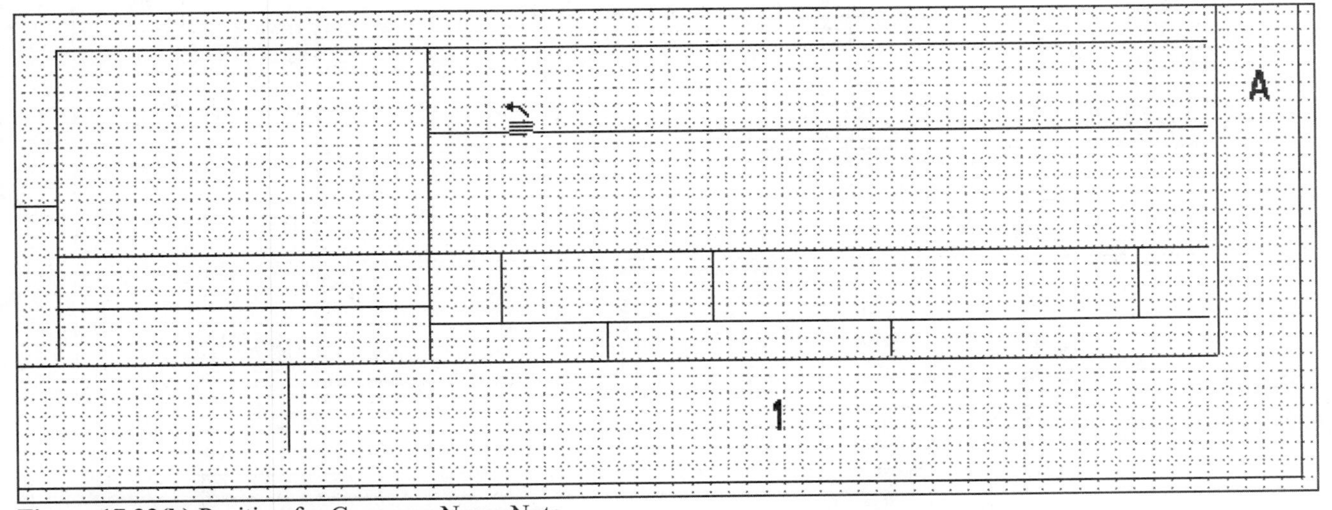

Figure 17.22(b) Position for Company Name Note

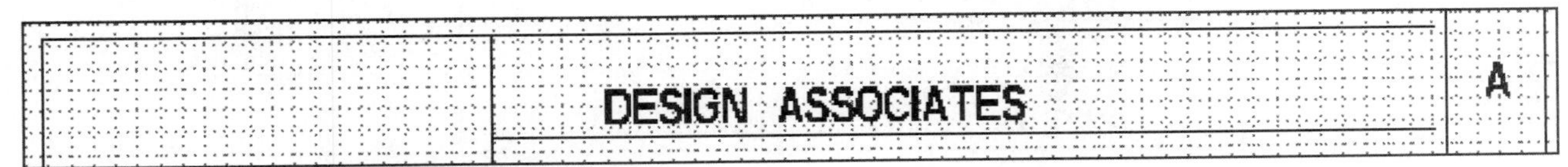

Figure 17.22(c) DESIGN ASSOCIATES Note

Click: **Make Note** ⇒ pick a position for the note [Fig. 17.22(d)] ⇒ **Caps Lock** on ⇒ type: **TOLERANCES** ⇒ **Enter** ⇒ **Enter** ⇒ **Make Note** ⇒ pick a position for the note [Fig. 17.22(e)] ⇒ type: **DESCRIPTION** ⇒ **Enter** ⇒ **Enter** ⇒ **Make Note** ⇒ pick a position for the note [Fig. 17.22(f)] ⇒ type **DWG NO.** ⇒ **Enter** ⇒ **Enter** [Fig. 17.22(g)] ⇒ **Make Note** ⇒ pick a position for the note on the lower left-hand side of the drawing ⇒ **GENERAL SURFACE** ⇒ **Enter** ⇒ **FINISH UNLESS** ⇒ **Enter** ⇒ **OTHERWISE SPECIFIED** ⇒ **Enter** ⇒ **Enter** [Fig. 17.22(h)] ⇒ **Done/Return**

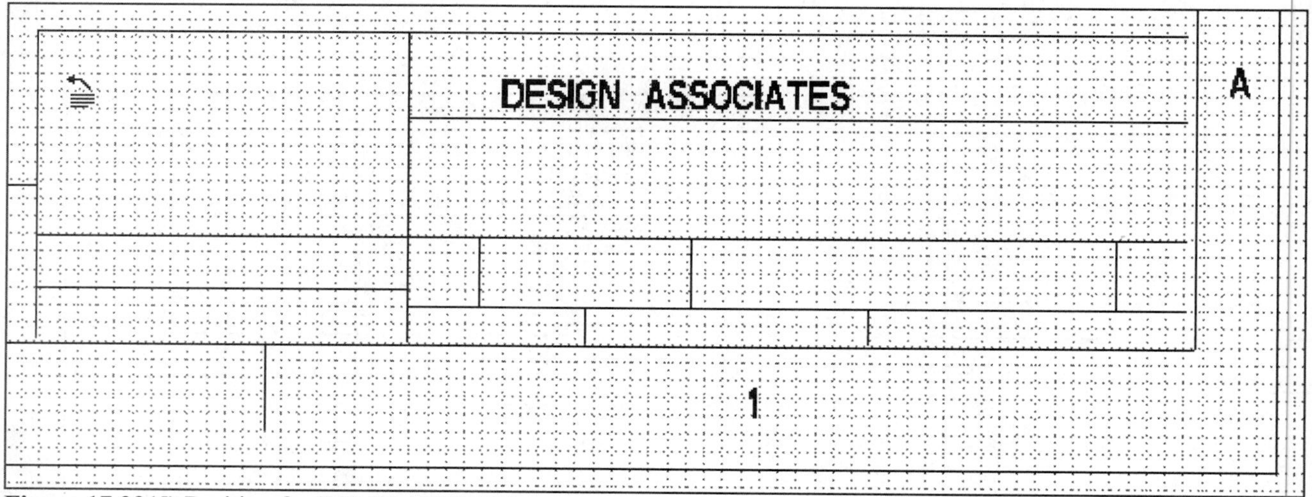

Figure 17.22(d) Position for TOLERANCES Note

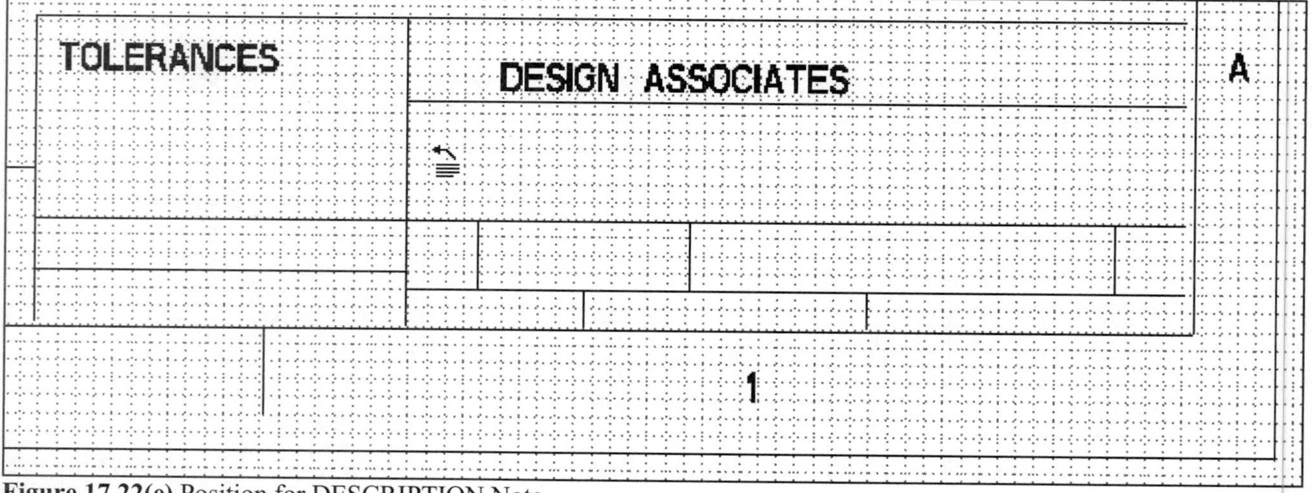

Figure 17.22(e) Position for DESCRIPTION Note

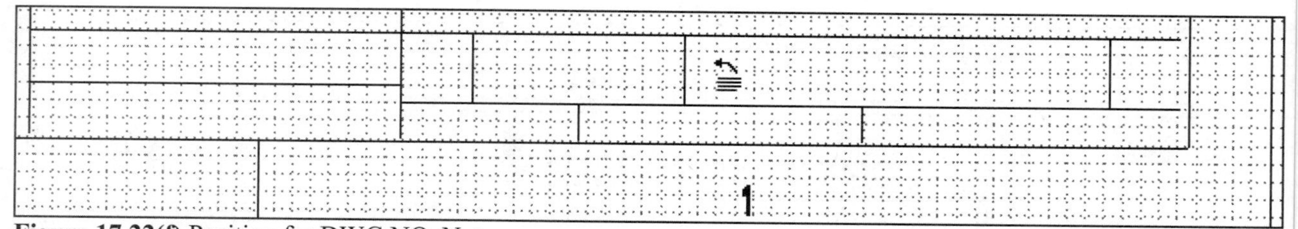

Figure 17.22(f) Position for DWG NO. Note

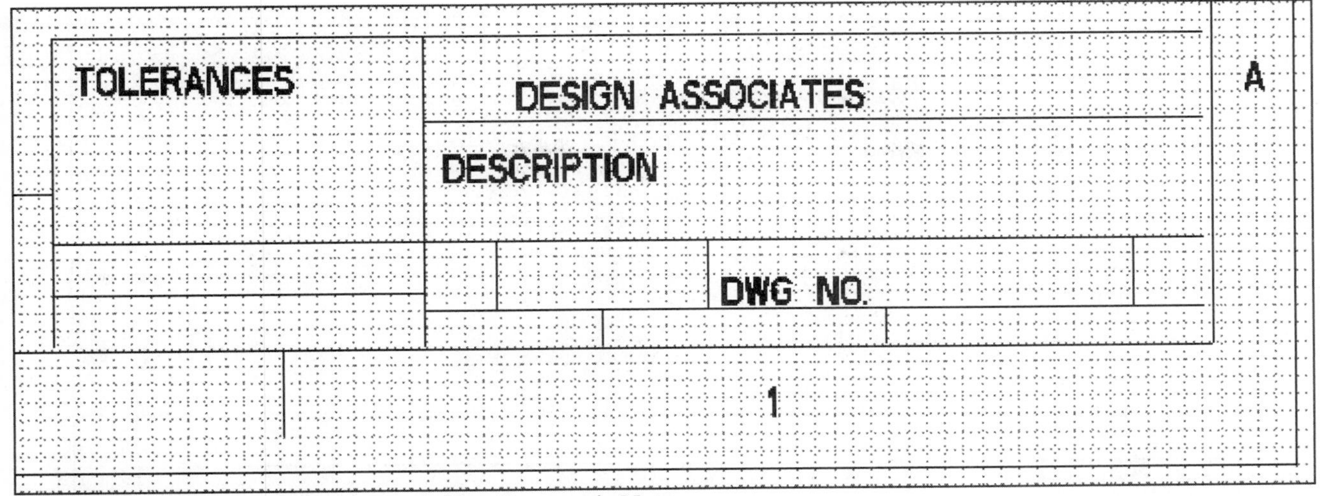

Figure 17.22(g) Completed Title Block Non-Parametric Notes

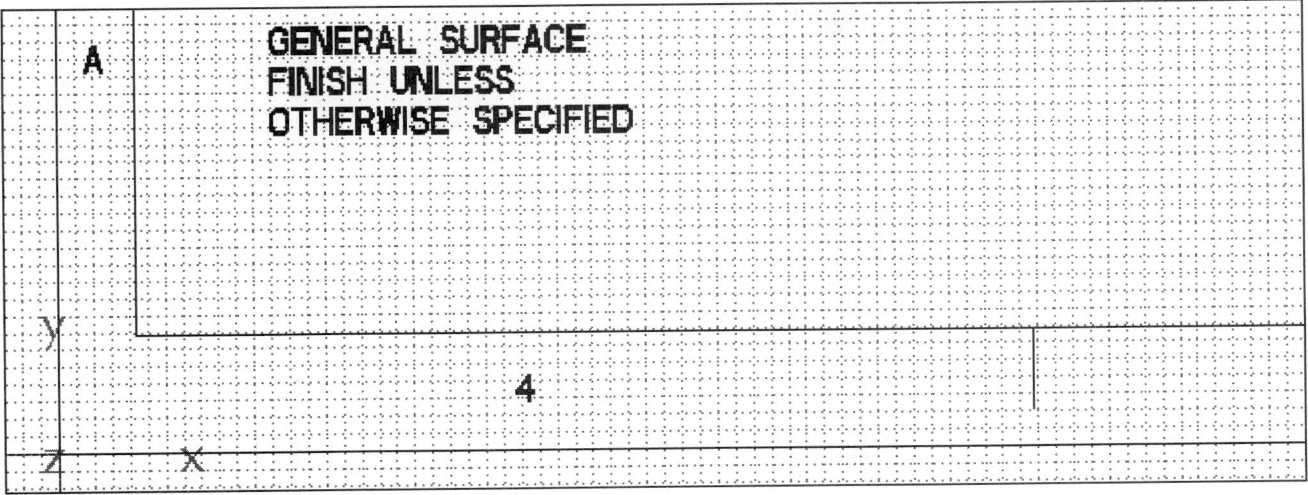

Figure 17.22(h) General Note

Modifying Model Notes

After you create the text, you often need to modify the text parameters or the note itself. The most common modification is to the text height. You can also use the Angle option to rotate text. This allows you to write text upside down along the top of a format or at 90° or 270° to make it parallel to the right or left border. The Mirror option allows you to change text so that it can be read from the backside of the vellum (when you plot it). Modify the text height of the notes and place them at the proper location on the format.

When you modify the parameters of text, Pro/E treats each text segment as a separate entity. This allows you to set the text height, for example, of different entity sets within a text string.

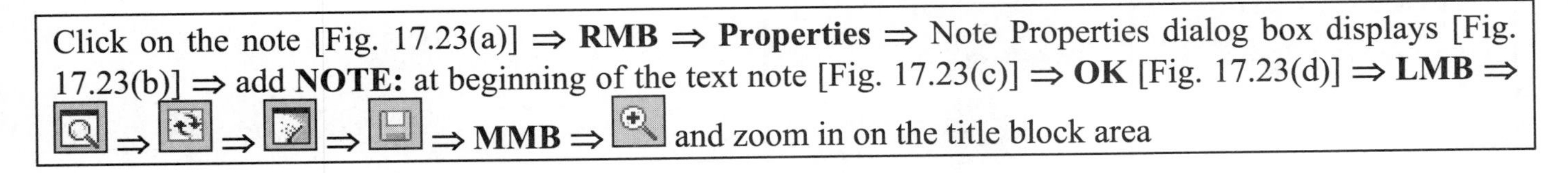

Click on the note [Fig. 17.23(a)] ⇒ **RMB** ⇒ **Properties** ⇒ Note Properties dialog box displays [Fig. 17.23(b)] ⇒ add **NOTE:** at beginning of the text note [Fig. 17.23(c)] ⇒ **OK** [Fig. 17.23(d)] ⇒ **LMB** ⇒ ⇒ ⇒ ⇒ ⇒ **MMB** ⇒ and zoom in on the title block area

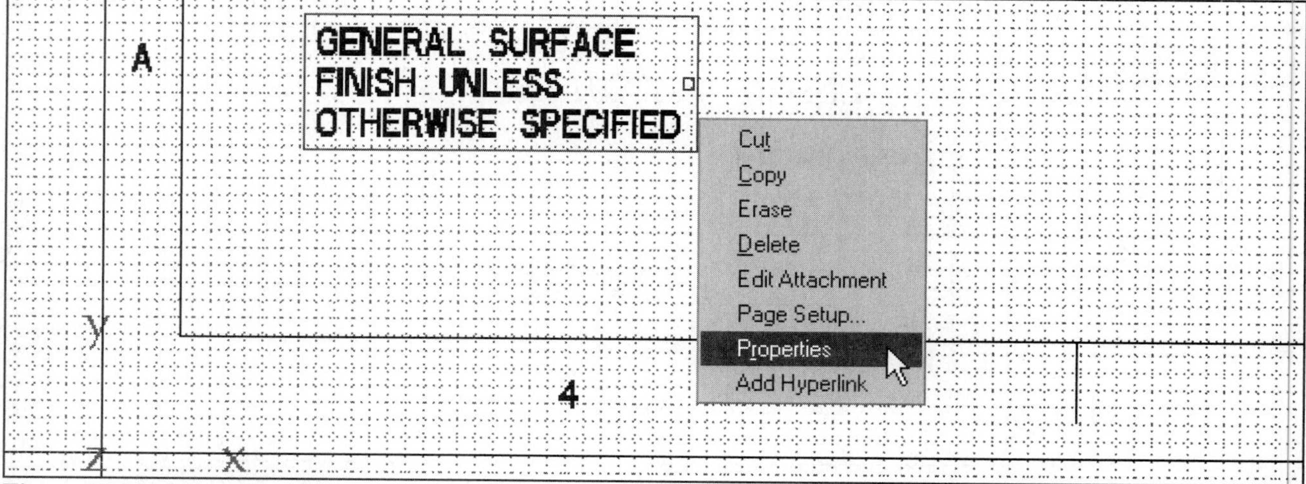

Figure 17.23(a) Note Properties

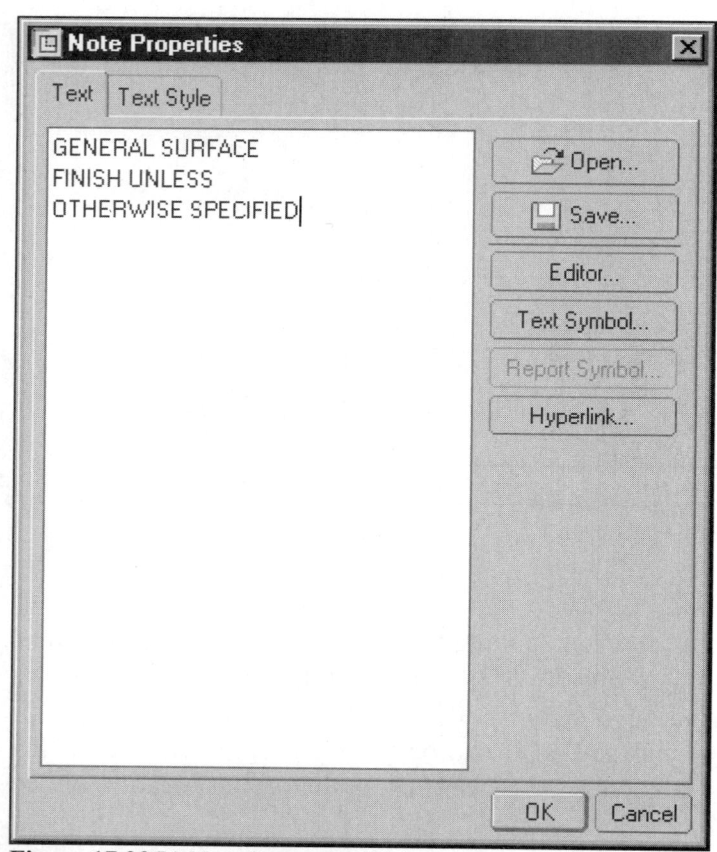

Figure 17.23(b) Note Properties Dialog Box

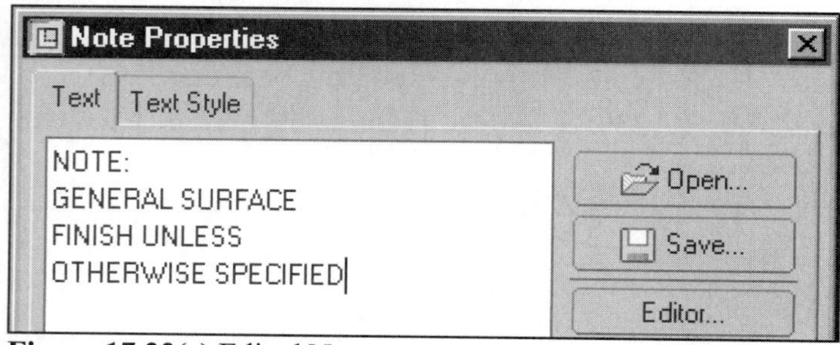

Figure 17.23(c) Edited Note

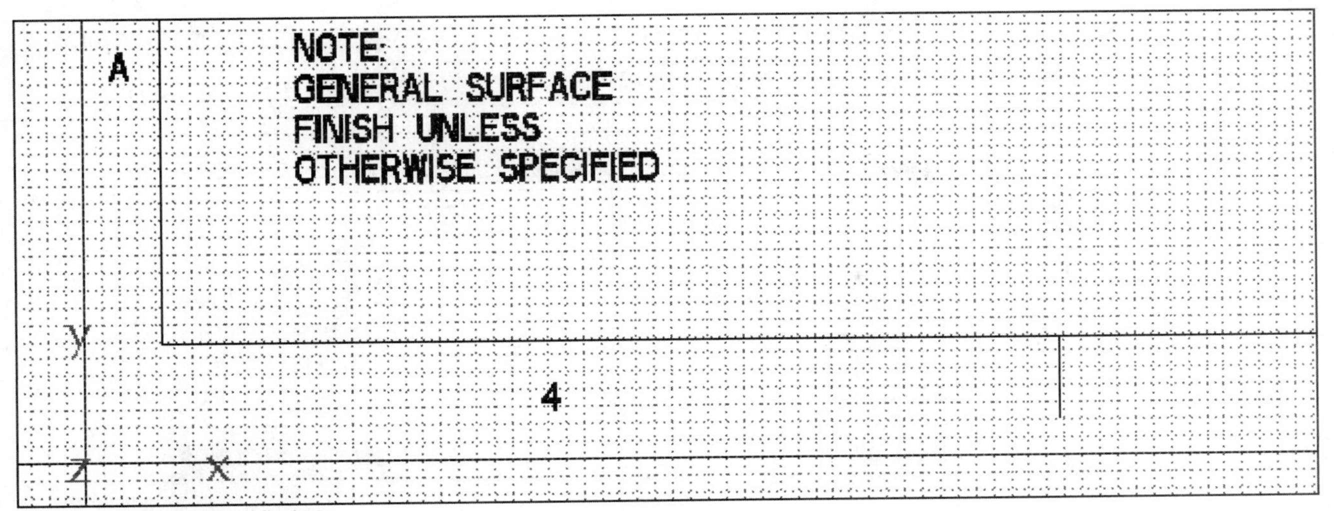

Figure 17.23(d) Modified Note

Click on the note **DESIGN ASSOCIATES** [Fig. 17.24(a)] ⇒ drag and drop the note to be centered in the block [Fig. 17.24(b)] ⇒ do the same with the **TOLERANCES** note ⇒ ⇒ **MMB** ⇒ **File** ⇒ **Delete** ⇒ **Old Versions** ⇒ **MMB**

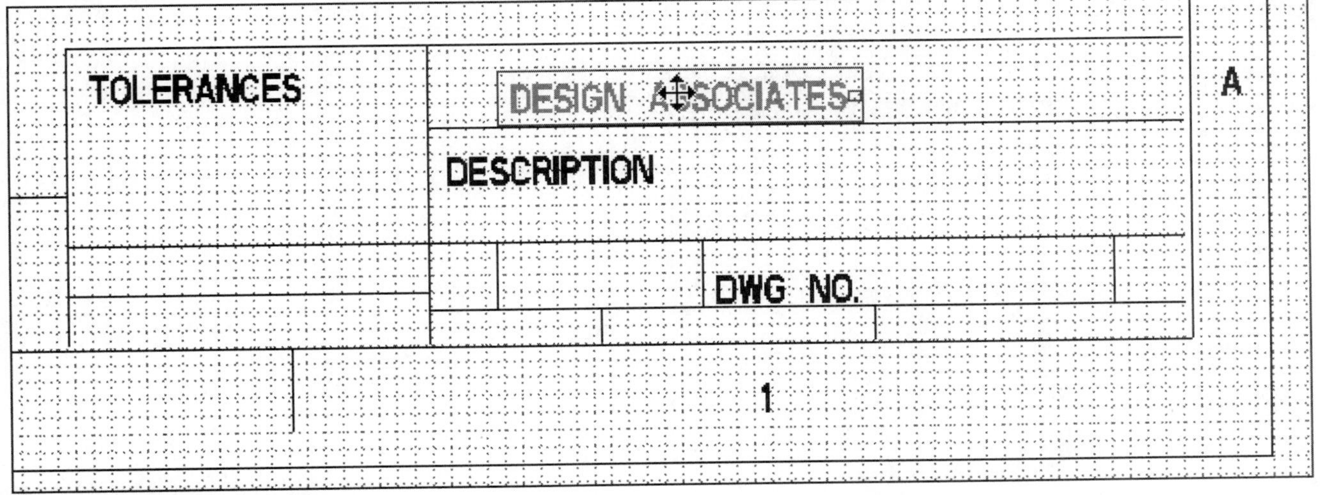

Figure 17.24(a) Click on Note

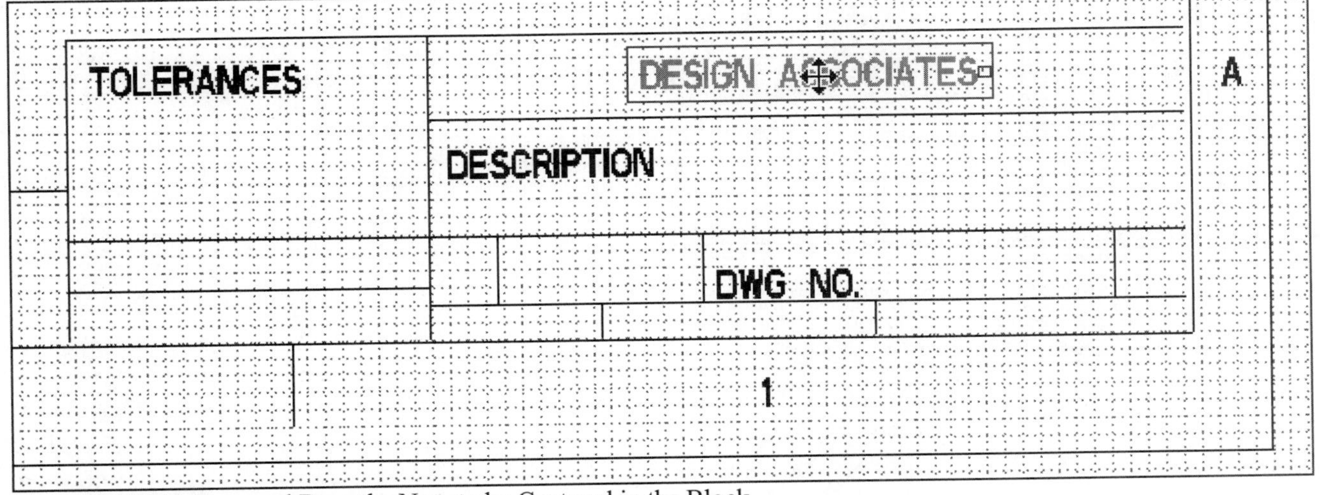

Figure 17.24(b) Drag and Drop the Note to be Centered in the Block

Click on the note **DESIGN ASSOCIATES** [Fig. 17.25(a)] ⇒ **RMB** ⇒ **Properties** ⇒ **Text Style** tab [Fig. 17.25(b)] ⇒ Font [Fig. 17.25(c)] ⇒ CG Century Schbk Bold [Fig. 17.25(d)] ⇒ **OK**

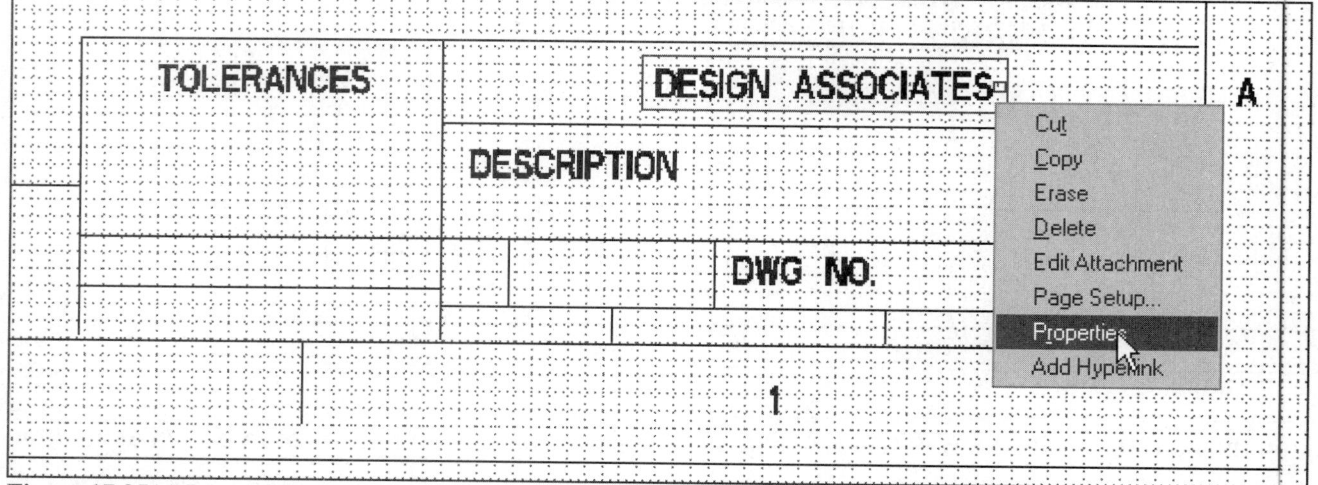

Figure 17.25(a) Properties

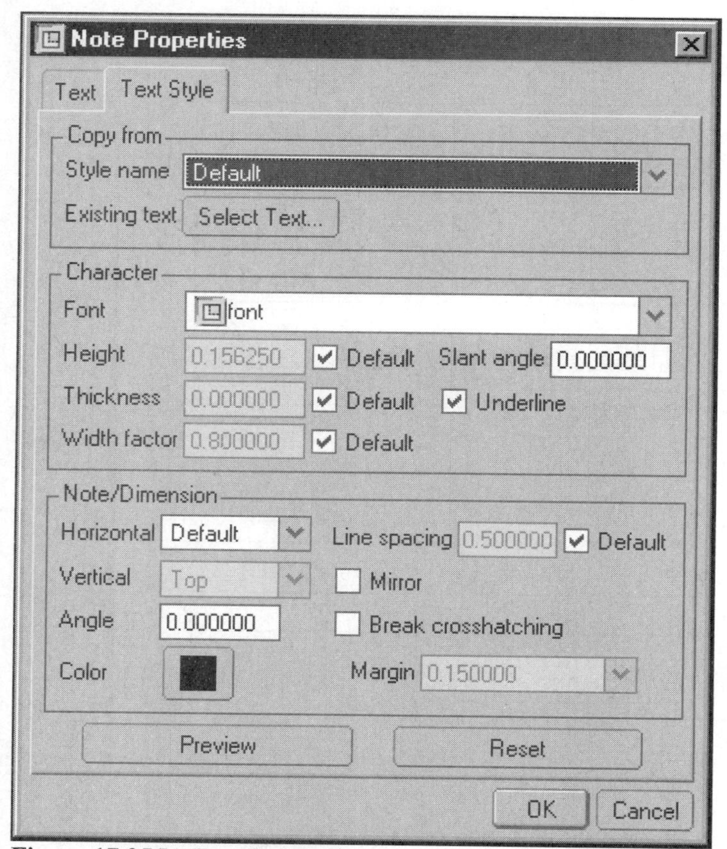

Figure 17.25(b) Text Style Tab

Character
Font filled
Height 0.156250 Default Slant angle 0.000000
Thickness 0.000000 Default Underline
Width factor 0.800000 Default

Figure 17.25(c) Font

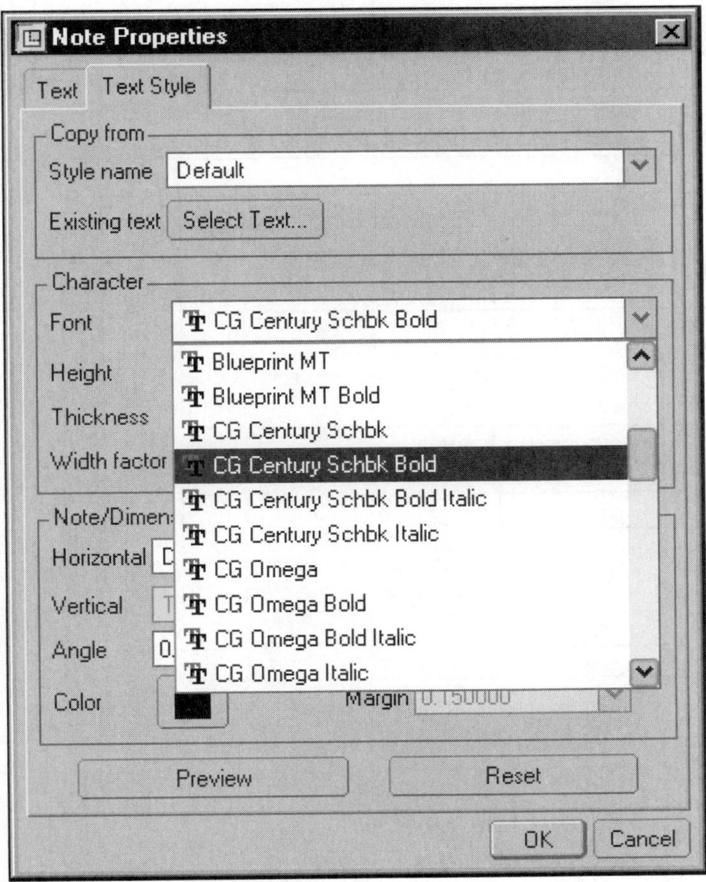

Figure 17.25(d) New Font

Center the note if necessary ⇒ [icon] ⇒ **MMB** [Fig. 17.25(e)]

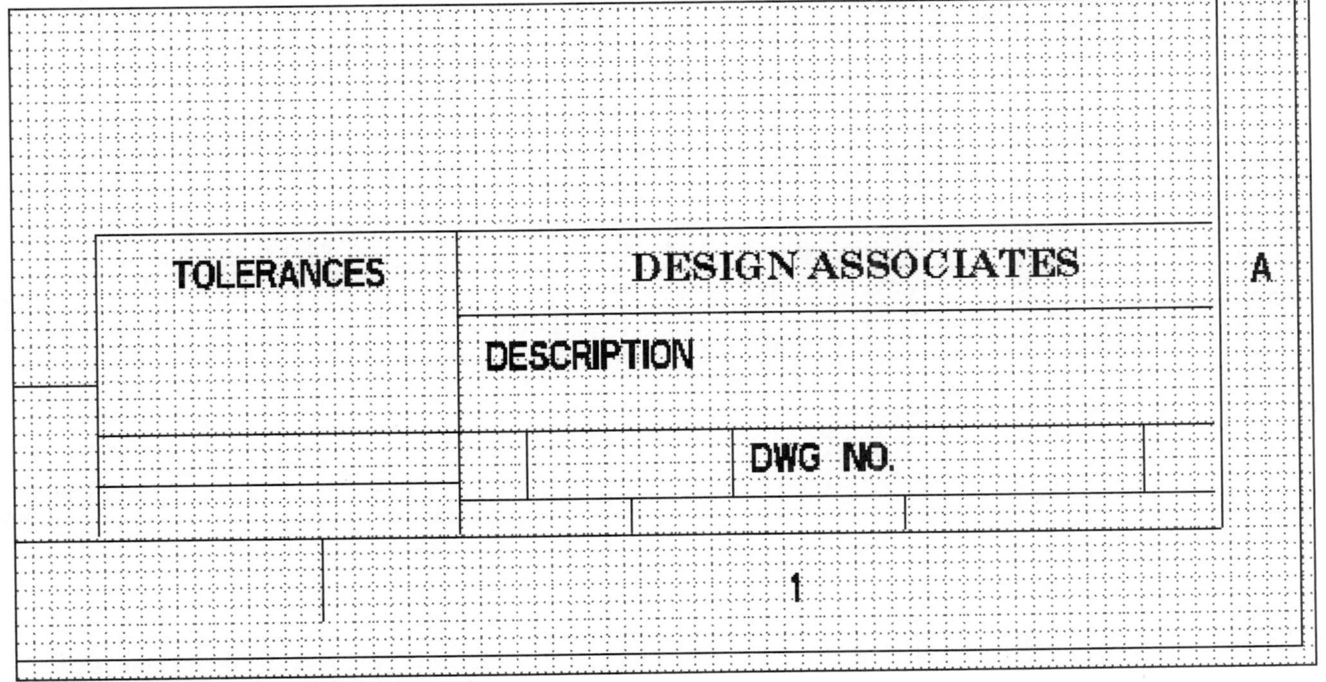

Figure 17.25(e) New Font for Company Name

Press and hold the **Ctrl** key and click on **TOLERANCES**, **DESCRIPTION**, and **DWG NO.** ⇒ **RMB** [Fig. 17.26(a)] ⇒ **Text Style** ⇒ Height [0.156250 ☐ Default] ⇒ Height [.100 ☐ Default] ⇒ **Apply** [Fig. 17.26(b)] ⇒ **OK** ⇒ move the notes if necessary ⇒ [save icon] ⇒ **MMB** ⇒ **Caps Lock** off

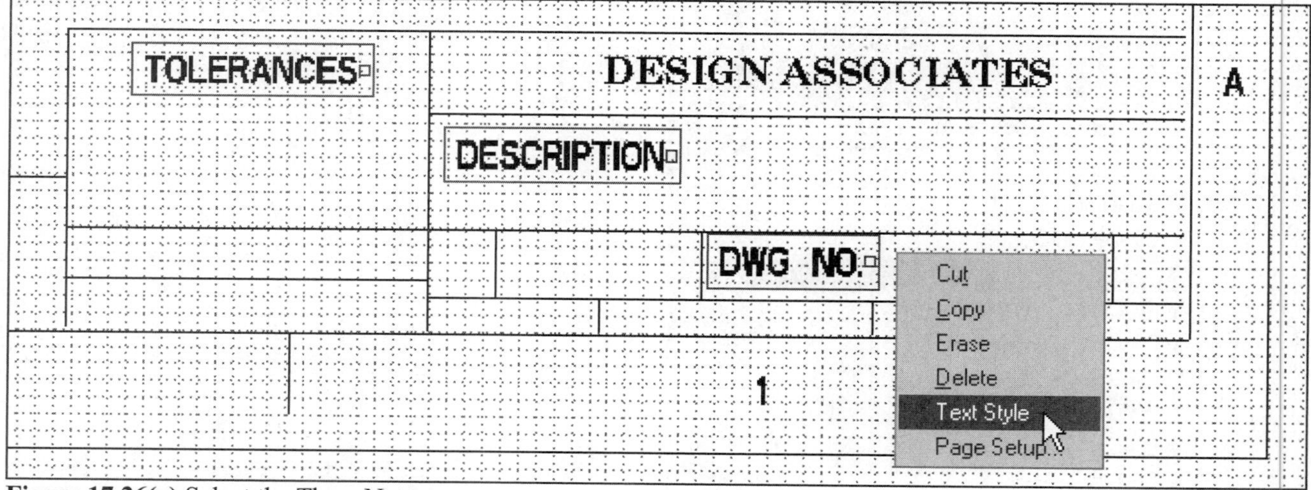

Figure 17.26(a) Select the Three Notes

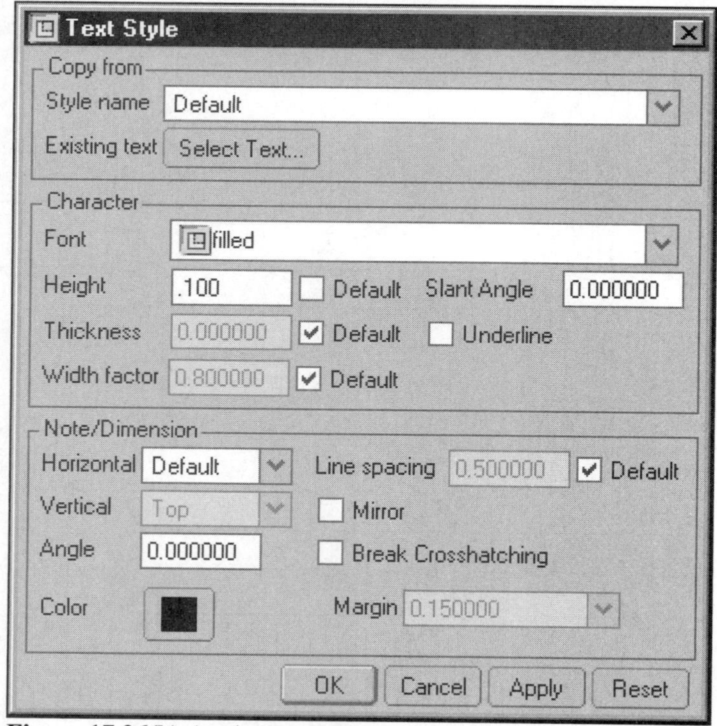

Figure 17.26(b) Apply New Height

Click on the note **DESIGN ASSOCIATES** ⇒ **RMB** [Fig. 17.27(a)] ⇒ **Add Hyperlink** ⇒ Type the URL or internal link: **www.cad-resources.com** [Fig. 17.27(b)] ⇒ **ScreenTip** ⇒ Screen tip text: type **CAD-RESOURCES WEB SITE** [Fig. 17.27(c)] ⇒ **OK** ⇒ **OK** ⇒ place the cursor on **DESIGN ASSOCIATES** to see Tool Tip and one line Help lower left of graphics window CAD-RESOURCES WEB SITE CTRL+click to follow link ⇒ press **Ctrl** and click on the note to open web site in Browser window [Fig. 17.27(d)] ⇒ ⇒ **MMB**

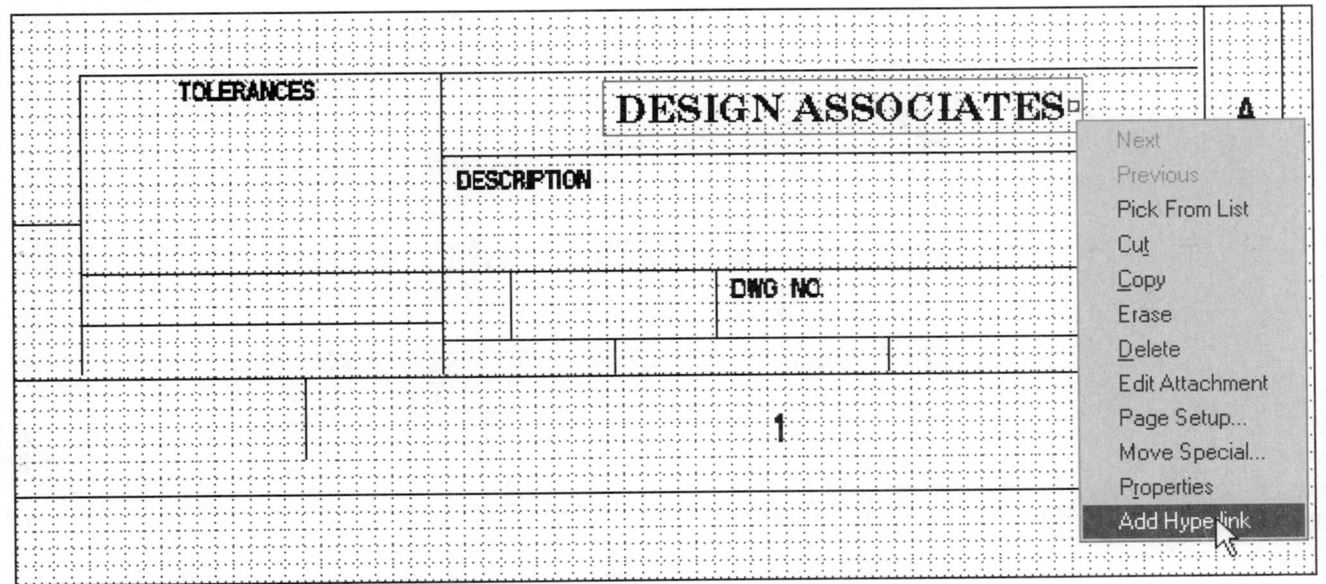

Figure 17.27(a) Add Hyperlink

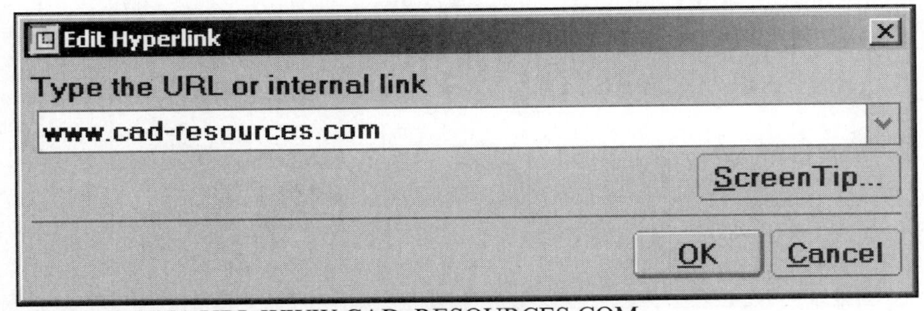

Figure 17.27(b) URL WWW.CAD_RESOURCES.COM

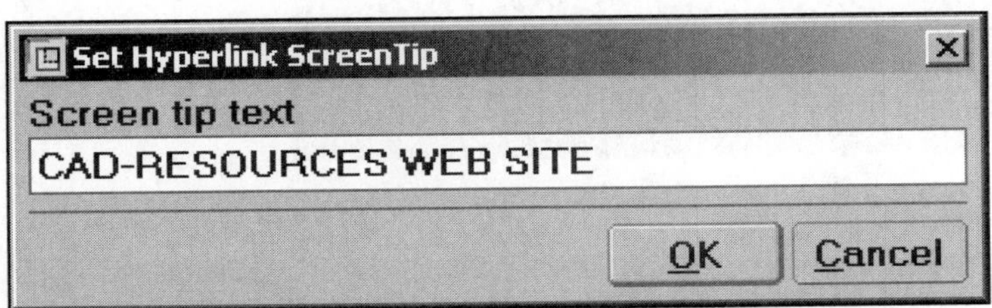

Figure 17.27(c) Screen Tip CAD-RESOURCES WEB SITE

Figure 17.27(d) CAD-RESOURCES WEB SITE

Parametric Notes in Formats and Tables

When you place a parametric note in a format, the note in the format acquires the appropriate value when you use it in a drawing. For example, if you create the note *&model_name* in a format, Pro/E displays the actual model name in the note in the drawing. For Pro/E to update these parameters when you add the format to a drawing, you can include in parametric notes only the drawing labels listed in the topic titled *System Parameters for Drawings* (except *&todays_date*). You must include all user-defined model and drawing parameters in format table cells in the form of *¶m* in order for Pro/E to update them in a drawing. Make sure your *Caps Lock* is off when typing parametric text. *You cannot enter the name of a Pro/E parameter (such as &linear_tol_0_0) in capital letters. Pro/E will not recognize it as a parameter.*

A variety of Pro/E standard drawing and format parametric notes are available including:

- ***&dwg_name***
- ***&todays_date***
- ***&scale***
- ***&type*** (drawing model type)
- ***&format*** (format size)
- ***&linear_tol_0_0*** through ***&linear_tol_0_000000***
- ***&angular_tol_0_0*** through ***&angular_tol_0_000000***
- ***¤t_sheet***
- ***&total_sheets***
- ***&det_scale*** (detail scale)

Format mode interprets the following types of parametric notes:

- Notes with symbol instances
- Notes with standard system symbols
- Notes with drawing labels
- Notes with a default tolerance

The following rules apply to including parameters as labels in format tables:

- Pro/E correctly evaluates parametric labels that you include in a format table *only* if you create the drawing first, add the model, and then add the format.
- When you add a format to a drawing with more than one model present, parametric notes can reference *only* the active drawing model.
- When you add the format to a drawing, Pro/E parses any and all that are present of the standard parametric symbols that it supports and displays the correct values in the table.

PARAMETER NAME	DEFINITION
&<param_name>	Displays a user-defined parameter value in a drawing note
&<param_name>:att_cmp	An object parameter that indicates the parameters of the component to which a note is attached
&angular_tol_0_0	Specifies angular tolerance values from one to six decimal places
&dwg_name	Displays a drawing label indicating the name of the drawing
&linear_tol_0_0	Specifies the format of dimensional tolerance values in a note
&model_name	Displays a drawing label indicating the name of the model used for the drawing
&scale	Displays a drawing label indicating the scale of the drawing
&total_sheets	Displays a drawing label indicating the total number of sheets

Create the tolerances note [Fig. 17.28(a)]. The *Spacebar* counts as a character on the current line and makes Pro/E think that you have entered actual text. Therefore, "Enter" is not taken to mean that you are finished making the note.

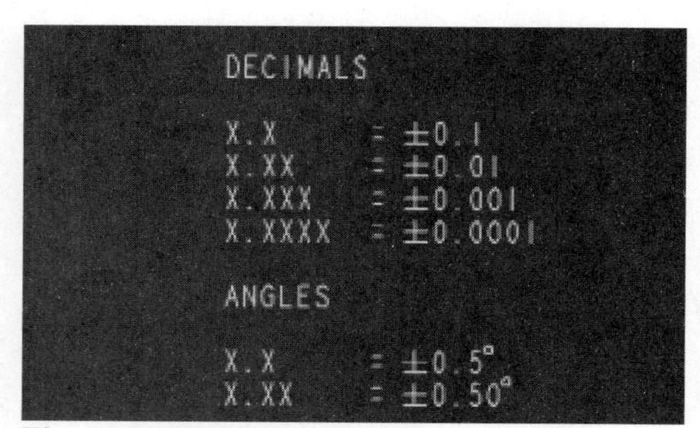

Figure 17.28(a) Tolerance Note

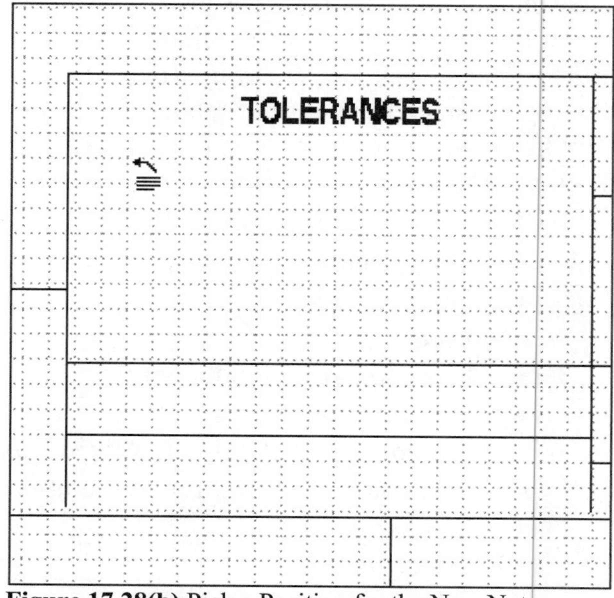

Figure 17.28(b) Pick a Position for the New Note

Click: **Create a note** ⇒ **Make Note** ⇒ pick a position for the note below the TOLERANCES note [Fig. 17.28(b)] ⇒ type **DECIMALS** ⇒ **Enter** ⇒

You now need a blank line. If you just press Enter again, Pro/E will assume that you are finished making the note, which you are not.

Press the **Spacebar** ⇒ **Enter** ⇒

Now you must input the parameterized variables.

Type **X.X** ⇒ **Spacebar** ⇒ **Spacebar** ⇒ **Spacebar** ⇒ **Spacebar** ⇒ **Spacebar** ⇒ type = ⇒ **Spacebar** ⇒ choose the **plus and minus** symbol ± from the palette [Fig. 17.28(c)] ⇒ type **&linear_tol_0_0** Enter NOTE: X.X = ±&linear_tol_0_0 ⇒ **Enter** ⇒ type **X.XX** ⇒ **Spacebar** ⇒ **Spacebar** ⇒ **Spacebar** ⇒ **Spacebar** ⇒ type = ⇒ **Spacebar** ⇒ ± from the palette ⇒ type **&linear_tol_0_00** ⇒ **Enter** ⇒ type **X.XXX** ⇒ **Spacebar** ⇒ **Spacebar** ⇒ **Spacebar** ⇒ type = ⇒ **Spacebar** ⇒ ± from the palette ⇒ type **&linear_tol_0_000** ⇒ **Enter** ⇒ type **X.XXXX** ⇒ **Spacebar** ⇒ **Spacebar** ⇒ type = ⇒ **Spacebar** ⇒ ± from the palette ⇒ type **&linear_tol_0_0000** ⇒ **Enter** ⇒

You now need another blank line. If you press Enter again, Pro/E will assume that you are finished making the note, and you are not.

Press the **Spacebar** ⇒ **Enter** ⇒ type **ANGLES** ⇒ **Enter** ⇒ **Spacebar** ⇒ **Enter** ⇒ type **X.X** ⇒ **Spacebar** ⇒ **Spacebar** ⇒ **Spacebar** ⇒ **Spacebar** ⇒ **Spacebar** ⇒ type = ⇒ **Spacebar** ⇒ ± from the palette ⇒ type **&angular_tol_0_0** Enter NOTE: X.X = ±&angular_tol_0_0 ⇒ **Enter** ⇒ type **X.XX** ⇒ **Spacebar** ⇒ **Spacebar** ⇒ **Spacebar** ⇒ **Spacebar** ⇒ type = ⇒ **Spacebar** ⇒ ± from the palette ⇒ type **&angular_tol_0_00** ⇒ **Enter** ⇒ **Enter** to finish creating the note ⇒ **Done/Return** [Fig. 17.28(d)]

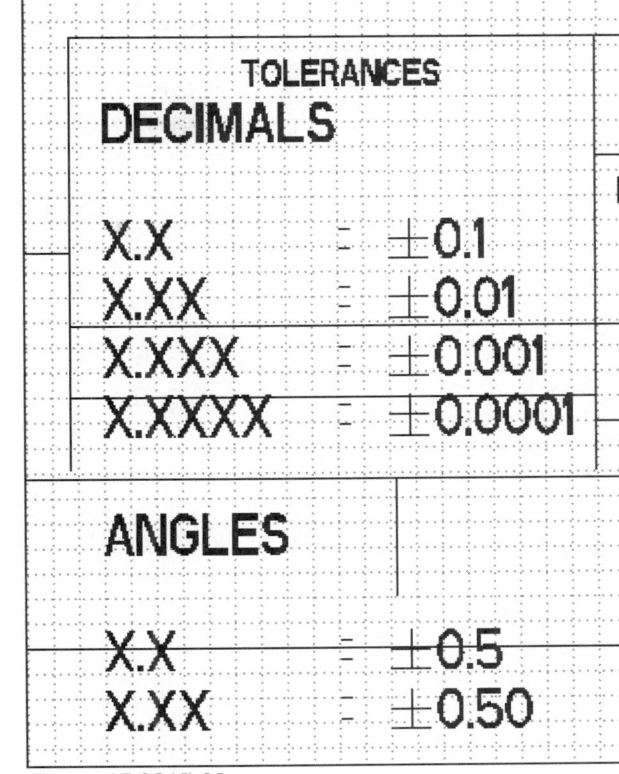

Figure 17.28(d) Note

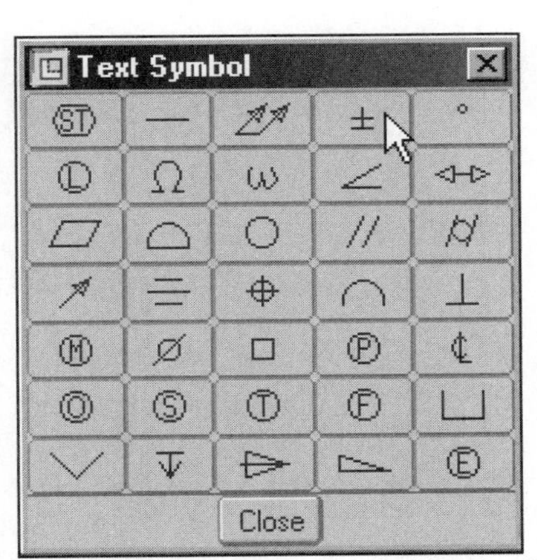

Figure 17.28(c) Text Symbol Dialog Box

Click: **LMB** in the graphics window to deselect ⇒ **Format** [Fig. 17.28(e)] ⇒ **Text Style** [Fig. 17.28(f)] ⇒ select all of the note elements [Fig. 17.28(g)] *region selection is available* ⇒ **MMB** [Fig. 17.28(h)]

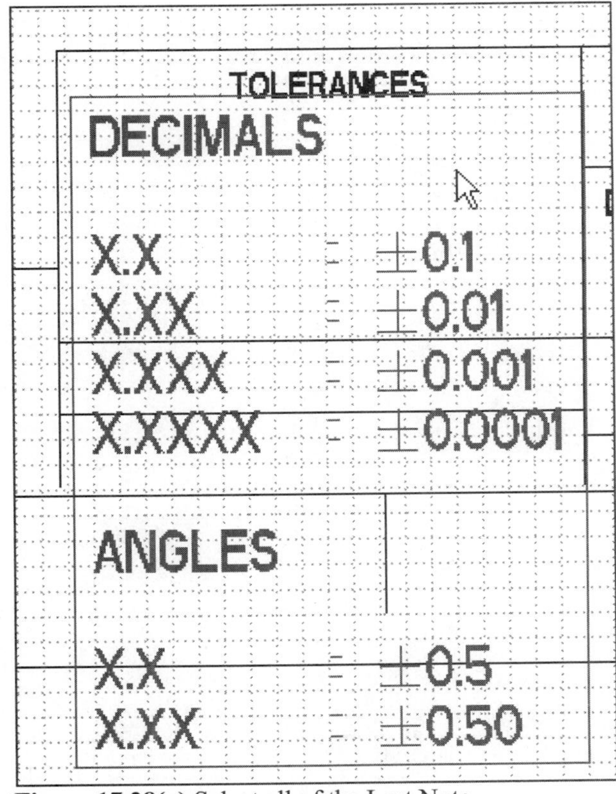

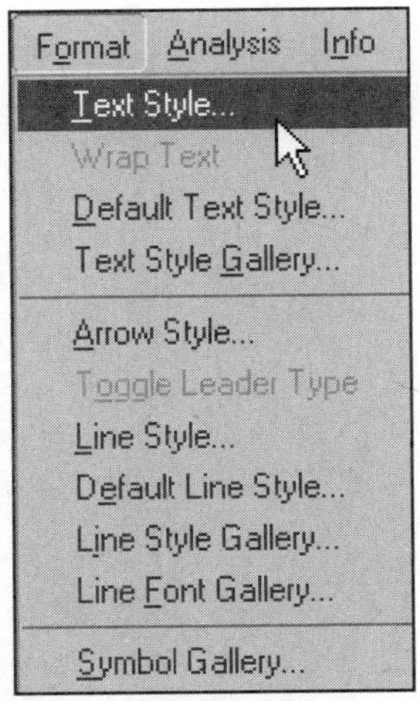

Figure 17.28(e) Format

Figure 17.28(f) Select

Figure 17.28(g) Select all of the Last Note

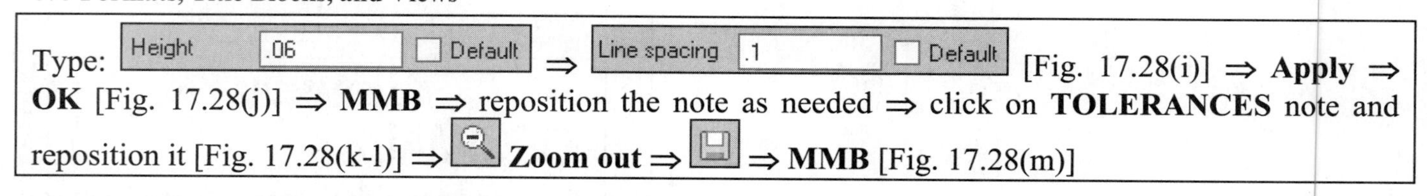

Type: Height .06 ☐ Default ⇒ Line spacing .1 ☐ Default [Fig. 17.28(i)] ⇒ **Apply** ⇒ **OK** [Fig. 17.28(j)] ⇒ **MMB** ⇒ reposition the note as needed ⇒ click on **TOLERANCES** note and reposition it [Fig. 17.28(k-l)] ⇒ **Zoom out** ⇒ ⇒ **MMB** [Fig. 17.28(m)]

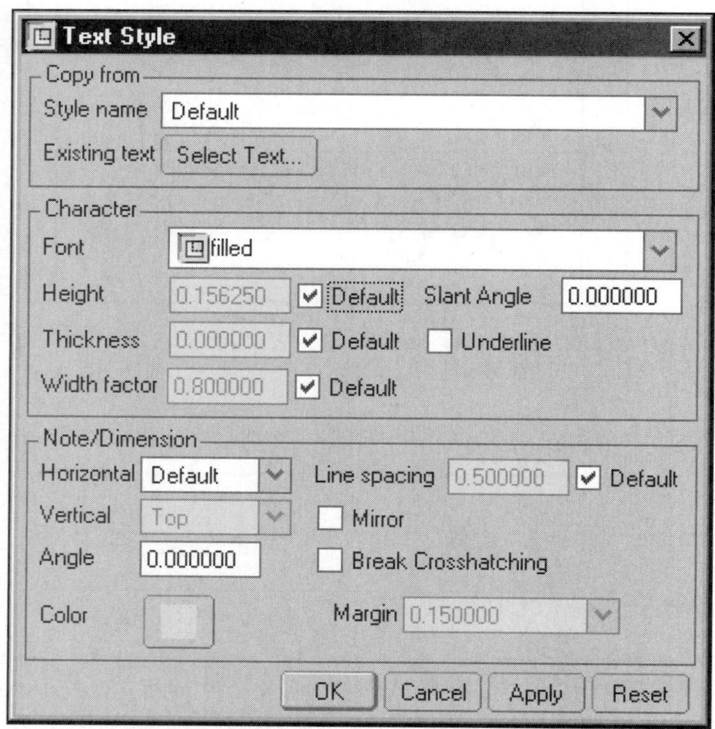

Figure 17.28(h) Text Style Dialog Box Defaults

Figure 17.28(i) Text Style Dialog Box New Height and Line Spacing

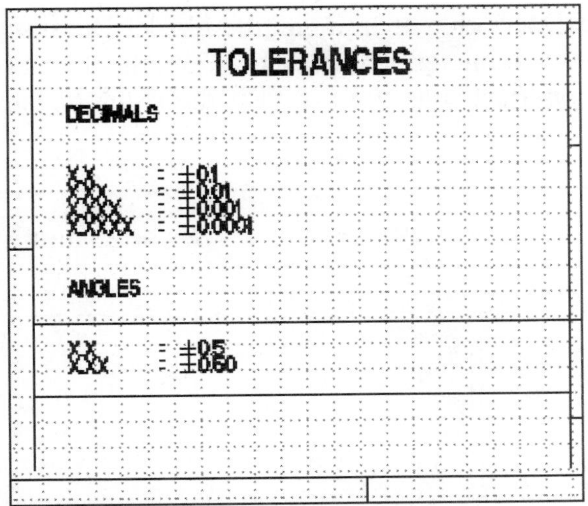

Figure 17.28(j) Note with a Smaller Font

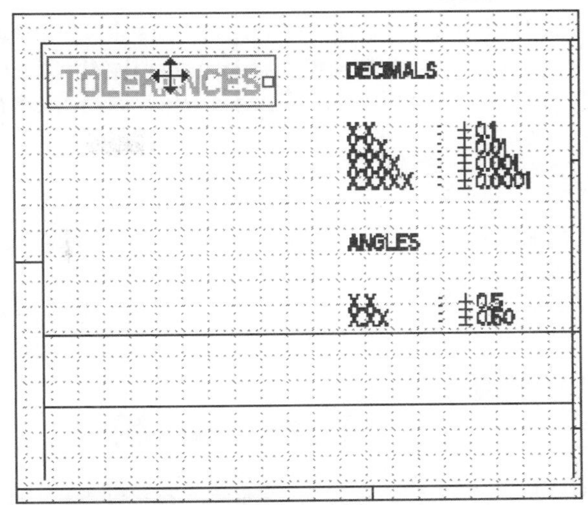

Figure 17.28(k) Repositioned Notes

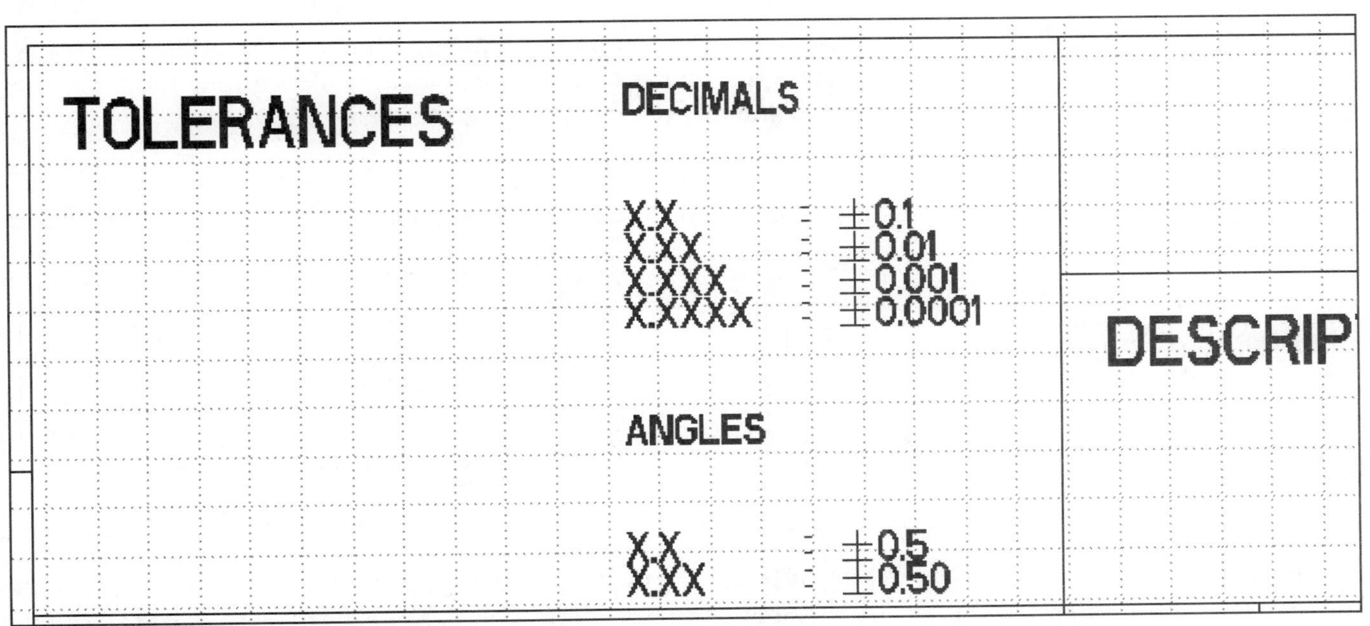

Figure 17.28(l) Tolerance Block

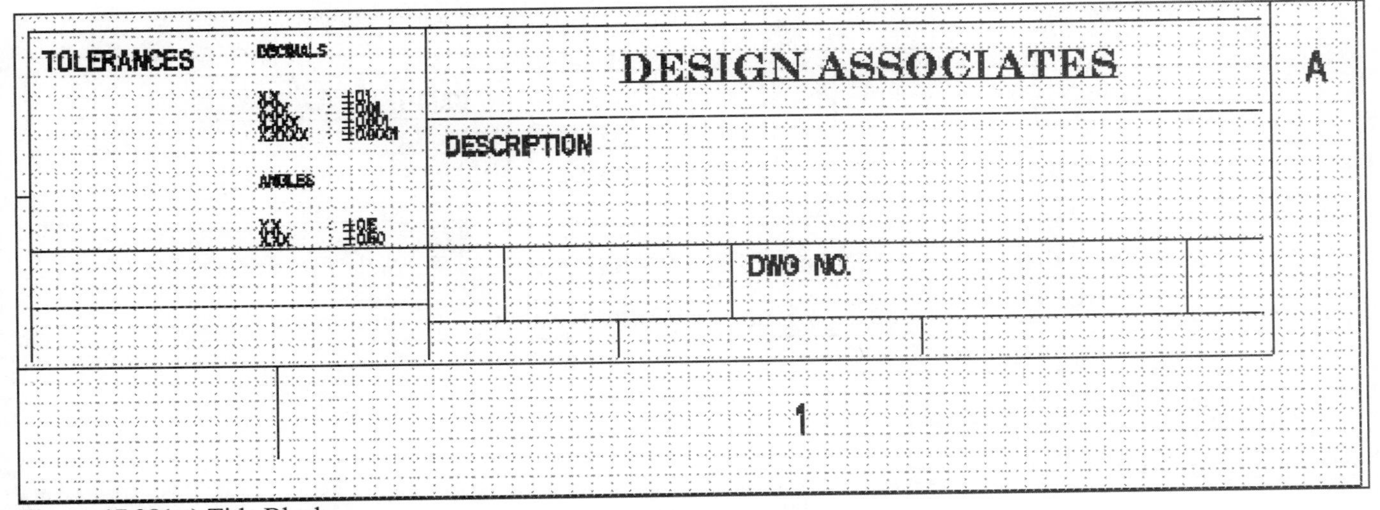

Figure 17.28(m) Title Block

Create the note that will display the model name in the description area in the title block. Because there is no model now, therefore only the text of the name of the parameter is displayed. When this format is placed on a drawing and a view is added (thus adding a model), this text will reflect the name of the model.

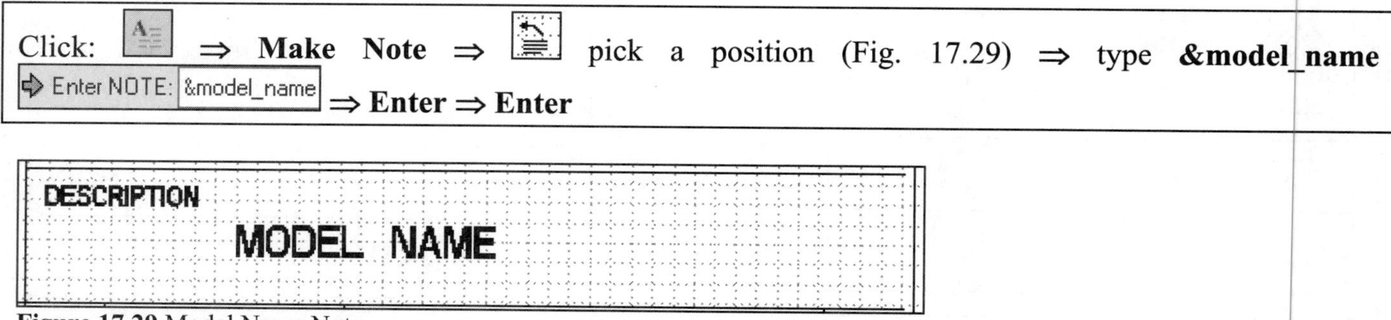

Click: ⇒ **Make Note** ⇒ pick a position (Fig. 17.29) ⇒ type **&model_name** Enter NOTE: &model_name ⇒ **Enter** ⇒ **Enter**

Figure 17.29 Model Name Note

Create the note that will display the drawing name in the drawing number area in the title block. Your version of this note will actually show the name of the format that is the current format name. When this format is placed on a drawing, the drawing name will replace this text.

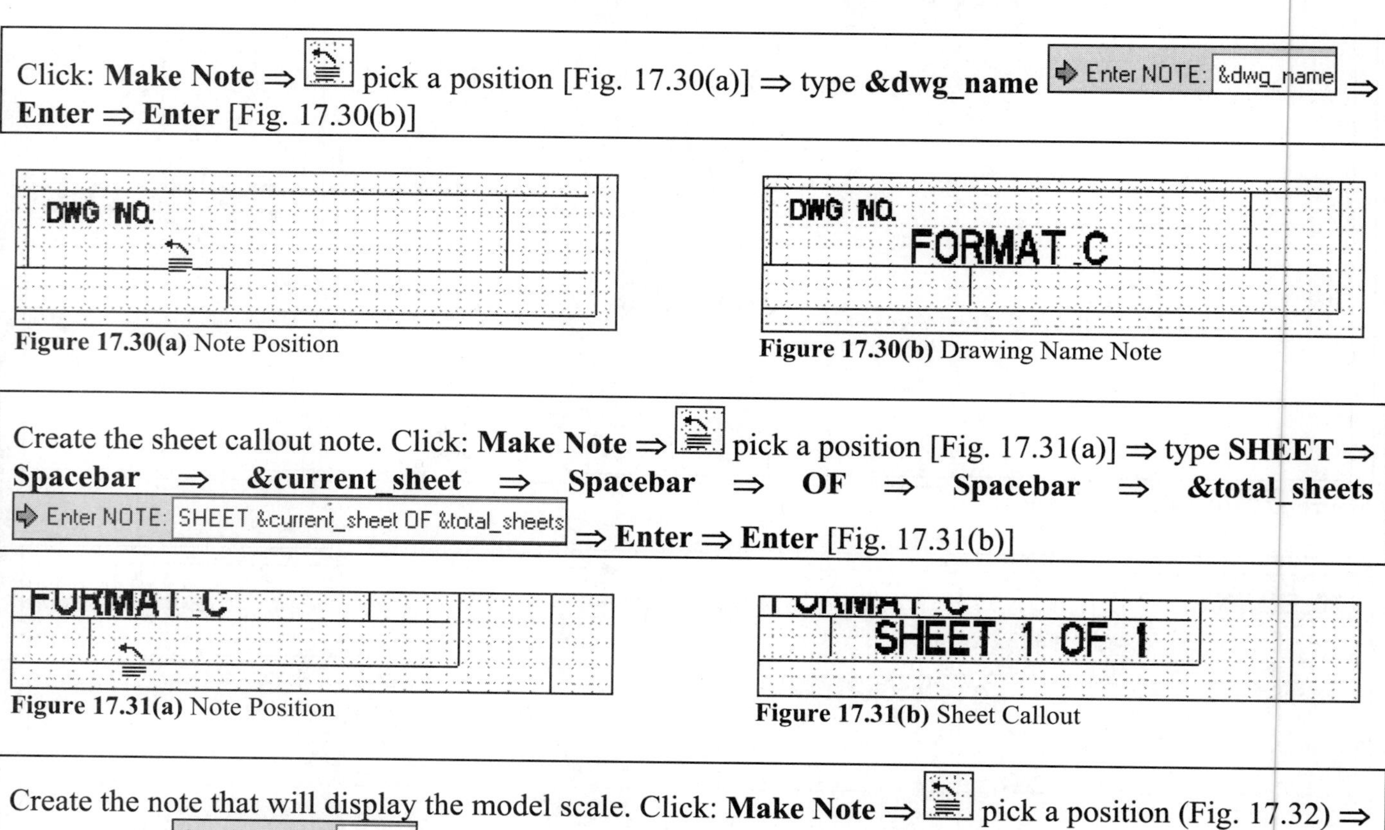

Click: **Make Note** ⇒ pick a position [Fig. 17.30(a)] ⇒ type **&dwg_name** Enter NOTE: &dwg_name ⇒ **Enter** ⇒ **Enter** [Fig. 17.30(b)]

Figure 17.30(a) Note Position

Figure 17.30(b) Drawing Name Note

Create the sheet callout note. Click: **Make Note** ⇒ pick a position [Fig. 17.31(a)] ⇒ type **SHEET** ⇒ **Spacebar** ⇒ **¤t_sheet** ⇒ **Spacebar** ⇒ **OF** ⇒ **Spacebar** ⇒ **&total_sheets** Enter NOTE: SHEET ¤t_sheet OF &total_sheets ⇒ **Enter** ⇒ **Enter** [Fig. 17.31(b)]

Figure 17.31(a) Note Position

Figure 17.31(b) Sheet Callout

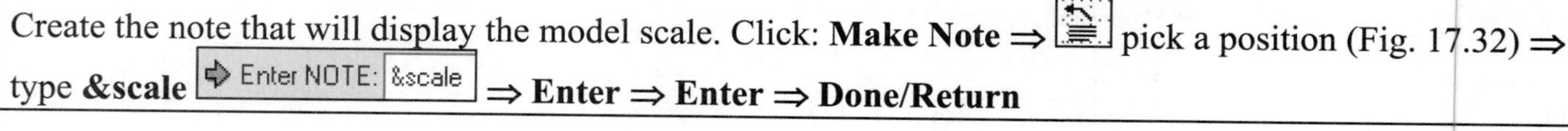

Create the note that will display the model scale. Click: **Make Note** ⇒ pick a position (Fig. 17.32) ⇒ type **&scale** Enter NOTE: &scale ⇒ **Enter** ⇒ **Enter** ⇒ **Done/Return**

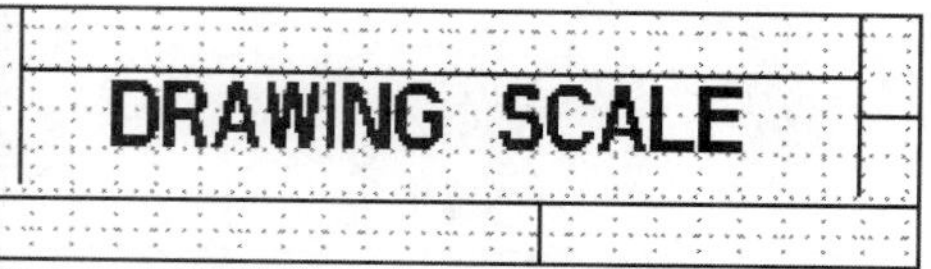

Figure 17.32 Drawing Scale Note

Add a note to display the format size, click: ⇒ **Make Note** ⇒ pick a position (Fig. 17.33) ⇒ type **&format** Enter NOTE: &format ⇒ **Enter** ⇒ **Enter** ⇒ **Done/Return** (Fig. 17.34) ⇒ **LMB** to deselect

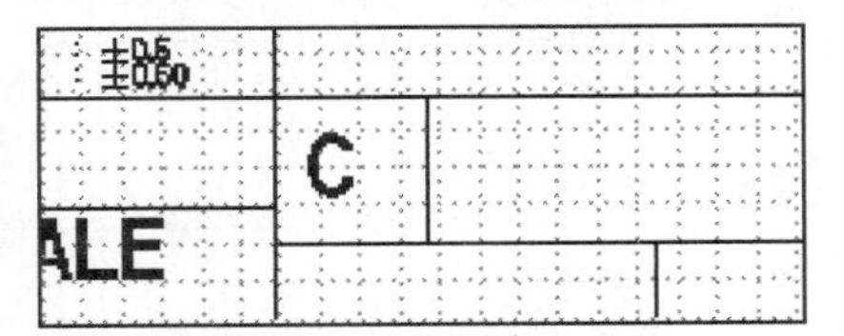

Figure 17.33 Format Size Note

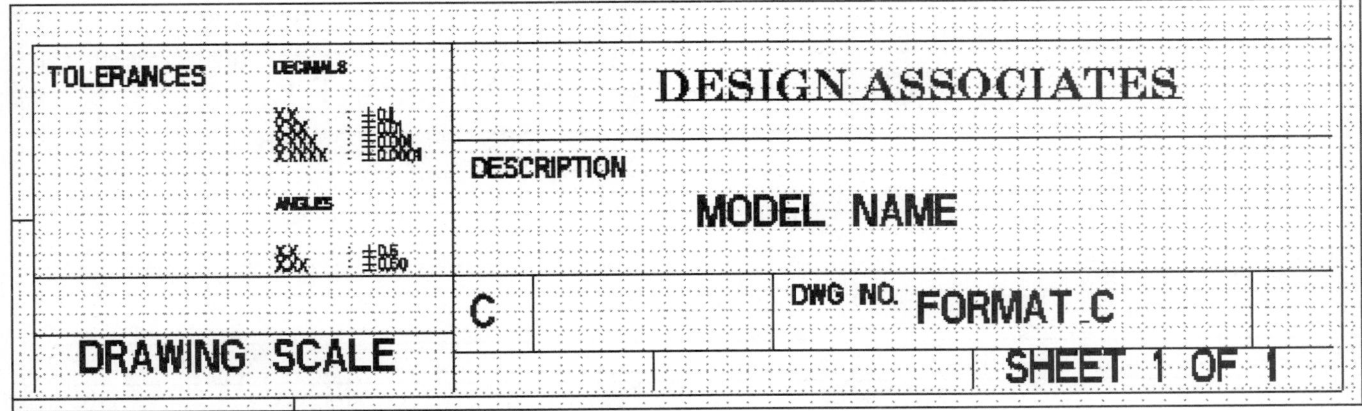

Figure 17.34 Title Block with Parametric Notes

Double-click on **SHEET 1 OF 1** note [Fig. 17.35(a)] ⇒ **Text Style** tab ⇒ Height **.1** [Fig. 17.35(b)] ⇒ **Preview** ⇒ **OK** ⇒ click on the note and reposition it as needed

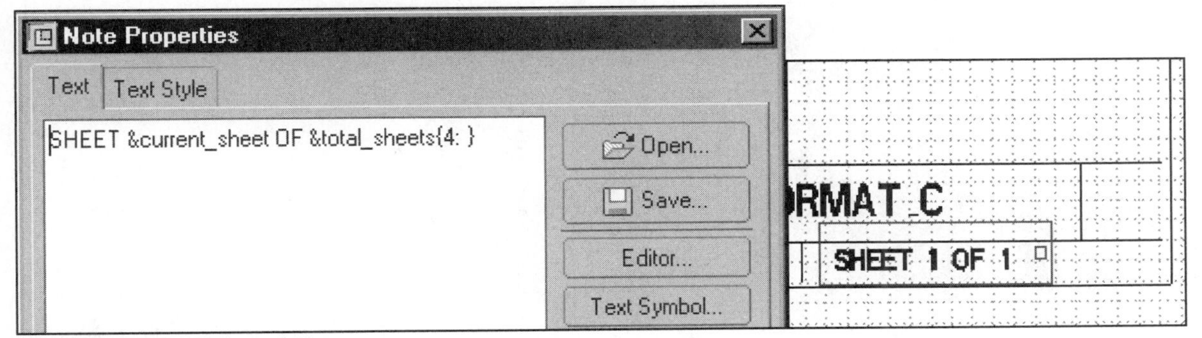

Figure 17.35(a) Selected Note

Text Style
Copy from
Style name: Default
Existing text: Select Text...
Character
Font: filled
Height: .1 ☐ Default Slant Angle: 0.000000
Thickness: 0.000000 ☑ Default ☐ Underline
Width factor: 0.800000 ☑ Default

Figure 17.35(b) Text Style, Height **.1**

Modify the text height (to **.10**) and reposition the drawing scale note as needed (Fig. 17.36) ⇒ **LMB** to deselect ⇒ [icon] ⇒ [icon] **Update the display of all views** ⇒ [icon] ⇒ [icon] ⇒ **MMB** (Fig. 17.37)

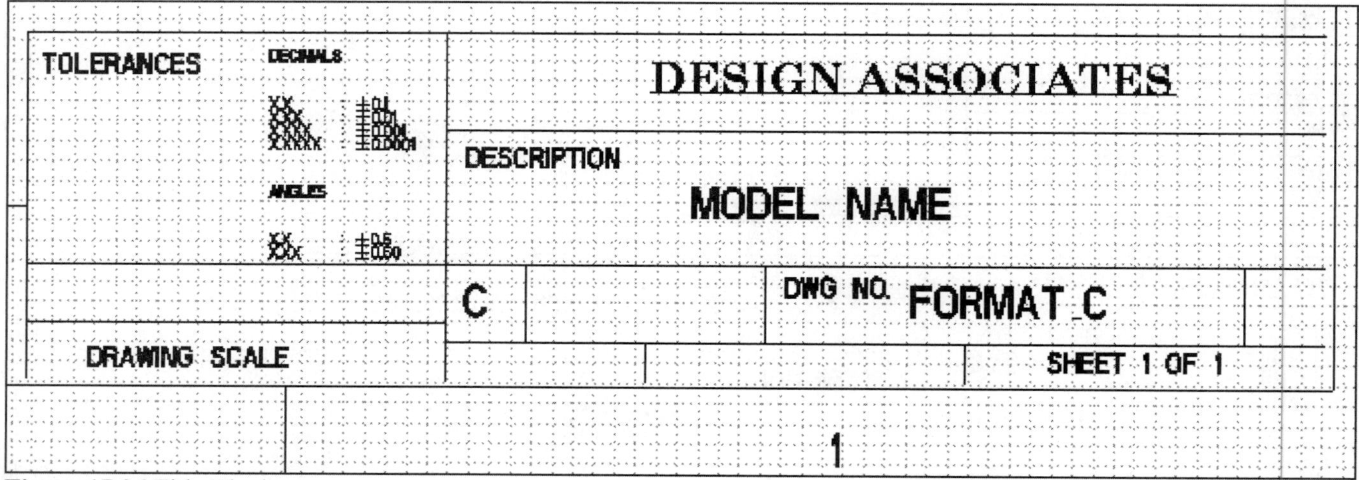

Figure 17.36 Title Block

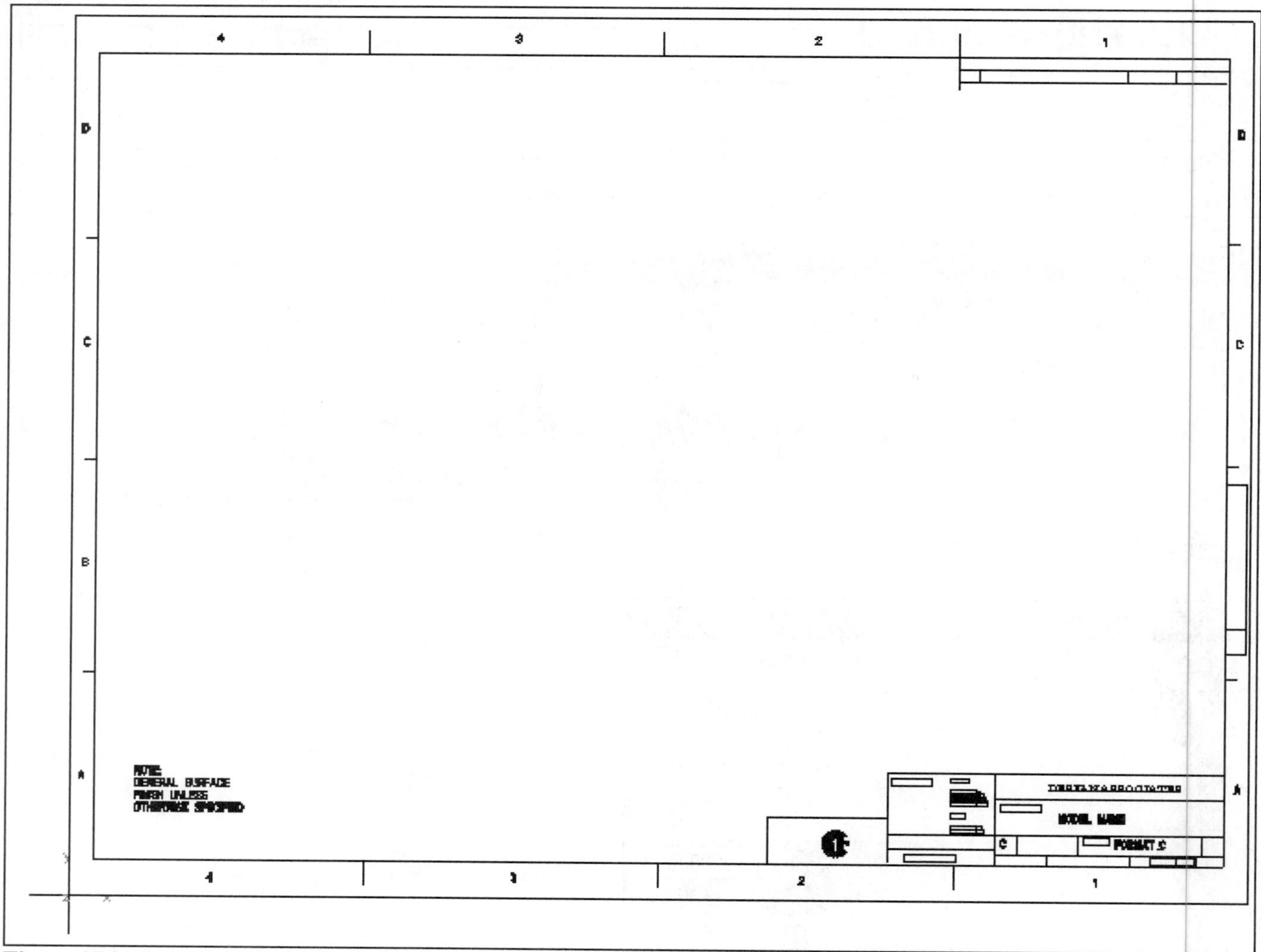

Figure 17.37 Format

Add a finish symbol to go with the note you created earlier.

Click: **Zoom In** on the drawing note [Fig. 17.38(a)] ⇒ **Insert** ⇒ **Drawing Symbol** [Fig. 17.38(b)] ⇒ **Custom** [Fig. 17.38(c)] ⇒ **Browse** ⇒ User Syms ⇒ **System Syms** [Fig. 17.38(d)] ⇒ **surftextsymlib** [Fig. 17.38(e)] ⇒ **Open** ⇒ **surftexture.sym** [Fig. 17.38(f)] ⇒ **Open** [Fig. 17.38(g)]

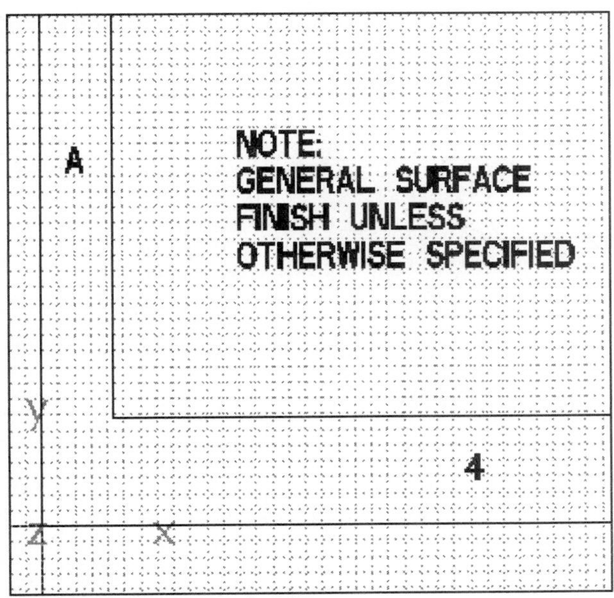

Figure 17.38(a) Drawing Note

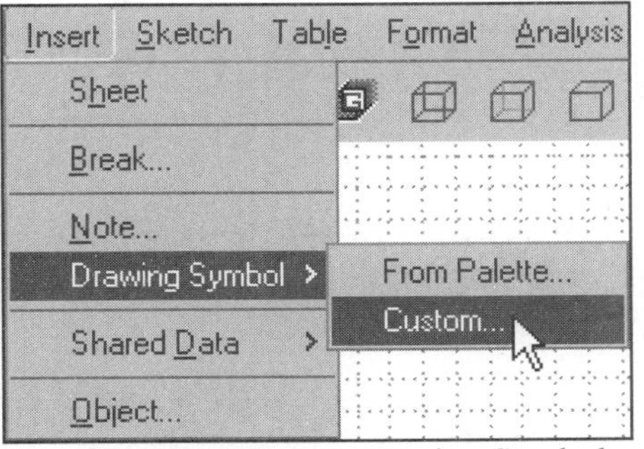

Figure 17.38(b) Insert Drawing Symbol

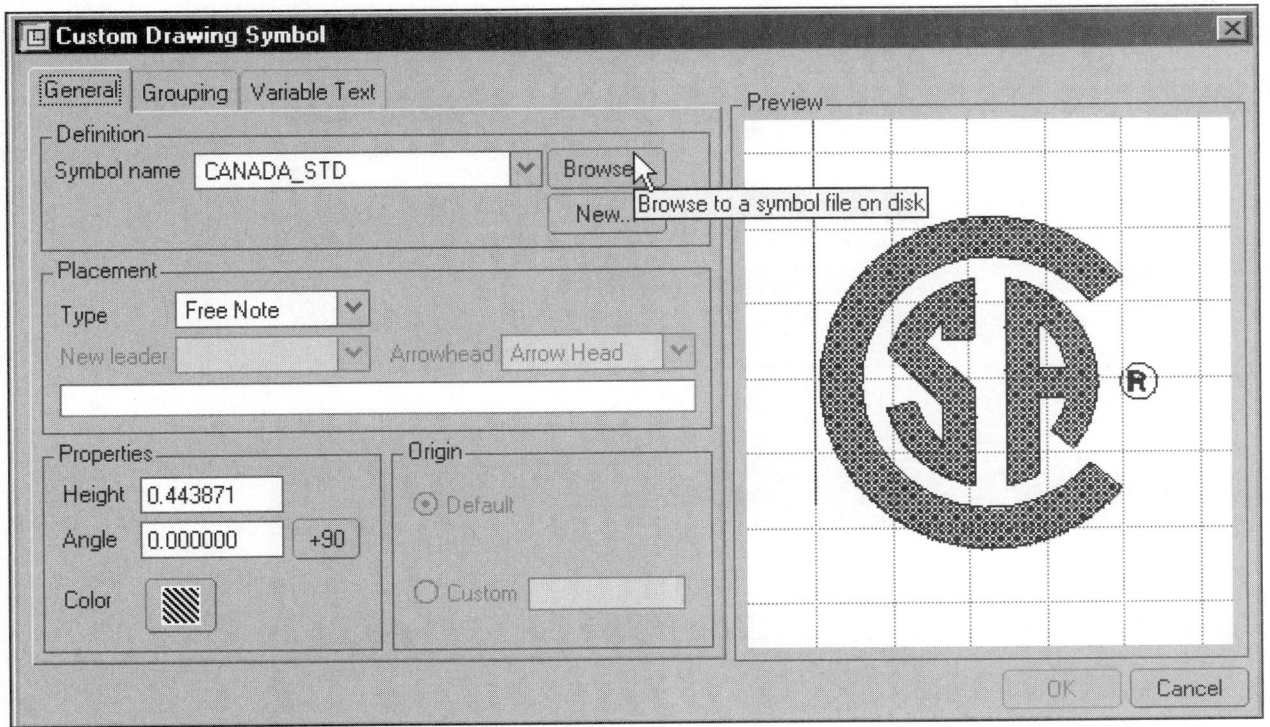

Figure 17.38(c) Custom Drawing Symbol Dialog Box

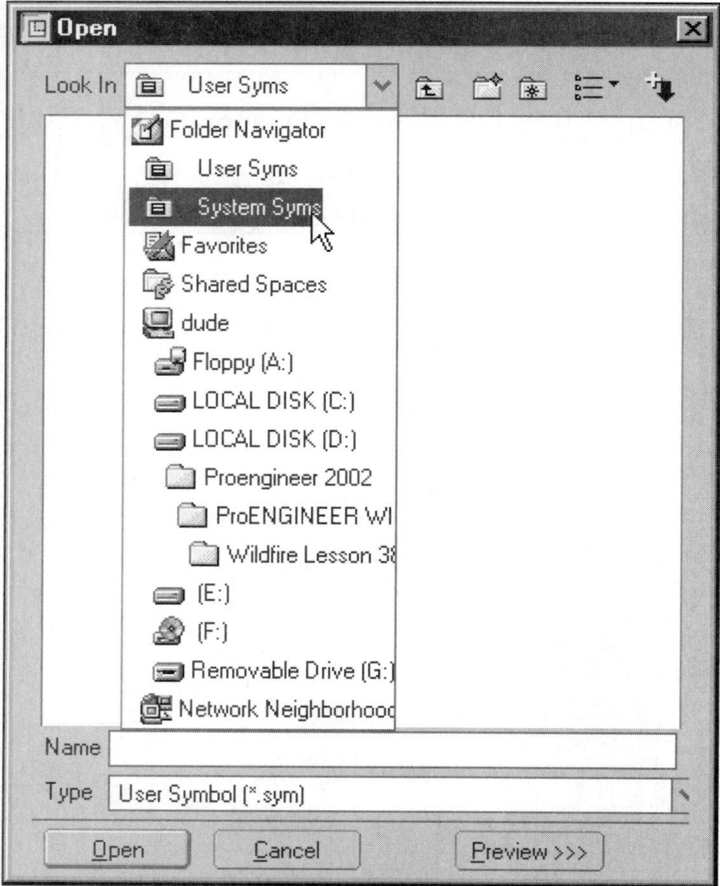

Figure 17.38(d) System Syms

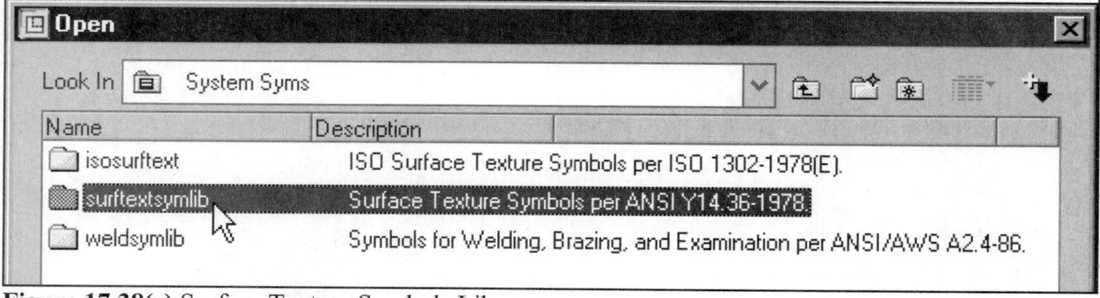

Figure 17.38(e) Surface Texture Symbols Library

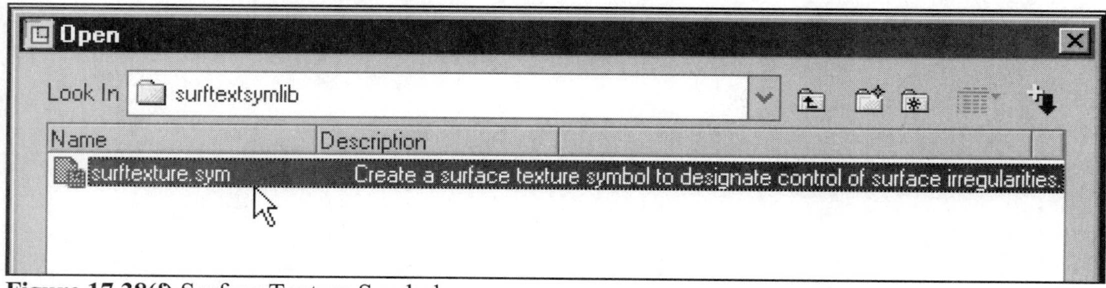

Figure 17.38(f) Surface Texture Symbol

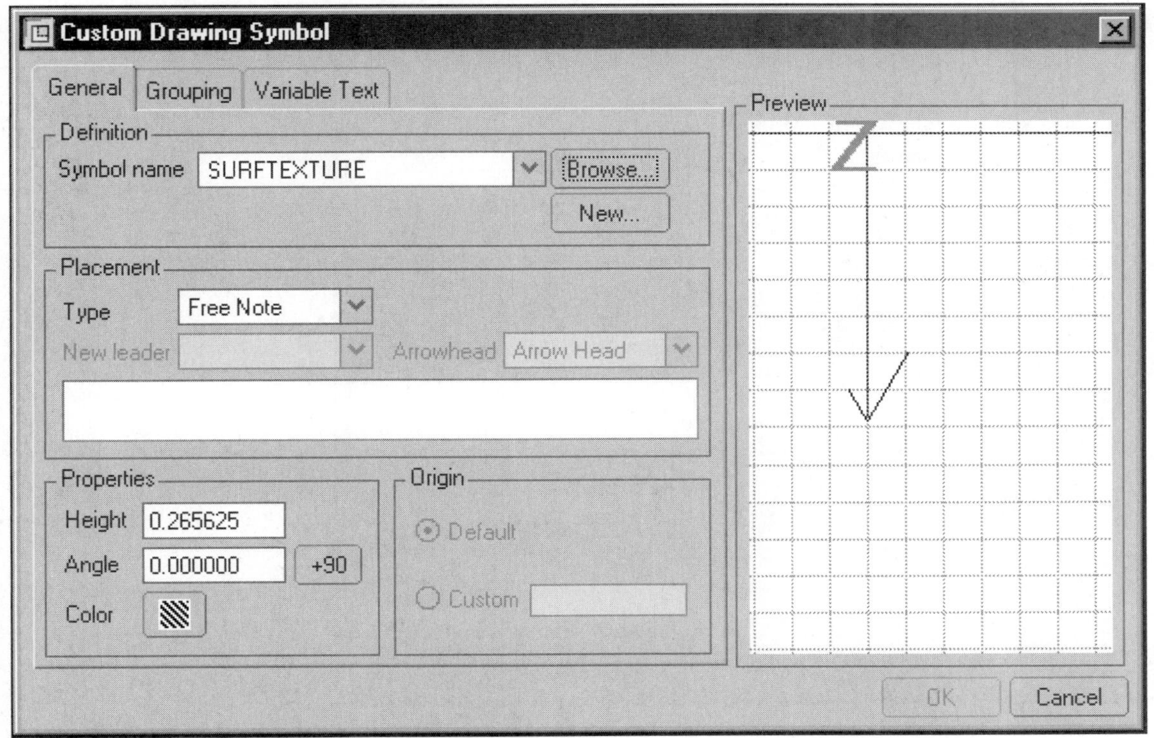

Figure 17.38(g) Custom Drawing Symbol Dialog Box

Properties Height **.5** [Fig. 17.38(h)] ⇒ **Grouping** tab [Fig. 17.38(i)]

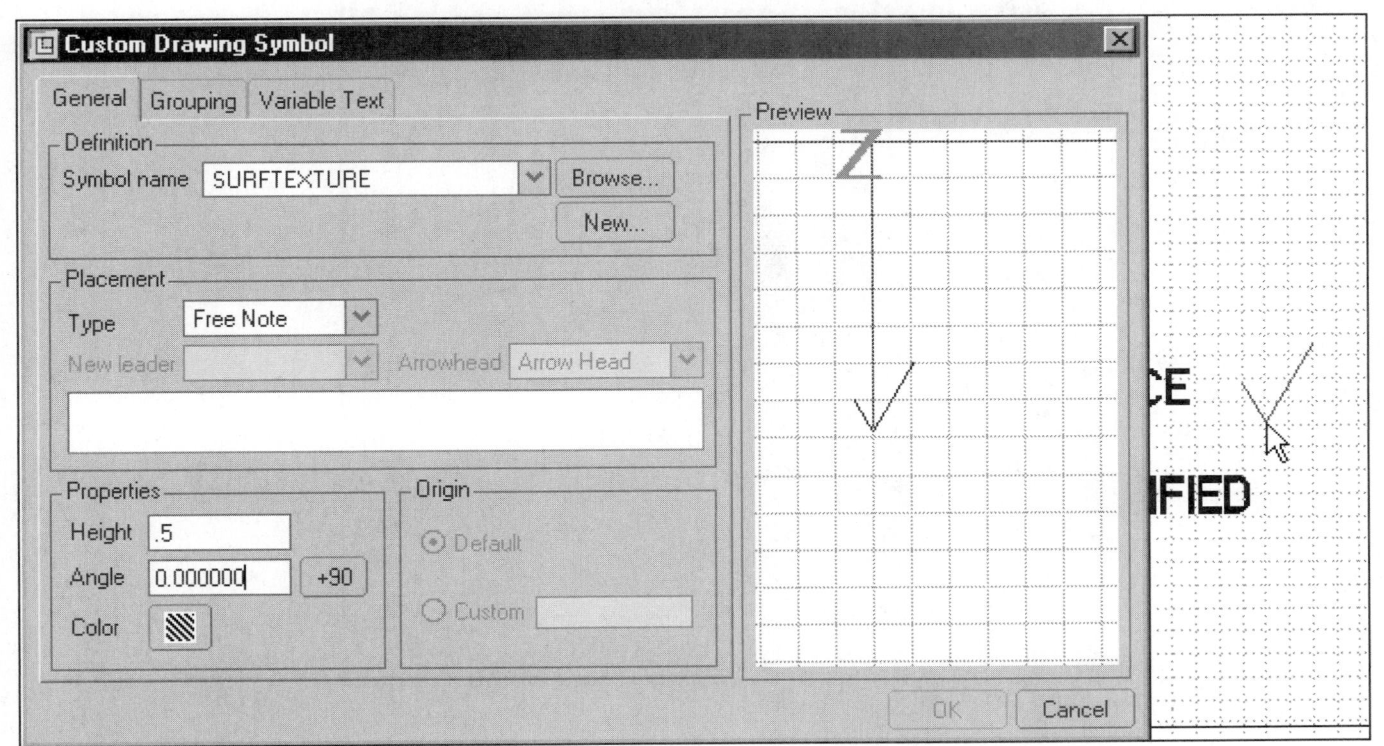

Figure 17.38(h) Height **.5**

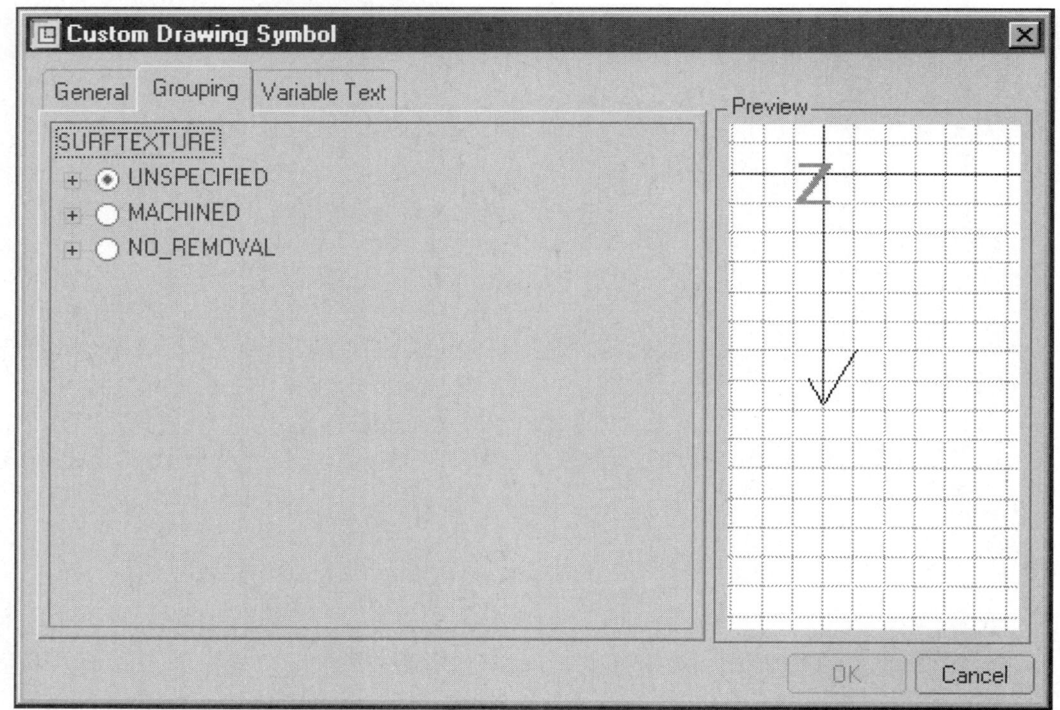

Figure 17.38(i) Grouping Tab

Click on the ⊞ just to the left of the text **UNSPECIFIED** [Fig. 17.38(j)] to expand

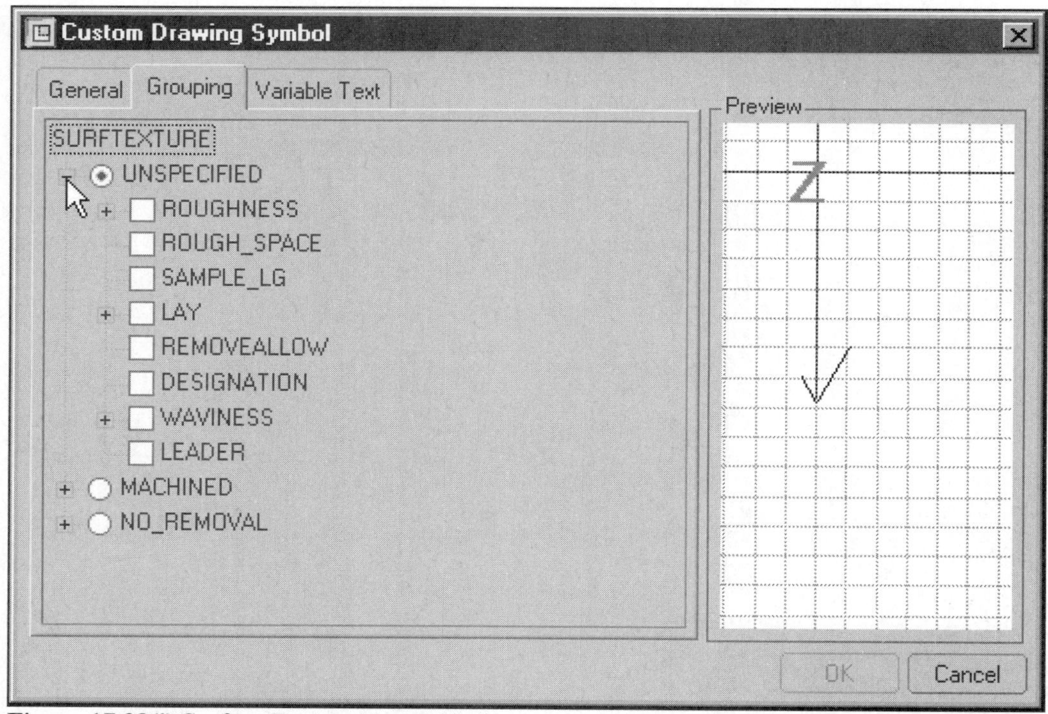

Figure 17.38(j) Surface Texture Unspecified

Check ☑ROUGHNESS [Fig. 17.38(k)] ⇒ **Variable Text** tab ⇒ average roughness, type **32** [Fig. 17.38(l)]

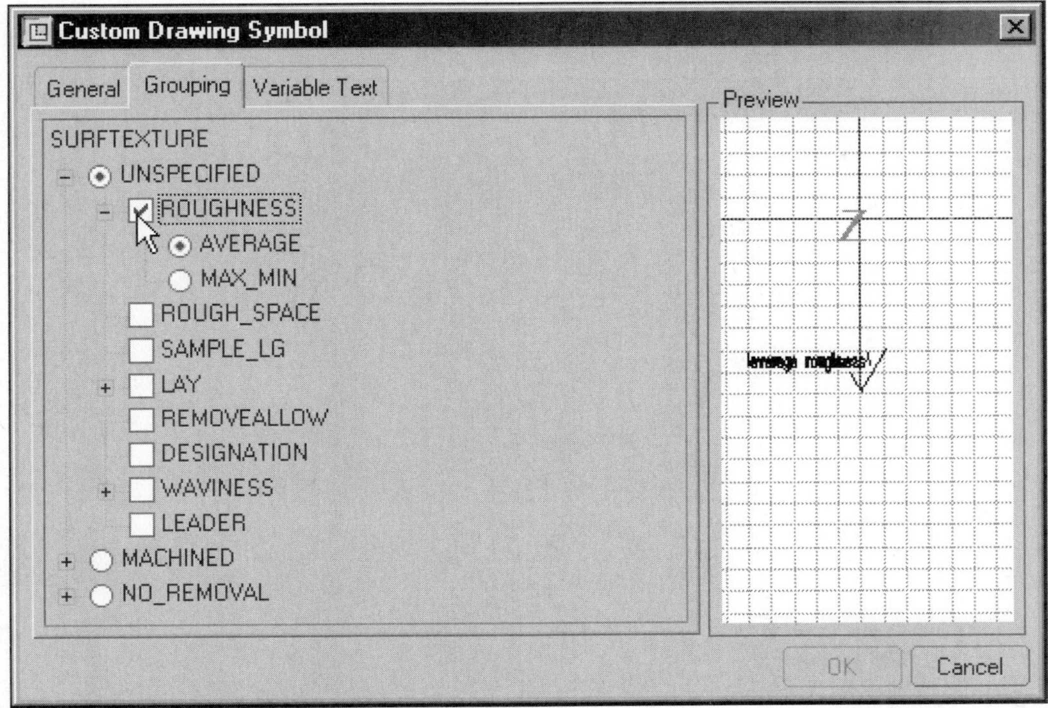

Figure 17.38(k) Roughness

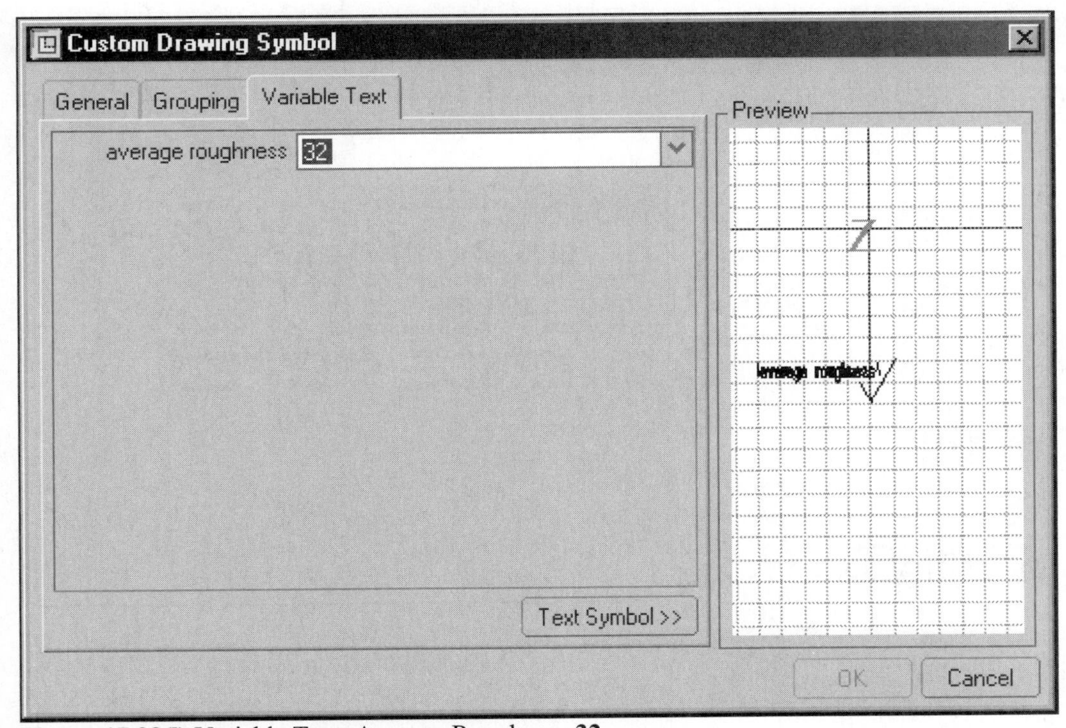

Figure 17.38(l) Variable Text, Average Roughness **32**

Click: **General** tab ⇒ Type ⇒ **Free Note** [Fig. 17.38(m)] ⇒ place the symbol [Fig. 17.38(n-o)] ⇒ **OK** ⇒ **LMB** ⇒ ⇒ ⇒ ⇒ ⇒ **MMB** ⇒ **File** ⇒ **Delete** ⇒ **Old Versions** ⇒ **MMB** (Fig. 17.39) ⇒ **File** ⇒ **Close Window**

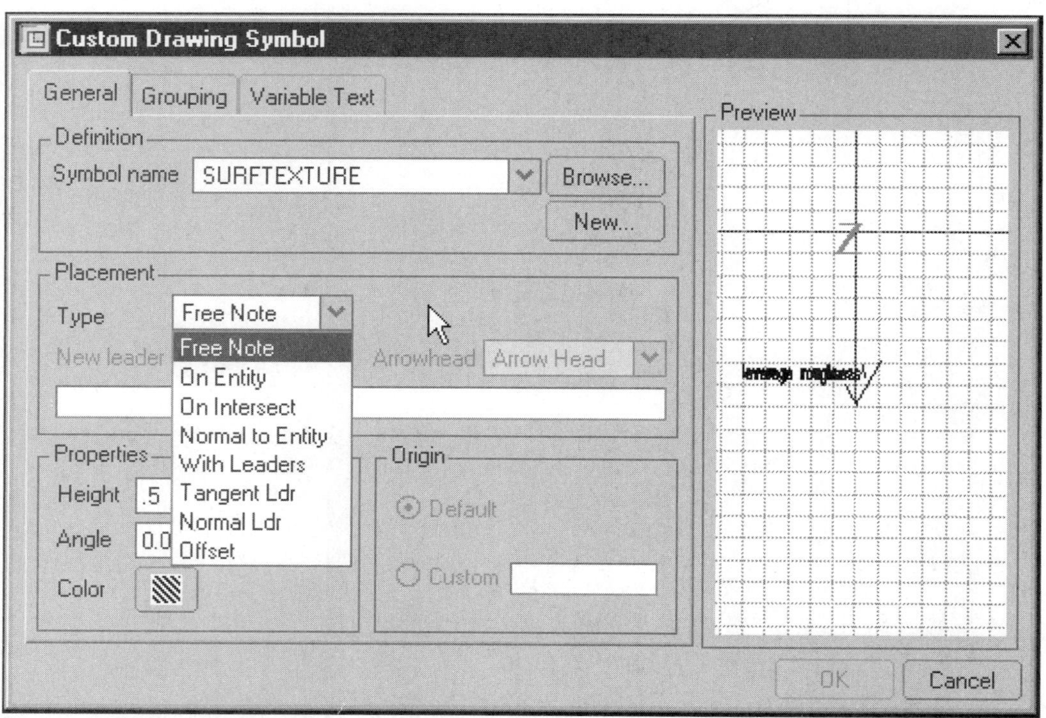

Figure 17.38(m) Free Note

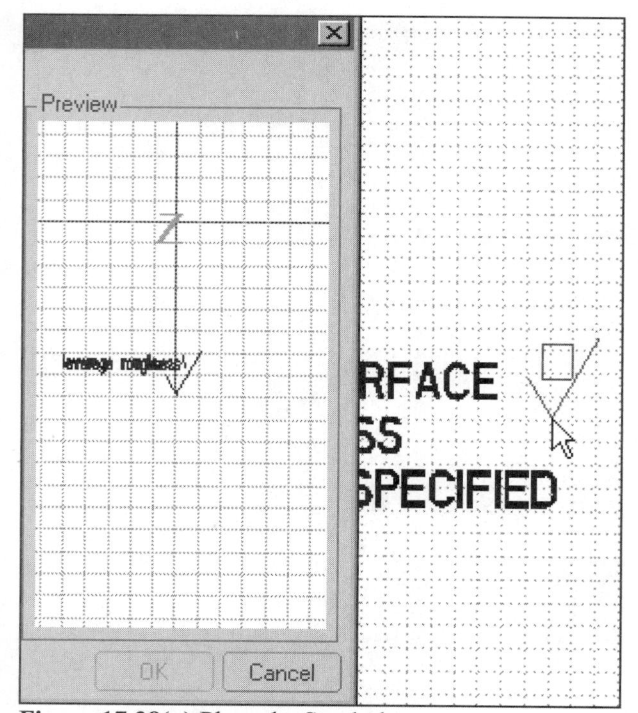

Figure 17.38(n) Place the Symbol

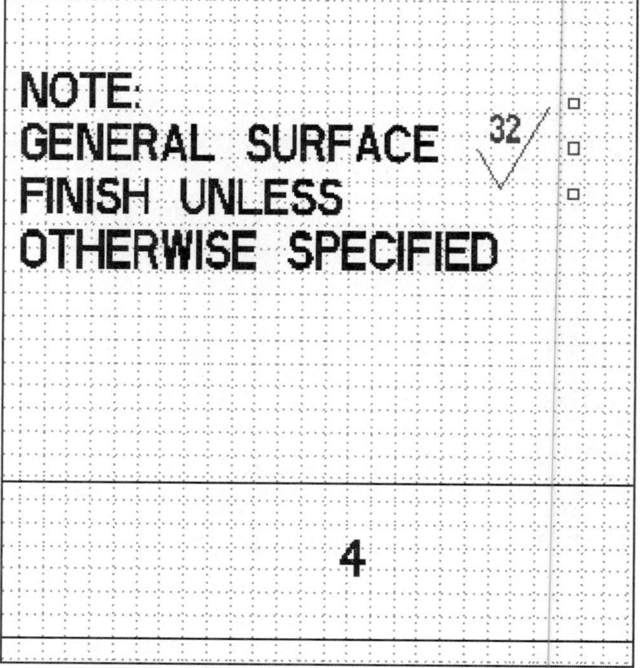

Figure 17.38(o) Symbol Placed Near the Note

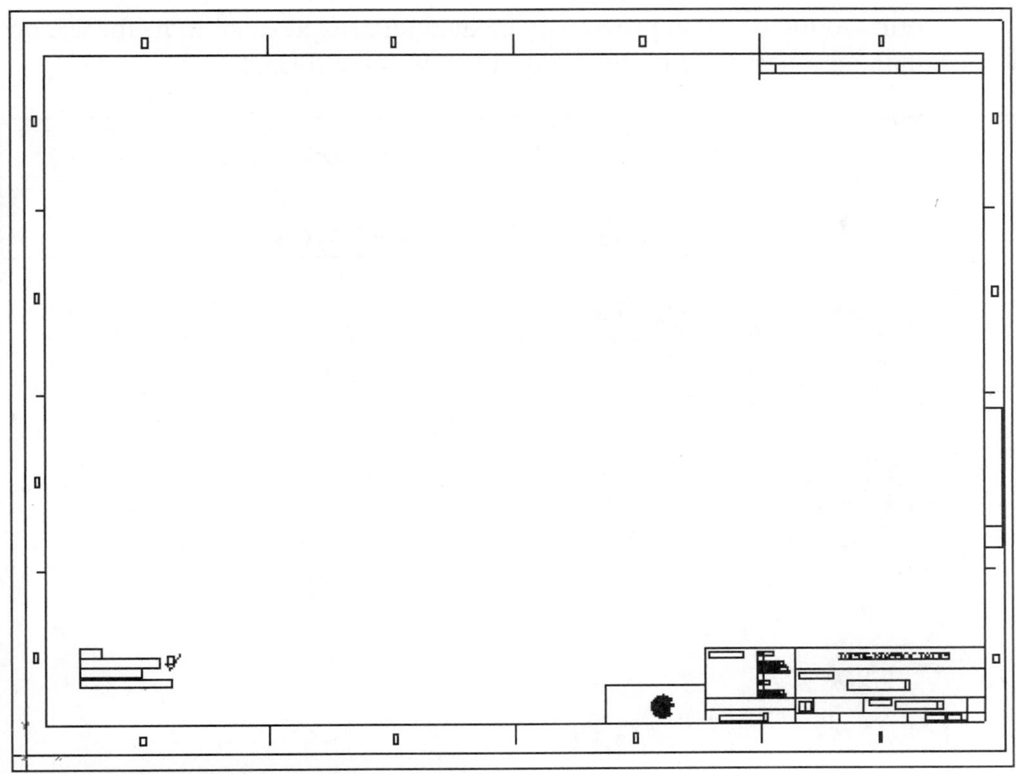

Figure 17.39 Completed FORMAT_C

Click: **Window ⇒ BASE_ANGLE.DRW ⇒ Activate** *or File ⇒ Open ⇒ BASE_ANGLE.DRW if it is not in session* (Fig. 17.40)

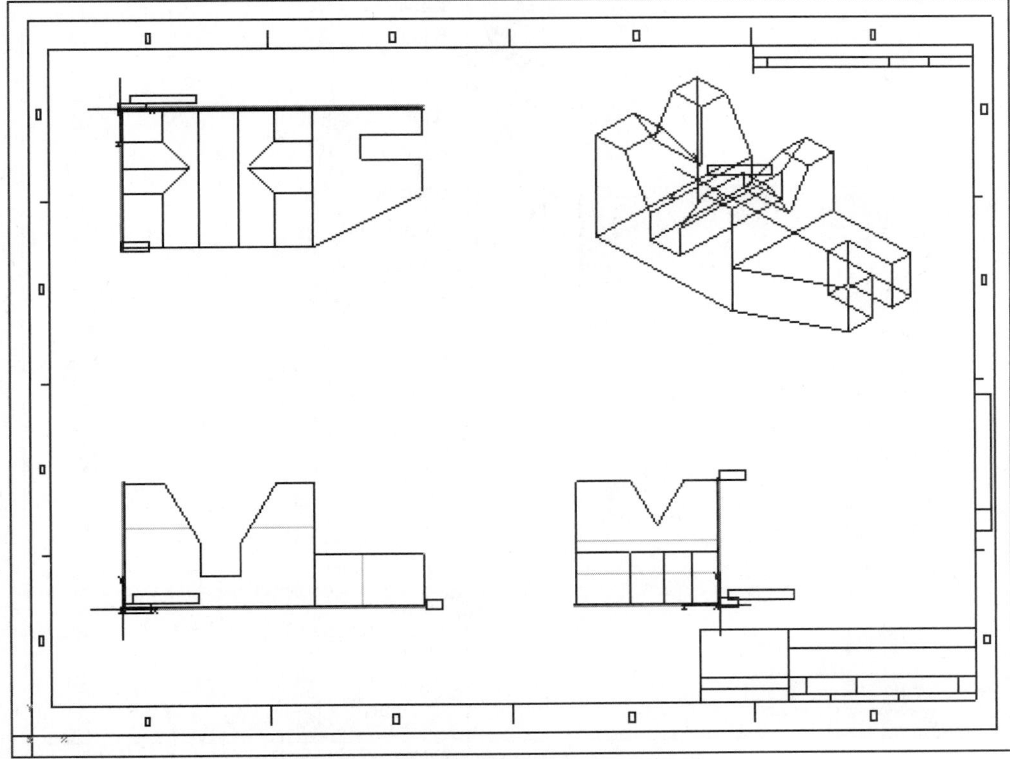

Figure 17.40 BASE_ANGLE Drawing with Standard Drawing Format

Now change the drawing format to the one you previously created and saved as an **.frm** for use on your projects. This format will contain parameters you set up and saved with the format.

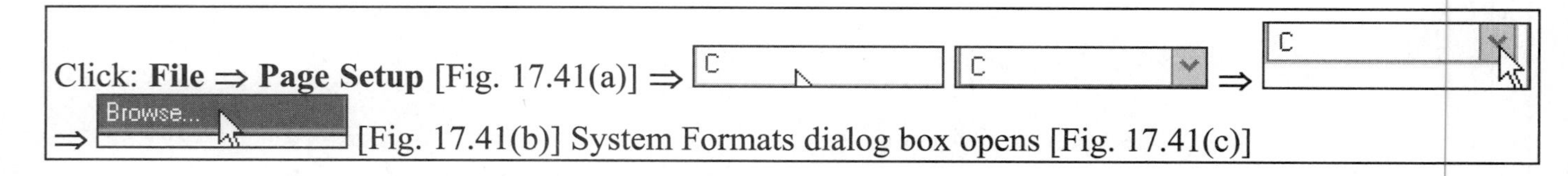

Click: **File** ⇒ **Page Setup** [Fig. 17.41(a)] ⇒ ⇒ ⇒ [Fig. 17.41(b)] System Formats dialog box opens [Fig. 17.41(c)]

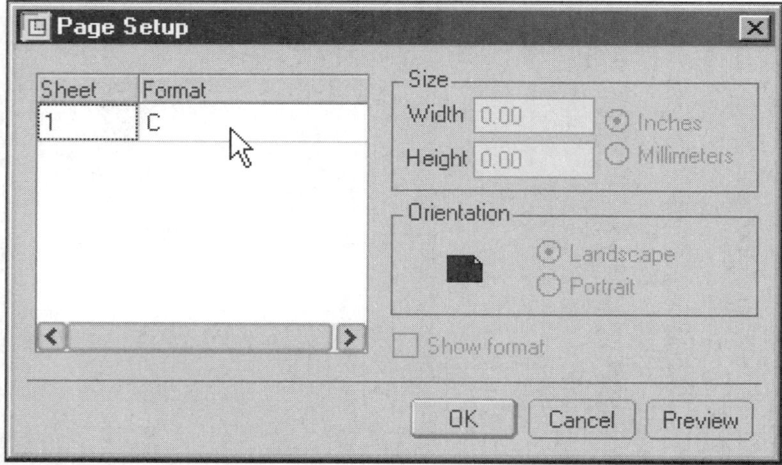

Figure 17.41(a) Page Setup Dialog Box

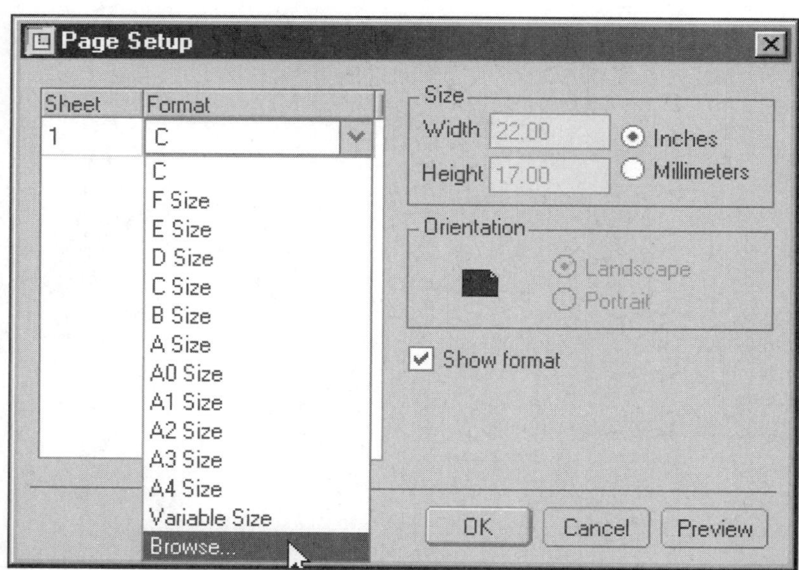

Figure 17.41(b) Browse

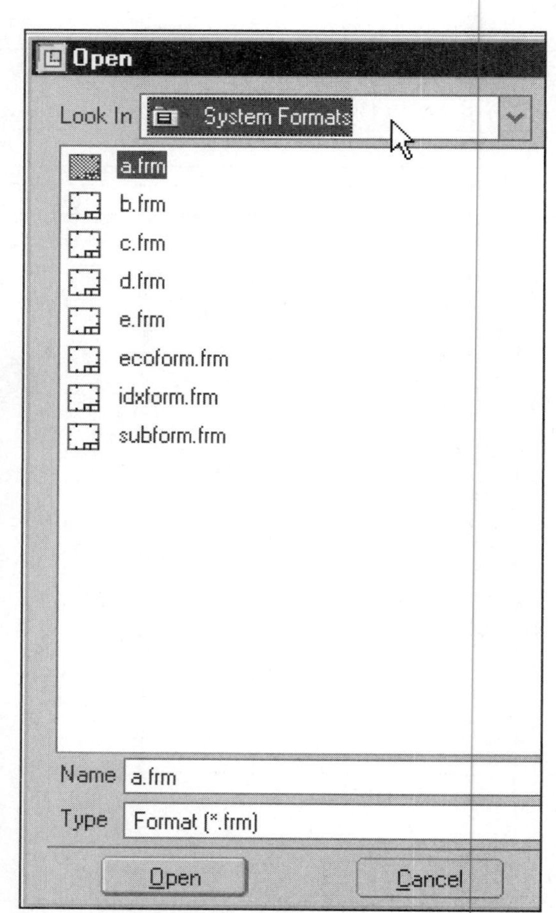

Figure 17.41(c) System Formats

Navigate your file directory to your current project file folder [Fig. 17.41(d)] ⇒ select **format_c.frm** [Fig. 17.41(e)] ⇒ **Open** [Fig. 17.41(f)]

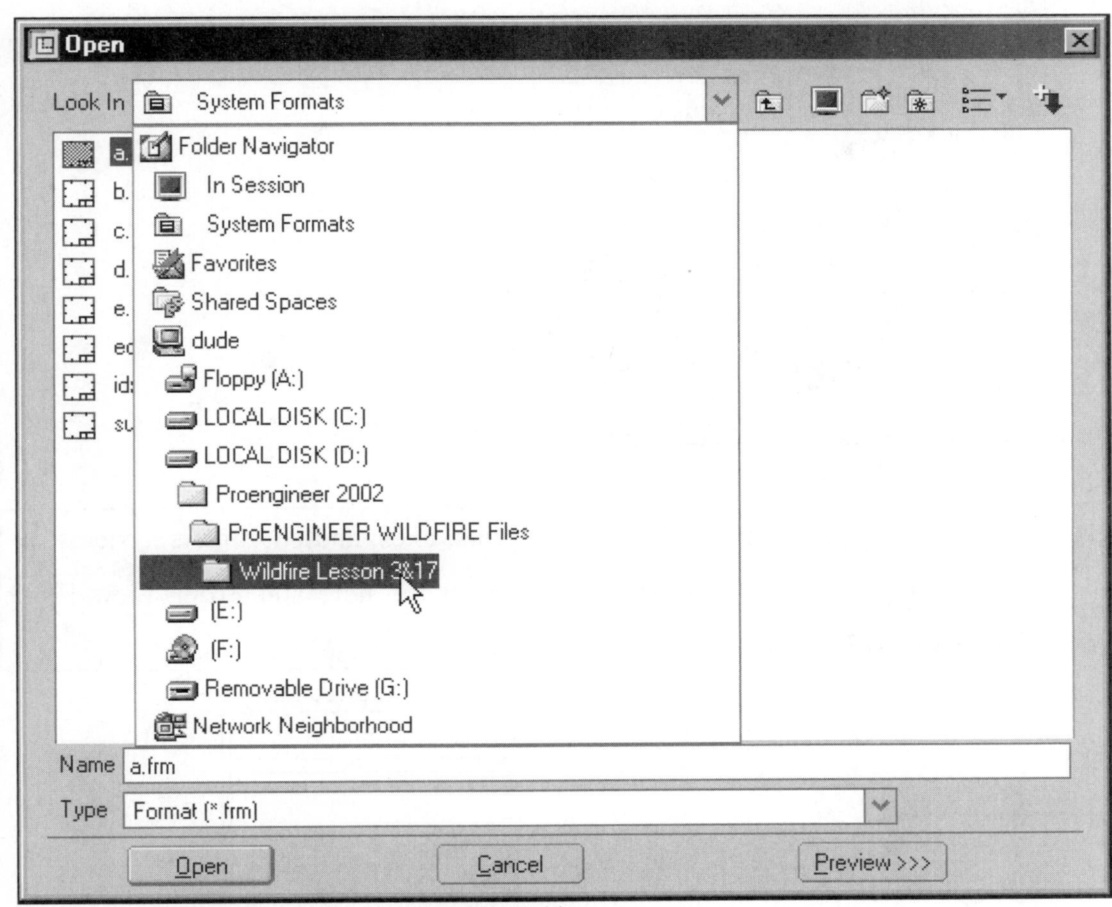

Figure 17.41(d) Navigate to your Project File Folder

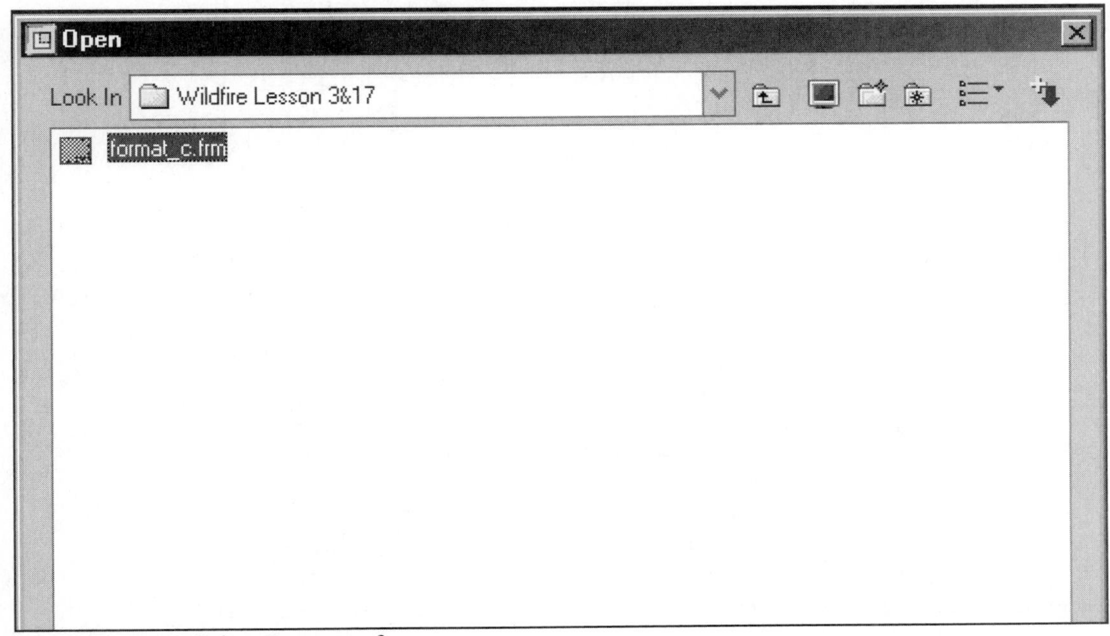

Figure 17.41(e) Select Format_c.frm

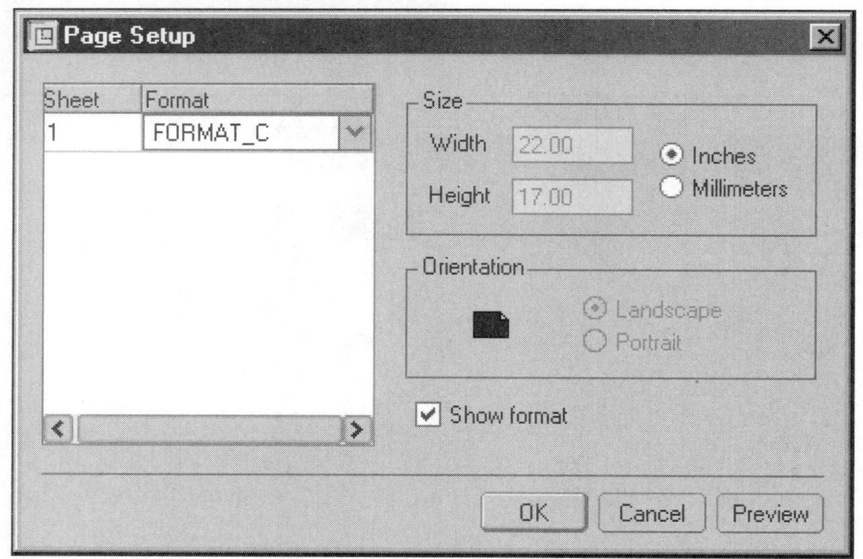

Figure 17.41(f) Format_C Added to Sheet One

Click: **OK** [Fig. 17.41(g)] ⇒ if on, turn off your datum planes, points, axes, and coordinate systems ⇒ **Zoom in** on the title block [Fig. 17.41(h)] ⇒ **Zoom in** on the note and Front View [Fig. 17.41(i)] ⇒ ⇒ ⇒ ⇒ **MMB** ⇒ **File** ⇒ **Delete** ⇒ **Old Versions** ⇒ **MMB**

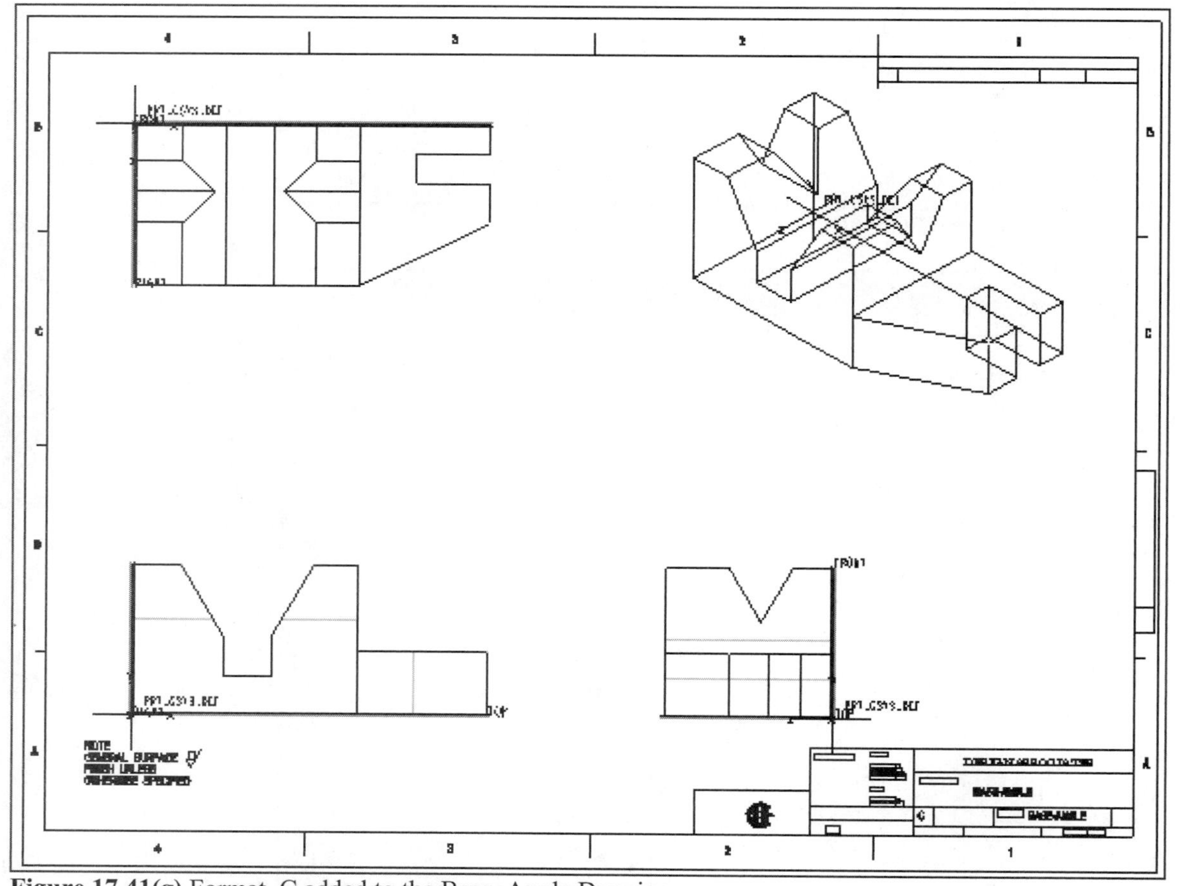

Figure 17.41(g) Format_C added to the Base_Angle Drawing

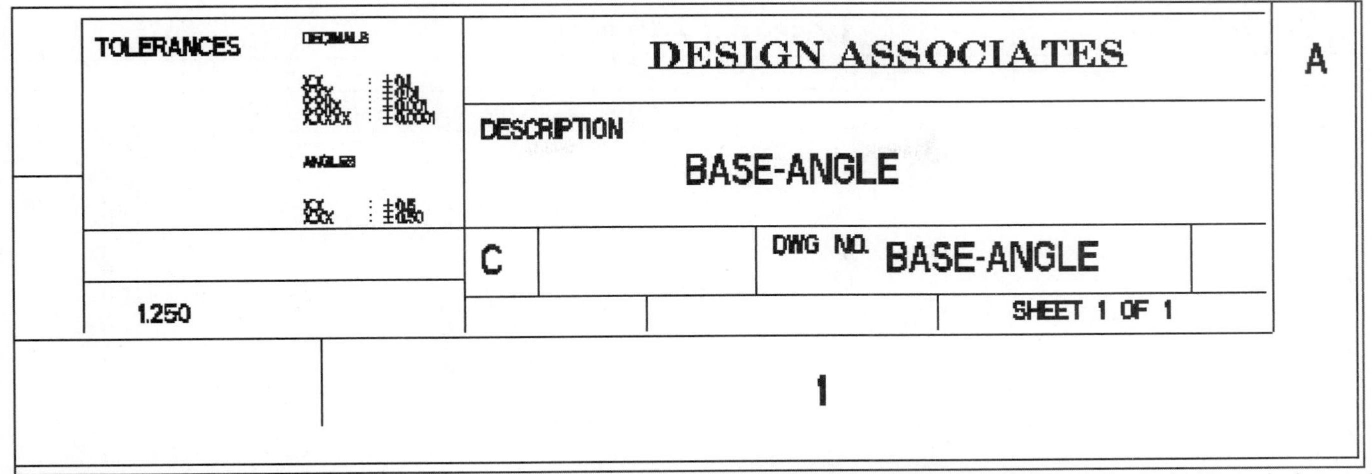

Figure 17.41(h) Title Block. Since the Part and the Drawing have the same Name, it will Display in both the Description and the Drawing Number Blocks.

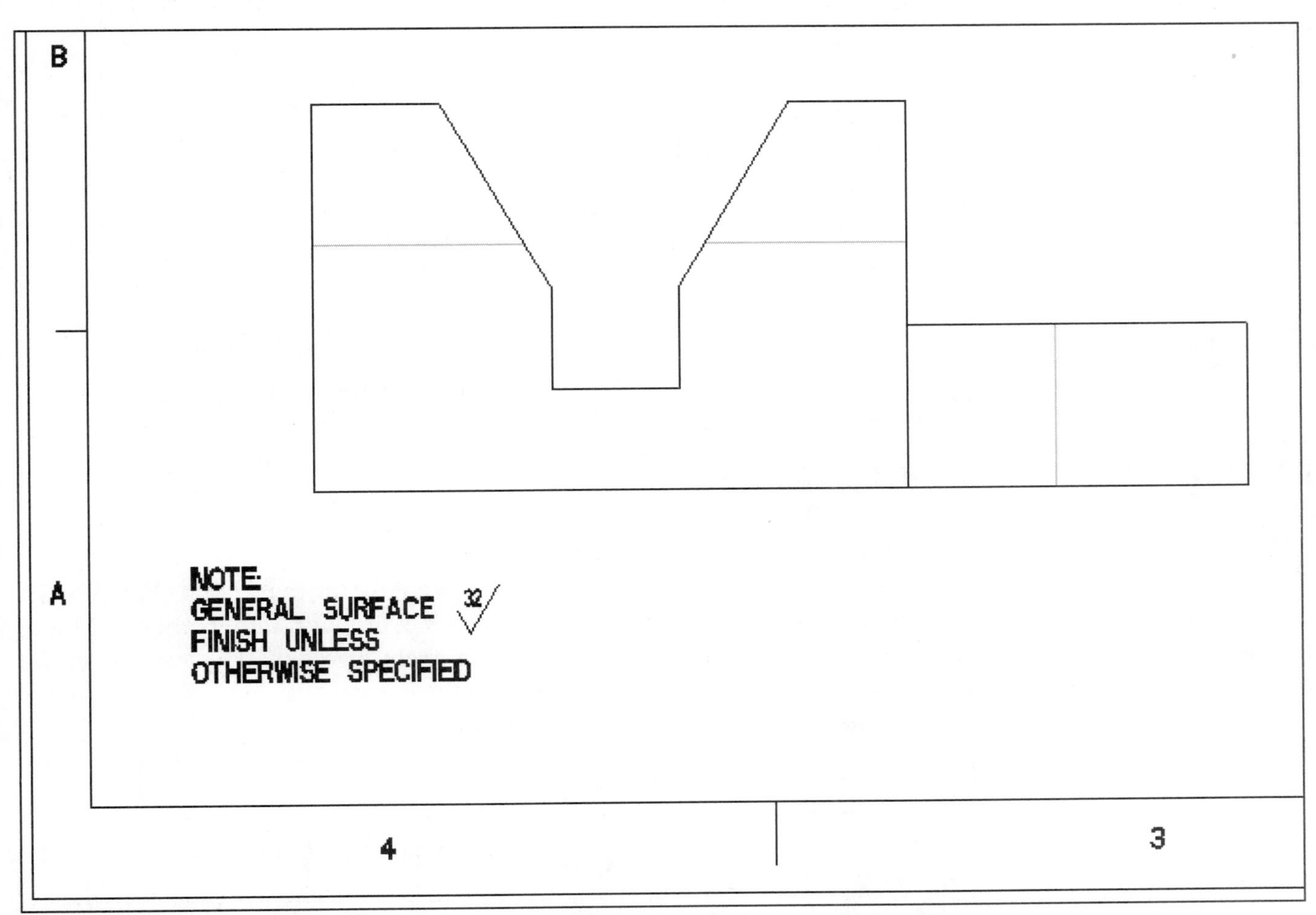

Figure 17.41(i) Note and Front View

You may edit the format, creating more blocks, adding the note **SCALE** before the scale parameter, adding notes for **DRAWN BY**, **DATE**, **CHECKED BY**, **REV**, **ISSUED**, removing the Canadian Standard symbol, and so forth.

You have completed this lesson. Continue with the lesson project.

Lesson 17 Project

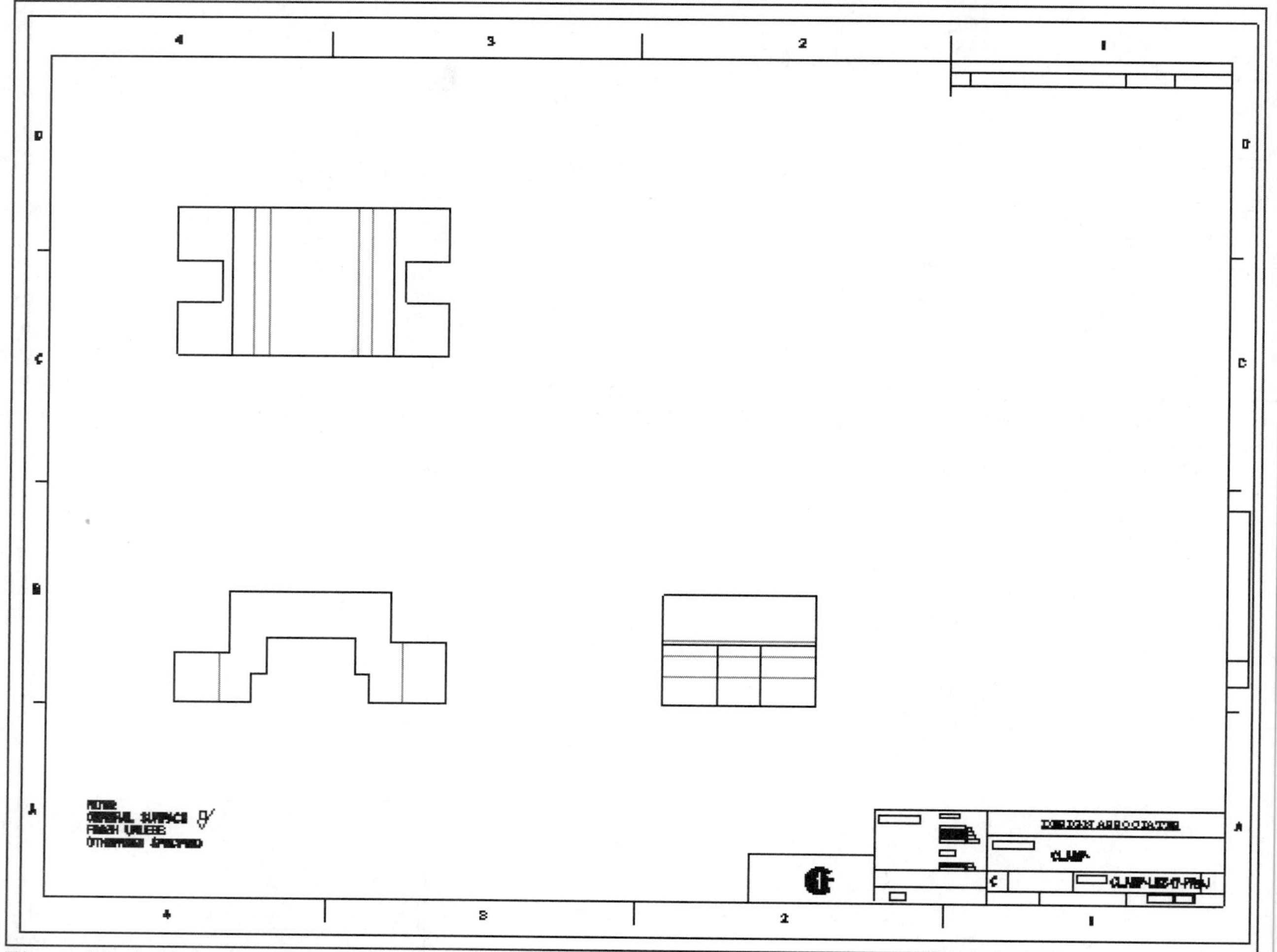

Figure 17.42 Clamp Drawing Using a Custom Format

Clamp Drawing

This lesson project will use the project modeled in Lesson 2. Create a drawing for the Clamp (Fig. 17.42). This is just one of many lesson parts and lesson projects that can be brought into Drawing Mode, formatted (Fig. 17.43) and detailed. Analyze the part, and plan the sheet size and the drawing views required to display its features for detailing. Use the format created in this lesson.

Create a personalized format for an A, B, D, E, and F size drawing.

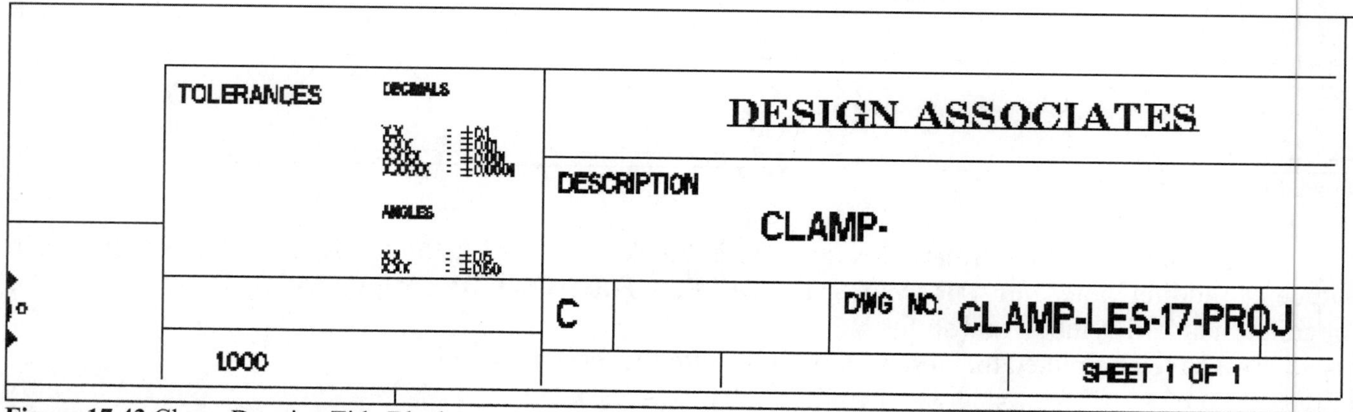

Figure 17.43 Clamp Drawing Title Block

Lesson 18 Detailing

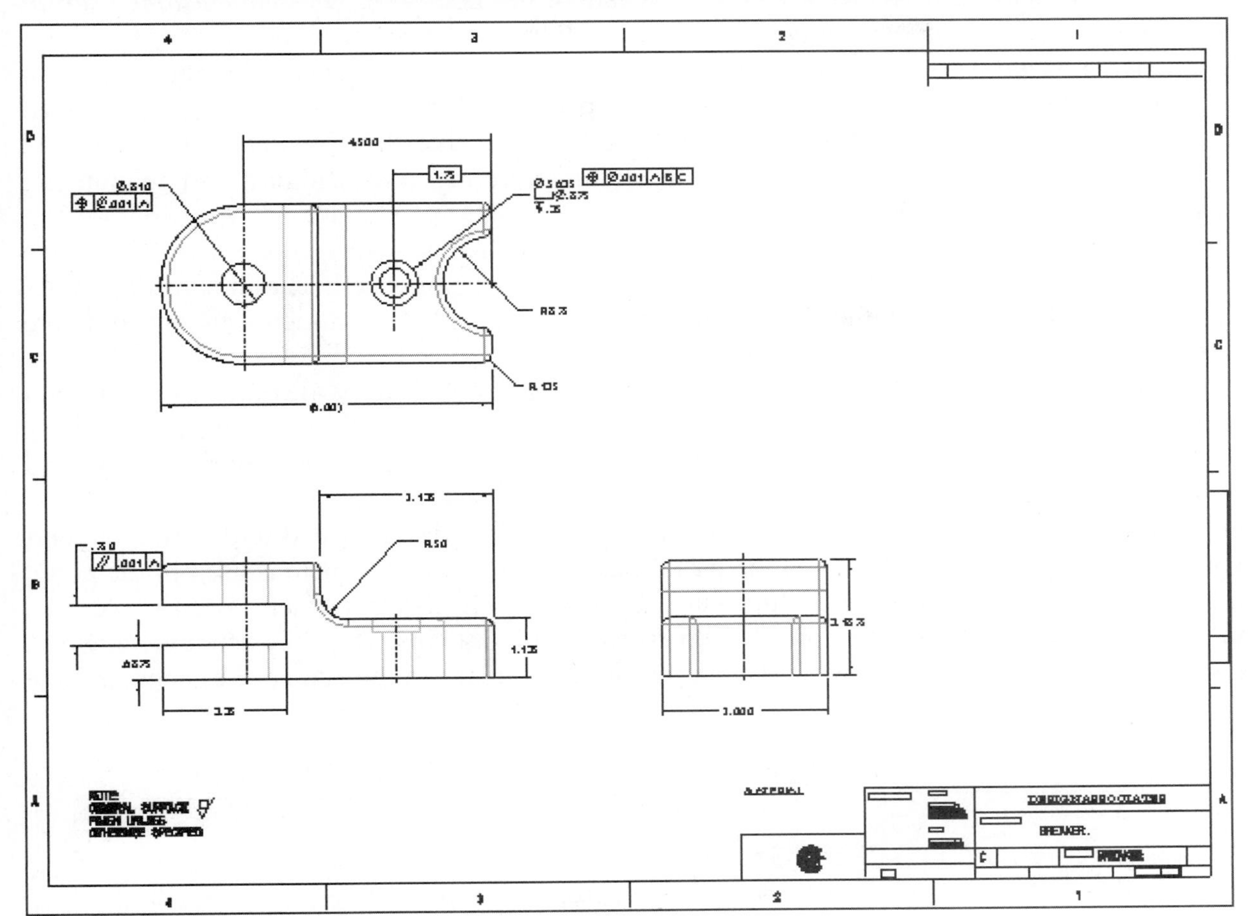

Figure 18.1 Breaker Detail Drawing

OBJECTIVES

- Detail and **Dimension** a part
- Create and save **Drawing Options**
- Add **Geometric Tolerancing** information to a drawing
- Use **Pro/MARKUP** to create checker changes
- **Move** and **Modify** dimensions
- Add a **URL** to a drawing feature

DETAILING

The purpose of an engineering drawing is to convey information so that the part can be manufactured correctly. Engineering drawings use dimensions and notes to convey this information (Fig. 18.1). Knowledge of the methods and practices of dimensioning and tolerancing is essential to the engineer or designer. The multiview projections of a model (part or assembly) provide a graphical representation of its shape (*shape description*). However, the drawing must also contain information that specifies size and other requirements.

Drawings are *annotated* with dimensions and notes. Dimensions must be provided between points, lines, or surfaces that are functionally related or to control relationships with other parts. With Pro/E, the *design intent* used in the original sequence of feature creation and the selection of dimensions used on the feature's sketch will determine the dimensions shown on the drawing (Fig. 18.2).

You need not create dimensions (unless desired), because the dimensions were established during the modeling of the part. Dimensions created in Drawing mode, to describe drafting features, will not be *driving dimensions* (they will be *driven dimensions* that cannot be modified). *Only dimensions used in modeling the part drive the part feature database.*

Each dimension on a drawing has a tolerance, implied or specified. The general tolerance, given in the title block, is called a general or sheet tolerance. Specific tolerances are provided with each appropriate dimension. Together, the views, dimensions, and notes give the complete shape and size description of the part. Uniform practices for stating and interpreting dimensioning and tolerancing requirements are maintained in **ASME Y14.5 1994**.

Views of a model (part or assembly) may be dimensioned in Drawing mode. Pro/E displays dimensions in a view based on the way the part (or the assembly) was modeled. The dimension type is selected from options before showing the dimensions on the drawing. Linear dimensions and ordinate dimensions are two of these options.

After a part's features are sketched, aligned, and dimensioned, you modify the dimension values to the exact sizes required for the design. The dimensioning scheme, the controlling features, the parent-child relationships, and the datums used to define and control the part features are determined as you design and model with Pro/E.

When detailing, you simply choose Pro/E commands to display views needed to describe the part and then display the dimensions used to model the part. These are the same dimensions used in the part's design. You cannot under-dimension or over-dimension, because Pro/E displays exactly what is required to model the part. Pro/E will not duplicate driving dimensions on a drawing. If a dimension is shown in one view, it will not be shown in another view. The dimension, however, can be switched to another view via detailing options.

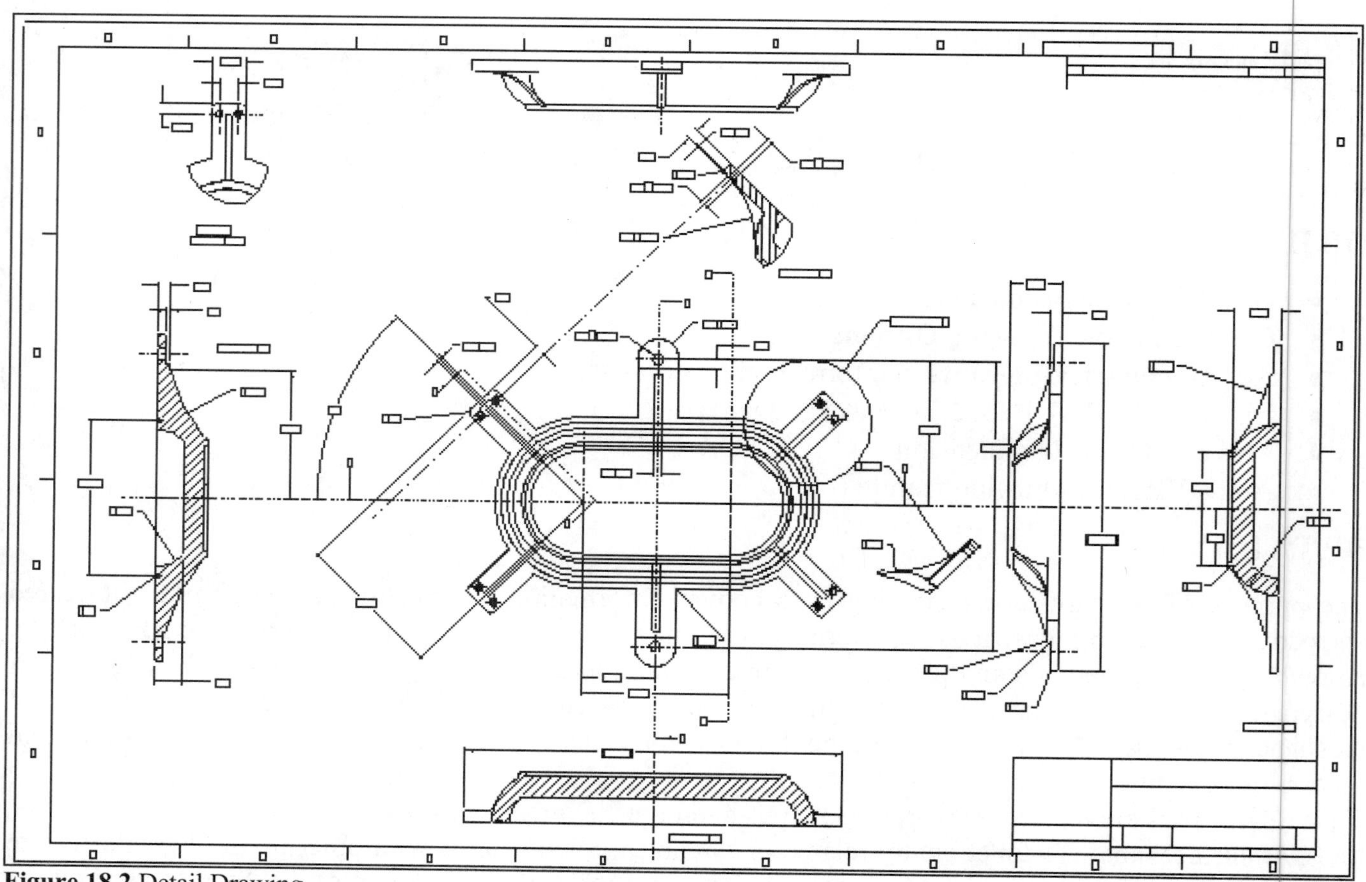

Figure 18.2 Detail Drawing

Dimensioning

Applying dimensions to drawing views in Pro/E is a different process from that of any other drawing and dimensioning programs you may have used. The difference is the Pro/E associativity factor; instead of using the drafting program to add dimensions to views, you *show* dimensions (Fig. 18.3) that are already imported from the 3D model. These dimensions are actively linked to the 3D model, so you can directly edit the 3D model through the dimensions in the drawing. When shown in the drawing, these dimensions are called *shown* or ***driving dimensions***, because you can use them to drive the shape of the model through the drawing. Of course, there will be instances where you need additional dimensions to show the same value for the same object, for example, a view repeated on another sheet. To add these, use the command: Insert ⇒ Dimension. These inserted dimensions are called *added* or ***driven dimensions***, because their association is only one-way, from the model to the drawing. If dimensions are changed in the model, all edited dimension values and the drawing are updated, but you can't use these driven dimensions to edit the 3D model. This concept has consequences that a drafter must bear in mind when making drawings in Pro/E:

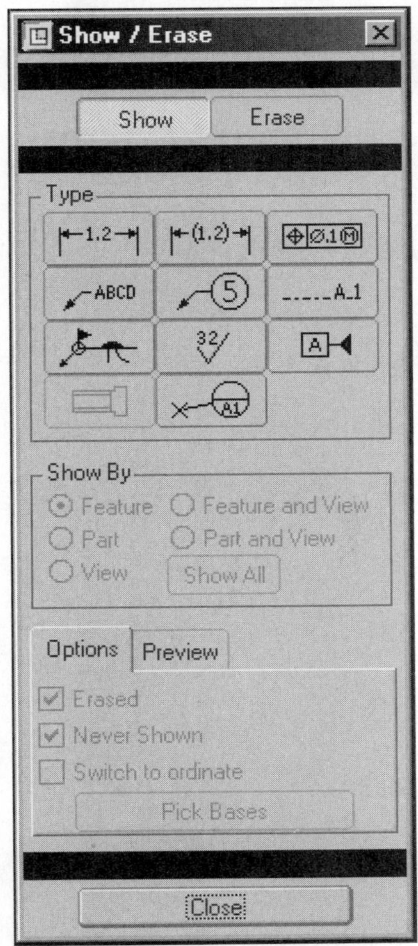

Figure 18.3 Show/Erase Dialog Box

- Only one driving dimension for each model dimension may exist in a drawing. A drawing may have several views of the same object, but only one driving dimension for each feature of the model may be shown. You can move a driving dimension from one view to another, for example, from a general view to a detailed view where it is more appropriate. To provide dimensioning for views when driving dimensions are "used up" in other views, you use the added or driven dimensions.
- It is possible to unintentionally edit the model. If a driving dimension is edited, it changes color as a warning that there is a change or discrepancy between the drawing and the model. When you regenerate the model, the drawing accepts the new dimension. It is possible, using configuration options, to break the link between model and drawing, but this is not the usual Pro/E usage.

The Concepts of Showing and Erasing

Understanding the concepts of showing and erasing detail items is fundamental to preparing Pro/E drawings. The underlying idea is that all items displayed in a drawing *originate* from the 3D model. The dimensions, geometric tolerances, cross sections etc., that exist in the Pro/E 3D model are imported to the 2D drawing, and maintain associativity with the 3D model. By default, they are imported as invisible. The detailer selectively *shows* these dimensions or other detail items where they belong on a particular view.

Items that are not shown are referred to as *erased*. They are still within the 3D file database, and can be shown in the drawing at any time. They cannot be deleted from the drawing, unless they are deleted from the 3D model. To control whether an item is shown or erased, use the View ⇒ Show and Erase dialog box. Drawing views can also be shown and erased using View ⇒ Drawing Display ⇒ Drawing View Visibility. Again, this does not delete them from the drawing. Erasing views can help in repainting large drawing files. For short-term invisibility, you can use the Hide and Unhide commands available as right mouse button shortcut commands from selected parts in the model tree.

Your menu bar will display a variety of buttons for quickly accessing drawing mode functions. *You can add the missing buttons to your Menu bar, or commands and options can be accessed from the menu bar* Edit View Insert Sketch Table Format.

Click: **Tools** ⇒ **Customize Screen** ⇒ Categories **Drawing** ⇒ drag and drop the command buttons to the Tool Bar ⇒ **OK**

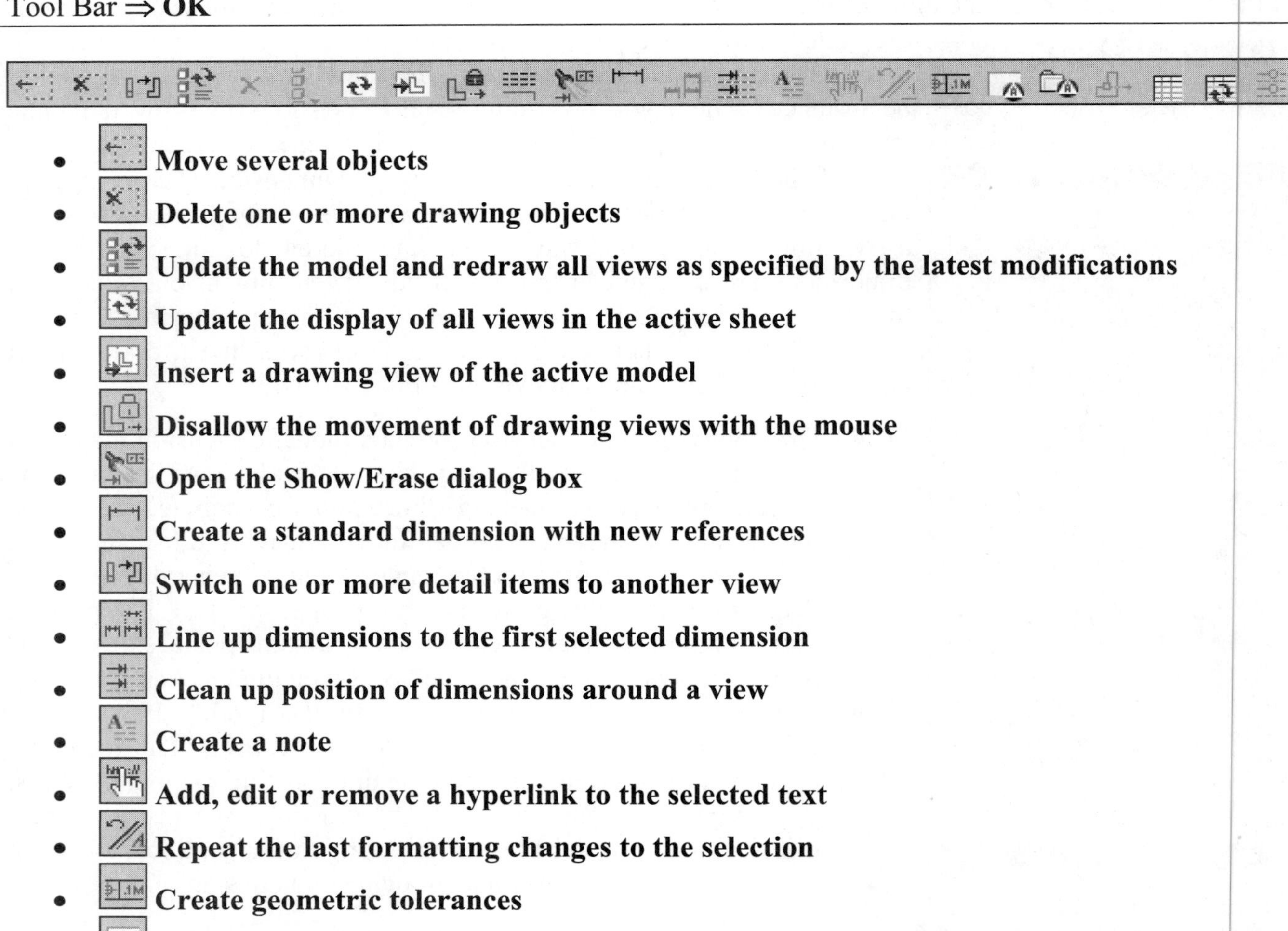

- **Move several objects**
- **Delete one or more drawing objects**
- **Update the model and redraw all views as specified by the latest modifications**
- **Update the display of all views in the active sheet**
- **Insert a drawing view of the active model**
- **Disallow the movement of drawing views with the mouse**
- **Open the Show/Erase dialog box**
- **Create a standard dimension with new references**
- **Switch one or more detail items to another view**
- **Line up dimensions to the first selected dimension**
- **Clean up position of dimensions around a view**
- **Create a note**
- **Add, edit or remove a hyperlink to the selected text**
- **Repeat the last formatting changes to the selection**
- **Create geometric tolerances**
- **Insert a drawing symbol instance from a standard palette**
- **Insert a customized instance of a drawing symbol**

The Show/Erase Dialog box also has a number of radio buttons to toggle selections on and off:

Show **Show detail items** Erase **Erase detail items**

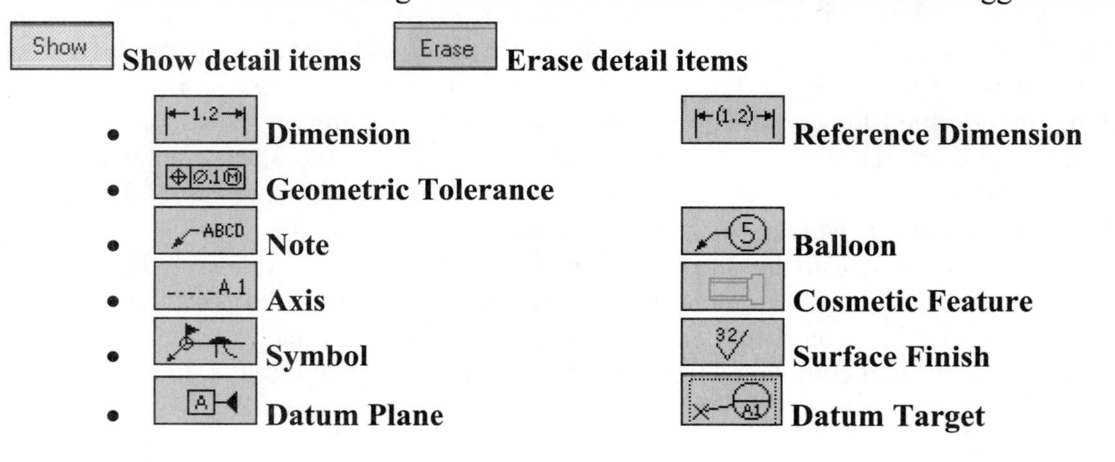

- **Dimension**
- **Reference Dimension**
- **Geometric Tolerance**
- **Note**
- **Balloon**
- **Axis**
- **Cosmetic Feature**
- **Symbol**
- **Surface Finish**
- **Datum Plane**
- **Datum Target**

Tolerancing

You can set the tolerance standard as **ANSI** or **ISO**, and set the tolerance display on or off. You can drive dimensional tolerances using a set of tolerance tables. Pro/E assigns each model a tolerance standard of either ANSI or ISO:

- When you switch from ISO to ANSI, Pro/E assigns the ANSI tolerances based on the nominal dimension's number of digits and deletes the tolerance table reference
- When you switch from ANSI to ISO, a set of tolerance tables drives the ISO-standard tolerances

The manufacturing of parts and assemblies uses a degree of precision determined by tolerances. A typical parametric design system supports three types of tolerances:

- **Dimensional** Specifies allowable variation of size
- **Geometric** Controls form, profile, orientation, and runout
- **Surface finish** Controls the deviation of a part surface from its nominal value

When you design a part, you specify dimensional tolerance, which means the *allowable variations in size*. All dimensions are controlled by tolerances, except "basic" dimensions, which for the purpose of reference are considered exact.

Dimensional tolerances on a drawing can be expressed in two forms:

- As **general tolerances** presented in a tolerance table. These apply to those dimensions that are displayed in nominal format--that is, without tolerances.

X.X	**±0.1**
X.XX	**±0.01**
X.XXX	**±0.001**
ANG.	**±0.5**

X.X	+-0.1
X.XX	+-0.01
X.XXX	+-0.001
ANG.	+-0.5

- As **individual tolerances** specified for individual dimensions. $4.500^{+.001}_{-.000}$ $\varnothing^{.812}_{.810}$

You can use general tolerances given as defaults in a table or set individual tolerances by modifying default values of selected dimensions. Default tolerance values are used when you start to create a model; therefore, *default tolerances must be set prior to creating geometry.*

When you start to create a part, the table at the bottom of the window will display the current defaults for tolerances. If you have not specified tolerances, Pro/E defaults are assumed.

You have a choice of displaying or blanking tolerances. If tolerances are not displayed, Pro/E still stores dimensions with their default tolerances. You can specify geometric tolerances, create "basic" dimensions, and set selected datums as reference datums for geometric tolerancing. ISO tolerances are generated from a table.

The available tolerance formats are:

- **Nominal** Dimensions displayed without tolerances.
- **Limits** Tolerances displayed as upper and lower limits.
- **Plus-Minus** Tolerances displayed as nominal with plus-minus tolerance. The positive and negative values are independent.
- **±Symmetric** Tolerances displayed as nominal with a single value for both the positive and the negative tolerance.
- **As Is** Tolerances as is.

Geometric Tolerances

Geometric tolerances provide a comprehensive method of specifying where on a part the critical surfaces are, how they relate to one another, and how the part should be inspected to determine if it is acceptable [Fig. 18.4(a)]. In manufacturing, you can use geometric tolerances to specify the maximum allowable deviation from the exact size and shape specified by designers. You can create, modify, display, and delete geometric tolerances (gtols) in Drawing mode. When you store a Pro/E geometric tolerance in a solid model, it contains parametric references to the geometry or feature it controls, its *reference entity*, and parametric references to referenced datums and axes. As a result, Pro/E updates a gtol's display when you rename a referenced datum.

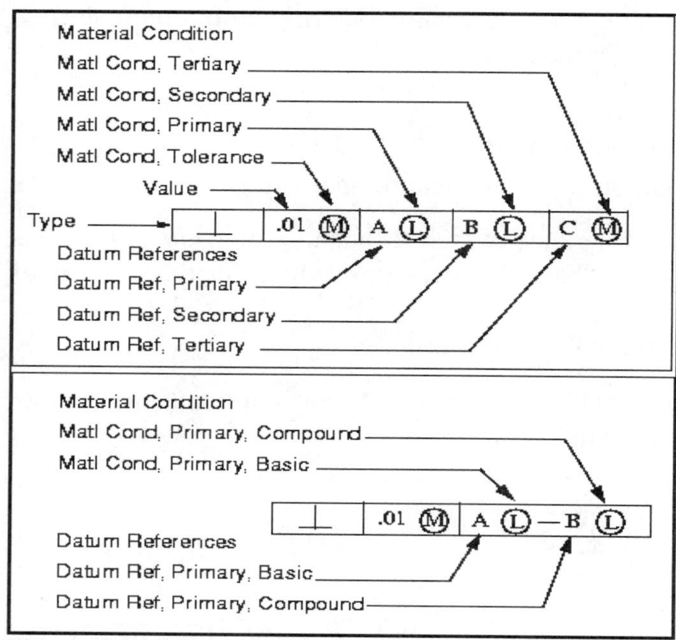

Figure 18.4(a) Geometric Tolerancing

In Assembly mode, you can create a geometric tolerance in a subassembly or a part. A geometric tolerance that you create in Part or Assembly mode automatically belongs to the part or assembly that occupies the graphics window; however, it can refer only to set datums belonging to that model itself, or to components within it. It cannot refer to datums *outside* of its model in some encompassing assembly, unlike assembly-created features.

Geometric Tolerance Dialog Box

To access the geometric tolerance functionality in Drawing mode, use Insert ⇒ Geometric Tolerance or **Create geometric tolerances**. The left side of the dialog box displays the geometric tolerance symbols available for selection. The next table shows the geometric tolerance types and the appropriate entities that you can reference. When you place the cursor on one of the geometric tolerance symbols, the type appears in the dialog box as balloon help and in the message status line. Using the four tabbed pages, you can perform the following tasks:

- Specify the model and the reference entity to which to add the geometric tolerance, as well as place the geometric tolerance on the drawing [Fig. 18.4(b)]
- Specify the datum references and material conditions for a geometric tolerance, as well as the value and datum reference of a composite tolerance [Fig. 18.4(c)]
- Specify the tolerance value and the material condition [Fig. 18.4(d)]
- Specify the symbols and modifiers, and projected tolerance zone [Fig. 18.4(e)]

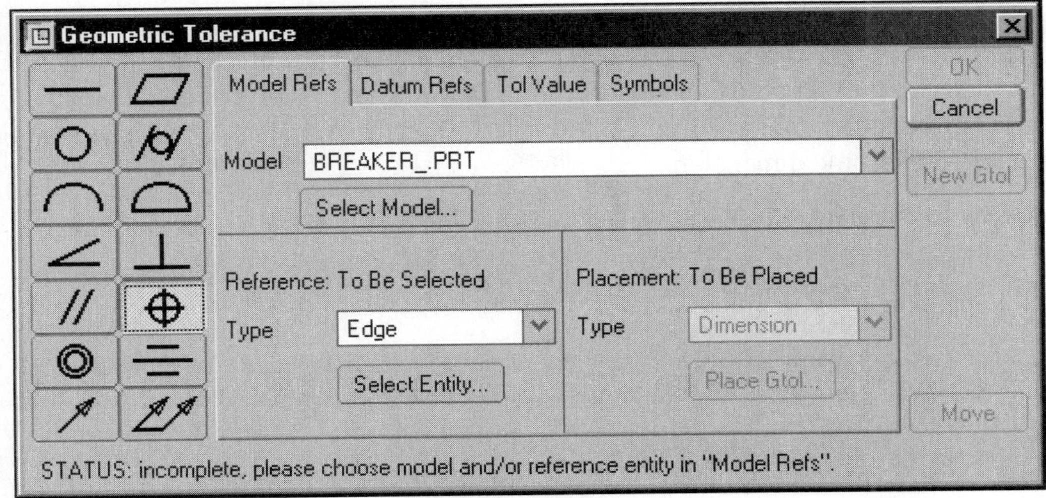

Figure 18.4(b) Geometric Tolerance Model Refs Tab

Model Refs | Datum Refs | Tol Value | Symbols
Datum References
Unordered
Primary | Secondary | Tertiary
Basic none RFS(no symbol)
Compound none RFS(no symbol)
Composite Tolerance
Value 0.001 Datum Reference None

Figure 18.4(c) Geometric Tolerance Datum Refs Tab

Model Refs | Datum Refs | Tol Value | Symbols
Tolerance Value
Overall Tolerance 0.001
Name gp0
Per Unit Tolerance
Value/Unit 0.001
Unit Length 0.001
Material Condition RFS(no symbol)

Figure 18.4(d) Geometric Tolerance Tol Value Tab

Model Refs | Datum Refs | Tol Value | Symbols
Symbols and Modifiers
Statistical Tolerance
Diameter Symbol
Free State
All Around Symbol
Tangent Plane
Set Boundary
Additional Text on right
Projected Tolerance Zone
None
Below Gtol
Inside Gtol
Zone Height
0.001

Figure 18.4(e) Geometric Tolerance Symbols Tab

You can attach a geometric tolerance to a dimension, datum, single or multiple edges, or another geometric tolerance. You can also create a *driven* dimension to which you can attach the geometric tolerance, place it free (anywhere on the drawing), or relate it to dimension text. You can attach stacked (multiple) geometric tolerances to another tolerance; or, if the first tolerance in a stack is attached to a dimension, you can attach them to the same dimension.

Tolerance Value and Material Condition

Using the buttons on the Tol Value tab of the Geometric Tolerances dialog box, you can specify the tolerance value and material condition.

- **Tolerance Value** Specifies the tolerance value. Type a value in the Overall Tolerance box, or select the Per Unit Tolerance check box and type a value in the Value/Unit and Unit Length boxes (or Unit Area for some geometric tolerance types).
- **Material Condition** Specifies the material condition

LMC	Ⓛ	Least material condition
MMC	Ⓜ	Maximum material condition
RFS	Ⓢ	Regardless of feature size
RFS/Default		RFS, but does not show a symbol in frame

Symbols, Modifiers, and Geometric Tolerance Classes and Types

Using the buttons on the Symbols tab of the Geometric Tolerance dialog box, you can specify the gtol's symbols, modifiers, and projected tolerance zone.

- **Symbols and Modifiers** Specifies the geometric tolerance symbols and modifiers. Select Statistical Tolerance, Diameter Symbol, Free State, All Around Symbol, Tangent Plane, or Set Boundary.
- **Projected Tolerance Zone** Specifies the location of the projected tolerance zone. Select None, Below Gtol, or Inside Gtol. You can also specify the height of the projected tolerance zone by typing a value in the Zone Height box.
- **Profile Boundary** (available for Profile gtols only) Specifies unilateral direction, bilateral direction, or both.

Class	Type	Symbol	Reference Entity
Form	**Straightness** **Flatness** **Circularity** **Cylindricity**	— ⏥ ○ ⌭	Surface of revolution, axis, straight edge Plane surface (not datum plane) Cylinder, cone, sphere Cylindrical surface
Profile	**Line** **Surface**	⌒ ⌓	Edge Surface (not datum plane)
Orientation	**Angularity** **Parallelism** **Perpendicularity**	∠ // ⊥	Plane, surface, axis Cylindrical, surface, axis Planar surface
Location	**Position** **Concentricity** **Symmetry**	⌖ ◎ ⌯	Any Axis, surface of revolution Any
Runout	**Circular** **Total**	↗ ⌰	Cone, cylinder, sphere, plane Cone, cylinder, sphere, plane

Geometric tolerances (Fig. 18.5) provide a method for controlling the *location*, *form*, *profile*, *orientation*, and *runout* of features. You add geometric tolerances to the model from Part mode or Drawing mode. The geometric tolerances are treated by Pro/E as annotations, and they are always associated with the model. *Unlike dimensional tolerances, geometric tolerances do not have any effect on part geometry.* When adding a geometric tolerance to the model, you can attach it to existing dimensions, edges, and existing geometric tolerances, or you can display it as a note without a leader.

Before you can reference a datum in a geometric tolerance, you must first indicate your intention by setting the datum. Once a datum is set, hyphens are added before and after the datum name, and it is enclosed in a rectangle (using the old standards) [Fig. 18.6(a)]. The ASME Y14.5M 1994 standards display the datum name [Fig. 18.6(b)]. You can change the name of a datum, either before or after it has been set, by using the Name option in the SetUp menu from Part or Assembly mode.

Geometric Tolerances

Geometric tolerances (gtols) provide a comprehensive method of specifying where on a part the critical surfaces are, how they relate to one another, and how the part should be inspected to determine if it is acceptable. They provide a method for controlling the location, form, profile, orientation, and runout of features. When you store a Pro/ENGINEER gtol in a solid model, it contains parametric references to the geometry or feature it controls--its *reference entity*--and parametric references to referenced datums and axes. As a result, the system updates a gtol's display when you rename a referenced datum.

In Assembly mode, you can create a gtol in a subassembly or a part. A gtol that you create in Part or Assembly mode automatically belongs to the part or assembly that occupies the window; however, it can refer only to set datums belonging to that model itself, or to components within it. It cannot refer to datums *outside* of its model in some encompassing assembly, unlike assembly-created features. You can add gtols in Part or Drawing mode, but they are reflected in all other modes. The system treats them as annotations, and they are always associated with the model. Unlike dimensional tolerances, though, gtols do *not* affect part geometry.

The Layout of a Gtol

Pro/ENGINEER specifies a gtol for an individual feature by means of a feature control frame (a rectangle) divided into compartments containing the gtol symbol followed by the tolerance value. Where applicable, it also follows the tolerance with a material condition symbol. If a gtol is related to a datum, it places the reference datum name in a compartment following the tolerance value. Where applicable, it follows the datum reference letter with a material condition symbol.

Geometric Tolerance Layout

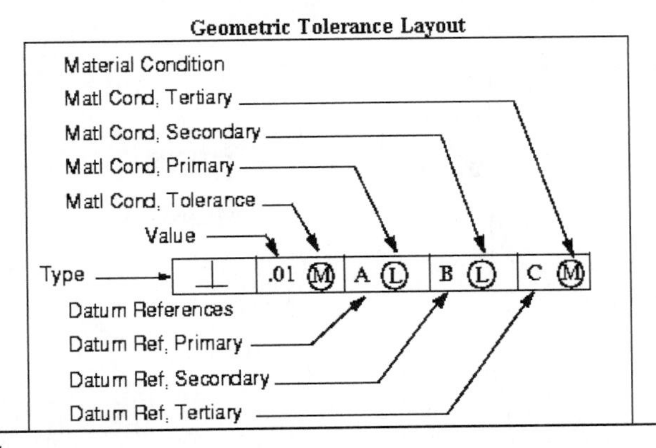

Figure 18.5 Geometric Tolerances

A geometric tolerance for individual features is specified by means of a **feature control frame** (a rectangle) divided into compartments containing the geometric tolerance symbol followed by the tolerance value. Where applicable, the tolerance is followed by a **material condition symbol**. Where a geometric tolerance is related to a datum, the reference datum name is placed in a compartment following the tolerance value. Where applicable, the reference datum name is followed by a material condition symbol. For each class of tolerance, the types of tolerances available and the appropriate types of entities can be referenced. The available material condition symbols are shown in the dialog box.

You are guided in the building of a geometric tolerance by Pro/E requests for each piece of required information. You respond by making menu choices, entering a tolerance value, and selecting entities and datums. As the tolerance is built, the choices are limited to those items that make sense in the context of the information you have already provided. For example, if the geometric characteristic is one that does not require a datum reference, you will not be prompted for one. Other checks are made to help prevent mistakes in the selection of entities and datums.

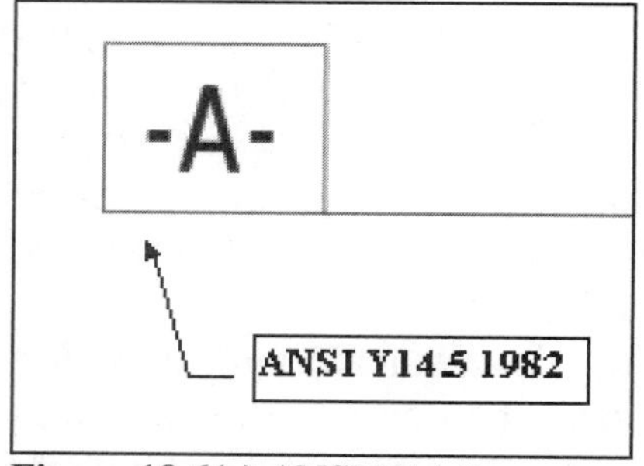

Figure 18.6(a) ANSI Y14.5

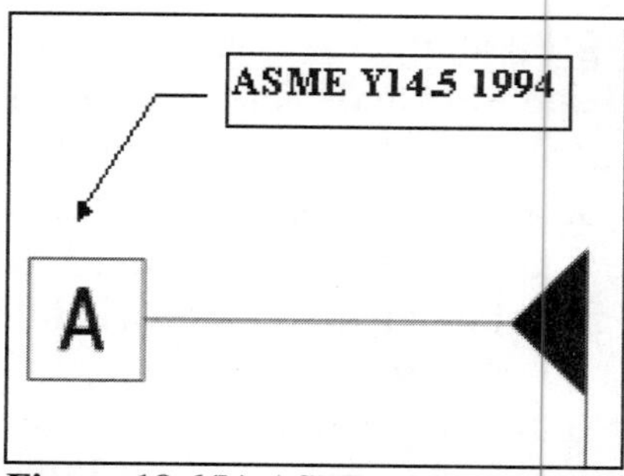

Figure 18.6(b) ASME Y14.5

To display set datums with the correct ASME Y14.5 1994 symbology, the .dtl file must be changed, use: File ⇒ Properties ⇒ Drawing Options, select *gtol_datums* and change *std_ansi* to *std_asme* (Fig. 18.7).

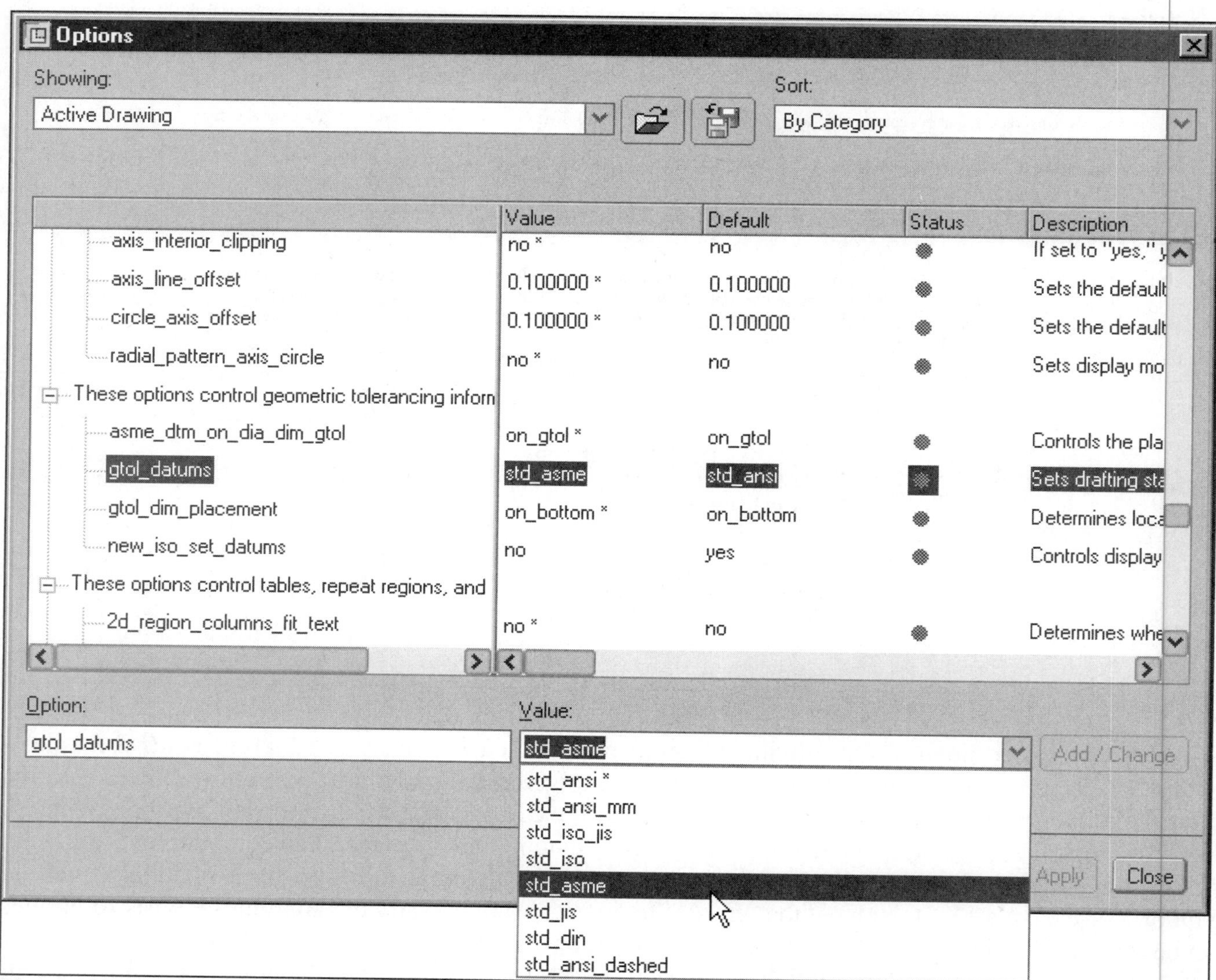

Figure 18.7 Setting Gtol Standards with Drawing Options (.dtl file)

Lesson 18 STEPS

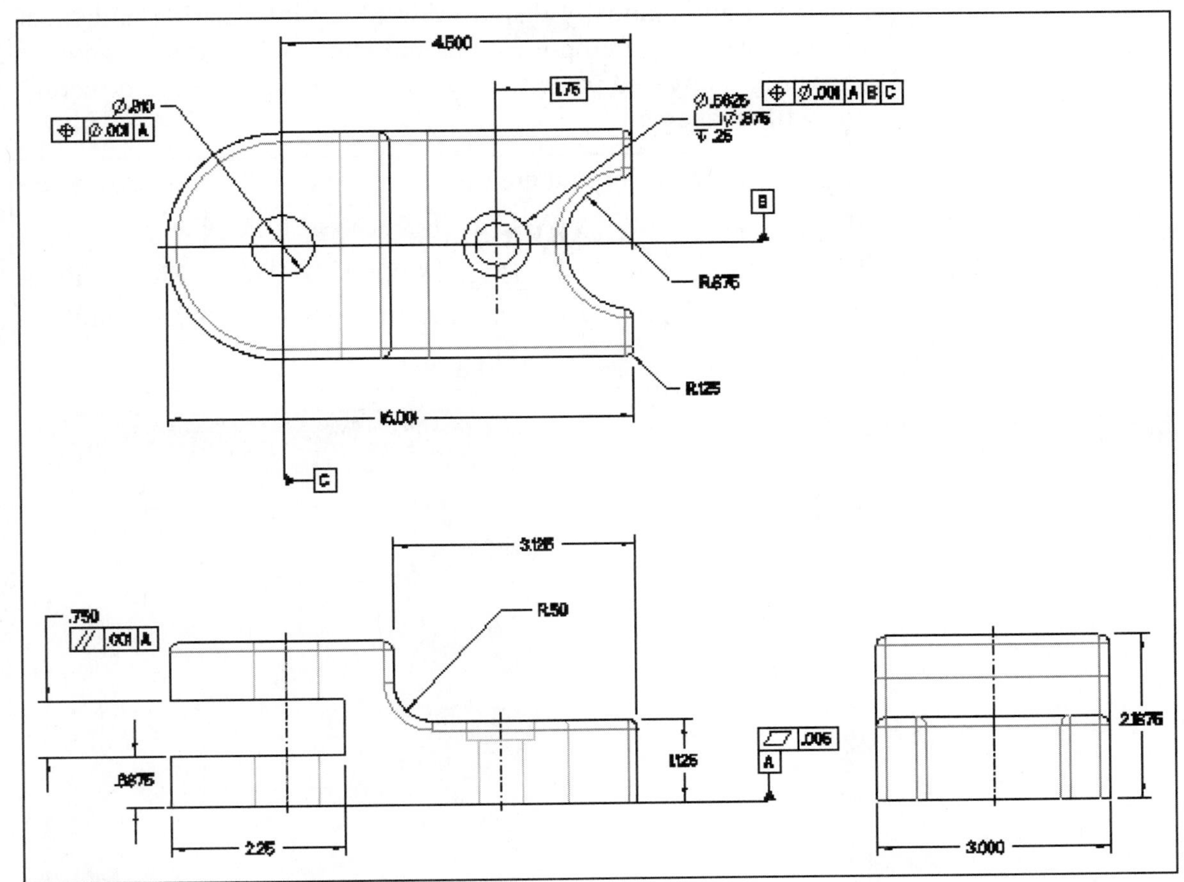

Figure 18.8 Dimensioned Breaker Drawing

Breaker Drawing

The Breaker (Fig. 18.8) from Lesson 4 will be detailed in this lesson. Dimensioning, centerlines (axes), annotation, and other drawing requirements will be used to complete the detailing of the model.

Pro/E allows you to dimension a drawing automatically based on the design intent dimensioning scheme used to create the part. Using this method, you are virtually assured that the part will be fully dimensioned without being overdimensioned. In addition to automatic dimensioning, Pro/E can automatically display centerlines based on datum axes and radial patterns that were used to model the part.

After the dimensions and axes are displayed, the only remaining detailing work involves cosmetically cleaning up their display and adding other annotations, such as general notes, title block information, and required tolerancing information.

Pro/E also allows you to create associative dimensions that are based on existing feature constraints. These dimensions are created using the same techniques you used to create sketch dimensions. These dimensions are referred to as *driven* dimensions, because their display is "driven" by changes to the model. Driven dimensions cannot, however, be used to make parametric changes to the model.

In this lesson, you will learn how to display dimensions automatically using the original constraints, move dimensions to another view, move dimensions to more appropriate locations, display centerlines on your drawing, erase dimensions, modify extension lines to show the proper gap to the appropriate edges, add annotation to a drawing, alter decimal places, and add text to dimension text.

Upon entering the Drawing mode, a template is generally required prior to selecting the model to be detailed. The template can be an empty format, a Pro/E-provided ANSI standard format, or a user-defined format created with parameters that will automatically display title block information and sheet callouts, as in Lesson 17. For the default template, use an empty format and later replace it with another format. If you are using a new working directory, copy and paste the *format_c.frm* from the old directory to the directory with the breaker part, in your file manager.

Click: **File** ⇒ **Set Working Directory** ⇒ select the directory where the Breaker.prt was saved ⇒ **OK** ⇒ **Create a new object** ⇒ Drawing ⇒ Name **BREAKER** ⇒ Use default template [Fig. 18.9(a)] ⇒ **OK** ⇒ Default Model **Browse** ⇒ pick **breaker.prt** ⇒ **Open** [Fig. 18.9(b)] ⇒ **OK** ⇒ **File** ⇒ **Page Setup** ⇒ C Size ⇒ ⇒ Format **Browse** [Fig. 18.9(c)] ⇒ navigate to your project folder ⇒ select **format_c.frm** ⇒ **Open** ⇒ **OK** [Figs. 18.9(d-e)] ⇒ ⇒ **MMB**

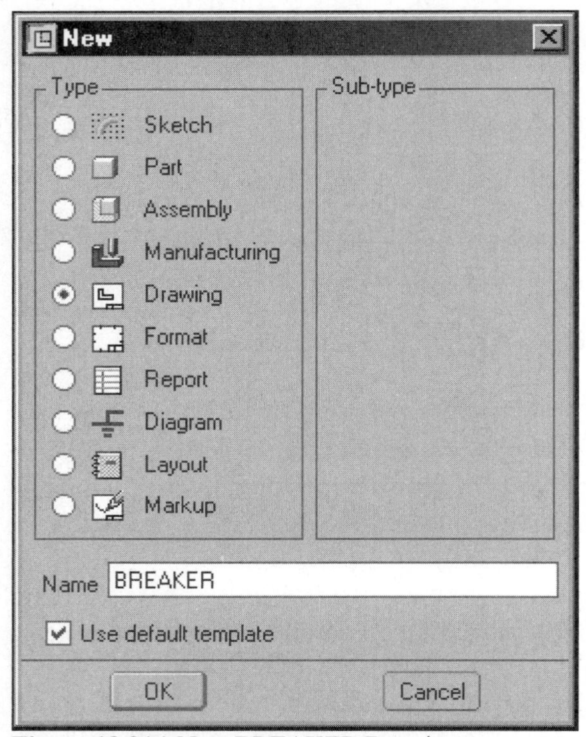

Figure 18.9(a) New BREAKER Drawing

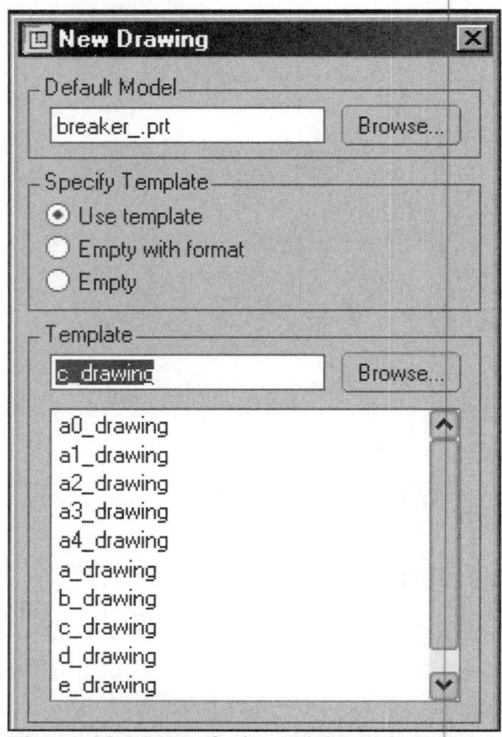

Figure 18.9(b) Default Model Breaker.prt

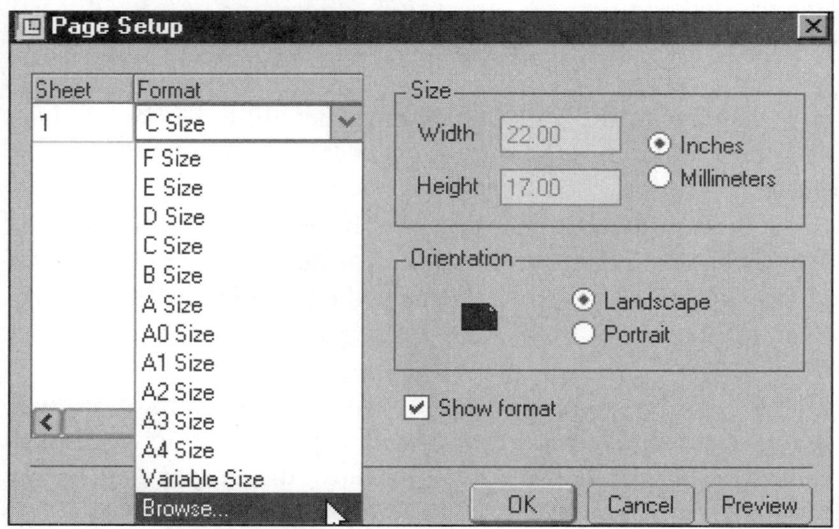

Figure 18.9(c) Browse for Format_C

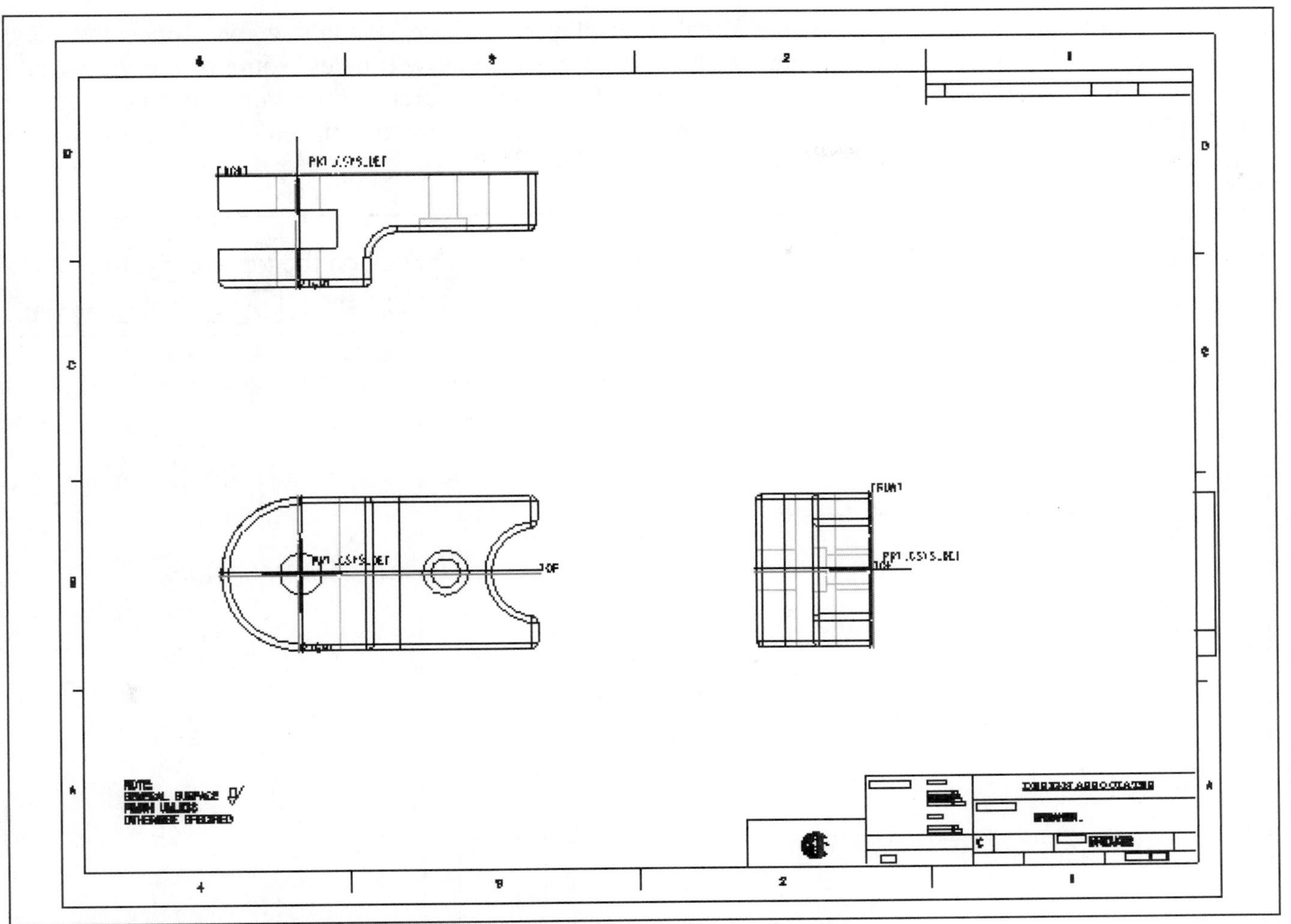

Figure 18.9(d) Breaker Displayed with Three Views on Drawing using Format_C

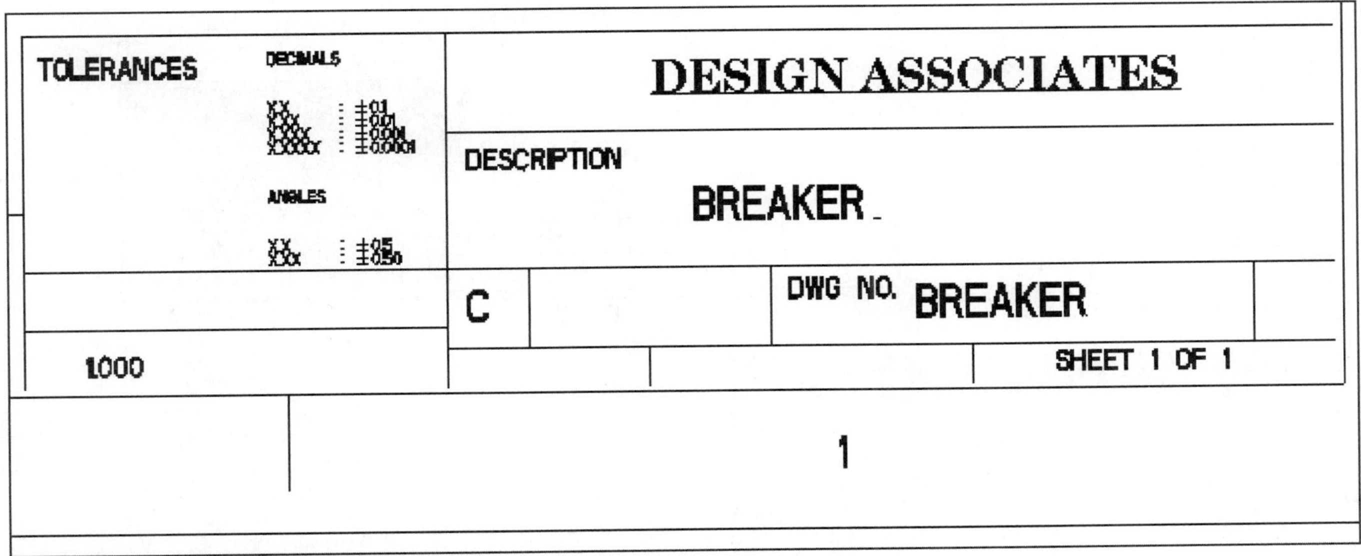

Figure 18.9(e) Breaker Drawing Title Block

You now have a drawing with the Breaker displayed in three standard views. Since the view selection may not be the most appropriate for the part geometry, you will delete some of the views, add new views, and then relocate them to better positions for detailing. Remember, you can reposition the views at any time in the detailing process, even after the dimensions are placed. Turn off the datum planes, axes, points, and coordinate system first.

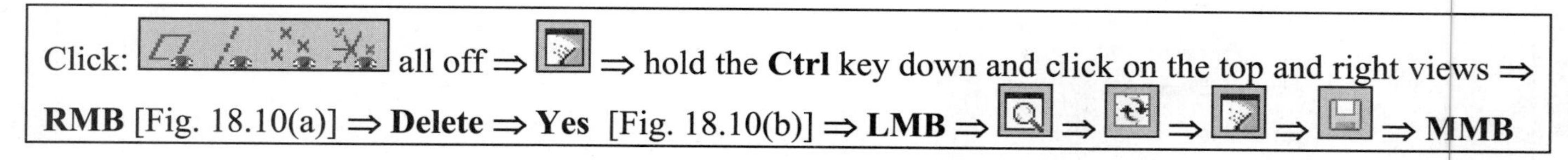

Click: all off ⇒ ⇒ hold the **Ctrl** key down and click on the top and right views ⇒ **RMB** [Fig. 18.10(a)] ⇒ **Delete** ⇒ **Yes** [Fig. 18.10(b)] ⇒ **LMB** ⇒ ⇒ ⇒ ⇒ ⇒ **MMB**

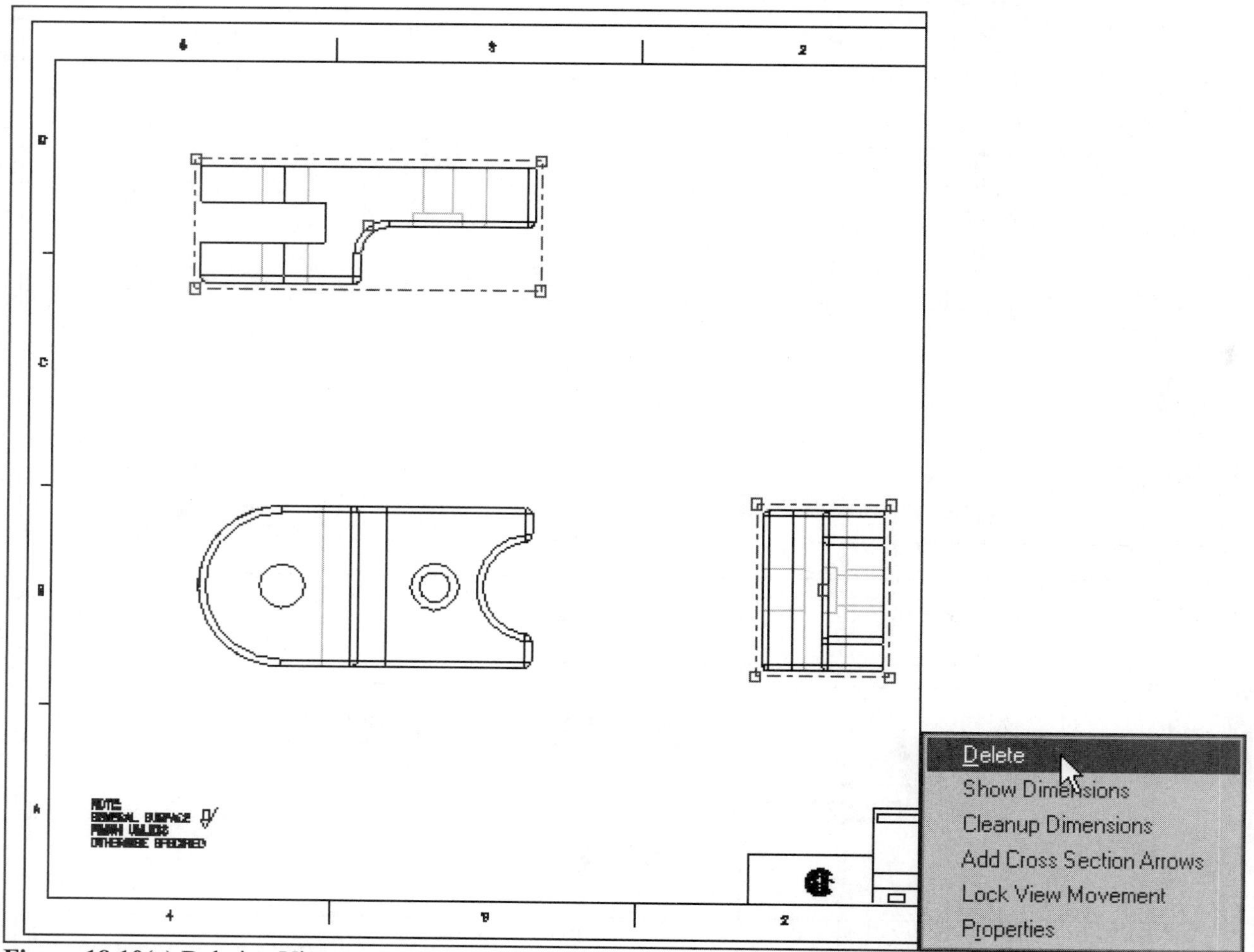

Figure 18.10(a) Deleting Views

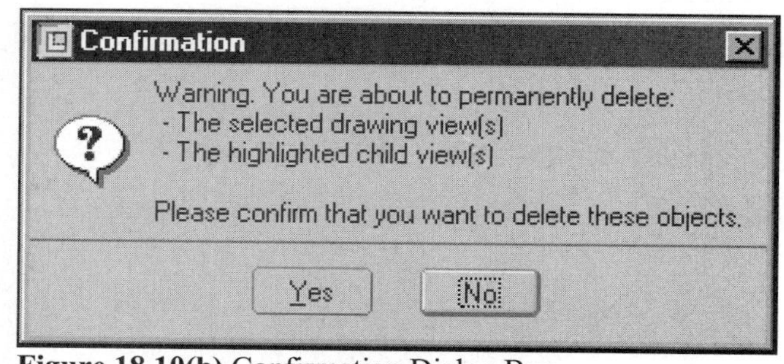

Figure 18.10(b) Confirmation Dialog Box

Click: unlock ⇒ click on the front view ⇒ place the cursor in the middle of the view and press and hold **LMB** [Fig. 18.10(c)] ⇒ drag and drop the view [Fig. 18.10(d)]

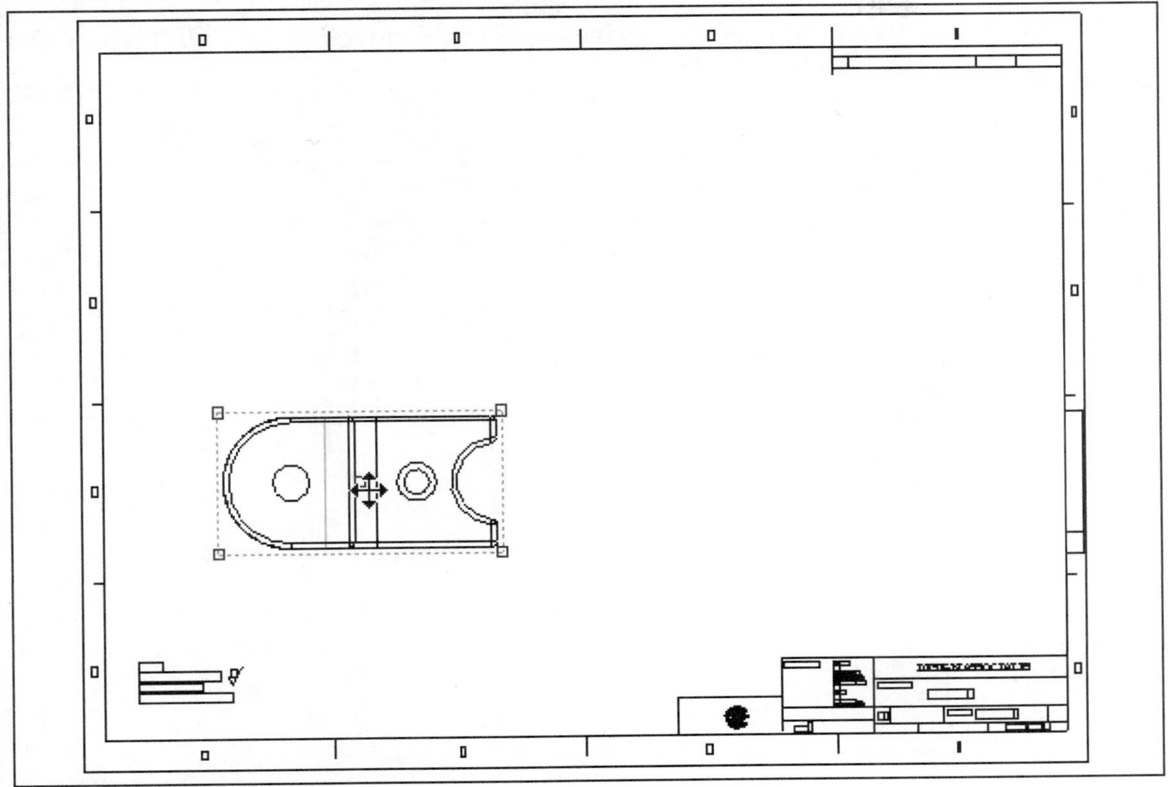

Figure 18.10(c) Click on View

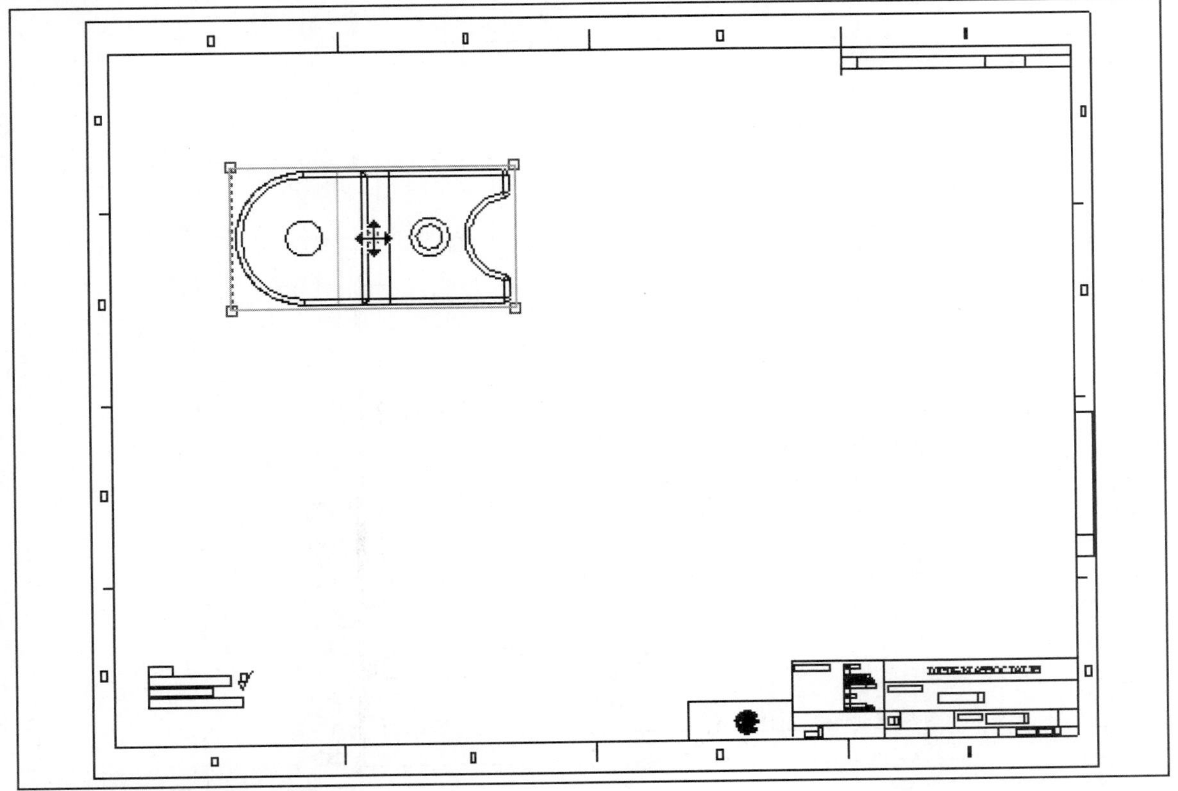

Figure 18.10(d) Drag and Drop

Add two new views, click: **Insert a drawing view of the active model ⇒ Projection ⇒ Full View ⇒ No Xsec ⇒ No Scale ⇒ MMB** to accept defaults ⇒ Select CENTER POINT for drawing view. pick a position [Fig. 18.10(e)] ⇒ **Insert a drawing view of the active model ⇒ Projection ⇒ Full View ⇒ No Xsec ⇒ No Scale ⇒ MMB ⇒** Select CENTER POINT for drawing view. pick a position [Fig. 18.10(f)] ⇒ reposition the views as necessary ⇒ **LMB** to deselect ⇒ ⇒ ⇒ ⇒ ⇒ **MMB**

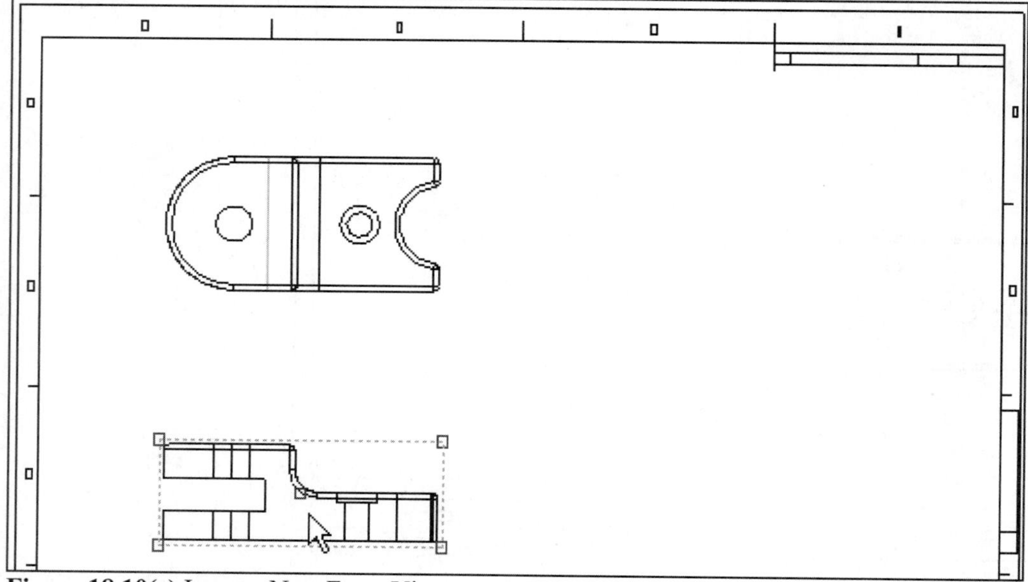

Figure 18.10(e) Insert a New Front View

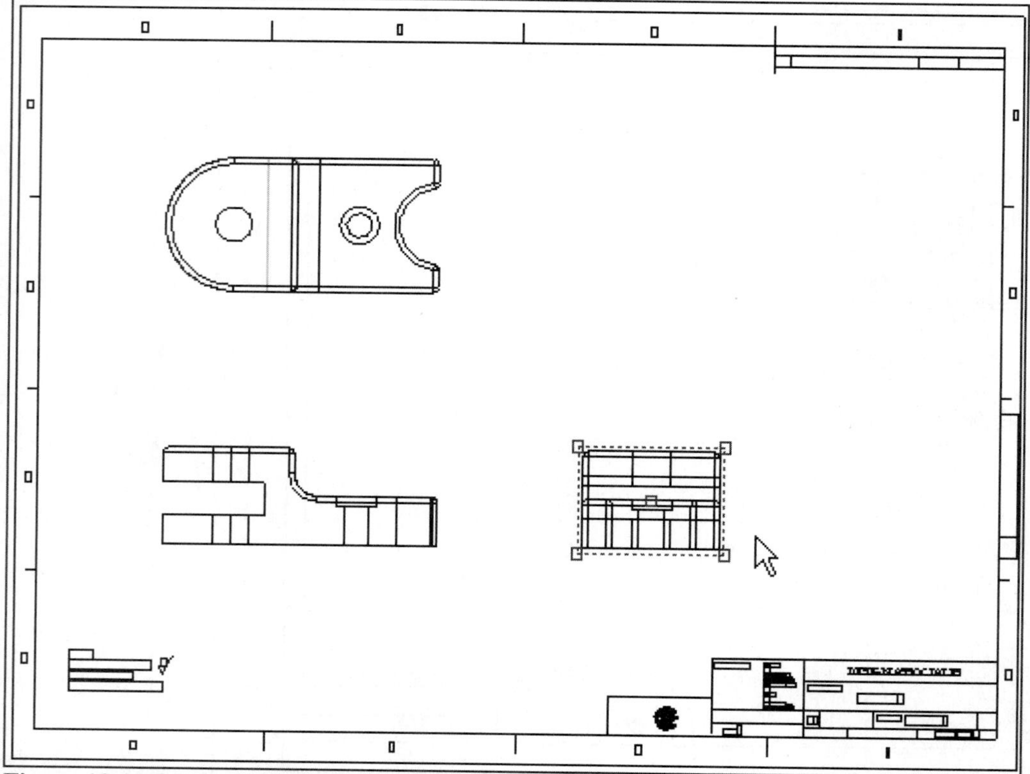

Figure 18.10(f) Insert a New Right Side View

Modifying Individual View Display

The display for a view, edge, or assembly member can be set individually. Once you have set the display mode for a specific view, it remains set regardless of the setting in the Environment dialog box, unless you choose Default in the View Disp menu. Options include:

- **Wireframe** Sets the display mode to wireframe
- **Hidden Line** Sets the display mode to hidden line
- **No Hidden** Sets the display mode to no hidden
- **Default** Sets the display mode as set in the Environment dialog box
- **Qlt HLR** Includes quilts in the hidden line removal process (except cross-sectional views)
- **No Qlt HLR** Excludes quilts from the hidden line removal process (except cross-sectional views)
- **Tan Solid** Displays tangent edges as solid lines
- **No Disp Tan** Turns off the display of tangent edges
- **Tan Ctrln** Displays tangent edges in centerline font
- **Tan Phantom** Displays tangent edges in phantom font
- **Tan Dimmed** Displays tangent edges in dimmed color
- **Tan Default** Sets the display of tangent edges as set in the Environment dialog box

Click on the right side view ⇒ **RMB** [Fig. 18.11(a)] ⇒ **Properties** ⇒ **View Disp** ⇒ **No Hidden** ⇒ **Tan Dimmed** ⇒ **MMB** ⇒ **MMB** [Fig. 18.11(b)] ⇒ **LMB** ⇒ click on the front and top views (press and hold **Ctrl** during selection) ⇒ **RMB** ⇒ **Properties** [Fig. 18.11(c)] ⇒ **View Disp** ⇒ **Hidden Line** ⇒ **Tan Dimmed** ⇒ **MMB** ⇒ **MMB** [Fig. 18.11(d)] ⇒ **LMB**

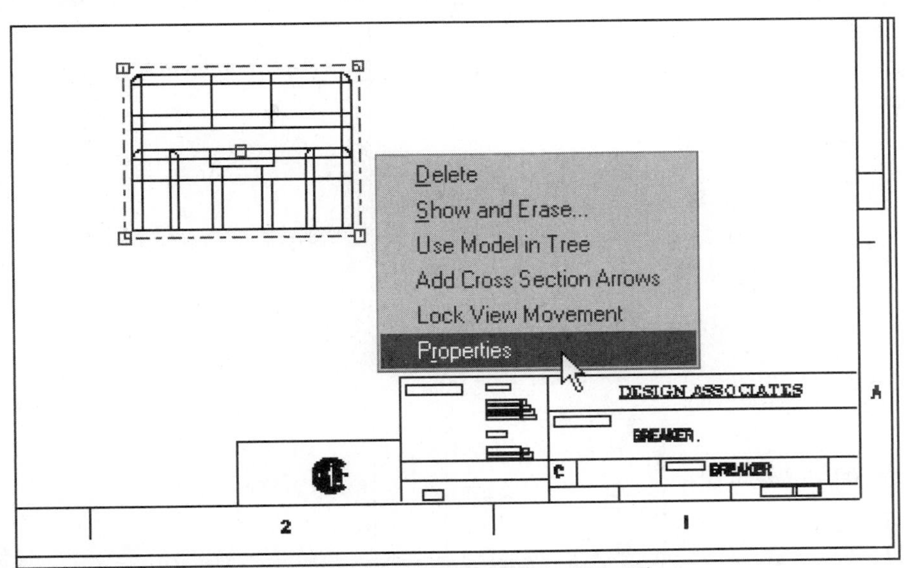

Figure 18.11(a) Change the Properties of the Right Side View

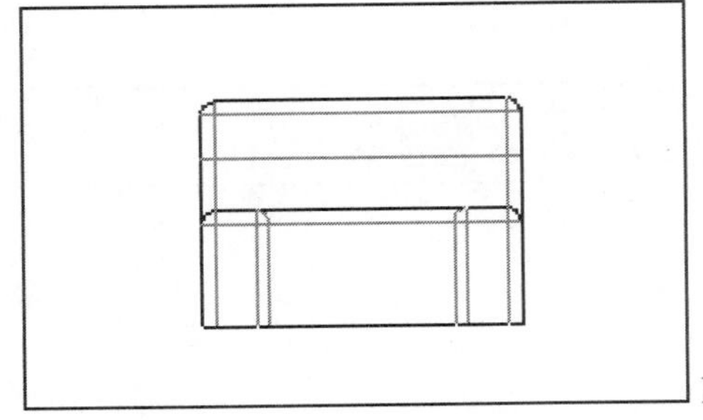

Figure 18.11(b) No Hidden, Tan Dimmed

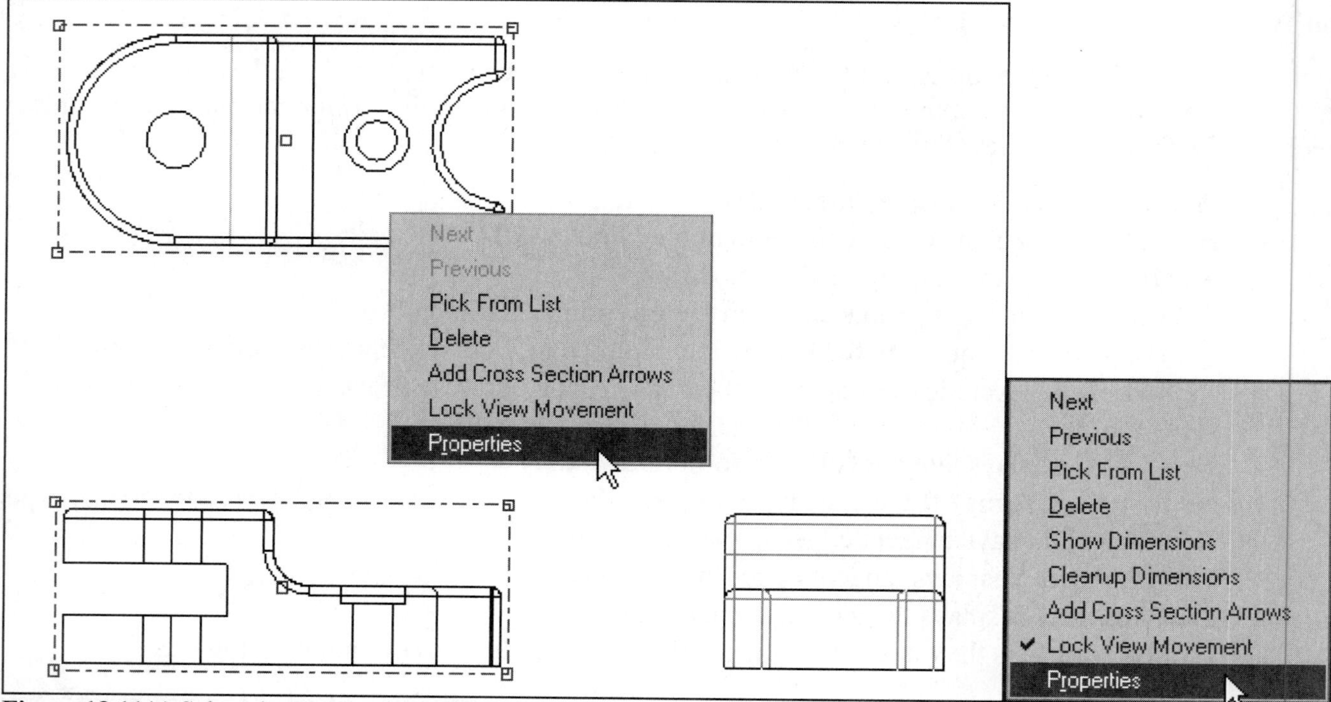

Figure 18.11(c) Select the Front and Top Views, RMB off the View
(If you **RMB** on the view itself, you will get more pop-up options)

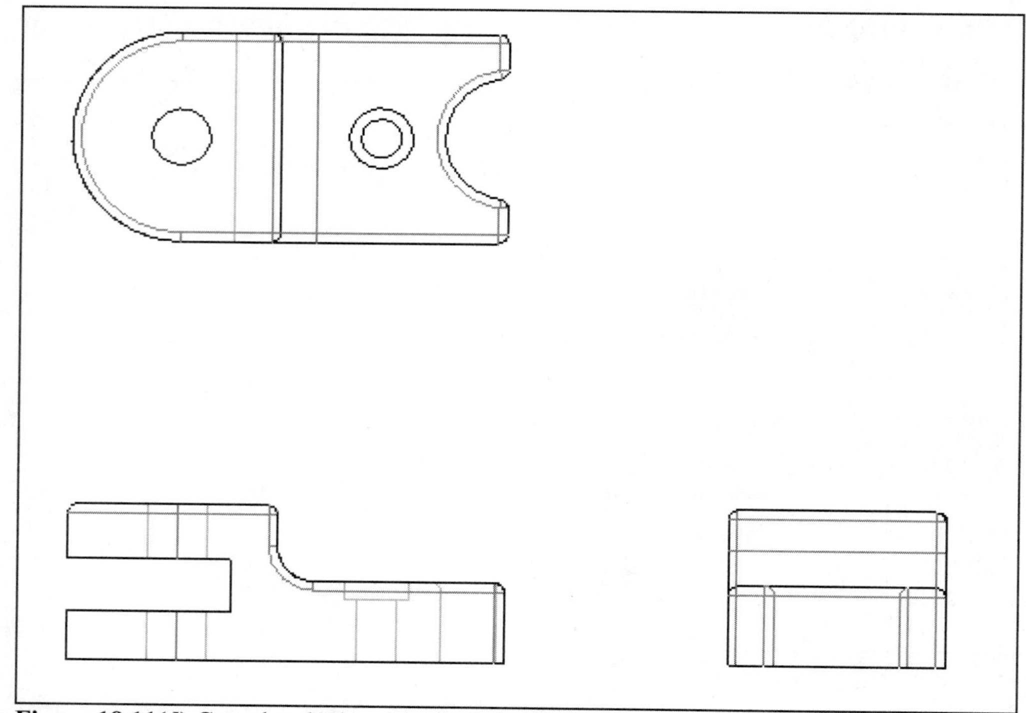

Figure 18.11(d) Completed View Display Changes

Click: **Update the display of all views in the active sheet ⇒ ⇒ ⇒ ⇒ MMB**

Drawing Options

These options determine characteristics like height of dimension and note text, text orientation, geometric tolerance standards, font properties, drafting standards and arrow lengths. Pro/E assigns default values to the setup file options, but you can modify the values (to customize a drawing), and save them to use again. Pro/E saves the default values in a setup file named filename **.dtl**. Pro/E saves drawing setup options with each individual drawing file.

Each *format* will have a ***.dtl*** file associated with it, and each *drawing* will have a different ***.dtl*** file associated with it. *They are separate .dtl files and are independent of each other.* When you activate a drawing and then add a format, the .dtl for the format controls the font etc., for the format only. If you modified the .dtl file in the format to have a filled font, the format will show as filled font when used in a drawing. The dimensions and text of the drawing will *not* show as filled because it has a separate .dtl file.

The file that you specify in the configuration option *drawing_setup_file* establishes the default drawing setup options for any drawing that you create during a Pro/E session. If you do not set this configuration option, Pro/E uses the default setup file. You can install sample drawing setup files for DIN, ISO, and JIS from the loadpoint/text directory with the following names:

- **din.dtl**
- **iso.dtl**
- **jis.dtl**

Retrieve these setup files to set the desired environment in your drawing, using: File ⇒ Properties ⇒ Drawing Options to create, retrieve, and modify a drawing setup file. The drawing .dtl file still needs to be established.

Click: **File ⇒ Properties ⇒ Drawing Options** *(or RMB anywhere in graphics window ⇒ Properties ⇒ Drawing Options)* ⇒ Sort **By Category** [Fig. 18.12(a)] ⇒ make the changes: ***default_font filled*** ⇒ **Add/Change** ⇒ **Apply** ⇒ ***draw_arrow_style filled*** ⇒ **Add/Change** ⇒ ***gtol_datums std_asme*** ⇒ **Add/Change ⇒ Apply** ⇒ Sort **As Set** [Fig. 18.12(b)] ⇒ **Close ⇒ MMB** ⇒ [icon] ⇒ [icon] ⇒ **MMB**

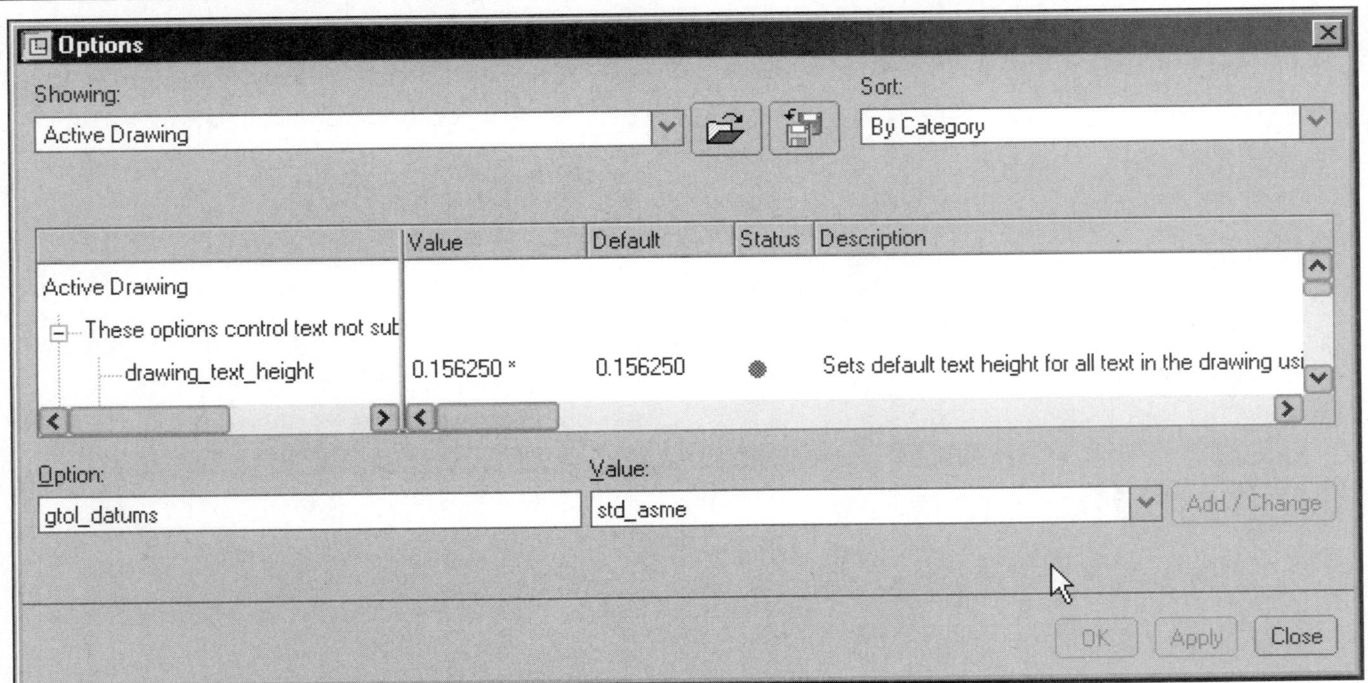

Figure 18.12(a) Drawing Options Sorted Using By Category

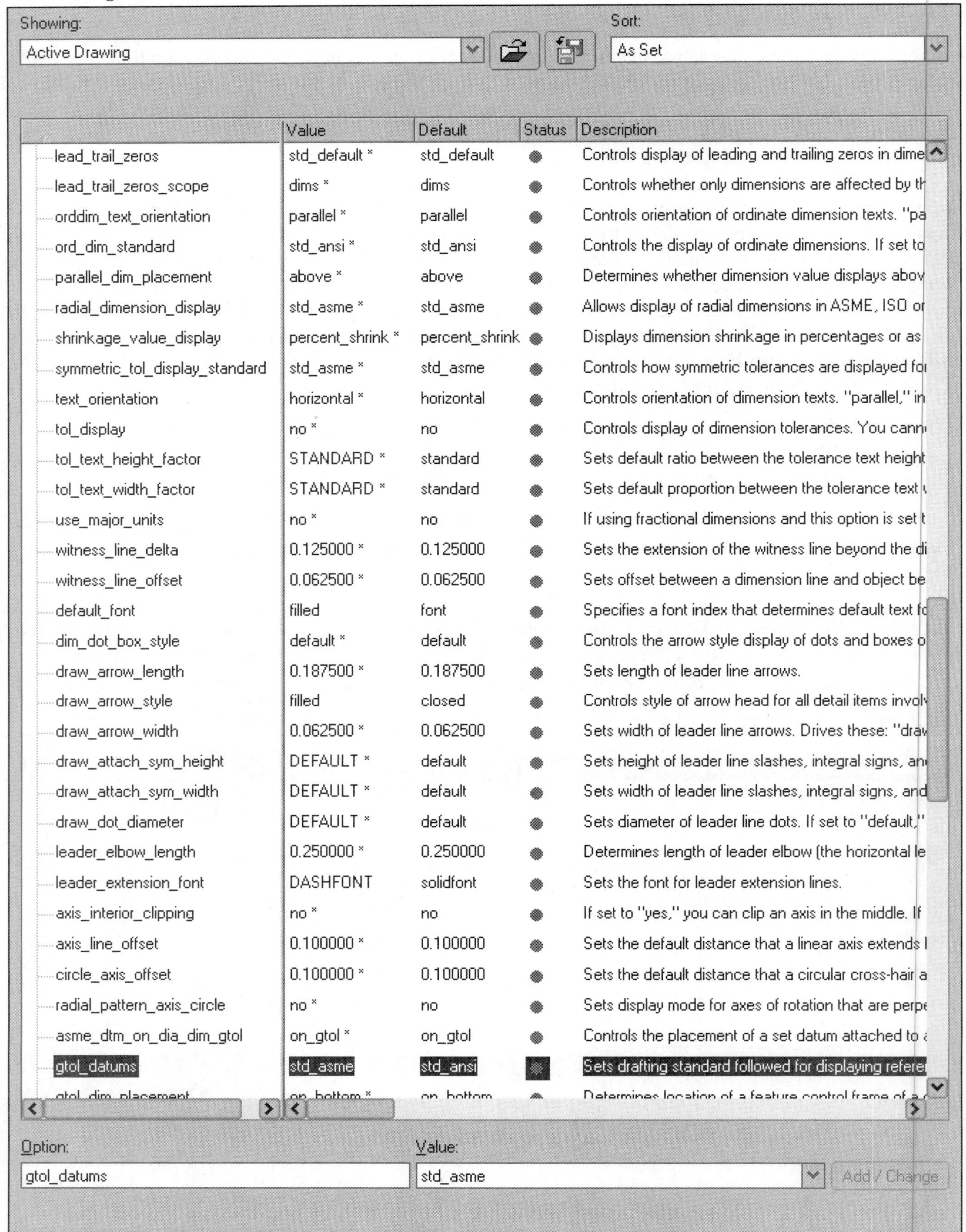

Figure 18.12(b) Drawing Options Sorted Using As Set

Showing Axes

Before dimensions are added, it is a good idea to add centerlines and adjust them to the drawing and view requirements. Axes can be displayed at the same time dimensions are shown, but in many cases, the drawing views become cluttered, and clean up becomes more time consuming.

Pro/E can automatically display centerlines at each location where there is a Datum Axis. The axis must be either parallel or perpendicular to the screen to be displayed. If the axis is perpendicular to the screen, Pro/E displays a "crosshairs" centerline. If the axis is parallel to the screen, Pro/E displays a linear centerline. If the Environment option Datum Axes is off, the name of the axis is not displayed on the drawing. A crosshairs axis actually consists of four centerline segments. Each segment makes up one "leg" of the crosshairs. A linear axis actually consists of two centerline segments. Each segment makes up one "half" of the centerline. This distinction means very little, unless you need to trim the centerline segments because of interference with other drafting objects or entities.

There are several Drawing Options (**.dtl**) controlling the display of centerlines. The option *circle_axis_offset* controls the distance that centerlines extend past the circle on which they lie. The option *axis_line_offset* controls the distance that centerlines extend beyond its associated feature. The option *axis_interior_clipping* controls whether or not you can trim the interior segments of a centerline. If this option value is set to *no*, you can shorten or lengthen the ends of only linear and crosshairs centerlines. If the option value is set to *yes*, you can shorten or lengthen any end of any centerline segment. The option *radial_pattern_axis_circle* will display a bolt circle centerline.

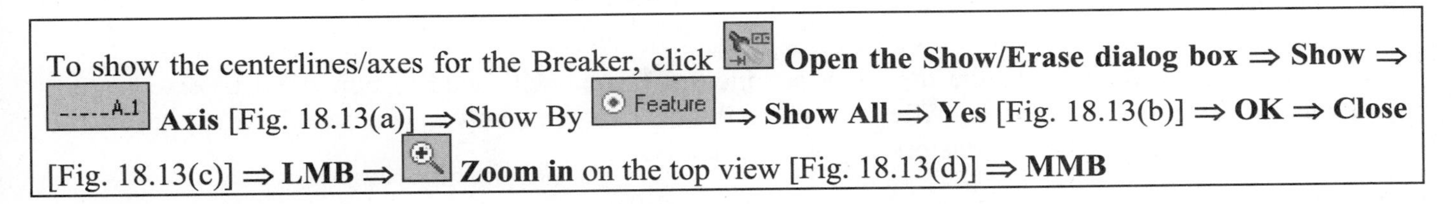
To show the centerlines/axes for the Breaker, click **Open the Show/Erase dialog box ⇒ Show ⇒** Axis [Fig. 18.13(a)] ⇒ Show By Feature **⇒ Show All ⇒ Yes** [Fig. 18.13(b)] **⇒ OK ⇒ Close** [Fig. 18.13(c)] **⇒ LMB ⇒ Zoom in** on the top view [Fig. 18.13(d)] **⇒ MMB**

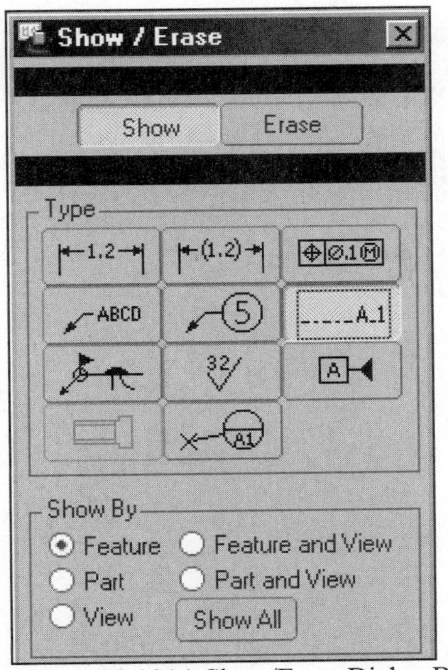

Figure 18.13(a) Show/Erase Dialog Box

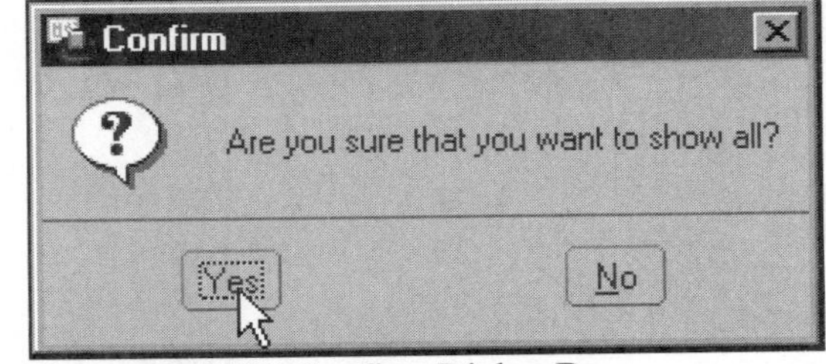

Figure 18.13(b) Confirm Dialog Box

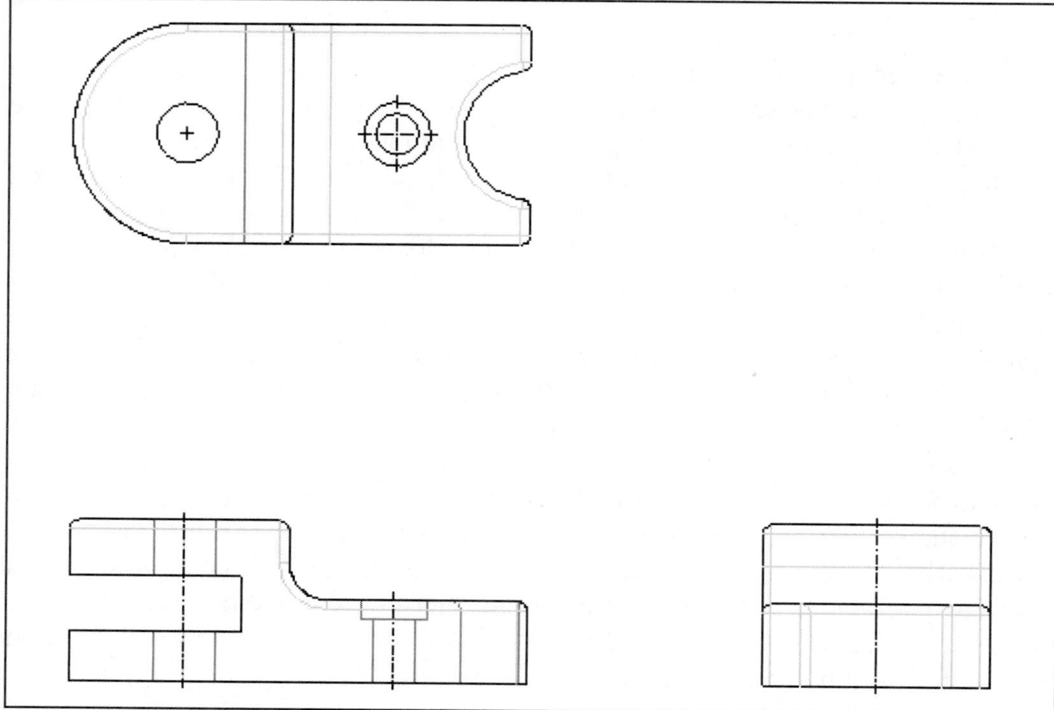

Figure 18.13(c) Displayed Axes

Since the centerlines in the top view are incorrect. You will need to edit them to extend their segments to the proper positions. After the parts dimensions are shown, you will have to reedit the centerlines and the dimension extension lines appropriately.

Click: **Sketch** ⇒ **Sketcher Preferences** ⇒ Grid intersection off ⇒ **Close** ⇒ click on the centerpoint, representing the *axis crosshair* for the hole [Fig. 18.13(d)] ⇒ click on a drag handle and extend the axis/centerline [Fig. 18.13(e)] ⇒ repeat for each handle of the centerpoint [Fig. 18.13(f)] ⇒ do the same for the counterbore hole [Figs. 18.13(g-h)] ⇒ **LMB** to deselect ⇒ ⇒ ⇒ ⇒ ⇒ **MMB**

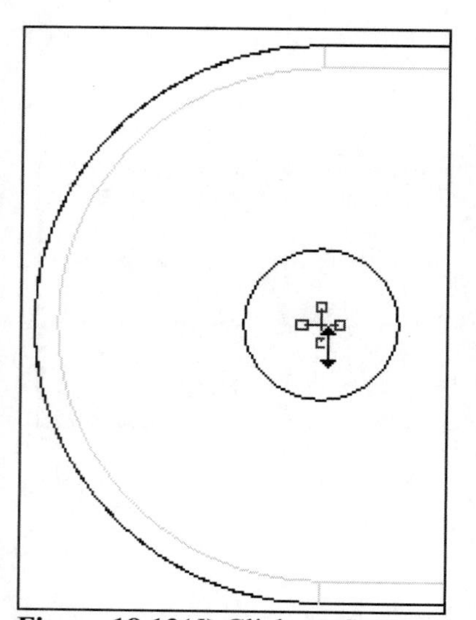

Figure 18.13(d) Click on Center

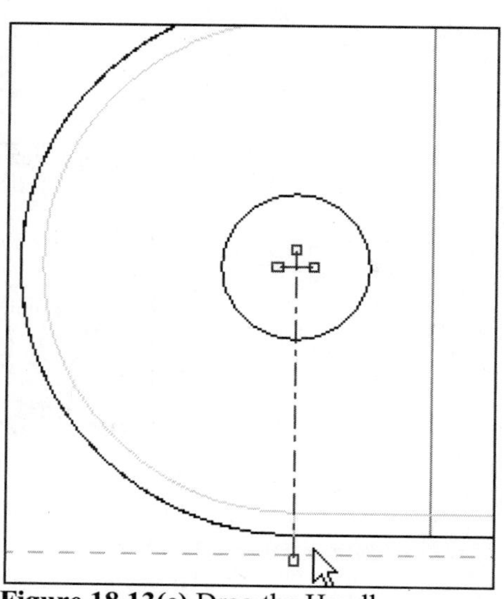

Figure 18.13(e) Drag the Handle

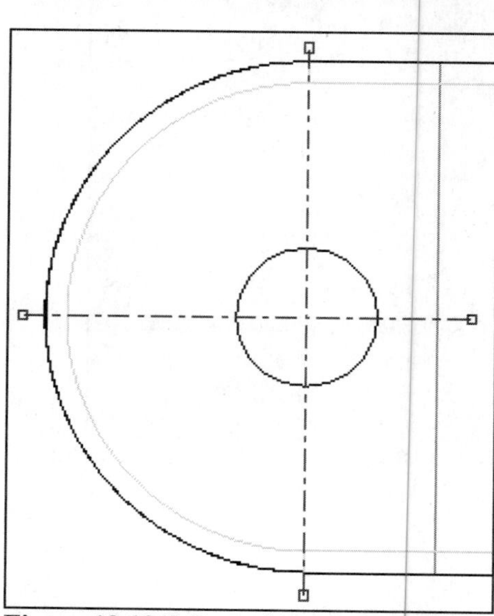

Figure 18.13(f) Reposition the Handles

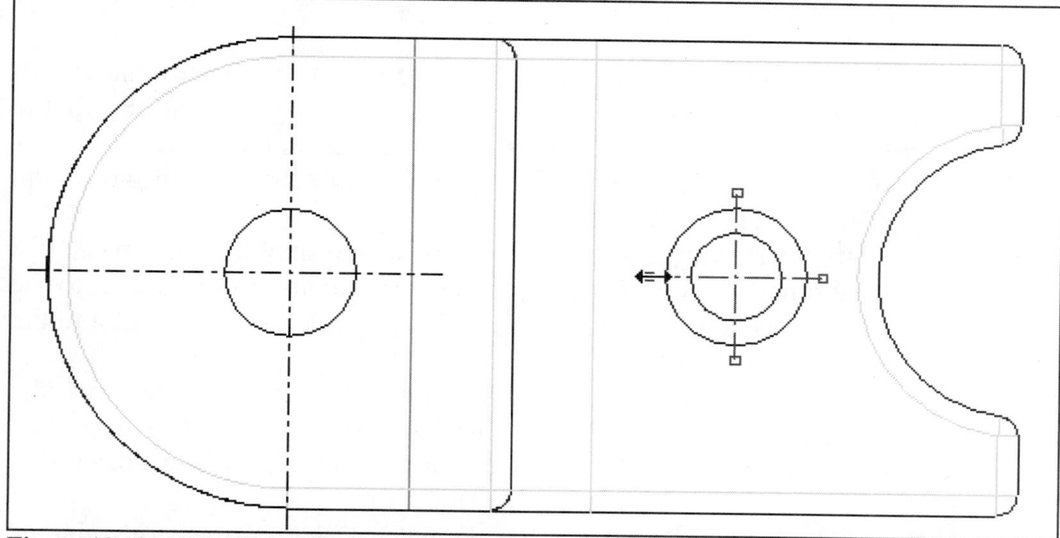

Figure 18.13(g) Modifying the Counterbore Hole Axes/Centerlines

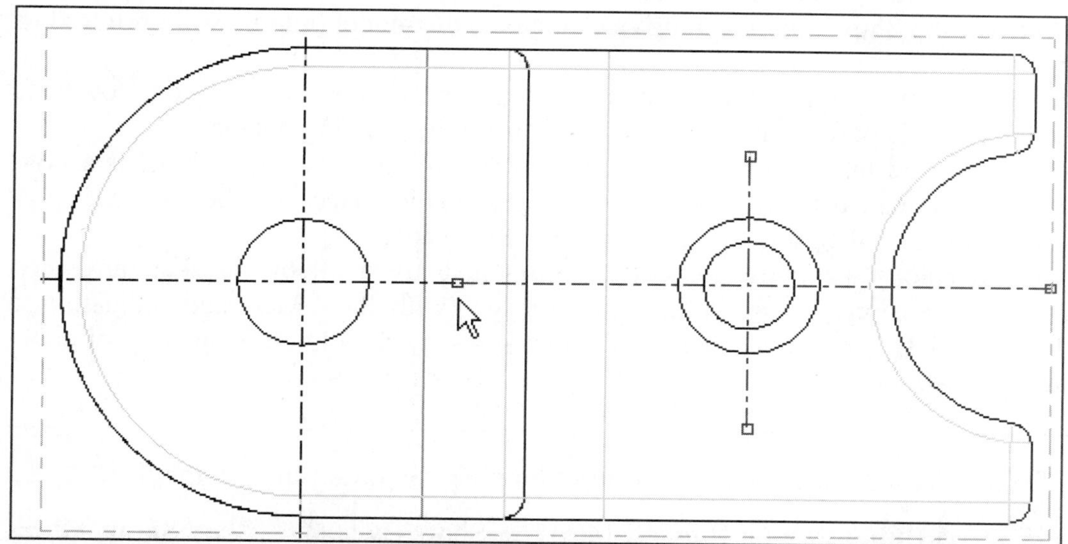

Figure 18.13(h) Modified Counterbore Hole Axes/Centerlines

The front and right side views show the centerlines correctly [Fig. 18.13(i)]. The side view of a hole will display the axis as linear and it can be modified with the same technique as used for centerpoints.

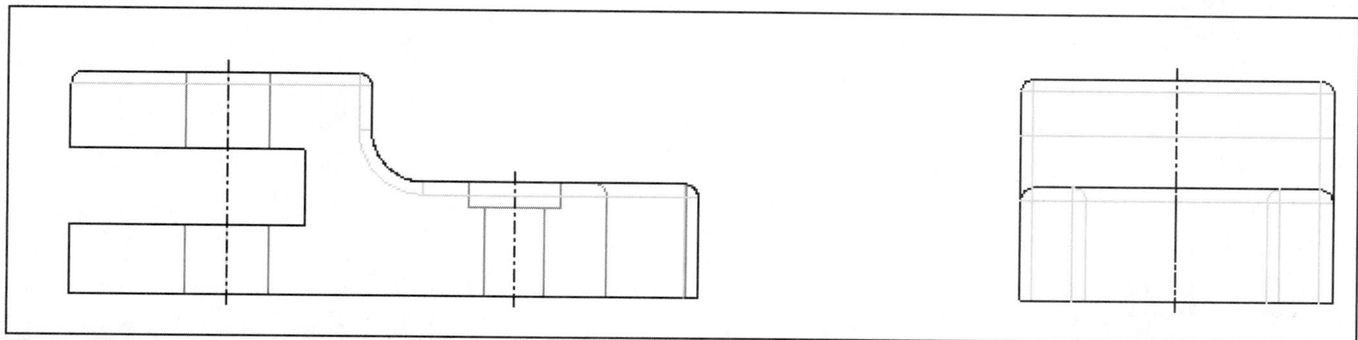

Figure 18.13(i) Front and Right Views Show the Axes as Linear Centerlines

Dimensions

The **Show** option also allows you to display dimensions based on the dimensional references and sketch dimensions that were used to create the model. You can show the dimensions of a feature, all the dimensions of all features in a particular view, or all the dimensions of all features on the model.

If you show dimensions in more than one view, and a feature can be dimensioned in more than one view, Pro/E attempts to decide which view is most appropriate.

Diameter dimensions are displayed differently based on the view where they are displayed. If a dimension is shown in a view where the cylinder axis is normal to the screen, the arrows are drawn to the circular edge of the cylinder. If the dimension is shown in a view where the cylinder axis is parallel to the screen, extension lines are added along the silhouette of the cylinder.

You can show dimensions, axes, datums, and so on at the same time, but for a complex part, the views quickly become cluttered. Showing axes first, then the dimensions, and then other less important detail items gives you an opportunity to modify each item separately and to see the view requirements more clearly.

Dimensions that you add "manually" are called driven dimensions. Driven dimensions change when the model changes, but they cannot be used to make changes to the model. Only dimensions that you display in the drawing using the Show option can drive changes to the model (and they are referred to as driving dimensions).

Driven dimensions can be erased in the same manner as driving dimensions. In addition, however, driven dimensions can be deleted; this permanently removes them from the drawing database.

You can also create notes and labels to add to the annotation on your drawing. A label is a note that has a leader. Leaders can be attached to edges or drawn to positions in space. It is better practice to modify an existing diameter dimension than it is to create a note with a leader.

You create driven dimensions using the same techniques that are available in sketching. For example, you create parallel dimensions by selecting a linear edge (with the LMB) and indicating a placement location (with the MMB). You create diameter dimensions by double-clicking on a circular edge and indicating a placement location.

To show the dimensions for the Breaker, click: **Open the Show/Erase dialog box** ⇒ toggle -----A_1 **Axis** radio button off ⇒ 1.2 **Dimension** on [Fig. 18.14(a)] ⇒ Show By Feature ⇒ **Show All** ⇒ **Yes** [Fig. 18.14(b)] ⇒ **OK** ⇒ **Close** ⇒ **LMB** to deselect

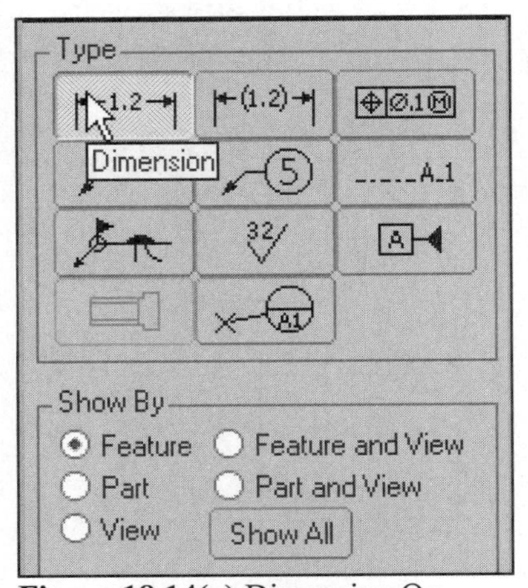

Figure 18.14(a) Dimension On

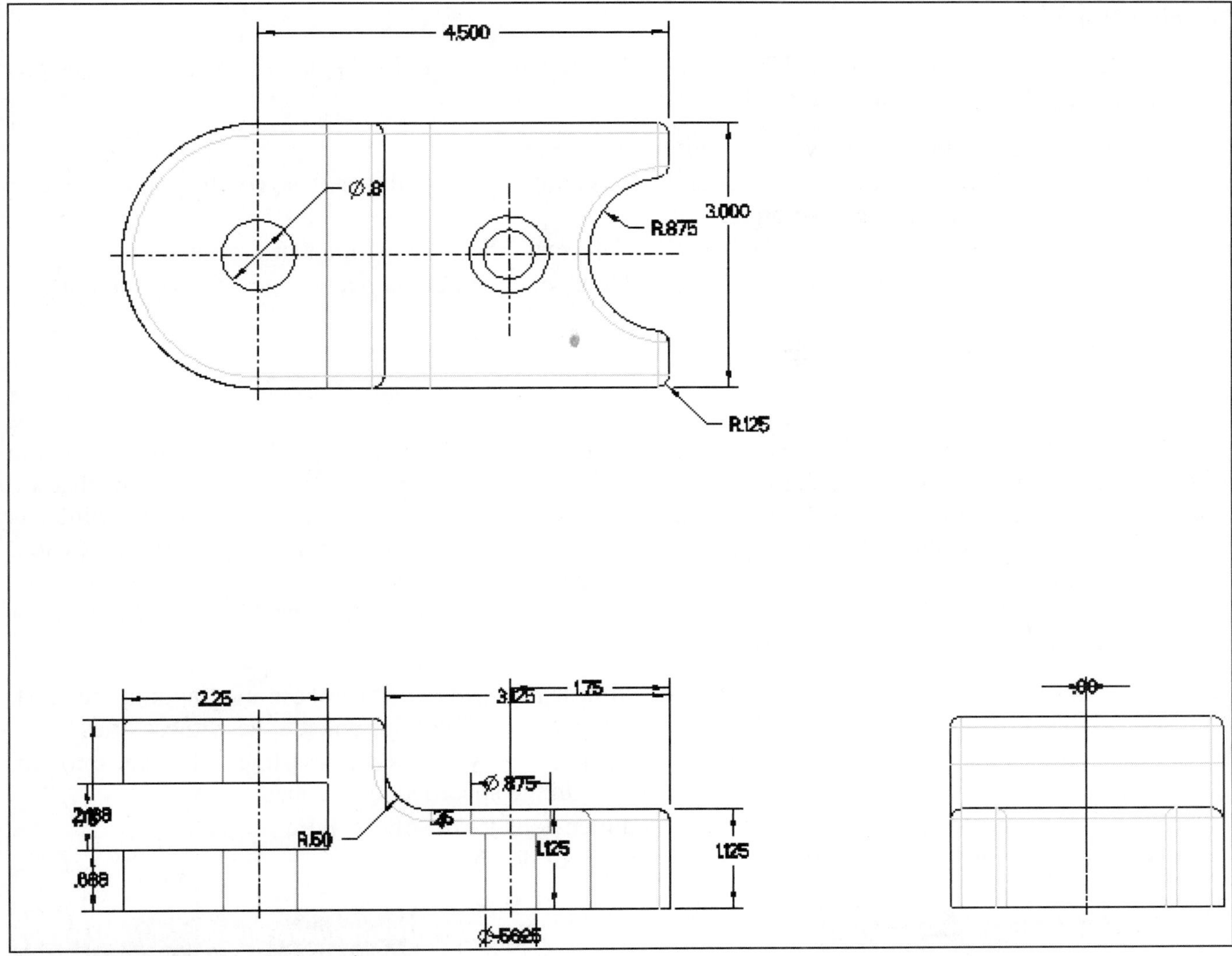

Figure 18.14(b) Displayed Dimensions

Reference dimensions can also be added to a drawing. Reference dimensions are *driven* dimensions, not to be used in manufacturing the part. Create a reference dimension to show the total width of the part.

Click: **Insert** ⇒ **Reference Dimension** ⇒ **New References** ⇒ [icon] pick both ends of the part in the top view [Fig. 18.14(c)] ⇒ place the dimension below the view by clicking **MMB** [Fig. 18.14(d)] ⇒ **Tangent** ⇒ **MMB** ⇒ **LMB**

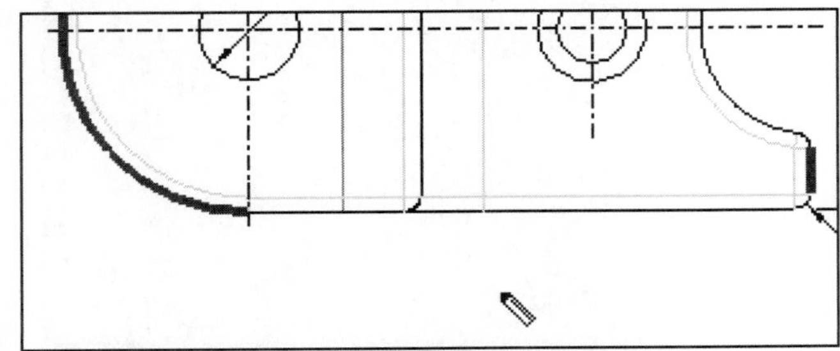

Figure 18.14(c) Reference Dimension References

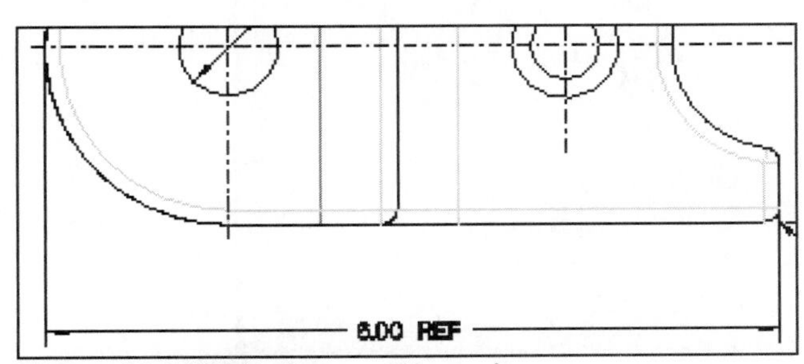

Figure 18.14(d) Reference Dimension

Cleaning Dimensions

Using Edit ⇒ Cleanup ⇒ Dimensions [Fig. 18.14(e)], you can arrange the display of selected dimensions automatically by doing any or all of the following:

- Clean dimensions by view, or by individual dimensions
- Center dimensions between witness lines (including the entire text box with gtols, diameter symbols, tolerances, and so on)
- Create breaks in witness lines where they intersect other witness lines or draft entities
- Place all dimensions on one side of a model edge, datum plane, view edge, axis, or snap line
- Flip arrows
- Offset dimensions from edges or view boundaries
- Create snap lines

On the Placement tab, Create Snap Lines creates snap lines under all moved dimensions (if the destination does not already have a snap line in the vicinity). They appear only under dimensions that are parallel to baselines (if you have selected a baseline) or parallel to a view border (if the view outline has been selected). Once you clear this check box, it remains cleared for the remainder of your Pro/E session. In general, you should clear this selection for most projects.

On the Cosmetic tab [Fig. 18.14(f)], by default, Pro/E selects all of the check boxes. Clear any or all of them, as necessary; then click Apply:

- **Flip Arrows** Flips the arrows inside the witness lines if they fit (without overlapping the text); if they do not fit or they would be overlapping the text, they flip outside the witness lines.
- **Center Text** Centers the text of each dimension between its witness lines. If it does not fit, Pro/E moves the text outside the witness lines in the specified direction.
- **Horizontal** Moves the text of horizontal dimensions to the left or right.
- **Vertical** Moves text of vertical dimensions up or down.

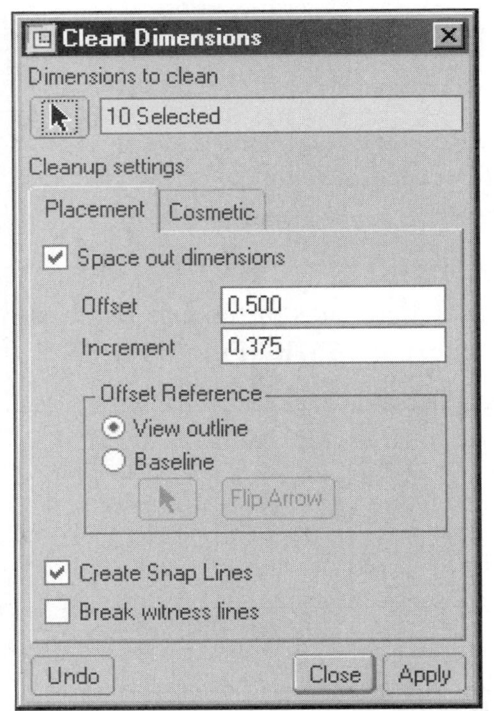

Figure 18.14(e) Clean Dimensions Dialog Box

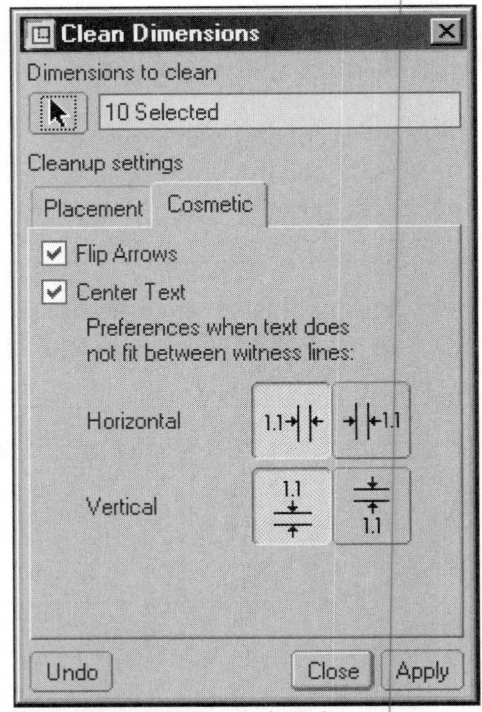

Figure 18.14(f) Cosmetic Tab

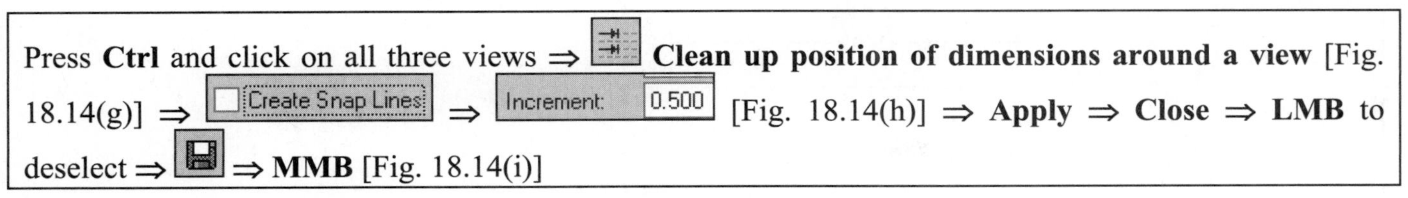

Press **Ctrl** and click on all three views ⇒ **Clean up position of dimensions around a view** [Fig. 18.14(g)] ⇒ Create Snap Lines ⇒ Increment: 0.500 [Fig. 18.14(h)] ⇒ **Apply** ⇒ **Close** ⇒ **LMB** to deselect ⇒ ⇒ **MMB** [Fig. 18.14(i)]

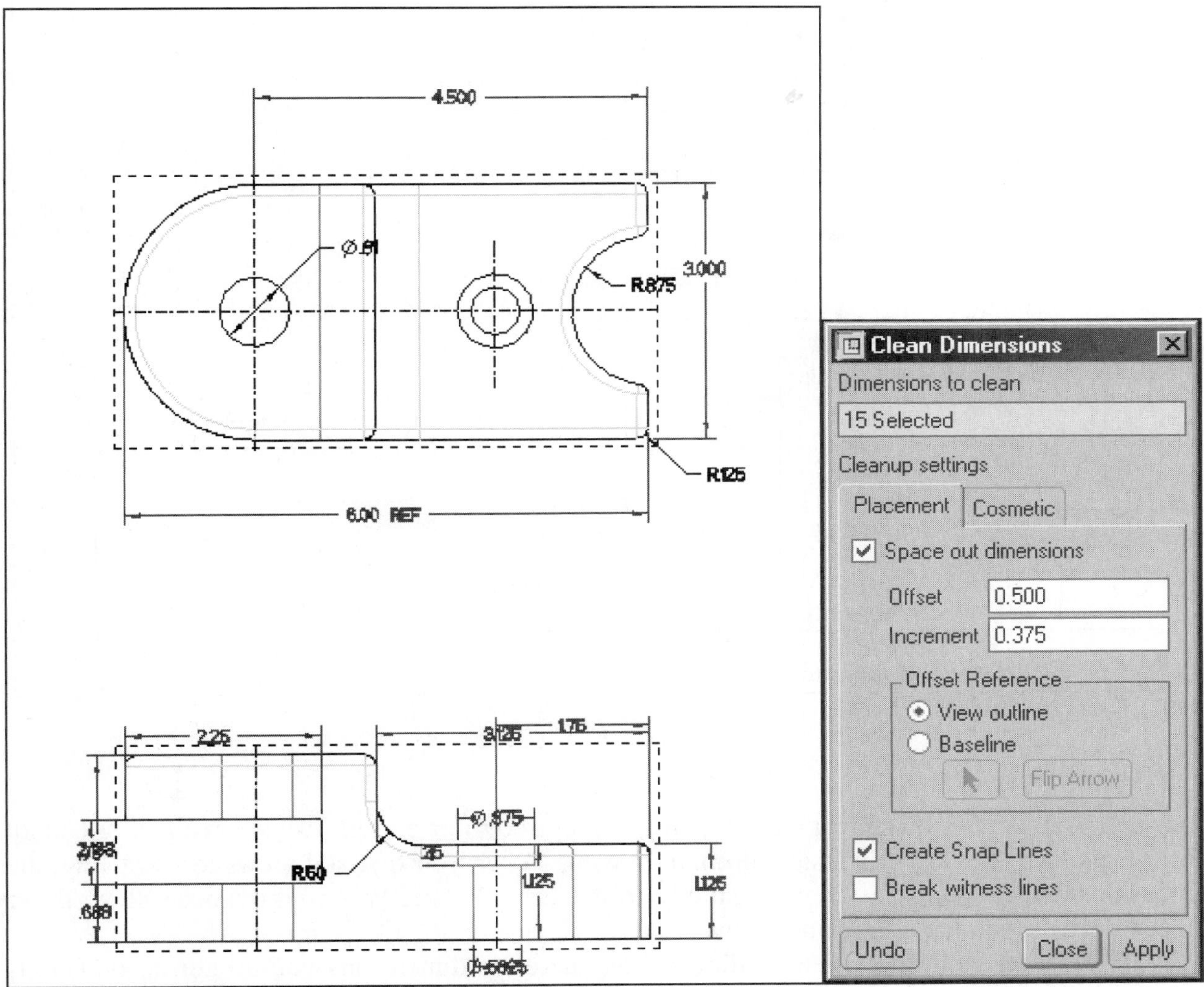

Figure 18.14(g) Highlighting Dimensions to be Cleaned

Space Dimensions
Offset: 0.500
Increment: 0.500
Offset Reference
View Outline
Baseline
Flip Arrow
Create Snap Lines
Break witness lines

Figure 18.14(h) Placement Tab Selections and Values

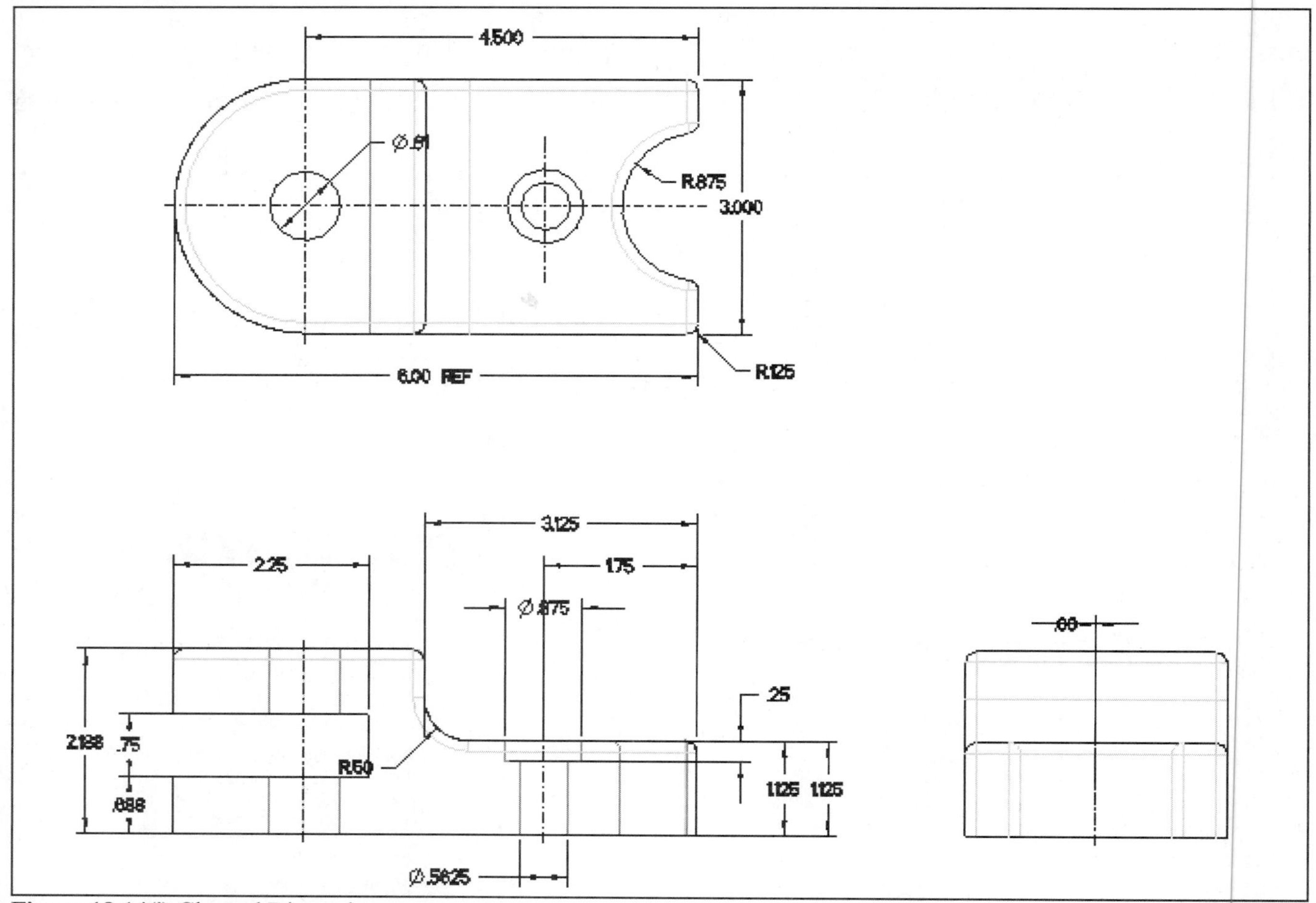

Figure 18.14(i) Cleaned Dimensions

Erasing

Feature dimensions can either be shown on a drawing (using Show) or erased (using Erase). You cannot delete feature (*true part-driving*) dimensions from a drawing. When you first add views to a drawing, the feature dimensions are also present; they are simply in the Erased state. When you choose Show/Erase and the desired options, you are un-erasing them.

Many times, when you Show feature dimensions, there are dimensions you do not need. This is especially true of dimensions of thickness and pattern number callouts. In addition, when you locate a hole from a datum at a distance of **.00** you will create what we call "*zero*" dimensions. These dimensions should be erased in the Drawing mode. If you accidentally erase desired dimensions, you can redisplay them using Show. Show does not know which dimensions you want to redisplay (from the set of dimensions that you have erased and not shown). Therefore, you may have to redisplay several dimensions and re-erase those you really do not need. Using the Feature & View option with Show (instead of Show All) will minimize the number of dimensions you need to redisplay. Use Preview to give you a choice of what to keep or remove.

Erase gives you a great deal of control over how you select the items to erase. You can choose to select a type of drafting object individually or erase all objects of a type, in the drawing, in a view, and/or by feature. If you choose to Erase All of a type, Pro/E prompts you to confirm the erasure. If you pick No, the erasure operation is aborted. If you pick Yes, the items are all erased.

The simplest way of erasing unwanted dimensions (notes and other entities) requires the selection of the item or items, RMB and selecting Erase. Selected items will highlight. Picking Erase will make the items *gray out*. Choose LMB to complete the removal of the items.

Erase the dimension in the right view, click on the **.00** dimension ⇒ **RMB** [Fig. 18.14(j)] ⇒ **Erase** ⇒ **LMB** ⇒ ⇒ **Update the display of all views in the active sheet** [Fig. 18.14(k)] ⇒

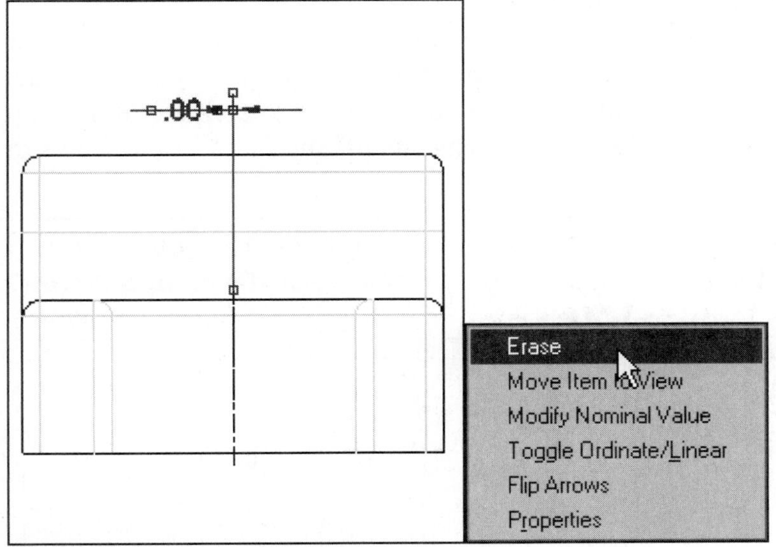

Figure 18.14(j) Erase **.00** Dimension

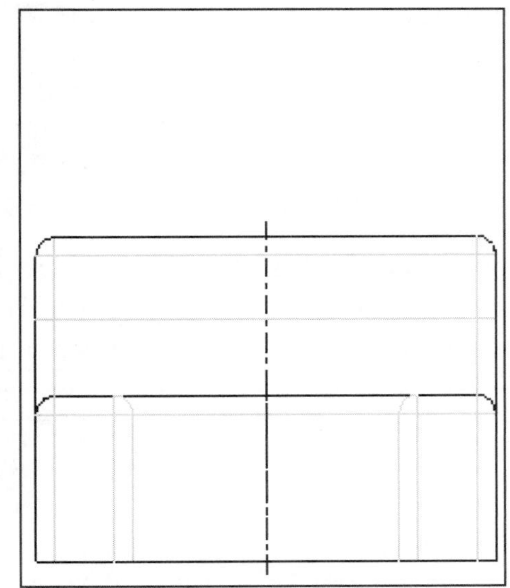

Figure 18.14(k) Dimension Erased

Click on the extra **1.125** dimension (if you have two) in the front view ⇒ **RMB** [Fig. 18.14(l)] ⇒ **Erase LMB** ⇒ **Update the display of all views in the active sheet** [Fig. 18.14(m)] ⇒ ⇒ ⇒ ⇒ **MMB**

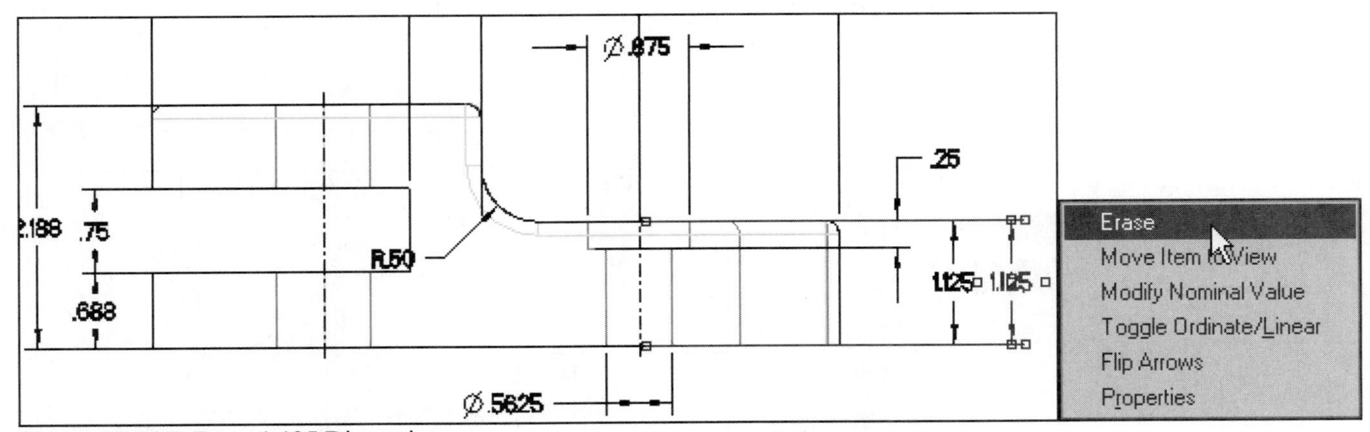

Figure 18.14(l) Erase **1.125** Dimension

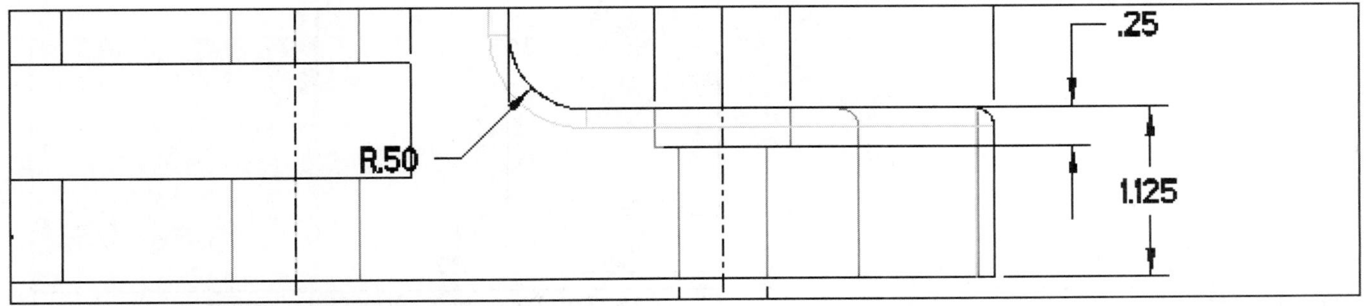

Figure 18.14(m) Extra Dimension Erased

Dimension Modifications

Dimension modifications include: changing the position of the dimension value, switching a dimensions view, altering the arrow direction and position, and controlling the number of digits displayed. Most of these can be accomplished using the RMB and the pop-up menu.

The **Flip Arrows** option is used to change the dimension arrows from inside to outside arrows or from outside to inside arrows. Small dimension values will create arrows on top of the dimension value, and radial dimensions sometimes point to the wrong side of the arc. You may have to pick the dimension more than once to obtain the required result. To start, move the dimensions off the faces of the views in accordance with ASME standards.

Click on the **∅.81** note in the top view [Fig. 18.15(a)] ⇒ **LMB** to drag and drop the note [Fig. 18.15(b)] ⇒ click on the **R.875** note [Fig. 18.15(c)] ⇒ **LMB** to drag and drop the note ⇒ **RMB** ⇒ **Flip Arrows** (you may need to flip more than once) [Fig. 18.15(d)] ⇒ **LMB**

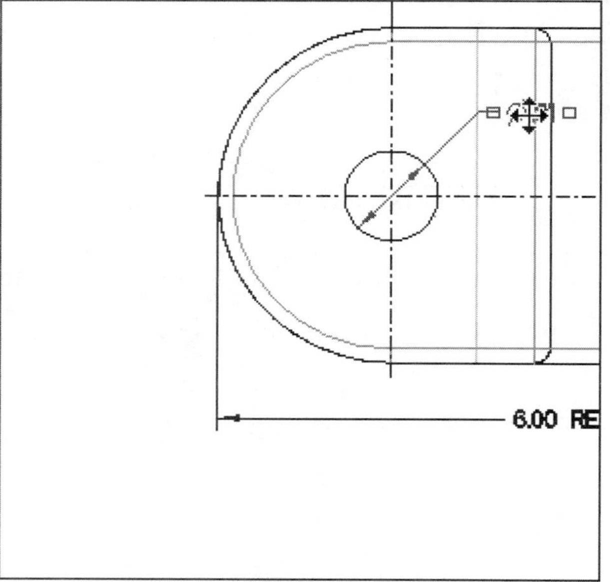

Figure 18.15(a) Click on the Diameter Dimension Note

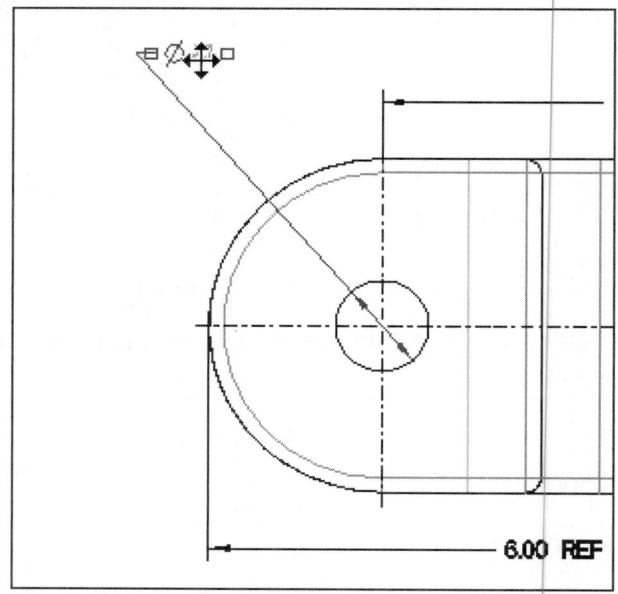

Figure 18.15(b) Drag and Drop at a New Location

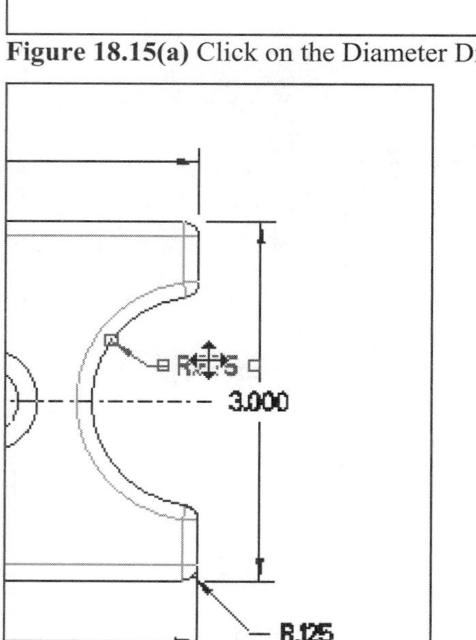

Figure 18.15(c) Click on the **R.875** Note

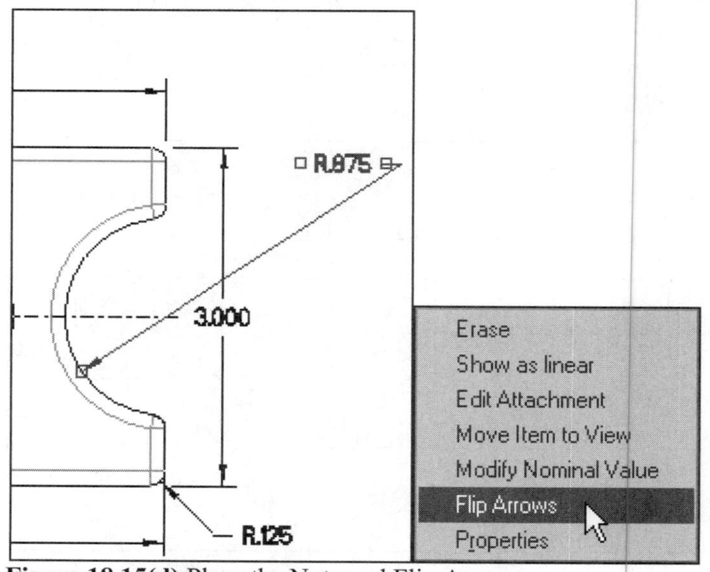

Figure 18.15(d) Place the Note and Flip Arrows

If there is a dimension appearing in the wrong view, you do *not* have to erase it. The **Move Item to View** option allows you to move that dimension to the view where you want it displayed. If, however, it is a dimension that you do not need, you can Erase it.

Click on the **Ø.875** dimension in the front view [Fig. 18.15(e)] ⇒ **RMB** ⇒ **Move Item to View** ⇒ click inside the top view [Fig. 18.15(f)] ⇒ click on the **Ø.5625** dimension in the front view [Fig. 18.15(g)] ⇒ **RMB** ⇒ **Move Item to View** ⇒ click inside the top view [Fig. 18.15(h)]

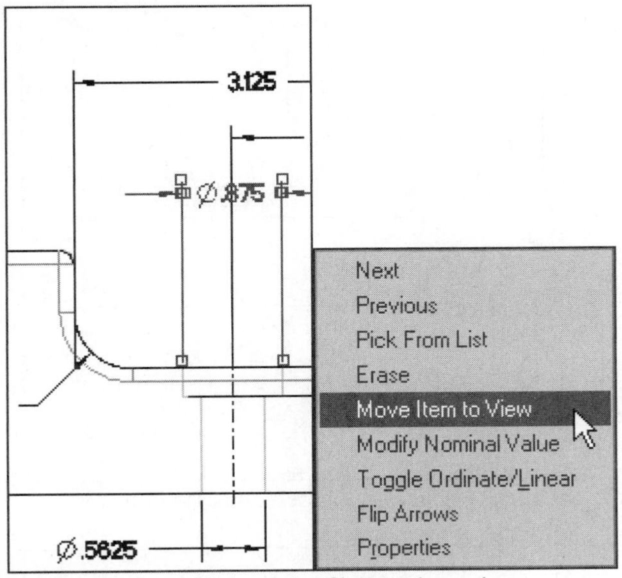

Figure 18.15(e) Click on the **Ø.875** Dimension

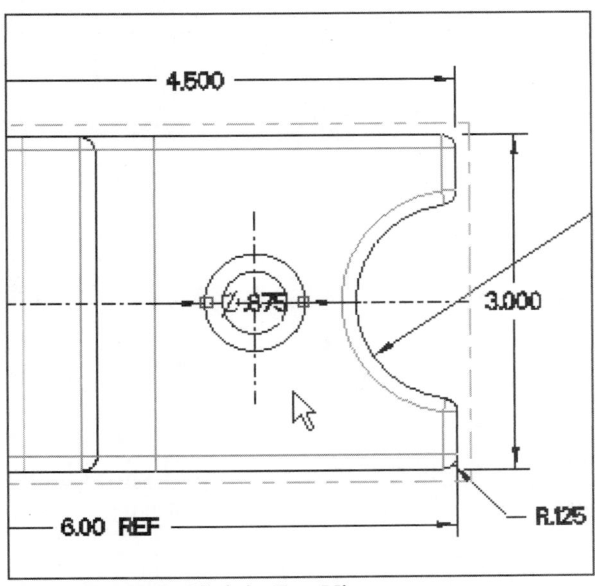

Figure 18.15(f) Click in Top View

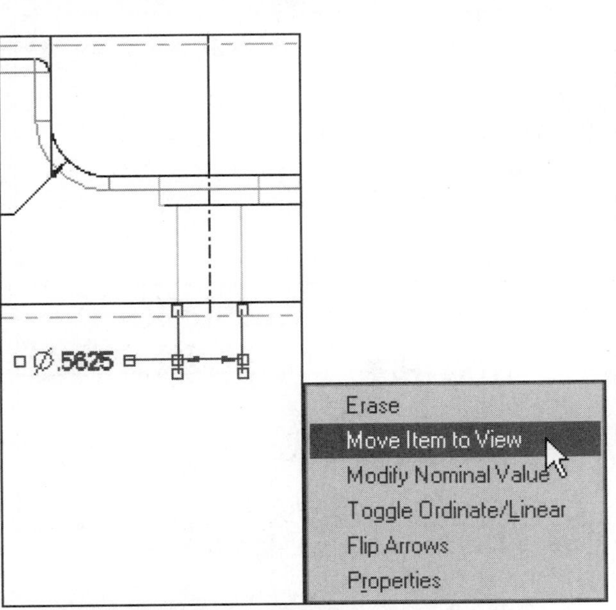

Figure 18.15(g) Click on the **Ø.5625** Dimension

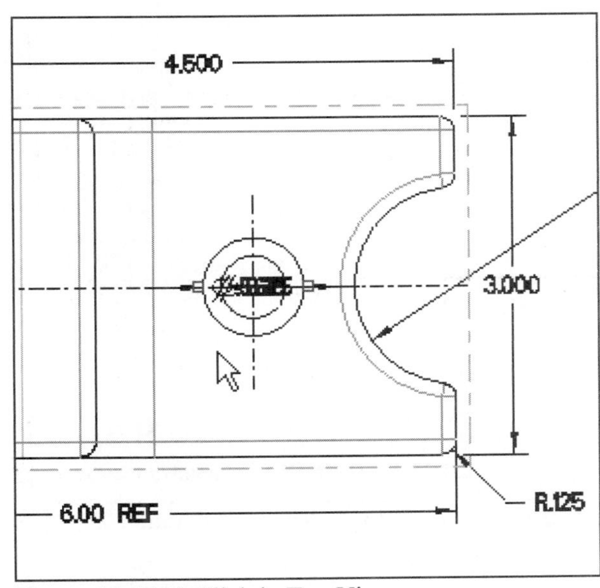

Figure 18.15(h) Click in Top View

The two diameter dimensions display in the same spot, over the top of each other. You must move them to separate locations off the face of the object.

Click on **∅.875** in the top view [Fig. 18.15(i)] ⇒ **LMB** *dimension changes from a dimension to a note outline* ⇒ drag and drop ⇒ click on the left handle of the **∅.875** note [Fig. 18.15(j)] ⇒ drag the note to the right side [Fig. 18.15(k)] ⇒ **LMB** ⇒ repeat the process to reposition the **∅.5625** dimension [Fig. 18.15(l)]

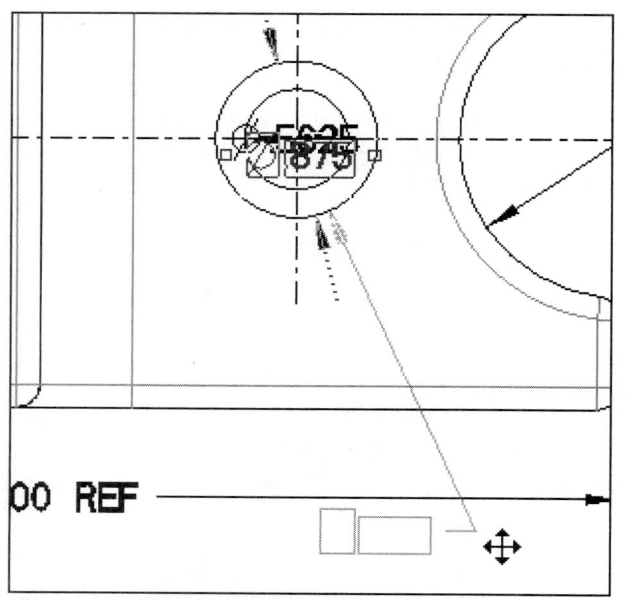

Figure 18.15(i) Click on the **∅.875** Dimension

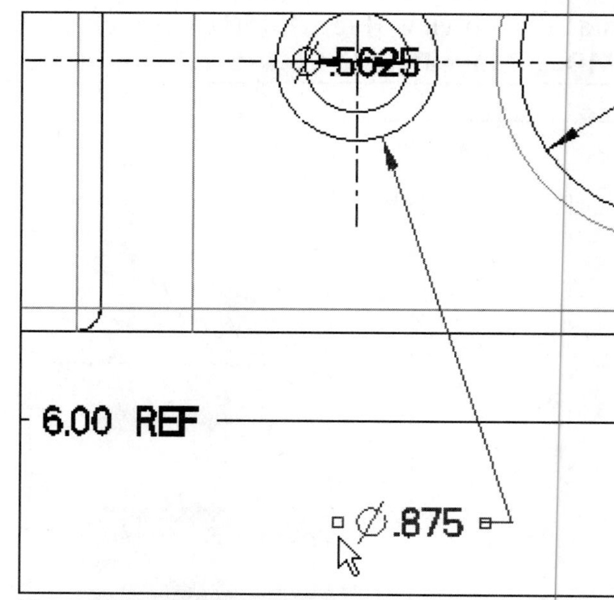

Figure 18.15(j) Click on the Left Drag Handle

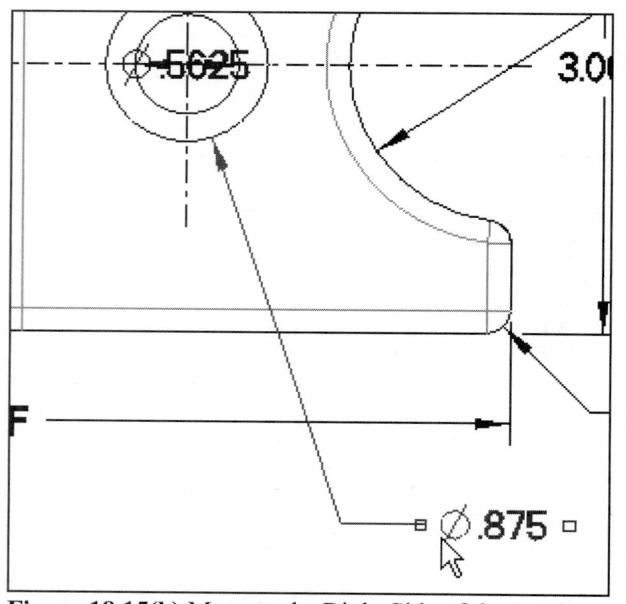

Figure 18.15(k) Move to the Right Side of the Leader

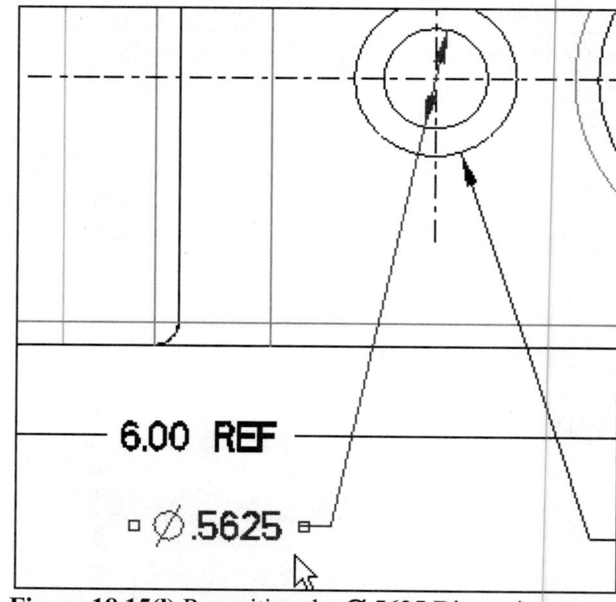

Figure 18.15(l) Reposition the **∅.5625** Dimension

Move the **1.75** dimension to the top view using **RMB** options [Fig. 18.15(m)] ⇒ reposition and flip the arrows on the **R.50** dimension [Fig. 18.15(n)] ⇒ move the **3.00** dimension from the top view to the right side view [Figs. 18.15(o-p)]

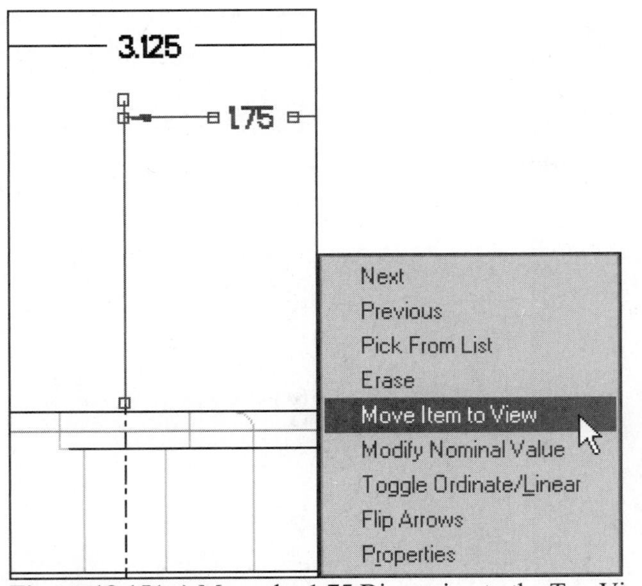

Figure 18.15(m) Move the **1.75** Dimension to the Top View

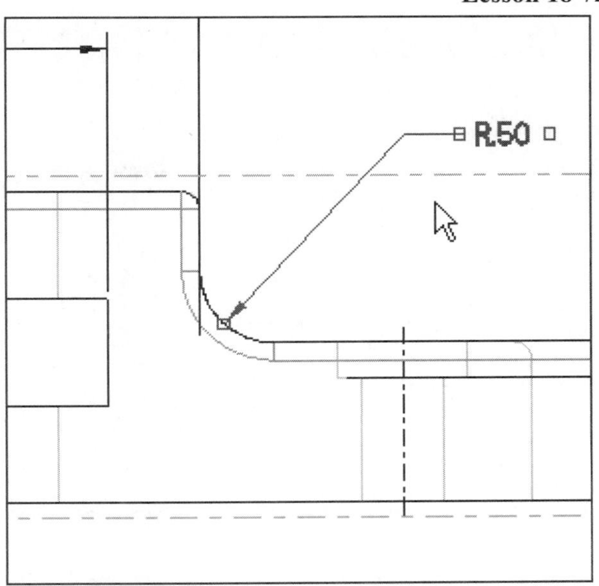

Figure 18.15(n) Reposition and Flip Arrows

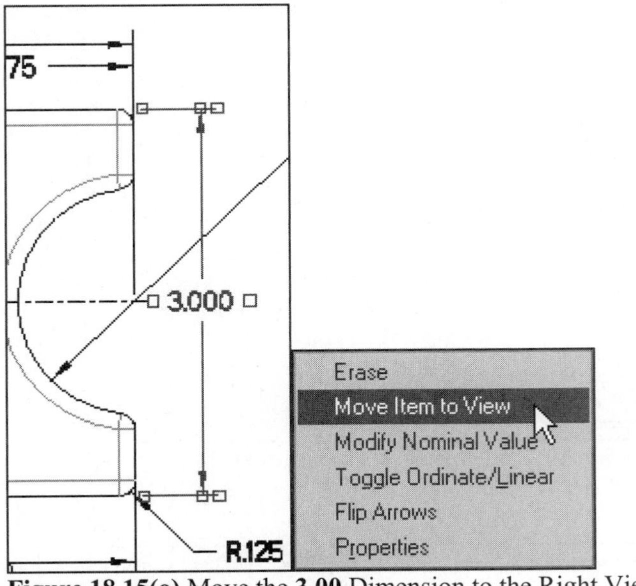

Figure 18.15(o) Move the **3.00** Dimension to the Right View

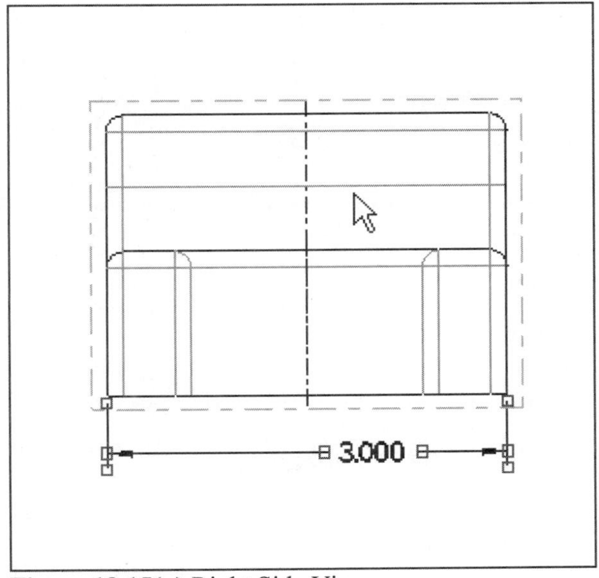

Figure 18.15(p) Right Side View

Another aspect to change on the drawing is the number of digits displayed for individual dimensions. In most cases, the number of digits is dependent on the tolerance for the dimension. Change the number of digits for the dimension (**.75**) displayed with two decimal places to three (**.750**). *In order to illustrate alternative techniques, you will be changing to four decimal places using one method and then use another method to modify to the required three decimal places.*

Click on the **.75** dimension ⇒ **RMB** ⇒ **Properties** ⇒ Number of decimal places-- type **4** [Fig.18.16(a)] ⇒ **OK** ⇒ **LMB** ⇒ double-click on **.7500** ⇒ Number of decimal places-- type **3** ⇒ **OK** ⇒ **LMB** ⇒ [Fig. 18.16(b)] ⇒ change the **Ø.81** to **Ø.810** ⇒ [save icon] ⇒ **MMB**

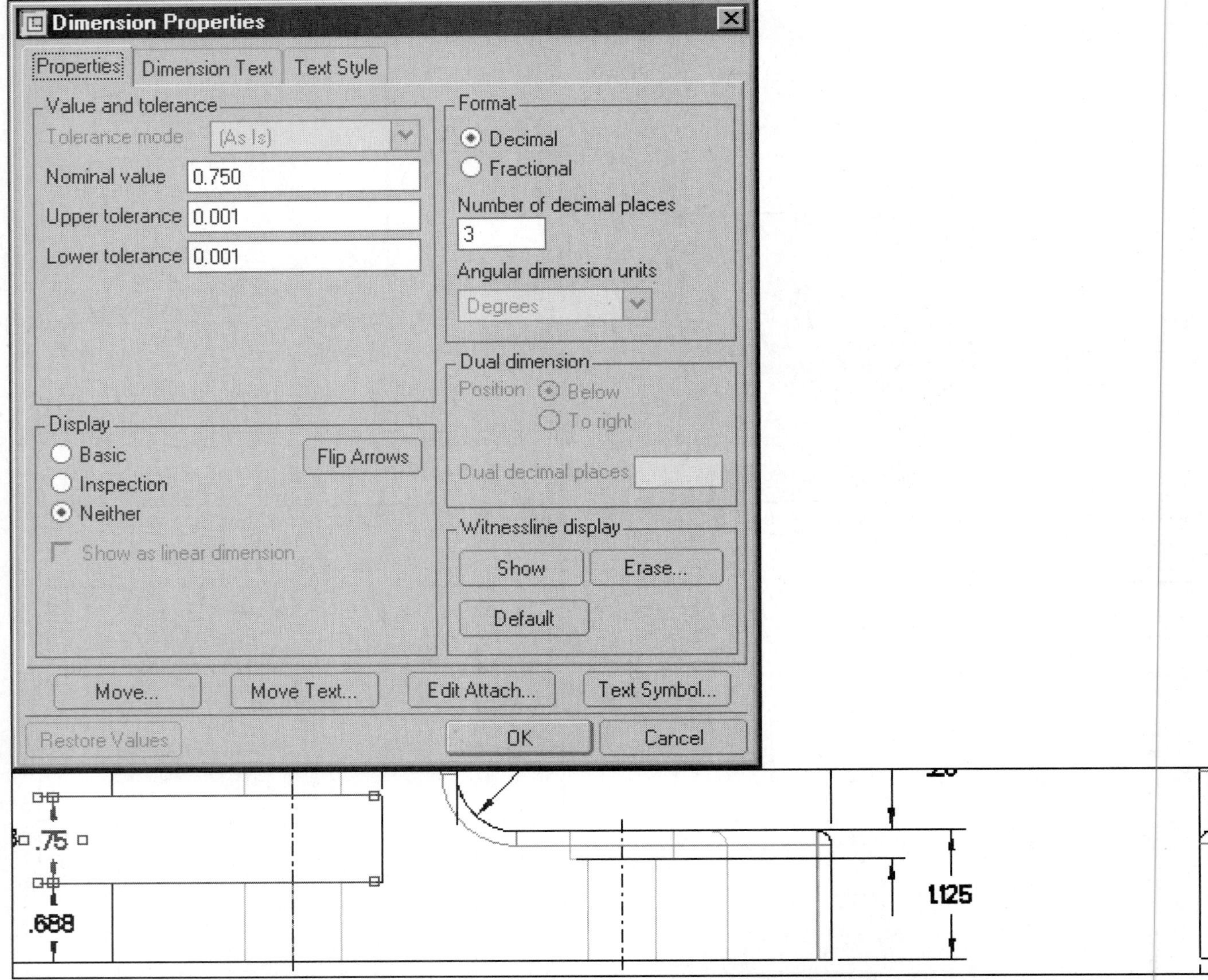

Figure 18.16(a) Change Dimension **.75** Number of Decimal Places to **3** using the Dimension Properties Dialog Box

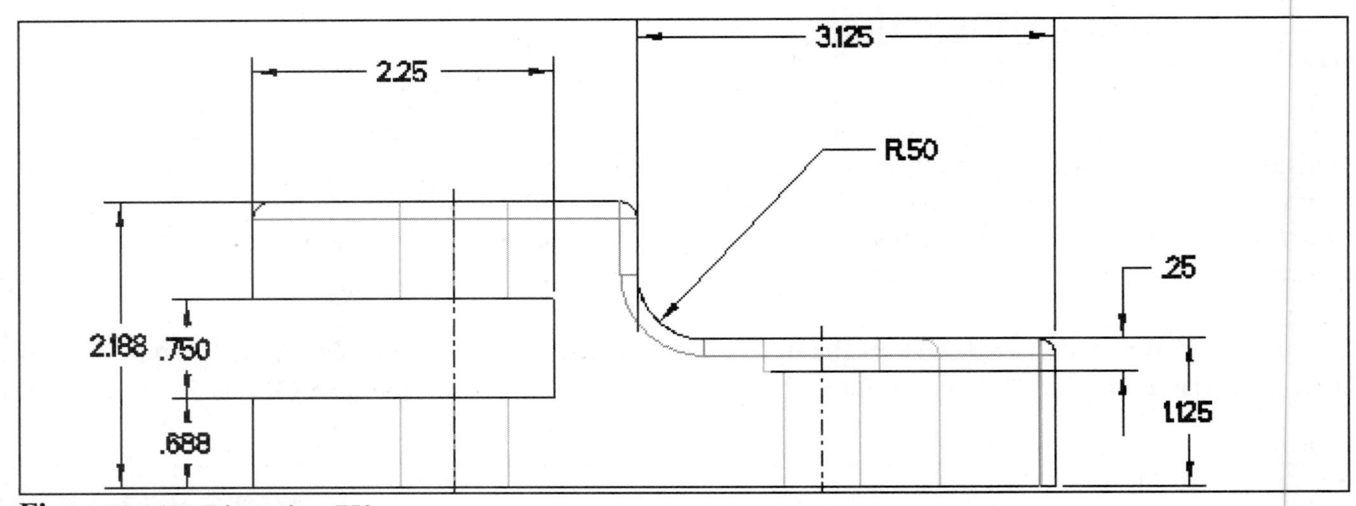

Figure 18.16(b) Dimension **.750**

Clipping Drafting Objects

You can dynamically move (clip) the endpoints of many drafting object elements, such as extension lines and centerlines. This allows you to "pick up" the end of an extension line and drag it to a new length. The most common application of this option involves "re-gaping" extension lines. (You have already experienced this capability when you edited the length of the centerpoint and linear axes.)

When you Show feature dimensions, the extension lines are often gapped to a point on the part that is not appropriate for the view in which the dimension is displayed. Clipping allows you to drag the extension line end to a location that is visually more pleasing (and consistent with drafting standards). When you clip the extension line, Pro/E remembers the clip distance and continues to apply it even if the model changes or if you Erase and redisplay the dimension. The grid snap status will affect the positioning of the dimensions.

Click: **Zoom in** on the front view ⇒ click on the **3.125** dimension [Fig. 18.17(a)] ⇒ hold down the **LMB** on the extension line drag handle [Fig. 18.17(b)] ⇒ move the mouse to adjust "clip" the extension line beyond the parts outline [Fig. 18.17(c)] ⇒ release the **LMB** ⇒ **LMB** to deselect

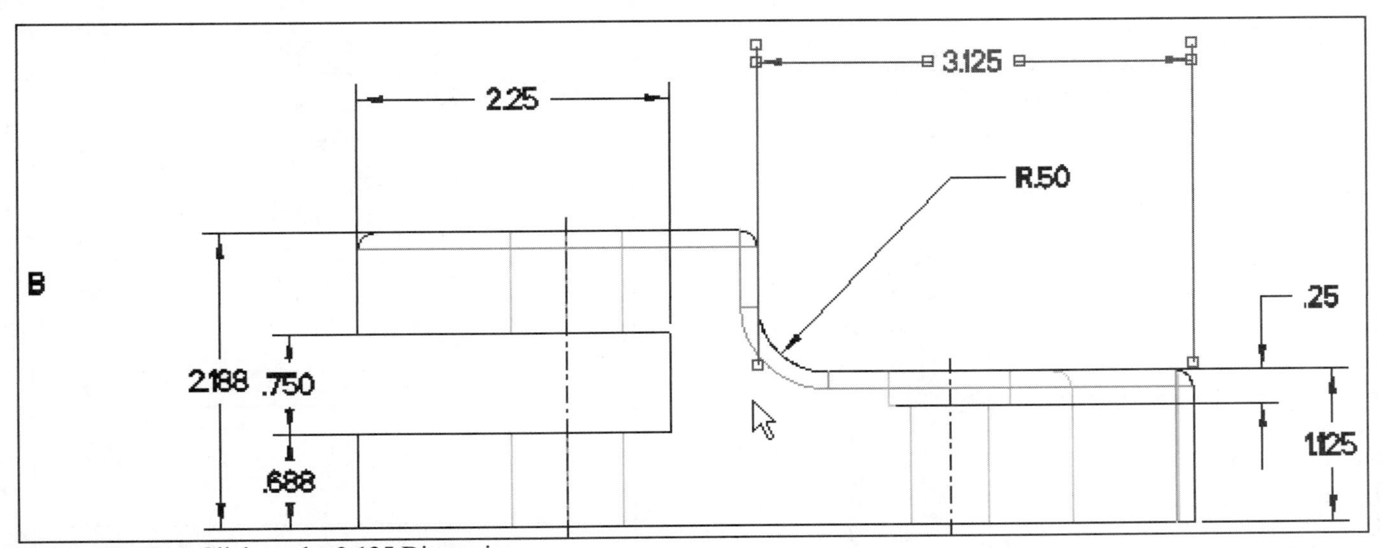

Figure 18.17(a) Click on the **3.125** Dimension

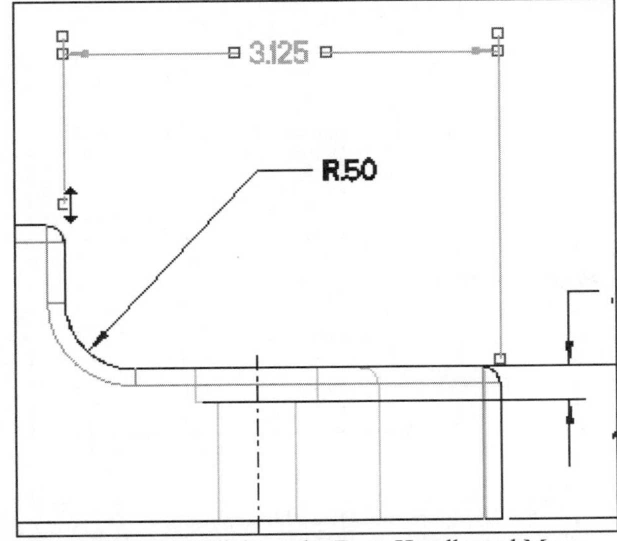

Figure 18.17(b) Click on the Drag Handle and Move

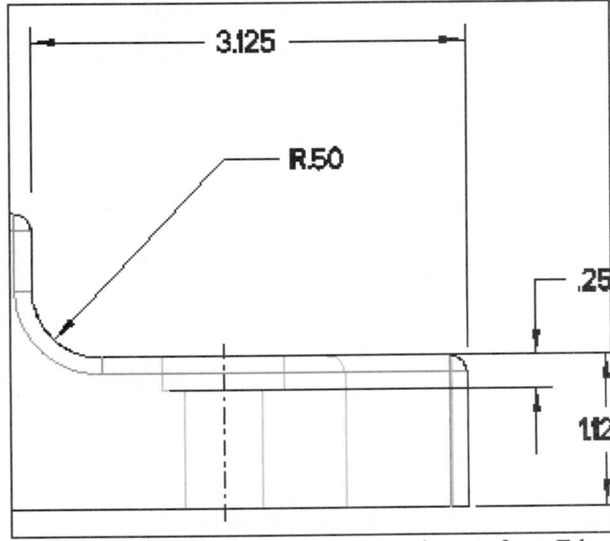

Figure 18.17(c) Extension Line moved away from Edge

Press and hold the **Ctrl** key and click on the **.25** and the **1.125** dimensions [Fig. 18.17(d)] ⇒ hold down **LMB** on the drag handle ⇒ clip the extension line beyond the counterbore edge [Fig. 18.17(e)] ⇒ click on the drag handle [Fig. 18.17(f)] ⇒ clip the extension line beyond the top edge of the part [Fig. 18.17(g)] ⇒ **LMB** ⇒ clip the extension lines for the **2.188**, **.750**, and the **.688** dimensions [Fig. 18.17(h-i)] ⇒ **LMB** ⇒ move and clip the dimensions and extension lines for the top view [Fig. 18.17(j)] ⇒ **LMB**

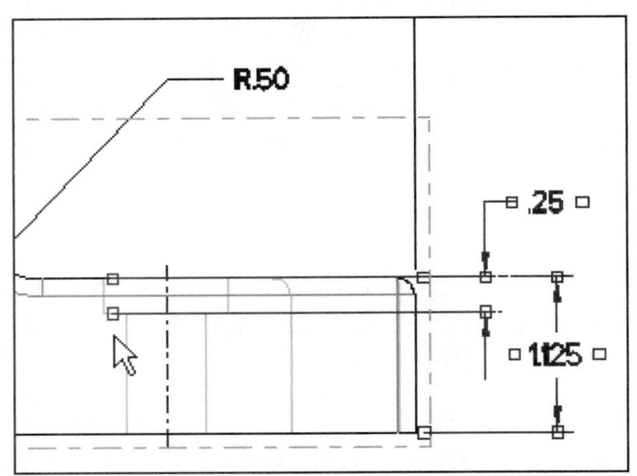

Figure 18.17(d) Click and Slide the Drag Handle

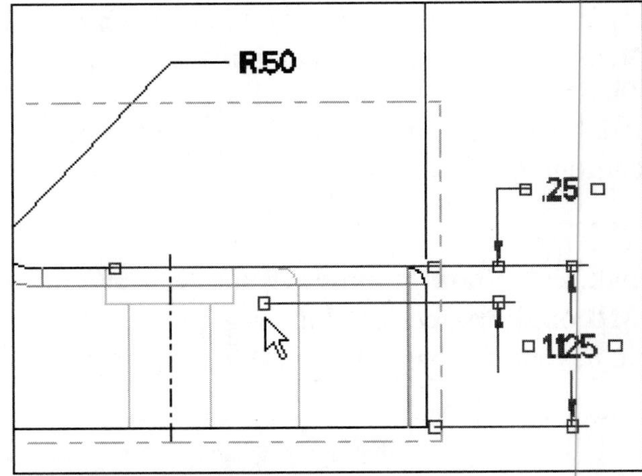

Figure 18.17(e) Move

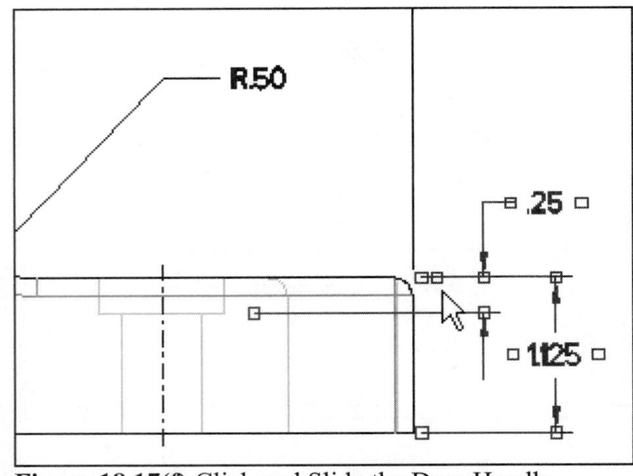

Figure 18.17(f) Click and Slide the Drag Handle

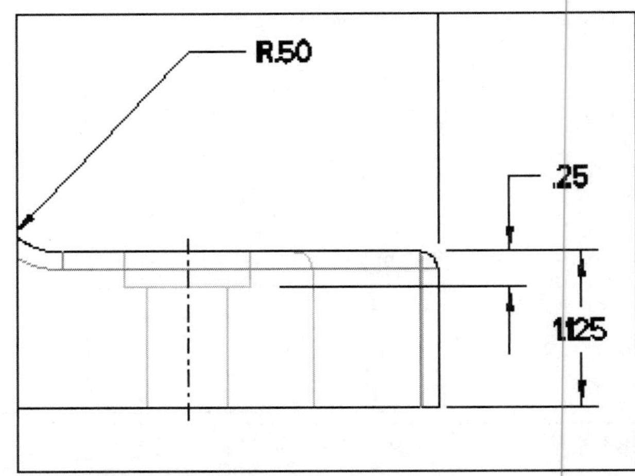

Figure 18.17(g) Move Extension Line

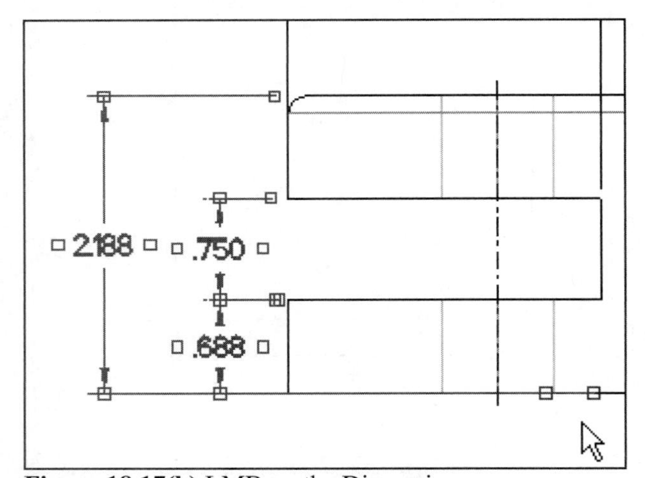

Figure 18.17(h) LMB on the Dimensions

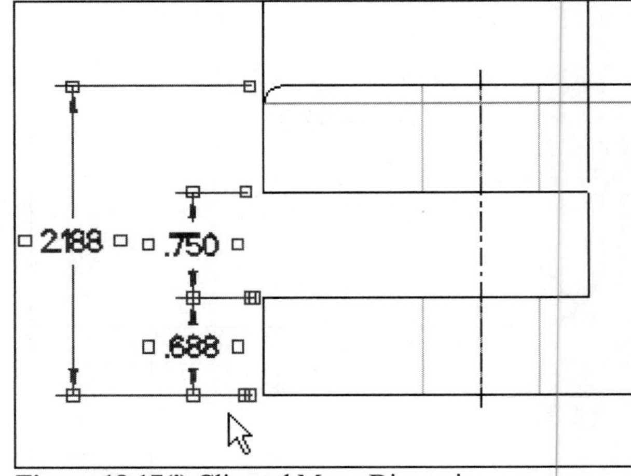

Figure 18.17(i) Clip and Move Dimensions

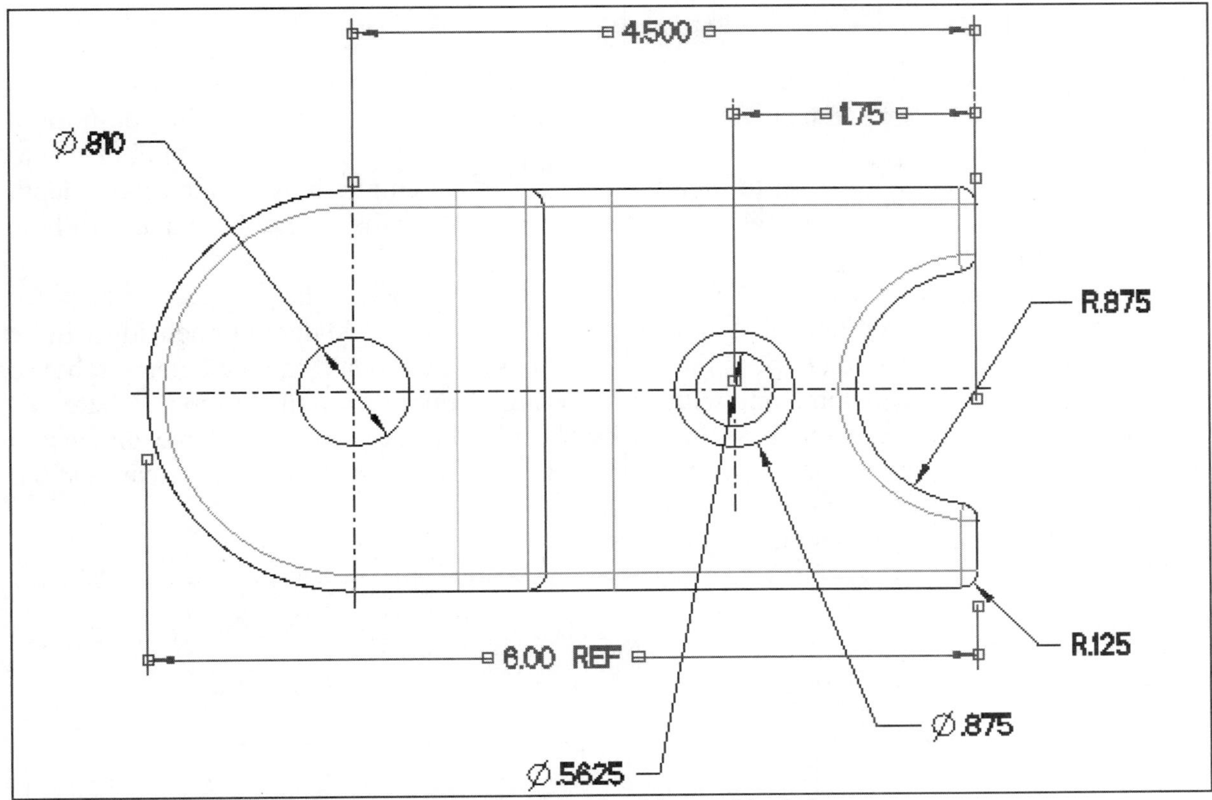

Figure 18.17(j) Move Dimensions and Clip the Extension Lines as Needed

Click on the **6.00 REF** dimension [Fig. 18.18(a)] ⇒ **RMB** ⇒ **Properties** ⇒ **Dimension Text** tab [Fig. 18.18(b)] ⇒ remove the **REF** and add *parentheses*- **(@D)** [Fig. 18.18(c)] ⇒ **OK** ⇒ [save icon] ⇒ **MMB**

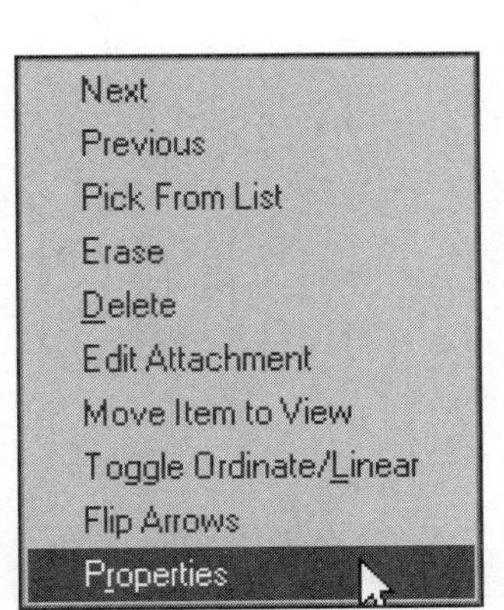

Figure 18.18(a) Properties

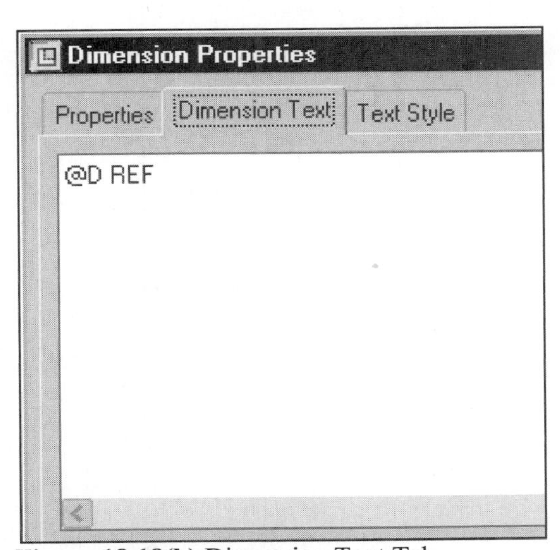

Figure 18.18(b) Dimension Text Tab

Dimension Properties

Properties | Dimension Text | Text Style

(@D)

Figure 18.18(c) Edit Text to (**@D**)

Notes

There are times when simply showing the dimensions is not enough to annotate a part completely. You can add to a drawing additional dimensions, labels, and notes that were not a part of its original definition. As an example, the counterbore depth, the thru hole, and the counterbore hole need to be combined into one note using the thru hole as the dimension to modify. The individual counterbore diameter and depth dimensions are then erased from the drawing. The diameter dimensions for the counterbore and thru hole were already switched to the top view.

The counterbore was created with a Revolved Cut instead of using the Hole command for demonstrating this process. The **.5625** diameter thru hole dimension and the **.250** depth dimension will be combined with the **.875** counterbore diameter into one note. You will need to add the diameter symbol to the value as a prefix to the **.5625** dimension along with a counterbore symbol from the Symbol Palette.

To make the note parametric, enter **&** followed by the dim symbols instead of the dimension values (e.g., **d32** and **d33**); the dimensions are then parameters that will reflect any modifications and changes to your model. Your **d** symbols will probably be different:
[**Ø.875 (&d32)**, **Ø.5625 (&d33)**, and the depth **.250 (&d35)**].

To see the dim symbols in Drawing mode [Fig. 18.19(a)], click **Info** ⇒ Switch Dimensions [Fig. 18.19(b-c)] ⇒ **Info** ⇒ Switch Dimensions ⇒ click on **Ø.875** ⇒ **RMB** [Fig. 18.19(d)] ⇒ **Properties**

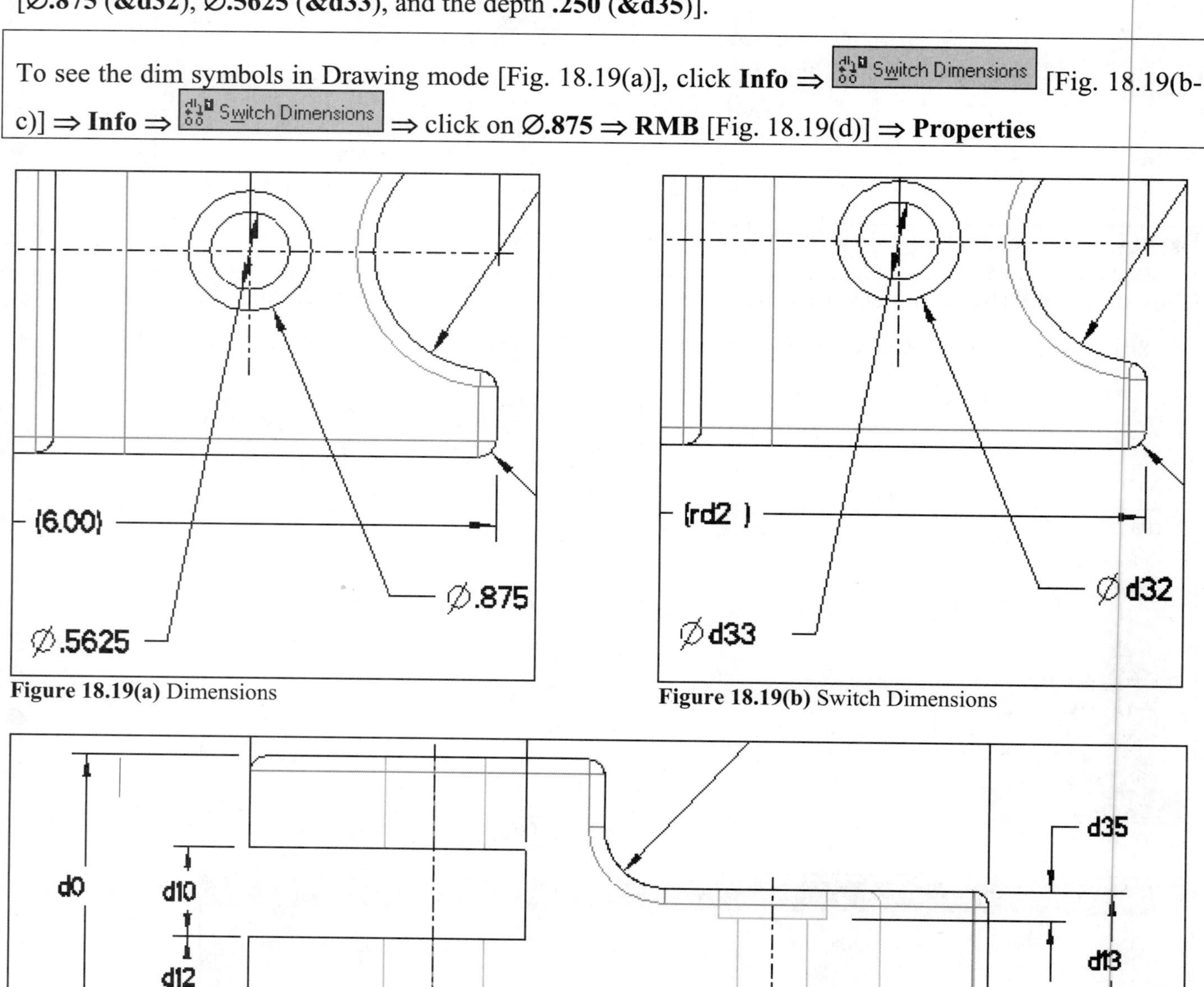

Figure 18.19(a) Dimensions

Figure 18.19(b) Switch Dimensions

Figure 18.19(c) Switch Dimensions Front View

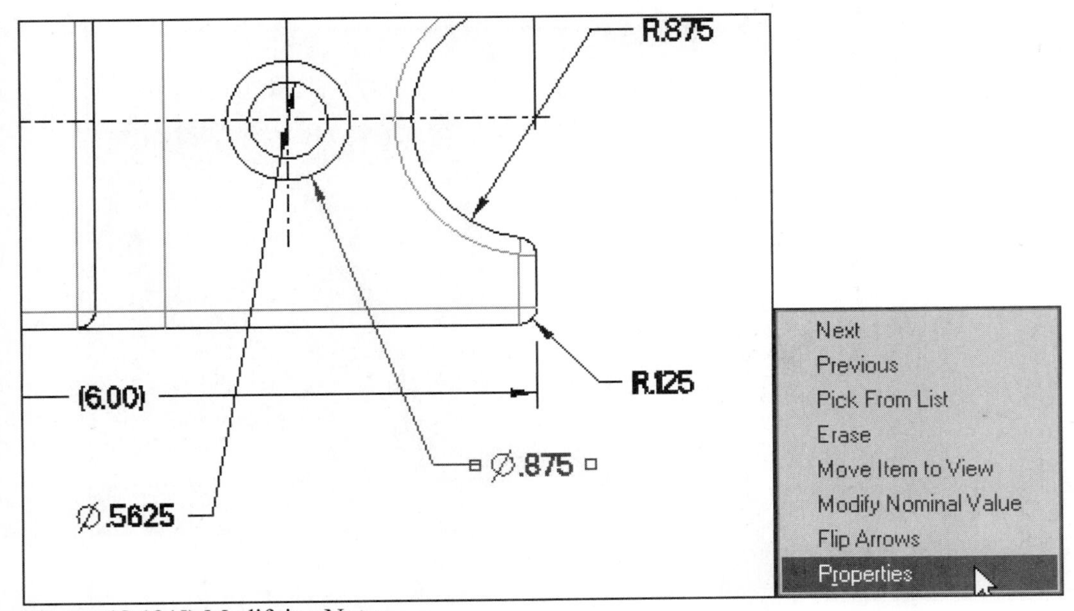

Figure 18.19(d) Modifying Notes

Dimension Properties dialog box displays [Fig. 18.19(e)] ⇒ **Dimension Text** tab [Fig. 18.19(f)] ⇒ edit the note [Fig. 18.19(g)] ⇒ beginning of first line, click **Text Symbol** ⇒ ∅ ⇒ type **&d33** [∅&d33] ⇒ **Enter** ⇒ second line [⌴] [⌴∅@D] ⇒ **Enter** ⇒ third line [↧]**&d35** [↧&d35] ⇒ **OK** ⇒ **LMB**

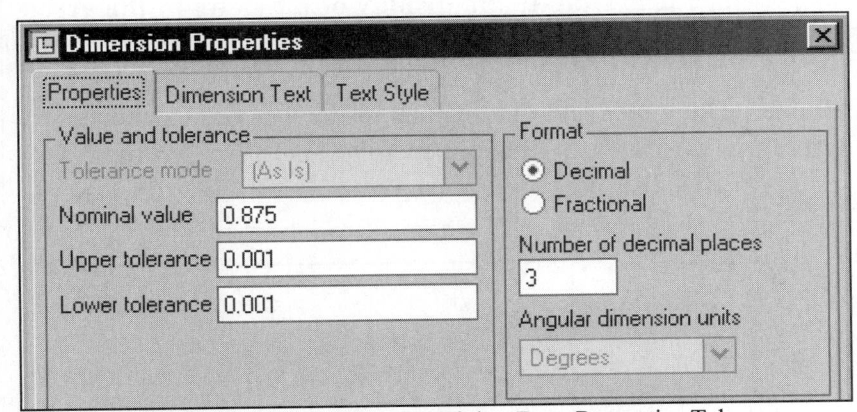

Figure 18.19(e) Dimension Properties Dialog Box- Properties Tab

Dimension Properties

Properties | Dimension Text | Text Style

∅@D

Figure 18.19(f) Dimension Properties Dialog Box- Dimension Text Tab

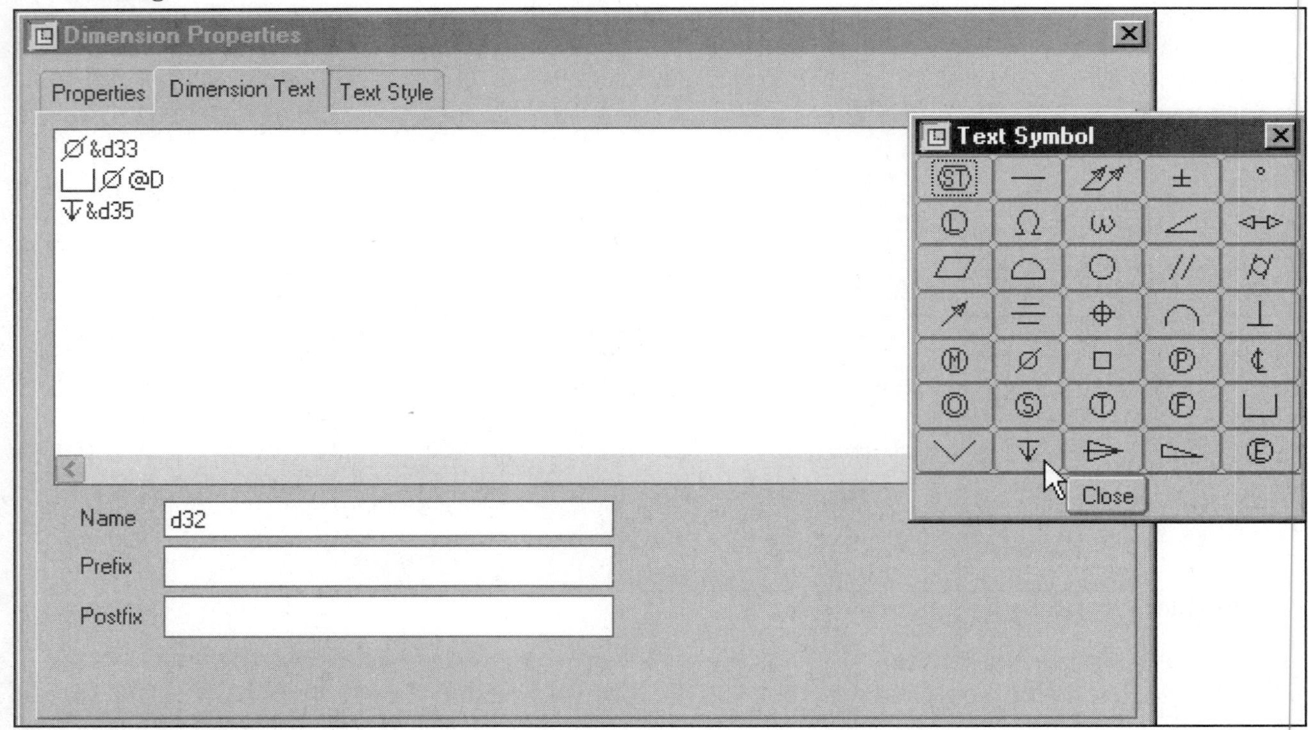

Figure 18.19(g) Edit the Note

Click: **Info ⇒ Switch Dimensions** [Fig. 18.19(h)] ⇒ **Update the display of all views in the active sheet ⇒ Info ⇒ Switch Dimensions ⇒** [Fig. 18.19(i)] ⇒ erase the **Ø.5625** and **.250** dimensions from the drawing ⇒ ⇒ move the positions of the notes and dimensions in the top view as necessary [Fig. 18.20(a)] ⇒ in the front view, move the positions of the dimensions and **Flip Arrows** [Fig. 18.20(b)] ⇒ ⇒ ⇒ ⇒ ⇒ **MMB**

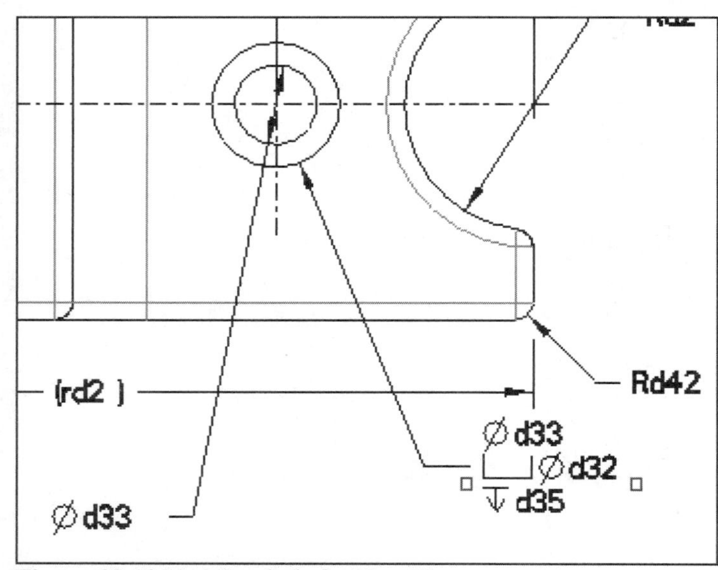

Figure 18.19(h) d Symbols for New Note

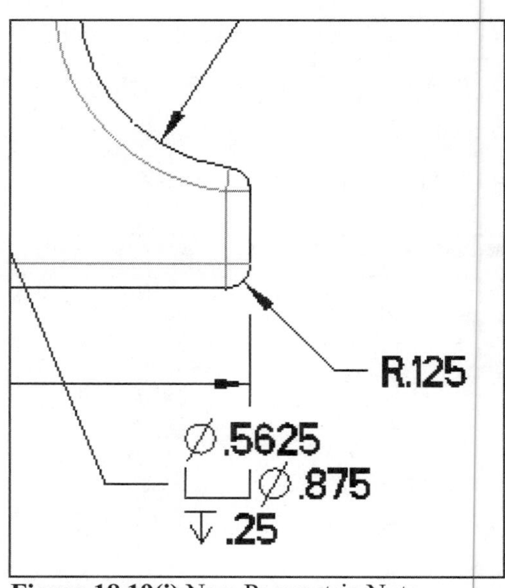

Figure 18.19(i) New Parametric Note

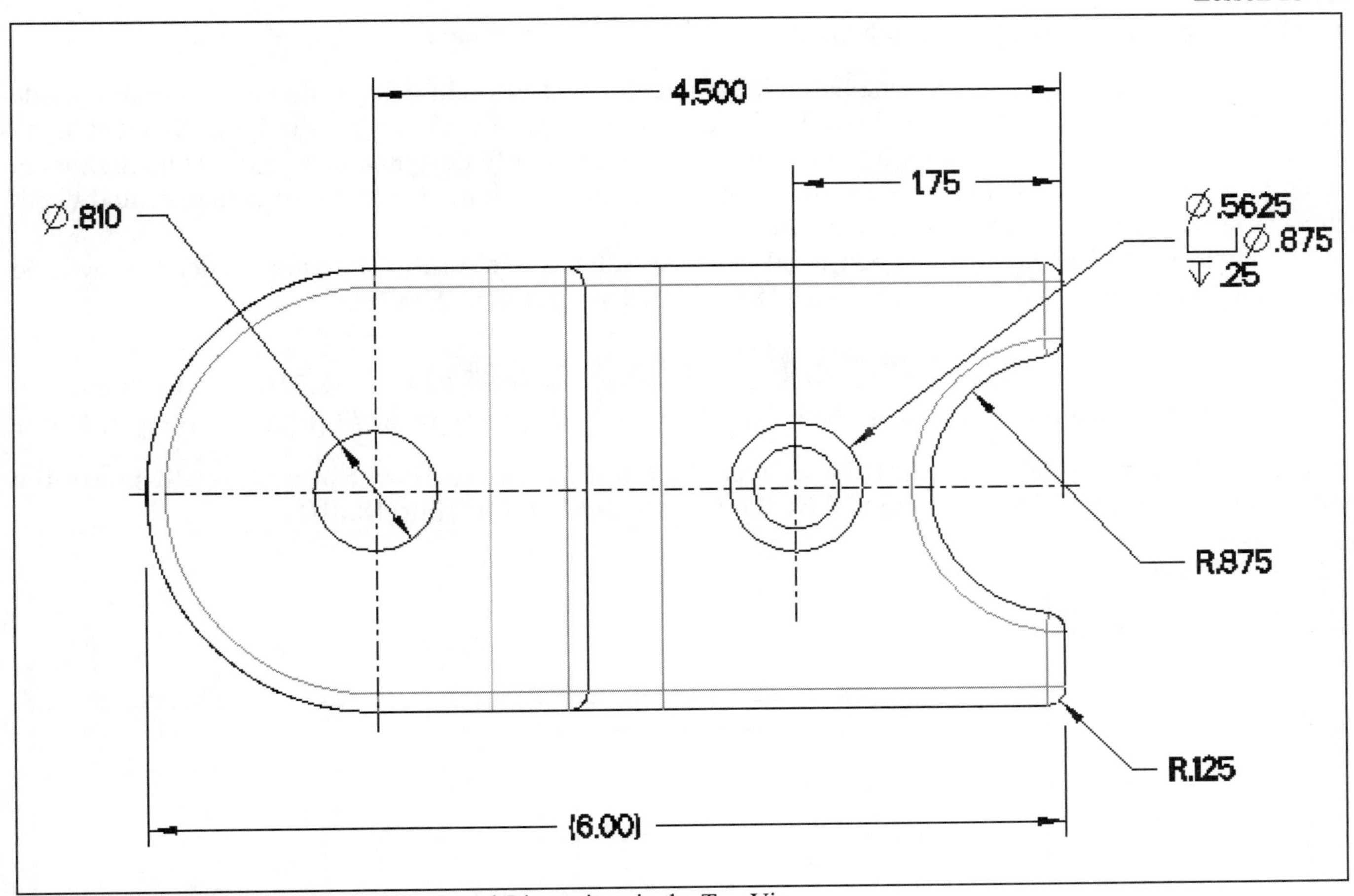

Figure 18.20(a) Adjust Position of Notes and Dimensions in the Top View

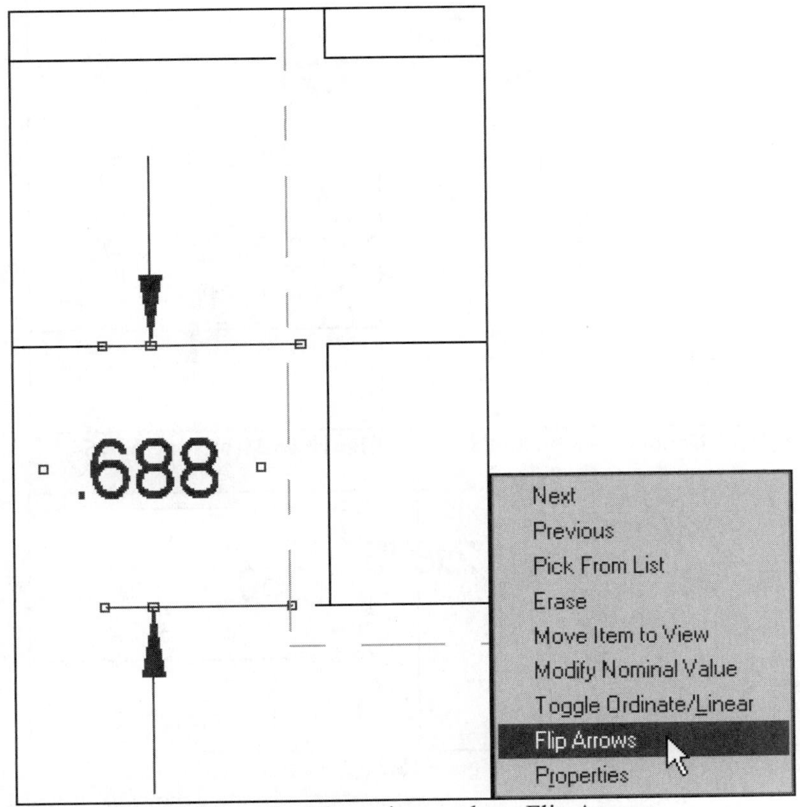

Figure 18.20(b) Move the Dimensions and use Flip Arrows

Modifying the Part in the Drawing Mode

Since the part, assembly, and drawings are all parametric, you can edit design dimensions in any mode and the features size will update throughout. In general, because of design conflicts that could be generated in the part and assembly, it is not good practice to modify design values in the Drawing mode. Since you created a parametric note for the counterbore; the thru hole, counterbore diameter, and depth dimension can be modified directly in the note.

Now we will make small dimensional changes to the model in the Drawing Mode. You will be using Edit to change the design dimensions (**.688** to **.6875** and **2.188** to **2.1875**).

Click in Selection Filter Drawing Item and View ⇒ Note and Dimension [Fig. 18.21(a)] ⇒ click once on **.688** to highlight, then double-click on **.688** again [Fig. 18.21(b)] ⇒ type **.6875** [Fig. 18.21(c)] ⇒ **Enter** ⇒ double-click on **2.188** [Fig. 18.21(d)] ⇒ type **2.1875** [Fig. 18.21(e)] ⇒ **Enter** ⇒ **Update the model and redraw all views as specified by the latest modifications** [Fig. 18.21(f)]

Figure 18.21(a) Set Filter Selection

Figure 18.21(b) Double-click on **0.688**

Figure 18.21(c) Type **.6875**

Figure 18.21(d) Double-click on **2.188**

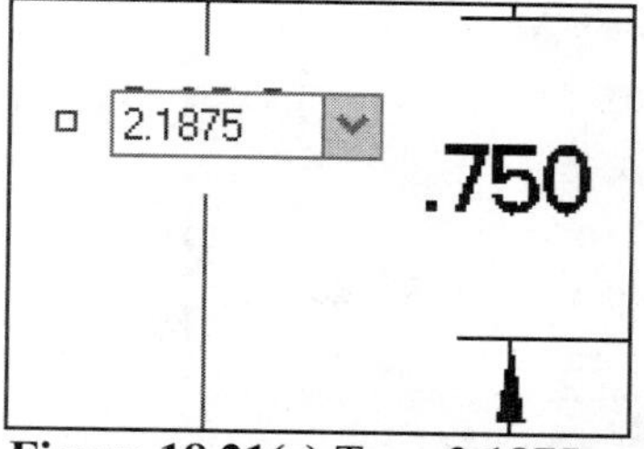

Figure 18.21(e) Type **2.1875**

Figure 18.21(f) Regenerated Dimensions

URL's on Notes and Dimensions

A URL was added to the Company Name in the title block of the format created in the last lesson and used on the drawing for the breaker in this lesson. If you try to launch the site from the drawing, it will not activate the URL. Therefore, we will create another URL in the drawing. You must have an existing note to attach the URL. To activate a Hyperlink, press and hold down the Ctrl button and click on the note. The embedded browser in Pro/E will go to the desired URL.

Click: **Create a note** ⇒ **No leader** ⇒ **Make Note** ⇒ pick a position for the note above the Symbol box near the title block [Fig. 18.22(a)] ⇒ type **MATERIAL** ⇒ **Enter** ⇒ **Enter** [Fig. 18.21(b)] ⇒ **Done/Return** ⇒ **Add, edit or remove hyperlink to the selected text** ⇒ Type the URL or internal link **www.matweb.com** [Fig. 18.22(c)] ⇒ **ScreenTip** ⇒ Screen tip text **MatWeb Property Data** [Fig. 18.21(d)] ⇒ **OK** ⇒ **OK** ⇒ place curser over **MATERIAL** [Figs. 18.22(e-f)] *(optional- Ctrl ⇒ LMB to launch website)* ⇒ ⇒ **MMB**

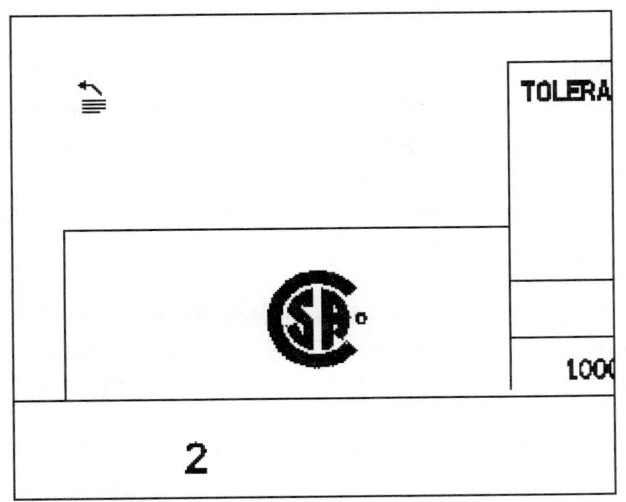

Figure 18.22(a) Locate Note

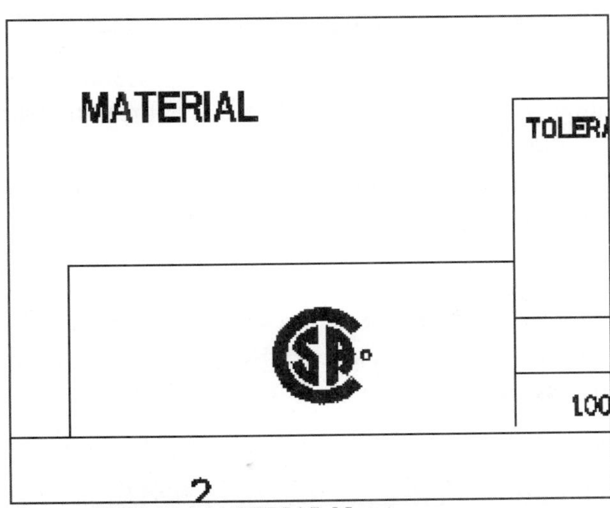

Figure 18.22(b) MATERIAL Note

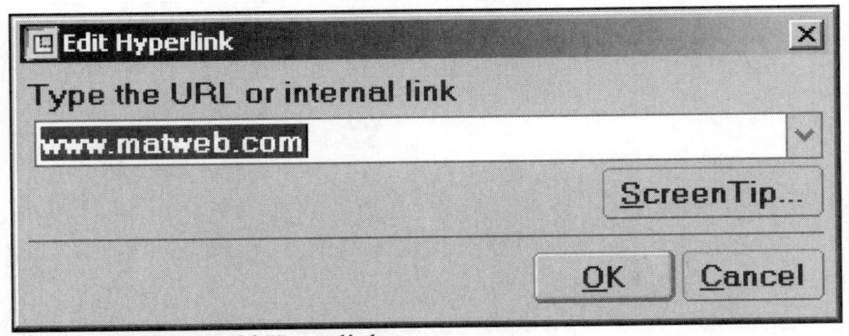

Figure 18.22(c) Add Hyperlink

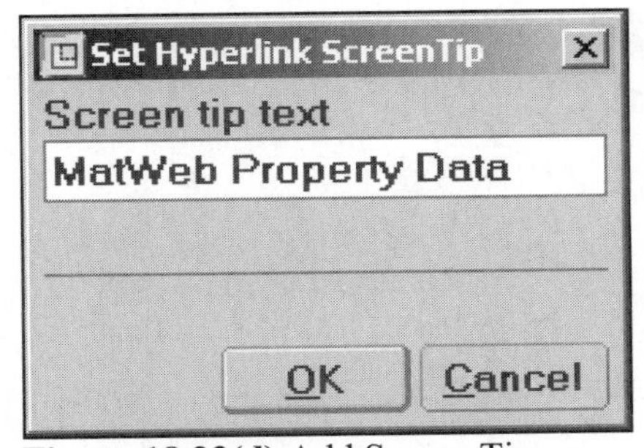

Figure 18.22(d) Add Screen Tip

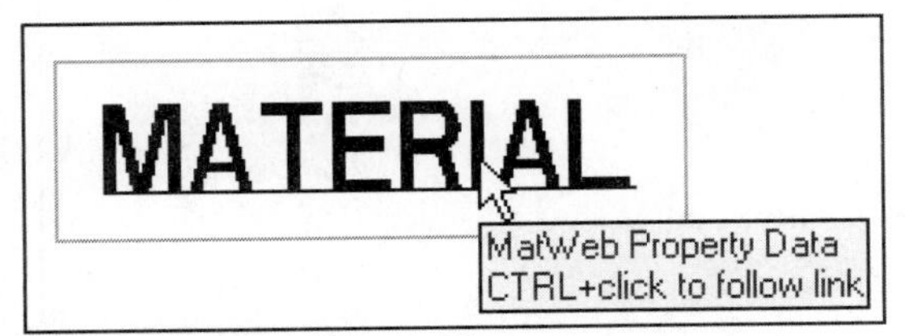

Figure 18.22(e) Screen Tip

MatWeb Property Data CTRL+click to follow link

Figure 18.22(f) One Line Help in Message

Geometric Tolerancing

Pro/DETAIL enables you to add geometric tolerances to the model from Drawing mode. Note that geometric tolerances can be added in Part, Assembly or Drawing mode but are reflected in all other modes.

Geometric tolerances are treated by Pro/E as *annotations*, and they are always associated with the model. Unlike dimensional tolerances, *geometric tolerances do not have any effect on part geometry*. We will add a few simple geometric tolerances to the model so you can experience this process. For in depth coverage of geometric tolerancing using Pro/E Wildfire, consult the Help Center.

To access the geometric tolerance functionality from the Part mode, click Edit ⇒ Setup ⇒ Geom Tol. The menu displays the following options:

- **Set Datum** Sets a datum to be used as a reference for geometric tolerances
- **Basic Dim** Creates basic dimensions
- **Inspect Dim** Creates inspection dimensions
- **Specify Tol** Creates a geometric tolerance
- **Clear** Deletes a tolerance
- **Make Target** Sets a datum target

When you add a geometric tolerance (**gtol**) to a drawing, you may choose the model in which to place the gtol, or instead create a nonparametric graphical gtol within the drawing itself. However, before you can add a gtol to a drawing, you must have set datums. It is easier to set datums in the Part mode, so the first step will be to open the drawings' part. If your part is already open, you merely activate its window. Regardless, if your part is open or not, it is *in session* because the drawing is referencing it.

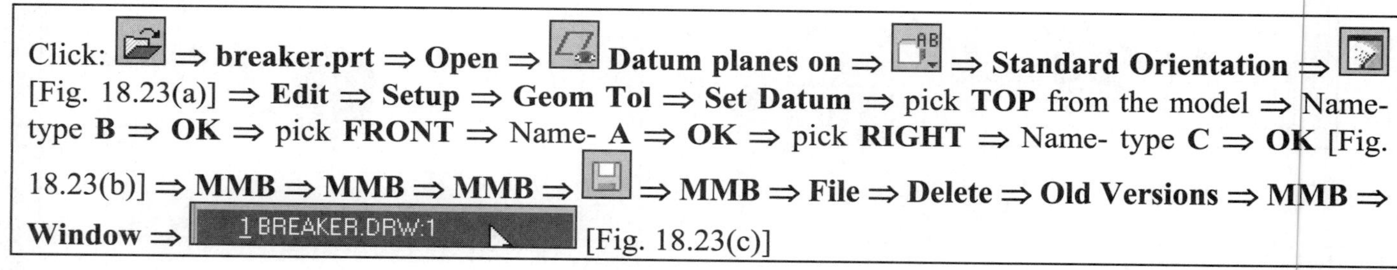

Click: ⇒ **breaker.prt** ⇒ **Open** ⇒ **Datum planes on** ⇒ ⇒ **Standard Orientation** ⇒ [Fig. 18.23(a)] ⇒ **Edit** ⇒ **Setup** ⇒ **Geom Tol** ⇒ **Set Datum** ⇒ pick **TOP** from the model ⇒ Name- type **B** ⇒ **OK** ⇒ pick **FRONT** ⇒ Name- **A** ⇒ **OK** ⇒ pick **RIGHT** ⇒ Name- type **C** ⇒ **OK** [Fig. 18.23(b)] ⇒ **MMB** ⇒ **MMB** ⇒ **MMB** ⇒ ⇒ **MMB** ⇒ **File** ⇒ **Delete** ⇒ **Old Versions** ⇒ **MMB** ⇒ **Window** ⇒ 1 BREAKER.DRW:1 [Fig. 18.23(c)]

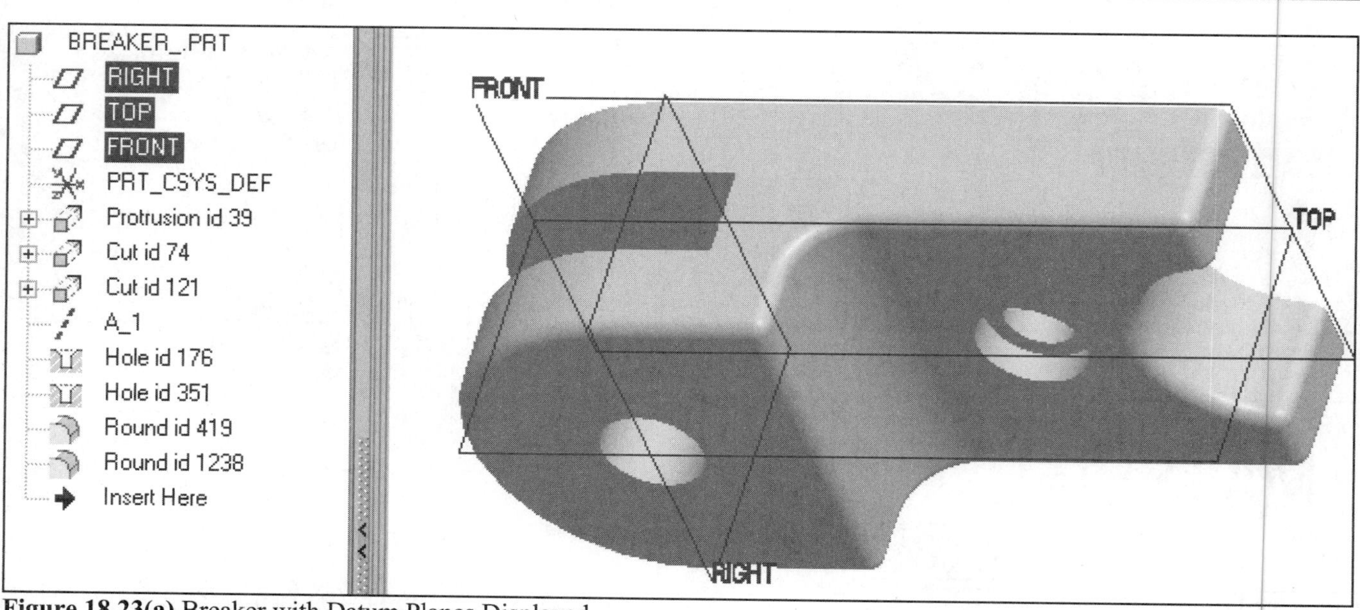

Figure 18.23(a) Breaker with Datum Planes Displayed

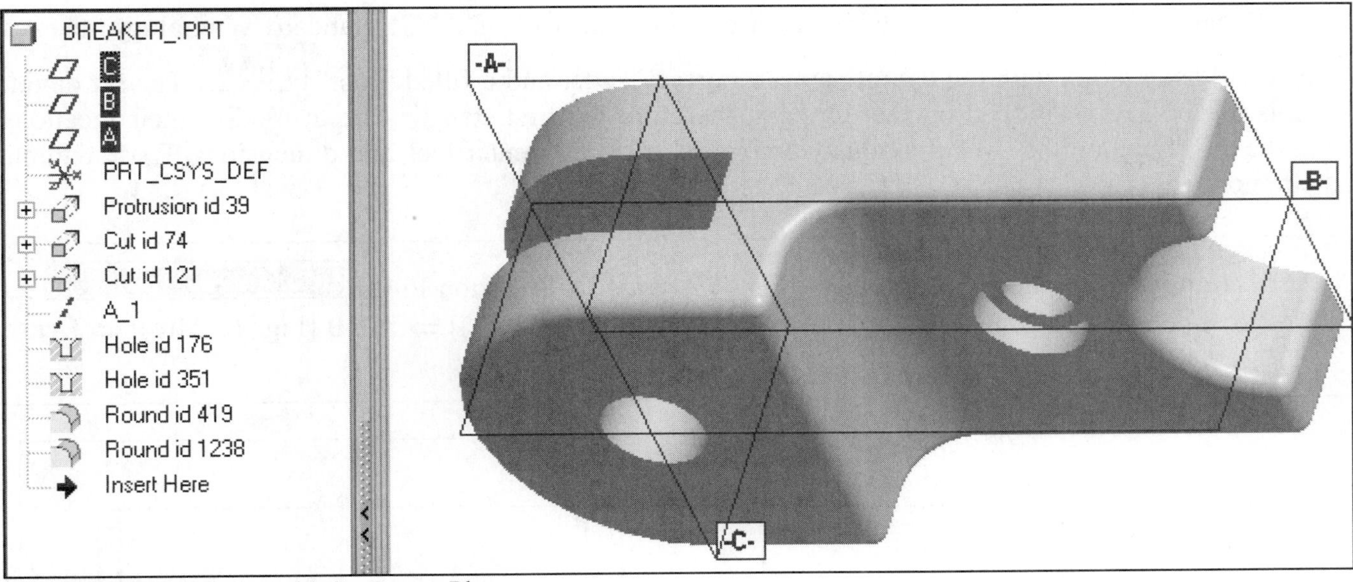

Figure 18.23(b) Breaker with Set Datum Planes

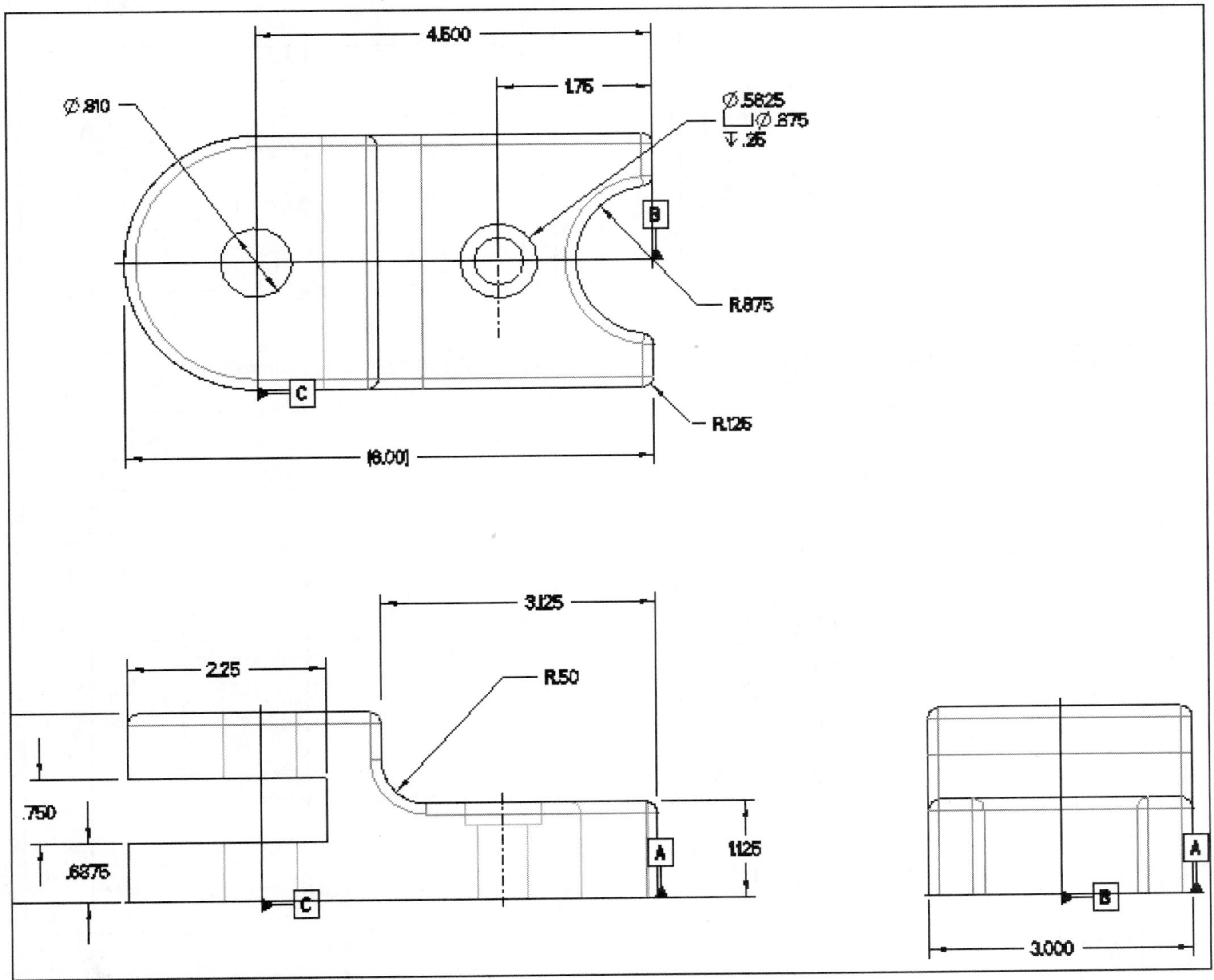

Figure 18.23(c) Breaker Drawing Displaying Geom Tol Set Datums

Because a *.dtl* option was altered to use ASME instead of ANSI standard symbology, the set datums will have a box with the datum name, a normal leader and a filled "foot" [icon]. The set datum symbols are poorly positioned on the drawing, therefore the first step will be to modify their locations using the clip function that was previously employed to adjust centerlines and dimensions. Erase datums that are not needed.

Click: [Zoom icon] **Zoom in** on the right side view [Fig. 18.24(a)] ⇒ Selection Filter [Datum Plane] ⇒ ⇒ Press and hold **Ctrl** key and click on the two datums [Fig. 18.24(b)] ⇒ **RMB** [Fig. 18.24(c)] ⇒ **Erase** ⇒ **LMB** [Fig. 18.24(d)] ⇒ [Save icon] ⇒ **MMB**

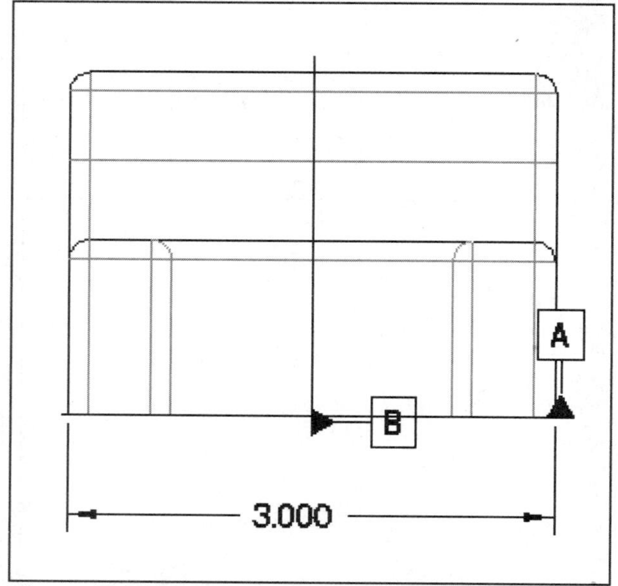

Figure 18.24(a) Zoom in on Right Side View

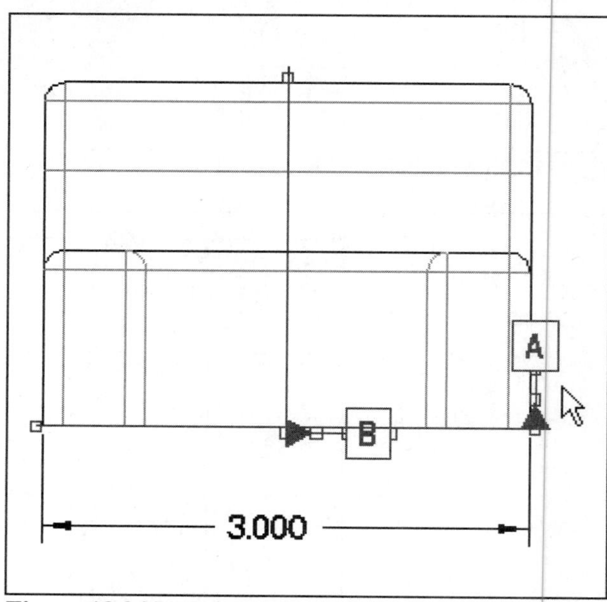

Figure 18.24(b) Selected Datums

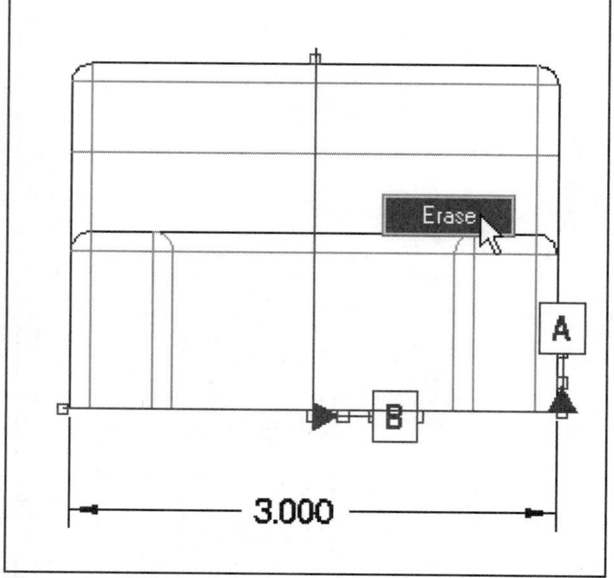

Figure 18.24(c) Erase

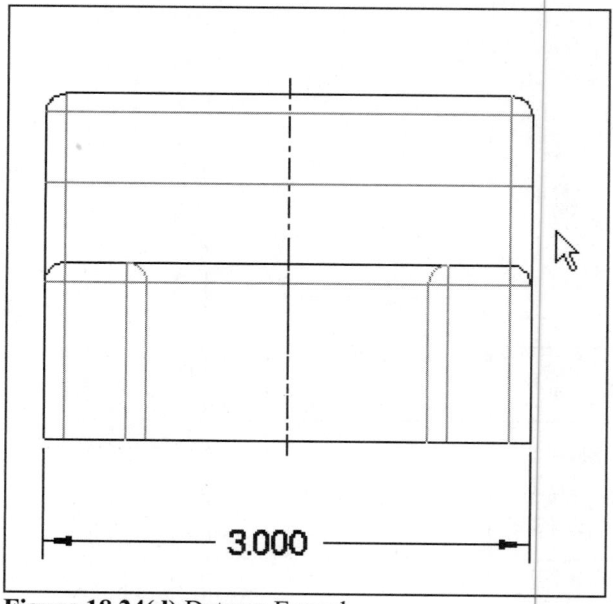

Figure 18.24(d) Datums Erased

Click: **Zoom in** on the front view ⇒ click on the datum **C** ⇒ **RMB** [Fig. 18.24(e)] ⇒ **Erase** ⇒ **LMB** ⇒ ⇒ click on the datum **A** in the front view [Fig. 18.24(f)] and move it to the right [Fig. 18.24(g)]

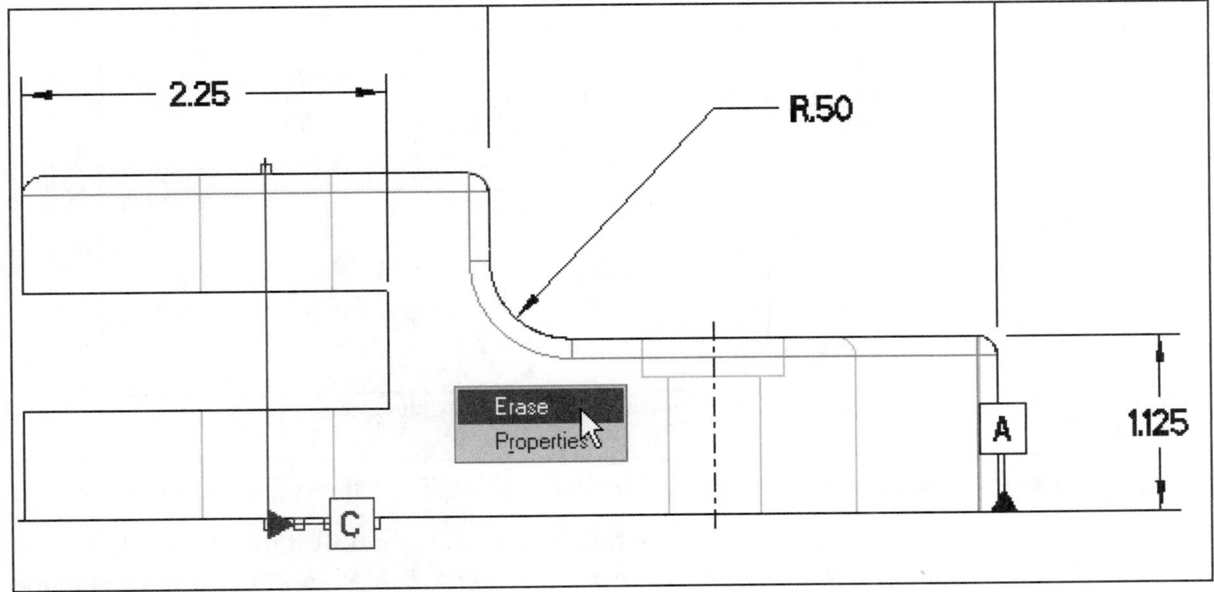

Figure 18.24(e) Erase Datum C

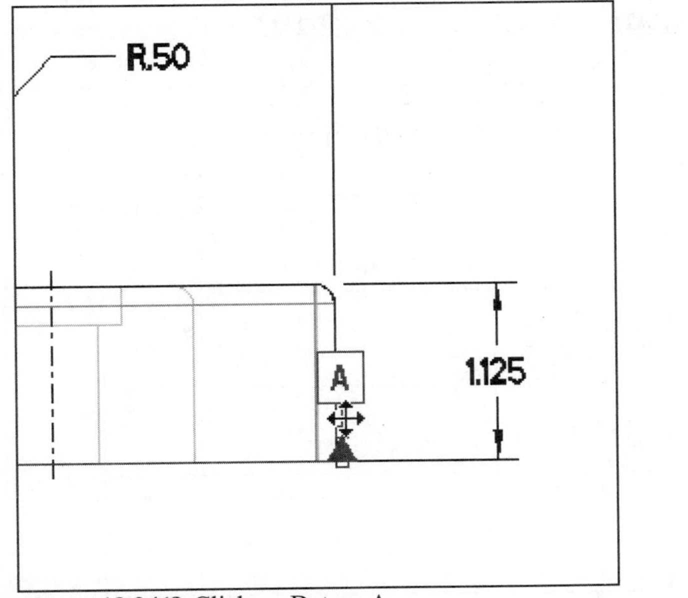

Figure 18.24(f) Click on Datum A

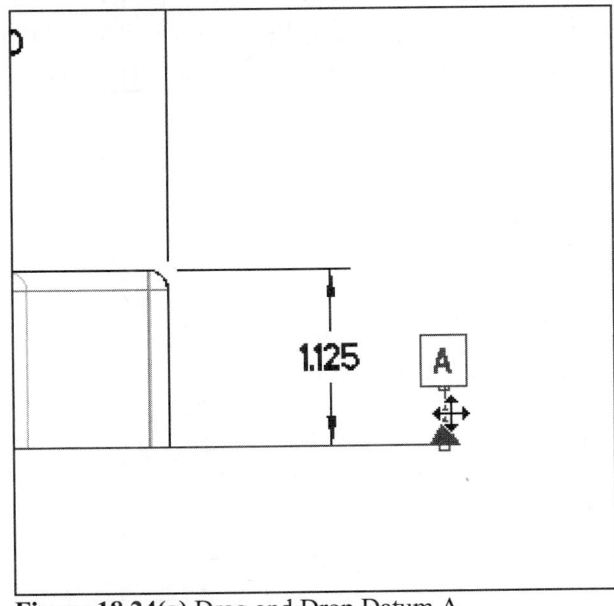

Figure 18.24(g) Drag and Drop Datum A

Click on the datum **B** in the top view and move it to the right [Fig. 18.24(h)] ⇒ click on the datum **C** in the top view and move down [Fig. 18.24(i)] ⇒ **LMB** ⇒ ⇒ ⇒ ⇒ ⇒ **MMB**

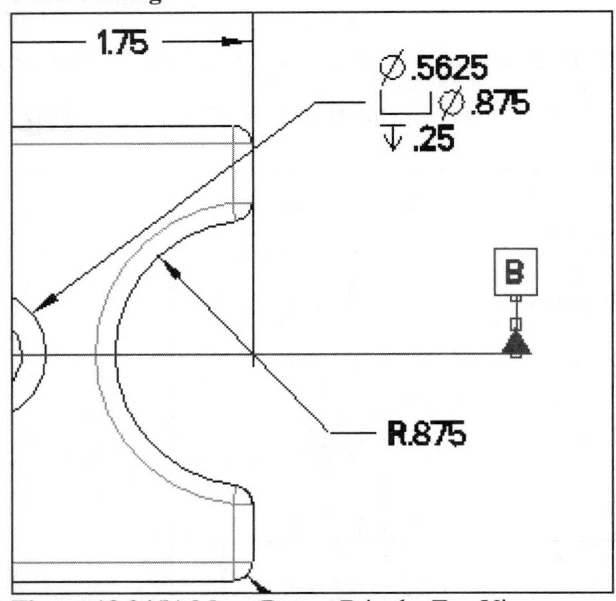

Figure 18.24(h) Move Datum B in the Top View

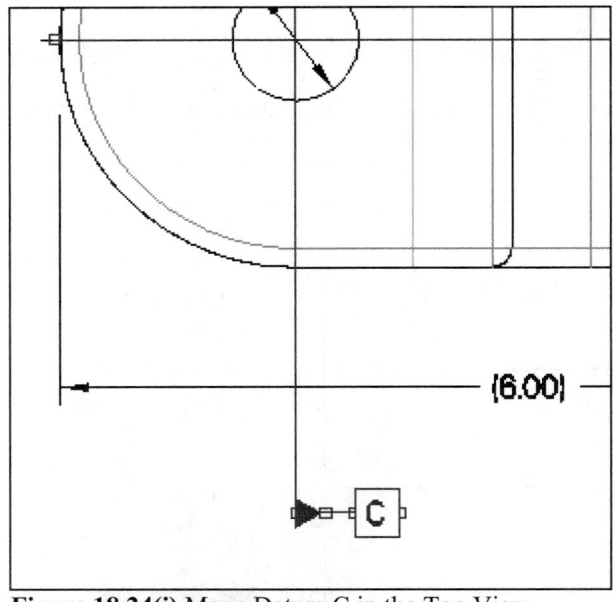

Figure 18.24(i) Move Datum C in the Top View

Before inserting geometric tolerances, move the height dimension (**2.1875**) to the right side view, and the slot dimension (**2.25**) below the front view, click: **Switch one or more detail items to another view** ⇒ click on **2.1875** in the front view *(or RMB ⇒ Move Item to View)* ⇒ **MMB** ⇒ click in the right side view [Fig. 18.25(a)] ⇒ **MMB** ⇒ **Move several objects** ⇒ click on the **2.25** dimension ⇒ **MMB** ⇒ drag to a position below the view [Fig. 18.25(b)] ⇒ **LMB** ⇒ **MMB** ⇒ click the **2.25** dimension ⇒ clip extension lines [Fig. 18.25(c)] ⇒ **LMB**

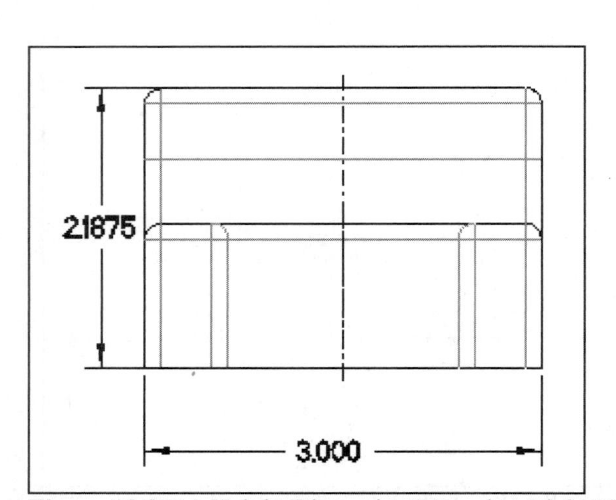

Figure 18.25(a) Height Dimension Moved to Right View

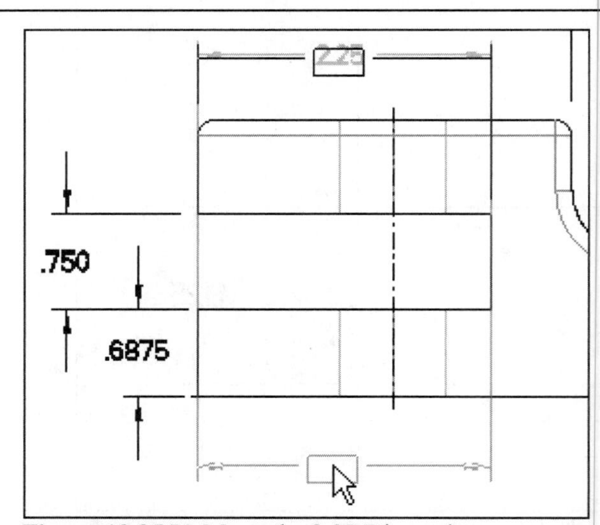

Figure 18.25(b) Move the **2.25** Dimension

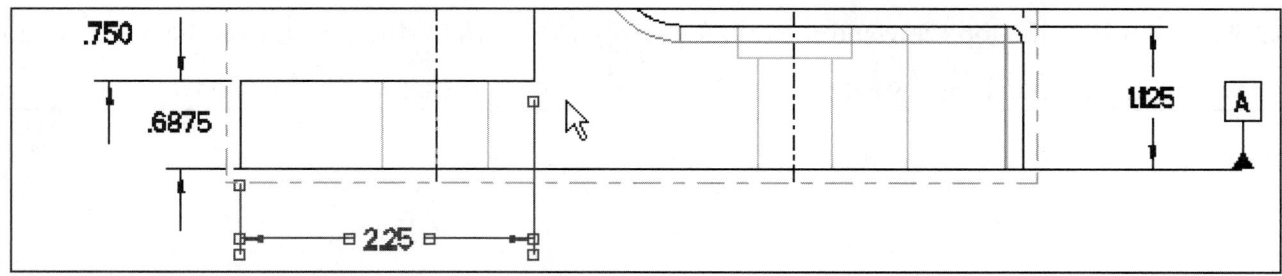

Figure 18.25(c) Clip the **2.25** Dimension Extension Lines

Click: **Create geometric tolerances** *(or Insert ⇒ Geometric Tolerance)* ⇒ Reference: Type ⇒ **Feature** [Fig. 18.26(a)] Select Entity... is now selected ⇒ click on **Ø.810** hole [Fig. 18.26(b)] ⇒ **Datum Refs** tab ⇒ **Primary** tab- Basic [Fig. 18.26(c)] ⇒ click on datum **A** in the front view as the Primary Datum Reference [Fig. 18.26(d)]

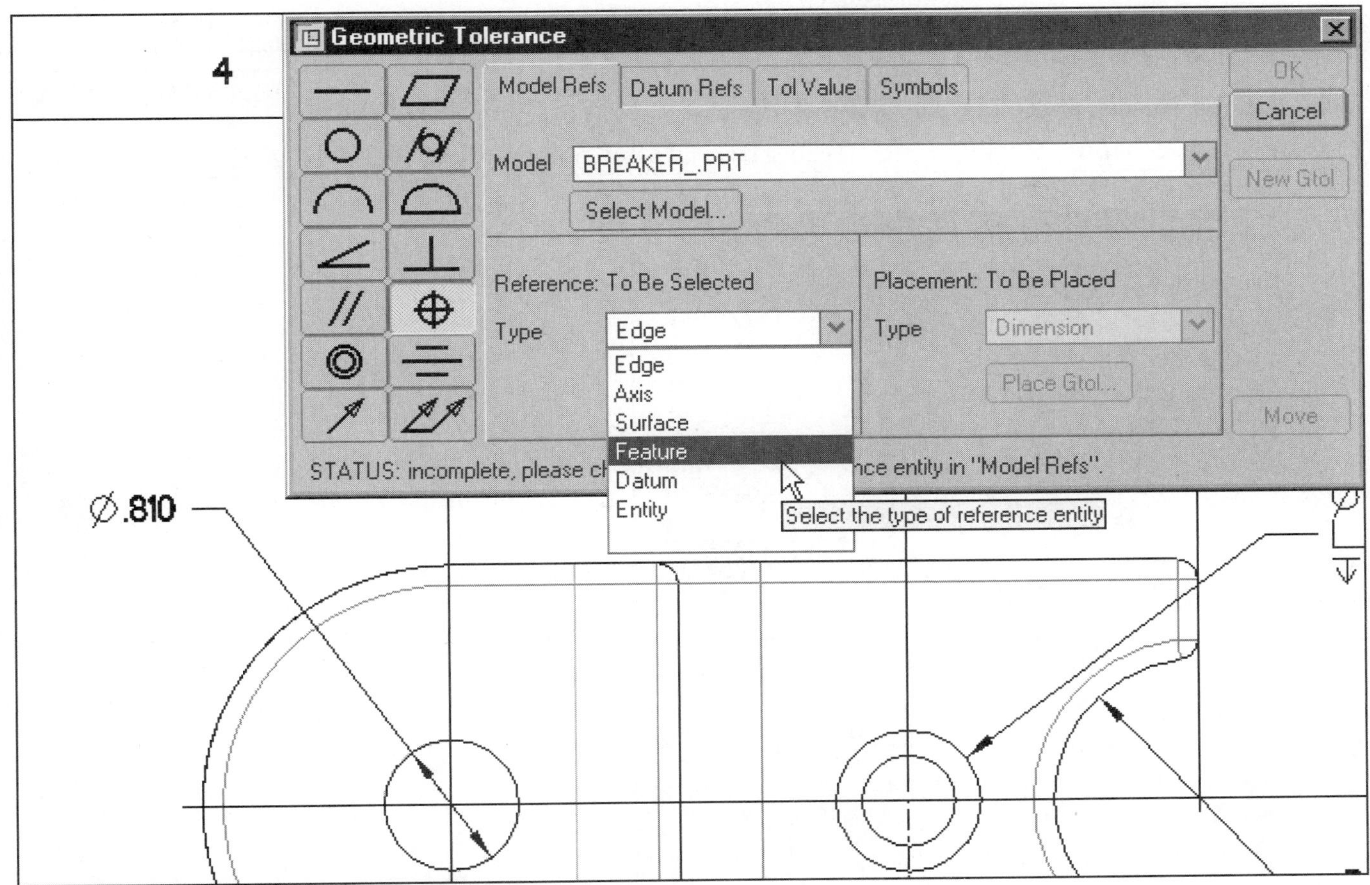

Figure 18.26(a) Geometric Tolerance Dialog Box

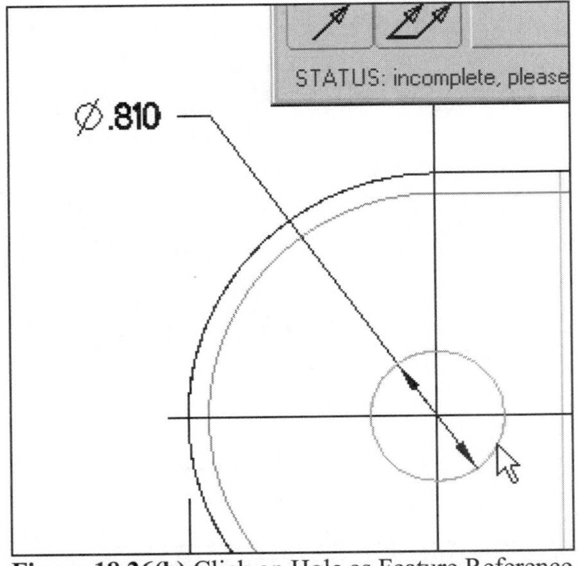

Figure 18.26(b) Click on Hole as Feature Reference

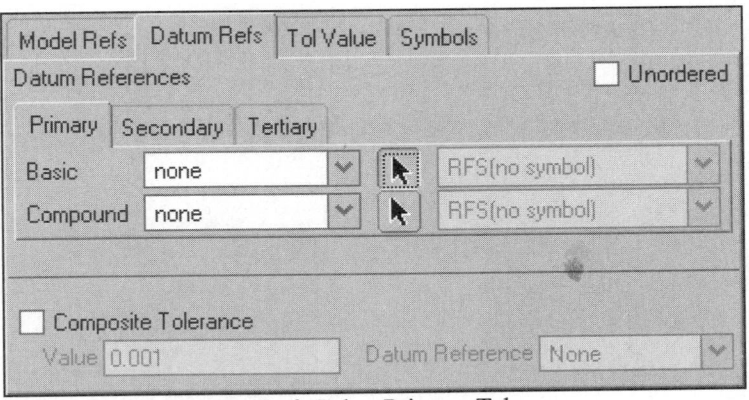

Figure 18.26(c) Datum Refs Tab – Primary Tab

Datum References Unordered
Primary Secondary Tertiary
Basic A RFS(no symbol)
Compound none RFS(no symbol)

Figure 18.26(d) Primary Datum Reference- Datum A

Click: **Tol Value** tab ⇒ Overall Tolerance **0.001** [Fig. 18.26(e)] ⇒ **Symbols** tab ⇒ [✓ ⌀ Diameter Symbol] [Fig. 18.26(f)] ⇒ **Model Refs** tab [Fig. 18.26(g)] ⇒ [Place Gtol...] ⇒ select **⌀.810** [Fig. 18.26(h)]

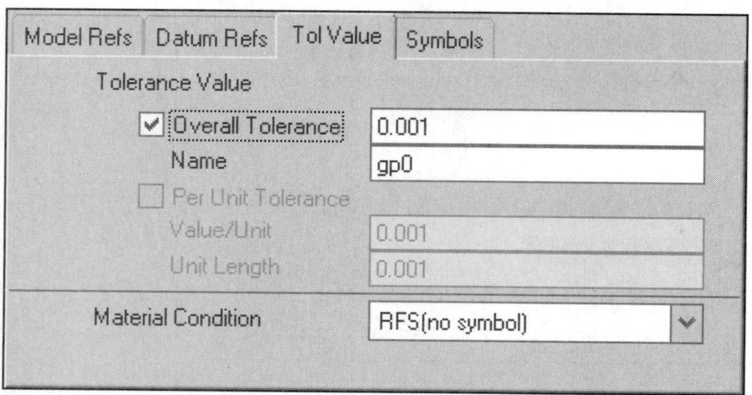

Figure 18.26(e) Tolerance Value **0.001**

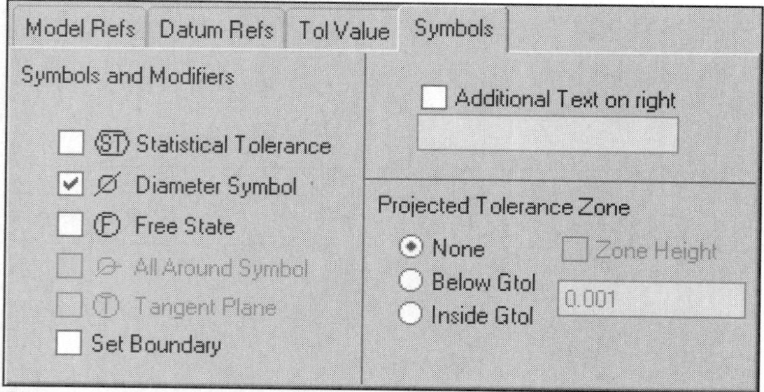

Figure 18.26(f) Symbols Tab

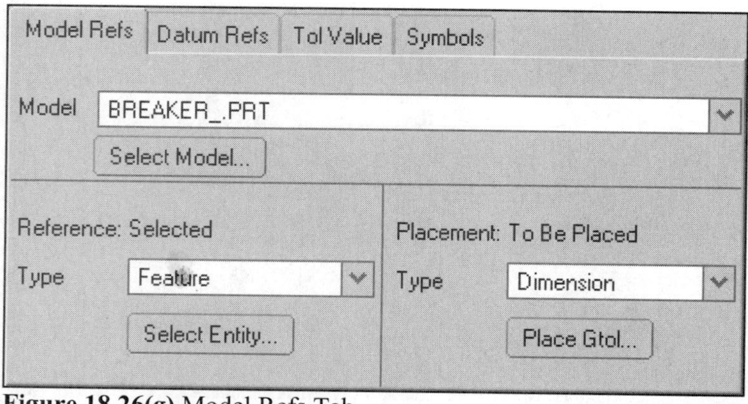

Figure 18.26(g) Model Refs Tab

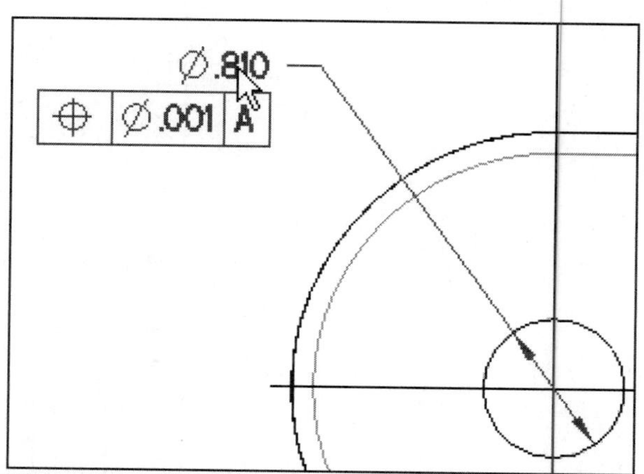

Figure 18.26(h) Click on Dimension to Place

Click: **New Gtol** ⇒ Select Entity... ⇒ select the counterbore ⇒ Placement: Type [▼] ⇒ **Free Note** [Fig. 18.26(i)] ⇒ click anywhere near the counterbore note [Fig. 18.26(j)] ⇒ **Datum Refs** tab ⇒ **Primary** tab-Basic ⇒ leave datum **A** as the Primary Datum Reference [Fig. 18.26(k)] ⇒ **Secondary** tab ⇒ [▼] ⇒ click on datum **B** as the Secondary Datum Reference [Fig. 18.26(l)] ⇒ **Tertiary** tab ⇒ [▼] ⇒ click on datum **C** as the Tertiary Datum Reference [Fig. 18.26(m)] ⇒ **Model Refs** tab ⇒ Move ⇒ move to a position to the right of **Ø.5625** [Fig. 18.26(n)] ⇒ **LMB** ⇒ **OK** ⇒ **No** ⇒ [icon] ⇒ [icon] ⇒ **MMB** [Fig. 18.26(o)]

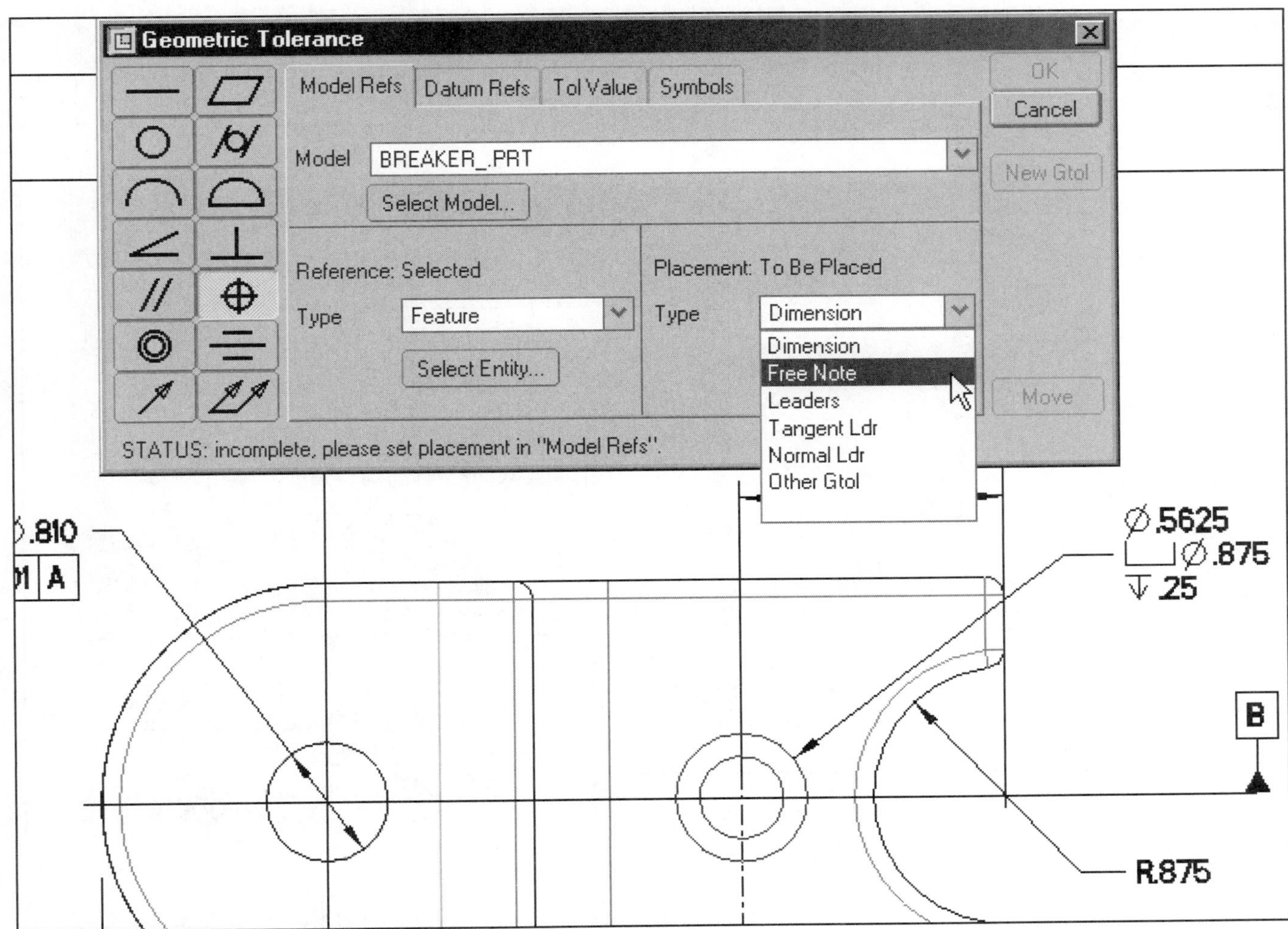

Figure 18.26(i) Click on Counterbore Feature

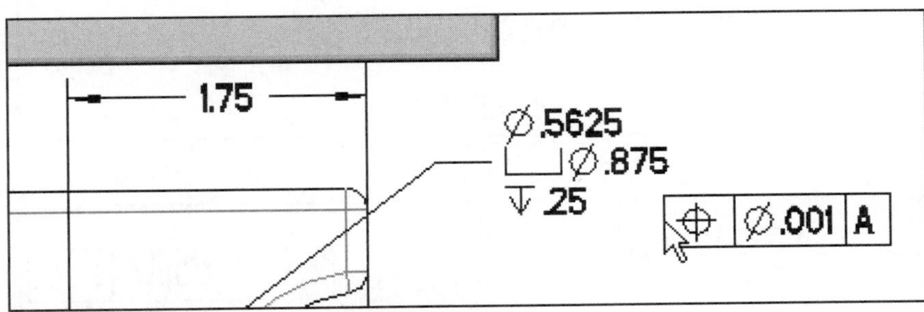

Figure 18.26(j) Click Near the Hole Note

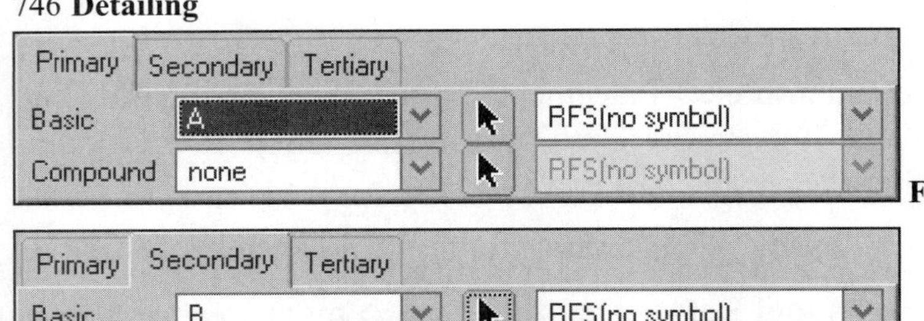

Figure 18.26(k) Primary Reference

Figure 18.26(l) Secondary Reference

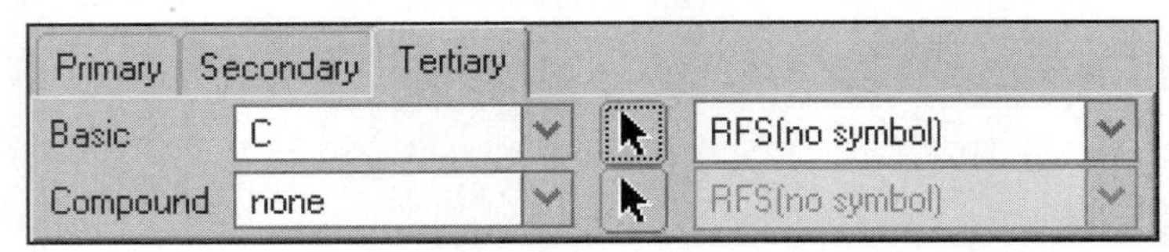

Figure 18.26(m) Tertiary Reference

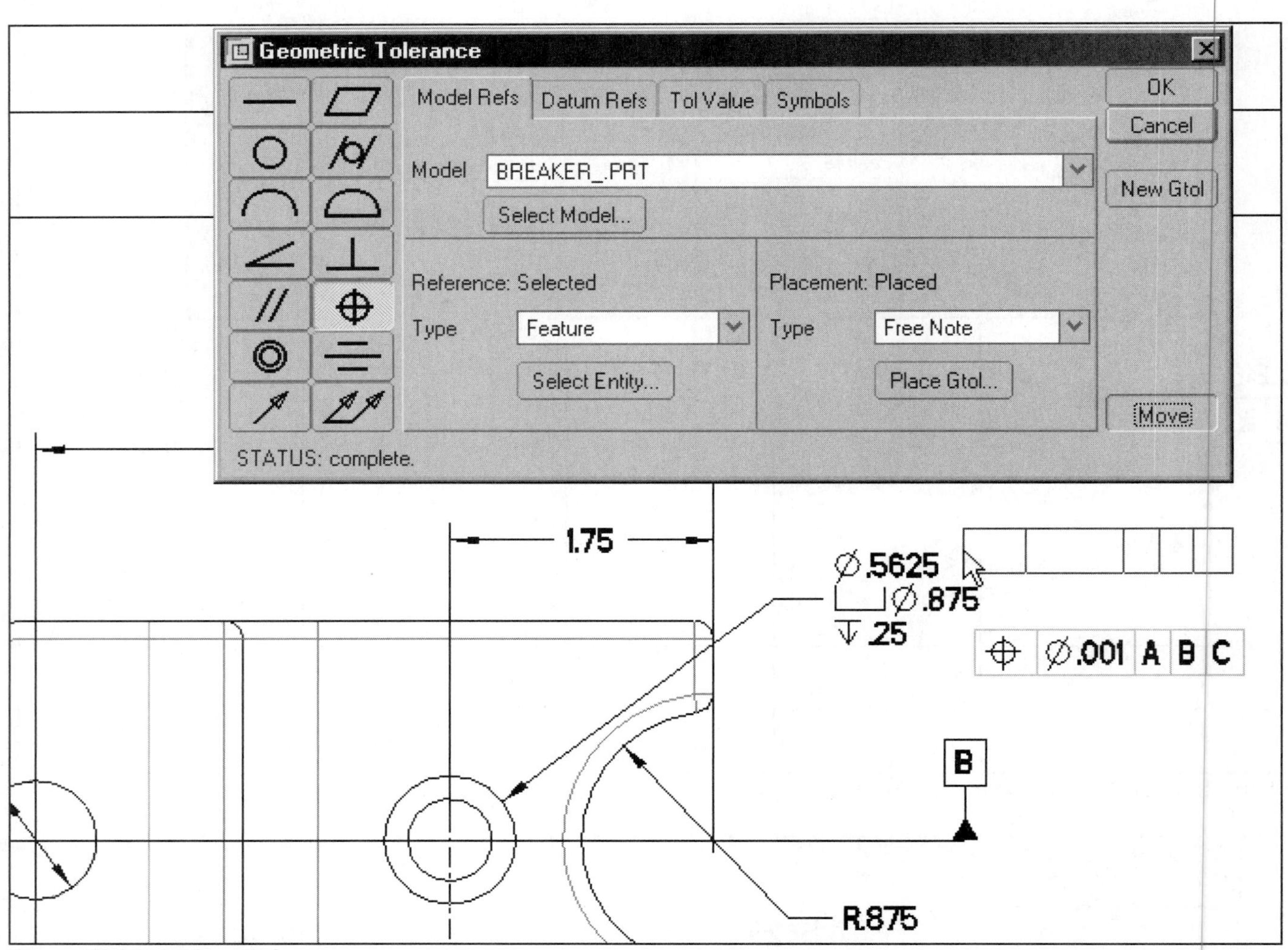

Figure 18.26(n) Move Tolerance

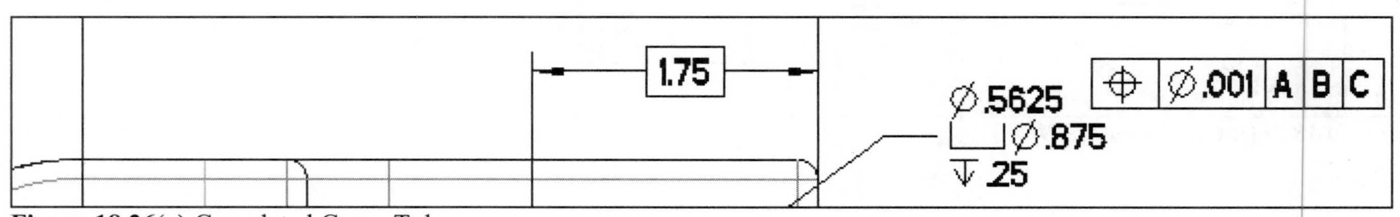

Figure 18.26(o) Completed Geom Tol

Click: **Insert** from the menu bar ⇒ **Geometric Tolerance** ⇒ [flatness icon] **Flatness** ⇒ Reference: Type [dropdown] ⇒ **Datum** [Fig. 18.27(a)] [Select Entity...] active ⇒ click on datum **A** [Fig. 18.27(b)] ⇒ **Tol Value** tab ⇒ [☑ Overall Tolerance 0.005] [Fig. 18.27(c)] ⇒ **OK** ⇒ **LMB** [Fig. 18.27(d)] ⇒ [repaint icon]

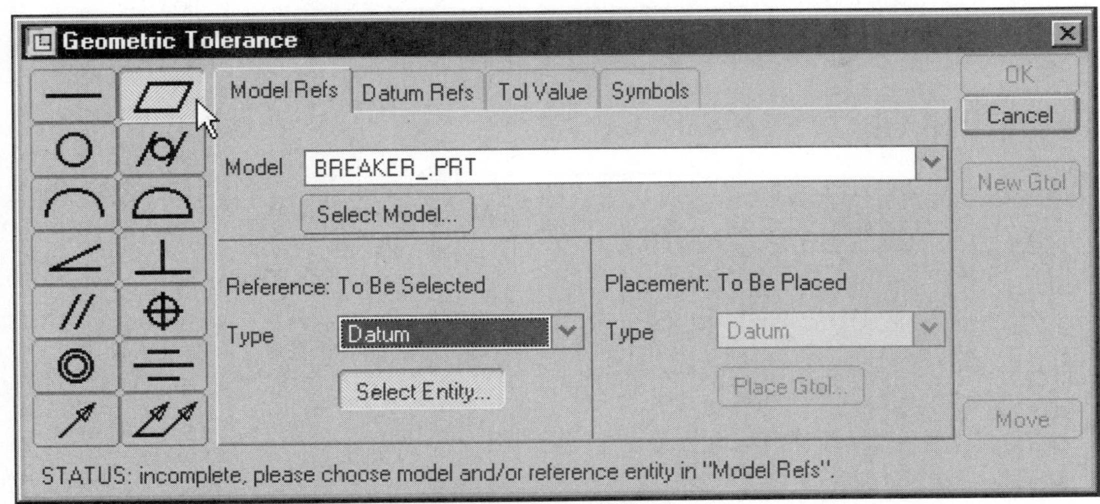

Figure 18.27(a) Select Flatness and Reference: Type- Datum

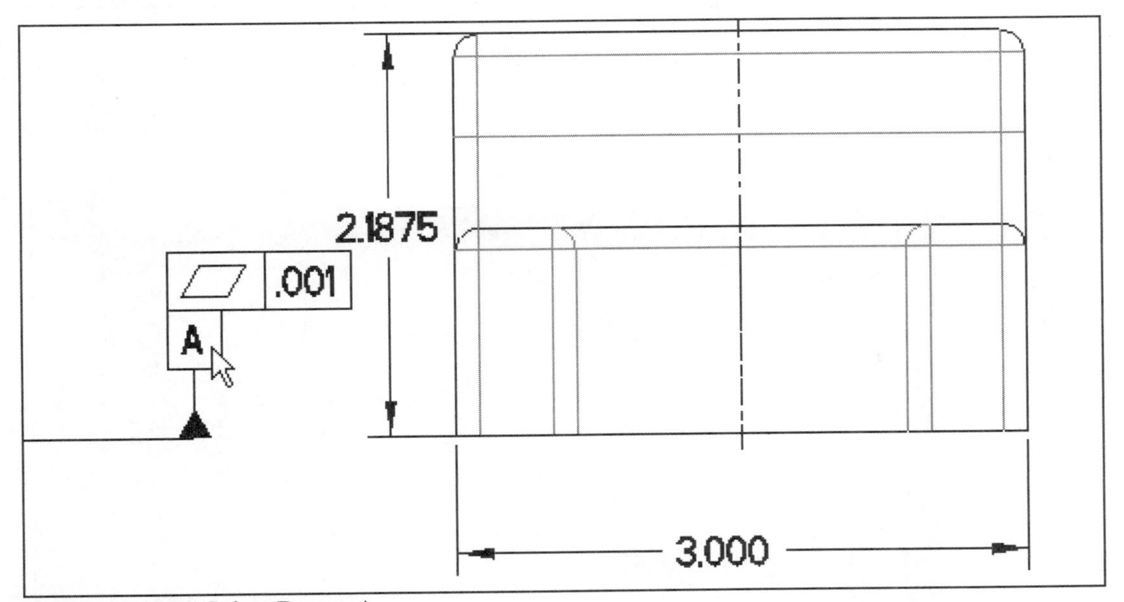

Figure 18.27(b) Select Datum A

Tolerance Value
☑ Overall Tolerance 0.005
Name gp5
☐ Per Unit Tolerance
Value/Unit 0.001
Unit Area 0.001
Material Condition RFS(no symbol)

Figure 18.27(c) Tolerance Value **0.005**

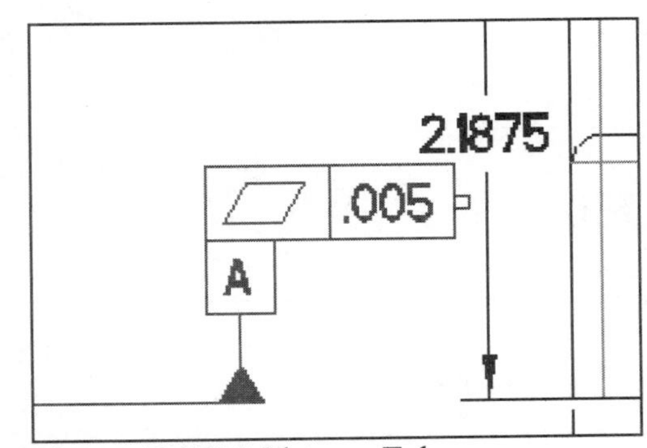

Figure 18.27(d) Flatness Tolerance

Move the **2.1875** height dimension to the opposite side of the right view and clip its extension lines as needed ⇒ [Save icon] ⇒ **MMB** [Fig. 18.27(e)]

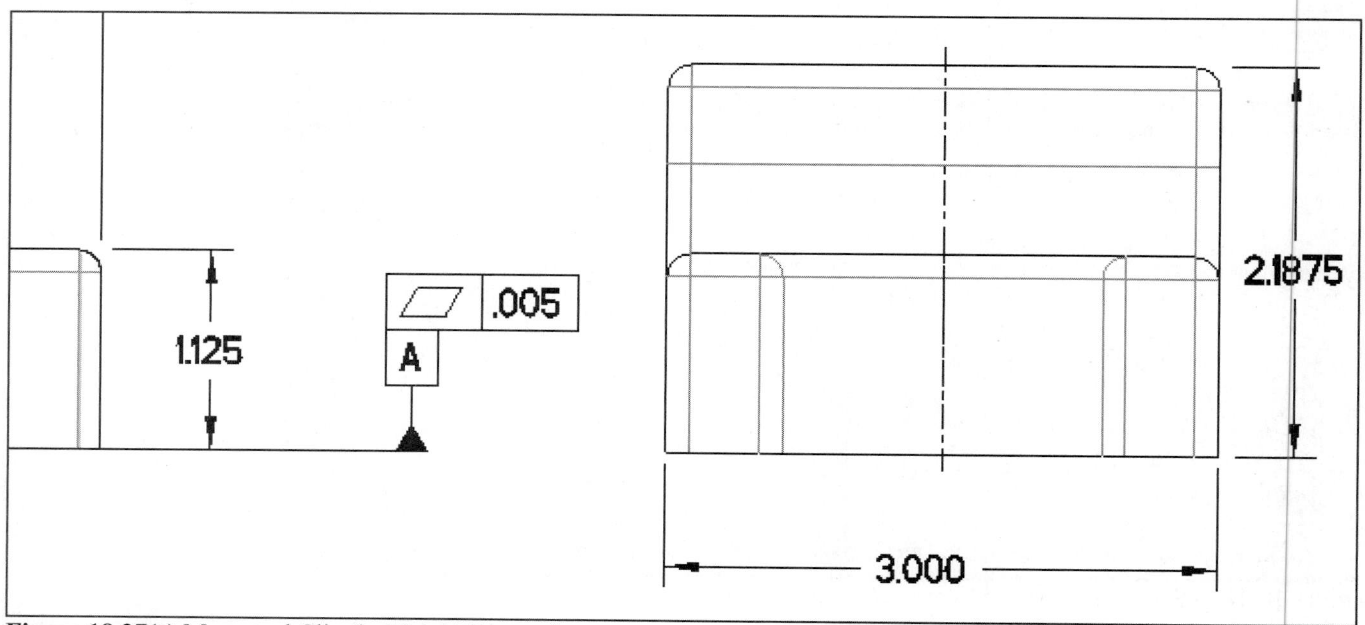

Figure 18.27(e) Move and Clip the Height Dimension

Click: [icon] **Create geometric tolerances** ⇒ [// icon] **Parallelism** ⇒ Reference: Type [▼] ⇒ **Surface** ⇒ select edge [Fig. 18.28(a)]

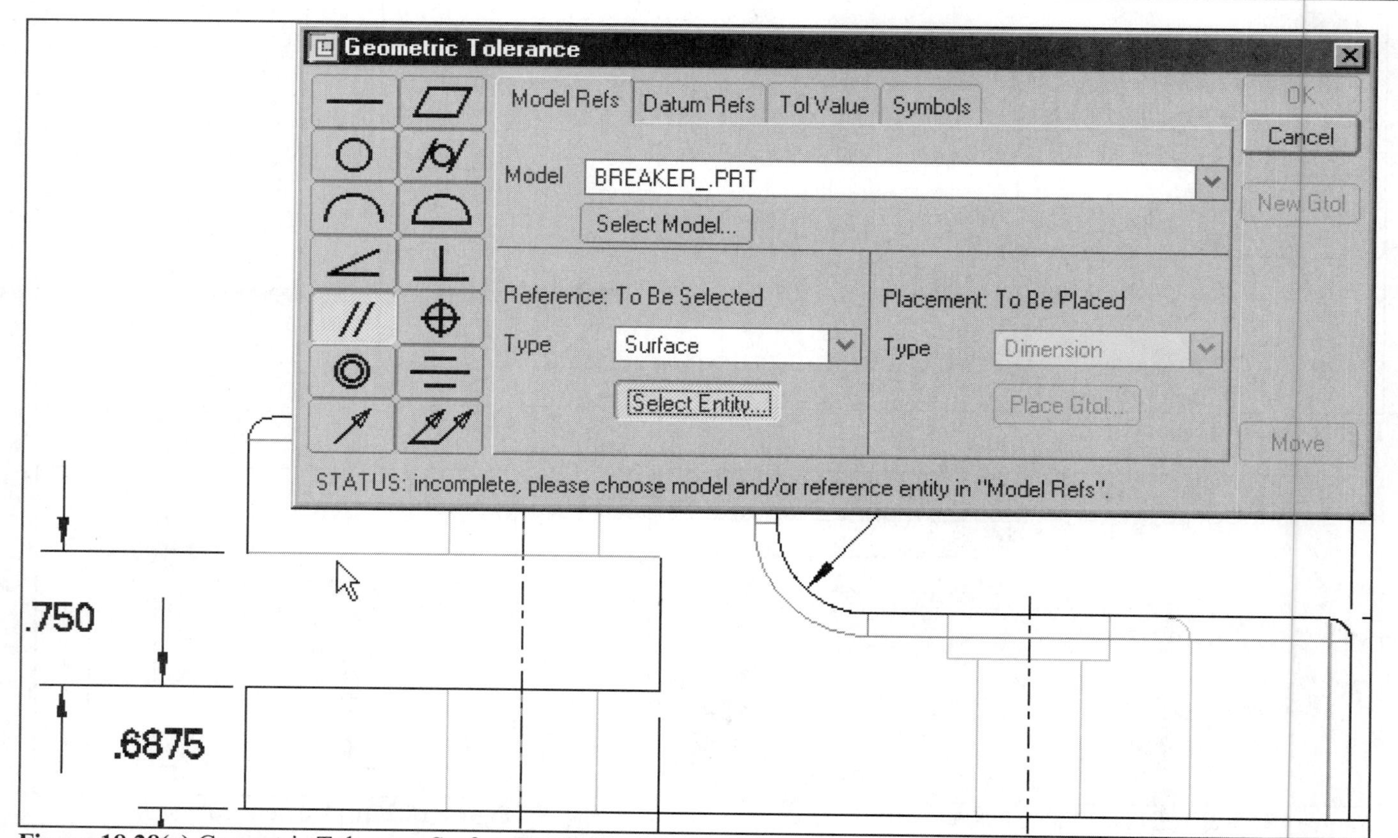

Figure 18.28(a) Geometric Tolerance Surface Parallelism

Click: **Datum Refs** tab ⇒ **Primary** tab- Basic ⇒ click on datum **A** [Fig. 18.28(b)] ⇒ **Tol Value** tab ⇒ **0.001** [Fig. 18.28(c)] ⇒ **Model Refs** tab ⇒ Placement: Type ⇒ **Dimension** Place Gtol... active ⇒ pick on **.750** dimension [Figs. 18.28(d-e)] ⇒ Move ⇒ move up [Fig. 18.28(f)] ⇒ **LMB** to place [Fig. 18.28(g-h)] ⇒ **OK** ⇒ ⇒ ⇒ ⇒ ⇒ **MMB** [Fig. 18.28(i)]

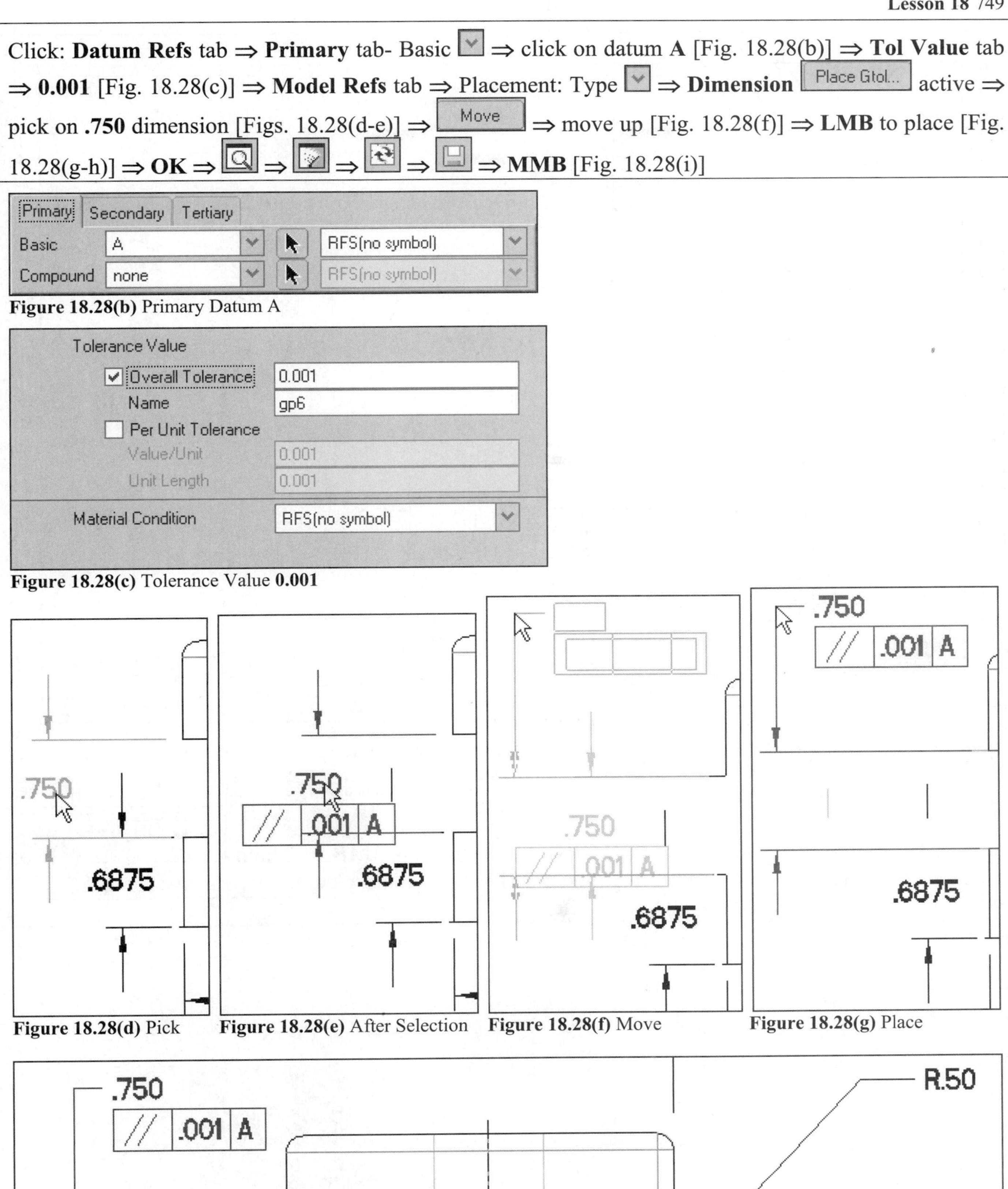

Figure 18.28(b) Primary Datum A

Figure 18.28(c) Tolerance Value **0.001**

Figure 18.28(d) Pick

Figure 18.28(e) After Selection

Figure 18.28(f) Move

Figure 18.28(g) Place

Figure 18.28(h) Parallelism Tolerance Created

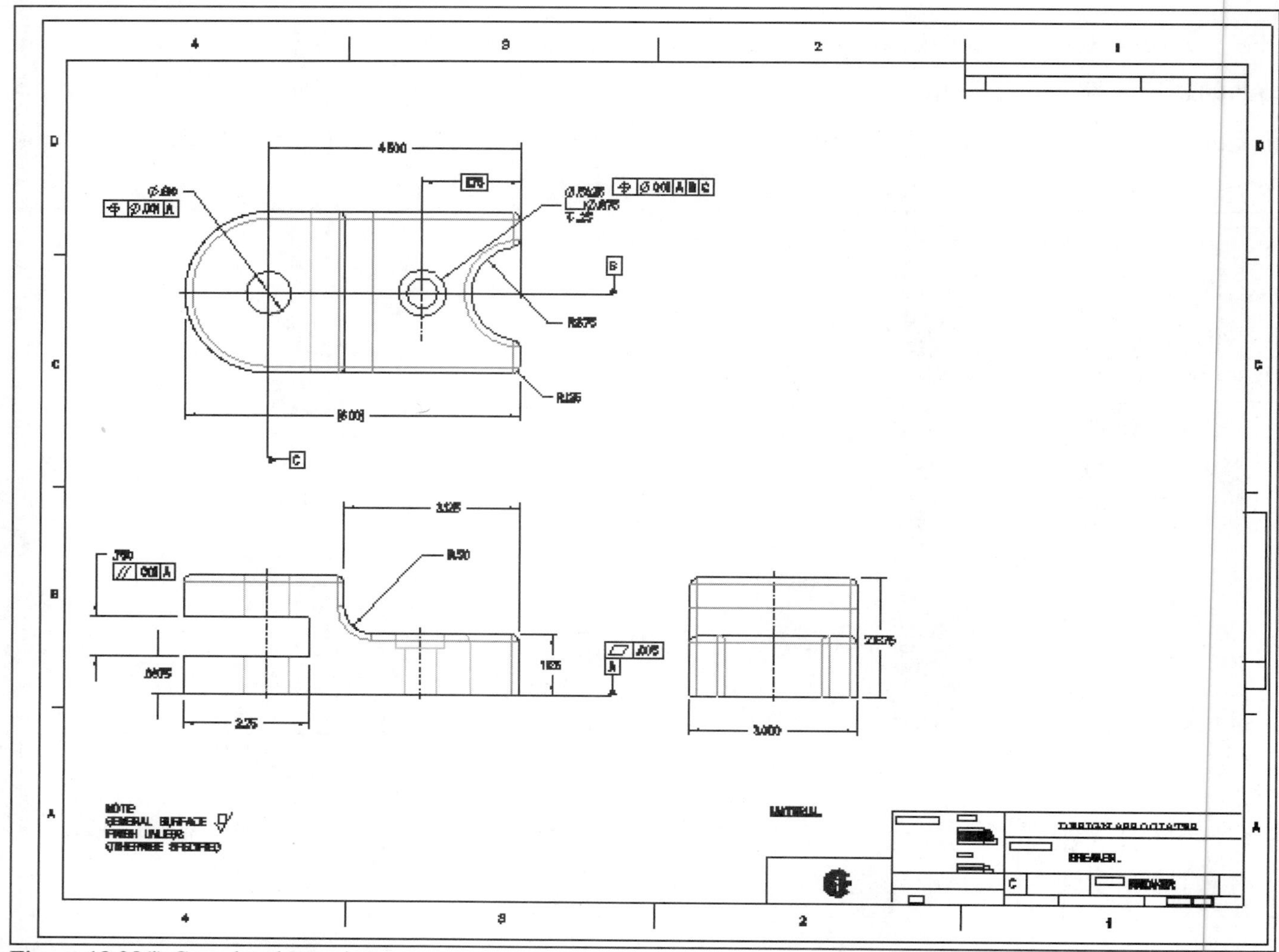

Figure 18.28(i) Completed Drawing

In the Model Tree of the Drawing, click **BREAKER.PRT** ⇒ **RMB** ⇒ **Open** (Fig. 18.29) ⇒ **Window** ⇒ **Close** ⇒ **Window** ⇒ **Activate** (will make the Drawing active again)

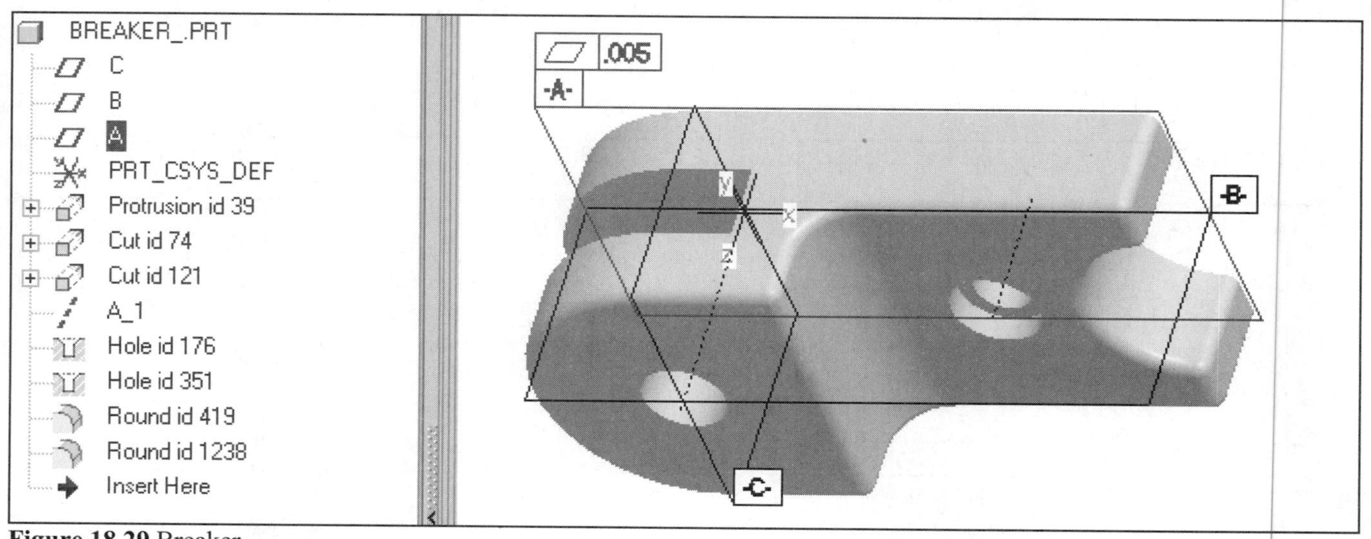

Figure 18.29 Breaker

Markup Mode

You believe the project is complete, so print or plot out a copy and submit it to the checker [teacher, boss, design checker (or yourself being checker)]. The checker will use Markup Mode to check and "mark up" your drawing. A markup is an object that, like a set of transparent sheets on top of a drawing sheet, enables you to superimpose text and sketched entities in a variety of colors to indicate where changes may be required. A markup is an informal sketch that you can create within Pro/E.

The basis for the markup is the object that you select to mark up. To create a markup, you use an object in the mode in which it was created:

- Part (includes all objects with a .prt extension)
- Assembly (includes all objects with a *.asm* extension)
- Drawing
- Manufacturing
- Layout
- Report
- Diagram

The object accompanies the markup file, which uses the file extension .mrk. The Orientation, Model Display and Advanced commands are not available in Markup mode. You must orient your three-dimensional object before you create the markup. Pro/E saves the orientation with the markup.

Each markup is as if a separate transparent sheet lay over the object containing sketched entities and/or notes in a single color. During the markup process, Pro/E does not change the original object in any way, and saves all sketched entities and notes created in Markup mode with the markup file, not with the original object. Since the driving object does not change, Pro/E does not store it when it stores the markup. The MARKUP menu includes options for:

- **Setup** Changes the color of the markup, toggles its display, or changes to another sheet of the markup and drawing. The menu displays the following commands:
  - **Color** Changes the color of the markup. The default markup color is red.
  - **Show** Lists the markups in the current working area along with their color. A check mark appears next to the markup's name and color if it is currently displayed. Choosing the markup name again removes the check mark and turns off its display.
  - **Switch Sht** lets you view and mark other sheets of the drawing.
  - **Text Height** Sets the text height before creating notes.
  - **Line Width** Sets the width of sketched entities before creating them.
- **Note** Places a note without a leader.
- **Arrow** Sketches an arrow. Use the LMB to select the point to which the arrow points, then select the point for the other end of the arrow. Press the MMB to abort arrow creation.
- **Curve** Selects the points of a curve to which a spline is to be fitted. Use the LMB to select the point, the MMB to stop the curve, and the RMB to abort curve creation.
- **Sketch** Sketches a curve without selecting points for it. Hold the LMB down to create a free curve. Release the LMB to end the curve.
- **Line** Creates a line using the LMB.
- **Move** Moves a note, curve, or line.
- **Modify** Changes the note text or line width of a markup item.
- **Delete** Deletes a markup item (note, arrow, curve, sketch, or line).

The checker (you?) will choose the following commands to enter Markup mode and show the ***checker changes*** he or she feels are necessary. The checker may also make design changes at this time.

To enter Markup mode, click: **Create a new object** [Fig. 18.30(a)] ⇒ Markup ⇒ Name **BREAKER** ⇒ **OK** ⇒ Type Drawing (*.drw) ⇒ select the object you want to markup--here select the drawing name you are presently working on breaker.drw [Fig. 18.30(b)] ⇒ **Open** a new window will open ⇒ **Setup** ⇒ **Text Height** ⇒ **.375** ⇒ **Enter** ⇒ **Line Width** ⇒ **.050** ⇒ **Enter** ⇒ **Show** active markup displays BREAKER ⇒ **Color** leave red Red ⇒ **Curve** sketch a spline curve (by picking free points) about the **.810** dimension [Fig. 18.30(c)] ⇒ **MMB** [Fig. 18.30(d)]

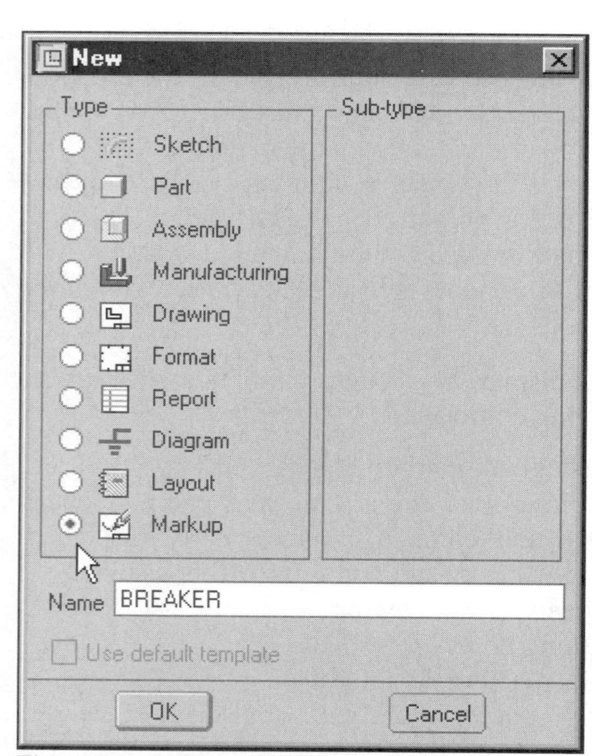

Figure 18.30(a) New Dialog Box- Markup

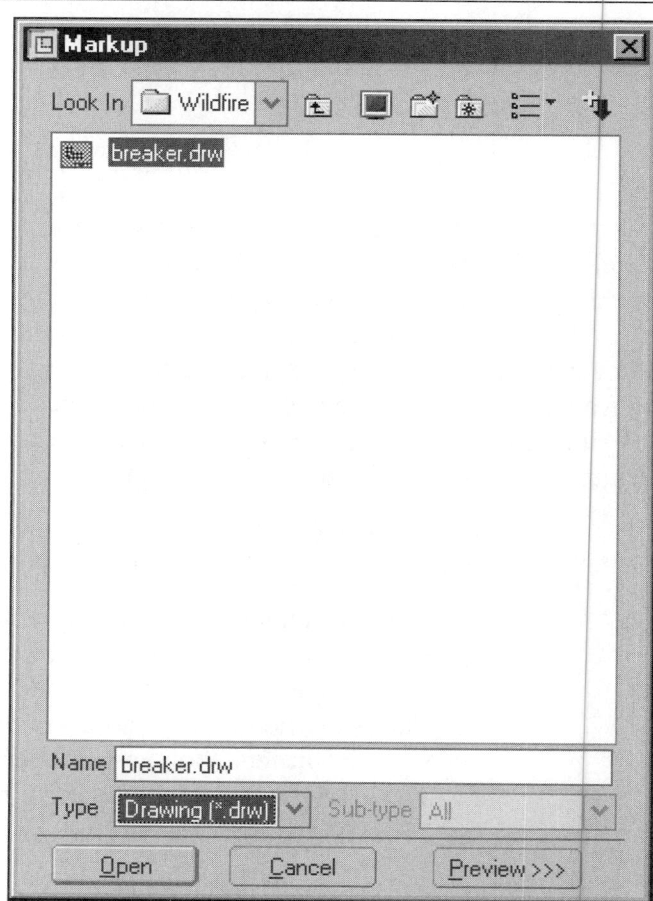

Figure 18.30(b) Markup Dialog Box- breaker.drw Selected

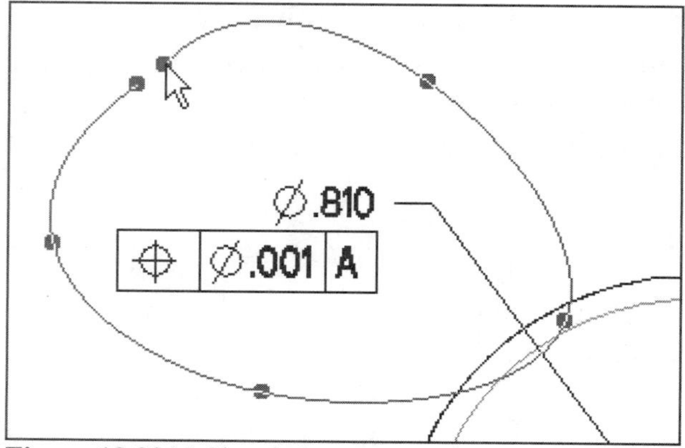

Figure 18.30(c) Sketch the curve

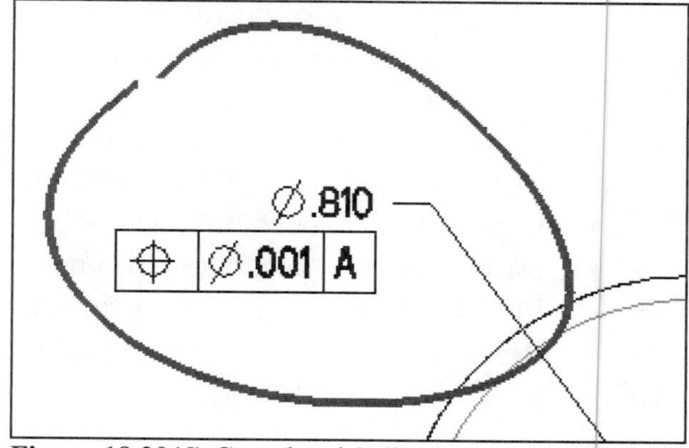

Figure 18.30(d) Completed Spline Curve

Click: **Note** ⇒ pick a position for the note [Fig. 18.30(e)] ⇒ type **1.000** ⇒ **Enter** ⇒ **Enter** [Fig. 18.30(f)] ⇒ **Modify** ⇒ **Note** ⇒ **Text Line** ⇒ pick the **1.000** note and the Text Symbol dialog box displays [Fig. 18.30(g)] ⇒ Edit line {0:1.000} ⇒ place cursor at beginning of Edit line ⇒ ∅ Edit line ∅{0:1.000} ⇒ **Enter** ⇒ **Enter** [Fig. 18.30(h)] ⇒ **Line** sketch two crossing lines ⇒ **Note** ⇒ pick a position for the note ⇒ type **Remove View** ⇒ **Enter** ⇒ **Enter** [Fig. 18.30(i)] ⇒ ⇒ ⇒ **MMB**

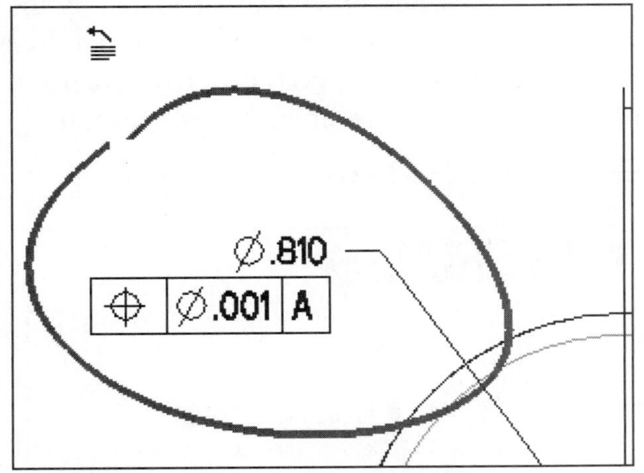

Figure 18.30(e) Pick a Position for the Note

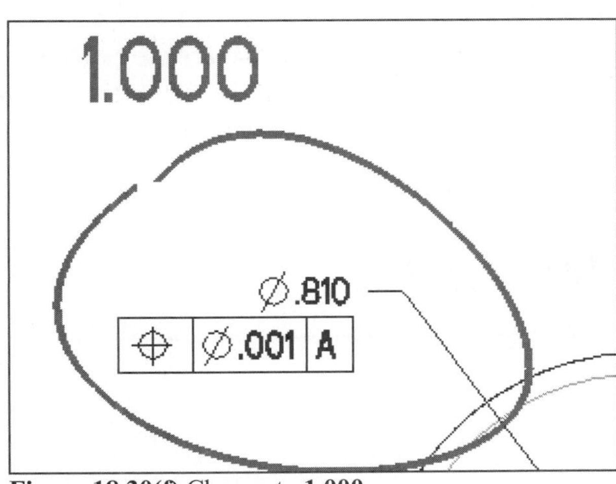

Figure 18.30(f) Change to **1.000**

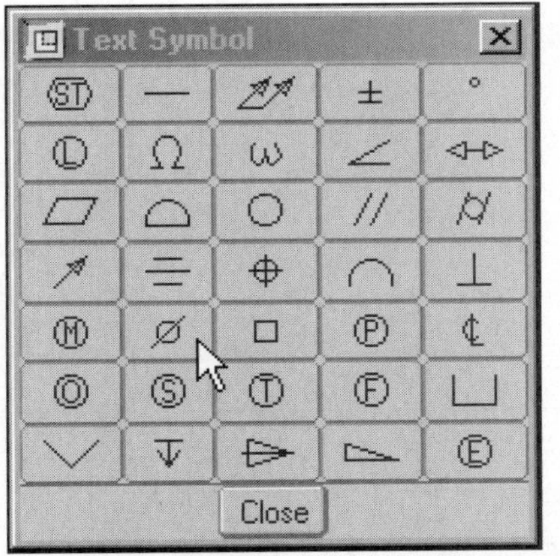

Figure 18.30(g) Text Symbol Dialog Box

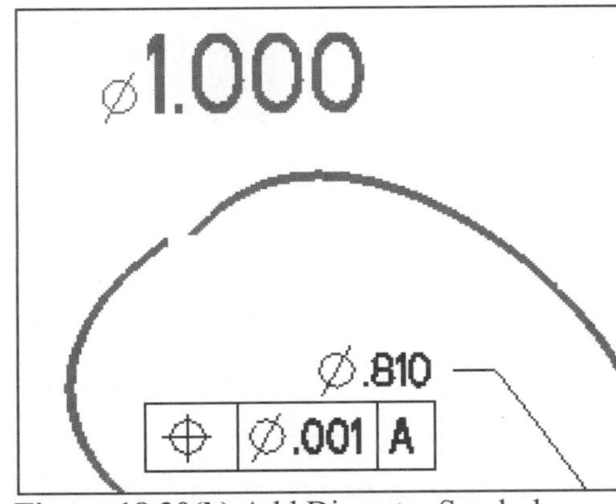

Figure 18.30(h) Add Diameter Symbol

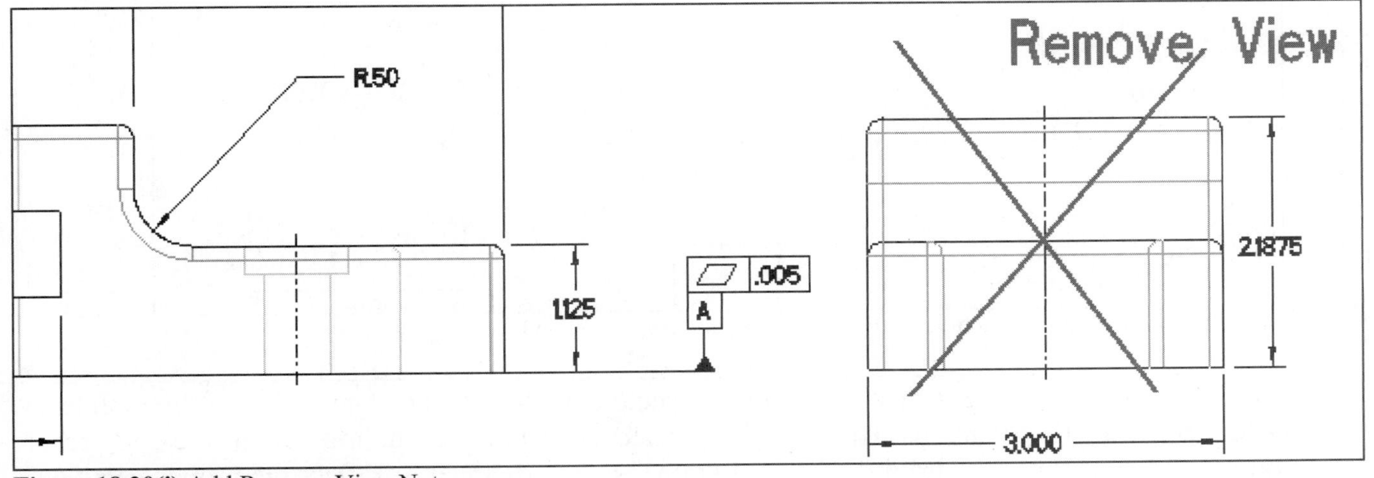

Figure 18.30(i) Add Remove View Note

Since the right view is to be removed (deleted), the height dimension will have to be moved to the front view and the width dimension to the top view when incorporating the checker changes. To delete the Canadian symbol and move the Material note inside the box you must activate the format and make the changes in the Format mode.

Click: **Line** ⇒ cross out the Canadian Standard symbol by sketching two lines [Fig. 18.30(j)] *(to delete the Canadian symbol, open and activate the format and delete the symbol)* ⇒ **Arrow** add an arrow to show the movement of the **MATERIAL** note to be inside the title block box previously occupied by the Canadian symbol. Pick the point of the arrow first then its tail [Fig. 18.30(k)]. (*To move the Material note inside the box you must activate the format and make the changes there.*) ⇒ **Sketch** draw a freeform curve about the **.750** dimension and geometric tolerance in front view ⇒ **Arrow** add an arrow pointing in the direction of the desired position [Fig. 18.30(l)] ⇒ ⇒ ⇒ ⇒ **MMB** [Fig. 18.30(m)]

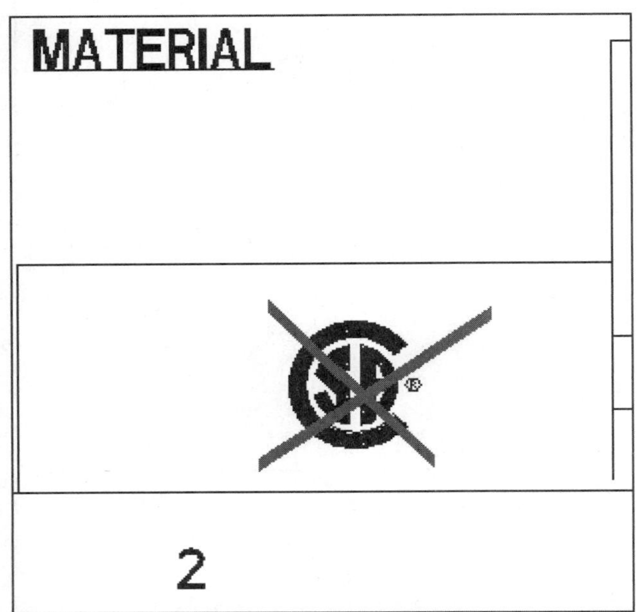

Figure 18.30(j) Cross out the Canadian Standard Symbol

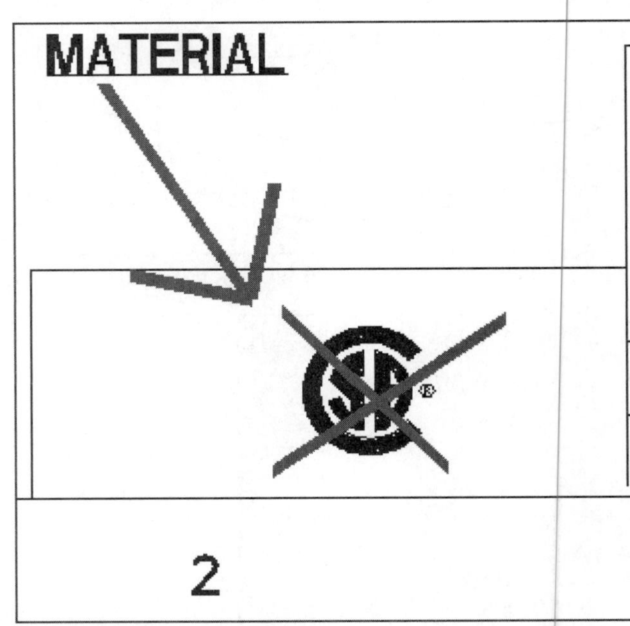

Figure 18.30(k) Add the Arrow

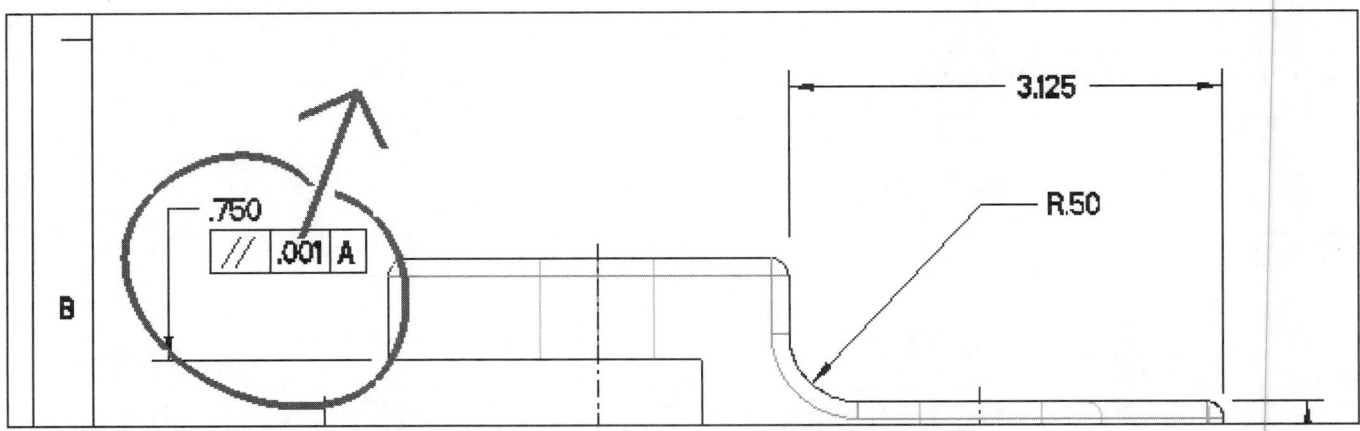

Figure 18.30(l) Add the Spline Sketch (drag the cursor to create the spline) and Arrow

You may add any other changes that you, your instructor, or the checker may see fit to complete the drawing to ASME Y14.5 Standards. (If you want the leader to point to the top line in the counterbore note, you will have to show the original thru hole dimension and edit it to reflect the proper parametric values. It will point to the inner circle in the top view instead of the outer circle.)

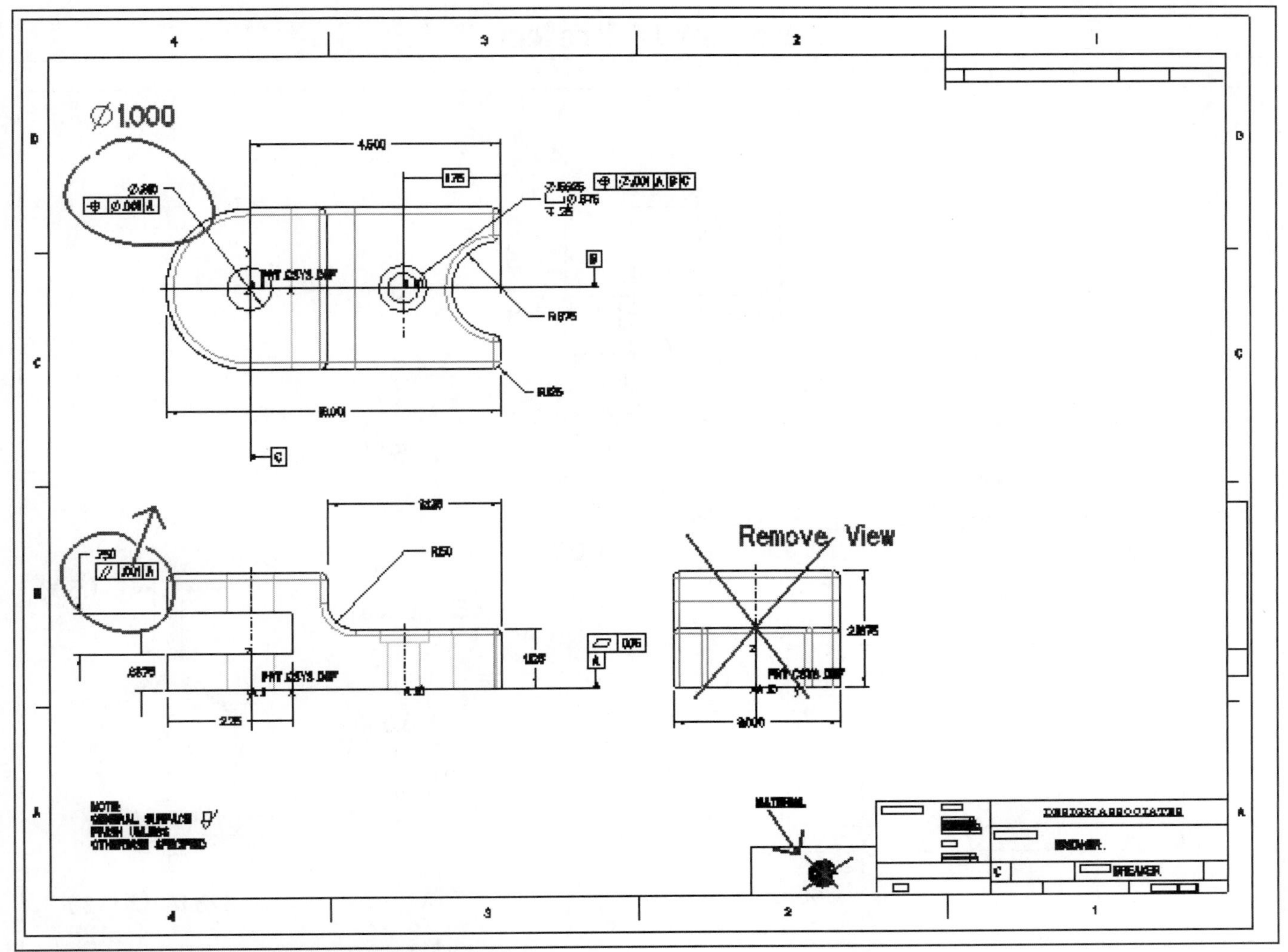

Figure 18.30(m) Completed Markup of the Breaker

After the checker saves the changes, bring up the Breaker markup and your Breaker detail drawing. By keeping the two files in session, you can work in your detail window and still see the markup. Because you will also be making some design changes, it is a good idea to have the part in a window to see the changes propagated throughout the model, for that matter, have the assembly (if the part is used in an assembly) open as well. All files will be updated with the new design changes after regeneration.

If you are starting a new session, use the following commands:

File ⇒ **Open** ⇒ Type **Part** ⇒ **breaker.prt** ⇒ **Open**

File ⇒ **Open** ⇒ Type **Markup** ⇒ **breaker.mrk** ⇒ **Open**

File ⇒ **Open** ⇒ Type **Format** ⇒ **c_format.frm** ⇒ **Open** ⇒ make the changes ⇒ ⇒ **MMB**

File ⇒ **Open** ⇒ Type **Drawing** ⇒ **breaker.drw** ⇒ **Open** ⇒ make the changes ⇒ ⇒ **MMB**

You have now completed this lesson. Continue with the Lesson Project.

Lesson 18 Project

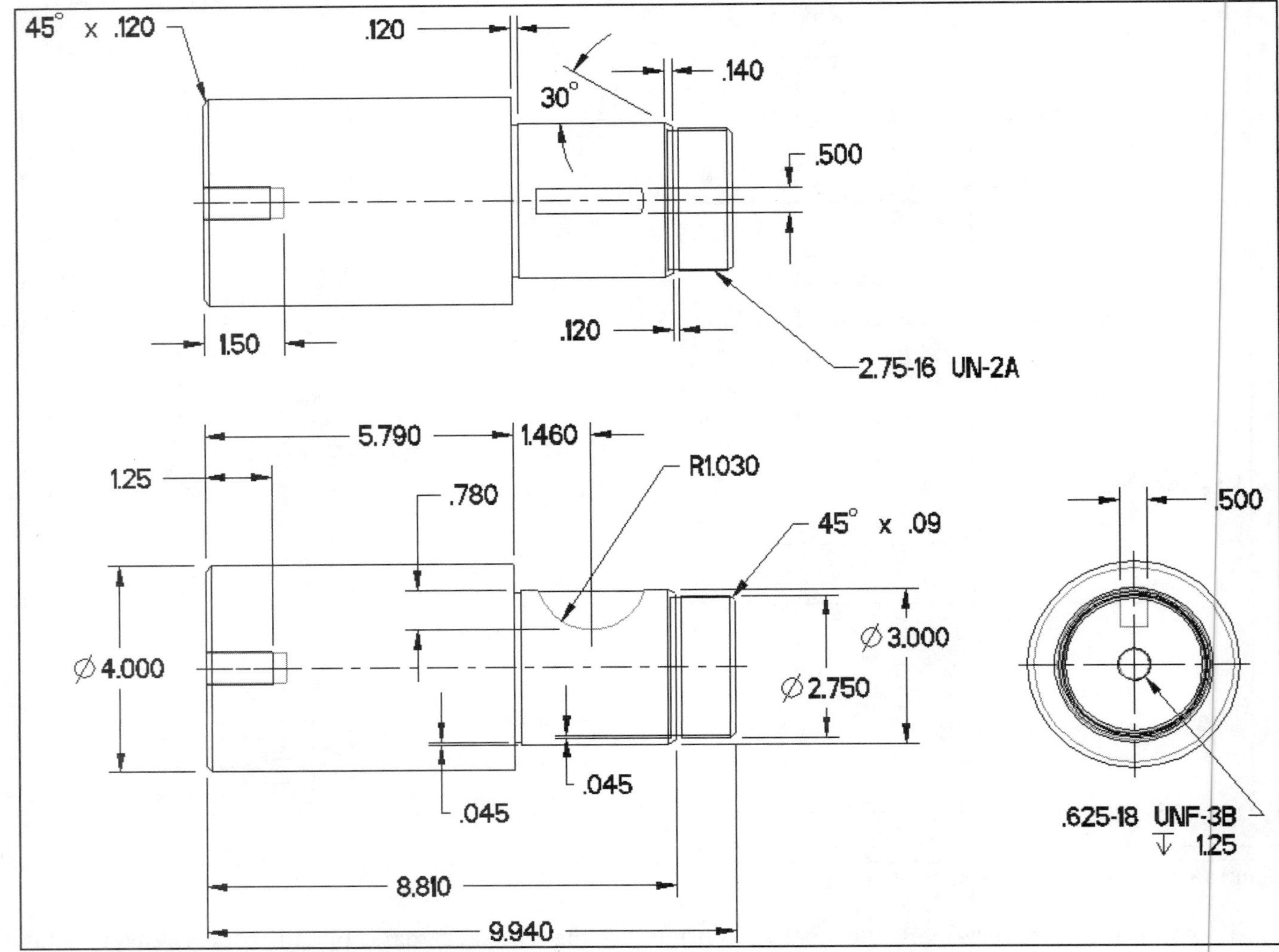

Figure 18.31 Cylinder Rod Drawing without a Format

Cylinder Rod Drawing

This lesson project uses the part modeled in Lesson 7. The drawing for the Cylinder Rod (Fig. 18.31) is just one of many lesson parts and lesson projects that can be brought into Drawing mode and detailed. Analyze the part and plan the sheet size and the drawing views required to display the model's features for detailing. Use the formats created in Lesson 17. Detail the part according to ASME Y14.5.

Lesson 19 Sections and Auxiliary Views

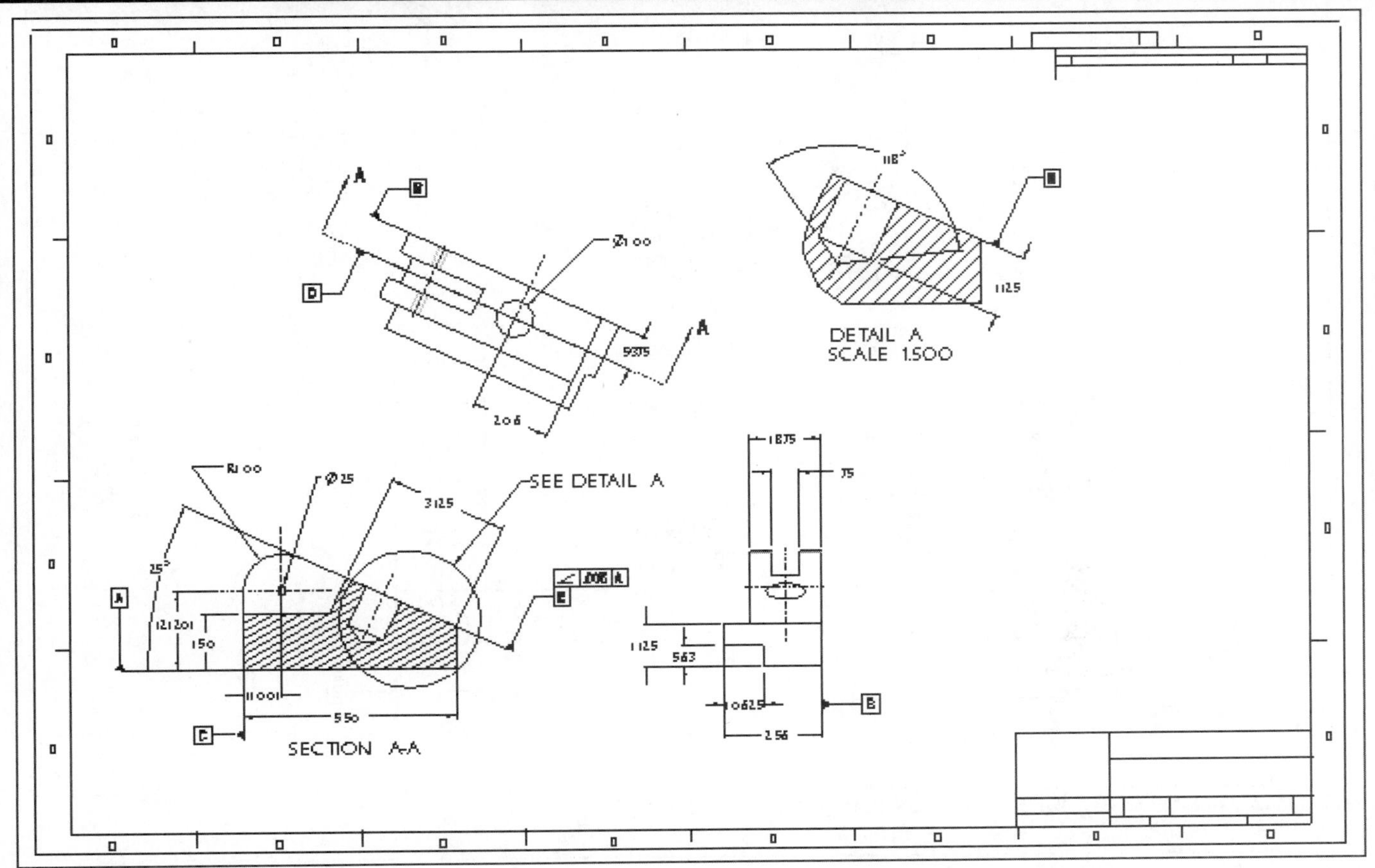

Figure 19.1 Anchor Drawing

OBJECTIVES

- Establish a **Drawing Options** file to use when detailing
- Identify the need for **sectional views** to clarify interior features of a part
- Identify **Cutting Planes** and the resulting views
- Create **Cross Sections** using datum planes
- Produce **Auxiliary Views**
- Create **Detail Views**
- Use **multiple drawing sheets**
- Apply **standard drafting conventions** and linetypes to illustrate interior features

SECTIONS AND AUXILIARY VIEWS

Designers and drafters use **sectional views**, also called **sections**, to clarify and dimension the internal construction of an object. Sections are needed for interior features that cannot be clearly described by hidden lines in conventional views (Figs. 19.1 and 19.2).

Auxiliary views are used to show the *true shape/size* of a feature or the relationship of features that are not parallel to any of the principal planes of projection. Many objects have inclined surfaces and features that cannot be adequately displayed and described by using principal views alone. To provide a clearer description of these features, it is necessary to draw a view that will show the *true shape/size*. Besides showing a feature's true size, auxiliary views are used to dimension features that are distorted in principal views and to solve a variety of engineering problems graphically (Fig. 19.3).

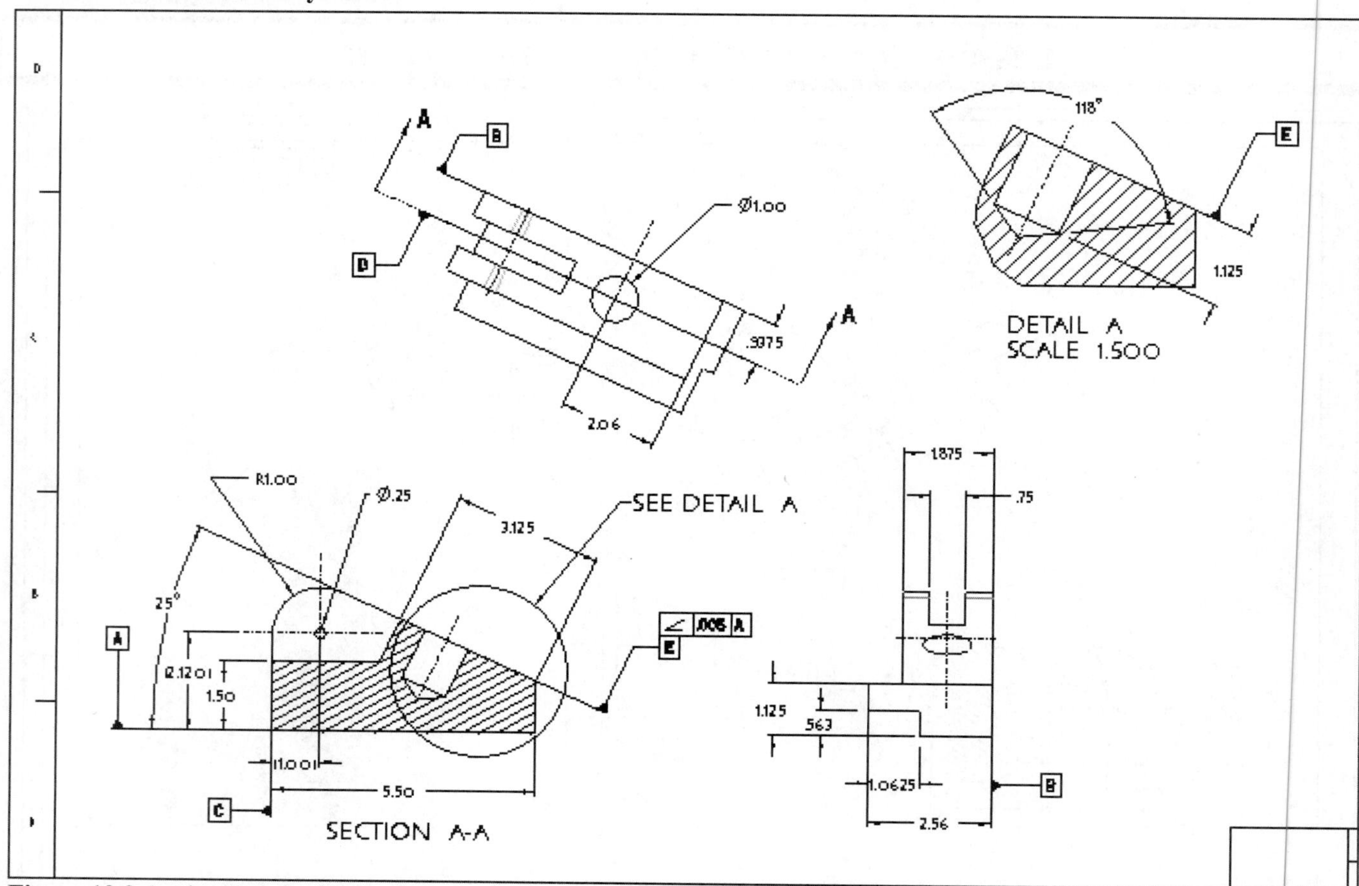

Figure 19.2 Anchor Detail with Section, Auxiliary and Detail Views

The angle is defined by a ***Hinge Line*** which you specify. The view is projected using the drawing projection angle (either First or Third Angle).

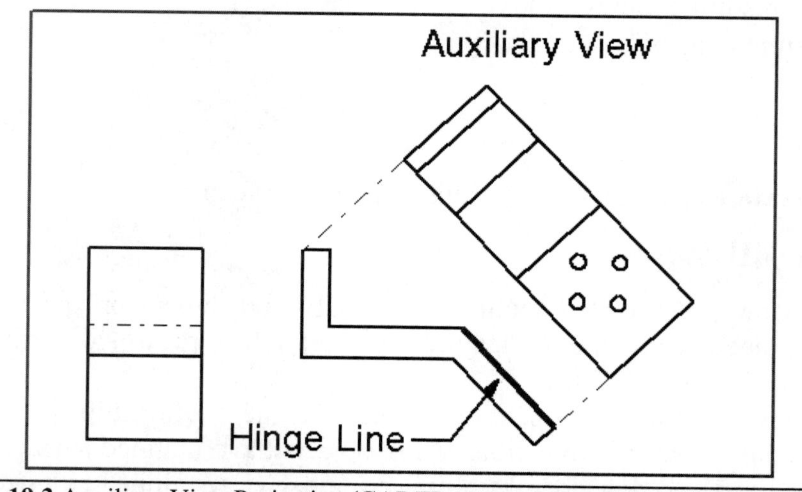

Figure 19.3 Auxiliary View Projection (CADTRAIN, COAch for Pro/E)

Sections

A sectional view (Fig. 19.4) is obtained by passing an imaginary *cutting plane* through the part (Fig. 19.5), perpendicular to the *line of sight*. The line of sight is the direction in which the part is viewed. The portion of the part between the cutting plane and the observer is "removed." The part's exposed solid surfaces are indicated by section lines. *Section lines* are uniformly spaced, angular lines drawn in proportion to the size of the drawing. There are many different types of section views.

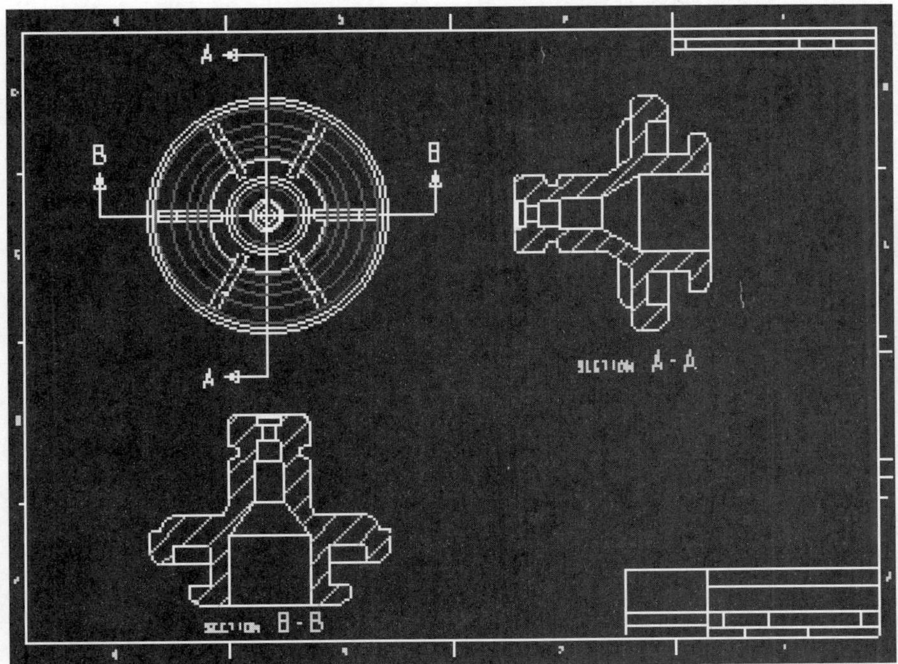

Figure 19.4 Planar Sections (CADTRAIN, COAch for Pro/E)

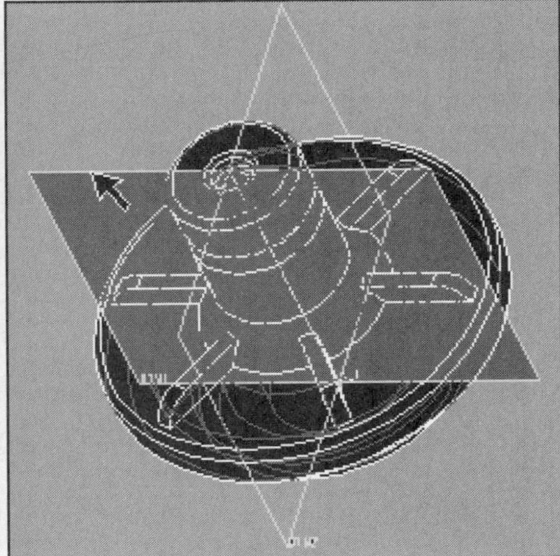

Figure 19.5 Planar Sections

Sectional views are slices through a part or assembly and are valuable for opening up the part or assembly for displaying features and detailing in Drawing mode. Part sectional views can also be used to calculate sectional view mass properties. Each sectional view has its own unique name within the part or assembly, allowing any number of sectional views to be created and then retrieved for use in a drawing. A variety of standard (ANSI) section lining symbols (cross-hatch patterns) representing the type of material can be generated and displayed.

You can create a variety of sectional view types:

- Standard planar sections of models (parts or assemblies)
- Offset sections of models (parts or assemblies)

Planar Sections

Planar sectional views are created along a datum plane. The datum may be established during the creation of the sectional view by defining a datum plane on the fly, or an existing plane may be selected.

An *offset* sectional view (Fig. 19.6) is created by extruding a 2D section perpendicular to the sketching plane, just like creating an extruded cut but without removing any material. This type of sectional view is valuable for opening up the object to display several features with a single section.

The sketched section must be an *open section*. The first and last segments of the open section must be straight lines.

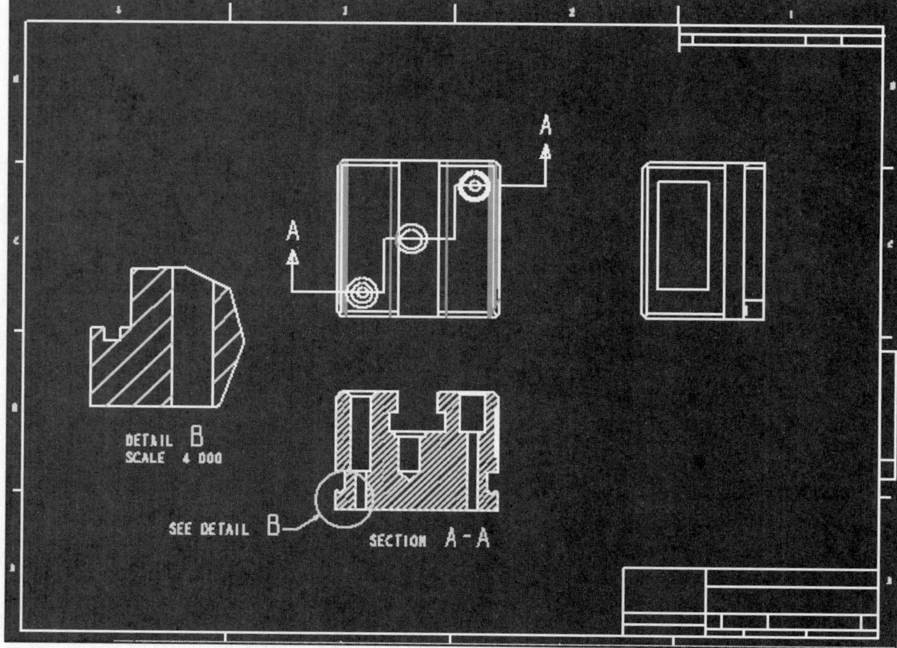

Figure 19.6 Offset Sections (CADTRAIN, COAch for Pro/E)

Auxiliary Views

An object's features and its natural or assembled position determine the proper selection of views, view orientation, and view alignment. Normally, the front view is the primary view and the top view is obvious, based on the position of the part in space or when assembled. The choice of additional views is determined by the part's features (Fig. 19.7), and the minimum number of views necessary to describe the object and display dimensions. Auxiliary views are created by making a projection of the model perpendicular to a selected edge. They are normally used to discern the true size and shape of a planar surface on an object. An auxiliary view can be created from any other type of view. Auxiliary views can have arrows created for them that point back at the view(s) from which they were created.

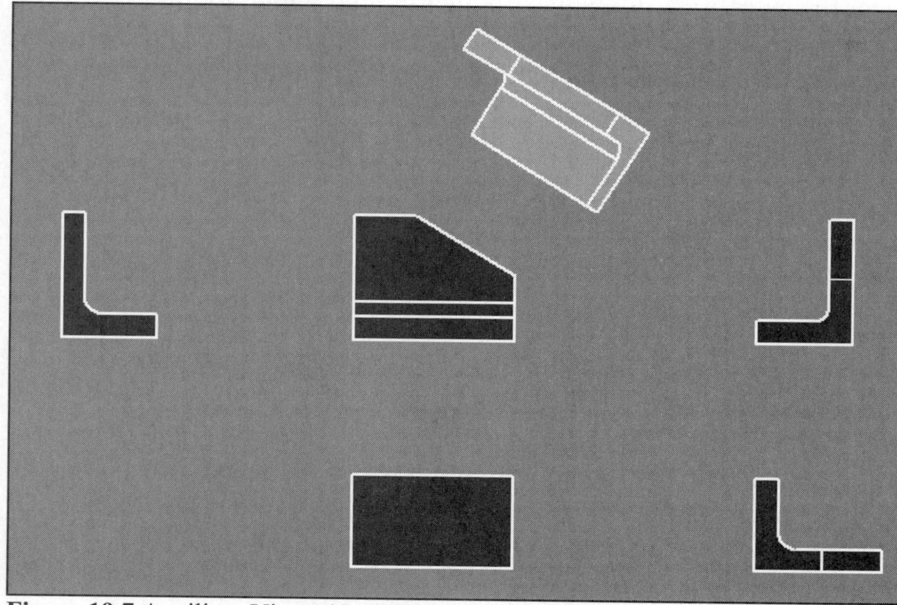

Figure 19.7 Auxiliary Views (CADTRAIN, COAch for Pro/E)

Views and Drawing Options

Auxiliary views are created, using Pro/E, by making a projection of the model perpendicular to a selected edge or along an axis. They are normally used to discern the true size and shape of a planar surface on a object, such as the Anchor, which has an angled surface with a hole machined perpendicular to that surface. Pro/E is able to create and display fully associative *section views* of solid models. You can create a section view by using an existing view on a drawing in Drawing mode or by creating a section in Part mode for retrieval in a drawing. As you were completing some of the lessons in this text, you were instructed to create sections in a number of lesson parts and lesson projects. You create a section view while looking at the drawing by sketching (with the help of Pro/E) a section line (*Offset*) or by selecting an existing Datum Plane for the section line (*Planar*) to pass through.

If you are going to create sections while you are in Drawing mode, it may require Datum Planes. Because most views in a drawing are orthographic, and because they cannot be rotated, so, you may wish to prepare the model with the necessary Datum Planes *before* you enter Drawing mode. You were asked to create a section of the Anchor [Fig. 19.8(a-b)] in Lesson 5. If you did not, you now need to create a section that passes vertically, lengthwise, through the model in Part mode or to create a section in Drawing mode, as described next.

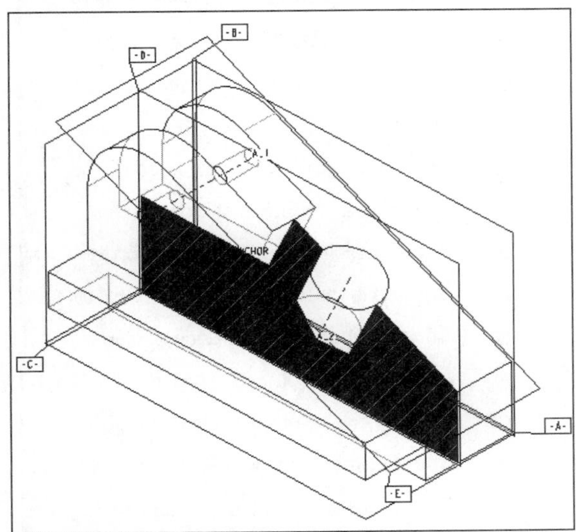

Figure 19.8(a) Sectioned Model

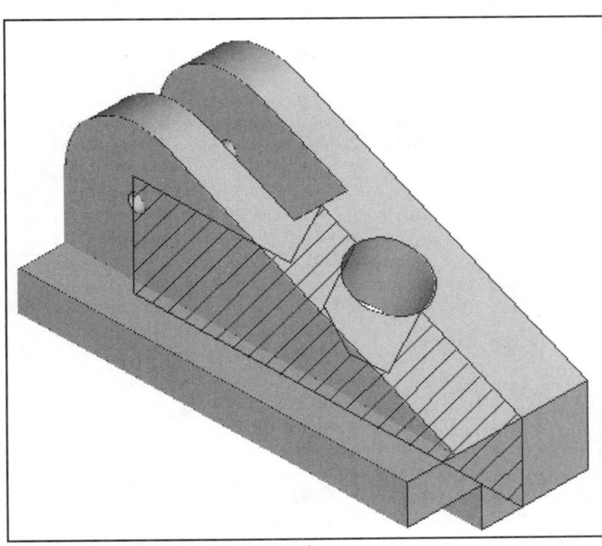

Figure 19.8(b) Shaded Section

The display of the *section line symbol, section cutting plane* and the *section view text* are controlled by the Option Values in the Drawing Options file (SETUP .dtl options file) [Fig. 19.9(a-c)].

The sectioning parameters apply to the two basic types of standards: the ANSI (ASME) Standards and the ISO/JIS/DIN Standards. Section line display and the manner in which the view titles (parameters) are created vary based on the standard and options you choose [Fig. 19.10(a-b)]. For a listing of section line and view parameters, see the Help Center for Pro/DETAIL.

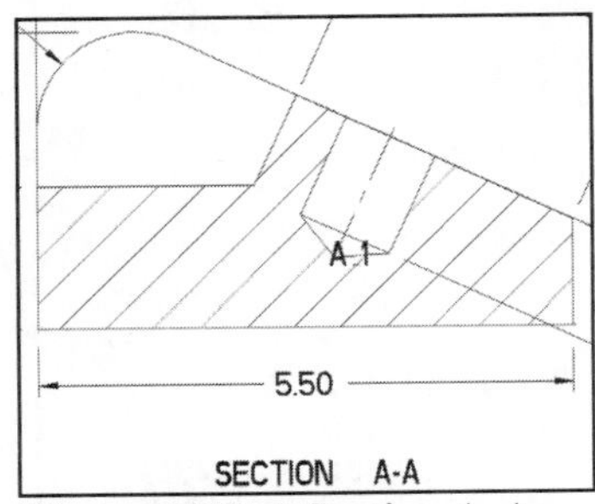

Figure 19.9(a) Section A-A

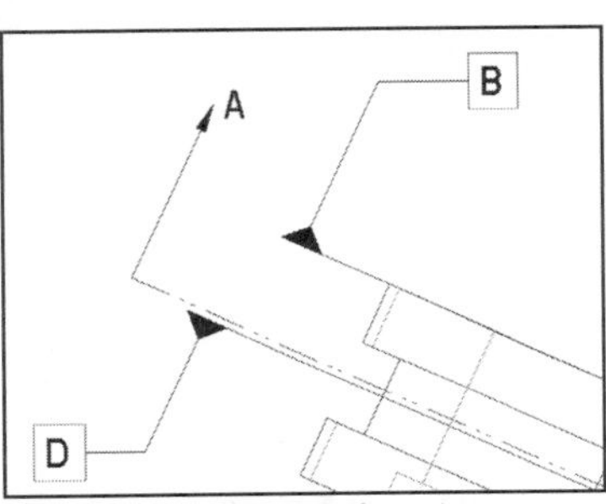

Figure 19.9(b) Section Arrows

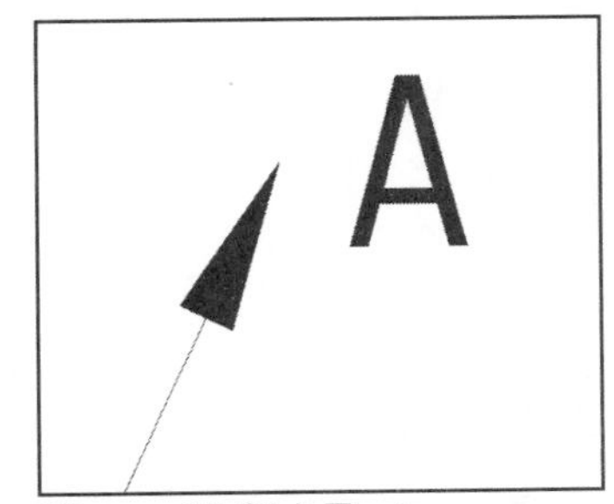

Figure 19.9(c) Text

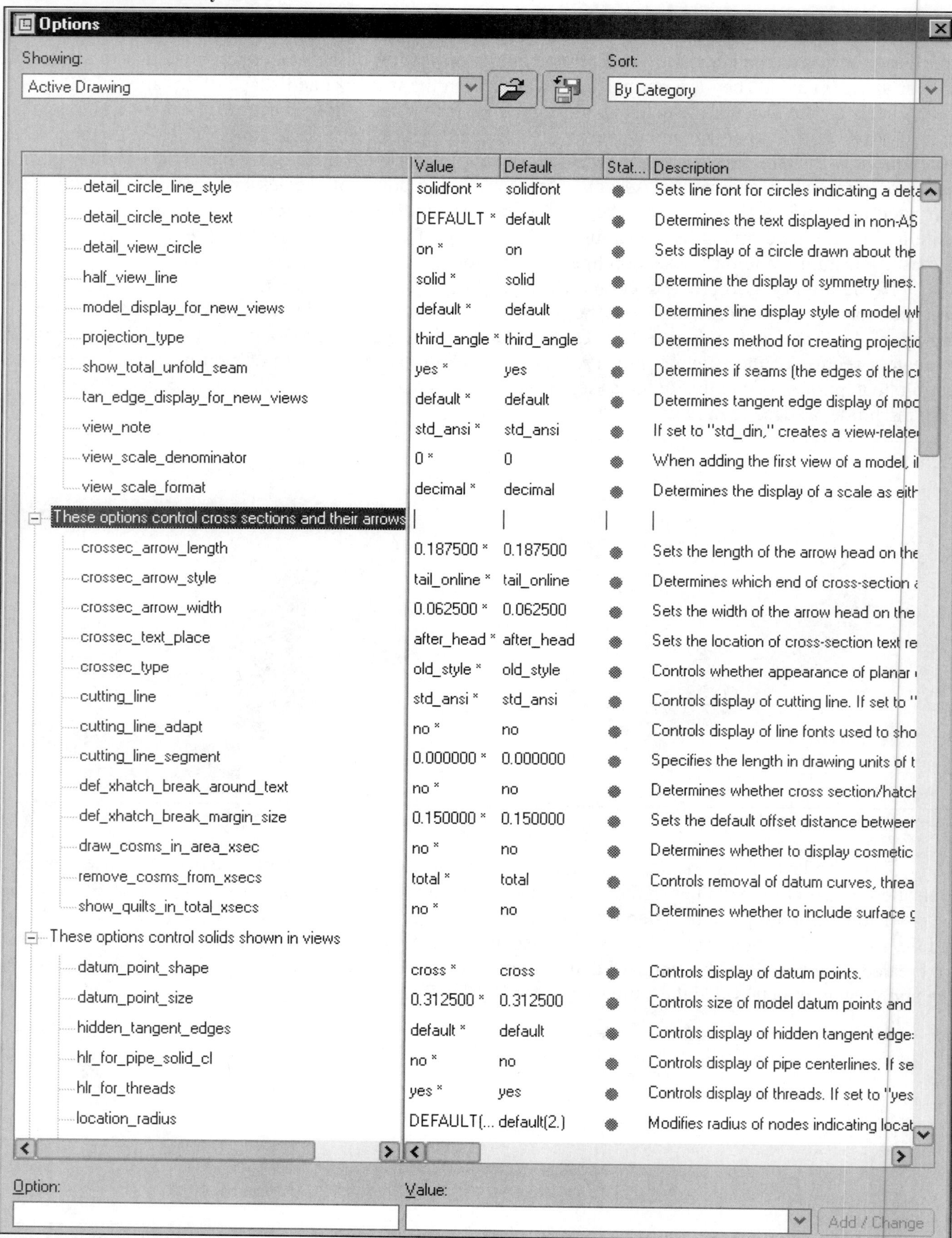

Figure 19.10(a) Drawing Options- Cross Sections

In Drawing mode, the **Planar** section option allows you to create a section that passes straight through a part without any "jogs" (steps) in the section line. Planar sections can be defined by retrieving a section, creating a datum plane, or selecting an existing datum plane to define the cut position.

When you want to create the section while in Drawing mode, you can select either a planar face or a Datum Plane as the "plane" the section cut passes through. Even though you can pick a planar face, this is generally not good practice. The section cut takes place *at* the plane you select. If you pick a face, this causes the section to be tangent to it. Not only is this poor drafting practice, but in some cases it also causes Pro/E to generate incomplete sections.

Parameter	Default	Hint
crossec_arrow_length	.187	
crossec_arrow_width	.0625	
crossec_arrow_style	tail_online	Tail Head
crossec_text_place	after_head	A after_head A before_tail A above_tail A above_line
cutting_line	std_ansi	std_ansi std_din std_jis std_iso std_ansi_dashed std_jis_alternate
cutting_line_segment	0	0 value
def_view_text_height	0	A
view_note	std_ansi	std_ansi std_din std_jis std_iso

Figure 19.10(b) Cutting Plane Options

When you create a Planar section on the drawing, Pro/E first asks you to define the location of the center of the section view. It then displays the orthographic projection to which the section edges (created by the cut) and the crosshatching will be added. Pro/E prompts you to enter a name for the section. This "name" is actually the section letter you wish to use. If you are creating your first section and you want it to have the letter **A** displayed at each arrow end and have the title **SECTION A-A**, you should enter **A** as the name of the section. It is a common mistake to type **AA**, thinking that you are establishing **SECTION A-A**, when in reality you are getting **SECTION AA-AA**. Type one letter only. You must then select a Datum Plane to define the section cutting plane. Pro/E then prompts you to pick a view in which the cutting plane is perpendicular to the screen. This is the view in which it will draw the section line and the view where the cut will actually take place. Pro/E displays the section line, with its arrows pointing in the direction it "thinks" you want to view the cut.

After the section has been created, pick the section cutting plane ⇒ RMB ⇒ Flip Material Removal Side (to flip the section identification arrows). The arrow direction does not affect the cross-sectioned area, which was defined when you indicated the location of the view. It does affect what you see in the "background," behind the cutting plane. You can change the name of a section by picking the section cutting plane ⇒ RMB ⇒ Rename, or you can accomplish the same thing in the Part mode.

Lesson 19 STEPS

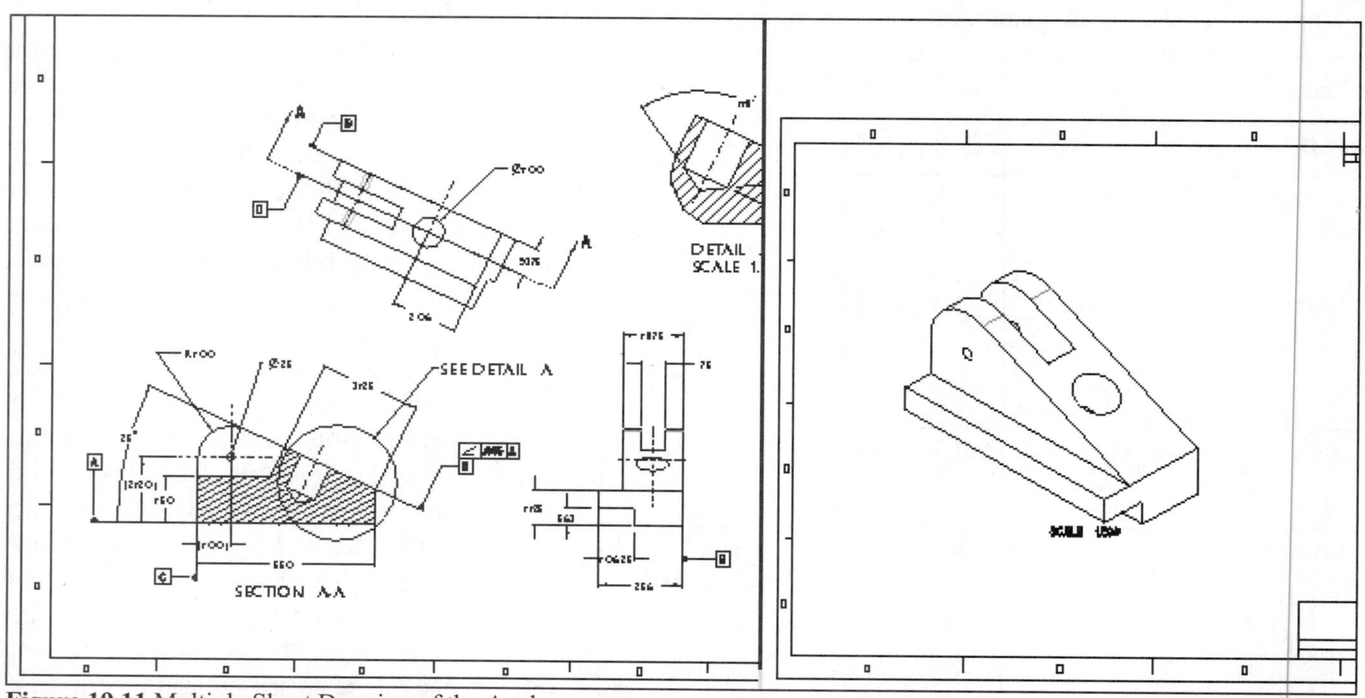

Figure 19.11 Multiple Sheet Drawing of the Anchor

Anchor Drawing

You will be creating a detail drawing of the Anchor (Fig. 19.11). The front view will be a full section. A right side view and an auxiliary view are required to detail the part. Views will be displayed according to visibility requirements per ANSI standards, such as no hidden lines in sections. The part is to be dimensioned according to ASME Y14.5M 1994. You will add the format created previously. Detailed views of other parts will be introduced to show the wide variety of view capabilities of Pro/E's Drawing mode.

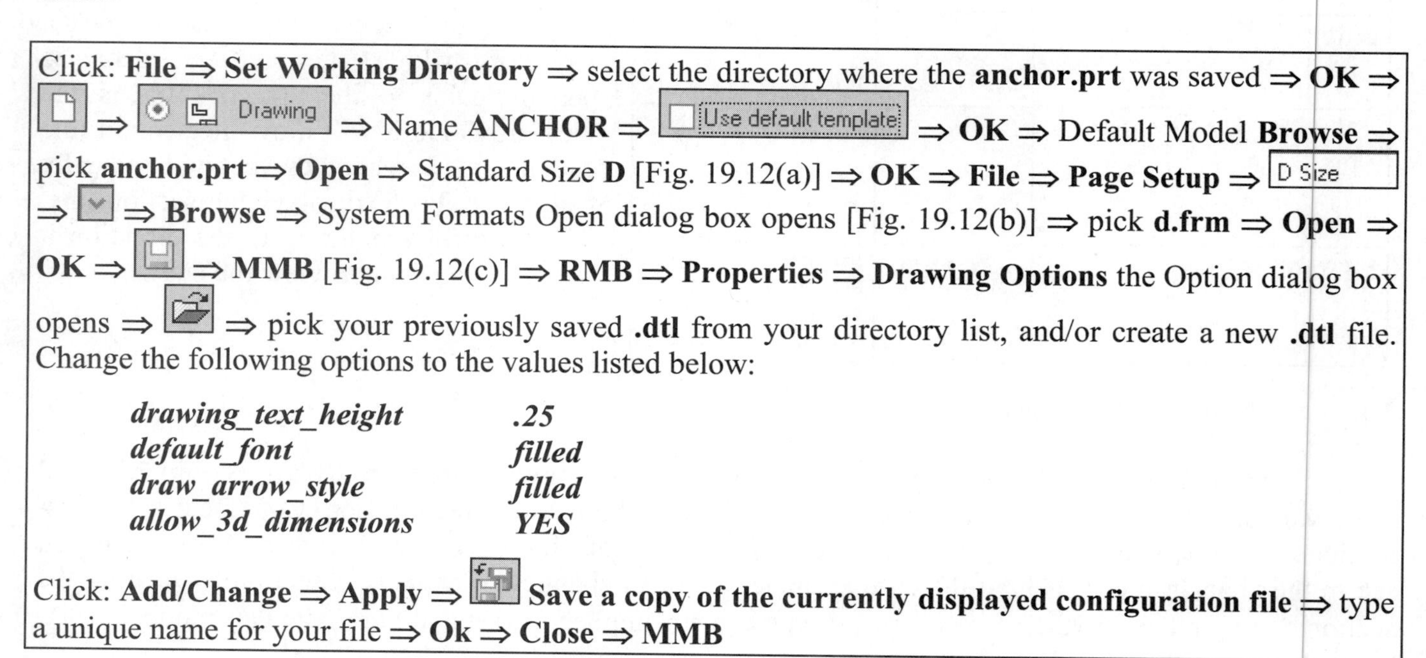

Click: **File** ⇒ **Set Working Directory** ⇒ select the directory where the **anchor.prt** was saved ⇒ **OK** ⇒ ⇒ Drawing ⇒ Name **ANCHOR** ⇒ Use default template ⇒ **OK** ⇒ Default Model **Browse** ⇒ pick **anchor.prt** ⇒ **Open** ⇒ Standard Size **D** [Fig. 19.12(a)] ⇒ **OK** ⇒ **File** ⇒ **Page Setup** ⇒ D Size ⇒ ⇒ **Browse** ⇒ System Formats Open dialog box opens [Fig. 19.12(b)] ⇒ pick **d.frm** ⇒ **Open** ⇒ **OK** ⇒ ⇒ **MMB** [Fig. 19.12(c)] ⇒ **RMB** ⇒ **Properties** ⇒ **Drawing Options** the Option dialog box opens ⇒ ⇒ pick your previously saved **.dtl** from your directory list, and/or create a new **.dtl** file. Change the following options to the values listed below:

drawing_text_height	***.25***
default_font	***filled***
draw_arrow_style	***filled***
allow_3d_dimensions	***YES***

Click: **Add/Change** ⇒ **Apply** ⇒ **Save a copy of the currently displayed configuration file** ⇒ type a unique name for your file ⇒ **Ok** ⇒ **Close** ⇒ **MMB**

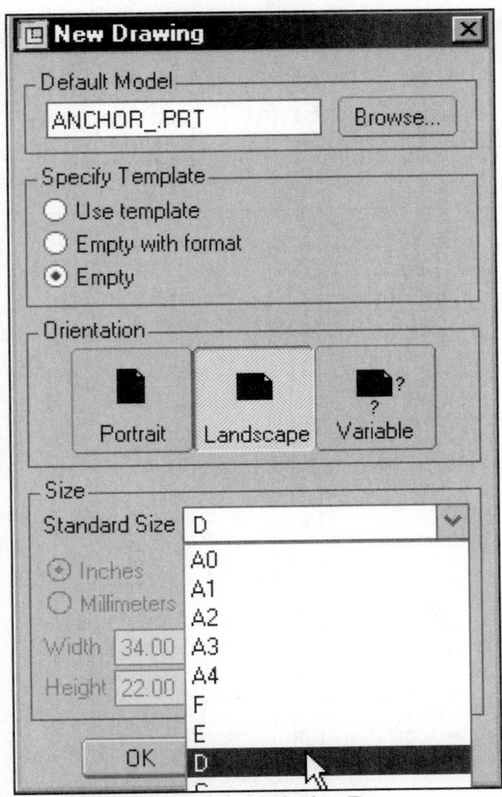

Figure 19.12(a) Standard Size D

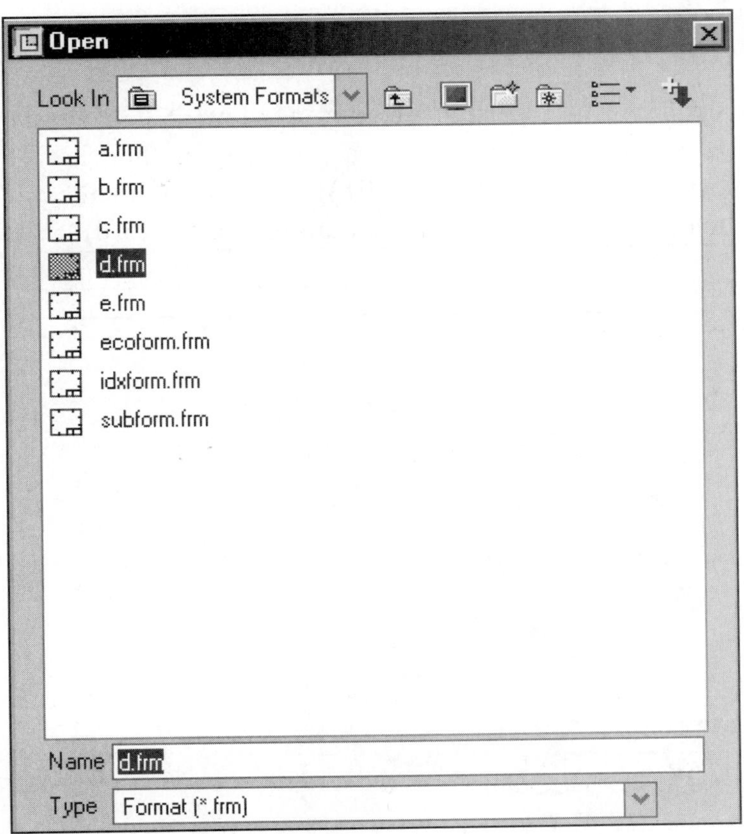

Figure 19.12(b) System Formats

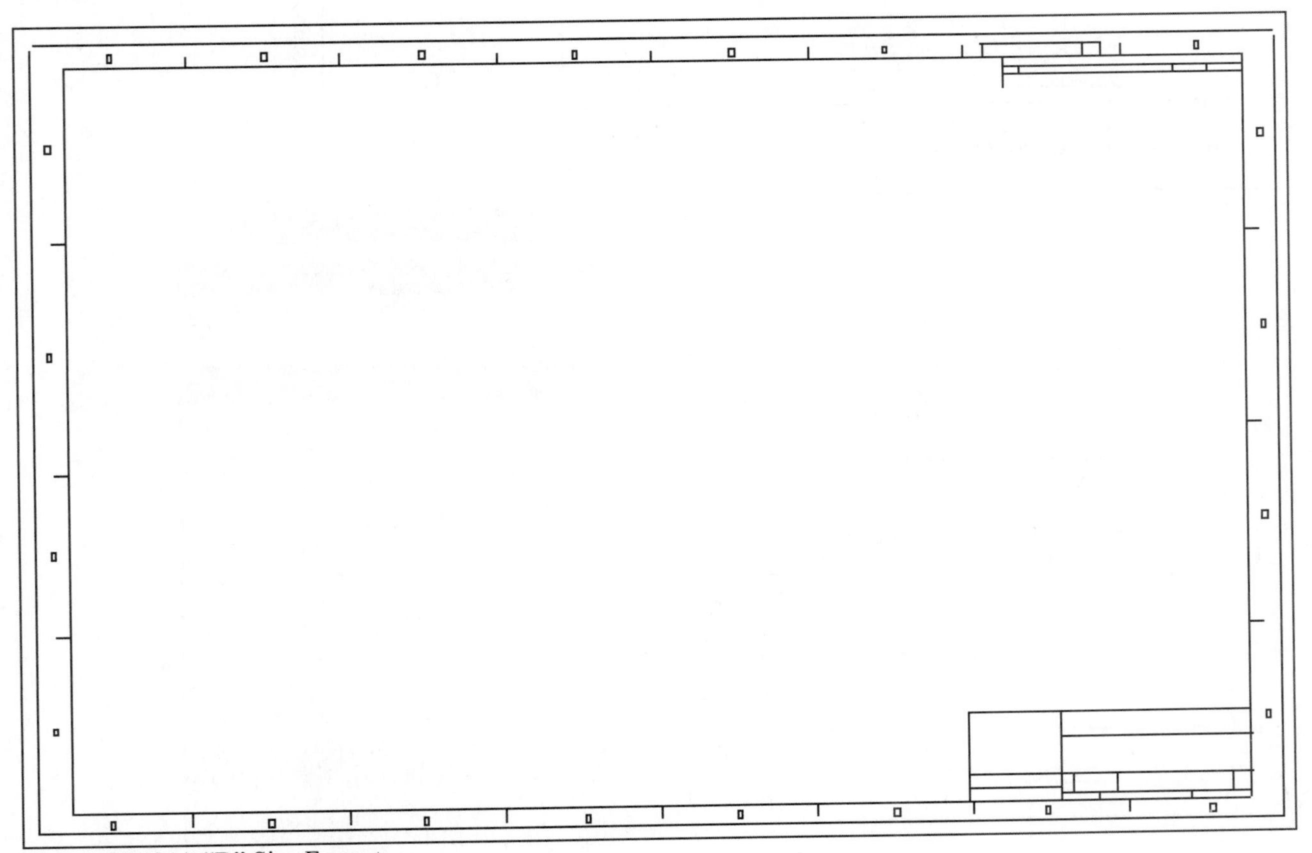

Figure 19.12(c) "D" Size Format

Click: **Datum planes off** ⇒ **Datum axes off** ⇒ **Datum points off** ⇒ **Coordinate systems off** ⇒ **Redraw the current view** ⇒ **Insert a drawing view of the active model** ⇒ **MMB** to accept defaults ⇒ pick a position for the view [Fig. 19.13(a)] ⇒ **Front** for Reference 1: Orientation dialog box [Fig. 19.13(b)] ⇒ pick datum **B** ⇒ **Top** for Reference 2 from the Orientation dialog box [Fig. 19.13(c)] ⇒ pick datum **A** [Fig. 19.13(d)] ⇒ **OK**

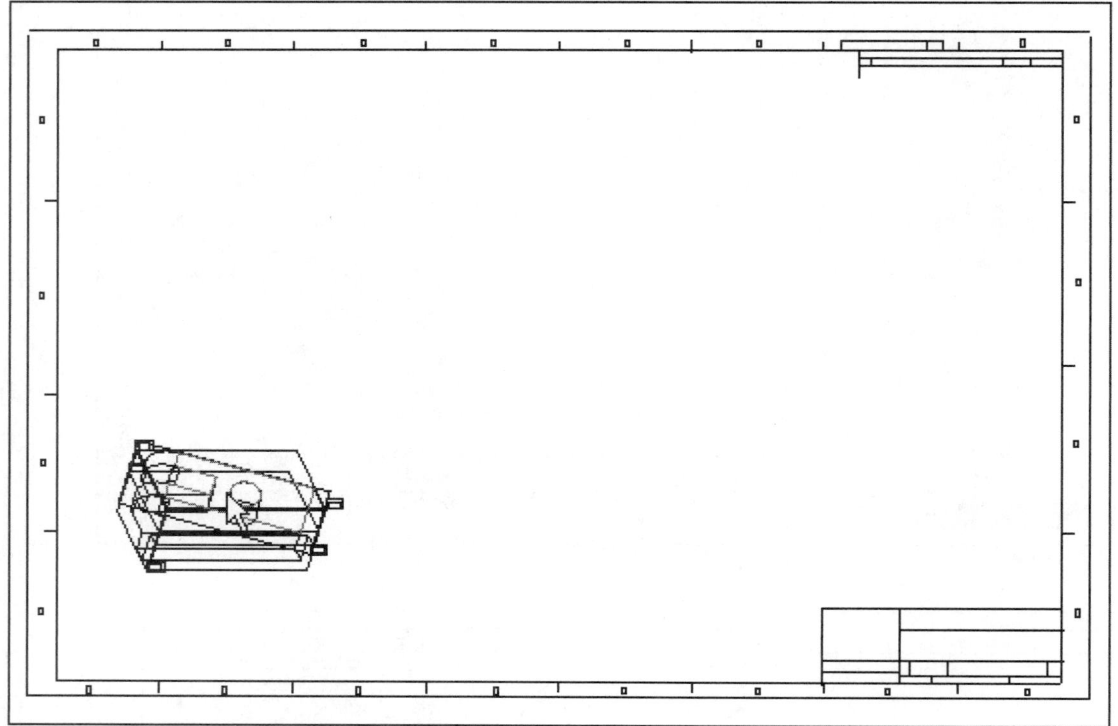

Figure 19.13(a) Pick a Position for the First View

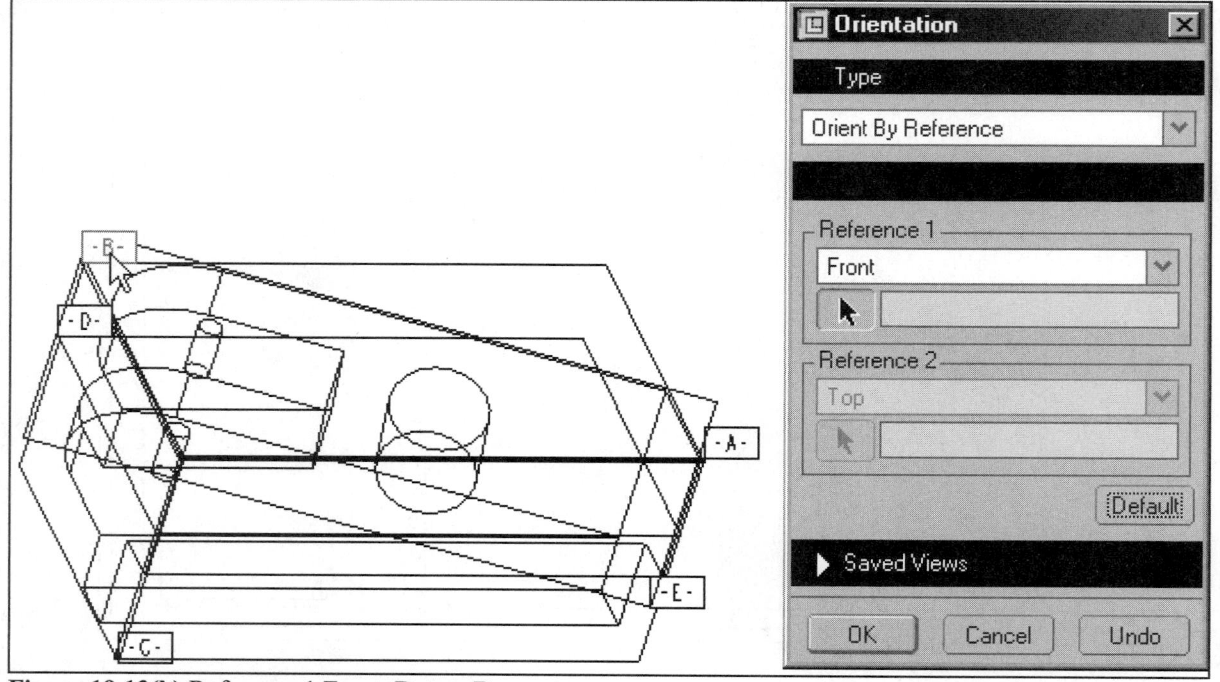

Figure 19.13(b) Reference 1 Front, Datum B

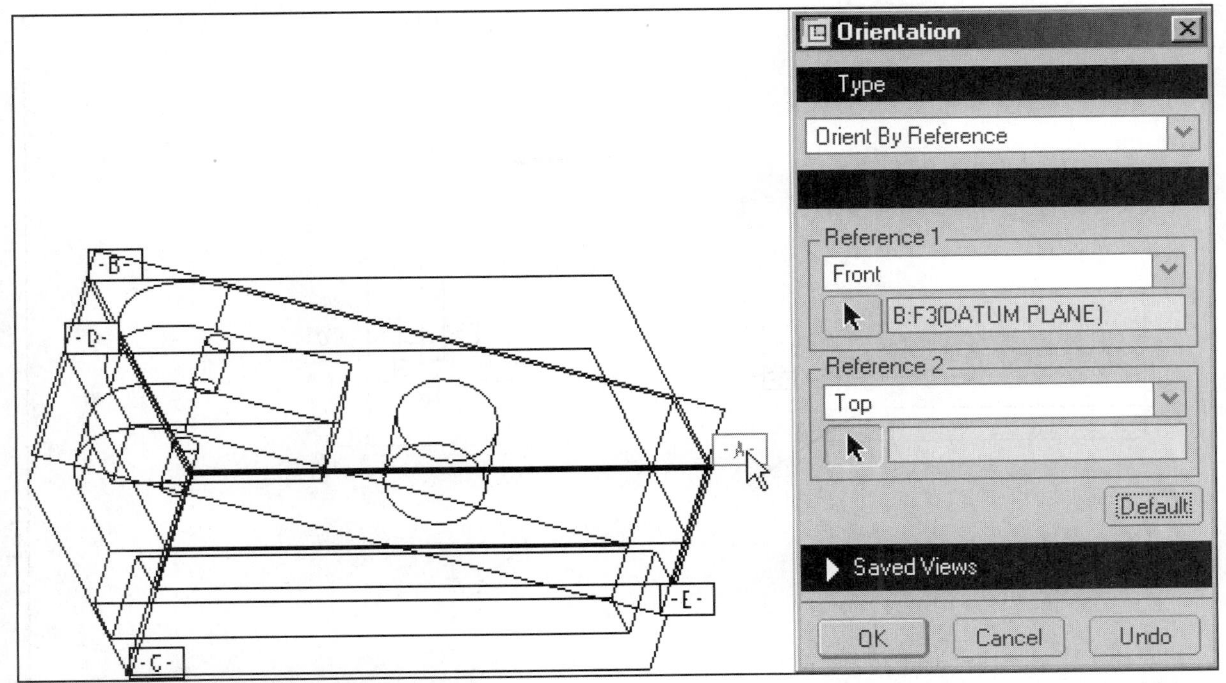

Figure 19.13(c) Reference 2 Top, Datum A

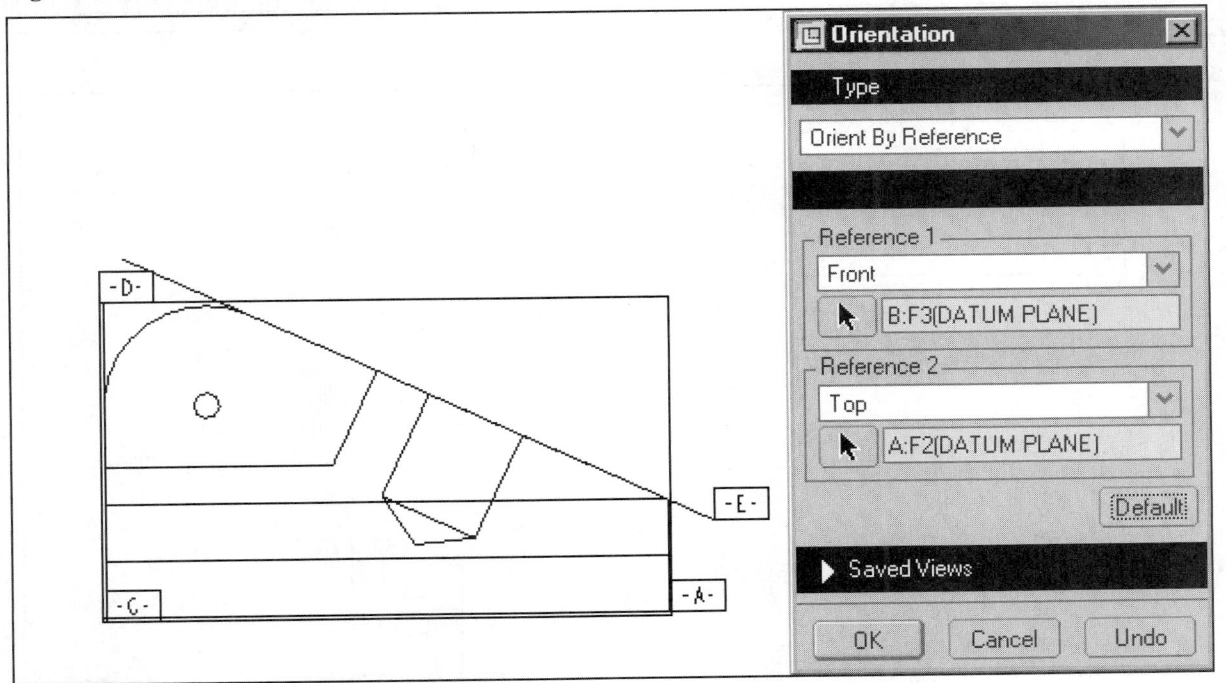

Figure 19.13(d) First View Established

Add two more views, click: **Insert a drawing view of the active model ⇒ Projection ⇒ Full View ⇒ No Xsec ⇒ No Scale ⇒ Done ⇒** Select CENTER POINT for drawing view. select a position for the view [Fig. 19.13(e)] ⇒ **Insert a drawing view of the active model ⇒ Projection ⇒ Full View ⇒ No Xsec ⇒ No Scale ⇒ MMB ⇒** Select CENTER POINT for drawing view. select a position for the view [Fig. 19.13(f)] ⇒ **LMB** ⇒ unlock ⇒ reposition the views ⇒ ⇒ ⇒ ⇒ **MMB** [Fig. 19.13(g)]

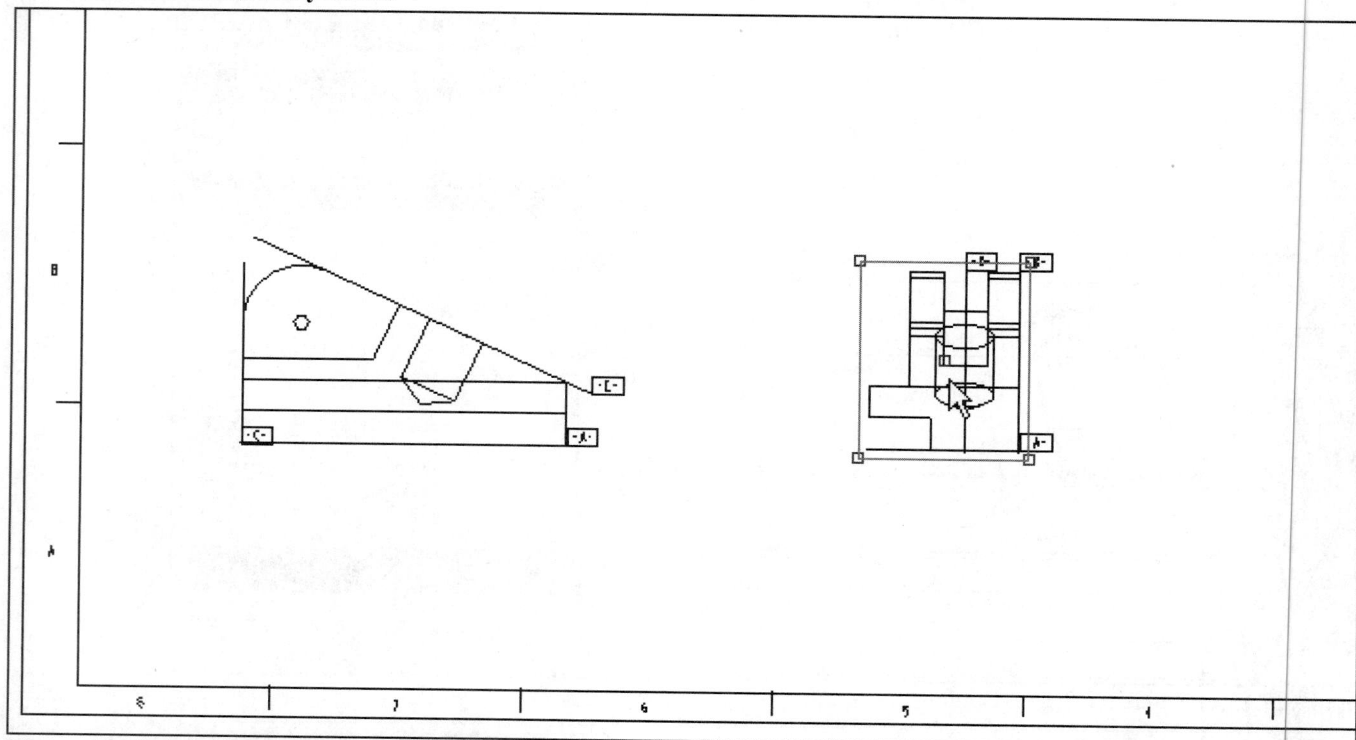

Figure 19.13(e) Add Projected View on Right Side

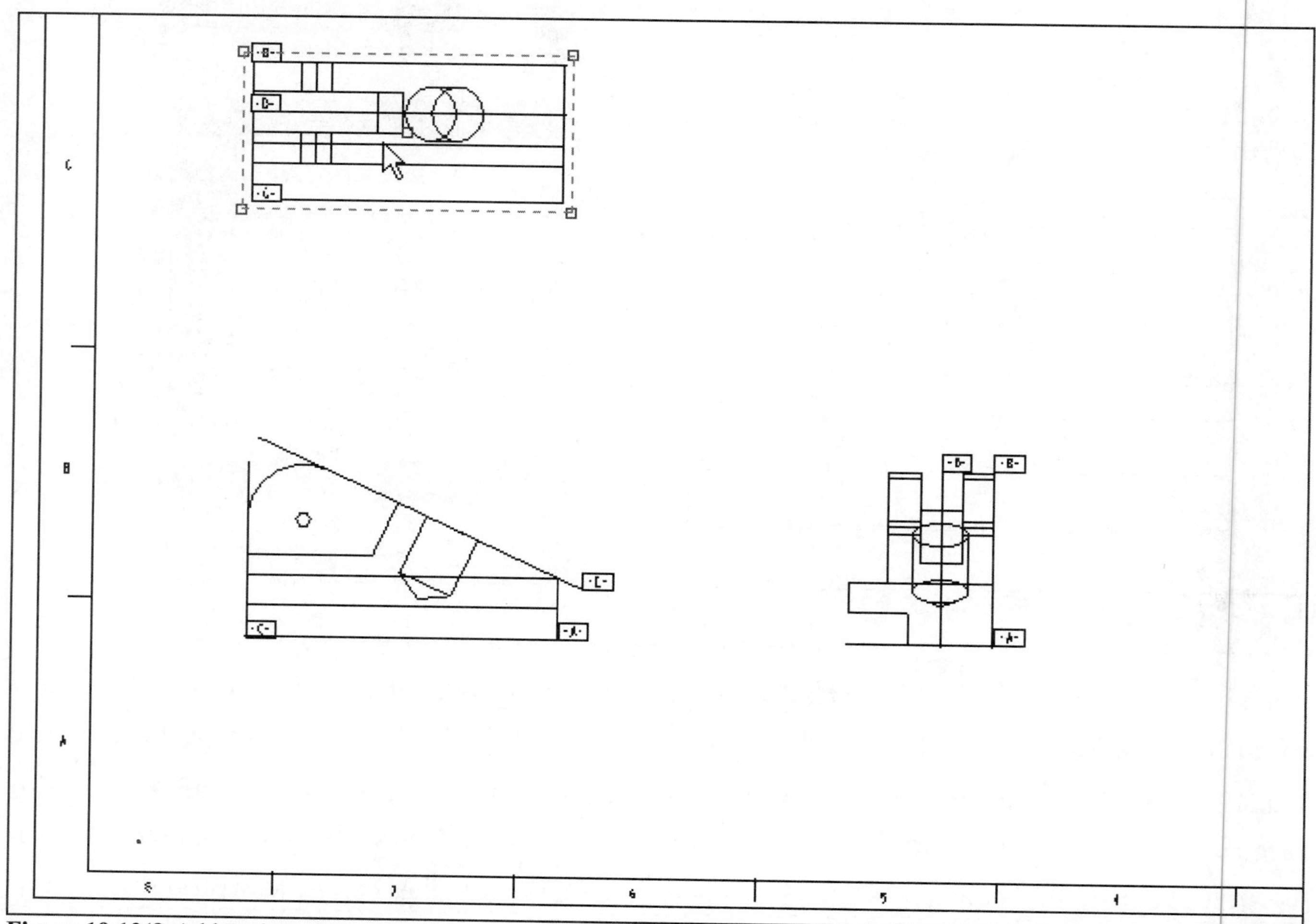

Figure 19.13(f) Add Projected View on Top

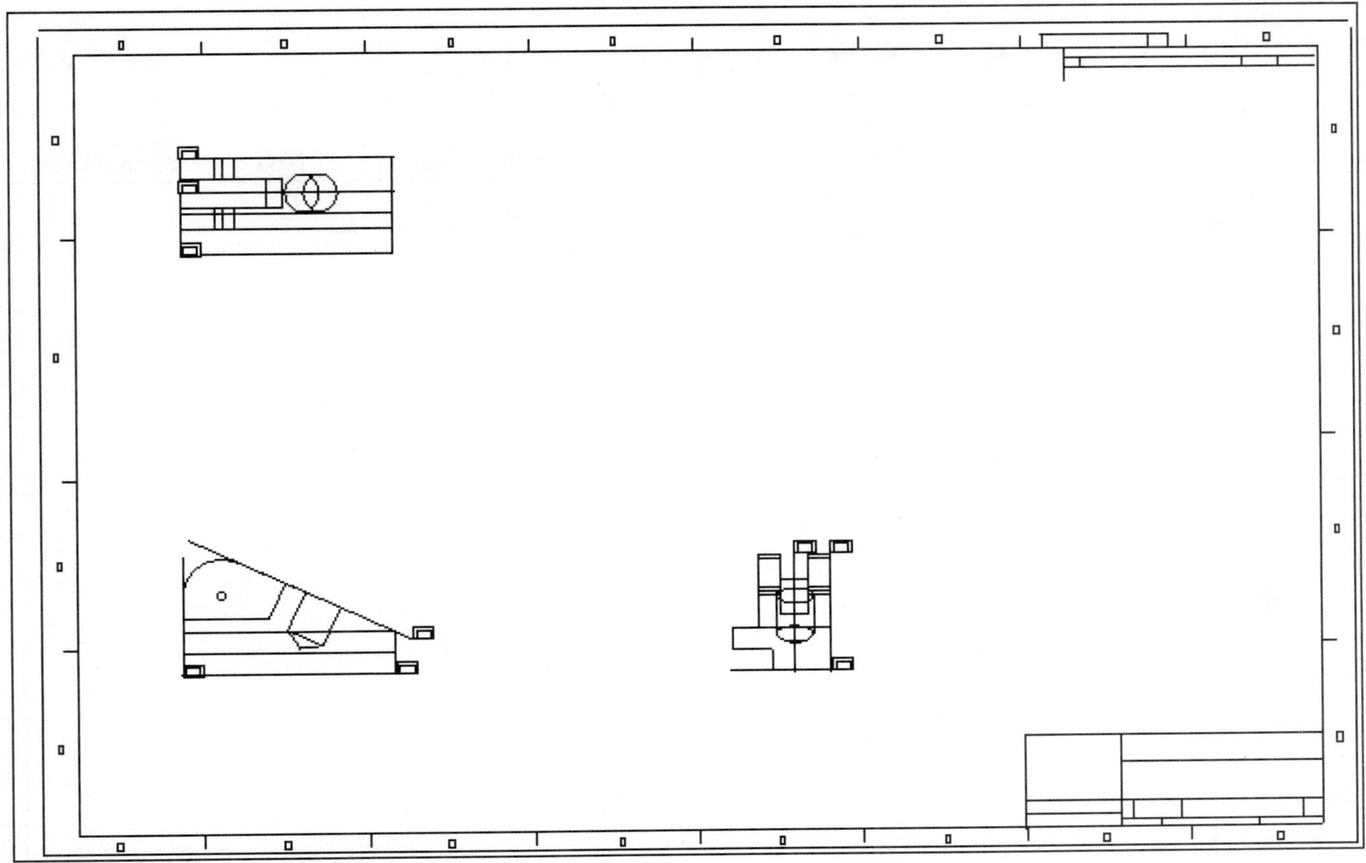

Figure 19.13(g) Repositioned Views

The top view is quite useless since it does not help in the description of the part's geometry. Delete the top view before adding an auxiliary view that will display the true shape of the angled surface.

Click on the top view ⇒ **RMB** ⇒ **Delete** (Fig. 19.14) ⇒ **Yes** ⇒ **LMB** ⇒ [icon] ⇒ [icon] ⇒ [icon] ⇒ **MMB**

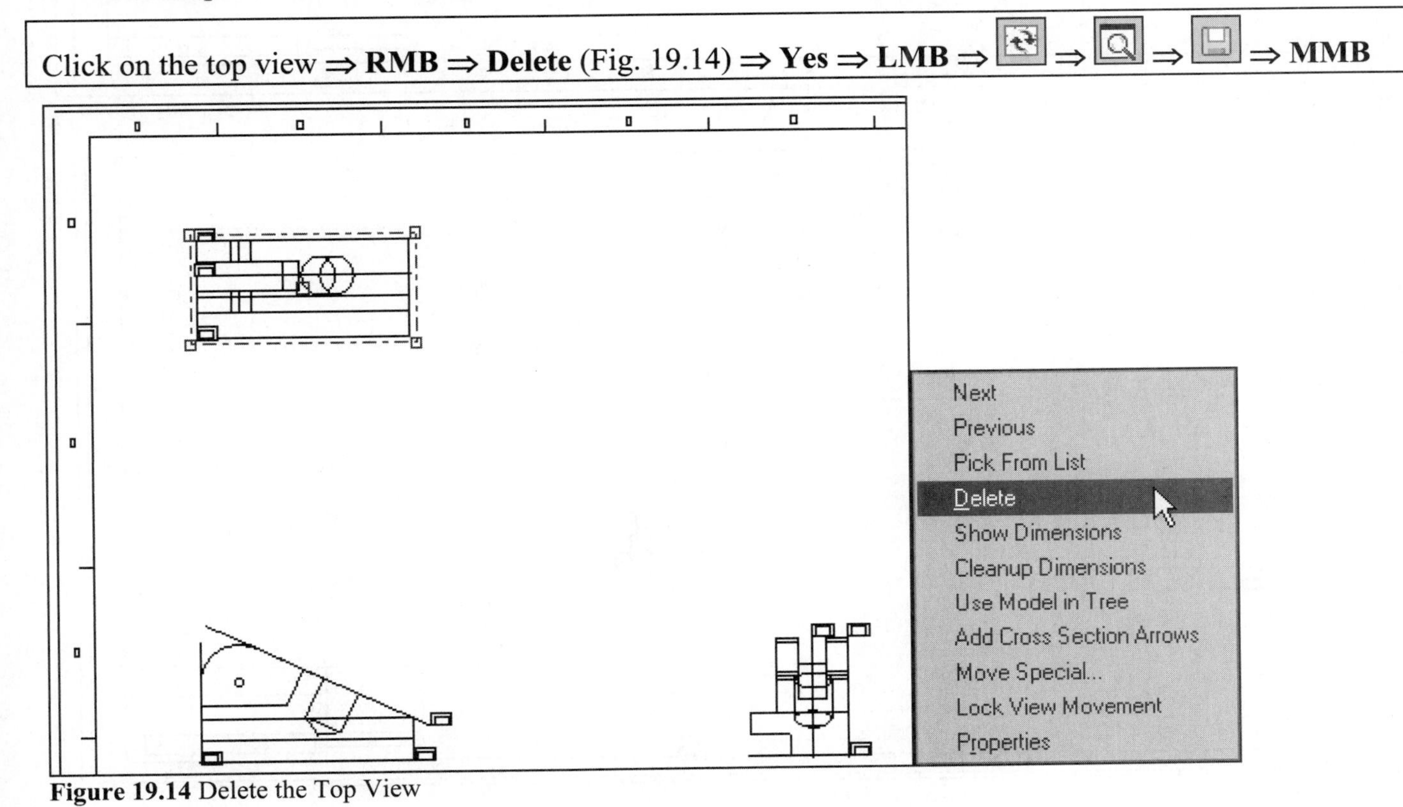

Figure 19.14 Delete the Top View

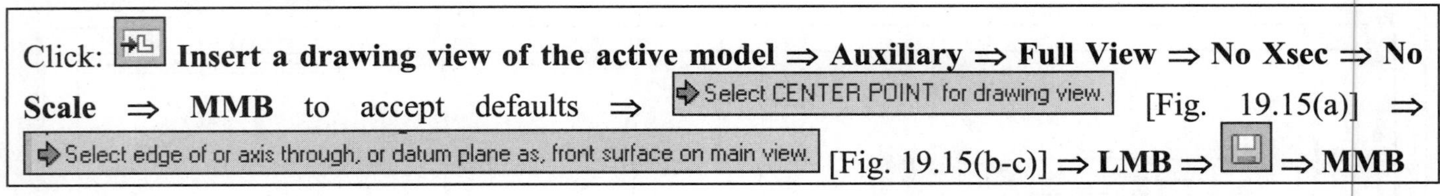

Click: **Insert a drawing view of the active model ⇒ Auxiliary ⇒ Full View ⇒ No Xsec ⇒ No Scale ⇒ MMB** to accept defaults ⇒ Select CENTER POINT for drawing view. [Fig. 19.15(a)] ⇒ Select edge of or axis through, or datum plane as, front surface on main view. [Fig. 19.15(b-c)] ⇒ **LMB** ⇒ ⇒ **MMB**

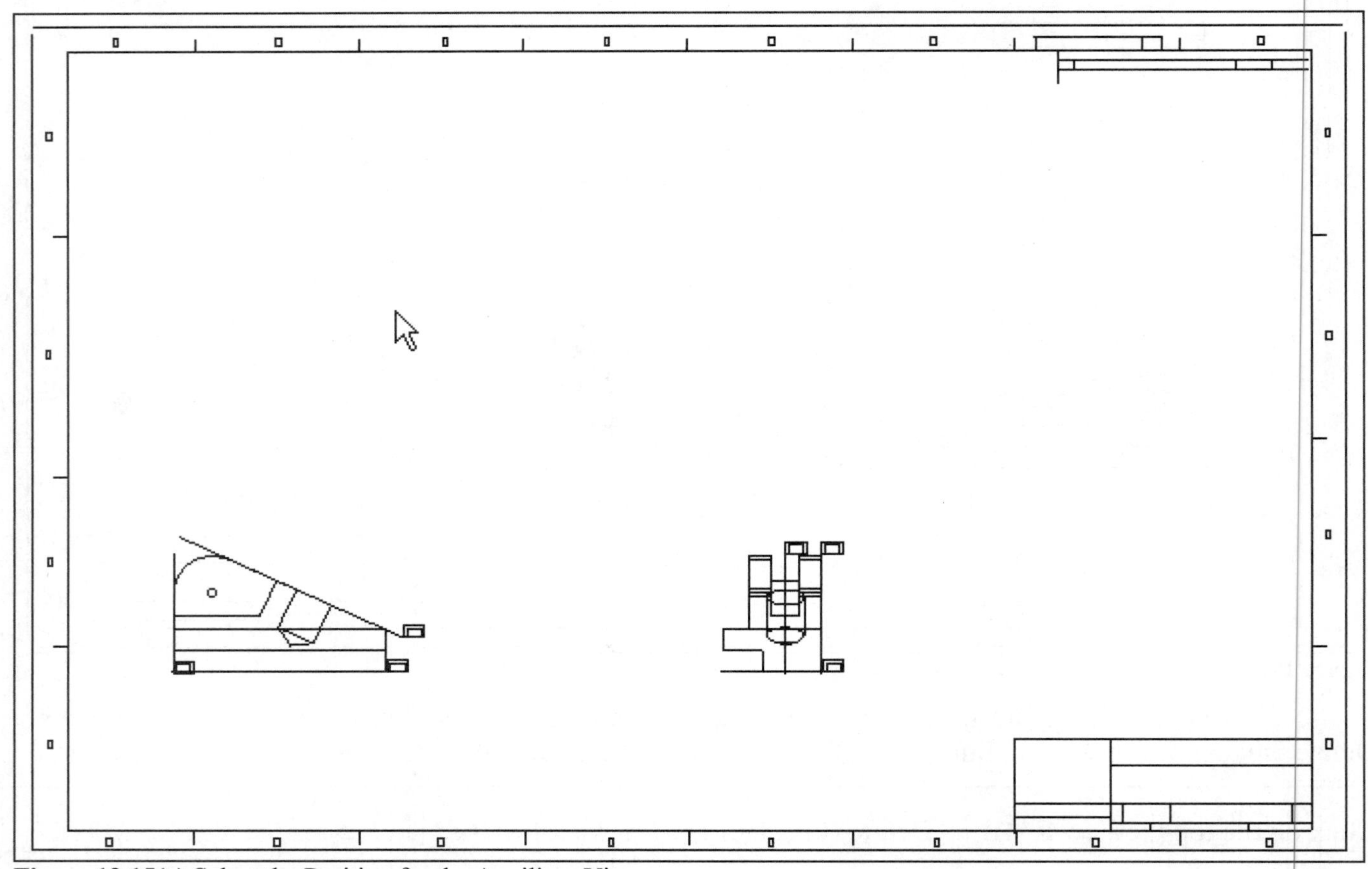

Figure 19.15(a) Select the Position for the Auxiliary View

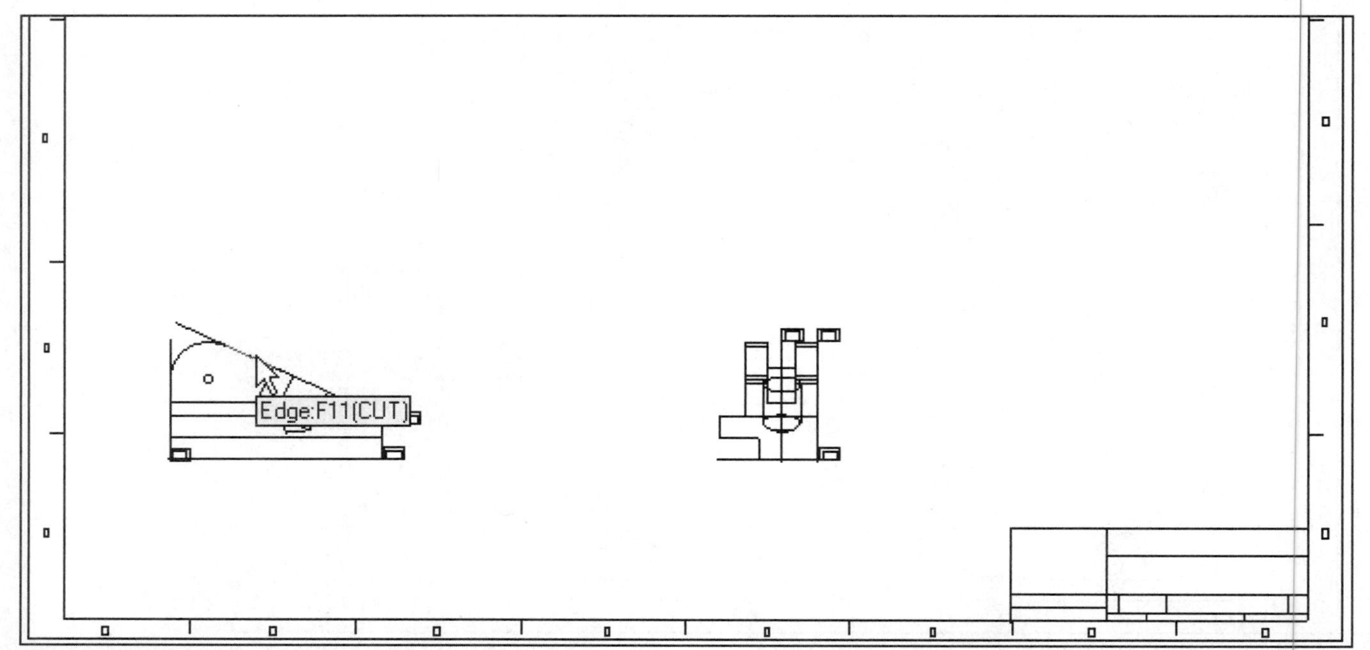

Figure 19.15(b) Select the Edge of the Angled Surface

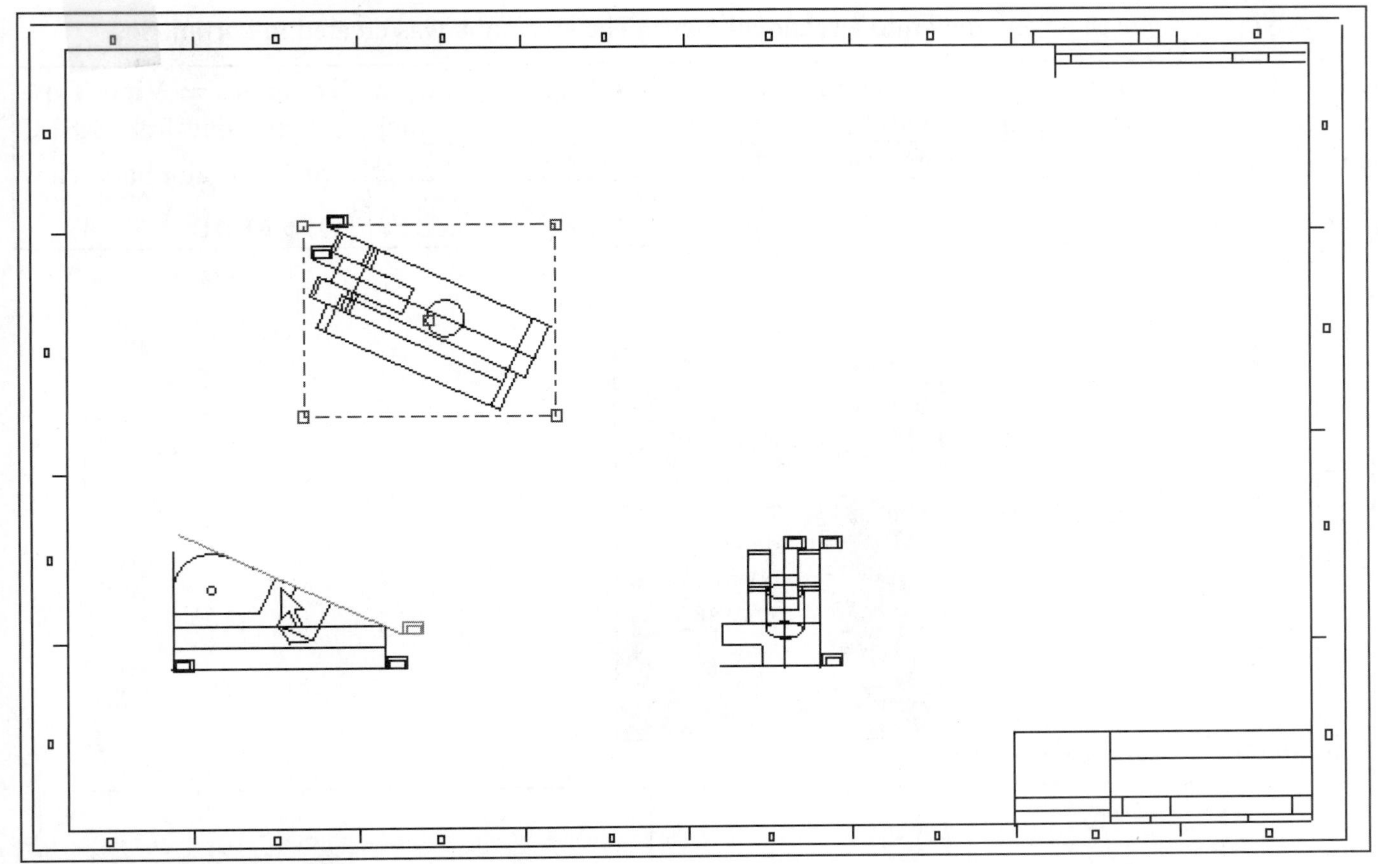

Figure 19.15(c) Auxiliary View

The proper standard was not selected when you made changes to the Drawing Options. ASME symbols for datum planes (gtol_datums) are the correct style standard used on drawings. The ANSI style [Fig. 19.16(a)] was discontinued in 1994, though retained by some companies as an "in house" standard. For outside vendors and for manufacturing internationally, the ISO-ASME standards should be applied to all manufacturing drawings.

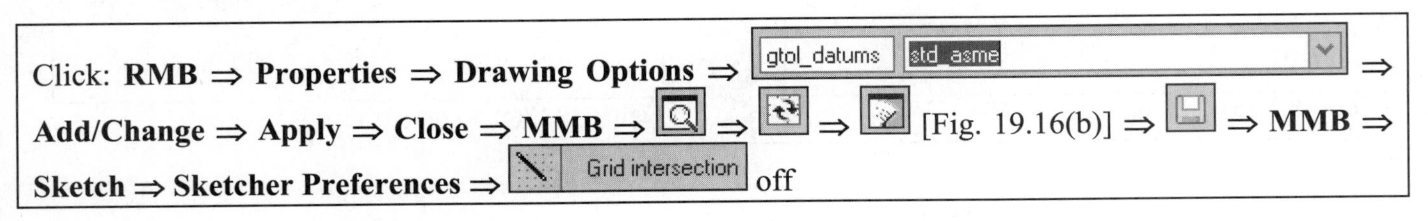

Click: **RMB** ⇒ **Properties** ⇒ **Drawing Options** ⇒ gtol_datums std_asme ⇒ **Add/Change** ⇒ **Apply** ⇒ **Close** ⇒ **MMB** ⇒ ⇒ ⇒ [Fig. 19.16(b)] ⇒ ⇒ **MMB** ⇒ **Sketch** ⇒ **Sketcher Preferences** ⇒ Grid intersection off

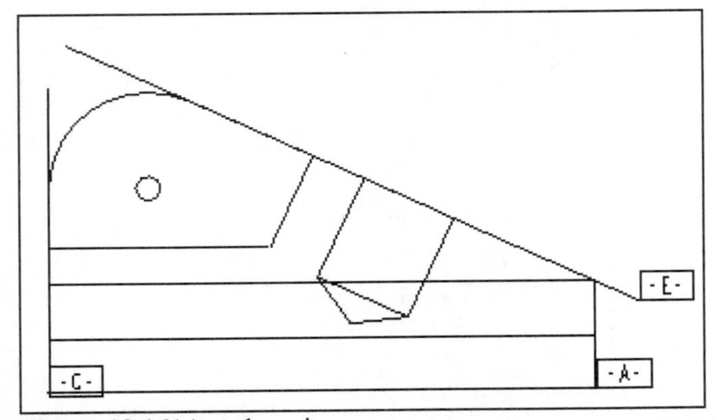

Figure 19.16(a) gtol_ansi

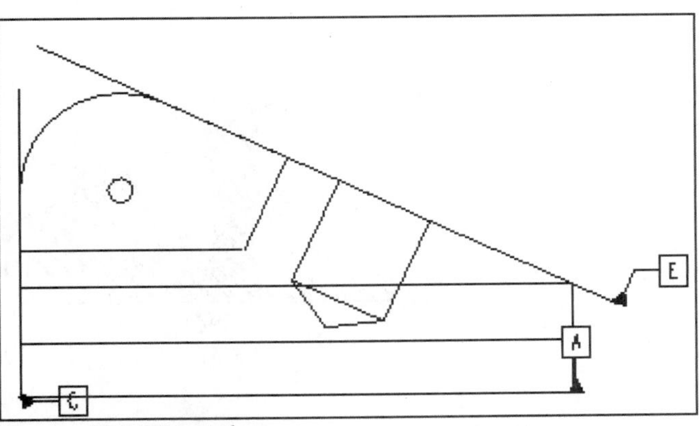

Figure 19.16(b) gtol_asme

Now, change the front view into a sectional view. The section **A** was created in Part mode.

Click on the front view as the view to be modified ⇒ **RMB** [Fig. 19.17(a)] ⇒ **Properties** ⇒ **View Type** ⇒ **Section** ⇒ **MMB** ⇒ **Full** ⇒ **Total Xsec** ⇒ **MMB** ⇒ **Retrieve** ⇒ pick section name from XSEC NAMES menu: **A** ⇒ Pick a view for arrows where the section is perp. MIDDLE button for none. pick the auxiliary view [Fig. 19.17(b)] ⇒ **MMB** [Figs. 19.17(c-d)] ⇒ **LMB** ⇒ ⇒ ⇒ ⇒ ⇒ **MMB**

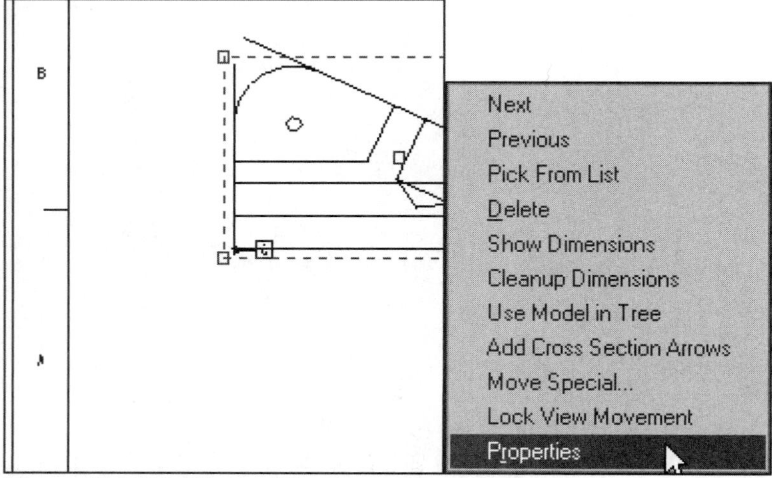

Figure 19.17(a) Change View Properties

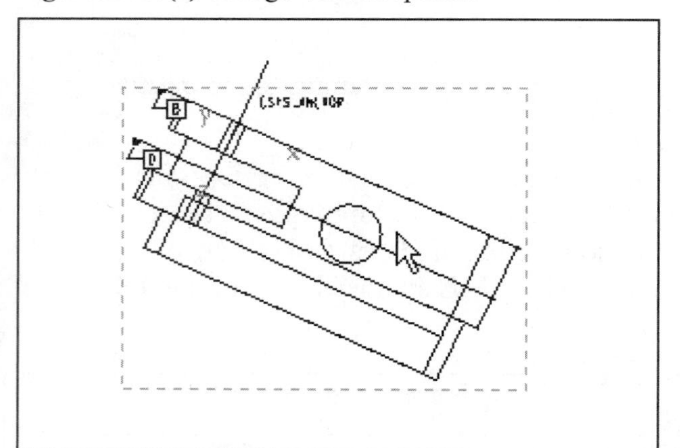

Figure 19.17(b) Click on Auxiliary View

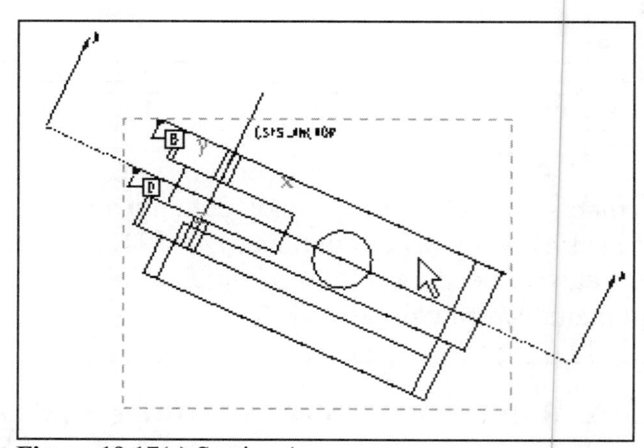

Figure 19.17(c) Section Arrows

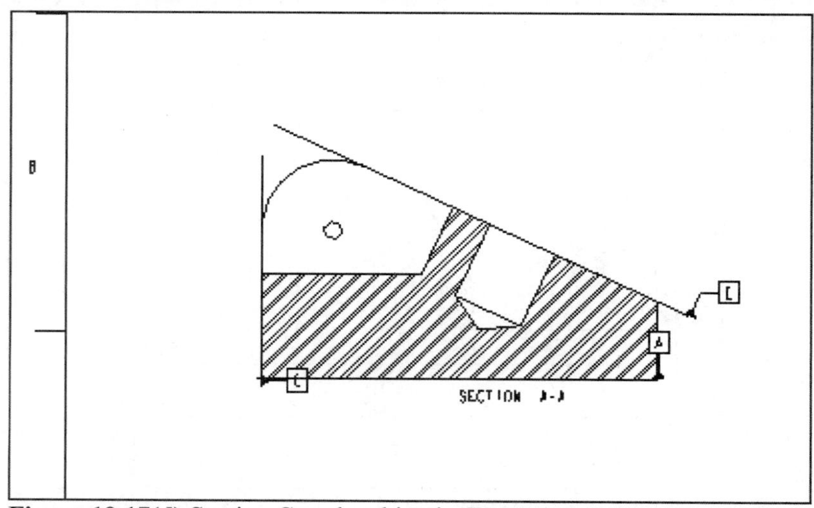

Figure 19.17(d) Section Crosshatching in SECTION A-A

Modify the visibility of the views to remove all hidden lines. While pressing and holding down the **Ctrl** key, click on all three views ⇒ click **RMB** with the cursor outside of the view outlines [Fig. 19.18(a)] ⇒ **Properties** ⇒**View Disp** ⇒ **No Hidden** ⇒ **Tan Dimmed** ⇒ **MMB** ⇒ **MMB** [Fig. 19.18(b)] ⇒ **LMB**

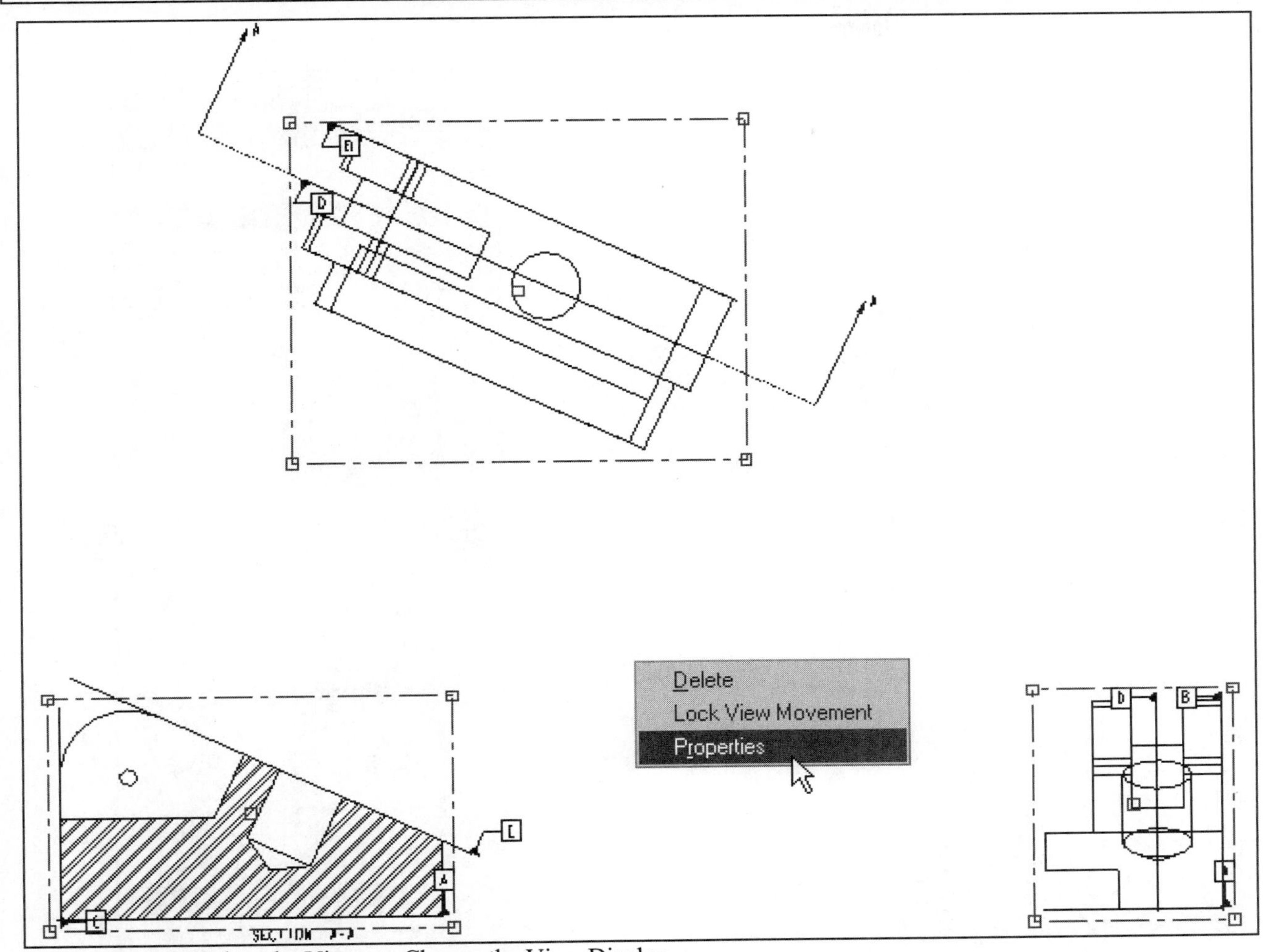

Figure 19.18(a) Select the Views to Change the View Display

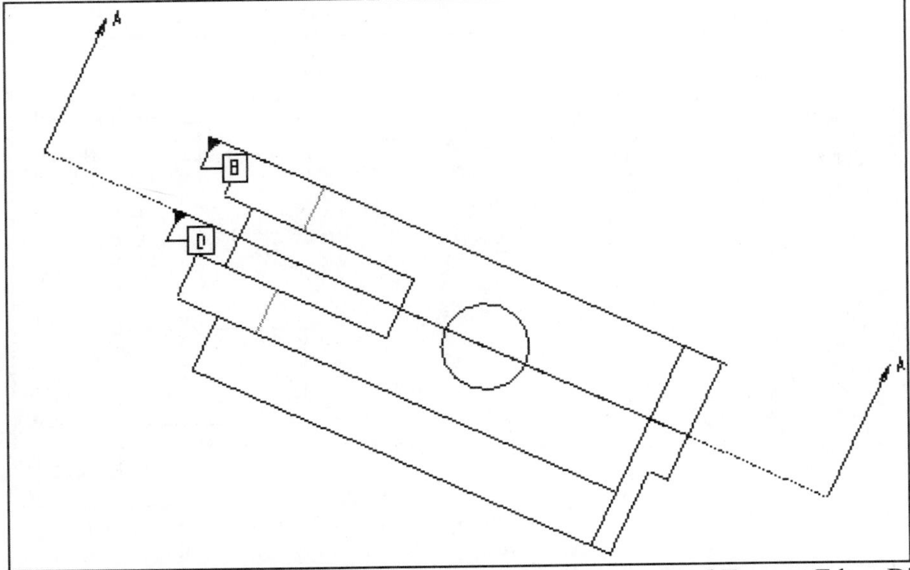

Figure 19.18(b) Auxiliary View with Hidden Lines Removed and Tangent Edges Dimmed

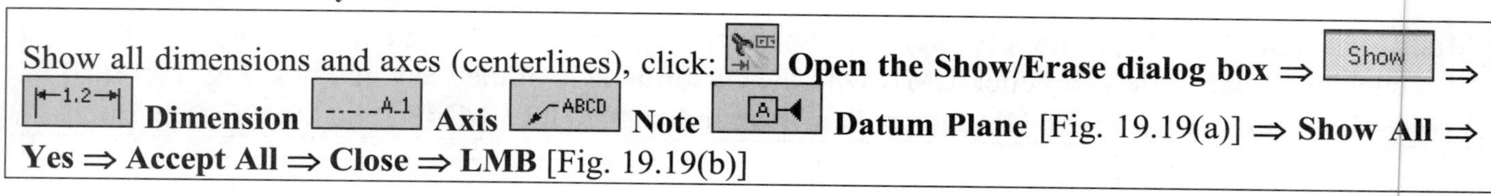

Show all dimensions and axes (centerlines), click: **Open the Show/Erase dialog box** ⇒ Show ⇒ **Dimension** **Axis** **Note** **Datum Plane** [Fig. 19.19(a)] ⇒ **Show All** ⇒ **Yes** ⇒ **Accept All** ⇒ **Close** ⇒ **LMB** [Fig. 19.19(b)]

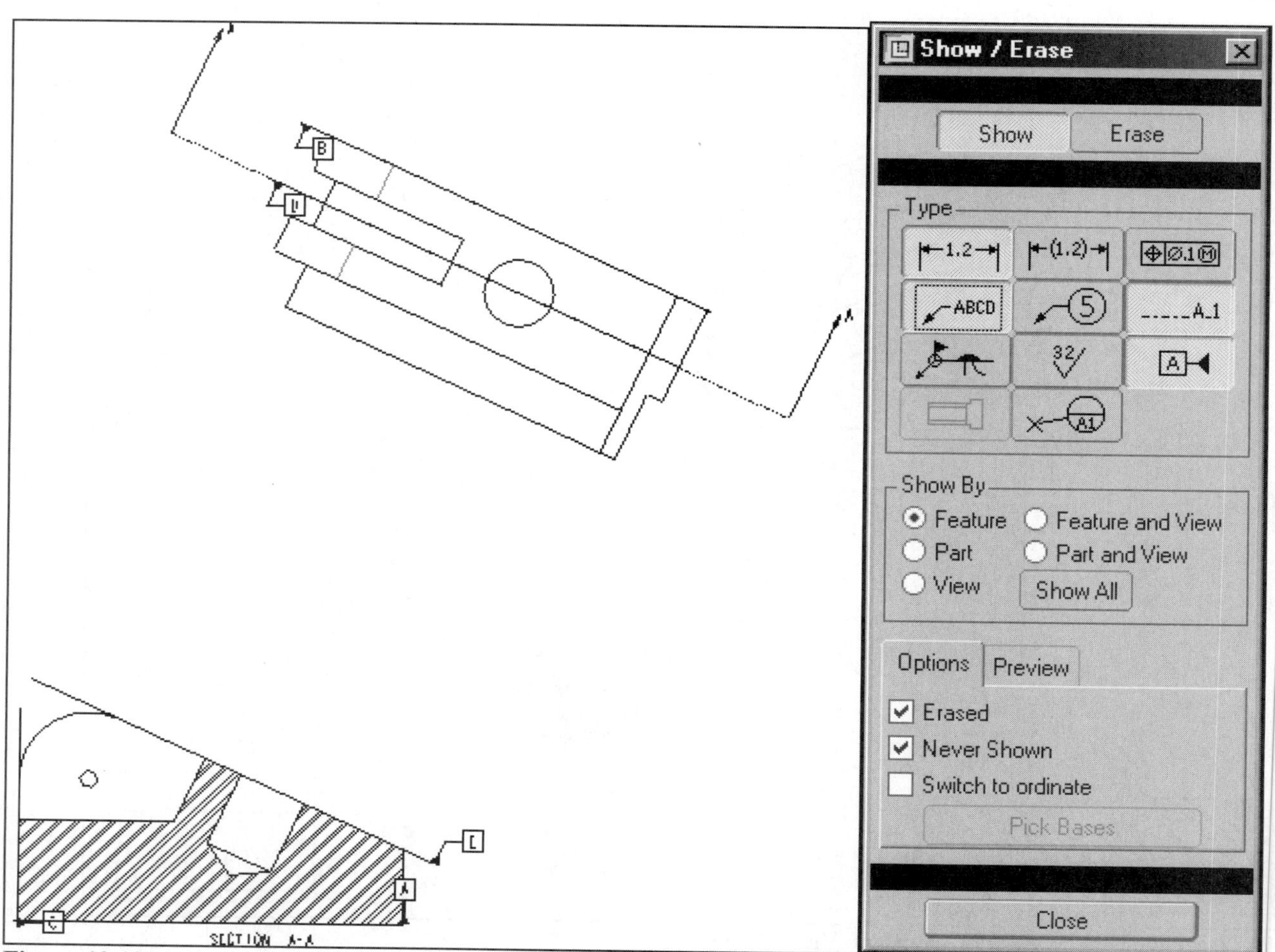

Figure 19.19(a) Show/Erase Dialog Box

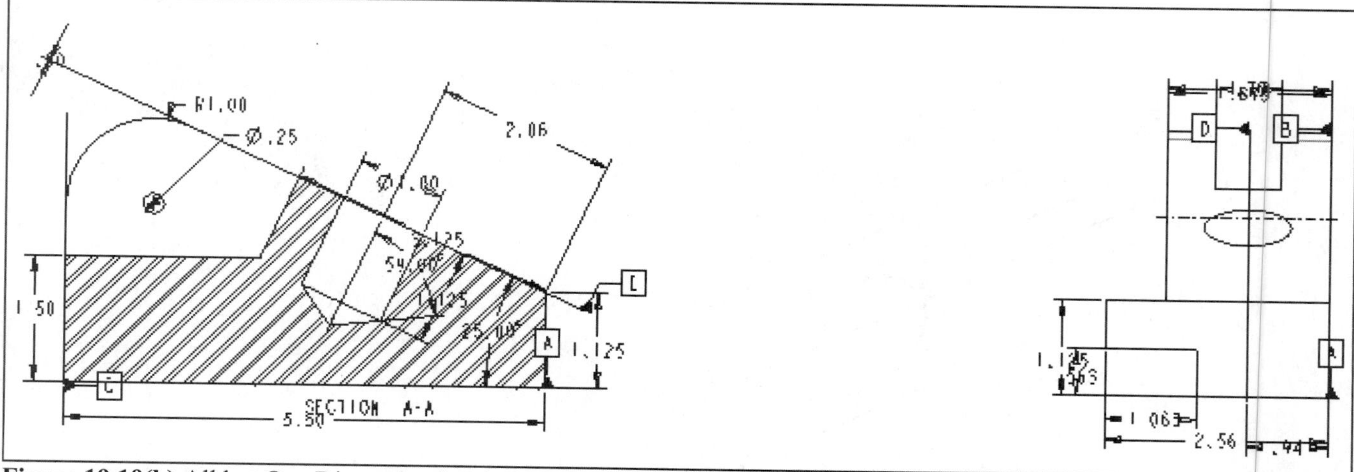

Figure 19.19(b) All but One Dimension is Displayed in the Front Section View and the Right Side View

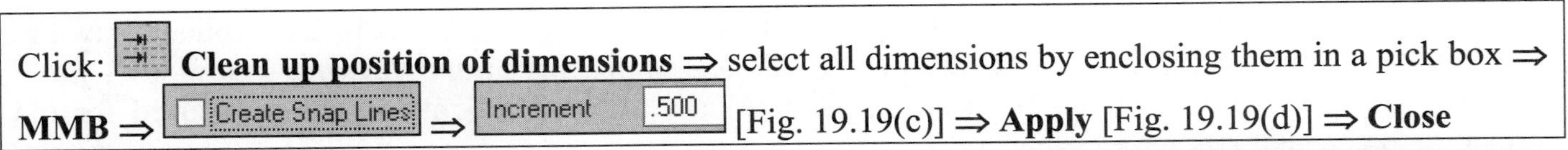

Click: **Clean up position of dimensions** ⇒ select all dimensions by enclosing them in a pick box ⇒ **MMB** ⇒ Create Snap Lines ⇒ Increment .500 [Fig. 19.19(c)] ⇒ **Apply** [Fig. 19.19(d)] ⇒ **Close**

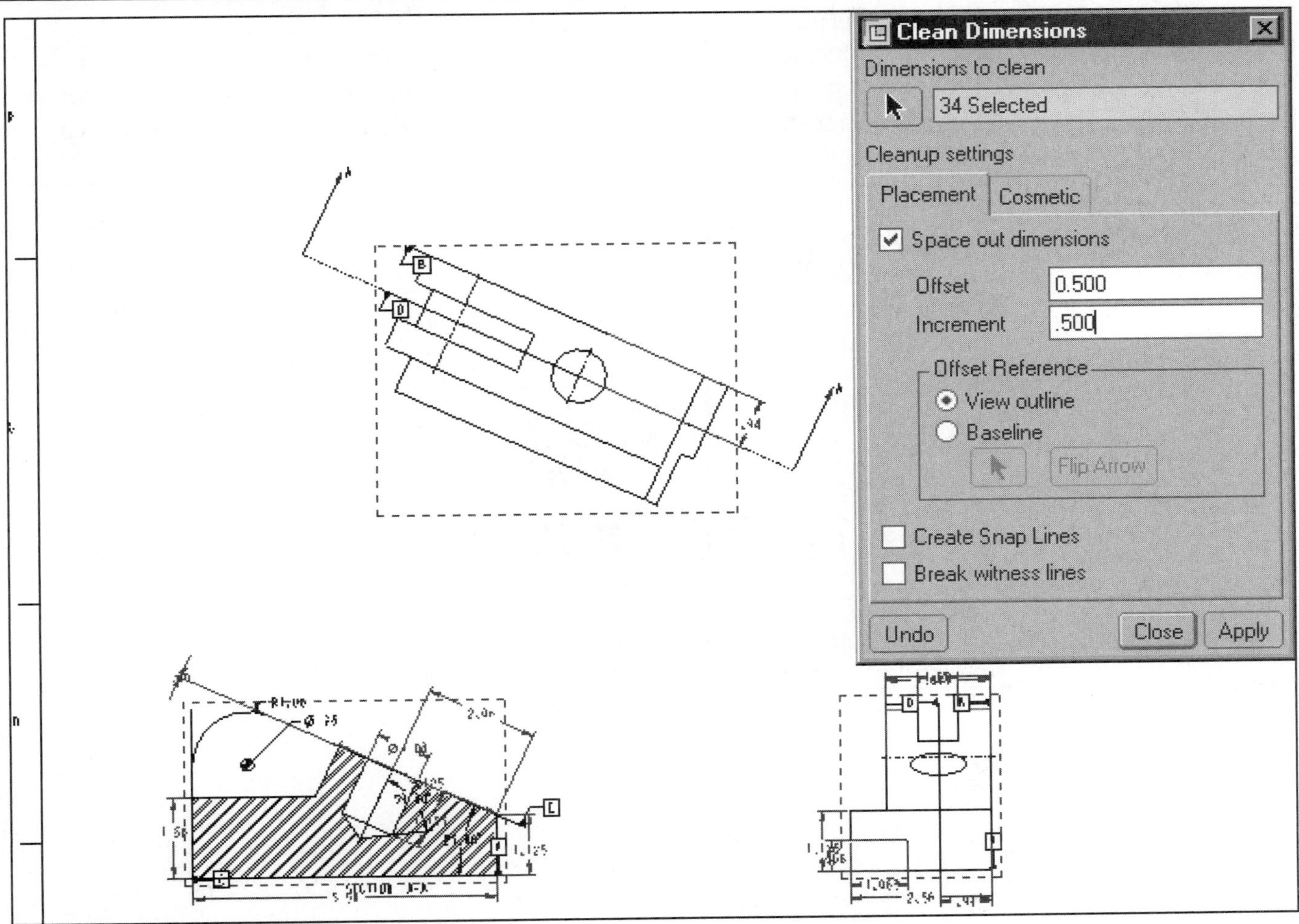

Figure 19.19(c) Clean Dimensions Dialog Box

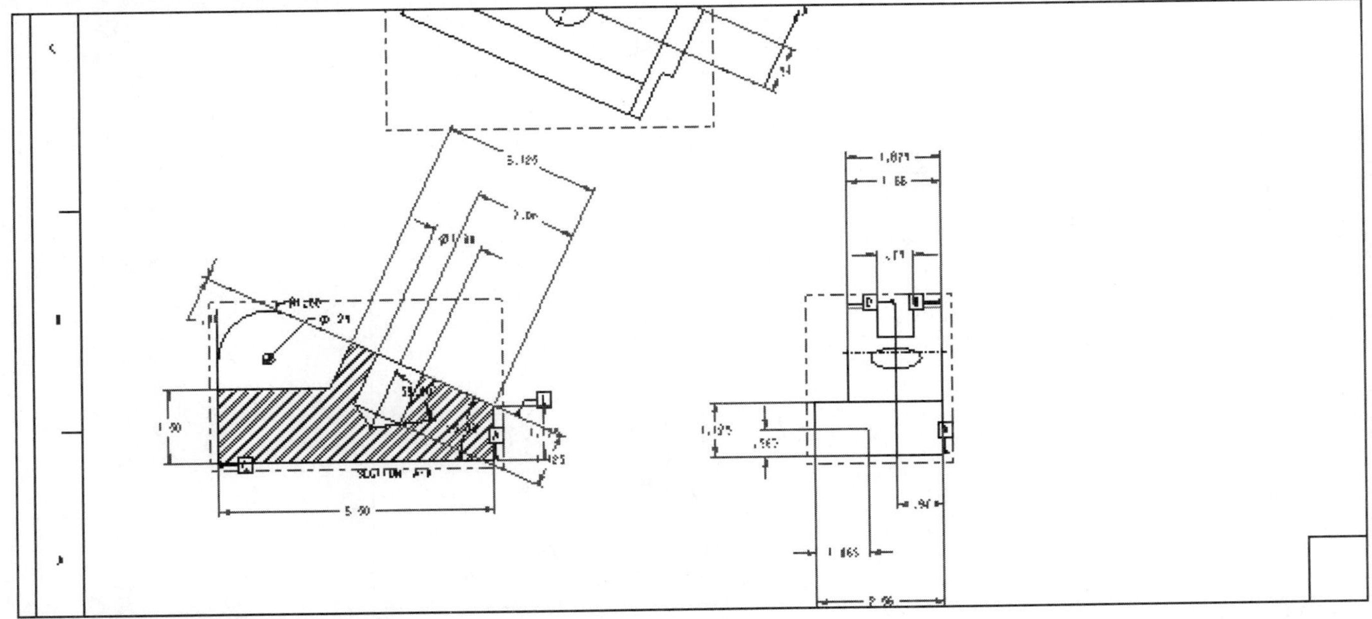

Figure 19.19(d) Cleaned Dimensions

Click on each individual drawing element and edit to create an ASME Y14.5 compliant drawing. Note that the Drawing Options: *draw_text_height .25*, *default_font filled* and *drawing_arrow_style filled* are now applied to the drawing. For sake of illustration clarity, these options were not used for the previous screen captures.

Click on the **1.00** diameter hole dimension ⇒ **RMB** [Fig. 19.20(a)]⇒ **Move Item to View** ⇒ click on auxiliary view [Fig. 19.20(b)] ⇒ reposition, erase or move dimensions [Fig. 19.20(c)], clip extension lines, and flip arrows where appropriate ⇒ move, erase and clip axes and datum planes as necessary ⇒ add a reference dimension to the small hole ⇒ change decimal place display as needed

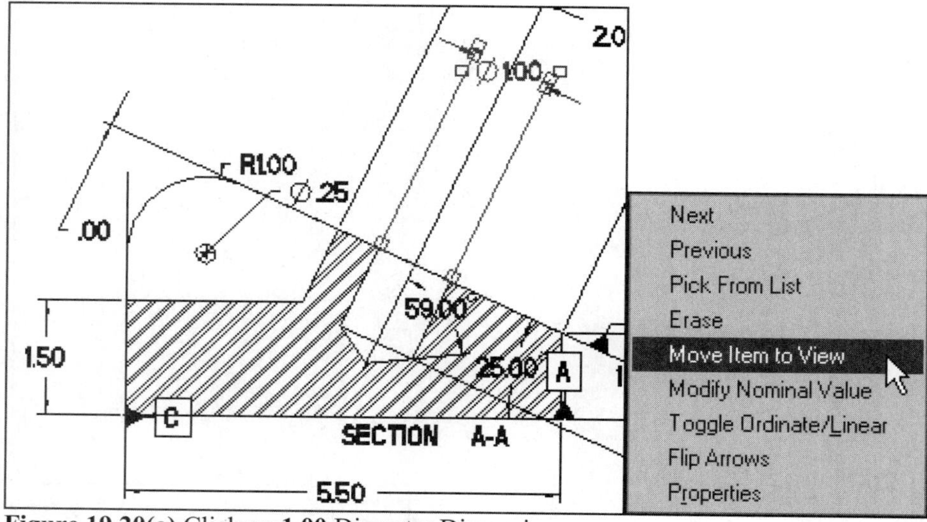

Figure 19.20(a) Click on **1.00** Diameter Dimension

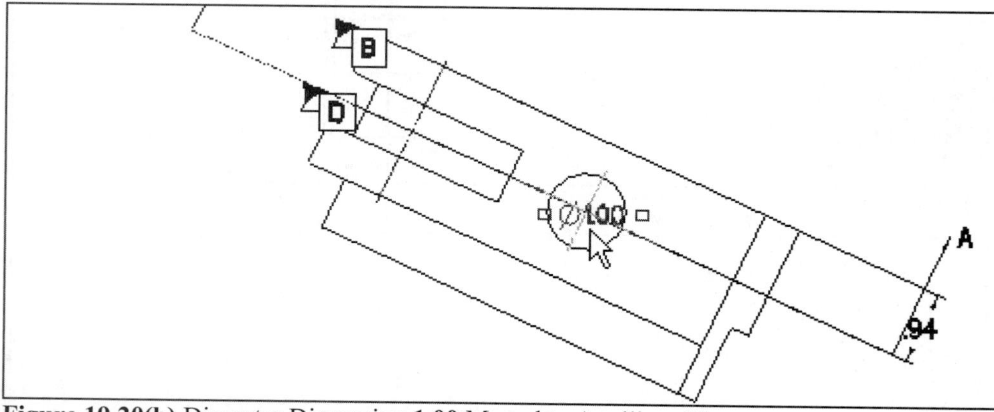

Figure 19.20(b) Diameter Dimension **1.00** Moved to Auxiliary View

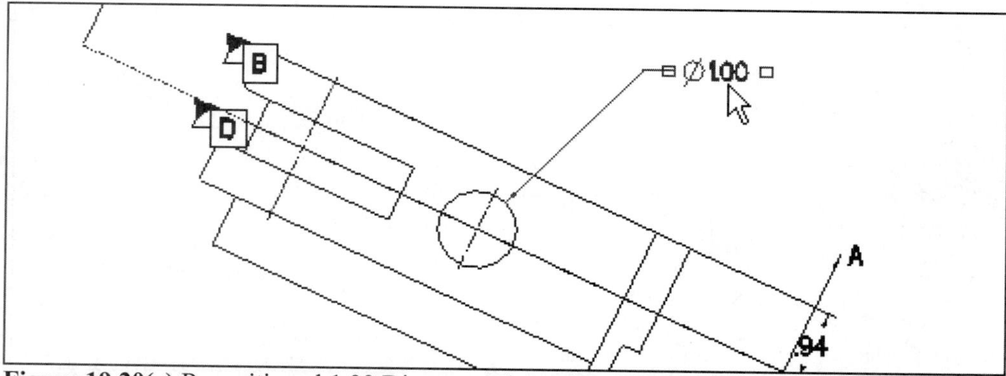

Figure 19.20(c) Repositioned **1.00** Diameter Dimension

Change the arrow length and width to be in proportion to the text height. Click **RMB ⇒ Properties ⇒ Drawing Options** opens an Options dialog box ⇒ ***draw_arrow_length .25***, ***draw_arrow_width .10*** ⇒ **Add/Change ⇒ Apply ⇒ Close ⇒ MMB** [Fig. 19.20(d-f)] ⇒ **Update the display of all views in the active sheet**

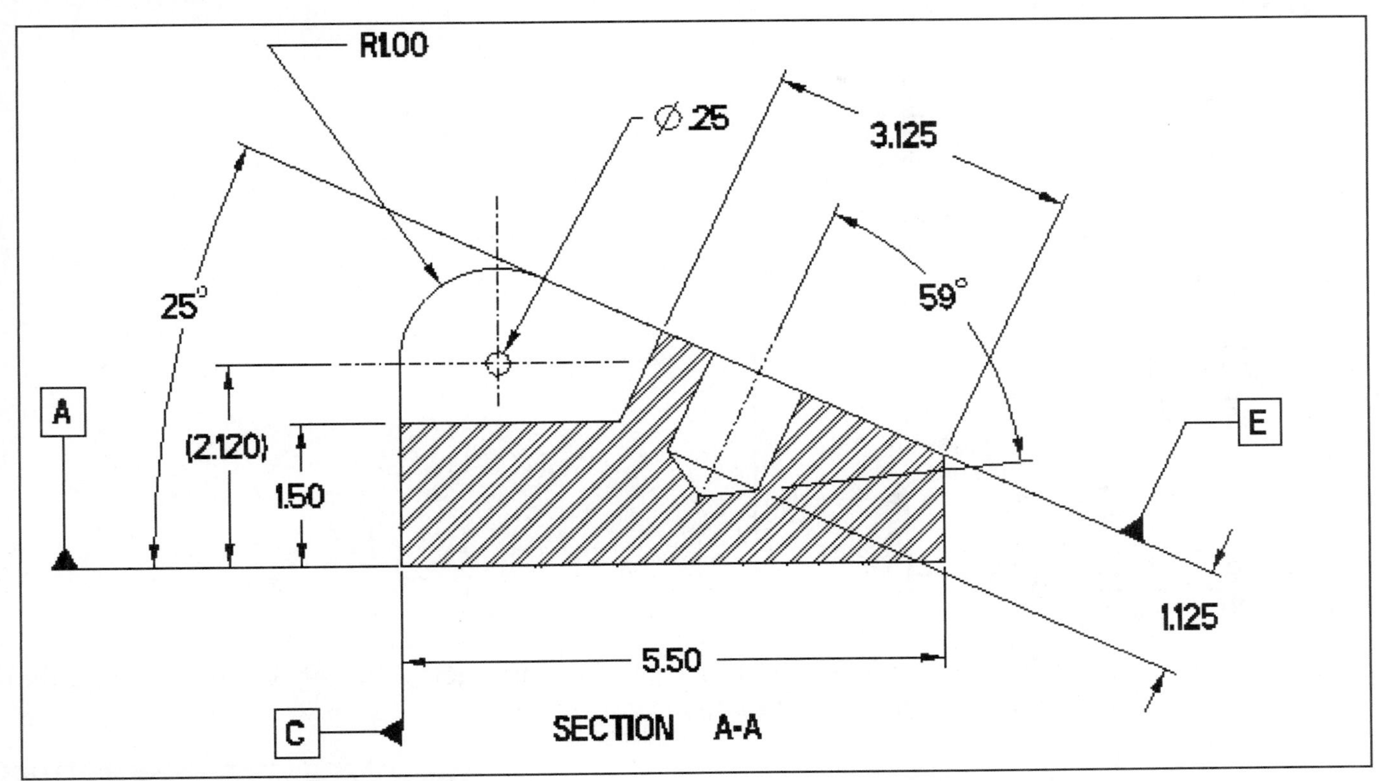

Figure 19.20(d) Edited Front Section View

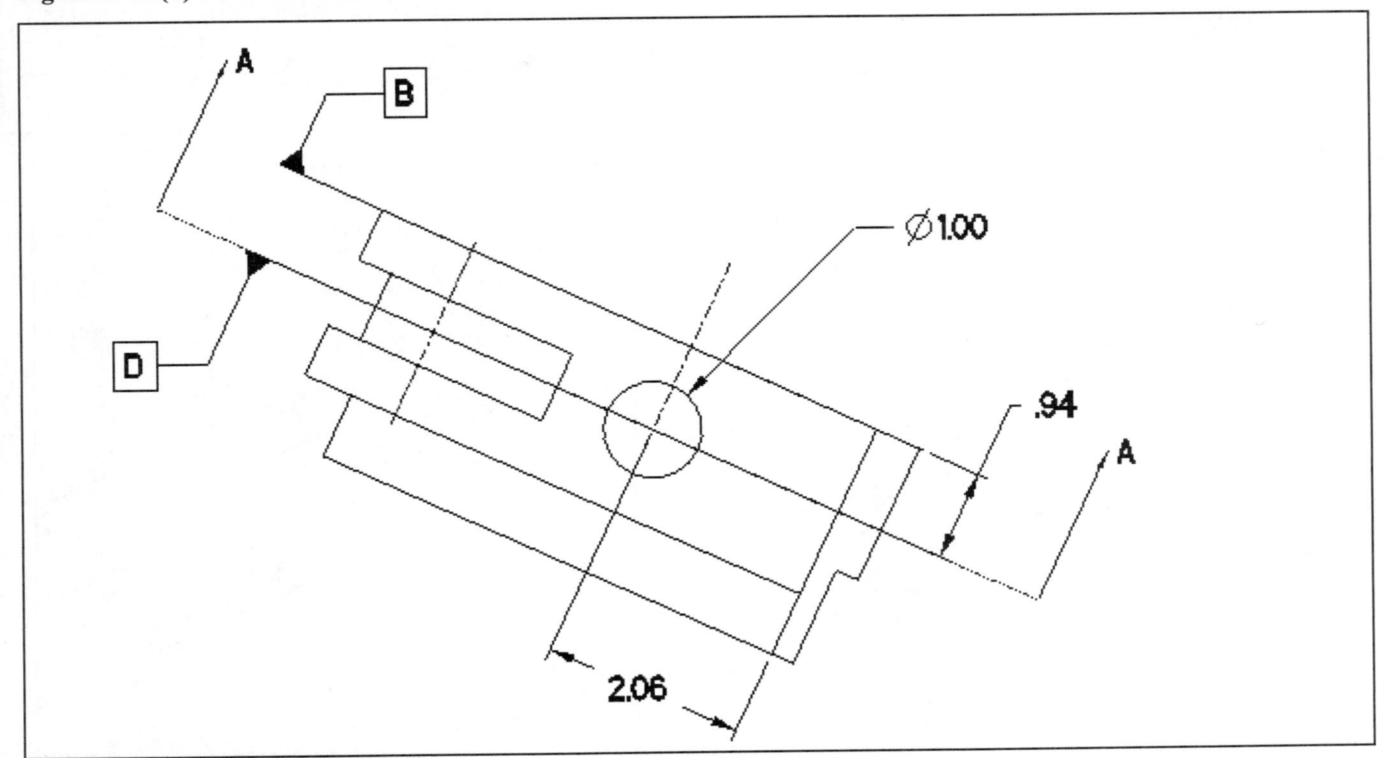

Figure 19.20(e) Edited Auxiliary View

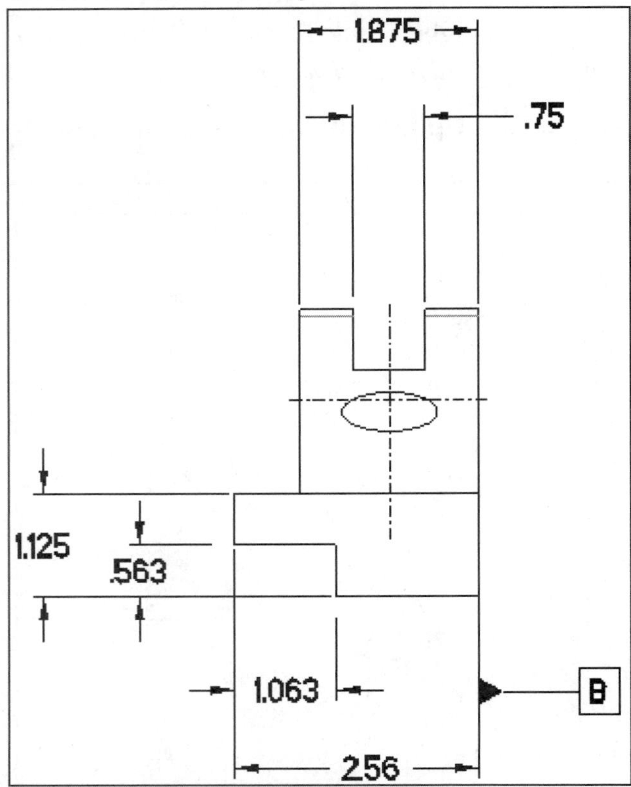

Figure 19.20(f) Edited Right Side View

Add the edges of the small thru hole back into the auxiliary view. The edges will display in the graphics window as light gray, but print as dashed on the drawing plot.

Click: **View** from Menu bar ⇒ **Drawing Display** ⇒ **Edge Display** ⇒ **Hidden Line** ⇒ press and hold the **Ctrl** key and pick near where the small thru hole would show as hidden in the auxiliary view [Fig. 19.21(a)] ⇒ pick the opposite edge [Fig. 19.21(b)] ⇒ **MMB** ⇒ **Done** ⇒ ⇒ ⇒ ⇒ ⇒ **MMB** [Fig. 19.21(c-d)]

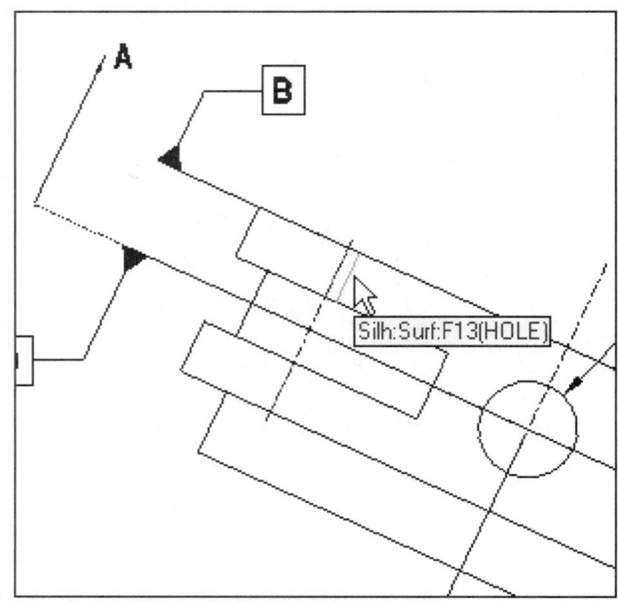

Figure 19.21(a) Click near the Hole Edge

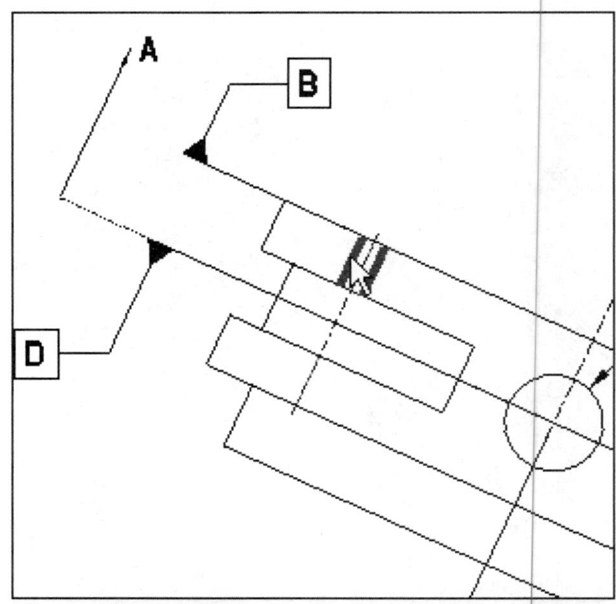

Figure 19.21(b) Click near the Opposite Edge of Hole

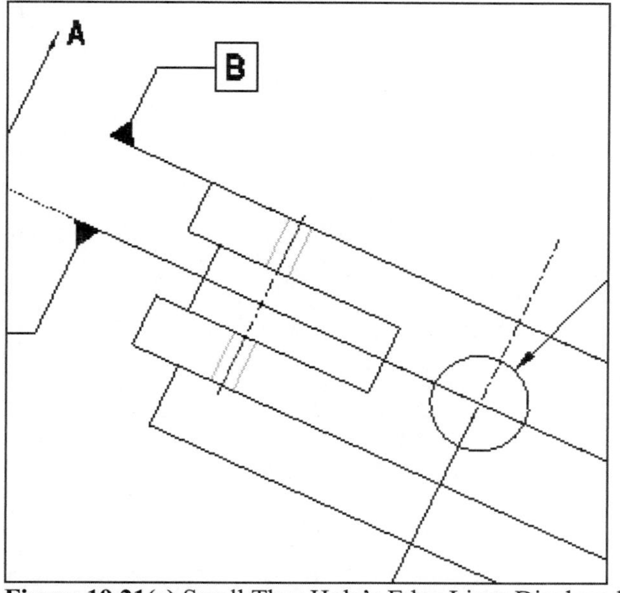

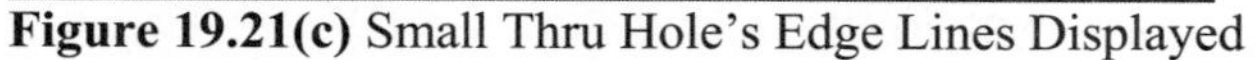

Figure 19.21(c) Small Thru Hole's Edge Lines Displayed

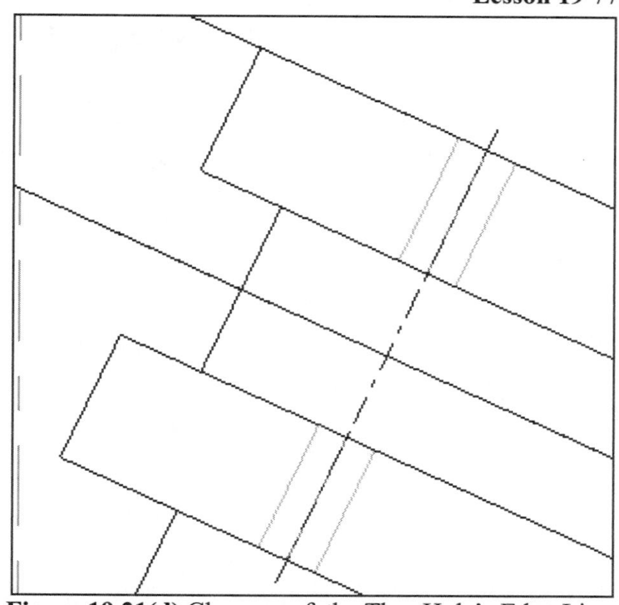

Figure 19.21(d) Close-up of the Thru Hole's Edge Lines

To increase the clarity of this drawing, you will need to master a number of capabilities. Partial views, detail views, using multiple sheets, and modifying section lining (crosshatch lines) are just a few of the many options available in Drawing mode. Create a Detail View.

Click: **Insert a drawing view of the active model** ⇒ **Detailed** ⇒ **MMB** ⇒ Select CENTER POINT for drawing view. click in the upper right of the drawing sheet ⇒ Enter scale for view 1.500 ⇒ **MMB** ⇒ select the center point for the detail on an existing view; pick the top edge of the hole in the front section view [Fig. 19.22(a)] ⇒ Sketch a spline, without intersecting other splines, to define an outline. **LMB** to sketch the spline and the **MMB** to end it [Fig. 19.22(b)] ⇒ Enter a name for the detailed view: A ⇒ **MMB** ⇒ **Circle** ⇒ select a location for the note [Fig. 19.22(c)] ⇒ **LMB** [Fig. 19.22(d)]

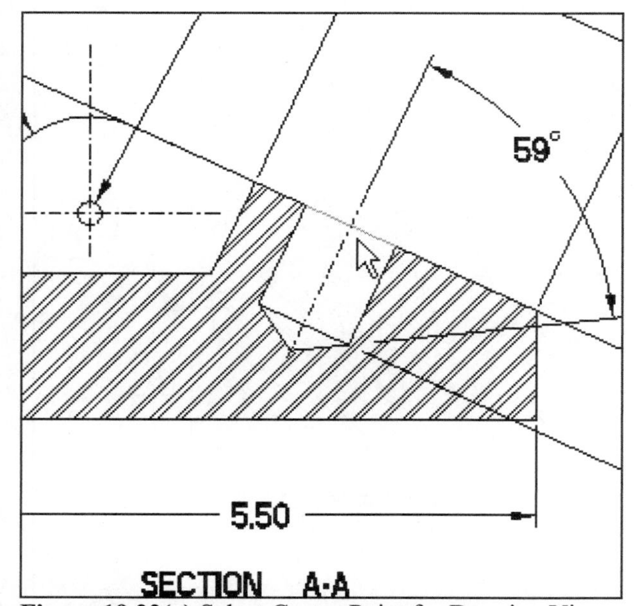

Figure 19.22(a) Select Center Point for Drawing View

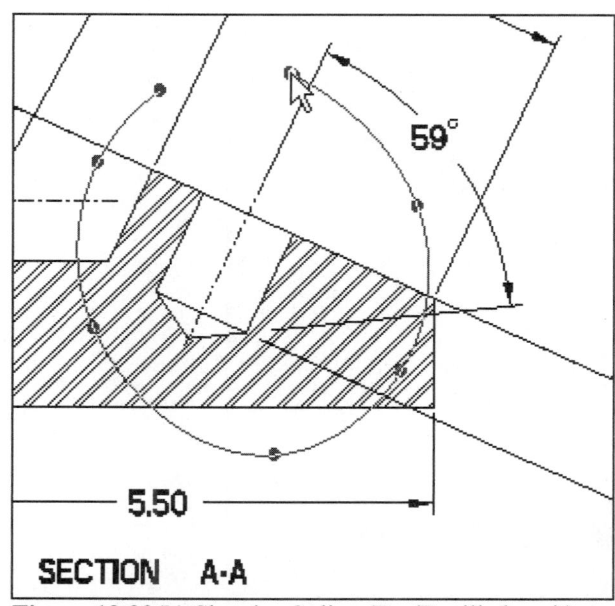

Figure 19.22(b) Sketch a Spline (Pro/E will close it)

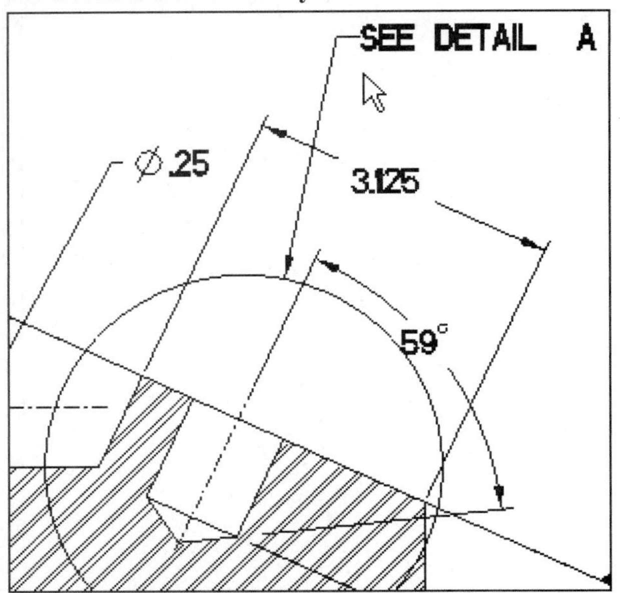

Figure 19.22(c) Select Location for Note

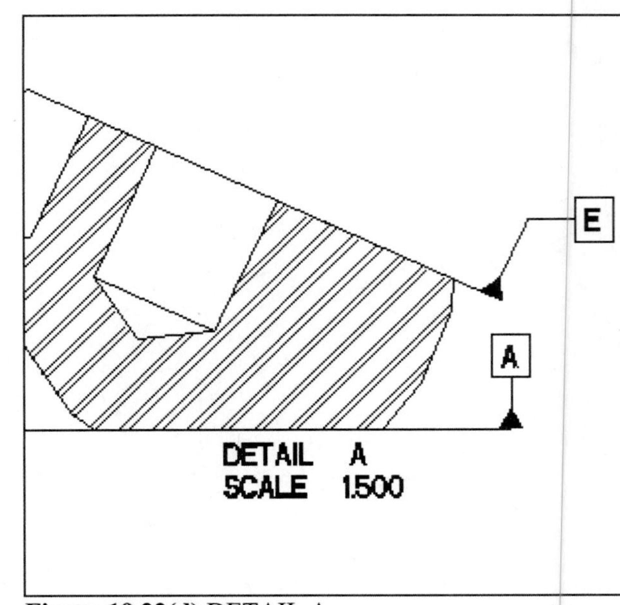

Figure 19.22(d) DETAIL A

Add an axis to the detail of the hole; click **View** from Menu bar ⇒ **Show and Erase** ⇒ Show ⇒ only ----A.1 **Axis** active ⇒ **Show All** ⇒ **Yes** ⇒ **Accept All** [Fig. 19.22(e)] ⇒ **Close** [Fig. 19.22(f)] ⇒ **LMB**

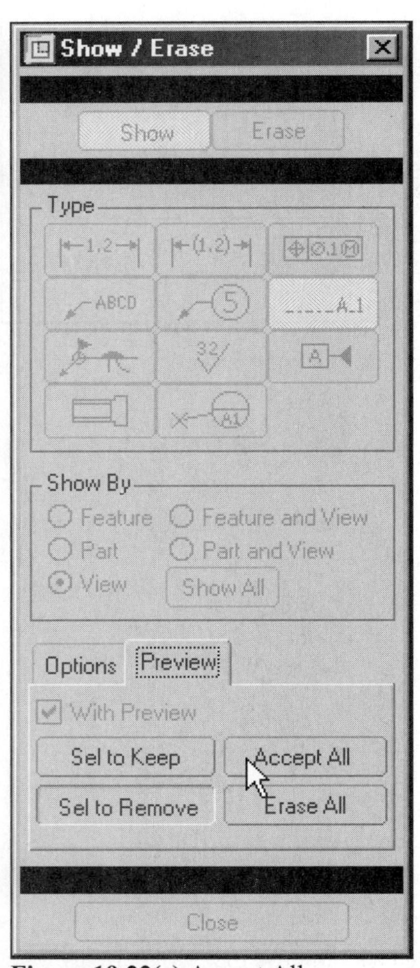

Figure 19.22(e) Accept All

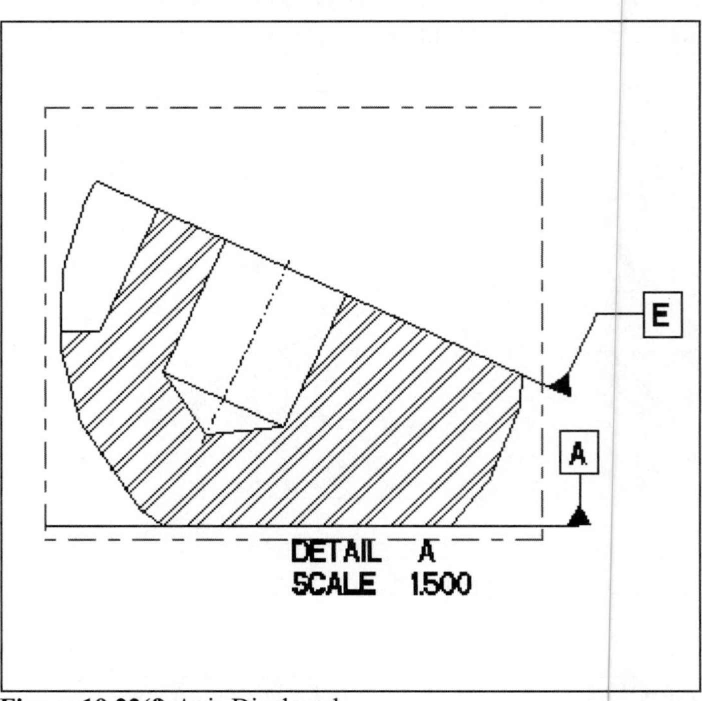

Figure 19.22(f) Axis Displayed

In DETAIL A, erase datum **A** and clip datum **E**. Also, clip the axis. ⇒ with the **Ctrl** key pressed, click on the text items: **SECTION A-A**, **SEE DETAIL A** and **DETAIL A SCALE 1.500** [Fig. 19.23(a)] ⇒ **RMB** ⇒ **Text Style** ⇒ Default ⇒ Height **.375** [Fig. 19.23(b)] ⇒ **Apply** ⇒ **OK** [Fig. 19.23(c)]

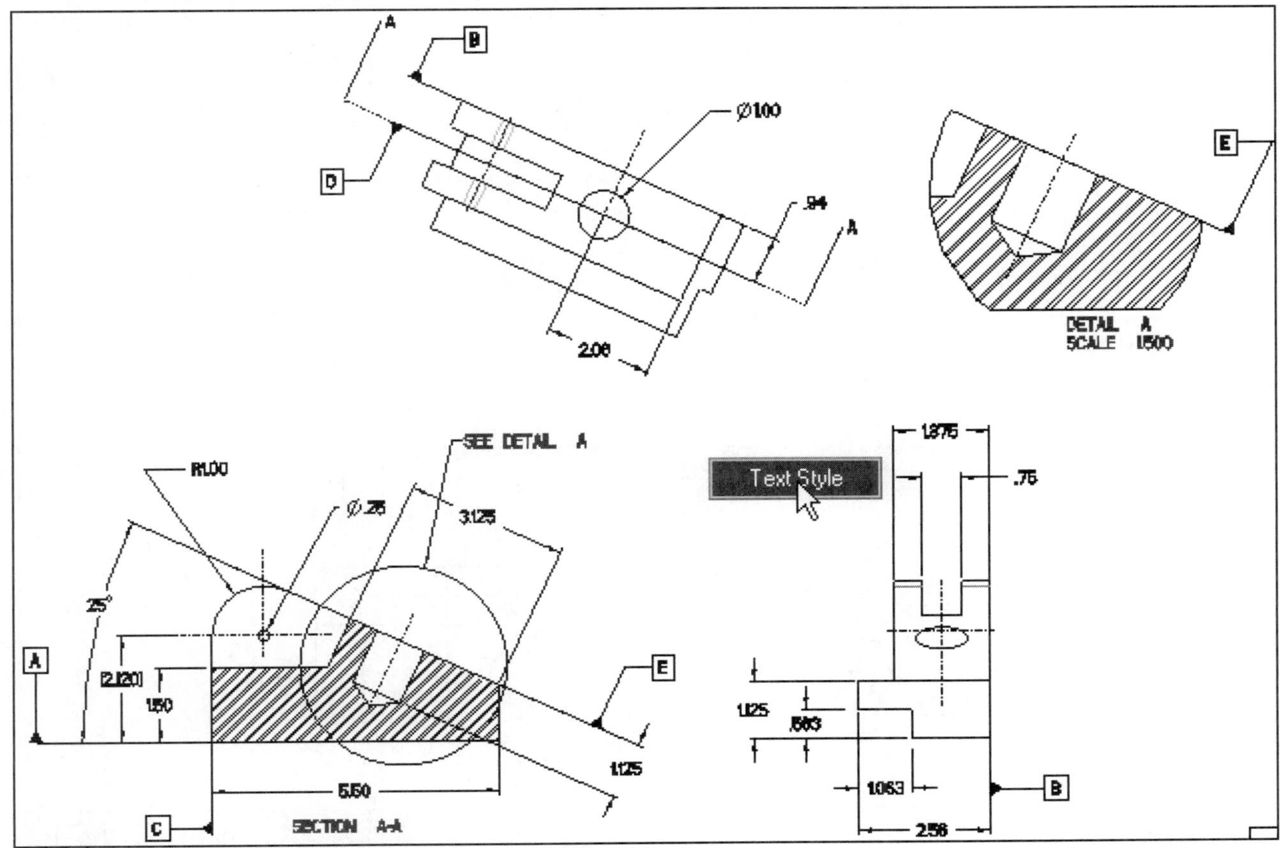

Figure 19.23(a) Select the Text Items and Change Their Height

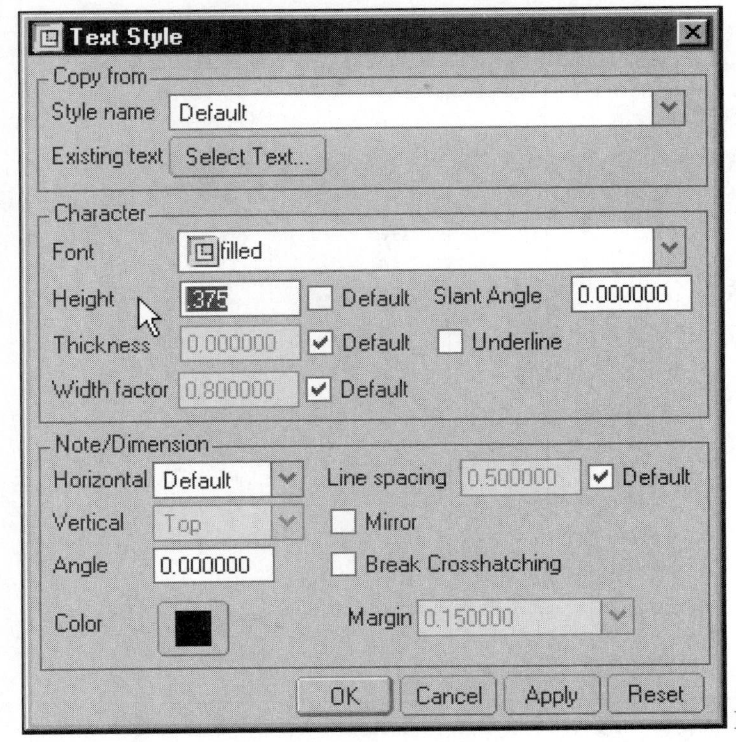

Figure 19.23(b) Text Style Dialog Box

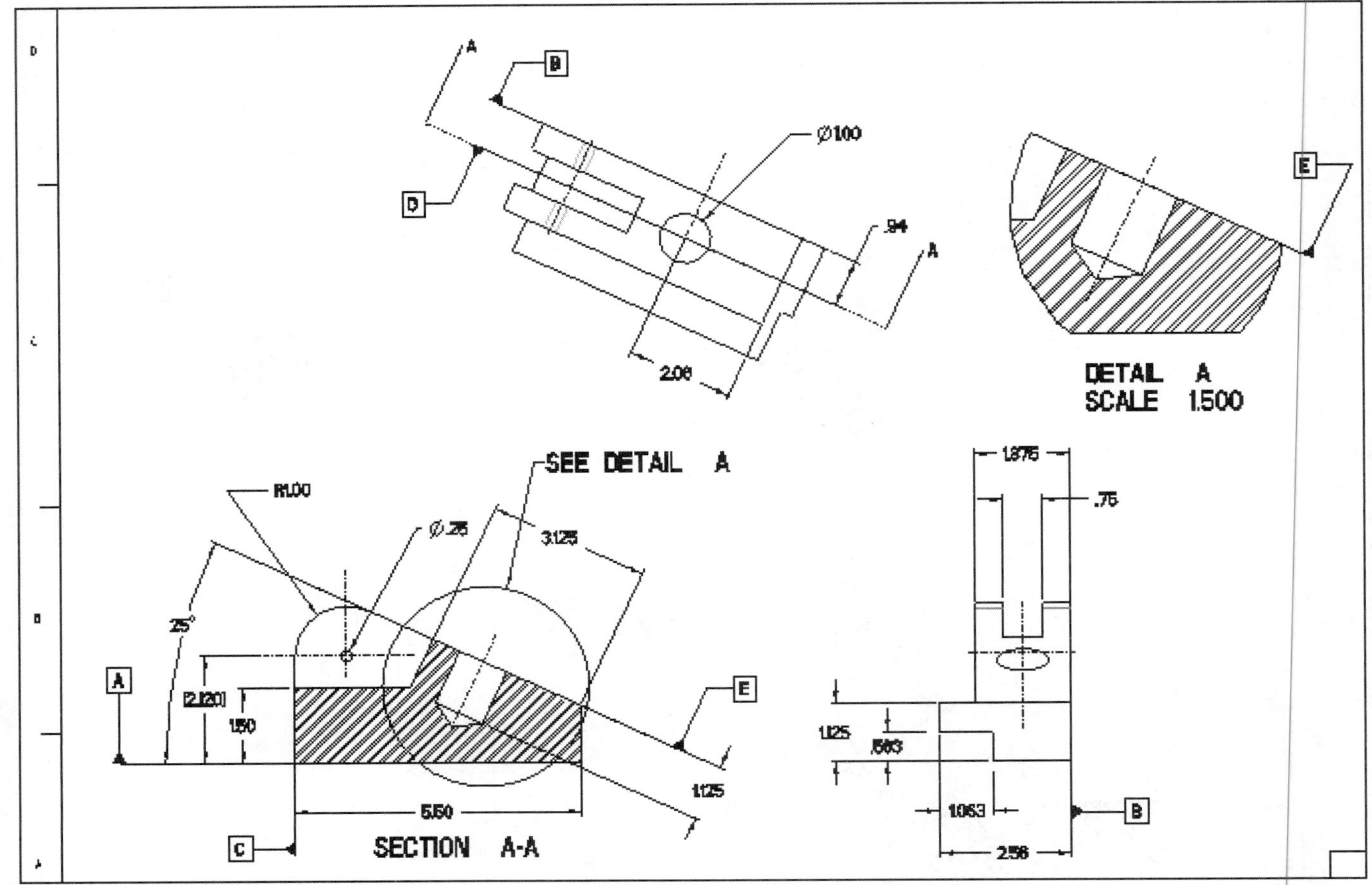

Figure 19.23(c) Changed Text Style

Change the height of the section identification lettering to **.375** [Fig. 19.23(d-e)] ⇒ **LMB** ⇒ ⇒ **MMB**

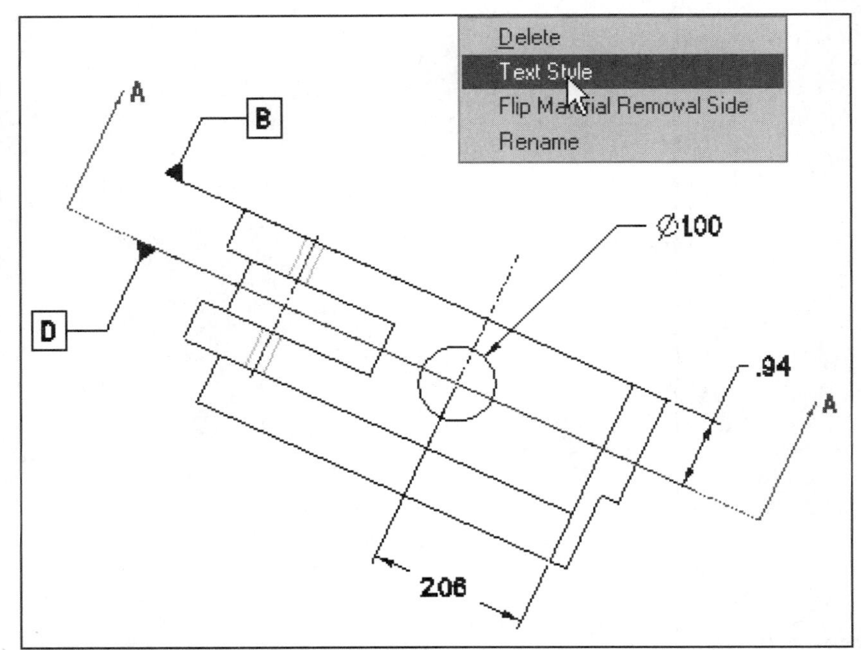

Figure 19.23(d) Text Style

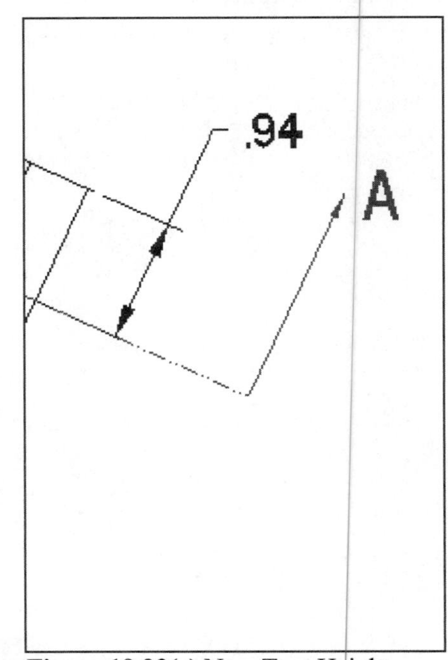

Figure 19.23(e) New Text Height

The Arrows for the cutting plane are now too small. These are controlled by the Drawing Options.

Click: **RMB ⇒ Properties ⇒ Drawing Options ⇒ *crossec_arrow_length .375*, *crossec_arrow_width .125*** [Fig. 19.24(a)] **⇒ Add/Change ⇒ Apply ⇒ Close ⇒ MMB ⇒** [Fig. 19.24(b)]

These options control cross sections and their arrows				
crossec_arrow_length	.375	0.187500	✱	Sets the length of the arrow head on the cros
crossec_arrow_style	tail_online *	tail_online	●	Determines which end of cross-section arrow
crossec_arrow_width	.125	0.062500	✱	Sets the width of the arrow head on the cros

Figure 19.24(a) Cross Section Drawing Options

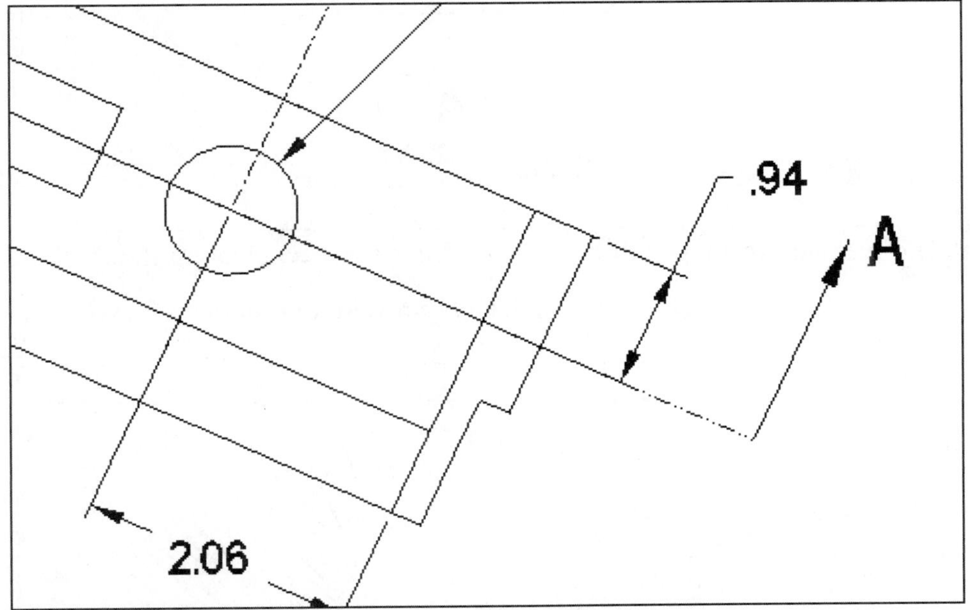

Figure 19.24(b) Section Arrows Length and Width Changed

Click on the **1.125** dimension in the front section view [Fig. 19.25(a)] ⇒ **RMB ⇒ Move Item to View** ⇒ click on view **DETAIL A** [Fig. 19.25(b)] ⇒ reposition and clip as needed ⇒ [Fig. 19.25(c)]

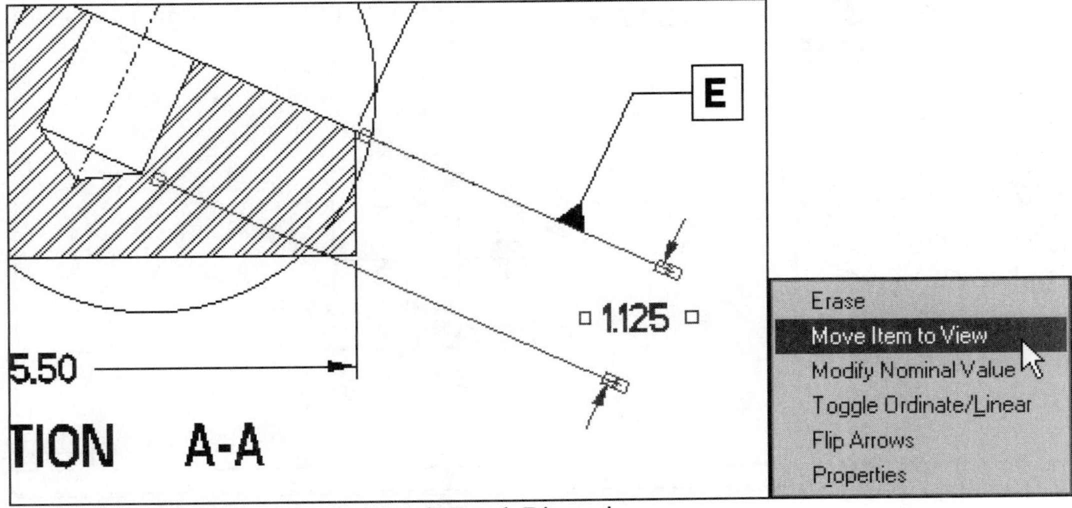

Figure 19.25(a) Click on the **1.125** Hole Depth Dimension

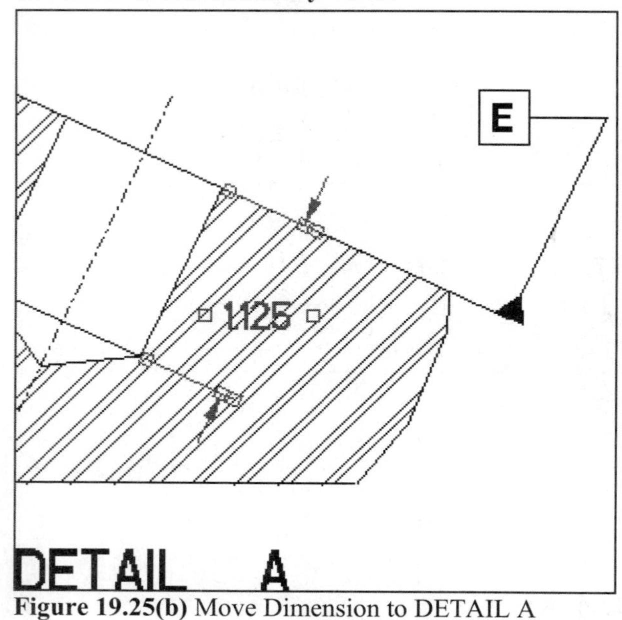

Figure 19.25(b) Move Dimension to DETAIL A

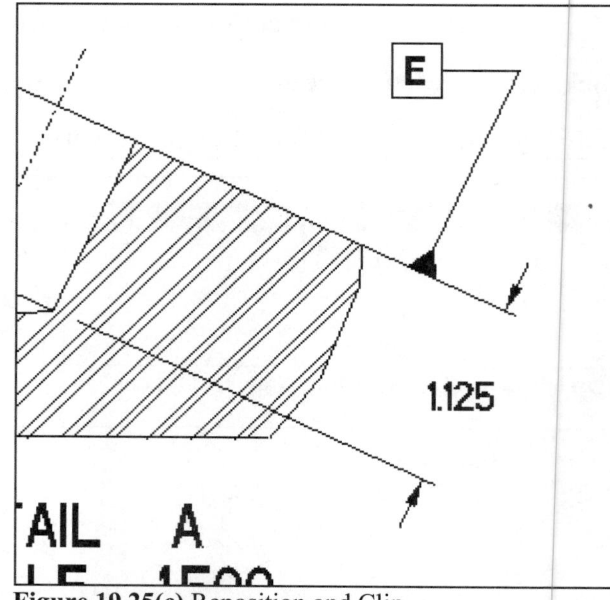

Figure 19.25(c) Reposition and Clip

Add another reference (horizontal) dimension to the small hole in the front view ⇒ cleanup the drawing as you see fit ⇒ ⇒ ⇒ ⇒ ⇒ **MMB** ⇒ **File** ⇒ **Delete** ⇒ **Old Versions** ⇒ **MMB** (Fig. 19.26)

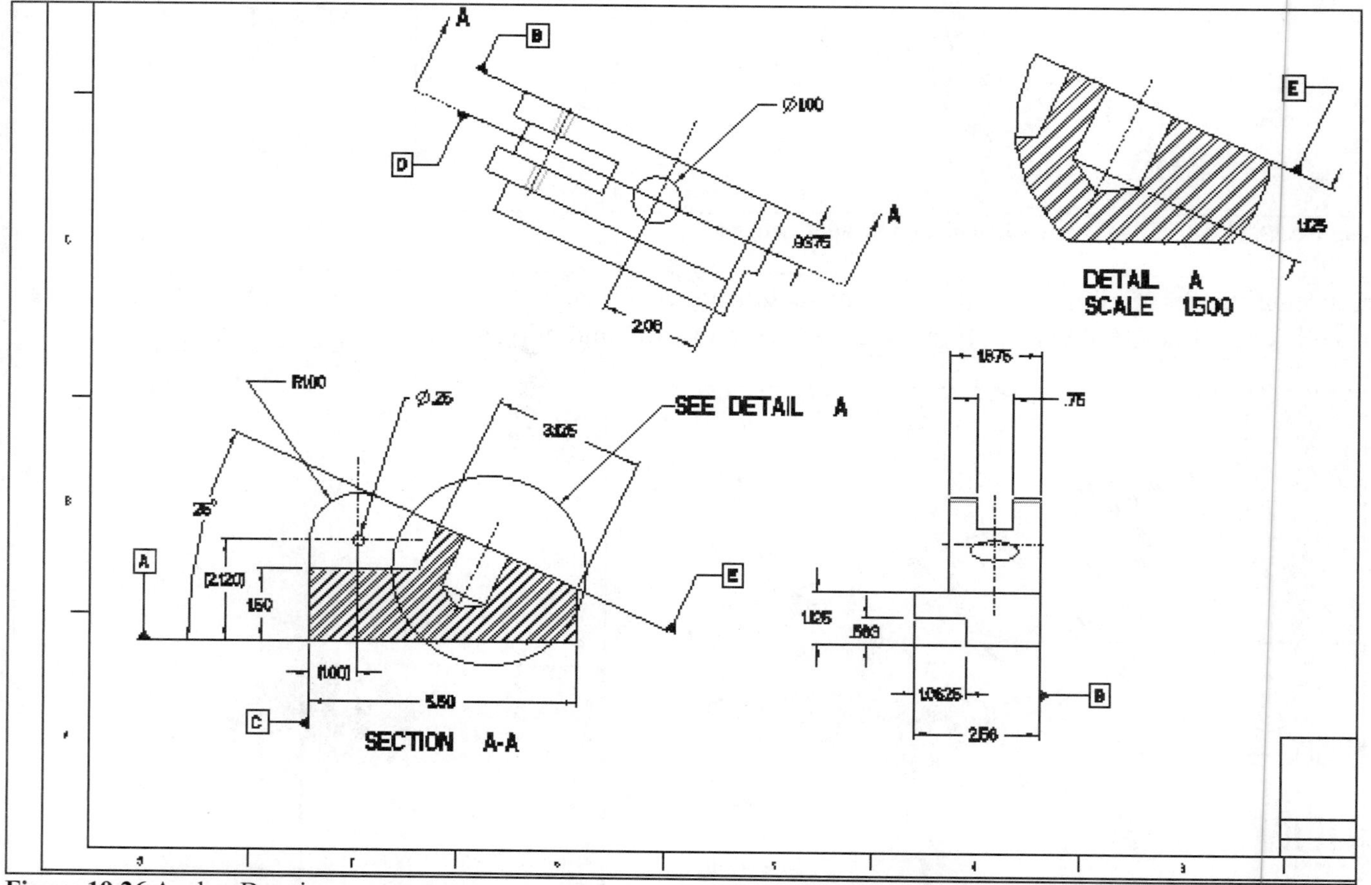

Figure 19.26 Anchor Drawing

Next, you will change the boundary of DETAIL A.

Click on **DETAIL A** ⇒ **RMB** if you RMB click outside the view, you get a pop-up list of options [Fig. 19.27(a)], whereas if you RMB click inside the view, there are more options [Fig. 19.27(b)] ⇒ **Properties** the same menu displays regardless of the option list ⇒ **Boundary** [Fig. 19.27(c)] ⇒ re-sketch the spline in the front (SECTION A-A) view [Fig. 19.27(d)] ⇒ **MMB** ⇒ **MMB** [Fig. 19.27(e)] ⇒ **MMB** ⇒ **LMB** [Fig. 19.27(f)] ⇒ [icon] ⇒ [icon] ⇒ [icon] ⇒ **MMB**

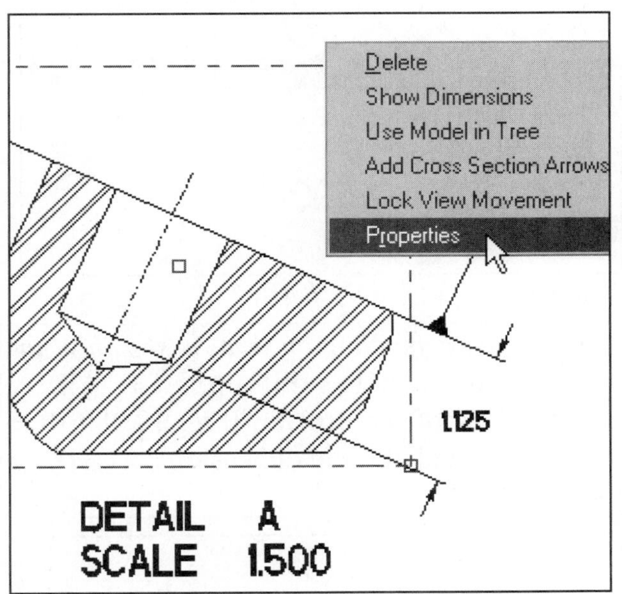

Figure 19.27(a) Options

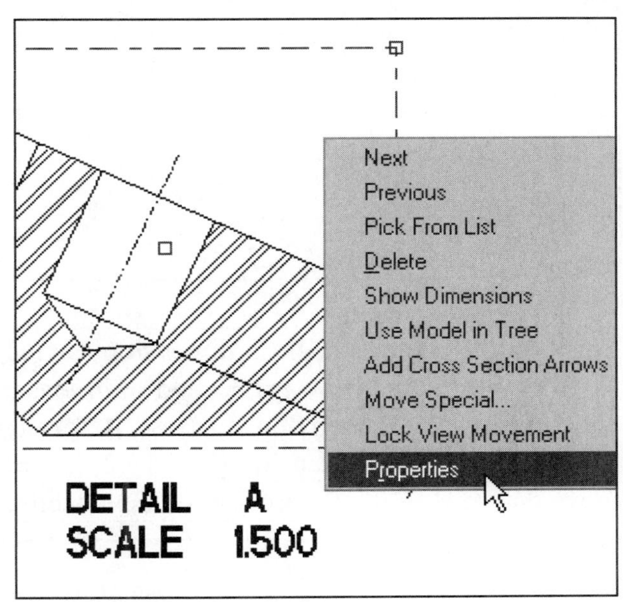

Figure 19.27(b) More Options

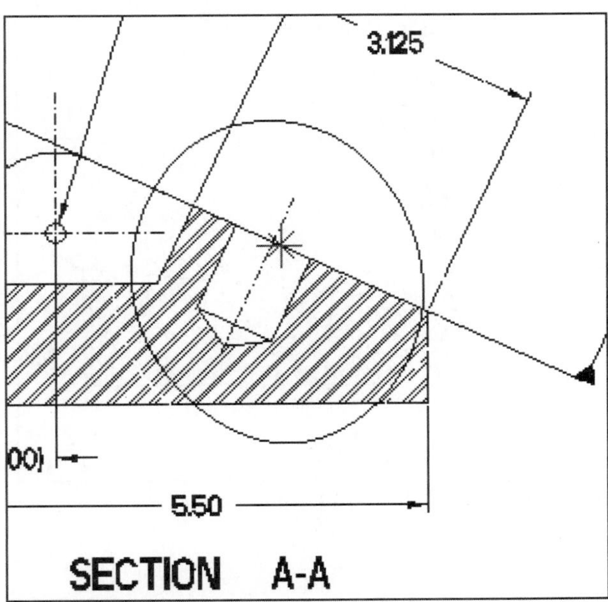

Figure 19.27(c) Old Spline

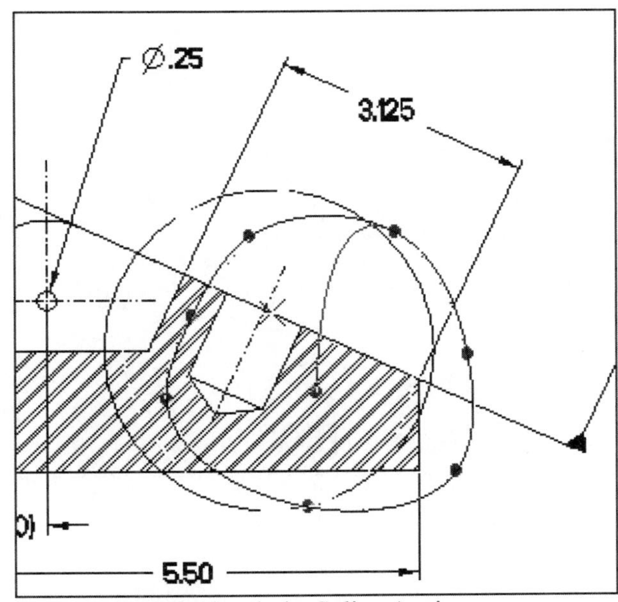

Figure 19.27(d) Sketch the Spline Again

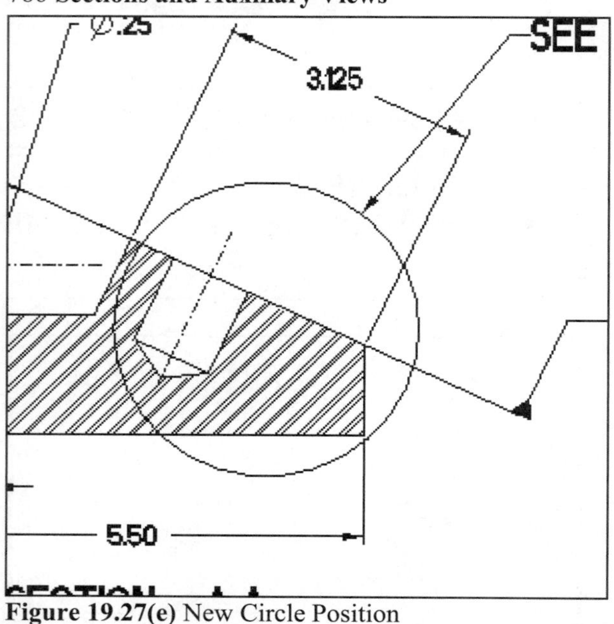

Figure 19.27(e) New Circle Position

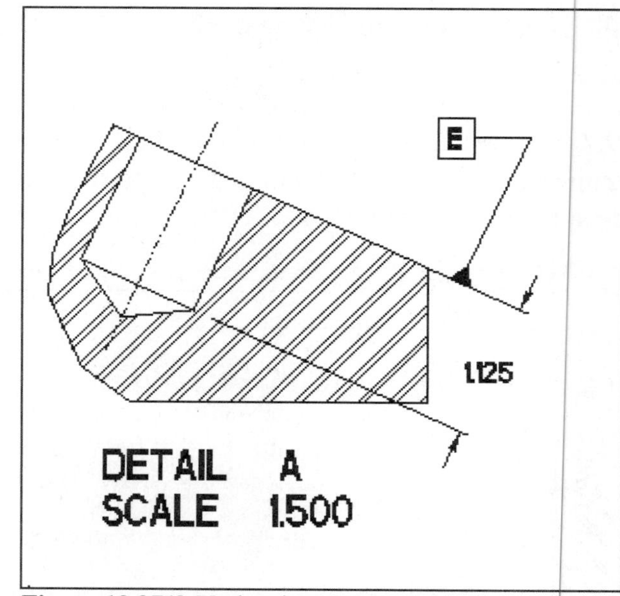

Figure 19.27(f) Updated DETAIL A

Erase the **118** degree drill tip dimension (*or the **59** degree dimension, depending on what is shown in the front view of your drawing*) ⇒ add a dimension for the full angle of the hole's drill tip in DETAIL A, click: **Insert** ⇒ **Dimension** ⇒ **New References** ⇒ pick both edges of the drill tip [Fig. 19.28(a)] ⇒ **MMB** to place the dimension [Fig. 19.28(b)] ⇒ **MMB** ⇒ reposition and clip the dimension ⇒ **LMB** ⇒ move datum **E** [Fig. 19.28(c)] ⇒ change decimal places to **0** [Fig. 19.28(d-e)]

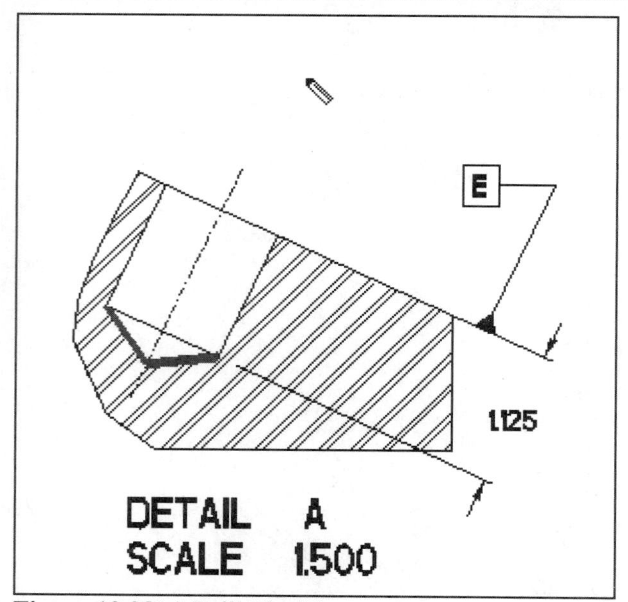

Figure 19.28(a) Drill Tip Edges

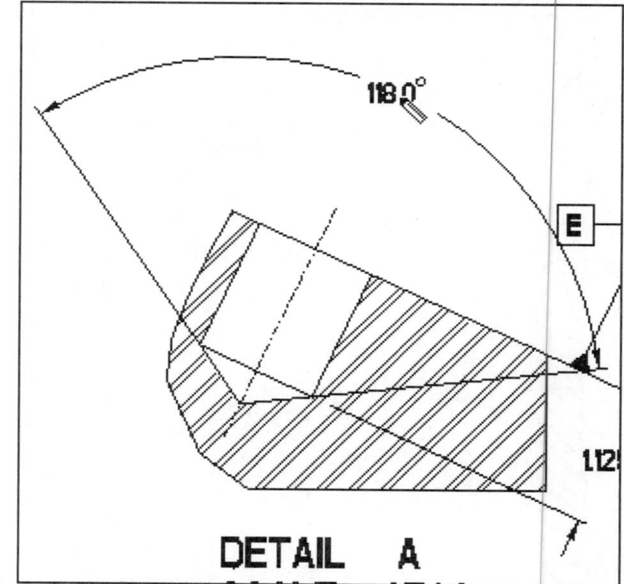

Figure 19.28(b) Place Dimension

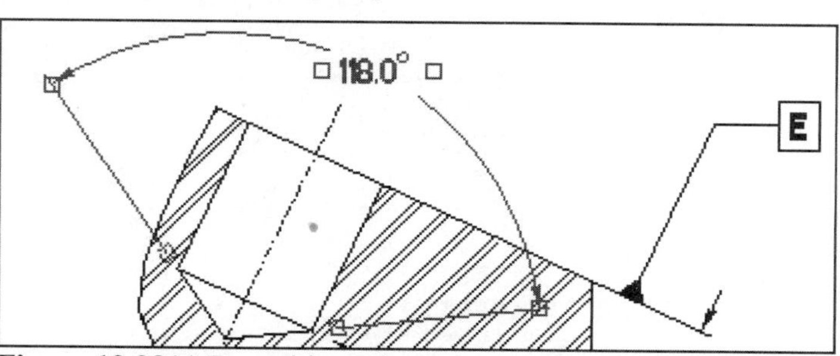

Figure 19.28(c) Reposition Dimension and Move Datum E

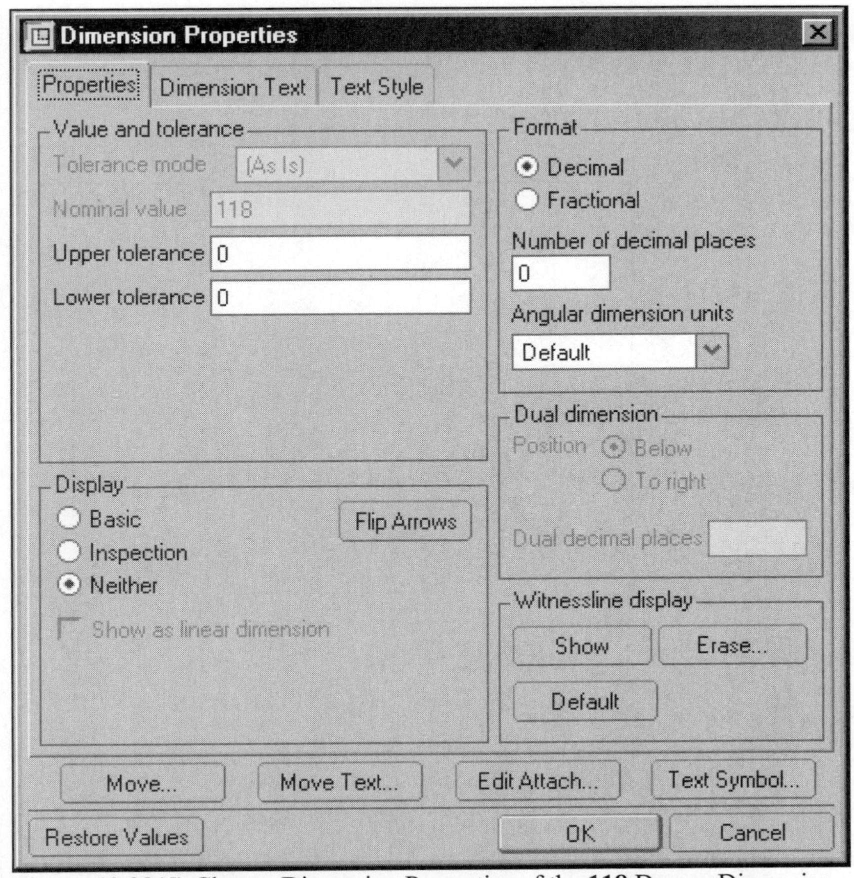

Figure 19.28(d) Change Dimension Properties of the **118** Degree Dimension

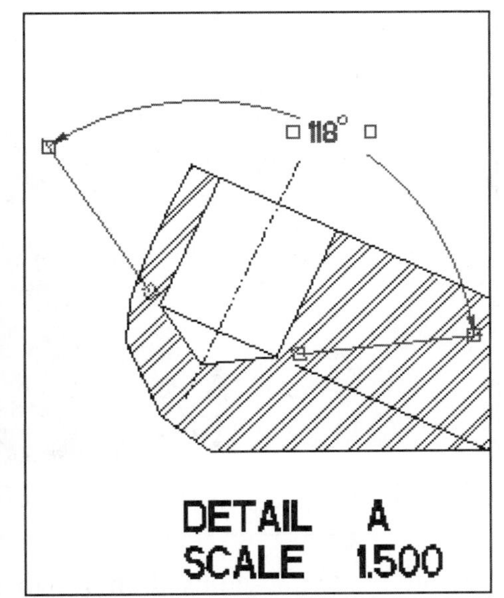

Figure 19.28(e) 118 Degrees

Change the spacing and the angle of the section lining. Click on the hatching in **SECTION A-A** [Fig. 19.29(a)] ⇒ **RMB** ⇒ **Properties** ⇒ **Delete Line** ⇒ **Angle** ⇒ **30** ⇒ **MMB** [Fig. 19.29(b)] ⇒ **LMB** ⇒ ⇒ ⇒ **MMB**

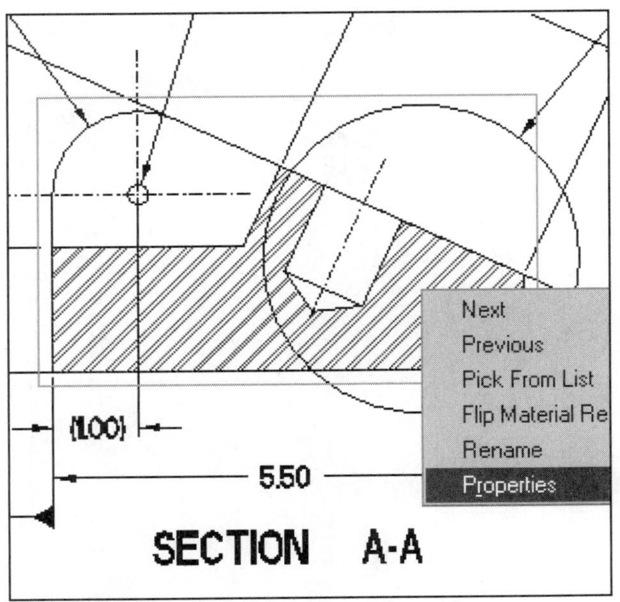

Figure 19.29(a) Xhatching Properties

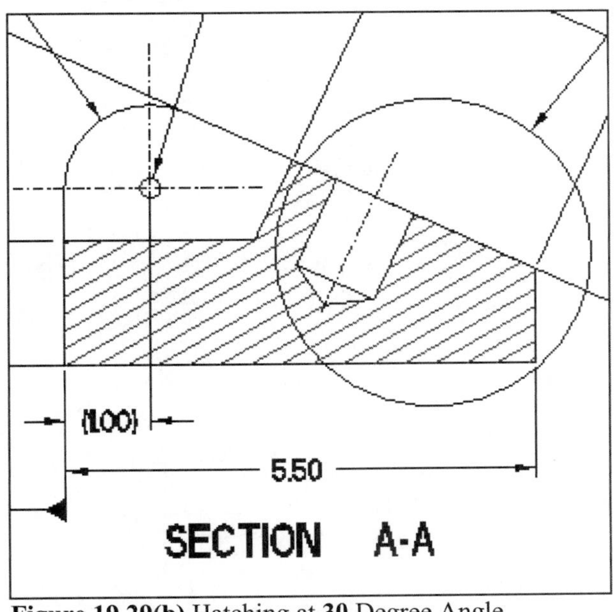

Figure 19.29(b) Hatching at **30** Degree Angle

Change the spacing and the angle of the section lining in DETAIL A. Click on the hatching in **DETAIL A** [Fig. 19.30(a)] ⇒ **RMB** ⇒ **Properties** ⇒ **Det Indep** ⇒ **Hatch** ⇒ **Spacing** ⇒ **Double** ⇒ **Angle** ⇒ **45** ⇒ **MMB** [Fig. 19.30(b)] ⇒ **LMB** ⇒ ⇒ ⇒ ⇒ **MMB**

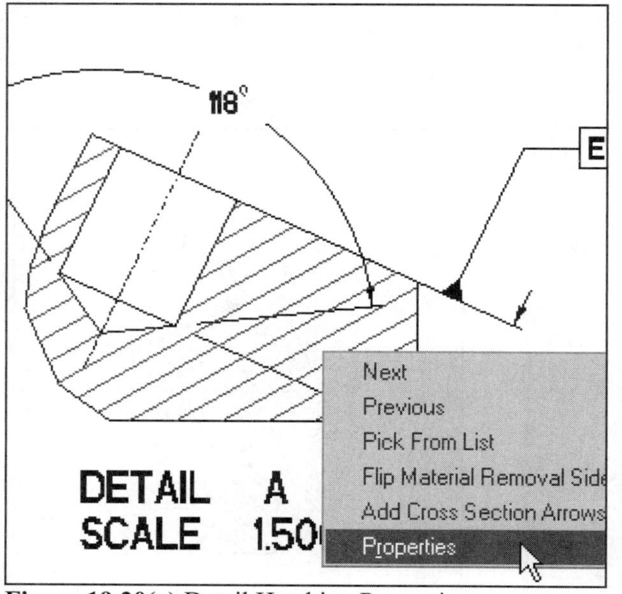

Figure 19.30(a) Detail Hatching Properties

Figure 19.30(b) Hatching at **45** Degree Angle

Change the text style used on drawing to a more "drafting" style. Click: ⇒ enclose all of the drawing text with a pick box [Fig. 19.31(a)] ⇒ **RMB** ⇒ **Text Style**

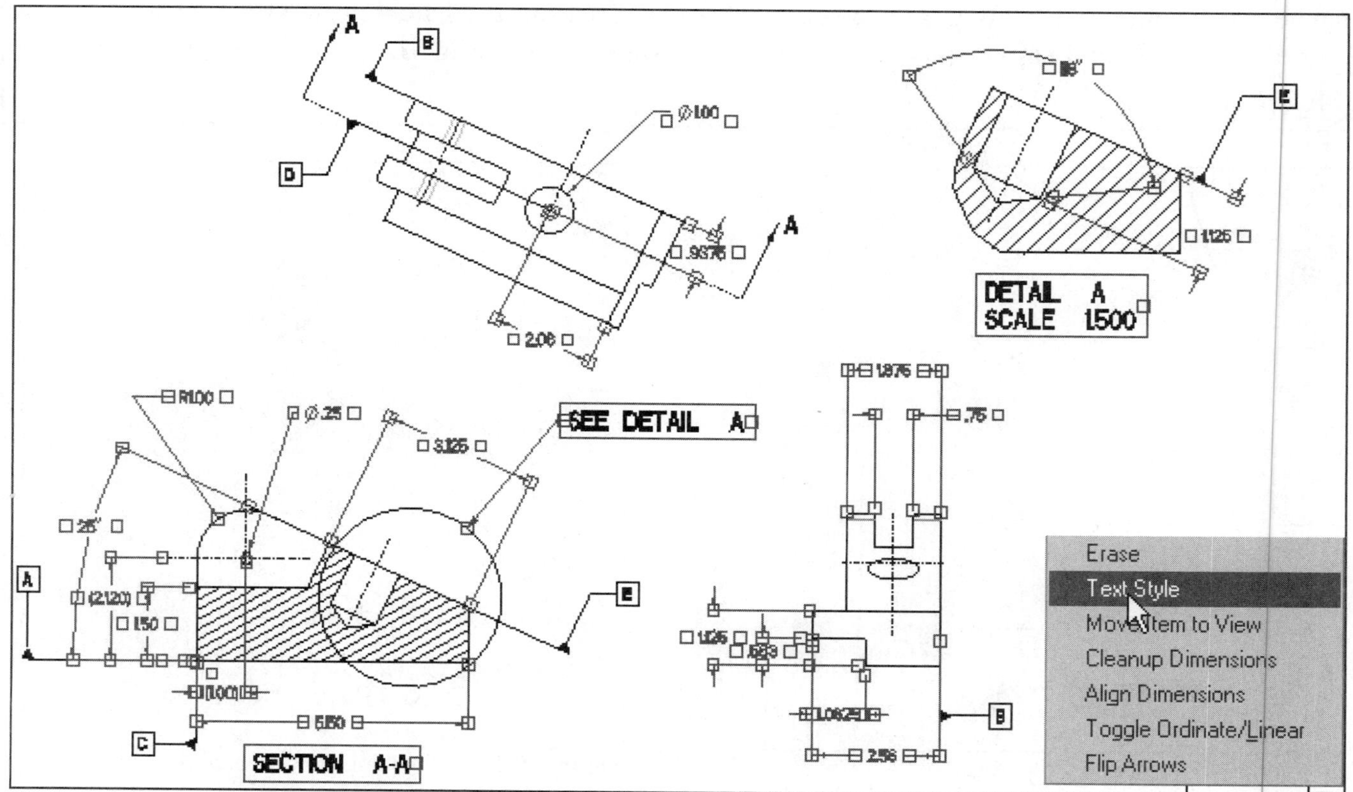

Figure 19.31(a) Select all Text

Font **Blueprint MT** [Fig. 19.31(b)] ⇒ **Apply** ⇒ **OK** ⇒ **LMB** [Fig. 19.31(c)] ⇒ ⇒ **MMB**

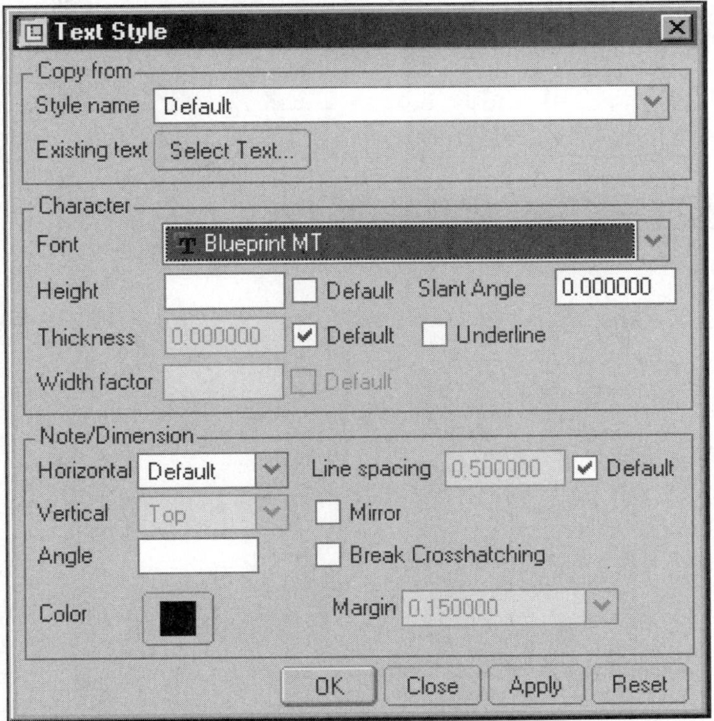

Figure 19.31(b) Text Style Dialog Box, Character- Font- Blueprint MT

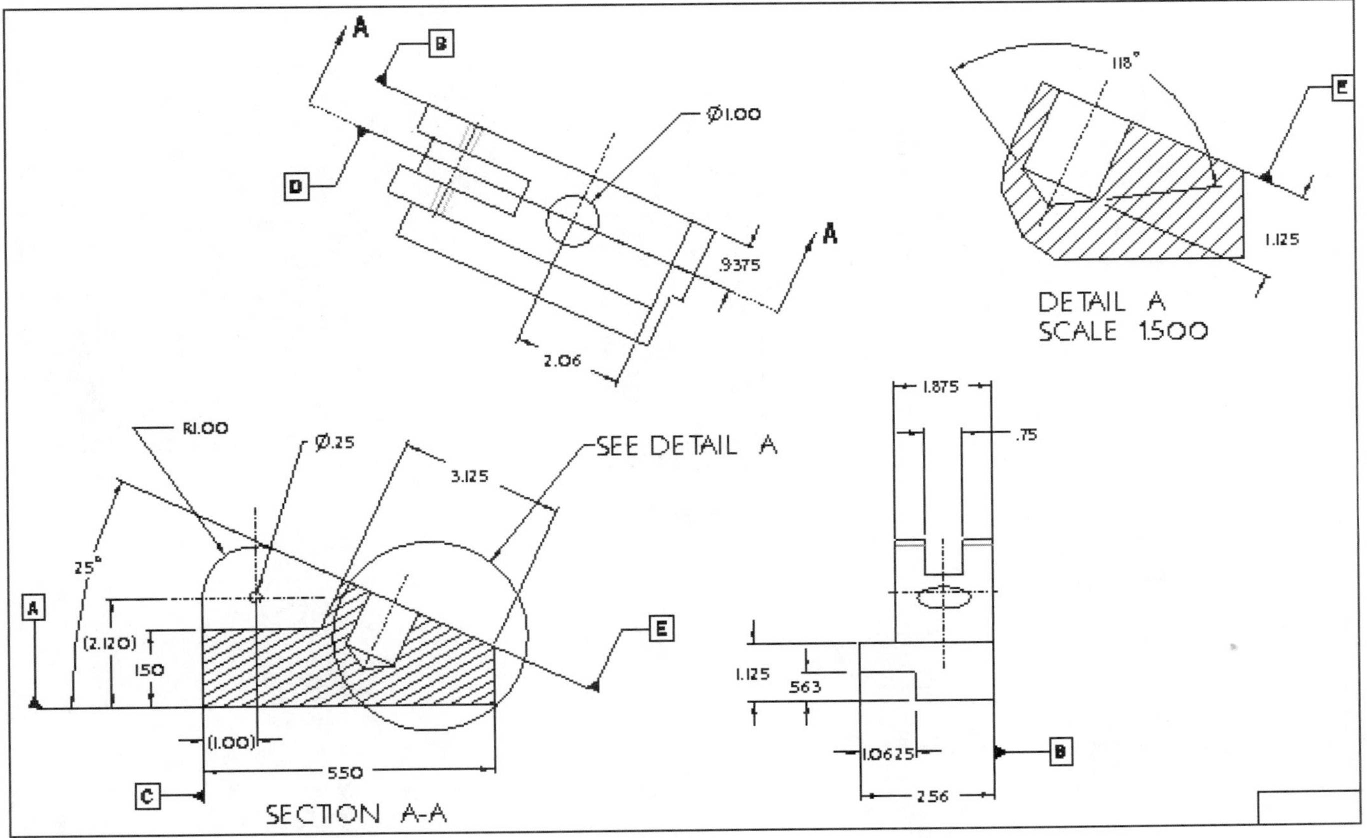

Figure 19.31(c) New Text Style

Add a geometric tolerance to the angled surface, click **Create geometric tolerances** (*or Insert ⇒ Geometric Tolerance*) ⇒ **Angularity** ⇒ Type **Datum** [Fig. 19.32(a)] ⇒ Select Entity... ⇒ click on datum **E** [Fig. 19.32(b)] ⇒ **Datum Refs** tab ⇒ **Primary** tab- Basic ⇒ click on datum **A** in the front view as the Primary Datum Reference [Fig. 19.32(c)] ⇒ **Tol Value** tab ⇒ Overall Tolerance 0.005 [Fig. 19.32(d)] ⇒ **OK** ⇒ **LMB** ⇒ **RMB** ⇒ **Update Sheet** [Fig. 19.32(e)]

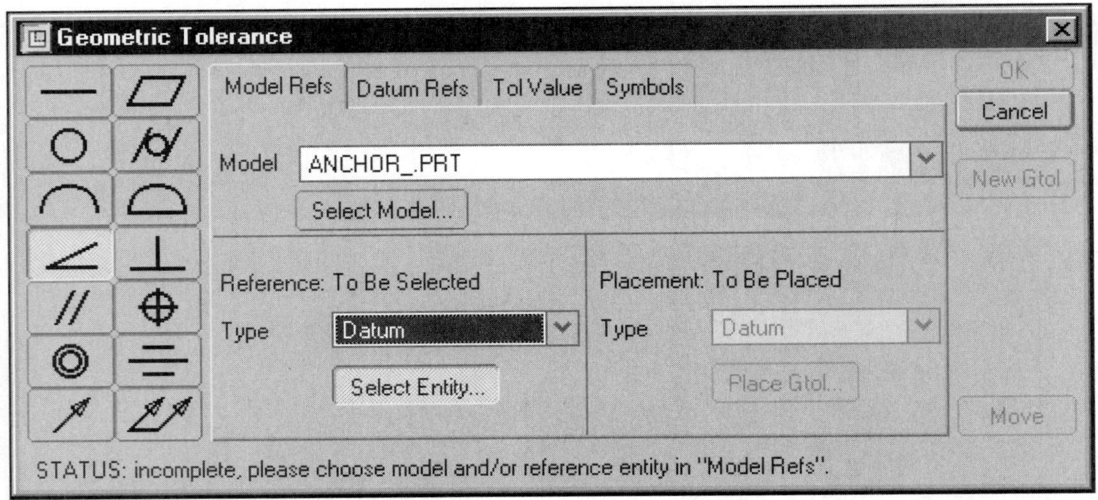

Figure 19.32(a) Angularity, Type Datum

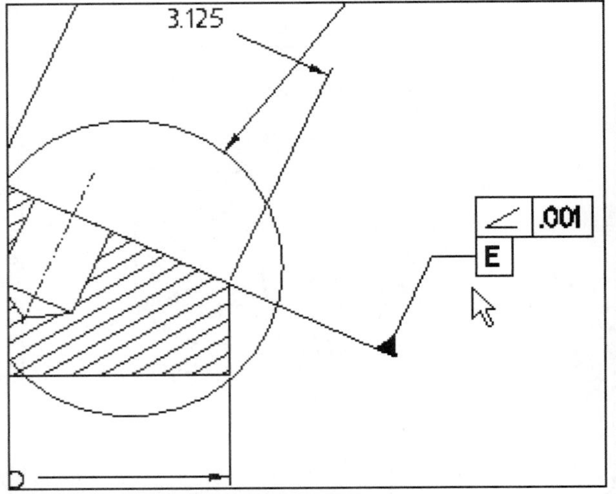

Figure 19.32(b) Pick Datum E

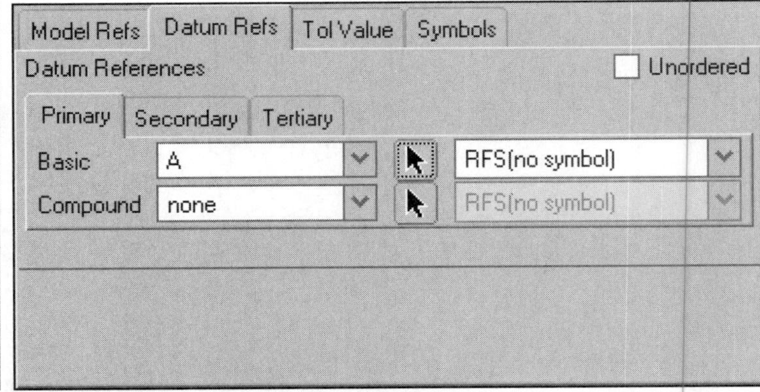

Figure 19.32(c) Primary Reference Datum A

Tolerance Value
Overall Tolerance 0.005
Name gp0
Per Unit Tolerance
Value/Unit 0.001
Unit Length 0.001
Material Condition RFS(no symbol)

Figure 19.32(d) Tolerance Value **.005**

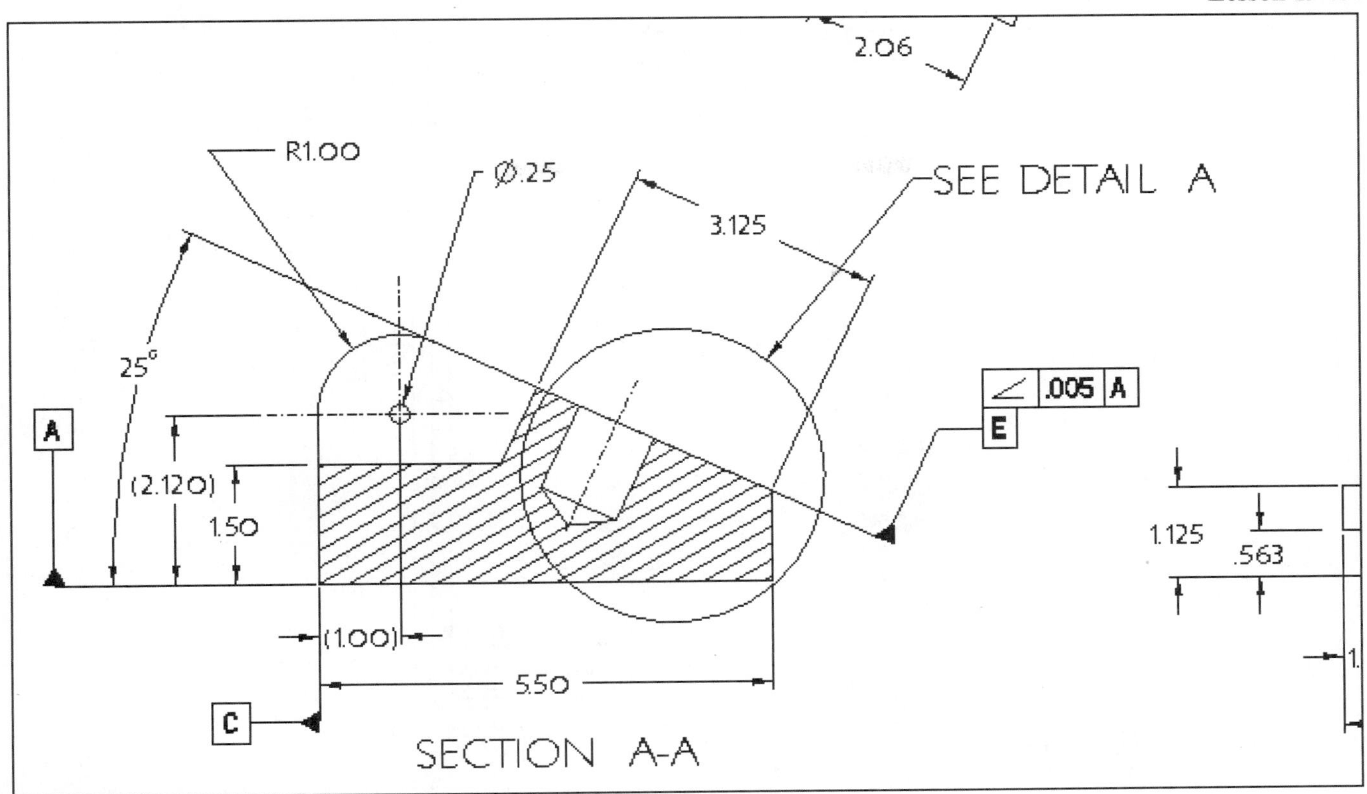

Figure 19.32(e) Added Geometric Tolerance

Create a *second sheet* with an isometric view of the Anchor by clicking: **Tools** ⇒ **Environment** ⇒ Standard Orient Isometric ⇒ **Apply** ⇒ **OK** ⇒ **Insert** ⇒ **Sheet** ⇒ **RMB** ⇒ **Page Setup** ⇒ ⇒ **Browse** [Fig. 19.33(a)] ⇒ System Formats **c.frm** ⇒ **Open** ⇒ **OK** ⇒ [Fig. 19.33(b)] ⇒ **Insert** ⇒ **Drawing View** ⇒ **Scale** leave other defaults ⇒ **MMB** ⇒ pick a center point for the view ⇒ type **1.5** as the scale ⇒ **Enter** [Fig. 19.33(c)] ⇒ **Saved Views** ⇒ **Standard Orientation** [Fig. 19.33(d)] ⇒ **Set** ⇒ **OK** ⇒ **LMB**

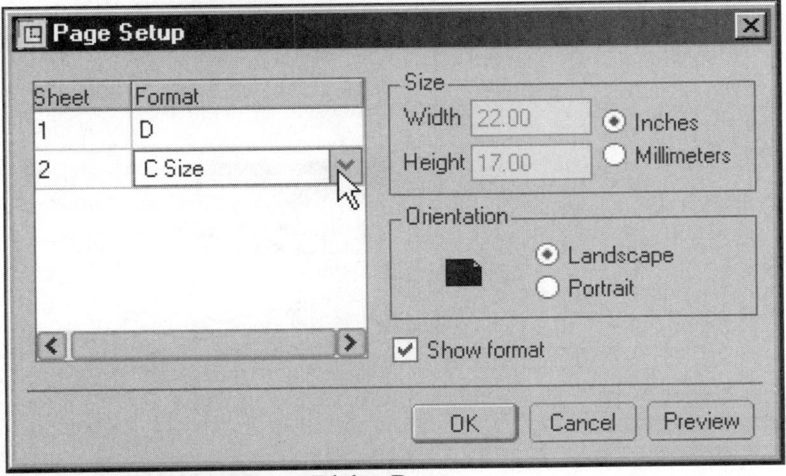

Figure 19.33(a) Page Setup Dialog Box

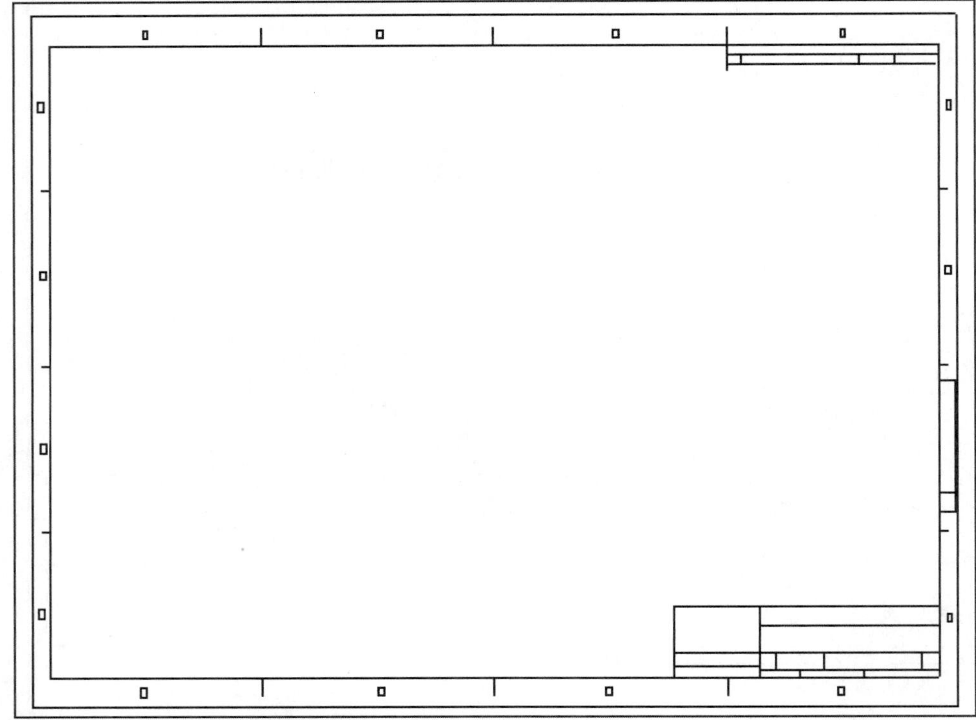

Figure 19.33(b) C Format for Page 2

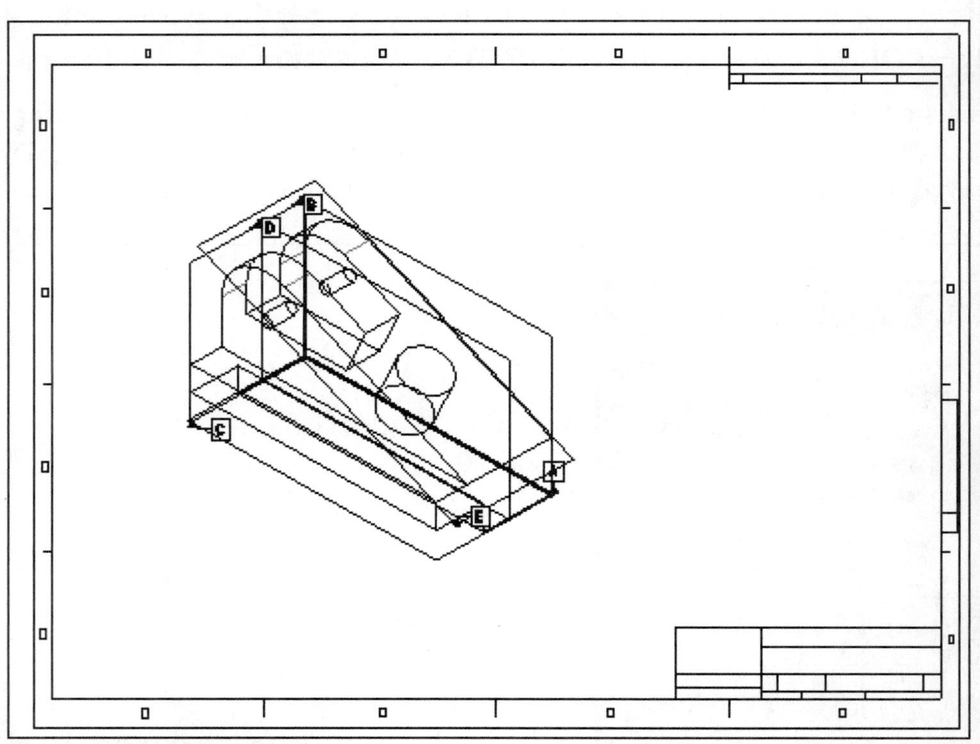

Figure 19.33(c) Pictorial View

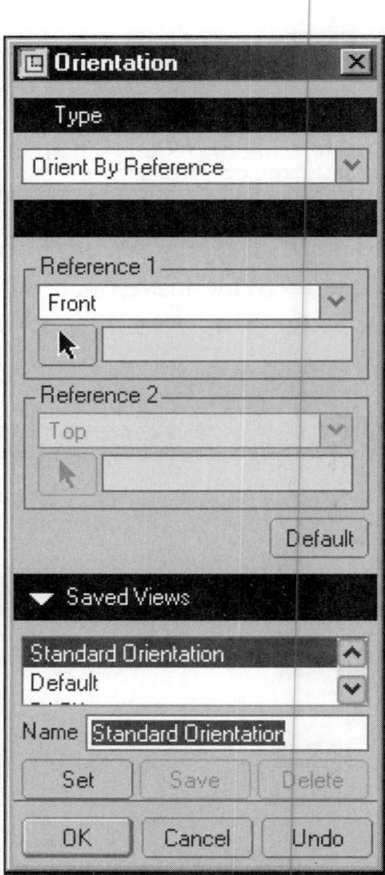

Figure 19.33(d) Orientation

Click on the pictorial view ⇒ **RMB** ⇒ **Properties** ⇒ **View Disp** ⇒ **No Hidden** ⇒ **Tan Dimmed** ⇒ **MMB** ⇒ **MMB** ⇒ **LMB** [Fig. 19.33(e)] ⇒ **File** ⇒ **Save** ⇒ **MMB**

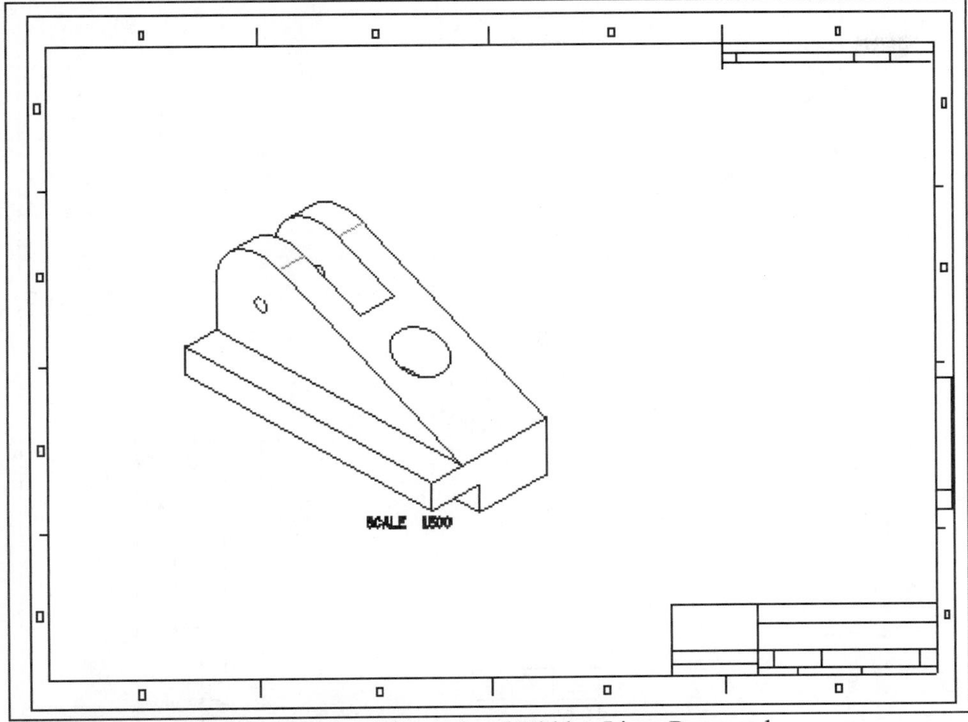

Figure 19.33(e) Sheet 2, Tangent Dimmed and Hidden Lines Removed

Click: 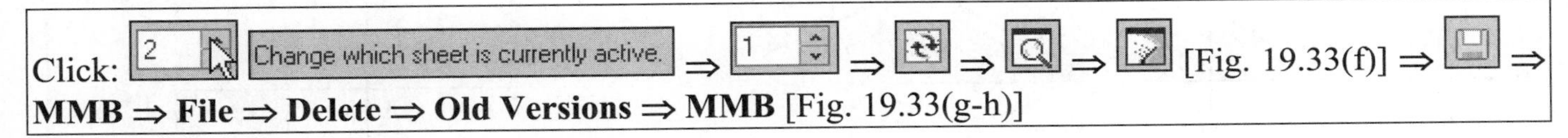

[Fig. 19.33(f)] ⇒ ⇒ **MMB** ⇒ **File** ⇒ **Delete** ⇒ **Old Versions** ⇒ **MMB** [Fig. 19.33(g-h)]

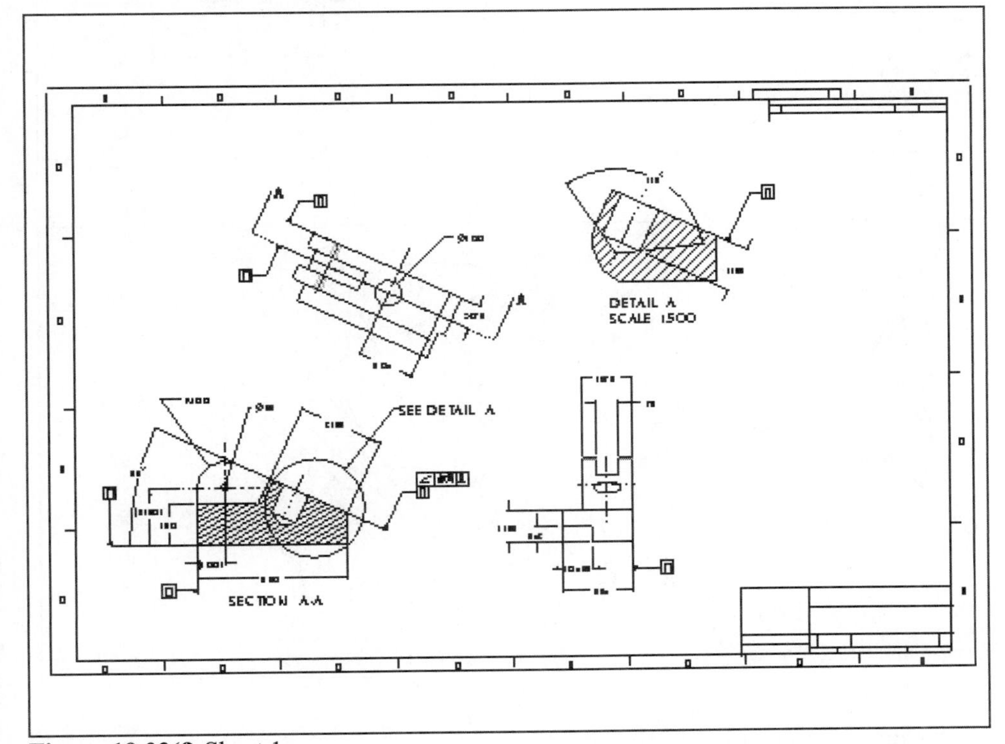

Figure 19.33(f) Sheet 1

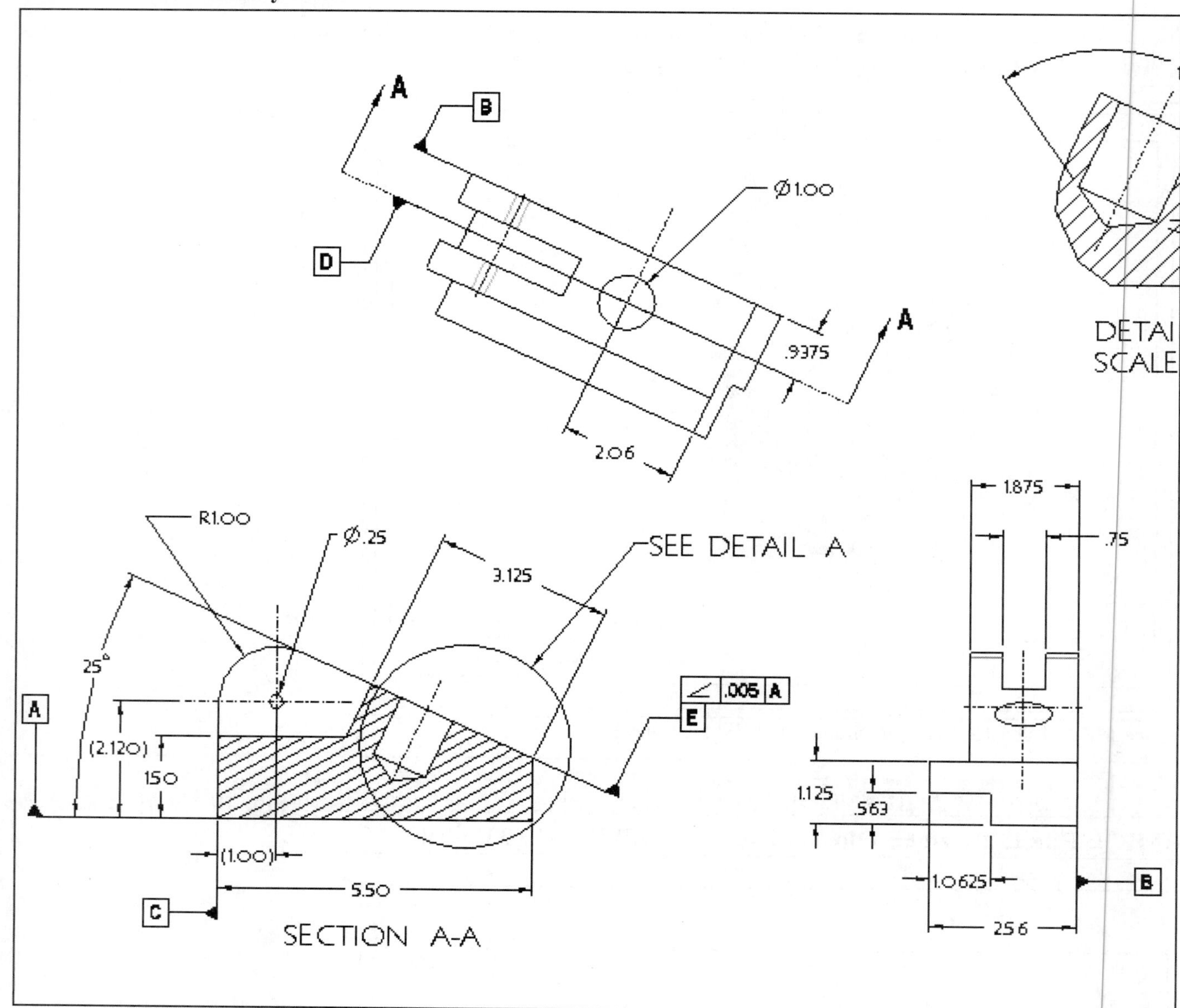

Figure 19.33(g) Sheet 1 Views

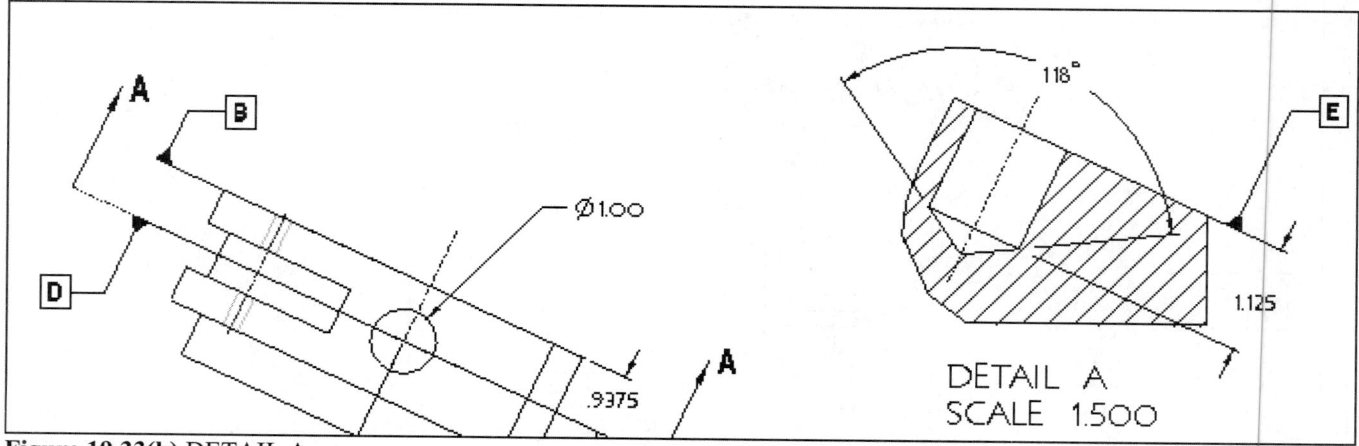

Figure 19.33(h) DETAIL A

You have now completed this lesson. Continue with the Lesson Project.

Lesson 19 Project

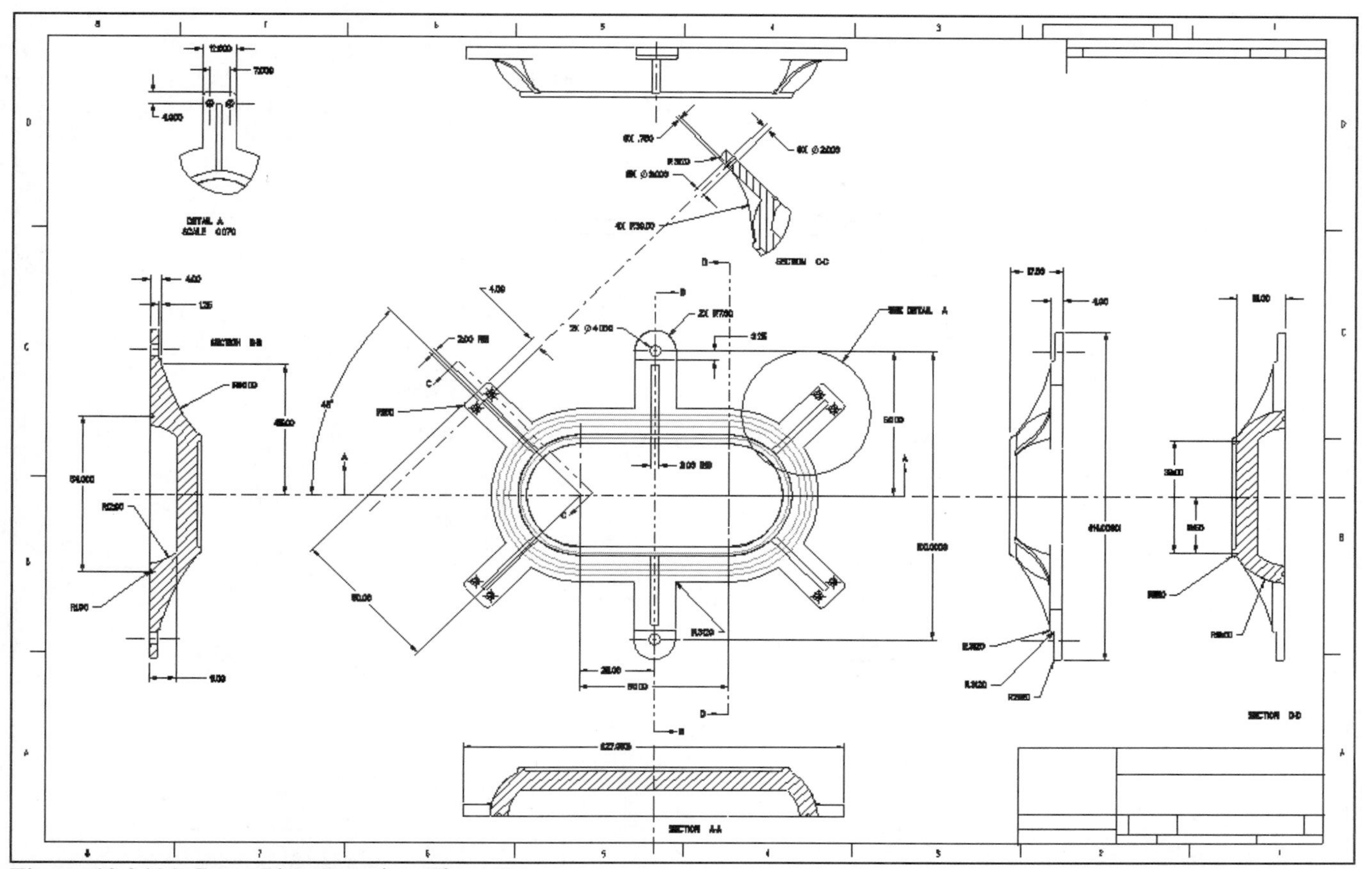

Figure 19.34(a) Cover Plate Drawing, Sheet One

Cover Plate

Detail the Cover Plate [Figs. 19.34(a-d)]. *See the complete set of drawings and views provided in the Lesson 12 Project.* Use your own format and title block. Analyze the Cover Plate for sheet size and view requirements. Create the required sections in Part mode to be used on the Drawing mode views. You may also detail any of the parts created in Lessons 2 through14 at this time.

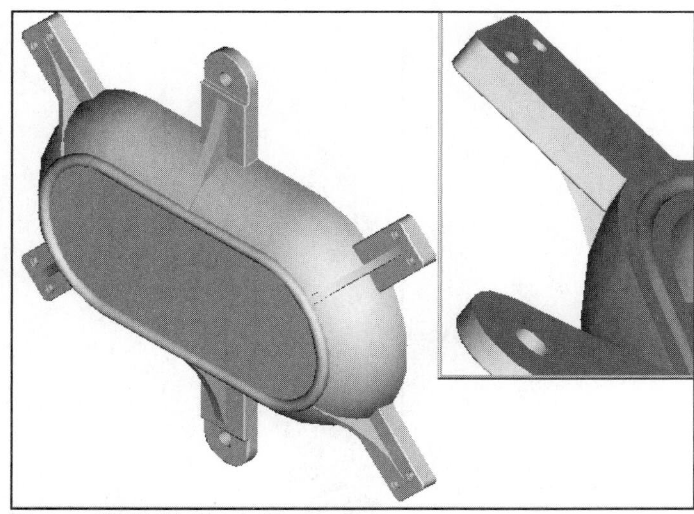

Figure 19.34(b) Cover Plate

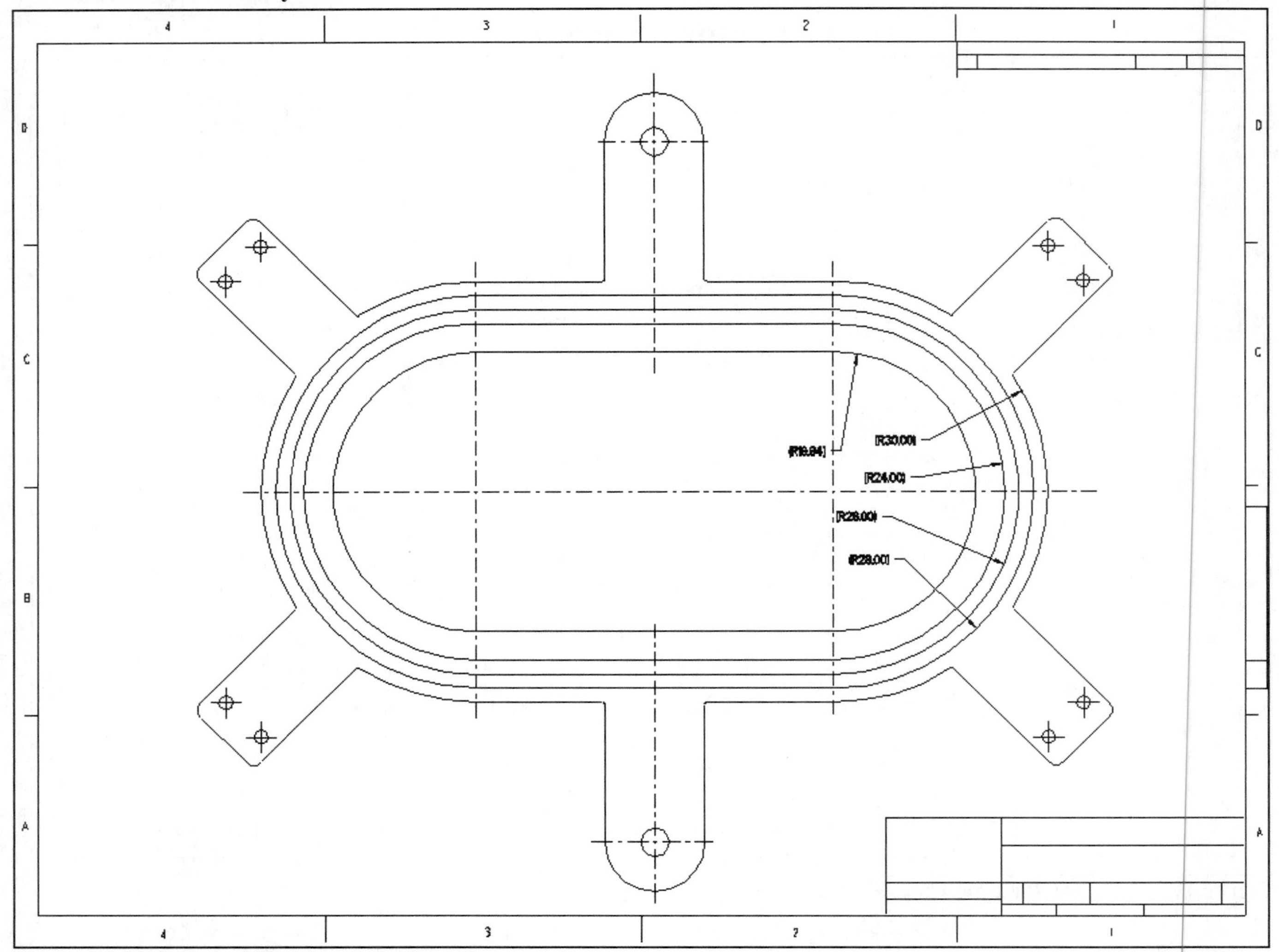

Figure 19.34(c) Cover Plate Drawing, Sheet Two

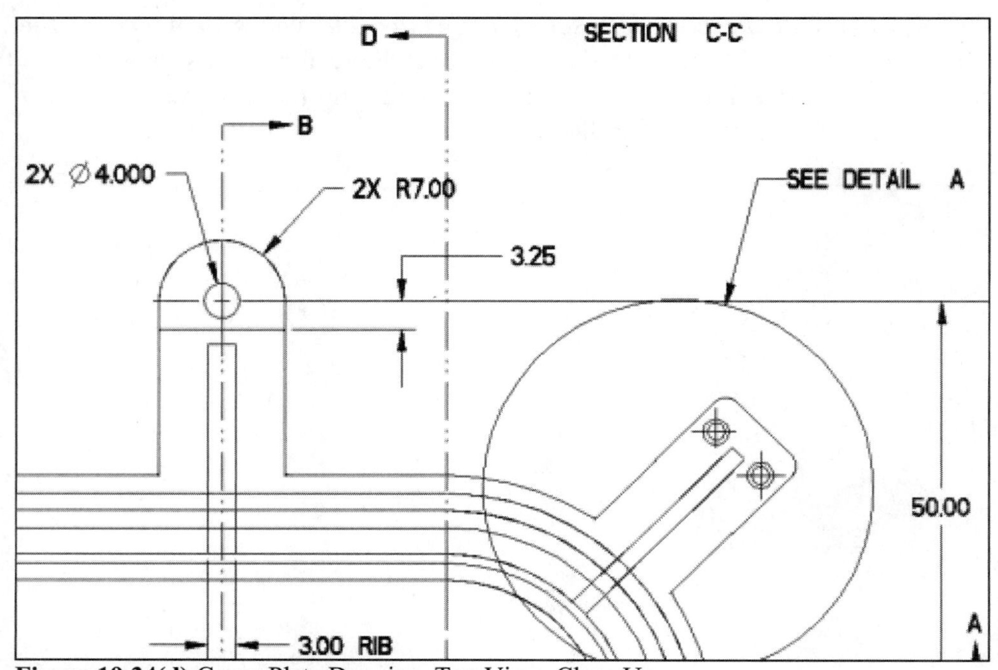

Figure 19.34(d) Cover Plate Drawing, Top View, Close Up

Lesson 20 Assembly Drawings

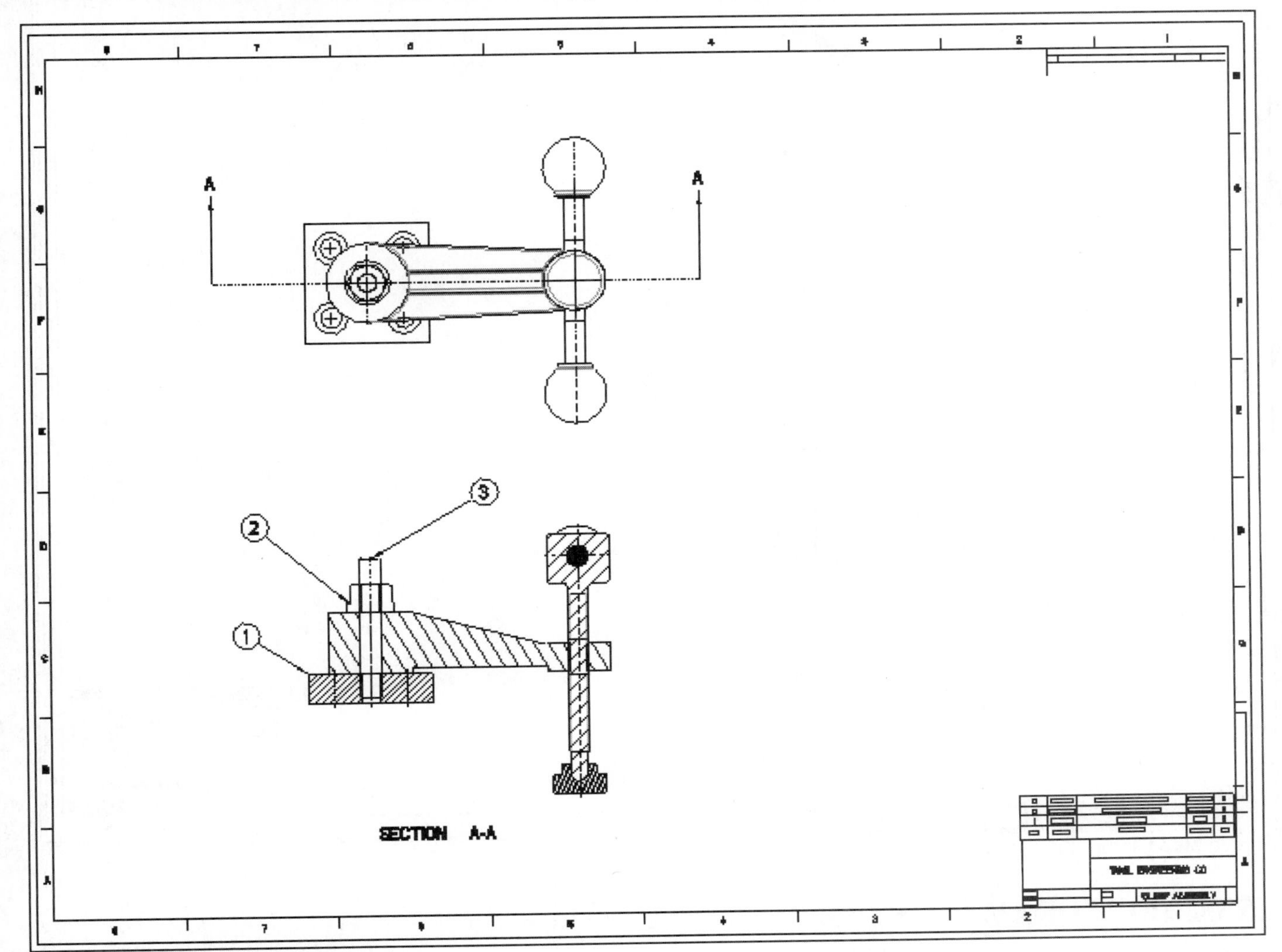

Figure 20.1 Clamp_Assembly Drawing

OBJECTIVES

- Create an **Assembly Drawing**
- Generate a **Parts List** from a bill of materials (**BOM**)
- **Balloon** an assembly drawing
- Create a **section assembly view** and change **component visibility**
- Add **Parameters** to parts
- Create a **Table** to generate a parts list automatically
- Create drawings with **Exploded Views**
- Use **Multiple Sheets**
- Make assembly **Drawing Sheets** with **multiple models**
- Create **Balloons** on exploded assembly views

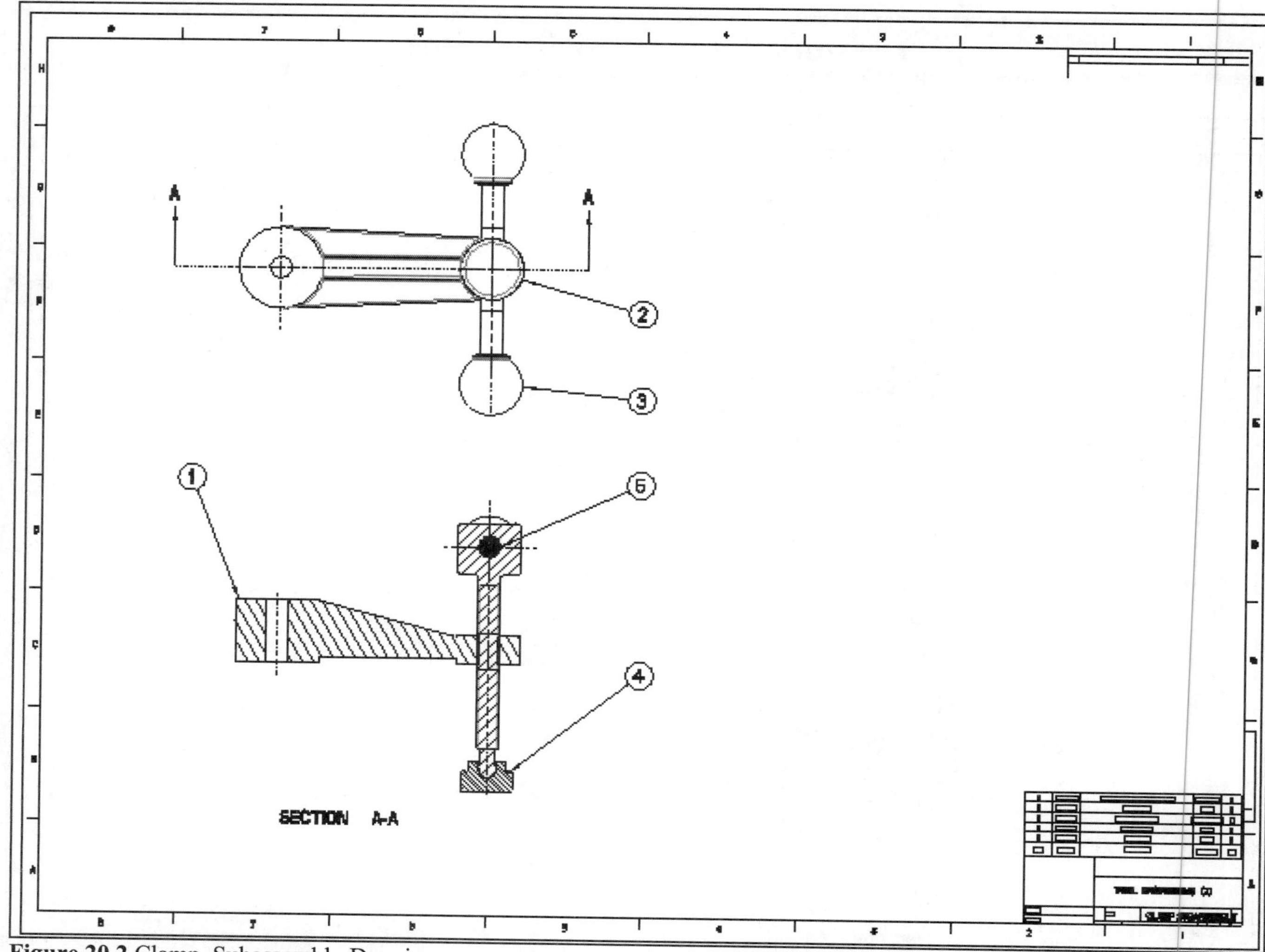

Figure 20.2 Clamp_Subassembly Drawing

ASSEMBLY DRAWINGS

Pro/E incorporates a great deal of functionality into drawings of assemblies (Fig. 20.1). You can assign parameters to parts in the assembly that can be displayed on a *parts list* in an assembly drawing (Fig. 20.2). Pro/E can also generate the item balloons for each component on standard orthographic views or on an exploded view.

In addition, a variety of specialized capabilities allow you to alter the manner in which individual components are displayed in views and in sections. The format for an assembly is usually different from the format used for detail drawings. The most significant difference is the presence of a *Parts List*.

As part of this lesson, you will create a set of assembly formats and place your standard parts list in them. A parts list is actually a *Drawing Table object* that is formatted to represent a bill of materials in a drawing. By defining *parameters* in the parts in your assembly that agree with the specific format of the parts list, you make it possible for Pro/E to add pertinent data to the assembly drawings parts list automatically as components are added to the assembly. After the parts list and parameters have been added, Pro/E can balloon the assembly drawing automatically.

Exploded Assembly Drawings

Exploding an assembly affects only the display of the assembly; it does not alter actual distances between components (Fig. 20.3). *Explode states* are created and saved to allow a clear visualization and understanding of the positional relationships of all of the components in an assembly. For each explode state, you can toggle the explode status of components, change the explode locations of components, and create explode offset lines. You can define multiple explode states for each assembly and then explode the assembly using any of these explode states at any time. You can also set an exploded state for each drawing view of an assembly.

If none of the exploded views you have created is exactly the exploded position you wish to use in this lesson, bring up the model (assembly), create additional exploded states, and save them for use in this lesson and lesson project.

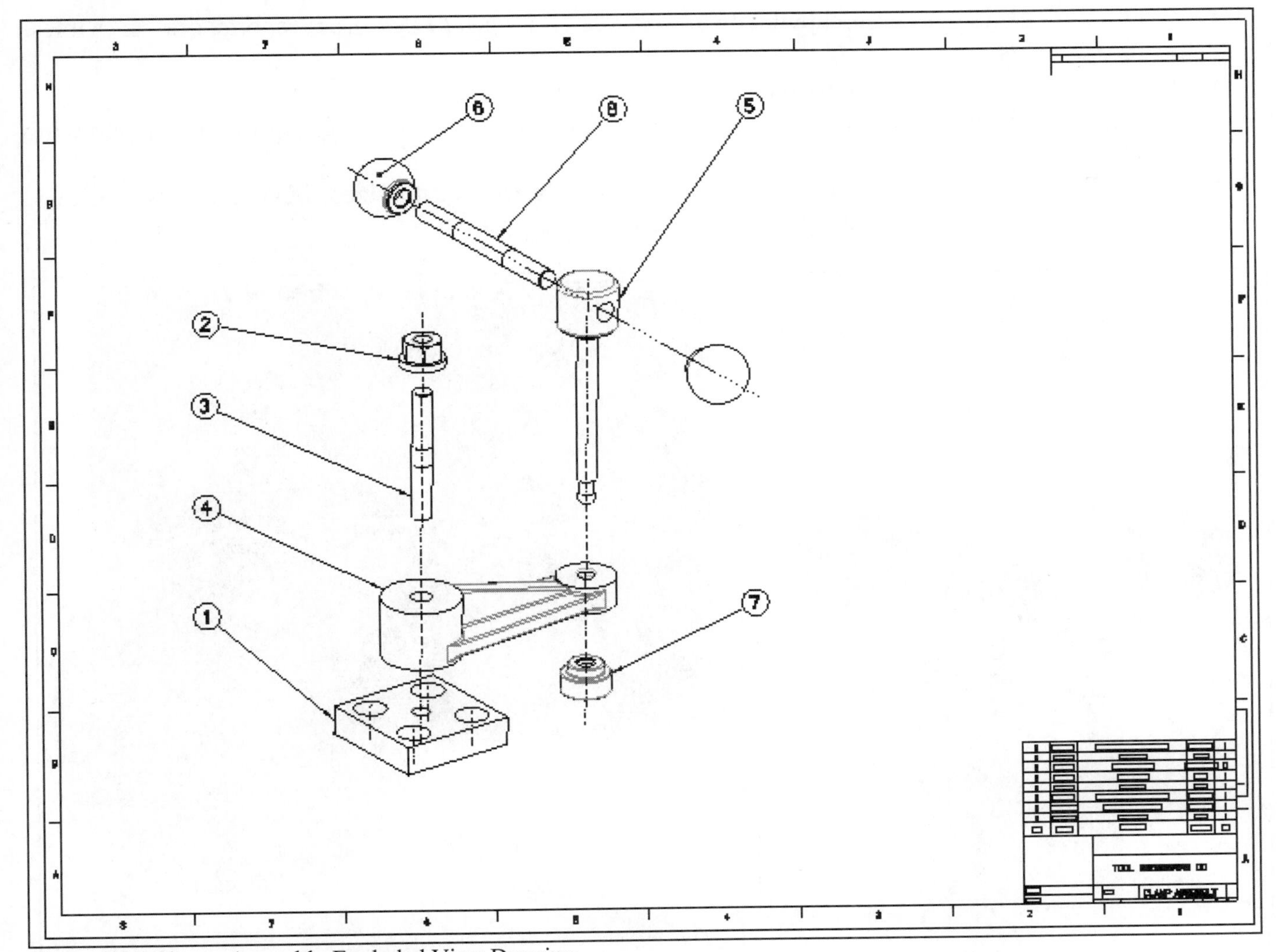

Figure 20.3 Clamp_Assembly Exploded View Drawing

BOM

An assembly drawing is created after the assembly is complete [Fig. 20.4(a)]. With **Pro/REPORT**, you can then generate a bill of materials or other tabular data as required for the project. Pro/REPORT introduces a formatting environment where text, graphics, tables, and data can be combined to create a dynamic report [Fig. 20.4(b)]. Specific tools enable you to generate customized **bill of materials (BOM)**, report tables, and other associative reports:

- Dynamic, customized reports with drawing views and graphics can be created.
- User-defined or predefined model data can be listed on reports, drawing tables, or layout tables. These reported data can be sorted by any requested data-type display.
- Regions in drawing tables, report tables, and layout tables can be defined to expand and shrink automatically with the amount of model information you have asked to display.
- Filters can be added to eliminate the display of specific types of data from reports, drawing tables, or layout tables.
- Recursive or top-level assembly data can be searched for display.
- Duplicate occurrences of model data can be listed individually or as a group in a report, drawing table, or layout table.
- Assembly component balloons can be linked directly to a customized BOM and automatically updated when assembly modifications are made.

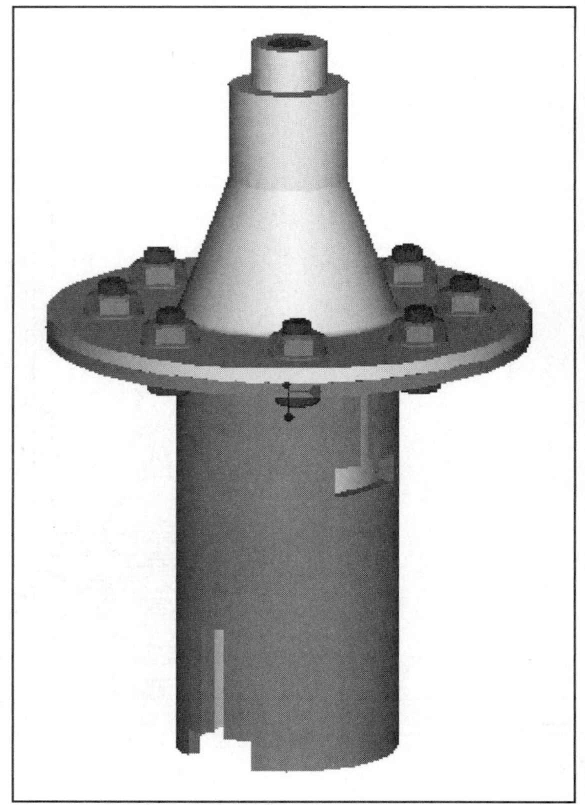

Figure 20.4(a) Assembly (CADTRAIN)

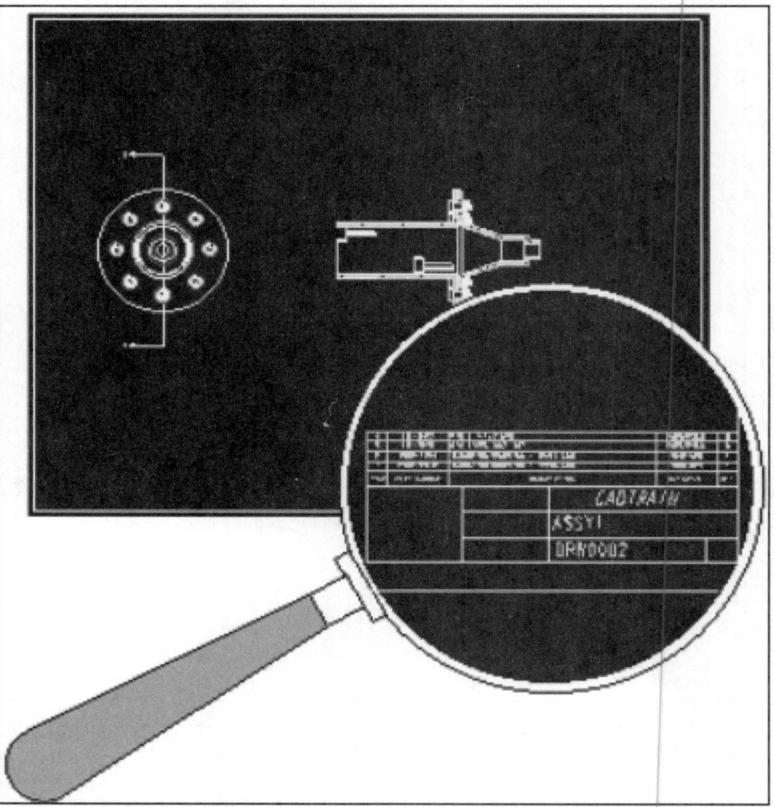

Figure 20.4(b) Assembly Drawing and BOM (CADTRAIN)

In **Report** mode, data can be displayed in a tabular form on reports, just as they are in drawing tables (Fig. 20.5). The data reported on the tables are taken directly from a selected model and update automatically when the model is modified or changed. A common example of a report is a bill of materials report.

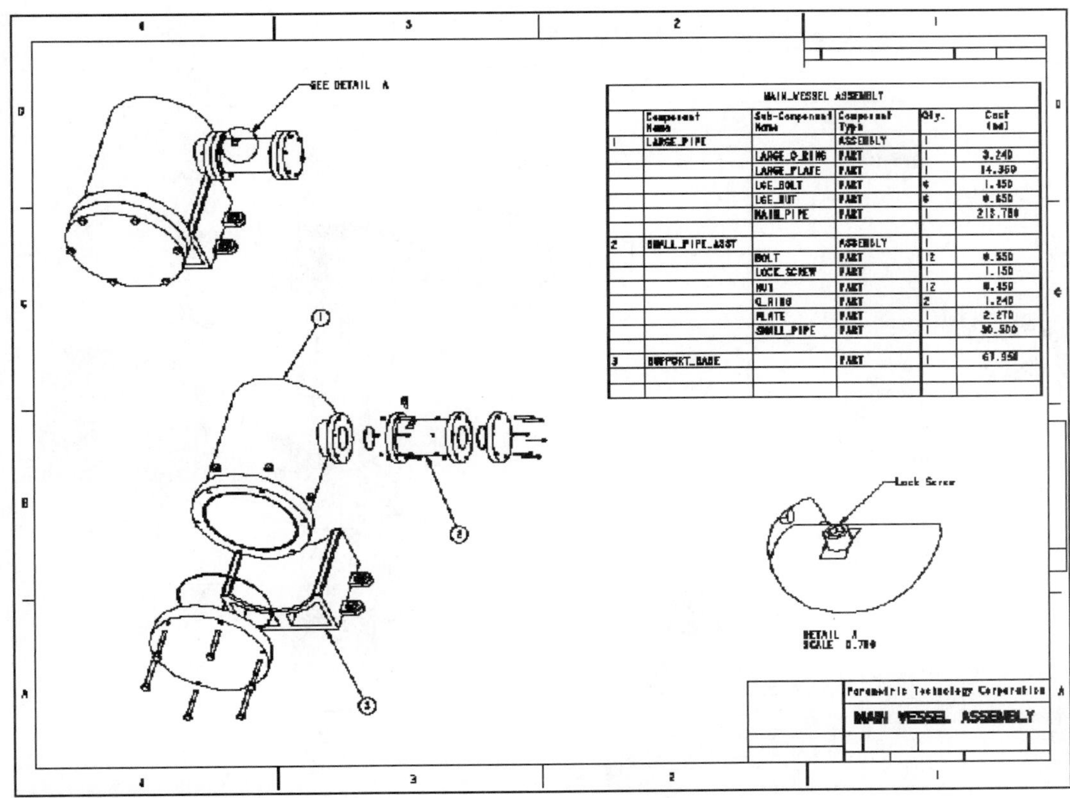

Figure 20.5 Swing Clamp

Including a Bill of Materials in a Drawing

If you do not have Pro/REPORT and want to add a bill of materials [Fig. 20.6(a)], add the BOM to the drawing as a note entered from a file. To format or arrange the information in the BOM, you must use a text editor. A BOM that is added to a drawing as a note [Fig. 20.6(b)] is not parametric with the BOM file that was used to create the note. If the composition of the assembly changes, you must create a new BOM file and add it to the drawing as a new note. You can edit this form of BOM displayed on drawings without affecting the original text file.

8	SW100-22FLN	.500-13 X HEX FLANGE NUT	PURCHASED	1
7	SW100-21ST	.500-13 X 3.50 DOUBLE END STUD	PURCHASED	1
6	SW100-20PL	SWING CLAMP PLATE	1020 CRS	1
5	SW101-95TL	.500-13 X 5.00 DOUBLE END STUD	PURCHASED	1
4	SW101-8FT	SWING CLAMP FOOT	NYLON	1
3	SW101-7BA	SWING CLAMP BALL	BLACK PLASTIC	2
2	SW101-6SW	SWING CLAMP SWIVEL	STEEL	1
1	SW101-5AR	SWING CLAMP ARM	STEEL	1
ITEM	PT NUM	DESCRIPTION	MATERIAL	QTY

TOOL ENGINEERING CO.

A

DRAWN
ISSUED

2.000 SW_CL_ASSEMBLY
SHEET 1 OF 1

2 1

Figure 20.6(a) Table BOM

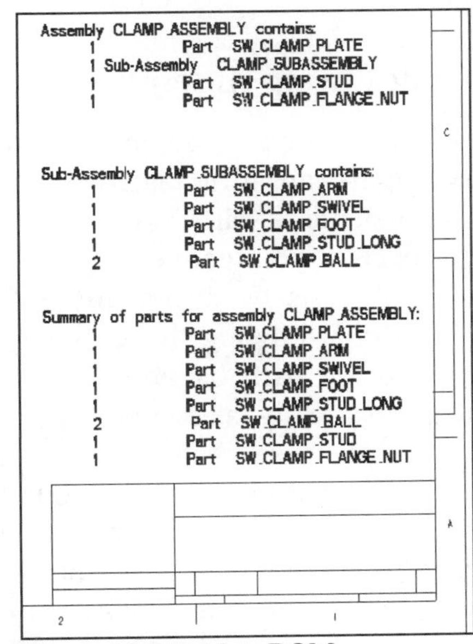

Assembly CLAMP_ASSEMBLY contains:
1 Part SW_CLAMP_PLATE
1 Sub-Assembly CLAMP_SUBASSEMBLY
1 Part SW_CLAMP_STUD
1 Part SW_CLAMP_FLANGE_NUT

Sub-Assembly CLAMP_SUBASSEMBLY contains:
1 Part SW_CLAMP_ARM
1 Part SW_CLAMP_SWIVEL
1 Part SW_CLAMP_FOOT
1 Part SW_CLAMP_STUD_LONG
2 Part SW_CLAMP_BALL

Summary of parts for assembly CLAMP_ASSEMBLY:
1 Part SW_CLAMP_PLATE
1 Part SW_CLAMP_ARM
1 Part SW_CLAMP_SWIVEL
1 Part SW_CLAMP_FOOT
1 Part SW_CLAMP_STUD_LONG
2 Part SW_CLAMP_BALL
1 Part SW_CLAMP_STUD
1 Part SW_CLAMP_FLANGE_NUT

Figure 20.6(b) Note BOM

Lesson 20 STEPS

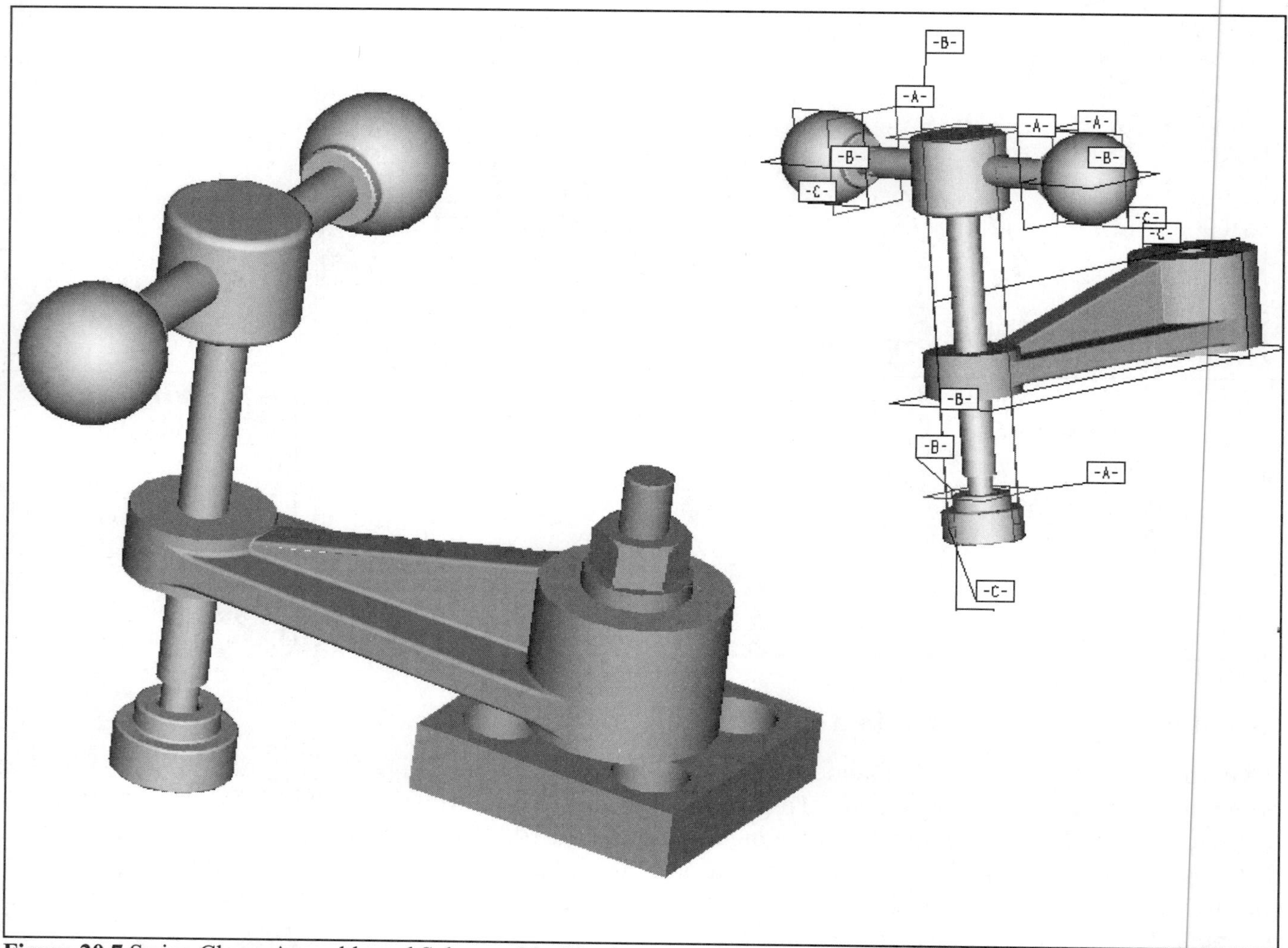

Figure 20.7 Swing Clamp Assembly and Subassembly (inset)

Swing Clamp Assembly Drawing

The format for an assembly drawing is usually different from the format used for detail drawings. The most significant difference is the presence of a parts list. We will create a standard "E" size format and place a standard parts list on it. You should create a set of assembly formats on "B", "C", "D", and "F" size sheets at your convenience.

A parts list is actually a *Drawing Table object* that is formatted to represent a bill of material (BOM) in a drawing. By defining parameters in the parts in your assembly that agree with the specific format of the parts list, you make it possible for Pro/E to add pertinent data to the assembly drawing parts list automatically, as components are added to the assembly.

After you create an "E" size format sheet with a parts list table, you will create two new drawings (each with two views) using your new assembly format. The Swing Clamp *subassembly* (Fig. 20.7), will be used in the first drawing. The second drawing will use the Swing Clamp *assembly* (Fig. 20.7). Both drawings use the "E" size format created in the first section of this lesson. The format will have a parameter-driven title block (as in Lesson 17) and an integral parts list. Using steps similar to those outlined in Lesson 17, where a "C" size format was created and saved, create an "E" size format.

Click: **File ⇒ Set Working Directory** select working directory ⇒ **OK** ⇒ **Open an existing object** ⇒ ⇒ System Formats ⇒ e.frm ⇒ **Open ⇒ File ⇒ Save a Copy** ⇒ type a unique name for your format: New Name **ASM_FORMAT_E ⇒ OK ⇒ Window ⇒ Close** ⇒ ⇒ pick **asm_format_e ⇒ Open**

To repeat what has been said before, and which cannot be stressed enough, the **format** will have a **.dtl** associated with it, and the **drawing** will have a different **.dtl** file associated with it. *They are separate **.dtl** files.* When you activate a drawing and then add a format, the **.dtl** for the format controls the font, etc. for the format only. The drawing **.dtl** file that controls items on the drawing needs to be established separately in the Drawing mode.

Click: **RMB** in the graphics window ⇒ **Properties** Options dialog box opens ⇒ ***drawing_text_height .25*** ⇒ **Add/Change** ⇒ ***default_font filled*** ⇒ **Add/Change** ⇒ ***draw_arrow_style filled*** ⇒ **Add/Change** ⇒ ***draw_arrow_length .25*** ⇒ **Add/Change** ⇒ ***draw_arrow_width .08*** ⇒ **Add/Change** (Fig. 20.8) ⇒ **Apply** option in Status column has default column value active, option in Status column shows pending value for this drawing ⇒ **Close** ⇒ ⇒ ⇒ **MMB**

	Value	Default	Status	Description
Active Drawing				
These options control text not subject to oth				
drawing_text_height	.25	0.156250	✳	Sets default text height for all text in the drawing usin
text_thickness	0.000000 *	0.000000	●	Sets default text thickness for new text after regener
text_width_factor	0.800000 *	0.800000	●	Sets default ratio between the text width and text hei
draft_scale	1.000000 *	1.000000	●	Determines value of draft dimensions relative to actu
These options control text and line fonts				
default_font	filled	font	✳	Specifies a font index that determines default text for
These options control leaders				
draw_arrow_length	.25	0.187500	✳	Sets length of leader line arrows.
draw_arrow_style	filled	closed	✳	Controls style of arrow head for all detail items involvi
draw_arrow_width	.08	0.062500	✳	Sets width of leader line arrows. Drives these: "draw
draw_attach_sym_height	DEFAULT *	default	●	Sets height of leader line slashes, integral signs, and
draw_attach_sym_width	DEFAULT *	default	●	Sets width of leader line slashes, integral signs, and b
draw_dot_diameter	DEFAULT *	default	●	Sets diameter of leader line dots. If set to "default," u
leader_elbow_length	0.250000 *	0.250000	●	Determines length of leader elbow (the horizontal leg
sort_method_in_region	delimited		●	Determines repeat regions sort mechanism. String_or
Miscellaneous options				
draft_scale	1.000000 *	1.000000	●	Determines value of draft dimensions relative to actu

Figure 20.8 Drawing Options Changes

Zoom into the title block region, and create notes for the title text and parameter text required to display the proper information.

Click: **Tools** ⇒ **Environment** ⇒ [Snap to Grid] ⇒ **Apply** ⇒ **OK** ⇒ **View** ⇒ **Draft Grid** ⇒ **Show Grid** ⇒ **Grid Params** ⇒ **X&Y Spacing** ⇒ type **.1** ⇒ **MMB** ⇒ **MMB** ⇒ **MMB** ⇒ **Create a note** ⇒

Make Note ⇒ pick point for the note in the largest area of the title block (Fig. 20.9) ⇒ type **TOOL ENGINEERING CO.** ⇒ **Enter** ⇒ **Enter** ⇒

Make Note ⇒ (Fig. 20.9) ⇒ type **DRAWN** ⇒ **Enter** ⇒ **Enter** ⇒

Make Note ⇒ (Fig. 20.9) ⇒ type **ISSUED** ⇒ **Enter** ⇒ **Enter** ⇒

Make Note ⇒ (Fig. 20.9) ⇒ type **&dwg_name** ⇒ **Enter** ⇒ **Enter** ⇒

Make Note ⇒ (Fig. 20.9) ⇒ type **&scale** ⇒ **Enter** ⇒ **Enter** ⇒

Make Note ⇒ (Fig. 20.9) ⇒ type **SHEET ¤t_sheet OF &total_sheets** ⇒ **Enter** ⇒ **Enter** ⇒ **Done/Return** ⇒ **LMB** ⇒ **Tools** ⇒ **Environment** ⇒ [Snap to Grid] ⇒ **Apply** ⇒ **OK** ⇒ modify the text height (**.10**) and the placement of the notes so that they are positioned correctly (Fig. 20.9) ⇒ ⇒ ⇒ ⇒ ⇒ **MMB** ⇒ **File** ⇒ **Delete** ⇒ **Old Versions** ⇒ **MMB**

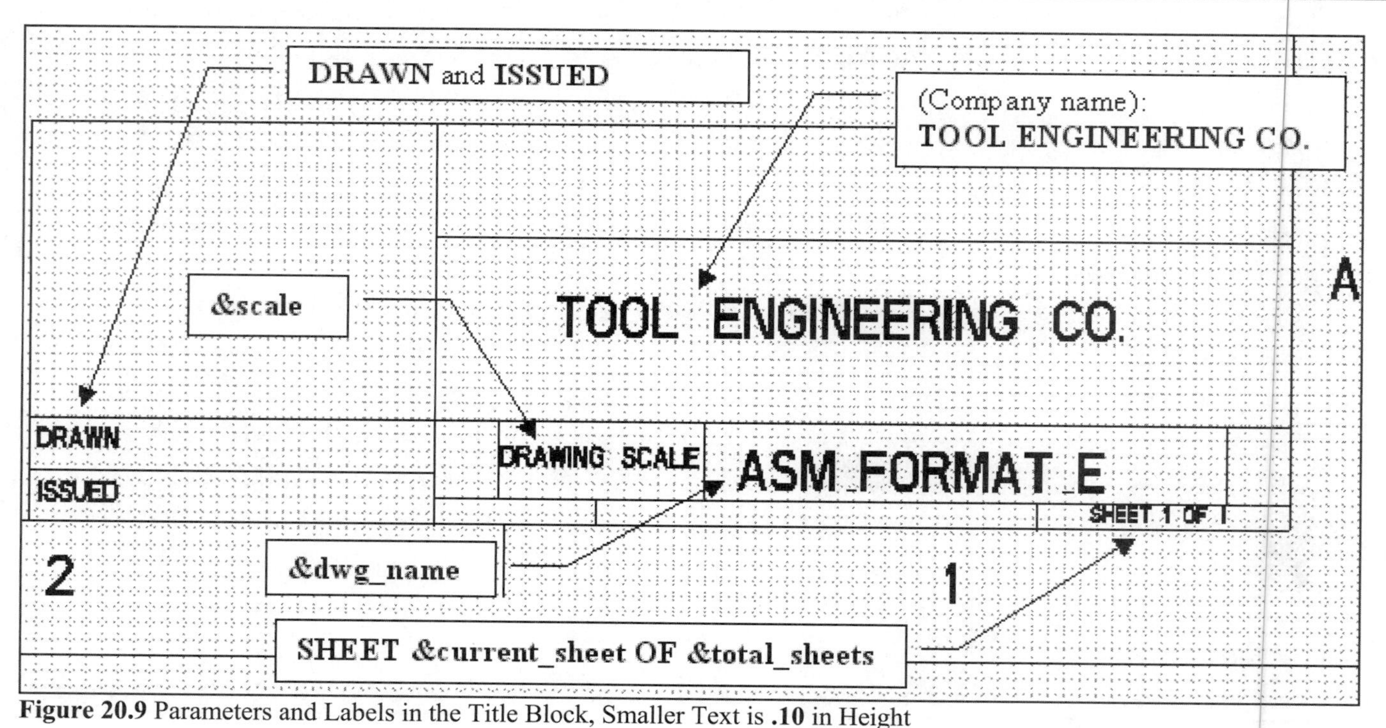

Figure 20.9 Parameters and Labels in the Title Block, Smaller Text is **.10** in Height

The parts list table can now be created and saved with this format. You can add and replace formats and still keep the table associated with the drawing. Start the parts list by creating a table.

Click: **View** ⇒ **Draft Grid** ⇒ **Hide Grid** ⇒ **MMB** ⇒ **Insert a table by specifying the columns and rows sizes** ⇒ **Ascending** ⇒ **Rightward** ⇒ **By Length** ⇒ pick a point at the upper left-hand corner of the title block [Figs. 20.10(a-b)] ⇒

Enter the width of the first column in drawing units (INCH) [Quit] 1.00 ⇒ **MMB** ⇒

Enter the width of the next column in drawing units (INCH) [Done] 1.00 ⇒ **MMB** ⇒

Enter the width of the next column in drawing units (INCH) [Done] 4.00 ⇒ **MMB** ⇒

Enter the width of the next column in drawing units (INCH) [Done] 1.00 ⇒ **MMB** ⇒

Enter the width of the next column in drawing units (INCH) [Done] .75 ⇒ **MMB** ⇒ **MMB** ⇒

Enter the height of the first row in drawing units(INCH) [Quit] .50 ⇒ **MMB** ⇒

Enter the height of the next row in drawing units (INCH) [Done] .375 ⇒ **MMB** ⇒ **MMB** [Fig. 20.10(b)] ⇒ **LMB**

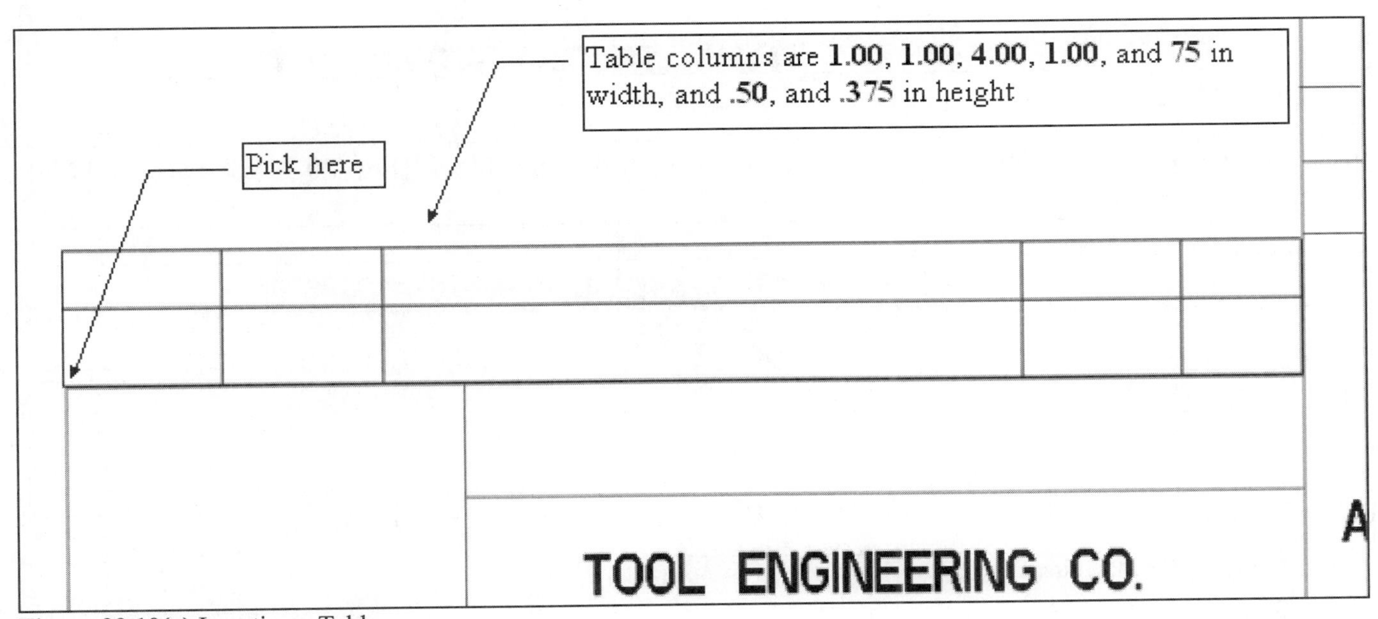

Figure 20.10(a) Inserting a Table

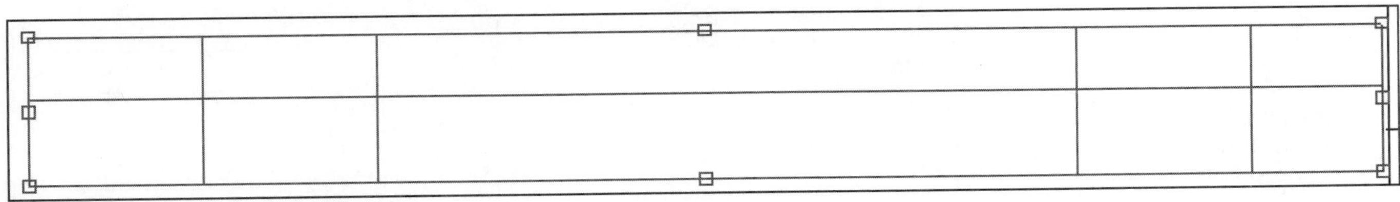

Figure 20.10(b) Highlighted Table

Click: **Table** from Menu bar ⇒ **Repeat Region** ⇒ **Add** ⇒ click in the left block [Fig. 20.10(c)] ⇒ click in the right block [Fig. 20.10(c)] ⇒ **Attributes** ⇒ select the Repeat Region just created ⇒ **No Duplicates** ⇒ **Recursive** ⇒ **MMB** ⇒ **MMB** ⇒ **MMB** ⇒ **Update the display of all views in the active sheet**

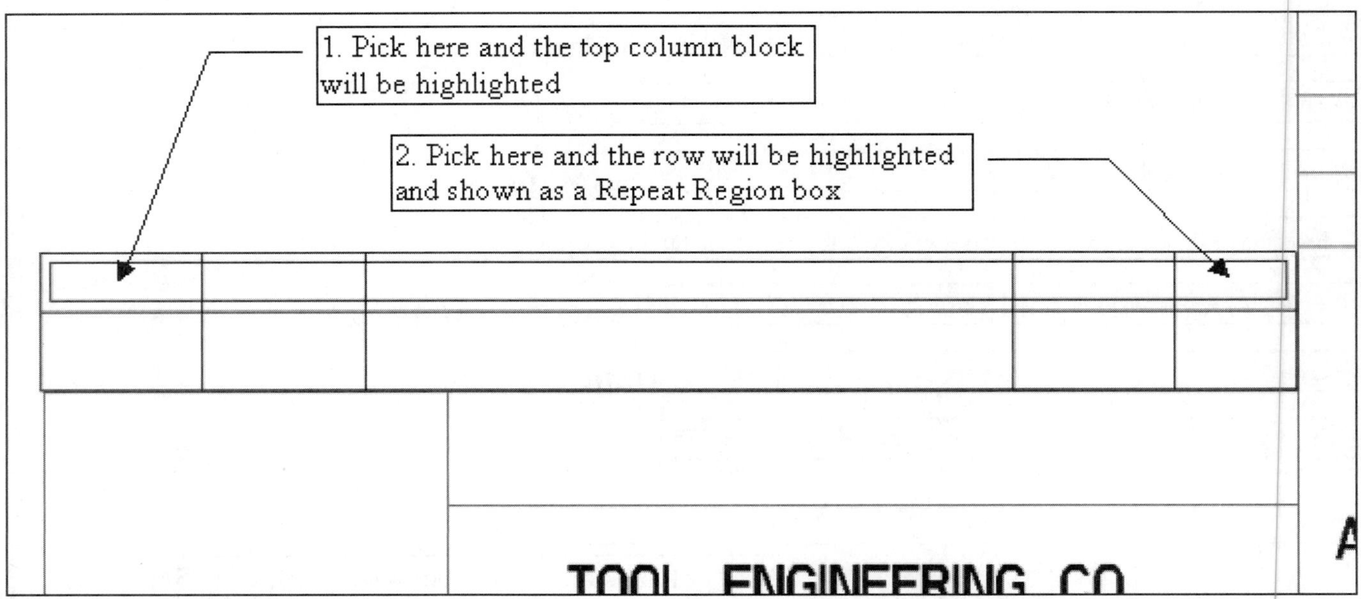

Figure 20.10(c) Repeat Region

Insert column headings using plain text: double-click on the first block [Fig. 20.11(a)] ⇒ type **ITEM** [Fig. 20.11(b)] ⇒ **OK**

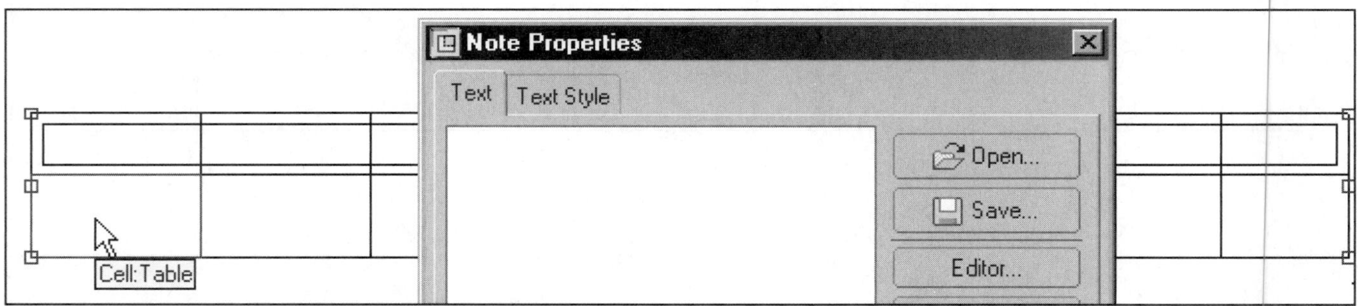

Figure 20.11(a) Select the First Block

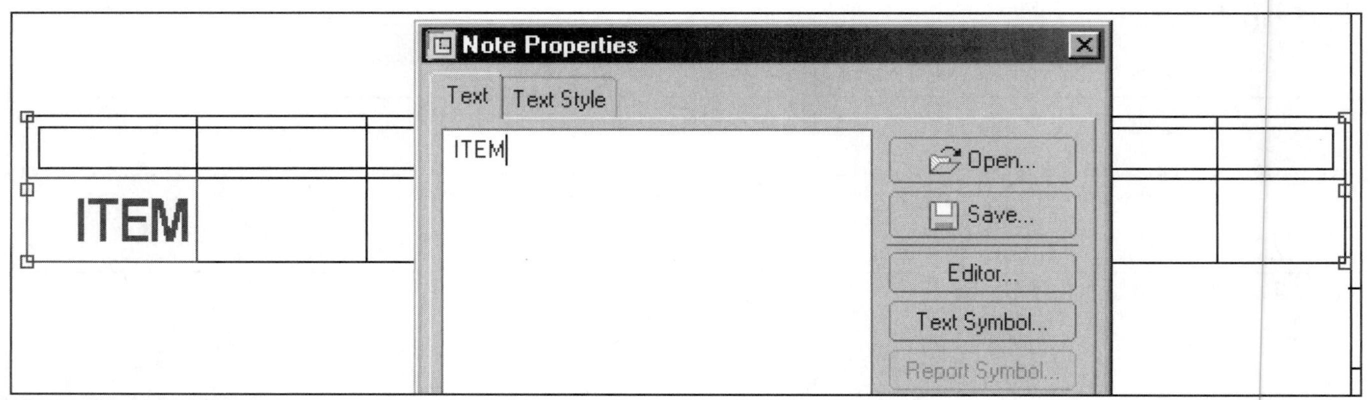

Figure 20.11(b) ITEM

The Repeat Region now needs to have some of its headings correspond to the parameters that will be created for each component model.

Double-click on the second block ⇒ type **PT NUM** [Fig. 20.11(c)] ⇒ **OK** ⇒ double-click on the third block ⇒ type **DESCRIPTION** ⇒ **OK** ⇒ double-click on the fourth block ⇒ type **MATERIAL** ⇒ **OK** ⇒ double-click on the fifth block ⇒ type **QTY** ⇒ **OK** [Fig. 20.11(d)]

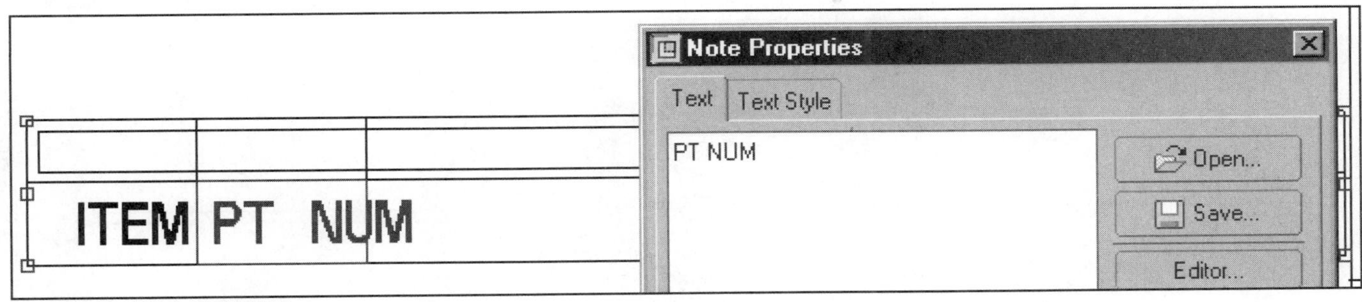

Figure 20.11(c) PT NUM

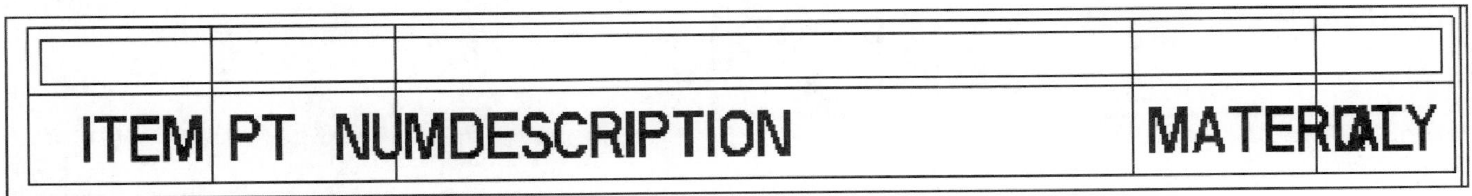

Figure 20.11(d) ITEM, PT MUM, DESCRIPTION, MATERIAL, QTY

Change the text height (**.125**), position (middle) and justification (centered) for all five titles.

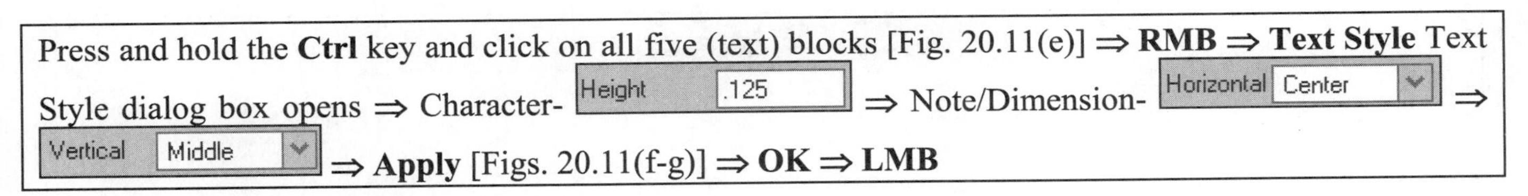

Press and hold the **Ctrl** key and click on all five (text) blocks [Fig. 20.11(e)] ⇒ **RMB** ⇒ **Text Style** Text Style dialog box opens ⇒ Character- Height .125 ⇒ Note/Dimension- Horizontal Center ⇒ Vertical Middle ⇒ **Apply** [Figs. 20.11(f-g)] ⇒ **OK** ⇒ **LMB**

Figure 20.11(e) Select All Five Blocks

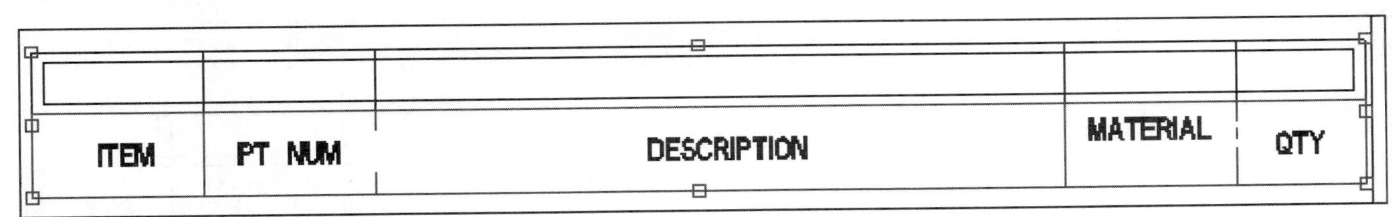

Figure 20.11(f) New Text Style

Text Style
Copy from
Style name Default
Existing text Select Text...
Character
Font filled
Height .125 Default Slant Angle 0.000000
Thickness 0.000000 Default Underline
Width factor 0.800000 Default
Note/Dimension
Horizontal Center Line spacing 0.500000 Default
Vertical Middle Mirror
Angle 0.000000 Break Crosshatching
Color Margin 0.150000
OK Cancel Apply Reset

Figure 20.11(g) Text Style Dialog Box

Insert parametric text into each appropriate repeat region block.

Double-click on the first table cell of the Repeat Region [Fig. 20.12(a)] ⇒ pick **rpt...** from the Report Symbol dialog box ⇒ pick **index** [Figs. 20.12(b-c)]

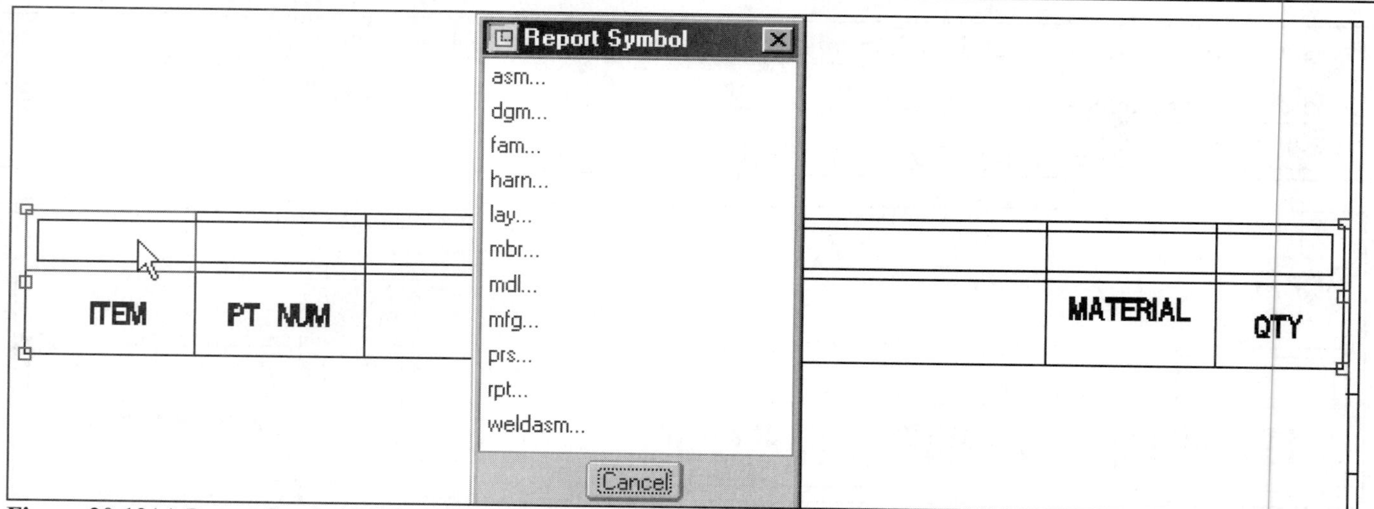

Figure 20.12(a) Report Symbol Dialog Box, Pick rpt…

Report Symbol
UP..
index
level
qty
rel...
ITEM
PT NUM
MATERIAL
QTY

Figure 20.12(b) Pick index

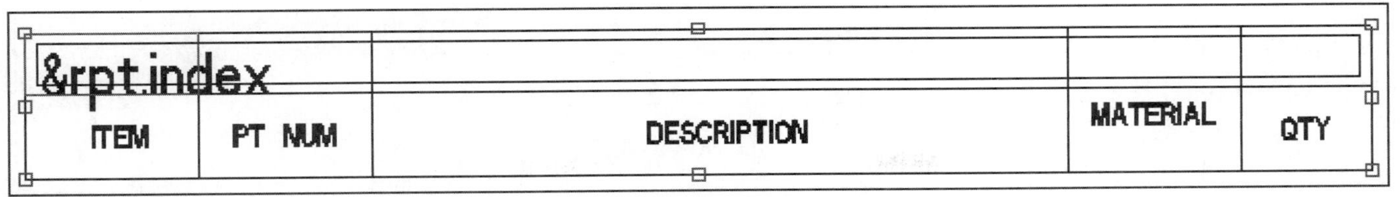

Figure 20.12(c) &rpt.index

Double-click on the fifth table cell of Repeat region ⇒ pick **rpt…** [Fig. 20.12(d)] ⇒ **qty** [Fig. 20.12(e-f)]

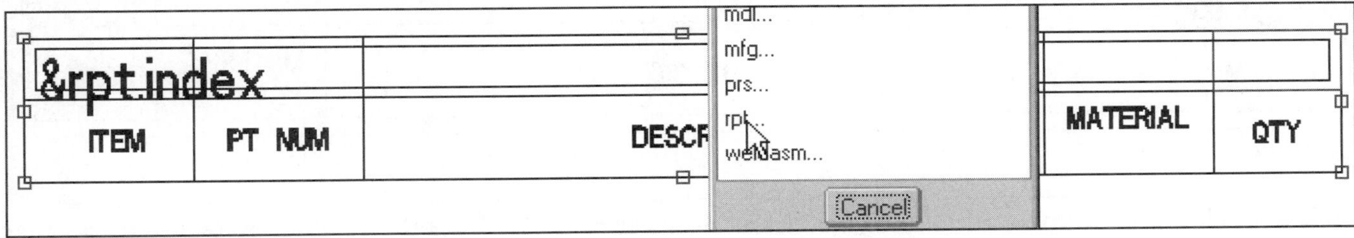

Figure 20.12(d) Pick rpt…

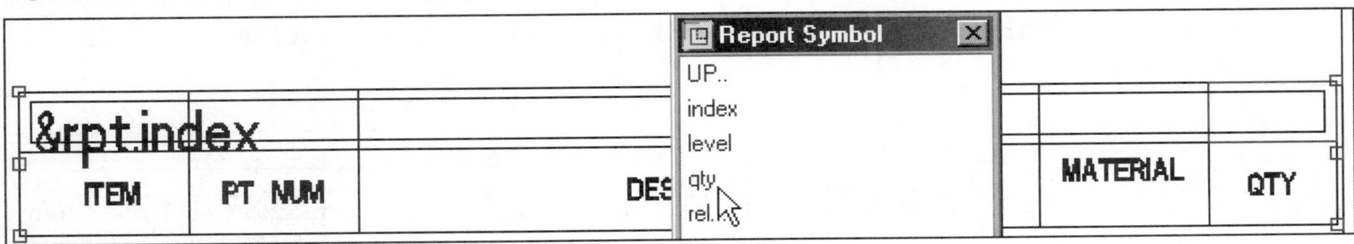

Figure 20.12(e) Pick qty

Figure 20.12(f) &rpt.qty

Double-click on the third (middle) table cell of the Repeat Region ⇒ **asm…** [Fig. 20.12(g)] ⇒ **mbr…** [Fig. 20.12(h)] ⇒ **User Defined** [Fig. 20.12(i)] ⇒ Enter symbol text: type **DSC** ⇒ **MMB** [Fig. 20.12(j)]

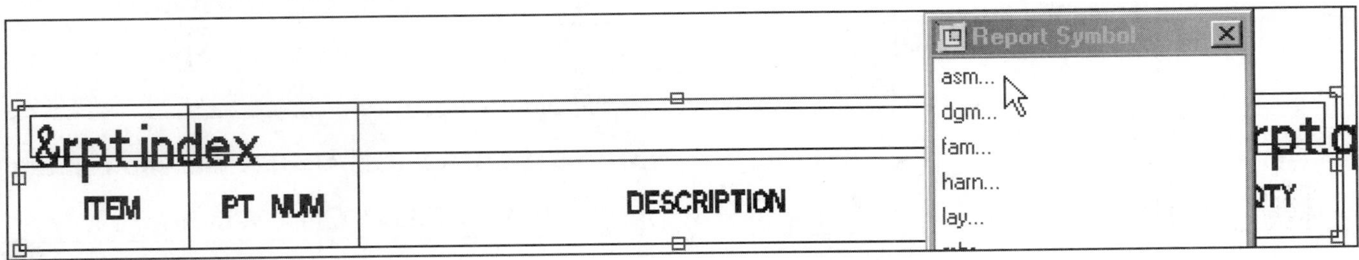

Figure 20.12(g) Pick asm…

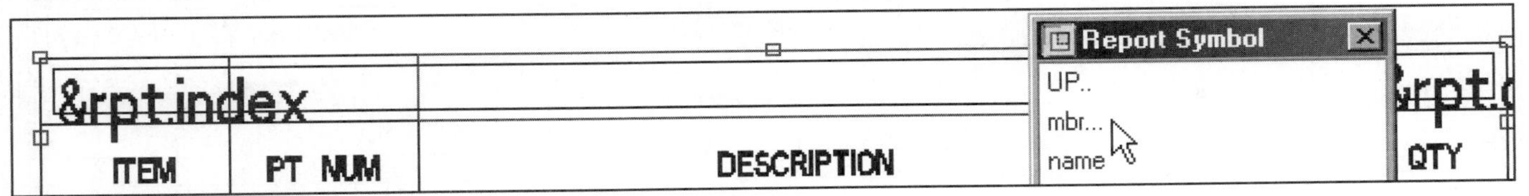

Figure 20.12(h) Pick mbr…

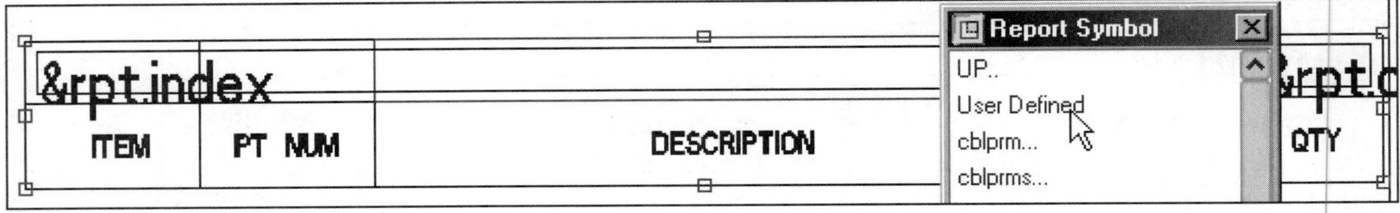

Figure 20.12(i) Pick User Defined

Figure 20.12(j) &asm.mbr.DSC

Double-click on the fourth table cell of the Repeat Region ⇒ **asm…** ⇒ **mbr…** ⇒ **User Defined** ⇒ Enter symbol text: type **MAT** ⇒ **MMB** [Fig. 20.12(k)] ⇒ double-click on the second table cell of the Repeat Region ⇒ **asm…** ⇒ **mbr…** ⇒ **User Defined** ⇒ Enter symbol text: type **PRTNO** ⇒ **MMB** [Fig. 20.12(l)] ⇒ **LMB**

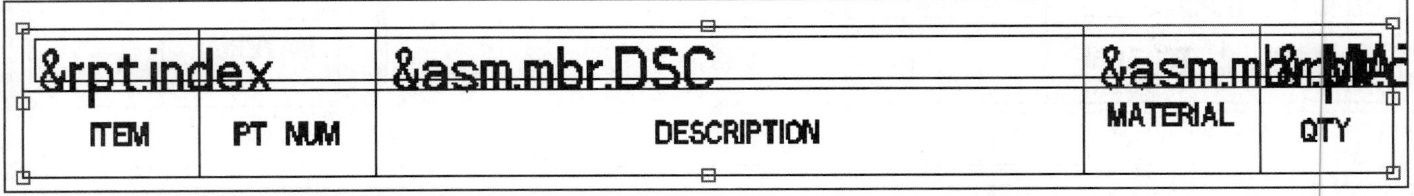

Figure 20.12(k) &asm.mbr.MAT

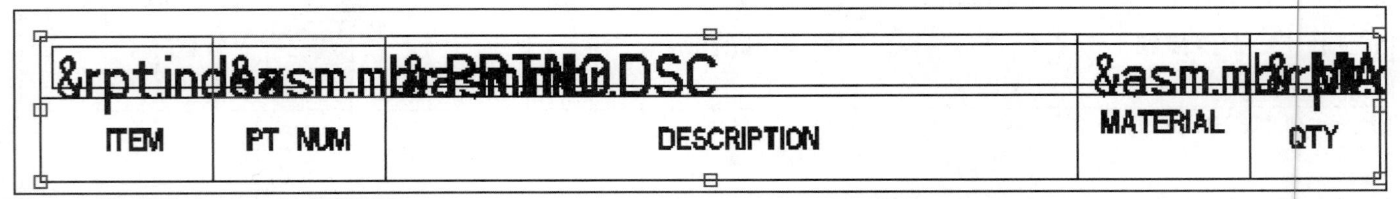

Figure 20.12(l) &asm.mbr.PRTNO

Change the text height of the Report Symbols in the table cells of the Repeat Region to **.125**.

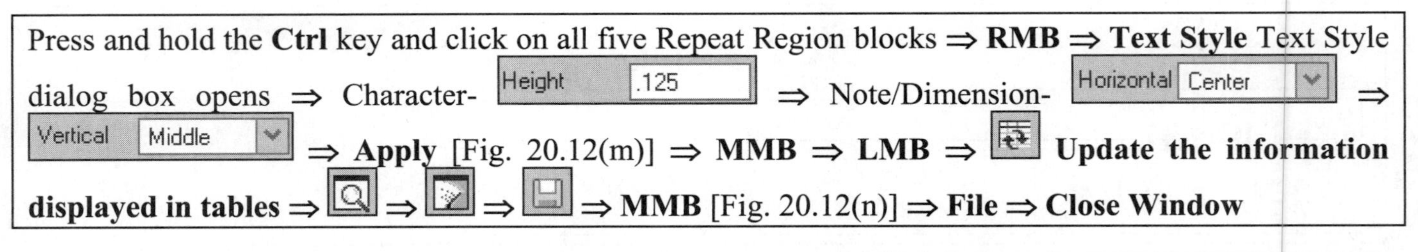

Press and hold the **Ctrl** key and click on all five Repeat Region blocks ⇒ **RMB** ⇒ **Text Style** Text Style dialog box opens ⇒ Character- Height .125 ⇒ Note/Dimension- Horizontal Center ⇒ Vertical Middle ⇒ **Apply** [Fig. 20.12(m)] ⇒ **MMB** ⇒ **LMB** ⇒ **Update the information displayed in tables** ⇒ ⇒ ⇒ ⇒ **MMB** [Fig. 20.12(n)] ⇒ **File** ⇒ **Close Window**

&rpt.index	&asm.mbr.PRTNO	&asm.mbr.DSC	&asm.mbr.MAT	&rpt.qty
ITEM	PT NUM	DESCRIPTION	MATERIAL	QTY

Figure 20.12(m) Completed BOM

&rpt.index	&asmmbr.PRTNO	&asmmbr.DSC	&asmmbr.MAT	&rpt.qty
ITEM	PT NUM	DESCRIPTION	MATERIAL	QTY

	TOOL ENGINEERING CO.		
DRAWN	DRAWING SCALE	ASM_FORMAT_E	
ISSUED			SHEET 1 OF 1

2 1

Figure 20.12(n) Title Block and BOM Table

Adding Parts List Data

When you save your standard assembly format, the Drawing Table that represents your standard parts list is now included. You must be aware of the titles of the parameters under which the data is stored, so that you can add them properly to your parts.

As you add components to an assembly, Pro/E reads the parameters from them and updates the parts list. You can also see the same effect by adding these parameters after the drawing has been created.

Pro/E also creates Item Balloons on the first view that was placed on the drawing. To improve their appearance, you can move these balloons to other views and alter the locations where they attach.

Retrieve the clamp arm, click: **File ⇒ Open ⇒ Part ⇒** select **clamp_arm.prt ⇒ Open** [Fig. 20.13(a)]

Figure 20.13(a) Clamp_Arm

Click: **Tools** from menu bar ⇒ **Parameters** [Fig. 20.13(b)] ⇒ **Parameters** [Fig. 20.13(c)] ⇒ **Add Parameter** from Parameters dialog box ⇒ in the Name field and type **PRTNO** [Figs. 20.13(d-e)]

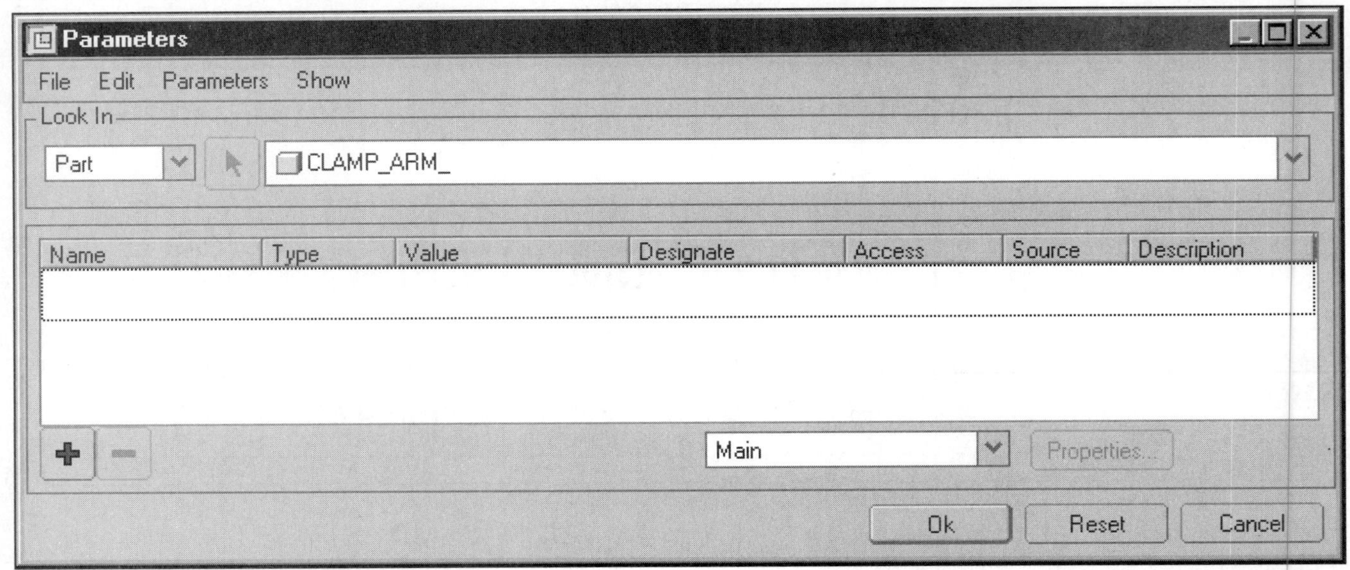

Figure 20.13(b) Parameters Dialog Box

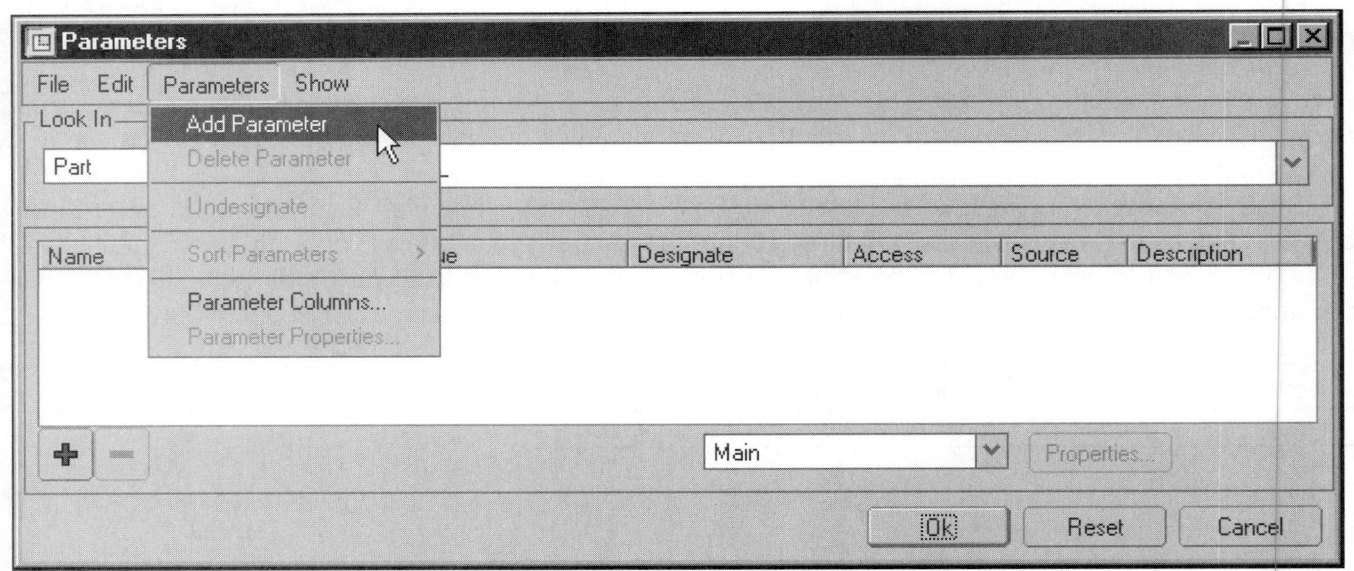

Figure 20.13(c) Add Parameter

Name	Type	Value	Designate	Access	Source	Description
PARAMETER_1	Real Num...	0.0	☐	Full	User-Defi...	

Main | Properties...

Figure 20.13(d) Adding Parameters

Name	Type	Value	Designate	Access	Source	Description
PRTNO	Real Num...	0.0	☐	Full	User-Defi...	

Figure 20.13(e) Name PRTNO

Click in Type field [Fig. 20.13(f)] ⇒ [dropdown arrow] [Fig. 20.13(g)] ⇒ String ⇒ click in Value field and type **SW101-5AR** [Fig. 20.13(h)] ⇒ click in Description field and type **part number** [Fig. 20.13(i)]

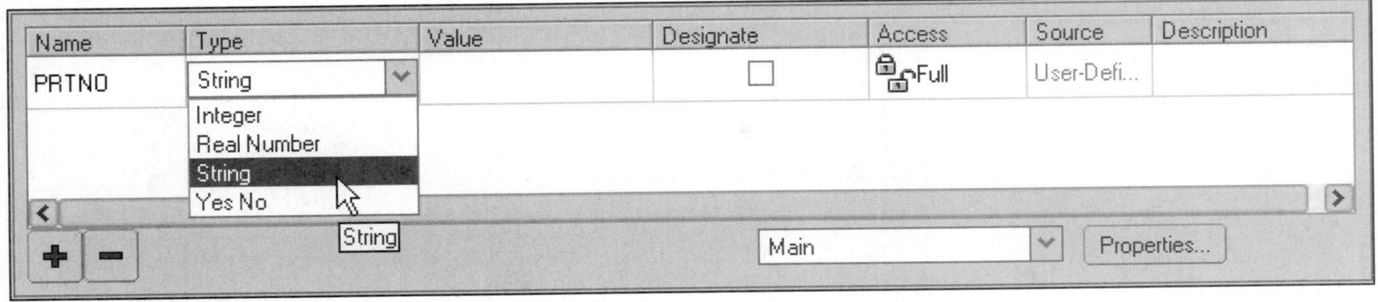

Figure 20.13(f) Click in Type Field- Real Number

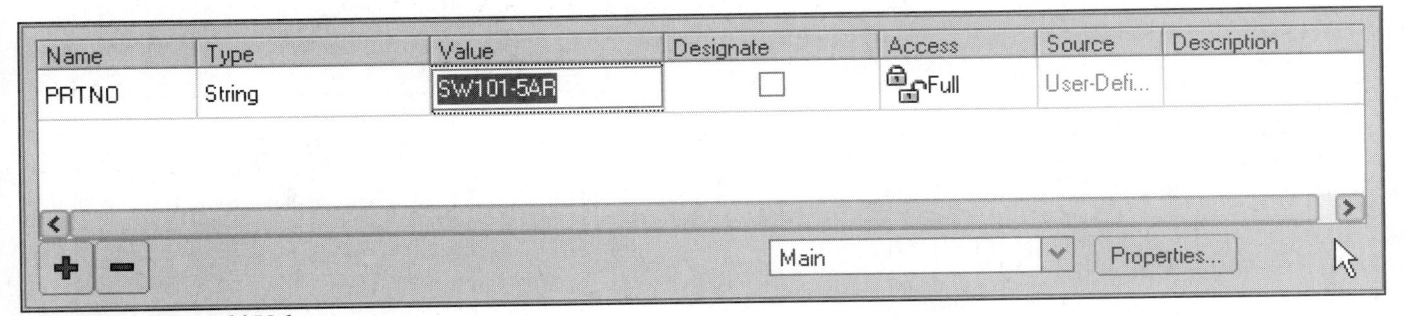

Figure 20.13(g) String

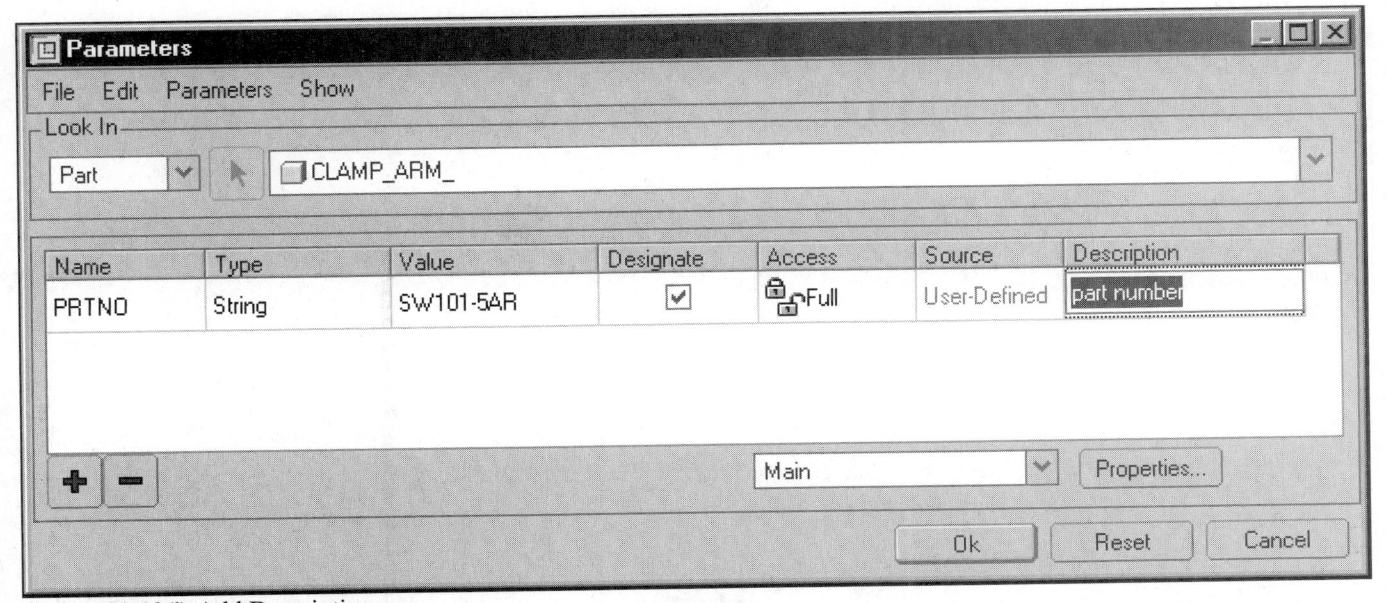

Figure 20.13(h) Add Value

Figure 20.13(i) Add Description

Click: **Add new Parameter** [Fig. 20.13(j)] ⇒ in the Name field, type **DSC** [Fig. 20.13(k)] ⇒ click in Type field ⇒ String ⇒ click in Value field and type **CLAMP ARM** ⇒ click in Description field and type **part name** [Fig. 20.13(l)] ⇒ click **Add new Parameter** ⇒ in the Name field, type **MAT** ⇒ click in Type field ⇒ String ⇒ click in Value field and type **STEEL** ⇒ click in Description field and type **material** [Fig. 20.13(m)] ⇒ **Ok** ⇒ **File** ⇒ **Save** ⇒ **MMB**

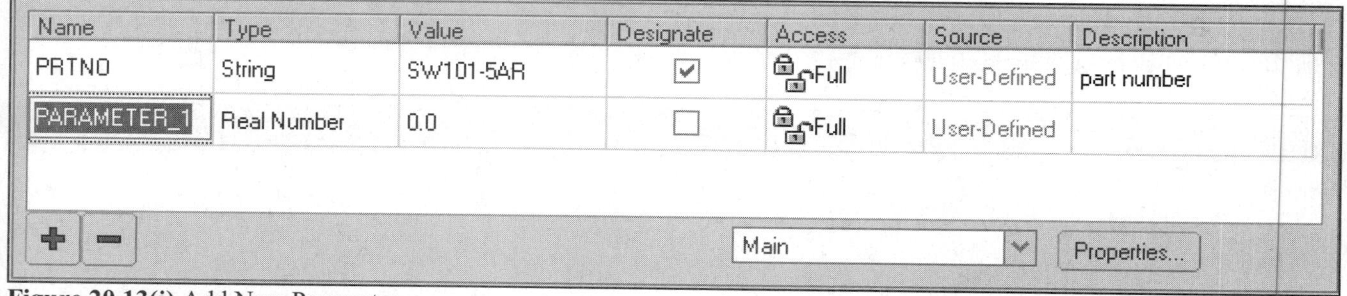

Figure 20.13(j) Add New Parameter

Name	Type	Value	Designate	Access	Source	Description
PRTNO	String	SW101-5AR	☑	Full	User-Defined	part number
DSC	String		☐	Full	User-Defined	

Figure 20.13(k) String

Name	Type	Value	Designate	Access	Source	Description
PRTNO	String	SW101-5AR	☑	Full	User-Defined	part number
DSC	String	CLAMP ARM	☑	Full	User-Defined	part name

Figure 20.13(l) Add Value and Description

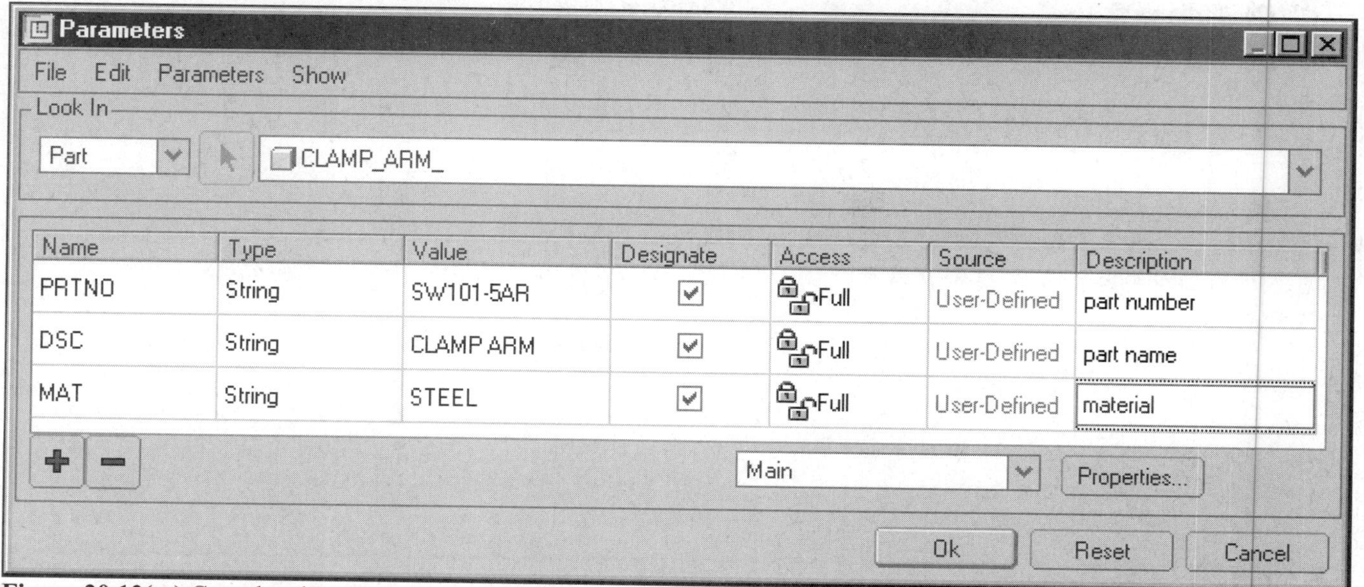

Figure 20.13(m) Completed Parameters

You can also access the parts parameters using the Relations dialog box. In the case of the Clamp_Arm, there was a relation created for controlling a features location.

Click: **Tools** ⇒ **Relations** ⇒ **Local Parameters** [Fig. 20.13(n)] ⇒ **Ok** ⇒ **File** ⇒ **Save** ⇒ **MMB** ⇒ **Window** ⇒ **Close**

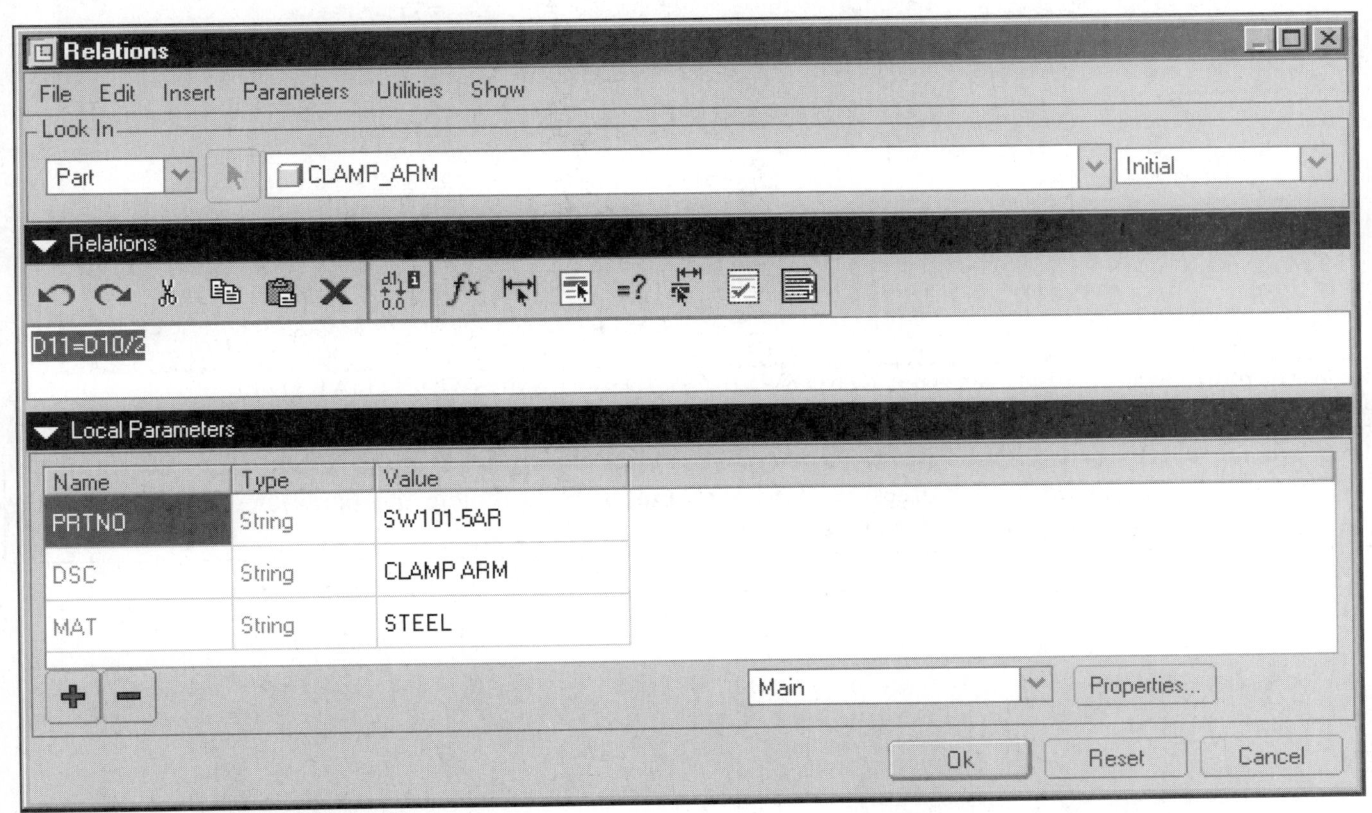

Figure 20.13(n) Relations Dialog Box

Retrieve the clamp swivel, click: **File** ⇒ **Open** ⇒ **Part** ⇒ **clamp_swivel.prt** ⇒ **Open** [Fig. 20.14(a)] ⇒ **Tools** ⇒ **Parameters** ⇒ **Parameters** ⇒ **Add Parameter** ⇒ complete the parameters as shown [Fig. 20.14(b)] ⇒ **Ok** ⇒ **File** ⇒ **Save** ⇒ **MMB** ⇒ **File** ⇒ **Close Window**

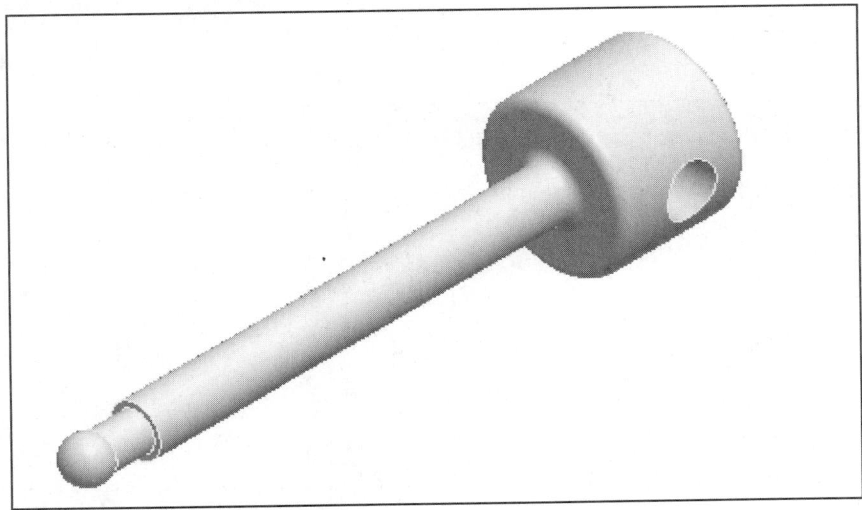

Figure 20.14(a) Clamp_Swivel

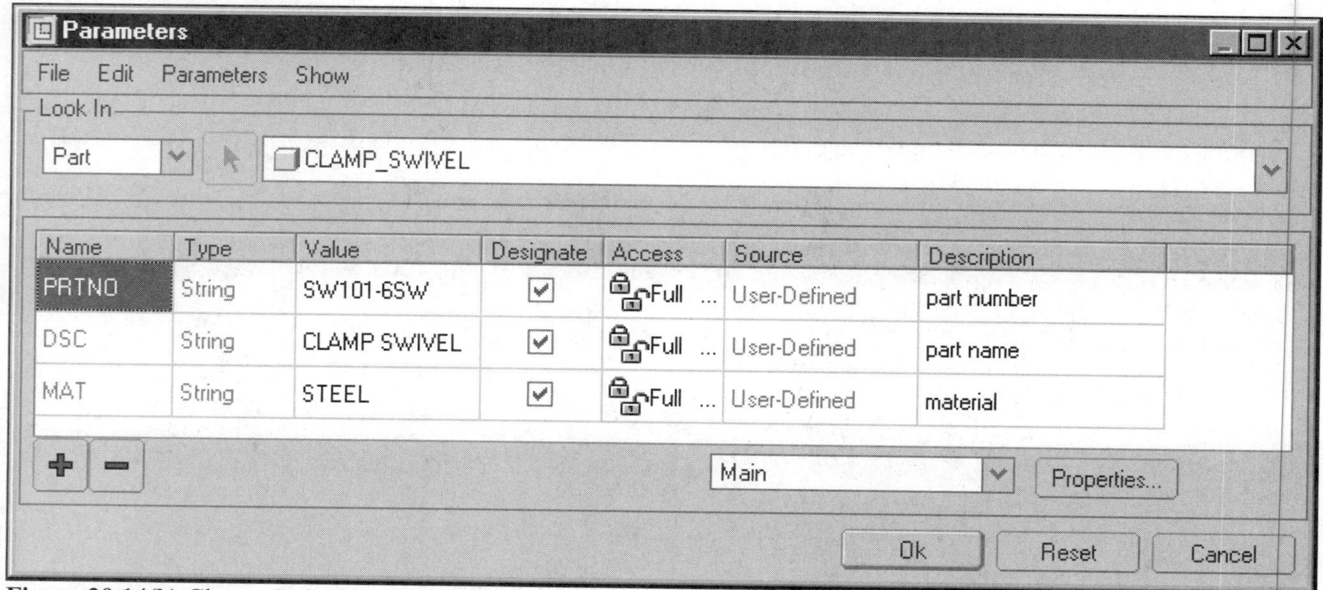

Name	Type	Value	Designate	Access	Source	Description
PRTNO	String	SW101-6SW	✓	Full ...	User-Defined	part number
DSC	String	CLAMP SWIVEL	✓	Full ...	User-Defined	part name
MAT	String	STEEL	✓	Full ...	User-Defined	material

Figure 20.14(b) Clamp_Swivel Parameters: PRTNO SW101-6SW, DSC CLAMP SWIVEL, MAT STEEL

Retrieve the clamp ball, click: **File** ⇒ **Open** ⇒ **Part** ⇒ **clamp_ball.prt** ⇒ **Open** [Fig. 20.15(a)] ⇒ **Tools** ⇒ **Parameters** ⇒ **Parameters** ⇒ **Add Parameter** ⇒ complete the parameters as shown [Fig. 20.15(b)] ⇒ **Ok** ⇒ **File** ⇒ **Save** ⇒ **MMB** ⇒ **File** ⇒ **Close Window**

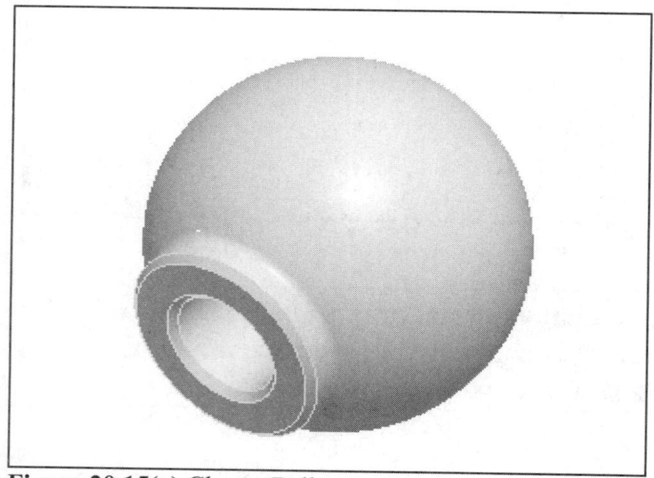

Figure 20.15(a) Clamp_Ball

Look In

Part | CLAMP_BALL

Name	Type	Value	Designate	Access	Source	Description
PRTNO	String	SW101-7BA	✓	Full ...	User-Defined	part number
DSC	String	CLAMP BALL	✓	Full ...	User-Defined	part name
MAT	String	BLACK PLASTIC	✓	Full ...	User-Defined	material

Figure 20.15(b) Clamp_Ball Parameters: PRTNO SW101-7BA, DSC CLAMP BALL, MAT BLACK PLASTIC

Parameters can be added, deleted, and modified in Part Mode, Drawing Mode, or Assembly Mode. You can also add *parameter columns* to the Model Tree in Assembly Mode and edit the parameter value. Use the following information to add parameters both to purchased components (standard parts) and to the remaining parts required for the subassembly and the assembly (Fig. 20.16). Use the Assembly Mode to input the information shown in Figures 20.17(a-e) to establish the part parameters.

Click: **File ⇒ Open ⇒ Assembly ⇒ clamp_assembly.asm ⇒ Open**
(see commands on the following pages)

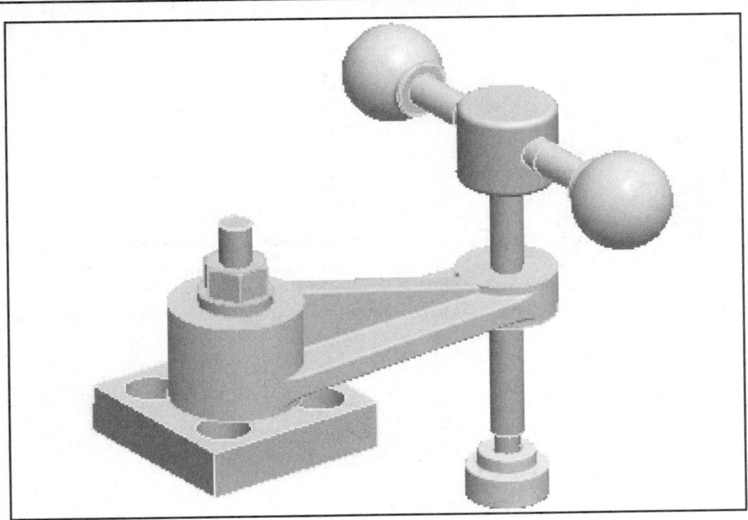

Figure 20.16 Clamp_Assembly

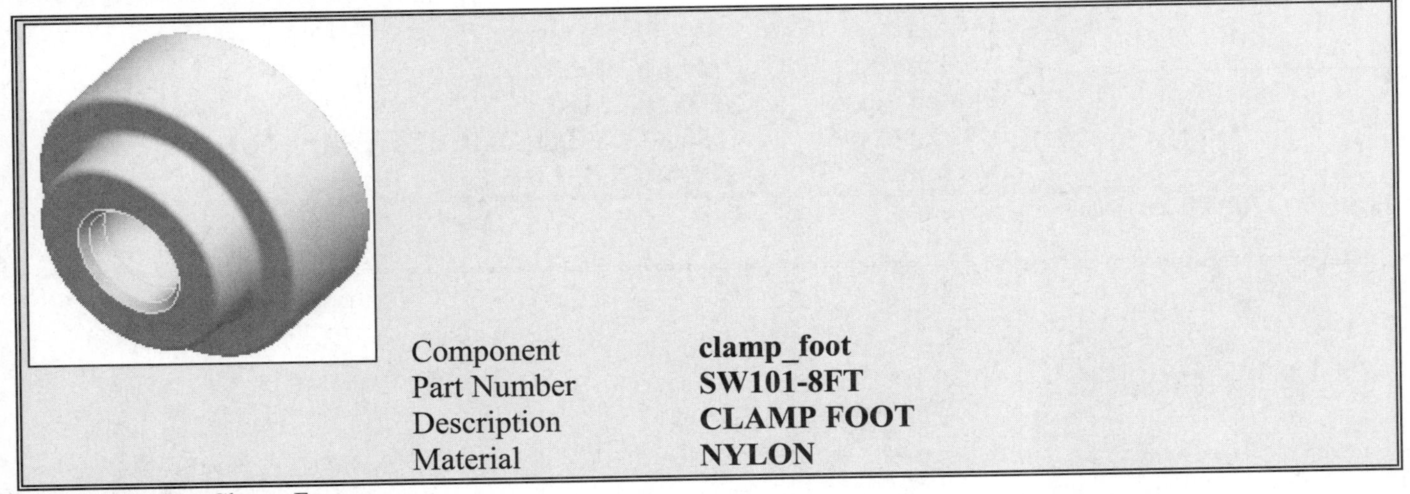

Component	**clamp_foot**
Part Number	**SW101-8FT**
Description	**CLAMP FOOT**
Material	**NYLON**

Figure 20.17(a) Clamp_Foot

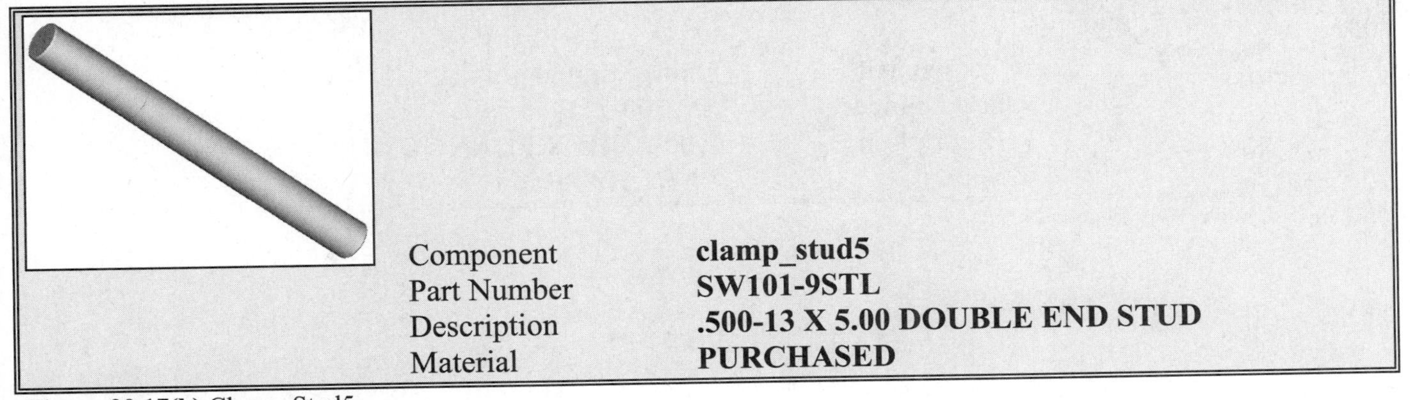

Component	**clamp_stud5**
Part Number	**SW101-9STL**
Description	**.500-13 X 5.00 DOUBLE END STUD**
Material	**PURCHASED**

Figure 20.17(b) Clamp_Stud5

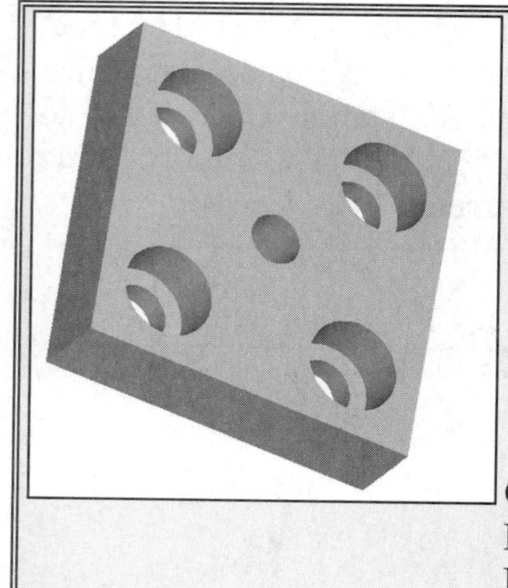

Component	**clamp_plate**
Part Number	**SW100-20PL**
Description	**CLAMP PLATE**
Material	**STEEL**

Figure 20.17(c) Clamp_Plate

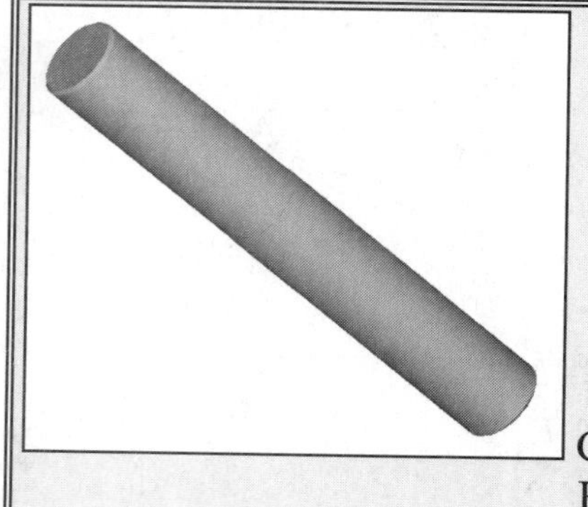

Component	**clamp_stud35**
Part Number	**SW100-21ST**
Description	**.500-13 X 3.50 DOUBLE END STUD**
Material	**PURCHASED**

Figure 20.17(d) Clamp_Stud35

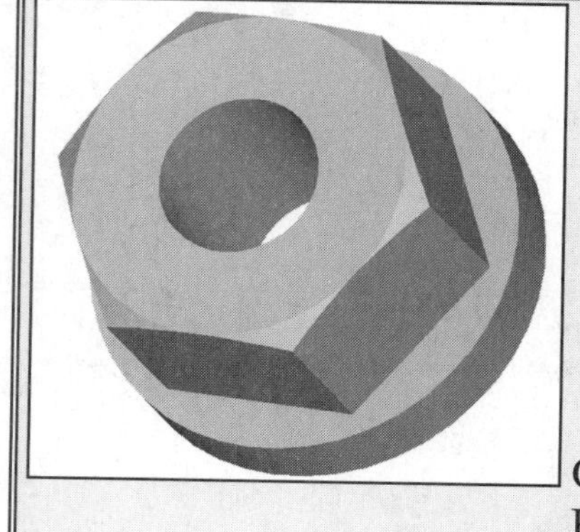

Component	**clamp_flange_nut**
Part Number	**SW100-22FLN**
Description	**.500-13 HEX FLANGE NUT**
Material	**PURCHASED**

Figure 20.17(e) Clamp_Flange_Nut

Click: **Settings** in the Model Tree ⇒ **Tree Columns** [Fig. 20.18(a)] ⇒ Type **Model Params** [Fig. 20.18(b)] ⇒ Name field, type **PRTNO** [Fig. 20.18(c)] ⇒ **Enter** ⇒ Name field, type **DSC** [Fig. 20.18(d)] ⇒ **Enter** [Fig. 20.18(e)] ⇒ Name field, type **MAT** [Fig. 20.18(f)] ⇒ **Enter** [Fig. 20.18(g)] ⇒ **Apply** ⇒ **OK**

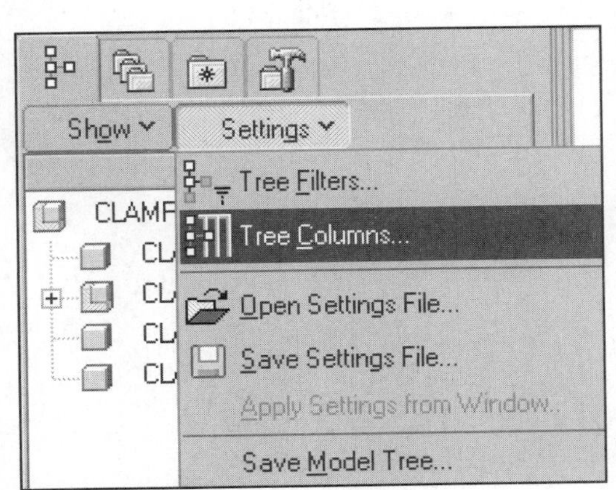

Figure 20.18(a) Tree Columns

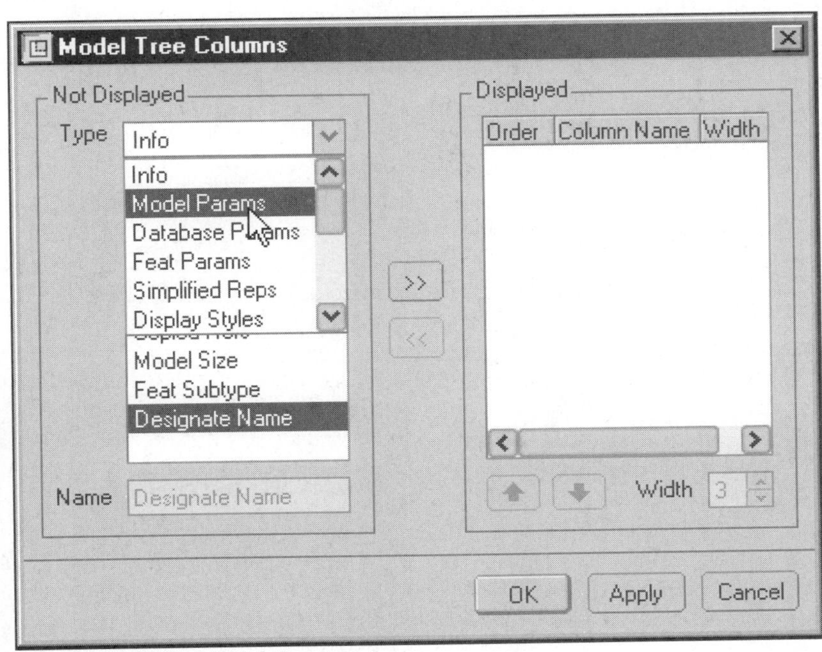

Figure 20.18(b) Model Params

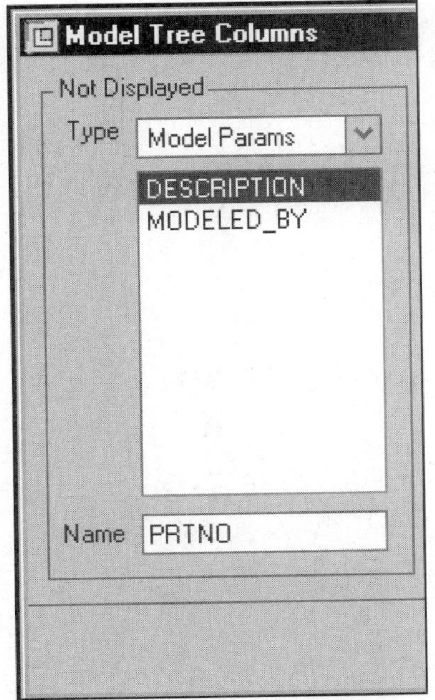

Figure 20.18(c) Name PRTNO

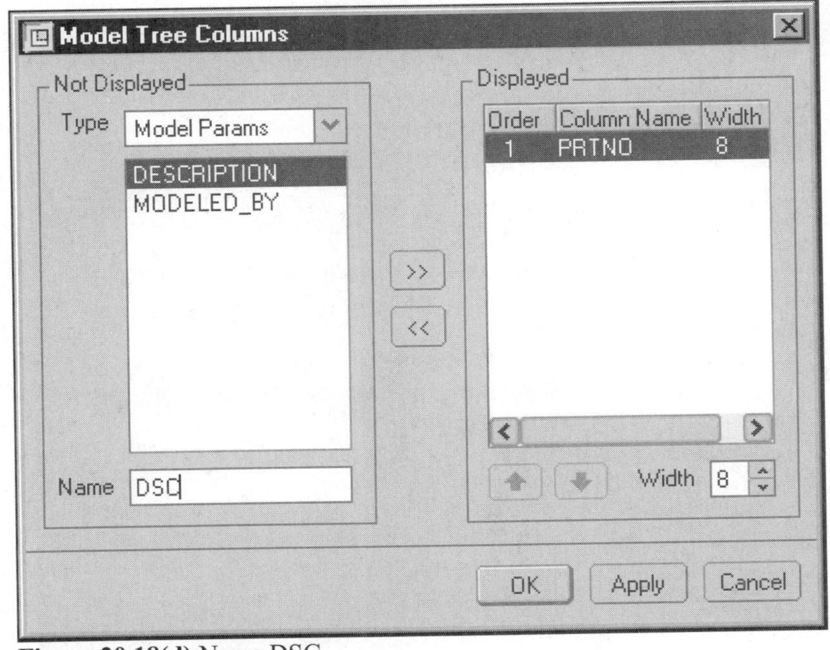

Figure 20.18(d) Name DSC

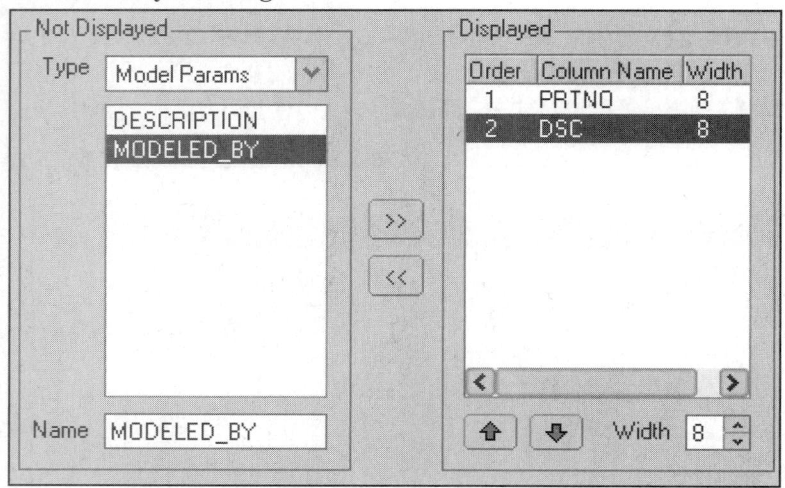

Figure 20.18(e) PRTNO and DSC

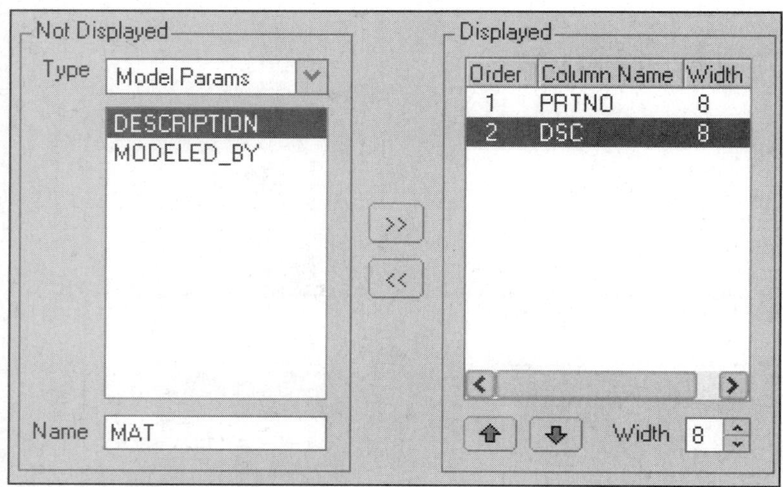

Figure 20.18(f) Name MAT

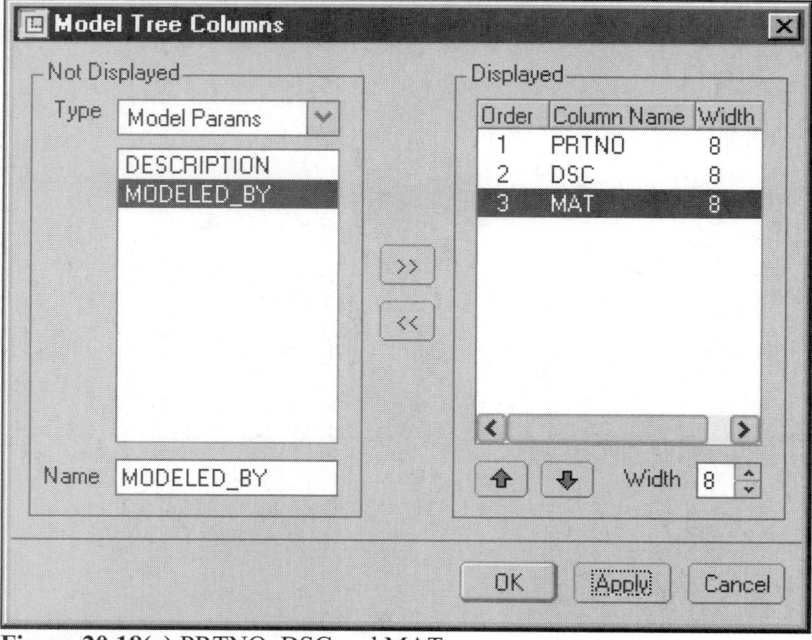

Figure 20.18(g) PRTNO, DSC and MAT

Resize the Model Tree and click on **CLAMP_PLATE.PRT** ⇒ click in **PRTNO** field [Fig. 20.19(a)] ⇒ String ⇒ **OK** ⇒ PRTNO type **SW100-20PL** [Fig. 20.19(b)] ⇒ **Enter** ⇒ click in **DSC** field ⇒ String [Fig. 20.19(c)] ⇒ **OK** ⇒ DSC type **CLAMP PLATE** [Fig. 20.19(d)] ⇒ **Enter** ⇒ click in **MAT** field ⇒ String ⇒ **OK** ⇒ MAT type **STEEL** ⇒ **Enter** [Fig. 20.19(e)] ⇒ **File** ⇒ **Save** ⇒ **MMB**

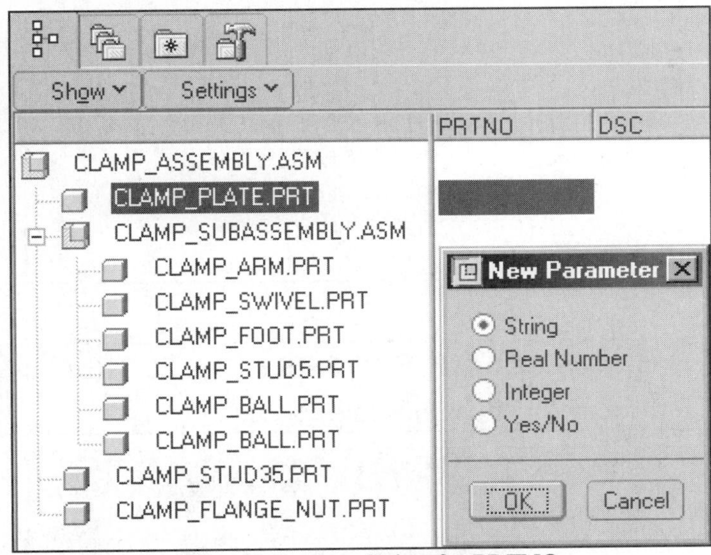

Figure 20.19(a) New Parameter String for PRTNO

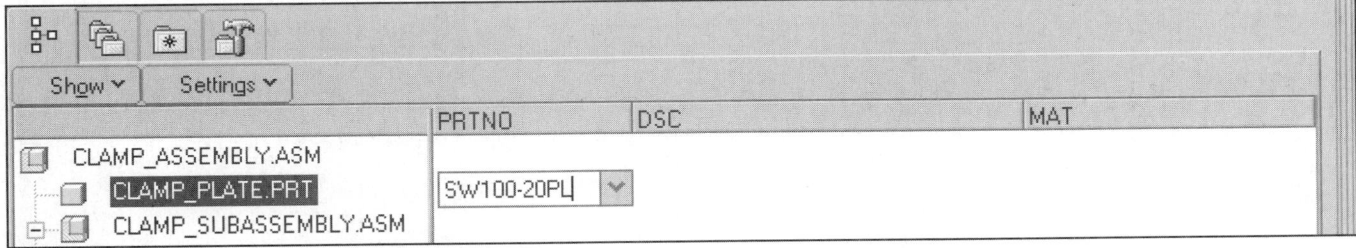

Figure 20.19(b) Part Number SW100-20PL

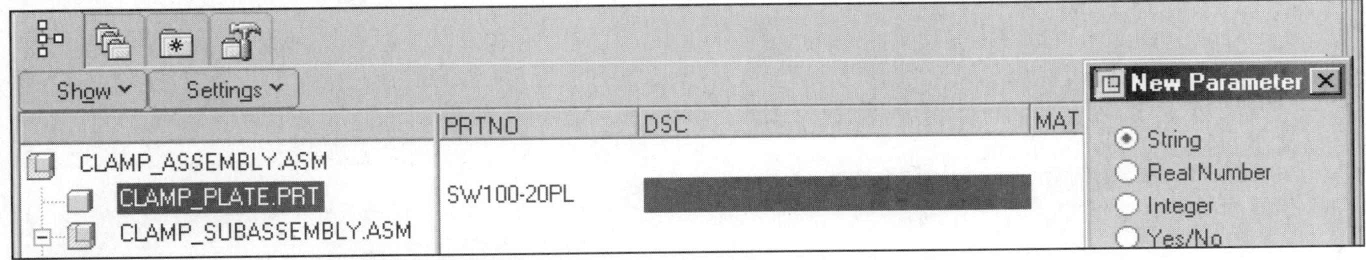

Figure 20.19(c) New Parameter String for DSC

Figure 20.19(d) Description CLAMP PLATE

Figure 20.19(e) Material STEEL

Complete the parameters for the remaining components (Fig. 20.20)

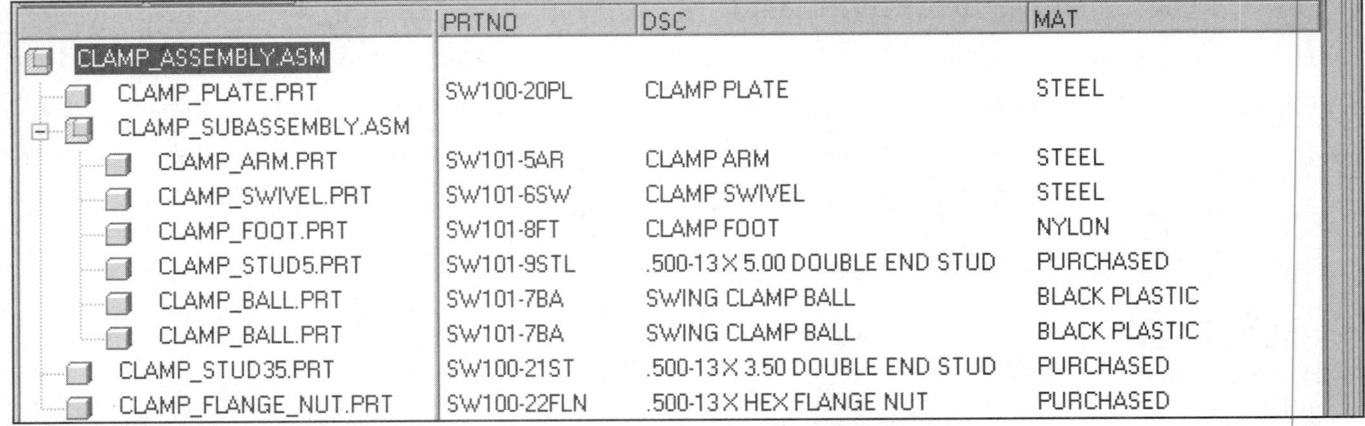

PRTNO
DSC
MAT
CLAMP_ASSEMBLY.ASM
CLAMP_PLATE.PRT
SW100-20PL
CLAMP PLATE
STEEL
CLAMP_SUBASSEMBLY.ASM
CLAMP_ARM.PRT
SW101-5AR
CLAMP ARM
STEEL
CLAMP_SWIVEL.PRT
SW101-6SW
CLAMP SWIVEL
STEEL
CLAMP_FOOT.PRT
SW101-8FT
CLAMP FOOT
NYLON
CLAMP_STUD5.PRT
SW101-9STL
.500-13 X 5.00 DOUBLE END STUD
PURCHASED
CLAMP_BALL.PRT
SW101-7BA
SWING CLAMP BALL
BLACK PLASTIC
CLAMP_BALL.PRT
SW101-7BA
SWING CLAMP BALL
BLACK PLASTIC
CLAMP_STUD35.PRT
SW100-21ST
.500-13 X 3.50 DOUBLE END STUD
PURCHASED
CLAMP_FLANGE_NUT.PRT
SW100-22FLN
.500-13 X HEX FLANGE NUT
PURCHASED

Figure 20.20 Assembly Model Tree and Parameters

Click: **Tools ⇒ Parameters ⇒ Part** [Fig. 20.21(a)] ⇒ click on **CLAMP_FOOT.PRT** [Fig. 20.21(b)] ⇒ Designate ☑ ⇒ add Parameter Descriptions ⇒ repeat for remaining components ⇒ **Ok**

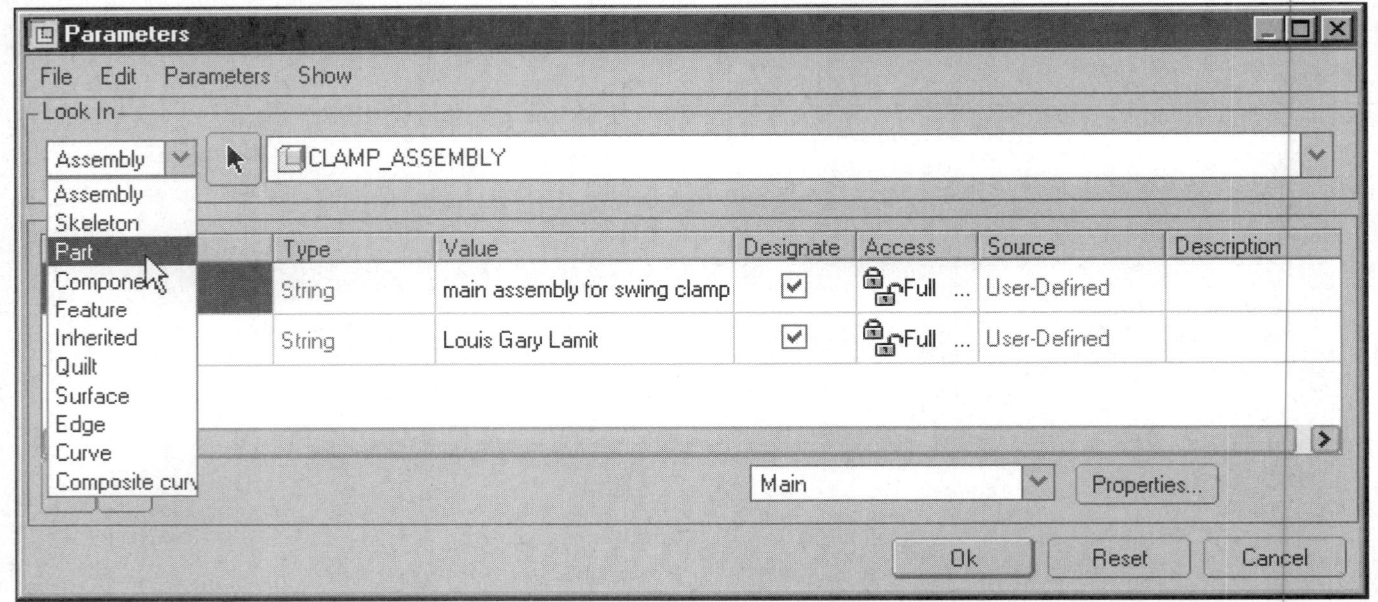

Figure 20.21(a) Extracting Part Parameters from the Assembly

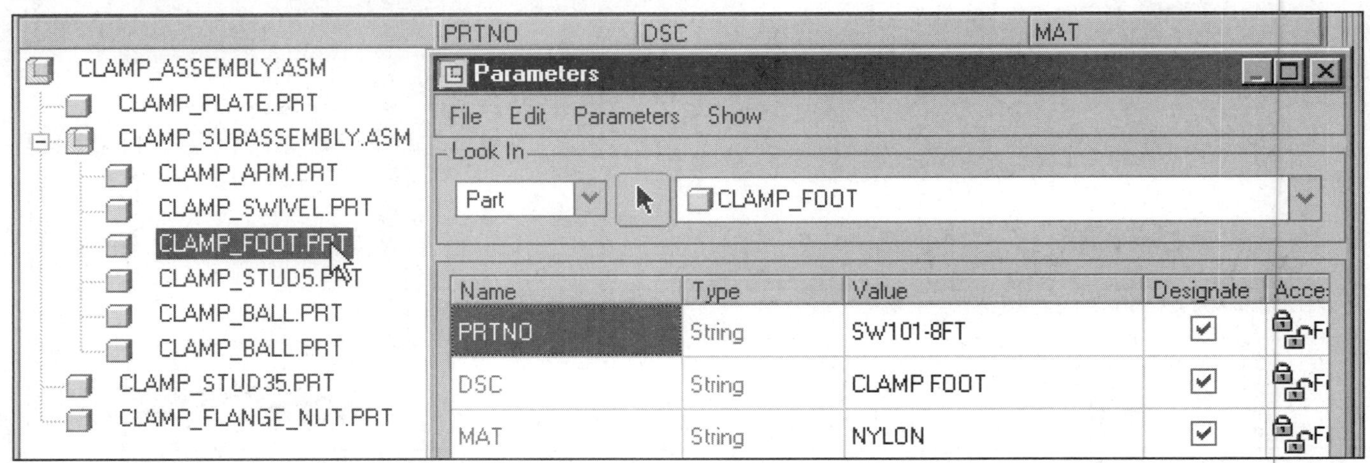

Figure 20.21(b) Part Parameters for Clamp_Foot

The Clamp_Flange_Nut, Clamp_Stud35 and the Clamp_Stud5 are standard parts that are copied from the PTC parts library, therefore they will have additional parameters [Fig. 20.22(a)] and relations [Figs. 20.22 (b-c)].

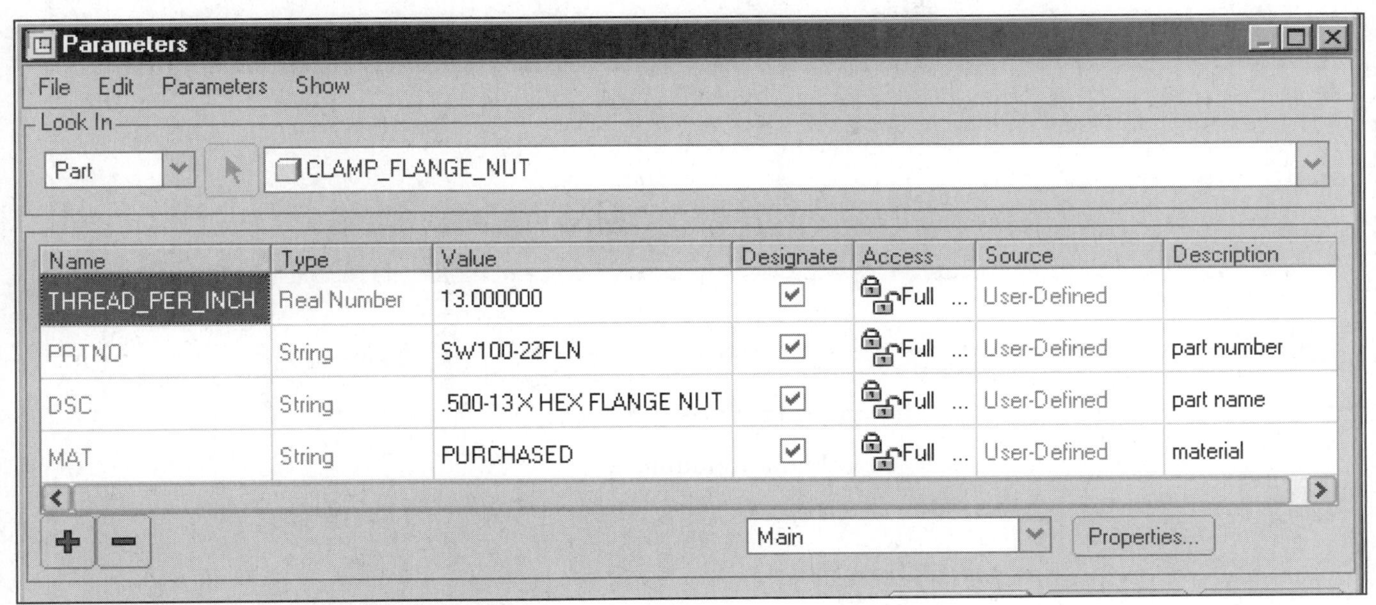

Name	Type	Value	Designate	Access	Source	Description
THREAD_PER_INCH	Real Number	13.000000	✓	Full ...	User-Defined	
PRTNO	String	SW100-22FLN	✓	Full ...	User-Defined	part number
DSC	String	.500-13 X HEX FLANGE NUT	✓	Full ...	User-Defined	part name
MAT	String	PURCHASED	✓	Full ...	User-Defined	material

Figure 20.22(a) Clamp_Flange_Nut Parameters

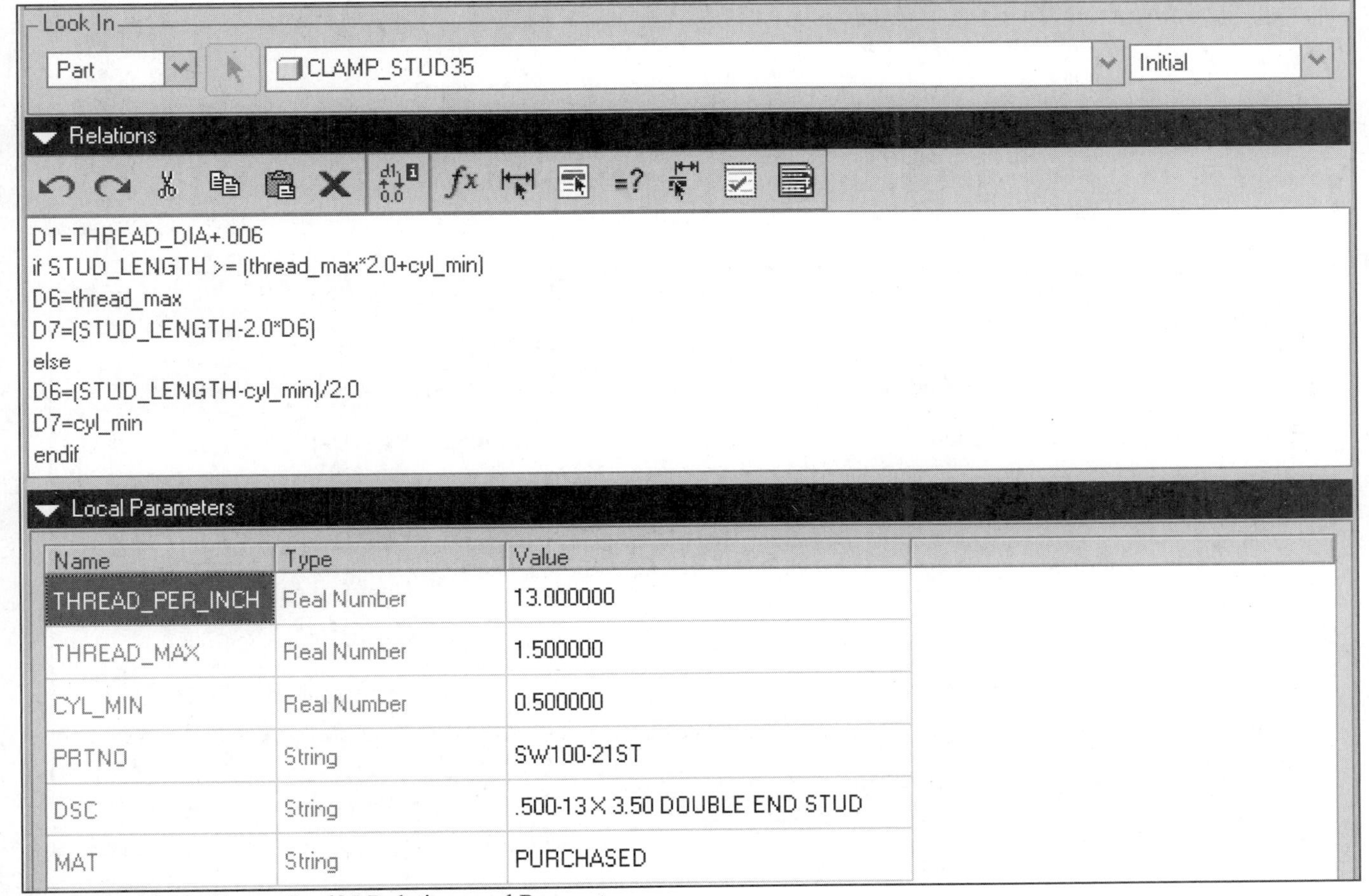


```
D1=THREAD_DIA+.006
if STUD_LENGTH >= (thread_max*2.0+cyl_min)
D6=thread_max
D7=(STUD_LENGTH-2.0*D6)
else
D6=(STUD_LENGTH-cyl_min)/2.0
D7=cyl_min
endif
```

Name	Type	Value
THREAD_PER_INCH	Real Number	13.000000
THREAD_MAX	Real Number	1.500000
CYL_MIN	Real Number	0.500000
PRTNO	String	SW100-21ST
DSC	String	.500-13 X 3.50 DOUBLE END STUD
MAT	String	PURCHASED

Figure 20.22(b) Clamp_Stud35 Relations and Parameters

Relations and parameters can also be displayed in the Part Mode by clicking: **Info ⇒ Relations and Parameters** [Fig. 20.22(c)].

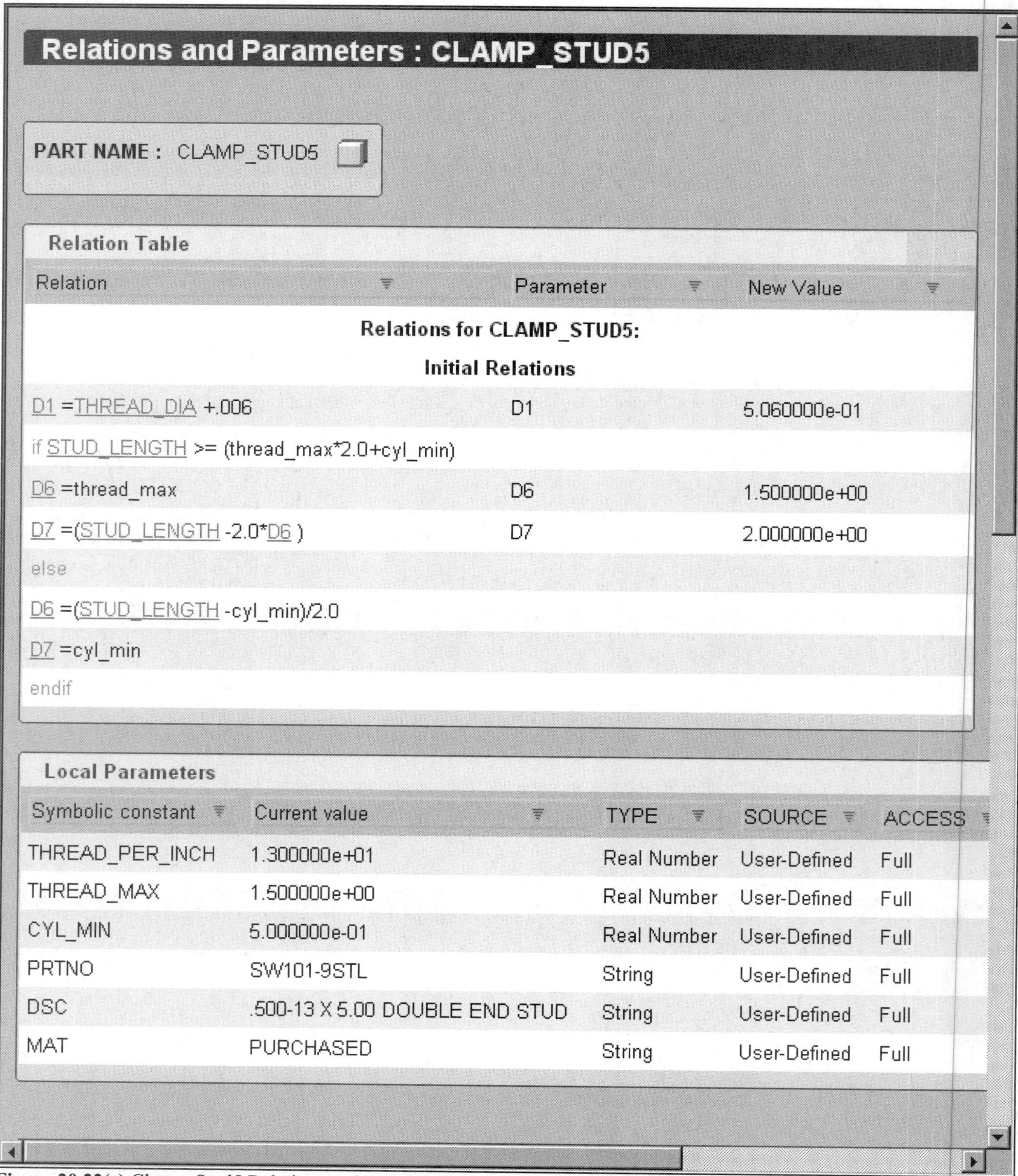

Relations and Parameters : CLAMP_STUD5

PART NAME : CLAMP_STUD5

Relation Table

Relation	Parameter	New Value
Relations for CLAMP_STUD5:		
Initial Relations		
D1 =THREAD_DIA +.006	D1	5.060000e-01
if STUD_LENGTH >= (thread_max*2.0+cyl_min)		
D6 =thread_max	D6	1.500000e+00
D7 =(STUD_LENGTH -2.0*D6)	D7	2.000000e+00
else		
D6 =(STUD_LENGTH -cyl_min)/2.0		
D7 =cyl_min		
endif		

Local Parameters

Symbolic constant	Current value	TYPE	SOURCE	ACCESS
THREAD_PER_INCH	1.300000e+01	Real Number	User-Defined	Full
THREAD_MAX	1.500000e+00	Real Number	User-Defined	Full
CYL_MIN	5.000000e-01	Real Number	User-Defined	Full
PRTNO	SW101-9STL	String	User-Defined	Full
DSC	.500-13 X 5.00 DOUBLE END STUD	String	User-Defined	Full
MAT	PURCHASED	String	User-Defined	Full

Figure 20.22(c) Clamp_Stud5 Relations and Parameters

Assembly Drawings

The parameters (and their values) have been established for each part. The assembly format with related parameters in a parts list table has been created and saved in your (format) directory. You can now create a drawing of the assembly, where the parts list will be generated automatically, and the assembly ballooned. The first assembly drawing will be of the Clamp_Subassembly [Figs. 20.23(a-b)].

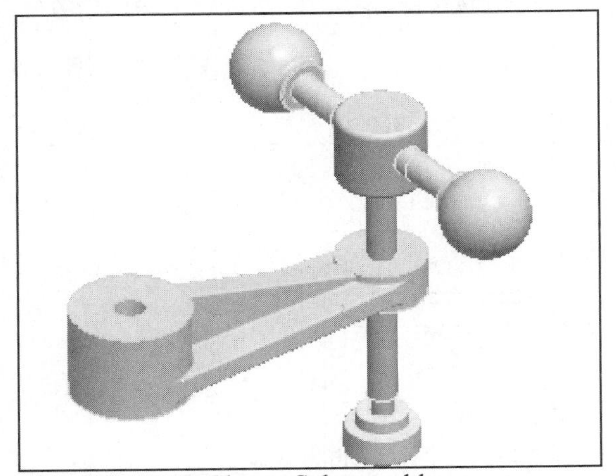

Figure 20.23(a) Clamp_Subassembly

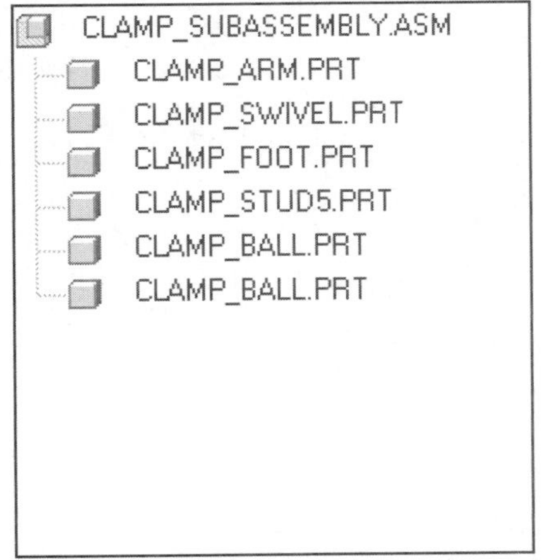

Figure 20.23(b) Subassembly Model Tree

Click: **File** ⇒ **Set Working Directory** ⇒ select the working directory ⇒ **OK** ⇒ [icon] **Create a new object** ⇒ [Drawing] ⇒ Name **CLAMP_SUBASSEMBLY** ⇒ [Use default template] [Fig. 20.24(a)] ⇒ **OK** ⇒ Default Model **Browse** ⇒ **clamp_subassembly.asm** ⇒ **Open** ⇒ [Empty with format] ⇒ **Browse** ⇒ Format **asm_format_e. frm** ⇒ **Open** [Fig. 20.24(b)] ⇒ **OK**

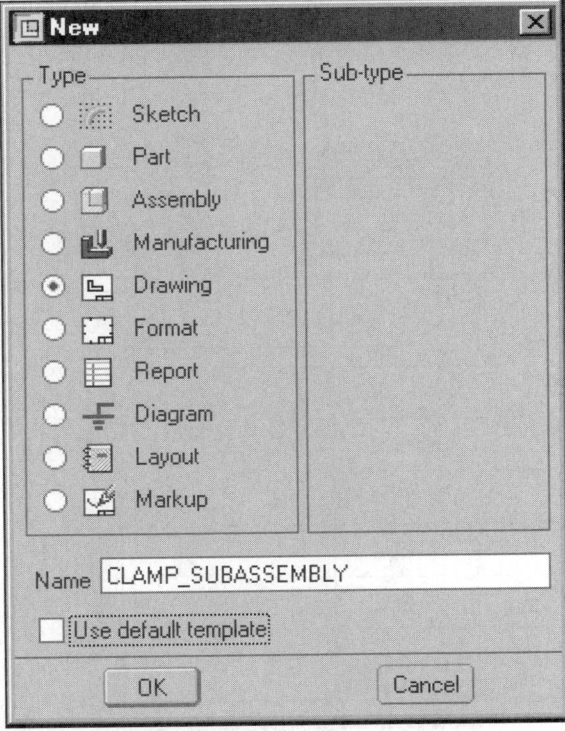

Figure 20.24(a) New Dialog Box

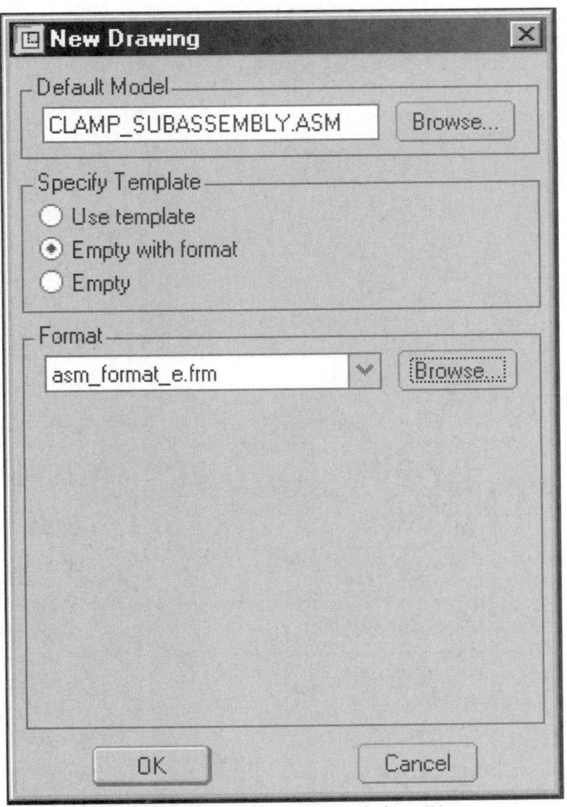

Figure 20.24(b) New Drawing Dialog Box

For the drawing to display with the correct style, modify the values of the Drawing Options.

Click: **Tools** ⇒ **Environment** ⇒ Snap to Grid ⇒ Display Style Hidden Line ⇒ Tangent Edges Dimmed ⇒ **OK** ⇒ **File** ⇒ **Properties** ⇒ **Drawing Options** ⇒ ***default_font filled*** ⇒ **Enter** ⇒ ***draw_arrow_style filled*** ⇒ **MMB** ⇒ ***drawing_text_height .50*** ⇒ **MMB** ⇒ ***crossec_arrow_length .50*** ⇒ **MMB** ⇒ ***crossec_arrow_width .17*** ⇒ **MMB** ⇒ ***max_balloon_radius .50*** ⇒ **MMB** ⇒ ***min_balloon_radius .50*** ⇒ **MMB** ⇒ **Save a copy of the currently displayed configuration file** ⇒ Name **CLAMP_ASM** ⇒ **Ok** ⇒ **Apply** ⇒ **Close** ⇒ **MMB** ⇒ ⇒ ⇒ ⇒ **MMB** [Fig. 20.24(c)]

Note that the Repeat Region of the BOM [Fig. 20.4(d)] has been automatically filled in with the parameters read from the individual components.

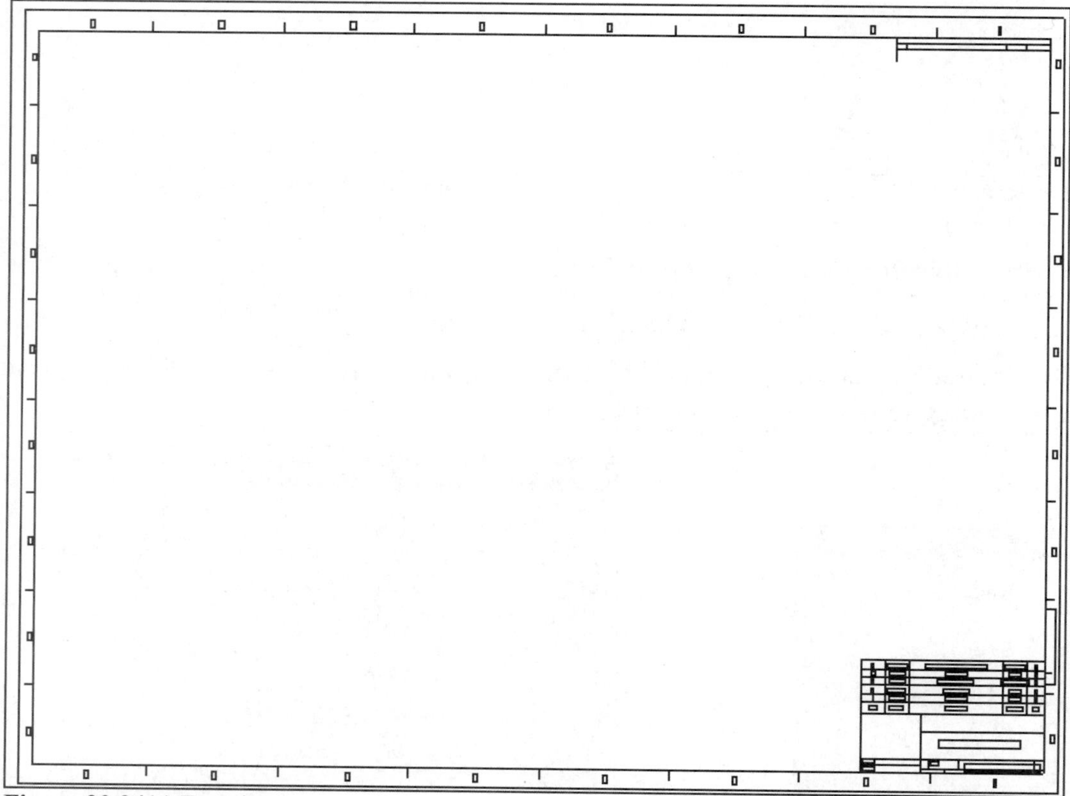

Figure 20.24(c) Drawing

5	SW101-9STL	.500-13 X 5.00 DOUBLE END STUD	PURCHASED	1
4	SW101-8FT	CLAMP FOOT	NYLON	1
3	SW101-7BA	SWING CLAMP BALL	BLACK PLASTIC	2
2	SW101-6SW	CLAMP SWIVEL	STEEL	1
1	SW101-5AR	CLAMP ARM	STEEL	1
ITEM	PT NUM	DESCRIPTION	MATERIAL	QTY

Figure 20.24(d) BOM

Click: **Datum axes off** ⇒ **Datum points off** ⇒ **Coordinate systems off** ⇒ double-click on **SCALE** in the lower left-hand of the graphics window SCALE : 1.000 TYPE : ASSEM ⇒ Enter value for scale Enter value for scale 1.000 ⇒ type **1.50** Enter value for scale 1.50 ⇒ **MMB** ⇒ **Disallow the movement of drawing views with the mouse** unlock ⇒ **Insert a drawing view of the active model** ⇒ **MMB** to accept all of the defaults ⇒ Select CENTER POINT for drawing view. (top view) [Figs. 20.25(a-b)] ⇒ select Reference 1 Front ASM_FRONT:F3(DATUM PLANE) [Fig. 20.25(c)]

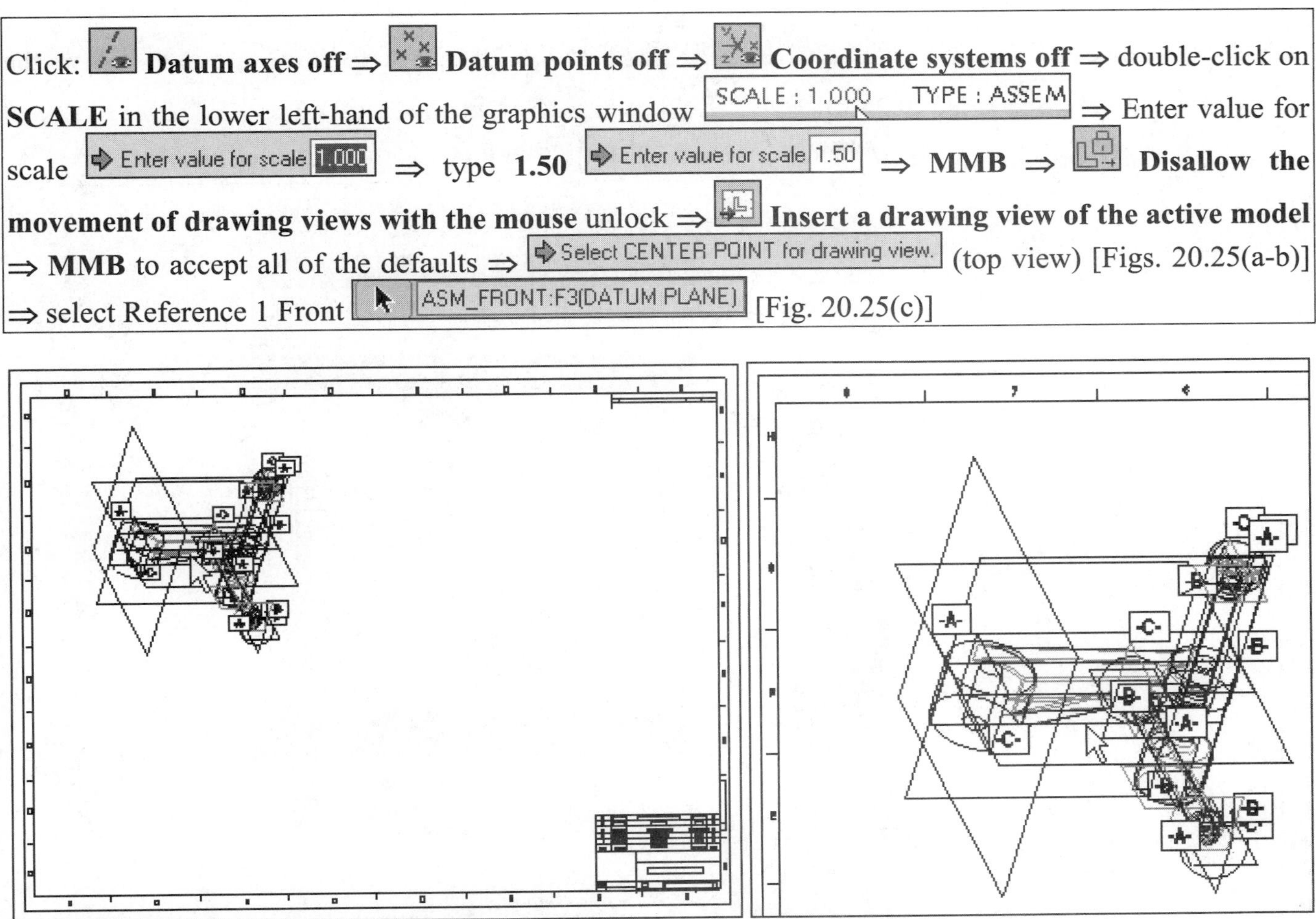

Figure 20.25(a) Select Centerpoint for First Drawing View

Figure 20.25(b) View Center

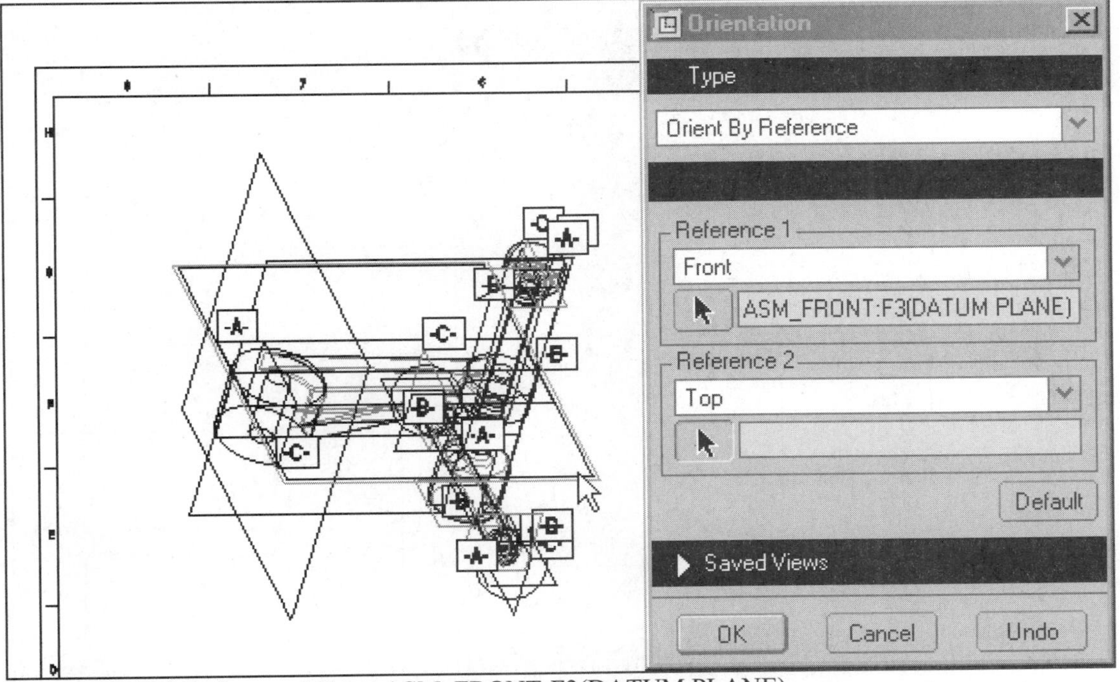

Figure 20.25(c) Reference 1 Front ASM_FRONT-F3(DATUM PLANE)

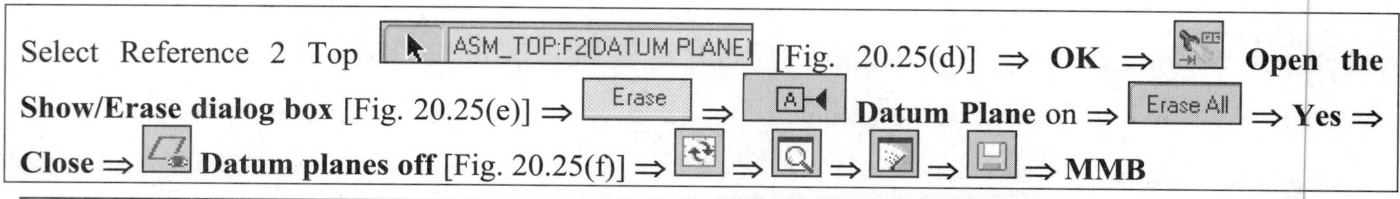
Select Reference 2 Top ASM_TOP:F2(DATUM PLANE) [Fig. 20.25(d)] ⇒ **OK** ⇒ **Open the Show/Erase dialog box** [Fig. 20.25(e)] ⇒ Erase ⇒ **Datum Plane** on ⇒ Erase All ⇒ **Yes** ⇒ **Close** ⇒ **Datum planes off** [Fig. 20.25(f)] ⇒ ⇒ ⇒ ⇒ ⇒ **MMB**

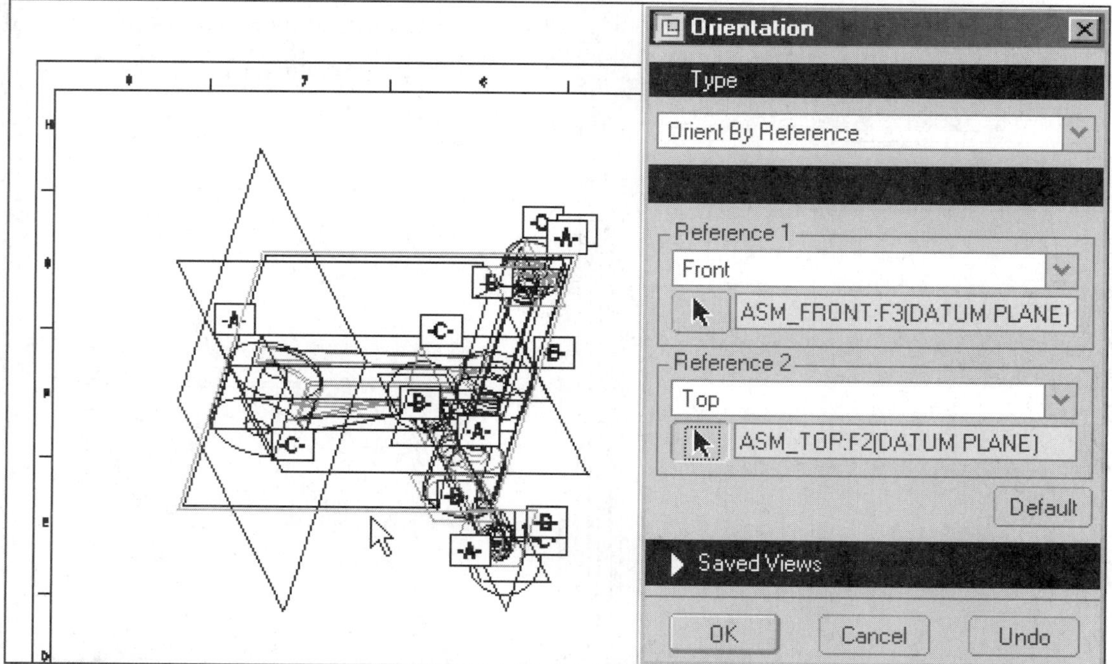

Figure 20.25(d) Reference 2 Top ASM_TOP:F2(DATUM PLANE)

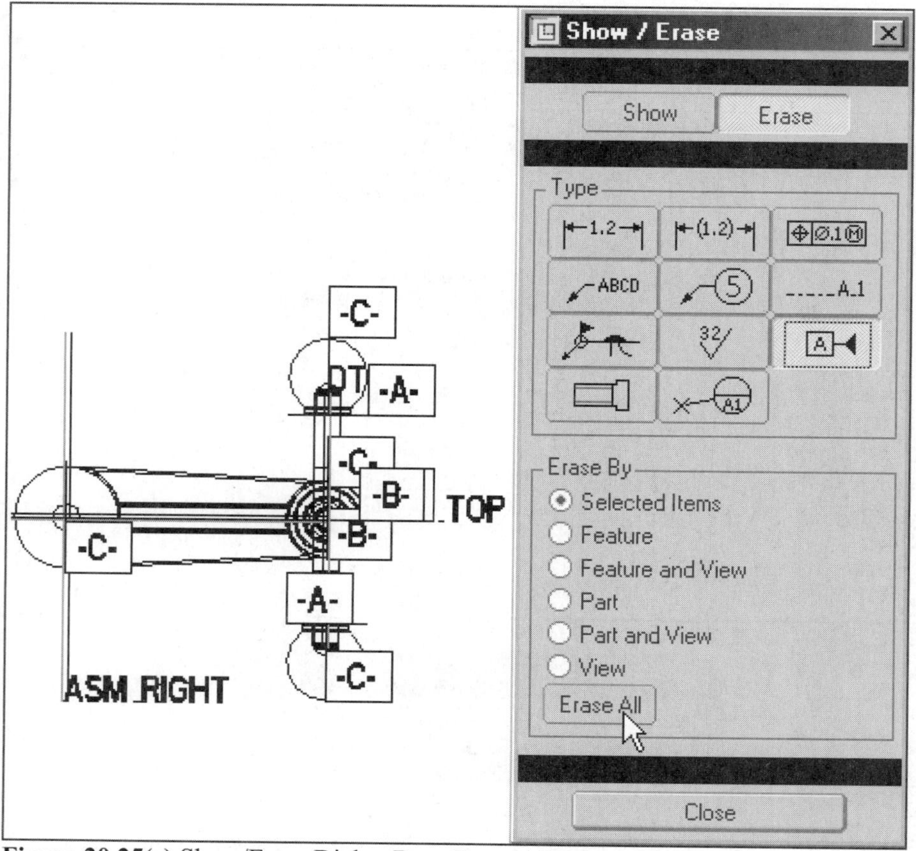

Figure 20.25(e) Show/Erase Dialog Box- Erase All Datum Planes

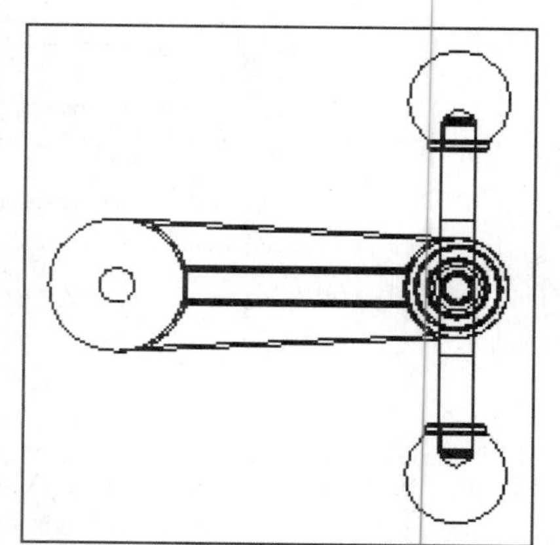

Figure 20.25(f) Datums Off

The only other view needed to show the subassembly is a front section view.

Click: **Insert a drawing view of the active model** ⇒ **Projection** ⇒ **Full View** ⇒ **Section** ⇒ **Unexploded** ⇒ **No Scale** ⇒ **MMB** ⇒ **Full** ⇒ **Total Xsec** ⇒ **MMB** ⇒ Select CENTER POINT for drawing view. select below the view previously created [Fig. 20.26(a)] ⇒ **Create** ⇒ **Planar** ⇒ **Single** ⇒ **MMB** ⇒ type the section name **A** Enter NAME for cross-section [QUIT]: A ⇒ **MMB** Select or create an assembly datum ⇒ **Datum planes on** ⇒ **Redraw the current view** ⇒ click on ASM_TOP:F2(DATUM PLANE) [Fig. 20.26(b)]

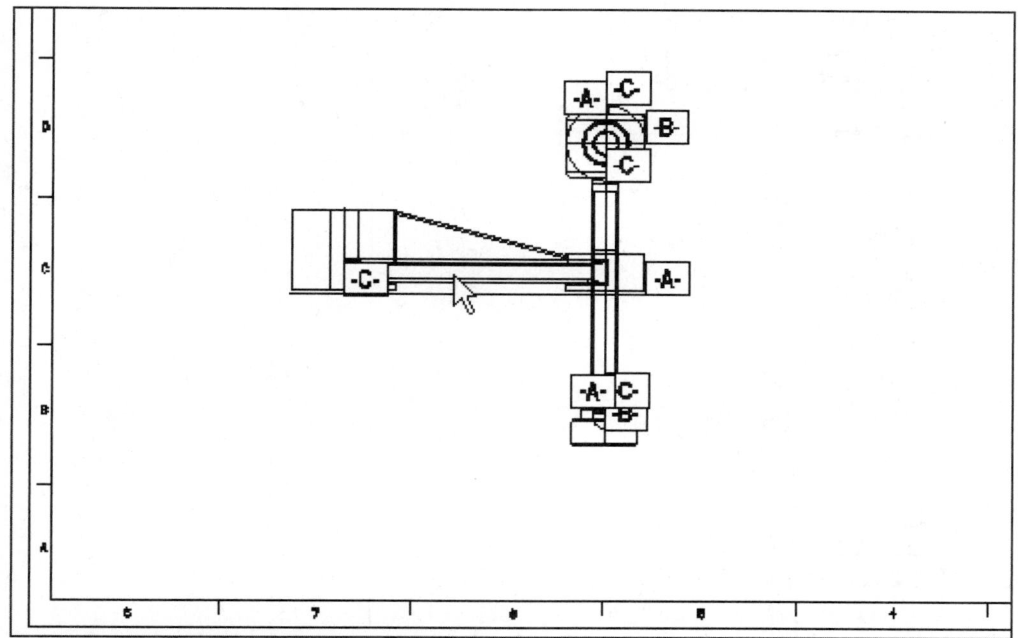

Figure 20.26(a) Place the Front View

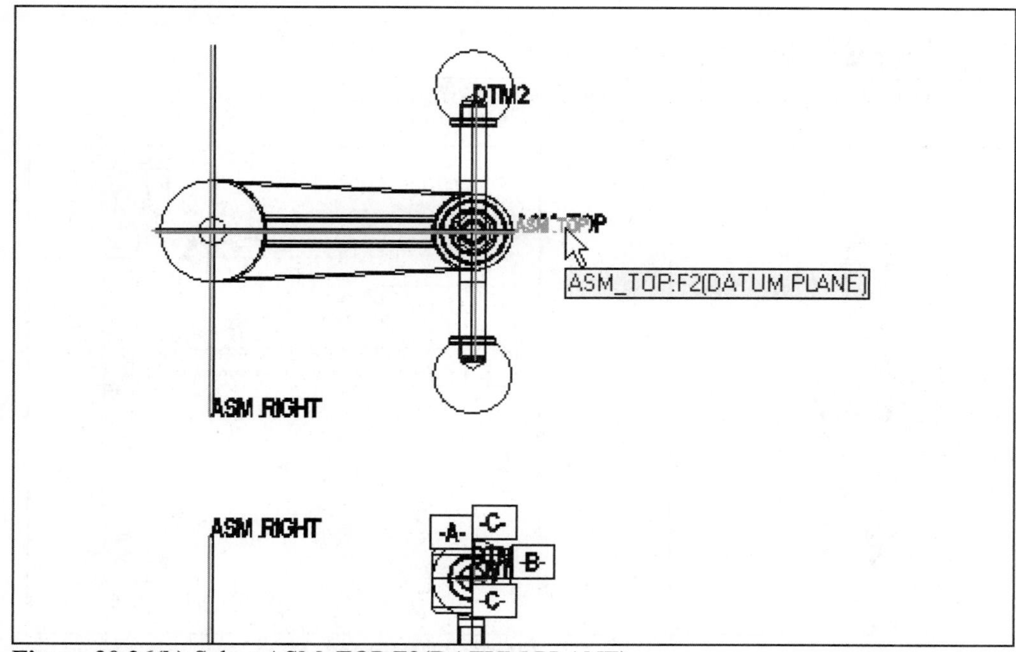

Figure 20.26(b) Select ASM_TOP:F2(DATUM PLANE)

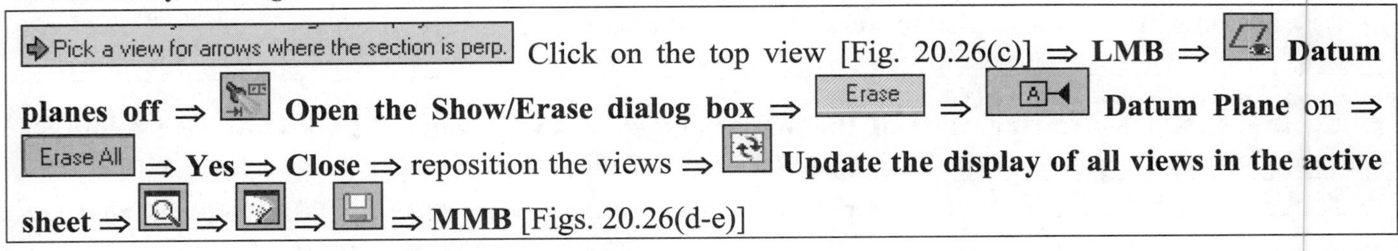

Pick a view for arrows where the section is perp. Click on the top view [Fig. 20.26(c)] ⇒ **LMB** ⇒ **Datum planes off** ⇒ **Open the Show/Erase dialog box** ⇒ Erase ⇒ **Datum Plane** on ⇒ Erase All ⇒ **Yes** ⇒ **Close** ⇒ reposition the views ⇒ **Update the display of all views in the active sheet** ⇒ ⇒ ⇒ ⇒ **MMB** [Figs. 20.26(d-e)]

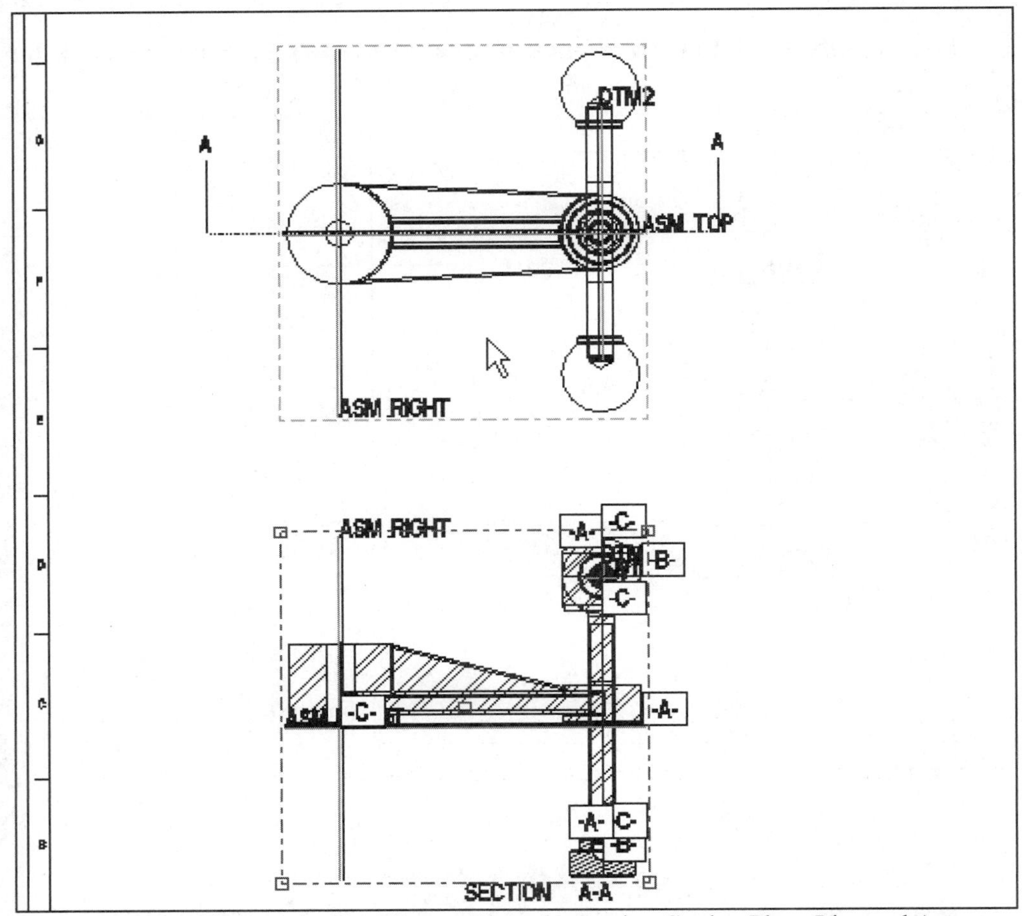

Figure 20.26(c) Click on the Top View to Place the Section Cutting Plane Line and Arrows

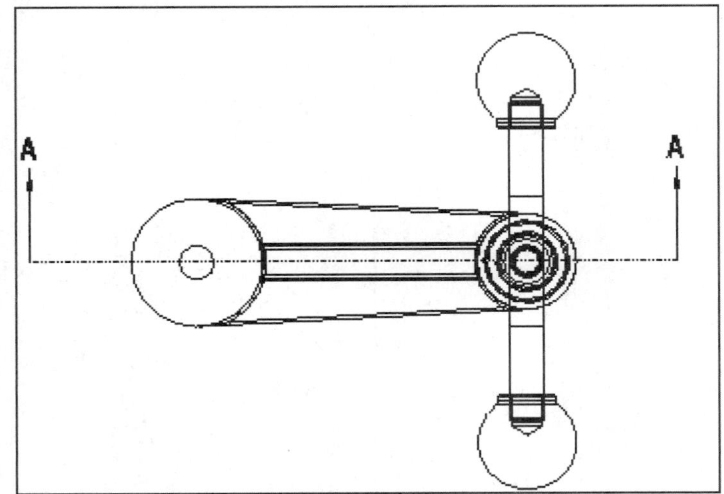

Figure 20.26(d) Top View with Section Cutting Plane

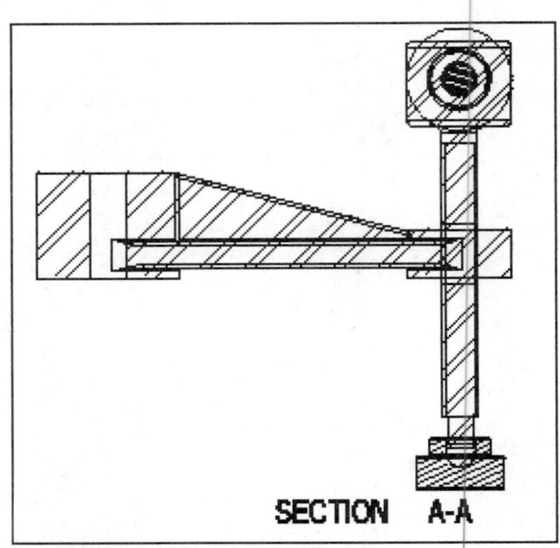

Figure 20.26(e) SECTION A-A

Pro/E provides tools to alter the display of the section views to comply with industry ASME standard practices. Most companies require that the crosshatching on parts in section views of assemblies be "clocked" such that parts that meet do not use the same section lining (crosshatching) spacing and angle. This makes the separation between parts more distinct. First, modify the visibility of the views to remove all hidden lines and make the tangent edges dimmed. Next, show all centerlines and clip as needed.

Press and hold down the **Ctrl** key and click on both views ⇒ click **RMB** with cursor outside of the view outlines ⇒ **Properties** ⇒ **View Disp** ⇒ **No Hidden** ⇒ **Tan Dimmed** ⇒ **MMB** ⇒ **MMB** ⇒ **LMB** ⇒ **Open the Show/Erase dialog box** ⇒ Show ⇒ **Datum Plane** off ⇒ A_1 **Axis** (on) ⇒ **Show All** ⇒ **Yes** ⇒ **Accept All** ⇒ **Close** ⇒ **LMB** [Fig. 20.27(a)] ⇒ ⇒ ⇒ ⇒ ⇒ **MMB** ⇒ **Sketch** ⇒ **Sketcher Preferences** ⇒ Grid intersection off

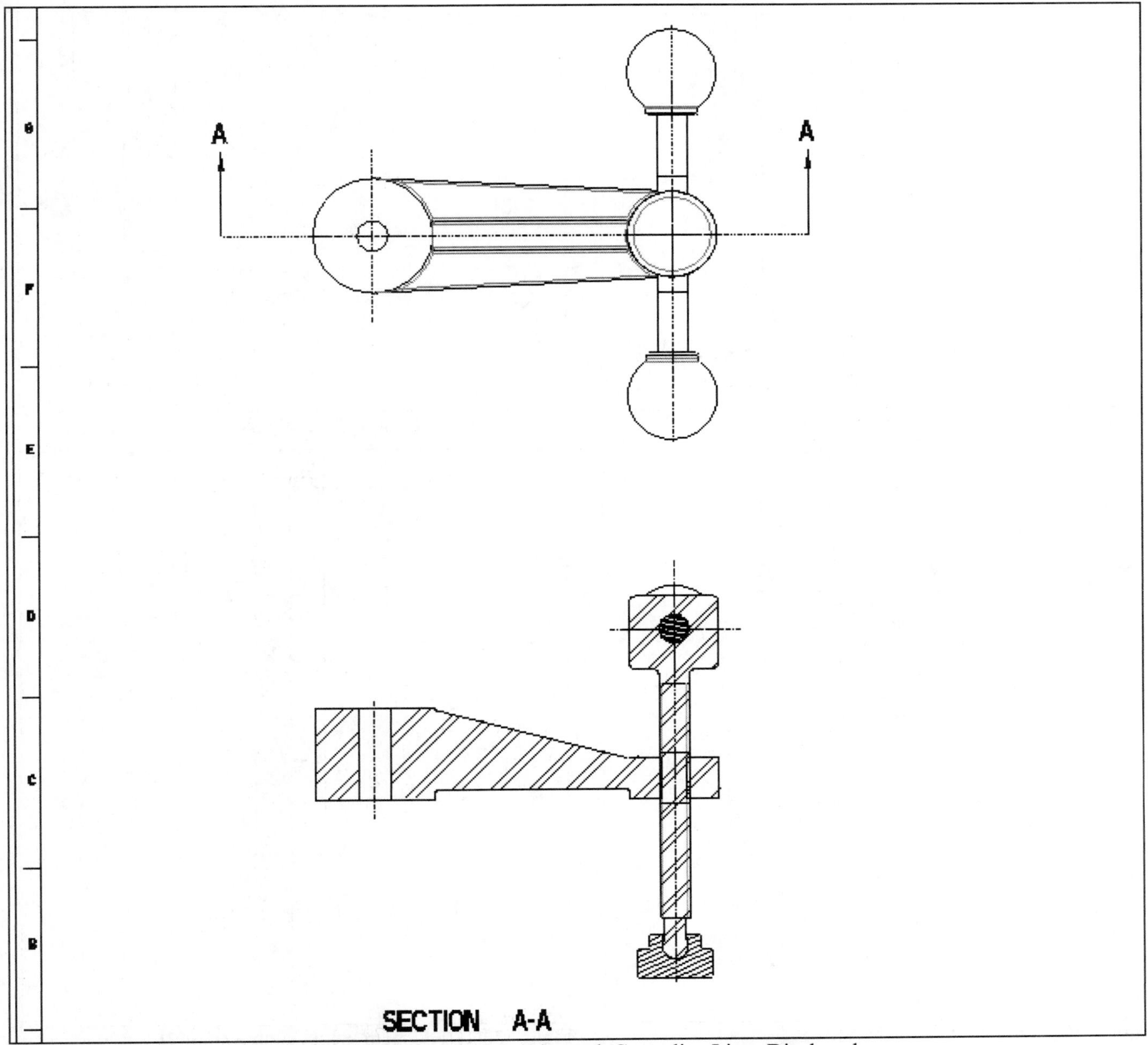

Figure 20.27(a) Hidden Lines Removed, Tangent Edges Dimmed, Centerline Lines Displayed

Double-click on the crosshatching in the front view (Clamp_Stud5 is now active) [Fig. 20.27(b)] ⇒ **Fill** [Fig. 20.27(c)] ⇒ click **Next Xsec** until Clamp_Foot is active [Fig. 20.27(d)] ⇒ **Hatch** ⇒ **Angle** ⇒ **135** [Fig. 20.27(e)]

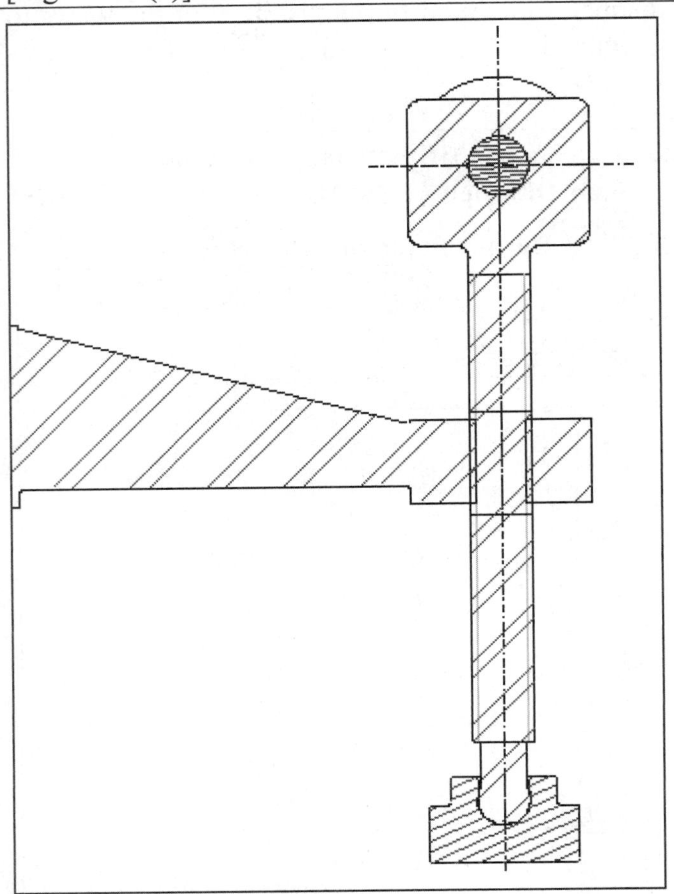

Figure 20.27(b) Clamp_Stud5 is Active

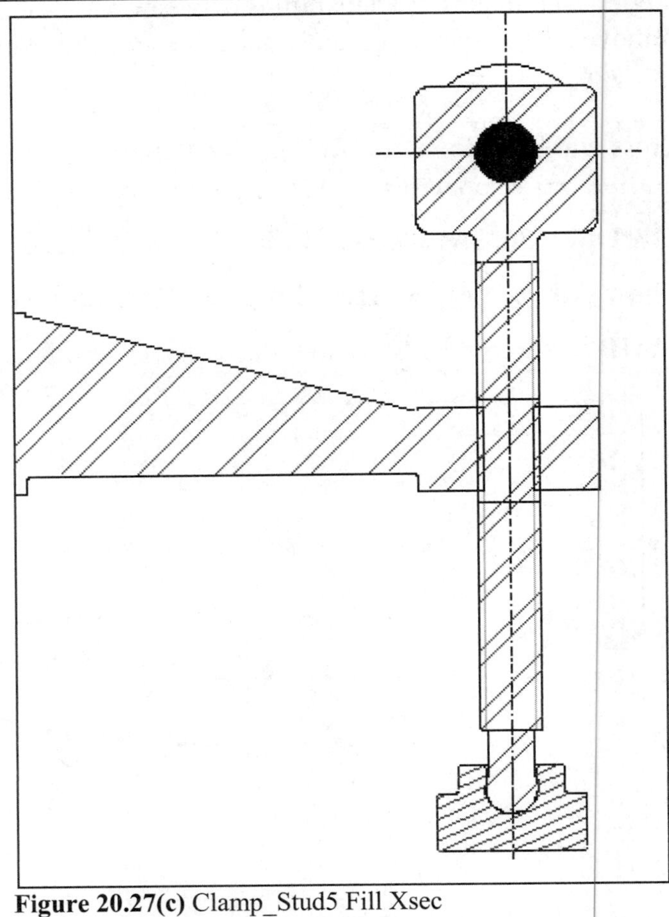

Figure 20.27(c) Clamp_Stud5 Fill Xsec

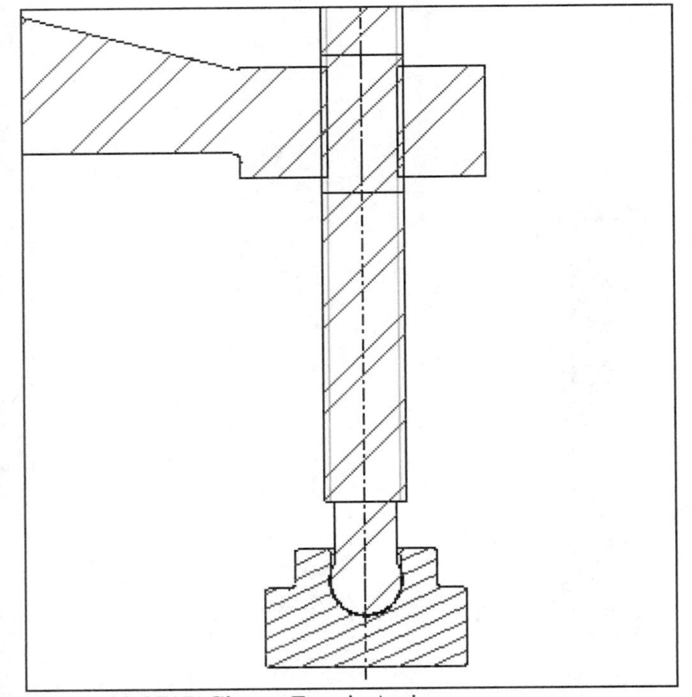

Figure 20.27(d) Clamp_Foot is Active

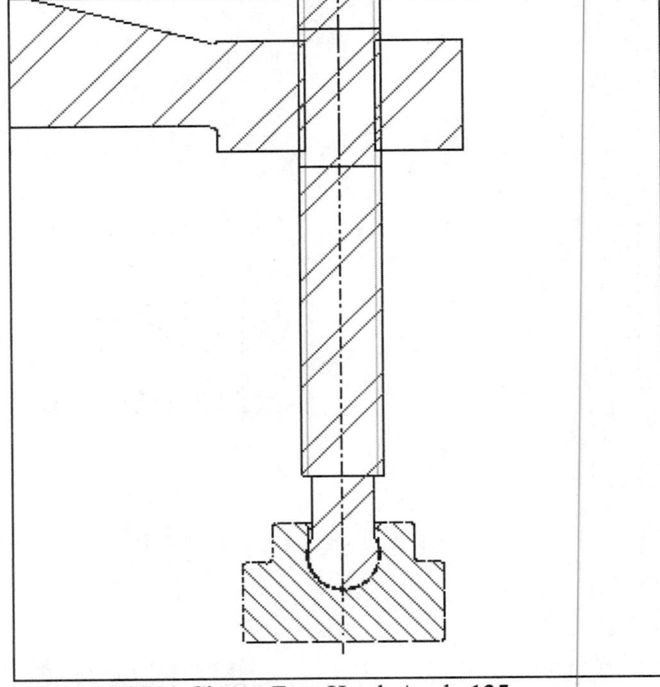

Figure 20.27(e) Clamp_Foot Hatch Angle **135**

Click: **Next Xsec** ⇒ **Next Xsec** Clamp_Arm is active ⇒ **Hatch** ⇒ **Delete line** ⇒ **Spacing** ⇒ **Half** ⇒ **Angle** ⇒ **120** [Fig. 20.27(f)] ⇒ **Prev Xsec** Clamp_Swivel is active ⇒ **Delete line** ⇒ **Spacing** ⇒ **Half** ⇒ ⇒ **MMB** ⇒ **LMB** ⇒ ⇒ ⇒ [Fig. 20.27(g)] ⇒ ⇒ **MMB**

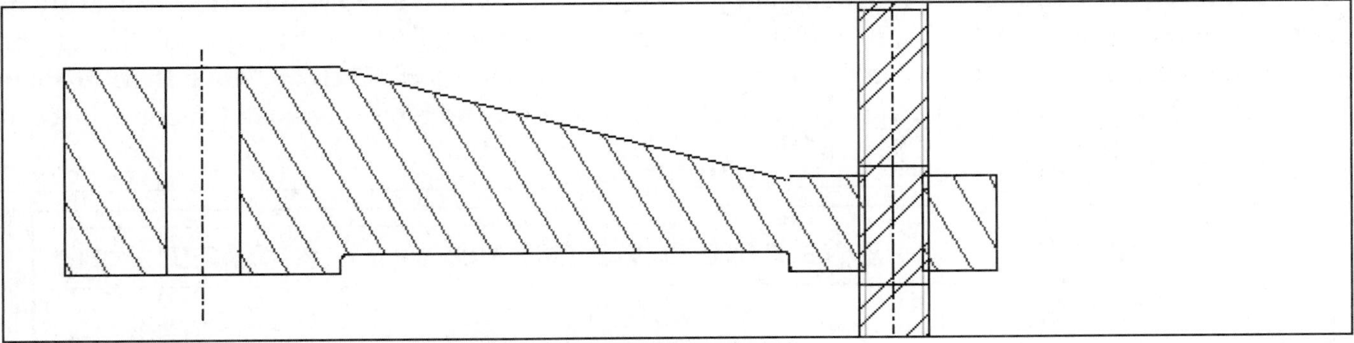

Figure 20.27(f) Clamp_Arm, Delete Line, Hatch Spacing Half, Angle **120**

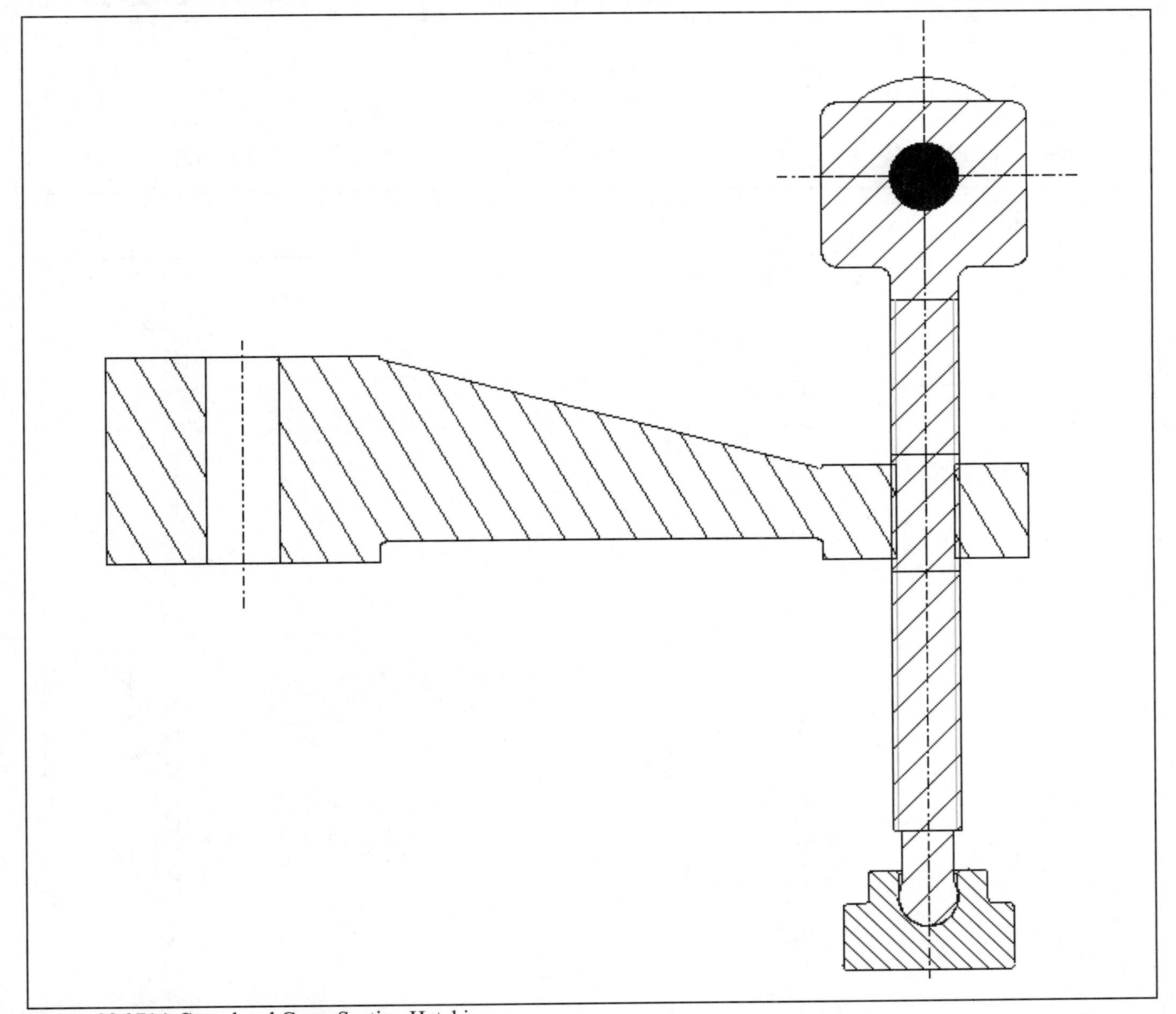

Figure 20.27(g) Completed Cross Section Hatching

To complete the subassembly drawing, you must display the balloons for each component. Show the **Item Balloons** on the drawing. Balloons are displayed in the top view, because it was the first view that was created.

Click: **Table** ⇒ **BOM Balloons** ⇒ **Set Region** ⇒ **Simple** ⇒ click in the BOM field [Fig. 20.28(a)] ⇒ **Create Balloon** ⇒ **Show All** [Fig. 20.28(b)] ⇒ **MMB** ⇒ press and hold the **Ctrl** key and click on the balloons for the Clamp_Arm (1), Clamp_Stud5 (5) and Clamp_Foot (4) ⇒ **RMB** ⇒ **Move Item to View** [Fig. 20.28(c)] ⇒ click in the front view [Fig. 20.28(d)] ⇒ click on and reposition each balloon as needed [Fig. 20.28(e)] ⇒ ⇒ ⇒ ⇒ ⇒ **MMB**

ITEM	PT NUM	DESCRIPTION	MATERIAL	QTY
5	SW101-9STL	.500-13 X 5.00 DOUBLE END STUD	PURCHASED	1
4	SW101-8FT	CLAMP FOOT	NYLON	1
3	SW101-7BA	SWING CLAMP BALL	BLACK PLASTIC	2
2	SW101-6SW	CLAMP SWIVEL	STEEL	1
1	SW101-5AR	CLAMP ARM	STEEL	1

Figure 20.28(a) Set Region

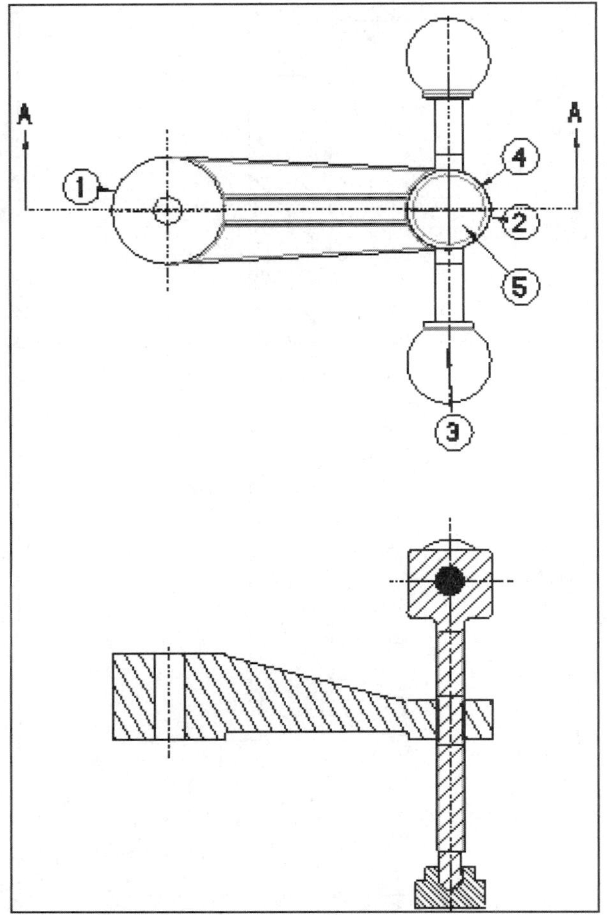

Figure 20.28(b) Show All Balloons

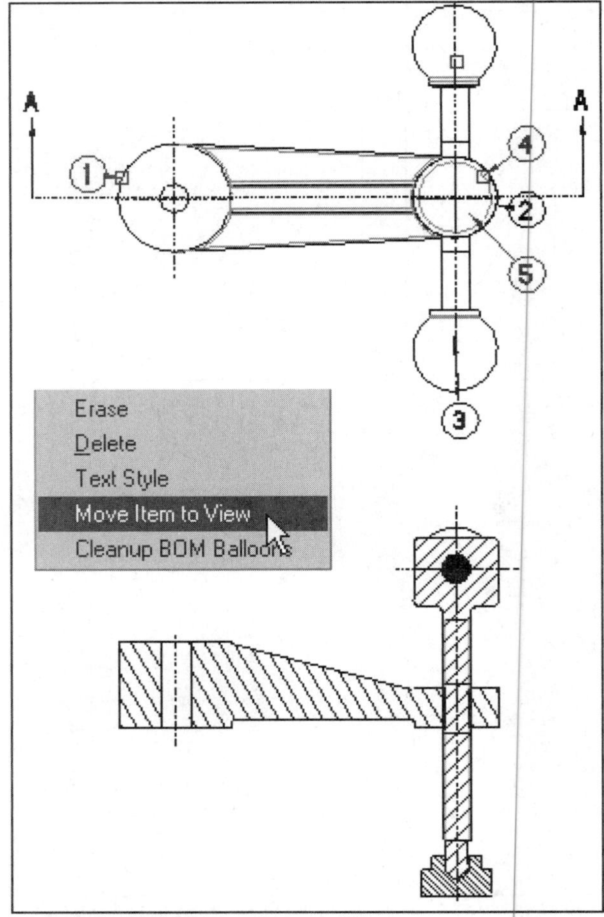

Figure 20.28(c) Move Item to View

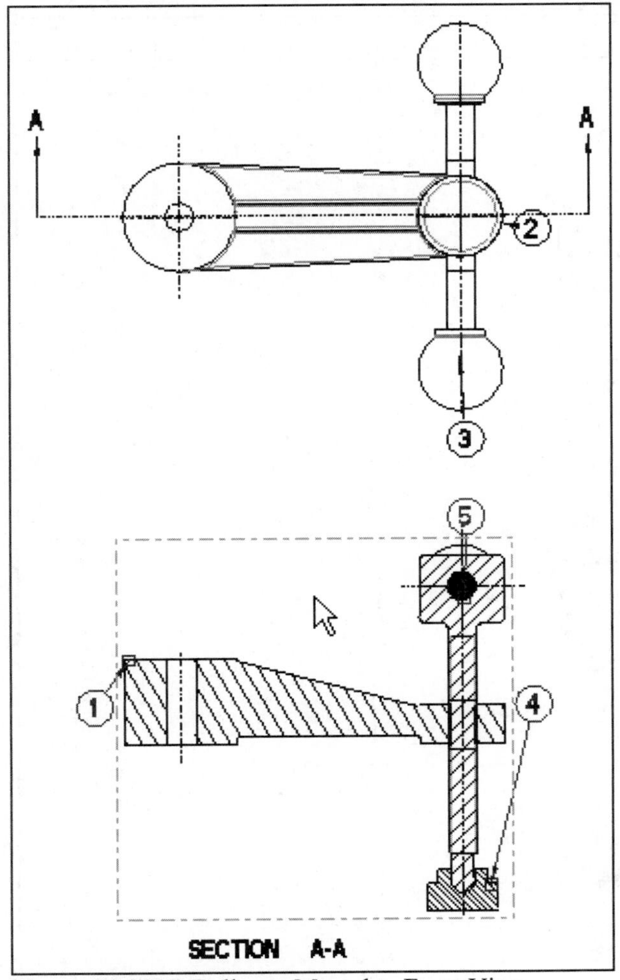

Figure 20.28(d) Balloons Moved to Front View

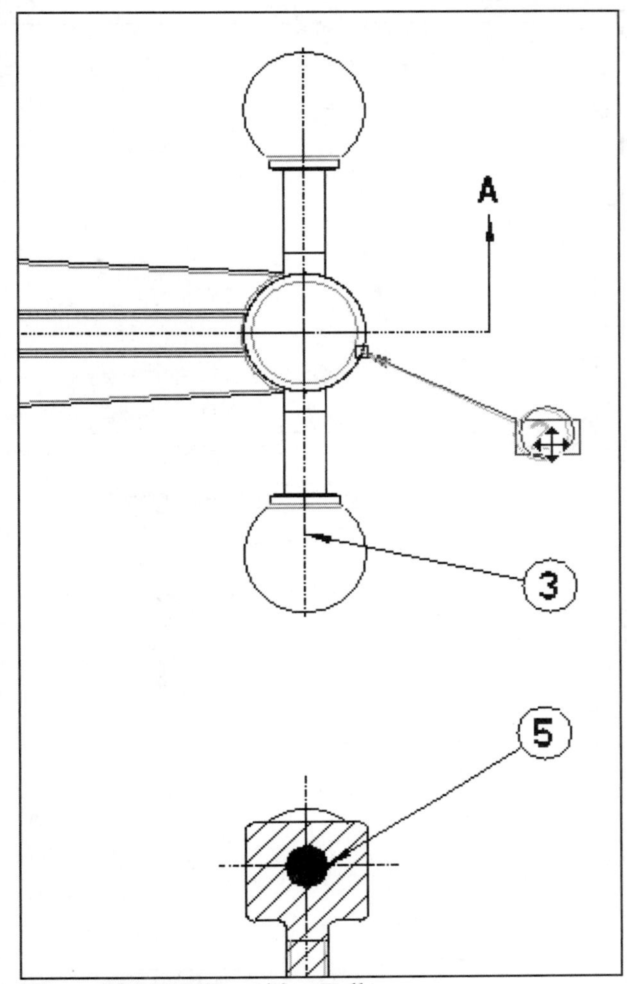

Figure 20.28(e) Reposition Balloons

Click on balloon **3** (Clamp_Ball) ⇒ **RMB** [Fig. 20.28(f)] ⇒ **Edit Attachment** ⇒ **On Entity** pick on the edge of the Clamp_Ball [Fig. 20.28(g)] ⇒ **MMB** ⇒ [icon] ⇒ [icon] ⇒ [icon] ⇒ [icon] ⇒ **MMB** ⇒ **File** ⇒ **Delete** ⇒ **Old Versions** ⇒ **MMB** [Fig. 20.28(h)] ⇒ **Window** ⇒ **Close**

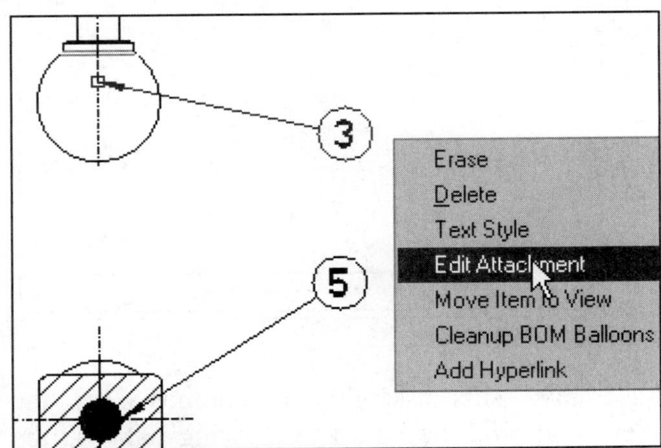

Figure 20.28(f) Edit Attachment

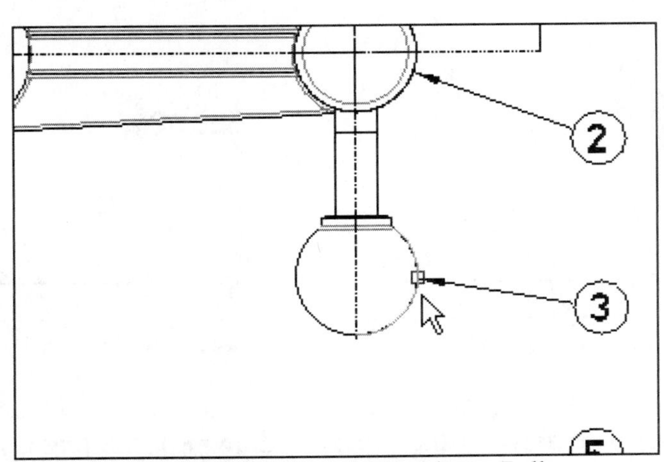

Figure 20.28(g) Pick the Edge of the Clamp_Ball

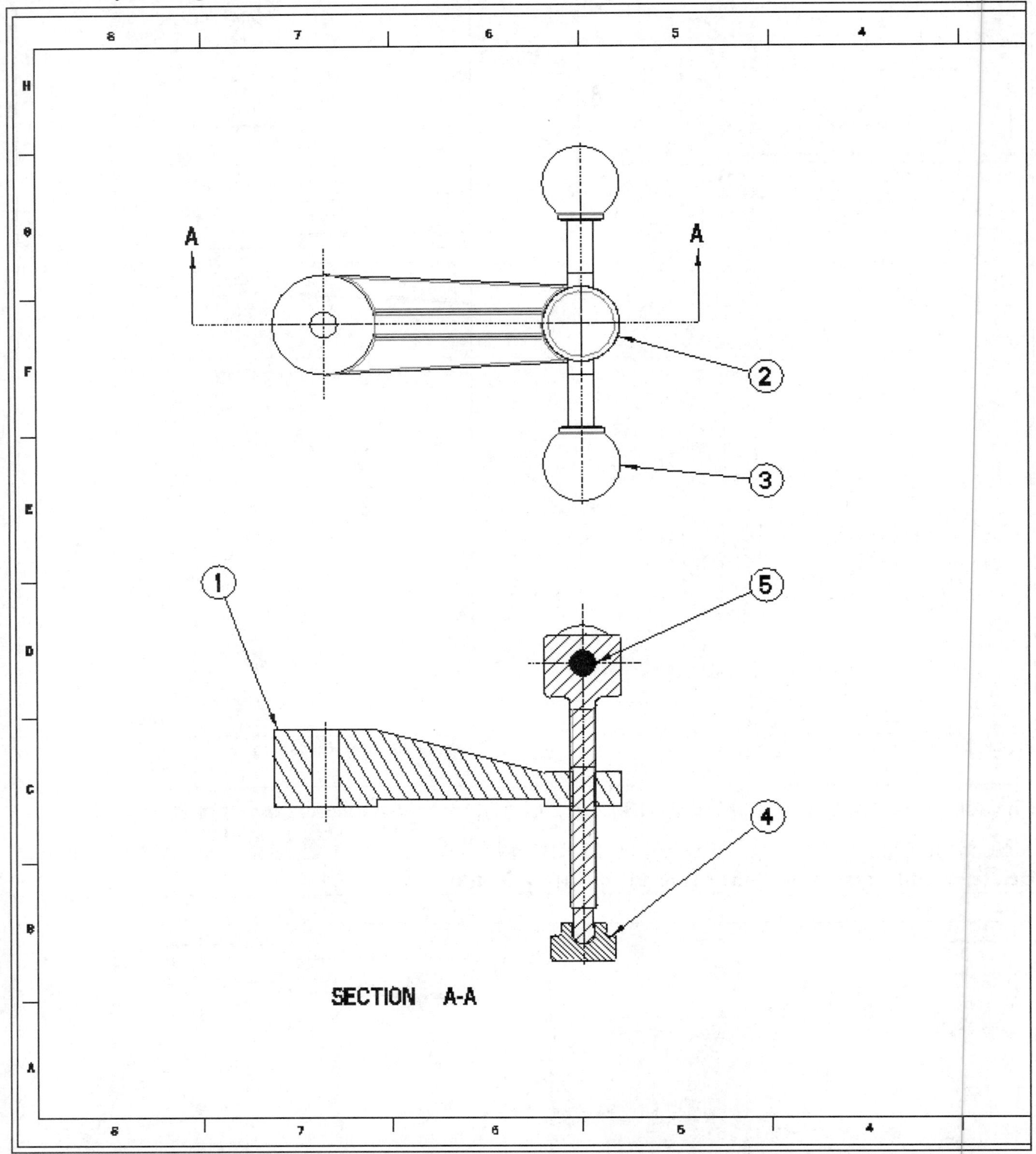

Figure 20.28(h) Completed Clamp_Subassembly Drawing

Create the drawing for the Swing Clamp Assembly. This assembly is composed of the subassembly (in the previous drawing), the plate, the short stud, and the nut. The drawing will use the same format created for the subassembly. Formats are read-only files that can be used as many times as needed.

Click: **Create a new object** ⇒ Drawing ⇒ Name **CLAMP_ASSEMBLY** ⇒ Use default template ⇒ **OK** ⇒ Default Model **Browse** ⇒ **clamp_assembly.asm** ⇒ **Open** ⇒ Empty with format ⇒ **Browse** ⇒ Format **asm_format_e. frm** ⇒ **Open** ⇒ **OK** ⇒ **Zoom in** on the title block [Fig. 20.29(a)] ⇒ ⇒ ⇒ **RMB** ⇒ **Properties** ⇒ **Drawing Options** ⇒ **Open a configuration file** ⇒ **clamp_asm.dtl** clamp_asm.dtl ⇒ **Open** ⇒ **Apply** ⇒ **Close** ⇒ **MMB**

ITEM	PT NUM	DESCRIPTION	MATERIAL	QTY
8	SW101-95TL	.500-13 X 5.00 DOUBLE END STUD	PURCHASED	1
7	SW101-8FT	CLAMP FOOT	NYLON	1
6	SW101-7BA	SWING CLAMP BALL	BLACK PLASTIC	2
5	SW101-6SW	CLAMP SWIVEL	STEEL	1
4	SW101-5AR	CLAMP ARM	STEEL	1
3	SW100-22FLN	.500-13 X HEX FLANGE NUT	PURCHASED	1
2	SW100-21ST	.500-13 X 3.50 DOUBLE END STUD	PURCHASED	1
1	SW100-20PL	CLAMP PLATE	STEEL	1

TOOL ENGINEERING CO.

DRAWN
ISSUED
1.000
CLAMP_ASSEMBLY
SHEET 1 OF 1
A

Figure 20.29(a) BOM and Title Block

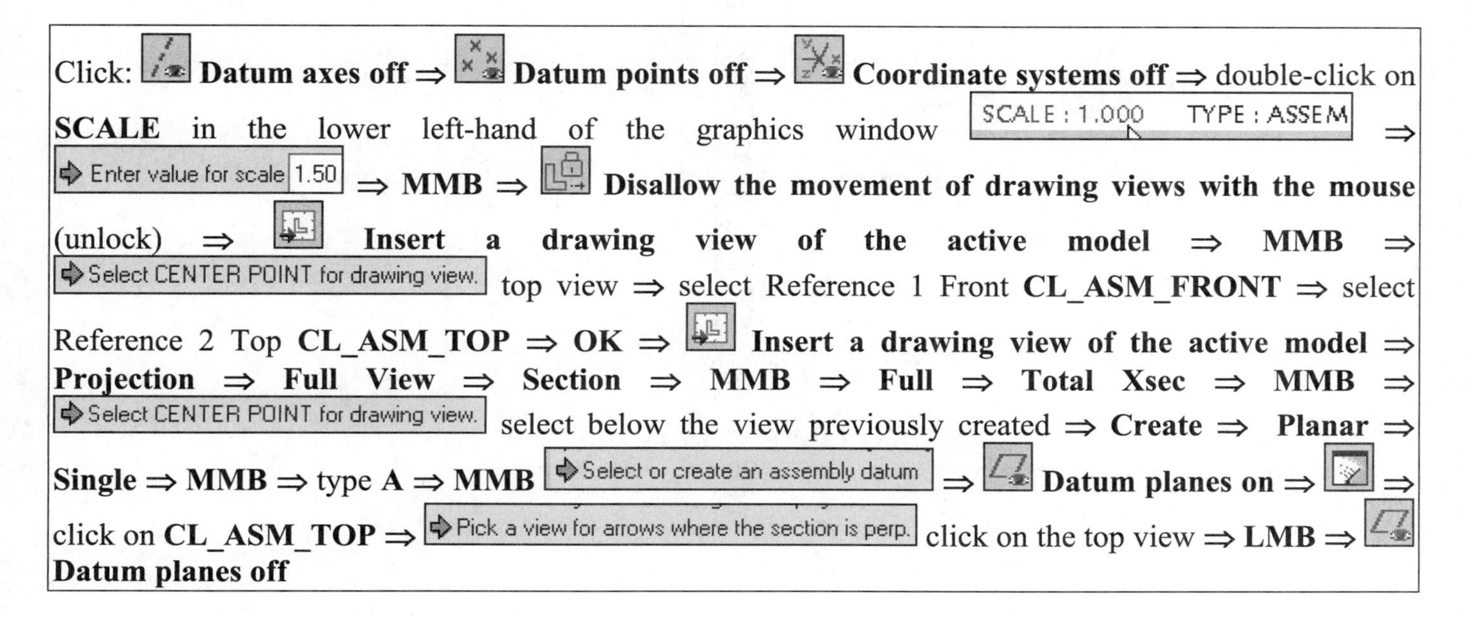

Click: **Datum axes off** ⇒ **Datum points off** ⇒ **Coordinate systems off** ⇒ double-click on **SCALE** in the lower left-hand of the graphics window SCALE : 1.000 TYPE : ASSEM ⇒ Enter value for scale 1.50 ⇒ **MMB** ⇒ **Disallow the movement of drawing views with the mouse** (unlock) ⇒ **Insert a drawing view of the active model** ⇒ **MMB** ⇒ Select CENTER POINT for drawing view. top view ⇒ select Reference 1 Front **CL_ASM_FRONT** ⇒ select Reference 2 Top **CL_ASM_TOP** ⇒ **OK** ⇒ **Insert a drawing view of the active model** ⇒ **Projection** ⇒ **Full View** ⇒ **Section** ⇒ **MMB** ⇒ **Full** ⇒ **Total Xsec** ⇒ **MMB** ⇒ Select CENTER POINT for drawing view. select below the view previously created ⇒ **Create** ⇒ **Planar** ⇒ **Single** ⇒ **MMB** ⇒ type **A** ⇒ **MMB** Select or create an assembly datum ⇒ **Datum planes on** ⇒ ⇒ click on **CL_ASM_TOP** ⇒ Pick a view for arrows where the section is perp. click on the top view ⇒ **LMB** ⇒ **Datum planes off**

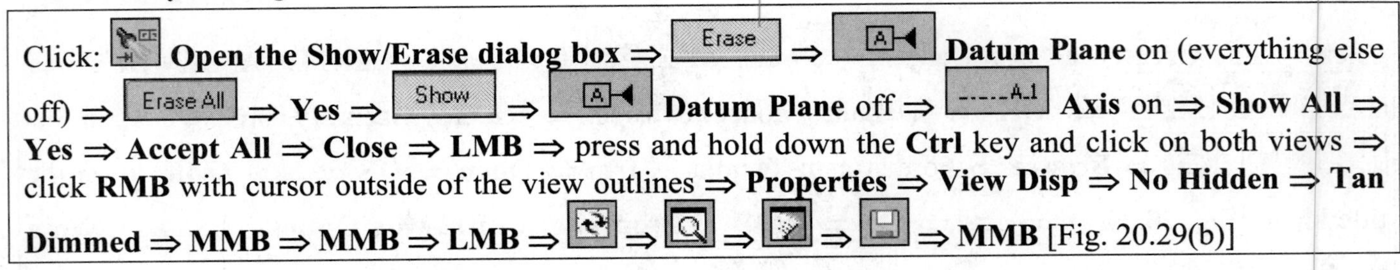

Click: **Open the Show/Erase dialog box** ⇒ Erase ⇒ **Datum Plane** on (everything else off) ⇒ Erase All ⇒ **Yes** ⇒ Show ⇒ **Datum Plane** off ⇒ **Axis** on ⇒ **Show All** ⇒ **Yes** ⇒ **Accept All** ⇒ **Close** ⇒ **LMB** ⇒ press and hold down the **Ctrl** key and click on both views ⇒ click **RMB** with cursor outside of the view outlines ⇒ **Properties** ⇒ **View Disp** ⇒ **No Hidden** ⇒ **Tan Dimmed** ⇒ **MMB** ⇒ **MMB** ⇒ **LMB** ⇒ ⇒ ⇒ ⇒ ⇒ **MMB** [Fig. 20.29(b)]

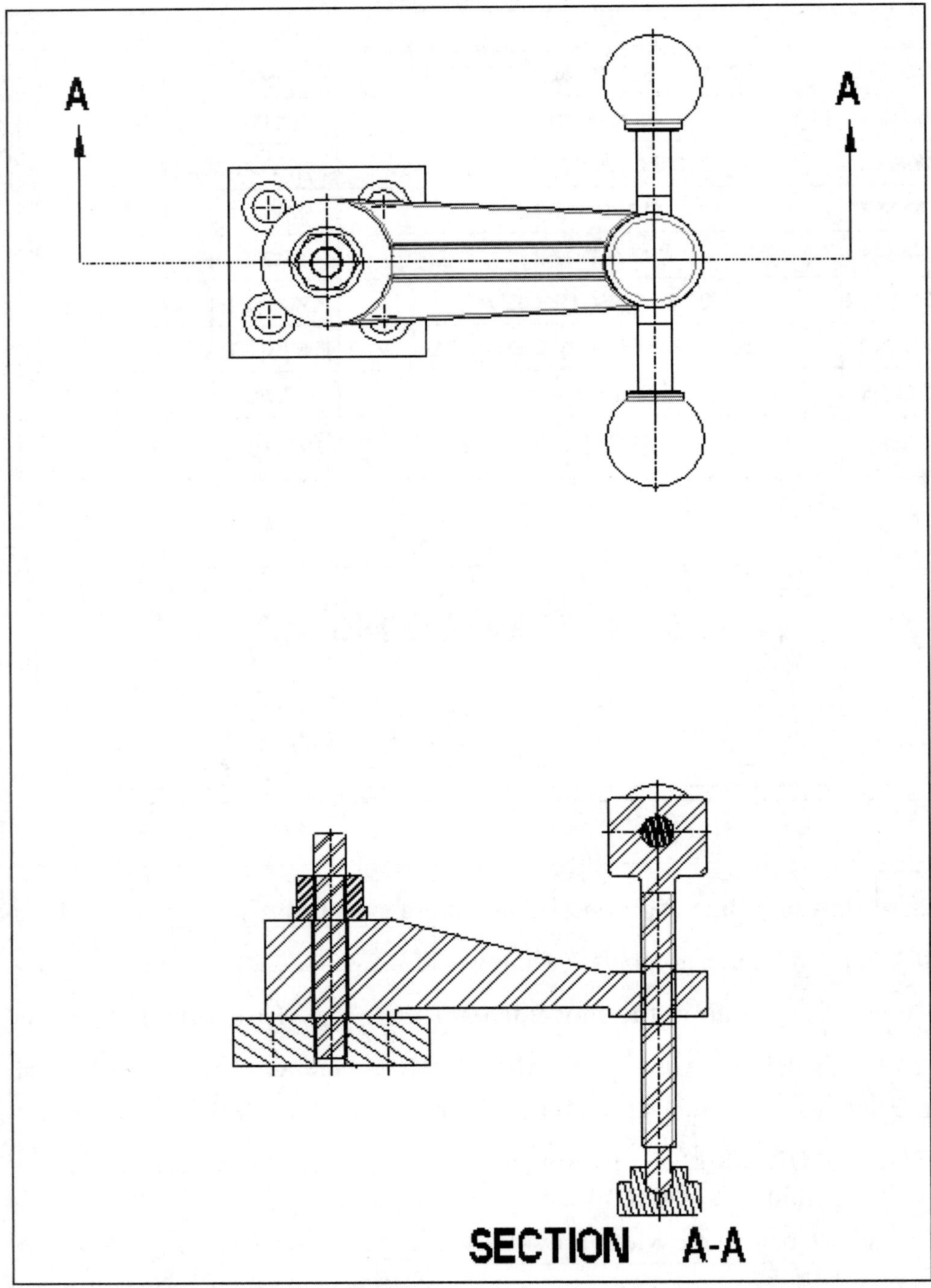

Figure 20.29(b) Assembly Drawing Views

Click: **Table** ⇒ **Repeat Region** ⇒ **Attributes** ⇒ click in the BOM field [Fig. 20.30(a)] ⇒ **Flat** ⇒ **MMB** ⇒ **MMB** ⇒ **MMB** [Fig. 20.30(b)] ⇒ **Table** ⇒ **BOM Balloons** ⇒ **Set Region** ⇒ click in the BOM field ⇒ **Create Balloon** ⇒ **Show All** ⇒ **MMB** [Fig. 20.30(c)]

8	SW101-9STL	.500-13 X 5.00 DOUBLE END STUD	PURCHASED	1
7	SW101-8FT	CLAMP FOOT	NYLON	1
6	SW101-7BA	SWING CLAMP BALL	BLACK PLASTIC	2
5	SW101-6SW	CLAMP SWIVEL	STEEL	1
4	SW101-5AR	CLAMP ARM	STEEL	1
3	SW100-22FLN	.500-13 X HEX FLANGE NUT	PURCHASED	1
2	SW100-21ST	.500-13 X 3.50 DOUBLE END STUD	PURCHASED	1
1	SW100-20PL	CLAMP PLATE	STEEL	1
ITEM	PT NUM	DESCRIPTION	MATERIAL	QTY

Figure 20.30(a) Showing with BOM Attribute Recursive

3	SW100-22FLN	.500-13 X HEX FLANGE NUT	PURCHASED	1
2	SW100-21ST	.500-13 X 3.50 DOUBLE END STUD	PURCHASED	1
1	SW100-20PL	CLAMP PLATE	STEEL	1
ITEM	PT NUM	DESCRIPTION	MATERIAL	QTY

TOOL ENGINEERING CO.

Figure 20.30(b) Showing with BOM Attribute Flat

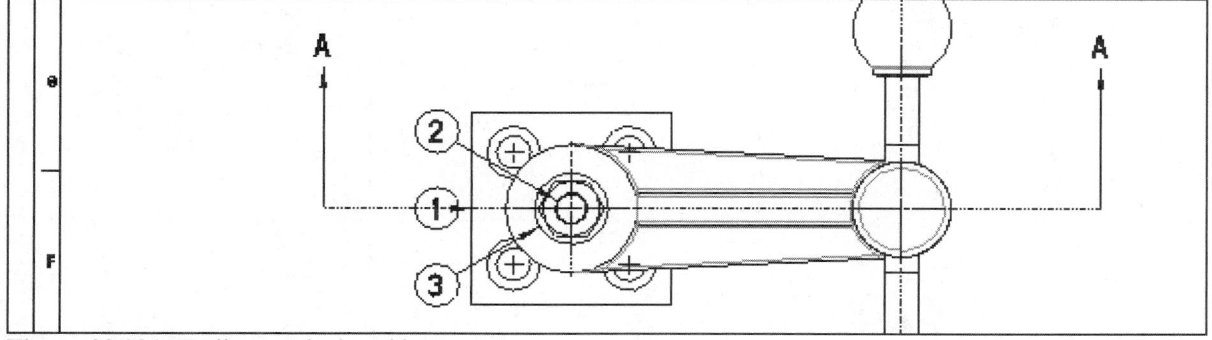

Figure 20.30(c) Balloons Displayed in Top View

While pressing the **Ctrl** key, click on all three balloons ⇒ **RMB** ⇒ **Move Item to View** ⇒ click in the front view [Fig. 20.30(d)] ⇒ click on and reposition each balloon as needed ⇒ click on balloon **2** [Fig. 20.30(e)] ⇒ **RMB** ⇒ **Edit Attachment** ⇒ **On Entity** pick on edge [Fig. 20.30(f)] ⇒ **MMB** ⇒ **LMB**

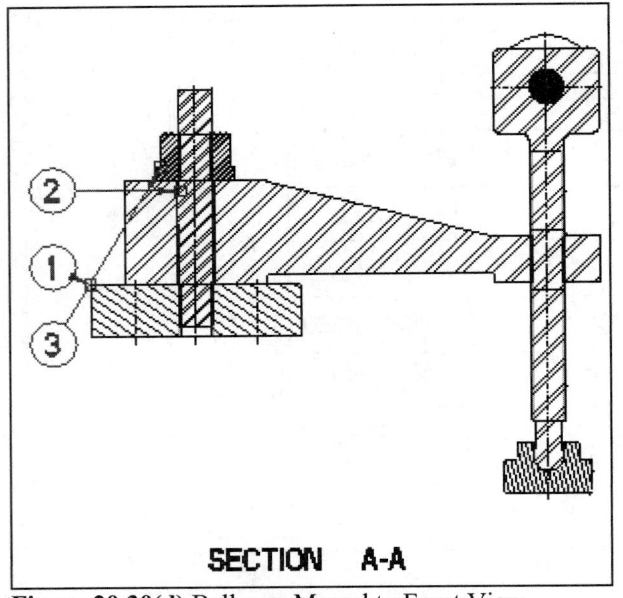

Figure 20.30(d) Balloons Moved to Front View

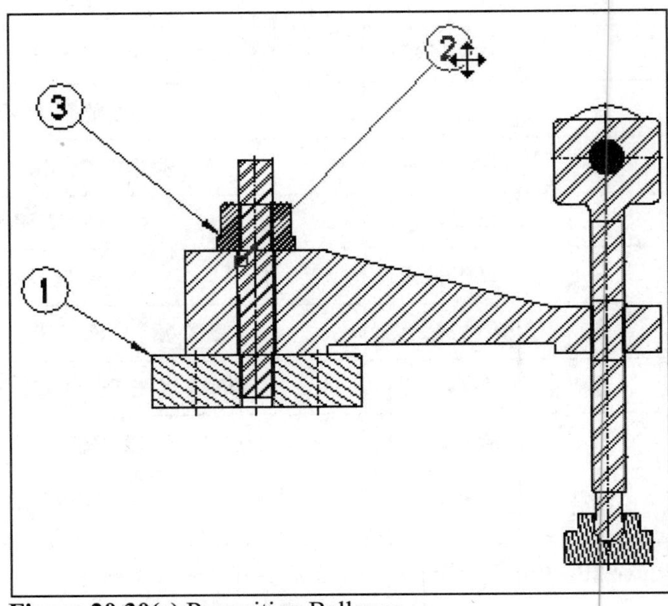

Figure 20.30(e) Reposition Balloons

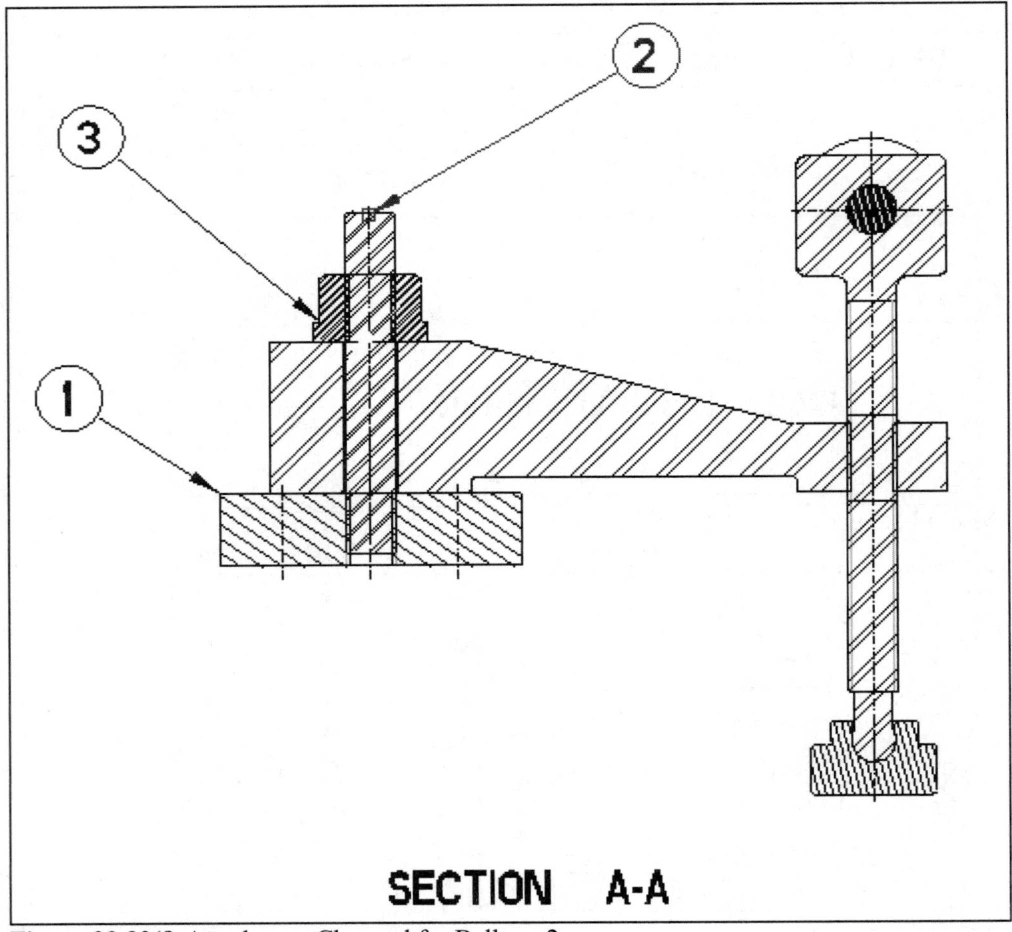

Figure 20.30(f) Attachment Changed for Balloon 2

Most companies (and as per drafting standards) require that round purchased items, such as nuts, bolts, studs, springs, and die pins be excluded from sectioning even when the section cutting plane passes through them. Remove the section lining (crosshatching) from the Clamp_Flange_Nut and the Clamp_Stud35 in the front section view.

Double-click on the crosshatching in the front view [Fig. 20.31(a)]. The Clamp_Flange_Nut is now active [Fig. 20.31(b)]. ⇒ **Excl Comp** to eliminate Xsec of Clamp_Flange_Nut [Fig. 20.31(c)] ⇒ **Next Xsec** Clamp_Stud35 is now active ⇒ **Excl Comp** to eliminate Xsec of Clamp_Stud35 [Fig. 20.31(d)] ⇒ **MMB** ⇒ **LMB** ⇒ [icon] ⇒ [icon] ⇒ [icon] ⇒ **MMB**

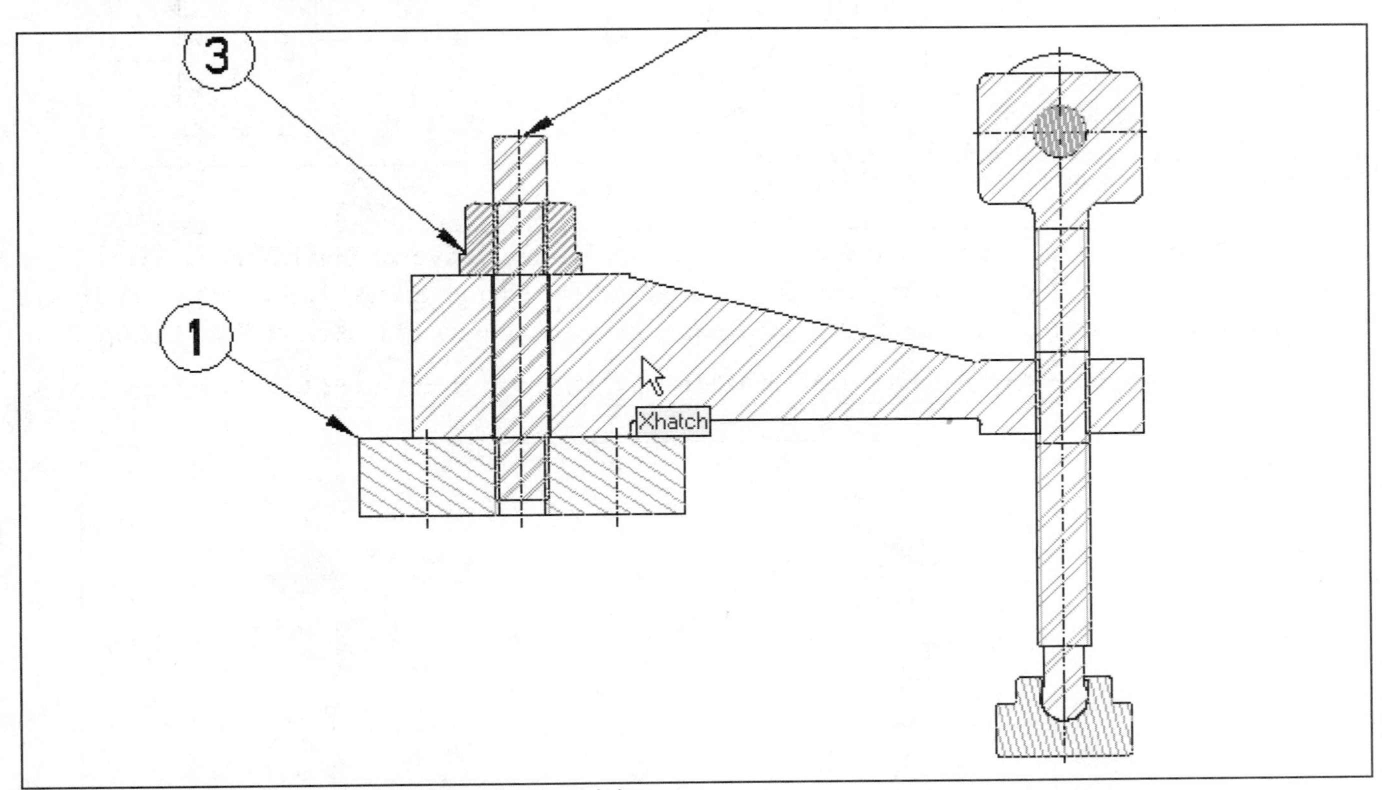

Figure 20.31(a) Double-Click on the Cross Section Lining

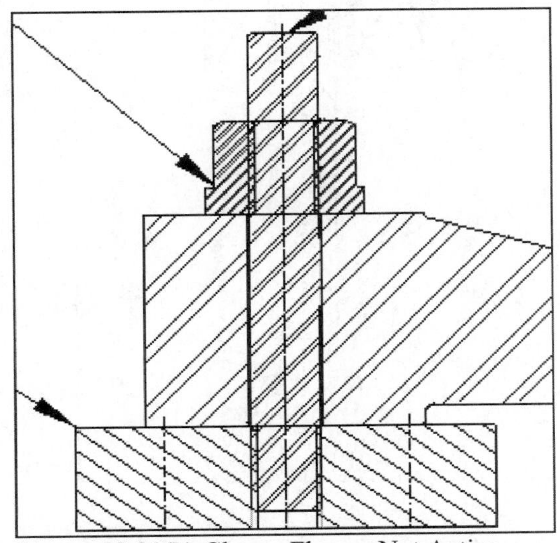

Figure 20.31(b) Clamp_Flange_Nut Active

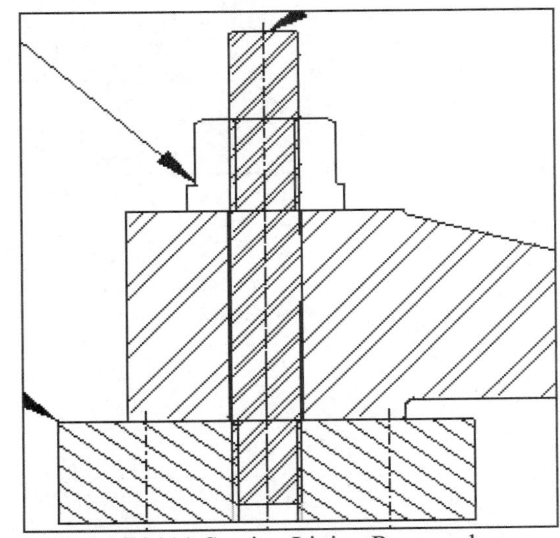

Figure 20.31(c) Section Lining Removed

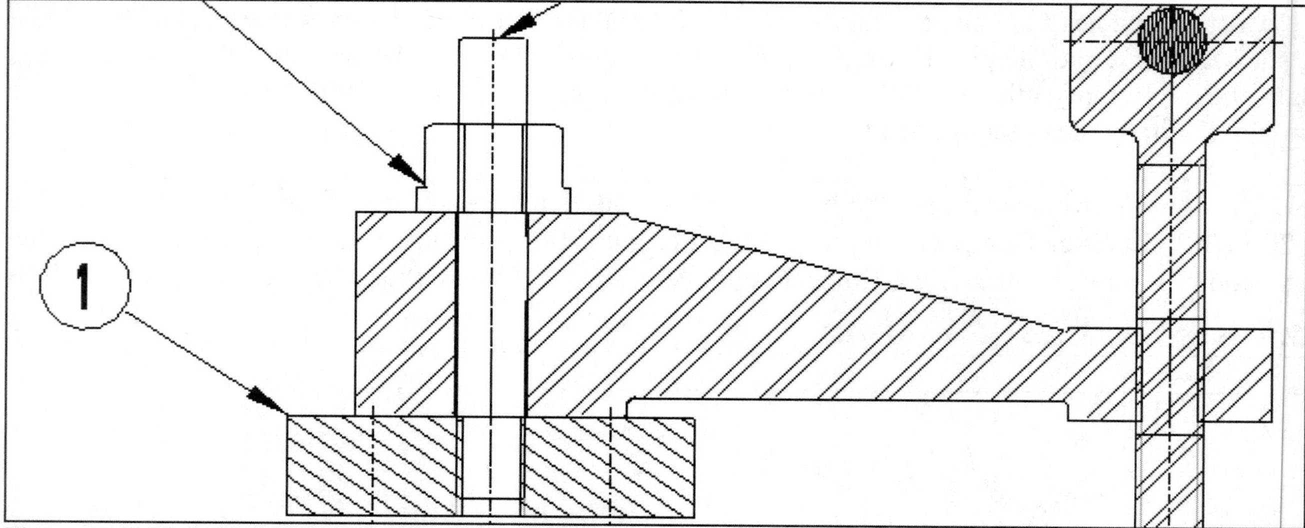

Figure 20.31(d) Section Lining Removed from Clamp_Stud35

Double-click on the crosshatching, the Clamp_Flange_Nut is now active ⇒ **Next Xsec** ⇒ **Next Xsec** Clamp_Stud5 is now active ⇒ **Fill** ⇒ **Next Xsec** ⇒ **Next Xsec** Clamp_Swivel is now active ⇒ **Delete line** ⇒ **Next Xsec** Clamp_Arm is now active ⇒ **Delete line** ⇒ **Angle** ⇒ **120** ⇒ **Next Xsec** Clamp_Plate is now active ⇒ **Angle** ⇒ **45** [Fig. 20.31(e)] ⇒ **MMB** ⇒ **LMB** ⇒ ⇒ ⇒ ⇒ ⇒ **MMB**

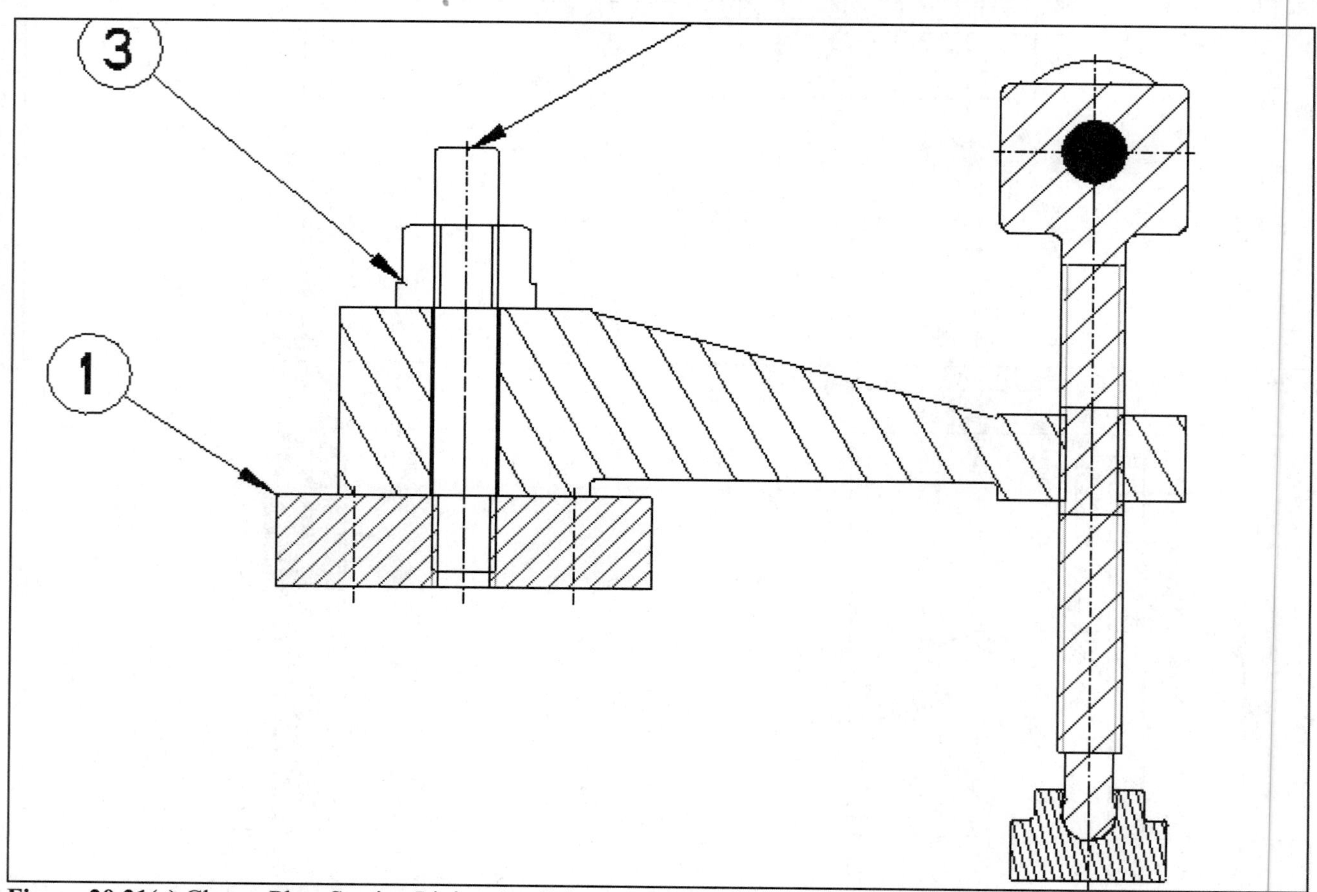

Figure 20.31(e) Clamp_Plate Section Lining Angle is now **45** Degrees

The numbering of the components in assemblies may need to be different from the default setting. To change the balloon numbering, you must use **Fix Index**.

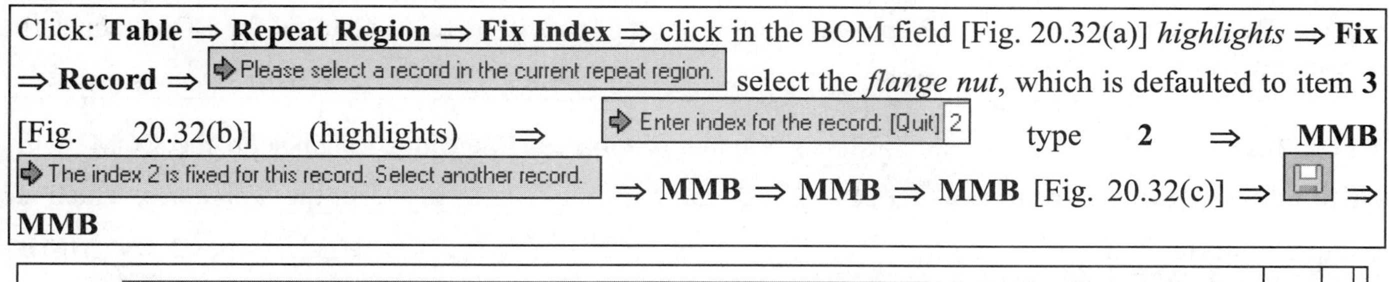
Click: **Table** ⇒ **Repeat Region** ⇒ **Fix Index** ⇒ click in the BOM field [Fig. 20.32(a)] *highlights* ⇒ **Fix** ⇒ **Record** ⇒ [Please select a record in the current repeat region.] select the *flange nut*, which is defaulted to item **3** [Fig. 20.32(b)] (highlights) ⇒ [Enter index for the record: [Quit] 2] type **2** ⇒ **MMB** [The index 2 is fixed for this record. Select another record.] ⇒ **MMB** ⇒ **MMB** ⇒ **MMB** [Fig. 20.32(c)] ⇒ [save icon] ⇒ **MMB**

3	SW100-22FLN	.500-13 X HEX FLANGE NUT	PURCHASED	1
2	SW100-21ST	.500-13 X 3.50 DOUBLE END STUD	PURCHASED	1
1	SW100-20PL	CLAMP PLATE	STEEL	1
ITEM	PT NUM	DESCRIPTION	MATERIAL	QTY

Figure 20.32(a) Click in the BOM Field

3	SW100-22FLN	.500-13 X HEX FLANGE NUT	PURCHASED	1
2	SW100-21ST	.500-13 X 3.50 DOUBLE END STUD	PURCHASED	1
1	SW100-20PL	CLAMP PLATE	STEEL	1
ITEM	PT NUM	DESCRIPTION	MATERIAL	QTY

Figure 20.32(b) Click in the Clamp_Flange_Nut Table Cell

3	SW100-21ST	.500-13 X 3.50 DOUBLE END STUD	PURCHASED	1
2	SW100-22FLN	.500-13 X HEX FLANGE NUT	PURCHASED	1
1	SW100-20PL	CLAMP PLATE	STEEL	1
ITEM	PT NUM	DESCRIPTION	MATERIAL	QTY

TOOL ENGINEERING CO.

DRAWN

ISSUED

1.500

CLAMP_ASSEMBLY

SHEET 1 OF 1

A

2

1

Figure 20.32(c) The Clamp_Flange_Nut is Now Listed Second

Exploded Swing Clamp Assembly Drawings

The process required to place an exploded view on a drawing is similar to adding assembly orthographic views. The BOM will display all components on this sheet. You will be required to fix the BOM sequence and manually create balloons.

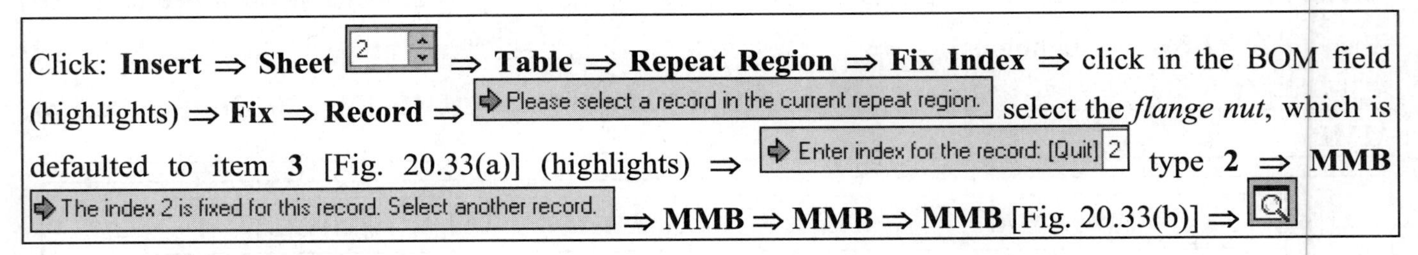

Click: **Insert** ⇒ **Sheet** 2 ⇒ **Table** ⇒ **Repeat Region** ⇒ **Fix Index** ⇒ click in the BOM field (highlights) ⇒ **Fix** ⇒ **Record** ⇒ Please select a record in the current repeat region. select the *flange nut*, which is defaulted to item **3** [Fig. 20.33(a)] (highlights) ⇒ Enter index for the record: [Quit] 2 type **2** ⇒ **MMB** The index 2 is fixed for this record. Select another record. ⇒ **MMB** ⇒ **MMB** ⇒ **MMB** [Fig. 20.33(b)] ⇒

8	SW101-9STL	.500-13 X 5.00 DOUBLE END STUD	PURCHASED	1
7	SW101-8FT	CLAMP FOOT	NYLON	1
6	SW101-7BA	SWING CLAMP BALL	BLACK PLASTIC	2
5	SW101-6SW	CLAMP SWIVEL	STEEL	1
4	SW101-5AR	CLAMP ARM	STEEL	1
3	SW100-22FLN	.500-13 X HEX FLANGE NUT	PURCHASED	1
2	SW100-21ST	.500-13 X 3.50 DOUBLE END STUD	PURCHASED	1
1	SW100-20PL	CLAMP PLATE	STEEL	1
ITEM	PT NUM	DESCRIPTION	MATERIAL	QTY

Figure 20.33(a) Click in the Flange Nut Table Cell on Sheet 2

8	SW101-9STL	.500-13 X 5.00 DOUBLE END STUD	PURCHASED	1
7	SW101-8FT	CLAMP FOOT	NYLON	1
6	SW101-7BA	SWING CLAMP BALL	BLACK PLASTIC	2
5	SW101-6SW	CLAMP SWIVEL	STEEL	1
4	SW101-5AR	CLAMP ARM	STEEL	1
3	SW100-21ST	.500-13 X 3.50 DOUBLE END STUD	PURCHASED	1
2	SW100-22FLN	.500-13 X HEX FLANGE NUT	PURCHASED	1
1	SW100-20PL	CLAMP PLATE	STEEL	1
ITEM	PT NUM	DESCRIPTION	MATERIAL	QTY

Figure 20.33(b) On Sheet 2 the Flange Nut is Now Listed Second

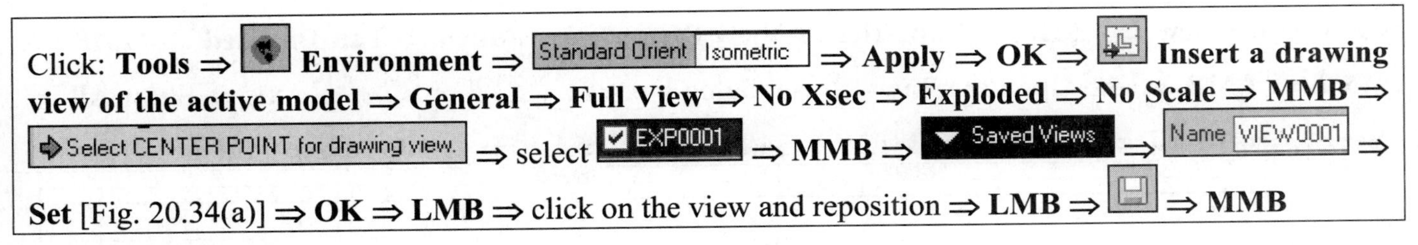

Click: **Tools** ⇒ **Environment** ⇒ Standard Orient Isometric ⇒ **Apply** ⇒ **OK** ⇒ **Insert a drawing view of the active model** ⇒ **General** ⇒ **Full View** ⇒ **No Xsec** ⇒ **Exploded** ⇒ **No Scale** ⇒ **MMB** ⇒ Select CENTER POINT for drawing view. ⇒ select EXP0001 ⇒ **MMB** ⇒ Saved Views ⇒ Name VIEW0001 ⇒ **Set** [Fig. 20.34(a)] ⇒ **OK** ⇒ **LMB** ⇒ click on the view and reposition ⇒ **LMB** ⇒ ⇒ **MMB**

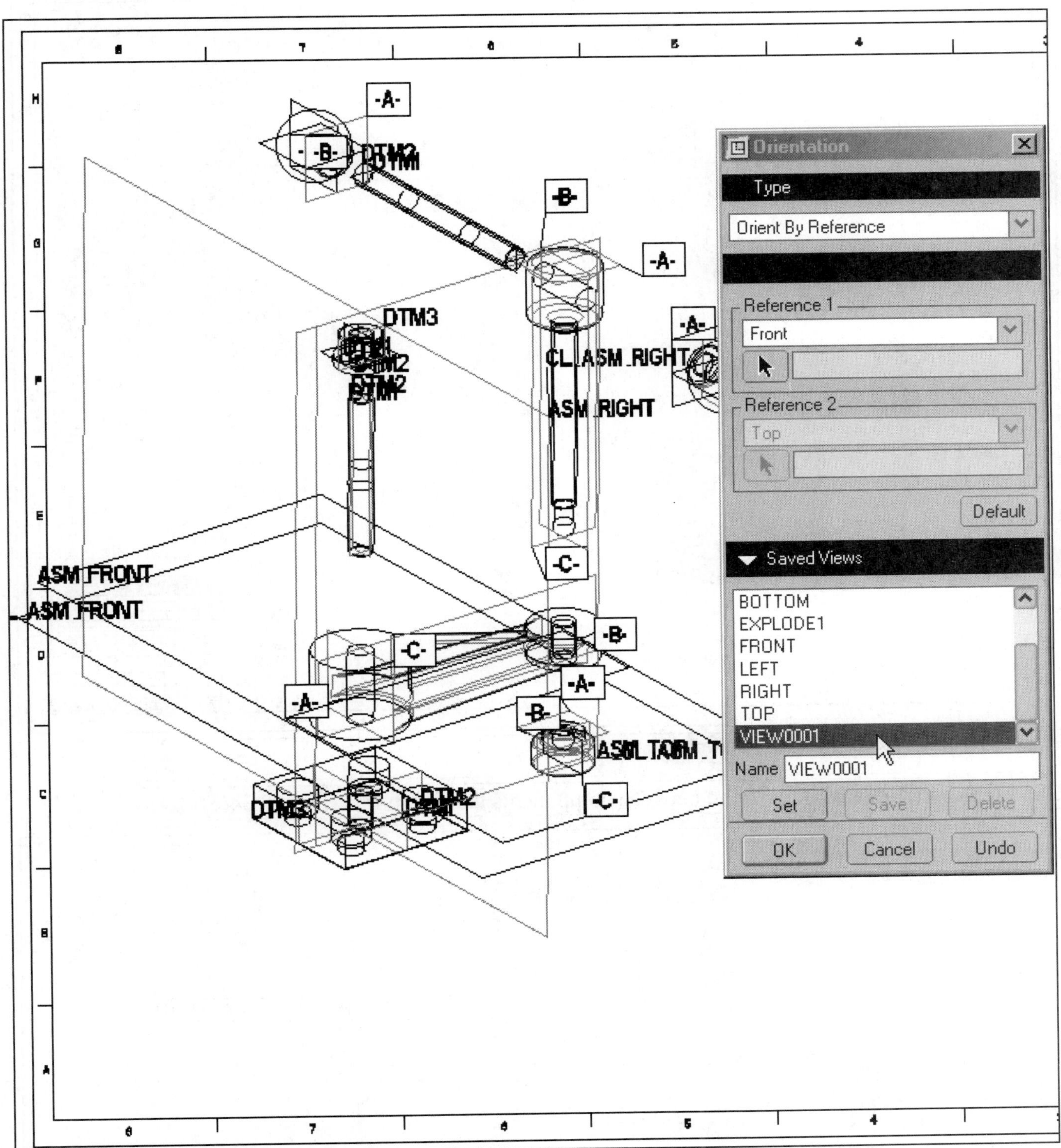

Figure 20.34(a) Placing the Exploded View

Click on the view ⇒ **RMB** ⇒ **Properties** ⇒ **View Disp** ⇒ **No Hidden** ⇒ **Tan Dimmed** ⇒ **MMB** ⇒ **MMB** ⇒ **LMB** ⇒ **Open the Show/Erase dialog box** ⇒ Show ⇒ ----A_1 **Axis** ⇒ **Show All** ⇒ **Yes** ⇒ **Accept All** ⇒ **Close** ⇒ **LMB** [Fig. 20.34(b)] ⇒ ⇒ ⇒ ⇒ ⇒ **MMB**

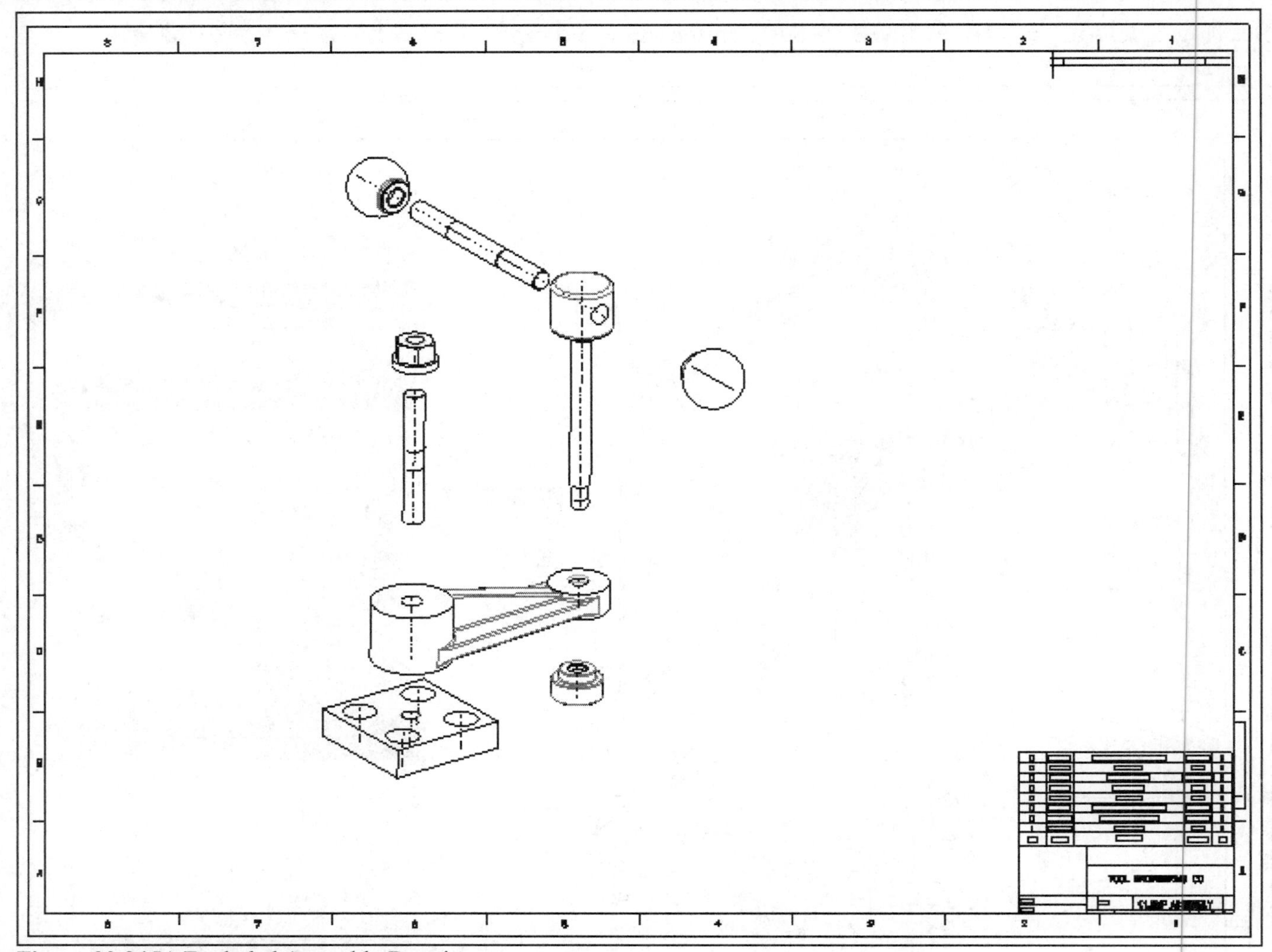

Figure 20.34(b) Exploded Assembly Drawing

Clip each centerline (axis) to extend between components that are in line [Fig. 20.34(c)]

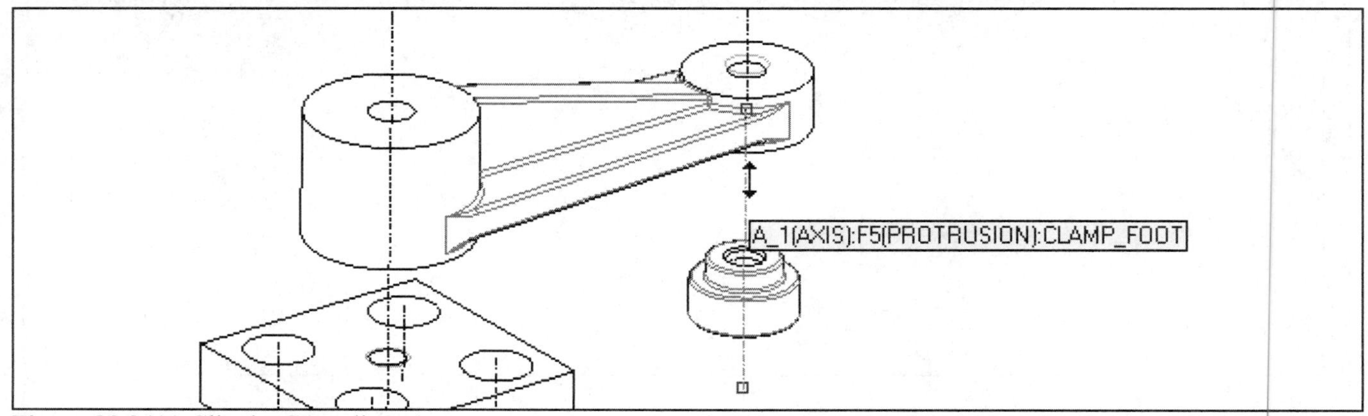

Figure 20.34(c) Clip the Centerlines Axes

To add balloons to a drawing sheet not using a parametric title block, you need to create each balloon separately. Create balloons for the components on the second sheet. The balloons added must correspond to the BOM.

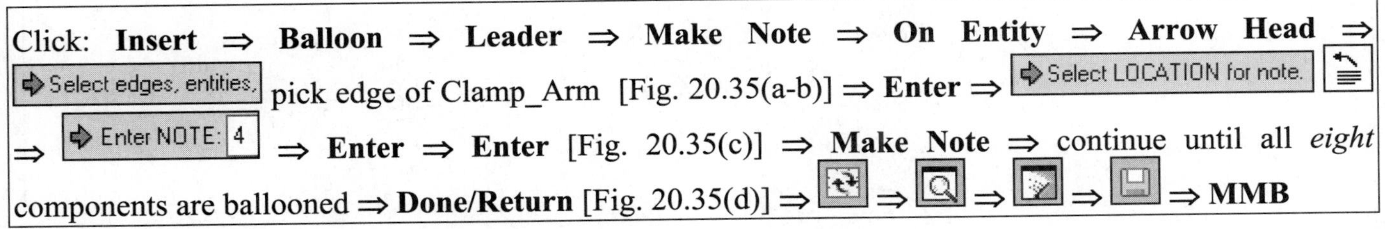

Click: **Insert** ⇒ **Balloon** ⇒ **Leader** ⇒ **Make Note** ⇒ **On Entity** ⇒ **Arrow Head** ⇒ [Select edges, entities,] pick edge of Clamp_Arm [Fig. 20.35(a-b)] ⇒ **Enter** ⇒ [Select LOCATION for note.] ⇒ [Enter NOTE: 4] ⇒ **Enter** ⇒ **Enter** [Fig. 20.35(c)] ⇒ **Make Note** ⇒ continue until all *eight* components are ballooned ⇒ **Done/Return** [Fig. 20.35(d)] ⇒ ⇒ ⇒ ⇒ ⇒ **MMB**

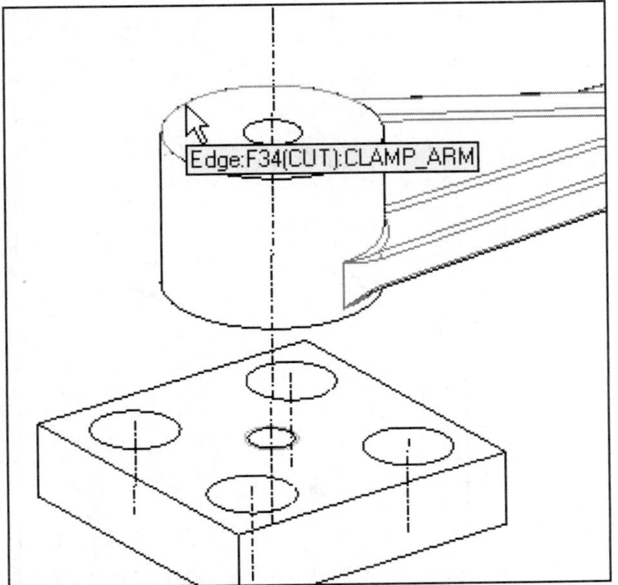

Figure 20.35(a) Highlight Edge of Clamp_Arm

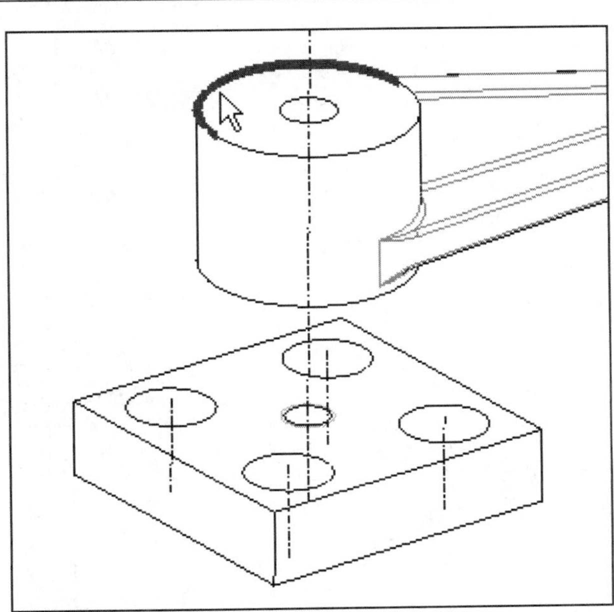

Figure 20.35(b) Select on the Edge

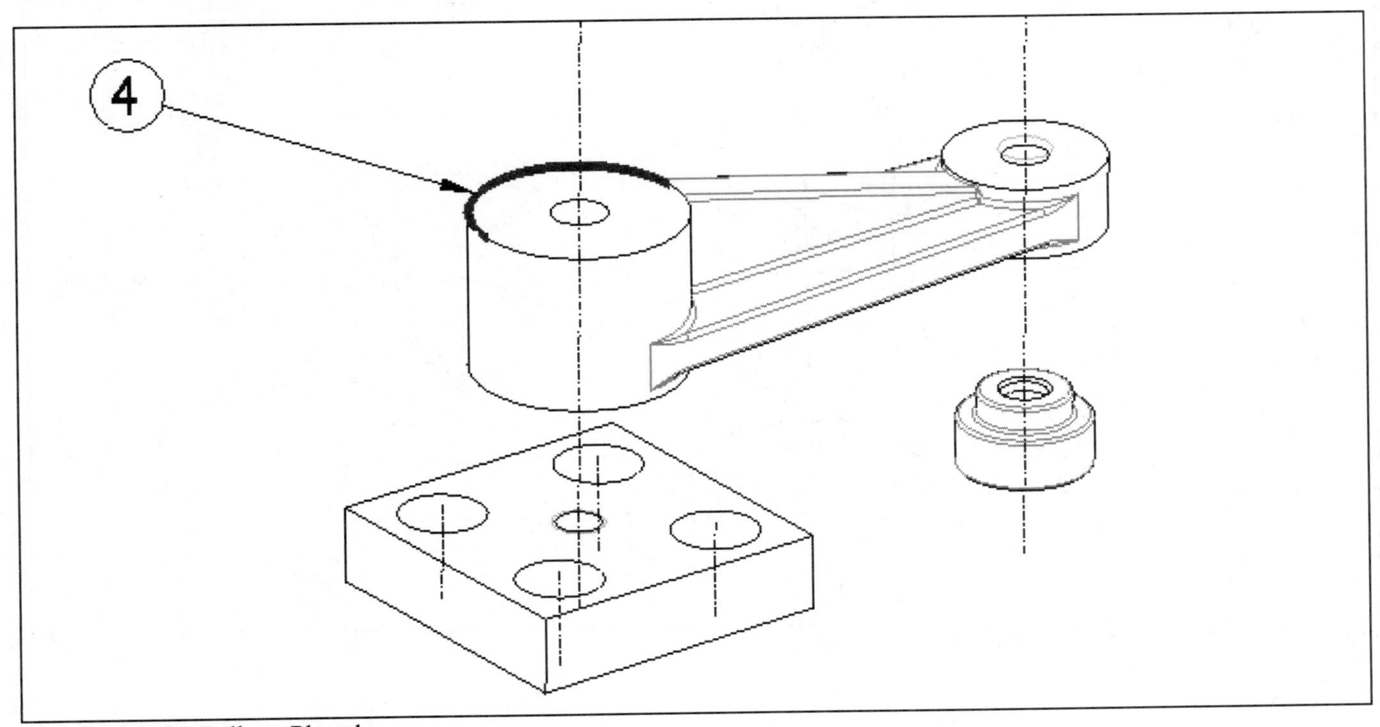

Figure 20.35(c) Balloon Placed

After the ballooning is complete, reposition the balloons and their attachment points as needed to clean up the drawing.

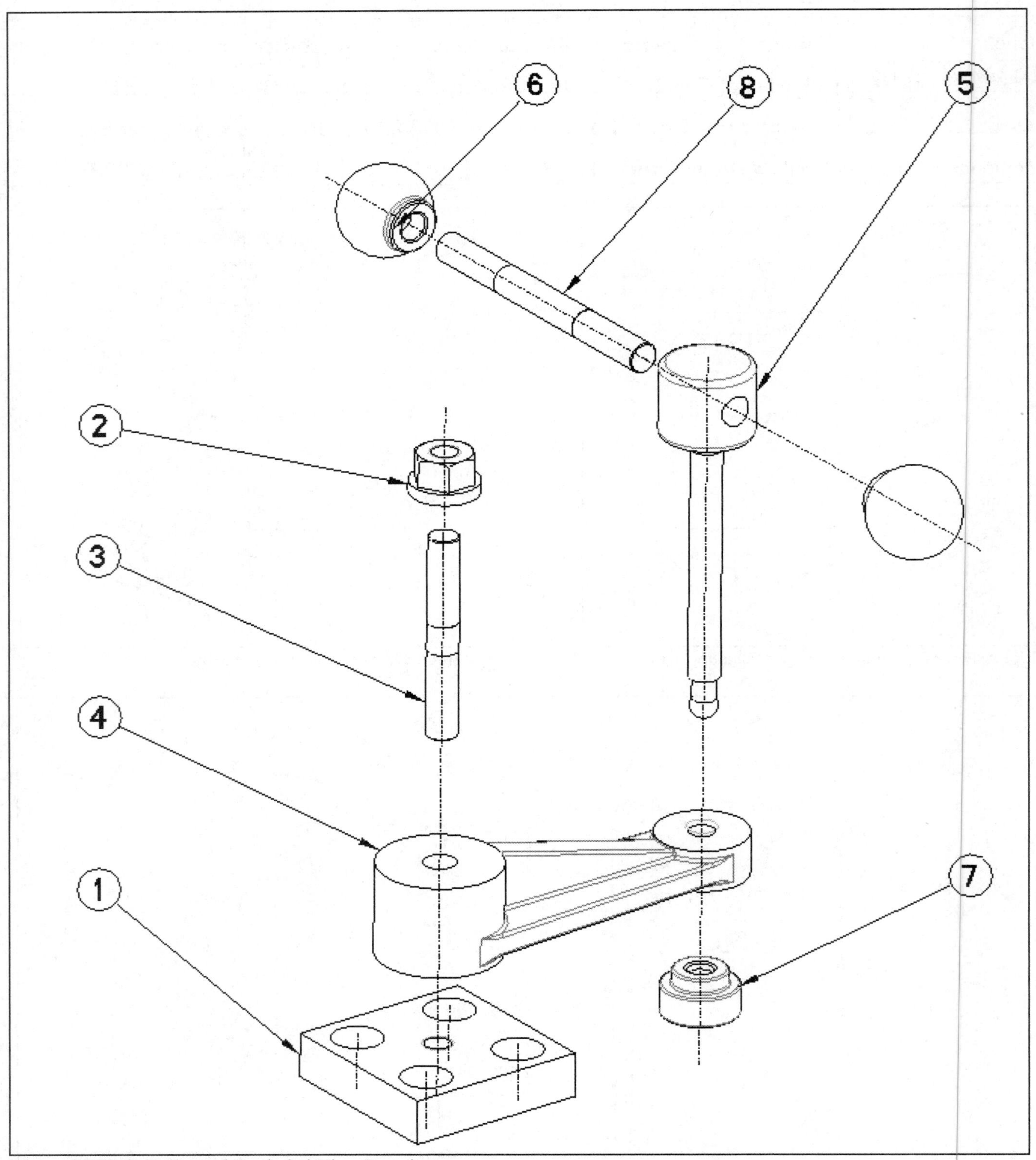

Figure 20.35(d) Ballooned Exploded View Drawing

Click on pick balloon **6** ⇒ **RMB** [Fig. 20.35(e)] ⇒ **Edit Attachment** ⇒ **On Surface** ⇒ **Filled Dot** [Fig. 20.35(f)] ⇒ pick a place on the Clamp_Ball's surface ⇒ **MMB** ⇒ **LMB** [Fig. 20.35(g)] ⇒ ⇒ ⇒ ⇒ ⇒ **MMB** ⇒ **File** ⇒ **Delete** ⇒ **Old Versions** ⇒ **MMB** [Fig. 20.35(h)]

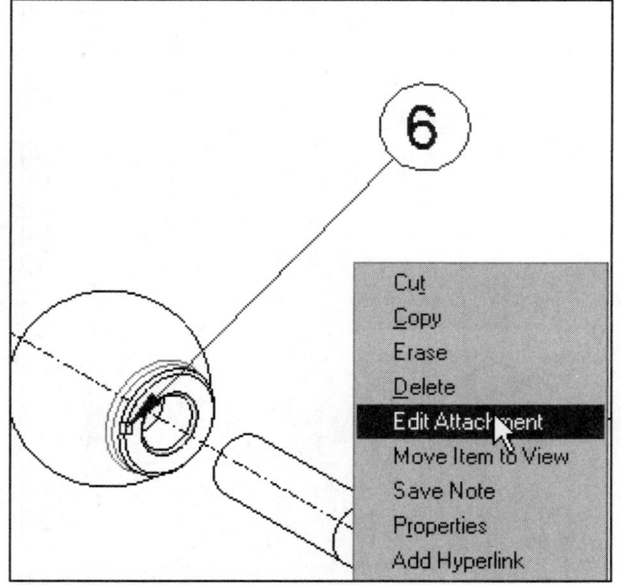

Figure 20.35(e) Edit Attachment

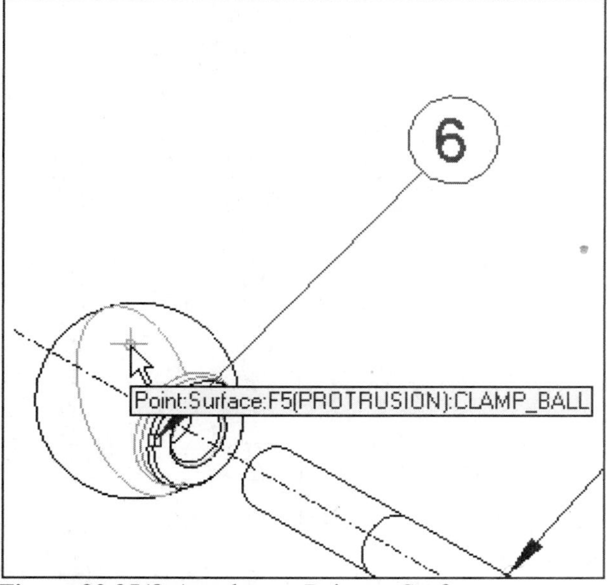

Figure 20.35(f) Attachment Point on Surface

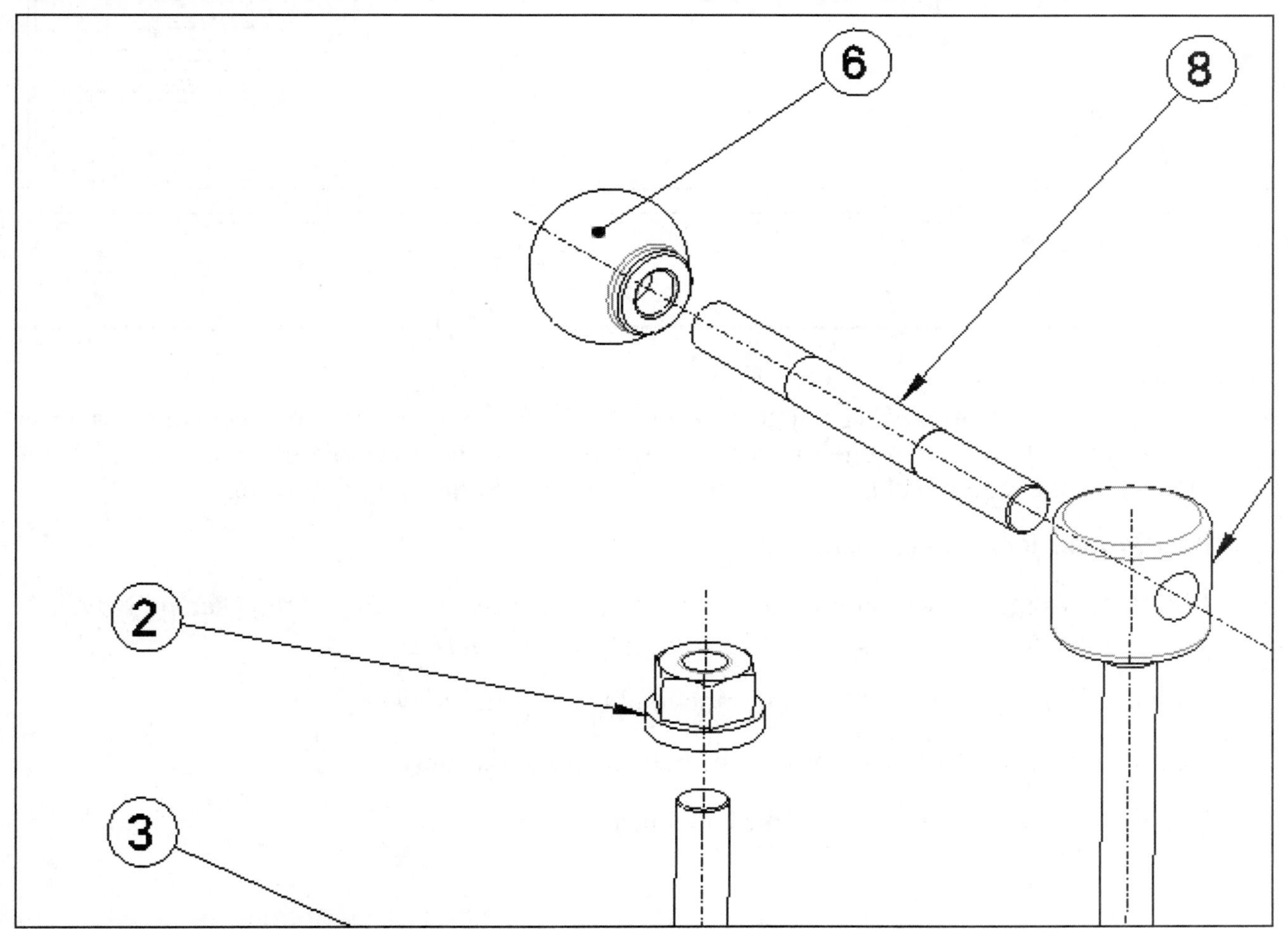

Figure 20.35(g) Filled Dot Surface Attachment

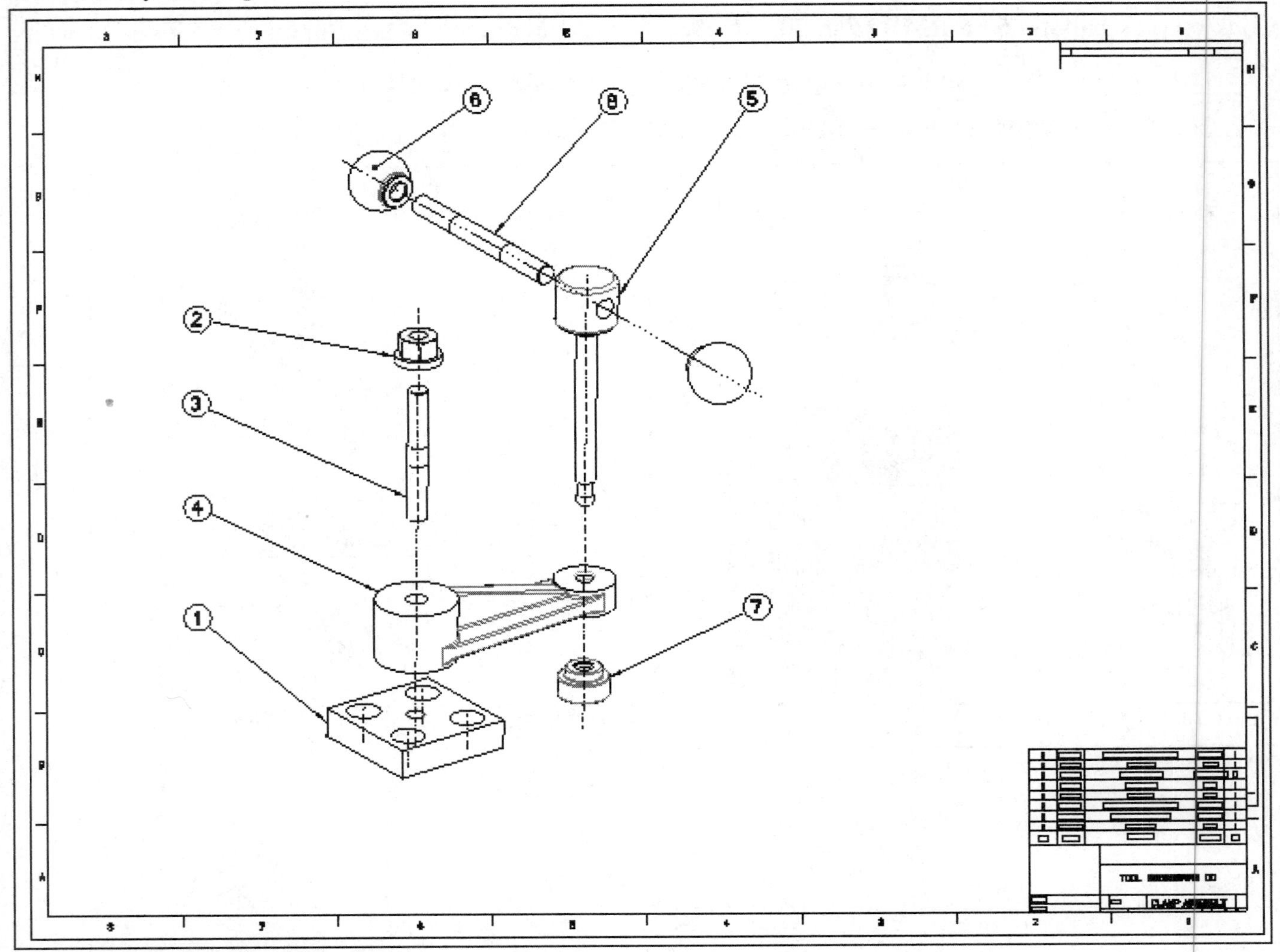

Figure 20.35(h) Completed Drawing

Create a *documentation package* for the Clamp_Assembly. A complete documentation package contains all models and drawings required to manufacture the parts and assemble the components. Your instructor may change the requirements, but in general, create and plot/print the following:

- **Part Models** for all components
- **Detail Drawings** for each nonstandard component, such as the Clamp_Arm, Clamp_Swivel, Clamp_Foot, and Clamp_Ball <u>*(do not detail the standard parts)*</u>
- **Assembly Drawings** using standard orthographic ballooned views
- **Exploded Subassembly Drawing** of the ballooned subassembly
- **Exploded Assembly Drawing** of the ballooned assembly

Lesson 20 Project

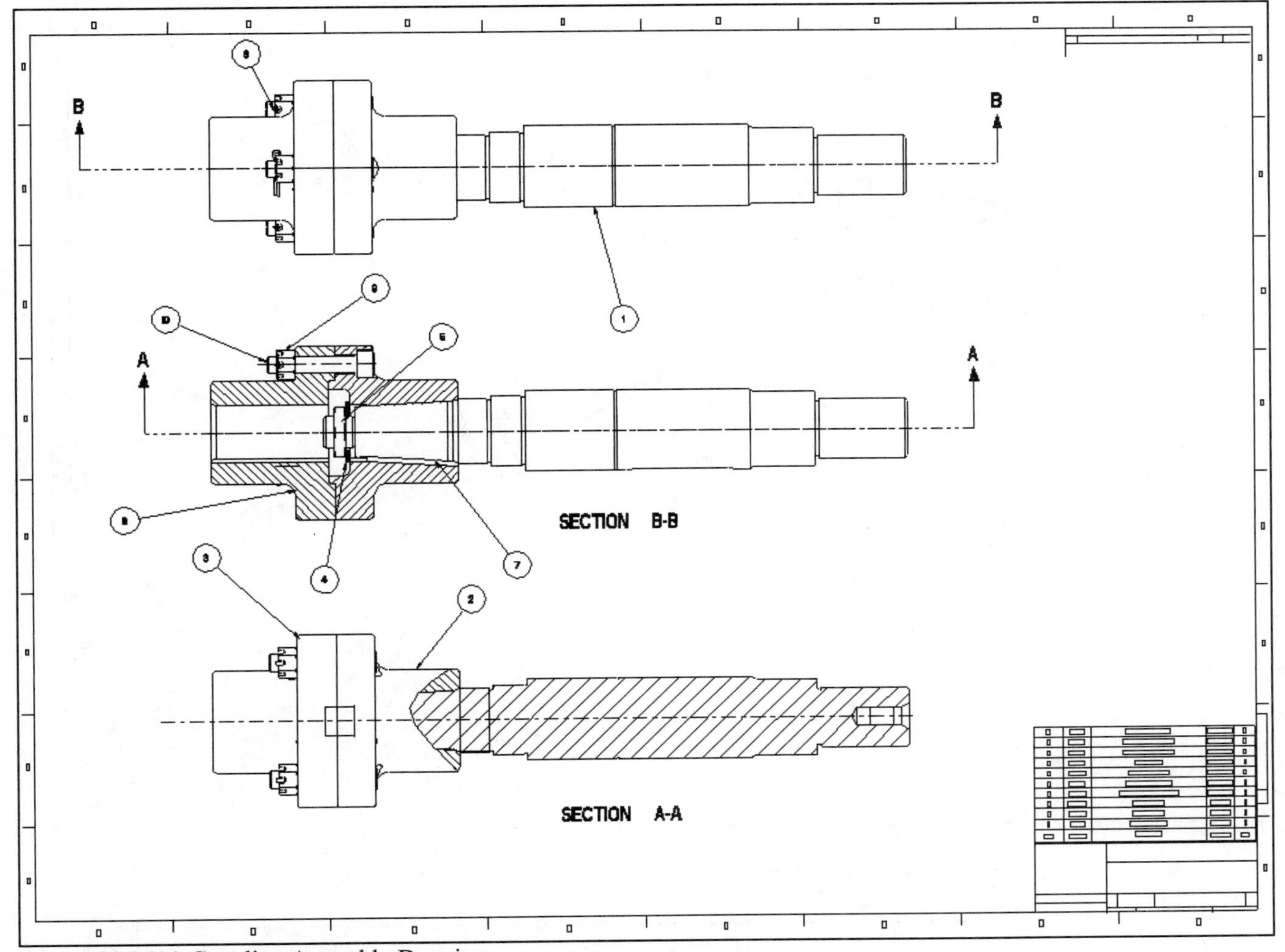

Figure 20.36(a) Coupling Assembly Drawing

Coupling Assembly Drawing and Exploded Coupling Assembly Drawing

Create a complete documentation package for the Coupling Assembly [Figs. 20.36(a-k)]. The ballooned assembly drawing will have three views and a parts list. Assign parameters to the parts in the assembly so that they can be displayed on a parts list in the assembly drawing. Some of the items listed here have been created in other lessons. Create or extract existing models and drawings, and plot/print the following:

- **Part Models** for all coupling assembly components
- **Detail Drawings** for each nonstandard component, for example, the Coupling Shaft
- **Assembly Drawing** and **Parts List (BOM)** using standard orthographic ballooned views
- **Exploded Assembly Drawing** of the ballooned assembly

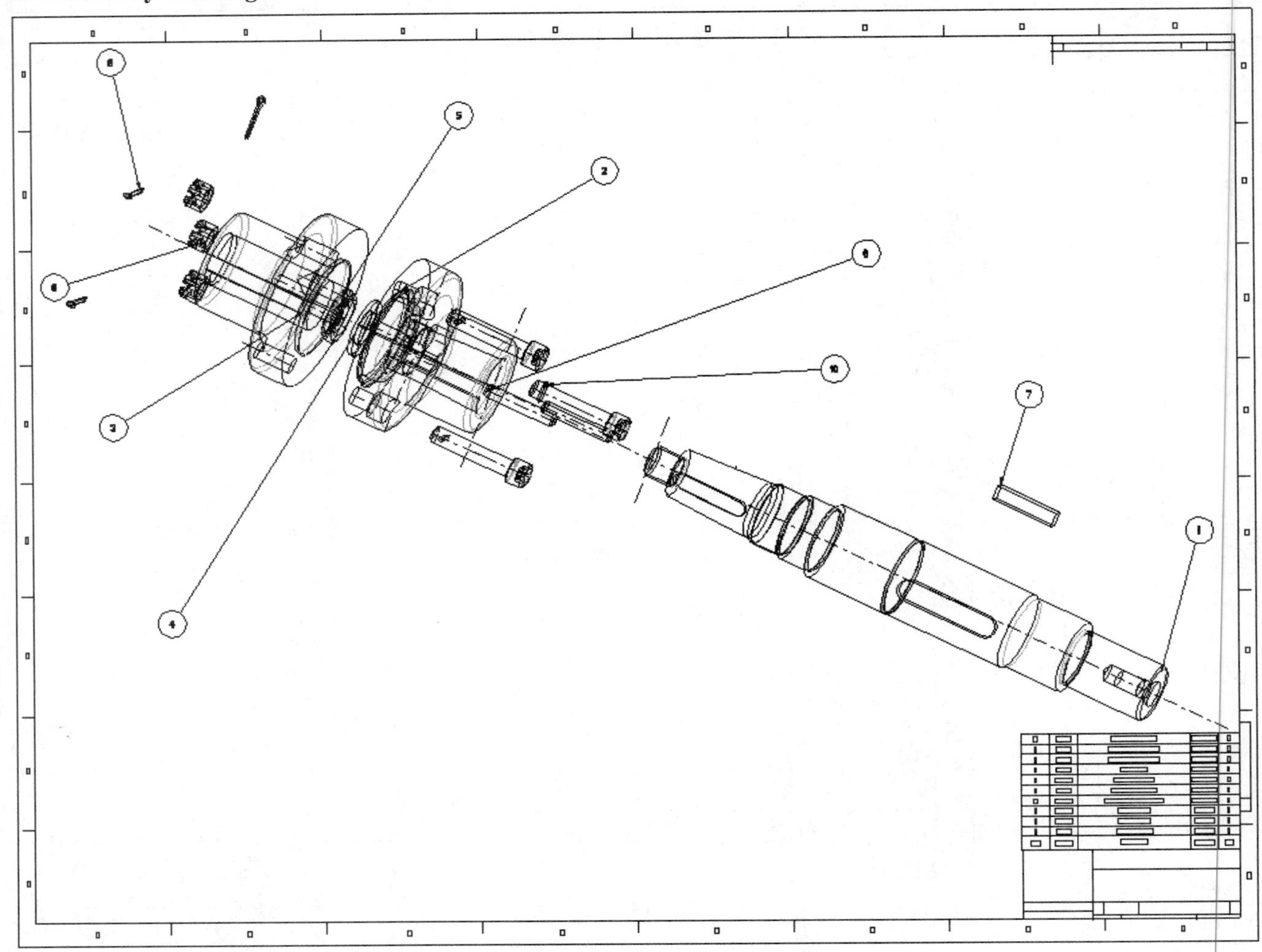

Figure 20.36(b) Exploded Coupling Assembly Drawing

ITEM	PT NUM	DESCRIPTION	MATERIAL	QTY
10	110-2CS	SOC HD CAP SCREW	PURCHASED	3
9	109-2SN	HEX SLOT NUT 16 X 2	PURCHASED	3
8	108-2CP	COTTER PIN .150 X 1.25	PURCHASED	3
7	107-2KY	KEY 14 X 61	PURCHASED	1
6	106-2DW	DOWEL 12OD X 70	PURCHASED	2
5	105-2HN	HEX NUT M30 X 3.5	PURCHASED	1
4	104-2WA	WASHER 33ID X 50OD X 4	PURCHASED	1
3	103-2CP2	COUPLING TWO	1040 CRS	1
2	102-2CP1	COUPLING ONE	1040 CRS	1
1	101-2SH	COUPLING SHAFT	1020 CRS	1

Figure 20.36(c) Coupling Assembly BOM

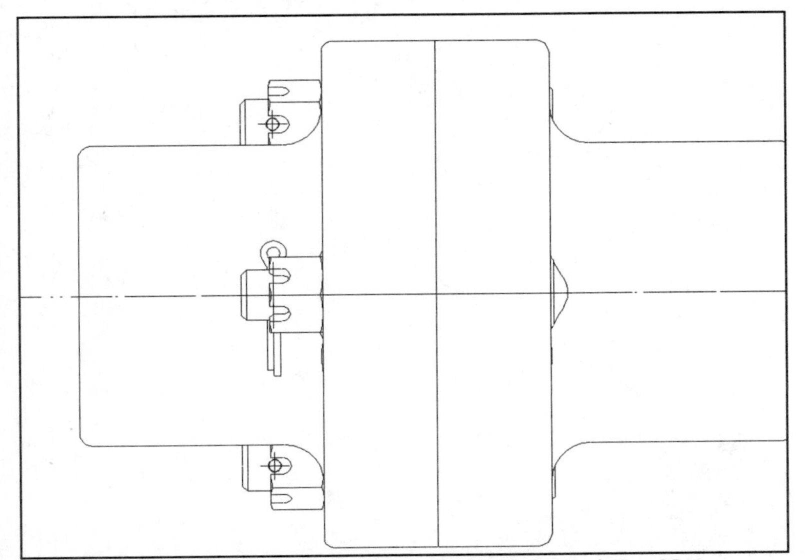

Figure 20.36(d) Coupling Assembly Drawing, Slotted Hex Nut

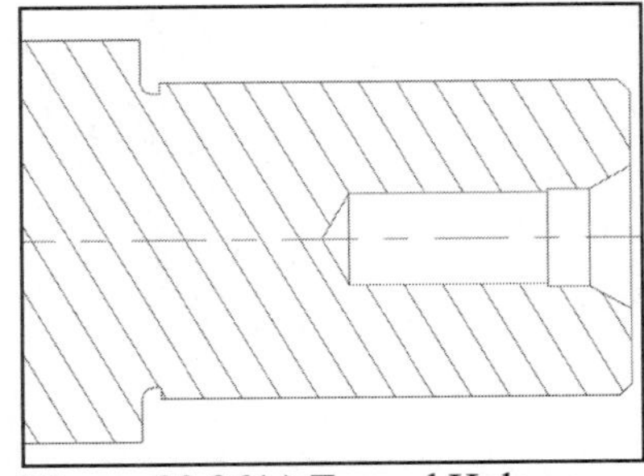

Figure 20.36(e) Tapped Hole

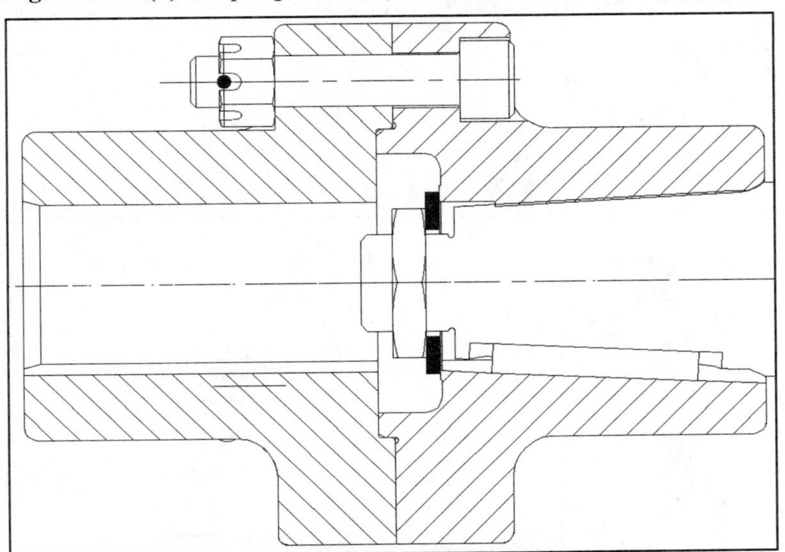

Figure 20.36(f) Coupling Assembly Drawing, Section Close-up

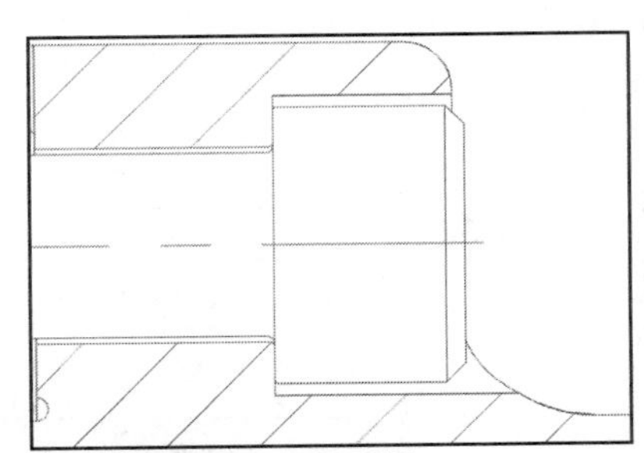

Figure 20.36(g) SHCS

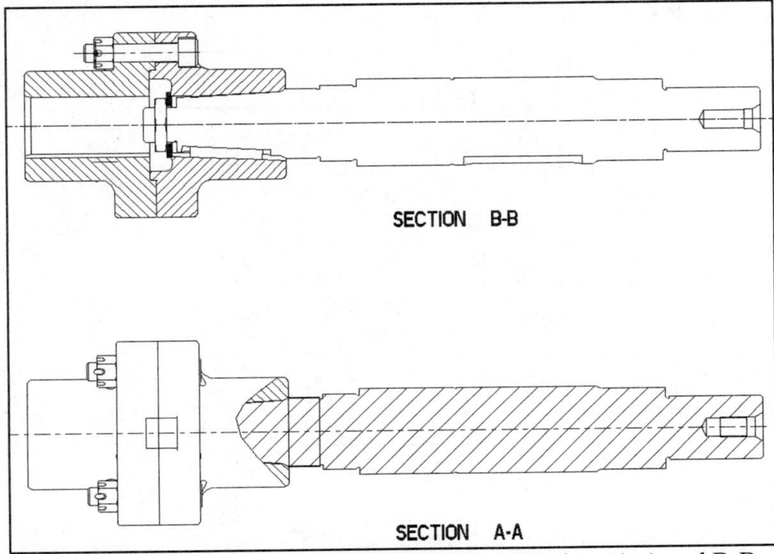

Figure 20.36(h) Coupling Assembly Drawing, Sections A-A and B-B

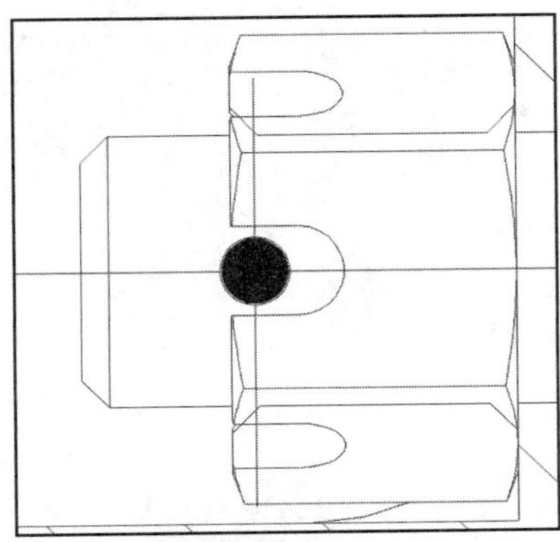

Figure 20.36(i) Nut

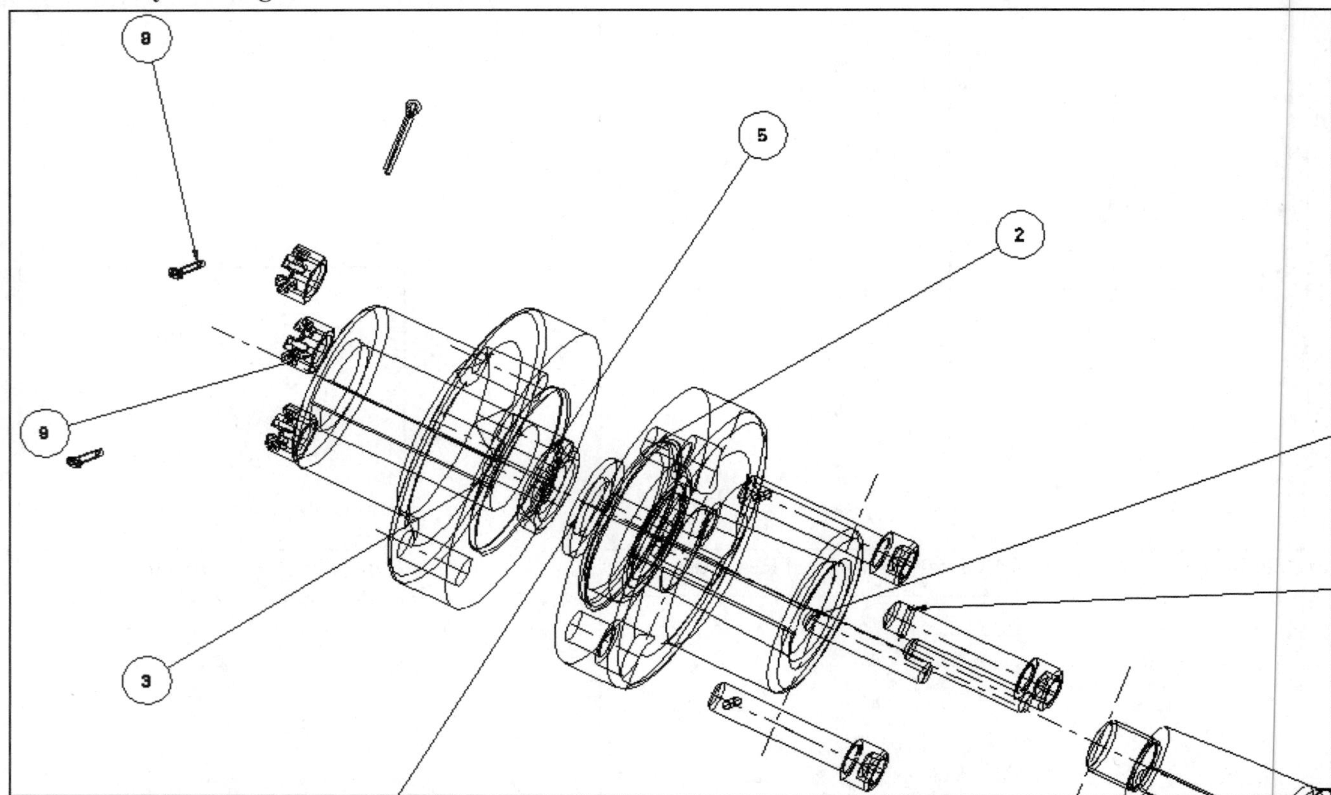

Figure 20.36(j) Coupling Assembly, Close-up

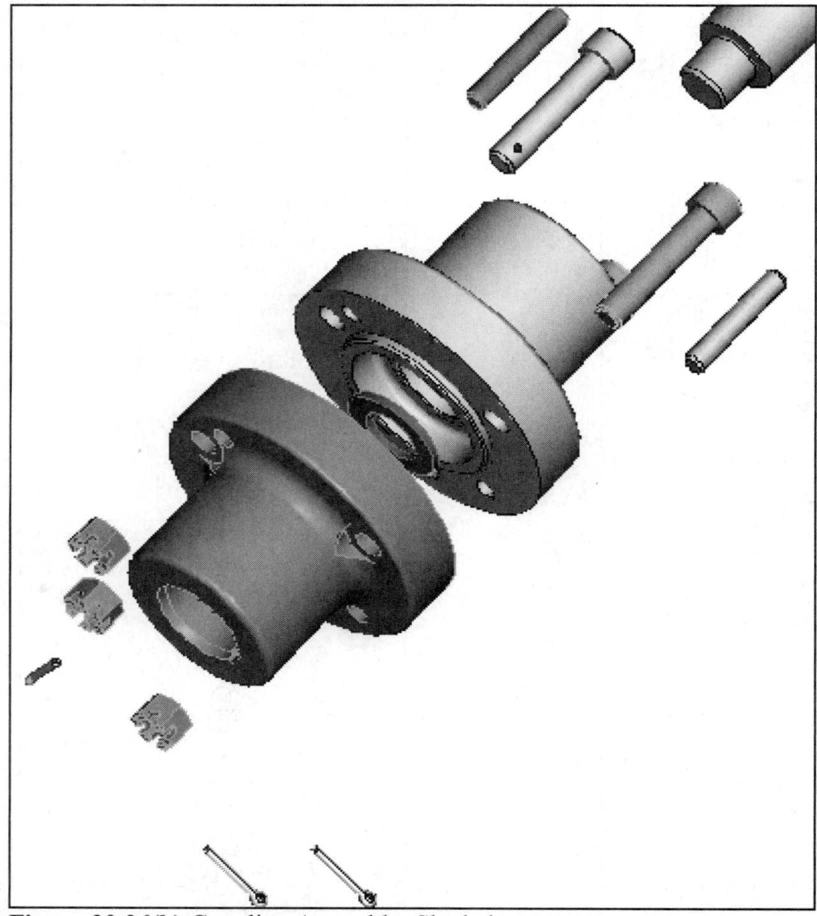

Figure 20.36(k) Coupling Assembly, Shaded

Index

S

T

U

V

W

X

Y

Z